A Man would do nothing,
if he waited until he could do it so well
that no one would find fault
with what he has done.

CARDINAL NEWMAN

ORGANIC

I. L. Finar, B.Sc., Ph.D(London), A.R.I.C.
Principal Lecturer in Organic Chemistry
The Polytechnic of North London, Holloway

CHEMISTRY

VOLUME 1

THE FUNDAMENTAL PRINCIPLES

SIXTH EDITION

Longman

LONGMAN GROUP LIMITED
London
Associated companies, branches and representatives throughout the world

© I. L. Finar 1959, 1967, 1973.

First published 1951
Second Edition 1954
Third Edition 1959
Fourth Edition 1963
Fifth Edition 1967
Sixth Edition 1973

ISBN 0 582 44221 4

Photoset by Oliver Burridge Filmsetting Limited
and printed in Hong Kong by The Continental Printing Co., Ltd.

List of journal abbreviations

ABBREVIATIONS	JOURNALS
Chem. Comm.	Chemical Communications
Chem. Rev.	Chemical Reviews
Chem. in Britain	Chemistry in Britain
Educ. in Chem.	Education in Chemistry
J. Amer. Chem. Soc.	Journal of the American Chemical Society
J. Chem. Educ.	Journal of Chemical Education
J. Chem. Soc.	Journal of the Chemical Society
Nature	Nature
Quart. Rev.	Quarterly Reviews
Research	Research
Roy. Inst. Chem.	Royal Institute of Chemistry
Tetrahedron	Tetrahedron

Contents

Preface to the sixth edition

In this book my aim has been to describe the fundamental principles of organic chemistry. Since I do not consider the chemistry of natural products to be fundamental chemistry but rather the application of fundamental principles, I have largely excluded the study of these compounds. Many types of natural products and a more detailed account of stereochemistry have been dealt with in Vol. 2. Those who prefer to deal differently with the subject-matter should find it fairly easy to use Vol. 2 in conjunction with this volume, and to help the reader I have included in the text references to further work described in Vol. 2.

The subject-matter described in this book covers much of the basic organic chemistry that is needed by a student who wishes to study chemistry as a main subject at degree level. How much a student learns depends on himself. Two types of print have been used and, in general, the material in small print contains more detailed and/or more advanced work. This 'extra' material may be omitted by those who do not require it but, even so, the student will find it worthwhile to at least read it.

To many beginners organic chemistry may seem to consist of a large variety of methods and reactions which appear isolated and, consequently, only to be learned by heart. However, the introduction of the electronic theories of organic chemistry as early as possible and a constant application of their principles give to organic chemistry a certain coherence that is soon appreciated by the beginner, and thus facilitate his study in this branch of chemistry. Even so, it is my belief that in order to learn organic chemistry it is necessary to acquire a comprehensive knowledge of chemical behaviour and then, with this as a background, to consider the principles and the theoretical basis of the subject. Thus, stress has been laid on structure and reactions of compounds. Nomenclature of organic compounds has also been stressed, since it is highly desirable that a beginner should be able to assign names to compounds.

The arrangement of the subject-matter is based on homologous series, and in general, descriptions of reactions are followed by discussions of their mechanisms, and these include an elementary account of the sort of evidence that led workers to suggest mechanisms that are acceptable at the present time.

Chapter 2 contains much of the theoretical aspects used throughout the book. I have expanded the elementary account of reaction kinetics and thermodynamic functions. This, I hope, will help those students who are unfamiliar with these topics (at this stage of the course) to better appreciate the theoretical matter discussed throughout the text.

It is my experience that only a fairly detailed study of the subject-matter enables the student to appreciate the problems involved. Too short an account usually leaves the impression that 'everything works according to plan'. This is undesirable for those who are expected to learn to think for themselves. I have therefore included detailed discussions on developments of a straightforward nature and also of a controversial nature, in the hope of encouraging the student to weigh up the evidence for himself. This will also give him an idea of some of the problems being investigated at the present time.

Many changes have been made in this new edition. The order of some topics has been changed, *e.g.*, the principles of conformational analysis, geometrical and optical isomerism have been dealt with earlier in this edition. Expanded sections include spectroscopic methods, addition and elimination reactions, reduction with complex metal hydrides, conformational analysis, the theory of aromaticity, the Hammett equation, etc. Also, more detailed accounts have been given on reactivity in terms of charge distribution and stability of carbonium ions and carbanions.

New subject-matter includes a description of SI units, enamines, the chemistry of organophosphorus, silicon, boron, aluminium, and tin compounds, etc. Mass spectrometry has been described and applied to many classes of compounds. A section on organic photochemistry has also been added, including the Woodward–Hoffmann principle of the conservation of orbital symmetry.

Only by reading original papers can the student hope to gain a more mature outlook. A selected number of reading references have therefore been given at the end of each chapter. Since summaries of many topics by workers in various fields are of great value in extending the student's knowledge, references of this type have also been included. A brief account of the literature of organic chemistry has been given in an appendix.

In describing methods of preparation of various compounds, I have given, wherever possible, actual percentage yields (taken mainly from *Organic Syntheses*). The student will thus be enabled to assess the value of a particular method. Where general methods of preparation have been described the yields have often been indicated according to the following (arbitrary) scheme:

0–15 per cent, very poor (*v.p.*); 16–30 per cent, poor (*p.*); 31–45 per cent, fair (*f.*); 46–60 per cent, fairly good (*f.g.*); 61–75 per cent, good (*g.*); 76–90 per cent, very good (*v.g.*); 91–100 per cent, excellent (*ex.*).

In this new edition, most of the questions at the end of each chapter are new and most are now in the form of problems. Another new feature is the inclusion of equations which have to be completed. In general, these will provide a rapid revision of 'book knowledge'. A book containing solutions to these problems is being prepared for publication.

It is impossible to express my indebtedness to those authors of monographs, articles, etc., from which I have gained so much information. I can only hope that some measure of my gratitude is expressed by the references I have given to their works.

I. L. FINAR

1971

1 Determination of structure

Historical introduction

Although organic substances such as sugar, starch, alcohol, resins, oils, indigo, etc., had been known from earliest times, very little progress in their chemistry was made until about the beginning of the eighteenth century. In 1675, Lemery published his famous *Cours de Chymie*, in which he divided compounds from natural sources into three classes: *mineral*, *vegetable* and *animal*. This classification was accepted very quickly, but it was Lavoisier who first showed, in 1784, that all compounds obtained from vegetable and animal sources always contained at least carbon and hydrogen, and frequently, nitrogen and phosphorus. Lavoisier, in spite of showing this close relationship between vegetable and animal products, still retained Lemery's classification. Lavoisier's analytical work, however, stimulated further research in this direction, and resulted in much-improved technique, due to which Lemery's classification had to be modified. Lemery had based his classification on the *origin* of the compound, but it was now found (undoubtedly due to the improved analytical methods) that in a number of cases the *same* compound could be obtained from both vegetable and animal sources. Thus, no difference existed between these two classes of compounds, and hence it was no longer justifiable to consider them under separate headings. This led to the reclassification of substances into two groups: all those which could be obtained from vegetables or animals, *i.e.*, substances that were produced by the *living organism*, were classified as *organic*; and all those substances which were not prepared from the living organism were classified as *inorganic*.

At this stage of the investigation of organic compounds it appeared that there were definite differences between organic and inorganic compounds, *e.g.*, complexity of composition and the combustibility of the former. Berzelius (1815) thought that organic compounds were produced from their elements by laws different from those governing the formation of inorganic compounds. This then led him to believe that organic compounds were produced under the influence of a *vital force*, and that they could not be prepared artificially.

In 1828, Wöhler converted ammonium cyanate into urea, a substance hitherto obtained only from animal sources. This synthesis weakened the distinction between organic and inorganic compounds, and this distinction was completely ended with the synthesis of acetic acid from its elements by Kolbe in 1845, and the synthesis of methane by Berthelot in 1856. A common belief appears to be that Wöhler's synthesis had little effect on the vital-force theory because it did not start with the elements. Wöhler had prepared his ammonium cyanate from ammonia and cyanic acid, both of which were of animal origin. Partington (1960), however, has pointed out that Priestley (1781) had obtained ammonia by reduction of nitric acid, which was later synthesised from its elements by Cavendish (1785). Also, potassium cyanide was obtained by Scheele (1783) by passing nitrogen over a strongly heated mixture of potassium carbonate and carbon, and since one form of carbon used was graphite, this reaction was therefore carried out with inorganic materials. Since potassium cyanide is readily converted into potassium cyanate, Wöhler's synthesis is one which starts from the elements.

Since the supposed differences between the two classes of compounds have been disproved, the terms organic and inorganic would appear to be no longer necessary. Nevertheless, they have been retained, but it should be appreciated that they have lost their original meaning. The retention of the terms organic and inorganic may be ascribed to several reasons: (i) all so-called organic compounds contain carbon; (ii) the compounds of carbon are far more numerous (over 1 000 000) than the known compounds of *all* the other elements put together; (iii) carbon has the power to combine with other carbon atoms to form long chains. This property, known as *catenation*, is not shown to such an extent by any other element.

Hence organic chemistry is the chemistry of the carbon compounds.

This definition includes compounds such as carbon monoxide, carbon dioxide, carbonates, carbon disulphide, etc. Since these occur chiefly in the inorganic kingdom (*original meaning*), they are usually described in text-books of inorganic chemistry.

Analysis of organic compounds

The following is an outline of the methods used in the study of organic compounds.

Purification. Before the properties and structure of an organic compound can be completely investigated, the compound must be pure. Common methods of purification are:

(i) Recrystallisation from suitable solvents.

(ii) Distillation: (a) at atmospheric pressure; (b) under reduced pressure or *in vacuo*; (c) under increased pressure.

(iii) Steam distillation.

(iv) Sublimation.

(v) Chromatography. This offers a means of concentrating a product that occurs naturally in great dilution, and is an extremely valuable method for the separation, isolation, purification and identification of the constituents of a mixture.

Chromatography depends on the distribution of the mixture between two phases, one fixed and the other moving. The mixture is dissolved in the moving phase and passed over a fixed phase which may be a column or a paper strip. The moving phase may be a liquid or a gas. When the fixed phase is a solid, *e.g.*, alumina, silica gel, etc., the process is usually called *adsorption chromatography*, and the solutes are adsorbed on *different* parts of the column. The adsorbed compounds are then eluted by the passage of a series of *eluants* (solvents). *Thin-layer chromatography* (TLC) uses a chromatoplate as the column, *i.e.*, a glass strip coated with the adsorbent. Alternatively, the fixed phase may consist of a liquid strongly adsorbed on a solid (as support). The solutes now become distributed between the 'fixed' liquid and the moving liquid (solvent). This technique is known as *partition chromatography*. *Paper chromatography* is a special case of partition chromatography; a paper strip is now the adsorbent column. When the moving phase is a mixture of gases, the usual method is to use a stationary solid phase (*gas-solid chromatography*, GSC) or a solid coated with a non-volatile liquid (*gas-liquid chromatography*, GLC; this was formerly called *vapour-phase chromatography*, VPC). Further-more, in GSC and GLC, a 'carrier' gas such as nitrogen replaces the solvent in column chromatography. In *ion-exchange chromatography*, the solid phase consists of a synthetic resin. Two types of exchange resins are used, cation and anion exchangers. In this way, mixtures of bases or mixtures of acids may be separated.

(vi) Zone melting. This is a most recent method of purification of organic compounds. A solid in a horizontal tube is slowly moved through a heating apparatus, or alternatively, the heater is moved with the tube fixed. In this way, a relatively narrow zone of solid is liquefied. When this zone is moved out of the heater, the liquid solidifies to a purer solid and the impurity is now in solution (in the heater portion). The basic principle is the redistribution of the components in a mixture during a phase change. A well-known example is freezing a salt solution, whereupon pure ice separates first.

There are various criteria for purity. The most common one for solids is m.p.; for liquids, b.p., density and refractive index are used; and more recently, the examination of the infra-red spectrum of a compound (whatever its normal physical state) has been used as a test for purity. In all cases, the process of purification is repeated until the physical constant or spectrum remains unchanged.

Before a natural product can be purified, it first has to be extracted from its source. Some of the earlier methods were solvent extraction, distillation and steam distillation. The methods used now are much superior, *e.g.*, counter-current separation (this has replaced solvent extraction), electrophoresis, dialysis, diffusion, molecular distillation, ultracentrifugation, etc. The most important method nowadays is chromatography in its various forms (see (v) above). Zone melting ((vi) above) is now also used for separating constituents in a mixture.

Qualitative Analysis

The elements commonly found in organic substances are: carbon (always: by definition), hydrogen, oxygen, nitrogen, halogens, sulphur, phosphorus and metals.

(i) **Carbon and hydrogen.** The compound is intimately mixed with dry cupric oxide and the mixture then heated in a tube. Carbon is oxidised to carbon dioxide (detected by means of calcium hydroxide solution), and hydrogen is oxidised to water (detected by condensation on the cooler parts of the tube).

(ii) **Nitrogen, halogens and sulphur.** These are all detected by the *Lassaigne method*. The compound is fused with metallic sodium, whereby nitrogen is converted into sodium cyanide, halogen into sodium halide and sulphur into sodium sulphide. The presence of these sodium salts is then detected by inorganic qualitative methods.

(iii) **Phosphorus.** The compound is heated with fusion mixture, whereby the phosphorus is converted into metallic phosphate, which is then detected by the usual inorganic tests.

(iv) **Metals.** When organic compounds containing metals are strongly heated, the organic part usually burns away, leaving behind a residue. This residue is usually the oxide of the metal, but in certain cases it may be the free metal, *e.g.*, silver, or the carbonate, *e.g.*, sodium carbonate.

As a rule, no attempt is made to carry out any test for oxygen: its presence is usually inferred from the chemical properties of the compound.

> The non-metallic elements which occur in *natural* organic compounds, in order of decreasing occurrence, are hydrogen, oxygen, nitrogen, sulphur, phosphorus, iodine, bromine and chlorine. Halogen compounds are essentially synthetic compounds, and are not found to any extent naturally. Some important exceptions are chloramphenicol (chlorine), Tyrian Purple (bromine) and thyroxine (iodine). In addition to these non-metallic elements, various metallic elements occur in combination with natural organic compounds, *e.g.*, sodium, potassium, calcium, iron, magnesium, copper.

Quantitative Analysis

The methods used in the determination of the composition by weight of an organic compound are based on simple principles.

(i) **Carbon and hydrogen** are estimated by burning a known weight of the substance in a current of dry oxygen, and weighing the carbon dioxide and water formed. If elements (non-metallic) other than carbon, hydrogen and oxygen are present, special precautions must be taken to prevent their interfering with the estimation of the carbon and hydrogen.

(ii) **Nitrogen** may be estimated in several ways, but only two are commonly used.

(*a*) *Dumas' method.* This consists in oxidising the compound with copper oxide, and measuring the volume of nitrogen formed. This method is applicable to all organic compounds containing nitrogen.

(*b*) *Kjeldahl's method.* This depends on the fact that when organic compounds containing nitrogen are heated with concentrated sulphuric acid, the organic nitrogen is converted into ammonium sulphate. This method however, has certain limitations.

(iii) **Oxygen** is usually estimated by difference. There are, however, some semi-micro methods, *e.g.*, the organic compound is subjected to pyrolysis in a stream of nitrogen, and all the oxygen in the pyrolysis products is converted into carbon monoxide by passage over carbon heated at 1 120°C. The carbon monoxide is then passed over iodine pentoxide, and the iodine liberated is estimated titrimetrically.

(iv) **Chlorine, bromine, iodine, sulphur** and **phosphorus** are usually estimated by the 'oxygen flask' method of combustion. A sample of the compound is wrapped in ashless filter paper and ignited electrically in a flask of oxygen containing the appropriate absorption liquid for the halogen or sulphur oxides produced. Suitable titration (or gravimetric determination) gives the amounts of these elements present. For example, sulphur oxides are absorbed in hydrogen peroxide and the sulphuric acid produced is titrated with standard alkali.

Fluorine is estimated by fusing with sodium or potassium in a nickel bomb, and the alkali fluoride is then determined gravimetrically or by titration.

Quantitative analysis falls into three groups according to the amount of material used for the estimation:

(i) **Macro-methods** which require about 0·1–0·5 g of material (actual amount depends on the element being estimated).

(ii) **Semi-micro methods** which require 20–50 mg of material.

(iii) **Micro-methods** which require 3–5 mg of material.

The micro-methods are almost always used now. Furthermore, with the increasing use of superior methods of separation, there is a trend to use still smaller amounts, the range being from 500 to 30 μg, and there is an increasing tendency to use instrumental techniques.

Empirical formula determination. The empirical formula indicates the *relative numbers* of each kind of atom in the molecule, and is calculated from the percentage composition of the compound. The following example illustrates the procedure.

0·202 g of a substance gave on combustion 0·361 g of CO_2 and 0·147 g of H_2O. What is the empirical formula of the substance?

$$\text{Wt. of C in sample} = 12/44 \times 0·361 = 0·0985 \text{ g}$$
$$\text{Wt. of H in sample} = 2/18 \times 0·147 = 0·0163 \text{ g}$$
$$\text{Wt. of 0} = 0·202 - (0·0985 + 0·0163) = 0·0872 \text{ g}$$

(i) *Percentage composition*

$$\% \text{ C} = \frac{0·0985}{0·202} \times 100 = 48·76$$

$$\% \text{ H} = \frac{0 \cdot 0163}{0 \cdot 202} \times 100 = 8 \cdot 07$$

In practice, the percentages of C and H are evaluated and the percentage of O is obtained by subtraction of their sum from 100, *i.e.*,

$$\% \text{ O} = 100 - (48 \cdot 76 + 8 \cdot 07) = 43 \cdot 17$$

(ii) *Empirical formula*

(*a*) If given the *weights* of sample and combustion products, divide the weights of C, H and O (by difference) by their respective atomic weights. Thus the ratio of atoms is:

$$\text{C} : \text{H} : \text{O} = \frac{0 \cdot 0985}{12} : \frac{0 \cdot 0163}{1} : \frac{0 \cdot 0872}{16}$$

$$= 0 \cdot 0082 : 0 \cdot 0163 : 0 \cdot 00545$$

To obtain this ratio as whole numbers, divide throughout by the smallest number.

$$\text{C} : \text{H} : \text{O} = 1 \cdot 50 : 2 \cdot 99 : 1$$

(Had these numbers been integers or very near to integers, *e.g.*, $1 \cdot 02 : 3 \cdot 04 : 1$, then the empirical formula would have been CH_3O.)

The procedure is to multiply the ratio by small numbers until a ratio of nearly whole numbers is obtained. Thus, multiplying by 2 gives the ratio $3 \cdot 00 : 5 \cdot 98 : 2$. The empirical formula is therefore $C_3H_6O_2$.

(*b*) If given the *percentage composition*, the procedure is similar to the above, *i.e.*

$$\text{C} : \text{H} : \text{O} = \frac{48 \cdot 76}{12} : \frac{8 \cdot 07}{1} : \frac{43 \cdot 17}{16}$$

$$= 4 \cdot 06 : 8 \cdot 07 : 2 \cdot 7$$

$$= 1 \cdot 50 : 2 \cdot 99 : 1$$

Multiplying by 2 gives $3 \cdot 00 : 5 \cdot 98 : 2$, and so the empirical formula is $C_3H_6O_2$.

It should be noted that the final ratio is the same as before. This is to be expected since, to obtain the percentage composition, each weight was multiplied by the *same constant*, $100/0 \cdot 202$.

Molecular weight determination. The molecular formula—this gives the *actual* number of atoms of each kind in the molecule—is obtained by multiplying the empirical formula by some whole number which is obtained from consideration of the molecular weight of the compound. In many cases this whole number is *one*.

The methods used for the determination of molecular weights are mainly physical. The standard physical methods are the determination of: (i) vapour density; (ii) elevation of boiling point; (iii) depression of freezing point. These methods are described fully in text-books of physical chemistry. In addition to these standard methods, which are used mainly for relatively simple molecules, there are other physical methods used for compounds having high molecular weights, *e.g.*, rate of diffusion, rate of sedimentation, viscosity of the solution, osmotic pressure. X-ray analysis and mass spectrometry are also used for molecular weight determination (see later).

Determination of structure, *i.e.*, the manner in which the atoms are arranged in the molecule. The usual procedure for elucidating the structure of an unknown compound is to make a detailed study of its chemical reactions. This procedure is known as the *analytical method*, and includes breaking down (*degrading*) the compound into smaller molecules of *known* structure.

When sufficient evidence has been accumulated, a tentative structure which best fits the facts is accepted. Sometimes two (or even more) structures fit the facts almost equally well, and it has been shown in certain cases that the compound exists in both forms which are in equilibrium. This phenomenon is known as *tautomerism*. Where tautomerism has not been shown to be present, one must accept (with reserve) the structure that has been chosen.

Until recently, structural determinations were based almost completely on purely chemical evidence. Nowadays, physical methods are considered as necessary tools for elucidating structures. They are used on the compound itself, and may also be used in the examination of the fragments obtained by degradative work. Of all the physical methods used, infra-red spectroscopy is probably the most widely applicable.

Spectroscopic methods. For light absorption spectra, the following equation is the starting-point for pure substances:

$$\log(I_0/I) = A = \varepsilon c l$$

I_0 is the intensity of the incident radiation and I the intensity of the transmitted radiation; A is the *absorbance* (*optical density*) and ε the *molar absorptivity* (*molar extinction coefficient*); c is the concentration of the solute (moles/litre) and l the length (in centimetres) of the medium traversed. Thus the units for ε are litre per mole per cm ($1 \text{ mol}^{-1} \text{ cm}^{-1}$), and if ε (sometimes $\log \varepsilon$; or the per cent absorption, *i.e.*, $100(1 - I/I_0)$) is plotted as ordinate against wavelength (or frequency) as abscissa, the *absorption curve* or *absorption spectrum* of the compound is obtained, and this is characteristic of a *pure* compound. On the other hand, the per cent trans-mission ($100 \, I/I_0$; the ratio I/I_0 is called the *transmittance*, T), or the absorbance, A, may be plotted against wavelength. Of particular importance are the values of the absorption maxima and their intensities. These are reported for ultra-violet and visible absorption spectra as, *e.g.*, λ_{max} (EtOH) 280 nm ($\varepsilon 4\,000$), which means that the ethanolic solution of the substance has a maximum absorption of $4\,000$ at a wavelength of 280 nm. Infra-red absorption spectra are reported as, *e.g.*, ν_{max} (CS$_2$) $1\,030 \text{ cm}^{-1}$ (s), which means that the substance, in carbon disulphide solution, has a strong absorption maximum at $1\,030 \text{ cm}^{-1}$. In recording infra-red spectra, it is customary to use the following letters to indicate intensity: m (medium), s (strong), w (weak), v (variable), vs (very strong).

The International System of Units (SI) has been used in this book, but a few recommended and non-SI units are also used, *e.g.*, the wave number is given in reciprocal centimetres (cm^{-1}) and not in reciprocal metres (m^{-1}), nor in reciprocal millimetres (mm^{-1}). In general, the SI unit, where used in the text for the first time, is followed (in brackets) by the previously accepted unit.

Table 1.1   Base-units

Physical quantity	Symbol	Name of base-unit	Symbol for unit
length	l	metre	m
mass	m	kilogramme	kg
time	t	second	s
thermodynamic temperature	T	kelvin	K
electric current	I	ampere	A
amount of substance	n	mole	mol

Table 1.2   Prefixes for SI units

Fraction	Prefix	Symbol	Multiple	Prefix	Symbol
10^{-1}	deci	d	10	deka	da
10^{-2}	centi	c	10^2	hecto	h
10^{-3}	milli	m	10^3	kilo	k
10^{-6}	micro	μ	10^6	mega	M
10^{-9}	nano	n	10^9	giga	G
10^{-12}	pico	p	10^{12}	tera	T
10^{-15}	femto	f			
10^{-18}	atto	a			

The symbol of the prefix is combined with the unit symbol, but the mass unit is an exception, in this case the prefix being attached to g and not to kg, *e.g.*, mg, *not* μkg (for 10^{-6} kg); Mg, *not* kkg (for 10^3 kg).

Table 1.3 Derived SI units

Quantity	Symbol	SI unit
activation energy	$E, E\ddagger$	$J\ mol^{-1}$
concentration	c	$mol\ m^{-3}$
density	ρ	$kg\ m^{-3}$
energy	E	J [joule]
enthalpy	H	J
entropy	S	JK^{-1}
force	F	N [newton]
frequency	v, f	Hz [hertz]
gas constant	R	$JK^{-1}\ mol^{-1}$
Gibbs function	G	J
kinetic energy	E_k, T, K	J
molar refraction	R_m	
potential energy	E_p, V, Φ	J
pressure	p, P	$N\ m^{-2}$
quantity of heat	q	J
quantum yield	Φ	
refractive index	n	
specific optical rotatory power	α_m	
thermodynamic energy	U	J
transmittance	T	
wavelength	λ	m
wave number	σ, v	m^{-1}
work	w, W	J

Table 1.4 Non-SI units still in use

Name of unit	Symbol	Definition
ångstrom	Å	10^{-8} cm; 10^{-10} m
atmosphere	atm	mm Hg
calorie:		
(i) international	cal_{IT}	4·1868J
(ii) thermochemical	cal	4·184J*
dyne	dyn	10^{-5}N
erg	erg	10^{-7}J
gauss	G	10^{-4}T [tesla]
litre	l	$10^{-3}\ m^3 = dm^3$
micron	μ	$10^{-6}\ m = \mu m$
millimicron	$m\mu$	$10^{-9}\ m = nm$

*This is the conversion factor used in the text.

Ultra-violet (200–400 nm) and **visible spectroscopy** (400–750 nm). A few functional groups may be detected, but this part of the spectrum is particularly useful for detecting the presence and nature of unsaturation, especially conjugated systems (see Ch. 31).

Infra-red spectroscopy (4 000–650 cm^{-1}). The study of infra-red spectra leads to a great deal of information, *e.g.*, the presence of various functional groups, hydrogen bonding (intra-molecular and intermolecular), the identification of *cis* and *trans* isomers, conformational orientations, orientation in aromatic compounds, etc.

The essential requirement for a substance to absorb in the infra-red region is that vibrations in the molecule must give rise to an unsymmetrical charge distribution. Thus it is not necessary for the molecule to possess a *permanent* dipole moment.

Just as electronic transitions are quantised (see Ch. 31), so are rotational and vibrational energy levels also quantised. Absorption in the near infra-red is due to changes in vibrational energy levels. A non-linear molecule can undergo a number of vibrational motions, the two main types being *stretching* (vibration along the bonds) and *deformation* (*bending*; displacements perpendicular to bonds). Figure 1.1 illustrates possible modes for a non-linear molecule (asym. = asymmetrical; def. = deformation; str. = stretching; sym. = symmetrical; and the plus and minus signs represent relative movement perpendicular to the page).

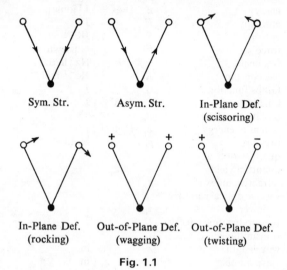

Sym. Str.	Asym. Str.	In-Plane Def. (scissoring)
In-Plane Def. (rocking)	Out-of-Plane Def. (wagging)	Out-of-Plane Def. (twisting)

Fig. 1.1

The stretching regions have higher frequencies (shorter wavelengths) than the deformation regions, and the intensities of the former are much greater than those of the latter. Although the masses of the bonded atoms predominantly influence the frequency of the absorption, other effects, *e.g.*, environment (*i.e.*, the nature of neighbouring atoms), steric effects, etc., also play a part. Thus, in general, a particular group will not have a fixed maximum absorption wavelength, but will have a *region* of absorption, the actual maximum in this region depending on the rest of the molecule. The spectrum also depends on the physical state of the compound: gas, liquid (as a thin film), solid (as a thin film or as a mull) or solution (preferably dilute; CCl_4, $CHCl_3$, CS_2).

The absorption regions of functional groups have been obtained empirically; many of these regions are described in the text (see also the Index under Infra-red Spectra). Most of the values have been taken from Cross *et al.* (see reading reference).

In the initial examination of the spectrum, the usual practice is to look for the presence of the various functional groups. In this way, it may be possible to assign the compound to some particular structural class (or classes). Knowledge of the molecular formula will often help to reject some of the alternatives, and chemical reactions of the compound will help further in this direction. Identification of a compound is carried out by comparison with published spectra (or with the spectrum of an authentic specimen). The region $1\,400{-}650$ cm^{-1} is known as the 'finger-print region'; this is the region usually checked for identification, since it is the region associated with vibrational (and rotational) energy changes of the *molecular skeleton*, and so is characteristic of the compound.

If a band has been found which corresponds to a particular group, the presence of this group should be confirmed by ascertaining the presence of another band which is also characteristic of the group, *e.g.*, saturated aliphatic esters show a strong band in the region $1\,750$–$1\,735\ \mathrm{cm}^{-1}$ (C=O str.) and another strong band in the region $1\,250$–$1\,170\ \mathrm{cm}^{-1}$ (C—O str.). Furthermore, the absence of a band which is characteristic of a particular group is not conclusive evidence that this group is not present in the molecule. One cause for this is that groups in a molecule may interact, and the result is that both regions are now different from the 'expected' individual regions. It is therefore always desirable to have chemical information about the compound and also spectroscopic data obtained from other methods (uv and NMR).

Various spectra (with wave numbers) have been given in the text. The reader will find it worth his while to make a list of infra-red absorption regions (described for different functional groups, etc.), and to examine the spectra with the legends covered. In this way, he will become familiar with the positions of some of the more important bands.

Raman spectroscopy. The information obtained is similar to that from infra-red spectroscopy.

Nuclear magnetic resonance (NMR) spectroscopy. In order to explain certain observations in rotational spectra, it was suggested that atomic nuclei spin about their axes. Now, since a rotating charged sphere has associated with it a magnetic moment, then all the charged particles in a nucleus will cause that nucleus to behave (to a first approximation) like a small bar magnet, with its magnetic moment along the axis of rotation. Nuclei are composed of protons and neutrons, the former carrying a unit positive charge, and the latter being electrically neutral (see Ch. 2). Magnetic properties occur with those nuclei which have (*a*) odd atomic and odd mass numbers, *e.g.*, $^1_1\mathrm{H}$, $^{15}_7\mathrm{N}$, $^{19}_9\mathrm{F}$, $^{31}_{15}\mathrm{P}$; (*b*) odd atomic number and even mass number, *e.g.*, $^2_1\mathrm{H}$ (D), $^{14}_7\mathrm{N}$; (*c*) even atomic number and odd mass number, *e.g.*, $^{13}_6\mathrm{C}$. Nuclei which have no magnetic moment are those with even atomic and even mass numbers, *e.g.*, $^{12}_6\mathrm{C}$, $^{16}_8\mathrm{O}$, $^{18}_8\mathrm{O}$, $^{32}_{16}\mathrm{S}$. It has been assumed that the particles in such nuclei are paired, *i.e.*, spinning in opposite directions, with the result that there is no resultant spin and consequently no magnetic moment (*cf.* covalent pairs of electrons, Ch. 2). In those nuclei where the magnitude of the spin is not zero, the *nuclear spin quantum number*, I, may assume any of the values $1/2$, 1, $3/2$, 2, etc. Nuclei possessing a resultant spin will thus behave as spinning magnets, and so will tend to orient themselves in an applied magnetic field, and the number of possible energy levels, *i.e.*, orientations with respect to the applied field, is given by $2I+1$. The simplest example of a spinning nucleus is that of the proton. Here, $I = 1/2$ (*cf.* the electron), and in this case there are only two orientations possible, lined up with or against the direction of the applied field. Since work must be done to turn a magnet against a magnetic field, each orientation corresponds to a different energy state of the nucleus. These levels are quantised (*cf.* ultra-violet and infra-red spectroscopy), and so it should be possible to find electromagnetic radiation of a definite frequency which will be absorbed, thereby changing the orientation of the proton from alignment with to against the field, the change being from a lower to a higher energy level. This electromagnetic radiation is supplied by an oscillator (with its *magnetic field* at right angles to the applied field), and since the position of the absorption peak, *i.e.*, where *resonance* occurs, depends on the frequency of the oscillator or the strength of the applied field (see below), it is possible to change from the lower to the higher energy level by using a variable frequency with a fixed applied magnetic field, or vice versa. In practice, it has been found easier to vary the field rather than the frequency. The result is that the NMR spectrum is usually a graph of signal intensity (ordinate) against magnetic field (abscissa); expressed in milligauss at a fixed frequency. For a given field, the strength of the signal depends on the magnetic moment of the nucleus, and since the proton has one of the largest moments, **proton magnetic resonance (PMR)** is of special importance. Other nuclei used in NMR spectroscopic studies in organic chemistry are $^{13}\mathrm{C}$, $^{19}\mathrm{F}$, $^{29}\mathrm{Si}$, and $^{31}\mathrm{P}$, but to study these nuclei it is necessary to modify the spectrometer.

The difference between the two energy levels, ΔE, for a proton is given by expression (a), where h is Planck's constant, H the strength of

$$(a)\ \Delta E = h\gamma H/2\pi \qquad\qquad (b)\ v = \gamma H/2\pi$$

the field *experienced* by the proton, and γ is the *magnetogyric (gyromagnetic)* ratio for the proton. The frequency of the radiation (v) which induces the transition is given by expression (b), since $\Delta E = hv$ (see p. 869). Thus, the position of the energy absorption is a function of both the frequency of the oscillator and the strength of the applied field. When radiation is absorbed, the proton changes from the lower to the higher energy state.

For an applied field of 9 400 gauss, the resonance frequency of a proton is about 40 MHz (megahertz = megacycles per second). The energy associated with this frequency is about $0.0167\,\mathrm{J\,mol^{-1}}$ (0.004 cal). Thus, ΔE is very small, and it is because of this that radiofrequency radiation can effect these transitions. Since ΔE is very small, the population in the lower state is only *slightly* greater than that in the higher state. This is the situation when the compound is placed in a magnetic field, and so there is net absorption when the radiofrequency is applied, and it is this absorption which is measured.

In order that a PMR signal be observed, the proton must be in a single state for 10^{-2} to 10^{-1} second. It has also been shown that the spectral line width is inversely proportional the average time the proton occupies the *higher* energy state. Hence, the longer the time spent in this state, the sharper is the line; and conversely, the shorter the time, the broader is the line.

From what has been said above, it might be expected that the resonance frequency for a given field depends only on the nature of the atomic nucleus concerned. This, however, is not the case. The applied field causes electrons round a nucleus to circulate in a plane perpendicular to the field, and these currents produce a field in opposition to the applied field. Thus, the effective magnetic field (H) experienced by the nucleus is smaller than the applied field (H_0), the relationship between the two being given by the expression

$$H = H_0(1-\sigma)$$

σ (which is non-dimensional) is called the shielding or screening constant, and has a positive value, but in certain circumstances it may be negative, *i.e.*, the effective field is larger than the applied field. In this case, the proton is said to be deshielded. Since the numerical value of σ depends on the *chemical environment* of a given nucleus, the shielding or deshielding of a nucleus varies with its environment. Shielding causes a shift of the resonance frequency to higher values of the applied field, *i.e.*, the shift is *upfield*. On the other hand, deshielding causes a shift of the resonance frequency to lower values of the applied field, *i.e.*, the shift is *downfield*. The magnitude of this shift is known as the *chemical shift*. Since the value of the field experienced by the sample cannot be determined accurately, chemical shifts are measured *relative* to some standard which contains the nucleus under consideration. Various reference compounds (which are usually added to the sample) for PMR have been used, but tetramethylsilane (TMS), $(CH_3)_4Si$, is particularly useful since it contains twelve *equivalent* protons. The PMR spectrum of this compound shows a single sharp line which occurs at a higher field than any protons in most of the common organic compounds, *i.e.*, most PMR signals occur downfield with respect to TMS (see also below).

The chemical shift may be reported in various ways. Since the resonance frequency is dependent on the strength of the applied field, the shift may be reported as field units (milligauss). However, because the field can be expressed in terms of frequency (see expression (b) above), the shift may also be expressed in terms of Hz (c.p.s.). This separation in Hz is also proportional to the frequency of the oscillator, *e.g.*, if the separation between a proton signal and TMS is 60 Hz at 40 MHz, the separation at 60 MHz becomes 90 Hz $[60 \times 60/40 = 90]$. Hence it is desirable to be able to report chemical shifts in units which are independent of the operating conditions of the spectrometer. This has been done by defining the chemical shift, δ, by the expression

$$\delta = \frac{\text{separation in Hz}}{\text{oscillator frequency}} \times 10^6$$

The factor 10^6 is introduced in order to record the chemical shift as a convenient value. This is usually in the range 1–10, and is quoted in parts per million (p.p.m.).

The independence of the chemical shift of the oscillator frequency is shown by the following example. A separation of 40 Hz at an oscillator frequency of 40 MHz becomes 90 Hz at an oscillator frequency of 60 MHz (see above). However, δ remains unchanged: $\delta = (10^6 \times 60)/(40 \times 10^6) = 1.5$ p.p.m.; $(10^6 \times 90)/(60 \times 10^6) = 1.5$ p.p.m.

It is now becoming common practice to express chemical shifts in τ (tau)-values, defined by the expression

$$\tau = 10 - \delta$$

where 10 p.p.m. is the value assigned to the line of TMS. Most protons have positive τ-values (*i.e.*, $\delta < 10$); strongly acidic protons, however, have negative τ-values (*i.e.*, $\delta > 10$). The greater the shielding of the nucleus, the larger is its τ-value (the smaller is δ). Since the degree of shielding depends on the electron density round the proton, any structural feature that decreases this density will cause a decrease in shielding, with consequent lowering of the τ-value (the chemical shift moves downfield). Halogens are electron-attracting (electronegative groups; p. 32), and when joined to a methyl group, the electron density round each proton is decreased. Consequently, the presence of halogens weakens the induced opposing field, *i.e.*, deshields these protons and so the τ-value will be expected to be lowered (the chemical shift is moved downfield). This prediction is observed in practice. The order of electronegativity of the halogens is $I < Br < Cl < F$, and the τ-values of the methyl protons are: CH_3I, 7·83; CH_3Br, 7·35; CH_3Cl, 6·98; CH_3F, 5·70 p.p.m.

Similarly, since the order of electronegativity of carbon, nitrogen, and oxygen is $C < N < O$, the τ-values of the methyl protons in CH_3—C, CH_3—N, CH_3—O are, respectively, 9·12, 7·85, 6·70 p.p.m. Silicon is less electronegative than carbon, and so methyl protons in TMS are more shielded than those in a methyl group attached to carbon, *i.e.*, the protons in TMS absorb upfield with respect to protons in most of the common organic compounds (see Table 3.2).

Figure 1.2 summarises the terminology described in the foregoing account of NMR spectroscopy.

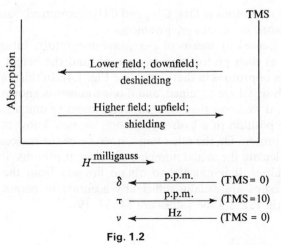

Fig. 1.2

Measurements of NMR spectra are normally carried out on liquids or solutions (5–20 per cent concentration). The best solvents are those which do not contain protons, *e.g.*, deuterio-chloroform and other deuterated compounds, carbon tetrachloride, etc. However, because of solubility problems, solvents containing protons may also be used, *e.g.*, chloroform. A difficulty with respect to solutions is that τ-values may change with the nature of the solvent, particularly aromatic solvents. Chemical shifts may also change with concentration in a given solvent. Protons attached to carbon are very little affected, but when attached to atoms such as O, N, S, the chemical shift is very much affected. In the latter case, the changes are due to changes in the degree of hydrogen-bonding, which causes a downfield shift relative to the unbonded state (see also later).

The study of NMR spectra of liquids and solutions is known as **high resolution NMR**. The study of solids, since they give spectra which consist of broad resonance lines, is referred to as **broad line resonance**.

Figure 1.3 is the NMR spectrum of ethanol, (liquid; 60 MHz), carried out at low resolution. The position of each peak is characteristic of the environment of a particular proton, and the areas under the curves of the peaks have been shown to be in the ratio $1:2:3$. This ratio

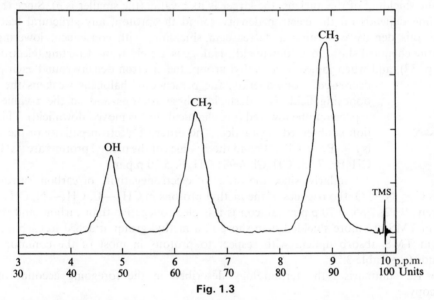

Fig. 1.3

corresponds to the number of protons in OH, CH_2 and CH_3, respectively, and so it is therefore possible to 'count' the protons in various environments.

These areas are now evaluated by means of electronic integrators. Integration produces a trace which rises in steps as each proton signal is passed, and the height (between steps) is proportional to the number of protons in that signal (see Fig. 1.4). In this way, the ratios of the numbers of protons in each signal are obtained, and if one number is known, all the others can be estimated. For example, if we know that the compound under examination is a monohydric alcohol then, knowing the position of a hydroxylic atom, we now know that the area of this signal is equivalent to one proton. On the other hand, if we know the molecular formula of the compound, we can also calculate the actual numbers of different protons. In practice, because of the experimental difficulties, it is unusual to obtain integers from the integration trace. However, the numerical results are usually sufficiently accurate to permit 'counting' of the different types of protons (in Fig. 1.4 the values are $54:37:19$).

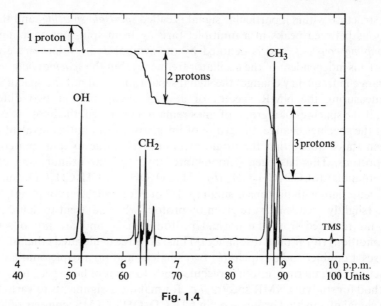

4   5   6   7   8   9   10 p.p.m.
40   50   60   70   80   90   100 Units

Fig. 1.4

Spin-spin coupling. Figure 1.4 is the high-resolution NMR spectrum of ethanol (20 per cent in CCl_4; 60 MHz), containing a trace of hydrochloric acid. Instead of the broad bands shown in Fig. 1.3, two have been split into multiplets. The *total* areas under the multiplets are still in the same ratio as before, *i.e.*, $2:3$. This fine structure has been explained as being due to shielding of protons by *protons on adjacent carbon atoms*.

First let us consider the influence of the methylene group, CH_2, on the methyl group, CH_3. Each proton in the methylene group may have its magnetic moment lined up with or against the applied field. If we represent the former alignment by an arrow pointing upwards and the latter alignment by an arrow pointing downwards, then there are the following four possible combinations:

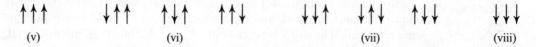

Thus the shielding effect will depend on the type of combination operating. However, combinations (ii) and (iii) will have the *same* shielding effect on the adjacent methyl group. Hence there are *three different* shielding combinations, and statistically it can be expected that at any given moment, 25 per cent of the methylene protons will be in combination (i), 25 per cent in (iv) and 50 per cent in (ii) and (iii). The net result is that the methyl proton signal is split into a triplet, the ratio of the areas being $1:2:1$.

If we apply the same argument to the effect of the protons of the methyl group on the methylene protons, then there are eight possible combinations:

↑↑↑   ↓↑↑   ↑↓↑   ↑↑↓   ↓↓↑   ↓↑↓   ↑↓↓   ↓↓↓
(v)   (vi)   (vii)   (viii)

As before, each combination in (vi) gives rise to the *same* shielding effect; this is also the case for (vii). The net result is that there are four different shielding combinations, resulting in the splitting of the methylene signal into a quartet, the ratio of the areas being $1:3:3:1$.

This fine structure within a particular signal is called *spin-spin splitting*, and the magnitude of the separations between peaks in a multiplet (arising from spin-spin couplings) is called the *spin-spin coupling constant*, and is denoted by the symbol J; its values are given in Hz. The magnitude of J is independent of the oscillator frequency, but the *spacings between signals* is not. Hence, a change in frequency changes the signal separations, but not the spacings in a multiplet. Thus, by measuring the NMR spectra of a given compound at two different oscillator frequencies, if the spacings of a group of lines remain unchanged, the lines are components of a multiplet. If the spacings change, the group of lines arises from non-equivalent protons.

It has been stated above that the origin of multiplets is due to spin-spin coupling between groups of protons. This has been demonstrated by, *e.g.*, the examination of the following deuterated ethanols: (*a*) CD_3CH_2OH, (*b*) CH_3CD_2OH, (*c*) CH_3CH_2OD. Since the coupling constant of deuterium with protons is small ($\sim 1/7$ of J for hydrogen coupling), the result is that single peaks (slightly broadened) are given by protons when adjacent to deuterium. Deuterium signals are far removed from the normal proton signals, and are not observed under the operating conditions (for proton resonance). Thus, all three deuterated ethanols give two signals only: (*a*) gives a doublet and a triplet, (*b*) two singlets, and (*c*) a triplet and a quartet. Because the introduction of deuterium into a molecule leads to a simplified spectrum, this is used as a general method for studying NMR spectra, *i.e.*, for making assignments to various spectral lines.

Since the signal of a methyl group, *e.g.*, in CH_3CD_2OH or TMS, consists of a single line, this means that there is apparently no magnetic coupling between the hydrogen atoms attached to the same carbon atom (*geminal* hydrogens). The three hydrogen atoms are equivalent, all three having the same chemical shift. This leads to the general rule that protons with the same chemical shift do not give rise to observable splitting (see also below). Only when protons are non-equivalent is splitting possible.

There are two types of equivalent protons, *chemically* equivalent and *magnetically* equivalent. *Chemically equivalent protons* are those which occupy chemically equivalent positions, *i.e.*, are in identical chemical environments. Such protons have the *same* chemical shift. A simple test for chemical equivalence of two (or more) protons is to replace each proton one at a time by substituent Z, and if by doing so, the *same* compound (or its mirror image) is obtained, then these protons are chemically equivalent. Let us consider ethane, CH_3CH_3, as an example. Replacement of each proton, one at a time, on one carbon atom gives CH_2ZCH_3, which is also obtained when the other carbon atom is treated in the same way (CH_3CH_2Z). Thus, the three hydrogens on *each* carbon atom are chemically equivalent, and *all six* hydrogens are also chemically equivalent. Hence, all have the same chemical shift, and consequently there is no splitting *within* a methyl group and none due to coupling *between* the groups. The observed spectrum of ethane consists of one signal which is a singlet.

A group of two (or more) protons are said to be *magnetically equivalent* when not only do they have the *same* chemical shift, *i.e.*, are chemically equivalent, but also the set of coupling constants to all other protons is identical for each member of the group. This may be illustrated with 1-bromo-2-chloroethane, which is represented by the Newman formula shown, (I) (see also p. 79). If we assume, for the moment, that this conformation is fixed, it can be seen that H_a and H_b are chemically equivalent, as also are H_c and H_d (for each pair, the chemical environments are identical). The coupling constants, J_{ad} and J_{bd}, however, are not equal, and hence H_a and H_b are *not* magnetically equivalent.

To understand non-magnetic equivalence, we must now consider the problem of the magnitude of the coupling constant. It has been found that the value of J decreases very rapidly with increase in the number of bonds connecting the interacting protons. Only two cases are important, the methylene type (two bonds), H—C—H (geminal hydrogens) and the vicinal type (three bonds), H—C—C—H (adjacent carbon atoms). Geminal hydrogens, *e.g.*, H_aH_b or H_cH_d, couple with splitting only if their environments are different (see also p. 81). If coupling occurs through four or more bonds, the coupling is referred to as *long range coupling*. This is usually negligible for *saturated* compounds, but may be large in *unsaturated* systems (see, *e.g.*, p. 122).

In addition to depending on the number of intervening bonds, J is also *angular dependent*, *i.e.*, its magnitude varies with the angle between the methylene protons (when these are not chemically equivalent), and the *dihedral angle* for vicinal hydrogens. In practice, rotation about a C—C single bond takes place quite readily (see p. 78). If we take the eclipsed conformation as the starting point, *i.e.*, in (I), the CH_2Br group is kept stationary and the CH_2Cl group is rotated until Cl and Br are in line, and consequently H_aH_c and H_bH_d are also in line (as pairs). In this *eclipsed* conformation, the dihedral angle is $0°$, and then keeping the CH_2Br group stationary, the other group can rotate through $360°$ before the molecule reaches its starting position. It has been found that the value of J, for a given pair of vicinal hydrogens, depends on the **dihedral angle (angle of rotation)**, being largest when the angle is $0°$ or $180°$, and very small (or zero) when the angle is $90°$. Thus, since the dihedral angle for H_aH_d in (I) is $60°$ and that for H_bH_d is $180°$, J for the latter will be larger than that for the former. Hence, these coupling constants for H_a and H_b with H_d are different, and consequently H_a and H_b are not magnetically equivalent (see also p. 81 for a further discussion).

> One theory for the transmission of coupling is as follows. A given proton, because of its magnetic moment (due to spin), affects the spins of the electrons forming the covalent C—H bond. This change affects the spins of the C—C electron pair, and this in turn affects the spins of the electron pair in the adjacent C—H bond. Thus, the spin effects of one proton are transmitted through the covalent bonds to the other spinning proton. Since these effects depend only on the structure and geometry of the molecule and are not due to the presence of the applied field, they are independent of the strength of the applied field.

Rules for determining multiplicity. Some simple rules for determining the multiplicity of a signal have been developed, but these usually apply only when $(\Delta v/J) \geqslant 6$ (where Δv in Hz is the separation of the signals of the interacting groups). Spectra of this type are said to be **first-order spectra**.

(i) Equivalent protons, when coupled, do not cause splitting.

(ii) The spacings (*i.e.*, J values) in a multiplet are equal and are also equal to the spacings of a multiplet arising from mutual coupling, *e.g.*, in the NMR spectrum of ethanol, the spacings in the methyl triplet *and* the methylene quartet are all equal.

(iii) The multiplicity of a group of equivalent protons is equal to $(n+1)$, where n is the number of equivalent protons in the group which are coupled to the first group, *e.g.*, in —O—CH_2—CH_3, the two equivalent methylene protons are coupled to *three* equivalent protons in the methyl group. Hence, $n = 3$, and so there will be *four* lines shown by the two methylene protons. Similarly, the three methyl protons give rise to *three* lines due to coupling with the *two* ($n = 2$) methylene protons. Furthermore, the relative intensities of the individual lines of a multiplet correspond (ideally) to the numerical coefficients of the terms in the binomial expansion $(1+x)^n$. Thus, the lines of the quartet ($n = 3$) have relative intensities $1:3:3:1$, and those of the triplet ($n = 2$), $1:2:1$. Also, the relative intensities of a multiplet are symmetrical (ideally) about the mid-point of the multiplet.

(iv) When a group of equivalent protons is coupled to groups of equivalent protons, n_a in one group, n_b in another, etc., the multiplicity of that group is equal to $(n_a+1)(n_b+1)$ . . . . To illustrate this rule, let us consider the NMR spectrum of *very pure* ethanol (see also below). The methyl group is coupled to the methylene group ($n = 2$), and so the methyl signal will be a triplet. The hydroxyl proton is also coupled to the methylene group, and its signal is therefore a triplet. The methylene group is coupled to *both* the methyl group ($n_a = 3$) and to the hydroxyl proton ($n_b = 1$). Hence, the multiplicity of the methylene signal is given by $(3+1)(1+1) = 8$, *i.e.*, an octet. This is the spectrum observed in practice (for very pure ethanol; see also below).

When dealing with first-order spectra, it is possible to construct them diagrammatically. Let us consider ethanol (plus a trace of acid) as our example. This spectrum is shown in Fig. 1.4; the diagrammatic spectrum is shown in Fig. 1.5. The signal of the methylene group is a quartet of relative intensities $1:3:3:1$, and so the *total* intensity of this signal $= 1+3+3+1 = 8$ units. Since these 8 units represent two protons, one unit is equivalent to 0·25 proton, and 3 units to 0·75 proton. Hence these four components of the quartet may be drawn as vertical lines, the lengths of which are proportional to the number of protons they represent. *e.g.*, using 5 mm to be equivalent to 0·25 proton (the weakest line of the multiplets in *all* signals), the quartet is then represented by four lines, 5 mm, 15 mm, 15 mm, 5 mm, equally spaced and centred about the τ-value of this signal. In the same way, the intensities of other multiplets in the spectrum may be estimated and represented as vertical lines whose lengths are also based on the arbitrarily chosen unit of 5 mm $= 0·25$ proton. For ethanol, the τ-values of the signals are known, but if the values are not known, it is usually possible to estimate them from tables of τ-values (see, *e.g.*, Table 3.2).

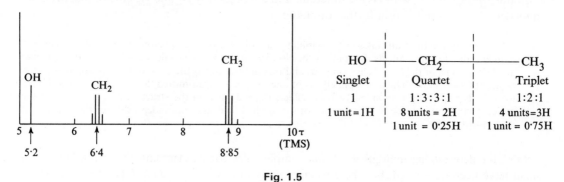

Fig. 1.5

The spectrum given in Fig. 1.5 is idealised. In practice, distortion of intensities may occur, inner lines increasing at the expense of outer lines (*cf.* Fig. 1.4). This becomes more pronounced as the ratio $\Delta v/J$ decreases. Since this distortion occurs between *coupled* signals, it may be used to show which groups of protons are coupled.

As we have seen, the simple rules of splitting are applicable only when $(\Delta v/J) \geqslant 6$. When $\Delta v \cong J$, the simple rules no longer hold good. The spectra are now complicated, and it is usually difficult to recognise the pattern of the lines. Mathematical techniques, however, have been developed for analysing complicated spectra. If the chemical shifts and coupling constants of all the magnetic nuclei present in the molecule are known, it is possible to predict the NMR spectrum of the compound. Alternatively, it may be possible to evaluate the chemical shifts and coupling constants from the observed spectrum.

One cause of the appearance of complicated spectra is that the components of a multiplet may overlap. For the purpose of analysing NMR spectra, a notation has been introduced for discussing spin-spin coupling. Protons which have similar chemical shifts are designated by the letters A, B, C, ..., and protons which have similar chemical shifts that are quite different from those of set A, B, C, ..., are designated by the letters ... X, Y, Z. When a proton has a chemical shift that lies between the above two sets and is well separated from both, it is designated by a middle letter of the alphabet, *e.g.*, M. The number of protons of the *same* type is then indicated by a subscript (an integer).

The earlier letters of the alphabet are usually chosen to represent those protons which absorb at lower fields, and the choice of letters depends on which protons are different, on how different they are, and whether they are coupled or not. One practice is to designate protons as A and X if $\Delta v/J \geqslant 6$, and if less than this, as A and B. The scheme described so far designates the same letter to protons with the same chemical shift, *i.e.*, to chemically *and* magnetically equivalent protons. One method of indicating that the protons in one group are *not* magnetically equivalent is to repeat the same letter with one of them primed, *e.g.*,

$$CH_3CH_2OH \qquad\qquad CH_3NHCHO \qquad\qquad ClCH_2CH_2Br$$
$$A_3M_2X \qquad\qquad\qquad A_3XY \qquad\qquad\quad AA'BB'$$

The term *spin system* is used to describe groups of nuclei that are spin-spin coupled among each other, but not with any nuclei outside the spin system. It is not necessary, however, for all the nuclei within the spin system to be coupled to *all* the other nuclei. In many cases, the spin system embraces the complete molecule, *e.g.*, $CH_3CH{=}CH_2$ is a six-spin system. On the other hand, a molecule may consist of two (or more) parts 'insulated' from each other, and thereby gives rise to two (or more) independent spin systems, *e.g.*, $CH_3CH_2{-}O{-}CH_2CH_2CH_3$. This contains two spin systems, the five-spin ethyl group and the seven-spin n-propyl group. The latter is an example of the case where not all nuclei in the spin system are coupled (coupling between OCH_2 and CH_3 is negligible).

Now let us analyse the spectrum of 1-bromo-3-chloropropane. The τ-values (at 40 MHz in $CDCl_3$) are: CH_2Br, 6·30; $-CH_2-$, 7·72; CH_2Cl, 6·45; J value for the first pair is $\sim$6·2 Hz, $\sim$6·2 Hz for the second pair, and zero for the first and third groups. Since $\delta(\text{p.p.m.}) = (\text{Hz/r.f.}) \times 10^6$, $\text{Hz} = \delta \times \text{r.f.} \times 10^{-6}$. Therefore, for the first pair, $\Delta v = 1\cdot42 \times (40 \times 10^6) \times 10^{-6} = 57$. Hence, $\Delta v/J = 57/6\cdot2 = \sim9\cdot2$. This value is greater than 6; and this also is the case for the second pair $(51/6\cdot2 = \sim8\cdot2)$. This spectrum is therefore first-order.

Analysis of this spectrum may now be carried out in a similar manner to that used for ethanol (p. 13), but here we shall deal with an alternative method, the graphical method. Each component of a multiplet may be represented by a line whose length is proportional to its intensity (*cf.* Fig. 1.5), but here we shall indicate relative intensities by numbers (see Fig. 1.6). The single lines in (i) represent each group

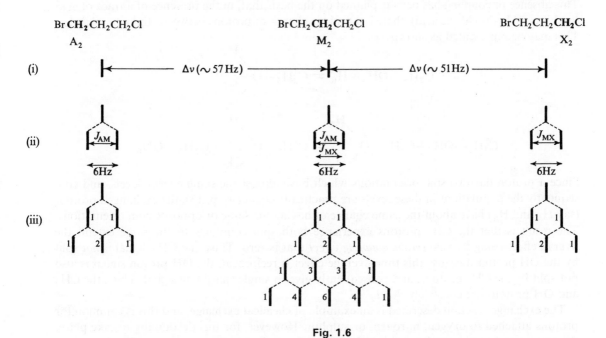

Fig. 1.6

of equivalent protons, and their spacings are proportional to the differences in chemical shifts. If there were *no* coupling, then each methylene group would have given a single line signal, all three signals being of equal intensity. Since, however, A_2 and M_2 are coupled, as are also M_2 and X_2, splitting *does* occur. Let us first consider the signal given by the A_2 group of protons, always bearing in mind the *two* possible orientations of a coupled proton. In the absence of coupling, the signal is the single line shown in (i). If we now consider the coupling with *one* proton in the M_2 group, this produces *two* lines of equal intensity, and they are separated by J_{AM} (Fig. 6ii). However, since A_2 is also coupled with the *other* proton in M_2, each component of the doublet is split into a *second* doublet, each pair being separated by J_{AM}. Hence, the 'inner' line of one doublet overlaps with that of the other doublet, thereby producing a triplet with equally spaced lines and with relative intensities 1:2:1 (Fig. 1.6iii). In the same way, it can be shown that couplings of the protons X_2 with M_2 also produce a triplet. In both cases, using the simple rules of multiplicity, with $n = 2$, the number of lines in the multiplet is $3 (= n+1)$, and their relative intensities are 1:2:1 (the coefficients of the expansion $(1+x)^2$).

Now let us consider the signal of M_2. According to the simple rules, their signal would be expected to be a multiplet of *nine* lines $[(2+1)(2+1) = 9]$. However, by using the scheme outlined for the coupling between A_2 and M_2, the diagram for the signal of M_2 is a multiplet of *five* lines (Fig. 1.6iii). It should be emphasised that the pattern of this multiplet is the same whether the M protons are coupled first to the A group, followed by coupling with the X group, or vice versa.

Because all J values are (almost) equal, overlapping of 'inner' components occurs, and consequently the observed number of lines is smaller than that calculated from the simple rules. This situation is generally the case when the J values between different groups of coupled protons are all the same or very nearly the same. Also, in these cases, the total number of lines of the multiplet is given by $N = (n_A+n_X+1)$. In the example discussed, $n_A = n_X = 2$; hence $N = 5$. Furthermore, the relative intensities of the lines are given by the coefficients of the expansion $(1+x)^{(n_A+n_X)}$, and the lines are equally spaced.

It can be seen from Fig. 1.4 (p. 13) that the signal from the proton of the hydroxyl group is not split, *i.e.*, there is no evidence of spin-spin coupling between the OH proton and the CH_2 protons, since had there been coupling, the OH proton signal would be expected to be a triplet. This absence of coupling has been explained on the basis that, in the presence of a trace of acid (in this case, hydrochloric acid), there is a rapid exchange of protons between ethanol molecules. This may be represented as shown:

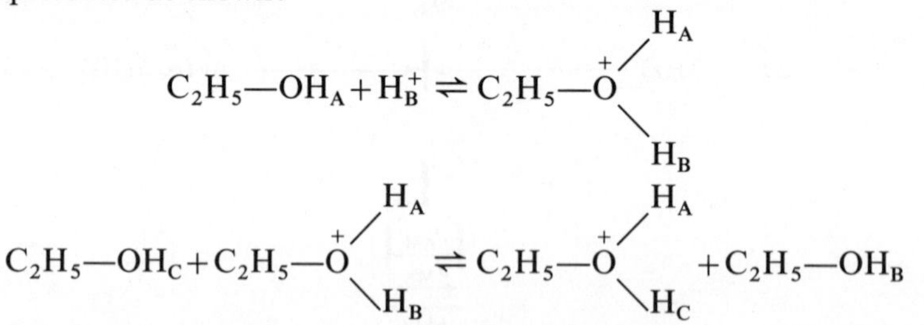

$$C_2H_5{-}OH_A + H_B^+ \rightleftharpoons C_2H_5{-}\overset{+}{O}\overset{\displaystyle H_A}{\underset{\displaystyle H_B}{\big<}}$$

$$C_2H_5{-}OH_C + C_2H_5{-}\overset{+}{O}\overset{\displaystyle H_A}{\underset{\displaystyle H_B}{\big<}} \rightleftharpoons C_2H_5{-}\overset{+}{O}\overset{\displaystyle H_A}{\underset{\displaystyle H_C}{\big<}} + C_2H_5{-}OH_B$$

Since a proton has two spin orientations which have almost the same energy levels, and consequently the populations in these levels are practically equal (see p. 13), the exchanged protons (say H_A and H_B) have about the same chance of having the same or opposite spin orientations. The result is that the CH_2 protons experience both spin couplings to the same extent, the overall effect being a *time-average coupling effect* that is zero. Thus, the CH_2 signal is *not* split by the OH proton. Because this time-average effect is reciprocal, the OH proton signal is also not split by the CH_2 protons, and consequently gives a single sharp line signal. Thus, the CH_2 and OH protons are effectively *decoupled*.

The exchange reaction described is an example of **chemical exchange,** and this is common for protons attached to oxygen, nitrogen, or sulphur. However, for this decoupling to take place, the rate of exchange must be rapid. Calculations have shown that if the rate of exchange per second is much greater than the separation (in Hz) of the two separate signals (had there been no exchange), then decoupling occurs. For pure ethanol the rate of exchange is very much slower than Δv (the separation of the two signals), and so coupling occurs, the CH_2 appearing as an octet and the OH as a doublet (see p. 16). Since the rate of chemical exchange is accelerated by a rise in temperature, *spin decoupling* may sometimes be observed by raising the temperature at which the NMR spectrum is measured. Similarly, *spin coupling* may be observed by lowering the temperature. It also might be noted here that complications arising from a proton attached to oxygen (or nitrogen) may be removed by deuterium exchange. This is readily carried out by shaking a solution of the compound in deuteriochloroform with a small amount of D_2O. Signals arising from OH (or NH) protons will no longer be observed.

From what has been said, it can be seen that in exchange reactions, the shape of a band depends on the rate of exchange. When the rate is very fast, the signal of the exchanged proton is a single sharp line, and as the rate becomes slower, a point is reached when this begins to broaden, and finally splits into the number of components required for coupling with the vicinal protons. Thus, analysis of the shape of the band affords a means of evaluating the rate of exchange.

As we have seen above, the simple rules for determining multiplicity usually apply only when $(\Delta v/J) \geqslant 6$. We have also seen that Δv (the spacing *between* the signals) is dependent on the oscillator frequency, whereas J is independent of it. Hence, by increasing the oscillator frequency, Δv can be increased, and if increased sufficiently the conditions for obtaining a first-order spectrum $(\Delta v/J) \geqslant 6$, may be fulfilled, thereby resulting in a simplified spectrum. Oscillator frequencies that have been used are 100 MHz, and better still, 220 MHz.

Double nuclear resonance. It has been pointed out above that spin-spin coupling may be annihilated by rapid exchange. This annihilation may also be effected as follows. Suppose we have two protons, A and B, in different environments and are coupled. Now suppose the resonance frequency of A is being measured. Then, at the same time as this is being done, a strong radiofrequency field whose frequency is the resonance frequency of B is also applied. This latter field produces both absorption and emission by B many times a second, and consequently coupling of B with A is now prevented, *i.e.*, B is decoupled with respect to A, and the time average of coupling between A and B is zero. Splitting of A by B has been prevented, but it is not possible under the conditions of the experiment to record the resonance of B. By applying double nuclear resonance to each type of proton, each signal can be made to collapse into a singlet line. In this way, the double resonance method produces a much simpler spectrum, and hence makes interpretation easier (*cf.* spectra of deuterated ethanol, above).

Uses of NMR spectroscopy. NMR absorption spectroscopy has found great use in the study of chemical structure, configurations and conformations of molecules, and the study of rates of chemical exchange. It has also been used in quantitative work, *e.g.*, the determination of enol content in keto-enol mixtures. Barcza (1963) has shown that NMR spectroscopy may be used to determine molecular weights with greater accuracy than the usual methods such as cryoscopic, ebullioscopic, etc.

Mass spectrometry. When a compound, in a high vacuum, is bombarded with electrons in a mass spectrometer, it is converted into positive ions by loss of an electron (positive ions may also be produced by other methods).

$$M + e \longrightarrow M^+ + 2e$$

The positive ion, M^+ (or P^+), is known as the **molecular ion** (or the **parent ion**), and is formed when the energy of the electrons is equal to that of the ionisation potential (usually 10–15 electronvolts). In practice, the energy of the electrons is 50–70 eV, and under these conditions the molecular ion is formed with an excess of energy, large enough for it to break down into a mixture of neutral and positively charged fragments, and the latter, if they have an excess of energy, also undergo further fragmentation, etc.

Most ions carry a unit positive charge, but some may carry a double (or greater) positive charge (and some ions may be negatively charged by electron *capture*). All *positive* ions are accelerated in an electric field and separated by their passage through an electric field and then a magnetic field. In this way, ions which have the same mass/charge (m/e) ratio are collected into beams, and fall on a collector plate. Thus, the ions are sorted out according to their mass/charge ratios, and so the masses of the ions can be determined. Since most ions have a unit positive charge, m/e is equivalent to m. To be of value, however, the instrument must be capable of at least separating, *i.e.*, *resolving*, adjacent beams of m/e and $(m+1)/e$. Mass spectrometers are now available which are capable of very high resolution, being able to differentiate between ions whose masses differ in the third decimal place.

If a positive ion is doubly-charged, it will behave, as far as collection in a beam is concerned, in the same way as a singly-charged ion of half the mass. There is also the problem of isotopes, *e.g.*, at low resolution, $^{12}C_2H_2$ and $^{12}C^{13}CH$ both have m/e of 26; but if these differ in the third decimal place, it is possible to differentiate between them by high resolution. Since *individual* ions are collected in the mass spectrometer, then the molecular ion will also give rise to a number of peaks, *e.g.*, bromine exists as ^{79}Br and ^{81}Br. Thus, the molecular ion of methyl bromide will appear (low resolution) at peaks of m/e 94 ($^{12}CH_3^{79}Br$), 95 ($^{13}CH_3^{79}Br$), 96 ($^{12}CH_3^{81}Br$), and 97 ($^{13}CH_3^{81}Br$). The relative intensities of these M^+, $(M+1)^+$, . . . , peaks depend on the abundance ratios of the isotopes in the molecule (see Table 1.5). It can therefore be seen that mass spectrometry affords a means of determining accurate isotopic abundance ratios, molecular weights and consequently molecular formulae (these are obtained from tables).

Table 1.5

Element	Mass	%	Element	Mass	%
H	1	99·985	F	19	100
D	2	0·015	S	32	95·0
C	12	98·89	S	33	0·74
C	13	1·11	S	34	4·24
N	14	99·63	Cl	35	75·4
N	15	0·37	Cl	37	24·6
O	16	99·76	Br	79	50·6
O	17	0·04	Br	81	49·4
O	18	0·20	I	127	100

The mass spectrum is a plot of ion beam intensities (ordinate) against mass/charge (m/e), and for any pure compound is characteristic of that compound, *i.e.*, a pure compound is characterised by its **cracking pattern** (and its molecular ion, if it has one). It should be noted that by cracking pattern is meant not only the fragmentation pattern, but also that the relative abundance of the peaks are fixed ratios. Since mass spectra are dependent on the conditions of the experiment, these spectra are reproducible only if the conditions are identical. Hence it is best to use the same instrument and the same conditions for the purpose of comparisons.

The largest peak in a mass spectrum corresponds to the most abundant ion, and is known as the **base peak**. The base peak is used to report mass spectra in a standardised form; its intensity is arbitrarily given the value of 100, and all other peaks are reported as percentages of the base peak. This series of calculated values is the cracking pattern, and it may be described in the form of a line diagram (bar graph; see Fig. 1.7), or may be reported in tabular form in which mass numbers and relative abundances are listed.

The interpretation of a mass spectrum is difficult, complications arising from various sources. The main source of difficulty is that the molecular ion may undergo rearrangement to give fragmentation patterns not anticipated from the structure of the compound, *e.g.*,

$$ABC + e \longrightarrow [ABC]^+ + 2e$$

$$[ABC]^+ \longrightarrow [AB]^+ + C \cdot \ or \ AB \cdot + C^+ \ or \ [BC]^+ + A \cdot \ or \ [AC]^+ + B \cdot, \ etc.$$

Metastable ions. If an ion (molecular or fragment), m_1^+, is accelerated *before* it breaks down, then, when it decomposes into m_2^+ and m_3, part of the kinetic energy of m_1^+ is lost to the neutral fragment m_3, and m_2^+ continues to be accelerated and is then collected.

$$m_1^+ \longrightarrow m_2^+ + m_3$$

Ion m_2^+, *produced in this way*, is *not* recorded as mass m_2, but as mass m^*, where $m^* = m_2^2/m_1$. This ion, known as a *metastable ion*, is usually recorded as a weak broad peak, and is not (usually) an integral value. It is evaluated from the *recorded* masses m_1 and m_2, which arise from m_1^+ ions undergoing decomposition and acceleration in the normal way. The presence of metastable peaks is very useful for deducing fragmentation mechanisms, since they indicate the conversion of m_1^+ into m_2^+ in *one* step; *e.g.*, three ions, *m/e* 32, 31, and 30 were recorded. This suggests loss of a hydrogen atom one at a time. A broad peak (metastable peak), however, was also recorded at 28·1, and since $28·1 = 30^2/32$, this means that ion *m/e* 32 was converted into ion *m/e* 30 in *one* step. It should be emphasised, however, that the absence of a metastable peak does not indicate that ion *m/e* 32 $\longrightarrow$ ion *m/e* 30 does not occur, and it is also possible that ion *m/e* 31 $\longrightarrow$ ion *m/e* 30 occurs.

Mass spectrometry may be used with gases, liquids, and solids, and only very small amounts of material are necessary (a few μg). It has proved extremely valuable for the determination of accurate molecular weights, obtaining molecular formulae, elucidation of structure, quantitative analysis of mixtures, ionisation potentials, and bond strengths.

Cracking patterns are largely dependent of the relative labilities of bonds, relative stabilities of possible fragment ions and neutral molecules, etc. Many examples are discussed in the text (see Index: Mass spectrometry), but some general principles are mentioned here. When alternative fragmentations are possible, splitting usually occurs in all the alternative ways, but the direction leading to the most stable carbonium ion and/or free radical is the one that predominates. Most molecules show a peak for the molecular ion, the stability of which is usually in the order: aromatics > conjugated acyclic polyenes > alicyclics > n-hydrocarbons > ketones > ethers > branched-chain hydrocarbons > alcohols. The ease of formation of a molecular ion depends on the type of electron removed; the order is usually n (lone-pair electron) > π (pi-electron) > σ (sigma-electron) (see Ch. 2). However, since the bombardment electron energy is 50–70 eV, *all* types of electrons may be removed, the one most easily removed being that from the atom with the lowest ionisation potential. Rearrangements occur most readily when a hydrogen atom is involved in a six-membered cyclic transition state or when 1,2-shifts are involved.

As an example of some of the principles discussed, we shall describe the mass spectrum of ethanol. Figure 1.7 is the line diagram (bar graph), and shows the most intense lines. Since there are many decomposition paths, only some of these will be considered. Single headed arrows are used to denote the transfer of one electron (*homolytic fission*), and when the position of the positive charge is known, a plus sign is placed above that atom. When, however, the position of the positive charge is uncertain, the ion is enclosed in square brackets with the symbol $\stackrel{+}{}$ as a superscript. Since the oxygen atom (with two lone pairs) in ethanol is the one with the lowest ionisation potential, one form of the molecular ion (M^+; *m/e* 46) will have the positive charge on the oxygen atom. However, in order to propose various paths for the decompositions, it will be necessary to consider other forms of the molecular ion in which the

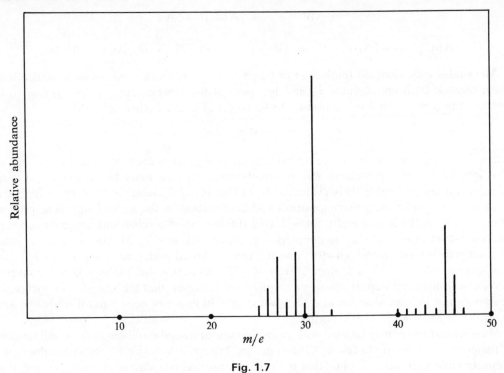

Fig. 1.7

position of the positive charge is uncertain (as we saw above, all types of electrons, n, π, and σ may be removed). It should be also noted that in some cases a path may be postulated which involves the transfer of *two* electrons (*heterolytic fission*), but this appears to be uncommon

$$CH_3—CH_2—\overset{..}{O}H + e \longrightarrow CH_3—CH_2—\overset{+}{O}H + 2e \quad m/e\ 46\ (M^+)$$

$$CH_3—CH_2—\overset{+}{O}H \longrightarrow CH_3\cdot + CH_2=\overset{+}{O}H \quad m/e\ 31\ (M\text{-}15)$$

$$CH_3—\underset{\underset{H}{|}}{CH}—\overset{+}{O}H \longrightarrow H\cdot + CH_3—CH=\overset{+}{O}H \quad m/e\ 45\ (M\text{-}1)$$

$$CH_3—CH_2—OH + e \longrightarrow [CH_3—CH_2—OH]^+ + 2e \quad (M^+)$$

$$[CH_3—CH_2—OH]^+ \longrightarrow HO\cdot + CH_3—CH_2^+ \quad m/e\ 29\ (M\text{-}17)$$

Although it may often be possible to predict the mass spectrum of a known compound, it is usually much more difficult, if not impossible in many cases, to elucidate the structure of a compound from its observed mass spectrum. Hence, in general, other information is required, and this is usually spectroscopic data: ir, uv, and NMR. High resolution mass spectrometry permits elucidation of molecular formulae, and this knowledge then enables the double bond equivalents (D.B.E.), *i.e.*, the number of double bonds and/or rings, in the compound to be ascertained. If the general formula of the compound is $C_aH_bN_cO_d$, then

$$\text{D.B.E.} = a + 1 - (b - c)/2$$

e.g., (i) Benzene is C_6H_6.

   D.B.E. $= 6+1-6/2 = 4$ (3 double bonds; 1 ring).

(ii) Allylamine is C_3H_7N.

   D.B.E. $= 3+1-(7-1)/2 = 1$ (1 double bond).

Univalent elements, such as halogens, may be replaced by one hydrogen atom, and bivalent elements may be ignored (*cf.* oxygen, above).

The major peaks and peaks near the molecular ion are examined, and m/e values are listed in terms of $M-m/e$. By checking m/e and $M-m/e$ values against a list of common fragments (see Table 1.6), and knowing the fragmentation patterns of compounds containing various functional groups, it is possible to make a great deal of progress towards solving the structure of an unknown compound. It should be noted that it is not necessary to identify every peak.

Table 1.6 lists some of the more common fragments which appear as ions and/or lost as radicals or molecules (only the lowest isotopic values are given).

Table 1.6

m/e	Fragment	m/e	Fragment	m/e	Fragment
1	H	41	C_3H_5	71	C_5H_{11}
2	H_2	42	C_3H_6	72	$C_4H_8NH_2$
14	CH_2	43	C_3H_7, CH_3CO	73	$CO_2C_2H_5$
16	O, NH_2	44	$CO_2, CH_2{=}CHOH,$	74	$CH_2{=}C(OH)OCH_3$
17	OH		$C_2H_4NH_2$	77	C_6H_5
18	H_2O, NH_4	45	$C_2H_5O, CH_3CHOH,$	78	C_6H_6
19	F		CO_2H	79	C_6H_7, Br
20	HF	46	NO_2	80	HBr
26	C_2H_2, CN	51	C_4H_3	83	C_6H_{11}
27	C_2H_3, HCN	53	C_4H_5	85	C_6H_{13}
28	C_2H_4, CO, N_2	55	C_4H_7	88	$CH_2{=}CH(OH)OC_2H_5$
29	C_2H_5, CHO	56	C_4H_8	91	C_7H_7
30	CH_2NH_2, NO	57	C_4H_9, C_2H_5CO	93	C_6H_5O
31	CH_2OH, CH_3O	58	$C_3H_6NH_2, CH_2{=}C(OH)CH_3$	94	C_6H_6O
32	CH_3OH	59	$CO_2CH_3, C_2H_5CHOH,$	97	C_7H_{13}
33	SH		$CH_2{=}C(OH)NH_2$	99	C_7H_{15}
34	H_2S	60	$CH_2 \quad CO_2H_2$	105	C_6H_5CO
35	Cl	65	C_5H_5	127	I
36	HCl	66	C_5H_6	128	HI
39	C_3H_3	69	C_5H_9		
40	CH_2CN	70	C_5H_{10}		

Electron spin resonance (ESR) is particularly useful for the detection and estimation of free radicals.

Diffraction methods.

(*a*) *X-ray diffraction* offers a means of determining the complete structure (molecular and spatial) of any crystalline compound. With the advent of computers, calculations from X-ray data can now be quickly performed, and so the use of this method for structural determination will become common. This method will therefore, in principle, render chemical methods superfluous. A particularly interesting example is the alkaloid thelepogine, $C_{20}H_{31}NO$; its structure has been determined entirely by X-ray analysis, no chemical work being carried out at all (Fridrichsons *et al.*, 1960).

X-ray analysis may also be used for the determination of molecular weights.

(*b*) *Electron diffraction* offers a means of determining the complete structure of fairly simple molecules, its main use being the examination of gases or compounds in the vapour state.

Optical properties. *Optical rotation* (rotation at a single wavelength) and *rotary dispersion* (rotation at different wavelengths) offer a means of determining relative and absolute configurations, and are also very useful in the study of conformations.

Dipole moments. This method gives information on configuration and conformation of fairly simple molecules, and information on electronic displacements.

Synthesis of the compound. The term *synthesis* means the building up of a compound, step by step, from a simpler substance of *known* structure. The term *complete synthesis* means the building up of a compound, step by step, *starting from its elements* (and any others that may be necessary). In either case (synthesis or complete synthesis), *the structure of each intermediate compound is taken as proved by its synthesis from the compound that preceded it.*

The synthesis of a compound is necessary to establish its structure beyond doubt. There is always the possibility of one or more steps not proceeding 'according to plan'. Hence the larger the number of syntheses of a compound by *different* routes, the more reliable will be the structure assigned to that compound. However, as has been pointed out above, physical methods—especially X-ray analysis—make synthesis *as a means of structure determination* less important than previously. Nevertheless, synthesis will still be an extremely important problem in the production of organic compounds, both natural and synthetic.

Structural formulae and isomerism

In 1857, Kekulé postulated the *constant* quadrivalency (tetravalency) of carbon. From 1900 onwards, however, compounds containing *tervalent* carbon have been prepared, and their number is increasing rapidly. These compounds are free radicals, carbonium ions and carbanions, and are usually reactive intermediates in many reactions. More recently, compounds containing bivalent carbon (*carbenes*) are believed to be formed as intermediates during certain reactions. Hence, unless there is definite evidence to the contrary, carbon is always assumed to be quadrivalent.

If 'valency units' or 'valency bonds' (see Ch. 2) are represented by lines, then the number of lines drawn from the symbol shows the valency of that atom, *e.g.*,

$$-\overset{\displaystyle |}{\underset{\displaystyle |}{C}}- ; \quad -O- ; \quad -\overset{\displaystyle |}{N}- ; \quad H-$$

The molecular formula shows the number of each kind of atom present in the molecule, but does not indicate their arrangement. In organic chemistry there are many cases where a given formula represents two or more compounds that differ in physical and chemical properties, *e.g.*, there are at least seven compounds having the same molecular formula $C_4H_{10}O$. Such compounds, having the same molecular formula, but differing in physical and chemical properties, are known as *isomers* or *isomerides*, the phenomenon itself being known as **isomerism**. The existence of isomerism may be explained by assuming that the atoms are arranged in a definite manner in a molecule, and that there is a different arrangement in each isomer, *i.e.*, the isomers differ in *structure* or *constitution*. This type of isomerism is known as **structural isomerism**.

Obviously, then, from what has been said above, it is always desirable to show the arrangement (if known) of the atoms in the molecule, and this is done by means of *structural formulae* or *bond-diagrams*; *e.g.*, the molecular formula of ethanol is C_2H_6O; its structural formula is

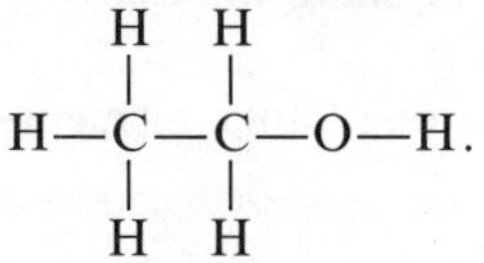

$$\begin{array}{cc} H & H \\ | & | \\ H-C-C-O-H. \\ | & | \\ H & H \end{array}$$

So far nothing has been said about the *spatial* disposition of the four valencies of carbon. Later (Ch. 2) it will be shown that when carbon is joined to four univalent atoms or groups, the four valencies are directed towards the four corners of a tetrahedron. Thus the above plane-structural formula does not show the disposition of the atoms in space; a three-dimensional formula is necessary for this. Usually the plane-formula is satisfactory. Since the actual *spatial* arrangements of a given structure may differ, this gives rise to isomerism of the type known as **stereoisomerism**. **Stereoisomers** have different **configurations**, *i.e.*, the spatial arrangements are different but not their structures.

A structural formula is really a short-hand description of the properties of the compound. Hence the study of organic chemistry is facilitated by mastering the structural formula of every compound the reader meets. An organic molecule, however, is only completely described when the following facts are known: *structure* or *constitution* (this includes a knowledge of the electron distribution; see resonance, p. 34), *configuration* (p. 98), and *conformation* (p. 78).

Saturated and unsaturated compounds

If, in an organic compound containing two or more carbon atoms, there are only *single* bonds linking any two adjacent carbon atoms, then that compound is said to be *saturated*, *e.g.*, ethane, C_2H_6 (I), normal propanol, C_3H_8O (II), acetaldehyde, C_2H_4O (III).

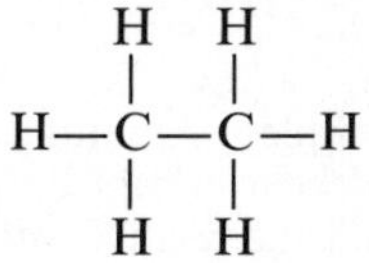

$$\begin{array}{ccc} \begin{array}{cc} H & H \\ | & | \\ H-C-C-H \\ | & | \\ H & H \end{array} & \begin{array}{ccc} H & H & H \\ | & | & | \\ H-C-C-C-O-H \\ | & | & | \\ H & H & H \end{array} & \begin{array}{cc} H & O \\ | & \nearrow\nearrow \\ H-C-C \\ | & \searrow \\ H & H \end{array} \\ (I) & (II) & (III) \end{array}$$

On the other hand, if the compound contains at least one pair of adjacent carbon atoms linked by a *multiple* bond, then that compound is said to be *unsaturated*, *e.g.*, ethylene, C_2H_4 (IV); this compound contains a *double* bond. Acetylene, C_2H_2 (V); this contains a *triple* bond. Acraldehyde, C_3H_4O (VI); this contains a double bond. *The double bond between the carbon and oxygen atoms is not a sign of unsaturation (cf. acetaldehyde above).*

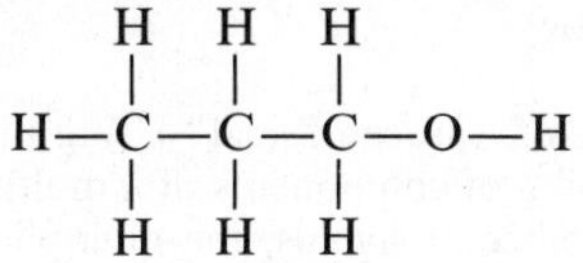

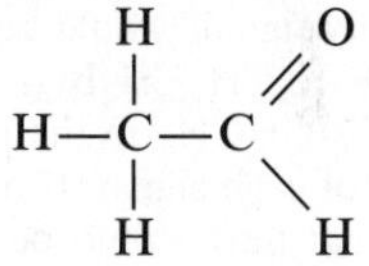

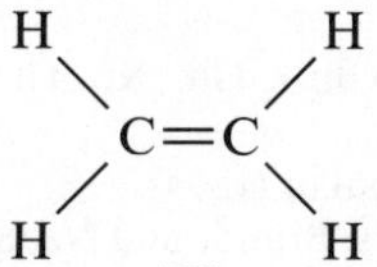

$$\begin{array}{ccc} \begin{array}{c} H \\ \searrow \quad H \\ C=C \\ \nearrow \quad \searrow \\ H \qquad H \end{array} & H-C\equiv C-H & \begin{array}{c} H \qquad H \quad O \\ \searrow \quad | \quad \nearrow\nearrow \\ C=C-C \\ \nearrow \qquad \searrow \\ H \qquad \qquad H \end{array} \\ (IV) & (V) & (VI) \end{array}$$

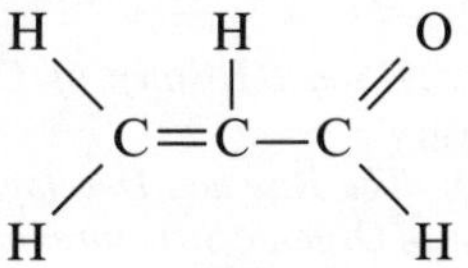

Classification of organic compounds

Organic compounds are classified into three major groups:

1. (a) **Aliphatic, open-chain** or **acyclic** compounds.
   (b) **Alicyclic** compounds. These are carbocyclic or ring compounds which resemble aliphatic compounds in many ways.

2. **Aromatic** compounds. These are carbocyclic or ring compounds *containing at least one benzene ring*. There is also the group of *non-benzenoid aromatic compounds*.

3. **Heterocyclic** compounds. These are cyclic (ring) compounds containing other elements besides carbon in the ring. In a few cases *no* carbon atom is in the ring.

PROBLEMS

1. Calculate the percentage composition of the elements of $C_{14}H_{14}N_2O_4S$; C_8H_9NO; $C_8H_8ClNO_2S$; $C_{15}H_9BrO_2$.

2. Calculate the empirical formulæ of the following compounds from the given elemental analyses:

	C%	H%	Br%	Cl%	N%	S%
(i)	40·1	6·7				
(ii)	42·6	3·5			20	
(iii)	73·8	5·3			16·4	
(iv)	60·6	4·8			11·8	9·0
(v)	51·2	4·0	21·35		14·95	
(vi)	24·3	4·0		71·7		

3. The sulphur content of cystine is 26·7 per cent. Given that cystine contains *two* sulphur atoms, what is the molecular weight of cystine?

4. How many proton signals would be expected in the NMR spectra of:
   (i) $ClCH_2CH_2Cl$; (ii) CH_3OCH_3; (iii) $CH_3OC_2H_5$; (iv) $CH_3CHClC_2H_5$; (v) CH_3COCH_3;
   (vi) CH_3CH_2CHO; (vii) $CH_3CO_2CH_3$; (viii) $(CH_3)_2CHCH_2Br$; (ix) $CH_3CH(OCH_3)CH_2CH_2OH$?

5. Give, for each of the following compounds: (a) the number of signals and their relative intensities, (b) the multiplicity of each signal and the relative intensities of the components, (c) the relative field positions of the signals with respect to TMS:
   (i) CH_3OCH_2Cl; (ii) CH_3CHO; (iii) $BrCH_2CHBr_2$; (iv) $CHCl_2CHBr_2$; (v) $CD_3CH_2CH_3$.

6. How many signals would be given by
   (i) $CH_3OCH_2CH_2CH_2Br$;
   (ii) $CH_3CHOHCH_3$ in the presence of a trace of acid? Using the simple rules of splitting, calculate the multiplicity of each signal. If overlapping of components of a multiplet can occur, in which signals will this happen? What now would be the number of signals, the multiplicities, and the relative intensities of the components in each signal? What ratios would be indicated by the integration trace?

7. Suggest molecular formulæ, in terms of ^1H and ^{12}C, for the following fragment ions obtained from ethanol: m/e 25, 26, 28, 42.

8. The ion with the highest mass obtained from $CBrCl_3$ is 167. Discuss.

9. Calculate the double bond equivalents of the following compounds:
   (i) $C_7H_6O_2$; (ii) C_3H_7N; (iii) $C_{10}H_7Cl$; (iv) $C_{12}H_{16}N_2O$; (v) $C_{13}H_9BrS$.

REFERENCES

PARTINGTON, *A Short History of Chemistry*, Macmillan (1967, 3rd edn.). Ch. X. 'The Beginnings of Organic Chemistry'.

SCHORLEMMER, *The Rise and Development of Organic Chemistry*, Macmillan (1894).

Determination of Organic Structures by Physical Methods, Academic Press, Braude and Nachod (eds.), Vol. 1 (1955). Nachod and Phillips (eds.), Vol. 2 (1962).

CROSS and JONES, *Introduction to Practical Infra-Red Spectroscopy*, Butterworths (1969, 3rd edn.).

DYER, *Applications of Absorption Spectroscopy of Organic Compounds*, Prentice-Hall (1965).

WHIFFEN, *Spectroscopy*, Longmans, Green (1966).

BIBLE, *Interpretation of NMR Spectra*, Plenum Press (1965).

JACKMAN and STERNHELL, *Applications of Nuclear Magnetic Resonance Spectroscopy in Organic Chemistry*, Pergamon Press (1969, 2nd edn.).

WILLIAMS and FLEMING, *Spectroscopic Methods in Organic Chemistry*, McGraw-Hill (1966).

HILL, *Introduction to Mass Spectrometry*, Heyden and Son (1966).

BIEMANN, *Mass Spectrometry. Applications to Organic Chemistry*, McGraw-Hill (1962).

REED, *Applications of Mass Spectrometry to Organic Chemistry*, Academic Press (1966).

BEYNON, SAUNDERS, and WILLIAMS, *The Mass Spectra of Organic Molecules*, Elsevier (1968).

2 Properties of molecules

Structure of the atom

According to modern theory, an atom consists of a *nucleus* which contains *protons* and *neutrons*, and which is surrounded by *electrons*. The mass of a proton is almost the same as that of a neutron, but whereas the proton carries a unit of *positive* charge, the neutron is electrically *neutral*. The electron has about $\frac{1}{1850}$th of the mass of a proton, and carries a unit of *negative* charge. The electrons are arranged in shells around the nucleus, each shell being able to contain up to a maximum number of electrons, this maximum depending on the number of the shell, n. n is known as the *principal quantum number*, and indicates the main energy level of the electrons in that shell. n has *whole* number values, 1, 2, 3, 4 . . ., the shells corresponding to which are also denoted by the letters K, L, M, N . . . respectively. In every principal quantum shell there are n energy sublevels, and these are indicated by l, the *orbital quantum number* (also known as the *azimuthal* or *serial* quantum number). Just as the principal quantum number n can have values 1, 2, 3 . . ., so can l have values 0, 1, 2, 3 . . ., $n-1$. The energy state corresponding to $l = 0$ is called the *s* state; $l = 1$, the *p* state; $l = 2$, the *d* state; etc. As we shall see later (p. 42), these *s*, *p* and *d* states are subdivided into a number of *orbitals*. The total number of orbitals that a principal quantum shell can contain is given by n^2. Thus, when the principal quantum number is 1 (*i.e.*, the first or K shell), then $l = 0$, *i.e.*, there is a single orbital in this K shell and is of the *s* type and is known as the 1*s* orbital. When $n = 2$ (*i.e.*, the second or L shell), then $l = 0$ or 1. This means there are *two* energy sublevels in the L shell. As pointed out above, the total number of orbitals in a given quantum shell is equal to n^2. Thus, when $n = 2$, there are four possible orbitals. When $l = 0$, this corresponds to the 2*s* orbital. When $l = 1$, then there are *three equivalent* orbitals; these are the three 2*p* orbitals. When $n = 3$ (*i.e.*, the third or M shell), then the total number of possible orbitals is 9 (3^2). These correspond to one 3*s* orbital ($l = 0$), three 3*p* orbitals ($l = 1$) and five 3*d* orbitals ($l = 2$). The existence of one *s* level, three *p*, five *d* and seven *f* levels of energy was used to explain the existence of spectral lines observed in the spectra of atoms and molecules. It should also be noted that the farther an electron is from the nucleus, the greater is its potential energy. Owing to the phenomenon of *penetration* of orbitals, *i.e.*, outer electrons can penetrate into the shell of inner electrons, the energy level of an electron is thus not completely determined by its principal quantum number, but also depends on the orbital quantum number (*i.e.*, on the *shape* of the orbital). Thus there is the following order of increasing energy: 1*s*, 2*s*, 2*p*, 3*s*, 3*p*, 4*s*, 3*d*, 4*p*, 5*s*, 4*d*, 5*p* . . . (see also p. 42).

In addition to the energy levels of an electron described by quantum numbers n and l, electrons have spin about their axis, some spinning in one direction and others in the opposite direction. This is indicated by the *spin quantum number* (s), and can have values of $+\frac{1}{2}$ and $-\frac{1}{2}$. Finally, an electron also has a *magnetic quantum number* (m), and this gives the allowed orientations of the orbitals in an external magnetic field. Thus an electron is described by *four* quantum numbers, n, l, s and m.

By the fundamental *Pauli Exclusion Principle* (1925), no two electrons, in any system, can be assigned the same set of four quantum numbers. Hence there can be only *two* electrons in any one orbital, and these must be differentiated from each other by their spins, which *must* be *antiparallel*, *i.e.*, in the opposite sense. Such electrons are said to be *paired*, and a pair of electrons with anti-parallel spins in the same orbital is represented by the symbol ↓↑. Since a moving charge is accompanied by a magnetic field, a spinning electron behaves as a small bar-magnet, and consequently two paired electrons will give a zero resultant magnetic field.

The hydrogen atom consists of one proton and one electron. When the hydrogen atom is in the 'ground' state, *i.e.*, the state of lowest energy, its electron will be in the lowest energy level, *i.e.*, the $1s$ level, and is represented by ($1s$). When hydrogen is in an 'excited' state, its electron will occupy a higher energy level, the actual level depending on the amount of 'excitation'.

Helium has two electrons; hence its electron configuration in the ground state is represented as $(1s)^2$. Lithium has three electrons. Since the maximum number of electrons in the K shell ($n = 1$, $l = 0$) is two, the third electron must start the L shell ($n = 2$, $l = 0, 1$). Electrons occupy lowest energy levels first. Thus this third electron occupies the $2s$ orbital, and not the $2p$, because the $2p$ is a higher energy level than the $2s$. Hence the electron configuration of lithium is $(1s)^2(2s)$. Thus the K shell is filled first. Then the electrons enter the L shell until that is filled. In this shell the s level is filled before the p. In fitting electrons into shells containing orbitals of *equivalent* energy, Hund's rules are used to assign the electrons to their orbitals. These rules are: (i) electrons tend to avoid being in the same orbital as far as possible; (ii) two electrons, each singly occupying a given pair of *equivalent* orbitals, tend to have their spins parallel when the atom is in the ground state. Thus carbon, with six electrons, may be represented as $(1s)^2(2s)^2(2p)^2$. The K shell is filled first; the L shell is filled next, the $2s$ orbital being doubly filled before a higher level is used; then singly two of the $2p$ orbitals. Nitrogen, with seven electrons is $(1s)^2(2s)^2(2p)^3$: all three $2p$ orbitals each contain one electron. Oxygen, with eight electrons, is $(1s)^2(2s)^2(2p)^4$: here one of the $2p$ orbitals is doubly filled.

The electronic theory of valency

The electronic theory of valency starts with the assumption that valency involves the electrons in the outer shells: in some cases only those in the highest sublevel in the outermost shell; in other cases those in the highest and penultimate sublevels, even though the penultimate sublevel may be in a lower quantum shell. Lewis (1916) assumed that the electron configuration in the rare gases was particularly stable (since these gases are chemically inert), and that chemical combination between atoms took place by achieving this configuration. The outermost shell of the rare gases always contains an *octet* of two s and six p electrons. Since both the s and p sublevels are completely filled, the octet will be a stable configuration. In the case of helium, however, an octet is impossible; here the stable arrangement is the *duplet*, the two $1s$ electrons of which completely fill the first quantum shell.

The octet rule applies only to atoms with $2s$ and $2p$ electrons, *i.e.*, to elements in the second period (Li—F). With the other elements, d orbitals may also be used in bond formation, and hence higher covalencies (*i.e.*, expansions beyond an octet) are possible, *e.g.*, PBr_5 (10 electrons),

SF_6 (12 electrons) and IF_7 (14 electrons). Since elements in period 2 have only $2s$ and $2p$ orbitals, the maximum covalency they can exhibit is 4.

Lewis also suggested that there was a definite tendency for electrons in a molecule to form pairs. This rule of 2, as we have seen, became established by the developments of quantum mechanics. There are also molecules that contain an odd number of valency electrons: where such *odd electron molecules* do exist, unusual properties are found to be associated with them (see free radicals).

There are three general *extreme* types of chemical bonds: electrovalent, covalent and metallic bonds. In addition to these extreme types, there are also bonds of *intermediate* types.

1. Electrovalency is manifested by the *transfer* of electrons, and gives rise to the *ionic* bond. Consider sodium chloride. Sodium is $(1s)^2(2s)^2(2p)^6(3s)$: chlorine is $(1s)^2(2s)^2(2p)^6(3s)^2(3p)^5$. Sodium has completed K and L shells, and is starting the M shell with one electron. This electron (the $3s$ electron) is the valency electron of sodium. Chlorine has completed K and L shells, and has seven electrons in the M shell. These M electrons are the valency electrons of chlorine. If the sodium completely transfers its valency electron to the chlorine atom, then each atom will have eight electrons in its outermost shell, and this, as we have seen, is a stable arrangement. Since both atoms were originally electrically neutral, the sodium atom, in losing one electron, will now have a single positive charge, *i.e.*, the neutral atom has become a positive ion. Similarly, the neutral chlorine atom, in gaining one electron, has become a negative ion. In the sodium chloride crystal these ions are held together by electrostatic forces. If the symbol of an element is used to represent the nucleus of an atom and all the electrons other than the valency electrons, and dots are used to represent the valency electrons, then the combination of the sodium and chlorine atoms to form sodium chloride may be represented as follows:

$$Na\cdot + \overset{\cdot\cdot}{\underset{\cdot\cdot}{:Cl\cdot}} \longrightarrow Na^+ \overset{\cdot\cdot}{\underset{\cdot\cdot}{:Cl:}}{}^-$$

2. Covalency. This type of bonding involves a *sharing* of electrons in pairs, each atom contributing one electron to form a shared pair, each pair of electrons having their spins antiparallel. This method of completing an octet (or any of the other possible values) gives rise to the *covalent* bond.

Hydrogen is usually *unicovalent*: occasionally it is unielectrovalent, *e.g.*, in sodium hydride, hydrogen exists as the hydride anion, formed by accepting an electron from the sodium:

$$Na\cdot + H\cdot \longrightarrow Na^+ H:^-$$

Carbon almost invariably forms covalent compounds. The electron configuration of carbon is $(1s)^2(2s)^2(2p)^2$. Since the two $2s$ electrons are paired, it would appear that carbon is bivalent, only the two single $2p$ electrons being involved in compound formation. As pointed out previously, carbon is almost always quadrivalent; thus the $2s$ and $2p$ electrons must be involved. Just how these four electrons give quadrivalent carbon will be described later; at this stage we shall assume it done, and write quadrivalent carbon as $\cdot \overset{\cdot}{\underset{\cdot}{C}} \cdot$.

In methane the four hydrogen atoms each contribute one electron and the carbon atom four electrons towards the formation of four shared pairs:

$$4H\cdot + \cdot\overset{\cdot}{\underset{\cdot}{C}}\cdot \longrightarrow H\overset{\overset{\textstyle H}{\cdot\cdot}}{\underset{\underset{\textstyle H}{\cdot\cdot}}{:C:}}H$$

Each hydrogen atom has its duplet (as in helium), and the carbon atom has an octet.

Each pair of shared electrons is equivalent to the ordinary 'valency-bond', and so electronic formulae are readily transformed into the usual structural formulae, each bond representing a shared pair, *e.g.*,

$$H:\overset{..}{\underset{..}{O}}:H \text{ or } H-O-H; \quad H:C\vdots C:H \text{ or } H-C\equiv C-H; \quad H:\overset{..}{N}:H \text{ or } H-N-H$$
$$\overset{\displaystyle H}{} \qquad\qquad\qquad\qquad \overset{\displaystyle |}{\underset{\displaystyle H}{}}$$

From these examples it can be seen that there is a very important difference between an electronic formula and its equivalent structural formula. In the former, all valency electrons are shown whether they are used to form covalent bonds or not; in the latter, only those electrons which are actually used to form covalent bonds are indicated. This is a limitation of the usual structural formula. A widely used scheme is to represent structures with ordinary valency bonds and to indicate lone pairs by pairs of dots (see below).

3. Co-ordinate valency is a special type of covalency. Its distinguishing feature is that *both* of the shared electrons forming the bond are supplied by only **one** of the two atoms linked together, *e.g.*, when ammonia combines with boron trifluoride, it is the lone pair of the nitrogen atom that is involved in the formation of the new bond. In boron trifluoride, the boron has only six electrons in its valency shell; hence it can accommodate two more to complete its octet. Thus, if the nitrogen atom uses its lone pair, the combination of ammonia with boron trifluoride may be shown as (I) or (II). In the latter, a co-ordinate bond is represented by an arrow pointing *away* from the atom supplying the lone pair (Sidgwick, 1927).

$$\begin{array}{ccc}
\text{H} \quad \text{F} & \text{H F} & \text{H} \qquad \text{F} \\
| \quad\ | & |\ \ | & |\qquad\ | \\
\text{H}-\text{N:}+\text{B}-\text{F} \longrightarrow \text{H}-\text{N:B}-\text{F} & or & \text{H}-\text{N}\longrightarrow\text{B}-\text{F} \\
| \quad\ | & |\ \ | & |\qquad\ | \\
\text{H} \quad \text{F} & \text{H F} & \text{H} \qquad \text{F}
\end{array}$$

$$\text{(I)} \qquad\qquad\qquad \text{(II)}$$

The atom that supplies the lone pair is known as the *donor*, and the atom that receives a share is the *acceptor*. Since it is one atom that donates the lone pair, the co-ordinate bond is also known as the *dative* bond (Sidgwick, 1927).

Before combination, both donor and acceptor are electrically neutral: after combination, the donor has lost a share in the lone pair, and the acceptor has gained a share. Therefore the donor acquires a positive charge and the acceptor a negative charge, and the presence of these charges may be indicated by writing the formula $\overset{+}{H_3N}-\overset{-}{BF_3}$.

Once the co-ordinate bond has been formed, there may be no way of distinguishing it from a covalent bond, but since one atom has supplied the pair of shared electrons, charges are produced in the molecule. When a covalent bond is formed, charges may also be produced in the molecule, giving rise to a dipole (*q.v.*). Hence the co-ordinate bond is effectively a covalent bond. The extent of the charge on each atom in a dative (or covalent) bond may be found as follows. Add the number of unshared electrons to one half of the shared electrons, and compare the result with the number of valency electrons of the neutral atom, *e.g.*, (i) methane, CH_4. Here there are 8 shared electrons; $\frac{1}{2}\times 8 = 4 =$ number of electrons in the neutral carbon atom; therefore methane is uncharged. (ii) $\overset{+}{H_3N}-\overset{-}{BF_3}$. For the nitrogen atom we have $\frac{1}{2}\times 8 = 4$, but since the neutral nitrogen atom has 5 valency electrons, in the compound $\overset{+}{H_3N}-\overset{-}{BF_3}$ the nitrogen has a charge of $+1$. For boron we have $\frac{1}{2}\times 8 = 4$; but since the neutral boron atom has 3 electrons, in this molecular compound the boron has a charge of -1.

Electrovalent compounds are good electrical conductors in the fused state or in solution. They are generally non-volatile, and are usually insoluble in hydrocarbons and allied solvents. Covalent compounds are non-electrical conductors, are generally volatile, and are usually soluble in hydrocarbons and allied solvents. Since the covalent bond is directional, *stereo-isomerism* (space-isomerism) is possible (see p. 472). Co-ordinated compounds behave very much like covalent compounds, but they are usually less volatile than pure covalent compounds.

Two important atomic quantities are:

(i) *Ionisation potential* (I.P.): this is the amount of energy *absorbed* when one electron is *removed* from a neutral (or charged) atom. It is measured in electronvolts ($1 \text{ eV} = 96\cdot44 \text{ kJ mol}^{-1} = 23\cdot05 \text{ kcal mol}^{-1}$), and the higher its value, the more difficult it is to remove the electron. The I.P.s of carbon are: $C \longrightarrow C^+ + e \ (+11\cdot27 \text{ eV})$; $C^+ \longrightarrow C^{2+} + e \ (24\cdot38 \text{ eV})$; $C^{2+} \longrightarrow C^{3+} + e \ (47\cdot87 \text{ eV})$. Even the first I.P. is relatively high, and hence carbon shows little tendency to form electrovalent bonds.

(ii) *Electron affinity* (E.A.): this is the amount of energy (in eV) which is *evolved* when a neutral (or charged) atom *adds* one electron. Most non-metals have relatively high (negative) E.A.s, and consequently easily form negative ions, *e.g.*, $Cl + e \longrightarrow Cl^- \ (-13.0)$; $Br + e \longrightarrow Br^- \ (-11\cdot84)$; $I + e \longrightarrow I^- \ (-10\cdot44 \text{ eV})$.

Now let us reconsider the co-ordinate bond. This may be imagined to be formed in two steps:

$$H_3N\text{:} + BF_3 \longrightarrow H_3\overset{+}{N}\cdot + \cdot\overset{-}{B}F_3 \longrightarrow H_3\overset{+}{N}\!-\!\overset{-}{B}F_3$$

Thus, the lower the I.P. of the donor atom and the higher the E.A. of the acceptor atom, the more readily will the co-ordinate bond be formed. Also, when the bond has been formed, since the donor atom now has a positive charge, it will be more difficult for it to form a second co-ordinate bond (provided that the atom has more than one unshared pair), *e.g.*, the I.P.s of O are $O \longrightarrow O^+ \ (13\cdot62)$ and $O^+ \longrightarrow O^{2+} \ (35\cdot15 \text{ eV})$. The latter is too great for its formation under normal conditions, and consequently the hydronium ion, H_3O^+, is readily formed (one co-ordinate bond), but not H_4O^{2+} (two co-ordinate bonds).

Dipole moments

When a covalent bond is formed between two identical atoms, *e.g.*, H—H, Cl—Cl, etc., the two electrons forming the covalent bond may be regarded as being symmetrically disposed between the two atoms. The centres of gravity of the electrons and nuclei therefore coincide. With two *dissimilar* atoms the two electrons are no longer symmetrically disposed, because each atom has a different *electronegativity*, *i.e.*, attraction for electrons. Chlorine has a much greater electronegativity than hydrogen that when chlorine and hydrogen combine to form covalent hydrogen chloride, the electrons forming the covalent bond are displaced towards the chlorine atom without any separation of the nuclei:

$$H\cdot + \text{:}\overset{..}{\underset{..}{C}l}\cdot \longrightarrow H\text{:}\overset{..}{\underset{..}{C}l}\text{:} \qquad or \qquad \overset{\delta+}{H}\!-\!\overset{\delta-}{Cl}$$

The hydrogen atoms will, therefore, be slightly positively charged, and the chlorine atom slightly negatively charged. Thus, owing to the greater electronegativity of the chlorine atom, the covalent bond in hydrogen chloride is characterised by the separation of small charges in the bond. A covalent bond such as this, in which one atom has a larger share of the electron-pair, is said to possess *partial ionic character*.

In analogy with a magnet, such a molecule is called a *dipole*, and the product of the electronic charge, e, and the distance d, between the charges (positive and negative centres) is called the *dipole moment*, μ; *i.e.*, $\mu = e \times d$; e is of the order of 10^{-10} e.s.u.; d, 10^{-8} cm. Therefore μ is of the order 10^{-18} e.s.u., and this unit is known as the Debye (D), in honour of Debye, who did a large amount of work on dipole moments.

The dipole moment is a vector quantity, and its direction is often indicated by an arrow parallel to the line joining the points of charge, and pointing towards the negative end, e.g., $\overrightarrow{H-Cl}$. The greater the value of the dipole moment, the greater is the *polarity* of the bond. The terms *polar* and *non-polar* are used to describe bonds, molecules and groups, and the reader is advised to make sure he appreciates how the terms are applied in each case under consideration.

> Not only do polar bonds contribute to the dipole moment of a molecule, but so do lone pairs, *e.g.*, the lone pair in ammonia makes a large contribution to the dipole moment.

Table 2.1 gives some electronegativity values, and it will be seen that electronegativity increases from left to right and decreases from top to bottom of the periodic table. The values are not absolute values; they are *relative* values and this is a satisfactory scheme since their use involves differences in electronegativities (see p. 272 for the measurement of relative electronegativities). Furthermore, it should be noted that electronegativity deals with electron attraction *within* a molecule, whereas electron affinity deals with the attraction of an electron from outside the atom (and has an *absolute* value).

Table 2.1

| | | | | H | | | |
				2·1			
Li	Be	B	C		N	O	F
1·0	1·5	2·0	2·5		3·0	3·5	4·0
Na	Mg	Al	Si		P	S	Cl
0·9	1·2	1·5	1·8		2·1	2·5	3·0
K	Ca				As	Se	Br
0·8	1·0				2·0	2·4	2·8
			Sn		Sb		I
			1·7		1·8		2·5

Electron displacements in a molecule

1. Inductive effect. Consider a carbon chain in which one terminal carbon atom is joined to a chlorine atom: $-C_3-C_2-C_1-Cl$. Chlorine has a greater electronegativity than carbon; therefore the electron pair forming the covalent bond between the chlorine atom and C_1 will be displaced towards the chlorine atom. This causes the chlorine atom to acquire a small negative charge, and C_1 a small positive charge. Since C_1 is positively charged, it will attract towards itself the electron pair forming the covalent bond between C_1 and C_2. This will cause C_2 to acquire a small positive charge, but the charge will be smaller than that on C_1 because the effect of the chlorine atom has been transmitted through C_1 to C_2. Similarly, C_3 acquires a positive charge which will be smaller than that on C_2. This type of electron displacement along a chain is known as the *inductive effect*; it is *permanent*, and decreases rapidly as the distance from the source increases. From the practical point of view, it may be ignored after the second carbon atom. It is important to note that the electron pairs, although permanently displaced, *remain in the same valency shells.*

> This inductive effect is sometimes referred to as a *transmission effect*, since it takes place by a displacement of the intervening electrons in the molecule. There is also another effect possible, the *direct* or *field effect*, which results from the electrostatic interaction across space or through a solvent of two charged centres in the same molecule, *i.e.*, the direct effect takes place independently of the electronic system in the molecule (Ingold, 1934). Apparently it has not been possible to separate these two modes of inductive effect in practice.

The inductive effect may be represented in several ways. The following will be adopted in this book: $—C \rightarrow C \rightarrow C \rightarrow Cl$.

Inductive effects may be due to atoms or groups, and the following is the order of decreasing inductive effects:

$+I$ O^-, CO_2^-, $(CH_3)_3C$, $(CH_3)_2CH$, $(CH_3)CH_2$, CH_3

$-I$ NR_3^+, SR_2^+, NH_3^+, NO_2, SO_2R, CN, CO_2H, F, Cl, Br, I, OR, OH

For measurement of relative inductive effects, hydrogen is chosen as reference in the molecule $CR_3—H$ as standard. If, when the H atom in this molecule is replaced by Z (an atom or group), the electron density in the CR_3 part of the molecule is *less* in this part than in $CR_3—H$, then Z is said to have a $-I$ effect (electron-attracting or electron-withdrawing). If the electron density in the CR_3 part is *greater* than in $CR_3—H$, then Z is said to have a $+I$ effect (electron-repelling or electron-releasing) *e.g.*, Br is $-I$; C_2H_5 is $+I$. This terminology is due to Ingold (1926); Robinson suggests the opposite signs for I, *i.e.*, Br is $+I$; C_2H_5, $-I$. Ingold's terminology will be used in this book. The inductive effect may be measured with respect to hydrogen by the effect of substituents on the change in acid strength of substituted acetic acids (p. 272).

2. Electromeric effect. This is a *temporary* effect involving the *complete transfer* of a shared pair of electrons to one or other atom joined by a multiple bond, *i.e.*, a double or triple bond. The electromeric effect is brought into play only at the requirements of the attacking reagent, and because of this, *the direction of the electromeric effect is always that which facilitates reaction.*

The electromeric effect is represented as follows:

$$A \overset{\frown}{=} B \; \rightleftharpoons \; \overset{+}{A} — \overset{-}{B}$$

The curved arrow shows the displacement of the shared electron pair, beginning at the position where the pair was originally, and ending where the pair has migrated. Since A has lost its share in the electron pair and B has gained this share, A acquires a positive charge and B a negative charge.

The electromeric effect is represented by the symbol E, and is said to be $+E$ when the displacement is *away* from the atom or group, and $-E$ when *towards* the atom or group (*cf.* the I effect).

> The displacement of the electron pair forming a covalent bond when a unit charge is brought up is a measure of the *polarisability* of that bond. It is not a permanent polarisation since, when the charge is removed, the electron displacement disappears. Thus, the electromeric effect is a polarisability effect and operates in the excited state.

3. Mesomerism or Resonance. The theory of mesomerism was developed on chemical grounds. It was found that no structural formula could satisfactorily explain all the properties of certain compounds, *e.g.*, benzene. This led to the idea that such compounds exist in a state which is some combination of two or more electronic structures, all of which seem equally capable of describing most of the properties of the compound, but none of describing *all* the properties. Ingold (1933) called this phenomenon *mesomerism* ('between the parts', *i.e.*, an intermediate structure). Heisenberg (1926), from quantum mechanics, supplied a theoretical background for mesomerism; he called it *resonance*, and this is the name which is widely used.

The chief conditions for resonance are:

(i) The positions of the nuclei in each structure must be the same or nearly the same.

(ii) The number of unpaired electrons in each structure must be the same.

(iii) Each structure must have about the same internal energy, *i.e.*, the various structures have approximately the same stability.

Let us consider carbon dioxide as an example. The electronic structure of carbon dioxide may be represented by at least three possible electronic arrangements which satisfy the above conditions:

$$\ddot{O}\!:\!C\!:\!\ddot{O} \qquad :\!\overset{-}{\ddot{O}}\!:\!C\!:\!\overset{+}{\ddot{O}}: \qquad :\!\overset{+}{\ddot{O}}\!:\!C\!:\!\overset{-}{\ddot{O}}:$$

(I)   (II)   (III)

Structures (II) and (III) are identical *as a whole*, since both oxygen atoms are the same.* Each structure, however, shows a given oxygen atom to be in a *different* state, *e.g.*, the oxygen atom on the left in (II) is negative, whereas in (III) it is positive. Although two (or more) of the electronic structures may be the same when the molecule is considered as a whole, each one must be treated as a separate individual which makes its own contribution to the resonance state. Structures (I), (II) and (III) are called the *resonating, unperturbed* or *canonical* structures of carbon dioxide, and carbon dioxide is said to be a *resonance hybrid* of these structures, or in the *mesomeric state*.

It is hoped that the following crude analogy will help the reader to grasp the concept of resonance. Most readers will be familiar with the rotating disc experiment that shows the composite nature of white light. When stationary, the disc is seen to be coloured with the seven colours of the rainbow. When rotating quickly, the disc appears to be white. The resonating structures of a resonance hybrid may be compared to the seven colours, and the actual state of the resonance hybrid to the 'white'; *i.e.*, the resonating structures may be regarded as superimposed on one another, the final result being *one kind of molecule*. **In a resonance hybrid all the molecules are the same; a resonance hybrid cannot be expressed by any single structure.**

In a resonance hybrid the molecules have, to some extent, the properties of each resonating structure. The greater the contribution of any one structure, the more closely does the actual state approach to that structure. At the same time, however, a number of properties differ from those of any one structure. The observed enthalpy of formation of carbon dioxide is greater than the calculated value by $132 \cdot 2$ kJ mol^{-1} ($31 \cdot 6$ kcal). In other words, carbon dioxide requires $132 \cdot 2$ kJ more energy than expected to break it up into its elements, *i.e.*, carbon dioxide is more stable than anticipated on the structure $O{=}C{=}O$. How can this be explained? Arguments based on quantum mechanics show that a resonance hybrid would be more stable than any single resonating structure, *i.e.*, the internal energy of a resonance hybrid is less than that calculated for any one of the resonating structures. The difference between the enthalpy of formation of the *actual* compound, *i.e.*, the *observed* value, and that of the resonating structure which has the *lowest* internal energy (obtained by *calculation*) is called the **resonance energy**. Thus the value of the resonance energy of any resonance hybrid is *not an absolute value*; it is a *relative* value, the resonating structure containing the least internal energy being chosen as the arbitrary standard for the resonance hybrid. The greater the resonance energy, the greater is the stabilisation. The resonance energy is a maximum when the resonating structures have equal energy content, and the more resonating structures there are, the greater is the resonance energy, provided that all contributing structures have similar stabilities.

Bond energies

There are two types of bond energy: (i) Dissociation energy (designated by D). (ii) Bond energy (designated by E).

(i) **Dissociation energy** (D). This is the energy required to break a *particular* bond in a polyatomic molecule, in the gaseous phase, into neutral fragments (free radicals), also in the gaseous phase, *i.e.*, D is the energy required for the reaction:

$$Y{-}Z(g) \longrightarrow Y{\cdot}(g) + Z{\cdot}(g)$$

*If the two oxygen atoms are not the same but one is isotope ^{16}O and the other isotope ^{18}O, then clearly structures (II) and (III) are different.

(ii) **Bond energy** (E). In polyatomic molecules, *e.g.*, CH_4, the four bonds are equivalent, but the energy required to break the first bond ($CH_4 \longrightarrow CH_3 \cdot + H \cdot$) is not the same as that for the second bond ($CH_3 \cdot \longrightarrow :CH_2 + H \cdot$); etc. Each individual value is, as we have seen, the bond dissociation energy of that particular bond. It can also be seen from Table 2.2 that the values of the dissociation energies of the C—H bond vary in the first four compounds. Thus, the energy of a bond depends on the nature of the rest of the molecule. In practice, it is usual to take the average of all the different values, and this *average value* is called the bond energy. For diatomic molecules, D and E are identical. Various factors are responsible for the differences in energy of a given bond in different compounds, *e.g.*, steric effects, angular strain (see also below). Table 2.2 lists some bond energies (but see also Table 4.1). P 131.

Table 2.2

Bond	Energy		Bond	Energy		Bond	Energy	
	kJ	(kcal)		kJ	(kcal)		kJ	(kcal)
CH_3—H	426·8	(102)	C≡N	866·1	(207)	O=O	497·9	(119)
$MeCH_2$—H	405·8	(97)	C—F	447·7	(107)	O—O	146·4	(35)
Me_2CH—H	393·3	(94)	C—Cl	326·4	(78)	O—H	464·4	(111)
Me_3C—H	374·5	(89·5)	C—Br	284·5	(68)	S–S	225·9	(54)
C—H (av.)	414·2	(99)	C—I	213·4	(51)	S—H	347·3	(83)
C—C	347·3	(83)	C—S	272·0	(65)	S=O	497·9	(119)
C=C	606·7	(145)	H—H	431·0	(103)	F—F	150·6	(36)
C≡C	803·3	(192)	H—N	389·1	(93)	H—F	560·7	(134)
C—O	334·7	(80)	N—N	163·2	(39)	Cl—Cl	242·7	(58)
C=O	694·5	(166)	N=N	418·4	(100)	H—Cl	426·8	(102)
O=C=O	803·3	(192)	N≡N	945·6	(226)	Br—Br	188·3	(45)
C—N	284·5	(68)	N—O	200·8	(48)	H—Br	364·0	(87)
C=N	615·1	(147)	N=O	606·7	(145)	I—I	150·6	(36)
						H—I	297·1	(71)

The resonance energy of a molecule is a property of the molecule in the ground state. Most measurements of resonance energies have been obtained from heats of combustion, but some measurements have also been obtained from heats of hydrogenation; the latter method is more accurate than the former. The heat of combustion of the most stable classical structure (*i.e.*, the resonating structure with the lowest internal energy) is, as mentioned above, obtained by *calculation*. This presupposes that *accurate* values for bond-energies are known. If these values are *not* accurate, then one cannot expect to obtain accurate resonance energies. In addition to this problem, there are other difficulties involved with the measurement of resonance energies. The reference molecule, not being identical with the actual molecule (resonance hybrid molecule), will therefore differ in a number of physical properties, *e.g.*, bond lengths (see also below). Although these differences may be small, nevertheless energy changes are involved in bond compression and bond stretching, and hence these energy terms should also be considered when comparing the reference molecule to the actual molecule. There are also energy terms involving hybridisation changes (see later), steric factors, etc. Now, the list of bond energies has been obtained for compounds in which these factors (steric, resonance, etc.) are absent. In general, the total chemical binding energy of a molecule may be regarded as being comprised of three factors:

(i) the sum of the bond energies; (ii) stabilising effects which increase the total binding energy, *e.g.*, resonance; (iii) destabilising effects which decrease the total binding energy, *e.g.*, steric effects. It can therefore be seen that the difference between the energy of the reference molecule (calculated from bond energies) and the actual molecule is not necessarily a measure of the resonance energy. The difficulty here is that there appears to be no unambiguous way of evaluating the individual contributions of the above three factors. Because of this, the term **stabilisation energy** is often used instead of *resonance energy* to designate the energy difference between the reference molecule and the resonance hybrid molecule. In this sense, stabilisation energy is the sum of the stabilising and destabilising effects. However, when the S.E. is relatively large, the R.E. greatly outweighs the destabilising effects, and so the term resonance energy is used in this book to represent the overall stabilising effect due to (predominantly) resonance (see p. 576 for a further discussion).

Another property of the resonance hybrid which differs from that of any of the resonating structures is the **bond length**, *i.e.*, the equilibrium distance between atoms joined by a covalent bond. The normal length of the carbonyl double bond ($C\!=\!O$) in ketones is about 1·20 Å; the value found in carbon dioxide is 1·15 Å. For a given pair of atoms, the length of a single bond is greater than that of a double bond, which, in turn, is greater than that of a triple bond. Resonance, therefore, accounts for the carbonyl bond in carbon dioxide not being single, double or triple.

Table 2.3 gives some bond lengths (but see also Table 4.1).

In a resonance hybrid, the electronic arrangement and bond lengths will be different from those of the resonating structures. Consequently the observed dipole moment may differ from that calculated for any one structure (but see also p. 333).

Table 2.3

Bond	Length (Å)	Bond	Length (Å)	Bond	Length (Å)
C—C	1·54	C—S	1·82	C—F	1·42
C=C	1·40	C—O	1·43	C—Cl	1·77
C≡C	1·21	C=O	1·20	C—Br	1·91
C—H	1·12	O—H	0·97	C—I	2·13
C—N	1·47	N—H	1·03		

As we have seen above, in a resonance hybrid all the molecules have the same structure. A difficulty that arises with the resonance theory is the representation of a resonance hybrid. The molecules corresponding to the structures chosen as the resonating structures do not necessarily have an actual existence. Thus, if these resonating structures are fictitious, what fictitious structures are we to choose? The normal way of solving this problem is first to ascertain the structure of the molecule by the usual methods, and then describe it by means of the classical valency-bond formula. Let us consider again the case of carbon dioxide. The classical structure is (I) (see above). As we have also seen, it has been found that not all the properties of carbon dioxide are described by this classical formula. Thus the classical structure is an approximation, *and it is in this sense that classical structures are fictitious*. By *postulating* other electronic structures (II) and (III), wave functions can then also be obtained for these fictitious structures. By a linear combination of all three functions, a 'structure' is obtained which describes the properties of carbon dioxide. This 'structure' is called the resonance hybrid of the classical (I) and the two postulated electronic structures (II) and (III).

It is very important to note here that wave-mechanics offers a theoretical method of studying the electron distribution in a molecule, but starts with a knowledge of the relative positions of all the nuclei concerned, *i.e.*, with the 'classical structure'. Theoretically, it is possible to start from a molecular formula, and then solve the structure. The number of possibilities and mathematical difficulties, however, are far too great at present, and so it seems that the chemist, who arrives at the classical structures by chemical and physical methods, will still be 'in business' for a long time to come. Since, however, by means of wave-mechanics one can calculate the density of electronic charge at all points in a molecule (of known classical structure), it is possible from this information to deduce charge distributions, bond lengths and bond angles, and consequently the size and shape of a molecule.

The question that now arises is: Starting with the classical structure, what other electronic structures are we justified in postulating? A very important point in this connection is that *resonance can occur only when all the atoms involved lie in the same plane* (*or nearly in the same plane*). Thus, any change in structure which prevents planarity will diminish or inhibit resonance. This phenomenon is known as *steric inhibition of resonance* (p. 656).

In practice, then, the conditions described above must be considered when choosing canonical structures. At the same time, the following observations will be a useful guide:

(i) Elements of the first two rows never violate the octet rule (hydrogen, of course, can never have more than a duplet).

(ii) The more stable a structure, then the larger will be its contribution to the resonance state. The stability of a molecule can be found from its bond energies. Generally, the structure with the largest number of bonds is the most stable. This is because the sum of the bond energies in a molecule gives a measure of the stability of the compound.

(iii) If the different resonating structures have the same number of bonds, but some structures are charged, then the charged molecules will be less stable than the uncharged. On the other hand, if charged forms have more covalent bonds than uncharged forms, then these charged forms may make a significant contribution to the resonance. The high energy content of a charged molecule is due to the work put into the molecule to separate unlike charges, and the greater the distance of charge separation, the less stable is that structure. In the same way, structures in which adjacent atoms carry like charges are relatively unstable, since work must also be put into the molecule to bring like charges together. However. because charged structures are stabilised by solvation (p. 159), the greater the polarity of the solvent the greater will be the contributions of charged structures. It should also be noted that the most stable-charged structures are those in which the negative charge is on the most electronegative atom and the positive charge on the least electronegative atom. Furthermore, an electron-releasing group at one end of a molecule and an electron-withdrawing group at the other end tend to stabilise (and hence increase the contribution of) charged structures, *e.g.*, the charged contribution in (V) will be much greater than that in (IV):

(IV) $H_2\overset{..}{N}$—CH=CH—CH=CH_2 $H_2\overset{+}{N}$=CH—CH=CH—$\overset{-}{C}H_2$

(V) $H_2\overset{..}{N}$—CH=CH—CH=O $H_2\overset{+}{N}$=CH—CH=CH—$\overset{-}{O}$

When the different resonating structures contain the same number of bonds, the phenomenon is called *isovalent* resonance, and when a different number of bonds, *heterovalent* resonance (Mulliken, 1959).

The final problem is the method of representing a resonance hybrid. Various methods have been used, and the one used in this book is that introduced by Bury (1935). This consists of writing down the resonating structures with a double-headed arrow between each pair:

$$O=C=O \longleftrightarrow \overset{-}{O}-C\equiv\overset{+}{O} \longleftrightarrow \overset{+}{O}\equiv C-\overset{-}{O}$$

Inductive and resonance (mesomeric) effects are permanently operating in the 'real' molecule; collectively they are known as the *polarisation effects*. On the other hand, there are also two temporary (time-variable) effects, the *electromeric effect* and the *inductomeric effect* (which operates by an inductive mechanism). Both of these are brought into play by the attacking reagent, and collectively they are known as the *polarisability effects*. Remick (1943) has suggested the use of subscripts s and d to represent the *static* (permanent) and *dynamic* (time-variable) effects. Thus the inductive effect may be represented by the symbol I_s, and the inductomeric effect by I_d. Since polarisability effects are brought into play only by the approach of the attacking reagent, they will therefore *aid* and never *inhibit* a reaction.

Strictly speaking, the term *resonance effect* (R) is not the same as the *mesomeric effect* (M). The mesomeric effect is a permanent polarisation, and the mechanism of electron transfer is the same as that in the electromeric effect, *i.e.*, the mesomeric effect is a permanent displacement of electron pairs which occurs in a system of the type $Z-C=C$; *e.g.*, $Z=R_2N$, Cl:

$$R_2N-C=C; \qquad :\overset{..}{Cl}-C=C-$$

Thus the essential requirement for mesomerism is the presence of a *multiple* bond in the molecule. On the other hand, the resonance effect embraces *all* permanent electron displacements in the molecule in the ground state, *e.g.*, the hydrogen chloride molecule is a resonance hybrid of two resonating structures:

$$H-Cl \longleftrightarrow H^+Cl^-$$

Since there is no multiple bond in this molecule, the mesomeric effect is not possible.

When the electronic displacement is *away* from the group the mesomeric (resonance) effect is said to be $+M$ $(+R)$, and when *towards* the group, $-M$ $(-R)$.

The mesomeric effect is particularly important in conjugated systems (p. 125), and the *combined* mesomeric and electromeric effects are known as the *conjugative effect*. This term is also used in the same sense as the resonance effect. Also, since this combined effect was first recognised in connection with tautomerism, it has also been called the *tautomeric effect* $(\pm T)$.

It might be noted here that when a $+M$ effect is present, then a strong electromeric effect will operate if a $+E$ effect is the requirement of the attacking reagent. If, however, the attacking reagent requires a $-E$ effect, then with the permanent $+M$ effect present, this $-E$ effect will be extremely weak or absent. Similarly, for a $-M$ effect, a $-E$ effect, if required, is strong, whereas a $+E$ effect, if required, is extremly weak or absent.

The possible polar influences of groups are shown in the following table:

Electronic mechanism	Polarisation effect (permanent)	Polarisability effect (temporary)
Inductive $(\pm I)$	Inductive (I or I_s)	Inductomeric (I_d)
Conjugative (Tautomeric, $\pm T$;	Mesomeric (M) or	Electromeric (E)
or Resonance, $\pm R$)	Resonance (R)	
Fields in which operative	Physical properties	—
	Reaction equilibria	—
	Reaction rates	Reaction rates only

As has been pointed out above, resonance describes *all* permanent electron displacements in the molecule in the ground state. It therefore follows that the I-effect can be described as due to resonance. Thus resonance is the combination of I- and M-effects. It is more convenient, however, from the point of view of the organic chemist, to consider a molecule with respect to its I-effect and 'resonance' (mesomeric) effect separately. Hence, from this point of view, resonance is the *additional permanent* electronic displacements to the I-effect, and it is customary to ignore the latter effect when discussing 'resonance'. In other words, resonance, in this context, is concerned only with the part of the molecule containing multiple bonds and is therefore, strictly speaking, *π-electron resonance*. This is the sense in which the term resonance will be used in this book.

Effect of structure on reactivity. The *type* of reaction of an organic compound is largely dependent on the nature of the functional group present (p. 74). It has been found that various structural changes, *e.g.*, the introduction of a given group into different positions in a molecule containing a given functional group, usually affect the rate of a given type of reaction and also the equilibrium position, and may even change the type of mechanism of the reaction. Much work has been done to try to correlate structure and reactivity (*i.e.*, rate of reaction), and as an outcome of this work, it appears that some sort of quantitative correlation can be made on the basis of consideration of independent contributions of inductive, resonance (mesomeric) and steric effects. When each effect has been assessed, then all three may be combined, and in this way there is obtained a relationship between structure and reactivity. In the text are discussed many cases of the effects on reaction rates and mechanism by polar (I and R) and steric effects.

Reactions in organic chemistry may be classified into the following main types: (i) substitution; (ii) replacement or displacement; (iii) addition; (iv) elimination; (v) isomerisation (rearrangement).

The hydrogen bond

Compounds containing OH or NH groups often exhibit unexpected properties such as relatively high boiling points, and it was soon felt necessary to assume that the elements oxygen or nitrogen were linked by means of hydrogen, thereby producing the *hydrogen bond*. Detailed study has shown that the unexpected properties were exhibited only when the atoms participating in the bond had high electronegativity—fluorine, oxygen and nitrogen (decreasing in this order), and to a less extent, chlorine and sulphur. Thus the hydrogen bond explained, for example, the association of hydroxylic compounds such as water, alcohols, etc., and the association of ammonia.

The exact nature of the hydrogen bond has been the subject of much discussion. The difficulty lies mainly in the fact that the energy of a hydrogen bond varies between that of the 'van der Waals forces' ($4 \cdot 184$ kJ mol^{-1}) and that of a chemical bond. Some values obtained are: H—F$\cdots$H, $41 \cdot 84$ kJ mol^{-1}; H—O$\cdots$H, $29 \cdot 29$ kJ; H—N$\cdots$H, $8 \cdot 37$ kJ. These are 'weak' hydrogen bonds, and for these the geometry of Z—H and Y is little changed when the hydrogen bond produces the complex Z—H$\cdots$Y. Hence, the bond length of Z—H is almost the same in both Z—H and Z—H$\cdots$Y. It is accepted that a number of factors contribute, but it appears that the most important one is electrostatic. In bond Z—H, if Z has high electronegativity, there will be a relatively large amount of polarity, *i.e.*, the state of affairs will be $\overset{\delta-}{Z}—\overset{\delta+}{H}$, where $\delta+$ is relatively large. Since the hydrogen atom has a tiny volume, the $\overset{\delta+}{H}$ will exert a large electrostatic force and so can attract atoms with a relatively large $\delta-$ — charge, providing these atoms have a small atomic radius. Fluorine, oxygen and nitrogen are of this character. If the atom has a greater radius the electrostatic forces are weaker; thus chlorine, although it has about the same electronegativity as nitrogen, forms very weak hydrogen bonds since its atomic radius is greater. C—H bonds do not normally form hydrogen bonds, but when the electronegativity of carbon is increased by *sp* and *sp²* hybridisation (see p. 139), then apparently this group can form hydrogen bonds.

Hydrogen bond formation *intramolecularly*, *i.e.*, involving one molecule only, gives rise to *ring formation* or *chelation*, and this usually when the formation of a 5-, 6- or 7-membered ring is possible. Hydrogen bonding *intermolecularly*, *i.e.*, between two or more molecules, gives rise to *association*. Many examples of hydrogen bonding will be found in the text, and this is represented by a dotted line between the hydrogen and other atom involved (as shown above).

Hydrogen bonding affects all physico-chemical properties such as m.p., b.p., solubility, spectra, etc., *e.g.*, association produces a higher boiling point than expected (*e.g.*, from the molecular weight of the compound). On the other hand, chelation usually produces a lower boiling point than expected, *e.g.*, a nitro-compound usually has a higher boiling point than its parent compound, but if chelation is possible in the nitro-compound, the boiling point is lowered (see, *e.g.*, nitrophenols, p. 710).

Atomic and molecular orbitals

So far, we have discussed the structure of molecules in terms of valency bonds. There is an alternative method of investigating the structure of molecules, and to appreciate this approach—and to extend the other—it is necessary to consider the structure of matter from the point of view of wave-mechanics. Classical physics (*i.e.*, the laws of mechanics, etc.) is satisfactory when dealing with large masses. These laws are approximations, but deviations become significant only when dealing with very small particles such as electrons and nuclei. The behaviour of these small particles, however, may be satisfactorily studied by *wave (quantum) mechanics*. This uses the idea of the particle-wave duality of matter. It has already been pointed out that the electron may be regarded as a tiny mass carrying a negative charge. In 1923, de Broglie proposed that every moving particle has wave properties associated with it. This was first experimentally verified in the case of the electron (Davison and Germer, 1927; G. P. Thomson, 1928). Thus an electron has a dual nature, particle and wave, but it behaves as one or the other according to the nature of the experiment; *it cannot at the same time behave as both*. According to wave-mechanics, a moving particle is represented by a wave function ψ such that $\psi^2 d\tau$ is the *probability* of finding the particle in the element of volume $d\tau$. The greater the value of ψ^2, the greater is the probability of finding the electron in that volume $d\tau$. Theoretically, ψ has a finite value at a large distance (compared with atomic dimensions) from the nucleus, but in practice there is very little probability of finding the electron beyond a distance of 2–3 Å. This spatial behaviour of an electron, in terms of probability, is known as an *orbital*. When associated with one nucleus, the electron is said to be in an *atomic orbital* (A.O.), and when with a number of nuclei, the electron is said to be in a *molecular orbital* (M.O.). The behaviour of an electron may be represented by a *charge-cloud*, the density of the cloud at any point being equal to the value of ψ^2 at that point.

In 1926, Schrödinger developed the wave-equation, which connected the wave function ψ of an electron with its energy, E. This equation has an infinite number of solutions, but very few of these solutions describe the *known* behaviour of electrons. Thus only certain values for E are permissible, since certain conditions must be satisfied. The permitted solutions for ψ are called the *eigenfunctions*, and the corresponding values of E are called the *eigenvalues*. A number of eigenfunctions exist, the simplest being those which possess *spherical symmetry* (ψ_s function), and the next simplest being those which possess an *axis of symmetry* (ψ_p function). Atomic orbitals are thus classified as $s, p, d, f, \ldots$ orbitals, and the energy of any A.O. is the eigenvalue (energy value) corresponding to that wave function ψ. We shall here be concerned with only s and p orbitals.

In addition to its wave function ψ, an electron also has spin. Two electrons can have the same wave function, *i.e.*, can occupy the same orbital, *provided their spins are opposite* (Pauli exclusion principle). In this case the electrons are said to be paired.

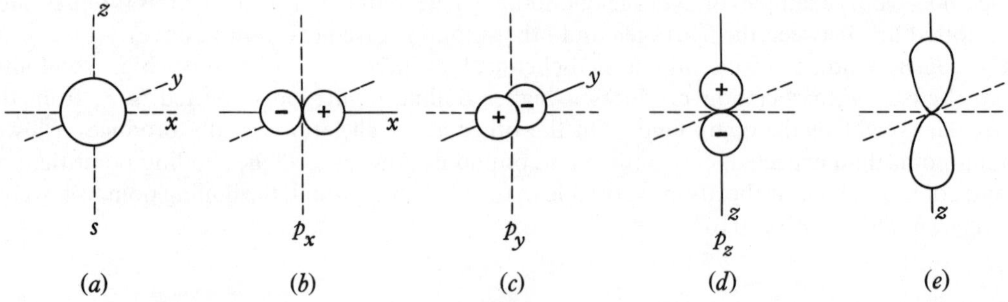

(a)   (b)   (c)   (d)   (e)

Fig. 2.1

The wave function ψ is a function of the co-ordinates of the electron and consists of two parts, the distance of the electron from the nucleus, *i.e.*, the *radial distribution function*, and the orientation of the electron with respect to the nucleus, *i.e.*, the *angular distribution function*. The two common ways of representing orbitals omit the radial distribution function and draw polar graphs in three dimensions of the angular function or its square, and the orbitals are depicted as sections through these polar (three-dimensional) graphs. The plot of the angular function for an s-orbital consists of a sphere, and the orbital is thus represented as a circle. This is also the case for the plot of the square of the angular function (Fig. 2.1a). On the other hand, the plot of the angular function for a p-orbital consists of two spheres in contact at the origin, and hence this orbital is represented as two circles in contact (Figs. 2.1 b, c, d). If, however, the polar plot is of the square of the angular function, then the p-orbital consists of two lobes in contact at the origin, and hence is represented, *e.g.*, the p_z-orbital, as shown in Fig. 2.1e. It can be seen in Fig. 2.1 that p-orbitals have a marked directional character, each orbital having an axis at right angles to those of the other two, and hence they are known as the p_x, p_y, p_z orbitals, respectively. These orbitals are entirely equivalent except for their directional property. It should also be noted that one circle of a p-orbital is marked with a positive sign and the other with a negative sign. When the plot of the square of the angular function is used, the resultant values are always positive and so both lobes are positive. In the same way, an s-orbital can be $+$ or $-$ for the plot of the angular function (but not its square). These signs arise from the fact that a wave function can have positive and negative regions, but the signs have no physical significance. Thus, it does not matter which circle in a p-orbital is given the positive sign; the other is then given the $-$ sign. Where ψ is zero, there is a node, *i.e.*, there is no likelihood of finding an electron in a nodal plane. These relative signs are useful when considering the overlap of atomic orbitals (see p. 48), and lobes are often used instead of circles.

The order of orbital energies is $1s < 2s < 2p < 3s < 3p \ldots$. When an electron absorbs energy, it is driven into an orbital of higher energy. The atom is then said to be 'excited' and is more reactive. On returning to its normal orbital, the electron loses this extra energy. When all the electrons in an atom are in their normal orbitals, *i.e.*, orbitals of lowest energy, the atom is said to be in the 'ground' state.

So far, we have dealt only with atoms, *i.e.*, with electrons associated with *one* nucleus. The wave-equations for molecules cannot be solved without making some approximations. Two types of approximations have been made, one set giving rise to the *valence-bond* method (V.B.); and the other set to the *molecular orbital* (M.O.). The V.B. method—due mainly to the work of Heitler, London, Slater and Pauling—considers the molecule as being made up of atoms *with electrons in atomic orbitals on each atom*. Thus a molecule is treated as if it were composed of atoms which, to some extent, retain their individual character when linked to other atoms. The M.O. method—due mainly to the work of Hund, Lennard-Jones and Mulliken—treats a molecule in the same way as an atom, except that in the molecule an electron moves in the field of *more than one nucleus, i.e., molecular orbitals* are *polycentric*. Thus each electron in a molecule is described by a certain wave function, the *molecular* orbital, for which various shapes can be drawn as for A.O.s, but differing in that the former are polycentric and the latter monocentric. In general, the greater the freedom (*i.e.*, the larger the region for movement) allowed to an electron, the lower will be its energy. Hence atoms combine to form a molecule because, owing to the overlap of the A.O.s when the atoms are brought together, the electrons acquire a greater freedom, and the energy of the system is lowered below that of the separate atoms. Energy would therefore have to be supplied to separate the atoms in the molecule, and the greater the amount of energy necessary, the stronger are the bonds formed between the various atoms.

Let us now consider the case of the hydrogen molecule. A hydrogen atom has one $1s$ electron. When the bond is formed between two hydrogen atoms to form the hydrogen molecule, these two $1s$ electrons become *paired* to form *molecular electrons, i.e.*, both occupy the same M.O., a state of affairs which is possible provided their spins are antiparallel. A very important principle for obtaining the M.O. is that the bond energy is greatest when the component A.O.s overlap one another as much as possible. To get the maximum amount of overlap of orbitals, the orbitals should be in the same plane. Thus the M.O. is considered as being a *linear combination of atomic orbitals with maximum overlap* (L.C.A.O.). Furthermore, according to L.C.A.O. theory, the binding energy is greater the more nearly equal are the energies of the component A.O.s. If these energies differ very much, then there will be no significant combination between the two atoms concerned.

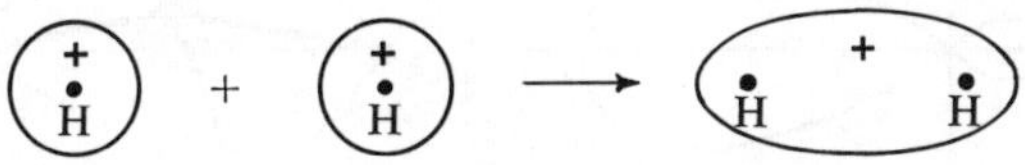

Fig. 2.2

Since the hydrogen molecule is composed of two identical atoms, the probability of finding both electrons simultaneously near the same nucleus is very small. Hence one might expect the M.O. to be symmetrical with respect to the two hydrogen nuclei, *i.e.*, the M.O. in the hydrogen molecule will be 'plum-shaped' (Fig. 2.2).

Although the probability of finding the two electrons simultaneously near the same nucleus is very small, nevertheless this probability exists, and gives rise to the two *ionic* structures H^+H^- and H^-H^+. Thus the hydrogen molecule will be a resonance hybrid of three resonating structures, one purely covalent (*i.e.*, the two electrons are *equally* shared), and two ionic (*i.e.*, the pair of electrons are associated with *one* nucleus all the time):

$$\text{H:H} \longleftrightarrow \overset{+}{\text{H}}\text{:}\overset{-}{\text{H}} \longleftrightarrow \overset{-}{\text{H}}\text{:}\overset{+}{\text{H}}$$

Calculation has shown that the ionic structures contribute very little to the actual state of the hydrogen molecule, and the bond between the two hydrogen atoms is described as a *covalent bond with partial ionic character.* It should here be noted that when the single bond is formed between the two hydrogen atoms, the probability of finding the electrons is greatest in the region *between* the two nuclei. It is this concentration of the negatively charged electrons between the two positive hydrogen nuclei that binds the nuclei together. Since electrons are negatively charged, they will repel each other and so tend to keep out of the region between the two nuclei. On the other hand, since the spins of the two electrons are antiparallel, this produces attraction between the two electrons, thereby tending to concentrate them in the internuclear region. The net result is that the electron density for *paired electrons* is greatest *between* the two nuclei. Such a bond is said to be a *localised* M.O., and preserves the idea of a bond connecting the two atoms. This localisation (in a covalent bond) gives rise to the properties of bond lengths, dipole moments, polarisability and force constants.

Hybridisation of bond orbitals

The electron configuration of carbon is $(1s)^2(2s)^2(2p_x,2p_y)$. It therefore appears to be bivalent. To be quadrivalent, the $(2s)^2$ and the $(2p_x,2p_y)$ electrons must be involved. One way is to *uncouple* the paired $2s$ electrons, and then *promote* one of them to the empty $2p_z$ orbital. Should this be done, four valencies would be obtained, since each of the electrons could now be paired with an electron of another atom. The resulting bonds, however, would not all be equivalent, since we would now have the component A.O.s $2s, 2p_x, 2p_y, 2p_z$. All work on *saturated* carbon compounds of the type Ca_4 indicates that the four valencies of carbon are equivalent. In order to get four equivalent valencies, the four 'pure' A.O.s must be 'mixed' or *hybridised*. It is possible, however, to hybridise these four 'pure' A.O.s in a number of ways to give four valencies which may, or may not, be equivalent. Three methods of hybridisation are important: (i) *tetrahedral* (sp^3 bond), (ii) *trigonal* (sp^2 bond), (iii) *digonal* (sp bond). The following discussion applies to molecules of the type CH_4 (sp^3), $CH_2{=}CH_2$ (sp^2), and $CH{\equiv}CH$ (sp).

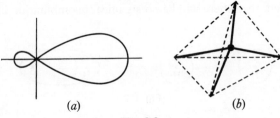

(a)   (b)

Fig. 2.3

sp^3 (i) In **tetrahedral hybridisation**, the $(2s)$ and $(2p_x, 2p_y, 2p_z)$ electrons are *all* hybridised, resulting in **four equivalent orbitals arranged tetrahedrally**, *i.e.,* pointing towards the four corners of a *regular* tetrahedron (Fig. 2.3b) The orbitals are greatly concentrated along these four directions (Fig. *a* shows the shape along one of these directions). Then, by linear combination with the $1s$ orbitals of four hydrogen atoms, four equivalent M.O.s are obtained for methane. Because of the large amount of overlapping between the hybridised A.O.s of the carbon and the s A.O. of the hydrogen atom, there will be strong binding between the nuclei. As in the case of the hydrogen molecule, each M.O. is almost completely confined to the region between the two nuclei concerned, *i.e.,* in methane are four *localised* molecular orbitals. This scheme of localised M.O.s may be satisfactorily applied to all compounds containing single covalent bonds. Bond orbitals which are symmetrical about the line joining the two nuclei concerned are known as **σ-bonds** (sigma-bonds).

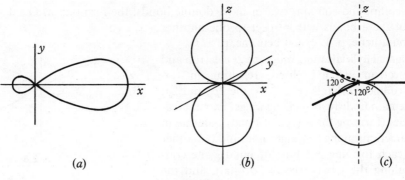

Fig. 2.4

sp^2

(ii) In **trigonal hybridisation**, the $2s$, $2p_x$ and $2p_y$ orbitals are hybridised, resulting in **three equivalent coplanar orbitals pointing at angles of 120° in the xy plane** (Fig. 2.4a). The remaining orbital is the undisturbed $2p_z$ (Fig. 2.4b). Thus there will be three equivalent valencies in one plane and a fourth pointing at right angles to this plane (Fig. 2.4c). The three coplanar valencies form σ-bonds, and the $2p_z$ valency forms the so-called π-bond (pi-bond). The $2p_z$ electrons are known as π-electrons, *mobile* electrons or *unsaturation* electrons when they form the π-bond. The trigonal arrangement occurs in compounds containing a *double* bond, which is regarded as being made up of a strong bond (σ-bond) between two trigonal hybrid A.O.s of carbon, and a weaker bond (π-bond) due to the relatively small overlap of the two pure p_z orbitals in a plane *at right angles* to the trigonal hybrids. Fig. 2.5(a) shows the plan, and (b) the elevation of ethylene, $CH_2=CH_2$.

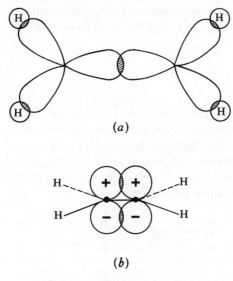

Fig. 2.5

The H—C—H angle in ethylene has been measured spectroscopically, and it has been found to be 119° 55′ (Gallaway *et al.*, 1942). This is in agreement with the value expected for trigonal hybridisation.

It is the π-electrons which are involved in the electromeric and resonance effects.

When a compound contains two or more double bonds, the resulting M.O.s depend on the positions of these bonds with respect to one another (see, *e.g.*, butadiene, p. 129, and benzene, p. 577).

(iii) In **digonal hybridisation**, only one 2s electron and the $2p_x$ electron are hybridised, resulting in **two equivalent collinear orbitals** (Fig. 2.6); the $2p_y$ and $2p_z$ electrons remain undisturbed. Thus we get two equivalent valencies (forming the σ-type of bond) pointing in opposite directions along a straight line, and two other valencies (each forming a π-type of bond), one concentrated along the *y*-axis (the $2p_y$ orbital), and the other along the *z*-axis (the $2p_z$ orbital). The digonal arrangement occurs in compounds containing a *triple* bond, *e.g.*, acetylene.

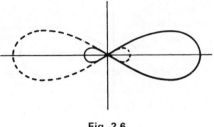

Fig. 2.6

From the above discussion, it can be seen that the carbon valency angles follow the order sp^3 (109° 28′) < sp^2 (120°) < sp (180°). Thus, increasing the *p*-character of a bond decreases the bond angle, or increasing the *s*-character of a bond increases the bond angle. At first sight, it might have been anticipated that the bond angle is therefore fixed, *i.e.*, it depends only on the nature of the hybridisation. In practice, it is found that bond angles are also dependent on the size (which produces a steric effect) and the electronegativity (which changes the amount of *s*-character in the bonds; see p. 139) of the atoms (or groups) attached to the carbon atom, *e.g.*, in $CHCl_3$, the Cl—C—Cl angles are increased from the normal angle of 109° 28′ to about 111° and the Cl—C—H angles decreased to about 108° (see also below).

> When the electrons of any atom have been placed in hybridised orbitals, that atom is said to be in a 'valence state'. The atom on its own cannot exist in a valence state; energy is required to promote the atom to this condition. This energy is obtained through the formation of bonds which are stronger with the hybridised orbitals than with the 'pure' orbitals, *i.e.*, more energy is released with the former than with the latter.

Shapes of molecules

As we have seen, in *saturated* carbon compounds the four sp^3 orbitals point towards the corners of a tetrahedron. In ethylene we have used sp^2 trigonal hybridisation (one σ- and one π-bond) to describe the double bond. It is possible, however, to still use sp^3 hybridisation to describe ethylene. In this case two electrons are in one orbital of 'banana' shape ('bent' bond), and the other two electrons in a second 'banana' orbital, equivalent to the first but the mirror image of it. Thus ethylene may be represented (Fig. 2.7):

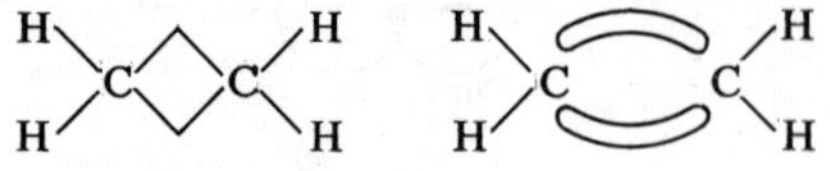

Fig. 2.7

It is interesting to note that this 'bent' bond method of representing ethylene is equivalent to Baeyer's description of a double bond (p. 539).

In the same way, the triple bond in acetylene (previously described in terms of one σ- and two π-bonds) may also be regarded as made up of three equivalent sp^3 hybrids symmetrically disposed round the C—C axis.

Quantum mechanical arguments show that both methods of representing these multiple bonds are equal to each other, each method having certain advantages. The σ-π bond method is more convenient for describing transitions from one state into another (*e.g.*, in electronic spectra), whereas the 'bent' bond method is more convenient for describing electron distribution in a molecule.

Now let us consider molecules in which the central atom has lone pairs. Consider nitrogen, with electron configuration $(1s)^2(2s)^2(2p)^3$. This has three $2p$ orbitals, and if each combines with a hydrogen atom, then the molecule of ammonia is $:NH_3$. In this molecule there is a lone pair $(2s)^2$, and the three hydrogen atoms, bound by the overlap with $2p$ orbitals, will therefore have a valency angle of 90°. The nitrogen atom is tercovalent, and by virtue of its lone pair, can act as a donor to become quadrivalent uni-electrovalent, *e.g.*,

$$:NH_3 + HCl \longrightarrow NH_4{}^+Cl^-$$

At first sight it might appear that the four hydrogen atoms in the ammonium ion are not equivalent; three bonds are formed from $2p$ electrons, and one by the lone $2s^2$. However, the fact is that all four hydrogen atoms in $NH_4{}^+$ are equivalent, and also the valency angle is about 109·5° (the tetrahedral value). Moreover, the valency angle in ammonia is about 107° and not 90° (the expected value from $2p$ bonding).

Now consider oxygen $(1s)^2(2s)^2(2p)^4$. One $2p$ orbital is doubly filled, and so, in water, one would expect that the two single $2p$ electrons would combine with hydrogen to form water in which the bond angle HOH is 90°. Actually the bond angle is about 104·5°. Also, since the water molecule has two lone pairs, it can act as a donor to form the hydroxonium ion, $H_3O:^+$, and it is found that all the three hydrogen atoms are equivalent. The valency angles are nearly those expected of tetrahedral configurations, and the difference from 90° is far too large to be accounted for by repulsion between hydrogen atoms. A satisfactory explanation is as follows. Sidgwick *et al.* (1940) assumed that lone pairs of electrons and bonding pairs were of equal importance, and that they arranged themselves symmetrically so as to minimise the repulsions between them. Thus pairs of electrons in a valency shell, whether a bonding pair or a lone pair, are always arranged in the *same* way, and this depends only on the total number of pairs. Thus two pairs are arranged linearly (*e.g.*, $HgCl_2$), three pairs in the form of an equilateral triangle (*e.g.*, BCl_3), four pairs tetrahedrally (*e.g.*, CH_4), five pairs in the form of trigonal bipyramid and six pairs octahedrally. In all cases it is assumed that all the pairs of electrons occupy hybridised orbitals and only σ-bonds are present (see also p. 402).

Now let us consider ammonia and water from this point of view. In each of the valency shells of these central atoms there are four pairs of electrons (nitrogen with one lone pair, and oxygen with two lone pairs). If these four pairs occupy four tetrahedrally hybridised orbitals, then the valency angles should be about 109·5°. As we have seen above, in ammonia the angle is about 107°, and in water about 104·5°. The problem then is to account for these deviations from the anticipated 'regular' shapes. This may be explained by assuming electrostatic repulsions between electron pairs in the valency shell (both bonding and lone pairs) are in the following order:

lone pair—lone pair > lone pair—bond pair > bond pair—bond pair.

This order can be explained on the basis that lone pairs are more concentrated than bonding pairs (the latter are 'stretched' in the formation of covalent bonds). Thus lone-pair electrons exert greater electrostatic repulsions on other lone pairs than on bonding pairs. Consequently, when lone pairs occupy the valency shell, bonding pairs are 'forced together'. In CH_4 there are four bonding pairs, and so the distribution is symmetrical with a valency angle of 109° 28'. In NH_3, there are three bonding pairs and one lone pair, and since the latter has a bigger repulsive force, the bonding pairs are forced closer together (the bond angle is about 107°). It should be noted that the lone pair is also in a hybridised orbital, and so when the ammonia molecule is converted into the ammonium ion the four bonding pairs are in the same state of hybridisation, and consequently all four hydrogen atoms are equivalent.

In the water molecule (bond angle 104·5°), there are *two* lone pairs, and consequently the bonding pairs are 'closed up' more than in ammonia.

Finally, it should be noted that when deciding the shape of a molecule, it is the number of σ-electrons and lone pairs around the bonded atom that are counted; π-bonds are fitted in *afterwards* (see, *e.g.*, p. 402). The effect of π-bonds is to *shorten* bond lengths without affecting the shape of the molecule.

Bonding and anti-bonding orbitals. When two atomic orbitals overlap, the nature of the M.O. depends on whether the overlap occurs between two regions having the same sign or opposite signs. When the former occurs, the result is a *bonding orbital*, whereas when the latter occurs, the result is an *anti-bonding orbital*. In some cases, two orbitals of opposite sign cancel out each other; the result is *non-bonding*. In bonding orbitals, the electron charge is concentrated in the region between the two nuclei, thereby holding the two nuclei together. On the other hand, in anti-bonding orbitals electron charge is withdrawn from the region between the two nuclei, thereby resulting in increased repulsion between the two nuclei.

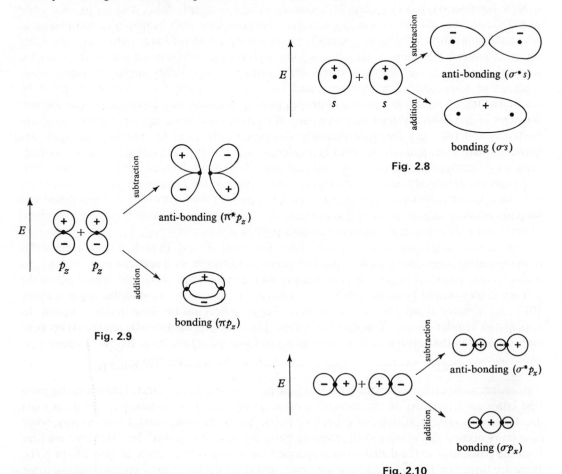

Fig. 2.8

Fig. 2.9

Fig. 2.10

The normal symbols, σ and π, used to designate bonding orbitals, are starred to designate the corresponding anti-bonding orbitals, *i.e.*, σ^*, π^*. Figures 2.8, 2.9, and 2.10 illustrate combinations of various A.O.s, the anti-bonding orbital involving the reversal of the sign (or signs) in one of the orbitals (arbitrarily the one on the right-hand side). In general, if n A.O.s are combined, then there are n M.O.s. Some of these are bonding, other anti-bonding, and in fewer cases, some are non-bonding (see, *e.g.*, Fig. 20.2, p. 577).

The general nature of organic reactions

Much work has been done on reaction mechanisms, *i.e.*, the actual steps by which a reaction takes place. A chemical equation indicates the initial and final products of a reaction; rarely does it indicate *how* the reaction proceeds. Many reactions take place via intermediates which may or may not have been isolated. When the products of a reaction are formed by a single collision of the reactant molecules, *i.e.*, the reaction proceeds without any intermediates, the reaction is said to be a *one-step* (or *elementary*) reaction. Most reactions, however, are *complex*, *i.e.*, they occur via a number of reaction steps.

Mechanisms are, in general, theories that have been devised to explain the facts which have been obtained experimentally. Mechanistic studies also include the nature of the *transition state* leading to intermediates (real or postulated) and to the products (see below). The interpretation of experimental data may not be clear-cut; several mechanisms may appear to fit equally well. In any case, the acceptability is strengthened when a particular mechanism can be used to make predictions which are borne out in practice. Many methods are used to elucidate mechanisms, some of the commoner ones being:

1. Kinetics. Kinetic studies are concerned with rates of reactions and provide the most general method for determining reaction mechanisms.
2. The identification of *all* the products of a reaction.
3. The detection, or better still (if it is possible), the isolation of intermediates. A special method of detection of intermediates is the *trapping experiment*.
4. The effect on reaction rates of changing the structure of the reactants.
5. The effect on reaction rates of changing the solvent.
6. Stereochemical evidence.
7. The use of isotopes. This method is particularly useful for tracing the part played by a particular atom in a reaction.
8. The use of crossed-experiments.

Applications of these methods are discussed in the text, but before ending this discussion, there are two other points of interest. One is the *principle of microscopic reversibility*. According to this principle, the mechanism of any reaction, under a given set of conditions, is identical in microscopic detail to that of the reverse reaction under the same conditions, except that it proceeds in the opposite way. With this principle it has been possible to deduce mechanisms where the forward or backward reactions do not lend themselves to kinetic studies (see, *e.g.*, p. 108).

The other point is that when the various possible products of a reaction are *not* interconvertible under the conditions of the reaction, then the product formed most rapidly will be the one that predominates in the products. Such reactions are called *kinetically controlled reactions*, and the most rapidly formed product is known as the *kinetically controlled* product. If, however, the possible products of the reaction are interconvertible under the reaction conditions, then the most stable product will predominate in the final products provided that enough time is given for equilibrium to be established. Reactions such as these are called *equilibrium controlled reactions*, and the most stable product is known as the *thermodynamically controlled* product (see p. 801 for a detailed discussion). Kinetically controlled reactions are the more common ones.

Now let us examine in more detail what happens when molecules containing *covalent* bonds undergo chemical reaction. Consider the reaction

$$Y + R - X \longrightarrow Y - R + X$$

where RX and RY are both covalent molecules. It can be seen that in this reaction, bond R—X has been broken and the new bond Y—R has been formed. The mechanism of the reaction depends on the way in which these bonds are broken. There are three possible ways in which this may occur, and the result of much work has shown that the actual way in which the break occurs depends on the nature of R, X and Y, and the experimental conditions.

(i) Each atom (forming the X—R bond) retains one electron of the shared pair, *i.e.*,

$$R - X \longrightarrow R \cdot + X \cdot \qquad or \qquad Y \cdot + R - X \longrightarrow Y - R + X \cdot$$

To make the representation of this type of bond breaking and making similar to those described below, one could use a half arrow head (*cf.* mass spectrometry, p. 21):

$$R \overset{\frown}{-} X \longrightarrow R \cdot + X \cdot \qquad or \qquad \overset{\frown}{Y} \; R \overset{\frown}{-} X \longrightarrow Y - R + X \cdot$$

This gives rise to *free radicals*, and the breaking of the bond in this manner is known as *homolytic* fission (*homolysis*). Free radicals are *odd electron molecules*, *e.g.*, methyl radical $CH_3 \cdot$, triphenylmethyl radical $(C_6H_5)_3C \cdot$, etc. The majority are electrically neutral (some free radical ions are known). All possess addition properties, and are extremly reactive; when a free radical is stable, its stability is believed to be due to resonance. Free radicals are *paramagnetic*, *i.e.*, possess a small permanent magnetic moment, due to the presence of the odd (unpaired) electron. This property is used to detect the presence of free radicals. Diradicals are also known; these have an even number of electrons, but *two are unpaired* (see, *e.g.*, methylene, anthracene). In general, free-radical reactions are catalysed or initiated by compounds which generate free radicals on decomposition, or by heat or light. Furthermore, a reaction which proceeds by a free-radical mechanism can be inhibited by the presence of compounds that are known to combine with free radicals. Another important characteristic is that a free-radical mechanism leads to abnormal orientation in aromatic substitution.

(ii) Atom (or group) R retains the shared pair. This may be represented as:

$$R \overset{\frown}{-} X \longrightarrow R^- + X^+ \qquad or \qquad Y \; \overset{\frown}{R} \overset{\frown}{-} X \longrightarrow Y - R + X^+$$

This is known as *heterolytic* fission (*heterolysis*), and Y is said to be an *electrophilic* (electron-seeking) or *cationoid* reagent, since it *gains* a share in the two electrons retained by R. Also an electrophilic reagent is most likely to attack a molecule at the point of highest electron density. When an electrophilic reagent is involved in a *substitution* or a *replacement* reaction, that reaction is represented by S_E (S referring to substitution, and E to the electrophilic reagent). When $\bar{R}:$ is a negative group in which the carbon atom carries the negative charge, *i.e.*, has an unshared pair of electrons, the group is known as a **carbanion** (but see below). There is a great deal of evidence to show that the configuration of the carbanion is tetrahedral; it is sp^3 hybridised, with the unshared pair of electrons occupying one orbital.

(iii) Atom (or group) R loses the shared pair, *i.e.*, the shared pair remains with X. This may be represented as:

$$R \overset{\frown}{-} X \longrightarrow R^+ + X^- \qquad or \qquad \overset{\frown}{Y} \; R \overset{\frown}{-} X \longrightarrow Y - R + X^-$$

This also is heterolytic fission (heterolysis), and Y is said to be a *nucleophilic* (nucleus-seeking) or *anionoid* reagent, since it *supplies* the electron pair. A nucleophilic reagent is most likely to attack a molecule at the point of lowest electron density. When a nucleophilic reagent is involved in a substitution or a replacement reaction, that reaction is represented by S_N. When R^+ is a positive group in which the carbon atom carries the positive charge, *i.e.*, lacks a pair of electrons in its valency shell, the group is known as a **carbonium ion**, and is in a trigonal state of hybridisation (but see later). Such a carbonium ion is said to have a 'classical' structure. There are, however, various cases where the ion is better represented as a *bridged carbonium ion*, and in these cases the ions are said to have a 'non-classical' structure (see, *e.g.*, p. 147).

Because of their charge, carbonium ions and carbanions are very reactive. In many cases they are also very unstable, but they may be stabilised by delocalisation (*i.e.*, by spreading) of the charge by means of solvation. Alternatively, the ion may be stabilised by delocalisation of the charge *within* the molecule by inductive and/or resonance effects. It should be noted that if there is spreading of charge, the carbon atom may, in fact, carry very little charge in the carbonium ion or carbanion (see, *e.g.*, p. 215).

Transition state theory of reactions. According to the *collision theory* of reactions, before molecules can enter into chemical reaction, they must collide and they must be *activated*, *i.e., they must attain a certain amount of energy* (E) *above the average value*. However, the rate of a reaction depends not only on the frequency of collisions in which the energy of activation is reached but also on whether the colliding molecules are suitably oriented with respect to each other for effective reaction to occur. This limitation is known as the *probability* or *steric* factor, and depends, for a given type of reaction, on the geometry of the reacting molecules. A simple example of the steric factor is that in the reaction

$$2HI \longrightarrow H_2 + I_2$$

If hydrogen iodide decomposes on collision, then activated molecules can collide in one of two ways, the 'right' way leading to decomposition, and the 'wrong' way leading to merely a 'change in partners'.

$$
\begin{array}{cccc}
& H \cdots\!\!\rightarrow H & & H\!-\!H \\
\text{'Right way'} & | \quad\quad | & \longrightarrow & + \\
& I \cdots\!\!\rightarrow I & & I-I
\end{array}
$$

$$
\begin{array}{cccc}
& H \cdots\!\!\rightarrow I & & H\!-\!I \\
\text{'Wrong way'} & | \quad\quad | & \longrightarrow & + \\
& I \cdots\!\!\rightarrow H & & I-H
\end{array}
$$

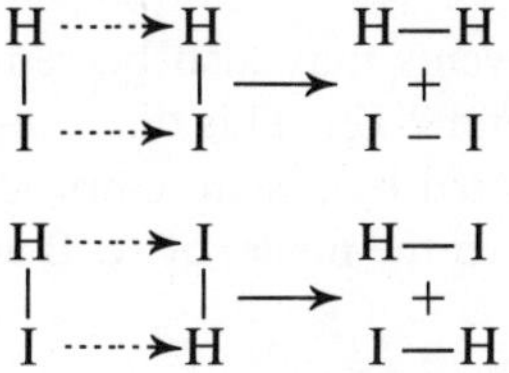

One can expect that when various paths are possible for a given reaction under given conditions, then the path actually followed will be the one requiring the lowest energy of activation. The problem therefore is to try to work out the path that requires the minimum energy of activation.

The *transition state theory* of reactions does not use the simple idea of collision, but considers how the potential energy of a system of atoms and/or molecules varies as the molecules are brought together. Consider the reaction

$$A + BC \longrightarrow AB + C$$

This is known as a *three-centre reaction*. London (1929), by making certain approximations, showed that the minimum energy required in a three-centre reaction is when the reaction proceeds by an *end-on approach*, *i.e.*, in the above reaction, the approach of A to BC requiring the minimum activation energy is for A to approach BC along the bonding line of BC and on the side remote from C:

$$A \cdots\!\!\!> \cdots B - C \longrightarrow A - B + C$$

In this three-centre reaction, the value of the activation energy depends on four factors: (i) The strength of the B—C bond. The stronger this is, the greater will be E. (ii) The repulsion between A and BC. The greater this repulsion, the greater will be E. (iii) The repulsion between AB and C. The greater this repulsion, the greater will be E. (iv) The strength of the A—B bond. The greater the strength of this bond, the lower will be E.

Since most reactions are carried out in solution, another factor affecting the value of E is solvation of molecules and ions.

When we consider the mechanism of activation of this three-centre reaction, we can imagine that there are two extreme cases possible: (i) A is forced up against the repulsion of BC until it is close enough to compete with B on equal terms with C, which is finally expelled. (ii) BC acquires so much energy that the bond B—C is broken, and then A and C combine without any opposition.

Polanyi *et al.* (1931–38) amplified London's ideas into the *transition state theory*. These authors showed by mathematical treatment that the lowest value of E is obtained when the reaction proceeds through a compromise between the two extremes (i) and (ii) mentioned above. A approaches BC along the bonding line of BC remote from C, and is forced against the repulsion of BC, and at the same time bond BC stretches until A and C can compete on equal terms for B. Thus a point is reached when the distances A—B and B—C are such that the forces between each pair are the same. This condition is the *transition state* (*activated complex*); in this state neither molecule AB nor BC exists independently. The system can now proceed in either direction to form A and BC or AB and C. This sequence of events may be represented by the following equation (T.S. = transition state):

$$A + BC \longrightarrow A \cdots B \cdots C \longrightarrow AB + C$$
$$\text{T.S.}$$

The above sequence of events may also be represented graphically by means of an *energy profile diagram* (Figs. 2.11 and 2.12). This diagram is obtained by plotting the potential energy (P.E.) of the system (calculated by a semi-empirical method) against the reaction co-ordinate (the various distances between the nuclei of A, B and C).

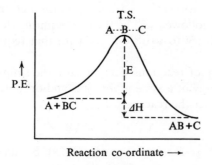

Fig. 2.11

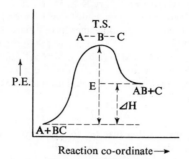

Reaction co-ordinate $\longrightarrow$

Fig. 2.12

E is the activation energy, and ΔH is the heat of reaction at constant pressure. It is assumed that the reaction rate is given by the rate at which the reactant molecules pass through the transition state. For a given shaped 'hump', the lower it lies (*i.e.*, the lower the *energy barrier* is), the easier it is for the reactant molecules to enter the transition state. Also, the wider the hump for a given height, the easier it is for the reactant molecules to enter the transition state, since there is now a wider latitude in nuclei positions for the activated complex. Fig. 2.11 is the energy profile of an exothermic reaction, whereas Fig. 2.12 is that of an endo-thermic reaction (see also below).

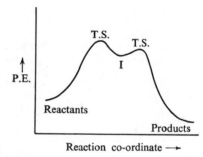

Reaction co-ordinate $\longrightarrow$

Fig. 2.13

The activated complex is *not* a true molecule; it contains *partial bonds*, and the energy content of the system is a maximum. Its life is extremely short, and hence it cannot be isolated; it is always decomposing into reactants or products. The reaction, however, if complex, will proceed through true intermediates which possess some measure of stability, and if this is great enough, the inter-mediates may be isolated. If the reaction proceeds through a true intermediate (I) (Fig. 2.13), there will be a minimum in the energy profile diagram. The greater the dip, the more stable will be the intermediate, and conversely, the shallower the dip, the less stable will be the intermediate. In the extreme case, the dip may be so shallow that the intermediate is indistinguishable from the transition state. It should be noted here that each intermediate has its own transition state.

Since E is the difference in energy content between the T.S. and the reactants, any factor which stabilises the T.S. (relative to the reactants), *i.e.*, lowers the energy content of the T.S., will therefore tend to lower E. Conversely, if any factor stabilises the reactants relative to the T.S., then E will tend to be raised. Resonance and steric effects may operate in these ways (see text).

Structure of the transition state. From Figs. 2.11 and 2.12, it can be seen that the structure of the T.S. must be somewhere between that of the reactants and that of the products, and it may be closer to the former or to the latter. If the reaction is strongly exothermic, then generally the T.S. structure resembles that of the reactants, *i.e.*, existing bonds are almost intact and new bonds are very little formed. The reverse is generally true, *i.e.*, the T.S. structure resembles that of the products when the reaction is strongly endothermic. These generalisations also hold good when the reaction proceeds through inter-mediates (Fig. 2.13), but in this case the intermediate is the 'product' for the preceding T.S. and is the 'reactant' for the T.S. that follows. These generalisations are known as the **Hammond principle** (1955).

As pointed out above, as the reactant molecules approach, the P.E. of the system increases because of the increasing repulsion between them. This repulsion is overcome by the kinetic energy of the particles, and the higher the K.E. required the higher is E. However, for a given value of E, the larger the number of molecules possessing the necessary K.E. for reaction (*i.e.*, to reach E), the faster will be the reaction. If we consider a gaseous system, then it can be shown that at a given temperature (the system then has a fixed energy content), the molecules have different velocities, and the total K.E. (for 1 mole) is given by K.E. $= N_A (1/2\ mu^2)$, where N_A is the Avogadro constant, m the mass of the molecule and u is the root-mean-square (RMS) velocity. Maxwell (1860), using the theory of probability, calculated the distribution of molecular velocities, *i.e.*, the fraction of molecules (ΔN_A) having a particular velocity. Since K.E. is a function of velocity, the equation for the distribution of molecular velocities can be written in the form

$$N_A = B\,e^{-(K.E.)/RT}$$

where B is a proportionality constant. Because the relationship between ΔN_A and T is exponential, a small change in T has a large effect on ΔN_A. The results discussed may be represented graphically (Fig. 2.14), where $T_2 > T_1$. It can be seen that as T is raised, the maximum is flattened and the curve is shifted to the right. Since the maximum represents the fraction of molecules with average K.E., as T is raised, the fraction of molecules with a K.E. higher than average increases (*e.g.*, AC > AB).

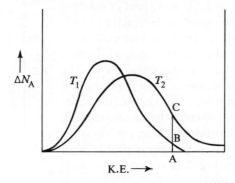

Fig. 2.14

Arrhenius (1889) investigated the effect of temperature on rates of reaction and derived the empirical relationship

$$k = Ae^{-E/RT} \text{ or } k = A\exp(-E/RT)$$

where A is the frequency factor. It can be seen that this equation has the same form as the K.E. distribution described above, and so k and E are related in the same way as are ΔN_A and K.E., *i.e.*, a small change in E will have a large effect on k, and as T is raised, k will increase rapidly. This follows from the fact that the repulsion between approaching reactant molecules is overcome by the K.E. of the molecules, and since raising T increases the number of molecules possessing the necessary K.E., the rate of the reaction increases. These arguments are based on the assumption that E is independent of T. Although this is not strictly true, it is satisfactory for limited variations of T (up to about 300–400° C).

The Arrhenius equation was later modified to

$$k = PZ\exp(-E/RT)$$

where Z is the total number of collisions per second of the reactant molecules (in unit volume) and P is the *probability* or *steric factor* (p. 51).

Summary. Activation energy is the excess of energy over the average energy that the reactant molecules must acquire for reaction to occur. Values of E lie in the range $41\cdot84$–$188\cdot3$ kJ mol^{-1} (10–45 kcal mol^{-1}) for most reactions. The existence of an energy barrier arises mainly from the fact that there is repulsion between reacting molecules and because bond breaking requires an input of energy. Repulsion is overcome by the kinetic energy of the approaching molecules, and the greater their velocity the more easily the molecules approach each other (see below).

Also, there is an output of energy when new bonds are formed, and this is utilised in assisting the breaking of bonds, since both making and breaking of bonds occur simultaneously in the formation of the T.S. Because of this, E is usually less than the bond energy (of the bond being broken). In addition to these factors, there is a steric factor which may have to be satisfied for reaction to occur. In solution, the nature of the solvent affects rates (and mechanisms) of reaction.

Reaction kinetics. Since arguments based on reaction kinetics are used throughout the book, it is worth while to consider this problem briefly at this point. According to the law of mass action, the rate of a chemical reaction is proportional to the product of the active masses of the reacting substances, the molar concentration generally being taken as a measure of the active mass of a substance. The rate of a reaction is defined as the amount of reactant that is consumed in unit time or, alternatively, the amount of product formed in unit time. The equation relating reaction rate and molar concentrations (represented by square brackets) is called a *rate law*. The constant k is the *rate constant* or *specific rate* (this has a definite value at a given temperature), the exponent of a concentration factor is the reaction order for that substance, and the *sum* of the exponents of the concentrations is the *order of the reaction*. A rate law must be established *experimentally*.

Consider the reaction

$$A + B \longrightarrow products$$

If it were found experimentally that the rate was proportional to the concentration of A and to the square of the concentration of B, the rate law would be written as

$$rate = k[A][B]^2$$

This reaction is first order with respect to A, second order with respect to B, the reaction itself being therefore a third-order reaction.

The units of k depend on the *order* of the reaction, *e.g.*, for first- and second-order reactions we have k (s^{-1}) and k $(1 \ mol^{-1} \ s^{-1})$. It can be seen from the above rate laws that k is numerically equal to the rate of the reaction when all the concentrations of the reactants are unity. It is thus not equal to the rate (except under the above conditions); it is a *measure* of the rate.

Cases also arise where the rate law does not contain the concentration of one of the reactants, *e.g.*, in the above reaction, it might be found that

$$rate = k[B]^2$$

In this case, the reaction is *zero order* with respect to A. When a reaction occurs via a number of steps, the rate of the overall reaction is determined by the slowest step, provided the others are relatively rapid. Thus the kinetics and order of such a reaction are basically those of the *slowest* step; this is called the *rate-determining step*. Hence, when the reaction is zero order with respect to one reactant, it means that the mechanism consists of two (or more) steps, the step in which this reactant is involved being relatively fast (see also below).

When two (or more) steps of a complex reaction are fairly slow, the rate law will then likely be complicated.

The *order* of a reaction has already been defined above, but there is another term, the *molecularity* of a reaction, that is also used to define the *number of reactant molecules participating in the rate-determining step*. Order and molecularity may or may not be the same (see, *e.g.*, p. 155).

A rate law may also be derived *mathematically* on the basis of a *postulated* mechanism. By comparing the experimentally observed rate law with various derived rate laws, it may be possible to formulate a reasonably acceptable mechanism. Additional evidence to support this mechanism may then be sought by the application of other methods.

Now let us consider the *reversible* reaction

$$A+B \underset{k_{-1}}{\overset{k_1}{\rightleftharpoons}} C+D$$

and let us suppose that forward and reverse reactions are all first order with respect to each reactant. Then, at equilibrium, *i.e.*, when the forward rate is equal to the reverse rate,

$$k_1[A][B] = k_{-1}[C][D]$$
$$k_1/k_{-1} = [C][D]/[A][B] = K$$

The *equilibrium constant, K*, can thus be determined experimentally from this equation.

When a reaction is complex, the step with the highest activation energy will be rate-determining (r/d), *i.e.*, the step leading to the highest point in the energy profile. The concentrations of reactants up to this point are involved in the rate law, but the rate law is independent of concentrations after this point. Let us consider the two-step reaction with reactants A, B, and C to form product D and involving an intermediate, I. The most common reaction of this type is:

$$\text{(i) } A+B \underset{k_{-1}}{\overset{k_1}{\rightleftharpoons}} I \qquad \text{(ii) } I+C \overset{k_2}{\longrightarrow} D$$

Equilibrium involving the formation of I is reached slowly. Since the *first* step is r/d, the overall rate of reaction is given by the rate at which A and B are consumed, *i.e.*,

$$\text{rate} = k_1[A][B] - k_{-1}[I]$$

To solve this equation, it is necessary to know $[I]$. This cannot be measured, but since the rate of reaction of I to form the products is very much faster than the reversible formation of A and B (*i.e.*, $k_2 \gg k_1$ and k_{-1}), then I is used up too quickly for the reversible reaction to occur to any appreciable extent (*i.e.*, $[I] = 0$). In these circumstances, the rate law will be

$$\text{rate} = k_1[A][B]$$

Thus, the rate law is independent of the concentration of C when the first step is rate-determining.

Equilibrium involving the formation of I is reached rapidly. Since it is now the *second* step which is r/d, the overall rate of reaction is given by

$$\text{rate} = k_2[I][C]$$

Also, because the equilibrium is reached rapidly (*i.e.*, k_1 and $k_{-1} \gg k_2$), $[I]$ is given by

$$K = [I]/([A][B]); \textit{ i.e., } [I] = K[A][B]$$

Hence, the rate of the reaction is

$$\text{rate} = k_2K[A][B][C] = k[A][B][C]$$

Thus, the rate law involves the concentrations of all three reactants when the second step is r/d. It should also be noted that the rate constant k is not k_2 (it is the product of k_2 and K).

Summary. (i) If a reactant does not appear in the rate law, it is involved in reaction after the r/d step. (ii) If all reactants appear in the rate law, then (*a*) all reactants are involved in the r/d step (a termolecular reaction), or (*b*) some are involved in a rapid equilibrium step which precedes the r/d step. Unless other information is available, it is not possible to distinguish between (*a*) and (*b*). (iii) Any proposed mechanism must give the observed rate law.

The energy profiles of the two reactions discussed above will be of the types shown in Figs. 2.15(*a*) and (*b*), respectively.

In both profiles, the activation energies for the separate steps are E_1 and E_2, but the activation for the *overall* reaction is E. This is the extra energy over the average energy of the reactants which must be acquired before the reactants can surmount the *highest* energy barrier that separates them from the products. It should also be noted that in Fig. 2.15(*a*), $E_3 > E_2$, whereas in Fig. 2.15(*b*), $E_3 < E_2$ (where E_3 is the activation energy of the *reverse* reaction: I $\longrightarrow$ A+B).

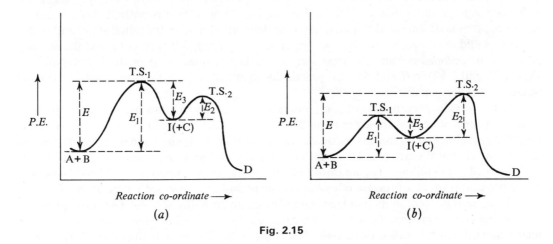

Fig. 2.15

Since $k = PZ \exp(-E/RT)$, it follows that the rate of the reaction: $A + B \longrightarrow C$ is given by

$$\text{rate} = k[A][B] = PZ \exp(-E/RT)[A][B]$$

Thus, the rate of a reaction is dependent on the concentrations of the reactants, P, Z, T, and E. As pointed out above, for a given value of E, small changes in T cause large changes in the number of molecules reaching E. On the other hand, P and Z are almost independent of T. Consequently, for a given reaction which can proceed by only one mechanistic path, the rate is predominantly affected by the $\exp(-E/RT)$ factor. When a reaction can proceed by different mechanisms, the path followed (exclusively or predominantly) will be the one with the lowest activation energy. However, since rate is dependent on concentration, changes in concentration of a reactant may speed up the slow mechanistic path to such an extent that this is now the path through which the reaction proceeds.

In addition to depending largely on T, E and concentration of reactants, the rate of a chemical reaction also depends on solvent and on pressure. Pressure effects on reactions in solution, however, are appreciable only at very high pressures (several thousand atmospheres). The effect of pressure on rate arises chiefly from volume changes that may occur when reactants form the transition state, and hence this relationship can often provide a great deal of information on the nature of the T.S.

Thermodynamic properties. Thermodynamics is the study of energy changes and equilibrium positions involved in physical and chemical transformations. The method is independent of any theory of the structure of matter and is not concerned with the time taken for equilibrium to be reached. Thus, chemical thermodynamics is mainly concerned with the position of equilibrium in chemical reactions and is independent of mechanism, whereas chemical kinetics is concerned with rates and mechanisms of reactions. The study of thermodynamics involves the use of various functions (properties), *e.g.*, P, V, T, U, H, G, S (see below). These are variables whose values depend only on the state of the system and are independent of the path by which this is reached. Functions such as these are known as **state functions.**

Thermodynamic properties are of two types: (i) *extensive properties*; these depend on quantity, *e.g.*, mass, V, U, H, G, S. (ii) *Intensive properties*; these are independent of quantity, *e.g.*, P, T.

Let us now consider some thermodynamic applications. If U is the internal energy of a system at pressure P (atmospheres) and volume V (litres), then H, the *enthalpy (heat content)* of the system, is defined by the equation

$$H = U + PV$$

Thus, when a reaction is carried out *at constant temperature*, the enthalpy change, ΔH, is the difference between the enthalpy of the final and initial states. If heat is evolved, ΔH is negative. In this case the final state of the system has a smaller enthalpy than the initial state, and such a reaction is said to be *exothermic*. If, however, heat is absorbed, ΔH is positive, and the reaction is said to be *endothermic* and the final state has a larger enthalpy than the initial state (see Figs. 2.11 and 2.12). ΔH and ΔU are practically identical for reactions involving solids and liquids (ΔV is very small).

In order to evaluate enthalpies of compounds, it is necessary to know the absolute enthalpies of the elements present in these compounds. Since, in practice, it is not possible to measure absolute enthalpies, the difficulty is overcome by dealing with the changes in energy which accompany chemical reactions. This treatment requires an *arbitrary* choice of zero enthalpy, and is carried out as follows. The *standard state* of an element or compound is defined as its most stable physical form at a pressure of one atmosphere and at a specified temperature. Then, *by convention*, each element in its standard state is given an enthalpy of zero. The enthalpy which accompanies the formation of one mole of a compound in its standard state from its elements in their standard states is called the *standard enthalpy of formation* at temperature T K. T K is *usually* 298 K (*i.e.*, 25° C), and if the temperature is not specified, then it is assumed to be 298 K. Properties of elements and compounds in their standard states are indicated by the superscript symbol $^{\ominus}$, *e.g.*,

$$C(s) + O_2(g, 1 \text{ atm}) \longrightarrow CO_2(g, 1 \text{ atm})$$
$$0 \qquad 0 \qquad\qquad\qquad -393 \cdot 3 \text{ kJ}$$

Thus the standard enthalpy of formation, $\Delta H^{\ominus}_f$, is $-393 \cdot 3$ kJ at 25° C. (N.B. s, g, l stand for solid, gas and liquid, respectively.) It follows from what has been said above that compounds for which $\Delta H^{\ominus}_f$ is negative are thermally stable (with respect to their elements); and vice versa. Although this is generally true, there are exceptions, *i.e.*, some endothermic compounds are thermally stable with respect to their elements.

In reactions that proceed in steps, the enthalpy of the overall reaction is given by the sum of the enthalpies of each step (Hess's law), *e.g.*,

$$A \xrightarrow{\Delta H_1} B \xrightarrow{\Delta H_2} C \xrightarrow{\Delta H_3} D; \Delta H = \Delta H_1 + \Delta H_2 + \Delta H_3$$

By means of Hess's law it is possible to calculate one ΔH if all except this one are known.

Another property used in thermodynamics is *entropy*, S ($\text{JK}^{-1} \text{ mol}^{-1}$; cal $\text{K}^{-1} \text{ mol}^{-1}$). The physical interpretation of entropy is not easy, but one way is to associate entropy with the *disorder* (or *randomness*) of the system. In a crystal, the molecules (or ions) have fixed positions, and the entropy of the system (at 25° C) is small. When the solid is liquefied, the molecules can now move everywhere. The entropy is therefore greater than in the solid state, and is very much greater when the liquid is vaporised. This idea of disorder can also be applied to molecules, *e.g.*, n-hexane, a straight-chain compound, can take up many conformations (shapes), whereas cyclohexane, a ring compound, can take up much fewer conformations. Thus, the entropy of n-hexane ($388 \cdot 3$ $\text{JK}^{-1} \text{ mol}^{-1}$) is greater than that of cyclohexane ($289 \cdot 3$ $\text{JK}^{-1} \text{ mol}^{-1}$). Hence, in general terms, a system changing from order to disorder is accompanied by an increase in entropy (ΔS is positive); and vice versa. A particularly important example of this general statement is that in chemical reactions in which the number of product molecules is greater than the number of reactant molecules, there will be an increase in entropy (the number of degrees of freedom has increased); and vice versa.

The *Gibbs free energy*, G, is defined by the equation

$$G = H - TS$$

Free energy is the energy available from the system at *constant temperature and pressure* for useful work, whereas the product TS gives the energy unavailable for useful work. The *change* in free energy, ΔG, of a system is given by

$$\Delta G = \Delta H - T\Delta S$$

When a system at constant pressure and temperature is in equilibrium, its free energy is a *minimum* (at equilibrium, ΔG, *i.e.*, the change in free energy, is zero). Thus a system not in equilibrium will tend to change irreversibly, thereby lowering its free energy so as to reach equilibrium.

Entropies have been assigned absolute values, and free energies standard values:

$$\Delta S^\ominus = \Sigma S^\ominus \text{ (products)} - \Sigma S^\ominus \text{ (reactants)}$$

$$\Delta G^\ominus = \Sigma G^\ominus \text{ (products)} - \Sigma G^\ominus \text{ (reactants)}$$

Here, the entropy of a perfect crystal of any element or compound is zero at 0 K, and the free energy of any element in its standard state is zero. The standard state for free energy is a *solution at molar concentration* or a *gas at one atmosphere pressure*, at a specified temperature.

H, S, and G are not temperature-independent, but their variation is very small over a limited range of temperature (*cf.* activation energy). Hence, at different temperatures, the main change in $\Delta G^\ominus$ arises from the term $T\Delta S^\ominus$, *i.e.*, to a first approximation:

$$\Delta G^\ominus_T \cong \Delta H^\ominus_{298} - T\Delta S^\ominus_{298}$$

If $\Delta G^\ominus$ is the standard free energy change of the reaction (at temperature T) whose equilibrium constant is K (at temperature T), it can be shown that ($\ln = \log e$; $R = 8{\cdot}314\,\mathrm{JK}^{-1}\mathrm{mol}^{-1}$ $= 1{\cdot}987\,\mathrm{cal}\,\mathrm{K}^{-1}\,\mathrm{mol}^{-1}$).

$$\Delta G^\ominus = -RT \ln K = -2{\cdot}3\,RT \log K$$

Thus, if $\Delta G^\ominus$ is negative, K is greater than unity, and the more negative $\Delta G^\ominus$ is, the greater K is. Here the reaction is thermodynamically favourable. If $\Delta G^\ominus$ is positive, K is less than unity, and the reaction is thermodynamically unfavourable.

Although a chemical reaction can proceed spontaneously (*i.e.*, without the assistance of any external agency) only in the direction of loss of free energy (*i.e.*, $\Delta G^\ominus$ is negative), it does not necessarily mean that it will proceed at a measurable rate. The presence of catalyst does not increase the yield, but increases the rate to the equilibrium position of a reaction which is thermodynamically favourable. As we have seen above, rates of reaction depend on the activation energy, E, and the frequency factor, A. Since catalysts provide alternative routes in which they generally participate to form intermediates, this means that either E is less and/or A is greater for the catalysed than for the uncatalysed reaction. It appears that, in general, both E and A vary. However, for closely related reactions, A will very likely be little changed, the changes in rates then being due to changes in E.

Now let us consider the equation

$$\Delta G^\ominus = \Delta H^\ominus - T\Delta S^\ominus$$

When $\Delta H^\ominus$ is negative and $\Delta S^\ominus$ is positive, $\Delta G^\ominus$ is negative and the reaction will be thermodynamically favourable. If, however, both $\Delta H^\ominus$ and $\Delta S^\ominus$ are negative, the term $T\Delta S^\ominus$ may be larger or smaller than $\Delta H^\ominus$, and so $\Delta G^\ominus$ will be positive or negative, respectively. In the former case, the reaction will not be thermodynamically favourable. In the same way, if $\Delta H^\ominus$ is positive, then the sign of $\Delta G^\ominus$ will depend on the sign and magnitude of $\Delta S^\ominus$. For very closely similar reactions, if it be *assumed* that $\Delta S^\ominus$ remains sensibly constant, then

$$\Delta G^\ominus = \Delta H^\ominus + \text{constant}$$

In these circumstances, $\Delta G^\ominus$ is a function of enthalpy. We may therefore write

$$\Delta H^\ominus = \Delta G^\ominus = -RT \ln K$$

Thus $\Delta H^\ominus$ may be used (but with caution) as a criterion for deciding whether a reaction is thermodynamically favourable. Usually, reasonable estimates of $\Delta H^\ominus$ may be obtained from a knowledge of bond energies in the reactants and products:

$$\Delta H^\ominus = \Sigma H^\ominus(\text{bonds formed}) - \Sigma H^\ominus(\text{bonds broken})$$
$$\text{(negative)} \qquad\qquad\qquad \text{(positive)}$$

This arbitrary approach of equating $\Delta G^\ominus$ with $\Delta H^\ominus$ is made necessary by the fact that the relevant values of $\Delta S^\ominus$ are very often unavailable. It is also difficult to arrive at a reasonable estimate for these values. One way out of the difficulty has been to make assumptions about entropy changes, *e.g.*, as was done above for closely similar reactions. This may be a very bad estimate of the situation under consideration, and if so, it usually manifests itself by giving an answer in disagreement with experiment. On the other hand, where we get a good answer, we may be too hasty in accepting that our assumptions were sound.

What has been said above applies to estimating (qualitatively) $\Delta G^\ominus$ in order to arrive at some idea about the position of equilibrium of various reactions. For rates of reaction, we need to know the relevant energies of activation. E may, in many cases, be determined experimentally from the equation:

$$2 \cdot 3 \log k = 2 \cdot 3 \log A - E/RT$$

If k is measured at several temperatures, the graph of $\log k$ against $1/T$ is usually a 'good' straight line whose slope is $-E/2\cdot3R$ and intercept is $2\cdot3 \log A$. Furthermore, since the slope is either zero or negative, the value of E is *zero* or *positive*.

Now let us consider the following reaction in terms of transition-state theory, where AB represents the transition state:

$$A + B \rightleftharpoons AB^\ddagger \rightleftharpoons \text{products}$$

Treating the T.S. as a molecular species, we have:

$$K^\ddagger = [AB^\ddagger]/[A][B]$$

and

$$\Delta G^{\ominus\ddagger} = \Delta H^{\ominus\ddagger} - T\Delta S^{\ominus\ddagger} = -RT \ln K^\ddagger$$

The quantities $\Delta G^{\ominus\ddagger}$, $\Delta H^{\ominus\ddagger}$ and $\Delta S^{\ominus\ddagger}$ are called the *standard free energy of activation, heat* (or *enthalpy*) and *entropy of activation*. Now, it can be shown from transition-state theory that

$$k_{\text{T.S.}} = \frac{kT}{h} K^\ddagger = \frac{kT}{h} \exp(-\Delta G^{\ominus\ddagger}/RT)$$

$$= \frac{kT}{h} \exp(\Delta S^{\ominus\ddagger}/R) \, . \, \exp(-\Delta H^{\ominus\ddagger}/RT)$$

where $k_{T.S.}$ is the rate constant for the formation of the T.S., k is Boltzmann's constant $(= R/N_A)$, and h is Planck's constant. Also, since

$$k = A \exp(-E/RT)$$

we may therefore equate $\Delta S^{\ominus\ddagger}$ with A and $-\Delta H^{\ominus\ddagger}$ with $-E$. It has been found that $\Delta H^{\ominus\ddagger}$ is very close to the value of E for liquid and solid systems, and $\Delta S^{\ominus\ddagger}$ can be calculated from the experimental rate constant and activation energy.

If the T.S. is formed very early, its structure will closely resemble that of the reactants and so $\Delta S^{\ominus\ddagger} \cong 0$. On the other hand, if the T.S. is formed fairly late, since it is more organised than the reactants, $\Delta S^{\ominus\ddagger}$ would be expected to be negative, and the more organised the T.S., the more negative will be $\Delta S^{\ominus\ddagger}$. Hence, an evaluation of $\Delta S^{\ominus\ddagger}$ will give information on the structure of the T.S.

Because the rate constant involves entropy of activation in T.S. theory, the rate constant can be interpreted in terms of entropy. Furthermore, since the entropy factor may outweigh the enthalpy factor, some authors prefer to use the free-energy profile rather than the P.E. profile. In this case, $\Delta G^{\ominus\ddagger}$ replaces the activation energy and $\Delta G^{\ominus}$ replaces $\Delta H^{\ominus}$.

One other problem will be discussed here, *viz.*, the effect of resonance on chemical equilibria. It has been assumed that for very closely similar reactions, ΔH is a measure of ΔG, and also that, *in solution*, ΔH and ΔU are practically identical. We may therefore conclude that, under these conditions, ΔU is a measure of ΔG, *i.e.*,

$$\Delta U = \Delta G^{\ominus} = -RT \ln K$$

Since
$$\Delta U = \Sigma U(\text{products}) - \Sigma U(\text{reactants})$$

the lower the value of $\Sigma U(\text{products})$, the more negative will be ΔU. Thus $\Delta G^{\ominus}$ will be more negative, and consequently the greater will be K. As we have seen, a resonance hybrid has lower internal energy than any of its contributing resonating structures. Thus, if the products of a reaction have greater resonance stabilisation than the reactants, the position of equilibrium will be displaced towards the formation of the products; and vice versa. In other words, resonance can be a driving force or an inhibiting force in chemical reactions. This argument has considered resonance only. However, reactions in solution involve *solvated* molecules and ions, and solvation energies play an important part in equilibria positions (see pp. 160, 272).

Use of isotopes in organic chemistry. In recent years the use of isotopes has been extremely helpful in the study of reaction mechanisms and rearrangements, in the elucidation of structures, and also in quantitative analysis. The application of isotopes in biochemistry has also been particularly fruitful, since they offer a means of identifying intermediates and the 'brickwork' of the final products. The common isotopes that have been used in organic chemistry are: deuterium (^{2}H, D; stable), tritium (^{3}H, T; radioactive), ^{13}C (stable), ^{14}C (radioactive), ^{15}N (stable), ^{18}O (stable), ^{32}P (radioactive), ^{35}S (radioactive), ^{37}Cl (stable), ^{82}Br (radioactive), ^{131}I (radioactive).

Various methods of analysis are used. Radioactive isotopes are usually analysed with the Geiger-Müller counter, and the stable isotopes by means of the mass spectrograph. Deuterium is often determined by means of infra-red spectroscopy; and there are also the older methods for deuterium and ^{18}O of density, or refractive index measurements (of the water produced after combustion of the compound). Nuclear magnetic resonance is now also used, being applicable only to those isotopes having nuclear magnetic moments.

Isotopes are usually used as *tracers*, *i.e.*, the starting material is labelled at some particular position, and after reaction the labelled atom is then located in the product. This does not mean that labelled compounds contain 100 per cent of the isotope, but that they usually contain an abnormal amount of the isotope. Many examples of the use of isotopic indicators will be found in the text (see Index, Isotopic indicators).

The use of isotopes, stable or radioactive, is based on the fact that the chemical behaviour of any particular isotope is the same as that of the other atoms isotopic with it (the chemical properties of an element depend on the nuclear positive charge and the number of electrons surrounding the nucleus, and not on the number of neutrons in the nucleus). This identity in chemical behaviour is essentially true for the heavy atoms, but in the case of the lightest elements, reactions involving heavier isotopes are slower, but so long as identical paths are followed, the final result is unaffected. This difference in rates of reaction of isotopes is known as the *kinetic isotope effect*, and the magnitude of such effects depends on the weight ratio of the isotopes involved. Thus the kinetic isotope effect is greatest with H and D (and T). The kinetic effect has been widely used to study reaction mechanisms, since this difference in rate is significant when the bond attaching the isotopic atom is stretched in the activated state, *i.e.*, if substitution by the isotope changes the rate, then breaking of that bond is involved in the rate-determining step.

It was generally assumed at first that differences in rate (and equilibrium) constants would arise only in reactions involving making and breaking of bonds with respect to isotopic atoms. It has now been found, however, that the presence of an isotope may affect rate or equilibrium constants even though 'isotopic bonds' are neither broken nor formed during the reaction. This phenomenon is known as the *secondary isotope effect* (see p. 703), and is much weaker than the kinetic isotope effect.

Ion accelerators and nuclear reactors produce, as by-products, artificial isotopes, particularly those which are radioactive. These isotopes are then supplied in the form of some compound from which a labelled compound can be synthesised, *e.g.*, ^{14}C is usually supplied as $Ba^{14}CO_3$, ^{15}N as $^{15}NH_4Cl$, etc. A simple example of the synthesis of a labelled compound is that of acetic acid (in the following equations, which involve a Grignard reagent, $\overset{*}{C}$ is ^{14}C; this is a common method of representing a tracer atom, provided its nature has been specified).

$$Ba\overset{*}{C}O_3 + 2HCl \longrightarrow BaCl_2 + H_2O + \overset{*}{C}O_2$$

(i) $\overset{*}{C}O_2 + CH_3MgI \longrightarrow CH_3\overset{*}{C}O_2H$

(ii) $\overset{*}{C}O_2 + 3H_2 \xrightarrow{\text{Catalyst}} H_2O + \overset{*}{C}H_3OH \xrightarrow{I_2/P} \overset{*}{C}H_3I \xrightarrow{Mg} \overset{*}{C}H_3MgI$

$$\overset{CO_2}{\underset{}{\longleftarrow}} \quad \overset{\downarrow}{\overset{*}{C}O_2}$$

$$\overset{*}{C}H_3CO_2H \qquad \overset{*}{C}H_3\overset{*}{C}O_2H$$

In a number of cases, an exchange reaction is a very simple means of preparing a labelled compound, *e.g.*, dissolving a carboxylic acid in water enriched with deuterium.

$$RCO_2H + D_2O \rightleftharpoons RCO_2D + DHO$$

Because of the possibility of this exchange reaction, it is often necessary to carry out control experiments.

Isotopes are also very useful in the analysis of mixtures, particularly for the determination of the yield of products in a chemical reaction when isolation is difficult. A simple method is that of *isotopic dilution*. The labelled compound is prepared, and a known amount is then added to the mixture to be analysed. A portion of the substance is now taken and analysed for its isotopic content. From a knowledge of the isotopic content of the labelled compound added and recovered, and the weight of the labelled compound added, it is thus possible to calculate the weight of the labelled compound in the mixture. This method can only be used as long as there is no isotopic exchange during the isolation.

Nomenclature of labelled compounds. Several methods have been used, *e.g.*, the positions of isotopes are indicated by arabic numerals placed within square brackets:

[1—^2H]ethanol, CH_3CH^2HOH; [1—^{14}C,2—^{13}C]acetaldehyde, $^{13}CH_3\,^{14}CHO$.

Acids and bases. The study of the acidity and basicity of organic compounds offers a very good means of examining the theories relating structure with reactivity.

There are various definitions of acids and bases. According to Arrhenius (1884), acids are compounds which ionise in aqueous solution to produce hydrogen ions, and bases ionise to produce hydroxide ions. According to the Brønsted–Lowry definition (1923), an *acid* is a *proton donor*, *i.e.*, is *protogenic*, and a *base* is a *proton acceptor*, *i.e.*, is *protophilic*. Thus:

$$B + H^+ \rightleftharpoons BH^+$$
$$\text{base} \quad \text{proton} \quad \text{acid}$$

Proton donation and acceptance are reversible, and the proton transfer, which is extremely fast, is known as *protonation*. The acid and base in the above equation are said to be *conjugate* with respect to each other; B is the conjugate base (cB) of the acid BH^+, and BH^+ is the conjugate acid (cA) of the base B. Thus every acid must have its conjugate base, and every base its conjugate acid. The Brønsted–Lowry definition is more general than the Arrhenius definition, since the former applies to *any* type of solvent, whereas the latter is restricted to aqueous solutions.

The *free* proton is encountered only in a vacuum or in a very dilute gas. Thus, since a free proton cannot exist in solution (in measurable concentrations), all acid-base reactions are of the type:

$$A_1 + B_2 \rightleftharpoons B_1 + A_2$$

where A_1—B_1 and A_2—B_2 are conjugate acid-base pairs, *i.e.*, in order that an acid may exhibit its acidic properties, there must be a proton acceptor (base) present.

Some examples are:

Acid (A_1)	+ *Base* (B_2)	*Acid* (A_2)	+ *Base* (B_1)
HCl	+ H_2O	H_3O^+	+ Cl^-
$HSO_4{}^-$	+ NH_3	$NH_4{}^+$	+ $SO_4{}^=$
$NH_4{}^+$	+ H_2O	H_3O^+	+ NH_3
Ph_3CH	+ $NH_2{}^-$	NH_3	+ Ph_3C^-

Let us now consider the case of water. This is the most important solvent for acids and bases, and acts as an acid or a base according as the other compound acts as a base or an acid. The 'other' compound, however, can be water itself, and this produces the following equilibrium:

$$H_2O + H_2O \rightleftharpoons H_3O^+ + OH^-$$
$$(A_1) \quad (B_2) \quad (A_2) \quad (B_1)$$

This 'self-ionisation' is known as *autoprotolysis*. Many organic compounds, particularly hydroxy compounds, can undergo autoprotolysis, *e.g.*,

$$C_2H_5OH + C_2H_5OH \rightleftharpoons C_2H_5OH_2{}^+ + C_2H_5O^-$$

We can summarise the situation so far as follows: (i) acids and bases may be neutral or charged; (ii) solvents such as water, which can undergo autoprotolysis, are called *amphiprotic* solvents; the cA (of the solvent) is often called the *lyonium* ion, and the cB the *lyate* ion; (iii) solvents which are neither proton donors nor acceptors are called *aprotic* solvents.

A third definition of acids and bases is that due to Lewis (1938). *Lewis acids* are molecules or ions which are capable of co-ordinating with unshared electron pairs, and *Lewis bases* are molecules or ions having unshared electron pairs available for sharing with acids. Since a reagent must have an unshared electron pair to accept a proton, some reagents may act as bases both in the Brønsted–Lowry and Lewis sense. On the other hand, *all* Brønsted–Lowry acids are Lewis acids, since a proton from any proton donor can co-ordinate with an unshared electron pair. The Lewis definition of an acid, however, also includes many reagents such as BF_3, $AlCl_3$, etc., since these can also co-ordinate with unshared electron pairs (to form co-ordination compounds), *e.g.*,

$$:NH_3 + BF_3 \longrightarrow H_3N^+{-}BF_3^-$$

It might be noted here that Lewis acids and bases are respectively electrophilic and nucleophilic reagents (p. 50).

Strengths of acids and bases. The 'strength' of an acid is a measure of its tendency to lose a proton, and the 'strength' of a base is a measure of its tendency to take up a proton. If we consider the equilibrium between the conjugate acid-base pairs, A_1—B_1 and A_2—B_2, *i.e.*,

$$A_1 + B_2 \rightleftharpoons B_1 + A_2$$

then the equilibrium will be the more to the right the stronger A_1 and B_2 are, and the weaker B_1 and A_2 are, *i.e.*, the equilibrium always shifts in the direction of the formation of the weaker acid and weaker base. Thus, it is not possible to speak of *absolute* strengths of acids and bases; the strength of acid A_1 will be *relative* to the strength of base B_2. B_2 is usually the solvent, and water is used for the purpose of measuring relative strengths. There is much evidence to show that the 'aqueous' proton is firmly 'bound' to four water molecules:

$$HA + 4H_2O \rightleftharpoons A^- + H_3O^+ \cdot 3H_2O \ \textit{or} \ H^+(H_2O)_4$$

However, usual practice is to write this equation as involving the formation of the *hydroxonium* (*hydronium*) ion, H_3O^+:

$$HA + H_2O \rightleftharpoons A^- + H_3O^+$$

Ionisation of an acid is also often written:

$$HA \rightleftharpoons H^+ + A^-$$

it being understood that there are no free protons but protonated solvent molecules.

From what has been said above, it follows that very strong acids have very weak conjugate bases, and vice versa, *e.g.*, if HY is a stronger acid than HZ, then Z^- is a stronger base than Y^-. Similarly, very strong bases have very weak conjugate acids, and vice versa.

Ionisation constants (K_a and K_b). These measure the strengths of acids and bases, and they are derived by the application of the law of mass action.

$$\textit{Acids} \qquad HA + H_2O \rightleftharpoons H_3O^+ + A^-$$
$$\therefore \quad K' = [H_3O^+][A^-]/[HA][H_2O]$$
$$\therefore \quad K_a = [H_3O^+][A^-]/[HA]$$

$[H_2O]$ is effectively constant. Also

$$pK_a = \log_{10}(1/K_a) = -\log_{10} K_a$$

Thus, the stronger an acid is, the larger is its K_a and consequently the smaller is its pK_a.

$$\textit{Bases} \qquad B + H_2O \rightleftharpoons BH^+ + OH^-$$
$$\therefore \quad K_b = [BH^+][OH^-]/[B][H_2O]$$
$$\therefore \quad K_b = [BH^+][OH^-]/[B]$$

Thus, the stronger a base is, the larger is its K_b and consequently the smaller is its pK_b.

Since in the Brønsted–Lowry system, acid-base systems involve proton transfer, there is a tendency to replace base ionisation constants, K_b, by the acid ionisation constants, K_a of the *conjugate acid*, BH^+.

Thus:

$$BH^+ + H_2O \rightleftharpoons B + H_3O^+$$

$$\therefore \quad K_a = [B][H_3O^+]/[BH^+]$$

It therefore follows that for a base and its conjugate acid,

$$K_b . K_a = \frac{[BH^+][OH^-]}{[B]} . \frac{[B][H_3O^+]}{[BH^+]} = [H_3O^+][OH^-] = K_w$$

where K_w is the autoprotolysis constant of water ($= 10^{-14}$, or $pK_w = 14$; at 25° C). Finally, since

$$pK_a + pK_b = pK_w$$

$$\therefore \quad pK_a = pK_w - pK_b$$

Thus, the *stronger* a base is (*i.e.*, the lower its pK_b), the *larger* is the pK_a of its conjugate acid.

An important point to note is that the preceding discussion has dealt with *thermodynamic acidity*, *i.e.*, the position of equilibrium of the ionisation of an acid is a thermodynamically controlled reaction (and is measured by K). On the other hand, there is also *kinetic acidity*, which deals with the rate at which an acid transfers its proton to a base. This is a kinetically controlled reaction (and measured by k).

In the text will be found the pK_a values of various acids and bases. Most of these values have been taken from Albert and Serjeant (see References).

Levelling effect. Most carboxylic acids behave as weak acids in water, but their pK_a values are different. When dissolved in liquid ammonia, however, all these acids form ammonium salts:

$$RCO_2H + NH_3 \rightleftharpoons RCO_2^- + NH_4^+$$

Since the acids are completely ionised (by proton transfer), *all* the acids now behave as *strong* acids due to the strong basic property of ammonia (relative to water). The basicity of ammonia is said to exert a *levelling effect* on the 'strengths' of acids dissolved in it. Thus, by using very strong bases, very weak acids may be converted into their conjugate bases. One very important example of this is the conversion, in ethanol solution, of ethyl malonate (a very weak acid) into its conjugate base (the ethyl malonate carbanion) by the action of sodium ethoxide (the ethoxide ion is an extremely strong base; it is the conjugate base of the extremely weak acid, ethanol):

$$CH_2(CO_2C_2H_5)_2 + C_2H_5O^- \rightleftharpoons \bar{:}CH(CO_2C_2H_5)_2 + C_2H_5OH$$

It should be noted that many compounds may therefore behave as acids (or bases) in suitable solvents, whereas in water they behave as 'neutral' substances, *e.g.*, methanol (pK_a 15·5). Such pK_a values have been estimated by indirect methods and then related to water.

Many factors control the strengths of organic acids and bases: inductive, resonance and steric effects, solvation, entropy, hydrogen bonding and conformational effects. Some of these problems are discussed in the text.

An important point in this connection is that although the bond strength of H—Y may be greater than that of H—Z, nevertheless the former may be a stronger acid than the latter. Bond strengths are dissociation energies (of the gas into radicals, *i.e.*, homolytic fission occurs), whereas when we are dealing with acid strengths, proton release occurs by *heterolytic* fission. This depends not only on bond strengths but also on solvation energies, entropy changes, etc. Thus a high bond energy may be more than compensated for by, *e.g.*, the high solvation energy of an ion (see pp. 160, 272).

PROBLEMS

1. Draw the electronic configurations (of the valency electrons) in:

(i) the carbon atom; (ii) CH_4; (iii) $CH_2=CH_2$; (iv) $CH\equiv CH$; (v) oxygen atom; (vi) H_2O; (vii) nitrogen atom; (viii) NH_3; (ix) $NH_4{}^+$.

Where bonds are involved, name their types, and describe the geometry of the molecules.

2. Draw the V.B. and M.O. structures of:

(i) CH_3COCH_3; (ii) CH_3CONH_2; (iii) $CH_2=CHCOCH_3$; (iv) $CH_3CO_2C_2H_5$; (v) CH_3NO_2.

3. Which of the following compounds have a dipole moment? Comment. Br_2, HI, CH_2Cl_2, $CHCl_3$, CCl_4, CH_3OH, CH_3OCH_3, $(CH_3)_2NH$, H_2O, CH_4, CO_2, BrCl, BF_3, CH_3COCH_3.

4. Draw the energy profiles for a three-step reaction in which (i) the first step is slowest and the last step fastest; (ii) the first step fastest and the last step slowest; (iii) second step slowest and last step fastest.

5. The order of bond lengths in the halogen acids is: HI > HBr > HCl > HF; the order of their dipole moments is: HF > HCl > HBr > HI. Offer an explanation.

6. Arrange the four halide ions in order of decreasing basicity. Explain your order.

7. $CH_4(g) + 2O_2(g) \longrightarrow CO_2(g) + 2H_2O(g)$

$\Delta H^\ominus = -802.5$ kJ mol^{-1} ($-191\cdot 8$ kcal mol^{-1});

$\Delta G^\ominus = -800\cdot 8$ kJ mol^{-1} ($-191\cdot 4$ kcal mol^{-1}).

What is $\Delta S^\ominus$? Comment on its value.

8. Consider the reaction:

$A \rightleftharpoons B$		$\Delta H^\ominus_f$ kJ	$S^\ominus$ JK^{-1}
	A	$-24\cdot 27$	246·4
	B	$-25\cdot 44$	252·3

What is the composition of the equilibrium mixture at 25° C?

9. Arrange the acids with the following pK_a values in order of increasing strength: 6·3, 4·7, 9·5, $-2\cdot 1$, 15·8. Do the same for bases with pK_a values: 8·1, 4·2, 16, 10, 25; and for bases with pK_b values: 7, 11·4, 9·2, 4·7, 12·8.

REFERENCES*

HAMMETT, *Physical Organic Chemistry*, McGraw-Hill (1970, 2nd edn.).
COULSON, *Valence*, Oxford University Press (1961, 2nd edn.).
WHELAND, *Advanced Organic Chemistry*, Wiley (1960, 3rd edn.).
FERGUSON, *The Modern Structural Theory of Organic Chemistry*, Prentice-Hall (1963).
INGOLD, *Structure and Mechanism in Organic Chemistry*, Bell and Sons (1969, 2nd edn.).
HINE, *Physical Organic Chemistry*, McGraw-Hill (1962, 2nd edn.).
GOULD, *Mechanism and Structure in Organic Chemistry*, Holt and Co. (1959).
DE MAYO (ed.), *Molecular Rearrangements*, Interscience, Part 1 (1963); Part 2 (1964).
MACWOOD and VERHOEK, 'How Can You Tell Whether a Reaction Will Occur?', *J. Chem. Educ.*, 1961, **38**, 334.
SANDERSON, 'Principles of Chemical Reaction', *J. Chem. Educ.*, 1964, **41**, 13.
PIMENTAL and MCCLELLAN, *The Hydrogen Bond*, Freeman and Co. (1960).

---

*Books which deal with the electronic theories of organic chemistry will not be given as references in subsequent chapters in this book. The reader should always be prepared to refer to them on any matter dealing with mechanisms.

LEFFLER, *The Reactive Intermediates of Organic Chemistry*, Interscience (1956).

LINNETT, *The Electronic Structure of Molecules, A New Approach*, Methuen (1964).

GILLESPIE, 'The Electron-Pair Repulsion Model for Molecular Geometry', *J. Chem. Educ.,* 1970, **47**, 18.

SANDERSON, 'Principles of Chemical Bonding', *J. Chem. Educ.,* 1961, **38**, 382.

BRAUDE and NACHOD (eds.), *Determination of Organic Structures by Physical Methods*, Academic Press (1955), Chs. 14, 15.

NEWMAN (ed.), *Steric Effects in Organic Chemistry*, Wiley (1956), Ch. 9.

ALBERT and SERJEANT, *Ionisation Constants of Acids and Bases*, Methuen (1962).

CLEVER, 'The Hydrated Hydronium Ion', *J. Chem. Educ.,* 1963, **40**, 637.

HULETT, 'Deviations From the Arrhenius Equation', *Quart. Rev.,* 1964, **18**, 227.

LE NOBLE, 'Reactions in Solution under Pressure', *J. Chem. Educ.,* 1967, **44**, 729.

KAYE and LABY, *Tables of Physical and Chemical Constants*, Longmans, Green (1966, 13th edn.).

3 Alkanes

Aliphatic compounds are *open-chain* or *acylic* compounds, and the name *aliphatic* arises from the fact that the first compounds of this class to be studied were the fatty acids (Greek: *aliphos*, fat).

Carbon forms a large number of compounds with hydrogen only and these are known collectively as *hydrocarbons*. There are two groups of hydrocarbons: (i) *saturated* hydrocarbons; (ii) *unsaturated* hydrocarbons.

The *alkanes* or the *paraffins* are the saturated hydrocarbons. Many occur naturally, and the chief source of the alkanes is *mineral oil* or *petroleum*, which occurs in many parts of the world.

Structural formulæ. The simplest alkane is methane, and its molecular formula is CH_4. Assuming the quadrivalency of carbon and the univalency of hydrogen, we find that there is only one structure possible for methane, *viz.*, (I).

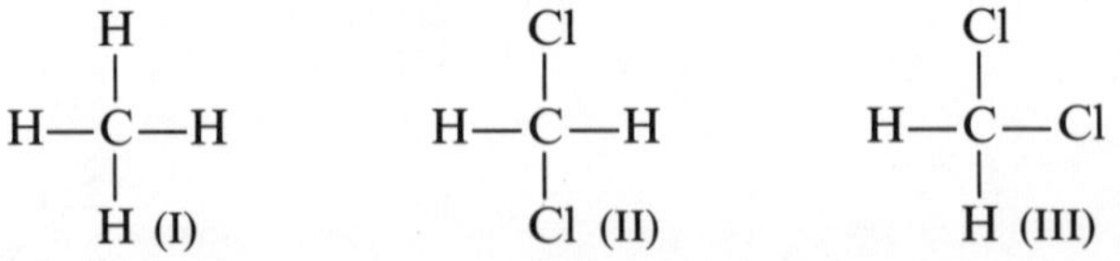

$$
\begin{array}{ccc}
\text{H} & \text{Cl} & \text{Cl} \\
| & | & | \\
\text{H—C—H} & \text{H—C—H} & \text{H—C—Cl} \\
| & | & | \\
\text{H (I)} & \text{Cl (II)} & \text{H (III)}
\end{array}
$$

Study of the reactions of methane shows that all four hydrogen atoms are equivalent, *e.g.*, methylene dichloride, CH_2Cl_2, prepared by totally different methods, is always the same. Thus (II) and (III) are different ways of writing the *same* structure. At first sight it may appear that these two structural formulæ are different. They *are* different if the molecule is two-dimensional, but, as we have seen (p. 44), in saturated compounds the four valencies of carbon are arranged tetrahedrally. Examination of the two structures as tetrahedral figures shows that they are identical.* The chief disadvantage of the plane-structural formula is that it does not show the spatial arrangement of the atoms. On the other hand, the three-dimensional structural formula is cumbersome, and for many complicated molecules cannot easily be drawn on paper. Hence we usually adopt the plane-formulæ when dealing with compounds from the point of view of their structure; only when we wish to stress the spatial arrangement of the atoms or groups in a molecule do we resort to solid diagrams (see, *e.g.*, Ch. 17). How we show the relative positions of the various atoms or groups in the plane-structural formula of a given compound usually depends on ourselves, but where possible the simplest method of writing the structure should be chosen.

*'Atomic models' are very useful to the organic chemist. Sets of these models may be bought.

Consider 1,5-dichloropentane, $C_5H_{10}Cl_2$. This is a 'straight' chain compound, and its structural formula is usually written $ClCH_2CH_2CH_2CH_2CH_2Cl$. Now, 1,5-dichloropentane gives cyclopentane when treated with zinc. Cyclopentane is a ring compound, and to show its formation from 1,5-dichloropentane, we write the structure of the latter as follows:

$$
\begin{array}{c}
CH_2CH_2Cl \\
\diagup \\
CH_2 \qquad + Zn \longrightarrow \\
\diagdown \\
CH_2CH_2Cl
\end{array}
\qquad
\begin{array}{c}
CH_2 \\
\diagup \quad \diagdown \\
CH_2 \qquad\quad CH_2 \\
\mid \\
\diagdown \quad CH_2 \\
\diagup \\
CH_2
\end{array}
\ + ZnCl_2
$$

Thus 'straight' chains may be 'bent' to stress a particular point we may have in mind.

Substitution reactions of methane. Chlorine has no action on methane in the dark. In bright sunlight the reaction is explosive, and hydrogen chloride and carbon are formed:

$$CH_4 + 2Cl_2 \longrightarrow C + 4HCl$$

In diffused sunlight no explosion occurs, but a series of reactions takes place whereby the four hydrogen atoms in methane are successively replaced by chlorine atoms:

$$
\begin{array}{ll}
CH_4 + Cl_2 \longrightarrow CH_3Cl + HCl & \text{methyl chloride} \\
CH_3Cl + Cl_2 \longrightarrow CH_2Cl_2 + HCl & \text{methylene dichloride} \\
CH_2Cl_2 + Cl_2 \longrightarrow CHCl_3 + HCl & \text{chloroform} \\
CHCl_3 + Cl_2 \longrightarrow CCl_4 + HCl & \text{carbon tetrachloride}
\end{array}
$$

In methane the four carbon valencies are satisfied by combination with four hydrogen atoms. Carbon never exhibits a valency of more than four, and so cannot combine with more than four hydrogen atoms or four other univalent atoms or groups. Hence in the reaction with chlorine, the hydrogen atoms are displaced, and chlorine atoms take their place. This type of reaction is known as *substitution*, and is the *direct replacement of hydrogen* by some other atom or group. The products so formed are known as *substitution products*. The atom or group that has replaced the hydrogen atom is called the *substituent*, and when a substituent atom or group is replaced by some other atom or group, the reaction is referred to as a *replacement* (or *displacement*) reaction. It should be noted that in substitution or replacement reactions there is *no change in structure*. The *spatial* arrangement of the molecule, however, may have changed (see p. 485).

Structural isomerism in alkanes. All the alkanes can be derived formally from methane by substituting hydrogen atoms by methyl groups (CH_3). When the alkane contains three or more carbon atoms, substitution now gives rise to isomerism, *e.g.*, (i) Propane, C_3H_8, can give rise to two butanes, C_4H_{10}; (I) by substitution at a terminal carbon atom, and (II) by substitution at the central carbon atom:

$$
\begin{array}{c}
\quad H\ \ H\ \ H\ \ H \\
\quad \mid\ \ \mid\ \ \mid\ \ \mid \\
H-C-C-C-C-H \\
\quad \mid\ \ \mid\ \ \mid\ \ \mid \\
\quad H\ \ H\ \ H\ \ H
\end{array}
\qquad\qquad
\begin{array}{c}
\qquad\qquad H \\
\qquad\qquad \mid \\
H\quad H-C-H\quad H \\
\mid\qquad\quad \mid\qquad\quad \mid \\
H-C\text{———}C\text{———}C-H \\
\mid\qquad\quad \mid\qquad\quad \mid \\
H\qquad\quad H\qquad\quad H \\
\qquad\quad CH_3 \\
\qquad\quad \mid
\end{array}
$$

$$
CH_3CH_2CH_2CH_3 \qquad\qquad CH_3CHCH_3 \quad \text{or} \quad (CH_3)_3CH
$$

$$
\text{(I)} \qquad\qquad\qquad\qquad \text{(II)}
$$

(I) has a *straight* chain, and (II) a *branched* chain. Both isomers are known, and thus 'butane' is the first alkane to exhibit structural isomerism. This is an example of *chain* (or *nuclear*) isomerism, and is characterised by the manner of linking of the *carbon chain*. (I) is known as normal butane, and (II) as isobutane.

(ii) Three pentanes, C_5H_{12}, may be derived from the butanes:

$$CH_3CH_2CH_2CH_2CH_3 \qquad (CH_3)_2CHCH_2CH_3 \qquad (CH_3)_4C$$
$$\text{normal pentane} \qquad\qquad \text{isopentane} \qquad\qquad \text{neopentane}$$

The bond-diagram way of writing structural formulæ uses up a lot of space. Hence it is general practice to use 'contracted' structural formulæ; the method is to write the formula in a 'straight' line and enclosing in parentheses any side-chain or substituent group. The atoms or groups in parentheses are joined directly to the *preceding* carbon atom in the chain not placed in parentheses, *e.g.*,

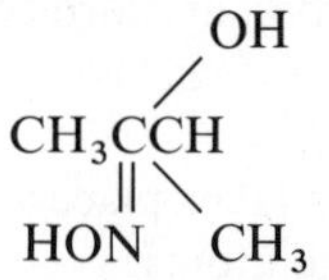

$$CH_3\overset{\overset{\displaystyle OH}{\diagup}}{\underset{\underset{\displaystyle HON \quad CH_3}{\| \diagdown}}{C}}CH \qquad \text{may be written} \quad CH_3C(=NOH)CH(OH)CH_3$$

In many cases where there is no ambiguity, the parentheses may be omitted, *e.g.*, $CH_3CH(OH)CH_3$ is often written $CH_3CHOHCH_3$.

As the number of carbon atoms in the alkane increases, the number of possible isomers increases rapidly, *e.g.*, the alkane $C_{15}H_{32}$ can exist in 4 347 isomeric forms. The number of isomers of a given alkane may be calculated by means of mathematical formulæ; in most cases very few have actually been prepared.

Examination of the two butane structures shows that all the carbon atoms are not equivalent, and also that all the hydrogen atoms are not equivalent. A *primary* carbon atom is one that is joined to *one* other carbon atom; a *secondary* carbon atom to *two* other carbon atoms; and a *tertiary* carbon atom to *three* other carbon atoms. Hydrogen atoms joined to primary, secondary and tertiary carbon atoms are known as primary, secondary, or tertiary hydrogen atoms, respectively. Thus normal butane contains two primary carbon atoms, two secondary carbon atoms, six primary and four secondary hydrogen atoms. Isobutane contains three primary carbon atoms, one tertiary carbon atom, nine primary hydrogen atoms, and one tertiary hydrogen atom. It will also be seen that normal pentane contains two primary and three secondary carbon atoms, six primary and six secondary hydrogen atoms; and isopentane contains one secondary, one tertiary, and three primary carbon atoms, and one tertiary, two secondary, and nine primary hydrogen atoms. Neopentane, however, contains four primary carbon atoms and one *quaternary* carbon atom, *i.e.*, a carbon atom joined to *four* other carbon atoms; also, all the twelve hydrogen atoms are primary. As we shall see later, the behaviour of these various types of hydrogen atoms differs considerably.

Nomenclature of the alkanes

Whenever a new branch of knowledge is opened up, there is always the problem of introducing a system of nomenclature. The early chemists usually named a compound on the basis of its history, *e.g.*, methane. This is the parent hydrocarbon of methyl alcohol, CH_3OH. Methyl alcohol was originally obtained by the destructive distillation of wood, and was named 'wood-spirit'. From this arose the word methyl, which is a combination of two Greek words, *methu* (wine) and *hule* (wood). Other examples of this way of naming compounds are acetic acid, which is the chief constituent of vinegar (Latin: *acetum*, vinegar); malic acid, which was first isolated from apples (Latin: *malum*, apple), and so on. Thus grew up a system of *common* or *trivial* names, but in many cases the origin of the name has been forgotten. One advantage of the trivial system is that the names are usually short and easily remembered, but a disadvantage is that a particular compound may have a number of names.

As the number of known organic compounds increased, it became apparent that it was necessary to systematise the method of nomenclature. The most satisfactory system is one which indicates the structure of the compound. This task was originally begun in 1892 by an international committee of chemists at Geneva, and hence is referred to as the *Geneva system of nomenclature*. The work was carried on by the *International Union of Chemists* (I.U.C.) by a committee appointed in 1922, and in 1931 these drew up a report which is often referred to as the I.U.C. system. Nomenclature is always undergoing revision, and the latest rules are those recommended in 1957 by the Commission on the Nomenclature of Organic Chemistry of the International Union of Pure and Applied Chemistry (I.U.P.A.C.). The reader should always consult the *Chemical Society Handbook* if he wishes to publish work in the *Journal*, and any mistakes he makes in nomenclature will soon be put right by the Editors!

There are at least three systems in use for naming alkanes, and in all three the *class-suffix*, i.e., the ending of the name which indicates the particular homologous series (see later), is *-ane*.

(1) In the **trivial system** of nomenclature the straight-chain compounds are always designated as *normal* compounds, and the word normal is usually abbreviated to n-. If the compound contains the grouping $(CH_3)_2CH-$, it is known as the iso-compound; if it contains a quaternary carbon atom, the compound is known as the neo-compound. It is impossible to name many of the more complex alkanes by the trivial system (see later for examples of the trivial system).

The first four alkanes have special names (related to their history); from the fifth member onwards, Latin or Greek numerals are used to indicate the number of carbon atoms in the molecule.

Name	Formula	Name	Formula
methane	CH_4	hexadecane	$C_{16}H_{34}$
ethane	C_2H_6	heptadecane	$C_{17}H_{36}$
propane	C_3H_8	octadecane	$C_{18}H_{38}$
butane	C_4H_{10}	nonadecane	$C_{19}H_{40}$
pentane	C_5H_{12}	eicosane	$C_{20}H_{42}$
hexane	C_6H_{14}	heneicosane	$C_{21}H_{44}$
heptane	C_7H_{16}	docosane	$C_{22}H_{46}$
octane	C_8H_{18}	tricosane, etc.	$C_{23}H_{48}$, etc.
nonane	C_9H_{20}	triacontane	$C_{30}H_{62}$
decane	$C_{10}H_{22}$	hexatriacontane	$C_{36}H_{74}$
undecane	$C_{11}H_{24}$	tetracontane	$C_{40}H_{82}$
hendecane		pentacontane	$C_{50}H_{102}$
dodecane	$C_{12}H_{26}$	hexacontane	$C_{60}H_{122}$
tridecane	$C_{13}H_{28}$	heptacontane	$C_{70}H_{142}$
tetradecane	$C_{14}H_{30}$	octacontane	$C_{80}H_{162}$
pentadecane	$C_{15}H_{32}$		

Univalent groups that are formed by the removal of one hydrogen atom from an alkane are known as *alkyl* or *alphyl* groups. The name of each individual group is obtained by changing the suffix -ane of the parent hydrocarbon into -yl. The first four alkyl groups are often represented by a shorthand notation.

The group derived from pentane was named amyl. This name is now abandoned and pentyl and isopentyl used.

The *alkanes*, originally and still known as the *paraffins*, derived this name from the fact that an alkyl group plus one hydrogen atom gives a paraffin, *i.e.*, *alkyl* + H = *alkane*.

In chemical equations, if we are dealing with alkyl compounds as a group and we do not wish to specify any particular member, we use the symbol R to represent the unspecified alkyl group, *e.g.*, RCl represents *any* alkyl chloride.

Alkane	Group	Short-hand notation
methane	methyl CH_3—	Me
ethane	ethyl C_2H_5—	Et
propane	n-propyl $CH_3CH_2CH_2$—	n-Pr, Pr^α, or Pr
	isopropyl CH_3CHCH_3	isoPr, Pr^β, or Pr^i
butane	n-butyl $CH_3CH_2CH_2CH_2$—	n-Bu, Bu^α, or Bu
	s-butyl $CH_3CH_2CHCH_3$	s-Bu, Bu^β, Bu^s
	isobutyl $(CH_3)_2CHCH_2$—	isoBu or Bu^i
	t-butyl $(CH_3)_3C$—	t-Bu, Bu^t

(2) In this system of nomenclature the hydrocarbon, except the n-compound, is regarded as a substitution product of methane. The most highly branched carbon atom in the compound is named as the methane nucleus, and the alkyl groups attached to this carbon atom are named in alphabetical order. Hydrogen atoms, if joined to the carbon atom chosen as the methane nucleus, are not named.

This system of nomenclature is fairly good, since the name indicates the structure of the compound. It is impossible, however, to name the complex alkanes by this system.

(3) In the **I.U.P.A.C. system** of nomenclature the longest chain possible is chosen, and the compound is named as a derivative of this n-hydrocarbon. The carbon chain is numbered from one end to the other by arabic numerals, and the positions of *side-chains* are indicated by numbers, the direction of numbering being so chosen as to give the lowest numbers possible to the side-chains. When series of locants containing the same number of terms are compared term by term, that series is 'lowest' which contains the lowest number on occasion of the first difference, and this principle is used irrespective of the nature of the substituents, *e.g.*,

$$\overset{10}{C}H_3\overset{9}{C}H_2\overset{8}{C}H-\overset{7}{C}H\overset{6}{C}H_2\overset{5}{C}H_2\overset{4}{C}H_2\overset{3}{C}H_2\overset{2}{C}H\overset{1}{C}H_3$$
$$\qquad\qquad |\qquad\; |\qquad\qquad\qquad\qquad |$$
$$\qquad\qquad CH_3\;\; CH_3\qquad\qquad\qquad CH_3$$

This is named 2,7,8-trimethyldecane and *not* 3,4,9-trimethyldecane; the first set is 'lower' than the second set because at the first difference 2 is less than 3. When two sets of numbers are equally possible, then the order of the prefixes in the name decides which shall be used, *e.g.*, 1-bromo-3-chloropropane and *not* 3-bromo-1-chloropropane. It should also be noted that the names of prefixes are arranged alphabetically, regardless of the number of each, *e.g.*, 5-ethyl-2,3-dimethyl-octane.

The I.U.P.A.C. system of nomenclature is undoubtedly superior to the other two, since it permits the naming of any alkane on sight.

The following are examples of the three systems of nomenclature:

	1	2	3
$CH_3CH_2CH_2CH_3$	n-butane	n-butane	n-butane
$\overset{1}{C}H_3\overset{2}{C}H\overset{3}{C}H_2\overset{4}{C}H_3$ $\quad\|$ $\quad CH_3$	isopentane	ethyldimethylmethane $\overset{2}{(C}$ is most highly branched)	isopentane
$\quad CH_3$ $\overset{1}{C}H_3\overset{2}{\underset{\|}{C}}\overset{3}{C}H_2\overset{4}{C}H_3$ $\quad\|$ $\quad CH_3$	neohexane	ethyltrimethylmethane $\overset{2}{(C}$ is most highly branched)	2,2-dimethylbutane
$CH_3CH_2\overset{5}{C}H\overset{6}{C}H_2\overset{7}{C}H_2\overset{8}{C}H_3$ $\quad\|$ $\quad ^4CH_2$ $\quad\|$ $\quad ^3CH{-}CH_3$ $\quad\|$ $\quad ^2CH{-}CH_3$ $\quad\|$ $\quad ^1CH_3$	—	—	5-ethyl-2,3-dimethyl-octane

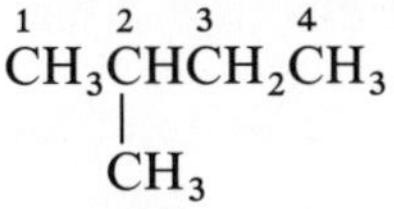

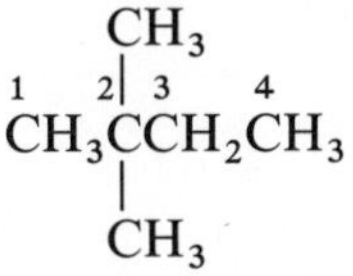

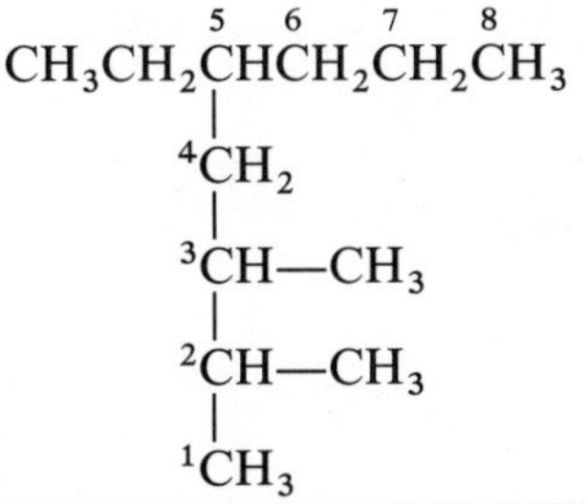

N.B. The following names are retained for *unsubstituted alkanes only*: isobutane, isopentane, neopentane, isohexane.

When several chains are of equal length, that chain chosen goes in series to: (*a*) the chain which has the greatest number of side-chains; (*b*) the chain whose side-chains have the lowest-numbered locants; (*c*) the chain having the greatest number of carbon atoms in the smaller side-chains; (*d*) the chain having the least-branched side-chains. Also, where there is a side-chain within a side-chain, the latter is also numbered, and the name of the complex group is considered to begin with the *first* letter of its complete name, *e.g.*,

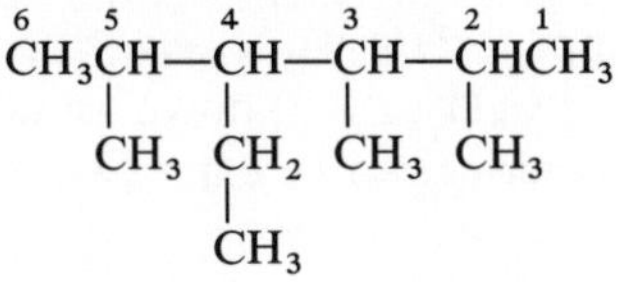

$$\overset{6}{C}H_3\overset{5}{C}H{-}\overset{4}{C}H{-}\overset{3}{C}H{-}\overset{2}{C}H\overset{1}{C}H_3$$

4-ethyl-2,3,5-trimethylhexane

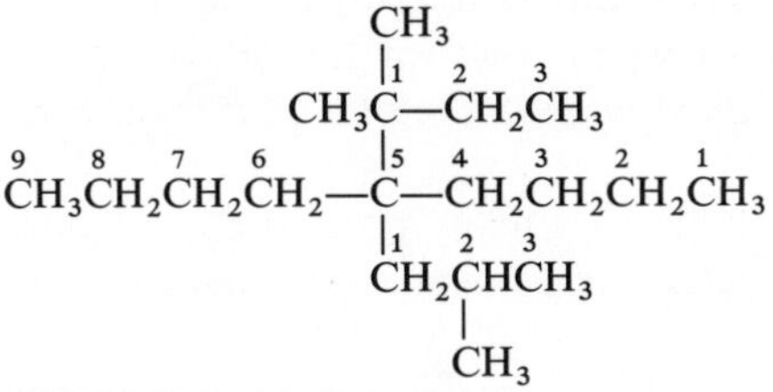

$$\overset{9}{C}H_3\overset{8}{C}H_2\overset{7}{C}H_2\overset{6}{C}H_2{-}\overset{5}{\underset{\|}{C}}{-}\overset{4}{C}H_2\overset{3}{C}H_2\overset{2}{C}H_2\overset{1}{C}H_3$$

5-(1,1-dimethylpropyl)-5-(2-methylpropyl)nonane

Homologous series

If we examine the formulæ of the various alkanes, we find that the formula of each individual differs from that of its 'neighbour' by CH_2, *e.g.*, CH_4, C_2H_6, C_3H_8, C_4H_{10}, C_5H_{12}, ... A set of compounds, such as the alkanes, in which the members differ in composition from one another by CH_2, is known as an *homologous series*, the individual members being known as *homologues*.

Throughout organic chemistry we find homologous series, each series being characterised by the presence of a *functional group*. The functional group is an atom or a group of atoms that causes a compound to behave in a particular way, *i.e.*, it is the functional group that gives rise to homologous series. Some of the more important functional groups and the classes of compounds to which they give rise are shown in Table 3.1.

Table 3.1

Class of Compound	Functional Group	
	Formula	Name*
Alcohols	—OH	Hydroxyl group
Aldehydes and ketones	$\diagdown \diagup C{=}O$	Carbonyl group
Carboxylic acids	$-C\diagup\diagdown \overset{O}{\underset{OH}{}}$	Carboxyl group
Cyanides	—C≡N	Cyano group
Nitro-compounds	—NO$_2$	Nitro group
Amines	—NH$_2$	Amino group
Mercaptans	—SH	Mercapto group
Sulphonic acids	—SO$_3$H	Sulphonic acid group

*Many functional groups are known by more than one name. Nomenclature is dealt with in each homologous series described in the text.

It is also possible for a compound to contain two (or more) identical or different functional groups and this gives rise to *polyfunctional* compounds (see text).

If we examine the formulæ of the various alkanes, we find that the formula C_nH_{2n+2} will represent any particular homologue when n is given the appropriate value, *e.g.*, for pentane n is 5; therefore the formula of pentane is C_5H_{12}. The formula C_nH_{2n+2} is known as the *general formula* of the alkanes. The composition of any homologous series can be expressed by means of a general formula.

When we study the methods of preparation of the different alkanes, we find that several methods are common to all, *i.e.*, similar methods may be used for the preparation of all the homologues. This gives rise to the *general methods of preparation* of a particular homologous series.

Examination of the properties of the alkanes shows that many properties are, more or less, common to all the alkane homologues. This gives rise to the *general properties* of an homologous series.

The occurrence of homologous series facilitates the study of organic chemistry, since it groups together compounds having many resemblances. If we know the properties of several of the lower homologues, we can obtain a fair idea of the properties of higher homologues, *i.e.*, we can forecast (within limits) the properties of a compound that we have not yet prepared. The reason for these limitations is that the behaviour of a functional group in an homologous series depends on structural differences. We can now, however, explain many of these differences in behaviour in terms of polar and steric effects, and we are therefore in a much stronger position to make our predictions.

Only when *one* functional group is present do the properties of the compounds containing it show many similarities. If two (or more) functional groups are present, the properties depend on the 'distance' between them. When the groups are widely separated, each generally behaves in the normal way, but when they are close to each other, the properties of the compound may be very different from those of the individual functional groups (see text).

There are various ways of studying organic chemistry, *e.g.*, as homologous series, or treating as a class compounds containing a particular functional group, irrespective of the 'rest' of the molecule. In this book, the treatment is the discussion of homologous series.

The general methods of preparation of the alkanes

1. By the catalytic reduction of unsaturated hydrocarbons, *e.g.*, reduction of ethylene using nickel. Formerly, the nickel catalyst was usually prepared by reducing with hydrogen nickel

$$C_2H_4 + H_2 \xrightarrow{\text{Ni}} C_2H_6 \quad (ex.)*$$

oxide deposited on a suitable inert, porous support, *e.g.*, kieselguhr. The support is impregnated with a nickel salt, treated with sodium hydroxide, washed and dried, and the resulting nickel oxide reduced with hydrogen at 300–450° C. Many organic compounds may be reduced by passing their vapours mixed with hydrogen over nickel heated at 200–300° C. Any reduction that is carried out in this manner is referred to as the *Sabatier–Senderens reduction*, in honour of the workers who first introduced this method. It is quite a common feature in organic chemistry to name a reaction after its discoverer or, in certain cases, after a worker who investigated the reaction and extended its application. The reader should always make himself familiar with the reaction *associated* with a particular name.

A common nickel catalyst is that prepared by the method introduced by Raney (1927). An alloy containing equal amounts of nickel and aluminium is digested with sodium hydroxide; the aluminium is dissolved away, and the residual very finely divided nickel is washed and stored under water, ethanol, or any other suitable liquid. Raney nickel is more reactive than the supported nickel catalyst, and is usually effective at lower temperatures, often at room temperature (see also Table 6.1).

Alkenes may also be reduced by a variety of chemical reagents (see text).

2. Reduction of alkyl halides. This may be done in different ways:

(i) Reduction by dissolving metals, *e.g.*, zinc and acetic or hydrochloric acid, zinc and sodium hydroxide, zinc-copper couple and ethanol, etc. It is no longer considered that 'nascent' hydrogen is the reducing agent. The belief now is that there is an electron-transfer from the metal to the substrate, followed by the addition of protons from the solvent. Thus, reduction with a zinc-copper couple may be formulated as (where *e* is an electron):

$$Zn \longrightarrow Zn^{2+} + 2e$$

$$RX + e \longrightarrow X^- + R\cdot \xrightarrow{e} R:^-$$

$$R:^- + C_2H_5OH \longrightarrow R\!-\!H + {}^-OC_2H_5 \quad (g.\text{-}v.g.)$$

(ii) Primary and secondary alkyl halides are readily reduced to alkanes by lithium aluminium hydride ($LiAlH_4$); tertiary halides give mainly alkenes. On the other hand, sodium borohydride ($NaBH_4$) reduces secondary and tertiary halides, but not primary, whereas triphenyltin hydride (Ph_3SnH) reduces all three types of halides.

*See preface for the significance of these terms in parentheses.

(iii) Catalytic hydrogenolysis (see p. 161).

(iv) *Iodides* may be reduced by heating with concentrated hydriodic acid at $150°$ C.

Reduction with concentrated hydriodic acid is usually carried out in the presence of a small amount of red phosphorus which regenerates the hydriodic acid from the iodine formed. The hydriodic acid-red phosphorus mixture is one of the most powerful reducing agents used in organic chemistry.

Alcohols, aldehydes, ketones, and carboxylic acids are also reduced to alkanes by hydrogen iodide and red phosphorus. On the other hand, aldehydes and ketones may be reduced to alkanes by the Clemmensen (p. 210) and Wolff-Kishner (p. 213) reductions.

3. Grignard reagents. Alkyl halides in ether react with magnesium to form alkylmagnesium halides or *Grignard reagents* which, on treatment with water or dilute acid, are decomposed to alkanes (see Ch. 15).

$$RI + Mg \xrightarrow{\text{ether}} R\text{---}Mg\text{---}I \xrightarrow{\text{H}^+} RH \quad (v.g.)$$

4. By heating a mixture of the sodium salt of a carboxylic acid and soda-lime:

$$RCO_2Na + NaOH(CaO) \longrightarrow RH + Na_2CO_3$$

This process of eliminating carbon dioxide from a carboxylic acid is known as *decarboxylation*. Soda-lime is a very useful reagent for this purpose, but various other reagents may also be used.

Oakwood *et al.* (1950) have shown that only sodium acetate decomposes according to the equation given above. In all of the other cases tested—propionate, butyrate and caproate—various products were obtained, *e.g.*, with sodium propionate:

$$C_2H_5CO_2Na \xrightarrow{\text{NaOH}} C_2H_6 + CH_4 + H_2 + \text{Unsaturated compounds.}$$
$$\quad\quad\quad\quad (44\%) \quad (20\%) \ (33\%)$$

This is therefore not a useful general method for the preparation of alkanes, since the separation of the products is usually difficult.

When dealing with a preparation, there are at least three important points that must be considered: (i) the yield of crude product; (ii) the yield of pure product; (iii) the *accessibility* of the starting materials. In certain cases the yield of crude material is high, but the nature of the impurities is such that purification causes a large loss of material, resulting in a poor yield of pure product. On the other hand, it often happens that the product of one reaction is to be used as the starting material for some other compound which can readily be freed (*i.e.*, purified with very little loss) from the original impurity. Provided, then, that this impurity does not interfere with the second reaction, the crude material of the first step can be used as the starting material for the second. Thus the yield alone of a particular reaction cannot decide the usefulness of that method of preparation; the subsequent history of the product must also be taken into consideration. Furthermore, all things being equal, the more accessible materials, *i.e.*, readily prepared or purchased, are used as the starting materials.

5. Wurtz reaction (1854). An ethereal solution of an alkyl halide (preferably the bromide or iodide) is treated with sodium, *e.g.*,

$$R^1X + R^2X + 2Na \longrightarrow R^1\text{---}R^2 + 2NaX$$

As previously pointed out, when we do not wish to specify a particular alkyl group, we use the symbol R. When we deal with two unspecified alkyl groups which may, or may not, be the same, we can indicate this by R^1 and R^2: also, when dealing with compounds containing a halogen atom, and we do not wish to specify the halogen, we can indicate the presence of the unspecified halogen atom by means of X.

Consideration of the equation given above shows that in addition to the desired alkane R^1—R^2, there will also be present the alkanes R^1—R^1 and R^2—R^2. Unsaturated hydrocarbons are also obtained. Obviously, then, the best yield of an alkane will be obtained when R^1 and R^2 are the same, *i.e.*, when the alkane contains an even number of carbon atoms and is symmetrical. It has been found that the Wurtz reaction gives good yields only for 'even carbon' alkanes of high molecular weight, and that the reaction generally fails with tertiary alkyl halides (*q.v.*).

Sodium is used in the Wurtz reaction. Other metals, however, in a finely divided state, may also be used, *e.g.*, Ag, Cu (see text).

Two mechanisms have been suggested for the Wurtz reaction.

(i) The intermediate formation of an *organo-metallic* compound, *e.g.*, the formation of n-butane from ethyl bromide:

$$C_2H_5—Br + 2Na\cdot \longrightarrow C_2H_5^- Na^+ + NaBr$$
$$C_2H_5^- Na^+ + C_2H_5Br \longrightarrow C_2H_5—C_2H_5 + NaBr$$

(ii) The intermediate formation of free radicals, *e.g.*,

$$C_2H_5—Br + Na\cdot \longrightarrow C_2H_5\cdot + NaBr$$
$$C_2H_5\cdot + C_2H_5\cdot \longrightarrow C_2H_5—C_2H_5$$

One of the properties of free radicals is *disproportionation*, *i.e.*, intermolecular hydrogenation, one molecule acquiring hydrogen at the expense of the other, *e.g.*,

$$C_2H_5\cdot + C_2H_5\cdot \longrightarrow C_2H_6 + C_2H_4$$

This would account for the presence of ethane and ethylene in the products. According to Morton *et al.* (1942), however, ethane and ethylene may be produced as follows:

$$Na^+CH_3CH_2^- \quad H—CH_2—CH_2—Br \longrightarrow CH_3CH_3 + CH_2{=}CH_2 + NaBr$$

This is a bimolecular elimination mechanism. On the other hand, the mechanism of alkane formation is generally accepted as usually being a bimolecular nucleophilic substitution:

$$Na^+CH_3CH_2^- \overset{\overset{\displaystyle Br}{|}}{CH_2CH_3} \longrightarrow CH_3CH_2CH_2CH_3 + NaBr$$

The free-radical mechanism, however, has been shown to operate in some cases (Bryce-Smith, 1963).

6. Kolbe's electrolytic method (1849). A concentrated solution of the sodium or potassium salt of a carboxylic acid or mixture of carboxylic acids is electrolysed, *e.g.*,

$$R^1CO_2K + R^2CO_2K + 2H_2O \longrightarrow R^1—R^2 + 2CO_2 + H_2 + 2KOH$$

If R^1 and R^2 are different, then hydrocarbons R^1—R^1 and R^2—R^2 are also obtained (*cf.* Wurtz reaction). Such mixtures can often be separated readily. Yields of 50–90 per cent have been obtained with straight-chain acids containing 2–18 carbon atoms. Alkyl groups in the α-position decrease the yield (usually below 10 per cent). The by-products are alkenes, alcohols (particularly in alkaline solution) and esters. It is also interesting to note that the yields of the alkanes are increased when dimethylformamide is used as solvent (Finkelstein *et al.*, 1960).

The Kolbe electrolytic method now has application in the synthesis of natural compounds, particularly lipids.

Several mechanisms have been proposed for the Kolbe reaction. The free-radical theory is the one now favoured, and strong support for it has been provided by Lippincott *et al.* (1956), *e.g.*, when sodium propionate is electrolysed, n-butane, ethane, ethylene and ethyl propionate are obtained. The propionate ion discharges at the anode to form a free radical:

$$C_2H_5CO_2:^- \longrightarrow C_2H_5CO_2\cdot + e$$

This propionate free radical then breaks up into the ethyl free radical and carbon dioxide:

$$C_2H_5CO_2\cdot \longrightarrow C_2H_5\cdot + CO_2$$

Then:

(i) $2C_2H_5\cdot \longrightarrow C_4H_{10}$

(ii) $C_2H_5\cdot + C_2H_5\cdot \longrightarrow C_2H_6 + C_2H_4$

(iii) $C_2H_5\cdot + C_2H_5CO_2\cdot \longrightarrow C_2H_5CO_2C_2H_5$

Reaction (i) gives n-butane; (ii) gives ethane and ethylene by disproportionation (*cf.* Wurtz reaction); and (iii) gives ethyl propionate.

7. A very useful method for preparing long-chain alkanes involves the coupling of alkylboranes (see p. 114) by means of silver nitrate in the presence of sodium hydroxide at 25°C, *e.g.*, 2-methylpent-1-ene gives 4,7-dimethyldecane (Brown *et al.*, 1961):

$$6CH_3(CH_2)_2C(CH_3)=CH_2 \xrightarrow{2B_2H_6} [2CH_3(CH_2)_2CH(CH_3)CH_2-]_3B \xrightarrow[\text{NaOH}]{AgNO_3}$$

$$3CH_3(CH_2)_2CH(CH_3)CH_2CH_2CH(CH_3)(CH_2)_2CH_3 \quad (75–80\%)$$

Coupling may also be effected between *unlike* boranes.

Conformation. It has already been pointed out that the valency angle at the central carbon atom of a regular tetrahedron is 109° 28′. It was originally postulated (van't Hoff) that this angle was fixed, but it is now known that it may deviate from this value ($\sim \pm 5°$). Another postulate was the **principle of free rotation about a single bond.** Let us consider the ethane molecule, CH_3—CH_3, and let us imagine that one methyl group is rotated about the C—C bond as axis with the other group at rest. Suppose we use, as the starting point, the position in which two C—H bonds are parallel, *i.e.*, these four atoms lie in a plane (Fig. 3.1*c*). In this position, the *dihedral angle* (*angle of rotation* or *angle of torsion*) is zero, and if the rotation is free, then as the dihedral angle changes, the energy content of the molecule will remain constant (the plot of energy content against the dihedral angle will be a horizontal line). In this situation, the two 'halves' can assume, with complete freedom, an infinite number of positions relative to each other. Thus, the entropy of the molecule will be a maximum. Pitzer *et al.* (1936) calculated the entropy of ethane based on the assumption that the *internal rotation* was free, and found that

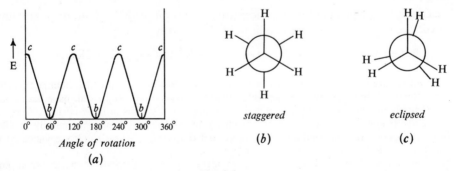

Angle of rotation

(*a*)

staggered

(*b*)

eclipsed

(*c*)

Fig. 3.1

calculated value was greater than the observed value. This means that there is less freedom to assume all possible dihedral angles than was expected on the principle of free rotation. Pitzer therefore assumed that the internal rotation is hindered by a potential energy barrier, and calculated the change in entropy with increasing barrier height. When he assumed a barrier of $12{\cdot}55\,kJ\,mol^{-1}$ (3 kcal), the calculated and observed entropies were brought into good agreement. The potential energy curve obtained for ethane is shown in Fig. 3.1a and, by convention, the potential energy is measured relative to the energy of the most stable form. Because of the existence of potential energy barriers, the result is *mutual oscillation* (*libration*) about the conformations with minimum potential energy, thereby producing 'more order' (more restriction) than had there been no barriers.

Figure 3.1b is the Newman projection formula. This is obtained by viewing the molecule along the bonding line of the two carbon atoms, with the carbon atom nearer to the eye being designated by equally spaced radii, and the carbon atom further from the eye by a circle with three equally spaced radial extensions. Figure 3.1b represents the *staggered* (or *transoid*) conformation (in which the hydrogen atoms are as far apart as possible), and 1c the *eclipsed* (or *cisoid*) conformation (in which the hydrogen atoms are as close together as possible). It can be seen from Fig. 3.1a that the energy of the eclipsed conformation is greater than that of the staggered conformation. The potential energy barrier ($11{\cdot}92\,kJ\,mol^{-1}$) is much too small for either conformation to remain stable, *i.e.*, the eclipsed and staggered forms are readily interconvertible and hence neither can be isolated as such. However, the staggered conformation is the preferred form, *i.e.*, its population is greater than that of the eclipsed form (see below).

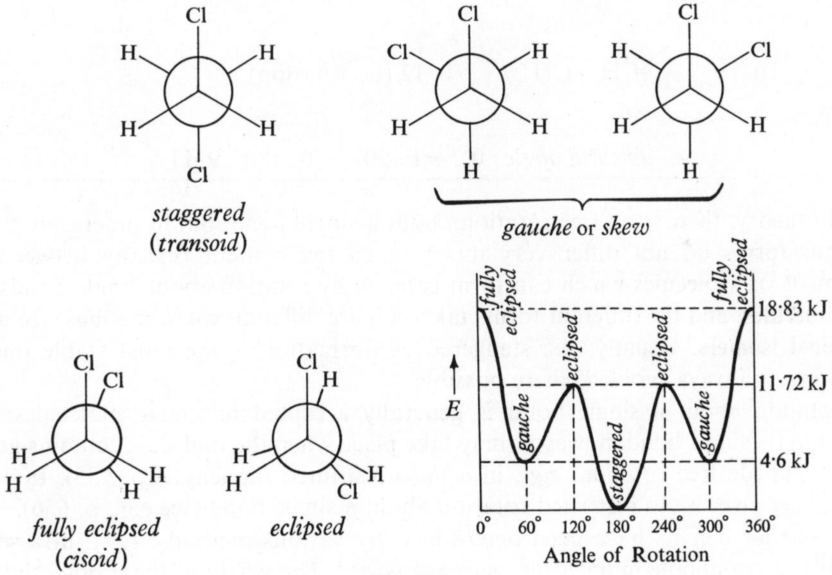

staggered
(transoid)

gauche or skew

fully eclipsed
(cisoid)

eclipsed

Fig. 3.2

Now let us consider ethylene dichloride. According to Bernstein (1949), the potential energy undergoes the changes shown in Fig. 3.2. There are two positions of minimum energy, but the staggered conformation is the more preferred one, *i.e.*, the one in which the molecule largely remains. Dipole moment studies show that this is so in practice, and according to Mizushima *et al.* (1938), only the staggered form is present at low temperatures.

Table 3.2

Z	CH₃Z	R'CH₂Z	R'₂CHZ	Z	CH₃Z	R'CH₂Z	R'₂CHZ
	τ	τ	τ		τ	τ	τ
—R	9·12	8·75	8·50	—I	7·83	6·85	5·78
—CO₂R	8·00	7·90	—	—Br	7·35	6·70	5·97
—CN	8·00	7·52	7·3	—Cl	6·98	6·56	5·98
—CONH₂	7·98	7·95	—	—OR	6·70	6·64	6·20
—CO₂H	7·93	7·65	7·43	—$\overset{+}{N}R_3$	6·67	6·60	6·50
—COR	7·90	7·60	7·5	—OH	6·62	6·42	6·15
—SH, —SR	7·90	7·60	6·90	—OCOR	6·35	5·89	4·95
—NH₂, —NR₂	7·85	7·50	7·13	—F	5·70	5·66	5·40
—CHO	7·83	7·80	7·60	—NO₂	5·67	5·60	5·40

	τ
R—OH	9·5–6·0 (lower for enols; −1 to −6)
Ar—OH	∼5·5
R—SH	9–8
Ar—SH	∼6·5
RNH₂	8·5
ArNH₂	6·6–5·0
RCO₂H	0 to −3

J values (Hz)

$$\begin{array}{c} \diagdown \quad \diagup H \\ C \\ \diagup \quad \diagdown H \end{array} \quad 10\text{--}18 \qquad >CH\text{---}CH< \quad 0\text{--}12 \text{ (no rotation)}$$

dihedral angle: 0°, ∼8; 90°, ∼0; 180°, 9–11

Thus, in theory, there is no free rotation about a single bond, but in practice it may occur if the various forms do not differ very much in energy content (usually between 4·18 and 41·84 kJ mol⁻¹). Molecules which can form isomers by rotation about single bonds are called **flexible molecules**, and the different forms taken up are different **conformations** (see also p. 550) or **rotational isomers**. Usually, the staggered conformation is the most stable one, and the eclipsed form is always avoided where possible.

Free rotation about a single bond is generally accepted in *simple* molecules. *Restricted rotation* about a single bond, however, may take place when the molecule contains groups large enough to impede free rotation, *e.g.*, in *ortho*-substituted biphenyls (p. 785). In some cases, resonance may give rise to restricted rotation about a single bond (see *e.g.*, p. 656).

Energy barrier heights have been determined by various methods, *e.g.*, infra-red, Raman and NMR spectra, dipole moment measurements, etc. The origin of these potential barriers is complex; several types of contributing interactions operate simultaneously, their relative importance depending on the structure of the molecule in question. Four factors are: (i) **Dipole-dipole interactions**; poles of like sign will tend to be as far apart as possible. (ii) **Bond opposition strain.** Covalent bonds, because they 'consist' of electron pairs, will repel other bonds. (iii) **Steric strain**, *i.e.*, repulsion between non-bonded atoms or groups. When molecules approach closely together, forces of repulsion come into play, and it has been found that the distance of closest approach is greater than the sum of the covalent radii of the atoms concerned. This distance of closest approach for each atom (or group) is known as the *van der Waals radius*, *e.g.*, H, $1 \cdot 2$ Å; O, $1 \cdot 4$ Å; Cl, $1 \cdot 8$ Å; Br, $1 \cdot 95$ Å; I, $2 \cdot 13$ Å (*cf.* Table 2.3).

(iv) **Bond angle strain**, *i.e.*, distortion of bond angles.

When the stability of a molecule is decreased by internal forces produced by interaction between constituent parts, that molecule is said to be under strain. The study of the existence of preferred conformations in molecules, and the relating of physical and chemical properties of a molecule to its preferred conformation, is known as **conformational analysis** (see also p. 550).

A particularly useful method for studying the conformations of molecules is NMR spectroscopy. However, before dealing with this, let us first consider the chemical shifts of the protons of CH_3, CH_2, and the CH group in acyclic compounds and the chemical shift of a proton attached to O, N, etc. As we have seen (p. 11), the more electronegative Z is in the groups —CH—Z and —Z—H, the more is the proton deshielded and consequently the lower is the τ-value. Table 3.2 lists the τ-values of some of the more important groupings, and also lists some coupling constants.

We return now to the use of NMR spectroscopy in conformational analysis. Let us first consider ethyl chloride, and since the most stable conformation is the staggered one, the molecule may be represented as (I), (II), and (III). In (I), the environments of H_a and H_c are identical, but differ from that of H_b. This can be seen to be the case by replacing each proton, one at a time, by Z. This procedure for H_a and H_c produces mirror images, and so the two protons are chemically equivalent,

(I)   (II)   (III)

but for H_b the substituted product is different, and therefore H_b is not chemically equivalent to H_a and H_c. The coupling constants of H_a with H_d and H_e are different because of the different dihedral angles, and similarly for H_c. Thus, H_a and H_c are chemically but not magnetically equivalent. By using similar arguments, it can be shown that H_d and H_e are also chemically but not magnetically equivalent. Hence, if the population of ethyl chloride conformation were *completely* represented by (I), protons H_a and H_c would give one signal and H_b another signal, provided that the chemical shifts were sufficiently different ($0 \cdot 1$–$0 \cdot 2$ p.p.m.), *i.e.*, the methyl group would give *two* signals. In practice, the methyl group gives only *one* signal (a triplet), and therefore protons H_a, H_b, and H_c must be equivalent. This is explained on the basis that there is free rotation about the carbon-carbon single bond. When the methyl group rotates (with respect to the CH_2Cl group), it can take up the other two stable staggered conformations, (II) and (III), the result being that each proton has an average environment. Since these average environments are identical, rotation results in the equivalence of H_a, H_b, and H_c.

This equivalence may be demonstrated as follows. Let P_I, P_{II}, P_{III} be the populations in conformations (I), (II), and (III), respectively, where $P_I + P_{II} + P_{III} = 1$. Also, let δ (the chemical shift) of any proton *trans* (*anti*) to another proton be x, and y if *trans* to Cl. Now, the observed δ of a given proton is the sum of the δ-contributions in each conformation, and so it follows that:

$$\delta(H_a) = P_I x + P_{II} y + P_{III} x$$
$$\delta(H_b) = P_I y + P_{II} x + P_{III} x$$
$$\delta(H_c) = P_I x + P_{II} x + P_{III} y$$

Since all three populations are equal (all three are indistinguishable because all protons, as such, are identical), *i.e.*, $P_I = P_{II} = P_{III} = 1/3$, therefore

$$\delta(H_a) = \delta(H_b) = \delta(H_c) = 2/3x + 1/3y$$

Hence, all three protons have the same (average) chemical shift (in the freely rotating state) and so there is no observable splitting for the methyl group.

Application of this method to protons H_d and H_e gives:

$$\delta(H_d) = P_I x + P_{II} x + P_{III} x = x$$
$$\delta(H_e) = P_I x + P_{II} x + P_{III} x = x$$

From the above discussion, it can be seen that if ethyl chloride were undergoing slow rotation, it would be an ABB'CC' spin system, but when undergoing fast rotation, it is an A_3B_2 spin system (see also below).

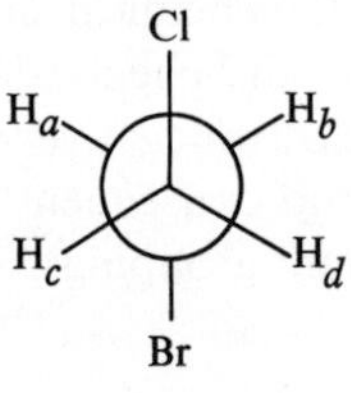

(IV) 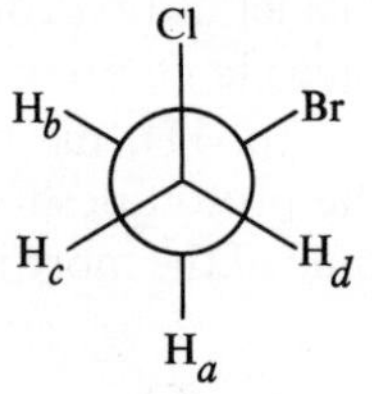   (V) 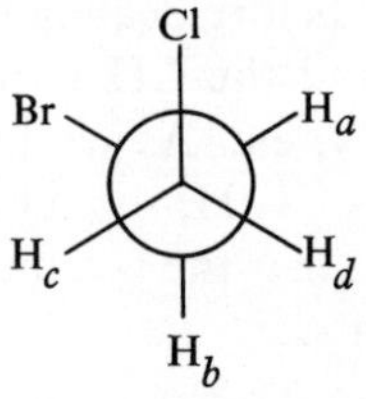   (VI)

Now let us consider the case of 1-bromo-2-chloroethane, and if we use the same arguments as before, then in (IV), H_a and H_b are chemically but not magnetically equivalent, and this is also true for H_c and H_d. Also, if δ for a proton *trans* to a proton is x, *trans* to Cl is y, and *trans* to Br is z, then:

$$\delta(H_a) = P_{IV} x + P_V y + P_{VI} x$$
$$\delta(H_b) = P_{IV} x + P_V x + P_{VI} y$$

In this molecule, although $P_V = P_{VI} \neq P_{IV}$, it still follows that $\delta(H_a) = \delta(H_b)$. In the same way, it can be shown that $\delta(H_c) = \delta(H_d)$.

We can extend the argument by also considering the possible coupling constants. If J_t and J_s represent, respectively, the vicinal *trans* and skew coupling, then:

$$J_{ad} = P_{IV} J_t + P_V J_s + P_{VI} J_s$$
$$J_{bd} = P_{IV} J_s + P_V J_t + P_{VI} J_s$$

Since $P_V = P_{VI} \neq P_{IV}$, therefore $J_{ad} \neq J_{bd}$. Hence, H_a and H_b (and also H_c and H_d) are chemically but not 'magnetically' equivalent. Application of this method will show that the three protons in the methyl group in ethyl chloride are magnetically equivalent, as are also the two protons in the CH_2Cl group, for the freely rotating molecule.

Finally, let us consider a molecule of the type $R^1 R^2 CHCHR^3 R^4$. Each carbon atom is asymmetric, and the environments of H_a are different in all three conformations, and this is also true for H_b.

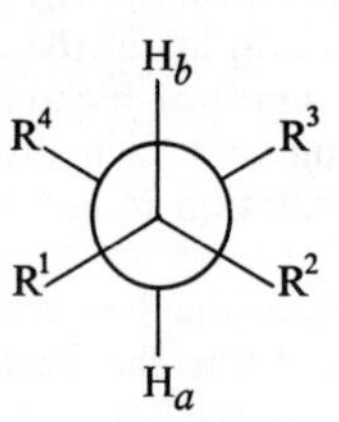

(VII) 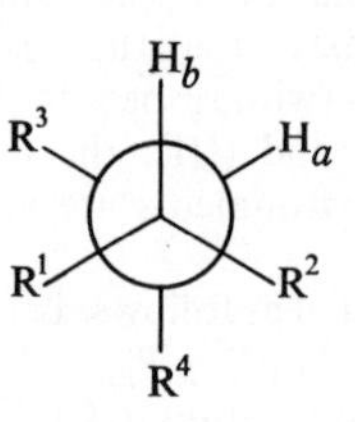   (VIII) 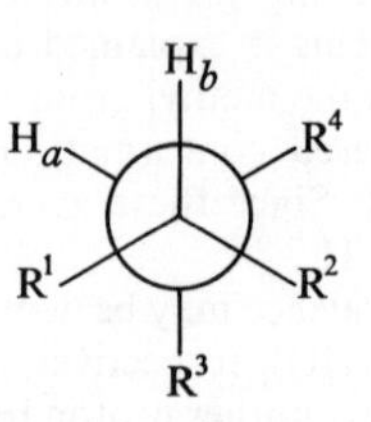   (IX)

Furthermore, because of the different steric effects, the three populations will be different (*i.e.*, $P_{VII} \neq P_{VIII} \neq P_{IX}$). Hence, the chemical shifts of H_a (and those of H_b) are different in each conformation. If rotation were slow, the NMR spectrum would be a composite of all three spectra (three AB spin systems for each pair of enantiomers). If the temperature is raised, the rate of rotation is increased, and the result is an average of chemical shifts and coupling constants to give one AB spin system (for each pair of enantiomers). These averages, however, will be weighted in favour of the most stable conformation, *i.e.*, the one with the highest population.

General properties of the alkanes

The name paraffin arose through contracting the two Latin words *parum affinis*, which means 'little affinity'. This name was suggested because these hydrocarbons were apparently very unreactive. It is difficult to define the terms 'reactive' and 'unreactive', since a compound may be reactive under one set of conditions and unreactive under another. Under 'ordinary' conditions, the alkanes are inert towards reagents such as acids, alkalis, oxidising reagents, reducing reagents, etc., but are reactive if the 'right' conditions are used (see below). One possible reason for this lack of reactivity is that the C—H bond is very strong (see Table 2.2).

General physical properties of the alkanes

The normal alkanes from C_1 to C_4 are colourless gases; C_5 to C_{17}, colourless liquids; and from C_{18} onwards, colourless solids. The b.p.s. rise fairly regularly as the number of carbon atoms in the compound increases. This holds good only for the normal compounds, and the difference in b.p.s. decreases as the higher homologues are reached. Other physical properties, such as m.p., density, viscosity, also increase in the same way as the b.p.s. (of the normal alkanes), *e.g.*, the density of the normal alkane increases fairly steadily for the lower members, and eventually tends to a maximum value of about 0·79.

Physical properties and structure. A great deal of work has been done to derive quantitative relationships between physical properties and chemical constitution. It is believed that variation in b.p.s. of compounds is due to different *intermolecular* forces such as hydrogen bonding, dipole moments, etc. Hydrogen bonding may produce association, and this will cause the b.p. to be higher than anticipated (see, *e.g.*, alcohols). The greater the dipole moment of the compound, the higher is the b.p., since, owing to the charges, more work is required to separate the molecules, *e.g.*, nitro-compounds, RNO_2, which have large dipole moments, have much higher b.p.s. than the alkanes in which the dipole moment is absent or very small. As far as the alkanes are concerned, the only attractive forces are the weak van der Waals forces.

Observation has shown that in a group of isomeric compounds (acyclic), the normal compound *always* has the highest b.p. and m.p., and generally, the greater the branching, the lower is the b.p. In the n-compound, owing to its large surface area due to the linear nature of the chain, the van der Waals attractive forces are relatively large. As branching increases, the molecule tends to become spherical, thereby resulting in a decreased surface area available for intermolecular attractions. Furthermore, because intermolecular forces are small in alkanes, the b.p.s. of these compounds are used as reference values for given molecular weights, *e.g.*, propane (M.W. 44; b.p. −44·5°C); ethanol (M.W. 46; b.p. 78·1°C). The high b.p. of the latter is attributed to hydrogen bonding.

The alkanes are almost insoluble in water, but readily soluble in ethanol and ether, the solubility diminishing with increase in molecular weight.

It is believed that solubility depends on the following intermolecular forces: solvent/solute; solute/solute; solvent/solvent. If the solvent/solute attractive forces are greater than either of the other two, solution will be expected to occur fairly readily. A non-electrolyte dissolves readily in water only if it can form hydrogen bonds with the water. Thus alkanes are insoluble, or almost insoluble, in water. Methane is more soluble than any of its homologues; hydrogen bonding with the water is unlikely, and so other factors—possibly molecular size—must also play a part. A useful rule in organic chemistry with respect to solubility is that 'like dissolves like', *e.g.*, if a compound contains a hydroxyl group, then the best solvents usually contain hydroxyl groups.

X-ray analysis of *solid* alkanes has shown that the carbon chains are fully extended, *i.e.*, zigzag. In the *liquid* state this extended form is also one of the stable conformations provided that the carbon chain is not very long. When the number of carbon atoms is sixteen or more, the extended form is no longer present in the liquid state.

Since the dipole moment of methane is zero, the dipole moment of the methyl group is equal to that of the fourth C—H bond (in methane) and is directed along this axis. Thus replacement of hydrogen by a methyl group will not be expected to change the dipole moment, *i.e.*, the dipole moment of all alkanes, whether straight- or branched-chain, will be zero. This has been found to be so in practice. This will always hold good whatever conformation is taken up by the alkane provided that no deformation of the normal carbon valency angle (of 109° 28′) is produced in the twisting, since all methyl groups will be balanced by a C—H bond. It therefore follows that the electronegativity of *all* alkyl groups is equal to that of hydrogen, namely zero (p. 34). As soon as one hydrogen atom is replaced by another atom or group (other than alkyl), the resultant molecule will now be found to possess a dipole moment (see also p. 153).

The position of the infra-red absorption region of the CH group (stretch) in alkanes depends on whether the carbon atom is primary, secondary or tertiary. Thus we have the following regions: —CH_3, 2975–2950 and 2885–2860 cm^{-1} (m); —CH_2—, 2940–2915 and 2870–2845 cm^{-1} (m); $\overset{\backslash}{\underset{/}{-}}CH$, 2900–2880 cm^{-1} (w). Some C—H deformation absorption frequencies are: —CH_3, 1470–1435 (m) and 1385–1370 cm^{-1} (s); —CH_2—, 1480–1440 cm^{-1} (m). Two useful skeletal vibrations are: iso, Me_2CH—, 1175–1165 cm^{-1} (s); neo, Me_3C—, 1255–1245 cm^{-1} (s). It is thus possible to detect the presence of these groups in a molecule.

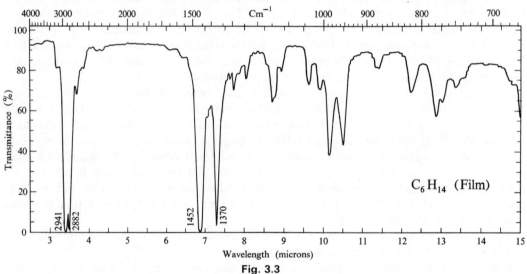

Fig. 3.3

2941 and 2882 cm^{-1}: C—H (str.) in Me and/or CH_2. 1452 cm^{-1}: C—H (def.) in Me and/or CH_2. 1370 cm^{-1}: C—H (def.) in Me.

Figure 3.3 is the infra-red spectrum of 3-methylpentane (as a film). These bands are characteristic of all types of alkanes containing methyl and methylene groups.

Mass spectra. Since the stability of carbonium ions is in the order t > s > prim., fission of bonds in alkanes occurs preferentially at branched carbon atoms. When alternative fissions can occur, it is the heaviest side-chain that is eliminated preferentially. Since alkyl carbonium ions are formed, all of them (with ${}^1H$ and ${}^{12}C$) will give peaks of odd masses. In particular, n-alkanes give a series of peaks separated by 14 mass units (CH_2). The relative abundance of these peaks is usually greatest for $C_3H_7^+$ (43), $C_4H_9^+$ (57), and $C_5H_{11}^+$ (71), and decreases fairly regularly for the larger masses. Furthermore, each peak is generally accompanied by peaks of mass 1 and 2 units lower, corresponding to the loss of 1 and 2 hydrogen atoms, respectively.

The molecular ion is always present for n-alkanes, its intensity decreasing with increasing molecular weight. On the other hand, the greater the branching in the alkane, the less is the likelihood of the appearance of the molecular ion and, if this does appear, its intensity is usually low.

General chemical properties of the alkanes

1. Halogenation. Chlorination may be brought about by light, heat or catalysts, and the extent of chlorination depends largely on the amount of chlorine used. A mixture of all possible isomeric monochlorides is obtained, but the isomers are formed in unequal amounts, due to the difference of the reactivity of primary, secondary and tertiary hydrogen atoms. Markownikoff (1875) found experimentally that the order of ease of substitution is tertiary hydrogen > secondary > primary; e.g., chlorination of isobutane at 300°C gives a mixture of two isomeric monochlorides:

$$(I) \quad \overset{\overset{\textstyle CH_3}{|}}{CH_3CHCH_2Cl} \quad \text{and} \quad \overset{\overset{\textstyle CH_3}{|}}{CH_3CClCH_3} \quad (II)$$

Isobutane has one tertiary and nine primary hydrogen atoms. On the basis of equal rates of substitution, the statistical results would be 90 per cent of (I) and 10 per cent of (II). Actually, the results obtained were 67 per cent of (I) and 33 per cent of (II). Thus tertiary hydrogen is replaced about 4·5 times as fast as primary hydrogen (see also p. 149).

Bromination is similar to chlorination, but not so vigorous. Iodination is reversible, but it may be carried out in the presence of an oxidising agent, such as HIO_3, HNO_3, HgO, etc., which destroys the hydrogen iodide as it is formed and so drives the reaction to the right, e.g.,

$$CH_4 + I_2 \rightleftharpoons CH_3I + HI$$
$$5HI + HIO_3 \longrightarrow 3I_2 + 3H_2O.$$

Iodides are more conveniently prepared by treating the chloro- or bromo-derivative with sodium iodide in methanol or acetone solution, e.g.,

$$RCl + NaI \xrightarrow{\text{acetone}} RI + NaCl$$

This reaction is possible because sodium iodide is soluble in methanol or acetone, whereas sodium chloride and sodium bromide are not. This reaction is known as the **Finkelstein** or **Conant–Finkelstein reaction** (1910, 1924).

Direct fluorination is usually explosive; special conditions are necessary for the preparation of the fluorine derivatives of the alkanes (see p. 172).

2. Nitration (see also p. 362). Under certain conditions, alkanes react with nitric acid, a hydrogen atom being replaced by a *nitro-group*, NO_2. This process is known as *nitration*. Nitration of the alkanes may be carried out in the vapour phase between 150° and 475°C, whereupon a complex mixture of mononitroalkanes is obtained. The mixture consists of all the possible mononitroderivatives and the nitro-compounds formed by every possibility of *chain fission* of the alkane; *e.g.*, propane gives a mixture of 1-nitropropane, 2-nitropropane, nitroethane and nitromethane:

$$CH_3CH_2CH_3 \xrightarrow[400°C]{HNO_3} CH_3CH_2CH_2NO_2 + CH_3\overset{\overset{\displaystyle NO_2}{|}}{C}HCH_3 + C_2H_5NO_2 + CH_3NO_2$$

As in the case of halogenation, the various hydrogen atoms in propane are not replaced with equal ease.

3. Sulphonation (see also p. 689) is the process of replacing a hydrogen atom by a *sulphonic acid group*, SO_3H. Sulphonation of a normal alkane from hexane onwards may be carried out by treating the alkane with oleum (fuming sulphuric acid). The ease of replacement of hydrogen atoms is: Tertiary very much greater than secondary, and secondary greater than primary; replacement of a primary hydrogen atom in sulphonation is very slow indeed. Isobutane, which contains a tertiary hydrogen atom, is readily sulphonated to give t-butylsulphonic acid:

$$(CH_3)_3CH + H_2SO_4/SO_3 \longrightarrow (CH_3)_3CSO_3H + H_2SO_4$$

Sulphuryl chloride, in the presence of light and a catalyst, converts hydrocarbons into sulphonyl chlorides (p. 405).

It has been shown that in concentrated sulphuric acid, hydrocarbons containing a tertiary hydrogen atom undergo hydrogen exchange (Ingold *et al.*, 1936). The mechanism is believed to occur via a carbonium ion:

$$R_3CH + 2H_2SO_4 \longrightarrow R_3C^+ + HSO_4^- + SO_2 + 2H_2O$$
$$R_3C^+ + R_3CH \longrightarrow R_3CH + R_3C^+, \text{ etc.}$$

This reaction is of particular interest since optically active hydrocarbons have been racemised in sulphuric acid (see p. 484); *e.g.*, Burwell *et al.* (1948) have shown that optically active 3-methylheptane is racemised in sulphuric acid.

4. Oxidation. All alkanes readily burn in excess of air or oxygen to form carbon dioxide and water. Controlled oxidation, on the other hand, under various conditions, leads to different products. Intensive oxidation gives a mixture of acids consisting of the complete range of C_1 to C_n carbon atoms. Less extensive oxidation gives a mixture of products in which no chain fission has occurred. Under moderate conditions, mixed ketones are the major products, and oxidation in the presence of boric acid produces a mixture of secondary alcohols. The oxidation of alkanes in the vapour state occurs via free radicals, *e.g.*, alkyl (R·), alkylperoxy (RCOO·), and alkoxy (RO·).

Oxidising reagents such as potassium permanganate readily oxidise a *tertiary* hydrogen atom to a hydroxyl group, *e.g.*, isobutane is oxidised to t-butanol:

$$(CH_3)_3CH + [O] \xrightarrow{KMnO_4} (CH_3)_3COH$$

5. Isomerisation of n-alkanes into branched-chain alkanes in which the side-chain is a methyl group, may be brought about by heating the n-alkane with aluminium chloride at 300°C, *e.g.*, n-butane isomerises into isobutane.

According to Pines *et al.* (1946), this isomerisation does not occur unless a trace of water is present (to form HCl from the $AlCl_3$) together with a trace of alkyl halide or an alkene. The isomerisation is believed to be an ionic chain reaction. The alkene 'impurity' is converted into a carbonium ion (by the $AlCl_3$ and HCl), and this initiates the chain reaction, isomerisation then occurring by a 1,2-shift involving a hydride ion in the second step (p. 146).

$$RCH\!=\!CH_2 + H^+ \rightleftharpoons R\overset{+}{C}HCH_3$$

$$R\overset{+}{C}HCH_3 + CH_3CH_2CH_2CH_3 \longrightarrow RCH_2CH_3 + \underset{CH_2CHCH_3}{\overset{CH_3}{\underset{+}{\diagup}}} \rightleftharpoons \underset{\overset{+}{C}H_2C(CH_3)_2}{\overset{H}{\diagup}} \rightleftharpoons (CH_3)_3C^+$$

$$(CH_3)_3C^+ + CH_3CH_2CH_2CH_3 \rightleftharpoons (CH_3)_3CH + CH_3CH_2\overset{+}{C}HCH_3; \text{ etc.}$$

The formation of an equilibrium mixture of n- and isobutane from n-butane with the iso-compound predominating may be explained by the fact that the t-carbonium ion is more stable than the s-carbonium ion (see also p. 104).

6. The thermal decomposition of the alkanes (see *Cracking*, below).

Some individual alkanes

Methane. This occurs in 'natural gas' (*q.v.*) and the gases from oil-wells. Methane is the principal product of organic decay in swamps and marshes, the gas being set free by the action of bacteria; this method of formation in nature has given rise to the name 'marsh-gas' for methane. Sewage sludge which has been fermented by bacteria yields a gas containing about 70 per cent methane, and this is used as a liquid fuel. Methane also forms about 40 per cent by volume of coal-gas.

> Methane may be synthesised by striking an electric arc between carbon electrodes in an atmosphere of hydrogen, or by heating a mixture of carbon and reduced nickel at 475°C in the presence of hydrogen.
> Sabatier and Senderens (1897) synthesised methane by passing a mixture of hydrogen and carbon monoxide or dioxide over finely divided nickel heated at about 300°C, *e.g.*,

$$CO + 3H_2 \longrightarrow CH_4 + H_2O.$$

Methane may conveniently be prepared by heating a mixture of anhydrous sodium acetate and soda-lime (p. 76), or by reduction of methyl iodide with dissolving metals or with lithium aluminium hydride or lithium hydride (p. 75). Other methods of preparation are by the action of water on aluminium carbide or methylmagnesium iodide (a Grignard reagent):

$$Al_4C_3 + 12H_2O \longrightarrow 3CH_4 + 4Al(OH)_3$$
$$CH_3MgI + H_2O \longrightarrow CH_4 + MgI(OH)$$

Methane is obtained in vast quantities from natural gas, gas from the oil-wells, and from cracked petroleum (*q.v.*).

Properties of methane. Methane is a colourless, odourless, non-poisonous gas; its b.p. is $-164°C/760\,mm$, and m.p. $-184°C$. It is somewhat soluble in water, 100 ml of water dissolving about 5 ml of methane at 20°C; but is quite soluble in ethanol and ether. It burns with a non-luminous flame in air or oxygen, forming carbon dioxide and water:

$$CH_4 + 2O_2 \longrightarrow CO_2 + 2H_2O$$

It explodes violently when mixed with air (or oxygen) and ignited, and this is believed to be the cause of explosions in coal-mines, where methane is known as *fire-damp*. Methane may be *catalytically* oxidised to methanol and formaldehyde.

Uses of methane. Methane, by heating to 1 000°C or by incomplete combustion, produces carbon in a very finely divided state. Carbon prepared this way is known as *carbon black*, and

$$C + 2H_2O \xleftarrow{\;O_2\;} CH_4 \xrightarrow{\;1000°C\;} C + 2H_2$$

is used to make paints and printers' ink; it is also used in the rubber industry for motor tyres, etc.

Methane is used as a source of **synthesis gas**. A mixture of methane and steam is passed over heated nickel (800°C) supported on alumina.

$$CH_4 + H_2O \longrightarrow CO + 3H_2; \qquad CH_4 + 2H_2O \longrightarrow CO_2 + 4H_2$$

The reaction to give carbon dioxide occurs to a much less extent than that to give carbon monoxide, and the carbon dioxide is removed from the mixture by washing with water under pressure.

Methane is also used as a fuel and as the starting point of many chemicals (see text).

Ethane, C_2H_6, occurs with methane in natural gas and the gases from the oil-wells. It may be prepared by the action of dissolving metals on ethyl iodide, by a Wurtz reaction with MeI (p. 76), by the electrolysis of sodium acetate (p. 77), or by the action of water on EtMgI. The best way of preparing pure ethane in large quantity is by the catalytic reduction of ethylene (p. 75).

Properties of ethane. Ethane is a colourless gas, b.p. $-89°C$, sparingly soluble in water but readily soluble in ethanol. It burns in air or oxygen to form carbon dioxide and water. It reacts with halogens in a similar manner to methane to form substitution products, but a much greater number of products are possible due, firstly to the presence of six hydrogen atoms in ethane compared with four in methane, and secondly, to the fact that isomerism is possible in the substitution products of ethane, and not in those of methane. Thus, for example, two dichloro-ethanes are possible: CH_3CHCl_2 and $ClCH_2CH_2Cl$. Ethane is mainly used as a fuel.

Propane, C_3H_8, is a constituent of natural gas and gas from the oil-wells. It may be prepared by methods used for ethane, using the appropriate starting materials. It is a colourless gas, b.p., $-44\cdot5°C$, and is used as a fuel.

Butanes, C_4H_{10}. Both n- and isobutane occur in natural gas and petroleum gas, and they may be separated by fractional distillation under pressure. The butanes are used as a liquid fuel and for the preparation of various compounds (see text).

n-Butane, b.p., $-0\cdot5°C$, may be prepared by the Wurtz reaction:

$$2CH_3CH_2I + 2Na \longrightarrow CH_3CH_2CH_2CH_3 + 2NaI$$

Isobutane, b.p., $-10\cdot2°C$, may be prepared by treating t-butyl iodide with a zinc-copper couple in ethanol:

$$(CH_3)_3CI \xrightarrow{\;e;\;H^+\;} (CH_3)_3CH + HI$$

Pentanes, C_5H_{12}. All three pentanes occur in natural gas and petroleum gas and are used for preparing chloropentanes and the pentanols.

n-Pentane, $CH_3CH_2CH_2CH_2CH_3$, b.p., $36°C$; isopentane, $(CH_3)_2CHCH_2CH_3$, b.p., $28°C$; neopentane, $(CH_3)_4C$, b.p., $9\cdot4°C$.

Petroleum and natural gas

Crude **petroleum** (*mineral oil*) is the term usually applied to the gases occurring naturally in the oilfields, the liquid from the wells, and the solids which are dissolved in, or have been separated from, the liquid. The composition of crude petroleum varies with the locality of occurrence, but all contain alkanes (straight- and branched-chain from about C_1 to C_{40}), cycloalkanes or naphthenes, and aromatic hydrocarbons. The low-boiling fractions of almost all petroleums are composed of alkanes; it is the composition of the higher-boiling fractions which differs according to the source of the petroleum. In addition to hydrocarbons, there are also present compounds containing oxygen, nitrogen, sulphur and metallic constituents.

Natural gas is the term applied to the large quantities of gas associated with or unassociated with liquid petroleum. The composition of natural gas varies with the source, and consists chiefly of the first six alkanes, the percentage of each decreasing with increasing molecular weight. Other gases such as water vapour, hydrogen, nitrogen, carbon dioxide and hydrogen sulphide may be present in amounts that vary with the locality of occurrence.

Distribution and general composition of crude petroleum. If the residue of petroleum, after removal of volatile compounds, contains a large amount of alkanes or wax, the petroleum is classified as *paraffinic* or *paraffin base oil*. If naphthenes predominate, the petroleum is classified as *asphaltic* or *asphalt base oil*. The crudes from the wells in Pennsylvania, Iran, Irak and Rumania are paraffinic; those from Baku and Venezuela are asphaltic; and those from Oklahoma, Texas and Mexico are intermediate in composition, and may be classified as paraffinic and asphaltic.

Distillation of petroleum. The crude oil is nearly always associated with water and sand; hence the crude petroleum discharged from the top of the well contains water and sand in suspension. The mixture is passed, under pressure, into cylindrical tanks, and the gas, oil and solids are drawn off separately.

Except for the low-boiling hydrocarbons, no attempt is made to separate the individual hydrocarbons. The crude oil is fractionated by continuous distillation into four main fractions: **petrol** (*gasoline*), **kerosene** (*kerosine, paraffin oil*), **gas oil** (*heavy oil*) and **lubricating oil.** The residue may be fractionated by means of vacuum-distillation to give light, medium and heavy lubricating oils, paraffin wax, and asphaltic bitumen. Each of the four main fractions may be further split up by batch distillation into fractions of narrow boiling range. Recently, it has been possible to isolate individuals by 'superfractionation'. The *final* number of fractions taken depends on the purpose in view.

Table 3.3 shows one set of fractions that may be obtained.

Table 3.3

Name	B.p. °C	Approximate composition	Uses
Light petrol	20–100	C_5H_{12}–C_7H_{16}	Solvent
Benzine	70–90	C_6–C_7	Dry cleaning
Ligroin	80–120	C_6–C_8	Solvent
Petroleum (gasoline)	70–200	C_6–C_{11}	Motor fuel
Kerosene (paraffin oil)	200–300	C_{12}–C_{16}	Lighting
Gas oil (heavy oil)	above 300	C_{13}–C_{18}	Fuel oil
Lubricating oil (mineral oil)	,,	C_{16}–C_{20}	Lubricants
Greases, vaseline, petrolatum	,,	C_{18}–C_{22}	Pharmaceutical preparations
Paraffin wax (hard wax)	,,	C_{20}–C_{30}	Candles, waxed paper, etc.
Residue (asphaltic bitumen)	,,	C_{30}–C_{40}	Asphalt tar; petroleum coke

Refining of the various fractions. It appears that refining was originally introduced to remove the bad colour and objectionable odour of petrol. Today it is realised that it is more important to remove sulphur compounds which lower the response of petrol to added tetraethyl-lead.

An internal-combustion engine, *i.e.*, one which burns fuel within the working cylinder, is more efficient the higher the compression ratio. Petrol engines use 'spark ignition', and as the compression ratio increases, a point is reached when 'knocking' is observed, *i.e.*, after passage of the firing spark, instead of all the fuel gas burning smoothly, the end portion burns with explosive violence, giving rise to a metallic rattle. The phenomenon of knocking is still not fully understood, but it has been found that, among other factors, the tendency to knock depends on the nature of the petrol. n-Alkanes tend to produce knocking far more than branched-chain alkanes. Edgar (1927) introduced 2,2,4-trimethylpentane (incorrectly known as iso-octane), which has higher anti-knock properties, and n-heptane, which has lower anti-knock properties than any commercial petrol, as standards for rating fuels. 'Iso-octane' is arbitrarily given the value of 100 and n-heptane, O, and the *octane number* of any fuel is the per cent of 'iso-octane' in a mixture of this compound and n-heptane which will knock under the same conditions as the fuel being tested.

Alkenes and aromatic compounds have high octane numbers. Tetraethyl-lead also raises the octane number of a given petrol, but if sulphur compounds are present, the response to this 'dope' is lowered. Hence it is very important to remove sulphur compounds from petrol. The method of refining depends on the particular fraction concerned, and it is not practicable to refine before distillation (of the petroleum).

Cracking

The thermal decomposition of organic compounds is known as *pyrolysis*; pyrolysis, when applied to alkanes, is known as *cracking*.

When heated to about 500–600°C, alkanes are decomposed into smaller molecules, and the products obtained from a given alkane depend on: (i) the structure of the alkane; (ii) the pressure under which cracking is carried out; and (iii) the presence or absence of catalysts such as silica–alumina, silica–alumina–thoria, silica–alumina–zirconia.

The mechanism of cracking is still obscure. Many theories have been suggested, and one that is highly favoured is a free-radical mechanism, evidence for which has been obtained from the observation that at cracking temperatures many hydrocarbons produce alkyl free radicals.

When petroleum is cracked, of all the compounds produced, the most important are those containing up to four carbon atoms: methane, ethane, ethylene, propane, propene, butane, butene and isobutene. All of these have found wide application as the materials for the preparation of a large number of chemicals (see text).

By using suitable catalysts, alkanes containing six or more carbon atoms may be catalytically cyclised, *e.g.*, n-hexane, under pressure, passed over chromic oxide carried on an alumina support and heated at 480–550°C, gives benzene (see also p. 569):

$$C_6H_{14} \longrightarrow C_6H_6 + 4H_2$$

There are two main types of cracking: (i) liquid phase, and (ii) vapour phase.

(i) **Liquid phase cracking.** Heavy oil (from the petroleum distillation) is cracked by heating at a suitable temperature (475–530°C) and under pressure (100–1 000 lb/sq. in.), by means of which the cracked material is maintained in the liquid condition. The heavy oil is converted into gasoline to the extent of 60–65 per cent of the oil (by volume), and has an octane number of 65–70. If attempts are made to increase the yield of gasoline, the octane number decreases.

(ii) **In vapour phase cracking** the cracking temperature is 600°C and the pressure is $3 \cdot 5$–$10 \cdot 5$ kg/cm². The cracking stock may be gasoline, kerosene, gas oils, but not the heavy oils, since these cannot be completely vaporised under the above conditions.

Reforming is the process whereby straight-run gasoline is cracked in order to raise the octane number. The gasoline is heated to about 600°C and under pressure of 28–52·5 kg/cm² and the yield varies from 60 to 90 per cent; the greater the yield, the lower is the octane number. Catalysts—the oxides of silicon and aluminium, plus small amounts of other oxides such as magnesia, zirconia, etc.—are usually employed in the reforming process. The chief chemical changes that occur are isomerisation, dehydration and cyclisation.

Catalytic cracking is also increasing in use, since it has been found that catalytically cracked gasoline contains few alkenes that readily polymerise. Gum formation in cracked gasolines is prevented by the addition of *inhibitors*, which are mainly phenols or aromatic amines, *e.g.*, catechol, *p*-benzyl-aminophenol, naphthylamine.

Treatment of natural gas. When natural gas does not contain hydrocarbons above ethane, it is said to be 'lean' or 'dry'; when it contains the higher hydrocarbons (up to hexane) it is said to be 'rich' or 'wet'. The alkanes may be separated by fractional distillation under increased pressure, thereby giving methane, ethane, propane, n- and isobutanes, n-, iso- and neopentanes, and hexane. These gases are used for various purposes (see text). On the other hand, natural gas itself may be used in the manufacture of various compounds. Oxidation of natural gas under carefully controlled conditions produces a complex mixture of compounds, among which are formaldehyde, acetaldehyde, acetic acid, acetone, methanol, ethanol, propanols and butanols. These are separated by distillation, solvent extraction, etc.

Wet gas is also used as a source of gasoline. The vapours of the liquid hydrocarbons (pentanes and hexane) in the wet gas are removed by various methods, *e.g.*, compression, and cooling. The liquid product obtained from wet gas is known as 'natural' or 'casinghead' gas, and is 'wild' because of the dissolved gases in it. These gases may be removed by distillation under pressure, and the resulting liquid is known as 'stabilised natural gasoline'; this has very high anti-knock properties.

Synthetic Fuels. Fischer–Tropsch Gasoline Synthesis or Synthine Process. Synthesis gas, at about 200–300°C and a pressure of 1–200 atm, is passed over a catalyst.

$$x\text{CO} + y\text{H}_2 \longrightarrow \text{mixture of hydrocarbons} + \text{water}$$

Various metals and oxides have been used as catalysts. One of the best is: cobalt (100 parts), thoria (5 parts), magnesia (8 parts) and kieselguhr (200 parts). When synthesis gas is passed over the catalyst at moderate pressure (9–11 atm), more of the high-boiling fraction is obtained than when the process is carried out at atmospheric pressure. The liquid products are fractionally distilled, and refined in the same way as are the petroleum fractions; furthermore the higher-boiling fractions are cracked.

Gasoline from the synthine process costs more than that from petroleum. The Fischer–Tropsch oils appear to be more valuable as chemical raw materials than as fuels, and have been used for the production of higher alkenes, monocarboxylic acids, detergents and in the *oxo*-process (see p. 180).

PROBLEMS

1. $\Delta G_f^{\ominus}$ for the formation of n-butane is $-15\cdot69\,\text{kJ mol}^{-1}$, and for isobutane is $-17\cdot99\,\text{kJ mol}^{-1}$. Starting with 1 mole of n-butane, calculate the moles of n- and isobutane in the mixture after heating (in the presence of a catalyst) until equilibrium is obtained.

2. Write out all the possible structures for the formula C_6H_{14}. State how many primary, secondary, tertiary and quaternary carbon atoms there are in each isomer, and name the isomers by the I.U.P.A.C. system.

3. Complete the following equations:

(i) $\text{Me}_2\text{CO} \xrightarrow{?} \text{Me}_2\text{CH}_2$; (ii) $\text{PrBr} \xrightarrow{?} \text{PrH}$; (iii) $\text{Me}_3\text{CH} \xrightarrow{?} \text{Me}_3\text{COH}$; (iv) $\text{MeCH}{=}\text{CHMe}$ $\xrightarrow{?} \text{MeCH}_2\text{CH}_2\text{Me}$; (v) $\text{MeI} + \text{EtI} \xrightarrow{\text{Na}} ?$; (vi) $\text{EtMgI} \xrightarrow{\text{H}_2\text{O}} ?$; (vii) $\text{EtCO}_2^- + e \longrightarrow ?$; (viii) $\text{EtCO}_2^-\text{Na}^+ \xrightarrow{\text{heat}} ?$; (ix) $\text{EtCH}{=}\text{CH}_2 \xrightarrow{\text{B}_2\text{H}_6} ? \xrightarrow[\text{NaOH}]{\text{AgNO}_3} ?$

4. An alkane has a M.W. of 72, and monochlorination produces one product only. What is the compound?

5. Draw Newman projection formulæ for the eclipsed and staggered forms of: (*a*) ethanol; (*b*) 1-bromo-2-chloroethane; (*c*) propane; (*d*) 1,2-dibromopropane.

6. The table lists a number of fragment ions and relative abundances for A, B and C, which are isomers of molecular formula C_5H_{12}. What are A, B, and C? Give reasons for your answer.

m/e	15	27	29	39	41	42	43	44	56	57	58	72
A (%)	7	41	46	21	68	87	100	33	17	55	24	6
B (%)	—	34	25	14	41	59	100	34	—	13	6	9
C (%)	5·3	15	—	13	42	—	—	—	—	100	44	—

7. How many signals can be expected in the NMR spectrum of $\text{MeCHBrCH}_2\text{Br}$? Give reasons for your answer.

N.B. The reader should test himself on the general methods of preparation and general properties of any homologous series.

REFERENCES

Handbook for Chemical Society Authors. Special Publication No. 14. (The Chemical Society, 1960.)
BEEZER, 'Latin and Greek Roots in Chemical Terminology', *J. Chem. Educ.*, 1940, **17,** 63.
Advances in Organic Chemistry, Vol. 1 (1960), Weeden, The Kolbe Electrolytic Synthesis, p. 1.
MEINSCHEIN *et al.*, 'Origins of Natural Gas and Petroleum', *Nature*, 1968, **220,** 1185.
SAMUEL, 'Industrial Chemistry: Organic', No. XI, *Roy. Inst. Chem.*, Monographs for Teachers (1966).
GAIT, *Heavy Organic Chemicals*, Pergamon Press (1967).
GOLDSTEIN and WADDAMS, *The Petroleum Chemicals Industry*, Spon (1967, 3rd edn.).

4 Alkenes and alkynes

Alkenes or olefins

The alkenes are the unsaturated hydrocarbons that contain one double bond. They have the general formula C_nH_{2n}, and the double bond is known as the 'olefinic bond' or 'ethylenic bond'.

Nomenclature

The name olefin arose from the fact that ethylene was called 'olefiant gas' (oil-forming gas), since it formed oily liquids when treated with chlorine or bromine. The original name given to this homologous series was olefine; but it was later decided to reserve the suffix *-ine* for *basic* substances only. Since the name olefine had gained wide usage, it was decided to compromise and call the series the olefins.

One method of nomenclature is to name the olefin from the corresponding alkane by changing the suffix *-ane* of the latter into *-ylene*, *e.g.*, (methylene), ethylene, propylene, etc.

Isomers differing only in the position of the double bond are prefixed by Greek letters or numbers which indicate the position of the double bond. The lowest number is usually given to the double bond (see below), and the number (or Greek letter) indicates the first of the two carbon atoms that are joined together by the double bond.

Another method of nomenclature is to consider ethylene as the parent substance and the higher members as derivatives of ethylene. If the compound is a monosubstituted derivative of ethylene, then no difficulty is encountered in naming it; if the compound is a disubstituted derivative of ethylene isomerism is possible, since the alkyl groups can be attached to the same or different carbon atoms. When the groups are attached to the same carbon atom the olefin is named as the *asymmetrical* or *unsymmetrical* compound (abbreviated to *as-* or *unsym-*); when attached to different carbon atoms the olefin is named as the *symmetrical* (*sym-* or *s-*) compound.

According to the I.U.P.A.C. system of nomenclature, the class suffix of the olefins is *-ene*, and so the series becomes the **alkene series**. The longest carbon chain containing the double bond is chosen as the parent alkene, the name of which is obtained by changing the suffix *-ane* of the corresponding alkane into *-ene*. The positions of the double bond and side-chains are indicated by numbers, the lowest number possible being given to the double bond, and this is placed before the suffix. The I.U.P.A.C. system is now the one generally used.

$$CH_3CH_2CH{=}CH_2 \qquad \text{but-1-ene (1-butylene, } \alpha\text{-butylene, ethylethylene)}$$

$$\overset{3}{C}H_3{-}\overset{2}{C}{=}\overset{1}{C}H_2 \qquad \text{2-methylprop-1-ene (isobutene, isobutylene,}$$
$$\underset{\displaystyle CH_3}{|} \qquad\qquad\qquad \textit{as}\text{-dimethylethylene)}$$

$$\overset{1}{C}H_3\overset{2}{C}H{=}\overset{3}{C}(CH_3)\overset{4}{C}H_2\overset{5}{C}H_3 \qquad \text{3-methylpent-2-ene (3-methyl-2-pentylene)}$$

$$CH_3CH_2\overset{2}{C}{-}\!\!-\overset{3}{C}H\overset{4}{C}H_3 \qquad \text{2-ethyl-3-methylbut-1-ene}$$
$$\underset{\displaystyle CH_2}{\|}\;\;\underset{\displaystyle CH_3}{|}$$

When there are several chains of equal length containing the double bond, then the same principles apply as for the alkanes (see last example).

The names of univalent groups derived from alkenes have the ending -enyl, and the carbon atom with the free valency is numbered 1. There are three exceptions (see examples):

$$\overset{3}{C}H_3\overset{2}{C}H{=}\overset{1}{C}H{-} \qquad\qquad \overset{5}{C}H_3\overset{4}{C}H(CH_3)\overset{3}{C}H{=}\overset{2}{C}H\overset{1}{C}H_2{-}$$

prop-1-enyl 4-methylpent-2-enyl

$$CH_2{=}CH{-} \qquad CH_3{=}CHCH_2{-} \qquad CH_2{=}C(CH_3){-}$$

vinyl allyl isopropenyl

(for ethenyl) (for propen-2-yl) (for 1-methylvinyl)

A double or triple bond is regarded as a functional group. When there are several functional groups in the molecule, then the lowest number possible is given, in order of preference: (i) to the *principal* functional group of the compound; (ii) to the double or triple bond; and (iii) to atoms or groups designated by prefixes.

General methods of preparation of the alkenes

1. By the action of concentrated sulphuric acid, at 160–170°C, on primary alcohols. The acid acts as a dehydrating agent, *e.g.*, ethylene from ethanol:

$$C_2H_5OH+H^+ \xrightarrow{\text{H}_2\text{SO}_4} C_2H_5OH_2^+ \longrightarrow H_2O+C_2H_5^+ \longrightarrow C_2H_4+H^+ \quad (f.\text{-}g.)$$

Dehydration of secondary and tertiary alcohols is best carried out using *dilute* sulphuric acid, since the alkenes produced from these alcohols (particularly tertiary alcohols) tend to polymerise under the influence of the concentrated acid. The yields of alkenes from secondary and tertiary alcohols are very good. Also, the ease of dehydration of alcohols is t > s > primary (see p. 183).

Acid-catalysed dehydration of primary alcohols normally gives 1-alkenes, but s- and t-alcohols give mixtures of alkenes which arise from removal of a proton from either adjacent carbon atom (to the COH group) and also from rearrangements undergone by the various carbonium ion intermediates (see p. 184). Rearrangements, however, may be avoided by dehydrating the alcohol over heated alumina in the presence of a little pyridine (see p. 183).

2. Cyclic eliminations. Most eliminations occur by a polar mechanism (see below), whereas cyclic eliminations are unimolecular non-polar reactions which take place in *one* step. Most occur when the compound is subjected to pyrolysis, and proceed via a cyclic T.S. This mechanism is supported by the fact that these reactions show a *negative* entropy of activation (a cyclic structure has less freedom than an open-chain structure). In general, *trans*-elimination occurs in the polar mechanism (see p. 164), whereas in pyrolytic eliminations, the elimination is *cis*. This is a consequence of the formation of a cyclic T.S.; both the eliminated proton and 'leaving group' are in the *cis*-position.

An advantage of cyclic eliminations is that no carbon skeleton rearrangement occurs (*cf.* dehydration of alcohols, method 1). Also, in the pyrolysis of esters and xanthates of s- and t-alcohols, mixtures of alkenes are produced, but the terminal alkene is favoured over the non-terminal alkene (*cf.* Saytzeff's rule, p. 163).

(*a*) **Pyrolysis of esters.** The esters are usually acetates:

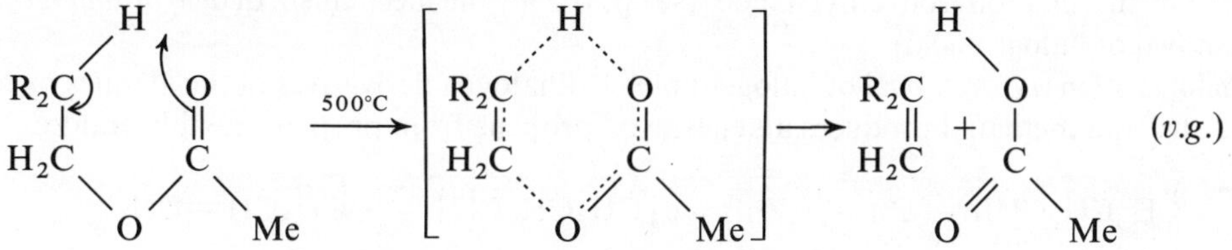

In many cases, pyrolysis of the alcohol with acetic anhydride is simpler than starting with the pre-formed ester (Aubrey *et al.*, 1965).

(*b*) **Pyrolysis of xanthates** (p. 407). This is known as the *Tschugaev (Chugaev) reaction* (1899).

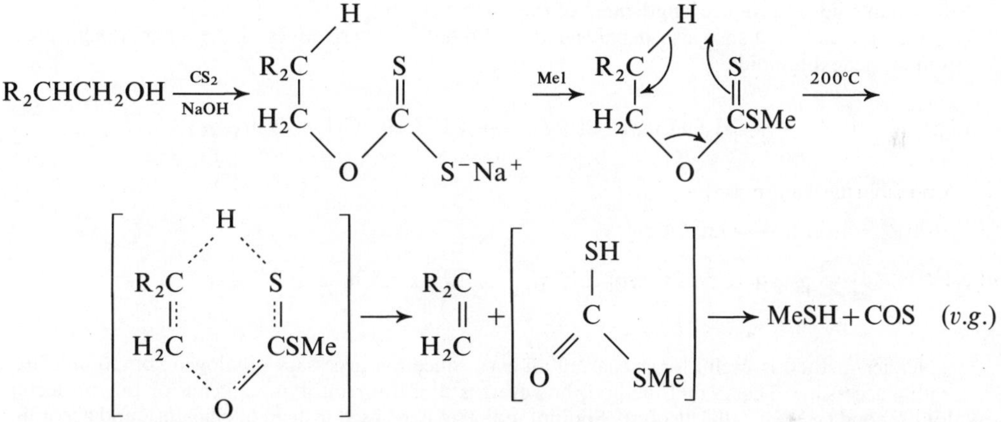

The yield of alkene is increased by heating the xanthate with, *e.g.*, chloroacetic acid ($-CS^-Na^+ \to -CSCH_2CO_2H$), and then refluxing this in feebly acid solution. Alternatively, heating the xanthate derivative in the presence of a Lewis acid catalyst, *e.g.*, BF_3, shortens the reaction time and the yield is increased.

(*c*) **Cope reaction** (1949). This is the reaction in which alkenes are formed when amine oxides (p. 378) are heated:

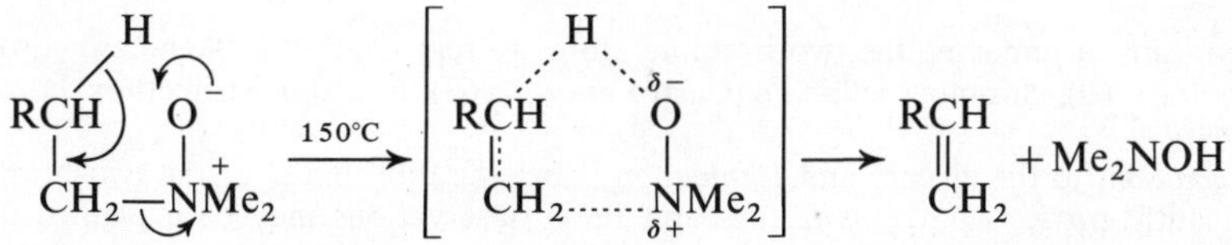

The reaction may also be carried out in dimethyl sulphoxide or tetrahydrofuran at room temperature (Cram *et al.*, 1962).

3. By the action of *ethanolic* potassium hydroxide on alkyl halides, *e.g.*, propene from propyl bromide:

$$CH_3CH_2CH_2Br + KOH \xrightarrow{\text{ethanol}} CH_3CH{=}CH_2 + KBr + H_2O$$

This is not a very important method for the preparation of the lower alkenes, since these may be prepared directly from the corresponding alcohols, which are readily accessible. The reaction, however, is very important, since by means of it a double bond can be introduced into an organic compound (see text). The yield of alkene depends on the nature of the alkyl halide used; it is fair with primary, and very good for secondary and tertiary alkyl halides. Ethylene cannot be prepared by this method from ethyl halides (see p. 162 for the mechanism of *dehydrohalogenation*, *i.e.*, removal of halogen acid).

Dehalogenation (*i.e.*, removal of halogen) of 1,1-dihalogen derivatives of the alkanes by means of zinc dust and methanol produces alkenes, *e.g.*, propene from propylidene dibromide:

$$CH_3CH_2CHBr_2 + Zn \longrightarrow ZnBr_2 + [CH_3CH_2\overset{/}{\underset{\backslash}{C}H}] \longrightarrow CH_3CH{=}CH_2$$

If sodium is used instead of zinc, and the reaction is carried out preferably in ether solution, comparatively little propene is formed, the main product being hex-3-ene:

$$2CH_3CH_2CHBr_2 + 4Na \longrightarrow CH_3CH_2CH{=}CHCH_2CH_3 + 4NaBr$$

This reaction is really an extension of the Wurtz synthesis, and the important point to note is that the use of sodium tends to produce lengthening of the carbon chain.

Zinc dust and methanol also dehalogenate 1,2-dihalogen derivatives of the alkanes, *e.g.*, propene from propene dibromide:

$$CH_3CHBrCH_2Br + Zn \longrightarrow CH_3CH{=}CH_2 + ZnBr_2$$

A possible mechanism is:

$$Zn \longrightarrow Zn^{2+} + 2e$$

$$2e + Br{-}\overset{|}{\underset{|}{C}}{-}\overset{|}{\underset{|}{C}}{-}Br \longrightarrow Br^- \ + {-}\overset{|}{\underset{|}{\ddot{C}}}{-}\overset{|}{\underset{|}{C}}{-}Br \longrightarrow \overset{\backslash}{\underset{/}{C}}{=}\overset{/}{\underset{\backslash}{C}} \ + Br^-$$

Neither method is useful for preparing alkenes, since the necessary dihalogen compounds are not readily accessible. The second method, however, is useful for purifying alkenes or for 'protecting' a double bond (see, *e.g.*, allyl alcohol). Sodium iodide may be used instead of zinc dust, and according to Mulders *et al.* (1963), the mechanism is:

$$\overset{\backslash}{\underset{/}{C}}{-}\overset{/}{\underset{\backslash}{C}} \underset{Br:}{\overset{Br}{\rightleftharpoons}} \ \overset{fast}{\rightleftharpoons} \ Br^- + \overset{\backslash}{\underset{/}{C}}{\cdots}\overset{/}{\underset{\backslash}{C}} \ \overset{slow}{\longrightarrow} \ \overset{\backslash}{\underset{/}{C}}{=}\overset{/}{\underset{\backslash}{C}} \ + IBr$$

Other means of removing the two bromine atoms to regenerate the alkene are chromous chloride (Allen *et al.*, 1960), chromous sulphate (Castro *et al.*, 1963), or sodium trimethoxyborohydride (King *et al.*, 1968).

Both *addition* (to the alkene) and *elimination* (from the dibromide) of the two bromine atoms are predominantly *trans*, *i.e.*, *stereoselective*, and these observations may be explained in terms of the formation of a bridged intermediate (see p. 102). It also follows from this that the configuration of the 'protected' alkene remains unchanged.

4. By heating a quaternary ammonium hydroxide, *e.g.*, ethylene from tetraethylammonium hydroxide (see p. 379):

$$(C_2H_5)_4N^+OH^- \longrightarrow C_2H_4 + (C_2H_5)_3N + H_2O$$

5. Boord *et al.* (1930–33) have prepared alkenes by conversion of an aldehyde into its chloro-ether, treating this with bromine followed by a Grignard reagent, and finally treating the product with zinc and n-butanol.

$$R^1CH_2CHO \xrightarrow[HCl]{C_2H_5OH} R^1CH_2CHClOC_2H_5 \xrightarrow{Br_2} R^1CHBrCHBrOC_2H_5 \xrightarrow{R^2MgBr}$$

$$R^1CHBrCHR^2OC_2H_5 \xrightarrow[C_4H_9OH]{Zn} R^1CH{=}CHR^2$$

This method is very useful for preparing alkenes of definite structure, and an interesting point about it is the replacement of the α-chlorine atom by bromine when the α-chloro-ether undergoes bromination in the β-position. The greater reactivity of the α-bromine atom may be explained as follows:

$$RCHBr{-}CH{-}\overset{..}{\underset{..}{O}}{-}C_2H_5 \longrightarrow RCHBrCH{=}\overset{+}{\underset{..}{O}}{-}C_2H_5 + Br^-$$
$$\underset{Br}{|}$$

6. The Wittig reaction. This is also a means of preparing alkenes where the position of the double bond is definite (see p. 410).

7. A number of alkenes are prepared by the cracking of petroleum (p. 90), *e.g.*, ethylene, propene, butenes, etc. (see also the individuals). For the production of the lower alkenes the most suitable starting material is gas oil, whereas for the higher alkenes it is best to use paraffin wax or Fischer–Tropsch wax (p. 91). The lower alkenes (C_2—C_5) are also prepared by the catalytic dehydrogenation of saturated hydrocarbons, the most satisfactory catalysts being those of the chromium oxide-alumina type.

General properties of the alkenes

The members containing two to four carbon atoms are gases; five to seventeen, liquids; eighteen onwards, solids at room temperature and they burn in air with a luminous smoky flame. In general, the physical properties of the alkenes are similar to those of the alkanes, since the alkenes are also subject only to weak van der Waals attractive forces. Geometrical isomers (*cis* and *trans*), however, behave differently because of their different geometry.

Owing to the presence of a double bond, the alkenes undergo a large number of *addition* reactions, but under special conditions they also undergo substitution reactions. The high reactivity of the olefinic bond is due to the presence of the two π-electrons, and when addition occurs, the trigonal arrangement in the alkene changes to the tetrahedral arrangement in the saturated compound produced. Alkene addition reactions involve an *electrophile* in the *first* step. This is the rate-determining step, and the intermediate, a carbonium ion, then reacts rapidly with a *nucleophile*. The energy profile of a two-step reaction of this type is shown in Fig. 2.15(*a*) (p. 57). Also, the more stable the carbonium ion, the faster is the addition (see p. 105).

If E and Nu represent, respectively, the electrophilic and nucleophilic 'fragments' of the addendum E—Nu, the addition may be generalised as follows:

$$\overset{\backslash}{\underset{/}{C}}{=}\overset{/}{\underset{\backslash}{C}} \quad E{-}Nu \xrightarrow[(slow)]{r/d} \overset{\backslash}{\underset{/}{C}}{-}\overset{+}{\underset{\backslash}{C}}E \xrightarrow[(fast)]{Nu-} \overset{\backslash}{\underset{/}{C}}Nu{-}\overset{/}{\underset{\backslash}{C}}E$$

Geometrical isomerism. Many compounds that contain one double bond exist in two forms which differ in most of their physical and in many of their chemical properties. van't Hoff suggested that if we assume there is *no free rotation about a double bond*, two spatial arrangements are possible for the molecule (*cf.* rotation about a single bond, p. 78). If we assume that the two unsaturated carbon atoms are sp^3 hybridised, then they are joined by two 'banana' bonds (see Fig. 2.7, p. 46). The two banana bonds are formed by the maximum overlap possible of the sp^3 orbitals (in pairs), and to decrease this overlap (in either bond), energy must be supplied to break the bond. Thus there is resistance to rotation about a double bond. On the other hand, if we assume sp^2 hybridisation for the two unsaturated carbon atoms, then in this case it is the maximum overlap of the two π-orbitals which causes resistance to the rotation about the double bond (see Fig. 2.5, p. 45). Furthermore, either way of representing the molecule produces a *planar* spatial arrangement.

Isomers which have the same structure but differ in their spatial arrangement, *i.e.*, have different *configurations*, are said to be *stereoisomers*. This type of isomerism is called **stereoisomerism**, and is of two kinds: **optical isomerism** and **geometrical isomerism** (or *cis-trans* **isomerism**). In optical isomerism, all (or at least some) of the isomers are optically active (see p. 154). On the other hand, geometrical isomerism is characterised by the existence of different configurations for a given structure which may or may not be optically active, *i.e.*, optical activity is *not* a criterion for geometrical isomerism but is, *by definition*, a criterion for optical isomerism (see also p. 496).

Several types of structures are capable of exhibiting geometrical isomerism, but we shall here discuss the one in which the molecule contains a double bond. Such compounds may be of various types, *e.g.*, $Ca_2{=}Cb_2$, $Ca_2{=}Cbd$, $Cab{=}Cab$, $Cab{=}Cad$. These are drawn conventionally as (I)–(IV).

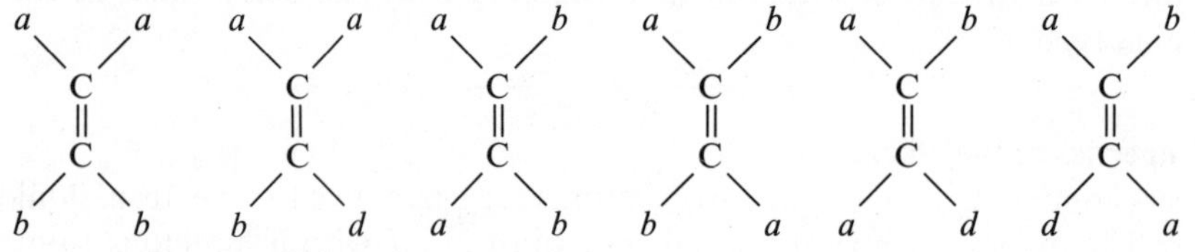

| (II) | (IIIa) | (IIIb) | (IVa) | (IVb) |

Inspection of these formulæ shows that geometrical isomerism is not possible for (I) and (II) but is possible for (III) and (IV). Thus, a double bond is not the only condition for geometrical isomerism; the groups attached to the unsaturated carbon atoms must also be taken into consideration.

Geometrical isomers are named as *cis* and *trans* compounds, respectively, according as identical (or similar) atoms or groups are on the *same* side or *opposite* sides, *e.g.*, but-2-enes:

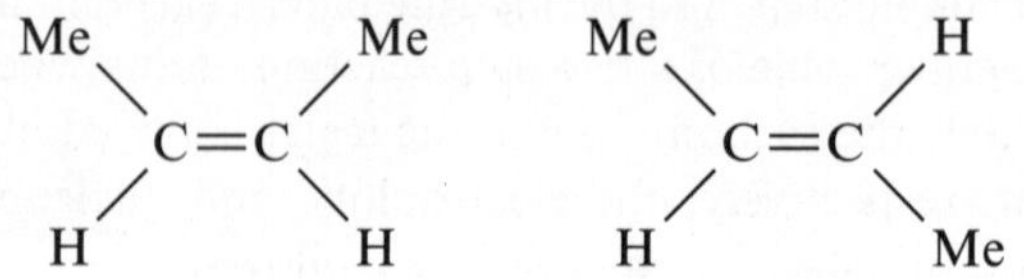

cis   *trans*

The two terms *conformation* (p. 80) and *configuration* often lead to confusion, since both are concerned with spatial arrangements. They may be differentiated as follows: change from one configuration to another involves breaking and making of bonds, whereas change from one conformation to another involves rotations (or twisting) about a single bond of one part of the molecule with respect to another part.

Stabilities of alkenes. One way of measuring the stability of an alkene is the determination of its heat of hydrogenation, e.g., (ΔH in kJ mol^{-1}):

$$CH_2{=}CH_2 \qquad MeCH{=}CH_2 \qquad MeCH_2CH{=}CH_2$$
$$-137\cdot2 \qquad\qquad -125\cdot9 \qquad\qquad -126\cdot8$$
$$MeCH{=}CHMe \qquad\qquad Me_2C{=}CH_2$$
$$cis,\ -119\cdot7;\ trans,\ -115\cdot5 \qquad\qquad -118\cdot8$$

Since the reaction is exothermic, the smaller ΔH is (numerically), the more stable is the alkene *relative* to its parent alkane. Thus, it is only possible to compare the stabilities of different alkenes which produce the *same* alkane on hydrogenation. This arises from the fact that the enthalpy of formation of alkanes is not a purely additive property; it also depends on, *e.g.*, steric effects, and these tend to vary from molecule to molecule. Since the three n-butenes all give n-butane on reduction, it follows that the order of their stabilities is: *trans* 2-ene > *cis* 2-ene > 1-ene.

This order may be explained in terms of steric effects and hyperconjugation. In but-1-ene, steric repulsion is virtually absent. In the but-2-enes, the two methyl groups in the *cis* isomer, being closer together than in the *trans* isomer, experience greater steric repulsion (p. 157) and consequently the *cis* form is under greater strain than the *trans*. Thus steric repulsion destabilises a molecule. On the other hand, hyperconjugation stabilises a molecule (see p. 332), and is small in but-1-ene but much larger in the but-2-enes. Since *trans*-but-2-ene is the most stable isomer, it follows that hyperconjugation has a greater stabilising effect than steric repulsion a destabilising effect (in these three butenes).

Stabilities of alkenes may also be compared by the determination of their heats of combustion (exothermic reaction), *e.g.*, (ΔH in kJ mol^{-1}):

$$MeCH_2CH{=}CH_2 \qquad\qquad MeCH{=}CHMe \qquad\qquad Me_2C{=}CH_2$$
$$-2719 \qquad\qquad cis,\ -2712;\ trans,\ -2707 \qquad\qquad -2703$$

In this case all four butenes may be compared, since all give the *same* products on combustion, viz., $4CO_2 + 4H_2O$. The order of stabilities is thus: iso > *trans* 2-ene > *cis* 2-ene > 1-ene.

In general, the order of stability of alkenes is:

$$R_2C{=}CR_2 > R_2C{=}CHR > R_2C{=}CH_2 \sim RCH{=}CHR\ (trans > cis) >$$
$$RCH{=}CH_2 > CH_2{=}CH_2$$

Reactions of alkenes

1. Alkenes are readily hydrogenated under pressure in the presence of a catalyst. Finely divided platinum and palladium are effective at room temperature; nickel on a support (Sabatier–Senderens reduction) requires a temperature between 200° and 300°C; Raney nickel is effective at room temperature and atmospheric pressure:

$$C_2H_4 + H_2 \xrightarrow{\text{catalyst}} C_2H_6$$

Platinum and palladium-black; *i.e.*, the metals in a very finely divided state, may be prepared by reducing their soluble salts with formaldehyde. Adams' platinum–platinum oxide catalyst is prepared by reducing platinum oxide with hydrogen before the addition of the compound being hydrogenated, or it may be added to the compound, reduction of the oxide taking place during hydrogenation.

Brown *et al.* (1962) have described the preparation of a 'highly active' platinum catalyst for alkene hydrogenation, and Eaborn *et al.* (1962) have prepared an even more active platinum catalyst.

One molecule of hydrogen is absorbed for each double bond present in the unsaturated compound.

The rate of hydrogenation of olefinic bonds (at room temperature and atmospheric pressure) is:

$$-CH\!=\!CH_2 > -CH\!=\!CH- \quad \text{or a ring double bond.}$$

When the alkene is of the type $R_2C\!=\!CR_2$ or $R_2C\!=\!CHR$, then hydrogenation under the above conditions is difficult.

The mechanism of catalytic hydrogenation is still not completely understood. It is widely accepted that the hydrogen is adsorbed on the metal surface and is present as hydrogen atoms ($H_2 \longrightarrow 2H\cdot$), and that the alkene is also adsorbed on the metal surface. Although the nature of the bonding to the metal in both cases is uncertain, it appears that it is more chemical than physical, and so is described as a *chemisorptive bond*. Linstead *et al.* (1942), from their studies on catalytic hydrogenation, proposed that the less hindered side of an unsaturated molecule is adsorbed on the metal surface, and that this is then followed by the *simultaneous* addition of two hydrogen atoms. In this way, the addition to form the *cis*-product was explained (attack must be from one side only). Furthermore, since chemisorption has converted hydrogen into hydrogen atoms and has broken one of the multiple bonds, the activation energy of catalytic hydrogenation is considerably lower than that in the uncatalysed reaction.

Later work has shown that this mechanism is unsatisfactory. Isomerisation of alkenes may occur during hydrogenation, *e.g.*, Smith *et al.* (1962) isolated, in addition to n-butane, *cis*- and *trans*-but-2-ene from the products of the incomplete catalytic hydrogenation of but-1-ene. Also, although *cis*-addition usually predominates (*i.e.*, the addition is stereoselective), *trans*-addition may occur and even predominate. These results and hydrogen–deuterium exchange experiments lead to the conclusion that addition of the adsorbed hydrogen atoms occurs one at a time and that the reaction is reversible (*cf.* catalytic dehydrogenation), *e.g.* (asterisks indicate metallic sites):

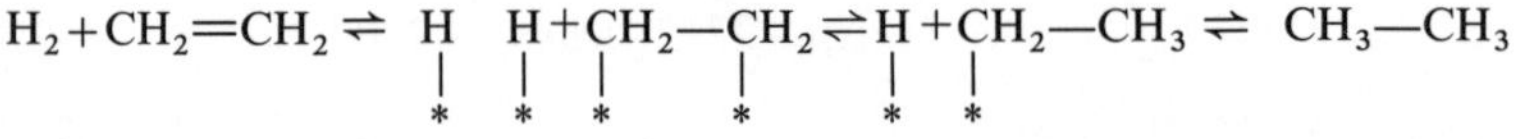

$$H_2 + CH_2\!=\!CH_2 \rightleftharpoons \underset{*\ \ \ *\ \ \ *}{H\ \ \ H} + CH_2\!-\!CH_2 \rightleftharpoons \underset{*\ \ \ \ \ \ *}{H} + CH_2\!-\!CH_3 \rightleftharpoons CH_3\!-\!CH_3$$

When reducible functional groups are also present, *e.g.*, $C\!=\!O$, $C\!\equiv\!N$, CO_2R, etc., it is usually possible to find conditions to selectively reduce the olefinic bond (see text).

Hydrogenation of alkenes is also of interest from the thermodynamic point of view. Let us consider the hydrogenation of ethylene:

$$CH_2\!=\!CH_2(g) + H_2(g) \longrightarrow CH_3CH_3(g)$$

In this reaction, one C—C double bond and one H—H bond are broken (energy is *supplied* to the system), and one C—C single bond and two C—H bonds are formed (energy is *given out* by the system). Hence, from bond-energy Table 2.2:

$\Delta H^{\ominus} = +(606\cdot7 + 431\cdot0) - (347\cdot3 + 2 \times 414\cdot2) = -138\cdot0$ kJ (the observed value is $-137\cdot2$ kJ). Also, the entropy values S, 298 K) of ethylene, hydrogen and ethane are $219\cdot7$, $130\cdot5$, and $230\cdot1$ J K^{-1}, respectively. $\therefore \Delta S^{\ominus} = 230\cdot1 - (219\cdot7 + 130\cdot5) = -120\cdot1$ J K^{-1}. Thus there is a *decrease* in entropy.

$$\Delta G^{\ominus} = \Delta H^{\ominus} - T\Delta S^{\ominus} = -138\cdot0 - (298 \times -120\cdot1/1\,000) = -138\cdot0 + 35\cdot8 = -102\cdot2 \text{ kJ}$$

Since $\Delta G^{\ominus}$ (at 25°C) has a large negative value, this means that the hydrogenation of ethylene is thermodynamically highly favourable. ($\Delta G^{\ominus} = -RT \ln K = -2\cdot3\,RT \log K$; $\therefore \log K = -\Delta G^{\ominus}/2\cdot3\,RT = 102\cdot2/5\cdot7 = \sim18$, *i.e.*, $K = 10^{18}$). Even so, the reaction does not proceed at any measurable rate at room temperature. As we have seen, the rate of a reaction depends on activation energies (and frequency factors). The activation energy of the above reaction is far too high to be attained at room temperature, but in the presence of a catalyst a new path with a much lower activation energy is now available, and so the reaction can therefore proceed at a fast rate.

The olefinic bond is not reduced by metal and acid, sodium and ethanol, lithium aluminium hydride, etc. unless it is α, β with respect to certain groups, *e.g.*, C=O (see Ch. 12).

Some chemical reagents, however, do reduce olefinic bonds, *e.g.*, alkanes are reduced to *cis* products by di-imide, and this stereospecificity can be explained by the formation of a cyclic T.S.

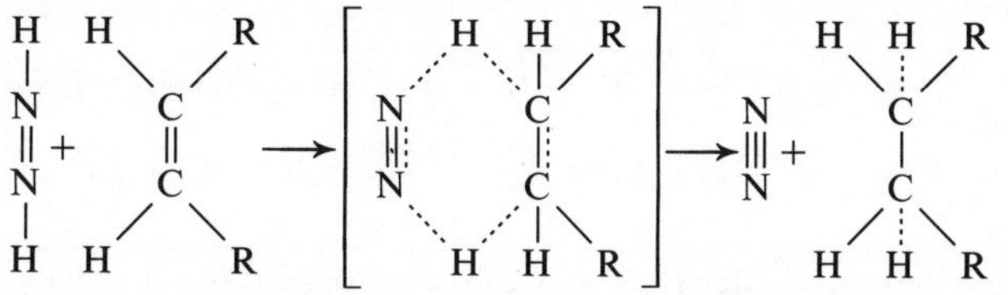

Di-imide is an unstable solid (at low temperature), and so is prepared *in situ*, *e.g.*, by the oxidation hydrazine with hydrogen peroxide, etc. Di-imide is selective in that it reduces carbon—carbon and nitrogen—nitrogen multiple bonds, but does not usually reduce C=O, NO_2, C≡N, etc.

Reduction by di-imide is an example of *transfer hydrogenation* (hydrogen is supplied by a donor molecule which is itself oxidised). Another donor molecule is cyclohexene (which is oxidised to benzene):

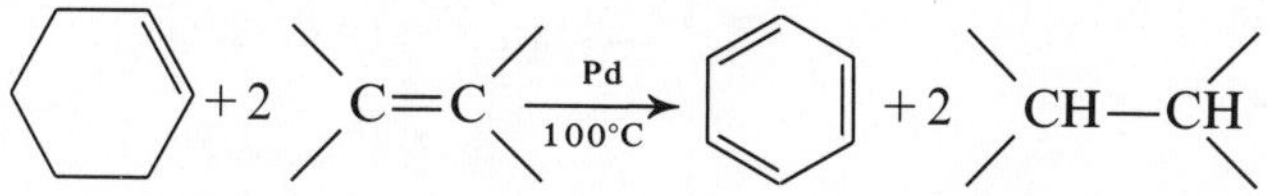

This system reduces many types of functional groups, *e.g.*, olefinic and acetylenic bonds, NO_2, —N=N—, etc.

A *terminal* double bond may be reduced by sodium in liquid ammonia *in the presence of an alcohol* (MeOH or EtOH; alcohols are stronger acids than ammonia). This method is known as the **Birch reduction,** and is believed to proceed stepwise via an anionic free radical:

$$RCH{=}CH_2 \xrightarrow[e]{Na} R\dot{C}H\bar{C}H_2 \xrightarrow{EtOH} R\dot{C}HCH_3 \xrightarrow[e]{Na}$$

$$R\overset{..}{\bar{C}}HCH_3 \xrightarrow{EtOH} RCH_2CH_3$$

The double bond is also reduced in excellent yield by $NaBH_4$—$PtCl_2$ (Brown *et al.*, 1962).

2. Alkenes form addition compounds with chlorine or bromine, *e.g.*, ethylene adds bromine to form ethylene dibromide:

$$CH_2{=}CH_2 + Br_2 \longrightarrow BrCH_2CH_2Br \quad (85\%)$$

Halogen addition can take place either by a heterolytic (polar) or a free-radical mechanism.

Heterolytic (polar) mechanism. Halogen addition readily occurs in solution, in the absence of light or peroxides, and is catalysed by inorganic halides, *e.g.*, aluminium chloride, or by polar surfaces (of the reaction vessel). These facts lead to the conclusion that reaction occurs by a polar mechanism. Further evidence for the polar mechanism comes from the work of Francis (1925), who showed that when ethylene reacts with bromine in aqueous sodium chloride solution, the products are ethylene dibromide and 1-bromo-2-chloroethane; *no ethylene dichloride is obtained.*

$$Br_2 + CH_2{=}CH_2 \xrightarrow[H_2O]{NaCl} BrCH_2CH_2Br + BrCH_2CH_2Cl$$

Francis also showed:

$$Br_2 + CH_2{=}CH_2 \begin{cases} \xrightarrow{\text{aq. NaI}} BrCH_2CH_2Br + BrCH_2CH_2I \\ \xrightarrow{\text{aq. NaNO}_3} BrCH_2CH_2Br + BrCH_2CH_2ONO \end{cases}$$

This work demonstrates that these addition reactions are at least a *two-stage* process, but does not demonstrate the order of the addition, *i.e.*, whether halogens are electrophilic or nucleophilic reagents. However, coupled with the fact that ethylene does not react with these solutions in the absence of bromine, this is a strong indication for electrophilic attack by the halogen. A mechanism consistent with these facts is:

$$CH_2{=}CH_2 \quad Br{-}Br \xrightarrow{\text{slow}} \overset{+}{C}H_2{-}CH_2Br + Br^-$$

$$BrCH_2CH_2Br \underset{\text{fast}}{\overset{Br-|}{\longleftarrow}} \quad \underset{\text{fast}}{\overset{|Cl-}{\longrightarrow}} ClCH_2CH_2Br$$

In addition to the evidence described above for electrophilic attack by halogens, there is the argument that negative ions (nucleophilic reagents) are hindered from attacking ethylenic carbon atoms by the screen of π-electrons. In fact, π-electrons are very susceptible to attack by electrophilic reagents. Thus alkenes are themselves nucleophilic reagents; *e.g.*, many form addition complexes with the silver ion as, *e.g.*, perchlorate. These addition compounds are known as **π-complexes**; the silver atom is not directly joined to either carbon atom, but is bound by overlap of the π-orbital with the vacant silver ion orbital.

$$Me_2C{=}CMe_2 + Ag^+ \longrightarrow Me_2C{=}CMe_2 \left.\right\} ClO_4^-$$
$$\qquad\qquad\qquad\qquad\qquad\quad \underset{Ag^+}{\downarrow}$$

Stereochemistry of halogen addition to alkenes (and alkynes). Addition of halogen is usually predominantly *trans*, *i.e.*, the addition is *stereoselective* (one stereoisomer predominates over the other), *e.g.*, the addition of bromine to maleic acid gives $(\pm)$-dibromosuccinic acid, which can result only from *trans* addition. To account for this, Roberts *et al.* (1937) suggested that the first step is the formation of a *bridged* (or *non-classical*) carbonium ion, followed by attack by the bromide ion.

The bromide ion attacks from *behind and along* the bonding C—Br$^+$ line, and this results in a Walden inversion at this carbon atom (p. 154). Since the *brominium* (*bromonium*) ion is symmetrical, each carbon atom is attacked equally well, and consequently results in the formation of (I) and (II) in *equal* amounts. (I) and (II) are a pair of enantiomers and hence the product is a racemic modification (see p. 154).

The existence of this bridged ion has been demonstrated by various workers (see p. 487). It should be noted, however, that although addition is usually *trans*, *cis* addition may also occur in certain circumstances (see p. 487).

Free-radical mechanism. It has generally been accepted that addition of halogen to alkenes, *in the absence of light*, is polar. Thus, for example, Stewart *et al.* (1935) showed that the addition of chlorine to ethylene is accelerated by light. This suggests a free-radical mechanism:

$$Cl_2 \xrightarrow{hv} 2Cl\cdot$$

$$CH_2{=}CH_2 + Cl\cdot \longrightarrow CH_2Cl{-}CH_2\cdot \xrightarrow{Cl_2} CH_2Cl{-}CH_2Cl + Cl\cdot\,; \text{ etc.}$$

Poutsma (1965), however, has shown that chlorine adds, *e.g.*, to but-2-ene at 25°C in the *absence* of light, and proposed a *spontaneous* free-radical mechanism:

$$MeCH{=}CHMe + Cl_2 \xrightarrow{\text{spont.}} MeCHClCHMe \xrightarrow{Cl_2} MeCHClCHClMe + Cl\cdot\,; \text{ etc.}$$

This mechanism is supported by the fact that the reaction was much slower in the presence of oxygen. Under these conditions, free-radical addition is inhibited and the raction proceeds by the slower polar mechanism.

Instead of addition reactions with the halogens, the alkenes may undergo substitution provided the right conditions are used. Thus, when straight-chain alkenes are treated with chlorine at a high temperature, they form mainly monochlorides of the allyl type, *i.e.*, in the chain $-C{-}\overset{\times}{C}{-}C{=}C{-}$, it is the hydrogen of C that is substituted. This high temperature substitution is strongly presumptive of a free-radical mechanism, *e.g.*,

$$Cl_2 \xrightarrow{400{-}600°C} 2Cl\cdot$$

$$CH_3CH{=}CH_2 + Cl\cdot \longrightarrow HCl + \cdot CH_2CH{=}CH_2 \xrightarrow{Cl_2} ClCH_2CH{=}CH_2 + Cl\cdot\,; \text{ etc.}$$

The reason for allyl substitution occurring is not certain, but one possibility is that, of the three types of free radicals that may be formed, $CH_3CH{=}CH\cdot$, $CH_3\overset{\cdot}{C}{=}CH_2$, and $\cdot CH_2CH{=}CH_2$, it is the allyl one that will have the greatest resonance stabilisation (the contributing structures are identical):

$$\cdot CH_2CH{=}CH_2 \longleftrightarrow CH_2{=}CHCH_2\cdot$$

From the M.O. point of view, in the allyl free radical, each carbon atom is associated with a p_z electron. Thus an M.O. is formed covering all three carbon atoms, with consequent delocalisation. Since the formation of the free radical with maximum stability means that the reaction path through this intermediate will be accompanied by the greatest decrease in free energy, the reaction will therefore proceed along this route (see p. 59).

An alternative explanation is that of the two types of C—H bonds present, *viz.*, —C—H and =C—H, the bond energy of the latter is greater than that of the former, and so the methyl group will be attacked preferentially (see Table 4.1, p. 131; see also below).

Allyl substitution in alkenes is also readily carried out with *N*-bromosuccinimide (p. 452).

The action of fluorine on alkenes usually results in the formation of carbon tetrafluoride, but addition to the double bond may be effected by treating the alkene with hydrogen fluoride in the presence of lead dioxide (Henne *et al.*, 1945); the fluorinating agent is lead tetrafluoride:

$$PbO_2 + 4HF \longrightarrow PbF_4 + 2H_2O$$

$$\overset{\diagdown}{\underset{\diagup}{}}C{=}C\overset{\diagup}{\underset{\diagdown}{}} + PbF_4 \longrightarrow \overset{\diagdown}{\underset{\diagup}{}}CFCF\overset{\diagup}{\underset{\diagdown}{}} + PbF_2$$

Oxidation of alkenes is readily effected in the allyl position by selenium dioxide, *e.g.*,

$$MeCH_2CH{=}CHMe \xrightarrow{SeO_2} MeCHOHCH{=}CHMe$$

It should be noted that the CH_2 group is preferred to the 1—Me group. Alkenes of the type $R^1CH{=}CHR^2$ are oxidised to α-diketones, R^1COCOR^2.

3. Alkenes form addition compounds with the halogen acids, *e.g.*, ethylene adds hydrogen bromide to form ethyl bromide:

$$C_2H_4 + HBr \longrightarrow C_2H_5Br$$

The order of reactivity of the addition of the halogen acids is hydrogen iodide > hydrogen bromide > hydrogen chloride > hydrogen fluoride (this is also the order of acid strength). The conditions for the addition are similar to those for the halogens; the addition of hydrogen fluoride, however, is effected only under pressure.

In the case of unsymmetrical alkenes it is possible for the addition of the halogen acid to take place in two different ways, *e.g.*, propene might add on hydrogen iodide to form propyl iodide;

$$CH_3CH = CH_2 + HI \longrightarrow CH_3CH_2CH_2I$$

or it might form isopropyl iodide:

$$CH_3CH = CH_2 + HI \longrightarrow CH_3CHICH_3$$

Markownikoff studied many reactions of this kind, and as a result of his work, formulated the following rule: *the negative part of the addendum adds on to the carbon atom that is joined to the least number of hydrogen atoms.* In the case of the halogen acids the halogen atom is the negative part, and so isopropyl halide is obtained.

Markownikoff's rule is empirical, but may be explained theoretically on the basis that the addition occurs by a polar mechanism. As with halogens, the addition of halogen acid is an electrophilic reaction, the proton adding first, followed by the halide ion. Also, the addition is predominantly *trans*, and this may be explained in terms of the formation of a bridged carbonium ion (p. 102):

$$CH_2{=}CH_2 \quad H{-}Cl \xrightarrow{slow} CH_2 \overset{\overset{\displaystyle H}{\underset{+}{\cdots\cdots}}}{} CH_2 + Cl^- \xrightarrow{fast} CH_3CH_2Cl$$

Now consider the case of propene. Since the methyl group has a $+I$ effect, the π-electrons are displaced towards the terminal carbon atom which, in consequence, acquires a negative charge. Thus, the proton adds on to the carbon *farthest* from the methyl group, and the halide ion then adds to the carbonium ion:

$$CH_3 {\rightarrow} CH{=}CH_2 \quad H{-}I \rightleftharpoons CH_3\overset{+}{C}HCH_3 + I^- \longrightarrow CH_3CHICH_3$$

An alternative explanation for Markownikoff's rule is in terms of stabilities of carbonium ions. (I), (II) and (III) represent a primary, secondary and tertiary carbonium ion, respectively.

$$\underset{\text{(I)}}{MeCH_2{}^+} \qquad \underset{\text{(II)}}{Me_2CH^+} \qquad \underset{\text{(III)}}{Me_3C^+} \qquad \underset{\text{(IV)}}{\overset{\delta+}{Me} \rightarrow \overset{\Delta+}{CH} \leftarrow \overset{\delta+}{Me}}$$

The positive charge on the central carbon atom makes this atom strongly electron-attracting, and this direction of electron movement is assisted by the $+I$ effect of the Me group, and consequently the charge on the central carbon atom is partially neutralised (represented by $\Delta+$ in IV) and, at the same time, Me acquires a small positive charge. Thus the net result is spreading of the charge (as in IV), and this produces increased stability. The reaction will therefore proceed through the intermediate formation of this ion.

Still another explanation for the order of the stability of the carbonium ions is *no bond resonance* (see also p. 332), *e.g.*,

$$H-\underset{\underset{H}{|}}{\overset{\overset{H}{|}}{C}}-\underset{\underset{H}{|}}{\overset{\overset{H}{|}}{C^+}} \longleftrightarrow H-\underset{\underset{H}{|}}{\overset{\overset{H^+}{|}}{\bar{C}}}-\underset{\underset{H}{|}}{\overset{\overset{H}{|}}{C^+}} \longleftrightarrow H-\underset{\underset{H}{|}}{\overset{\overset{H^+}{|}}{C}}=\underset{\underset{H}{|}}{\overset{\overset{H}{|}}{C}} \longleftrightarrow \cdots$$

Thus the number of hydrogen atoms available for no bond resonance is 3 for (I), 6 for (II) and 9 for (III). Consequently, (III) would be expected to be the most stable. At the same time, the charge is spread (over the methyl hydrogens).

The direction of addition of ICl and BrCl to propene is explained in the same way:

$$CH_3CHClCH_2I \overset{ICl}{\longleftarrow} CH_3CH=CH_2 \overset{BrCl}{\longrightarrow} CH_3CHClCH_2Br$$

Chlorine is more electronegative than either iodine or bromine, and so is the negative part of the addendum and adds in the *second step*.

Any factor that lowers the activation energy of a reaction will increase the rate of that reaction. Hence, the more stable the carbonium ion the faster will be the addition. Since the order of stability is (III) > (II) > (I), it can be anticipated that the rate of addition of HX will be

$$Me_2C=CH_2 > MeCH=CH_2 > CH_2=CH_2$$

This order is observed in practice.

In certain cases (according to the nature of the carbonium ion), these addition reactions may be accompanied by rearrangement (see, *e.g.*, p. 147).

Relative reactivities of various positions in a molecule. The addition of halogen acid to unsymmetrical alkenes is an example of the prediction of the relative reactivities of various positions in a given molecule. Two different methods were used, and these are widely applied in the study of reaction mechanisms.

(i) *Predictions based on charge distribution.* This approach assumes that the reactivity of a position in a polar reaction is determined by the charge density at that position, *i.e.*, an electrophilic reagent will attack preferentially at the site of highest electron density and a nucleophilic reagent will attack at the site of lowest electron density. The charge distribution is ascertained by the examination of the consequences of inductive and/or resonance effects. This approach is referred to as the **isolated molecule method**. Two disadvantages are the neglect of changes in charge distribution as the reagent approaches the substrate and the neglect of electrostatic forces between the substrate molecule and a charged reagent. Finar (1968) has shown that these electrostatic forces may be the predominant factor in predicting the site of attack.

(ii) *Predictions based on stabilities of intermediates.* This method assumes that the reactivities of different positions in a given molecule are determined only by their relative activation energies. Since the activation energy depends on the structure of the transition state, a difficulty here is the nature of the model that is to be used for the T.S. For reactions that are believed to involve intermediate carbonium ions, carbanions and free radicals, it is these that are chosen to represent the structures of the transition states, the stabilities of which are then examined in terms of inductive and/or resonance effects. The assumption is then made that the more stable the intermediate (*i.e.*, the lower its energy), the lower is the energy of the T.S., and consequently the lower is the energy of activation leading to that intermediate. This approach is referred to as the **localisation energy method.** It also has disadvantages, *e.g.*, electrostatic forces (*cf.* above) and the nature of the reagent are neglected. Since the latter factor plays a great part in deciding the structure of the T.S. (p. 53), the use of the intermediate as the model for all types of reagents may lead to wrong predictions.

In this book, many reaction mechanisms are dealt with by both methods, since it is instructive to compare them.

The Peroxide Effect (Kharasch, 1933). The presence of oxygen or peroxides that are formed when the alkene stands exposed to the air, or added peroxides such as benzoyl peroxide, causes the addition of hydrogen bromide to take place in the direction opposite to that predicted by Markownikoff's rule. This departure from the rule is known as the 'abnormal' reaction, and was shown to be due to the 'peroxide effect' (Kharasch *et al.*, 1933). Hydrogen chloride, hydrogen iodide and hydrogen fluoride do not exhibit the abnormal reaction. It has been found that the addition of hydrogen bromide is 'abnormally' effected photochemically as well as by peroxide catalysts.

The mechanism of the peroxide effect is a free-radical chain reaction, the peroxide generating the free radical $R^1\cdot$ (*cf.* polymerisation, below):

$$(R^1CO_2)_2 \longrightarrow 2R^1CO_2\cdot \longrightarrow 2R^1\cdot + 2CO_2$$

$$R^1\cdot + HBr \longrightarrow R^1H + Br\cdot$$

$$R^2CH{=}CH_2 + Br\cdot \longrightarrow R^2\dot{C}HCH_2Br \xrightarrow{HBr} R^2CH_2CH_2Br + Br\cdot, \text{ etc.}$$

In the photochemical addition, the bromine atom is produced by absorption of a quantum of light:

$$HBr \xrightarrow{h\nu} H\cdot + Br\cdot$$

Abell *et al.* (1962), using ESR spectroscopy, showed that the photochemical addition of HBr produces free radicals which appear to be best represented as bridged structures. They also showed, using DBr, that the bromine atom is the initial event in the attack.

The reason for the addition being contrary to Markownikoff's rule is not certain. A favoured theory is that the order of stability of free radicals is the same as that for carbonium ions, *i.e.*, t > s > primary. One explanation for this order is no bond resonance. Thus the reaction proceeds as shown in the equations above, the primary free-radical, $R^2CHBrCH_2\cdot$, being much less favoured energetically than the secondary, $R^2\dot{C}HCH_2Br$.

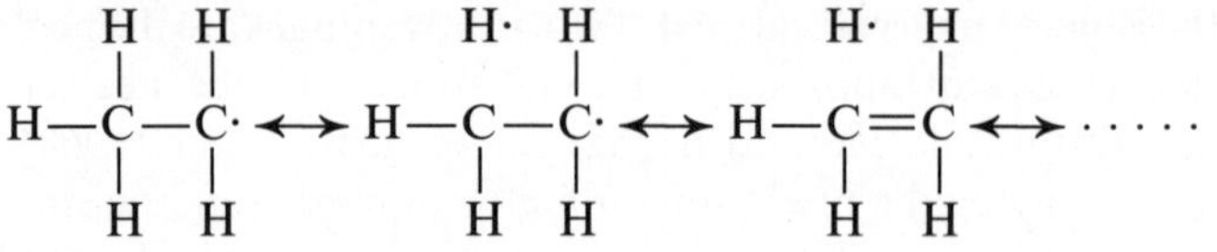

$$\begin{array}{ccc}
\underset{H}{\overset{H}{\mid}}\ \underset{H}{\overset{H}{\mid}} & \underset{H}{\overset{H\cdot}{\mid}}\ \underset{H}{\overset{H}{\mid}} & \underset{H}{\overset{H\cdot}{\mid}}\ \underset{H}{\overset{H}{\mid}} \\
H{-}C{-}C\cdot \longleftrightarrow & H{-}C{-}C\cdot \longleftrightarrow & H{-}C{=}C \longleftrightarrow \cdots \cdots
\end{array}$$

It has previously been pointed out (p. 52) that, among other factors, the energy of activation in a given reaction is lower the greater the strength of the new bond formed. Now, the bond broken is the π-bond, and so this energy change will not affect the activation energy whichever way the free-radical addition occurs. Also, the strength of a C—H bond is in the order p > s > t (Table 4.1, p. 131); this order also applies to bonds other than C—H. Hence, on this basis, it can be expected that the intermediate free radical will be preferably $R\dot{C}HCH_2X$ rather than $RCHXCH_2\cdot$. This, then, is another possible explanation for free-radical addition occurring contrary to Markownikoff's rule.

The abnormal reaction in the presence of peroxides can be prevented by the addition, in sufficient amount, of an *inhibitor* such as diphenylamine, catechol, etc. Inhibitors combine with free radicals, and so prevent propagation of the chain-reaction. Thus an inhibitor produces (if present in small amount only) an *induction period*, *i.e.*, a period when apparently no reaction is occurring. When the inhibitor is used up, it is then possible for the chain reaction to proceed in the normal fashion.

Another problem is why only hydrogen bromide exhibits the peroxide effect. A reasonable explanation may be offered in terms of the energetics of the chain steps in radical additions of HX to ethylene (kJ mol^{-1} at 25°C).

H—X	$\Delta H_1^{\ominus}$ $X\cdot + CH_2{=}CH_2 \rightarrow$ $XCH_2{-}CH_2\cdot$	$\Delta H_2^{\ominus}$ $XCH_2{-}CH_2\cdot + H{-}X \rightarrow$ $XCH_2{-}CH_3 + X\cdot$
H—Cl	$-67\cdot0$	$+12\cdot6$
H—Br	$-25\cdot1$	$-50\cdot2$
H—I	$+46\cdot0$	$-117\cdot1$

If we assume that $\Delta S^{\ominus}$ is the same for the corresponding steps in the three reactions, then $-RT \ln K = \Delta G^{\ominus} = \Delta H^{\ominus} + \text{constant}$. Thus, $\Delta G^{\ominus}$ may be equated with $\Delta H^{\ominus}$, and hence the more negative $\Delta H^{\ominus}$ is, the more thermodynamically favourable will be the reaction. Only for HBr are both steps exothermic and consequently only for HBr is each step thermodynamically favourable.

Now let us consider the following reaction:

$$H\cdot + CH_2{=}CH_2 \longrightarrow CH_3{-}CH_2\cdot \xrightarrow{\text{HBr}} CH_3CH_2Br + H\cdot$$

$$\Delta H^{\ominus} = -167\cdot4 \text{ kJ} \qquad \Delta H^{\ominus} = +100\cdot4 \text{ kJ}$$

Although the first step is highly thermodynamically favourable, the reverse is true for the second step. Hence, the free-radical addition of HBr occurs first by addition of Br·, followed by addition of H·.

Tri- and tetra-halogenated methanes also add on to a *terminal* double bond in the presence of peroxides (Kharasch *et al.*, 1945–1950). Here again the mechanism is believed to be a free-radical chain reaction, *e.g.*,

$$(RCO_2)_2 \longrightarrow 2R\cdot + 2CO_2$$

$$R\cdot + CHCl_3 \longrightarrow RH + \cdot CCl_3$$

$$RCH{=}CH_2 + \cdot CCl_3 \longrightarrow R\dot{C}HCH_2CCl_3 \xrightarrow{\text{CHCl}_3} RCH_2CH_2CCl_3 + \cdot CCl_3; \text{ etc.}$$

4. Hypohalous acids add on to alkenes to form halohydrins. Usually the reaction is carried out by treating the alkene with chlorine- or bromine-water; alkene dihalide is also formed.

The mechanism is believed to be (*cf.* p. 102):

$$CH_2{=}CH_2 + Br_2 \rightleftharpoons Br^- + BrCH_2\overset{+}{C}H_2 \xrightarrow{H_2O} BrCH_2CH_2\overset{+}{O}H_2 \longrightarrow H^+ + BrCH_2CH_2OH$$

With unsymmetrical alkenes, the hydroxyl group adds on to the carbon atom joined to the least number of hydrogen atoms:

$$RCH{=}CH_2 \quad Cl{-}Cl \longrightarrow Cl^- + R\overset{+}{C}HCH_2Cl \xrightarrow{H_2O} RCHOHCH_2Cl + H^+$$

Aqueous solutions of hypohalous acid also add on to alkenes in the presence of strong acids to form halohydrins, *e.g.*, the addition of hypochlorous acid to ethylene. The mechanism is probably via the formation of the *chlorinium ion*:

$$ClOH + H^+ \rightleftharpoons Cl\overset{+}{O}H_2 \rightleftharpoons Cl^+ + H_2O$$

$$CH_2{=}CH_2 + Cl^+ \longrightarrow ClCH_2\overset{+}{C}H_2 \xrightarrow{H_2O} ClCH_2CH_2\overset{+}{O}H_2 \underset{}{\overset{-H^+}{\rightleftharpoons}} ClCH_2CH_2OH$$

Halohydrins regenerate the alkene when treated with zinc dust in acetic acid (*cf.* p. 95, 3).

The addition of 'HOX' involves the addition of the solvent, in this case, water. Water is a nucleophilic solvent and competes with the other nucleophile, Cl^-, which is present. All other nucleophilic solvents behave in the same way as water, *e.g.*, propene reacts with methanol in the presence of sulphuric acid as follows:

$$MeCH{=}CH_2 \xrightarrow{H_2SO_4} Me_2CH^+ \xrightarrow{MeOH} Me_2C\overset{+}{H}\overset{\displaystyle H}{\underset{\displaystyle Me}{O}} \xrightarrow{-H^+} Me_2CHOMe$$

Various sulphenyl halides form adducts with alkenes; 2,4-dinitrobenzenesulphenyl chloride in particular has been found extremely useful for identifying alkenes (Karasch *et al.*, 1949–1952).

$$Ar\overset{\delta+}{S}{-}\overset{\delta-}{Cl}+CH_2{=}CH_2 \longrightarrow CH_2\underset{\underset{\displaystyle SAr}{+}}{\diagdown\underline{\qquad}\diagup}CH_2+Cl^- \longrightarrow ArSCH_2CH_2Cl$$

Two products are obtained with unsymmetrical alkenes, the predominating adduct being accepted as the one formed in accordance with Markownikoff's rule. However, Calo *et al.* (1968) have shown that the direction of addition depends on the nature of the solvent.

Compounds containing triple bonds also form adducts with one molecule of the sulphenyl chloride.

5. Alkenes are absorbed by concentrated sulphuric acid to form alkyl hydrogen sulphates. Addition takes place according to Markownikoff's rule, *e.g.*, propene reacts with sulphuric acid to form isopropyl hydrogen sulphate:

$$CH_3CH{=}CH_2 \quad H{-}OSO_2OH \longrightarrow CH_3\overset{+}{C}HCH_3+\overset{-}{O}SO_2OH \longrightarrow (CH_3)_2CHOSO_2OH$$

Alkanes are not absorbed by cold concentrated sulphuric acid, and hence may be separated from alkenes (see also ethers and alcohols).

6. Hydration of alkenes. Alkenes may be hydrated to alcohols by absorption in concentrated sulphuric acid followed by hydrolysis of the alkyl sulphate (see above section and p. 189). Alkenes, however, may also be catalytically hydrated in dilute acid solution, *e.g.*, Lucas *et al.* (1934) showed that isobutene forms t-butanol in dilute acid solution:

$$(CH_3)_2C{=}CH_2 + H_2O \xrightarrow{H^+} (CH_3)_3COH$$

The mechanism of alkene hydration has been the subject of much discussion. The hydration reaction, as such, cannot be examined kinetically, but since the reaction is reversible, the principle of microscopic reversibility may be applied (p. 49). Hughes and Ingold (1941) have examined the elimination of water from t-alcohols, and on the basis of their results have proposed the following mechanism for hydration:

$$Me_2C{=}CH_2 \quad H{-}OH_2^+ \underset{\text{slow}}{\rightleftharpoons} Me_2\overset{+}{C}{-}Me + H_2O \rightleftharpoons \underset{\underset{\displaystyle {}^+OH_2}{|}}{Me_2C{-}Me} \overset{-H^+}{\rightleftharpoons} \underset{\underset{\displaystyle OH}{|}}{Me_2CMe}$$

Since hydration cannot be carried out in alkaline solution, this is very strong evidence that the first step is proton addition in acid-catalysed hydration, *i.e.*, the reaction is electrophilic. Further support for this mechanism is the fact that only those alkenes which form t-carbonium ions are readily hydrated under these conditions. Such ions are more stable than s- or prim.-carbonium ions, and consequently will be produced more readily than the latter.

It appears the hydration is not stereoselective, as is the case for halogen acid addition (Collins *et al.*, 1960). This may be explained on the basis that a classical carbonium ion is formed (as shown in the mechanism).

A further point of interest is that water is not a sufficiently strong acid to transfer a proton to an alkene; hence the use of dilute acid solution. In concentrated acid, the nucleophilic anion of the acid adds on (see **5** above). In sufficiently diluted acid, $[H_2O] \gg [HSO_4^-]$ and consequently water is involved in the second step.

Since hydration of alkenes and dehydration of alcohols are reversible, both proceed through the same intermediate carbonium ions, and if these are capable of undergoing rearrangement, 'unexpected' alcohols may be obtained from the alkene (*cf.* p. 184). Rearrangement, however, may be avoided by treatment of the alkene with mercuric acetate (in tetrahydrofuran), followed by reduction in aqueous sodium hydroxide with sodium borohydride (Brown *et al.*, 1967),

$$Me_3CCH=CH_2 \xrightarrow[\text{THF}]{\text{Hg(OOCMe)}_2} Me_3CCHHgOOCMe \underset{\text{OOCMe}}{\Big|} \xrightarrow[\text{NaOH}]{\text{NaBH}_4} Me_3CCHOHMe + Hg$$

This method is referred to as *oxymercuration-demercuration*, and the final result is hydration in accordance with Markownikoff's rule (*cf.* hydroboration, below).

7. Alkenes add on nitrosyl chloride, nitrosyl bromide and oxides of nitrogen; *e.g.*, ethylene nitrosochloride with nitrosyl chloride:

$$CH_2=CH_2 + NOCl \longrightarrow ClCH_2CH_2NO$$

Since the X atom is the negative end of the dipole in NOX, it will add on to the carbon atom joined to the least number of hydrogen atoms (Markownikoff's rule), *e.g.*, trimethylethylene adds on nitrosyl bromide to form the following trimethylethylene nitrosobromide (see also p. 366):

$$(CH_3)_2C\!=\!CHCH_3 + NO\!-\!Br \longrightarrow (CH_3)_2CBrCH(NO)CH_3$$

The addition of oxides of nitrogen to alkenes is complicated, the products formed depending on the structure of the alkene and the nature of the 'nitrous fumes' (see p. 362).

8. Alkenes are readily hydroxylated, *i.e.*, add on hydroxyl groups, to form dihydroxy-compounds known as glycols (*q.v.*). Hydroxylation may be effected:

(i) By cold dilute alkaline permanganate solution (*cis*-hydroxylation); *e.g.*, ethylene is converted into ethylene glycol:

$$CH_2=CH_2 + H_2O + [O] \xrightarrow{\text{KMnO}_4} HOCH_2CH_2OH$$

The mechanism of hydroxylation with permanganate is believed to proceed via a cyclic intermediate (*cf.* osmium tetroxide, below).

This mechanism accounts for *cis*-hydroxylation and is supported by the work of Wiberg *et al.* (1957), who used potassium permanganate labelled with ^{18}O and showed that *both* glycol oxygen atoms come from the oxidising agent.

Similarly, chromic acid converts alkenes into glycols (or epoxides). The mechanism again involves a cyclic intermediate ($HCrO_4^-$ replaces MnO_4^-, and $>Cr(O^-)OH$ replaces $>Mn(=O)O^-$).

(ii) By 90 per cent hydrogen peroxide in glacial acetic acid or better, in formic acid; *e.g.*, oleic acid is converted into 9,10-dihydroxystearic acid (see also reaction **9**):

$$CH_3(CH_2)_7CH{=}CH(CH_2)_7CO_2H + H_2O_2 \longrightarrow CH_3(CH_2)_7CH(OH)CH(OH)(CH_2)_7CO_2H$$

The addition of hydrogen peroxide may be catalysed by various oxides, *e.g.*, osmium tetroxide in t-butanol (*cis*-addition), selenium dioxide in t-butanol or acetone (*trans*-addition; see p. 487).

(iii) **By means of osmium tetroxide.** This compound adds very readily to an ethylenic double bond at room temperature (Criegee, 1936):

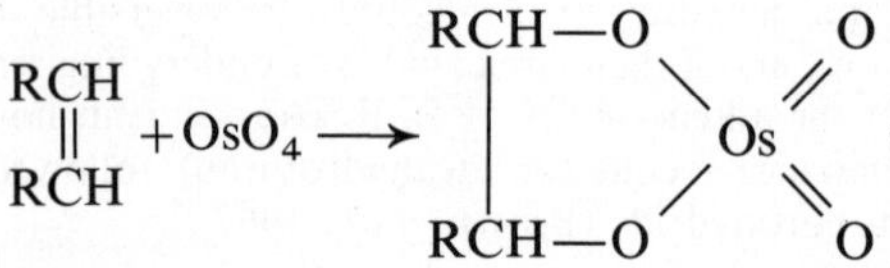

$$\begin{array}{c} RCH \\ \parallel \\ RCH \end{array} + OsO_4 \longrightarrow \begin{array}{c} RCH{-}O \\ | \qquad\qquad Os \\ RCH{-}O \end{array} \begin{array}{c} O \\ \diagup\diagdown \\ \diagdown\diagup \\ O \end{array} \quad (v.g.{-}ex.)$$

These cyclic compounds (*osmic esters*), on refluxing with aqueous ethanolic sodium hydrogen sulphite, are hydrolysed to 1,2-glycols (*cis*-glycols).

If the addition of osmium tetroxide is carried out in the presence of pyridine, coloured crystalline compounds are obtained, usually in theoretical yield:

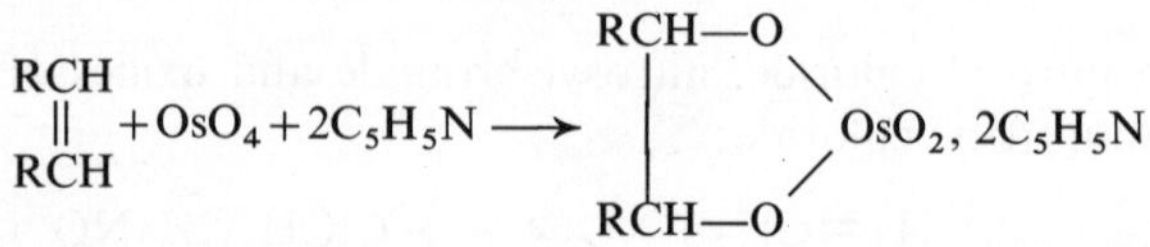

$$\begin{array}{c} RCH \\ \parallel \\ RCH \end{array} + OsO_4 + 2C_5H_5N \longrightarrow \begin{array}{c} RCH{-}O \\ | \qquad\qquad OsO_2, 2C_5H_5N \\ RCH{-}O \end{array}$$

Berkowitz *et al.* (1958) have shown that ruthenium tetroxide is more convenient to use than osmium tetroxide; it is less toxic.

cis-Hydroxylation of a double bond may also be effected by treating the alkene with iodine and silver acetate in wet acetic acid, and then hydrolising the mixed mono- and di-acetates with alkali (Barkley *et al.*, 1954):

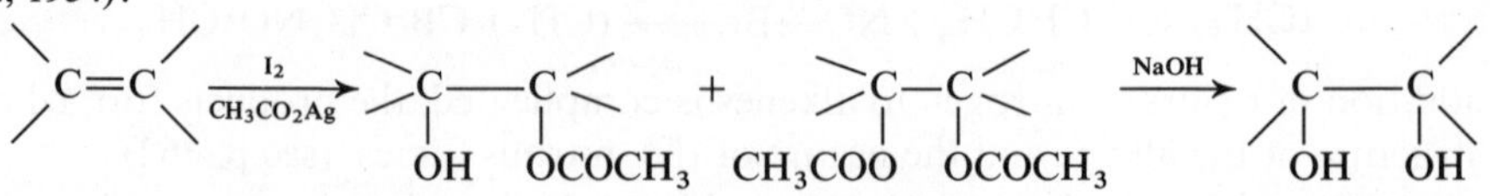

$$\begin{array}{c} \diagdown \quad \diagup \\ C{=}C \\ \diagup \quad \diagdown \end{array} \xrightarrow[CH_3CO_2Ag]{I_2} \begin{array}{c} \diagdown \quad \diagup \\ C{-}C \\ | \quad | \\ OH \quad OCOCH_3 \end{array} + \begin{array}{c} \diagdown \quad \diagup \\ C{-}C \\ | \quad | \\ CH_3COO \quad OCOCH_3 \end{array} \xrightarrow{NaOH} \begin{array}{c} \diagdown \quad \diagup \\ C{-}C \\ \diagup | \quad | \diagdown \\ OH \quad OH \end{array}$$

If the above reaction is carried out in *dry* acetic acid, the di-acetate and diol have the *trans*-configuration.

Glycols (or the alkenes themselves) are readily oxidised to acids or ketones by means of acid permanganate or acid dichromate, the nature of the products being determined by the structure of the glycol, *e.g.*,

(*a*) propene glycol gives acetic and formic acid:

$$CH_3CH(OH)CH_2OH \xrightarrow{[O]} CH_3CO_2H + HCO_2H$$

(*b*) Isobutene glycol gives acetone and formic acid:

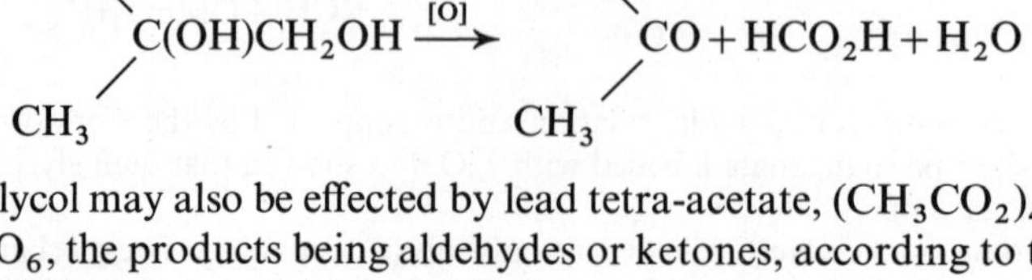

$$\begin{array}{c} CH_3 \\ \diagdown \\ C(OH)CH_2OH \\ \diagup \\ CH_3 \end{array} \xrightarrow{[O]} \begin{array}{c} CH_3 \\ \diagdown \\ CO + HCO_2H + H_2O \\ \diagup \\ CH_3 \end{array}$$

Oxidation of a glycol may also be effected by lead tetra-acetate, $(CH_3CO_2)_4Pb$, or by periodic acid, HIO_4 or H_5IO_6, the products being aldehydes or ketones, according to the structure of the glycol (see also p. 312), *e.g.*,

(a) ethylene glycol gives two molecules of formaldehyde:

$$HOCH_2CH_2OH \xrightarrow[(CH_3CO_2)_4Pb]{[O]} HCHO + HCHO$$

(b) Isobutene glycol gives actone and formaldehyde:

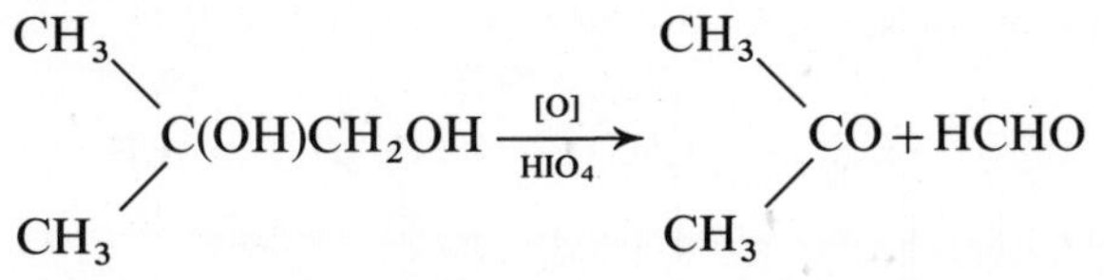

$$\begin{array}{c} CH_3 \\ \diagdown \\ C(OH)CH_2OH \\ \diagup \\ CH_3 \end{array} \xrightarrow[HIO_4]{[O]} \begin{array}{c} CH_3 \\ \diagdown \\ CO + HCHO \\ \diagup \\ CH_3 \end{array}$$

It can be seen that whatever oxidising agent is used the glycol is split into two fragments, the rupture of the carbon chain occurring between the two carbon atoms joined to the hydroxyl groups. Since these two carbon atoms were linked together by the double bond in the original alkene, identification of the two fragments which may be acids, aldehydes or ketones, will indicate the position of the double bond in the alkene, e.g.,

$$R^1CH{=}CHR^2 \xrightarrow{H_2O_2} R^1CH(OH)CH(OH)R^2 \xrightarrow{HIO_4} R^1CHO + R^2CHO$$

The method of oxidative cleavage of the olefinic bond described above has now been replaced by the use of the **Lemieux reagent** (1955). This is better both for determining the position of a double bond and for preparing carbonyl compounds. The reagent is an aqueous solution of sodium periodate and a trace of potassium permanganate. The alkene is oxidised to the cis-diol, which is then cleaved by the periodate to aldehydes and/or ketones. Aldehydes are further oxidised by the permanganate to acids (formaldehyde is usually obtained from a terminal alkene). The reaction proceeds at room temperature, and the lower valent form of Mn is re-oxidised to the higher valent form by the periodate (hence only a catalytic trace of permanganate is needed).

Instead of permanganate, a trace of osmium tetroxide has also been used by Lemieux et al. In this case, aldehyde is produced (it is not oxidised further), e.g.,

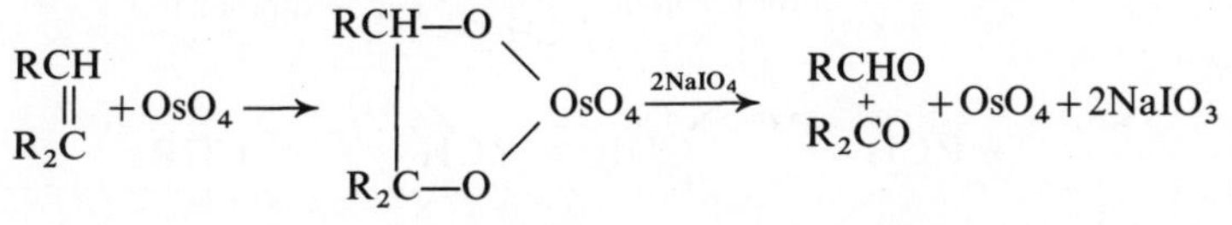

$$\begin{array}{c} RCH \\ \| \\ R_2C \end{array} + OsO_4 \longrightarrow \begin{array}{c} RCH{-}O \\ \diagdown \\ \diagup \\ R_2C{-}O \end{array} OsO_4 \xrightarrow{2NaIO_4} \begin{array}{c} RCHO \\ + \\ R_2CO \end{array} + OsO_4 + 2NaIO_3$$

9. Prileschaiev's reaction (1912). By means of per-acids, the double bond in alkenes is converted into the *epoxide* (alkene oxide). Perbenzoic acid, $C_6H_5COO_2H$, monoperphthalic acid $HO_2CC_6H_4COO_2H$, and p-nitroperbenzoic acid, have been used; e.g.,

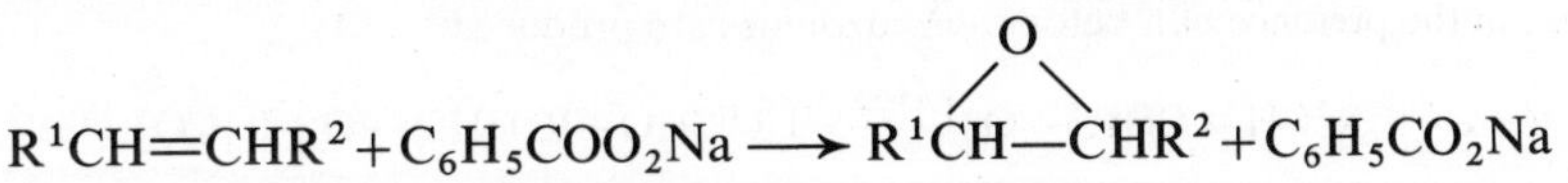

$$R^1CH{=}CHR^2 + C_6H_5COO_2Na \longrightarrow R^1CH\overset{\displaystyle O}{\overbrace{}}CHR^2 + C_6H_5CO_2Na$$

Emmons et al. (1954, 1955) have found that peroxytrifluoroacetic acid (CF_3COO_2H) is a very good reagent for epoxidation and hydroxylation.

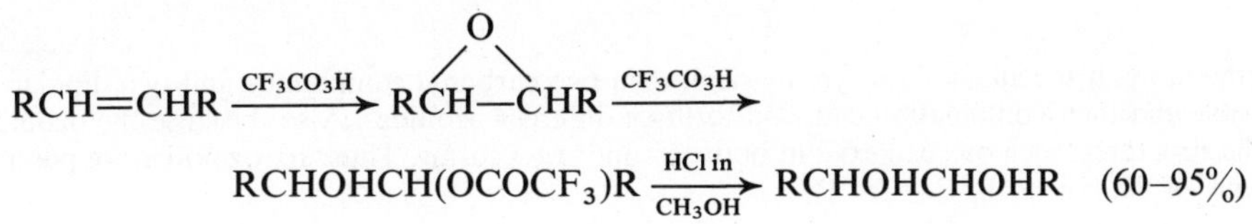

$$RCH{=}CHR \xrightarrow{CF_3CO_3H} RCH\overset{\displaystyle O}{\overbrace{}}CHR \xrightarrow{CF_3CO_3H}$$

$$RCHOHCH(OCOCF_3)R \xrightarrow[CH_3OH]{HCl\ in} RCHOHCHOHR \quad (60\text{--}95\%)$$

This method is particularly useful for high molecular-weight alkenes with a terminal double bond (these are only slowly hydroxylated by other per-acids). Furthermore, peroxytrifluoroacetic acid may be used to hydroxylate α,β-unsaturated carbonyl compounds, *e.g.*, ethyl acrylate, $CH_2{=}CHCO_2C_2H_5$ (see p. 340).

Many mechanisms have been proposed for epoxidation, but none is certain. According to Pausacker *et al.* (1955), the mechanism is:

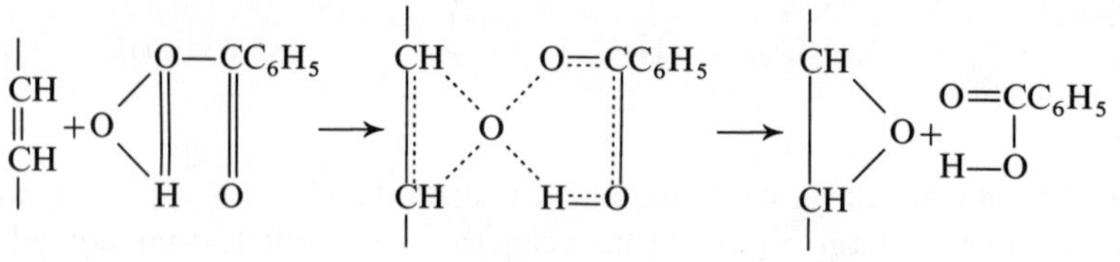

transition state

This mechanism accounts for the fact that epoxidation is usually stereospecific, *i.e.*, one product is formed from each geometrical isomer. Furthermore, the diol produced has the *trans* configuration, and this is in keeping with attack on the intermediate epoxide (see p. 314).

10. Alkenes add on ozone to form **ozonides**. These are usually explosive in the free state, and their structure and mechanism of formation have been the subject of a great deal of work. Staudinger (1922) suggested that the *molozonide* is formed first, and this then rearranges to the ozonide, some dimerising (to the diperoxide); polymers may also be formed. A widely accepted mechanism involves a 1,3-dipolar cyclo-addition reaction (p. 394; this accounts for the stereospecific *cis*-addition to form the molozonide); ozone is considered to behave as

$$\overset{+}{O}{=}O{-}\overset{-}{O} \longleftrightarrow \overset{+}{O}{-}O{-}\overset{-}{O}:$$

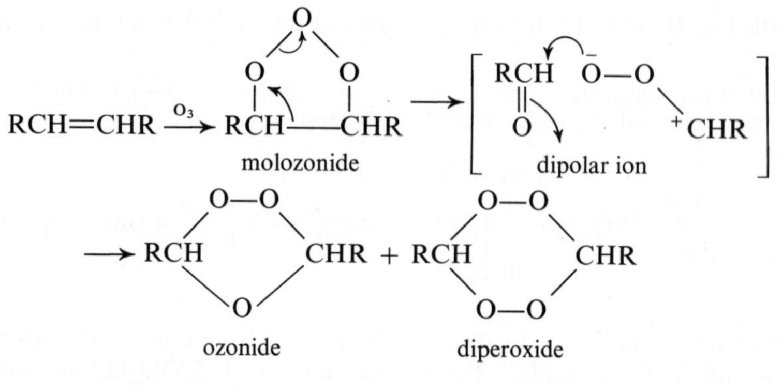

This mechanism of fission followed by recombination is supported by the fact that if the reaction is carried out in the presence of a ketone, *two* ozonides are produced.

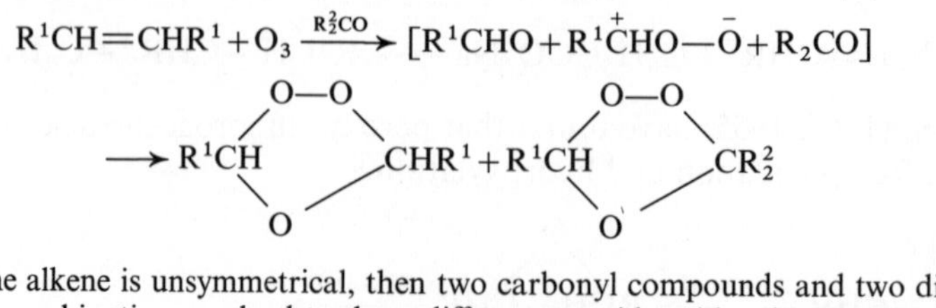

Furthermore, if the alkene is unsymmetrical, then two carbonyl compounds and two dipolar ions are possible and their combination can lead to three different ozonides. Also, because the ozonides have cyclic structures, each one can exist in both *cis*- and *trans*-forms. Thus, *six* ozonides are possible from alkenes of the type $R^1R^2C{=}CR^3R^4$.

The ozonide is prepared by dissolving the olefinic compound in a solvent that is unaffected by ozone, *e.g.*, chloroform, carbon tetrachloride, glacial acetic acid, light petrol, etc., and a stream of ozonised oxygen is passed through. Subsequent treatment may be one of the following procedures.

(i) The ozonide is oxidised by means of silver oxide, hydrogen peroxide or per-acids, thereby producing acids and/or ketones.

(ii) (*a*) Reduction of the ozonide with zinc-acid, H_2-Raney nickel, triphenylphosphine, etc., gives aldehydes and/or ketones. (*b*) Reduction may also be carried out with lithium aluminium hydride or sodium borohydride; the products are the corresponding alcohols of the carbonyl compounds formed in (*a*).

$$
R_2^1C \underset{O}{\overset{O-O}{\diagdown\diagup}} CHR^2
\begin{array}{l}
\xrightarrow{\text{(i)}} R_2^1CO + R^2CO_2H \\
\overset{(a)}{\longrightarrow} R_2^1CO + R^2CHO \\
\underset{(b)}{\searrow} R_2^1CHOH + R^2CH_2OH
\end{array}
$$

Unlike mono-, di-, and tri-substituted ethylenes, tetra-alkyl-substituted ethylenes do not give normal ozonides on ozonolysis by the usual procedures; the product is a polymeric peroxide, $(-CR_2-O-O-)_n$. It would therefore appear that the dipolar ion, $R_2\overset{+}{C}O_2^-$, cannot combine with a ketone. However, Nebel (1968) has shown that ketones may be isolated in high yield by the addition of water to the ozonolysis products of a tetra-alkyl-substituted ethylene in an alcohol at low temperature.

$$
R_2^1\overset{+}{C}-O-\overset{-}{O} + R^2OH \longrightarrow R_2^1C(OR^2)OOH \xrightarrow{H_2O}
$$
$$
R_2^1CO + R^2OH + H_2O_2
$$

The complete process of preparing the ozonide and decomposing it (and identifying the products formed) is known as **ozonolysis**, and is used for determining the position of a double bond in any olefinic compound. Ozonolysis is also used as a means of preparing aldehydes, ketones and acids, and chromatography, particularly GLC and TLC, has been used to isolate the products. There is, however, an increasing tendency to use the Lemieux reagent instead of ozonolysis (see p. 111).

11. Alkenes isomerise when heated at high temperature (500–700°C), or at a lower temperature (200–300°C) in the presence of various catalysts, *e.g.*, aluminium chloride. Isomerisation may be due (i) to the change in position of the double bond, which always tends to move towards the centre of the chain, *e.g.*, pent-1-ene isomerises to pent-2-ene:

$$
CH_3CH_2CH_2CH\!=\!CH_2 \overset{\frown}{} AlCl_3 \rightleftharpoons CH_3CH_2CH_2\overset{+}{C}HCH_2-\overset{-}{A}lCl_3 \underset{+H^+}{\overset{-H^+}{\rightleftharpoons}}
$$
$$
CH_3CH_2CH\!=\!CHCH_2-\overset{-}{A}lCl_3 \underset{-H^+}{\overset{H^+}{\rightleftharpoons}} CH_3CH_2CH\!=\!CHCH_3 + AlCl_3
$$

(ii) To the migration of a methyl group, *e.g.*, but-1-ene isomerises to isobutene:

$$
\overset{\overset{\displaystyle Me}{|}}{CH_2}-CH\!=\!CH_2 \overset{\frown}{} AlCl_3 \rightleftharpoons \overset{\overset{\displaystyle Me}{|}}{CH_2}-\overset{+}{C}H-CH_2-\overset{-}{A}lCl_3
$$
$$
\rightleftharpoons \overset{+}{C}H_2-\overset{\overset{\displaystyle Me}{|}}{\underset{\underset{\displaystyle H}{}}{C}}-CH_2-\overset{-}{A}lCl_3 \rightleftharpoons CH_2\!=\!CMe_2 + AlCl_3
$$

(i) and (ii) may, or may not, occur together.

12. Hydroboronation (hydroboration). Alkenes react rapidly with diborane in, *e.g.*, ether at room temperature to form trialkylboranes. Terminal alkenes give primary alkylboranes which, on oxidation with alkaline hydrogen peroxide, produce primary alcohols (Brown *et al.*, 1957). Also, the overall hydration of the alkene has been shown to occur by *cis*-addition.

$$RCH{=}CH_2 \xrightarrow{B_2H_6} (RCH_2CH_2{-})_3B \xrightarrow[NaOH]{H_2O_2} 3RCH_2CH_2OH + H_3BO_3$$

The anti-Markownikoff direction of addition to form the primary trialkylborane may be explained by the fact that borane is an electron-deficient molecule and so behaves as an electrophile, tending to attack at the point of highest electron density. To explain the overall *cis*-hydration, however, is not so simple. A possible mechanism is that hydroboronation is stereospecifically *cis*-addition, and this would be explained by the reaction proceeding stepwise through an intermediate cyclic T.S.

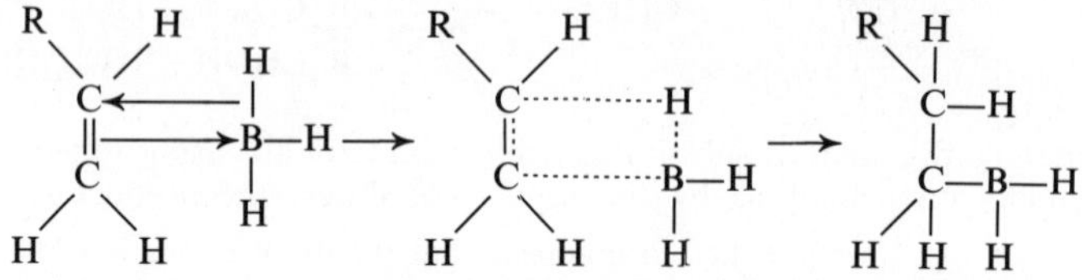

$$\xrightarrow{RCH=CH_2} (RCH_2CH_2{-})_2BH \xrightarrow{RCH=CH_2} (RCH_2CH_2{-})_3B$$

The oxidation must therefore occur by conversion of the C—B bond into C—OH *without inversion*. A possibility is one which involves a 1,2-*shift* (see p. 146).

$$H{-}O{-}O{-}H + OH^- \rightleftharpoons H_2O + \overset{-}{O}OH$$

$$\longrightarrow R_2BOH + RO^- \xrightarrow{H_2O} R_2BOH + ROH + OH^-$$

Since R migrates *with its bonding pair*, the overall hydration is *cis*.

The Markownikoff direction of hydration may be carried out by the oxymercuration-demercuration method (Reaction **6**).

If the trialkylborane is treated with propionic acid, the corresponding alkane is obtained by protolysis:

$$(RCH_2CH_2{-})_3B \xrightarrow{H^+} 3RCH_2CH_3$$

Thus the final product is formed by reduction of an alkene by a *non-catalytic* method.

In the same way, alkynes undergo monohydroboronation and protolysis to give almost pure *cis*-alkene:

$$-C{\equiv}C- \xrightarrow{B_2H_6} (-CH{=}\overset{|}{C}{-})_3B \xrightarrow{H^+} -CH{=}CH-$$

Addition of diborane to a non-terminal double bond is at the least hindered carbon atom, and the nature of the product depends on the amount of hindering at the double bond, *e.g.*,

MeCH=CHMe	Me$_2$C=CHMe	Me$_2$C=CMe$_2$
↓	↓	↓
(MeCH$_2$CHMe—)$_3$B	(Me$_2$CHCHMe—)$_2$BH	Me$_2$CHCMe$_2$—BH$_2$

These s- and t-alkylboranes undergo isomerisation to primary alkylboranes when heated, *e.g.*,

$$(MeCH_2CHMe{-})_3B \longrightarrow (MeCH_2CH_2CH_2)_3B$$

Bis-3-methyl-2-butylborane (*disiamylborane*; second example above) is particularly useful for selective hydroboronation; this borane is often written as Siα_2BH. Thus:

$$C_6H_5CH{=}CH_2 \qquad C_6H_5CH{=}CH_2$$

$$20\% \uparrow \quad \uparrow 80\% \qquad 2\% \uparrow \quad \uparrow 98\%$$

$$(B_2H_6) \qquad\qquad (Si\alpha_2BH)$$

In this way, a compound containing more than one double bond can be converted into one particular alcohol, *e.g.*, Brown *et al.* (1961):

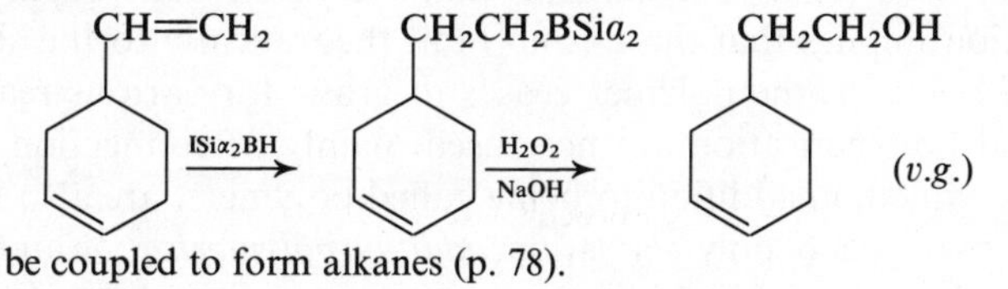

$$(v.g.)$$

Organoboranes may also be coupled to form alkanes (p. 78).

13. Alkenes condense with acetic anhydride in the presence of a catalyst, *e.g.*, zinc chloride, to form unsaturated ketones, *e.g.*, ethylene forms methyl vinyl ketone:

$$CH_2{=}CH_2 + (CH_3CO)_2O \xrightarrow{\text{ZnCl}_2} CH_2{=}CHCOCH_3 + CH_3CO_2H$$

Acid chlorides, alkyl chlorides and α-halogenated ethers also combine with alkenes in the presence of aluminium chloride, *e.g.*,

$$CH_3COCl + CH_2{=}CH_2 \xrightarrow{\text{AlCl}_3} CH_3COCH_2CH_2Cl$$

$$(CH_3)_3CCl + CH_2{=}CH_2 \xrightarrow{\text{AlCl}_3} (CH_3)_3CCH_2CH_2Cl$$

$$CH_3OCH_2Cl + CH_2{=}CH_2 \xrightarrow{\text{AlCl}_3} CH_3OCH_2CH_2CH_2Cl$$

All of these are examples of the Friedel–Crafts reaction with aliphatic compounds (see p. 609).
14. In the presence of a catalyst, *e.g.*, concentrated sulphuric acid, alkenes alkylate alkanes containing a tertiary hydrogen atom. This reaction is particularly useful for preparing 2,2,4-trimethylpentane ('iso-octane; p. 89) by treating isobutane (4 parts) with isobutene (1 part) in the presence of concentrated sulphuric acid.

$$(CH_3)_2C{=}CH_2 + H^+ \longrightarrow (CH_3)_3C^+$$

$$(CH_3)_3C^+ \overset{\frown}{} CH_2{=}C(CH_3)_2 \longrightarrow (CH_3)_3CCH_2\overset{+}{C}(CH_3)_2 \xrightarrow{(CH_3)_3CH}$$

$$(CH_3)_3CCH_2CH(CH_3)_2 + (CH_3)_3C^+\,;\ \text{etc.}$$

On the other hand, various alkenes may dimerise under the influence of suitable catalysts, *e.g.*,

$$Me_2C{=}CH_2 \xrightarrow[\text{(H}_2\text{SO}_4)]{\text{H}^+} Me_3C^+ \overset{\frown}{} CH_2{=}CMe_2 \longrightarrow$$

$$Me_3CCH_2\overset{+}{C}Me_2 \xrightarrow{-\text{H}^+} Me_3CCH{=}CMe_2 + Me_3CCH_2CMe{=}CH_2$$

Under different conditions, alkenes polymerise, *e.g.*, isobutene gives a polymer:

$$nC_4H_8 \xrightarrow{\text{H}^+} (C_4H_8)_n$$

When two compounds have the same empirical formula but differ in molecular weight, the more complicated compound is called a *polymer* of the simpler one. The term *polymerisation* was used originally to indicate the process that took place when a single substance—the *monomer*—gave products having the same empirical formula but different molecular weights, each of these being a multiple of that of the monomer. As the investigation of polymerisation reactions progressed, it was found that many compounds of high molecular weight, although they produced a large number of monomer molecules on suitable treatment, did not always have exactly the same empirical formula as the parent monomer. This led to a modification of the definitions of the terms polymer and polymerisation. According to Carothers (1931) polymerisation is best defined as *intermolecular combinations that are functionally capable of proceeding indefinitely*. This definition implies that there is no limit theoretically to the size of the polymer molecule. In practice, however, the polymer ceases to grow, for various reasons (see below). The terms polymer and polymerisation are now used mainly in connection with high molecular weight compounds, which, in addition to being called polymers, are also known as *macromolecules*.

There are two types of polymerisation, *addition polymerisation* and *condensation polymerisation*.

Addition polymerisation. Addition polymerisation occurs between molecules containing double or triple bonds; but in certain cases it can also occur between bifunctional compounds that result from the opening of ring structures (see, *e.g.*, ethylene oxide). *There is no liberation of small molecules during addition polymerisation.*

A very important group of olefinic compounds that undergo addition polymerisation is of the type $CH_2=CHY$, where Y may be H, X, CO_2R, CN, etc.:

$$nCH_2=CHY \longrightarrow (-CH_2CHY-)_n$$

There are three possible ways in which this polymerisation can occur:

(i) Head to tail: $-CH_2CHY-CH_2CHY-$

(ii) Head to head and tail to tail: $-CHYCH_2-CH_2CHY-CHYCH_2-CH_2CHY-$

(iii) A random arrangement involving (i) and (ii).

Experimental work indicates that (i) is favoured.

Most polymerisations are carried out in the presence of catalysts, and polymerisation of alkenes can be accelerated by ionic-type catalysts or radical-type catalysts. Both types of reaction consist of a number of steps which follow one another consecutively and rapidly, and take place in three principal stages:

(i) The initiation or activation. (ii) The growth or propagation. (iii) The termination or cessation.

The ionic mechanism of catalysis. Ionic catalysts are usually Lewis acids, *e.g.*, H_2SO_4, HF, $AlCl_3$, BF_3, etc. The ionic mechanism may be illustrated by the polymerisation of isobutene in the presence of sulphuric acid (*cf.* above).

(i) H^+ $CH_2=CMe_2 \longrightarrow Me-\overset{+}{C}Me_2$

(ii) Me_3C^+ $CH_2=CMe_2$ $CH_2=CMe_2 \cdots \longrightarrow Me_3C-(CH_2-CMe_2)_n-CH_2-\overset{+}{C}Me_2$

(iii) $Me_3C-(CH_2-CMe_2)_n-CH_2-\overset{+}{C}Me_2 \xrightarrow{-H^+} Me_3C-(CH_2-CMe_2)_n-CH=CMe_2$

This polymerisation, carried out in the presence of a small amount of butadiene, produces *butyl rubber*. The most important ionic catalysts are those introduced by Ziegler and Natta (see below).

The free-radical mechanism of catalysis. Addition polymerisation may take place by chain reactions brought about by catalysts that are known to generate free radicals. The most widely used catalysts are the organic and inorganic peroxides and the salts of the per-acids, *e.g.*, benzoyl peroxide, acetyl peroxide, hydrogen peroxide, potassium perborate, etc.

This type of mechanism is illustrated for polymerisations in the presence of organic peroxides.

(i) $(RCO)_2O_2 \longrightarrow 2RCO_2\cdot \longrightarrow 2R\cdot + 2CO_2$

$R\cdot + CH_2=CH_2 \longrightarrow R-CH_2-CH_2\cdot$

(ii) $R-CH_2-CH_2\cdot + nCH_2=CH_2 \longrightarrow R-(CH_2-CH_2)_n-CH_2-CH_2\cdot$

(iii) The termination reaction may take place in various ways. One obvious way is by the combination of two growing chains, *i.e.*,

$$2R—(CH_2—CH_2)_n· \longrightarrow R—(CH_2—CH_2)_{2n}—R$$

On the other hand, termination may also occur by disproportionation:

$$2R—(CH_2—CH_2)_n—CH_2—CH_2· \longrightarrow R—(CH_2—CH_2)_n—CH_2CH_3$$
$$+R—(CH_2—CH_2)_n—CH=CH_2$$

Still another way for termination to occur is by reaction between the growing chain and another species that has been added in relatively small amount before polymerisation.

$$R—(CH_2—CH_2)_{\overline{n}}+YZ \longrightarrow R—(CH_2—CH_2)_n—Y+Z·$$

$$Z·+CH_2=CH_2 \longrightarrow Z—CH_2—CH_2· \cdots\rightarrow Z—(CH_2—CH_2)_{\overline{n}}$$

$$Z—(CH_2—CH_2)_{\overline{n}}+YZ \longrightarrow Z—(CH_2—CH_2)_n—Y+Z·$$

Polymers of the type $Z—M_n—Y$, *i.e.*, those which contain a 'foreign' molecule in addition to the recurring unit, are known as *telomers*, and the process of their formation is called *telomerisation*.

In 'uncatalysed' polymerisation, *i.e.*, in the absence of foreign substances, initiation may begin by *dimerisation* of the monomer:

$$2CH_2=CH_2 \longrightarrow ·CH_2—CH_2—CH_2—CH_2·$$

Not only can addition polymerisation take place between molecules of one kind, but it can also take place between molecules of two kinds, when the phenomenon is known as *copolymerisation* or *interpolymerisation*.

Condensation polymerisation or polycondensation. In condensation polymerisation, bi- or poly-functional molecules condense with one another, and in doing so repeatedly eliminate a small molecule such as water, ammonia, hydrogen chloride, etc., as the reaction proceeds. This type of polymerisation takes place by a series of steps, and is discussed in various parts of the text (see, for example, the aldol condensation, the esterification of glycols with dibasic acids, etc.).

Polymers usually possess a certain amount of crystallinity, and their tensile strength increases with molecular weight. Also, the greater the crystallinity, the greater is the tensile strength, the lower is the solubility and the higher is the m.p.

Polymers may be classified into three groups:

(i) *natural*, *e.g.*, rubber, proteins, cellulose; (ii) *semi-synthetic*, *e.g.*, nitrocellulose, cellulose acetate; and (iii) *synthetic*, *e.g.*, nylon, bakelite, perspex.

Plastics form a group of high polymers which have a fair range of deformability and mouldability, particularly at high temperatures. In plastics the polymers formed do not all have the same molecular weight, and since the polymers are not amenable to the ordinary methods of separation, the molecular weight of a 'polymer' is the *average* molecular weight. Polymerisation is carried out with the object of building up compounds with predicted properties, and since the properties of a plastic depend on the degree of polymerisation it is necessary to stop polymerisation when the desired average molecular weight is reached. This may be done by various means, *e.g.*, variation of the concentration of the catalyst. The average molecular weight of plastics varies from about 20,000 (*e.g.*, nylon) to several hundred thousand (*e.g.*, polyvinyl chloride, 250,000).

Plastics are generally tough, resistant to the action of acids and alkalis, and not very much affected over a fair range of temperature. They can be moulded to any desired shape or form.

Plastics are of two main types, *thermoplastic* and *thermosetting*. Thermoplastics are linear polymers which are soluble in many organic solvents, and which soften on heating and become rigid on cooling. The process of heat-softening, moulding and cooling can be repeated as often as desired, and hardly affects the properties of the plastic. Typical thermoplastics are cellulose acetate, nitrocellulose and vinyl polymers such as polythene, perspex, etc.

Thermosetting plastics are three-dimensional polymers which are insoluble in any kind of solvent, and which can be heat-treated only once before they set, *i.e.*, their formation, after which heating results in chemical decomposition, and hence they cannot be 'reworked'. Typical thermosetting plastics are phenol-formaldehyde, urea-formaldehyde, melamine-formaldehyde, silicones, etc.

In thermoplastics the chains are, more or less, free chemically, but are held together by van der Waals' forces. It is possible, however, to link together these linear molecules (*cf.* the rungs of a ladder) and the cross-linking agent converts the thermoplastic into a thermosetting plastic, *e.g.*, in the vulcanisation of rubber the sulphur cross-links the long chains. Furthermore, such thermosetting plastics may be reconverted into thermoplastics by opening the cross-links, *e.g.*, the reclaiming of rubber. Most thermosetting plastics may be regarded as cross-linked polymers.

Those plastics which do not soften very much with rise in temperature are made soft and readily workable by the addition of certain compounds known as *plasticisers*; *e.g.*, polyvinyl chloride is extremely stiff and hard, but addition of tricresyl phosphate makes it soft and rubber-like.

Stereochemistry of polymers. If we consider the polymerisation of the monomer, CH_2=CYZ, then there are two extreme stereoisomers possible: (i) all the Y groups lie on one side of the chain, and all the Z goups on the other side; this is the *isotactic* polymer; (ii) the Y and Z groups lie alternately on each side of the chain; this is the *syndiotactic* polymer.

isotactic   syndiotactic

Between these two extremes, there are those polymers in which the two groups Y and Z are arranged in a random fashion along the two sides of the chain; these are the *atactic* polymers.

Ziegler-Natta catalysts are stereospecific; they are metal compounds, *e.g.*, $TiCl_4$, $Zr(OR)_4$, etc., together with co-catalysts such as R_3Al, $LiAlH_4$, etc.; *e.g.*, a typical Ziegler-Natta catalyst is the triethylaluminium-titanium chloride complex.

$$Et_3Al + CH_2{=}CH_2 \longrightarrow EtCH_2CH_2AlEt_2 \xrightarrow{n(CH_2=CH_2)}$$
$$Et(CH_2CH_2)_nCH_2CH_2AlEt_2 \xrightarrow{H^+} Et(CH_2CH_2)_nCH_2CH_3$$

It should be noted, however, that a stereochemically pure polymer is not usually obtained, the product being a mixture of predominantly the stereospecific product together with some atactic compound.

Some individual alkenes

Diradicals (biradicals). Electrons have a spin number of $\pm 1/2$. When all the electrons in a molecule are paired, each pair has the spins anti-parallel ($+1/2$ and $-1/2$) and so cancel each other, the result being a diamagnetic molecule, *i.e.*, one which does not possess a permanent magnetic moment. Now suppose one of the paired electrons is raised by, *e.g.*, electromagnetic radiation, to a higher electronic level. In the excited molecule the spins may still be antiparallel and so cancel out. The molecule is therefore still diamagnetic. The spins may, however, be parallel (this is possible because the two electrons are in different orbitals), and the molecule will now be paramagentic.

If S is the algebraic sum of the spins of the electrons in a molecule, then M, the *multiplicity*, is given by the expression

$$M = 2S + 1.$$

In covalent pairs,

$$S = (+1/2) + (-1/2) = 0.$$

Therefore $M = 1$, and the molecule is said to be in a *singlet* state. If the molecule has *one odd electron*, $S = 1/2$ and therefore $M = 2$; the molecule is now in a *doublet* state. If the molecule has two electrons in different orbitals then, if the spins are antiparallel, $S = 0$, $M = 1$, and the molecule is in a singlet state. If the spins are parallel, $S = (1/2) + (+1/2) = 1$, $M = 3$ and the molecule is in a *triplet* state; in this state the two electrons interact strongly with each other. There is also the case where the interaction (coupling) is very weak or absent, *i.e.*, the unpaired electrons act independently of each other; in this case, the molecule is a *diradical* (one might call it a 'double doublet'). A molecule in the triplet state can behave as a diradical in chemical reactions (see also p. 870).

Methylene (carbene), CH_2. This is the first member of the alkenes, but it has only a very short life, and is a 'bivalent' carbon compound. It is formed by the photolysis (photochemical decomposition) or pyrolysis of diazomethane or keten:

$$CH_2N_2 \longrightarrow CH_2 + N_2$$
$$CH_2{=}C{=}O \longrightarrow CH_2 + CO$$

Methylene undergoes two types of reaction, insertion and addition. Insertion reactions occur mainly in the C—H bond, but can also occur in O—H and C—Cl bonds, *e.g.* (Strachan *et al.*, 1954; Bradley *et al.*, 1961):

$$CH_3CHOHCH_3 \xrightarrow{\text{CH}_2} (CH_3)_3COH + CH_3CH_2CHOHCH_3 + (CH_3)_2CHOCH_3$$
$$CH_3CH_2CH_2Cl \xrightarrow{\text{CH}_2} CH_3(CH_2)_2CH_2Cl + CH_3CH_2CHClCH_3$$

The usual mechanism for insertion is believed to be via a cyclic transition state, *e.g.*,

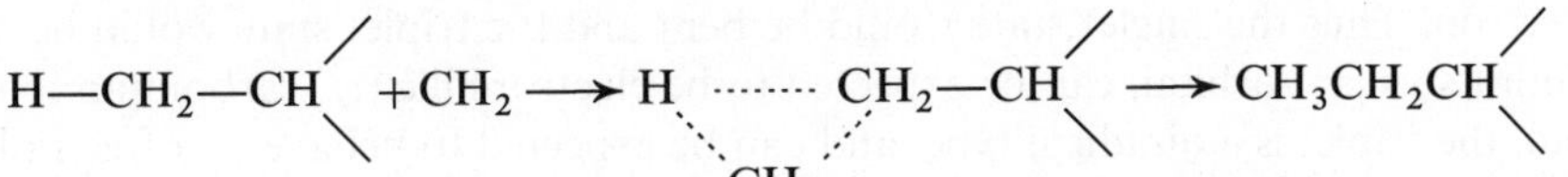

Methylene adds across double bonds to form cyclopropanes; at the same time, insertion reactions also occur. Skell *et al.* (1956, 1959) showed the addition of methylene (from diazomethane) to *cis*- and *trans*-but-2-ene is a *cis*-addition. Thus the addition is *stereospecific*, *i.e.*, each geometrical isomer forms one product, and the configurations of the two products are different.

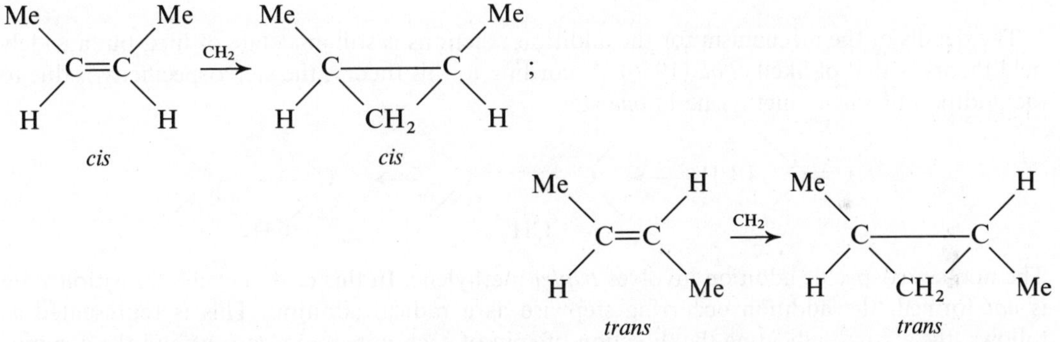

Anet *et al.* (1960), however, showed that this stereospecific addition is lost when the reaction is carried out in the presence of an inert gas (nitrogen), *i.e.*, each substrate (reactant) now gives a mixture of the *cis*- and *trans*-products. On the other hand, Duncan *et al.* (1962) showed that methylene formed by the photolysis of keten added to the above substrates in a non-stereospecific manner.

To explain these results, let us first consider the problem of the *relationship between reactivity and selectivity*. A general (empirical) principle is that the more reactive the reagent is, the less selective it is in its reactions, *e.g.*, in the chlorination of alkanes, the rate of hydrogen substitution is t > s > primary (see p. 149). If the temperature is raised above 300°C, chlorine becomes more reactive and the substitution now becomes less selective, the rates of substitution being the same for primary-, s- and t-hydrogen. A possible basis for this *reactivity-selectivity principle* is that the more reactive the reagent is, the more likely it will react at every collision, *i.e.*, there would be less discrimination between the various positions. The less reactive the reagent, the more will the electronic distribution in the substrate play a part, thereby resulting in more discrimination at the various positions.

The above non-stereospecific additions are an indication that in an inert gas, methylene becomes less reactive; it also follows that methylene generated from keten is less reactive than that from diazomethane. The latter conclusion had already been reached by Frey (1958), who found that methylene from the photolysis of diazomethane was less selective in its insertion in primary and secondary C—H bonds than was methylene from keten. It therefore appears that there are two forms of methylene. Herzberg *et al.* (1961) showed, from spectroscopic evidence, that methylene (from diazomethane) was initially formed in the singlet state (this is the excited state) and some then rapidly changes to the triplet state (this is the ground state).

The electronic distribution in methylene has been the subject of much discussion. It appears that the general feeling now is that in the singlet state, the carbon atom is sp^2 hybridised with two electrons in each of these three orbitals, whereas in the triplet state, the carbon is sp hybridised with two electrons in each of these orbitals, and the two p-orbitals each occupied by one electron. Thus the singlet state would be bent and the triplet state would be linear. The singlet, with its vacant orbital, can be expected to be electrophilic (*cf.* carbonium ions). On the other hand, the triplet is a diradical type, and can be expected to behave as a free radical.

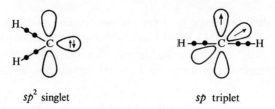

sp^2 singlet sp triplet

The details of the mechanism for the addition reactions is still in a state of flux, but a widely held theory is that of Skell *et al.* (1956). According to this theory, the stereospecificity is due to the addition of *singlet* methylene in *one* step:

$$\begin{array}{ccc} \diagup\!\!\!\diagdown C=C \diagdown\!\!\!\diagup + CH_2 \longrightarrow & \diagup\!\!\!\diagdown C \cdots\cdots\cdots C \diagdown\!\!\!\diagup & \longrightarrow \quad \diagup\!\!\!\diagdown C \text{——} C \diagdown\!\!\!\diagup \\ & CH_2 & CH_2 \end{array}$$

The non-stereospecific addition involves *triplet* methylene. In this case, a cyclic transition state is *not* formed, the addition occurring stepwise as a radical addition. This is represented as follows, the arrows indicating the direction of spin of each unpaired electron; and the assumption is made that rotation about the single bond is more rapid than spin inversion:

$$\begin{array}{c}
\begin{array}{ccccccc}
R\diagdown \quad \diagup R & & R\diagdown \quad \diagup R & & R\diagdown \quad \diagup R & & R\diagdown \qquad \diagup R \\
C=C & + CH_2 \longrightarrow & C{-}C & \longrightarrow & C{-}C & \longrightarrow & C\text{———}C \\
H\diagup \quad \diagdown H & & H\diagup \;\downarrow\; \diagdown H & & H\diagup \;\uparrow\; \diagdown H & & H\diagup \quad CH_2 \quad \diagdown H \\
& & \downarrow\!\cdot CH_2 & & \downarrow\!\cdot CH_2 & &
\end{array}
\end{array}$$

$$\Updownarrow$$

$$\begin{array}{ccccccc}
H\diagdown \quad \diagup R & & H\diagdown \quad \diagup R & & H\diagdown \qquad \diagup R \\
C{-}C & \longrightarrow & C{-}C & \longrightarrow & C\text{———}C \\
R\diagup \;\downarrow\; \diagdown H & & R\diagup \;\uparrow\; \diagdown H & & R\diagup \quad CH_2 \quad \diagdown H \\
\downarrow\!\cdot CH_2 & & \downarrow\!\cdot CH_2 & &
\end{array}$$

Ethylene may be prepared by most of the general methods of preparation, but the most convenient laboratory method is to heat ethanol with excess of concentrated sulphuric acid (p. 94). It is obtained in huge quantities as a by-product in the cracking of crude oil and of ethane and propane.

Ethylene is a colourless gas, b.p. $-105°C$, sparingly soluble in water. It burns with a smoky luminous flame. When ethylene is heated with chlorine at 350–450°C, vinyl chloride is obtained:

$$CH_2{=}CH_2 + Cl_2 \longrightarrow CH_2{=}CHCl + HCl$$

Ethylene may be oxidised to ethylene oxide (p. 313), and polymerises under high pressure and high temperature to form *polyethylene* or *polythene*:

$$nCH_2{=}CH_2 \longrightarrow -(-CH_2-CH_2-)_n-$$

This polymerisation is catalysed by traces of oxygen (which produces the free radicals). Polythene is very resistant to acids, bases, and most of the usual organic solvents.

> Polythene is also manufactured by the Ziegler process. Ethylene is passed, under pressure, into a hydrocarbon solvent containing a suspension of, *e.g.*, $Et_3Al + TiCl_4$ as catalyst at about 70°C.

Ethylene is used for ripening fruit. Unripe fruit may be transported easily without damage, and ripens on exposure to ethylene gas for a few days, the product being apparently indistinguishable from the natural ripened fruit. It appears that all fruits give off ethylene (Burg *et al.*, 1962). Ethylene is also used as an anæsthetic, in the manufacture of mustard gas and plastics (polythene, polystyrene), and in the preparation of various solvents such as glycol, dioxan, cellosolves, etc.

Structure of ethylene. The molecular formula of ethylene is C_2H_4. Two carbon atoms have the power to combine with six univalent atoms or groups, as in ethane, neopentane, etc. There are only four univalent hydrogen atoms present in ethylene: therefore ethylene is said to be unsaturated, and should be capable of adding on two univalent atoms or groups, and this, as we have seen above, is observed in practice. Thus the structure of ethylene must be such as to be capable of undergoing addition reactions. Assuming carbon to be quadrivalent and hydrogen univalent, three structures are possible for ethylene:

$$CH_3-\ddot{C}H \qquad CH_2{=}CH_2 \qquad \cdot CH_2-CH_2\cdot$$

$$\text{(I)} \qquad\qquad \text{(II)} \qquad\qquad \text{(III)}$$

Two isomeric compounds of molecular formula $C_2H_4Cl_2$ are possible: CH_3CHCl_2 and and CH_2ClCH_2Cl. Both isomers are known, one (ethylene dichloride) being formed by the direct combination between ethylene and chlorine, and the other (ethylidene dichloride) by the action of phosphorus pentachloride on acetaldehyde. The structure of ethylidene dichloride is CH_3CHCl_2 (see p. 169); hence the structure of ethylene dichloride is CH_2ClCH_2Cl. If (I) were the structure of ethylene, then the addition of chlorine should give ethylidene dichloride, and not ethylene dichloride. We may, therefore, reject structure (I). Furthermore, since (I) is unsymmetrical it would have a fairly large dipole moment; actually ethylene has a zero dipole moment.

Structure (II) represents ethylene as possessing a *double bond*. Such a bond would prevent free rotation and would therefore explain geometrical isomerism. Hence (II) was acceptable to the classical chemists. The modern theory of a double bond is that it consists of one σ- and one π-bond (p. 45), or two 'bent' bonds (p. 46). Structure (III) represents ethylene as a free diradical, but since ethylene does not exhibit the usual properties of a diradical, we must reject (III).

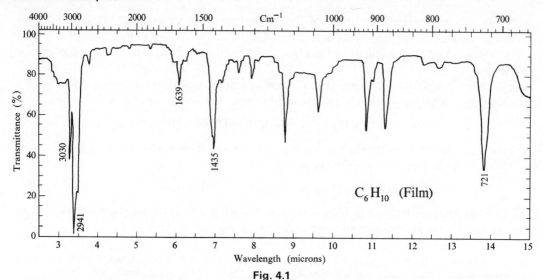

Fig. 4.1

1639 cm^{-1}: C=C (str.). 3030 cm^{-1}: =C—H (str.). 721 cm^{-1}: =C—H (def.); *cis* isomer. 2941 cm^{-1}: C—H (str.) in (Me and/or) CH$_2$. 1435 cm^{-1}: C—H (def.) in (Me and/or) CH$_2$.

The presence of a double (or triple bond) in an organic compound may be found readily by means of bromine water, bromine in chloroform solution, or dilute alkaline permangante. If the compound under investigation is unsaturated, then the above reagents are decolorised. Perbenzoic acid or monoperphthalic acid can be used to detect the presence of a double bond, and also to estimate the number of double bonds (see also iodine value, p. 323). The presence of an isolated double bond is readily detected by its infra-red absorption spectrum (C=C stretch), 1680–1620 cm^{-1} (v). *cis-* and *trans*-Isomers of the type RCH=CHR may be distinguished by the C—H deformation frequencies: *cis*, 630–665 cm^{-1} (s); *trans*, 970–960 cm^{-1} (s). Also, there are several regions of absorption due to the =C—H stretch. The actual positions depend on the nature of the alkene: mono-, *s*-di- (*both cis and trans*) and trisubstituted ethylenes all show a band at 3040–3010 cm^{-1} (m), and *as*-disubstituted ethylenes shown a band at 3095–3075 cm^{-1} (m) [this bond is also shown by monosubstituted ethylenes].

Figure 4.1 is the infra-red spectrum of cyclohexene (as a film).

NMR spectra of alkenes. Since the chemical shift of an olefinic proton ($\tau = 4\cdot3$–$5\cdot3$ p.p.m.) is very different from that of a methyl, methylene, or methine proton ($\tau = 7\cdot0$–$9\cdot1$) in hydrocarbons, and from aromatic (and heterocyclic) protons ($\tau = 1\cdot0$–$3\cdot0$), it is relatively simple to detect the presence of a double bond in a compound (olefinic) containing the =CH group (see also p. 132). Figure 4.2 is the NMR spectrum of cyclohexene (in CCl$_4$ at 60 MHz).

It has been pointed out that the spin-spin coupling constant depends on the number of covalent bonds between the two interacting protons and on their geometry. The former factor is the same for a pair of *cis-trans* 1,2-disubstituted ethylenes. The latter factor, however, is different, and this produces different *J* values and so offers a means of distinguishing between *cis* and *trans* isomers.

cis   *trans*   *geminal*

J   7–14 Hz   12–19 Hz   0–3 Hz

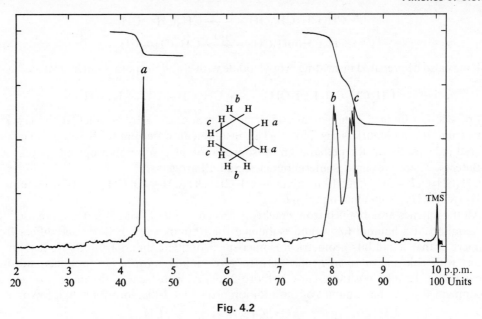

Fig. 4.2

J values for other groupings involving a double bond are, *e.g.* (note the long range coupling):

CH₂=CH—CH₂—R (structures)

$$\text{vicinal}\qquad\qquad\text{allylic}\qquad\qquad\text{homo-allylic}$$
$$J\quad 4\text{–}10\ \text{Hz}\qquad (cis\ \text{or}\ trans)\qquad\qquad 0\text{–}2\ \text{Hz}$$
$$0\text{–}2\ \text{Hz}$$

Mass spectra. The molecular ion of mono-alkenes is usually present and tends to undergo allylic cleavage (*i.e.*, at the β-bond with respect to the double bond), with the positive charge usually remaining with the fragment containing the double bond, since the allyl ion is stabilised by resonance.

$$\left[\text{CH}_2\!=\!\text{CH}\!-\!\underset{\alpha}{\text{CH}}_2\!-\!\underset{\beta}{\text{R}}\right]^{\dot{+}} \longrightarrow \text{R}\cdot + \text{CH}_2\!=\!\text{CH}\!-\!\text{CH}_2{}^+ \longleftrightarrow \overset{+}{\text{CH}}_2\!-\!\text{CH}\!=\!\text{CH}_2$$

If the formation of a six-membered cyclic T.S. involving a γ-hydrogen is possible, a **McLafferty rearrangement** usually occurs to give two alkenes, and the positive charge may reside with either alkene, *e.g.*,

$$\left[\begin{array}{c}\text{R}^1\ \ \ \ \ \ \text{H}\\ {}_\gamma\text{CH}\ \ \ \ \text{CHR}^2\\ {}_\beta\text{CH}_2\ \ \ \ \text{CH}\\ {}_\alpha\text{CH}_2\end{array}\right]^{\dot{+}} \longrightarrow \left[\begin{array}{c}\text{R}^1\text{CH}\\ \ \ \text{CH}_2\end{array} + \begin{array}{c}\text{CH}_2\text{R}^2\\ \text{CH}\\ \text{H}_2\text{C}\end{array}\right]^{\dot{+}}$$

Thus, the mass spectra of mono-alkenes are characterised by the presence of peaks $C_nH_{2n}-1$ (27, 41, 55, 69, etc.) and peaks C_nH_{2n} (28, 42, 56, etc.).

A difficulty encountered with alkenes is that, because of the ready migration of the double bond, fragmentation of isomeric alkenes are often similar.

Propene (propylene) may be prepared by heating propanol or isopropanol with sulphuric acid (mechanism as for ethylene from ethanol):

$$CH_3CH_2CH_2OH \xrightarrow{-H_2O} CH_3CH=CH_2$$

$$CH_3CH(OH)CH_3 \xrightarrow{-H_2O} CH_3CH=CH_2$$

It may also be prepared by heating propyl iodide with ethanolic potassium hydroxide:

$$CH_3CH_2CH_2I + KOH \xrightarrow{ethanol} CH_3CH=CH_2 + KI + H_2O$$

Propene is obtained commercially in huge quantities as a by-product in the cracking of petroleum. It is a colourless gas, b.p. $-48°C$, insoluble in water but fairly soluble in ethanol. It is used industrially for the preparation of isopropanol, glycerol, polypropylene (a plastic), etc.

Butenes. There are three isomeric butenes, and all are gases:

$CH_3CH_2CH=CH_2$, but-1-ene (b.p. $-6·1°C$); $CH_3CH=CHCH_3$, but-2-ene (b.p. $1°C$); $(CH_3)_2C=CH_2$, isobutene (b.p. $-6·6°C$).

All the butenes are obtained from cracked petroleum. The 1- and 2- butenes are used for the preparation of s-butanol (*q.v.*), and isobutene for t-butanol (*q.v.*). But-2-ene differs from its isomers in that it exhibits geometrical isomerism. Isobutene differs from its isomers in that it reacts with chlorine *at room temperature* to give mainly *substitution* products, substitution occurring in the allyl position. Thus 3-chloro-2-methylprop-1-ene is the main product, and is accompanied by a small amount of the addition product 1,2-dichloro-2-methylpropane:

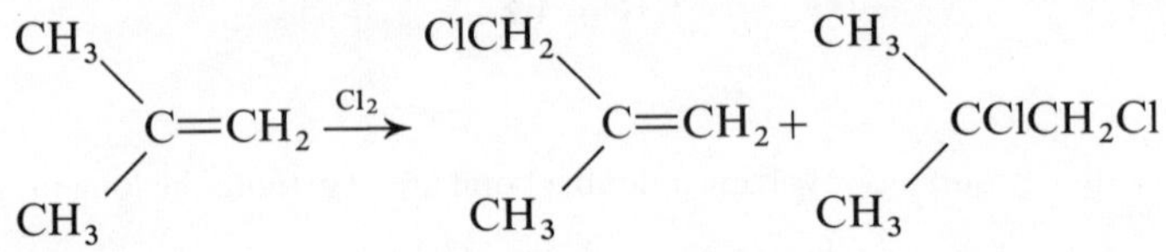

Mechanistic studies have shown that the substitution product is not obtained by *direct* substitution. Reeve (1952), using isobutene labelled at position 1 (^{14}C) did not obtain labelled formaldehyde on ozonolysis of the monochloroisobutene. Since the dichloride is stable under the conditions of the experiment, a mechanism which is in accord with these facts is:

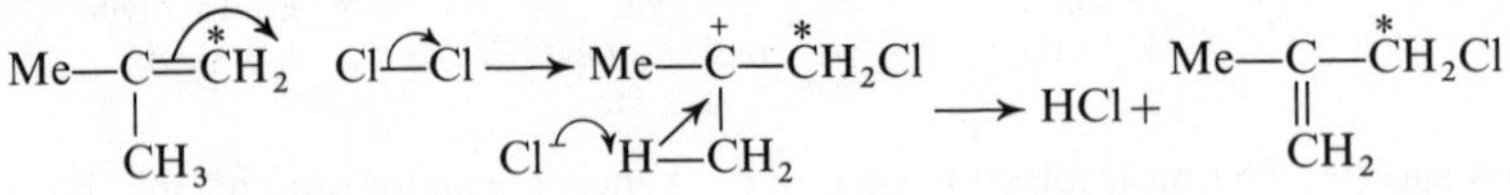

Unsaturated compounds with two or more double bonds

When the compound contains two double bonds, it is known as a diolefin or alkadiene, and has the general formula C_nH_{2n-2}; when there are three double bonds present, the compound is known as a triolefin or alkatriene, and has the general formula C_nH_{2n-4}; etc.

Nomenclature

The longest carbon chain containing the maximum number of double bonds is chosen as the parent hydrocarbon, and the chain is so numbered as to give the lowest possible numbers to the double bonds, *e.g.*,

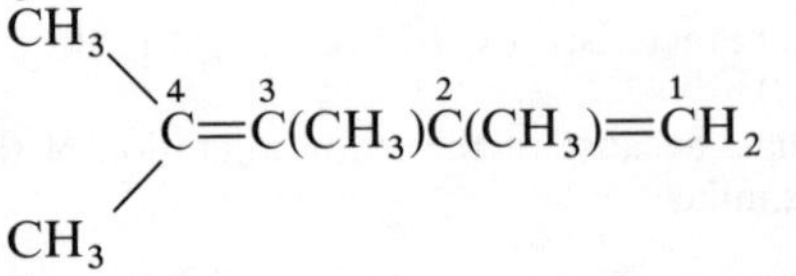

2,3,4-trimethylpenta-1,3-diene

There are three different types of compounds with two double bonds.

1. Hydrocarbons with isolated double bonds contain the arrangement $\diagdown C\!\!=\!\!CH(CH_2)_nCH\!\!=\!\!C\diagup$,

where $n > 0$. One of the simplest compounds of this type is **diallyl** or **hexa-1,5-diene,** which may be prepared by the Wurtz reaction with allyl chloride:

$$2CH_2\!\!=\!\!CHCH_2Cl+Mg \xrightarrow{\text{ether}} CH_2\!\!=\!\!CH(CH_2)_2CH\!\!=\!\!CH_2+MgCl_2 \quad (56\text{–}65\%)$$

Hexa-1,5-diene is a liquid, b.p. 59·6°C. Alkadienes with *isolated* double bonds react like the alkenes each olefinic bond acting independently of the other.

2. Hydrocarbons with cumulated double bonds contain the arrangement $\diagdown C\!\!=\!\!C\!\!=\!\!C\diagup$. The simplest compound of this type is **allene** or **propadiene**, which may be prepared by heating 1,2,3-tribromopropane with solid potassium hydroxide, and then treating the resulting 2,3-dibromopropene with zinc dust in methanol solution:

$$BrCH_2CHBrCH_2Br \xrightarrow{KOH} BrCH_2CBr\!\!=\!\!CH_2 \xrightarrow{Zn/CH_2OH} CH_2\!\!=\!\!C\!\!=\!\!CH_2$$

Allene is a gas, b.p. $-32°C$. With bromine it forms 1,2,2,3-tetrabromopropane; with sulphuric acid it forms acetone; and when treated with sodium in ether, the sodium derivative of propyne is produced.

(a) $CH_2\!\!=\!\!C\!\!=\!\!CH_2 \xrightarrow{H^+} CH_3\overset{+}{C}\!\!=\!\!CH_2 \xrightarrow{H_2O}$

$CH_3\overset{+}{C}(OH_2)\!\!=\!\!CH_2 \xrightarrow{-H^+} CH_3C(OH)\!\!=\!\!CH_2 \longrightarrow CH_3COCH_3$

(b) $CH_2\!\!=\!\!C\!\!=\!\!CH_2 \xrightarrow[(-H^+)]{Na} CH_2\!\!=\!\!C\!\!=\!\!\ddot{C}H^- \longleftrightarrow {}^-\ddot{C}H_2\!\!-\!\!C\!\!\equiv\!\!CH$

$\xrightarrow{H^+} CH_3C\!\!\equiv\!\!CH \xrightarrow[(-H^+)]{Na} CH_3C\!\!\equiv\!\!C^-Na^+$

Allenes may be prepared from alkynes by reversal of the above rearrangement, *e.g.,*

$$CH\!\!\equiv\!\!CCH_2CO_2H \xrightarrow[40°C]{\text{aq. }K_2CO_3} CH_2\!\!=\!\!C\!\!=\!\!CHCO_2H$$

A novel way of preparing allenes is by treating the cyclopropane derivative formed from an alkene and bromoform and alkali with magnesium in ether (Doering *et al.*, 1958). Dibromomethylene is an intermediate (p. 170).

$$R_2C\!\!=\!\!CR_2+:CBr_2 \longrightarrow R_2C\!\!\underset{\underset{CBr_2}{\diagdown\;\diagup}}{\rule{1.5em}{0.4pt}}\!\!CR_2 \xrightarrow{Mg} R_2C\!\!=\!\!C\!\!=\!\!CR_2$$

An extended allene type of linkage gives the *cumulene* system, the simplest member of which is **butatriene,** and this has been prepared by debrominating 1,4-dibromobut-2-yne with zinc (Schubert *et al.*, 1952, 1954):

$$BrCH_2C\!\!\equiv\!\!CCH_2Br \xrightarrow{Zn} CH_2\!\!=\!\!C\!\!=\!\!C\!\!=\!\!CH_2$$

Allenes are very important from the stereochemical point of view (see p. 498).

3. Hydrocarbons with conjugated double bonds contain single and double bonds arranged *alternately, i.e.,* they contain the arrangement $-C\!=\!CH\!\!-\!\!CH\!\!=\!\!CH\!\!-\!\!CH\!\!=\!\!CH\!-$. The simplest member of this group of compounds is buta-1,3-diene, which may be prepared by passing cyclohexene over a heated nichrome wire (an alloy of nickel, chromium and iron):

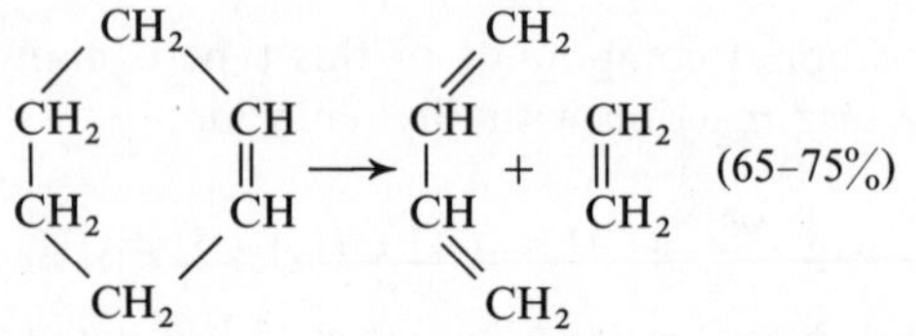

$$\begin{array}{ccccc} & CH_2 & & CH_2 & \\ CH_2 & CH & CH & CH_2 & \\ CH_2 & \parallel & \longrightarrow & | & + & \parallel & (65-75\%) \\ CH_2 & CH & CH & CH_2 & \\ & CH_2 & & CH_2 & \end{array}$$

Butadiene is prepared technically in various ways, *e.g.* (i) By dehydrogenating n-butane (from natural gas or petroleum gas) over a heated chromium oxide-alumina catalyst. (ii) By dehydrogenating but-1-ene and but-2-ene (from cracked petroleum) over a heated mixture of nickel and calcium phosphates (plus a little chromic oxide) as catalyst.

Butadiene is a gas, b.p. $-4 \cdot 4°C$. Under the influence of sodium as catalyst, butadiene readily polymerises to a product which has been used as a rubber substitute known as *buna* (*bu*tadiene + *Na*).

A very important alkadiene is **isoprene** or **2-methylbut-1,3-diene,** $CH_2{=}CMeCH{=}CH_2$, which may be obtained, in poor yield, by the slow distillation of rubber.

Isoprene is manufactured in various ways, *e.g.* (i) By passing isopentane or isopentene over heated Cr_2O_3—Al_2O_3. (ii) Propene is dimerised by heating at 200°C under pressure in the presence of tri-isopropylaluminium as catalyst, the dimer, 2-methylpent-1-ene, then isomerised to 2-methylpent-2-ene by heating to about 300°C in the presence of a catalyst, and this compound, mixed with steam and a small amount of hydrogen bromide, is heated to about 800°C:

$$2CH_2{=}CHCH_3 \longrightarrow CH_2{=}C(CH_3)CH_2CH_2CH_3 \longrightarrow$$
$$CH_3C(CH_3){=}CHCH_2CH_3 \longrightarrow CH_2{=}C(CH_3)CH{=}CH_2 + CH_4$$

Isoprene is a liquid, b.p. 35°C, and when heated with sodium at 60°C it polymerises to a substance resembling natural rubber.

$$CH_2{=}C(CH_3)—CH{=}CH_2 + Na\cdot \longrightarrow \overset{+}{Na}\overset{..}{C}H_2—C(CH_3){=}CH—CH_2\cdot$$
$$\Big\downarrow CH_2{=}C(CH_3)CH{=}CH_2$$

$$\overset{+}{Na}\overset{..}{C}H_2—C(CH_3){=}CH—CH_2—CH_2—C(CH_3){=}CH—CH_2\cdot, \text{etc.}$$

Compounds containing conjugated double bonds have physical and chemical properties that are not usually shown by compounds containing isolated or cumulated double bonds, *e.g.*, they undergo abnormal addition reactions, readily polymerise, and undergo the Diels–Alder reaction (p. 536). Butadiene also exhibits a special case of the Diels–Alder reaction in that it dimerises to 4-vinylcyclohexene, etc. (see p. 916).

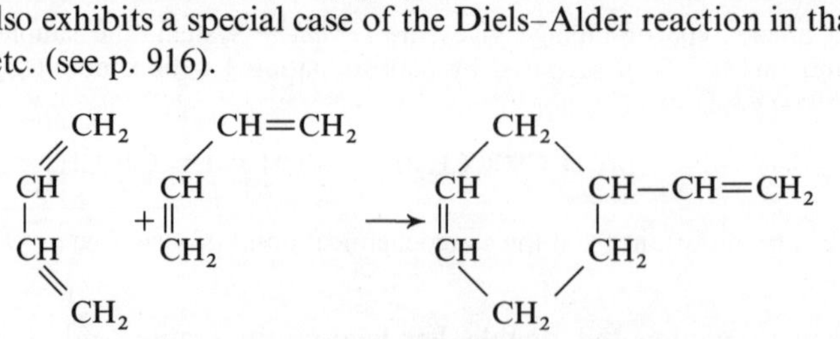

$$\begin{array}{ccccc} CH_2 & CH{=}CH_2 & & CH_2 & \\ CH & CH & CH & CH—CH{=}CH_2 \\ | & +\parallel & \longrightarrow & \parallel & | \\ CH & CH_2 & CH & CH_2 \\ CH_2 & & & CH_2 & \end{array}$$

This compound regenerates butadiene on heating. Another typical reaction of conjugated dienes is their combination with sulphur dioxide to form a cyclic sulphone, *e.g.*,

$$CH_2{=}CH{-}CH{=}CH_2 + SO_2 \longrightarrow \begin{matrix} CH{-}CH_2 \\ \parallel \qquad\qquad SO_2 \\ CH{-}CH_2 \end{matrix}$$

sulpholene

Although an isolated double bond is not usually reduced by dissolving metals, conjugated systems are, *e.g.*, isoprene and sodium in liquid ammonia:

$$CH_2{=}CMeCH{=}CH_2 + 2e \xrightarrow{2Na} [\bar{C}H_2{-}CMe{=}CH{-}\bar{C}H_2] \xrightarrow{2NH_3}$$
$$CH_3CMe{=}CHCH_3 + 2NH_2{}^-$$

Conjugated systems also show optical exaltation, and their absorption maximum in the infra-red ($C{=}C$ stretch) is shifted from 1680 to 1620 cm^{-1} for an isolated double bond to 1660–1580 cm^{-1} (s).

Thiele's Theory of Partial Valencies. Conjugated compounds undergo abnormal addition reactions, *e.g.*, when butadiene is treated with bromine (one molecule), two dibromo-derivatives are obtained, the 'expected' 3,4-dibromobut-1-ene (1,2-addition), and the 'unexpected' 1,4-dibromobut-2-ene (1,4-addition):

$$CH_2{=}CH{-}CH{=}CH_2 \xrightarrow{Br_2} BrCH_2CHBrCH{=}CH_2 + BrCH_2CH{=}CHCH_2Br$$

It has been found that 1,2- and 1,4-additions usually take place together, and the relative amount of each generally depends on the nature of the addendum and the conditions of the experiment, *e.g.*, type of solvent, temperature.

Thiele (1899) suggested his theory of *partial valencies* to account for 1,4-addition. According to Thiele, a single bond is sufficient to hold two carbon atoms together, and the two valencies of the double bond are not used completely to link the two carbon atoms, but only one valency and *part* of the other, leaving a surplus on each carbon atom. Thiele called this surplus valency the *residual* or *partial* valency, and if we represent it by a broken line, the formula of butadiene (and similarly for any other conjugated compound) may be written $CH_2{-}CH{-}CH{-}CH_2$. Thiele thought that the two middle partial valencies mutually satisfied each other rather than remain free. Thus the actual state of butadiene is

$$CH_2{-}CH{-}CH{-}CH_2 \text{ or } CH_2{-}CH{-}CH{-}CH_2.$$

The ends of this molecule are therefore the most active parts, and so addition of, *e.g.*, bromine will occur at these ends, first by attachment through the partial valencies, and then by each bromine atom acquiring a full valency, causing the two middle carbon atoms to utilise completely the two valencies left:

$$CH_2{-}CH{-}CH{-}CH_2 + Br_2 \longrightarrow \underset{Br}{CH_2}{-}CH{-}CH{-}\underset{Br}{CH_2} \longrightarrow \underset{Br}{CH_2}{-}CH{=}CH{-}\underset{Br}{CH_2}$$

Thiele's theory explains 1,4-addition so well that it does not account at all for 1,2-addition!

One explanation for 1,2- and 1,4-additions is obtained from a consideration of the stabilities of the various carbonium ion intermediates (and assumes electrophilic addition as for mono-alkenes).

$$CH_2{=}CH{-}CH{=}CH_2 \xrightarrow{Cl_2} CH_2{=}CH{-}\overset{+}{C}H{-}CH_2Cl \text{ or } CH_2{=}CH{-}CHCl{-}\overset{+}{C}H_2$$
(A)   (B)

(A) is a secondary and (B) is a primary carbonium ion, and since (A) is more stable than (B), the former is preferred. (A) is also a resonance hybrid of (A) and (A') and may be represented as (C). This extended conjugation (of the allyl group) is absent in (B), and consequently (A) is further stabilised with respect to (B) and hence (A) is formed, *i.e.*, the *first* step involves addition

$$CH_2{=}CH{-}\overset{+}{C}H{-}CH_2Cl \longleftrightarrow \overset{+}{C}H_2{-}CH{=}CH{-}CH_2Cl; \; \overset{\delta +}{CH_2}{=\!=\!=}CH{=\!=\!=}\overset{\delta +}{CH}{-}CH_2Cl$$
$$\text{(A)} \qquad\qquad\qquad \text{(A')} \qquad\qquad\qquad \text{(C)}$$

at the *terminal* carbon atom. The single representation of the resonance hybrid, (C), shows the presence of two positive centres. An ion such as this, *i.e.*, a cation with two positively charged centres, is called an *ambident cation*. Because of these two positive centres, addition of the chloride ion in the second step can occur at either centre to give 1,2- and 1,4-dichlorides:

$$\overset{+}{CH_2}{=\!=\!=}CH{=\!=\!=}\overset{+}{CH}{-}CH_2Cl + Cl^- \longrightarrow CH_2{=}CHCHClCH_2Cl + ClCH_2CH{=}CHCH_2Cl$$

However, (A) will contribute more than (A') to (C), *i.e.*, the resonance hybrid will carry more positive charge on the 'inner' carbon atom than on the terminal one. Hence, 1,2-addition will be expected to predominate over 1,4-, *i.e.*, the 1,2-dichloride is the kinetically controlled product (provided the two products are not intervonvertible under the conditions of the experiment; see p. 49). This has been shown to be so in practice (see also below).

An alternative explanation for the course of addition of chlorine to butadiene proposes that attack occurs at a terminal carbon atom because this becomes electron-rich through the $+E$ effect of the vinyl group:

$$CH_2{=}CH{-}CH{=}CH_2 \;\; Cl{-}Cl \longrightarrow CH_2{=}CH{-}\overset{+}{CH}{-}CH_2Cl + Cl^-$$

The argument to explain the formation of both 1,2- and 1,4- dichlorides is the same as above, but differs in that it does not predict which product predominates. This problem is solved by consideration of the relative stabilities of the products. If the conditions permit interconversion, then the thermodynamically controlled product will predominate. If the conditions do not permit interconversion, then the kinetically controlled product will predominate (p. 49). Consider the following:

$$-C{=}C{-}C{-}X$$

In this type of structure, halogen hyperconjugation is possible (p. 332), and this makes the molecule more stable than one in which such hyperconjugation is not possible. Inspection of the two butadiene dichlorides shows that the 1,2-product has one chlorine-hyperconjugated system whereas the 1,4-product has two. Thus, the latter is more stable and so should be the thermodynamically controlled product if the two forms are interconvertible, and consequently will be the predominant product under these conditions. These predictions have been observed experimentally, *e.g.*, Muskat *et al.* (1930) treated butadiene with chlorine under conditions where the two isomers were not interconvertible, and obtained about 60 per cent of the 1,2- and 40 per cent of the 1,4-product. This implies that the 1,2-compound is the kinetically controlled product. Pudovic (1949) heated each isomer at 200°C and obtained the same equilibrium mixture containing about 30 per cent 1,2- and 70 per cent 1,4-. Thus, the 1,4-isomer is the thermodynamically controlled product, and predominates under conditions of interconvertibility.

The addition of halogen acid to butadiene also produces two products, the 1,2- and the 1,4-. Furthermore, since the proton adds on first, this adds always to the *terminal* carbon, the halogen then adding at position 2 or 4. Hydrogen bromide adds to butadiene in the absence of oxygen to form a mixture of 3-bromobut-1-ene (80 per cent) and 1-bromobut-2-ene. In the presence of air or peroxides, these two products are again formed, but now the yields are reversed (Kharasch *et al.*, 1936; *cf.* the peroxide effect, p. 106).

In the foregoing account of the reactions of butadiene, we have assumed that the molecule has the structure $CH_2=CH—CH=CH_2$. There are, however, alternative electronic structures which are *charged*. Hence butadiene is a resonance hybrid of a number of resonating structures:

$$CH_2=CH—CH=CH_2 \longleftrightarrow CH_2=CH—\overset{+}{C}H—\overset{..}{C}H_2 \longleftrightarrow \overset{+}{C}H_2—CH=CH—\overset{..}{C}H_2, \text{ etc.}$$

There is still, however, another contributing structure of butadiene, *viz.*, (I). The two electrons have antiparallel spins, since the number of unpaired electrons in each resonating structure must be the same (p. 35). Thus these paired electrons would, in the ordinary way, form a covalent bond. The distance between them, however, is too great for them for form an *effective* bond. Consequently this bond is referred to as a *formal bond*, and may be represented by a dotted line (II). Because of the 'long bond' in (II), this will make very little contribution to the resonance hybrid. Since the charged structures have, apart from the charges, fewer covalent bonds than (III), they will contribute relatively little (*cf.* p. 38). Thus, the most important contribution is made by (III) and hence the resonance energy can be expected to be small. Actually, calculation has shown it to be about $14 \cdot 64 \, \text{kJ mol}^{-1}$. From this it follows that butadiene is best represented

$$\overset{.}{C}H_2—CH=CH—\overset{.}{C}H_2 \qquad CH_2\overset{\cdots\cdots\cdots}{—}CH=CH—CH_2 \qquad CH_2=CH—CH=CH_2$$
$$\text{(I)} \qquad\qquad\qquad\qquad \text{(II)} \qquad\qquad\qquad\qquad \text{(III)}$$

as a resonance hybrid whose structure is very close to that of the classical one. This means that the two terminal 'double' bonds have a very small amount of single bond character and that the two terminal 'double' bonds have a very small amount of single bond character and that the central 'single' bond has a very small amount of double bond character (*cf.* Thiele's formula; see also below).

So far we have considered the structure of conjugated compounds from V.B. theory. When we consider their structure from M.O. theory, we get a different picture. Each carbon atom in butadiene has the trigonal arrangement, and Fig. 4.3 (*a*) shows the p_z electrons associated with each carbon atom. If the molecule is planar, the p_z electron of C_2 overlaps that of C_1 as much as it does that of C_3, etc. Therefore, all four p_z orbitals can be treated as forming an *M.O. covering all four carbon atoms* (*b*). In this condition, a pair of electrons are no longer mainly confined to the region between two nuclei, *i.e.*, the bond formed is no longer a localised bond. The bonds produced are therefore called *delocalised* bonds. The wave equation for butadiene may therefore be written as: $\psi = c_1\psi_1 + c_2\psi_2 + c_3\psi_3 + c_4\psi_4$,

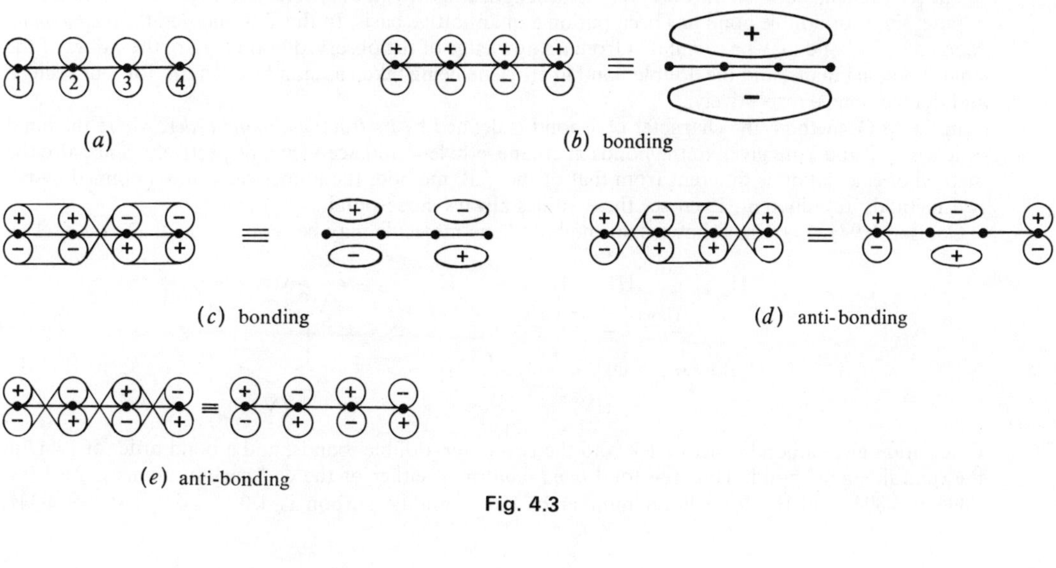

Fig. 4.3

where ψ_1 to ψ_4 represent the wave functions of the four p_z orbitals and c_1 to c_4 are their respective coefficients which indicate the individual contributions of the orbitals to the molecular orbital. This equation has four solutions, the values of the coefficients for each M.O. being: (b) 0·37, 0·60, 0·60, 0·37; (c) 0·60, 0·37, −0·37, −0·60; (d) 0·60, −0·37, −0·37, 0·60; (e) 0·37, −0·60, 0·60, −0·37 (b to e correspond to Figs 3b to 3e; the more conventional diagrams are also given). In these diagrams the values of the coefficients have been ignored; only their signs are used, and it should be remembered that change in sign means that there is a node between the adjacent pair of atoms (this is indicated where the lines cross; see also p. 42). As pointed out on p. 48, as the number of nodes in an orbital increases, so does the energy associated with that orbital. Furthermore, according to the Pauli exclusion principle, no more than *two* electrons can occupy the same M.O. Therefore in the ground state of butadiene, two of the π-electrons will occupy the M.O. in (b), and the other two the M.O. with the next higher energy level, *i.e.*, (c). In any excited state of butadiene, electrons will occupy orbitals (d) or (e).

In delocalised bonds, the electrons have greater freedom of movement than in localised bonds. Thus the total energy of the system is lowered, *i.e.*, delocalisation of bonds makes the molecule more stable. This energy of stabilisation is also called the *delocalisation energy* (Coulson), but it is also usual to call it the resonance energy. It should here be noted that *delocalisation of bonds* is in M.O. theory what *resonance* is in V.B. theory. Furthermore, since delocalisation of electrons stabilises a system, the localisation of electrons will make a system less stable (see, *e.g.*, acids and bases, p. 242, and aromatic substitution, Ch. 20).

It can be seen from the foregoing discussion that the M.O. treatment of conjugated systems does away with the idea of 'bonds' between atoms (this applies to the π-bonds, and not the to σ-bonds). Also the term *conjugation* is used in M.O. theory to indicate the existence in any part of a molecule of *molecular orbitals which embrace three or more nuclei.*

(T)   (C)

If we consider the problem of rotation about the central 'single' bond, then there are only two conformations in which all four p-orbitals are parallel (and therefore give maximum overlap). (T) is the *s-trans* (*transoid*) form and (C) the *s-cis* (*cisoid*); s refers to the fact that there is geometrical isomerism with respect to a *single* bond. According to Sheppard *et al.* (1962), butadiene exists only in the *transoid* form. This is the more stable conformation owing to greater steric repulsion (between the two H) in (C) than in (T).

The relationship between the observed bond-length and the value expected on the assumption that it is a 'pure' single or double bond has been put on a quantitative basis. In the V.B. method the *double-bond character* of a bond may be calculated from a knowledge of the observed bond length, the values of the single bond in ethane, and the double bond in ethylene being taken as standard lengths for 'pure' single and double bonds, respectively.

In the M.O. method, the character of a bond is defined by its *fractional bond order*, where the bond orders of 1, 2 and 3 are given to the bonds in ethane, ethylene and acetylene, respectively. Since also the method of calculation is different from that of the V.B. method, the numerical values obtained by the two methods are different. Even so, these values always correspond.

Coulson (1941, 1947) has shown that the butadiene molecule may be represented as shown in (IV).

(IV)   (V)

Calculation gives a bond order of 1·894 to the two outer 'double' bonds, and a bond order of 1·447 to the central 'single' bond. Thus the total *bond number* of either of the end carbon atoms is $2 \times 1·0 + 1·894 = 3·894$, and the total bond number of either middle carbon is $1·0 + 1·447 + 1·894 = 4·341$.

Furthermore, since calculation has shown that the maximum bond number for a carbon atom is 4·732, it follows that each carbon atom in butadiene has the 'free valency' shown in (IV) (Coulson has suggested that free valency be represented by an arrow). On the other hand, if a structure containing fractional double bonds is written with single bonds labelled with the bond order, and charges are placed on the atoms, then the resulting diagram is known as a *molecular diagram*, e.g., the molecular diagram for benzene is (V).

More recently, it has been suggested that the relative stability of butadiene is not due to resonance stabilisation but is due to the variations in bond energies that arise from differences in the hybrid state of carbon (see p. 139). Table 4.1 gives some *experimental* bond lengths, and bond-energy values *calculated* by Bloor and Gartside (1959).

Table 4.1

Bond type	Bond length (Å)	Bond energy kJ mol^{-1}	(kcal mol^{-1})
$C(sp^3)$—$C(sp^3)$	1·54	346·3	(82.76)
$C(sp^3)$—$C(sp^2)$	1·50	357·6	(85·48)
$C(sp^3)$—$C(sp)$	1·46	382·5	(91·42)
$C(sp^2)$—$C(sp^2)$	1·48	383·2	(91·58)
$C(sp^2)$—$C(sp)$	1·43	403·7	(96·48)
$C(sp)$—$C(sp)$	1·38	433·5	(103·6)
$C(sp^3)$—H	1·11	412·8	(98·67)
$C(sp^2)$—H	1·10	412·9	(98·69)
$C(sp)$—H	1·08	428·4	(102·38)
$C(sp^2)$=$C(sp^2)$	1·34	598·7	(143·1)

Alkynes or acetylenes

The alkynes are unsaturated hydrocarbons that contain one triple bond. They have the general formula C_nH_{2n-2} and the triple bond is known as the 'acetylenic bond'. Many alkynes have been found in nature.

Nomenclature

One method is to name higher homologues as derivatives of acetylene, the first member of the series. In the I.U.P.A.C. system of nomenclature the class suffix is *-yne*, and the rules for numbering are as for the alkenes (p. 93), *e.g.*,

CH≡CH	CH_3C≡CH	$(CH_3)_2CHC$≡CCH_3
ethyne	propyne	4-methylpent-2-yne
(acetylene)	(methylacetylene)	(isopropylmethylacetylene)

Univalent groups have the endings ynyl, *e.g.*, ethynyl (CH≡C—), propynyl (CH_3C≡C—), prop-2-ynyl (CH≡CCH_2—).

Acetylene or **ethyne** is the most important member of this series, and it may be prepared by any of the following methods:

(1) By the action of water on calcium carbide:

$$CaC_2 + 2H_2O \longrightarrow C_2H_2 + Ca(OH)_2$$

This method of preparation is used industrially. Other industrial methods are: (i) The cracking of methane-ethane mixtures; (ii) Heating to 1000–1300°C a mixture of ethane or propane and steam; natural gas is also used as the starting material.

(2) By the action of ethanolic potassium hydroxide on ethylene dibromide. The reaction proceeds in two steps, and under suitable conditions the intermediate product vinyl bromide may be isolated:

$$BrCH_2CH_2Br + KOH \xrightarrow{\text{ethanol}} CH_2{=}CHBr + KBr + H_2O$$

$$CH_2{=}CHBr + KOH \xrightarrow{\text{ethanol}} CH{\equiv}CH + KBr + H_2O$$

Sodamide in liquid ammonia can be used instead of ethanolic potassium hydroxide, and the yields are usually better since there is less tendency to form by-products, e.g.,

$$-CHBrCHBr- + 2NaNH_2 \longrightarrow -C{\equiv}C- + 2NaBr + 2NH_3$$

(3) The following methods may also be used to prepare acetylene:

(a) By the action of ethanolic potassium hydroxide on ethylidene dichloride:

$$CH_3CHCl_2 \xrightarrow{\text{KOH}} CH_2{=}CHCl \xrightarrow{\text{KOH}} CH{\equiv}CH$$

(b) By heating a haloform with silver powder, e.g., iodoform:

$$2CHI_3 + 6Ag \longrightarrow C_2H_2 + 6AgI \quad (p)$$

(c) By the electrolysis of a concentrated solution of the sodium (or potassium) salt of maleic or fumaric acid (cf. Kolbe's method):

$$NaO_2CCH{=}CHCO_2Na \xrightarrow{2H_2O} C_2H_2 + 2CO_2 + 2NaOH + H_2 \quad (p)$$

Acetylene may be synthesised from its elements by striking an electric arc between carbon rods in an atmosphere of hydrogen. It is also formed by the incomplete combustion of hydrocarbons, e.g., when a bunsen burner 'strikes back'.

Acetylene is a colourless gas, b.p. $-84°C$, and has an ethereal smell when pure. It is sparingly soluble in water but readily soluble in acetone. When compressed or liquefied acetylene is explosive, but its solution under pressure (10 atm) in acetone adsorbed on some suitable porous material can be handled with safety. Acetylene burns with a luminous smoky flame (due to the high carbon content), and hence is used for lighting purposes. It is also used in the oxy-acetylene blow-pipe, a temperature above 3000°C being reached. Acetylene is used for the preparation of a large number of compounds, e.g., acetaldehyde, ethanol, acetic acid, etc. (see text).

The infra-red absorption region of acetylenic compounds depends on whether they contain acetylenic hydrogen or not. Thus, in $RC{\equiv}CH$, there is one region 3310–3300 cm^{-1} (m) due to the $C{\equiv}H$ stretch, and another region, 2140–2100 cm^{-1} (w), due to the $C{\equiv}C$ stretch. In compounds $RC{\equiv}CR$, the region for the $C{\equiv}C$ group is 2260–2190 cm^{-1} (v).

On p. 139 it is shown that the electronegativities of the carbon atoms in the following compounds are in the order $C_2H_2 > C_2H_4 > C_2H_6$. It therefore follows that the chemical shifts of the protons should also be in this order. In actual practice, the order is $C_2H_4(\delta \sim 5 \text{ p.p.m.}) > C_2H_2(\delta \sim 2.5) > C_2H_6(\delta \sim 0.9)$. This can be explained on the assumption that a *shielding* effect is operating in acetylene, and is very much smaller in ethylene. The cause of this shielding effect is believed to be as follows. If acetylene is placed in a magnetic field with its molecular axis parallel to the field (Fig. 4.4a), the π-electrons circulate in the annular π-molecular orbital (cf. Fig. 4.5b), thereby producing an induced field in opposition to the applied field. Thus, protons *in line* with the triple bond are *shielded*, resulting in a decreased chemical shift (increased τ-value). Protons which lie *above* or *below* the bonding line, however, are *deshielded*. The overall result is that there are cones within which shielding is experienced, and outside which deshielding is experienced by protons. This is represented by Fig. 4.4(*b*), *shielding* being indicated by a *positive* sign, and *deshielding* by a *negative* sign.

If the molecular axis of acetylene is perpendicular to the applied field (Fig. 4.4*c*), no π-electron circulation is produced, and consequently no shielding or deshielding occurs. The effect of these π-electron circulations, averaged out over all possible orientations in the applied field, must therefore produce some shielding (in line with the bond) of acetylenic protons, and this results in a relatively high τ-value (~ 7.5).

In a similar way, a double bond is associated with shielding and deshielding (Figs. 4.4*d* and 4*e*), but in this case the induced current is produced only when the molecular axis lies *perpendicular* to the applied field. Also, shielding and deshielding effects are weaker for a double bond than for a triple bond (hence the higher τ-value for the latter). In both types of multiple bonds, double and triple, the magnitude of the induced magnetic field depends on the angle of the molecular axis with respect to the applied magnetic field. Because of this, compounds with multiple bonds are said to be *magnetically anisotropic*.

In addition to this shielding effect in acetylenes, there is also the existence of long range coupling, *e.g.*,

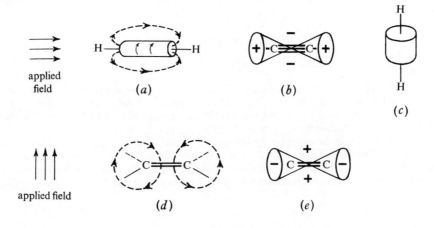

applied field

(*a*)   (*b*)   (*c*)

applied field

(*d*)   (*e*)

Fig. 4.4

in methylacetylene, the methyl group gives a doublet and the acetylenic hydrogen a quartet. This coupling extends over *four* bonds (which includes one triple bond); coupling is absent or negligible over four *single* bonds. This coupling also extends over four bonds when one is a double bond. In both cases, *J* is small (1–2 Hz).

Owing to the presence of a triple bond, acetylene is more unsaturated than ethylene, and forms addition products with two or four univalent atoms or groups, never one or three (*cf.* ethylene). A triple bond consists of one σ-bond and two π-bonds; this produces a *linear* molecule. When two univalent atoms add on to a triple bond the digonal arrangement changes into the trigonal, and the further addition of two univalent atoms changes the trigonal into the tetrahedral arrangement. Under suitable conditions it is possible to isolate the intermediate alkene.

Acetylene is *less* reactive than ethylene towards electrophilic reagents. This is unexpected in view of the fact that the π-electron density in a triple bond is higher than that in a double bond. The reason for this behaviour is not clear. Arguments based on bond strengths would lead to the conclusion that acetylene should be more reactive than ethylene: $C{\equiv}C$ (803·3) $\longrightarrow C{=}C$ (606·7); $\Delta H = +196·6$ kJ mol^{-1}; $C{=}C$ (606·7) $\longrightarrow C{-}C$ (347·3); $\Delta H = +259·4$ kJ mol^{-1}.

This order of reactivity is observed for catalytic hydrogenation (see Reaction 1, below). The mechanism of this reaction, however, does not involve electrophiles. A possible contributing factor for decreased reactivity of a triple bond to electrophiles may be the fact that the bridged carbonium ion from acetylene (assuming its formation) is more strained than the bridged carbonium ion from ethylene (the broken lines represent the bonds actually present in alkenes and alkanes):

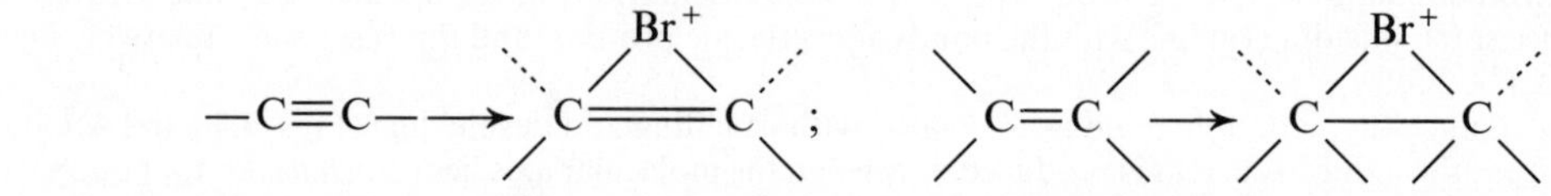

Justification for the assumption of these intermediate bridged ions being involved in additions to acetylene by the polar mechanism lies in the fact that the additions are stereoselectively *trans* (*cf.* ethylene, p. 102).

Reactions of acetylene

1. Acetylene adds on hydrogen in the presence of a catalyst, the reaction proceeding in two stages:

$$C_2H_2 \xrightarrow[\text{cat.}]{H_2} C_2H_4 \xrightarrow[\text{cat.}]{H_2} C_2H_6$$

Most catalysts effect hydrogenation of a triple bond, and it is possible, by using suitable conditions, to isolate the intermediate alkene. One method uses **Lindlar's catalyst** (1952), which consists of a Pd—CaCO$_3$ catalyst partially poisoned with lead acetate (see p. 178). A better catalyst is Pd—BaSO$_4$ partially poisoned with quinoline (Cram *et al.*, 1956; see also Table 6.1).

Dialkylacetylenes may be catalytically reduced to a mixture of *cis*- and *trans*-alkenes, the former predominating. On the other hand, reduction with sodium in liquid ammonia produces the *trans*-alkene, which is also produced by reduction with lithium aluminium hydride (Slaugh, 1966). *cis*-Reduction may be carried out with diborane (p. 114), or with di-isobutylaluminium hydride (Wilke *et al.*, 1960).

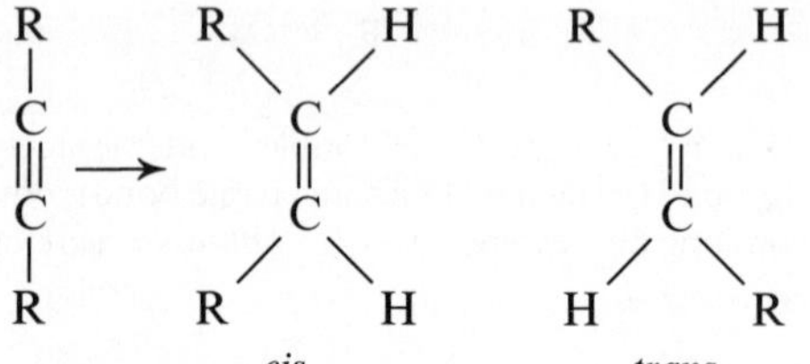

2. Acetylene adds on gaseous chlorine or bromine in the dark to form acetylene di- and tetra-halides; the addition is catalysed by light and metallic halides (*cf.* alkenes):

$$C_2H_2 \xrightarrow{Cl_2} C_2H_2Cl_2 \xrightarrow{Cl_2} C_2H_2Cl_4$$

Direct combination of acetylene with chlorine may be accompanied by explosions, but this is prevented by the presence of a metal chloride catalyst.

Acetylene reacts with dilute bromine water to produce acetylene dibromide:

$$CH\equiv CH + Br_2 \xrightarrow{H_2O} CHBr=CHBr$$

With liquid bromine and in the absence of a solvent, acetylene forms acetylene tetrabromide:

$$C_2H_2 + 2Br_2 \longrightarrow C_2H_2Br_4$$

Acetylene adds on iodine with difficulty, but if the reaction is carried out in ethanolic solution, acetylene di-iodide is obtained:

$$CH{\equiv}CH + I_2 \xrightarrow{\text{ethanol}} CHI{=}CHI$$

The addition of halogens to acetylene is stereoselective; the predominant product is the *trans* isomer (*cf.* the alkenes).

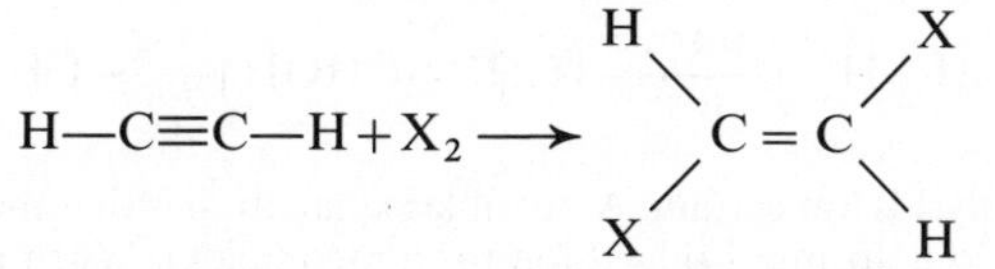

$$H{-}C{\equiv}C{-}H + X_2 \longrightarrow$$

Acetylene also undergoes substitution with halogen provided the right conditions are used, *e.g.*, dichloroacetylene is formed when acetylene is passed into sodium hypochlorite solution at 0°C in the absence of air and light:

$$H{-}C{\equiv}C{-}H \quad O{-}Cl \rightleftharpoons H{-}C{\equiv}C \quad Cl{-}O{-}H \longrightarrow OH^- + H{-}C{\equiv}C{-}Cl \xrightarrow{\text{OCl}^-} Cl{-}C{\equiv}C{-}Cl$$

Since ethylene does not behave in a similar manner, it is reasonable to assume that it is the greater acidity of acetylene that enables this reaction to occur (see p. 139).

Similarly, if acetylene is passed into a solution of iodine in liquid ammonia, di-iodoacetylene is formed (Vaughan and Nieuland, 1932):

$$C_2H_2 + 2I_2 + 2NH_3 \longrightarrow C_2I_2 + 2NH_4I$$

3. Acetylene can add on the halogen acids, their order of reactivity being HI > HBr > HCl > HF; HF adds on only under pressure (*cf.* ethylene). The addition of the halogen acids can take place in the dark, but is catalysed by light or metallic halides. The addition is in accordance with Markownikoff's rule, *e.g.*, acetylene combines with hydrogen bromide to form first vinyl bromide, and then ethylidene dibromide:

$$CH{\equiv}CH + HBr \longrightarrow CH_2{=}CHBr \xrightarrow{\text{HBr}} CH_3CHBr_2$$

Peroxides have the same effect on the addition of hydrogen bromide to acetylene as they have on alkenes (p. 106).

The mechanism of the addition of halogens and halogen acids is the same as that for the alkenes, *e.g.*, the addition of hydrogen bromide

$$CH{\equiv}CH \quad H{-}Br \longrightarrow Br^- \; \overset{+}{C}H{=}CH_2 \longrightarrow CHBr{=}CH_2$$

Addition of another molecule of hydrogen bromide could give either the intermediate carbonium ion, $CH_3\overset{+}{C}HBr$ (a secondary ion), or $\overset{+}{C}H_2CH_2Br$ (a primary ion). Since the former is more stable than the latter, the reaction proceeds via the former to form ethylidene dibromide. Thus:

$$Br{-}H \quad CH_2{=}CHBr \longrightarrow CH_3\overset{+}{C}HBr \; \overset{-}{Br} \longrightarrow CH_3CHBr_2$$

Because of the $-I$ effect of the bromine atom, the availability of the π-electrons in the first step is decreased and the addition is much slower than that for ethylene.

$$\overset{+}{C}H_2-\overset{-}{C}HBr \xleftarrow{-I} CH_2{=}CH{-}Br \xleftrightarrow{+R} \overset{-}{C}H_2-CH{=}\overset{+}{B}r$$

It might also be noted here that the contribution of the charged structure to the resonance hybrid, vinyl bromide, is not sufficient to give the bromine atom a positive charge. The direction of the dipole is still $\overrightarrow{C{-}Br}$, but its value is less than that in ethyl bromide. This observation, however, can also be explained in terms of the sp^2 hybridisation of the carbon atom ($={C}HBr$); its electronegativity is greater than that of an sp^3 carbon atom (p. 139).

4. When passed into dilute sulphuric acid at 60°C in the presence of mercuric sulphate as catalyst, acetylene adds on one molecule of water to form acetaldehyde. The mechanism of this hydration probably takes place via the formation of vinyl alcohol as an intermediate (*cf.* p. 108).

$$CH{\equiv}CH+H_2O \xrightarrow[Hg^{2+}]{H_2SO_4} [CH_2{=}CHOH] \longrightarrow CH_3CHO$$

The function of the catalyst is not certain. A possible explanation is that the mercuric ion, because of its large size, can fairly readily form a bridged ion (or a π-complex). When this is formed, reaction then proceeds as follows:

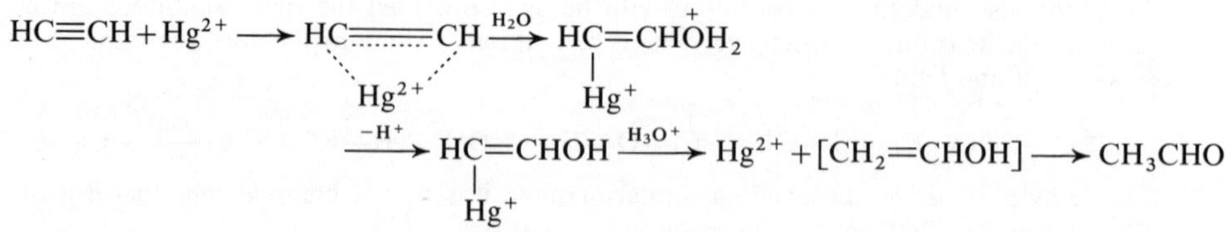

$$HC{\equiv}CH+Hg^{2+} \longrightarrow \underset{Hg^{2+}}{HC{=\!=\!=}CH} \xrightarrow{H_2O} \underset{Hg^+}{HC{=}\overset{+}{C}HOH_2}$$

$$\xrightarrow{-H^+} \underset{Hg^+}{HC{=}CHOH} \xrightarrow{H_3O^+} Hg^{2+}+[CH_2{=}CHOH] \longrightarrow CH_3CHO$$

The homologues of acetylene form ketones when hydrated, *e.g.*, propyne gives acetone:

$$CH_3C{\equiv}CH+H_2O \xrightarrow[Hg^{2+}]{H_2SO_4} [CH_3C(OH){=}CH_2] \longrightarrow CH_3COCH_3$$

5. When acetylene is passed into dilute hydrochloric acid at 65°C in the presence of mercuric ions as catalyst, vinyl chloride is formed:

$$CH{\equiv}CH+HCl \xrightarrow{Hg^{2+}} CH_2{=}CHCl$$

Acetylene adds on hydrogen cyanide in the presence of cuprous chloride in hydrochloric acid as catalyst to form vinyl cyanide:

$$CH{\equiv}CH+HCN \longrightarrow CH_2{=}CHCN$$

Vinyl cyanide is manufactured by passing a mixture of propene, ammonia, steam and air over a catalyst, *e.g.*, a mixture of oxides of molybdenum, cobalt and aluminium.

$$2CH_2{=}CHCH_3+2NH_3+3O_2 \longrightarrow 3CH_2{=}CHCN+6H_2O$$

Vinyl cyanide is used in the manufacture of *Buna N* synthetic rubber, which is a copolymer of vinyl cyanide and butadiene.

When acetylene is passed into warm acetic acid in the presence of mercuric ions as catalyst, vinyl acetate and ethylidene diacetate are formed:

$$CH{\equiv}CH+CH_3CO_2H \xrightarrow{Hg^{2+}} CH_2{=}CHOOCCH_3$$

$$CH_2{=}CHOOCCH_3+CH_3CO_2H \xrightarrow{Hg^{2+}} CH_3CH(OOCCH_3)_2$$

Vinyl acetate is manufactured by passing a mixture of acetylene and acetic acid vapour over zinc acetate on charcoal at about 170°C, or by the reaction between ethylene, palladium chloride and sodium acetate in aqueous solution in the presence of cupric chloride

$$CH_2{=}CH_2 + PdCl_2 + 2CH_3CO_2Na \xrightarrow{Cu^{2+}} CH_2{=}CHOOCCH_3 + 2NaCl + Pd + CH_3CO_2H$$

Vinyl acetate (liquid) is used in the plastics industry. Ethylidene diacetate (liquid), when heated rapidly to 300–400°C, gives acetic anhydride and acetaldehyde.

Acetylene reacts with nitric acid in the presence of mercuric ions to form nitroform, $CH(NO_3)_3$, and combines with arsenic trichloride to form Lewisite (p. 440). When acetylene is passed into methanol at 160–200°C in the presence of a small amount (1–2 per cent) of potassium methoxide and under pressure just high enough to prevent boiling, methyl vinyl ether is formed. The mechanism is believed to involve nucleophilic attack in the first step.

$$CH{\equiv}CH + CH_3O^- \longrightarrow \bar{C}H{=}CHOCH_3 \xrightarrow{CH_3OH} CH_2{=}CHOCH_3 + CH_3O^-$$

Methyl vinyl ether is used for making the polyvinyl ether plastics.

This process whereby acetylene adds on to compounds containing an active hydrogen atom (p. 425) to form vinyl compounds is known as **vinylation**.

Acetylene and formaldehyde interact in the presence of sodium alkoxide as catalyst to form butynediol, together with smaller amounts of propargyl alcohol:

$$HC{\equiv}CH + CH_3\bar{O} \rightleftharpoons HC{\equiv}C^- + CH_3OH$$

$$O{=}CH_2 \quad {}^-C{\equiv}CH \longrightarrow {}^-O{-}CH_2{-}C{\equiv}CH \xrightarrow{CH_3OH}$$

$$HO{-}CH_2{-}C{\equiv}CH \xrightarrow[CH_2O]{CH_3O^-} HOCH_2C{\equiv}CCH_2OH$$

This reaction in which acetylene (or any compound containing the $\equiv$CH group, *i.e.*, a methyne hydrogen atom) adds on to certain unsaturated links (such as in the carbonyl group), or eliminates a molecule of water by reaction with certain hydroxy-compounds, is known as **ethinylation**. Thus the above reactions with formaldehyde are examples of ethinylation; another example is the following:

$$R_2NCH_2OH + CH{\equiv}CH \longrightarrow R_2NCH_2C{\equiv}CH + H_2O$$

Alkyl bromo- and chloro-methyl ethers (p. 430) add to acetylenes in the presence of the corresponding aluminium halide to form alkenes (Bindácz *et al.*, 1960); *e.g.*,

$$EtOCH_2Cl + CH{\equiv}CH \xrightarrow{AlCl_3} EtOCH_2CH{=}CHCl$$

Under the same conditions acyl chlorides behave similarly:

$$RCOCl + CH{\equiv}CH \xrightarrow{AlCl_3} RCOCH{=}CHCl$$

6. When acetylene is passed into hypochlorous acid solution, dichloroacetaldehyde is formed:

$$CH{\equiv}CH + HOCl \longrightarrow [CHCl{=}CHOH] \xrightarrow{HOCl} [Cl_2CHCH(OH)_2] \longrightarrow Cl_2CHCHO + H_2O$$

Dichloroacetic acid, Cl_2CHCO_2H, is also formed by the oxidation of dichloroacetaldehyde by the hypochlorous acid.

7. Acetylene and its homologues form ozonides with ozone, and these compounds are decomposed by water to form diketones, which are then oxidised to acids by the hydrogen peroxide formed in the reaction:

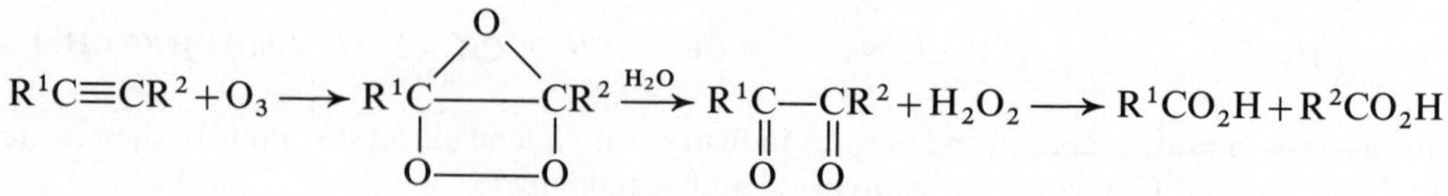

$$R^1C\equiv CR^2+O_3 \longrightarrow R^1C\underset{\underset{O-\!\!-\!\!-O}{|\qquad|}}{\overset{\overset{O}{\diagup\,\diagdown}}{-\!\!-\!\!-}}CR^2 \xrightarrow{H_2O} R^1C\underset{\overset{\|}{O}}{-}CR^2+H_2O_2 \longrightarrow R^1CO_2H+R^2CO_2H$$

Acetylene is exceptional in that it gives glyoxal as well as formic acid (Hurd and Christ, 1936):

$$CH\equiv CH \xrightarrow[\text{(ii) } H_2O]{\text{(i) } O_3} OCHCHO+HCO_2H$$

Since oxidation (ozonide formation being one example) is very much slower with alkynes than with alkenes (*cf.* p. 133), by choosing a suitable oxidising agent, a triple bond may be left intact in enynes, *e.g.*,

$$R^1C\equiv C(CH_2)_nCH=CHR^2 \xrightarrow[\text{AcOH}]{CrO_3} R^1C\equiv C(CH_2)_nCO_2H+R^2CO_2H$$

Hydrogen peroxide-formic acid mixture also preferentially hydroxylates the double bond in enynes. On the other hand, permanganate attacks a triple bond, *e.g.*,

$$CH_3(CH_2)_7C\equiv C(CH_2)_7CO_2H \xrightarrow{KMnO_4} CH_3(CH_2)_7COCO(CH_2)_7CO_2H$$

8. When passed through a heated tube acetylene polymerises, to a small extent, to benzene.

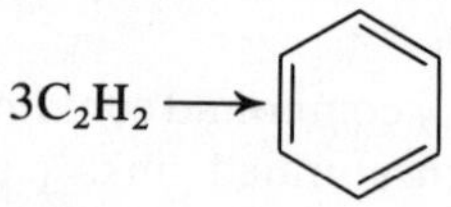

$$3C_2H_2 \longrightarrow \text{(benzene)}$$

Homologues of acetylene behave in a similar manner, *e.g.*, propyne polymerises to 1,3,5-trimethylbenzene, and but-2-yne to hexamethylbenzene:

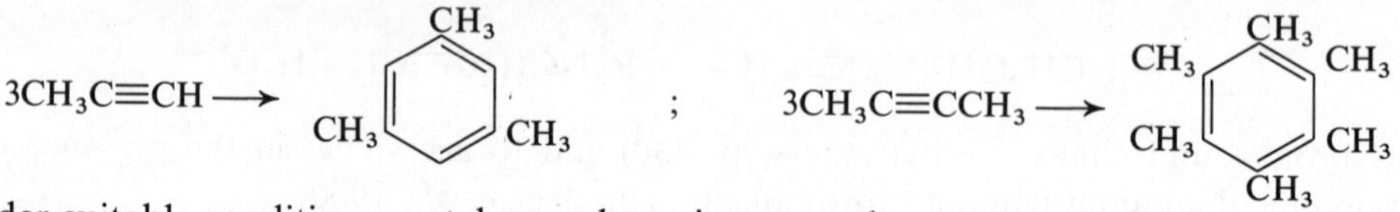

$$3CH_3C\equiv CH \longrightarrow \text{(1,3,5-trimethylbenzene)} \quad ; \quad 3CH_3C\equiv CCH_3 \longrightarrow \text{(hexamethylbenzene)}$$

Under suitable conditions acetylene polymerises to cyclooctatetraene (*q.v.*):

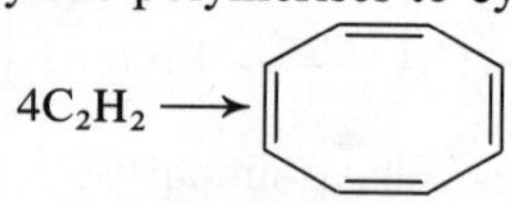

$$4C_2H_2 \longrightarrow \text{(cyclooctatetraene)}$$

In addition to the above type of *cyclic* polymerisation, acetylene undergoes *linear* polymerisation when passed into a solution of cuprous chloride in ammonium chloride to give vinylacetylene and divinylacetylene:

$$CH\equiv CH+CH\equiv CH \longrightarrow CH_2=CH-C\equiv CH \xrightarrow{C_2H_2} CH_2=CH-C\equiv C-CH=CH_2$$

Compounds containing both a double and triple bond are named systematically as *alkenynes*. The double bond is always expressed first in the name, and numbers as low as possible are given to the double and triple bonds, even though this may give 'yne' the lower number. Thus vinylacetylene is but-1-en-3-yne, and divinylacetylene is hexa-1,5-dien-3-yne. The following compound, $CH_3CH{=}CHC{\equiv}CH$, however, is pent-3-en-1-yne.

Chloroprene (2-chlorobuta-1,3-diene; b.p. 60°C) is manufactured by passing vinylacetylene into concentrated hydrochloric acid in the presence of cuprous and ammonium chlorides. There is a great deal of evidence to show that the reaction proceeds by 1,4-addition followed by rearrangement under the influence of the catalyst.

$$CH_2{=}CH{-}C{\equiv}CH + HCl \longrightarrow [CH_2Cl{-}CH{=}C{=}CH_2] \longrightarrow CH_2{=}CCl{-}CH{=}CH_2$$

Chloroprene readily polymerises to a rubber-like substance known as *neoprene*.

9. Acetylene forms metallic derivatives by replacement of one or both hydrogen atoms, *e.g.*, if acetylene is passed over heated sodium, both the monosodium and disodium acetylides are formed:

$$CH{\equiv}CH \xrightarrow{\text{Na}} CH{\equiv}CNa \xrightarrow{\text{Na}} NaC{\equiv}CNa$$

By using a large excess of acetylene the main product is monosodium acetylide, which is also obtained by passing acetylene into a solution of sodium in liquid ammonia until the blue colour disappears. By treatment of a fine dispersion of sodium in xylene at 100–105°C with acetylene, sodium acetylide can be obtained in 98 per cent yield (Rutledge, 1957). The monosodium derivative possesses the interesting property of being able to absorb dry carbon dioxide to form the sodium salt of propiolic acid (*q.v.*):

$$CH{\equiv}CNa + CO_2 \longrightarrow CH{\equiv}CCO_2Na$$

The alkali metal acetylides react with carbonyl compounds to form acetylenic alcohols:

$$\diagdown CO + NaC{\equiv}CR \longrightarrow \diagdown C(OH)C{\equiv}CR$$

Acetylene also forms Grignard reagents by reaction with alkylmagnesium halides (p. 426).

When acetylene is passed into an ammoniacal solution of cuprous chloride or silver nitrate, cuprous acetylide Cu_2C_2 (red) or silver acetylide Ag_2C_2 (white) is precipitated. Both these compounds, when dry, explode when struck or heated. Acetylenic copper compounds undergo oxidative coupling (see p. 141).

The acidic nature of hydrogen in acetylene is characteristic of hydrogen in the group $\equiv CH$, and it has been suggested that this is because the $C{\equiv}H$ bond has considerable ionic character due to resonance.

$$H{-}C{\equiv}C{-}H \leftrightarrow H{-}C{\equiv}\overset{-}{\underset{+}{C}}H \leftrightarrow \overset{+}{\underset{-}{H}}C{\equiv}C{-}H \leftrightarrow \overset{+}{\underset{-}{H}}C{\equiv}\overset{-}{\underset{+}{C}}H$$

There is, however, evidence to show that the electronegativity of a carbon atom depends on the number of bonds by which it is joined to its neighbouring carbon atom (Walsh, 1947). Since π-electrons are more weakly bound than σ-electrons, the electron density round a carbon atom with π-bonds is less than that when only σ-bonds are present. Thus, a carbon atom having one π-bond has a slight positive charge compared with a carbon atom which has only σ-bonds. Hence, the electronegativity of an sp^2 hybridised carbon atom is greater than that of an sp^3 hybridised carbon atom. Similarly, a carbon atom which has two π-bonds carries a small positive charge which is greater than that carried by a carbon atom with only one π-bond. Thus the electronegativity of an sp hybridised carbon atom is greater than that of an sp^2 hybridised carbon atom. It therefore follows that the more s character a bond has, the more electronegative is that carbon atom. Thus the attraction for electrons by hybridised carbon will be $sp > sp^2 > sp^3$. Hence proton release (by heterolytic fission of the C—H bond) will occur most readily in acetylene, less readily in ethylene, and least readily in methane (*cf.* Table 4.1). Alternatively, since the greater the electronegativity of an atom, the more readily it can accommodate a negative charge, stabilities of the following carbanions follow the order: $CH{\equiv}C^- > CH_2{=}CH^- > CH_3^-$.

Acidities of hydrocarbons have been studied by determining the position of the following equilibrium in inert solvents:

$$R^{1-}Na^+ + R^2H \rightleftharpoons R^1H + R^{2-}Na^+$$

The more the equilibrium is displaced to the right, the stronger is R^2H as an acid. The position of equilibrium may be determined in several ways, *e.g.*, (i) By passing carbon dioxide through the solution; only the sodium carboxylate of the more acidic hydrocarbon is formed. Since carboxylation upsets the equilibrium, this method actually gives information on which is the stronger acid. (ii) By ultraviolet or visible spectroscopy; the anions usually differ from their parent hydrocarbons in these regions of the spectrum.

The structure of metallic carbides, *i.e.*, compounds formed between carbon and metals, is still a matter of dispute. It appears certain, so far, that the carbides of the strongly electropositive metals: Na, K, Ca, Sr, Ba, are ionic—X-ray crystal analysis has shown the lattice of these carbides to be ionic, containing the ion $\bar{C}{\equiv}\bar{C}$. These carbides react with water to produce acetylene. Copper and silver carbides are not affected by water, and are explosive when dry. On account of these differences it seems likely that the carbides of these two metals are covalent, *e.g.*, Cu—C≡C—Cu. Thus these compounds may be regarded as *acetylides*.

In addition to the above carbides, there are a number of carbides which react with water or dilute acids to produce methane, *e.g.*, aluminium carbide; or a mixture of hydrocarbons, *e.g.*, uranium carbide which gives acetylene and other unsaturated hydrocarbons; iron carbide which gives methane and hydrogen. Many authors believe these carbides to be ionic, and include carbides of copper and silver in this group.

There is also one other group of carbides which are highly refractory, and which are extremely stable chemically; *e.g.*, vanadium carbide is not attacked by water or hydrochloric acid even at 600°C. These carbides are believed to be *interstitial* compounds, *i.e.*, their lattice is not composed of ions, but resembles a metallic or atomic lattice.

Structure of acetylene. By reasoning similar to that used for ethylene, the structure of acetylene is shown to be H—C≡C—H, and may be represented as Fig. 4.5 (*a*). This representation, however, appears to be inadequate. According to Coulson (1952), the two π-bonds form a charge cloud which has cylindrical symmetry about the carbon-carbon axis (Fig. 4.5*b*; but see also p. 46).

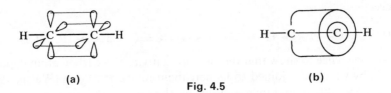

(a)   Fig. 4.5   (b)

Homologues of acetylene

Homologues of acetylene may be prepared by any of the following methods:
1. By the action of ethanolic potassium hydroxide on 1,2-dihalogen derivatives of the alkanes (*cf.* preparation of acetylene, method 2), *e.g.*, propyne from propene dibromide:

$$CH_3CHBrCH_2Br + 2KOH \xrightarrow{\text{ethanol}} CH_3C{\equiv}CH + 2KBr + 2H_2O$$

Sodamide in liquid ammonia usually gives better yields (see 3 below).

This method affords a simple means of introducing a triple bond into an organic compound, *e.g.*, n-butanol is catalytically dehydrated to but-1-ene, which on treatment with bromine gives 1,2-dibromo-butane, and this, when heated with ethanolic potassium hydroxide, yields but-1-yne:

$$CH_3CH_2CH_2CH_2OH \xrightarrow[350°C]{Al_2O_3} CH_3CH_2CH{=}CH_2 \xrightarrow{Br_2}$$

$$CH_3CH_2CHBrCH_2Br \xrightarrow[ethanol]{KOH} CH_3CH_2C{\equiv}CH$$

An interesting point in this connection is that open-chain 1,2-dihalides give alkynes ¬ather than con-jugated dienes on treatment with strong bases:

$$R^1CH{=}CHCH{=}CHR^2 \xleftarrow{\ \ \ //\ \ \ } R^1CH_2CHXCHXCH_2R^2 \xrightarrow{-2HX} R^1CH_2C{\equiv}CCH_2R^2$$

2. Monosodium acetylide (see reaction 9 of acetylene (p. 139)) is treated with an alkyl halide, preferably a bromide, whereupon an acetylene homologue is produced:

$$CH{\equiv}CH + Na \xrightarrow[NH_3]{liquid} CH{\equiv}CNa \xrightarrow{RX} CH{\equiv}CR + NaX$$

This reaction may be carried further as follows:

$$R^1C{\equiv}CH + Na \xrightarrow[NH_3]{liquid} R^1C{\equiv}CNa \xrightarrow{R^2X} R^1C{\equiv}CR^2 + NaX$$

In practice this method is limited to the use of primary alkyl halides, since higher s- and t-halides give mainly alkenes when they react with the monosodium derivatives of acetylene or its homologues.

3. By the action of acetylene on a Grignard reagent (*q.v.*) and then treating the resulting magnesium complex with an alkyl halide:

$$CH{\equiv}CH + R{-}Mg{-}Br \longrightarrow RH + CH{\equiv}C{-}Mg{-}Br \xrightarrow{RBr} CH{\equiv}CR$$

The properties of the homologues of acetylene are very similar to those of acetylene, particularly when they are of the type $RC{\equiv}CH$, *i.e.*, contain the $\equiv CH$ group. Acetylene homologues isomerise when heated with ethanolic potassium hydroxide, the triple bond moving towards the centre of the chain, *e.g.*, but-1-yne isomerises to but-2-yne:

$$CH_3CH_2C{\equiv}CH \longrightarrow [CH_3CH{=}C{=}CH_2] \longrightarrow CH_3C{\equiv}CCH_3$$

There is a great deal of evidence to show that an allene is formed as an intermediate.

On the other hand, when alkynes are heated with sodamide in an inert solvent, *e.g.*, paraffin, the triple bond moves towards the end of the chain; *e.g.*, but-2-yne gives the sodium derivative of but-1-yne, which is converted into but-1-yne by the action of dilute acid:

$$CH_3C{\equiv}CCH_3 + NaNH_2 \xrightarrow{paraffin} NH_3 + CH_3CH_2C{\equiv}CNa \xrightarrow{H^+} CH_3CH_2C{\equiv}CH$$

This reaction affords a means of stepping up the alkyne series by method 2 above.

Of the two ynes, it is the 2-isomer which is the thermodynamically controlled product, but in the presence of the strong base sodamide, the formation of the *insoluble sodium salt* of the 1-isomer shifts the equilibrium in the direction of this product (*cf.* p. 125).

A very useful acetylene derivative for synthetic work is ethoxyacetylene. This may conveniently be prepared by the action of sodamide on chloroacetaldehyde diethyl acetal (Jones *et al.*, 1954):

$$ClCH_2CH(OC_2H_5)_2 \xrightarrow{NaNH_2} CH{\equiv}COC_2H_5$$

Lithium derivatives of alkynes can be prepared by interaction of the alkyne and lithium amide in dioxan (Schlubach *et al.*, 1958).

Cuprous acetylides undergo oxidative coupling when heated in acetic acid in the presence of air (Weedon *et al.*, 1958):

$$2RC{\equiv}CCu \xrightarrow{O_2} RC{\equiv}C{-}C{\equiv}CR$$

Eglington *et al.* (1959) have used cupric acetate in pyridine instead of oxygen (see p. 561).

A variation of the method has been introduced by Chodkiewicz (1957) who used a 1-alkyne and the bromo derivative of a 1-alkyne (prepared by the action of sodium hybromite on a 1-alkyne; p. 135). The method has also been extended to 'crossed' oxidative coupling, e.g.,

$$MeCH{=}CMeC{\equiv}CBr + CH{\equiv}C(CH_2)_8CO_2H \xrightarrow{Cu^+} MeCH = CMeC{\equiv}C{-}C{\equiv}C(CH_2)_8CO_2H$$

$$(98\%)$$

PROBLEMS

1. Write out the structure (ignoring stereochemistry) of the isomeric pentenes, and name them by the IUPAC system. Give the structures of the products formed from each on ozonolysis.

2. Draw the energy profiles for the reactions:

(i) $MeCH{=}CH_2 + HX \longrightarrow MeCH_2CH_2^+ + X^-$

(ii) $MeCH{=}CH_2 + HX \longrightarrow Me\overset{+}{C}HMe + X^-$

Give reasons for your answers.

3. Using the Board synthesis, how would you prepare:

(i) but-2-ene;

(ii) hex-3-ene?

4. Draw only the geometrical isomeric forms of: (i) n-pentenes; (ii) isopentenes; (iii) monobromopropenes.

5. Can you compare the stabilities of pent-1-ene, cis- and trans-pent-2-enes and 3-methylbut-1-ene by measuring their heats of hydrogenation? If not, why not? What method could you use?

6. Arrange the following carbonium ions in order of increasing stability, and give your reasons.

$$Me\overset{+}{C}HMe \qquad CH_2{=}CHCH_2^+ \qquad MeCH_2CH_2^+$$

7. What alkenes could you expect from the following compounds? Give the experimental conditions:

(i) Me_3CCl;

(ii) $Me_2C(OH)CMe_3$;

(iii) Me_2CHCH_2OAC;

(iv) $MeCH_2CH(OAc)Me$.

8. Addition of HCl to 3,3-dimethylbut-1-ene gives two isomeric alkyl chlorides. Explain.

9. Using bond energies (Table 2.2, p. 00), estimate $\Delta H^{\ominus}$ for the following reactions:

(i) $MeCH{=}CH_2 + HBr \longrightarrow MeCH_2CH_2Br$

(ii) $MeCH{=}CH_2 + HBr \longrightarrow MeCHBrMe$

What would you predict from the results? Are these predictions correct? If not, why not? Under what experimental conditions are the equations correct?

10. Arrange the following alkenes in order of increasing stability, and give reasons for your answer.

$$CH_2{=}CH_2 \qquad CH_2{=}CMe_2 \qquad MeCH{=}CHMe$$
$$MeCH{=}CH_2 \qquad Me_2C{=}CHMe \qquad Me_2C{=}CMe_2$$

11. Draw the energy profiles for: (i) the hydrogenation of ethylene with and without a catalyst; (ii) the catalytic hydrogenation of cis- and trans-but-2-enes.

12. Isobutene, in the presence of H_2SO_4, forms a mixture of two isomeric alkenes, C_8H_{16}. Formulate the reaction. How could you distinguish between the isomers chemically?

13. Formulate the reactions between but-1-ene, in the presence of a small amount of benzoyl peroxide, and (i) CCl_4; (ii) $CBrCl_3$. Give your reasons.

14. Draw the energy profile for the addition of HI to propene. Explain your diagram.

15. Compound A (C, 88·9%; H, 11·1%), M.W. 54, on catalytic hydrogenation, gives B (C, 82·8%; H, 17·2%). The i.r. spectrum of A shows a band in the region 2 240 cm^{-1}. What is A?

16. Discuss the mechanisms of the following reactions:

(i) $MeCH{=}CH_2 + HOBr \longrightarrow MeCHOHCH_2Br$

(ii) $CCl_3CH{=}CH_2 + HOBr \longrightarrow CCl_3CHBrCH_2OH$

17. $\Delta G_f^\ominus$ for *cis*-but-2-ene is $65 \cdot 86 \, kJ \, mol^{-1}$ and that for *trans*-but-2-ene is $62 \cdot 97 \, kJ \, mol^{-1}$. What is the value of K at 25°C for the equilibrium mixture? Does either isomer form this equilibrium mixture at 25°C? Discuss.

18. Explain the course of the following reaction:

$$CH_2{=}CHCHOHCH{=}CHMe \xrightarrow[H_2SO_4]{MeOH} CH_2{=}CHCH{=}CHCH(OMe)Me$$

19. Compound A (C, 87·1%; H, 12·8%), on catalytic reduction, gives B (C, 84·1%; H, 15·9%). Ozonolysis of A gives acetic acid, acetone, and pyruvic acid ($MeCOCO_2H$). What are A and B?

20. Write out the structures of the isomeric hexynes, and name them by the I.U.P.A.C. system.

21. Show how you could distinguish between ethane, ethylene, and acetylene by (i) chemical methods; (ii) i.r. spectroscopy; (iii) NMR spectroscopy.

22. Complete the following equations:

$$CH_3COCH_3 \xrightarrow{NaNH_2} ? \qquad \xrightarrow{C_2H_2} ? \qquad \xrightarrow{H^+} (CH_3)_2C(OH)C{\equiv}CH \xrightarrow[Pd]{1H_2} ? \qquad \xrightarrow[400°C]{Al_2O_3} ?$$

23. What are the geometries of:

(i) but-1-yne; (ii) but-2-yne; (iii) hept-2-en-5-yne?

24. Complete the following equations, and comment where necessary:

(i) $MeCH_2CH_2CMe{=}CHMe \xrightarrow{SeO_2} ?$

(ii) $MeCH_2C{\equiv}CBr + CH{\equiv}CMe \xrightarrow{Cu^+} ?$

(iii) $Me_2CHCH{=}CH_2 + HBr \longrightarrow ?$

(iv) $MeCH_2C{\equiv}CH \xrightarrow{Na/NH_3} ? \xrightarrow{EtBr} ?$

(v) $BrCH_2CH_2Br \xrightarrow{?} C_2H_2$

(vi) $MeCH_2CH_2OH \xrightarrow{?} MeCH{=}CH_2$

(vii) $MeCHBrCH_2Br \xrightarrow[MeOH]{Zn} ?$

(viii) 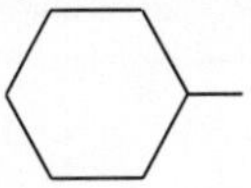 $C{\equiv}CH \xrightarrow[H_2SO_4]{Hg^{2+}} ?$

25. Some of the mass spectra peaks given by the three methylpent-1-enes, A, B, and C are listed in the table. What are A, B, and C? Give reasons for your answer.

m/e	27	28	29	41	42	43	55	56	57	69	84
A (%)	48	7	40	79	21	7·5	100	39	—	71	30
B (%)	35	8	28	80	29	8	43	100	44	36	31
C (%)	33	4	6	72	33	100	8	44	20	13	11

26. How many NMR signals would you expect for the following compounds? If a signal is other than a singlet indicate this by (m):

(i)

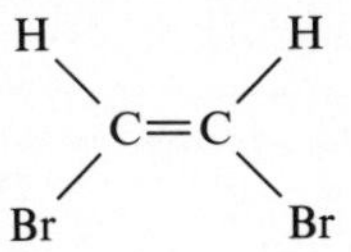

(ii)

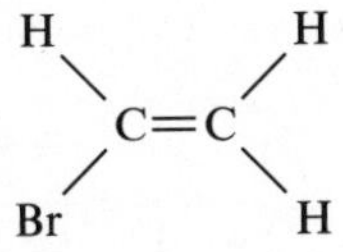

(iii) $CH{\equiv}CCH_2Br$

(iv) $CH_2{=}C{=}CH_2$

(v) $CH_3C{\equiv}CH$

(vi)

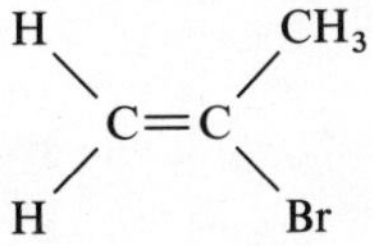

REFERENCES

Organic Reactions. Wiley, Vol. XII (1962), Ch. 2. 'The Preparation of Olefins by the Pyrolysis of Esters.'

BROWN, *Hydroboration*, Benjamin (1962).

PATAI (ed.), *The Chemistry of the Alkenes*, Interscience (1964).

DE LA MARE and BOLTON, *Electrophilic Additions to Unsaturated Systems*, Elsevier (1966).

HOUSE, *Modern Synthetic Reactions*, Benjamin (1965).

KIRMSE, *Carbene Chemistry*, Academic Press (1964).

HINE, *Divalent Carbon*, Ronald Press (1964).

LEDWITH, 'The Chemistry of Carbenes', *Roy. Inst. Chem., Lecture Series*, No. 4 (1964).

MILLER, 'Hydrogenation with Di-imide', *J. Chem. Educ.*, 1965, **42,** 254.

TURNEY, *Oxidation Mechanisms*, Butterworths (1965).

SKLARZ, 'Organic Chemistry of Periodates', *Quart. Rev.*, 1967, **21,** 3.

LEE and UFF, 'Organic Reactions Involving Electrophilic Oxygen', *Quart. Rev.,* 1967, **21,** 429.

GUNSTONE, 'Uses of Ozone in Organic Chemistry', *Educ. in Chem.*, 1968, **5,** 166.

5 Halogen derivatives of the alkanes

Halogen derivatives of the alkanes are divided into mono-, di-, tri-, etc., substitution products according to the number of halogen atoms in the molecule.

Nomenclature.

Monohalogen derivatives are usually named as the halide of the corresponding alkyl group, *e.g.*, C_2H_5Cl ethyl chloride; $CH_3CHBrCH_3$ isopropyl bromide; $(CH_3)_3CCl$ t-butyl chloride.

 Dihalogen derivatives. (i) When both halogen atoms are attached to the *same* carbon atom, they are said to be in the *geminal* (*gem-*) position. Since the loss of two hydrogen atoms from the same carbon atom gives the *alkylidene* group, *gem*-dihalides are named as the alkylidene dihalides, *e.g.*, CH_3CHBr_2 ethylidene dibromide; $CH_3CCl_2CH_3$ isopropylidene dichloride.

 (ii) When the two halogen atoms are on *adjacent* carbon atoms they are said to be in the *vicinal* (*vic-*) position, and these dihalides are named as the dihalide of the alkene from which they may be prepared by the addition of halogen, *e.g.*, CH_2ClCH_2Cl ethylene dichloride; $(CH_3)_2CBrCH_2Br$ isobutene dibromide.

 (iii) When there is a halogen atom on each of the *terminal* carbon atoms of the chain, *i.e.*, in the α,ω-position, the compound is named as the polymethylene dihalide, *e.g.*, $CH_2ClCH_2CH_2CH_2Cl$ tetramethylene dichloride.

 (iv) When the two halogen atoms occupy positions other than those mentioned above, the compounds are named as dihalogen derivatives of the parent hydrocarbon, the positions of the halogen atoms being indicated by numbers (use principle of lowest numbers), *e.g.*,

$$CH_3CHClCH_2CHClCH_2CH_3 \qquad \text{2,4-dichlorohexane}$$

 The alkyl halides, *gem-*, *vic-* and α,ω-dihalides may also be named by this (I.U.P.A.C.) system, *e.g.*, $CH_3CH_2CHBrCH_3$ 2-bromobutane; $CH_2ClCH_2CH_2Cl$ 1,3-dichloropropane.

 Polyhalogen derivatives are best named by the I.U.P.A.C. system (method iv), and the names of the halogens (and any other substituents present) are arranged alphabetically:

$CH_2BrCHClCHClCH_3$ 1-bromo-2,3-dichlorobutane

$CH_2ClCHICH(CH_3)CH_2CH_2Br$ 5-bromo-1-chloro-2-iodo-3-methylpentane

Alkyl halides

The alkyl halides have the general formula $C_nH_{2n+1}X$ or RX, where X denotes chlorine, bromine or iodine. Fluorine is not included, since fluorides do not behave like the other halides.

General methods of preparation.

1. The method most widely used is the replacement of the hydroxyl group of an alcohol by an X atom. Halogen acid is the most common reagent used with alcohols, but the conditions depend on both the structure of the alcohol and the particular acid.

$$ROH + HX \xrightarrow{\text{heat}} RX + H_2O$$

Hydrogen chloride is used with zinc chloride (*Groves' process*) for primary and secondary alcohols, but tertiary alcohols readily react with concentrated hydrochloric acid in the absence of zinc chloride.

Zinc chloride is a Lewis acid and consequently can co-ordinate with the alcohol:

$$ROH + ZnCl_2 \longrightarrow R\overset{+}{\underset{\underset{H}{|}}{O}} - \bar{Z}nCl_2 \xrightarrow{S_N1} R^+ + [HO-ZnCl_2]^-$$

The R—O bond is weakened and so the complex readily forms a carbonium ion (S_N1 mechanism; see p. 155). If the nature of R^+ is such that it can undergo rearrangement, the product will be a mixture of isomeric alkyl chlorides (see also p. 181). When the concentration of zinc chloride is low, the reaction is still catalysed, but no rearrangement occurs (S_N2 mechanism).

$$ZnCl_2 + HCl + ROH \longrightarrow ZnCl_3^- + ROH_2^+ \xrightarrow{S_N2} ZnCl_2 + RCl + H_2O$$

Pyridine and dimethylamine also catalyse the reaction between alcohols and hydrochloric acid without rearrangement.

Constant boiling hydrobromic acid (48 per cent) is used for preparing alkyl bromides. The reaction is carried out in the presence of a little concentrated sulphuric acid (as catalyst) for primary alcohols, but not for secondary and tertiary alcohols, since these tend to be dehydrated to alkene. The presence of sulphuric acid does not cause any significant rearrangement in the products.

Constant boiling hydriodic acid (57 per cent) is used without sulphuric acid to prepare alkyl iodides. Alternatively, good yields of alkyl iodides may be obtained by heating alcohols (or ethers and alkenes) with sodium or potassium iodide in 95 per cent phosphoric acid (Stone *et al.*, 1950).

The formation of the various halides by rearrangement is one example of **1,2-shifts**. In this type of rearrangement a group migrates from a *carbon* atom (the *migration origin*) to an *adjacent* atom (the *migration terminus*), which is usually carbon or nitrogen. The rearrangement may be expressed in the following general form:

$$\overset{Z}{\underset{|}{A}}-B\overset{\frown}{-}\bar{Y} \rightarrow \bar{Y} + \overset{Z}{\underset{|}{A}}-\overset{+}{B} \rightarrow \overset{Z}{\underset{|}{A}}-B \xrightarrow[\text{reagent}]{\text{nucleophilic}} \text{products}$$

When Y ionises, B is left with only six valency electrons (open sextet), and consequently carries a positive charge. Z migrates *with its bonding pair of electrons*, and A now has an open sextet.

Attack now takes place at A by Y^- or some other nucleophilic reagent to form the products. This mechanism was first suggested by Whitmore (1932) and is often referred to as the *Whitmore mechanism*.

The nature of Z is quite varied; the 'key' atom (*i.e.*, the atom joined to the bonding pair which migrates) may be carbon, halogen, oxygen, etc. If Z has a lone pair of electrons, *e.g.*, halogen, then this pair may be used in the migration to give a cyclic intermediate, *i.e.*, a **bridged** or **non-classical carbonium ion.**

$$
\begin{array}{ccc}
Z & & \overset{+}{Z} \\
| & & \diagup\;\diagdown \\
A\!-\!B^+ & \longrightarrow & A\!\!-\!\!\!-\!\!\!-\!\!B \longrightarrow \text{products}
\end{array}
$$

If Z is an aryl group, then this group can supply an electron pair to form a bridged ion (see, *e.g.*, p. 230). If Z is an alkyl group, then there are no lone pairs or multiple bonds, but again it is believed that a bridged ion is possible, *i.e.*,

$$
\begin{array}{ccc}
R & & R \\
| \quad | & & \diagup \overset{+}{\cdots} \diagdown \\
-\!\overset{|}{C}\!-\!\overset{|}{C}{}^+ & \longrightarrow & \diagdown C\!\cdots\!\cdots\!C\diagup \\
| \quad | & & \diagup \qquad \diagdown
\end{array}
$$

This type of bridged ion contains three *partial* bonds formed from *one pair of electrons*. The formation of these bridged ions in 1,2-shifts is an example of *neighbouring group participation*, and when the rate is increased because of this effect, the rearrangement is said to be *anchimerically assisted* (Winstein *et al.*, 1953; see also p. 487).

A very important point about 1,2-shifts is that they are *intramolecular*, *i.e.*, Z is never actually free. If the mechanism were such that Z was free during its migration, then the rearrangement would be called *intermolecular*. The intramolecular nature of the 1,2-shifts has been established by using a group Z which contains an asymmetric carbon atom attached to A, *i.e.*, have type *abc*C—A. If Z actually separated during the migration, then it means that the free migrating unit is a carbanion, *i.e.*, *abc*C:⁻. Examination of such carbanions has shown that none ever retains its configuration; it racemises, *i.e.*, half the molecules have their configuration inverted, and so the product is no longer optically active. Thus if Z moves with its bonding pair and the configuration is *retained* in the product, then Z can never have been free, *i.e.*, the rearrangement is intramolecular.

We can now illustrate the foregoing discussion with neopentyl alcohol as our example. Whitmore *et al.* (1932) showed that the reactions undergone by neopentyl alcohol and neopentyl halides are of two types: (i) replacement reactions, which occur very slowly (if at all) and produce neopentyl compounds; (ii) replacement (and elimination) reactions, which occur very fast and produce t-amyl compounds.

Hughes and Ingold (1946) have shown that the slow reactions take place by the S_N2 mechanism (p. 155) and consequently without rearrangement, *e.g.*, neopentyl bromide reacts with ethanolic sodium ethoxide to give ethyl neopentyl ether:

$$
\begin{array}{ccccc}
Me_3C & & Me_3C & & Me_3C \\
| & +OEt^- \rightleftharpoons & \underset{\delta-}{\;}| & \xrightarrow{-Br^-} & | \\
BrCH_2 & & Br\cdot\cdot CH_2\cdot\cdot OEt^{\delta-} & & CH_2OEt
\end{array}
$$

On the other hand, when neopentyl alcohol reacts with anhydrous hydrogen bromide, the main product is t-amyl bromide (72 per cent), together with primary and secondary bromides. The rearranged bromides are the result of 1,2-shifts (methyl and hydride):

$$Me_3CCH_2OH \underset{}{\overset{HBr}{\rightleftharpoons}} Me_3CCH_2\overset{+}{-}OH_2^+ \rightleftharpoons Me_3C-CH_2^+ + H_2O$$

$$Me_2CBrCH_2Me \xleftarrow{Br^-} Me_2\overset{+}{C}CH_2Me \xleftarrow{Me\ shift} Me_3CCH_2Br$$

$$Me_2CHCHBrMe \xleftarrow{Br^-} Me_2CH\overset{+}{C}HMe$$

(with H^- shift)

The 'driving force' of the rearrangement is probably due to the stabilities of carbonium ions being tertiary > secondary > primary.

Any alkyl halide may be prepared by the action of a phosphorus halide on the alcohol. Phosphorus pentachloride gives variable yields depending on the alcohol:

$$ROH + PCl_5 \longrightarrow RCl + HCl + POCl_3$$

Straight-chain primary alcohols react with phosphorus trihalides to give unrearranged alkyl halides but many secondary alcohols undergo rearrangement (see also p. 413).

$$PX_3 + 3ROH \longrightarrow (RO)_3P + 3HX \longrightarrow (RO)_3\overset{+}{P}H \xrightarrow[S_N2]{X^-} RX + (RO)_2P(H)O \xrightarrow[]{HX}$$

$$RX + (RO)P(H)(OH)O \xrightarrow[S_N2]{HX} RX + H_3PO_3$$

Thionyl chloride reacts with straight-chain primary alcohols without rearrangement in the presence or absence of pyridine (**Darzens procedure**, 1911); the reaction proceeds via a chloro-sulphite (see also p. 156).

$$ROH + SOCl_2 \longrightarrow HCl + ROSOCl \longrightarrow Cl^- + ROSO^+ \xrightarrow{S_N2} RCl + SO_2$$

2. By the addition of halogen acids to an alkene.

3. By direct halogenation, *i.e.*, substitution reactions of the alkanes with halogen, and this may be brought about by light, catalysts or heat.

Photohalogenation is carried out by treating the alkane with chlorine (or bromine) at ordinary temperature in the presence of light. The reaction takes place by a free-radical chain mechanism, the initiation being brought about by the formation of chlorine atoms by absorption of light:

$$Cl_2 \xrightarrow{h\nu} 2Cl\cdot$$

$$Cl\cdot + H-CH_3 \longrightarrow Cl\cdots H\cdots CH_3 \longrightarrow HCl + CH_3\cdot$$

$$CH_3\cdot + Cl-Cl \longrightarrow CH_3\cdots Cl\cdots Cl \longrightarrow CH_3Cl + Cl\cdot$$

$$Cl\cdot + H-CH_2Cl \longrightarrow Cl\cdots H\cdots CH_2Cl \longrightarrow HCl + CH_2Cl\cdot$$

$$ClCH_2\cdot + Cl-Cl \longrightarrow ClCH_2\cdots Cl\cdots Cl \longrightarrow CH_2Cl_2 + Cl\cdot; \quad etc.$$

The termination of the chain reaction may take place by adsorption of the chlorine atoms on the walls of the containing vessel, or by two chlorine atoms combining with each other to form a chlorine molecule. There appears to be no evidence to show that $CH_3\cdot$, $\cdot CH_2Cl$, etc., radicals combine to terminate the chain reaction, but another termination step is the combination of $R\cdot$ with $Cl\cdot$ (to give RCl).

Thermal halogenation. Thermal chlorination has been studied in great detail by Hass, McBee and their co-workers (1935 onwards), and as a result of their work they suggested a number of rules for chlorination:

(i) If high temperature is avoided, no carbon skeleton rearrangements occur in either thermal or photochemical chlorination.

(ii) Every possible monochloride is formed, and over-chlorination, *i.e.*, chlorination beyond mono-substitution, may be suppressed by controlling the ratio of chlorine to alkane (this rule also holds good for photochemical chlorination).

(iii) The order of ease of substitution is tertiary hydrogen > secondary > primary. At 300°C, with reaction in the vapour phase, the relative rates of substitution of primary, secondary and tertiary hydrogen atoms are $1\cdot00:3\cdot25:4\cdot43$.

(iv) As the temperature rises above 300°C the relative rates of substitution tend to become equal, *i.e.*, $1:1:1$. Increased pressure causes an increase in the relative rate of primary substitution.

(v) According to Tedder *et al.* (1960), the effect of the halogen atom already present in the molecule is to retard substitution at a β-carbon atom, and also affects substitution at an α-carbon atom.

It is of interest here to consider the method of calculation of rates of substitution. Let r_p, r_s and r_t be the *relative* rates of substitution of a primary, secondary and tertiary hydrogen atom, respectively. Hass found that the thermal chlorination of propane at 300°C gave 48 per cent of the 1-chloro- and 52 per cent of the 2-chloro-isomer. Since there are six H_p atoms, and statistically each one has an equal chance of being replaced, their total rate of replacement is $6r_p$. Similarly, for two H_s atoms, their total rate of replacement is $2r_s$. Since the amount of each isomer formed depends on its rate of formation, it therefore follows that

$$\frac{2r_s}{6r_p} = \frac{52}{48}; \qquad \therefore \ \frac{r_s}{r_p} = \frac{3\cdot25}{1}$$

Hass also found that chlorination of isobutane gave 67 per cent of the 1-chloro- and 33 per cent of the 2-chloro-isomer. Since isobutane contains $9H_p$ atoms and $1H_t$ atom,

$$\frac{r_t}{9r_p} = \frac{33}{67}; \qquad \therefore \ \frac{r_t}{r_p} = \frac{4\cdot43}{1}$$

Thus the relative rates of substitution of H_p, H_s and H_t are $1:3\cdot25:4\cdot43$ (these are the figures given in (iii) above).

With these values it is possible to calculate the proportions of the various monochloro isomers for any alkane, *e.g.*, isopentane.

	%	1	2	3	4
calc.		30	22	33	14
obs.		33·5	22	28	16·5

$$\overset{1}{C}H_3$$

$$\overset{2}{C}H\overset{3}{C}H_2\overset{4}{C}H_3$$

$$\overset{1}{C}H_3$$

There are $9H_p$, $2H_s$ and $1H_t$. Thus the total rate is:

$$9 \times 1 + 2 \times 3\cdot25 + 1 \times 4\cdot43 = 19\cdot93$$

$$\therefore \ r_p \times 100 = \frac{9 \times 100}{19\cdot93} = 45\%; \qquad r_s \times 100 = \frac{6\cdot5 \times 100}{19\cdot93} = 33\%;$$

$$r_t \times 100 = \ = \frac{4\cdot43 \times 100}{19\cdot93} = 15\%$$

Since there are two CH_3 groups labelled 1 and one CH_3 group labelled 4, the rate of formation of the 1-chloro- derivative will be twice that of the 4-chloro-. Thus 30 per cent will be the 1- and 15 per cent the 4-chloro-isomer. These calculated values compare favourably with the values observed by Hass (see also later).

Bromination takes place with greater difficulty than chlorination; and there is less tendency for polysubstitution. Furthermore, Roberts *et al.* (1952) have shown that bromination is more selective; bromine has a much greater tendency than chlorine to replace tertiary hydrogen in preference to secondary or primary. It is the study of reactions such as these where the more reactive reagent shows less selectivity that has given rise to the empirical *reactivity-selectivity principle* (p. 119).

Pease *et al.* (1931) investigated the thermal chlorination of methane and found that the reaction was retarded by the presence of oxygen. This is strong evidence for a free-radical chain reaction, and the authors proposed the following mechanism (*cf.* photohalogenation; see also below):

$$Cl_2 \xrightarrow{\text{heat}} Cl\cdot + Cl\cdot$$
$$Cl\cdot + CH_4 \longrightarrow HCl + CH_3\cdot, \qquad \text{etc.}$$

Estimations have shown that in thermal chlorination of some alkanes, one chlorine atom may propagate a chain reaction involving about 10^7 hydrocarbon molecules (see also p. 901).

Pease *et al.* also had proposed an alternative mechanism:

$$Cl\cdot + CH_4 \longrightarrow CH_3Cl + H\cdot$$
$$H\cdot + Cl_2 \longrightarrow HCl + Cl\cdot, \qquad \text{etc.}$$

When we consider the enthalpy changes in these two chain reactions, we find (using bond energies in Table 2.2, p. 36):

(i) (a) $Cl\cdot + Me\text{—}H\ (+414\cdot2) \longrightarrow H\text{—}Cl\ (-426\cdot8) + Me\cdot$ $\Delta H^\ominus = -12\cdot6$ kJ

    (b) $Me\cdot + Cl\text{—}Cl\ (242\cdot7) \longrightarrow Me\text{—}Cl\ (-326\cdot4) + Cl\cdot$ $\Delta H^\ominus = -83\cdot7$ kJ

(ii) (a) $Cl\cdot + Me\text{—}H\ (+414\cdot2) \longrightarrow Me\text{—}Cl\ (-326\cdot4) + H\cdot$ $\Delta H^\ominus = +87\cdot8$ kJ

    (b) $H\cdot + Cl\text{—}Cl\ (+242\cdot7) \longrightarrow H\text{—}Cl\ (-426\cdot8) + Cl\cdot$ $\Delta H^\ominus = -184\cdot1$ kJ

The overall enthalpy change for both mechanisms is $-96\cdot3$ kJ. This, of course, is to be expected, since the overall equation for both mechanisms (obtained by adding the steps of the reaction) is the same (Hess's law), *viz.*,

$$Cl\text{—}Cl\ (+242\cdot7) + Me\text{—}H\ (+414\cdot2) \longrightarrow Me\text{—}Cl\ (-326\cdot4) + H\text{—}Cl\ (-426\cdot8)$$

$$\Delta H^\ominus = -96\cdot3 \text{ kJ}$$

Now let us consider the entropy change for the overall reaction. $S^\ominus$ values (JK^{-1} mol^{-1}) are: Cl_2, $223\cdot0$; CH_4, $186\cdot2$; CH_3Cl, $234\cdot3$; HCl, $186\cdot6$.

$$\therefore \quad \Delta S^\ominus = (234\cdot3 + 186\cdot6) - (223\cdot0 + 186\cdot2) = 11\cdot7 \text{ kJ}$$

Hence there is a very small increase in entropy.

$$\Delta G^\ominus = \Delta H^\ominus - T\Delta S^\ominus = -96\cdot3 - (298 \times 11\cdot7/1000) = -96\cdot3 - 3\cdot49 = -99\cdot79 \text{ kJ}$$
$$\therefore \quad RT\ln K = 99\cdot79; \log K = 99\cdot79/2\cdot3\ RT = 99\cdot79/5\cdot7 = 17\cdot4; K = \sim 10^{17} \text{ at } 25°C.$$

Thus, the overall reaction (which is independent of mechanism) is thermodynamically highly favourable.

In the foregoing account, $\Delta H^\ominus$ has been *estimated* from bond energies. This usually gives a reasonably good estimate. Entropy changes have been evaluated from a knowledge of absolute entropy values of the reactants and products. Unfortunately, many of these values are not available. Since their estimation is very difficult, one can assume that the entropy changes are small at $25°C$ and so may be neglected in comparison with $\Delta H^\ominus$ values if these are reasonably high (say, 40 kJ or more). In this case, then, $\Delta H^\ominus$ may be taken as a measure of $\Delta G^\ominus$ (*cf.* p. 60).

Although the above reaction is thermodynamically highly favourable, it is immeasurably slow in the dark at room temperature. As we have seen, rates depend on activation energy values. Since the dissociation of chlorine into chlorine atoms is the first step in both mechanisms, energy (as heat or light) must be supplied from outside the system because the value required, $242\cdot7$ kJ, is far too high to be reached through thermal collision at room temperature. In mechanism (i), both (a) and (b) are exothermic, and so both are thermodynamically favourable. Calculation has shown that E for (a) is $16\cdot11$ kJ and for (b) is < 21 kJ. These low values are therefore readily reached. Reaction (iia) is strongly endothermic and so is thermodynamically highly unfavourable. Now, in an endothermic reaction, E must be at least equal to $\Delta H^\ominus$ (*cf.* Fig. 2.12). Thus a *minimum* value of E for (iia) is $87\cdot86$ kJ.

This step is therefore energetically highly unfavourable, and consequently, for a chain mechanism, (i) is involved rather than (ii). There is, however, one further point that should also be considered. It can be shown, from thermodynamic considerations, that the value of the equilibrium constant, K, decreases with rise in temperature for exothermic reactions and increases for endothermic reactions. This, of course, follows from Le Chatelier's principle. Thus, exothermic reactions become thermodynamically less favourable and endothermic reactions more favourable as the temperature rises, but in both cases the equilibrium position is reached much more quickly. Hence endothermic reactions can give reasonable yields at a fast rate when the temperature is raised sufficiently. From this it follows that at some particular temperature, mechanism (ii) could compete with mechanism (i).

It is also of interest to consider the other halogens, fluorine, bromine and iodine.

	F	Cl	Br	I
$X_2 \longrightarrow 2X\cdot$	$+150\cdot6$	$+242\cdot7$	$+188\cdot3$	$+150\cdot6$
(ia) $X\cdot + MeH \longrightarrow HX + Me\cdot$	$-146\cdot4$	$-12\cdot55$	$+50\cdot21$	$+117\cdot2$
(ib) $Me\cdot + X_2 \longrightarrow MeX + X\cdot$	$-334\cdot7$	$-83\cdot68$	$-96\cdot23$	$-62\cdot76$
$\Delta H^{\ominus}$ kJ (overall)	$-481\cdot1$	$-96\cdot23$	$-46\cdot02$	$+54\cdot44$

The rate of each step will depend on the relevant activation energy. E for fluorination (ia) is apparently less than that for chlorination, and because the reaction is so strongly exothermic, fluorination is often explosive. E for bromination (ia) is about $74\cdot48$ kJ, and so the rate is considerably slower than that of chlorination. E for iodination (ia) must be at least $117\cdot2$ kJ, and so the reaction will be much slower still. Also, because iodination (overall reaction) is thermodynamically unfavourable, the equilibrium will lie very much to the left.

If we consider the different C—H bond energies in various alkanes, we have (Table 2.2, p. 36): primary, $405\cdot8$; secondary, $393\cdot3$; tertiary, $374\cdot5$ kJ. The weaker the bond (that is broken), the lower is the necessary activation energy (p. 52), and so the rate of substitution will be t > s > p. For bromination, the activation energy is reached for t-hydrogen, and so this is replaced almost exclusively. It might also be noted that order of replacement, t > s > p, is what we might expect from the consideration of the stabilities of the free radicals produced in step (ia); (cf. p. 103). Finally, as the temperature rises above 300°C (in thermal chlorination), the chlorine and alkane molecules acquire so much energy that reaction occurs at every collision irrespective of the type of C—H bond. Under these conditions, the replacement rates will be the same (cf. (iv), p. 149).

4. Direct chlorination of alkanes may be effected by means of sulphuryl chloride. Sulphuryl chloride in the absence of light and catalysts does not react with alkanes even at their boiling points, but in the presence of light and a trace of an organic peroxide the reaction is fast. The mechanism is probably (cf. p. 116):

$$(R^1CO_2)_2 \longrightarrow 2R^1CO_2\cdot \longrightarrow 2R^1\cdot + 2CO_2$$
$$R^1\cdot + SO_2Cl_2 \longrightarrow R^1Cl + \cdot SO_2Cl \longrightarrow SO_2 + Cl\cdot$$
$$Cl\cdot + R^2H \longrightarrow HCl + R^2\cdot \xrightarrow{SO_2Cl_2} R^2Cl + \cdot SO_2Cl, \quad \text{etc.}$$

In practice, chlorination with sulphuryl chloride is best carried out in the presence of α,α-azobis-isobutyronitrile (this readily produces free radicals):

$$\underset{\underset{CN}{|}}{Me_2C} - N{=}N - \underset{\underset{CN}{|}}{CMe_2} \longrightarrow N_2 + 2Me_2C(CN)\cdot$$

$$Me_2C(CN)\cdot + SO_2Cl_2 \longrightarrow Me_2C(CN)Cl + \cdot SO_2Cl.$$

Sulphuryl chloride is less selective than chlorine, and Russell (1958) has shown this is due to the nature of the solvent (when one is used).

5. Hunsdiecker or **Borodine-Hunsdiecker reaction.** Hunsdiecker *et al.* (1935) found that silver salts of the carboxylic acids in carbon tetrachloride solution are decomposed by chlorine or bromine to form the alkyl halide, *e.g.*,

$$RCO_2Ag + Br_2 \longrightarrow RBr + CO_2 + AgBr$$

The yield of halide is primary > secondary > tertiary, and bromine is generally used, chlorine giving a poorer yield of alkyl chloride. The mechanism is uncertain, but a favoured theory is that the first step is the formation of an acyl hypohalite which then decomposes into free radicals. The details of the later steps are less certain:

$$RCO_2Ag + Br_2 \longrightarrow RCOOBr + AgBr$$
$$RCOOBr \longrightarrow Br\cdot + RCOO\cdot \longrightarrow R\cdot + CO_2$$
$$R\cdot + Br_2 \longrightarrow RBr + Br\cdot$$
$$R\cdot + RCOOBr \longrightarrow RBr + RCOO\cdot, \qquad etc.$$

Cristol *et al.* (1961) have obtained very good yields of alkyl bromide by adding bromine to a refluxing carbon tetrachloride solution of the acid in the presence of an excess of red mercuric oxide.

$$2RCO_2H + 2Br_2 + HgO \longrightarrow 2RBr + 2CO_2 + HgBr_2$$

Iodine forms esters with the silver salts; this is known as the **Birnbaum-Simonini reaction** (1892):

$$2RCO_2Ag + I_2 \longrightarrow RCO_2R + CO_2 + 2AgI$$

Cristol *et al.* (1964), however, have shown that acids form alkyl iodides if iodine and red mercuric oxide are used.

On the other hand, Bachman *et al.* (1963) have shown that aliphatic carboxylic acids may be converted into esters and alkyl halides as follows:

$$RCO_2H \longrightarrow (RCO_2)_4Pb \xrightarrow{X_2} RCO_2R + 2RX$$

Alkyl chlorides may be prepared without simultaneous ester formation by the action of lead tetra-acetate and lithium chloride on the carboxylic acid in benzene solution (Kochi, 1965). This reaction is applicable to primary, secondary and tertiary acids, and occurs without rearrangement, *e.g.*,

$$Me_3CCH_2CO_2H \longrightarrow Me_3CCH_2Cl$$

The reaction is believed to proceed by a free-radical mechanism.

6. Alkyl halides may be prepared by methods involving the use of phosphorus compounds (see p. 413).

General properties of the alkyl halides.

The lower members methyl chloride, methyl bromide and ethyl chloride are gases; methyl iodide and the majority of the higher members are sweet-smelling liquids. The order of values of the poiling points (and densities) of the alkyl halides is iodide > bromide > chloride > fluoride. In a group of isomeric alkyl halides, the order of the boiling points is primary > secondary > tertiary. Many of the alkyl halides burn with a green-edged flame. The chemical reactions of the alkyl halides are similar, but they are not equally reactive, the order of reactivity being iodide > bromide > chloride; the reactivity of alkyl fluorides depends on the nature of the fluoride (p. 174). Alkyl iodides are sufficiently reactive to be decomposed by light, the iodide darkening due to the liberation of iodine:

$$2RI \longrightarrow R-R + I_2$$

Alkyl halides (and the polyhalides) are covalent compounds, insoluble in water, with which they cannot form a hydrogen bond, but soluble in organic solvents.

The alkyl halides are classified as primary, secondary and tertiary, according as the halogen atom is present in the respective groups $-CH_2X$, $=CHX$ and $\equiv CX$, *i.e.*, according as the halogen atom is joined to a primary, secondary or tertiary carbon atom.

The absorption region of alkyl halides (C—X stretch) depends on the nature of the halogen atom: C—F, 1100–1000 cm^{-1} (s); C–Cl, 750–700 cm^{-1} (s); C—Br, 600–500 cm^{-1} (s); C—I, ~ 500 cm^{-1} (s).

NMR spectroscopy of alkyl halides has been previously discussed (p. 11; see also Table 3.2, p. 80). Here, we shall mention the problem of fluorine. Apart from being the most electronegative atom of the halogen group, fluorine has nuclear spin, with $I = +1/2$ and $-1/2$ (as for hydrogen). It can therefore spin-spin couple with protons and so give rise to splitting, *e.g.*, CH_3F gives a doublet for the methyl group (τ 5·70), whereas CH_3Cl gives a singlet (τ 6·98) and CH_3Br also a singlet (τ 7·35). The signal for fluorine is not observed under the conditions used for proton resonance, but with the conditions fulfilled, the fluorine spectrum will be a quartet.

Mass spectra. Loss of one electron from a lone pair on the halogen atom occurs most readily to form the molecular ion, and this then undergoes fragmentation in a way which depends on the nature of the halogen atom and on the nature of the alkyl group. Since Cl and Br exist as isotopes, the molecular ions appear as doublets of mass M and $M + 2$. However, because the abundance ratios are quite different (see Table 1.5, p. 20), alkyl chlorides and bromides may be readily distinguished. Some of the fragmentation patterns are:

$$RCH_2CH_2^+ + X \cdot \xleftarrow{\ Br, I\ } [RCH_2CH_2\text{—}X]^+ \xrightarrow{\ F, Cl\ } [RCH=CH_2]^+ + HX$$
$$\downarrow$$
$$RCH_2 \cdot + CH_2 = X^+$$

These are the predominant paths and the intensity of M^+ is in the order I > Br > Cl > F, and decreases with increase in the size of the alkyl group. The $RCH_2CH_2^+$ and $[RCH=CH_2]^+$ fragments undergo further fragmentation typical of alkanes and alkenes, respectively. Alkyl halides (predominantly for Cl and Br) containing six or more carbon atoms in a straight chain also fragment to give the ion $C_4H_8X^+$:

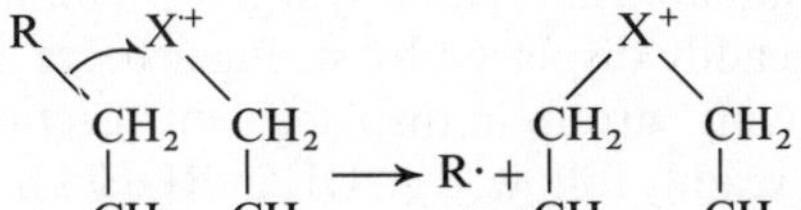

It has already been pointed out that alkyl groups have a $+I$ effect. Before considering the reactions of the alkyl halides, it is important to explain why an alkyl group is electron-repelling. Several explanations have been offered; the following is a highly favoured one. As we have seen, methane and ethane are non-polar (p. 84). Thus the methyl group has a zero inductive effect; and this is true for all alkyl groups since all alkanes, whether straight- or branched-chain, have a zero dipole moment. When, however, one hydrogen atom in an alkane is replaced by some polar atom (or group), the alkyl group now exerts a polar effect which is produced by the presence of the polar atom. Thus, in an alkyl halide, the alkyl group possesses an inductive effect, but it is one which is produced mainly by the mechanism of *interaction polarisation* (*cf.* p. 33). It therefore follows that the alkyl group is more polarisable than a hydrogen atom, and since most of the groups attached to the alkyl group are electron-attracting groups, the alkyl groups thus usually becomes an electron-repelling group, *i.e.*, alkyl groups normally have a $+I$ effect.

Optical isomerism. *Optical activity* is the name given to the phenomenon exhibited by compounds which, when placed in the path of a beam of plane-polarised light, are capable of rotating the plane of polarisation to the left or right. Such compounds are said to be *optically active*, and are dextrorotatory if the substance rotates the plane of polarisation to the right, and lævorotatory if to the left. The symbols $(+)$- and $(-)$- are used to indicate the sign of rotation, *e.g.*, $(+)$-glucose (dextrorotatory), $(-)$-fructose (lævorotatory).

Optical activity may be due to the nature of the crystalline structure of the compound (see p. 474), but for the present brief discussion we shall consider optical activity that is due to the nature of the *molecular* structure of the compound. When a molecular structure is not superimposable on its mirror image, that molecule is said to possess *molecular asymmetry*, and as a result, exists in two forms, one being the $(+)$-form and the other the $(-)$-form, the rotations being numerically equal. These two forms constitute a pair of **enantiomorphs** or **enantiomers**, and a mixture of a pair of enantiomers in equimolecular amounts is called a **racemic modification**. This is optically inactive (by external compensation) and is represented as DL- or $(\pm)$-.

From the chemical point of view it is less important to know the signs of rotation than to know *stereochemical relationships*, *i.e.*, the relationships between the configurations of various optically active compounds. These relationships are indicated by the symbols D and L, *e.g.*, D$(+)$-glucose and D$(-)$-fructose are stereochemically related, *i.e.*, their configurations are related (both are members of the D-series) even though their rotations happen to be of opposite sign.

A more detailed account of optical isomerism is given in Ch. 17, but it should be noted here that any molecule that contains an asymmetric carbon atom, *i.e.*, a carbon atom joined to four *different* groups, possesses molecular asymmetry. Furthermore, substitution reactions occurring at an asymmetric carbon atom may be accompanied by *optical inversion* (*Walden inversion*), *i.e.*, a D-substrate, on undergoing reaction, produces the L-product, or vice versa.

General reactions of the alkyl halides.

The alkyl halides are extremely important reagents because they undergo a large variety of reactions that make them valuable in organic syntheses. This arises from the fact that the halide ion is a weak base and is readily displaced by stronger bases. In general, the stronger the base, the less facile is it as a 'leaving' group in displacement reactions, but if protonation can occur first, the displacement may readily follow, *e.g.*, OH in ROH is not displaced but OH_2^+ is in ROH_2^+.

The order of ease of displacement of groups is not fixed; it depends on the nature of the R group and on the conditions (solvent). However, it appears that, in general, the order is:

$$OTs > I > Br > Cl > OH_2^+ > F > OAc > NR_3^+ > OR > NR_2$$

The order of *nucleophilicity* (nucleophilic reactivity) is also dependent on the nature of the solvent and, in general, is in the following order:

$$PhS^- > CN^- > I^- > EtO^- > OH^- > Br^- > PhO^- > Cl^- > Me_3N$$

An important point in this connection is that *nucleophilicity* is the ability to form bonds to carbon atoms, whereas *basicity* is the affinity for protons. Also, nucleophilicity is a measure of *rate of reaction*, whereas basicity is a measure of the *value of the equilibrium constant* (see p. 64). Even so, it might have been expected that there would be some sort of parallelism between the two. Although this is often the case, deviations do occur, particularly with elements in the same periodic group, *e.g.*, RS^- is much more reactive as a nucleophile but is much less basic than RO^-. The reasons for this behaviour are not clear (see also p. 399).

1. Alkyl halides are hydrolysed to alcohols very slowly by water, but rapidly by silver oxide suspended in boiling water, or by boiling aqueous alkalis (see also below):

$$RX + KOH \longrightarrow ROH + KX$$

This type of reaction is an example of nucleophilic substitution (S_N) since the attacking reagent is nucleophilic reagent (p. 51). It has been assumed that this type of heterolytic reaction in solution can take place by two different mechanisms, unimolecular or bimolecular.

Unimolecular mechanism. This is a *two-stage process*, the first stage consisting of *slow* heterolysis of the compound to form a carbonium ion, followed by *rapid* combination between the carbonium ion and the substituting nucleophilic reagent. Since the rate-determining step is the first one, and since in this step only *one* molecule is undergoing a covalency change, this type of mechanism is called *unimolecular*, and is labelled S_N1 (Ingold *et al.*, 1928, 1933). Thus, if the hydrolysis of an alkyl halide is an S_N1 reaction, it may be written:

$$R\frown X \underset{\text{fast}}{\overset{\text{slow}}{\rightleftharpoons}} X^- + R^+ \xrightarrow{\text{OH}^-} ROH$$

Heavy metal ions, particularly silver ions, catalyse S_N1 reactions and, according to Kevill *et al.* (1967), the mechanism involves a fast pre-equilibrium step (the silver ion has an empty orbital):

$$R-X + Ag^+ \overset{\text{fast}}{\rightleftharpoons} [R-X-Ag]^+ \xrightarrow{\text{slow}} R^+ + AgX$$

Bimolecular mechanism. This is a *one-stage process*, two molecules simultaneously undergoing covalency change in the rate-determining step. This type of mechanism is called *bimolecular*, and is labelled S_N2. Since the rate-determining step in this reaction is the formation of the transition state, the hydrolysis of an alkyl halide by an S_N2 reaction may be written:

$$HO^\frown R-X \xrightarrow{\text{slow}} \overset{\delta-}{HO}\cdots R\cdots\overset{\delta-}{X} \xrightarrow{\text{fast}} HO-R + X^-$$

A number of differences exist between S_N2 and S_N1 reactions, *e.g.* (i) When both reactants are present in small and controllable concentrations, S_N2 reactions are second-order and S_N1 reactions are first-order. If one reactant is present in large excess, *e.g.*, one of the reactants is the solvent, then an S_N2 mechanism will lead to first-order kinetics. (ii) The S_N2 mechanism always leads to inversion of the product, whereas with the S_N1 mechanism the configuration of the product depends on various factors. (iii) No rearrangement is possible for the S_N2 mechanism, but is possible (and usually occurs) for the S_N1. (iv) The rate constant of an S_N2 reaction with a given substrate depends on the nature of the nucleophile and for a given nucleophile, on the nature of the leaving group in the substrate (see above).

Stereochemistry of substitution reactions.

(i) *The S_N2 mechanism.*

$$HO^-\frown R-X \longrightarrow \overset{\delta-}{HO}\cdots R\cdots\overset{\delta-}{X} \longrightarrow HO-R + X^-$$

In the transition state, the groups OH and X are collinear and on opposite sides of the attacked carbon atom. Furthermore, the line joining OH and X is perpendicular to the plane containing the other three groups *a*, *b* and *d*. In the original molecule *CabdX*, the four groups are arranged tetrahedrally. Hence, to achieve a planar configuration of *Cabd* in the transition state, the carbon atom changes from tetrahedral to trigonal hybridisation, the remaining p_z orbital being used (by means of its two lobes) to hold the groups OH and X by 'half-bonds'. When X is ejected, the carbon atom returns to its state of tetrahedral hybridisation (see diagram).

The above reaction is a three-centre reaction, proceeding by an end-on approach.

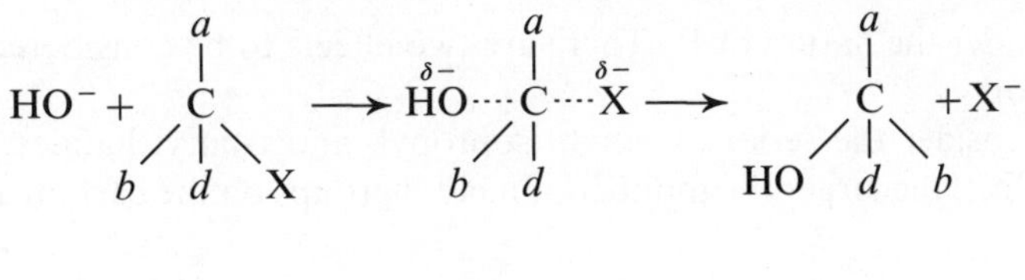

Hughes *et al.* (1935) studied: (*a*) the interchange reaction of (+)-2-iodo-octane with radio-active iodine (as NaI*) in acetone solution, and (*b*) the racemisation of (+)-2-iodo-octane by ordinary NaI under the same conditions. These reactions were shown to occur by the S_N2 mechanism, and the results showed that every halide-halide* displacement was always accompanied by inversion. Thus, this experiment leads to the *assumption that an* S_N2 *reaction always gives inversion*; this is fully supported by other experimental work. Hence, if it can be shown that the S_N2 mechanism is operating, then inversion must have occurred.

(ii) *The* S_N1 *mechanism*

$$R\overset{\frown}{-}X \underset{\text{fast}}{\overset{\text{slow}}{\rightleftharpoons}} X^- + R^+ \xrightarrow{\text{OH}^-} ROH$$

When the reaction proceeds by this mechanism, then inversion and retention (racemisation) will occur, the amount of each depending on various factors. The carbonium ion is flat (trigonal hybridisation), and hence attack by nucleophilic reagents can take place equally well on either side, *i.e.*, equal amounts of (+)- and (−)-forms are produced; this is racemisation. Complete racemisation, however, can be expected only if the carbonium ion is sufficiently long-lived; this is favoured by low reactivity of the carbonium ion and low concentration of the nucleophilic reagent. On the other hand, during the actual ionisation, the retiring negative group will shield attack on that side; this encourages end-on attack *on the other side*, thereby leading to inversion. In general, the S_N1 mechanism is accompanied by inversion and racemisation, but in some cases there may be complete inversion.

It will be seen from the above discussion that if heterolysis of RX is the rate-determining step, then the mechanism is S_N1. However, circumstances might arise where the *second* step is the rate-determining one. In this case, Hughes and Ingold (1954) have labelled the reaction $S_N2(C^+)$ [substitution, nucleophilic, bimolecular, with a rapidly formed carbonium ion].

Another S_N reaction is the S_Ni type (substitution, nucleophilic, internal), *e.g.*, Hughes, Ingold *et al.* (1937) have shown that optically active α-phenylethanol (C_6H_5CHMe- = R) reacts with thionyl chloride to give α-phenylethyl chloride with almost complete retention of configuration. The mechanism proposed is S_Ni via a chlorosulphite:

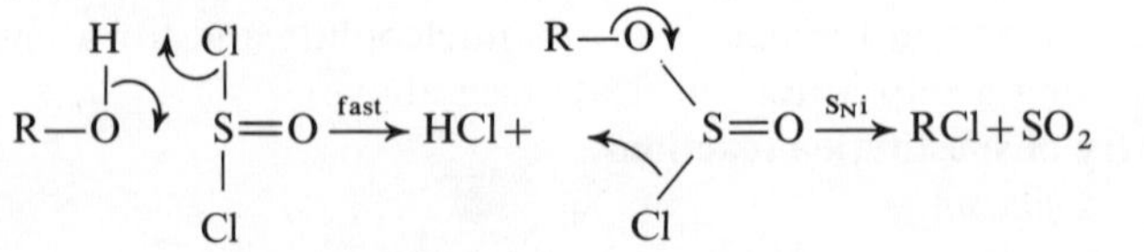

No inversion occurs in either stage (both stages are a four-centre type). This situation arises with *arylmethanols*, but in the presence of pyridine, the chloride now has the *inverted* configuration. This has been explained on the basis that the S_N2 mechanism is operating, but now involves a pyridine complex.

$$ROSOCl + C_5H_5N \xrightarrow{\text{fast}} \overset{+}{ROSONC_5H_5} + Cl^- \underset{S_N2}{\xrightarrow{\text{slow}}} \overset{\delta-}{Cl}\cdots R\cdots \overset{\delta-}{O}\overset{+}{SONC_5H_5}$$
$$\xrightarrow{\text{fast}} Cl-R + SO_2 + C_5H_5N$$

Any factor that affects the energy of activation of a given type of reaction will affect the rate and/or the mechanism. The following discussion is largely qualitative, and because of this, one cannot be sure which are the predominant factors in deciding the value of the energy of activation (see also later).

Let us first consider the nature of R. There are two effects to be considered, the polar factor and the steric factor.

Polar factor. Consider the series of ethyl, isopropyl, and t-butyl halides. Since the methyl group has a $+I$ effect, the larger the number of methyl groups on the carbon atom of the C—X, the greater will be the electron density on this carbon. If we use one arrow-head to represent (qualitatively) the electron-repelling effect of a methyl group, then we have the following state of affairs:

$$Me{\rightarrow}CH_2{\rightarrow}X \qquad \begin{matrix} Me\searrow \\ \quad CH{\rightarrow}{\rightarrow}X \\ Me\nearrow \end{matrix} \qquad \begin{matrix} Me\searrow \\ Me{\rightarrow}C{\rightarrow}{\rightarrow}{\rightarrow}X \\ Me\nearrow \end{matrix}$$

This increasing electron density on the central carbon atom increasingly opposes attack at this carbon atom by a negatively charged nucleophilic reagent. Thus, the formation of the transition state for the S_N2 mechanism becomes increasingly difficult. It can therefore be anticipated that the S_N2 mechanism is made more difficult in passing from EtX to t-BuX. On the other hand, since the S_N1 mechanism involves ionisation as the first step, then it can be expected that as the electron density increases on the central carbon atom, the bonding pair in C—X becomes more and more displaced towards the X atom, and consequently ionisation of X as a negative ion will become easier. It therefore follows that the tendency for the S_N1 mechanism should increase from EtX to t-BuX. These anticipated results have been realised in practice. Hughes, Ingold *et al.* (1935–1940) examined the hydrolysis of alkyl bromides in alkaline aqueous ethanol and showed that MeBr and EtBr undergo hydrolysis by the S_N2 mechanism, isoPrBr by *both* S_N2 and S_N1 mechanisms, and t-BuBr by the S_N1 mechanism only. It has also been shown, however, that the actual position where the mechanism changes over from S_N2 to S_N1 in a given graded polar series (such as the one above) is not fixed, but also depends on other factors such as the concentration of the nucleophilic reagent and the nature of the solvent (see below).

An alternative explanation for the effect of the nature of R is as follows. When the S_N1 mechanism is favoured, it means that the activation energy for this path is less than that for the S_N2; and vice versa. Since the S_N1 reaction proceeds through a carbonium ion, the more stable this ion, the lower is the activation energy and the more favoured will be this mechanism. Now, the order of stabilities of carbonium ions is p < s < t, and hence the tendency for the S_N1 mechanism to operate will be t-BuX > isoPrX > EtX, and the reverse tendency for the S_N2 mechanism.

Steric effects. The steric effect was originally thought to be a spatial effect brought into play by mechanical interference between groups, and was described as *steric hindrance* (see p. 771). The term steric hindrance considered the geometry of the *reactant* molecule. When, however, a molecule undergoes chemical reaction, it does so via a transition state. Consequently the geometry of both the initial and transition state must be taken into consideration. Thus steric factors may affect the speed and/or the mechanism of a reaction. It is, however, often very difficult to distinguish between steric and polar factors. Nevertheless, the effects of the steric factor can often be assessed in some sort of qualitative manner. When steric effects slow down a reaction, that reaction is said to be subject to **steric hindrance** (*or* **retardation**), and when they speed up a reaction, that reaction is said to be subject to **steric acceleration**.

In an S_N2 reaction there will be five groups 'attached' to the carbon atom at which reaction occurs (p. 155). Thus there will be 'crowding' in the transition state, and the bulkier the groups, the greater will be the *compression energy*, and consequently the reaction will be hindered sterically.

Let us consider the following example, Hughes *et al.* (1946) examined the rate of exchange of iodide ions in acetone solution with MeBr, EtBr, isoPrBr and t-BuBr under conditions where only the S_N2 mechanism operated. The relative reactivities were found to be 10,000; 65; 0·50; 0·039.

$$\overset{-}{I} \overset{\frown}{R} \overset{\frown}{\underset{}{Br}} \longrightarrow \overset{\delta-}{I} \cdots R \cdots \overset{\delta-}{Br} \longrightarrow I-R+Br^-$$

Thus, increasing the number of methyl groups on the central carbon atom increases the steric retardation.

The problem is somewhat different for the S_N1 mechanism. Here the transition state does not contain more than four groups attached to the central carbon atom, and hence one would expect steric hindrance to be less important in the S_N1 mechanism. If, however, the molecule contains bulky groups, then by ionising, the molecule can relieve the steric strain, since the carbonium ion produced is flat (trigonal hybridisation) and so there is more room to accommodate the three alkyl groups. Thus, in such S_N1 reactions, there will be steric acceleration. Brown *et al* (1949) showed that the solvolysis of tertiary halides is subject to steric acceleration (*solvolysis* is a nucleophilic substitution reaction in which the *solvent* is the nucleophilic reagent).

$$R_3C\overset{\frown}{-}X \overset{H_2O}{\longrightarrow} R_3C^+ + X^-$$

(tetrahedral;   (trigonal; planar;
large strain)   small strain)

These authors showed that as R increases in size, the rate of solvolysis increases. However, the larger the R groups are, the more slowly will the carbonium ion be expected to react with solvent molecules, since steric strain will be reintroduced into the molecule. In cases like this, the carbonium ion tends to undergo an elimination reaction to form an alkene, and Brown *et al.* (1950) have shown that this elimination process increases as the alkyl groups become larger (see p. 165).

It should be noted that a fundamental part of the S_N1 mechanism is the postulate of the transient existence of carbonium ions. Triarylmethyl carbonium ions have been isolated as their salts, *e.g.*, triphenylmethyl carbonium ion, Ph_3C^+, as its perchlorate and borofluoride (Dauben *et al.*, 1960). The stability of ions such as these is attributed to resonance (*cf.* p. 788). The order of stability of alkyl carbonium ions is $t > s > p$, and so one is more likely to expect success in the isolation of the t-ions. Olah *et al.* (1964) have prepared, *e.g.*, $Me_3C^+ SbF_6^-$, and their infra-red spectra studies have substantiated the planar sp^2 hybridised structure of simple alkyl carbonium ions.

Now let us consider the effect of the nature of the halogen atom. Experimental work has shown that the nature of X has very little effect, if any, on *mechanism*, but it does affect the *rate* of reaction for a given mechanism, *e.g.*, it has been found that in both S_N2 and S_N1 reactions, the rate follows the order $RI > RBr > RCl$. Experimental work has shown that many S_N2 reactions of alkyl chlorides and bromides are catalysed by the presence of iodide ions. This may be explained on the basis that the activation energies of *both* steps in reaction (i) are lower than the activation energy of the one-step reaction (ii):

(i) $I^- + R-Cl \longrightarrow I-R+Cl^- \overset{OH^-}{\longrightarrow} ROH + I^- + Cl^-$

(ii) $HO^- + R-Cl \longrightarrow HO-R + Cl^-$

The more pronounced the nucleophilic activity of the attacking reagent, then the more the S_N2 mechanism will be favoured, since in the S_N1 mechanism the reagent does not enter into the rate-determining step of ionisation. However, it can also be expected that as the nucleophilic activity of the reagent decreases, a point will be reached where the nucleophilic activity is so low that the mechanism will change from S_N2 to S_N1. Hughes, Ingold, *et al.* (1935) examined the mechanism and rate of decomposition of various trimethylsulphonium salts ($Me_3S^+X^-$; see also p. 402) and showed that for X = OH, the mechanism is S_N2, and for X = I, Cl or Br, the mechanism is S_N1.

Mechanism and rate of reaction are very much affected by the nature of the solvent. It has been found that the ionising power of a solvent depends on its dielectric constant and its power of solvation, and it appears that the latter is more important than the former. Solvation is due to the 'attachment' between solvent and solute molecules, and one important contributing factor is attraction of a charge for a dipole. Since electrostatic work is done in the process of solvation, energy is lost by the system, and consequently the system is more stable. Furthermore, the greater the polarity (dipole moment) of the solvent, the greater is the attraction between solute and solvent molecules and consequently the more stable is the system. Thus, increasing the dielectric constant of the solvent increases the ionising potentiality of the solute molecules, and the higher the polarity of the solvent the greater is the amount of solvation. Solvation, however, may also be partly due to hydrogen bonding of the solvent (if this is protic) with the solute (if this can form hydrogen bonds). Some common solvents and their dielectric constants and dipole moments (D) are: water (81·1; 2–3); formic acid (48·0; 1·5); nitrobenzene (35·7; 4·0); ethanol (25·8; 1·7); acetone (21·3; 3·0); acetic acid (7·1; 1–1·5); chloroform (4·6; 1·1); ether (4·3; 1·25); benzene (2.3; 0); carbon tetrachloride (2·2; 0).

When solute molecules have ionised, the oppositely charged pair of ions may become enclosed in a 'cage' of the surrounding solvent molecules, and may therefore recombine before they can escape from the cage. Such a complex is known as an *ion-pair*, and their recombination is known as *internal return*. It has now been shown that many organic reactions proceed via ion-pairs rather than dissociated ions (see p. 335). Thus we have the possibility of the following steps:

$$RX \rightleftharpoons \underset{\text{ion-pair}}{R^+X^-} \rightleftharpoons \underset{\text{dissociated ions}}{R^+ + X^-}$$

Many attempts have been made to correlate reaction rates and nature of the solvent. Hughes and Ingold (1935, 1948) proposed the following qualitative theory of solvent effects:

(i) Ions and polar molecules, when dissolved in polar solvents, tend to become solvated.

(ii) For a given solvent, solvation tends to increase with increasing magnitude of charge on the solute molecules or ions.

(iii) For a given solute, solvation tends to increase with the increasing dipole moment of the solvent.

(iv) For a given magnitude of charge, solvation decreases as the charge is spread over a larger volume.

(v) The decrease of solvation due to the dispersal of charge will be less than that due to its destruction.

Let us consider the following S_N1 reaction:

$$R\!-\!X \underset{\text{slow}}{\rightleftharpoons} \overset{\delta+}{R} \cdots \overset{\delta-}{X} \xrightarrow{\text{fast}} R^+ + X^- \xrightarrow[\text{fast}]{OH^-} R\!-\!OH$$

Since the transition state has a larger charge than the reactant molecule, the former will be more solvated than the latter (rule ii). Thus the transition state is more stabilised than the reactant molecule, *i.e.*, the energy of activation is lowered, and so the reaction proceeds faster than had there been no (or less) solvation.

The rates of S_N2 reactions are also affected by the polarity of the solvent:

$$\overset{\frown}{HO} \, R\!-\!X \rightleftharpoons \overset{\delta-}{HO} \cdots R \cdots \overset{\delta-}{X} \longrightarrow HO\!-\!R + X^-$$

A solvent with a high dipole moment will solvate both the reactant ion (nucleophilic reagent) and the transition state, but more so the former than the latter, since in the latter the charge, although unchanged in magnitude ($\delta- = -1/2$), is more dispersed than in the former. Therefore solvation tends to stabilise the reactants more than the transition state (rule iv). Thus the activation energy is increased, and so the reaction is retarded.

The rate of the following S_N2 reaction (**Menschutkin reaction**, 1890) is increased as the polarity of the solvent increases.

$$R_3\overset{\frown}{N} \quad R-X \rightleftharpoons R_3-\overset{\delta-}{N}\cdots R\cdots \overset{\delta-}{X} \longrightarrow R_4N^+X^-$$

Here the charge on the transition state is greater than that on the reactants. Hence the transition state is more solvated than the reactants and consequently stabilised, and so the activation energy is lowered.

The polarity of the solvent may *change* the mechanism of a reaction, *e.g.*, Olivier (1934) showed that the alkaline hydrolysis of benzyl chloride in 50 per cent. aqueous acetone proceeds by both S_N2 and S_N1 mechanisms. When water was used as solvent, the mechanism was now mainly S_N1. The dielectric constant of water is greater than that of aqueous acetone, and so ionisation of the benzyl chloride is facilitated (see also Vol. 2, Ch. 3).

The above discussion has been purely qualitative, but it is also of interest to consider the problem in a quantitative way. This involves the use of the following thermochemical cycle:

$$
\begin{array}{ccc}
\text{RX (solvated)} & \xrightarrow[\text{(ionisation)}]{1} & \text{R}^+\text{ (solvated)} + \text{X}^-\text{ (solvated)} \\
\;2 \downarrow \text{ (desolvation)} & & \;6\uparrow\text{(solvation)} \qquad 7\uparrow\text{(solvation)} \\
& & \text{R}^+\text{ (vapour)} \qquad \text{X}^-\text{ (vapour)} \\
& & \;4\uparrow(-e) \qquad\quad 5\uparrow(+e) \\
\text{RX (vapour)} & \xrightarrow[\text{(dissociation)}]{3} & \text{R}\cdot\text{ (vapour)} + \text{X}\cdot\text{ (vapour)}
\end{array}
$$

Then, if the enthalpy change for each step is represented by the number (of the step) as a subscript, it follows from Hess's law that

$$\Delta H_1 = \Delta H_2 + \Delta H_3 + \Delta H_4 + \Delta H_5 + \Delta H_6 + \Delta H_7$$

where ΔH_4 is the ionisation potential of $R\cdot$ and ΔH_5 is the electron affinity of $X\cdot$.

All of these values except ΔH_2 have been calculated from suitable experimental data, and the value of ΔH_2 has been estimated. The values given in the following discussion are those used by Frazer and Singer (see References).

First let us consider the problem of reactions in the gas phase and in solution. In the gas phase, if the reaction is homolytic, then ΔH for step 3 in the above cycle is ΔH_3. If we use methyl chloride as our example, then ΔH_3 is $+334\cdot7$ kJ. If, however, the reaction is heterolytic, then

$$\Delta H = \Delta H_3 + \Delta H_4 + \Delta H_5 = +334\cdot7 + 970\cdot7 - 364\cdot0 = 941\cdot4 \text{ kJ}$$

It can be seen that heterolytic reaction in the gas phase is very unfavourable energetically. If the reaction is carried out in water, then

$$\Delta H_1 = +33\cdot47 + 334\cdot7 + 970\cdot7 - 364\cdot0 - 343\cdot1 - 288\cdot7 = 343\cdot1 \text{ kJ}$$

This much more favourable value is due to the large contribution of the heats of solvation (ΔH_6 and ΔH_7).

As we have already seen, in the S_N1 mechanism (with all ions and molecules solvated):

$$RX_{(aq)} \overset{\text{slow}}{\rightleftharpoons} R^+_{(aq)} + X^-_{(aq)} \xrightarrow[\text{fast}]{H_2O} ROH_{(aq)} + H^+_{(aq)}$$

the rate order is t-Bu > isoPr > Et > Me. The following Table gives the values of ΔH_1 (heat of ionisation in kJ mol^{-1}) for various alkyl halides.

Group	Cl	Br	I
Me	343·1	343·1	364·0
Et	238·5	213·4	238·5
isoPr	104·6	96·23	115·1
t-Bu	71·13	62·76	75·31

It can be seen that in going in a downward direction for a given halogen atom, ΔH_1 decreases rapidly. Moreover, calculations have shown that ΔH_1 and E (activation energy) are almost identical. Hence the rate of the S_N1 reaction will increase in the downward direction, *i.e.*, t-Bu > isoPr > Et > Me.

It has been shown experimentally that the rate of the S_N1 reaction (and that of the S_N2) for a given alkyl group is RI > RBr > RCl. The above Table indicates the order RBr > RCl > RI. This again illustrates the point that even though a quantitative approach to a problem is always desirable, unless all the necessary values are known (reasonably) accurately, estimates may lead to some wrong conclusions. This is more the case with reactions in solution than with gas-phase reactions.

2. Alkyl halides are reduced by dissolving metals to alkanes:

$$RX \xrightarrow{e; H^+} RH + HX$$

Primary and secondary alkyl chlorides and bromides are not affected in neutral or acidic media, whereas prim., s- and t-iodides, t-chlorides and t-bromides are converted into alkanes by *catalytic hydrogenolysis*, *i.e.*, hydrogenation in the presence of a catalyst. The best catalyst is Pd—C, but Raney nickel is also effective provided it is used in large amounts.

$$RX + H_2 \xrightarrow{Pd-C} RH + HX$$

Reduction to alkane can also be carried out with lithium aluminium hydride or sodium borohydride.

Johnson *et al.* (1964) have shown that primary and secondary alkyl iodides are oxidised by dimethyl sulphoxide to aldehydes or ketones, respectively. The mechanism is uncertain, but a possibility is:

$$RCH_2I + Me_2S{=}O \longrightarrow I^- + R{-}CH{-}O{-}\overset{+}{S}Me_2 \longrightarrow RCHO + Me_2S + H^+$$
$$\underset{H}{|}$$

3. Alkyl halides undergo the Wurtz reaction to form alkanes (p. 76). On the other hand alkyl halides, when treated with certain metallic alloys, form *organometallic* compounds, *e.g.*, ethyl chloride heated with a sodium lead alloy under pressure gives tetraethyl-lead:

$$4C_2H_5Cl + 4Na/Pb \longrightarrow (C_2H_5)_4Pb + 4NaCl + 3Pb$$

4. When an alkyl halide is heated at about 300°C, and at a lower temperature in the presence of aluminium chloride as catalyst, the alkyl halide undergoes rearrangement, *e.g.*,

$$CH_3CH_2CH_2Cl \xrightarrow{-Cl^-} CH_3CH_2CH_2{}^+ \longrightarrow CH_3\overset{+}{C}HCH_3 \xrightarrow{Cl^-} CH_3CHClCH_3$$

If there is no hydrogen atom on the carbon adjacent to the C—X group, an alkyl group migrates; *e.g.*, neopentyl chloride rearranges to 2-chloro-2-methylbutane (see also p. 147).

$$(CH_3)_3CCH_2Cl \longrightarrow (CH_3)_2CClCH_2CH_3$$

5. When alkyl halides are boiled with ethanolic potassium hydroxide, alkenes are obtained, *e.g.*, propyl bromide gives propene:

$$CH_3CH_2CH_2Br + KOH \xrightarrow{ethanol} CH_3CH{=}CH_2 + KBr + H_2O$$

This is an example of the **1,2-** or **β-elimination reaction**, in which two groups are lost from *neighbouring* atoms to form a double bond. Other elimination types (α, γ, etc.) are also known, but these are much less common (see pp. 170 and 532). There are two important mechanisms which operate with alkyl halides, E1 and E2, both of which are *polar* (*cf.* cyclic eliminations, p. 94).

E1 mechanism (unimolecular elimination). This mechanism is parallel to the S_N1 mechanism in that the rate-determining step is ionisation to a carbonium ion, but the second step is now elimination of a β-proton to form an alkene.

$$HCR_2-CR_2\!-\!Z \underset{}{\overset{slow}{\rightleftharpoons}} HCR_2-CR_2^{+} + Z^{-}$$
$$B^{\,}H\!-\!CR_2\!-\!CR_2^{+} \xrightarrow{fast} BH + CR_2=CR_2$$

The reaction is carried out in the presence of a base and the proton is transferred to this, *e.g.*, t-butyl bromide in aqueous ethanolic potassium hydroxide gives isobutene:

$$(CH_3)_3CBr \overset{slow}{\rightleftharpoons} (CH_3)_3C^{+} + Br^{-}$$
$$HO^{-}\;\;H\!-\!CH_2\!-\!\overset{+}{C}(CH_3)_2 \xrightarrow{fast} H_2O + CH_2=C(CH_3)_2$$

Evidence for the E1 mechanism has been obtained in several ways, *e.g.*, the rate law for many eliminations is first-order (alkyl halide only), *i.e.*, rate $= k[RX]$. This is consistent with the mechanism given.

E2 mechanism (bimolecular elimination). The general reaction may be written:

$$B^{\,}H\!-\!CR_2\!-\!CR_2\!-\!Z \longrightarrow BH + CR_2=CR_2 + Z^{-}$$

An example of this type is the action of ethanolic potassium hydroxide on isopropyl bromide.

$$EtO^{-}\,H\!-\!CH_2\!-\!CH(CH_3)\!-\!Br \longrightarrow EtOH + CH_2=CHCH_3 + Br^{-}$$

Evidence for the E2 mechanism comes from the fact that the rate law for many eliminations is second-order, *i.e.*, rate $= k[RX][B]$. There is, however, a difficulty here. The second-order rate law is consistent with the one-step mechanism given above, but it is also consistent with the following two-step mechanism (Ph $= C_6H_5$):

$$EtO^{-} + PhCH_2CH_2Br \overset{fast}{\rightleftharpoons} EtOH + Ph\overset{-}{C}HCH_2Br$$
$$Ph\overset{-}{C}H\!-\!CH_2\!-\!Br \xrightarrow{slow} PhCH=CH_2 + Br^{-}$$

A simplified derivation of the rate law for this reaction is as follows (*cf.* p. 56). Since the reaction is carried out in ethanol, the concentration of the *ethanol produced* in the first step may be neglected. Hence:

$$K = [Ph\overset{-}{C}HCH_2Br]/([PhCH_2CH_2Br][OEt^{-}])$$

$$\therefore \quad rate = k[Ph\overset{-}{C}HCH_2Br] = kK[PhCH_2CH_2Br][OEt^{-}]$$

Since k and K are both constants, their product may be replaced by a constant k', and the rate law is:

$$rate = k'[PhCH_2CH_2Br][OEt^{-}]$$

i.e., the rate law is second-order. Thus, kinetics cannot distinguish between these two possible elimination mechanisms.

In the second mechanism, a slow unimolecular elimination occurs in the conjugate base (cB) of the reactant, and hence this mechanism is called the **E1cB** or **carbanion mechanism**. Since the first step must be reversible (acid-conjugate base equilibrium), if ethanol containing EtOD is used as solvent, it would be expected that the original bromide would incorporate deuterium. Skell *et al.* (1945) examined this reaction and found that after the reaction was half-completed, the recovered bromide did not contain deuterium. Hence, *this* mechanism is not E1cB, but is E2.

It appears that definite examples of the E1cB mechanism are rare.

Which mechanism, E1 or E2, operates depends on the nature of the alkyl halide and the conditions used. Before discussing this, let us consider the situation where *two* alkenes may be formed by dehydrohalogenation of an alkyl halide. Originally, the direction of elimination was described empirically by **Saytzeff's rule** (1875). This rule may be stated in two ways: The predominant product is the most substituted alkene, *i.e.*, the one carrying the largest number of alkyl substituents; *or*, Hydrogen is eliminated preferentially from the carbon atom joined to the least number of hydrogen atoms. There is also another rule which describes empirically the direction of elimination in quaternary ammonium hydroxides. This rule, the **Hofmann rule** (1881), is discussed later (p. 379), but may be stated here as follows: The predominant alkene (from an 'onium hydroxide) is the least substituted alkene. An illustration of these rules is the dehydrobromination of 2-bromobutane.

$$\text{(I)} \quad CH_3CH{=}CHCH_3 \longleftarrow CH_3CH_2CHBrCH_3 \longrightarrow CH_3CH_2CH{=}CH_2 \quad \text{(II)}$$

(I) is the Saytzeff product and (II) is the Hofmann product, and since we are dealing with elimination from an alkyl halide, (I) will be expected to be the predominant product. Ingold *et al.* (1941) have offered the following explanation for the Saytzeff rule. If (I) and (II) are produced by an E2 mechanism, their formation may be written as follows:

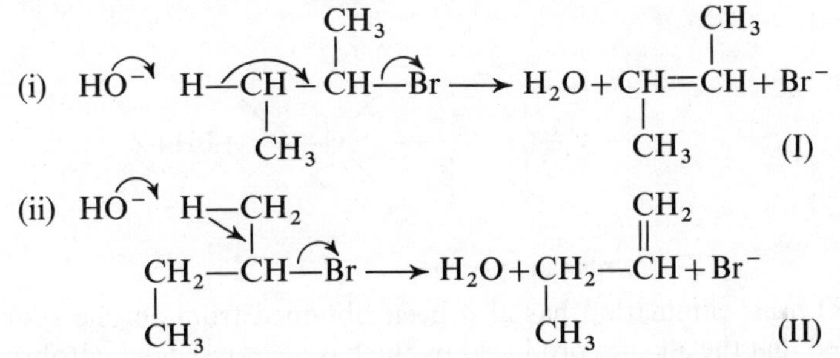

In the transition state of (i), the CH_3 group can enter into hyperconjugation with the partly formed double bond (p. 332), thereby lowering the energy of the transition state. In (ii), however, the CH_3 group cannot enter into hyperconjugation with the partly formed double bond, and so the energy of this transition state is higher than that of (i), and consequently the latter path is favoured.

On the basis that the hyperconjugative effect determines the stability of the alkene produced, it can be deduced that the ease of formation of alkene from an alkyl halide by the E2 mechanism should be t—R > s—R > p—R, *e.g.*, t—Bu (2 Me groups) > isoPr (1 Me group) > Et (no Me groups):

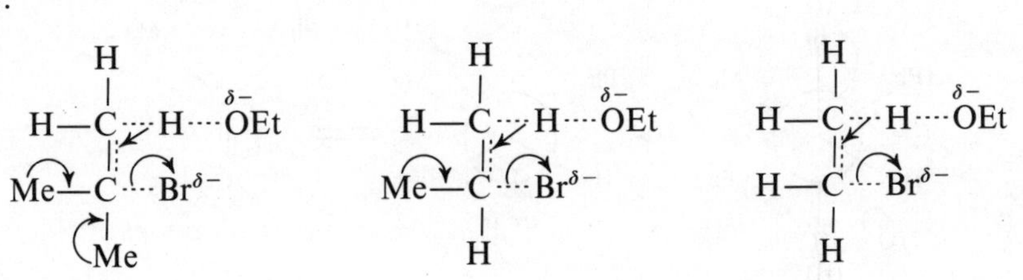

These predictions are borne out in practice. Also, since the most substituted alkene is generally the most stable isomer (p. 99), it therefore follows that Saytzeff's rule leads to the predominance of the most stable product. A further point in this connection is that when the alkene is capable of existing as geometrical isomers, the *trans* -isomer generally predominates over the *cis*.

Saytzeff's rule also applies when the mechanism is E1. In this mechanism the formation of the carbonium ion is the rate-determining step (as for the S_N1 mechanism), and then the carbonium ion eliminates a proton to form an alkene, the stability of which will be largely determined by the hyperconjugative effect.

Stereochemistry of elimination reactions. Many examples in the literature show that *trans* elimination occurs more readily than *cis*, e.g., chlorofumaric acid undergoes dehydrochlorination about fifty times as fast as chloromaleic acid:

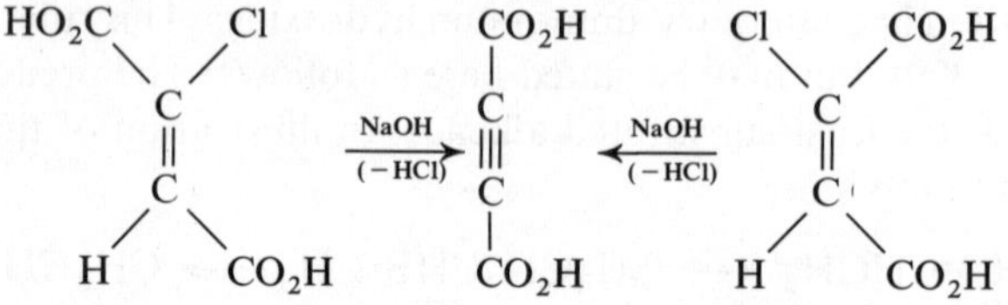

Experiments of this type (involving 'rigid' systems) led Hughes and Ingold to propose that bimolecular elimination reactions take place when the two groups to be eliminated are *trans* and lie in one plane with the two carbon atoms to which they are attached, *i.e.*, E2 reactions are stereoselectively *trans*:

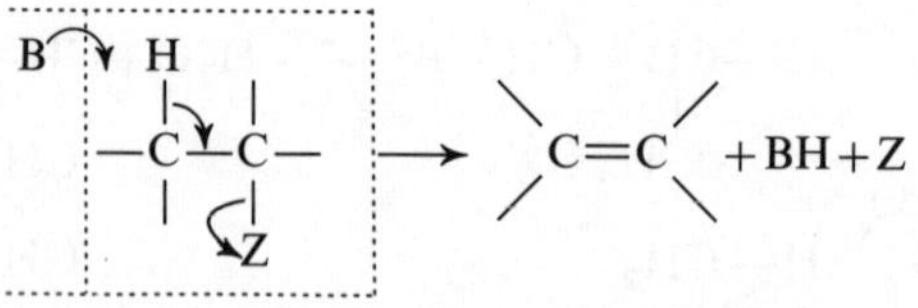

trans and planar.

Evidence for E2 *trans*-elimination has also been obtained from acyclic systems, *e.g.*, Cram *et al.* (1952) showed that the alkenes produced by the base-catalysed dehydrobromination of the diastereoisomeric 1-bromo-1,2-diphenylpropanes (I and II) can only arise by *trans*-elimination (see also p. 381).

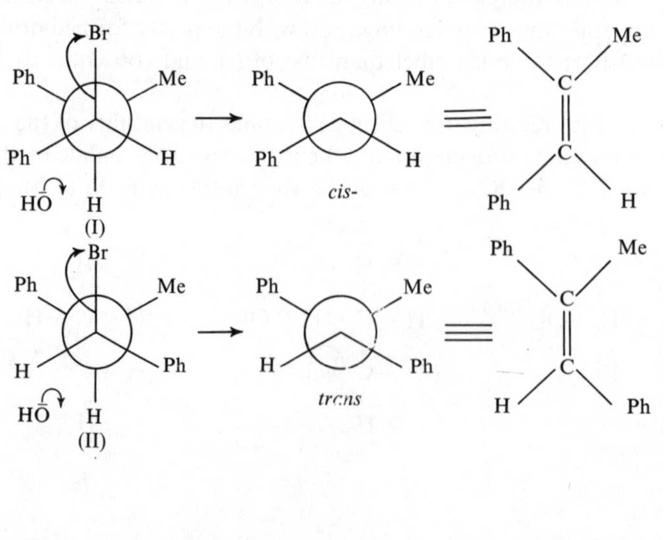

Later work has shown that bimolecular eliminations may, in certain circumstances, proceed by *cis*-elimination (see p. 488).

Steric effects in elimination reactions. Let us consider the E1 reaction with RCH_2CMe_2Br. Accepting that elimination of HBr is selectively *trans*, then we may represent the two directions of elimination as follows (see also Ch. 17).

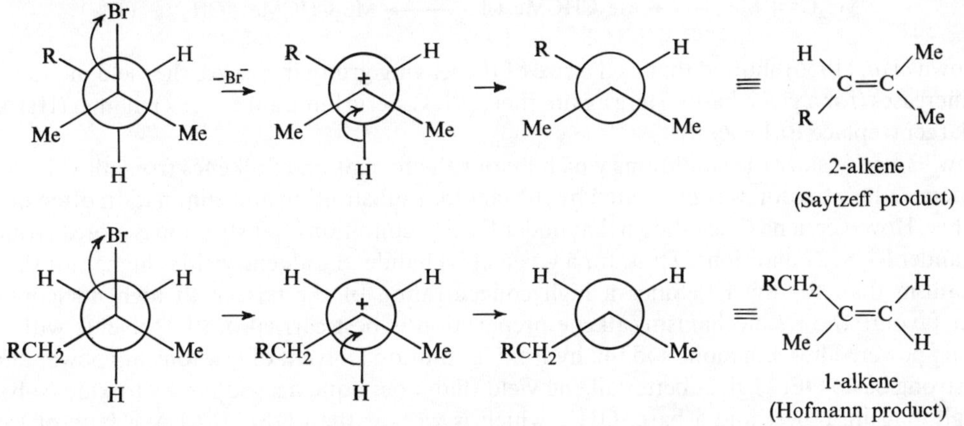

2-alkene
(Saytzeff product)

1-alkene
(Hofmann product)

On the basis of hyperconjugation, the 2-alkene should predominate (Saytzeff's rule). However, Brown *et al.* (1955) showed that as R increased in size, the ratio 1-alkene/2-alkene also increased. As R gets bigger, the steric repulsion in the transition state leading to the 2-alkene increases (due to eclipsing of R and Me), whereas steric repulsion is minimised in the T.S. leading to the 1-alkene (Hofmann rule). Thus, in this case, the 1-alkene is more stable than the 2-isomer (*cf.* p. 99). Hence, in the E1 mechanism, the structure of the alkene is determined by hyperconjugation and steric effects.

The E2 mechanism is more complicated, since not only does the size of R play a part, but also do the sizes of the attacking base and the group ejected. The eclipsing effect of R groups was shown to be the same as that in the E1 mechanism. Brown *et al.* (1956, 1968) also showed that 2-halogenoalkanes gave an increasing ratio of 1-alkene/2-alkene with the potassium alkoxides $EtO^- K^+$, $Me_3CO^- K^+$, $Et_3CO^- K^+$. These results are attributed to the increasing steric requirements of the attacking base. As the base increases in size, attack is less hindered on a primary hydrogen atom and increasingly more hindered at a secondary or tertiary hydrogen atom. This may be illustrated with the E2 mechanism using $Me_2CHCBrMe_2$ as substrate.

(III) 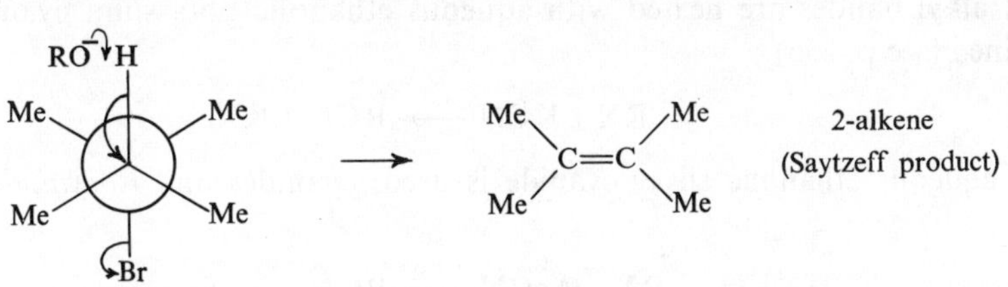   2-alkene
(Saytzeff product)

(IV) 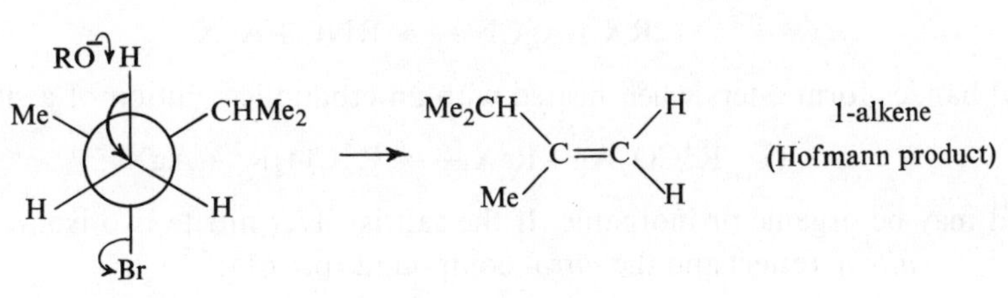   1-alkene
(Hofmann product)

As R becomes larger, the steric effect increases more in (III) than in (IV) (RO$^-$ approaches from the front), and so 1-alkene formation increases.

It might be noted here that this change in direction of elimination may be used to shift a double bond in an alkene, *e.g.*,

$$Me_2C{=}CMe_2 \xrightarrow{\text{HCl}} Me_2CHCMe_2Cl \xrightarrow{\text{Et}_3CO^-K^+} Me_2CHCMe{=}CH_2 \quad (97\%)$$

Brown *et al.* (1956) showed that as the size of the leaving group increases, the yield of 1-alkene also increases (for a given base). Here again, there is less steric hindrance in (IV) than in (III) as Z gets larger (replace Br by Z).

Now let us consider the conditions which favour the formation of alkenes from alkyl halides. In practice, the situation is complicated by the fact that substitution and elimination often occur together. However, it has been shown that under E1/S_N1 conditions, substitution is more favoured than under E2/S_N2 conditions. Thus, for a given alkyl halide, the alkene yield is higher for the E2 mechanism than for the E1. Since a high concentration of the base with high nucleophilic power favours the E2 mechanism, alkene preparation is best carried out in solvents with *low* ionising power. Thus, ethanolic sodium hydroxide solution (solvent of low ionising power and a very strong base, OEt$^-$) gives better alkene yield than does aqueous sodium hydroxide (solvent of high ionising power and a base, OH$^-$, which is weaker than OEt$^-$). Also, a base of *weak* nucleophilic power will have little affinity for hydrogen, and consequently alkene formation will be decreased and substitution encouraged. The acetate ion is a very weak base, and hence its use will give an ester as the main product rather than alkene. Thus, the conversion of an alkyl halide into the alcohol is often best carried out via the ester, which is then hydrolysed.

A point of interest here is that even under the most favourable conditions, ethyl halides give only 1 per cent. of ethylene; the main reaction leads to substitution products.

The eliminations discussed above have dealt with only alkyl halides as substrates. However, many other types of alkyl compounds also undergo E1 and E2 reactions, *e.g.*, in RZ, Z may be R$_3$N$^+$, R$_2$S$^+$, R$_3$P$^+$, OTs, OCOR, etc., and the base may be OH$^-$, OR$^-$, NH$_2^-$, etc. (see text).

6. When alkyl halides are heated with ethanolic ammonia under pressure, a mixture of *amines*, *i.e.*, substituted ammonias, is obtained, *e.g.* (see also p. 367):

$$RX + NH_3 \longrightarrow RNH_2 + HX \longrightarrow RNH_2 \cdot HX \quad \text{or} \quad [RNH_3]^+X^-$$

7. When alkyl halides are heated with aqueous ethanolic potassium cyanide, alkyl *cyanides* are obtained (see p. 355):

$$RX + KCN \longrightarrow RCN + KX$$

When aqueous ethanolic silver cyanide is used, cyanides and *isocyanides* are formed (see p. 367):

$$RX + AgCN \longrightarrow RCN + AgX$$
$$RX + AgCN \longrightarrow RNC + AgX$$

8. Alkyl halides form esters when heated with an ethanolic solution of a silver salt.

$$R^1CO_2Ag + R^2X \longrightarrow R^1CO_2R^2 + AgX$$

The acid may be organic or inorganic. If the salt is silver nitrite two isomeric compounds are obtained, the *nitrite* (ester) and the *nitro-* compound (p. 361):

$$R—O—N{=}O \longleftarrow RX + Ag—O—N{=}O \longrightarrow R—\overset{+}{N}\underset{O^-}{\overset{\displaystyle O}{\diagdown}}$$

9. When alkyl halides are heated with sodium alkoxides, *ethers* are obtained (see p. 196):

$$R^1ONa + R^2X \longrightarrow R^1OR^2 + NaX$$

10. Alkyl halides heated with aqueous ethanolic sodium hydrogen sulphide form *thioalcohols* (p. 398), and *thioethers* (p. 401) when mercaptides are used.

$$RSH \xleftarrow{\text{NaSH}} RX \xrightarrow{\text{RSNa}} RSR$$

With sodium sulphite, a *sulphonate* is formed (p. 405):

$$RX + Na_2SO_3 \longrightarrow RSO_3Na + NaX$$

11. Alkyl halides may be used in the *Friedel-Crafts* reaction (p. 609) and to prepare Grignard reagents (p. 424).

Methyl chloride is prepared industrially by chlorinating methane with chlorine diluted with nitrogen, the ratio by volume of $CH_4{:}Cl_2{:}N_2$ being 8:1:80. The reaction is carried out in the presence of partly reduced cupric chloride as catalyst. All four chloromethanes are obtained, the methyl chloride comprising 90 per cent of the chlorine used:

$$CH_4 + Cl_2 + (N_2) \longrightarrow CH_3Cl + HCl + (N_2) + (CH_2Cl_2 + CHCl_3 + CCl_4)$$

By adjusting the ratio of chlorine to methane, each chloromethane can be obtained as the main product.

Methyl chloride, b.p. $-24°C$, is used in the manufacture of aniline dyes, silicone rubbers, tetramethyl-lead, methylation of cellulose, as a refrigerating agent, local anæsthetic and as a fire extinguisher.

Methyl bromide, b.p. $4.5°C$, is manufactured by passing the vapours of methanol and hydrobromic acid over a catalyst; it is used as a fumigating reagent.

Methyl iodide is prepared industrially:

(i) By warming a mixture of methanol and red phosphorus with iodine:

$$6CH_3OH + 2P + 3I_3 \longrightarrow 6CH_3I + 2H_3PO_3$$

(ii) By the action of methyl sulphate on potassium iodide solution in the presence of calcium carbonate:

$$KI + (CH_3)_2SO_4 \longrightarrow CH_3I + K(CH_3)SO_4 \qquad (90\text{–}94\%)$$

Methyl iodide is a sweet-smelling liquid, b.p. $42.5°C$. Since it is a liquid, it is easier to handle than methyl chloride, and so is used a great deal as a methylating agent in laboratory organic syntheses, but the chloride is used industrially because it is cheaper.

Ethyl chloride is prepared industrially:

(i) By the chlorination of ethane at about 400°C.

$$C_2H_6 + Cl_2 \longrightarrow C_2H_5Cl + HCl$$

The hydrogen chloride is used for method (ii).

(ii) By the addition of hydrogen chloride to ethylene (from cracked petroleum) in the presence of aluminium chloride as catalyst:

$$C_2H_4 + HCl \xrightarrow{\text{AlCl}_3} C_2H_5Cl$$

Ethyl chloride, b.p. $12.5°C$, is used in the preparation of tetraethyl-lead, etc., and as a refrigerating agent.

Ethyl bromide, b.p. 38·5°C, is manufactured by irradiating a mixture of ethylene and hydrogen bromide with γ-radiation from a cobalt-60 source.

$$HBr \xrightarrow{\gamma\text{-rays}} H\cdot + Br\cdot$$

$$C_2H_4 + Br\cdot \longrightarrow \cdot CH_2-CH_2Br \xrightarrow{HBr} CH_3CH_2Br + Br\cdot; \quad \text{etc.}$$

It is used for preparing barbiturate drugs.

Dihalogen derivatives

General methods of preparation.

1,1-Dihalides may be prepared:

(i) By the action of phosphorus pentahalides on aldehydes or ketones; *e.g.*, acetone gives isopropylidene dichloride when treated with phosphorus pentachloride:

$$CH_3COCH_3 + PCl_5 \longrightarrow CH_3CCl_2CH_3 + POCl_3$$

(ii) By the addition of halogen acids to alkynes, *e.g.*, ethylidene dibromide from acetylene and hydrogen bromide:

$$CH{\equiv}CH + HBr \longrightarrow CH_2{=}CHBr \xrightarrow{HBr} CH_3CHBr_2$$

1,2-Dihalides may be prepared by the addition of halogen to alkenes, *e.g.*, propene dibromide from propene and bromine:

$$CH_3CH{=}CH_2 + Br_2 \longrightarrow CH_3CHBrCH_2Br$$

The method of preparing an $\alpha\omega$-**dihalide** is special to the particular halide, *e.g.*, trimethylene dibromide may be prepared by the addition of hydrogen bromide to allyl bromide at low temperatures:

$$CH_2{=}CHCH_2Br + HBr \longrightarrow CH_2BrCH_2CH_2Br$$

Oldham (1950), however, has found that $\alpha\omega$-dibromides may be prepared by the action of bromine in carbon tetrachloride on the silver salt of a dicarboxylic acid (*cf.* p. 151):

$$(CH_2)_n(CO_2Ag)_2 + 2Br_2 \longrightarrow Br(CH_2)_nBr + 2AgBr + 2CO_2$$

The yields are very good provided n is 5 or more.

General properties and reactions of the dihalides.

The dihalides are sweet-smelling, colourless liquids.

1,1-Dihalides are hydrolysed by aqueous alkalis to the corresponding carbonyl compound (aldehyde or ketone), *e.g.*, isopropylidene dichloride gives acetone:

$$CH_3CCl_2CH_3 + 2KOH \longrightarrow 2KCl + [CH_3C(OH)_2CH_3] \longrightarrow CH_3COCH_3 + H_2O$$

It has generally been found that a compound containing two (or more) hydroxyl groups attached to the *same* carbon atom is unstable, and readily eliminates a molecule of water (see p. 228).

1,1-Dihalides gives alkenes when treated with zinc dust and methanol (p. 96), and alkynes when treated with ethanolic potassium hydroxide (p. 132).

1,2- and α,ω-Dihalides are just as reactive as the alkyl halides. When heated with zinc and methanol, 1,2-dihalides give alkenes (p. 96), but α,ω-dihalides in which the two halogen atoms are in the 1,3 to the 1,6 positions give cyclic compounds, *e.g.*, trimethylene dibromide gives cyclopropane:

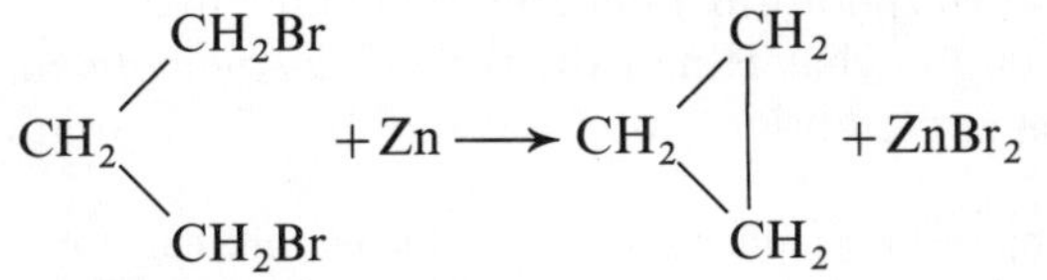

$$
\begin{array}{c}
CH_2Br \\
\diagup \\
CH_2 \qquad\qquad +Zn \longrightarrow \\
\diagdown \\
CH_2Br
\end{array}
\qquad
\begin{array}{c}
CH_2 \\
\diagup \quad | \\
CH_2 \quad | \qquad +ZnBr_2 \\
\diagdown \quad | \\
CH_2
\end{array}
$$

1,2-Di-iodides tend to eliminate iodine, particularly at raised temperatures, to form alkenes, *e.g.*, propene di-iodide gives propene:

$$CH_3CHICH_2I \xrightarrow{\text{heat}} CH_3CH{=}CH_2 + I_2$$

Methylene dichloride, b.p. 40°C, is prepared industrially by the direct chlorination of methane (see methyl chloride). It is used as an industrial solvent.

Methylene dibromide, b.p. 97°C, and **methylene di-iodide**, b.p. 181°C, are prepared by the partial reduction of bromoform and iodoform, respectively, with sodium arsenite in alkaline solution (the yield of CH_2Br_2 is 88–90 per cent; CH_2I_2, 90–97 per cent), *e.g.*,

$$CHI_3 + Na_3AsO_3 + NaOH \longrightarrow CH_2I_2 + NaI + Na_3AsO_4$$

All the methylene dihalides are used in organic syntheses.

Ethylene dichloride, b.p. 84°C, and **ethylidene dichloride**, b.p. 57°C, are isomers; the former is prepared from ethylene and chlorine, and the latter by the action of phosphorus pentachloride on acetaldehyde (*q.v.*).

Ethylene dibromide, b.p. 131°C, is made industrially by passing ethylene into cooled bromine; it is used as a component in anti-knock fluids.

Trihalogen derivatives

The most important trihalogen derivatives are those of methane, and they are usually known by their trivial names: chloroform, $CHCl_3$, bromoform, $CHBr_3$ and iodoform, CHI_3.

Chloroform may be prepared in the laboratory or industrially by heating ethanol or acetone with bleaching powder, or with chlorine and alkali (yield about 40 per cent). The reaction is extremely complicated, and the mechanism is obscure (see chloral, p. 227).

The equations usually given for the action of bleaching powder on ethanol are: (i) oxidation of ethanol to acetaldehyde; (ii) chlorination of acetaldehyde to trichloroacetaldehyde; (iii) decomposition of trichloroacetaldehyde (choral) by free calcium hydroxide (present in the bleaching powder) into chloroform and formic acid:

(i) $CH_3CH_2OH + Cl_2 \longrightarrow CH_3CHO + 2HCl$
(ii) $CH_3CHO + 3Cl_2 \longrightarrow CCl_3CHO + 3HCl$
(iii) $2CCl_3CHO + Ca(OH)_2 \longrightarrow 2CHCl_3 + (HCO_2)_2Ca$

When acetone is used, the first product given is trichloroacetone, which is then decomposed by the calcium hydroxide into chloroform and acetic acid:

(i) $CH_3COCH_3 + 3Cl_2 \longrightarrow CCl_3COCH_3 + 3HCl$
(ii) $2CCl_3COCH_3 + Ca(OH)_2 \longrightarrow 2CHCl_3 + (CH_3CO_2)_2Ca$

Chloroform is also prepared industrially by the chlorination of methane (see methyl chloride).

Chloroform is a sickly, sweet-smelling, colourless liquid, b.p. 61°C. It is sparingly soluble in water but readily soluble in ethanol and ether. It does not burn in air under usual conditions, but its vapour may be ignited, when it burns with a green-edged flame. According to Hine (1950, 1954), chloroform (and other haloforms) undergoes alkaline hydrolysis to produce the formate ion and carbon monoxide by what Hine calls the *alpha-elimination mechanism*: this involves the removal of hydrogen and chloride ions from the *same* carbon atom.

The mechanism proposed is as follows, involving the intermediate formation of dichloromethylene:

$$CHCl_3 + OH^- \rightleftharpoons :CCl_3^- + H_2O$$
$$:CCl_3^- \longrightarrow :CCl_2 + Cl^-$$
$$:CCl_2 \xrightarrow[H_2O]{OH^-} CO + HCO_2^- + 2Cl^-$$

The removal of a proton is facilitated by the $-I$ effect of the chlorine atoms. However, this cannot be the complete story since, although fluorine has a much stronger $-I$ effect than chlorine, $CHClF_2$ behaves differently. Hine *et al.* (1954–57) have shown that $CHCl_3$, $CHBr_3$, $CHBr_2F$, etc., readily undergo deuterium exchange in alkaline solution:

$$CHCl_3 + OH^- \rightleftharpoons H_2O + :CCl_3^- \xrightarrow{D_2O} CDCl_3 + OD^-$$

$CHClF_2$ was found to react more slowly. To account for this, it has been suggested that the CCl_3^- ion is stabilised by resonance. Cl can expand its shell to a decet because it has *d*-orbitals, but F has no *d*-orbitals and so this expansion cannot take place:

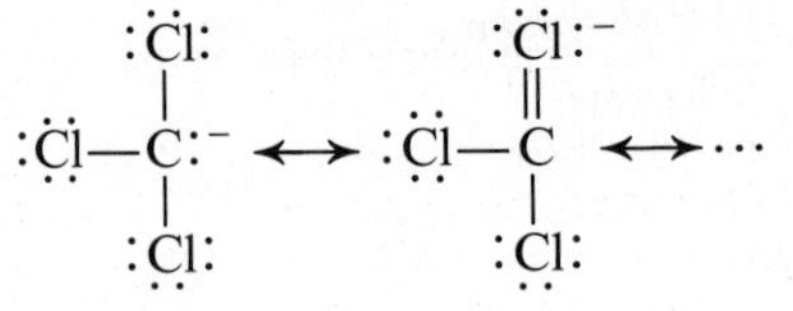

This type of resonance, which requires expansion of an octet, is known as **d orbital resonance**.

When chloroform is treated with zinc and hydrogen chloride in ethanolic solution, methylene dichloride (*q.v.*) is obtained; when treated with zinc and water, methane is obtained:

$$CHCl_3 \xrightarrow{Zn/H_2O} CH_4 + 3HCl$$

When chloroform is warmed with silver powder, acetylene is obtained:

$$2CHCl_3 + 6Ag \longrightarrow C_2H_2 + 6AgCl$$

When treated with concentrated nitric acid, chloroform forms *chloropicrin*:

$$CHCl_3 + HNO_3 \longrightarrow CCl_3NO_2 + H_2O$$

Chloropicrin or nitrochloroform (liquid, b.p. 112°C) is used as an insecticide, and has been used as a war-gas. Chloroform adds on to the carbonyl group of ketones in the presence of potassium hydroxide, *e.g.*, with acetone it forms *chloretone* (colourless needles, m.p. 97°C), which is used as a drug. The mechanism is possibly:

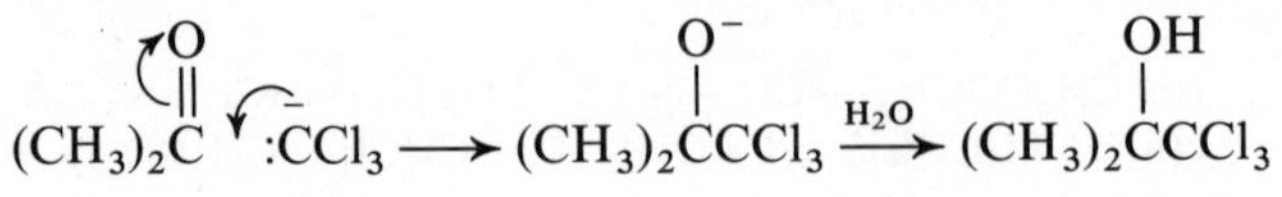

Chloroform has been used in surgery as an anæsthetic, and for this purpose it had to be pure. In the presence of air and light, chloroform slowly forms carbonyl chloride, which is extremely poisonous:

$$CHCl_3 + \tfrac{1}{2}O_2 \longrightarrow COCl_2 + HCl$$

Chlorine, water and carbon dioxide are also produced. Anæsthetic chloroform was therefore kept in well-stoppered dark-brown or blue bottles. Ethanol was also added (1 per cent), but its function is not quite clear. According to some authors, it *retards* the decomposition of the chloroform. This is supported by the fact that infra-red measurements of such mixtures show the absence of the carbonyl frequency.

A delicate test for chloroform is the 'isocyanide test'. This is carried out by heating chloroform with ethanolic potassium hydroxide and aniline, whereby phenyl isocyanide is formed, and is readily detected by its nauseating odour (see p. 357):

$$CHCl_3 + 3KOH + C_6H_5NH_2 \longrightarrow C_6H_5NC + 3KCl + 3H_2O$$

Chloroform is widely used in industry as a solvent for fats, waxes, resins, rubber, etc.

Bromoform may be prepared by methods similar to those used for chloroform, but it is prepared industrially by the electrolysis of an aqueous solution of acetone or ethanol containing sodium carbonate and potassium bromide (acetone gives a better yield than ethanol). The solution is maintained at about 20°C, and hydrobromic acid is run in to neutralise the sodium hydroxide produced during the electrolysis. Bromine is set free at the anode, and probably reacts in the same way as does chlorine in the preparation of chloroform.

Bromoform is a liquid, b.p. 149·5°C, and smells like chloroform, which it closely resembles chemically.

Iodoform is prepared industrially by the electrolysis of an aqueous solution of ethanol or acetone containing sodium carbonate and potassium iodide (ethanol gives a better yield than acetone). The solution is maintained at 60–70°C, and a current of carbon dioxide is passed through the solution to neutralise the sodium hydroxide formed.

Iodoform crystallises in yellow hexagonal plates, m.p. 119°C. It is insoluble in water, but is readily soluble in ethanol and ether. It is used as an antiseptic, but its antiseptic properties are due to the liberation of free iodine, and not to iodoform itself. Iodoform chemically resembles chloroform and bromoform.

Polyhalogen derivatives

Carbon tetrachloride, CCl_4, is prepared industrially in several ways:

(i) By the action of chlorine on carbon disulphide in the presence of aluminium chloride as catalyst:

$$CS_2 + 3Cl_2 \xrightarrow{\text{AlCl}_3} CCl_4 + S_2Cl_2$$

The sulphur monochloride is removed by fractional distillation, and the carbon tetrachloride is then shaken with sodium hydroxide, and finally distilled.

(ii) By the chlorination of methane (see methyl chloride).

(iii) By *chlorinolysis*. This term describes the process of chlorinating an organic compound under conditions which rupture the carbon–carbon bond to yield chloro-compounds with fewer carbon atoms than the original compound. Chlorinolysis may be effected with or without a catalyst, *e.g.*, the hydrocarbon and chlorine are heated at high temperature (300–400°C) and under high pressure (about 70 kg/cm^2). The product is usually a mixture, *e.g.*, propane gives both carbon tetrachloride and hexachloroethane:

$$C_3H_8 \xrightarrow{Cl_2} CCl_4 + C_2Cl_6$$

Carbon tetrachloride is a colourless liquid, b.p. 77°C, which has a sickly smell. It is insoluble in water but readily soluble in ethanol and ether. Since its vapour is non-inflammable, carbon tetrachloride is widely used as an industrial solvent (for fats, oils, resins, lacquers, etc.). It is also used as a fire-extinguisher under the name of *Pyrene*.

Carbon tetrachloride is stable at red heat (about 500°C), but when its vapour comes into contact with water vapour at this temperature, some carbonyl chloride is formed:

$$CCl_4 + H_2O \longrightarrow COCl_2 + 2HCl$$

Hence after using Pyrene to extinguish a fire, the room should be well ventilated.

Carbon tetrachloride is reduced by moist iron filings to chloroform (*q.v.*). The alkaline hydrolysis of carbon tetrachloride gives the same products (formate and carbon monoxide) as chloroform, but the rate of reaction is slower (Hine, 1954).

1,1,2,2-Tetrachloroethane (acetylene tetrachloride), b.p. 146°C, is manufactured by the reaction between acetylene and chlorine in the presence of ferric chloride at 80°C. It is non-inflammable, and hence is widely used as a solvent for oils, fats, paints, varnishes, rubber, etc.

Tetrachloroethane is the starting material for the industrial preparation of **trichloroethylene** (b.p. 88°C) and **tetrachloroethylene** (b.p. 121°C); both are used as solvents.

$$CHCl_2CHCl_2 \xrightarrow[\text{cat.}]{300-500°C} CCl_2{=}CHCl \xrightarrow[\text{light}]{Cl_2} CCl_3CHCl_2 \xrightarrow[\text{cat.}]{300°C} CCl_2{=}CCl_2$$

1,2-Dichloroethylene may be prepared by the action of zinc dust on 1,1,2,2-tetrachloroethane.

$$CHCl_2CHCl_2 \xrightarrow{Zn} CHCl{=}CHCl$$

It is manufactured by the action of iron borings on hexachloroethane. It is a liquid, and exists as two geometrical isomers, *cis*, b.p. 60°C; *trans*, b.p. 48°C, and is used as a rubber solvent, etc.

Hexachloroethane (*perchloroethane*), m.p. 187°C, is manufactured by passing a mixture of s-tetrachloroethane and chlorine over aluminium chloride, and by the chlorinolysis of propane (see carbon tetrachloride, above). It is used as a moth repellent.

Fluorine derivatives of the alkanes

Most organic compounds burn or explode when treated with fluorine gas. This great reactivity is partly due to the low dissociation energy of the fluorine molecule (150·6 kJ mol^{-1}). Also, one of the termination steps is R· + F· → RF. This is very strongly exothermic (447·7 kJ), and the energy liberated is larger than that required to break a C—C bond (347·3 kJ), thereby resulting in fission of the organic molecule.

Aliphatic fluorine compounds may be obtained in several ways:

(i) Direct fluorination of hydrocarbons may be carried out successfully by diluting the fluorine with nitrogen, and carrying out the reaction in a metal tube packed with copper gauze at a temperature of 150–350°C. It is very difficult to control the fluorination, and the product is usually a complex mixture, *e.g.*, methane gives CH_3F, CH_2F_2, CHF_3, CF_4, C_2F_6 and C_3F_8; ethane gives CF_4, C_2F_6, CH_3CHF_2, CH_2FCHF_2; no mono- or s-difluoroethane is obtained.

When catalysts other than copper (actually CuF_2) are used, *e.g.*, AgF, CoF_2, CeF_3, MnF_2, perfluoro-compounds are obtained, *e.g.*, n-heptane gives perfluoroheptane:

$$2AgF + F_2 \longrightarrow 2AgF_2$$
$$C_7H_{16} \xrightarrow{AgF_2} C_7F_{16}$$

It appears that when a catalyst is used, the perfluoro-compound obtained usually has the same number of carbon atoms as the original compound; if no catalyst is used, fluoro-compounds with fewer carbon atoms are usually obtained.

The mechanism of direct fluorination is still obscure, but it appears that the first step is the conversion of the catalyst into a higher fluoride, *e.g.*, AgF into AgF_2; CoF_2 into CoF_3, etc. Some of these higher fluorides have been isolated, *e.g.*, AgF_2.

(ii) Alkenes and alkynes add on hydrogen fluoride under pressure to form fluoro-derivatives of the alkanes, *e.g.*,

$$C_2H_4 + HF \longrightarrow C_2H_5F \quad (f.g.\text{-}g.)$$
$$CH_3C{\equiv}CH + 2HF \longrightarrow CH_3CF_2CH_3 \quad (f.g.\text{-}g.)$$

If the unsaturated compound contains a halogen atom (other than fluorine), this atom may be replaced by fluorine, *e.g.*,

$$RCCl{=}CH_2 + HF \longrightarrow RCFClCH_3$$
$$RCCl{=}CH_2 + 2HF \longrightarrow RCF_2CH_3 + HCl$$

Lead tetrafluoride (from lead dioxide and hydrogen fluoride) is particularly useful for the addition of two fluorine atoms to an alkene containing chlorine, *e.g.*,

$$CCl_2{=}CCl_2 + PbF_4 \xrightarrow{\text{25°C}} CFCl_2CFCl_2 + PbF_2$$

It is also possible to add fluorine directly without a catalyst to highly halogenated alkenes, *e.g.*,

$$CFCl{=}CFCl + F_2 \longrightarrow CF_2ClCF_2Cl$$

(iii) Fluorine compounds may be prepared indirectly by heating organic halides with inorganic fluorides such as AsF_3, SbF_3, AgF, Hg_2F_2, etc., *e.g.*,

$$C_2H_5Cl + AgF \longrightarrow C_2H_5F + AgCl \quad (ex.)$$

This method was first used by Swarts (1898), and so is known as the **Swarts reaction**.

When the organic halide contains two or three halogen atoms attached to the same carbon atom, the best yield of fluoride is obtained when CoF_3 is used, but SbF_3 gives yields almost as good (and is more accessible), *e.g.*,

$$3CH_3CCl_2CH_3 + 2SbF_3 \longrightarrow 3CH_3CF_2CH_3 + 2SbCl_3 \quad (v.g.)$$

Alternatively, hydrogen fluoride may be used under pressure in the presence of a catalyst, *e.g.*,

$$CCl_4 \xrightarrow[\text{(C + FeCl}_3)]{\text{HF: 300°C}} CF_2Cl_2 + CFCl_3$$

(iv) Another method of fluorination is the direct electrolytic method. Nickel electrodes are used, and electrochemical fluorination takes place at the anode, the reaction being carried out by the electrolysis of a solution of the organic compound in anhydrous hydrogen fluoride, *e.g.*,

$$CH_3CO_2H \longrightarrow CF_3COF$$
$$C_2H_5OC_2H_5 \longrightarrow C_2H_5OC_2F_5$$

The particular merit of this method is that it usually leaves untouched many types of functional groups.

Sulphur tetrafluoride (Smith, 1962) and phenylsulphur trifluoride, $PhSF_3$ (Sheppard, 1962) are capable of replacing *oxygen* atoms by fluorine; *e.g.*,

$$RCO_2H \xrightarrow{SF_4} RCF_3; \qquad R_2CO \longrightarrow R_2CF_2; \qquad ROH \longrightarrow RF$$

Bromine trifluoride is also useful for replacing oxygen atoms (Stevens, 1961).

The lower n-alkyl fluorides are gases. The first four members are stable, and the higher members tend to decompose spontaneously into alkene and hydrogen fluoride, *e.g.*,

$$CH_3CH_2CH_2CH_2CH_2F \longrightarrow CH_3CH_2CH_2CH{=}CH_2 + HF$$

Secondary and tertiary alkyl fluorides are so unstable that it is impossible to prepare them free from alkene.

1,2-Difluorides are also usually unstable, *e.g.*, ethylene difluoride decomposes spontaneously at 0°C into hydrogen fluoride and butadiene:

$$2CH_2FCH_2F \longrightarrow 4HF + CH_2{=}CH{-}CH{=}CH_2$$

Alkyl fluorides are readily hydrolysed by strong acids to the corresponding alcohols; alkalis have no effect. On the other hand, ethylene difluoride is immediately hydrolysed by water to glycol:

$$CH_2FCH_2F + 2H_2O \longrightarrow HOCH_2CH_2OH + 2HF$$

Fluorides with two or three fluorine atoms on the same carbon atom are stable to water and strong acids, *e.g.*, CHF_3, CHF_2CHF_2, etc.

Alkyl fluorides do not react with sodium, *i.e.*, do not undergo the Wurtz reaction, and do not form Grignard reagents. An interesting compound is trifluoromethyl iodide, CF_3I. It is converted into fluoroform, CHF_3, by potassium hydroxide, and it combines directly with many non-metals such as P, As, Sb, S, Se, to give, *e.g.*, with phosphorus, $(CF_3)_3P$, $(CF_3)_2PI$ and CF_3PI_2.

Chlorofluoro-derivatives of methane and ethane are used as refrigerants and for air-conditioning under the name of *Freons*, which are prepared by the action of hydrogen fluoride on carbon tetrachloride, chloroform and hexachloroethane.

Tetrafluoroethylene, b.p. $-76°C$, is prepared by the action of antimony trifluoride and hydrogen fluoride on chloroform, and then heating the chlorodifluoromethane so produced at 800°C:

$$CHCl_3 \xrightarrow[HF]{SbF_3} CHF_2Cl \xrightarrow{800°C} C_2F_4 + 2HCl + \text{other products}$$

When tetrafluoroethylene is polymerised, the plastic *Teflon* is produced. Teflon is difficult to work, but is inert to chemical reagents, even to boiling aqua regia.

Polychlorofluoroethylenes are valuable as oils and greases. Perfluoroheptane is useful in a process for the separation of uranium isotopes by gaseous diffusion.

The C—F bond is much stronger than any other C—X bond (see Table 2.2, p. 36) and increases in strength as the number of fluorine atoms attached to the carbon atom increases (CH_3F, 447·7; CF_4, 485·3 kJ). It is largely because of this that *perfluoro* compounds are stable to heat and very resistant to attack by chemical reagents.

PROBLEMS

1. Write out the structures and IUPAC names of all the dichloro-derivatives of:
   (i) butane; (ii) isobutane.
2. Distinguish between ethylene dichloride and ethylidene dichloride:
   (i) chemically; (ii) by one physical method.
3. Draw all possible stereoisomers for:
   (i) $MeCHBrCH{=}CH_2$; (ii) $MeCHOHCH{=}CHMe$; (iii) $MeCHClCO_2H$; (iv) $BrCH_2CH_2CO_2H$;
   (v) $Me_2C(OH)CO_2H$.
4. Explain the following reactions:

   (i) $n\text{-BuBr} + KCN \xrightarrow{EtOH-H_2O} n\text{-BuCN}$

   (ii) $t\text{-BuBr} + KCN \xrightarrow{EtOH-H_2O} Me_2C{=}CH_2$
5. Explain the following reactions:
   (i) $EtOH + HI \longrightarrow EtI$ \qquad (ii) $EtOH + HCN \longrightarrow$ no reaction
6. Calculate $\Delta H^\ominus$ for the reaction:

$$C_2H_6 + Br_2 \longrightarrow C_2H_5Br + HBr$$

Is the reaction thermodynamically favourable? If so, will it proceed in the dark at room temperature? If not, why not?

7. Treatment of $Me_3CCH{=}CH_2$ and $Me_3CCHOHMe$ with concentrated hydrochloric acid gives the same two isomeric alkyl chlorides. What are these two products? Explain.

8. Complete the following equations:

(i) $ClCH_2CH_2CH_2Br + KCN \xrightarrow[\text{heat}]{\text{EtOH--H}_2\text{O}}$?

(ii) $PhCHO \xrightarrow{\text{SF}_4}$?

(iii) $BrCH_2CH{=}CHCO_2Me \xrightarrow{\text{AgF}}$?

(iv) $isoPrBr \xrightarrow{?} PrH$

(v) $Me_2C{=}CH_2 + HCl \xrightarrow{\text{peroxide}}$?

9. $EtCH_2CHBrMe \xrightarrow[(-\text{HBr})]{\text{EtOK}} \underset{A}{1\text{-ene}} + \underset{B}{2\text{-ene}} + \underset{C}{2\text{-ene}}$

What are A, B, and C, and which one predominates? (Assume the mechanism is E2.)

10. Discuss the reaction: $ClCH_2CHCl_2 \xrightarrow{\text{OH}^-} CH_2{=}CCl_2$

11. The hydrolysis of n-BuCl in aqueous ethanol is accelerated in the presence of NaI. Explain, and draw the energy profiles for the uncatalysed and catalysed reactions.

12. What is the simplest alkane that is optically active?

13. The NMR spectrum of the compound $C_3H_5Cl_3$ showed two signals, one a doublet and the other a quintet. What is the structure of the compound?

14. State the number of signals and their multiplicity for the NMR spectra of:

(i) CH_2F_2; (ii) Me_3CCH_2Cl; (iii) Me_3CCl; (iv) $ClCH_2CHClCCl_3$; (v) $ClCH_2CCl_2CH_3$.

15. The table gives some of the peaks in the mass spectra of two isomeric straight-chain alkyl halides, A and B. One is a primary halide. What is this one? Give your reasons.

m/e	41	43	57	85	135	164	166
A(%)	38	100	22	49	50	2·3	2·3
B(%)	38	100	12	57	1	1	1

REFERENCES

HUNTRESS, *Organic Chlorine Compounds*, Wiley (1949).

Organic Reactions, Wiley. Vol. IX (1957), Ch. 5. 'The Reaction of Halogens with Silver Salts of Carboxylic Acids.'

BUNTON, *Nucleophilic Substitution at a Saturated Carbon Atom*, Elsevier (1963).

BANTHORPE, *Elimination Reactions*, Elsevier (1963).

FRAZER and SINGER, 'Thermochemical Cycles', *Educ. in Chem.*, 1964, **1**, 39.

IBNE-RASA, 'Equations for Correlation of Nucleophilic Reactivity', *J. Chem. Educ.*, 1967, **44**, 89.

PHELAN et al., 'A Molecular Orbital Description of the Non-Classical Ion in 1,2-Rearrangements', *J. Chem. Educ.*, 1967, **44**, 626.

GERRARD and HUDSON, 'Rearrangement in Alkyl Groups During Substitution Reactions', *Chem. Rev.*, 1965, **65**, 697.

MCLENNAN, 'The Carbanion Mechanism of Olefin-forming Elimination', *Quart. Rev.*, 1967, **21**, 490.

DAVIDSON, 'Hydrogen Abstraction in the Liquid Phase by Free Radicals', *Quart. Rev.*, 1967, **21**, 249.

STACEY et al. (eds.), *Advances in Fluorine Chemistry*, Butterworths. Vols. 1–5 (1960–1965).

SHARTS, 'Organic Fluorine Chemistry', *J. Chem. Educ.*, 1968, **45**, 185.

6

Monohydric alcohols

An alcohol is a compound that contains one or more *hydroxyl* groups, *i.e.*, alcohols are hydroxy-derivatives of the alkanes. They are classified according to the number of hydroxyl groups present. Monohydric alcohols contain one hyroxyl group; dihydric, two; trihydric, three; etc. When the alcohols contain four or more hydroxyl groups, they are usually called polyhydric alcohols.

The monohydric alcohols form an homologous series with the general formula $C_nH_{2n+2}O$, but, since their functional group is the hydroxyl group, their general formula is more satisfactorily written as $C_nH_{2n+1}OH$ or ROH.

Nomenclature.

The simpler alcohols are commonly known by their trivial names, which are obtained by naming the alcohol as a derivative of the alkyl group attached to the hydroxyl group, *e.g.*, CH_3OH, methyl alcohol; $CH_3CH_2CH_2OH$. n-propyl alcohol; $CH_3CH(OH)CH_3$, isopropyl alcohol; $(CH_3)_3COH$, t-butyl alcohol.

Another system of nomenclature considers the alcohols as derivatives of methyl alcohol, which is named *carbinol* or *methanol*, *e.g.*, $CH_3CH_2CHOHCH_3$, ethylmethylcarbinol or ethyl-methylmethanol (both methods have been used in this book).

In the I.U.P.A.C. system of nomenclature, the longest carbon chain containing the hydroxyl group is chosen as the parent hydrocarbon. The class suffix is *-ol*, and the positions of side-chains and the hydroxyl group are indicated by numbers, the lowest possible number being given to the hydroxyl group (p. 94), *e.g.*, CH_3OH, methanol; C_2H_5OH, ethanol; $CH_3CH_2CH_2OH$, propan-1-ol; $(CH_3)_2CHCHOHCH_3$, 3-methylbutan-2-ol. According to this system the alcohols are referred to as the **alkanols**.

Monohydric alcohols are subdivided into primary, secondary and tertiary alcohols according as the alkyl group attached to the hydroxyl group is a primary, secondary or tertiary group, respectively. Primary alcohols contain the *primary alcoholic group*—CH_2OH, *e.g.*, ethanol, CH_3CH_2OH; secondary alcohols, the *secondary alcoholic group* $\cdot CH(OH)\cdot$, *e.g.*, isopropanol, $(CH_3)_2CHOH$; and tertiary alcohols the *tertiary alcoholic group* $\equiv C(OH)$, *e.g.*, t-butanol, $(CH_3)_3COH$.

General methods of preparation.

1. By the hydrolysis of an alkyl halide with aqueous alkali or silver oxide suspended in water:

$$RX + \text{'AgOH'} \longrightarrow ROH + AgX$$

2. By the hydrolysis of esters with alkali:

$$R^1CO_2R^2 + KOH \longrightarrow R^1CO_2K + R^2OH$$

This method is important industrially for preparing certain alcohols that occur naturally as esters. Since tertiary alkyl halides give large amounts of alkene on hydrolysis, they are best converted into alcohols as follows:

$$R_3CX + CH_3CO_2Ag \xrightarrow[\text{heat}]{\text{EtOH}} AgX + CH_3CO_2CR_3 \xrightarrow{\text{NaOH}} CH_3CO_2Na + R_3COH$$

3. By heating ethers with dilute sulphuric acid under pressure, *e.g.*, diethyl ether forms ethanol:

$$(C_2H_5)_2O + H_2O \xrightarrow{H_2SO_4} 2C_2H_5OH$$

This method is important industrially, since ethers are formed as by-products in the preparation of certain alcohols (see ethanol and propanols).

4. By the reduction of aldehydes, ketones or esters by means of excess of sodium and ethanol or n-butanol as the reducing agent (*Bouveault-Blanc reduction*, 1903), *e.g.*,

(i) *Aldehydes*: $RCHO \xrightarrow{e; H^+} RCH_2OH$ (*g.-v.g.*)

(ii) *Esters*: $R^1CO_2R^2 \xrightarrow{e; H^+} R^1CH_2OH + R^2OH$ (*g.*)

(iii) *Ketones*: $R_2CO \xrightarrow{e; H^+} R_2CHOH$ (*g.*)

The Bouveault-Blanc reduction is believed to occur in steps involving the transfer of one electron at a time; *e.g.*,

$$RC\underset{\underset{O}{\|}}{-}OEt \xrightarrow{Na} R\overset{..}{C}\underset{\underset{O\cdot}{|}}{-}OEt \xrightarrow{EtOH} RCH\underset{\underset{O\cdot}{|}}{-}OEt \xrightarrow{Na} RCH\overset{\frown}{-}OEt \longrightarrow$$

$$OEt^- + RCH{=}O \xrightarrow{Na} R\overset{..}{C}H{-}O\cdot \xrightarrow{EtOH} RCH_2{-}O\cdot \xrightarrow{Na} RCH_2{-}O^- \xrightarrow{EtOH} RCH_2OH$$

Reduction with metallic hydrides. Many complex metallic hydrides reduce various functional groups, and their number is increasing rapidly. The most versatile reagent is **lithium aluminium hydride** (LAH). This reduces most functional groups (see Table 6.1), but does not normally reduce the olefinic bond (see also p. 774). An unusual feature of this reagent is its reduction of the *carboxyl group* to primary alcohol.

Reductions with lithium aluminium hydride are usually carried out in ethereal solutions, the compound in ether being added to the lithium aluminium hydride solution. In certain cases the reverse addition is necessary, *i.e.*, the hydride solution is added to the solution of the compound to be reduced.

Sodium borohydride, which is insoluble in ether, is used in *ethanolic* solution to reduce *carbonyl* compounds, the only important exception being the carboxyl group. It also does not normally reduce esters, but reduction to primary alcohol can often be effected by use of a large excess of reagent in methanol.

The reduction of the carbonyl group by LAH or sodium borohydride occurs in a stepwise manner, each step involving hydride ion transfer, *e.g.*,

$$R_2C{=}O \quad H{-}\bar{A}lH_3 \longrightarrow \left[R_2\overset{O^-}{CH} + AlH_3 \right] \longrightarrow R_2CH{-}O{-}\bar{A}lH_3 \xrightarrow{R_2CO}$$

$$(R_2CHO)_2\bar{A}lH_2 \xrightarrow{R_2CO} (R_2CHO)_3\bar{A}lH \xrightarrow{R_2CO} (R_2CHO)_4Al^- \xrightarrow{H^+} 4R_2CHOH$$

Experimental work has shown that the rate of reduction is decreased as the size of R increases. One contributing factor is steric hindrance, and this is made use of by employing reducing complexes containing large groups, *e.g.*, when the substrate contains two carbonyl groups and one is sterically hindered, lithium tri-t-butoxyaluminium hydride (Table 6.1) reduces only the non-hindered carbonyl group. This reagent is generally used in tetrahydrofuran (THF) solution. In addition to steric hindrance, there is also the inductive effect in the anion, *e.g.*,

$$(R{-}O{\twoheadrightarrow})_3Al \twoheadleftarrow H \quad \overset{\diagdown}{\underset{\diagup}{C}}{=}O$$

The electronegative oxygen atom exerts a strong $-I$ effect and consequently makes hydride ion transfer more difficult. Hence, the reactivity of lithium tri-t-butoxyaluminium hydride is lower than that of LAH.

Many other metallic hydrides are also reducing reagents, and because of their different reactivities, they offer a means of *selective* reduction.

Lithium borohydride (in ether or THF) is more reactive than sodium borohydride but less than LAH (Table 6.1).

Lithium tri-t-butoxyaluminium hydride (in THF) is useful for reducing acid chlorides to aldehydes (Table 6.1).

Sodium aluminium hydride (in THF) is similar to LAH, but it reduces esters to aldehydes.

Diborane (in THF) reduces many functional groups (Table 6.1).

Aluminium hydride (in ether or THF) is similar to LAH, but is better for reducing α,β-unsaturated carbonyl compounds to unsaturated alcohols.

The reducing power of a *given* metallic hydride is affected by the nature of the solvent and the presence of certain other compounds, *e.g.*, **LAH and aluminium chloride** in ether form AlH_3, AlH_2Cl, or $AlHCl_2$ according to the proportion of reagents used:

$$3LiAlH_4 + AlCl_3 \longrightarrow 4AlH_3 + 3LiCl$$
$$LiAlH_4 + AlCl_3 \longrightarrow 2AlH_2Cl + LiCl$$
$$LiAlH_4 + 3AlCl_3 \longrightarrow 4AlHCl_2 + LiCl$$

Hence, the reducing power of the reagent will depend on which one is actually present; all are milder reducing reagents than LAH, *e.g.*, LAH—$AlCl_3$ does not reduce alkyl halides. It has also been shown that a solution of LAH in pyridine which has been allowed to stand will reduce a carbonyl group but not a carboxyl or carbalkoxyl group. Thus, LAH has been converted into a milder reducing reagent, the structure of which appears to be uncertain.

Sodium borohydride-aluminium chloride (in diethylene glycol dimethyl ether, *i.e.*, diglyme, p. 314) is more reactive than sodium borohydride itself (Table 6.1).

Catalytic hydrogenation. Many functional groups are reduced catalytically (Table 6.1), the most common catalysts being nickel, platinum, palladium, rhodium, and ruthenium; other catalysts used are copper chromite and copper-barium-chromium oxide. The catalytic activity of a given metal is dependent on its method of preparation, *e.g.*, nickel (p. 75). It also depends on other factors: (*a*) the presence of other compounds which may either increase (*promoters*) or decrease or inhibit (*poisons*) catalytic activity; (*b*) the nature of the solvent: neutral, basic, or acidic.

In addition to having different reactivities, catalysts have different selectivities, *e.g.*, carboxylic acids can be hydrogenated to alcohols with a Ru—C catalyst but not with a Pt catalyst. In general, catalysts are deposited on the surface of a support such as charcoal, alumina, calcium carbonate, etc., and the activity of the catalyst may be changed by changing the support.

Table 6.1

Group	Product	$LiAlH_4$ in ether	$LiBH_4$ in THF	$LiAlH(OBu^t)_3$ in THF	$NaBH_4$ in EtOH	$NaBH_4 + AlCl_3$ in diglyme	B_2H_6 in THF	H_2 + catalyst
—CHO	—CH$_2$OH	+	+	+	+	+	+	+
>CO	>CHOH	+	+	+	+	+	+	+
—CO$_2$H	—CH$_2$OH	+	−	−	−	+	+	+[1]
—CO$_2$R	—CH$_2$OH	+	+	−	−[2]	+	+	
—COCl	—CH$_2$OH	+	+	[3]	+	+	−	+[4]
—CONH$_2$	—CH$_2$NH$_2$	+			−	+	+	+
(RCO)$_2$O	RCH$_2$OH	+			−	+	+	+
lactone	diol	+	+		−	+	+	+
epoxide	alcohol	+	+		−	+	+	+
—CN	—CH$_2$NH$_2$	+[5]	−	−	−	+	+	+
C=NOH	—CH$_2$NH$_2$	+						+
RNO$_2$	RNH$_2$	+	−		−	−	−	+
ArNO$_2$	ArN=NAr	+	−		−	−	−	+[6]
azoxy	azo	+						+[6]
C=C	CH—CH	−		−	−	+	+	+
—C≡C—	—CH=CH—	−[7]					+	+
RX	RH	+			+		−	+
ArX	ArH					−	−	+

1. Only Ru—C and Cu—Ba—Cr oxide. 2. Often reduces if use large excess of $NaBH_4$ in MeOH. 3. Gives aldehyde at −78°C. 4. Gives aldehyde with Pd—BaSO$_4$—S—quinoline. 5. Reverse addition gives aldehyde. 6. Also to amine. 7. LiAlH$_4$—AlCl$_3$ reduces RC≡CH (and ArC≡CH).

5. Primary, secondary and tertiary alcohols may be prepared by means of a Grignard reagent and the appropriate carbonyl compound.

6. A number of alcohols are obtained by fermentation processes (see later).

7. In recent years synthetic methods have become very important for preparing various alcohols commercially:

(i) By the hydration of alkenes (pp. 108, 114).

(ii) By heating a mixture of carbon monoxide and hydrogen under pressure in the presence of a catalyst, *e.g.*, zinc chromite plus small amounts of alkali metal or iron salts. A mixture of alcohols containing methyl, ethyl, n-propyl, isobutyl and higher-branched alcohols is obtained, the individuals being separated by fractional distillation (see below).

(iii) The **Oxo process** (*Oxo synthesis, carbonylation* or *hydroformylation reaction*). A mixture of alkene, carbon monoxide and hydrogen, under pressure and elevated temperature, in the presence of a catalyst, forms aldehydes. A common catalyst is cobalt carbonyl hydride, $[CoH(CO)_4]$, and the product is a mixture of isomeric straight-chain and branched-chain aldehydes (the former predominating). These are reduced catalytically to the corresponding alcohols, *e.g.*, propene gives a mixture of n- and isobutanols:

$$2MeCH{=}CH_2 + 2CO + 2H_2 \longrightarrow Me(CH_2)_2CHO + Me_2CHCHO \xrightarrow[Cu-Zn]{H_2}$$

$$Me(CH_2)_2CH_2OH + Me_2CHCH_2OH$$

The aldehydes are first separated by fractional distillation.

(iv) Methanol, ethanol, propanols and butanols are prepared industrially by the oxidation of natural gas (p. 91).

Most of the methods given above can be used for the preparation of any particular class of alcohol: it is only a question of starting with the appropriate compound, *e.g.*, all three types may be prepared via Grignard reagents (see also p. 427): *primary* alcohols from formaldehyde ($R^1 = R^2 = H$), *secondary* from aldehydes ($R^1 = R; R^2 = H$), and *tertiary* from ketones (R^1 and R^2 = alkyl groups).

$$R^1C\!\!\overset{O}{\underset{R^2}{\diagdown}} + R^3MgX \longrightarrow R^1C\!\!\overset{OMgX}{\underset{R^2}{\diagup}}\!\!R^3 \xrightarrow{H^+} R^1R^2R^3COH$$

General properties of the alcohols.

The alcohols are neutral substances: the lower members are liquids, and have a distinctive smell and a burning taste; the higher members are solids and are almost odourless.

In a group of isomeric alcohols, the primary alcohol has the highest boiling point and the tertiary the lowest, with the secondary having an intermediate value. The lower members are far less volatile than is to be expected from their molecular weight, and this is due to association through hydrogen bonding *extending over a chain of molecules*, thus giving rise to a 'large molecule' the volatility of which would be expected to be low:

$$\cdots O\!\!-\!\!H\underset{|}{\overset{R}{\cdots}}O\!\!-\!\!H\underset{|}{\overset{R}{\cdots}}O\!\!-\!\!H\underset{|}{\overset{R}{\cdots}}O\!\!-\!\!H\underset{|}{\overset{R}{\cdots}}\cdots$$

The lower alcohols are very soluble in water, and the solubility diminishes as the molecular weight increases. Their solubility in water is to be expected, since the oxygen atom of the hydroxyl group in alcohols can form hydrogen bonds with the water molecules. In the lower alcohols the hydroxyl group constitutes a large part of the molecule, whereas as the molecular weight of the alcohol increases the hydrocarbon character of the molecule increases, and hence the solubility in water decreases. This, however, is not the complete story; the structure of the carbon chain also plays a part, *e.g.*, n-butanol is fairly soluble in water, but t-butanol is miscible with water in all proportions.

General reactions of the alcohols.

1. Alcohols react with organic and inorganic acids to form *esters*:

$$R^1CO_2H + R^2OH \longrightarrow R^1CO_2R^2 + H_2O$$

Esters of the halogen acids are, as we have seen (p. 146), the alkyl halides.

The order of reactivity of an alcohol with a given acid is primary alcohol > secondary > tertiary, provided the mechanism is bimolecular for all cases (see p. 248). With a given alcohol, the order of reactivity of the halogen acids in the bimolecular mechanism is HI > HBr > HCl.

Let us consider the following example:

$$EtOH + H^+ + X^- \underset{}{\overset{fast}{\rightleftharpoons}} \ X^- + EtOH_2^+ \ \rightleftharpoons \ \overset{\delta-}{X} \cdots Et \cdots \overset{\delta+}{OH_2} \overset{fast}{\longrightarrow} X - Et + H_2O$$

The order of reactivity of the halogen acids is attributed to the variation in polarisability of the anion; the greater the size of the anion, the greater is its polarisability (p. 34). What this amounts to is that the larger the anion, the more readily it can donate a lone pair to form a covalent bond. Also, the nature of the alkyl group will influence the mechanism of halide formation (in aqueous solution) in the same way as it does their hydrolysis. Protonated t-alcohols readily eliminate a molecule of water to form a carbonium ion and so react by the S_N1 mechanism. Straight-chain primary alcohols will favour the S_N2 mechanism, but if branched-chain, then the S_N1 mechanism may operate (see neopentyl alcohol, p. 147). Secondary alcohols would be expected to react by both mechanisms.

2. Alcohols react with phosphorus halides to form alkyl halides (p. 148).

3. Alcohols combine with phenyl isocyanate to form phenyl-substituted urethans (p. 459):

$$C_6H_5NCO + ROH \longrightarrow C_6H_5NHCO_2R$$

4. Strongly electropositive metals (K, Na, Mg, Al, Zn) liberate hydrogen from alcohols to form *alkoxides*, *e.g.*, sodium reacts with ethanol to form sodium ethoxide:

$$2C_2H_5OH + 2Na \longrightarrow 2C_2H_5O^-Na^+ + H_2$$

Alkoxides are white deliquescent solids, readily soluble in water with hydrolysis:

$$RO^-Na^+ + H_2O \ \rightleftharpoons \ ROH + NaOH$$

Alkoxides react with carbon disulphide to form xanthates:

$$RONa + C\overset{\displaystyle S}{\underset{\displaystyle S}{\big\|}} \longrightarrow RO-C\overset{\displaystyle S}{\underset{\displaystyle SNa^+}{\big\|}}$$

The fact that alcohols liberate hydrogen by the action of metals shows that alcohols can behave as acids, but since they do not affect the pH of water, they are weaker acids than water.

This accounts for alkoxides being hydrolysed by water and for the fact that the ethoxide ion (the cB of EtOH) is a stronger base than the hydroxide ion (the cB of H_2O; see p. 64).

Since alkyl groups have a $+I$ effect, there will be an increased electron displacement towards the oxygen atom in going from primary to secondary to tertiary alcohol. This may be represented (qualitatively) as follows:

$$Me \rightarrow CH_2 \overset{\delta-}{\rightarrow} O \rightarrow H \qquad \underset{Me}{\overset{Me}{\diagdown}}CH \overset{2\delta-}{\twoheadrightarrow} O \twoheadrightarrow H \qquad \underset{Me}{\overset{Me}{\diagdown}}Me \rightarrow C \overset{3\delta-}{\Rrightarrow} O \Rrightarrow H$$

The greater the negative charge on the oxygen atom, the closer is the covalent pair in the O—H bond driven to the hydrogen atom and consequently separation of a proton becomes increasingly difficult. Thus the acid strengths of alcohols will be in the order: prim. $>$ s $>$ t. This is the order of reactivity of alcohols towards metals, and with t-alcohols the reaction with sodium is so slow that it is better to use the more electropositive potassium for these alcohols, e.g., potassium t-butoxide is the usual salt prepared from t-butanol.

From what has been said above, it can be seen that the tendency for the C—O bond to break will be the reverse of that for the O—H bond, i.e., reactions involving the breaking of the C—O bond will follow the order of reactivity: t $>$ s $>$ prim. This order would be expected from the consideration of the stabilities of the carbonium ions produced (p. 105).

If we now consider β-substituted alcohols containing strong $-I$ groups, the acidity of the alcohol is increased, e.g., pK_a of ethanol is 15·5 whereas that of trifluoroethanol is 12·4.

$$Me \rightarrow CH_2 \rightarrow O—H \qquad CF_3 \lll CH_2 \lll O—H$$

The Me group is weakly $+I$, whereas the CF_3 group is strongly $-I$. Thus the covalent pair is pulled away from the oxygen atom in the C—O bond in the latter compound, and consequently the covalent pair in the O—H bond is drawn towards the O atom, thereby facilitating release of the hydrogen as a proton.

A number of alkoxides are important as synthetic reagents; e.g., sodium ethoxide, C_2H_5ONa; aluminium ethoxide, $(C_2H_5O)_3Al$; aluminium t-butoxide $[(CH_3)_3CO]_3Al$ (see text for their uses). The aluminium alkoxides may be conveniently prepared by the action of aluminium amalgam or aluminium shavings on the alcohol.

5. Primary and secondary alcohols may be acetylated with acetyl chloride, e.g., ethanol gives ethyl acetate:

$$CH_3COCl + C_2H_5OH \longrightarrow CH_3CO_2C_2H_5 + HCl$$

With tertiary alcohols the reaction is often accompanied by dehydration of the alcohol to alkene or by the formation of a tertiary alkyl chloride; e.g., t-butanol gives a good yield of t-butyl chloride:

$$(CH_3)_3COH + CH_3COCl \longrightarrow (CH_3)_3CCl + CH_3CO_2H$$

However, in the presence of a base such as dimethylaniline, the ester is produced in good yield:

$$CH_3COCl + (CH_3)_3COH \xrightarrow{PhNMe_2} CH_3CO_2C(CH_3)_3 + PhNMe_2 \cdot HCl \quad (63–68\%)$$

A possible explanation for this is that t-butyl esters are very readily decomposed by hydrochloric acid as follows:

$$MeCOCl + Me_3COH \longrightarrow MeCOOCMe_3 + HCl \longrightarrow MeCO_2H + Me_3CCl$$

In the presence of a base, the HCl is removed as base · HCl, and consequently the second step cannot occur.

Acetylation can also be carried out with acetic anhydride in the presence of a catalyst, *e.g.*, pyridine. This method is successful for prim.- and s-alcohols, but fails with t-alcohols, *e.g.*, t-butanol is not acetylated. However, t-alcohols can be acetylated if *p*-toluenesulphonic acid is used as catalyst.

6. Alcohols may be oxidised, and the products of oxidation depend on the class of the alcohol and on the nature of the oxidising agent (see also p. 185).

Halogens, in *aqueous* solution, oxidise alcohols to carbonyl compounds. According to Swain *et al.* (1961), the mechanism involves hydride ion removal:

$$Br-Br \quad H-CHMe-\ddot{O}-H \longrightarrow Br^- + HBr + MeCH=\overset{+}{O}H \xrightarrow{-H^+} MeCHO$$

In *alkaline* solution, the mechanism proposed is:

$$MeCH_2O-H \quad OH \rightleftharpoons H_2O + MeCH_2-O^-$$

$$MeCH_2-\bar{O} + Br_2 \longrightarrow MeCH=O + HBr + Br^-$$

In both cases, the carbonyl compound is then halogenated under the catalytic influence of acid or alkali (see p. 215).

On the other hand, alcohols are oxidised by hydrogen peroxide and ferrous sulphate to diols which, according to Coffman *et al.* (1960), are produced by dimerisation of free-radical intermediates, *e.g.*, t-butanol forms, 2,5-dimethylhexane-2,5-diol.

$$Fe^{2+} + H_2O_2 \longrightarrow Fe^{3+} + HO^- + HO\cdot$$

$$(CH_3)_2C(OH)CH_3 \xrightarrow[-H_2O]{\cdot OH} (CH_3)_2C(OH)CH_2\cdot \xrightarrow{2} (CH_3)_2C(OH)CH_2CH_2C(OH)(CH_3)_2$$

t-Butanol is particularly useful since all the alkyl hydrogens are equivalent and so only one free radical is produced.

7. Alcohols may be dehydrated to alkenes, the ease of dehydration being t > s > prim. Dehydration may be effected by heat alone (400–800°C), but in the presence of a catalyst, lower temperatures may be used, *e.g.*, when passed over heated alumina, t-alcohols are dehydrated at about 150°C, s-alcohols at 250°C and prim.-alcohols at 350°C. Also, primary alcohols are dehydrated by concentrated sulphuric acid at about 170°C, and secondary and tertiary alcohols by boiling dilute sulphuric acid (this is used to avoid polymerisation of the alkene). Isomerisation usually occurs with dehydration by acid or alumina, but in the latter case, isomerisation is suppressed by the addition of a small amount of pyridine.

The mechanism of dehydration with acid is described on p. 108 and, on the basis that a carbonium ion is formed as an intermediate, it can be seen why the ease of dehydration is t > s > prim.; this is the order of stabilities of the carbonium ions.

With secondary and tertiary alcohols, dehydration may occur in two ways, *e.g.*,

$$CH_3CH_2CH(OH)CH_3 \xrightarrow[(1:1H_2SO_4)]{-H_2O} \begin{cases} CH_3CH_2CH=CH_2 \\ CH_3CH=CHCH_3 \end{cases}$$

Experiment shows that hydrogen attached to the adjacent carbon atom joined to the least number of hydrogen atoms is eliminated most easily. Thus, in the above reaction, the main product is but-2-ene (65–80 per cent). This elimination therefore occurs in accordance with Saytzeff's rule for the dehydrohalogenation of alkyl halides, and the reason is the same (see p. 163). However, when thorium oxide or lanthanide metal oxides are used as the dehydrating agents, the yield of 1-alkene from 2-alcohols is generally above 98 per cent (Lundeen *et al.*, 1963).

As mentioned above, rearrangement often occurs with acid-catalysed dehydration. All three types of alcohol may behave in this way via a carbonium ion that may undergo methyl and/or hydride ion 1,2-shift (p. 146). In the following examples, only methyl shifts are shown (all the possible hydride ion shifts lead finally to the same products resulting from some of the competing methyl shifts).

(i) $Me_2CHCH_2OH \xrightarrow[\text{(ii) } -H_2O]{\text{(i) } H^+} Me_2CHCH_2^+ \xrightarrow{-H^+} Me_2C{=}CH_2$

Me shift

$MeCH{=}CHMe \xleftarrow{-H^+} Me\overset{+}{C}HCH_2Me \xrightarrow{-H^+} CH_2{=}CHCH_2Me$
(major product)

N.B. An illustration of hydride ion shift is:

$Me_2CHCH_2^+ \longrightarrow Me_2\overset{+}{C}CH_3 \xrightarrow{-H^+} Me_2C{=}CH_2$

(ii) $Me_3CCHOHMe \xrightarrow[\text{(ii) } -H_2O]{\text{(i) } H^+} Me_3C\overset{+}{C}HMe \xrightarrow{-H^+} Me_3CCH{=}CH_2$
(very little)

Me shift

$Me_2C{=}CMe_2 \xleftarrow{-H^+} Me_2\overset{+}{C}CHMe_2 \xrightarrow{-H^+} CH_2{=}CMeCHMe_2$
(major product)   (moderate amount)

(iii) $Me_3CCH_2OH \xrightarrow[\text{(ii) } -H_2O]{\text{(i) } H^+} Me_3CCH_2^+$

Me shift

$Me_2C{=}CHMe \xleftarrow{-H^+} Me_2\overset{+}{C}CH_2Me \xrightarrow{-H^+} CH_2{=}CMeCH_2Me$
(major product)   (small amount)

In each case, the major product is in accordance with Saytzeff's rule. Another point to note is that the rearrangements occur extremely rapidly. In fact, in the examples given, rearrangement is faster than proton elimination. Furthermore, Phelan *et al.* (1967) have shown that ease of anion migration is Ph ≫ Me > H.

When a double bond is produced in the product and is accompanied by a 1,2-shift, the reaction is said to be a *retropinacol rearrangement*. This type of rearrangement, when occurring in *open-chain* compounds, is also sometimes called the *Wagner rearrangement*.

Alcohols may also be converted into alkenes via their esters or xanthates, but in these cases no rearrangements occur (p. 95), *e.g.*, $Me_3CCH(OAc)Me \longrightarrow Me_3CCH{=}CH_2$ (*cf.* the action of acid on the alcohol, above).

Dehydration may be used as the first step in the conversion of a primary alcohol of suitable structure into a secondary or tertiary alcohol, or a s-alcohol of suitable structure into a t-alcohol, *e.g.* (*N.B.* HI additions are in accordance with Markownikoff's rule):

(i) $MeCH_2CH_2OH \xrightarrow[\text{350°C}]{Al_2O_3} MeCH{=}CH_2 \xrightarrow[\text{(ii) AgOH}]{\text{(i) HI}} MeCHOHMe$

(ii) $Me_2CHCH_2OH \xrightarrow[\text{350°C}]{Al_2O_3} Me_2C{=}CH_2 \xrightarrow[\text{(ii) AgOH}]{\text{(i) HI}} Me_3COH$

(iii) $Me_2CHCHOHMe \xrightarrow[250°C]{Al_2O_3} Me_2C{=}CHMe \xrightarrow[\text{(ii) AgOH}]{\text{(i) HI}} Me_2C(OH)CH_2Me$

The 'reverse' procedure, *i.e.*, the conversion t → s → prim. alcohol can be carried out via hydroboronation (see also p. 114), *e.g.*,

$$Me_2C(OH)CH_2Me \xrightarrow{-H_2O} Me_2C{=}CHMe \xrightarrow{B_2H_6}$$

$$(Me_2CHCHMe{-})_2BH \xrightarrow{heat} (Me_2CHCH_2CH_2{-})_2BH$$

$$\downarrow{H_2O_2;\ OH^-} \qquad\qquad\qquad \downarrow{H_2O_2;\ OH^-}$$

$$Me_2CHCHOHMe \qquad\qquad\qquad Me_2CHCH_2CH_2OH$$

It is also possible to step down the alcohol series by first dehydrating to the alkene, which is then subjected to ozonolysis, *e.g.*,

(*a*) $RCH_2CH_2OH \xrightarrow[350°C]{Al_2O_3} RCH{=}CH_2 \xrightarrow[\text{(ii) Zn/H}^+]{\text{(i) O}_3} RCHO \xrightarrow[cat.]{H_2} RCH_2OH$

(*b*) $RCHOHCH_3 \xrightarrow[300°C]{ThO_2} RCH{=}CH_2 \dashrightarrow RCH_2OH$

8. Alcohols combine with acetylene in the presence of mercury compounds as catalyst to form acetals:

$$2ROH + CH{\equiv}CH \xrightarrow{Hg^{2+}} CH_3CH(OR)_2$$

If, however, the reaction is carried out in the presence of potassium alkoxides at high temperature and under pressure, vinyl ethers are obtained (p. 137).

Methods of distinguishing between the three classes of alcohols.

1. By means of *oxidation*. The nature of the oxidation products of an alcohol depends on whether the alcohol is primary, secondary or tertiary.

(i) A *primary* alcohol on oxidation first gives an *aldehyde*, and this, on further oxidation, gives an acid. *Both aldehyde and acid contain the same number of carbon atoms as the original alcohol. e.g.*,

$$CH_3CH_2OH \xrightarrow{[O]} CH_3CHO \xrightarrow{[O]} CH_3CO_2H$$

(ii) A *secondary* alcohol, on oxidation, first gives a *ketone with the same number of carbon atoms as the original alcohol.* Ketones are fairly difficult to oxidise, but prolonged action of the oxidising agents produces a mixture of acids, each containing *fewer* carbon atoms than the original alcohol, *e.g.*, methyl-n-propylmethanol gives first pentan-2-one, and then a mixture of acetic and propionic acids:

$$CH_3CHOHCH_2CH_2CH_3 \xrightarrow{[O]} CH_3COCH_2CH_2CH_3 \xrightarrow{[O]} CH_3CO_2H + CH_3CH_2CO_2H$$

(iii) *Tertiary* alcohols are resistant to oxidation in neutral or alkaline solution, but are readily oxidised by acid oxidising agents to a mixture of *ketone and acid, each containing fewer carbon atoms than the original alcohol.*

$$(CH_3)_2C(OH)CH_2CH_3 \xrightarrow{[O]} (CH_3)_2CO + CH_3CO_2H$$

The oxidising agents usually used for oxidising alcohols are: acid dichromate, acid or alkaline potassium permanganate, and dilute nitric acid.

The accepted mechanism for the oxidation of primary and secondary alcohols by dichromate and permanganate involves an ester intermediate, e.g., with acid dichromate:

$$Cr_2O_7^{2-} + H_2O \; \rightleftharpoons \; 2HCrO_4^-$$

$$Me_2CHOH + HCrO_4^- + 2H^+ \; \rightleftharpoons \; Me_2CH\!-\!O\!-\!CrO_3H_2^+ + H_2O$$

$$Me_2C\!-\!O\!-\!CrO_3H_2^+ \xrightarrow{\text{slow}} Me_2CO + H_3O^+ + H_2CrO_3$$

The rate-determining step is the removal of hydrogen from the intermediate ester. This has been demonstrated by Westheimer *et al.* (1949), who showed that Me_2CDOH is oxidised at about 1/7 the rate of either Me_2CHOH or $(CD_3)_2CHOH$. Thus, the r/d step involves the fission of the C—H bond (this accounts for the kinetic isotope effect).

This mechanism cannot operate with t-alcohols, since there is no H—COH group present. According to Sager (1956), the mechanism involves dehydration, e.g.,

$$Me_3COH \xrightarrow[\text{(ii) } -H_2O]{\text{(i) } H^+} Me_3C^+ \xrightarrow{-H^+} Me_2C\!=\!CH_2 \xrightarrow{[O]} Me_2C\overset{O}{\overbrace{}}CH_2 \xrightarrow{[O]} Me_2CO + HCO_2H$$

2. The three classes of alcohols differ in their behaviour when the vapour is passed over copper at 300°C:

(i) A primary alcohol is dehydrogenated to an aldehyde, e.g.,

$$CH_3CH_2OH \xrightarrow[300°C]{Cu} CH_3CHO + H_2$$

(ii) A secondary alcohol is dehydrogenated to a ketone, e.g.,

$$CH_3CH(OH)CH_3 \xrightarrow[300°C]{Cu} CH_3COCH_3 + H_2$$

(iii) A tertiary alcohol is dehydrated to an alkene, e.g.,

$$(CH_3)_2C(OH)CH_2CH_3 \xrightarrow[300°C]{Cu} (CH_3)_2C\!=\!CHCH_3 + H_2O$$

Methyl alcohol, methanol (*carbinol*) is prepared industrially by several methods. The earliest method was the destructive distillation of wood, whereby tar and an aqueous fraction known as *pyroligneous acid* are obtained. Pyroligneous acid contains methanol, acetone and acetic acid, and all three compounds may be obtained by suitable treatment (see acetic acid, p. 241). It was this method which gave rise to the name 'wood spirit' for methanol. The modern methods are synthetic.

(i) Synthesis gas (p. 88) is passed at a pressure of 200 atmospheres over a catalyst containing the oxides of copper, zinc and chromium at 300°C:

$$CO + 2H_2 \longrightarrow CH_3OH$$

If the proper precautions are taken, the yield of methanol is almost 100 per cent, and its purity is above 99 per cent. By changing the catalyst and the ratio of carbon monoxide to hydrogen, methanol and a variety of higher alcohols are produced (p. 180).

(ii) By the catalytic oxidation of methane. A mixture of methane and oxygen (ratio by volume of 9:1) at a pressure of 100 atmospheres is passed through a copper tube at 200°C:

$$CH_4 + \tfrac{1}{2}O_2 \longrightarrow CH_3OH$$

Methanol is a colourless, inflammable liquid, b.p. 64°C, and is poisonous. It is miscible with water in all proportions, and is also miscible with most organic solvents. It burns with a faintly luminous flame, and its vapour forms explosive mixtures with air or oxygen when ignited. It combines with calcium chloride to form $CaCl_2 \cdot 4CH_3OH$, and hence cannot be dried this way (*cf*. ethanol).

Methanol is used as a solvent for paints, varnishes, shellac, celluloid cements, etc.; in the manufacture of dyes, perfumes, formaldehyde, etc. It is also used for making methylated spirit and automobile antifreeze mixtures.

Structure of methanol. Analysis and molecular-weight determinations show that the molecular formula of methanol is CH_4O. Assuming that carbon is quadrivalent, oxygen bivalent and hydrogen univalent, only one structure is possible:

$$
\begin{array}{c}
\text{H} \\
| \\
\text{H---C---O---H} \qquad or \qquad CH_3OH \\
| \\
\text{H}
\end{array}
$$

This is supported by all the chemical reactions of methanol, *e.g.*, (i) only one hydrogen atom in methanol is replaceable by sodium; this suggests that one hydrogen atom is in a different state of combination from the other three.

(ii) Methanol is formed from methyl chloride by hydrolysis with sodium hydroxide. Methyl chloride can have only the structure CH_3Cl. It is reasonable to suppose that the methyl group in methyl chloride is unchanged by the action of dilute alkali, and that the reaction takes place by the replacement of the chlorine atom by a hydroxyl group.

(iii) The presence of the hydroxyl group is confirmed, for example, by the reaction between methanol and phosphorus pentachloride, when methyl chloride, hydrogen chloride and phosphoryl chloride are formed. Thus one oxygen atom (bivalent) and one hydrogen atom (univalent) have been replaced by one chlorine atom (univalent). This implies that the oxygen and hydrogen atoms exist as a univalent group in methanol: the only possibility is as a hydroxyl group, OH. It is the hydrogen of the hydroxyl group which is displaced by sodium.

Infra-red spectroscopic studies of hydroxy-compounds show that they absorb (O—H stretch) in the region of 3650–3580 cm^{-1} (v). This is true only if there is no hydrogen bonding. Intermolecular hydrogen bonding produces absorption (broad band) in the region 3550–3230 cm^{-1} (v), whereas intramolecular hydrogen bonding absorbs in the region 3590–3420 cm^{-1} (v). Moreover, the former shows changes in intensity and frequency shifts on dilution, whereas the latter is unaffected by dilution; thus the two may be distinguished.

In addition to O—H stretching vibrations, there are also C—O stretching vibrations which are characteristic of the type of alcohol: primary, close to 1050 (s); secondary, close to 1100 (s); tertiary, close to 1150 cm^{-1} (s). Thus the classes of alcohols may be identified in these regions.

Two infra-red spectra of ethanol are shown in Fig. 6.1 (as a film) and in Fig. 6.2 (as a 1 per cent solution in carbon tetrachloride). In the former, all the OH groups are associated, but in the latter, some OH groups are free.

The NMR spectrum of ethanol has been given in Figs. 1.3 and 1.4 (pp. 12, 13). However, a further point that may be mentioned here is the τ-value for methylene protons in the —CH_2—O— group is between 6·5–6·65 p.p.m., and when the alcohol is acetylated, the τ-value changes to $\sim 5 \cdot 9$. This change may be used to identify a primary alcohol. Secondary alcohols can also be identified; $>$ CH—O— ($\tau = \sim 6 \cdot 1$) and the acetylated alcohol ($\tau = \sim 5 \cdot 0$). These downfield shifts are due to the increased deshielding effect of the acetoxyl group ($OOCCH_3$) compared to hydroxyl oxygen.

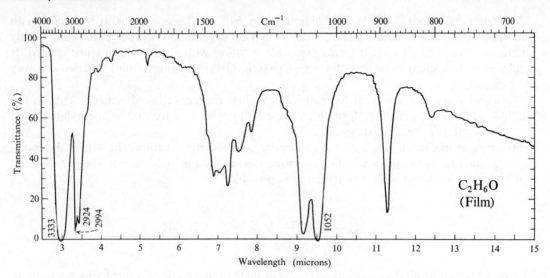

Fig. 6.1.

3 333 cm⁻¹: O—H (str.); association (through hydrogen bonding). 2 994 and 2 924 cm⁻¹: C—H (str.) in Me and/or CH_2. 1052 cm⁻¹: C—O (str.); characteristic of primary alcohols.

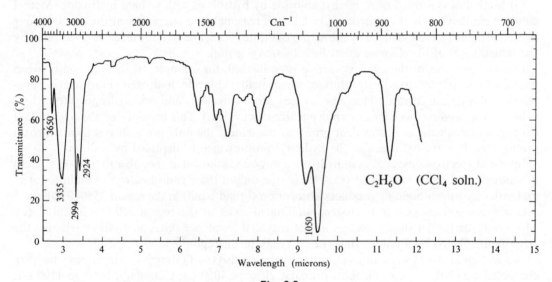

Fig. 6.2.

3 650 cm⁻¹: O—H (str.); free OH. 3 333, 2 994, 2 924, and 1050 cm⁻¹; As for Fig. 6.1.

Mass spectra. The mass spectrum of ethanol is shown in Fig. 1.7 (p. 22), but here we shall describe the mass spectra of alcohols from a general point of view. Usually, strong peaks are shown by fragmentation involving β-cleavage (*i.e.*, the C—COH bond), and the fragment containing the hydroxyl group is stabilised by resonance:

$$\left[\begin{array}{c} R^2 \\ | \\ R^1 - \overset{.+}{C} - \overset{..}{O}H \\ | \\ R^3 \end{array}\right] \longrightarrow R^1 \cdot + \left[\begin{array}{c} R^2 \\ \diagdown \\ \overset{.}{C} - \overset{..+}{O}H \\ \diagup \\ R^3 \end{array} \longleftrightarrow \begin{array}{c} R^2 \\ \diagdown \\ C = \overset{+}{O}H \\ \diagup \\ R^3 \end{array}\right]$$

The alkyl group with the heaviest mass is eliminated preferentially, but ions also appear for alternative eliminations. Further fragmentation may also occur as follows ($R^2 = H$, $R^3 = C_2H_5$):

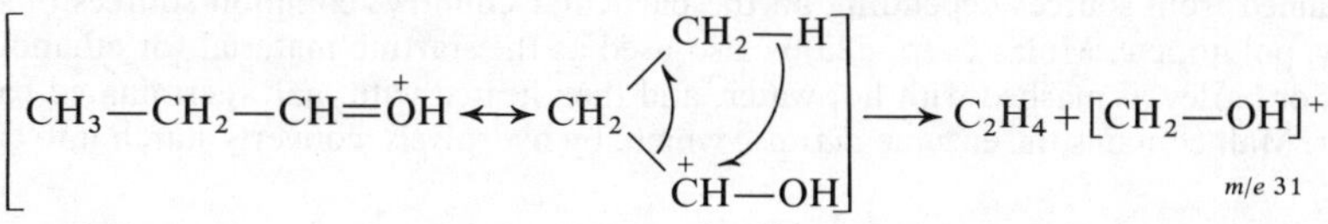

$$\left[CH_3-CH_2-CH=\overset{+}{O}H \longleftrightarrow CH_2 \underset{\overset{+}{CH}-OH}{\overset{CH_2-H}{\Big)}} \right] \longrightarrow C_2H_4 + \underset{m/e\ 31}{[CH_2-OH]^+}$$

Hence, both s- and t-alcohols can also show peaks at m/e 31 (the most characteristic ion of alcohols).

Primary and secondary alcohols usually give weak molecular ions, and for tertiary alcohols the molecular ion is very weak or absent. Long-chain alcohols show peaks at M-18 due to the loss of a molecule of water and formation of a cyclic ion (cf. alkyl halides):

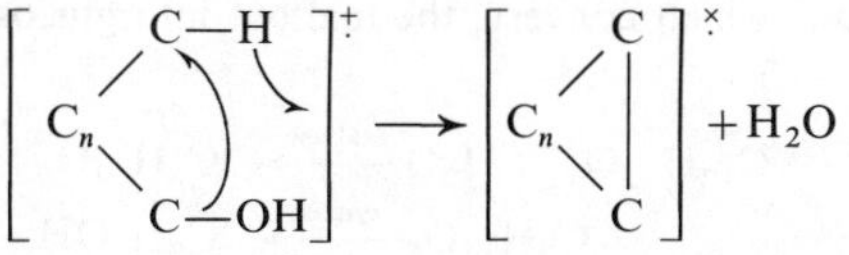

$$\left[C_n \underset{C-OH}{\overset{C-H}{\Big)}} \right]^{+} \longrightarrow \left[C_n \overset{C}{\underset{C}{\Big|}} \right]^{x} + H_2O$$

On the other hand, a McLafferty rearrangement can also occur with loss of water:

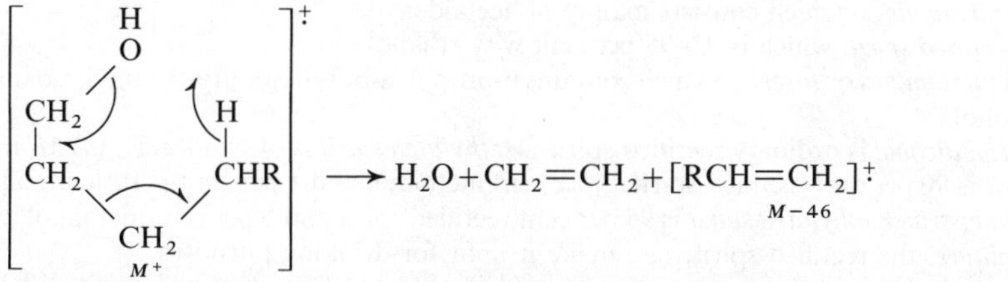

$$\left[\begin{array}{c} H \\ O \\ CH_2 \quad\quad H \\ | \quad\quad\quad | \\ CH_2 \quad\quad CHR \\ \diagdown \quad\quad \diagup \\ CH_2 \\ M^+ \end{array} \right]^{+} \longrightarrow H_2O + CH_2{=}CH_2 + \underset{M-46}{[RCH{=}CH_2]^{+}}$$

In general, long-chain alcohols also show the fragmentation patterns of alkanes and alkenes.

Ethyl alcohol, ethanol, is prepared industrially by several methods, e.g., ethylene (from cracked petroleum) is absorbed in concentrated sulphuric acid (98 per cent) at 75–80°C, under pressure (17·5–35 kg/cm^2). Ethyl hydrogen sulphate and ethyl sulphate are formed:

$$C_2H_4 + (HO)_2SO_2 \longrightarrow C_2H_5OSO_2OH$$
$$C_2H_5OSO_2OH + C_2H_4 \longrightarrow (C_2H_5O)_2SO_2$$

The reaction mixture is then diluted with about an equal volume of water, and warmed. Hydrolysis takes place and ethanol together with some diethyl ether is formed:

$$C_2H_5OSO_2OH + H_2O \longrightarrow C_2H_5OH + H_2SO_4$$
$$(C_2H_5O)_2SO_2 + 2H_2O \longrightarrow 2C_2H_5OH + H_2SO_4$$
$$(C_2H_5O)_2SO_2 + C_2H_5OH \longrightarrow (C_2H_5)_2O + C_2H_5OSO_2OH$$

The ether is kept to a minimum by separating the ethyl sulphate from the reaction products, and hydrolysing it separately. The hydrolysed liquids are distilled, and the aqueous ethanol distillate is concentrated by fractional distillation (see also below).

Ethanol is also manufactured by the *direct* hydration of ethylene with steam under pressure in the presence of a suitable catalyst, e.g., phosphoric acid on a support.

$$C_2H_4 + H_2O \longrightarrow C_2H_5OH$$

Some ether ($\sim$ 5 per cent) is formed as a by-product.

The earliest method of preparing ethanol is by fermentation, and this is still used for the manufacture of beer, wine, brandy, etc., and also as a source of ethanol. The starting material is starch, which is obtained from sources depending on the particular country: common sources of starch are wheat, barley, potato, etc. Molasses (p. 523) is also used as the starting material for ethanol. The grain, *e.g.*, wheat or barley, is mashed with hot water, and then heated with malt (germinated barley) at 50°C for 1 hour. Malt contains the enzyme *diastase* which, by hydrolysis, converts starch into the sugar, maltose (*q.v.*):

$$(C_6H_{10}O_5)_n + \frac{n}{2}H_2O \xrightarrow{\text{diastase}} \frac{n}{2}C_{12}H_{22}O_{11}$$

If molasses is used, then this step is unnecessary, since it contains carbohydrates already present as sugars which can be fermented.

The liquid is cooled to 30°C and fermented with yeast for 1–3 days. Yeast contains various enzymes, among which are *maltase*, which converts the maltose into glucose, and *zymase*, which converts the glucose into ethanol:

$$C_{12}H_{22}O_{11} + H_2O \xrightarrow{\text{maltase}} 2C_6H_{12}O_6$$
$$C_6H_{12}O_6 \xrightarrow{\text{zymase}} 2C_2H_5OH + 2CO_2$$

The carbon dioxide is recovered and sold as a by-product. The fermented liquor or 'wort', which contains 6–10 per cent ethanol and some other compounds, is fractioned into three fractions:

(i) *First runnings*, which consists mainly of acetaldehyde.

(ii) *Rectified spirit*, which is 93–95 per cent w/w ethanol.

(iii) *Final runnings* or *fusel oil*, which contains n-propyl, n-butyl, isobutyl, n-amyl, isoamyl and 'active' amyl alcohol.

Industrial alcohol is ordinary rectified spirit. *Methylated spirit* is of two kinds: (*a*) *Mineralised methylated spirit* is 90 per cent rectified spirit, 9 per cent methanol and 1 per cent petroleum oil, and a purple dye. (*b*) *Industrial methylated spirit* is 95 per cent rectified spirit and 5 per cent methanol, whose purpose is to '*denature*' the rectified spirit, *i.e.*, make it unfit for drinking purposes.

Absolute alcohol is 99·5 per cent ethanol, and is obtained from rectified spirit. When an aqueous solution of ethanol is fractionated, it forms a constant-boiling mixture containing 96 per cent ethanol from which 100 per cent ethanol may be obtained by adding a small amount of benzene, and then distilling. The first fraction is the ternary azeotrope, *i.e.*, a constant-boiling mixture containing three constituents, b.p. 64·8°C (water, 7·4 per cent; ethanol, 18·5 per cent; benzene, 74·1 per cent). After all the water has been removed, the second fraction that distils over is the binary azeotrope, b.p. 68·2°C (ethanol, 32·4 per cent; benzene, 67·6 per cent). After all the benzene has been removed, pure ethanol, b.p. 78·1°C, distils over.

Ethanol cannot be dried by means of calcium chloride, since a compound (an *alcoholate*) is formed, *e.g.*, $CaCl_2 \cdot 3C_2H_5OH$ (*cf.* methanol). Distillation of rectified spirit over calcium oxide, and then over calcium, gives absolute alcohol. This method is often used in the laboratory, and was formerly used industrially.

Ethanol is a colourless, inflammable liquid, b.p. 78·1°C. It is miscible with water in all proportions, and is also miscible with most organic solvents. Ethanol and methanol resemble each other very closely, but they may be distinguished (i) by the fact that ethanol gives the haloform reaction (*q.v.*), whereas methanol does not; and (ii) ethanol gives acetic acid on oxidation; methanol gives formic acid. These two acids are readily distinguished from each other (p. 240).

Ethanol is used for the preparation of esters, ether, chloral, chloroform, etc. It is also used as a solvent for gums, resins, paints, varnishes, etc., and as a fuel.

Structure of ethanol. Analysis and molecular-weight determinations show that the molecular formula of ethanol is C_2H_6O. Assuming that carbon is quadrivalent, oxygen bivalent, and hydrogen univalent, two structures are possible:

$$\underset{\text{(I)}}{CH_3-CH_2-OH} \qquad \underset{\text{(II)}}{CH_3-O-CH_3}$$

(i) Only one hydrogen atom in ethanol is replaceable by sodium or potassium. This indicates that one hydrogen atom is in a different state of combination from the other five. In (I), one hydrogen atom differs from the other five, but in (II) all the hydrogen atoms are equivalent.

(ii) When ethanol is treated with hydrochloric acid or phosphorus pentachloride, one oxygen atom (bivalent) and one hydrogen atom (univalent) are replaced by one chlorine atom (univalent) to give ethyl chloride, C_2H_5Cl. This implies the presence of a hydroxyl group (*cf.* methanol).

(iii) When ethyl chloride is hydrolysed with dilute alkali, ethanol is obtained. This reaction also indicates the presence of a hydroxyl group in ethanol.

(iv) Ethanol may be prepared as follows:

$$C_2H_6 \xrightarrow{Cl_2} C_2H_5Cl \xrightarrow{NaOH} C_2H_5OH$$

The arrangement of the six hydrogen atoms in ethane is known, and it is reasonable to suppose that five retain their original arrangement in ethyl chloride and ethanol, since these five hydrogen atoms do not enter (presumably) into the above reactions. Thus there is an ethyl group C_2H_5— in ethanol. This is so in (I), but not in (II).

(v) Structure (II) is definitely eliminated, since it can be shown that it is the structure of dimethyl ether (*q.v.*), a compound that has very little resemblance, physically or chemically, to ethanol.

Hence (I) is accepted as the structure of ethanol, and it accounts for all the known properties of ethanol.

It can be seen from the various examples given on structure determination, *e.g.*, methane, butanes, ethylene, methanol and ethanol, that the method of approach follows certain definite lines. First the molecular formula is obtained. Then, if the compound is simple—in the sense that it contains a small number of unlike atoms, and that the total number of atoms is also small—the valencies of the atoms present are assumed, and various possible structures are written down. If the compound is 'simple' the number of possible structures will not be large (four or five at the most). Then by considering the chemical properties of the compound in question, the structure which best fits the observed facts is accepted as the correct one. If the compound is not 'simple', the procedure is to detect the presence of as many functional groups as possible; to degrade the compound into simpler substances whose structures are already known or which may be determined by further degradation; to build up structural formulæ based on the facts obtained; and then to choose that structure which best fits the facts. Finally a synthesis is attempted, and if successful, will usually give proof of the correctness of the structure suggested.

It should be noted that the arguments used for structure determination by purely chemical methods are based on the assumption that no rearrangements have occurred in the series of reactions performed. Thus, the possibility of rearrangement must always be borne in mind when carrying out chemical reactions.

As was pointed out in Ch. 1, physical methods are being used nowadays for structural determination. Once the molecular formula is obtained, functional groups may be identified, etc. In this way, the structures of many compounds have now been ascertained with a minimum of chemical evidence.

Propyl alcohols. Two isomeric propyl alcohols are possible, and both are known.

n-Propyl alcohol, propan-1-ol, n-propanol, was originally obtained from fusel oil (see above), but it is now produced by the hydrogenation of carbon monoxide. A more recent method is by the catalytic reduction of propargyl alcohol (from acetylene and formaldehyde):

$$CH{\equiv}CCH_2OH + 2H_2 \xrightarrow{Ni} 2CH_3CH_2CH_2OH$$

n-Propanol, b.p. 97·4°C, is miscible with water, ethanol and ether. It is used in the preparation of propionic acid, toilet preparations such as lotions, etc.

Isopropyl alcohol, *propan-2-ol*, *isopropanol*, is prepared industrially by passing propene (from cracked petroleum) into concentrated sulphuric acid, then diluting with water, and distilling off the isopropanol. Isopropyl ether is obtained as a by-product (*cf.* ethanol):

$$CH_3CH{=}CH_2 + (HO)_2SO_2 \longrightarrow (CH_3)_2CHOSO_2OH$$

$$(CH_3)_2CHOSO_2OH + CH_3CH{=}CH_2 \longrightarrow [(CH_3)_2CHO]_2SO_2$$

$$(CH_3)_2CHOSO_2OH + H_2O \longrightarrow (CH_3)_2CHOH + H_2SO_4$$

$$[(CH_3)_2CHO]_2SO_2 + 2H_2O \longrightarrow 2(CH_3)_2CHOH + H_2SO_4$$

$$[(CH_3)_2CHO]_2SO_2 + (CH_3)_2CHOH \longrightarrow [(CH_3)_2CH]_2O + (CH_3)_2CHOSO_2OH$$

Isopropanol is also prepared by *direct* hydration by passing a mixture of propene and steam at 220–250°C, and under pressure (220 atm), over a catalyst of tungsten oxide plus zinc oxide, on a silica carrier.

Isopropanol, b.p. 82·4°C, is soluble in water, ethanol and ether. It is used for preparing esters, acetone, keten, as a solvent, and for high-octane fuel.

Butyl alcohols. Four isomers are possible, and all are known.

n-Butyl alcohol, *butan-1-ol*, *n-butanol*, b.p. 117·4°C, is prepared industrially from propene by the Oxo process (p. 180). It is widely used as a solvent.

Isobutyl alcohol, *2-methylpropan-1-ol*, *isobutanol*, b.p. 108°C, is also obtained by the Oxo process. It behaves as a primary alcohol, but it readily rearranges due to the presence of a branched chain near the COH group, *e.g.*, when it is treated with hydrochloric acid, isobutyl chloride and t-butyl chloride are obtained (*cf.* p. 147):

$$(CH_3)_2CHCH_2OH \xrightarrow{HCl} (CH_3)_2CHCH_2Cl + (CH_3)_3CCl$$

s-Butyl alcohol, *butan-2-ol*, *s-butanol*, b.p. 100°C, is prepared industrially by the hydration of 1- or 2-butene (from cracked petroleum) by means of concentrated sulphuric acid (*cf.* ethanol and isopropanol). It is used for the preparation of butanone, esters, and as a lacquer solvent.

t-Butyl alcohol, *t-butanol* (*trimethylmethanol*), m.p. 25·5°C, b.p. 83°C, is prepared synthetically by the hydration of isobutene (from cracked petroleum). It is mainly used as an alkylating agent.

Amyl alcohols. Eight isomers are possible, and all are known:

1. $CH_3CH_2CH_2CH_2CH_2OH$, n-amyl alcohol, pentan-1-ol, n-pentanol, b.p. 138°C.
2. $(CH_3)_2CHCH_2CH_2OH$, isoamyl alcohol, isopentanol, b.p. 130°C.
3. $CH_3CH_2CH(CH_3)CH_2OH$, 'active' amyl alcohol, 2-methylbutan-1-ol, b.p. 128°C.
4. $(CH_3)_3CCH_2OH$, neopentyl alcohol, t-butylcarbinol, b.p. 113°C.
5. $CH_3CH_2CH_2CH(OH)CH_3$, pentan-2-ol, methyl-n-propylcarbinol, b.p. 120°C.
6. $CH_3CH_2CH(OH)CH_2CH_3$, pentan-3-ol, diethylcarbinol, b.p. 117°C.
7. $(CH_3)_2CHCH(OH)CH_3$, 3-methylbutan-2-ol, isopropylmethylcarbinol, b.p. 114°C.
8. $(CH_3)_2C(OH)CH_2CH_3$, 2-methylbutan-2-ol, t-amyl alcohol, t-pentanol, ethyldimethylcarbinol, b.p. 102°C.

Three amyl alcohols, *viz.*, n-pentanol, isopentanol and 'active' amyl alcohol, have been isolated from fusel oil (*q.v.*). The last two are the chief constituents of fusel oil, and all three are produced by the fermentation of protein matter associated with the carbohydrates in starch. This mixture of amyl alcohols (from fusel oil) is used for the preparation of esters (for artificial essences), scents, and as a laboratory reducing agent with sodium it is better than ethanol owing to its higher boiling point (this mixture of amyl alcohols will be referred to in future as isopentanol).

A mixture of amyl alcohols, known as *pentasol*, is prepared industrially by chlorinating at 200°C, and in the dark, a mixture of n- and isopentanes (from petroleum) to the amyl chlorides which are hydrolysed with dilute sodium hydroxide solution plus a little sodium oleate for emulsification, to the amyl alcohols. Seven isomeric amyl chlorides are theoretically possible, but in practice six are obtained. 2-Chloro-2-methylbutane (the only t-chloride produced) readily eliminates hydrogen chloride to form trimethylethylene. Pentasol (the mixture of six amyl alcohols) finds great use as a solvent in the lacquer industry.

Commercial 's-amyl alcohol' (80 per cent 2- and 20 per cent 3-pentanol) is made by hydrating 1- and 2-pentenes (from cracked petroleum).

A number of higher alcohols occur as esters in waxes, *e.g.*, cetyl (palmityl) alcohol (hexadecan-1-ol), $C_{16}H_{33}OH$, m.p. 49°C, occurs as the palmitate in spermaceti (obtained from the oil of the sperm whale); carnaubyl alcohol (tetracosan-1-ol), $C_{24}H_{49}OH$, m.p. 69°C, as esters in wool-grease; ceryl alcohol (hexacosan-1-ol), $C_{26}H_{53}OH$, m.p. 79°C, as the cerotate in chinese wax; myricyl (melissyl) alcohol (triacontan-1-ol), $C_{30}H_{61}OH$, m.p. 88°C, as esters in bees-wax.

A number of higher alcohols are now prepared industrially by the catalytic reduction of the ethyl (or glyceryl) esters of the higher monocarboxylic acids, particularly the alcohols lauryl (dodecan-1-ol), $C_{12}H_{25}OH$, m.p. 24°C; myristyl (tetradecan-1-ol), $C_{14}H_{29}OH$, m.p. 38°C; palmityl (hexadecan-1-ol), $C_{16}H_{33}OH$, m.p. 49°C; and stearyl (octadecan-1-ol), $C_{18}H_{37}OH$, m.p. 59°C. These alcohols are used in the form of sodium alkyl sulphates, $ROSO_2ONa$, as detergents, emulsifying agents, wetting agents, insecticides, fungicides, etc. These sodium salts lather well, and are not affected by hard water, and hence can be used as a soap substitute.

PROBLEMS

1. Write out the structures and I.U.P.A.C. names of the isomeric secondary heptanols containing one methyl side-chain.

2. The dehydration of butan-2-ol with acid gives a mixture of 1- and 2-butenes. Write out the equations, state which predominates, and give your reasons. Draw the energy profiles of the reactions.

3. Convert ethanol into (i) n-propanol; (ii) n-butanol.

4. Arrange EtOH, CF_3CH_2OH, and CCl_3CH_2OH in order of increasing strength as acids and give your reasons.

5. The dehydration of n-BuOH with acid gives 1- and 2-butene. Explain this, state which product predominates, and why.

6. Convert propene into (i) n-PrOH; (ii) isoPrOH.

7. How could you show the existence of the following equilibrium?

$$ROH + H^+ \overset{H_2SO_4 - H_2O}{\rightleftharpoons} R^+ + H_2O$$

8. What are the disadvantages of preparing alkenes by:
(i) RX + base; ROH + acid?

9.

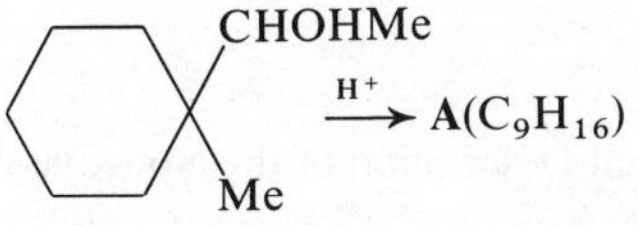

A, on ozonolysis, gives nonane-2,8-dione. What is **A** and how is it formed?

10. Complete the following equation and comment:

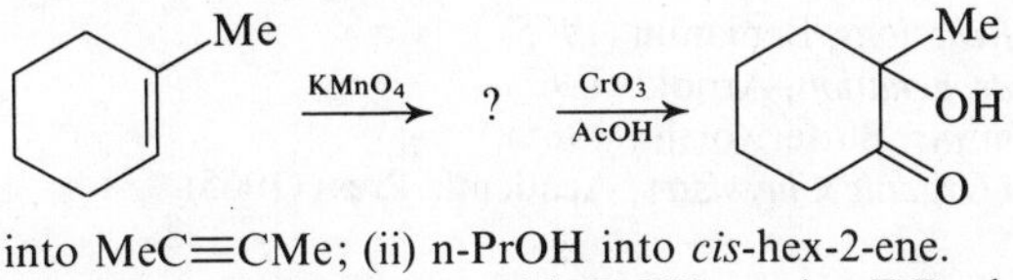

11. Convert (i) $n\text{-}C_4H_9OH$ into $MeC\equiv CMe$; (ii) n-PrOH into *cis*-hex-2-ene.

12. A concentrated aqueous solution of HBr reacts with EtOH to give EtBr, but a concentrated aqueous solution of NaBr does not. Explain.

13. What are **A** to **E** in the following reactions?

$$(\textbf{A})\ C_4H_{10}O \xrightarrow{SOCl_2} C_4H_9Cl\ (\textbf{B})$$

$$\xrightarrow[]{H_2SO_4}$$

$$(\textbf{D})\ C_3H_6O_2 \xleftarrow[\text{(ii) }H_2O_2]{\text{(i) }O_3} C_4H_8(\textbf{C}) \xrightarrow{HCl} C_4H_9Cl\ (\textbf{E})$$

14. Complete the following equations:

(i) $n\text{-}C_3H_7CO_2H \xrightarrow{?} n\text{-}C_4H_9OH$

(ii) $Me_2CO + EtMgI \longrightarrow ? \xrightarrow{H^+} ?$

(iii) $EtCO_2Et + 2MeMgI \longrightarrow ? \xrightarrow{H_2O} ?$

(iv) 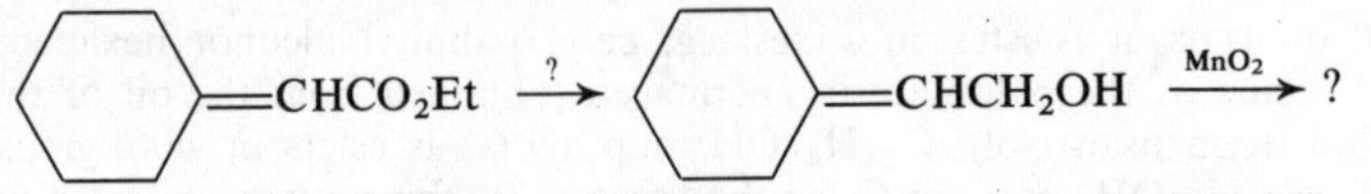   [structure] $=CHCO_2Et \xrightarrow{?}$ [structure] $=CHCH_2OH \xrightarrow{MnO_2}$?

(v) $CH_3CH_2CH_2OH \longleftarrow BrCH_2CH_2COCl \longrightarrow BrCH_2CH_2CH_2OH$

15. How many NMR signals would be given by the following alcohols in the presence of a trace of acid? Indicate all signals other than singlets by (m).

(i) $CH_3CH_2CH_2OH$; (ii) Me_2CHCH_2OH; (iii) Me_3COH; (iv) Me_3CCH_2OH; (v) $MeCH_2CMe_2OH$.

16. The table lists some peaks in the mass spectra of the four isomeric butyl alcohols, **A, B, C, D**. Assign each alcohol to a letter, and give your reasons.

m/e	18	29	31	41	43	45	56	59	74
A(%)	–	34	100	63	60	7	86	–	1
B(%)	4	24	67	55	100	52	3	5	9
C(%)	–	15	22	11	10	100	1	18	0·2
D(%)	5	13	34	20	15	–	1·4	100	0·1

17. What reagents could you use for the following conversions?

(i) $MeCO(CH_2)_2CO_2Et \longrightarrow MeCHOH(CH_2)_2CO_2Et$
(ii) $HO_2C(CH_2)_4COCl \longrightarrow HO_2C(CH_2)_4CH_2OH$
(iii) $O_2N(CH_2)_2CN \longrightarrow O_2N(CH_2)_2CH_2NH_2$
(iv) $O_2N(CH_2)_2CH=CH_2 \longrightarrow H_2N(CH_2)_2CH=CH_2$
(v) $O_2N(CH_2)_2CH=CH_2 \longrightarrow O_2N(CH_2)_3CH_3$
(vi) $Me_2CHCOCl \longrightarrow Me_2CHCHO$
(vii) $O_2N(CH_2)_3CHO \longrightarrow O_2N(CH_2)_3CH_2OH$

REFERENCES

PHELAN et al., 'A Molecular Orbital Description of the Non-Classical Ion in 1,2-Rearrangements', *J. Chem. Educ.*, 1967, **44**, 626.

GAYLORD, *Reduction with Complex Metal Hydrides*, Interscience (1956).

BROWN, *Hydroboration*, Benjamin (1962).

HOUSE, *Modern Synthetic Reactions*, Benjamin (1965).

AUGUSTINE, *Catalytic Hydrogenation*, Arnold (1965).

TURNEY, *Oxidation Mechanisms*, Butterworths (1965).

WIBERG (ed.), *Oxidation in Organic Chemistry*, Academic Press (1965).

7

Ethers

The general formula of the ethers is $C_nH_{2n+2}O$ (which is the same as that for the monohydric alcohols), and since their general structure is R—O—R, they may be regarded as alkyl oxides or the anhydrides of the alcohols (see below).

When the two alkyl groups in an ether are the same, the ether is said to be symmetrical or simple, e.g., diethyl ether, C_2H_5—O—C_2H_5. When the two alkyl groups are different, the ether is said to be unsymmetrical or mixed, e.g., ethyl methyl ether, CH_3—O—C_2H_5.

Nomenclature.

1. In this system of nomenclature all the members are known as ethers, and the individuals are named according to the alkyl groups attached to the oxygen atom, e.g., CH_3—O—CH_3, dimethyl ether; C_2H_5—O—$CH(CH_3)_2$, ethyl isopropyl ether.

2. According to the I.U.P.A.C. system of nomenclature, the ethers are regarded as hydro-carbons in which a hydrogen atom is replaced by an alkoxyl group, —OR, the larger group being chosen as the alkane. For symmetrical ethers, method 1 is to be used, e.g., $C_2H_5OC_2H_5$, diethyl ether; $CH_3OC_2H_5$, methoxyethane.

General methods of preparation.

1. Ethers may be prepared from alcohols in acid media, and the generally accepted mechanisms are that straight-chain primary alcohols react by the bimolecular mechanism (S_N2), t-alcohols by the unimolecular mechanism (S_N1), and s-alcohols by either of these mechanisms (the alcohol is the nucleophilic reagent):

S_N2

$$ROH + H^+ \underset{\text{fast}}{\rightleftharpoons} ROH_2^+$$

$$R-\overset{}{\underset{H}{\ddot{O}}} \quad R-OH_2^+ \underset{\text{slow}}{\rightleftharpoons} R-\overset{+}{\underset{H}{O}}-R + H_2O \xrightarrow{\text{fast}} ROR + H_3O^+$$

S_N1

$$ROH + H^+ \underset{\text{fast}}{\rightleftharpoons} ROH_2^+ \underset{}{\overset{\text{slow}}{\rightleftharpoons}} R^+ + H_2O$$

$$R^+ \quad \overset{}{\underset{H}{\ddot{O}}}-R \xrightarrow{\text{fast}} R-\overset{+}{\underset{H}{O}}-R \xrightarrow[\text{fast}]{H_2O} ROR + H_3O^+$$

On the other hand, it still appears to be uncertain whether the preparation of ethers from primary alcohols and *concentrated* sulphuric acid proceeds by the S_N2 mechanism or via an intermediate alkyl hydrogen sulphate (*cf*. p. 189).

Dehydration of alcohols to ethers may also be carried out by passing the alcohol vapour over a heated catalyst such as alumina, aluminium phosphate, etc.

2. Ethers may be prepared by the addition of alcohols to alkenes in the presence of acid, *e.g.*, ethanol and isobutene give t-butyl ethyl ether:

$$Me_2C{=}CH_2 + H^+ \underset{H_2SO_4}{\rightleftharpoons} Me_3C^+ \underset{EtOH}{\rightleftharpoons} Me_3C{-}\overset{+}{O}(H)Et \xrightarrow{-H^+} Me_3COEt$$

This reaction gives very good yields with alkenes which can produce t-carbonium ions, and is very useful for preparing mixed ethers.

3. By **Williamson's synthesis**, in which sodium or potassium alkoxide is heated with an alkyl halide (S_N2):

$$R^1O^- \quad R^2{-}X \rightleftharpoons R^1\overset{\delta-}{O}{\cdots}R^2{\cdots}\overset{\delta-}{X} \longrightarrow R^1OR^2 + X^- \ (g.)$$

This method is particularly useful for preparing mixed ethers, and it is best to use the alkoxide of the secondary or tertiary alcohol, and primary alkyl halide because secondary and tertiary alkyl halides readily undergo E2 elimination in the presence of a strong base to form alkenes. Williamson's synthesis proves the structure of the ethers.

When R^2 is a methyl or ethyl group, methyl or ethyl sulphate, respectively, can be used instead of the corresponding alkyl halide, *e.g.*,

$$Na^+\{C_2H_5O \quad CH_3{-}O{-}SO_3CH_3 \longrightarrow C_2H_5OCH_3 + CH_3OSO_3^- Na^+$$

This has been modified to a *general* method for preparing ethers by refluxing an alcohol with an ester of toluene-*p*-sulphonic acid in the presence of sodium:

$$R^1O \quad R^2{-}O{-}SO_2 {-\!\!\!\bigcirc\!\!\!-} Me \xrightarrow{Na} R^1OR^2 + Me {-\!\!\!\bigcirc\!\!\!-} SO_3^-$$

4. From halogeno-ethers and Grignard reagents (p. 430).

5. Methyl ethers are readily prepared by the action of diazomethane on alcohols (p. 393).

General properties of the ethers.

The lower members are gases or volatile liquids, and their vapours are highly inflammable. Their boiling points are much lower than those of the alcohols containing the same number of carbon atoms, and this is due to the fact that ethers cannot associate through hydrogen bonding. All the ethers are less dense than water in which they are not very soluble, but their solubility is very much increased in the presence of small amounts of alcohol.

The C—O group (stretch) in acyclic ethers containing two primary alkyl groups, *i.e.*, those containing the group —CH_2—O—CH_2—, absorbs in the region 1150–1060 cm^{-1} (s). Fig. 7.1 is the infra-red spectrum of di-n-butyl ether (as a film).

The NMR spectra of a number of methyl ethers have been examined, and it appears that the methoxyl group in alkyl ethers shows one band with a τ-value of 6·60–6·80 p.p.m. Hence the presence of this band is very strong evidence that there is a methoxyl group in the molecule. In aromatic ethers, the τ-value is around 6.30 p.p.m.

Mass spectra. The molecular ion of ethers is weak, and the principal modes of fission occur through α- and β-cleavage:

$$R^1_3C^+ + R^2O\cdot \xleftarrow{\;\alpha\;} [R^1\overset{\beta}{-}CR^1_2\overset{\alpha}{-}OR^2]^{\ddagger} \xrightarrow{\;\beta\;} R^1\cdot + CR^1_2 = \overset{+}{O}R^2$$

Ethers containing α-substituted alkyl groups also undergo *double* cleavage and rearrangement, and the resulting oxygen-containing fragment ions have high intensity and may even be the base peaks:

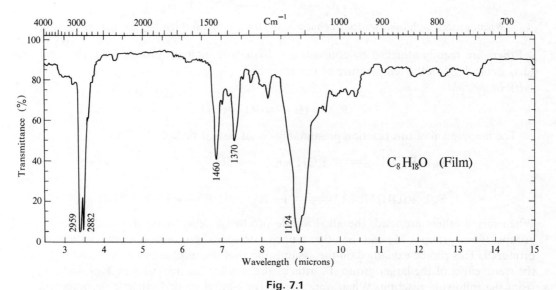

$$R^1_2C\overset{\cdot+}{-}\overset{}{O} \cdots CHR^2 \longrightarrow R^1\cdot + CR^1_2 = \overset{+}{O}H + R^2CH = CH_2$$

This is the mechanism for the formation of ions *m/e* 45, 59, etc.

Fig. 7.1

$1\,124\,\mathrm{cm}^{-1}$: C—O (str.); this is characteristic of the group —CH_2—O—CH_2—. $2\,959$ and $2\,882\,\mathrm{cm}^{-1}$: C—H (str.) in Me and/or CH_2. $1\,460\,\mathrm{cm}^{-1}$: C—H (def.) in Me and/or CH_2. $1\,370\,\mathrm{cm}^{-1}$: C—H (def.) in Me.

(within figure: $C_8H_{18}O$ (Film))

General reactions.

1. Ethers dissolve in concentrated solutions of strong inorganic acids to form oxonium salts, *i.e.*, ethers behave as Brønsted-Lowry bases, *e.g.*,

$$R_2O + H_2SO_4 \rightleftharpoons [R_2OH]^+HSO_4^-$$

It is the presence of the lone-pair electrons on the oxygen atom which predominantly characterises the reactions of the ethers. Thus, ethers also readily form co-ordination complexes (*etherates*) with Lewis acids, and treatment of an etherate with an alkyl fluoride produces a tertiary oxonium salt, *e.g.*,

$$R_2^1O + BF_3 \longrightarrow R_2^1\overset{+}{O}-\overset{-}{BF_3} \overset{R^2F}{\longrightarrow} [R_2^1OR^2]^+ BF_4^-$$

These trialkyloxonium fluoroborates are powerful alkylating reagents for nucleophilic sub-strates such as carboxylic acids, phenols, thioethers, etc., *e.g.*, thioethers are readily converted into trialkylsulphonium fluoroborates (see also p. 402):

$$R_2^2\overset{\cdot\cdot}{S} \,\, R^1 - \overset{+}{O}R_2^1\} BF_4^- \longrightarrow [R_2^2S-R^1]^+ BF_4^- + R_2^1O$$

2. When heated with dilute sulphuric acid under pressure, ethers form the corresponding alcohols:

$$R_2O + H_2O \overset{H_2SO_4}{\longrightarrow} 2ROH$$

The mechanism of this reaction is believed to be bimolecular (S_N2):

$$
\begin{array}{c}
\text{H} \\
| \\
R-\overset{+}{O}-R \quad \overset{\cdot\cdot}{O}H_2 \longrightarrow ROH + ROH_2^+
\end{array}
$$

When warmed with *concentrated* sulphuric acid, ethers form alkyl hydrogen sulphates.

Ethers are readily attacked by concentrated hydriodic or hydrobromic acid, the final pro-ducts depending on the temperature of the reaction.

(i) *In the cold:*

$$R_2O + HI \longrightarrow RI + ROH$$

The mechanism of this reaction presumably could be S_N1 or S_N2.

$$S_N1 \quad R_2OH^+ \underset{}{\overset{slow}{\rightleftharpoons}} ROH + R^+ \overset{I^-}{\underset{fast}{\longrightarrow}} RI$$

$$S_N2 \quad R(OH)^+R + I^- \rightleftharpoons \overset{\delta-}{I} \cdots R \cdots \overset{\delta+}{(OH)R} \longrightarrow I-R + ROH$$

When mixed ethers are used, the alkyl iodide produced depends on the nature of the alkyl groups. If one group is Me and the other a prim.- or s-alkyl group, it is methyl iodide which is produced. This can be explained on the assumption that the mechanism is S_N2, and because of the steric effect of the larger group, I^- attacks the smaller (methyl) group. This finds support from the following reaction. When optically active s-butyl methyl ether is decomposed with hydrogen iodide, methyl iodide and optically active s-butanol are obtained. Thus the s-Bu-O bond is *not* broken in this reaction (see p. 485).

When the substrate is a methyl t-alkyl ether, the products are t-RI and MeOH. This can be explained by an S_N1 mechanism, the carbonium ion produced being the t-R since this type is more stable than a prim.- or s-R carbonium ion.

(ii) *When heated:*

$$R_2O + 2HI \longrightarrow 2RI + H_2O$$

In this case, (i) is followed by:

$$ROH + HI \longrightarrow RI + H_2O$$

Reactions (i) and (ii) are very useful for identifying the groups present in an ether. Further-more, since (i) occurs very easily, it is possible to 'protect' a hydroxyl group in a polyfunctional compound by converting the hydroxyl group into an ether, which is later treated with concentrated hydriodic acid to regenerate the hydroxyl group (see, *e.g.*, p. 317).

Decomposition of ethers by concentrated hydriodic acid is the basis of the *Zeisel* method for estimating methoxyl and ethoxyl groups (p. 374).

Ethers are readily split by Lewis acids (*cf.* p. 197), *e.g.*,

$$R_2O + BCl_3 \xrightarrow{\text{heat}} RCl + ROBCl_2 \xrightarrow{R_2O} RCl + (RO)_2BCl \xrightarrow{R_2O} RCl + (RO)_3B$$

3. When treated with chlorine or bromine, ethers undergo substitution, the extent of which depends on the conditions. Hydrogen joined to the carbon directly attached to the oxygen atom is most readily replaced, *e.g.*, diethyl ether reacts with chlorine in the dark to form 1,1'-dichlorodiethyl ether:

$$CH_3CH_2OCH_2CH_3 \xrightarrow{Cl_2} CH_3CHClOCH_2CH_3 \xrightarrow{Cl_2} CH_3CHClOCHClCH_3$$

In the presence of light, perchlorodiethyl ether is obtained:

$$(C_2H_5)_2O \xrightarrow{Cl_2} (C_2Cl_5)_2O$$

The reaction proceeds by a free-radical mechanism, and α-substitution occurs readily because of resonance stabilisation of the intermediate radical.

$$R^1\ddot{O}{-}CH_2R^2 \xrightarrow{Cl\cdot} R^1\ddot{O}{-}\dot{C}HR^2 \longleftrightarrow R^1\dot{O}{=}CHR^2 \xrightarrow{Cl_2} R^1\ddot{O}{-}CHClR^2 + Cl\cdot; \quad \text{etc.}$$

4. Acid chlorides react with ethers when heated in the presence of anhydrous zinc chloride, aluminium chloride, etc.; *e.g.*,

$$MeCOCl + AlCl_3 \rightleftharpoons MeCO^+ + AlCl_4^-$$

$$R_2O + MeCO^+ \longrightarrow R{-}\overset{+}{\underset{R}{O}}{-}COMe \longrightarrow MeCO_2R + R^+ \xrightarrow{AlCl_4^-} RCl + AlCl_3$$

Acid anhydrides also split ethers to form esters:

$$R_2O + (CH_3CO)_2O \xrightarrow{ZnCl_2} 2CH_3CO_2R$$

5. Ethers react with carbon monoxide at 125–180°C and at a pressure of 500 atmospheres, in the presence of boron trifluoride plus a little water:

$$R_2O + CO \longrightarrow RCO_2R$$

6. Many ethers are attacked by alkyl derivatives of the alkali metals (and sodamide), *e.g.*,

$$Na^+C_2H_5^- \overset{\frown}{} H{-}CH_2{-}CH_2{-}O{-}C_2H_5 \longrightarrow C_2H_6 + C_2H_4 + C_2H_5O^-Na^+$$

Only very strong bases are capable of abstracting hydrogen from ethers. The alkanes are extremely weak acids, but their conjugate bases are extremely strong bases (p. 64).

Dimethyl ether (*methyl ether*) is prepared industrially by passing methanol vapour at 350–400°C, and at a pressure of 15 atmospheres, over aluminium phosphate as catalyst (p. 196). It is a gas, b.p. −23·6°C, and is used as a refrigerating agent.

Diethyl ether (*ether*, '*sulphuric ether*') is prepared in the laboratory and industrially by the '*continuous etherification process*', *i.e.*, heating excess of ethanol with concentrated sulphuric acid. It is also obtained industrially: (i) as a by-product in the preparation of ethanol from ethylene and sulphuric acid (p. 189); (ii) by passing ethanol vapour, under pressure, over heated alumina or aluminium phosphate (*cf.* dimethyl ether).

Diethyl ether is a colourless liquid, b.p. 34·5°C. It is fairly soluble in water, and is miscible with ethanol in all proportions. It is highly inflammable and forms explosive mixtures with air; this is a great disadvantage in its use as an industrial solvent for oils, fats, gums, resins, etc., and as an extracting solvent. It is also used in surgery as an anaesthetic, and is the usual solvent for carrying out Grignard reactions. In the presence of air and light, ether forms ether peroxide, $CH_3CH(OOH)OC_2H_5$, which is a heavy, pungent, oily liquid, and is explosive. Since its boiling point is higher than that of ether, it is left in the residue after ether distillations, and may cause explosions. Addition of a small amount of a cuprous compound, *e.g.*, cuprous oxide, has been recommended for avoiding the formation of ether peroxide. The chief impurity in ether is ethanol, and this has the property of preventing the formation of ether peroxide.

An important derivative of ether is 2,2′-dichlorodiethyl ether, which is prepared by heating ethylene chlorohydrin with concentrated sulphuric acid at 100°C:

$$2ClCH_2CH_2OH \xrightarrow[-H_2O]{H_2SO_4} ClCH_2CH_2OCH_2CH_2Cl$$

It may also be prepared by passing a mixture of ethylene and chlorine into ethylene chlorohydrin at 80°C:

$$HOCH_2CH_2Cl + C_2H_4 + Cl_2 \longrightarrow (ClCH_2CH_2)_2O + HCl$$

2,2′-Dichlorodiethyl ether is used as a solvent, as a soil fumigant and also as the starting point of many chemicals.

2,2′-Dichlorodiethyl ether is also named as bis-(2-chloroethyl) ether. The prefix bis (Latin, twice) indicates that there are *two identical groups* attached to a given atom; it is generally used for complex groups.

Di-isopropyl ether is obtained industrially as a by-product in the preparation of isopropanol from propene and sulphuric acid (p. 192). It is also prepared by passing propene into 75 per cent sulphuric acid at 75-125°C under a pressure of 3-7 atmospheres; very little isopropanol is formed under these conditions.

Di-isopropyl ether is a colourless liquid, b.p. 69°C. It is used as an industrial solvent for extraction operations, and for decreasing the knocking properties of petrol which, mixed with di-isopropyl ether, acquires a high octane number.

Di-n-butyl ether, b.p. 142°C, is manufactured by heating n-butanol with concentrated sulphuric acid. It is used as an industrial solvent and as a solvent in Grignard reactions for which higher temperatures are required than can be obtained by using diethyl ether.

Di-isoamyl ether, $[(CH_3)_2CHCH_2CH_2—]_2O$, is prepared by the action of concentrated sulphuric acid on isopentanol. It is a colourless liquid, b.p. 172°C, and has a pear-like odour. It is used as an industrial solvent, and as a solvent in Grignard reactions.

Mixed ethers of the following types have been prepared: primary–primary; primary–secondary; primary–tertiary; secondary–secondary; secondary–tertiary; tertiary–tertiary.

PROBLEMS

1. The starting material is an optically active form of s-BuOH, and is subjected to the following reactions:

$$s\text{-BuOH} \xrightarrow{Na} ? \xrightarrow{MeI} ?$$

Complete the equations and state the configuration of the s-Bu group in each product. Explain your answer.

2. Which of the two reactions would you use to prepare Me_3COMe, and why?

(i) $Me_3CBr + MeO^-K^+ \longrightarrow$;

(ii) $Me_3CO^-K^+ + MeBr \longrightarrow$.

3. Complete the following equations and comment:

(i) $MeOEt \xrightarrow{\ HI\ } ?$ (ii) $Et_2O \xrightarrow{\ Na\ } ?$

(iii) $Me_2C{=}CH_2 \xrightarrow[\text{pressure}]{H_2SO_4 - H_2O} ? \xrightarrow[\text{pressure}]{Me_3COH} ?$

4. The compound $C_6H_{14}O$ showed two NMR signals. What is the compound?

5. The reaction between HI and C_2H_4 in EtOH gives predominantly EtI, whereas the reaction with HCl under the same conditions gives predominantly Et_2O. Explain.

6. Show the fragmentation paths whereby the ether, Et—O—CHMeEt, could produce ions m/e: 102, 87, 59, 57, 45, 31, 29.

REFERENCE

PATAI (ed.), *The Chemistry of the Ether Linkage*, Interscience (1967).

8

Aldehydes and ketones

Aldehydes and ketones both have the general formula $C_nH_{2n}O$ and both contain the **oxo (carbonyl) group**, $\diagdown$C=O. In **aldehydes**, the functional group is —CHO, *i.e.*, one of the available valencies of the carbonyl group is attached to hydrogen, and so the aldehyde group occurs at the *end* of a chain. In **ketones** both available valencies are attached to carbon atoms, and so the keto group occurs *within* a chain (*cf.* ketens, p. 349).

Nomenclature

Aldehydes. The lower members are commonly named after the acids that they form on oxidation. The suffix of the names of acids is *-ic* (the names of the trivial system are used, see p. 233); this suffix is deleted and replaced by *aldehyde*, *e.g.*,

$$HCHO \xrightarrow{[O]} HCO_2H$$
form*aldehyde*   form*ic* acid

$$(CH_3)_2CHCHO \xrightarrow{[O]} (CH_3)_2CHCO_2H$$
isobutyr*aldehyde*   isobutyr*ic* acid

The positions of side-chains or substituents are indicated by Greek letters, the α carbon atom being the one *adjacent* to the aldehyde group, *e.g.*,

$$\overset{\gamma}{C}H_3\overset{\beta}{C}H(OH)\overset{\alpha}{C}H_2CHO \qquad \beta\text{-hydroxybutyraldehyde}$$

According to the I.U.P.A.C. system of nomenclature, aldehydes are designated by the suffix *-al*, which is added to the name of the hydrocarbon from which they are derived. The longest carbon chain containing the aldehyde group is chosen as the parent hydrocarbon; the positions of side-chains or substituents are indicated by numbers, and the aldehyde group is given the number 1, which may be omitted from the name if it is the only functional group present in the compound, *e.g.*,

CH$_3$CHO	ethanal
CH$_3$CH$_2$CHCH(CH$_3$)CH$_2$CH$_3$	2-ethyl-3-methylpentanal
$\quad\quad\quad$|	
$\quad\quad$CHO	

Ketones. The lower members are commonly named according to the alkyl groups attached to the keto group, *e.g.*,

CH$_3$COCH$_3$	dimethyl ketone
CH$_3$CH$_2$COCH(CH$_3$)$_2$	ethyl isopropyl ketone

The positions of side-chains or substituents are indicated by Greek letters, the α-carbon atom being the one *adjacent* to the keto group, *e.g.*,

CH$_3$CHClCOCH$_2$CH$_2$Cl $\quad\quad$ α, β'-dichlorodiethyl ketone

If the two alkyl groups in a ketone are the same, the ketone is said to be simple or symmetrical; if unlike, mixed or unsymmetrical (*cf.* ethers).

According to the I.U.P.A.C. system of nomenclature, ketones are designated by the suffix *-one*, which is added to the name of the hydrocarbons from which they are derived. The longest carbon chain containing the keto group is chosen as the parent hydrocarbon; the positions of side-chains or substituents are indicated by numbers, and the keto group is given the lowest number possible, *e.g.*,

CH$_3$COCH$_2$CH$_2$CH$_3$	pentan-2-one
(CH$_3$)$_2$CHCOCH(CH$_3$)CH$_2$CH$_3$	2,4-dimethylhexan-3-one

The carbonyl or oxo group, $\diagdown$CO, is prefixed in aldehydes as the oxo, aldo (for aldehydic O), $\diagup$

or formyl (for CHO), and in ketones is prefixed as oxo or keto.

Since aldehydes and ketones both contain the carbonyl group, it might be expected that they would resemble one another. It is therefore instructive to compare their general methods of preparation and their general properties.

General methods of preparation of aldehydes and ketones

1. *Aldehydes.* By the oxidation or dehydrogenation of a *primary* alcohol (pp. 185, 186).

> t-Butyl chromate (prepared by adding chromium trioxide to t-butanol) oxidises primary alcohols to aldehydes almost quantitatively (Oppenauer *et al.*, 1949).

Ketones. By the oxidation or dehydrogenation of a *secondary* alcohol.

There is, however, a specific reagent for oxidising secondary alcohols to ketones, *viz.*, aluminium t-butoxide, [(CH$_3$)$_3$CO]$_3$Al (the **Oppenauer oxidation**, 1937). The secondary alcohol is refluxed with the reagent, and then acetone (or cyclohexanone, etc.) is added. The mechanism of the reaction is the *reverse* of that of the Meerwein–Ponndorf–Verley reduction (p. 210):

$$R^1CHOHR^2 + Me_3COAl/3 \rightleftharpoons R^1R^2CHOAl/3 + Me_3COH$$
$$R^1R^2CHOAl/3 + Me_2CO \rightleftharpoons R^1R^2CO + Me_2CHOAl/3$$

t-Butyl alkoxide is used as the reagent, since the t-butyl alcohol produced is *not* oxidised under these conditions.

This reagent is particularly useful for oxidising *unsaturated* secondary alcohols because it does not affect the double bond. On the other hand, *primary* alcohols (particularly unsaturated alcohols) may also be oxidised to aldehydes if acetone is replaced by *p*-benzoquinone. In general, quinones and aromatic ketones are better hydrogen acceptors than acetone.

> Primary and secondary alcohols, both saturated and unsaturated, may be oxidised to the corresponding carbonyl compound by means of manganese dioxide in acetone solution (*inter alia*, Bharucha, 1956). Ruthenium tetroxide also oxidises saturated primary and secondary alcohols to the corresponding carbonyl compound (Berkowitz *et al.*, 1958). On the other hand, *p*-toluenesulphonates (tosylates) of n-alcohols can be oxidised to aldehydes by dimethyl sulphoxide in the presence of sodium hydrogen carbonate at 100°C (Kornblum *et al.*, 1959).

$$RCH_2OSO_2C_6H_4CH_3 + (CH_3)_2SO \xrightarrow{NaHCO_3} RCHO + (CH_3)_2S + CH_3C_6H_4SO_3Na \quad (65\text{--}85\%)$$

> Another method of oxidising primary and secondary alcohols to the corresponding carbonyl compound is that of Barton *et al.* (1964). The alcohol is converted into its alkyl chloroformate, which is then dissolved in dimethyl sulphoxide, and finally triethylamine is added:

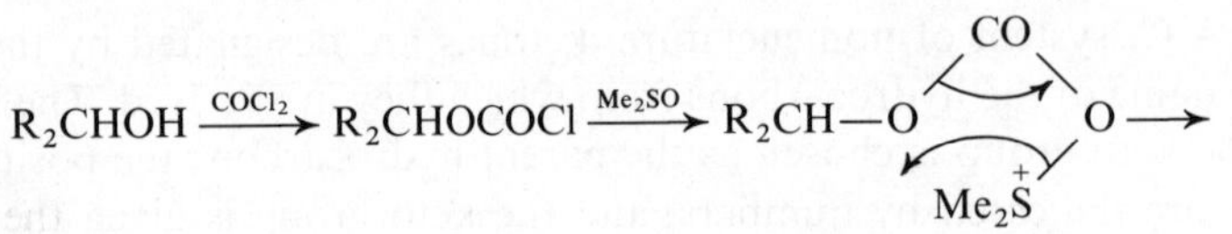

$$R_2CHOH \xrightarrow{COCl_2} R_2CHOCOCl \xrightarrow{Me_2SO} R_2CH\!-\!O \overset{CO}{\underset{Me_2S}{\rightleftarrows}} O \longrightarrow$$

$$CO_2 + R_2CH\!-\!O\!-\!\overset{+}{S}Me_2 \xrightarrow[-H^+]{Et_3N} R_2CO + Me_2S \quad (57\text{--}80\%)$$

2. *Aldehydes.* By heating a mixture of the calcium salts of formic acid and any one of its homologues (yield: variable due to many side reactions):

$$(RCO_2)_2Ca + (HCO_2)Ca \longrightarrow 2RCHO + 2CaCO_3$$

Ketones. By heating the calcium salt of any monocarboxylic acid other than formic acid.

$$(RCO_2)_2Ca \longrightarrow R_2CO + CaCO_3$$

The yields are usually low, but are high when the iron salts of n-acids are heated (Schultz *et al.*, 1962). The mechanism of the reaction is uncertain, but a possibility is one via the formation of an aldol and then a β-keto-acid (M = metal):

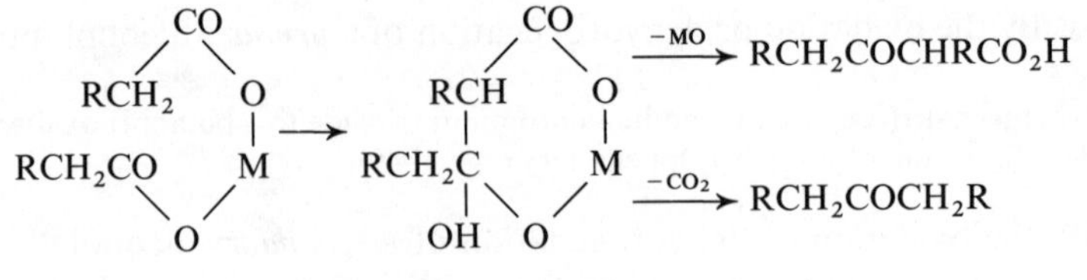

$$\begin{array}{ccc} & & \xrightarrow{-MO} RCH_2COCHRCO_2H \\ \text{[aldol/β-keto-acid scheme]} & & \\ & & \xrightarrow{-CO_2} RCH_2COCH_2R \end{array}$$

> No aldol formation is possible unless the acid contains an α-hydrogen atom. Similarly, the anhydrides of acids with an α-hydrogen also form ketones on heating:

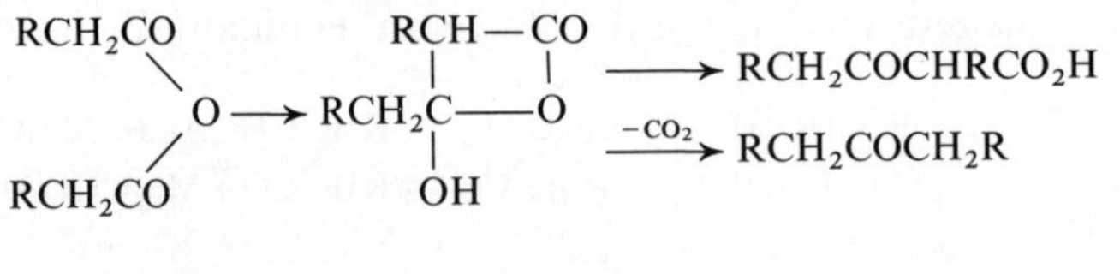

$$\text{[anhydride scheme]} \longrightarrow RCH_2C\!\!-\!\!O \xrightarrow{} RCH_2COCHRCO_2H$$
$$\xrightarrow{-CO_2} RCH_2COCH_2R$$

If a mixture of calcium salts is used, mixed ketones are obtained:

$$(R^1CO_2)_2Ca + (R^2CO_2)_2Ca \longrightarrow 2R^1COR^2 + 2CaCO_3$$

3. *Aldehydes.* By passing a mixture of the vapours of formic acid and any one of its homologues over manganous oxide as catalyst at 300°C:

$$RCO_2H + HCO_2H \xrightarrow{\text{MnO}} RCHO + CO_2 + H_2O \quad (f.g.-g.)$$

R_2CO and $RCHO$ are obtained as by-products, and the reaction probably proceeds via the manganous salt.

Ketones. By passing the vapour of any monocarboxylic acid other than formic acid over manganous oxide at 300°C:

$$2RCO_2H \xrightarrow{\text{MnO}} R_2CO + CO_2 + H_2O \quad (g.)$$

A mixture of monocarboxylic acids gives mixed ketones:

$$R^1CO_2H + R^2CO_2H \longrightarrow R^1COR^2 + CO_2 + H_2O$$

R_2^1CO and R_2^2CO are obtained as by-products.

4. *Aldehydes.* By the reduction of an acid chloride with hydrogen in boiling xylene using a palladium catalyst supported on barium sulphate (**Rosenmund's reduction**, 1918):

$$RCOCl + H_2 \longrightarrow RCHO + HCl \quad (g.-v.g.)$$

Aldehydes are more readily reduced than are acid chlorides, and therefore one would expect to obtain the alcohol as the final product. It is the barium sulphate that prevents the aldehyde from being reduced, acting as a poison (to the palladium catalyst) in this reaction. Generally, when the Rosenmund reduction is carried out, a small amount of quinoline and sulphur is added; these are very effective in poisoning the catalyst in the aldehyde reduction.

Acid chlorides are readily reduced to aldehydes by lithium tri-t-butoxyaluminium hydride (Table 6.1, p. 179) and by tri-n-butyltin hydride. Esters also give aldehydes with sodium aluminium hydride.

Ketones. There is no analogous method for the preparation of ketones.

5. *Aldehydes.* By the ozonolysis of alkenes of the type $R^1CH{=}CHR^2$.

Ketones. By the ozonolysis or oxidation with the Lemieux reagent (p. 111) of alkenes of the type $R_2^1C{=}CR_2^2$.

6. *Aldehydes.* By the oxidation of 1,2-glycols of the type $R^1CH(OH)CH(OH)R^2$ with lead tetra-acetate or periodic acid (see p. 110).

Ketones. By the oxidation of 1,2-glycols of the type $R_2^1C(OH)C(OH)R_2^2$ with lead tetra-acetate or periodic acid.

7. *Aldehydes.* Acetylene, when passed into hot dilute sulphuric acid in the presence of mercuric sulphate as catalyst, is converted into acetaldehyde (p. 136).

On the other hand, 1-alkynes give aldehydes by hydroboronation with disiamylborane followed by oxidation (Brown *et al.*, 1961; see also p. 114).

$$RC{\equiv}CH + Si\alpha_2BH \longrightarrow RCH{\equiv}CHBSi\alpha_2 \xrightarrow[\text{OH}^-]{\text{H}_2\text{O}_2} [RCH{=}CHOH] \longrightarrow RCH_2CHO \quad (g.)$$

The addition to the *terminal* carbon atom is due to the fact that the borane is sterically hindered to addition at C-2.

Ketones. All homologues of acetylene, treated in the same way as acetylene, form ketones.

Hydroboronation of non-terminal alkynes, followed by oxidation, gives ketones:

$$RC{\equiv}CR \xrightarrow{\text{Sia}_2\text{BH}} RCH{=}CRBSi\alpha_2 \xrightarrow[\text{OH}^-]{\text{H}_2\text{O}_2} [RCH{=}CR(OH)] \longrightarrow RCH_2COR$$

8. *Aldehydes.* By **Stephen's method** (1925). An alkyl cyanide is dissolved in ether, or better, in ethyl formate or ethyl acetate (Stephen *et al.*, 1956), and reduced with stannous chloride and hydrochloric acid, and then steam distilled.

Acording to Hantzsch (1931), Stephen's reaction proceeds via the iminochloride or aldimine hydrochloride (which is present as the stannichloride):

$$RC{\equiv}N \xrightarrow{\text{HCl}} [RC{\equiv}NH]^+Cl^- \xrightarrow[\text{HCl}]{\text{SnCl}_2} [RCH{=}NH_2]_2{}^+SnCl_6{}^{2-} \xrightarrow{\text{H}_2\text{O}} RCHO$$

Tolbert *et al.* (1963) have isolated these aldimine stannichlorides.

Cyanides and t-amides ($R^1CONR_2^2$) are reduced to aldehydes by lithium triethoxyaluminium hydride.

Ketones. There is no analogous method for the preparation of ketones.
9. *Aldehydes.* By means of a Grignard reagent and formic ester (see p. 430).
Ketones. By means of a Grignard reagent and, for example, an alkyl cyanide (see p. 430). Ketones are also formed by reaction between an alkyl cyanide and lithium-alkyls (p. 436).
10. *Ketones.* By the ketonic hydrolysis of the alkyl derivatives of acetoacetic ester (see p. 290), and also from ethyl malonate derivatives (p. 296).
Aldehydes. There is no analogous method for the preparation of aldehydes. However, the reduction of acylmalonic esters with sodium borohydride gives aldehydes (Muxfeldt *et al.*, 1965).

$$RCOCH(CO_2Et)_2 \xrightarrow{\text{NaBH}_4} RCHO + CH_2(CO_2Et)_2$$

11. *Aldehydes.* Many aldehydes may be prepared by the *oxo* process (p. 180), *e.g.*, propionaldehyde (*q.v.*).
Ketones. There is no analogous method for the preparation of ketones.
12. *Aldehydes.* Aldehydes may be prepared (in yields up to 50 percent) by means of the modified **Sommelet reaction** (Angyal *et al.*, 1953; see benzaldehyde, p. 735); *e.g.*, n-hexanal is produced by adding an aqueous solution of n-hexylamine hydrochloride to a solution of hexamine in acetic acid through which steam is passed.

$$CH_3(CH_2)_5NH_2{\cdot}HCl + (CH_2)_6N_4 \longrightarrow CH_3(CH_2)_4CHO \quad (17\%)$$

Ketones. There is no analogous method for the preparation of ketones.
13. *Ketones.* Methyl ketones may be prepared from esters and dimethyl sulphoxide (Corey *et al.*, 1964). The latter compound is converted into a methylide by means of sodium hydride, and this is then treated as follows (THF = tetrahydrofuran):

$$2CH_3SOCH_3 \xrightarrow{2\text{NaH}} 2CH_3SOCH_2{}^- \xrightarrow{\text{RCO}_2\text{Me}} MeO^- + CH_3SOCH_3 + [RCOCHSOCH_3]$$

$$\xrightarrow{\text{H}^+} RCOCH_2SOCH_3 \xrightarrow[\text{THF}-\text{H}_2\text{O}]{\text{Al}-\text{Hg}} RCOCH_3 \quad (70{-}98\%)$$

14. *Aldehydes and ketones.* See the Darzens Glycidic Ester condensation (p. 219).

General properties of aldehydes and ketones

Dipole moment measurements of aldehydes and ketones have shown that the values are larger than can be accounted for by the inductive effect of the oxygen atom, but can be accounted for if carbonyl compounds are resonance hybrids:

$$\diagup\!\!\!\!\diagdown C=O \longleftrightarrow \diagup\!\!\!\!\diagdown C^+ - O^-$$

Thus the carbon atom has a positive charge and consequently can be attacked by nucleophilic reagents. The carbonyl group also exhibits basic properties; it is readily protonated by strong acids to form oxonium salts. Since oxygen is more electronegative than carbon, the second resonating structure will make a larger contribution than the first.

$$\diagup\!\!\!\!\diagdown C=O + H^+ \rightleftharpoons \diagup\!\!\!\!\diagdown C=O^+ - H \longleftrightarrow \diagup\!\!\!\!\diagdown C^+ - O - H$$

Hence, protonation increases the electrophilic character of the carbonyl group and so it can be expected that nucleophilic additions will be catalysed by acids. It should also be noted that, because of the positive charge on the carbon atom, the CO group has a strong $-I$ effect.

Many addition reactions of carbonyl compounds may be represented by the general equation

$$
\begin{array}{c}
R^1 \\ \diagdown \\ \diagup \\ R^2
\end{array} C=O + H-Z \rightleftharpoons
\begin{array}{c}
R^1 \quad OH \\ \diagdown \diagup \\ C \\ \diagup \diagdown \\ R^2 \quad Z
\end{array}
$$

Thus, the hybridisation of the carbon atom changes from sp^2 to sp^3. This results in increased crowding and so reaction can be expected to be increasingly sterically hindered with increasing size of R^1 and R^2. For aldehydes, $R^2 = H$ (for formaldehyde, $R^1 = R^2 = H$), and hence it can be anticipated that aldehydes will be more reactive than ketones. This is observed in practice, reaction being faster and the equilibrium shifted more to the right.

The $+I$ effect of an alkyl group also decreases the reactivity of the carbonyl group towards nucleophilic reagents, this being due to the partial neutralisation of the positive charge on the carbonyl carbon atom. Hence, ketones will be less reactive than aldehydes due to both steric and inductive factors (see also p. 228).

If we consider the energetics of these additional reactions:

$$
\diagup\!\!\!\!\diagdown C=O + H-Z \rightleftharpoons
\begin{array}{c}
OH \\ \diagdown \diagup \\ C \\ \diagup \diagdown \\ Z
\end{array}
$$

$$(+694{\cdot}5) \quad (+x) \qquad (-334{\cdot}7 - 464{\cdot}4 - y)$$

$$\text{then } \Delta H^{\ominus} = (x - y - 104{\cdot}6)\,\text{kJ}$$

where x and y are the bond energies of the H—Z and C—Z bonds, respectively. Suppose Z is OH, i.e., H—Z is H−OH. In this case, $\Delta H^{\ominus} = (464{\cdot}4 - 334{\cdot}7 - 104{\cdot}6) = 25{\cdot}1\,\text{kJ}$. Now suppose Z is CN, i.e., HZ is H—CN. Then

$$\Delta H^{\ominus} = (414{\cdot}2 - 347{\cdot}3 - 104{\cdot}6) = -37{\cdot}7\,\text{kJ}$$

Since $\Delta G^{\ominus} = \Delta H^{\ominus} - T\Delta S^{\ominus} = -RT\ln K$ and since the number of product molecules in both cases is less than the number of reactant molecules, there will be a decrease of entropy (p. 58), *i.e.*, the term $-T\Delta S^{\ominus}$ will be positive. Hence, with water, $\Delta G^{\ominus}$ must be positive and the reaction is therefore thermodynamically unfavourable. For hydrogen cyanide, $\Delta G^{\ominus}$ will be considerably less positive (it may actually be negative if $\Delta H^{\ominus} > T\Delta S^{\ominus}$) than for the hydration reaction and consequently, relative to the latter, will be highly thermodynamically favourable (see p. 211).

Aldehydes and ketones have low boiling points compared with those of the alcohols, and this is probably due to their inability to form intermolecular hydrogen bonds. On the other hand, carbonyl compounds have higher boiling points than the corresponding alkanes, and this is due to dipole-dipole interactions in the former. Since carbonyl compounds can form hydrogen bonds with water and also form hydrates (to various extents), they are soluble (or partially soluble) in water.

The infra-red absorption region (C=O stretch) for saturated aldehydes is 1740–1720 cm^{-1} (s) and for acyclic ketones is 1725–1700 cm^{-1} (s). It is therefore possible to distinguish between aldehydes and ketones. Once the presence of a carbonyl group has been established, aldehydes and ketones may be differentiated by the fact that there is the C—H (str.) in the aldehyde group that absorbs in the region 2880–2665 cm^{-1} (w-m). There are also bands due to C—H (def.) which are characteristic of a methyl or a methylene group directly attached to the carbonyl group in ketones: CH$_3$CO—, 1365–1355 cm^{-1} (s); —CH$_2$CO—, 1435–1405 cm^{-1} (s).

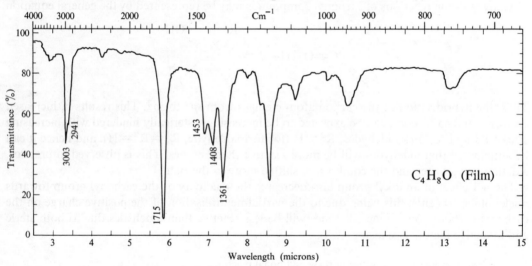

Fig. 8.1

1715 cm^{-1}: C=O (str.) in saturated ketones (and aldehydes?).   3003 and 2941 cm^{-1}: C—H (str.) in Me and/or CH$_2$.   1453 cm^{-1}: C—H (def.) in Me and/or CH$_2$.   1408 cm^{-1}: C—H (def.) in CH$_2$ in —CH$_2$CO—. 1361 cm^{-1}: C—H (def.) in Me in MeCO—.

Figure 8.1 is the infra-red spectrum of ethyl methyl ketone (as a film). The region of the CO absorption is in that for saturated ketones, but is very close to the lower frequency region for aldehydes. However, the absence of a band at 2880–2665 cm^{-1} is strongly in favour of the absence of an aldehyde.

The NMR spectra of carbonyl compounds are affected by the strong inductive effect ($-I$) of the carbonyl group and also by the magnetic anisotropy of the carbon-oxygen double bond (see ethylene, Figs. 4.4(*d*) and (*e*), p. 133; replace one CH$_2$ group by oxygen). Both effects deshield an *aldehydic* proton, the result being a very low τ-value 0·1–0·7 p.p.m. On the other hand, protons in *ketones* are deshielded mainly by the $-I$ effect (the protons are at the 'edge' of the shielding cone) and consequently the shift downfield is much less (than for the aldehydic proton), *e.g.*, aliphatic ketones containing the MeCO group have a τ-value 7·8–8·1 for the methyl protons. Thus, infra-red spectroscopy will show the presence of a carbonyl group, and the NMR spectrum will distinguish between aldehydes and ketones. This is clearly shown by the NMR spectra of acetaldehyde (Fig. 8.2) and that of ethyl methyl ketone (Fig. 8.3; both in CCl_4 at 60 MHz).

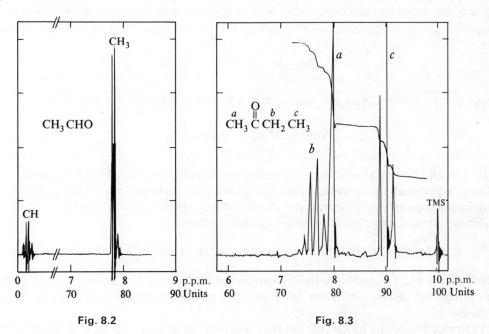

Fig. 8.2   Fig. 8.3

Mass spectra. Aldehydes give molecular ions of low intensity and readily undergo α-cleavage to produce acylium ions:

$$H\cdot + RC\equiv O^+ \longleftarrow RCH=O^+ \longrightarrow R\cdot + HC\equiv O^+ \quad (m/e\ 29)$$

The presence of ions M–1 and m/e 29 are usually characteristic of aldehydes (R^+ is also formed). It should be noted that the ion m/e 29 could also be $C_2H_5^+$, which is given by the higher aldehydes (the two ions may be distinguished by high resolution).

Aldehydes also undergo β-cleavage (the ion m/e 43 could also be $C_3H_7^+$):

$$[R-CH_2-CH=O]^+ \longrightarrow R\cdot(M-43) + [CH_2=CH-O]^+ \quad (m/e\ 43)$$

When a γ-hydrogen atom is present, the McLafferty rearrangement also occurs:

Ketones undergo fragmentation patterns similar to those of aldehydes, but the intensity of the molecular ion is very strong, and loss of the group with the heavier mass occurs predominantly. Hence, for methyl ketones, the acylium ion, $CH_3C\equiv O^+$ (m/e 43; also equivalent to $C_3H_7{}^+$) is often the base peak. Alkyl ions are also produced, as well as alkenes and $CH_2=CR-\overset{+}{O}H$ by the McLafferty rearrangement.

Reactions common to aldehydes and ketones

1. Catalytic hydrogenation (see Table 6.1, p. 179) readily converts aldehydes and ketones into primary and secondary alcohols, respectively. Reduction with dissolving metals also produces alcohols:

$$\underset{/}{\overset{\backslash}{\,}}C=O+2e \longrightarrow \underset{/}{\overset{\backslash}{\,}}C^--O^- \xrightarrow{2H^+} \underset{/}{\overset{\backslash}{\,}}CHOH$$

This reaction, however, may also lead to the formation of a 1,2-glycol (see p. 224):

$$2\underset{/}{\overset{\backslash}{\,}}C=O \xrightarrow{e;\,H^+} \underset{\backslash}{\overset{\backslash}{\,}}C(OH)C(OH)\underset{\backslash}{\overset{/}{\,}}$$

Reduction of both aldehydes and ketones to alcohols may be effected by means of the **Meerwein-Ponndorf-Verley reduction** (1925, 1926). The carbonyl compound is heated with aluminium isopropoxide in isopropanol solution; the isopropoxide is oxidised to acetone, which is removed from the equilibrium mixture by slow distillation:

$$R_2CO+(CH_3)_2CHOAl/_3 \rightleftharpoons R_2CHOAl/_3+CH_3COCH_3 \xrightarrow[H_2SO_4]{dilute} R_2CHOH \quad (g.)$$

This reducing agent is specific for the carbonyl group, and so may be used for reducing aldehydes and ketones containing some other functional group that is reducible, *e.g.*, a double bond or a nitro-group.

The reduction occurs by hydride ion transfer from aluminium isopropoxide to the carbonyl compound via a cyclic transition state.

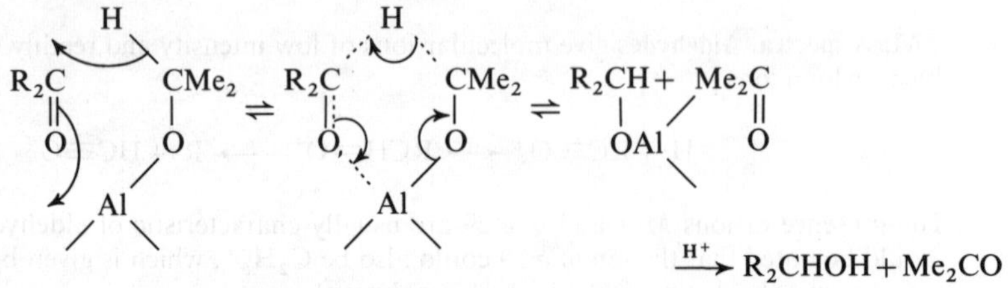

$$\xrightarrow{H^+} R_2CHOH+Me_2CO$$

The equilibrium is displaced to the right by distillation of the acetone (which is the component with the lowest b.p. in the system). Of all the metal alkoxides that may be used for this reaction, *e.g.*, K, Na, etc., aluminium alkoxide is the best, probably because of its great ability to form co-ordination compounds.

Aldehydes and ketones are readily reduced to alcohols by various metallic hydrides (see Table 6.1, p. 179).

Clemmensen reduction (1913): the carbonyl compound is reduced with *amalgamated zinc* and concentrated hydrochloric acid:

$$R^1COR^2 \xrightarrow{e;\,H^+} R^1CH_2R^2$$

The Clemmensen reduction does not appear to work well for aldehydes, but is reasonably good for many ketones.

Several mechanisms have been proposed for the Clemmensen reduction, *e.g.*,

$$R-\underset{\displaystyle \overset{\displaystyle O}{\|}}{C}-R + Zn + Cl^- \rightleftharpoons R-\underset{\displaystyle \underset{\displaystyle ZnCl}{|}}{\overset{\displaystyle \overset{\displaystyle O^-}{|}}{C}}-R \xrightarrow{H^+} R-\underset{\displaystyle \underset{\displaystyle ZnCl}{|}}{\overset{\displaystyle \overset{\displaystyle OH}{|}}{C}}-R \xrightarrow[-H_2O]{H^+}$$

$$R-\underset{\displaystyle \underset{\displaystyle ZnCl}{|}}{\overset{\displaystyle +}{C}}-R \xrightarrow{Zn} {}^+ZnCl + R-\overset{\displaystyle +}{C}{}^+-R \xrightarrow[Cl^-]{Zn} R-\underset{\displaystyle \underset{\displaystyle ZnCl}{|}}{\overset{\displaystyle \overset{\displaystyle Zn^-}{|}}{C}}-R$$

$$\xrightarrow{H^+} R-\underset{\displaystyle \underset{\displaystyle ZnCl}{|}}{C}H-R \xrightarrow{H^+} R-CH_2-R$$

2. Aldehydes and ketones add on sodium hydrogen sulphite to form *bisulphite* compounds:

$$\underset{/}{\overset{\backslash}{}}C{=}O + NaHSO_3 \rightleftharpoons \underset{\underset{SO_3Na}{\diagdown}}{\overset{\overset{OH}{\diagup}}{C}} \quad or \quad \left[\underset{\underset{SO_3}{\diagdown}}{\overset{\overset{OH}{\diagup}}{C}}\right]^- Na^+$$

These bisulphite compounds are *hydroxysulphonic acid salts*, since the sulphur atom is *directly* attached to the carbon atom. This structure is supported by work with isotope ^{34}S (Sheppard *et al.*, 1954).

> The position of equilibrium lies largely to the right for most aldehydes and to the left for most ketones, *e.g.*, Hooper (1967) showed from NMR studies of solutions of ketones in aqueous sodium hydrogen sulphite that the amounts of *unreacted* ketone were: Me_2CO, 24·6; MeCOEt, 71·8; MeCOisoPr, 100 per cent. These results may be partly explained in terms of steric effects.

Bisulphite compounds are usually crystalline solids, insoluble in sodium hydrogen sulphite solution. Since they regenerate the carbonyl compound when heated with dilute acid or sodium carbonate solution, their formation affords a convenient means of separating carbonyl compounds from non-carbonyl compounds.

> The mechanism of bisulphite formation has been the subject of much discussion. Hinshelwood *et al.* (1958), from a study of the *reverse* reaction (see principle of microscopic reversibility, p. 48), have proposed the following mechanism:
>
> $$HSO_3^- \rightleftharpoons H^+ + SO_3^{2-}$$
>
> $$\underset{/}{\overset{\backslash}{}}C{=}O + SO_3^{2-} \rightleftharpoons \underset{\underset{SO_3^-}{\diagdown}}{\overset{\overset{O^-}{\diagup}}{C}} \underset{\longleftarrow}{\overset{HSO_3^-}{\rightleftharpoons}} \underset{\underset{SO_3^-}{\diagdown}}{\overset{\overset{OH}{\diagup}}{C}} + SO_3^{2-}$$

3. Aldehydes and ketones add on hydrogen cyanide to form *cyanohydrins*. The carbonyl compound is treated with sodium cyanide and dilute sulphuric acid:

$$\underset{/}{\overset{\backslash}{}}C{=}O + HCN \rightleftharpoons \underset{\underset{CN}{\diagdown}}{\overset{\overset{OH}{\diagup}}{C}} \quad (g.\text{--}v.g.)$$

Lapworth (1903,1904) showed that the addition of hydrogen cyanide to carbonyl compounds is accelerated by bases and retarded by acids, and he concluded that the addendum was the cyanide ion, and proposed the following mechanism:

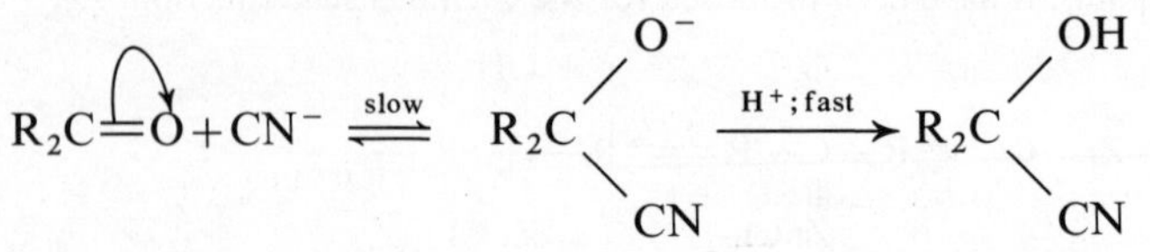

$$R_2C{=}O + CN^- \underset{slow}{\rightleftharpoons} R_2C\!\!\begin{array}{c} O^- \\ \diagup \\ \diagdown \\ CN \end{array} \xrightarrow{H^+;\,fast} R_2C\!\!\begin{array}{c} OH \\ \diagup \\ \diagdown \\ CN \end{array}$$

Cyanohydrins are important compounds in organic synthesis since they are readily hydrolysed to α-hydroxy-acids:

$$RCH(OH)CN \xrightarrow{H^+} RCH(OH)CO_2H$$

All aldehydes form cyanohydrins; only the ketones acetone, butanone, pentan-3-one, and pinacolone form cyanohydrins.

4. Aldehydes and ketones combine with a variety of compounds of the type Z—NH₂ (Z = OH, NH₂, etc.; see below). The general reaction may be written

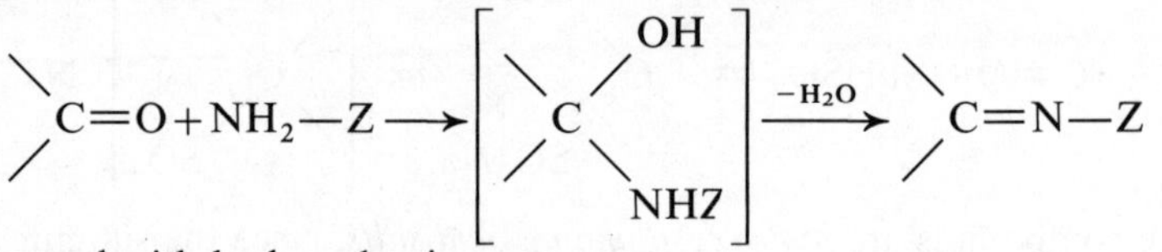

$$\diagdown\!\!\underset{\diagup}{C}{=}O + NH_2{-}Z \longrightarrow \left[\,\underset{\diagup\;\diagdown}{\overset{\diagup\;\diagup}{C}}\begin{array}{c}OH\\ \\NHZ\end{array}\right] \xrightarrow{-H_2O} \diagdown\!\!\underset{\diagup}{C}{=}N{-}Z$$

Oximes are formed with hydroxylamine:

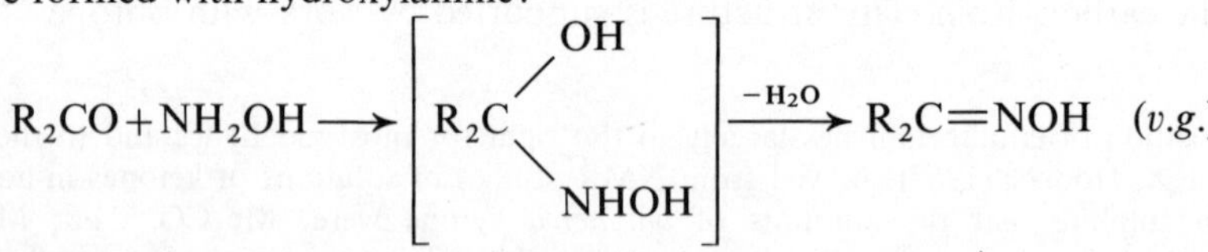

$$R_2CO + NH_2OH \longrightarrow \left[\,R_2C\begin{array}{c}OH\\ \\NHOH\end{array}\right] \xrightarrow{-H_2O} R_2C{=}NOH \quad (v.g.)$$

Oximes are usually well-defined crystalline solids, and may be used to identify carbonyl compounds.

The absorption region of oximes (C=N stretch) is 1690–1630 cm⁻¹ (v) and that for the O—H stretch is 3650–3500 cm⁻¹ (s) (*cf.* alcohols).

Aldoximes form cyanides when boiled with acetic anhydride, whereas ketoximes form the acetyl derivative of the oxime:

$$RCH{=}NOH \xrightarrow[-H_2O]{(CH_3CO)_2O} RCN$$

$$R_2C{=}NOH \xrightarrow{(CH_3CO)_2O} R_2C{=}NOCOCH_3$$

Aldehydes and ketones combine with *N*-substituted hydroxylamines to form **nitrones**:

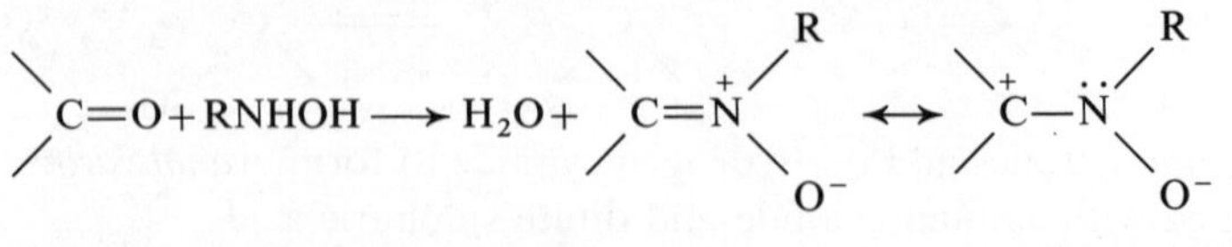

$$\diagdown\!\!\underset{\diagup}{C}{=}O + RNHOH \longrightarrow H_2O + \diagdown\!\!\underset{\diagup}{C}{=}\overset{+}{N}\begin{array}{c}R\\ \diagup\\ \diagdown\\ O^-\end{array} \longleftrightarrow \diagdown\!\!\underset{\diagup}{C}{-}\overset{+}{\underset{\diagdown}{\ddot{N}}}\begin{array}{c}R\\ \diagup\\ \\ O^-\end{array}$$

Nitrones readily undergo 1,3- addition reactions, *e.g.*,

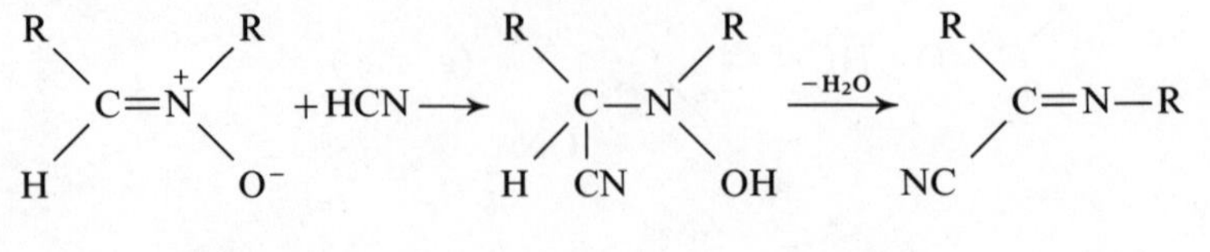

$$\underset{H}{\overset{R}{\diagdown}}C{=}\overset{+}{N}\underset{O^-}{\overset{R}{\diagup}} + HCN \longrightarrow \underset{H\;\;CN}{\overset{R}{\diagdown}}C{-}N\underset{OH}{\overset{R}{\diagdown}} \xrightarrow{-H_2O} \underset{NC}{\overset{R}{\diagdown}}C{=}N{-}R \qquad \text{cyano-imine}$$

Hydrazones and **azines** are formed with hydrazine:

$$\diagdown C=O+NH_2NH_2 \longrightarrow \left[\diagdown C \diagup^{NHNH_2}_{\diagdown OH} \right] \xrightarrow{-H_2O} \diagdown C=NNH_2 \xrightarrow{CO} \diagdown C=NN=C \diagup$$

By using suitable derivatives of hydrazine, more well-defined crystalline products are obtained (and azine formation is avoided), e.g., phenylhydrazine forms **phenylhydrazones**:

$$C_6H_5NHNH_2 + O=C\diagup \longrightarrow H_2O + C_6H_5NHN=C\diagup$$

Semicarbazide forms **semicarbazones**:

$$NH_2CONHNH_2 + O=C\diagup \longrightarrow NH_2CONHN=C\diagup + H_2O$$

The mechanism of the formation of semicarbazones and oximes has been studied in great detail.

These reactions (as are most reactions of this type, i.e., $\diagdown CO + NH_2—Z$) are catalysed by acids, and it has been found that the rates of formation of the products reach a maximum at some particular pH. According to Jencks et al. (1959, 1962), the pH optima observed are due to transitions in the rate-determining steps with changing concentration of acid. Let us consider the following mechanism (Z = —NHCONH$_2$ or —OH):

$$R_2C=O+NH_2—Z \longrightarrow R_2C\diagup^{O^-}_{\diagdown \overset{+}{N}H_2—Z} \longrightarrow R_2C\diagup^{OH}_{\diagdown NHZ} \underset{(I)}{}$$

$$\longrightarrow R_2C=NZ+H_2O$$
$$\text{(II)}$$

Jencks showed, by means of ultraviolet and infra-red studies, that at pH 7 the formation of (I) is fast, and that as the acidity increases, the overall rate of the reaction increases, since the dehydration step, which is the r/d step, also increases (through protonation of the OH, etc.). However, as the acidity increases, the addition step becomes progressively slower because Z—NH$_2$ is decreasing in concentration due to its conversion into its conjugate acid, Z—NH$_3{}^+$. The latter no longer has a lone pair of electrons on the N atom and so is not a nucleophile. Hence, at a sufficiently high acidity, the addition step is slowed down to such an extent that this now becomes the r/d step in the reaction.

Oximes and hydrazones regenerate the carbonyl compound when refluxed with dilute hydrochloric acid. Regeneration from phenylhydrazones by this method is usually difficult. One fairly successful method is by exchange of the phenylhydrazino group with another oxo compound, e.g., pyruvic acid. The most satisfactory method, however, appears to be the use of acetylacetone in faintly acid solution (Ried et al., 1962); the product is 3,5-dimethyl-1-phenylpyrazole.

$$\diagdown C=NNHPh + MeCOCH_2COMe \longrightarrow \diagdown C=O + \underset{\underset{Ph}{N}}{N}\diagdown^{Me}_{Me}$$

Wolff–Kishner reduction (1912). When hydrazones (or semicarbazones) are heated with sodium ethoxide at 180°C, nitrogen is eliminated, and a hydrocarbon is obtained, i.e., by this means the carbonyl group is converted into the methylene group:

$$\diagdown C=NNH_2 \longrightarrow \diagdown CH_2 + N_2 \quad (g.\text{--}v.g.)$$

The yield of hydrocarbon is better than that obtained by the Clemmensen reduction, but both the Wolff–Kishner and the Clemmensen reduction are generally unsuccessful with sterically hindered ketones. Barton *et al.* (1954, 1955), however, have overcome this problem by using a modification of the improved technique of the Wolff–Kishner reduction introduced by Huang-Minlon (1946). All the methods described use elevated temperatures (180–200°C), but Cram *et al.* (1962) have shown that the reaction proceeds at room temperature if dimethyl sulphoxide is used as solvent.

According to Szmant *et al.* (1952), the mechanism of the Wolff–Kishner reaction is (B is the base):

$$R_2C{=}NNH_2 + B \rightleftharpoons R_2C{=}N\bar{N}H + BH^+ \longrightarrow R_2\bar{C}H + N_2 \xrightarrow{BH^+} R_2CH_2$$

With α,β-unsaturated carbonyl compounds, the Wolff–Kishner reduction may lead to migration of the double bond. In the above mechanism, the first step is the abstraction of a proton. Hence the use of a very strong base should enable the reaction to be carried out at a lower temperature. Henbest *et al.* (1963), using potassium t-butoxide, found that the reduction proceeded normally with certain compounds that had previously been reported to give abnormal products.

Girard's reagents for carbonyl compounds (1934, 1936). Girard introduced two reagents for carbonyl compounds: Girard's reagents 'T' and 'P', which are respectively trimethylaminoaceto-hydrazide chloride: $[(CH_3)_3NCH_2CONHNH_2]^+Cl^-$, and pyridinium-acetohydrazide chloride, $[C_5H_5NCH_2CONHNH_2]^+Cl^-$ (see p. 268). These reagents react with carbonyl compounds to form

derivatives of the type $\left[(CH_3)_3NCH_2CONHN{=}C\diagdown\right]^+Cl^-$. These compounds are *soluble* in water.

and have been found particularly useful for isolating certain ketonic hormones.

Aldimines, Schiff bases, or **azomethines,** are formed when aldehydes react with aliphatic *primary* amines:

$$R^1CHO + R^2NH_2 \longrightarrow R^1CH(OH)NHR^2 \xrightarrow{-H_2O} R^1CH{=}NR^2$$

In some cases, the product is a polymer of the aldimine (see p. 221).

Aldehydes and ketones with α-hydrogen form *enamines* with cyclic secondary amines (see p. 383).

5. Aldehydes and ketones condense with thioalcohols (mercaptans), in the presence of hydrochloric acid, to form *mercaptals* and *mercaptols*, respectively (see also p. 400):

$$\diagup^\diagdown C{=}O + 2RSH \longrightarrow \diagup^\diagdown C(SR)_2 + H_2O$$

6. Aldehydes and ketones add on a molecule of a Grignard reagent, and the complex formed, when decomposed with acid, gives a secondary alcohol from an aldehyde (except formaldehyde, which gives a primary alcohol), and a tertiary alcohol from a ketone (see p. 428):

$$\diagup^\diagdown C{=}O + RMgX \longrightarrow \overset{OMgX}{\underset{R}{\diagup C \diagdown}} \xrightarrow{H^+} \overset{OH}{\underset{R}{\diagup C \diagdown}}$$

7. Both aldehydes and ketones undergo the Wittig reaction to form alkenes (p. 410).

8. Phosphorus pentachloride reacts with *simple* carbonyl compounds to form 1,1-dichlorides:

$$\diagup^\diagdown C{=}O + PCl_5 \longrightarrow \diagup^\diagdown CCl_2 + POCl_3$$

The mechanism of this reaction is uncertain, but a possibility is attack by PCl_4^+ (*solid* PCl_5 is $PCl_4^+PCl_6^-$):

$$\text{C=O} + PCl_4^+ \rightleftharpoons \overset{+}{\text{C}}\text{—}OPCl_4 \xrightarrow{Cl^-} \underset{OPCl_4}{\overset{Cl}{\text{C}}} \xrightarrow{-OPCl_4^-} \overset{+}{\text{C}}\text{—}Cl \xrightarrow{Cl^-} \underset{Cl}{\overset{Cl}{\text{C}}}$$

The chloride ions are probably derived from PCl_6^- (Newman *et al.*, 1959).

Good yields of 1,1-dichlorides are obtained by reaction between carbonyl compounds and 1,1-dichlorodimethyl ether (Rieche *et al.*, 1960).

Very little 1,1-dibromides are obtained with phosphorus pentabromide; the main product from any carbonyl compound is the α-bromo-derivative. Phosphorus pentabromide dissociates more readily then phosphorus pentachloride, and it is probable that the halogenation in the α-position is brought about by the free halogen (see below).

9. Chlorine or bromine replaces one or more α-hydrogen atoms in aldehydes and ketones, *e.g.*, acetone may be brominated in glacial acetic acid to give monobromoacetone:

$$CH_3COCH_3 + Br_2 \longrightarrow CH_3COCH_2Br + HBr \quad (43\text{–}44\%)$$

The halogenation of carbonyl compounds is catalysed by acids and bases. Let us consider the case of acetone. In alkaline solution, tribromoacetone and bromoform are isolated. Thus, the introduction of a second and a third bromine atom is more rapid than the first. In aqueous sodium hydroxide, the rate has been shown to be independent of the bromine concentration, but first order with respect to both acetone and base (Lapworth, 1904), *i.e.*,

$$\text{rate} = k[\text{acetone}][OH^-]$$

Thus the reaction is zero order in bromine, and this suggests at least a two-stage process, the first being the rate-determining step of formation of an intermediate which then rapidly reacts with bromine to give the products. The problem is now to formulate this intermediate. A most likely one is the enolate ion, since it is known that both acids and bases catalyse enolisation of a ketone (p. 281) and that enols react extremely rapidly with bromine. The following mechanism is consistent with the facts mentioned (see also p. 55):

$$CH_3COCH_3 + OH^- \underset{\text{slow}}{\rightleftharpoons} H_2O + CH_3\overset{O}{\overset{\|}{C}}CH_2^- \longleftrightarrow CH_3\overset{O^-}{\overset{|}{C}}=CH_2 \xrightarrow[\text{fast}]{Br_2}$$

$$CH_3COCH_2Br + Br^-$$

There is much evidence to support this mechanism, *e.g.*, iodination of acetone proceeds at the same rate as bromination and is zero order in iodine (Bartlett, 1934). The above mechanism is independent of both the nature of the halogen and its concentration.

In the acid-catalysed bromination of acetone, it is possible to isolate the mono-, di-, and tribromo-derivatives. Hence, in contrast with the base-catalysed reaction, the introduction of a second and a third bromine atom is less rapid than the first.

Another point that we shall discuss here is: How is it possible for a C—H bond to be broken so readily? This has been explained as follows. Owing to the inductive effect of the carbonyl group, the electrons of the C—H bonds on the α-carbon atom are displaced towards the carbon atom, thus facilitating the release of a proton from C_α. At the same time, the anion produced has greater resonance stabilisation than the parent carbonyl compound, *i.e.*, the conjugate base is stabilised with respect to its acid (see above equation).

The inductive effect is very much weaker on a β-carbon atom since the I effect falls off rapidly from the source. Hence proton release is far less easy than on α-carbon. Also, if a proton were released from the β—C, this negative atom would be 'insulated' from the CO group by the intervening saturated carbon atom, and consequently such an anion would not have increased resonance stabilisation. Thus, because of the increased reactivity of hydrogen in a methylene group (—CH_2—) or in a methyne group (=CH—) when *adjacent* to a carbonyl group or any other strongly electron-attracting group, these groups are referred to as the 'active' methylene or methyne group.

Aldehydes and ketones with α-hydrogen atoms readily react with sulphuryl chloride at room temperature in the absence of a catalyst, to replace α-hydrogen atoms only, *e.g.*,

$$CH_3COCH_3 \xrightarrow{SO_2Cl_2} CH_3COCH_2Cl$$

A general method for preparing α-bromoaldehydes is as follows (see also Perkin reaction, p. 739):

$$RCH_2CHO + (MeCO)_2O \xrightarrow{MeCO_2^-K^+} RCH=CHOOCMe \xrightarrow{Br_2}$$

$$RCHBrCHBrOOCMe \xrightarrow{EtOH} RCHBrCH(OEt)_2 \xrightarrow{HCl} RCHBrCHO \quad (v.g.)$$

The halogen atom in the group CXCO is very reactive by the S_N2 mechanism, but not by the S_N1. The reason for this is not clear, but a contributing factor is the $-I$ effect of the carbonyl group which draws the covalent pair in the C—Cl bond away from the Cl and consequently makes its ionisation difficult.

10. Aldehydes and ketones with a methyl or methylene group adjacent to the carbonyl group are oxidised by selenium dioxide at room temperature to dicarbonyl compounds; *e.g.*, acetaldehyde forms glyoxal, and acetone forms methylglyoxal:

$$CH_3CHO + SeO_2 \longrightarrow OCHCHO + Se + H_2O$$

$$CH_3COCH_3 + SeO_2 \longrightarrow CH_3COCHO + Se + H_2O$$

The reaction is usually carried out in acetic acid, and the actual reagent is selenous acid. The mechanism of the oxidation appears to be far from settled, but a possibility is:

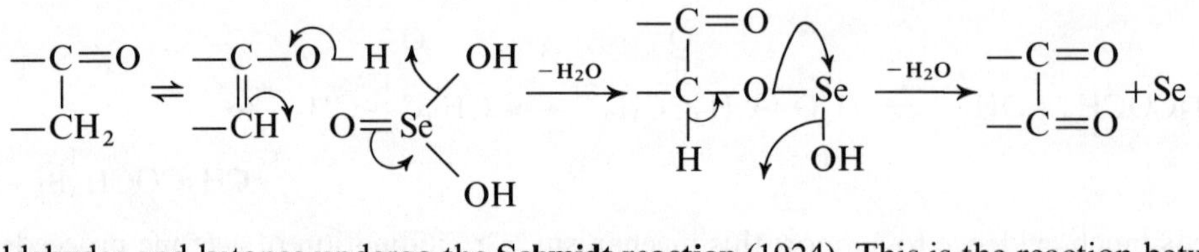

11. Aldehydes and ketones undergo the **Schmidt reaction** (1924). This is the reaction between a carbonyl compound and hydrazoic acid in the presence of, *e.g.*, concentrated sulphuric acid. Aldehydes give a mixture of cyanide and formyl derivatives of primary amines, whereas ketones give amides:

$$RCHO + HN_3 \xrightarrow{H_2SO_4} RCN + RNHCHO + N_2$$

$$RCOR + HN_3 \xrightarrow{H_2SO_4} RCONHR + N_2$$

The mechanism of the reaction is uncertain. It has been shown to be intramolecular, and Smith (1948) has proposed the following mechanism, which is an example of the 1,2-shift (from carbon to nitrogen); for ketones:

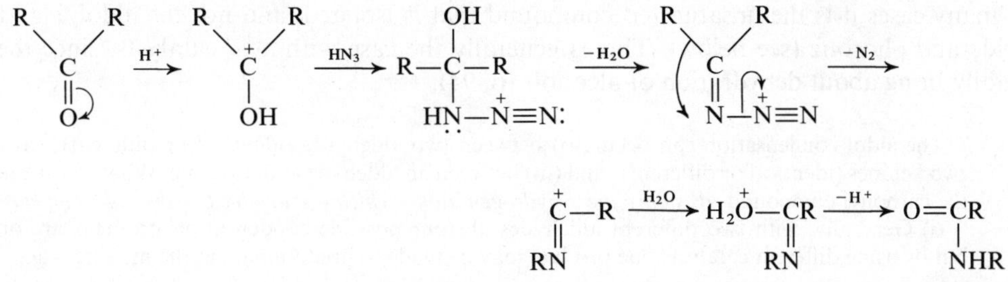

$$\underset{\underset{O}{\overset{R}{\underset{\shortparallel}{C}}}}{\overset{R}{\diagup}} \xrightarrow{H^+} \underset{OH}{\overset{R}{\underset{\shortparallel}{\overset{+}{C}}}\overset{R}{\diagup}} \xrightarrow{HN_3} \underset{\underset{HN-N\equiv N:}{\shortparallel}}{R-\overset{OH}{\underset{\shortparallel}{C}}-R} \xrightarrow{-H_2O} \underset{N-N\equiv N}{\overset{R}{\diagup}\overset{R}{\diagdown}}{\overset{R}{\diagup}C} \xrightarrow{-N_2}$$

$$\underset{RN}{\overset{\shortparallel}{\overset{+}{C}}-R} \xrightarrow{H_2O} \underset{RN}{H_2\overset{+}{O}-\underset{\shortparallel}{C}R} \xrightarrow{-H^+} \underset{NHR}{O=\underset{\shortparallel}{C}R}$$

In ketones, if the two groups are not identical, then two geometrical isomers of (I) are possible. It is also reasonable to suppose that the *anti* group (to the diazonium nitrogen) is the group that migrates (*cf.* the Beckmann rearrangement, p. 755). In this way it is possible to explain how steric factors may influence the isomer ratio of amides formed:

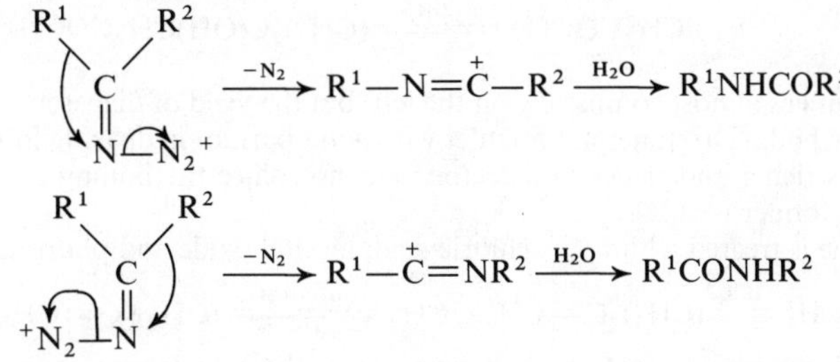

$$\underset{N-N_2{}^+}{\overset{R^1}{\diagup}\overset{R^2}{\diagdown}C} \xrightarrow{-N_2} R^1-N=\overset{+}{C}-R^2 \xrightarrow{H_2O} R^1NHCOR^2$$

$$\underset{{}^+N_2-N}{\overset{R^1}{\diagup}\overset{R^2}{\diagdown}C} \xrightarrow{-N_2} R^1-\overset{+}{C}=NR^2 \xrightarrow{H_2O} R^1CONHR^2$$

For aldehydes, $R^1=H$, and so the reaction may be formulated:

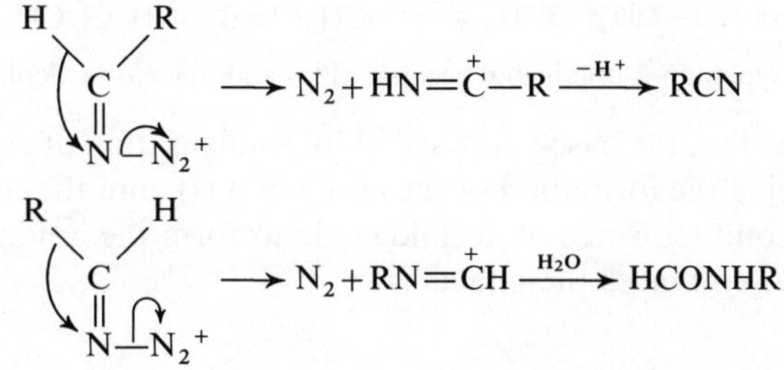

$$\underset{N-N_2{}^+}{\overset{H}{\diagup}\overset{R}{\diagdown}C} \longrightarrow N_2+HN=\overset{+}{C}-R \xrightarrow{-H^+} RCN$$

$$\underset{N-N_2{}^+}{\overset{R}{\diagup}\overset{H}{\diagdown}C} \longrightarrow N_2+RN=\overset{+}{C}H \xrightarrow{H_2O} HCONHR$$

12. Aldehydes and ketones undergo cyclic additions with certain substrates to form heterocyclic compounds, *e.g.*, formaldehyde forms β-propiolactone with keten (p. 361).

13. Aldehydes and ketones undergo condensation reactions, *i.e.*, two or more (identical or different) molecules unite with, or without, the elimination of water (or any other simple molecule). There is very little difference between the terms condensation and condensation polymerisation or polycondensation (p. 117); generally, condensation is used for those reactions in which the resulting compound is made up of a small number of the reacting molecules.

Aldol condensation. Acetaldehyde, in the presence of dilute sodium hydroxide, potassium carbonate or hydrochloric acid, undergoes condensation to form a syrupy liquid known as *aldol*:

$$2CH_3CHO \xrightarrow{NaOH} CH_3CH(OH)CH_2CHO \quad (50\%)$$

On heating, aldols eliminate water to form unsaturated compounds, *e.g.*, aldol forms crotonaldehyde:

$$CH_3CH(OH)CH_2CHO \longrightarrow CH_3CH=CHCHO+H_2O$$

In many cases it is the unsaturated compound that is isolated, and not the aldol, *e.g.*, mesityl oxide and phorone (see below). This is generally the case with *acid* catalysts, since these can readily bring about dehydration of alcohols (p. 94).

The aldol condensation can occur: (i) between two aldehydes (identical or different); (ii) between two ketones (identical or different); and (iii) between an aldehyde and a ketone. Whatever the nature of the carbonyl compound, *it is only the α-hydrogen atoms which are involved in the aldol condensation.*

(i) Generally, with two different aldehydes all four possible condensation products are obtained; but by using different catalysts one product may be made to predominate in the mixture, *e.g.*,

$$CH_3CHO + CH_3CH_2CHO \xrightleftharpoons{NaOH} CH_3CH_2CH(OH)CH_2CHO$$

$$CH_3CHO + CH_3CH_2CHO \xrightleftharpoons{HCl} CH_3CH(OH)CH(CH_3)CHO$$

(ii) Acetone, in the presence of barium hydroxide, gives diacetone alcohol:

$$2CH_3COCH_3 \xrightleftharpoons{Ba,OH.2} (CH_3)_2C(OH)CH_2COCH_3$$

This equilibrium lies almost completely on the left, but the yield of diacetone alcohol may be increased (71 per cent) by boiling acetone in a Soxhlet with solid barium hydroxide in the thimble. The acetone in the flask gets richer and richer in diacetone alcohol, since the boiling point of the latter is 164°C, and that of the former is 56°C.

When acetone is treated with hydrochloric acid, mesityl oxide and phorone are formed:

$$2CH_3COCH_3 \xrightleftharpoons{HCl} (CH_3)_2C{=}CHCOCH_3 \xrightleftharpoons{CH_3COCH_3} (CH_3)_2C{=}CHCOCH{=}C(CH_3)_2$$

(iii) When aldehydes condense with ketones, it is the α-hydrogen atom of the *ketone* which is involved in the condensation, *e.g.*,

$$CH_3CHO + CH_3COCH_3 \xrightleftharpoons{NaOH} CH_3CH(OH)CH_2COCH_3 \quad (25\%)$$

The yield of 4-hydroxypentan-2-one is low because aldol and diacetone alcohol are also formed.

It is generally accepted that the base-catalysed aldol condensation of acetaldehyde takes place in two steps, the first being the formation of the carbanion (I), and the second the combination of this anion with a second molecule of acetaldehyde to form the anion (II) of the aldol. The simplest mechanism that embraces these facts is:

$$HO^{\frown} H{-}CH_2CHO \rightleftharpoons H_2O + \bar{C}H_2CHO$$

(I)

$$\overset{\frown O}{\underset{CH_3CH}{\parallel}} \overset{\frown}{CH_2CHO} \rightleftharpoons CH_3\overset{O^-}{\underset{|}{C}}HCH_2CHO \xrightleftharpoons{H_2O} CH_3\overset{OH}{\underset{|}{C}}HCH_2CHO + OH^-$$

(II)

On the basis of analogy with acetaldehyde, the mechanism of the base-catalysed condensation of acetone will be:

$$MeCOMe + OH^- \rightleftharpoons H_2O + \bar{C}H_2COMe$$

$$\overset{\frown O}{\underset{Me_2C}{\parallel}} \overset{\frown}{CH_2COMe} \rightleftharpoons Me_2\overset{O^-}{\underset{|}{C}}CH_2COMe \xrightleftharpoons{H_2O} Me_2\overset{OH}{\underset{|}{C}}CH_2COMe + OH^-$$

There appear to be no detailed kinetic investigations on acid-catalysed aldol condensations, but it is generally assumed that the condensation proceeds by reaction between conjugate acid and the *enol form* of the carbonyl compound; *e.g.*, the formation of mesityl oxide:

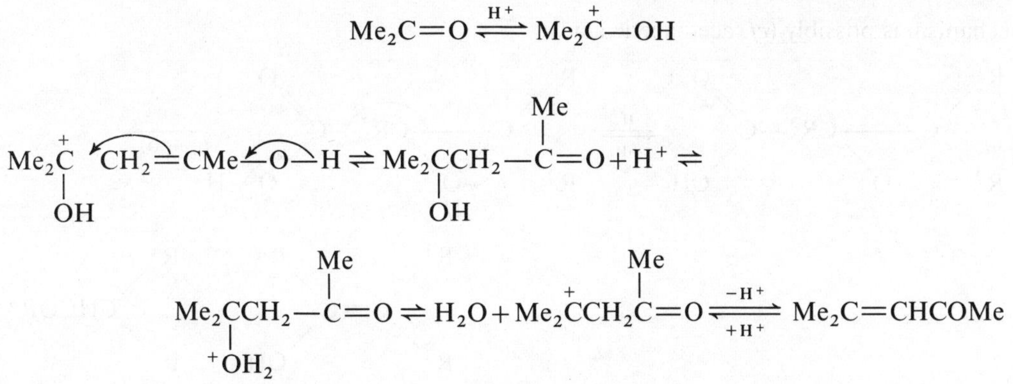

$$Me_2C=O \underset{}{\overset{H^+}{\rightleftharpoons}} Me_2\overset{+}{C}-OH$$

$$\underset{\underset{OH}{|}}{Me_2\overset{+}{C}} \quad CH_2=CMe-O-H \rightleftharpoons Me_2CCH_2-\underset{\underset{OH}{|}}{\overset{\overset{Me}{|}}{C}}=O+H^+ \rightleftharpoons$$

$$Me_2CCH_2-\underset{\underset{+OH_2}{|}}{\overset{\overset{Me}{|}}{C}}=O \rightleftharpoons H_2O + Me_2\overset{+}{C}CH_2\overset{\overset{Me}{|}}{C}=O \underset{+H^+}{\overset{-H^+}{\rightleftharpoons}} Me_2C=CHCOMe$$

Claisen condensation (1881). This is the condensation between an ester and another molecule of an ester or ketone (see p. 285).

Claisen–Schmidt reaction (also known as the **Claisen reaction**) is the condensation between an *aromatic* aldehyde (or ketone) and an aldehyde or ketone, *in the presence of dilute alkali* to form an α,β-unsaturated compound. This reaction is similar to the aldol condensation, and is illustrated by the formation of cinnamaldehyde from benzaldehyde and acetaldehyde:

$$C_6H_5CHO + CH_3CHO \xrightarrow{NaOH} C_6H_5CH=CHCHO + H_2O$$

The mechanism of the Claisen–Schmidt reaction is (Noyce *et al.*, 1959; Stiles *et al.*, 1959):

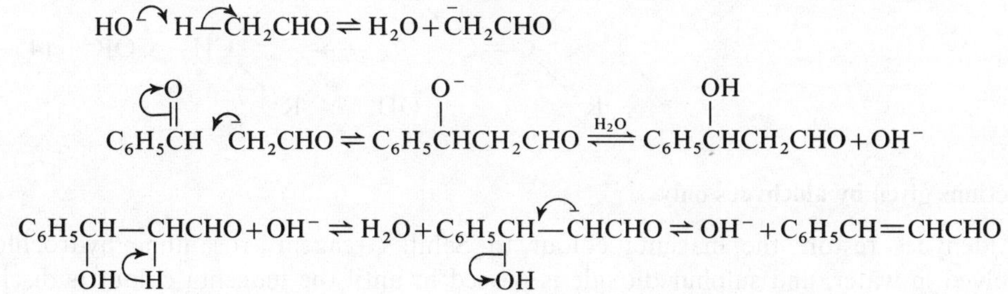

$$HO \quad H-CH_2CHO \rightleftharpoons H_2O + \bar{C}H_2CHO$$

$$\underset{}{C_6H_5CH} \quad CH_2CHO \rightleftharpoons C_6H_5\underset{\underset{O^-}{|}}{CH}CH_2CHO \underset{}{\overset{H_2O}{\rightleftharpoons}} C_6H_5\underset{\underset{OH}{|}}{CH}CH_2CHO + OH^-$$

$$C_6H_5\underset{\underset{OH}{|}}{CH}-\underset{\underset{H}{|}}{CH}CHO + OH^- \rightleftharpoons H_2O + C_6H_5CH-CHCHO \rightleftharpoons OH^- + C_6H_5CH=CHCHO$$

Other condensations involving aldehydes or ketones are the *Knoevenagel* (p. 342) and *Perkin* (p. 739) reactions.

14. Darzens Glycidic Ester Condensation (1904). This reaction involves the condensation of an aldehyde or ketone with an α-halogeno-ester to produce an α,β-epoxy-ester (*glycidic ester*); the condensing agent is usually sodium ethoxide or sodamide:

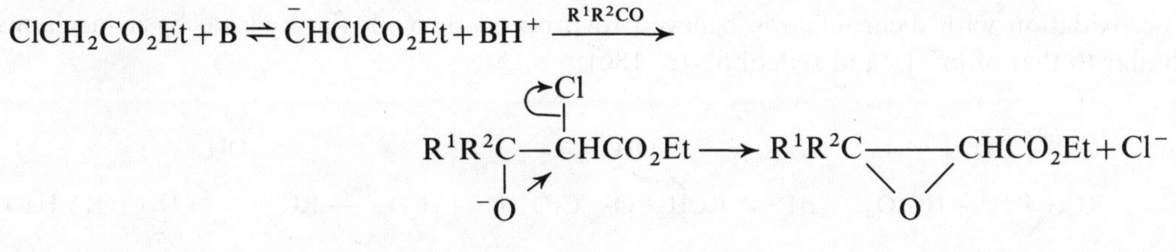

$$ClCH_2CO_2Et + B \rightleftharpoons \bar{C}HClCO_2Et + BH^+ \xrightarrow{R^1R^2CO}$$

$$R^1R^2C-\underset{\underset{-O}{|}}{\overset{\overset{Cl}{|}}{C}}HCO_2Et \longrightarrow R^1R^2C\underset{O}{\diagdown\diagup}CHCO_2Et + Cl^-$$

This is an example of neighbouring group participation.

On alkaline hydrolysis, the ester gives the salt of the epoxy-acid, and when this is warmed with dilute acid, the product is an aldehyde if R^3=H, and a ketone if R^3=alkyl group. The mechanism is possibly (*cf.* acetoacetic acid):

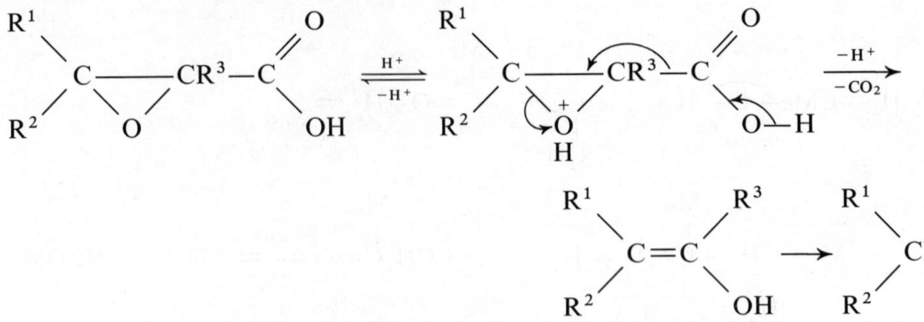

$$R^1 \overset{R^2}{\diagdown}C\diagup\!\!\diagdown O \diagup\!\!\diagdown CR^3\!\!-\!\!C\overset{O}{\diagdown}_{OH} \underset{-H^+}{\overset{H^+}{\rightleftharpoons}} R^1 \overset{R^2}{\diagdown}C\diagup\!\!\diagdown \overset{+}{O}\!\!-\!\!H \diagup\!\!\diagdown CR^3\!\!-\!\!C\overset{O}{\diagdown}_{O\!-\!H} \overset{-H^+}{\underset{-CO_2}{\longrightarrow}}$$

$$R^1 \overset{R^2}{\diagdown}C\!\!=\!\!C\overset{R^3}{\diagup}_{OH} \longrightarrow R^1 \overset{R^2}{\diagdown}CHCOR^3$$

This synthesis can be shortened by using t-butyl glycidates, since the ester, on pyrolysis, eliminates isobutene to give an aldehyde or ketone (Blanchard *et al.*, 1963):

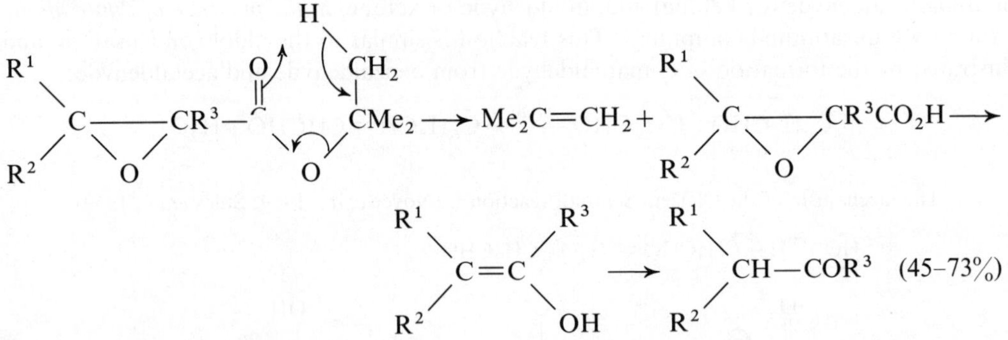

$$R^1 \overset{R^2}{\diagdown}C\diagup\!\!\diagdown O \diagup\!\!\diagdown CR^3\!\!-\!\!C\overset{O}{\diagdown}_{O} \overset{H}{\diagdown}\overset{CH_2}{\underset{CMe_2}{|}} \longrightarrow Me_2C\!\!=\!\!CH_2 + R^1 \overset{R^2}{\diagdown}C\diagup\!\!\diagdown O \diagup\!\!\diagdown CR^3CO_2H \longrightarrow$$

$$R^1 \overset{R^2}{\diagdown}C\!\!=\!\!C\overset{R^3}{\diagup}_{OH} \longrightarrow R^1 \overset{R^2}{\diagdown}CH\!\!-\!\!COR^3 \quad (45\text{--}73\%)$$

Reactions given by aldehydes only

1. Aldehydes restore the magenta colour to Schiff's reagent (rosaniline hydrochloride is dissolved in water, and sulphur dioxide is passed in until the magenta colour is discharged). The mechanism of this reaction is uncertain.

2. Aldehydes are very easily oxidised, and hence are powerful reducing agents. Aldehydes reduce Fehling's solution (an alkaline solution containing a complex of copper tartrate) to red cuprous oxide; and Tollens' reagent (ammoniacal silver nitrate solution) to metallic silver, *e.g.*,

$$RCHO + 2Ag(NH_3)_2OH \longrightarrow RCO_2NH_4 + 2Ag + 3NH_3 + H_2O$$

Aldehydes are readily oxidised by acid dichromate, permanganate, etc. (*cf.* ketones, below) The oxidation with dichromate is believed to proceed through the hydrate by a mechanism similar to that of prim.- and s-alcohols (p. 186):

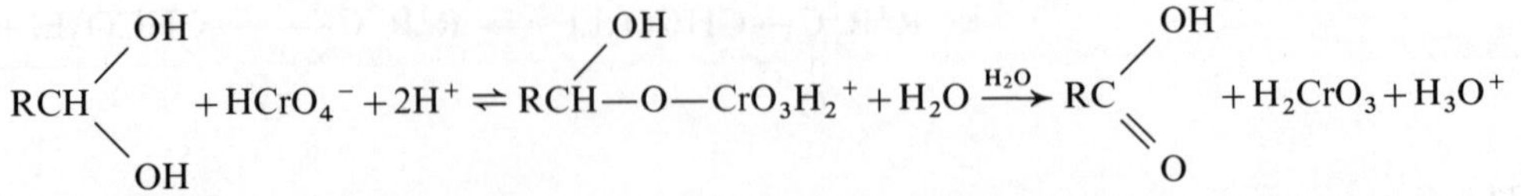

$$RCH\overset{OH}{\underset{OH}{\diagup}} + HCrO_4^- + 2H^+ \rightleftharpoons RCH\!\!-\!\!O\!\!-\!\!CrO_3H_2^+ + H_2O \overset{H_2O}{\longrightarrow} RC\overset{OH}{\underset{O}{\diagup\!\!\!\parallel}} + H_2CrO_3 + H_3O^+$$

3. All aldehydes except formaldehyde form resinous products when warmed with concentrated sodium hydroxide solution. The resin is probably formed via a series of condensations, *e.g.*,

$$2CH_3CHO \longrightarrow CH_3CH(OH)CH_2CHO \xrightarrow{-H_2O} CH_3CH=CHCHO \xrightarrow{CH_3CHO}$$

$$CH_3CH=CHCH(OH)CH_2CHO \xrightarrow{-H_2O} CH_3CH=CHCH=CHCHO, \text{ etc.}$$

Aldehydes that have no α-hydrogen atoms undergo the **Cannizzaro reaction** (1853) in which two molecules of the aldehyde are involved, one molecule being converted into the corresponding alcohol, and the other into the acid. The usual reagent for bringing about the Cannizzaro reaction is 50 per cent aqueous or ethanolic alkali, *e.g.*,

$$2HCHO + NaOH \longrightarrow HCO_2Na + CH_3OH$$

The Cannizzaro reaction is mainly applicable to aromatic aldehydes (see p. 737).

Although the Cannizzaro reaction is characteristic of aldehydes having no α-hydrogen atoms, it is not confined to them, *e.g.*, certain aliphatic α-monoalkylated aldehydes undergo quantitative disproportionation when heated with aqueous sodium hydroxide at 170–200°C (Häusermann, 1951):

$$2(CH_3)_2CHCHO + NaOH \xrightarrow{200°C} (CH_3)_2CHCO_2Na + (CH_3)_2CHCH_2OH \quad (100\%)$$

The Cannizzaro reaction can take place between two different aldehydes, and is then known as a 'crossed' Cannizzaro reaction (see formaldehyde, p. 226).

All aldehydes can be made to undergo the Cannizzaro reaction by treatment with aluminium ethoxide. Under these conditions the acid and alcohol are combined as the ester, and the reaction is then known as the **Tischenko reaction** (1906); *e.g.*, acetaldehyde gives ethyl acetate, and propionaldehyde gives propyl propionate.

The mechanism is uncertain, but a possibility is one that involves a hydride-ion shift as in the MPV reduction (p. 210):

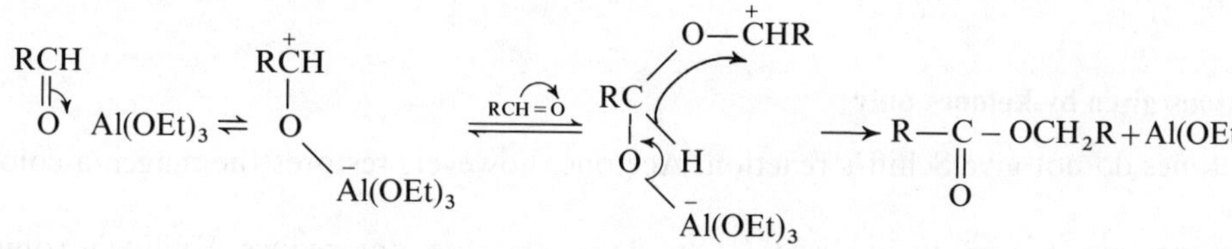

4. Aldehydes (except formaldehyde) react with ammonia in ethereal solution to give a precipitate of *aldehyde-ammonia*, *e.g.*,

$$CH_3CHO + NH_3 \longrightarrow CH_3CH(OH)NH_2 \quad (50\%)$$

The structure is actually more complicated than that given, and X-ray analysis has led to the suggestion that it is the trihydrate of trimethylhexahydrotriazine, which may be prepared anhydrous by allowing 'acetaldehyde ammonia' to stand over sulphuric acid. Acetaldimine is formed first and then trimerises.

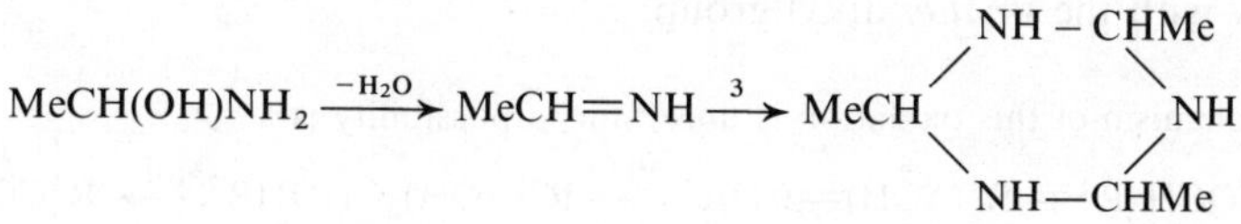

5. Aldehydes combine with alcohols in the presence of *e.g.*, dry hydrogen chloride, to form first the *hemi-acetal*, and then the *acetal*:

$$R^1CHO + R^2OH \overset{HCl}{\rightleftharpoons} R^1CH(OH)OR^2 \overset{R^2OH}{\rightleftharpoons} RCH(OR^2)_2 + H_2O \quad (g.)$$

The hemi-acetal is rarely isolated since it very readily forms the acetal. Acetals are diethers of the unstable 1,1-dihydroxyalcohols, and may be named as 1,1-dialkoxyalkanes. Unlike the parent dihydroxyalcohols these acetals are stable. They are also stable in the presence of alkali, but are converted into the aldehyde by acid. Thus acetal formation may be used to protect the aldehyde group against *alkaline* oxidising agents. On the other hand, the aldehyde group can be protected in *acid* solution by mercaptal formation (p. 401).

> The stability of acetals to *alkaline* media may be attributed to the fact that the alkoxyl group is a very poor leaving group, whereas the protonated group is a very good leaving group.
> The mechanism of acetal formation has been elucidated by use of the principle of microscopic reversibility on acetals:

> Both aldehydes and ketones form 1,3-dioxalans with ethylene glycol (p. 311).

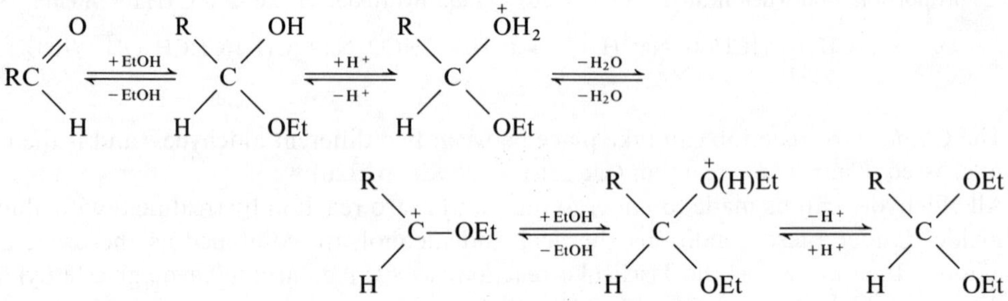

6. The lower aldehydes polymerise with great ease (see formaldehyde and acetaldehyde, below).

Reactions given by ketones only

1. Ketones do not give Schiff's reaction. Acetone, however, restores the magenta colour *very slowly*.

2. Ketones are not easily oxidised (*cf.* p. 185); they do not reduce Fehling's solution or ammoniacal silver nitrate.

α-Hydroxyketones, *i.e.*, compounds containing the group CH(OH)CO, readily reduce Fehling's solution and ammoniacal silver nitrate.

Strong oxidising agents, *e.g.*, acid dichromate, nitric acid, etc., oxidise ketones, only the carbon atoms adjacent to the carbonyl group being attacked, and the carbon atom joined to the smaller number of hydrogen atoms is preferentially oxidised:

$$CH_3COCH_2CH_3 \overset{[O]}{\longrightarrow} 2CH_3CO_2H$$

If the adjacent carbon atoms have the same number of hydrogen atoms, the carbonyl group remains chiefly with the *smaller* alkyl group.

> The mechanism of this oxidation is uncertain; a possibility is:
> $$R^1COCH_2R^2 \rightleftharpoons R^1C(OH)=CHR^2 \overset{[O]}{\longrightarrow} R^1C(OH)_2CHOHR^2 \overset{[O]}{\longrightarrow} R^1CO_2H + R^2CO_2H$$

All ketones containing the acetyl group, CH_3CO, undergo the **haloform reaction**. This reaction is best carried out by dissolving the compound in dioxan, adding dilute sodium hydroxide, then a slight excess of iodine in potassium iodide solution, warming, and finally adding water. If the compound contains the acetyl group, iodoform is precipitated.

$$RCOCH_3 \xrightarrow{I_2, OH^-} RCOCI_3 \xrightarrow{OH^-} RCO_2{}^- + CHI_3$$

The haloform reaction is very useful in organic degradations. A positive iodoform test is given by most compounds containing the acetyl group attached to either carbon or hydrogen, or by compounds which are oxidised under the conditions of the test to derivatives containing the acetyl group, *e.g.*, ethanol, isopropanol, etc.

Booth and Saunders (1950) have shown that a number of other compounds besides those mentioned above also give the haloform reaction, *e.g.*, certain quinones, quinols and *m*-dihydric phenols. On the other hand, acetoacetic ester, for example, does *not* give the test although it contains an acetyl group. The reason is that it contains an active methylene group, and it is this group that is halogenated and not the Me of the acetyl group.

An alternative method to the haloform reaction is to treat the monobrominated methyl ketone with pyridine, and then decompose the precipitated pyridinium salt with dilute ethanolic sodium hydroxide (Krohnke, 1933; King, 1944):

$$RCOCH_3 \xrightarrow{Br_2} RCOCH_2Br \xrightarrow{C_5H_5N} RCOCH_2\overset{+}{-}NC_5H_5Br^- \xrightarrow{NaOH/EtOH}$$

$$RCO_2H + C_5H_5\overset{+}{N}Me\}Br^-$$

The advantage of this method is that the reaction can be carried out in non-aqueous solution.

Aliphatic ketones undergo the **Baeyer–Villiger oxidation** (1899–1900) to form esters or their hydrolysed products:

$$RCOR \longrightarrow RCO_2R \xrightarrow{H^+} RCO_2H + ROH$$

The authors used Caro's acid (permonosulphuric acid, H_2SO_5), but the reaction is now carried out with various organic per-acids, *e.g.*, perbenzoic, peracetic, or monoperphthalic acid. The yields are reasonable, and so this reaction is useful in structure determination, and may also be used as a method of preparing acids. According to Mateos *et al.* (1964), the mechanism is:

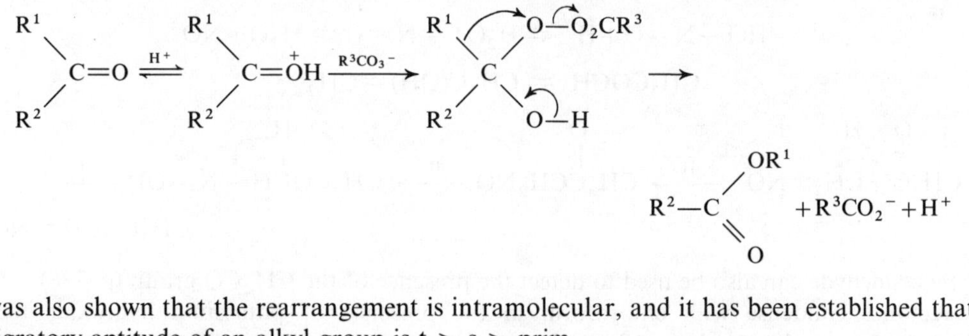

It was also shown that the rearrangement is intramolecular, and it has been established that the migratory aptitude of an alkyl group is t > s > prim.

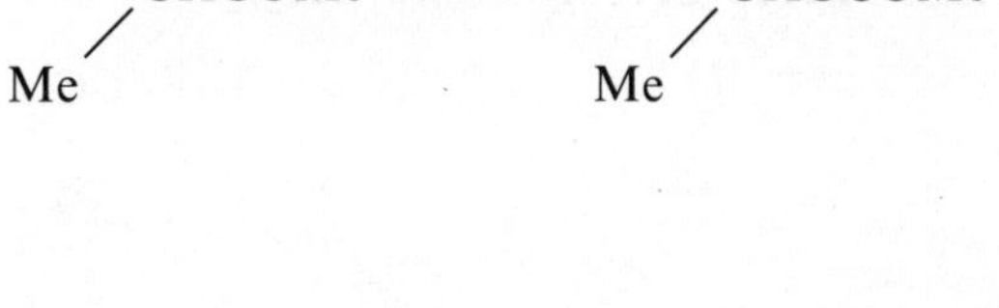

3. Ketones react with ammonia to form complex condensation products, *e.g.*, if acetone is treated with ammonia and then followed by acidification of the reaction products, *diacetonamine* and *triacetonamine* are formed (see p. 340).

4. Ketones do not readily form *ketals* when treated with alcohols in the presence of hydrogen chloride (*cf.* acetals, above). Ketals may, however, be prepared by treating the ketone with ethyl orthoformate (Helferich and Hauser, 1924):

$$R_2CO + HC(OC_2H_5)_3 \longrightarrow R_2C(OC_2H_5)_2 + HCO_2C_2H_5 \quad (g.)$$

In the presence of acids, however, the product is an enol ether.

$$RCOCH_2R + H(COEt)_3 \longrightarrow RC(OEt)_2CH_2R \xrightarrow{HCl} RC(OEt)=CHR + EtOH$$

Ketones readily react with 1,2-glycols to form dioxolans (see p. 311).

5. Ketones are reduced catalytically or by dissolving metals in alkaline solution to alcohols, but reduction by the latter reagent in neutral or acid solution gives 1,2-glycols as the main product; *e.g.*, with magnesium amalgam, acetone forms pinacol (*cf.* p. 210). This reaction is believed to proceed via free radicals:

$$2Me_2C{=}O + 2e \xrightarrow{Mg} [2Me_2\dot{C}{-}\bar{O}]Mg^{2+} \longrightarrow$$

$$\begin{bmatrix} Me_2C{-}CMe_2 \\ | \quad\quad | \\ O^- \quad O^- \end{bmatrix} Mg^{2+} \xrightarrow[H^+]{H_2O} \begin{matrix} Me_2C{-}CMe_2 \\ | \quad\quad | \\ OH \; OH \end{matrix} \quad (43{-}50\%)$$

6. When ketones are treated with nitrous acid, the 'half oxime' of the α-dicarbonyl compound is formed, *e.g.*, acetone gives oximinoacetone (isonitrosoacetone; see p. 366):

$$CH_3COCH_3 + HNO_2 \xrightarrow{C_2H_5ONO/HCl} CH_3COCH{=}NOH + H_2O$$

All compounds containing the CH_2CO group form the oximino-derivative with nitrous acid, and this has been used to detect the presence of the CH_2CO group.

The mechanism of this reaction is uncertain. A possibility is via attack by the nitrosonium ion (*cf.* nitration, p. 639); the nitroso compound first formed then rearranges to the oxime (p. 366):

$$HO{-}N{=}O + H^+ \rightleftharpoons H_2O^+{-}N{=}O \rightleftharpoons H_2O + NO^+$$

$$CH_3COCH_3 \rightleftharpoons CH_3C(OH){=}CH_2$$

$$CH_3\overset{O-H}{\underset{}{C}}={CH_2} \quad NO^+ \xrightarrow{-H^+} CH_3\overset{O}{\overset{\|}{C}}CH_2NO \xrightarrow{+H^+} CH_3COCH{-}\overset{H}{\underset{}{N}}{=}\overset{+}{O}H \xrightarrow{-H^+}$$

$$CH_3COCH{=}NOH$$

Benzaldehyde can also be used to detect the presence of the CH_2CO group (p. 738):

7. Ketones condense with chloroform in the presence of potassium hydroxide to form chlorohydroxy-compounds (p. 170):

$$CH_3COCH_3 + CHCl_3 \xrightarrow{KOH} (CH_3)_2C{\overset{OH}{\underset{CCl_3}{<}}}$$

8. Ketones form *sodio-derivatives* when treated with sodium or sodamide in ethereal solution *e.g.*,

$$CH_3COCH_3 + NaNH_2 \longrightarrow \left[\begin{array}{c} O^- \\ | \\ CH_3C{=}CH_2 \end{array} \right] \overset{+}{Na} + NH_3$$

Formaldehyde (*methanal*) is prepared industrially:

(i) By the dehydrogenation or air oxidation of methanol in the presence of a silver catalyst at 400°C:

$$CH_2O + H_2 \overset{Ag}{\longleftarrow} CH_3OH \underset{Ag}{\overset{air}{\longrightarrow}} CH_2O + H_2O$$

In the oxidation method, the condensate obtained is a mixture of formaldehyde, methanol and water. It is freed from excess of methanol by distillation, and the resulting mixture is known as *formalin* (40 per cent formaldehyde, 8 per cent methanol, 52 per cent water).

(ii) By the air oxidation of methane or propane-butane mixtures (from natural gas) in the presence of various metallic oxides.

Formaldehyde is a colourless, pungent-smelling gas, b.p. $-21°C$, extremely soluble in water. It is a powerful disinfectant and antiseptic, and so is used for preserving anatomical specimens.

Its main uses are in the manufacture of dyes, the hardening of casein and gelatin, and for making plastics.

Formaldehyde undergoes many of the general reactions of aldehydes, but differs in certain respects. When treated with ammonia, it does not form an aldehyde-ammonia, but gives instead *hexamethylenetetramine* (m.p. 260°C).

$$6HCHO + 4NH_3 \longrightarrow (CH_2)_6N_4 + 6H_2O \quad (80\%)$$

The structure of hexamethylenetetramine consists of three fused chair conformations (p. 550).

Formaldehyde, since it has no α-hydrogen atoms, readily undergoes the Cannizzaro reaction (see p. 221).

Primary and secondary amines are methylated by heating with formaldehyde (formalin solution) and an excess of formic acid at 100°C, *e.g.*, ethylmethylamine from ethylamine:

$$C_2H_5NH_2 + CH_2O + HCO_2H \longrightarrow C_2H_5NHCH_3 + CO_2 + H_2O$$

This reaction is known as the **Eschweiler–Clarke methylation** (1905, 1933).

Polymers of formaldehyde. (i) In dilute aqueous solution, formaldehyde is almost 100 per cent hydrated to form methylene glycol (Bieber *et al.*, 1947). This is believed to be the reason for the stability of dilute formaldehyde solutions (see also p. 228):

$$CH_2O + H_2O \rightleftharpoons CH_2(OH)_2$$

(ii) When a formaldehyde solution is evaporated to dryness, a white crystalline solid, m.p. 121–123°C, is obtained. This is known as *paraformaldehyde*, $(CH_2O)_n \cdot H_2O$, and it appears to be a mixture of polymers, n having values between 6 and 50. Paraformaldehyde reforms formaldehyde when heated.

Formaldehyde cannot be separated from methanol (in formalin) by fractionation; pure aqueous formaldehyde may be obtained by refluxing paraformaldehyde with water until solution is complete.

(iii) When a formaldehyde solution is treated with concentrated sulphuric acid, *polyoxymethylenes*, $(CH_2O)_n \cdot H_2O$ — n is greater than 100 — are formed. Polyoxymethylenes are white solids, insoluble in water, and reform formaldehyde when heated.

(iv) When allowed to stand at room temperature, formaldehyde gas slowly polymerises to a white solid, *trioxymethylene* (*metaformaldehyde, trioxan*) $(CH_2O)_3$, m.p. 61–62°C. Trioxan is prepared by distilling formaldehyde solution (60 per cent) containing a little sulphuric acid. This trimer is soluble in water, and does not show any reducing properties. Hence it is believed to have a cyclic structure (in which there is no free aldehyde group).

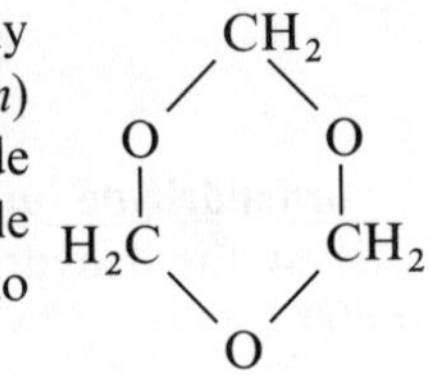

Trioxan is very useful for generating formaldehyde since: (*a*) it is an anhydrous form of formaldehyde; (*b*) the rate of depolymerisation can be controlled; and (*c*) it is soluble in organic solvents.

(v) Formaldehyde polymerises in the presence of weak alkalis, *e.g.*, calcium hydroxide, to a mixture of sugars of formula $C_6H_{12}O_6$, which is known as *formose* or *α-acrose*.

Condensation reactions of formaldehyde. Formaldehyde can participate in the 'crossed' Cannizzaro reaction, and the nature of the final product depends on the structure of the other aldehyde. Aldehydes with *no* α-hydrogen atoms readily undergo the crossed Cannizzaro reaction; *e.g.*, benzaldehyde forms benzyl alcohol:

$$C_6H_5CHO + HCHO + NaOH \longrightarrow C_6H_5CH_2OH + HCO_2Na$$

Aldehydes with *one* α-hydrogen atom react as follows:

$$R^1R^2CHCHO + HCHO \xrightarrow{NaOH} R^1R^2C\begin{array}{c} CH_2OH \\ \diagup \\ \diagdown \\ CHO \end{array}$$

This β-hydroxyaldehyde in the presence of excess of formaldehyde forms a substituted trimethylene glycol:

$$R^1R^2C\begin{array}{c} CH_2OH \\ \diagup \\ \diagdown \\ CHO \end{array} + HCHO + NaOH \longrightarrow R^1R^2C\begin{array}{c} CH_2OH \\ \diagup \\ \diagdown \\ CH_2OH \end{array} + HCO_2Na$$

Thus the first step in the above reaction is the replacement of the α-hydrogen atom by a *hydroxymethyl* group, CH_2OH, and the second step is the crossed Cannizzaro reaction.

In a similar manner, aldehydes with *two* α-hydrogen atoms are converted first into the hydroxymethyl, then into the bishydroxymethyl, and finally into the trishydroxymethyl compound:

$$RCH_2CHO + HCHO \xrightarrow{NaOH} RCH\begin{array}{c} CHO \\ \diagup \\ \diagdown \\ CH_2OH \end{array} \xrightarrow{HCHO}$$

$$RC\begin{array}{c} CHO \\ \diagup \\ -CH_2OH \\ \diagdown \\ CH_2OH \end{array} \xrightarrow[NaOH]{HCHO} RC\begin{array}{c} CH_2OH \\ \diagup \\ -CH_2OH \\ \diagdown \\ CH_2OH \end{array} + HCO_2Na$$

A special case is acetaldehyde, which has *three* α-hydrogen atoms. This reaction is best carried out by adding powdered calcium oxide to a suspension of paraformaldehyde in water containing acetaldehyde. Tetrakishydroxymethylmethane (tetramethylolmethane) or *pentaerythritol* is formed:

$$CH_3CHO + 4HCHO \xrightarrow{Ca(OH)_2} C(CH_2OH)_4 + HCO_2Ca/2 \quad (55–57\%)$$

Pentaerythritol is important industrially since its tetranitrate is a powerful explosive.

Acetaldehyde (*ethanal*) is prepared industrially:

(i) By the dehydrogenation or air oxidation of ethanol in the presence of a silver catalyst at 300°C (*cf.* formaldehyde):

$$CH_3CHO + H_2 \xrightarrow{Ag} CH_3CH_2OH \xrightarrow[Ag]{air} CH_3CHO + H_2O$$

(ii) By the hydration of acetylene (p. 136). This method is decreasing in importance.

(iii) By passing a mixture of ethylene and oxygen, under pressure, into an aqueous solution of palladium and cupric chlorides at 50°C.

$$CH_2=CH_2 + PdCl_2 + H_2O \longrightarrow CH_3CHO + Pd + 2HCl$$
$$2CuCl_2 + Pd \longrightarrow 2CuCl + PdCl_2$$
$$2CuCl + 2HCl + \tfrac{1}{2}O_2 \longrightarrow 2CuCl_2 + H_2O$$

Acetaldehyde is a colourless, pungent-smelling liquid, b.p. 21°C, miscible with water, ethanol and ether in all proportions. It is used in the preparation of acetic acid, ethanol, paraldehyde, rubber accelerators, phenolic resins, synthetic drugs, etc.

Polymers of acetaldehyde. (i) When acetaldehyde is treated with a few drops of concentrated sulphuric acid, a vigorous reaction takes place and the trimer *paraldehyde*, $(CH_3CHO)_3$, is formed. This is pleasant-smelling liquid, b.p. 128°C, and is used in medicine as an hypnotic. When paraldehyde is distilled with dilute sulphuric acid, acetaldehyde is regenerated. Paraldehyde has no reducing properties, and its structure is believed to be (I):

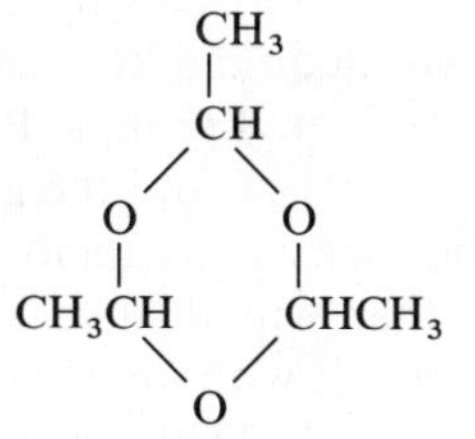

(I) 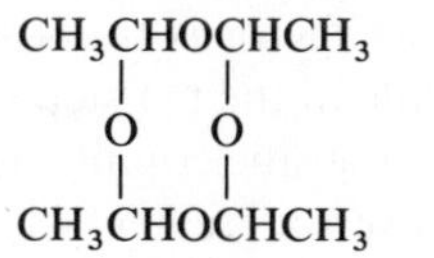 (II)

(ii) When acetaldehyde is treated with a few drops of concentrated sulphuric acid at 0°C, the tetramer *metaldehyde* $(CH_3CHO)_4$, is formed. This is a white solid, m.p. 246°C, and regenerates acetaldehyde when distilled with dilute sulphuric acid. Its structure may be (II).

Chloral (*trichloroacetaldehyde*) is prepared industrially by the chlorination of ethanol. Chlorine is passed into cooled ethanol, and then at 60°C, until no further absorption of chlorine takes place. The final product is chloral alcoholate, $CCl_3CH(OH)OC_2H_5$, which separates out as a crystalline solid which, on distillation with concentrated sulphuric acid, gives chloral.

Chloral is a colourless, oily, pungent-smelling liquid, b.p. 98°C, soluble in water, ethanol and ether. When heated with concentrated potassium hydroxide, it yields pure chloroform:

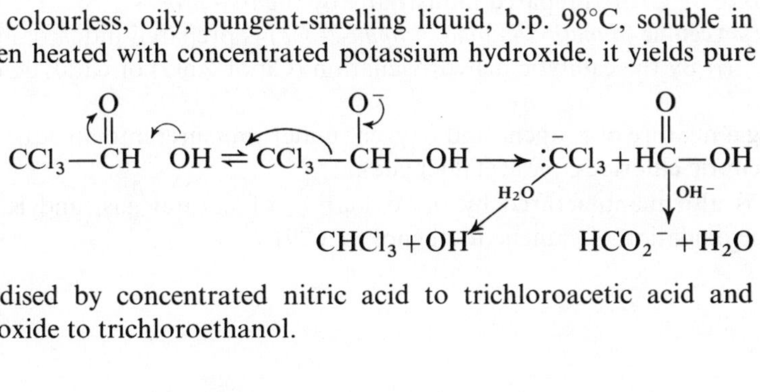

Chloral is oxidised by concentrated nitric acid to trichloroacetic acid and is reduced by aluminium ethoxide to trichloroethanol.

$$CCl_3CO_2H \xleftarrow{\text{HNO}_3} CCl_3CHO \xrightarrow{\text{Al(OC}_2\text{H}_5)_3} CCl_3CH_2OH$$

Chloral undergoes the usual reactions of an aldehyde, but its behaviour towards water and ethanol is unusual. When chloral is treated with water or ethanol, combination takes place with the evolution of heat, and a crystalline solid is formed, chloral hydrate, m.p. 57°C, or chloral alcoholate, m.p. 46°C, respectively. *These compounds are stable*, and the water or ethanol can only be removed by treatment with concentrated sulphuric acid. Thus, in chloral hydrate the water is present as *water of constitution*, i.e., the structure of chloral hydrate is $CCl_3CH(OH)_2$; similarly, the structure of chloral alcoholate is $CCl_3CH(OH)OC_2H_5$.

Carbonyl compounds are hydrated by water, the reaction being catalysed by acids and bases. The mechanisms are believed to be:

Acid

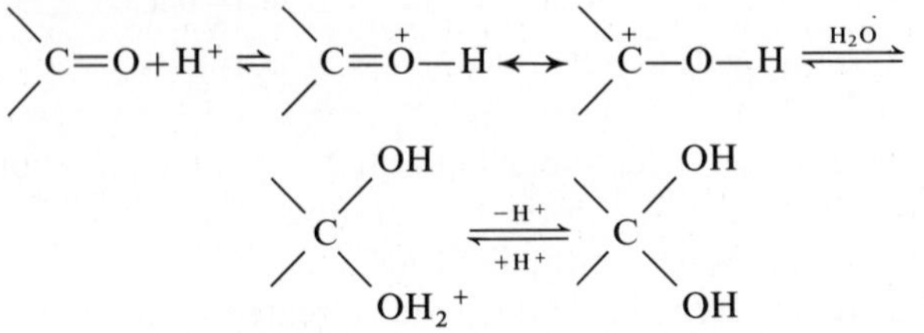

$$\backslash C{=}O + H^+ \rightleftharpoons \quad \backslash C{=}\overset{+}{O}{-}H \longleftrightarrow \quad \overset{+}{\backslash C}{-}O{-}H \xrightarrow{\text{H}_2\text{O}}$$

$$\backslash C \overset{OH}{\underset{OH_2^+}{}} \underset{+H^+}{\overset{-H^+}{\rightleftharpoons}} \backslash C \overset{OH}{\underset{OH}{}}$$

Base

$$\backslash C{=}O + OH^- \rightleftharpoons \backslash C \overset{O^-}{\underset{OH}{}} \xrightarrow{\text{H}_2\text{O}} \backslash C \overset{OH}{\underset{OH}{}} + OH^-$$

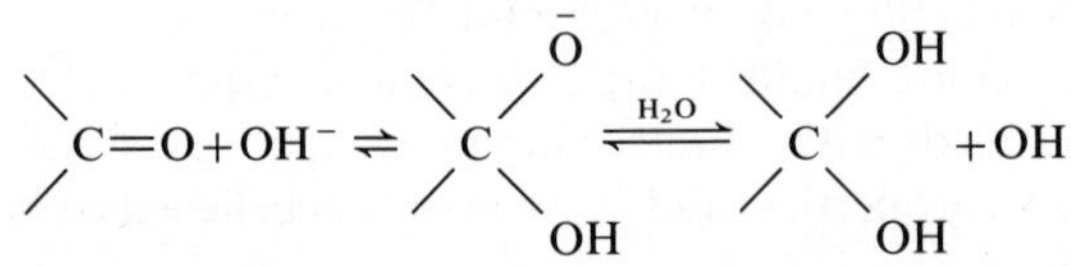

Since the reactions are reversible, any factor that stabilises the hydrate will shift the equilibrium to the right. In practice, aldehydes are more hydrated than ketones. One contributing factor is the steric effect. Another factor is the +I effect of alkyl groups. Both mechanisms involve the addition of a nucleophile in the r/d step (H_2O or OH^-), and the greater the +I effect, the less easy is the addition of the nucleophile and the easier is its removal. Ketones, with two alkyl groups, will therefore exert a greater +I effect than will aldehydes (with one alkyl group), and consequently the equilibrium will be more to the left with ketones than with aldehydes.

Hooper (1967) has examined these equilibria by N.M.R. spectroscopy and found that the amounts of *unreacted* compound in dilute aqueous solution were: MeCHO, 58; EtCHO, 63; Me_3CCHO, 100; Me_2CO, 100; $MeCOCH_2Cl$, 66 per cent.

The unusual stability of chloral hydrate has been attributed to the −I effect of chlorine and

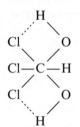

to the formation of intramolecular hydrogen bonds, the presence of which has been shown by Davies (1940) from infra-red studies. At the same time, the combination with water as *water of constitution* has also been shown, from i.r. and u.v. spectroscopy, by the presence of the carbonyl group in chloral itself and its absence in the hydrate.

Propionaldehyde, b.p. 49°C, **butyraldehyde**, b.p. 75°C, and **isobutyraldehyde**, b.p. 61°C, are prepared industrially by the *oxo process*.

Acetone (*dimethyl ketone, propan-2-one*) is prepared industrially:

(i) By the catalytic dehydrogenation (Cu or ZnO) or catalytic oxidation (Ag) of isopropanol.

(ii) By passing a mixture of propene and oxygen, under pressure, into an aqueous solution of palladium and cupric chlorides (*cf.* acetaldehyde).

(iii) Acetone is also manufactured by the oxidation of natural gas, and is obtained as a by-product in the oxidation of cumene to phenol (p. 699).

Acetone is a colourless, pleasant-smelling liquid, b.p. 56°C, miscible with water, ethanol and ether in all proportions. Acetone is used as a solvent for acetylene, cellulose acetate and nitrate, celluloid, lacquers, etc., and for the preparation of keten, sulphonal, etc.

> Although acetone does not exist as a hydrate in water (see above), it nevertheless undergoes reversible hydration. On the other hand, Wilson *et al.* (1963) have shown that acetone hydrate exists in the *solid* state.

Ketones do not polymerise, but readily undergo condensation reactions. Acetone readily forms mesityl oxide, phorone and diacetone alcohol (p. 299), but in addition to these condensations, acetone forms mesitylene when distilled with concentrated sulphuric acid (see p. 584).

> **Ethyl methyl ketone, butan-2-one,** is prepared industrially by the catalytic dehydrogenation or air oxidation of s-butanol, and from but-1-ene and palladium chloride, etc. (*cf.* acetone).
> Butanone is a pleasant-smelling liquid, b.p. 80°C, and is widely used as a solvent for vinyl resins, synthetic rubber, etc.
> **Isopropyl methyl ketone, 3-methylbutan-2-one,** b.p. 94°C, may be prepared in several ways, but the best method is by treating t-pentanol with bromine and hydrolysing the dibromo-derivative, whereupon a molecular rearrangement takes place (see pinacone, below):

$$(CH_3)_2C(OH)CH_2CH_3 \xrightarrow{Br_2} (CH_3)_2CBrCHBrCH_3 \xrightarrow{H_2O}$$

$$(CH_3)_2C(OH)CH(OH)CH_3 \xrightarrow[\text{rearranges}]{-H_2O} (CH_3)_2CHCOCH_3$$

t-Butyl methyl ketone, pinacolone, pinacone, b.p. 119°C, may be prepared by distilling pinacol or pinacol hydrate with sulphuric acid:

$$(CH_3)_2C(OH)C(OH)(CH_3)_2 \xrightarrow{H^+} CH_3COC(CH_3)_3 \quad (65\text{--}72\%)$$

This conversion of pinacol into pinacone is an example of the 1,2-shift (p. 146), and is known as the *pinacol-pinacone rearrangement*.

Pinacolone is oxidised by alkaline sodium hypobromite to trimethylacetic acid:

$$CH_3COC(CH_3)_3 \xrightarrow{Br_2/NaOH} (CH_3)_3CCO_2H \quad (71\text{--}74\%)$$

The pinacol-pinacone rearrangement is general for 1,2-glycols under acid conditions; the migrating group may be alkyl or aryl. The detailed mechanism of this arrangement is not certain, but it is generally accepted that the first stage is the addition of a proton to one of the hydroxy-groups, followed by loss of this as water. This leaves a carbonium ion, and it is the nature of this ion and the steps leading to the product that are uncertain. According to Bunton *et al.* (1959), the mechanism for the rearrangement of pinacol itself is:

$$Me_2C-CMe_2 + H^+ \rightleftharpoons Me_2C——CMe_2 \rightleftharpoons$$
$$\quad | \quad | \qquad\qquad\qquad | \quad\; {}^+OH_2$$
$$\quad OH\; OH \qquad\qquad\quad OH$$

(I)

$$\qquad\qquad Me$$
$$\qquad\qquad |$$
$$H_2O + Me-\overset{+}{C}-CMe_2 \longrightarrow Me-\overset{+}{C}-CMe_3 \longrightarrow H^+ + MeCOCMe_3$$
$$\qquad\qquad |\qquad\qquad\qquad\quad |$$
$$\qquad\qquad OH \qquad\qquad\qquad H-O$$

(II)

Kinetic measurements have shown that the conversion of (I) into (II) is the rate-determining step.

Whitmore *et al.* (1939), using a related pinacolic deamination, showed that when an optically active amine containing one asymmetric centre was used, this asymmetric centre was predominantly inverted in the product (*cf.* Walden inversion, p. 485; $Ph = C_6H_5$):

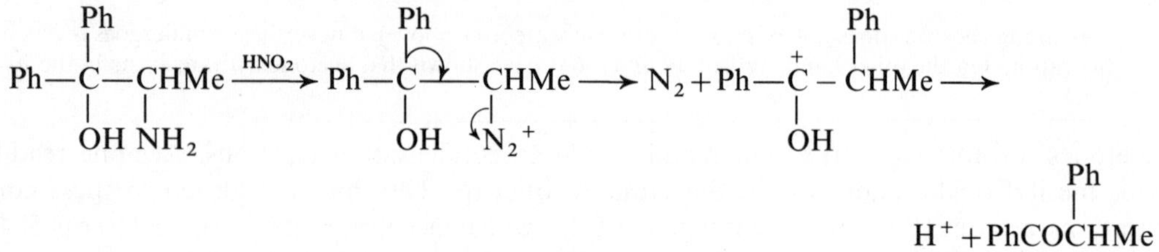

$$Ph-\underset{\underset{OH\ NH_2}{|\ \ |}}{C}-CHMe \xrightarrow{HNO_2} Ph-\underset{\underset{OH\ N_2^+}{|\ \ |}}{C}-CHMe \longrightarrow N_2 + Ph-\overset{+}{\underset{\underset{OH}{|}}{C}}-CHMe \longrightarrow$$

$$H^+ + Ph\underset{\underset{}{}}{COCHMe}\ \ (Ph)$$

These results mean that the classical carbonium ion, which is flat, if formed at all, must rearrange so fast that racemisation cannot occur (see p. 484). On the other hand, the inversion may be explained by a synchronous migration of a phenyl group and breaking of the C—N bond. In this case, the intermediate could be the *phenonium ion, i.e.,* a bridged ion in which the configuration of the asymmetric carbon atom is held in the inverted position:

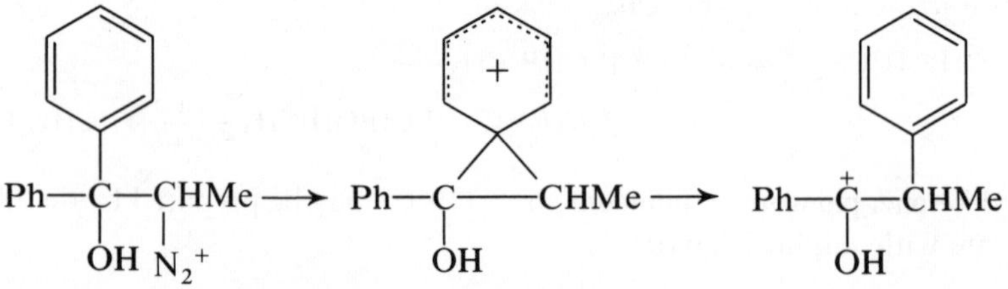

$$Ph-\underset{\underset{OH\ N_2^+}{|\ \ |}}{C}-CHMe \longrightarrow Ph-\underset{\underset{OH}{|}}{C}\!\!-\!\!-\!\!CHMe \longrightarrow Ph-\overset{+}{\underset{\underset{OH}{|}}{C}}-CHMe$$

A pair of electrons of the benzene ring is used to form the bridge, and the rest of the benzene molecule is represented as a pentadienyl cation (see p. 592).

When dealing with unsymmetrical pinacols, there are several other points to be considered, *e.g.,* which hydroxyl group is removed. Since the OH group is removed as water, any polar effect in the molecule which weakens one C—O bond more than the other will facilitate the release of this OH group. Consider the two groups phenyl and methyl. The phenyl group has a powerful conjugative effect, whereas the methyl group has a weak inductive effect in comparison:

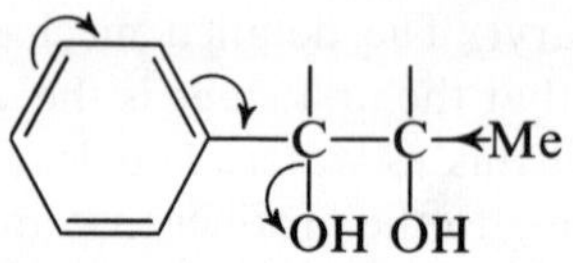

Therefore the OH group lost should be the one on the phenyl side, and consequently it will be the methyl group that migrates. This is borne out in practice, *e.g.,*

$$Ph_2C(OH)C(OH)Me_2 \xrightarrow{H_2SO_4} Ph_2MeCCOMe$$

Another way of explaining which hydroxyl group is removed is based on the relative stabilities of the alternative carbonium ions, the more stable of the two being the preferred one. In the example given here, (III) is stabilised by resonance, whereas (IV) is stabilised by the $+I$ effect of the methyl groups. Since phenyl-conjugation is a much greater stabilising factor than simply spreading the charge by the inductive effect, (III) is far more stable than (IV).

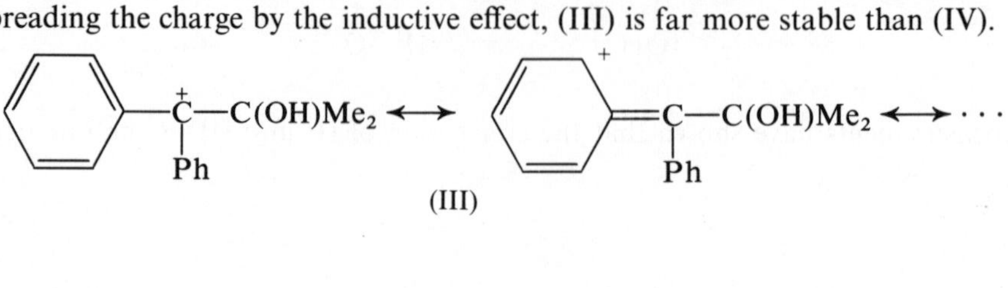

$$\underset{\underset{Ph}{|}}{\overset{+}{C}}-C(OH)Me_2 \longleftrightarrow \underset{\underset{Ph}{|}}{\overset{+}{=}C}-C(OH)Me_2 \longleftrightarrow \cdots$$

$$(III)$$

$$Ph_2C(OH)—\overset{+}{C}\overset{Me}{\underset{Me}{<}}$$

(IV)

Another problem is that of 'migration aptitude', *i.e.*, which of the two groups migrates when the groups are different. A great deal of work has been done with pinacols of the type (Ar = aryl) $Ar^1Ar^2C(OH)C(OH)Ar^1Ar^2$, and it has been shown that the aryl group with the greater electron-releasing property is the one that migrates (Bachmann *et al.*, 1932–34). This is understandable on the basis of the formation of an intermediate phenonium ion. Any polar factor that helps to release a pair of electrons from a benzene ring will therefore facilitate the formation of the bridged ion with *this* benzene ring. It has also been found that migration aptitudes depend on steric factors.

PROBLEMS

1. Show, by equations, how acetaldehyde may be converted into:

   (i) EtOH; (ii) CH_3COCl; (iii) CHI_3; (iv) MeCOEt; (v) CH_3CO_2Et; (vi) $CH_3CHOHCO_2H$; (vii) $EtNH_2$; (viii) $CH_3CHBrCHBrCO_2H$.

2. Convert $ClCH_2CH_2CHO$ into $HOCH_2CHOHCHO$.
3. Draw the energy profile for the following, and give your reasons.

$$CH_3COCH_3 + NaOH + Br_2 \longrightarrow CH_3COCH_2Br + H_2O + NaBr$$

4. Acetone, in dilute aqueous solution is 100 per cent unhydrated. When acetone is dissolved in water enriched with ^{18}O, recovered acetone contains ^{18}O. Explain.
5. Convert: (i) n-butanol into 2-ethylhexan-1-ol; (ii) bromocyclohexane into cyclopentanecarboxylic acid.
6. Calculate ΔH for each reaction (use Table 2.2, p. 36):

$$CH_3CHO + OH^- \rightleftharpoons \bar{C}H_2CHO + H_2O$$

$$CH_3COCH_3 + O\bar{H} \rightleftharpoons \bar{C}H_2COCH_3 + H_2O$$

Are these reactions equally thermodynamically favourable? Are the experimental results in agreement with your calculations? If not, why not?

7. Compound $C_5H_{10}O$ had a band at 1715 cm^{-1} in its i.r. spectrum, and showed two signals, a triplet and a quartet, in its N.M.R. spectrum. What is the compound?

8. One of the i.r. bands of compound **A**, C_3H_6O, shifts to a higher frequency when its solution in CCl_4 is progressively diluted. Write out possible acyclic structures [4] for the formula C_3H_6O, and suggest which is most likely for **A**. Using i.r. and N.M.R. spectroscopy, how would you distinguish between the four structures?

9. The bromination of acetone is catalysed by acids, and the rate law is zero order with respect to bromine. Discuss this statement.

10. Suggest what compounds the letters stand for in the following equations:

   (i) $MeCOMe + CH\equiv CH \xrightarrow[\text{(ii) A}]{\text{(i) NaNH}_2} \textbf{B}$

   (ii) $PhCOMe + ClCH_2CO_2Et \xrightarrow{\text{NaNH}_2} \textbf{C}$

   (iii)

   $$\text{[cyclohexanone]} + ClCH_2CO_2Et \xrightarrow[\text{t-BuOH}]{\text{t-BuOK}} \textbf{D}$$

   (iv) $CH_3COC_2H_5 + EtONO \xrightarrow[\text{gas}]{\text{HCl}} \textbf{E}$

   (v) $Et_2CO + NH_4Cl + NaCN \xrightarrow{\text{NH}_4\text{OH}} \textbf{F}$

11. Complete the following equations and give reasons for your answer.

(i) $Me_2C(OH)CH_2OH \xrightarrow{H^+}$?   (ii) $PhMeC(OH)C(OH)MePh \xrightarrow{H^+}$?

12. Describe paths that could lead to the formation of the following fragment ions:

(i)

m/e	29	43	44	71	72
n-Butyraldehyde (%)	54	79	100	5	73
Isobutyraldehyde (%)	23	100	—	1	53

(ii)

m/e	29	43	57	58	71	86
Pentan-2-one (%)	16	100	26	9	9	25
Pentan-3-one (%)	99	—	100	—	1	17

REFERENCES

Organic Reactions (Wiley). Vol. I (1942), Ch. 7, 'The Clemmensen Reduction': Vol. II (1944), Ch. 3, 'The Cannizzaro Reaction'; Ch. 5, 'Reduction with Aluminium Alkoxides': Vol. III (1946), Ch. 8, 'The Schmidt Reaction': Vol. IV (1948), Ch. 7, 'The Rosenmund Reduction of Acid Chlorides to Aldehydes'; Ch. 8, 'The Wolff-Kishner Reduction': Vol. V (1949), Ch. 7, 'The Leuckart Reaction'; Ch. 8, 'Selenium Dioxide Oxidation'; Ch. 10, 'The Darzens Glycidic Ester Condensation': Vol. VI (1951), Ch. 5, 'The Oppenauer Oxidation': Vol. VII (1953), Ch. 6, 'The Nitrosation of Aliphatic Carbon Atoms': Vol. VIII (1954), Ch. 5, 'The Synthesis of Aldehydes from Carboxylic Acids': Vol. IX (1957), Ch. 3, 'The Baeyer-Villiger Oxidation of Aldehydes and Ketones': Vol. XVI (1968), Ch. 1, 'The Aldol Condensation'.

COLLINS, 'The Pinacol Rearrangement', *Quart. Rev.,* 1960, **14,** 357.

PATAI (ed.), *The Chemistry of the Carbonyl Group*, Interscience (1966).

BUCHANAN and WOODGATE, 'The Clemmensen Reduction of Difunctional Ketones', *Quart. Rev.,* 1969, **23,** 522.

CARNDUFF, 'Recent Advances in Aldehyde Synthesis', *Quart. Rev.,* 1966, **20,** 169.

WHEELER, 'The Girard Reagents', *J. Chem. Educ.,* 1968, **45,** 435.

DELPIERRE and LAMCHEN, 'Nitrones', *Quart. Rev.,* 1965, **19,** 329.

LEE and UFF, 'Organic Reactions Involving Electrophilic Oxygen', *Quart. Rev.,* 1967, **21,** 449.

9

Saturated monocarboxylic acids and their derivatives

The **saturated monocarboxylic acids** are also referred to as the **fatty acids**, a name which arose from the fact that some of the higher members, particularly palmitic and stearic acids, occur in natural fats. Their general formula is $C_nH_{2n}O_2$ but, because their functional group is the carboxyl group, $-CO_2H$, they are more conveniently expressed as $C_nH_{2n+1}CO_2H$ or RCO_2H, since these show the nature of the functional group. The structure of the carboxyl group is written as

$$-C\overset{\displaystyle O}{\underset{\displaystyle OH}{\diagup}}$$

(or $-COOH$), but, as we shall see later, it does not represent accurately the behaviour of this group. Also, since only the carboxyl group hydrogen atom is replaceable by a metal, the monocarboxylic acids are *monobasic*.

Nomenclature

The monocarboxylic acids are commonly known by the trivial names, which have been derived from the source of the particular acid; *e.g.*, formic acid, HCO_2H, was so named because it was first obtained by the distillation of ants; the Latin word for ant is *formica*. Acetic acid, CH_3CO_2H, is the chief constituent of vinegar, the Latin word for which is *acetum*; etc. (see below).

Another system of nomenclature considers the acids, except formic acid, as derivatives of acetic acid, *e.g.*,

$CH_3CH_2CO_2H$	methylacetic acid
$(CH_3)_3CCO_2H$	trimethylacetic acid
$(CH_3)_2CHCH_2CO_2H$	isopropylacetic acid

In the above two systems the positions of substituents in the chain are indicated by Greek letters, the α-carbon atom being the one joined to the carboxyl group, *e.g.*,

$CH_3CH(OH)CH_2CH_2CO_2H$	γ-hydroxyvaleric acid
$CH_3CHClCH(CH_3)CO_2H$	β-chloro-α-methylbutyric acid

According to the I.U.P.A.C. system of nomenclature, the suffix of the monocarboxylic acids is *-oic*, which is added to the name of the alkane corresponding to the longest carbon chain containing the carboxyl group, *e.g.*,

$$HCO_2H \qquad\qquad \text{methanoic acid}$$
$$CH_3CH_2CO_2H \qquad\qquad \text{propanoic acid}$$

The positions of side-chains (or substituents) are indicated by numbers, *the carboxyl group always being given number* 1 (see p. 94), *e.g.*,

$$CH_3CH(\dot{C}H_3)CH(CH_3)CH_2CO_2H \qquad \text{3,4-dimethylpentanoic acid}$$

Alternatively, the carboxyl group is regarded as a substituent, and is denoted by the suffix carboxylic acid, *e.g.*,

$$\overset{4}{C}H_3\overset{3}{C}H_2\overset{2}{C}H(CH_3)\overset{1}{C}H_2CO_2H \qquad \text{2-methylbutane-1-carboxylic acid}$$

General methods of preparation

1. Oxidation of alcohols, aldehydes or ketones with acid dichromate yields acids (see also p. 186):

$$RCH_2OH \xrightarrow{[O]} RCHO \xrightarrow{[O]} RCO_2H \quad (g.\text{-}v.g.)$$
$$R_2CHOH \xrightarrow{[O]} R_2CO \xrightarrow{[O]} R^1CO_2H + R^2CO_2H \quad (g.)$$

Many *primary* alcohols, on oxidation with acid dichromate, form esters in addition to acids, *e.g.*, n-butanol gives n-butyric acid and n-butyl n-butyrate (41–47 per cent). Several mechanisms have been proposed, but the generally accepted one involves hemiacetal formation:

$$RCH_2OH \xrightarrow{[O]} RCHO \xrightarrow{RCH_2OH} RCH(OH)OCH_2R \xrightarrow{[O]} RCO_2CH_2R$$

2. A very good synthetic method is the hydrolysis of cyanides with acid or alkali:

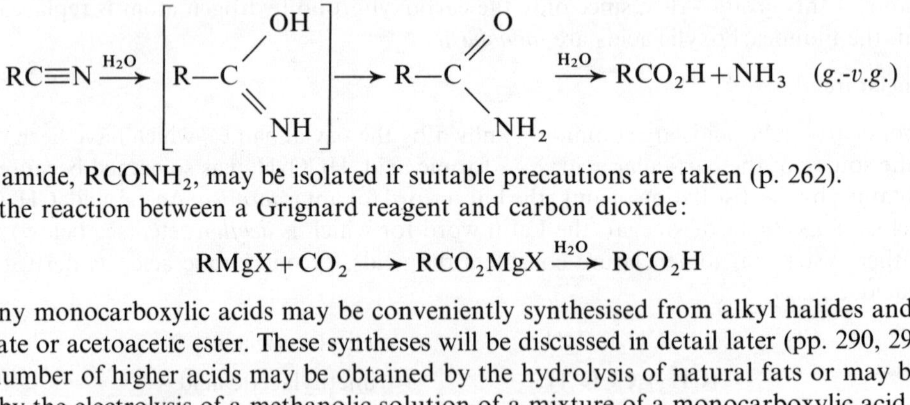

$$RC{\equiv}N \xrightarrow{H_2O} \left[R-C \overset{OH}{\underset{NH}{\Big\langle}} \right] \longrightarrow R-C \overset{O}{\underset{NH_2}{\Big\langle}} \xrightarrow{H_2O} RCO_2H + NH_3 \quad (g.\text{-}v.g.)$$

The amide, $RCONH_2$, may be isolated if suitable precautions are taken (p. 262).

3. By the reaction between a Grignard reagent and carbon dioxide:

$$RMgX + CO_2 \longrightarrow RCO_2MgX \xrightarrow{H_2O} RCO_2H$$

4. Many monocarboxylic acids may be conveniently synthesised from alkyl halides and ethyl malonate or acetoacetic ester. These syntheses will be discussed in detail later (pp. 290, 294).

5. A number of higher acids may be obtained by the hydrolysis of natural fats or may be prepared by the electrolysis of a methanolic solution of a mixture of a monocarboxylic acid and a half-ester of a dicarboxylic acid (Linstead *et al.*, 1950; *cf.* p. 77):

$$RCO_2H + HO_2C(CH_2)_nCO_2CH_3 \longrightarrow H_2 + 2CO_2 + R(CH_2)_nCO_2CH_3$$

The chain of a monocarboxylic acid can be extended by five or six carbon atoms in a synthesis involving enamines (p. 384; see also p. 307).

6. Monocarboxylic acids may be obtained by heating a sodium alkoxide with carbon monoxide under pressure:

$$R\overset{-}{O}\overset{+}{Na} + CO \longrightarrow RCO_2^-\overset{+}{Na}$$

Acids may also be obtained by heating the alcohol with carbon monoxide under pressure in the presence of a catalyst (see p. 241).

The sodium salts of acetic and propionic acids are formed by the interaction of methylsodium or ethylsodium, respectively, and carbon dioxide, *e.g.*,

$$C_2\bar{H}_5\overset{+}{Na} + CO_2 \longrightarrow C_2H_5CO_2Na$$

Monocarboxylic acids may be prepared by heating an alkene with carbon monoxide and steam under pressure at 300–400°C in the presence of a catalyst, *e.g.*, phosphoric acid:

$$CH_2{=}CH_2 + CO + H_2O \longrightarrow CH_3CH_2CO_2H$$
$$CH_3CH{=}CH_2 + CO + H_2O \longrightarrow (CH_3)_2CHCO_2H$$

General properties of the monocarboxylic acids

The first three acids are colourless, pungent-smelling liquids; the acids from butyric, $C_4H_8O_2$, to nonanoic, $C_9H_{18}O_2$, are oils which smell like goats' butter; and those higher than decanoic acid, $C_{10}H_{20}O_2$, are odourless solids. The lower members are far less volatile than is to be expected from their molecular weight. This can be explained by hydrogen bonding, and electron diffraction studies (Pauling, 1934) have shown that an eight-membered ring is present, *i.e.*, the

$$
\begin{array}{ccc}
 & O \cdots H{-}O & \\
 & \diagup\!\!/ \qquad \diagdown & \\
R{-}C & & C{-}R \\
 & \diagdown \qquad /\!\!\diagup & \\
 & O{-}H \cdots O &
\end{array}
$$

acids exist as cyclic dimers. The lower members exist as dimers in the vapour phase and in aqueous solution, but in the liquid phase they exist as 'polymers'.

The melting point of the n-monocarboxylic acids show alternation or oscillation from one member to the next, the melting point of an 'even' acid being higher than that of the 'odd' acid immediately below and above it in the series (see the physical constants of the individuals). A number of homologous series follow this oscillation or 'saw-tooth' rule.

The first four members are very soluble in water, and the solubility decreases as the molecular weight increases. This solubility is due to the acids being capable of forming hydrogen bonds with water.

The infra-red absorption region of the carboxyl group contains a number of bands. The carbonyl group (stretch) absorbs in the same region as that of acyclic ketones. 1725–1700 cm^{-1} (s), but the (free) hydroxyl group (stretch) absorbs at 3560–3500 cm^{-1} (m), which is in the lower frequency region than for alcohols (3670–3580 cm^{-1}).

Figure 9.1 is the infra-red spectrum of stearic acid (as a 1 per cent solution in carbon tetrachloride). It should be noted that the band for the bonded O—H (str.) is hidden by the strong C—H (str.))

On the other hand, Fig. 9.2 is the spectrum of stearic acid as a Nujol mull. In this case, the stearic acid is in the solid (dispersed) state, and it can be seen that there is an extremely large difference in the region 1300–1180 cm^{-1}. In this region there are eight maxima, and this pattern is characteristic of the spectra of long-chain n-alkyl compounds (in the solid state), the number of peaks depending on the chain length. Nujol (a petroleum fraction), since it is a hydrocarbon mixture, has the bands characteristic of alkanes (*cf.* Fig. 3.2). The O—H (str.) is hidden (as for Fig. 9.1). These two spectra illustrate the influence of the physical state of a compound on its infra-red spectrum.

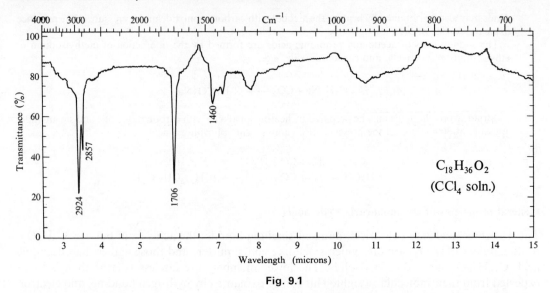

Fig. 9.1

1 706 cm^{-1}: C=O (str.) in (saturated ketones and) saturated aliphatic acids.   2 942 and 2 857 cm^{-1}: C—H (str.) in Me and/or CH$_2$. 1 460 cm^{-1}: C—H (def.) in Me and/or CH$_2$.

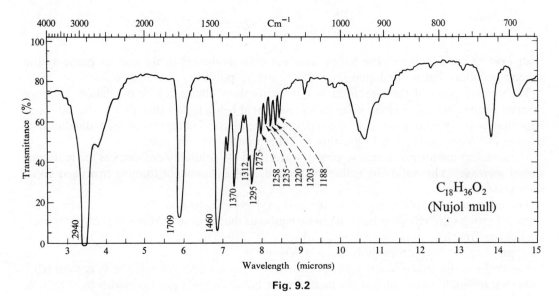

Fig. 9.2

2 940, 1 460 and 1 370 cm^{-1}: Bands for Nujol. 1 709 cm^{-1}: C=O (str.) as for Fig. 9.1. 1 312, 1 295, 1 279, 1 258, 1 235, 1 220, 1 203 and 1 188 cm^{-1}: These bands are characteristic of long-chain n-alkyl compounds (in the solid state).

Because of hydrogen bonding (intermolecular) and exchange reactions, the proton peak of the carboxyl group can be found in many different positions in the NMR spectrum. This applies to hydroxyl protons in general (see p. 12). However, it has been found that in non-polar solvents, the τ-value of the carboxyl proton in aliphatic acids is around $-1 \cdot 0$ p.p.m. (*i.e.*, there is a great shift to lower field due to strong hydrogen bonding).

Mass spectra. Acids, esters, and amides undergo similar fragmentation patterns, *e.g.*, α-cleavage for lower members (Z=OH, OR, NH_2):

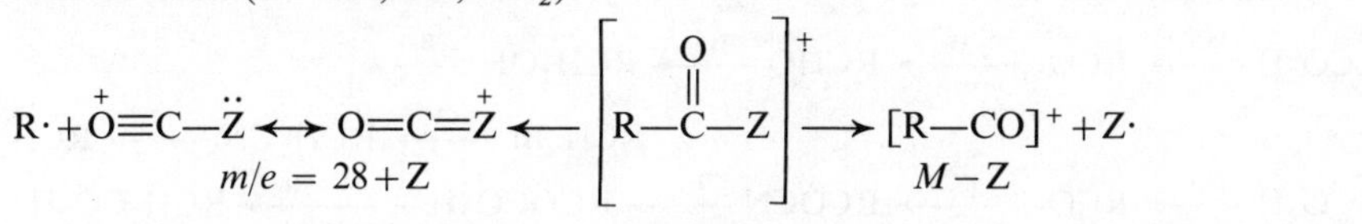

$$R\cdot + \overset{+}{O}\equiv C - \overset{..}{Z} \longleftrightarrow O = C = \overset{+}{Z} \longleftarrow \left[R - \overset{\overset{\displaystyle O}{\|}}{C} - Z \right]^{+} \longrightarrow [R-CO]^{+} + Z\cdot$$
$$m/e = 28+Z \qquad\qquad\qquad\qquad M-Z$$

Also, R and Z may carry the positive charge. Hence, peaks are obtained at $M-17$ and m/e 45 (Z = OH), $M-31$, $M-45$, and m/e 59, 73 (Z = OMe and OEt, respectively), and $M-16$ and m/e 44 (Z = NH_2). Since they are more volatile than their corresponding acids, esters are used preferentially. All can undergo the McLafferty rearrangement if they contain a γ-hydrogen atom:

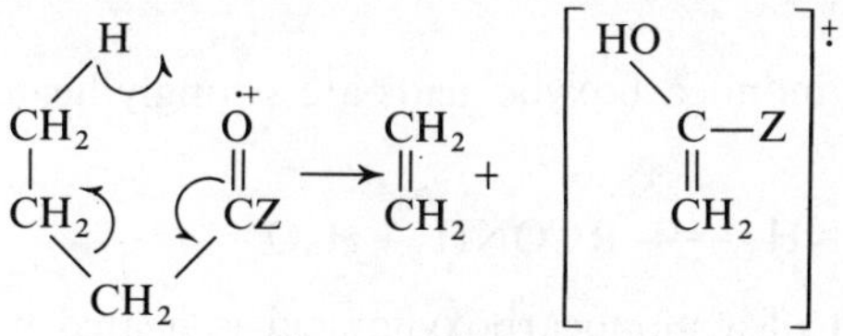

m/e 60 (Z = OH);

74 (Z = OMe);

88 (Z = OEt);

59 (Z = NH_2)

When the alkyl group in OR of esters is higher than methyl, esters can also undergo the McLafferty rearrangement to give alkene and acid, *e.g.*,

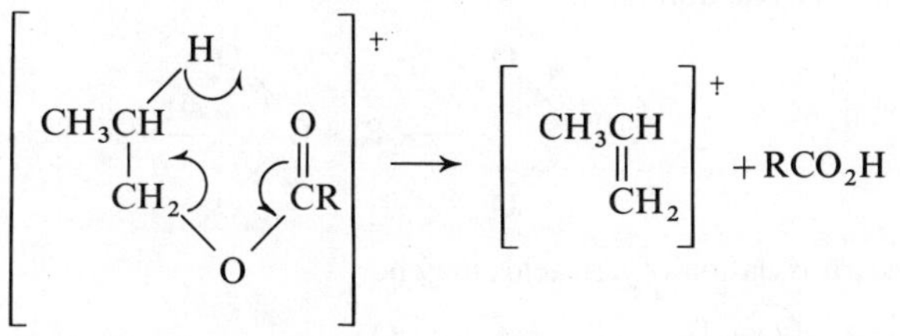

General reactions of the monocarboxylic acids

1. The acids are acted upon by the strongly electropositive metals with the liberation of hydrogen and formation of a salt:

$$RCO_2H + Na \longrightarrow RCO_2^- Na^+ + \tfrac{1}{2}H_2$$

Salts are also formed by the reaction between an acid and an alkali:

$$RCO_2H + NaOH \longrightarrow RCO_2^- Na^+ + H_2O$$

2. Monocarboxylic acids react with alcohols to form *esters*:

$$R^1CO_2H + R^2OH \rightleftharpoons R^1CO_2R^2 + H_2O$$

3. Phosphorus trichloride, phosphorus pentachloride or thionyl chloride act upon the monocarboxylic acids to form *acid chlorides*:

$$3RCO_2H + PCl_3 \longrightarrow 3RCOCl + H_3PO_3$$
$$RCO_2H + PCl_5 \longrightarrow RCOCl + HCl + POCl_3$$
$$RCO_2H + SOCl_2 \longrightarrow RCOCl + HCl + SO_2$$

The acid series may be 'stepped up' via the acid chloride, but since the process involves many steps, the yield of the higher acid homologue is usually only f.-f.g.:

(a) $RCO_2H \xrightarrow{PCl_5} RCOCl \xrightarrow[\text{catalyst}]{H_2} RCHO \xrightarrow[\text{catalyst}]{H_2} RCH_2OH \xrightarrow{PBr_3}$

$$RCH_2Br \xrightarrow{KCN} RCH_2CN \xrightarrow{H^+} RCH_2CO_2H$$

(b) $RCO_2H \xrightarrow{PCl_5} RCOCl \xrightarrow{KCN} RCOCN \xrightarrow{H^+} RCOCO_2H \xrightarrow[\text{reduction}]{\text{Clemmensen}} RCH_2CO_2H$

Although method (b) involves fewer steps than (a), the final yield of acid by route (a) is higher than that by (b), since the yields in each step of (b) are generally low. Furthermore, the yield from (a) can be improved by direct reduction of acid to alcohol by means of various metallic hydrides (see Table 6.1, p. 179).

4. When the ammonium salts of the monocarboxylic acids are strongly heated, the *acid amide* is formed by the elimination of water:

$$RCO_2NH_4 \longrightarrow RCONH_2 + H_2O$$

5. When the anhydrous sodium salt of a monocarboxylic acid is heated with soda-lime, an alkane and other products are formed (p. 76).

The mechanism of this decarboxylation is uncertain, but there is much evidence to show that *salts* decompose by an S_E1 reaction (*cf.* p. 274):

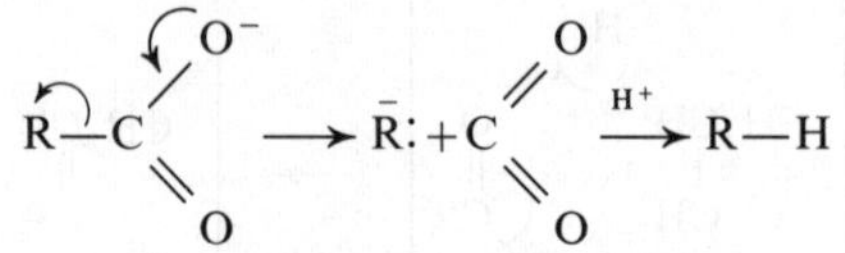

The thermal decarboxylation of *free* acids may be:

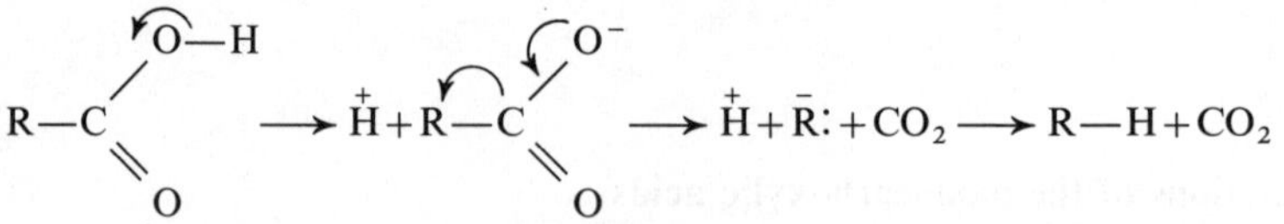

Monocarboxylic acids may be 'stepped down' by heating the calcium salt with calcium acetate (see p. 204), and the methyl ketone produced is oxidised with acid dichromate:

$$RCH_2CO_2Ca/_2 + CH_3CO_2Ca/_2 \longrightarrow RCH_2COCH_3 \xrightarrow{[O]} RCO_2H + CH_3CO_2H$$

It should be noted that the second step is based on the general rule that when an unsymmetrical ketone is oxidised, the carbonyl group remains chiefly with the smaller alkyl group. Since monocarboxylic acids containing an odd number of carbon atoms occur only in small amounts, this method of descending the series affords a useful means of obtaining 'odd' acids from 'even'.

When a mixture of the calcium salt (or any of the others mentioned above) and calcium formate is heated, an aldehyde is obtained (p. 204). The pyrolysis of iron salts gives very good yields of symmetrical ketones (p. 205).

Metallic salts, particularly the silver salts, are converted by bromine into the alkyl bromide. This reaction may also be used to 'step down' the acid series (p. 152).

6. When a concentrated aqueous solution of the sodium or potassium salt of a monocarboxylic acid is electrolysed, an alkane is obtained:

$$2RCO_2K + 2H_2O \longrightarrow R\!-\!R + 2CO_2 + 2KOH + H_2$$

7. Monocarboxylic acids react slowly with chlorine or bromine in the cold, but at higher temperatures and in the presence of a small amount of phosphorus, reaction proceeds smoothly to give α-halogeno-acids (see also p. 269); *e.g.*,

$$RCH_2CO_2H \xrightarrow{P/Br_2} RCHBrCO_2H \xrightarrow{P/Br_2} RCBr_2CO_2H$$

8. All the acids, except formic acid, are extremely resistant to oxidation, but prolonged heating with oxidising agents ultimately produces carbon dioxide and water.

9. The reduction products of monocarboxylic acids depend on the nature of the reducing agent. Thus, heating with hydrogen iodide-red phosphorus under pressure, or with hydrogen under pressure at elevated temperature in the presence of a nickel catalyst, produces an alkane.

$$RCO_2H + 3H_2 \xrightarrow{Ni} RCH_3 + 2H_2O$$

If the catalyst is Ru—C, the alcohol is formed. Alcohols are also produced by reduction with various metallic hydrides (see Table 6.1, p. 179).

10. The acids undergo the *Schmidt reaction* (p. 216) to form a primary amine:

$$RCO_2H + HN_3 \xrightarrow{H_2SO_4} RNH_2 + CO_2 + N_2$$

Schmidt's reaction with acids is a modification of the Curtius reaction (p. 268). The following mechanism has been proposed (*cf.* p. 217):

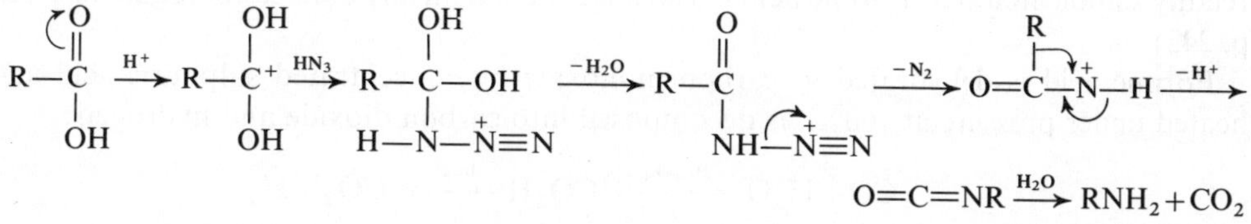

$$O=C=NR \xrightarrow{H_2O} RNH_2 + CO_2$$

The rearrangement (1,2-shift) has been shown to be intramolecular, *e.g.*, Kenyon *et al.* (1939) showed retention of optical activity when α-phenylpropionic acid underwent the Schmidt reaction:

$$C_6H_5CH(CH_3)CO_2H \longrightarrow C_6H_5CH(CH_3)NH_2$$

Formic acid (*methanoic acid*) is prepared industrially by heating sodium hydroxide with carbon monoxide at 210°C, and at a pressure of 6–10 atmospheres:

$$NaOH + CO \longrightarrow HCO_2Na$$

An aqueous solution of formic acid is obtained by distilling the sodium salt with dilute sulphuric acid:

$$2HCO_2Na + H_2SO_4 \longrightarrow 2HCO_2H + Na_2SO_4$$

Anhydrous formic acid is obtained from the aqueous solution (70–77 per cent) by the addition of butyl formate followed by distillation. The first fraction is an azeotrope of ester and water, and then the excess of ester is removed from the formic acid by fractionation.

The most convenient laboratory preparation of formic acid is to heat glycerol with oxalic acid at 100–110°C. Glyceryl monoxalate is produced, and decomposes into glyceryl monoformate (monoformin) and carbon dioxide. When the evolution of carbon dioxide ceases, more oxalic acid is added, whereupon formic acid is produced:

$$
\begin{array}{l}
\text{CH}_2\text{OH} \\
| \qquad\qquad \text{COOH} \\
\text{CHOH} + | \qquad\qquad\quad\; \longrightarrow \\
| \qquad\qquad \text{COOH} \\
\text{CH}_2\text{OH}
\end{array}
\quad
\begin{array}{l}
\text{CH}_2\text{OCOCOOH} \\
| \\
\text{CHOH} \\
| \\
\text{CH}_2\text{OH}
\end{array}
\quad + \text{H}_2\text{O} \xrightarrow{\;100\text{--}110°\text{C}\;}
$$

$$
\text{CO}_2 +
\begin{array}{l}
\text{CH}_2\text{OCHO} \\
| \\
\text{CHOH} \\
| \\
\text{CH}_2\text{OH}
\end{array}
\xrightarrow{(\text{COOH})_2}
\begin{array}{l}
\text{CH}_2\text{OCOCOOH} \\
| \\
\text{CHOH} \\
| \\
\text{CH}_2\text{OH}
\end{array}
\quad + \text{HCO}_2\text{H}
$$

The distillate contains formic acid and water. The aqueous formic acid solution cannot be fractionated to give anhydrous formic acid because the boiling point of the acid is 100·5°C. The procedure adopted is to neutralise the aqueous acid solution with lead carbonate, and concentrate the solution until lead formate crystallises out. The precipitate is then recrystallised, dried, and heated at 100°C in a current of hydrogen sulphide:

$$(\text{HCO}_2)_2\text{Pb} + \text{H}_2\text{S} \longrightarrow 2\text{HCO}_2\text{H} + \text{PbS}$$

The anhydrous formic acid which distils over contains a small amount of hydrogen sulphide, and may be freed from the latter by adding some dry lead formate and redistilling. This procedure for obtaining the anhydrous acid from its aqueous solution can only be used for volatile acids.

Formic acid is a pungent corrosive liquid, m.p. 8·4°C, b.p. 100·5°C, miscible in all proportions with water, ethanol and ether. It forms salts which, except for the lead and silver salts, are readily soluble in water. Formic acid is a stronger acid than any of its homologues (see Table 9.1, p. 242).

Formic acid is dehydrated to carbon monoxide by concentrated sulphuric acid and when heated under pressure at 160°C, is decomposed into carbon dioxide and hydrogen:

$$\text{CO} + \text{H}_2\text{O} \xleftarrow{\text{H}_2\text{SO}_4} \text{HCO}_2\text{H} \xrightarrow{\text{heat}} \text{CO}_2 + \text{H}_2$$

The same decomposition takes place at room temperature in the presence of a catalyst such as iridium, rhodium, etc.

When metallic formates are heated with an alkali, hydrogen is evolved.

$$\text{HCO}_2\text{Na} + \text{NaOH} \longrightarrow \text{H}_2 + \text{Na}_2\text{CO}_3$$

When calcium or zinc formate is strongly heated, formaldehyde is produced:

$$(\text{HCO}_2)_2\text{Ca} \longrightarrow \text{HCHO} + \text{CaCO}_3$$

When sodium or potassium formate is rapidly heated to 360°C, hydrogen is evolved and the oxalate is formed:

$$2\text{HCO}_2\text{Na} \longrightarrow (\text{COONa})_2 + \text{H}_2$$

Formic acid forms esters, but since it is a relatively strong acid it is not necessary to use a catalyst; refluxing 90 per cent formic acid with the alcohol is usually sufficient.

Formic acid differs from the rest of the members of the monocarboxylic acid series in being a powerful reducing agent; it reduces ammoniacal silver nitrate and the salts of many of the heavy metals, e.g., it converts mercuric chloride into mercurous chloride.

Acetic acid (*ethanoic acid*) is prepared industrially by various methods, e.g., (i) the air oxidation of acetaldehyde in acetic acid in the presence of manganous ion as catalyst; (ii) the air oxidation of n-butane in the liquid phase under pressure and at 130–230°C in the presence of a suitable catalyst, *e.g.*, manganous ion; (iii) by reaction between carbon monoxide and methanol under pressure in the presence of cobalt octacarbonyl at about 210°C:

$$CO + CH_3OH \longrightarrow CH_3CO_2H.$$

One of the earliest methods for preparing acetic acid was by the destructive distillation of wood to give pyroligneous acid. This contains about 10 per cent acetic acid, and was originally treated by neutralising with lime and then distilling off the volatile compounds (these are mainly methanol and acetone). On distillation with dilute sulphuric acid, the residue gives dilute acetic acid. More recently, the acetic acid is extracted by means of solvents, *e.g.*, isopropyl ether.

Vinegar, which is a 6–10 per cent aqueous solution of acetic acid, may be made in several ways. Malt vinegar is prepared by the oxidation of wort (p. 190) by means of the bacteria *Mycoderma aceti*:

$$CH_3CH_2OH + O_2 \longrightarrow CH_3CO_2H + H_2O$$

In the 'quick vinegar process', beech shavings, contained in barrels, are moistened with strong vinegar containing the bacteria. A 10 per cent aqueous solution of ethanol containing phosphates and inorganic salts (which are necessary for the fermentation) is then poured through the barrels, and the ethanol is thereby oxidised to acetic acid. A plentiful supply of air is necessary, otherwise the oxidation is incomplete and acetaldehyde is produced.

Acetic acid is a pungent corrosive liquid, m.p. 16·6°C, b.p. 118°C, miscible in all proportions with water, ethanol and ether. It is stable towards oxidising agents, and so is a useful solvent for chromium trioxide oxidations. Acetic acid is commonly used as a solvent, and in the preparation of acetates, acetone, acetic anhydride, etc.

Strengths of the monocarboxylic acids. If the structure of the carboxyl group were simply (I), then, because of the $-I$ effect of the carbonyl group (as shown in (II)), proton release will be facilitated as compared with alcohols.

$$\underset{\text{(I)}}{\overset{\overset{\displaystyle O}{\|}}{-C-OH}} \qquad \underset{\text{(II)}}{\overset{\overset{\displaystyle O}{\|}}{R \rightarrow C \leftarrow O \leftarrow H}}$$

Thus carboxylic acids will be stronger acids than alcohols. However, the inductive effect alone cannot account for the very large difference in acid strength. One explanation is that the carboxyl group is a resonance hybrid and because of the positive charge on the oxygen atom

$$RC \overset{\overset{\displaystyle OH}{\diagup}}{\underset{\diagdown}{\diagdown O}} \longleftrightarrow RC \overset{\overset{\displaystyle \overset{+}{O}H}{\|}}{\underset{\diagdown}{\diagdown \overset{-}{O}}} + H_2O \rightleftharpoons H_3O^+ + RC \overset{\overset{\displaystyle O}{\|}}{\underset{\diagdown}{\diagdown \overset{-}{O}}} \longleftrightarrow RC \overset{\overset{\displaystyle \overset{-}{O}}{\diagup}}{\underset{\diagdown}{\diagdown O}}$$

of the OH group, proton release is facilitated; this positive charge is absent in alcohols since they are not resonance hybrids.

This argument can be extended as follows. When the acid ionises, *i.e.*, donates its proton to water (which is the solvent), the conjugate base of the acid, *i.e.*, the carboxylate ion, is a resonance hybrid of two *equivalent* resonating structures. At the same time, since the negative charge is spread, the resonance energy of this anion will be greater than that of the un-ionised acid, which is a resonance hybrid of two different resonating structures, one with separation of *unlike* charges. Therefore the internal energy of the anion is lower than that of the un-ionised acid. In the case of alcohols, there is no resonance stabilisation in the alkoxide ion. It therefore follows that $\Delta G^{\ominus}$ for the ionisation of monocarboxylic acids will be more negative than that of alcohols (see p. 59). Hence the equilibrium constant of the former will be greater than that of the latter, *i.e.*, the pK_a values of the acids will be lower than those of the alcohols. Thus alcohols are weaker acids than the monocarboxylic acids.

X-ray and electron-diffraction measurements of the carboxylate ion have shown that the two C—O bonds are of equal length. This is in keeping with the ion being a resonance hybrid.

Table 9.1

Acid	pK_a	Acid	pK_a
Formic	3·75	Chloroacetic	2·85
Acetic	4·76	Dichloroacetic	1·25
Propionic	4·87	Trichloroacetic	0·66
n-Butyric	4·82	Trifluoroacetic	0·23
Isobutyric	4·86	Bromoacetic	2·9
n-Valeric	4·86	Iodoacetic	3·16
Isovaleric	4·78	α-Chloropropionic	2·80
Trimethylacetic	5·05	β-Chloropropionic	4·08
Hexanoic	4·88	Nitroacetic	1·68
Heptanoic	4·89	Hydroxyacetic	3·83
Octanoic	4·90	Hydrochloric	−7·00

The treatment of the carboxyl group and the carboxylate ion from the M.O. point of view also leads to the same results as those obtained by the V.B. method, the increased stability now being due to *delocalisation*. Let us first consider the *undissociated* carboxylic acid (Fig. 9.3a). The carbon atom is linked to the hydroxyl group by a σ-bond, and to the oxygen of the carbonyl group by one σ- and one π-bond. This is shown in (Fig. 9.3b). If the oxygen atom of the hydroxyl group has sp^2 hybridisation, this would leave a *p* orbital perpendicular to the plane of the carboxyl group, and it would contain a pair of electrons (Fig. 9.3c). This oribtal could now overlap with both the *p* orbital of the carbon atom and that of the oxygen atom of the carbonyl group. The combination of these three orbitals would give rise to M.O.s embracing *three* nuclei (Fig. 9.3d). Since *four* electrons are involved, they must be accommodated, in the ground state, in the lowest two M.O.s. The M.O. with the lowest energy level has *one* node (Fig. 9.3e), and the next energy level has *two* nodes (Fig. 9.3f). Thus (Fig. 9.3g) represents the ground state of the undissociated carboxyl group. Because of delocalisation, this group has become more stable. Since a hydroxyl group is more electron-attracting than an oxygen atom, the delocalisation of the pair of electrons in the *p* oribtal is not very large, *i.e.*, although this lone pair embraces three nuclei, this pair is far more likely to be found in the region of the donor atom than anywhere else. Since the oxygen atom of the hydroxyl group has lost 'full-control' of this lone pair, it will acquire a small positive charge, and since the oxygen atom of the CO group has acquired a small share of this lone pair, this oxygen atom acquires a small negative charge (actually this oxygen atom already has a small negative charge due to the inductive effect). The greater electronegativity of oxygen in a hydroxyl group over that of an oxygen atom alone is due to the hydrogen atom sharing one pair of electrons in forming the bond, and thereby decreasing, to some extent, the electron density on that oxygen atom.

Now let us consider the *carboxylate ion* (Fig. 9.4a). The oxygen atom of the original hydroxyl group retains the σ-electrons when the hydrogen atom is removed as a proton. As in the case of the *carboxyl group*, one lone pair enters into conjugation with the π-bond of the C=O group (Fig. 9.4b). In the latter case, however, delocalisation is complete, *i.e.*, the lone pair is now likely to be found equally well on either oxygen atom (each atom will have a charge of $-\frac{1}{2}$). Since delocalisation is complete, the resonance (delocalisation) energy is much greater than in the case of the undissociated carboxyl group. Because the carbon atom is sp^2 hybridised in both the carboxyl group and carboxylate ion, the O—C—O angle in both would be expected to be about 120°. This has been observed in practice.

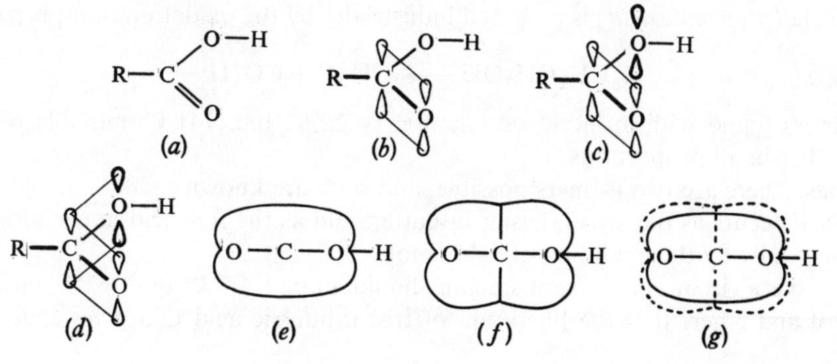

Fig. 9.3

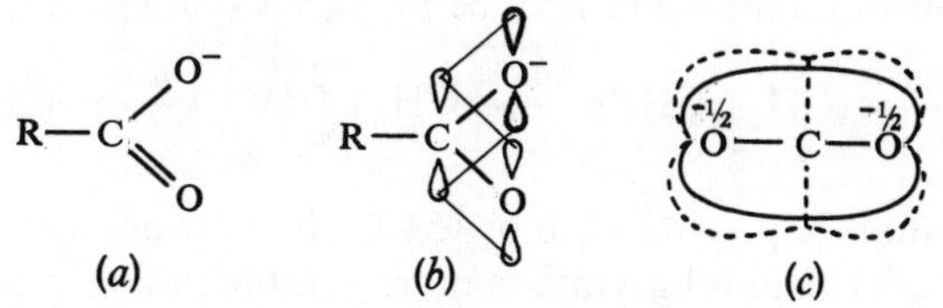

Fig. 9.4

In Figs. 9.3 and 9.4, the orbitals drawn in thicker type indicate that the atoms concerned contain *two* electrons in these orbitals. It should also be noted that lobes and not spheres (with + and − signs) have been used in these diagrams (see p. 42), and that overlapping orbitals have been drawn in an alternative way.

Peracetic (peroxyacetic) acid, $CH_3C\!\!\begin{smallmatrix}O\\ \\O-OH\end{smallmatrix}$, may be prepared by treating acetic anhydride with concentrated hydrogen peroxide, and then distilling under reduced pressure.

Swern *et al.* (1962) have prepared per-acids (85–97 per cent yield) by treating a solution or slurry of an aliphatic (or aromatic) acid in methanesulphonic acid with 90–95 per cent hydrogen peroxide. This appears to be the best method so far.

Peracetic acid is an unpleasant-smelling liquid, f.p. $+0\cdot1°C$, soluble in water, ethanol and ether. It explodes violently when heated above 110°C. It is a powerful oxidising agent; it oxidises the olefinic bond to the oxide.

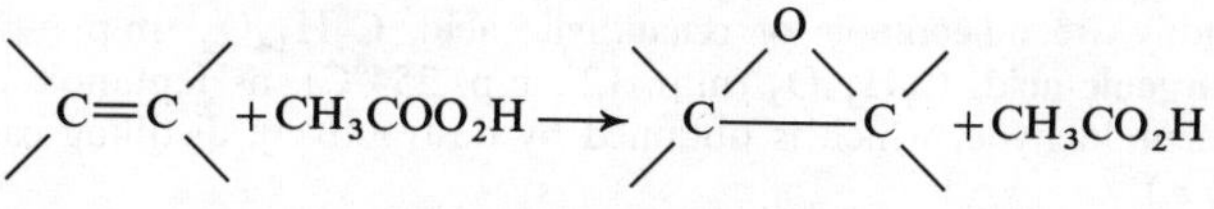

$$\begin{array}{c}\diagdown\\ \diagup\end{array}C{=}C\begin{array}{c}\diagup\\ \diagdown\end{array} + CH_3COO_2H \longrightarrow \begin{array}{c}\diagdown\\ \diagup\end{array}C\overset{\displaystyle O}{\underset{\textstyle\diagdown}{\diagup\!\!\!\!\diagdown}}C\begin{array}{c}\diagup\\ \diagdown\end{array} + CH_3CO_2H$$

It also oxidises primary aromatic amines to nitroso-compounds, *e.g.*, aniline is converted into nitroso-benzene:

$$C_6H_5NH_2 + 2CH_3COO_2H \longrightarrow C_6H_5NO + 2CH_3CO_2H + H_2O$$

Infra-red absorption spectra measurements of per-acids show that these acids exist in solution very largely in the monomeric, intramolecularly hydrogen-bonded form (*inter alia*, Minkoff, 1954).

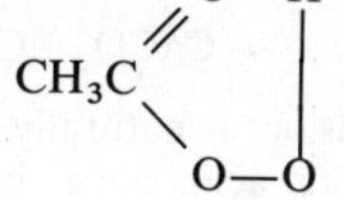

$$CH_3C\!\!\begin{smallmatrix}O\cdots H\\ \;\;\;\;\;|\\ O-O\end{smallmatrix}$$

Propionic acid (*propanoic acid*) is prepared industrially by the oxidation of n-propanol:

$$CH_3CH_2CH_2OH \xrightarrow{[O]} CH_3CH_2CO_2H$$

It is a colourless liquid with an acrid odour, m.p. $-22°C$, b.p. $141°C$, miscible with water, ethanol and ether in all proportions.

Butyric acids. There are two isomers possible, and both are known.

n-Butyric acid occurs as the glyceryl ester in butter, and as the free acid in perspiration. It is prepared industrially by the oxidation of n-butanol.

n-Butyric acid is a viscous unpleasant-smelling liquid, m.p. $-4\cdot7°C$, b.p. $162°C$, miscible with water, ethanol and ether. It is the liberation of free n-butyric acid that gives stale butter its rancid odour.

Isobutyric acid occurs in the free state and as its esters in many plants. It is prepared industrially by the oxidation of isobutanol and may be prepared synthetically as follows:

$$(CH_3)_2CHOH \xrightarrow{PBr_3} (CH_3)_2CHBr \xrightarrow{KCN} (CH_3)_2CHCN \xrightarrow[acid]{H_2O} (CH_3)_2CHCO_2H$$

Isobutyric acid is a liquid, m.p. $-47°C$, b.p. $154°C$. Its calcium salt is more soluble in hot water than in cold, whereas calcium butyrate is more soluble in cold water than in hot.

Valeric acids. There are four isomers possible, and all are known:

n-Valeric acid, $CH_3CH_2CH_2CH_2CO_2H$, m.p. $-34\cdot5°C$, b.p. $187°C$.

Isovaleric acid, $(CH_3)_2CHCH_2CO_2H$, m.p. $-51°$, b.p. $175°C$.

Ethylmethylacetic acid or *active valeric acid*, $CH_3CH_2CH(CH_3)CO_2H$, b.p. $175°C$.

Trimethylacetic acid or *pivalic acid*, $(CH_3)_3CCO_2H$, m.p. $35\cdot5°$, b.p. $164°C$.

The higher monocarboxylic acids which occur in nature usually have straight chains, and usually contain an *even* number of carbon atoms. **Caproic** (*hexanoic*), $C_6H_{12}O_2$ (m.p. $-9\cdot5°$, b.p. $205°C$), **caprylic** (*octanoic*), $C_8H_{16}O_2$ (m.p. $16°$, b.p. $237°C$) and *capric* (*decanoic*) acid, $C_{10}H_{20}O_2$ (m.p. $31\cdot5°$ b.p. $270°C$), are present as glyceryl esters in goats' butter. *Lauric* (*dodecanoic*), $C_{12}H_{24}O_2$ (m.p. $44°C$) and **myristic** (*tetradecanoic*) acid, $C_{14}H_{28}O_2$ (m.p. $58°C$), occur as their glyceryl esters in certain vegetable oils. The most important higher acids are *palmitic* (*hexadecanoic*), $C_{16}H_{32}O_2$ (m.p. $64°C$), and **stearic** (*octadecanoic*), $C_{18}H_{36}O_2$ (m.p. $72°C$), which are very widely distributed as their glyceryl esters (together with oleic acid) in most animal and vegetable oils and fats. The sodium and potassium salts of palmitic and stearic acids are the constituents of ordinary soaps.

Some still higher acids are found in waxes: **arachidic** (*eicosanoic*), $C_{20}H_{40}O_2$ (m.p. $77°C$), **behenic** (*docosanoic*), $C_{22}H_{44}O_2$ (m.p. $82°C$), **lignoceric** (*tetracosanoic*), $C_{24}H_{48}O_2$ (m.p. $83\cdot5°C$), **cerotic** (*hexacosanoic*), $C_{26}H_{52}O_2$ (m.p. $87\cdot7°C$), and **melissic** acid (*triacontanoic*), $C_{30}H_{60}O_2$ (m.p. $90°C$).

The odd fatty acids may be obtained by degrading an even acid (see below). Two odd acids which may be prepared readily are **n-heptanoic** or **œnanthylic** acid, $C_7H_{14}O_2$ (m.p. $-10°$, b.p. $223\cdot5°C$), and **nonanoic** or **pelargonic** acid, $C_9H_{18}O_2$ (m.p. $12°$, b.p. $254°C$). n-Heptanoic acid is prepared by the oxidation of n-heptaldehyde, which is obtained by destructively distilling castor oil which contains ricinoleic acid (*q.v.*):

$$CH_3(CH_2)_5CHO + [O] \xrightarrow[H_2SO_4]{KMnO_4} CH_3(CH_2)_5CO_2H \quad (68\text{–}70\%)$$

Nonanoic acid is obtained by the oxidation of aleic acid (*q.v.*).

Margaric acid (*heptadecanoic acid*), $C_{16}H_{33}CO_2H$, m.p. $61°C$, is prepared by heating a mixture of the calcium salts of stearic and acetic acids, and then oxidising the heptadecyl methyl ketone so produced.

$$C_{17}H_{35}CO_2ca + CH_3CO_2ca \longrightarrow CaCO_3 + C_{17}H_{35}COCH_3 \xrightarrow{[O]} C_{16}H_{33}CO_2H$$

Odd-numbered monocarboxylic acids occur naturally in very small amounts up to $n = 23$, and their main source is butter fat. Branched-chain acids have also been isolated from butter fat.

General reactions of the monocarboxylic acids and their derivatives

Acids and their derivatives may be represented as follows, where $Z = OH, OR, X, NH_2$, etc.

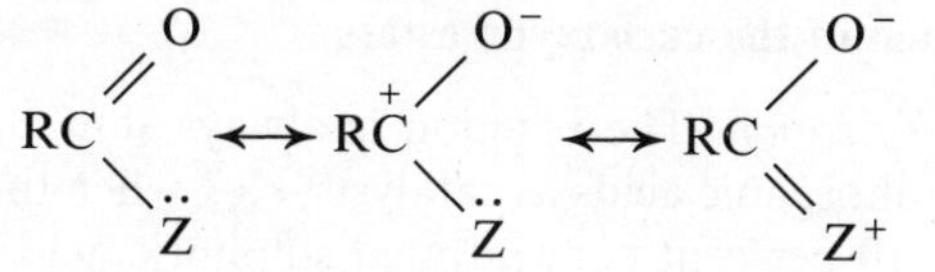

$$RC\underset{\overset{..}{Z}}{\overset{O}{\diagup\hspace{-0.3em}\diagdown}} \longleftrightarrow R\overset{+}{C}\underset{\overset{..}{Z}}{\overset{O^-}{\diagup}} \longleftrightarrow RC\underset{Z^+}{\overset{O^-}{\diagdown}}$$

Thus, all are resonance hybrids and the behaviour of the 'carbonyl' group depends on the nature of Z. Z may have a $-I$ and a $+R$ effect and consequently the actual weights of the canonical forms will depend on the relative contributions of these two electronic effects, *e.g.*, for $Z = NH_2$ (acid amide), $+R > -I$, whereas for $Z = Cl$ (acid chloride), $-I > +R$. Hence, the rates of reaction of the carbonyl group in these compounds can be expected to be different (see below).

Many of the reactions of the acids and their derivatives may be generalised as:

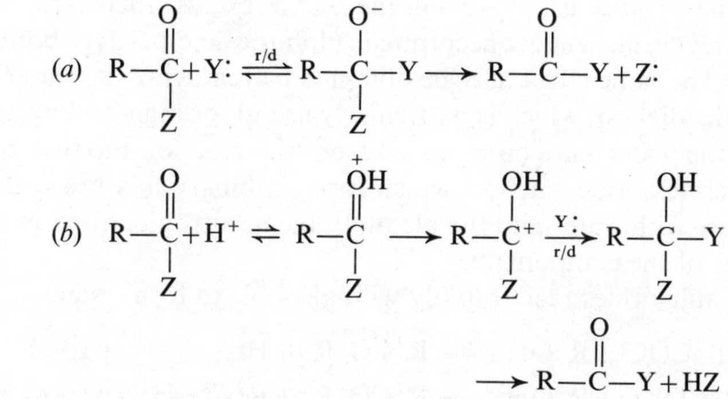

$$(a)\quad \underset{Z}{\overset{O}{R-C}}+Y: \underset{r/d}{\rightleftharpoons} \underset{Z}{\overset{O^-}{R-C-Y}} \longrightarrow \underset{Z}{\overset{O}{R-C-Y}}+Z:$$

$$(b)\quad \underset{Z}{\overset{O}{R-C}}+H^+ \rightleftharpoons \underset{Z}{\overset{\overset{+}{O}H}{R-C}} \longrightarrow \underset{Z}{\overset{OH}{R-\overset{+}{C}}} \xrightarrow[r/d]{Y:} \underset{Z}{\overset{OH}{R-C-Y}}$$

$$\longrightarrow \overset{O}{R-C-Y}+HZ$$

These reactions are typical of the behaviour of the carbonyl group (in aldehydes and ketones), and the greater the positive charge on the carbonyl atom, the more easily is this carbon atom attacked by a nucleophile (in the r/d step). Since the $+R$ effect tends to neutralise the positive charge, the greater the $+R$ effect of Z, the slower will be the reaction. Experimental work has shown the rates of hydrolysis are in the order $RCONH_2 < RCO_2R < RCOCl$, and this is the reverse order of the $+R$ effects of the groups, *i.e.*, $NH_2 > OR > Cl$.

Esters

Esters are compounds which are formed when the hydroxylic hydrogen atom in oxygen acids is replaced by an alkyl group; the acid may be organic or inorganic. The most important esters are derived from the carboxylic acids. The general formula of the carboxylic esters is $C_nH_{2n}O_2$, which is the same as that of the carboxylic acids, and they are named as the alkyl salts of the acid, *e.g.*,

$$CH_3COOC_2H_5 \qquad \text{ethyl acetate}$$
$$(CH_3)_2CHCOOCH(CH_3)_2 \qquad \text{isopropyl isobutyrate}$$

Carboxylic esters are formed by the action of the acid on an alcohol:

$$\text{acid} + \text{alcohol} \rightleftharpoons \text{ester} + \text{water}$$

The reaction is reversible, the *forward* reaction being known as *esterification*, and the *backward* reaction as *hydrolysis* (see p. 248 for mechanisms).

General methods of preparation of the carboxylic esters

1. The usual method is *esterification*. The reaction is always slow, but is speeded up by the presence of small amounts of inorganic acids as catalysts, *e.g.*, when the acid is refluxed with the alcohol in the presence of 5–10 per cent concentrated sulphuric acid:

$$R^1COOH + R^2OH \xrightarrow{\text{H}_2\text{SO}_4} R^1COOR^2 + H_2O \quad (v.g.)$$

Alternatively, hydrogen chloride is passed into the mixture of alcohol and acid until there is a 3 per cent increase in weight, and the mixture is refluxed (the yields are very good). This is known as the *Fischer–Speier method* (1895), and is more satisfactory for secondary and tertiary alcohols than the sulphuric acid method, which tends to dehydrate the alcohol to alkene.

Esterification without the use of catalysts, and starting with one mole of acid and one mole of alcohol, gives rise to about $\frac{2}{3}$ mole of ester. The yield of ester may be increased by using excess of acid or alcohol, the cheaper usually being the one in excess. Increased yields may also be effected by dehydrating agents, *e.g.*, concentrated sulphuric acid behaves both as a catalyst and a dehydrating agent. The same effect may be obtained by removing the water or ester from the reaction mixture by distillation, which is particularly useful for high-boiling acids and alcohols. On the other hand, the water may be removed from the reaction mixture by the addition of benzene or carbon tetrachloride, each of which forms a binary mixture with water (and may form a ternary mixture with water and the alcohol), the azeotropic mixtures boiling at a lower temperature than any of the components.

2. Acid chlorides or anhydrides react rapidly with alcohols to form esters:

$$R^1COCl + R^2OH \longrightarrow R^1CO_2R^2 + HCl \qquad (v.g.-ex.)$$
$$(R^1CO)_2O + R^2OH \longrightarrow R^1CO_2R^2 + R^1CO_2H \quad (v.g.-ex.)$$

The reaction with tertiary alcohols is very slow, and is often accompanied by side reactions (alkene or alkyl halide formation), but by using the appropriate conditions, esters are produced (see p. 182). Esters of tertiary alcohols may also be conveniently prepared by means of a Grignard reagent (see p. 432).

3. Esters may be prepared by refluxing the silver salt of an acid with an alkyl halide in ethanolic solution:

$$R^1CO_2Ag + R^2Br \longrightarrow R^1CO_2R^2 + AgBr \quad (v.g.-ex.)$$

This method is useful where direct esterification is difficult (*cf.* **2** above).

4. *Methyl* esters are very conveniently prepared by treating an acid with an ethereal solution of diazomethane (*q.v.*):

$$RCO_2H + CH_2N_2 \longrightarrow RCO_2CH_3 + N_2 \quad (ex.)$$

According to Morrison *et al.* (1961), the BF_3—MeOH reagent is better than diazomethane for the methylation of carboxylic acids.

5. The interaction between an ether and carbon monoxide at 125–180°C, under 500 atmospheres pressure, in the presence of boron trifluoride plus a little water, produces esters:

$$R_2O + CO \longrightarrow RCO_2R$$

Esters may also be prepared by the Tischenko reaction (with aldehydes; p. 221); ethyl acetate is made commercially this way.

General properties of the esters

The carboxylic esters are pleasant-smelling liquids or solids. The boiling points of the straight-chain isomers are higher than those of the branched-chain isomers. The boiling points of the methyl and ethyl esters are lower than those of the corresponding acid, and this is due to the fact that the esters are not associated since they cannot form intermolecular hydrogen bonds. The esters of low molecular weight are fairly soluble in water—hydrogen bonding between ester and water is possible—and the solubility decreases as the series is ascended; all esters are soluble in most organic solvents.

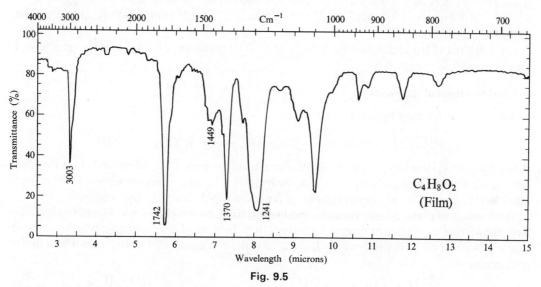

Fig. 9.5

1742 cm^{-1}: C=O (str.) in saturated aliphatic esters.   1241 cm^{-1}: C—O (str.) in *acetates*.   3003 cm^{-1}: C—H (str.) in Me and/or CH$_2$.   1449 cm^{-1}: C—H (def.) in Me and/or CH$_2$. 1370 cm^{-1}: C—H (def.) in Me.

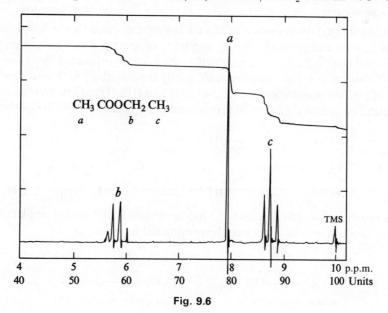

Fig. 9.6

The structure of the esters, like their parent acids, is best represented as a resonance hybrid:

$$R^1-C\overset{\displaystyle O}{\underset{\displaystyle OR^2}{\Big\backslash}} \longleftrightarrow R^1-C\overset{\displaystyle O^-}{\underset{\displaystyle \overset{+}{O}R^2}{\Big\backslash}}$$

The infra-red absorption region of the carbonyl group (stretch) in saturated aliphatic esters is $1750-1735 \ cm^{-1}$ (s). This is a higher frequency than that of the carbonyl group in ketones and acids, and so esters may usually be identified. Characteristic bands are also shown by the C—O (stretch) in formates, $1200-1180 \ cm^{-1}$ (s); acetates, $1250-1230 \ cm^{-1}$ (s); propionates, etc., $1200-1170 \ cm^{-1}$ (s).

The infra-red (film) and NMR (in $CDCl_3$ at 60 MHz) spectra of ethyl acetate are shown in Figs. 9.5 and 9.6, respectively.

General reactions of the esters

1. Esters are hydrolysed by acids or alkalis:

$$R^1CO_2H + R^2OH \xleftarrow{H_3O^+} R^1CO_2R^2 \xrightarrow{OH^-} R^1CO_2^- + R^2OH$$

When hydrolysis is carried out with alkali, the salt of the acid is obtained, and since the alkali salts of the higher acids are soaps, alkaline hydrolysis is known as *saponification* (derived from Latin word meaning soap); saponification is far more rapid than acid hydrolysis.

Mechanisms of carboxyl esterification and hydrolysis of carboxylic esters. Mechanistic studies have shown that in alkaline or neutral hydrolysis, it is the neutral ester molecule which undergoes reaction, whereas in acid-catalysed hydrolysis, it is the conjugate acid of the ester which undergoes reaction.

$$\underset{\textit{neutral ester}}{R^1-\overset{\displaystyle O}{\overset{\displaystyle \|}{C}}-OR^2} \qquad R^1-\overset{\displaystyle \overset{+}{O}H}{\overset{\displaystyle \|}{C}}-OR^2 \quad \text{rather than} \quad \underset{\textit{conjugate acid}}{R^1-\overset{\displaystyle O}{\overset{\displaystyle \|}{C}}-\overset{\displaystyle H}{\overset{\displaystyle |}{O^+}}-R^2}$$

The protonated carbonyl oxygen form of the cA is preferred, and this is based on the work of Fraenkel (1961), who examined the NMR spectrum of methyl formate in, *e.g.*, 100 per cent sulphuric acid and concluded from his results that the ester is protonated chiefly on the carbonyl oxygen. This formulation is also used for *acids*, partly from analogy with esters and partly from the fact that of the two possible conjugate acids, (I) and (II), (I) may be written as a resonance hybrid to which two identical canonical forms contribute, whereas for (II) the two

$$R-C\overset{\displaystyle \overset{+}{O}H}{\underset{\displaystyle OH}{\Big\backslash}} \longleftrightarrow R-C\overset{\displaystyle OH}{\underset{\displaystyle \overset{+}{O}H}{\Big\backslash}} \qquad R-C\overset{\displaystyle O}{\underset{\displaystyle \overset{+}{O}H_2}{\Big\backslash}} \qquad R-C\overset{\displaystyle \overset{-}{O}}{\underset{\displaystyle \overset{+}{O}H_2}{\Big\backslash}}$$
$$\qquad\qquad (I) \qquad\qquad\qquad\qquad\qquad (II)$$

contributing structures are different and one has fewer bonds and carries unlike charges.

The hydrolysis of carboxylic esters may be formulated in two ways:

$$\underset{\textit{acyl-oxygen heterolysis}}{R^1-\overset{\displaystyle O}{\overset{\displaystyle \|}{C}}\frown OR^2} \qquad\qquad \underset{\textit{alkyl-oxygen heterolysis}}{R^1-\overset{\displaystyle O}{\overset{\displaystyle \|}{C}}-O\frown R^2}$$

Both modes of fission have been demonstrated experimentally, and kinetic studies have shown that two types of mechanism may operate for each mode, *unimolecular* and *bimolecular*. Hence, *eight* mechanisms are possible for ester hydrolysis but, so far, *six* have been observed (Table 9.2). Also, since in principle, the *acid-catalysed* reactions (esterification and hydrolysis) are reversible, but not *alkaline* hydrolysis, then *four* mechanisms are possible for esterification. So far, *three* have been observed (Table 9.2). The eight possible mechanisms are indicated by the symbols shown in Table 9.2, where A represents the cA of the ester or acid (reaction in *acid* solution), B represents the unprotonated ester (reaction in *basic* or *neutral* solution), subscripts AC and AL, respectively, denote acyl- and alkyl-heterolysis, and numbers 1 and 2 represent the molecularity of the rate-determining step (Ingold, 1941).

Table 9.2

Type of mechanism	Hydrolysis	Esterification
$B_{AC}1$	—	—
$B_{AC}2$	Very common	—
$A_{AC}1$	Special cases	Special cases
$A_{AC}2$	Very common	Very common
$B_{AL}1$	Special cases	—
$B_{AL}2$	Rare	—
$A_{AL}1$	Very common for t-alcohols	Very common for t-alcohols
$A_{AL}2$	—	—

Most of the experimental work has been carried out on hydrolysis because this process can be studied in acidic, basic, or neutral media, whereas esterification can be studied only in acidic media. It might appear at first sight that esterification could also be studied in neutral media, but since carboxylic acids are acids, reaction in the absence of added acid (as catalyst) is still effectively in an acid medium. Because of this, mechanisms of esterification are partly based on the principle of microscopic reversibility (p. 49).

Bimolecular basic hydrolysis with acyl-oxygen heterolysis ($B_{AC}2$ mechanism). It has long been known that the alkaline hydrolysis of esters is a second-order reaction, *i.e.*, rate = k [ester] [OH$^-$]. Evidence for acyl-oxygen heterolysis has been obtained in several ways, *e.g.*, Polanyi *et al.* (1934) showed that the alkaline hydrolysis of n-pentyl acetate in water enriched with ^{18}O gave n-pentyl alcohol containing *no* ^{18}O. Therefore acyl-oxygen heterolysis must have occurred:

$$CH_3CO—OC_5H_{11} + \overset{*}{O}H^- \longrightarrow CH_3CO\overset{*}{O}H + C_5H_{11}O^- \longrightarrow CH_3CO\overset{*}{O}^- + C_5H_{11}OH$$

Another method uses optical activity as a diagnostic test for demonstrating acyl-oxygen heterolysis. If the alcohol group is optically active, and if the $B_{AC}2$ mechanism operates, then the ion RO$^-$ will be liberated and will retain its optical activity since the bond R—O is never broken. Various examples of retention are known when the reaction is bimolecular.

A mechanism consistent with the above facts (and with the work of Bender (1951), who used esters labelled with ^{18}O in the carbonyl group) is:

Any R^1 group with a $-I$ effect should accelerate hydrolysis by creating a positive charge on the attached carbon atom, thereby facilitating attack by the hydroxide ion. On the other hand, because the r/d step is bimolecular, the larger R^1 is the larger can be expected to be the steric retardation, since the hybridisation of the carbon atom of the carbonyl group changes from sp^2 to sp^3 (with consequent increased crowding). These anticipated results have been observed in practice, e.g.,

$-I$ effect			
Ester	$MeCO_2Me$	$ClCH_2CO_2Me$	$CHCl_2CO_2Me$
Relative rates	1	761	16 000

Steric effect			
Ester	$MeCO_2Et$	$EtCO_2Et$	$isoPrCO_2Et$
Relative rates	1·00	0·47	0·10

Bimolecular acid-catalysed hydrolysis and esterification with acyl-oxygen heterolysis ($A_{AC}2$ mechanism). Acid-catalysed hydrolysis and acid-catalysed esterification are the reverse of each other, and so, according to the principle of microscopic reversibility the mechanism of acid-catalysed hydrolysis will be that of acid-catalysed esterification, but in reverse.

The rate law for acid-catalysed hydrolysis has been shown to be second-order, first-order in both ester and hydrogen ion concentration, i.e., rate = k [ester] [H_3O^+]. Also, evidence for acyl-oxygen heterolysis has been obtained in several ways, e.g., Ingold et al. (1939) showed that the acid-catalysed hydrolysis of methyl hydrogen succinate in water enriched with ^{18}O gave methanol with no extra ^{18}O. Thus, acyl-oxygen heterolysis must have occurred:

$$HO_2CCH_2CH_2CO\frown OMe + H_2\overset{*}{O} \xrightarrow{H^+} HO_2CCH_2CH_2CO\overset{*}{O}H + MeOH$$

Furthermore, Roberts et al. (1938) esterified benzoic acid with methanol enriched with ^{18}O and obtained water not enriched with ^{18}O. Thus, acyl-oxygen heterolysis must have occurred:

$$PhCO\frown OH + Me\overset{*}{O}H \longrightarrow PhCO\overset{*}{O}Me + H_2O$$

A mechanism that fits these facts is:

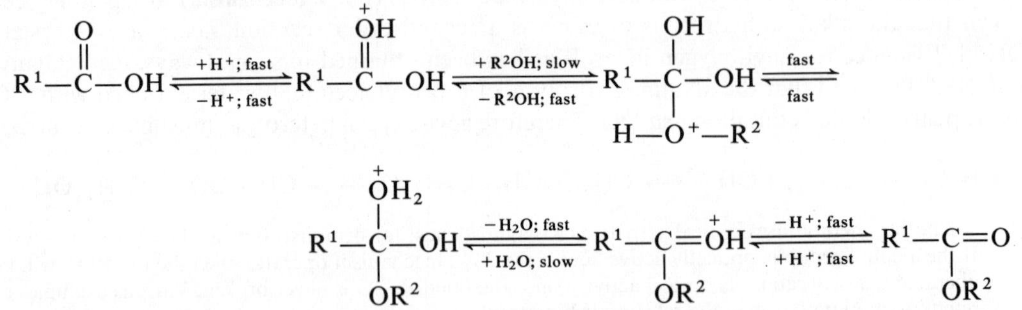

It should be noted that the r/d step for esterification is the addition of alcohol, whereas that for hydrolysis is the addition of water. In both cases, the r/d step involves change of hybridisation of the carbon atom of the carbonyl group from sp^2 to sp^3, and consequently steric retardation can be expected to increase as R^1 increases in size (cf. the $B_{AC}2$ mechanism), e.g. (esterification with methanol):

Acid	$MeCO_2H$	$n\text{-}PrCO_2H$	Me_3CCO_2H	Et_3CCO_2H
Relative rates	1·0	0·51	0·037	0·00016

Unimolecular acid-catalysed hydrolysis and esterification with acyl-oxygen heterolysis ($A_{AC}1$ mechanism). Let us consider the following *observed* results (M represents the mesityl group, 1,3,5-trimethylphenyl, in *mesitoic acid*, $Me_3C_6H_2CO_2H$):

$$PhCO_2H + MeOH \xrightarrow{HCl} PhCO_2Me \qquad \textit{occurs readily}$$

$$PhCO_2Me + OH^- \longrightarrow PhCO_2^- + MeOH \quad \textit{occurs readily}$$

$$MCO_2H + MeOH \xrightarrow{HCl} \textit{No reaction}$$

$$MCO_2Me + OH^- \longrightarrow MCO_2^- + MeOH \quad \textit{very slow}$$

Hammett *et al.* (1937) showed that when methyl benzoate is dissolved in concentrated sulphuric acid and the solution then poured into ice/water, almost all of the ester is recovered. On the other hand, methyl mesitoate, on the same treatment, gave a quantitative yield of mesitoic acid. Then Newman (1941) showed that when a solution of mesitoic acid in concentrated sulphuric acid is poured into cold methanol, methyl mesitoate is formed. This different behaviour of mesitoyl derivatives from benzoyl derivatives suggests that different mechanisms are operating in esterification and hydrolysis for these two sets of compounds. Now, Hammett (1937) had shown that when mesitoic acid is dissolved in concentrated sulphuric acid, the van't Hoff factor i is 4 (measured by depression of freezing point). Under the same conditions, benzoic acid gives $i = 2$. Hammett therefore proposed that the mesitoyl reactions proceed through an intermediate *acylium* (*oxocarbonium*) ion (*mesitoylium* ion in this case) formed by acyl-oxygen heterolysis:

(i) $MCO_2H + H_2SO_4 \underset{}{\overset{fast}{\rightleftharpoons}} M\overset{+}{C}\begin{smallmatrix}OH\\\\OH\end{smallmatrix} + HSO_4^-$

(ii) $M\overset{+}{C}\begin{smallmatrix}OH\\\\OH\end{smallmatrix} + H_2SO_4 \overset{slow}{\rightleftharpoons} M\overset{+}{C}{=}O + H_3O^+ + HSO_4^-$

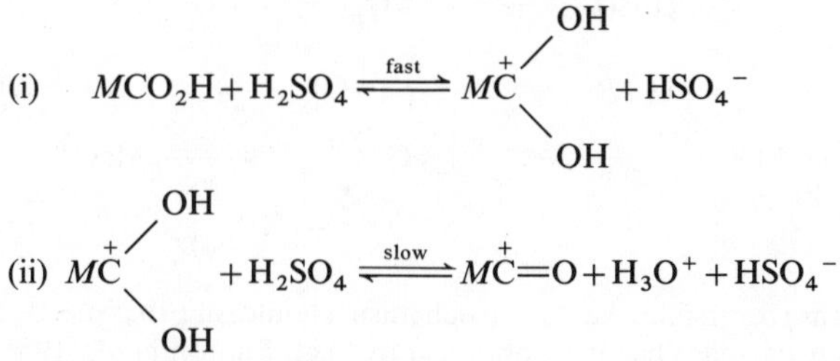

The overall equation may be written:

$$MCO_2H + 2H_2SO_4 \rightleftharpoons M\overset{+}{C}{=}O + H_3O^+ + 2HSO_4^-$$

This accounts for $i = 4$ for mesitoic acid, and for $i = 2$ for benzoic acid if only reaction (i) is involved. Thus, the reactions of mesitoic acid and its ester may be formulated as:

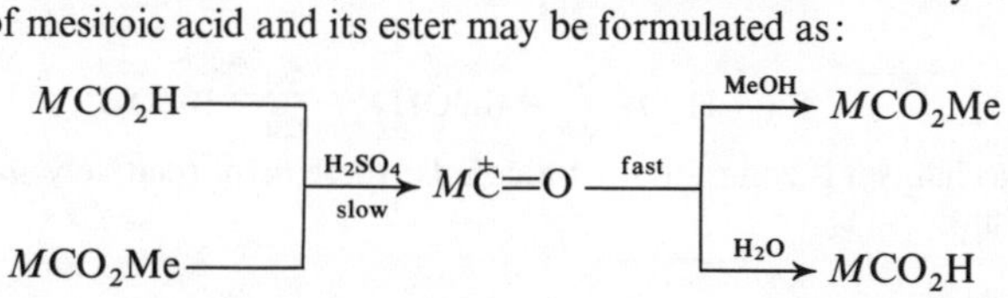

This type of reaction occurs with highly sterically hindered acids and esters, and so steric hindrance must be an important factor.

Further support for this mechanism has been provided by Olah *et al.* (1967), who showed that acids (and esters), when dissolved in FSO_3H—SbF_5—SO_2 (a very strongly acidic medium), at 60°C, lose a molecule of water at 0°C to form the acylium ion:

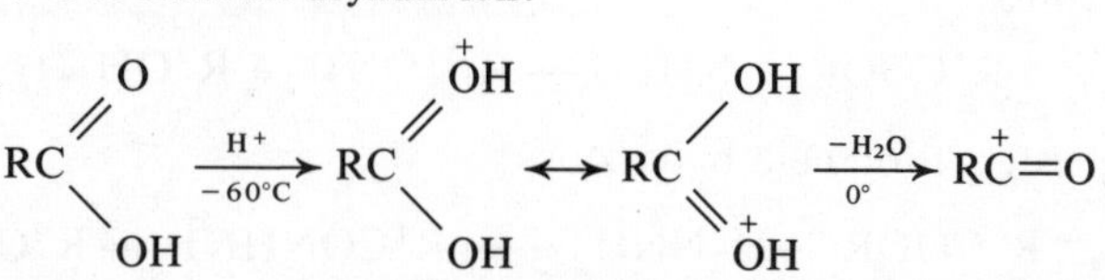

Unimolecular basic hydrolysis with alkyl-oxygen heterolysis ($B_{AL}1$ mechanism). This mechanism has been shown to occur when the alkyl group of the alcohol is capable of forming a relatively stable carbonium ion, the solvent has high ionising power (thereby encouraging alkyl-oxygen heterolysis), and the hydroxide ion concentration is low (thereby suppressing the bimolecular mechanism). An example of the $B_{AL}1$ mechanism is the hydrolysis of optically active α-phenylethyl hydrogen phthalate in *faintly* alkaline solution. Kenyon *et al.* (1936) showed that the product was racemised α-phenylethanol (phenylmethylmethanol); this is in keeping with the formation of an intermediate carbonium ion (sp^2 hybridised) [R = $o\text{-}HO_2CC_6H_4$—]:

$$\text{RCO—O—CHMePh} \underset{}{\overset{slow}{\rightleftharpoons}} \text{RCO}_2{}^- + \text{PhMeCH}^+ \xrightarrow[fast]{H_2O} \text{PhMeCHOH}_2{}^+ \xrightarrow[fast]{RCO_2{}^-}$$
$$(\pm)\text{-PhMeCHOH} + \text{RCO}_2\text{H}$$

Bimolecular basic hydrolysis with alkyl-oxygen heterolysis ($B_{AL}2$ mechanism). This mechanism is rare. One example is the reaction between methyl benzoate and sodium methoxide in methanol (Bunnett *et al.*, 1950). The products were dimethyl ether and sodium benzoate; these results are readily explained on the basis of the $B_{AL}2$ mechanism:

$$\text{MeO}^- \quad \text{Me—O—OCPh} \longrightarrow \text{Me}_2\text{O} + \text{PhCO}_2{}^-$$

Unimolecular acid-catalysed hydrolysis and esterification with alkyl-oxygen heterolysis ($A_{AL}1$ mechanism). Definite evidence for this mechanism was obtained by Hughes, Ingold *et al.* (1939), who showed that the esterification of acetic acid with optically active octan-2-ol in the presence of sulphuric acid gave a large amount of racemised ester. Racemisation is to be expected if an intermediate carbonium ion is formed, and so the reaction may be formulated as $A_{AL}1$ (R = $n\text{-}C_6H_{11}\text{CHMe}$—):

$$\text{ROH} + \text{H}^+ \underset{fast}{\overset{fast}{\rightleftharpoons}} \text{ROH}_2{}^+ \underset{fast}{\overset{slow}{\rightleftharpoons}} \text{H}_2\text{O} + \text{R}^+$$

$$\underset{\text{OH}}{\overset{\ddot{\text{O}} \quad \text{R}^+}{\text{MeC}}} \underset{slow}{\overset{fast}{\rightleftharpoons}} \underset{\text{O—H}}{\overset{\overset{+}{\text{OR}}}{\text{MeC}}} \underset{fast}{\overset{fast}{\rightleftharpoons}} \underset{\text{O}}{\overset{\text{OR}}{\text{MeC}}} + \text{H}^+$$

Since all steps are reversible, the $A_{AL}1$ mechanism should, in principle, be possible for ester hydrolysis. Evidence for its occurrence has been obtained by, *e.g.*, Bunton *et al.* (1951), who carried out the acid-catalysed hydrolysis of t-butyl acetate in water enriched with ^{18}O and obtained t-butanol containing ^{18}O.

$$\underset{}{\overset{\text{O}}{\text{MeC—O—Bu}^t}} + \text{H}^+ \underset{fast}{\overset{fast}{\rightleftharpoons}} \text{Me—C—O—Bu}^t \underset{fast}{\overset{slow}{\rightleftharpoons}} \underset{\text{O}}{\overset{\text{OH}}{\text{MeC}}} + \text{Bu}^t$$

$$\overset{+}{\text{Bu}^t} + \text{H}_2\overset{*}{\text{O}} \underset{slow}{\overset{fast}{\rightleftharpoons}} \text{Bu}^t\overset{*}{\text{OH}}_2{}^+ \underset{fast}{\overset{fast}{\rightleftharpoons}} \text{Bu}^t\overset{*}{\text{OH}} + \text{H}^+$$

The $A_{AL}1$ mechanism is common for t-alcohols; these form relatively stable carbonium ions (*cf.* the $B_{AL}1$ mechanism).

2. Esters are converted into alcohols by the Bouveault-Blanc reduction (p. 177), catalytic hydrogenation, and by metallic hydrides (see Table 6.1, p. 179).

3. Esters react with ammonia to form amides. This reaction is an example of *ammonolysis* (which means, literally, splitting by ammonia; see also p. 261):

$$\text{R}^1\text{COOR}^2 + \text{NH}_3 \longrightarrow \text{R}^1\text{CONH}_2 + \text{R}^2\text{OH} \quad (g.)$$

With hydrazine, esters form acid hydrazides:

$$\text{R}^1\text{COOR}^2 + \text{H}_2\text{NNH}_2 \longrightarrow \text{R}^1\text{CONHNH}_2 + \text{R}^2\text{OH}$$

Esters react with phosphorus pentachloride or thionyl chloride to form acyl chlorides, *e.g.*,

$$\text{R}^1\text{CO}_2\text{R}^2 + \text{PCl}_5 \longrightarrow \text{R}^1\text{COCl} + \text{POCl}_3 + \text{R}^2\text{Cl}$$

4. By means of *alcoholysis* (splitting by alcohol), an alcohol residue in an ester can be replaced by another alcohol residue. Alcoholysis is carried out by refluxing the ester with a large excess of alcohol, preferably in the presence of a small amount of acid or sodium alkoxide as catalyst. Alcoholysis is usually effective in replacing a higher alcohol by a lower one, *e.g.*,

$$CH_3CO_2C_4H_9 + C_2H_5OH \underset{}{\overset{C_2H_5ONa}{\rightleftharpoons}} CH_3CO_2C_2H_5 + C_4H_9OH$$

Alcoholysis of esters is also known as *transesterification*.

The mechanism of alcoholysis is uncertain; a possibility is (in basic media):

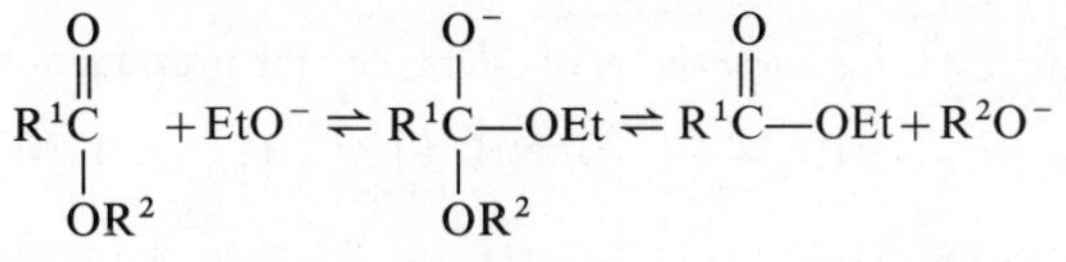

$$\underset{\underset{OR^2}{|}}{\overset{\overset{O}{\|}}{R^1C}} + EtO^- \rightleftharpoons \underset{\underset{OR^2}{|}}{\overset{\overset{O^-}{|}}{R^1C}}-OEt \rightleftharpoons \overset{\overset{O}{\|}}{R^1C}-OEt + R^2O^-$$

$$R^2O^- + EtOH \rightleftharpoons R^2OH + EtO^-$$

In *acidolysis*, the acid residue is displaced from its ester by another acid residue, *e.g.*,

$$CH_3CO_2C_2H_5 + C_5H_{11}CO_2H \rightleftharpoons C_5H_{11}CO_2C_2H_5 + CH_3CO_2H$$

Acidolysis is a useful reaction for converting the neutral ester of a dibasic acid into its acid ester.

According to Parker *et al.* (1968), who used ^{18}O-labelled toluic acids and their esters, the mechanism of acidiolysis is:

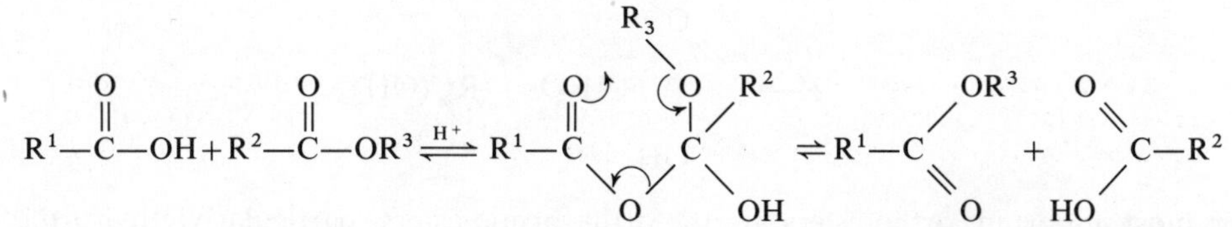

5. When an ester—preferably the methyl or ethyl ester— is treated with sodium in an inert solvent, *e.g.*, ether, benzene or toluene, and subsequently with acid, an **acyloin** is formed (50–70 per cent yield). It is important that the reaction be carried out *in the absence of any free alcohol*. Acyloins are α,β-keto-alcohols, and according to Kharasch and his co-workers (1939), their formation takes place via a free-radical mechanism, *e.g.*, propionin from ethyl propionate:

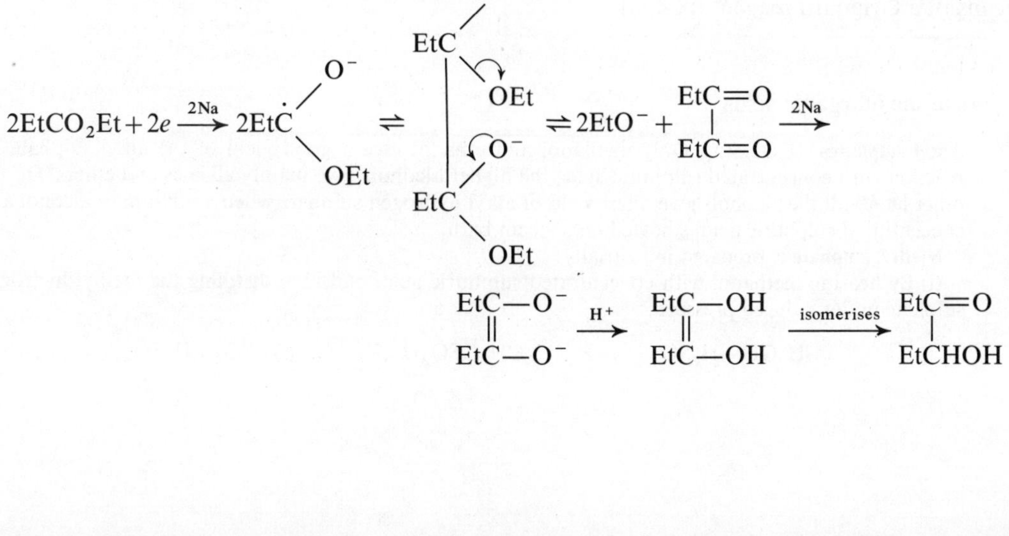

Acyloins are reduced by the Clemmensen method to alkanes.

6. Carboxylic esters which contain α-hydrogen atoms react with sodamide in liquid ammonia solution to form the acid amide and a condensation product involving two molecules of the ester, *e.g.*, ethyl acetate gives acetamide and acetoacetic ester (see also p. 285):

$$CH_3CO_2C_2H_5 + NaNH_2 \longrightarrow CH_3CONH_2 + C_2H_5ONa$$

$$CH_3CO_2C_2H_5 + NaNH_2 \longrightarrow NH_3 + [\overset{..}{C}H_2CO_2C_2H_5]^-Na^+ \xrightarrow{CH_3CO_2C_2H_5}$$
$$CH_3COCH_2CO_2C_2H_5 + C_2H_5OH$$

Of particular interest is the *carbonation* of esters, *i.e.*, the introduction of the carboxyl group.

$$R^1CH_2CO_2R^2 + NaNH_2 \xrightarrow{\text{liquid } NH_3} NH_3 + [R^1\overset{..}{C}HCO_2R^2]^-Na^+ \xrightarrow[\text{ether}]{CO_2} R^1\underset{\underset{CO_2Na}{|}}{C}HCO_2R^2$$

The yields of these *malonic acid derivatives* are 54–60 per cent with acetates, and becomes progressively lower as the molecular weight of the acid increases.

7. The prolysis of esters, particularly acetates, gives alkenes (see p. 95).

Esters are used as solvents for cellulose, oils, gums, resins, etc., and as plasticisers. They are also used for making artificial flavours and essences, *e.g.*, isoamyl acetate—banana oil; amyl butyrate—apricot; isoamyl isovalerate—apple; methyl butyrate—pineapple; etc.

Ortho-esters are compounds of the type $R^1C(OR^2)_3$. They are derived from the ortho-acids, $RC(OH)_3$, which have not yet been isolated, but which are possibly present in aqueous solution:

$$R-C\overset{\displaystyle O}{\underset{\displaystyle OH}{\diagup\!\!\!\!\diagdown}} \quad + H_2O \quad RC(OH)_3$$

The most important ortho-esters are the orthoformic esters, particularly ethyl orthoformate. This may be prepared by running ethanol and chloroform into sodium covered with ether:

$$2CHCl_3 + 6C_2H_5OH + 6Na \longrightarrow 2HC(OC_2H_5)_3 + 6NaCl + 3H_2 \quad (70\%)$$

A possible mechanism is via the formation of dichloromethylene (p. 170):

$$CHCl_3 \xrightarrow{EtO^-} :CCl_2 \xrightarrow{EtOH} EtOCHCl_2 \xrightarrow{2EtO^-} HC(OEt)_3 + 2Cl^-$$

Ethyl orthoformate may be used for preparing ketals (p. 224), and for preparing aldehydes by means of a Grignard reagent (p. 430).

Esters of the inorganic acids

Alkyl sulphates, $(RO)_2SO_2$. Only methanol and ethanol give a good yield of the alkyl sulphate by reaction with concentrated sulphuric acid; the higher alcohols give mainly alkenes and ethers. On the other hand, all the alcohols give a fair yield of alkyl hydrogen sulphate when a mixture of alcohol and concentrated sulphuric acid is heated on a steam bath.

Methyl sulphate is prepared industrially:

(i) By heating methanol with concentrated sulphuric acid, and then distilling the methyl hydrogen sulphate under reduced pressure:

$$CH_3OH + H_2SO_4 \longrightarrow H_2O + CH_3OSO_3H \xrightarrow{2} (CH_3O)_2SO_2 + H_2SO_4$$

(ii) By treating methanol with sulphur trioxide at low temperatures:

$$2SO_3 + 2CH_3OH \longrightarrow (CH_3)_2SO_4 + H_2SO_4$$

Ethyl sulphate may be prepared by the same methods as methyl sulphate, but in addition, there is the industrial preparation by passing ethylene in excess into cold concentrated sulphuric acid:

$$2C_2H_4 + H_2SO_4 \longrightarrow (C_2H_5)_2SO_4$$

Sulphates of the higher alcohols are conveniently prepared by the oxidation of sulphites as follows:

$$2ROH + SOCl_2 \longrightarrow 2HCl + (RO)_2SO \xrightarrow{KMnO_4} (RO)_2SO_2$$

Methyl sulphate, b.p. 188°C, and ethyl sulphate, b.p. 208°C, are heavy poisonous liquids. They are largely used as alkylating agents since the alkyl group will replace the hydrogen atom of the groups —OH, —NH— or —SH. The alkylation may be carried out by treating the compound with the alkyl sulphate in sodium hydroxide solution. Usually only one of the alkyl groups takes part in the reaction, *e.g.*, methylation of a primary amine:

$$RNH_2 + (CH_3)_2SO_4 + NaOH \longrightarrow RNHCH_3 + CH_3NaSO_4 + H_2O$$

The methyl and ethyl esters of the carboxylic acids may be conveniently prepared by treating the sodium salt of the acid with respectively methyl or ethyl sulphate:

$$RCO_2Na + (C_2H_5)_2SO_4 \longrightarrow RCO_2C_2H_5 + C_2H_5NaSO_4$$

The sodium alkyl sulphates of the higher alcohols are used as detergents, *e.g.*, sodium lauryl sulphate:
Alkyl nitrates, $RONO_2$. The only important alkyl nitrate is ethyl nitrate, $C_2H_5ONO_2$. This may be prepared by heating ethyl iodide with silver nitrate in ethanolic solution:

$$C_2H_5I + AgNO_3 \longrightarrow C_2H_5ONO_2 + AgI$$

When concentrated nitric acid is added to ethanol, the reaction is usually violent; part of the ethanol is oxidised, and some of the nitric acid is reduced to nitrous acid. Apparently it is the presence of the nitrous acid which produces the violent oxidation of the alcohol by the nitric acid. This danger may be avoided by first boiling the nitric acid with urea, which destroys any nitrous acid present, and then adding this mixture to cool ethanol, any nitrous acid produced being immediately destroyed by the urea:

$$CO(NH_2)_2 + 2HNO_2 \longrightarrow CO_2 + 2N_2 + 3H_2O$$

Ethyl nitrate is a pleasant-smelling liquid, b.p. 87·5°C. When reduced with tin and hydrochloric acid, it forms hydroxylamine and ethanol:

$$C_2H_5ONO_2 \xrightarrow{Sn/HCl} C_2H_5OH + NH_2OH + H_2O$$

Alkyl nitrites, RONO. Alkyl nitrites are isomeric with the nitroalkanes (p. 361). The only important alkyl nitrites are the ethyl and amyl nitrites; the latter is actually mainly isoamyl nitrite, since the amyl alcohol used is the isopentanol from fusel oil (p. 190). Ethyl and amyl nitrites are prepared by adding concentrated hydrochloric acid or sulphuric acid to aqueous sodium nitrite and the alcohol (see also p. 224):

$$R-\overset{..}{\underset{..}{O}}-H + NO^+ \longrightarrow R-\overset{+}{O}\overset{\diagup H}{\diagdown NO} \xrightarrow{-H^+} RONO$$

Ethyl nitrite, b.p. 17°C, and amyl nitrite, b.p. 99°C, are pleasant-smelling liquids, and are used as a means of preparing nitrous acid in anhydrous media; *e.g.*, an ethanolic solution of nitrous acid may be prepared by passing dry hydrogen chloride into amyl nitrite dissolved in ethanol.

Acid or acyl chlorides

The general formula of the *acyl* groups is $R—\overset{\overset{\displaystyle O}{\|}}{C}—$. Acid chlorides, which may be prepared by the replacement of the hydroxyl in the carboxyl group by chlorine, are also known as *acyl chlorides* because they contain the acyl group.

Nomenclature

According to the I.U.P.A.C. system the class suffix of the acyl chlorides is *-oyl*; the common names are formed by changing the suffix *-ic* of the *trivial* name of the acid into *-yl*, *e.g.*,

CH_3COCl $(CH_3)_2CHCOCl$

ethanoyl chloride   2-methylpropanoyl chloride

acetyl chloride   isobutyryl chloride

If the carboxyl group is considered as a substituent, then according to the I.U.P.A.C. system, the nomenclature of all substances containing acyl groups is in all cases based on the name 'carbonyl' for the CO group, *e.g.*,

$CH_3CH_2CH_2COCl$ propane-1-carbonyl chloride

When naming amides (see below), the 'yl' is elided before the suffix amide, *e.g.*,

$CH_3CH_2CH_2CONH_2$ propane-1-carbonamide

General methods of preparation

1. The acid is heated with phosphorus trichloride or pentachloride, *e.g.*,

$$3RCOOH + PCl_3 \longrightarrow 3RCOCl + H_3PO_3 \quad (g.)$$

$$RCOOH + PCl_5 \longrightarrow RCOCl + HCl + POCl_3 \quad (v.g.)$$

The reaction with phosphorus trichloride is accompanied by the formation of small amounts of volatile phosphorus compounds, *e.g.*,

$$RCO_2H + PCl_3 \longrightarrow RCO_2PCl_2 + HCl$$

Thionyl chloride may be used instead of the phosphorus chlorides:

$$RCO_2H + SOCl_2 \longrightarrow RCOCl + SO_2 + HCl \quad (v.g.)$$

The inorganic chloride is chosen according to the boiling point of the acyl chloride formed. Phosphorous acid decomposes at 200°C; the boiling point of phosphoryl chloride is 107°C, and that of thionyl chloride is 76°C. Since acetyl chloride boils at 52°C, any of the three inorganic halides may be used, but it is difficult to separate acetyl chloride from thionyl chloride (which is generally used in excess) by fractionation. n-Butyryl chloride boils at 102°C, and so phosphorus pentachloride cannot be used. Usually thionyl chloride is the most convenient.

The mechanism of the formation of acid chlorides is uncertain; it may be similar to that when an alcohol is the substrate (p. 148).

Acyl halides undergo the following exchange reaction with acids (probably via the acid anhydride):

$$R^1COCl + R^2CO_2H \rightleftharpoons [R^1COOOCR^2 + HCl] \rightleftharpoons R^1CO_2H + R^2COCl$$

This therefore offers a means of preparing acyl halides, but the success depends on displacement of the equilibrium to the right. This is readily achieved by using oxalyl chloride (or bromide) as the acyl halide.

The half-acid chloride of oxalic acid produced is decomposed to carbon monoxide and carbon dioxide under the conditions of the experiment (p. 448).

2. By distilling the salt of the acid with either phosphorus trichloride, phosphoryl chloride or sulphuryl chloride, *e.g.*,

$$3CH_3CO_2Na + PCl_3 \longrightarrow 3CH_3COCl + Na_3PO_3$$
$$2CH_3CO_2Na + POCl_3 \longrightarrow 2CH_3COCl + NaCl + NaPO_3$$
$$(CH_3CO_2)_2Ca + SO_2Cl_2 \longrightarrow 2CH_3COCl + CaSO_4$$

This method is used industrially since the salts are cheaper than the acid.

Acyl bromides may be prepared by the action of phosphorus tribromide or pentabromide (red phosphorus and bromine) on the acid, or by the action of excess of hydrogen bromide on the acyl chloride:

$$RCOCl + HBr \longrightarrow RCOBr + HCl$$

Acyl iodides may be prepared by the action of phosphorus tri-iodide on acid anhydrides. *Acyl fluorides* are also known, and may be prepared by the action of hydrogen fluoride on acid anhydrides:

$$(RCO)_2O + HF \longrightarrow RCOF + RCO_2H$$

General properties and reactions of the acyl chlorides. The lower acyl chlorides are colourless liquids with irritating odours; the higher members are colourless solids. The chlorine atom is very reactive and so the acid chlorides are important reagents.

The infra-red absorption region of the carbonyl group (stretch) in acyl chlorides is 1815–1785 cm^{-1} (s).

1. The acyl chlorides are readily hydrolysed by water, the lower members reacting vigorously:

$$RCOCl + H_2O \longrightarrow RCO_2H + HCl$$

Acyl chlorides usually react rapidly with compounds containing 'active' hydrogen atoms, *i.e.* hydrogen attached to oxygen, nitrogen or sulphur; *e.g.*, esters are formed with alcohols:

$$R^1COCl + R^2OH \longrightarrow R^1COOR^2 + HCl$$

Amides are formed with ammonia, and *N*-substituted amides with primary and secondary amines:

$$RCOCl + 2NH_3 \longrightarrow RCONH_2 + NH_4Cl$$
$$R^1COCl + R^2NH_2 \longrightarrow R^1CONHR^2 + HCl$$

Hydrazides are formed with hydrazine, and hydroxamic acids with hydroxylamine:

$$RCOCl + H_2NNH_2 \longrightarrow RCONHNH_2 + HCl$$
$$RCOCl + NH_2OH \longrightarrow RCONHOH + HCl$$

The mechanisms of all of these reactions are uncertain; a possibility is:

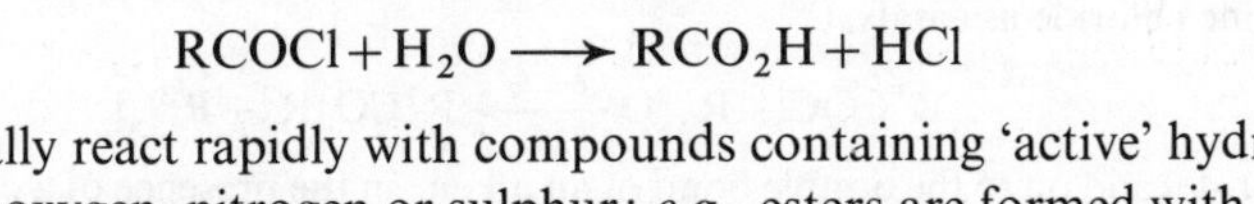

$$
\underset{R^1-\overset{\overset{\displaystyle O}{\|}}{C}-Cl + R^2OH}{}
\underset{slow}{\rightleftharpoons}
\underset{R^1-\overset{\overset{\displaystyle O^-}{|}}{\underset{\underset{\displaystyle H-O^+-R^2}{|}}{C}}-Cl}{}
\xrightarrow{fast}
\underset{R^1-\overset{\overset{\displaystyle O}{\|}}{\underset{\underset{\displaystyle OR^2}{|}}{C}} + HCl}{}
$$

2. Acyl chlorides may be reduced catalytically to aldehydes or to alcohols:

$$RCOCl \xrightarrow[Pd]{H_2} RCHO \xrightarrow{H_2} RCH_2OH$$

Acid chlorides are reduced to esters of an enediol by sodium amalgam in an inert solvent, *e.g.*, ether (see also p. 791):

$$4RCOCl + 4Na \longrightarrow \begin{array}{c} R \\ \diagdown \\ RCOO \end{array} C = C \begin{array}{c} R \\ \diagup \\ OOCR \end{array} + 4NaCl$$

Under suitable conditions, small yields of α-diketones have been obtained. Acid chlorides are reduced to alcohols by lithium aluminium hydride or sodium borohydride, and to aldehydes by lithium hydride or lithium tri-t-butoxyaluminium hydride (see Table 6.1, p. 179).

3. Acyl chlorides react with salts of the acids to form *acid anhydrides* (*q.v.*):

$$R^1COCl + R^2COONa \longrightarrow R^1CO \cdot O \cdot OCR^2 + NaCl$$

4. Acyl chlorides may be used in the Friedel–Crafts reaction to produce an aromatic ketone; *e.g.*, acetyl chloride reacts with benzene in the presence of anhydrous aluminium chloride to form acetophenone:

$$C_6H_6 + CH_3COCl \xrightarrow{AlCl_3} C_6H_5COCH_3 + HCl$$

5. Acyl chlorides react with Grignard reagents to produce ketones or tertiary alcohols, according to the conditions (see p. 431); *e.g.*, acetyl chloride forms butanone with ethylmagnesium iodide:

$$CH_3COCl + C_2H_5MgI \longrightarrow CH_3COC_2H_5 + MgClI$$

6. Acyl chlorides are readily halogenated in the α-position (see p. 269).

7. Acyl chlorides form esters (and other products) when heated with an ether in the presence of anhydrous zinc chloride as catalyst:

$$R^1COCl + R_2^2O \xrightarrow{ZnCl_2} R^1CO_2R^2 + R^2Cl$$

8. Acyl chlorides add on to the double bond of an alkene in the presence of a catalyst, *e.g.*, zinc chloride or aluminium chloride, to form a chloroketone which, on heating, eliminates a molecule of hydrogen chloride to form an unsaturated ketone:

$$(CH_3)_2C{=}CH_2 + CH_3COCl \xrightarrow{ZnCl_2} (CH_3)_2CClCH_2COCH_3 \xrightarrow{heat} (CH_3)_2C{=}CHCOCH_3 + HCl$$

Formyl fluoride, HCOF, has been isolated by Olah *et al.* (1961). Formyl chloride, HCOCl, has been obtained in solution by the action of hydrogen chloride on a solution of 1-formylimidazole (p. 847) in chloroform at $-10°C$ (Staab *et al.*, 1964). A mixture of carbon monoxide and hydrogen chloride behaves as if it were formyl chloride in the Gattermann–Koch aldehyde synthesis (p. 735).

Acetyl chloride is the most important acyl halide. It is a colourless fuming liquid, b.p. 52°C, and is soluble in ether and chloroform. It is largely used as an *acetylating* agent, *i.e.*, as a means of introducing the acetyl group, into compounds containing 'active' hydrogen atoms. It is also used to detect the presence of hydroxyl groups in organic compounds, and to estimate their number.

The accepted abbreviation for the acetyl group is Ac; *e.g.*, AcOH is acetic acid; Ac_2O, acetic anhydride; AcCl, acetyl chloride; C_6H_5NHAc, acetanilide; etc.

Acid anhydrides

The acid anhydrides may be regarded theoretically as being derived from an acid by the removal of one molecule of water from two molecules of the acid:

$$2RCO_2H \xrightarrow{-H_2O} \overset{O \quad O}{\underset{}{\overset{\parallel \quad \parallel}{RCOCR}}}$$

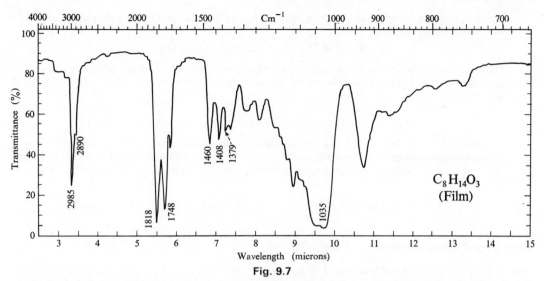

Fig. 9.7

1818 and 1748 cm^{-1}: C=O (str.) in acyclic anhydrides. 1035 cm^{-1}: C—O (str.) in —C—O—C— (acyclic anhydrides). 2985 and 2890 cm^{-1}: C—H (str.) in Me and/or CH$_2$. 1460 and 1408 cm^{-1}: C—H (def.) in Me and/or CH$_2$. 1379 cm^{-1}: C—H (def.) in Me.

It is possible in practice to prepare many acid anhydrides by direct dehydration, but this method is usually confined to the anhydrides of the higher members of the acid series (see below).

The infra-red absorption regions of the carbonyl group (stretch) in acyclic anhydrides are 1840–1800 (s) and 1780–1740 cm^{-1} (s). There is also a band in the region 1180–1030 cm^{-1} (s) due to the C—O (stretch) in the grouping —C—O—C—.

Figure 9.7 is the infra-red spectrum of n-butyric anhydride (as a film).

Nomenclature

The acid anhydrides are named as the anhydride of the acid groups present, the *trivial* name of the acid being used, *e.g.*,

$$(CH_3CO)_2O \qquad \text{acetic anhydride}$$
$$(CH_3CH_2CO)_2O \quad \text{propionic anhydride}$$

Acetic anhydride may be conveniently prepared by distilling a mixture of anhydrous sodium acetate and acetyl chloride:

$$CH_3\overset{O}{\underset{Cl}{C}} + CH_3\overset{O}{C}{-}O^- \rightleftharpoons CH_3\overset{\overset{-}{O}\ \ O}{\underset{Cl}{C}}OCCH_3 \longrightarrow CH_3\overset{O\ \ O}{C}OCCH_3 + Cl^-$$

Acetic anhydride is prepared industrially in a number of ways, *e.g.*,

(i) By passing acetylene into glacial acetic acid in the presence of mercuric ions as catalyst, and distilling the resulting ethylidene diacetate:

$$C_2H_2 + 2CH_3CO_2H \xrightarrow{Hg^{2+}} CH_3CH(OCOCH_3)_2 \xrightarrow{heat} (CH_3CO)_2O + CH_3CHO$$

(ii) By passing keten into glacial acetic acid:

$$CH_2{=}C{=}O + CH_3CO_2H \longrightarrow (CH_3CO)_2O$$

Anhydrides of the higher acids may be prepared by heating, and then fractionating, a mixture of the acid and acetic anhydride:

$$2RCO_2H + (CH_3CO)_2O \rightleftharpoons (RCO)_2O + 2CH_3CO_2H$$

This method is only satisfactory for anhydrides which have higher boiling points than acetic acid. A much better method of preparation is to treat the acid chloride with pyridine and benzene, add the acid, and then heat:

$$RCOCl + RCO_2H + C_5H_5N \longrightarrow (RCO)_2O + C_5H_5\overset{+}{N}HCl^- \quad (g.\text{-}ex.)$$

Suschitzky *et al.* (1964) have shown that acid anhydrides which are unaffected by cold water may be readily prepared in high yield and purity by shaking an aqueous solution of a metal salt of the carboxylic acid with an acyl halide at room temperature in the presence of a t-amine, *e.g.*, pyridine; *e.g.*, heptanoic anhydride:

$$CH_3(CH_2)_5CO_2Na + CH_3(CH_2)_5COCl \xrightarrow{\text{pyridine}} [CH_3(CH_2)_5CO]_2O \quad (68\%)$$

Properties of acetic anhydride

Acetic anhydride is a colourless liquid, b.p. 139·5°C, with an irritating smell. It is neutral when pure, and is slightly soluble in water, but readily soluble in ether and benzene. It is hydrolysed slowly in water, but rapidly by alkali:

$$(CH_3CO)_2O + H_2O \longrightarrow 2CH_3CO_2H$$

It undergoes reactions similar to those of acetyl chloride, but with less vigour; only half of the acetic anhydride molecule is used in acetylation, the other half being converted into acetic acid:

$$(CH_3CO)_2O + C_2H_5OH \longrightarrow CH_3CO_2C_2H_5 + CH_3CO_2H$$
$$(CH_3CO)_2O + NH_3 \longrightarrow CH_3CONH_2 + CH_3CO_2H$$

Acetylation with acetic anhydride is usually best carried out in pyridine solution or in the presence of a small amount of sodium acetate or concentrated sulphuric acid as catalyst. Acetic anhydride reacts with dry hydrogen chloride to form acetyl chloride:

$$(CH_3CO)_2O + HCl \longrightarrow CH_3COCl + CH_3CO_2H$$

It is readily halogenated, and may be used in the Friedel–Crafts reaction. It reacts with aldehydes to form alkylidene diacetates; *e.g.*, with acetaldehyde it forms ethylidene diacetate:

$$(CH_3CO)_2O + CH_3CHO \longrightarrow CH_3CH(OCOCH_3)_2$$

Acetic anhydride is reduced to ethanol by some metallic hydrides (Table 6.1, p. 179).

Acetic anhydride reacts with nitrogen pentoxide to form **acetyl nitrate**, b.p. 22°C/70 mm, which explodes violently if heated suddenly. It is hydrolysed by water to acetic and nitric acid:

$$CH_3COONO_2 + H_2O \longrightarrow CH_3CO_2H + HNO_3$$

Acetyl nitrate is very useful for preparing certain aromatic nitro-compounds, but is dangerous to handle (see p. 641).

Acylation of a hydroxy-compound to give an ester is generally a better procedure than direct esterification of the acid. It has already been pointed out that sterically hindered acids are difficult to esterify, but Parish *et al.* (1965) have shown that such acids may be readily converted into their esters by dissolving the acid in trifluoroacetic anhydride and adding the hydroxy-compound (yield: up to 97 per cent). According to the authors, the active agent is the mixed anhydride, $RCOOCOCF_3$, that is formed.

Acetyl peroxide, $CH_3COOOCOCH_3$, may be prepared by the action of barium peroxide on acetic anhydride. It is colourless, pungent-smelling liquid, b.p. 63°C/21 mm, and tends to explode on warming. It is a powerful oxidising agent.

Propionic anhydride, b.p. 169°C, is manufactured, *e.g.*,

$$CH_3CH_2CO_2H + C_2H_4 + CO \xrightarrow[\text{pressure}]{\text{heat;}} (CH_3CH_2CO)_2O$$

It is mainly used for the preparation of cellulose esters.

Formic anhydride may be prepared (about 90 per cent purity) by the distillation of a mixture of diethoxymethyl formate and n-butyric formic anhydride (Schijf *et al.*, 1965).

$$HCO_2CH(OEt)_2 + PrCOOOCH \longrightarrow (HCO)_2O + PrCO_2CH(OEt)_2$$

Mixed anhydrides containing the formic acid group have been known for some time and may be prepared as follows, *e.g.*,

$$CH_3COCl + HCO_2Na \longrightarrow CH_3COOOCH + NaCl$$

This behaves as a *formylating* reagent.

Acid amides

Acid amides are compounds in which the hydroxyl of the carboxyl group has been replaced by the amino-group, NH_2, to form the *amido-group*, $—CONH_2$.

There are three classes of amides: primary, $RCONH_2$; secondary, $(RCO)_2NH$; and tertiary, $(RCO)_3N$. Only the primary amides are important, and it can be seen from their formulae that all three classes may be regarded as the acyl derivatives of ammonia.

Nomenclature

According to the I.U.P.A.C. system of nomenclature, the suffix *-oic* of the parent acid is replaced by *amide* (see also acid chlorides, above), but they are commonly named by replacing the suffix *-ic* of the trivial name of the parent acid by *amide*.

$HCONH_2$ methanamide; formamide

CH_3CONH_2 ethanamide *or* methylcarbonamide; acetamide

General methods of preparation of the amides

1. By heating the ammonium salt of the acid:

$$RCOONH_4 \longrightarrow RCONH_2 + H_2O \quad (g.-v.g.)$$

Since the ammonium salts tend to dissociate on heating, the reaction is best carried out in the presence of some free acid, which represses the hydrolysis and the dissociation of the ammonium salt.

Amides may also be prepared by heating an acid with urea (Cherbuliez *et al.*, 1946):

$$RCO_2H + CO(NH_2)_2 \longrightarrow RCONH_2 + CO_2 + NH_3 \quad (g.-v.g.)$$

2. By *ammonolysis*, *i.e.*, the action of concentrated ammonia solution on acid chlorides, acid anhydrides or esters:

$$RCOCl + 2NH_3 \longrightarrow RCONH_2 + NH_4Cl \qquad (v.g.)$$
$$(RCO)_2O + 2NH_3 \longrightarrow RCONH_2 + RCO_2NH_4 \quad (v.g.)$$
$$R^1COOR^2 + NH_3 \longrightarrow R^1CONH_2 + R^2OH \qquad (g.)$$

The mechanisms of these reactions are uncertain, but a possibility is:

$$R-\overset{\overset{O}{\parallel}}{\underset{\underset{Z}{|}}{C}}:NH_3 \rightleftharpoons R-\overset{\overset{O^-}{|}}{\underset{\underset{Z}{|}}{C}}-\overset{+}{N}H_3 \xrightarrow{NH_3} R-\overset{\overset{O^-}{|}}{\underset{\underset{Z}{|}}{C}}-NH_2+NH_4{}^+ \longrightarrow R-\overset{\overset{O}{\parallel}}{C}-NH_2+ZH+NH_3$$

N-substituted amides may be prepared by using primary or secondary amines instead of ammonia, *e.g.*,

$$R^1COCl+R^2NH_2 \longrightarrow R^1CONHR^2+HCl$$

3. By the *graded* hydrolysis of alkyl cyanides. This may be carried out satisfactorily by dissolving the alkyl cyanide in concentrated sulphuric acid and then pouring the solution into cold water, or by shaking the alkyl cyanide with cold concentrated hydrochloric acid:

$$RC\equiv N+H_2O \xrightarrow{HCl} RCONH_2 \quad (g.)$$

Snyder *et al.* (1954) have shown that the hydrolysis of cyanides with polyphosphoric acid gives very good yields of amide.

A possible mechanism for this hydrolysis is:

$$RC\equiv N:+\overset{+}{H} \longrightarrow RC\equiv \overset{+}{N}-H \longleftrightarrow R\overset{+}{C}=\overset{..}{N}-H \xrightarrow{H_2O} R\underset{\underset{\overset{+}{O}H_2}{|}}{C}=\overset{..}{N}-H \xrightarrow{-H^+} R\underset{\underset{OH}{|}}{C}=\overset{..}{N}H \rightleftharpoons R\overset{\overset{O}{\parallel}}{C}-NH_2$$

Sometimes the conversion of an alkyl cyanide into the amide may be effected by means of alkaline hydrogen peroxide. A possible mechanism is (Wiberg, 1953):

$$HOOH+OH^- \longrightarrow H_2O+{}^-OOH$$

$$RC\equiv N+{}^-OOH \longrightarrow R\underset{\underset{OOH}{|}}{C}=N^- \xrightarrow{H_2O} R\underset{\underset{O-OH}{|}}{C}=NH \xrightarrow{H_2O} RCONH_2+O_2+H_2O$$

Other methods for preparing amides are the Ritter reaction (p. 370), the action of ammonia on diazoketones (p. 394), the Willgerodt reaction (p. 751), and the Beckmann rearrangement (p. 755).

General properties and reactions of the amides

Except for formamide, which is a liquid, all the amides are colourless crystalline solids, and those of low molecular weight are soluble in water (with which they can form hydrogen bonds). The lower amides have much higher melting points and boiling points than are to be expected from their molecular weights; this indicates association (through hydrogen bonding):

$$H_2N-CR{=}O\cdots H-NH-CR{=}O\cdots H-NH-CR{=}O\cdots$$

This association has been demonstrated from infra-red spectroscopic studies on amides in solution and in the solid state (Richards *et al.*, 1947).

Infra-red spectroscopic evidence has also indicated that amides are resonance hybrids, and this is supported by N.M.R. studies, *e.g.*, the two methyl groups in dimethylformamide give two separate signals at room temperature. This can be explained on the basis that, because of the partial double bond character of the C—N bond, the two methyl groups have *different* environments, one methyl group being *cis* and the other *trans* to the hydrogen atom. At room temperature the rate of rotation about the C—N bond is slow enough for the two methyl groups to be non-equivalent. As the temperature is raised, the signals broaden and finally collapse to a single line. At these higher temperatures, the molecule has absorbed sufficient energy to overcome the energy barrier to rotation to make the average environment of the two methyl groups the same, *i.e.*, the two methyl groups are *now* equivalent.

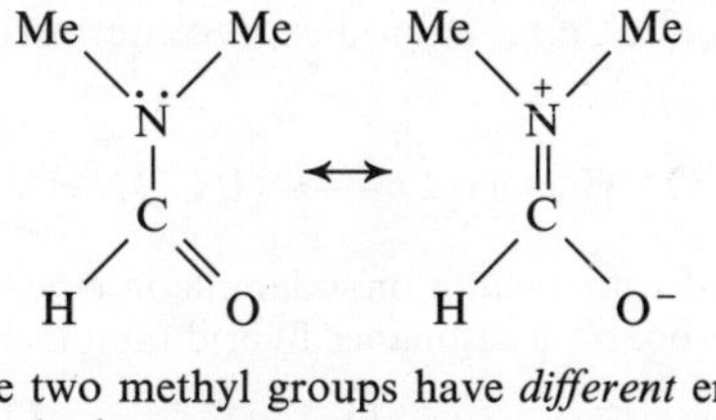

The infra-red absorption regions of the carbonyl group (stretch) in primary amides are near $1\,690$ (s) and $1\,650$ cm^{-1} (s), and those of the free NH group (stretch) are near $3\,500$ (s) and $3\,400$ cm^{-1} (s). For a hydrogen-bonded NH group, the regions are near $3\,350$ (m) and $3\,200$ cm^{-1} (m).

Dimethylformamide, b.p. $152°$C, is a very good solvent for polar and non-polar compounds, and is used as a formylating agent in conjunction with phosphoryl chloride (p. 744).

It is manufactured by reaction between methyl formate and dimethylamine:

$$HCO_2Me + Me_2NH \longrightarrow HCONMe_2 + MeOH$$

1. Amides are hydrolysed slowly by water, rapidly by acids, and far more rapidly by alkalis:

$$RCONH_2 + H_2O \longrightarrow RCO_2H + NH_3$$

Possible mechanisms are:

Base hydrolysis

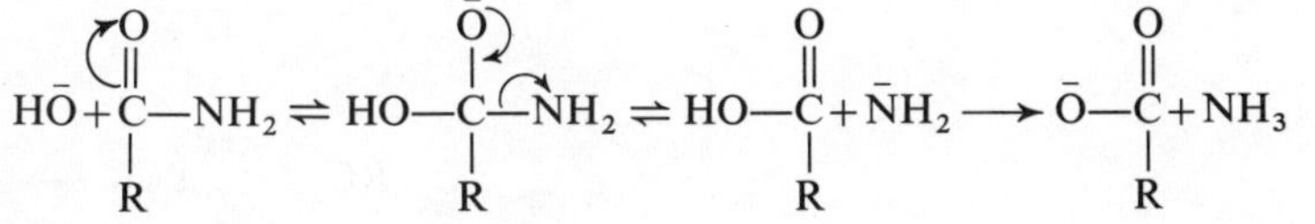

Acid hydrolysis

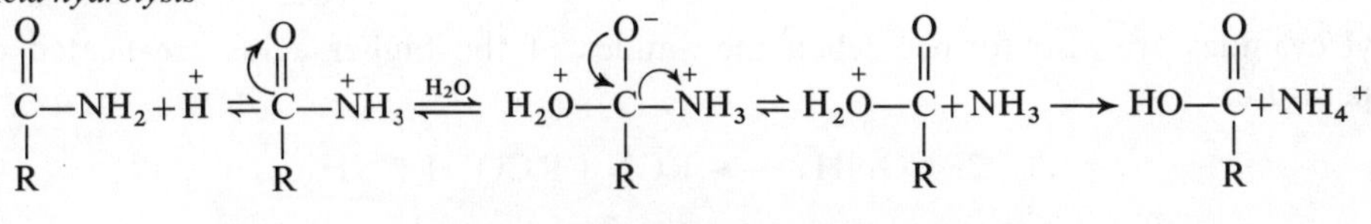

2. Amides are very feebly basic and form unstable salts with strong inorganic acids, *e.g.*, $RCONH_2 \cdot HCl$. The structure of these salts may be (I) or (II):

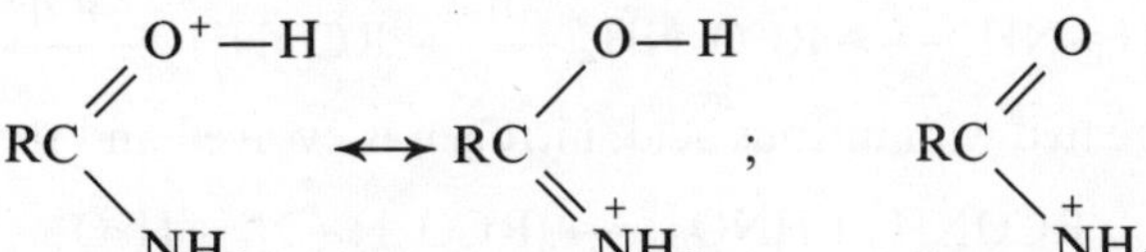

Kutzennigg *et al.* (1962), from their proton magnetic resonance studies, have provided evidence that the salts of both acetamide and the *N,N*-disubstituted acetamides are protonated on the oxygen atom. This would be in agreement with the fact that amides are resonance hybrids, the O carrying a negative charge and the N a positive charge.

3. Amides are also feebly acidic; *e.g.*, they dissolve mercuric oxide to form covalent mercury compounds:

$$2RCONH_2 + HgO \longrightarrow (RCONH)_2Hg + H_2O$$

When amides are treated with sodium or sodamide in ethereal solution, the sodium *salt* is formed, the structure of which is probably a resonance hybrid (ambident cation):

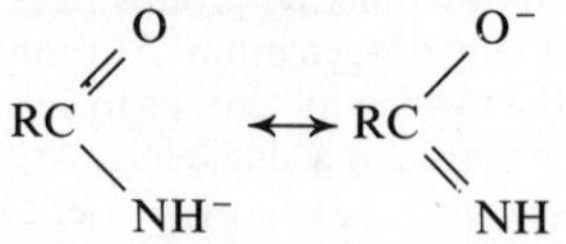

Treatment of the *sodium* salt with alkyl iodide gives the *N*-alkyl derivative, whereas with the silver salt, the *O*-alkyl derivative (*imido ester*, p. 267) is obtained. Imido esters are also obtained by alkylation of amides with alkyl sulphates or oxonium salts, *e.g.*, $R_3O^+BF_4^-$.

4. Amides are reduced by sodium and ethanol, catalytically, etc. (see Table 6.1, p. 179), *e.g.*,

$$RCONH_2 \xrightarrow{\text{Na/C}_2\text{H}_5\text{OH}} RCH_2NH_2 + H_2O$$

5. When heated with phosphorus pentoxide, or better, with PCl_5, $POCl_3$, or $SOCl_2$, amides are dehydrated to alkyl cyanides:

$$RCONH_2 \xrightarrow{-H_2O} RCN$$

The mechanisms of these dehydrations are not certain, but according to Kirsanov (1954) and Lapidot *et al.* (1962), the mechanism with phosphorus pentachloride is:

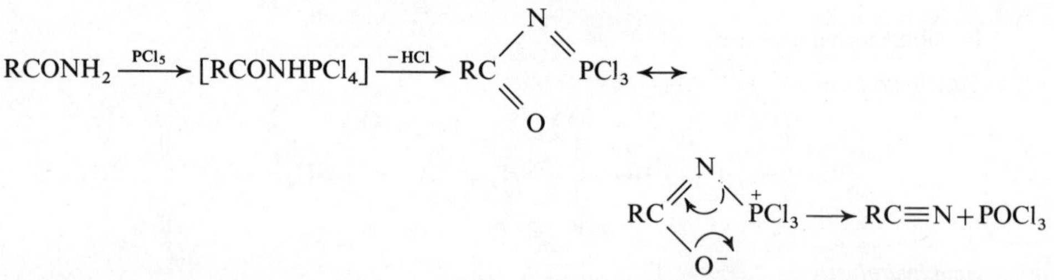

Alkyl cyanides are also formed when the amides of the higher acids are heated to a high temperature:

$$2RCONH_2 \longrightarrow RCN + RCO_2H + NH_3$$

If this reaction is carried out in the presence of excess of ammonia, then all the amide is converted into alkyl cyanide:

$$RCO_2H + NH_3 \longrightarrow RCO_2NH_4 \xrightarrow{-H_2O} RCONH_2 \xrightarrow{-H_2O} RCN$$

6. When amides are treated with nitrous acid, nitrogen is evolved and the acid is formed:

$$RCONH_2 + HNO_2 \longrightarrow RCO_2H + N_2 + H_2O$$

This reaction is usually slow; a better method is to use butyl nitrite:

$$RCONH_2 + BuONO \longrightarrow RCO_2H + N_2 + BuOH$$

7. Hofmann preparation of primary amines (1881). This is the conversion of an amide into a primary amine with one carbon atom less by means of bromine (or chlorine) and alkali. The overall equation for the reaction may be written:

$$RCONH_2 + Br_2 + 4KOH \longrightarrow RNH_2 + 2KBr + K_2CO_3 + 2H_2O$$

The mechanism of this reaction has been the subject of a great deal of work. A number of intermediates have been isolated: N-bromamides, RCONHBr; salts of these bromamides, $[RCONBr]^-K^+$; isocyanates, RNCO. On this basis, a possible mechanism is the following, which is an example of the 1,2-shift:

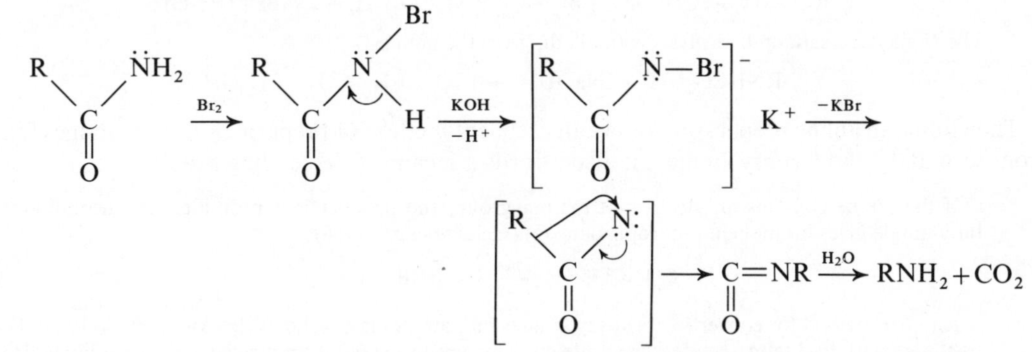

(I)

The uncertainty of this mechanism lies mainly in the formation of the acyl nitrene (I; see also p. 385). According to Wright (1968), who used ^{14}C- and ^{15}N-labelled compounds, the Hofmann, Curtius and Lossen rearrangements (see below) may not involve nitrenes as intermediates; he has proposed the following *concerted* mechanism:

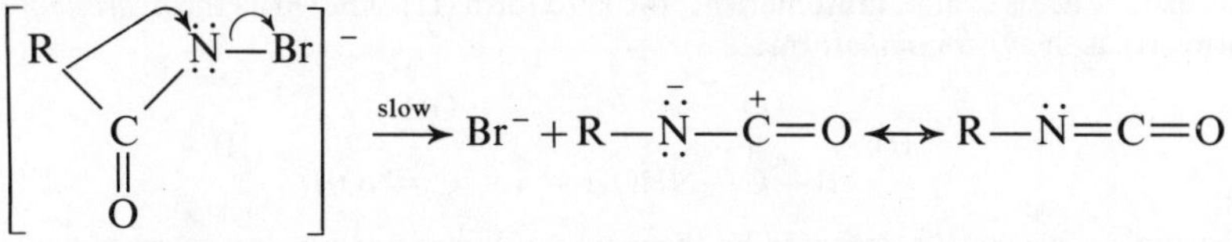

Hauser *et al.* (1937) have shown that electron-releasing groups in the *para*-position of a migrating aryl group accelerate the reaction and electron-withdrawing groups retard the reaction. This supports the formation of a bridged ion (a phenonium ion), (Ia), where the migrating group is aromatic, and by inference, a bridged ion (Ib) when the migrating group is

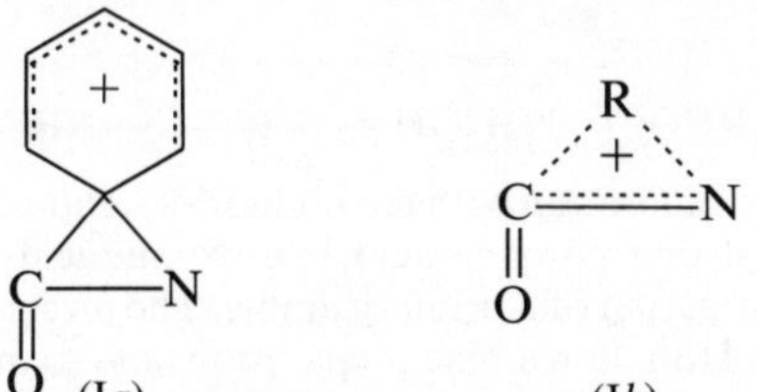

alkyl. This is further supported by the fact that the rearrangement is intramolecular, *e.g.*, Wallis *et al.* (1931), using the amide of optically active α-benzylpropionic acid, found that there was retention of configuration in the product, α-benzylethylamine; thus the migrating group is never free:

$$C_6H_5CH_2CHMeCONH_2 \longrightarrow C_6H_5CH_2CHMeNH_2$$

The Hofmann amine preparation can be carried out on all monocarboxylic acids, and the yields are 70–90 per cent for amides containing up to seven carbon atoms; amides with more than seven carbon atoms in the chain give mainly alkyl cyanide:

$$RCH_2CONH_2 \xrightarrow{Br_2/KOH} RCN$$

The alkyl cyanide, however, may readily be reduced to the primary amine RCH_2NH_2.

It is possible to obtain the amine in good yield from the long-chain amides by modifying the procedure as follows. Bromine is added rapidly to a methanolic solution of the amide containing sodium methoxide:

$$RCONH_2 + 2CH_3ONa + Br_2 \longrightarrow RNHCO_2CH_3 + 2NaBr + CH_3OH$$

The *N*-alkylurethan, on hydrolysis with alkali, gives the amine:

$$RNHCO_2CH_3 + 2NaOH \longrightarrow RNH_2 + Na_2CO_3 + CH_3OH$$

The Hofmann amine preparation offers an excellent method of preparing primary amines free from secondary and tertiary amines; it also affords a means of descending a series.

If the amide contains an electron-attracting group, the product is a bromide, *e.g.*, heptafluoro-butyramide gives bromoheptafluoropropane (Haszeldine *et al.*, 1956):

$$C_3F_7CONH_2 \xrightarrow[NaOH]{Br_2} C_3F_7Br \quad (85\%)$$

Another method for converting primary amides into amines is as follows (Beckwith *et al.*, 1965). On treatment with lead tetra-acetate, the amide gives an acylamine (40–81 per cent); this is readily hydrolysed to the amine.

$$RCONH_2 \xrightarrow[\text{benzene; reflux}]{Pb(OAc)_4} RNHAc$$

Hydroxamic acids

The hydroxamic acids exhibit tautomerism; the keto form (I) is known as the *hydroxamic* form, and the enol form (II) as the *hydroximic* form:

$$\text{(I)} \quad \underset{\substack{\|\\O}}{R-C}-NHOH \rightleftharpoons R-\underset{\substack{|\\OH}}{C}=NOH \quad \text{(II)}$$

Hydroxamic acids may be prepared by the action of hydroxylamine on esters or acid chlorides:

$$RCOCl + NH_2OH \longrightarrow R-\underset{\substack{\|\\O}}{C}-NHOH + HCl$$

Hydroxamic acids give a red coloration with ferric chloride solution, a reaction which is characteristic of enols. When treated with a strong inorganic acid, hydroxamic acids undergo the **Lossen rearrangement** (1875), which results in the formation of a primary amine. The mechanism of the Lossen rearrangement is closely related to that of the Hofmann amine preparation, and so may be formulated (see also above):

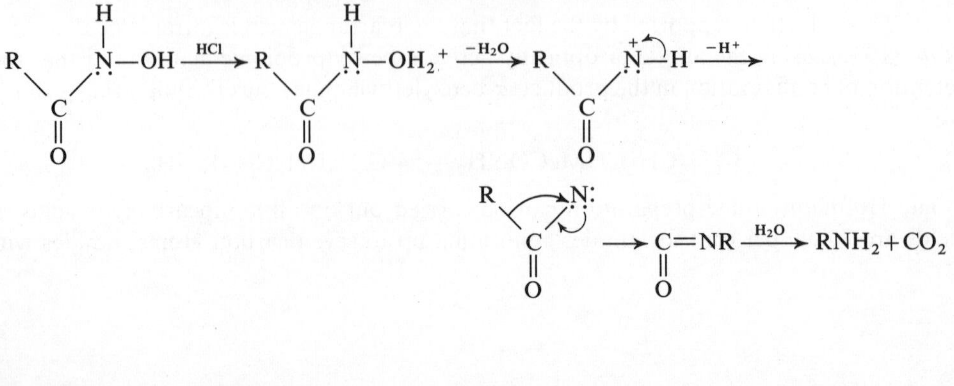

This rearrangement (1,2-shift) has been shown to be intramolecular, and electron-releasing groups in a migrating aryl group accelerate the reaction, thereby supporting the formation of a phenonium ion.

The amides of the hydroxamic acids are known as **amidoximes**. These are tautomeric substances:

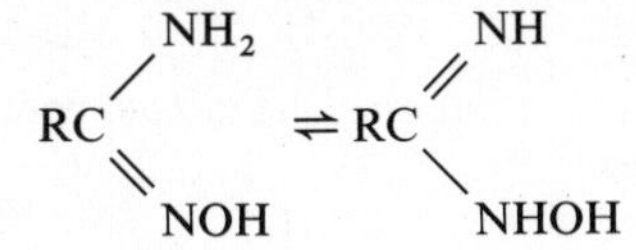

$$RC\underset{NOH}{\overset{NH_2}{<}} \rightleftharpoons RC\underset{NHOH}{\overset{NH}{<}}$$

Aliphatic amidoximes are best obtained by the action of hydroxylamine on an alkyl cyanide:

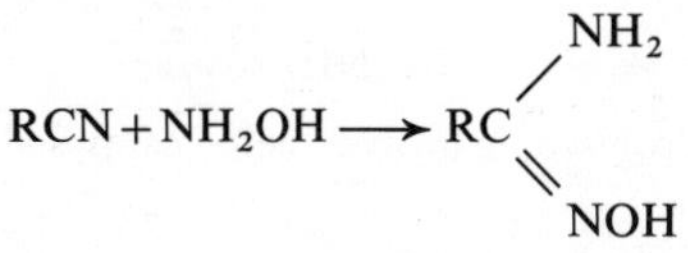

$$RCN + NH_2OH \longrightarrow RC\underset{NOH}{\overset{NH_2}{<}}$$

Imidic esters and amidines

Imidic esters, which are also known as *imino-ethers*, are best prepared by passing dry hydrogen chloride into a solution of an alkyl cyanide in anhydrous alcohol; the imidic ester hydrochloride is slowly deposited (Pinner, 1877):

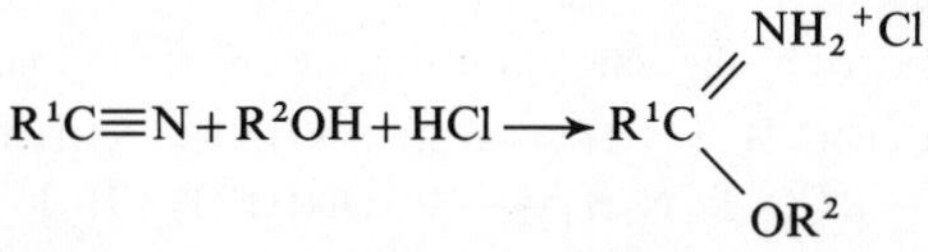

$$R^1C\equiv N + R^2OH + HCl \longrightarrow R^1C\underset{OR^2}{\overset{NH_2{}^+Cl^-}{<}} \quad (g.\text{-}v.g.)$$

Imidic ester hydrochlorides form ortho-esters when allowed to stand in an alcohol, and are readily hydrolysed to esters by water:

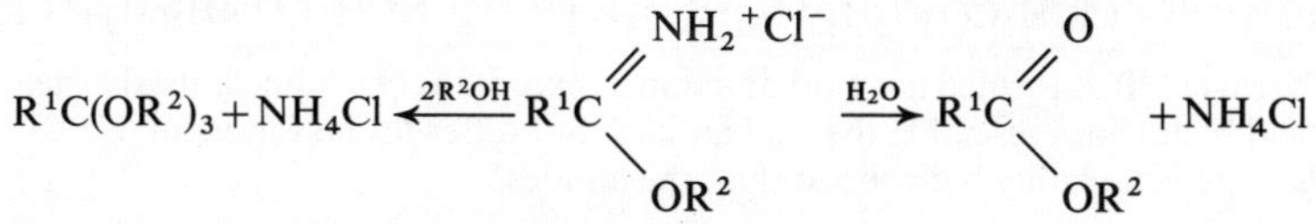

$$R^1C(OR^2)_3 + NH_4Cl \xleftarrow{2R^2OH} R^1C\underset{OR^2}{\overset{NH_2{}^+Cl^-}{<}} \xrightarrow{H_2O} R^1C\underset{OR^2}{\overset{O}{<}} + NH_4Cl$$

Imidic ester hydrochlorides are decomposed on heating:

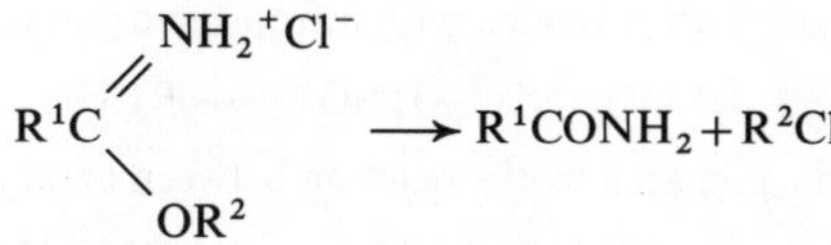

$$R^1C\underset{OR^2}{\overset{NH_2{}^+Cl^-}{<}} \longrightarrow R^1CONH_2 + R^2Cl$$

When an ethereal solution of an imidic ester hydrochloride is treated with potassium carbonate, the free ester is obtained. When treated with an ethanolic solution of ammonia, the imidic ester hydrochloride forms the **amidine**:

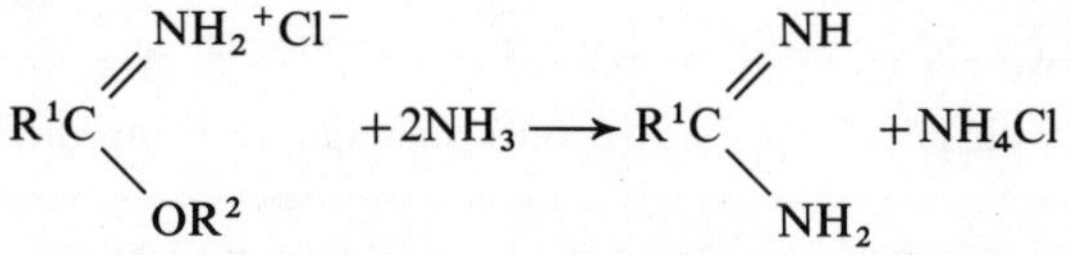

$$R^1C\underset{OR^2}{\overset{NH_2{}^+Cl^-}{<}} + 2NH_3 \longrightarrow R^1C\underset{NH_2}{\overset{NH}{<}} + NH_4Cl$$

Formamidine acetate and acetamidine acetate may be prepared from ortho-esters (R = H or Me) as follows (Taylor *et al.*, 1960):

$$RC(OEt)_3 + NH_3 + CH_3CO_2H \longrightarrow H_2NCR{=}NH_2{}^+CH_3CO_2{}^-$$

These acetates are better precursors of their respective amidines than are the corresponding hydrochlorides.

Amidines are strong monoacid bases, forming salts with strong acids. Their basic strength may be explained by resonance. In the amidine, the resonating structures are different and one carries *unlike* charges.

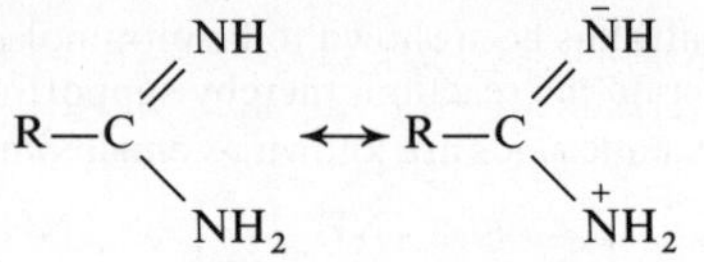

$$R-C \overset{NH}{\underset{NH_2}{\diagup}} \longleftrightarrow R-C \overset{\bar{N}H}{\underset{\overset{+}{N}H_2}{\diagup}}$$

In the amidine *ion*, the gain of a proton results in *both* contributing structures being equivalent and carrying only a positive charge.

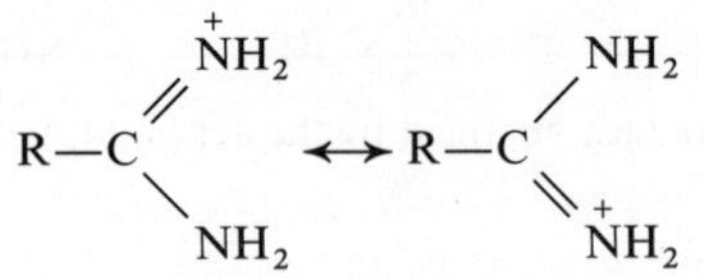

$$R-C \overset{\overset{+}{N}H_2}{\underset{NH_2}{\diagup}} \longleftrightarrow R-C \overset{NH_2}{\underset{\overset{+}{N}H_2}{\diagup}}$$

Amidines are readily hydrolysed, hydrolysis being catalysed by acids or alkalis:

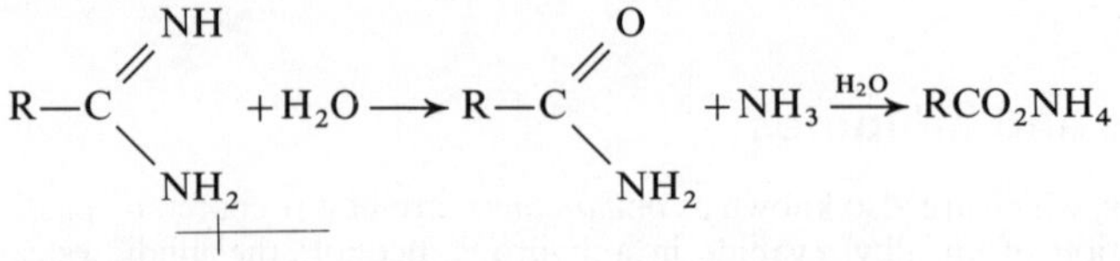

$$R-C \overset{NH}{\underset{NH_2}{\diagup}} + H_2O \longrightarrow R-C \overset{O}{\underset{NH_2}{\diagup}} + NH_3 \xrightarrow{H_2O} RCO_2NH_4$$

Acid hydrazides and acid azides

Acid hydrazides may be prepared by the action of hydrazine on esters or acyl chlorides:

$$RCOOC_2H_5 + N_2H_4 \longrightarrow RCONHNH_2 + C_2H_5OH$$
$$RCOCl + N_2H_4 \longrightarrow RCONHNH_2 + HCl$$

Girard's reagent 'T' (p. 214) is trimethylaminoacetohydrazide, which may be prepared by interaction between trimethylamine, ethyl chloroacetate and hydrazine:

$$(CH_3)_3N + ClCH_2CO_2C_2H_5 + N_2H_4 \longrightarrow [(CH_3)_3NCH_2CONHNH_2]^+Cl^- + C_2H_5OH$$

Girard's reagent 'P' is formed in a similar manner, except that pyridine is used instead of trimethylamine.
The acid hydrazides resemble the acid amides, but differ in the following ways:
(i) They are less readily hydrolysed than the amides:

$$RCONHNH_2 + H_2O \longrightarrow RCO_2H + N_2H_4$$

(ii) They are reducing agents—hydrazine is a powerful reducing agent.
(iii) They form acid azides when treated with nitrous acid; no nitrogen is evolved:

$$RCONHNH_2 + HNO_2 \longrightarrow RCON_3 + 2H_2O$$

Acid azides may also be prepared by the reaction between an acyl chloride and sodium azide:

$$RCOCl + NaN_3 \longrightarrow RCON_3 + NaCl$$

The structure of the acid azides is best represented as a resonance hybrid:

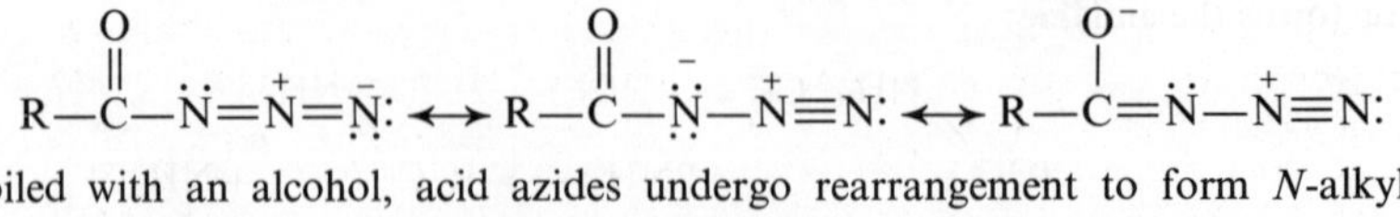

$$R-\overset{O}{\overset{||}{C}}-\overset{..}{N}=\overset{+}{N}=\overset{-}{\underset{..}{N}}: \longleftrightarrow R-\overset{O}{\overset{||}{C}}-\overset{-}{\underset{..}{N}}-\overset{+}{N}\equiv N: \longleftrightarrow R-\overset{O^-}{\overset{|}{C}}=\overset{..}{N}-\overset{+}{N}\equiv N:$$

When boiled with an alcohol, acid azides undergo rearrangement to form *N*-alkyl-substituted urethans:

$$R^1CON_3 + R^2OH \longrightarrow RNHCO_2R^2 + N_2 \quad (v.g.\text{-}ex.)$$

The mechanism of this rearrangement—the **Curtius rearrangement** (1894)—is similar to that of the Hofmann and Lossen rearrangements (see above), and so may be formulated:

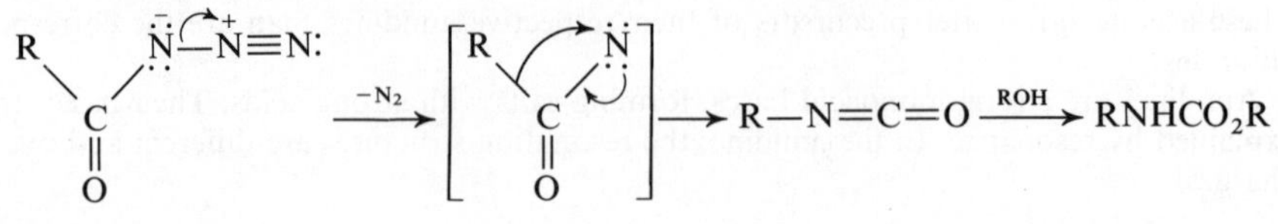

$$\underset{\overset{||}{O}}{\overset{R}{\diagdown}}\overset{..}{\underset{..}{N}}-\overset{+}{N}\equiv N: \xrightarrow{-N_2} \left[\underset{\overset{||}{O}}{\overset{R}{\diagdown}}\overset{..}{\underset{..}{N}} \right] \longrightarrow R-N=C=O \xrightarrow{ROH} RNHCO_2R$$

This mechanism is supported by the fact that when the terminal nitrogen atom is ^{15}N in 3,5-dinitro-benzazide, all of this tracer is found in the nitrogen eliminated in the reaction (Bothner—By et al., 1951). The Curtius reaction may be used to step down a series (cf. the Hofmann amine preparation):

$$RCO_2H \xrightarrow{C_2H_5OH} RCO_2C_2H_5 \xrightarrow{N_2H_4} RCONHNH_2 \xrightarrow{HNO_2}$$

$$RCON_3 \xrightarrow{CH_3OH} RNHCO_2CH_3 \xrightarrow{NaOH} RNH_2$$

The Curtius reactions offers a very good method for preparing primary amines free from secondary and tertiary amines, and may also be used for preparing isocyanates and urethans; e.g., when heated in benzene or chloroform solution, an acid azide rearranges to the alkyl isocyanate which can be isolated. If the isocyanate is warmed with an alcohol, an N-substituted urethan is formed:

$$R^1CON_3 \xrightarrow[\text{solution}]{\text{benzene}} R^1NCO \xrightarrow{R^2OH} R^1NHCO_2R^2$$

Halogen derivatives of the monocarboxylic acids

The halogen derivatives of the monocarboxylic acids are compounds in which one or more hydrogen atoms in the *carbon chain* have been replaced by halogen.

Nomenclature

The following examples illustrate the methods (cf. p. 233):

CH_2ClCO_2H	monochloroacetic acid
$CH_3CHClCO_2H$	α-chloropropionic acid or 2-chloropropanoic acid
$(CH_3)_2CClCHBrCO_2H$	α-bromo-β-chloroisovaleric acid or 2-bromo-3-chloro-3-methylbutanoic acid

α-Halogeno-acids. Although the monocarboxylic acids themselves are not readily halogenated, their acid chlorides and anhydrides may be halogenated very easily. Bromination takes place only at the α-position, but chlorination, although it occurs mainly at the α-position, may also take place further in the chain; e.g., chlorination of propionic acid results in the formation of the α- and β-chloro-derivatives (see also below):

$$CH_3CH_2CO_2H \xrightarrow{Cl_2} CH_3CHClCO_2H + ClCH_2CH_2CO_2H$$

Fluorination of the monocarboxylic acids has not yet been studied in great detail, but the work done so far seems to indicate that fluorine enters the chain somewhat indiscriminately; e.g., fluorination of butyric acid in carbon tetrachloride solution gives the β- and γ-fluoro-derivatives (see also below).

Iodo-derivatives are prepared from the chloro- or bromo-compound by the action of potassium iodide in methanolic or acetone solution (cf. alkyl iodides, p. 85).

The usual method for preparing α-chloro- or bromo-acids is by the **Hell–Volhard–Zelinsky reaction (H.V.Z. reaction)**, which is carried out by treating the acid with chlorine or bromine in the presence of a small amount of red phosphorus.

The mechanism of the H.V.Z. reaction is uncertain, but there is some evidence to show that it may proceed through an enol intermediate (Zimmerman et al., 1959):

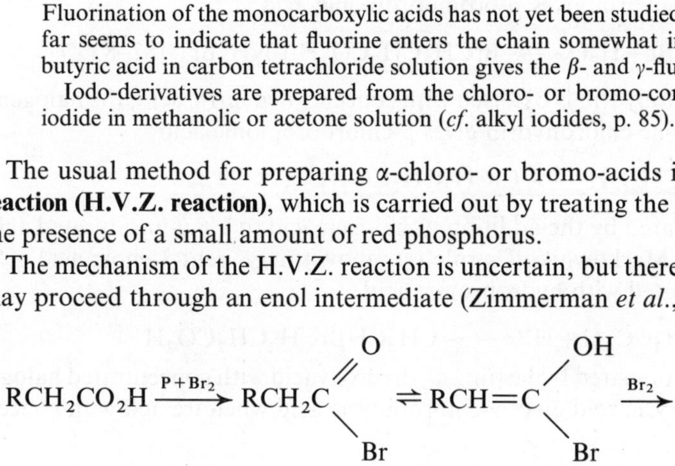

$$\text{HBr} + \text{RCHBrCOBr} \underset{}{\overset{\text{RCH}_2\text{CO}_2\text{H}}{\rightleftharpoons}} \text{RCHBrCO}_2\text{H} + \text{RCH}_2\text{COBr, etc.}$$

Enolisation of the acid bromide is far more rapid than that of the acid.

The second α-hydrogen atom may be replaced by chlorine or bromine by using excess of the halogen, but whereas bromination ceases when both α-hydrogen atoms have been replaced, chlorination proceeds further in the chain (see above). Since the H.V.Z. reaction with bromine is specific for α-hydrogen atoms, it can be used to detect the presence of α-hydrogen in an acid; e.g., trimethylacetic acid does not undergo the H.V.Z. reaction with bromine.

The H.V.Z. reaction with bromine is applicable to di- and polycarboxylic acids, all the α-positions being substituted if sufficient bromine is used.

Another convenient method of preparing α-bromo-acids is by brominating an alkylmalonic acid and heating the bromo-acid, whereupon it is decarboxylated to the monobasic acid:

$$\text{RCH(CO}_2\text{H)}_2 \xrightarrow{\text{Br}_2} \text{RCBr(CO}_2\text{H)}_2 \xrightarrow{\text{heat}} \text{RCHBrCO}_2\text{H} + \text{CO}_2$$

Sulphuryl chloride, in the presence of a small amount of iodine, chlorinates monocarboxylic acids in the α-position, but in the presence of organic peroxides (to give a free-radical reaction), the α-, β-, γ-, etc. positions are substituted. On the other hand, Louw (1966) has shown that cupric chloride (plus LiCl) chlorinates acid chlorides, anhydrides, and carboxylic acids (containing the acid chloride or anhydride as 'catalyst') at the α-position only (yields > 95 per cent). The reaction is carried out in a polar inert solvent, e.g., sulpholane, and only the α-monochloro derivative is obtained regardless of the number of α-hydrogen atoms in the compound.

β-Halogeno-acids may be prepared by treating an α,β-unsaturated acid with halogen acid; e.g., acrylic acid forms β-bromopropionic acid when treated with hydrogen bromide in acetic acid solution:

$$\text{CH}_2\text{=CHCO}_2\text{H} + \text{HBr} \longrightarrow \text{BrCH}_2\text{CH}_2\text{CO}_2\text{H}$$

β-Halogeno-acids may also be prepared by treating an α,β-unsaturated aldehyde with halogen acid, and oxidising the β-chloroaldehyde produced, e.g., acraldehyde gives β-chloropropionic acid when treated with concentrated hydrochloric acid, and the product then oxidised with concentrated nitric acid:

$$\text{CH}_2\text{=CHCHO} + \text{HCl} \longrightarrow \text{ClCH}_2\text{CH}_2\text{CHO} \xrightarrow{\text{[O]}} \text{ClCH}_2\text{CH}_2\text{CO}_2\text{H} \quad (60\text{--}65\%)$$

The addition of all the halogen acids to α,β-unsaturated carbonyl compounds takes place very readily, and in a direction contrary to Markownikoff's rule (see p. 344).

When an alkene cyanohydrin is refluxed with 40 per cent hydrobromic acid, the β-bromoacid is obtained; e.g., ethylene cyanohydrin gives β-bromopropionic acid:

$$\text{CH}_2\text{(OH)CH}_2\text{CN} + 2\text{HBr} + \text{H}_2\text{O} \longrightarrow \text{BrCH}_2\text{CH}_2\text{CO}_2\text{H} + \text{NH}_4\text{Br} \quad (82\text{--}83\%)$$

Alternatively, when an alkene halohydrin is oxidised with concentrated nitric acid, the halogeno-acid is produced; e.g., trimethylene chlorohydrin gives β-chloropropionic acid:

$$\text{ClCH}_2\text{CH}_2\text{CH}_2\text{OH} \longrightarrow \text{ClCH}_2\text{CH}_2\text{CO}_2\text{H} \quad (78\text{--}79\%)$$

γ-Halogeno-acids may be prepared by the addition of halogen acid to a β,γ-unsaturated acid; the addition occurs contrary to Markownikoff's rule (cf. above); e.g., pent-3-enoic acid gives γ-bromo-n-valeric acid when treated with hydrobromic acid:

$$\text{CH}_3\text{CH=CHCH}_2\text{CO}_2\text{H} + \text{HBr} \longrightarrow \text{CH}_3\text{CHBrCH}_2\text{CH}_2\text{CO}_2\text{H}$$

γ-Halogeno-acids may also be prepared by heating a γ-hydroxyacid with concentrated halogen acid solution; e.g., γ-hydroxybutyric acid gives γ-chlorobutyric acid when treated with concentrated hydrochloric acid:

$$HOCH_2CH_2CH_2CO_2H + HCl \longrightarrow ClCH_2CH_2CH_2CO_2H + H_2O$$

The methods described above apply to particular types of halogeno-acids. However, the following method may be used for preparing any halogeno-acid (*cf.* Hunsdiecker reaction, p. 151); *e.g.*, a δ-halogeno-acid:

$$MeO_2C(CH_2)_4CO_2H \xrightarrow[\text{KOH}]{\text{AgNO}_3} MeO_2C(CH_2)_4CO_2Ag \xrightarrow[\text{CCl}_4]{\text{Br}_2}$$

$$MeO_2C(CH_2)_3CH_2Br \quad (65–68\%)$$

Properties of the halogeno-acids

The α-halogeno-acids undergo most of the reactions of the alkyl halides, but the halogen atom in the acid is far more reactive than that in the alkyl halide; the enhanced reactivity is due to the $-I$ effect of the adjacent carbonyl group. The reactions of the carboxyl group are unchanged. β-, γ- and δ-halogeno-acids undergo some of the reactions of the alkyl halides, but tend to eliminate a molecule of hydrogen halide to form an unsaturated acid or a lactone (p. 469). They do not form Grignard reagents, but the halogeno-acid esters react with Grignard reagents to form halogen-substituted tertiary alcohols. The halogeno-acids are reduced to the corresponding acids by sodium amalgam, but the reduction of acids of the type $R^1R^2CClCO_2H$ with lithium aluminium hydride gives chlorohydrins (I), alcohols (II), aldehydes (III) and glycols (IV).

$R^1R^2CClCH_2OH$	$R^1R^2CHCH_2OH$	R^1R^2CHCHO	$R^1R^2C(OH)CH_2OH$
(I)	(II)	(III)	(IV)

The amount of each depends on the nature of R^1 and R^2, *e.g.*, very little glycol is obtained unless at least one of these groups is phenyl (Eliel *et al.*, 1956).

The behaviour of an halogeno-acid with alkali depends on the position of the halogen atom relative to the carboxyl group.

(*a*) α-Halogeno-acids are converted into the corresponding α-hydroxyacid:

$$RCHXCO_2H + H_2O \xrightarrow[\text{(ii) H}^+]{\text{(i) NaOH}} RCH(OH)CO_2H$$

On the other hand, if the α-halogeno-acid ester is heated with a tertiary amine, the α,β-unsaturated acid ester is formed by the elimination of a molecule of hydrogen halide; *e.g.*. ethyl α-bromobutyrate gives ethyl crotonate when heated with dimethylaniline, $C_6H_5N(CH_3)_2$:

$$CH_3CH_2CHBrCO_2C_2H_5 \xrightarrow{-\text{HBr}} CH_3CH=CHCO_2C_2H_5$$

(*b*) β-Halogeno-acids are converted into the corresponding β-hydroxyacid, which, on continued reflux with alkali, eliminates a molecule of water to form the α,β-unsaturated acid:

$$RCHBrCH_2CO_2H \xrightarrow{\text{NaOH}} RCH(OH)CH_2CO_2^- \xrightarrow[-\text{H}_2\text{O}]{\text{NaOH}} RCH=CHCO_2^-$$

(*c*) γ- and δ-Halogeno-acids are converted into lactones; *e.g.*, γ-chlorobutyric acid gives γ-butyrolactone:

This is an example of neighbouring group participation.

(d) ε-Halogeno-acids, etc., give the corresponding hydroxyacid; e.g., ε-bromocaproic acid gives ε-hydroxycaproic acid:

$$BrCH_2CH_2CH_2CH_2CH_2CO_2H \xrightarrow[\text{(ii) H}^+]{\text{(i) NaOH}} HOCH_2CH_2CH_2CH_2CH_2CO_2H$$

The most characteristic reaction of the α-halogeno-acid esters is the *Reformatsky reaction* (p. 437).

The halogeno-acids are all stronger acids than the parent acid (see Table 9.1, p. 242), and for a group of isomeric acids, the further the halogen is removed from the carboxyl group the weaker is the acid. The increase in acid strength may be explained by the high electronegativity of the halogen atom which exerts a strong inductive effect, thereby facilitating the release of proton from the carboxyl group.

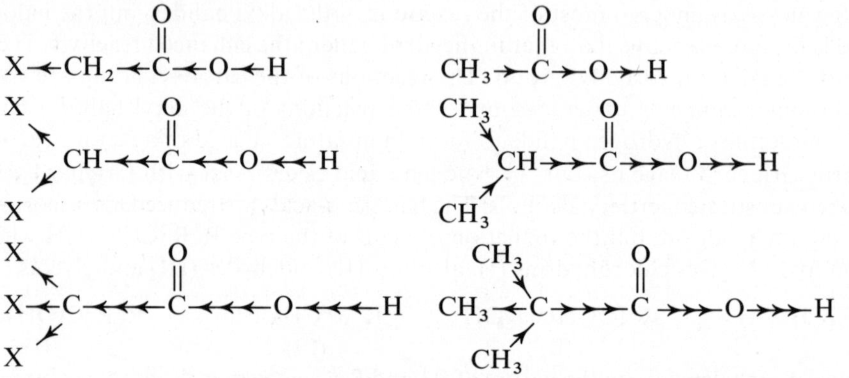

The larger the number of halogen atoms on the α-carbon atom, the stronger is the inductive effect, and consequently the stronger is the acid; also, the further removed the halogen atom is from the carboxyl group, the weaker is the inductive effect at the carboxyl group, and consequently the weaker is the acid. In the same way, since an alkyl group is electron-repelling, increasing their number on the α-carbon atom should *decrease* the strength of the acid since release of the proton will be hindered (see Table 9.1, p. 242). Since $-I$ groups in the α-position, e.g., X, NO_2, OH, increase the strength of the acid, and $+I$ groups, e.g., Me, weaken the acid, the measurement of the pK_a's of various substituted acetic acids affords a means of measuring the *relative* inductive effects of different groups.

Relative inductive effects (or electronegativities) may also be measured by NMR spectroscopy (p. 11). The results, however, must be accepted with reserve, since the applied field itself may affect the electron distribution in the molecule.

To study the strengths of the different members of the carboxylic acids quantitatively, we must use the equations:

(i) HA (solvated) $\rightleftharpoons$ H$^+$ (solvated) + A$^-$ (solvated)

(ii) $-RT \ln K_a = \Delta G^\theta$ (ionisation) $= \Delta H^\theta$ (ionisation $- T\Delta S^\theta$ (ionisation)

This, in theory, may be done by using the following thermochemical cycle (see also p. 160):

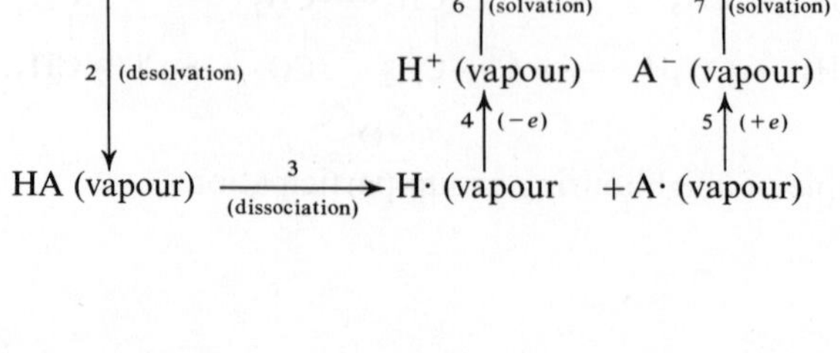

It then follows that:

$$\Delta H_1 = \Delta H_2 + \Delta H_3 + \Delta H_4 + \Delta H_5 + \Delta H_6 + \Delta H_7$$

and

$$\Delta S_1 = \Delta S_2 + \Delta S_3 + \Delta S_4 + \Delta S_5 + \Delta S_6 + \Delta S_7$$

Unfortunately, very few of these values are known for the various organic acids. However, it is possible to evaluate ΔH_1 from experimental values of K_a over a range of temperatures, and since ΔG is also obtained, ΔS_1 can thus be calculated. Some values of ΔH_1 (kJ) and ΔS_1 (JK^{-1}), respectively, are, in water: formic acid, -0.08368, -73.64; acetic, -0.3766, -92.47; propionic, -0.6694, -95.40; chloroacetic, -4.853, -71.13.

It can be seen that ΔH_1 is quite small, but ΔS_1 can vary a great deal. Thus the difference in acid strengths must be due to entropy effects. An interesting point to note is the smaller entropy loss for chloroacetic acid makes this a stronger acid than acetic. ΔS_1 for trichloroacetic acid is -8.368 JK^{-1}, and this acid is a very strong acid. These entropy changes may be explained as follows. Acetic acid is far less polar than the acetate ion, and so the latter is far more solvated than the former. Since solvation produces some sort of arrangement of the solvent molecules around the anion, this loss of randomness of solvent molecules means a decrease in entropy (i.e., ΔS is negative). Because trichloroacetic acid is highly polar, it is highly solvated both in the un-ionised form and in the anion, and so the loss in entropy in this case is very much smaller. It should also be noted that when ionisation occurs, the liberated proton is also solvated. This solvation causes the 'freezing' of four molecules of water (to form $H^+(H_2O)_4$), and so this contributes to the decrease in entropy when the acid ionises. However, since the solvated proton is the same for all acids, only the change in entropy of the anion need be considered.

Chloroacetic acid may be prepared by the H.V.Z. reaction; the reaction can be carried out in direct sunlight, and in the absence of phosphorus. Chloroacetic acid is prepared industrially: (i) by heating trichloroethylene with 74 per cent sulphuric acid:

$$CHCl{=}CCl_2 + 2H_2O \xrightarrow[140°C]{H_2SO_4} ClCH_2CO_2H + 2HCl$$

(ii) By the chlorination of acetic acid in the presence of acetic anhydride-sulphuric acid mixture.

Chloroacetic acid is a deliquescent solid, m.p. 61°C, soluble in water and ethanol. It finds many uses in organic syntheses, and is used in the industrial preparation of indigotin (q.v.). It is readily esterified with ethanol in the presence of sulphuric acid.

Ethyl chloroacetate is converted into chloroacetamide when shaken with aqueous ammonia:

$$ClCH_2COOC_2H_5 + NH_3 \longrightarrow ClCH_2CONH_2 + C_2H_5OH \quad (78–84\%)$$

Chloroacetamide is dehydrated to chloroacetonitrile (b.p. 123°C) by phosphorus pentoxide:

$$ClCH_2CONH_2 \xrightarrow[-H_2O]{P_2O_5} ClCH_2CN \quad (61–70\%)$$

Dichloroacetic acid may be prepared in the laboratory and industrially by adding calcium carbonate to a warm aqueous solution of chloral hydrate, then adding an aqueous solution of sodium cyanide, and finally heating the mixture:

$$2CCl_3CH(OH)_2 + 2CaCO_3 \xrightarrow{NaCN} (CHCl_2CO_2)_2Ca + 2CO_2 + CaCl_2 + 2H_2O \xrightarrow{HCl}$$
$$CHCl_2CO_2H \quad (88–92\%)$$

The action of the sodium cyanide is not understood.

Dichloroacetic acid, b.p. 194°C, gives glyoxylic acid (q.v.) on hydrolysis:

$$CHCl_2CO_2H + H_2O \xrightarrow[(ii)\ H^+]{(i)\ NaOH} OCHCO_2H + 2HCl$$

Vigorous hydrolysis with concentrated alkali gives oxalate and glycollate, due to the glyoxylic acid undergoing the Cannizzaro reaction (p. 221):

$$2OCHCO_2H + 3NaOH \longrightarrow (CO_2Na)_2 + CH_2(OH)CO_2Na + 2H_2O$$

Trichloroacetic acid is best prepared by oxidising chloral hydrate with concentrated nitric acid:

$$CCl_3CH(OH)_2 + [O] \xrightarrow{HNO_3} CCl_3CO_2H + H_2O$$

Trichloroacetic acid is a deliquescent solid, m.p. 58°C, and is one of the strongest organic acids.

The presence of three chlorine atoms on a carbon atom adjacent to a carbonyl group causes the C—C bond to break very easily. Thus, when trichloroacetic acid is boiled with dilute sodium hydroxide, or even water, chloroform is obtained:

$$CCl_3CO_2H \longrightarrow CHCl_3 + CO_2$$

The ready fission of the C—C bond may be attributed to the strong inductive effect of the chlorine atoms, and a possible mechanism is:

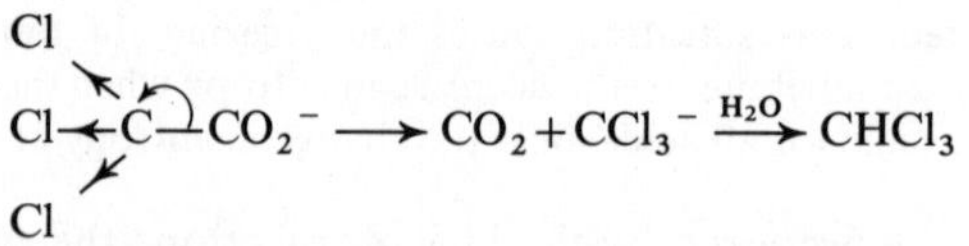

With *concentrated* alkali, formates are produced, and these probably result via the formation of dichloromethylene (see p. 170).

Trifluoroacetic acid is conveniently prepared by the oxidation of *p*-trifluoromethyltoluidine with chromic acid:

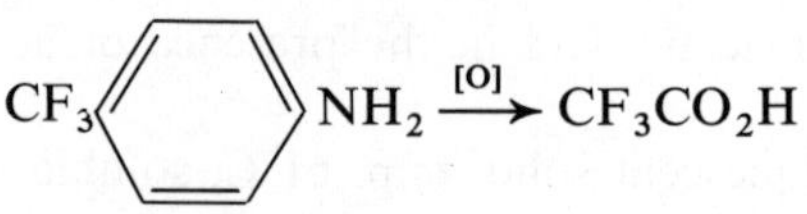

Trifluoroacetic acid is a liquid that fumes in the air, and is one of the strongest organic acids known. It does not form fluoroform when heated with alkali, and is reduced to trifluoroacetaldehyde by lithium aluminium hydride. Peroxytrifluoroacetic acid, CF_3CO_3H, is a useful oxidising agent (Emmons, 1954; *cf.* peroxyacetic acid).

Fluoroacetic acid may be prepared by heating a mixture of carbon monoxide, hydrogen fluoride and formaldehyde under pressure.

$$CO + HF + HCHO \longrightarrow FCH_2CO_2H$$

The simplest chloro-acid would be chloroformic acid, $ClCO_2H$, but it is unknown; chlorination of formic acid results in the formation of hydrogen chloride and carbon dioxide:

$$HCO_2H \xrightarrow{Cl_2} 2HCl + CO_2$$

On the other hand, esters of chloroformic acid are known (see p. 458).

PROBLEMS

1. Convert n-$C_5H_{11}CO_2H$ into n-$C_4H_9CO_2H$ by three methods.
2. Which of the following would be most and least readily hydrolysed with NaOH, and why?

   (i) $MeCO_2Me$; (ii) Me_2CHCO_2Me; (iii) $MeCO_2Bu^t$.

3. Using MeI, $\overset{*}{\text{MeI}}$, NaCN, Na$\overset{*}{\text{C}}$N ($\overset{*}{\text{C}}$=^{14}C) and any other chemicals, suggest a synthesis for each of the following:

(i) $\overset{*}{\text{CH}}_3\text{CO}_2\text{H}$; (ii) $\text{CH}_3\overset{*}{\text{CO}}_2\text{H}$; (iii) $\overset{*}{\text{CH}}_3\overset{*}{\text{CO}}_2\text{H}$; (iv) $\overset{*}{\text{CH}}_3\text{CO}\overset{*}{\text{CH}}_3$; (v) $\overset{*}{\text{CH}}_3\overset{*}{\text{CHO}}$.

4. Complete the equations and comment:

(i) $\text{CHCl}_2\text{COCl} \xrightarrow{\text{LAH}}$?   (ii) $\text{HOCH}_2\text{CH}_2\text{CN} \xrightarrow{\text{HBr}-\text{H}_2\text{O}}$?

(iii) $B_{\text{AC}}2$ mech. $\text{Me}_3\text{CCH}_2\text{OAc} + \text{OH}^- \longrightarrow$?   (iv) $B_{\text{AL}}1$ mech. $\text{Me}_3\text{CCH}_2\text{OAc} + \text{OH}^- \longrightarrow$?

(v) $\text{ClCH}_2\text{CO}_2\text{H} + \text{NaI} \xrightarrow{\text{acetone}}$?   (vi) $\text{Me}_2\text{CBrCONH}_2 \xrightarrow[\text{NaOH}]{\text{Br}_2}$?

5. Calculate $\Delta G^{\ominus}$ for the equilibrium:

$$\text{AcOH} + \text{EtOH} \rightleftharpoons \text{AcOEt} + \text{H}_2\text{O}$$
$$\text{moles} \quad \tfrac{1}{3} \qquad \tfrac{1}{3} \qquad \tfrac{2}{3} \qquad \tfrac{2}{3}$$

What information does the value of $\Delta G^{\ominus}$ give on the time required for this equilibrium to be reached?

6. Draw the energy profile for the $A_{\text{AC}}2$ esterification mechanism and comment.

7. A solution of $\text{Ph}_3\text{CCO}_2\text{H}$ in conc. H_2SO_4 gives MeOCPh_3 when poured into methanol. Discuss.

8. Convert: (i) n-BuOH into n-PrNH$_2$; (ii) $\text{Et}_2\text{CHCO}_2\text{H}$ into Et_2CHCHO.

9. Can MeCH_2COCl, $\text{MeCHClCO}_2\text{H}$ and $\text{ClCH}_2\text{CH}_2\text{CO}_2\text{H}$ be differentiated by NMR spectroscopy? Give reasons for your answer.

10. Draw the diagrammatic NMR spectra of: (i) MeCO_2Et; (ii) EtCO_2Me.

11. Give the fragmentation paths that could produce the following ions:

(i)

m/e	31	43	59	71	74	102
Me n-butyrate (%)	6	100	25	55	69	2

(ii)

m/e	16	43	44	59	71	87
n-Butanamide (%)	3	40	94	100	6	3

REFERENCES

HOPKINSON, 'Acid-catalysed Hydrolysis of Alkyl Benzoates', *J. Chem. Soc. (B)*, 1969, 203.

Organic Reactions, Wiley. Vol. III (1946), (i) Ch. 7, 'The Hofmann Reaction'. (ii) Ch. 8, 'The Schmidt Reaction'. (iii) Ch. 9, 'The Curtius Reaction'. Vol. IV (1948), Ch. 4, 'Acyloins'.

SIDGWICK, *The Organic Chemistry of Nitrogen*, Oxford University Press (1966, 3rd edn. by Millar and Springall).

GENSLER, 'Recent Developments in the Synthesis of Fatty Acids', *Chem. Rev.*, 1957, **57**, 191.

BENDER, 'Mechanisms of Catalysis of Nucleophilic Reactions of Carboxylic Acid Derivatives', *Chem. Rev.*, 1960, **60**, 53.

BELL, *The Proton in Chemistry*, Methuen (1959).

SATCHELL, 'An Outline of Acylation', *Quart. Rev.*, 1963, **17**, 160.

ELOY and LANAERS, 'The Chemistry of Amidoximes and Related Compounds', *Chem. Rev.*, 1962, **62**, 155.

WHEELER, 'The Girard Reagents', *Chem. Rev.*, 1962, **62**, 205.

10 Polycarbonyl compounds

Tautomerism

Acetoacetic ester or **ethyl acetoacetate** (E.A.A.) is the ethyl ester of acetoacetic acid, $CH_3COCH_2CO_2H$, a β-ketonic acid. Acetoacetic ester was first discovered by Geuther (1863), who prepared it by the action of sodium on ethyl acetate, and suggested the formula $CH_3C(OH)=CHCO_2C_2H_5$ (β-hydroxycrotonic ester). In 1865, Frankland and Duppa, who, independently of Geuther, also prepared acetoacetic ester by the action of sodium on ethyl acetate, proposed the formula $CH_3COCH_2CO_2C_2H_5$ (β-ketobutyric ester). These two formulæ immediately gave rise to two schools of thought, one upholding the Geuther formula, and the other the Frankland–Duppa formula, each school bringing forward evidence to prove its own claim, *e.g.*,

1. Evidence in favour of the Geuther formula (*reactions of an unsaturated alcohol*). (i) When acetoacetic ester is treated with sodium, hydrogen is evolved and the sodium derivative is formed. This indicates the presence of a hydroxyl group.

(ii) When acetoacetic ester is treated with an ethanolic solution of bromine, the colour of the latter is immediately discharged. This indicates the presence of an olefinic double bond.

(iii) When acetoacetic ester is treated with ferric chloride, a reddish-violet colour is produced.

This is characteristic of compounds containing the group $-C(OH)=C\diagup^{\diagup}_{\diagdown}$ (*cf.* phenols).

2. Evidence in favour of the Frankland–Duppa formula (*reactions of a ketone*). (i) Acetoacetic ester forms a bisulphite compound with sodium hydrogen sulphite.

(ii) Acetoacetic ester forms a cyanohydrin with hydrogen cyanide.

Thus the remarkable position arose where it was possible to show that a given compound had two different formulæ, each of which was based on a number of *particular* reactions. The controversy continued until about 1910, when chemists were coming to the conclusion that both formulæ were correct, and that the two compounds existed together in equilibrium in solution (or in the liquid state):

$$CH_3COCH_2CO_2C_2H_5 \rightleftharpoons CH_3\overset{\overset{\displaystyle OH}{|}}{C}=CHCO_2C_2H_5$$

When a reagent which reacts with ketones is added to acetoacetic ester, the ketone form is removed. This upsets the equilibrium, and in order to restore the equilibrium mixture, the hydroxy-form of acetoacetic ester changes into the ketone form. Thus, provided sufficient reagent is added, acetoacetic ester reacts completely as the ketone form. Similarly, when a reagent which reacts with alkenes or with hydroxy-compounds is added in sufficient quantity, acetoacetic ester reacts completely as the hydroxy-form.

The problem was finally settled by Knorr (1911), who succeeded in isolating both forms. He cooled a solution of acetoacetic ester in light petrol to $-78°C$, and obtained crystals which melted at $-39°C$. This substance gave no coloration with ferric chloride and did not combine with bromine, and was therefore the pure ketone form corresponding to the Frankland–Duppa formula. Knorr then suspended the sodium derivative of acetoacetic ester in light petrol cooled to $-78°C$, and treated this suspension with just enough hydrogen chloride to decompose the sodium salt. He now obtained a product which did not crystallise, but set to a glassy solid when cooled. This substance gave an intense coloration with ferric chloride, and was therefore the pure hydroxy-form corresponding to the Geuther formula.

Thus acetoacetic ester is a substance that does the duty of two structural isomers, each isomer being capable of changing rapidly into the other when the equilibrium is disturbed, *e.g.*, by the addition of certain reagents. This is a case of *dynamic isomerism*, and the name *tautomerism* (Greek: *same parts*) was given to this phenomenon by Laar (1885). The two forms are known as *tautomers* or *tautomerides*, the ketone isomer being called the *keto* form, and the hydroxy isomer the *enol* form. Hence this type of tautomerism is known as *keto-enol* tautomerism. (In the I.U.P.A.C. system of nomenclature the suffix *-en* indicates the presence of a double bond, and the suffix *-ol*, a hydroxyl group. The word *enol* is a combination of these suffixes and indicates the structure of this form.)

When one tautomer is more stable than the other under ordinary conditions, the former is known as the *stable* form, and the latter as the *labile* form. In practice, it is generally difficult to say which is the labile form, since very often a slight change in the conditions, *e.g.*, temperature or solvent, shifts the equilibrium from keto to enol or vice versa (see below). Tautomerism in the solid state is rare, and hence, in the solid state, one or other tautomer is normally stable, but in the liquid or gaseous state, or in solution, the two forms usually exist as an equilibrium mixture.

It has been found that the enol form is more volatile than the keto, and that the change from enol to keto is extremely sensitive to catalysts. Meyer *et al.* (1920, 1921) found that traces of basic compounds were very effective catalysts. Thus they found that soft glass vessels were unsuitable for the separation of the keto and enol forms, since, when fractionated, the more volatile enol form rapidly changed into the original keto-enol mixture under the catalytic influence of the walls of the containing vessel. Meyer, however, succeeded in separating the enol form from the keto by carrying out the fractional distillation under reduced pressure in silica apparatus which had been thoroughly cleaned (freed from dust, moisture, etc.). Distillation under these conditions is known as *aseptic distillation*.

In tautomerism in general there may be an equilibrium between two or more forms. One may predominate, or all may be present to about the same extent, the concentration of each form depending on the temperature and the solvent (if in solution). Although it may not be possible to separate tautomers owing to the ease and rapidity of their interconversion or, as in many cases, due to one form being almost completely absent, the presence of more than one compound may be shown by special properties of each isomer. Experiments using deuterium exchange reactions have also shown the presence of keto-enol mixtures, *e.g.*, Klar (1934) showed that hydrogen is exchanged slowly by deuterium when acetaldehyde is dissolved in D_2O. Since the

C—H bond in alkanes is stable under these conditions, the inference is that a hydroxyl group is present in acetaldehyde, *i.e.*, some enol form is present. Acetone was found to undergo this exchange more rapidly.

The methods used for the quantitative estimation of each form in any tautomeric equilibrium mixture, of which the keto-enol system is only one example, fall into two distinct groups, physical and chemical.

Physical methods

Physical methods do not disturb the equilibrium, for they do not depend on the removal of one form, and they should therefore be used wherever possible.

(i) The refractive index of the equilibrium mixture is determined experimentally. The refractive indices of both the keto and the enol forms are calculated (from a table of atomic refractions), and from these figures it is then possible to calculate the amount of each form present in the equilibrium mixture. In some cases the refractive index of each form may be obtained directly by isolating it, *e.g.*, acetoacetic ester.

(ii) If *one* form is an electrolyte, the electrical conductivity of the mixture is determined experimentally, and the amount of this form present may be calculated from the results, *e.g.*, nitromethane (p. 364).

(iii) The composition of the equilibrium mixture may be determined by means of optical rotation measurements (*cf.* mutarotation, p. 514).

(iv) The position of equilibrium in β-diketones and β-keto-esters has been determined by proton magnetic resonance (see later).

Chemical methods

Since chemical methods cause the removal of one form, it is necessary to use a reagent that reacts with this form faster than the rate of interconversion of the tautomers. Meyer (1911, 1912) found that in the case of keto-enol tautomerism, bromine reacts instantaneously with the enol form, and so slowly with the keto form in comparison with the enol, that the keto reaction may be ignored. Meyer introduced two procedures, the *direct* and the *indirect* method. In the *direct* method a weighed sample of the keto-enol mixture dissolved in ethanol is *rapidly* titrated with a dilute ethanolic solution of bromine at 0°C (to slow down the interconversion of the tautomers). The first appearance of excess of bromine indicates the end point.

The titration must be carried out rapidly; otherwise the keto form changes into the enol during the time taken for the titration. In any case, it has been found impossible to carry out the titration sufficiently rapidly to avoid the conversion of some keto into enol, and so this method always results in too high a value for the enol form.

More reliable results may be obtained by the *indirect* method. An *excess* of dilute ethanolic solution of bromine is added rapidly to the weighed sample dissolved in ethanol, and then an excess of 2-naphthol dissolved in ethanol is added *immediately*. By this means, the excess of bromine is removed almost instantaneously, and so the keto-enol equilibrium is not given time to be disturbed. Potassium iodide solution and hydrochloric acid are now added, and the liberated iodine is titrated with standard thiosulphate. The overall equation is:

$$CH_3COCHBrCO_2Et + 2I^- + H^+ \longrightarrow CH_3COCH_2CO_2Et + I_2 + Br^-$$

The mechanism of bromination of ethyl acetoacetate is probably (*cf.* acetone, p. 219):

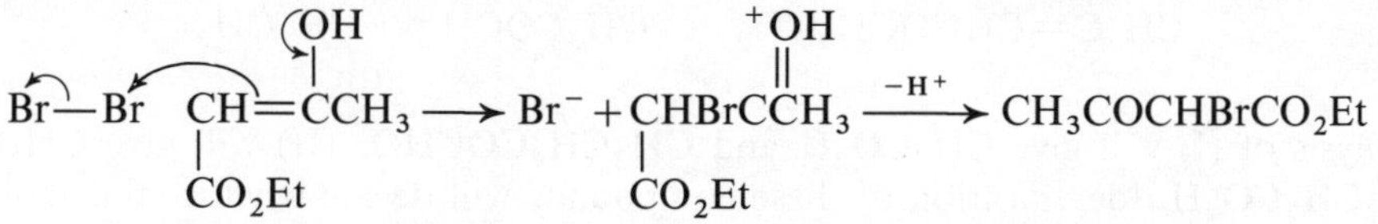

Gero (1960, 1961) has determined the enol content of monoketones by adding an excess of iodine monochloride in methanol to a mixture of the dry ketone and sodium hydrogen carbonate, then adding an excess of aqueous sodium iodide and titrating with standard thiosulphate.

Schwarzenbach and Wittwer (1947) have introduced the *flow-method* for the estimation of the enol content. The solution of the keto-enol mixture and an acidified bromide-bromate solution are simultaneously mixed and diluted in a mixing chamber, and the mixture made to flow past a platinum electrode. The relative amounts of the two solutions are adjusted so that the potential measured at the platinum electrode shows a sharp rise—this corresponds to the end-point in the titration of the enol form by the bromine. This method gives good results of enol contents as low as 10^{-5} per cent.; *e.g.*, the enol content of acetone in aqueous solution was found to be $2 \cdot 5 \times 10^{-4}$ per cent.

Table 10.1

Compound	Per cent enol (in ethanol)
$CH_3COCH_2CO_2CH_3$	4.8
$CH_3COCH_2CO_2C_2H_5$	7.5
$CH_3COCH_2COCH_3$	76
$CH_3COCH(CH_3)COCH_3$	31
$C_6H_5COCH_2COC_6H_5$	96
$CH_2(CO_2C_2H_5)_2$	trace
Aldehydes of type RCH_2CHO	trace
Ketones of type RCH_2COCH_2R	trace

The enol content in keto-enol mixtures (in dipropyl ether) has been determined by means of lithium aluminium hydride; the hydrogen liberated is estimated (Höfling *et al.*, 1952). The values obtained are about 10 per cent higher than those obtained by the bromine titration or by physical methods.

Enolisation

The phenomenon of enolisation is exhibited by compounds containing either a methylene group,

$-CH_2-$, or a methyne group, $\diagdown CH-$, adjacent to a carbonyl group. The actual amount of

enol form present depends on a number of factors; these are discussed later.

When the methylene or methyne group is attached to two or three carbonyl groups, the hydrogen atom might migrate equally well to one or other carbonyl group. This is not found to be so in practice for unsymmetrical compounds, one enol form being present exclusively, or largely predominating; *e.g.*, in acetoacetic ester the hydrogen atom migrates exclusively to the acetyl carbonyl group (see also below). When two or more enol forms are theoretically possible, ozonolysis may be used to ascertain the structure of the form present; *e.g.*, in hexane-2,4-dione, $CH_3COCH_2COCH_2CH_3$, the two possible enols are:

$$\underset{\text{(I)}}{CH_3\overset{\overset{\displaystyle OH}{|}}{C}=CHCOCH_2CH_3} \qquad \underset{\text{(II)}}{CH_3COCH=\overset{\overset{\displaystyle OH}{|}}{C}CH_2CH_3}$$

Ozonolysis of (I) will give CH_3CO_2H and CH_3CH_2COCHO; (II) will give CH_3COCHO and $CH_3CH_2CO_2H$. Identification of these compounds will decide whether the enol is (I) or (II), or both.

The type of tautomerism discussed above is known as the **keto–enol triad system**. (In this system a hydrogen atom migrates from atom 1 (*oxygen*) to atom 3 (*carbon*):

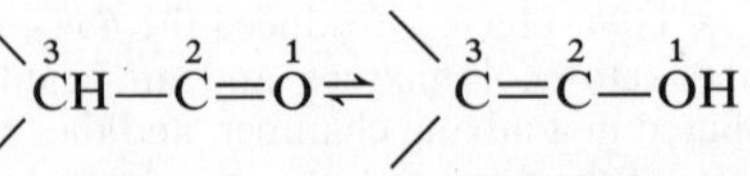

$$\overset{\diagdown}{\underset{\diagup}{}}\overset{3}{C}H-\overset{2}{C}=\overset{1}{O} \rightleftharpoons \overset{\diagdown}{\underset{\diagup}{}}\overset{3}{C}=\overset{2}{C}-\overset{1}{O}H$$

Enols resemble phenols in a number of ways; *e.g.*, both form soluble sodium salts; both give characteristic colorations with ferric chloride; and both couple with diazonium salts.

The keto–enol type of tautomerism is only one example of a triad system. The triad system is the most important class of tautomeric systems, and the following, which are exemplified in the text, are the commonest:

(i)   *Three-carbon systems:*
$$CH-C=C \rightleftharpoons C=C-CH$$

(ii)   *Nitro-acinitro (pseudonitro) system:*
$$CH-NO_2 \rightleftharpoons C=NOOH$$

(iii) *Nitroso-oximino system:*
$$CH-N=O \rightleftharpoons C=NOH$$

(iv)   *Amidine system:*
$$NH-C=N \rightleftharpoons N=C-NH$$

(v)   *Amido-imidol system:*
$$NH-C=O \rightleftharpoons N=C-OH$$

(vi)   *Azo-hydrazone system:*
$$N=N-CH \rightleftharpoons HN-N=C$$

(vii) *Diazo-amino (triazen) system:*
$$N=N-NH \rightleftharpoons NH-N=N$$

(viii) *Diazo-nitrosamine system:*
$$Ar-N=NOH \rightleftharpoons Ar-NH-N=O$$

In addition to the 'open' systems of tautomerism, there is also, *e.g.*, *ring-chain tautomerism* (see, *e.g.*, aldol, p. 298; carbohydrates, p. 514).

Theories of tautomerism

Ingold (1927) suggested the name *cationotropy* for all those cases of tautomerism which involve the separation of a *cation*; and the name *anionotropy* for those cases which involve the separation of an *anion*. Lowry (1923) suggested the name *prototropy* for those cases in which a proton separates, and called such systems *prototropic* systems. Using Ingold's generalised classification of tautomeric systems, it can be seen that prototropy is a special case of cationotropy. Braude and Jones (1944) have proposed the term *oxotropy* for anionotropic rearrangements involving only the migration of a *hydroxyl* group (see allylic rearrangement, p. 334).

The actual steps involved in keto–enol tautomerism are still the subject of much discussion. According to Hughes and Ingold, base-catalysed enolisation of a ketone proceeds through an enolate anion (I) whose formation is controlled largely by the inductive effects of the alkyl groups.

$$B + R_2CHCOR \longrightarrow BH^+ + R_2\overset{-}{C}-\overset{\overset{\displaystyle O}{\|}}{C}R \longleftrightarrow R_2C=\overset{\overset{\displaystyle O^-}{|}}{C}R$$
$$\text{(I)}$$

Acid-catalysed enolisation involves the removal of a proton from the conjugate acid of the ketone, (II), and this process is dependent mainly on the hyperconjugation (p. 332) by the alkyl groups in the transition state for the formation of the carbon–carbon double bond.

$$H^+ + R_2\overset{\overset{\displaystyle H}{|}}{C}-\overset{\overset{\displaystyle O}{\|}}{C}-R \longrightarrow R_2\overset{\overset{\displaystyle H}{|}}{C}-\overset{\overset{\displaystyle OH}{|}}{\overset{+}{C}}-R \longrightarrow H^+ + R_2C=\overset{\overset{\displaystyle OH}{|}}{C}-R$$
$$\text{(II)}$$

Evidence for these mechanisms is that alkyl groups depress base-catalysed reaction rates.

The mechanisms described above are step-wise mechanisms, and they appear to operate in certain cases. On the other hand, there is evidence that in other cases, acid- and base-catalysis of enolisation take place by a *concerted* mechanism, *i.e.*, the molecule undergoing change is attacked simultaneously at two places. Thus the enolisation of, *e.g.*, acetone, proceeds by the simultaneous removal of a proton from an α-carbon and the addition of a proton to the oxygen of the carbonyl group. This may be represented as follows (B is a base and HA is an acid):

$$\begin{array}{ccc}
B\cdots H-CH_2 & B\cdots H\cdots CH_2 & CH_2 \\
\quad\downarrow & \quad\vdots & \quad\overset{+}{\|} \\
C=O\cdots HA & C=O\cdots HA & BH + C-OH + A^- \\
\quad| & \quad| & \quad| \\
CH_3 & CH_3 & CH_3
\end{array}$$
$$\text{transition state}$$

In this type of mechanism, the solvent water molecules are believed to be involved, one acting as a proton acceptor and the other as a proton donor. Furthermore, the above mechanism is termolecular, and Swain (1950) has presented evidence to show that both the acid- and base-catalysed enolisation of acetone are termolecular reactions. Emmons *et al.* (1956) have obtained evidence supporting Swain's termolecular mechanism for ketone enolisation, but propose that the transition state for both acid- and base-catalysed enolisation may be represented by (III). Under these conditions in either acid- or base-catalysed reactions, the transition state will be very close to the enol and hence will be stabilised by hyperconjugation. In the acid-catalysed reaction, bond *b* is relatively tight and bond *a* is relatively loose. Hence in acid-catalysed enolisation, steric hindrance around bond *a* is not observed in most cases, and hyperconjugation is therefore a dominating factor in the enolisation of a ketone. In the base-catalysed reaction, bond *a* is relatively tight and bond *b* is relatively loose. Under these conditions a steric factor becomes important and is probably the reason why alkyl groups depress base-catalysed reaction rates.

$$\begin{array}{c}
\overset{\displaystyle b}{\overbrace{\qquad\qquad}} \\
RCH_2 \quad O\cdots HA \\
\quad\searrow\!\overset{\vdots}{\|} \\
\qquad C=CR \\
\quad\nearrow \quad \vdots \\
RCH_2\;\overset{-}{}\;\vdots \\
\qquad B\cdots H \\
\underbrace{\qquad\qquad}_{\displaystyle a} \\
\text{(III)}
\end{array}$$

The tremendous increase in speed by acid and base catalysts is well illustrated by the fact that the pure keto and enol forms of ethyl acetoacetate change very slowly into the equilibrium mixture (several weeks), whereas the latter is obtained rapidly (several minutes) by the addition of acid or base catalysts (Lowry et al., 1924).

Position of equilibrium in keto–enol tautomerism

Since

$$\Delta G^{\ominus} = \Delta H^{\ominus} - T\Delta S^{\ominus} = -RT \ln K$$

a knowledge of the values of the enthalpy and entropy changes would enable us to calculate $\Delta G^{\ominus}$, and from this, K, the equilibrium constant of the reaction:

$$\text{keto} \rightleftharpoons \text{enol}; \ K = [\text{enol}]/[\text{keto}]$$

The sum of the bond energies of the —CH$_2$CO— group is 1 870 and that of the —CH=C(OH)— group is 1 820 kJ (from Table 2.2):

$$2C—H(828\cdot4) + C—C(347\cdot3) + C=O(694\cdot5) = 1\,870\,\text{kJ}$$

$$C—H(414\cdot2) + C=C(606\cdot7) + C—O(334\cdot7) + O—H(464\cdot4) = 1\,820\,\text{kJ}$$

This means that the *enthalpy of formation* of the CH$_2$CO group is $-1\,870$ kJ and that of the CH=C(OH) group is $-1\,820$ kJ. Thus the keto form is more stable than the enol by 50 kJ mol^{-1}, and hence there must be some driving force to bring about enolisation.

If the equilibrium mixture contains 1 per cent enol, then

$$\Delta G^{\ominus} = -5\cdot7 \log 1/99 = 11\cdot42\,\text{kJ}$$

If the equilibrium mixture contains 99 per cent enol, then

$$\Delta G^{\ominus} = -5\cdot7 \log 99 = -11\cdot42\,\text{kJ}$$

Thus a difference of 22·84 kJ mol^{-1} in ΔG shifts the equilibrium from 1 per cent enol to 99 per cent enol. It can therefore be seen that small changes in ΔG will have large effects on the position of the equilibrium. Hence any attempt to calculate K from *estimated* values of ΔH and ΔS is almost certainly doomed to failure. However, a semi-quantitative approach is instructive. Any factors that affect $\Delta H^{\ominus}$ and $\Delta S^{\ominus}$ will affect the value of $\Delta G^{\ominus}$. Consider the following equilibria:

acetylacetone Me—C(=O)—CH$_2$—C(=O)—Me $\rightleftharpoons$ (enol form)

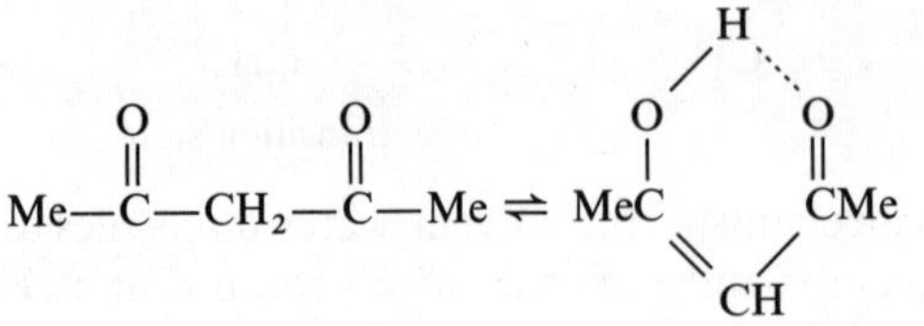

ethyl acetoacetate Me—C(=O)—CH$_2$—C(=O)—OEt $\rightleftharpoons$ (enol form)

ethyl malonate EtOC(=O)—CH$_2$—C(=O)—OEt $\rightleftharpoons$ (enol form)

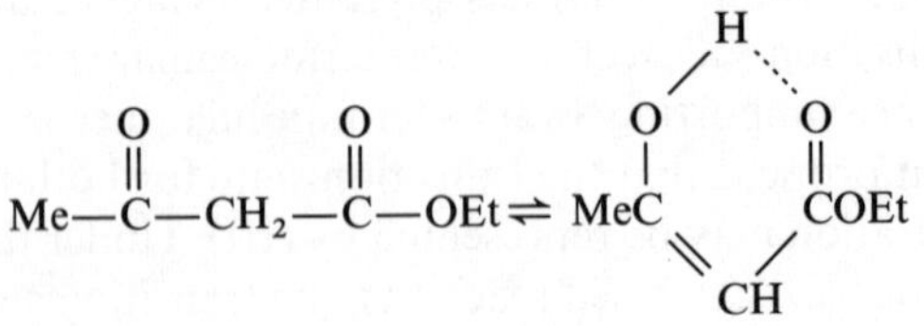

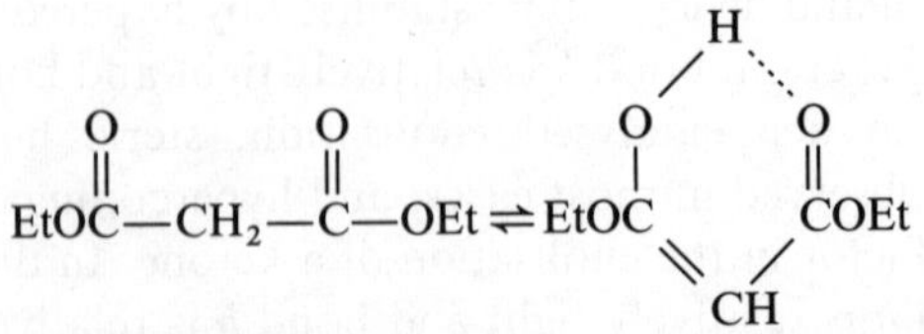

The enol form, in each case, is stabilised by intramolecular hydrogen bonding (say, 29·3 kJ mol^{-1}).

Thus, in the absence of any other factor affecting ΔH, we have:

$$\Delta H^{\ominus} = \Sigma H^{\ominus} \text{ (products)} - \Sigma H^{\ominus} \text{ (reactants)}$$

$$= -(1\,820+29\cdot3)-(-1\,870) = +20\cdot7\,\text{kJ}$$

$$\Delta G^{\ominus} = 20\cdot7 - T\Delta S^{\ominus} = -RT\ln K$$

Now let us consider the resonance energies of crotonaldehyde (I), ethyl acetate (II) and ethyl

$$\underset{\substack{\text{R.E. (kJ)} \\ \text{(I) } 10\cdot04}}{\text{MeCH}{=}\text{CHCHO}} \qquad \underset{\text{(II) } 75\cdot31}{\text{Me}\overset{\overset{\displaystyle O}{\|}}{\text{C}}{-}\text{OEt}} \qquad \underset{\text{(III) } 75\cdot31}{\text{CH}_2{=}\overset{\overset{\displaystyle Me}{|}}{\text{C}}{-}\overset{\overset{\displaystyle O}{/\!/}}{\text{C}}{-}\text{OEt}}$$

methacrylate (III). The important point here is that in (III) the ethylenic bond does not enter into resonance with the CO_2Et group. (III) contains crossed conjugation (see p. 724), and the contribution to resonance is only the path that leads to the greater R.E., which, in this case, is the carbethoxy group.

Inspection of the enol formulæ shows that the ring contains a conjugated system:

$$\text{H}{-}\text{O}{-}\overset{|}{\text{C}}{=}\text{CH}{-}\overset{|}{\text{C}}{=}\text{O} \leftrightarrow \text{H}{-}\overset{+}{\text{O}}{=}\overset{|}{\text{C}}{-}\text{CH}{=}\overset{|}{\text{C}}{-}\text{O}^-$$

The R.E. value of this is difficult to assess, but it can be expected to be considerably less than $75\cdot31$ kJ. Thus, in E.A.A., only the remaining group, $\text{HO}{-}\text{C}{=}\text{CH}{-}$, will make any contribution to the 'ring' resonance. It therefore follows that the enol form of E.A.A. is less stabilised by resonance than that of acetylacetone, and so will be present in smaller amount. It might also be noted here that if enolisation in E.A.A. occurred at the CO group of CO_2Et, then the R.E. of the system produced is far less. Thus, there is no tendency to enolise in this direction. From this it can be seen that D.E.M. will have far less enol form than E.A.A. In the case of simple carbonyl compounds, e.g., acetone, acetaldehyde, etc., there is no extended conjugation in the enol form, and, of far greater importance, there is no intramolecular hydrogen bonding to stabilise the enol. Thus there is very little tendency for these carbonyl compounds to enolise.

Just as it is difficult to estimate $\Delta H^{\ominus}$, so it is more difficult to estimate $\Delta S^{\ominus}$. In the diketo form, the molecule is acyclic and consequently can assume various conformations. In the enol form, however, the presence of a double bond prevents change in conformation at this part of the molecule and, at the same time, conformational change is further restricted by hydrogen bonding, since this tends to maintain a ring structure. Thus both of these factors will tend to cause a *decrease* in entropy when the diketo compound enolises. Some support for this argument is that Gero (1961) has found that cyclic monoketones contain more enol than the corresponding acyclic 2-one. In the cyclic ketones, change from keto to enol involves a relatively small change in strain in the ring due to the introduction of a double bond. For acyclic ketones, the introduction of the double bond has a much greater effect on the freedom to take up different conformations.

Another factor that determines the enol content is the steric factor. Let us consider the enol content of acetylacetone and α-methylacetylacetone; the values given are for the gas phase (Conant *et al.*, 1932). In the latter compound, there is much greater steric repulsion due to the presence of the

α-methyl group. Thus, the α-methyl enol form has greater internal strain than the enol of acetylacetone and so the former is less favourable energetically than the latter. Hence less enol will be present in the former equilibrium mixture than in the latter. A striking example of strain affecting the position of equilibrium is the case of triacylmethanes, $(RCO)_3CH$. Nonhebel (1967) has shown, by means of i.r., u.v., and NMR studies, that as the size of R increases, the amount of enol content decreases, e.g., $(Bu^tCO)_3CH$ is *completely* in the keto form.

Finally, let us consider the effect of the nature of the solvent on the enol content. If one form is more solvated than the other, then there may be considerable entropy effects connected with the nature of the solvent (*cf.* acids, p. 273). Any solvent that can form hydrogen bonds with the carbonyl group of the keto form will cause a decrease in entropy. The enol form, however, since it forms an intra-molecular bond, will be largely prevented from forming hydrogen bonds with the solvent, and conse-quently there will be a smaller decrease in entropy. Thus, the term $\Delta S^{\ominus}$ becomes important in deciding the position of keto–enol equilibrium, *e.g.*, solvents such as water, methanol, acetic acid, etc., tend to reduce the enol content. On the other hand, in solvents such as hexane, benzene, etc., the enol content will be larger, *e.g.*, the enol content of acetylacetone in hexane is 92 per cent (see also Table 10.1).

It has been shown that the enol content of E.A.A. (pure liquid) decreases as the pressure increases (le Noble, 1960). From this it follows that the molar volume of the enol form is greater than that of the keto (*cf.* p. 58).

The acidity of keto compounds

The proton transfer for these compounds may be written:

$$-CH_2-\overset{\overset{\textstyle O}{\|}}{C}-+B \rightleftharpoons BH^+ + -\overset{\cdot\cdot}{C}H-\overset{\overset{\textstyle O}{\|}}{C}- \longleftrightarrow -CH=\overset{\overset{\textstyle O^-}{|}}{C}-$$

The carbanion produced is a resonance hybrid, and the transference of charge to the oxygen atom which has a higher electronegativity than the carbon atom will make this resonating structure a greater contributor to the resonance hybrid; the result is an ambident anion (p. 128). The presence of the carbonyl group facilitates the release of a proton from the α-carbon atom, and the resonance hybrid nature of the conjugate base stabilises this with respect to its acid. Both factors are acting in the 'same' direction, and consequently such compounds will be acidic. However, the pK_a of the compound will depend on how strong these two driving forces are. As we have seen above, since enolisation involves a removal of a proton first, it follows that the acid strengths of these carbonyl compounds, in a proton-accepting solvent, will run parallel with the enol content of each compound, *e.g.*,

E.A.A.	$(MeCO)_2CH_2$
pK_a 10–68	8·24

Infra-red spectra of β-diketones

If β-diketones existed as such, one would expect to find absorption in the region of a single carbonyl group ($1\,725$–$1\,700\,cm^{-1}$), but with greater intensity because two such groups are present. In practice, those β-diketones which exist partly (or wholly) in the enol form (as indicated by the coloration with ferric chloride) show the carbonyl absorption in the region $1\,640$–$1\,540\,cm^{-1}$ (s); the enolic form of β-keto-esters absorbs in the region $1\,655$–$1\,635\,cm^{-1}$ (s). α,β-Unsaturated ketones, *i.e.*, those containing the grouping $C{=}C{-}C{=}O$ absorb in the region $1\,685$–$1\,665\,cm^{-1}$ (s). Enols contain this grouping, but the shift that occurs in these com-pounds is attributed to hydrogen bonding as follows: $-O{-}H\cdots O{=}C{-}$ (see chelated structures of the enols described above). At the same time, the true alcoholic hydroxyl band near $3\,700\,cm^{-1}$ is absent in enols, but there is a band near $2\,700\,cm^{-1}$ (s) which is attributed to the *chelated* hydroxyl group. Thus, infra-red spectral studies of β-dicarbonyl compounds may be used to ascertain their structures.

NMR spectra studies have been very useful in the examination of keto–enol tautomeric mixtures, *e.g.*, Fig. 10.1 is the NMR spectrum of ethyl acetoacetate (in CCl_4 at 60 MHz), and it

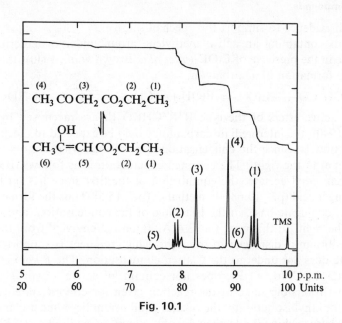

Fig. 10.1

can be seen that the CH_2 and $=CH$ protons give separate signals. Furthermore, very careful integration will give the proton ratio of these two groups and so it is possible to estimate the amount of enol content.

The preparation of ethyl acetoacetate

In the preparation of ethyl acetoacetate, condensation is effected between two molecules of the same ester. This is one example of the **Claisen condensation** (1887), in which a keto-ester is formed by the reaction between two molecules of an ester containing α-hydrogen atoms. The Claisen condensation may also take place between an ester and a ketone to form a 1,3-diketone. The condensation is brought about by sodium ethoxide, sodamide, triphenylmethylsodium, etc.

Many mechanisms have been proposed for the formation of ethyl acetoacetate. The one most widely accepted is (*cf.* aldol condensation):

$$EtO^- \quad H{-}CH_2CO_2Et \;\rightleftharpoons\; EtOH + \overset{-}{C}H_2CO_2Et$$

$$\underset{OEt}{\overset{O}{\underset{|}{\overset{||}{Me\overset{}{C}}}}} CH_2CO_2Et \;\rightleftharpoons\; \underset{OEt}{\overset{O^-}{\underset{|}{\overset{|}{Me\overset{}{C}}}}}{-}CH_2CO_2Et \;\rightleftharpoons\;$$

$$\overset{O}{\overset{||}{MeC}}{-}\underset{}{\overset{H}{\underset{|}{CHCO_2Et}}} + EtO^- \;\rightleftharpoons\; EtOH + \overset{\bar{O}}{\underset{}{MeC}}{=}CHCO_2Et \xrightarrow{\;MeCO_2H\;}$$

$$\underset{OH}{\overset{}{\underset{|}{MeC}}}{=}CHCO_2Et \;\rightleftharpoons\; \underset{}{\overset{O}{\overset{||}{MeCCH_2CO_2Et}}} \quad (28{-}29\%)$$

There is much evidence to support this mechanism, *e.g.*;

(i) Compounds containing an active methylene group undergo deuterium exchange with sodium ethoxide in the presence of EtOD (*inter alia*, Brown *et al.*, 1940); this can be explained by the *reversible* formation of a carbanion,

$$-CH_2CO- + EtO^- \rightleftharpoons EtOH + -\bar{C}HCO- \underset{EtOD}{\rightleftharpoons} -CHDCO- + EtO^-$$

(ii) Optically active esters of the type $R^1R^2CHCO_2Et$ are racemised by the ethoxide ion (Kenyon *et al.*, 1940); the intermediate carbanion would be expected to racemise (see p. 484).

The problem now is: Since the condensation is reversible, what factors favour shifting the overall equilibrium to the right? The accepted explanation is as follows. Because ethanol is a stronger acid than ethyl acetate, the equilibrium of the first stage lies on the left. However, E.A.A. is a stronger acid (pK_a 10·68) than ethanol (pK_a 15·5). Thus the former is present mainly as its anion, *i.e.*, conjugate base, in the last stage of the condensation, and so the equilibrium lies largely on the right. In this way, the E.A.A. anion is 'removed' from the reaction, and so the overall equilibrium is displaced to the right. This argument is supported by the fact that ethyl isobutyrate does *not* undergo the Claisen condensation. The β-keto ester that would be produced, $Me_2CHCOCMe_2CO_2Et$, does not contain an active α-hydrogen atom, and consequently the last reversible stage of the reaction given for ethyl acetate as substrate is now absent. Hence, for ethyl isobutyrate, the overall equilibrium lies almost completely on the left. If a very strong base such as the triphenylmethyl anion (as sodium salt), $Ph_3C^-Na^+$ (pK_a of Ph_3CH is about 33) is used, then the equilibrium of the first stage with ethyl isobutyrate lies on the right, and the Claisen condensation will proceed, since the position of equilibrium of the last stage will no longer matter.

Ethyl acetoacetate is also prepared industrially by polymerising keten in acetone solution to diketen, which is then treated with ethanol. A possible mechanism is (see also p. 352):

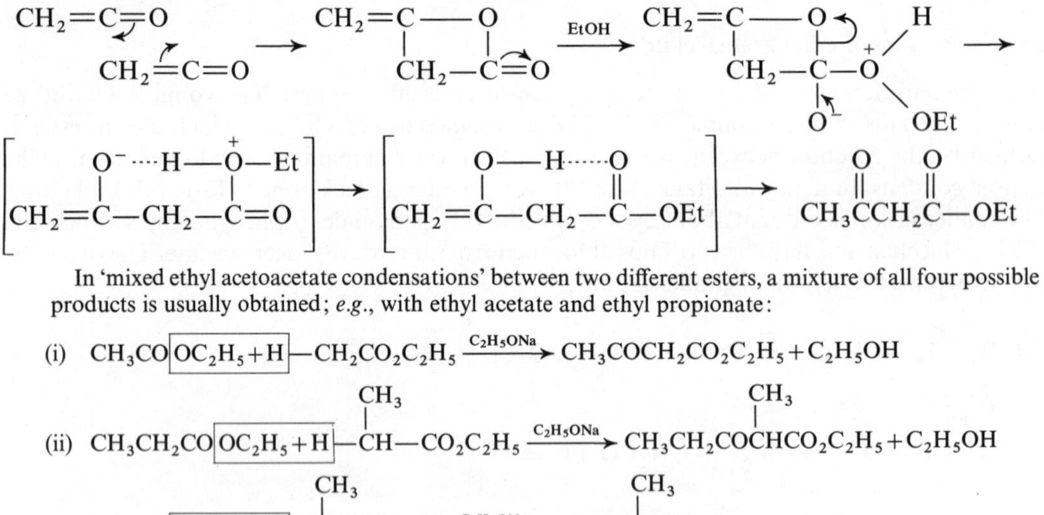

In 'mixed ethyl acetoacetate condensations' between two different esters, a mixture of all four possible products is usually obtained; *e.g.*, with ethyl acetate and ethyl propionate:

(i) $CH_3CO\boxed{OC_2H_5 + H}{-}CH_2CO_2C_2H_5 \xrightarrow{C_2H_5ONa} CH_3COCH_2CO_2C_2H_5 + C_2H_5OH$

(ii) $CH_3CH_2CO\boxed{OC_2H_5 + H}{-}\overset{\overset{\displaystyle CH_3}{|}}{CH}{-}CO_2C_2H_5 \xrightarrow{C_2H_5ONa} CH_3CH_2CO\overset{\overset{\displaystyle CH_3}{|}}{CH}CO_2C_2H_5 + C_2H_5OH$

(iii) $CH_3CO\boxed{OC_2H_5 + H}{-}\overset{\overset{\displaystyle CH_3}{|}}{CH}CO_2C_2H_5 \xrightarrow{C_2H_5ONa} CH_3CO\overset{\overset{\displaystyle CH_3}{|}}{CH}CO_2C_2H_5 + C_2H_5OH$

(iv) $CH_3CH_2CO\boxed{OC_2H_5 + H}{-}CH_2CO_2C_2H_5 \xrightarrow{C_2H_5ONa} CH_3CH_2COCH_2CO_2C_2H_5 + C_2H_5OH$

Esters with no α-hydrogen atoms will condense only 'one way' with compounds containing an active methylene group. Three important esters of this type are formic, benzoic and oxalic esters, and these give, respectively, the formyl, benzoyl and oxalyl derivatives. Ethyl carbonate also does not contain any α-hydrogen atoms, and is very useful for introducing a carbethoxyl group into compounds with an active methylene group.

β-Keto esters, other than E.A.A., may be prepared by the Claisen condensation between ketones and ethyl carbonate (see p. 458) or ethyl oxalate (*cf.* decarbonylation of ethyl pyruvate, p. 305):

$$\text{RCOMe} + \text{EtO}_2\text{CCO}_2\text{Et} \xrightarrow{\text{EtONa}} \text{RCOCH}_2\text{COCO}_2\text{Et} \xrightarrow{\text{heat}} \text{RCOCH}_2\text{CO}_2\text{Et} + \text{CO}$$

Properties of ethyl acetoacetate

Ethyl acetoacetate is a colourless, pleasant-smelling liquid, b.p. 181°C (with slight decomposition), sparingly soluble in water, but miscible with most organic solvents. It is readily reduced by sodium amalgam, or catalytically to β-hydroxybutyric ester:

$$\text{CH}_3\text{COCH}_2\text{CO}_2\text{C}_2\text{H}_5 \longrightarrow \text{CH}_3\text{CH(OH)CH}_2\text{CO}_2\text{C}_2\text{H}_5$$

This is also produced by using lithium aluminium hydride in pyridine (see p. 178). In the absence of pyridine, the product is butane-1,3-diol (Buchta *et al.*, 1951).

$$\text{CH}_3\text{COCH}_2\text{CO}_2\text{C}_2\text{H}_5 \xrightarrow{\text{LiAlH}_4} \text{CH}_3\text{CHOHCH}_2\text{CH}_2\text{OH} \quad (30\%)$$

Kent *et al.* (1963) have shown that sodium borohydride in boiling ethanol also produces the diol.

Ethyl acetoacetate is neutral to litmus, but is soluble in dilute sodium hydroxide solution: it is the enol form which dissolves to form the sodium salt. It does *not* give the haloform reaction (see p. 278). When ethyl acetoacetate is hydrolysed in the cold with dilute sodium hydroxide, the solution then acidified, extracted with ether and the ether removed in the cold under reduced pressure, free acetoacetic acid is obtained. Krueger (1952) has prepared acetoacetic acid as a crystalline solid, m.p. 36–37°C. As prepared in the usual way it is unstable, readily decomposing into acetone and carbon dioxide:

$$\text{CH}_3\text{COCH}_2\text{CO}_2\text{H} \longrightarrow \text{CH}_3\text{COCH}_3 + \text{CO}_2$$

All β-ketonic acids readily decompose in this manner, but it is interesting to note that trifluoroacetoacetic acid, $\text{CF}_3\text{COCH}_2\text{CO}_2\text{H}$, is quite stable and can be distilled without much decomposition. Ethyl acetoacetate forms α-oximinoacetoacetate when treated with nitrous acid:

$$\text{CH}_3\text{COCH}_2\text{CO}_2\text{C}_2\text{H}_5 \xrightarrow{\text{HNO}_2} \text{CH}_3\text{COC}(=\text{NOH})\text{CO}_2\text{C}_2\text{H}_5$$

On the other hand, monosubstituted derivatives of ethyl acetoacetate are split by nitrous acid; the nitroso-compound is formed first and then rearranges to the oximino form with fission of the molecule.

$$\text{CH}_3\text{COCHRCO}_2\text{C}_2\text{H}_5 \xrightarrow{\text{HNO}_2} \underset{\overset{|}{\text{NO}}}{\text{CH}_3\text{COCRCO}_2\text{C}_2\text{H}_5} \longrightarrow \underset{\overset{\|}{\text{NOH}}}{\text{RCCO}_2\text{C}_2\text{H}_5}$$

Ethyl acetoacetate forms a green copper compound which is soluble in organic solvents. This indicates that the copper compound is not an ionic but a chelate compound (the alternative is a π-delocalised structure). Bromination of E.A.A. is catalysed by the presence of cupric ions; this

is due to the stabilisation of the enol form by chelate formation.

E.A.A. reacts with ammonia (and primary amines) to form an enamine (p. 383):

$$Me-\overset{\overset{\displaystyle O}{\|}}{C}-CH_2CO_2Et + NH_3 \longrightarrow Me-\overset{\overset{\displaystyle OH}{|}}{\underset{\underset{\displaystyle NH_2}{|}}{C}}-CH_2CO_2Et \xrightarrow{-H_2O} Me-\overset{}{\underset{\underset{\displaystyle NH_2}{|}}{C}}=CHCO_2Et$$

Monoalkyl derivatives of E.A.A. also behave in a similar manner, but dialkyl derivatives form amides, $MeCOCR_2CONH_2$. In the latter, there is no α-hydrogen atom.

Ethyl acetoacetate reacts with hydroxylamine to form an isoxazolone.

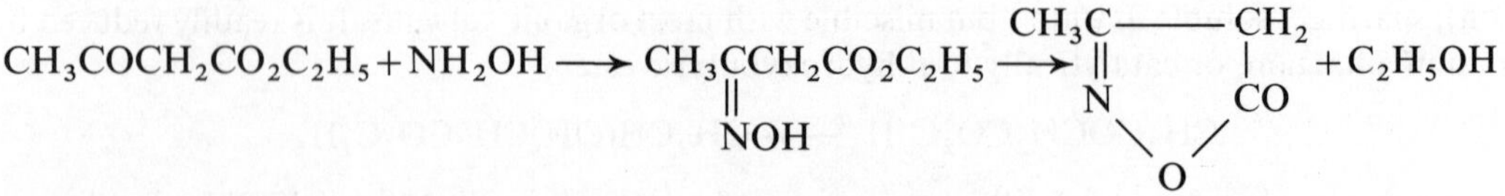

$$CH_3COCH_2CO_2C_2H_5 + NH_2OH \longrightarrow CH_3\underset{\underset{\displaystyle NOH}{\|}}{C}CH_2CO_2C_2H_5 \longrightarrow \overset{CH_3C \rule[0.5ex]{2em}{0.4pt} CH_2}{\underset{\displaystyle N \qquad CO}{\underset{\displaystyle \diagdown \; O \; \diagup}{}}} + C_2H_5OH$$

With phenylhydrazine, ethyl acetoacetate forms a pyrazolone (p. 846).

Acetoacetic ester undergoes the Knoevenagel reaction due to the presence of an 'active' methylene group (p. 216), and also couples with diazonium salts. It reacts with Grignard reagents to form the hydrocarbon, which indicates that it reacts in the enol form, e.g.,

$$CH_3\overset{\overset{\displaystyle OH}{|}}{C}=CHCO_2C_2H_5 + CH_3MgI \longrightarrow CH_3\overset{\overset{\displaystyle OMgI}{|}}{C}=CHCO_2C_2H_5 + CH_4$$

The O-methyl derivative is also produced by the action of diazomethane on ethyl acetoacetate. Thus the latter compound reacts in the enol form (see also p. 393).

When acetoacetic ester is heated under reflux with a trace of sodium hydrogen carbonate, **dehydroacetic acid**, m.p. 109°C, is obtained.

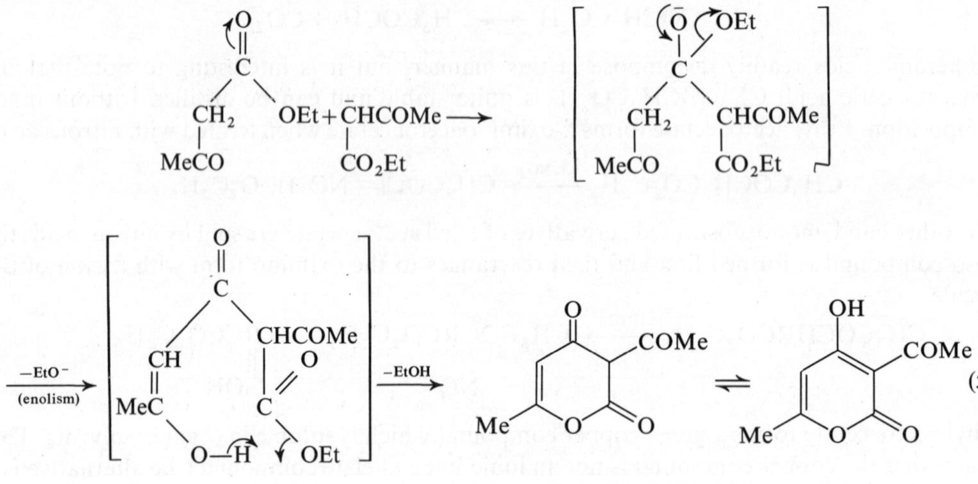

Dehydroacetic acid may also be prepared by the polymerisation of diketen:

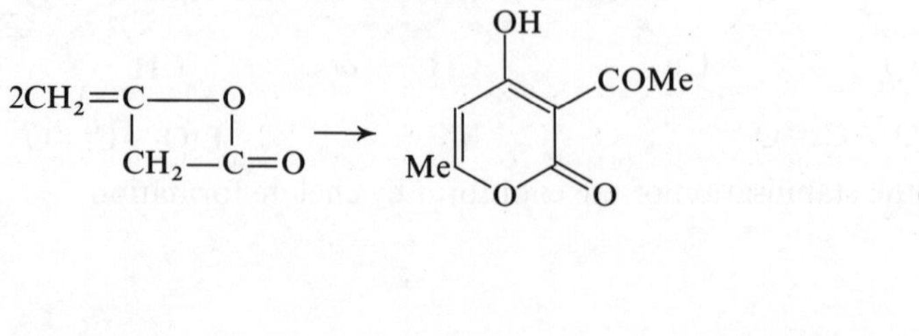

(60–80%)

When E.A.A. is allowed to stand in concentrated sulphuric acid (5–6 days), **isodehydroacetic acid**, m.p. 155°C, and its ester are formed:

$$EtO_2C \cdots \text{(structure, Me)} \quad (27\text{–}36\%) \quad \longleftarrow 2MeCOCH_2CO_2Et \longrightarrow \quad HO_2C \cdots \text{(structure, Me)} \quad (22\text{–}27\%)$$

Both acids are α-pyrone derivatives (p. 862).

The use of acetoacetic ester in the synthesis of ketones and monocarboxylic acids

The synthetic use of acetoacetic ester depends on two chemical properties:

1. (*a*) When treated with sodium ethoxide, acetoacetic ester forms sodioacetoacetic ester:

$$\overset{OH}{\underset{|}{CH_3C}}{=}CHCO_2C_2H_5 + C_2H_5ONa \longrightarrow \left[\overset{O^-}{\underset{|}{CH_3C}}{=}CHCO_2C_2H_5 \right] \overset{+}{Na} + C_2H_5OH \quad (ex.)$$

(*b*) Sodioacetoacetic ester readily reacts with primary and secondary alkyl halides (vinyl and aryl halides do *not* react) to produce alkyl derivatives of acetoacetic ester in which the alkyl group is attached to carbon. This fact has given rise to considerable speculation regarding the mechanism of the alkylation process. The problem is still not settled, but a highly favoured theory is that the negative ion is a resonance hybrid.

$$\overset{O^-}{\underset{|}{CH_3C}}{=}CHCO_2C_2H_5 \longleftrightarrow \overset{O}{\underset{||}{CH_3C}}{-}\overset{-}{CH}CO_2C_2H_5$$
$$\text{(I)} \qquad\qquad\qquad \text{(II)}$$

i.e., this ion is ambident. Kornblum *et al.* (1955) have pointed out that the tendency for alkylation at the more electronegative atom of an ambident anion usually increases with the S_N1 character of the reaction. As far as the alkylation of E.A.A. is concerned, the evidence is in favour of an S_N2 mechanism and consequently attack occurs at the carbon atom.

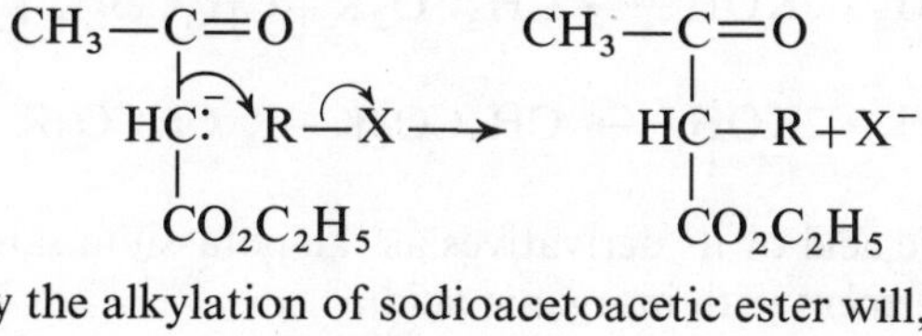

$$\begin{array}{cc} CH_3{-}C{=}O & CH_3{-}C{=}O \\ | & | \\ HC{:}\,{}^{\frown}R{-}X \longrightarrow & HC{-}R + X^- \\ | & | \\ CO_2C_2H_5 & CO_2C_2H_5 \end{array}$$

For the sake of simplicity the alkylation of sodioacetoacetic ester will, in future, be represented as:

$$[CH_3COCHCO_2C_2H_5]^- Na^+ + RX \longrightarrow CH_3COCHRCO_2C_2H_5 + NaX \quad (v.g.)$$

After one alkyl group has been introduced, the whole process may then be repeated to give the dialkyl derivative of acetoacetic ester:

$$CH_3COCHR^1CO_2C_2H_5 \xrightarrow{C_2H_5ONa} [CH_3COCR^1CO_2C_2H_5]^- Na^+ \xrightarrow{R^2X}$$
$$CH_3COCR^1R^2CO_2C_2H_5 \quad (g.\text{–}v.g.)$$

Formerly, it was not considered possible to prepare *disubstituted* derivatives of acetoacetic ester in *one* step. Sandberg (1957), however, has now carried out this one-step reaction in the preparation of ethyl β-acetotricarballylate from acetoacetic ester (0·5 mole), ethyl bromoacetate (1·2 mole) and sodium hydride (1·2 mole) in benzene solution.

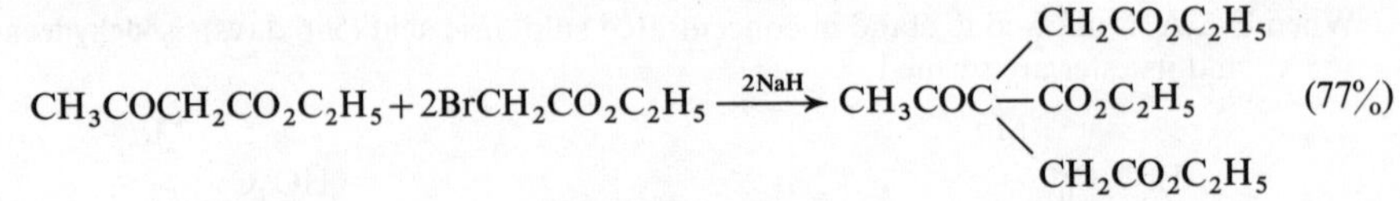

$$CH_3COCH_2CO_2C_2H_5 + 2BrCH_2CO_2C_2H_5 \xrightarrow{2NaH} CH_3COC\begin{smallmatrix} \diagup CH_2CO_2C_2H_5 \\ -CO_2C_2H_5 \\ \diagdown CH_2CO_2C_2H_5 \end{smallmatrix} \quad (77\%)$$

2. Acetoacetic ester and its alkyl derivatives can undergo two types of hydrolysis with potassium hydroxide:

(*a*) **Ketonic hydrolysis.** Ketonic hydrolysis, so called because a ketone is the chief product, is carried out by boiling with *dilute* aqueous or ethanolic potassium hydroxide solution, and is usually followed by acidification (dilute sulphuric acid) and warming:

$$CH_3COCHRCO_2C_2H_5 \xrightarrow[\text{(ii) H}_2\text{SO}_4]{\text{(i) KOH}} [CH_3COCHRCO_2H] \xrightarrow{-CO_2} CH_3COCH_2R \quad (g.)$$

The ketone obtained is acetone or its derivatives, and the latter *always contain the group* $CH_3CO—$.

Dehn and Jackson (1933) found that 85 per cent phosphoric acid was a very good catalyst for the ketonic hydrolysis of acetoacetic ester and its alkyl derivatives, the yield of ketone reaching 95 per cent.

The mechanism of the decarboxylation of E.A.A. has been the subject of much discussion. Evidence obtained by several workers has led to the proposal that decarboxylation occurs through a cyclic transition state, and that the product is originally produced in the enol form.

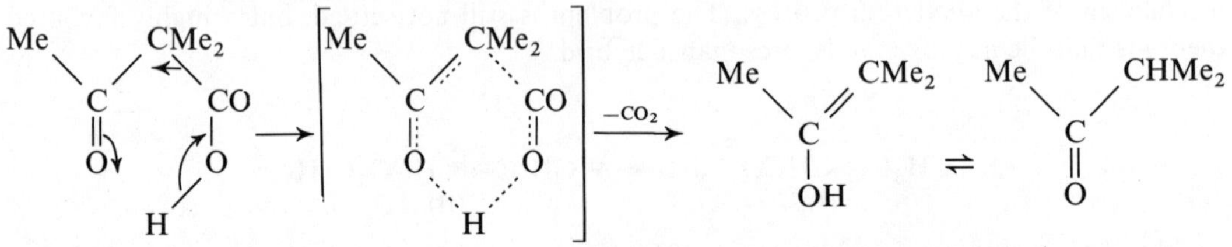

(*b*) **Acid hydrolysis.** Acid hydrolysis, so called because an acid is the chief product, is carried out by boiling with *concentrated* ethanolic potassium hydroxide solution, e.g.,

$$CH_3COCH_2CO_2C_2H_5 + 2KOH \longrightarrow CH_3CO_2K + CH_3CO_2K + C_2H_5OH \quad (f.g.)$$

$$CH_3COCR_2CO_2C_2H_5 + 2KOH \longrightarrow CH_3CO_2K + R_2CHCO_2K + C_2H_5OH \quad (f.g.)$$

The acid obtained is acetic acid or its derivatives as the potassium salt. From this the free acid is readily obtained by treatment with inorganic acid.

The mechanism of this cleavage is possibly:

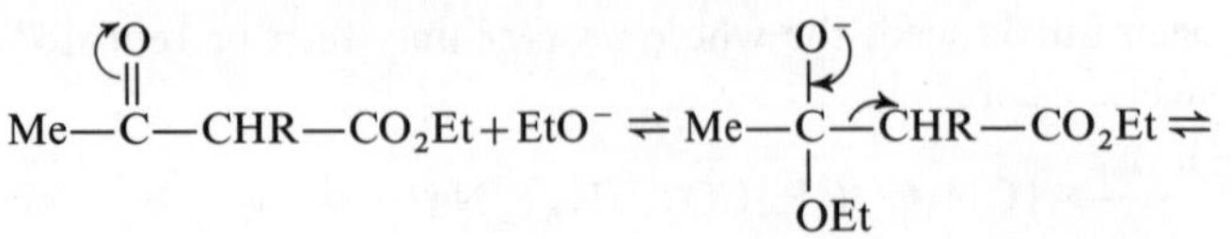

$$Me—\overset{\overset{\displaystyle O}{\|}}{C}—CHR—CO_2Et + EtO^- \rightleftharpoons Me—\overset{\overset{\displaystyle O^-}{|}}{\underset{\displaystyle OEt}{C}}—CHR—CO_2Et \rightleftharpoons$$

$$MeCO_2Et + {}^-CHRCO_2Et \xrightarrow{EtOH} CH_2RCO_2Et + EtO^-$$

This is the reverse of the Claisen condensation. However, the esters formed are hydrolysed by the potassium hydroxide to their salts, and since these do not undergo the condensation, the final equilibrium is driven completely to the right.

Examples of the synthesis of ketones

The formula of the ketone is written down, and *provided it contains the group* $CH_3CO—$, the ketone can be synthesised via acetoacetic ester as follows. The acetone nucleus is picked out, and the alkyl groups attached to it are then introduced into acetoacetic ester one at a time; this is followed by ketonic hydrolysis. It is usually better to introduce the larger group before the smaller (steric effect):

(i) *Butanone.* $\boxed{CH_3COCH_2}CH_3$

$$CH_3COCH_2CO_2C_2H_5 \xrightarrow{C_2H_5ONa} [CH_3COCHCO_2C_2H_5]^-Na^+ \xrightarrow{CH_3I}$$

$$CH_3COCH(CH_3)CO_2C_2H_5 \xrightarrow[\text{(ii) } H^+]{\text{(i) dil. KOH}} CH_3COCH_2CH_3$$

(ii) *3-Methylpentan-2-one.* $\boxed{CH_3COCH}^{\overset{\displaystyle CH_3}{\displaystyle |}}CH_2CH_3$

$$CH_3COCH_2CO_2C_2H_5 \xrightarrow{C_2H_5ONa} [CH_3COCHCO_2C_2H_5]^-Na^+ \xrightarrow{C_2H_5I}$$

$$CH_3COCH(C_2H_5)CO_2C_2H_5 \xrightarrow{C_2H_5ONa} [CH_3COC(C_2H_5)CO_2C_2H_5]^-Na^+ \xrightarrow{CH_3I}$$

$$CH_3COC(CH_3)(C_2H_5)CO_2C_2H_5 \xrightarrow[\text{(ii) } H^+]{\text{(i) dil. KOH}} CH_3COCHCH_2CH_3 \overset{}{\underset{\displaystyle CH_3}{|}}$$

Examples of the synthesis of monocarboxylic acids

The approach is similar to that for ketones except that the acetic acid nucleus is picked out, and the acetoacetic ester derivative is subjected to 'acid' hydrolysis. The acetoacetic ester acid synthesis is usually confined to the preparation of straight-chain acids or branched-chain acids where the branching occurs on the α-carbon atom:

(i) n-*Butyric acid.* $CH_3CH_2\boxed{CH_2CO_2H}$

$$CH_3COCH_2CO_2C_2H_5 \xrightarrow{C_2H_5ONa} [CH_3COCHCO_2C_2H_5]^-Na^+ \xrightarrow{C_2H_5I}$$

$$CH_3COCH(C_2H_5)CO_2C_2H_5 \xrightarrow[\text{(ii) } H^+]{\text{(i) c. KOH}} CH_3CH_2CH_2CO_2H$$

(ii) α-*Methyl-n-valeric acid.* $CH_3CH_2CH_2\boxed{CHCO_2H}^{\overset{\displaystyle CH_3}{\displaystyle |}}$

$$CH_3COCH_2CO_2C_2H_5 \xrightarrow{C_2H_5ONa} [CH_3COCHCO_2C_2H_5]^-Na^+ \xrightarrow{C_3H_7Br}$$

$$CH_3COCH(C_3H_7)CO_2C_2H_5 \xrightarrow{C_2H_5ONa} [CH_3COC(C_3H_7)CO_2C_2H_5]^-Na^+ \xrightarrow{CH_3I}$$

$$CH_3COC(CH_3)(C_3H_7)CO_2C_2H_5 \xrightarrow[\text{(ii) } H^+]{\text{(i) c. KOH}} CH_3CH_2CH_2CHCO_2H \overset{\displaystyle CH_3}{\underset{}{|}}$$

It has been found that 'ketonic' hydrolysis and 'acid' hydrolysis of acetoacetic ester *always take place simultaneously*, but one or the other can be made to predominate by adjusting the concentration of the potassium hydroxide. Thus, in the preparation of acids, there will always be some ketone formed as a by-product; and vice versa. For this reason it is better to use ethyl malonate to synthesise acids since the yields are greater. However, Ritter *et al.* (1962) have improved the preparation of acids via E.A.A. by heating the mono-alkyl derivatives with catalytic amounts of alkoxide in excess of absolute ethanol:

$$MeCOCHRCO_2Et + EtOH \xrightarrow{OEt^-} MeCO_2Et + RCH_2CO_2Et \quad (70\text{–}90\%)$$

In addition to the acetoacetic ester ketone synthesis, ketones may be prepared:

(i) by the alkylation of simpler ketones (sodamide followed by RI). The difficulty with this method is that over-alkylation usually occurs. This synthesis has been modified to prepare t-carboxylic acids from acetophenone.

$$CH_3COC_6H_5 \xrightarrow[\text{(ii) R}^1\text{I}]{\text{(i) NaNH}_2} R^1CH_2COC_6H_5 \xrightarrow[\text{(ii) R}^2\text{I}]{\text{(i) NaNH}_2} \underset{R^2}{\overset{R^1}{\diagdown}}CHCOC_6H_5 \xrightarrow[\text{(ii) R}^3\text{I}]{\text{(i) NaNH}_2}$$

$$\underset{R^3}{\overset{R^1}{\underset{|}{R^2-}}}CCOC_6H_5 \xrightarrow{NaNH_2} \underset{R^3}{\overset{R^1}{\underset{|}{R^2-}}}CCONH_2 \xrightarrow{HNO_2} \underset{R^3}{\overset{R^1}{\underset{|}{R^2-}}}CCO_2H$$

(ii) Stork *et al.* (1963) have introduced a new method of alkylating aldehydes and ketones, and so this affords a means of preparing more complicated aldehydes and ketones from simpler ones. An aliphatic primary amine is condensed with an aldehyde or ketone containing at least one active α-hydrogen atom, and the imine produced is treated with ethylmagnesium bromide in boiling tetrahydrofuran. The magnesium salt so produced readily reacts with alkyl halides, *e.g.*, α-benzylisobutyraldehyde:

$$Me_2CHCHO \xrightarrow{Me_3CNH_2} Me_2CHCH=NCMe_3 \xrightarrow{3EtMgBr} Me_2C=CHN \underset{CMe_3}{\overset{MgBr}{\diagup}} \xrightarrow{PhCH_2Cl}$$

$$Me_2CCH=NCMe_3 \xrightarrow{HCl} PhCH_2CMe_2CHO \quad (80\%)$$
$$\underset{CH_2Ph}{|}$$

(iii) β-Keto-acids, $R^1COCHR^2CO_2H$, can be prepared by α-carboxylation of ketones, $R^1COCH_2R^2$, with magnesium methyl carbonate. The intermediate magnesium salt can be alkylated *in situ* and then decarboxylated to give ketones, $R^1COCHR^2R^2$ (Stiles *et al.*, 1959).

Sodioacetoacetic ester reacts with many other halogen compounds besides alkyl halides, and so may be used to synthesise a variety of compounds.

(i) 1,3-*Diketones*. In the synthesis of 1,3-diketones the halogen compound used is an acid chloride. Since acid chlorides react with ethanol, the reaction cannot be carried out in this solvent in the usual way. The reaction, however, may conveniently be carried out by treating acetoacetic ester in benzene solution with magnesium and the acid chloride; *e.g.*, pentane-2,4-dione may be obtained by the ketonic hydrolysis of the intermediate product ethyl diacetylacetate:

$$CH_3COCH_2CO_2C_2H_5 + CH_3COCl \xrightarrow{Mg} (CH_3CO)_2CHCO_2C_2H_5 \xrightarrow[\text{(ii) H}^+]{\text{(i) dil. KOH}}$$

$$CH_3COCH_2COCH_3$$

If sodioacetoacetic ester or acetoacetic ester itself is treated with acetyl chloride in pyridine as solvent, the *O-acetyl* derivative of acetoacetic ester, acetoxycrotonic ester, is obtained, and not the carbon-linked compound (as above):

$$[CH_3COCHCO_2C_2H_5]^-Na^+ + CH_3COCl \xrightarrow{pyridine} CH_3\underset{|}{\overset{OCOCH_3}{C}}=CHCO_2C_2H_5 + NaCl$$

If we use Kornblum's generalisation with ambient ions (p. 289), then the first reaction presumably occurs by the S_N2 mechanism and the latter by the S_N1.

(ii) *Dicarboxylic acids.* Dicarboxylic acids may be prepared by interaction of sodioacetoacetic ester and a halogen derivative of an ester, *e.g.*, succinic acid from ethyl chloroacetate:

$$[CH_3COCHCO_2C_2H_5]^-Na^+ + ClCH_2CO_2C_2H_5 \longrightarrow$$

$$\begin{array}{c} CH_2CO_2C_2H_5 \\ | \\ CH_3COCHCO_2C_2H_5 \end{array} \xrightarrow[\text{(ii) H}^+]{\text{(i) c. KOH}} \begin{array}{c} CH_2CO_2H \\ | \\ CH_2CO_2H \end{array}$$

Ketonic hydrolysis of this acetoacetic ester derivative gives the γ-keto-acid ester (ethyl ester of lævulic acid):

$$\begin{array}{c} CH_2CO_2C_2H_5 \\ | \\ CH_3COCHCO_2C_2H_5 \end{array} \xrightarrow[\text{(ii) H}^+]{\text{(i) dil. KOH}} CH_3COCH_2CH_2CO_2C_2H_5$$

(iii) *Long-chain monocarboxylic acids.* This method involves a combination of methods (i) and (ii) described above (Mrs Robinson, 1930):

$$[CH_3COCHCO_2C_2H_5]^-Na^+ \xrightarrow{Br(CH_2)_xCO_2C_2H_5} \begin{array}{c} (CH_2)_xCO_2C_2H_5 \\ | \\ CH_3COCHCO_2C_2H_5 \end{array} \xrightarrow[\text{(ii) CH}_3\text{(CH}_2)_y\text{COCl}]{\text{(i) C}_2\text{H}_5\text{ONa}}$$

$$\begin{array}{c} (CH_2)_xCO_2C_2H_5 \\ | \\ CH_3COCCO_2C_2H_5 \\ | \\ CO(CH_2)_yCH_3 \end{array} \xrightarrow[\text{hydrolysis}]{\text{graded}} CH_3(CH_2)_yCOCH_2(CH_2)_xCO_2H$$

These keto-acids are readily reduced by means of the Clemmensen reduction (p. 210). Improved modifications of this method have now been developed (see, *e.g.*, p. 384).

(iv) The acetyl group in acetoacetic ester may be replaced by another acyl group under suitable conditions, *e.g.*, ethyl benzoylacetate:

$$MeCOCH_2CO_2Et + PhCOCl \xrightarrow{NaOH} PhCOCH(COMe)CO_2Et \xrightarrow{\text{aq. NH}_4\text{Cl}}$$

$$PhCOCH_2CO_2Et + MeCO_2NH_4$$

(v) Alicyclic compounds (p. 534) and heterocyclic compounds (Ch. 30) may be prepared by the use of E.A.A.

Malonic ester syntheses

Malonic ester, $CH_2(CO_2C_2H_5)_2$, which is the diethyl ester of malonic acid, $CH_2(CO_2H)_2$, is prepared by dissolving potassium cyanoacetate in ethanol, adding concentrated hydrochloric acid, and warming the mixture on the water-bath:

$$NCCH_2-CO_2K + 2C_2H_5OH + 2HCl \longrightarrow CH_2(CO_2C_2H_5)_2 + KCl + NH_4Cl \quad (v.g.)$$

It is a pleasant-smelling liquid, b.p. 199°C. Its use as a synthetic reagent depends on two chemical properties.

1. With sodium ethoxide it forms a sodium derivative, sodiomalonic ester, which reacts with compounds containing a reactive halogen atom, *e.g.*, alkyl halides, acid chlorides, halogen-substituted esters, etc. (*cf.* acetoacetic ester). In all cases the yields are g.-v.g.

$$C_2H_5O-\overset{\overset{\displaystyle O}{\|}}{C}-CH_2-\overset{\overset{\displaystyle O}{\|}}{C}-OC_2H_5 \xrightarrow{C_2H_5ONa} \left[C_2H_5O-\overset{\overset{\displaystyle \bar{O}}{|}}{C}=CH-\overset{\overset{\displaystyle O}{\|}}{C}-OC_2H_5 \right] Na^+ \longleftrightarrow$$

$$\left[C_2H_5O-\overset{\overset{\displaystyle O}{\|}}{C}-\overset{-}{C}H-\overset{\overset{\displaystyle O}{\|}}{C}-OC_2H_5 \right] Na^+ \xrightarrow{RX} C_2H_5O-\overset{\overset{\displaystyle O}{\|}}{C}-CHR-\overset{\overset{\displaystyle O}{\|}}{C}-OC_2H_5 + NaX$$

The process may then be repeated to produce the disubstituted derivative of malonic ester:

$$R^1CH(CO_2C_2H_5)_2 \xrightarrow{C_2H_5ONa} [R^1C(CO_2C_2H_5)_2]^-Na^+ \xrightarrow{R^2X} R^1R^2C(CO_2C_2H_5)_2$$

These disubstituted derivatives of malonic ester can readily be prepared in *one* step by treating the ester with *two* equivalents of sodium ethoxide and then with *two* equivalents of alkyl halide. This procedure is only used if it is required to introduce two identical alkyl groups.

$$CH_2(CO_2C_2H_5)_2 \xrightarrow[\text{(ii) 2RX}]{\text{(i) 2EtONa}} R_2C(CO_2C_2H_5)_2 + 2NaX$$

2. Malonic acid and its derivatives eliminate a molecule of carbon dioxide when heated just above the melting point of the acid (usually between 150° and 200°C) to form acetic acid or its derivatives (the yields are *v.g.-ex.*):

$$HO_2CCH_2CO_2H \longrightarrow CH_3CO_2H + CO_2$$
$$HO_2CCHRCO_2H \longrightarrow RCH_2CO_2H + CO_2$$

Decarboxylation may also be effected by refluxing malonic acid or its derivatives in sulphuric acid solution.

The mechanism of decarboxylation is probably similar to that of acetoacetic acid:

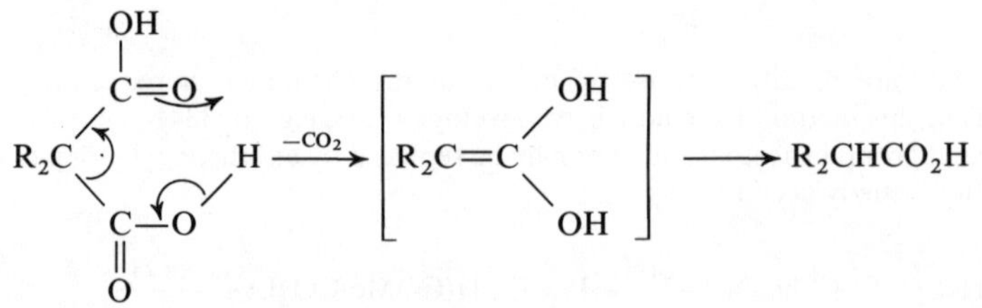

In a number of cases ethyl cyanoacetate (b.p. 207°C) may be used instead of malonic ester in many syntheses. The cyano-group is readily converted into the carboxyl group on hydrolysis.

Synthesis of monocarboxylic acids

Malonic ester is preferable to acetoacetic ester in synthesising acids, and should be used wherever possible. The structural formula of the acid required is written down, the acetic acid nucleus picked out, and the required alkyl groups introduced into sodiomalonic ester. The substituted ester is then refluxed with potassium hydroxide solution, acidified with hydrochloric acid, and the precipitated acid dried and then heated just above its melting point. Alternatively, the potassium salt may be refluxed with sulphuric acid:

(i) n-*Valeric acid.* $CH_3CH_2CH_2\boxed{CH_2CO_2H}$

$$CH_2(CO_2C_2H_5)_2 \xrightarrow{C_2H_5ONa} [CH(CO_2C_2H_5)_2]^-Na^+ \xrightarrow{C_3H_7Br} C_3H_7CH(CO_2C_2H_5)_2 \xrightarrow{KOH}$$

$$\text{C}_3\text{H}_7\text{CH(CO}_2\text{K)}_2 \xrightarrow{\text{HCl}} \text{C}_3\text{H}_7\text{CH(CO}_2\text{H)}_2 \xrightarrow{150-200°\text{C}} \text{CH}_3\text{CH}_2\text{CH}_2\text{CH}_2\text{CO}_2\text{H}$$

(ii) *Dimethylacetic acid.* $\text{CH}_3\boxed{\text{CHCO}_2\text{H}}$ with $\overset{\text{CH}_3}{\underset{|}{}}$ above

$$\text{CH}_2\text{(CO}_2\text{C}_2\text{H}_5)_2 \xrightarrow{\text{C}_2\text{H}_5\text{ONa}} [\text{CH(CO}_2\text{C}_2\text{H}_5)_2]^-\text{Na}^+ \xrightarrow{\text{CH}_3\text{I}} \text{CH}_3\text{CH(CO}_2\text{C}_2\text{H}_5)_2 \xrightarrow{\text{C}_2\text{H}_5\text{ONa}}$$

$$[\text{CH}_3\text{C(CO}_2\text{C}_2\text{H}_5)_2]^-\text{Na}^+ \xrightarrow{\text{CH}_3\text{I}} \text{(CH}_3)_2\text{C(CO}_2\text{C}_2\text{H}_5)_2 \xrightarrow{\text{KOH}}$$

$$\text{(CH}_3)_2\text{C(CO}_2\text{K)}_2 \xrightarrow{\text{HCl}} \text{(CH}_3)_2\text{C(CO}_2\text{H)}_2 \xrightarrow{150-200°\text{C}} \text{(CH}_3)_2\text{CHCO}_2\text{H}$$

Since two methyl groups are required for this synthesis, their introduction can be carried out in one step (see above):

$$\text{CH}_2\text{(CO}_2\text{C}_2\text{H}_5)_2 \xrightarrow[2\text{CH}_3\text{I}]{2\text{C}_2\text{H}_5\text{ONa}} \text{(CH}_3)_2\text{C(CO}_2\text{C}_2\text{H}_5)_2 \xrightarrow{\text{KOH, etc.}} \text{(CH}_3)_2\text{CHCO}_2\text{H}$$

Since aryl groups cannot be introduced directly into malonic ester, aryl-substituted derivatives must be prepared indirectly. A very good method uses the Claisen condensation followed by decarboxylation (*cf.* p. 287); *e.g.*, ethyl phenylmalonate:

$$\text{PhCH}_2\text{CO}_2\text{Et} + \text{(CO}_2\text{Et)}_2 \xrightarrow{\text{EtONa}} \text{EtOH} + \text{PhCH(CO}_2\text{Et)COCO}_2\text{Et} \xrightarrow{\text{heat}}$$

$$\text{PhCH(CO}_2\text{Et)}_2 + \text{CO} \quad (80-85\%)$$

This method may also be used for the preparation of long-chain alkylmalonic esters; the ester of a long-chain monocarboxylic acid with two α-hydrogen atoms is used instead of the ester of phenylacetic acid given above.

Synthesis of dicarboxylic acids

Dicarboxylic acids of the type $\text{R}^1\text{R}^2\text{C(CO}_2\text{H)}_2$ are readily prepared from malonic ester as shown above.

Dicarboxylic acids of the type $\text{HO}_2\text{C(CH}_2)_n\text{CO}_2\text{H}$ are very important, and the malonic ester synthesis is particularly useful for their preparation. The actual procedure depends mainly on the value of n, and the following examples illustrate this point:

(i) *Adipic acid.* $\boxed{\text{HO}_2\text{CCH}_2}\text{CH}_2\text{CH}_2\vdots\text{CH}_2\text{CO}_2\text{H}$

The acetic acid is blocked off at *one* end, and the remaining fragment is considered from the point of view of accessibility. In this example the fragment required is γ-bromobutyric ester. This is not readily accessible, and so the procedure now is to block off the acetic acid nucleus at the *other* end of the adipic acid molecule, and then to use *two* molecules of malonic ester and the α,ω-*dihalide* of the polymethylene fragment that joins together the two acetic acid nuclei. In this case it is ethylene dibromide, and since this is readily accessible, the synthesis of adipic acid may be carried out as follows:

$$[(\text{C}_2\text{H}_5\text{O}_2\text{C})_2\text{CH}]^-\text{Na}^+ + \text{BrCH}_2\text{CH}_2\text{Br} + \text{Na}^+[\text{CH(CO}_2\text{C}_2\text{H}_5)_2]^- \longrightarrow$$

$$\text{(C}_2\text{H}_5\text{O}_2\text{C})_2\text{CHCH}_2\text{CH}_2\text{CH(CO}_2\text{C}_2\text{H}_5)_2 \xrightarrow[\text{(ii) HCl}]{\text{(i) KOH}}$$

$$\text{(HO}_2\text{C})_2\text{CHCH}_2\text{CH}_2\text{CH(CO}_2\text{H)}_2 \xrightarrow{150-200°\text{C}} \text{HO}_2\text{CCH}_2\text{CH}_2\text{CH}_2\text{CH}_2\text{CO}_2\text{H}$$

Should the fragments required by either route by inaccessible, then a method not involving the use of malonic ester may be more satisfactory (see p. 445).

(ii) *Succinic acid.* $\boxed{\text{HO}_2\text{CCH}_2}\text{CH}_2\text{CO}_2\text{H}$

If *one* acetic acid nucleus is blocked off, the fragment required is ethyl chloroacetate. This is readily accessible and hence succinic acid may be synthesised from *one* molecule of malonic ester as follows:

$$[CH(CO_2C_2H_5)_2]^-Na^+ + ClCH_2CO_2C_2H_5 \longrightarrow (C_2H_5O_2C)_2CHCH_2CO_2C_2H_5 \xrightarrow[\text{(ii) HCl}]{\text{(i) KOH}}$$

$$(HO_2C)_2CHCH_2CO_2H \xrightarrow{150-200°C} HO_2CCH_2CH_2CO_2H$$

If the *other* acetic acid nucleus in succinic acid is blocked off, there is no intervening fragment left. However, the union of two malonic ester molecules may be effected by means of iodine. Thus:

$$[(C_2H_5O_2C)_2CH]^-Na^+ + I_2 + Na^+[CH(CO_2C_2H_5)_2]^- \longrightarrow$$

$$(C_2H_5O_2C)_2CHCH(CO_2C_2H_5)_2 \xrightarrow[\text{(ii) HCl}]{\text{(i) KOH}} (HO_2C)_2CHCH(CO_2H)_2$$

$$\xrightarrow{150-200°C} HO_2CCH_2CH_2CO_2H$$

Synthesis of ketones

Bowman *et al.* (1952) have introduced the following general synthesis of ketones and β-keto-esters ($R^2 = $ tetrahydropyran-2-ol):

$$R^1CH(CO_2R^2)_2 \xrightarrow[\text{(ii) } R^3COCl]{\text{(i) Na}} R^3COCR^1(CO_2R^2)_2 \xrightarrow[\text{reflux}]{CH_3CO_2H} R^3COCH_2R^1 \quad (50-92\%)$$

Johnson *et al.* (1952) have used t-butyl esters of malonic acid ($R^2 = $ t-butyl):

$$R^1CH(CO_2R^2)_2 \xrightarrow[\text{(ii) } R^3COCl]{\text{(i) NaH}} R^3COCR^1(CO_2R^2)_2 \xrightarrow[CH_3CO_2H]{CH_3 -\!\!\bigcirc\!\!- SO_3H} R^3COCH_2R^1 \quad (56-85\%)$$

Good yields of ketone are obtained by the above method only when anhydrous conditions are used throughout the preparation. Consequently, instead of the usual method of hydrolysis (aqueous alkali), acidolysis is used (p. 253). Acetic acid is satisfactory for this purpose, especially in the presence of a small amount of sulphuric acid or *p*-toluenesulphonic acid. In this way, the ester group is transferred from the malonic ester to the acetic acid. However, for the transfer to be successful, the ester group must be very sensitive to acidolysis; the above two esters are particularly sensitive in this way.

Synthesis of higher ketonic acids. Sodiomalonic ester is treated with the acid chloride-ester derivative of a dicarboxylic acid, *e.g.*, ε-ketoheptanoic acid:

$$[(C_2H_5O_2C)_2CH]^-Na^+ + ClCO(CH_2)_4CO_2C_2H_5 \longrightarrow$$

$$(C_2H_5O_2C)_2CHCO(CH_2)_4CO_2C_2H_5 \xrightarrow[\text{(ii) HCl}]{\text{(i) KOH}} (HO_2C)_2CHCO(CH_2)_4CO_2H \xrightarrow{150-200°C}$$

$$[HO_2CCH_2CO(CH_2)_4CO_2H] \xrightarrow{-CO_2} CH_3CO(CH_2)_4CO_2H$$

The intermediate β-keto-dicarboxylic acid is unstable and is readily decarboxylated. This might have been anticipated, since it may be regarded as a derivative of acetoacetic acid, which is readily decarboxylated on warming.

Synthesis of polycarboxylic acids

It has been pointed out previously (p. 270) that the monoalkyl derivatives of malonic acid are readily brominated, and that these bromo-derivatives give α-bromo-acids on decarboxylation. Malonic ester is also readily brominated, and this monobromo-derivative may be used to synthesise polycarboxylic acids, *e.g.*,

$$CH_2(CO_2C_2H_5)_2 \xrightarrow{Br_2} CHBr(CO_2C_2H_5)_2$$

$$CHBr(CO_2C_2H_5)_2 + [CH(CO_2C_2H_5)_2]^-Na^+ \longrightarrow NaBr + (C_2H_5O_2C)_2CHCH(CO_2C_2H_5)_2 \xrightarrow{Br_2}$$

$$\begin{array}{c} (C_2H_5O_2C)_2C-Br \\ | \\ (C_2H_5O_2C)_2C-Br \end{array} \xrightarrow{2[CH(CO_2C_2H_5)_2]^-Na^+} \begin{array}{c} (C_2H_5O_2C)_2C-CH(CO_2C_2H_5)_2 \\ | \\ (C_2H_5O_2C)_2C-CH(CO_2C_2H_5)_2 \end{array}$$

Malonic ester may also be used for the preparation of unsaturated acids (Knoevenagel reaction, p. 342), alicyclic compounds (p. 534), and heterocyclic compounds (p. 419).

Hydroxyaldehydes and hydroxyketones

When naming a compound which contains more than one functional group, it is necessary to choose one as the principal function. The compound is then named by using the suffix of the principal function and the prefixes of the other functions. The carboxylic and the sulphonic acid group are *always* chosen as principal functions, and the usual order for choosing the principal function is:

Carboxylic, sulphonic, acid halide, amide, imide, aldehyde, cyanide, isocyanide, ketone, alcohol, phenol, thioalcohol, amine, imine.

Examples. Acetoacetic acid contains a ketonic group and a carboxyl group, and since the latter is the principal function, acetoacetic acid is named as a ketone acid, *viz.*, β-ketobutyric acid, 3-ketobutanoic acid, or 3-oxobutanoic acid. Acetoacetic acid may also be named propan-2-one-1-carboxylic acid—each functional group is indicated by its appropriate suffix.

Table 10.2

Function	Prefix	Suffix
Acid	carboxy	carboxylic or -oic
Acid amide	carbamoyl	amide or carbonamide
Acid chloride	chloroformyl	oyl or carbonyl
Alcohol	hydroxy	ol
Aldehyde	oxo, aldo (for aldehydic O) or formyl (for CHO)	al
Amine	amino	amine
Azo-derivative	azo	—
Azoxy-derivative	azoxy	—
Carbonitrile (nitrile)	cyano	carbonitrile
Double bond	—	ene
Ester	alk(yl)oxycarbonyl	carboxylate
Ether	alkoxy	—
Ethylene oxide, etc.	epoxy	—
Halogenide (halide)	halogeno (halo)	—
Hydrazine	hydrazino	hydrazine
Ketone	oxo or keto	one
Nitro-derivative	nitro	—
Nitroso-derivative	nitroso	—
'Quinquevalent' nitrogen	—	onium, inium
Sulphide	alkylthio	—
Sulphinic derivative	sulphino	sulphinic
Sulphone	sulphonyl	—
Sulphonic derivative	sulpho	sulphonic
Sulphoxide	sulphinyl	—
Thioalcohol (Mercaptan)	mercapto	thiol
Triple bond	—	yne
Urea	ureido	urea

CH_2OHCH_2CHO. This is both an alcohol and an aldehyde, so that, bearing in mind the order of preference, the name of the compound will be β-hydroxypropionaldehyde or 3-hydroxypropanal.

Table 10.2 indicates the prefixes and suffixes used for designating the functions (in alphabetical order).

The simplest hydroxyaldehyde is **glycolaldehyde** (*hydroxyethanal*). It may be prepared by the oxidation of glycol with *Fenton's reagent* (hydrogen peroxide and ferrous sulphate):

$$HOCH_2CH_2OH + H_2O_2 \xrightarrow{FeSO_4} HOCH_2CHO + 2H_2O$$

It may also be prepared by ozonolysis of allyl alcohol:

$$CH_2=CHCH_2OH \xrightarrow{O_3} \begin{array}{c} CH_2-O-CHCH_2OH \\ | \qquad\qquad | \\ O \underline{\qquad\qquad} O \end{array} \xrightarrow[Zn]{H_2O} OCHCH_2OH$$

The most convenient means of preparing glycolaldehyde, however, is by heating dihydroxymaleic acid, which is obtained from the oxidation of tartaric acid (*q.v.*):

$$\begin{array}{c} HO_2C-C-OH \\ \| \\ HO_2C-C-OH \end{array} \longrightarrow 2CO_2 + \left[\begin{array}{c} H-C-OH \\ \| \\ H-C-OH \end{array}\right] \longrightarrow \begin{array}{c} CHO \\ | \\ CH_2OH \end{array}$$

Glycolaldehyde exists in the solid form as the dimer, m.p. 96°C; but in aqueous solution it exists as the monomer which forms the stable hydrate, $CH(OH)_2CH_2OH$ (*cf.* chloral, p. 228). Careful oxidation of glycolaldehyde with bromine water produces glycollic acid:

$$HOCH_2CHO + [O] \longrightarrow HOCH_2CO_2H$$

Glycolaldehyde is a powerful reducing agent, reducing ammoniacal silver nitrate and Fehling's solution at room temperature. With phenylhydrazine it forms the *osazone*. This osazone is identical with that formed from glyoxal (see p. 300).

Glycolaldehyde undergoes the aldol condensation in the presence of alkali; with sodium hydroxide solution a tetrose sugar is formed, and with sodium carbonate solution, a hexose sugar:

$$2HOCH_2CHO \xrightarrow{NaOH} HOCH_2CHOHCHOHCHO$$

$$3HOCH_2CHO \xrightarrow{Na_2CO_3} HOCH_2CHOHCHOHCHOHCHOHCHO$$

Glycolaldehyde is formed in small amounts when formaldehyde is allowed to stand in the presence of calcium carbonate:

$$2HCHO \xrightarrow{CaCO_3} HOCH_2CHO$$

Aldol (*acetaldol*, *β-hydroxybutyraldehyde*, *3-hydroxybutanal*), b.p. 83°C/20 mm., may be prepared by the aldol condensation of acetaldehyde (p. 217):

$$2CH_3CHO \xrightarrow{NaOH} CH_3CHOHCH_2CHO$$

When heated, aldol is dehydrated to crotonaldehyde:

$$CH_3CHOHCH_2CHO \longrightarrow CH_3CH=CHCHO + H_2O$$

There is some doubt about the structure of aldol. Recent work suggests it is an equilibrium mixture of β-hydroxybutyraldehyde and the cyclic hemiacetal:

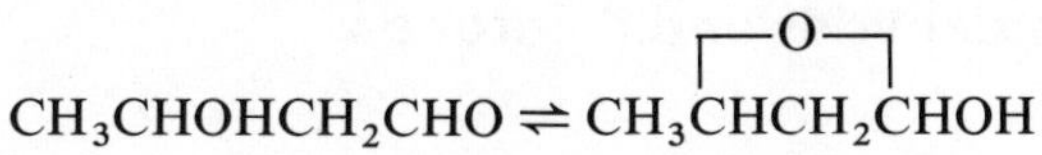

$$\overset{\displaystyle\lceil\!-\!O\!-\!\rceil}{CH_3CHOHCH_2CHO \rightleftharpoons CH_3CHCH_2CHOH}$$

This is an example of ring-chain tautomerism.

The simplest hydroxyketone is **hydroxyacetone** (*acetol, pyruvic alcohol*), b.p. 145°C. The best method of preparation is by heating bromoacetone with potassium hydroxide in methanolic solution, and adding ethyl formate:

$$HCO_2C_2H_5 + KOH \longrightarrow HCO_2K + C_2H_5OH$$
$$CH_3COCH_2Br + HCO_2K \longrightarrow CH_3COCH_2OOCH + KBr$$
$$CH_3COCH_2OOCH + CH_3OH \longrightarrow CH_3COCH_2OH + HCO_2CH_3 \quad (54-88\%)$$

This is an example of alcoholysis, hydroxyacetone being replaced by methanol.

Hydroxyacetone reduces ammoniacal silver nitrate, thereby being oxidised to DL-lactic acid, and reduces Fehling's solution, thereby being oxidised to a mixture of formic and acetic acids. Although ketones do not normally reduce ammoniacal silver nitrate and Fehling's solution, α-hydroxyketones are exceptions. Similarly, ketones form phenylhydrazones with phenylhydrazine, but α-hydroxyketones form osazones; thus, hydroxyacetone gives the same osazone as methylglyoxal (*cf.* glycolaldehyde):

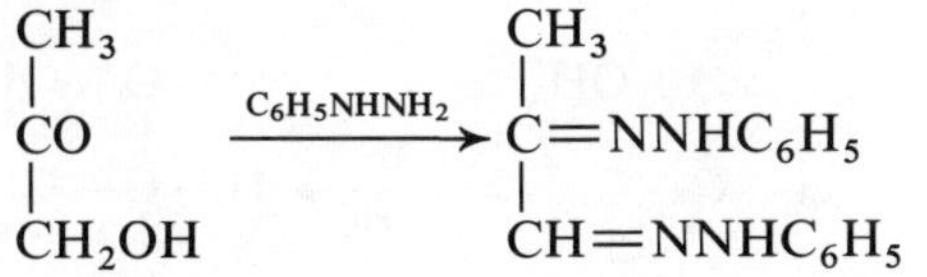

$$
\begin{array}{ccc}
CH_3 & & CH_3 \\
| & & | \\
CO & \xrightarrow{\;C_6H_5NHNH_2\;} & C{=}NNHC_6H_5 \\
| & & | \\
CH_2OH & & CH{=}NNHC_6H_5
\end{array}
$$

Hydroxyacetone appears to exist as an equilibrium mixture of hydroxyketone and cyclic hemiacetal forms (ring-chain tautomerism).

$$CH_3COCH_2OH \rightleftharpoons \underset{\underset{OH}{|}}{CH_3C}\!\!\overset{\diagdown}{\underset{\diagup}{}}\!\!\overset{O}{}\!\!CH_2$$

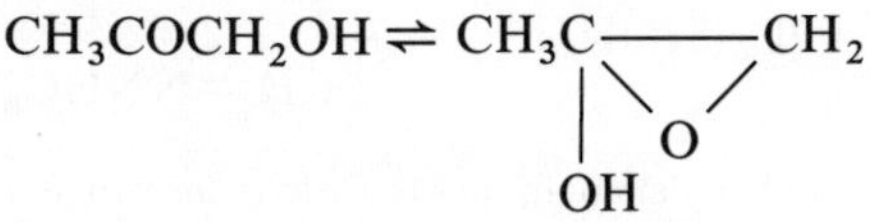

Acyloins are α-hydroxyketones, and may be readily prepared by reduction of esters (p. 253). Acetoin, b.p. 148°C, the simplest member, may be prepared from ethyl acetate.

Diacetone alcohol (4-*hydroxy*-4-*methylpentan*-2-*one*), b.p. 164°C, may be prepared by the condensation of acetone in the presence of barium hydroxide (p. 218):

$$2CH_3COCH_3 \xrightarrow{\;Ba(OH)_2\;} (CH_3)_2C(OH)CH_2COCH_3$$

When heated with a trace of acid or iodine, diacetone alcohol eliminates a molecule of water to form mesityl oxide:

$$(CH_3)_2C(OH)CH_2COCH_3 \xrightarrow{\;I_2\;} (CH_3)_2C{=}CHCOCH_3 + H_2O$$

Diacetone alcohol is oxidised by sodium hypobromite to β-hydroxyisovaleric acid (*cf.* haloform reaction):

$$(CH_3)_2C(OH)CH_2COCH_3 \xrightarrow{\;[O]\;} (CH_3)_2C(OH)CH_2CO_2H$$

Dialdehydes, ketoaldehydes and diketones

Dialdehydes

The simplest dialdehyde is **glyoxal** (*oxaldehyde, ethanedial*). It may be obtained by the oxidation of ethanol, acetaldehyde or glycol with nitric acid, but the yields are poor. It is most conveniently prepared, as the bisulphite compound, by refluxing a mixture of paraldehyde, 50 per cent aqueous acetic acid, dioxan and selenous acid (yield 72–74 per cent on the selenous acid):

$$(CH_3CHO)_3 + 3H_2SeO_3 \longrightarrow 3OCHCHO + 3Se + 6H_2O \xrightarrow{NaHSO_3} OCHCHO{\cdot}2NaHSO_3{\cdot}H_2O$$

Glyoxal is manufactured by the vapour-phase oxidation of glycol with air at 250–300°C in the presence of copper as catalyst.

Glyoxal exists as a colourless polymer giving, on distillation, the monomer. This is a green vapour, which condenses to yellow crystals (m.p. 15°C) which polymerise on standing to a colourless solid of unknown molecular weight. Glyoxal exists in aqueous solution as the dihydrate, $CH(OH)_2CH(OH)_2$, which has been isolated (*cf.* chloral hydrate, p. 228).

Glyoxal undergoes many of the reactions of a dialdehyde; *e.g.*, it reduces ammoniacal silver nitrate, and forms addition compounds with two molecules of hydrogen cyanide and sodium hydrogen sulphite. It does *not* reduce Fehling's solution. Since it has no α-hydrogen atom, glyoxal undergoes an intramolecular Cannizzaro reaction in the presence of alkali to form glycollic acid.

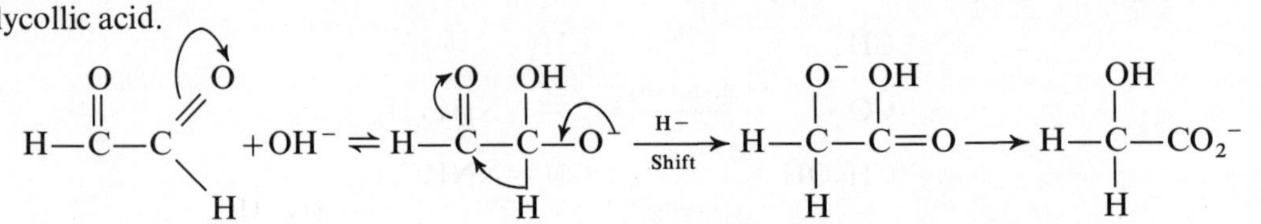

With phenylhydrazine, glyoxal forms the osazone (this is identical with that from glycolaldehyde):

$$\begin{matrix} CHO \\ | \\ CHO \end{matrix} + 2C_6H_5NHNH_2 \longrightarrow \begin{matrix} CH{=}NNHC_6H_5 \\ | \\ CH{=}NNHC_6H_5 \end{matrix} + 2H_2O$$

It also combines with *o*-phenylenediamines to form *quinoxalines* (heterocyclic compounds); *e.g.*, with *o*-phenylenediamine it forms quinoxaline itself:

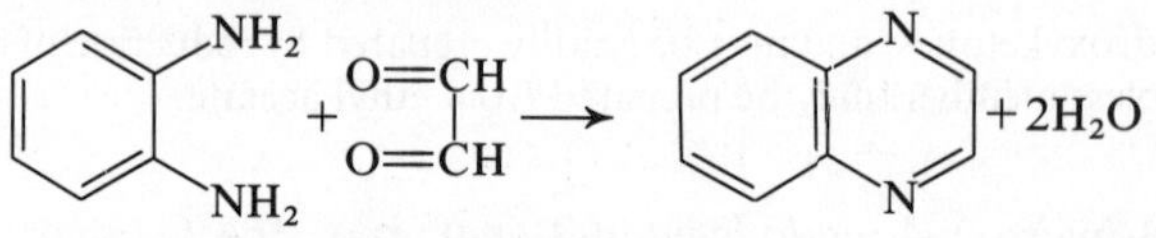

Methylglyoxal (pyruvaldehyde) may be prepared by the oxidation of hydroxyacetone with hydrogen peroxide, or of acetone with selenium dioxide, or by the hydrolysis of oximinoacetone:

$$\left.\begin{matrix} CH_3COCH_2OH \xrightarrow{H_2O_2} \\ CH_3COCH_3 \xrightarrow{SeO_2} \\ CH_3COCH{=}NOH \xrightarrow{H_3O^+} \end{matrix}\right\} \longrightarrow CH_3COCHO$$

It is manufactured by the vapour-phase oxidation of propene glycol with air at 250–300°C in the presence of copper as catalyst:

$$CH_3CHOHCH_2OH \xrightarrow[Cu]{O_2} CH_3COCHO + 2H_2O$$

It is a yellow oil with a pungent odour, and begins to boil at 72°C to give a light green vapour.

The liquid form of methylglyoxal is the dimer, and this slowly polymerises at room temperature to form a glassy mass of unknown molecular weight (*cf.* glyoxal).

> **Malondialdehyde** is prepared commercially as its tetra-acetal, 1,1,3,3-tetraethoxypropane, $(EtO)_2CHCH_2CH(OEt)_2$, b.p. 98°C/11 mm. The free aldehyde is liberated by the action of acid, but is very unstable (due to polymerisation).
> **Formylacetone (acetoacetaldehyde)**, b.p. 100°C, is formed when ethyl formate undergoes the Claisen condensation with acetone.

$$CH_3COCH_3 + HCO_2Et \xrightarrow{EtONa} CH_3COCH_2CHO$$

Formylacetone is strongly acidic, *e.g.*, it dissolves in sodium hydroxide solution to form the sodium salt, and gives a violet colour with ferric chloride. Thus it is enolised. Moreover, infra-red spectra studies by Boileau (1954) have shown that it is the aldehyde group that has enolised almost 100 per cent *i.e.*, formylacetone is hydroxymethyleneacetone, $MeCOCH=CHOH$. It appears that all β-keto-aldehydes behave in this way. The interesting point about their direction of enolisation is that they were the first exceptions to the *Erlenmeyer rule* (1880, 1892) that a terminal $=CHOH$ group (in acyclic compounds) is so unstable that it always rearranges to the aldehyde (*cf.* 'vinyl alcohol', p. 327). More recently, Garbisch, Jr. (1963) has shown, from his NMR spectra studies of cyclic β-keto-aldehydes, that the direction of enolisation depends on their structure.

Succinaldehyde (succindialdehyde, butanedial) may be prepared by the ozonolysis of hexa-1,5-diene:

$$CH_2=CHCH_2CH_2CH=CH_2 \xrightarrow{ozonolysis} OCHCH_2CH_2CHO$$

Another method of preparation is to allow pyrrole and hydroxylamine to interact, and to treat the product, succinaldoxime, with aqueous nitrous acid (see also p. 839):

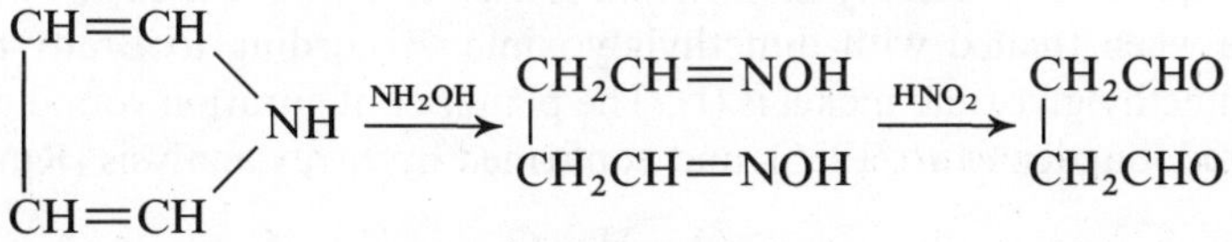

Barton *et al.* (1964) have obtained succinaldehyde (yield: 80 per cent) from butane-1,4-diol via the alkyl chloroformate (see p. 204).

Succinaldehyde is a colourless oil, b.p. 170°C, which readily polymerises. It undergoes the usual reactions of a (di)aldehyde; *e.g.*, it reduces ammoniacal silver nitrate and Fehling's solution, forms addition compounds with two equivalents of hydrogen cyanide and sodium hydrogen sulphite; etc. When treated with phosphorus pentoxide, ammonia and phosphorus pentasulphide, it gives respectively furan, pyrrole and thiophen:

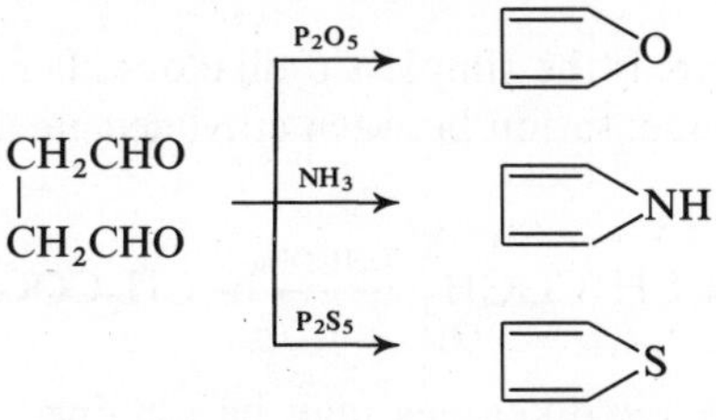

Diketones

These are classified as α, β, γ ... diketones according as the two carbonyl groups are in the 1,2, 1,3, 1,4, ... positions, respectively.

The infra-red absorption region of α-diketones is 1730–1710 cm^{-1} (s), and that of γ-diketones is 1725–1705 cm^{-1} (s). Both of these are in the absorption region of acyclic monoketones

(1725–1700 cm^{-1}) but of greater intensity, and since they are completely different from the region of enolisable β-diketones (p. 284), it therefore means that α- and γ-diketones are like acetone, *i.e.*, are in the keto form.

Butane-2,3-dione, dimethylglyoxal (*diacetyl*), is the simplest α-diketone. It may be prepared:

(i) By the hydrolysis of oximinobutan-2-one formed by the action of nitrous acid on the ketone:

$$CH_3COCH_2CH_3 \xrightarrow{HNO_2} CH_3COC(=NOH)CH_3 \xrightarrow{HCl} CH_3COCOCH_3$$

(ii) By the oxidation of butan-2-one with selenium dioxide:

$$CH_3COCH_2CH_3 + SeO_2 \longrightarrow CH_3COCOCH_3 + Se + H_2O$$

(iii) Acetoin may be readily oxidised by bismuth trioxide in acetic acid; the yield of dimethylglyoxal is almost quantitative (Rigby, 1951). This reagent is specific for this reaction, *i.e.*, for the oxidation of acyloins to α-diketones.

Butane-2,3-dione is a yellow oil, b.p. 88°C. It gives the usual reactions of a diketone; *e.g.*, it forms the monoxime and di-oxime with hydroxylamine, the bisulphite compound with sodium hydrogen sulphite, etc. It is oxidised by hydrogen peroxide to acetic acid:

$$CH_3COCOCH_3 \xrightarrow{H_2O_2} 2CH_3CO_2H$$

Reduction with lithium aluminium hydride produces the corresponding diol.

The dioxime of butane-2,3-dione, *i.e.*, *dimethylglyoxime*, $CH_3C(=NOH)C(=NOH)CH_3$, forms chelate compounds with many metals, and is used to estimate nickel, whose salts produce a red precipitate when treated with dimethylglyoxime. According to Brady *et al.* (1930), the structure of bisdimethylglyoxime nickel is (I). The planar configuration round the nickel ion has been demonstrated (Sugden *et al.*, 1935), and confirmed by X-ray analysis (Rundle *et al.*, 1953).

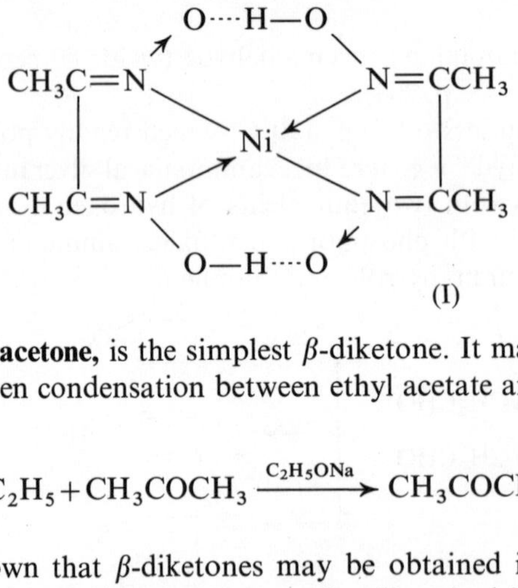

(I)

Pentane-2,4-dione, acetylacetone, is the simplest β-diketone. It may be prepared:

(i) By means of the Claisen condensation between ethyl acetate and acetone (yield 38–45 per cent on the acetone):

$$CH_3CO_2C_2H_5 + CH_3COCH_3 \xrightarrow{C_2H_5ONa} CH_3COCH_2COCH_3$$

Bloomfield (1962) has shown that β-diketones may be obtained in very good yields by this method by using dimethyl sulphoxide as solvent and sodium hydride as the base.

(ii) By the ketonic hydrolysis of acetylacetoacetic ester:

$$CH_3COCH_2CO_2C_2H_5 \xrightarrow[Mg]{CH_3COCl} CH_3COCH\overset{\displaystyle COCH_3}{|}CO_2C_2H_5 \xrightarrow[(ii)\ H^+]{(i)\ dil.\ KOH} CH_3COCH_2COCH_3$$

(iii) By the condensation of acetic anhydride with acetone in the presence of boron trifluoride as catalyst (yield 80–85 per cent on the acetone):

$$CH_3COCH_3 + (CH_3CO)_2O \xrightarrow{BF_3} CH_3COCH_2COCH_3 + CH_3CO_2H$$

Pentane-2,4-dione is a colourless liquid, b.p. 139°C/746 mm. It exhibits tautomerism, e.g., it gives a red colour with ferric chloride.

$$CH_3COCH_2COCH_3 \rightleftharpoons CH_3COCH=C(OH)CH_3$$

It is readily oxidised to acetic acid, and is converted into a mixture of acetone and acetic acid when heated with potassium hydroxide solution (cf. 'acid hydrolysis' of acetoacetic ester, p. 290). It forms pyrazoles when treated with hydrazine or its derivatives; e.g., with phenyl-hydrazine it forms 3,5-dimethyl-1-phenylpyrazole:

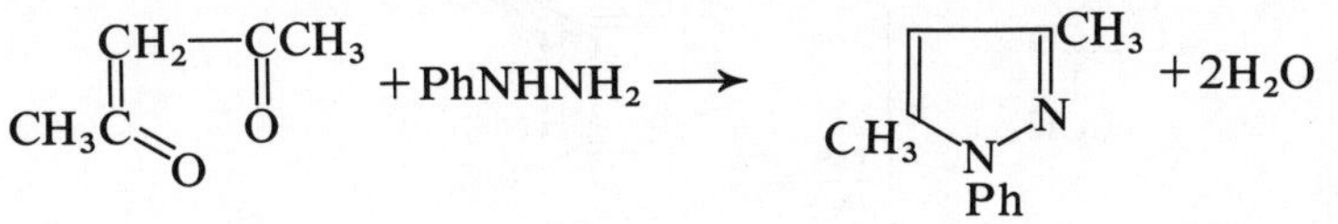

Pentane-2,4-dione forms chelated compounds with various metals, e.g., iron, aluminium, copper, etc. (cf. E.A.A.):

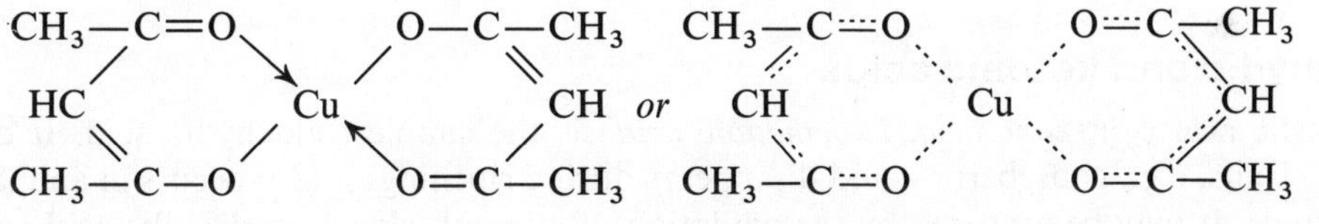

All β-diketones are split by aqueous alkali, and when they are unsymmetrical, fission can occur in two different ways:

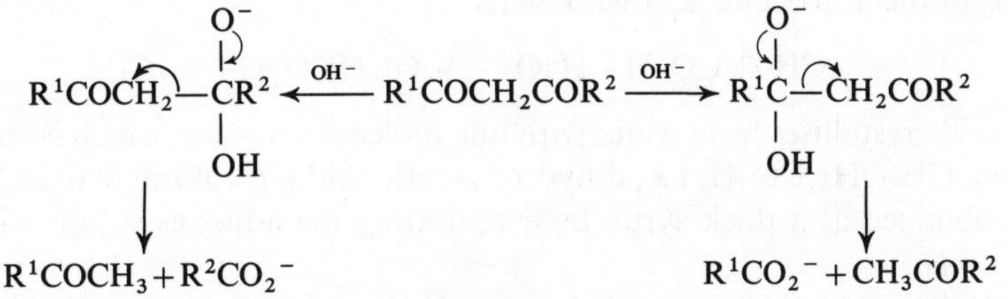

In general, fission occurs in the direction that produces the stronger acid in greater yield. Pentane-2,4-dione is oxidised by selenium dioxide to pentane-2,3,4-trione:

$$MeCOCH_2COMe \xrightarrow{SeO_2} MeCOCOCOMe \quad (29\%)$$

Hexane-2,5-dione, acetonylacetone, b.p. 188°C, is the simplest γ-diketone. It may be prepared by means of the acetoacetic ester synthesis:

$$[CH_3COCHCO_2C_2H_5]^-Na^+ + BrCH_2COCH_3 \longrightarrow$$

$$\overset{\displaystyle CH_2COCH_3}{\underset{\displaystyle CH_3COCHCO_2C_2H_5}{|}} \xrightarrow[\text{(ii) H}^+]{\text{(i) dil. KOH}} CH_3COCH_2CH_2COCH_3$$

An alternative and more convenient acetoacetic ester synthesis is as follows:

$$2[CH_3COCHCO_2C_2H_5]^-Na^+ \xrightarrow{I_2} \begin{array}{c} CH_3COCHCO_2C_2H_5 \\ | \\ CH_3COCHCO_2C_2H_5 \end{array} \xrightarrow[\text{(ii) HCl}]{\text{(i) KOH}}$$

$$\begin{array}{c} CH_3COCHCO_2H \\ | \\ CH_3COCHCO_2H \end{array} \xrightarrow{\text{heat}} \begin{array}{c} CH_3COCH_2 \\ | \\ CH_3COCH_2 \end{array} + 2CO_2$$

It readily forms five-membered rings: with phosphorus pentoxide, 2,5-dimethylfuran; with ammonia, 2,5-dimethylpyrrole; and with phosphorus pentasulphide, 2,5-dimethylthiophen (*cf.* succinaldehyde).

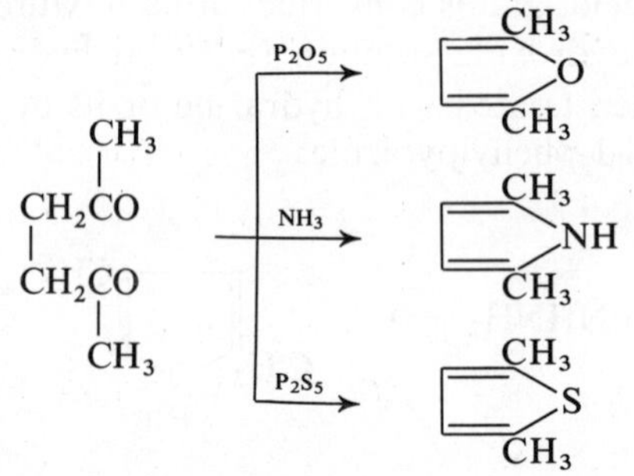

Hexane-2,5-dione is oxidised by selenium dioxide to hex-3-ene-2,5-dione (*cf.* pentane-2,4-dione):

$$MeCOCH_2CH_2COMe \xrightarrow{SeO_2} MeCOCH=CHCOMe$$

Aldehydic and ketonic acids

Glyoxylic acid (*glyoxalic acid, oxoethanoic acid*) is the simplest aldehydic acid. It occurs in unripe fruits, *e.g.*, gooseberries, and disappears during ripening; it also occurs in animal tissues and fluids. It may be prepared by the oxidation of ethanol, glycol or glycollic acid with nitric acid (yields are poor). It may also be prepared by the reduction of oxalic acid either electrolytically or by means of magnesium and sulphuric acid, but it is most conveniently prepared by the hydrolysis of dichloroacetic acid with water:

$$CHCl_2CO_2H + H_2O \longrightarrow OCHCO_2H + 2HCl$$

Glyoxylic acid crystallises from water with one molecule of water which is combined as water of constitution, $CH(OH)_2CO_2H$, *i.e.*, dihydroxyacetic acid (*cf.* chloral hydrate). The anhydrous acid may be obtained as a thick syrup by evaporating the aqueous solution over phosphorus pentoxide *in vacuo*.

Glyoxylic acid gives all the reactions of an aldehyde and an acid; *e.g.*, it reduces ammoniacal silver nitrate, forms a bisulphite compound, etc. Since it has no α-hydrogen atoms, glyoxylic acid undergoes the Cannizzaro reaction to form glycollic and oxalic acids:

$$2OCHCO_2H \xrightarrow{NaOH} HOCH_2CO_2Na + (CO_2Na)_2$$

When glyoxylic acid is reduced by means of metal and acid, tartaric acid is obtained as well as glyoxylic acid:

$$OCHCO_2H \xrightarrow{e;H^+} HOCH_2CO_2H + \begin{array}{c} CH(OH)CO_2H \\ | \\ CH(OH)CO_2H \end{array}$$

Pyruvic acid (*acetylformic acid, pyroracemic acid, α-ketopropionic acid, 2-oxopropanoic acid*), b.p. 165°C, is the simplest keto-acid. It may be prepared:

(i) By heating tartaric acid alone, or better, with potassium hydrogen sulphate at 210–220°C.

The reaction is believed to take place via the formation of hydroxymaleic acid (I), which rearranges to oxalacetic acid, (II):

$$\begin{array}{c} CH(OH)CO_2H \\ | \\ CH(OH)CO_2H \end{array} \xrightarrow{-H_2O} \begin{array}{c} CHCO_2H \\ \| \\ C(OH)CO_2H \\ (I) \end{array} \longrightarrow \begin{array}{c} CH_2CO_2H \\ | \\ COCO_2H \\ (II) \end{array} \longrightarrow CH_3COCO_2H + CO_2 \quad (50\text{--}55\%)$$

This is the best method for preparing pyruvic acid, and it was this method which gave rise to the name pyroracemic acid.

(ii) Other methods of preparation are the oxidation of lactic (α-hydroxypropionic) acid, the hydrolysis of α,α-dibromopropionic acid or acetyl cyanide, e.g.,

$$CH_3COCl \xrightarrow{KCN} CH_3COCN \xrightarrow{H^+} CH_3COCO_2H$$

A general method of preparing α-keto-acids uses the Claisen condensation between ethyl oxalate and monocarboxylic esters:

$$EtO_2CCO_2Et + RCH_2CO_2Et \xrightarrow{EtONa} EtO_2CCOCHRCO_2Et$$
$$\xrightarrow{10\% H_2SO_4} [HO_2CCOCHRCO_2H] \xrightarrow{-CO_2} RCH_2COCO_2H \quad (94\%)$$

Pyruvic acid behaves as a ketone and as an acid; it forms an oxime, hydrazone, etc. It is reduced by sodium amalgam to lactic acid and dimethyltartaric acid (cf. glyoxylic acid, above):

$$3CH_3COCO_2H \xrightarrow{e;\ H^+} CH_3CH(OH)CO_2H \quad + \quad \begin{array}{c} CH_3C(OH)CO_2H \\ | \\ CH_3C(OH)CO_2H \end{array}$$

Pyruvic acid reduces ammoniacal silver nitrate, itself being oxidised to acetic acid:

$$CH_3COCO_2H + [O] \longrightarrow CH_3CO_2H + CO_2$$

It is oxidised by warm nitric acid to oxalic acid:

$$CH_3COCO_2H + 4[O] \longrightarrow (CO_2H)_2 + CO_2 + H_2O$$

Pyruvic acid is easily decarboxylated with warm dilute sulphuric acid to give acetaldehyde:

$$CH_3COCO_2H \longrightarrow CH_3CHO + CO_2$$

The mechanism of this reaction is uncertain, but since it occurs only with α-keto-acids, it suggests that the -I effect of the α-carbonyl group plays an important part (cf. trichloroacetic acid). On this basis, the following mechanism could be postulated, involving a carbene intermediate.

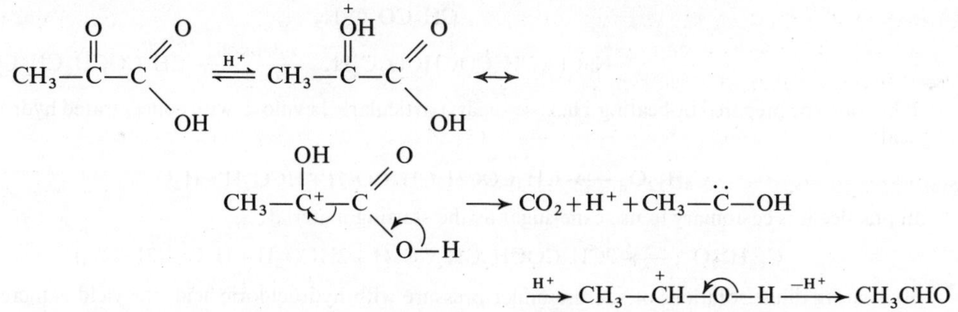

On the other hand, when warmed with *concentrated* sulphuric acid, pyruvic acid undergoes *decarbonylation*, i.e., eliminates a molecule of carbon monoxide, to form acetic acid:

$$CH_3\overset{O}{\overset{\|}{C}}-C\overset{O}{\overset{\diagup}{\diagdown}}_{OH} \xrightarrow{+2H^+} CH_3\overset{\overset{+}{O}H}{\overset{\|}{C}}-C^+\overset{OH}{\diagdown_{OH}} \xrightarrow[-H^+]{-H_2O}$$

$$CH_3\overset{O}{\overset{\|}{C}}-\overset{+}{C}{=}O \longrightarrow CO+CH_3\overset{+}{C}{=}O \xrightarrow[\text{(ii) }-H^+]{\text{(i) }H_2O} CH_3CO_2H$$

In this case, the possible mechanism proceeds through an intermediate acylium ion (*cf.* the $A_{AC}1$ mechanism, p. 251).

Both of these reactions with sulphuric acid are characteristic of α-keto-acids. α-Keto-esters are also decarbonylated when heated, and Calvin *et al.* (1940), using ethyl pyruvate labelled with ^{14}C, showed that the carbon monoxide came from the carbethoxyl group. In this case, the results may be explained by the formation of an epoxide intermediate:

$$CH_3-\underset{\underset{:OEt}{|}}{\overset{\overset{O}{\diagdown}}{C}}-\overset{O}{\overset{\|}{C}}_{*} \xrightarrow{110°C} CH_3-\underset{\underset{\underset{Et}{O}}{+\diagup}}{\overset{\overset{O}{\diagdown}}{C}}-C_{*}\overset{O}{\overset{\|}{}} \longrightarrow CH_3-\underset{OEt}{\overset{O}{\overset{\|}{C}}}_{*} +CO$$

The methyl group is made reactive by the adjacent carbonyl group, and so pyruvic acid undergoes many condensation reactions characteristic of a compound containing an active methylene group; *e.g.*, in the presence of dry hydrogen chloride, pyruvic acid forms α-keto-γ-valerolactone-γ-carboxylic acid:

$$CH_3C{=}O+CH_3COCO_2H \xrightarrow{HCl} \left[\underset{\underset{OH}{|}}{\overset{\overset{CO_2H}{|}}{CH_3C}}-CH_2-COCO_2H \right] \xrightarrow{-H_2O} \overset{\overset{CO_2H}{|}}{CH_3C}-CH_2-\underset{\underset{O}{\rule{2.5em}{0.4pt}}}{COCO}$$

Pyruvic acid is a very important substance biologically, since it is an intermediate product in the metabolism of carbohydrates and proteins.

Lævulic acid (*β-acetylpropionic acid, γ-ketovaleric acid, butan-3-one-1-carboxylic acid*), m.p. 34°C, is the simplest γ-keto-acid. It may be prepared via acetoacetic ester as follows:

$$[CH_3COCHCO_2C_2H_5]^-Na^+ +ClCH_2CO_2C_2H_5 \longrightarrow$$

$$NaCl+CH_3COCHCO_2C_2H_5 \overset{\overset{CH_2CO_2C_2H_5}{|}}{} \xrightarrow[\text{(ii) }H^+]{\text{(i) dil. KOH}} CH_3COCH_2CH_2CO_2H$$

It may also be prepared by heating a hexose sugar, particularly lævulose, with concentrated hydrochloric acid:

$$C_6H_{12}O_6 \longrightarrow CH_3COCH_2CH_2CO_2H+HCO_2H+H_2O$$

In practice it is customary to use cane-sugar as the starting material:

$$C_{12}H_{22}O_{11} \longrightarrow 2CH_3COCH_2CH_2CO_2H+2HCO_2H+H_2O \quad (21-22\%)$$

By heating dilute solutions of sucrose under pressure with hydrochloric acid, the yield is increased to 50 per cent.

Lævulic acid behaves as a ketone (forms oxime, etc.) and as an acid (forms esters, etc.). On the other hand, many of the reactions of lævulic acid indicate that it exists in the *lactol* form, *i.e.*, hydroxylactone (this is an example of ring-chain tautomerism), *e.g.*,

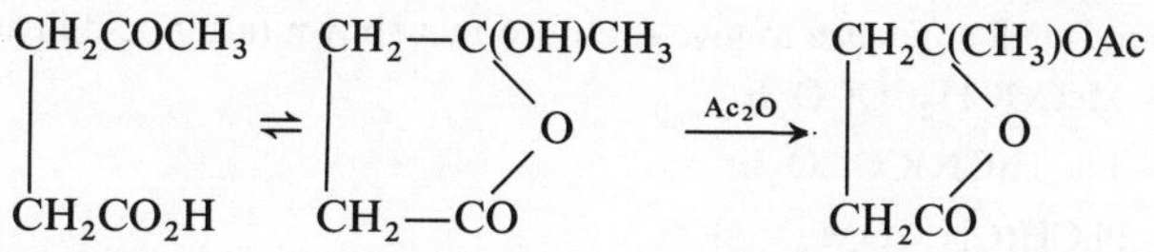

$$CH_2COCH_3 \quad CH_2\!-\!C(OH)CH_3 \qquad\qquad CH_2C(CH_3)OAc$$
$$\Big| \qquad\quad \Big| \qquad\qquad O \xrightarrow{\;Ac_2O\;} \qquad \Big| \qquad\qquad O$$
$$CH_2CO_2H \quad CH_2\!-\!CO \qquad\qquad CH_2CO$$

When heated for some time, lævulic acid is converted into α- and β-*angelica* lactones (derived from the lactol form):

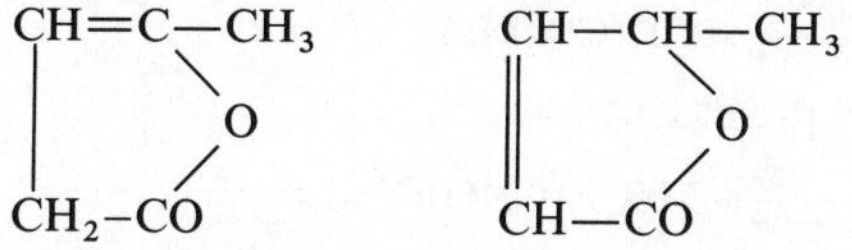

$$CH\!=\!C\!-\!CH_3 \qquad\qquad CH\!-\!CH\!-\!CH_3$$
$$\Big| \qquad\qquad O \qquad\qquad \| \qquad\qquad O$$
$$CH_2\!-\!CO \qquad\qquad CH\!-\!CO$$

α-   β-

A general method for preparing γ-keto-acids is to treat a ketone with lithium amide and lithium bromoacetate; the ketone undergoes α-carboxymethylation (Puterbaugh *et al.*, 1959):

$$R^1COCH_2R^2 \xrightarrow[\;CH_2BrCO_2Li\;]{\;LiNH_2\;} R^1COCHR^2CH_2CO_2H$$

γ-Keto-acids may also be synthesised by other methods (see, *e.g.*, p. 451).

Keto-acids with the oxo group in the δ- or a higher position are very conveniently prepared by the action of a dialkylcadmium on the half-ester acid chloride (see p. 437). On the other hand, δ-keto-acids may be prepared as follows from cyclohexane-1,3-dione (Stetter *et al.*, 1952, 1955); *e.g.*, 5-oxoheptanoic acid:

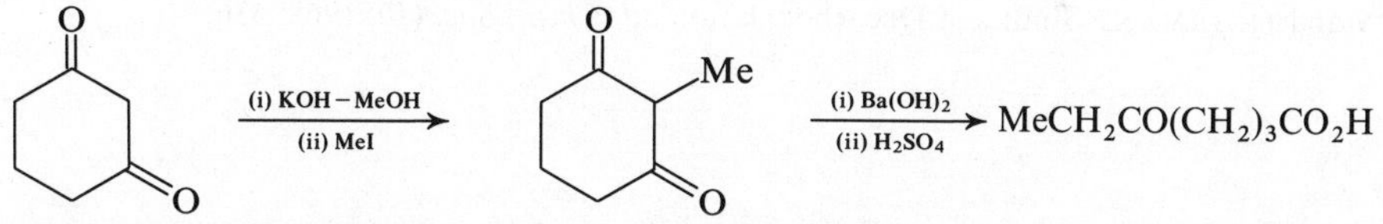

$$\text{(cyclohexane-1,3-dione)} \xrightarrow[\text{(ii) MeI}]{\text{(i) KOH}-\text{MeOH}} \text{(2-methyl-cyclohexane-1,3-dione)} \xrightarrow[\text{(ii) H}_2\text{SO}_4]{\text{(i) Ba(OH)}_2} MeCH_2CO(CH_2)_3CO_2H$$

The cyclic dione, since it contains an active methylene group, forms the potassium salt which can be alkylated, and the product readily undergoes fission, thus behaving like the acyclic β-diketones. The keto-acids may be reduced by the Wolff–Kishner method to the corresponding saturated acid; in the above example, to heptanoic acid. Alternatively, the alkylated ring may be opened and the carbonyl group reduced in one step by refluxing with hydrazine in alkaline glycol solution.

PROBLEMS

1. Draw the energy profiles for tautomerism between A and B and for resonance between A and B.

2. Explain the following equation:

$$MeCOCH_2CO_2Et + MeCOCHR^1CO_2Et \xrightarrow[\text{(ii) 1R}^2\text{X}]{\text{(i) 1EtONa}}$$

$$MeCOCHR^2CO_2Et + MeCOCHR^1CO_2Et$$

3. How could you differentiate, by i.r. and NMR spectroscopy, between **A**, $MeCOCH_2CH_2COMe$ and **B**, $MeCOCH_2COCH_2CH_3$?

4. Starting with E.A.A. or D.E.M., give *one* synthesis (using equations) for: (i) 4-methylpentan-2-one; (ii) 2,3-dimethylbutanoic acid; (iii) methylsuccinic acid; (iv) cyclopentane-1,2-dicarboxylic acid; (v) $PhCOCH_2CO_2Et$; (vi) 3-carboxy-2-methylpentanedioic acid.

5. The β-keto-acid shown is decarboxylated with difficulty. Suggest an explanation.

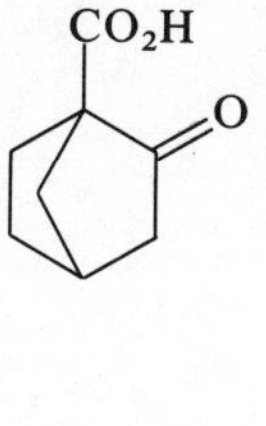

$$\text{CO}_2\text{H}$$

6. What *two* compounds will condense to give each product shown (after acidification)?

(i) $A + B \xrightarrow{\text{EtONa}} MeCOCH_2COCO_2Et$

(ii) $C + D \xrightarrow{\text{EtONa}} PhCH(CN)COCO_2Et$

(iii) $E + F \xrightarrow{\text{EtONa}} PhCH(CN)COMe$

7. Starting with E.A.A., synthesise 3-methylpentan-2-ol.

8. Complete the equations:

(i) $CH_2(CO_2Me)_2 + ? \xrightarrow[\text{(ii) AcOH}]{\text{(i) Na}} CH(CO_2Me)_3$

(ii) $MeCH(CO_2Et)COCO_2Et \xrightarrow{\text{heat}} ?$

(iii) $MeCH_2CO_2Et + ? \xrightarrow[\text{(ii) AcOH}]{\text{(i) EtONa}} MeCH(CHO)CO_2Et$

(iv) $MeCH_2CO_2Et + (CO_2Et)_2 \xrightarrow[\text{(ii) AcOH}]{\text{(i) EtONa}} ? \xrightarrow{\text{heat}} ?$

9. Suggest a synthesis for ethyl 4-keto-3-methylpentanoate.

10. A compound $C_4H_6O_2$, shows a very strong i.r. band at $1\,721$ cm^{-1} and its N.M.R. spectrum shows only one singlet signal. What is the compound?

REFERENCES

Organic Reactions, Wiley. Vol. I (1942), Ch. 9, 'The Acetoacetic Ester Condensation and Certain Related Reactions': Vol. VIII (1954), Ch. 3, 'The Acylation of Ketones to form β-Diketones or β-Ketoaldehydes': Vol. IX (1957), Ch. 1, 'The Cleavage of Nonenolisable Ketones with Sodium Amide'; Ch. 4, 'The Alkylation of Esters and Nitriles'.

BIGLEY and THURMAN, 'Studies in Decarboxylation', *J. Chem. Soc. (B)*, 1968, 436.

11 Polyhydric alcohols

Dihydric alcohols or glycols

Dihydric alcohols are compounds containing two hydroxyl groups. They are classified as α, β, γ . . . glycols, according to the relative positions of the two hydroxyl groups: α is the 1,2 glycol; β, 1,3; γ, 1,4 . . . Although it is unusual to find a compound with two hydroxyl groups attached to the same carbon atom, ether derivatives of these 1,1 glycols are stable, *e.g.*, acetals (p. 222). The commonest glycols are the α-glycols.

Nomenclature

The common names of the α-glycols are derived from the corresponding alkene from which they may be prepared by direct hydroxylation, *e.g.*,

$HOCH_2CH_2OH$	ethylene glycol
$(CH_3)_2COHCH_2OH$	isobutene glycol

β-, γ- . . . Glycols are named as the corresponding *polymethylene* glycols, *e.g.*,

$HOCH_2CH_2CH_2OH$	trimethylene glycol
$HOCH_2CH_2CH_2CH_2CH_2OH$	pentamethylene glycol

According to the I.U.P.A.C. system of nomenclature, the class suffix is *-diol*, and numbers are used to indicate the positions of side-chains and the two hydroxyl groups, *e.g.*,

$CH_3CHOHCH_2OH$ propane-1,2-diol

$$\begin{array}{c}CH_3CH_3\\ ||\\ HOCH_2CH_2CHCH_2CHCH_2OH\end{array}$$ 2,4-dimethylhexane-1,6-diol

Ethylene glycol, glycol (*ethane*-1,2-*diol*), is the simplest glycol, and may be prepared as follows:

(i) By passing ethylene into cold dilute alkaline permanganate solution (p. 109).

(ii) By passing ethylene into hypochlorous acid, and then hydrolysing the ethylene chlorohydrin by boiling with aqueous sodium hydrogen carbonate:

$$CH{=}CH_2 + HOCl \longrightarrow ClCH_2CH_2OH \xrightarrow{\text{aq. NaHCO}_3} HOCH_2CH_2OH + NaCl + CO_2$$

(iii) By boiling ethylene dibromide with aqueous sodium carbonate:

$$BrCH_2CH_2Br + NaCO_3 + H_2O \longrightarrow HOCH_2CH_2OH + 2NaBr + CO_2 \quad (50\%)$$

The low yield in this reaction is due to the conversion of some ethylene dibromide into vinyl bromide:

$$BrCH_2CH_2Br + Na_2CO_3 \longrightarrow CH_2{=}CHBr + NaBr + NaHCO_3$$

If aqueous sodium hydroxide is used instead of sodium carbonate, vinyl bromide is again obtained as a by-product. The best yield of glycol is obtained by heating ethylene dibromide with potassium acetate in glacial acetic acid, and subsequently hydrolysing the glycol diacetate with hydrogen chloride in methanolic solution:

$$BrCH_2CH_2Br + 2CH_3CO_2K \longrightarrow$$

$$CH_2(OCOCH_3)CH_2(OCOCH_3) + 2KBr \xrightarrow{\text{HCl}} HOCH_2CH_2OH$$
$$(90\%) \qquad\qquad\qquad\qquad (83\text{--}84\%)$$

(iv) By treating ethylene oxide with dilute hydrochloric acid:

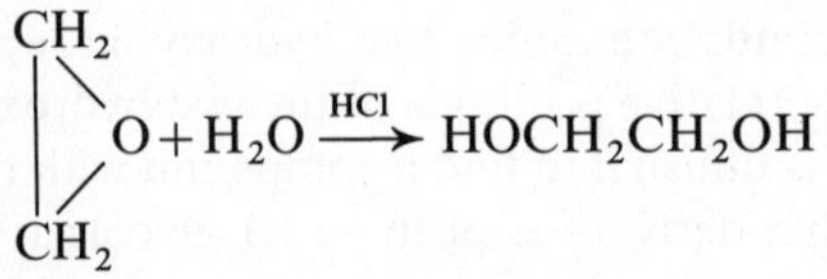

$$\begin{array}{c} CH_2 \\ | \;\backslash \\ | \quad O + H_2O \xrightarrow{\text{HCl}} HOCH_2CH_2OH \\ | \;/ \\ CH_2 \end{array}$$

Glycol is prepared industrially by method (iv), and by the catalytic reduction of methyl glycollate which is produced synthetically (see p. 471).

Glycol is a colourless viscous liquid, b.p. 197°C, and has a sweet taste (the prefix *glyc-* indicates that the compound has a sweet taste: Greek *glukus*, sweet). It is miscible in all proportions with water and ethanol, but is insoluble in ether. It is widely used as a solvent and as an antifreeze agent.

The high boiling points and high solubility in water of polyhydric alcohols are due to hydrogen bonding involving all hydroxyl groups.

The chemical reactions of glycol are those which might have been expected of a monohydric primary alcohol. One hydroxyl group, however, is almost always completely attacked before the other reacts.

(i) Sodium at 50°C forms the monosodium salt, and the di-sodium salt only when the temperature is raised to 160°C.

(ii) Hydrogen chloride forms ethylene chlorohydrin at 160°C, whereas at 200°C, ethylene dichloride is formed.

Ethylene chlorohydrin (β-chloroethyl alcohol, 2-chloroethanol) is a colourless liquid, b.p. 128·8°C. It is very useful in organic syntheses since it contains two different reactive groups; *e.g.*, by heating with aqueous sodium cyanide it may be converted into ethylene cyanohydrin, which, on hydrolysis, gives β-hydroxypropionic acid:

$$HOCH_2CH_2Cl + NaCN \longrightarrow HOCH_2CH_2CN + NaCl \quad (79\text{--}80\%)$$

$$HOCH_2CH_2CN \xrightarrow{\text{HCl}} HOCH_2CH_2CO_2H$$

(iii) When glycol is treated with phosphorus trichloride or phosphorus tribromide, the corresponding ethylene dihalide is obtained:

$$\begin{array}{c} CH_2OH \\ | \\ CH_2OH \end{array} \xrightarrow{PBr_3} \begin{array}{c} CH_2Br \\ | \\ CH_2Br \end{array}$$

Phosphorus tri-iodide, however, produces ethylene. Ethylene di-iodide is formed as an intermediate, but readily eliminates iodine (see p. 169):

$$\begin{array}{c} CH_2OH \\ | \\ CH_2OH \end{array} \xrightarrow{PI_3} \left[\begin{array}{c} CH_2I \\ | \\ CH_2I \end{array} \right] \longrightarrow I_2 + \begin{array}{c} CH_2 \\ \| \\ CH_2 \end{array}$$

(iv) When glycol is treated with organic acids or inorganic oxygen acids, the mono- or di-esters are obtained, depending on the relative amounts of glycol and acid; *e.g.*, glycol diacetate is obtained by heating glycol with acetic acid in the presence of a small amount of sulphuric acid as catalyst:

$$\begin{array}{c} CH_2OH \\ | \\ CH_2OH \end{array} + 2CH_3CO_2H \xrightarrow{H_2SO_4} \begin{array}{c} CH_2OCOCH_3 \\ | \\ CH_2OCOCH_3 \end{array} + 2H_2O$$

When glycol is heated with dicarboxylic acids, condensation polymers (polyesters) are obtained:

$$n\,HOCH_2CH_2OH + n\,HO_2C(CH_2)_xCO_2H \longrightarrow HOCH_2[CH_2OCO(CH_2)_xCOOCH_2]_nCH_2OH$$

(v) Glycol condenses with aldehydes or ketones in the presence of *p*-toluenesulphonic acid to yield respectively cyclic acetals or cyclic ketals (1,3-*dioxolans*):

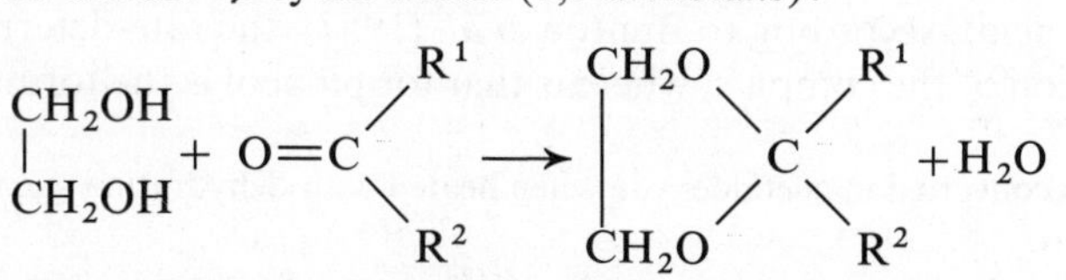

$$\begin{array}{c} CH_2OH \\ | \\ CH_2OH \end{array} + O{=}C\begin{array}{c} R^1 \\ \diagup \\ \diagdown \\ R^2 \end{array} \longrightarrow \begin{array}{c} CH_2O \\ | \qquad \diagdown \\ \qquad\quad C \\ | \qquad \diagup \\ CH_2O \end{array}\begin{array}{c} R^1 \\ \diagup \\ \diagdown \\ R^2 \end{array} + H_2O$$

Dioxolan formation offers a means of protecting a carbonyl group in reactions carried out in alkaline media. The carbonyl compound may be regenerated by the addition of periodic acid to an aqueous dioxan solution of the dioxolan:

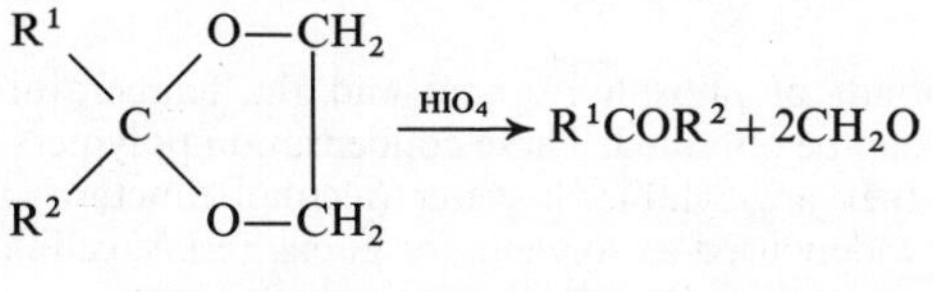

$$\begin{array}{c} R^1 \\ \diagdown \\ \quad C \\ \diagup \\ R^2 \end{array}\begin{array}{c} O{-}CH_2 \\ \diagdown \quad\; | \\ \quad\; | \\ \diagup \quad\; | \\ O{-}CH_2 \end{array} \xrightarrow{HIO_4} R^1COR^2 + 2CH_2O$$

(vi) When glycol is oxidised with nitric acid, glycollic and oxalic acids may be readily isolated. All theoretically possible oxidation products have been isolated, but since, except for glycollic and oxalic acids, they are more readily oxidised than glycol itself, they have only been obtained in very poor yields.

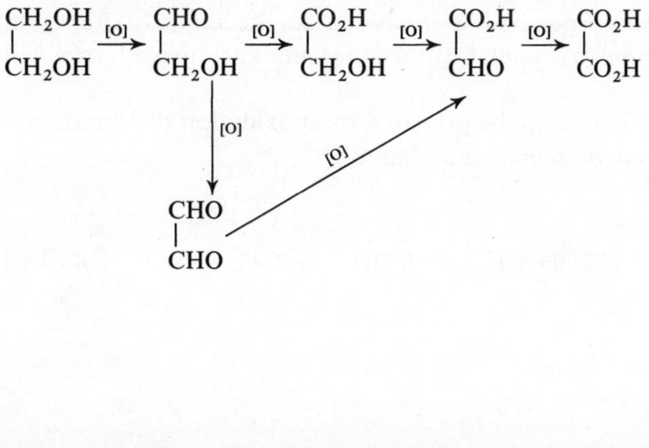

$$\begin{array}{c} CH_2OH \\ | \\ CH_2OH \end{array} \xrightarrow{[O]} \begin{array}{c} CHO \\ | \\ CH_2OH \end{array} \xrightarrow{[O]} \begin{array}{c} CO_2H \\ | \\ CH_2OH \end{array} \xrightarrow{[O]} \begin{array}{c} CO_2H \\ | \\ CHO \end{array} \xrightarrow{[O]} \begin{array}{c} CO_2H \\ | \\ CO_2H \end{array}$$

$$\Big\downarrow {[O]} \qquad \nearrow {[O]}$$

$$\begin{array}{c} CHO \\ | \\ CHO \end{array}$$

Oxidation of glycols with acid, alkaline or neutral permanganate, acid dichromate or chromic acid (chromium trioxide in acetic acid), lead tetra-acetate or periodic acid results in fission of the carbon–carbon bond to give aldehydes, acids and/or ketones.

Oxidation of 1,2-glycols occurs more rapidly with *cis*-glycols than with the corresponding *trans*-isomers, and this has been interpreted as evidence for the formation of cyclic intermediates, *e.g.*,

(*a*)

$$\begin{array}{c} R_2C-OH \\ | \\ R_2C-OH \end{array} + (AcO)_4Pb \underset{-AcOH}{\rightleftharpoons} \begin{array}{c} R_2C-O-Pb(OAc)_3 \\ | \\ R_2C-O-H \end{array} \underset{-AcOH}{\rightleftharpoons}$$

$$\begin{array}{c} R_2C-O \\ | \\ R_2C-O \end{array} Pb(OAc)_2 \longrightarrow \begin{array}{c} R_2C=O \\ + \\ R_2C=O \end{array} + (AcO)_2Pb$$

(*b*)

$$\begin{array}{c} R_2C-OH \\ | \\ R_2C-OH \end{array} + HIO_4 \rightleftharpoons \begin{array}{c} R_2C-O \\ | \\ R_2C-O \end{array} IO_3H \longrightarrow \begin{array}{c} R_2C=O \\ + \\ R_2C=O \end{array} + HIO_3$$

The fact that *trans*-isomers *are* oxidised suggests that the reaction proceeds through a non-cyclic intermediate for these compounds. This could be the non-cyclic ester formed in the first step.

These oxidations are also subject to steric hindrance, *e.g.*, glycol is oxidised much faster than pinacol by periodic acid. According to Bunton *et al.* (1957), the rate-determining step for glycol oxidation is the fission of the complex, whereas that for pinacol is the formation of the complex.

(vii) Glycol is converted in acetaldehyde when heated with dehydrating agents, *e.g.*,

$$HOCH_2CH_2OH \xrightarrow{ZnCl_2} CH_3CHO + H_2O$$

On the other hand, when glycol is heated with a dehydrating agent such as phosphoric acid, *polyethylene glycols* are obtained, *e.g.*, diethylene glycol:

$$2HOCH_2CH_2OH \xrightarrow{-H_2O} HOCH_2CH_2OCH_2CH_2OH$$

By varying the amount of phosphoric acid and the temperature, the polyethylene glycols up to decaethylene glycol can be obtained. These condensation polymers contain both the alcohol and ether functional groups; they are soluble in water (alcohol function) and are very good solvents (ether function). They are widely used as solvents for gums, resins, cellulose esters, etc.

Ethylene oxide, b.p. 10.7°C. According to the I.U.P.A.C. system of nomenclature, an oxygen atom linked to two of the carbon atoms in a carbon chain is denoted by the prefix *epoxy* in all cases other than those in which a substance is named as a cyclic compound; thus ethylene oxide is *epoxyethane*. Epoxy-compounds contain the *oxiran (oxirane)* ring, and ethylene oxide is also known as *oxiran (oxirane)*. Oxiran compounds are also referred to as *cyclic ethers* or *alkene oxides*.

Epoxides may, in general, be prepared by epoxidation of alkenes with peroxy acids (p. 111). Ethylene oxide may be prepared as follows:

$$HOCH_2CH_2Cl + KOH \longrightarrow \underset{O}{CH_2-CH_2} + KCl + H_2O$$

The mechanism is possibly:

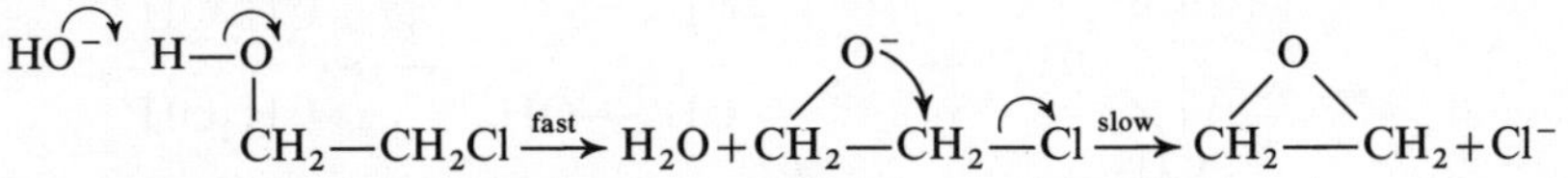

$$\overset{\curvearrowleft}{HO^-}\;\overset{\curvearrowleft}{H-O}$$
$$\underset{CH_2-CH_2Cl}{\mid}\xrightarrow{fast} H_2O + \overset{O^-}{CH_2-CH_2}\overset{\curvearrowleft}{-Cl}\xrightarrow{slow}\overset{O}{CH_2-CH_2}+Cl^-$$

This is an example of neighbouring group participation.

Ethylene oxide is manufactured by passing ethylene and air under pressure over a silver catalyst at 200–400°C:

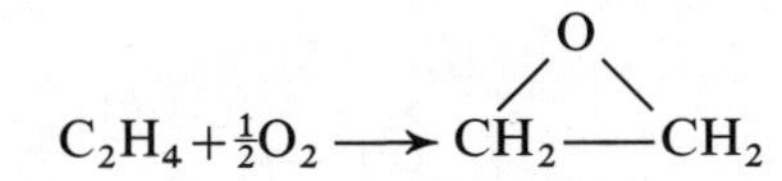

$$C_2H_4+\tfrac{1}{2}O_2\longrightarrow CH_2\overset{O}{-\!\!-}CH_2$$

Ethylene oxide undergoes molecular rearrangement on heating to form acetaldehyde:

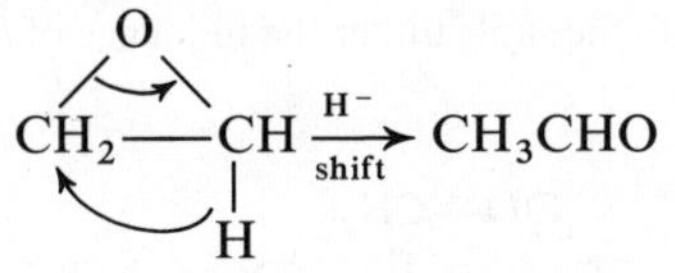

$$CH_2\overset{O}{-\!\!-}\underset{H}{\overset{\mid}{C}}H\xrightarrow[\text{shift}]{H^-}CH_3CHO$$

Epoxides are reduced by lithium aluminium hydride to alcohols, unsymmetrical epoxides giving as main product the more highly substituted alcohol; thus (I) is the main product:

$$RCH\overset{O}{-\!\!-}CH_2\xrightarrow{LiAlH_4}RCHOHCH_3+RCH_2CH_2OH$$
$$\qquad\qquad\qquad\qquad\quad (I)\qquad\qquad (II)$$

On the other hand, (II) is the main product if LAH—AlCl$_3$ is used (Eliel *et al.*, 1958; see also Table 6.1). Brown *et al.* (1968) have shown that (II) is the major product (> 90%) when BH$_3$—BF$_3$ is the reducing reagent.

Epoxides are readily reduced to the corresponding alkenes by t-phosphines, *e.g.*,

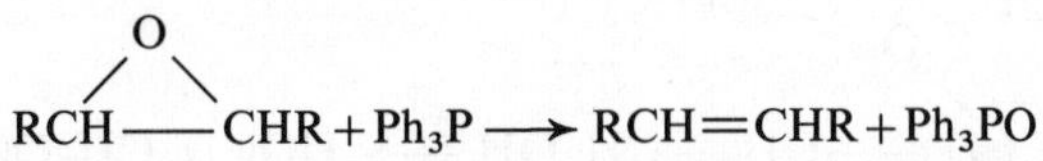

$$RCH\overset{O}{-\!\!-}CHR+Ph_3P\longrightarrow RCH\!=\!CHR+Ph_3PO$$

They are also oxidised by dimethyl sulphoxide to α-hydroxyketones (acyloins).

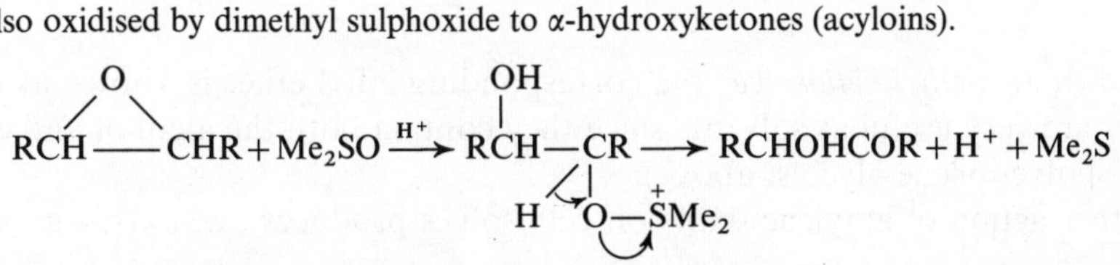

$$RCH\overset{O}{-\!\!-}CHR+Me_2SO\xrightarrow{H^+}RCH-\underset{H\;\;O-\overset{+}{S}Me_2}{\overset{OH}{\overset{\mid}{C}R}}\longrightarrow RCHOHCOR+H^++Me_2S$$

Ethylene oxide is converted into ethylene glycol in *dilute* acid solution and into ethylene halogenohydrins with *concentrated* halogen acid solutions. Ethylene oxide also forms mono-ethers with alcohols in the presence of a small amount of acid as catalyst. The mechanisms of these reactions in acid media probably follow the same pattern, *e.g.*, with *dilute* acid (with alcohols, replace H$_2$O by ROH):

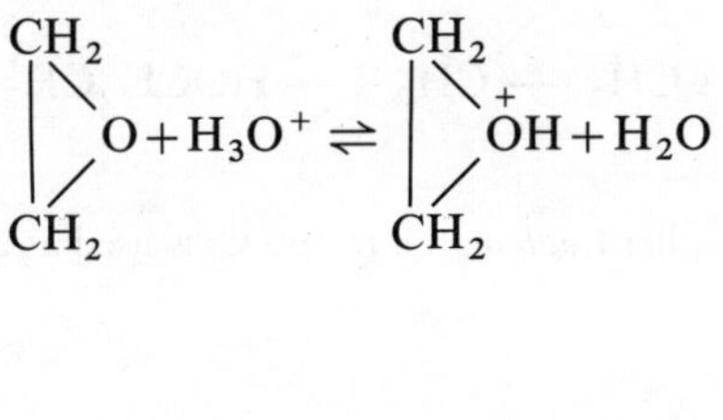

$$\underset{CH_2}{\overset{CH_2}{\mid}}\!\!>\!O+H_3O^+\rightleftharpoons\underset{CH_2}{\overset{CH_2}{\mid}}\!\!>\!\overset{+}{O}H+H_2O$$

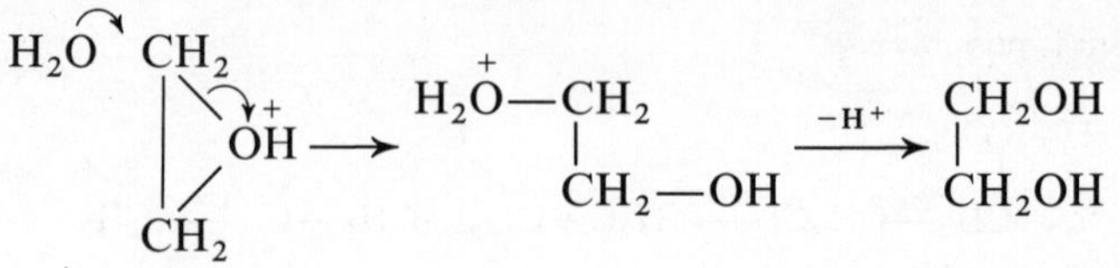

$$\text{H}_2\text{O} \quad \begin{matrix} \text{CH}_2 \\ \backslash \\ \overset{+}{\text{OH}} \\ / \\ \text{CH}_2 \end{matrix} \longrightarrow \quad \begin{matrix} \overset{+}{\text{H}_2\text{O}}\!\!-\!\!\text{CH}_2 \\ | \\ \text{CH}_2\!\!-\!\!\text{OH} \end{matrix} \quad \xrightarrow{-\text{H}^+} \quad \begin{matrix} \text{CH}_2\text{OH} \\ | \\ \text{CH}_2\text{OH} \end{matrix}$$

Attack occurs from the position remote from the $^+$OH, and the product is the *trans* diol. This cannot be demonstrated with ethylene oxide itself but has been, *e.g.*, with 1,2-epoxycyclohexane. With *concentrated* halogen acid:

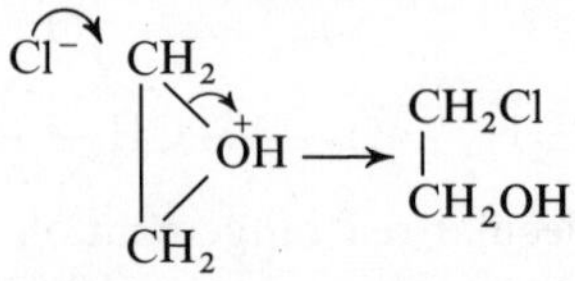

$$\text{Cl}^- \quad \begin{matrix} \text{CH}_2 \\ \backslash \\ \overset{+}{\text{OH}} \\ / \\ \text{CH}_2 \end{matrix} \longrightarrow \quad \begin{matrix} \text{CH}_2\text{Cl} \\ | \\ \text{CH}_2\text{OH} \end{matrix}$$

Ethylene oxide also reacts with alcohols under the influence of *basic* catalysts. A possible mechanism is:

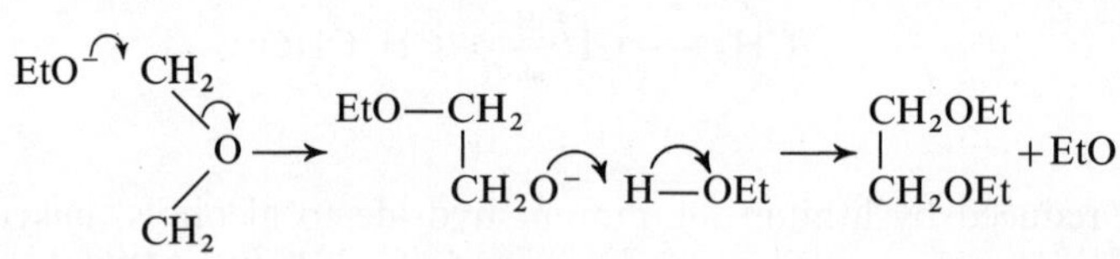

$$\text{EtO}^- \quad \begin{matrix} \text{CH}_2 \\ \backslash \\ \text{O} \\ / \\ \text{CH}_2 \end{matrix} \longrightarrow \quad \begin{matrix} \text{EtO}\!\!-\!\!\text{CH}_2 \\ | \\ \text{CH}_2\text{O}^- \quad \text{H}\!\!-\!\!\text{OEt} \end{matrix} \longrightarrow \quad \begin{matrix} \text{CH}_2\text{OEt} \\ | \\ \text{CH}_2\text{OEt} \end{matrix} + \text{EtO}^-$$

Of particular interest is the addition of hydrogen cyanide to form ethylene cyanohydrin:

$$\begin{matrix} \text{O} \\ / \quad \backslash \\ \text{CH}_2\!\!-\!\!\text{CH}_2 \end{matrix} + \text{HCN} \longrightarrow \text{HOCH}_2\text{CH}_2\text{CN}$$

The structure of ethylene oxide is uncertain; it may contain 'bent' bonds (see cyclopropane, p. 519).

When ethylene oxide is heated with methanol under pressure, the monomethyl ether of glycol is formed:

$$\begin{matrix} \text{O} \\ / \quad \backslash \\ \text{CH}_2\!\!-\!\!\text{CH}_2 \end{matrix} + \text{CH}_3\text{OH} \longrightarrow \text{HOCH}_2\text{CH}_2\text{OCH}_3$$

This is known as *methyl cellosolve*; the corresponding ethyl ether is known as *ethyl cellosolve*. Cellosolves are very useful as solvents since they contain both the alcohol and ether functional groups (*cf.* polyethylene glycols, above).

The further action of ethylene oxide on cellosolves produces *carbitols*, *e.g.*,

$$\text{HOCH}_2\text{CH}_2\text{OCH}_3 + \begin{matrix} \text{O} \\ / \quad \backslash \\ \text{CH}_2\!\!-\!\!\text{CH}_2 \end{matrix} \longrightarrow \text{HOCH}_2\text{CH}_2\text{OCH}_2\text{CH}_2\text{OCH}_3$$

When glycol is treated with ethylene oxide, diethylene glycol is formed:

$$\text{HOCH}_2\text{CH}_2\text{OH} + \begin{matrix} \text{O} \\ / \quad \backslash \\ \text{CH}_2\!\!-\!\!\text{CH}_2 \end{matrix} \longrightarrow \text{HOCH}_2\text{CH}_2\text{OCH}_2\text{CH}_2\text{OH}$$

Diethylene glycol dimethyl ether (*diglyme*), b.p. 160°C, is used as a solvent.

Ethylene oxide reacts with ammonia to form a mixture of three aminoalcohols which are usually referred to as the *ethanolamines*:

$$C_2H_4O + NH_3 \longrightarrow HOCH_2CH_2NH_2 \xrightarrow{C_2H_4O}$$

$$(HOCH_2CH_2-)_2NH \xrightarrow{C_2H_4O} (HOCH_2CH_2-)_3N$$

The ethanolamines are widely used as emulsifying agents.

The mechanism of these reactions is possibly:

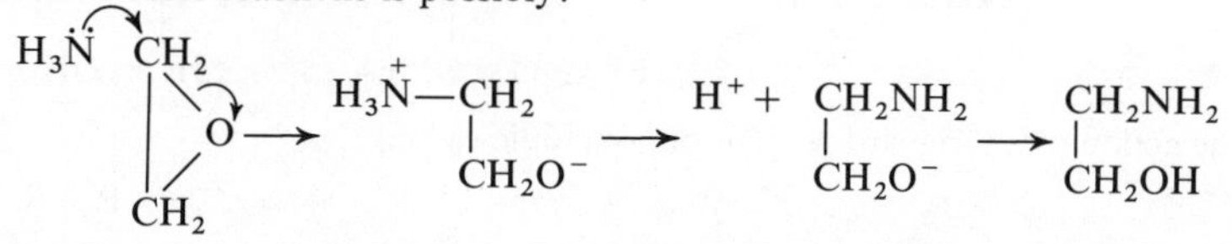

$$H_3\overset{+}{N}-CH_2 \longrightarrow H^+ + \underset{CH_2O^-}{\overset{CH_2NH_2}{|}} \longrightarrow \underset{CH_2OH}{\overset{CH_2NH_2}{|}}$$

In the presence of excess of ethylene oxide, the reaction proceeds with the ethanolamine now being the amino reagent, etc.

The C—O group (stretch) in epoxides absorbs in the region $1\,260$–$1\,240\,cm^{-1}$ (s), and so these compounds may be readily distinguished from ethers (p. 196).

The NMR spectra of epoxy compounds show a τ-value $\sim 7 \cdot 2$–$7 \cdot 6$ p.p.m. for methylene protons and $7 \cdot 0$ for methine protons. These ranges are sufficiently different from those of CH_2 and CH groups in other environments that epoxides may usually be recognised in this way (*cf.* Table 3.2, p. 80).

Dioxan (1,4-*dioxan, diethylene dioxide*), b.p. $101 \cdot 5°C$, is manufactured:

(i) By distilling glycol with fuming sulphuric acid:

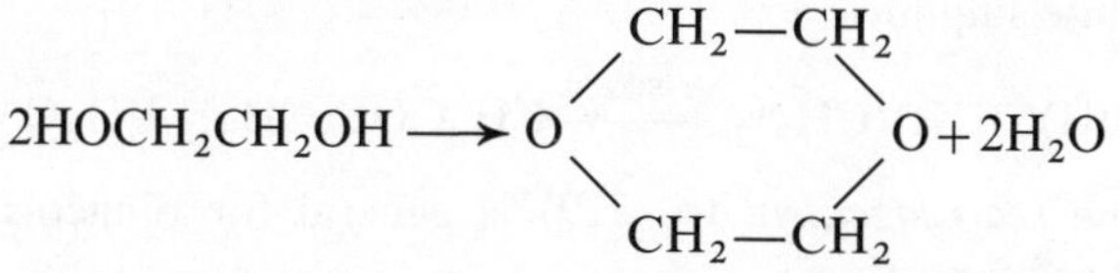

$$2HOCH_2CH_2OH \longrightarrow O\overset{CH_2-CH_2}{\underset{CH_2-CH_2}{<\quad>}}O + 2H_2O$$

(ii) By distilling ethylene oxide with dilute sulphuric acid:

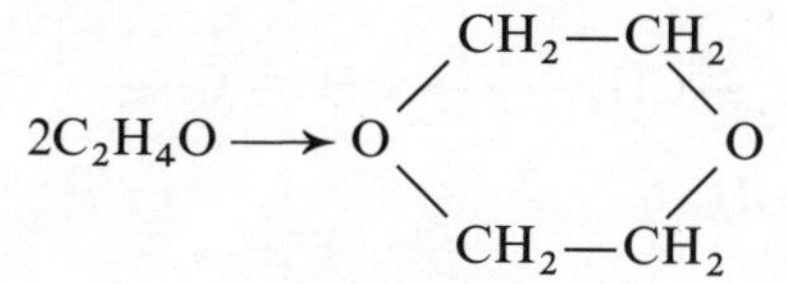

$$2C_2H_4O \longrightarrow O\overset{CH_2-CH_2}{\underset{CH_2-CH_2}{<\quad>}}O$$

Dioxan is a useful solvent for cryoscopic and ebullioscopic work.

Propene oxide, b.p. $35°C$, is manufactured as follows:

(i) $2CH_3CH=CH_2 \xrightarrow{2HOCl} CH_3CHOHCH_2Cl + CH_3CHClCH_2OH$

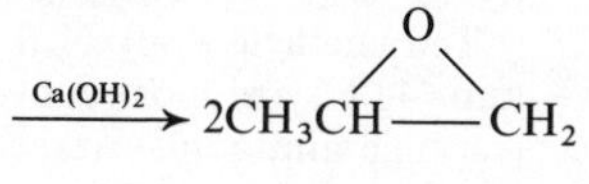

$$\xrightarrow{Ca(OH)_2} 2CH_3CH\overset{O}{\overset{/\backslash}{—}}CH_2$$

(ii) 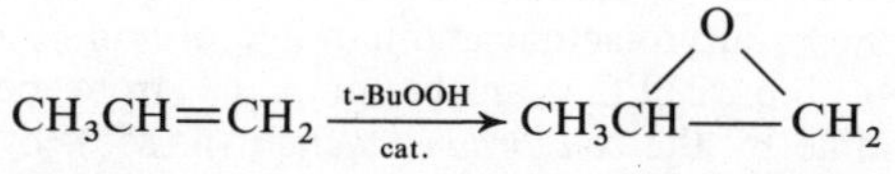 $CH_3CH=CH_2 \xrightarrow[\text{cat.}]{t\text{-BuOOH}} CH_3CH\overset{O}{\overset{/\backslash}{—}}CH_2$

Its reactions are similar to those of ethylene oxide. When heated with dilute hydrochloric acid, propene oxide forms propene glycol, b.p. $188°C$, which is used as a solvent, as a substitute for glycerol, etc.

General methods of preparation of 1,2-glycols other than glycol itself

1. By the reduction of ketones with magnesium amalgam, *e.g.*, pinacol (see p. 224).

$$2CH_3COCH_3 \xrightarrow[\text{(ii) } H_2O]{\text{(i) Mg-Hg}} (CH_3)_2C(OH)C(OH)(CH_3)_2 \quad (43–50\%)$$

2. By the action of a Grignard reagent on α-diketones, *e.g.*,

$$CH_3COCOCH_3 \xrightarrow{2RMgX} \underset{\underset{XMgO}{|}}{\overset{\overset{R}{|}}{CH_3C}} - \underset{\underset{OMgX}{|}}{\overset{\overset{R}{|}}{CCH_3}} \xrightarrow{H_2O} \underset{\underset{HO}{|}}{\overset{\overset{R}{|}}{CH_3C}} - \underset{\underset{OH}{|}}{\overset{\overset{R}{|}}{CCH_3}}$$

3. By the action of a Grignard reagent on α-ketonic esters, *e.g.*,

$$CH_3COCO_2C_2H_5 \xrightarrow{3RMgX} \underset{\underset{XMgO}{|}}{\overset{\overset{R}{|}}{CH_3C}} - \underset{\underset{OMgX}{|}}{\overset{\overset{R}{|}}{CR}} \xrightarrow{H_2O} \underset{\underset{HO}{|}}{\overset{\overset{R}{|}}{CH_3C}} - \underset{\underset{OH}{|}}{\overset{\overset{R}{|}}{CR}}$$

4. By the catalytic reduction of acyloins:

$$RCH(OH)COR \xrightarrow[\text{Ni}]{H_2} RCH(OH)CH(OH)R$$

5. By hydroxylation of unsaturated compounds (see p. 109).

Pinacol (*tetramethylethylene glycol, 2,3-dimethylbutane-2,3-diol*) is most conveniently prepared by reducing acetone with magnesium amalgam (see above). It crystallises out of solution as the hexahydrate. The most important reaction of pinacol is the rearrangement it undergoes when distilled with dilute sulphuric acid:

$$(CH_3)_2C(OH)C(OH)(CH_3)_2 \xrightarrow{H_2SO_4} CH_3COC(CH_3)_3 + H_2O \quad (65–72\%)$$

The *pinacol–pinacolone rearrangement* (p. 229) is general for pinacols (ditertiary alcohols). Some diene is formed as by-product:

$$\underset{\underset{OH}{|}}{\overset{\overset{CH_3}{|}}{CH_3-C}} - \underset{\underset{OH}{|}}{\overset{\overset{CH_3}{|}}{C-CH_3}} \xrightarrow{H_2SO_4} \overset{\overset{CH_3}{|}}{CH_2=C} - \overset{\overset{CH_3}{|}}{C=CH_2} + 2H_2O$$

Polymethylene glycols

A general method of preparing polymethylene glycols is to reduce α,ω-dicarboxylic esters with sodium and ethanol or lithium aluminium hydride, or catalytically. Individual polymethylene glycols are usually prepared by special methods.

Trimethylene glycol (propane-1,3-diol), b.p. 214°C (with decomposition), may be prepared by the hydrolysis of trimethylene dibromide.

Tetramethylene glycol (butane-1, 4-diol), b.p. 230°C, and **hexamethylene glycol (hexane-1, 6-diol),** m.p. 42°C, are most conveniently prepared by reducing the corresponding α,ω-dicarboxylic esters (succinic and adipic, respectively). Tetramethylene glycol is prepared industrially by hydrogenating butynediol. It is used for preparing butadiene, γ-butyrolactone and tetrahydrofuran.

Pentamethylene glycol (pentane-1,5-diol), b.p. 239°C, can be prepared from pentamethylene dibromide, which is obtained from piperidine by the *von Braun reaction* (p. 855), by heating with potassium acetate in glacial acetic acid, and subsequently hydrolysing the diacetate with alkali (*cf.* glycol):

$$BrCH_2(CH_2)_3CH_2Br \xrightarrow{CH_3CO_2K} CH_3COOCH_2(CH_2)_3CH_2OOCCH_3 \xrightarrow{NaOH}$$

$$HOCH_2(CH_2)_3CH_2OH \quad (v.g.)$$

Pentamethylene glycol may also be prepared by the catalytic hydrogenation (copper chromite) of tetrahydrofurfuryl alcohol (*cf.* p. 833) under pressure (40–47 per cent yield).

The polymethylene glycols are readily converted into the corresponding mono- or di-halogen derivative by halogen acid, according to the amount of halogen acid used. These polymethylene halides are useful reagents in organic syntheses since they contain two reactive halogen atoms, one or both of which may be made to undergo reaction. If the synthesis requires reaction with one halogen atom only, the most satisfactory procedure is to 'protect' the other halogen atom by ether formation and subsequently decompose the ether with concentrated hydrobromic acid, *e.g.*,

$$C_2H_5ONa + BrCH_2CH_2CH_2Br \longrightarrow NaBr + C_2H_5OCH_2CH_2CH_2Br \xrightarrow{\text{AgNO}_2}$$

$$C_2H_5OCH_2CH_2CH_2NO_2 \xrightarrow{\text{HBr}} BrCH_2CH_2CH_2NO_2$$

Then, *e.g.*,

$$BrCH_2CH_2CH_2NO_2 \xrightarrow{\text{KCN}} NCCH_2CH_2CH_2NO_2$$

Trihydric alcohols

The only important trihydric alcohol is **glycerol (propane-1,2,3-triol),** b.p. 290°C. It occurs in almost all animal and vegetable oils and fats as the glyceryl esters of mainly palmitic, stearic and oleic acids (see below).

Glycerol is obtained in large quantities as a by-product in the manufacture of soap, and this is still a commercial source of glycerol (see below). It is prepared synthetically as follows:

$$CH_3CH=CH_2 \xrightarrow[480-500°C]{\text{Cl}_2} ClCH_2CH=CH_2 \xrightarrow[150°C\ 12\ atm.]{\text{aq. Na}_2CO_3} HOCH_2CH=CH_2 \xrightarrow{\text{HOCl}}$$

allyl chloride   allyl alcohol

$$HOCH_2CHClCH_2OH \xrightarrow{\text{NaOH}} HOCH_2CHOHCH_2OH$$

glycerol β-monochlorohydrin

An alternative route that is used is:

$$ClCH_2CH=CH_2 \xrightarrow{\text{HOCl}} \left.\begin{array}{c} ClCH_2CHOHCH_2Cl \\ + \\ ClCH_2CHClCH_2OH \end{array}\right\} \xrightarrow{\text{CaO}}$$

$$\underset{\displaystyle CH_2\!-\!\!-\!\!CHCH_2Cl}{\overset{\displaystyle O}{\diagdown\diagup}} \xrightarrow{\text{NaOH}} HOCH_2CHOHCH_2OH$$

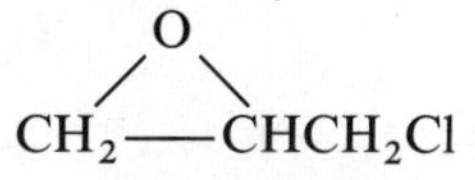

epichlorohydrin

Other methods of manufacture are:

(i) $$CH_2=CHCHO \xrightarrow[\text{cat.}]{\text{H}_2} CH_2=CHCH_2OH \xrightarrow{\text{H}_2O_2} HOCH_2CHOHCH_2OH$$

(ii) $$CH_2=CHCH_3 + O_2 \xrightarrow[300-400°C]{\text{Cu}_2O} CH_2=CHCHO$$

$$CH_2=CHCH_3 + H_2O \xrightarrow[\text{WO}_3-\text{ZnO}]{220-250°C} CH_3CHOHCH_3$$

$$CH_2=CHCHO + (CH_3)_2CHOH \xrightarrow[400°C]{\text{MgO}-\text{ZnO}} (CH_3)_2CO + CH_2=CHCH_2OH$$

$$(CH_3)_2CHOH + O_2 \xrightarrow{100-140°C} (CH_3)_2CO + H_2O_2$$

$$CH_2=CHCH_2OH + H_2O_2 \xrightarrow[\text{WO}_3]{60-70°C} HOCH_2CHOHCH_2OH$$

Acetone is obtained as a by-product.

Glycerol is used as an antifreeze, for making explosives, and, because of its hygroscopic properties, as a moistening agent for tobacco, shaving soaps, etc.

Glycerol contains one secondary and two primary alcoholic groups, and it undergoes many of the reactions to be expected of these types of alcohols. The carbon atoms in glycerol are indicated as shown: $\overset{\alpha}{C}H_2OH\overset{\beta}{C}HOH\overset{\alpha' \text{ or } \gamma}{C}H_2OH$.

(i) When glycerol is treated with sodium, one α-hydroxyl group is readily attacked, and the other α-group less readily; the β-hydroxyl groups is not attacked at all:

$$HOCH_2CHOHCH_2OH \xrightarrow{Na} NaOCH_2CHOHCH_2OH \xrightarrow{Na} NaOCH_2CHOHCH_2ONa$$

(ii) On passing hydrogen chloride into glycerol at 110°C until there is the theoretical increase in weight corresponding to the esterification of one hydroxyl group, both α- and β-glycerol monochlorohydrin are formed, the former predominating (66 per cent). Continued action of hydrogen chloride at 110°C, using 25 per cent of acid in excess required by theory for the esterification of two hydroxyl groups, produces glycerol α,α'-dichlorohydrin (α-dichlorohydrin) and glycerol α,β-dichlorohydrin (β-dichlorohydrin), the former predominating (55–57 per cent); some other products are also formed:

$$HOCH_2CHOHCH_2OH \xrightarrow{HCl} ClCH_2CHOHCH_2OH + HOCH_2CHClCH_2OH \xrightarrow{HCl}$$
$$ClCH_2CHOHCH_2Cl + ClCH_2CHClCH_2OH$$

When either of these dichlorohydrins or glycerol itself is treated with phosphorus pentachloride, glycerol trichlorohydrin (1,2,3-trichloropropane) is obtained. This is a liquid, b.p. 156–158°C.

When glycerol α,α'-dichlorohydrin is oxidised with sodium dichromate-sulphuric acid mixture, 1,3-dichloroacetone (1,3-dichloropropan-2-one), m.p. 45°C, is obtained:

$$ClCH_2CHOHCH_2Cl \xrightarrow{[O]} ClCH_2COCH_2Cl \quad (68\text{–}75\%)$$

When glycerol α,α'-dichlorohydrin is treated with powdered sodium hydroxide in ether solution, epichlorohydrin (3-chloro-1,2-epoxypropane), b.p. 117°C, is obtained (see p. 313 for the mechanism):

$$ClCH_2CHOHCH_2Cl + NaOH \longrightarrow ClCH_2CH\!\!\overset{O}{\overset{\diagup \diagdown}{—\!\!—}}\!\!CH_2 + NaCl + H_2O \quad (76\text{–}81\%)$$

Epichlorohydrin is also obtained by distilling an alkaline solution of the α-dichlorohydrin under reduced pressure (yield: 90 per cent). It is used for the manufacture of epoxy resins, plasticisers, etc.

Hydrogen bromide and the phosphorus bromides react with glycerol in the same way as the corresponding chlorine compounds, but the analogous iodine compounds behave differently, the products depending on the amount of reagent used. When glycerol is heated with a *small* amount of hydrogen iodide or phosphorus tri-iodide, *allyl iodide* is the main product:

$$HOCH_2CHOHCH_2OH \xrightarrow{PI_3} [ICH_2CHICH_2I] \longrightarrow I_2 + CH_2{=}CHCH_2I$$

When a *large* amount of phosphorus tri-iodide is used, the main product is *isopropyl iodide*:

$$CH_2{=}CHCH_2I \xrightarrow{HI} [CH_3CHICH_2I] \xrightarrow{-I_2} CH_3CH{=}CH_2 \xrightarrow{HI} CH_3CHICH_3 \quad (80\%)$$

(iii) When glycerol is treated with monocarboxylic acids, esters are obtained which may be mono-, di- or tri-esters, according to the amount of acid used; high temperature and an excess of acid favour the formation of the tri-ester (see the glycerides, below).

Nitroglycerine is manufactured by adding glycerol in a thin stream to a cold mixture of concentrated nitric and sulphuric acids:

$$
\begin{array}{l}
CH_2OH \\
| \\
CHOH + 3HNO_3 \longrightarrow \\
| \\
CH_2OH
\end{array}
\quad
\begin{array}{l}
CH_2ONO_2 \\
| \\
CHONO_2 + 3H_2O \\
| \\
CH_2ONO_2
\end{array}
$$

Nitroglycerine is an ester, not a nitro-compound; it is *glyceryl trinitrate*. The incorrect name appears to have been introduced due to the use of 'mixed acid' which is normally used for nitration (see p. 639).

Nitroglycerine is a poisonous, colourless, oily liquid, insoluble in water. It usually burns quietly when ignited, but when heated, rapidly struck or detonated, it explodes violently. Nobel (1867) found that nitroglycerine could be stabilised by absorbing it in kieselguhr. This was *dynamite*, which is now, however, usually manufactured by using wood pulp as the absorbent, and adding solid ammonium nitrate. *Blasting gelatin* or *gelignite* is made by mixing nitroglycerine with gun-cotton (cellulose nitrate). The smokeless powder, *cordite*, is a mixture of nitroglycerine, gun-cotton and vaseline. Nitroglycerine is also used in the treatment of angina pectoris.

When heated with formic acid or oxalic acid at 260°C, glycerol is converted into allyl alcohol (p. 329). With dibasic acids, glycerol forms condensation polymers known as *alkyd resins*, the commonest of which is *glyptal*, formed by heating glycerol with phthalic anhydride.

When boric acid is added to an aqueous solution of glycerol, a complex is produced which has a higher electrical conductivity than boric acid itself, *i.e.*, is a stronger acid. This complex is believed to be a borospiranic acid in which the two rings attached to the boron atom are perpendicular to each other (hence name spiran):

$$
\left[
\begin{array}{ccc}
C-O & & O-C \\
& B & \\
C-O & & O-C
\end{array}
\right]^{-} H^{+}
$$

This borospiranic acid is formed from α-glycols in which the two hydroxyl groups are in the *cis-*configuration. Ethylene glycol apparently does not form this complex; the reason is unknown.

(iv) Glycerol can theoretically give rise to a large variety of oxidation products. The actual product obtained in practice depends on the nature of the oxidising agent used:

$$
\begin{array}{l}
CH_2OH \\
| \\
CHOH \\
| \\
CH_2OH
\end{array}
\left<
\begin{array}{lll}
CH_2OH & CH_2OH & CO_2H \\
| & | & | \\
CHOH \longrightarrow & CHOH \longrightarrow & CHOH \\
| & | & | \\
CHO & CO_2H & CO_2H \\
\text{glyceraldehyde} & \text{glyceric acid} & \text{tartronic acid} \\
CH_2OH & CO_2H & \\
| & | & \\
CO \longrightarrow & CO \longleftarrow & \\
| & | & \\
CH_2OH & CO_2H & \\
\text{dihydroxy-} & \text{mesoxalic acid} & \\
\text{acetone} & &
\end{array}
\right.
$$

Dilute nitric acid converts glycerol into glyceric and tartronic acids; concentrated nitric acid oxidises it to mainly glyceric acid (80 per cent); bismuth nitrate produces mainly mesoxalic acid. Bromine water, sodium hypobromite and Fenton's reagent (hydrogen peroxide and ferrous sulphate) oxidise glycerol to a mixture of glyceraldehyde (predominantly) and dihydroxyacetone; this mixture is known as *glycerose*. These two compounds are interconvertible in the presence of anhydrous pyridine, this being known as the *Lobry de Bruyn–van Ekenstein rearrangement* (see also p. 514):

$$
\begin{array}{ccc}
\text{CHO} & \left[\begin{array}{c}\text{CHOH}\end{array}\right. & \text{CH}_2\text{OH} \\
| & \|\ & | \\
\text{CHOH} \rightleftharpoons & \text{COH} & \rightleftharpoons \text{CO} \\
| & | & | \\
\text{CH}_2\text{OH} & \left.\begin{array}{c}\text{CH}_2\text{OH}\end{array}\right] & \text{CH}_2\text{OH}
\end{array}
$$

(v) When heated with potassium hydrogen sulphate, glycerol is dehydrated to acraldehyde:

$$\text{HOCH}_2\text{CHOHCH}_2\text{OH} \xrightarrow{\text{KHSO}_4} \text{CH}_2{=}\text{CHCHO} + 2\text{H}_2\text{O}$$

(vi) The mixed ethers of glycerol may be conveniently prepared by the action of sodium alkoxide on a glycerol chlorohydrin, *e.g.*, *triethylin*, the triethyl ether of glycerol, is prepared by heating 1,2,3-trichloropropane with sodium ethoxide:

$$
\begin{array}{ccc}
\text{CH}_2\text{Cl} & & \text{CH}_2\text{OC}_2\text{H}_5 \\
| & & | \\
\text{CHCl} + 3\text{C}_2\text{H}_5\text{ONa} \longrightarrow & \text{CHOC}_2\text{H}_5 + 3\text{NaCl} \\
| & & | \\
\text{CH}_2\text{Cl} & & \text{CH}_2\text{OC}_2\text{H}_5
\end{array}
$$

Helferich and co-workers (1923) found that triphenylmethyl chloride (*trityl chloride*), $(\text{C}_6\text{H}_5)_3\text{CCl}$, usually formed ethers only with primary alcoholic groups. Thus the α-mono- and the α,α'-ditriphenylmethyl (trityl) ethers of glycerol can be prepared, the latter offering a means of preparing β-esters of glycerol, *e.g.*,

$$
\begin{array}{cccc}
\text{CH}_2\text{OH} & \text{CH}_2\text{OC}(\text{C}_6\text{H}_5)_3 & \text{CH}_2\text{OC}(\text{C}_6\text{H}_5)_3 & \text{CH}_2\text{OH} \\
| & | & | & | \\
\text{CHOH} \xrightarrow{(\text{C}_6\text{H}_5)_3\text{CCl}} & \text{CHOH} \xrightarrow{\text{RCOCl}} & \text{CHOCOR} \xrightarrow{\text{HBr}} & \text{CHOCOR} \\
| & | & | & | \\
\text{CH}_2\text{OH} & \text{CH}_2\text{OC}(\text{C}_6\text{H}_5)_3 & \text{CH}_2\text{OC}(\text{C}_6\text{H}_5)_3 & \text{CH}_2\text{OH}
\end{array}
$$

A better means of obtaining the β-ester is to protect the α,α'-hydroxyl groups by cyclic ether formation with benzaldehyde (Bergmann and Carter, 1930). This cyclic acetal (p. 311) is formed when glycerol and benzaldehyde are heated together, or the cool mixture treated with hydrogen chloride:

$$
\begin{array}{cc}
\text{CH}_2\text{OH} & \text{CH}_2\text{O} \\
| & | \quad \diagdown \\
\text{CHOH} + \text{C}_6\text{H}_5\text{CHO} \xrightarrow{\text{HCl}} & \text{CHOH} \quad \text{CHC}_6\text{H}_5 + \text{H}_2\text{O} \\
| & | \quad \diagup \\
\text{CH}_2\text{OH} & \text{CH}_2\text{O}
\end{array}
$$

(vii) Glycerol can be fermented to produce a variety of compounds, *e.g.*, propionic acid, succinic acid, acetic acid, n-butanol, trimethylene glycol, lactic acid, n-butyric acid, etc. Fermentation by means of a particular micro-organism usually produces more than one compound, *e.g.*, propionic acid bacteria produce propionic acid, succinic acid and acetic acid.

Polyhydric alcohols

Tetrahydric alcohols. Erythritol, $\text{HOCH}_2\text{CHOHCHOHCH}_2\text{OH}$, exists in three forms: dextrorotatory erythritol, m.p. 88°; lævorotatory erythritol, m.p. 88°; *meso*-erythritol, m.p. 121·5°C. The *meso*-form occurs in certain lichens and seaweeds. All three forms may be oxidised to tartaric acid,

$$\text{HO}_2\text{CCHOHCHOHCO}_2\text{H}$$

Pentaerythritol, $\text{C}(\text{CH}_2\text{OH})_4$, is prepared by the condensation of formaldehyde with acetaldehyde (see p. 226).

Pentahydric alcohols (*pentitols*). There are four pentitols with the structure $HOCH_2(CHOH)_3CH_2OH$, which may be prepared by reducing the corresponding aldopentoses (see p. 504). Two are optically active, forming a pair of enantiomers—D- and L-arabitol—and the other two are optically inactive, existing as *meso* forms—adonitol and xylitol.

Aldopentose	Pentitol
D- and L-ribose	adonitol (ribitol)
D- and L-xylose	xylitol
D-arabinose ⎱ D-lyxose ⎰	D-arabitol (D-lyxitol)
L-arabinose ⎱ L-lyxose ⎰	L-arabitol (L-lyxitol)

Hexahydric alcohols (*hexitols*). There are ten hexitols with the structure $HOCH_2(CHOH)_4CH_2OH$, which may be prepared by reducing the corresponding aldohexoses (see p. 505). Eight exist as four pairs of enantiomers, and the remaining two as *meso* forms.

Aldohexose	Pentitol
D-glucose and L-gulose	D-sorbitol (D-glucitol)
L-glucose and D-gulose	L-sorbitol (L-glucitol)
D-mannose	D-mannitol
L-mannose	L-mannitol
D-idose	D-iditol
L-idose	L-iditol
D-talose and D-altrose	D-talitol
L-talose and L-altrose	L-talitol
D- and L-galactose	dulcitol
D- and L-allose	allodulcitol (allitol)

D-*Sorbitol*, D-*mannitol* and *dulcitol* occur naturally.

Lipids

The term 'lipids' embraces a variety of naturally occurring compounds which have in common the property of being soluble in organic solvents, but sparingly soluble in water. Also, all lipids yield monocarboxylic acids (saturated and unsaturated) on hydrolysis. The lipids include oils, fats, waxes and phospholipids.

Oils and fats

This group of lipids is often referred to as the 'simple' lipids. They are compounds of glycerol and various organic acids, *i.e.*, they are glyceryl esters or *glycerides*. Oils, which are liquids at ordinary temperatures, contain a larger proportion of unsaturated acids than do the fats, which are solids at ordinary temperatures. The acids present in the glycerides are almost exclusively straight-chain acids, and almost always contain an even number of carbon atoms. The chief saturated acids are lauric, myristic, palmitic and stearic (see p. 244). The chief unsaturated acids are oleic, linoleic and linolenic, and these usually occur in the *cis*-form (see p. 348). Palmitic acid is the most abundant of the saturated acids, and acids containing less than eighteen carbon atoms are usually present only as minor constituents, but sometimes they are present in appreciable amounts in insect waxes and marine fats.

Glycerides are named according to the nature of the acids present, the suffix -*ic* of the common name of the acid being changed to -*in*. Glycerides are said to be 'simple' when all the acids are the same, and 'mixed' when the acids are different:

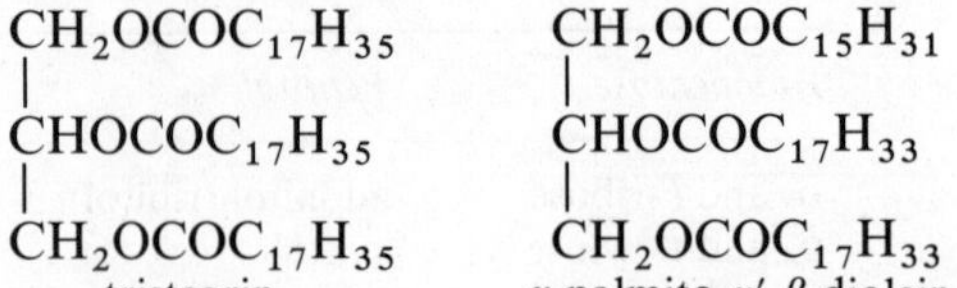

$$CH_2OCOC_{17}H_{35} \qquad CH_2OCOC_{15}H_{31}$$
$$CHOCOC_{17}H_{35} \qquad CHOCOC_{17}H_{33}$$
$$CH_2OCOC_{17}H_{35} \qquad CH_2OCOC_{17}H_{33}$$

tristearin   α-palmito-α′,β-diolein

(simple glyceride)   (mixed glyceride)

It is still not certain whether simple glycerides occur naturally; if they do, they are definitely not as common as the mixed glycerides.

According to Desnuelle *et al.* (1959), the structure of natural glycerides is not random; the position of the acid residue depends on the chain-length of the unsaturated acids and on the degree of their unsaturation. In vegetable oils the saturated acids occur mainly at the 1- and 3-positions and unsaturated acids at the 2-position. In animal fats the distribution is not so rigid.

Synthesis of glycerides

Simple triglycerides are readily prepared from glycerol and an excess of acyl chloride, or from 1,2,3-tribromopropane and the silver or potassium salts of the acids.

Mixed triglycerides are far more difficult to prepare. A number of methods have been developed, *e.g.*,

(i) The sodium salt of an acid is heated with glycerol monochlorohydrin, and the monoester so formed is acylated:

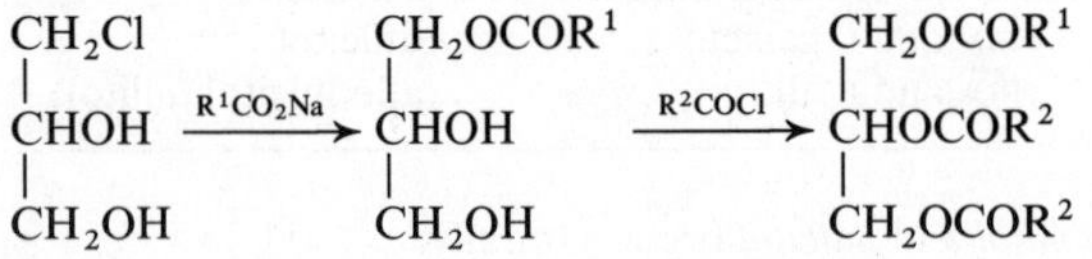

$$CH_2Cl \qquad\qquad CH_2OCOR^1 \qquad\qquad CH_2OCOR^1$$
$$CHOH \xrightarrow{R^1CO_2Na} CHOH \xrightarrow{R^2COCl} CHOCOR^2$$
$$CH_2OH \qquad\qquad CH_2OH \qquad\qquad CH_2OCOR^2$$

The preparation of 1-monoglycerides has been improved by Hartman (1960). The 2- and 3-positions in glycerol are protected by the formation of the isopropylidene derivative (*cf.* (ii) below; see also p. 506), and the protecting group is removed by boric acid in hot 2-methoxyethanol.

(ii) 1,3-Benzylidene-glycerol is treated with an acid chloride, the benzaldehyde residue is removed by hydrolysis, and the monoester so formed is then further acylated:

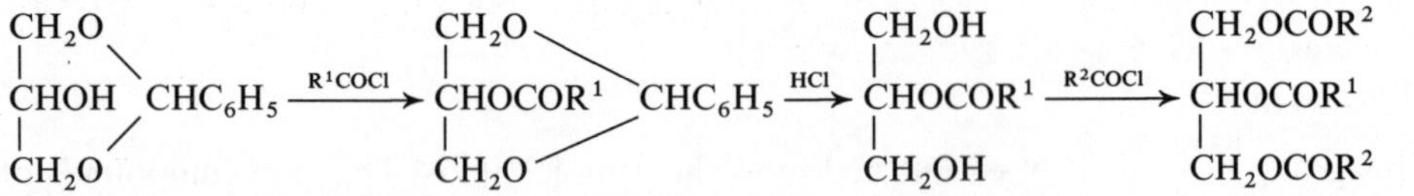

$$CH_2O \qquad\qquad CH_2O \qquad\qquad CH_2OH \qquad\qquad CH_2OCOR^2$$
$$CHOH \; CHC_6H_5 \xrightarrow{R^1COCl} CHOCOR^1 \; CHC_6H_5 \xrightarrow{HCl} CHOCOR^1 \xrightarrow{R^2COCl} CHOCOR^1$$
$$CH_2O \qquad\qquad CH_2O \qquad\qquad CH_2OH \qquad\qquad CH_2OCOR^2$$

In both methods (i) and (ii) there is, however, still difficulty in knowing with certainty the position of the acyl group introduced first, since the acyl group tends to migrate in the monoester, producing the following equilibrium:

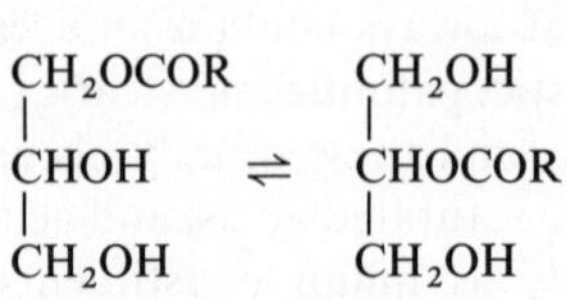

$$CH_2OCOR \qquad CH_2OH$$
$$CHOH \quad \rightleftharpoons \quad CHOCOR$$
$$CH_2OH \qquad CH_2OH$$

This isomerisation has been shown to be intramolecular.

Analysis of oils and fats

Oils and fats are characterised by means of physical as well as chemical tests. The usual physical constants that are determined are melting point, solidifying point, density and refractive index.

The chemical tests give an indication of the type of fatty acids present in the oil or fat.

The **acid value**, which is the number of milligrams of potassium hydroxide required to neutralise 1 gram of the oil or fat, indicates the amount of free acid present.

The **saponification value** is the number of milligrams of potassium hydroxide required to neutralise the acids resulting from the complete hydrolysis of 1 gram of the oil or fat.

The **iodine value**, which is the number of grams of iodine that combine with 100 grams of oil or fat, gives the degree of unsaturation of the acids in the substance. Several methods are used for determining the iodine value. In *Hubl's method*, a carbon tetrachloride solution of the substance is treated with a solution of iodine and mercuric chloride in ethanol; in *Wijs' method*, iodine monochloride in glacial acetic acid is used. Another method is to use a solution of glacial acetic acid containing pyridine, bromine and concentrated sulphuric acid (*Dam's solution*).

The **Reichert–Meissl value**, which is the number of ml of $0 \cdot 1$ N-potassium hydroxide required to neutralise the distillate of 5 grams of hydrolysed fat, indicates the amount of steam-volatile fatty acids (*i.e.*, acids up to lauric) present in the substance.

The **acetyl value**, which is the number of milligrams of potassium hydroxide required to neutralise the acetic acid obtained when 1 gram of an acetylated oil or fat is hydrolysed, indicates the number of free hydroxyl groups present in the substance.

Preparation of glycerol from oils and fats

Glycerol and some acids are prepared for industrial use by the hydrolysis of oils and fats with water under pressure at 220°C. The glycerol is recovered from the aqueous solution; the free fatty acids are used in the manufacture of candles.

$$
\begin{array}{l}
\text{CH}_2\text{OCOR} \\
| \\
\text{CHOCOR} + 3\text{H}_2\text{O} \longrightarrow \\
| \\
\text{CH}_2\text{OCOR}
\end{array}
\qquad
\begin{array}{l}
\text{CH}_2\text{OH} \\
| \\
\text{CHOH} + 3\text{RCO}_2\text{H} \\
| \\
\text{CH}_2\text{OH}
\end{array}
$$

If an oil is saponified, *i.e.*, hydrolysed with alkali, soaps are obtained. Any metallic salt of a fatty acid is a soap, but the term soap is usually applied to the water-soluble salts since only these have detergent properties. The saturated fats give hard soaps whereas the unsaturated fats, *i.e.*, the oils, give soft soaps. Ordinary soap is a mixture of the sodium salts of the even fatty acids from octanoic to stearic. The sodium salts of a given oil or fat are harder and less soluble than the corresponding potassium salts. Thus soft soaps are usually the potassium salts, particularly when they are derived from oils.

Hardening of oils

Glycerides of the unsaturated acids are liquid at room temperature and so are unsuitable for edible fats. By converting the unsaturated acids into saturated acids, oils are changed into fats. This introduction of hydrogen is known as the *hardening of oils*. The oil is heated to 150–200°C and hydrogen is passed in, under pressure, in the presence of a finely divided nickel catalyst. In the hydrogenation process only a proportion of the unsaturated acids are converted into the saturated acids, otherwise a fat as hard as tallow would be obtained; the hydrogenation is carried out until a fat of the desired consistency is obtained.

Phospholipids

These differ from the simple lipids in that they also contain phosphorus and nitrogen; they are glycerides in which one organic acid residue, the α- or β-, is replaced by a group containing phosphoric acid and a base. One subgroup of the phospholipids is the **phosphatides**. When the base is **cholamine**, $CH_2OHCH_2NH_2$, the phosphatide is known as a **kephalin**; when the base is **choline**, $CH_2OHCH_2\overset{+}{N}(CH_3)_3\}OH$, the phosphatide is known as a **lecithin**. These compounds have betaine structures (see p. 391), and it appears unlikely that β-kephalins or β-lecithins occur in nature.

$$
\begin{array}{ll}
CH_2OCOR^1 \\
\quad| \\
R^2COOCH \quad O^- \\
\quad\quad\quad\quad| \quad\quad | \\
\quad\quad CH_2OPOCH_2CH_2NH_3{}^+ \\
\quad\quad\quad\quad\| \\
\quad\quad\quad\quad O
\end{array}
\qquad\qquad
\begin{array}{ll}
CH_2OCOR^1 \\
\quad| \\
R^2COOCH \quad O^- \\
\quad\quad\quad\quad| \quad\quad | \\
\quad\quad CH_2OPOCH_2CH_2\overset{+}{N}(CH_3)_3 \\
\quad\quad\quad\quad\| \\
\quad\quad\quad\quad O
\end{array}
$$

α-kephalin \qquad\qquad\qquad\qquad\qquad α-lecithin

Kephalins may also contain L-serine instead of aminoethanol. These three types of phospholipids have been called *phosphoglycerides*, in order to distinguish them from another group of phospholipids, the *acetal phospholipids* (*plasmalogens*). The structures of these compounds appear to be uncertain; one proposal is (for an ethanolamine plasmalogen):

$$
\begin{array}{ll}
CH_2OCH{=}CHR^1 \\
\quad| \\
R^2COOCH \quad O^- \\
\quad\quad\quad\quad| \quad\quad | \\
\quad\quad CH_2OPOCH_2CH_2NH_3{}^+ \\
\quad\quad\quad\quad\| \\
\quad\quad\quad\quad O
\end{array}
$$

All of these phosphoglycerides occur in the brain and the spinal chord.

Only a few monocarboxylic acids have been isolated from phosphatides: stearic, oleic, linoleic and arachidonic from kephalins, and palmitic, stearic, oleic, linoleic, linolenic and arachidonic from lecithins. Enzymic studies of phospholipids have shown that saturated acids predominate at the 1-position and unsaturated acids at the 2-position (Hanahan *et al.*, 1960).

Another group of phospholipids is the **sphingolipids**. These are *not* glycerides, and yield, on hydrolysis, fatty acids, phosphoric acid, choline and sphingosine (which has the *trans*-configuration about the double bond). Sphingolipids are also known as **sphingomyelins**, and their structure is believed to be:

$$
\begin{array}{l}
CH_3(CH_2)_{12}C{-}H \quad\quad\quad\quad O^- \\
\quad\quad\quad\quad\| \quad\quad\quad\quad\quad\quad\quad | \\
\quad\quad H{-}C{-}CH{-}CHCH_2O\overset{+}{P}OCH_2CH_2\overset{+}{N}(CH_3)_3 \\
\quad\quad\quad\quad| \quad\quad | \quad\quad\quad\quad\| \\
\quad\quad\quad\quad OH \quad NH \quad\quad\quad O \\
\quad\quad\quad\quad\quad\quad\quad | \\
\quad\quad\quad\quad\quad\quad\quad RCO
\end{array}
$$

Glycolipids (cerebrosides) also contain sphingosine, but instead of the phosphocholine group, they have

$$
\cdots CH\underset{\underset{NH}{|}}{CH_2}OCH(CHOH)_3CHCH_2OH,
$$

i.e., they are glycosides, and apparently only two sugars have been isolated so far, glucose and galactose.

Drying oils

These are oils which, on exposure to air, change into hard solids, *e.g.*, linseed oil. All drying oils contain a large proportion of the unsaturated acids linoleic and linolenic, and it is this 'drying' property which makes these oils valuable in the paint industry. The mechanism of drying appears to be a complicated process involving oxidation, polymerisation and colloidal gel formation, and it has been found to be catalysed by various metallic oxides, particularly lead monoxide.

Waxes

These are esters of the higher homologues of both the fatty acids and monohydric alcohols, *e.g.*,

beeswax	myricyl palmitate	$C_{15}H_{31}CO_2C_{30}H_{61}$
spermaceti	cetyl palmitate	$C_{15}H_{31}CO_2C_{16}H_{33}$
carnauba wax	myricyl cerotate	$C_{25}H_{51}CO_2C_{30}H_{61}$

Some waxes are also esters of cholesterol, *e.g.*, cholesteryl esters occur in wool-wax. Also present in waxes are free acids, long-chain alcohols, ketones and alkanes.

PROBLEMS

1. Write out the structures of all the possible isomeric diols derived from the butanes, and indicate which can be oxidised with periodic acid.

2. Arrange the following in order of increasing b.p., and give your reasons. $(CH_2OH)_2$, $(CH_2OMe)_2$, $HOCH_2CH_2OMe$.

3. Complete the following equations:

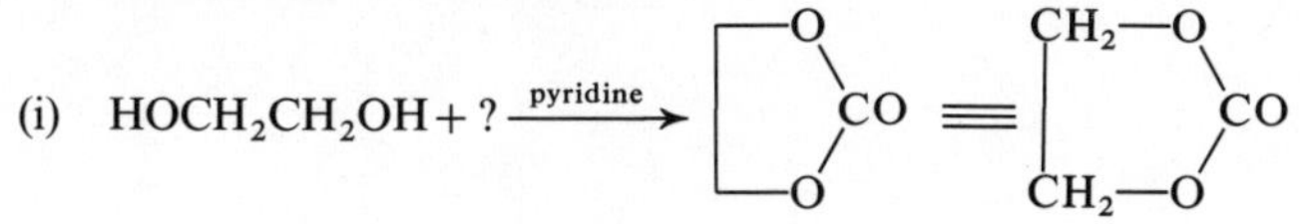

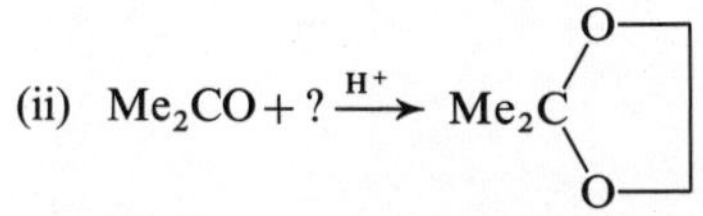

(i) $HOCH_2CH_2OH + ? \xrightarrow{\text{pyridine}}$

(ii) $Me_2CO + ? \xrightarrow{H^+}$

(iii) $MeCHO + HOCH_2CH_2SH \xrightarrow{H^+} ?$

(iv) $R_2C(OH)CH_2R \xleftarrow{?}$ $\xrightarrow{?} R_2CHCHOHR$

(v) $\xrightarrow[\text{(ii) } H^+]{\text{(i) } HCO_3H} ?$

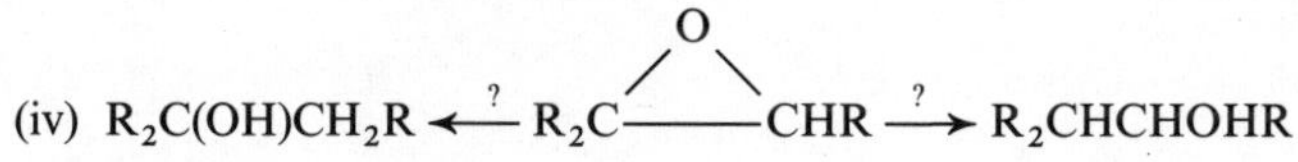

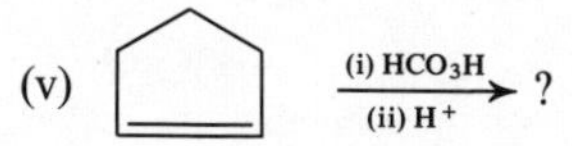

(vi) $CH_2{=}CH(CH_2)_2CH{=}CH_2 \xrightarrow[\text{(ii) } H_2O_2;\ OH^-]{\text{(i) } B_2H_6} ?$

(vii) $(CH_2)_8(CO_2Et)_2 \xrightarrow{?} (CH_2)_8(CH_2OH)_2$

4. Write out the order of rates of oxidation with periodic acid of the following diols, and give your reasons: $HOCH_2CH_2OH$, $MeCHOHCHOHMe$, Me_2COHCH_2OH, $Me_2C(OH)C(OH)Me_2$, $Me_2COHCHOHMe$.

5. Convert: into (*cis*) and (*trans*)

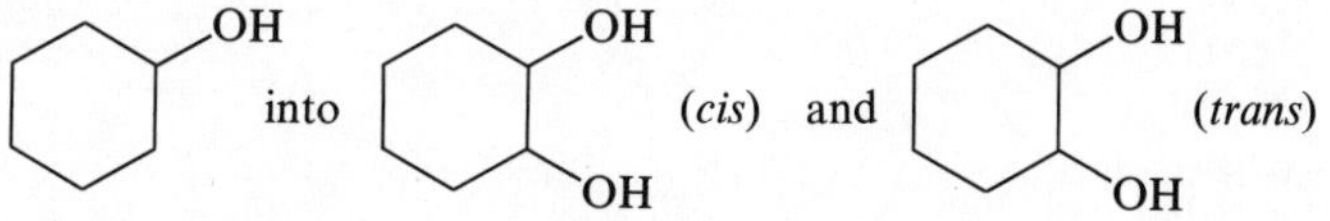

6. Suggest a mechanism for the conversion of ethylene glycol into dioxan by conc. H_2SO_4.

7. How many N.M.R. signals can be expected for:

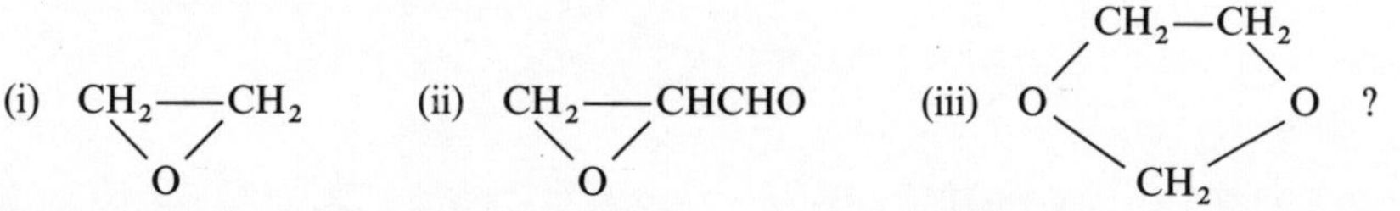

(i)   (ii)   (iii)   ?

8. Tetramethylethylene, on oxidation with acid dichromate, gives the compound $C_6H_{12}O$. This compound showed an i.r. band at about $1\,700\ cm^{-1}$, and undergoes the haloform reaction. Suggest a structure for this compound, and give your reasons.

REFERENCES

TURNEY, *Oxidation Mechanisms*, Butterworths (1965).

ELDERFIELD (ed.), *Heterocyclic Compounds*, Wiley, Vol. I (1949), Ch. 1, 'Ethylene and Trimethylene Oxides'.

ACHESON, *An Introduction to the Chemistry of Heterocyclic Compounds*, Interscience (1967, 2nd edn).

PARKER and ISAACS, Mechanisms of Epoxide Reactions, *Chem. Rev.*, 1959, **59,** 737.

HILDITCH, *Chemical Constitution of Natural Fats,* Wiley (1964, 4th edn).

12 Unsaturated alcohols, ethers, carbonyl compounds and acids

Unsaturated alcohols

The simplest unsaturated alcohol is **vinyl alcohol**, $CH_2=CHOH$. This is, however, unknown; all attempts to prepare it result in the formation of acetaldehyde (together with a small amount of ethylene oxide), *e.g.*, vinyl bromide on treatment with silver oxide in boiling water gives acetaldehyde (see the Erlenmeyer rule, p. 301):

$$CH_2=CHBr + \text{`AgOH'} \longrightarrow AgBr + [CH_2=CHOH] \longrightarrow CH_3CHO$$

Although vinyl alcohol itself is unknown, many of its derivatives have been prepared, and are quite stable. These derivatives may be prepared by the interaction between acetylene and the other reactant in the presence of a suitable catalyst (see also p. 136); *e.g.*, *vinyl chloride*:

$$C_2H_2 + HCl \xrightarrow{Hg^{2+}} CH_2=CHCl$$

Vinyl chloride is manufactured by the thermal decomposition of ethylene dichloride at 600–650°C:

$$ClCH_2CH_2Cl \longrightarrow CH_2=CHCl + HCl$$

Vinyl chloride and *bromide* are conveniently prepared in the laboratory by heating ethylene dichloride and dibromide, respectively, with ethanolic potassium hydroxide:

$$BrCH_2CH_2Br + KOH \xrightarrow{ethanol} CH_2=CHBr + KBr + H_2O$$

Many of the vinyl compounds are used to make plastics. *Vinyon*, which is a thermoplastic, is a copolymer of vinyl chloride and vinyl acetate.

The halogen atom in vinyl halides is not reactive; vinyl halides do not undergo the usual double decomposition reactions of the alkyl halides. Vinyl chlorides and bromides readily form Grignard reagents in *tetrahydrofuran* as solvent (Normant, 1954, 1957), and they react with lithium to form lithium compounds which undergo the usual reactions (Braude, 1950–1952; see p. 436), *e.g.*,

$$R_2^1C=CHBr \xrightarrow{Li} R_2^1C=CHLi \xrightarrow{R^2CHO} R_2^1C=CHCHOHR^2$$

The reason for the unreactivity of the halogen atom is not clear. Some believe it to be due to resonance through which the halogen atom acquires some double-bond character, and is thereby more strongly bound to the carbon atom due to the shortening of the C—Cl bond (p. 37):

$$CH_2{=}CH{-}\overset{..}{\underset{..}{C}}l\colon \longleftrightarrow {}^{-}\overset{..}{C}H_2{-}CH{=}\overset{+}{\underset{..}{C}}l\colon$$

An alternative explanation is that due to Walsh (1947). As we have seen (p. 139), an sp^2 hybridised carbon atom attracts electrons more than does an sp^3 hybridised carbon atom. Thus the Cl atom in the C—Cl group is more tightly bound in C=C—Cl than in C—C—Cl.

The non-reactivity of the chlorine atom in vinyl chloride may be explained from the M.O. point of view as follows. If the chlorine atom has sp^2 hybridisation, the C—Cl bond will be a σ-bond and the two lone pairs of electrons would occupy the other two sp^2 orbitals. This would leave a p orbital containing a lone pair, and this orbital could now conjugate with the π-bond of the ethylenic link (Fig. 12.1). Thus two M.O.s will be required to accommodate these four π-electrons (*cf.* the carboxyl group, p. 243). Furthermore, since chlorine is more electronegative than carbon, the electrons will tend to be found in

Fig. 12.1

the vicinity of the chlorine atom. Nevertheless, the chlorine atom has now lost 'full control' of the lone pair, and so is less negative than it would have been had there been no conjugation. Since two carbon atoms have acquired a share in the lone pair, each carbon atom acquires a small negative charge. Hence, owing to delocalisation of bonds (through conjugation), the vinyl chloride molecule has an increased stability. Before the chlorine atom can be displaced by some other group, the lone pair must be localised again on the chlorine atom. This requires energy, and so the chlorine is more 'firmly bound' than had no conjugation occurred.

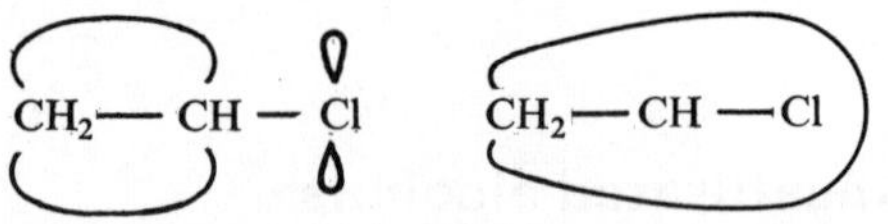

Vinylidene chlorides, $R_2C{=}CCl_2$. Dichloromethylene, formed by the action of potassium t-butoxide or an alkyl-lithium on chloroform (p. 170), may be trapped by triphenylphosphine to form the ylid, dichloromethylenetriphenylphosphorane (Speziale *et al.*, 1960; see also p. 410). This readily undergoes the Wittig reaction to form a vinylidene chloride.

$$CHCl_3 \longrightarrow [CCl_2] \xrightarrow{\text{Ph}_3\text{P}} Ph_3P{=}CCl_2 \xrightarrow{\text{R}_2\text{C}={\text{O}}} R_2C{=}CCl_2 + Ph_3P{=}O$$

Vinyl cyanide (*acrylonitrile*) is very useful for introducing the cyanoethyl group, $-CH_2CH_2CN$, by reaction with compounds containing an active methylene group. The best catalyst for **cyanoethylation** is benzyltrimethylammonium hydroxide, $C_6H_5CH_2\overset{+}{N}(CH_3)_3\}OH^-$, but in many cases an aqueous or ethanolic solution of alkalis is effective (Bruson, 1942, 1943). Cyanoethylation is one example of the Michael condensation (p. 340), *e.g.*,

$$MeCOCH_2Me + OH^- \rightleftharpoons Me\overset{-}{CO}CHMe + H_2O$$

$$Me\overset{-}{CO}CHMe + CH_2{=}CH{-}C{\equiv}N \underset{}{\overset{H_2O}{\rightleftharpoons}} \underset{\underset{CH_2CH{=}C{=}NH}{|}}{MeCOCHMe} \longrightarrow$$

$$\underset{\underset{CH_2CH_2CN}{|}}{MeCOCHMe} \xrightarrow[\text{(ii) CH}_2\,=\,\text{CHCN}]{\text{(i) OH}^-} \underset{\underset{CH_2CH_2CN}{|}}{Me\overset{\overset{CH_2CH_2CN}{|}}{CO}CMe} \qquad (90\%)$$

This type of reaction offers a means of preparing a variety of compounds. Vinyl cyanide normally causes di- or tri-cyanoethylation of compounds which have an active methylene or methyl group respectively, but Campbell *et al.* (1956) have described conditions for mono-cyanoethylation. In addition to active methylene and methyl groups, compounds such as primary and secondary amines, alcohols, phenols, etc., can also undergo cyanoethylation.

Vinyl cyanide is manufactured by the direct combination of hydrogen cyanide with acetylene (p. 136), and from propene as follows (acraldehyde is an intermediate):

$$2CH_2{=}CHCH_3 + 2NH_3 + 3O_2 \xrightarrow[450°C]{\text{cat.}} 2CH_2{=}CHCN + 6H_2O$$

Allyl alcohol (*prop*-2-*en*-1-*ol*), b.p. 97°C, may be prepared as follows:

(i) By boiling allyl chloride with dilute sodium hydroxide solution under pressure (this is a commercial method).

$$CH_2{=}CHCH_2Cl + NaOH \xrightarrow{150°C} CH_2{=}CHCH_2OH + NaCl$$

Another commercial method is the isomerisation of propene oxide in the presence of lithium phosphate as catalyst.

(ii) By heating glycerol with formic acid or oxalic acid at 260°C:

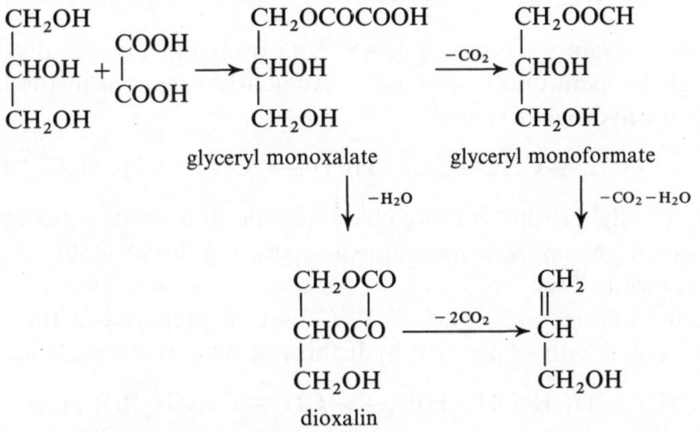

$$
\begin{array}{ccc}
CH_2OH & & CH_2OCOCOOH & & CH_2OOCH \\
| & COOH & | & & | \\
CHOH \ + \ | & \longrightarrow & CHOH & \xrightarrow{-CO_2} & CHOH \\
| & COOH & | & & | \\
CH_2OH & & CH_2OH & & CH_2OH \\
& & \text{glyceryl monoxalate} & & \text{glyceryl monoformate}
\end{array}
$$

glyceryl monoxalate: $\downarrow -H_2O$ glyceryl monoformate: $\downarrow -CO_2 - H_2O$

$$
\begin{array}{ccc}
CH_2OCO & & CH_2 \\
| \quad | & & \| \\
CHOCO & \xrightarrow{-2CO_2} & CH \\
| & & | \\
CH_2OH & & CH_2OH
\end{array}
$$

dioxalin

In practice it is better to use formic acid since this gives a higher yield (45–47 per cent).

A general synthesis of allylic alcohols is the selenium dioxide oxidation of alkenes of the type $R^1CH_2CH{=}CHR^2$ (see p. 103).

Allyl alcohol has the properties of an unsaturated compound and a primary alcohol. It is oxidised to glycerol by dilute alkaline permanganate:

$$CH_2{=}CHCH_2OH + H_2O + [O] \longrightarrow HOCH_2CHOHCH_2OH$$

Allyl alcohol (and other α,β-unsaturated primary and secondary alcohols) may be conveniently oxidised to the corresponding oxo compound with manganese dioxide or nickel peroxide, *e.g.*,

$$CH_2{=}CHCH_2OH \xrightarrow{MnO_2} CH_2{=}CHCHO$$

t-Butyl chromate (p. 203), since it does not attack a double bond, is also a very useful reagent for oxidising unsaturated primary alcohols to the corresponding unsaturated aldehydes.

In general, oxidising agents attack the double bond in allyl alcohol as well as the alcoholic group. Oxidation, however, may be carried out by 'protecting' the double bond by bromination and then oxidising and debrominating with zinc dust in methanolic solution, e.g.,

$$CH_2{=}CHCH_2OH \xrightarrow{Br_2} BrCH_2CHBrCH_2OH \xrightarrow{HNO_3}$$

$$BrCH_2CHBrCO_2H \xrightarrow[MeOH]{Zn} CH_2{=}CHCO_2H$$

Allyl alcohol adds on chlorine or bromine to form the corresponding 2,3-dihalogeno-propan-1-ol. Allyl alcohol also adds on halogen acids and hypohalous acids, but the addition takes place *contrary* to Markownikoff's rule, e.g., glycerol β-monochlorohydrin is formed with hypochlorous acid. This may be due to the presence of the oxygen atom which exerts a strong −I effect.

$$CH_2{=}CH{-}CH_2 \rightharpoonup OH + HO{-}Cl \longrightarrow HOCH_2CHClCH_2OH$$

If we consider this addition from the point of view of the stabilities of the intermediate carbonium ions, $^+CH_2CHClCH_2OH$ (primary) and $ClCH_2\overset{+}{C}HCH_2OH$ (secondary), it might have been anticipated that the product would have arisen via the more stable secondary ion to give $ClCH_2CHOHCH_2OH$ (the Markownikoff product). There are, however, various complications, e.g., the carbonium ion is probably not the classical one shown, but is a chlorinium (bridged) ion (it might also be the epoxide type?). If a chlorinium ion, the hydroxide ion might be expected to attack at the CH_2 carbon atom, which is less hindered than the other carbon atom (C of the $CHCH_2OH$).

Allyl chloride (3-*chloropropene*), b.p. 45°C, is prepared industrially by the chlorination of propene at high temperature. It may be conveniently prepared in the laboratory by warming allyl alcohol with hydrochloric acid:

$$CH_2{=}CHCH_2OH + HCl \longrightarrow CH_2{=}CHCH_2Cl + H_2O$$

This reactivity of allyl alcohol is more characteristic of a tertiary alcohol than a primary. This may be explained by an S_N1 mechanism operating because of the stability of the allyl carbonium ion (see below).

Allyl bromide (3-*bromopropene*), b.p. 70°C, is best prepared in the laboratory by distilling aqueous allyl alcohol with 48 per cent hydrobromic acid in the presence of sulphuric acid:

$$CH_2{=}CHCH_2OH + HBr \longrightarrow CH_2{=}CHCH_2Br + H_2O \quad (92{-}96\%)$$

Allyl iodide (3-*iodopropene*), b.p. 103·1°C, may be prepared by heating glycerol with a small amount of hydriodic acid (p. 318), or by the Finkelstein reaction (p. 85):

$$CH_2{=}CHCH_2Cl + NaI \xrightarrow{acetone} CH_2{=}CHCH_2I + NaCl$$

The halogen atom in the allyl halides is very reactive, and it has been found experimentally that the position of the double bond with respect to the halogen atom determines the reactivity of the halogen atom. In compounds of the type C=C—X, the halogen atom X is unreactive (see vinyl halides, above). In C=C—C—X, X is more reactive than in the alkyl halides. When the halogen atom in an unsaturated compound is further removed from the double bond than in the allyl position, it behaves similarly to the halogen atom in alkyl halides; e.g., 4-chlorobut-1-ene, $CH_2{=}CHCH_2CH_2Cl$, undergoes the usual reactions of the alkyl halides with reagents that do not affect the double bond.

The high reactivity of the allyl halides may be explained by the fact that when the halogen atom ionises, the allyl carbonium ion produced is a resonance hybrid, and consequently it is stabilised:

$$CH_2\!=\!CH\!-\!CH_2{}^+ \longleftrightarrow \overset{+}{C}H_2\!-\!CH\!=\!CH_2$$

The allyl halides add on the halogen acids to form a mixture of the 1,2- and 1,3-dihalides. The addition of hydrogen bromide to allyl bromide has been studied in great detail (Kharasch, 1933):

$$CH_2\!=\!CHCH_2Br + HBr \longrightarrow \underset{\text{1,3-dibromide}}{BrCH_2CH_2CH_2Br} + \underset{\text{1,2-dibromide}}{CH_3CHBrCH_2Br}$$

In the presence of peroxides, the 1,3-dibromide is obtained in 90 per cent yield. In the absence of peroxides, the 1,2-dibromide is obtained in 90 per cent yield. When allyl bromide, which has not been carefully purified, is treated with hydrogen bromide, without special precautions being taken to exclude all traces of air, a mixture of the 1,2- and 1,3-dibromides is obtained.

The explanation for these results is as follows. The 1,3-dibromide is formed by a free-radical mechanism:

$$(RCO_2)_2 \longrightarrow 2RCOO\cdot \longrightarrow 2R\cdot + 2CO_2$$
$$R\cdot + HBr \longrightarrow RH + Br\cdot$$

$$Br\cdot + CH_2\!=\!CHCH_2Br \longrightarrow BrCH_2\overset{\cdot}{C}HCH_2Br \overset{HBr}{\longrightarrow} BrCH_2CH_2CH_2Br + Br\cdot\,;\text{ etc.}$$

The alternative free radical, $\overset{\cdot}{C}H_2CHBrCH_2Br$, is a primary one, and since this is less stable than a secondary, the reaction proceeds via the latter.

The formation of the 1,2-dibromide proceeds by means of a polar mechanism, but now a proton is added in the first step to give the secondary carbonium ion, $CH_3CH^+CH_2Br$, which is more stable than the alternative primary carbonium ion, $^+CH_2CH_2CH_2Br$. Arguments based on the $-I$ effect of the bromine atom would have anticipated the formation of the 1,3-dibromide (*cf.* with allyl alcohol, above).

A completely different approach to this problem of 1,2-addition is the use of the **Baker–Nathan effect** (1935). As we have seen, the general inductive effect of alkyl groups is $Me_3C > Me_2CH > MeCH_2 > Me$. This inductive order has been used satisfactorily to explain various physical data, etc. In some reactions, however, the inductive order is reversed, *e.g.*, Baker and Nathan (1935) examined kinetically the reaction between *p*-substituted benzyl bromides and pyridine. The reaction was carried out in acetone, and was shown to be entirely S_N2. Thus the reaction may be formulated:

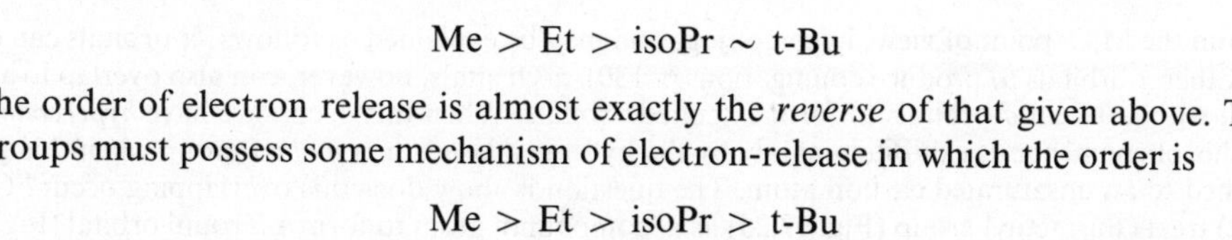

The greater the $+I$ effect of the R group, the faster should be the reaction, but it was found that the rate order for the R groups was

$$Me > Et > isoPr \sim t\text{-Bu}$$

Thus the order of electron release is almost exactly the *reverse* of that given above. Therefore alkyl groups must possess some mechanism of electron-release in which the order is

$$Me > Et > isoPr > t\text{-Bu}$$

Further work showed that in all cases where the general inductive order of alkyl groups was reversed, the alkyl group was attached to an aromatic nucleus. This led Baker and Nathan (1935, 1939) to suggest that when the H—C bond is attached to an unsaturated carbon atom, the σ-electrons of the H—C bond become less localised by entering into *partial* conjugation with the attached unsaturated system, *i.e.*, σ,π-conjugation:

$$H—\overset{\frown}{C}—\overset{\frown}{C}{=}C$$

Thus there is conjugation between electrons of *single* and those of multiple bonds. This type of conjugation is known as **hyperconjugation**, and is a *permanent* effect (this name was given by Mulliken, 1941).

There are various ways of looking at the **hyperconjugative effect**. A widely used one is to regard the H—C bond as possessing partial ionic character due to resonance, *e.g.*, propene may be written as a resonance hybrid as follows:

$$\overset{\displaystyle H}{\underset{\displaystyle H}{H—C—}}CH{=}CH_2 \longleftrightarrow \overset{\displaystyle H^+}{\underset{\displaystyle H}{H—\overset{..}{C}—}}CH{=}CH_2 \longleftrightarrow \overset{\displaystyle H^+}{\underset{\displaystyle H}{H—C{=}}}CH{—}\bar{C}H_2 \longleftrightarrow \cdots$$

From this point of view, hyperconjugation may be regarded as '*no bond resonance*'. The hydrogen atoms are not *free*; the effect is to increase the ionic character of the C—H bond, the electrons of which become partially delocalised through conjugation.

Various physical data have been explained by hyperconjugation, *e.g.*, bond lengths:

$$CH_3—CH_3 \qquad CH_3—C{\equiv}CH$$
$$\text{1·54Å} \qquad\qquad \text{1·46Å}$$

Because of hyperconjugation, the CH$_3$—C bond has some partial double bond character and consequently is shortened. Also, since hyperconjugation stabilises a molecule (through resonance), propene, for example, should be more stable than 'expected'. This has been found to be so; the observed heat of hydrogenation of propene is less than the calculated value.

Using this scheme of hyperconjugation, we are now in a position to explain the formation of the 1,2-dibromide from allyl bromide and hydrogen bromide by the polar mechanism.

$$\overset{\displaystyle H}{CH_2{=}CH—CHBr} \longleftrightarrow \overset{\displaystyle H^+}{CH_2{=}CH—\overset{..}{C}HBr} \longleftrightarrow \overset{\displaystyle H^+}{\bar{C}H_2—CH{=}CHBr} \xrightarrow{\text{HBr}}$$
$$CH_3CHBrCH_2Br$$

The inductive effect of the bromine atom would tend to give the 1,3-dibromide (see above). If, however, the hyperconjugative effect is operating, and is *stronger* than the −I effect of the bromine atom, then the 1,2-dibromide will be formed.

An interesting point here is that with allyl alcohol, the −I effect of the hydroxyl group is stronger than the hyperconjugative effect (p. 598).

From the M.O. point of view, hyperconjugation may be explained as follows: π-orbitals can overlap with other π-orbitals to produce conjugation (p. 130). π-Orbitals, however, can also overlap to a certain extent with adjacent σ-orbitals to form extended orbitals. When this occurs we have *hyperconjugation*, and this phenomenon is exhibited mainly by the hydrogen bonds in a *methyl* group when this group is attached to an unsaturated carbon atom. The question is: how does this overlapping occur? Coulson (1942) treats the methyl group (Fig. 12.2*a*) as a 'compound' atom to form a 'group' orbital (Fig. *b*). An alternative 'group' orbital may also be formed as in (*c*). In this arrangement, the methyl group behaves as a group with a π-orbital, and this can conjugate with an adjacent π-bond in, *e.g.*, propene (Fig. *d*). Hence, owing to partial delocalisation of bonds (σ and π) in this way, the propene molecule is more stable than 'expected'.

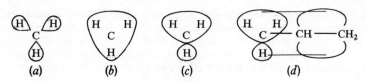

Fig. 12.2

The unusual reactivity of the halogen atom in allyl halides may be explained as follows. If the chlorine were to ionise, the carbon atom to which it was attached now has a positive charge and has only *six* electrons. Hence the π-bond covering the other two carbon atoms can extend to embrace this third carbon atom. The net result is an M.O. covering *three* carbon atoms, and so the delocalisation energy will be increased, *i.e.*, the new arrangement is stabilised, and behaves as if the chlorine atom is ionic (Fig. 12.3).

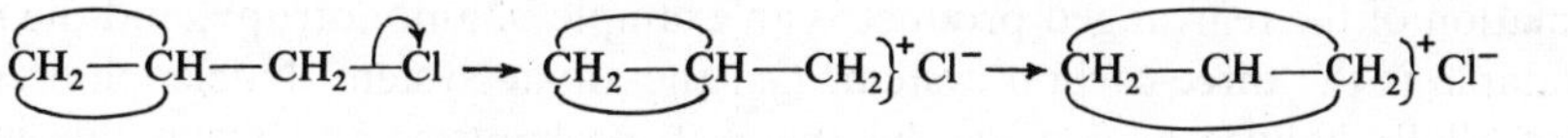

$$CH_2-CH-CH_2-Cl \rightarrow CH_2-CH-CH_2\}^+Cl^- \rightarrow CH_2-CH-CH_2\}^+Cl^-$$

Fig. 12.3

There is, however, a certain amount of doubt that dipole moment data are evidence for hyperconjugation. Since the electronegativity of a carbon atom depends on its state of hybridisation, there will be a bond moment when two differently hybridised carbon atoms are linked together. Petro (1958) has calculated bond moments for several types of carbon carbon single bonds:

Bond	Bond moment
$C(sp^3)^+ - C(sp^2)^-$	0·68D
$C(sp^2)^+ - C(sp)^-$	1·15D
$C(sp^3)^+ - C(sp)^-$	1·48D

Ferreira (1960) has calculated the dipole moments of various molecules, *e.g.*, toluene, propene, etc., on these principles, and has shown that the calculated values agree well with the experimental values. Thus it appears that dipole moment evidence for hyperconjugation is not valid.

It has also been found experimentally that all C—C single bonds are *not* the same length. This shortening has been attributed to hyperconjugation, but it can also be explained in terms of different electronegatives of the two carbon atoms. Bloor and Gartside (1959) have calculated bond energies of C—C and C—H bonds for differently hybridised carbon atoms (see Table 4.1, p. 131), and since these vary with the nature of the hybrid state of the carbon atom, this makes the evaluation of resonance energies from 'average' bond energies (Table 2.2, p. 36) very uncertain.

It is quite possible, however, since bond moments, bond lengths and bond energies are ground-state properties, that hyperconjugation exists in excited states and in transition states of molecules, and therefore is important for the interpretation of reactivity of molecules.

Mulliken (1959) has concluded that the explanation offered above is partly right, but does not believe the π-delocalisation shortenings are negligible. He has made some calculations and has attributed 40 per cent of observed shortenings to π-electron resonance and 60 per cent to hybridisation differences. On the other hand, Ehrenson (1964), from his calculations, supports the theory that hyperconjugation may be the major driving force in some reactions where carbonium ion products are formed.

Crotyl alcohol (*crotonyl alcohol*, *but-2-en-1-ol*), b.p. 118°C, may be prepared by the reduction of crotonaldehyde with aluminium isopropoxide (p. 210):

$$CH_3CH=CHCHO \xrightarrow{[(CH_3)_2CHO]_3Al} CH_3CH=CHCH_2OH \quad (85–90\%)$$

When treated with hydrogen bromide, crotyl alcohol produces a mixture of crotyl and methyl-vinylcarbinyl bromide:

$$CH_3CH=CHCH_2OH \xrightarrow{HBr} CH_3CH=CHCH_2Br + CH_3CHBrCH=CH_2$$

The formation of the rearranged product is an example of anionotropy, and the most widely studied example is the three-carbon system, *i.e.*, the allylic system. Nucleophilic substitution reactions of allylic halides may occur by the S_N2 mechanism, and when this is operating, substitution proceeds normally (see below). When, however, nucleophilic substitution reactions are carried out under conditions which favour the unimolecular mechanism, then the product may be a mixture of two isomers, the rearranged product being an example of the *allylic rearrangement*. The mechanism by which this rearrangement occurs is known as S_N1', and is believed to occur via a resonance hybrid (ambident) cation:

$$MeCH=CHCH_2OH \underset{}{\overset{H^+}{\rightleftharpoons}} MeCH=CHCH_2OH_2{}^+ \underset{+H_2O}{\overset{-H_2O}{\rightleftharpoons}}$$

$$MeCH=CHCH_2{}^+ \longleftrightarrow Me\overset{+}{C}HCH=CH_2$$

$$\downarrow{\scriptstyle Br^-} \qquad\qquad\qquad \downarrow{\scriptstyle Br^-}$$

$$MeCH=CHCH_2Br \qquad MeCHBrCH=CH_2$$

$$S_N1 \qquad\qquad\qquad S_N1'$$

There is much evidence to support this S_N1' mechanism, *e.g.*, Hughes *et al.* (1948) examined the conversion of α-methylallyl and crotyl chloride into α-methylallyl and crotyl ether by reaction with ethanolic sodium ethoxide. The concentration of the sodium ethoxide was made so small as to give first-order kinetics, or so large as to give second-order kinetics. The results were:

$$MeCHClCH=CH_2 \xrightarrow{\quad 1st\ order \quad} \qquad \xleftarrow{\quad 1st\ order \quad} MeCH=CHCH_2Cl$$

$$\downarrow{\scriptstyle 2nd\ order} \qquad\qquad\qquad\qquad\qquad\qquad\qquad\qquad \downarrow{\scriptstyle 2nd\ order}$$

$$MeCH(OEt)CH=CH_2\ (100\%) \quad \begin{cases} MeCH(OEt)CH=CH_2\ (13\%) \\ \qquad\qquad + \\ MeCH=CHCH_2OEt\ (87\%) \end{cases} \quad \begin{array}{c} MeCH=CHCH_2OEt \\ (100\%) \end{array}$$

With second-order substitution, each chloride gives its own *unrearranged ether* (thus the S_N2 mechanism). With first-order substitution, however, each chloride gives the *same* mixture of isomeric ethers. This latter result may be explained by assuming that *two* mechanisms are operating, S_N1 (without rearrangement) and S_N1' (with rearrangement), both proceeding through a *common* intermediate ambident cation.

In addition to the S_N1' mechanism, there is also the S_N2', *i.e.*, a bimolecular nucleophilic substitution with allylic rearrangement. This has been difficult to demonstrate, but a good example of it has been given by de la Mare *et al.* (1953). These authors have suppressed the S_N2 reaction by steric factors and reduced the S_N1 reaction by means of a reagent of high nucleophilic activity and a solvent of low ionising power. They found that the reaction between α,α-dimethylallyl chloride and sodium thiophenoxide in ethanol gives 62 per cent of rearranged product by a second-order reaction (*i.e.*, S_N2'):

$$Me_2C-CH=CH_2 \quad SPh \longrightarrow Me_2C=CHCH_2SPh + NaCl$$

One other problem that will be discussed here is the case of ion pairs in the allylic rearrangement (see also p. 159). Young *et al.* (1951) have shown that α,α-dimethylallyl chloride in acetic acid undergoes acetolysis accompanied by rearrangement to γ,γ-dimethylallyl chloride:

$$Me_2CClCH=CH_2 \xrightarrow{MeCO_2H} \text{acetolysis products} + Me_2C=CHCH_2Cl$$

The rate of the isomerisation was shown to be proportional to the concentration of the starting chloride only. Thus a possible mechanism is the S_N1'. If this were so, then the rate of isomerisation would be affected by addition of chloride ions (since the first step is ionisation). Experiment showed that the rate was unaffected by added chloride ions, and it was also shown that when an *excess* of radioactive chloride ion was added, the rate of isomerisation was far greater than the incorporation of Cl^{*-}. It thus appears that the chloride ion is set free during the isomerisation and is mainly the 'first ion' to recombine, even though the solution contains a large excess of Cl^{*-} ions. Young *et al.* explained this result by assuming that the starting chloride undergoes ionisation to form an ion-pair, and this is followed by internal return:

$$Me_2CClCH=CH_2 \rightleftharpoons [Me_2C\cdots CH\cdots CH_2]^+Cl^- \rightleftharpoons Me_2C=CHCH_2Cl$$
$$\text{ion pair}$$

Nevertheless, some ion pairs escape from the solvent cage, and these 'free' carbonium ions are capable of undergoing acetolysis and combination with Cl^{*-}.

A number of unsaturated alcohols occur in *essential oils*. The essential oils are volatile oils with pleasant odours, and are obtained from various plants to which they impart their characteristic odours. They are widely used in the perfume industry (see Vol. 2, Ch. 8).

The simplest acetylenic alcohol is **propargyl alcohol** (*prop*-2-*yn*-1-*ol*), b.p. 114°C. This may be prepared from 1,2,3-tribromopropane:

$$BrCH_2CHBrCH_2Br \xrightarrow[\text{KOH}]{\text{ethanolic}} CH_2=CBrCH_2Br \xrightarrow[\text{KOH}]{\text{aqueous}} CH_2=CBrCH_2OH$$

$$\xrightarrow[\text{KOH}]{\text{ethanolic}} CH\equiv CCH_2OH$$

Propargyl alcohol (together with butynediol) is prepared by the interaction of acetylene and formaldehyde in the presence of silver or cuprous acetylide as catalyst:

$$C_2H_2 + HCHO \xrightarrow{Ag_2C_2} CH\equiv CCH_2OH \xrightarrow[\text{HCHO}]{Ag_2C_2} HOCH_2C\equiv CCH_2OH$$

Propargyl alcohol behaves in many ways like acetylene, *e.g.*, it adds on two or four bromine atoms, and forms the silver or cuprous compounds when treated with an ammoniacal solution of silver nitrate or cuprous chloride. It may be catalytically reduced to allyl and n-propyl alcohols.

Unsaturated ethers

The simplest unsaturated ether is methyl vinyl ether, which is prepared industrially by passing acetylene into methanol at 160–200°C in the presence of 1–2 per cent potassium methoxide, and under pressure sufficient to prevent boiling:

$$C_2H_2 + CH_3OH \longrightarrow CH_3OCH{=}CH_2$$

The acetylene is diluted with nitrogen to prevent explosions.

Methyl vinyl ether is a very reactive gas, b.p. 5–6°C. It is hydrolysed rapidly by dilute acid at room temperature to give methanol and acetaldehyde (*cf.* the ethers), but is stable in alkaline

$$CH_3OCH{=}CH_2 + H_2O \xrightarrow{\ H^+\ } CH_3OH + CH_3CHO$$

solution. It undergoes many addition reactions at the double bond, *e.g.*,

$$CH_2{=}CHOCH_3 + HCl \xrightarrow{\ 0°C\ } CH_3CHCl{-}O{-}CH_3$$

$$CH_2{=}CHOCH_3 + CH_3OH \xrightarrow[\text{trace of HCl}]{25°C} CH_3CH(OCH_3)_2$$

It readily polymerises, and is used for making polyvinyl ether plastics.

Divinyl ether (*vinyl ether*), b.p. 39°C, may be prepared by heating ethylene chlorohydrin with sulphuric acid, and then passing the product, 2,2′-dichlorodiethyl ether, over potassium hydroxide at 200–240°C:

$$2ClCH_2CH_2OH \xrightarrow[(-H_2O)]{H_2SO_4} ClCH_2CH_2OCH_2CH_2Cl \xrightarrow[(-2HCl)]{KOH} CH_2{=}CHOCH{=}CH_2$$
$$\qquad\qquad\qquad\qquad\qquad (75\%) \qquad\qquad\qquad\qquad\qquad\qquad\qquad (25\%)$$

It is used as an anæsthetic, being better for this purpose than diethyl ether.

Unsaturated aldehydes

The carbonyl (C=O stretch) absorption band of **α,β-unsaturated aldehydes** occurs in the region 1 705–1 680 cm^{-1} (s). It is therefore readily distinguished from the carbonyl group in saturated (acyclic) aldehydes. The molecular formula will also help here.

The τ-values of both α- and β-protons in α,β-unsaturated carbonyl compounds occur at lower field than those in alkenes (4·3–5·3), and the value for the β-proton is also dependent on whether it is *cis* or *trans* with respect to the carbonyl group. The usual range for α-protons is τ ~ 3·5–4·2 and for β-protons, τ ~ 2·4–3·5. This downfield shift for both α- and β-protons may be attributed to the −I effect of the CO group, and the greater shift for the β-proton may be explained by conjugation which results in a small positive charge on the β-carbon atom, thereby increasing deshielding of a β-proton:

$$-CH{=}CH{-}C{=}O \longleftrightarrow -\overset{+}{C}H{-}CH{=}C{-}O^-$$

Since shifts for protons attached to an unsaturated carbon atom further along the chain are fairly close to those of protons in alkenes, this may be used to show whether a double bond is α,β with respect to the carbonyl group.

The simplest unsaturated aldehyde is **acraldehyde** (*acrolein, prop-2-en-1-al*), b.p. 52°C. This is conveniently prepared by heating glycerol with potassium hydrogen sulphate, which is the most satisfactory dehydrating agent for this reaction:

$$HOCH_2CHOHCH_2OH \xrightarrow{\ -2H_2O\ } CH{=}CHCHO \quad (33\text{–}48\%)$$

Acraldehyde may also be readily prepared by the oxidation of allyl alcohol with manganese dioxide or nickel peroxide (p. 329).

Acraldehyde is prepared industrially in many ways, *e.g.*,
(i) by passing a mixture of acetaldehyde and formaldehyde vapour over lithium phosphate as catalyst:

$$CH_3CHO + CH_2O \longrightarrow CH_2{=}CHCHO + H_2O$$

(ii) By the direct oxidation of propene over cuprous oxide on alumina:

$$CH_2{=}CHCH_3 + O_2 \longrightarrow CH_2{=}CHCHO + H_2O$$

Acraldehyde is unstable, readily polymerising to a white solid. It is an α,β-unsaturated aldehyde, and undergoes many of the usual reactions of an alkene and an aldehyde, *e.g.*, it adds on two halogen atoms or a molecule of halogen acid (contrary to Markownikoff's rule; *cf.* allyl alcohol):

$$CH_2{=}CHCHO + Br_2 \longrightarrow BrCH_2CHBrCHO$$
$$CH_2{=}CHCHO + HCl \longrightarrow ClCH_2CH_2CHO$$

The addition of hydrogen bromide is unaffected by the presence of peroxides. Acraldehyde reduces ammoniacal silver nitrate, forms a cyanohydrin with hydrogen cyanide, and a phenyl-hydrazone with phenylhydrazine:

$$CH_2{=}CHCHO + [O] \xrightarrow{AgNO_3} CH_2{=}CHCO_2H$$
$$CH_2{=}CHCHO + HCN \longrightarrow CH_2{=}CHCH(OH)CN$$
$$CH_2{=}CHCHO + C_6H_5NHNH_2 \longrightarrow CH_2{=}CHCH{=}NNHC_6H_5 + H_2O$$

Many of the reactions of α,β-unsaturated aldehydes are 1,2-additions, but α,β-unsaturated ketones tend to undergo 1,4-additions. Aldehydes are more reactive than ketones towards nucleophiles, and the kinetically controlled product with aldehydes is 1,2-addition and with ketones 1,4-addition. With the latter, rearrangement occurs if part of the addendum is a proton to give the thermodynamically controlled product (see p. 340).

Acraldehyde undergoes the Tischenko reaction (p. 221) to form allyl acrylate:

$$2CH_2{=}CHCHO \xrightarrow{(C_2H_5O)_3Al} CH_2{=}CHCO_2CH_2CH{=}CH_2$$

It does not, however, undergo the normal aldol condensation; instead, in the presence of alkali, cleavage of the molecule takes place:

$$CH_2{=}CHCHO + OH^- \rightleftharpoons HO{-}CH_2{-}\bar{C}HCHO \underset{H_2O}{\rightleftharpoons} H{-}OCH_2CH_2CHO + OH^- \rightleftharpoons$$

$$O^-{-}CH_2{-}CH_2CHO + H_2O \rightleftharpoons CH_2O + \bar{C}H_2CHO \underset{H_2O}{\rightleftharpoons} CH_3CHO + OH^-$$

The first part of the reaction is the reversal of dehydration, and the second part is the reversal of the aldol condensation. This type of reaction is characteristic of α,β-unsaturated aldehydes.

When acraldehyde is reduced with metal and acid, there are obtained an unsaturated alcohol, a saturated aldehyde, a saturated alcohol and a compound formed by bimolecular reduction:

$$CH_2{=}CHCHO \xrightarrow{e;H^+} CH_2{=}CHCH_2OH + CH_3CH_2CHO +$$
$$CH_3CH_2CH_2OH + CH_2{=}CHCHOHCHOHCH{=}CH_2$$

A good yield of the bimolecular product is obtained by reducing acraldehyde with magnesium amalgam (*cf.* pinacol, p. 316). Acraldehyde is reduced by sodium amalgam to n-propanol and by aluminium isopropoxide (Meerwein–Ponndorf–Verley reduction, p. 210) to allyl alcohol. It should be noted that metal and acid do not usually reduce a double bond; it is only when the double bond is in the α,β-position with respect to a carbonyl group that it is reduced in this way (see also below).

The product obtained by the catalytic hydrogenation of α,β-unsaturated carbonyl compounds depends largely on the nature of the catalyst. Saturated carbonyl compounds are obtained with Pd—C at room temperature, unsaturated alcohols with PtO_2—FeCl—$(CH_3CO_2)_2Zn$, and saturated alcohols with Ni, Pd—C, or PtO_2 at elevated temperatures and hydrogen under pressure.

α,β-Unsaturated carbonyl compounds may be reduced to unsaturated alcohols by lithium aluminium hydride (but see p. 774), sodium borohydride, or best of all, by aluminium hydride.

Acraldehyde adds on to butadiene to form a cyclic compound (see the *Diels–Alder* reaction, p. 536).

An interesting property of α,β-unsaturated aldehydes, ketones and acids is that they often undergo abnormal ozonolysis:

$$R_2^1C{=}CR^2CR^3{=}O \longrightarrow R_2^1CO + R^2CO_2H + R^3CO_2H$$
$$R_2^1C{=}CR^2CO_2H \longrightarrow R_2^1CO + R^2CO_2H + CO_2$$

Crotonaldehyde (*but-2-en-1-al*), b.p. 104°C, may be prepared by heating aldol alone, or better, with a dehydrating agent, *e.g.*, zinc chloride. The best yield is obtained by distilling aldol with acetic acid as catalyst:

$$CH_3CHOHCH_2CHO \xrightarrow{-H_2O} CH_3CH{=}CHCHO$$

Crotonaldehyde closely resembles acraldehyde in its chemical properties. It exists, however, in two geometrical isomeric forms, *cis* and *trans*:

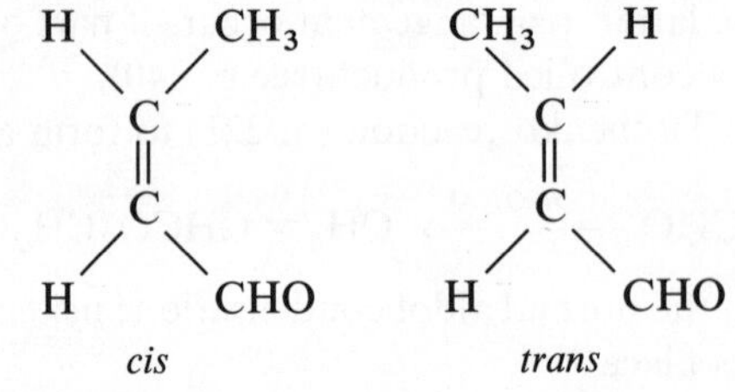

A number of unsaturated aldehydes occur naturally in the essential oils, many accompanying the corresponding unsaturated alcohol.

The simplest acetylenic aldehyde is **propargylaldehyde** (propiolaldehyde, propynal), b.p. 60°C. This may be prepared from acraldehyde, the aldehyde group of which is 'protected' by acetal formation:

$$CH_2{=}CHCHO + 2C_2H_5OH \underset{}{\overset{HCl}{\rightleftharpoons}} H_2O + CH_2{=}CHCH(OC_2H_5)_2 \xrightarrow{Br_2}$$
$$BrCH_2CHBrCH(OC_2H_5)_2 \xrightarrow[KOH]{ethanolic} CH{\equiv}CCH(OC_2H_5)_2 \xrightarrow[HCl]{dilute} CH{\equiv}CCHO$$

It may also be obtained by the controlled oxidation of propargyl alcohol.

Propargylaldehyde undergoes cleavage when treated with sodium hydroxide (*cf.* acraldehyde, above):

$$CH{\equiv}CCHO + NaOH \longrightarrow CH{\equiv}CH + HCO_2Na$$

Unsaturated ketones

The carbonyl infra-red absorption band (C=O stretch) in **α,β-unsaturated ketones** occurs in the region $1\,685$–$1\,665\,\text{cm}^{-1}$ (s). This distinguishes these compounds from α,β-unsaturated aldehydes (see also p. 336; NMR spectra).

The simplest unsaturated ketone is **methyl vinyl ketone** (*but-3-en-2-one*), b.p. 79°C. This may be prepared by means of an aldol condensation between formaldehyde and acetone, the product being dehydrated by heat.

$$HCHO + CH_3COCH_3 \xrightarrow{\text{NaOH}} HOCH_2CH_2COCH_3 \xrightarrow{-H_2O} CH_2{=}CHCOCH_3$$

It may also be prepared by the action of acetic anhydride on ethylene in the presence of zinc chloride as catalyst:

$$CH_2{=}CH_2 + (CH_3CO)_2O \xrightarrow{\text{ZnCl}_2} CH_2{=}CHCOCH_3 + CH_3CO_2H$$

> Methyl vinyl ketone can be prepared by the action of acetic anhydride on vinylmagnesium bromide in tetrahydrofuran at $-60°$ to $-70°C$ (Martin, 1957); this is a general method for preparing vinyl ketones:
>
> $$2CH_2{=}CHMgBr + 2(CH_3CO)_2O \longrightarrow$$
> $$2CH_2{=}CHCOCH_3 + MgBr_2 + (CH_3CO_2)_2Mg \quad (60\text{–}80\%)$$
>
> It is manufactured by hydrating vinylacetylene in the presence of dilute sulphuric acid and mercuric sulphate (*cf.* acetaldehyde, p. 136):
>
> $$CH_2{=}CHC{\equiv}CH + H_2O \longrightarrow CH_2{=}CHCOCH_3$$

Methyl vinyl ketone polymerises on standing and is used commercially as the starting material for plastics.

Mesityl oxide (4-methylpent-3-en-2-one), b.p. 130°C, may be prepared by distilling diacetone alcohol (*q.v.*) with a trace of iodine:

$$(CH_3)_2C(OH)CH_2COCH_3 \xrightarrow{-H_2O} (CH_3)_2C{=}CHCOCH_3 \quad (90\%)$$

Stross *et al.* (1947) have shown that mesityl oxide prepared this way is a mixture of mesityl oxide and *isomesityl oxide*, $CH_2{=}C(CH_3)CH_2COCH_3$.

Mesityl oxide may also be prepared as follows:

$$(CH_3)_2C{=}CH_2 + CH_3COCl \xrightarrow{\text{ZnCl}_2} (CH_3)_2CClCH_2COCH_3 \xrightarrow{-HCl} (CH_3)_2C{=}CHCOCH_3$$

It is used as a solvent for oils, gums, etc.

Phorone (2,6-dimethylhepta-2,5-dien-4-one), m.p. 28°C, may be prepared by the action of hydrochloric acid on acetone (see p. 218):

$$3CH_3COCH_3 \xrightarrow{\text{HCl}} (CH_3)_2C{=}CHCOCH{=}C(CH_3)_2 + 2H_2O$$

> **Isophorone** (3,5,5-trimethylcyclohex-2-enone), b.p. 215°C, is formed when acetone is distilled with aqueous sodium hydroxide under pressure.

$$3CH_3COCH_3 \xrightarrow{\text{OH}^-} \text{[structure]} + 2H_2O$$

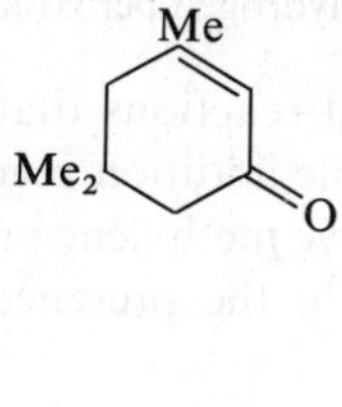

It is used as a solvent.

Catalytic reduction and reduction with metal hydrides of α,β-unsaturated ketones has been described on p. 338. Reduction with metal and acid, or better, with magnesium amalgam, gives bimolecular products (cf. acraldehyde, above):

$$2(CH_3)_2C{=}CHCOCH_3 \xrightarrow{Mg/Hg} \begin{array}{c} (CH_3)_2CCH_2COCH_3 \\ | \\ (CH_3)_2CCH_2COCH_3 \end{array}$$

The addition of halogen to α,β-unsaturated ketones occurs by the normal electrophilic mechanism across the 3,4-positions. The addition of halogen acids, however, occurs by nucleophilic attack by the 1,4-mechanism (see below).

The α,β-unsaturated ketones add on ammonia, primary amines and secondary amines to form β-amino-compounds; with ammonia, mesityl oxide forms diacetonamine and phorone forms triacetonamine:

$$(CH_3)_2C{=}CHCOCH_3 + NH_3 \longrightarrow \overset{\overset{\displaystyle NH_2}{\displaystyle |}}{(CH_3)_2CCH_2COCH_3}$$

$$(CH_3)_2C{=}CHCOCH{=}C(CH_3)_2 + 2NH_3 \longrightarrow$$

$$\left[\begin{array}{c} (CH_3)_2CCH_2COCH_2C(CH_3)_2 \\ | \qquad\qquad\qquad | \\ NH_2 \qquad\qquad\qquad NH_2 \end{array} \right] \xrightarrow{-NH_3} OC \overset{\displaystyle CH_2{-}C(CH_3)_2}{\underset{\displaystyle CH_2{-}C(CH_3)_2}{\diagup \diagdown}} NH$$

All the evidence is in favour of a 1,4-addition mechanism when the addendum is of the type H—Z, the product being an enol which then tautomerises to the stable keto form, the overall result being 3,4-addition, e.g., with halogen acids:

$$R^1{-}\overset{4}{C}H{=}\overset{3}{C}H{-}\overset{2}{C}R^2{=}\overset{1}{\overset{\displaystyle\cdot}{O}} \xrightarrow{HX} X^- + R^1{-}CH{=}CH{-}CR^2{=}\overset{+}{O}H \longleftrightarrow$$

$$\overset{+}{R^1{-}CH}{-}CH{=}CR^2{-}OH \xrightarrow{X^-} R^1{-}CHX{-}CH{=}CR^2{-}O{-}H \longrightarrow$$

$$R^1{-}CHX{-}CH_2{-}CR^2{=}O$$

The 1,4-addition intermediate is the kinetically controlled product and the 3,4-product is the thermodynamically controlled product (see p. 337).

Similarly, α,β-unsaturated ketones add on hydrogen cyanide to form β-cyano-compounds, $R^1CH(CN)CH_2COR^2$; and sodium hydrogen sulphite to form β-sulphonic acids, $R^1CH(SO_3Na)CH_2COR^2$. The reaction with hydroxylamine usually results in a mixture of the oxime, $R^1CH{=}CHC({=}NOH)R^2$, and the β-hydroxylaminoketone, $R^1CH(NHOH)CH_2COR^2$. Trialkylboranes also undergo 1,4-addition, e.g.,

$$R^1_3B + CH_2{=}CHCOR^2 \longrightarrow R^1CH_2CH{=}CR^2OBR^1_2 \xrightarrow{H_2O} R^1CH_2CH_2COR^2$$

A number of factors affect the direction of these addition reactions, an important one being steric hindrance. Steric hindrance at the carbonyl group favours 1,4-addition, whereas steric hindrance at the 3- and/or 4-positions favours 1,2-addition (see also p. 434).

α,β-Unsaturated ketones (and aldehydes) are more difficult to epoxidise than alkenes (p. 111), but are epoxidised by alkaline hydrogen peroxide.

One of the most important reactions that α,β-unsaturated ketones undergo is the **Michael condensation** (1887). This is the addition reaction between an α,β-unsaturated keto-compound and a compound with an active methylene group, e.g., malonic ester, acetoacetic ester, etc. The condensation is carried out in the presence of a base, e.g., sodium ethoxide, or secondary amines. The mechanism of the reaction is believed to be (using malonic ester and mesityl oxide as an example):

$$CH_2(CO_2Et)_2 + EtO^- \rightleftharpoons \bar{C}H(CO_2Et)_2 + EtOH$$

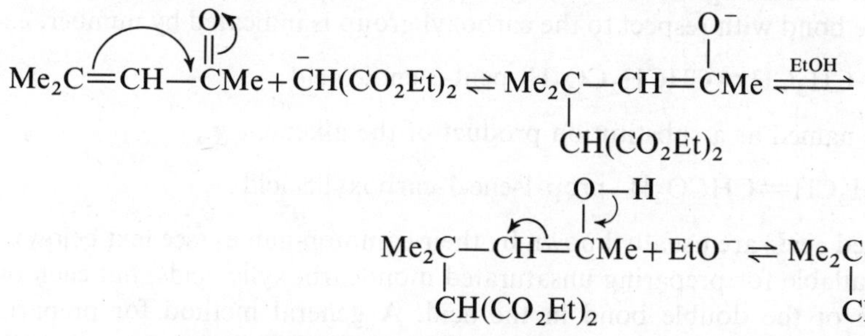

$$Me_2C\!\!=\!\!CH\!\!-\!\!\overset{\overset{\displaystyle O}{\|}}{C}Me + \bar{C}H(CO_2Et)_2 \rightleftharpoons Me_2C\!\!-\!\!CH\!\!=\!\!\overset{\overset{\displaystyle O^-}{|}}{C}Me \xrightarrow{EtOH}$$
$$\underset{CH(CO_2Et)_2}{|}$$

$$Me_2C\!\!-\!\!\overset{\overset{\displaystyle O\!-\!H}{|}}{CH}\!\!=\!\!CMe + EtO^- \rightleftharpoons Me_2C\!\!-\!\!CH_2COMe + EtO^-$$
$$\underset{CH(CO_2Et)_2}{|} \qquad\qquad \underset{CH(CO_2Et)_2}{|}$$

With this example, after the Michael condensation has occurred, it is followed by an internal Claisen condensation and leads to the formation of dimedone.

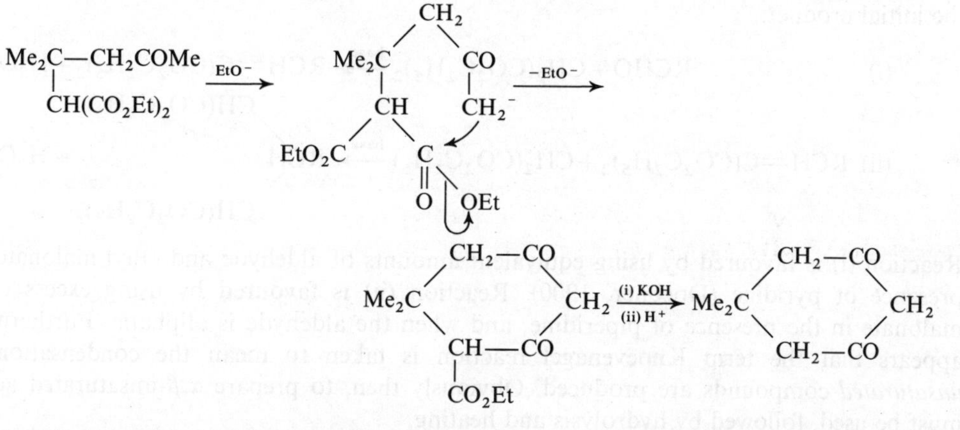

Me₂C—CH₂COMe   EtO⁻ → ...

The description of the Michael condensation given above is a limited one. The term now includes a large variety of addenda and acceptors; the α,β-unsaturated compounds may be ketones, esters, cyanides, etc., and the addenda may be malonic ester, E.A.A., cyanoacetic ester, nitroalkanes, etc.

The Michael condensation, as shown above, is reversible. Thus a Michael *retrogression, i.e.,* reversal, may occur to produce compounds different from the starting compounds, *e.g.,* starting with benzylidenemalonic ester and E.A.A. in the presence of a large amount of sodium ethoxide results in a mixture containing the reactants, the Michael condensation product, *and* benzylideneacetoacetate and malonic ester (produced by the retrogression). If the reaction is carried out with the last two compounds as starting materials, the *same* mixture is obtained. Thus:

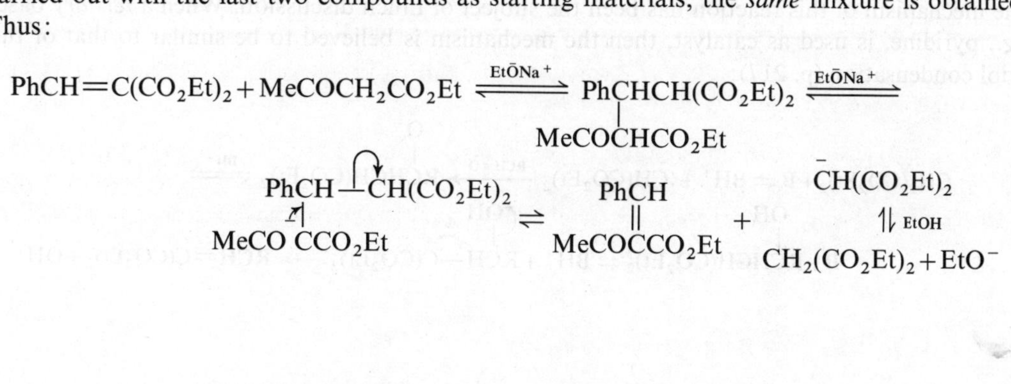

$$PhCH\!\!=\!\!C(CO_2Et)_2 + MeCOCH_2CO_2Et \underset{\overline{E}tONa^+}{\rightleftharpoons} PhCHCH(CO_2Et)_2 \underset{\overline{E}tONa^+}{\rightleftharpoons}$$
$$\underset{MeCOCHCO_2Et}{|}$$

$$PhCH\!\!-\!\!\bar{C}H(CO_2Et)_2 \rightleftharpoons PhCH + \bar{C}H(CO_2Et)_2$$
$$\underset{MeCO\ CCO_2Et}{|} \qquad\quad \underset{MeCOCCO_2Et}{\|} \qquad \underset{CH_2(CO_2Et)_2 + EtO^-}{\Updownarrow EtOH}$$

Unsaturated monocarboxylic acids

Nomenclature

The longest carbon chain containing the carboxyl group and the double bond is chosen, and the position of the double bond with respect to the carboxyl group is indicated by number, *e.g.*.

$$CH_3CH=CHCH_2CO_2H \quad \text{pent-3-enoic acid}$$

Alternatively, the acid is named as a substitution product of the alkene, *e.g.*,

$$CH_3CH=CHCO_2H \quad \text{prop-1-ene-1-carboxylic acid}$$

In practice the unsaturated acids are usually known by their common names (see text below).

Many methods are available for preparing unsaturated monocarboxylic acids, but each one depends on the position of the double bond in the acid. A general method for preparing α,β-unsaturated acids is by the **Knoevenagel reaction** (1898). This is the reaction between aldehydes and compounds with active methylene groups in the presence of an organic base. The reaction may stop at stage (i) or proceed to stage (ii) via a Michael condensation involving the initial product.

(i) $\qquad RCHO + CH_2(CO_2C_2H_5)_2 \xrightarrow{\text{base}} RCH=C(CO_2C_2H_5)_2 + H_2O$

(ii) $RCH=C(CO_2C_2H_5)_2 + CH_2(CO_2C_2H_5) \xrightarrow{\text{base}} RCH\begin{array}{l} \diagup CH(CO_2C_2H_5)_2 \\ \diagdown CH(CO_2C_2H_5)_2 \end{array} + H_2O$

Reaction (i) is favoured by using equivalent amounts of aldehyde and ethyl malonate in the presence of pyridine (Doebner, 1900). Reaction (ii) is favoured by using excess of ethyl malonate in the presence of piperidine, and when the aldehyde is aliphatic. Furthermore, it appears that the term Knoevenagel reaction is taken to mean the condensation when *unsaturated* compounds are produced. Obviously then, to prepare α,β-unsaturated acids, (i) must be used, followed by hydrolysis and heating.

$$RCHO + CH_2(CO_2C_2H_5)_2 \xrightarrow{\text{pyridine}} RCH=C(CO_2C_2H_5)_2 \xrightarrow[\text{(ii) HCl}]{\text{(i) KOH}}$$

$$RCH=C(CO_2H)_2 \xrightarrow{150-200°C} RCH=CHCO_2H + CO_2$$

In practice, it is usually sufficient to treat the aldehyde with malonic acid in the presence of pyridine; *e.g.*, acetaldehyde gives crotonic acid:

$$CH_3CHO + CH_2(CO_2H)_2 \xrightarrow{\text{pyridine}} CH_3CH=CHCO_2H + CO_2 + H_2O \quad (60\%)$$

The mechanism of this reaction has been the subject of much discussion. When a *tertiary* base, *e.g.*, pyridine, is used as catalyst, then the mechanism is believed to be similar to that of the aldol condensation (p. 217):

$$CH_2(CO_2Et)_2 + B \rightleftharpoons BH^+ + :\bar{C}H(CO_2Et)_2 \xrightarrow{RCH=O} \overset{\overset{O^-}{|}}{RCHCH(CO_2Et)_2} \xrightarrow{BH^+}$$

$$B + \overset{\overset{OH}{|}}{RCHCH(CO_2Et)_2} \rightleftharpoons BH^+ + RCH\overset{OH}{-}C(CO_2Et)_2 \longrightarrow RCH=C(CO_2Et)_2 + OH^-$$

When a primary or secondary base is used as catalyst, the mechanism may still possibly be as given above, but a complication arises from the fact that carbonyl compounds can form addition products with such bases and these addition products may therefore be intermediates in the reaction.

Ethyl malonate condenses with aldehydes only. On the other hand, ethyl cyanoacetate condenses with ketones in the presence of an amino-acid (as catalyst) such as 3-aminopropionic acid; e.g., acetone forms isopropylidenecyanoacetic ester:

$$(CH_3)_2CO + CH_2(CN)CO_2C_2H_5 \xrightarrow{-H_2O} (CH_3)_2C{=}C\genfrac{}{}{0pt}{}{CN}{CO_2C_2H_5}$$

Since cyano-compounds are readily hydrolysed to the corresponding acid, the above condensation may be used to prepare α,β-unsaturated acids of the type $R_2C{=}CHCO_2H$.

α,β-Unsaturated acids may also be prepared by heating α-bromo-acids with ethanolic potassium hydroxide, or better still, with potassium t-butoxide (Cason *et al.*, 1953); or by heating β-hydroxyacids with aqueous sodium hydroxide:

$$RCH_2CHBrCO_2H + KOH \xrightarrow{ethanol} RCH{=}CHCO_2H + KBr + H_2O$$

$$RCHOHCH_2CO_2H \xrightarrow{NaOH} RCH{=}CHCO_2H + H_2O$$

β,γ-Unsaturated acids may be prepared by heating an aldehyde with sodium succinate and acetic anhydride at 100°C. The product is a *γ-alkylparaconic acid*, and when this is heated at 150°C, it eliminates carbon dioxide to form the β,γ-unsaturated acid (Fittig *et al.*, 1885). This is a variation of the Perkin reaction, and succinic anhydride (formed by interaction of sodium succinate and acetic anhydride) is probably the reacting species (see p. 739).

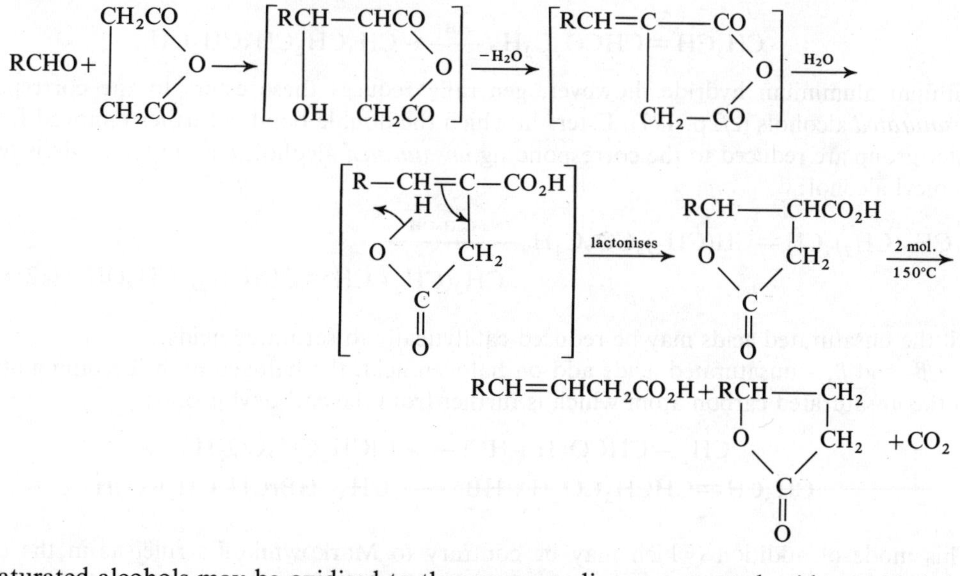

Unsaturated alcohols may be oxidised to the corresponding unsaturated acid, provided the double bond is protected, *e.g.*, acrylic acid from allyl alcohol (see p. 330).

Oxidation of unsaturated aldehydes produces unsaturated acids. The oxidation, however, cannot be carried out with the usual oxidising agents, such as acid or alkaline permanganate, acid dichromate, etc., since these will attack the double bond. A useful oxidising agent for unsaturated aldehydes is ammoniacal silver nitrate (see p. 220); *e.g.*, crotonaldehyde is oxidised to crotonic acid:

$$CH_3CH=CHCHO + [O] \xrightarrow[\text{AgNO}_3]{\text{ammoniacal}} CH_3CH=CHCO_2H$$

Unsaturated methyl ketones may be oxidised to unsaturated acids by sodium hypochlorite (*cf.* the haloform reaction), *e.g.*,

$$RCH=CHCOCH_3 \xrightarrow{\text{NaOCl}} RCH=CHCO_2H$$

The carbonyl (C=O stretch) infra-red absorption band in **α,β-unsaturated acids** occurs in the region $1\,715\text{–}1\,690\,\text{cm}^{-1}$ (s). This readily distinguishes these compounds from α,β-unsaturated ketones, but is not so clear cut with α,β-unsaturated aldehydes, and may also lead to confusion with the saturated monocarboxylic acids (p. 235). However, the absorption region of the double bond (C=C stretch), provided it is not conjugated, is $1\,650\text{–}1\,630\,\text{cm}^{-1}$ (v). Thus, although this does not indicate its position, it does show that a double bond is present. The molecular formula should distinguish unsaturated acids from the saturated acids and from α,β-unsaturated aldehydes (and ketones) [see also p. 336; N.M.R. spectra].

General reactions of the unsaturated acids

The esters of α,β-unsaturated acids undergo the same addition reactions (including the Michael condensation) at the double bond as the α,β-unsaturated ketones. When reduced by the Bouveault–Blanc method or with a large excess of sodium borohydride in methanol (see Table 6.1, p. 179), α,β-unsaturated esters are converted into the corresponding saturated alcohol:

$$CH_3CH=CHCO_2C_2H_5 \xrightarrow{e;\text{H}^+} CH_3CH_2CH_2CH_2OH$$

Lithium aluminium hydride, however, generally reduces these esters to the corresponding *unsaturated* alcohols (*cf.* p. 177). Esters in which the double bond is further removed from the ester group are reduced to the corresponding *unsaturated* alcohol; *e.g.*, butyl oleate is reduced to oleyl alcohol:

$$CH_3(CH_2)_7CH=CH(CH_2)_7CO_2C_4H_9 \xrightarrow{\text{Na/C}_4\text{H}_9\text{OH}}$$
$$CH_3(CH_2)_7CH=CH(CH_2)_7CH_2OH \quad (82\text{–}84\%)$$

All the unsaturated acids may be reduced catalytically to saturated acids.

α,β- and β,γ- unsaturated acids add on halogen acid, the halogen atom becoming attached to the unsaturated carbon atom which is further from the carboxyl group:

$$CH_2=CHCO_2H + HCl \longrightarrow ClCH_2CH_2CO_2H$$
$$CH_3CH=CHCH_2CO_2H + HBr \longrightarrow CH_3CHBrCH_2CH_2CO_2H$$

This mode of addition (which may be contrary to Markownikoff's rule, as in the case of acrylic acid) may be ascribed to the inductive effect of the carboxyl group (*cf.* allyl alcohol, etc.). On the other hand, addition of halogen acid to γ,δ-unsaturated acids of the type $CH_2=CHCH_2CH_2CO_2H$ takes place in accordance with Markownikoff's rule. This must be due to the fact that the inductive effect of the carboxyl group ceases to be felt beyond the β-carbon atom.

When unsaturated acids are boiled with alkali, the double bond tends to move so as to form the α,β-unsaturated acid:

$$CH_3CH=CHCH_2CO_2H \xrightarrow{NaOH} CH_3CH_2CH=CHCO_2H$$

An interesting example of this migration of the double bond is the hydrolysis of allyl cyanide with boiling alkali to produce *crotonic* acid. But-3-enoic acid is probably formed first, and this then rearranges to crotonic acid:

$$CH_2=CHCH_2Br \xrightarrow{CuCN} CH_2=CHCH_2CN \xrightarrow{NaOH}$$

$$[CH_2=CHCH_2CO_2H] \xrightarrow{NaOH} CH_3CH=CHCO_2H$$

The movement of a double bond to the α,β-position shows that the latter acid is thermodynamically more stable, *e.g.*, Linstead *et al.* (1943) showed that the migration is reversible and that the equilibrium between the sodium salts of vinylacetic acid and crotonic acid is 98 per cent on the side of the latter:

$$CH_2=CHCH_2CO_2^- \rightleftharpoons CH_3CH=CHCO_2^-$$

The mechanism of this shift is believed to be:

$$CH_2=CHCH_2CO_2^- \underset{OH^-;-H^+}{\overset{}{\rightleftharpoons}} CH_2=CH\bar{C}HCO_2^- \longleftrightarrow$$

$$\bar{C}H_2CH=CHCO_2^- \overset{H^+}{\rightleftharpoons} CH_3CH=CHCO_2^-$$

When unsaturated acids are fused with alkali, cleavage of the chain takes place with the formation of two acids, one of which is always acetic acid. This again indicates the migration of the double bond to the α,β-position; *e.g.*, oleic acid gives acetic and palmitic acids:

$$CH_3(CH_2)_7CH=CH(CH_2)_7CO_2H \xrightarrow{KOH} [CH_3(CH_2)_{14}CH=CHCO_2H] \xrightarrow{KOH}$$

$$CH_3(CH_2)_{14}CO_2H + CH_3CO_2H$$

α,β-Unsaturated esters add on aliphatic diazo-compounds to form *pyrazoline* derivatives; *e.g.*, acrylic ester reacts with diazomethane to give pyrazoline-5-carboxylic ester:

$$CH_2=CHCO_2C_2H_5 + CH_2N_2 \longrightarrow \underset{\underset{N}{\overset{\displaystyle HC \diagdown \diagup NH}{\mid \qquad \mid}}}{CH_2\!\!-\!\!CHCO_2C_2H_5}$$

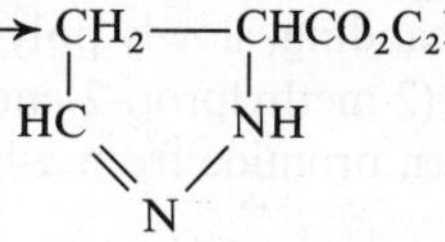

α,β-Unsaturated acids may be degraded by the Hofmann method (p. 264) using a modified procedure. The α,β-unsaturated acid amide in methanol is treated with an alkaline solution of sodium hypochlorite, the urethan produced being hydrolysed in acid solution to give the *aldehyde* in good yield:

$$RCH=CHCONH_2 \xrightarrow[MeOH]{NaOCl} RCH=CHNHCO_2Me \xrightarrow{H_3O^+}$$

$$RCH=CHNH_2 \rightleftharpoons RCH_2CH=NH \xrightarrow{H_3O^+} RCH_2CHO$$

On the other hand, β,γ- and γ,δ-unsaturated acid amides produce the corresponding unsaturated primary amine, but in poor yield:

$$RCH=CHCH_2CONH_2 \longrightarrow RCH=CHCH_2NH_2$$

Baizer *et al.* (1964) have shown that α,β-unsaturated acids may be *hydrodimerised* electrochemically in faintly aqueous alkaline solution ($X=CO_2R$, CN, etc.):

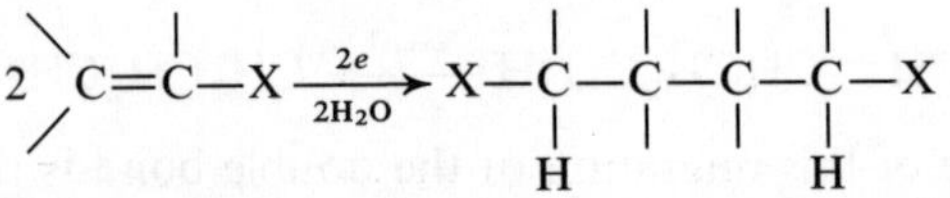

$$2 \quad \overset{\diagdown}{\underset{\diagup}{C}}=\overset{|}{C}-X \xrightarrow[2H_2O]{2e} X-\overset{|}{\underset{|}{C}}-\overset{|}{\underset{H}{C}}-\overset{|}{\underset{|}{C}}-\overset{|}{\underset{H}{C}}-X$$

This reaction can be carried out with two different substrates and so offers a method of preparing a variety of 1,4-bifunctional compounds (see p. 387).

γ,δ- and δ,ε-Unsaturated acids form lactones in the presence of concentrated sulphuric acid (p. 469).

Acrylic acid (*prop-2-enoic acid*), b.p. 141°C, may be prepared:
(i) By the oxidation of allyl alcohol or acraldehyde (see p. 330).
(ii) By heating β-hydroxypropionic acid with aqueous sodium hydroxide:

$$HOCH_2CH_2CO_2H \xrightarrow{NaOH} CH_2=CHCO_2H + H_2O$$

Alternatively, ethylene cyanohydrin may be heated with sulphuric acid:

$$HOCH_2CH_2CN \xrightarrow{H_2SO_4} HOCH_2CH_2CO_2H \xrightarrow{-H_2O} CH_2=CHCO_2H$$

Vinyl cyanide also gives acrylic acid on hydrolysis:

$$CH_2=CHCN \xrightarrow{H^+} CH_2=CHCO_2H$$

Acrylic acid is prepared industrially in many ways, and so are its esters, *e.g.*,

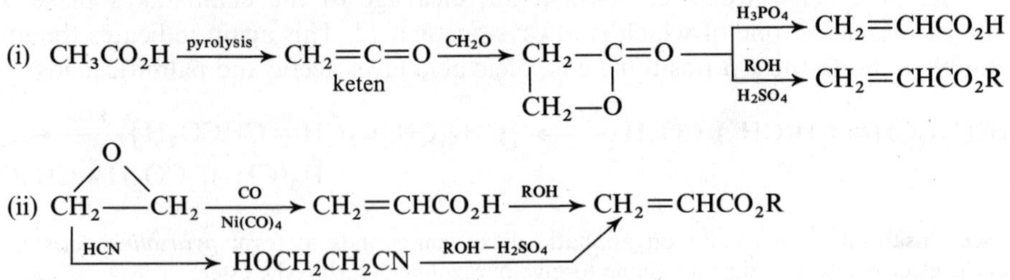

$$(i) \quad CH_3CO_2H \xrightarrow{pyrolysis} \underset{keten}{CH_2=C=O} \xrightarrow{CH_2O} \overset{|}{\underset{|}{CH_2}}-\overset{|}{\underset{O}{C}}=O \quad\left[\begin{array}{l} \xrightarrow{H_3PO_4} CH_2=CHCO_2H \\ \xrightarrow[H_2SO_4]{ROH} CH_2=CHCO_2R \end{array}\right.$$

$$(ii) \quad \overset{O}{\overset{\diagup\diagdown}{CH_2-CH_2}} \xrightarrow[Ni(CO)_4]{CO} CH_2=CHCO_2H \xrightarrow{ROH} CH_2=CHCO_2R$$
$$\downarrow{HCN} \qquad HOCH_2CH_2CN \xrightarrow{ROH-H_2SO_4} \nearrow$$

Acrylic acid, on standing, slowly polymerises to polyacrylic acid, which is a solid.

Methacrylic acid (2-methylprop-2-enoic acid), m.p. 15°C, may be prepared by removing a molecule of hydrogen bromide from α-bromoisobutyric acid:

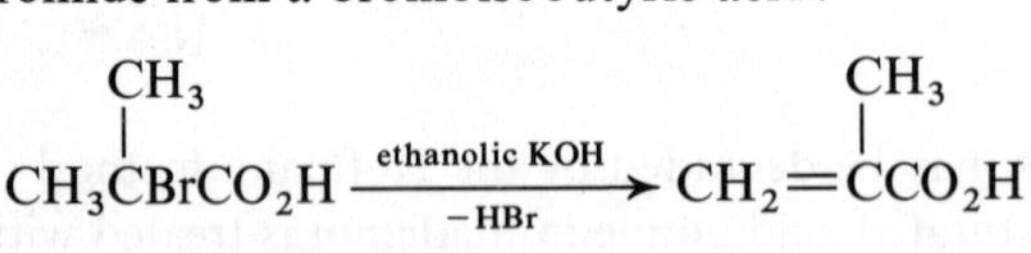

$$\overset{CH_3}{\underset{|}{CH_3CBrCO_2H}} \xrightarrow[-HBr]{ethanolic\ KOH} \overset{CH_3}{\underset{|}{CH_2=CCO_2H}}$$

Methyl methacrylate is very important commercially, since it polymerises to polymethyl methacrylate under the influence of heat. This polymer is tough, transparent, and can be moulded; one of its trade names is *perspex*. One industrial method of preparing methyl methacrylate is as follows:

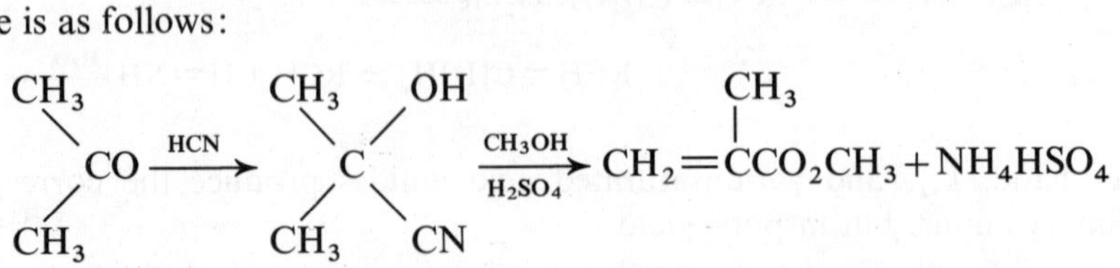

$$\overset{CH_3}{\underset{CH_3}{\diagdown\diagup}}CO \xrightarrow{HCN} \overset{CH_3 \quad OH}{\underset{CH_3 \quad CN}{\diagdown|\diagup}}C \xrightarrow[H_2SO_4]{CH_3OH} \overset{CH_3}{\underset{|}{CH_2=CCO_2CH_3}} + NH_4HSO_4$$

Methyl acrylate also polymerises, but this polymer is softer than that from the methacrylate.

Crotonic acid and **isocrotonic acid** are geometrical isomers, crotonic acid being the *trans* isomer, and isocrotonic the *cis*:

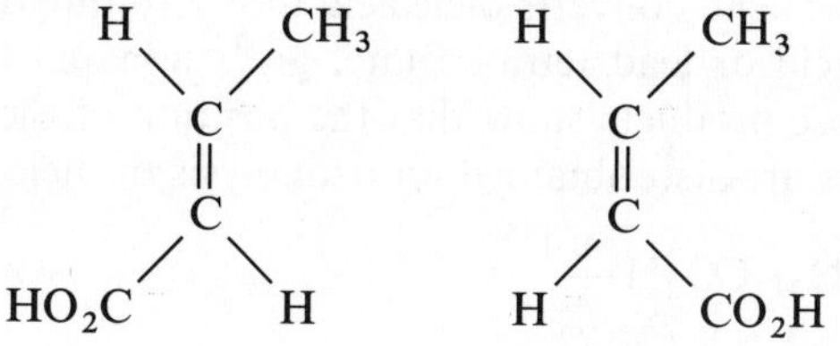

crotonic acid   isocrotonic acid

The more stable form is crotonic acid, m.p. 72°C. Isocrotonic acid, m.p. 15°C, slowly changes into crotonic acid when heated at 100°C.

Crotonic acid (*trans*-but-2-enoic acid) may be prepared by oxidising crotonaldehyde with ammoniacal silver nitrate; by heating β-hydroxybutyric acid with sodium hydroxide; or by the Knoevenagel reaction:

$$CH_3CHO + CH_2(CO_2H)_2 \xrightarrow[20°C]{\text{pyridine}} CH_3CH{=}CHCO_2H + CO_2 + H_2O \quad (60\%)$$

Isocrotonic acid (*cis*-but-2-enoic acid) may be prepared by the action of sodium amalgam on β-chloroisocrotonic ester which is obtained from acetoacetic ester by reaction with phosphorus pentachloride:

$$CH_3COCH_2CO_2C_2H_5 \xrightarrow{PCl_5} CH_3CCl_2CH_2CO_2C_2H_5 \xrightarrow{-HCl}$$
$$CH_3CCl{=}CHCO_2C_2H_5 \xrightarrow{Na/Hg} CH_3CH{=}CHCO_2C_2H_5$$

Angelic and **tiglic acids** are geometrical isomers with the structure $CH_3CH{=}C(CH_3)CO_2H$ (*2-methylbut-2-enoic acid*):

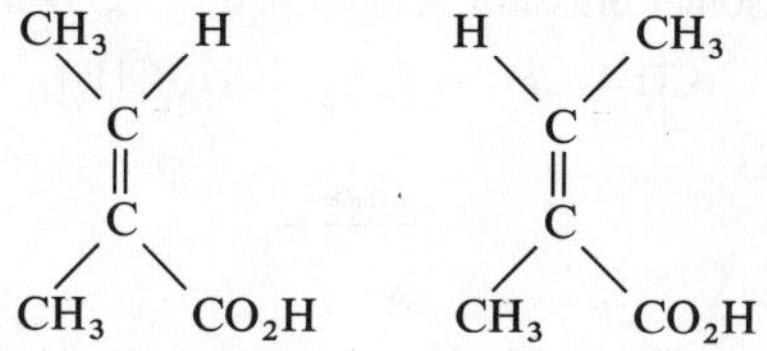

tiglic acid, m.p. 64°C   angelic acid, m.p. 45°C

Undecylenic acid (*dec-9-ene-1-carboxylic acid*), m.p. 24·5°C, may be obtained by the destructive distillation of *ricinoleic acid* (11-*hydroxyheptadec-8-ene-1-carboxylic acid*), which occurs as the glyceride ester in castor oil. Heptanal is the other product, 10 per cent yield of each being obtained:

$$CH_3(CH_2)_5CHOHCH_2CH{=}CH(CH_2)_7CO_2H \longrightarrow CH_3(CH_2)_5CHO + CH_2{=}CH(CH_2)_8CO_2H$$

When ricinoleic acid is heated with sodium hydroxide in air, octan-2-ol and sebacic acid are obtained:

$$CH_3(CH_2)_5CHOHCH_2CH{=}CH(CH_2)_7CO_2H \xrightarrow[\text{air}]{NaOH} CH_3(CH_2)_5CHOHCH_3 + HO_2C(CH_2)_8CO_2H$$

When ricinoleic acid (castor oil) is treated with concentrated sulphuric acid, it gives a complex mixture consisting of the hydrogen sulphate of ricinoleic acid in which the hydroxyl group is esterified, and a compound in which the sulphuric acid has added to the double bond: esterification and addition do not occur together in the same molecule of ricinoleic acid. The product, which is known as *Turkey-red oil*, is used as a wetting agent.

Oleic acid (*cis-heptadec-8-ene-1-carboxylic acid*), m.p. 16°C, occurs as the glyceryl ester in oils and fats. Catalytic reduction converts it into stearic acid. Cold dilute alkaline permanganate, or hydrogen peroxide in acetic acid, converts oleic acid into 9,10-dihydroxystearic acid, which, on oxidation with periodic acid or lead tetra-acetate, gives nonanal (perlargonic aldehyde) and azelaic half aldehyde. These products show that the position of the double bond in oleic acid is 9,10. The same products are also obtained on ozonolysis of oleic acid.

$$Me(CH_2)_7CH{=}CH(CH_2)_7CO_2H \xrightarrow[\text{AcOH}]{H_2O_2}$$

$$Me(CH_2)_7CHOHCHOH(CH_2)_7CO_2H \xrightarrow{HIO_4} Me(CH_2)_7CHO + OHC(CH_2)_7CO_2H$$

Oleic acid is the *cis* form. The *trans* form is **elaidic acid**, m.p. 51°C, which may be obtained by the action of nitrous acid on oleic acid:

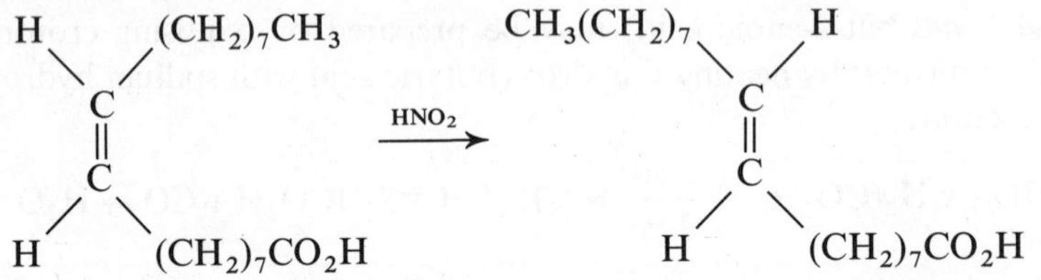

Linoleic acid (*heptadeca-8,11-diene-1-carboxylic acid*), $CH_3(CH_2)_4CH{=}CHCH_2CH{=}CH(CH_2)_7CO_2H$, b.p. 228°C/14 mm, occurs as the glyceryl ester in linseed oil, hemp oil, etc.; it has been synthesised by Raphael *et al.* (1950).

Linolenic acid (*heptadeca-8,11,14-triene-1-carboxylic acid*), $CH_3CH_2CH{=}CHCH_2CH{=}CHCH_2CH{=}CH(CH_2)_7CO_2H$, is a liquid which occurs as the glyceryl ester in, *e.g.*, poppy-seed oil. Oils containing linoleic and linolenic acids are drying oils (p. 324). Linolenic acid is the most abundant and widespread triene acid in Nature; it has been synthesised by Weedon *et al.* (1956).

Erucic acid, m.p. 34°C, is the *cis* form of *heneicos-12-ene-1-carboxylic acid*, $CH_3(CH_2)_7CH{=}CH(CH_2)_{11}CO_2H$. It occurs as the glyceryl ester in various oils, *e.g.*, rape oil, cod-liver oil, etc. It is converted into the *trans* isomer brassidic acid, m.p. 61·5°C, by the action of nitrous acid.

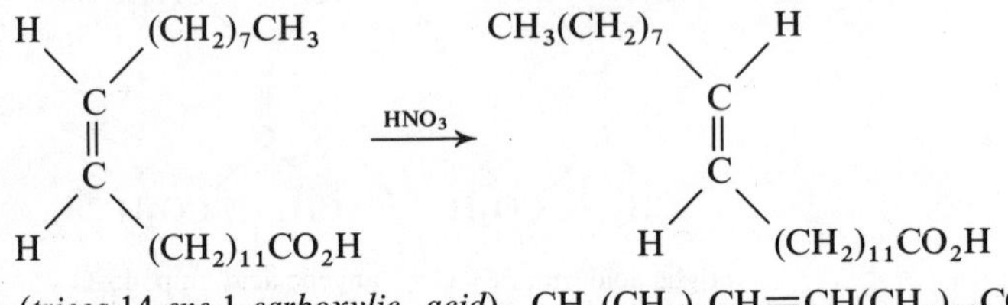

Nervonic acid (*tricos-14-ene-1-carboxylic acid*), $CH_3(CH_2)_7CH{=}CH(CH_2)_{13}CO_2H$, occurs in human brain-tissue and in fish oils. Both the *cis* and *trans* forms are known, but it appears to be the *cis* acid, m.p. 43°C, which occurs naturally.

Arachidonic acid (*nonadeca-4,7,10,13-tetraene-1-carboxylic acid*), $CH_3(CH_2)_3(CH_2CH{=}CH)_4CH_2)_3CO_2H$, has been isolated from phosphatides (p. 324).

Propiolic acid (*propargylic acid, prop-2-ynoic acid*), m.p. 9°C, is the simplest acetylenic carboxylic acid. It is conveniently prepared by the action of dry carbon dioxide on sodium acetylide:

$$CH{\equiv}CNa \xrightarrow{CO_2} CH{\equiv}CCO_2Na \xrightarrow[H_2SO_4]{\text{dilute}} CH{\equiv}CCO_2H$$

It is reduced by sodium amalgam to propionic acid and it adds on halogen acid to give the β-halogeno-acrylic acid. It forms the silver or cuprous salt when treated with an ammoniacal solution of silver nitrate or cuprous chloride, respectively. When exposed to sunlight, it polymerises to *trimesic acid* (benzene 1,3,5-tricarboxylic acid; *cf.* acetylene):

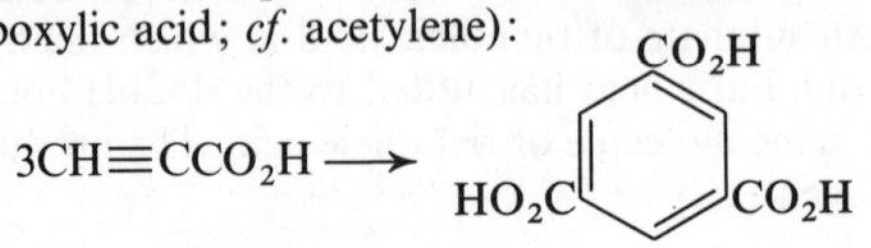

$$3CH{\equiv}CCO_2H \longrightarrow$$

α,β-Acetylenic acids are stronger acids than their ethylenic and saturated analogues due to the acetylene bond exerting a strong electron-attracting effect on the sp hybridised carbon atom.

The principle of vinylogy

According to Fuson (1935), the principle of vinylogy may be described as follows: If E_1 and E_2 represent non-metallic elements, then in the compound of the type $A-E_1=E_2$ or $A-E_1\equiv E_2$, if a structural unit of the type $\left(-\underset{|}{C}=\underset{|}{C}\right)_n$ is interposed between A and E_1, the function of E_2 remains

qualitatively unchanged, but that of E_1 may be usurped by the carbon atom attached to A. Thus

$$A-\left(\underset{|}{C}=\underset{|}{C}\right)_n-E_1=E_2 \text{ or } A-\left(\underset{|}{C}=\underset{|}{C}\right)_n-E_1\equiv E_2$$

form vinylogous series. This may be illustrated by the compounds ethyl acetate and ethyl crotonate. These are vinylogues (n being equal to 1):

$$CH_3-\underset{\underset{OC_2H_5}{|}}{C}=O \qquad CH_3-\overbrace{CH=CH}-\underset{\underset{OC_2H_5}{|}}{C}=O$$

A is equivalent to CH_3 and $E_1=E_2$ to $-\underset{\underset{OC_2H_5}{|}}{C}=O$

Ethyl acetate condenses with ethyl oxalate to form *oxalacetic ester* (this, of course, is an example of the Claisen condensation):

$$C_2H_5OOCCOOC_2H_5 + CH_3COOC_2H_5 \xrightarrow[\text{ether}]{Na} C_2H_5OOCCOCH_2COOC_2H_5 + C_2H_5OH$$

In the same way, ethyl crotonate also condenses with ethyl oxalate:

$$C_2H_5OOCCOOC_2H_5 + CH_3\ CH=CH\ COOC_2H_5 \xrightarrow[\text{ether}]{Na}$$
$$C_2H_5OOCCOCH_2\overbrace{CH=CH}COOC_2H_5 + C_2H_5OH$$

The principle of vinylogy is readily understood in terms of the inductive effect, *e.g.*,

$$CH_3CH=CHC\overset{\displaystyle O}{\underset{\displaystyle OEt}{\diagdown}}$$

Thus the methyl group behaves like an active methylene group due to the $-I$ effect of the carbethoxyl group, which is readily transmitted through the $-CH=CH-$ group. At the same time, the $-I$ effect is also assisted by the increased electronegativity of the sp^2 hybridised carbon atom adjacent to the methyl group.

Ketens (ketenes)

Ketens are compounds which are characterised by the presence of the grouping $\diagup\diagdown C=C=O$.

If the compound is of the type $RCH=C=O$, it is known as an *aldoketen*; and if $R_2C=C=O$, then a *ketoketen*.

Ketoketens are generally prepared by debrominating an α-bromoacyl bromide with zinc, *e.g.*, dimethylketen from α-bromoisobutyryl bromide:

$$(CH_3)_2CBrCOBr + Zn \longrightarrow (CH_3)_2C=C=O + ZnBr_2$$

Aldoketens are usually prepared by refluxing an acid chloride in pyridine solution, *e.g.*,

$$CH_3CH_2COCl + C_5H_5N \longrightarrow CH_3CH=C=O + C_5H_5NH^+Cl^-$$

The simplest member of the keten series is **keten** (*ketene*), $CH_2=C=O$, and may be prepared by debrominating bromoacetyl bromide with zinc:

$$CH_2BrCOBr + Zn \longrightarrow CH_2=C=O + ZnBr_2$$

It is, however, usually prepared by the thermal decomposition of acetone (ethyl acetate or acetic anhydride may also be used as the starting material):

$$CH_3COCH_3 \longrightarrow CH_2=C=CO + CH_4$$

The thermal decomposition of acetone has been shown to be a free-radical chain reaction.

Keten is a colourless, poisonous, pungent gas, b.p. $-41°C$, which oxidises in air to the unstable peroxide, the structure of which may be as shown. It rapidly polymerises to *diketen*, and undergoes photochemical decomposition to give methylene (p. 119).

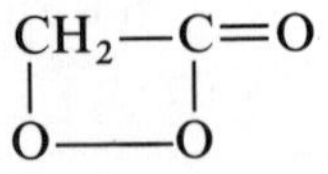

$$\begin{array}{cc} CH_2 & C=O \\ | & | \\ O & O \end{array}$$

The reactions of keten are generally those of an acid anhydride; it acetylates most compounds with an active hydrogen atom.

The mechanism of acetylations by keten can be generalised in terms of the addition of a nucleophile as the rate-determining step:

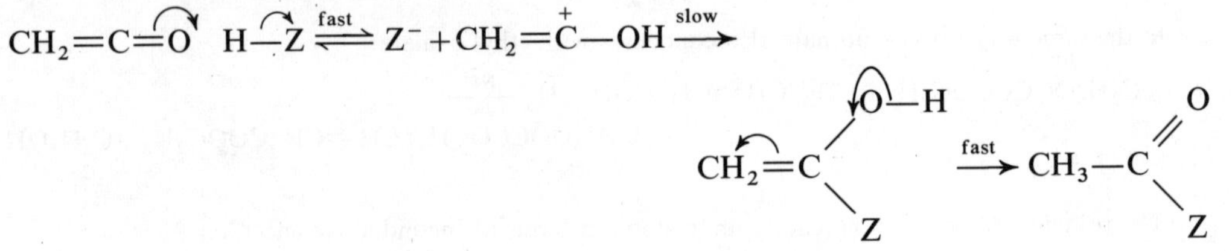

$$CH_2=C=O \quad H-Z \overset{fast}{\rightleftharpoons} Z^- + CH_2=\overset{+}{C}-OH \overset{slow}{\longrightarrow}$$

$$CH_2=C\overset{O-H}{\underset{Z}{\diagup}} \overset{fast}{\longrightarrow} CH_3-C\overset{O}{\underset{Z}{\diagup}}$$

(i) When keten is passed into water, acetic acid is formed slowly:

$$CH_2=CO + H_2O \longrightarrow CH_3CO_2H$$

(ii) When keten is passed into glacial acetic acid, acetic anhydride is formed:

$$CH_2=CO + CH_3CO_2H \longrightarrow (CH_3CO)_2O$$

By means of this reaction a mixed anhydride may be obtained, one group of which, of course, must be acetyl:

$$RCO_2H + CH_2=CO \longrightarrow RCOOCOCH_3$$

(iii) Keten reacts with aliphatic or aromatic hydroxy-compounds:

$$CH_2=CO + ROH \longrightarrow CH_3CO_2R$$

The reaction, however, is best carried out in the presence of a catalyst, *e.g.*, sulphuric acid. It also acetylates enolisable compounds, *e.g.*,

$$CH_3COCH_2CO_2C_2H_5 \xrightarrow[H_2SO_4]{CH_2=CO} CH_3C=CHCO_2C_2H_5 + CH_2=CCH_2CO_2C_2H_5$$
$$\underset{OCOCH_3}{\quad} \qquad \underset{OCOCH_3}{\quad}$$

(iv) Keten reacts with ammonia to form acetamide, and with primary or secondary amines to form *N*-alkyl-acetamides:

$$CH_3CONH_2 \xleftarrow{NH_3} CH_2{=}CO \xrightarrow{RNH_2} CH_3CONHR$$

Primary amino-groups are acetylated extremely readily, and hence it is possible to acetylate this group selectively in compounds containing both an amino and hydroxyl group, *e.g.*, *p*-aminophenol:

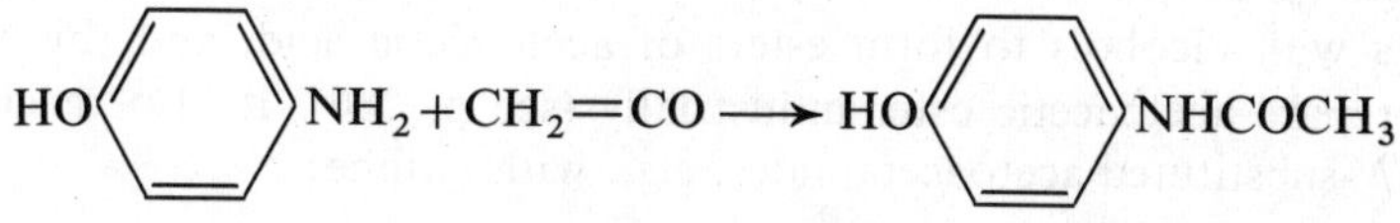

$$HO\text{—}\langle\!\!\!\bigcirc\!\!\!\rangle\text{—}NH_2 + CH_2{=}CO \longrightarrow HO\text{—}\langle\!\!\!\bigcirc\!\!\!\rangle\text{—}NHCOCH_3$$

A particularly useful reaction of this type is the acetylation of amino-acids:

$$RCH(NH_2)CO_2H + CH_2{=}CO \longrightarrow RCH(NHCOCH_3)CO_2H$$

Amides are also acetylated by keten in the presence of sulphuric acid, but at elevated temperatures (in the absence of sulphuric acid) the cyanide is formed.

$$RCONH_2 + CH_2{=}CO \longrightarrow RCN + CH_3CO_2H$$

(v) Thiols are acetylated (preferably in the presence of sulphuric acid as catalyst):

$$RSH + CH_2{=}CO \longrightarrow RSCOCH_3$$

(vi) Keten reacts with Grignard reagents to form methyl ketones:

$$RMgX + CH_2{=}CO \longrightarrow RCOCH_3$$

In addition to behaving as an acetylating reagent, keten behaves as an unsaturated compound, *e.g.*,

(i) Keten adds on bromine to form bromoacetyl bromide:

$$CH_2{=}CO + Br_2 \longrightarrow BrCH_2COBr$$

The halogeno-acetyl halide is also formed by the interaction of keten and phosphorus pentahalides:

$$CH_2{=}CO + PCl_5 \longrightarrow ClCH_2COCl + PCl_3$$

(ii) Keten adds on halogen acid to form the acetyl halide:

$$CH_2{=}CO + HX \longrightarrow CH_3COX$$

Alkyl halides also react with keten in the presence of charcoal at 100°C to form acid chlorides:

$$RX + CH_2{=}CO \longrightarrow RCH_2COX$$

This offers a means of stepping up a series.

(iii) Keten reacts with aldehydes to form β-lactones; *e.g.*, with formaldehyde, *β-propiolactone* is formed:

$$CH_2{=}CO + HCHO \longrightarrow \underset{\displaystyle \quad\;\;|\!\!\rule{0pt}{0pt}\text{O}\!\rule{0pt}{0pt}|}{CH_2CH_2CO}$$

Ketones also form β-lactones, but in this case a catalyst is usually required, *e.g.*, zinc chloride.

Diketen is readily formed by passing keten into acetone cooled in solid carbon dioxide (yield: 50–55 per cent). Diketen is a pungent-smelling liquid, b.p. 127°C, which, when strongly heated (550–600°C), is depolymerised to keten.

Various structures have been suggested for diketen, but based mainly on N.M.R. spectroscopy (which showed the presence of two methylene groups) and X-ray analysis, the structure accepted is that shown (vinylaceto-β-lactone). Diketen may be further polymerised to dehydroacetic acid (p. 288).

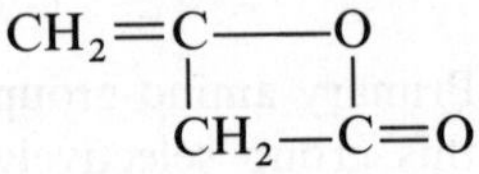

$$CH_2{=}C{-}\!\!-\!\!-O$$
$$CH_2{-}C{=}O$$

vinylaceto-β-lactone

Diketen reacts with alcohols to form esters of acetoacetic acid, and this reaction is now being used to prepare acetoacetic ester industrially (see p. 286). It also reacts with primary amines to form N-substituted acetoacetamides, $e.g.$, with aniline:

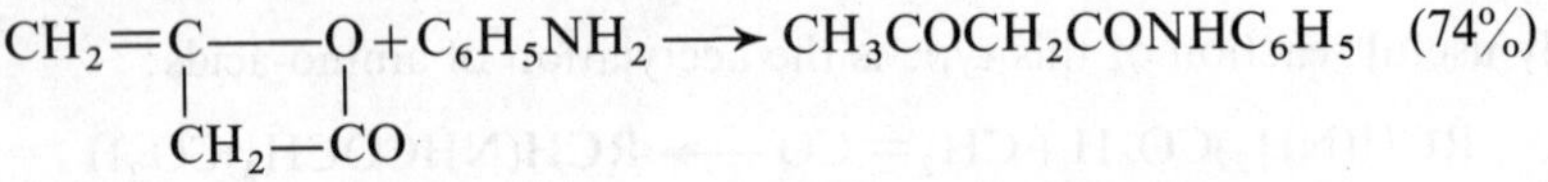

$$CH_2{=}C{-}\!\!-\!\!-O + C_6H_5NH_2 \longrightarrow CH_3COCH_2CONHC_6H_5 \quad (74\%)$$
$$CH_2{-}CO$$

Diketen also undergoes the Friedel–Crafts reaction with benzene to form benzoylacetone:

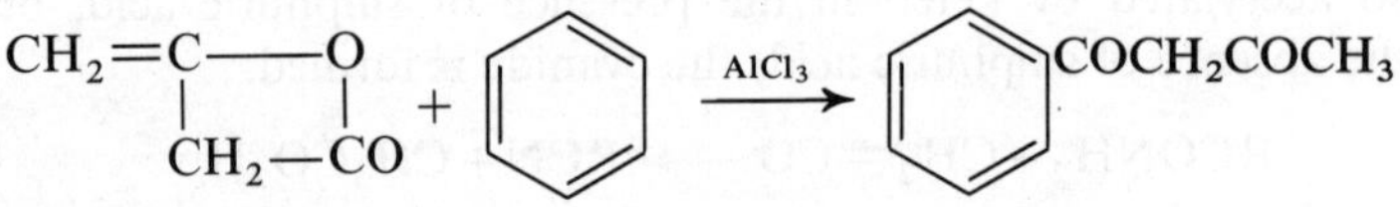

$$CH_2{=}C{-}\!\!-\!\!-O \quad + \quad \bigcirc \quad \xrightarrow{AlCl_3} \quad \bigcirc COCH_2COCH_3$$
$$CH_2{-}CO$$

PROBLEMS

1. Complete the following equations and comment where necessary:

(i) $\quad Me(C{\equiv}C)_3COCH{=}CH_2 \xrightarrow{NaBH_4} ? \xrightarrow{H^+} ?$

(ii) $\quad CH_2{-}\!\!\overset{O}{\triangle}\!\!{-}CH_2 + HCN \longrightarrow ? \xrightarrow[300°C]{Al_2O_3} ?$

(iii) $\quad CH_2{=}CHCl + Cl_2 \longrightarrow ? \xrightarrow[Ca(OH)_2]{aq.} ?$

(iv) $\quad CH_2{=}CHCN + EtOH \xrightarrow{H_2SO_4 - H_2O} ?$

(v) $\quad CH_2{=}CHCOCH_3 \xrightarrow{NaOCl} ?$

(vi) $\quad MeCH{=}CHCOMe \xrightarrow[H_2 \,(1\,atm)]{Pd - C} ?$

(vii) $\quad MeCH{=}CHCOEt \xrightarrow[press.]{Ni; H_2} ?$

(viii) $\quad MeCH{=}CHCO_2Et + HCN \longrightarrow ?$

(ix) $\quad CH_2{=}CHCO_2Et + EtOH \xrightarrow{EtONa} ?$

(x) $\quad CH_2{=}CHCOMe \xrightarrow{H_2O_2; \, OH^-} ?$

(xi) $\quad CH{\equiv}CNa + Cl(CH_2)_3I \xrightarrow[NH_3]{liq.} ? \xrightarrow[(ii)\,H^+]{(i)\,NaCN} ?$

(xii) $\quad MeCH{=}CHCOMe + NH_3 \longrightarrow ?$

(xiii) $\quad MeCH{=}CHCH_2Cl + ? \xrightarrow{?} MeCH{=}CHCH_2I$

(xiv) $\quad MeCH{=}CHCO_2Et + CH_3NO_2 \xrightarrow{EtONa} ?$

(xv) $\quad MeCH{=}CHCHO + HCN \xrightarrow{OH^-} ?$

(xvi) $\quad MeCH{=}CHCHO \xrightarrow{LAH} ?$

(xvii) $\quad PhCHO + ? \xrightarrow{pyridine} PhCH{=}CHCO_2H$

(xviii) $EtCHO + 2CH_2(CO_2Et)_2 \xrightarrow{\text{piperidine}}$?

(xix) $EtCHO + CH_2(CO_2H)_2 \xrightarrow{\text{pyridine}}$?

(xx) $EtCH{=}CHCH_2CONH_2 \xrightarrow{\text{NaOCl/NaOH}}$? $\xrightarrow{\text{HCl}}$?

(xxi) 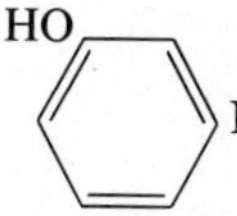
HO

⬡ $NH_2 + CH_2{=}CO \longrightarrow$?

2. A compound has structure $MeCH{=}CHCH_2CHO$ (**A**) or $MeCH_2CH{=}CHCHO$ (**B**). Distinguish between the two by (i) a chemical method; (ii) i.r.; (iii) NMR.

3. Explain the following:

(i) μ of acraldehyde $>$ μ of propionaldehyde; (ii) μ of ethyl bromide $>$ μ of vinyl bromide.

4. Complete the following equation for:

(i) $B_{AC}2$; (ii) $B_{AL}1$; (iii) $A_{AC}1$ mechanism.

$$CH_2{=}CHCHMeOOCCH_3 \xrightarrow{\text{OH}^- \text{ or H}^+} ?$$

5. How could you demonstrate the S_N1' mechanism in the reaction

$$CH_2{=}CHCH_2OH + HCl \longrightarrow CH_2{=}CHCH_2Cl ?$$

6. Calculate $\Delta H^\ominus$ for the 1,2- and 1,4-additions of HZ to $C{=}C{-}C{=}O$. Which addition is favoured?

7. Draw the M.O. for the group $C{=}C{-}C{=}O$, and suggest a reason for the 1,4-addition properties.

8. Convert $MeCH{=}CHCO_2H$ into $MeCH(CH_2CO_2Et)_2$.

9. Synthesise $EtO_2CCH{=}CHCH_2COCO_2Et$ starting from compounds not containing more than *four* carbon atoms.

10. Aldol may undergo dehydration in two possible ways:

$$CH_2{=}CHCH_2CHO \xleftarrow{-H_2O} MeCHOHCH_2CHO \xrightarrow{-H_2O} MeCH{=}CHCHO$$

One of these products is obtained almost exclusively. Which one and why?

11. 2,2-Dimethylbut-3-enoic acid, on warming, gives trimethylethylene. Suggest a mechanism.

12. Suggest a synthesis for:

(i)
$Me_2CCH_2CO_2H$
$|$
$Me_2CCH_2CO_2H$
; (ii) $Me_2C(CH_2CO_2H)_2$; (iii) $CH_3CH{=}CHCH_2CH_2OH$.

REFERENCES

BENT, 'Distribution of Atomic *s* Character in Molecules and its Chemical Implications', *J. Chem. Educ.*, 1960, **37**, 616.

Symposium. An Epistologue on the Effect of Environment on the Properties of Carbon Bonds, *Tetrahedron*, 1962, **17**, 123.

DEWAR, *Hyperconjugation*, Ronald Press (1962).

PATAI (ed.), *The Chemistry of Alkenes*, Interscience (1964), Ch. 10, 'Allylic Reactions'.

Organic Reactions, Wiley, Vol. V (1949), Ch. 2, 'Cyanoethylation': Vol. X (1959), Ch. 3, 'The Michael Reaction': Vol. XV (1967), Ch. 2, 'The Knoevenagel Condensation'.

LACEY, 'Ketene in Organic Syntheses', *Advances in Organic Chemistry*, Interscience, Vol. II (1960), p. 217.

YOUNG, 'Unexpected Rearrangement and Lack of Rearrangement in Allylic Systems', *J. Chem. Educ.*, 1962, **39**, 455.

LEE and UFF, 'Organic Reactions Involving Electrophilic Oxygen', *Quart. Rev.*, 1967, **21**, 429.

13 Nitrogen compounds

Some compounds containing nitrogen have been described in Chapter 9, since these were regarded primarily as acid derivatives. It is now proposed to deal with many other nitrogen compounds, most of which may be regarded as alkyl nitrogen compounds.

Hydrocyanic acid, **hydrogen cyanide** (*prussic acid*) b.p. 26°C, was discovered by Scheele (1782), who obtained it from bitter almonds which contain the glycoside *amygdalin*. Amygdalin, on hydrolysis with dilute acid, yields hydrogen cyanide, benzaldehyde and glucose:

$$C_{20}H_{27}O_{11}N + 2H_2O \xrightarrow{\text{acid}} HCN + C_6H_5CHO + 2C_6H_{12}O_6$$

Hydrogen cyanide also occurs in the leaves of certain plants, *e.g.*, laurel. It may be prepared by heating sodium cyanide with concentrated sulphuric acid, the gas being dried over calcium chloride:

$$NcCN + H_2SO_4 \longrightarrow HCN + NaHSO_4 \quad (93-97\%)$$

Hydrogen cyanide is prepared industrially in several ways, *e.g.*, by passing a mixture of ammonia, air and methane over a platinum–rhodium gauze catalyst at 1 000°C:

$$2NH_3 + 3O_2 + 2CH_4 \longrightarrow 2HCN + 6H_2O$$

Hydrogen cyanide is a very weak acid, and hydrolyses slowly in aqueous solution, and more rapidly in the presence of inorganic acids to form first formamide, and then ammonium formate:

$$H-C\equiv N \xrightarrow{H_2O} HCONH_2 \xrightarrow{H_2O} HCO_2NH_4$$

Two types of organic derivatives of hydrocyanic acid are known, alkyl cyanides, RCN, and alkyl isocyanides, RNC. Their formation from cyanides of the alkali metals may be explained as follows. Alkali cyanides are ionic ($:C\equiv N^-:$), and the cyanide ion is ambident and so can form a covalent bond either from the carbon or the nitrogen. Silver cyanide, $AgC\equiv N$, is essentially covalent, and hence the lone pair on the nitrogen is mainly available for covalent bond formation, resulting in the predominant formation of isocyanides.

Alkyl cyanides

These are also known as **nitriles** or **carbonitriles**.

Nomenclature

This group of compounds is usually named either as the *alkyl cyanides* (*i.e.*, the alkyl derivatives of hydrogen cyanide), or as the *nitrile* of the acid which is produced on hydrolysis, the suffix *-ic* of the trivial name being replaced by *-onitrile*, *e.g.*,

HCN	hydrogen cyanide or formonitrile
CH$_3$CN	methyl cyanide or acetonitrile
(CH$_3$)$_2$CHCN	isopropyl cyanide or isobutyronitrile

General methods of preparation

1. By the dehydration of acid amides with phosphorus pentoxide (see p. 264). High molecular weight acid amides are dehydrated to the corresponding cyanide by heat alone (p. 264).

Cyanides are prepared industrially by passing a mixture of carboxylic acid and ammonia over alumina at 500°C. This reaction probably occurs as follows:

$$RCO_2H + NH_3 \longrightarrow RCO_2NH_4 \xrightarrow{Al_2O_3} H_2O + RCONH_2 \xrightarrow{Al_2O_3} H_2O + RCN$$

Another industrial method is the *ammoxidation* of alkanes:

$$2RCH_3 + 2NH_3 + 3O_2 \xrightarrow[500-600°C]{cat.} 2RCN + 6H_2O$$

This term describes the process involving oxidation in the presence of ammonia.

2. By the dehydration of aldoximes with phosphorus pentoxide, or better, with acetic anhydride:

$$RCH{=}NOH \xrightarrow[-H_2O]{(CH_3CO)_2O} RCN \quad (g.)$$

Aldehydes may be converted directly to cyanides by refluxing a solution of an aldehyde in formic acid with hydroxylamine hydrochloride and sodium acetate (30–42 per cent; van Es, 1965).
3. The most convenient method is to heat an alkyl halide with potassium cyanide in aqueous ethanolic solution; a small amount of isocyanide is also obtained:

$$RX + KCN \longrightarrow RCN + KX \quad (g.)$$

This method is satisfactory only if R is a primary or secondary alkyl group. If it is a tertiary group very little cyanide is obtained, the tertiary halide being converted into the corresponding alkene (*cf.* p. 158). Tertiary alkyl cyanides are best prepared by method 4 below. Cyanide yield from primary chlorides has been much improved by using dimethyl sulphoxide or dimethylformamide as solvent (Smiley *et al.*, 1960; Friedman *et al.*, 1960).

Many cyanides, particularly the lower members, may be readily prepared by warming the potassium alkyl sulphate with potassium cyanide (some isocyanide is also obtained):

$$ROSO_2OK + KCN \longrightarrow RCN + K_2SO_4 \quad (g.-v.g.)$$

Tosylates, *i.e.*, *p*-toluenesulphonates (see p. 694), are also useful for cyanide preparation:

$$p\text{-}CH_3C_6H_4SO_3R + KCN \longrightarrow RCN + p\text{-}CH_3C_6H_4SO_3K$$

4. By the interaction of a Grignard reagent and cyanogen chloride:

$$RMgCl + ClCN \longrightarrow RCN + MgCl_2$$

This is the best method of preparing tertiary alkyl cyanides.

General properties

The alkyl cyanides are stable neutral substances with fairly pleasant smells, and are not as poisonous as hydrogen cyanide. The lower members are liquids which are soluble in water (with which they can form hydrogen bonds), but the solubility diminishes as the molecular weight increases. Since they cannot form intermolecular hydrogen bonds, alkyl cyanides are much more volatile than their corresponding acids.

The infra-red absorption band of the cyano group (C—N stretch) in alkyl cyanides is $2\,260$–$2\,240\ \text{cm}^{-1}$ (w–m). Figure 13.1 is the infra-red spectrum of methyl cyanide (as a film).

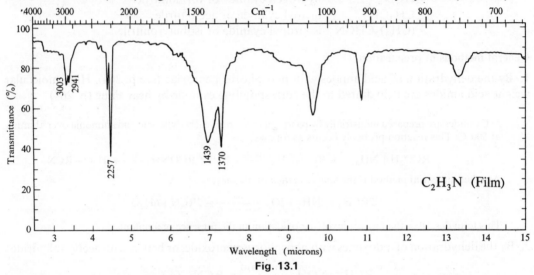

Fig. 13.1

$2257\ \text{cm}^{-1}$: C≡N (str.) in saturated alkyl cyanides. 3003 and $2941\ \text{cm}^{-1}$: C—H (str.) in Me (and/or CH$_2$). $1439\ \text{cm}^{-1}$: C—H (def.) in Me (and/or CH$_2$). $1370\ \text{cm}^{-1}$: C—H (def.) in Me.

Reactions

1. The alkyl cyanides are hydrolysed by acids or alkalis to the corresponding acid via the intermediate formation of an amide (see also p. 262):

$$RC{\equiv}N \xrightarrow{H_2O} \left[R-C \begin{smallmatrix} \diagup OH \\ \diagdown NH \end{smallmatrix} \right] \longrightarrow RCONH_2 \xrightarrow{H_2O} RCO_2H + NH_3 \quad (g.\text{-}v.g.)$$

When a solution of alkyl cyanide in an alcohol is heated with concentrated sulphuric acid or hydrochloric acid, the *ester* is obtained:

$$R^1CN + R^2OH + H_2O \longrightarrow R^1CO_2R^2 + NH_3$$

If dry hydrogen chloride is passed into the solution of an alkyl cyanide in anhydrous alcohol, the *imidic ester hydrochloride* is formed (p. 267), and when treated with *dry* hydrogen chloride, alkyl cyanides form the *imidochloride*:

$$RC{\equiv}N + 2HCl \longrightarrow RCCl{=}NH_2{}^+Cl^-$$

Alkyl cyanides combine with dry ammonia to form *amidines* (p. 267) and with acid anhydrides on heating to form *tertiary acid amides*.

$$RC(=NH)NH_2 \xleftarrow{NH_3} RCN \xrightarrow{(RCO)_2O} (RCO)_3N$$

2. Alkyl cyanides may be converted into aldehydes (p. 206), and are reduced by sodium and ethanol or metal hydrides (Table 6.1, p. 179) to primary amines.

$$RCN \longrightarrow RCH_2NH_2$$

Some secondary amine, $(RCH_2)_2NH$, is also formed. Catalytic reduction of an alkyl cyanide also produces a primary amine accompanied by the secondary amine, but in this case the amount of the latter is greater than when the reduction is carried out with dissolving metals. Von Braun and co-workers (1933) suggested that the mechanism of the formation of secondary amine is via the imine formed first:

$$RC{\equiv}N \xrightarrow[Ni]{H_2} RCH{=}NH \xrightarrow{H_2} RCH_2NH_2$$

$$RCH{=}NH + RCH_2NH_2 \longrightarrow \underset{\underset{NH_2}{|}}{RCHNHCH_2R} \longrightarrow NH_3 + RCH{=}NCH_2R \xrightarrow{H_2} RCH_2NHCH_2R$$

Support for this mechanism is to be found in the fact that the formation of secondary amine is prevented if the reduction is carried out with a Raney nickel catalyst *in the presence of ammonia*.

Alternatively, the formation of secondary amine may be prevented by carrying out the hydrogenation in acetic anhydride solution containing sodium acetate (Gould *et al.*, 1960). Under these conditions, the primary amine formed is acetylated; this acetyl derivative cannot react with the imine.

3. Alkyl cyanides react with Grignard reagents to form ketones:

$$R^1C{\equiv}N + R^2MgX \longrightarrow R^1R^2C{=}NMgX \xrightarrow{H^+} R^1COR^2$$

4. Alkyl cyanides undergo condensation reactions in the presence of sodium, only the α-hydrogen atoms being involved. These condensations are to be expected in view of the fact that the cyano group is strongly electron-attracting. When the reaction is carried out in ether, two molecules of cyanide condense:

$$CH_3CH_2C{\equiv}N + CH_3CH_2CN \xrightarrow[ether]{Na} CH_3CH_2\underset{\overset{||}{NH}}{C}{-}\underset{\overset{|}{CH_3}}{CH}{-}CN$$

This is known as the **Thorpe nitrile condensation** (1904).

On the other hand, alkyl cyanides with α-hydrogen atoms can condense with esters if the reaction is carried out in ether solution in the presence of sodamide (Levine and Hauser, 1946); *e.g.*, methyl cyanide condenses with ethyl propionate to form propionylmethyl cyanide:

$$CH_3CN + NaNH_2 \longrightarrow [:\bar{C}H_2CN]Na^+ + NH_3 \xrightarrow{CH_3CH_2CO_2C_2H_5}$$

$$CH_3CH_2COCH_2CN + C_2H_5OH \quad (40\%)$$

Alkyl isocyanides

These are also known as **isonitriles** or **carbylamines**.

General methods of preparation

1. By heating an alkyl iodide with silver cyanide in aqueous ethanolic solution; a small amount of cyanide is also formed (see p. 354):

$$RI + AgCN \longrightarrow RNC + AgI$$

2. By heating a mixture of a primary amine and chloroform with ethanolic potassium hydroxide:

$$RNH_2 + CHCl_3 + 3KOH \longrightarrow RNC + 3KCl + 3H_2O$$

The mechanism of the isocyanide reaction is not certain. Robinson (1961) has suggested that the reaction proceeds via the intermediate formation of dichloromethylene, which is produced from the chloroform in alkaline solution (p. 170):

$$\overset{\frown}{RN H_2} \;\; \overset{..}{C Cl_2} \longrightarrow R\overset{+}{N} H_2 - \overset{-}{C}Cl_2 \xrightarrow{-H^+} RNH - \overset{-}{C}Cl_2 \xrightarrow[-H^+]{-Cl^-} RN = \overset{-}{C}Cl \xrightarrow{-Cl^-} RN \equiv C$$

3. Dehydration of *N*-substituted formamides with phosphoryl chloride in pyridine produces isocyanides (Ugi *et al.*, 1960).

$$RNHCHO \xrightarrow{POCl_3} RNC \quad (g.)$$

General properties

The alkyl isocyanides are poisonous, most unpleasant-smelling liquids, with lower boiling points than the isomeric cyanides. They are not very soluble in water, the nitrogen atom not having a lone pair of electrons available for hydrogen bonding. Alkyl isocyanides absorb in the region 2185–$2120 \text{ cm}^{-1}(s)$.

Reactions

1. Alkyl isocyanides are hydrolysed to an amine and formic acid by dilute acids, *but are not hydrolysed by alkalis*:

$$RNC + 2H_2O \xrightarrow{acid} RNH_2 + HCO_2H$$

2. Alkyl isocyanides are reduced to secondary amines either by dissolving metals or by catalytic reduction:

$$RNC \longrightarrow RNHCH_3$$

3. Alkyl isocyanides add on halogen to form *alkyliminocarbonyl halides*, sulphur to form *alkyl isothiocyanates*, and are readily oxidised by mercuric oxide to *alkyl isocyanates*:

$$RNC + X_2 \longrightarrow RNCX_2$$
$$RNC + S \longrightarrow RNCS$$
$$RNC + HgO \longrightarrow RNCO + Hg$$

4. When alkyl isocyanides are heated for a long time, they rearrange to the cyanide:

$$RNC \longrightarrow RCN$$

Cyanogen (*ethanedinitrile*) may be prepared by heating the cyanides of mercury or silver:

$$Hg(CN)_2 \longrightarrow (CN)_2 + Hg$$

Cyanogen is a poisonous, colourless gas, b.p. $-21°C$, and is stable in spite of the fact that its heat of formation is $+292 \cdot 9 \text{ kJ mol}^{-1}$. It is hydrolysed by water, the main products being hydrocyanic acid and cyanic acid, and minor products being oxamide and ammonium formate.

$$(CN)_2 \xrightarrow{H_2O} HCN + HNCO + (CONH_2)_2 + HCO_2NH_4$$

When cyanogen is heated at $400°C$, it polymerises to *paracyanogen*, which, at $800°C$, regenerates cyanogen. The structure of cyanogen is probably best represented as a resonance hybrid, (I) being the most important:

$$\underset{\text{(I)}}{:N \equiv C - C \equiv N:} \longleftrightarrow \underset{\text{(II)}}{:\overset{+}{N} = C = C = \overset{-}{N}:} \longleftrightarrow \underset{\text{(III)}}{:\overset{-}{N} = C = C = \overset{+}{N}:}$$

Cyanogen halides, XCN (X=Cl) (b.p. 12°C), Br (m.p. 52°), or I (m.p. 146·5°C) are readily prepared by the action of halogen on sodium cyanide:

$$NaCN + X_2 \longrightarrow XCN + NaX$$

When pure, cyanogen halides are stable, but in the presence of impurities they readily trimerise to cyanuric halides, $X_3C_3N_3$ (cyclic compounds). They are stable to water, but are hydrolysed by alkalis:

$$XCN + 2KOH \longrightarrow KNCO + KX + H_2O$$

They readily react with amines to form alkyl cyanamides (see also p. 378), *e.g.*,

$$2RNH_2 + XCN \longrightarrow RNHCN + RNH_3{}^+X^-$$

Cyanic acid, HNCO. Urea, on dry distillation, gives *cyanuric acid*:

$$3CO(NH_2)_2 \longrightarrow H_3N_3C_3O_3 + 3NH_3$$

Cyanuric acid is a colourless, crystalline solid, not very soluble in water, and is strongly acid, reacting as a mono-, di- and tribasic acid. It has a cyclic structure (a *triazine* derivative) and X-ray analysis indicates that the acid is best represented as a resonance hybrid.

When cyanuric acid is heated, it does not melt but decomposes into cyanic acid vapour which, when condensed below 0°C, gives a colourless condensate:

$$H_3N_3C_3O_3 \longrightarrow 3HNCO$$

Cyanic acid is a colourless, volatile, strongly acid liquid which, above 0°C, readily polymerises to cyanuric acid and *cyamelide*, $(HNCO)_n$. Aqueous solutions of cyanic acid rapidly hydrolyse to give carbon dioxide and ammonia:

$$HNCO + H_2O \longrightarrow CO_2 + NH_3$$

There are two possible structures for cyanic acid, $H-O-C\equiv N$ and $H-N=C=O$. There is very little evidence for the existence of the former (cyanic acid), *e.g.*, infra-red absorption spectra indicate the latter (isocyanic acid).

When a solution of potassium isocyanate (usually called potassium cyanate) and ammonium sulphate is evaporated to dryness, urea and potassium sulphate are obtained, the urea being formed by the molecular rearrangement of the intermediate product ammonium isocyanate:

$$NH_4NCO \longrightarrow CO(NH_2)_2$$

Cyanic acid reacts with alcohols to form *urethans*; excess of cyanic acid converts urethans into *allophanates*, which are esters of the unknown acid, *allophanic acid*, $NH_2CONHCO_2H$.

$$ROH + HNCO \longrightarrow NH_2CO_2R \xrightarrow{\text{HNCO}} NH_2CONHCO_2R$$

N-substituted urethans are formed by the reaction between alkyl or aryl isocyanates and an alcohol (see p. 459). Phenyl isocyanate is also used to characterise primary and secondary amines, with which it forms substituted ureas (see p. 463).

Phenyl isocyanate is usually prepared by the action of carbonyl chloride on aniline:

$$C_6H_5NH_2 + COCl_2 \longrightarrow HCl + C_2H_5NHCOCl \xrightarrow{\text{heat}} HCl + C_6H_5NCO$$

This is a general method for preparing isocyanates; another general method is the Curtius reaction (p. 268).

Di-isocyanates, *e.g.*, hexamethylene di-isocyanate, $OCN(CH_2)_6NCO$, are used to prepare polyurethan plastics by reaction with di- and polyhydroxyalcohols, *e.g.*, tetramethylene glycol:

$$HO(CH_2)_4OH + OCN(CH_2)_6NCO + HO(CH_2)_4OH + \ldots \longrightarrow$$
$$HO(CH_2)_4OCONH(CH_2)_6NHCOO(CH_2)_4O- \ldots$$

Cyanamide, m.p. 42°C, may be prepared by the action of ammonia on cyanogen chloride:

$$ClCN + NH_3 \longrightarrow NH_2CN + HCl$$

It is also readily prepared by the action of water and carbon dioxide on calcium cyanamide (see below).

Cyanamide is converted into *guanidine* (*q.v.*) by ammonia, and into *thiourea* (*q.v.*) by hydrogen sulphide. When cyanamide is melted it forms the dimer *dicyandiamide*, $(NH_2)_2C=NCN$, and the trimer *melamine*, which is a cyclic compound. X-ray analysis has shown that all the C—N bond lengths are the same (Hughes, 1941). This can be explained on the assumption that melamine is a resonance hybrid of the amino-form (*cf.* cyanuric acid).

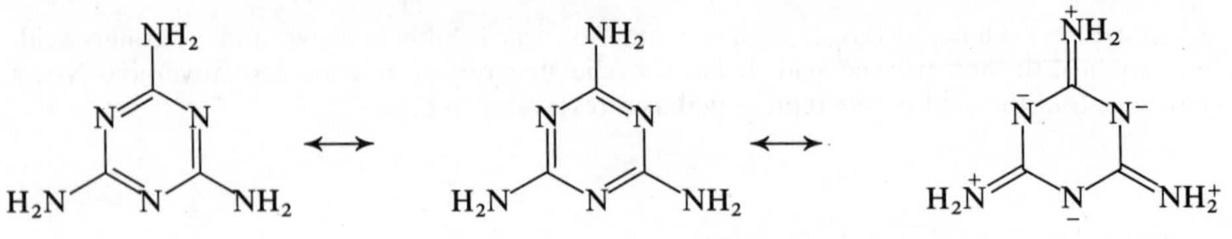

Melamine is manufactured by heating urea and passing the gaseous products over a heated catalyst:

$$CO(NH_2)_2 \xrightarrow{-NH_3} HNCO \xrightarrow{6 \text{ mol.}} C_3H_6N_6 + 3CO_2$$

Melamine is used for making melamine–formaldehyde plastics.

Cyanamide itself is a tautomeric compound (amidine system), and Raman spectra investigations have shown that two forms are present in equilibrium in the solid or fused state, or in solution:

$$H_2N—C\equiv N \rightleftharpoons HN=C=NH$$

Cyanamide forms salts, the most important of which is the calcium salt. This is manufactured by heating calcium carbide mixed with 10 per cent its weight of calcium chloride in a stream of nitrogen at 800°C:

$$CaC_2 + N_2 \longrightarrow CaNCN + C$$

The calcium cyanamide–carbon mixture is used as a fertiliser; it is hydrolysed in the soil to cyanamide, which is then hydrolysed to urea, which, in turn, is converted into ammonium carbonate by bacteria in the soil:

$$CaNCN \xrightarrow{H_2O} Ca(OH)_2 + NH_2CN \xrightarrow{H_2O} CO(NH_2)_2 \xrightarrow{H_2O} (NH_4)_2CO_3$$

Calcium cyanamide is also used to prepare urea industrially.

Carbodi-imides. These are most conveniently prepared by heating disubstituted thioureas with yellow mercuric oxide:

$$RNHCSNHR + HgO \longrightarrow RN=C=NR + HgS + H_2O$$

Carbodi-imides are very powerful dehydrating agents and are used for removing the elements of water in various reactions, the most common one used for this purpose being *dicyclohexylcarbodi-imide*, m.p. 20°C, *e.g.*,

$$2RCO_2H + C_6H_{11}N=C=NC_6H_{11} \longrightarrow (RCO)_2O + (C_6H_{11}NH)_2CO$$

Nitroalkanes

Nomenclature

The nitroalkanes are named as the nitro-derivative of the corresponding alkane, the positions of the nitro-groups being indicated by numbers, *e.g.*,

$$CH_3NO_2 \qquad \text{nitromethane}$$

$$\underset{CH_3CHCH_2CHCH_2CH_3}{\overset{NO_2 \quad NO_2}{|\qquad\;\;|}} \quad \text{2,4-dinitrohexane}$$

The nitroalkanes, the structure of which is RNO_2, are isomeric with the alkyl nitrites, RONO. The evidence that may be adduced for these respective formulæ is to be found in the study of the reactions of these two groups of compounds. The reactions of the nitrites have already been described (p. 255), and those of the nitroalkanes are described below. It is, however, worth while at this stage to mention the reaction which most clearly indicates their respective structures, *viz*., reduction. When alkyl nitrites are reduced, an alcohol and ammonia or hydroxylamine are formed. This shows that the alkyl group in nitrites is attached to an oxygen atom:

$$R—ONO \xrightarrow{e; H^+} ROH + NH_3 + H_2O$$

On the other hand, when nitroalkanes are reduced, a primary amine is formed. This shows that the alkyl group is attached to the nitrogen atom, since the structure of a primary amine is known to be $R—NH_2$ from its method of preparation (see below):

$$RNO_2 \xrightarrow{e; H^+} RNH_2 + 2H_2O$$

The structure of the nitro-compounds is best represented as a resonance hybrid:

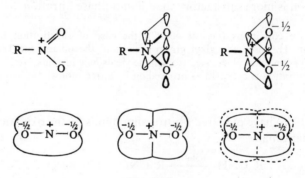

For most purposes, however, the non-committal formula RNO_2, is satisfactory.

From the M.O. point of view, the nitro-group is conjugated (Fig. 13.2), and delocalisation of bonds increases its stability (the two oxygen atoms are equivalent; *cf*. the carboxylate ion, p. 209).

Fig. 13.2

General methods of preparation

1. By heating an alkyl halide with silver nitrite in aqueous ethanolic solution:

$$RX + AgONO \longrightarrow RNO_2 + RONO + AgX$$

This method is only useful for the preparation of primary nitroalkanes. With *s*-halides the yield is about 15 per cent, and with *t*-halides 0–5 per cent; the yield of nitrite increases from primary to tertiary halides (Kornblum *et al*., 1954, 1955). Kornblum *et al*., (1956) have shown that sodium nitrite reacts with alkyl halides to give good yields of nitroalkane (55–62 per cent) together with alkyl nitrites (25–33 per cent). The success of the reaction depends on the use of dimethylformamide as solvent, and the addition of urea increases the solubility of sodium nitrite. Alkyl bromides and iodides are most satisfactory; the chlorides react too slowly to be useful. On the other hand, if dimethyl sulphoxide is used as solvent instead of D.M.F., the addition of urea is unnecessary (Smiley *et al.*, 1960; Friedman *et al.*, 1960). It should be noted that the nitrite ion is ambident.

Nitromethane may be prepared by boiling an aqueous solution of sodium nitrite with a halogeno-acetic acid, *e.g.*,

$$ClCH_2CO_2Na + NaNO_2 \longrightarrow NaCl + [CH_2(NO_2)CO_2H] \longrightarrow CH_3NO_2 + CO_2 \quad (35\text{–}38\%)$$

The ready decarboxylation of the intermediate nitroacetic acid may be explained on the assumption that the electron-attracting nitro-group facilitates the loss of carbon dioxide (*cf.* p. 274):

2. Nitro-compounds are prepared industrially by direct nitration, which is carried out in two ways.

In *liquid-phase nitration*, the hydrocarbon is heated with concentrated nitric acid under pressure at 140°C. Nitration under these conditions is always slow, and a large amount of polynitro-compounds is produced.

In *vapour-phase nitration*, the hydrocarbon is heated with nitric acid (or with oxides of nitrogen) at 150–475°C; each hydrocarbon has its optimum temperature, *e.g.*,

$$CH_3CH_2CH_3 \xrightarrow[400°C]{HNO_3} CH_3CH_2CH_2NO_2 + CH_3CH(NO_2)CH_3 + CH_3CH_2NO_2 + CH_3NO_2$$

Vapour-phase nitration is more satisfactory than liquid-phase nitration.

Under the above conditions (*i.e.*, at 400°C), the ease of substitution by the nitro group is t-H > s-H > prim. H. Also, any alkyl group present in the alkane can be replaced by a nitro-group, *i.e.*, chain fission takes place; *e.g.*, isopentane yields nine nitroalkanes.

Vapour-phase nitration is accepted as proceeding by a free-radical mechanism, but the details are uncertain.

3. Kornblum *et al.* (1956) have prepared t-nitro-compounds by the oxidation of t-carbinamines with potassium permanganate:

$$R_3CNH_2 \xrightarrow{KMnO_4} R_3CNO_2 \quad (70\text{–}80\%)$$

4. The addition of 'nitrous fumes' to alkenes is complicated, and much of the earlier work is of a doubtful nature. Stevens (1960) has carried a detailed investigation of the addition of pure dinitrogen tetroxide in ether to 1-alkenes in the presence of oxygen, and showed that the major products are the dinitro-compound (I), nitro-nitrite (II), nitro-nitrate (III) and nitro-ketone (IV).

$$RCH{=}CH_2 \xrightarrow[O_2]{N_2O_4} \underset{\underset{X}{|}}{RCHCH_2NO_2} + RCOCH_2NO_2$$
$$\qquad\qquad\qquad\qquad\qquad\qquad (IV)$$

(I), X=NO$_2$; (II), X=ONO; (III), X=ONO$_2$. It is the oxygen which is responsible for the formation of (III) and (IV), both being formed at the expense of (I). According to Schechter *et al.* (1952–55) and Stevens *et al.* (1958), (I) and (II) are formed by the addition of the nitronium free-radical:

$$N_2O_4 \rightleftharpoons 2 \cdot NO_2$$

$$RCH{=}CH_2 \xrightarrow{\cdot NO_2} R\dot{C}H{-}CH_2NO_2 \xrightarrow{\cdot NO_2} \underset{\underset{NO_2}{|}}{RCHCH_2NO_2} + \underset{\underset{ONO}{|}}{RCHCH_2NO_2}$$

General properties

The nitroalkanes are colourless (when pure), pleasant-smelling liquids which are sparingly soluble in water. Most of them can be distilled at atmospheric pressure.

There are two infra-red absorption bands of the nitro-group (NO_2 vibrations) in alkyl primary and secondary nitro-compounds, 1565–1545 (s) and 1385–1360 cm^{-1} (s), and for the tertiary nitro-compounds, 1545–1530 (s) and 1360–1340 cm^{-1} (s). These regions clearly distinguish nitro-compounds from alkyl nitrites, in which the nitroso-group (N=O stretch; R—O—N=O) absorbs in the region 1680–1610 cm^{-1} (v.s.).

> Witanowski *et al.* (1964), from their NMR studies, have shown that it is possible to identify prim.-, s- and t-nitroalkanes (*cf.* Table 3.2, p. 80).

The infra-red spectrum of nitroethane (as a film) is shown in Fig. 13.3.

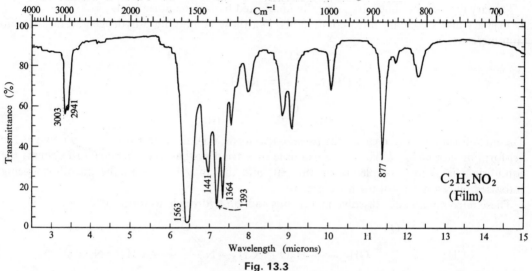

Fig. 13.3

1563 and 1393 cm^{-1}: NO_2 vibrations in alkyl nitro-compounds. 877 cm^{-1}: C—N (str.) or NO_2 (def.). 3003 and 2941 cm^{-1}: C—H (str.) in Me and/or CH_2. 1441 cm^{-1}: C—H (def.) in Me and/or CH_2. 1364 cm^{-1}: C—H (def.) in Me.

Reactions

1. Nitroalkanes are reduced catalytically, by lithium aluminium hydride, or by metal in *acid* solution to primary amines.

$$RNO_2 \longrightarrow RNH_2 + 2H_2O$$

When the reduction with metal is carried out in *neutral* solution, *e.g.*, with zinc dust and ammonium chloride solution, nitro-compounds are converted into an *N*-alkylhydroxylamine.

$$RNO_2 \xrightarrow{Zn/NH_4Cl} RNHOH + H_2O$$

When stannous chloride and hydrochloric acid are used as the reducing agent, nitro-compounds are converted into a mixture of *N*-alkylhydroxylamine and oxime:

$$RCH_2NO_2 \xrightarrow{SnCl_2/HCl} RCH_2NHOH + RCH=NOH$$

Since hydroxylamine can give rise to two types of derivatives, *e.g.*, RNHOH and NH_2OR, it is necessary to distinguish one from the other. A common method is to name the former as the

N-alkylhydroxylamine, and the latter as the *O*-alkylhydroxylamine, the capital letters *N* and *O*, respectively, indicating where the alkyl group is attached in the molecule.

2. Primary nitro-compounds are hydrolysed by boiling hydrochloric acid or by 85 per cent sulphuric acid to a carboxylic acid and hydroxylamine:

$$RCH_2NO_2 + H_2O \xrightarrow{\text{HCl}} RCO_2H + NH_2OH$$

This reaction is used for manufacturing hydroxylamine.

Secondary nitro-compounds are hydrolysed by boiling hydrochloric acid to ketones and nitrous oxide:

$$2R_2CHNO_2 \xrightarrow{\text{HCl}} 2R_2CO + N_2O + H_2O$$

Tertiary nitro-compounds are generally unaffected by hydrochloric acid.

3. Primary and secondary nitro-compounds, *i.e.*, those containing α-hydrogen atoms, exhibit tautomerism:

$$RCH_2\overset{+}{N}\underset{O^-}{\overset{O}{\diagup}} \rightleftharpoons RCH{=}\overset{+}{N}\underset{O^-}{\overset{OH}{\diagup}}$$

$$\text{(I)} \qquad\qquad\qquad \text{(II)}$$

The nitro-form (I) is often called the *pseudo-acid* form; (II), the isonitro-form, is known as the *aci*-form or *nitronic acid*. This is an example of a triad system, the *nitro-acinitro* system; the equilibrium is almost completely on the left, and this may be due to the nitro-form being stabilised by resonance (of the nitro-group).

These nitro-compounds dissolve in aqueous sodium hydroxide to form salts, *e.g.*,

$$CH_3{-}\overset{+}{N}\underset{O^-}{\overset{O}{\diagup}} + OH^- \longrightarrow H_2O + {^-}CH_2{-}\overset{+}{N}\underset{O^-}{\overset{O}{\diagup}} \longleftrightarrow CH_2{=}\overset{+}{N}\underset{O^-}{\overset{O^-}{\diagup}}$$

$$\text{(I)} \qquad\qquad\qquad\qquad \text{(I}a\text{)} \qquad\qquad \text{(II}a\text{)}$$

The removal of the proton is made possible by the strong $-I$ effect of the nitro-group, *i.e.*, the adjacent methylene group is an active methylene group. At the same time, the extended conjugation in the conjugate base stabilises it with respect to its acid.

When the sodium salt is acidified at low temperature, there is not always an immediate separation of oily drops (of the nitro-form). On standing, however, the acidified solution slowly deposits oily drops. This time factor can be explained by protonation of (II*a*) to produce the *aci*-form (soluble) which then slowly tautomerises to the insoluble nitro-form (see also p. 648).

When the sodium salt solution of the nitronic acid is acidified with 50 per cent sulphuric acid at room temperature, an aldehyde (from a primary nitro-compound) and a ketone (from a secondary nitro-compound) is obtained, *e.g.* (R² is either an alkyl group or a hydrogen atom):

$$2R^1R^2C{=}\overset{+}{N}\overset{-}{O_2}Na + 2H_2SO_4 \longrightarrow 2R^1R^2CO + N_2O + 2NaHSO_4 + H_2O \quad (85\%)$$

This reaction is known as the **Nef carbonyl synthesis** (1894).

When treated with stannous chloride and hydrochloric acid, the sodium salt of the nitronic acid is reduced to the aldoxime or ketoxime:

$$R^1R^2C{=}\overset{-}{N}\overset{+}{O_2}Na \xrightarrow{\text{SnCl}_2/\text{HCl}} R^1R^2C{=}NOH$$

4. Primary and secondary nitro-compounds are readily halogenated in alkaline solution in the α-position only:

$$R_2C{=}\overset{-}{\underset{+}{N}O_2}Na + Br_2 \xrightarrow{\text{NaOH}} R_2CBrNO_2 + NaBr$$

Primary nitro-compounds can form the mono- and dibromo-derivatives, whereas secondary form only the monobromo-derivative: nitromethane is exceptional in that it can form the tribromo-derivative. *Chloropicrin* (p. 170) is manufactured by the reaction between nitromethane, chlorine and sodium hydroxide.

5. Nitro-compounds react with nitrous acid, the product depending on the nature of the alkyl group.

Primary nitro-compounds form *nitrolic acids*; these are crystalline substances which dissolve in sodium hydroxide to give red solutions:

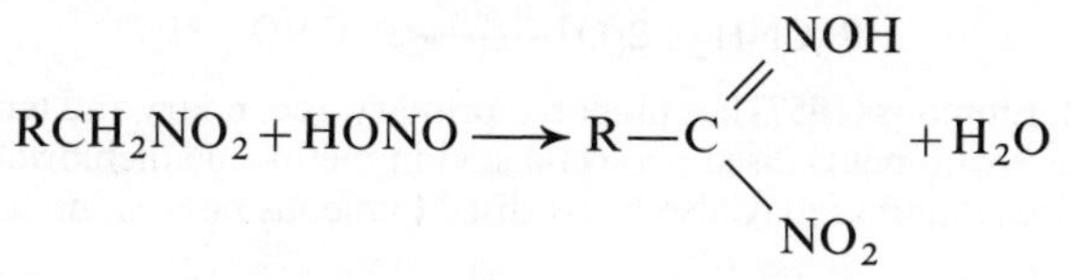

$$RCH_2NO_2 + HONO \longrightarrow R{-}C\overset{\displaystyle /\!/NOH}{\underset{\displaystyle \backslash NO_2}{}} + H_2O$$

Secondary nitro-compounds form *pseudonitroles* (ψ-nitroles); these are colourless crystalline substances which dissolve in sodium hydroxide to give blue solutions (the blue colour is probably due to the presence of the nitroso-group; see below):

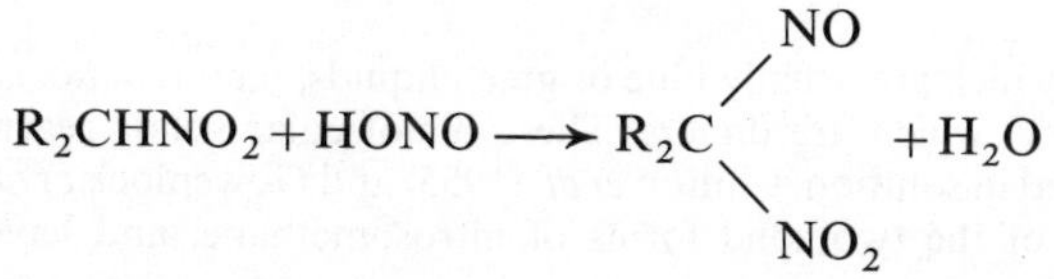

$$R_2CHNO_2 + HONO \longrightarrow R_2C\overset{\displaystyle /NO}{\underset{\displaystyle \backslash NO_2}{}} + H_2O$$

Tertiary nitro-compounds do not react with nitrous acid since they have no α-hydrogen atom.

These reactions with nitrous acid are the basis of the 'red, white and blue' test for the nature of monohydric alcohols (Victor Meyer *et al.*, 1874).

6. Owing to the presence of active α-hydrogen atoms, primary and secondary nitro-compounds undergo condensation with aldehydes; *e.g.*, nitromethane condenses with benzaldehyde in the presence of ethanolic potassium hydroxide to form *ω-nitrostyrene*:

$$CH_3NO_2 + OH^- \rightleftharpoons H_2O + {}^-CH_2NO_2 \xrightarrow{\text{PhCHO}}$$

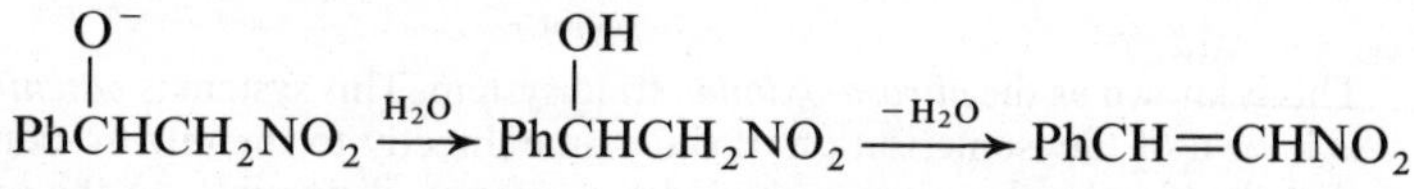

$$\underset{\displaystyle \overset{|}{O^-}}{Ph\overset{|}{C}HCH_2NO_2} \xrightarrow{H_2O} \underset{\displaystyle \overset{|}{OH}}{Ph\overset{|}{C}HCH_2NO_2} \xrightarrow{-H_2O} PhCH{=}CHNO_2$$

Nitroalkanes can act as addenda in the Michael condensation (p. 340), *e.g.*, with mesityl oxide nitromethane forms 4,4-dimethyl-5-nitropentan-2-one:

$$(CH_3)_2C{=}CHCOCH_3 + CH_2{=}NO_2{}^-Na^+ \longrightarrow$$

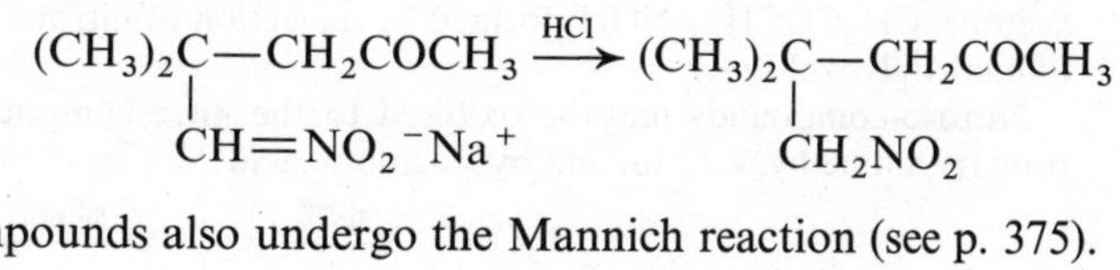

$$\underset{\displaystyle \overset{|}{CH{=}NO_2{}^-Na^+}}{(CH_3)_2\overset{|}{C}{-}CH_2COCH_3} \xrightarrow{\text{HCl}} \underset{\displaystyle \overset{|}{CH_2NO_2}}{(CH_3)_2\overset{|}{C}{-}CH_2COCH_3}$$

Primary and secondary nitro-compounds also undergo the Mannich reaction (see p. 375).

Nitrosoalkanes

The nitrosoalkanes contain a nitroso-group, $-N=O$, directly attached to a carbon atom. They are named as the nitroso-derivatives of the corresponding alkane, e.g.,

$$(CH_3)_3CNO \quad \text{2-methyl-2-nitrosopropane}$$

General methods of preparation

1. By the addition of nitrosyl chloride or bromide to alkenes, whereby alkene nitrosohalides are formed:

$$CH_2{=}CH_2 + NOCl \longrightarrow ClCH_2CH_2NO \xrightarrow{\text{2 mol.}} (ClCH_2CH_2NO)_2$$

2. By the action of nitrous acid on certain types of compounds, e.g., secondary nitroalkanes (see p. 365).

3. By the oxidation of primary amines containing a tertiary alkyl group with e.g., Caro's acid (peroxy-(mono)sulphuric acid):

$$R_3CNH_2 + 2[O] \xrightarrow{H_2SO_5} R_3CNO + H_2O$$

On the other hand, Emmons (1957) has prepared primary, secondary and tertiary nitroso-compounds by oxidation of amines with neutralised peracetic acid in methylene dichloride (yield: 33–80 per cent).

N-Substituted hydroxylamines may also be oxidised (aqueous potassium dichromate-sulphuric acid) to nitroso compounds:

$$RNHOH \xrightarrow{[O]} RNO$$

General properties

Nitroso-compounds, which are usually blue or green liquids, tend to associate to give colourless solids (smelling like camphor) which are dimers. These bimolecular solids regenerate the monomer when fused or when dissolved in solution. Chilton et al. (1955) and Gowenlock et al. (1955) have examined the absorption spectrum of the two solid forms of nitrosomethane, and have concluded that they are geometrical isomers.

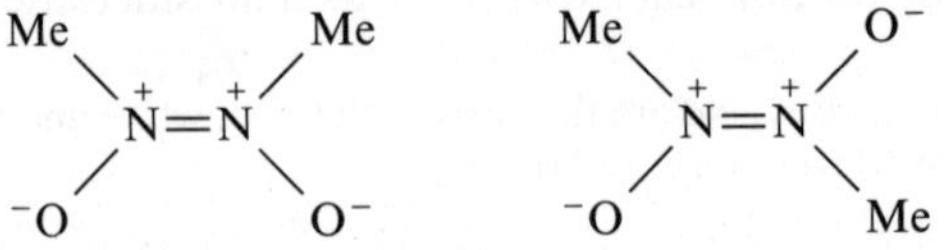

cis   trans

Nitroso-compounds exist as such only when the nitroso-group is attached to a tertiary carbon atom. If the nitroso-group is attached to a primary or secondary carbon atom, the nitroso-compound is generally unstable, tending to rearrange to the oxime:

$$R_2CH{-}NO \longrightarrow R_2C{=}NOH$$

blue and unstable   colourless and stable

This is known as the *nitroso-oximino* triad system. This system is *potentially* tautomeric, but no example is known of the isomerisation of the oxime directly to the nitroso-compound. It has also been found that the bimolecular nitroso-chlorides rearrange, if possible, to the oxime when treated with sodium hydroxide:

$$(ClCH_2CH_2NO)_2 \xrightarrow{OH^-} 2ClCH_2CH{=}NOH$$

The so-called isonitroso-compounds (oximino-compounds) are actually the oximes, e.g., isonitroso-acetone, $CH_3COCH{=}NOH$, formed by the action of nitrous acid on acetone, is really the half oxime of methylglyoxal (q.v.).

Nitroso-compounds may be oxidised to the nitro-compound by nitric acid, and reduced to the primary amine by, e.g., tin and hydrochloric acid:

$$R_3CNO_2 \xleftarrow{HNO_3} R_3CNO \xrightarrow{Sn/HCl} R_3CNH_2$$

Monoamines

Amines are derivatives of ammonia in which one or more hydrogen atoms have been replaced by alkyl groups. The amines are classified as primary, secondary or tertiary amines according as one, two or three hydrogen atoms in the ammonia molecule have been replaced by alkyl groups. Thus the general formulæ of primary, secondary and tertiary amines may be written RNH_2, R_2NH and R_3N, and each is characterised by the presence of the *amino-group* $-NH_2$, the *imino-group* $\diagdown NH$, and the *tertiary nitrogen atom* $\diagdown N-$, respectively.

In addition to the amines the tetra-alkyl derivatives of ammonium hydroxide, $[R_4N]^+OH^-$, are known; these are called the *quaternary ammonium hydroxides*.

Nomenclature

The suffix of the series if *amine*, and each member is named according to the alkyl groups attached to the nitrogen atom, *e.g.*,

CH_3NH_2	methylamine
$(CH_3)_2NH$	dimethylamine
$(CH_3)_3N$	trimethylamine
$(C_2H_5)_2NCH(CH_3)_2$	diethylisopropylamine

The amines are said to be 'simple' when all the alkyl groups are the same, and 'mixed' when the alkyl groups are different.

The method of naming a quaternary ammonium hydroxide is illustrated by the following examples:

$(CH_3)_4N^+OH^-$	tetramethylammonium hydroxide
$[(CH_3)_3\overset{+}{N}C_2H_5]OH^-$	ethyltrimethylammonium hydroxide

General methods of preparation

Many methods are available for preparing amines, but it is instructive to consider them in the following groups:

Methods for the three classes of amines

1. Hofmann's method (1850). This is by means of *ammonolysis*, an alkyl halide and an ethanolic solution of ammonia being heated in a sealed tube at 100°C. A mixture of all three classes of amines is obtained, together with some quaternary ammonium compound.

$$RX + NH_3 \longrightarrow RNH_3^+X^-$$
$$RNH_3^+X^- + NH_3 \rightleftharpoons RNH_2 + NH_4X$$
$$RNH_2 + RX \longrightarrow R_2NH_2^+X^- \overset{NH_3}{\rightleftharpoons} NH_4X + R_2NH$$
$$\overset{RX}{\longrightarrow} R_3NH^+X^- \overset{NH_3}{\rightleftharpoons} NH_4X + R_3N \overset{RX}{\longrightarrow} R_4N^+X^-$$

The alkylation of amines works well with primary alkyl halides, fairly well with many s-alkyl halides, but t-alkyl halides give alkenes, *e.g.*,

$$H_3\overset{\frown}{N} \quad H\overset{\frown}{-}CH_2\overset{\text{─}}{-}CMe_2\overset{\frown}{-}Cl \longrightarrow CH_2=CMe_2 + NH_4Cl$$

The order of reactivity of halides is $RI > RBr > RCl$, and the mechanism of the reaction is S_N2.

2. A mixture of the three types of amines may be prepared by the ammonolysis of alcohols; the alcohol and ammonia are heated under pressure in the presence of a catalyst, *e.g.*, copper chromite or alumina:

$$ROH \xrightarrow{\text{NH}_3} H_2O + RNH_2 \xrightarrow{\text{ROH}} H_2O + R_2NH \xrightarrow{\text{ROH}} H_2O + R_3N$$

Separation of amine mixtures

When the mixture contains the three amine salts and the quaternary salt, it is distilled with potassium hydroxide solution. The three amines distil, leaving the quaternary salt unchanged in solution, *e.g.*,

$$R_2NH_2{}^+X^- + KOH \longrightarrow R_2NH + KX + H_2O$$

The distillate of mixed amines may now be separated into the individual amines as follows:

Fractional distillation. This is the most satisfactory method and is now used industrially, its success being due to the high efficiency of industrial fractionation apparatus.

Hinsberg's method (1890). The mixture of amines is treated with an aromatic sulphonyl chloride. Benzenesulphonyl chloride, $C_6H_5SO_2Cl$, was used originally, but has now been replaced by *p*-toluenesulphonyl chloride, $CH_3C_6H_4SO_2Cl$. After treatment with this acid chloride, the solution is made alkaline with potassium hydroxide. Primary amines form the *N*-alkyl-sulphonamide, *e.g.*, ethylamine forms *N*-ethyl-*p*-toluenesulphonamide:

$$CH_3C_6H_4SO_2Cl + C_2H_5NH_2 \longrightarrow CH_3C_6H_4SO_2NHC_2H_5 + HCl$$

These sulphonamides are often soluble in potassium hydroxide solution due to the formation of the potassium salt. The toluenesulphonyl group is strongly electron-attracting, and thereby facilitates the release of the proton attached to the nitrogen atom:

$$CH_3C_6H_4SO_2NHC_2H_5 + KOH \longrightarrow [CH_3C_6H_4SO_2\bar{N}C_2H_5] \longleftrightarrow$$

$$\left[CH_3C_6H_4\overset{\overset{\displaystyle O^-}{|}}{\underset{\displaystyle \|}{\underset{\displaystyle O}{S}}}{=}NC_2H_5 \right] K^+ + H_2O$$

Secondary amines form *N,N*-dialkyl-sulphonamides, which are insoluble in potassium hydroxide solution because there is no α-hydrogen atom, *e.g.*, diethylamine forms *N,N*-diethyl-*p*-toluenesulphonamide:

$$CH_3C_6H_4SO_2Cl + (C_2H_5)_2NH \longrightarrow CH_3C_6H_4SO_2N(C_2H_5)_2 + HCl$$

Tertiary amines do not react with *p*-toluenesulphonyl chloride.

The alkaline solution is distilled, and the tertiary amine thereupon distils off. The residual liquid is filtered: the filtrate contains the primary amine derivative which, on acidification, gives the alkyl-sulphonamide; the residual solid is the dialkyl-sulphonamide. The amines are regenerated from the sulphonamides by refluxing with 70 per cent sulphuric acid or 25 per cent hydrochloric acid.

Unfortunately, the Hinsberg separation is not always successful because many monoalkyl-suphonamides are insoluble in alkali. This insolubility appears to be general for the higher primary amines, *e.g.*, the sulphonamide from n-heptylamine.

Preparation of primary amines

1. By the reduction of nitro-compounds with metal and acid, or with hydrogen and a nickel catalyst, or with LAH (see p. 363).

2. (i) By the reduction of alkyl cyanides (see p. 357).

(ii) By the reduction of oximes or amides with sodium and ethanol, LAH, or catalytically (see Table 6.1, p. 179).

On the other hand, oximes are reduced by diborane to N-substituted hydroxylamines (Feuer *et al.*, 1965):

$$R_2C{=}NOH \xrightarrow{\text{B}_2\text{H}_6} R_2CHNHOH \quad (50\text{--}90\%)$$

3. By passing a mixture of aldehyde or ketone and a *large excess* of ammonia and hydrogen under pressure (20–150 atm) over Raney nickel at 40–150°C. The reaction may also be carried out at 3 atmospheres in the presence of excess of ammonium chloride and Adams' platinum catalyst. A possible mechanism is via an imine (*cf.* p. 357):

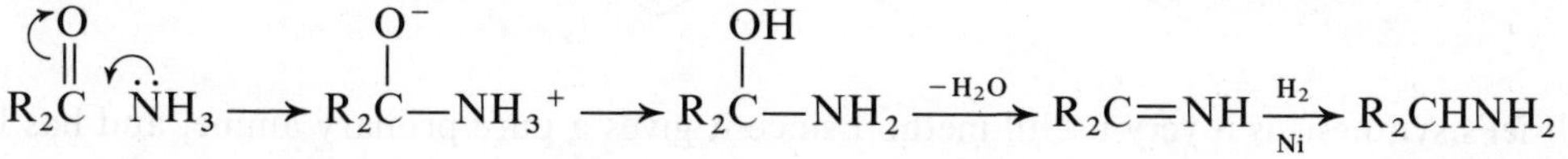

$$\underset{R_2C}{\overset{O}{\|}}\;\ddot{N}H_3 \longrightarrow R_2\overset{O^-}{\underset{|}{C}}{-}NH_3{}^+ \longrightarrow R_2\overset{OH}{\underset{|}{C}}{-}NH_2 \xrightarrow{-\text{H}_2\text{O}} R_2C{=}NH \xrightarrow[\text{Ni}]{\text{H}_2} R_2CHNH_2$$

Small amounts of secondary and tertiary amines are obtained as by-products. This process of introducing alkyl groups into ammonia or primary or secondary amines by means of an aldehyde or ketone in the presence of a reducing agent is known as *reductive alkylation*.

4. Aldehydes and ketones react with ammonium formate or with formamide to give formyl derivatives of primary amines (**Leuckart reaction**, 1885):

$$\underset{/}{\overset{\backslash}{C}}{=}O + 2HCO_2NH_4 \longrightarrow \underset{/}{\overset{\backslash}{C}}HNHCHO + 2H_2O + CO_2 + NH_3$$

$$\underset{/}{\overset{\backslash}{C}}{=}O + 2HCONH_2 \longrightarrow \underset{/}{\overset{\backslash}{C}}HNHCHO + CO_2 + NH_3$$

These formyl derivatives are readily hydrolysed by acid to the primary amine.

The mechanism of the Leuckart reaction is not settled, but it appears that an imine is formed first and this is then reduced to an amine by hydride ion transfer from formic acid:

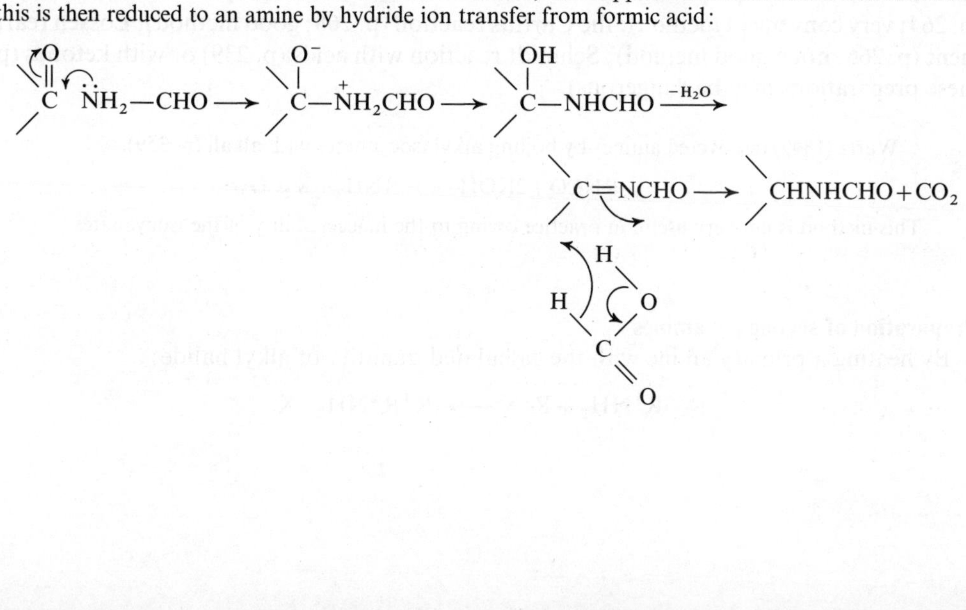

$$\underset{/}{\overset{O}{\overset{\|}{C}}}\;\ddot{N}H_2{-}CHO \longrightarrow \underset{/}{\overset{O^-}{\underset{|}{C}}}{-}\overset{+}{N}H_2CHO \longrightarrow \underset{/}{\overset{OH}{\underset{|}{C}}}{-}NHCHO \xrightarrow{-\text{H}_2\text{O}}$$

$$\underset{/}{\overset{\backslash}{C}}{=}NCHO \longrightarrow \underset{/}{\overset{\backslash}{C}}HNHCHO + CO_2$$

5. Gabriel's phthalimide synthesis (1887). In this method phthalimide is converted, by means of ethanolic potassium hydroxide, into its salt which, on heating with an alkyl halide, gives the N-alkylphthalimide. This is then hydrolysed to phthalic acid and a primary amine by heating with 20 per cent hydrochloric acid under pressure, or by refluxing with potassium hydroxide solution:

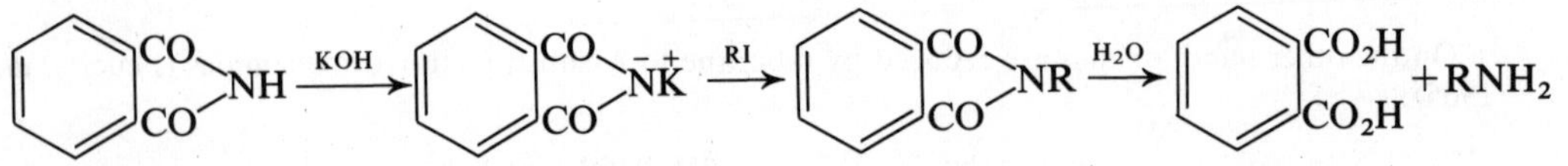

When hydrolysis is difficult, the alkylphthalimide can be treated with hydrazine to give the amine (Ing, 1926):

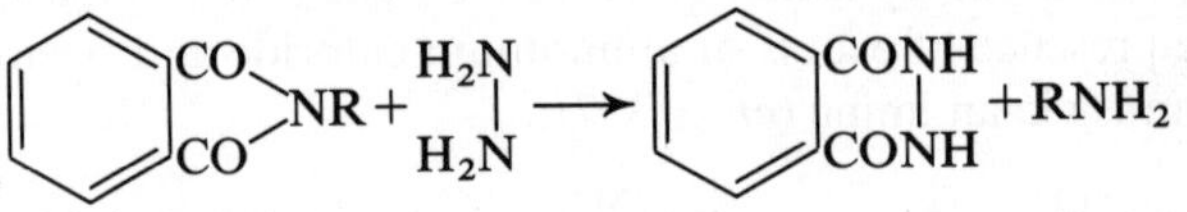

Gabriel's synthesis is a very useful method since it gives a pure primary amine, and has been improved by using dimethylformamide as solvent.

6. By means of a Grignard reagent (or a trialkylborane) and chloramine. The product is pure:

$$RMgX + ClNH_2 \longrightarrow RNH_2 + MgXCl$$

Alternatively, a primary amine may be prepared by interaction of a Grignard reagent and O-methylhydroxylamine (see p. 390).

7. A good method for preparing primary amines containing a t-alkyl group is by the **Ritter reaction** (1948). This reaction depends on the formation of intermediate carbonium ions, and hence is mainly used with t-alcohols (or alkenes that can produce t-carbonium ions). The reaction occurs between alcohol (or alkene) and alkyl cyanide (or hydrocyanic acid) in the presence of sulphuric acid to produce an amide, e.g.,

$$R_3COH \xrightarrow{H^+} H_2O + R_3C^+ \xrightarrow{NaCN} R_3C\overset{+}{N}\equiv CH \xrightarrow{H_2O}$$
$$R_3CNHCHO \xrightarrow{OH^-} R_3CNH_2 + HCO_2^-$$

8. Other methods for preparing primary amines are: Hofmann's preparation of primary amines (p. 264; very convenient method); the Curtius reaction (p. 268; good method); Lossen rearrangement (p. 266; not a good method); Schmidt reaction with acids (p. 239) or with ketones (p. 216; these preparations may be dangerous).

Wurtz (1849) discovered amines by boiling alkyl isocyanates with alkali (p. 359).

$$RNCO + 2KOH \longrightarrow RNH_2 + K_2CO_3$$

This method is not very useful in practice owing to the inaccessibility of the isocyanates.

Preparation of secondary amines

1. By heating a primary amine with the calculated quantity of alkyl halide:

$$R^1NH_2 + R^2X \longrightarrow R^1R^2NH_2^+X^-$$

This reaction usually results in some over-alkylation, but this may be avoided as follows (see also p. 368).

$$R^1NH_2 \xrightarrow[OH^-]{TsCl} Ts\bar{N}R^1 \xrightarrow{R^2X} TsNR^1R^2 \xrightarrow{H^+} TsOH + R^1R^2NH$$

2. Aniline is heated with an alkyl halide, and the product, dialkylaniline, is treated with nitrous acid; the p-nitroso-dialkylaniline so formed is then boiled with sodium hydroxide solution. A pure secondary amine and p-nitrosophenol are produced:

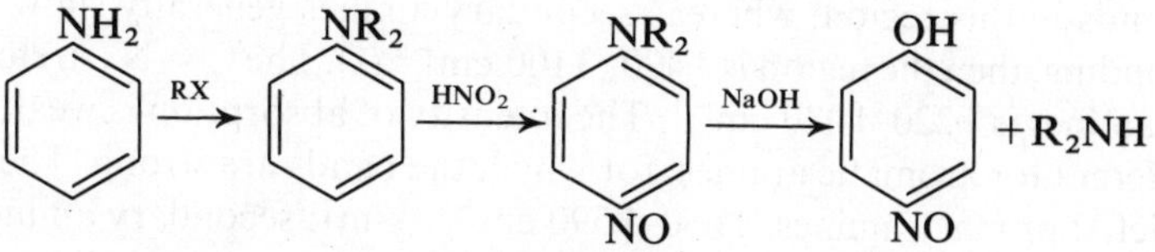

This is one of the best methods of preparing pure secondary amines.

3. Pure secondary amines may be prepared from calcium cyanamide as follows:

$$CaNCN \xrightarrow{NaOH} Na_2NCN \xrightarrow{2RX} R_2NCN \xrightarrow[OH^-]{H^+ \ or} R_2NH + CO_2 + NH_3$$

The catalytic reduction of imines may be used to prepare secondary amines, but this method is usually confined to amines containing a phenyl group (since the aldehyde is commonly benzaldehyde). This may be avoided as follows:

$$PhCHO + R^1NH_2 \longrightarrow PhCH=NR^1 \xrightarrow[Ni]{H_2} PhCH_2NHR^1$$
$$\downarrow R^2X$$
$$[PhCH=\overset{+}{N}R^1R^2]X^- \xrightarrow{H^+} PhCHO + R^1NHR^2$$

On the other hand, reductive alkylation (p. 369) of any carbonyl compound in the presence of a *primary* amine gives secondary amines.

Preparation of tertiary amines

1. The best method is to heat an ethanolic solution of ammonia with alkyl halide which is used in slight excess required by the equation:

$$3RX + NH_3 \longrightarrow R_3NH^+X^- + 2HX$$

2. By the reductive alkylation of a carbonyl compound (in excess) in the presence of a secondary amine, *e.g.*,

$$R_2^1NH + R^2CHO + H_2 \xrightarrow{Ni} R_2^1NCH_2R^2 + H_2O$$

Preparation of quaternary compounds

There is only one satisfactory method of preparing quaternary compounds, *viz.*, by heating ammonia with a very large excess of alkyl halide; a primary, secondary or tertiary amine may be used instead of ammonia. The starting materials will depend on the nature of the desired quaternary compound, *e.g.*, ethyltrimethylammonium iodide may be prepared by heating trimethylamine with ethyl iodide:

$$(CH_3)_3N + C_2H_5I \longrightarrow [(CH_3)_3NC_2H_5]^+I^-$$

General properties of the amines

The lower members are gases, soluble in water (with which they can form hydrogen bonds). These are followed by members which are liquids, and finally by members which are solids. The solubility in water decreases as their molecular weight increases. All the volatile members have a powerful fishy smell, and are combustible.

The reactions of an amine depend very largely on the class of that amine.

Primary and secondary amines both contain the NH group; both types (N—H stretch) absorb in the region $3\,500$–$3\,300$ cm^{-1} (v), but the two are distinguished by the fact that primary amines show *two* bands in this region, whereas secondary amines generally show only one. If there is hydrogen bonding, then the region is $3\,400$–$3\,100$ cm^{-1} (s). The C—N (stretch) absorption region of aliphatic amines is $1\,220$–$1\,020$ cm^{-1}. The intensity of absorption is weak to medium, and this region is different for aromatic amines, for which the bands are strong. There is also a band due to N—H (def.): primary amines, $1\,650$–$1\,590$ cm^{-1} (s-m); secondary amines, $1\,650$–$1\,550$ cm^{-1} (s). Because of the overlap of these regions, it is not usually possible to distinguish the two types of amines in this way.

Figure 13.4 is the infra-red spectrum of cyclohexylamine (as a film).

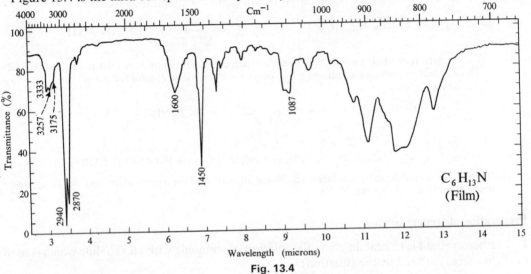

Fig. 13.4

3333 and 3257 cm^{-1}: N—H (str.) in primary amines. 3175 cm^{-1}: N—H (str.); association? 1600 cm^{-1}: N—H (def.) in primary (or secondary) amines. 1087 cm^{-1}: C—N (str.) in primary (or secondary) amines. 2940 and 2870 cm^{-1}: C—H (str.) in (Me and/or) CH$_2$. 1450 cm^{-1}: C—H (def.) in (Me and/or) CH$_2$.

Protons attached to nitrogen usually give rise to a single absorption line in the NMR spectrum. The reason for this is the same as that for alcohols, *i.e.*, there is a rapid chemical exchange of the amino hydrogens (see p. 18). The following τ-values are characteristic of the NH proton: aliphatic and cyclic amines, 8–9 p.p.m.; aromatic amines, 5·3–7·4; amides, 2·0–5·0. NMR studies have also been used to detect, in particular, *N*-methyl groups (τ-value 7·8–8·0).

A problem in connection with nitrogen compounds is that of restricted rotation, *e.g.*,

In both examples, the environments of the methyl groups are different and so these will give different signals (see p. 263).

Mass spectra

When a compound contains an *odd* number of nitrogen atoms, its molecular weight is an *odd* number. The molecular ion of amines is very weak or absent, and its most characteristic fragmentation pattern is via β-cleavage, *e.g.*,

$$R\overset{\frown}{}CH_2\!-\!\overset{.+}{N}H_2 \longrightarrow R\cdot + CH_2\!=\!\overset{+}{N}H_2 \quad (m/e\ 30)$$

Thus, the base peak of primary amines is *m/e* 30 provided no branching occurs at the α-carbon atom. The presence of this peak, however, does not necessarily mean that the amine is primary, since s- and t-amines can also give rise to this peak via McLafferty rearrangements. The behaviour of these amines is similar to that of the ethers (p. 197); β-cleavage occurs preferentially to eliminate the largest alkyl group, *e.g.*, ($R^1 > R^2$ and R^3):

$$R^1\!-\!CH_2\!-\!\overset{.+}{N}R^2R^3 \longrightarrow R^1\!\cdot + CH_2\!=\!\overset{+}{N}R^2R^3$$

If, *e.g.*, $R^2 = $ Me and $R^3 = $ Et, then a McLafferty rearrangement can now occur:

$$\underset{\underset{CH_3}{|}}{CH_2\!=\!\overset{+}{N}}\!\overset{\overset{\displaystyle H\!\!\overset{\frown}{}\!CH_2}{\overset{+}{|}\quad\overset{\frown}{|}}}{\!-\!CH_2} \longrightarrow CH_2\!=\!CH_2 + \underset{\underset{CH_3}{|}}{CH_2\!=\!\overset{+}{N}H} \quad (m/e\ 44)$$

When both amino and hydroxyl groups are present in the compound, because the ionisation potential of nitrogen is lower than that of oxygen, fission occurs to give preferentially the positive fragment which contains the nitrogen atom, *e.g.*, the intensity of $CH_2\!=\!\overset{+}{N}H_2$ (*m/e* 30) is about ten times that of $CH_2\!=\!\overset{+}{O}H$ (*m/e* 31).

$$\underset{\underset{OH}{|}}{\cdot CH_2} + \underset{\underset{^+NH_2}{|}}{\overset{\displaystyle CH_2}{\|}} \longleftarrow \left[\underset{\underset{OH}{|}}{CH_2}\!-\!\underset{\underset{NH_2}{|}}{CH_2}\right]^{\ddagger} \longrightarrow \underset{\underset{^+OH}{|}}{\overset{\displaystyle CH_2}{\|}} + \underset{\underset{NH_2}{|}}{\cdot CH_2}$$

The order of stability of fragments of the above type is:

$$CH_2\!=\!\overset{+}{N}H_2 > CH_2\!=\!\overset{+}{S}H > CH_2\!=\!\overset{+}{O}H.$$

Reactions given by all three classes of amines

1. All the amines are basic, and they are stronger bases than ammonia.

$$RNH_2 + H_2O \rightleftharpoons RNH_3{}^+ + OH^-$$

One factor that controls the basicity of amines will be the availability of the lone pair for protonation. At first sight, it would appear that as the number of alkyl groups increases, then because of their $+I$ effect, the lone pair will become more available, thereby increasing the basicity of the amine.

Inspection of the pK_a's in Table 13.1 shows that the introduction of one alkyl group into ammonia strengthens the base (remember: $pK_a = pK_w - pK_b$; p. 65), a second alkyl group further strengthens the base, but the third alkyl group *decreases* the strength, *e.g.*,

$$NH_3 < Me_3N < MeNH_2 < Me_2NH$$

The reason for this order is not clear. It has been suggested that a steric factor operates. Addition of the proton increases crowding and so sets up strain, which will be greatest in the tertiary amine, and consequently the stability of this molecule is decreased, *i.e.*, its basicity is reduced. However, when we consider that the three bonding pairs and the lone pair all occupy sp^3 orbitals, and that a lone pair causes crowding of bonding pairs (p. 47), it would seem that protonation should *relieve* crowding by transforming the lone pair into a bonding pair.

Table 13.1

Base	pK_a	Base	pK_a
NH_3	9·27	n-$PrNH_2$	10·58
$MeNH_2$	10·62	iso$PrNH_2$	10·63
Me_2NH	10·77	n-$BuNH_2$	10·61
Me_3N	9·80	t-$BuNH_2$	10·45
$EtNH_2$	10·63	n-$AmNH_2$	10·63
Et_2NH	10·93	$NCCH_2NH_2$	5·34
Et_3N	10·87		

All the 'onium ions carry a positive charge and so are solvated. The greater the ion, the less will be the solvation and so the less stabilised is the ion. It seems possible that in going from Me_2NH to Me_3N, although the availability of the lone pair is increased, the ion has now grown sufficiently in size at this stage to become less solvated and so the stabilisation lost due to this decrease in solvation is greater than the inductive effect making the lone pair available for protonation.

That the inductive effects are important is shown by the fact that, *e.g.*, in cyanomethylamine, in which the cyano-group has a strong $-I$ effect, the pK_a value is much lower than for methylamine.

2. All the amines combine with acids to form salts, *e.g.*, methylamine combines with hydrochloric acid to form *methylammonium chloride*:

$$CH_3NH_2 + HCl \longrightarrow [CH_3NH_3]^+Cl^-$$

The nitrogen is quadricovalent unielectrovalent in amine salts; but to show their relationship to the amines, their salts are often written as, *e.g.*, $CH_3NH_2 \cdot HCl$, methylamine hydrochloride; $[(C_2H_5)_2NH]_2 \cdot H_2SO_4$, diethylamine sulphate.

3. All the amines combine with alkyl halides to form alkyl-substituted ammonium halides with more alkyl groups than the amine used.

4. When an amine salt is heated at high temperature, a molecule of alkyl halide is eliminated (*i.e.*, reverse of reaction 3, above):

$$R_3NH^+Cl^- \longrightarrow RCl + R_2NH \xrightarrow{\text{HCl}} RCl + RNH_2 \xrightarrow{\text{HCl}} RCl + NH_4Cl$$

If a t-alkyl group is present, this is eliminated as alkene (via pyrolysis of the alkyl halide). If a methyl group is present, this is generally eliminated preferentially.

When the methyl ether of a high boiling hydroxy compound is heated with concentrated hydriodic acid at 100°C, methyl iodide is formed (*cf.* p. 198):

$$ArOCH_3 + HI \xrightarrow{100°C} ArOH + CH_3I$$

This is the basis of the **Zeisel method** for the quantitative estimation of methoxyl groups (p. 718).

When the temperature is raised to 150°C, *N*-methyl groups are converted into methyl iodide:

$$ArNHCH_3 + HI \xrightarrow{150°C} ArNH_2 + CH_3I$$

This is the basis of the **Herzig–Meyer method** for the quantitative estimation of methylimino groups. Thus methoxyl and methylimino groups can be estimated separately when both are present in the same compound.

Reactions given by primary and secondary amines

1. Primary and secondary amines react with acid chlorides or anhydrides to form the *N*-alkyl acid amide, *e.g.*,

$$RNH_2 + (CH_3CO)_2O \longrightarrow CH_3CONHR + CH_3CO_2H$$

These monoacetyl derivatives are easily prepared and, since they are usually well-defined crystalline solids, are used to characterise primary amines.

It is difficult to prepare the diacetyl derivative; excess of the acetylating agent and high temperature, however, often yields the diacetyl derivative:

$$RNHCOCH_3 + (CH_3CO)_2O \longrightarrow RN(COCH_3)_2 + CH_3CO_2H$$

Secondary amines form only the monoacyl derivative:

$$R_2NH + CH_3COCl \longrightarrow CH_3CONR_2 + HCl$$

Sulphonyl chlorides also react with primary and secondary amines to form *N*-alkyl-sulphonamides (see the Hinsberg separation, p. 368).

2. Primary and secondary amines form substituted ureas with phenyl isocyanate, and may be characterised by these derivatives (see p. 463).

3. Primary and secondary amines can participate in the **Mannich reaction** (1917). This is the condensation between formaldehyde (but other aldehydes may be used), ammonia or a primary or secondary amine (preferably as the hydrochloride) and a compound containing at least one active hydrogen atom, *e.g.*, ketones, β-keto-esters, β-cyano-esters, nitroalkanes, alkynes (with $C \equiv H$), etc. The reaction is usually carried out in acid solution (but may also be base-catalysed).

The mechanism of the Mannich reaction is uncertain. According to Masui *et al.* (1970), the 'Mannich intermediate' is formed as follows, *e.g.*, for diethylamine in acid solution:

$$2Et_2NH + CH_2O \rightleftharpoons H_2O + Et_2N{-}CH_2{-}NEt_2 \xrightarrow{H^+} Et_2NH + Et_2\overset{+}{N}{=}CH_2 \longleftrightarrow Et_2N{-}\overset{+}{C}H_2$$

The subsequent steps are possibly, for acetophenone (the enol form appears to be the reactive species in all cases):

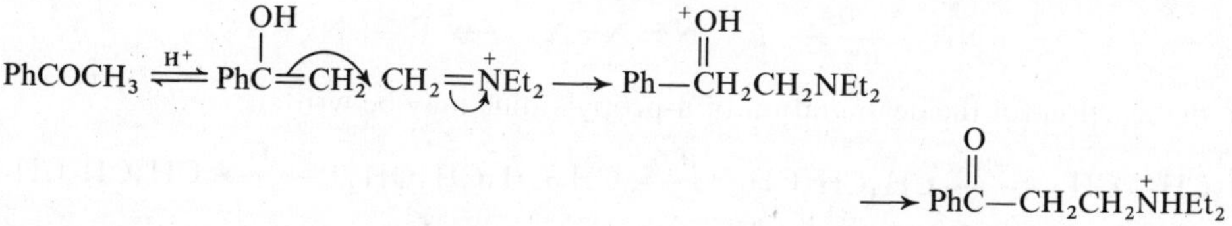

The free base is obtained by basifying the solution at the end of the reaction.

The products of the Mannich reaction are known as **Mannich bases**, and many are useful as intermediates in synthesis, *e.g.*,

(i) $PhCOCH_2CH_2NHR_2{}^+Cl^- \xrightarrow{heat} PhCOCH{=}CH_2 \xrightarrow[cat.]{H_2} PhCOCH_2CH_3$

(ii) $PhCOCH_2CH_2NR_2 \xrightarrow{CN^-} PhCOCH_2CH_2CN$

4. Primary and secondary amines are methylated by heating with formaldehyde and an excess of formic acid at 100°C (see p. 225).

5. Primary and secondary amines form *N*-halogeno-amines with halogen in weakly alkaline solution, e.g.,

$$RNH_2 \xrightarrow[\text{aq. Na}_2\text{CO}_3]{Br_2} RNHBr + RNBr_2$$

Reactions given by primary amines

1. Primary amines form isocyanides when heated with chloroform and ethanolic potassium hydroxide (p. 357):

$$RNH_2 + CHCl_3 + 3KOH \longrightarrow RNC + 3KCl + 3H_2O$$

This reaction is used as a test for primary amines (or for chloroform).

2. When warmed with carbon disulphide, primary amines form a *dithiocarbamic acid*, which is decomposed by mercuric chloride to the *alkyl isothiocyanate*:

$$RNH_2 + CS_2 \longrightarrow S{=}C{\overset{\displaystyle NHR}{\underset{\displaystyle SH}{\diagup\diagdown}}} \xrightarrow{HgCl_2} RNCS + HgS + 2HCl$$

This is known as the *Hofmann mustard oil reaction*, and may be used as a test for primary amines.

3. Primary amines combine with aldehydes to form imines (p. 214):

$$C_6H_5CHO + RNH_2 \longrightarrow C_6H_5CH{=}NR + H_2O$$

4. Primary amines react with nitrous acid with the evolution of nitrogen. The equation is usually written:

$$RNH_2 + HNO_2 \longrightarrow ROH + N_2 + H_2O$$

The reaction, however, is far more complicated than this equation indicates. When the nature of R permits it, the products are mixtures of isomeric alcohols, isomeric alkenes, and other compounds, e.g., n-propylamine gives n-propanol (7 per cent), isopropanol (32 per cent), propene (28 per cent), and propyl and isopropyl halides (when the nitrous acid is generated from $NaNO_2$ and HX). These products can be explained on the basis of the formation of an intermediate carbonium ion:

$$RNH_2 \xrightarrow[\text{HX}]{NaNO_2} R{-}N{\equiv}N + X^- \longrightarrow R^+ + N_2 + X^-$$

Thus, the reaction for the deamination of n-propylamine may be written:

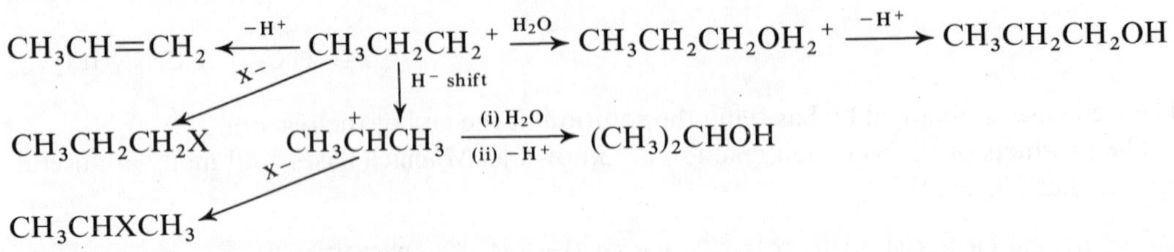

$$CH_3CH{=}CH_2 \xleftarrow[X^-]{-H^+} CH_3CH_2CH_2^+ \xrightarrow{H_2O} CH_3CH_2CH_2OH_2^+ \xrightarrow{-H^+} CH_3CH_2CH_2OH$$

$$\Big\downarrow \text{H}^- \text{ shift}$$

$$CH_3CH_2CH_2X \qquad CH_3\overset{+}{C}HCH_3 \xrightarrow[\text{(ii) }-H^+]{\text{(i) }H_2O} (CH_3)_2CHOH$$

$$CH_3CHXCH_3$$

Whatever are the products formed by the action of nitrous acid on a primary amine, nitrogen is always evolved. Thus this reaction may be used as a test for primary amines, since none of the other classes of amines liberate nitrogen.

The mechanism given above does not explain many other experimental results, e.g., n-propylamine gives some cyclopropane (Silver, 1961; Skell *et al.*, 1962); s-butylamine gives a small amount of methylcyclopropane (Lee *et al.*, 1965); and if the starting amine is an optically active form, then the corresponding alcohol produced is largely inverted:

$$\text{D-MeCH}_2\text{CHMe} \xrightarrow{\text{HNO}_2} \triangleright\!\!-\text{Me} + \text{D-MeCH}_2\text{CHOHMe} + \text{L-MeCH}_2\text{CHOHMe}$$
$$\overset{|}{\underset{\text{NH}_2}{}}$$

The formation of inverted alcohol *without* rearrangement suggests that the diazonium ion is attacked directly:

$$\text{H}_2\overset{\curvearrowright}{\text{O}}\ \ \text{R}\overset{\curvearrowleft}{\underset{}{-\text{N}_2}}{}^+ \longrightarrow \text{N}_2 + \text{H}_2\overset{+}{\text{O}}\!-\!\text{R} \xrightarrow{-\text{H}^+} \text{HO}\!-\!\text{R}$$

On the other hand, some amines lead to a large amount of retention, *e.g.*, α-phenyl-s-butylamine (Streitwieser *et al.*, 1959). It must therefore be concluded that a number of paths are followed during deamination. One suggestion (Streitwieser, 1957) is that all acyclic products can be explained in terms of *both* the carbonium ion and the diazonium cation undergoing reaction. Also, the formation of cyclic products can be explained by internal electrophilic attack in a carbonium ion, *e.g.*,

$$\text{H}\!\!-\!\!\overset{\curvearrowright}{\underset{\underset{\overset{+}{\text{CH}_2}}{\curvearrowright}}{\text{CH}_2\!\!-\!\!\text{CH}_2}} \longrightarrow \text{H}^+ + \overset{\text{CH}_2\!\!-\!\!\text{CH}_2}{\underset{\text{CH}_2}{\diagdown\ \diagup}}$$

5. *Oxidation of primary amines.* The products obtained depend on the oxidising agent used, and on the nature of the alkyl group, *e.g.*,

(i) with potassium permanganate:

$$\text{RCH}_2\text{NH}_2 \xrightarrow{[\text{O}]} \underset{\text{aldimine}}{\text{RCH}\!=\!\text{NH}} \xrightarrow{\text{H}^+} \text{RCHO} + \text{NH}_3$$

$$\text{R}_2\text{CHNH}_2 \xrightarrow{[\text{O}]} \underset{\text{ketimine}}{\text{R}_2\text{C}\!=\!\text{NH}} \xrightarrow{\text{H}^+} \text{R}_2\text{CO} + \text{NH}_3$$

$$\text{R}_3\text{CNH}_2 \xrightarrow{[\text{O}]} \text{R}_3\text{CNO}_2$$

(ii) with Caro's acid (H_2SO_5):

$$\text{RCH}_2\text{NH}_2 \xrightarrow{[\text{O}]} \text{RCH}_2\!-\!\text{NHOH} + \text{RCH}\!=\!\text{NOH} + \text{RC}\overset{\diagup\text{OH}}{\underset{\diagdown\text{NOH}}{}}$$

$$\underset{\substack{\textit{N}\text{-alkyl-}\\\text{hydroxylamine}}}{} \qquad \underset{\text{aldoxime}}{} \qquad \underset{\substack{\text{hydroxamic}\\\text{acid}}}{}$$

$$\text{R}_2\text{CHNH}_2 \xrightarrow{[\text{O}]} \text{R}_2\text{C}\!=\!\text{NOH}$$

$$\text{R}_3\text{CNH}_2 \xrightarrow{[\text{O}]} \text{R}_3\text{CNO}$$

Reactions given by secondary amines

1. Secondary amines react with nitrous acid to form insoluble oily *nitrosamines*; *nitrogen is not evolved*:

$$\text{HO}\!-\!\text{NO} + \text{H}^+ \rightleftharpoons \text{H}_2\overset{+}{\text{O}}\!-\!\text{NO} \rightleftharpoons \text{H}_2\text{O} + \text{NO}^+$$

$$\text{R}_2\text{NH} + \text{NO}^+ \rightleftharpoons \text{R}_2\overset{+}{\text{N}}\text{HNO} \longrightarrow \text{R}_2\text{NNO} + \text{H}^+$$

Nitrosamines are yellow neutral oils which are steam-volatile. When warmed with a crystal of phenol and a few drops of concentrated sulphuric acid, nitrosamines form a green solution which, when made alkaline with aqueous sodium hydroxide, turns deep blue. This procedure may be used as a test for secondary amines; it is known as *Liebermann's nitroso reaction*.

Nitrosamines are readily hydrolysed to the amine by boiling with dilute hydrochloric acid:

$$\text{R}_2\text{NNO} + \text{H}_2\text{O} \xrightarrow{\text{HCl}} \text{R}_2\text{NH} + \text{HNO}_2$$

Peroxytrifluoroacetic acid converts secondary nitrosamines into nitramines (Emmons, 1954):

$$R_2NNO \xrightarrow{CF_3CO_3H} R_2NNO_2 \quad (73\text{–}95\%)$$

2. When warmed with carbon disulphide, secondary amines form a *dithiocarbamic acid*, which is *not* decomposed by mercuric chloride to the alkyl isothiocyanate (*cf.* primary amines):

$$S{=}C{=}S + R_2NH \longrightarrow S{=}C\underset{SH}{\overset{NR_2}{\big<}}$$

3. Secondary amines readily form *enamines* with aldehydes containing α-hydrogen (see p. 383).

4. Secondary amines may be oxidised, the product depending on the oxidising agent used, *e.g.*,

$$R_2NH \xrightarrow{KMnO_4} R_2N{-}NR_2; \quad R_2NH \xrightarrow[\text{or } H_2O_2]{H_2SO_5} R_2NOH$$

tetra-alkylhydrazine

Reactions given by tertiary amines

1. Tertiary amines dissolve in *cold* nitrous acid to form the nitrite salt, $R_3N\cdot HNO_2$ or $[R_3NH]^+NO_2{}^-$. If the amine is of the type $(RCH_2)_3N$ or $(R_2CH)_3N$, then, when the solution is warmed, the former produces an aldehyde, the latter a ketone, and in both cases a nitrosamine and nitrous oxide.

2. Tertiary amines are not affected by potassium permanganate, but are oxidised to the *amine oxide* by Caro's acid, ozone, or by hydrogen peroxide:

$$R_3N + [O] \longrightarrow R_3\overset{+}{N}{-}\overset{-}{O}$$

These amine oxides are basic and exist as $[R_3NOH]^+OH^-$ in solution. Amine oxides form addition compounds with gaseous hydrogen halide, *e.g.*, $[R_3NOH]^+Cl^-$, and with alkyl halides, *e.g.*, $[R_3NOR]^+Cl^-$.

Tertiary amines are oxidised at the α-hydrogen atoms by potassium permanganate, manganese dioxide, or mercuric acetate; the product is an enamine or its hydrolytic products (which may be further oxidised), *e.g.*,

$$R^1CH_2CH_2NR_2^2 \xrightarrow{Hg(OAc)_2} R^1CH{=}CHNR_2^2 \xrightarrow{H_2O} R^1CH_2CHO + R_2^2NH$$

3. Tertiary amines react with cyanogen bromide to form a *dialkylcyanamide* which, on hydrolysis, gives a secondary amine (p. 371):

$$R_3N + BrCN \longrightarrow [R_3NCN]^+Br^- \longrightarrow RBr + R_2NCN \xrightarrow[OH^-]{H^+ \text{ or}} R_2NH$$

This method therefore offers a means of converting a tertiary amine into a secondary, but is mainly used to open the ring of a cyclic amine containing a tertiary nitrogen atom (see p. 855).

Quaternary ammonium compounds

The quaternary ammonium salts are white crystalline solids, soluble in water, and completely dissociated in solution (the nitrogen is quadricovalent unielectrovalent). When a quaternary ammonium halide is heated *in vacuo*, it gives the tertiary amine:

$$[R_4N]^+X^- \longrightarrow R_3N + RX \quad (g.\text{-}v.g.)$$

Quaternary ammonium halides are converted into their corresponding hydroxides by moist silver oxide, methanolic potassium hydroxide, or by passing the halide over a basic ion-exchange resin, e.g.,

$$[R_4N]^+X^- + {}'AgOH' \longrightarrow [R_4N]^+OH^- + AgX$$

The quaternary ammonium hydroxides are white deliquescent crystalline solids which are as strongly basic as sodium and potassium hydroxides (all exist as ions in the solid state). Their solutions absorb carbon dioxide, and will liberate ammonia from its salts.

Quaternary ammonium salts containing a long-chain alkyl group have detergent properties, and are known as *invert soaps* because their surface activity is due to the presence of the *positive* ion and not to a negative ion as in alkyl hydrogen sulphate detergents.

Elimination reactions

Quaternary ammonium hydroxides (and their salts in the presence of alkoxide ions) undergo thermal decomposition to form alkenes, and do so by the E2 mechanism (the reader should revise the section on elimination reactions undergone by alkyl halides, pp. 162–6, before proceeding further). The early work was done by Hofmann (1881), and he believed that only tetramethylammonium hydroxide gave an alcohol (methanol) as one product, but Musker (1964) has shown that the major decomposition products are trimethylamine and dimethyl ether; the minor products are methanol and some other compounds:

$$2(CH_3)_4N^+OH^- \longrightarrow 2(CH_3)_3N + (CH_3)_2O + CH_3OH$$

Musker worked with the solid substrate, but Tanaka et al. (1966) showed that when this hydroxide is heated in aqueous solution under pressure, the decomposition proceeds according to the equation:

$$(CH_3)_4N^+OH^- \longrightarrow (CH_3)_3N + CH_3OH$$

As an outcome of his work, Hofmann proposed an empirical rule (the **Hofmann rule**) for the decomposition of 'onium hydroxides: *If an ethyl group is present, the predominant olefinic product is ethylene.* However, a detailed study of all cases where 'onium hydroxides give alkenes on heating, has led to a general statement of Hofmann's rule: *Elimination of hydrogen occurs preferentially from that β-carbon atom which is joined to the largest number of hydrogen atoms, e.g.,*

$$\underset{\substack{|\\ \underset{\beta}{CH_2CH_3}}}{\overset{\substack{Me\\|}}{Me-\overset{+}{N}-CH_2\overset{\beta}{CH_2}CH_3}} \xrightarrow[\text{heat}]{OH^-} \underset{(98\%)}{C_2H_4} + \underset{(2\%)}{MeCH=CH_2}$$

Sulphonium salts also behave in a similar manner, e.g.,

$$\underset{\substack{|\\ \beta CH_3}}{\overset{\beta}{Me}CH_2CH\overset{+}{S}Me_2} \xrightarrow[\text{heat}]{EtO^-} \underset{(74\%)}{MeCH_2CH=CH_2} + \underset{(2\%)}{MeCH=CHMe} + Me_2S$$

From these examples it can be seen that the Hofmann generalised rule may be stated in an alternative way: *The predominant alkene is the least substituted ethylene, i.e., the one carrying the smallest number of alkyl groups.*

Summary. For alkene-forming reactions of 'onium hydroxides, the mechanism has been shown to be E2 (see below). Since the most substituted alkene is generally the most stable isomer, Saytzeff's rule leads to the predominance of the most stable product and the Hofmann rule to the least stable one. Also, if the alkene produced is capable of geometrical isomerism, the *trans* isomer usually predominates over the *cis* isomer (the former is generally more stable than the latter; see p. 99).

Mechanism of elimination in 'onium ions. Kinetic studies have shown that the elimination is a second-order reaction; it is first-order in ammonium ion and in hydroxide ion. Thus, the general equation for the E2 mechanism may be written as follows:

$$HO^{\curvearrowleft} \quad H \overset{\frown}{-} CR_2 \overset{\frown}{-} CR_2 \overset{+}{-} \overset{\curvearrowright}{N}R_3 \longrightarrow H_2O + CR_2 = CR_2 + R_3N$$

Since elimination proceeds by removal of a β-hydrogen, any factor that affects its ease of removal will affect the rate of reaction. Hence, when alternative β-hydrogens are present, the one most readily removed will decide the structure of the predominating alkene product. One factor affecting the ease of removal is the acidity of the β-hydrogen, and the more acidic it is, the easier it is to remove it. Ingold *et al.* (1927) showed that the inductive effect due to the positive charge on the nitrogen atom is an important factor in deciding the strength of this acidity, and so it follows that the Hofmann generalised rule is a consequence of the $-I$ effect operating in 'onium ions (*cf.* the hyperconjugative effect for the Saytzeff rule). The C—H bonds are weakened sufficiently for a β-proton to be eliminated by the E2 mechanism.

$$CH_3 \rightarrow \overset{\overset{\displaystyle H}{|}}{CH} \rightarrow CH_2 \rightarrow \overset{+}{N} \overset{\curvearrowleft}{\rightharpoondown} CH_2 \overset{\curvearrowleft}{\leftharpoondown} CH_2 - H \overset{OH^-}{\longrightarrow}$$

$$\underset{CH_3 \quad CH_3}{}$$

$$CH_3CH_2CH_2N(CH_3)_2 + CH_2 = CH_2 + H_2O \quad (98\%)$$

Since the terminal methyl group of the n-propyl group is electron-releasing, the positive charge induced on the β-carbon atom (of the propyl group) is partially neutralised. This 'tightens' the bonding of these β-hydrogen atoms, and so a less tightly bound β-hydrogen atom in the ethyl group is eliminated preferentially with the formation of ethylene.

In some cases, the Hofmann generalised rule is not followed, *e.g.*, styrene and not ethylene is formed in the following elimination:

$$PhCH_2CH_2\overset{+}{N}Me_2Et \xrightarrow[heat]{OH^-} PhCH = CH_2 + Me_2NEt + H_2O$$

This may be explained on the basis that the β-hydrogen adjacent to the benzene ring (*benzylic* hydrogen) is more acidic than the β-hydrogen of the ethyl group because of stabilisation of the transition state by conjugation (which is absent for the ethyl group).

It might also be noted that alcohols may be obtained in addition to alkenes, *e.g.*,

$$RCH_2CH_2\overset{+}{N}Me_3OH^- \longrightarrow RCH = CH_2 + RCH_2CH_2OH + Me_3N + H_2O$$

In some cases, no alkene is obtained at all, *e.g.*, quaternised tetrahydroquinoline:

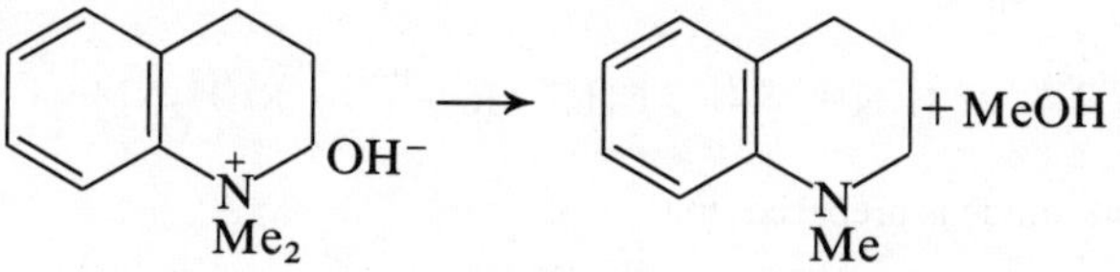

Stereochemistry of the elimination reaction. Just as it was shown that *trans* elimination occurs more readily than *cis* for alkyl halides, so experimental work has shown the same rule applies to 'onium salts, *e.g.*, Cram *et al.* (1956) examined the following reaction:

$$PhCHMeCHPh\overset{+}{N}Me_3I^- + EtO^- \longrightarrow PhMeC = CHPh + Me_3N + EtOH$$

The reaction was shown to be second-order (hence the bimolecular mechanism), and it was also shown that the *erythro* form of the ion gave the *cis* alkene, whereas the *threo* form gave the *trans* alkene (see p. 477 for the meanings of these prefixes). These results are in keeping with *trans* elimination.

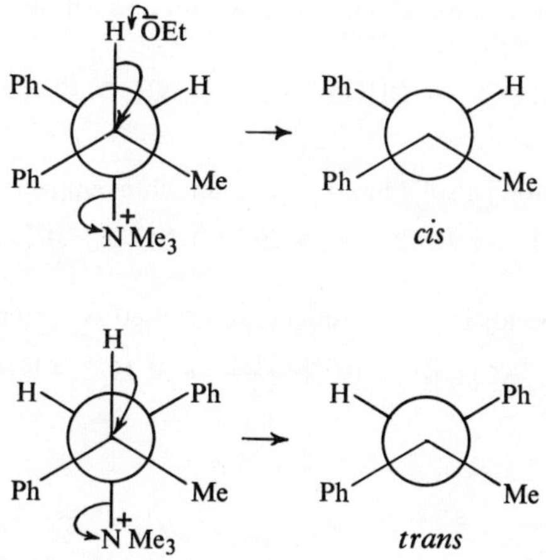

The rates of reaction were very different, the *threo* diastereoisomer undergoing elimination about fifty times as fast as the *erythro*. In the *cis* product, the phenyl groups become eclipsed and consequently, due to the large steric repulsion, the energy of activation for the formation of the *cis* product is larger than that for the *trans*.

cis Eliminations have been shown to occur with various 'onium hydroxides and are far more common than was originally believed (see p. 488).

The elimination reaction of 'onium hydroxides is the basis of the **Hofmann exhaustive methylation method** (1851, 1881). This method may be used to prepare alkenes, but it is largely used to ascertain the nature of the carbon skeleton of cyclic compounds containing a nitrogen in the ring (see p. 856).

An alternative to heating quaternary ammonium hydroxides to prepare alkenes is the **Cope reaction**, in which an amine oxide is the starting material. This is an example of a *cyclic* elimination mechanism (see p. 95).

Baumgarten *et al.* (1957) have applied the Cope reaction starting from a carboxylic acid containing an α-methylene group. The first step is the conversion of the acid to an amine by the Schmidt reaction (p. 239):

$$RCH_2CO_2H \xrightarrow{HN_3} RCH_2NH_2 \xrightarrow{MeI} RCH_2NMe_2 \xrightarrow{H_2O_2} RCH_2\overset{+}{N}Me_2-O^- \xrightarrow{heat} R^1CH=CH_2$$

Alternatively, the amide is prepared first:

$$RCH_2CO_2H \xrightarrow[\text{(ii) NH}_3]{\text{(i) SOCl}_2} RCH_2CONH_2 \xrightarrow{LiAlH_4} RCH_2CH_2NH_2 \cdots\cdots\rightarrow RCH=CH_2$$

In the former case, oxidation of the alkene gives an acid with two carbon atoms less than the original acid, and in the latter case, one carbon atom less. Thus either reaction may be used to step down the acid series.

Methylamine, dimethylamine and *trimethylamine* may be prepared by any of the general methods, but are conveniently prepared by special methods.

Methylamine: By heating ammonium chloride with two equivalents of formaldehyde (in formalin soln); the yield is 45–51 per cent based on the ammonium chloride (*cf.* p. 225):

$$2HCHO + NH_4Cl \longrightarrow CH_3NH_3{}^+Cl^- + HCO_2H$$

It is prepared industrially by passing a mixture of methanol and ammonia over alumina at 450°C under pressure and separating the mixture of amines by fractional distillation (see p. 368).

Methylamine is a gas, b.p. −7·6°C, which is used as a refrigerant.

Dimethylamine: By heating ammonium chloride with about four equivalents of formaldehyde (in formalin solution):

$$NH_4Cl + 4HCHO \longrightarrow (CH_3)_2NH_2{}^+Cl^- + 2HCO_2H$$

It is a gas, b.p. 7°C.

Trimethylamine: By heating a solid mixture of ammonium chloride and paraformaldehyde:

$$2NH_4Cl + 9HCHO \longrightarrow 2(CH_3)_3NH^+Cl^- + 3CO_2 \quad (89\%)$$

It is a gas, b.p. 3·5°C. It occurs in sugar residues, and is used as a source of methyl chloride:

$$(CH_3)_3N + 4HCl \xrightarrow[\text{pressure}]{heat} 3CH_3Cl + NH_4Cl$$

Nitrogen ylids (ylides)

A general method for preparing nitrogen ylids is by the reaction between quaternary ammonium halides and phenyl- or butyl-lithium, *e.g.*, tetramethylammonium chloride gives trimethylammonium methylide:

$$Me_4N^+Cl^- + PhLi \xrightarrow{Et_2O} PhH + [Me_3NCH_2Li]^+Cl^- \xrightarrow{-LiCl} Me_3\overset{+}{N}CH_2{}^-$$

Many ylids are of the type

$$\overset{+}{Z}-\overset{..}{\underset{}{C}}\diagup \longleftrightarrow Z=C\diagup$$

where Z is a heteroatom such as N, P, S, etc., attached to various groups. However, on the basis that N cannot expand its octet, N ylids differ from the other ylids in that they are not stabilised by *d*-orbital conjugation, *i.e.*, the second contributing resonance structure is absent for N ylids.

Ylids undergo reaction with a variety of compounds to form products, many of which may not be readily prepared by other methods, *e.g.*,

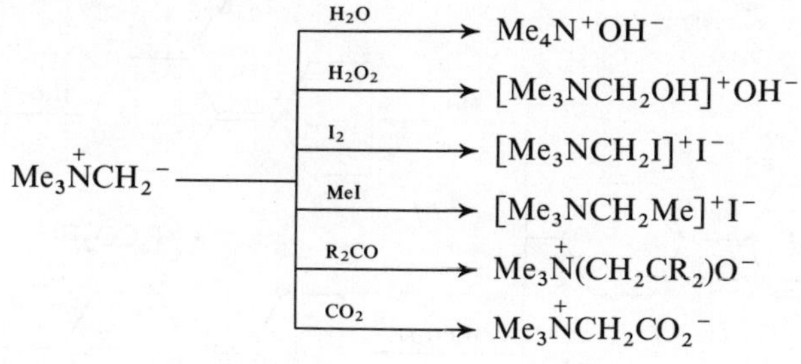

$$Me_3\overset{+}{N}CH_2{}^-$$

- $\xrightarrow{H_2O}$ $Me_4N^+OH^-$
- $\xrightarrow{H_2O_2}$ $[Me_3NCH_2OH]^+OH^-$
- $\xrightarrow{I_2}$ $[Me_3NCH_2I]^+I^-$
- $\xrightarrow{MeI}$ $[Me_3NCH_2Me]^+I^-$
- $\xrightarrow{R_2CO}$ $Me_3\overset{+}{N}(CH_2CR_2)O^-$
- $\xrightarrow{CO_2}$ $Me_3\overset{+}{N}CH_2CO_2{}^-$

Enamines

This name usually refers to α,β-unsaturated amines, and their most common method of preparation is by reaction between a carbonyl compound which contains at least one α-hydrogen atom and a secondary amine. This may be represented in general terms as:

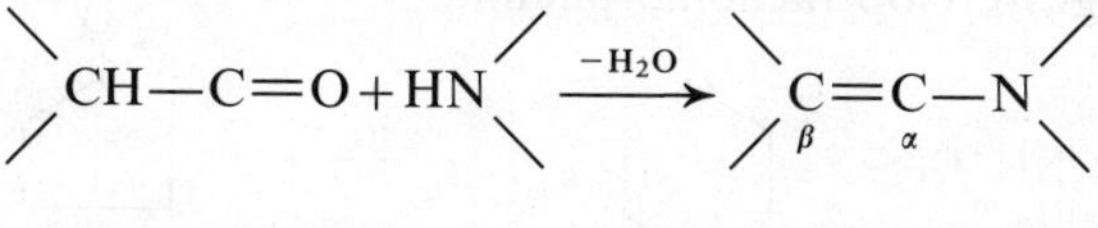

$$\begin{array}{c}\diagdown\\ \diagup\end{array}CH{-}C{=}O + HN\begin{array}{c}\diagup\\ \diagdown\end{array} \xrightarrow{-H_2O} \begin{array}{c}\diagdown\\ \diagup\end{array}\underset{\beta}{C}{=}\underset{\alpha}{C}{-}N\begin{array}{c}\diagup\\ \diagdown\end{array}$$

Enamines derived from aldehydes are less stable than those from ketones, and both types derived from acyclic secondary amines are far less stable than those from cyclic secondary amines, the most common of which are pyrrolidine, piperidine, and morpholine, e.g., cyclohexanone and pyrrolidine form *N*-cyclohexenopyrrolidine:

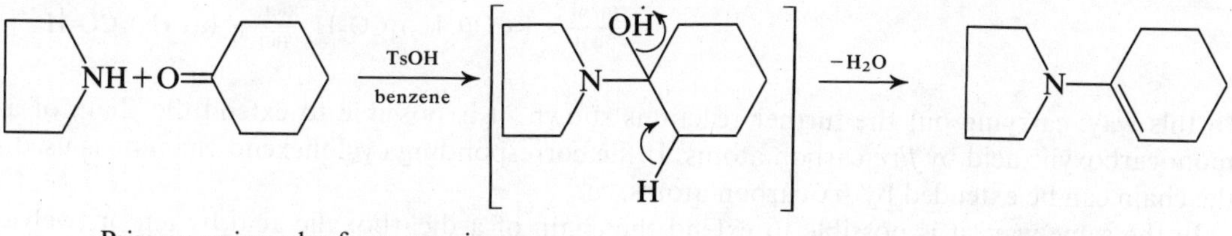

Primary amines also form enamines:

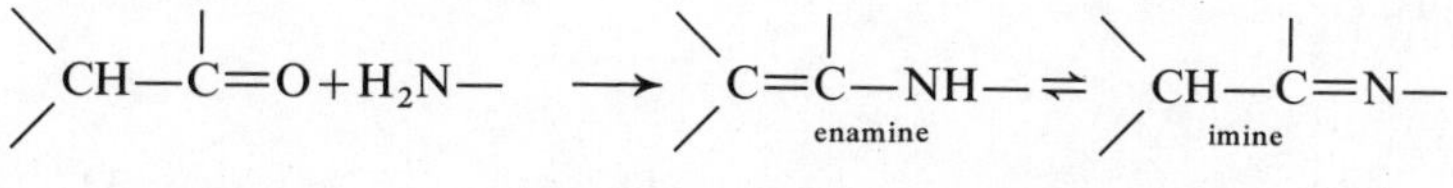

$$\begin{array}{c}\diagdown\\ \diagup\end{array}CH{-}\overset{|}{C}{=}O + H_2N{-} \longrightarrow \begin{array}{c}\diagdown\\ \diagup\end{array}C{=}\overset{|}{C}{-}\underset{enamine}{NH}{-} \rightleftharpoons \begin{array}{c}\diagdown\\ \diagup\end{array}CH{-}\overset{|}{\underset{imine}{C}}{=}N{-}$$

However, in this case, enamine-imine tautomerism is possible and the equilibrium lies virtually completely on the imine side, and consequently the enamine cannot be isolated.

Enamines may be represented as resonance hybrids of two canonical structures:

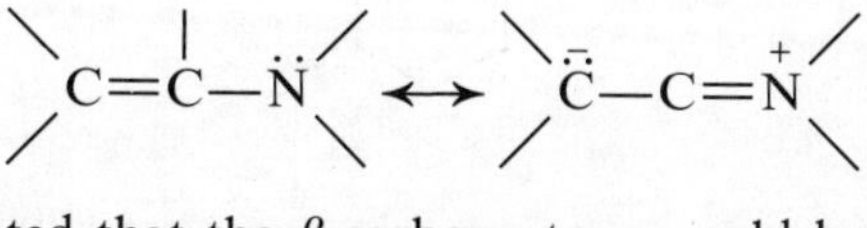

$$\begin{array}{c}\diagdown\\ \diagup\end{array}C{=}\overset{|}{C}{-}\ddot{N}\begin{array}{c}\diagup\\ \diagdown\end{array} \longleftrightarrow \begin{array}{c}\diagdown\\ \diagup\end{array}\overset{\bar{\bar{}}}{C}{-}C{=}\overset{+}{N}\begin{array}{c}\diagup\\ \diagdown\end{array}$$

Hence, it might be anticipated that the β-carbon atom would be susceptible to electrophilic attack, or alternatively, the β-carbon atom is a nucleophilic centre. This is so in practice, and consequently enamines are very useful as intermediates in organic syntheses, e.g.,

(i) **Alkylation.** Enamines are alkylated at the β-carbon atom by reactive alkyl halides, and react with α-bromoacetic esters, e.g.,

MeI
H₃O⁺
Me
I⁻
BrCH₂CO₂Et
CH₂CO₂Et
Br⁻

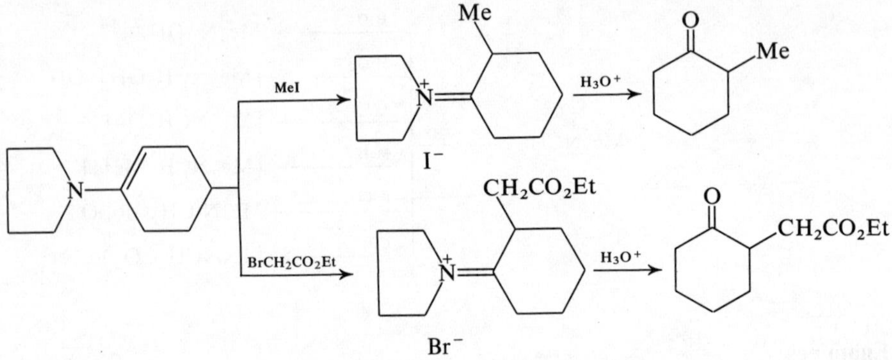

One advantage of this method over that of the direct action on the ketone (after treatment with sodium ethoxide) is that monosubstituted enamines are relatively difficult to alkylate further, and so can be isolated.
(ii) **Acylation**, *e.g.*, using *N*-cyclopentenomorpholine:

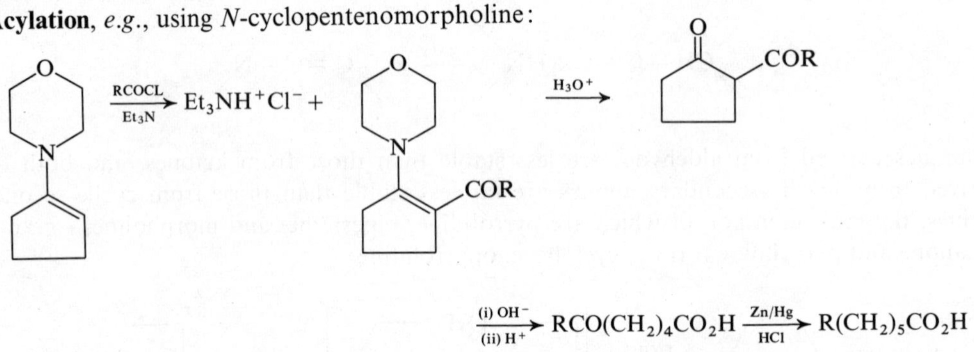

RCOCL
Et₃N
Et₃NH⁺Cl⁻ +
COR
H₃O⁺
(i) OH⁻
(ii) H⁺
RCO(CH₂)₄CO₂H
Zn/Hg
HCl
R(CH₂)₅CO₂H

In this way, carrying out the further reactions shown, it is possible to extend the chain of a monocarboxylic acid by *five* carbon atoms. If the corresponding cyclohexeno-enamine is used, the chain can be extended by *six* carbon atoms.

In the same way, it is possible to extend the chain of a dicarboxylic acid by ten or twelve carbon atoms, *e.g.*,

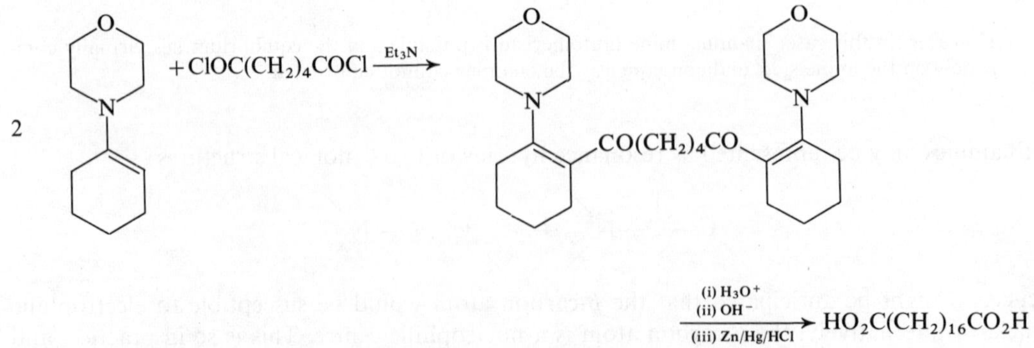

2
+ClOC(CH₂)₄COCl
Et₃N
CO(CH₂)₄CO
(i) H₃O⁺
(ii) OH⁻
(iii) Zn/Hg/HCl
HO₂C(CH₂)₁₆CO₂H

It might be noted that triethylamine is used with acyl halides because *acylated* enamines are too weakly basic to absorb the HX produced in the reaction.

Formylation of enamines can be carried out with dimethylformamide as follows:

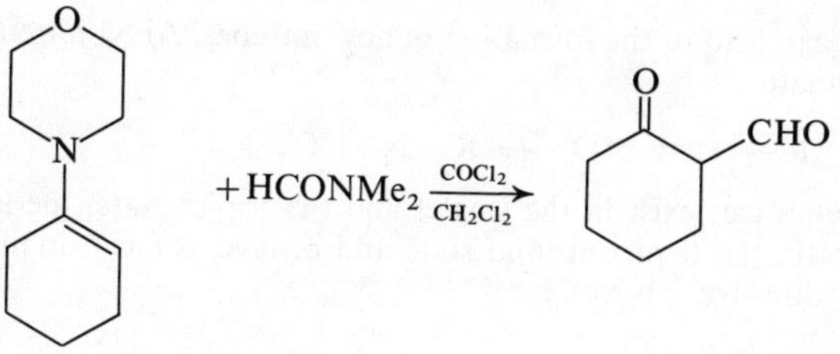

(iii) **Michael-type of condensation.** Enamines behave as addenda to α,β-unsaturated ketones, esters, cyanides, etc. (*cf.* p. 340). *e.g.*,

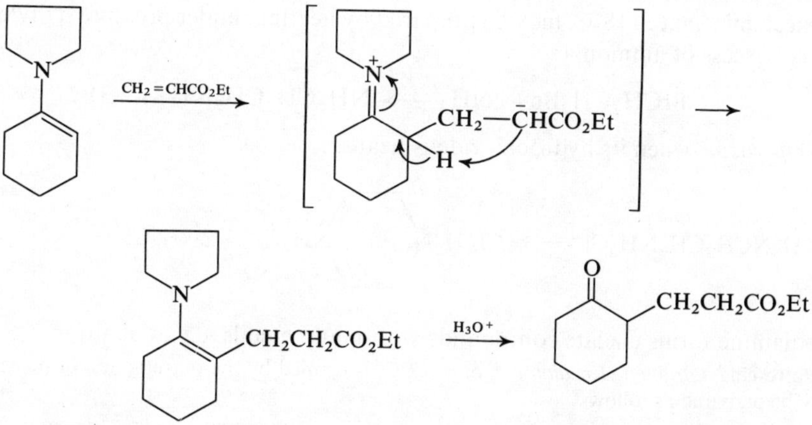

(iv) **Introduction of a cyano group.** Enamines are treated with cyanogen chloride in the presence of Et_3N:

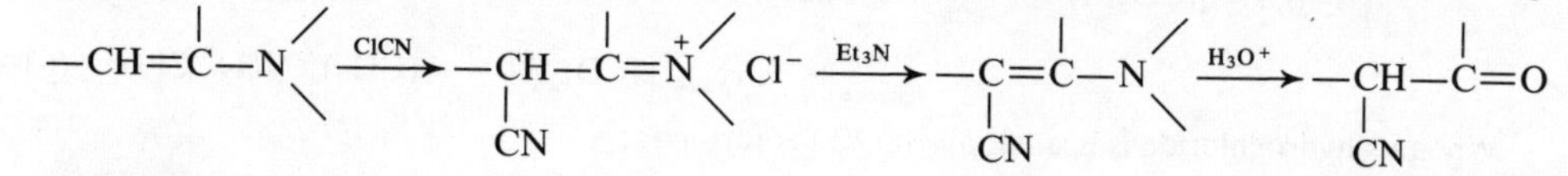

(v) **Reduction.** Enamines are reduced catalytically or by sodium borohydride (but not by LAH) to saturated t-amines:

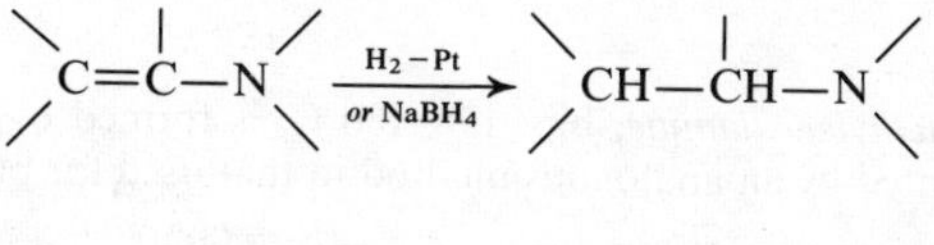

Nitrenes

The parent compound is nitrene, NH (also known as *imidogen*, *azene*, or *imene*), and is formed when hydrazoic acid is irradiated with ultraviolet light. In the presence of ethylene, nitrene is trapped to form ethylenimine (p. 387).

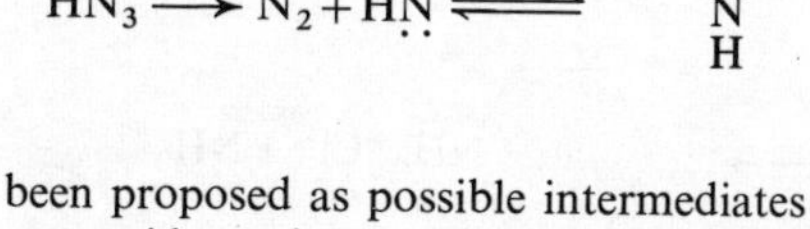

Acyl nitrenes, $RCO\ddot{N}$, have been proposed as possible intermediates in the Hofmann, Curtius, and Lossen rearrangements, but recent evidence does not favour this (see p. 265). However, there is evidence that the *photolysis* of acyl azides can lead to the formation of acyl nitrenes. Alkyl nitrenes may be prepared by the photolysis of isocyanates:

$$R{-}N{=}C{=}O \xrightarrow{h\nu} R{-}\ddot{N}+CO$$

Like methylene (p. 119), nitrenes can exist in the singlet and the triplet state, the latter being the ground state. Most nitrenes exist in the triplet ground state and behave as electrophiles, undergoing cyclo-addition and insertion reactions (see above).

Diamines

Diamines may be prepared by methods similar to those for the monoamines, using polymethylene halides instead of alkyl halides.

Ethylenediamine. b.p. 118°C, may be prepared by heating, under pressure, ethylene dibromide with a large excess of ammonia:

$$BrCH_2CH_2Br + 2NH_3 \longrightarrow NH_2CH_2CH_2NH_2 + 2HBr$$

It forms *piperazine* when its hydrochloride is heated:

$$Cl^-H_3\overset{+}{N}CH_2CH_2\overset{+}{N}H_3Cl^- \longrightarrow Cl^-H_2\overset{+}{N}\hspace{3em}\overset{+}{N}H_2Cl^- + 2NH_4Cl$$

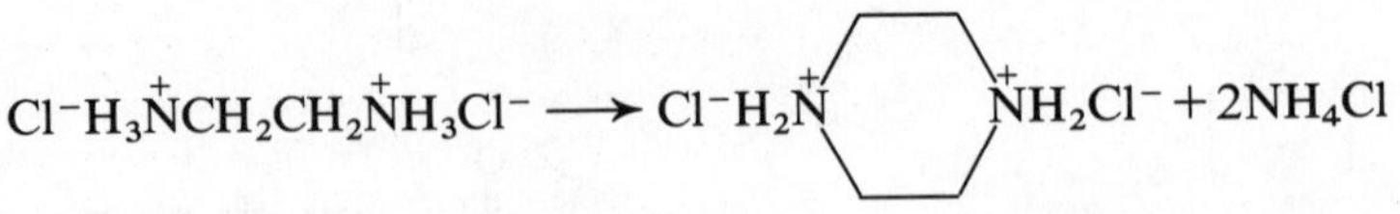

Ethylenediamine forms chelate compounds with many metals, *e.g.*, cobalt.

Putrescine, *tetramethylenediamine*, m.p. 27°C, is formed by the putrefaction of proteins (in flesh). It may be prepared as follows:

$$BrCH_2CH_2Br \xrightarrow{KCN} NCCH_2CH_2CN \xrightarrow{Na/C_2H_5OH} NH_2(CH_2)_4NH_2$$

A convenient synthesis starts with ethyl adipate and uses the Curtius rearrangement (see p. 268):

$$(CH_2)_4(CO_2Et)_2 \xrightarrow[\text{reflux}]{N_2H_4/EtOH} (CH_2)_4(CONHNH_2)_2 \xrightarrow{HNO_2} (CH_2)_4(CON_3)_2$$

$$\xrightarrow[\text{benzene}]{\text{heat in}} (CH_2)_4(NCO)_2 \xrightarrow{aq.\ HCl} (CH_3)_4(NH_3{}^+Cl^-)_2 \quad (63\text{--}73\%)$$

When its hydrochloride is heated, *pyrrolidine* is formed:

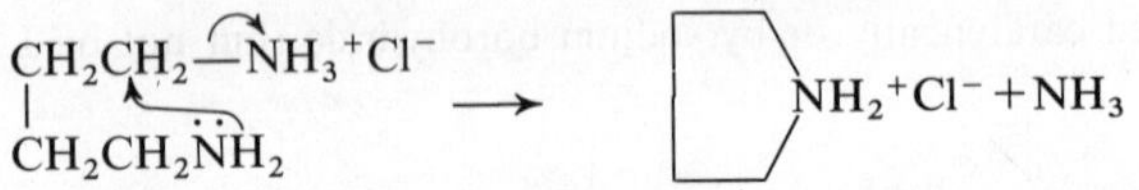

$$\begin{array}{l} CH_2CH_2{-}NH_3{}^+Cl^- \\ | \\ CH_2CH_2\ddot{N}H_2 \end{array} \longrightarrow \left[\quad\right]NH_2{}^+Cl^- + NH_3$$

Cadaverine, *pentamethylenediamine*, b.p. 178–180°C, is formed by the putrefaction of proteins (in flesh). It may be prepared by an analogous method to that used for putrescine:

$$Br(CH_2)_3Br \xrightarrow{KCN} NC(CH_2)_3CN \xrightarrow{Na/C_2H_5OH} NH_2(CH_2)_5NH_2$$

It is, however, prepared more conveniently by heating pentamethylene dibromide with an excess of ammonia:

$$Br(CH_2)_5Br + 2NH_3 \longrightarrow NH_2(CH_2)_5NH_2 + 2HBr^-$$

The starting material pentamethylene dibromide is readily obtained from piperidine (see p. 855). When its hydrochloride is heated, piperidine is formed:

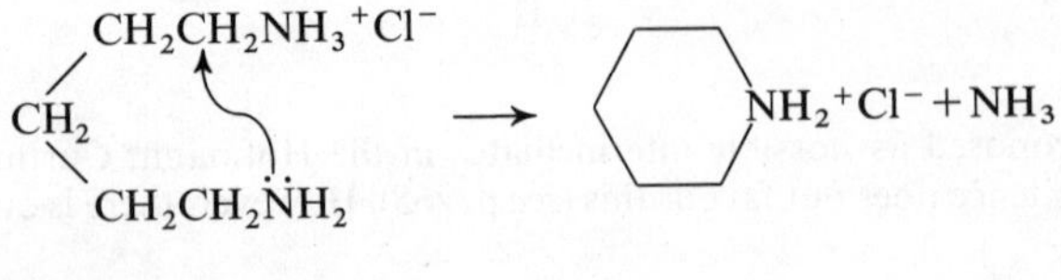

$$\begin{array}{l} CH_2CH_2NH_3{}^+Cl^- \\ CH_2 \\ CH_2CH_2\ddot{N}H_2 \end{array} \longrightarrow \langle\quad\rangle NH_2{}^+Cl^- + NH_3$$

Hexamethylenediamine is manufactured by the catalytic hydrogenation of adiponitrile prepared from adipic acid:

$$HO_2C(CH_2)_4CO_2H \xrightarrow[\text{catalyst}]{NH_3} NC(CH_2)_4CN \xrightarrow[\text{catalyst}]{H_2} NH_2(CH_2)_6NH_2$$

Adiponitrile is also manufactured by the electrolytic hydrodimerisation of vinyl cyanide (p. 346).

$$2CH_2{=}CHCN \xrightarrow[2H_2O]{2e} NC(CH_2)_4CN$$

Hexamethylenediamine, m.p. 39°C, is used in the manufacture of *nylon* (p. 456).

Spermine, a deliquescent crystalline solid which is isolated from human sperm, is a tetramine:

$$NH_2(CH_2)_3NH(CH_2)_4NH(CH_2)_3NH_2.$$

Diamines with two amino-groups attached to the same carbon atom are unknown (*cf.* the group $C(OH)_2$, p. 228). On the other hand, *N*-substituted derivatives have been prepared, *e.g.*, tetraethylmethylenediamine:

$$(C_2H_5)_2NH + CH_2O + NH(C_2H_5)_2 \longrightarrow (C_2H_5)_2NCH_2N(C_2H_5)_2 + H_2O$$

Unsaturated amines

The simplest unsaturated amine would be vinylamine (ethenylamine) if it existed. All attempts to prepare it result in the formation of the cyclic compound **ethylenimine (aziridine)**, and a good method is to treat 2-aminoethanol with sulphuric acid and then to heat the product with aqueous sodium hydroxide.

$$NH_2CH_2CH_2OH \xrightarrow{H_2SO_4} \overset{+}{N}H_3CH_2CH_2OSO_3{}^- \xrightarrow{NaOH} \underset{CH_2{-}CH_2}{\overset{NH}{\diagup\diagdown}}$$

Thus this compound is a three-membered heterocyclic ring. Ethylenimine is a syrupy liquid, b.p. 56°C (see also p. 386). It combines with sulphurous acid to form *taurine* (2-aminoethanesulphonic acid), which occurs in human bile:

$$\underset{NH}{\overset{CH_2{-}CH_2}{\diagdown\diagup}} + H_2SO_3 \longrightarrow \overset{+}{N}H_3CH_2CH_2SO_3{}^-$$

Neurine (*trimethylvinylammonium hydroxide*) is found in the brain. It may be prepared:
(i) By boiling choline with barium hydroxide solution:

$$[(CH_3)_3NCH_2CH_2OH]^+OH^- \xrightarrow{Ba(OH)_2} [(CH_3)_3NCH{=}CH_2]^+OH^- + H_2O$$

(ii) By heating trimethylamine and ethylene dibromide (one molecule of each), and subsequently heating the product with silver oxide in water:

$$(CH_3)_3N + BrCH_2CH_2Br \longrightarrow [(CH_3)_3NCH_2CH_2Br]^+Br^- \xrightarrow{AgOH} [(CH_3)_3NCH{=}CH_2]^+OH^-$$

Neurine is a very poisonous syrupy liquid; it is a quaternary enamine.

Allylamine (2-*propenylamine*), b.p. 53°C, may be prepared by heating allyl iodide with ammonia:

$$CH_2{=}CHCH_2I + NH_3 \longrightarrow CH_2{=}CHCH_2NH_2 + HI$$

It is more conveniently prepared by boiling allyl isothiocyanate with dilute hydrochloric acid:

$$CH_2{=}CHCH_2NCS + H_2O \xrightarrow{HCl} CH_2{=}CHCH_2NH_2 + COS \quad (70\text{--}73\%)$$

Aminoalcohols

The simplest aminoalcohol is 2-**aminoethanol** (*cholamine, ethanolamine, 2-hydroxyethylamine*). It occurs in kephalins (p. 324), and is best prepared by the action of ethylene oxide on excess of ammonia.

$$\underset{CH_2{-}CH_2}{\overset{O}{\diagup\diagdown}} + NH_3 \longrightarrow NH_2CH_2CH_2OH$$

It is a viscous liquid, b.p. 171°C, miscible with water, and is strongly basic. It readily reacts with concentrated hydrobromic acid to form β-bromoethylamine hydrobromide (83 per cent).

Choline (*2-hydroxyethyltrimethylammonium hydroxide*) occurs in lecithins (p. 324); it is best prepared by the action of ethylene oxide on trimethylamine in aqueous solution:

$$CH_2\!\!-\!\!-\!\!CH_2 + (CH_3)_3N + H_2O \longrightarrow [(CH_3)_3NCH_2CH_2OH]^+OH^-$$

It is a colourless viscous liquid, soluble in water, and is strongly basic. It forms neurine when boiled with barium hydroxide solution (see above). It is present in the vitamin B complex; it is a growth factor in chicks.

Amino-acids

Amino-acids are derivatives of the carboxylic acids in which a hydrogen atom in the carbon chain has been replaced by an amino-group. The amino-group may occupy the α- or β- or γ- . . . position; there may also be two or more amino-groups present in the chain.

The three basic classes of foods are: proteins, fats and carbohydrates. The proteins are nitrogenous substances which occur in most cells of the animal body; they also occur in plants. When hydrolysed by strong inorganic acids or by enzymes, proteins yield a mixture of amino-acids, all of which are α-amino-acids. The number of amino-acids obtained from proteins is about twenty-five, of which about ten are essential, *i.e.*, a deficiency in any one prevents growth in young animals, and may even cause death.

The amino-acids are classified in several ways; Table 13.2 shows a convenient classification; the letters *g*, *l* and *e* which follow the name of the acid indicate that the acid is respectively of general occurrence, lesser occurrence and essential (in man).

In this book we shall consider in detail only the simplest amino-acid, **glycine** (*aminoacetic acid, glycocoll*), $CH_2(NH_2)CO_2H$ (see also Vol. 2, Ch. 13). This acid is found in many proteins, and occurs in certain animal excretions, usually in combination, *e.g.*, *hippuric acid* (in horses' urine), $C_6H_5CONHCH_2CO_2H$. Glycine may be readily prepared by the action of concentrated ammonium hydroxide solution on chloroacetic acid:

$$ClCH_2CO_2H + 2NH_3 \longrightarrow CH_2(NH_2)CO_2H + NH_4Cl \quad (64\text{--}65\%)$$

It may also be prepared pure by Gabriel's phthalimide synthesis (p. 370):

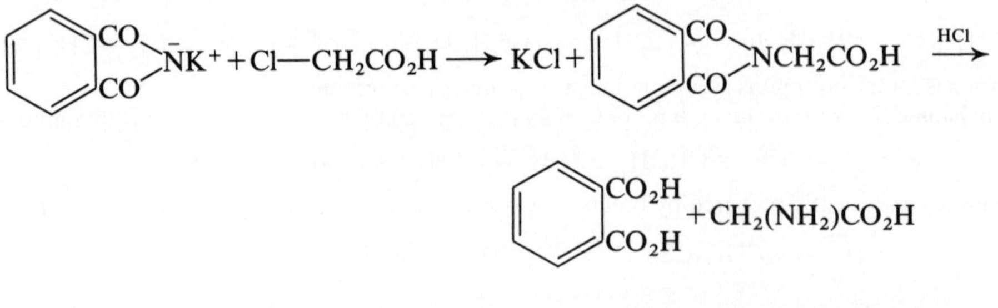

Both methods may also be used to prepare alanine, $CH_3CH(NH_2)CO_2H$, from α-bromopropionic acid. Alanine may also be prepared by the **Strecker synthesis** (1850); this is the reaction between an aldehyde-ammonia and hydrogen cyanide, and the resulting amino-nitrile is then hydrolysed. In practice, the reaction is carried out as follows:

$$MeCHO \xrightarrow[\text{KCN}]{\text{NH}_4\text{Cl}} MeCH(NH_2)CN \xrightarrow[\text{H}_2\text{O}]{\text{HCl}} MeCH(NH_2)CO_2H$$

Table 13.2

Name	Formula

Neutral amino-acids (one amino-group and one carboxyl group)

1. Glycine (g) $\quad$ $CH_2(NH_2)CO_2H$
2. Alanine (g) $\quad$ $CH_3CH(NH_2)CO_2H$
3. Valine (g, e) $\quad$ $(CH_3)_2CHCH(NH_2)CO_2H$
4. Leucine (g, e) $\quad$ $(CH_3)_2CHCH_2CH(NH_2)CO_2H$
5. Isoleucine (g, e) $\quad$ $(C_2H_5)(CH_3)CHCH(NH_2)CO_2H$
6. Norleucine (l) $\quad$ $CH_3(CH_2)_3CH(NH_2)CO_2H$

7. Phenylalanine (g, e)

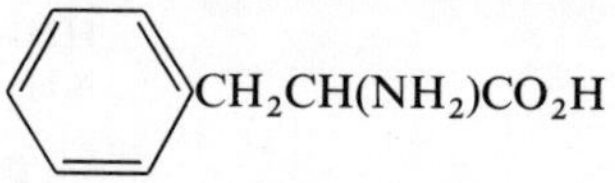

8. Tyrosine (g)

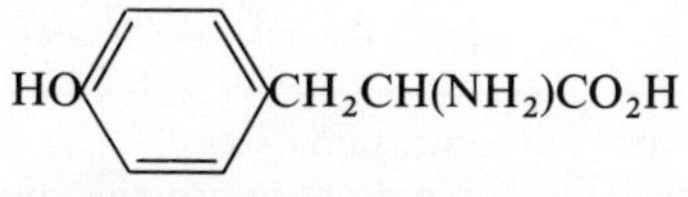

9. Serine (g) $\quad$ $HOCH_2CH(NH_2)CO_2H$
10. Cysteine (g) $\quad$ $HSCH_2CH(NH_2)CO_2H$
11. Cystine (g) $\quad$ $(-SCH_2CH(NH_2)CO_2H)_2$
12. Threonine (g, e) $\quad$ $CH_3CHOHCH(NH_2)CO_2H$
13. Methionine (g, e) $\quad$ $CH_3SCH_2CH_2CH(NH_2)CO_2H$

14. Di-iodotyrosine or iodogorgic acid (l)

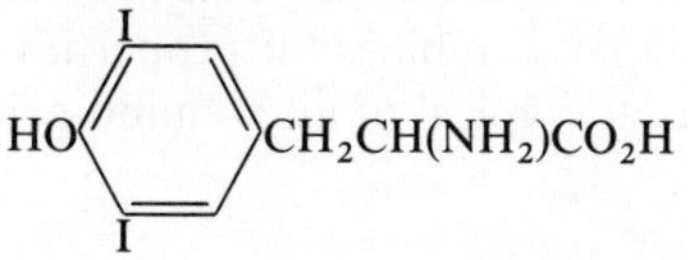

15. Thyroxine (l)

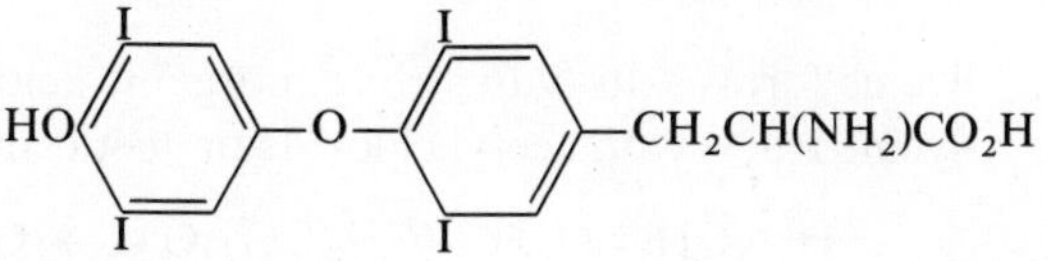

16. Dibromotyrosine (l)

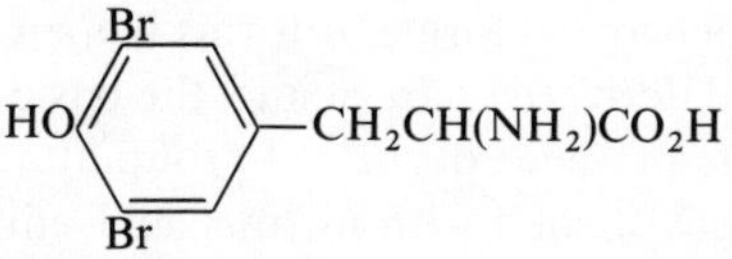

17. Tryptophan (g, e)

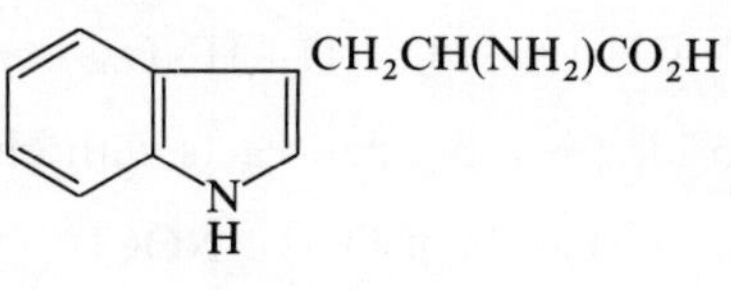

18. Proline (g)

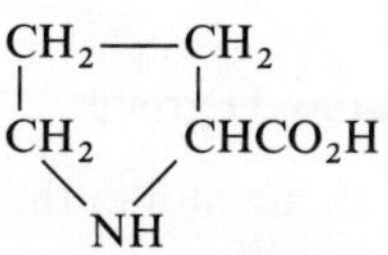

19. Hydroxyproline (l)

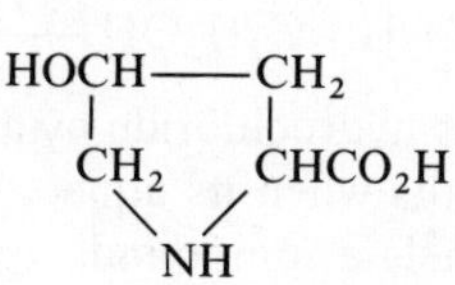

Table 13.2 *(continued)*

Name	Formula
Acidic amino-acids (one amino-group and two carboxyl groups)	
20. Aspartic acid (*g*)	$HO_2CCH_2CH(NH_2)CO_2H$
21. Glutamic acid (*g*)	$HO_2CCH_2CH_2CH(NH_2)CO_2H$
22. β-Hydroxyglutamic acid (*l*)*	$HO_2CCH_2CHOHCH(NH_2)CO_2H$
Basic amino-acids (two amino-groups and one carboxyl group)	
23. Ornithine†	$NH_2CH_2CH_2CH_2CH(NH_2)CO_2H$
24. Arginine (*g, e*)	$\overset{\displaystyle NH_2}{\overset{\displaystyle \vert}{HN{=}C}}{-}NHCH_2CH_2CH_2CH(NH_2)CO_2H$
25. Lysine (*g, e*)	$NH_2CH_2CH_2CH_2CH_2CH(NH_2)CO_2H$
26. Histidine (*g, e*)	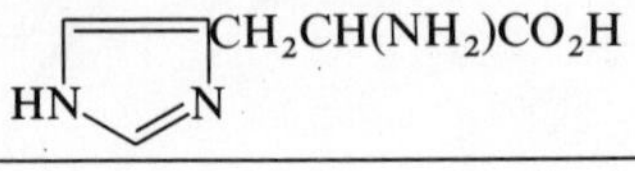 $\underset{HN\diagdown\!\!\diagup N}{\boxed{}}CH_2CH(NH_2)CO_2H$

*Occurrence in proteins uncertain.
†Ornithine is probably not present in proteins, but is formed by the hydrolysis of arginine.

Glycine exists as white prisms which melt, with decomposition, at 289–292°C. It has a sweet taste, and is soluble in water but insoluble in ethanol and ether. Since it contains an amino-group and a carboxyl group, it combines the properties of a base and an acid, *i.e.*, it is *amphoteric*. The following reactions are typical of all α-amino-acids.

Reactions characteristic of the amino-group

1. Glycine forms salts with strong inorganic acids, *e.g.*, $Cl\{\overset{+}{H_3}\overset{-}{N}CH_2CO_2H}$.
2. Glycine reacts with acetyl chloride or acetic anhydride to give the acetyl derivative:

$$H_2NCH_2CO_2H+(CH_3CO)_2O \longrightarrow CH_3CONHCH_2CO_2H+CH_3CO_2H$$

Similarly, with benzoyl chloride, it forms benzoylglycine (hippuric acid; see above).

These acylated derivatives are acidic, the basic character of the amino-group being effectively eliminated by the presence of the $-I$ group attached to the nitrogen (see p. 659).
3. When glycine is treated with nitrous acid, nitrogen is evolved and glycollic acid is formed:

$$CH_2(NH_2)CO_2H+HNO_2 \longrightarrow HOCH_2CO_2H+N_2+H_2O$$

4. Nitrosyl chloride (or bromide) reacts with glycine to form chloro- (or bromo-) acetic acid:

$$CH_2(NH_2)CO_2H+NOCl \longrightarrow ClCH_2CO_2H+N_2+H_2O$$

Reactions characteristic of the carboxyl group

1. Glycine may be esterified by an alcohol in the presence of inorganic acid, *e.g.*,

$$CH_2(NH_2)CO_2H+C_2H_5OH \xrightarrow{\text{HCl}} \overset{+}{Cl}\overset{-}{H_3}NCH_2CO_2C_2H_5+H_2O$$

The ester is liberated from its hydrochloride by alkali.
2. Glycine forms metallic salts when its aqueous solution is warmed with a metallic oxide or hydroxide. These salts are chelate compounds, *e.g.*, the copper salt (deep blue needles) is:

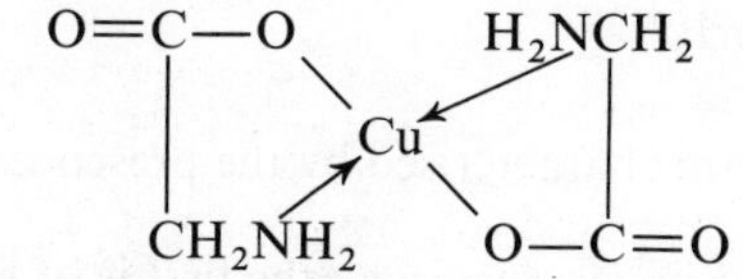

$$\begin{array}{ccc}
O{=}C{-}O & & H_2NCH_2 \\
| & \diagdown & | \\
& Cu & \\
| & \diagup \quad \diagdown & | \\
CH_2NH_2 & & O{-}C{=}O
\end{array}$$

3. Amino-acids may be reduced to aminoalcohols by lithium aluminium hydride (Vogel *et al.* 1952).

$$RCH(NH_2)CO_2H \xrightarrow{\text{LiAlH}_4} RCH(NH_2)CH_2OH$$

Reactions due to both the amino and carboxyl groups

1. When measured in aqueous solution, the dipole moment of glycine is found to have a large value. To account for this large value it has been suggested that glycine exists, in solution, as an *inner salt*:

$$H_2NCH_2CO_2H + H_2O \rightleftharpoons H_3\overset{+}{N}CH_2CO_2{}^- + H_2O$$

Such a doubly charged ion is known, in addition to an inner salt, as a *zwitterion*, *ampholyte* or a *dipolar ion*. This dipolar ion structure also accounts for the absence of acidic and basic properties of an amino-acid (the carboxyl and amino-groups of the *same* molecule neutralising each other to form a salt). The properties of crystalline glycine, *e.g.*, its high melting point and its insolubility in hydrocarbon solvents, also indicate that it exists as the inner salt in the solid state. The presence of the $NH_3{}^+$ group in glycine has now been confirmed by i.r. and NMR spectra studies.

Owing to its amphoteric character, glycine cannot be titrated directly with alkali. When formaline solution is added to glycine, methyleneglycine is formed (reaction probably more complex):

$$H_2NCH_2CO_2H + HCHO \longrightarrow CH_2{=}NCH_2CO_2H + H_2O$$

This is a strong acid (the basic character of the amino-group being now suppressed), and can be titrated with sodium hydroxide. This method is known as the *Sörensen titration*.

2. When heated, glycine forms *diketopiperazine*; glycine esters give a better yield:

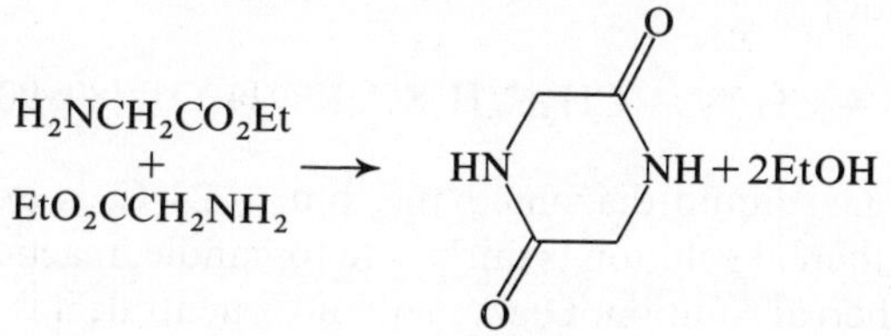

$$\begin{array}{c}
H_2NCH_2CO_2Et \\
+ \\
EtO_2CCH_2NH_2
\end{array}
\longrightarrow
\quad HN\bigg\langle\!\!\!\!\!\!\!\!\!\!\!\!\!\!\! \text{(ring)} \!\!\!\!\!\!\!\!\!\!\!\!\!\!\!\bigg\rangle NH + 2EtOH$$

Betaines. These are the trialkyl derivatives of glycine, which exist as dipolar ions of formula $R_3\overset{+}{N}CH_2\overset{-}{C}O_2$. Betaine itself is the trimethyl derivative, and may be prepared by heating glycine with methyl iodide in methanolic solution:

$$H_3\overset{+}{N}CH_2\overset{-}{C}O_2 + 3CH_3I \longrightarrow (CH_3)_3\overset{+}{N}CH_2\overset{-}{C}O_2 + 3HI$$

It is more conveniently prepared by warming an aqueous solution of chloroacetic acid with trimethylamine:

$$(CH_3)_3N + ClCH_2CO_2H \longrightarrow (CH_3)_3\overset{+}{N}CH_2\overset{-}{C}O_2 + HCl$$

Betaine is a solid, m.p. 300°C (with decomposition). It occurs in nature, especially in plant juices. It behaves as a base, *e.g.*, with hydrochloric acid it forms the stable crystalline hydrochloride, $Cl^-(CH_3)_3\overset{+}{N}CH_2CO_2H$.

Aliphatic diazo-compounds

The aliphatic diazo-compounds are characterised by the presence of the group $\diagdown \diagup CN_2$.

Diazomethane may be prepared in various ways; the first is of historical importance, and the others are very convenient methods.

Method of Von Pechmann (1894). Methylamine is treated with ethyl chloroformate to give *N*-methylurethan which, on treatment with nitrous acid in ethereal solution, forms *N*-methyl-*N*-nitroso-urethan. This, on warming with methanolic potassium hydroxide, decomposes into diazomethane, which is collected in cooled ether:

$$CH_3NH_2 + ClCO_2C_2H_5 \longrightarrow CH_3NHCO_2C_2H_5 \xrightarrow{HNO_2}$$

$$CH_3N(NO)CO_2C_2H_5 \xrightarrow{KOH} CH_2N_2 + CO_2 + C_2H_5OH$$

An improved method starts with methylurea:

$$MeNHCONH_2 \xrightarrow{HNO_2} MeN(NO)CONH_2 \xrightarrow{KOH} CH_2N_2 + KNCO + 2H_2O$$

Method of McKay (1948). Methylamine hydrochloride and nitroguanidine are allowed to react in potassium hydroxide solution, the product *N*-methyl-*N'*-nitroguanidine treated with nitrous acid, and the *N*-methyl-*N*-nitroso-*N'*-nitroguanidine so produced is then warmed with potassium hydroxide:

$$CH_3NH_3{}^+Cl^- + NH_2\overset{\overset{\displaystyle NH}{\|}}{C}NHNO_2 + KOH \longrightarrow$$

$$NH_3 + KCl + CH_3NH\overset{\overset{\displaystyle NH}{\|}}{C}NHNO_2 \xrightarrow{HNO_2} CH_3N(NO)\overset{\overset{\displaystyle NH}{\|}}{C}NHNO_2 \xrightarrow{KOH} CH_2N_2$$

Method of Backer et al. (1951). The nitroso-derivative of *p*-toluene-*N*-methylsulphonamide is distilled with ethanolic potassium hydroxide:

$$CH_3C_6H_4SO_2N(CH_3)NO + KOH \longrightarrow CH_2N_2 + CH_3C_6H_4SO_3K + H_2O \quad (80\text{--}90\%)$$

Diazomethane is a yellow, poisonous gas; liquid diazomethane, b.p. $-24°C$, is explosive. The gas is soluble in ether, and since the ethereal solution is fairly safe to handle, reactions with diazomethane are usually carried out in ethereal solution. Diazomethane is neutral; it is reduced by sodium amalgam to methylhydrazine.

Diazomethane is best represented as a resonance hybrid derived from linear resonating structures with opposing dipoles:

$$CH_2{=}\overset{+}{N}{=}\overset{-}{\underset{..}{N}}: \longleftrightarrow \overset{+}{CH_2}{-}\overset{..}{N}{=}\overset{-}{\underset{..}{N}}: \longleftrightarrow \overset{-}{CH_2}{-}\overset{+}{N}{\equiv}N: \longleftrightarrow \overset{-}{CH_2}{-}\overset{..}{N}{=}\overset{+}{N}:$$
$$\text{(I)} \qquad\qquad \text{(II)} \qquad\qquad \text{(III)} \qquad\qquad \text{(IV)}$$

Under appropriate conditions, diazomethane can behave as an electrophile (II) or a nucleophile (III), as a 1,3-dipole (IV), or as a source of methylene (p. 119).

Reactions

Most of the reactions of diazomethane are readily explained in terms of nucleophilic attack (structure III) on the substrate, followed by loss of nitrogen.

Halogen acids react with diazomethane to form methyl halides.

$$\overset{\frown}{Cl-H} \ \overset{\frown}{CH_2} - \overset{+}{N} \equiv N \longrightarrow Cl^{\frown} \ CH_3 - \overset{+}{N} \equiv N \longrightarrow Cl-CH_3 + N_2$$

Diazomethane methylates *acidic* hydroxyl groups very readily: carboxylic acids, sulphonic acids, phenols and enols. A possible mechanism is:

$$RCO-\overset{\frown}{O-H} \ \overset{\frown}{CH_2} - \overset{+}{N} \equiv N \longrightarrow RCO-O^- \ ^{\frown}CH_3 - \overset{+}{N} \equiv N \xrightarrow{-N_2} RCOOCH_3$$

$$\downarrow -N_2$$

$$RCO-O^- + CH_3^+$$

Alcohols can also be methylated by diazomethane in the presence of a suitable catalyst, *e.g.*, an aluminium alkoxide. Fluoroboric acid is also a very good catalyst for primary alcohols (Müller, 1958).

$$R^1OH + Al(OR^2)_3 \rightleftharpoons \quad \underset{H}{\overset{R^1}{\diagdown}} \overset{+}{O} - \bar{A}l(OR^2)_3 \xrightarrow{CH_2N_2}$$

$$N_2 + \quad \underset{H_3C}{\overset{R^1}{\diagdown}} \overset{+}{O} - \bar{A}l(OR^2)_3 \rightleftharpoons R^1OCH_3 + Al(OR^2)_3$$

Catalysts are necessary with alcohols because of the low acidity of the hydroxyl hydrogen.

Diazomethane reacts with aldehydes to form methyl ketones and with ketones to form the higher homologue. In both cases, an epoxide is formed and may be the predominant product.

$$R_2C\overset{\frown}{\underset{O}{\overset{\parallel}{C}}}{}^{\frown}CH_2 - \overset{+}{N} \equiv N \longrightarrow R-\underset{R}{\overset{O^-}{\underset{|}{\overset{|}{C}}}}-CH_2 - \overset{+}{N} \equiv N \xrightarrow{-N_2} R-\overset{O}{\overset{\parallel}{C}}-CH_2R$$

$$\longrightarrow R_2\overset{O^-}{\underset{|}{C}}-CH_2 - \overset{+}{N} \equiv N \xrightarrow{-N_2} R_2C \overset{O}{\overbrace{\qquad}} CH_2$$

Certain cyclic ketones undergo ring expansion when treated with diazomethane (p. 544).

Diazomethane adds on to ethylenic compounds to form *pyrazoline* derivatives (*cf.* α,β-unsaturated esters, p. 345); with ethylene, *pyrazoline* is formed. The addition may be formulated (see also below):

$$\begin{array}{c} CH_2{=}CH_2 \\ \overset{+}{CH_2}-N{=}N^- \end{array} \longrightarrow \left[\begin{array}{c} CH_2{-\!-\!-}\overset{+}{CH_2} \\ | \quad\quad \nwarrow \\ CH_2 \quad N^- \\ \diagdown \ \ /\!/ \\ N \end{array} \right] \longrightarrow \left[\begin{array}{c} CH_2{-\!-\!-}CH_2 \\ | \quad\quad | \\ CH_2 \quad N \\ \diagdown \ \ /\!/ \\ N \end{array} \right] \longrightarrow \begin{array}{c} CH_2{-\!-\!-}CH \\ | \quad\quad \| \\ CH_2 \quad N \\ \diagdown \quad / \\ NH \end{array}$$

Diazomethane also adds on to acetylenic compounds, in this case to form *pyrazole* derivatives; with acetylene, *pyrazole* is formed:

$$\begin{array}{c} CH{\equiv}CH \\ \overset{+}{CH_2}-N{=}N^- \end{array} \longrightarrow \left[\begin{array}{c} CH{=\!=}\overset{+}{CH} \\ | \quad\quad \nwarrow \\ CH_2 \quad N^- \\ \diagdown \ \ /\!/ \\ N \end{array} \right] \longrightarrow \left[\begin{array}{c} CH{=\!=}CH \\ | \quad\quad | \\ CH_2 \quad N \\ \diagdown \ \ /\!/ \\ N \end{array} \right] \longrightarrow \begin{array}{c} CH{-\!-\!-}CH \\ \| \quad\quad \| \\ CH \quad N \\ \diagdown \quad / \\ NH \end{array}$$

Additions such as these, which involve a molecule containing a multiple bond and a molecule containing three atoms in a chain with its terminal atoms carrying a small positive and negative charge, respectively, are referred to as 1,3-*dipolar additions*. It is quite possible that the cyclo-addition occurs in one step rather than two.

Diazomethane is used in the **Arndt–Eistert synthesis** (1935). This is a means of converting an acid (aliphatic, aromatic, alicyclic, or heterocyclic) into the next higher homologue. The acid chloride is treated with diazomethane (2 molecules); the product is a diazoketone, and this loses nitrogen to form a keten by rearrangement when warmed in the presence of silver oxide as catalyst.

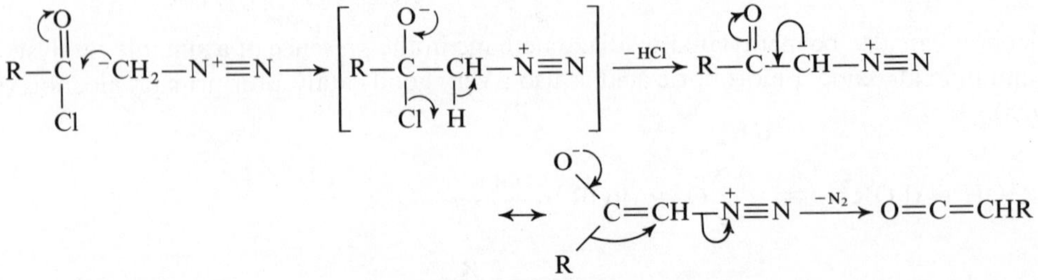

This rearrangement of the diazoketone is usually referred to as the **Wolff rearrangement** (1912).

When the rearrangement of the diazoketone is carried out in the presence of water, a carboxylic acid is formed (p. 350):

$$RCH=C=O + H_2O \longrightarrow RCH_2CO_2H$$

Diazoketones react with alcohols and ammonia in the presence of silver oxide as catalyst to form esters and amides, respectively (via the keten):

$$R^1CH_2CO_2R^2 \xleftarrow{R^2OH} R^1COCHN_2 \xrightarrow{NH_3} R^1CH_2CONH_2$$

In the absence of a catalyst and in the presence of water and formic acid, diazoketones form hydroxymethyl ketones:

$$RCOCHN_2 + H_2O \xrightarrow{HCO_2H} RCOCH_2OH + N_2$$

Diazoketones also react with hydrogen chloride to form chloromethyl ketones:

$$RCOCHN_2 + HCl \longrightarrow RCOCH_2Cl + N_2$$

Diazoacetic ester (*ethyl diazoacetate*), b.p. 141°/720 mm, may be readily prepared by treating a cooled solution of the hydrochloride of ethyl glycine ester with cold sodium nitrite solution:

$$Cl^- \overset{+}{H_3N}CH_2CO_2C_2H_5 + NaNO_2 \longrightarrow N_2CHCO_2C_2H_5 + NaCl + 2H_2O \quad (85\%)$$

The stability of diazoacetic ester may be attributed to extended resonance. Thus there will be larger resonance energy than for diazomethane, and so the latter is less stable than diazoacetic ester.

The reactions of diazoacetic ester are similar to those of diazomethane. It is reduced by zinc dust and acetic acid to ammonia and glycine. When boiled with dilute halogen acid, it eliminates nitrogen to form glycollic ester:

$$N_2CHCO_2C_2H_5 + H_2O \longrightarrow HOCH_2CO_2C_2H_5 + N_2$$

When, however, diazoacetic ester is warmed with *concentrated* halogen acid, ethyl halogeno-acetate is formed, *e.g.*,

$$N_2CHCO_2C_2H_5 + HCl \longrightarrow ClCH_2CO_2C_2H_5 + N_2$$

Diazoacetic ester reacts with compounds containing an active hydrogen atom, *e.g.*, it forms acetylglycollic ester with acetic acid, and the ethyl ether of glycollic ester with ethanol:

$$CH_3CO_2H + N_2CHCO_2C_2H_5 \longrightarrow CH_3COOCH_2CO_2C_2H_5 + N_2$$
$$C_2H_5OH + N_2CHCO_2C_2H_5 \longrightarrow CH_2(OC_2H_5)CO_2C_2H_5 + N_2$$

It reacts with ethylenic compounds to form *pyrazoline* derivatives, *e.g.*, with ethylene it forms pyrazoline-3-carboxylic ester:

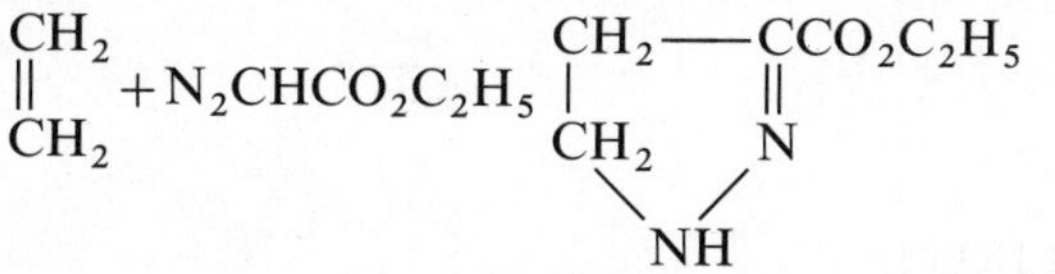

$$\begin{matrix} CH_2 \\ \| \\ CH_2 \end{matrix} + N_2CHCO_2C_2H_5 \quad \begin{matrix} CH_2 — CCO_2C_2H_5 \\ | \qquad \| \\ CH_2 \quad N \\ \diagdown \diagup \\ NH \end{matrix}$$

With acetylenic compounds it forms *pyrazole* derivatives, *e.g.*, with acetylene, it gives pyrazole-3-carboxylic ester:

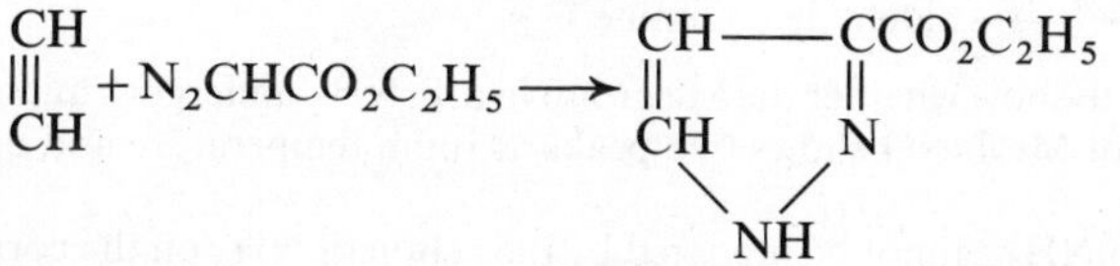

$$\begin{matrix} CH \\ \||| \\ CH \end{matrix} + N_2CHCO_2C_2H_5 \longrightarrow \begin{matrix} CH —— CCO_2C_2H_5 \\ \| \qquad \| \\ CH \quad N \\ \diagdown \diagup \\ NH \end{matrix}$$

In some of its reactions, diazoacetic ester decomposes to carbethoxymethylene, *e.g.*, in the presence of benzene, cycloheptatrienecarboxylic ester is produced (Doering *et al.*, 1956; *cf.* p. 549):

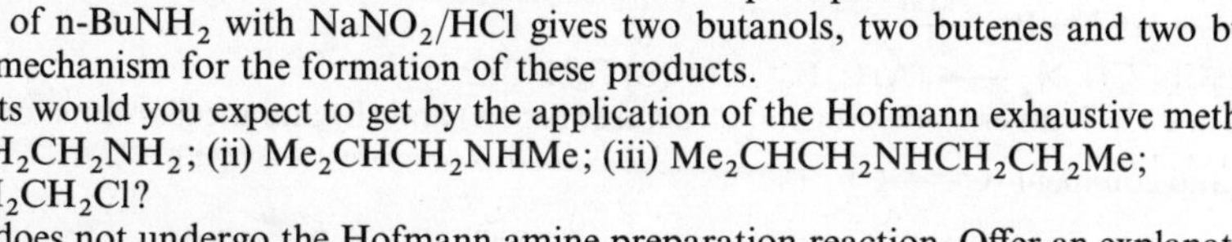

$$N_2CHCO_2Et \xrightarrow{h\nu} N_2 + H—C—CO_2Et \longrightarrow \bigcirc\!\!\!\!\bigcirc — CO_2Et$$

PROBLEMS

1. Write out the structures and names of the isomeric amines of formula $C_4H_{11}N$.

2. Of a group of isomeric amines, the t-amine has the lowest b.p. Explain.

3. Deamination of n-$BuNH_2$ with $NaNO_2$/HCl gives two butanols, two butenes and two butyl chlorides. Give a possible mechanism for the formation of these products.

4. What products would you expect to get by the application of the Hofmann exhaustive methylation to:
(i) $Me_2CHCH_2CH_2NH_2$; (ii) Me_2CHCH_2NHMe; (iii) $Me_2CHCH_2NHCH_2CH_2Me$;
(iv) $EtNHCH_2CH_2Cl$?

5. RCONHMe does not undergo the Hofmann amine preparation reaction. Offer an explanation.

6. Complete the following equations:

(i) $CH_2=CHCN + PhCH_2OH \xrightarrow{H_2SO_4}$?

(ii) $PhCH_2COCHN_2 + HCl(g) \xrightarrow{Et_2O}$?

(iii) $NH_2CONH_2 + n\text{-}BuOH \xrightarrow{heat}$?

(iv) $MeNH_2 + ? \longrightarrow MeNHCO_2Et$

(v) $HOCH_2CH_2CN \xrightarrow[\text{heat}]{\text{aq. HBr}}$?

(vi) 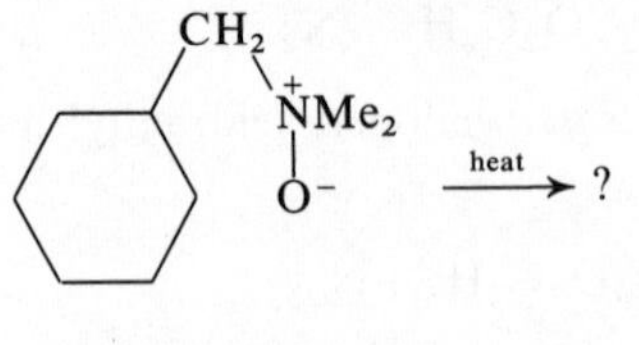 $\xrightarrow{\text{heat}}$?

(vii) $PhCH(CN)CH(CN)CO_2Et \xrightarrow[\text{heat}]{\text{aq. HCl}}$?

(viii) $+ ClCO_2Et \xrightarrow{Et_3N}$? $\xrightarrow{H^+}$?

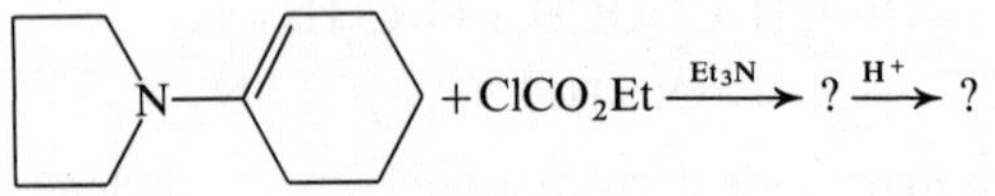

(ix) 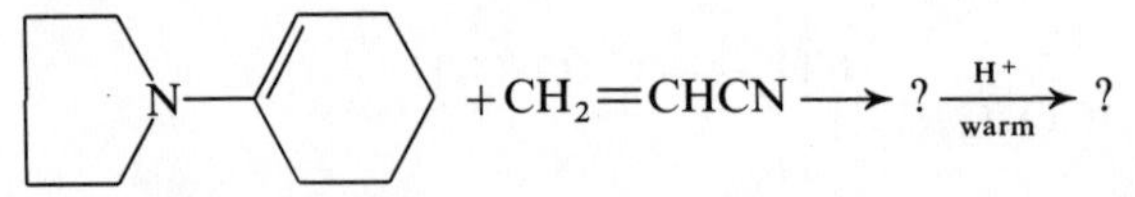 $+ CH_2{=}CHCN \longrightarrow$? $\xrightarrow[\text{warm}]{H^+}$?

(x) $Me_2CHNO_2 + HCHO + EtNH_2 \xrightarrow{HCl}$?

7. Discuss the reactions:

(i) $R_2C(OH)CH_2CR_2NH_2 \xrightarrow{HNO_2} R_2CO + R_2C{=}CH_2$

(ii) $PhCH{=}NR \xrightarrow{MeI}$? $\xrightarrow{H^+}$ aldehyde + s-amine

8. Devise an experiment to show whether the Me group in n-PrNH$_2$ undergoes migration during deamination.

9. The NMR spectrum of Me$_2$NNO shows two peaks at room temperature and one peak at $\sim 200°C$. Offer an explanation.

10. t-BuNH$_2$ and neoPenNH$_2$ cannot be prepared by the action of NH$_3$ on the corresponding alkyl bromide. Why not? Each can be prepared from a carboxylic acid. How?

11. Complete the following equations:

(i) $Me_2C(CN)NHNHC(CN)Me_2 \xrightarrow{HOCl}$?

(ii) $MeNHNHMe \xrightarrow{HgO}$?

(iii) $Me(CH_2)_2Me \xrightarrow[\sim 500°C]{HNO_3}$?

(iv) $RCH{=}CH_2 \xrightarrow{B_2H_6}$? $\xrightarrow{ClNH_2}$?

(v) $Me_3N + D\text{-}EtMeCHCl \longrightarrow$?

(vi) $PhNCO \xrightarrow{H_2O}$? $\xrightarrow{-CO_2}$? $\xrightarrow{PhNCO}$?

(vii) $CH_2(CO_2Et)_2 \xrightarrow[\text{(ii) BrCN}]{\text{(i) EtONa}}$?

(viii) $EtNH_2 + KCN + Br_2 \xrightarrow{KOH}$?

12. $MeCOCH_2Cl + CH_2N_2 \longrightarrow$ (A) C_4H_7ClO + (B) C_4H_7ClO
(A) is a ketone and (B) is an epoxide. Suggest structures for (A) and (B) and mechanisms for their formation.

13. Complete and comment ($\overset{*}{C} = {}^{13}C$):

$$RMgBr + \overset{*}{C}O_2 \longrightarrow \text{?} \xrightarrow{H^+} \text{?} \xrightarrow[\text{(ii) CH}_2\text{N}_2]{\text{(i) SOCl}_2} \text{?} \xrightarrow{Ag_2O} \text{?} \xrightarrow{H_2O} \text{?}$$

14. Amine (A), $C_6H_{15}N$, on treatment with MeI and then KOH, gives (B), $C_8H_{20}N^+OH^-$. This, on heating, produces isobutene and amine (C), $C_4H_{11}N$. What is (A)?

15. Complete the following equations:

(i) $R_2CO + Ph_3P{=}NH \longrightarrow$?

(ii) $Ph_3CCl + RNH_2 \longrightarrow$?

(iii) $EtNO_2 + 2CH_2O \xrightarrow{KHCO_3}$?

(iv) $RCH_2NH_2 + 2CH_2O + 2HCO_2H \longrightarrow$?

(v) $CH_2{=}CO + HCN \longrightarrow$? $\xrightarrow{CH_2=CO}$?

(vi) $Me_2C{=}CHCOMe + RCH_2NH_2 \longrightarrow$? $\xrightarrow{HNO_2}$? $\xrightarrow{EtONa}$?

(vii) $EtCHO + CH_2N_2 \longrightarrow$?

(viii) $MeCOEt + CH_2N_2 \longrightarrow$?

(ix)   ? $\xleftarrow{MeBr}$ $Et_3\overset{+}{N}{-}CH_2{}^-$ $\xrightarrow{Me_2CO}$?

(x) 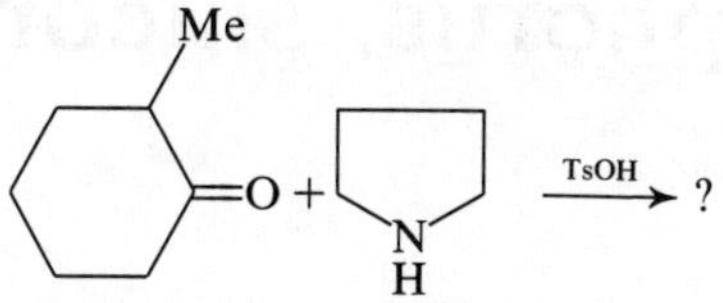 $\xrightarrow{TsOH}$?

(xi) $CH_2{=}CHCOMe + CH_2N_2 \longrightarrow$?

16. Discuss the NMR spectrum of $MeCH_2CH_2NO_2$ and give the relative positions of the signals.

17. Convert: (i) Me_2CHCO_2H into $Me_2CH(CH_2)_6CO_2H$; (ii) $(CH_2)_3Br_2$ into $Et_2N(CH_2)_3NH_2$; (iii) $O_2NCH_2CH_2CN$ into $(CH_2)_3(NH_2)_2$; (iv) $O_2NCH_2CH_2CN$ into $O_2NCH_2CH_2CH_2NH_2$.

18. How many N.M.R. signals would you expect for:
(i) $MeCONHCH_2CO_2H$; (ii) $MeCH{=}CHCN$; (iii) $MeCH_2CH(NH_2)Me$; (iv) $(MeCO_2)_2NMe$?

19. Show the fragmentation paths that would account for the ions m/e:
(i) 30 and 58 from ethyl-n-propylamine; (ii) 30, 58, and 86 from triethylamine; (iii) 41 from n-butyl cyanide.

REFERENCES

SIDGWICK, *The Organic Chemistry of Nitrogen*, Oxford University Press (1966, 3rd edn by Millar and Springall).

SMITH, *The Chemistry of Open-Chain Organic Nitrogen Compounds*, Benjamin, Vol. I (1965); Vol. II (1966).

PATAI (ed.), *The Chemistry of the Amino Group*, Interscience (1968).

FEUER (ed.), *The Chemistry of the Nitro and Nitroso Groups*, Interscience. Part I (1969).

JOHNSON, *Ylid Chemistry*, Academic Press (1966).

GILCHRIST and REES, *Carbenes, Nitrenes and Arynes*, Nelson (1969).

Organic Reactions, Wiley, Vol. I (1942), Ch. 10, 'The Mannich Reaction'; Vol. I (1942), Ch. 2, 'The Arndt–Eistert Synthesis': Vol. III (1946), Ch. 8, 'The Schmidt Reaction': Vol. IV (1948), Ch. 3, 'The Preparation of Amines by Reductive Alkylation': Vol. VII (1953), Ch. 3, 'Carbon–Carbon Alkylations with Amines and Ammonium Salts'; Ch. 6, 'The Nitrosation of Aliphatic Carbon Atoms': Vol. VIII (1954), Ch. 8, 'The Reaction of Diazomethane and its Derivatives with Aldehydes and Ketones': Vol. XI (1960), Ch. 5, 'Olefins from Amines: The Hofmann Elimination Reaction and Amine Oxide Pyrolysis': Vol. XII (1962), Ch. 3, 'The Synthesis of Aliphatic and Alicyclic Nitro-compounds': Vol. XVII (1969), Ch. 3, 'The Ritter Reaction'.

FINAR, *Organic Chemistry*, Longman, Vol. 2 (1968, 4th edn), Ch. 13, 'Amino-Acids and Proteins'.

Advances in Organic Chemistry; Methods and Results, Interscience. Vol. 4 (1963), Ch. 1, 'Enamines'.

COOK (ed.), *Enamines: Synthesis, Structure and Reactions*, Dekker (1969).

COWELL and LEDWITH, 'Developments in the Chemistry of Diazoalkanes', *Quart. Rev.*, 1970, **24**, 119.

14 Aliphatic compounds of sulphur, phosphorus, silicon and boron

Sulphur compounds

Mercaptans, thiols, or thioalcohols, RSH

These compounds occur in petroleum and give rise to 'sour petrol'.

Nomenclature

One method is to name them as alkyl *mercaptans*, the $-SH$ group being known as the *mercapto* or *sulph-hydryl group*. On the other hand, according to the I.U.P.A.C. system of nomenclature, the $-SH$ group is known as the *thiol* group, and the suffix of the series is *thiol*. This method of naming the mercaptans arises from the fact that they are the sulphur analogues of the alcohols, the usual procedure of showing that an oxygen atom has been replaced by a sulphur atom being indicated by the prefix *thio*.

CH_3SH	methyl mercaptan	methanethiol
$(CH_3)_2CHCH_2SH$	isobutyl mercaptan	2-methylpropanethiol

General methods of preparation

Thiols may be prepared by heating potassium or sodium hydrogen sulphide with an alkyl halide, a sodium alkyl sulphate, or with a tosylate:

$$RSH \xleftarrow{RX} KSH \xrightarrow[\text{or TsOR}]{ROSO_2ONa} RSH$$

Oxygen in an alcohol may be directly replaced by sulphur, *e.g.*, with phosphorus pentasulphide, but a more satisfactory method is to pass a mixture of alcohol vapour and hydrogen sulphide over a thoria catalyst at 400°C:

$$ROH + H_2S \xrightarrow{ThO_2} RSH + H_2O \quad (f.-g.)$$

Alternatively, alkenes may be used instead of alcohols, and the catalyst is nickel sulphide, *e.g.*,

$$MeCH{=}CH_2 + H_2S \xrightarrow[300°C]{NiS} MeCH(SH)Me + MeCH_2CH_2SH$$

The major product is the s-thiol (Markownikoff product; polar mechanism). Addition of hydrogen sulphide can also be induced by ultraviolet light to give predominantly the anti-Markownikoff product (free-radical mechanism).

Compounds containing the thiol group are known as *thiyl compounds*, and the addition of these to multiple bonds is called *thiylation* (see above examples).

The best method for preparing thiols is to decompose an *S-alkylisothiouronium salt* with alkali; these salts may be prepared by interaction of an alkyl halide (preferably the bromide or iodide) and thiourea (cyanamide polymers are also produced):

$$RBr + S{=}C(NH_2)_2 \longrightarrow [RSC({=}NH_2)NH_2]^+ Br^- \xrightarrow{\text{NaOH}}$$
$$RSH + NH_2CN + NaBr + H_2O \quad (80\text{–}90\%)$$

General properties

The thiols, except methanethiol, which is a gas, are colourless volatile liquids with disagreeable smells. Their boiling points are lower than those of the corresponding alcohols; this is due to the fact that they are very little associated, hydrogen bonding not readily taking place between hydrogen and sulphur, which is less electronegative than oxygen. The thiols are also less soluble in water than the corresponding alcohols, again, no doubt, due to their inability to form hydrogen bonds with water.

Thiols are stronger acids than alcohols, *e.g.*, EtSH, pK_a 10·5; EtOH, pK_a 15·5. This is unexpected in view of the fact that sulphur is less electronegative than oxygen. One possible explanation is that the S—H bond (347·3 kJ) is weaker than the O—H bond (464·4 kJ) and, because of its larger size, the sulphide ion can accommodate the negative charge more easily than can the smaller oxygen atom in the hydroxide ion.

The infra-red absorption region of the thiol group (S—H stretch) is $2\,590\text{–}2\,550\,cm^{-1}$ (w), and that of the C—S (stretch) is $705\text{–}570\,cm^{-1}$ (w).

Reactions

The thiols resemble the alcohols in many ways (sulphur and oxygen occur in the same periodic group); the main difference is their behaviour towards oxidising agents.

1. Thiols form mercaptides with the evolution of hydrogen when treated with alkali metals:

$$2RSH + 2Na \longrightarrow 2RS^- Na^+ + H_2$$

These alkali mercaptides are salts, and are decomposed by water:

$$R\bar{S}Na^+ + H_2O \longrightarrow RSH + NaOH$$

2. Thiols precipitate mercaptides when treated with an aqueous solution of the salt of a heavy metal, *e.g.*,

$$2RSH + (CH_3CO_2)_2Pb \longrightarrow (RS)_2Pb + 2CH_3CO_2H$$

Thiols also attack mercuric oxide in aqueous solution to form the mercury mercaptide:

$$2RSH + HgO \longrightarrow (RS)_2Hg + H_2O$$

It is this reaction which is the origin of the name mercaptans (*mercurius*, mercury; *captans*, seizing). These heavy metal mercaptides are *covalent* compounds.

3. Thiols react with carboxylic acids, preferably in the presence of inorganic acid, to form a *thiolester*:

$$R^1CO_2H + R^2SH \underset{}{\overset{H^+}{\rightleftharpoons}} R^1COSR^2 + H_2O$$

The equilibrium lies mainly to the left, but a good yield of ester may be obtained by using an acid chloride instead of acid.

4. Thiols may be oxidised, the nature of the product depending on the oxidising agent used. Mild oxidation with, *e.g.*, air, hydrogen peroxide, cupric chloride or sodium hypochlorite results in the formation of *dialkyl-disulphides*:

$$2RSH + H_2O_2 \longrightarrow R-S-S-R + 2H_2O$$

An alkaline solution of sodium plumbite containing sulphur ('doctor solution') converts thiols into disulphides:

$$2RSH + Na_2PbO_2 + S \longrightarrow R-S-S-R + PbS + 2NaOH$$

Dialkyl-disulphides may also be prepared by the action of iodine on sodium mercaptides:

$$2R\overset{-}{S}Na^+ + I_2 \longrightarrow R_2S_2 + 2NaI$$

These disulphides have an unpleasant smell (not as unpleasant as that of thiols). Diallyl disulphide, $(CH_2=CHCH_2-)_2S_2$, occurs in garlic. Disulphides are reduced to thiols by lithium aluminium hydride or by zinc and acid.

When oxidised with, *e.g.*, nitric or periodic acid, thiols are converted into *sulphonic acids*:

$$RSH + 3[O] \overset{HNO_3}{\longrightarrow} RSO_3H$$

5. Thiols readily undergo desulphurisation with Raney nickel:

$$RSH \overset{Ni}{\longrightarrow} RH + NiS$$

Desulphurisation, in general, involves the breaking of a C—S bond in an organic compound, and the source of the hydrogen is usually the hydrogen adsorbed by the Raney nickel prepared in the ordinary way. The first step is believed to be chemisorption of the S atom on the surface of the catalyst, and this is then followed by fission of the C—S bond to give free radicals and finally hydrogenation (*cf.* p. 100).

6. Thiols readily combine with aldehydes and ketones, in the presence of hydrochloric acid, to form *mercaptals* and *mercaptols* respectively; *e.g.*, ethanethiol forms diethylmethyl mercaptal with acetaldehyde, and diethyldimethyl mercaptol with acetone (see also sulphones, below):

$$CH_3CHO + 2C_2H_5SH \overset{HCl}{\longrightarrow} CH_3CH(SC_2H_5)_2 + H_2O$$

$$(CH_3)_2CO + 2C_2H_5SH \overset{HCl}{\longrightarrow} (CH_3)_2C(SC_2H_5)_2 + H_2O$$

Refluxing mercaptals (and mercaptols) with ethanol in the presence of freshly prepared Raney nickel replaces the thiol group by hydrogen (Wolfrom *et al.*, 1944):

$$\diagdown\!\!\!\underset{\diagup}{C}(SR)_2 + 4[H] \longrightarrow \diagdown\!\!\!\underset{\diagup}{C}H_2 + 2RSH$$

Thus a carbonyl group can be converted into a methylene group (*cf.* the Clemmensen and Wolff–Kishner reductions).

The aldehyde group may be 'protected' in *acid* solution by conversion into a mercaptal, and can be regenerated by treatment of the latter with mercuric chloride in the presence of cadmium carbonate (*cf.* acetals, p. 222).

An alternative way of protecting a carbonyl group makes use of β-mercaptoethanol:

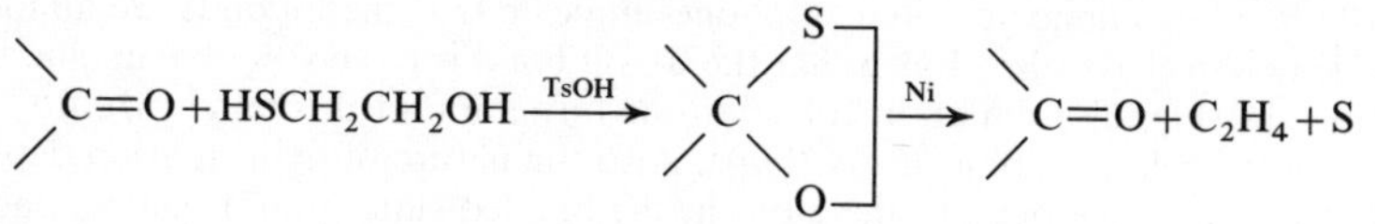

$$\diagdown\!\!\!\diagup C{=}O + HSCH_2CH_2OH \xrightarrow{\text{TsOH}} \diagdown\!\!\!\diagup C\!\!\left[\begin{array}{c}S\\O\end{array}\right] \xrightarrow{\text{Ni}} \diagdown\!\!\!\diagup C{=}O + C_2H_4 + S$$

Thioethers or alkyl sulphides, R_2S. These are the sulphur analogues of the ethers, from which they differ considerably in a number of ways.

General methods of preparation

1. By heating potassium sulphide with an alkyl halide, potassium alkyl sulphate, or a tosylate; *e.g.*,

$$S^= R{-}X \longrightarrow X^- + R{-}S^-$$
$$R{-}S \quad R{-}X \longrightarrow R_2S + X^- \quad (f.g.\text{-}g.)$$

2. Heating an ether with phosphorus pentasulphide:

$$5R_2O + P_2S_5 \longrightarrow 5R_2S + P_2O_5 \quad (p.)$$

3. By heating an alkyl halide with a sodium mercaptide (*cf.* Williamson's synthesis, p. 196):

$$R^1X + R^2\overset{-}{S}Na^+ \longrightarrow R^1{-}S{-}R^2 + NaX \quad (g.\text{-}v.g.)$$

4. By passing a thiol over a mixture of alumina and zinc sulphide at 300°C:

$$2RSH \longrightarrow R_2S + H_2S$$

5. By the addition of a thiol to an alkene in the presence of peroxides; in the absence of the latter very little reaction occurs (Kharasch *et al.*, 1939):

$$(R^1CO_2)_2 \longrightarrow 2R^1{\cdot} + 2CO_2$$
$$R^1{\cdot} + R^2SH \longrightarrow R^1H + R^2S{\cdot}$$
$$R^3CH{=}CH_2 + R^2S{\cdot} \longrightarrow R^3\dot{C}HCH_2SR^2 \xrightarrow{R^2SH} R^3CH_2CH_2SR^2 + R^2S{\cdot}: \text{etc.}$$

General properties and reactions

The thioethers are unpleasant-smelling oils, insoluble in water but soluble in organic solvents. Like the ethers, they behave as weak bases, *e.g.*, they dissolve in 100 per cent sulphuric acid to form sulphonium salts, $[R_2SH]^+HSO_4^-$. They may be oxidised to *sulphoxides* which, on further oxidation, are converted into *sulphones*: *e.g.*, ethyl sulphide, on oxidation with hydrogen peroxide in glacial acetic acid, gives first diethyl sulphoxide and then diethyl sulphone:

$$(C_2H_5)_2S \xrightarrow{H_2O_2} (C_2H_5)_2S{=}O \xrightarrow{H_2O_2} (C_2H_5)_2S\overset{\displaystyle O}{\underset{\displaystyle O}{\lessgtr}}$$

Only sulphoxide is obtained with periodic acid at 0°C; at higher temperatures, the sulphone is produced (Leonard *et al.*, 1962). Sulphoxide is also obtained by oxidation of sulphides with 1-chlorobenzotriazole (Johnson *et al.*, 1969).

Originally, sulphur was thought to be quadrivalent in sulphoxides and sexavalent in sulphones. Then, to conform with the 'octet' theory, the valencies were changed to 3 and 4, respectively, the oxygen atoms now being linked by co-ordinate bonds. More recently, however, bond length measurements of the S—O bond in sulphoxides and sulphones indicate that these bonds are almost all double bonds. M.O. calculations have also shown that the S—O bond is largely double in character and that the $3d$ orbitals of the sulphur atom are involved in the formation of this bond (Moffitt, 1950).

The valency shell of sulphur is $(3s)^2(3p)^4$. Also the nature of hybridisation depends on the number of σ-electrons and lone pairs in the atom in the bonded state (p. 47). Unless one wishes to consider multiple bonds as 'bent' bonds, electrons in π-bonds (double or triple) are not counted when deciding the type of hybridisation. In sulphides, each R group supplies one electron, and since the S atom has 6 valency electrons, the bond orbitals are sp^3, two being occupied by bonding pairs, and two by lone pairs, *i.e.*, R_2S. In sulphonium salts, the halide ion takes its bonding pair, and the R group uses one of the lone pairs of the sulphur atom, *i.e.*, $R_3\overset{..}{S}{}^+X^-$. In sulphoxides, $R_2S{=}O$, the orbitals are still sp^3, but here the R groups are joined by σ-bonds, the oxygen atom by a σ and π_{d-p} bond (one of the sulphur electrons has been promoted to $3d$), and a lone pair remains in the fourth sp^3 orbital. In

sulphones, $R_2S\!\!\!\begin{smallmatrix}\nearrow O \\ \searrow O\end{smallmatrix}$, there are four σ-bonds and two π_{d-p} bonds (one to each oxygen atom).

Alkyl sulphides readily undergo desulphurisation with Raney nickel:

$$R^1{-}S{-}R^2 + H_2 \xrightarrow{\text{Ni}} R^1H + R^2H + NiS$$

The alkyl sulphides form various addition products, *e.g.*, with bromine the alkyl sulphide dibromide is formed, R_2SBr_2. Alkyl sulphides also combine with a molecule of alkyl halide to form *sulphonium salts*, in which the sulphur is tercovalent unielectrovalent. When a sulphonium salt is heated, it decomposes into alkyl sulphide and alkyl halide (*cf.* quaternary ammonium salts, p. 378):

$$[R_3S]^+I^- \longrightarrow R_2S + RI$$

When they are treated with moist silver oxide, the *sulphonium hydroxide* is formed:

$$[R_3S]^+I^- + \text{`AgOH'} \longrightarrow [R_3S]^+OH^- + AgI$$

Sulphonium hydroxides are strongly basic, and on heating form alkyl sulphide and alkene, *e.g.*,

$$[(C_2H_5)_3S]^+OH^- \longrightarrow (C_2H_5)_2S + C_2H_4 + H_2O$$

This is believed to occur by the E2 mechanism:

$$HO^-\;\;H{-}CH_2{-}CH_2{-}\overset{+}{S}(C_2H_5)_2 \longrightarrow H_2O + CH_2{=}CH_2 + (C_2H_5)_2S$$

This reaction can also occur by the E1 mechanism provided the alkyl groups are of a suitable type, *e.g.*,

$$Me_3C{-}\overset{+}{S}{}^+Me_2 \xrightarrow{\text{OH}^-} Me_2S + Me_3C^+ \longrightarrow Me_2C{=}CH_2 + H_2O$$

Sulphur ylids (see also p. 382). Many bases remove α—H from sulphonium salts to form ylids, *e.g.*, dimethylsulphonium methylide:

$$Me_3S^+X^- \xrightarrow{\text{RLi}} RH + LiX + Me_2\overset{+}{S}{-}\overset{-}{C}H_2 \longleftrightarrow Me_2S{=}CH_2$$

The most satisfactory base appears to be the methylsulphinyl carbanion, $MeSOCH_2^-$, produced by the action of NaH, NaNH$_2$, etc. on dimethyl sulphoxide (see also below):

$$\text{MeSOMe} \xrightarrow{\text{NaH}} \text{MeSOCH}_2^- \xrightarrow{\text{Me}_3\text{S}^+\text{I}^-} \text{MeSOMe} + \text{Me}_2\text{S}{=}\text{CH}_2$$

Sulphonium ylids react with ketones to form epoxides:

$$\text{Me}_2\text{S}{=}\text{CH}_2 \quad \text{CR}_2{=}\overset{+}{\text{O}} \longrightarrow \text{Me}_2\overset{+}{\text{S}}{-}\overset{-}{\text{CH}_2}{-}\text{CR}_2 \longrightarrow \text{Me}_2\text{S} + \text{CH}_2{-}\text{CR}_2$$

Compounds containing activated double bonds readily form cyclopropanes, *e.g.*,

$$\text{Ph}_2\text{C}{=}\text{CH}_2 \quad \text{CH}_2{=}\text{SMe}_2 \longrightarrow \text{Ph}_2\text{C}{-}\text{CH}_2 \longrightarrow \text{Ph}_2\text{C}{-}\text{CH}_2 + \text{Me}_2\text{S}$$

Dimethyl sulphoxide also forms an ylid, *dimethylsulphoxonium methylide*, which reacts with ketones to form epoxides:

$$\text{Me}_2\text{SO} \xrightarrow{\text{MeI}} \text{Me}_3\overset{+}{\text{S}}\text{OI}^- \xrightarrow{\text{NaNH}_2} \text{Me}_2\overset{\text{O}}{\underset{}{\text{S}}}{=}\text{CH}_2 \xrightarrow{\text{R}_2\text{CO}} \text{R}_2\text{C}{-}\text{CH}_2 + \text{Me}_2\text{SO}$$

Mustard gas, 2,2′-*dichlorodiethyl sulphide, bis*(2-*chloroethyl*) *sulphide*, b.p. 215°C, may be prepared by the action of sulphur monochloride on ethylene:

$$2\text{C}_2\text{H}_4 + \text{S}_2\text{Cl}_2 \longrightarrow (\text{ClCH}_2\text{CH}_2{-})_2\text{S} + \text{S}$$

It may also be prepared, in a purer state, by heating ethylene chlorohydrin with sodium sulphide, and treating the product with hydrochloric acid:

$$2\text{HOCH}_2\text{CH}_2\text{Cl} + \text{Na}_2\text{S} \longrightarrow 2\text{NaCl} + (\text{HOCH}_2\text{CH}_2{-})_2\text{S} \xrightarrow{\text{HCl}} (\text{ClCH}_2\text{CH}_2{-})_2\text{S}$$

Mustard gas is a poison and a vesicant.

Sulphinium salts. Diaryl sulphides or sulphoxides produce sulphur radical ions when dissolved in concentrated sulphuric acid:

$$\text{Ar}_2\text{S} \xrightarrow{-e} \text{Ar}_2\text{S}^{\cdot+}$$

Solutions of this radical are coloured (red or blue), and sulphinium salts have been isolated as their hexachlorostannates.

Thiocyanic acid, isothiocyanic acid and their derivatives. Thiocyanic acid was believed to be a tautomeric equilibrium mixture of thiocyanic acid, HSCN, and isothiocyanic acid, HNCS. Spectroscopic studies of thiocyanic acid, however, indicate the structure HNCS (Beard *et al.*, 1947), and this has been confirmed by microwave studies (Dousmanis *et al.*, 1953).

Alkyl derivatives of both forms are known (*cf.* cyanic acid, p. 359).

Thiocyanic acid may be prepared by heating a mixture of potassium thiocyanate and potassium hydrogen sulphate:

$$\text{KNCS} + \text{KHSO}_4 \longrightarrow \text{HNCS} + \text{K}_2\text{SO}_4$$

Its dilute aqueous solutions are fairly stable, the concentrated solutions decomposing to form carbonyl sulphide and ammonia:

$$\text{HNCS} + \text{H}_2\text{O} \longrightarrow \text{COS} + \text{NH}_3$$

Alkyl thiocyanates may be prepared by heating potassium thiocyanate with an alkyl halide or tosyl ester:

$$\text{RI} + \text{KNCS} \longrightarrow \text{RSCN} + \text{KI}$$

Since the thiocyanate is obtained rather than the isothiocyanate, this indicates that sulphur is a stronger nucleophile than nitrogen (see also below).

Thiocyanates are very conveniently prepared from sodium thiosulphate:

$$\text{NaSSO}_3\text{Na} \xrightarrow{\text{RI}} \text{RSSO}_3\text{Na} \xrightarrow{\text{NaCN}} \text{RSCN} + \text{Na}_2\text{SO}_3$$

The alkyl thiocyanates are fairly stable volatile oils, with a slight odour of garlic. They are oxidised to sulphonic acids by concentrated nitric acid, and reduced to thiols by, *e.g.*, zinc and sulphuric acid:

$$RSO_3H \xleftarrow{HNO_3} RSCN \xrightarrow{Zn/H_2SO_4} RSH$$

They are converted into sulphonyl chlorides by chlorine water.

$$RSCN + Cl_2 \xrightarrow{H_2O} RSO_2Cl + ClCN$$

Alkyl thiocyanates rearrange on heating to the more thermodynamically stable isothiocyanate:

$$R\text{—}SCN \rightleftharpoons [R^+ + SCN^-] \rightleftharpoons RNCS$$

Alkyl isothiocyanates or **mustard oils** may be prepared by heating alkyl thiocyanates, but a much more satisfactory method is the **Hofmann mustard-oil reaction** (1868), which is carried out by heating a mixture of a primary amine, carbon disulphide and mercuric chloride (see p. 465 for further details).

Another convenient method is to add an aqueous solution of a primary amine to carbon disulphide in sodium hydroxide solution, and then ethyl chloroformate:

$$RNH_2 + CS_2 + NaOH \longrightarrow RNHCS_2Na + H_2O \xrightarrow{ClCO_2C_2H_5}$$
$$NaCl + RNHCS_2CO_2C_2H_5 \longrightarrow RNCS + COS + C_2H_5OH \quad (60\text{--}70\%)$$

The alkyl isothiocyanates are liquids with a powerful mustard smell; they are lachrymatory and vesicatory. They are hydrolysed to primary amines when heated with hydrochloric acid, and reduced to primary amines and thioformaldehyde by, *e.g.*, zinc and sulphuric acid:

$$RNH_2 + CO_2 + H_2S \xleftarrow[2H_2O]{HCl} RNCS \xrightarrow{Zn/H_2SO_4} RNH_2 + HCHS$$

Allyl isothiocyanate (*allyl mustard oil*), $CH_2=CHCH_2NCS$. When the glucoside *sinigrin* (which occurs in mustard seed) is hydrolysed by acid or by the enzyme *myrosin* (which is found in mustard seeds), allyl isothiocyanate, glucose and potassium hydrogen sulphate are obtained. Allyl isothiocyanate is a colourless oil, b.p. 151°C, and is the substance which gives mustard its characteristic odour and taste. It is lachrymatory and vesicatory, and is a convenient starting material for the preparation of allylamine.

Thiocyanogen, $(SCN)_2$, may be prepared by treating lead thiocyanate with bromine in ethereal solution at 0°C.

$$Pb(SCN)_2 + Br_2 \longrightarrow (SCN)_2 + PbBr_2$$

It is a gas and resembles the halogens in that it adds on to double bonds (*thiocyanation*) to form dithiocyanates.

Alkyl sulphoxides (*alkylsulphinylalkanes*) may be prepared by oxidising alkyl sulphides with the *theoretical* amount of hydrogen peroxide (in acetic acid) or with *dilute* nitric acid (see also p. 401):

$$R_2S + [O] \longrightarrow R_2S=O$$

The sulphoxides are odourless, relatively unstable solids, soluble in water, ethanol and ether, and are feebly basic, *e.g.*, they form salts with hydrochloric acid, $[R_2S\text{—}OH]^+Cl^-$. Sulphoxides are reduced to sulphides by zinc and acetic acid or by LAH.

The infra-red absorption region of the sulphoxide group (S=O stretch) is $1\,070\text{--}1\,030\,cm^{-1}$ (s).

Dimethyl sulphoxide, DMSO, b.p. 188°C, is prepared by the oxidation of dimethyl sulphide by air. It is an extremely valuable solvent and may be used as an oxidising agent for certain compounds, *e.g.*, phenacyl bromide (see p. 750). A solution of sodium dimethyl sulphoxide appears to be very useful for the estimation of all forms of active hydrogen. Titration is carried out in the presence of triphenylmethane as indicator; the blood-red colour of the triphenyl-methyl anion is formed immediately in the presence of excess of the reagent:

$$MeSOCH_3 \xrightarrow{NaH} MeSOCH_2^- \xrightarrow{Ph_3CH} MeSOCH_3 + Ph_3C^-$$

Alkyl sulphones may be prepared by oxidising alkyl sulphides with hydrogen peroxide *in excess* (in acetic acid) or with *concentrated* nitric acid:

$$R_2S + 2[O] \longrightarrow R_2SO_2$$

They are colourless, odourless, very stable solids, soluble in water; they absorb in the region 1 160–1 120 cm^{-1} (vs). They are very resistant to reduction, but some are reduced to sulphide by lithium aluminium hydride. The α-hydrogen atom in sulphones is very reactive (*cf.* the active methylene group) and so sulphones can enter into the Claisen-type of condensation. Many sulphones produce sulphinic acids when fused with potassium hydroxide at 200°C, *e.g.*,

$$(C_2H_5)_2SO_2 + KOH \longrightarrow C_2H_4 + C_2H_5SO_2K + H_2O \quad (60\%)$$

Aliphatic sulphinic acids are viscous oils unstable to heat, but their salts are stable.

Sulphenic acids (RSOH) are known only as their derivatives, *e.g.*, sulphenyl halides, RSX (see also p. 108).

Sulphens, $R^1R^2C{=}SO_2$. These may be generated by various methods, but because of their reactivity, they polymerise (*cf.* ketens), or they may be trapped. Conclusive evidence for the intermediate formation of sulphens has been obtained by Truce *et al.* (1964):

$$CH_3SO_3Me \xleftarrow{\text{MeOD}} CH_3SO_2Cl + Et_3N \longrightarrow Et_3N{\cdot}HCl + CH_2{=}SO_2 \xrightarrow{\text{MeOD}} CH_2DSO_3Me$$

The methyl methanesulphonate was shown, by NMR spectroscopy and mass spectrometry, to be a mixture of monodeuterated and undeuterated ester.

Sulphines, $R^1R^2C{=}S{=}O$, have been obtained, *e.g.*, by the action of triethylamine on sulphinyl chlorides, RSOCl. Sulphines are stable enough to be isolated.

Sulphonic acids. The aliphatic sulphonic acids are named either as alkylsulphonic acids or as alkanesulphonic acids; in the latter case the sulphonic acid group is considered as a substituent group, *e.g.*,

CH_3SO_3H methylsulphonic acid or methanesulphonic acid

$(CH_3)_2CHSO_3H$ isopropylsulphonic acid or propane-2-sulphonic acid

General methods of preparation

1. By the action of sulphuric acid, oleum, or chlorosulphonic acid on an alkane. The sulphonic acid group enters the chain, probably at the second carbon atom, *e.g.*,

$$CH_3(CH_2)_4CH_3 + SO_3 \xrightarrow{\text{H}_2\text{SO}_4} CH_3(CH_2)_3CH(SO_3H)CH_3 \quad (p.-f.)$$

2. By the action of sulphuryl chloride, or a mixture of chlorine and sulphur dioxide, on a hydrocarbon in the presence of light or peroxide at 40–60°C. The sulphonyl chloride is obtained, often in high yield; the sulphonyl chloride group appears to enter mainly at the second carbon atom, but varying amounts of product with this group at the first carbon are also obtained:

$$RH + SO_2Cl_2 \longrightarrow RSO_2Cl + HCl$$

3. By the oxidation of a thiol with nitric acid, permanganate, sodium hypobromite, etc., *e.g.*,

$$RSH + 3[O] \xrightarrow{\text{HNO}_3} RSO_3H \quad (g.-v.g.)$$

4. The Strecker reaction (1868). This is carried out by heating an alkyl halide with sodium sulphite.

$$RX + Na_2SO_3 \longrightarrow RSO_3Na + NaX \quad (g.-v.g.)$$

5. Sodium hydrogen sulphite adds on to alkenes in the presence of peroxides to form sulphonic acids (Kharasch *et al.*, 1939).

$$RCH{=}CH_2 + NaHSO_3 \longrightarrow RCH_2CH_2SO_3Na \quad (ex.)$$

6. By the oxidation of *S*-alkylisothiouronium salts (p. 465).

General properties and reactions

The sulphonic acids are generally thick liquids, soluble in water. They are isomeric with the alkyl hydrogen sulphites:

RSO₂OH   ROSOOH

alkanesulphonic acid   alkyl hydrogen sulphite

Sulphonic acids have two absorption regions (S$=$O stretch), 1 260–1 150 (s) and 1 080–1 010 cm^{-1} (s), whereas sulphites (S$=$O stretch) have only one absorption region, 1 220–1 170 cm^{-1} (s).

The sulphonic acids are strong acids, forming salts with metallic hydroxides or carbonates; the lead and barium salts are very soluble in water. They form the acid chloride, the sulphonyl chloride, when treated with phosphorus pentachloride (see p. 692):

$$RSO_3H + PCl_5 \longrightarrow RSO_2Cl + HCl + POCl_3$$

These sulphonyl chlorides are only very slowly hydrolysed by water (*cf.* acyl chlorides); they react readily with concentrated aqueous ammonia to form *sulphonamides*, and with alkoxides to form esters:

$$RSO_2NH_2 \xleftarrow{\text{NH}_3} RSO_2Cl \xrightarrow{\text{RONa}} RSO_3R$$

Sulphonyl chlorides are reduced by lithium aluminium hydride to thiols (Marvel *et al.*, 1950).

Thioacids may be prepared by the action of phosphorus pentasulphide on a carboxylic acid:

$$5RCO_2H + P_2S_5 \longrightarrow 5RCOSH + P_2O_5$$

The thioacids have a most disagreeable odour, and slowly decompose in air. In most of their reactions, the thioacids and their salts behave as if they contained the mercapto-group, but in a few reactions they behave as if they contained a hydroxyl group. This may be accounted for by tautomerism for the acid and resonance for the salts:

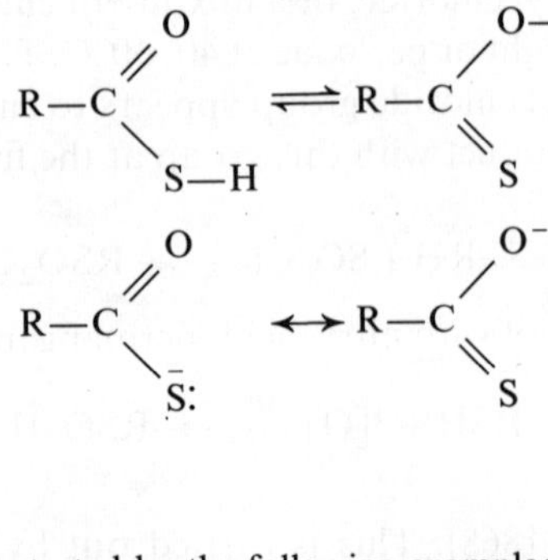

Nomenclature

The methods of nomenclature are illustrated by the following example:

CH₃COSH   CH₃CSOH

(I)   (II)

According to the trivial system of nomenclature, both (I) and (II) are named as thioacetic acid. According to the I.U.P.A.C. rules, the suffix *-oic* of the corresponding oxygen acid is changed to *thioic*. Thus both (I) and (II) are named either as ethanethioic acid or methanecarbothioic acid. On the other hand, the suffix *-thiolic* is used if it is certain that the oxygen of the hydroxyl group is replaced by sulphur, and the suffix *thionic* if it is the oxygen of the carbonyl group. Thus (I) is ethanethiolic acid or methane-carbothiolic acid, and (II) is ethanethionic or methanecarbothionic acid.

The most characteristic reaction of the thioacids is their extreme readiness to acylate alcohols and amines:

$$R^1COOR^2 \xleftarrow[(-H_2S)]{R^2OH} R^1COSH \xrightarrow[(-H_2S)]{R^2NH_2} R^1CONHR^2$$

Dithioacids may be prepared by the action of a Grignard reagent on carbon disulphide:

$$RMgX + C\overset{S}{\underset{S}{\diagdown}} \longrightarrow R-C\overset{S}{\underset{SMgX}{\diagdown}} \xrightarrow{acid} RCS_2H$$

Nomenclature

The methods of nomenclature are illustrated by the following example:

CH_3CS_2H is named as dithioacetic acid, ethanethionthiolic acid or methanecarbodithioic acid.

A very important dithioacid is **dithiocarbonic acid,** $HOCS_2H$. The free acid is unknown, but many of its derivatives have been prepared, *e.g.*, *potassium xanthate* may be prepared by the reaction between potassium hydroxide, ethanol and carbon disulphide (see also p. 95):

$$EtOH \xrightarrow{KOH} EtO^-K^+ \xrightarrow{CS_2} EtOC(=S)S^-K^+$$

Thiocarbonyl compounds. Aliphatic thioaldehydes (*thials*) are unknown; attempts to prepare them result in polymers. On the other hand, some thioketones (*thiones*) have been isolated as monomers. The reason for this instability is believed to be due to the difficulty with which sulphur forms a π_{d-p} bond with carbon (overlapping of the S$3d_\pi$ with the C$2p_\pi$ orbital is very difficult), and the result is ready polymerisation (to dimer, trimer, etc.), *e.g.* (6-ring):

$$R_2C\overset{\frown}{=}S \quad CR_2\overset{\frown}{=}S \quad CR_2\overset{\frown}{=}S \longrightarrow R_2C-S-CR_2-S-CR_2-S$$

Vulcanisation of rubber. This process is carried out by heating crude rubber with 4–5 per cent sulphur and certain organic compounds which accelerate the reaction between the rubber and sulphur. These organic compounds are known as *accelerators*, and all contain sulphur or nitrogen, or both. Vulcanising rubber causes the rubber to lose its stickiness, makes it no longer sensitive to temperature changes, causes it to retain its elasticity over a wide temperature range, and increases its tensile strength. The function of the sulphur appears to be to cross-link the long hydrocarbon chains in crude rubber.

Phosphorus compounds

Both nitrogen and phosphorus occur in Group V, and their valence shells are similar: N, $(2s)^2 (2p)^3$; P, $(3s)^2 (3p)^3$. The two elements, however, differ in some important respects. Nitrogen exhibits a maximum covalency of 4 (there are no available *d*-orbitals), whereas phosphorus can exhibit a covalency of 5, *e.g.*, PCl_5 (sp^3d; trigonal bipyramid), and 6, *e.g.*, PF_6^- (sp^3d^2; octahedron). The majority of P compounds are sp^3 hybridised, PX_4^+ (*cf.* N), but although both PX_3 and NH_3 are pyramidal structures, the angle in P compounds is less than that in N compounds, *e.g.*, $\angle$ HPH (PH_3) is 93°; $\angle$ HNH (NH_3) is 107°. It therefore appears that the lone pair $(3s^2)$ is less involved in tricovalent P compounds than the lone pair $(2s^2)$ in the corresponding N compounds.

Nomenclature of organophosphorus compounds

There is a very large variety of organophosphorus compounds; some are discussed here. For compounds containing *one* phosphorus atom, three parent structures have been used (whether they have an independent existence or not. *N.B.* In the following structures, **PO is P=O**).

Hydrides	Examples
Phosphine, H_3P	Me_2PH, dimethylphosphine
Phosphine oxide, H_3PO	Et_3PO, triethylphosphine oxide
Phosphorane, H_5P	Ph_5P, pentaphenylphosphorus

Tervalent acids	Examples
Phosphorous acid, H_3PO_3	$(EtO)_3P$, triethyl phosphite
Phosphonous acid, $HP(OH)_2$	$HP(OH)OMe$, methyl hydrogen phosphonite
Phosphinous acid, H_2POH	H_2POEt, ethyl phosphinite

Structures with carbon–phosphorus bonds formed by replacement of H joined to P by an alkyl, aryl, of a heterocyclic group, are named by prefixing to the parent name the name of the replacing group. Compounds with P—X or P—N bonds are named respectively as acid halides or amides if *all* OH groups have been replaced. If, however, at least one OH group is still present, then affixes are used and the name acid is retained, *e.g.*,

Parent	Examples
$HP(OH)_2$	$MePCl_2$, methylphosphonous dichloride
	$HP(OH)NH_2$, phosphonamidous acid
$(HO)_3P$	$HOPCl_2$, phosphorodichloridous acid
	$P(NH_2)_3$, phosphorous triamide

Quinquevalent acids	Examples
Phosphoric acid, $(HO)_3PO$	$(EtO)_3PO$, triethyl phosphate
	$HOPO(NH_2)_2$, phosphorodiamidic acid
	$(HO)_2PO(Cl)$, phosphorochloridic acid
Phosphonic acid, $HPO(OH)_2$	$HPO(Cl_2)$, phosphonic dichloride
	$MePO(OH)NH_2$, methylphosphonamidic acid
	$MePO(Cl)(NH_2)$, methylphosphonamidic chloride
Phosphinic acid, $H_2PO(OH)$	$MeHPO(Cl)$, methylphosphinic chloride
	$H_2PO(OEt)$, ethyl phosphinate

Preparation and reactions of some organophosphorus compounds

Since the phosphorus atom in tercovalent compounds has a lone pair of electrons, these compounds behave as nucleophiles in many reactions. It is because of this lone pair that phosphines are good reducing agents. Also, the C—P bond is very strong, and is resistant to oxidation and hydrolysis.

A number of organophosphorus compounds are used as nerve gases, insecticides and fungicides.

Phosphines and phosphonium salts. Trialkylphosphines are prepared most conveniently by the action of excess of Grignard reagent on phosphorus trihalide:

$$3RMgBr + PCl_3 \longrightarrow R_3P + 3MgClBr \quad (g.-v.g.)$$

This method may also be used to prepare triarylphosphines (from $ArMgBr$), and another useful method is:

$$3ArBr + PCl_3 + 6Na \longrightarrow Ar_3P + 3NaCl + 3NaBr$$

The three classes of phosphines may be prepared by interaction of alkyl halide and metal phosphides or metal hydrogen phosphides, *e.g.*,

$$3Na \xrightarrow[\text{liq. NH}_3]{\text{PH}_3} Na_3P \xrightarrow{\text{3RX}} R_3P + 3NaX$$

$$Na \xrightarrow[\text{liq. NH}_3]{\text{PH}_3} NaPH_2 \xrightarrow{\text{RX}} RPH_2 + NaX$$

$$Na \xrightarrow[\text{liq. NH}_3]{\text{R}^1\text{PH}_2} NaPHR^1 \xrightarrow{\text{R}^2\text{X}} R^1R^2PH + NaX$$

Mono- and di-alkylphosphines are formed when an alkyl halide is heated with phosphonium iodide in the presence of zinc oxide (to neutralise the HI formed in the reaction):

$$PH_4{}^+I^- + EtI + ZnO \longrightarrow EtPH_2 + ZnI_2 + H_2O$$
$$2EtPH_2 + 2EtI + ZnO \longrightarrow Et_2PH + ZnI_2 + H_2O$$

General properties and reactions

Except for methylphosphine (a gas), all the alkylphosphines are colourless, unpleasant-smelling liquids or low-melting solids. One of their most characteristic reactions is the ease with which they combine with atmospheric oxygen, often undergoing spontaneous combustion in the process. On the other hand, controlled oxidation is readily carried out with nitric acid:

$$EtPH_2 \xrightarrow{\text{HNO}_3} EtPO(OH)_2 \qquad \text{ethylphosphonic acid}$$

$$Et_2PH \xrightarrow{\text{HNO}_3} Et_2PO(OH) \qquad \text{diethylphosphinic acid}$$

$$Et_3P \xrightarrow{\text{HNO}_3} Et_3PO \qquad \text{triethylphosphine oxide}$$

t-Phosphines are also oxidised to phosphine oxide by ozone:

$$R_3P + O_3 \longrightarrow R_3PO + O_2$$

All the alkylphosphines are reducing agents, but trialkylphosphines appear to be the most useful; they react with a variety of oxygen-containing compounds:

$$R_3P + ZO \longrightarrow R_3PO + Z$$

Thus, *e.g.*, triphenylphosphine reduces epoxides to alkenes and amine oxides to amines. The major driving force in these reactions is the unusually large strength of the P=O bond ($502 \cdot 1$–$627 \cdot 6 \, \text{kJ mol}^{-1}$).

Phosphines are weaker bases than the corresponding amines, and the order of basicity in phosphines is t > s > prim. This is not quite parallel with that of the amines (p. 373), and a possible explanation is that solvation effects are less important in phosphines than in amines because of the larger size of the phosphorus atom (and consequently the charge is more diffuse than in amines).

Trialkylphosphines readily form phosphonium salts with alkyl halides (preferably iodides); the mechanism is S_N2

$$R_3^1\overset{\frown}{P:} \quad R^2{-}X \longrightarrow R_3^1\overset{+}{P}R^2X^-$$

(*cf.* the Menshutkin reaction). t-Phosphines are stronger nucleophiles than t-amines; this is illustrated by the fact that aminophosphines combine with alkyl halides to give phosphonium salts and not ammonium salts:

$$R_2^1NPR_2^1 + R^2X \longrightarrow R_2^1NPR_2^1\overset{+}{R^2}X^-$$

This greater nucleophilicity has been explained on the basis that the lone pair of electrons of phosphorus has a greater polarisability than that of nitrogen, due to the larger size of the phosphorus atom. However, according to Henderson *et al.* (1960), this greater nucleophilicity, *i.e.*, greater rate of reaction, is accounted for by the entropy of activation term in the two reactions.

When phosphonium halides are treated with moist silver oxide, the phosphonium hydroxide is produced:

$$(C_2H_5)_4P^+I^- + \text{‘AgOH’} \longrightarrow (C_2H_5)_4P^+OH^- + AgI$$

The quaternary phosphonium hydroxides are strongly basic, comparable in strength with the quaternary ammonium hydroxides and sodium hydroxide. When heated they form the trialkylphosphine oxide and a hydrocarbon (*cf.* R_4NOH, p. 379).

$$R_4P^+OH^- \longrightarrow R_3PO + RH$$

Some alkene, however, is formed when R contains β-hydrogen (with respect to the phosphorus):

$$R^1CH_2CH_2\overset{+}{P}R_3^2OH^- \longrightarrow R^1CH{=}CH_2 + R_3^2P + H_2O$$

Pentaphenylphosphorus (m.p. 124°C) has been prepared, and by means of X-ray analysis and dipole moment measurements, has been shown to have a trigonal bipyramidal structure.

Phosphine oxides. Phosphine oxides may be prepared by oxidation of a trialkylphosphine with, *e.g.*, nitric acid or hydrogen peroxide; by heating a phosphonium hydroxide (see above); or by reaction between a Grignard reagent and phosphoryl chloride:

$$3RMgBr + POCl_3 \longrightarrow R_3PO + 3MgBrCl$$

Tertiary phosphine oxides are usually solids, and form one of the most stable groups of organophosphorus compounds.

Primary (RH_2PO) and secondary (R_2HPO) phosphine oxides are also known, and may be prepared by oxidation of the corresponding phosphine with hydrogen peroxide.

Phosphorus ylids (see also p. 382). **Phosphoranes** may be prepared from phosphonium halides by reaction with a strong base. Triphenylphosphine is usually the starting point; it is commercially available because of its stability in air; *e.g.*, methylenetriphenylphosphorane:

$$Ph_3P + CH_3Br \longrightarrow Ph_3PCH_3{}^+Br^- \xrightarrow{\text{PhLi}} Ph_3P{=}CH_2 \longleftrightarrow Ph_3\overset{+}{P}{-}CH_2{}^- + PhH + LiBr$$

In general, the bases commonly used are butyl- and phenyl-lithium in ether or tetrahydrofuran, sodium or lithium ethoxide in ethanol or dimethylformamide, and dimethyl sulphoxide and sodium hydride. The last reagent is the best (see also p. 402).

These alkylidenephosphoranes are ylids, and because they readily react with oxygen (see later), reactions involving them are carried out in an atmosphere of nitrogen. Their reaction with carbonyl compounds leads to alkenes, and this alkene synthesis is one example of the **Wittig reaction** (1953, 1956). The mechanism of the Wittig reaction has been the subject of much discussion, but the evidence is now strongly in favour of the formation of an intermediate betaine, followed by ring closure and then fission:

Support for this mechanism comes from the observation that the entropy of activation is negative. This is in keeping with the formation of a cyclic intermediate. Furthermore, cyclic intermediates have been isolated in certain cases (see also below).

The reactivity of these ylids is associated with the carbanionic centre in the dipolar structure, and is decreased when the negative charge is delocalised. Delocalisation of charge may be brought about through resonance due to the presence of a suitable electron-attracting group (or groups) attached to the methylene carbon atom, e.g.,

$$Ph_3P\text{=}CH\text{—}CH\text{=}O \leftrightarrow Ph_3\overset{+}{P}\text{—}\overset{-}{C}H\text{—}CH\text{=}O \leftrightarrow Ph_3\overset{+}{P}\text{—}CH\text{=}CH\text{—}\overset{-}{O}$$

These ylids react readily with aldehydes, but react slowly (if at all) with ketones. In some cases, the ylid is so stable that it does not react even with aldehydes. Because of this variation in stability, the rate-determining step in the Wittig reaction may be either the formation of the betaine (for stable ylids) or betaine decomposition (for reactive ylids).

Since in the formation of the ylid only *one* α-hydrogen is lost, the structure of the ylid is definite and so is the structure of the alkene produced.

When the alkene product is capable of geometrical isomerism, it has been found that the *trans* isomer is favoured by non-polar solvents and the *cis* isomer by highly polar solvents. Thus, for the solvents in which the order of polarity is DMF > EtOH > THF > Et₂O > PhH, the yield of *trans* isomer increases from right to left, and left to right for the *cis* isomer. Alternatively, by carrying out the reaction in the presence of a Lewis base, e.g., amines, I⁻, there is a very large predominance of the *cis* isomer, e.g., Shemyakin et al. (1963) carried out the following stereospecific reactions:

$$Ph_3P\text{=}CHPh + EtCHO \longrightarrow$$

(structures)

Ph, H / C=C / Et 96% H
(DMF+LiI)

Ph, H / C=C / H 100% Et
(PhH)

The explanation for these results is still the subject of discussion.

Many alkylidenephosphoranes are hydrolysed by water, whereas others require heating in alkaline solution, e.g.,

$$Ph_3P\text{=}CHMe \xrightarrow{H_2O} Ph_3\overset{+}{P}EtOH^- \longrightarrow Ph_2EtPO + PhH$$

These ylids are also oxidised by a small amount of oxygen (or air) to alkenes, but with excess of oxygen the carbonyl compound is obtained:

$$Ph_3P\text{=}CR_2 \xrightarrow{O_2} \underset{\underset{O\text{—}O}{|\ \ \ |}}{Ph_3P\text{—}CR_2} \longrightarrow Ph_3PO + R_2CO \xrightarrow{Ph_3P\text{=}CR_2} Ph_3PO + R_2C\text{=}CR_2$$

With *excess* of oxygen, all of the ylid is oxidised and consequently the second step cannot occur.

Ylids containing groups other than alkylidene may also be used in the Wittig reaction. Some examples of their preparation and uses are:

(i)   Dichloromethylenetriphenylphosphorane (see p. 328).

(ii) Carbonyl compounds may be stepped up to aldehydes containing one more carbon atom:

$$Ph_3P + ClCH_2OMe \longrightarrow [Ph_3\overset{+}{P}CH_2OMe]Cl^- \xrightarrow{EtONa} Ph_3P\text{=}CHOMe \xrightarrow{PhCHO}$$

$$Ph_3PO + PhCH\text{=}CHOMe \xrightarrow{H^+} [PhCH\text{=}CHOH] \longrightarrow PhCH_2CHO$$

(iii) Keto-ylids may be produced as follows:

$$Ph_3P\text{=}CH_2 \xrightarrow{RCOCl} [Ph_3\overset{+}{P}CH_2COR]Cl^- \xrightarrow{Ph_3P\text{=}CH_2} Ph_3P\text{=}CHCOR + [Ph_3PMe]^+Cl^-$$

Alkylphosphonous dihalides, RPX_2, and **dialkylphosphinous halides**, R_2PX. Phosphonous dichlorides may be prepared as follows:

$$R_2Cd + 2PCl_3 \longrightarrow 2RPCl_2 + CdCl_2$$
$$R_3Al + 3PCl_3 \longrightarrow 3RPCl_2 + AlCl_3$$

The phosphinous chlorides may be prepared by partial alkylation of phosphorus trichloride with lead tetra-alkyls.

The chlorine in both types of halides is very reactive (these halides are acid chlorides), *e.g.*,

$$RPCl_2 + 2H_2O \longrightarrow RHPO(OH) + 2HCl$$
$$R_2PCl + H_2O \longrightarrow R_2HPO + HCl$$

The alkylphosphonous dihalides and dialkylphosphinous halides readily combine with chlorine to form chlorophosphoranes, and oxygen also adds on to give the phosphonic dichloride or phosphinic chloride, *e.g.*,

$$RPCl_4 \xleftarrow{Cl_2} RPCl_2 \xrightarrow{O_2} RPOCl_2$$

Alkylphosphonic dihalides, $RPOX_2$, and **dialkylphosphinic halides**, R_2POX. These may be prepared as described above, but a better method is to treat the corresponding halogenophosphorane with sulphur dioxide, *e.g.*,

$$R_2PCl + Cl_2 \longrightarrow R_2PCl_3 \xrightarrow{SO_2} R_2POCl + SOCl_2$$

Alternatively, phosphonic and phosphinic acids or their esters give the corresponding chloride when treated with thionyl chloride or phosphorus pentachloride, *e.g.*,

$$R^1PO(OR^2)_2 + 2PCl_5 \longrightarrow R^1POCl_2 + 2POCl_3 + 2R^2Cl$$
$$R_2PO(OH) + SOCl_2 \longrightarrow R_2POCl + SO_2 + HCl$$

Both types of chlorides react with water (*i.e.*, behave as acid chlorides), *e.g.*,

$$RPOCl_2 + 2H_2O \longrightarrow RPO(OH)_2 + 2HCl$$

Esters of phosphorous acid. These are readily prepared by the reaction between an alcohol and phosphorus trichloride in the presence of a tertiary base.

$$3ROH + PCl_3 \xrightarrow{3PhNMe_2} (RO)_3P + 3PhNMe_2 \cdot HCl$$

In the *absence* of a base, the phosphite is attacked by the hydrogen chloride produced in the reaction to give alkyl chlorides (see p. 148).

Phosphites usually oxidise in the air and are readily oxidised by ozone to phosphates (see later). They are readily hydrolysed by dilute acids, but only slowly by dilute alkali.

Triphenyl phosphite reacts with alcohols and halogen to produce alkyl halide in good yield (Rydon *et al.*, 1954).

$$(PhO)_3P + ROH + X_2 \longrightarrow RX + XPO(OPh)_2 + PhOH$$

Instead of halogen, a reactive halide may be used, *e.g.*, methyl iodide, benzyl chloride or bromide. The reaction does not take place through a carbonium-ion intermediate, since no rearrangement occurs (see also the Arbusov reaction, below), *e.g.*, neopentyl alcohol is converted into neopentyl bromide (*cf.* p. 147):

$$(PhO)_3P + Me_3CCH_2OH + PhCH_2Br \longrightarrow Me_3CCH_2Br + PhCH_2PO(OPh)_2 + PhOH$$

An interesting point about phosphites is their structures. Free phosphorous acid is (I), which is

(I)   (II)

phosphonic acid. Mono- and di-alkyl phosphites have structures derived from (I), *i.e.*, they are phosphonic esters. It has been shown, however, that dialkylphosphonates, (I*a*), are in equilibrium with the dialkyl phosphites, (II*a*), and that the equilibrium lies almost entirely on the phosphonate side.

$$\begin{array}{ccc}
\mathrm{RO} \quad \mathrm{H} & & \mathrm{RO} \\
\diagdown \nearrow & & \diagdown \\
\mathrm{P} & & \mathrm{P{-}OH} \\
\diagup \diagdown\!\!\!\diagdown & & \diagup \\
\mathrm{RO} \quad \mathrm{O} & & \mathrm{RO} \\
(\mathrm{I}a) & & (\mathrm{II}a)
\end{array}$$

A study of the ΔH values (obtained from bond energies) has shown that conversion of phosphonate to phosphite is an endothermic reaction, and so (I*a*) $\longrightarrow$ (II*a*) is thermodynamically unfavourable. Thus, the phosphoryl form (I*a*) is the more stable one.

Esters of phosphonous and phosphinous acids. Esters of the 'alkyl-acid' may be prepared by reaction between an alkylphosphonous dichloride or dialkylphosphinous chloride and an alcohol in the presence of a tertiary base, *e.g.*,

$$\mathrm{R^1_2PCl + R^2OH + C_5H_5N \longrightarrow R^1_2POR^2 + C_5H_5N{\cdot}HCl}$$

Esters of phosphoric acid. As with phosphites, the preparation of alkyl phosphates from alcohol and phosphoryl chloride requires the presence of a tertiary base. Also, as with phosphites, phosphates are readily hydrolysed by dilute acids, but only very slowly (if at all) by dilute alkali.

$$\mathrm{3ROH + POCl_3 + 3C_5H_5N \longrightarrow (RO)_3PO + 3C_5H_5N{\cdot}HCl}$$

Phosphate esters are very important biologically, and most naturally occurring phosphorus compounds contain a terminal unsubstituted $\mathrm{-PO(OH)_2}$ group. Introduction of this group into a molecule is known as *phosphorylation*. One important method uses a protected phosphorylating agent, *e.g.*,

$$\mathrm{(PhO)_2POCl + ROH + C_5H_5N \longrightarrow C_5H_5N{\cdot}HCl + (PhO)_2PO(OR)}$$
$$\xrightarrow{\mathrm{H_2-Pd}} \mathrm{(HO)_2PO(OR) + 2PhMe}$$

If the protected group is not used, then polymers containing the $\mathrm{P-O-P}$ linkage are obtained.

Esters of phosphonic and phosphinic acids. Alkylphosphonates and dialkylphosphinates may be prepared as follows:

$$\mathrm{R^1PO(Cl_2) + 2R^2ONa \longrightarrow R^1PO(OR^2)_2 + 2NaCl}$$
$$\mathrm{R^1_2POCl + R^2ONa \longrightarrow R^1_2PO(OR^2) + NaCl}$$

Their most useful method of preparation is the **Arbusov reaction** (1906). This is a good means of forming carbon–phosphorus bonds, and in its simplest form, is the reaction between a trialkyl phosphite and an alkyl halide to form a dialkyl alkylphosphonate.

$$\mathrm{P(OR^1)_3 + R^2X \longrightarrow R^2PO(OR^1)_2 + R^1X}$$

The Arbusov reaction also includes the formation of alkyl dialkylphosphinates from dialkyl alkylphosphonites.

$$\mathrm{R^1P(OR^2)_2 + R^3X \longrightarrow R^1R^3PO(OR^2) + R^2X}$$

The Arbusov reaction occurs in *two* stages, each by the S_N2 mechanism, and the reaction works best with primary alkyl halides, *e.g.*,

$$\mathrm{X{\frown}R^2 \;\;\ddot{P}(OR^1)_3 \longrightarrow R^2{-}\overset{+}{P}(OR^1)_2{\frown}O{-}R^1 \;\; X^- \longrightarrow R^2PO(OR^1)_2 + R^1X}$$

It should be noted that the product of the Arbusov reaction is a phosphonate (*i.e.*, the phosphoryl form) which, as we have seen above, is more stable than the phosphite form.

Radical reactions at phosphorus

The majority of reactions undergone by phosphorus compounds are heterolytic, but free-radical reactions may be initiated in the presence of, *e.g.*, acyl peroxides or by means of irradiation with ultra-violet light. Most free-radical reactions involve tervalent phosphorus compounds; these give a seven-electron species, the *phosphino* free radicals. On the other hand, there is also the less common nine-electron species, the *phosphoranyl* free radicals. Examples of their formation are (where $\mathrm{R^2{\cdot}}$ is generated from an acyl peroxide):

$$R_2^1PX + R^{2\cdot} \longrightarrow R_2^1P\cdot + R^2X$$
$$R_3^1P + R^{2\cdot} \longrightarrow R^2R_3^1P\cdot$$
$$R_2P\!-\!PR_2 \xrightarrow{hv} 2R_2P\cdot$$

Oxidation. t-Phosphines are oxidised by oxygen to phosphine oxide and phosphinate (together with small amounts of other products). According to Buckler (1962), the mechanism is:

$$R\cdot + O_2 \longrightarrow ROO\cdot \xrightarrow{R_3P} R_3PO + RO\cdot \xrightarrow{R_3P} R_3\dot{P}OR \longrightarrow R_2POR + R\cdot\,; \text{ etc.}$$

Trialkyl phosphites are usually oxidised by oxygen to phosphates very slowly, but in the presence of acyl peroxides or ultraviolet light, the oxidation is fast. Plumb *et al.* (1963) have proposed the following mechanism:

$$R\cdot + O_2 \longrightarrow RO_2\cdot \xrightarrow{(RO)_3P} (RO)_3PO + RO\cdot \xrightarrow{(RO)_3P} (RO)_4P\cdot \longrightarrow (RO)_3PO + R\cdot\,; \text{ etc.}$$

t-Phosphines and trialkyl phosphites are also oxidised by di-t-butyl peroxide (R^1 = t-Bu), *e.g.*,

$$R^1\!-\!O\!-\!O\!-\!R^1 \longrightarrow 2R^1O\cdot$$
$$R_3^2P + R^1O\cdot \longrightarrow R_3^2\dot{P}OR^1 \longrightarrow R_3^2PO + R^1\cdot\,; \text{ etc.}$$

Addition to alkenes. Phosphorus compounds containing a P—H bond add to alkenes (anti-Markownikoff) in the presence of acyl peroxides:

$$R_2^1PH + R^{2\cdot} \longrightarrow R^2H + R_2^1P\cdot \xrightarrow{R^3CH=CH_2} R_2^1PCH_2CHR^3\cdot \xrightarrow{R_2^1PH} R_2^1PCH_2CH_2R^3 + R_2^1P\cdot\,; \text{ etc.}$$

Reactions with thiyl radicals

Phosphines and trialkyl phosphites react with thiols in the presence of acyl peroxides (or on irradiation with ultraviolet light) to form phosphorus sulphides, *e.g.*,

$$R^2SH + R^{1\cdot} \longrightarrow R^2S\cdot + R^1H$$
$$(R^3O)_3P \xrightarrow{R^2S\cdot} (R^3O)_3\dot{P}SR^2 \longrightarrow (R^3O)_3PS + R^2\cdot\,; \text{ etc}$$

Reaction with polyhalogenoalkanes

t-Phosphines and trialkyl phosphites react with, *e.g.*, carbon tetrachloride in the presence of acyl peroxides:

$$CCl_4 + R^{2\cdot} \longrightarrow R^2Cl + \cdot CCl_3$$
$$(R^1O)_3P + \cdot CCl_3 \longrightarrow (R^1O)_3\dot{P}CCl_3 \xrightarrow{CCl_4} \cdot CCl_3 + (R^1O)_3\overset{+}{C}Cl_3Cl^- \longrightarrow (R^1O)_2PO(CCl_3) + R^1Cl$$

Silicon compounds

Both carbon and silicon are Group IV elements, and their valence electrons are similar: C, $2s^2$, $2p^2$; Si, $3s^2$, $3p^2$. Hence, it can be anticipated that sp^3 hybridisation would operate in quadrivalent silicon compounds (as in quadrivalent carbon compounds), *e.g.*, X-ray analysis of Me_4Si shows a tetrahedral arrangement. However, the chemical properties of these two elements differ in some respects. One reason is the difference in bond strengths for the same ligand, *e.g.*, some bond-strength orders are: Si—H < C—H; Si—O > C—O; Si—Si < C—C. Thus, enthalpies of similar reactions of Si and C may be quite different, and so the free energies will be different. Silicon, unlike carbon, does not form multiple bonds such as occur in acetylene (sp), ethylene (sp^2), carbonyl compounds (sp^2), etc. The reason for this is not certain, but as far as $p_\pi\!-\!p_\pi$ bonding is concerned, *e.g.*, in Si=O, the two orbitals have very different energies, and consequently there is very little overlap (p. 43). Connected with this is the fact that silicon free radicals, *e.g.*, $Ph_3Si\cdot$, are much more reactive than the corresponding carbon compounds, $Ph_3C\cdot$. The latter are stabilised by resonance involving $C=C$ bonds (see p. 788); there is no resonance stabilisation in the silicon compounds since the $Si=C$ bond is absent. However, silicon can form multiple bonds, particularly with strongly electronegative ligands. In these compounds silicon has a covalency of *five*, and this is achieved by overlap of a lone pair ($2p^2$) of the donor atom with the *empty* $3d$ orbitals of silicon to form a d_π—d_π bond. A consequence of the existence of five-co-ordinate silicon (sp^3d), for which there is no carbon analogue, is, *e.g.*, the difference in acidity between Me_3SiOH and Me_3COH. Carbon is more electronegative than silicon (Table 2.1, p. 33), and therefore the carbinol would be expected to be more acidic than the silanol. In actual fact the reverse is true, and this may be explained on the basis that the $(p \rightarrow d)$ π-bond in the silanol reduces the negative charge on the oxygen atom, thereby increasing the acidity (*cf.* p. 182).

$$Me_3Si \Leftarrow OH \quad \textit{or} \quad Me_3Si—\overset{..}{O}H \longleftrightarrow Me_3\overset{-}{Si}=\overset{+}{O}H$$

Nomenclature of organosilicon compounds

The following examples illustrate the classification and nomenclature of silicon compounds.

Silanes is the generic name of compounds of the general formula Si_nH_{2n+2}, and prefixes di, tri, etc., indicate the number of silicon atoms present.

$$Si_nH_{2n+2} \qquad SiH_4 \qquad H_3Si—SiH_3 \qquad H_3SiSiH_2SiH_3$$
$$\textit{silanes} \qquad \text{silane} \qquad \text{disilane} \qquad \text{trisilane}$$

Various substituted silanes are named by use of the appropriate prefixes or suffixes, *e.g.*,

$$EtSiH_3 \qquad BrSiH_3 \qquad H_3SiNH_2 \qquad H_2Si(NH_2)_2$$
$$\text{ethylsilane} \quad \text{bromosilane} \quad \text{silanamine} \quad \text{silanediamine}$$
$$MeSiH_2OH \qquad HOSiH_2SiH_2OH$$
$$\text{methylsilanol} \qquad \text{disilane-1,2-diol}$$

Compounds containing the Si—O—Si group are **siloxanes**, and those with the Si—N—Si group, **silazanes**, *e.g.*,

$$MeOSiH_2—O—SiH_3 \qquad \overset{5}{Cl}\overset{}{Si}H_2\overset{4}{N}H\overset{3}{Si}H_2\overset{2}{N}H\overset{1}{Si}H_2OH$$
$$\text{methoxydisiloxane} \qquad \text{5-chlorotrisilazanol}$$

Alkylsilanes. Many methods are available for the preparation of alkylsilanes, *e.g.*,

(i) $\quad SiCl_4 + 4MeMgI \overset{\text{heat}}{\longrightarrow} Me_4Si + 4MgClI$

(ii) $\quad SiCl_4 + 2EtMgBr \longrightarrow Et_2SiCl_2 \overset{\text{LAH}}{\longrightarrow} Et_2SiH_2$

(iii) $\quad Cl_3SiH + 3EtLi \longrightarrow Et_3SiH + 3LiCl$

(iv) $\quad 2Et_3SiCl + 2Na \longrightarrow Et_3Si—SiEt_3 + 2NaCl$

The alkylsilanes are colourless oils and, except for the lower monoalkylsilanes, are stable in air. The alkylsilicon hydrides are readily oxidised to silanols by oxygen in the presence of benzoyl peroxide. Tetra-alkylsilanes are usually oxidised at the alkyl group, and are also readily attacked at the alkyl group by electrophilic reagents, *e.g.*,

$$(C_2H_5)_4Si + Cl_2 \longrightarrow (C_2H_5)_3SiC_2H_4Cl + HCl$$

On the other hand, silicon hydrides give alkylsilicon chlorides, and form silanols with alkali (see below).

Chlorinated alkylsilanes undergo β-elimination when warmed with alkali (*cf.* the E2 mechanism), *e.g.*,

$$HO^{\frown} \; SiEt_3 \overset{\frown}{-}CH_2 \overset{\frown}{-}CH_2 \overset{\frown}{-}Cl \longrightarrow HOSiEt_3 + CH_2{=}CH_2 + Cl^-$$

The corresponding γ-chloro-compounds eliminate cyclopropane:

$$HO^{\frown} \; SiR_3 - CH_2 - CH_2 \\ \qquad\qquad | \qquad\qquad \longrightarrow HOSiR_3 + CH_2 - CH_2 + Cl^- \\ \qquad\qquad CH_2 - Cl \qquad\qquad\qquad\qquad CH_2$$

Silenes. These correspond to carbenes; dimethylsilene, $Me_2Si\colon$, has been generated by reaction between the vapours of dimethyldichlorosilane and sodium or potassium in an atmosphere of helium (Skell *et al.*, 1964).

Alkylsilicon chlorides. These may be prepared, *e.g.*:

(i) by partial alkylation of silicon chloride with a Grignard reagent (see above).

(ii) Silicon hydrides are directly halogenated at the silicon atom. The mechanism is uncertain, but since reaction occurs

$$R_3SiH + Cl_2 \longrightarrow R_3SiCl + HCl$$

with retention of configuration, the mechanism cannot be S_N2. There is also evidence against a *four-centre reaction*, *e.g.*, BrCl gives R_3SiBr but, on the basis of such a reaction, with the silicon atom being subject to nucleophilic attack (see below), R_3SiCl would be expected to be the predominant product.

$$\begin{array}{cc} \overset{\delta+}{R_3Si} \cdots\cdots \overset{\delta-}{H} \\ \vdots \qquad\qquad \vdots \\ \overset{\delta-}{Cl} \cdots\cdots \overset{\delta+}{Br} \end{array} \longrightarrow R_3SiCl + HBr$$

(iii) Methylsilicon chlorides are prepared commerically by passing methyl chloride over a mixture of silicon and copper at 350°C; the product is a mixture of $SiCl_4$, $MeSiHCl_2$, Me_2SiCl_2, Me_3SiCl, and disilanes, of which the dimethyl dichloride predominates. This is a free-radical reaction, and is believed to proceed via methylcopper (Me—Cu). Arylsilicons may also be prepared this way from aryl chlorides.

The alkylsilicon chlorides are very reactive, the chlorine being readily displaced by various nucleophilic reagents, *e.g.*, they react with alcohols to form alkoxysilanes (see also below):

$$MeOH^{\frown} \; SiEt_3 \overset{\frown}{-}Cl \longrightarrow Cl^- + MeO\overset{\overset{\displaystyle H}{\overset{|}{+}}}{-}SiEt_3 \xrightarrow{\;-H^+\;} MeOSiEt_3$$

They also add to alkenes when heated under pressure, and the reaction is catalysed by platinum catalysts; the mechanism is probably ionic.

$$\begin{array}{c} \diagdown \qquad \diagup \\ C{=}C \\ \diagup \qquad \diagdown \end{array} + Cl_3\overset{+}{Si}H^- \xrightarrow{\;H_2PtCl_6\;} \begin{array}{c} \diagdown \qquad | \\ CH - C - SiCl_3 \\ \diagup \qquad | \end{array}$$

This is an example of **hydrosilation**. On the other hand, this addition can be effected in the presence of acyl peroxides or by ultraviolet light radiation; the mechanism is now free-radical:

$$Cl_3SiH \xrightarrow[\text{or } h\nu]{R\cdot} Cl_3Si\cdot \xrightarrow{>C=C<} Cl_3Si - \overset{|}{\underset{|}{C}} - \overset{|}{\underset{|}{C}}\cdot \xrightarrow{Cl_3SiH} Cl_3Si - \overset{|}{\underset{|}{C}} - \overset{|}{\underset{|}{C}}H + Cl_3Si\cdot \; ; \text{ etc.}$$

Silanols. These are usually prepared by the hydrolysis of the corresponding chlorides; the mechanism is S_N2, *e.g.*,

$$H_2\overset{\frown}{O} \; SiMe_2Cl \overset{\frown}{-}Cl \longrightarrow Cl^- + H_2\overset{+}{O}{-}SiMe_2Cl \xrightarrow{\;-H^+\;} HO{-}SiMe_2Cl \xrightarrow{H_2O} HO{-}SiMe_2{-}OH$$

Because of the ease with which these hydroxysilanes lose water to form siloxanes, hydrolysis is best carried out in buffered solution to maintain a neutral solution.

Silanols are also formed by the action of alkali on silicon hydrides; hydrogen is liberated:

$$R_3SiH + OH^- \xrightarrow{H_2O} R_3SiOH + H_2$$

Siloxanes. Disiloxanes are formed when silanols are heated alone or with acid.

$$Me_3SiOH \xrightarrow{H_3O^+} Me_3Si{-}OH_2{}^+$$

$$Me_3Si\overset{\frown}{O}H \quad SiMe_3{-}OH_2{}^+ \longrightarrow H_2O + Me_3Si{-}OH^+{-}SiMe_3 \xrightarrow{-H^+} Me_3Si{-}O{-}SiMe_3$$

They are also formed when alkoxysilanes are hydrolysed by dilute acid:

$$2Me_3SiOEt \xrightarrow{H_3O^+} Me_3Si{-}O{-}SiMe_3 + 2EtOH$$

Silicones. Hydrolysis of methylsilicon dichlorides without special precautions (see above) results first in the formation of diols, which then combine with elimination of water to form polymers (polysiloxanes) of general formula $(R_2SiO)_n$. These are oils and are called *silicones*, and normally consist of mixtures of linear and cyclic polymers, *e.g.*,

$$HOSiMe_2(OSiMe_2)_nOH \qquad Me_2Si(O{-}SiMe_2)_2O$$

By using a mixture of chlorides, Me_2SiCl_2 and $MeSiCl_3$, cross-linked (linear) polymers are obtained. The main groups in silicones are methyl and phenyl, and according to the degree of polymerisation and cross-linking, silicone rubbers, greases, resins, etc. are obtained. These are used for various purposes, *e.g.*, water-repellents, lubricants, electrical insulators.

Silanamines (silylamines). These may be prepared by reaction between a monochloride and an amine, *e.g.*, *N*-methyltriethylsilanamine.

$$Et_3SiCl + Me_2NH \longrightarrow Et_3SiNMe_2 + Me_2NH{\cdot}HCl$$

If a primary amine is used, a silazane is also formed (see below).

Silazanes. Disilazanes may be prepared as follows:

$$Me_3SiNHEt + Me_3SiCl \longrightarrow Me_3SiNEtSiMe_3 + HCl$$

Silicon dichlorides react with ammonia to give linear and cyclic silazanes (*cf*. silanediols), *e.g.*,

$$Me_2SiCl_2 + 2NH_3 \longrightarrow [Me_2Si(NH_2)_2] \xrightarrow{3\ mol.} Me_2Si(NHSiMe_2)_2NH$$

Boron compounds

The electronic configuration of boron is $1s^2, 2s^2, 2p$, and most boron compounds are tricovalent (sp^2) and contain only three electron pairs (open sextet), *e.g.*, BF_3, R_3B, etc. These have a planar structure and the Z—B—Z angle is 120°. In six-membered rings, the third $2p$ orbital may be used to form a π-bond which becomes part of the delocalised M.O. (see later). On the other hand, because of the open sextet, boron compounds readily accept a lone pair of electrons from a donor atom, thereby completing their octet, *e.g.*, $Na^+BF_4{}^-$, $F_3\overset{-}{B}{-}\overset{+}{N}H_3$ (see p. 31). In these compounds, the boron becomes tetrahedrally hybridised (sp^3).

A difficult problem associated with boron chemistry is the structure of boron hydrides. The simplest hydride is diborane (often called borane), B_2H_6, and is a dimer of borane, BH_3, which has no free existence. The structure of diborane is accepted as having a bridged configuration, which may be as shown. The molecule contains 12 valency electrons. Each H—B bond is 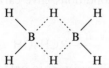 covalent (*i.e.*, contains an electron-pair) and each B····H····B bridge is considered to be covered by a three-centre M.O. containing *one pair of electrons*. Thus, in effect, each B····H bond may be regarded as a 'half-bond'. This molecule does not have a sufficient number of electrons for all ligands to be bound covalently by electron-pairs. Thus, diborane is an example of an *electron-deficient* molecule. Monomers, *e.g.*, R_3B, and carbonium ions, R_3C^+, *i.e.*, compounds containing an open sextet, have also been referred to as electron-deficient molecules, but it is clear that these and diborane differ in that in the former all linked atoms are covalently bound by electron-pairs.

Nomenclature of boron compounds

The following examples illustrate two methods of nomenclature.

$$BH_3 \qquad BCl_3 \qquad Et_3B \qquad MeBCl_2$$

borane $\qquad$ boron trichloride; $\qquad$ triethylborane $\qquad$ methylboron dichloride;
(borine) $\qquad$ trichloroborane $\qquad\qquad\qquad\qquad$ methyldichloroborane

$$MeB(OH)_2 \qquad Et_2BOH \qquad B(OH)_3$$

methylboronic acid; $\qquad$ diethylborinic acid; $\qquad$ orthoboric acid
methyldihydroxyborane $\qquad$ diethylhydroxyborane

The nomenclature of some heterocyclic boron compounds is described in the text.

Other ways of naming have also been used, but recent literature uses the two described here.

Trialkylboranes. Some convenient preparations are:

(i) Hydroboronation (see p. 114).

(ii) Reaction between a Grignard reagent and boron trifluoride or its etherate:

$$3RMgX + BF_3 \cdot (Et_2O)_2 \longrightarrow R_3B + 3MgFX + 2Et_2O$$

(iii) $\qquad\qquad B(OMe)_3 + 3RMgX \longrightarrow R_3B + 3MgXOMe$

(iv) $\qquad\qquad B(OR^1)_3 + R_3^2Al \longrightarrow R_3^2B + Al(OR^1)_3$

Trialkylboranes are gases or liquids, and exist as monomers, and this monomeric form is attributed to steric effects preventing dimerisation. They readily form co-ordination complexes, *e.g.*, (Z = N, P, etc.):

$$R_3^1B + ZR_3^2 \longrightarrow R_3^1\overset{-}{B}-\overset{+}{Z}R_3^2$$

The stability of these complexes depends on several factors: I effect of the R groups, electronegativity of Z, steric effects of the R groups, etc. Trimethylamine complexes, in particular, have been used to store volatile boranes, which are regenerated with hydrogen chloride.

Trialkylboranes react with halogens (the lower members inflame), hydrogen halides, and boron halides on heating to give mono- and di-alkylboron halides, *e.g.*,

$$R_2BCl + RCl \xleftarrow{Cl_2} R_3B \xrightarrow{HBr} R_2BBr + RH$$

All volatile trialkylboranes are spontaneously inflammable, but controlled oxidation with dry oxygen gives a dialkyl alkylboronate or an alkyl dialkylborinate with wet oxygen:

$$RB(OR)_2 \xleftarrow{O_2} R_3B \xrightarrow[H_2O]{O_2} R_2BOR$$

Peroxides, R_2BOOR, are intermediates, and undergo rapid rearrangement. Trialkylboranes are oxidised by alkaline hydrogen peroxide to alcohols (p. 114), and they readily react with carbon monoxide, and these carbonylated products, under suitable conditions, give carbonyl compounds, alcohols, etc., *e.g.*,

$$R_3B + CO \rightleftharpoons R_3\overset{-}{B}-\overset{+}{C}O \longrightarrow R_2BCOR \xrightarrow{LiBH_4} R_2BCHRO^- \xrightarrow{KOH} RCH_2OH$$

Triarylboranes are prepared by method (ii). They are solids, relatively resistant to oxidation, and react with sodium or potassium to form highly coloured anion-radicals, *e.g.*,

$$Ph_3B + Na \longrightarrow Ph_3B^{\cdot-} + Na^+$$

Alkylboron hydrides. The bridging hydrogens of diborane cannot be replaced by alkyl groups. Thus, *five* derivatives are theoretically possible; all five methyl compounds have been prepared. One method of preparing alkyldiboranes is by equilibrium between volatile trialkylboranes and diborane:

$$2Me_3B + B_2H_6 \rightleftharpoons 2Me_2BHBH_2Me$$

Alkyldiboranes are readily attacked by water, oxygen, etc., and may be prepared in monomeric form as co-ordination compounds, *e.g.*, $\overset{-}{R}BH_2—\overset{+}{N}Me_3$.

Alkylboron chlorides. These may be prepared in various ways, *e.g.*,
(i) By reaction between trialkylboranes and Cl_2, HCl, or BCl_3 (see above).
(ii) By reaction between boron trichloride and the appropriate amount of Grignard reagent (see above).
(iii) $$R_4Sn + 4BCl_3 \longrightarrow 4RBCl_2 + SnCl_4$$

Orthoborates may be prepared by reaction between boric acid or boron trichloride and an alcohol, or by transesterification (alcoholysis) with the appropriate alcohol.

$$3R^1OH + B(OH)_3 \longrightarrow (R^1O)_3B + 3H_2O$$
$$(R^1O)_3B + 3R^2OH \rightleftharpoons (R^2O)_3B + 3R^1OH$$

Orthoborates are generally readily hydrolysed by water:

$$(RO)_3B + 3H_2O \longrightarrow 3ROH + B(OH)_3$$

It appears to be uncertain whether this reaction is stepwise or not.

Alkylboronic acids, $RB(OH)_2$, and **dialkylborinic acids,** R_2BOH. These are obtained on hydrolysis of the corresponding alkylboron chloride, and their esters are produced if an alcohol is used:

$$R_2^1BOH \xleftarrow{H_2O} R_2^1BCl \xrightarrow{R^2OH} R_2^1BOR^2$$

The acids and their esters may also be prepared by reaction between an orthoborate and the appropriate amount of Grignard reagent, followed by hydrolysis (to give the acid):

$$(R^1O)_3B \xrightarrow{R^2MgX} R^2B(OR^1)_2 \xrightarrow{H_2O} R^2B(OH)_2$$

The boronic and borinic acids are weak acids, and are less sensitive to oxidation than the trialkylboranes, and the former acids are less sensitive than the latter. Both acids are readily dehydrated on heating to *boroxines* (*boroxoles*), $(RBO)_3$, and *borinic anhydrides* $(R_2B)_2O$, respectively. Boroxines are cyclic anhydrides; the alkylboroxines are stable, but the parent compound (R=H) is not. Although this structure could exhibit aromatic character (p. 582), calculations have shown that it is very small in boroxines (*cf.* borazines, below).

Organoboron–nitrogen compounds. It has already been mentioned that alkylboranes form co-ordination compounds with amines. On the other hand, dialkylboron chlorides react with secondary amines to form complexes which, on treatment with Et_3N, eliminate a molecule of HCl to form a *borazene*, R_2BNR_2. When they exist as monomers (aminoboranes), their structure is (I) but, unless the R groups are large (steric effect), the

borazenes usually exist as cyclic trimers (II; this is of the cyclohexane type), and are called *cycloborazanes*.

Borazines (borazoles). These are six-membered rings, the parent compound being (IIIa or IIIb), which may be prepared by heating diborane with ammonia. If borazine has substituents on the N atoms, this is indicated by prefixing the substituent with N-; if on the B atoms, by B-. B-Trichloro-N-trialkylborazines are conveniently prepared by heating boron trichloride with amines, e.g.,

$$3BCl_3 + 3RNH_2 \longrightarrow B_3Cl_3N_3R_3 + 6HCl$$

Borazines are of special interest in that they exhibit aromatic character; they contain a closed shell of six electrons (see IIIb; this is a flat hexagon; see also p. 582). Although borazines are 'aromatic', this aromaticity is not very pronounced, e.g., they are far more reactive than benzene; they readily add on three molecules of HX, H_2O, etc., the negative part of the addendum becoming attached to the boron atoms.

Complex anions of boron. These are of the type $M^+BR_4^-$. Sodium and lithium borohydrides (R=H), also called hydroborates, are useful reducing agents in organic chemistry (see Table 6.1, p. 179). Tetra-alkyl and -arylborates (R=R or Ar) may be prepared in several ways, e.g.,

$$R_3^1B + R^2Li \longrightarrow LiBR_3^1R^2$$

Sodium tetraphenylborate is used as a reagent for estimating potassium, nitrogen bases, etc. It may be prepared as follows:

$$BF_3 + 4PhMgBr \longrightarrow 3MgClF + Ph_4B^-MgCl^+ \xrightarrow{\text{aq. NaCl}} Na^+BPh_4^-$$

PROBLEMS

1. The reaction $RX + RS^-Na^+ \longrightarrow R_2S + NaX$ is faster than the reaction $RX + RO^-Na^+ \longrightarrow R_2O + NaX$. Explain.

2. Complete the following equations:

(i) $CH_2=CHCMe=CH_2 + SO_2 \longrightarrow$?

(ii) $MeNH_2 + CS_2 + NaOH \xrightarrow[\text{heat}]{H_2O}$?

(iii) $MeCONH_2 + P_2S_5 \xrightarrow{C_6H_6}$?

(iv) $CH_2=CHCN + HOCH_2CH_2SH \xrightarrow{\text{heat}}$?

(v) $CH_2\!\!-\!\!CH_2 + H_2S \longrightarrow$?
$\diagdown\!\!O\!\!\diagup$

(vi) $MeCH=CHCH_2SCN \xrightarrow{\text{heat}}$ $-CH(NCS)-$ (part of product)?

(vii) $CH_2=CH_2 + (SCN)_2 \longrightarrow$?

(viii) $Me_2C(SEt)_2 \xrightarrow{KMnO_4}$?

(ix) $PhSO_2Me \xrightarrow{EtONa}$? $\xrightarrow{PhCO_2Me}$? $\xrightarrow{-OMe^-}$?

(x) $MeCOCl + EtSH \longrightarrow$?

(xi) $(RS)_2Hg \xrightarrow{\text{heat}}$?

(xii) $Et_2S + H_2SO_4 \longrightarrow$? $\xrightarrow{H_2O}$?

(xiii) $Et_2S + BrCN \longrightarrow$?

3. Name the following compounds:
(i) $PhPO(OH)_2$; (ii) $HOP(OEt)_2$; (iii) $EtPCl_2$; (iv) $MePOCl_2$; (v) $H_2PCH_2CH_2Cl$; (vi) $H_3P=NH$; (vii) $(HO)POCl_2$; (viii) $POCl_2(NH_2)$; (ix) $HPOCl(OH)$; (x) H_2POCl; (xi) $MeHPO(NH_2)$; (xii) Et_2PCl.

4. Suggest a mechanism for the removal of oxygen from epoxides by Ph_3P to form alkenes (use ethylene oxide as your example).

5. Suggest a preparation for each of the following:
(i) Ph_3P; (ii) $MeEtPH$; (iii) Et_3PO; (iv) diethylphosphinous chloride; (v) diethyl methylphosphonate; (vi) dimethylphosphonic chloride; (vii) ethylphosphonic diamide.

6. Complete the following equations:

(i) $PhCH=CHCH_2PPh_3{}^+Cl^- + PhCHO \xrightarrow{EtOLi}$?

(ii) 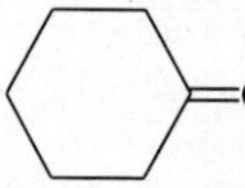 $=O + Ph_3P=CH_2 \longrightarrow$?

(iii) $Ph_3P=CCl_2 + p\text{-}Me_2NC_6H_4CHO \longrightarrow$?

(iv) $Ph_3P + CHCl_3 \xrightarrow{t\text{-BuOK}}$?

(v) $Ph_3P + ClCH_2COMe \longrightarrow ? \xrightarrow{Na_2CO_3} ? \xrightarrow{PhCHO}$?

(vi) $PhCO(CH_2)_4Br \xrightarrow{Ph_3P} ? \xrightarrow[EtOH]{EtONa} PhCO(CH_2)_3CH=PPh_3 \longrightarrow$?

(vii) $PhNO + Ph_3P=CHPh \longrightarrow$?

(viii) $Et_2PCl + Et_2PNa \longrightarrow$?

(ix) $Ph_3P + Et_3\overset{+}{N}-\overset{-}{O} \longrightarrow$?

(x) $R_3^1P + R_2^2SO \longrightarrow$?

(xi) $Ph_3P=CH_2 + MeCH=CHCOMe \longrightarrow$?

(xii) $(EtO)_3P + Cl_2 \longrightarrow$?

(xiii) $(EtO)_3P + Me_3\overset{+}{N}-\overset{-}{O} \longrightarrow$?

(xiv) $Me_2SO + Ph_3P \longrightarrow$?

(xv) $Ph_4P^+I^- + PhLi \longrightarrow$?

7. Name the following compounds:
(i) $H_3Si(SiH_2)_2SiH_3$; (ii) $MeHSi(NH_2)_2$; (iii) $EtSi(OH)_3$; (iv) Me_2SiCl_2; (v) $Et_2SiHOOCMe$; (vi) $EtOSiH_2(OSiH_2)_3SiH_2NH_2$; (vii) $SiH_3NMeSiH(OEt)NHSiH_3$.

8. Suggest a synthesis for each of the following:
(i) $n\text{-}Pr_4Si$; (ii) $MeSiH_3$; (iii) $Et_3SiSiEt_3$; (iv) Me_3SiCH_2Cl; (v) Me_3SiNEt_2; (vi) $Me_3SiNMeSiMe_3$; (vii) Me_3SiEt.

9. Complete the following equations:

(i) $MeCH=CHMe + Me_2SiHCl \xrightarrow{cat.}$?

(ii) $2CH_2=CHCH_2Cl + Si \xrightarrow{Cu}$?

(iii) $Me_3SiCl + BrCl \longrightarrow$?

(iv) $Me_3SiCl \xrightarrow{H_2O}$?

(v) $Et_3SiCl + Me_3SiOH \longrightarrow$?

(vi) $Me_3SiH + NaOH \xrightarrow{H_2O}$?

(vii) $Me_3SiCH_2CHClCH_3 + OH^- \longrightarrow$?

10. Write out the structures of:
(i) B-trichloroborazine; (ii) dimethoxyborane; (iii) methyl dimethylborinate; (iv) diethyl methyl-boronate; (v) trimethyl borate; (vi) ethoxy(methoxy)borane; (vii) acetoxyborane.

11. Write out the structures of the *five* possible methyldiboranes (mono, di, etc.). Comment.

12. Complete the following equations:

(i) $B(OMe)_3 + 3EtMgBr \longrightarrow$?

(ii) $3PhMgBr + BF_3 \longrightarrow$?

(iii) $Et_3B + Br_2 \longrightarrow$?

(iv) $2(MeO)_3B + NaH \longrightarrow$?

(v) $Me_3B + HCl \longrightarrow$?

(vi) $EtBCl_2 + 2MeOH \longrightarrow$?

(vii) $B(OMe)_3 + 2EtMgBr \longrightarrow$? $\xrightarrow{H_2O}$?

(viii) $MeB(OH)_2 \xrightarrow{heat}$?

13. Suggest mechanisms for the following rearrangements:

(i) $EtSCHMeCH_2OH \xrightarrow{HCl} EtSCH_2CHClMe$

(ii) $Ph_3P + ROOR \longrightarrow Ph_3PO + R_2O$ [*Not* free radical]

14. Convert:

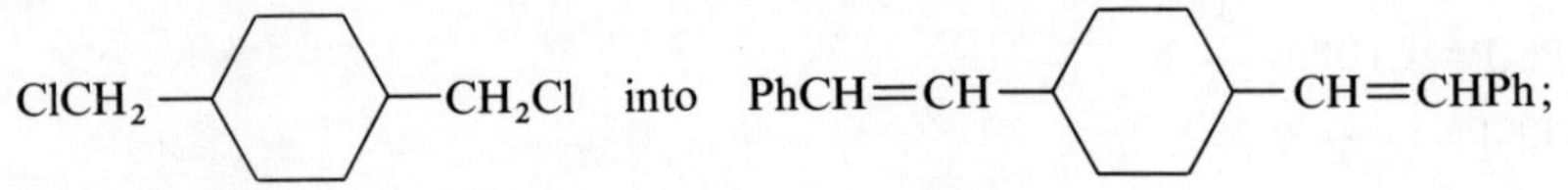

(i) $ClCH_2$—⟨ ⟩—CH_2Cl into $PhCH{=}CH$—⟨ ⟩—$CH{=}CHPh$;

(ii) $EtMeCHBr$ into $EtMeC{=}CHPr^n$.

15. Complete the following equations:

(i) $Ph_2CO + Ph_3P{=}CHOMe \longrightarrow$? $\xrightarrow{H^+}$?

(ii) $PhCN_2 + SO_2 \longrightarrow$? $\xrightarrow{ROH}$?

(iii) $MeSC{\equiv}CH + HCl \longrightarrow$?

(iv) $(R^1O)_3P + R^2COCl \longrightarrow$?

(v) $Et_3B + Et_3N \longrightarrow$?

(vi) $Et_3B + O_2(H_2O) \longrightarrow$?

(vii) $B(OMe)_3 + 2EtMgBr \longrightarrow$? $\xrightarrow{H_2O}$?

(viii) $EtB(OH)_2 \xrightarrow{heat}$?

(ix) ? $\xleftarrow{EtOH}$ CH_2—CH_2 $\xrightarrow{Me_2NH}$?
          \   /
           S

(x) $(CH_2SH)_2 + R_2CO \xrightarrow[BF_3.Et_2O]{AcOH}$? $\xrightarrow[H_2]{Ni}$?

(xi) $MeP(OEt)_2 + EtI \longrightarrow$?

(xii) $Et_2PNEt_2 + MeI \longrightarrow$?

(xiii) $R^1SSR^2 \xrightarrow[Ni]{Raney}$?

(xiv) $3BCl_3 + 3EtNH_2 \xrightarrow{heat}$?

16. Each of the following compounds showed only one singlet line in its NMR spectrum: (i) $C_2H_3F_3O_2S$; (ii) $C_2H_4O_4S$. Suggest a structure for each compound.

REFERENCES

KHARASCH (ed.), *Organic Sulphur Compounds*, Pergamon (1961).
PRYOR, *Mechanisms of Sulphur Reactions*, McGraw-Hill (1962).
GILBERT, *Sulphonation and Related Reactions*, Interscience (1965).
JOHNSON, *Ylid Chemistry*, Academic Press (1966).
WALLACE, 'The Chemistry of Sulphen Intermediates', *Quart. Rev.*, 1966, **20**, 67.
JANSSEN, *Organosulphur Chemistry*, Interscience (1967).
Handbook for Chemical Society Authors, Special Publication No. 14 (1960), pp. 115, 154 (Nomenclature).

HUDSON, *Structure and Mechanism in Organo-Phosphorus Chemistry*, Academic Press (1965).

KIRBY and WARREN, *The Organic Chemistry of Phosphorus*, Elsevier (1967).

EABORN, *Organosilicon Compounds*, Butterworths (1960).

SOMMER, *Stereochemistry, Mechanism and Silicon*, McGraw-Hill (1965).

COATES, GREEN, POWELL, and WADE, *Principles of Organometallic Chemistry*, Methuen (1968).

MUETTERTIES (ed.), *The Chemistry of Boron and its Compounds*, Wiley (1966).

CRAGG, 'Recent Developments in the Use of Organoboranes in Organic Synthesis', *J. Chem. Educ.*, 1969, **46**, 794.

15

Organometallic compounds

Generally speaking, organometallic compounds are those organic compounds in which a metal is directly joined to carbon. However, the ionic character of the carbon–metal bond depends on the nature of the metal, *e.g.*, the ionic character order is: Na > Li > Mg > Al > Zn > Cd > Hg; sodium and potassium alkyls are *salts* (see later). All of these metals are less electronegative than carbon (see Table 2.1, p. 33), and the larger the difference, the greater is the ionic character of the carbon–metal bond (the carbon being the *negative* end of the dipole). Also, the greater the ionic character, the more reactive is the metal alkyl.

The Grignard reagents

The alkylmagnesium halides, R—Mg—X, or Grignard reagents, introduced by Grignard in 1900, are extremely valuable in organic syntheses. A Grignard reagent is generally prepared by reaction between magnesium (1 atom) and alkyl halide (1 molecule) in dry, alcohol-free ether.

$$RX + Mg \longrightarrow RMgX \quad (v.g.-ex.)$$

The ethereal solution of the Grignard reagent is generally used in all reactions. Other solvents besides ether may be used, *e.g.*, tertiary amines, tetrahydrofuran and the dimethyl ether of ethylene glycol. Tetrahydrofuran is being used increasingly as a solvent, since it has been found that it increases the reactivity of organic halides towards magnesium. Thus, Normant (1953, 1957) has prepared Grignard reagents of vinyl halides and aryl chlorides in THF. In many cases where the preparation of the Grignard reagent is difficult, the addition of some methyl iodide may result in the formation of the desired Grignard reagent; this is referred to as the *entrainment method*. However, the best entrainment technique is to add ethylene dibromide (3 molecules or more) and to use magnesium in excess (6 molecules or more; Pearson *et al.*, 1959).

The ease with which an alkyl halide forms a Grignard reagent depends on a number of factors. It has been found that for a given alkyl group the ease of formation is alkyl iodide > bromide > chloride. It has also been found that the formation of a Grignard reagent becomes increasingly difficult as the number of carbon atoms in the alkyl group increases, *i.e.*, the ease of formation is $CH_3X > C_2H_5X > C_3H_7X > \cdots$. Since tertiary alkyl iodides readily eliminate hydrogen iodide with the formation of an alkene, tertiary alkyl chlorides are used. Also, in the preparation of the Grignard reagent, a common side-reaction is coupling between two R groups to give R—R (see p. 427).

Structure of Grignard reagents

Grignard *et al.* (1901), in their attempt to isolate the Grignard reagent, showed that the compound contained ether (which they called 'ether of crystallisation'). Although a great deal of work has been done, the structure of the Grignard reagent is still not settled. According to Ashby (1967), the composition of Grignard reagents *in ether* may be represented by the following equilibria:

$$\text{etc.} \rightleftharpoons \text{Dimer} \rightleftharpoons 2RMgX \rightleftharpoons R_2Mg + MgX_2 \rightleftharpoons \text{Dimer} \rightleftharpoons \text{etc.}$$

The extent of association depends on the concentration, and is extensive above about 0·3 M. Each molecule of Grignard reagent co-ordinates with one molecule of ether, and the halogen atom of one molecule co-ordinates with the magnesium atom of another molecule, *e.g.*, the monomer is (I) and the trimer is (II).

Crystalline Grignard reagents have been isolated (from ether) as 'di-etherates', *e.g.*, $EtMgBr.(OEt_2)_2$, and X-ray analysis has shown that the four groups are arranged tetrahedrally about the magnesium.

In this book, the Grignard reagent will be designated as RMgX, but as far as ether solutions are concerned, it will occasionally (and more accurately) be designated as $RMgX(OEt_2)_2$.

Reactions of the Grignard reagents

When working with Grignard reagents, it is usual to add the other reactant (often in ethereal solution) slowly to the Grignard solution or vice versa, and after a short time, decompose the magnesium complex with water, dilute acid, or aqueous ammonium chloride; the yields are usually *g.–v.g.*

The majority of Grignard reactions fall into two groups.

(i) *Addition of the Grignard reagent to a compound containing a multiple-bond group, e.g.,*

$$\diagdown\!\!\diagup C{=}O; \quad {-}C{\equiv}N; \quad \diagdown\!\!\diagup C{=}S; \quad {-}N{=}O; \quad \diagdown\!\!\diagup S{=}O$$

Addition of the R group of RMgX occurs at the less electronegative and that of the MgX group at the more electronegative atom. This is understandable on the basis that R and MgX are respectively the negative and positive ends of the dipole in RMgX.

(ii) *Double decomposition with compounds containing an active hydrogen atom or a reactive halogen atom.* We shall consider only the former at this stage (see p. 430 for an example of the latter). As we have seen, an active hydrogen atom is one joined to oxygen, nitrogen or sulphur. When such compounds react with a Grignard reagent, the alkyl group is converted into the alkane, *e.g.*,

$$R^1H + Mg(OH)X \xleftarrow{\text{H}_2\text{O}} R^1MgX \xrightarrow{\text{R}^2\text{OH}} R^1H + Mg(OR^2)X$$

Since reactions with compounds containing an active hydrogen atom result in the quantitative yield of hydrocarbon, this type of reaction is valuable for the determination of the number of active hydrogen atoms in a compound. The procedure is known as the **Zerewitinoff active hydrogen determination** (1907), and methylmagnesium iodide is normally used as the Grignard reagent. The methane which is liberated is measured (by volume), one molecule of methane being equivalent to one active hydrogen atom, *e.g.*,

$$RNH_2 + CH_3MgI \longrightarrow CH_4 + RNHMgI$$
$$R_2NH + CH_3MgI \longrightarrow CH_4 + R_2NMgI$$

Only one hydrogen atom in a primary amine reacts at room temperature. At a sufficiently high temperature, the active hydrogen atom in the magnesium derivative of the primary amine will react with a further molecule of methylmagnesium iodide:

$$RNHMgI + CH_3MgI \longrightarrow CH_4 + RN(MgI)_2$$

It is therefore possible to estimate the number of amino- and imino-groups in compounds containing both. It is not possible to get a high enough temperature for the second reaction with ether; a satisfactory solvent for the complete Zerewitinoff determination is pyridine (Lehman and Basch, 1945).

Lithium aluminium hydride also reacts with compounds containing active hydrogen, and so may be used for analytical determinations, *e.g.*,

$$LiAlH_4 + 4ROH \longrightarrow (RO)_4LiAl + 4H_2$$
$$LiAlH_4 + 4RNH_2 \longrightarrow 4H_2 + (RNH)_4LiAl \xrightarrow{LiAlH_4} 2(RN)_2LiAl + 4H_2$$

The enolic form of a compound, since it contains an active hydrogen atom, reacts with a Grignard reagent, *e.g.*, acetoacetic ester and nitroethane:

$$CH_3COCH_2CO_2C_2H_5 \rightleftharpoons CH_3\overset{\overset{\displaystyle OH}{|}}{C}{=}CHCO_2C_2H_5 \xrightarrow{CH_3MgI} CH_3\overset{\overset{\displaystyle OMgI}{|}}{C}{=}CHCO_2C_2H_5 + CH_4$$

$$CH_3CH_2NO_2 \rightleftharpoons CH_3CH{=}NO_2H \xrightarrow{CH_3MgI} CH_3CH{=}NO_2MgI + CH_4$$

In both cases the methane will not be liberated immediately, but at a rate depending on the speed of the conversion of keto into enol.

The hydrogen atom in the $\equiv$CH group is also active with respect to a Grignard reagent (to be expected, since it is acidic, p. 139). Thus, when acetylene is passed through an ether solution of ethylmagnesium bromide, ethynylbis(magnesium bromide) is formed:

$$C_2H_5MgBr + C_2H_2 \longrightarrow HC{\equiv}CMgBr + C_2H_6 \xrightarrow{C_2H_5MgBr} BrMgC{\equiv}CMgBr$$

The second step can be avoided by adding the Grignard solution to tetrahydrofuran saturated with acetylene (Jones *et al.*, 1956).

Order of reactivity of functional groups. In most cases of syntheses in which a Grignard reagent is used, the other compound has only one functional group, and consequently the reaction takes place in one direction only. Occasionally it is necessary to carry out a synthesis with a compound containing two (or possibly more) functional groups. If an excess of Grignard reagent is used, then both groups react as would be expected. It has been found experimentally that the reactivity of different groups is not equal, and hence when one equivalent of Grignard reagent is added, two competitive reactions take place simultaneously, but at different rates, resulting in two products in unequal amounts. Experiments have shown that an active hydrogen reacts very much faster than any other group; so much so, in fact, that a compound containing an active hydrogen and another group, reacts with one equivalent of a Grignard reagent as if it had only one reactive group, the active hydrogen. In general, the order of reactivity of the oxo group is as shown, and in all cases it is more reactive than halogen of the alkyl halide type.

$$-CHO > {\Large\diagdown}CO{\diagup} > -COCl > -CO_2R > -CH_2X$$

Synthetic uses of the Grignard reagents

Hydrocarbons. When a Grignard reagent is treated with any compound containing active hydrogen, a hydrocarbon is produced; in practice, water or dilute acid is used:

$$RMgBr + H_2O \xrightarrow{H^+} RH + Mg(OH)Br$$

Since alkyl halides are readily prepared from alcohols, it becomes a relatively simple matter to convert an alcohol (saturated or unsaturated) into the corresponding hydrocarbon.

It has already been mentioned (p. 424) that coupling to give R_2 usually occurs during the preparation of the Grignard reagent. On the other hand, Kharasch *et al.* (1941) showed that the addition of a small amount of cobaltous chloride to the Grignard reagent increases the yield of coupling product. According to Wilds *et al.* (1949) and Hey *et al.* (1965, 1969), the reaction proceeds via dialkylcobalt intermediates:

$$2RMgX + CoCl_2 \longrightarrow 2MgXCl + R_2Co \longrightarrow R_2 + Co$$
$$2RMgX + Co \longrightarrow Mg + MgBr_2 + R_2Co \longrightarrow R_2 + Co; \text{ etc.}$$

Many transition-metal halides may be used besides cobaltous chloride, *e.g.*, AgBr, $CuCl_2$, $CrCl_3$, $MnCl_2$, $FeCl_3$, etc.

Coupling between a Grignard reagent and alkyl halide containing a reactive halogen atom can be effected directly; this reaction is probably S_N2:

$$XMg \overset{\frown}{\diagup} R^1 \quad R^2 \overset{\frown}{\diagdown} X \longrightarrow R^1 - R^2 + MgX_2$$

The yield is very good if R^2 is allyl, t-butyl, or benzyl. This reaction also works well for α-halogeno-ethers (see later). A good yield $R^1 - R^2$ can be obtained for any R^2 by using its tosyl ester.

$$R^1MgI + TsOR^2 \longrightarrow R^1 - R^2 + TsOMgI$$

Unreactive halides react with Grignard reagents in the presence of cobaltous chloride, etc., preferably in benzene solution. The main products are R_2^1 and R_2^2; the crossed product is usually obtained in poor yield. The formation of R_2^1 (from R^1MgBr) has been described above; the other products are possibly formed as follows:

$$R_2^1Co + 2R^2Br \longrightarrow R_2^1 + R_2^2 + CoBr_2$$
$$R^2Br + Mg \longrightarrow R^2MgBr$$
$$R^1MgBr + R^2MgBr + CoCl_2 \longrightarrow 2MgBrCl + R^1CoR^2 \longrightarrow R^1 - R^2 + Co$$

Primary alcohols. A Grignard reagent may be used to synthesise an alcohol by treating it with dry oxygen and decomposing the product with acid:

$$RMgX \xrightarrow{O_2} RO_2MgX \xrightarrow{RMgX} 2ROMgX \xrightarrow{H_3O^+} 2ROH \quad (g.-v.g.)$$

This method (which can be used for all three classes of alcohols) is little used in practice since an alkyl halide may be converted into the corresponding alcohol by simpler means (p. 155). The method, however, is useful for converting aryl halides into phenols.

When a Grignard reagent (in ethereal solution) is treated with formaldehyde gas, or when a Grignard reagent in, *e.g.*, di-n-butyl ether (b.p. 141°C), is refluxed with paraformaldehyde, a primary alcohol is obtained by decomposing the magnesium complex with dilute acid. In view of the fact that mechanisms of reactions involving Grignard reagents are not known with certainty, and what is known leads to a complicated sequence (see tertiary alcohols, below), it is customary to use *simplified* equations for Grignard reactions. Thus:

$$\begin{array}{c} R \diagdown MgI \\ CH_2 \!\!=\!\! O \end{array} \longrightarrow \begin{array}{c} H \quad R \\ \diagdown \diagup \\ C \\ \diagup \diagdown \\ H \quad OMgI \end{array} \xrightarrow{H_3O^+} RCH_2OH \quad (g.)$$

A primary alcohol containing *two* carbon atoms more than the Grignard alkyl group can be prepared by adding one molecule of ethylene chlorohydrin to two molecules of Grignard reagent.

$$RMgBr + ClCH_2CH_2OH \longrightarrow RH + ClCH_2CH_2OMgBr \xrightarrow{RMgBr}$$

$$RCH_2CH_2OMgBr \xrightarrow{H_3O^+} RCH_2CH_2OH \quad (g.\text{-}v.g.)$$

Two molecules of Grignard reagent are not necessary if ethylene oxide is used instead of ethylene chlorohydrin.

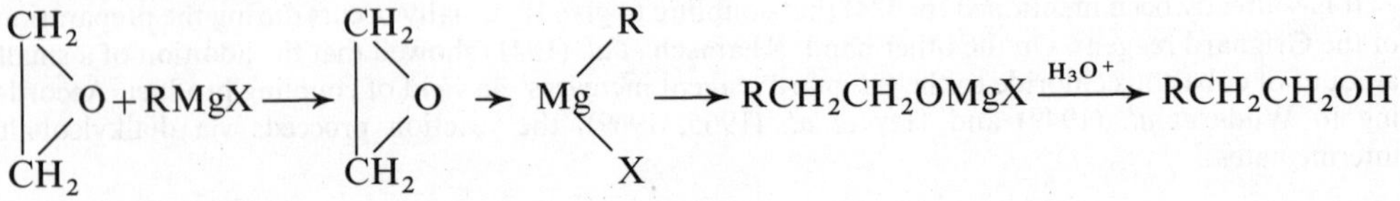

$$\xrightarrow{H_3O^+} RCH_2CH_2OH \quad (g.)$$

Similarly, a primary alcohol containing *three* more carbon atoms can be prepared from the Grignard reagent by using trimethylene oxide.

Secondary alcohols. When a Grignard reagent is treated with any aldehyde other than formaldehyde, a secondary alcohol is formed:

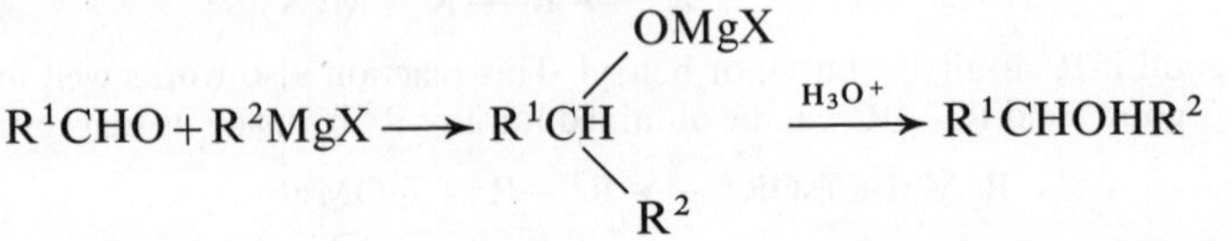

$$R^1CHO + R^2MgX \longrightarrow R^1CH \overset{OMgX}{\underset{R^2}{\diagdown}} \xrightarrow{H_3O^+} R^1CHOHR^2 \quad (f.g.\text{-}g.)$$

It can be seen that the secondary alcohol, R^1CHOHR^2, is obtained whether we start with R^1CHO and R^2MgX, or R^2CHO and R^1MgX. Which pair we use is generally a matter of their relative accessibility.

Secondary alcohols may also be prepared by interaction of a Grignard reagent (2 molecules) and ethyl formate (1 molecule) (see also t-alcohols, below):

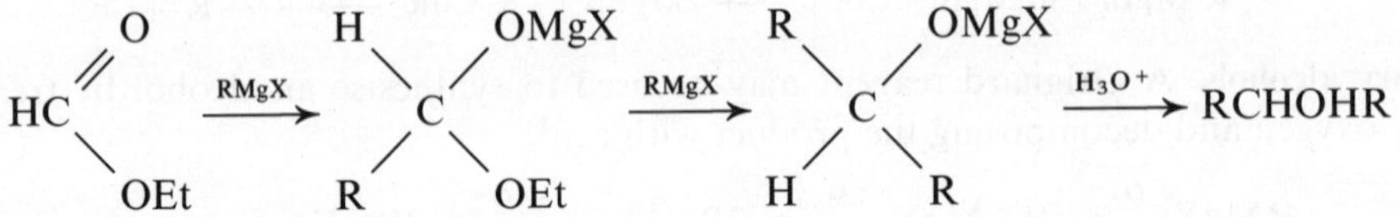

Tertiary alcohols. A tertiary alcohol may be prepared by the action of a Grignard reagent on a ketone:

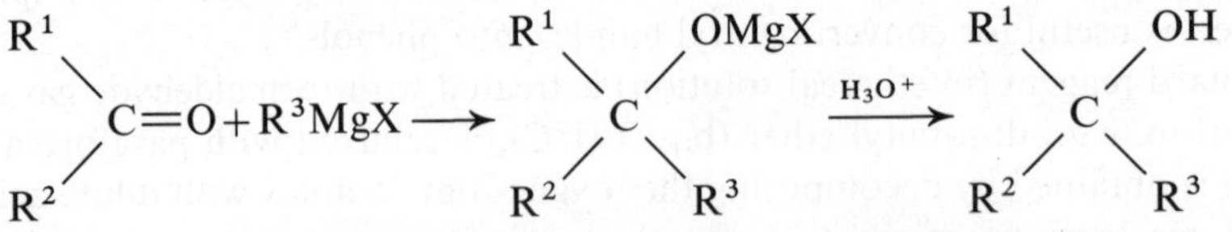

By this means a tertiary alcohol with three different alkyl groups may be prepared, and the starting materials may be any of the following pairs of compounds: R^1COR^2 and R^3MgX; R^1COR^3 and R^1MgX, or R^2COR^3 and R^1MgX (*cf.* secondary alcohols). It might be noted here that because t-alcohols are readily dehydrated to alkenes by acids, the complex is often broken up with aqueous ammonium chloride or buffered acid solution.

The mechanism of the reaction between a Grignard reagent and a carbonyl compound has been studied in great detail. The following mechanism is that proposed by Ashby (1967), who examined the reaction between phenylmagnesium bromide and benzophenone (R=Ph):

$$R_2C{=}O + RMgX(OEt_2)_2 \rightleftharpoons R_2C{=}O \rightarrow MgRX(OEt_2)$$

This is followed by combination with another molecule of Grignard reagent to form an intermediate alkoxymagnesium alkyl, which then reacts as shown:

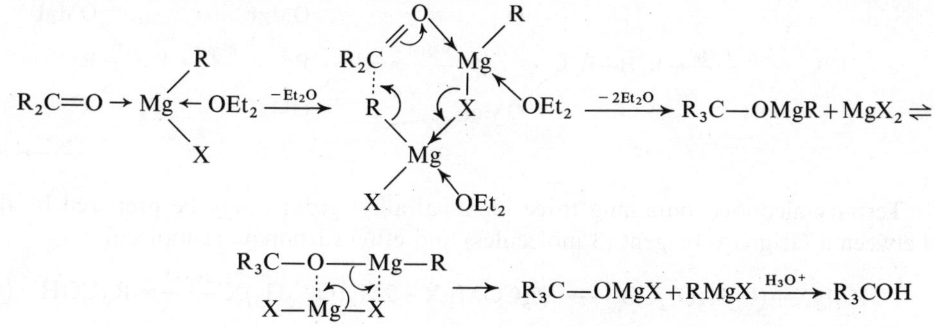

This mechanism is in keeping with many of the observed facts and, in particular, explains the observed decrease in rate of reaction when half of the Grignard reagent has been used up. Further support comes from the fact that a mixture of alkoxymagnesium alkyl and magnesium bromide produces the alkoxymagnesium halide (Ashby *et al.*, 1967).

Tertiary alcohols containing at least two identical alkyl groups may be prepared by the reaction between a Grignard reagent (2 molecules) and any ester (1 molecule) other than formic ester.

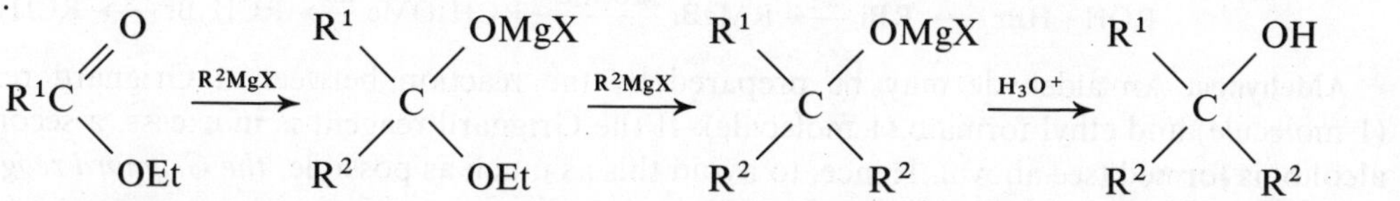

There is very little evidence for the mechanism of this reaction, but by analogy with that for Grignard/ketone reactions (see above), the steps may be (ether molecules are omitted):

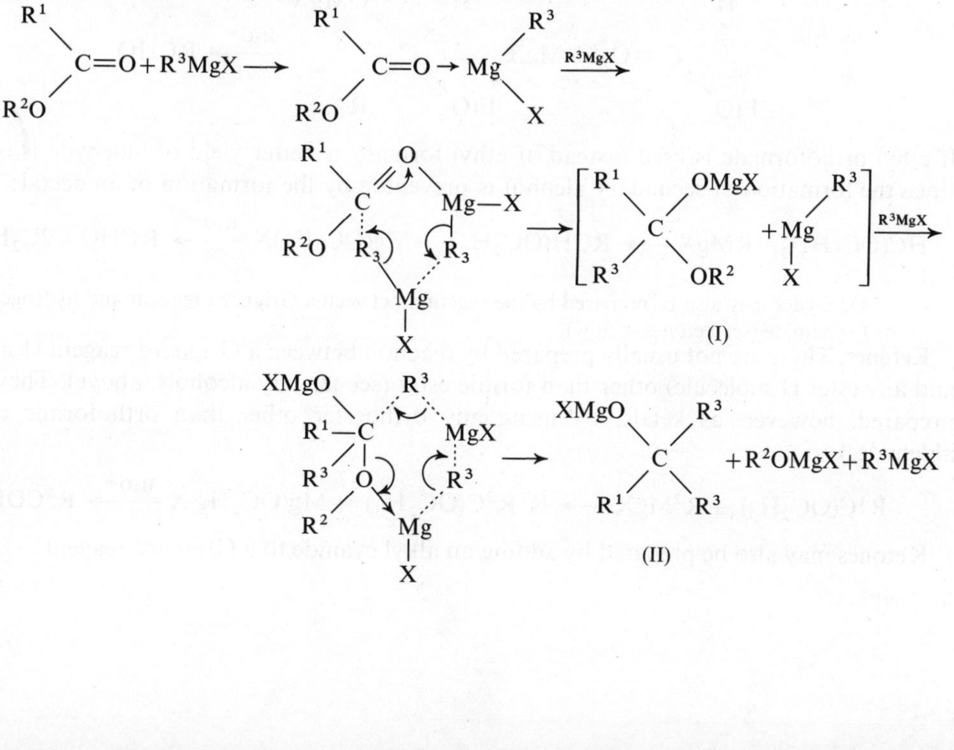

For most esters, it appears that the step leading to (II) is much faster than that leading to (I). Thus, treatment of an ester with one equivalent of a Grignard reagent generally gives t-alcohol and *recovery of half of the ester*. If, however, the Grignard is added to ester (*reverse addition*) and the reaction is carried out at low temperatures, then reasonable yields of ketone may be obtained. If the Grignard reagent (or the ester) is sterically hindered, ketone and not alcohol is the major product.

Tertiary alcohols may also be prepared by using an acid and a slight excess of Grignard reagent, *i.e.*, more than 3 molecules (Huston *et al.*, 1946):

$$R^1C{\overset{O}{\underset{OH}{\diagup\!\!\!\!\diagdown}}} \xrightarrow{R^2MgI} R^2H + R^1C{\overset{O}{\underset{OMgI}{\diagup\!\!\!\!\diagdown}}} \xrightarrow{R^2MgI} R^1C{\overset{OMgI}{\underset{OMgI}{\diagup\!\!\!\!\diagdown}}}R^2 \xrightarrow{R^2MgI} R^1C{\overset{OMgI}{\underset{R^2}{\diagup\!\!\!\!\diagdown}}}R^2$$

$$\xrightarrow{H_3O^+} R^1R_2^2COH$$

Tertiary alcohols containing three identical alkyl groups may be prepared by the reaction between a Grignard reagent (3 molecules) and ethyl carbonate (1 molecule):

$$3RMgX + (C_2H_5O)_2CO \longrightarrow R_3COMgX + 2Mg(OC_2H_5)X \xrightarrow{H_3O^+} R_3COH \quad (f.g.-g.)$$

Ethers. Ethers may be prepared by reaction between a Grignard reagent and an α-monochloroether (this is an example of double decomposition):

$$R^1OCH_2Cl + R^2MgX \longrightarrow R^1OCH_2R^2 + MgXCl \quad (g.)$$

Chloroethers of the above type are readily prepared, and using monochlorodimethyl ether with Grignard reagents, it is possible to step up both the ether and alcohol series, *e.g.*,

$$MeOH + CH_2O + HCl \longrightarrow MeOCH_2Cl + H_2O$$

$$ROH + HBr \longrightarrow RBr \xrightarrow{Mg} RMgBr \xrightarrow{MeOCH_2Cl} RCH_2OMe \xrightarrow{HBr} RCH_2Br \xrightarrow{OH^-} RCH_2OH$$

Aldehydes. An aldehyde may be prepared by the reaction between a Grignard reagent (1 molecule) and ethyl formate (1 molecule). If the Grignard reagent is in excess, a secondary alcohol is formed (see above). Hence, to avoid this as much as possible, *the Grignard reagent is added to the ester*:

$$\underset{EtO}{\overset{H}{\diagdown\!\!\!/}}C{=}O + RMgX \longrightarrow \underset{EtO}{\overset{H}{\diagdown}}\underset{R}{\overset{OMgX}{\diagup}}C \xrightarrow{H_3O^+} RCHO$$

If ethyl orthoformate is used instead of ethyl formate, a better yield of aldehyde is obtained, since the formation of secondary alcohol is prevented by the formation of an acetal:

$$HC(OC_2H_5)_3 + RMgX \longrightarrow RCH(OC_2H_5)_2 + Mg(OC_2H_5)X \xrightarrow{H_3O^+} RCHO + 2C_2H_5OH$$

Aldehydes may also be prepared by the reaction between a Grignard reagent and hydrogen cyanide or formamide (see ketones, below).

Ketones. These are not usually prepared by reaction between a Grignard reagent (1 molecule) and any ester (1 molecule) other than formic ester (see tertiary alcohols, above). They may be prepared, however, as ketals, by using any orthoester other than orthoformic ester (*cf.* aldehydes):

$$R^1C(OC_2H_5)_3 + R^2MgX \longrightarrow R^1R^2C(OC_2H_5)_2 + Mg(OC_2H_5)X \xrightarrow{H_3O^+} R^1COR^2$$

Ketones may also be prepared by adding an alkyl cyanide to a Grignard reagent:

$$R^1CN + R^2MgX \longrightarrow R^1R^2C\!\!=\!\!NMgX \xrightarrow{H_3O^+} [R^1R^2C\!\!=\!\!NH] \xrightarrow{H_3O^+} R^1COR^2 + NH_3$$
$$\text{ketimine}$$

The starting materials may be either R^1CN and R^2MgX or R^2CN and R^1MgX. If hydrogen cyanide is used instead of an alkyl cyanide, an aldehyde is formed.

Acyl chlorides (1 molecule) react rapidly with Grignard reagents (1 molecule) to form ketones:

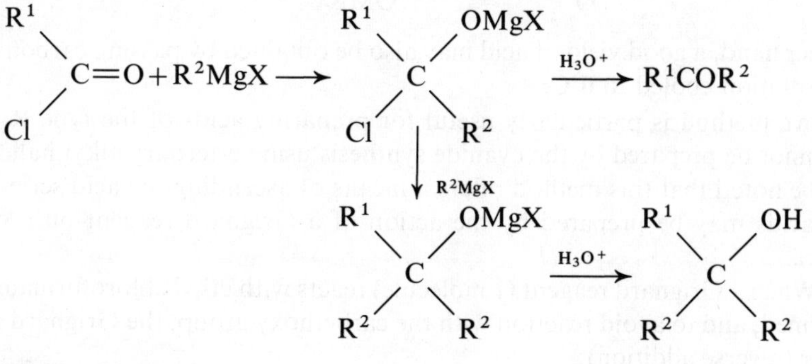

Tertiary alcohol is also formed.

By analogy with that for Grignard/ketone reactions, the mechanism may be:

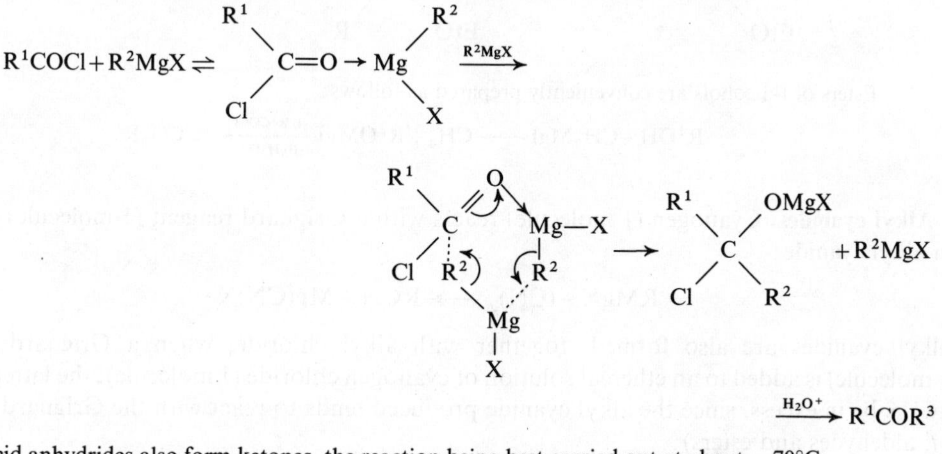

Acid anhydrides also form ketones, the reaction being best carried out at about $-70°C$:

$$(R^1CO)_2O + R^2MgX \xrightarrow{-70°C} R^1COR^2 + R^1CO_2MgX \quad (g.\text{-}v.g.)$$

Amides and N-substituted amides may also be used:

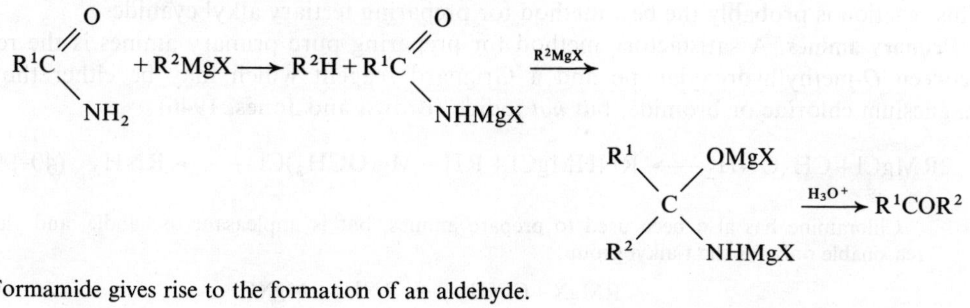

Formamide gives rise to the formation of an aldehyde.

Acids. When a Grignard reagent is treated with solid carbon dioxide and the complex decomposed with dilute acid, a monocarboxylic acid is obtained:

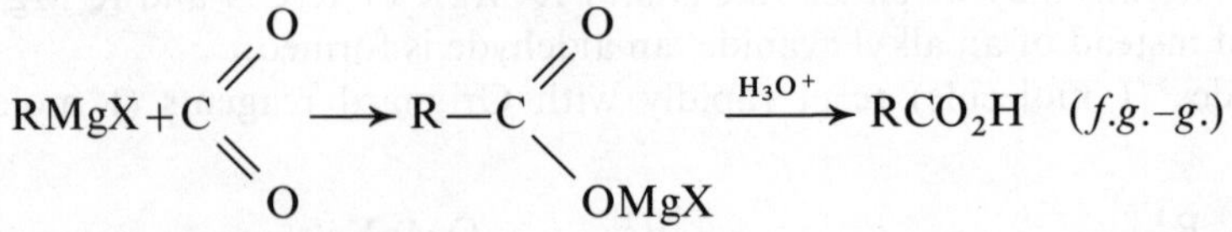

$$RMgX + C\underset{O}{\overset{O}{\lozenge}} \longrightarrow R-C\underset{OMgX}{\overset{O}{\lozenge}} \xrightarrow{H_3O^+} RCO_2H \quad (f.g.\text{--}g.)$$

On the other hand, a good yield of acid may also be obtained by passing carbon dioxide into the Grignard solution cooled to 0°C.

The above method is particularly useful for preparing acids of the type R_3CCO_2H, which usually cannot be prepared by the cyanide synthesis using a tertiary alkyl halide (*cf.* p. 612). It may also be noted that this method offers a means of ascending the acid series.

γ-Keto-acids may be prepared by the action of a Grignard reagent on a cyclic anhydride (see p. 451).

Esters. When a Grignard reagent (1 molecule) reacts with ethyl chloroformate (1 molecule) an ester is formed, and to avoid reaction with the carbethoxy group, the Grignard reagent is added to the ester (reverse addition):

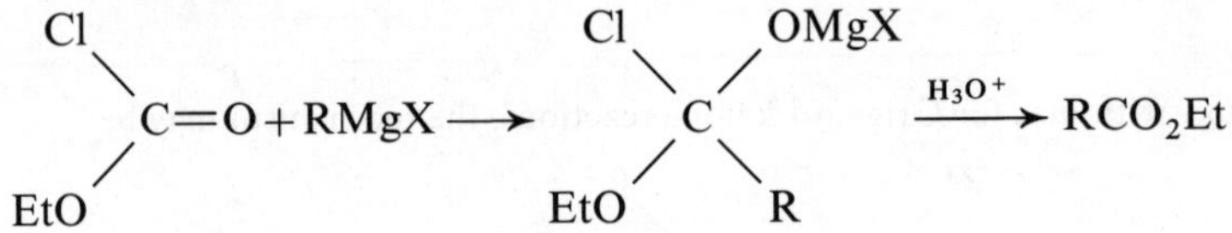

$$\underset{EtO}{\overset{Cl}{>}}C{=}O + RMgX \longrightarrow \underset{EtO}{\overset{Cl}{>}}C\underset{R}{\overset{OMgX}{<}} \xrightarrow{H_3O^+} RCO_2Et$$

Esters of t-alcohols are conveniently prepared as follows:

$$R^1OH + CH_3MgI \longrightarrow CH_4 + R^1OMgI \xrightarrow[\text{(ii) } H_3O^+]{\text{(i) } R^2COCl} R^2CO_2R^1$$

Alkyl cyanides. Cyanogen (1 molecule) reacts with a Grignard reagent (1 molecule) to form an alkyl cyanide:

$$RMgX + (CN)_2 \longrightarrow RCN + Mg(CN)X$$

Alkyl cyanides are also formed, together with alkyl chloride, when a Grignard reagent (1 molecule) is added to an ethereal solution of cyanogen chloride (1 molecule); the latter should always be in excess, since the alkyl cyanide produced tends to react with the Grignard reagent (*cf.* aldehydes and esters):

$$RMgCl + ClCN \longrightarrow RCN + MgCl_2$$
$$RMgCl + ClCN \longrightarrow RCl + Mg(CN)Cl$$

This reaction is probably the best method for preparing tertiary alkyl cyanides.

Primary amines. A satisfactory method for preparing pure primary amines is the reaction between *O*-methylhydroxylamine and a Grignard reagent which may be either the alkyl-magnesium chloride or bromide, but *not* iodide (Brown and Jones, 1946):

$$2RMgCl + CH_3ONH_2 \longrightarrow RNHMgCl + RH + Mg(OCH_3)Cl \xrightarrow{H_3O^+} RNH_2 \quad (40\text{--}90\%)$$

Chloramine has also been used to prepare amines, but is unpleasant to handle, and yields are reasonable only for the t-alkyl group.

$$RMgX + ClNH_2 \longrightarrow RNH_2 + MgXCl$$

Organometallic and organo-non-metallic compounds. These may be prepared by interaction of a Grignard reagent and an inorganic halide (see Ch. 14 and pp. 434–440).

Alkyl halides. Bromides and iodides are formed when a Grignard reagent is treated with bromine or iodine:

$$RMgX + I_2 \longrightarrow RI + MgXI$$

Thioalcohols. These may be prepared by the action of sulphur on a Grignard reagent (*cf.* alcohols):

$$RMgX + S \longrightarrow RSMgX \xrightarrow{H_3O^+} RSH$$

Sulphinic acids. When sulphur dioxide is passed into a well-cooled Grignard solution, a sulphinic acid (as its magnesium complex) is formed:

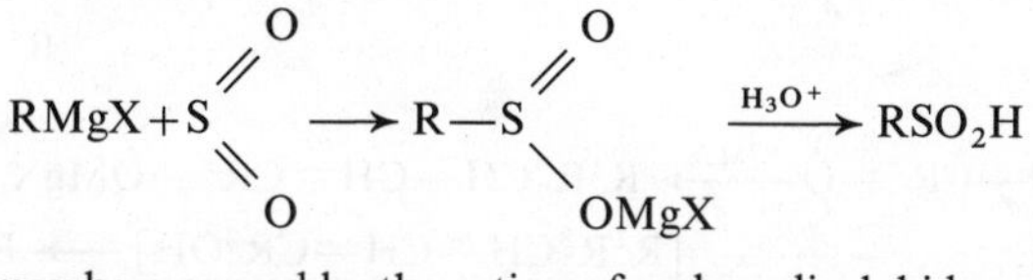

$$RMgX + S\overset{\displaystyle O}{\underset{\displaystyle O}{}} \longrightarrow R\!-\!S\overset{\displaystyle O}{\underset{\displaystyle OMgX}{}} \xrightarrow{H_3O^+} RSO_2H$$

Dithioic acids. These may be prepared by the action of carbon disulphide on a Grignard reagent (*cf.* carboxylic and sulphinic acids):

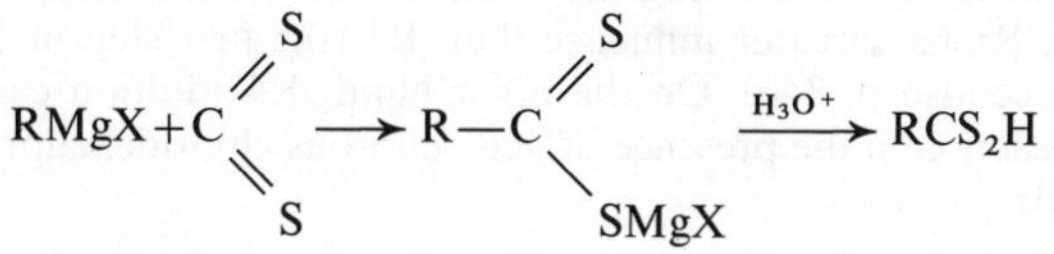

$$RMgX + C\overset{\displaystyle S}{\underset{\displaystyle S}{}} \longrightarrow R\!-\!C\overset{\displaystyle S}{\underset{\displaystyle SMgX}{}} \xrightarrow{H_3O^+} RCS_2H$$

Abnormal behaviour of Grignard reagents

In certain cases a Grignard reagent does not react with compounds containing a functional group which is normally capable of reaction. Generally, branching of the carbon chain near the functional group prevents reaction; the cause is probably the steric effect, *e.g.*, methylmagnesium bromide or iodide does not react with hexamethylacetone, $(CH_3)_3CCOC(CH_3)_3$. It has also been found that if the Grignard reagent contains large alkyl groups, reaction may be *prevented*; *e.g.*, isopropyl methyl ketone reacts with methylmagnesium iodide but not with t-butyl-magnesium chloride. In other cases, *abnormal reaction* may take place, *e.g.*, when isopropyl-magnesium bromide is added to di-isopropyl ketone, the expected tertiary alcohol is *not* formed; instead, the secondary alcohol, di-isopropylcarbinol, is obtained, resulting from the *reduction* of the ketone (see also p. 436):

$$(CH_3)_2CHCOCH(CH_3)_2 \xrightarrow{(CH_3)_2CHMgBr} (CH_3)_2CHCHOHCH(CH_3)_2 + CH_3CH{=}CH_2$$

This abnormal reaction may be explained by the transfer of a hydride ion from the Grignard reagent via a cyclic transition state (*cf.* the M.P.V. reduction):

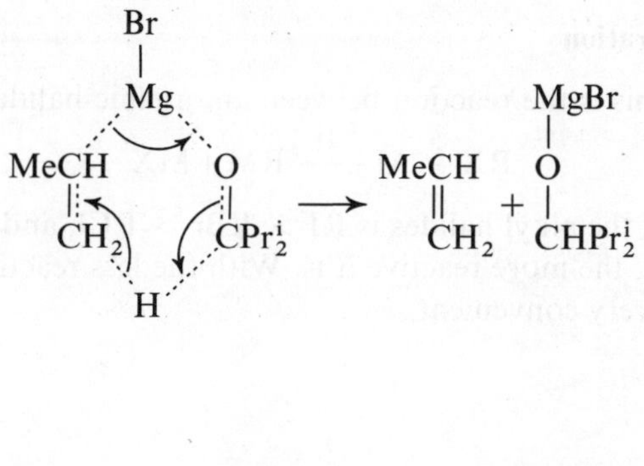

In some cases, aldehyde and the Grignard reagent interact to form ketone and a primary alcohol, *e.g.*,

$$R^1CHO + R^2MgBr \longrightarrow R^1R^2CHOMgBr \xrightarrow{R^1CHO} R^1R^2CO + R^1CH_2OMgBr$$

This also occurs by hydride ion transfer via a cyclic T.S.

Grignard reagents often reduce nitro groups (to nitroso or hydroxylamine), and hence mixtures of products may be obtained. However, in certain cases, normal reaction will proceed if carried out at low temperatures.

Compounds which are capable of enolisation liberate hydrogen with Grignard reagents (see p. 426).

α,β-Unsaturated carbonyl compounds react with a Grignard reagent in the 1,2- or 1,4-positions.

1,2-addition

$$R^1 \overset{4}{-}CH \overset{3}{=} CH \overset{2}{-} \overset{1}{C}R^2 \overset{}{=} \underset{}{O} \xrightarrow{R^3MgX} R^1 - CH = CH - CR^2R^3 - OMgX \xrightarrow{H_3O^+}$$
$$R^1 - CH = CH - R^2R^3 - OH$$

1,4-addition

$$R^1 \overset{4}{-}CH \overset{3}{=} CH \overset{2}{-} \overset{1}{C}R^2 \overset{}{=} \underset{}{O} \xrightarrow{R^3MgX} R^1R^3CH - CH = CR^2 - OMgX \xrightarrow{H_3O^+}$$
$$[R^1R^3CH - CH = CR^2OH] \longrightarrow R^1R^3CH - CH_2 - CR^2 = O$$
$$\qquad\qquad\text{enol} \qquad\qquad\qquad\qquad\qquad\qquad \text{keto}$$

The predominant factor in deciding the course of addition is steric hindrance, and when both R^1 and R^2 are large, R^2 has greater influence than R^1 (the first step in 1,2-addition is addition to the carbonyl group; see also p. 340). On the other hand, 1,4-addition can be made to predominate by carrying out the reaction in the presence of, *e.g.*, cuprous chloride, cupric acetate, etc. (the mechanism is now free-radical).

The reaction of dihalides of the type $Br(CH_2)_nBr$ with magnesium depends on the value of n. For $n = 1$–3, no Grignard reagent is formed; *e.g.*, ethylene dibromide ($n = 2$) gives ethylene; 1,3-dibromopropane ($n = 3$) gives propene, cyclopropane, and other products. When $n \geqslant 4$, the dimagnesium compound is obtained, but the yields are usually of the order of about 30 per cent., *e.g.*, $n = 5$:

$$Br(CH_2)_5Br + 2Mg \longrightarrow BrMg(CH_2)_5MgBr$$

The reactions of these di-Grignards depend on the nature of the added substrate (see p. 536).

Metal alkyls

Alkyl derivatives of practically all the metals have been prepared; they are named as the alkyl metal, *e.g.*, $(CH_3)_2Hg$, dimethylmercury, $(C_2H_5)_4Pb$, tetraethyl-lead.

Since metal displacement reactions depend on the standard electrode potentials of the metals concerned, it is useful to refer to the electrochemical series (*i.e.*, the arrangement of metals in order of their standard electrode potentials):

$$Li > K > Na > Mg > Al > Zn > Cd > Sn^{2+} (\longrightarrow Sn) > Pb^{2+} (\longrightarrow Pb) > H > Sn^{4+}$$
$$(\longrightarrow Sn^{2+}) > Bi > Cu > Ag^+ (\longrightarrow Ag) > Hg > Pb^{4+} (\longrightarrow Pb^{2+})$$

General methods of preparation

1. Direct displacement. This is the reaction between an organic halide and a metal:

$$RX + 2M \longrightarrow RM + MX$$

The order of reactivity of the alkyl halides is $RI > RBr > RCl$, and the higher the metal is in the electrochemical series, the more reactive it is. With the less reactive metals, use of an alloy (usually with sodium) is very convenient.

2. Exchange between organometallic compounds and metal halides. This is the most widely used method; it takes place when M^1 is higher in the electrochemical series than M^2. Thus, with M^1

$$RM^1 + M^2X \longrightarrow RM^2 + M^1X$$

as magnesium, *i.e.*, using a Grignard reagent, only a few metals (those which occur above Mg in the electrochemical series) fail to react.

3. Halogen-metal exchange. This is the reaction between an organic halide and an organo-

$$R^1X + R^2M \longrightarrow R^1M + R^2X$$

metallic compound. This method is particularly useful for preparing organolithium compounds.

4. Metal exchange (transmetalation). This is the reaction between a metal which is higher in the electrochemical series and an organo-compound of a metal which is lower in the series:

$$M^1 + RM^2 \longrightarrow RM^1 + M^2$$

GROUP IA METALS

The alkyl derivatives of these metals are generally best prepared by the metal exchange reaction (method 4, above) using a dialkylmercury:

$$2Na + Me_2Hg \longrightarrow 2Me^-Na^+ + Hg$$

Primary alkylsodium compounds are best prepared by reaction between an alkyl-lithium and sodium t-butoxide (Lochmann *et al.*, 1966):

$$Bu^tONa + RLi \longrightarrow Bu^tOLi + R^-Na^+$$

Alkyl-lithium compounds may be prepared by direct displacement (method 1, above); the chlorides give best yields, *e.g.*,

$$BuCl + 2Li \xrightarrow[\text{hexane}]{N_2} BuLi + LiCl$$

When the lithium compounds cannot be prepared this way, then the halogen-metal exchange reaction is used (method 3, above); n-butyl-lithium is normally used, *e.g.*,

$$BuLi + RX \longrightarrow RLi + BuX$$

When alkyl-lithium compounds are prepared in solution, hydrocarbon solvents are generally used. The use of ether encourages the Wurtz reaction, and, more important, lithium alkyls attack ethers. This reaction probably proceeds via metalation:

$$CH_3CH_2OC_2H_5 + BuLi \longrightarrow BuH + CH_3CHLiOC_2H_5 \longrightarrow CH_2{=}CH_2 + C_2H_5OLi$$

The sodium and potassium alkyls are insoluble, colourless, highly reactive solids. They are electrovalent compounds, *i.e.*, salts. The lithium compounds are colourless liquids or readily fusible solids, and are covalent, *e.g.*, CH_3Li, but, according to the nature of the alkyl group, they may have partial ionic character, *e.g.*, n-butyl-lithium. The large covalent character of the C—Li bond is shown by the fact that when R is an optically active group, RLi racemises slowly in non-polar solvents.

Lithium alkyls are associated in the liquid state and some are associated in the vapour state. These 'polymers' are believed to be electron-deficient molecules (p. 417).

From the practical point of view, lithium compounds are the most satisfactory to use in organic synthesis, since they are very easily prepared and dissolve in organic solvents. Generally speaking the lithium compounds (and those of sodium and potassium) behave like the Grignard reagents, but the lithium compounds are usually more reactive and the yield of product is often better. Because of their sensitivity to oxygen and to water, reactions with lithium alkyls are best carried out in an atmosphere of dry nitrogen. Some reactions of the lithium compounds are as follows. With carbon dioxide a good yield of ketone is obtained:

$$RLi + CO_2 \longrightarrow RC\!\!\begin{array}{c}O\\ \diagdown\\ OLi\end{array} \xrightarrow{RLi} R_2C\!\!\begin{array}{c}OLi\\ \diagup\\ \diagdown\\ OLi\end{array} \xrightarrow{H_2O} R_2CO$$

A ketone is also the main product when a carboxylic acid is used as starting material:

$$R^1C\!\!\begin{array}{c}O\\ \diagdown\\ OH\end{array} \xrightarrow{R^2Li} R^1C\!\!\begin{array}{c}O\\ \diagdown\\ OLi\end{array} \xrightarrow{R^2Li} R^1C\!\!\begin{array}{c}OLi\\ \diagup\\ |\\ \diagdown\\ OLi\end{array}\!\!-R^2 \xrightarrow{H_2O} R^1COR^2$$

Lithium compounds also react with cyanides to give ketones, and with aldehydes and ketones to give alcohols in better yields than with Grignard reagents.

It has been pointed out on p. 433 that highly sterically hindered t-alcohols cannot be prepared by the Grignard reaction. On the other hand, many of these alcohols can be prepared by means of lithium alkyls, e.g.,

$$(Me_2CH)_2CO + Me_2CHLi \longrightarrow (Me_2CH)_3COLi \xrightarrow{H_3O^+} (Me_2CH)_3COH \quad (53\%)$$

Thus, organolithium compounds are less susceptible to steric hindrance than Grignard reagents. This is also illustrated by the fact that addition of organolithium compounds to α,β-unsaturated carbonyl compounds is usually almost exclusively 1,2 (see p. 434). The main difference between the two reagents is that lithium compounds add to an ethylenic bond, whereas Grignard reagents do not, e.g.,

$$Bu^tLi + C_2H_4 \longrightarrow Bu^tCH_2CH_2Li$$

Apart from the synthetic uses described above, organolithium compounds are used in the preparation of carbenes, Wittig reagents, arynes, etc. The lithium alkyls, particularly the n-butyl compound, mixed with titanium chloride, are used as catalysts for the polymerisation of propene and isoprene (Ziegler–Natta catalyst; see p. 118).

GROUP IIB METALS

Organozinc compounds. Zinc alkyls may be prepared in several ways, e.g.,

(i) By heating an alkyl iodide with zinc, or better, a zinc-copper couple. Alkylzinc iodide is formed first, and this then disproportionates on distillation:

$$2RI + 2Zn \longrightarrow 2RZnI \longrightarrow R_2Zn + ZnI_2$$

(ii) By reaction between zinc chloride and an aluminium alkyl (method 2, p. 435), e.g.,

$$ZnCl_2 + 2Et_3Al \longrightarrow Et_2Zn + 2Et_2AlCl$$

The zinc dialkyls (collinear molecules) are volatile liquids, spontaneously inflammable in air and readily attacked by water. Their reactions are similar to those of the Grignard reagents, but the zinc compounds are less reactive; they do not react with carbon dioxide and react only slowly with cyanides, ketones and esters (see also below).

Alkylzinc halides. These may be intermediates in the preparation of zinc dialkyls (see above) and can be obtained as polymers as follows:

$$2ZnCl_2 + 2Et_2Zn \longrightarrow [EtZnCl]_4$$

Iodomethylzinc iodide, ICH_2ZnI, is an alkylzinc halide, and reacts with alkenes to form cyclopropanes (see p. 541).

The Reformatsky reaction (1887). The Reformatsky reaction is the reaction between an α-bromoacid ester and a carbonyl compound (aldehyde, ketone) in the presence of zinc to form a β-hydroxy-ester.

$$R^1CHBrCO_2Et + Zn \longrightarrow BrZnCHR^1CO_2Et \xrightarrow{R_2^2CO}$$

$$R_2^2C(OZnBr)CHR^1CO_2Et \xrightarrow{H_3O^+} R_2^2C(OH)CHR^1CO_2Et$$

The intermediate organozinc bromides have been isolated (Curé *et al.*, 1966).

The use of zinc instead of magnesium in the Reformatsky reaction is based on the fact that the Grignard reagent formed immediately attacks the ester group of a second molecule.

A difficulty with the Reformatsky reaction is the conversion of the β-hydroxy-ester into the corresponding acid; in general, hydrolysis leads to the α,β-unsaturated ester and acid (see p. 469). However, Read *et al.* (1969) have shown that by refluxing the adduct (Zn + carbonyl compound + ester) in benzene, and then followed by acid, good yields of hydroxy-acid are obtained.

Organocadmium compounds. Cadmium dialkyls are readily prepared by method 2 (p. 435):

$$2RMgX + CdCl_2 \longrightarrow R_2Cd + 2MgClX$$

Cadmium dialkyls (collinear molecules) are volatile liquids; they are chemically similar to the zinc compounds, but are less reactive.

Dialkylcadmium compounds are used to prepare ketones from acid chlorides; they are prepared *in situ*:

$$2R^1COCl + R_2^2Cd \longrightarrow 2R^1COR^2 + CdCl_2$$

This ketone synthesis may be used to prepare long-chain monocarboxylic acids as follows:

$$R(CH_2)_xOH \xrightarrow[\text{(ii) Mg}]{\text{(i) HBr}} [R(CH_2)_x]_2Cd \xrightarrow{ClOC(CH_2)_yCO_2C_2H_5} R(CH_2)_xCO(CH_2)_yCO_2C_2H_5$$
$$\text{(iii) CdCl}_2$$

$$\xrightarrow[\text{HCl}]{\text{Zn/Hg}} R(CH_2)_xCH_2(CH_2)_yCO_2H$$

An interesting point in connection with syntheses involving cadmium dialkyls is that Kollonitsch (1960) has shown that the *pure* compounds react very slowly (if at all) with acid chlorides. In the presence of magnesium (and other ether-soluble halides), reaction is fast.

Organomercury compounds. Mercury dialkyls are best prepared by the same method used for the cadmium compounds:

$$2RMgX + HgCl_2 \longrightarrow R_2Hg + 2MgClX$$

When X = Br or I, the main product is an alkylmercury bromide or iodide, *e.g.*,

$$RMgBr + HgCl_2 \longrightarrow RHgBr + MgBrCl$$

The mercury dialkyls (collinear molecules) are poisonous liquids, not spontaneously inflammable in air; they are not decomposed by water but are by dilute acids. With halogen acids, the alkylmercury halide is obtained.

$$R_2Hg + HCl \longrightarrow RHgCl + RH$$

Trihalogenomethyl derivatives of mercury have also been prepared, *e.g.*,

$$PhHgCl + CHBrCl_2 + Bu^tOK \longrightarrow PhHgCBrCl_2 + KCl + Bu^tOH$$

These compounds are a useful source of dihalogenomethylenes, *e.g.*,

$$PhMgCBrCl_2 \longrightarrow PhHgBr + [CCl_2]$$

Mercuration, *i.e.*, the introduction of a mercuri-acid group, has important applications in aromatic chemistry. The commonest two groups are *chloromercuri* (— HgCl) and *acetoxymercuri* (—HgOOCH$_3$). These are discussed elsewhere (see, *e.g.*, p. 617); *oxymercuration* has been discussed on p. 109.

GROUP IIIB METALS

Organoaluminium compounds. *Aluminium alkyls.* A number of trialkyls are prepared commercially by heating aluminium powder, hydrogen, a 1-alkene, and an aluminium trialkyl under pressure, *e.g.*,

$$Al + 3/2H_2 + 2Et_2Al \longrightarrow 3Et_2AlH \xrightarrow{3C_2H_4} 3Et_3Al$$

Dialkylaluminium hydrides are also prepared by the above method. On the other hand, mono-, di-, and tri-alkylaluminium compounds may be prepared by reaction between aluminium chloride and the appropriate amount of Grignard reagent (method 2, p. 435).

The aluminium n-trialkyls are dimers, but branched-chain compounds are monomers (due to the steric effect), *e.g.*, Pr$_3^i$Al. The dimers are electron-deficient molecules and have alkyl-bridges (*cf.* alkyl-

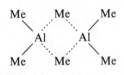

boranes, p. 418), *e.g.*, trimethylaluminium dimer is (I) and the bridging carbon atoms are five-co-ordinate. These dimers dissociate to a small extent in solution; in the gaseous phase dissociation is much greater, *e.g.*, Me$_6$Al$_2$ at 215°C and under low pressure is 100 per cent monomer (Haaland *et alt*, 1969). The N.M.R. spectroscopy of Me$_6$Al$_2$ (at room temperature) has shown that all the methyl groups are equivalent. This has been explained on the basis that the terminal and bridging methyl groups in (I) are exchanging rapidly (Muller *et al.*, 1960). Molecules such as this, which exhibit site exchange, are called *fluxional molecules.*

Dialkylaluminium hydrides (which exist as trimers) are formed when trialkyls are heated; alkene is eliminated, *e.g.*,

$$Bu_3^t Al \xrightarrow{150°C} Bu_2^t AlH + Me_2C{=}CH_2$$

This reaction is *reversible*, and because of the ease with which isobutene is lost, tri-t-butylaluminium is used to prepare other aluminium alkyls by exchange with alkenes:

$$Bu_3^t Al + 3RCH{=}CH_2 \longrightarrow (RCH_2CH_2)_3Al + 3Me_2C{=}CH_2$$

Since aluminium trialkyls are decomposed by water, the overall reaction is reduction of an alkene.

$$(RCH_2CH_2)_3Al \xrightarrow{H_2O} 3RCH_2CH_3 + Al(OH)_3$$

Di-isobutylaluminium hydride is a very useful reducing reagent for various functional groups, *e.g.*, aldehydes and ketones give alcohols; esters give alcohols, but at low temperatures, aldehydes are obtained; cyanides give aldehydes or amines (see also p. 134). The mechanism is similar to that with lithium aluminium hydride, *i.e.*, involves hydride ion transfer (see p. 178):

$$R_2CO + Bu_2^i AlH \longrightarrow R_2CHOAlBu_2^i \xrightarrow{H_2O} R_2CHOH$$

There is, however, an added complication; if only the $\diagdown$Al—H bond were involved, then one molecule of the hydride will supply one hydride ion. In practice, particularly with aldehydes, *both* isobutyl groups are also involved; *three* molecules of aldehyde are reduced by *one* molecule of Bu$_2^i$AlH, and two molecules of isobutene are formed. This could be explained by direct β-hydride ion transfer from the isobutyl groups (*cf.* p. 433).

Many other aluminium compounds are used as reducing reagents, particularly LAH (see Table 6.1, p. 179).

Aluminium trialkyls, like the alkylboranes, form co-ordination compounds, *e.g.*,

$$Me_6Al_2 + 2Me_3N \longrightarrow 2Me_3\bar{Al}{-}\overset{+}{N}Me_3$$

Triethylaluminium is the starting point for the commercial preparation of long straight-chain alcohols. Long chains are produced by insertion of ethylene when heated with this under pressure at about 100°C. The length of each individual chain is usually different; for simplicity, the reaction is written as follows:

$$AlEt_3 \xrightarrow{3nC_2H_4} Al[(C_2H_4)_nEt]_3 \xrightarrow{O_2} Al[O(C_2H_4)_nEt]_3 \xrightarrow{H_2O} 3Et(C_2H_4)_nOH + Al(OH)_3$$

The most important use of aluminium trialkyls is as a catalyst for the low temperature polymerisation of alkenes (p. 118).

GROUP IVB METALS

Organotin compounds. *Tetra-alkylstannanes* are conveniently prepared as follows (methods 1, 2, and 3; p. 434):

$$4RX + 4Na/Sn \longrightarrow R_4Sn + 3Sn + 4NaX$$
$$SnCl_4 + 4RMgX \longrightarrow R_4Sn + 4MgClX$$
$$3SnCl_4 + 4R_3Al + 4NaCl \longrightarrow 3R_4Sn + 4NaAlCl_4$$

The tetra-alkylstannanes are stable liquids; they are unaffected by air or water. They react with halogens to form alkyltin halides:

$$R_4Sn \xrightarrow{I_2} R_3SnI + RI$$

Tetravinyltin, b.p. 160°C, may be prepared as follows:

$$CH_2{=}CHBr + Mg \xrightarrow{THF} CH_2{=}CHMgBr \xrightarrow{SnCl_4} (CH_2{=}CH)_4Sn$$

This compound is of particular interest in that the vinyl group is very labile, and hence is used to prepare vinyl derivatives of other metals, *e.g.*,

$$(CH_2{=}CH)_4Sn + 4PhLi \longrightarrow 4CH_2{=}CHLi + Ph_4Sn$$

Organotin halides may be prepared, *e.g.*,

$$R_4Sn + SnCl_4 \xrightarrow{20°C} R_3SnCl + RSnCl_3 \xrightarrow{180°C} 2R_2SnCl_2$$

They are reduced by LAH to the corresponding tin hydride, *e.g.*,

$$R_2SnCl_2 \xrightarrow{LiAlH_4} R_2SnH_2$$

These hydrides selectively reduce the carbonyl group in α,β-unsaturated aldehydes and ketones (*cf.* p. 337).

Alkenes (and alkynes) undergo *hydrostannation* when heated with organotin hydrides, *e.g.*,

$$R_3^1SnH \longrightarrow H\cdot + R_3^1Sn\cdot \xrightarrow{R^2CH{=}CH_2} R_3^1SnCH_2CHR^2 \xrightarrow{R_3^1SnH} R_3^1SnCH_2CH_2R^2 + R_3^1Sn\cdot \text{ ; etc.}$$

Many organotin compounds are used as polymerisation catalysts, bacteriocides, etc.

Organolead compounds. *Lead tetra-alkyls* may be prepared by methods used for the tin compounds (see above). Tetramethyl- and tetraethyl-lead, which are used as anti-knock additives to petrol, are manufactured by the reaction between the alkyl chloride and a lead-sodium alloy:

$$4EtCl + 4Na/Pb \longrightarrow Et_4Pb + 3Pb + 4NaCl$$

Another industrial method is the electrolysis of an ethereal solution of methyl- or ethyl-magnesium chloride using a lead anode.

$$4RMgCl + Pb \longrightarrow R_4Pb + 2Mg + 2MgCl_2$$

A point of interest is that lead(II) halides generally give lead tetra-alkyls, *e.g.*,

$$2PbCl_2 + 4RMgX \longrightarrow R_4Pb + Pb + 2MgX_2 + 2MgCl_2$$

$$2PbCl_2 + 4RLi \longrightarrow R_4Pb + Pb + 4LiCl$$

The lead tetra-alkyls are stable liquids; they are very toxic, and are unaffected by air or water.

Alkyl-lead hydrides and *halides* are known; trialkyl-lead hydrides are sufficiently reactive to add to alkenes (or alkynes) at 0°C in the absence of a catalyst.

Organoarsenic compounds. In many ways, the chemistry of the organic compounds of arsenic (a *metalloid*) is similar to that of the phosphorus compounds; the methods of nomenclature of the two series are also similar (see p. 408).

Trialkylarsines may be prepared as follows:

$$3RMgX + AsBr_3 \longrightarrow R_3As + 3MgBrX$$

Alkylarsenic hydrides and *halides* can be prepared in various ways. One of the starting compounds is (i) an alkylarsonic acid, or (ii) a dialkylarsinic acid:

(i) $\quad RX + Na_3AsO_3 \longrightarrow NaX + RAsO(ONa)_2 \xrightarrow{H_3O^+} RAsO(OH)_2 \xrightarrow{Zn/HCl} RAsH_2$

(ii) $\quad RAsO(OH)_2 \xrightarrow[\text{(ii) } H_3O^+]{\text{(i) } RX-NaOH} R_2AsO(OH) \xrightarrow[HCl]{Zn} R_2AsH$

The arsines are oxidised by air; $RAsH_2$ gives an alkylarsonic acid, R_2AsH a dialkylarsinic acid, and R_3As an arsine oxide. Arsonium salts may be prepared by reaction between an alkyl halide and a trialkylarsine or arsenic:

$$2As + 4MeI \longrightarrow Me_4As^+I^- + AsI_3$$

Arsonium salts readily form *alkylidenearsoranes* (*arsenic ylids*)

$$Ph_3AsMe^+Br^- \xrightarrow{PhLi} Ph_3As{=}CH_2$$

With moist silver oxide, they give arsonium hydroxides, $R_4As^+OH^-$, which are strong bases.

Pentaphenylarsenic has been prepared as a crystalline solid, and is isomorphous with the corresponding phosphorus compound (p. 410). Pentavalent arsenic compounds of the type R_3AsX_2 are also known (prepared by the addition of X_2 to R_3As). Lewisite is $ClCH{=}CH-AsCl_2$ (see p. 137).

Organoantimony compounds. *Trialkylstibines* have been prepared, *e.g.*,

$$3MeMgI + SbCl_3 \longrightarrow Me_3Sb + 3MgClI$$
$$SbF_3 + 3Et_3Al \longrightarrow Et_3Sb + 3Et_2AlF$$

They are stable liquids, but the alkylstibine hydrides are unstable volatile liquids. Trimethylstibine combines with methyl iodide to form tetramethylstibonium iodide, $Me_4Sb^+I^-$, which, with moist silver oxide, gives the corresponding hydroxide, a strong base. Pentaphenylantimony has been prepared, and its structure has been shown to be a square pyramid, with the antimony atom lying inside the pyramid (*cf.* the analogous phosphorus compound). Me_5Sb is also known, and pentavalent compounds of the type R_3SbX_2 have been prepared.

Organobismuth compounds. *Trialkylbismuthines* are prepared as follows:

$$3RMgX + BiCl_3 \longrightarrow R_3Bi + 3MgClX$$

These compounds do *not* add a molecule of alkyl halide to form bismuthonium salts, of which only a few are known, *e.g.*, bismuthonium tetraphenylborate:

$$Ph_5Bi + Ph_3B \longrightarrow Ph_4Bi^+Ph_4B^-$$

Pentaphenylbismuth is an unstable solid; other pentacovalent compounds of the type Ph_3BiX_2 ($X{=}Cl$, Br) are also known.

Alkyl free radicals

A characteristic property of many metal alkyls is the generation of free radicals when they are subjected to pyrolysis or photolysis, *e.g.*,

$$Et_4Pb \xrightarrow{Heat} 4Et\cdot + Pb$$
$$Me_2Hg \xrightarrow{h\nu} 2Me\cdot + Hg$$

Because of the presence of an unpaired electron, free radicals may be detected by means of paramagnetic susceptibility measurements and electron spin resonance spectroscopy (see also p. 50). Chemical methods have also been used to detect free radicals (see p. 789).

Alkyl free radicals are formed as reactive intermediates in many types of reactions (see Index), and a common method of generating them in solution is to use acyl peroxides, but better reagents are azonitriles (p. 151).

The configuration of alkyl free radicals is uncertain. The methyl free radical has been shown, from ultraviolet and Raman spectra studies (Herzberg *et al.*, 1956), to have high symmetry. Thus, this radical is planar. On the other hand, infra-red and ESR studies indicate that $CF_3\cdot$ is tetrahedral.

In general, reactions involving free radicals which are asymmetric (at the free-radical centre) result in racemised product. However, in certain trapping experiments, optically active products have been obtained (*inter alia*, Jacobus *et al.*, 1969). Thus, there is chemical evidence to show that *some* free radicals are transiently non-planar.

PROBLEMS

1. Comment on the following equations:

$$HOCH_2CH_2COCH_2CH_2CO_2Et \xrightarrow{1RMgX} ? \xrightarrow{1RMgX} ? \xrightarrow{2RMgX} ?$$

2. Complete the following equations and comment where necessary:

(i) $n\text{-}C_4H_9MgBr + ? \xrightarrow[\text{(ii) H}^+]{\text{(i) Et}_2\text{O}} n\text{-}C_7H_{15}OH$

(ii) $3EtMgBr + (EtO)_2CO \xrightarrow[\text{(ii) aq. NH}_4\text{Cl}]{\text{(i) Et}_2\text{O}} ?$

(iii) $PhMgBr + MeOCH_2CN \xrightarrow[\text{(ii) H}^+]{\text{(i) Et}_2\text{O}} ?$

(iv) $C_6H_{11}MgBr + BrCH_2CBr{=}CH_2 \xrightarrow{\text{Et}_2\text{O}} ?$

(v) $CH_2{=}CO + EtMgBr \xrightarrow[\text{(ii) H}^+]{\text{(i) Et}_2\text{O}} ?$

(iv) $\overset{\displaystyle O}{MeCH\!\!\overset{\diagup\ \ \diagdown}{-\!\!-\!\!-}\!\!CH_2} + Me_2CHMgBr \xrightarrow[\text{(ii) H}^+]{\text{(i) Et}_2\text{O}} ?$

(vii) $EtCH{=}CHCOEt \xrightarrow[\text{(ii) H}^+]{\text{(i) EtMgBr + CuBr}} ?$

(viii) $PrMgBr + MeC(OMe)_3 \xrightarrow[\text{(ii) H}^+]{\text{(i) C}_6\text{H}_6} ?$

(ix) $ClCH_2CO_2Et + 2RMgX \xrightarrow[\text{(ii) H}^+]{\text{(i) Et}_2\text{O}} ?$

(x) $\overset{\displaystyle O}{MeCH\!\!\overset{\diagup\ \ \diagdown}{-\!\!-\!\!-}\!\!CH_2} + EtMgBr \xrightarrow[\text{(ii) H}^+]{\text{(i) Et}_2\text{O}} ?$

(xi) $(CH_2)_5Br_2 \xrightarrow[\text{Et}_2\text{O}]{\text{2Mg}} ? \xrightarrow[\text{(ii) H}^+]{\text{(i) 2MeONH}_2} ?$

(xii) $MeCOCH_2CO_2Et + Me_2Cd \xrightarrow[\text{(ii) H}^+]{\text{(i) Et}_2\text{O}} ?$

(xiii) $Et_2Cd + CCl_3CHO \xrightarrow[\text{(ii) H}^+]{\text{(i) Et}_2\text{O}} ?$

3. Starting with EtOH and n-PrOH, prepare pent-2-ene via a Grignard reagent. Is there more than one way? If so, which is the best and why?

4. Convert $n\text{-}C_5H_{11}CO_2H$ into $n\text{-}C_4H_9CO_2H$ by *two* methods.

5. $R^1CH{=}CHCOR^2 + R^3MgX \longrightarrow$ 1,2- + 1,4-products

How would the ratio of the products change if R^2 = Et, isoPr, t-Bu? Explain.

6. Convert CHD_2I into CHD_2CH_2OH. What alkenes would you expect to get by the acid-catalysed dehydration of the alcohol? Explain.

7. Carry out the following conversions via Grignard reagents, and use any other compounds you like:
   (i) $MeCO_2H \longrightarrow MeCOEt$;   (ii) $EtBr \longrightarrow EtCOEt$;   (iii) $Me_2CHOH \longrightarrow Me_2CHNH_2$;
   (iv) $MeCOEt \longrightarrow EtCH(Me)CHBrMe$.

8. Complete the following equations:

   (i) $4PhLi + (CH_2=CH)_4Sn \longrightarrow$?

   (ii) $2EtNa + Hg \longrightarrow$?

   (iii) $BuLi + MeONH_2 \longrightarrow ? \xrightarrow{H_2O}$?

   (iv) $Me_2CHLi + C_2H_4 \longrightarrow ? \xrightarrow[\text{(ii) } H_2O]{\text{(i) } CO_2}$?

   (v) $Me_3CCl + Me_2Zn \longrightarrow$?

   (vi) $2EtI + Na_2/Hg \longrightarrow$?

   (vii) $Et_2Hg + 2K \longrightarrow$?

   (viii) $4RMgCl + SnCl_4 \longrightarrow$?

   (ix) $Me_2Hg + Zn \longrightarrow$?

   (x) $2CH_2(CN)COCl + Et_2Cd \longrightarrow$?

   (xi) $3Ph_2Hg + 2Al \xrightarrow{\text{heat}}$?

   (xii) $CH_2=CHCOMe \xrightarrow{Me_2SnH_2}$?

   (xiii) $R_2CO + BrCH_2CH=CHCO_2Et \xrightarrow[\text{(ii) } H^+]{\text{(i) } Zn}$?

   (xiv) $EtMgX \xrightarrow{PbCl_2}$?

   (xv) $(PhCH_2)_4Sn \xrightarrow{\text{heat}}$?

   (xvi) $2MeI + Zn/Cu \longrightarrow$?

   (xvii) $EtLi + Br_2 \longrightarrow$?

9. $Me_2CHNO_2 \xrightarrow[Br_2]{NaOH} \mathbf{A} \xrightarrow{Me_2Zn} \mathbf{B}$

What are **A** and **B**? Formulate the reaction. Could you use MeMgI instead of Me_2Zn? Explain.

10. Give the structures of **A–E** in the reactions:

$$EtMeCHCOPh \xrightarrow[\text{(ii) aq. } NH_4Cl]{\text{(i) } PhMgBr} \mathbf{A} \xrightarrow{H^+} \mathbf{B} \xrightarrow{O_3} \mathbf{C} + \mathbf{D}$$

$$\mathbf{E} \uparrow$$

$$EtCHO$$

11. Suggest a mechanism for the reaction:

$$R_2CO + Et_3Al \longrightarrow R_2CHOAlEt_2 + C_2H_4$$

12. Using any compounds you like, convert:
   (i) $Me_2CO \longrightarrow Me_2C(OH)CH_2CO_2H$;   (ii) $EtOH \longrightarrow Me_2CHCHO$;   (iii) $Me_3COH \longrightarrow$
   $Me_3CCH_2CH=CH_2$;   (iv) $MeCHO \longrightarrow EtMeCHCO_2H$;   (v) $HOCH_2(CH_2)_2CH_2OH \longrightarrow$
   $Me_2C(OH)(CH_2)_4C(OH)Me_2$.

REFERENCES

Organic Reactions, Wiley (i) Vol. I (1942), Ch. 1, 'The Reformatsky Reaction': Vol. VI (1951), Ch. 7, 'The Halogen–Metal Interconversion with Organolithium Compounds': Vol. VIII (1954), Ch. 2, 'The Synthesis of Ketones from Acid Halides and Organo-metallic Compounds of Magnesium, Zinc, and Cadmium'; Ch. 6; 'The Metalation Reaction with Organo-lithium Compounds'.

KHARASCH and REINMUTH, *Grignard Reactions of Non-metallic Substances,* Constable (1954).

ASHBY, 'Grignard Reagents. Compositions and Mechanisms of Reaction', *Quart. Rev.,* 1967, **21,** 259.

COATES, GREEN and WADE, *Organometallic Compounds,* Methuen (3rd edn). Vol. I (1967); Vol. II (1968).

COATES, GREEN, POWELL and WADE, *Principles of Organometallic Chemistry,* Methuen (1968).

Journal of Organometallic Chemistry. Vol. I (1963); etc.

Advances in Organometallic Chemistry. Vol. I (1964); etc.

SEYFERTH and KING, *Annual Surveys of Organometallic Chemistry.* Vol. I (1965); etc.

WALLING, *Free Radicals in Solution,* Wiley (1957).

KERR and LLOYD, 'Decomposition Reactions of Radicals', *Quart. Rev.,* 1968, **22,** 549.

16 Saturated dicarboxylic acids

The general formula of the saturated dicarboxylic acids is $C_nH_{2n}(CO_2H)_2$ ($n = 0$ for oxalic acid), and the best-known examples are those which have the two carboxyl groups at the opposite ends of the carbon chain.

Nomenclature

The dicarboxylic acids are commonly known by names which indicate their source, *e.g.*, HO_2CCO_2H oxalic acid; this occurs in plants of the *oxalis* group (for further examples, see the individual acids).

In this trivial system of nomenclature, the positions of side-chains or substituents are indicated by Greek letters, *e.g.*,

$$HO_2CCH(CH_3)CH_2CH_2CHClCO_2H \quad \alpha\text{-chloro-}\alpha'\text{-methyladipic acid}$$

According to the I.U.P.A.C. system of nomenclature, the class suffix is *-dioic*, *e.g.*,

$$HO_2CCH_2CO_2H \quad \text{propanedioic acid}$$
$$HO_2CCH(CH_3)CH(CH_3)CH_2CO_2H \quad \text{2,3-dimethylpentanedioic acid}$$

When this method leads to cumbrous names, the alternative scheme is to regard the carboxyl group as a substituent, and the name of the acid is then obtained by adding the suffix *carboxylic acid*, *e.g.*,

$$HO_2CCH_2CH_2CO_2H \quad \text{1,2-ethanedicarboxylic acid or ethane-1,2-dicarboxylic acid}$$

General methods of preparation

1. The *cyanide synthesis* of dicarboxylic acids is a very useful method: the starting material may be either a halogeno-acid or a polymethylene dibromide:

(i) $\quad ClCH_2CO_2H \xrightarrow{\text{KCN}} CH_2(CN)CO_2H \xrightarrow{\text{H}_3\text{O}^+} CH_2(CO_2H)_2$

(ii) $Br(CH_2)_2Br \xrightarrow{\text{KCN}} NC(CH_2)_2CN \xrightarrow{\text{H}_3\text{O}^+} HO_2C(CH_2)_2CO_2H$

2. The **Crum-Brown and Walker electrolytic method** (1891, 1893). This is the electrolysis of an aqueous solution of the potassium alkyl esters of the dicarboxylic acids (*cf.* Kolbe's method):

$$K\vdots O_2C\vdots(CH_2)_n—CO_2C_2H_5$$
$$K\vdots O_2C\vdots(CH_2)_n—CO_2C_2H_5$$
$$+2H_2O \longrightarrow$$

$$\begin{array}{l}(CH_2)_nCO_2C_2H_5\\|\\(CH_2)_nCO_2C_2H_5\end{array}+2CO_2+2KOH+H_2 \quad (f.g.\text{-}g.)$$

It is obvious that this method can be used to prepare only *even* homologues.

Dicarboxylic acids may also be prepared by the electrochemical hydrodimerisation of suitable α,β-unsaturated compounds (see p. 346).

3. Dicarboxylic acids may be prepared by the acetoacetic ester synthesis (p. 293) and better, by the malonic ester synthesis (p. 295).

4. Cyclic ketones may be oxidised to dicarboxylic acids, *e.g.*, cyclohexanone gives adipic acid:

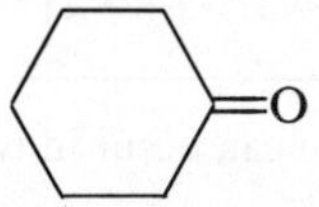

Alternatively, lactones can be converted into dicarboxylic acids (see p. 470).

5. The preparation of higher homologues from lower homologues may be carried out in several ways; which method is used depends on the homologue desired. Many *even* higher homologues can be prepared by the Crum-Brown and Walker method (method 2). Any dicarboxylic acid can be stepped up by two carbon atoms as follows:

$$(CH_2)_n(CO_2H)_2 \xrightarrow{LAH} (CH_2)_n(CH_2OH)_2 \xrightarrow{HBr} (CH_2)_n(CH_2Br)_2 \xrightarrow{KCN}$$
$$(CH_2)_n(CH_2CN)_2 \xrightarrow{H_3O^+} (CH_2)_n(CH_2CO_2H)_2$$

By using malonic ester instead of potassium cyanide, the acid may be stepped up by four carbon atoms:

$$(CH_2)_n(CH_2Br)_2 + 2[CH(CO_2C_2H_5)_2]^-Na^+ \longrightarrow$$
$$2NaBr + (CH_2)_n[CH_2CH(CO_2C_2H_5)_2]_2 \xrightarrow[\text{(ii) HCl}]{\text{(i) KOH}}$$
$$(CH_2)_n[CH_2CH(CO_2H)_2]_2 \xrightarrow{150-200°C} (CH_2)_n(CH_2CH_2CO_2H)_2 + 2CO_2$$

Dicarboxylic acids can be stepped up by *ten* carbon atoms by starting with a di-acyl chloride and an enamine (see p. 384).

Dicarboxylic acids may also be prepared from α-bromo-esters as follows (*cf.* p. 307).

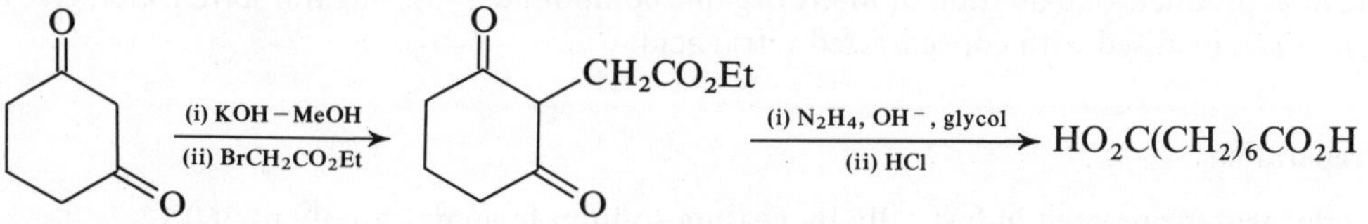

General properties

All the dicarboxylic acids are crystalline solids, the lower members being soluble in water, the solubility decreasing with increase in molecular weight; the odd acids are more soluble than the even. None is steam volatile, and except for oxalic acid, the dicarboxylic acids are stable towards oxidising agents. Their melting points follow the oscillation rule (p. 235), the even acids having higher melting points than the odd. This alternation is due to the relative positions of the carboxyl groups in the crystal structures. These acids dissociate in two steps, the dissociation constant of the first being much greater than that of the second. Furthermore, the acid strength of the discarboxylic acids decreases as the series is ascended (see Table 16.1).

Table 16.1

Acid	pK_a^1	pK_a^2	K_a^1/K_a^2
Oxalic	1·271	4·266	989
Malonic	2·86	5·70	692
Succinic	4·21	5·64	26·9
Glutaric	4·34	5·27	8·51
Adipic	4·41	5·28	7·41

Since these acids possess two carboxyl groups, ionisation can occur in two steps:

$$HO_2C(CH_2)_nCO_2H + H_2O \rightleftharpoons HO_2C(CH_2)_nCO_2^- + H_3O^+ \quad (1)$$
$$HO_2C(CH_2)_nCO_2^- + H_2O \rightleftharpoons {}^-O_2C(CH_2)_nCO_2^- + H_3O^+ \quad (2)$$

In the first-stage ionisation (1), the acid can lose a proton from *two* positions, but the 'half-ion' (conjugate base, cB_1) can add a proton to only *one* position. Hence, if no other factors operated, it would be anticipated that the acid in equilibrium (1) is twice as strong as a monocarboxylic acid, RCO_2H, *i.e.*, $K_a^1 = 2 \times K_a$ of RCO_2H. In the second-stage ionisation (2), the half-ion can lose a proton from only *one* position, but the dianion (cB_2) can add a proton to *two* positions. Therefore for equilibrium (2), $K_a^2 = 1/2 K_a$ of RCO_2H. Thus equilibrium (1) is statistically four times as favoured as equilibrium (2). Hence, on the assumption above, *i.e.*, that no other factors operate, the value of K_a^1 would be expected to be four times K_a^2. In practice, the two constants differ from this factor of 4 (see Table 16.1).

An explanation for the experimental results may be as follows. Bjerrum (1923) assumed that the electrostatic effect of the negatively charged carboxylate ion, cB_1, would require additional energy to separate the proton of the carboxyl group, *i.e.*, there is a field effect through the solvent (p. 33) and consequently increases the ratio K_a^1/K_a^2 above 4. At the same time, the carboxylate anion exerts a $+I$ effect with resultant weakening of the acid (*cf.* p. 182). It would therefore follow that as the chain increases in length, *i.e.*, as n increases, both the field and $+I$ effects would decrease, and so the ratio would approach 4. This is borne out in practice (see Table 16.1: see also conformational effects, p. 495).

The reactions of the dicarboxylic acids depend, to a large extent, on the length of the carbon chain. The dicarboxylic acids, therefore, will be described individually.

Oxalic acid (*ethanedioic acid*) is one of the most important dicarboxylic acids. It occurs in rhubarb, in sorrel and other plants of the *oxalis* group (hence its name). Oxalic acid is one of the final products of oxidation of many organic compounds, *e.g.*, sugars, starch, etc., give oxalic acid when oxidised with concentrated nitric acid.

Preparation

Oxalic acid is prepared industrially by heating sodium formate rapidly to 360°C.

$$2HCO_2Na \longrightarrow (CO_2Na)_2 + H_2$$

The free acid is obtained from its sodium salt by adding calcium hydroxide solution, then treating the precipitated calcium oxalate with the calculated amount of dilute sulphuric acid, removing the precipitated calcium sulphate, and finally evaporating the filtrate to crystallisation.

The usual laboratory method for preparing oxalic acid is to oxidise sucrose or molasses with concentrated nitric acid in the presence of vanadium pentoxide as catalyst:

$$C_{12}H_{22}O_{11} + 18[O] \xrightarrow[V_2O_5]{HNO_3} 6(CO_2H)_2 + 5H_2O \quad (25\%)$$

Properties and reactions

Oxalic acid crystallises from water as colourless crystals with two molecules of water of crystallisation; the melting point of the hydrate is 101·5°C; that of the anhydrous acid is 189·5°C. Oxalic acid is poisonous, soluble in water and ethanol but almost insoluble in ether. The dihydrate loses water when heated at 100–105°C, and when heated at about 200°C, oxalic acid decomposes into carbon dioxide, carbon monoxide, formic acid and water:

$$(CO_2H)_2 \longrightarrow CO_2 + HCO_2H \longrightarrow CO + H_2O$$

The anhydrous acid is conveniently obtained by heating the hydrate with carbon tetrachloride.

When heated with concentrated sulphuric acid at 90°C, oxalic acid is decomposed:

$$(CO_2H)_2 \xrightarrow{H_2SO_4} CO + CO_2 + H_2O$$

It is oxidised by permanganate to carbon dioxide:

$$(CO_2H)_2 + [O] \longrightarrow 2CO_2 + H_2O$$

It is only very slowly oxidised by concentrated nitric acid. When fused with potassium hydroxide, it evolves hydrogen:

$$(CO_2K)_2 + 2KOH \longrightarrow 2K_2CO_3 + H_2$$

The anhydride of oxalic acid is unknown. When *anhydrous* oxalic acid is refluxed with ethanol, ethyl oxalate is formed:

$$(CO_2H)_2 + 2C_2H_5OH \longrightarrow (CO_2C_2H_5)_2 + 2H_2O \quad (80–90\%)$$

A more general method for preparing di-esters is to heat a mixture of the dicarboxylic acid, alcohol, toluene and concentrated sulphuric acid. Sulphuric acid is the catalyst, and the toluene removes the water by forming a ternary azeotrope of alcohol, water and toluene; the yield of di-ester is 94–98 per cent from oxalic to sebacic acid. The esterification may be carried out in the absence of toluene; a larger amount of sulphuric acid is required and the yields are lower (70–90 per cent).

The ethyl and propyl esters of oxalic acid are liquid; the methyl ester is solid, and hence has been used to prepare pure methanol by hydrolysis with sodium hydroxide solution. Ethyl oxalate undergoes the Claisen condensation with esters containing two α-hydrogen atoms to form keto-esters, *e.g.*, with ethyl acetate, *oxalacetic ester* is formed:

$$C_2H_5O_2CCO_2C_2H_5 + CH_3CO_2C_2H_5 \xrightarrow[\text{ethanol}]{C_2H_5ONa \text{ in}}$$

$$\begin{array}{c} \overset{\displaystyle CHCO_2C_2H_5}{\underset{\displaystyle C(ONa)CO_2C_2H_5}{\|}} \end{array} \xrightarrow{CH_3CO_2H} \begin{array}{c} \overset{\displaystyle CH_2CO_2C_2H_5}{\underset{\displaystyle COCO_2C_2H_5}{|}} \end{array} \quad (60–70\%)$$

When oxalic acid is heated with ethylene glycol, the *cyclic* compound, *ethylene oxalate*, is obtained:

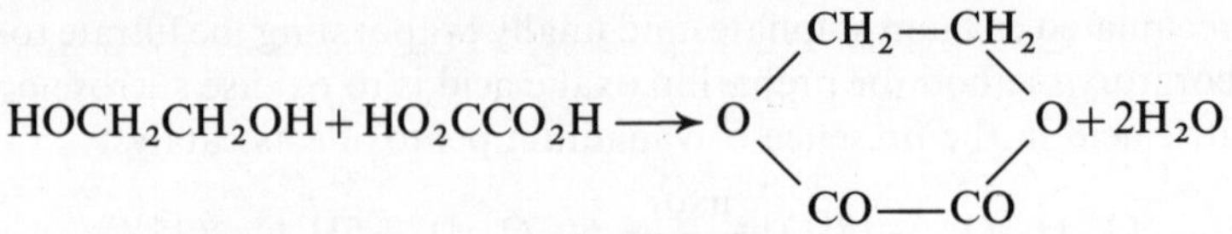

$$HOCH_2CH_2OH + HO_2CCO_2H \longrightarrow O \underset{\substack{\diagdown \\ CO-CO}}{\overset{\substack{CH_2-CH_2 \\ \diagup \qquad \diagdown}}{}} O + 2H_2O$$

This reaction is characteristic of oxalic acid; the other dicarboxylic acids usually react with glycol to form polyesters (see p. 450). When oxalic acid is heated with glycerol, formic acid (p. 239) or allyl alcohol (p. 329) is obtained according to the conditions.

Oxalic acid forms the diamide, *oxamide*. This may be prepared by shaking ethyl oxalate with concentrated ammonium hydroxide solution:

$$C_2H_5O_2CCO_2C_2H_5 + 2NH_3 \longrightarrow NH_2COCONH_2 + 2C_2H_5OH$$

It may also be prepared by passing cyanogen into *cold* concentrated hydrochloric acid:

$$(CN)_2 + 2H_2O \xrightarrow{HCl} (CONH_2)_2 \quad (50\%)$$

Oxamide (decomp. 350°C), in aqueous solution slowly changes into ammonium oxalate, a change which is brought about rapidly by hydrochloric acid (on warming):

$$(CONH_2)_2 \xrightarrow{H_3O^+} (CO_2NH_4)_2$$

When heated with phosphorus pentoxide, oxamide is dehydrated to cyanogen:

$$(CONH_2)_2 \xrightarrow{P_2O_5} (CN)_2 + 2H_2O$$

The monoamide of oxalic acid is also known. It is called *oxamic acid*; the suffix of the names of all the monoamides of the dicarboxylic acids is *-amic*. Oxamic acid, m.p. 210°C, may be prepared by the action of concentrated ammonium hydroxide solution on ethyl hydrogen oxalate, or by heating ammonium hydrogen oxalate.

When reduced with zinc and sulphuric acid, oxalic acid forms *glycollic acid*, $HOCH_2CO_2H$. Electrolytic reduction with a lead cathode gives glycollic and glyoxylic acids, the latter being obtained in better yield by reducing oxalic acid with magnesium and sulphuric acid.

When oxalic acid is treated with *excess* of phosphorus pentachloride, *oxalyl chloride*, b.p. 64°C, is formed:

$$(CO_2H)_2 \xrightarrow{PCl_5} (COCl)_2 \quad (f.-g.)$$

If an excess of phosphorus pentachloride is not used, oxalic acid is decomposed, possibly via the intermediate formation of the half-acid chloride (see also p. 256).

$$(CO_2H)_2 \xrightarrow{PCl_5} [HO_2CCOCl] \longrightarrow CO_2 + CO + HCl$$

Alkanes can be carboxylated by oxalyl chloride under the influence of light (Kharasch and Brown, 1942):

$$RH + (COCl)_2 \longrightarrow RCOCl + CO + HCl$$

Oxalic acid is used for the manufacture of ink and for bleaching straw. Its antimony salts are used as mordants in printing and dyeing.

Malonic acid (*propanedioic acid*), m.p. 136°C, was first obtained by the oxidation of malic acid (hence its name):

$$HO_2CCHOHCH_2CO_2H + [O] \xrightarrow{K_2Cr_2O_7/H_2SO_4} HO_2CCOCH_2CO_2H$$
oxalacetic acid

$$\xrightarrow{[O]} HO_2CCH_2CO_2H + CO_2$$

Malonic acid may be prepared by heating potassium chloroacetate with aqueous potassium cyanide and hydrolysing the product, potassium cyanoacetate, with hydrochloric acid:

$$\mathrm{ClCH_2CO_2K} \xrightarrow{\text{KCN}} \mathrm{CH_2(CN)CO_2K} \xrightarrow{\text{H}_3\text{O}^+} \mathrm{CH_2(CO_2H)_2} \quad (84\%)$$

When heated to 140–150°C, or when refluxed in sulphuric acid solution, it eliminates carbon dioxide (p. 294).

$$\mathrm{HO_2CCH_2CO_2H} \longrightarrow \mathrm{CH_3CO_2H + CO_2}$$

All dicarboxylic acids which have both carboxyl groups attached to the *same* carbon atom are decomposed in a similar manner. When malonic acid is heated with phosphorus pentoxide, a small amount of *carbon suboxide* is obtained:

$$\mathrm{HO_2CCH_2CO_2H} \xrightarrow{\text{P}_2\text{O}_5} \mathrm{O{=}C{=}C{=}C{=}O + 2H_2O}$$

Carbon suboxide, b.p. 7°C, may be regarded as a diketen; it combines with water to form malonic acid. Malonic acid does not form a cyclic anhydride, but dimethylmalonic acid does (*cf.* p. 561).

When malonic acid is treated with nitrous acid and the product hydrolysed, **mesoxalic acid** (*keto-malonic acid*) is obtained:

$$\mathrm{(HO_2C)_2CH_2} \xrightarrow{\text{HNO}_2} \mathrm{(HO_2C)_2C{=}NOH} \xrightarrow{\text{H}_3\text{O}^+} \mathrm{HO_2CCOCO_2H}$$

Mesoxalic acid crystallises from water with one molecule of water which is held firmly, and hence is believed to be water of constitution: $\mathrm{(HO)_2C(CO_2H)_2}$ (*cf.* glyoxylic acid). Ethyl malonate may be oxidised directly to ethyl mesoxalate by selenium dioxide (yield: 23 per cent), or with dinitrogen trioxide ($\mathrm{HNO_3 + As_2O_3}$; yield: 74–76 per cent).

Ethyl malonate is far more important than the acid because of its synthetic uses; the acid contains an active methylene group, the reactivity of which is more pronounced in the ester. Both the acid and ester may be readily brominated, *e.g.*, monobromomalonic acid, $\mathrm{CHBr(CO_2H)_2}$, is formed when a suspension of malonic acid in ether is treated with bromine; owing to the high reactivity of the methylene group, no red phosphorus is required as catalyst (*cf.* p. 270).

Ethyl malonate does not form the diamide, *malonamide*, when shaken with concentrated ammonium hydroxide solution; the dimethyl ester, however, gives a good yield of malonamide. Ethyl malonate forms *barbituric acid* when heated with urea in the presence of sodium ethoxide (see p. 462).

Succinic acid (*butanedioic acid*), m.p. 185°C, was originally obtained by the distillation of amber (Latin: *succinum*, amber). It may be synthesised by the following methods:

(i) From ethylene dibromide:

$$\begin{array}{c} \mathrm{CH_2Br} \\ | \\ \mathrm{CH_2Br} \end{array} \xrightarrow{\text{KCN}} \begin{array}{c} \mathrm{CH_2CN} \\ | \\ \mathrm{CH_2CN} \end{array} \xrightarrow{\text{H}_3\text{O}^+} \begin{array}{c} \mathrm{CH_2CO_2H} \\ | \\ \mathrm{CH_2CO_2H} \end{array} \quad (80\%)$$

(ii) By the reaction between malonic ester (1 molecule) and ethyl chloroacetate, or between malonic ester (2 molecules) and iodine (see p. 295).

Alkyl-substituted succinic acids may be prepared by using monoalkylmalonic ester and α-halogeno-acid esters or iodine. Alternatively, they may be prepared by the addition of hydrogen cyanide to α,β-unsaturated esters and hydrolysing the β-cyano-complex produced (see p. 341).

(iii) By heating malic acid, in a sealed tube, with constant boiling hydriodic acid and red phosphorus:

$$\begin{array}{c} \text{CHOHCO}_2\text{H} \\ | \\ \text{CH}_2\text{CO}_2\text{H} \end{array} + 2\text{HI} \longrightarrow \begin{array}{c} \text{CH}_2\text{CO}_2\text{H} \\ | \\ \text{CH}_2\text{CO}_2\text{H} \end{array} + \text{I}_2 + \text{H}_2\text{O} \quad (60\%)$$

Succinic acid is prepared industrially by the catalytic (or by the electrolytic) reduction of maleic acid:

$$\begin{array}{c} \text{CHCO}_2\text{H} \\ \| \\ \text{CHCO}_2\text{H} \end{array} + \text{H}_2 \xrightarrow{\text{Ni}} \begin{array}{c} \text{CH}_2\text{CO}_2\text{H} \\ | \\ \text{CH}_2\text{CO}_2\text{H} \end{array}$$

When heated, a large amount sublimes, the rest being converted into the *cyclic* anhydride, succinic anhydride:

$$\begin{array}{c} \text{CH}_2\text{CO}_2\text{H} \\ | \\ \text{CH}_2\text{CO}_2\text{H} \end{array} \longrightarrow \begin{array}{c} \text{CH}_2\text{CO} \\ | \qquad\diagdown \\ \qquad\qquad \text{O} + \text{H}_2\text{O} \\ | \qquad\diagup \\ \text{CH}_2\text{CO} \end{array}$$

When heated with excess of glycol, succinic acid forms high-polymer esters (*polyesters*, which belong to the group known as the *alkyl resins*—see p. 319). These esters are acidic, the end groups being succinic acid residues:

$$\text{HO}_2\text{C(CH}_2)_2\text{CO}_2\text{H} + \text{HOCH}_2\text{CH}_2\text{OH} + \text{HO}_2\text{C(CH}_2)_2\text{CO}_2\text{H} + \text{HOCH}_2\text{CH}_2\text{OH} + \ldots$$
$$\longrightarrow \text{HO}_2\text{C(CH}_2)_2\text{CO}-[-\text{O(CH}_2)_2\text{OCO(CH}_2)_2\text{CO}-]-\text{OH}$$

Dicarboxylic acids from malonic to adipic acid form these polyesters with ethylene glycol. These esters are linear polymers, but if glycerol is used instead of glycol, three dimensional polymeric esters are obtained.

Succinic anhydride (*butanedioic anhydride*) is obtained in excellent yield by distilling succinic acid with acetic anhydride, acetyl chloride, or phosphoryl chloride. The anhydride, and not the acid chloride, is obtained when succinic acid is heated with thionyl chloride.

Succinic anhydride, m.p. 119°C, when boiled with water or alkalis, is converted into succinic acid:

$$\begin{array}{c} \text{CH}_2\text{CO} \\ | \qquad\diagdown \\ \qquad\qquad \text{O} + \text{H}_2\text{O} \longrightarrow \\ | \qquad\diagup \\ \text{CH}_2\text{CO} \end{array} \begin{array}{c} \text{CH}_2\text{CO}_2\text{H} \\ | \\ \text{CH}_2\text{CO}_2\text{H} \end{array}$$

When reduced with sodium and ethanol, succinic anhydride is converted first into γ-butyrolactone, and finally into tetramethylene glycol:

$$\begin{array}{c} \text{CH}_2\text{CO} \\ | \qquad\diagdown \\ \qquad\qquad \text{O} \\ | \qquad\diagup \\ \text{CH}_2\text{CO} \end{array} \xrightarrow{e;\,\text{H}^+} \begin{array}{c} \text{CH}_2\text{CH}_2 \\ | \qquad\diagdown \\ \qquad\qquad \text{O} \\ | \qquad\diagup \\ \text{CH}_2\text{CO} \end{array} \xrightarrow{e;\,\text{H}^+} \begin{array}{c} \text{CH}_2\text{CH}_2\text{OH} \\ | \\ \text{CH}_2\text{CH}_2\text{OH} \end{array}$$

If the reduction is carried out with sodium amalgam in *acid* solution, the lactone is obtained in good yield; some butyric acid is also formed.

Succinic anhydride is very useful for preparing the 'half-derivatives' of succinic acid, *e.g.*, with alcohols it forms the acid-ester:

$$\begin{array}{c} CH_2CO \\ | \qquad\qquad O + ROH \longrightarrow \\ CH_2CO \end{array} \qquad \begin{array}{c} CH_2CO_2R \\ | \\ CH_2CO_2H \end{array}$$

With Grignard reagents, succinic anhydride forms γ-keto-acids:

$$\begin{array}{c} CH_2CO \\ | \qquad\qquad O + RMgX \longrightarrow RCOCH_2CH_2CO_2H \\ CH_2CO \end{array}$$

The carbonyl groups (C=O stretch) in *five*-membered ring anhydrides absorb in the regions 1 870–1 830 (s) and 1 800–1 760 cm^{-1} (s). Thus these anhydrides can be distinguished from acyclic anhydrides (p. 259). The molecular formula should also help here for all cyclic anhydrides.

Figure 16.1 is the infra-red spectrum of succinic anhydride (as a Nujol mull).

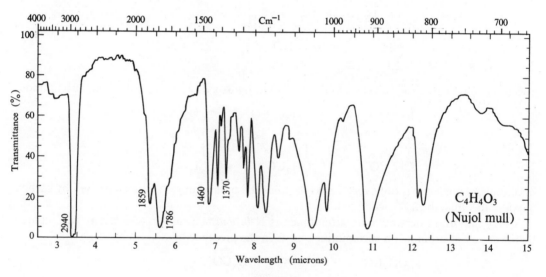

Fig 16.1

2940, 1460 and 1370 cm^{-1}: Bands for Nujol.   1859 and 1786 cm^{-1}: C=O (str.) in 5-ring anhydrides.

Succinimide (*butanimide*) is formed when succinic anhydride is heated in a current of dry ammonia:

$$\begin{array}{c} CH_2CO \\ | \qquad\qquad O + NH_3 \longrightarrow \\ CH_2CO \end{array} \qquad \begin{array}{c} CH_2CO \\ | \qquad\qquad NH + H_2O \quad (v.g.) \\ CH_2CO \end{array}$$

Succinamic acid (m.p. 157°C), (I), and *succinamide* (m.p. 243°C), (II), are both readily converted into succinimide when heated:

$$\begin{array}{c} \text{CH}_2\text{CONH}_2 \\ | \\ \text{CH}_2\text{CO}_2\text{H} \end{array} \xrightarrow{\ -\text{H}_2\text{O}\ } \begin{array}{c} \text{CH}_2\text{CO} \\ \diagdown \\ \diagup \\ \text{CH}_2\text{CO} \end{array}\!\!\text{NH} \xleftarrow{\ -\text{NH}_3\ } \begin{array}{c} \text{CH}_2\text{CONH}_2 \\ | \\ \text{CH}_2\text{CONH}_2 \end{array}$$

(I)   (II)

Succinimide, m.p. 125°C, when boiled with water or alkalis is converted into succinic acid:

$$\begin{array}{c} \text{CH}_2\text{CO} \\ \diagdown \\ \diagup \\ \text{CH}_2\text{CO} \end{array}\!\!\text{NH} + 2\text{H}_2\text{O} \longrightarrow \begin{array}{c} \text{CH}_2\text{CO}_2\text{H} \\ | \\ \text{CH}_2\text{CO}_2\text{H} \end{array} + \text{NH}_3$$

When succinimide is distilled with zinc dust, *pyrrole*, (III), is obtained. When reduced with sodium and ethanol, succinimide forms *pyrrolidine*, (IV), and when reduced electrolytically, it forms *pyrrolidone*, (V):

$$\begin{array}{c} \text{CH}\!=\!=\!\text{CH} \\ || \qquad || \\ \text{CH} \qquad \text{CH} \\ \diagdown \diagup \\ \text{NH} \end{array} \qquad \begin{array}{c} \text{CH}_2\!\!-\!\!\text{CH}_2 \\ | \qquad | \\ \text{CH}_2 \qquad \text{CH}_2 \\ \diagdown \diagup \\ \text{NH} \end{array} \qquad \begin{array}{c} \text{CH}_2\!\!-\!\!\text{CH}_2 \\ | \qquad | \\ \text{CH}_2 \qquad \text{CO} \\ \diagdown \diagup \\ \text{NH} \end{array}$$

(III)   (IV)   (V)

Succinimide is acidic, *e.g.*, it reacts with potassium hydroxide to form potassium succinimide:

$$\begin{array}{c} \text{CH}_2\text{CO} \\ | \qquad\quad \diagdown \\ \qquad\qquad \text{N}^- \\ | \qquad\quad \diagup \\ \text{CH}_2\text{CO} \end{array} \longleftrightarrow \begin{array}{c} \text{CH}_2\text{C}\!\!-\!\!\text{O}^- \\ | \qquad\quad \diagdown\!\!\diagdown \\ \qquad\qquad \text{N} \\ | \qquad\quad \diagup \\ \text{CH}_2\text{CO} \end{array} \longleftrightarrow \begin{array}{c} \text{CH}_2\text{CO} \\ | \qquad\quad \diagdown \\ \qquad\qquad \text{N} \\ | \qquad\quad \diagup\!\!\diagup \\ \text{CH}_2\text{C}\!\!-\!\!\text{O}^- \end{array}$$

Succinimide, however, is only very weakly acidic; the potassium salt is decomposed by carbon dioxide, the imide being regenerated.

A very important derivative of succinimide is N-*bromosuccinimide*, which may be prepared by the action of bromine on succinimide at 0°C in the presence of sodium hydroxide:

$$\begin{array}{c} \text{CH}_2\text{CO} \\ \diagdown \\ \diagup \\ \text{CH}_2\text{CO} \end{array}\!\!\text{NH} + \text{Br}_2 \xrightarrow{\ \text{NaOH}\ } \begin{array}{c} \text{CH}_2\text{CO} \\ \diagdown \\ \diagup \\ \text{CH}_2\text{CO} \end{array}\!\!\text{NBr} + \text{HBr}$$

N-bromosuccinimide (NBS) is a valuable reagent for brominating olefinic compounds in the *allyl* position (Ziegler, 1942).

$$-\text{CH}_2\!\!-\!\!\text{CH}\!\!=\!\!\text{CH}_2 + \begin{array}{c} \text{CH}_2\text{CO} \\ \diagdown \\ \diagup \\ \text{CH}_2\text{CO} \end{array}\!\!\text{NBr} \longrightarrow -\text{CHBrCH}\!\!=\!\!\text{CH}_2 + \begin{array}{c} \text{CH}_2\text{CO} \\ \diagdown \\ \diagup \\ \text{CH}_2\text{CO} \end{array}\!\!\text{NH}$$

This reaction offers a means of splitting off *three* carbon atoms from a compound by oxidative degradation, provided the compound contains one double bond or can have one produced, *e.g.*,

$$RCH_2CH_2CH_2CH_2OH \xrightarrow[350°C]{Al_2O_3} RCH_2CH_2CH=CH_2 \xrightarrow{\text{N-bromosuccinimide}}$$

$$RCH_2CHBrCH=CH_2 \xrightarrow[\text{ethanol}]{KOH} RCH=CHCH=CH_2 \xrightarrow{\text{ozonolysis}} RCHO$$

This method has found great use in steroid chemistry.

Normally, *N*-bromosuccinimide *substitutes* alkenes in the allyl position, and this reaction is believed to take place by a free-radical mechanism; it is catalysed by peroxides and is promoted by light (both are free-radical producing agents). If there are two allylic positions, then two monobromo-derivatives may be obtained. Since a secondary hydrogen is substituted more readily than a primary hydrogen atom, the former gives rise to the predominant product.

The details of the mechanism are uncertain. According to Goldfinger *et al.* (1953, 1956), the function of the *N*-bromosuccinimide is to supply a *low concentration of molecular bromine* (see also p. 103):

$$\diagup N{-}Br + HBr \longrightarrow \diagup N{-}H + Br_2$$

$$Br_2 \rightleftharpoons 2Br\cdot$$

$$Br\cdot + \diagup CHCH=CH_2 \longrightarrow \diagup \dot{C}CH=CH_2 + HBr$$

$$Br_2 + \diagup \dot{C}CH=CH_2 \longrightarrow \diagup CBrCH=CH_2 + Br\cdot$$

N-bromosuccinimide also brominates unsaturated esters, *e.g.*,

$$CH_3CH=CHCO_2CH_3 \xrightarrow{\diagdown N{-}Br} BrCH_2CH=CHCO_2CH_3$$

In addition to substitution, *N*-bromosuccinimide may also produce *addition* compounds, but these are usually formed only in small amount. Braude *et al.* (1952), however, have shown that the *addition* reaction is catalysed by tetraalkylammonium salts, *e.g.*, *N*-bromosuccinimide and cyclohexene give 3-bromocyclohexene (Ziegler *et al.*, 1942), but in the presence of, *e.g.*, tetraethylammonium bromide, 1,2-dibromocyclohexane is the main product.

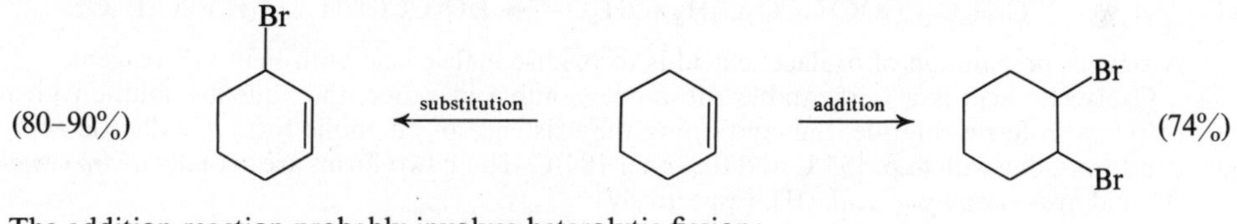

$$(80\text{–}90\%) \quad \xleftarrow{\text{substitution}} \quad \xrightarrow{\text{addition}} \quad (74\%)$$

The addition reaction probably involves heterolytic fission:

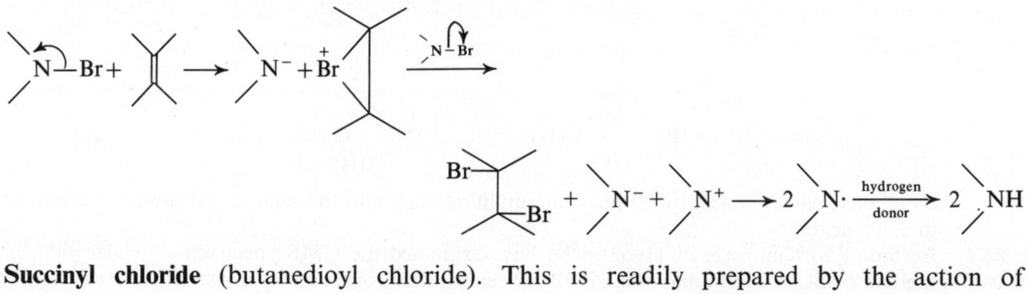

$$\diagdown N{\overset{\frown}{-}}Br + \diagup\!\!\!\diagdown \longrightarrow \diagdown N^- + \overset{+}{Br} \xrightarrow{\diagdown N{-}Br}$$

$$Br{-}\diagup\!\!\!\diagdown{-}Br + \diagdown N^- + \diagdown N^+ \longrightarrow 2\diagdown N\cdot \xrightarrow[\text{donor}]{\text{hydrogen}} 2\diagdown NH$$

Succinyl chloride (butanedioyl chloride). This is readily prepared by the action of phosphorus pentachloride on succinic acid. It was formerly believed that succinyl chloride existed in two forms, the chain isomer, (I), and the ring isomer, (II). However, all recent work indicates structure (I), *e.g.*, Schmid (1965, 1966), from NMR studies, showed that the molecule is symmetrical, and from infra-red studies that a γ-lactone oxo group (in II) is absent.

$$\begin{array}{cc} & CH_2CCl_2 \\ & | \quad \searrow O \\ (CH_2COCl)_2 & CH_2CO \nearrow \\ (I) & (II) \end{array}$$

Isomeric with succinic acid is *methylmalonic acid* or *isosuccinic acid*, m.p. 130°C. It may be readily prepared from sodiomalonic ester and methyl iodide, or as follows:

$$CH_3CH_2CO_2H \xrightarrow{Br_2/P} CH_3CHBrCO_2H \xrightarrow{KCN} CH_3CH(CN)CO_2H \xrightarrow{H^+} CH_3CH(CO_2H)_2$$

Oxalacetic ester is the diethyl ester of ketosuccinic acid, and may be prepared by the Claisen condensation between ethyl oxalate and ethyl acetate (see p. 447). It is a colourless liquid which can be distilled under reduced pressure (b.p. 132°C/24 mm), but at atmospheric pressure it eliminates a molecule of carbon monoxide to form malonic ester:

$$C_2H_5O_2CCOCH_2CO_2C_2H_5 \longrightarrow CH_2(CO_2C_2H_5)_2 + CO$$

Hydrogen of the methylene group adjacent to the carbonyl group may be replaced by alkyl groups by reactions similar to those used for acetoacetic ester, *e.g.*,

$$C_2H_5O_2CCOCH_2CO_2C_2H_5 \xrightarrow{C_2H_5ONa} [C_2H_5O_2CCOCHCO_2C_2H_5]^- Na^+ \xrightarrow{CH_3I}$$

$$C_2H_5O_2CCOCH(CH_3)CO_2C_2H_5 \xrightarrow[\text{(ii) } C_2H_5I]{\text{(i) } C_2H_5ONa} C_2H_5O_2CCOC(CH_3)(C_2H_5)CO_2C_2H_5$$

Oxalacetic ester and its alkyl derivatives undergo 'acid hydrolysis' when boiled with alkalis, and 'ketonic hydrolysis' when boiled with dilute sulphuric acid.

'*Acid hydrolysis*'
$$C_2H_5O_2CCOCHRCO_2C_2H_5 \xrightarrow[\text{(ii) } H^+]{\text{(i) NaOH}} (CO_2H)_2 + RCH_2CO_2H + 2C_2H_5OH$$

'*Ketonic hydrolysis*'
$$C_2H_5O_2CCOCHRCO_2C_2H_5 \xrightarrow{H_2SO_4} [HO_2CCOCHRCO_2H] \longrightarrow CO_2 + RCH_2COCO_2H$$

Hydrolysis of oxalacetic ester with concentrated hydrochloric acid in the cold gives *oxalacetic acid* (*ketosuccinic acid*):

$$C_2H_5O_2CCOCH_2CO_2C_2H_5 + 2H_2O \xrightarrow{HCl} HO_2CCOCH_2CO_2H + 2C_2H_5OH$$

A simpler preparation of oxalacetic acid is to oxidise maleic acid with Fenton's reagent.

Oxalacetic acid is a fairly stable substance, soluble in water, the aqueous solution giving a red colour with ferric chloride, thus indicating the existence of an enolic form. Oxalacetic acid exists in two forms, one with m.p. 155°C and the other 184°C. These two forms are actually *hydroxymaleic acid*, (I), and *hydroxyfumaric acid*, (II), respectively:

$$\begin{array}{cc} \begin{array}{c} H \quad\quad CO_2H \\ \diagdown \diagup \\ C \\ \| \\ C \\ \diagup \diagdown \\ HO \quad\quad CO_2H \end{array} & \begin{array}{c} HO_2C \quad\quad H \\ \diagdown \diagup \\ C \\ \| \\ C \\ \diagup \diagdown \\ HO \quad\quad CO_2H \end{array} \\ (I) & (II) \end{array}$$

(I) is convertible into (II) by 30 per cent sulphuric acid, and the salts of (II) give (I) on treatment with dilute acid.

According to Kumler *et al.* (1962), who have examined the NMR spectrum of oxalacetic acid in deuterium oxide, at equilibrium there is 50 per cent of keto-form and 8 per cent of enol, the rest being present in some other form.

Glutaric acid (*pentanedioic acid*), m.p. 97°C (relationship to *glu*tamic acid and tar*taric* acid gave rise to its name), may be prepared by a number of methods, *e.g.* (see also p. 470):

(i) By refluxing trimethylene dicyanide with concentrated hydrochloric acid:

$$NC(CH_2)_3CN + 4H_2O \xrightarrow{HCl} HO_2C(CH_2)_3CO_2H + 2NH_3 \quad (83\text{--}85\%)$$

(ii) By the action of methylene di-iodide on sodiomalonic ester:

$$2[CH(CO_2C_2H_5)_2]^- Na^+ + CH_2I_2 \longrightarrow 2NaI + CH_2[CH(CO_2C_2H_5)_2]_2 \xrightarrow[\text{(ii) HCl}]{\text{(i) KOH}}$$

$$CH_2[CH(CO_2H)_2]_2 \xrightarrow{150-200°C} HO_2C(CH_2)_3CO_2H + 2CO_2$$

(iii) By condensing formaldehyde with malonic ester in the presence of diethylamine:

$$2(C_2H_5O_2C)_2CH_2 + CH_2O \xrightarrow[\text{reflux}]{(C_2H_5)_2NH} (C_2H_5O_2C)_2CHCH_2CH(CO_2C_2H_5)_2 \quad (61\%)$$

This tetracarboxylic ester may now be treated as in (ii), or may be converted into glutaric acid directly by refluxing with concentrated hydrochloric acid:

$$(C_2H_5O_2C)_2CHCH_2CH(CO_2C_2H_5)_2 \xrightarrow{HCl} HO_2C(CH_2)_3CO_2H \quad (76\text{--}80\%)$$

When heated with acetic anhydride or thionyl chloride, glutaric acid is converted into glutaric anhydride:

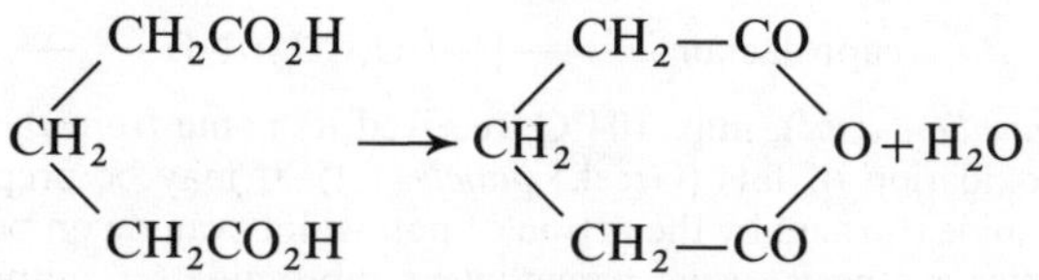

Succinic and glutaric acids differ from their higher homologues in that they readily form the cyclic anhydride when heated with acetic anhydride.

Adipic acid (*hexanedioic acid*), m.p. 150°C, received its name from the fact that it was first obtained by the oxidation of fats (Latin: *adeps*, fat). It may be synthesised from sodiomalonic ester and ethylene dibromide (see p. 295), but is prepared industrially by the oxidation of cyclohexane (from the hydrogenation of benzene) by air in the presence of a catalyst, *e.g.*, cobalt acetate:

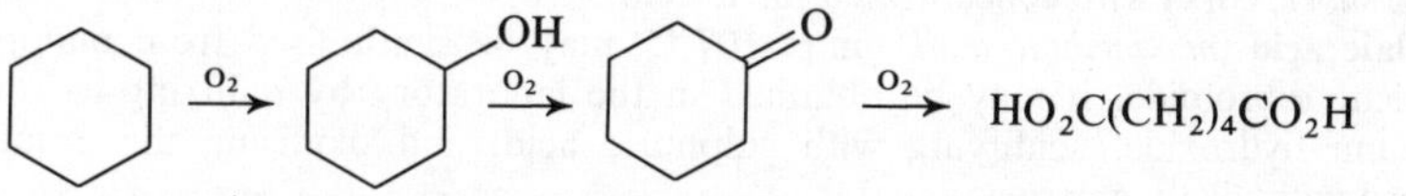

When adipic and pimelic acids are heated with acetic anhydride and the product distilled at 300°C, a cyclic *ketone* is obtained in each case:

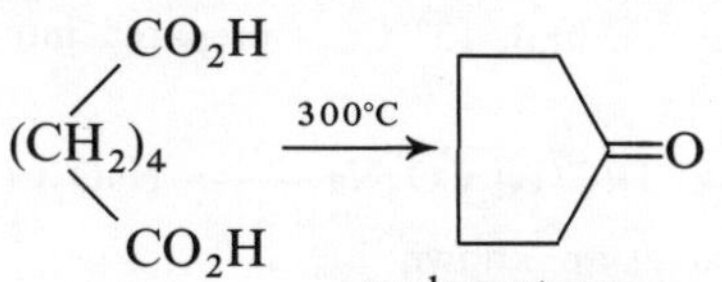

cyclopentanone

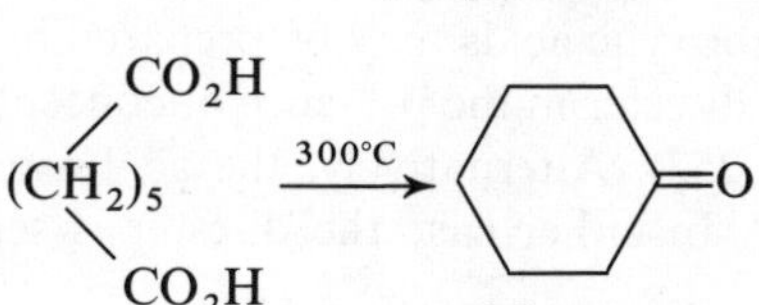

cyclohexanone

Adipic acid is used for the preparation of resins (poly-esters), and is an intermediate in the manufacture of *nylon*. Nylon is used as the generic name for all synthetic fibre-forming polymeric amides having a protein-like structure. Nylon yarns and fabrics are practically non-inflammable. Nylon '66' is formed from adipic acid and hexamethylenediamine, and polymerisation is effected by melting the mixture in the presence of various catalysts.

$$HO_2C(CH_2)_4CO_2H + H_2N(CH_2)_6NH_2 + HO_2C(CH_2)_4CO_2H + H_2N \text{---} \cdots \text{---} \longrightarrow$$
$$HO_2C(CH_2)_4CO \text{---} [\text{---}NH(CH_2)_6NHCO(CH_2)_4CO\text{---}]_n \text{---}NH(CH_2)_6NH_2$$

Nylon '6' is made from *caprolactam*, which is manufactured in several ways starting from cyclohexane:

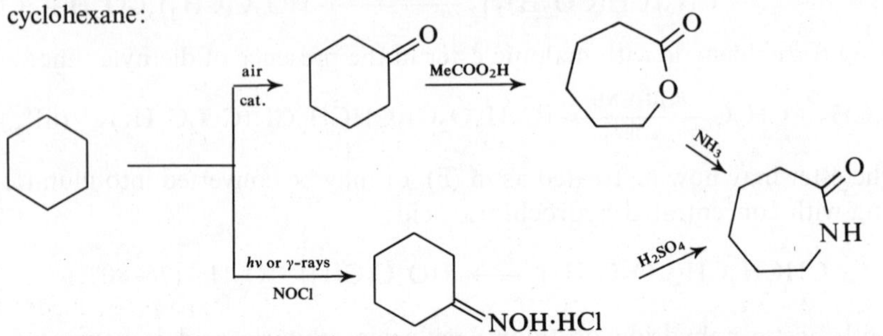

The conversion of the oxime to caprolactam occurs via the Beckmann rearrangement (p. 755), and the caprolactam is polymerised in the presence of a catalyst:

$$\text{caprolactam} \longrightarrow \text{---}[\text{---}CO(CH_2)_5NH\text{---}]_n\text{---}$$

Pimelic acid (*heptanedioic acid*), m.p. 104°C, received its name from the fact that it was obtained originally from the oxidation of fats (Greek: *pimele*, fat). It may be prepared by the hydrolysis of pentamethylene dicyanide (formed by the action of potassium cyanide on pentamethylene dibromide), or by the malonic ester synthesis, using trimethylene dibromide (*cf.* adipic acid). Pimelic acid may also be prepared by the reduction of salicylic acid with sodium and isopentanol, followed by the addition of water and then hydrochloric acid (see p. 766):

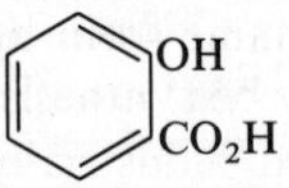

$+ H_2O \xrightarrow{e;\,H^+} HO_2C(CH_2)_5CO_2H$ (43–50%)

Suberic acid (*octanedioic acid*), $HO_2C(CH_2)_6CO_2H$, m.p. 144°C, may be synthesised by the electrolysis of potassium ethyl glutarate (*cf.* p. 445). It is prepared industrially by the oxidation of cork (Latin: *suber*, cork) with concentrated nitric acid.

Azelaic acid (*nonanedioic acid*), m.p. 107°C, may be synthesised from malonic ester and pentamethylene dibromide. It may be obtained in the laboratory by refluxing castor oil with ethanolic potassium hydroxide, acidifying with sulphuric acid, and oxidising the crude ricinoleic acid so obtained with alkaline permanganate:

$$CH_3(CH_2)_5CHOHCH_2CH{=}CH(CH_2)_7CO_2H \xrightarrow{[O]} HO_2C(CH_2)_7CO_2H \quad (32\text{–}36\%)$$

Azelaic acid is prepared industrially by the ozonolysis of oleic acid.

Sebacic acid (*decanedioic acid*), m.p. 133°C, is prepared industrially by heating castor oil with sodium hydroxide.

$$CH_3(CH_2)_5CHOHCH_2CH{=}CH(CH_2)_7CO_2Na \xrightarrow{NaOH} NaO_2C(CH_2)_8CO_2Na + C_6H_{13}CHOHCH_3$$

Many higher dicarboxylic acids are known.

The acid esters of the dicarboxylic acids may be prepared by refluxing a mixture of the acid with half its equivalent of the di-ester in the presence of concentrated hydrochloric acid. This is an example of acidolysis (p. 253). Alternatively, the acid ester may be obtained by the half saponification of the di-ester. In either case the di-ester is separated from the acid ester by fractional distillation.

Another method of preparing a half-ester is to heat equimolecular amounts of dicarboxylic acid and alcohol; the products are unchanged acid, half-ester and di-ester, which are then separated. If the dicarboxylic acid is substituted at one α-carbon atom, then the other carboxyl group is readily esterified by the Fischer–Speier method; this is probably a question of steric hindrance.

When a cyclic anhydride is available, this is a better way to prepare the half-ester (see p. 451). Here again, if one α-carbon atom is substituted, it is the half-ester of the other carboxyl group that is formed.

Half-esters are useful starting materials in various syntheses, e.g.,

(i) $\begin{matrix} CH_2CH_2CO_2H \\ | \\ CH_2CH_2CO_2Et \end{matrix} \xrightarrow[\text{(ii) Br}_2]{\text{(i) SOCl}_2} \begin{matrix} CH_2CHBrCOCl \\ | \\ CH_2CH_2CO_2Et \end{matrix} \xrightarrow{\text{ROH}} \begin{matrix} CH_2CHBrCO_2R \\ | \\ CH_2CH_2CO_2Et \end{matrix}$

(ii) $HO_2C(CH_2)_nCO_2Et \xrightarrow[\text{Br}_2/\text{CCl}_4]{\text{HgO}} Br(CH_2)_nCO_2Et$

(iii) $HO_2C(CH_2)_nCO_2Et \xrightarrow{\text{Na}-\text{EtOH}} HO_2C(CH_2)_nCH_2OH$

When the calcium, barium, or better still, the thorium salts of the carboxylic acids from adipic to nonadecanedioic are distilled, varying yields of cyclic ketones are obtained; e.g., adipic acid gives cyclopentanone (see also p. 558):

$$(CH_2)_4 \begin{smallmatrix} \diagup CO_2 \\ \diagdown CO_2 \end{smallmatrix} Ca \longrightarrow \bigcirc\!\!=\!O + CaCO_3$$

Based on the experimental results of the ease of formation of cyclic anhydrides and cyclic ketones, is **Blanc's rule** (1905). Blanc found that dicarboxylic acids, on heating with acetic anhydride, and then distilling at 300°C (or distilling directly at 300°C), gave cyclic anhydrides or cyclic ketones according to the relative positions of the two carboxyl groups. 1,4- and 1,5-dicarboxylic acids gave cyclic anhydrides; 1,6- and 1,7- gave cyclic ketones: this is Blanc's rule. By using Blanc's rule it is possible to determine the size of rings. A double bond is introduced into the ring, and the ring opened by oxidation to the corresponding dicarboxylic acid. This is then heated with acetic anhydride and distilled. If a cyclic anhydride is obtained, the acid is either 1,4- or 1,5-; if a cyclic ketone, either 1,6- or 1,7-; if there is no change, the acid is 1,8- or more. In naturally occurring compounds the rings are usually five- or six-membered. Hence the formation of the anhydride is taken to mean a five-membered ring, and of the ketone a six-membered ring. Blanc's rule, however, is only satisfactory for simple cyclic compounds. For substituted rings, the results may be misleading, since substituents in the chain bring about ring closure much more easily; e.g., substituted adipic acids may give rise to the anhydride and not to the cyclic ketone.

Carbonic acid and its derivatives

Orthocarbonic acid, $C(OH)_4$, is unknown in the free state, but its esters have been prepared. They may be obtained by the reaction between sodium alkoxide and nitrochloroform:

$$4RONa + CCl_3NO_2 \longrightarrow C(OR)_4 + 3NaCl + NaNO_2 \quad (f.g.)$$

The alkyl orthocarbonates are ethereal smelling liquids.

Carbonic acid (*metacarbonic acid*), $O\!=\!C(OH)_2$, is unknown in the free state, but its salts and esters have been prepared. Alkyl carbonates may be readily formed by heating alcohols with carbonyl chloride in the presence of pyridine:

$$COCl_2 + 2ROH \longrightarrow CO(OR)_2 + 2HCl$$

They may also be prepared by heating silver carbonate with alkyl iodide:

$$Ag_2CO_3 + 2RI \longrightarrow CO(OR)_2 + 2AgI$$

The alkyl carbonates are ethereal smelling liquids, readily soluble in water. In recent years they have found various uses in organic synthesis, *e.g.*,

(i) They may be used to introduce the carbalkoxy group, CO_2R, into ketones to produce β-ketoesters. The reaction is carried out by heating a ketone with sodium or sodium alkoxide in the presence of a large excess of alkyl carbonate:

$$R^2COCH_3 + (R^1O)_2CO + NaOR^1 \longrightarrow [R^2COCHCO_2R^1]^- Na^+ + 2R^1OH \xrightarrow{H^+} R^2COCH_2CO_2R^1$$

(ii) Alkyl carbonates react with primary alkyl cyanides in the presence of sodium alkoxide to form α-cyanoesters:

$$R^2CH_2CN + (R^1O)_2CO + NaOR^1 \longrightarrow$$
$$[R^2C(CN)CO_2R^1]^- Na^+ + 2R^1OH \xrightarrow{H^+} R^2CH(CN)CO_2R^1$$

(iii) Alkyl carbonates may be used to alkylate sodiomalonic esters:

$$[R^2C(CO_2C_2H_5)_2]^- Na^+ + (R^1O)_2CO \longrightarrow R^1R^2C(CO_2C_2H_5)_2 + R^1OCOONa$$

The acid chloride of carbonic acid is *carbonyl chloride* (*phosgene*). It may be obtained by the action of chlorine on carbon monoxide under the influence of sunlight, or in the presence of heated charcoal (200°C) as catalyst; the latter is the commercial method.

Carbonyl chloride is a colourless liquid, b.p. 8°C, and has been used as a toxic gas in warfare. It is used in various organic syntheses, behaving as an acid chloride. It is very slowly decomposed by water:

$$COCl_2 + H_2O \longrightarrow CO_2 + 2HCl$$

When treated with slight excess of one equivalent of alcohol in the cold, carbonyl chloride forms chloroformic esters:

$$COCl_2 + ROH \longrightarrow ClCO_2R + HCl$$

When treated with an excess of alcohol in the presence of pyridine, alkyl carbonates are formed (see above). Carbonyl chloride reacts with ammonia to form urea:

$$COCl_2 + 2NH_3 \longrightarrow CO(NH_2)_2 + 2HCl$$

With primary or secondary amines, substituted ureas are formed, *e.g.*,

$$COCl_2 + 2RNH_2 \longrightarrow CO(NHR)_2 + 2HCl$$

Chloroformic acid (*chlorocarbonic acid*), $ClCO_2H$, is not known in the free state, but its esters have been prepared. These may be readily obtained by treating carbonyl chloride with slight excess of one equivalent of alcohol in the cold (see above). Chloroformic esters are acid-chloride esters, and the chlorine atom reacts with compounds containing active hydrogen; thus ethyl chloroformate, b.p. 94°C, reacts with water, alcohols, ammonia, primary and secondary amines; *e.g.*, urethan is formed with ammonia:

$$ClCO_2C_2H_5 + NH_3 \longrightarrow NH_2CO_2C_2H_5 + HCl$$

Ethyl chloroformate is useful for introducing the carbethoxy group on a nitrogen atom; *e.g.*, ethyl *N*-tricarboxylate is formed when ethyl chloroformate is added to a mixture of urethan, ether and sodium, this mixture having been previously heated:

$$NH_2CO_2C_2H_5 + 2ClCO_2C_2H_5 + 2Na \longrightarrow N(CO_2C_2H_5)_3 + 2NaCl + H_2 \quad (51\text{--}57\%)$$

Ethyl chloroformate may also be used to prepare ethyl esters by reaction with Grignard reagents (p. 432).

Amides of carbonic acid. Since carbonic acid is dibasic, two amides are possible: the mono- and diamide. These are known, respectively, as *carbamic acid*, (I), and *urea*, (II):

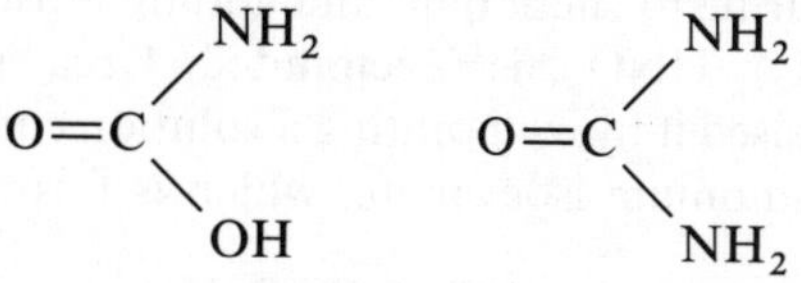

$$O=C\begin{matrix} \diagup NH_2 \\ \diagdown OH \end{matrix} \qquad O=C\begin{matrix} \diagup NH_2 \\ \diagdown NH_2 \end{matrix}$$

(I)   (II)

Carbamic acid is not known in the free state, but its salts and esters have been prepared. Ammonium carbamate is formed when dry ammonia reacts with dry carbon dioxide:

$$2NH_3 + CO_2 \longrightarrow NH_2COONH_4$$

It is a white crystalline solid, very soluble in water. When its aqueous solution is warmed to 60°C, ammonium carbamate is hydrolysed to ammonium carbonate:

$$NH_2COONH_4 + H_2O \longrightarrow CO(ONH_4)_2$$

Esters of carbamic acid are known as **urethans**. These may be prepared:

(i) By treating a chloroformic ester with ammonia, *e.g.*, ethyl chloroformate produces ethyl carbamate:

$$ClCO_2C_2H_5 + NH_3 \longrightarrow NH_2CO_2C_2H_5 + HCl$$

(ii) By refluxing urea in an alcohol, *e.g.*, n-butanol gives n-butyl carbamate when refluxed for 30 hours:

$$CO(NH_2)_2 \longrightarrow NH_3 + HNCO$$
$$HNCO + CH_3CH_2CH_2CH_2OH \longrightarrow NH_2CO_2CH_2CH_2CH_2CH_3 \quad (75\text{--}76\%)$$

(iii) *N*-substituted urethans may be prepared by the Curtius reaction (p. 268). The acid azide is refluxed in benzene solution and then an alcohol is added:

$$R^1CON_3 \longrightarrow N_2 + R^1NCO \xrightarrow{R^2OH} R^1NHCO_2R^2$$

N-phenylurethans may be prepared by reaction between phenyl isocyanate and alcohol:

$$C_6H_5NCO + ROH \longrightarrow C_6H_5NHCO_2R$$

These are crystalline solids and may be used to characterise the alcohols.

Readily dehydrated alcohols, particularly t-alcohols, do not form urethans; alkenes and diphenyl-urea are produced:

$$R_2C(OH)CH_2CH_3 \longrightarrow R_2C{=}CHCH_3 + H_2O$$
$$PhNCO + H_2O \longrightarrow CO_2 + PhNH_2 \xrightarrow{PhNCO} (PhNH)_2CO$$

Ethyl carbamate, usually known as *urethan*, is a crystalline solid, m.p. 50°C. It reacts with ammonia to form urea:

$$NH_2CO_2C_2H_5 + NH_3 \longrightarrow CO(NH_2)_2 + C_2H_5OH$$

It is decomposed by aqueous sodium hydroxide on warming:

$$NH_2CO_2C_2H_5 + 2NaOH \longrightarrow Na_2CO_3 + NH_3 + C_2H_5OH$$

Ethyl carbamate has hypnotic properties; many urethans are valuable hypnotics, *e.g.*, *aponal* (t-amyl carbamate, t-amylurethan), $NH_2COOC(CH_3)_2(C_2H_5)$.

Urea (*carbamide*) is very important physiologically. It is the chief nitrogenous product of protein metabolism; adults excrete about 30 g per day in the urine, from which it can be extracted by evaporating the urine to small bulk and adding nitric acid, whereupon the slightly soluble urea nitrate, $CO(NH_2)_2 \cdot HNO_3$, is precipitated. Urea is historically very important because Wöhler (1828) synthesised it by evaporating a solution containing potassium isocyanate and ammonium sulphate; ammonium isocyanate, which is formed first, undergoes molecular rearrangement:

$$NH_4NCO \rightleftharpoons CO(NH_2)_2$$

The reaction is reversible; the solution contains about 5 per cent ammonium isocyanate.

The mechanism of this rearrangement is uncertain. One possibility is as follows (via dissociation):

$$NH_4NCO \rightleftharpoons NH_3 + HNCO$$

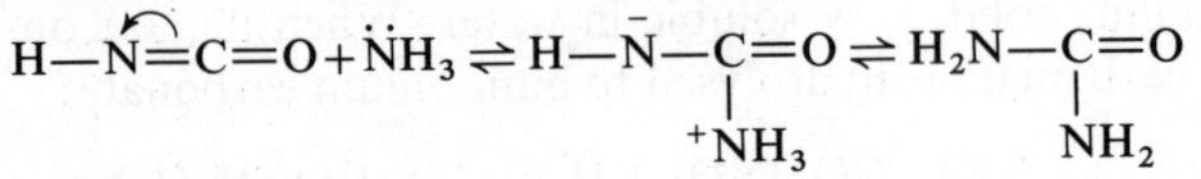

$$H-N{=}C{=}O + \dot{N}H_3 \rightleftharpoons H-\bar{N}-C{=}O \rightleftharpoons H_2N-C{=}O$$
$$\qquad\qquad\qquad\qquad\quad \overset{+}{N}H_3 \qquad\qquad NH_2$$

Urea may be prepared in the laboratory by the action of ammonia on carbonyl chloride, alkyl carbonates, chloroformates or urethans, *e.g.*,

$$COCl_2 + 2NH_3 \longrightarrow CO(NH_2)_2 + 2HCl \quad (f.)$$
$$(C_2H_5O)_2CO + 2NH_3 \longrightarrow CO(NH_2)_2 + 2C_2H_5OH \quad (f.)$$

Industrially, urea is prepared by allowing liquid carbon dioxide and liquid ammonia to interact, and heating the ammonium carbamate so formed to 130–150°C under about 35 atm pressure:

$$2NH_3 + CO_2 \longrightarrow NH_2COONH_4 \longrightarrow CO(NH_2)_2 + H_2O$$

Structure of urea. Although urea appears to be a simple molecule, it is only recently that its structure has been ascertained with any degree of certainty. The diamide structure, $CO(NH_2)_2$, seemed to be indicated by its synthesis from carbonyl chloride and ammonia:

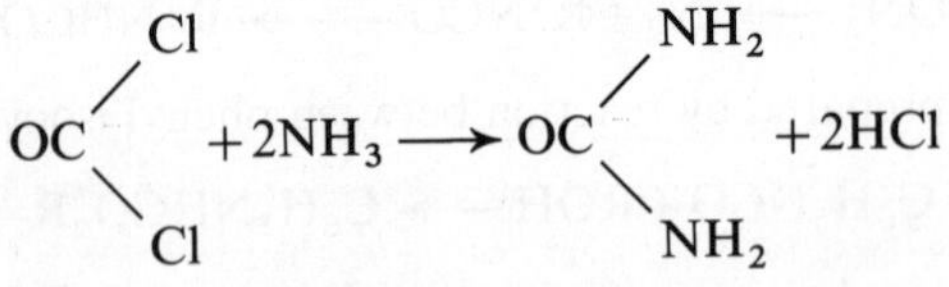

$$OC\overset{\displaystyle Cl}{\underset{\displaystyle Cl}{\big<}} + 2NH_3 \longrightarrow OC\overset{\displaystyle NH_2}{\underset{\displaystyle NH_2}{\big<}} + 2HCl$$

This structure, however, did not appear to explain all the reactions of urea and so gave rise to a great deal of controversy. Physical methods of structure determination have now shown that the diamide structure is correct, at least in the solid state. Crystal structure studies have shown that in solid urea both nitrogen atoms are identical. Bond-length measurements in urea give the C—N distance as 1·37 Å. In aliphatic amines the C—N bond length is 1·47 Å. This indicates that the C—N bond in urea has some double bond character (about 28 per cent); this can be explained by resonance:

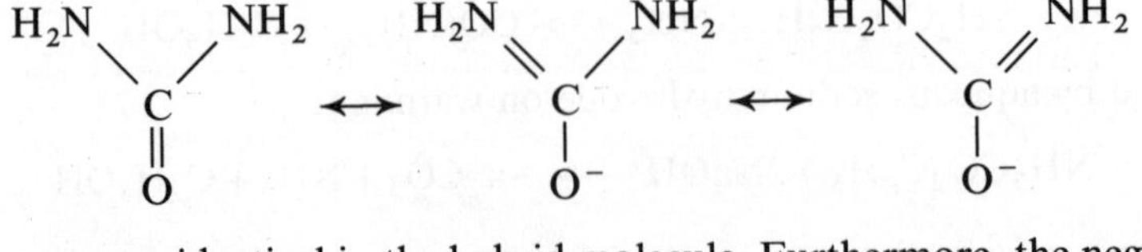

Both nitrogen atoms are identical in the hybrid molecule. Furthermore, the negatively charged oxygen atom is capable of co-ordinating with *one* proton (therefore urea will be a 'monoacidic'

base), and thus the salt may be formulated as a resonance hybrid:

$$
\underset{\substack{|\\ OH}}{\overset{\substack{H_2\overset{+}{N} \diagdown \qquad \diagup NH_2 \\ \diagdown \quad \diagup}}{C}}
\quad\longleftrightarrow\quad
\underset{\substack{|\\ OH}}{\overset{\substack{H_2N \diagdown \qquad \diagup \overset{+}{N}H_2 \\ \diagdown \quad \diagup}}{C}}
\quad\longleftrightarrow\quad
\underset{\substack{\|\\ {}^{+}OH}}{\overset{\substack{H_2N \diagdown \qquad \diagup NH_2 \\ \diagdown \quad \diagup}}{C}}
$$

NMR spectra of crystalline salts of urea (and thiourea) are consistent with protonation at the oxygen atom (or sulphur atom; Redpath *et al.*, 1962).

Properties and reactions of urea

Urea is a white crystalline solid, m.p. 132°C, soluble in water and ethanol, but insoluble in ether. It is used for preparing formaldehyde-urea plastics, barbiturates, as a fertiliser, etc. A recent use is for the manufacture of hydrazine; urea is treated with alkaline sodium hypochlorite (in effect, the Hofmann preparation of primary amines is applied to urea):

$$NH_2CONH_2 + NaOCl + 2NaOH \longrightarrow N_2H_4 + NaCl + Na_2CO_3 + H_2O$$

(i) Urea behaves as a 'monoacidic' base; the nitrate and oxalate are the most important, since neither is very soluble in water.

When urea nitrate is added to cold concentrated sulphuric acid, *nitrourea* is formed:

$$NH_2CONH_2{\cdot}HNO_3 \xrightarrow{\text{H}_2\text{SO}_4} NH_2CONHNO_2 + H_2O \quad (70\text{–}87\%)$$

(ii) Urea is hydrolysed by boiling with acids or alkalis:

$$CO(NH_2)_2 + H_2O \longrightarrow CO_2 + 2NH_3$$

The enzyme urease (which occurs in soyabeans) produces the same change.

(iii) When gently heated, urea loses ammonia to form *biuret*:

$$(NH_2)_2CO \rightleftharpoons NH_3 + HNCO \xrightarrow{(NH_2)_2CO} NH_2CONHCONH_2$$

When an aqueous biuret solution is treated with sodium hydroxide solution and a drop of copper sulphate solution, a violet colour is produced. This is known as the *biuret reaction*, which is characteristic of all compounds containing the grouping —CONH—, *e.g.*, proteins.

When heated rapidly, urea evolves ammonia and forms cyanic acid which rapidly polymerises to cyanuric acid (p. 359). When refluxed with alcohols, urea forms urethans (see above).

(iv) Nitrous acid reacts with urea with the liberation of nitrogen which, however, is not evolved quantitatively:

$$CO(NH_2)_2 + 2HNO_2 \longrightarrow CO_2 + 3H_2O + 2N_2$$

Nitrogen is also evolved, again not quantitatively, when urea is treated with *excess* of alkaline hypobromite:

$$CO(NH_2)_2 + 3NaOBr + 2NaOH \longrightarrow N_2 + Na_2CO_3 + 3NaBr + 3H_2O$$

(v) Acid chlorides and acid anhydrides react with urea to form *ureides*, *e.g.*, acetyl chloride forms acetylurea:

$$CH_3COCl + NH_2CONH_2 \longrightarrow CH_3CONHCONH_2 + HCl$$

Many of these ureides are useful drugs, particularly when the acid group has a branched chain, *e.g.*, *bromural* (α-bromoisovalerylurea), $(CH_3)_2CHCHBrCONHCONH_2$.

Dicarboxylic acids react with urea in the presence of phosphoryl chloride to form *cyclic* ureides; *e.g.*, oxalic acid forms *parabanic acid* (oxalylurea):

$$\begin{array}{c} CO_2H \\ | \\ CO_2H \end{array} + CO(NH_2)_2 \xrightarrow{POCl_3} 2H_2O + \begin{array}{c} CO—NH \\ \diagdown \\ CO \\ \diagup \\ CO—NH \end{array}$$

Cyclic ureides may also be prepared by refluxing a di-ester with urea in ethanolic solution containing sodium ethoxide; *e.g.*, malonic ester forms *barbituric acid* (*malonylurea*):

$$CH_2 \begin{array}{c} CO_2C_2H_5 \\ \diagup \\ \diagdown \\ CO_2C_2H_5 \end{array} + \begin{array}{c} H_2N \\ \diagdown \\ CO \\ \diagup \\ H_2N \end{array} \xrightarrow{C_2H_5ONa} \text{barbituric acid} + 2C_2H_5OH \quad (72–78\%)$$

Barbituric acid and its 5- or 5,5-derivatives are used in medicine as hypnotics and sedatives, *e.g.*, *barbitone* (5,5-diethylbarbituric acid), and *phenobarbitone* (5-phenyl-5-ethylbarbituric acid).

Inclusion complexes

Several types of inclusion complexes are known, and in all of them molecules of one component, the 'guest' molecules, are physically imprisoned in the cavities of the crystalline structure of the second component, the 'host' molecules. Two important types of inclusion complexes are the *channel* (*canal*) *complexes* and the *clathrate* (*cage*) *complexes* (see p. 716 for a discussion of the latter).

Channel complexes are those cases where the host crystallises in a form with parallel, approximately cylindrical channels in which molecules of the guest are enclosed lengthwise. Urea normally forms a crystal structure which is closely packed, but in the presence of various *straight-chain* molecules, *e.g.*, n-alkanes, n-alcohols, n-acids, n-esters, etc., urea crystallises in a more open structure which contains long channels enclosing the guest molecules. These channels contain a number of guest molecules and, in general, the number is inversely proportional to the length of the guest molecule. Branched-chain and cyclic structures cannot 'fit' into these channels, and so this property affords a means of separating straight-chain from branched-chain compounds (but see below). The complexes are decomposed by melting or by dissolving away the urea with water. The formula of the channel complexes is usually $A_n B$ (A is urea) where n, usually not a whole number, has values of 4 or more and increases as the length of B increases. Furthermore, since channel complexes are characterised by the fact that guest molecules bear some structural resemblance to each other, it is not possible to separate homologues by channel complexes. For each homologous series there is a certain minimum chain-length for channel complex formation, *e.g.*, six carbons for alkanes, seven for alcohols, five for acids, etc.

Channel complex formation has been used to resolve racemic modifications, and also to separate geometrical isomers. It is also interesting to note that if a molecule is sufficiently long, then 'small' side-chains will not prevent such a molecule behaving as a guest, *e.g.*, Schiessler *et al.* (1957) have shown that 2-methyldecane forms a channel complex with urea.

X-ray analysis of these channel complexes has shown that the diameter of the channel is about 5 Å (Smith, 1952; see also thiourea, p. 464).

Substituted ureas may be prepared by a reaction similar to Wöhler's synthesis of urea; the hydrochloride or sulphate of a primary or secondary amine is heated with potassium isocyanate, *e.g.*, methylamine hydrochloride (1 molecule) forms methylurea:

$$CH_3NH_2 \cdot HCl + KNCO \longrightarrow KCl + CH_3NH_2 \cdot HNCO \longrightarrow CH_3NHCONH_2$$

If *excess* of amine salt is used, *N,N'*-disubstituted ureas are obtained, *e.g.*, excess of aniline hydrochloride forms *N,N'*-diphenylurea:

$$C_6H_5NH_2 \cdot HCl + KNCO \longrightarrow KCl + C_6H_5NH_2 \cdot HNCO \longrightarrow$$

$$C_6H_5NHCONH_2 \xrightarrow{C_6H_5NH_2 \cdot HCl} C_6H_5NHCONHC_6H_5 + NH_4Cl$$

Alternatively, *N,N'*-diphenylurea may be obtained by refluxing an aqueous solution of aniline hydrochloride with urea; the urea produces ammonia and isocyanic acid, and the reaction proceeds as shown above. Both phenylurea (38–40 per cent) and *N,N'*-diphenylurea (52–55 per cent) may be isolated.

N,N'-Substituted ureas may be obtained by the action of carbonyl chloride on a primary or secondary amine:

$$COCl_2 + 2RNH_2 \longrightarrow CO(NHR)_2 + 2HCl$$

The reaction between phenyl isocyanate and a primary or secondary amine also produces an *s*-disubstituted urea, *e.g.*, with ethylamine, *N*-ethyl-*N'*-phenylurea is formed:

$$C_6H_5NCO + C_2H_5NH_2 \longrightarrow C_6H_5NHCONHC_2H_5$$

This reaction is used to characterise amines.

A very convenient method of preparing alkylureas is to evaporate an aqueous or ethanolic solution of an amine and nitrourea (see above); the latter decomposes into isocyanic acid and nitramide, NH_2NO_2, which readily decomposes into nitrous oxide:

$$NH_2CONHNO_2 \longrightarrow HNCO + NH_2NO_2$$
$$HNCO + R_2NH \longrightarrow R_2NCONH_2$$
$$NH_2NO_2 \longrightarrow N_2O + H_2O$$

When urea is treated with methyl sulphate in faintly alkaline solution, *methylisourea* is obtained:

$$NH_2CONH_2 \xrightarrow{Me_2SO_4} NH{=}C(OMe)NH_2$$

Compounds related to urea

Semicarbazide (*aminourea*), m.p. 96°C, may be prepared by treating hydrazine sulphate with potassium cyanate:

$$H_2NNH_2 \cdot HNCO \longrightarrow NH_2CONHNH_2$$

It is manufactured by heating urea with hydrazine hydrate.

$$NH_2CONH_2 + N_2H_4 \cdot H_2O \longrightarrow NH_2CONHNH_2 + NH_3 + H_2O$$

A more recent method is the electrolytic reduction of nitrourea in sulphuric acid solution using a lead anode:

$$NH_2CONHNO_2 \xrightarrow{e;\, H^+} NH_2CONHNH_2 + 2H_2O \quad (61–69\%)$$

Semicarbazide is an important reagent for the identification of carbonyl compounds, with which it forms *semicarbazones*; it is also used in the Wolff–Kishner reduction (p. 213).

Guanidine (*aminomethanamidine*), m.p. 50°C, is found in beet juice, and is one of the degradation products of the purines. It may be prepared by heating cyanamide with ammonium chloride:

$$NH_2CN + NH_4Cl \longrightarrow (NH_2)_2C{=}NH{\cdot}HCl$$

Guanidine is a *strong* 'monoacid' base, even forming a carbonate. Its strength as a base may be explained by resonance. There are unlike charges in the neutral molecule, whereas the ion has only one positive charge which is spread, and at the same time all the resonating structures are equivalent (*cf.* urea).

Neutral molecule

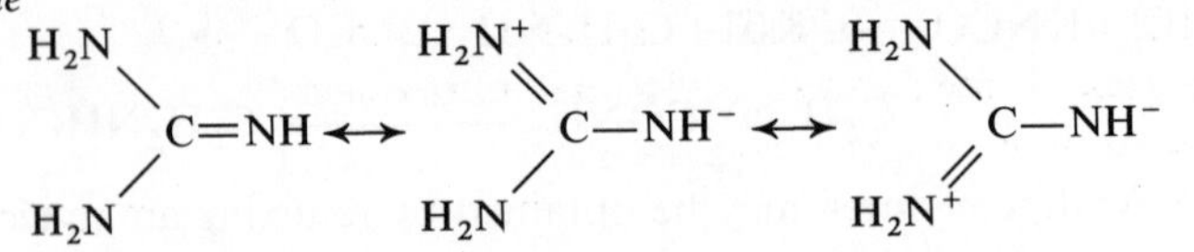

Ion

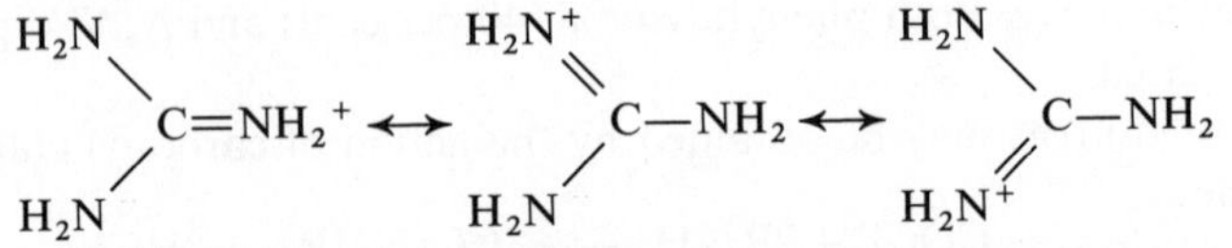

X-ray analysis of guanidinium iodide shows that the three nitrogen atoms are symmetrically placed round the carbon atom. Furthermore, the C—N distance has been found to be 1·38 Å (Theilacker, 1935; *cf.* urea). These facts are in keeping with the assumption that the guanidinium ion is a resonance hybrid.

Careful hydrolysis with barium hydroxide solution converts guanidine into urea:

$$(NH_2)_2C{=}NH + H_2O \longrightarrow (NH_2)_2C{=}O + NH_3$$

Guanidine nitrate may be prepared by heating dicyandiamide with ammonium nitrate:

$$(NH_2)_2C{=}NCN + 2NH_4NO_3 \longrightarrow 2(NH_2)_2C{=}NH{\cdot}HNO_3 \quad (85\%)$$

When treated with concentrated sulphuric acid, guanidine nitrate is converted into nitroguanidine, $(NH_2)_2C{=}NNO_2$ (*cf.* urea nitrate). Nitroguanidine is used for making flashless powders.

Thiourea (*thiocarbamide*), m.p. 180°C, may be prepared by heating ammonium thiocyanate at 170°C for some time:

$$NH_4NCS \longrightarrow S{=}C(NH_2)_2 \quad (14\text{--}16\%)$$

The structure of thiourea is similar to that of urea; it behaves as a 'monoacidic' base and, like urea, forms channel complexes. However, in this case, the diameter of the channel is about 7 Å, and so can accommodate larger guest molecules (than can urea). Thus thiourea readily encloses branched-chain aliphatic compounds and even cycloalkanes. Apparently the cross-section of straight-chain alkanes is too small to support the channels in thiourea and consequently these compounds do not form channel complexes. When heated with alkalis, thiourea is hydrolysed:

$$S{=}C(NH_2)_2 + 2H_2O \xrightarrow{\text{NaOH}} CO_2 + H_2S + 2NH_3$$

Oxidation with *alkaline* permanganate or alkaline hydrogen peroxide converts thiourea into urea:

$$S{=}C(NH_2)_2 + [O] \longrightarrow O{=}C(NH_2)_2 + S$$

Oxides of lead, silver or mercury remove a molecule of hydrogen sulphide from thiourea at room temperature to form cyanamide:

$$(NH_2)_2CS + HgO \longrightarrow NH_2CN + HgS + H_2O$$

When treated with alkyl halide, thiourea forms S-*alkyl-ψ-thiouronium salts* (S-*alkyliso-thiouronium salts*):

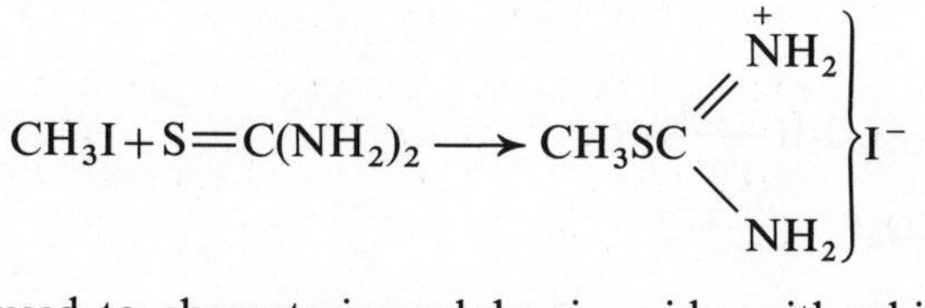

$$CH_3I + S{=}C(NH_2)_2 \longrightarrow CH_3SC{\overset{\overset{+}{N}H_2}{\underset{NH_2}{\Big\}}}}I^-$$

These compounds are used to characterise sulphonic acids, with which they form insoluble salts. They may also be used to prepare thioalcohols (p. 399), and also to prepare sulphonyl chlorides by oxidation with chlorine-water.

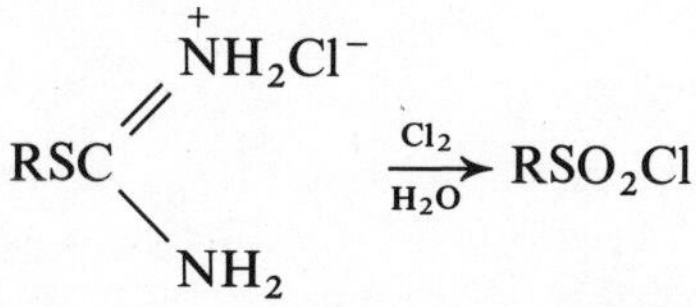

$$RSC{\overset{\overset{+}{N}H_2Cl^-}{\underset{NH_2}{}}} \xrightarrow[H_2O]{Cl_2} RSO_2Cl$$

s-*Diphenylthiourea* or *diphenylthiocarbanilide*, $(C_6H_5NH)_2CS$, which is used as a rubber accelerator, may be prepared by heating aniline with carbon disulphide in an ethanolic solution containing potassium hydroxide.

Dithiocarbamic acid, $S{=}C(NH_2)SH$, although unstable in the free state, gives rise to stable salts. These may be prepared by heating a primary or secondary amine with carbon disulphide. This reaction is complicated and the mechanism is uncertain. A possibility is via the formation of the unstable *N*-alkyldithiocarbamic acid which forms the stable dithiocarbamic acid salt with another molecule of amine:

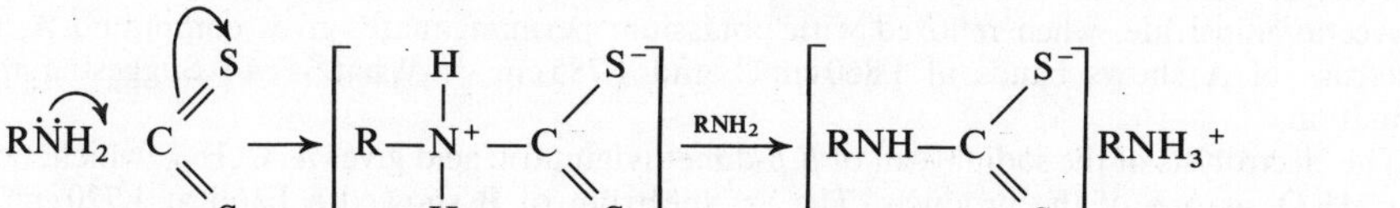

$$R\overset{\frown}{N}H_2 \quad C\overset{S}{\underset{S}{\Big\backslash}} \longrightarrow \left[R{-}\overset{\overset{H}{|}}{\underset{\underset{H}{|}}{N^+}}{-}C\overset{S^-}{\underset{S}{\Big\backslash}} \right] \xrightarrow{RNH_2} \left[RNH{-}C\overset{S^-}{\underset{S}{\Big\backslash}} \right] RNH_3^+$$

The *Hofmann mustard oil reaction* (p. 404) is believed to take place via the formation of a dithiocarbamic acid salt, which is then decomposed by the mercuric chloride, possibly as follows:

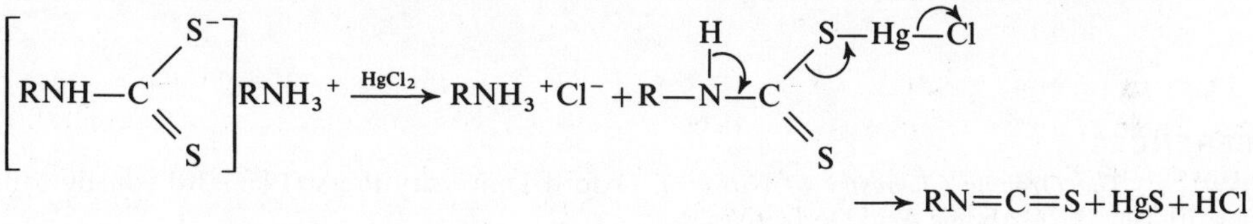

$$\left[RNH{-}C\overset{S^-}{\underset{S}{\Big\backslash}} \right] RNH_3^+ \xrightarrow{HgCl_2} RNH_3^+Cl^- + R{-}\overset{\overset{H}{|}}{N}{-}C\overset{S{-}Hg{-}Cl}{\underset{S}{\Big\backslash}}$$

$$\longrightarrow RN{=}C{=}S + HgS + HCl$$

PROBLEMS

1. Write out all the possible isomeric dicarboxylic acids of molecular formula $C_6H_{10}O_4$ and name them by the I.U.P.A.C. system.

2. Complete the following equations and comment where necessary:

(i) 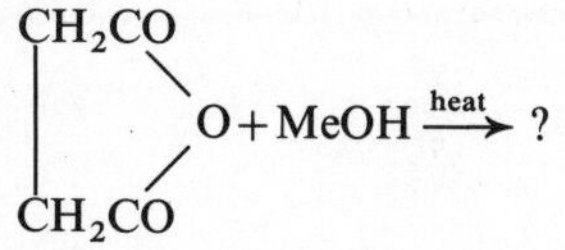 $\begin{array}{l} CH_2CO \\ \quad\quad\quad\searrow \\ \quad\quad\quad\quad O + MeOH \xrightarrow{heat} ? \\ \quad\quad\quad\nearrow \\ CH_2CO \end{array}$

(ii) $CH_2(CO_2H)_2 + 2Me_2C{=}CH_2 \xrightarrow[Et_2O]{H_2SO_4} ?$

(iii) $HO_2C(CH_2)_4CO_2Me \xrightarrow{AgNO_3} ? \xrightarrow[CCl_4]{Br_2} ?$

(iv)

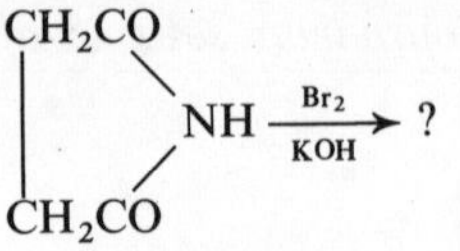

$$\begin{array}{c} CH_2CO \\ | \\ | \\ CH_2CO \end{array} NH \xrightarrow[KOH]{Br_2} ?$$

(v)　$PhCH(CO_2H)(CH_2)_2CO_2H \xrightarrow[heat]{Ac_2O} ?$

(vi)　$CH_2{=}CH(CH_2)_nCO_2R^1 \xrightarrow[R^2OH]{O_3} ?$

(vii)

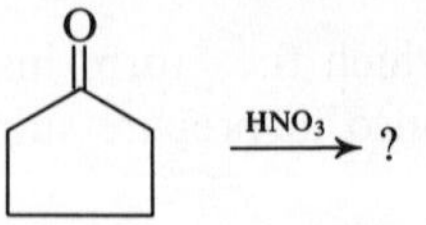

$$\xrightarrow{HNO_3} ?$$

(viii)　$ButOH + NH_2CONH_2 \xrightarrow{H_2SO_4} ?$

(ix)　$KNCS + NH_2NH_2 \xrightarrow[H_2O]{H_2SO_4} ?$

(x)　$CH_3CH{=}CHCH_2CH_3 \xrightarrow{NBS} ?$

(xi)　$HOCH_2(CH_2)_nCH_2OH \xleftarrow{?} HO_2C(CH_2)_nCO_2Et \xrightarrow{?} HO_2C(CH_2)_nCH_2OH$

3. Synthesise from readily accessible materials:
　(i) $EtCH(CN)CO_2Et$; (ii) $EtN(CO_2Et)_2$; (iii) $EtCH(CO_2H)CH_2CO_2H$.

4. Suggest a mechanism for the reaction:

$$NH_2CONH_2 \cdot HNO_3 \xrightarrow{H_2SO_4} NH_2CONHNO_2 + H_2O$$

5. Convert succinic acid into cyclopentanone.

6. Acetic anhydride, when refluxed with potassium permanganate, gives compound **A**, $C_4H_4O_3$. The i.r. spectrum of **A** shows bands at $1\,860\,cm^{-1}$ and $1\,785\,cm^{-1}$. What is **A**? Suggest a mechanism for its formation.

7. The electrolysis of the sodium salt of β,β-dimethylglutaric acid gives **A**, C_5H_{10}, which, on ozonolysis, gives **B**, C_4H_8O, as one of the products. The i.r. spectrum of **B** showed a band at $1\,720\,cm^{-1}$ and the NMR spectrum showed three signals, $\tau\,9{\cdot}1,\ 7{\cdot}8,\ 7{\cdot}6$. What are **A** and **B**? Give your reasons.

8. The N.M.R. spectrum of the compound $C_8H_{14}O_4$ shows three signals: triplet, $\tau\,8{\cdot}74$; quartet, $\tau\,5{\cdot}82$; singlet, $\tau\,7{\cdot}4$. Suggest a structure for the compound.

REFERENCES

SIDGWICK, *The Organic Chemistry of Nitrogen*, Oxford University Press (1966, 3rd edn by Millar and Springall). Ch. 12, 'Carbonic Acid Derivatives'.

FILLER, 'Oxidations and Dehydrogenations with *N*-Bromosuccinimide and Related *N*-Haloimides', *Chem. Rev.*, 1963, **63**, 21.

DIAPER and KUKSIS, 'Synthesis of Alkylated Alkanedioic Acids', *Chem. Rev.*, 1959, **59**, 89.

TRUTER, 'Sorting Molecules by Size and Shape', *Research*, 1963, **6**, 320.

HAGAN, *Clathrate Inclusion Compounds*, Reinhold (1962).

17

Hydroxyacids, stereochemistry, unsaturated dicarboxylic acids

Monobasic hydroxyacids

Monobasic hydroxyacids are monocarboxylic acids which contain one or more hydroxyl groups in the carbon chain.

Nomenclature

The usual method is to name the hydroxyacid as a derivative of the parent acid (named according to the trivial system), the position of the hydroxyl group being indicated by a Greek letter, *e.g.*,

$$CH_3CHOHCH_2CO_2H \quad \beta\text{-hydroxybutyric acid}$$

According to the I.U.P.A.C. system of nomenclature, the position of the hydroxyl group is indicated by a number, *e.g.*,

$$CH_3CH_2CHOHCO_2H \quad \text{2-hydroxybutanoic acid or}$$
$$\text{1-hydroxypropane-1-carboxylic acid}$$

Many hydroxyacids which occur in nature are given special names indicating the source, *e.g.*, $CH_3CHOHCO_2H$, lactic acid, occurs in sour milk (Latin: *lac*, milk).

General methods of preparation

The methods used depend on the position of the hydroxyl group in the chain.

α-Hydroxyacids may be prepared by the hydrolysis of aldehyde or ketone cyanohydrins (p. 212) or the hydrolysis or α-bromo-acids (p. 271).

> Less useful methods are the controlled oxidation of 1,2-glycols with dilute nitric acid (p. 311) or the deamination of α-amino-acids (p. 390).

β-Hydroxyacids are usually best prepared by means of the Reformatsky reaction (p. 437) or by the oxidation of aldols with Tollens' reagent (p. 298). The hydrolysis of ethylene cyanohydrin is useful for the preparation of β-hydroxypropionic acid (p. 310).

α-, β-, γ-, or δ-**Hydroxyacids** may be prepared by the catalytic reduction of the corresponding keto-esters, *e.g.*,

$$CH_3COCH_2CO_2Et \xrightarrow[\text{Press.; 100°C}]{H_2-Ni} CH_3CHOHCH_2CO_2Et$$

$$\xrightarrow[\text{(ii) } H_3O^+]{\text{(i) KOH}} CH_3CHOHCH_2CO_2H$$

Alternatively, hydrolysis of the corresponding halogeno-acid gives the hydroxyacid or lactone (see below). The reduction of half-esters of α,ω-dicarboxylic acids is particularly useful for preparing ω-hydroxyacids (p. 457).

General properties and reactions

Glycollic acid, the first member of the series, is a solid; the higher members are liquids. All are soluble in water, generally more so than are either the corresponding acid or alcohol. This is to be expected, since hydroxyacids have *two* functional groups which can form hydrogen bonds with water.

1. Hydroxyacids behave both as acids and alcohols; in many reactions the hydroxyl and carboxyl groups do not interfere with each other, particularly when they are far apart. Furthermore, by esterifying the carboxyl group, the ester then behaves predominantly as a hydroxy-compound, *i.e.*, esterification masks, to a large extent, the presence of the carboxyl group. The carboxyl group may be converted into the ester, amide, nitrile, acyl chloride, etc. The hydroxyl group (when the carboxyl group has been esterified) may be converted into the ester, ether, etc.; *e.g.*, glycollic acid reacts with acetyl chloride to form acetylglycollic acid (behaving as a hydroxy-compound):

$$HOCH_2CO_2H + CH_3COCl \longrightarrow CH_3COOCH_2CO_2H + HCl$$

Glycollic acid reacts with phosphorus pentachloride or thionyl chloride to form chloroacetyl chloride (behaving both as an alcohol and an acid); the chloroacetyl chloride is readily hydrolysed by water to chloroacetic acid:

$$HOCH_2CO_2H \xrightarrow{PCl_5} ClCH_2COCl \xrightarrow{H_2O} ClCH_2CO_2H$$

2. When heated with dilute sulphuric acid or acid permanganate, α-hydroxyacids are converted into aldehydes or ketones:

$$RCHOHCO_2H \xrightarrow{H_2SO_4} RCHO + HCO_2H \quad (v.g.)$$

$$R_2C(OH)CO_2H + [O] \xrightarrow{KMnO_4} R_2CO + CO_2 + H_2O \quad (v.g.)$$

These reactions offer a very good means of stepping down the monocarboxylic acid series (via the H.V.Z. reaction), one carbon atom at a time.

β-Hydroxyacids, on oxidation with alkaline permanganate, give ketones:

$$RCHOHCH_2CO_2H \longrightarrow [RCOCH_2CO_2H] \xrightarrow{-CO_2} RCOCH_3$$

3. When heated with concentrated hydriodic acid, hydroxyacids are reduced to the corresponding carboxylic acid:

$$CH_3CHOHCO_2H + 2HI \longrightarrow CH_3CH_2CO_2H + H_2O + I_2$$

4. When hydroxyacids are heated, the product formed depends on the relative positions of the hydroxyl and carboxyl groups.

(i) α-Hydroxyacids form **lactides**; these are six-membered ring compounds formed by reaction between *two* molecules of the hydroxyacid, and are named systematically as 3,6-dialkyl-1,4-dioxan-2,5-dione:

$$\begin{array}{c} \text{RCHO}\boxed{\text{H \quad HO}}\text{OC} \\ | \qquad\qquad | \\ \text{CO}\ \boxed{\text{OH \quad H}}\text{OCHR} \end{array} \longrightarrow \begin{array}{c} \text{RCH—O—CO} \\ | \qquad\quad | \\ \text{CO—O—CHR} \end{array} +2\text{H}_2\text{O}$$

The tendency to form lactides is very pronounced, in many cases the lactide being formed by allowing the α-hydroxyacid to stand in a desiccator over concentrated sulphuric acid. Lactides are readily hydrolysed to the acid by alkali.

The distillation of α-hydroxyacids produces aldehydes via the lactide:

$$2\text{RCHOHCO}_2\text{H} \longrightarrow \begin{array}{c} \text{RCHOCO} \\ | \qquad | \\ \text{COOCHR} \end{array} \longrightarrow 2\text{RCHO}+2\text{CO}$$

When heated with a trace of zinc chloride, lactides are converted into linear polyesters, $\text{HO—(—CHRCOO—)}_n\text{—H}$, which regenerate the lactide on distillation under reduced pressure.

(ii) When β-hydroxyacids are heated, they eliminate a molecule of water to form mainly the α,β-unsaturated acid and a very small amount of the β,γ-unsaturated acid. The reaction is best carried out by refluxing with dilute sulphuric acid or dilute alkali.

$$\text{RCHOHCH}_2\text{CO}_2\text{H} \longrightarrow \text{RCH}{=}\text{CHCO}_2\text{H}+\text{H}_2\text{O}$$

(iii) On heating, γ- and δ-hydroxyacids readily form *internal* esters known as **lactones**:

$$\begin{array}{c} \text{OH} \qquad\quad \text{HO} \\ | \qquad\qquad\quad | \\ \text{RCHCH}_2\text{CH}_2\text{CO} \end{array} \longrightarrow \begin{array}{c} \overset{\displaystyle \ulcorner\text{—O—}\urcorner}{\text{RCHCH}_2\text{CH}_2\text{CO}} \end{array} +\text{H}_2\text{O}$$

$$\gamma\text{-acid} \qquad\qquad \gamma\text{-lactone}$$

$$\begin{array}{c} \text{OH} \qquad\qquad\quad \text{HO} \\ | \qquad\qquad\qquad | \\ \text{RCHCH}_2\text{CH}_2\text{CH}_2\text{CO} \end{array} \longrightarrow \begin{array}{c} \overset{\displaystyle \ulcorner\text{——O——}\urcorner}{\text{RCHCH}_2\text{CH}_2\text{CH}_2\text{CO}} \end{array} +\text{H}_2\text{O}$$

$$\delta\text{-acid} \qquad\qquad\qquad \delta\text{-lactone}$$

Lactone formation often occurs spontaneously when the sodium salts of γ- or δ-hydroxyacids are acidified, particularly with the former.

γ- and δ-Lactones are also formed when γ,δ- or δ,ε-unsaturated acids are treated with concentrated sulphuric acid:

$$\text{CH}_2{=}\text{CHCH}_2\text{CH}_2\text{C}\!\!\begin{array}{c}\nearrow^{\text{O}}\\\searrow_{\text{OH}}\end{array} \xrightarrow{\text{H}_3\text{O}^+} \begin{array}{c}\text{CH}_3\text{—}^+\text{CH} \quad \text{CO} \\ \text{H}{-}\ddot{\text{O}} \end{array} \xrightarrow{-\text{H}^+} \begin{array}{c} \text{CH}_2\text{—CH}_2 \\ | \qquad\quad | \\ \text{CH}_3\text{CH} \quad\ \text{CO} \\ \searrow_{\text{O}}\nearrow \end{array}$$

β-Lactones from β-hydroxyacids can only be obtained under special conditions; in practice β-lactones may be prepared by shaking an aqueous solution of the sodium salt of the β-chloroacid with chloroform:

$$\text{RCHClCH}_2\text{CO}_2\text{Na} \longrightarrow \overset{\displaystyle \ulcorner\text{—O—}\urcorner}{\text{RCHCH}_2\text{CO}}+\text{NaCl}$$

They may, however, be prepared more readily by reactions between keten and a carbonyl compound; *e.g.*, β-propiolactone from keten and formaldehyde (p. 361). β-Lactones are more reactive than other

$$CH_2{=}C{=}O + HCHO \longrightarrow \overset{\lceil\quad O\quad \rceil}{CH_2CH_2CO}$$

types of lactones; *e.g.*, β-propiolactone readily reacts with nucleophilic reagents to form β-substituted propionic acid derivatives.

According to the I.U.P.A.C. system of nomenclature, lactones are known as *-olides*, *e.g.*,

$\overset{\lceil\qquad O\qquad\rceil}{CH_2CH_2CH_2CH_2CO}$, δ-valerolactone or 1,5-pentanolide. The systematic name of this lactone is 4-*hydroxybutane*-1-*carboxylic acid lactone*.

γ-Butyrolactone is manufactured by the oxidation of tetramethylene glycol over a copper catalyst:

$$HOCH_2CH_2CH_2CH_2OH \xrightarrow[Cu]{O_2} \overset{\lceil\quad O\quad\rceil}{CH_2CH_2CH_2CO}$$

It is an important intermediate in the preparation of polyamides.

Lactones are converted into alkali salts when refluxed with excess of alkali:

$$\overset{\lceil\quad O\quad\rceil}{RCHCH_2CH_2CO} + NaOH \longrightarrow RCHOHCH_2CH_2CO_2Na$$

They are reduced by sodium amalgam in acid solution to the corresponding acid:

$$\overset{\lceil\qquad O\qquad\rceil}{RCHCH_2CH_2CH_2CO} \xrightarrow{e;\ H^+} RCH_2CH_2CH_2CH_2CO_2H$$

On the other hand, lactones are reduced to diols by LAH or by catalytic hydrogenation (see also Table 6.1, p. 179).

When treated with concentrated halogen acid, lactones form the corresponding halogeno-acid:

$$\overset{\lceil\quad O\quad\rceil}{RCHCH_2CH_2CO} + HX \rightleftarrows RCHXCH_2CH_2CO_2H$$

With concentrated ammonium hydroxide solution, the hydroxyamide is formed; in some cases the product may be the γ-lactam:

$$\overset{\lceil\ O\ \rceil}{RCH(CH_2)_2CO} \xrightarrow{NH_3} RCHOH(CH_2)_2CONH_2 \quad or \quad \overset{\lceil\ NH\ \rceil}{RCH(CH_2)_2CO}$$

Lactones readily react with potassium cyanide to form cyano-acids, *e.g.*, γ-butyrolactone forms glutaric acid as follows:

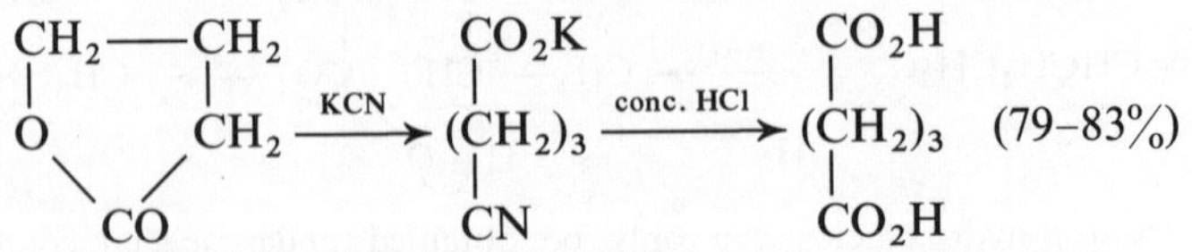

$$\begin{array}{c} CH_2{-}CH_2 \\ | \qquad\quad | \\ O \qquad CH_2 \\ \diagdown\quad\diagup \\ CO \end{array} \xrightarrow{KCN} \begin{array}{c} CO_2K \\ | \\ (CH_2)_3 \\ | \\ CN \end{array} \xrightarrow{conc.\ HCl} \begin{array}{c} CO_2H \\ | \\ (CH_2)_3 \\ | \\ CO_2H \end{array} \quad (79{-}83\%)$$

Lactones also react with Grignard reagents to form diols, *i.e.*, they behave as esters; *e.g.*,

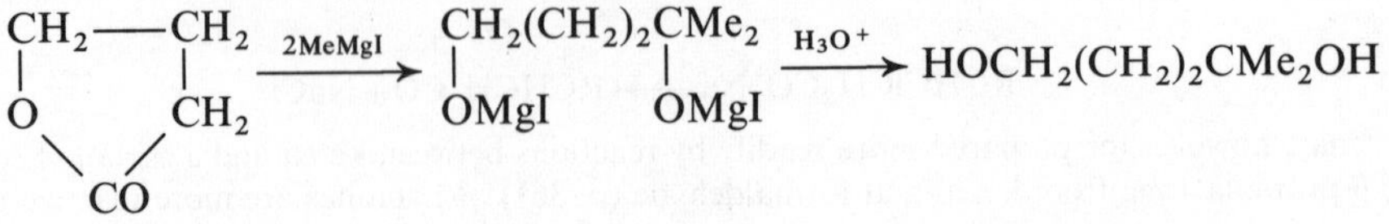

$$\begin{array}{c} CH_2{-}CH_2 \\ | \qquad\quad | \\ O \qquad CH_2 \\ \diagdown\quad\diagup \\ CO \end{array} \xrightarrow{2MeMgI} \begin{array}{c} CH_2(CH_2)_2CMe_2 \\ | \qquad\qquad\ | \\ OMgI \qquad OMgI \end{array} \xrightarrow{H_3O^+} HOCH_2(CH_2)_2CMe_2OH$$

The infra-red absorption region of the carbonyl group (C=O stretch) in γ-lactones is $1\,780$–$1\,760$ cm^{-1} (s), whereas that for δ-lactones is $1\,750$–$1\,735$ cm^{-1} (s). Thus the two may be distinguished, but the latter absorb in the same region as esters. The molecular formula should differentiate between saturated esters and lactones.

(iv) ε-Hydroxyacids, in certain cases, may form the lactone on heating. Usually they either eliminate a molecule of water to form two unsaturated acids, δ,ε- and ε,ζ-, or form linear poly-esters.

(v) Hydroxyacids with the hydroxyl group further removed than the ε-position, on heating, either eliminate a molecule of water to form unsaturated acids (of two types; cf. ε-hydroxy-acids), or form linear esters.

Large ring lactones (fourteen- to eighteen-membered rings) have been prepared by the oxidation of cyclic ketones with Caro's acid (Ruzicka and Stoll, 1928). This is an example of the Baeyer–Villiger oxidation (p. 223):

$$\begin{array}{c}
\text{(CH}_2)_n \\
\text{CH}_2 \quad \text{CO} \\
\text{(CH}_2)_n
\end{array}
+ [\text{O}] \xrightarrow{\text{H}_2\text{SO}_5}
\begin{array}{c}
\text{(CH}_2)_n\text{—CO} \\
\text{CH}_2 \quad | \\
\text{(CH}_2)_n\text{—O}
\end{array}$$

By using the **high dilution principle** of Ruggli (1912), large ring lactones have also been prepared from hydroxyacids in which the hydroxyl group is far removed from the carboxyl group. According to this principle, by using sufficiently dilute solutions of a hydroxyacid, the distance between *different* molecules can be made greater than the distance between the hydroxyl and carboxyl groups of the *same* molecule. Thus the cyclic compound (lactone) is formed instead of linear condensation taking place; e.g., Stoll and his co-workers (1934) found that ω-hydroxypentadecanoic acid gave a high yield of lactone in very dilute solution.

Hunsdiecker and Erlbach (1947) have also prepared large ring lactones by the dilution principle. These workers cyclised ω-bromo-aliphatic acids by boiling dilute solutions in butanone with excess of potassium carbonate; they obtained lactones in yields varying from $56 \cdot 3$ to $96 \cdot 8$ per cent, the yield increasing with the size of the ring.

Glycollic acid (*hydroxyacetic acid, hydroxyethanoic acid*), m.p. 80°C, is the simplest hydroxyacid, and occurs in the juice of beet and sugar-cane, and in unripe grapes. It may be prepared by refluxing an aqueous solution of potassium chloroacetate with sodium carbonate:

$$\text{ClCH}_2\text{CO}_2\text{K} + \text{H}_2\text{O} \xrightarrow[\text{(ii) HCl}]{\text{(i) Na}_2\text{CO}_3} \text{HOCH}_2\text{CO}_2\text{H} + \text{KCl} \quad (80\%)$$

It may also be prepared by warming a solution of formalin with potassium cyanide:

$$\text{HCHO} + \text{KCN} + 2\text{H}_2\text{O} \longrightarrow \text{HOCH}_2\text{CO}_2\text{K} + \text{NH}_3 \xrightarrow{\text{HCl}} \text{HOCH}_2\text{CO}_2\text{H} \quad (70\%)$$

Glycollic acid is prepared industrially by the electrolytic reduction of oxalic acid and by heating at 160–170°C and under pressure, a mixture of formaldehyde, carbon monoxide and water in acetic acid with sulphuric acid as catalyst:

$$\text{HCHO} + \text{CO} + \text{H}_2\text{O} \longrightarrow \text{HOCH}_2\text{CO}_2\text{H}$$

If methanol is used instead of water, methyl glycollate is obtained.

Glycollic acid is oxidised to oxalic acid by nitric acid. Its lactide is known as *glycollide*.

Lactic acid (*α-hydroxypropionic acid, 2-hydroxypropanoic acid*) may be prepared by any of the general methods used for α-hydroxyacids, e.g., the hydrolysis of acetaldehyde cyanohydrin, or by heating α-bromopropionic acid with aqueous sodium hydroxide.

One industrial preparation of lactic acid is the fermentation of lactose (the sugar in milk) by *Bacillus acidi lactiti*:

$$C_{12}H_{22}O_{11} + H_2O \longrightarrow 4CH_3CHOHCO_2H$$

Another method is the fermentation of sucrose by *Rhizopus oryzæ*.

Lactic acid undergoes all the general reactions of α-hydroxyacids; oxidation with Fenton's reagent converts it into pyruvic acid:

$$CH_3CHOHCO_2H \xrightarrow{\text{H}_2\text{O}_2/\text{Fe}^{2+}} CH_3COCO_2H$$

It is oxidised to acetic acid by permanganate. Lactic acid is used in the tanning and dyeing industries, and ethyl lactate is used as a solvent for cellulose nitrate.

Since lactic acid contains one asymmetric carbon atom, it can (theoretically) exhibit optical activity (see p. 153). Three forms are known: L(+)-lactic acid, m.p. 26°C; D(−)-lactic acid, m.p. 26°C; and DL or (±)-lactic acid, m.p. 18°C. The lactic acid prepared by the above methods is (±)-. (+)-Lactic acid may be obtained from meat extract; this acid is also known as *sarcolactic acid* (Greek: *sarkos*, flesh). (−)-Lactic acid may be obtained by the fermentation of sucrose by *Bacillus acidi lævolactiti*.

Stereochemistry

A description of conformational analysis (p. 81) and brief accounts of optical (p. 153) and geometrical isomerism (p. 97) have already been given, and these should be revised before reading the following discussion.

Isomerism consists of three types:

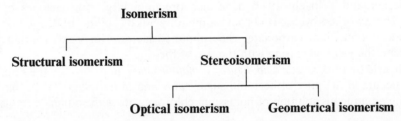

Structural isomerism is due to the difference in structure, and is exhibited in three different ways.

(i) *Chain* or *nuclear isomerism* is exhibited by compounds which differ in the arrangement of the *carbon atoms*, e.g., n- and isobutane.

(ii) *Position isomerism* is exhibited by compounds having the *same* carbon skeleton but differing in the position occupied by a substituent group, e.g., n- and isopropyl alcohols; α-, β- and γ-hydroxybutyric acids; *ortho-*, *meta-* and *para-*nitrophenols.

(iii) *Functional group isomerism* is exhibited by compounds having different functional groups, i.e., compounds with the same molecular formula but belonging to different homologous series, e.g., ethanol and dimethyl ether; acetone and propionaldehyde.

Tautomerism may be regarded as a special case of functional group isomerism.

Stereoisomerism is exhibited by isomers having the *same* structure but different *configurations*. Different configurations are possible because carbon forms mainly covalent bonds and these have direction in space (see also Vol. 2, Ch. 2).

Optical isomerism is characterised by compounds having the *same* structure but different configurations, and because of their *molecular asymmetry* these compounds are optically active.

Geometrical isomerism or *cis-trans isomerism* is characterised by compounds having the *same structure but different configurations*, and because of their *molecular symmetry* these compounds are *not* optically active. They can also exhibit optical isomerism if the structure of the molecule, apart from giving rise to geometrical isomerism, also satisfies the requirements for optical isomerism.

The instrument used for measuring the rotatory power of a substance is the *polarimeter*.* Essentially it consists of two Nicol prisms, one the polariser (P) and the other, the analyser (A), and between them a tube (T) which contains the substance (a liquid or a solution) to be examined (Fig. 17.1). S is a source of monochromatic light.

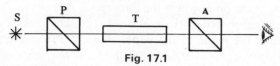

Fig. 17.1

If the substance rotates the plane of polarisation to the right, *i.e.*, if the analyser has to be turned to the right (clockwise) to restore the original field, the substance is said to be *dextrorotatory*; if to the left (anticlockwise), *lævorotatory*.

It has been found that the amount of the rotation depends, for a given substance, on a number of factors, *e.g.*, (i) the thickness of the layer traversed, (ii) the nature of the solvent (if in solution), (iii) the temperature and (iv) the wavelength of the light used. If $[\alpha]$ represents the *specific rotation*, l the thickness of the layer in decimetres, c the number of grams of substance (per 100 ml of solution), and the determination is carried out at temperature $t°C$ using sodium light (the D line), then if α is the *observed* rotation (+ or −),

$$[\alpha]\, t_D = \frac{100\alpha}{lc}$$

Since the value of the rotation depends on the solvent, this should also be stated.

Designation of optical isomers. Originally, the dextrorotatory (+) isomer was prefixed by d and the lævorotatory (−) isomer by l. It soon became apparent that *stereochemical relationships* were far more important than the actual direction of rotation, and so the symbols D and L are now used to indicate relationships between configurations. If such relationships are unknown, the symbols (+) and (−) are used to indicate the *sign* of rotation. The prefixes *dextro* and *lævo* (with hyphens) are also used.

In order to develop a system of stereochemical relationships it is necessary to choose a standard to which all compounds may be referred. Rosanoff (1906) chose the (+)- and (−)-forms of glyceraldehyde as the *arbitrary* standard and proposed that the formula of (+)-glyceraldehyde should be written as shown in Fig. 17.2(*a*). The tetrahedron is drawn so that three corners are

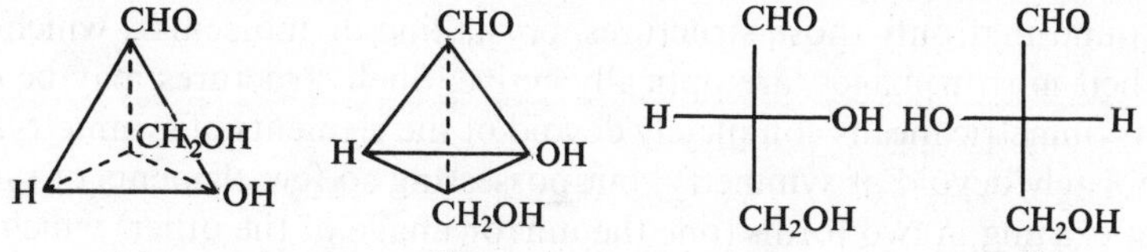

CHO   CHO   CHO   CHO

(*a*)   (*b*)   (*c*)   (*d*)

Fig. 17.2

*All standard text-books of Practical Physical Chemistry describe the construction and operation of polarimeters.

imagined to be *above* the plane of the paper, and the fourth *below* this plane. Furthermore, the spatial arrangement of the four groups joined to the central carbon atom *must be placed as shown in (a)*, *i.e.*, the accepted convention for drawing D(+)-glyceraldehyde places the *hydrogen atom at the left and the hydroxyl group at the right, with the aldehyde group at the top corner*. Now imagine the tetrahedron to rotate about the horizontal line joining H and OH until it takes up position (*b*). This is the conventional position for a tetrahedron, groups joined to *full horizontal* lines being *above* the plane of the paper, and those joined to *broken vertical* lines being *below* the plane of the paper. The *conventional plane-diagram is* obtained by drawing the full horizontal and broken vertical lines of (*b*) as full lines, placing the groups as they appear in (*b*), and taking the asymmetric carbon atom to be at the point where the lines cross. Although (*c*) is a plane-diagram, it is most important to remember that horizontal lines represent groups above the plane, and vertical lines groups below the plane of the paper. Figure (*d*) represents the plane-diagram formula of L(−)-glyceraldehyde; here *the hydrogen atom is to the right and the hydroxyl group to the left*. Another way of drawing (*c*) and (*d*) is to use *broken vertical* lines (instead of the full lines shown). Thus, any compound that can be prepared from, or converted into, D(+)-glyceraldehyde will belong to the D-series. Similarly, any compound that can be prepared from, or converted into, L(−)-glyceraldehyde will belong to the L-series. When representing relative configurational relationships of molecules containing more than one asymmetric carbon atom, *the asymmetric carbon atom of glyceraldehyde is always drawn at the bottom*, the rest of the molecule being built up from this unit.

$$H\!-\!\overset{|}{\underset{|}{C}}\!-\!OH$$
$$CH_2OH$$
D-series

$$HO\!-\!\overset{|}{\underset{|}{C}}\!-\!H$$
$$CH_2OH$$
L-series

$$\begin{array}{c} CO_2H \\ H\!-\!\!-\!\!-\!OH \\ HO\!-\!\!-\!\!-\!H \\ CO_2H \end{array}$$
(+)-tartaric acid

Thus we have a scheme of classification of *relative* configurations based on D(+)-glyceraldehyde as *arbitrary* standard. Until recently there was no way of determining, with certainty, the *absolute* configuration of molecules. *Arbitrary choice* makes the configuration of D(+)-glyceraldehyde have the hydrogen to the left and the hydroxyl to the right. Bijvoet *et al.* (1951), however, have shown by X-ray analysis that dextrorotatory tartaric acid has the configuration assigned to it by E. Fischer. Hence tartaric acid can be used as an *absolute* standard (see p. 490).

In 1848, Pasteur separated sodium ammonium racemate (p. 492) into two kinds of crystals by hand, and found that the specific rotation of each kind of crystal was the same, but one was dextrorotatory and the other, lævorotatory. Pasteur was able to separate the crystals by hand because he observed that they had hemihedral facets, one set of crystals being the mirror images of the other set. Such sets of crystals are said to be *enantiomorphous*.

It has been found that only those structures, crystalline or molecular, which are *not* superimposable on their mirror images, are optically active. Such structures may be *asymmetric*, or *dissymmetric*. Asymmetric means completely devoid of the elements of symmetry. Dissymmetric means not completely devoid of symmetry, but possessing so few elements of symmetry as still to be capable of existing in two forms (one the mirror image of the other) which are not superimposable. To avoid unnecessary complications, we shall use the term asymmetric to cover both cases (of asymmetry and dissymmetry). If a compound is asymmetric, then it is to be expected that the original molecule and its mirror image might differ in some properties although their structures are identical. Experience shows that the most marked difference is their action on polarised light.

Optical activity may be due entirely to the *crystal structure being asymmetric*, *e.g.*, quartz. In such cases the substance is optically active only so long as it remains solid, the optical activity being lost when the solid is fused or dissolved in a solvent. Quartz crystals exist in hemihedral forms. Hemihedral faces are those faces not symmetrically placed with respect to other faces; they occur only in *half* the positions where they might be expected to occur, and thus give the crystal an asymmetric structure (actually dissymmetric). The (+)- and (−)-forms of quartz are mirror images; but it should be noticed that many optically active crystals do not possess hemihedral faces.

On the other hand, optical activity may be due entirely to *molecular* structure. In this case the *molecular structure is asymmetric*, *i.e.*, the compounds have molecules in which the *atoms* are arranged spatially so that the original molecule is not superimposable on its mirror image. Such compounds are optically active in the solid, fused, dissolved or gaseous state, *e.g.*, sucrose, lactic acid, limonene, etc.

A molecule and its mirror image, when they are not superimposable, are known as *enantiomorphs* (this name taken from crystallography), *enantiomers* or *optical antipodes*. It appears that enantiomers are identical physically except (i) in their manner of rotating polarised light; the rotations are equal but opposite; (ii) they absorb dextro- and lævo-circularly polarised light unequally (circular dichroism). The crystal forms of enantiomers may be mirror images of each other, *i.e.*, the crystals themselves may be enantiomorphous, but this is unusual. Enantiomers are similar chemically, but their rates of reaction with other optically active substances are usually different. They may also be different physiologically, *e.g.*, (+)-histidine is sweet, (−)- tasteless; (−)-nicotine is more poisonous than (+)-; L(+)-ascorbic acid (vitamin C) is more efficient than D(−)-.

In 1874, van't Hoff and Le Bel, independently, gave the solution to the problem of optical isomerism. Van't Hoff proposed the theory that if the four valencies of the carbon atom are arranged tetrahedrally with the carbon atom at the centre, then all the cases of isomerism known are accounted for. Le Bel's theory is substantially the same as van't Hoff's, but differs in that whereas van't Hoff believed that the valency distribution was definitely tetrahedral and fixed as such, Le Bel believed that the valency directions were not rigidly fixed, and did not specify the tetrahedral arrangement, but thought that *whatever* the spatial arrangement, the molecule *Cabde* would be *asymmetric*. Van't Hoff's theory is more in keeping with later work, *e.g.*, in recent years physico-chemical evidence—X-ray and dipole moment studies—has shown that saturated carbon compounds exhibit a tetrahedral structure and that the carbon atom is situated inside the tetrahedron at the centre. Before the tetrahedral theory was suggested, it was believed that the four valencies of carbon were planar.

The simplest type of asymmetric structure is that which contains one carbon atom joined to four different atoms or groups, *i.e.*, a molecule of the type *Cabde*, in which the groups, *a*, *b*, *d* or *e* may or may not contain carbon. The carbon atom in *Cabde* is said to be asymmetric (actually, of course, it is the *group* which is asymmetric; a *carbon atom* cannot be asymmetric). Up to the present time, compounds of the type Ca_4 (*e.g.*, CCl_4), Ca_3b (*e.g.*, $CHCl_3$), Ca_2b_2 (*e.g.*, CH_2Cl_2), Ca_2bd (*e.g.*, $HOCH_2CO_2H$) have never been observed to exist as optical isomers. Only one form of each is known. This agrees with the tetrahedral configuration, *e.g.*, Ca_2bd (Fig. 17.3) (II), the mirror image of (I), is superimposable on (I); however the four groups are arranged in the tetrahedron, (II) is always superimposable on (I) (see footnote, p. 68). Thus there is only one form of Ca_2bd. Similarly, there is only one form of Ca_4, Ca_3b or Ca_2b_2. On the other hand, the tetrahedral structure of *Cabde* gives two forms (no more), one related to the other as object and mirror image, which are *not* superimposable (Fig. 17.4). Thus a molecule of the type *Cabde* should exist in two forms; and these should be detectable if the difference between them (physical or chemical) is sufficiently great. This would account for the structural identity and optical activity of

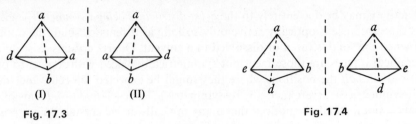

Fig. 17.3   Fig. 17.4

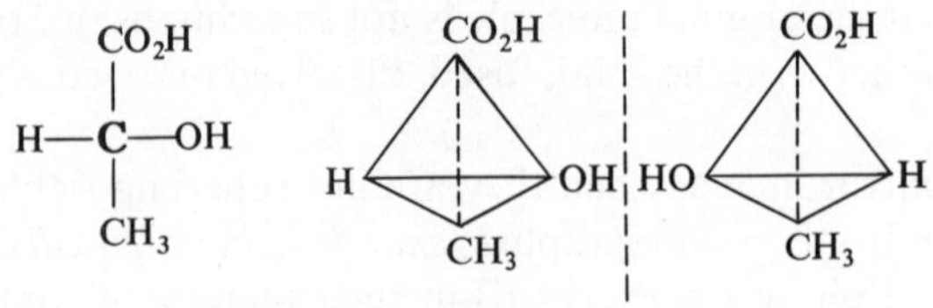

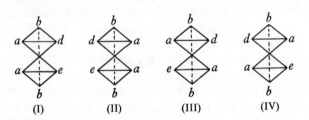

Fig. 17.5

molecules of the type $Cabde$, *e.g.*, the lactic acids, $CH_3CHOHCO_2H$ (Fig. 17.5). (III) and (IV) are mirror images and cannot be superimposed. Further evidence that optical activity is due to this arrangement is shown by the fact that if lactic acid is reduced to propionic acid, optical activity disappears; propionic acid, $CH_3CH_2CO_2H$, is a molecule of the type Ca_2bd, which is superimposable on its mirror image.

> Groups a, b, d, e are all different, but two or more may be *structural* isomers, *e.g.*, isopropylpropyl-methanol is optically active. The substitution of hydrogen by deuterium has also been investigated in recent years to ascertain whether these two atoms are sufficiently different to give rise to optical iso-merism. The earlier work gave conflicting results, but later work is conclusive in favour of optical activity, *e.g.*, Eliel (1949) prepared optically active methylphenyldeuteromethane, $CH_3CHDC_6H_5$, by reducing optically active methylphenylmethyl chloride with lithium aluminium deuteride.

If we examine an equimolecular mixture of a pair of enantiomers, we shall find that the mixture is optically inactive. This is to be expected, since enantiomers have equal but opposite rotatory power. Such a mixture (of equimolecular amounts) is said to be **optically inactive by external compensation**, and is known as a **racemic modification**. A racemic modification may be a purely mechanical mixture, a compound, or a solid solution, and is designated by the prefixes ($\pm$)- or DL-, *e.g.* ($\pm$)- or DL-lactic acid. Now let us examine the stereochemistry of a molecule containing two asymmetric carbon atoms. First let us consider the case of a compound containing two structurally dissimilar carbon atoms, i.e., compounds of the type $CabdCabe$, *e.g.*, $CH_3CHBrCHBrCO_2H$. There are four possible spatial arrangements for this type of structure (Fig. 17.6). (I) and (II) are enantiomers, and an equimolecular mixture of them forms a racemic

Fig. 17.6

modification; similarly for (III) and (IV). Thus there are four optically active forms. In general, a compound containing n different asymmetric carbon atoms exists in 2^n optically active forms (see also Vol. 2, Ch. 2).

(I) and (III) are not identical in configuration and are not mirror images; they are known as **diastereoisomers**, *i.e.*, they are optical isomers but not mirror images (not enantiomers). Thus a compound of the type $CabdCabe$ exists in six forms: two pairs of enantiomers, and two racemic modifications. Diastereoisomers differ in physical properties, such as melting point, solubility, specific rotation, etc. Chemically they are similar, but their rates of reaction with other optically active compounds are different (see also geometrical isomerism, p. 496).

The plane-diagrams of the molecules (I–IV) in Fig. 17.6 will be (V–VIII), respectively (see also below).

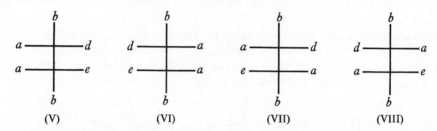

Pairs of enantiomers of the type $CabdCabe$ are distinguished by the prefixes *erythro* and *threo*. In the former, the two identical groups a and the two identical groups b can eclipse each other (a,a and b,b) in one conformation, whereas in the latter this cannot be done. Thus (I) and (II) are the *erythro*-forms, and (III) and (IV) are the *threo*-forms. These names are derived from erythrose and threose (see p. 519).

Instead of writing down all the possible configurations, the number of optical isomers for a compound of the type $CabdCabe$ may be obtained by indicating the *configuration* of each asymmetric carbon atom by the symbol $+$ or $-$, or by D or L; thus:

$$
\begin{array}{c|cccc}
Cabd & + \quad - & + \quad - & D_1 \ L_1 & D_1 \ L_1 \\
\mid & & & & \\
Cabe & + \quad - & - \quad + & D_2 \ L_2 & L_2 \ D_2 \\
& \underbrace{} & \underbrace{} & \underbrace{} & \underbrace{} \\
& (\pm) & (\pm) & \text{DL} & \text{DL}
\end{array}
$$

Now let us consider the case of a compound containing two asymmetric carbon atoms which are structurally the same, *i.e.*, compounds of the type $CabdCabd$, *e.g.*, tartaric acid, $HO_2CCHOHCHOHCO_2H$. In compounds of this type, it is obvious that $D_1 = D_2$. If we ignore the problem of the different conformations possible for this molecule (but see below), we can proceed as follows.

$$
\begin{array}{c|cccc}
Cabd & D & L & D & L \\
\mid & & & & \\
Cabd & D & L & L & D \\
& (\text{IX}) & (\text{X}) & (\text{XI}) & (\text{XII})
\end{array}
$$

In molecules (IX) and (X), the upper and lower halves reinforce each other; hence (IX), as a whole, has the dextro-, and (X), the lævo-configuration, *i.e.*, (IX) and (X) are optically active, and enantiomeric. On the other hand, in (XI) the two halves are in opposition, and hence the molecule, *as a whole*, will not show optical activity. It is also obvious that (XI) and (XII) are identical, *i.e.*, there is only *one* optically inactive form of $CabdCabd$. Molecule (XI) is said to be

optically inactive by internal compensation; it is known as the *meso*-form, and is a diastereoisomer of (IX) and (X). Thus there are four possible forms: D-, L-, DL- and *meso*-.

Molecule (XI) is an example of a *compound having two asymmetric carbon atoms, but is optically inactive* (by internal compensation). It is therefore obvious that inspection of the usual structural formula, which contains two (or more) asymmetric carbon atoms, is not sufficient to decide whether the molecule is optically active or not. *The molecule as a whole must be asymmetric.* The test of superimposing the original formula (tetrahedral) on its mirror image definitely indicates whether the molecule is symmetrical or not. The most satisfactory way in which superimposability may be ascertained is to build up models of the molecule and its mirror image. Usually, this is not convenient. A simple device to decide whether a molecule is symmetrical or not is to ascertain whether it contains a *plane of symmetry*, a *centre of symmetry*, or an *alternating axis of symmetry*. If any one of these is present the molecule is symmetrical, *i.e.*, superimposable on its mirror image.

A **plane of symmetry** divides a molecule in such a way that points (atoms or groups of atoms) on the one side of the plane form mirror images of those on the other side. This test may be applied to both solid and plane-formulæ, *e.g.*, the plane-formula of the *meso* form of *CabdCabd* possesses a plane of symmetry; the other two, D- and L-, do not:

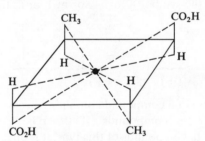

A **centre of symmetry** is a point from which lines, when drawn on one side and produced an equal distance on the other side, will meet identical points in the molecule. This test can be satisfactorily applied only to three-dimensional formulæ, particularly those of ring systems, *e.g.*,

2,4-dimethylcyclobutane-1,3-dicarboxylic acid. The form shown possesses a centre of symmetry, which is the centre of the ring. This form is therefore optically inactive.

(ii) Dimethyldiketopiperazine exists in two geometrical isomeric forms, *cis* and *trans*:

The *cis* isomer has neither a plane nor a centre of symmetry. It can therefore exist in two enantiomeric forms; both are known. The *trans* isomer has a centre of symmetry and is therefore optically inactive.

It is important to note that only *even-numbered* rings can possibly possess a centre of symmetry.

Up to the present, all optically active *natural* compounds have been found to consist of molecules which owe their asymmetry to the absence of both a plane and centre of symmetry. It is possible, however, for these elements of symmetry to be *absent* and the molecule to be superimposable on its mirror image and hence not be optically active. Such a molecule will possess an alternating axis of symmetry. This may be defined as follows: a molecule possesses an *n*-fold **alternating axis of symmetry** if, when rotated through an angle of $360°/n$ about this axis and then followed by reflection in a plane perpendicular to the axis, the molecule is the same as it was in the starting position.

(XIII)   (XIV)

McCasland and Proskow (1956) synthesised a compound which owes its symmetry to the presence of an alternating axis of symmetry only (this compound possesses neither a plane nor a centre of symmetry). It is the *N*-spiro-compound (XIII) (as the *p*-toluenesulphonate). If (XIII) is rotated through 90° about the co-axis of the two rings, (XIV) is obtained. Reflection of (XIV) through a central plane perpendicular to this axis gives a molecule identical and coincident with (XIII).

The application of the elements of symmetry is always to the Fischer plane projection formulæ, *which are always those of the eclipsed conformations*. These molecules, however, may have other conformations which are devoid of any element of symmetry. Nevertheless, such compounds are not optically active. Although such conformations are *individually* optically active, statistically it can be expected that for every optically active conformation present, so will also be present its mirror image conformation, *i.e.*, each conformation will be present as a pair of enantiomers. Thus, the whole collection of molecules will be optically inactive by 'external compensation'. It therefore follows that when a molecule can exist in a number of conformations, then provided that at least one conformation (whether preferred or not) is superimposable on its mirror image, the compound will be optically inactive by 'external compensation'. Such 'racemic modifications' cannot be resolved because of the low energy of interconversion of the enantiomers.

Chirality. As we have seen, the structures of enantiomers differ only in 'handedness', one being left-handed and the other being right-handed. Any molecule which is not superimposable on its mirror image is said to possess **chirality** (from the Greek *kheir*, hand). Thus, the term chirality means 'having handedness'. This term was first introduced by Kelvin (1884), and has been used by Cahn *et al.* in their system for the specification of absolute configuration (see below). Chirality expresses the necessary and sufficient condition for the existence of enantiomers (*i.e.*, chirality is now equivalent to asymmetry or dissymmetry). The adjective *chiral* is equivalent to being left- or right-handed, and so a **chiral centre** is one which can be left- or right-handed. On the other hand, when a molecule is superimposable on its mirror image, that molecule does not possess 'handedness' (it is not asymmetric), and is said to be *achiral*. The commonest cause of optical activity is the presence of one or more chiral centres, which in organic chemistry are usually asymmetric carbon atoms.

Specification of absolute configurations. Since the configuration of (+)-tartaric acid has been related to that of (+)-glyceraldehyde (p. 490), and since the *absolute* configuration of (+)-tartaric acid has been determined (p. 474), it is now possible to assign *absolute configurations* to many compounds whose relative configurations to (+)-glyceraldehyde are known. This raised the problem of using one system of specifying absolute configurations. Cahn *et al.* (1956, 1964) have proposed such a system and this is now widely used. Let us first consider the procedure for a molecule containing one asymmetric carbon atom, *i.e.*, containing one chiral centre.

(1) The four groups are first ordered according to the *sequence rule*. According to this rule, the groups are arranged in *decreasing atomic number* of the atoms by which they are bound to the asymmetric carbon atom. If two or more of these atoms have the same atomic number, then the relative priority of the groups is determined by a similar comparison of the atomic numbers of the *next* atoms in the groups. If this fails, then the next atoms in the groups are considered. Thus one works *outwards* from the asymmetric carbon atom until a selection can be made for the sequence of the groups.

(2) It is next determined whether the sequence describes a right- or left-handed pattern on the molecular model as viewed according to the *conversion rule*. When the four groups in the molecule C*abcd* have been ordered in the priority *a*, *b*, *c*, *d*, the conversion rule states that their spatial pattern shall be described as right- or left-handed according as the sequence $a \rightarrow b \rightarrow c$ is clockwise or anticlockwise when viewed from an external point on the side *remote* from *d* (the atom with the lowest priority), *e.g.*, (I) shows a right-handed (*i.e.*, clockwise) arrangement.

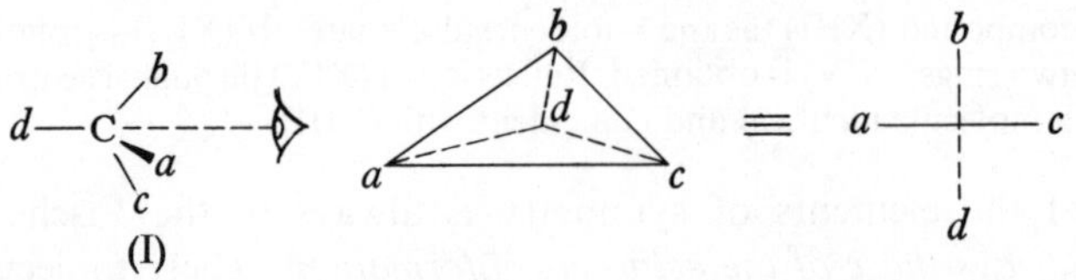

(I)

(3) Absolute-configuration labels are then assigned. The asymmetry leading under the sequence and conversion rules to a right- and left-handed pattern, is indicated by *R* and *S* respectively (*R*; *rectus*, right; *S*, *sinister*, left).

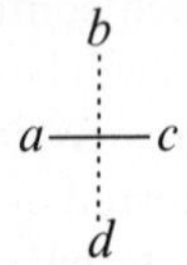

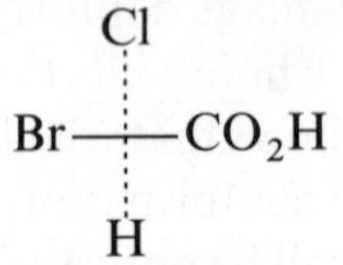

(II)

Consider bromochloroacetic acid (II). Priority of the groups according to the sequence rule is Br (*a*), Cl (*b*), C of CO_2H (*c*), and H (*d*). Hence by the conversion rule, (II) is the *R*-form.

When multiple bonds or rings are present, the procedure for determining priority is as follows. Both atoms attached to the multiple bond are considered to be duplicated (for a double bond) or triplicated (for a triple bond), *e.g.*,

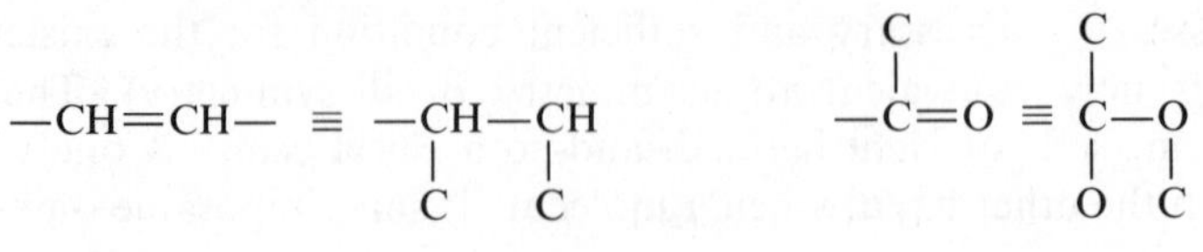

$$-\overset{\overset{\displaystyle H}{|}}{C}=O \equiv -\overset{\overset{\displaystyle H}{|}}{\underset{\underset{\displaystyle C}{|}}{C}}-\overset{O}{\underset{}{}}\qquad -C\equiv N \equiv -\overset{\overset{\displaystyle N \quad C}{|\quad|}}{\underset{\underset{\displaystyle N \quad C}{|\quad|}}{C}}-N$$

$$\bigcirc \equiv -CH\overset{\displaystyle CH_2-}{\underset{\displaystyle CH_2-}{}}\qquad \bigcirc \longleftrightarrow \bigcirc \equiv -\overset{\overset{\displaystyle C\quad C}{|\quad|}}{\underset{\underset{\displaystyle C-}{|}}{C}}-\overset{\overset{\displaystyle C-C-}{}}{\underset{}{C}}$$

The priority sequence is then determined by the consideration of the duplicated or triplicated 'structure' in which there are *phantom atoms, e.g.,* —CHO is —CH(O)—O(C) and —CH(OH)$_2$ is —CHOH—OH (the phantom atoms in the former are those in parentheses). Both groups contain a carbon atom joined to two oxygen atoms, but since C precedes H, —CHO precedes —CH(OH)$_2$ in priority.

Ring systems are treated as branched chains, and if unsaturated, then duplication is used for a double bond or triplication for a triple bond. By using these rules, it can be shown that the order of priority sequence (for some of the common substituents) is: I, Br, Cl, SO$_3$H, SH, F, OCOR, OR, OH, NO$_2$, NR$_2$, NHR, NH$_2$, CO$_2$R, CO$_2$H, COR, CHO, CH$_2$OH, CN, Ph, CR$_3$, CHR$_2$, CH$_2$R, CH$_3$, D, H.

Now let us consider D(+)-glyceraldehyde. By *convention* it is drawn as (III); (this is also the *absolute* configuration). Reference to the

$$\begin{array}{ccc}
\text{CHO} & \text{CHO} & \text{CHO} \\
\text{H}\!-\!\!-\!\text{OH} & \text{CH}_2\text{OH}\!-\!\!-\!\text{OH} & \text{HO}\!-\!\!-\!\text{CH}_2\text{OH} \\
\text{CH}_2\text{OH} & \text{H} & \text{H} \\
\text{D}(+) & \text{L}(-) & \text{D}(+) \\
\text{(III)} & \text{(IV)} & \text{(V)}
\end{array}$$

sequence list gives the priority sequence: OH (*a*), CHO (*b*), CH$_2$OH (*c*), and H(*d*). Since the interchanging of two groups inverts the configuration, the sequence (III) ⟶ (IV) ⟶ (V) gives the *original* configuration, and because (V) corresponds to *a* → *b* → *c* in a clockwise fashion, D(+)-glyceraldehyde is therefore (*R*)-glyceraldehyde.

When a molecule contains two or more asymmetric carbon atoms, each asymmetric carbon atom (chiral centre) is assigned a configuration according to the sequence and conversion rules and is then specified with *R* or *S, e.g.,* (+)-tartaric acid.

$$\begin{array}{ccc}
\text{CO}_2\text{H} & \text{CO}_2\text{H} & \overset{2}{\text{C}}\text{HOHCO}_2\text{H} \\
\text{H}\!-\!\!\underset{2}{|}\!\!-\!\text{OH} \equiv & \text{H}\!-\!\!\underset{2}{|}\!\!-\!\text{OH} \equiv & \text{HO}\!-\!\!\underset{1}{|}\!\!-\!\text{H} \\
{}^1\text{CHOHCO}_2\text{H} & \text{HO}\!-\!\!\underset{1}{|}\!\!-\!\text{H} & \text{CO}_2\text{H} \\
& \text{CO}_2\text{H} &
\end{array}$$

$$\begin{array}{ccc}
\|\text{(two} & \|\text{(two} & \|\text{(two} \\
\text{interchanges)} & \text{interchanges)} & \text{interchanges)}
\end{array}$$

$$\begin{array}{ccc}
\text{CO}_2\text{H} & b & \text{CO}_2\text{H} \\
\text{HO}\!-\!\!\underset{2}{|}\!\!-\!\text{CHOHCO}_2\text{H} & a\!-\!\!|\!\!-\!c & \text{HO}\!-\!\!\underset{1}{|}\!\!-\!\text{CHOHCO}_2\text{H} \\
\text{H} & d & \text{H}
\end{array}$$

Thus, (+)-tartaric acid is (*RR*)-tartaric acid.

The sequence rule was designed to relate the symbols D and L with the symbols R and S. However, D and L are obtained by means of chemical transformations, whereas R and S are derived from geometrical models and are independent of correlations. Because of this, R and S must be applied only to compounds whose absolute stereochemistry has been determined; they do not necessarily correlate chemical families, e.g., (+)-tartaric acid, whether it be D or L (according to the method of correlation, p. 490), has the absolute configuration specified by (RR).

Resolution of racemic modifications

When optically active compounds are prepared by synthetic methods, the usual result is a racemic modification; e.g., bromination of propionic acid results in the formation of DL-α-bromopropionic acid:

$$CH_3CH_2CO_2H \xrightarrow{\text{Br}_2/\text{P}} CH_3CHBrCO_2H$$

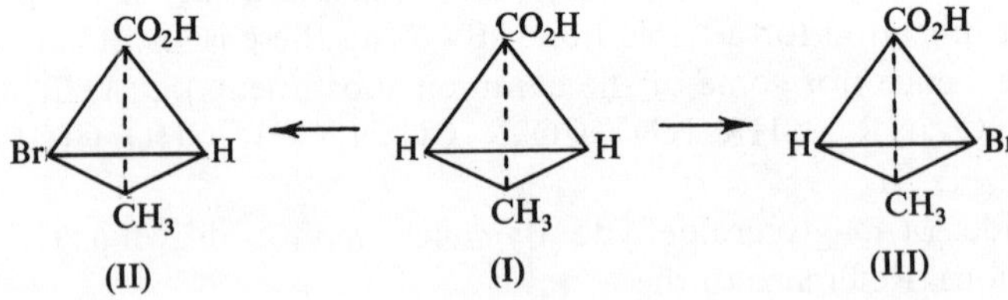

Fig. 17.7

(II) and (III) (Fig. 17.7) are enantiomers, and since molecule (I) is symmetrical about its vertical axis, it can be anticipated from the theory of probability that either hydrogen atom should be replaced equally well to give DL-α-bromopropionic acid. This actually does occur in practice.

From what has been said above, it would appear that the two hydrogen atoms in (I) are alike. This is true for their behaviour with bromine, but the point to note is that replacement of one or other hydrogen does not produce the *same* molecule; a pair of enantiomers is produced. These two hydrogens are therefore said to be **enantiotopic**. This term may be defined as follows: Two atoms or groups in a molecule are enantiotopic if replacement of each in turn by some other group leads to a pair of enantiomers. It should be noted that corresponding groups in a *pair of enantiomers* are also said to be enantiotopic, e.g., the methyl groups or bromine atoms in (II) and (III) are enantiotopic. If separate replacements produce the *same* molecule, the groups are said to be **homotopic**.

Enantiotopic groups react with non-asymmetric (achiral) reagents at the *same* rate, but react with asymmetric (chiral) reagents at different rates (*cf.* bromination, above).

In addition to referring to groups as being enantiotopic, *faces of double bonds* are also said to be enantiotopic if stereoisomers are produced by addition reactions, e.g., the addition of hydrogen cyanide to the carbonyl double bond (attack at the front face and at the back face).

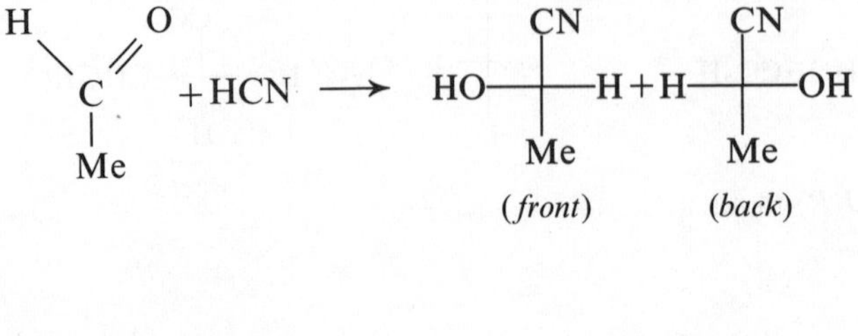

If we now consider the bromination of (IV), it can be seen that the two α-hydrogen atoms are *not* symmetrically placed with respect to the lower half (which is asymmetric). Hence it might be expected that the two hydrogen atoms would be replaced by bromine (an achiral reagent) at *different* rates. This is the case in practice; the product is a pair of diastereoisomers in *unequal* amounts. These two hydrogens are therefore said to be **diastereotopic**. This term may be defined as follows: Two atoms or groups in a molecule are diastereotopic if replacement of each in turn by some other group leads to a pair of diastereoisomers. Also, corresponding groups in a *pair of diastereoisomers* are said to be diastereotopic (*cf.* enantiotopic groups above).

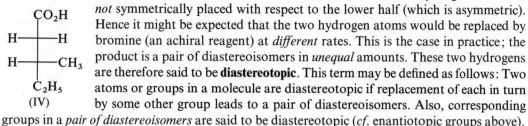

The term *diastereotopic faces* may be used with respect to the *faces of a double bond* when one of the groups attached to the unsaturated carbon atom contains a chiral centre.

The process of separating a racemic modification into its enantiomers is known as *resolution*. Various methods have been introduced (see also Vol. 2, Ch. 2).

1. Mechanical separation (Pasteur, 1848). In this method the crystals are, if sufficiently well defined, actually separated by hand; the crystals must be enantiomorphous. This method is applicable only to racemic *mixtures*, and is mainly of historical interest.

2. Biochemical separation (Pasteur, 1858). Certain bacteria and moulds, when they grow in a dilute solution of a racemic modification, destroy one enantiomer more rapidly than the other; *e.g.*, *Penicillium glaucum* (a mould), when grown in a solution of ammonium racemate, attacks the D-form leaving the L-.

3. By means of salt-formation (Pasteur, 1858). This method, which is the best of all the methods of resolution, consists in converting the enantiomers in a racemic modification into *diastereoisomers*; *e.g.*, if an optically active base is combined with a racemic acid, two diastereoisomers are obtained:

$$(D_{acid} + L_{acid}) + 2D_{base} \longrightarrow (D_{acid}D_{base}) + (L_{acid}D_{base})$$

Because of their different solubilities, these diastereoisomers may be separated by fractional crystallisation. After separation, the acids may be regenerated by hydrolysis with inorganic acids or with alkalis.

Bases which are used for the resolution of racemic acids are mainly alkaloids, *e.g.*, quinine, strychnine, brucine, cinchonine, morphine, etc., and acids which are used for racemic bases are, *e.g.*, tartaric acid, camphorsulphonic acid, bromocamphorsulphonic acid.

The method of salt-formation has been extended to compounds other than acids and bases, *e.g.*,

(i) Racemic alcohols are converted into the acid ester derivative with phthalic anhydride:

$$\text{(phthalic anhydride)} \quad CO{-}O + ROH \longrightarrow \quad CO_2R,\ CO_2H$$

The acid ester, consisting of equimolecular amounts of the D- and L-forms, may now be resolved by the method used for acids.

(ii) Racemic aldehydes and racemic ketones may be resolved by means of optically active derivatives of hydrazine, *e.g.*, (−)-menthylhydrazine.

Resolution of racemic modifications by means of salt formation may be complicated by the phenomenon of *asymmetric transformation*. This is exhibited by compounds that are optically unstable, *i.e.*, the enantiomers are readily interconvertible:

$$(+)\text{-C} \rightleftharpoons (-)\text{-C}$$

It is possible to get *complete* conversion of C into the form (as salt) that crystallises from solution. The form may be $(+)$ or $(-)$, depending on the nature of the base (used for resolving acids) and the solvent.

4. Selective adsorption. Optically active substances may be selectively adsorbed by some optically active adsorbent, *e.g.*, Henderson and Rule (1939) resolved a racemic modification of a camphor derivative on D-lactose as adsorbent; Prelog and Wieland (1944) resolved Tröger's base on D-lactose (p. 499).

5. Channel complex formation. This method is possible because the complexes of each enantiomer have different solubilities; *e.g.*, Schlenk (1952) has resolved $(\pm)$-2-chloro-octane by means of channel complex formation with urea (see p. 462).

Racemisation

By using suitable conditions, it is possible to cause most optically active compounds to lose their optical activity without changing their structure. This means that the $(+)$- and $(-)$-forms of most optically active compounds are convertible one into the other, the final result being a racemic modification. Such a transformation is known as *racemisation*. The method of effecting racemisation depends on the nature of the compound in question; generally, heat, light or chemical reagents may be used. Thus, if the starting material is the $(+)$-form, then after treatment half will be converted into the $(-)$-form; similarly, when starting with the $(-)$-form, half will be converted into the $(+)$-form.

In some cases racemisation occurs spontaneously at room temperature; it is then known as *autoracemisation*, *e.g.*, dimethyl bromosuccinate autoracemises.

Many different types of compounds can racemise, and a number of mechanisms have been proposed, each mechanism applying to a particular type of compound. One important type of compound that readily racemises is that in which the asymmetric carbon atom is joined to a hydrogen atom and can undergo tautomeric change; the mechanism proposed for this racemisation is one via enolisation, *e.g.*, the acid-catalysed racemisation of 2-butyl phenyl ketone (*cf.* p. 281):

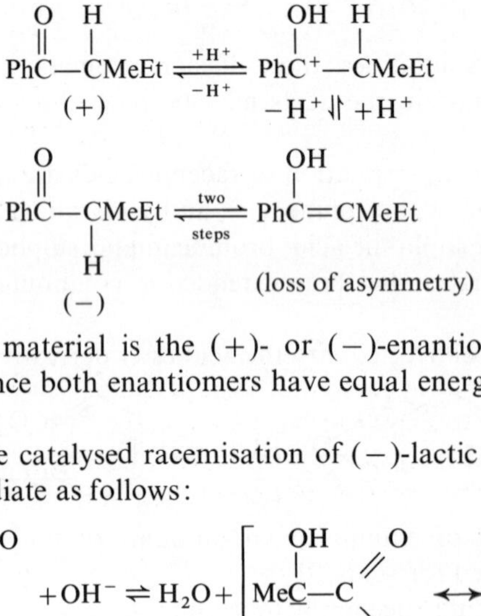

Hence, whether the starting material is the $(+)$- or $(-)$-enantiomer, the final product is $(\pm)$-2-butyl phenyl ketone, since both enantiomers have equal energy content and so both are formed in equal amounts.

Another example is the base catalysed racemisation of $(-)$-lactic acid. This may be formulated via a carbanion intermediate as follows:

Here again, whether we start with the (−)- or (+)-enantiomer, the final product is (±)-lactic acid.

Some compounds which cannot undergo tautomeric change can, nevertheless, be racemised, *e.g.*, (−)-limonene, some biphenyl compounds. The mechanism of some of these racemisations is uncertain (see Vol. 2, Ch. 2).

The Walden inversion (optical inversion)

By a series of reactions, Walden (1893) was able to transform an optically active compound into its enantiomer. In some cases the product is 100 per cent optically pure, *i.e.*, the *inversion* is quantitative; in other cases the product is a mixture of the (+)- and (−)-forms, but in unequal quantities, *i.e.*, a partial inversion has taken place (see also Vol. 2, Ch. 3).

The phenomenon was first discovered by Walden with the following reactions:

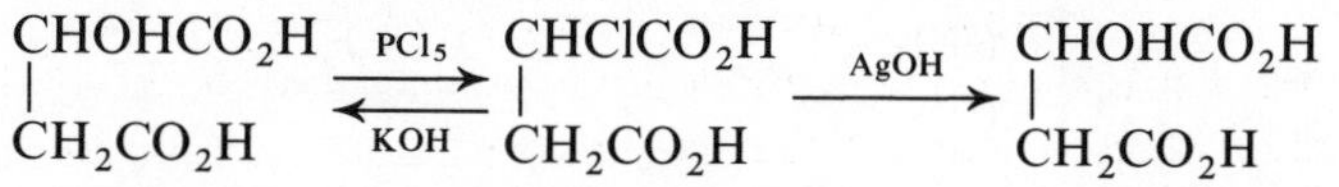

$$\underset{\substack{\text{L-malic acid}\\(I)}}{\overset{\text{CHOHCO}_2\text{H}}{\underset{\text{CH}_2\text{CO}_2\text{H}}{|}}} \xrightarrow[\text{KOH}]{\text{PCl}_5} \underset{\substack{\text{D-chlorosuccinic acid}\\(II)}}{\overset{\text{CHClCO}_2\text{H}}{\underset{\text{CH}_2\text{CO}_2\text{H}}{|}}} \xrightarrow{\text{AgOH}} \underset{\substack{\text{D-malic acid}\\(III)}}{\overset{\text{CHOHCO}_2\text{H}}{\underset{\text{CH}_2\text{CO}_2\text{H}}{|}}}$$

In one of the two reactions there must be an interchange of position between two groups; *e.g.*, if the configuration of (I) corresponds with that of (II), the inversion must have taken place between (II) and (III). *The step in which the inversion occurs constitutes a* **Walden inversion**.

As the above experiment stands, there is no way of telling at which stage inversion has occurred. Change in the sign of rotation does not necessarily indicate an inversion of configuration; *e.g.*, when D(+)-glyceraldehyde is oxidised to glyceric acid, the acid obtained is D(−)-glyceric acid.

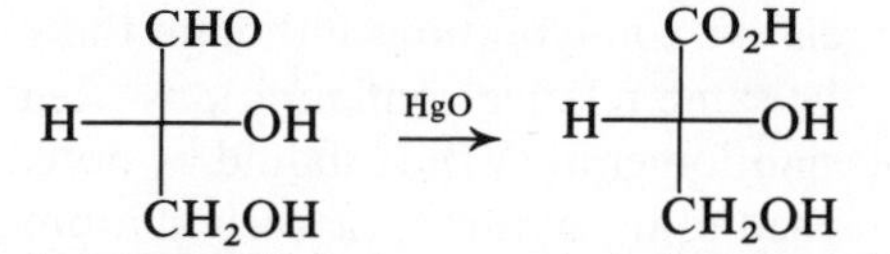

D(+)-glyceraldehyde → D(−)-glyceric acid

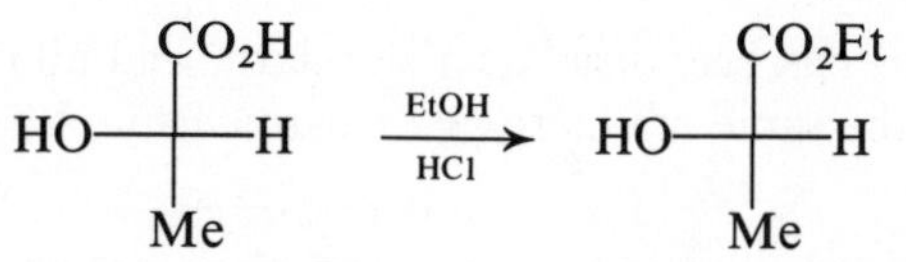

L(+)-lactic acid → ethyl L(−)-lactate

In each case, there is no change in configuration, but the signs of rotation are opposite. Thus it is necessary to have methods to determine relative configurations. The above examples are easy to solve; starting material and product both have the same relative configurations, since the asymmetric carbon atom is not affected in the reaction, and so inversion is not possible. At the same time, because the asymmetric carbon atom is not involved, starting with optically pure compounds will lead to optically pure products. In this way, it is easy to measure the **optical purity** of a sample of such products. Optical purity is expressed as a percentage, *e.g.*, if the (maximum) specific rotation of compound A is $+5°$ and an impure sample has a rotation $+3°$, this sample is 60 *per cent optically pure*.

When the asymmetric carbon atom is involved in various reactions, the problem is far more difficult. Kenyon *et al.* (1925) established a basis for the determination of relative configurations as follows. These authors carried out a series of reactions on optically active hydroxy compounds. Now it has been established that in the esterification of a monocarboxylic acid by an alcohol under ordinary conditions, acyl-oxygen fission occurs (see p. 248); *i.e.*,

$$R^1CO\!-\!OH + R^2O\!-\!H \longrightarrow R^1COOR^2 + H_2O$$

Kenyon assumed that in all reactions of this type the R^2—O bond remained intact, and consequently no inversion of the alcohol is possible. The following chart shows a series of reactions carried out on ethyl (+)-lactate; Ts = *p*-tosyl group, *i.e.*, *p*-toluenesulphonyl group, $CH_3C_6H_4SO_2$—, Ac = CH_3CO, and the symbol $\longrightarrow_{\!\!\!\!\!\sigma}\!\!\rightarrow$ is used to represent inversion in that step.

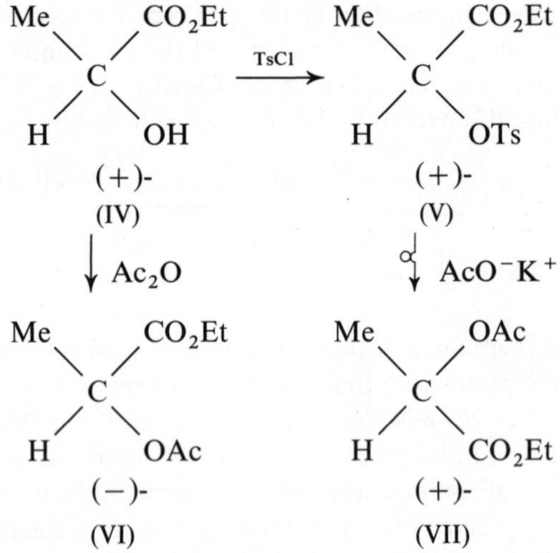

(IV) and (VI) have the same relative configurations although the sign of rotation has changed. Similarly, (IV) and (V) have the same relative configurations. Reaction of (V) with AcO^-K^+, however, produces (VII), the enantiomer of (VI). It should be noted that if inversion is going to occur, the *complete group* attached to the asymmetric carbon atom must be removed. The converse, however, is not necessarily true, *i.e.*, removal of a complete group does not invariably result in inversion (see below).

The above set of reactions has been used as a standard, and all closely analogous reactions are assumed to proceed in the same manner, *e.g.*, treatment of (V) with lithium chloride gives (VIII) with inversion.

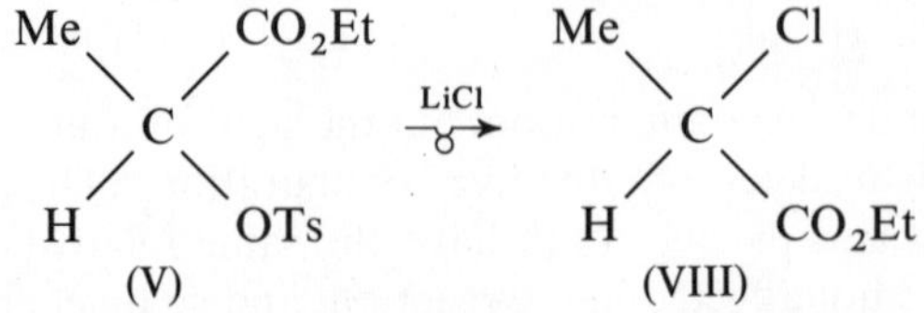

Also, since (+)-(IV) is obtained from the esterification of (−)-lactic acid, and (−)-lactic acid has been correlated with D(+)-glyceraldehyde, the stereochemical relationships to D(+)-glyceraldehyde are therefore established.

Participation of neighbouring groups in nucleophilic substitution. So far, we have discussed polar and steric effects on rates and mechanisms of reactions (for solvent effects, see p. 159). We have already seen that the phenomenon of *neighbouring group participation* may operate in various rearrangements (p. 147). In the same way, this effect may also operate in nucleophilic substitution reactions. Here we have a group attached to the carbon atom *adjacent* to the carbon atom where nucleophilic substitution occurs and, during the course of the reaction, becomes bonded or partially bonded to the reaction centre. Neighbouring group participation, however, may also involve a group further removed than the carbon atom adjacent to the reactive centre.

Whether a group Z can enter into neighbouring group participation depends on the nature of Z, and for a given Z it usually depends on the size of the ring that can be formed (by n.g.p.).

When the reaction is accelerated by neighbouring group participation, that reaction is said to be *anchimerically assisted*. As we have seen, various atoms exhibit this phenomenon of neighbouring group participation, *e.g.*, halogen. Brominium (bromonium) ions were first proposed by Roberts *et al.* (1937) as intermediates in the addition of bromine to alkenes (see p. 102). The existence of this cyclic brominium ion (bridged ion) has been demonstrated by Winstein *et al.* (1939), who found that the action of fuming hydrobromic acid on $(-)$-*threo*-3-bromobutan-2-ol gave $(\pm)$-2,3-dibromobutane.

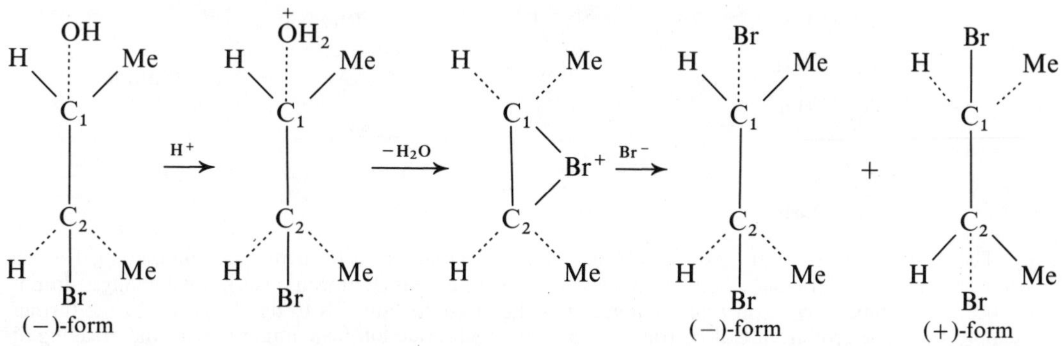

Had no neighbouring group participation occurred, then if the reaction were S_N2, complete inversion would have occurred only at C_1. If the reaction were S_N1, C_1 would have formed a classical carbonium ion, and so racemisation would have occurred at C_1 only. Since retention or inversion of *both* C_1 and C_2 occurs, the results are explained as shown above. If the Br^- ion attacks C_1 from the back, the $(-)$-form is produced; if it attacks C_2 from the back, the $(+)$-form is produced (see also p. 553, and Vol. 2, Ch. 3).

Further evidence for the existence of the brominium ion has been obtained by Olah *et al.* (1967), who examined the N.M.R. spectrum of the solution of the product formed from the addition of bromine to tetramethylethylene in SbF_5—SO_2 at $-60°C$.

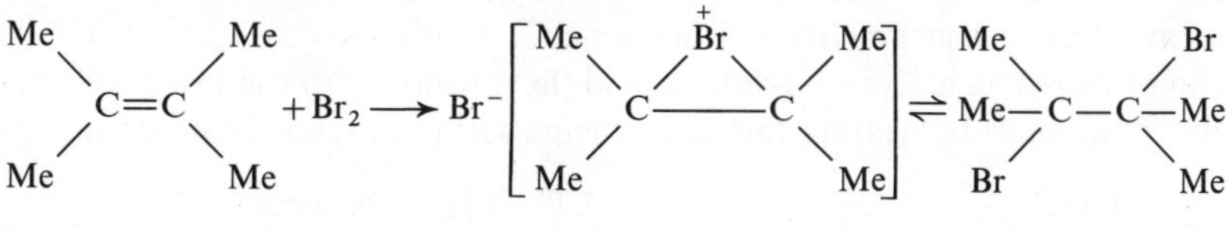

The authors have interpreted the results as being consistent with the formation of the brominium bromide salt, and also showed that chlorine and iodine formed bridged ions under these conditions, but not fluorine. On the other hand, Wynberg *et al.* (1969) believe that they have actually isolated a reasonably stable brominium ion as its bromide, $RBr^+Br_3^-$.

The problem of addition reactions of alkenes is more complicated than that given in the account above. In some cases, addition may be stereospecifically *trans*, *e.g.*, the addition of chlorine to *trans*-but-2-ene at $-90°C$ in the *absence* of solvent (Poutsma, 1965). On the other hand, addition of, *e.g.*, DBr to *cis*-1-phenylpropene (I) is stereoselectively *cis* (88 per cent) at $0°C$ in methylene dichloride (Dewar *et al.*, 1963). If addition through bridged ions were the only route, and the bridged ions were stable, *trans* addition would be stereospecific. However, if the classical carbonium ion becomes increasingly stable, *cis* addition will become increasingly favoured. One factor that increases the stability of the bridged ion is the nature of the addendum, *e.g.*, the order for the halogens (as shown experimentally) is $I > Br > Cl > F$. Hence stereospecific *trans* addition operates for iodine. On the other hand, when the addendum is not an iodine atom (in the first step) and the classical ion is stabilised, *e.g.*, by resonance, then the rate-determining step is the formation of a classical carbonium ion as an ion-pair (II), where the halide ion is held on the *same side* of the original double bond as the entering deuteron. Collapse of this ion-pair thus gives the *cis*-adduct but, at the same time, *rearrangement* of (II) to the isomeric ion-pair (III), gives the *trans*-adduct. It has also been found that the nature of the solvent affects the relative amounts of *cis* and *trans* products.

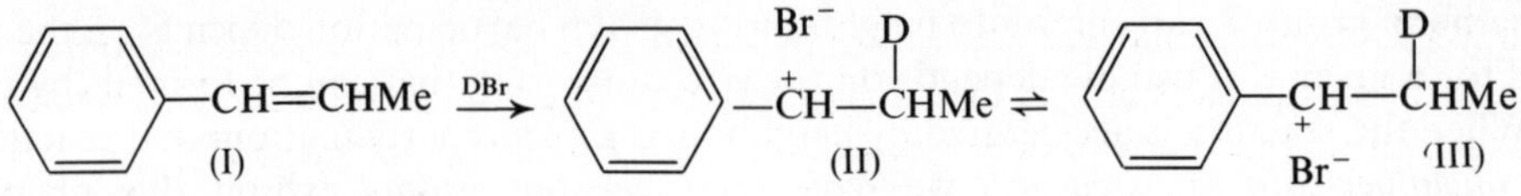

Elimination reactions can also be stereoselectively or stereospecifically *cis* (see also pp. 164, 381). An example of the former is (Cristol *et al.*, 1960):

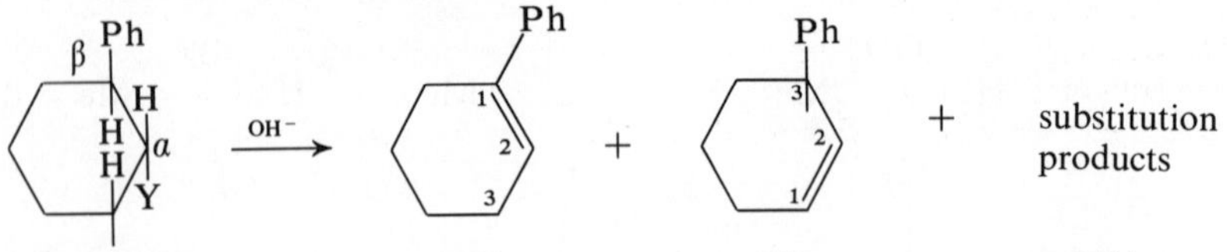

$Y = SMe_2^+$	22%	2%	61%
$Y = NMe_3^+$	64%	2%	8%

The reaction was shown to be E2, and hence 1-phenylcyclohexene is formed by *cis*-elimination. This has been explained by Ingold (1963) as follows. In *cis*-elimination (E2 mechanism), bond changes do not occur simultaneously (as in *trans*-elimination). The more difficult it is to detach group Y, the further ahead will be the proton on C-β be transferred to the hydroxide ion (as compared with the breaking of the C—Y bond). Thus charge builds up on C-β, and the electron-pair becomes increasingly available when C-α is "free". In this way, phenyl conjugation controls the orientation, and so *cis*-elimination becomes increasingly favoured the more difficult it is to remove Y. Since it has been shown (from other experiments) that it is more difficult to detach the NMe_3^+ group than the SMe_2^+, the former leads to more *cis*-elimination than the latter.

Asymmetric synthesis

In ordinary laboratory preparations of optically active compounds, the racemic modification is always obtained. By special means, however, it is possible to prepare optically active compounds from symmetrical compounds (*i.e.*, not optically active) without the necessity of resolution. The method involves the use of optically active compounds, and is known as *partial asymmetric synthesis*. The first partial asymmetric synthesis was carried out by Marckwald (1904), who prepared an active (−)-valeric acid (lævorotatory to the extent of about 10 per cent of the pure compound) by heating the half brucine salt of ethylmethylmalonic acid at 170°C.

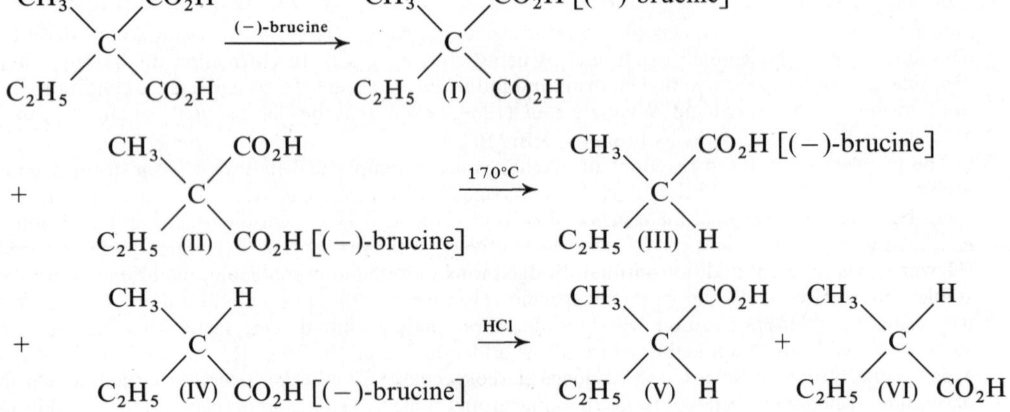

(I) and (II) are diastereoisomers; so are (III) and (IV). (V) and (VI) are enantiomers, and since the mixture is optically active, they must be present in unequal amounts. This was believed to be due to the different rates of decomposition of diastereoisomers (I) and (II).

This reaction was reinvestigated by Eisenlohr and Meier (1938), and they believed that the half-brucine salts were not present in *equal* amounts in the solid form (as thought by Marckwald). These authors suggested that as the *less* soluble salt crystallised out (during evaporation of the solution), some of the *more* soluble salt spontaneously changed into the *less* soluble salt to restore the equilibrium between the two; thus the final result was a mixture of the half-brucine salt containing a larger proportion of the *less soluble* diastereoisomer. If this be the case, then we are dealing with an example of asymmetric transformation (p. 483). Kenyon and Ross (1952) have also reinvestigated this reaction, and their work appears to show that the above reaction is a true asymmetric synthesis. According to these authors, when the half-brucine salt is heated, decarboxylation occurs with the intermediate formation of a *carbanion*. (I) and (II) both produce the *same* carbanion:

$$(I) \longrightarrow \quad \begin{array}{c} CH_3 \\ \diagdown \\ \bar{C}CO_2H[(-)\text{-brucine}] \longleftarrow (II) \\ \diagup \\ C_2H_5 \end{array}$$

Combination of this carbanion with a proton would produce diastereoisomers (III) and (IV) in different amounts, since, in general, diastereoisomers are formed at different rates (see also p. 482).

McKenzie (1905) reduced with aluminium amalgam pyruvic esters in which the alcohol was optically active, *e.g.*, (−)-amyl alcohol, (−)-menthol, etc. When the product, lactic acid ester, was hydrolysed, the resulting lactic acid was found to be slightly lævorotatory (see also p. 482).

$$CH_3COCO_2R \xrightarrow{Al/Hg} CH_3CHOHCO_2R \xrightarrow{H^+} CH_3CHOHCO_2H$$

Prelog *et al.* (1953) have studied, by conformational analysis, the steric course of the addition of Grignard reagents to benzoylformic esters of asymmetric alcohols, and have found that the configuration of the

$$C_6H_5COCO_2R^1 \xrightarrow[\text{(ii) hydrolysis}]{\text{(i) } R^2MgX} C_6H_5CHOHCO_2H$$

asymmetric carbon atom in the stereoisomer that predominated in this reaction could be correlated with the asymmetric carbon atom in R^1, *e.g.*, (−)-menthol and (−)-borneol are both configurationally related to (−)-glyceraldehyde, and both lead to a predominance of the (−)-hydroxyacid. This method has been used to correlate configurations.

The other type of asymmetric synthesis is *absolute asymmetric synthesis*. This is the preparation of an optically active compound *without* the intermediate use of optically active reagents. The first conclusive evidence for an absolute asymmetric synthesis was obtained by Kuhn and Knopf (1930), who irradiated (±)-α-azidopropionic dimethylamide, $CH_3CH(N_3)CON(CH_3)_2$, with right circularly polarised light and obtained a product that was slightly dextrorotatory. When the amide was irradiated with left circularly polarised light, the product was slightly lævorotatory (see also Vol. 2, Ch. 3).

Hydroxy-dibasic and polybasic acids

Tartronic acid (hydroxymalonic acid) may be prepared by heating bromomalonic acid with silver oxide suspended in water:

$$CHBr(CO_2H)_2 + \text{`AgOH'} \longrightarrow CHOH(CO_2H)_2 + AgBr$$

It is a crystalline solid which, heated to 160°C, melts with the evolution of carbon dioxide and the formation of polyglycollide, $(C_2H_2O_2)_n$.

Malic acid (*hydroxysuccinic acid, hydroxybutanedioic acid*), m.p. 100°C, occurs in sour apples (Latin: *malum*, apple), fruits, berries, etc. Malic acid is made synthetically by heating maleic acid with dilute sulphuric acid under pressure:

$$\begin{array}{c}CHCO_2H \\ \| \\ CHCO_2H\end{array} + H_2O \xrightarrow{H_2SO_4} \begin{array}{c}CHOHCO_2H \\ | \\ CH_2CO_2H\end{array} \quad (100\%)$$

It may be conveniently prepared in the laboratory by heating bromosuccinic acid with silver oxide suspended in water:

$$\begin{array}{c}CHBrCO_2H \\ | \\ CH_2CO_2H\end{array} + \text{'AgOH'} \longrightarrow \begin{array}{c}CHOHCO_2H \\ | \\ CH_2CO_2H\end{array} + AgBr$$

Malic acid contains one asymmetric carbon atom, and can therefore exist in the D-, L- and DL-forms. Malic acid from natural sources is L($-$); synthetic malic acid is DL; D($+$)-malic acid may be obtained by the careful reduction of D($+$)-tartaric acid with concentrated hydriodic acid:

$$\begin{array}{c}CO_2H \\ | \\ H-C-OH \\ | \\ HO-C-H \\ | \\ CO_2H\end{array} + 2HI \longrightarrow \begin{array}{c}CO_2H \\ | \\ H-C-OH \\ | \\ CH_2 \\ | \\ CO_2H\end{array} + I_2 + H_2O$$

Malic acid behaves both as an alcohol and acid. Inspection of the formula of malic acid shows it to be an α-hydroxyacid with respect to one carboxyl group, and a β-hydroxyacid with respect to the other. Such acids, when heated, undergo the reaction characteristic of the β-acid, *i.e.*, they eliminate a molecule of water to form an unsaturated acid (not the lactide). Thus, on heating, malic acid forms maleic anhydride and fumaric acid (*q.v.*):

$$\begin{array}{c}CHCO \\ \| \quad \diagdown \\ \quad \quad O \\ \| \quad \diagup \\ CHCO\end{array} \longleftarrow \begin{array}{c}CHOHCO_2H \\ | \\ CH_2CO_2H\end{array} \longrightarrow \begin{array}{c} H \diagdown \quad \diagup CO_2H \\ C \\ \| \\ C \\ \diagup \quad \diagdown \\ HO_2C \quad \quad H\end{array}$$

Tartaric acid (*α,α′ dihydroxysuccinic acid, 2,3-dihydroxybutanedioic acid*) contains two structurally identical carbon atoms, and can therefore exist in the D-, L-, DL- and *meso*-forms; all of these are known (Fig. 17.8).

The configurations of the tartaric acids are a troublesome problem. Fischer wrote the configuration of the natural dextrorotatory acid (*i.e.*, the ($+$)-acid) as shown. It is possible to synthesise ($-$)-tartaric acid from D($+$)-glyceraldehyde, and on this basis the ($+$)-acid would be L($+$)-tartaric acid. This is in agreement with Rosanoff's scheme of building *up* from D($+$)-glyceraldehyde. It is also possible, however, to degrade ($+$)-tartaric acid into D($-$)-glyceric acid, and hence ($+$)-tartaric acid will be D($+$)-tartaric acid. The designation L($+$)-tartaric acid is used by those chemists who regard this acid as a carbohydrate derivative. This difficulty may be overcome by specifying ($+$)-tartaric acid as (*RR*)-tartaric acid (p. 481).

When fumaric acid is treated with dilute alkaline permanganate, DL-tartaric acid is formed; maleic acid, under the same conditions, forms *meso*-tartaric acid.

$$HO_2CCH=CHCO_2H + [O] \xrightarrow{KMnO_4} HO_2CCHOHCHOHCO_2H$$

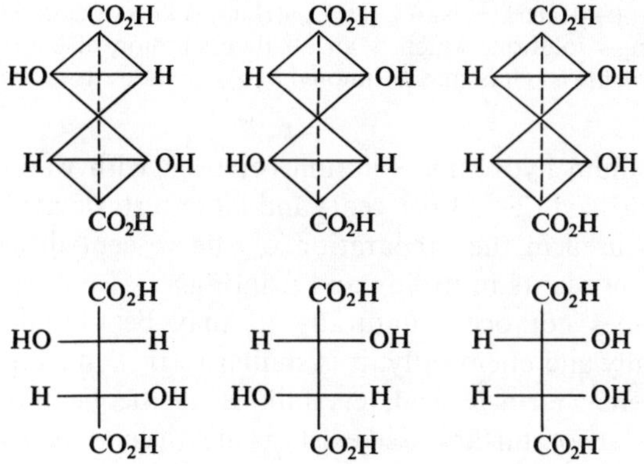

L(−)-tartaric
acid

D(+)-tartaric
acid

meso-tartaric
acid

Fig. 17.8

Both DL- and *meso*-tartaric acid are formed when α,α'-dibromosuccinic acid is boiled with silver oxide suspended in water:

$$\begin{array}{c}\mathrm{CHBrCO_2H}\\|\\\mathrm{CHBrCO_2H}\end{array} + 2\text{'AgOH'} \longrightarrow \begin{array}{c}\mathrm{CHOHCO_2H}\\|\\\mathrm{CHOHCO_2H}\end{array} + 2\mathrm{AgBr}$$

DL- and *meso*-Tartaric acids are also formed by the hydrolysis of glyoxal cyanohydrin:

$$\begin{array}{c}\mathrm{CHO}\\|\\\mathrm{CHO}\end{array} + 2\mathrm{HCN} \longrightarrow \begin{array}{c}\mathrm{CHOHCN}\\|\\\mathrm{CHOHCN}\end{array} \xrightarrow{\mathrm{H_2O}} \begin{array}{c}\mathrm{CHOHCO_2H}\\|\\\mathrm{CHOHCO_2H}\end{array}$$

All the foregoing synthetic preparations clearly show the structure of tartaric acid.

D(+)-**Tartaric acid**, occurs in the free state and as potassium hydrogen tartrate in the juice of grapes. During the fermentation of grapes, the acid potassium salt separates as a reddish-brown crystalline mass which is known as *argol*. When recrystallised, argol is converted into the purer substance (white), which is known as *cream of tartar*, from which D(+)-tartaric acid is obtained by dissolving it in water, adding calcium hydroxide, then calcium chloride to the filtrate, and decomposing the collected batches of precipitated calcium tartrate with the calculated quantity of dilute sulphuric acid:

$$2\mathrm{KHC_4H_4O_6} \xrightarrow{\mathrm{Ca(OH)_2}} \mathrm{CaC_4H_4O_6} + 2\mathrm{H_2O} + \mathrm{K_2C_4H_4O_6} \xrightarrow{\mathrm{CaCl_2}}$$

$$2\mathrm{KCl} + \mathrm{CaC_4H_4O_6} \xrightarrow{\mathrm{H_2SO_4}} \mathrm{H_2C_4H_4O_6} + \mathrm{CaSO_4}$$

Anhydrous D(+)-tartaric acid, m.p. 170°C, is obtained from the filtrate by evaporation to crystallisation.

When heated, tartaric acid is converted into pyruvic acid, and is

$$\begin{array}{c}\mathrm{CHOHCO_2H}\\|\\\mathrm{CHOHCO_2H}\end{array} \longrightarrow \mathrm{CH_3COCO_2H} + \mathrm{CO_2} + \mathrm{H_2O}$$

reduced by hydriodic acid, first to malic acid and then to succinic acid.

Sodium potassium D(+)-tartrate, $NaKC_4H_4O_6 \cdot 4H_2O$, is known as *Rochelle salt*, and is used in the preparation of Fehling's solution, which is an alkaline solution of cupric ion complexed with two tartrate ions. *Tartar emetic* is potassium antimonyl D(+)-tartrate; its structure is uncertain.

When heated with sodium hydroxide solution (or even with water), D(+)-tartaric acid is converted into DL-tartaric acid (29–33 per cent) and *meso*-tartaric acid (13–17 per cent).

D(+)-Tartaric acid is used in the preparation of effervescent drinks. The acid and tartar emetic are both used as mordants in dyeing and printing.

L(−)-**Tartaric acid**, does not occur naturally. It may be obtained by the resolution of DL-tartaric acid. Physically and chemically, it is similar to D(+)-tartaric acid.

DL-**Tartaric acid, racemic tartaric acid**, crystallises as the hemihydrate $(C_4H_6O_6)_2 \cdot H_2O$, m.p. 206°C. In the solid state DL-tartaric acid exists as the *racemic compound*, but dissociates into the D- and L-forms in solution. It is manufactured by the oxidation of fumaric acid with dilute alkaline permanganate.

meso-**Tartaric acid** (*i-tartaric acid*) crystallises as the monohydrate; the melting point of the anhydrous acid is 140°C. *meso*-Tartaric acid is optically inactive by *internal* compensation, and may be obtained (together with DL-) by heating D(+)-tartaric acid with alkali. It may also be prepared by the oxidation of maleic acid with dilute alkaline permanganate.

Citric acid (*β-hydroxytricarballylic acid*, 2-*hydroxypropane*-1,2,3-*tricarboxylic acid*) occurs in many fruits, especially unripe fruits of the *citrus* group, *e.g.*, lemon juice contains about 6–10 per cent citric acid. It is manufactured by the fermentation of solutions of glucose, sucrose or purified cane-molasses in the presence of certain inorganic salts, by various moulds or fungi, *e.g.*, *Citromyces pfefferianus*, *Aspergillus wentii*.

Citric acid may be synthesised from glycerol by the following reactions; this synthesis shows its structure:

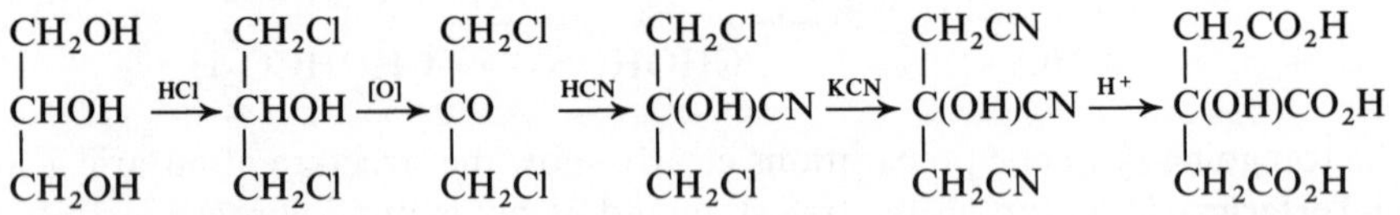

$$
\begin{array}{ccccccc}
CH_2OH & CH_2Cl & CH_2Cl & CH_2Cl & CH_2CN & CH_2CO_2H \\
| & | & | & | & | & | \\
CHOH \xrightarrow{HCl} & CHOH \xrightarrow{[O]} & CO \xrightarrow{HCN} & C(OH)CN \xrightarrow{KCN} & C(OH)CN \xrightarrow{H^+} & C(OH)CO_2H \\
| & | & | & | & | & | \\
CH_2OH & CH_2Cl & CH_2Cl & CH_2Cl & CH_2CN & CH_2CO_2H
\end{array}
$$

Lawrence (1897) synthesised citric acid by means of the Reformatsky reaction (p. 437), starting with ethyl bromoacetate and oxalacetic ester:

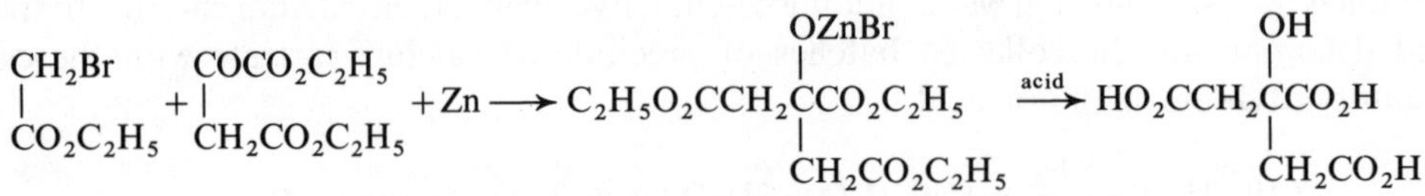

$$
\begin{array}{cc}
CH_2Br & COCO_2C_2H_5 \\
| & + | \\
CO_2C_2H_5 & CH_2CO_2C_2H_5
\end{array}
+ Zn \longrightarrow
\begin{array}{c}
OZnBr \\
| \\
C_2H_5O_2CCH_2CCO_2C_2H_5 \\
| \\
CH_2CO_2C_2H_5
\end{array}
\xrightarrow{acid}
\begin{array}{c}
OH \\
| \\
HO_2CCH_2CCO_2H \\
| \\
CH_2CO_2H
\end{array}
$$

The monohydrate of citric acid, obtained by crystallisation from water, loses its water of crystallisation when heated at 130°C, and melts at 153°C. Citric acid behaves as an alcohol and a tribasic acid, *e.g.*, it forms the acetyl derivative and three series of salts.

Citric acid is used for making beverages and as a mordant in dyeing.

Citric acid is both an α- and β-hydroxy-acid; when heated to 150°C, it eliminates a molecule of water to form *aconitic* acid. On pyrolysis, citric acid gives a number of products among which have been isolated *aconitic acid, citraconic* (*mesaconic*) and *itaconic anhydrides*, and acetone (I–VII).

(I) is acetonedicarboxylic acid; (II) acetoacetic acid; (III) aconitic acid; (IV) citraconic and mesaconic acid (*cis-trans* isomers); (V) itaconic acid; (VI) citraconic (mesaconic) anhydride; (VII) itaconic anhydride.

$$
\begin{array}{c}
\text{CHCO}_2\text{H} \\
\parallel \\
\text{CCO}_2\text{H} \\
\text{(III) CH}_2\text{CO}_2\text{H}
\end{array}
\xleftarrow{-\text{H}_2\text{O}}
\begin{array}{c}
\text{CH}_2\text{CO}_2\text{H} \\
| \\
\text{C(OH)CO}_2\text{H} \\
| \\
\text{CH}_2\text{CO}_2\text{H}
\end{array}
\xrightarrow{-\text{HCO}_2\text{H}}
\begin{array}{c}
\text{CH}_2\text{CO}_2\text{H} \\
| \\
\text{CO} \\
\text{(I) CH}_2\text{CO}_2\text{H}
\end{array}
\xrightarrow{-\text{CO}_2}
\begin{array}{c}
\text{CH}_3 \\
| \\
\text{CO} \\
\text{(II) CH}_2\text{CO}_2\text{H}
\end{array}
$$

(III) $\xrightarrow{-\text{CO}_2}$

(II) $\xrightarrow{-\text{CO}_2}$

$$
\begin{array}{c}
\text{CHCO} \\
\parallel \quad\diagdown \\
\qquad\quad \text{O} \\
\text{CCO} \quad\diagup \\
| \\
\text{CH}_3 \text{ (VI)}
\end{array}
\xleftarrow{-\text{H}_2\text{O}}
\begin{array}{c}
\text{CHCO}_2\text{H} \\
\parallel \\
\text{CCO}_2\text{H} \\
| \\
\text{CH}_3 \text{ (IV)}
\end{array}
\qquad
\begin{array}{c}
\text{CH}_2 \\
\parallel \\
\text{CCO}_2\text{H} \\
| \\
\text{CH}_2\text{CO}_2\text{H (V)}
\end{array}
\xrightarrow{-\text{H}_2\text{O}}
\begin{array}{c}
\text{CH}_2 \\
\parallel \\
\text{CCO} \diagdown \\
\qquad\quad \text{O} \\
\text{CH}_2\text{CO} \diagup \text{(VII)}
\end{array}
\qquad
\begin{array}{c}
\text{CH}_3 \\
| \\
\text{CO} \\
| \\
\text{CH}_3
\end{array}
$$

When citric acid is heated with concentrated sulphuric acid, aconitic acid is obtained (41–44 per cent). When treated with *fuming* sulphuric acid, citric acid forms acetonedicarboxylic acid (85–90 per cent), a reaction which is characteristic of α-hydroxyacids.

Tricarballylic acid (*propane*-1,2,3 *tricarboxylic acid*), m.p. 166°C, occurs in unripe beet-roots. It may be prepared by the reduction of aconitic acid, or from 1,2,3-tribromopropane as follows:

$$
\begin{array}{c}
\text{CH}_2\text{Br} \\
| \\
\text{CHBr} \\
| \\
\text{CH}_2\text{Br}
\end{array}
\xrightarrow{\text{KCN}}
\begin{array}{c}
\text{CH}_2\text{CN} \\
| \\
\text{CHCN} \\
| \\
\text{CH}_2\text{CN}
\end{array}
\xrightarrow[\text{(ii) HCl}]{\text{(i) aq. KOH}}
\begin{array}{c}
\text{CH}_2\text{CO}_2\text{H} \\
| \\
\text{CHCO}_2\text{H} \\
| \\
\text{CH}_2\text{CO}_2\text{H}
\end{array}
\quad (70\%)
$$

Tricarballylic acid may also be synthesised by the Michael condensation (p. 340) between diethyl fumarate and malonic ester, and heating the product, ethyl propane-1,1,2,3-tetracarboxylate, with concentrated hydrochloric acid:

$$
\text{C}_2\text{H}_5\text{O}_2\text{CCH}{=}\text{CHCO}_2\text{C}_2\text{H}_5 + \text{CH}_2(\text{CO}_2\text{C}_2\text{H}_5)_2 \xrightarrow{\text{C}_2\text{H}_5\text{ONa}}
$$

$$
\begin{array}{c}
\text{C}_2\text{H}_5\text{O}_2\text{CCHCH}_2\text{CO}_2\text{C}_2\text{H}_5 \\
| \\
\text{CH}(\text{CO}_2\text{C}_2\text{H}_5)_2
\end{array}
\xrightarrow{\text{HCl}}
\begin{array}{c}
\text{CH}_2\text{CO}_2\text{H} \\
| \\
\text{CHCO}_2\text{H} \\
| \\
\text{CH}_2\text{CO}_2\text{H}
\end{array}
\quad (88\text{–}90\%)
$$

Tricarballylic esters have been used as plasticisers.

Unsaturated dicarboxylic acids

The formula of the simplest unsaturated dicarboxylic acid is $\text{HO}_2\text{CCH}{=}\text{CHCO}_2\text{H}$. This formula actually represents two isomers: maleic acid and fumaric acid.

Maleic acid may be prepared:

(i) By heating malic acid at about 250°C:

$$
\begin{array}{c}
\text{CHOHCO}_2\text{H} \\
| \\
\text{CH}_2\text{CO}_2\text{H}
\end{array}
\xrightarrow{-\text{H}_2\text{O}}
\begin{array}{c}
\text{CHCO}_2\text{H} \\
\parallel \\
\text{CHCO}_2\text{H}
\end{array}
\xrightarrow{-\text{H}_2\text{O}}
\begin{array}{c}
\text{CHCO} \diagdown \\
\parallel \qquad\quad \text{O} \\
\text{CHCO} \diagup
\end{array}
\xrightarrow[\text{(ii) HCl}]{\text{(i) NaOH}}
\begin{array}{c}
\text{CHCO}_2\text{H} \\
\parallel \\
\text{CHCO}_2\text{H}
\end{array}
$$

(ii) By heating bromosuccinic acid with aqueous alkali; some fumaric acid is also obtained:

$$
\begin{array}{c}
\text{CHBrCO}_2\text{H} \\
| \\
\text{CH}_2\text{CO}_2\text{H}
\end{array}
+ \text{KOH} \longrightarrow
\begin{array}{c}
\text{CHCO}_2\text{H} \\
\parallel \\
\text{CHCO}_2\text{H}
\end{array}
+ \text{KBr} + \text{H}_2\text{O}
$$

Maleic anhydride is prepared industrially:

(a) By the oxidation of benzene with air in the presence of vanadium pentoxide as catalyst at 410–430°C:

$$2 \bigcirc + 9O_2 \xrightarrow{V_9O_5} 2 \begin{matrix} CHCO \\ \| \quad \quad O \\ CHCO \end{matrix} + 4CO_2 + 4H_2O$$

(b) As a by-product in the manufacture of phthalic anhydride from naphthalene (p. 776):

(c) By the oxidation of but-2-ene (from cracked petroleum) or crotonaldehyde with air in the presence of vanadium pentoxide as catalyst at 450°C, e.g.,

$$CH_3CH=CHCH_3 + 3O_2 \xrightarrow{V_2O_5} \begin{matrix} CHCO \\ \| \quad \quad O \\ CHCO \end{matrix} + 3H_2O$$

Maleic acid is a synthetic compound. It is a crystalline solid, m.p. 130°C, soluble in water (79 g per 100 ml at 25°C). When heated, some distils unchanged; the rest is converted into maleic anhydride. A much better yield of maleic anhydride is obtained by heating the acid with acetic anhydride. Maleic acid may be reduced catalytically or electrolytically to succinic acid. It is oxidised by dilute alkaline permanganate to *meso*-tartaric acid; this may be obtained in excellent yield by replacing the permanganate by potassium chlorate and osmium tetroxide. Prolonged heating of maleic acid at 150°C converts it into fumaric acid. Maleic acid and its anhydride are used in the Diels–Alder synthesis (p. 536).

Maleic acid inhibits rancidity in milk powders, oils and fats. Maleic anhydride is used for making varnishes and lacquers.

Fumaric acid may be prepared:

(i) By heating maleic acid for some time at 150°C.

(ii) By heating bromosuccinic acid with alkali; maleic acid is also formed.

(iii) By the Knoevenagel reaction (p. 342); malonic acid is condensed with glyoxylic acid in the presence of pyridine:

$$HO_2CCHO + CH_2(CO_2H)_2 \xrightarrow{pyridine} HO_2CCH=CHCO_2H + H_2O + CO_2$$

(iv) By the oxidation of furfural with sodium chlorate:

$$\begin{matrix} CH——CH \\ \| \quad\quad \| \\ CH \quad\quad CCHO \\ \diagdown \quad \diagup \\ O \end{matrix} + 4[O] \xrightarrow{NaClO_3} \begin{matrix} HO_2CCH \\ \| \\ CHCO_2H \end{matrix} + H_2O + CO_2 \quad (50-88\%)$$

Fumaric acid is prepared industrially by boiling maleic acid with hydrochloric acid or sodium hydroxide. Another industrial preparation is the fermentation of glucose (and other carbohydrates) by e.g., *Rhizopus nigricans*.

Fumaric acid occurs in nature in many plants. It is a crystalline solid, m.p. 287°C, slightly soluble in water (0·7 g per 100 ml at 25°C). It does not form an anhydride of its own, but gives maleic anhydride when heated at 230°C. It may be reduced to succinic acid, and is oxidised by alkaline permanganate to DL-tartaric acid; the latter is obtained in excellent yield if potassium chlorate and osmium tetroxide are used as the oxidising agent (cf. maleic acid). Fumaroyl chloride is formed when maleic anhydride is heated with phthaloyl chloride in the presence of zinc chloride.

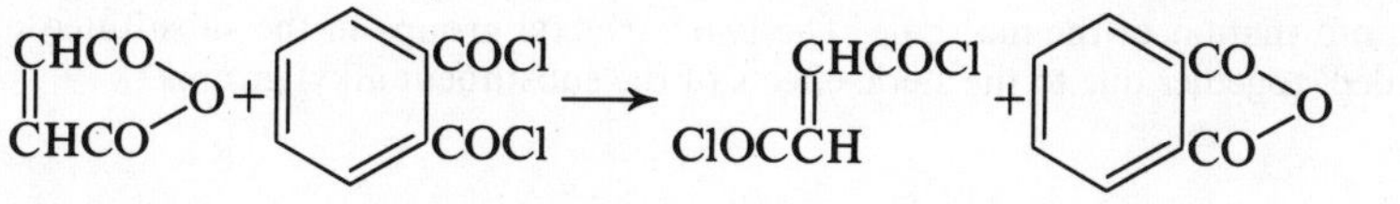

$$(82\text{--}95\%)$$

Many acid chlorides can be obtained in 95 per cent yield by this method.

	pK_a^1	pK_a^2	K_a^1/K_a^2
Maleic acid	1·92	6·23	20,420
Fumaric acid	3·02	4·38	22·9

At first sight, since the structures are identical, it might have been expected that the ionisation constants would be the same. As can be seen, however, maleic acid is stronger for the first ionisation but weaker for the second. The reason for this is believed to be due to the fact that hydrogen bonding can occur in the maleate anion, whereas this is not possible for the corresponding fumarate anion. Thus the former anion (cB) is stabilised with respect to its acid, and so there is a driving force for ionisation which is absent in the case of fumaric acid. Furthermore, since the maleate anion is stabilised by hydrogen bonding, the fumarate anion ionises to the dianion more readily than the former.

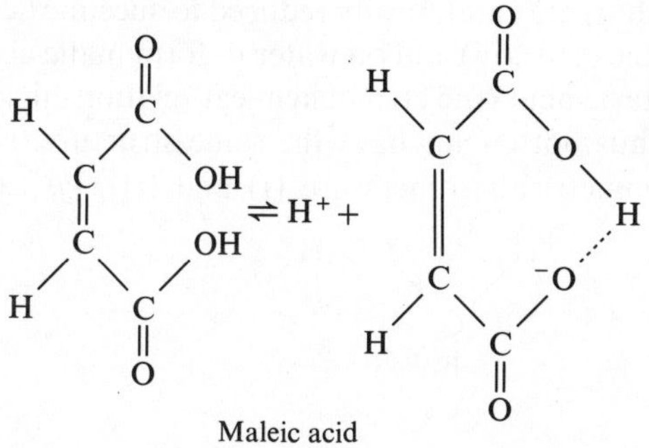

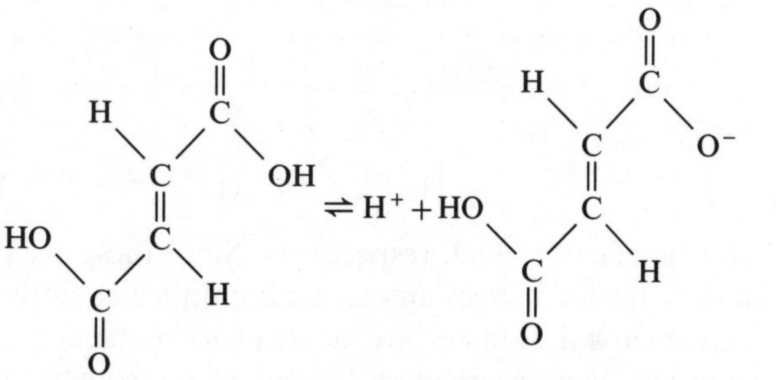

Not only can configuration affect acid strength but so also can conformation. Table 17.1 gives the pK_a values of some substituted malonic acids (in water). These differences cannot be explained by the polar effects of the alkyl groups alone. As we have seen (p. 446), there are the inductive and field effects operating in the dicarboxylic acids. The field effect, since it is a 'distance' effect, will depend very much

Table 17.1

Acid	pK_a^1	pK_a^2	K_a^1/K_a^2
$CH_2(CO_2H)_2$	2·86	5·7	692
$MeCH(CO_2H)_2$	3·05	5·76	510
$EtCH(CO_2H)_2$	2·99	5·83	702
$Me_2C(CO_2H)_2$	3·17	6·06	783
$MeEtC(CO_2H)_2$	2·86	6·41	3,580
$Et_2C(CO_2H)_2$	2·21	7·29	121,000

on the conformation of the molecule. The two carboxyl groups in the substituted malonic acids will be 'crowded' together due to the steric effects of the substituent alkyl groups.

$$
\begin{array}{ccc}
& CH_2 & \\
& / \quad \backslash & \\
HO_2C & & CO_2H
\end{array}
\qquad\qquad
\begin{array}{ccc}
R & & R \\
\backslash & & / \\
& C & \\
/ & & \backslash \\
HO_2C & & CO_2H
\end{array}
$$

The closer together the two carboxyl groups are, then when one is converted into the carboxylate anion, the more difficult it will be for the other carboxyl group to lose a proton (because of the increased field effect). Thus the K_a^{1}/K_a^{2} ratio will increase, the more so the greater the crowding.

Geometrical isomerism

The basic principles of geometrical isomerism have already been discussed (p. 97). The classical example of this phenomenon is the case of maleic and fumaric acids. Both have the same molecular formula $C_4H_4O_4$, but differ in most of their physical and in many of their chemical properties, and neither is optically active. It was originally thought that they were structural isomers, and because of this, different names were assigned to each form (this applies to many other geometrical isomers—see text). It was subsequently shown however, that maleic and fumaric acids were not structural isomers, e.g., both (i) are catalytically reduced to succinic acid; (ii) add on hydrogen bromide to form bromosuccinic acid; (iii) add on water to form malic acid; (iv) are oxidised by alkaline permanganate to tartaric acid (the stereochemical relationships in reactions (ii), (iii) and (iv) have been ignored). Thus both acids have the same structure, viz., $HO_2CCH{=}CHCO_2H$, and the two possible geometrical isomers are (I) and (II), cis- and

$$
\begin{array}{ccc}
H & & CO_2H \\
\backslash & & / \\
& C & \\
& \| & \\
& C & \\
/ & & \backslash \\
H & & CO_2H \\
& (I) &
\end{array}
\qquad\qquad
\begin{array}{ccc}
H & & CO_2H \\
\backslash & & / \\
& C & \\
& \| & \\
& C & \\
/ & & \backslash \\
HO_2C & & H \\
& (II) &
\end{array}
$$

trans-butenedioic acid, respectively. Since these are planar molecules, they are superimposable on their mirror images and so are not optically active.

Geometrical isomers are now also classified as diastereoisomers. According to the new definition, diastereoisomers are any stereoisomers which are not enantiomers of each other, i.e., the restriction that diastereoisomers are optical isomers has been dropped (see also p. 477). Thus, this type of isomerism (geometrical or cis-trans) is a sub-class of the general phenomenon of diastereoisomerism.

The problem now is to decide which formula represents maleic acid, and which fumaric acid. There is no general method for determining the configuration of geometrical isomers; the method used depends on the nature of the compound in question. The methods of cyclisation, dipole-moment measurements and spectroscopy may be used to determine the configurations of maleic and fumaric acids.

Method of cyclisation. Wislicenus was the first to suggest the principle that *intramolecular* reactions are more likely to occur the closer together the reacting groups are in the molecule. This principle is generally true, but has led to incorrect results (as shown by other work) in certain cases, e.g., the aromatic oximes (p. 755).

Of the two acids, only maleic acid forms the anhydride when heated. Fumaric acid does not form an anhydride of its own, but when strongly heated, gives maleic anhydride. If we accept the principle of cyclisation, then (I) is maleic acid, and (II) fumaric acid.

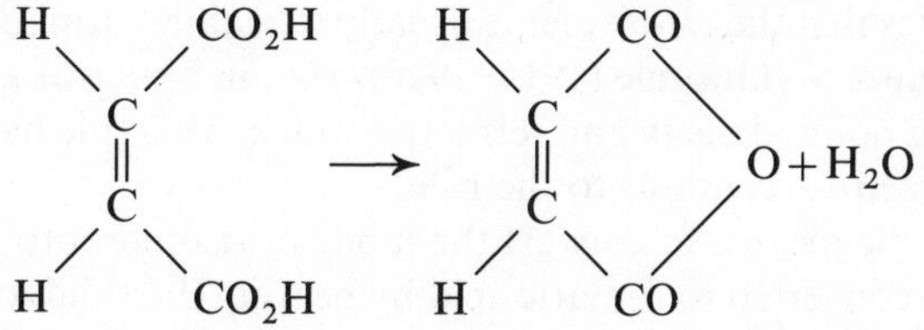

In fumaric acid, the two carboxyl groups are too far apart to react with each other.

Method of dipole-moment measurements. The use of dipole moments to assign configurations to geometrical isomers must be used with caution. Since dipole moments are vector quantities the sum of two equal and opposite group moments will be zero only if the two vectors are collinear or parallel. When the group moment is directed along the axis of the bond formed by the 'key' atom of the group and the carbon atom to which it is joined, then that group is said to have a *linear* moment. Such groups are H, halogen, Me, CN, NO_2, etc. On the other hand, groups which have *non-linear* moments are OH, OR, CO_2H, NH_2, etc. Thus for *cis-* and *trans*-olefinic compounds, the dipole moments of the latter will be zero only if the groups attached have linear moments, *e.g.*, $CH_2=CH_2$, $CHCl=CHCl$, $CHMe=CHMe$. When the groups have non-linear moments, their vector sum is *not* zero, and the difference between the dipole moments of the *cis-* and *trans*-forms may be too small to assign configuration with any confidence, *e.g.*, the dipole moment of diethyl maleate is 2·54D and that of diethyl fumarate is 2·38D.

Spectroscopic methods. Geometrical isomers may often be differentiated by examination of their infra-red and NMR spectra (p. 122), and by their ultra-violet spectra (p. 875).

Geometrical isomerism is also possible in *cyclic* compounds, the ring structure being comparable to the double bond in olefinic compounds (in giving rise to a more or less rigid structure), *e.g.*, hexahydroterephthalic acids:

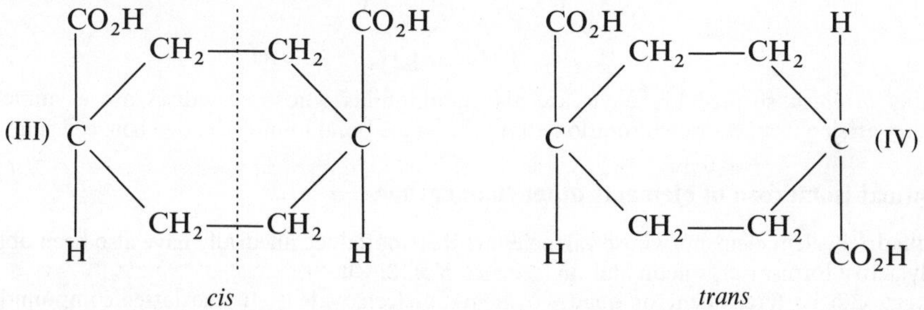

Both hexahydroterephthalic acids are optically inactive; (III) has a plane, and (IV) a centre of symmetry.

Compounds with a triple bond cannot exhibit geometrical isomerism, *e.g.*, acetylenedicarboxylic acid, $HO_2CC\equiv CCO_2H$. Using either sp^3 (tetrahedral) hybridisation (Fig. 17.9) or sp (digonal) hybridisation (*cf.* Fig. 4.5), the molecule is linear and consequently cannot exhibit geometrical (diastereoisomerism) (nor optical) isomerism.

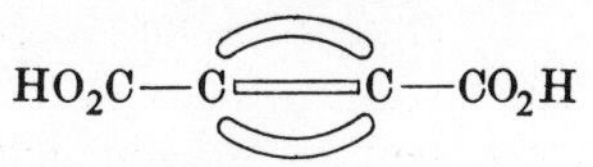

Fig. 17.9

Properties of *cis-trans* isomers

Comparison of the properties of *cis-* and *trans*-isomers of known configurations shows certain regularities, *e.g.*, the melting point and stability of the *cis-* are *lower* than those of the *trans*-isomer; the density, refractive index, solubility, dipole moment, heat of combustion and the dissociation constant (if an acid) of the *cis-* are *greater* than those of the *trans*-isomer. It can be seen from these properties that the *cis*-isomer is usually the labile form. Based on certain of these generalisations is the **Auwers–Skita rule** (1915, 1920), *viz.*, in a pair of geometrical isomers, the *cis* has the higher boiling point, density and refractive index. This rule has been used to elucidate configurations, but there are exceptions to the rule.

It is possible, by suitable means, to convert the labile *cis*-isomer into the stable *trans*-isomer, *e.g.*, maleic acid may be converted to fumaric acid by heating the solid, or a solution of the solid in water or benzene, to a temperature above its melting point (130°C). It is more difficult to convert the *trans*-isomer into the *cis*-. Usually the best method is to irradiate the *trans*-isomer with ultra-violet light in the presence of a trace of bromine; the product is generally an equilibrium mixture of both isomers (see also p. 906).

The stereochemistry of carbon compounds not containing an asymmetric carbon atom

As pointed out previously (p. 478), the presence of an asymmetric carbon atom is not essential for optical activity; the essential requirement is *the asymmetry of the molecule as a whole*. Allenes are compounds whose *structures* are asymmetric, and so should be resolvable. Several compounds of this type have been obtained in optically active forms, *e.g.*, where $R^1 = C_6H_5$ (phenyl) and $R^2 = 1\text{-}C_{10}H_7$ (1-naphthyl) [Mills and Maitland, 1936].

If both double bonds of allene are replaced by rings, *spirans* are obtained in which the rings are at right angles to each other. Hence by suitable substitution, it should be possible to obtain optically active *spiro*-compounds, *e.g.*, Backer (1928) resolved the following *spiro*heptane derivative.

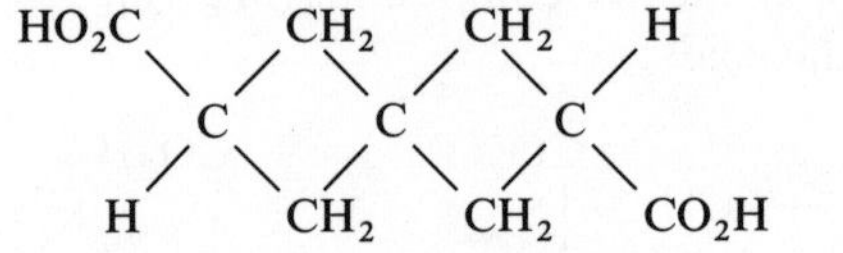

Suitably *ortho*-substituted biphenyls are also compounds whose structures are asymmetric, the asymmetry arising from restricted rotation about the single bond joining the two benzene rings (p. 785).

The optical isomerism of elements other than carbon

Many quadricovalent elements whose valencies are distributed tetrahedrally have also been obtained in optically active forms, *e.g.*, silicon and tin (see also Vol. 2, Ch. 6).

Nitrogen can be tercovalent or quadricovalent unielectrovalent. In the latter compounds, if the charge on the nitrogen atom is ignored, the molecule then closely resembles carbon compounds. Thus, for example, the following compounds have been obtained in optically active forms:

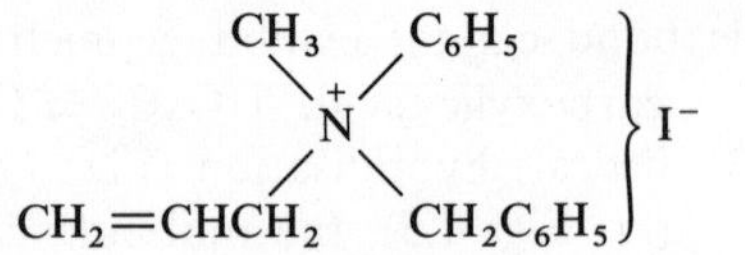

allylbenzylmethylphenylammonium iodide
(Pope and Peachey, 1899)

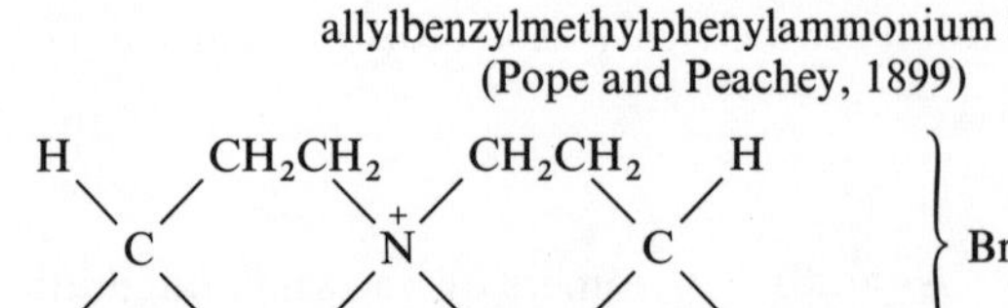

4-carbethoxy-4'-phenylbispiperidinium-1,1'-spiran   ethylmethylphenylamine
bromide (Mills and Warren, 1925)   oxide (Meisenheimer, 1908)

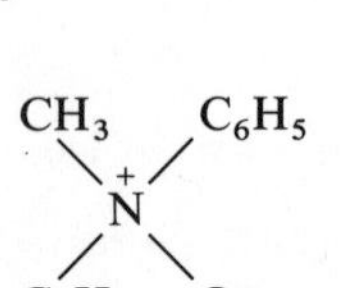

Racemisation of compounds of the type $\overset{+}{\text{N}}abde\,\text{X}^-$ is effected far more readily than with the carbon compounds, $Cabde$. The mechanism is believed to be due to the ready dissociation:

$$\overset{+}{\text{N}}abde\} \text{ X}^- \rightleftharpoons \text{N}abd + e\text{X}$$

The amine, $\text{N}abd$ readily racemises (see below), and so the quaternary compound will racemise.

Tercovalent nitrogen offers a very interesting problem from the point of view of stereoisomerism. No tertiary amine, $\text{N}abd$, has yet been resolved. It was therefore suggested that such molecules were planar, but physico-chemical evidence, $e.g.$, dipole moment measurements, absorption spectra, etc., shows that the configuration of ammonia and amines is tetrahedral, the nitrogen atom being at one corner of the tetrahedron with a valency angle of about 109°. This failure to resolve tertiary amines has been explained by rapid oscillation of the nitrogen atom through the plane of the three groups a, b, and d (Fig. 17.10). This oscillation is regarded as an 'umbrella' switch of bonds, $i.e.$, the bond lengths remain

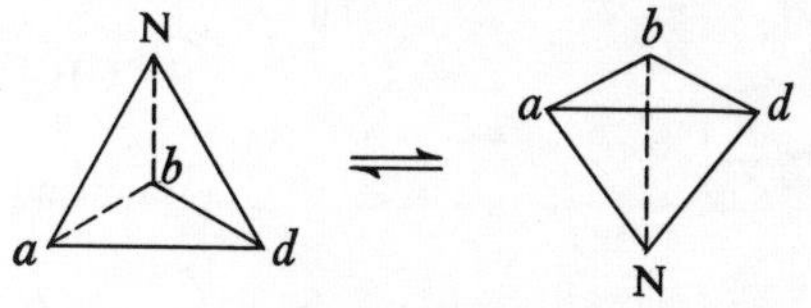

Fig. 17.10

unaltered and only the nitrogen valency angles change. Theoretical considerations have shown that if the nitrogen atom were 'anchored' by forming part of a ring, then the inversion (due to oscillation) would be inhibited. This has been confirmed by the resolution of Tröger's base.

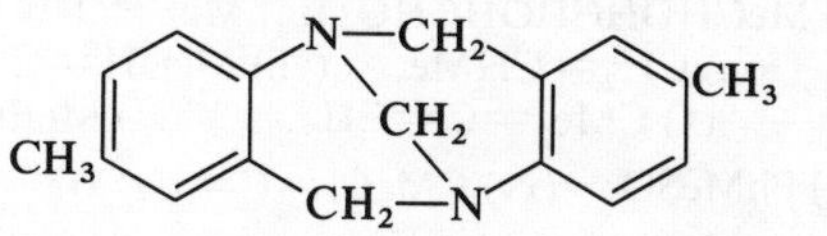

Oximes are also compounds of tercovalent nitrogen. They exhibit geometrical isomerism, and some have also been obtained in optically active forms, $e.g.$, the oxime of cyclohexanone-4-carboxylic acid (Mills and Bain, 1910).

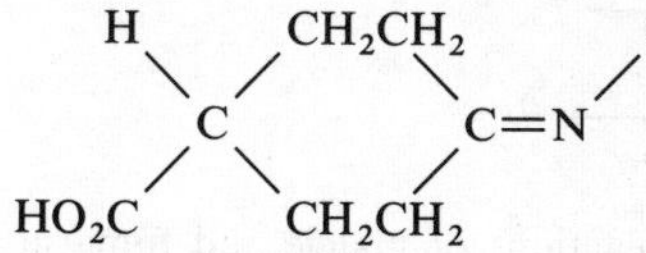

Several tertiary phosphines have been resolved, but many quadricovalent phosphorus compounds have been obtained in optically active forms, $e.g.$,

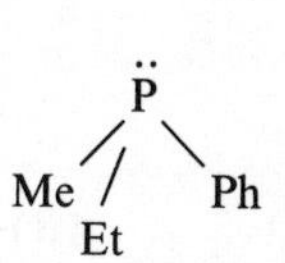

(Horner, 1961) 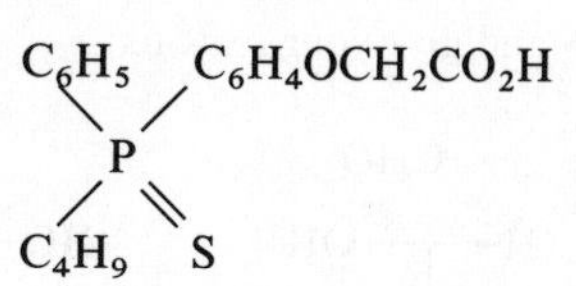   (Davies and Mann, 1944) 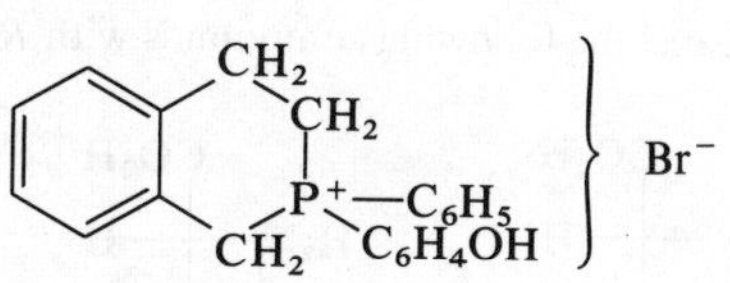   (Holliman and Mann, 1947)

Tercovalent and quadricovalent arsenic compounds have been resolved, $e.g.$,

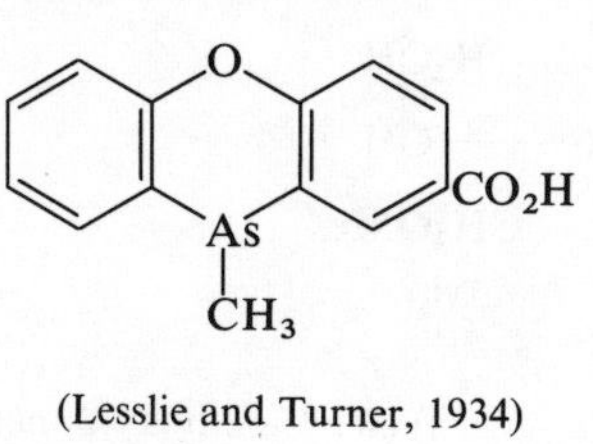

(Lesslie and Turner, 1934) 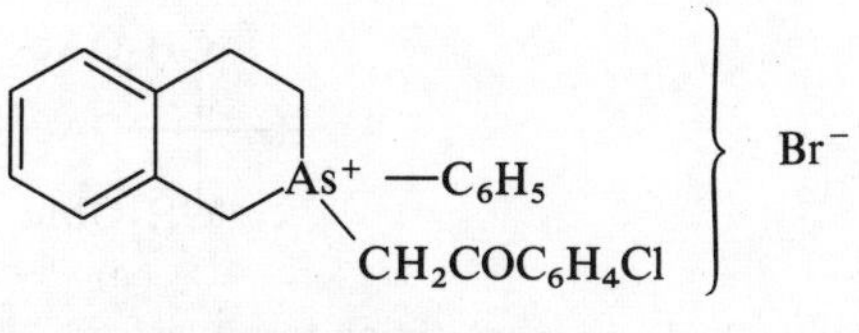   (Holliman and Mann, 1943)

Various sulphur compounds have also been resolved, *e.g.*,

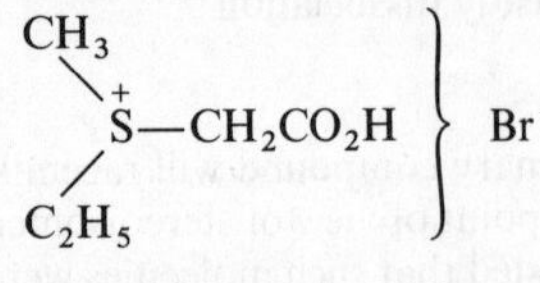

$$
\left.
\begin{array}{c}
CH_3 \\
\overset{+}{S}{-}CH_2CO_2H \\
C_2H_5
\end{array}
\right\} \ Br^-
$$

(Pope and Peachey, 1900)

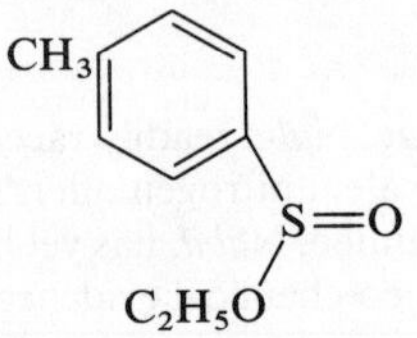

(Phillips, 1925)

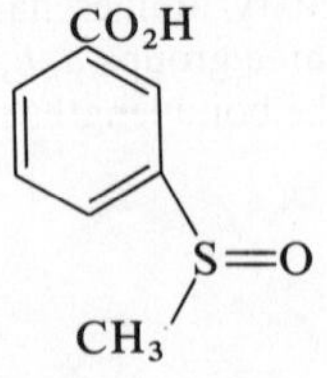

(Phillips, 1926)

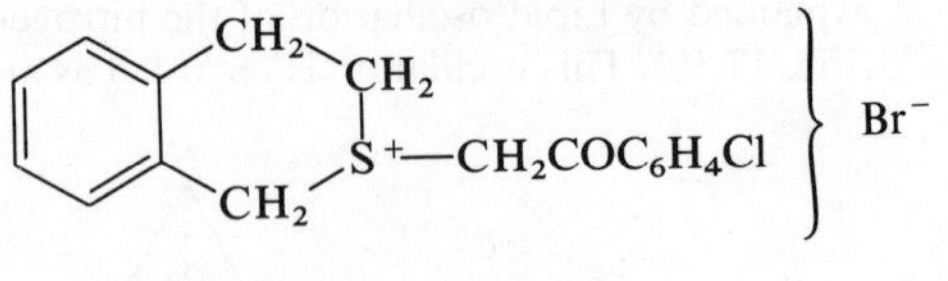

$$
\left.
\begin{array}{c}
CH_2\text{---}CH_2 \\
\quad\quad\quad S^+{-}CH_2COC_6H_4Cl \\
CH_2
\end{array}
\right\} \ Br^-
$$

(Mann and Holliman, 1946)

PROBLEMS

1. Draw all possible stereoisomers for the following and state which are optically active, inactive, and which are *meso* forms:

(i) MeCHOHCHOHMe; (ii) MeCHOHCHOHCHO; (iii) MeCHOHCOCHOHMe; (iv) MeCHClCO$_2$H;
(v) HO$_2$CCHBrCH$_2$CO$_2$Et;   (vi) CH$_2$=CHMe;   (vii) MeCH=CHEt;   (viii) MeCHBrCH=CH$_2$;
(ix) MeCHOHCH=CHMe;   (x) CMe$_2$=C=CH$_2$;   (xi) MeEtPrN;   (xii) MeEtPrPr^{i}N$^+$Br$^-$;
(xiii) PhMeEtN$^+$—Ō;   (xiv) PhMeSO;   (xv) CMeCl=C=CHMe.

2. Complete the following equations and state the relations between the configurations of the reactant and product. Give your reasons.

(i)   D-MeCHOHEt+Ac$_2$O ⟶ ?
(ii) L-MeCH(OAc)Et+NaOH ⟶ ?
(iii) L-MeCHOHEt+HBr ⟶ ?
(iv) D-MeCHClCO$_2$H+2NaOH ⟶ ?
(v)   L-MeEtPrCOMe+HI ⟶ ?

3. Complete the equations of the reactions of maleic and fumaric acids with the following reagents and comment:

(i) bromine; (ii) dilute alkaline permanganate solution; (iii) perbenzoic acid; (iv) catalytic hydrogenation.

4. Label the following compounds with *R* or *S* and give your reasons:

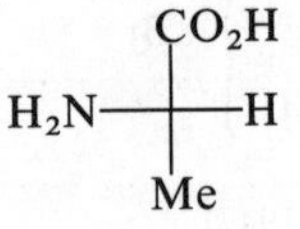

(i) 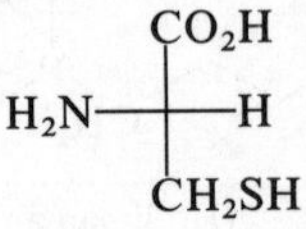   (ii) 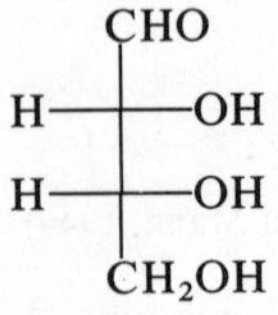   (iii) 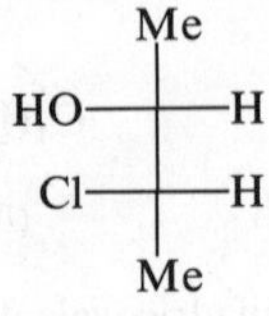   (iv) 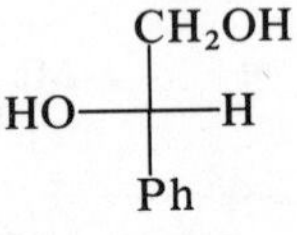   (v)

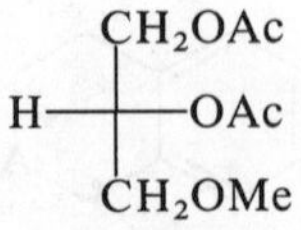

(vi) 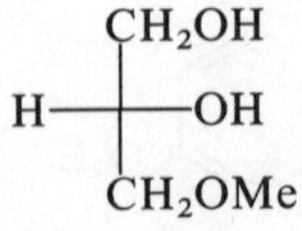   (vii)

5. Draw the Fischer projection formulæ for:

(i) (*R*)-3-methylpentan-1-ol;   (ii) (*S*)-2,3-dimethylhexane;   (iii) (*R*)-3-methoxy-2-methylpropan-1-ol;
(iv) (2*S*, 3*R*)-2,3-dichloropentane.

6. Write out the structures and I.U.P.A.C. names of the isomeric hydroxyacids with the molecular formula $C_4H_8O_3$. How would each acid behave towards:

(i) heat; (ii) dilute sulphuric acid on warming?

7. Complete the equations, and assuming all but one of the steps are S_N2, label each product with D or L. Comment.

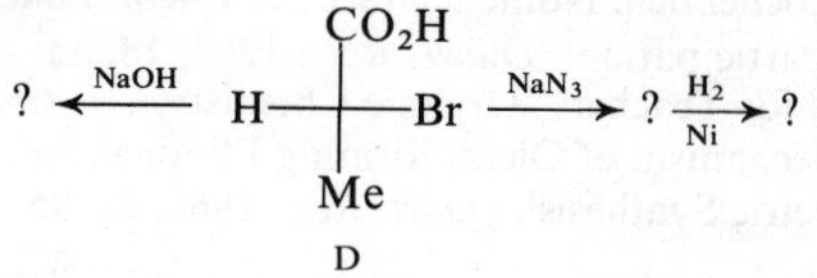

$$? \xleftarrow{\text{NaOH}} \underset{D}{\text{H}\!-\!\!\!\overset{\displaystyle CO_2H}{\underset{\displaystyle Me}{|}}\!\!\!-\!Br} \xrightarrow{\text{NaN}_3} ? \xrightarrow[\text{Ni}]{\text{H}_2} ?$$

8. Complete and comment:

(i) $CH_2CH_2\overset{\displaystyle\frown O\frown}{CO} + NaCl \xrightarrow{\text{H}_2\text{O}} ?$

(ii) $(HO_2CCH_2)_2C(OH)CO_2H \xrightarrow[\text{H}_2\text{SO}_4]{\text{fuming}} ? \xrightarrow[\text{HCl}]{\text{EtOH}} ?$

(iii) $EtO_2CC\!\equiv\!CCO_2Et + CH_2(CO_2Et)_2 \xrightarrow[\text{EtOH}]{\text{EtONa}} ?$

(iv) $CH_2(CH_2)_2CH_2\overset{\displaystyle\frown O\frown}{CO} \xrightarrow{\text{LAH}} ?$

9. Discuss the following data:

	$\Delta H_f^{\ominus}$ (kJ)	$S^{\ominus}$ (JK^{-1})
Maleic acid	$-787\cdot85$	$159\cdot4$
Fumaric acid	$-810\cdot9$	$166\cdot1$

10. At which steps do inversions occur? Explain.

$$PhCH_2CHMeOH \xrightarrow[\text{(i)}]{\text{TsCl}} PhCH_2CHMeOTs \xrightarrow[\text{(ii)}]{\text{AcONa}} PhCH_2CHMeOAc \xrightarrow[\text{(iii)}]{\text{OH}^-} PhCH_2CHMeOH$$

11. *meso*-2,3-Dibromobutane reacts with iodide ion more rapidly than does the ($\pm$)-compound. What are the respective alkenes formed? Explain.

12. Complete the equations:

(i) $n\text{-}C_5H_{11}CH(CH_2)_2\overset{\displaystyle\frown O\frown}{CO} + 2MeMgBr \xrightarrow[\text{(ii) aq. NH}_4\text{Cl}]{\text{(i) Et}_2\text{O}} ?$

(ii) $CH_2(CH_2)_2\overset{\displaystyle\frown O\frown}{CO} + EtOH + HBr(g) \longrightarrow ?$

(iii) $CH_3CHOHCO_2H \xrightarrow{\text{Na}} ?$

13. Mustard gas, $(ClCH_2CH_2\!\!-\!)_2S$, is hydrolysed by water to $ClCH_2CH_2SCH_2CH_2OH$ *much faster* than expected for a primary alkyl halide. Offer an explanation.

14. An aliphatic ketone, M.W. 86, when converted into its oxime which is then reduced, gives an amine that can be resolved. What is the ketone? If more than one is possible, how could they be distinguished by a physical method?

15. What are **A–D** in the following reactions?

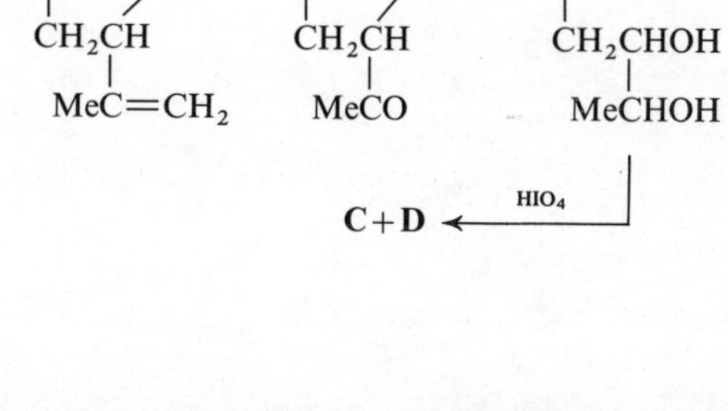

REFERENCES

FINAR, *Organic Chemistry*, Longman, Vol. 2 (1968, 4th edn). Ch. 2–6.

ELIEL, *Stereochemistry of Carbon Compounds*, McGraw-Hill (1962).

KLYNE (ed.), *Progress in Stereochemistry*, Butterworth (1954).

CAHN, 'An Introduction to the Sequence Rule', *J. Chem. Educ.*, 1964, **41**, 116.

ELIEL, 'Recent Advances in Stereochemical Nomenclature', *J. Chem. Educ.*, 1971, **48**, 1963.

CAPON, 'Neighbouring Group Participation', *Quart. Rev.*, 1964, **18**, 45.

DOLBIER, 'Research Summaries for Teachers; Organic Chemistry', *J. Chem. Educ.*, 1969, **46**, 342.

MCLENNAN, 'The Carbanion Mechanism of Olefin-forming Elimination', *Quart. Rev.*, 1967, **21**, 490.

BOYD and MCKERVEY, 'Asymmetric Synthesis', *Quart. Rev.*, 1968, **22**, 95.

18 Carbohydrates

Carbohydrates are substances with the general formula $C_x(H_2O)_y$, and were called carbohydrates (hydrates of carbon) because they contained hydrogen and oxygen in the same proportion as in water. However, a number of compounds have been discovered which are carbohydrates by chemical behaviour, but do not conform to the formula $C_x(H_2O)_y$, *e.g.*, 2-deoxyribose, $C_5H_{10}O_4$. It is also important to note that all compounds conforming to the formula $C_x(H_2O)_y$ are not necessarily carbohydrates, *e.g.*, formaldehyde, CH_2O; acetic acid, $C_2H_4O_2$; etc.

All carbohydrates are polyhydroxy-aldehydes or ketones, or substances that yield these on hydrolysis. Some partially methylated sugars occur naturally, and one natural nitro-sugar is known.

Nomenclature

The names of the simpler carbohydrates end in *-ose*; carbohydrates with an aldehydic structure are known as *aldoses*, and those with ketonic, *ketoses*. The number of carbon atoms in the molecule is indicated by a Greek prefix, *e.g.*, a *tetrose* contains four carbon atoms, a *pentose* five, a *hexose* six, etc.

Carbohydrates are divided into two main classes, **sugars and polysaccharides** (*polysaccharoses*). Sugars are crystalline substances with a sweet taste and soluble in water. Polysaccharides are more complex than the sugars, their molecular weights being far greater. Most of them are non-crystalline substances which are not sweet, and are insoluble or less soluble in water, than the sugars.

Sugars are subdivided into two major groups:

1. Monosaccharides (*monosaccharoses*). These are sugars which cannot be hydrolysed into smaller molecules. Their general formula is $C_nH_{2n}O_n$ (there are exceptions), where n is 2–10 (but see later), and the most important are the pentoses and hexoses.

2. Oligosaccharides. These yield 2–9 monosaccharide molecules on hydrolysis, *e.g.*, *disaccharides* (two monosaccharide molecules which may, or may not, be the same); *trisaccharides* (three monosaccharide molecules); etc.

Polysaccharides (*polysaccharoses*) are carbohydrates which yield a large number of monosaccharide molecules on hydrolysis. The most widely spread polysaccharides have the general formula $(C_6H_{10}O_5)_n$, *e.g.*, starch, cellulose, etc.; a group of polysaccharides which are not so widely spread in nature is the *pentosans*, $(C_5H_8O_4)_n$.

Monosaccharides

The simplest monosaccharide is glycolaldehyde, $HOCH_2CHO$. This does not contain an asymmetric carbon atom and is therefore not optically active. Since all naturally occurring sugars are optically active, it is more satisfactory to exclude glycolaldehyde from the group of sugars, and to define sugars as optically active polyhydroxy-aldehydes or ketones.

Triose, $C_3H_6O_3$. *Glyceraldehyde*, $HOCH_2CHOHCHO$, is an *aldotriose*. It contains one asymmetric carbon atom, and can therefore exist in two optically active forms, D- and L-; both are known. Glyceraldehyde has been chosen as the standard configuration in sugar chemistry (p. 473).

Dihydroxyacetone, $HOCH_2COCH_2OH$, is not optically active, and hence, by the definition given above, is not a sugar. If the definition be rejected, *i.e.*, the proviso that a sugar is always optically active is rejected, then dihydroxyacetone will be a *ketotriose*.

Tetroses, $C_4H_8O_4$. The structure of an aldotetrose is $HOCH_2CHOHCHOHCHO$ (see p. 519).

There is only one structure possible for a *ketotetrose*, *viz.*, $HOCH_2COCHOHCH_2OH$. This contains one asymmetric carbon atom, and corresponds to D- and L-*erythrulose*.

Pentoses, $C_5H_{10}O_5$. The *aldopentoses* are an important group of sugars (monosaccharides), and their structure has been elucidated by methods similar to those used for glucose and fructose (see later). Their structure is:

$$OCHCHOHCHOHCHOHCH_2OH$$

This contains three structurally different asymmetric carbon atoms, and can therefore exist in eight optically active forms. All are known, and correspond to the D- and L-forms of *arabinose*, *xylose*, *ribose* and *lyxose* (see p. 519).

L(+)-*Arabinose*, m.p. 158°C, occurs naturally as pentosans (arabans) in various gums, *e.g.*, gum arabic. It is usually obtained by the hydrolysis of cherry-gum with dilute sulphuric acid.

D(−)-*Arabinose* occurs in certain glucosides.

D(+)-*Xylose* (wood-sugar), m.p. 145°C, occurs as pentosans (xylans) in wood gums, bran and straw, from which it may be obtained by hydrolysis with dilute sulphuric acid.

D(+)-*Ribose*, m.p. 95 C, occurs in plant nucleic acids, and in liver and pancreas nucleic acids. All the other aldopentoses are synthetic compounds.

The chemical properties of the aldopentoses are similar to those of the aldohexoses (see glucose, below); pentoses do *not* undergo fermentation. All the aldopentoses are converted quantitatively into *furfural* when warmed with dilute acid:

$$C_5H_{10}O_5 \xrightarrow{HCl} \underset{O}{\boxed{}}CHO + 3H_2O$$

A very important aldopentose is 2-deoxyribose, $HOCH_2CHOHCHOHCH_2CHO$ (the prefix 'deoxy' indicates the replacement of a hydroxyl group by hydrogen; see also p. 791). It occurs in nucleic acids.

Ketopentoses with the structure $HOCH_2COCHOHCHOHCH_2OH$ have been prepared. The molecule contains two structurally different asymmetric carbon atoms, and can therefore exist in four optically active forms. All are known, and correspond to D- and L-forms of *ribulose* and *xylulose*.

Hexoses, $C_6H_{12}O_6$. The *aldohexoses* are another important group of monosaccharides, and their structure has been elucidated as follows:

(i) Analysis and molecular-weight determinations show that the molecular formula of the aldohexoses is $C_6H_{12}O_6$.

(ii) When treated with acetic anhydride, aldohexoses form the penta-acetate. This indicates the presence of five hydroxyl groups.

(iii) Aldohexoses form an oxime when treated with hydroxylamine, and therefore contain a carbonyl group.

(iv) When an aldohexose is oxidised with bromine-water, a pentahydroxy-acid of formula $C_6H_{12}O_7$ is obtained. This indicates that the carbonyl group is present in an *aldehydic* group.

(v) When reduced with concentrated hydriodic acid and red phosphorus at 100°C, aldo-hexoses give a mixture of 2-iodohexane and n-hexane. This indicates that the six carbon atoms in an aldohexose are in a straight chain. This conclusion is supported by ascending the series by the *Kiliani reaction* (p. 510) and reducing the product, a polyhydroxyacid, with hydriodic acid; n-heptanoic acid is produced.

The foregoing reactions show that the structure of the aldohexoses is:

$$OCHCHOHCHOHCHOHCHOHCH_2OH$$

This contains four structurally different asymmetric carbon atoms, and can therefore exist in sixteen optically active forms. All are known, and correspond to the D- and L-forms of *glucose, mannose, galactose, allose, altrose, gulose, idose* and *talose* (see p. 520).

Although the foregoing evidence indicates an open-chain structure, there is further evidence which shows that the sugars actually exist as *six-membered rings* in the *solid* state. The ring structure is hexagonal (I), but it is convenient to use the planar formula (II), *e.g.*, α-D-glucose, (II), corresponds to the open-chain formula (III) (see also p. 516), and it should be noted that the way of drawing the conventional planar formula of an aldose is with the carbon chain vertical and the aldehyde group at the top. In the D-series the hydroxyl group is always to the right on the bottom asymmetric carbon atom. The L-series is then the enantiomers of the D-series.

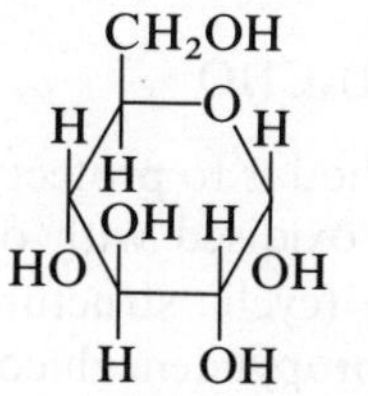

(I) 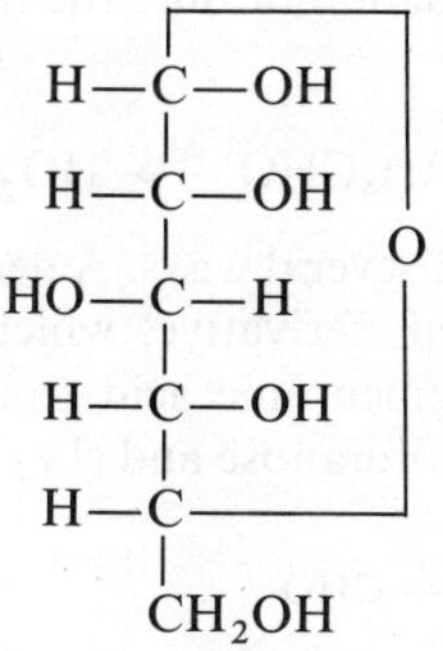   (II) 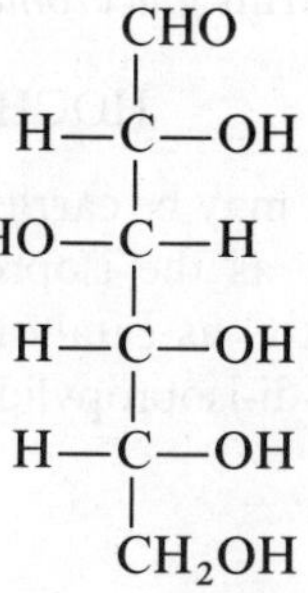   (III)

D(+)-**Glucose, dextrose** (*grape-sugar*), is found in ripe grapes, honey and most sweet fruits; it is also a normal constituent of blood, and occurs in the urine of diabetics. Commercially, pure D(+)-glucose is manufactured by heating starch with dilute hydrochloric acid under pressure:

$$(C_6H_{10}O_5)_n + nH_2O \xrightarrow{HCl} nC_6H_{12}O_6$$

D(+)-Glucose, m.p. 146°C, has a sweet taste, but is not as sweet as cane-sugar. Naturally-occurring glucose is dextrorotatory (hence name *dextrose*). It is a strong reducing agent, reducing both Fehling's solution and ammoniacal silver nitrate. When heated with sodium hydroxide, an aqueous solution of glucose turns brown (see also p. 514). Glucose forms a cyanohydrin with hydrogen cyanide (see also the Kiliani reaction, p. 510), and an oxime with hydroxylamine. All of these reactions are typical of aldehydes, but heating with phenyl-hydrazine in acetic acid does not give the phenylhydrazone; an *osazone* is formed (see pp. 298, 508).

Oxidation of sugars. The nature of the product formed by the oxidation of aldoses depends on the oxidising agent used. Mild oxidising reagents such as bromine-water or an alkaline solution of iodine oxidise only the aldehyde group to give **aldonic acids.** Thus, glucose gives *gluconic acid*, m.p. 131°C.

$$HOCH_2(CHOH)_4CHO + [O] \xrightarrow{Br_2/H_2O} HOCH_2(CHOH)_4CO_2H \quad (50\%)$$

This type of oxidation may also be carried out with mercuric oxide in the presence of calcium carbonate (calcium salt is obtained). Aldonic acids (also as calcium salt), however, are best prepared by the electrolysis of a solution of the aldose in the presence of calcium bromide and calcium carbonate. Gluconic acid is manufactured by the oxidation of glucose by the enzyme glucose oxidase, which is present in *Aspergillus niger*, *Penicillium chrysogenum*, etc.

The method of naming aldonic acids is to add prefixes to -onic to indicate the parent aldose, *e.g.*, galactonic and arabonic acids.

Oxidation of aldoses with *strong* oxidising reagents, *e.g.*, nitric acid (and under suitable conditions), produces α,ω-dicarboxylic acids known as **saccharic** or **aldaric acids**, *e.g.*, glucose gives glucosaccharic (still often called saccharic acid) or glucaric acid (some oxalic acid is also obtained):

$$HOCH_2(CHOH)_4CHO \xrightarrow[HNO_3]{[O]} HO_2C(CHOH)_4CO_2H \quad (40\text{--}46\%)$$

The general method of naming these acids is to indicate the parent aldose by a prefix, *e.g.*, mannosaccharic or mannaric acid, ribosaccharic or ribaric acid.

Aldaric acids readily form monolactones, but in some cases the dilactone may be produced.

Uronic acids are produced by oxidation of only the terminal —CH_2OH group in an aldose, *e.g.*, glucose forms *glucuronic acid*:

$$HOCH_2(CHOH)_4CHO \xrightarrow{[O]} HO_2C(CHOH)_4CHO$$

This oxidation may be carried out in several ways. A general method is to protect the aldehyde group, usually as the isopropylidene derivative, which is then oxidised with oxygen in the presence of Pt-C as catalyst, *e.g.*, glucuronic acid from glucose (cyclic structures are used). (III) is 1,2:5,6-di-isopropylideneglucofuranose and (IV) is 1,2-isopropylideneglucofuranose.

(III)   (IV)

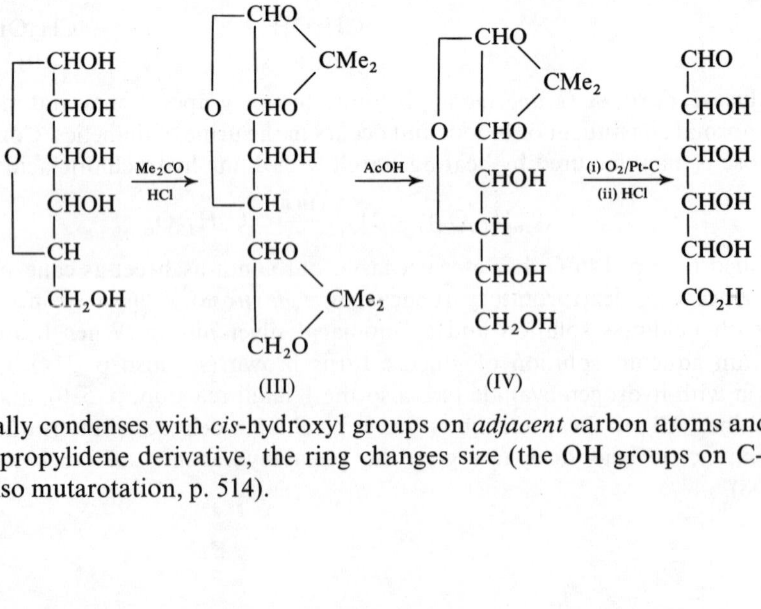

Acetone usually condenses with *cis*-hydroxyl groups on *adjacent* carbon atoms and hence, to form the di-isopropylidene derivative, the ring changes size (the OH groups on C-3 and C-4 are *trans*; see also mutarotation, p. 514).

Aldoses (and ketoses) are oxidised by periodic acid with complete breakdown of the carbon chain (see p. 522).

Reduction of sugars. As with oxidation, the product of reduction depends on the nature of the reagent. Sodium amalgam (see also p. 514), sodium borohydride, high-pressure catalytic hydrogenation ($Ni—H_2$), or electrolytic reduction in *acid* solution all convert aldoses into the corresponding alcohols, *e.g.*, glucose is converted into *sorbitol*:

$$HOCH_2(CHOH)_4CHO \xrightarrow{e; H^+} HOCH_2(CHOH)_4CH_2OH \quad (50\%)$$

Reduction of glucose with concentrated hydriodic acid and red phosphorus at 100°C produces 2-iodohexane; prolonged heating finally gives n-hexane.

Sugar derivatives. The sugars form a variety of derivatives, *e.g.*, ethers (p. 522), esters (p. 504), isopropylidene compounds (see above), etc. Because of the easy removal of the iso-propylidene group (dilute acid), these derivatives are very useful for protecting alcoholic groups (they are stable to alkali).

Aldohexoses, when heated with concentrated hydrochloric acid, first give hydroxymethyl-furfural, which then decomposes into lævulic acid and other products.

$$\begin{array}{l} CHO \\ (CHOH)_4 \\ CH_2OH \end{array} \longrightarrow 3H_2O + \; HOH_2C \underset{O}{\overset{}{\bigcirc}} CHO \longrightarrow$$

$$CH_3COCH_2CH_2CO_2H + HCO_2H + CO + CO_2$$

Glucose is fermented by yeast to ethanol (p. 190), and forms *glucosates* with various metallic hydroxides, *e.g.*, with calcium hydroxide it forms calcium glucosate, $C_6H_{12}O_6 \cdot CaO$; the structure of these glucosates is uncertain.

Aldohexoses (and aldopentoses) have a hydroxyl group at C-1 in the ring structure. This does not behave as an alcoholic group, *e.g.*, glucose combines with monohydric alcohols, in the presence of hydrochloric acid, to form *glucosides* (see p. 518).

D(+)-**Mannose**, m.p. 132°C, may be obtained by carefully oxidising mannitol with nitric acid:

$$HOCH_2(CHOH)_4CH_2OH + [O] \xrightarrow{HNO_3} HOCH_2(CHOH)_4CHO + H_2O$$

Commercially, mannose is prepared by hydrolysing the polysaccharide *seminine* with boiling dilute sulphuric acid. Seminine (a *mannan*) occurs in many plants, particularly in the shell of the ivory nut, which is used as the starting material for D(+)-mannose.

D(+)-**Galactose** (m.p. of monohydrate, 118°C) is found in several polysaccharides (*galactans*), and is combined with glucose in the disaccharide *lactose*. Galactose may be prepared by hydrolysing lactose, and separating it from glucose by fractional crystallisation from water, in which glucose is more soluble than galactose.

Rhamnose, $CH_3(CHOH)_4CHO$, is a 6-deoxyhexose; it occurs in several glycosides.

Ketohexoses. The only important ketohexose is D(−)-fructose, m.p. 102°C, the structure of which has been elucidated as follows:

(i) Analysis and molecular-weight determinations show that the molecular formula of fructose is $C_6H_{12}O_6$.

(ii) When treated with acetic anhydride, fructose forms the penta-acetate. This indicates the presence of five hydroxyl groups.

(iii) Fructose forms an oxime when treated with hydroxylamine, and therefore contains a carbonyl group.

(iv) When oxidised with nitric acid, fructose is converted into a mixture of trihydroxyglutaric, tartaric and glycollic acids. Since a mixture of acids each containing fewer carbon atoms than fructose is obtained, the carbonyl group in fructose must be present in a *ketonic* group.

(v) Fructose may be reduced to a hexahydric alcohol, *sorbitol*, which, on reduction with hydriodic acid and red phosphorus at 100°C, gives a mixture of 2-iodohexane and n-hexane. The formation of the latter two compounds indicates that the six carbon atoms in fructose are in a straight chain.

(vi) On ascending the series by the *Kiliani reaction*, and reducing the product with hydriodic acid, 2-methylhexanoic acid is obtained. This shows that the ketonic group in fructose is adjacent to one of the terminal carbon atoms.

The foregoing reactions show that the structure of fructose is:

$$HOCH_2CHOHCHOHCHOHCOCH_2OH$$

This contains three structurally different asymmetric carbon atoms, and can therefore exist in eight optically active forms. Of these the following six are known: D(−)- and L(+)-*fructose*, D(+)- and L(−)-*sorbose*, D(+)-*tagatose* and L(−)-*psicose* (see also p. 521).

Naturally-occurring fructose (*fruit-sugar*) is lævorotatory, and is therefore also known as lævulose (*cf.* dextrose). D(−)-Fructose is found in fruits and honey, and occurs combined with glucose in the disaccharide *cane-sugar*, from which it may be obtained by hydrolysis and fractional crystallisation. Fructose is prepared commercially by hydrolysis of *inulin*, a polysaccharide which occurs in dahlia tubers and Jerusalem artichokes:

$$(C_6H_{10}O_5)_n + nH_2O \xrightarrow{HCl} nC_6H_{12}O_6$$

The chemical reactions of fructose are similar to those of glucose, any differences being due to the fact that the former is a ketone and the latter an aldehyde. Thus, the most noticeable difference is the behaviour towards oxidising agents; fructose is not oxidised by bromine-water or an alkaline solution of iodine, and oxidation with nitric acid gives a mixture of acids, each containing less than six carbon atoms (see above).

Reaction of glucose and fructose with phenylhydrazine. Fischer (1884, 1887) showed that both aldoses and ketones react with excess of phenylhydrazine to form **osazones** (these contain two phenylhydrazino groups), aniline and ammonia. The overall equation may be written:

$$\begin{array}{c} CHO \\ | \\ CHOH \\ \vdots \end{array} \xrightarrow{3PhNHNH_2} \begin{array}{c} CH{=}NNHPh \\ | \\ C{=}NNHPh \\ \vdots \end{array} + PhNH_2 + NH_3 \xleftarrow{3PhNHNH_2} \begin{array}{c} CH_2OH \\ | \\ CO \\ \vdots \end{array}$$

Fischer also suggested a mechanism for this reaction, but it has been rejected in favour of that proposed by Weygand (1940). According to this mechanism, the first step is the formation of a phenylhydrazone, which then undergoes the **Amadori rearrangement** (1926). Weygand had originally proposed two mechanisms, but later (1958–1963) concluded that both were operating. According to Shemyakin *et al.* (1965), however, only one of these mechanisms is operating, *viz.*,

Glucose

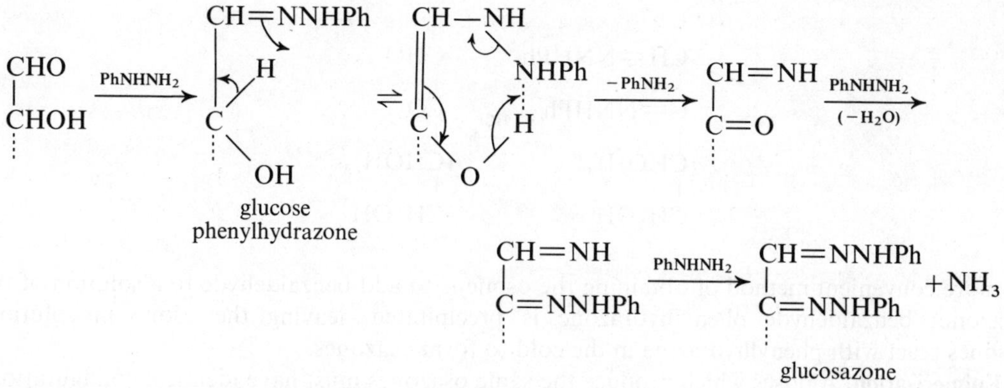

glucose phenylhydrazone

glucosazone

Fructose

fructose phenylhydrazone

fructosazone

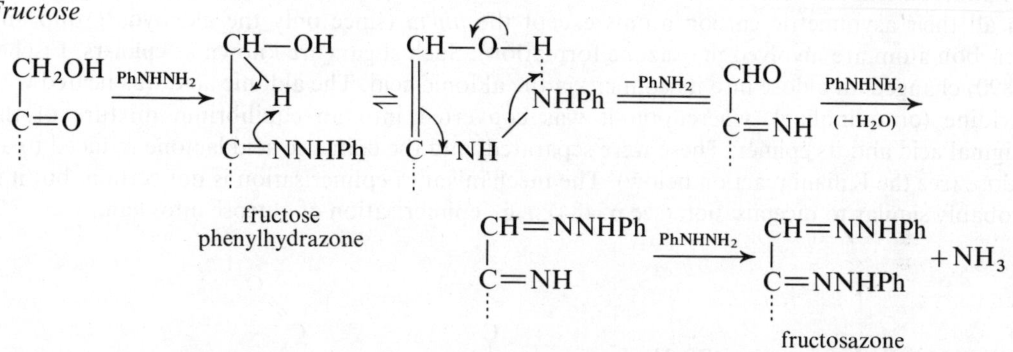

Weygand's mechanism does not explain why only the first two carbon atoms should be involved. At first sight, it might have been expected that the Amadori rearrangement could carry on down the chain to produce a hexaphenylhydrazone. The failure to undergo further reaction has been explained by stabilisation of the osazone by chelation, *i.e.*, phenylosazones have *ring* structures (Fieser and Fieser, 1944). Two alternatives (I) and (II) were proposed.

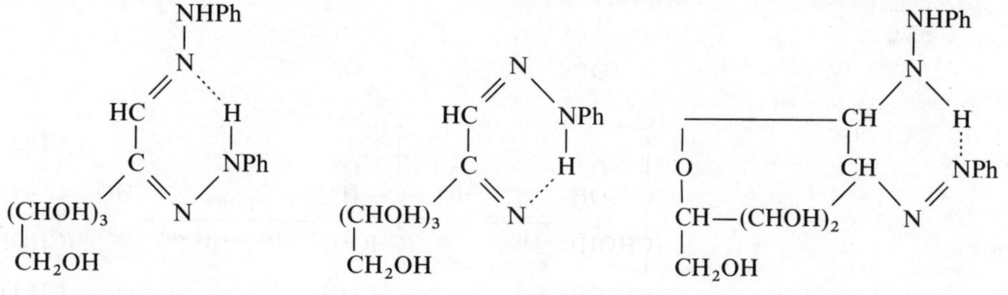

(I)   (II)   (III)

Mester (1955, 1965) believed that (I) was the correct structure, but Blair *et al.* (1969), from their chemical and i.r. and u.v. spectra studies, believe that neither is correct, and have proposed (III), the *phenylazo* structure (at C-2) for the compound in the *solid state*, and that *in solution*, (III) is in equilibrium with the open-chain form.

Osazones are yellow crystalline solids and are used to characterise the sugars. They may be hydrolysed with hydrochloric acid, both phenylhydrazino groups being eliminated; the dicarbonyl compound formed is known as an **osone**, *e.g.*, *glucosazone* forms *glucosone*;

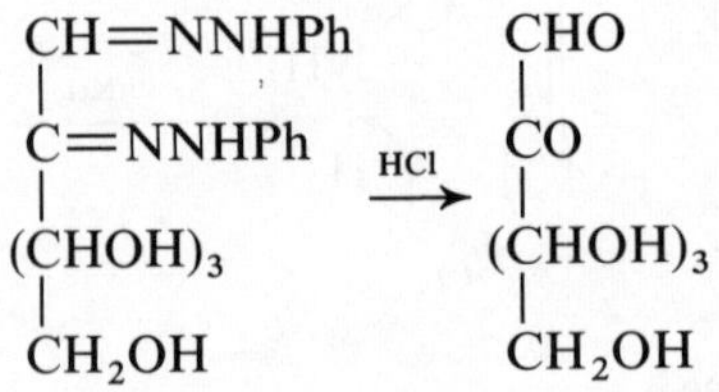

$$
\begin{array}{ccc}
\text{CH}=\text{NNHPh} & & \text{CHO} \\
| & & | \\
\text{C}=\text{NNHPh} & \xrightarrow{\text{HCl}} & \text{CO} \\
| & & | \\
(\text{CHOH})_3 & & (\text{CHOH})_3 \\
| & & | \\
\text{CH}_2\text{OH} & & \text{CH}_2\text{OH}
\end{array}
$$

A more convenient method of obtaining the osone is to add benzaldehyde to a solution of the osazone; benzaldehyde phenylhydrazone is precipitated, leaving the osone in solution. Osones react with phenylhydrazine in the cold to form osazones.

Epimerisation. Aldoses which produce the same osazones must have identical configurations on all their asymmetric carbon atoms except the *alpha* (since only the aldehyde group and α-carbon atom are involved in osazone formation). Such sugars are known as **epimers**. Fischer (1890) changed an aldose into its epimer via the aldonic acid. The aldonic acid was heated with pyridine (or quinoline), whereupon it was converted into an equilibrium mixture of the original acid and its epimer. These were separated, and the epimeric acid lactone reduced to an aldose (see the Kiliani reaction below). The mechanism of epimerisation is not certain, but it is probably similar to racemisation (see p. 484), *e.g.*, epimerisation of glucose into mannose:

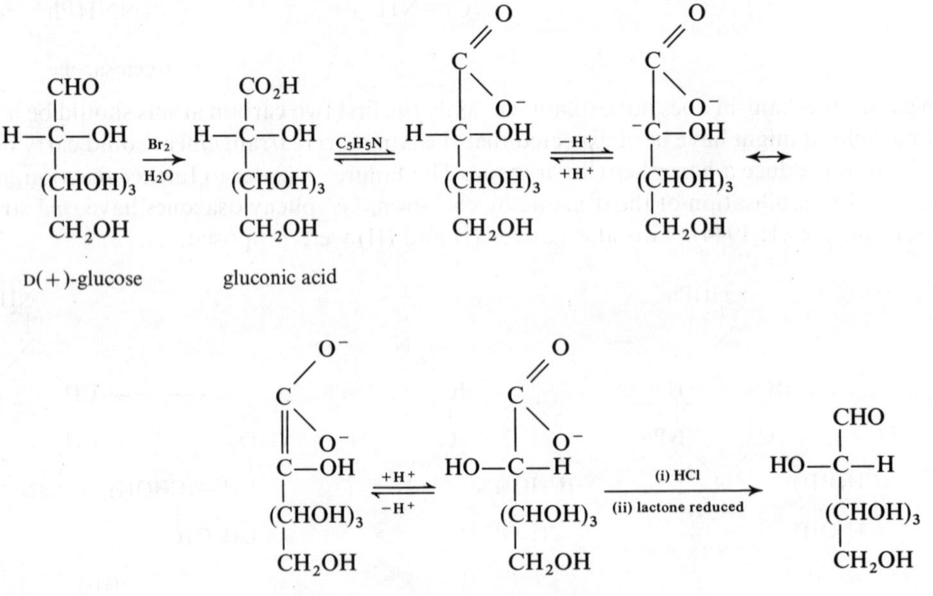

This change of configuration of *one* asymmetric carbon atom in a compound containing two or more asymmetric carbon atoms is known as **epimerisation**.

Method of ascending the sugar series. An aldose may be converted into its next higher aldose, *e.g.*, an aldopentose into an aldohexose, by means of the **Kiliani reaction** (1886).

CHO	CN	CO$_2$H	CO⎯	CHO
CHOH	CHOH	CHOH	CHOH	CHOH
CHOH	CHOH	CHOH	CHOH ⟩O	CHOH
CHOH	CHOH	CHOH	CH⎯	CHOH
CHOH	CHOH	CHOH	CHOH	CHOH
CH$_2$OH	CHOH	CH$_2$OH	CH$_2$OH	CH$_2$OH
	CH$_2$OH			

$\xrightarrow[\text{aq.}]{\text{HCN}}$ $\xrightarrow[\text{(ii) H}_2\text{SO}_4]{\text{(i) Ba(OH)}_2}$ $\xrightarrow{\text{heat}}$ $\xrightarrow[\text{H}_2\text{SO}_4]{\text{Na/Hg}}$

γ-lactone

Theoretically, two lactones are possible, since two cyanohydrins may be formed when hydrogen cyanide adds on to the aldopentose (a new asymmetric carbon is produced), *viz.*,

$$\text{CHO} \xrightarrow{\text{HCN}} \underset{\text{H}-\overset{\displaystyle\text{CN}}{\underset{\displaystyle}{\text{C}}}-\text{OH}}{} + \underset{\text{HO}-\overset{\displaystyle\text{CN}}{\underset{\displaystyle}{\text{C}}}-\text{H}}{}$$

Thus two epimeric aldohexoses should be obtained. In practice, one cyanohydrin predominates because the asymmetry present in the molecule produces diastereotopic faces at the double bond (see p. 483). Hence the final product will be mainly one aldohexose, and very little of its epimer.

1,4-Aldonolactones (γ-lactones) may also be reduced to the aldoses by sodium borohydride or lithium aluminium hydride in a mixture of tetrahydrofuran and pyridine.

There are now many other methods available for stepping up the sugars. Sowden *et al.* (1947) used nitromethane as follows (see also p. 364), and the proportions of the epimers formed by this method are different from those produced by the Kiliani reaction.

$$\text{CHO}+\text{CH}_3\text{NO}_2 \xrightarrow{\text{CH}_3\text{ONa}} \begin{array}{c}\text{CH}{=}\text{NO}_2\text{Na}\\|\\\text{CHOH}\end{array} \xrightarrow{\text{H}_2\text{SO}_4} \begin{array}{c}\text{CHO}\\|\\\text{H}{-}\text{C}{-}\text{OH}\end{array} + \begin{array}{c}\text{CHO}\\|\\\text{HO}{-}\text{C}{-}\text{H}\end{array}$$

Sowden (1950) has also stepped up an aldose to a *ketose* containing *two* additional carbon atoms using the above method except that 2-nitroethanol is used instead of nitromethane.

$$\begin{array}{c}\text{CHO}\\+\\\begin{array}{c}\text{CH}_2\text{NO}_2\\|\\\text{CH}_2\text{OH}\end{array}\end{array} \xrightarrow{\text{CH}_3\text{ONa}} \begin{array}{c}\text{CH}_2\text{OH}\\|\\\text{C}{=}\text{NO}_2\text{Na}\\|\\\text{CHOH}\end{array} \xrightarrow{\text{H}_2\text{SO}_4} \begin{array}{c}\text{CH}_2\text{OH}\\|\\\text{CO}\\|\\\text{H}{-}\text{C}{-}\text{OH}\end{array} + \begin{array}{c}\text{CH}_2\text{OH}\\|\\\text{CO}\\|\\\text{HO}{-}\text{C}{-}\text{H}\end{array}$$

On the other hand, Wolfrom *et al.* (1946) have stepped up an aldose to a ketose with *one* more carbon atom by a modified Arndt–Eistert reaction (p. 394).

$$\begin{array}{c}\text{CHO}\\|\\(\text{CHOH})_3\\|\\\text{CH}_2\text{OH}\end{array} \xrightarrow{\underset{\text{H}_2\text{O}}{\text{Br}_2}} \begin{array}{c}\text{CO}_2\text{H}\\|\\(\text{CHOH})_3\\|\\\text{CH}_2\text{OH}\end{array} \xrightarrow[\text{(ii) SOCl}_2]{\text{(i) Ac}_2\text{O}} \begin{array}{c}\text{COCl}\\|\\(\text{CHOAc})_3\\|\\\text{CH}_2\text{OAc}\end{array} \xrightarrow{\text{CH}_2\text{N}_2} \begin{array}{c}\text{CHN}_2\\|\\\text{CO}\\|\\(\text{CHOAc})_3\\|\\\text{CH}_2\text{OAc}\end{array} \xrightarrow[\text{(ii) Ba(OH)}_2]{\text{(i) AcOH}} \begin{array}{c}\text{CH}_2\text{OH}\\|\\\text{CO}\\|\\(\text{CHOH})_3\\|\\\text{CH}_2\text{OH}\end{array}$$

More recently, Kochetkov *et al.* (1965) have stepped up an aldose to an aldose containing two additional carbon atoms, *e.g.*, the aldose is condensed with carbethoxymethylenephosphorane (see Wittig reaction), and the product is then treated as follows (see also p. 410):

$$RCHOHCHO + Ph_3P\!=\!CHCO_2Et \longrightarrow Ph_3PO + RCHOHCH\!=\!CHCO_2Et \xrightarrow[HClO_4]{OsO_4}$$

$$[RCHOHCHOHCHOHCO_2H] \longrightarrow RCH(CHOH)_2\overset{\overset{\displaystyle O}{\frown}}{C}O \xrightarrow[\text{or } NaBH_4]{Na-Hg} RCHOHCHOHCHO$$

Method of descending the sugar series. There are various methods of converting a sugar into its next lower sugar. All of these methods start with the aldose, and hence, in order to convert a ketose into the lower aldose, it is first necessary to transform it into an aldose (see later).

Wohl's method (1893) starts with the aldose oxime, which is then treated as shown (for an aldohexose).

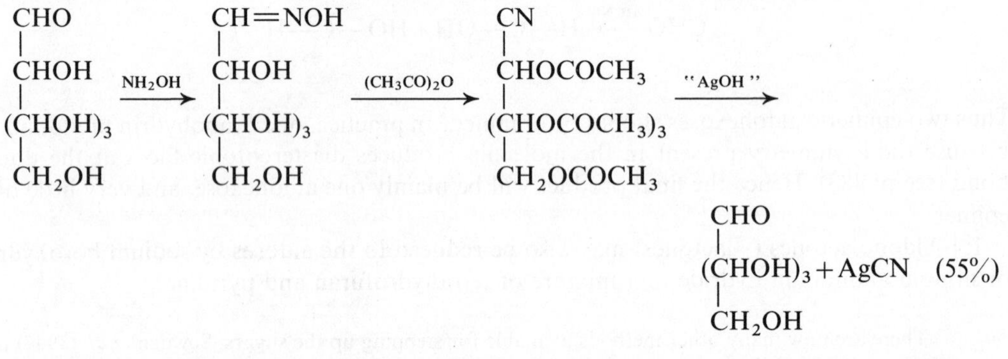

Zemplén (1917) modified Wohl's method by using a solution of sodium methoxide in chloroform instead of an aqueous solution of ammoniacal silver nitrate, to remove hydrogen cyanide and the acetyl groups, and thereby increased the yield of pentose to 60–70 per cent. Weygand *et al.* (1950) have treated the *oxime* with 1-fluoro-2,4-dinitrobenzene in aqueous sodium hydrogen carbonate; the products are the lower aldose (yield: 50–60%), hydrogen cyanide and 2,4-dinitrophenol.

Ruff's method (1898). The aldose is oxidised (by bromine-water) to the corresponding aldonic acid; when the calcium salt of this acid is treated with Fenton's reagent it is converted into the lower aldose (*cf.* oxidation of α-hydroxyacids, p. 468):

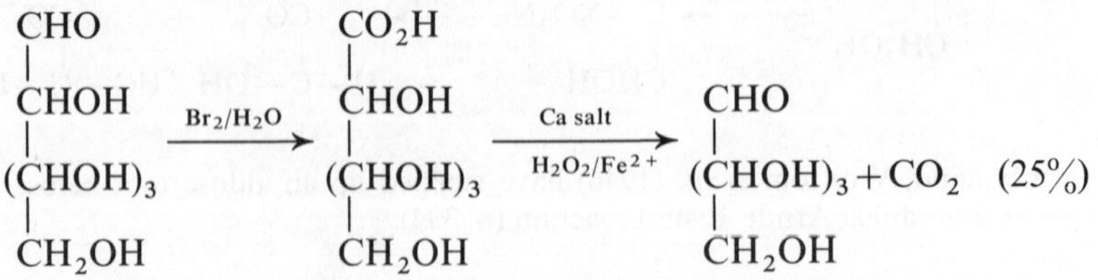

Berezovski *et al.* (1949) have shown that calcium gluconate can be oxidised with hydrogen peroxide in the presence of ferric sulphate and barium acetate to give D-arabinose (44%).

Weerman's reaction (1913). This is the reaction whereby an α-hydroxy- or α-methoxy-amide is degraded by means of a cold solution of sodium hypochlorite (*cf.* Hofmann amine preparation, p. 264).

α-Hydroxyamides

$$\begin{array}{c} CONH_2 \\ | \\ CHOH \\ | \\ R \end{array} \xrightarrow{\text{NaOH/NaOCl}} \left[\begin{array}{c} NCO \\ | \\ CHOH \\ | \\ R \end{array}\right] \xrightarrow{\text{NaOH}} \begin{array}{c} CHO \\ | \\ R \end{array} + NaNCO$$

α-Methoxyamides

$$\begin{array}{c} CONH_2 \\ | \\ CHOCH_3 \\ | \\ R \end{array} \xrightarrow{\text{NaOH/NaOCl}} \left[\begin{array}{c} NCO \\ | \\ CHOCH_3 \\ | \\ R \end{array}\right] \xrightarrow{\text{NaOH}} \begin{array}{c} CHO \\ | \\ R \end{array} + CO_2 + NH_3 + CH_3OH$$

More recent methods of descending the sugars are also available. Thus, Macdonald *et al.* (1953) have stepped down an aldose as follows:

$$\begin{array}{c} CHO \\ | \\ CHOH \\ | \\ R \end{array} \xrightarrow[\text{HCl}]{C_2H_5SH} \underset{\text{mercaptal}}{\begin{array}{c} CH(SC_2H_5)_2 \\ | \\ CHOH \\ | \\ R \end{array}} \xrightarrow{C_2H_5CO_3H} \underset{\text{disulphone}}{\begin{array}{c} CH(SO_2C_2H_5)_2 \\ | \\ CHOH \\ | \\ R \end{array}} \xrightarrow{NH_4OH} \begin{array}{c} CHO \\ | \\ R \end{array} + CH_2(SO_2C_2H_5)_2$$

On the other hand, Perlin (1954) has shown that aldoses may be stepped down by direct oxidation with lead tetra-acetate or sodium bismuthate, *e.g.*, D-mannose gives D-arabinose (35%).

Conversion of an aldose into a ketose. The aldose is converted into its osazone, which is then hydrolysed with hydrochloric acid to the osone. On reduction with zinc and acetic acid, the osone is converted into the ketose (an aldehyde group is reduced more readily than a ketonic group):

$$\begin{array}{c} CH{=}NNHC_6H_5 \\ | \\ C{=}NNHC_6H_5 \end{array} \xrightarrow{HCl} \begin{array}{c} CHO \\ | \\ CO \\ \vdots \end{array} \xrightarrow{e;\ H^+} \begin{array}{c} CH_2OH \\ | \\ CO \\ \vdots \end{array}$$

Conversion of a ketose into an aldose. The ketose is reduced (preferably by catalytic reduction, p. 210) to the corresponding polyhydric alcohol, which is then oxidised to a monocarboxylic acid (only one of the terminal CH_2OH groups being oxidised). On warming, the acid is converted into the γ-lactone which, on reduction with sodium amalgam in faintly acid solution, is converted into the aldose:

$$\begin{array}{c} CH_2OH \\ | \\ CO \\ | \\ CHOH \\ | \\ CHOH \\ | \\ CHOH \\ | \\ CH_2OH \end{array} \xrightarrow{H_2/Ni} \begin{array}{c} CH_2OH \\ | \\ CHOH \\ | \\ CHOH \\ | \\ CHOH \\ | \\ CHOH \\ | \\ CH_2OH \end{array} \xrightarrow[HNO_3]{[O]} \begin{array}{c} CO_2H \\ | \\ CHOH \\ | \\ CHOH \\ | \\ CHOH \\ | \\ CHOH \\ | \\ CH_2OH \end{array} \xrightarrow{heat} \begin{array}{c} CO{-}\!\!\!\!\\ | \\ CHOH \\ | \\ CHOH \\ | \\ CH{-} \\ | \\ CHOH \\ | \\ CH_2OH \end{array}\!\!\!O \xrightarrow{e;\ H^+} \begin{array}{c} CHO \\ | \\ CHOH \\ | \\ CHOH \\ | \\ CHOH \\ | \\ CHOH \\ | \\ CH_2OH \end{array}$$

Theoretically, two polyhydric alcohols may be formed on reduction of the ketose, due to the formation of a new asymmetric carbon atom:

$$\begin{array}{c} CH_2OH \\ | \\ CO \\ \vdots \end{array} \xrightarrow{H_2/Ni} \begin{array}{c} CH_2OH \\ | \\ H{-}C{-}OH \\ \vdots \end{array} + \begin{array}{c} CH_2OH \\ | \\ HO{-}C{-}H \\ \vdots \end{array}$$

In practice, however, one predominates (*cf*. the Kiliani reaction). Furthermore, when these two alcohols are oxidised, oxidation may take place at *either* end of the chain, and hence the final product will be a mixture of *four* aldoses, but these will not be present to the same extent.

Lobry de Bruyn–van Ekenstein rearrangement (1890). When warmed with *concentrated* alkali, sugars first turn yellow, then brown and finally resinify (*cf*. aldehydes, p. 220). In the presence of *dilute* alkali or amines, sugars undergo rearrangement; *e.g.*, a dilute solution of glucose, in the presence of sodium hydroxide, is converted into an almost optically inactive solution from which have been isolated D(+)-glucose, D(+)-mannose, D(−)-fructose. The same mixture is obtained if the starting material is D(−)-fructose or D(+)-mannose. The details of the mechanism are uncertain, but it is accepted that the rearrangement occurs through an enediol:

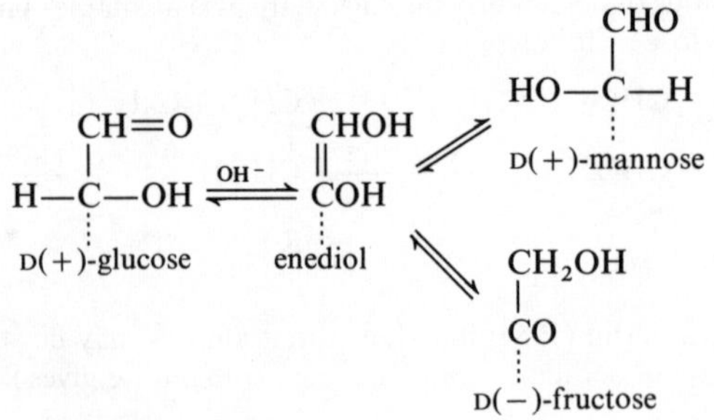

Since the Lobry de Bruyn–van Ekenstein rearrangement takes place in alkaline media, it is best to carry out reactions with the sugars in neutral or acid media.

Cyclic structures of monosaccharides.

When a monosaccharide is dissolved in water, the optical rotatory power of the solution gradually changes until it reaches a constant value, *e.g.*, a freshly prepared solution of glucose has a specific rotation of $+111°$; when this solution is allowed to stand, the specific rotation falls to $+52.5°$, and remains constant at this value. The final stage can be reached more rapidly either by heating the solution or by adding some catalyst which may be an acid or a base. This change in value of the specific rotation is known as **mutarotation.** All *reducing* sugars (except some ketoses) undergo mutarotation. To account for mutarotation, Tollens (1883) suggested an oxide ring structure for D(+)-glucose, whereby *two* forms would be produced since, in the formation of the ring, another asymmetric carbon atom (which can exist in *two* configurations) is produced (*cf*. the Kiliani reaction). Tollens assumed that a five-membered ring (the furanose form) was produced, (I) and (II). The difficulty of this suggestion was that there was no experimental evidence for the existence of these two forms. Tanret (1895), however, isolated two isomeric forms of D(+)-glucose, thus apparently verifying Tollens' supposition (see later). The two forms are called α- and β-D(+)-glucofuranose; (I) is the α-form, and (II) the β-.

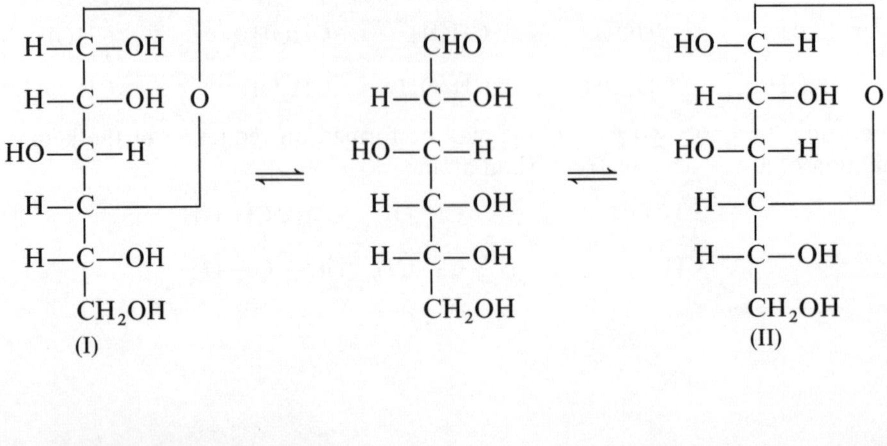

Ring formation of a sugar is really hemiacetal formation, one alcohol group of the sugar forming a hemiacetal with the aldehyde group of the *same* molecule, thus producing a ring structure. This equilibrium between the open and ring forms is an example of ring-chain tautomerism. The carbon atoms in the chain tend to form a loop, and this results in the OH on C_4 or C_5 being very close to the CHO group. With ketoses, it is the OH group on C_5 or C_6 that is close to the CO group. In both aldoses and ketoses, therefore, the geometry of the molecule is very favourable for hemi-acetal formation.

Later work by Haworth, Hirst and their co-workers (1926 onwards) has shown that glucose (and other sugars) exists, not as a five-membered ring but as a six-membered ring, the two forms being α- and β-D(+)-glucopyranose. Such *pairs* of stereoisomeric ring-forms of any sugar are known as **anomers**.

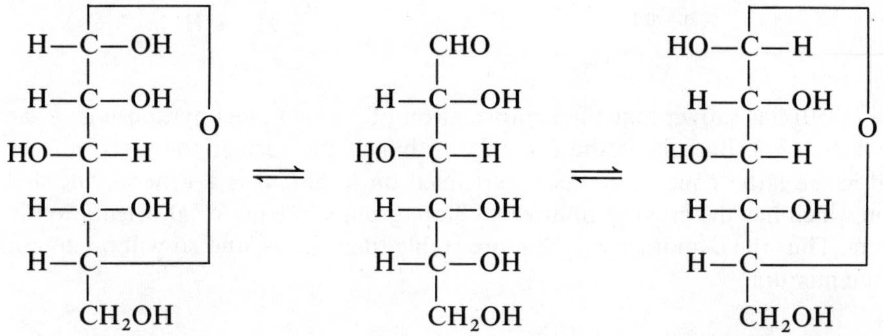

α-D(+)-glucopyranose   β-D(+)-glucopyranose

The widely accepted view is that monosaccharides in solution exist mainly as an equilibrium mixture of the α- and β-anomeric pyranoses, a small amount of the open-chain form together with small amounts of the α- and β-anomeric furanoses. The presence of the furanose forms has been inferred from the fact that monosaccharides undergo some reactions which lead to the formation of furanose derivatives (see p. 506, isopropylidene derivatives).

The mechanism of mutarotation has been the subject of a great deal of research, but it is still uncertain. A widely accepted mechanism involves a concerted attack by base and acid (water is an amphiprotic solvent) to produce the *open-chain structure*, and then this recloses to the anomeric form:

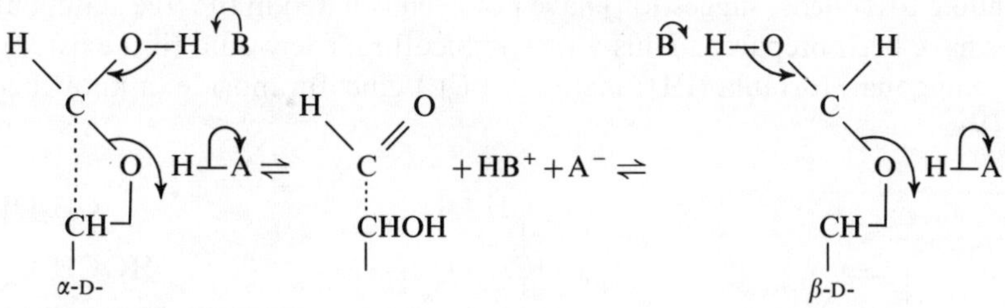

The ordinary form of D(+)-glucose is the α-anomer, m.p. 146°C, specific rotation +111°, and may be prepared by crystallisation of glucose from cold aqueous solution. The β-anomer, m.p. 148–150°C, specific rotation +19·2°, may be obtained by crystallising glucose from hot saturated aqueous solution. Both anomers show mutarotation, the final value of the specific rotation being 52·5°. This corresponds to about 36 per cent of the α-anomer and 64 per cent of the β-. It is therefore assumed that whenever a sugar is formed in solution, it immediately changes into a mixture of the two pyranose anomers, the open chain isomer (and furanoses) being present in extremely small amount. The cyclic structure of the sugars accounts for the following facts: (i) the existence of two isomers, *e.g.*, α- and β-glucose; (ii) mutarotation;

(iii) glucose and other aldoses do not give certain characteristic reactions of aldehydes, *e.g.*, Schiff's reaction, do not form a bisulphite or an aldehyde-ammonia compound.

Haworth (1926) proposed a six-membered ring formula (hexagonal formula) based on the *pyran* ring which is almost planar. The **pyranose** structure is applicable to nearly all the sugars, and is supported by X-ray crystal analysis (of the sugars). Thus α-D(+)-*glucopyranose* is (*a*).

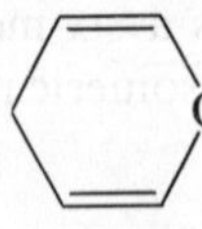

pyran

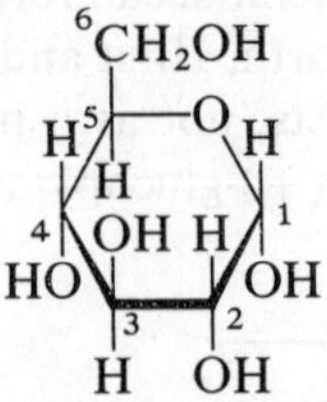

(*a*)

Reeves (1950) has shown that the conformation of α-D(+)-glucopyranose is (*b*) and that of the β-anomer is (*c*). Both have the chair form, but in the former the glycosidic hydroxyl is axial and in the latter equatorial. As is explained on p. 551, it is a general rule that the conformation which has the greatest number of large groups in equatorial orientation is the most stable form. Thus the β-anomer will be more stable than the α-, and so will predominate in the equilibrium mixture.

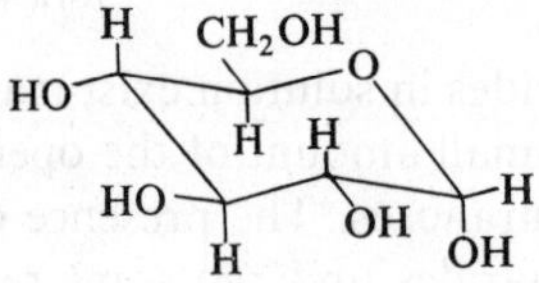

α-anomer

(*b*)

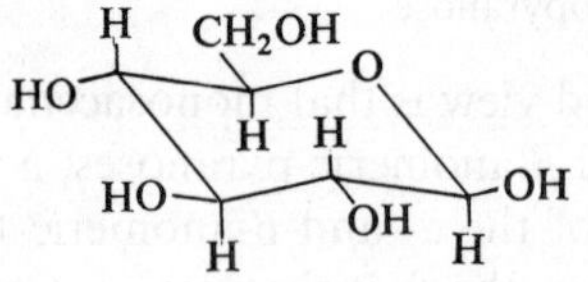

β-anomer

(*c*)

It has also been shown (Irvine, Haworth *et al.*) that glucose, fructose, etc. can exist as five-membered rings, which may be regarded as derivatives of *furan*. So far, the **furanose** sugars (corresponding to Tollens' suggestion) have not been isolated in the free state, but some of their derivatives have been prepared. Thus α-D(+)-glucofuranose would (if it existed) have perspective (*i.e.*, pentagonal) formula (III); methyl α-D(+)-glucofuranoside (a known compound) has formula (IV):

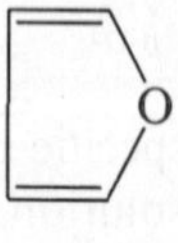

furan

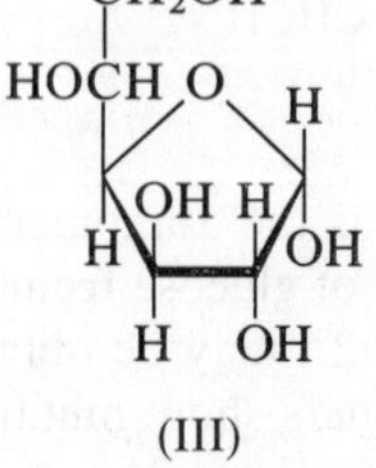

(III)

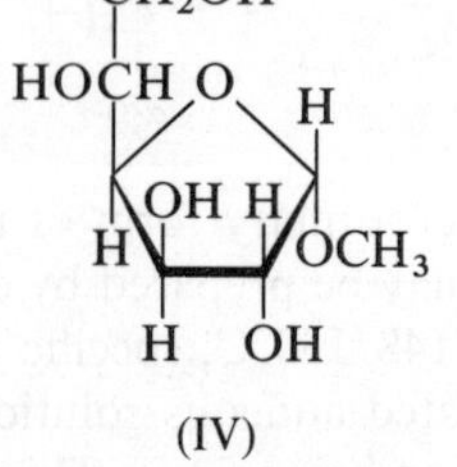

(IV)

Pentoses also normally exist in the pyranose form, and derivatives of the furanose form have been prepared.

Conversion of the plane-diagrams into the perspective formulæ may be done as follows. (V) is α-D-glucopyranose, and if the H on C_5 is interchanged with the group CH_2OH, then a Walden inversion has been effected; and if the H is now interchanged with the point of attachment of the oxygen ring, another Walden inversion is effected, and so the original configuration of (V) is retained; thus we now have (VI) (with no change in configuration). Since all horizontal bonds indicate groups lying above the plane of the paper, and vertical bonds groups lying behind this plane (see p. 473), then by twisting (VI) so that the oxide ring is perpendicular to the plane of the paper, and placing the oxygen atom as shown, (VII) is obtained. Thus, to change from (VI) to (VII), first draw the hexagon (as shown in (VII)), and then place all the groups on the left-hand side in (VI) above the plane of the ring in (VII), and all those on the right-hand side in (VI) below the plane of the ring in (VII).

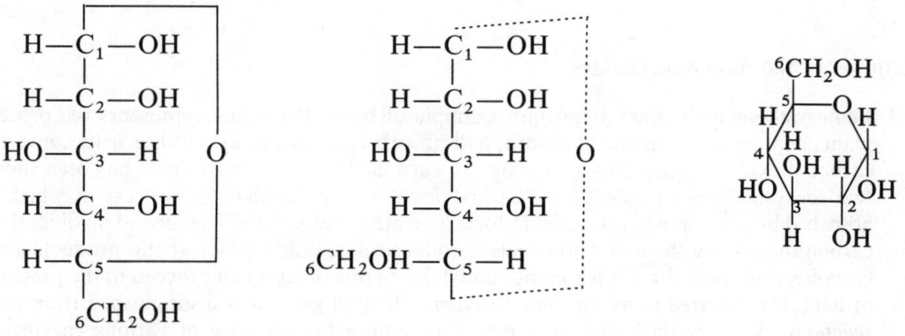

(V) (VI) (VII)

In a similar way, perspective formulæ may be obtained for the furanose sugars, *e.g.*, methyl β-D(+)-fructofuranoside (in this case, *i.e.*, for ketoses, treat the CH_2OH group adjacent to the keto group as H in the aldoses).

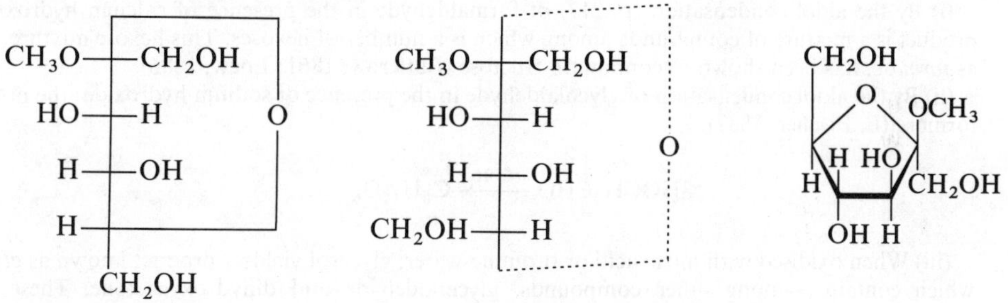

NMR spectroscopic studies of monosaccharides and their derivatives have led to several generalisations (see p. 551 for the meanings of *axial* and *equatorial*).

(i) Anomeric protons (*i.e.*, H on C-1) occur at lower field than any others (C-1 is linked to *two* oxygen atoms).

(ii) Axial hydrogens appear upfield with respect to equatorial hydrogens.

(iii) For *acetylated* pyranosides, the methyl group on axial hydroxyl groups occurs downfield with respect to those on equatorial hydroxyl groups.

(iv) For *methylated* pyranosides, the methyl group on axial hydroxyl groups occurs upfield with respect to those on equatorial hydroxyl groups.

These generalisations are useful, but may lead to wrong conclusions in rings which have been deformed by, *e.g.*, steric effects.

Just as simple hemiacetals react with another molecule of an alcohol to form acetals (p. 222), so can the hemiacetal form of a sugar react with a molecule of an alcohol to form the acetal derivative, which is known under the generic name of **glycoside**; those of glucose are known as *glucosides*; of fructose, *fructosides*, etc. For example, ethyl α-D(+)-glucopyranoside is prepared by refluxing glucose in excess of ethanol in the presence of a small amount of hydrochloric acid (see p. 521). These glycosides are stable compounds, and do not undergo many of the reactions of the sugars, *e.g.*, they show no reducing properties, they do not mutarotate, etc. The non-sugar part of a glycoside is known as the *aglycon*, and in most of the glycosides which occur naturally the aglycon is a phenolic compound, *e.g.*, the aglycon in *salicin* is *salicyl alcohol*; in *indican*, *indoxyl*.

Synthesis of the monosaccharides

Photosynthesis is the most important example of biosynthesis and represents the processes whereby plants containing the pigment chlorophyll absorb light energy and utilise it to convert atmospheric carbon dioxide, in the presence of water, to carbohydrates. Photosynthesis has been shown to involve two separate types of reactions. The first involves a *photochemical* process in which light energy, absorbed by chlorophyll, is utilised to form activated compounds. The second involves the reduction of carbon dioxide by the active molecules produced in the first process; the products are oxygen and carbohydrates (and some other compounds). Since this process can proceed in the presence or absence of light, it is referred to as the *dark reactions*; the first process is distinguished from this as the *light reactions*. Both the dark and light reactions require the presence of various enzymes. The overall equation for photosynthesis may be written as:

$$6CO_2 + 6H_2O \longrightarrow 6(CH_2O) + 6O_2$$

Calvin *et al.* using $^{14}CO_2$ as tracer, have worked out the pathway of carbon dioxide in photosynthesis.

In the laboratory, the sugars have been synthesised in various ways:

(i) By the aldol condensation (p. 217) of formaldehyde in the presence of calcium hydroxide; the product is a mixture of compounds among which is a number of hexoses. This hexose mixture, known as *formose*, has been shown to contain DL-fructose (Butlerow, 1861; Loew, 1886).

(ii) By the aldol condensation of glycolaldehyde in the presence of sodium hydroxide; the product is formose (E. Fischer, 1887):

$$3HOCH_2CHO \xrightarrow{NaOH} C_6H_{12}O_6$$

(iii) When oxidised with nitric acid or bromine-water, glycerol yields a product known as *glycerose*, which contains, among other compounds, glyceraldehyde and dihydroxyacetone. These, in the presence of barium hydroxide, are converted into a mixture of α- and *β-acrose* (E. Fischer, 1887); α-acrose is DL-fructose and β-acrose is DL-sorbose:

$$HOCH_2CHOHCHO + HOCH_2COCH_2OH \longrightarrow HOCH_2CO(CHOH)_3CH_2OH$$

Starting with α-acrose, Fischer isolated DL-fructosazone (DL-glucosazone), and making use of reduction, oxidation and epimerisation, he converted DL-fructosazone into D(−)-fructose, D(+)-glucose, D(+)-mannose and other aldohexoses.

(iv) When hydrolysed with barium hydroxide solution, dibromoacraldehyde forms a mixture of α- and *β-acrose* (E. Fischer, 1887):

$$BrCH_2CHBrCHO \xrightarrow{Ba(OH)_2} HOCH_2CHOHCHO \xrightarrow{dimerises} HOCH_2CO(CHOH)_3CH_2OH$$

Configurations of the monosaccharides

Aldotrioses. Glyceraldehyde is the only aldotriose.

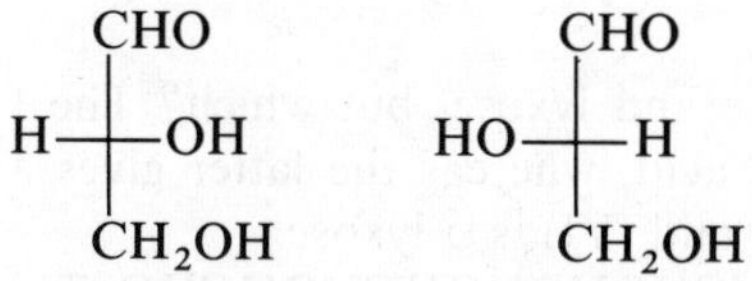

$$
\begin{array}{cc}
\text{CHO} & \text{CHO} \\
\text{H}-\!\!\!-\text{OH} & \text{HO}-\!\!\!-\text{H} \\
\text{CH}_2\text{OH} & \text{CH}_2\text{OH}
\end{array}
$$

D(+)-glyceraldehyde   L(−)-glyceraldehyde

Aldotetroses, $HOCH_2CHOHCHOHCHO$. This structure contains two asymmetric carbon atoms, and so there are four optical isomers (two pairs of enantiomers). All are known, and correspond to D- and L-threose and D- and L-erythrose. D(+)-Glyceraldehyde may be stepped up by the Kiliani reaction to give D(−)-threose and D(−)-erythrose. The question now is: which is which?

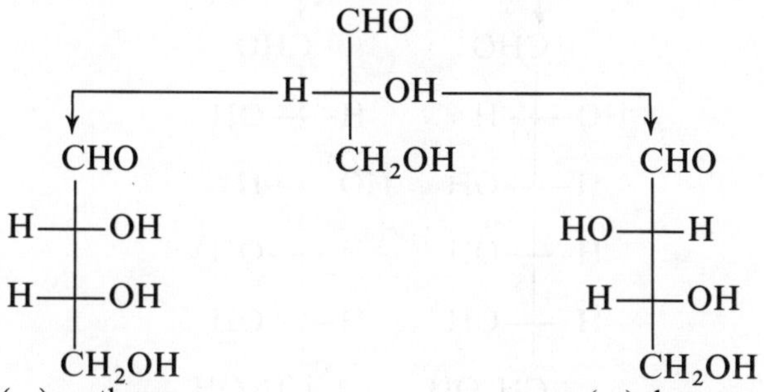

$$
\begin{array}{c}
\text{CHO} \\
\text{H}-\!\!\!-\text{OH} \\
\text{CH}_2\text{OH}
\end{array}
$$

$$
\begin{array}{cc}
\text{CHO} & \text{CHO} \\
\text{H}-\!\!\!-\text{OH} & \text{HO}-\!\!\!-\text{H} \\
\text{H}-\!\!\!-\text{OH} & \text{H}-\!\!\!-\text{OH} \\
\text{CH}_2\text{OH} & \text{CH}_2\text{OH}
\end{array}
$$

D(−)-erythrose   D(−)-threose
(I)   (II)

On oxidation, D-erythrose forms *meso*-tartaric acid. Therefore D-erythrose must be (I), and consequently (II) must be D-threose.

Aldopentoses, $HOCH_2CHOHCHOHCHOHCHO$. This structure contains three asymmetric carbon atoms, and so there are eight optical isomers (four pairs of enantiomers). All are known. D-Erythrose, when stepped up by the Kiliani reaction, gives D(−)-ribose and D(−)-arabinose. Similarly, D-threose gives D(+)-xylose and D(−)-lyxose.

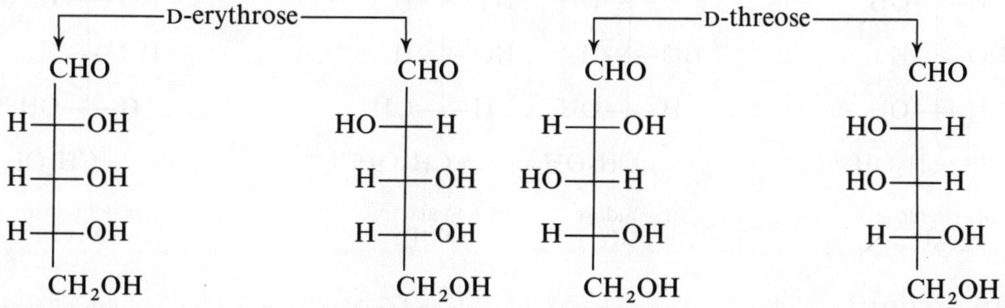

D-erythrose   D-threose

$$
\begin{array}{cccc}
\text{CHO} & \text{CHO} & \text{CHO} & \text{CHO} \\
\text{H}-\!\!\!-\text{OH} & \text{HO}-\!\!\!-\text{H} & \text{H}-\!\!\!-\text{OH} & \text{HO}-\!\!\!-\text{H} \\
\text{H}-\!\!\!-\text{OH} & \text{H}-\!\!\!-\text{OH} & \text{HO}-\!\!\!-\text{H} & \text{HO}-\!\!\!-\text{H} \\
\text{H}-\!\!\!-\text{OH} & \text{H}-\!\!\!-\text{OH} & \text{H}-\!\!\!-\text{OH} & \text{H}-\!\!\!-\text{OH} \\
\text{CH}_2\text{OH} & \text{CH}_2\text{OH} & \text{CH}_2\text{OH} & \text{CH}_2\text{OH}
\end{array}
$$

D(−)-ribose   D(−)-arabinose   D(+)-xylose   D(−)-lyxose
(III)   (IV)   (V)   (VI)

(III) and (IV) must be ribose and arabinose, but which is which? On oxidation with nitric acid, arabinose gives an optically active dicarboxylic acid (a trihydroxyglutaric acid), whereas ribose gives an optically inactive dicarboxylic acid. When the terminal groups (*i.e.*, CHO and

CH₂OH) of (III) are oxidised to carboxyl groups, the molecule produced possesses a plane of symmetry, and so this acid is inactive. The dicarboxylic acid produced from (IV), however, has no plane (or any other element) of symmetry, and so is optically active. Thus (III) is D-ribose and (IV) is D-arabinose.

(V) and (VI) must be xylose and lyxose, but which? The former, on oxidation, gives an optically inactive dicarboxylic acid, whereas the latter gives an optically active dicarboxylic acid. Therefore (V) is D-xylose and (VI) is D-lyxose.

Aldohexoses, HOCH₂CHOHCHOHCHOHCHOHCHO. This structure contains four asymmetric carbon atoms, and so there are sixteen optical isomers (eight pairs of enantiomers). All are known, and may be prepared by stepping up the aldopentoses. D-ribose gives D(+)-allose and D(+)-altrose; D-arabinose gives D(+)-glucose and D(+)-mannose; D-xylose gives D(−)-gulose and D(−)-idose, and D-lyxose gives D(+)-galactose and D(+)-talose.

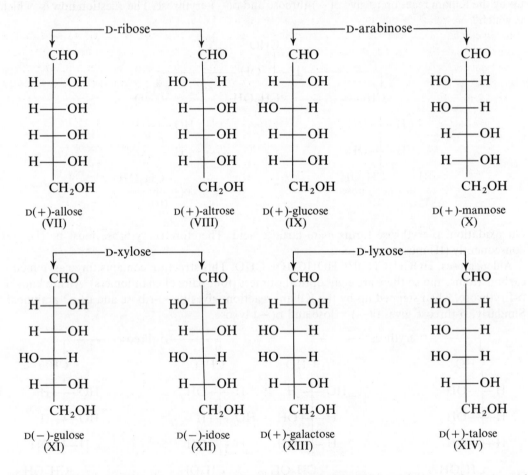

(VII) and (VIII) must be allose and altrose, but which is which? On oxidation with nitric acid, the former gives an optically inactive (allaric) and the latter an optically active (talaric) dicarboxylic acid. Therefore (VII) is allose and (VIII) is altrose.

(XIII) and (XIV) must be galactose and talose, but which is which? On oxidation with nitric acid, the former gives an optically inactive (galactaric) and the latter an optically active (talaric) dicarboxylic acid. Therefore (XIII) is galactose and (XIV) is talose.

The elucidation of the configurations of the remaining four aldohexoses is not quite so simple since, on oxidation with nitric acid, glucose and mannose *both* give optically active dicarboxylic acids, as also do gulose and idose; in all four configurations (IX, X, XI, XII), replacement of the two terminal groups (CHO and CH_2OH) by carboxyl groups leads to dicarboxylic acid whose structures have no plane (or any other element) of symmetry. It has been found, however, that the dicarboxylic acid from glucose (glucaric acid) is the same as that produced from gulose. Actually the two glucaric acids obtained are enantiomers, D-glucose giving D-glucaric acid and D-gulose L-glucaric acid. Since glucaric acid, $HO_2C(CHOH)_4CO_2H$, is produced by the oxidation of the terminal groups with the rest of the molecule unaffected, it therefore follows that the 'rest of the molecule' must be the same for both glucose and gulose. Inspection of formulæ (IX, X, XI and XII) shows that only (IX) and (XI) have the 'rest of the molecule' the same; by interchanging the CHO and CH_2OH groups of (IX), (XI) is obtained. Therefore (IX) must be glucose (since we know that glucose is obtained from arabinose), and (XI) must be gulose. Consequently (X) is mannose and (XII) is idose.

Ketohexoses. Fructose is a 2-ketohexose, and natural fructose is lævorotatory. Since D-glucose gives the *same* osazone as natural fructose, the latter must be D(−)-fructose. Furthermore, since osazone formation involves only the first two carbon atoms in a sugar, it therefore follows that the configuration of the rest of the molecules in glucose and fructose must be the same. Hence the configuration of D(−)-fructose is (XV).

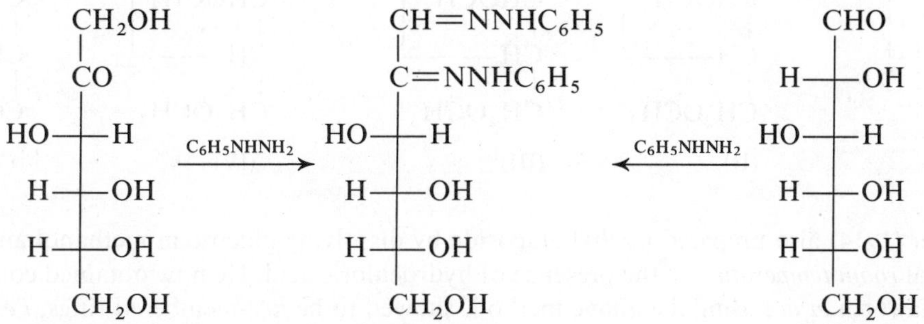

D(−)-fructose (XV) osazone D(+)-glucose

The configurations of the other 2-ketohexoses are:

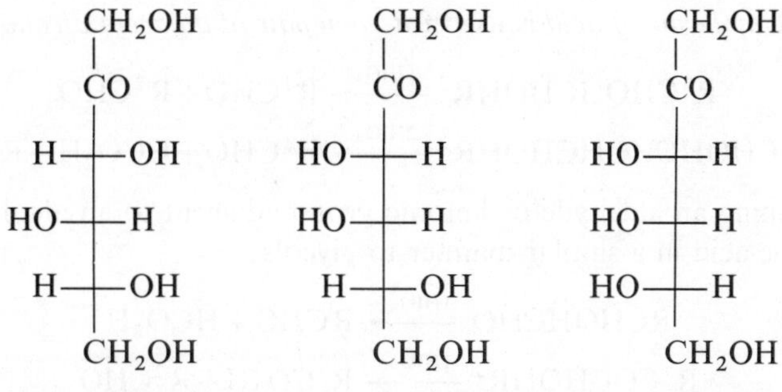

D(+)-sorbose D(+)-tagatose L(−)-psicose

Determination of the size of sugar-rings

E. Fischer (1893) refluxed glucose with excess of methanol in the presence of a small amount of hydrochloric acid and obtained a white crystalline solid containing *one* methyl group. This compound was no longer reducing, and also did not form an osazone. Thus Fischer had prepared methyl glucoside. Ekenstein (1894) isolated a second isomer from the same reaction, and Fischer explained the existence of these two isomers by suggesting a *ring* structure for them, and followed Tollens in believing that they were five-membered rings. There was no experimental evidence for the existence of a five-membered ring, and it was shown to be incorrect by Haworth, Hirst and their co-workers (1926 onwards). They proved that the above glucosides were *six-membered* rings, *i.e.*, glucopyranosides. Their method was to fully methylate the glucoside (I), hydrolyse the product, methyl tetramethyl-α-D-glucoside (II) to tetramethyl-α-D-glucose (III), oxidise this with bromine-water at 90°C to the lactone (IV), and then to oxidise this with nitric acid. The product obtained was shown to be xylotrimethoxy-glutaric acid (V; this can be obtained directly by the oxidation of methylated xylose). The most reasonable interpretation of these results is in accordance with the assumption that methyl glucoside is a six- and *not* a five-membered ring. Thus:

$$
\begin{array}{cccc}
\begin{array}{l}
\text{CHOCH}_3\\
\text{CHOH}\\
\text{CHOH}\\
\text{CHOH}\\
\text{CH}\\
\text{CH}_2\text{OH}
\end{array}\text{O}
&
\xrightarrow[\text{NaOH}]{(\text{CH}_3)_2\text{SO}_4}
\begin{array}{l}
\text{CHOCH}_3\\
\text{CHOCH}_3\\
\text{CHOCH}_3\\
\text{CHOCH}_3\\
\text{CH}\\
\text{CH}_2\text{OCH}_3
\end{array}\text{O}
&
\xrightarrow{\text{HCl}}
\begin{array}{l}
\text{CHOH}\\
\text{CHOCH}_3\\
\text{CHOCH}_3\\
\text{CHOCH}_3\\
\text{CH}\\
\text{CH}_2\text{OCH}_3
\end{array}\text{O}
&
\xrightarrow{\text{Br}_2/\text{H}_2\text{O}}
\begin{array}{l}
\text{CO}\\
\text{CHOCH}_3\\
\text{CHOCH}_3\\
\text{CHOCH}_3\\
\text{CH}\\
\text{CH}_2\text{OCH}_3
\end{array}\text{O}
&
\xrightarrow{\text{HNO}_3}
\begin{array}{l}
\text{CO}_2\text{H}\\
\text{CHOCH}_3\\
\text{CHOCH}_3\\
\text{CHOCH}_3\\
\text{CO}_2\text{H}
\end{array}
\\
\text{(I)} & \text{(II)} & \text{(III)} & \text{(IV)} & \text{(V)}
\end{array}
$$

Fischer (1914) also prepared methyl glucoside by dissolving glucose in methanol and letting it stand at *room temperature* in the presence of hydrochloric acid. He now obtained compounds which Haworth *et al.*, using the above method, showed to be *five*-membered rings, *i.e.*, methyl D-glucofuranosides.

Hudson (1937, 1939) has also determined the size of sugar-rings, but his method was to oxidise methyl glycosides with periodic acid. As we have seen (p. 312), periodic acid splits 1,2-glycols, and *one molecule of acid* is used for *each pair of adjacent alcoholic groups*, e.g.,

$$\text{R}^1\text{CHOHCHOHR}^2 \xrightarrow{\text{1HIO}_4} \text{R}^1\text{CHO} + \text{R}^2\text{CHO}$$

$$\text{R}^1\text{CHOHCHOHCHOHR}^2 \xrightarrow{\text{2HIO}_4} \text{R}^1\text{CHO} + \text{HCO}_2\text{H} + \text{R}^2\text{CHO}$$

Compounds containing an aldehyde or ketonic group adjacent to an alcoholic group are also attacked by periodic acid in a similar manner to glycols:

$$\text{RCHOHCHO} \xrightarrow{\text{1HIO}_4} \text{RCHO} + \text{HCO}_2\text{H}$$

$$\text{R}^1\text{COCHOHR}^2 \xrightarrow{\text{1HIO}_4} \text{R}^1\text{CO}_2\text{H} + \text{R}^2\text{CHO}$$

Aldoses and pentoses are *completely* broken down by periodic acid, *i.e.*, *free* sugars behave as open-chain compounds in this reaction, *e.g.*, an aldopentose consumes *four* molecules of periodic acid, and the products are four molecules of formic acid and one molecule of formaldehyde (from the terminal CH_2OH group).

$$\text{OCHCHOHCHOHCHOHCH}_2\text{OH} \xrightarrow{\text{4HIO}_4} 4\text{HCO}_2\text{H} + \text{CH}_2\text{O}$$

Thus, estimating the periodic acid used and the formic acid and/or formaldehyde produced will indicate the number of *free* adjacent oxidisable groups (CHOH, CHO or CO). On the other hand, the ring in *glycosides* is stable in the presence of periodic acid. Thus, the oxidation of methyl glucoside (produced under reflux conditions; see above) uses *two* molecules of periodic acid and produces *one* molecule of formic acid. Only one structure fits these facts, namely, a *six-membered* ring:

$$
\begin{array}{ccc}
\text{CHOCH}_3 & & \text{CHOCH}_3 \\
| & & | \\
\text{CHOH} & & \text{CHO} \\
| & & | \\
\text{CHOH} \quad \text{O} \xrightarrow{2\text{HIO}_4} & & \text{CHO} \quad \text{O} + \text{HCO}_2\text{H} \\
| & & | \\
\text{CHOH} & & \\
| & & \\
\text{CH}\!-\!\!\!-\!\! & & \text{CH}\!-\!\!\!-\!\! \\
| & & | \\
\text{CH}_2\text{OH} & & \text{CH}_2\text{OH}
\end{array}
$$

The other product of the oxidation has also been isolated and characterised by Hudson. This periodic acid oxidation method is now very widely used for structure investigation of carbohydrates.

Since methylation of sugars under different conditions produces different-sized rings, the question is: what is the size of the ring in the original sugar? Various sources of evidence, *e.g.*, X-ray analysis, show that the normal sugars are *pyranoses*.

One other point will be mentioned here, and that is the configuration of the *first* carbon atom. When the open-chain structure is closed to form the ring, two configurations of the new asymmetric carbon atom are possible, the one with the hydrogen to the left (the α-anomer), and the one with the hydrogen to the right (the β-). This is an *arbitrary* arrangement, but it has now been shown that the α- and β-anomers actually have the configurations originally arbitrarily assigned to them, *e.g.*, X-ray analysis of α-D-glucose has shown that the 1,2-hydroxyl groups are in the *cis*-position (McDonald *et al.*, 1950) (see also Vol. 2, Ch. 7).

Since the L-series are the enantiomers of the D-series, in the former the α-anomer has the hydrogen on the first carbon atom to the *right* and the β-anomer has the hydrogen to the *left*. These are *opposite* to those of the D-series, *i.e.*, the configuration of C-1 in an α-D-aldose is identical with that of the β-L-aldose.

Disaccharides

All the disaccharides are crystalline solids, soluble in water, and fall into two classes, the *reducing* sugars and the *non-reducing* sugars.

Just as methanol forms methyl glycosides with the monosaccharides, so can other hydroxy-compounds form similar unions with the monosaccharides. Since the latter are themselves hydroxy-compounds, it is possible that they can link up with themselves to form acetals, *i.e.*, glycosides in which the aglycon is another sugar molecule. A number of such compounds occur in nature, *e.g.*, *sucrose, maltose* and *lactose*.

Sucrose, cane-sugar, $C_{12}H_{22}O_{11}$, is one of the most important compounds commercially, and is obtained from the sugar-cane and sugar-beet. In addition to crystalline sucrose, a syrup is always obtained which will not crystallise. This syrup, known as *molasses*, is also a commercial product.

Sucrose is a white crystalline solid, m.p. 180°C, soluble in water. When heated above its melting point, it forms a brown substance known as caramel. Concentrated sulphuric acid chars sucrose, the product being almost pure carbon. Sucrose is dextrorotatory, its specific rotation being $+66 \cdot 5°$. On hydrolysis with dilute acids sucrose yields an equimolecular mixture of D($+$)-glucose and D($-$)-fructose:

$$C_{12}H_{22}O_{11} + H_2O \xrightarrow{\text{HCl}} \underset{\text{glucose}}{C_6H_{12}O_6} + \underset{\text{fructose}}{C_6H_{12}O_6}$$

Since D($-$)-fructose has a greater specific rotation than D($+$)-glucose, the resulting mixture is lævorotatory. Because of this, the hydrolysis of cane-sugar is known as *the inversion of cane-sugar* (this is not to be confused with the *Walden inversion*), and the mixture is known as *invert sugar*. The inversion (*i.e.*, hydrolysis) of cane-sugar may also be effected by the enzyme *invertase* which is found in yeast.

Controlled oxidation of sucrose in alkaline solution with air gives D-arabonic acid. Oxidation of sucrose with nitric acid under different conditions gives either oxalic acid (80%), tartaric acid (40%), or glucaric acid (30%). Hydrogenation of sucrose under controlled conditions gives a mixture of mannitol and sorbitol (these may be separated by fractional crystallisation).

Sucrose is *not* a reducing sugar, *e.g.*, it will not reduce Fehling's solution; it does not form an oxime or an osazone, and does not undergo mutarotation. This indicates that neither the aldehyde group of glucose nor the ketonic group of fructose is free in sucrose. Thus a tentative structure of sucrose is one in which two molecules, glucose and fructose, are linked by the aldehyde group of the former and ketonic group of the latter. This has been amply confirmed by further work, and sucrose has been shown to be α-D-glucopyranosyl-β-D-fructofuranoside (*i.e.*, α-glucose is linked to β-fructose).

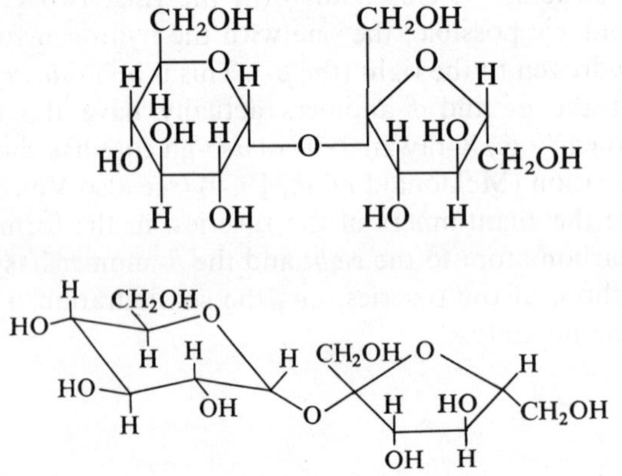

($+$)-**sucrose**

It should be noted that the fructose molecule in sucrose exists as the furanose form, and that when sucrose is hydrolysed, it is the pyranose form of fructose which is isolated.

Maltose (*malt sugar*), $C_{12}H_{22}O_{11}$, is produced by the action of malt (which contains the enzyme *diastase*) on starch:

$$(C_6H_{10}O_6)_n + \frac{n}{2}H_2O \xrightarrow{\text{diastase}} \frac{n}{2}C_{12}H_{22}O_{11}$$

Maltose is a white crystalline solid, m.p. 160–165°C, soluble in water, and is dextrorotatory.

When it is hydrolysed with dilute acids or by the enzyme *maltase*, maltose yields two molecules of D(+)-glucose. Maltose is a reducing sugar, *e.g.*, it reduces Fehling's solution; it forms an oxime and an osazone, and undergoes mutarotation. This indicates that at least one aldehyde group (of the two glucose molecules) is free in maltose. Further work has shown that maltose is 4-*O*-α-D-glucopyranosyl-D-glucopyranose (*i.e.*, the reducing half is linked to the non-reducing half by an α-link). It should also be noted that the enzyme maltase splits only α-glycosidic links, and hence is used to ascertain the presence of this bond in disaccharides and polysaccharides.

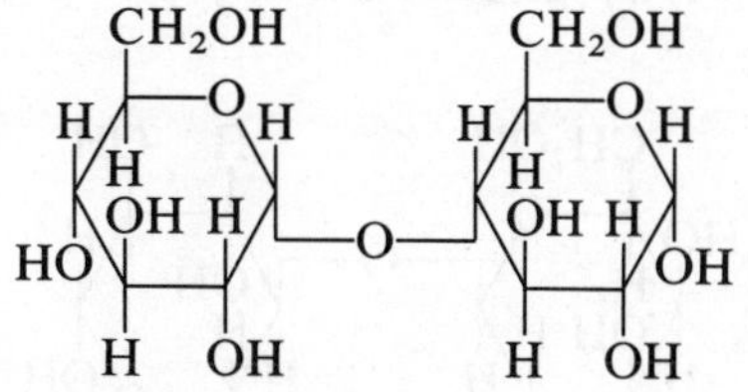

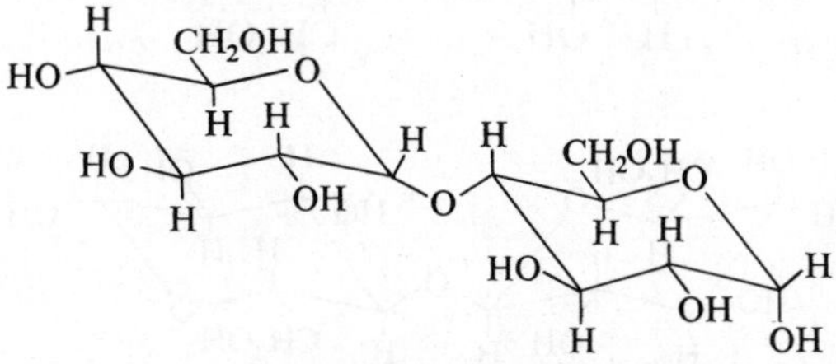

(+)-**maltose**; α-**anomer**

Cellobiose, $C_{12}H_{22}O_{11}$, may be obtained from cellulose by acetylating good filter paper (which is almost pure cellulose) with acetic anhydride in the presence of concentrated sulphuric acid. The octa-acetate of cellobiose so obtained is hydrolysed with potassium hydroxide or with sodium ethoxide, whereupon cellobiose is produced.

Cellobiose is a white crystalline solid, m.p. 225°C, soluble in water, and dextrorotatory. When hydrolysed with dilute acids or by the enzyme *emulsin*, it yields two molecules of D(+)-glucose. It is a reducing sugar, forms an oxime and osazone, and undergoes mutarotation. Cellobiose is 4-*O*-β-D-glucopyranosyl-D-glucopyranose. Thus, the only difference between cellobiose and maltose is that in the former the glycosidic link is β, whereas in the latter it is α. Furthermore, the enzyme emulsin splits only β-glycosidic links (*cf.* maltase).

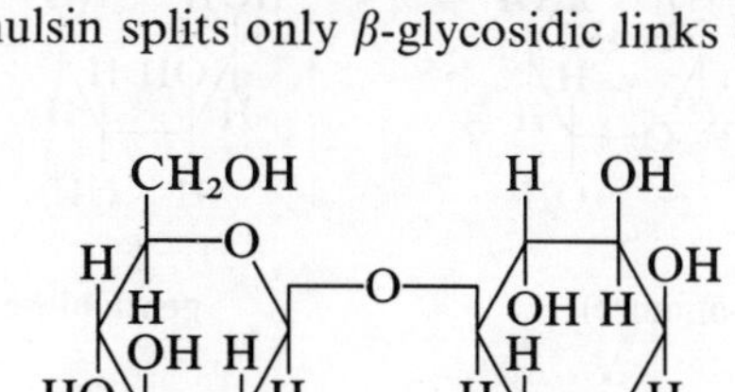

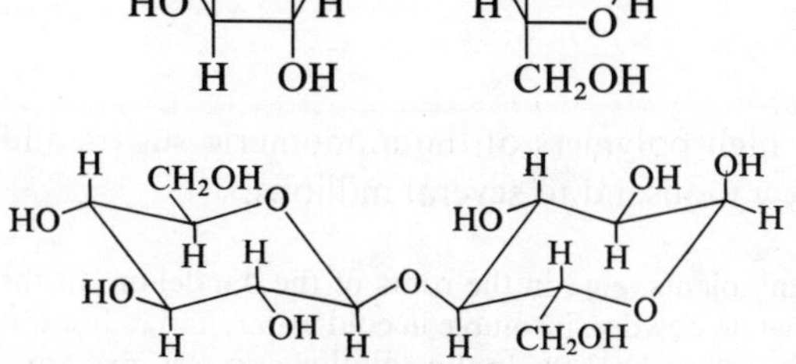

(+)-**cellobiose**; α-**anomer**

Lactose (*milk-sugar*), $C_{12}H_{22}O_{11}$, occurs in the milk of all animals, and is prepared commercially from whey by evaporation to crystallisation; whey is obtained as a by-product in the manufacture of cheese.

Lactose is a white crystalline solid, m.p. 203°C (with decomposition), soluble in water, and is dextrorotatory. It is hydrolysed by dilute acids or by the enzyme *lactase*, to an equimolecular mixture of $D(+)$-glucose and $D(+)$-galactose. Lactose is a reducing sugar, forms an oxime and osazone and undergoes mutarotation. It is 4-*O*-β-D-galactopyranosyl-D-glucopyranose. It should also be noted that lactase is a β-glycosidase, *i.e.*, splits β-glycosides (it has been shown to be identical with emulsin).

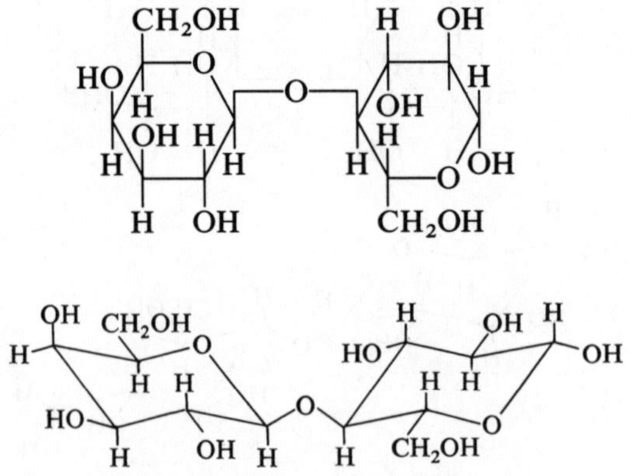

(+)-**lactose**; **β-anomer**

Sucrose, maltose and lactose are three disaccharides which occur naturally. Cellobiose may be prepared from cellulose. Two other disaccharides which have been prepared are **melibiose** (from the trisaccharide *raffinose*), and **gentiobiose** (from the trisaccharide *gentianose*). These differ from the other disaccharides in that the two monosaccharide molecules are linked by the *sixth* carbon atom (the aldehyde carbon atom being number one) of the reducing monosaccharide (see also Vol. 2, Ch. 7):

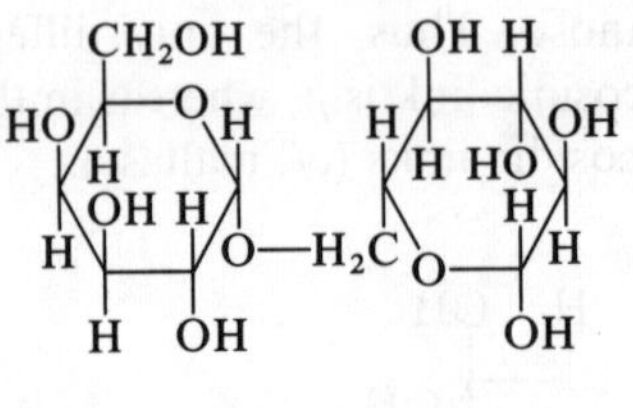

melibiose (*β-anomer*) 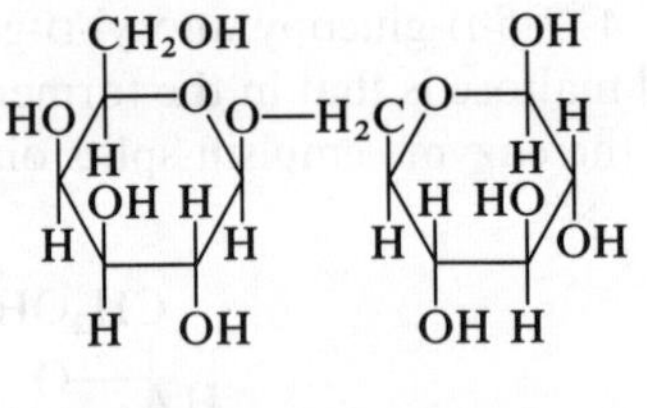   **gentiobiose** (*β-anomer*)

Polysaccharides

The polysaccharides are high polymers of the monomeric sugars and have molecular weights that may range from a few thousand to several millions.

Inulin occurs in many plants, *e.g.*, in the roots of the dandelion, in the tubers of the dahlia and in certain lichens. It is a white powder, insoluble in cold water, and in hot water forms a colloidal solution which does not form a gel on cooling. Inulin solutions do not give any colour with iodine. Inulin is hydrolysed by dilute acids to $D(-)$-fructose. Thus, inulin is $(C_6H_{10}O_5)_n$, and n has a value of about 30 (obtained by chemical methods).

Starch, $(C_6H_{10}O_5)_n$, occurs in all green plants; the commercial sources of starch are maize, wheat, barley, rice, potatoes and sorghum.

Starch consists of two fractions, one being known as α-**amylose** (the 'A' fraction), and the other as β-**amylose** or **amylopectin** (the 'B' fraction); the former comprises 10–20 per cent of starch, and the latter 80–90 per cent. α-Amylose is soluble in water, and the solution gives a blue colour with iodine. This blue colour is believed to be due to the formation of an inclusion complex. An aqueous solution of α-amylose slowly forms a precipitate, since α-amylose has a strong tendency to 'revert' to the insoluble state in solution. Amylopectin is insoluble in water, is stable in contact with water, and gives a violet colour with iodine. α-Amylose and amylopectin are both hydrolysed to maltose by the enzyme diastase, and to D(+)-glucose by dilute acids (amylopectin gives about 50 per cent of maltose). Thus, both contain D-glucopyranose units joined by the α-glucosidic linkage.

α-amylose

α-Amylose consists of an *unbranched chain*, with a molecular weight varying between 10 000 ($n = \sim 60$) and 1 000 000 ($n = \sim 6000$). The value of n depends on the source and treatment of α-amylose. Amylopectin differs from α-amylose in that it contains *branched chains*, the branching occurring through 1,6-linkages (and other linkages) and the length of the unbranched sections being about 24–30 glucose units. The molecular weights recorded for amylopectin vary between 50 000 and 10 000 000.

The **dextrins,** $(C_6H_{10}O_5)_n$, are produced by the partial hydrolysis of starch by boiling with water under pressure at about 250°. They are white powders, and are used for making adhesives and confectionery, for sizing paper, etc.

Glycogen, $(C_6H_{10}O_5)_n$, is found in nearly all animal cells, occurring mainly in the liver; it is the reserve carbohydrate of animals, and so is often known as *'animal starch'*. It has also been isolated from plant sources.

Glycogen is a white powder, soluble in water, the solution giving a purplish-red colour with iodine. On hydrolysis with dilute acid, glycogen gives D(+)-glucose. The molecular weight of glycogen has been given as 1 000 000 to 5 000 000, and glycogen contains highly branched chains.

Pectins are found in plant and fruit juices. Their characteristic property is the ability of their solutions to gelate, *i.e.*, form jellies. They have a high molecular weight, and are polygalacturonic acids (linked 1,4) with the carboxyl groups partially esterified with methanol.

Alginic acids occur in marine algæ, and are polymannuronic acids (linked 1,4).

Cellulose, $(C_6H_{10}O_5)_n$, is the main constituent of the cell-wall of plants, and also occurs in certain animal tissues. It is the most widely distributed organic compound, and its main source is cotton (almost pure cellulose) and wood (which also contains *lignin*, which is not a polysaccharide).

Cellulose is a white solid, insoluble in water but soluble in ammoniacal copper hydroxide solution (*Schweitzer's reagent*). Careful hydrolysis of cellulose gives cellobiose; it is also possible to isolate *cellotriose* (trisaccharide) and *cellotetrose* (tetrasaccharide). All of these saccharides, on further hydrolysis, yield only D(+)-glucose which exists in the β-form in cellulose (*cf.* starch).

The molecular weight of cellulose varies between 20 000 and 500 000, and the compound consists of an unbranched chain.

cellulose

Artificial silk. The term **rayon** is used collectively to cover all synthetic or manufactured fibres from cellulose, but it is usually most often applied to viscose yarns.

Cellulose nitrates (*nitrocellulose*). The highest nitrate of cellulose theoretically possible is the trinitrate, $(C_6H_7N_3O_{11})_n$, which contains 14·4 per cent nitrogen (each glucose unit contains *three* hydroxyl groups). The nitrates are prepared by the reaction of cellulose with a mixture of nitric and sulphuric acids, and the degree of 'nitration' depends on the concentrations of the acids and the time of the reaction. Cellulose trinitrate (12·2–13·2% N) is known as **gun-cotton** and is used in the manufacture of blasting explosives and smokeless powders. The lower nitrates, in the solid state, are known as **pyroxylin,** and their use depends on the nitrogen content, *e.g.*, (i) pyroxylin (10·8–11·0% N) mixed with camphor as plasticiser is converted into **celluloid**; (ii) *nitrocellulose lacquers* are made by dissolving pyroxylin (11·5–12·2% N), resins, pigments and plasticisers in a mixture of solvents.

Cellulose acetate (*celanese silk*). When acetylated with acetic anhydride in the presence of sulphuric acid, cellulose is converted into cellulose triacetate. When the reaction is complete, water is added to decompose the triacetate into the diacetate (approximately). The diacetate is washed, dried and dissolved in a mixture of organic solvents (of which acetone is generally the main constituent). The solution is then forced through a spinneret into a warm chamber, whereupon the solvent evaporates leaving behind fine threads of cellulose acetate.

Cellulose acetate is also used for making non-inflammable photographic and motion-picture films, non-shatterable glass, lacquers and varnishes.

Esters other than the acetate are also used, *e.g.*, cellulose acetate butyrate (prepared by the esterification of cellulose with a mixture of acetic anhydride, acetic acid, and butyric anhydride) mixed with plasticisers forms a thermoplastic which is used for many purposes, *e.g.*, film and sheeting.

Viscose rayon. In the viscose process, cellulose is digested with sodium hydroxide solution, and then carbon disulphide is passed into the solution. A mixture of sodium cellulose xanthates, soluble in sodium hydroxide, is formed (*cf.* p. 407): R = cellulose:

$$R-OH + CS_2 + NaOH \longrightarrow S{=}C{\begin{smallmatrix} \diagup OR \\ \diagdown SNa \end{smallmatrix}} + H_2O$$

This alkaline solution has a high viscosity, and hence the silk obtained by this process was named *viscose* rayon. The viscose solution is forced through a spinneret into a sulphuric acid bath, whereupon cellulose is precipitated as fine threads.

Cellophane is made by extruding a viscose solution through a long narrow slit into an acid bath, whereupon cellulose is precipitated as very thin sheets. These sheets are made moisture-proof by coating with a transparent nitrocellulose lacquer.

Some cellulose ethers, particularly the ethyl ether, are used in the manufacture of plastics.

When agitated with about 20 per cent aqueous sodium hydroxide, cellulose swells; apparently cellulose combines with the sodium hydroxide to form a sodio-cellulose. This is unstable, and is readily decomposed into cellulose by the addition of water, but the cellulose has a number of its physical properties changed, *e.g.*, it absorbs dyes more readily than untreated cellulose. This 'reactive' form of cellulose is known as *mercerised cellulose*.

Cellulose may be oxidised by nitrogen dioxide to give a product which retains its original form and much of the original tensile strength. It appears that about half of the CH_2OH groups (alternately) are oxidised to carboxyl, thereby giving a compound readily soluble in sodium hydroxide. This oxidised cellulose is useful in medicine, *e.g.*, it possesses hæmostatic properties, and is useful as sterile gauze.

PROBLEMS

1. Outline the evidence to justify the structure $HOCH_2(CHOH)_3CHO$ for arabinose. Is this a fair question? Comment.

2. Formulate the equations for the action of periodic acid on:

(i) $CH_3CHOHCH_2OH$;   (ii) $HOCH_2COCH_2OH$;   (iii) $HOCH_2CHOHCHOHCH_2OH$;
(iv) $HOCH_2CH_2CHOHCO_2H$; (v) $HO_2CCHOHCHOHCHO$; (vi) $OCHCH_2CHOMeCHOHCH_2OH$;
(vii) $OCHCHOMeCHOHCH_2OEt$.

3. Convert:
(i) D-erythrose to the next higher aldose; (ii) D-mannose to the next lower aldose; (iii) D-arabinose to the next higher ketose; (iv) D-fructose to the next lower aldose.

4. Designate, using R and S symbols, D(+)-glucose.

5. Complete and comment:

(i)   D-glucose $\xrightarrow[\text{HCl}]{\text{EtSH}}$? $\xrightarrow[\text{NaOH}]{\text{Me}_2\text{SO}_4}$? $\xrightarrow[\text{CdCO}_3]{\text{HgCl}_2}$?

(ii)   D-glyceraldehyde $\xrightarrow[\text{HCl}]{\text{Me}_2\text{CO}}$? $\xrightarrow{\text{CH}_2=\text{CHMgCl}}$? $\xrightarrow{\text{O}_3}$?

(iii)   D-glucose $\xrightarrow[\text{EtONa}]{\text{MeNO}_2}$? $\xrightarrow{\text{H}_2\text{SO}_4}$? $\xrightarrow[\text{Ni}]{\text{H}_2}$?

6. Draw the conformations of the α- and β-forms of:
(i) D-arabinopyranose; (ii) D-galactopyranose. State in each case which is the more stable form.

7. α-Glucopyranose is oxidised by HIO_4 more rapidly than the β-anomer at the 1,2-bond. Suggest a reason.

8. Complete the equations and comment.

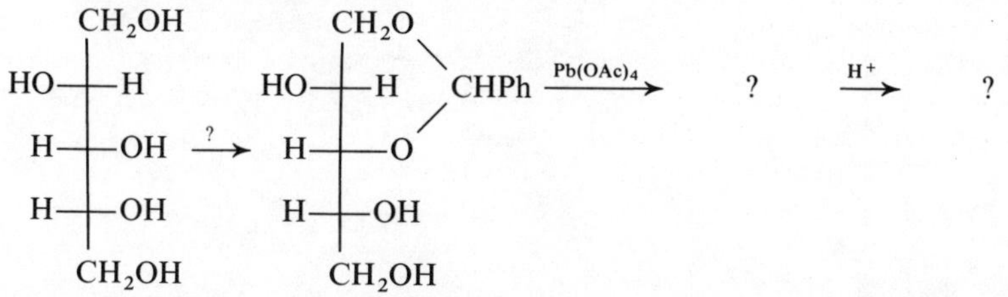

$$
\begin{array}{c}
CH_2OH \\
HO\!-\!\!-\!H \\
H\!-\!\!-\!OH \\
H\!-\!\!-\!OH \\
CH_2OH
\end{array}
\xrightarrow{\;?\;}
\begin{array}{c}
CH_2O \\
HO\!-\!\!-\!H \\
H\!-\!\!-\!O \\
H\!-\!\!-\!OH \\
CH_2OH
\end{array}
\!\!CHPh
\xrightarrow{\text{Pb(OAc)}_4}
\;?\;
\xrightarrow{\text{H}^+}
\;?
$$

9. Ruthenium tetroxide is known to oxidise a *single* CHOH group to ketone. How could you convert $HOCH_2(CHOH)_4CHO$ into $HOCH_2(CHOH)_2COCHOHCHO$?

10. What sugars are obtained by epimerising, respectively. C-2, C-3, C-4 of D-glucose?

11. Glucose, mannose and fructose give identical osazones. Explain.

12. Compound A, $C_5H_{10}O_5$, gives a tetra-acetate with Ac_2O, and oxidation of A with Br_2—H_2O gives an acid, $C_5H_{10}O_6$. Reduction of A with HI gives isopentane. What are possible structures of A and how could you distinguish between the possibilities?

13. Compound A, $C_5H_{10}O_4$, is oxidised by Br_2—H_2O to the acid, $C_5H_{10}O_5$, which readily forms a lactone. A forms a triacetate with Ac_2O and an osazone with $PhNHNH_2$. A is oxidised by HIO_4, only one molecule of which is consumed. What is the structure of A?

14. Compound A, $C_5H_{10}O_4$, on oxidation with Br_2—H_2O, gives acid B, $C_5H_{10}O_5$. A forms a triacetate (Ac_2O) and is reduced by HI to n-pentane. Oxidation of A with HIO_4 gives, among other products, 1 molecule of CH_2O and 1 molecule of HCO_2H. What are the possible structures of A and how could you distinguish between them?

REFERENCES

PERCIVAL, *Structural Carbohydrate Chemistry,* Miller (1962, 2nd edn.).

GUTHRIE and HONEYMAN, *An Introduction to the Chemistry of Carbohydrates,* Clarendon Press (1964, 2nd edn.).

DAVIDSON, *Carbohydrate Chemistry,* Holt, Rinehart and Winston (1967).

MCILROY, *Introduction to Carbohydrate Chemistry,* Butterworths (1967).

FINAR, *Organic Chemistry,* Vol. 2, Longman (1968, 4th edn.), Ch. 7, 'Carbohydrates'.

CAPON, 'Mechanism in Carbohydrate Chemistry', *Chem. Rev.,* 1969, **69**, 407.

FERRIER and OVEREND, 'Newer Aspects of the Stereochemistry of Carbohydrates', *Quart. Rev.,* 1959, **13**, 265.

MAYO (ed.), *Molecular Rearrangements,* Part II, Interscience (1964), Ch. 12, 'Rearrangements and Isomerisations in Carbohydrate Chemistry'.

SUNDERWITH and OLSON, 'Conformational Analysis of the Pyranoside Ring', *J. Chem. Educ.,* 1962, **39**, 410.

RODERICK, 'Structural Variety of Natural Products', *J. Chem. Educ.,* 1962, **39**, 2.

19

Alicyclic compounds

There is a large number of compounds which contain closed rings comprised of carbon atoms only. These compounds are known collectively as **carbocyclic** or **homocyclic compounds**. In this group falls a class of compounds which resemble the aliphatic compounds in many ways, and hence they are often called **alicyclic compounds** (*ali*phatic *cyclic* compounds). The saturated alicyclic hydrocarbons have the general formula C_nH_{2n} (the same as that of the alkenes); they do not contain a double bond but possess a ring structure.

When the molecular formula of a *saturated* hydrocarbon corresponds to the general formula C_nH_{2n-2}, then the compound contains two rings; if to C_nH_{2n-4}, three rings, etc

Nomenclature

Since the saturated alicyclic hydrocarbons contain a number of methylene groups joined together to form a ring, they are known as the **polymethylenes**, the number of carbon atoms in the ring being indicated by a Greek or Latin prefix, *e.g.*,

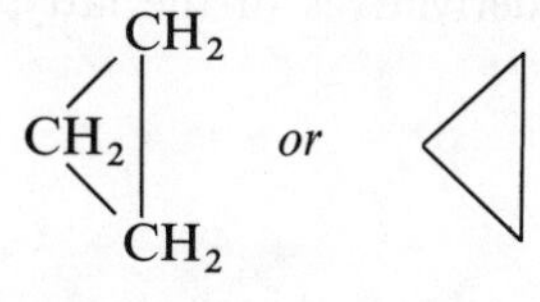

trimethylene 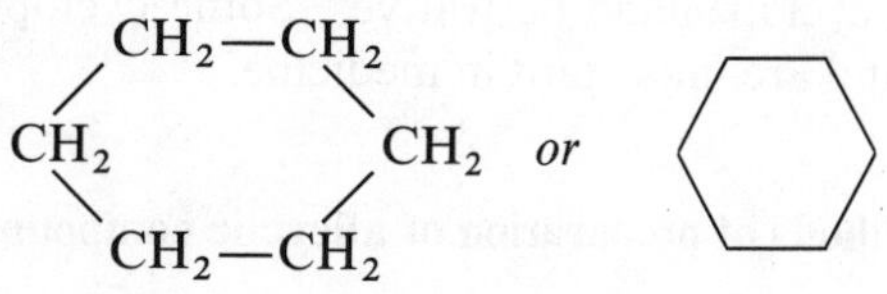   hexamethylene

According to the I.U.P.A.C. system, the saturated monocyclic hydrocarbons take the names of the corresponding open-chain saturated hydrocarbons, preceded by the prefix cyclo- and they are known collectively as the **cycloparaffins** or **cycloalkanes**, and if the alicyclic hydrocarbon is unsaturated, the rules applied to the alkenes are used, *e.g.*,

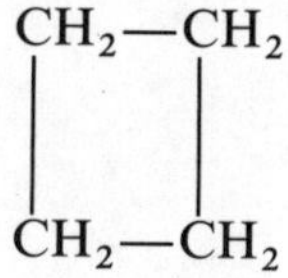

cyclobutane 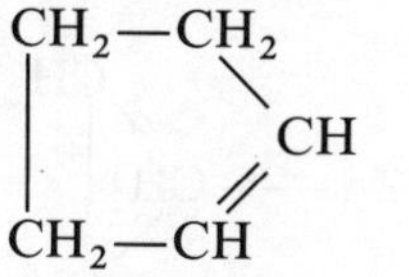   cyclopentene 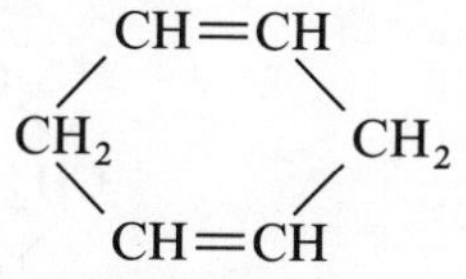   cyclohexa-1,4-diene

In addition to the simple monocyclic compounds, there are more complicated compounds with bridges linked across the ring, *e.g.*,

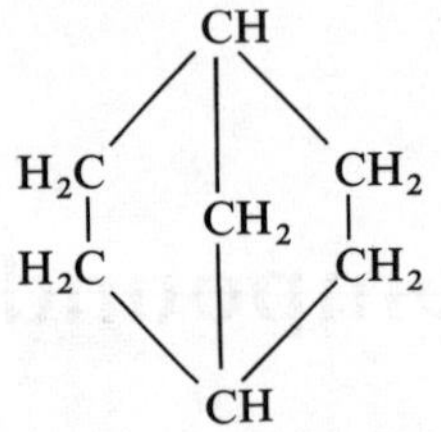

bicyclo[2,2,1]heptane 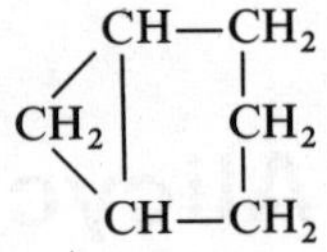   bicyclo[3,1,0]hexane

According to the I.U.P.A.C. system, cycloalkanes consisting of two rings only and having two or more atoms in common, take the prefix *bicyclo* followed by the name of the alkane *containing the same total number of carbon atoms*. The number of carbon atoms in each of the three bridges connecting the two tertiary carbon atoms is indicated in brackets in descending order. Numbering begins with one of the bridgeheads and proceeds by the longest possible path to the second bridgehead; numbering is then continued from this atom by the longer unnumbered path back to the first bridgehead and is completed by the shortest path, *e.g.*,

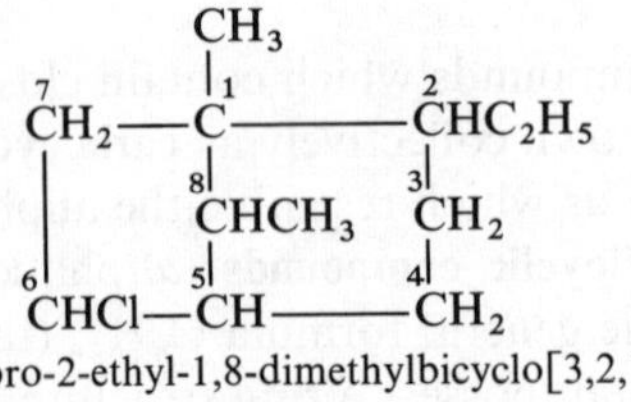

6-chloro-2-ethyl-1,8-dimethylbicyclo[3,2,1]octane

N.B. A bridged system is considered to have a number of rings equal to the number of scissions required to convert the system into an acyclic compound.

Cycloalkanes and cycloalkenes

Five- and six-membered cycloalkanes occur in petroleum (the *naphthenes*; see p. 89); three-, four- and five-membered rings occur in terpenes, the most important class of alicyclic compounds. Many cyclic acids also occur in petroleum; these are known as the *naphthenic acids* and are mainly cyclopentane derivatives. Some cyclopentene derivatives of the fatty acids occur naturally, and are important in medicine.

General methods of preparation of alicyclic compounds

It is interesting to note that up to about 1880, organic chemists thought that there were only two classes of compounds, aliphatic and aromatic; *e.g.*, V. Meyer (1876) believed that rings smaller than six carbon atoms were never likely to be obtained. Since 1882, however, many methods have been introduced to prepare various-sized rings. The following methods are typical.

1. When α,ω-dihalogen derivatives of the alkanes are treated with sodium or zinc, the corresponding cycloalkane is formed (Freund, 1882), *e.g.*, 1,3-dibromopropane forms cyclopropane:

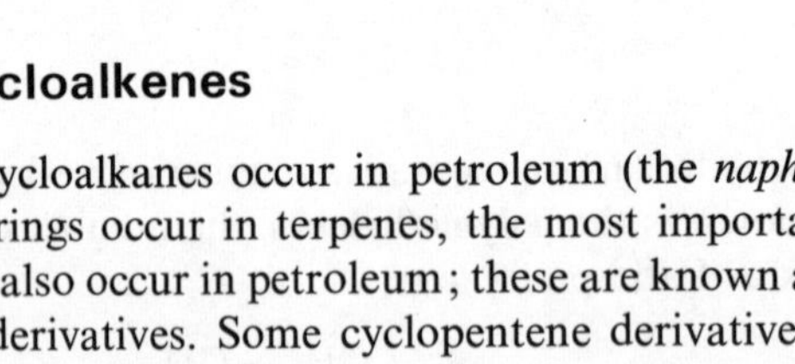

α,ω-Dihalogen derivatives of the alkanes in which the two halogen atoms are further apart than the 1,6 positions, do not form ring compounds but undergo the Wurtz reaction, e.g., $Br(CH_2)_nBr$ $(n > 6)$ gives, with sodium, $Br(CH_2)_{2n}Br$, $Br(CH_2)_{3n}Br$, etc.

2. When the calcium or barium salt of a dicarboxylic acid is distilled, a cyclic ketone is formed (Wislicenus, 1893), e.g., barium adipate gives cyclopentanone. The mechanism is uncertain, but a possibility is (see also p. 204):

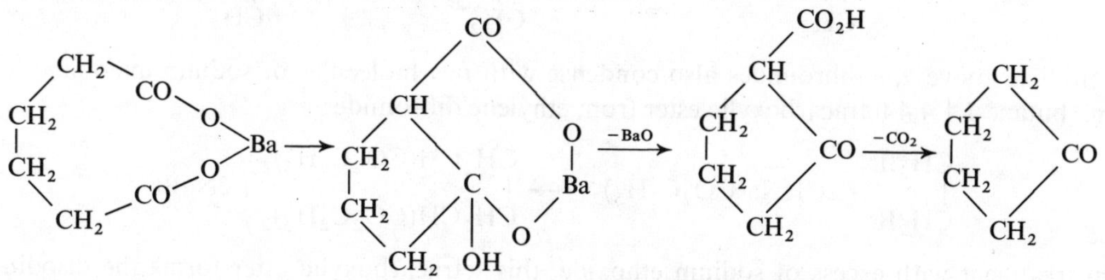

This method is useful for the preparation of only five-, six-, and seven-ring ketones (see also p. 558).

Cyclic ketones may readily be converted into the corresponding cycloalkanes by means of the Clemmensen reduction, e.g.,

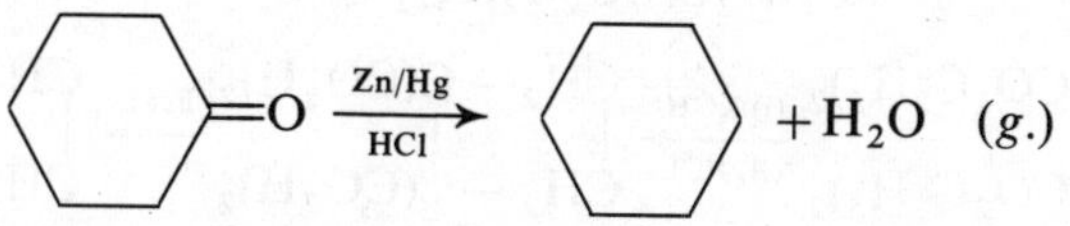

Alternatively, the conversion may be effected by either of the following two methods:

(i)

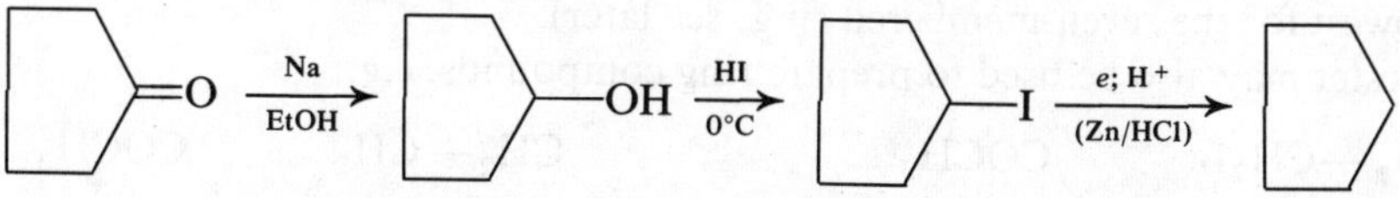

(ii)

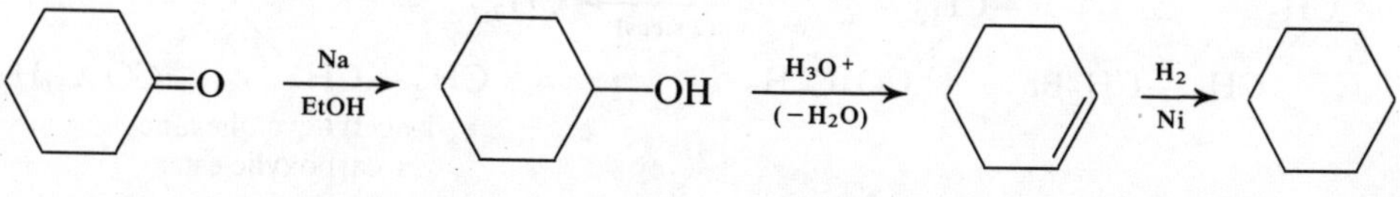

3. Six-membered alicyclic compounds may very conveniently be prepared by the reduction of benzene and its derivatives. Catalytic reduction under pressure using nickel is the most satisfactory, e.g., phenol is almost quantitatively converted into cyclohexanol:

$$C_6H_5OH + 3H_2 \xrightarrow[200°C]{Ni} C_6H_{11}OH$$

4. Various alicyclic compounds may be prepared by the condensation between certain dihalogen derivatives of the alkanes and sodium malonic ester or sodium acetoacetic ester (Perkin junior, 1883), e.g.,

(i) α,ω-Dibromides, $Br(CH_2)_nBr(n = 2–5)$, condense with *one* molecule of malonic ester in the presence of *two* molecules of sodium ethoxide to form cycloalkane-1,1-dicarboxylic esters. These may be converted into the monocarboxylic acid by the usual procedure used in malonic ester syntheses.

$$\underset{\underset{CH_2Br}{|}}{CH_2Br} \xrightarrow{Na^{+\ -}CH(CO_2C_2H_5)_2} \underset{\underset{CH_2Br}{|}}{CH_2CH(CO_2C_2H_5)_2} \xrightarrow{EtONa} \left[\underset{\underset{CH_2\overset{|}{\underset{}{}}Br}{|}}{CH_2\bar{C}(CO_2C_2H_5)_2} \right] \longrightarrow$$

$$\underset{CH_2}{\overset{CH_2}{<}}C(CO_2C_2H_5)_2 \xrightarrow[(ii)\ HCl]{(i)\ KOH} \underset{CH_2}{\overset{CH_2}{<}}C(CO_2H)_2 \xrightarrow{heat} \underset{CH_2}{\overset{CH_2}{<}}CHCO_2H + CO_2$$

(ii) The above α,ω-dibromides also condense with *two* molecules of sodium malonic ester; *e.g.*, butane-1,1,4,4-tetracarboxylic ester from ethylene dibromide:

$$\underset{\underset{CH_2Br}{|}}{CH_2Br} + 2CHNa(CO_2C_2H_5)_2 \longrightarrow \underset{\underset{CH_2CH(CO_2C_2H_5)_2}{|}}{CH_2CH(CO_2C_2H_5)_2} + 2NaBr$$

On treatment with excess of sodium ethoxide, this tetracarboxylic ester forms the disodium derivative which, when treated with iodine, is converted into a cyclobutane derivative. If methylene di-iodide is used instead of iodine, the cyclopentane derivative is obtained:

$$\underset{\underset{CH_2CH(CO_2C_2H_5)_2}{|}}{CH_2CH(CO_2C_2H_5)_2} \xrightarrow{2NaOC_2H_5} \underset{\underset{CH_2\bar{C}(CO_2C_2H_5)_2}{|}}{CH_2\bar{C}(CO_2C_2H_5)_2} \xrightarrow{I_2}$$

$$\underset{\underset{CH_2C(CO_2C_2H_5)_2}{|}}{CH_2C(CO_2C_2H_5)_2} \xrightarrow[(ii)\ HCl]{(i)\ KOH} \underset{\underset{CH_2-C(CO_2H)_2}{|}}{CH_2-C(CO_2H)_2} \xrightarrow{heat} \underset{\underset{CH_2-CHCO_2H}{|}}{CH_2-CHCO_2H} + 2CO_2$$

Thus, by using the appropriate dihalogen derivatives of the alkanes under suitable conditions, it is possible to prepare rings containing three to seven carbon atoms (the yield being highest for the five- and lowest for the seven-membered ring; see later).

Acetoacetic ester may also be used to prepare ring compounds, *e.g.*,

$$\underset{CH_2-CH_2Br}{\overset{CH_2-CH_2Br}{CH_2<}} + \underset{CO_2C_2H_5}{\overset{COCH_3}{CH_2<}} \xrightarrow[(2\ steps)]{2NaOC_2H_5} \underset{CH_2-CH_2}{\overset{CH_2-CH_2}{CH_2<}}\underset{CO_2C_2H_5}{\overset{COCH_3}{>C<}}$$

1-acetyl-cyclohexane-
1-carboxylic ester

Hydrolysis and decarboxylation of this compound produce cyclohexyl methyl ketone.

By using acetoacetic ester, it is possible to prepare rings containing three, five, six and seven carbon atoms, but not four; all attempts to prepare a four-membered ring result in the formation of a dihydropyran derivative (p. 862):

$$(CH_2)_3\underset{Br}{\overset{Br}{<}} + Na^+\bar{C}\underset{COCH_3}{\overset{CO_2C_2H_5}{<}} \longrightarrow \underset{\underset{H_2C}{|}}{\overset{CH_2}{H_2C}}\underset{\underset{CCH_3}{\underset{Br\ O}{|}}}{\overset{CCO_2C_2H_5}{\|}} \longrightarrow \underset{\underset{H_2C}{|}}{\overset{CH_2}{H_2C}}\underset{\underset{CCH_3}{O}}{\overset{CCO_2C_2H_5}{\|}}$$

5. Certain cyclic ketones can be obtained by the **Dieckmann reaction** (1901). This reaction, which is really an intramolecular Claisen condensation, is carried out by treating the esters of adipic, pimelic or suberic acids with sodium, or better, with sodium ethoxide (Mayer *et al.*, 1959), whereupon five-, six- or seven-membered rings, respectively, are obtained; *e.g.*, adipic ester forms cyclopentanone:

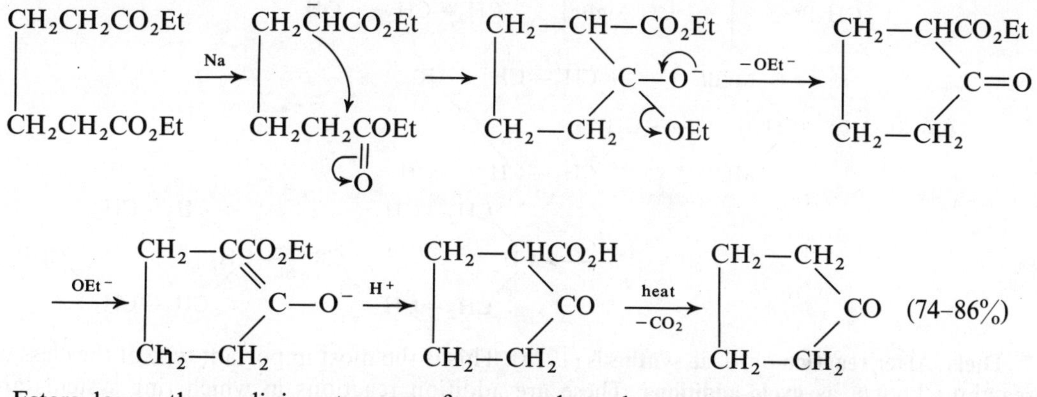

Esters lower than adipic ester may form products by *intermolecular* condensation and cyclisation, *e.g.*, in the presence of sodium or sodium ethoxide, ethyl succinate forms *succino-succinic ester* (cyclohexane-2,5-dione-1,4-dicarboxylic ester):

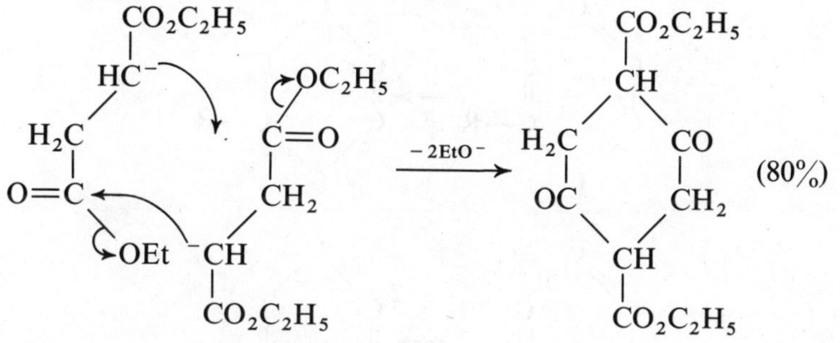

On the other hand, five-membered ring compounds may be prepared by the intermolecular condensation between oxalic and glutaric esters:

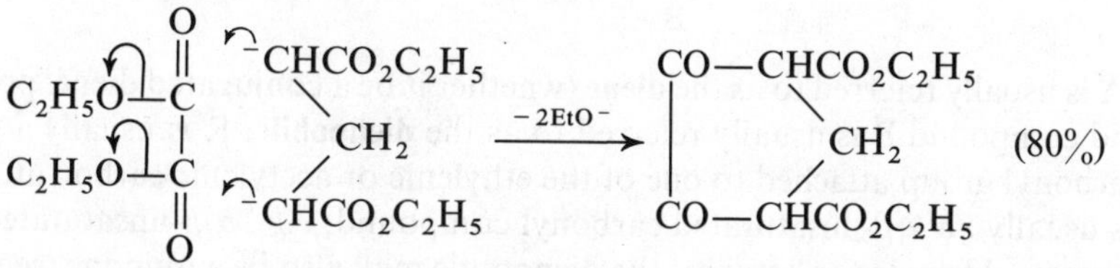

Certain ketones may also be used with ethyl oxalate to give five-membered rings, *e.g.*, acetone gives cyclopentane-1,2,4-trione.

6. Various types of bifunctional compounds undergo intramolecular reactions under suitable conditions to form cyclic compounds, particularly when the product is a five- or six-membered ring, *e.g.*,

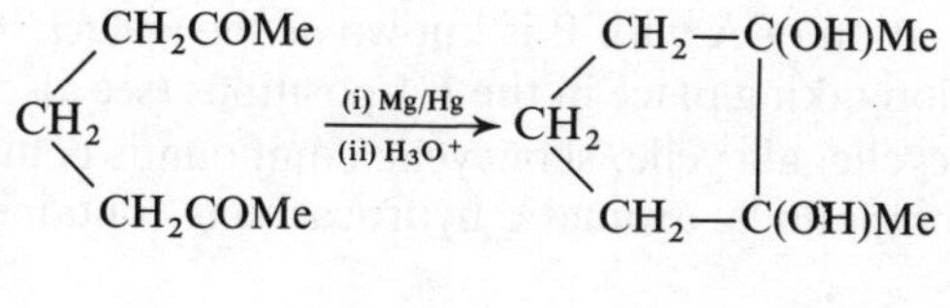

(ii) 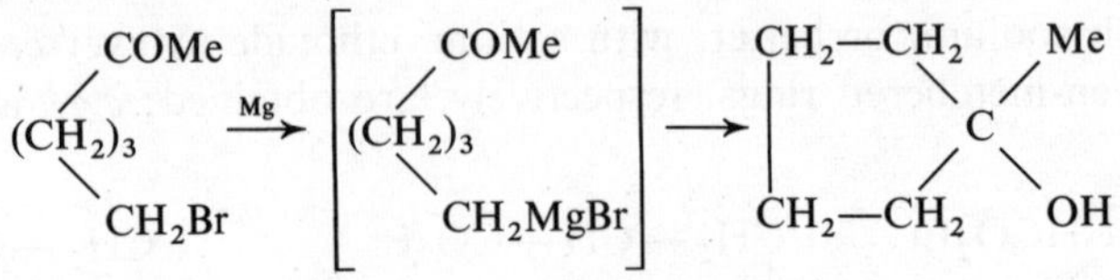 $(CH_2)_3$ with COMe and CH$_2$Br $\xrightarrow{\text{Mg}}$ $\left[(CH_2)_3 \text{ with COMe and } CH_2MgBr \right]$ $\longrightarrow$ $\begin{array}{c} CH_2—CH_2 \\ | \quad\quad | \\ CH_2—CH_2 \end{array} C \begin{array}{c} Me \\ OH \end{array}$

(iii) 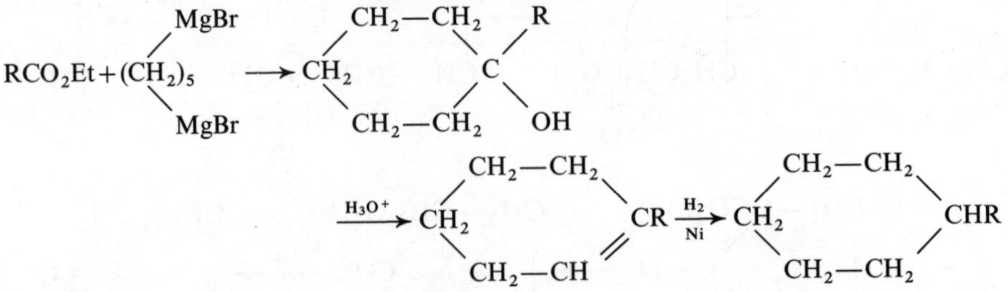 $RCO_2Et + (CH_2)_5$ with MgBr and MgBr $\longrightarrow$ $CH_2 \begin{array}{c} CH_2—CH_2 \\ | \quad\quad | \\ CH_2—CH_2 \end{array} C \begin{array}{c} R \\ OH \end{array}$

$\xrightarrow{H_3O^+}$ $CH_2 \begin{array}{c} CH_2—CH_2 \\ | \quad\quad | \\ CH_2—CH \end{array} CR \xrightarrow[\text{Ni}]{H_2}$ $CH_2 \begin{array}{c} CH_2—CH_2 \\ | \quad\quad | \\ CH_2—CH_2 \end{array} CHR$

7. Diels–Alder reaction or **diene synthesis** (1928). This is the most important type of the class of reactions known as **cyclo-additions**. These are addition reactions in which ring systems are formed without elimination of any compounds. In general terms, the Diels–Alder reaction may be written:

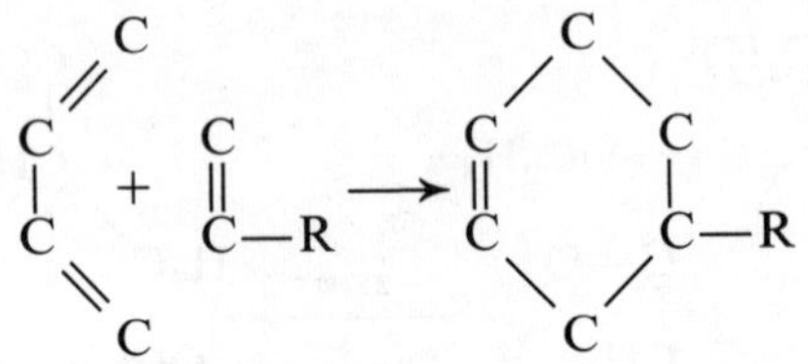

and

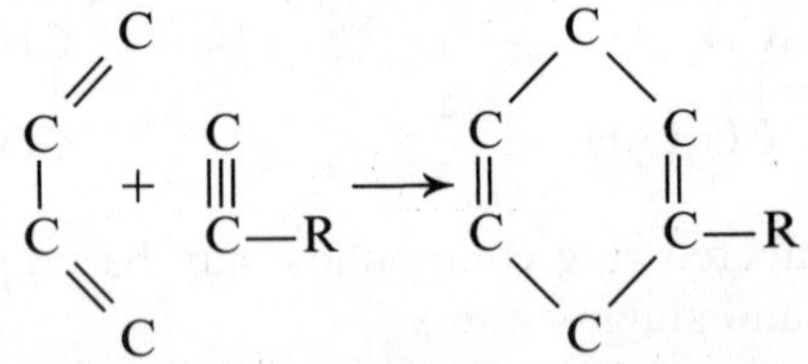

Compound A is usually referred to as the **diene** (whether it be a conjugated diene, polyene, enyne or diyne), and compound B is usually referred to as the **dienophile**. R is usually a group which contains a carbonyl group attached to one of the ethylenic or acetylenic carbon atoms, *e.g.*, the dienophile is usually an α,β-unsaturated carbonyl compound, *e.g.*, α,β-unsaturated acids, acid anhydrides, esters, aldehydes or ketones; the dienophile may also be a quinone (see p. 813). The presence of a carbonyl group (a −I group) in the dienophile is not essential: compounds which contain other −I groups such as nitro- or cyano-, can also behave as dienophiles. In certain cases, the dienophile may even be an unsaturated hydrocarbon (see p. 126). Nevertheless, the diene synthesis takes place most readily when the dienophile contains a carbonyl group, and the most useful dienophile is *maleic anhydride*. Tetracyanoethylene, $C(CN)_2{=}C(CN)_2$ m.p. 200°C, however, appears to be the most reactive dienophile discovered so far.

The compound formed by the condensation of A with B is known as the **adduct**. The adduct is usually a six-membered ring, the addition taking place in the 1,4-positions (see also below).

The diene may be of various types: acyclic, alicyclic, semicyclic compounds containing two double bonds in conjugation, bicyclic compounds, aromatic hydrocarbons containing at least three *linear* benzene rings (*e.g.*, anthracene; see p. 814), and certain heterocyclic compounds (*e.g.*, furan; see p. 830). It is important to note that only a cisoid and *not* a transoid diene (p. 130) enters into the Diels–Alder reaction, *e.g.*, in example (v), the conformation given is cisoid; if this could be 'frozen' in the transoid conformation, no reaction would occur.

The Diels–Alder reaction requires no catalyst; the two compounds are heated together or heated in some solvent, *e.g.*, benzene. Some examples are:

(i)

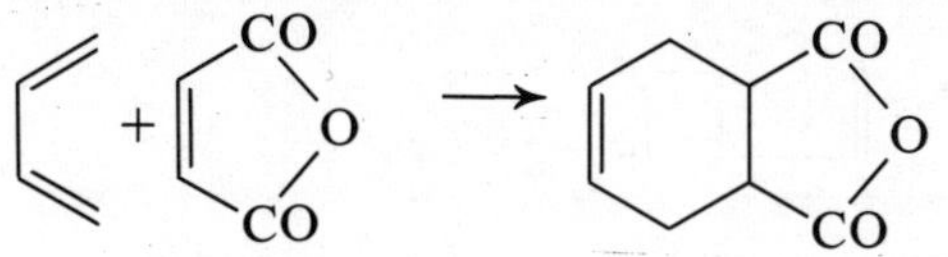

(ii)

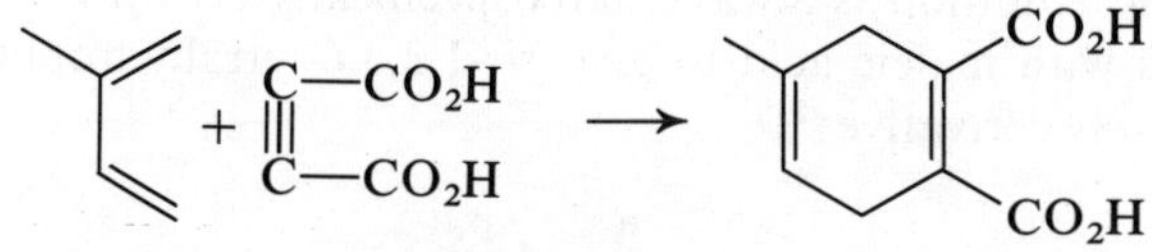

(iii)

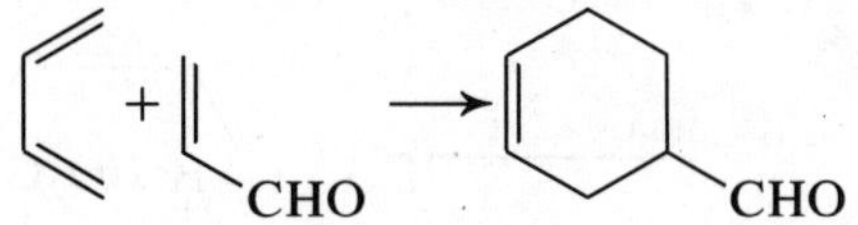

(iv)

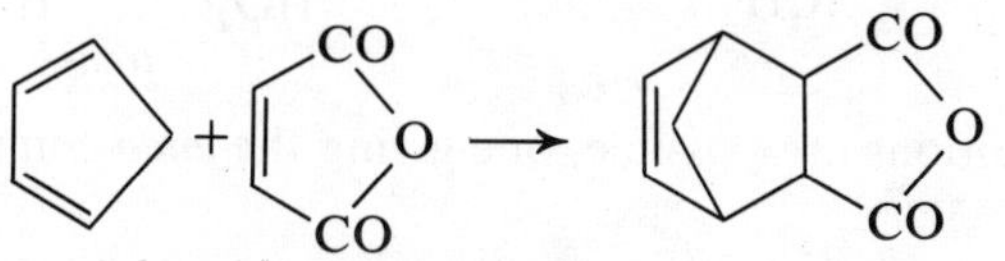

(v)

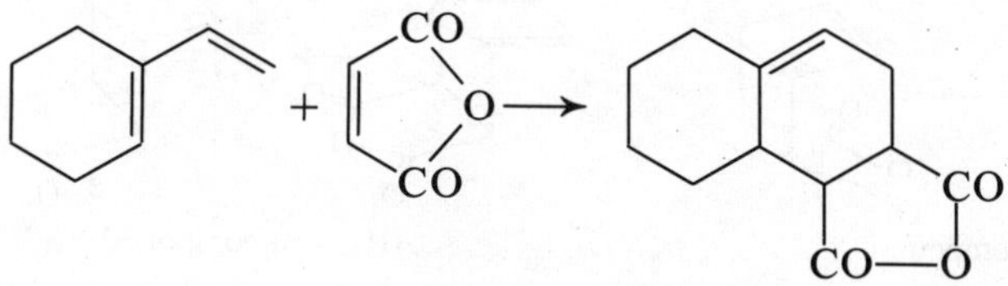

The reason why only polynuclear aromatics containing at least three linear benzene rings undergo the Diels–Alder reaction is uncertain. One possible explanation is as follows. The reaction is exothermic ($\Delta H = 50\cdot2$–$75\cdot3$ kJ mol^{-1}), and in formation of the adduct, resonance energy of the diene is lost. If ΔH−R.E. (loss) is still positive, the reaction is thermodynamically favourable, and so may proceed (see p. 59). In the examples described here (benzene, naphthalene, and anthracene), the R.E. of the adduct is given by $150\cdot6 \, n$kJ mol^{-1}; where n is the number of benzene rings remaining intact (the R.E. of benzene is $150\cdot6$ kJ mol^{-1}); R.E.$_{obs.}$ is the experimental value of each aromatic compound (see also p. 576).

	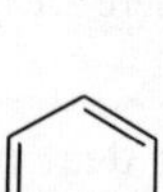	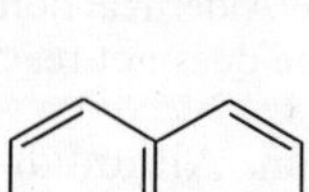	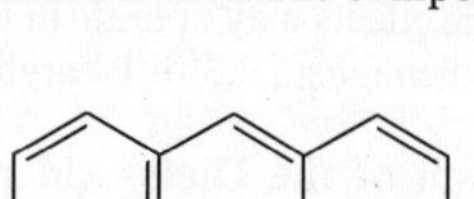
R.E. (*obs.*) kJ mol^{-1}	150·6	255·2	351·5
R.E. (adduct)	0	150·6	301·3
R.E. loss	150·6	104·6	50·2

Only for anthracene is ΔH−R.E. (loss) positive (see p. 814).

In a number of cases, addition in the Diels–Alder reaction is 1,2 to give a *four-membered ring*, and this is particularly the case when tetrafluorethylene is the dienophile, *e.g.*,

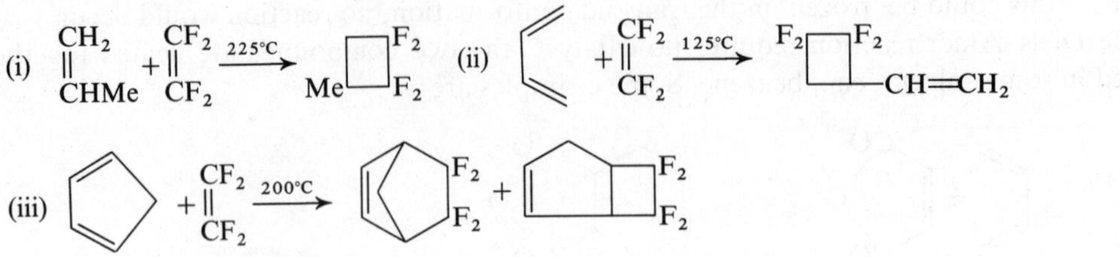

Stereochemistry of the addition. Addition is always stereospecifically *cis with respect to the dienophile*, *e.g.*, butadiene reacts with maleic acid to give *cis*-1,2,3,6-tetrahydrophthalic acid, whereas fumaric acid gives the *trans* derivative:

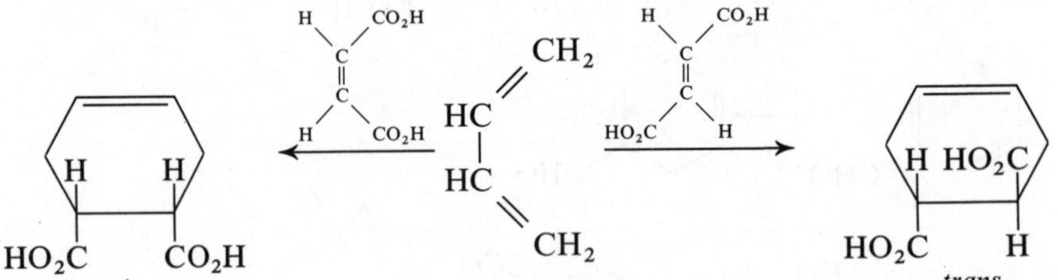

When the diene is cyclic, *two cis*-additions are possible, one giving the *endo*-compound and the other the *exo*-compound, *e.g.*,

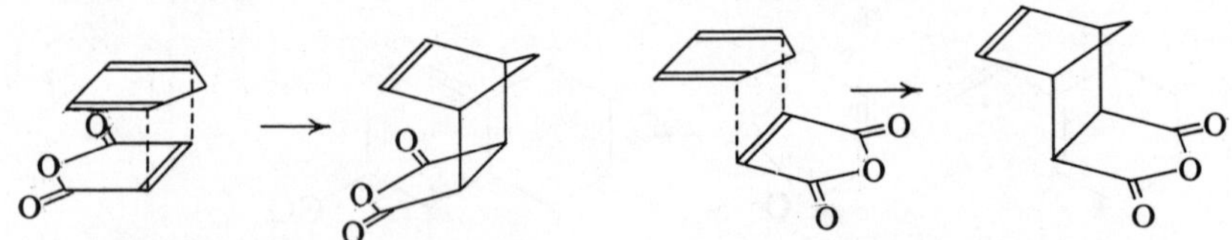

(I); *endo*-compound   (II); *exo*-compound

It has been found experimentally that the *endo*-product is usually obtained exclusively. It has been shown, however, that the diene synthesis is reversible and exothermic, and therefore raising the temperature shifts the equilibrium to the reactants (Wassermann, 1938). It has also been shown that the *endo*-compound is the kinetically controlled product and the *exo*-compound the thermodynamically controlled product. Thus raising the temperature of the reaction favours the formation of the *exo*-compound.

Steric effects may operate in the Diels–Alder reaction when there are large groups at positions 2 and 3 in the diene, *e.g.*, 2,3-di-t-butylbutadiene does not react.

Mechanism of the Diels–Alder reaction. Although a great deal of work has been done, the mechanism of this reaction is still uncertain. Wassermann (1950) proposed that ring formation occurs by *both* new bonds (of the addition) being formed *simultaneously*. This is imagined to proceed through a *non-planar* transition state, and is produced by bonds being formed simultaneously between diene and dienophile, which are in parallel planes, by overlap of the *p*-orbitals in an *endwise* fashion instead of the usual sideways overlap (see (I) and (II); the broken lines represent the overlapping orbitals). This mechanism accounts for stereospecific addition and also for steric effects (see above), but it does not account for polar effects, *e.g.*, acraldehyde reacts faster with isoprene than with butadiene. The presence of the methyl group (+I effect) would not be expected to affect the rate with the above mechanism. It has been argued, however, that the less symmetrical the diene and dienophile are, the less symmetrical are the charge distributions in each, and consequently the formation of the transition state is facilitated. At the same time, this polarisation would account for the formation of the predominant product when two orientations are possible, *e.g.*, isoprene and acraldehyde react to give predominantly the 4-methyl adduct (4-methyl-1,2,5,6-tetrahydrobenzaldehyde).

(major product)

These mechanisms are also discussed on p. 916.

Classification of monocyclic systems

Monocyclic systems have been classified according to the number of carbon atoms in the ring: **small rings**, 3–4; **common rings**, 5–7; **medium rings**, 8–11; **large rings**, 12–. As we shall see, many chemical properties depend on the class of the cycloalkane, and these differences in behaviour have been explained largely in terms of steric strain.

Baeyer's strain theory. Baeyer (1885) was the first to point out that the angle subtended by the corners and centre of a regular tetrahedron—109° 28′—lies between the values of the angles in a regular pentagon (108°) and a regular hexagon (120°). On this observation was based the **Baeyer strain theory**. According to the strain theory, the valency angle can be altered from this normal value (109° 28′), but when altered, a strain is set up in the molecule, and the greater the deviation from the normal angle, the greater is the strain. Thus, according to Baeyer, five- and six-membered rings form most readily, and are the most stable because they involve the least strain (or distortion) from the normal valency angle. If all the rings are *assumed* to be planar, the distortions for each ring-size can be readily calculated, *e.g.*, in cyclopropane the three carbon atoms each occupy a corner of an equilateral triangle. Since the angles of an equilateral triangle are 60°, the distortion in cyclopropane will be $\frac{1}{2}$ (109° 28′–60°) = +24° 44′ (Fig. 19.1). It should be noted that the distortion of the bond angle has been assumed to be *equally* shared between the two bonds. Table 19.1 gives a list of distortions, and it can be seen that these decrease rapidly from ethylene (counting this as a 'two-membered' ring) to cyclopentane and then increase more slowly. Thus, the smaller rings (including ethylene) suffer large distortion and this was believed to be the cause of their enhanced reactivity.

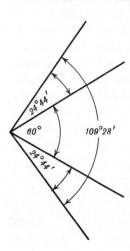

Fig. 19.1

Baeyer's strain theory is based on a mechanical concept of valency and on the assumption that *all* the rings are planar. This has led to conclusions which have now been shown to be wrong. In the first place, physical methods have shown that all rings, except cyclopropane, are not planar (cyclopropane is a three-point system and consequently *must* be planar). In the second place, quantum mechanical calculations do not permit very large distortions of bond angles. If cyclopropane were an equilateral triangle with bonds directed *along* the lines joining the carbon atoms, then the ring valency angle of each carbon atom would be 60°. This value is impossible, since the carbon valency angle can never be less than 90° (when they are pure *p*-orbitals). Furthermore, mixture of *p* with *s*-orbitals *opens* the valency angle. According to Coulson *et al.* (1949), calculation has shown that the smallest carbon valency angle that one can reasonably expect to have is 104°. Coulson has therefore suggested that in cyclopropane, the carbon hybridised orbitals are not pointing towards one another in the same straight line, and consequently there is a loss of overlap (Fig. 19.2*a*). It is this loss of overlap that gives rise to instability, the cyclopropane molecule being in a state of 'strain' due to 'bent' bonds. Applying this argument to cyclobutane (Fig. 19.2*b*), we see that this molecule also has 'bent' bonds, but loss of overlap is less in this case than for cyclopropane, and so the former will be more stable than the latter.

Since strain affects chemical stability, measurement of the latter will indicate the presence of the former. Chemical stability may be measured in various ways, *e.g.*, by the heat of formation, heat of combustion, dipole moment, absorption spectra, etc. One of the most convenient to work with in respect to hydrocarbons is the heat of combustion; this gives a measure of **thermochemical stability** and is the *total* strain, *i.e.*, the sum of the *Baeyer strain* (bond angle deformation only) and other sources of strain (steric repulsions, etc.). If the strain in a ring changes with the size of the ring, then this should be observed by changes in the heat of combustion.

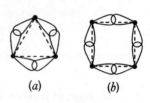

(a)   (b)

Fig. 19.2

The *total strain* in the molecule may be calculated as follows:

(number of C atoms in ring) × (observed heat of combustion/CH_2

− observed heat of combustion/CH_2 for *n*-alkanes)

According to the results shown in Table 19.1, stability increases up to the six-membered ring, then decreases from seven to eleven-membered rings, and from the twelve-membered onwards attains the stability of the six-membered ring. These conclusions have been reached from heats of combustion obtained experimentally, and included in Table 19.1 are large rings which, according to Baeyer's strain theory are difficult, if at all possible, to prepare (see later).

Table 19.1

Number of carbon atoms in the ring	Angle between valency bonds	Distortion	Heat of combustion in kJ/CH_2	Total strain (kJ)
2	0°	54° 44′	711	108
3	60°	24° 44′	697	120
4	90°	9° 44′	685	112
5	108°	0° 44′	664	35
6	120°	−5° 16′	659	12
7	128° 34′	−9° 33′	662	35
8–11	135°–147° 16′	−12° 46′−−18° 54′	661–665	32–88
12–	150°–	−20° 16′	657–661	0–48
n-alkanes	109° 28′	0°	657	0

General properties of the cycloalkanes

In many chemical respects the cycloalkanes are similar to the alkanes, but in others they are different in that the lower members form addition products with *ring fission*. Chlorine and bromine, in the presence of a catalyst, form halogenocycloalkanes, but with cyclopropane, the product is a mixture of 1,1-, 1,2-, and 1,3-dihalogenopropanes. Concentrated hydrogen iodide has no action on cycloalkanes except cyclopropane (at room temperature), which gives *n*-propyl iodide, and cyclobutane (on heating) which gives *n*-butyl iodide. Hydrogen in the presence of nickel at elevated temperatures does not attack cycloalkanes except cyclopropane (at 80°C to give propane) and cyclobutane (at 200°C to give *n*-butane).

Depending on the nature of the compound and on the reagents used, various rings may be opened or changed in size (see text).

The infra-red absorption region of the CH_2 group (C—H stretch) in cycloalkanes (see Fig. 13.4) is the same as that for alkanes (p. 84) except in the case of cyclopropane where the region is $3\,080$–$3\,040$ cm^{-1} (v). On the other hand, the absorption region of the carbonyl group (C=O stretch) in cycloalkanones depends on the size of the ring: 4-ring, $1\,780$–$1\,760$ cm^{-1} (s); 5-ring, $1\,750$–$1\,740$ cm^{-1} (s); 6- and 7-rings, $1\,720$–$1\,700$ cm^{-1} (s). Thus the first two rings can be distinguished from acyclic ketones (p. 208), but not the last two. However, the molecular formula will help here.

Small- and common-ring compounds

Cyclopropane b.p. $-34°C$, may be prepared by the action of zinc on trimethylene dibromide (p. 532).

Cyclopropane is one of the best anæsthetics known. When heated to a high temperature, preferably in the presence of a catalyst, *e.g.*, platinum, cyclopropane is converted into propene.

Cyclopropanes are readily prepared by addition reactions involving alkenes and methylenes (see also pp. 119, 125):

(i) $CH_3CH_2CH{=}CH_2 + :CCl_2 \xrightarrow[\text{Bu}^t\text{OK}]{\text{CHCl}_3} CH_3CH_2CH{-\!\!-}CH_2$ with CCl_2 bridging

(ii) $Me_2C{=}CMe_2 + :CHCl \xrightarrow[\text{BuLi}]{\text{CH}_2\text{Cl}_2} Me_2C{-\!\!-}CMe_2$ with $CHCl$ bridging

Cyclopropanes may also be produced by reaction between an alkene, methylene di-iodide and a zinc-copper couple. Since this method does not result in insertion reactions (which can occur with alkenes), the reaction is believed to proceed via iodomethylzinc iodide (Hoberg, 1962).

$$
\begin{array}{ccc}
\underset{C}{\overset{C}{\|}} \;+\; \underset{CH_2I}{ZnI} & \longrightarrow & \underset{-C-CH_2I}{-C-ZnI} \longrightarrow \; \underset{C}{\overset{C}{\diagdown}} CH_2 + ZnI_2
\end{array}
$$

Cyclopropanes may also be prepared by addition of diazomethane to activated double bonds, and then decomposing the pyrazoline by heat in the presence of copper powder or by photochemical decomposition.

$$
\begin{array}{c}
\text{CHCO}_2\text{Et} \\
\| \qquad\qquad + \text{CH}_2\text{N}_2 \\
\text{CHCO}_2\text{Et}
\end{array}
\longrightarrow
\begin{array}{c}
\text{CH}_2\!\!-\!\!\text{CHCO}_2\text{Et} \\
\mid \qquad\quad \mid \\
\text{HN} \qquad \text{CCO}_2\text{Et} \\
\diagdown \quad \diagup \\
\text{N}
\end{array}
\xrightarrow[\text{or } h\nu]{\text{Cu; heat}}
\begin{array}{c}
\text{CHCO}_2\text{Et} \\
\diagup \bullet \\
\text{CH}_2 \\
\diagdown \bullet \\
\text{CHCO}_2\text{Et}
\end{array}
+ \text{N}_2 \longrightarrow
\begin{array}{c}
\text{CHCO}_2\text{Et} \\
\diagup \mid \\
\text{CH}_2 \mid \\
\diagdown \mid \\
\text{CHCO}_2\text{Et}
\end{array}
$$

Many cyclopropanecarboxylic acids are known; some exhibit geometrical and optical isomerism, *e.g.*, cyclopropane-1,2-dicarboxylic acid exists in the *cis*-form (optically inactive; it has a plane of symmetry) and the *trans*-form which is optically active (it has neither a plane nor a centre of symmetry).

cis-acid m.p. 139°C   *trans*-acid m.p. 175 °C

Cyclopropane, b.p. −36°C, has been prepared from cyclopropylamine via the Hofmann exhaustive methylation reaction (see below).

Cyclopropylamine, b.p. 50°C, on treatment with nitrous acid forms allyl alcohol instead of the expected cyclopropanol.

$$
\begin{array}{c}
\text{CH}_2 \\
\mid \diagdown \\
\mid \quad \text{CHNH}_2 + \text{HNO}_2 \longrightarrow \text{CH}_2\!\!=\!\!\text{CHCH}_2\text{OH} + \text{N}_2 + \text{H}_2\text{O} \\
\mid \diagup \\
\text{CH}_2
\end{array}
$$

Cyclopropanol has not yet been prepared in a pure form; it readily rearranges to propionaldehyde. It has been prepared in a crude form by several methods (Roberts *et al.*, 1951), *e.g.*,

$$
\begin{array}{c}
\text{CH}_2 \\
\mid \diagdown \\
\mid \quad \text{CHCl} \\
\mid \diagup \\
\text{CH}_2
\end{array}
\xrightarrow{\text{Mg}}
\begin{array}{c}
\text{CH}_2 \\
\mid \diagdown \\
\mid \quad \text{CHMgCl} \\
\mid \diagup \\
\text{CH}_2
\end{array}
\xrightarrow[\text{(ii) H}_2\text{O}]{\text{(i) O}_2}
\begin{array}{c}
\text{CH}_2 \\
\mid \diagdown \\
\mid \quad \text{CHOH} \\
\mid \diagup \\
\text{CH}_2
\end{array}
$$

Cyclobutane, b.p. −15°C. Cyclobutane derivatives may be prepared by means of the malonic ester synthesis. The yield of cyclobutane by ring closure of 1,4-dihalogen derivatives of *n*-butane is very poor. Many derivatives of cyclobutane were known long before the parent hydrocarbon was obtained. This was first prepared by Willstätter (1907), using the *Hofmann exhaustive methylation reaction* (p. 856):

$$
\begin{array}{c}
\text{CH}_2\text{Br} \\
\mid \\
\text{CH}_2\text{CH}_2\text{Br}
\end{array}
+ \text{CH}_2(\text{CO}_2\text{C}_2\text{H}_5)_2
\xrightarrow{2\text{NaOC}_2\text{H}_5}
\begin{array}{c}
\text{CH}_2\!\!-\!\!\text{C}(\text{CO}_2\text{C}_2\text{H}_5)_2 \\
\mid \qquad\qquad \mid \\
\text{CH}_2\!\!-\!\!\text{CH}_2
\end{array}
\xrightarrow{\text{hydrolysis}}
$$

$$
\begin{array}{c}
\text{CH}_2\!\!-\!\!\text{C}(\text{CO}_2\text{H})_2 \\
\mid \qquad\quad \mid \\
\text{CH}_2\!\!-\!\!\text{CH}_2
\end{array}
\xrightarrow{\text{heat}}
\begin{array}{c}
\text{CH}_2\!\!-\!\!\text{CHCO}_2\text{H} \\
\mid \qquad\quad \mid \\
\text{CH}_2\!\!-\!\!\text{CH}_2
\end{array}
\xrightarrow[\text{(ii) NH}_3]{\text{(i) PCl}_5}
\begin{array}{c}
\text{CH}_2\!\!-\!\!\text{CHCONH}_2 \\
\mid \qquad\quad \mid \\
\text{CH}_2\!\!-\!\!\text{CH}_2
\end{array}
\xrightarrow{\text{Br}_2/\text{KOH}}
$$

$$
\begin{array}{c}
\text{CH}_2\!\!-\!\!\text{CHNH}_2 \\
\mid \qquad\quad \mid \\
\text{CH}_2\!\!-\!\!\text{CH}_2
\end{array}
\xrightarrow[\text{(ii) `AgOH}]{\text{(i) CH}_3\text{I}}
\begin{array}{c}
\text{CH}_2\!\!-\!\!\text{CHN(CH}_3)_3\text{OH}^- \\
\mid \qquad\quad \mid \\
\text{CH}_2\!\!-\!\!\text{CH}_2
\end{array}
\xrightarrow{\text{heat}}
\begin{array}{c}
\text{CH}_2\!\!-\!\!\text{CH} \\
\mid \qquad \| \\
\text{CH}_2\!\!-\!\!\text{CH}
\end{array}
\xrightarrow[\text{100°C}]{\text{H}_2/\text{Ni}}
\begin{array}{c}
\text{CH}_2\!\!-\!\!\text{CH}_2 \\
\mid \qquad \mid \\
\text{CH}_2\!\!-\!\!\text{CH}_2
\end{array}
$$

cyclobutene

It appears that attempts to decarboxylate cyclobutanecarboxylic acid (by heating) result in opening of the ring; hence the necessity for the steps such as given in the equation (see also below).

Cyclobutanecarboxylic acid may also be converted into cyclobutane by the following routes:

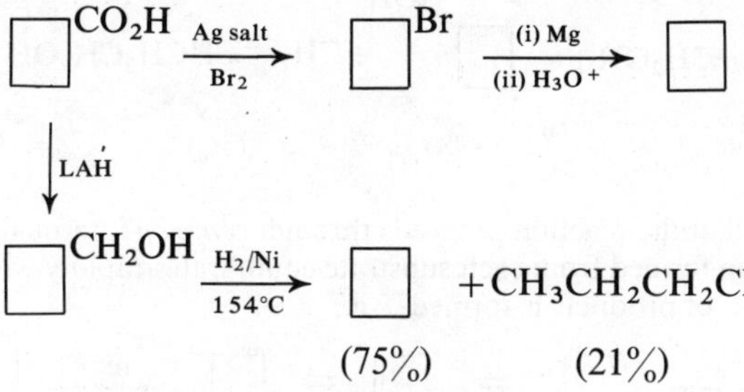
CO₂H
Ag salt
Br₂
Br
(i) Mg
(ii) H₃O⁺
LAH
CH₂OH
H₂/Ni
154°C
+CH₃CH₂CH₂CH₃
(75%)
(21%)

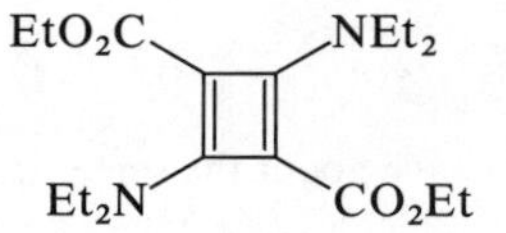
EtO₂C
NEt₂
Et₂N
CO₂Et

Many cyclobutane derivatives may be prepared by means of the Diels–Alder reaction (see p. 538).

Cyclobutene, b.p. 2°C, shows the ordinary olefinic reactions. **Cyclobutadiene** has not yet been isolated, but the tetra-substituted derivative shown has been prepared (Gompper *et al.*, 1968; see also p. 579).

Cyclobutylamine, b.p. 82°C, when treated with nitrous acid, gives a mixture of cyclobutanol and cyclopropylcarbinol, the latter being produced by ring contraction:

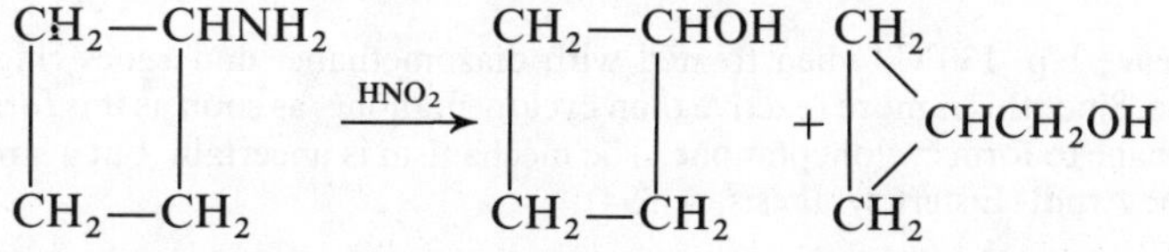
CH₂—CHNH₂
CH₂—CH₂
HNO₂
CH₂—CHOH
CH₂—CH₂
+
CH₂
CHCH₂OH
CH₂

This rearrangement of cyclic amines is known as the **Demjanov rearrangement** (1903).

If cyclobutylmethylamine is treated with nitrous acid, ring expansion by the Demjanov rearrangement takes place; four products are obtained: cyclobutylcarbinol, methylenecyclobutane, cyclopentanol and cyclopentene:

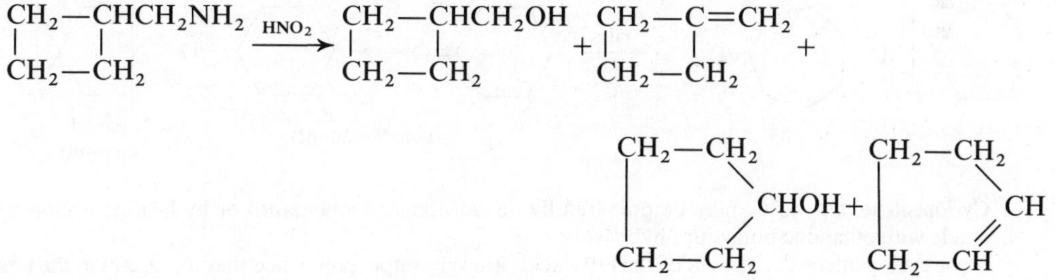
CH₂—CHCH₂NH₂
CH₂—CH₂
HNO₂
CH₂—CHCH₂OH
CH₂—CH₂
CH₂—C=CH₂
CH₂—CH₂
CH₂—CH₂
CHOH+
CH₂—CH₂
CH₂—CH₂
CH
CH₂—CH

All cycloalkylmethylamines from three- to eight-membered rings behave in this manner, and the yield of expanded products depends largely on the strain in the ring compound produced. Thus, highest yields are obtained when the products are five-, six-, or seven-membered rings.

The mechanism of the Demjanov rearrangement (a 1,2-shift) is still uncertain. Whatever the mechanism is, it must explain why, *e.g.*, both cyclopropylmethylamine and cyclobutylamine, on treatment with nitrous acid, give the same mixture of products:

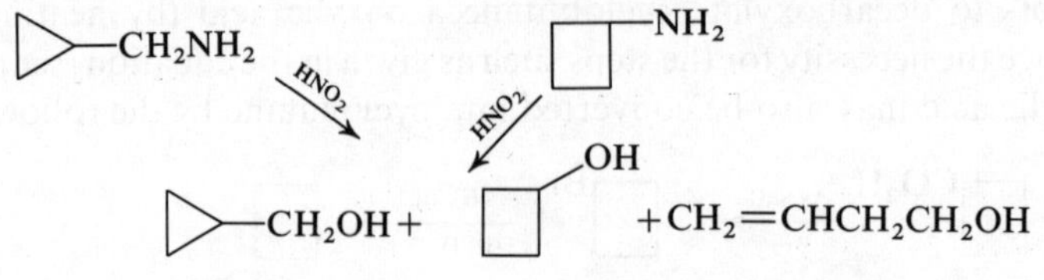

(48%)   (47%)   (5%)

These results indicate that the reaction proceeds through *common* intermediates, and can be explained on the basis that the ion formed from each substrate equilibrates rapidly with other structures through which the same mixture of products is formed, *e.g.*,

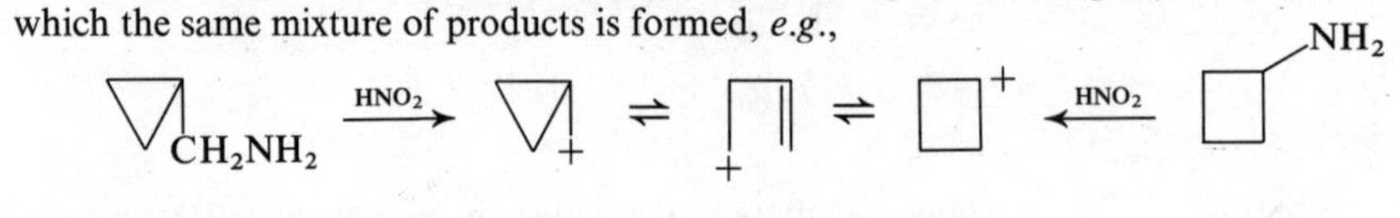

(I)   (II)   (III)

From the analysis of the products, it would appear that (I) and (III) have about the same stability, both being much more stable than (II).

Cyclopentane, b.p. 50°C, may be prepared by cyclising 1,5-dibromopentane with zinc or, better, by reducing cyclopentanone by the Clemmensen method.

Cyclopentanone, b.p. 130°C, when treated with diazomethane, undergoes ring expansion to form cyclohexanone. Since this is more reactive than cyclopentanone, as soon as it is formed most of it reacts with diazomethane to form cycloheptanone. The mechanism is uncertain, but a strong possibility is the 1,2-shift (*cf.* the Arndt–Eistert synthesis, p. 394):

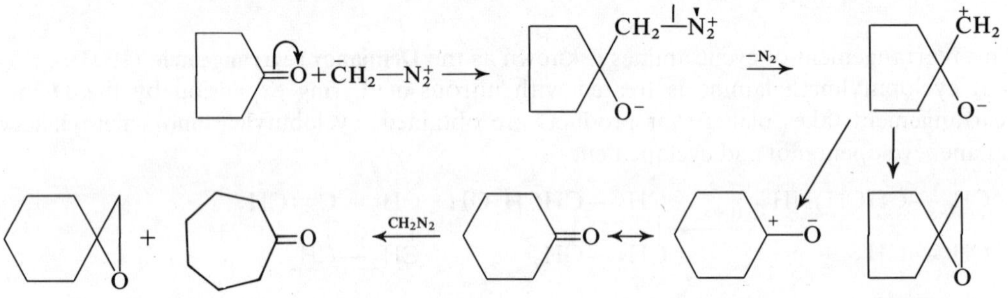

(small amount)   (small amount)

Cyclopentene, b.p. 45°C, may be prepared by dehydrating cyclopentanol or by heating cyclopentyl bromide with ethanolic potassium hydroxide.

Some cyclopentene derivatives of the fatty acids are very important, since they are useful in the treatment of leprosy and tuberculosis. Two acids, which occur in chaulmoogra oil, have been used from early times in the treatment of leprosy; both acids are optically active, the naturally occurring form being the (+);

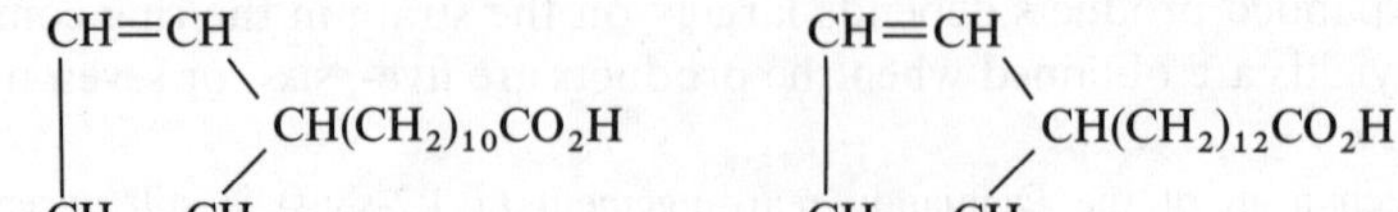

$$CH=CH$$
$$CH_2-CH_2 \quad CH(CH_2)_{10}CO_2H$$

hydnocarpic acid, m.p. 60°C

$$CH=CH$$
$$CH_2-CH_2 \quad CH(CH_2)_{12}CO_2H$$

chaulmoogric acid, m.p. 68°C

Cyclopentadiene, b.p. 41°C, is found in the crude benzene that is obtained from coal tar. The hydrogen atoms of the methylene group are very reactive, *e.g.*, cyclopentadiene reacts with Grignard reagents to form a hydrocarbon and cyclopentadienylmagnesium halide:

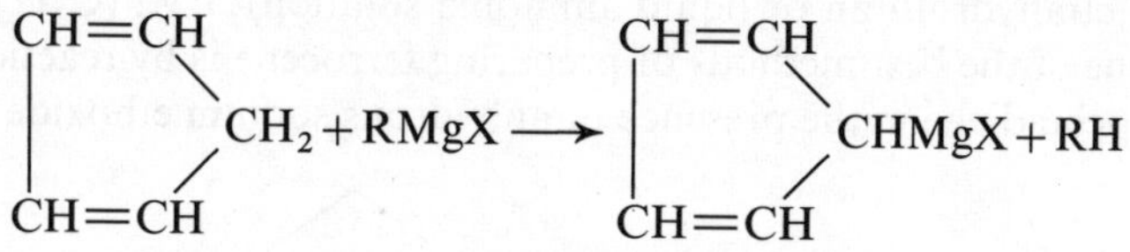

$$\begin{matrix} CH=CH \\ | \qquad\qquad\quad CH_2+RMgX \longrightarrow \\ CH=CH \end{matrix} \qquad \begin{matrix} CH=CH \\ | \qquad\qquad\quad CHMgX+RH \\ CH=CH \end{matrix}$$

Borg *et al* (1958) have treated cyclopentadienylsodium with chloroform and obtained chloro-benzene. The reaction is believed to proceed via the formation of dichloromethylene (p. 170):

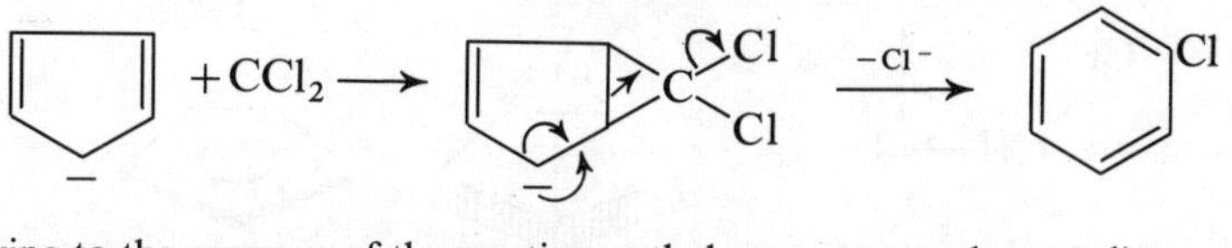

Owing to the presence of the reactive methylene group, cyclopentadiene condenses with aldehydes or ketones in the presence of sodium ethoxide to form *fulvenes*:

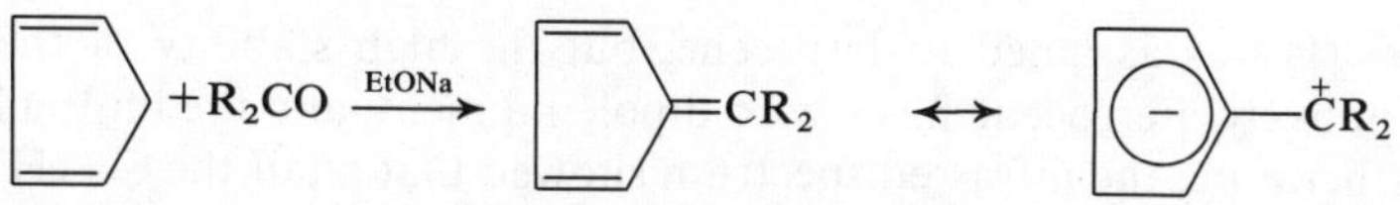

The parent compound, **fulvene** (R = H), is unstable. The fulvenes are yellow or red compounds, and dipole moment measurements indicate a large contribution of the dipolar structure (see also p. 581). Chemically, fulvenes behave predominantly as olefinic compounds.

Hydrogen of the methylene group in cyclopentadiene is acidic, *e.g.*, treatment with sodium produces sodium cyclopentadienide:

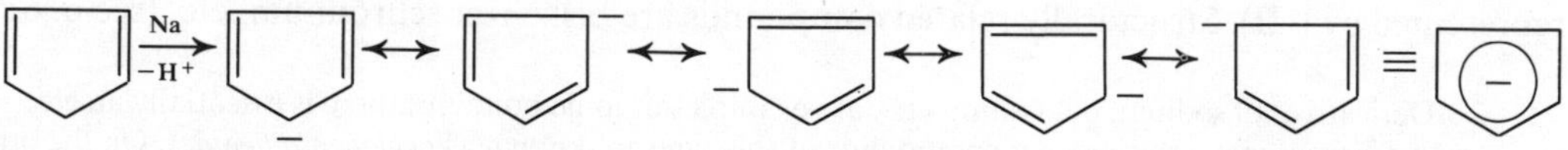

Since the carbanion is a resonance hybrid, it is this which stabilises the conjugate base with respect to its acid, *i.e.*, resonance (with the resulting delocalisation of the negative charge) is the driving force in the reaction. This cyclopentadienyl anion exhibits aromatic properties (see p. 579).

On standing, cyclopentadiene dimerises (it undergoes the Diels–Alder reaction) to form dicyclopentadiene, which regenerates the monomer on heating.

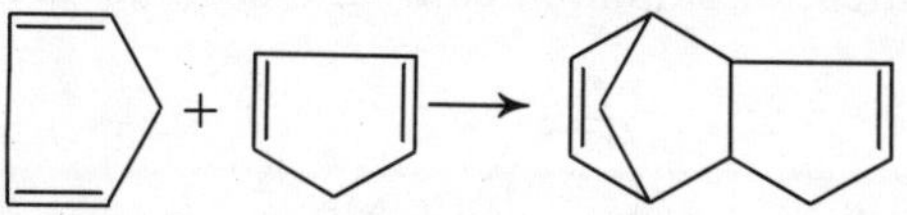

Cyclopentadiene complexes. Pauson *et al.* (1951) treated cyclopentadienylmagnesium bromide with ferric chloride and isolated dicyclopentadienyliron.

$$2C_5H_5MgBr \xrightarrow{\text{FeCl}_3} (C_5H_5)_2Fe$$

This iron (Fe II) complex was named **ferrocene** by Woodward *et al.* (1952) to indicate its aromatic character.

Many dicyclopentadienyls have now been prepared, the methods being by the use of the Grignard reaction (as shown above), by direct reaction between cyclopentadiene vapour and a heated metal or metal carbonyl, or by reaction between the sodium salt of cyclopentadiene and a metal halide (in tetrahydrofuran or liquid ammonia solution); *e.g.*, $(C_5H_5)_2Cr$, $(C_5H_5)_2Mn$, $(C_5H_5)_2Co$, etc. One of the best methods of preparing ferrocene is by reaction between ferrous chloride and cyclopentadiene in the presence of anhydrous sodium ethoxide (yield: 90%).

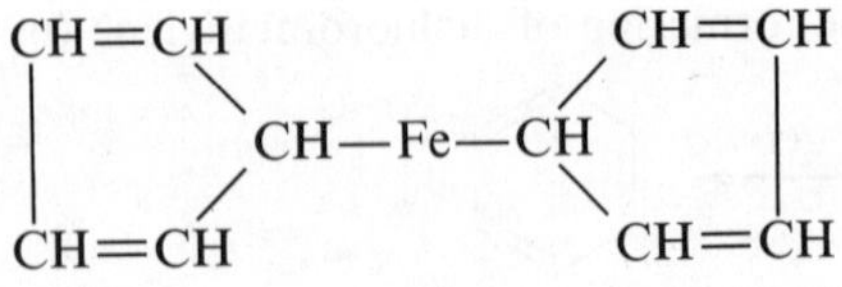

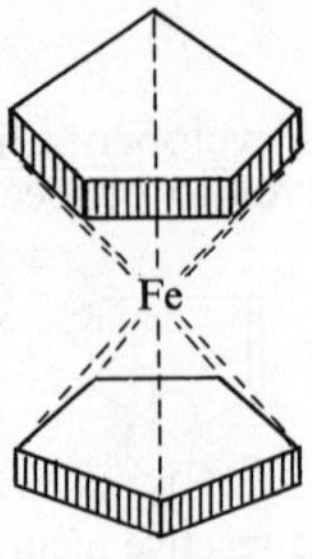

Originally, structure (I) was assigned to ferrocene, but the high stability of the molecule showed that this was unlikely. Ferrocene has a zero dipole moment, and the molecule is therefore symmetrical. Furthermore, the infra-red spectrum showed that all of the C—H bonds are equivalent, and so, on this evidence, structure (II) (this is a 'sandwich' structure, and is the anti-prism one) was proposed and has been confirmed by X-ray analysis, *i.e.*, that the two five-membered rings lie in parallel planes with the iron atom placed symmetrically between the two. In ferrocene and other cyclopentadienyls of the transition metals (Cr, Mn, Co, etc.), the entire ring is bonded uniformly to the metal atom; the bonding occurs by overlap of the sextet of π-electrons of the ring with the *d*-orbitals of the metal, thereby giving a *delocalised* covalent bond between the metal atom and the cyclopentadienyl ring as a whole. This, in ferrocene, is represented by (III). Structurally related compounds are bisbenzenechromium, etc. (see p. 619).

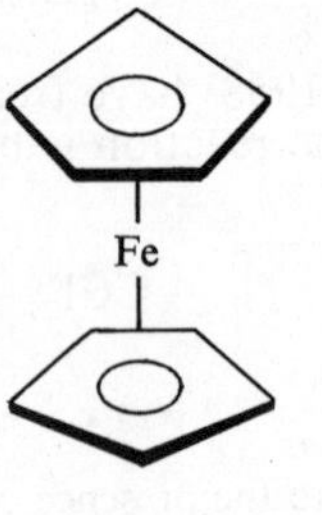

Derivatives of sodium, potassium, etc., are examples of ionic bonding; there is essentially an electron transfer from metal to ring, and compounds of this type are known as *cyclopentadienides*. On the other hand, there are cases where the key atom, *e.g.*, mercury, and *one* carbon atom of the ring each contribute one electron to form essentially a σ-bond.

Ferrocene is an orange solid, m.p. 173°C, and its reactions are aromatic, *e.g.*, it gives a mono- and diacetyl derivative by means of the Friedel–Crafts reaction under suitable conditions. Ferrocene forms an aldehyde by the *N*-methylformanilide method, and this undergoes the Cannizzaro reaction, the azlactone synthesis, etc. Ferrocene undergoes mercuration and metalation but cannot be directly halogenated, nitrated, or sulphonated by the usual methods. These

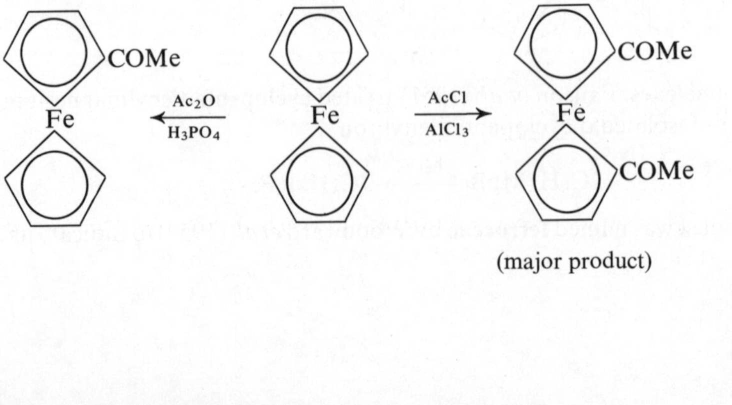

(major product)

failures are due to oxidation of the iron atom to give the blue *ferricinium* (Fe III) *cation*, $(C_5H_5)_2Fe^+$, which is resistant to electrophilic attack. Halogenoferrocenes, however, may be prepared via mercuration (p. 617), and mono- and disulphonic acids are obtained by sulphonation in acetic anhydride as solvent. Nitroferrocene is prepared by first metalating with butyllithium and then treating the lithium derivative with dinitrogen tetroxide. Metalation of ferrocene to give mono- and dilithiumferrocene (one lithium atom in each ring) is the starting point of a large variety of ferrocene derivatives (*cf.* p. 436). The aromatic character of ferrocene is further illustrated by the fact that it does *not* condense with maleic anhydride, and hence the conjugated system is absent in the rings.

Cyclohexane, b.p. 81°C, occurs in petroleum. When reduced with hydrogen iodide at 250°C, benzene is converted into a mixture of cyclohexane and methylcyclopentane. Pure cyclohexane may be conveniently prepared by the Clemmensen reduction of cyclohexanone, and is prepared industrially by the hydrogenation of benzene in the presence of nickel at 200°C. Another industrial method is the hydrogenation of phenol, dehydrating the product cyclohexanol to cyclohexene, which is then catalytically hydrogenated to cyclohexane; the process is carried out without isolating the intermediate compounds:

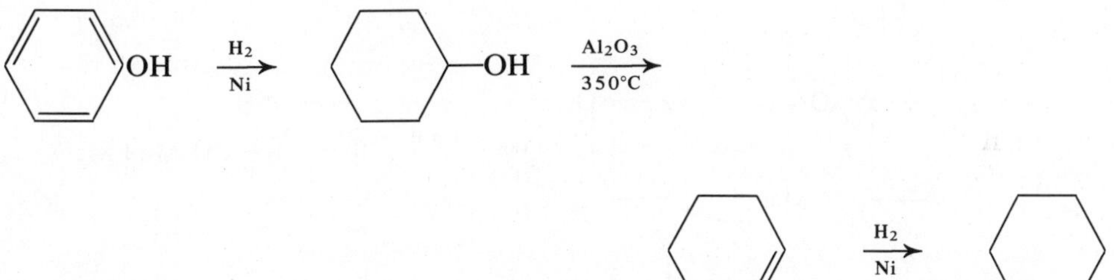

Hot concentrated nitric acid oxidises cyclohexane to adipic acid, and fuming sulphuric acid converts it into benzenesulphonic acid:

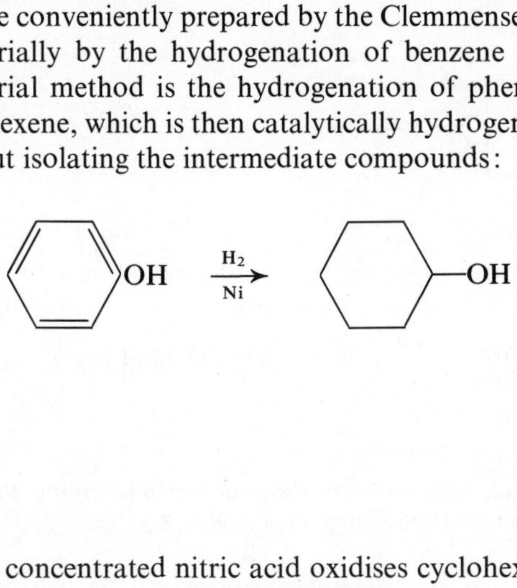

For a consideration of the spatial arrangement of the carbon atoms in cyclohexane, see p. 550.

Cyclohexane can be catalytically (Pt or Pd) dehydrogenated to benzene. Provided the ring contains at least one double bond, then dehydrogenation can be readily effected with sulphur of selenium. In general, this dehydrogenation is confined to six-membered rings.

Cyclohexanol, m.p. 24°C, is prepared industrially by the catalytic hydrogenation (Ni) of phenol. It undergoes the general reactions of an aliphatic secondary alcohol. It is converted by gentle oxidation (dilute nitric acid) into cyclohexanone; vigorous oxidation (concentrated nitric acid) produces adipic acid.

Cyclohexanone, b.p. 157°C, may be prepared by the oxidation of cyclohexanol; it is manufactured by the oxidation of cyclohexane with oxygen under pressure in the presence of a cobalt catalyst (cyclohexanol is also obtained). α-Halogenocyclohexanones undergo the **Favorsky rearrangement** (1894) when treated with alkali, *e.g.*, the six-membered ring changes to a five-membered ring carboxylic acid:

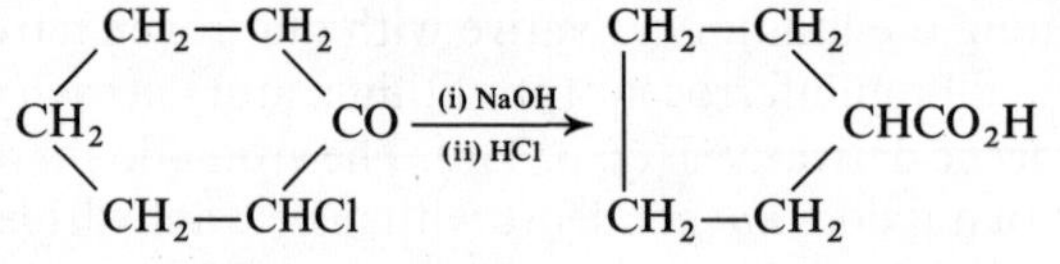

$$
\begin{array}{c}
CH_2{-}CH_2 \\
CH_2 \qquad CO \\
CH_2{-}CHCl
\end{array}
\xrightarrow[\text{(ii) HCl}]{\text{(i) NaOH}}
\begin{array}{c}
CH_2{-}CH_2 \\
\qquad\quad CHCO_2H \\
CH_2{-}CH_2
\end{array}
$$

The mechanism of the Favorsky rearrangement is uncertain, but a highly favoured one is via a cyclopropanone intermediate; *e.g.*, in the presence of sodium methoxide:

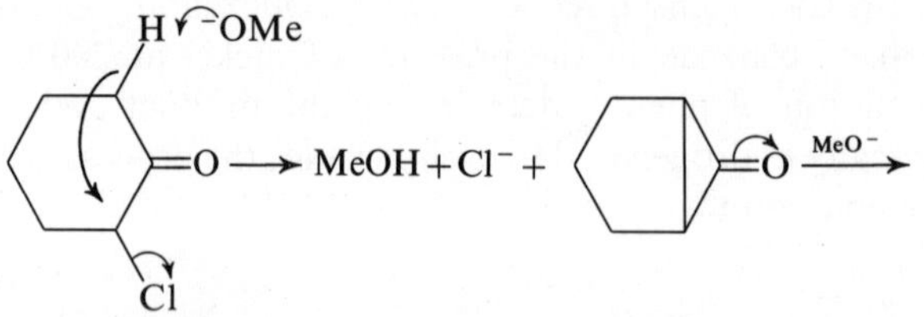

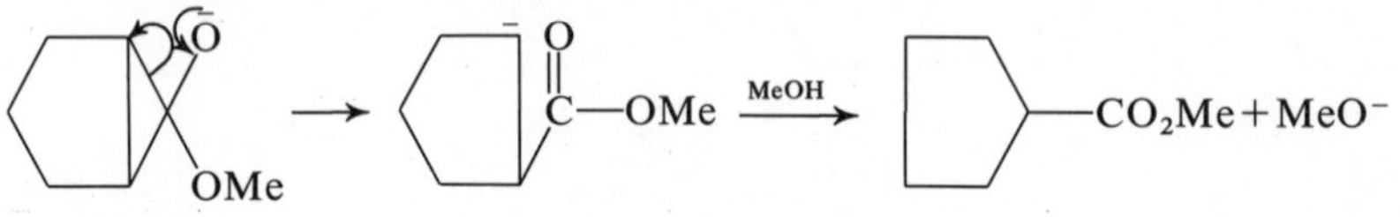

Treatment of acyclic α-halogenoketones with sodium methoxide also results in the Favorsky rearrangement (α-methoxyketones have also been isolated; Turro *et al.*, 1969), *e.g.*,

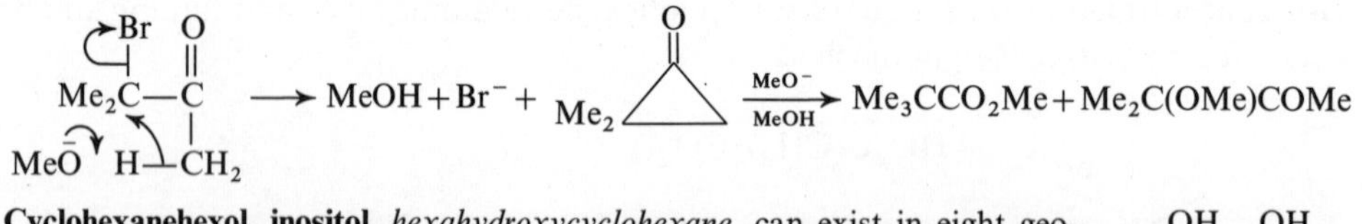

Cyclohexanehexol, **inositol**, *hexahydroxycyclohexane*, can exist in eight geometrical isomeric forms, of which only one is optically active (see formula). The (+)- and (−)-forms of this isomer (m.p. 248°C) occur in plants as the hexaphosphoric ester, which is known as *phytin*. Some of the optically inactive forms also occur as their hexaphosphoric esters. One of the optically inactive forms is present in the vitamin B complex; it appears to have growth-promoting properties in chicks, and is an anti-alopecia factor (anti-baldness) in mice.

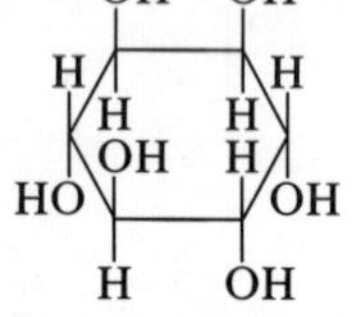

Hexachlorocyclohexane, $C_6H_6Cl_6$, exists in a number of geometrical isomeric forms, one of which is a powerful insecticide (see p. 622).

Cyclohexene (*tetrahydrobenzene*), b.p. 83°C, may conveniently be prepared by dehydrating cyclohexanol with sulphuric acid. It has the usual properties of an alkene.

Many cyclohexene derivatives may be prepared by means of the Diels–Alder reaction.

Cyclohexadienes (*dihydrobenzenes*). There are two isomeric cyclohexadienes, the 1,3- (b.p. 81°C) and the 1,4- (b.p. 86°C). They readily polymerise and undergo the usual reactions of an alkadiene. Since cyclohexa-1,3-diene contains a conjugated system of double bonds, it can undergo both the 1,2- and 1,4-addition reactions (see p. 127).

Cyclohexatriene is **benzene**, and differs enormously in its chemical properties from the cycloalkenes and cycloalkadienes. Benzene has '*aromatic properties*' (see benzene, p. 574).

5,5-Dimethylcyclohexane-1,3-dione, **dimedone**, m.p. 148°C, is a very sensitive reagent for formaldehyde, which may also be estimated gravimetrically with this reagent (see also p. 341).

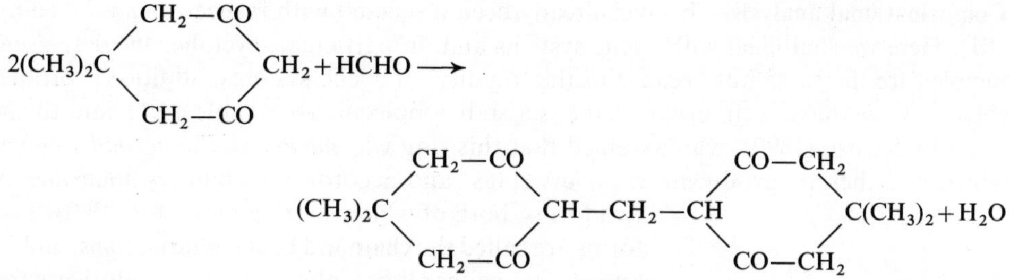

$$2(CH_3)_2C \overset{\displaystyle CH_2-CO}{\underset{\displaystyle CH_2-CO}{\diagdown\diagup}} CH_2 + HCHO \longrightarrow$$

$$(CH_3)_2C \overset{\displaystyle CH_2-CO}{\underset{\displaystyle CH_2-CO}{\diagdown\diagup}} CH-CH_2-CH \overset{\displaystyle CO-CH_2}{\underset{\displaystyle CO-CH_2}{\diagup\diagdown}} C(CH_3)_2 + H_2O$$

Cycloheptane, b.p. 118°C, occurs in petroleum; it may be prepared by the Clemmensen reduction of cycloheptanone.

Cycloheptatriene (tropilidene), b.p. 115·5°C, may be prepared by the photolysis of diazomethane in the presence of benzene (note the ring expansion):

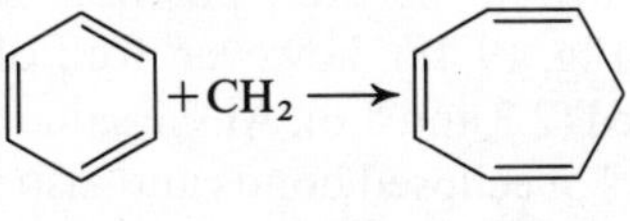

Two derivatives of cycloheptatriene, **tropone** (*cycloheptatrienone*) and **tropolone** (*cycloheptatrienolone*), are particularly interesting because they possess aromatic character, *e.g.*, although their classical structures contain a keto group, both compounds show a lack of ketonic properties (see p. 579 for a further discussion). Tropone may be prepared from anisole by the following sequence of reactions (Birch *et al.*, 1962):

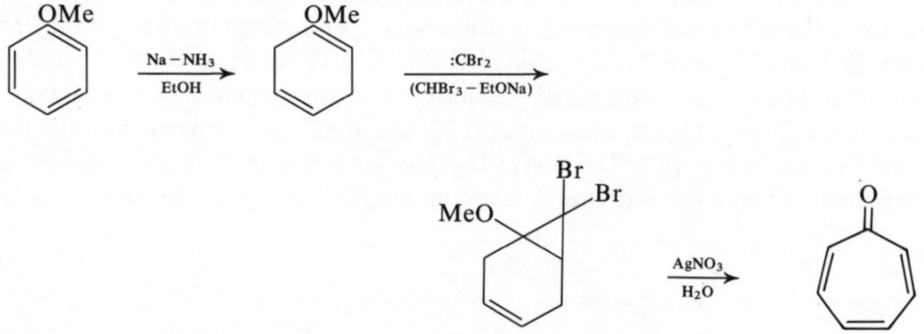

Tropolone may be prepared from (i) cycloheptatriene, and (ii) cyclopentadiene:

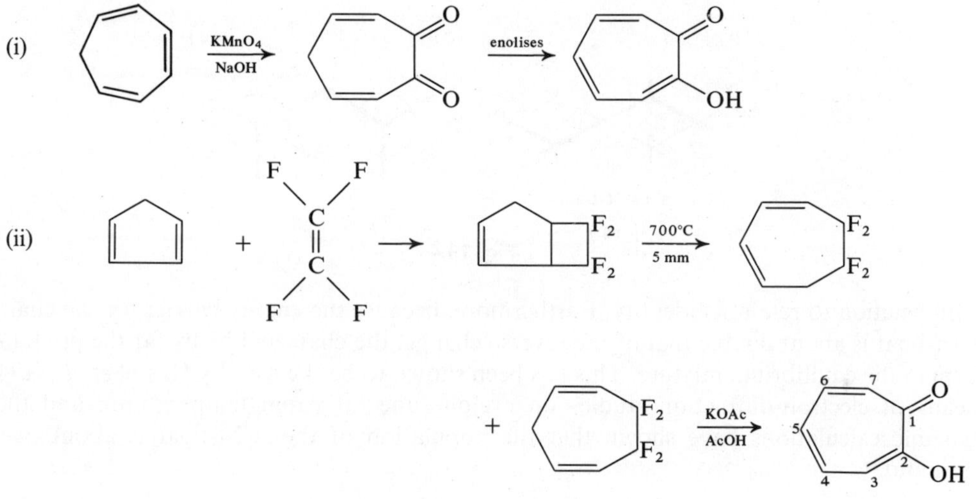

(i)

(ii)

Conformational analysis. This has already been discussed with respect to *acyclic* compounds (p. 81). Here we shall deal with cyclic systems and, in particular, cyclohexane derivatives. The principles are the same but because of the 'rigidity' of cyclic systems, additional problems are involved. As we have seen, cyclohexane is a stable molecule (see Table 19.1), and this was explained by Sachse (1890), who assumed that this ring was *puckered*, the *normal* valency being retained and thereby producing *strainless* rings; and according to him, cyclohexane exists in two forms, both of which are strainless (Fig. 19.3). These two forms are called the **chair** and **boat conformations**, and it is now known that neither is completely strainless. Both are free from angle strain, but because of differences in steric strain and bond opposition strain, the two forms differ in energy content. In the chair form (Fig. 19.4a) all the C—H bonds on adjacent carbon atoms are in the skew position (see Fig. 19.4e). In the boat form (Fig. 19.4c), however, four of the C—H bonds are skew

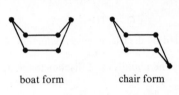

boat form   chair form

Fig. 19.3

(1,2; 3,4; 4,5 and 6,1), and two are eclipsed (2,3 and 5,6). At the same time, there will also be some bond opposition strain for these two pairs of eclipsed bonds and also steric repulsion between the hydrogens pointing towards each other at 1 and 4 (Fig. 19.4c). Hence the total strain in the boat conformation is larger than that in the chair conformation, and consequently the former is less stable than the latter. The boat form, however, is flexible and can be readily distorted into many shapes, and in these the hydrogen eclipsings are reduced (Fig. 19.4d).

The chair form is 'rigid' (in the sense that it resists distortion), and when it is changed into the boat form, some angular deformation is necessary. The energy barrier for this transformation is about 37·7–46·0 kJ mol^{-1} (determined from N.M.R. spectral data). Calculations have also shown that the twist-boat contains 6·7 kJ mol^{-1} less energy than the classical boat conformation, and that the chair conformation contains 22·2 kJ mol^{-1} less energy than the twist-boat. These data are shown in Fig. 19.5. The energy barriers are not large enough to prevent interconversion of the chair and boat conformations (at room temperature), but are large enough to permit each

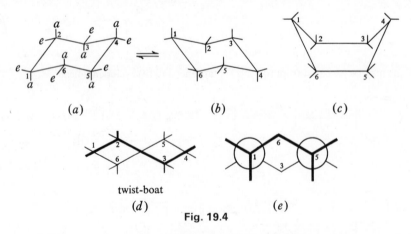

(a)   (b)   (c)

twist-boat

(d)   (e)

Fig. 19.4

conformation to retain its identity. Furthermore, because the energy barrier for the chair ⟶ twist-boat is about double that of the reverse change, the chair will be by far the predominant form in the equilibrium mixture. This has been shown to be the case by Hassel *et al.*, (1943) by means of electron-diffraction studies on cyclohexane (at room temperature), and thermodynamic calculations have shown that the population of the twist-boat is about one in a thousand.

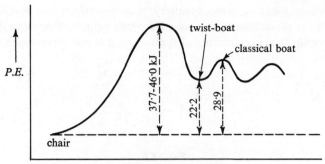

Fig. 19.5

Since the boat conformation occurs in relatively few cases, we shall confine our attention to the chair form. Consideration of the C—H bonds in this form shows that there are two sets of six. In one set the six C—H bonds are parallel to the axis of the ring (Fig. 19.4a); these are the **axial** (a) **bonds**. In the other set the six C—H bonds make an angle of 109° 28′ with the axis (or $\pm 19°$ 28′ with the horizontal plane of the ring; Fig. 19.4a); these are the **equatorial** (e) **bonds**. Each carbon atom has one axial and one equatorial bond, and because of the mobility of the chair conformation, one form (Fig. 19.4a) can readily change into the other (Fig. 19.4b), and when this occurs all hydrogens originally axial now become equatorial, and vice versa. The two forms are identical.

Calculations have shown that in the chair conformation the distances between pairs of hydrogen atoms are (Angyal and Mills, 1952):

1e,2e, 2·49 Å; 1e,2a, 2·49 Å; 1a,2a, 3·06 Å; 1a,3a, 2·51 Å.

The nearest four hydrogens to the equatorial hydrogen at 1 are those on 2 (a and e) and 6 (a and e), whereas for the axial hydrogen at 1 these four hydrogens are the axial hydrogens on 3 and 5 and the equatorial on 2 and 6 (see Fig. 19.4a). Thus, for hydrogen atoms, the interactions 1e,2e, 1e,2a and 1a,3a are about the same.

Now let us consider the problem when one hydrogen atom in cyclohexane is replaced by some substituent which, of necessity, must be larger than a hydrogen atom. A study of accurate scale models has shown that in monosubstituted cyclohexanes, an axial substituent at 1 is closer to the two axial hydrogens at 3 and 5 than an equatorial substituent at 1 is to the four hydrogens at 2 and 6. Thus 1a,3a interactions will be greater than 1e,2e or 1e,2a interactions, and so, in general, a monosubstituted cyclohexane will assume the conformation in which the substituent occupies an equatorial position. This has been confirmed experimentally, e.g., Hassel (1950) has shown from electron-diffraction studies that the methyl group in methylcyclohexane and the chlorine in chlorocyclohexane are equatorial (predominantly).

a-methyl   e-methyl

Let us now consider 1,2-disubstituted cyclohexanes in which the two substituents are identical.

According to the classical ideas of stereochemistry, there are two forms possible, *cis* and *trans*. The conformation of the *cis* configuration *must* have one axial and one equatorial substituent (Fig. 19.6a), and since this has no elements of symmetry, it is not superimposable on its mirror

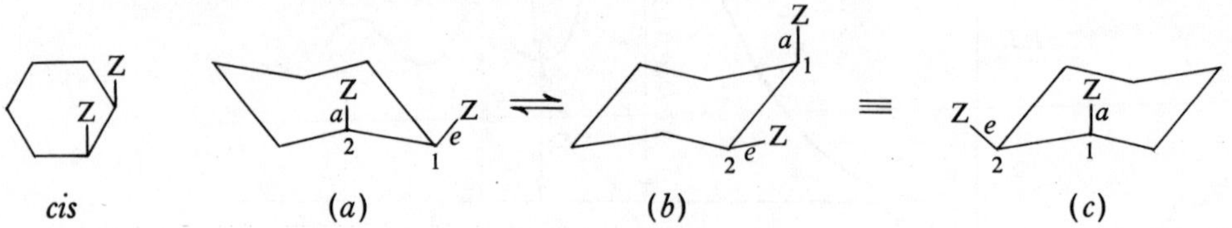

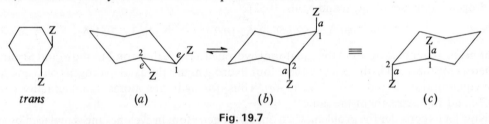

cis   (*a*)   (*b*)   (*c*)

Fig. 19.6

image (Fig. 19.6c). When Fig. 19.6(a) flips to give Fig. 19.6(b), axial hydrogens ⟶ equatorial and vice versa, and therefore Fig. 19.6(c), which is equivalent of Fig. 19.6(b), is the enantiomer of Fig. 19.6(a). Furthermore, since strain is identical in both forms, their populations are equal, and consequently this 1,2-disubstituted cyclohexane is optically inactive by *external* compensation. However, this type of compound (*i.e.*, Z = Z) has never yet been resolved, and this is due to the fact that the two enantiomers are readily interconvertible. In effect, each enantiomer undergoes very rapid autoracemisation (p. 484).

The *trans* configuration can exist in two different conformations, 1a,2a and 1e,2e (Fig. 19.7), but from what has been said above, the *trans*-1e,2e form will be more stable than the *trans*-1a,2a, and this *trans* form will be more stable than the *cis*. Also, neither conformation has any element of symmetry, and neither conformation can be converted into its mirror image by flipping. *Hence*, trans-1,2-disubstituted cyclohexanes exist as pairs of enantiomers.

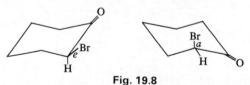

trans   (*a*)   (*b*)   (*c*)

Fig. 19.7

The above arguments can be applied to the 1,3- and 1,4-disubstituted cyclohexanes, and to the polysubstituted derivatives, and it will be found that the chair conformation with the maximum number of equatorial substituents will be the preferred form. This generalisation, however, is only true when other forces due to, *e.g.*, dipole interactions, hydrogen bonding, are absent. When these 'disturbing' factors are present, they may be large enough to outweigh the 1,3-interactions, *e.g.*, infra-red spectra studies have shown that the bromine in 2-bromocyclohexanone is predominantly *axial* (Fig. 19.8). The C—Br and C=O bonds are both strongly polar, and when the bromine is equatorial the dipolar repulsion is a maximum, and is a minimum when the bromine is axial. Since the axial form predominates, the equatorial dipolar repulsion must be much larger than the 1,3-interactions. In *cis*-1,3-dihydroxycyclohexane, the diaxial form predominates, since this form is stabilised by hydrogen bonding.

Fig. 19.8

It is also instructive to consider one example of a 1,2-disubstituted cyclohexane in which the two substituents are different, *e.g.*, *cis*-2-methylcyclohexanol. Since 1,3-interactions will be most powerful when the larger group is axial, the preferred form will therefore be the one in which the larger group is equatorial. Thus, since the methyl group is larger than hydroxyl, the preferred form of *cis*-2-methylcyclohexanol is $1a$-OH:$2e$-CH$_3$.

Effect of conformation on the course and rate of reactions. Because the environments of axial and equatorial groups are different, it may be expected that the reactivity of a given group will therefore depend on its orientation. Since S$_N$2 reactions always occur with inversion, it can be seen that the approach of the attacking group (Z) along the bonding line remote from the group to be expelled (Y) is less hindered in (I; aY) than in (II; eY). Thus, S$_N$2 reactions take place more readily with an axial substituent than with an equatorial.

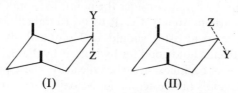

(I)   (II)

It can be expected that an S$_N$1 reaction will be sterically accelerated for an axial substituent, since the formation of the carbonium ion will relieve the steric strain due to the 1,3-interactions. As these interactions are absent for an equatorial substituent, in this case there will be no steric acceleration.

In elimination reactions, for the E2 mechanism to occur readily, the two groups concerned must be in the *trans*-position (p. 488). In cyclohexane derivatives, this requirement is found only in *trans*-1,2-diaxial compounds, and these thus readily undergo elimination reactions.

E1 reactions are less easy to study because of the difficulty of ascertaining the geometry of the intermediates involved.

Let us consider the case of 4-t-butylcyclohexyl-*p*-toluenesulphonate (Winstein *et al.*, 1955). Two forms, *cis* and *trans*, are possible, but because of the large bulk of the t-butyl group, this group is *always* equatorial, *i.e.*, the conformation of the two stereoisomers are made rigid. Under the same conditions (sodium ethoxide in ethanol at 70°C), the *cis* form readily undergoes bimolecular elimination, but the *trans* does not.

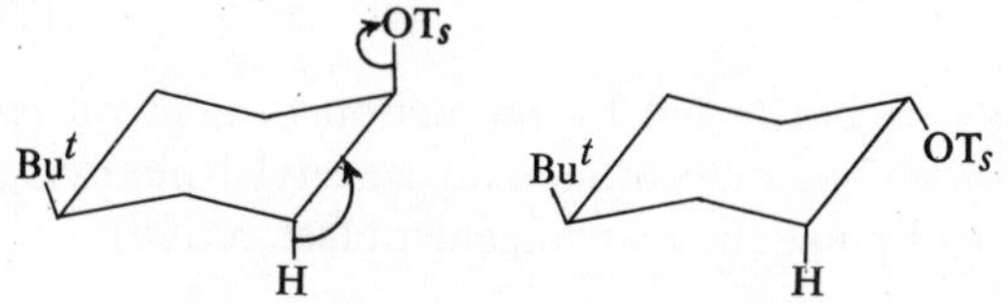

OT$_s$ Bu^t OT$_s$
Bu^t H
H
cis   *trans*

The following example is of particular interest in that it illustrates the phenomenon of anchimeric assistance in neighbouring group participation (p. 486). Bartlett (1935) has shown that alkali converts *trans*-2-chlorocyclohexanol into cyclohexene oxide, and proposed the following mechanism:

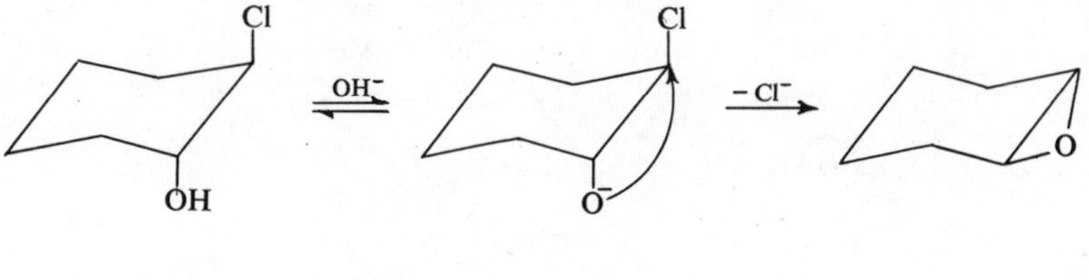

Cl   Cl
$\xrightleftharpoons{\text{OH}^-}$ $\xrightarrow{-\text{Cl}^-}$
OH   O$^-$ O

Bergkvist (1947) showed that this reaction proceeds more than 100 times as fast as that when the *cis*-compound is used. This may be explained as follows. The *cis*-compound probably forms the alkoxide ion just as readily as does the *trans*-compound, but the *cis*-ion cannot attack at the rear of the adjacent carbon atom (where the chlorine is being ejected) without great distortion of the molecule.

So far, we have examined reactions in which the bond linking an atom to the ring is broken. Now let us consider reactions in which it does not occur, *e.g.*, esterification and ester hydrolysis by the $A_{AC}2$ mechanism (p. 250). In the esterification of cyclohexanol, all stages involve groups larger than hydrogen becoming attached to the oxygen of the OH group. This will therefore increase non-bonded interactions, more so for an axial than for an equatorial OH. Hence the former should esterify more slowly than the latter. Hydrolysis by the $A_{AC}2$ mechanism is the reverse of esterification, but if we now consider the point that some steps involve change of sp^2 hybridisation (flat) of the carbon atom of CO_2R to sp^3 (tetrahedral), non-bonded interactions increase. Consequently, the axial ester should undergo hydrolysis more slowly than the equatorial one. These arguments hold good for hydrolysis by the $B_{AC}2$ mechanism, and also for esterification and hydrolysis of cyclohexanecarboxylic acid. These predictions are borne out in practice, *i.e.*, axial cyclohexanols (and esters) and cycloalkanecarboxylic acids (and esters) are more slowly esterified (and hydrolysed) than the corresponding equatorial isomers.

Rearrangements involving cyclohexanes. Many cyclohexane derivatives can undergo 1,2-shifts, but the course of the reaction depends on a number of factors. If the leaving group Z is *axial* and there is a group Y attached to an *adjacent* carbon atom which is also axial (*i.e.*, Z and Y are *trans-a,a*) and which is capable of migrating (*e.g.*, methyl, hydrogen), then the spatial requirements are satisfied and so a 1,2-shift can occur with formation of a *cyclohexane derivative* (see III). If, however, the leaving group Z is equatorial and Y is axial (*i.e.*, Z and Y are *cis*), then these two groups no longer satisfy the spatial requirements of a 1,2-shift. On the other hand, *an adjacent carbon atom of the ring system* does satisfy the spatial requirements of the 1,2-shift, and this occurs with the formation of a *cyclopentane derivative* (see IV).

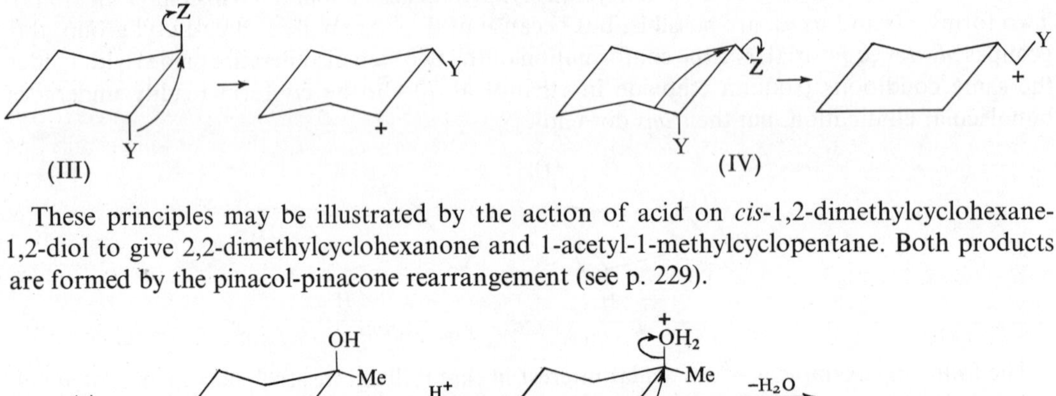

(III)   (IV)

These principles may be illustrated by the action of acid on *cis*-1,2-dimethylcyclohexane-1,2-diol to give 2,2-dimethylcyclohexanone and 1-acetyl-1-methylcyclopentane. Both products are formed by the pinacol-pinacone rearrangement (see p. 229).

(a)

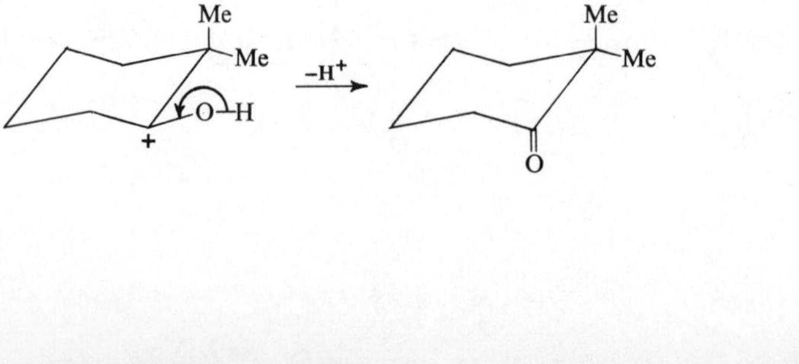

(b)

$$\text{OH} \quad \text{Me} \quad \text{OH} \quad \text{Me} \xrightarrow{H^+} \quad \text{OH} \quad \text{Me} \quad \overset{+}{O}H_2 \quad \text{Me} \xrightarrow{-H_2O}$$

$$\overset{+}{C}\text{—Me} \quad \text{O—H} \quad \text{Me} \xrightarrow{-H^+} \quad \text{COMe} \quad \text{Me}$$

NMR spectroscopy of cyclohexanes. The NMR spectra of unsubstituted cycloalkanes, except cyclopropane, show a τ-value around 8·5 p.p.m. due to the methylene protons (*cf*. acyclic methylene value, 8·7). As we have seen, cyclohexane exists in two stable chair conformations, (V) and (VI) which are undergoing rapid interconversion at room temperature. Let us suppose

$$H_a \quad H_b \quad (V) \rightleftharpoons H_a \quad H_b \quad (VI) \qquad H \quad Cl \quad H \quad (VII) \rightleftharpoons Cl \quad H \quad H \quad (VIII)$$

that cyclohexane is a rigid molecule, *i.e.*, no interconversion occurs. Then, because the environments of axial and equatorial hydrogens are different, their chemical shifts will be different. The general rule is that axial protons absorb upfield with respect to equatorial protons. At the same time, coupling occurs between axial and equatorial protons attached to the same carbon atom, and coupling also occurs between vicinal protons. Since the coupling constant depends on the dihedral angle, J for vicinal axial—equatorial hydrogens (skew; dihedral angle of 60°) will be smaller (2–4 Hz) than that (5–12 Hz) for vicinal axial–axial hydrogens (*trans*; dihedral angle of 180°). The overall result would be an NMR spectrum in which multiplets would have different J values. At room temperature, cyclohexane is *not* a rigid molecule, and since the interconversion rate is rapid, both axial and equatorial have the same average environment, thereby giving rise to a single sharp line. When the spectrum of cyclohexane is measured at about $-100°C$, two signals are observed, one due to axial and the other to equatorial hydrogens. Under these conditions, the interconversion rate is sufficiently slowed down for each type of proton to show its own chemical shift.

Now let us consider a monosubstituted cyclohexane, *e.g.*, chlorocyclohexane (VII and VIII). Since the equatorial chlorine conformation (VII) is favoured, the populations of (VII) and (VIII) are different, and consequently the time spent by a given hydrogen atom (axial or equatorial) will be longer in conformation (VII) than in (VIII). Even so, because of rapid interconversion at room temperature, only one *broad* proton signal is observed. When the temperature is cooled below $-55°C$, two broad peaks are observed.

In the case of fused systems, *e.g.*, the decalins (see below), *cis*-decalin shows one proton peak, whereas *trans*-decalin shows two. The former isomer is capable of interconversion, but the latter is not. The latter is a rigid system, and so equatorial and axial protons are held in different environments, the result being two different chemical shifts.

Cycloalkenes. The presence of a double bond in cyclohexene causes the ring to assume the half-chair conformation (atoms 1,2,3, and 6 lie in one plane):

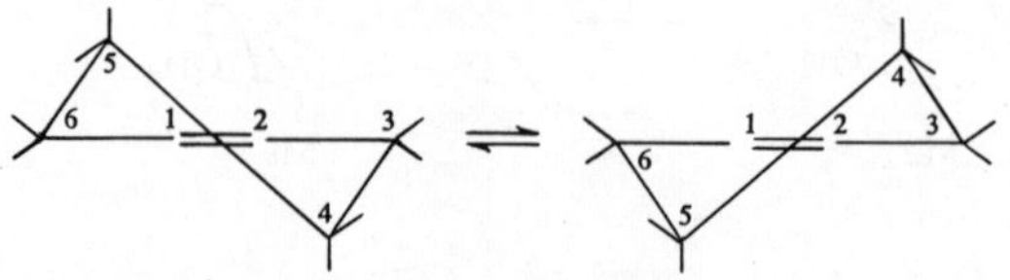

cyclohexene

Fused systems. Since the boat and chair forms of cyclohexane are readily interconvertible, neither form can be isolated, Mohr (1918), however, elaborated Sachse's theory and predicted that the fusion of two cyclohexane rings, *e.g.*, in decalin, should produce *cis* and *trans* forms which should be stable enough to retain their identities. Both forms have now been prepared. Several conventions have been introduced to represent these isomers. One uses *full* lines to represent groups *above* the plane of the molecule, and *broken* lines to represent those *below*. Another convention uses a black dot to represent a hydrogen atom above the plane, and the ordinary lines of the formula to indicate a hydrogen atom below the plane. Thus the decalins may be drawn:

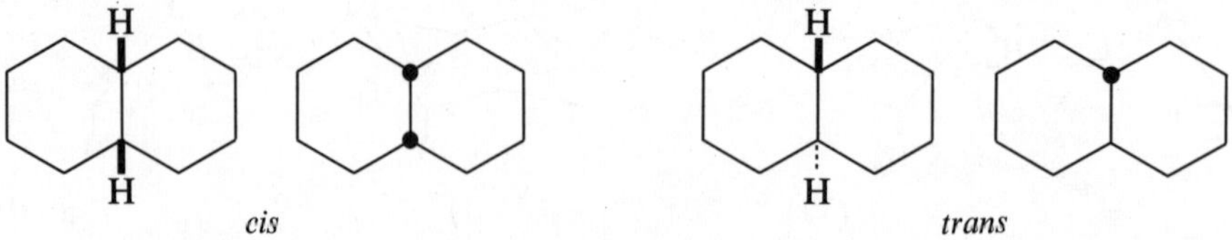

The configurations of decalin are complicated by the fact that a number of strainless modifications are possible which differ in the type of 'locking', *i.e.*, whether axial or equatorial bonds are used to fuse the rings. According to Hassel *et al.* (1946), *cis*- and *trans*-decalin are as shown in Fig. 19.9. In both cases the cyclohexane rings are *all chair forms*;

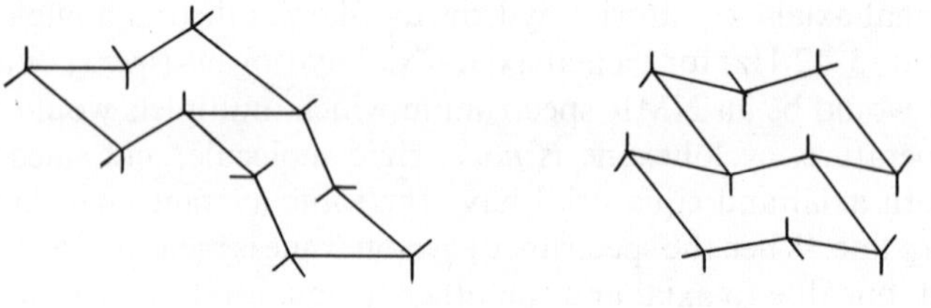

cis-decalin   trans-decalin

Fig. 19.9

the *cis* form is produced by joining one axial and one equatorial bond of each ring, whereas the *trans* form is produced by joining the two rings by equatorial bonds only. Calculation has shown that the *cis* isomer has 10·0 kJ energy content more than the *trans*. It is also of interest to note that if the decalins are regarded as 1,2-disubstituted cyclohexanes, then the *trans* form (1e,2e) would be expected to be more stable than the *cis* (1e,2a) (see also Vol. 2, Ch. 4).

Medium-ring compounds

Cyclo-octane, m.p. 11·8°C, may be obtained by the catalytic reduction of cyclo-octene, cyclo-octadiene or cyclo-octatetraene (Willstätter 1911).

(1,3,5,7)-Cyclo-octatetraene, b.p. 142–143°C, was first obtained by Willstätter (1911, 1913), who prepared it from the alkaloid *pseudo*-pelletierine (ψ-pelletierine) by means of the Hofmann exhaustive methylation method (p. 856):

ψ-pelletierine   cyclo-octatetraene

Reppe (1940) prepared cyclo-octatetraene in large quantities by the polymerisation of acetylene under pressure in the presence of a nickel compound, *e.g.*, nickel cyanide, as catalyst, in tetrahydrofuran solution:

$$4C_2H_2 \longrightarrow$$

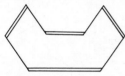

Infra-red and Raman spectra studies have shown that cyclo-octatetraene contains a conjugated system of double bonds, and this is supported by X-ray analysis, which has also shown that the ring is non-planar. According to Treibs (1950), cyclo-octatetraene is 'tub-shaped', there being comparatively small strain in this conformation. If the double bonds were delocalised, the molecule would be flat and consequently would experience large strain, too large to be compensated by the gain in resonance energy (see also p. 580).

Cyclo-octatetraene is a reactive compound and many of its reactions are accompanied by rearrangement to benzene and bicyclo[4,2,0]-ring compounds. These are the result of transannular reactions (see later).

Cyclononane, cyclodecane, and **cyclo-undecane** may be prepared by the methods used for large rings (see below).

Medium-ring compounds have many properties which are quite different from those of the other classes of ring compounds. Reference to Table 19.1 shows that medium rings have large strain relative to all other rings except the small rings. The contributing factors to strain in cycloalkanes are bond-angle deformation, steric repulsion, and bond-opposition forces. In medium rings, however, there is *another* source of strain. X-ray analysis studies on some cyclodecane derivatives have shown that the conformation is (I). This is derived from two chair conformations of cyclohexane joined by 1,3-axial bonds. Groups in 1,2-positions are partially eclipsed, thereby producing steric strain, but a more important source of strain is the steric repulsion between atoms on opposite sides of the ring (see I). This type of interaction has been designated **transannular interaction**, and this produces **transannular strain**. At the same time, it is this close proximity of *opposite* sides of the ring that gives rise to **transannular reactions**, *e.g.*, the formolysis of epoxycyclodecane (II), followed by hydrolysis gives, among other products, cyclodecane-1,6-diol (these structures are drawn in the conventional manner):

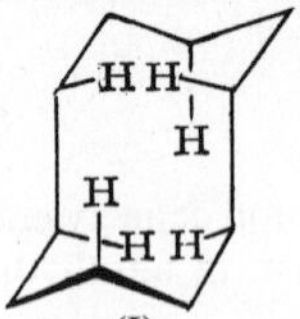

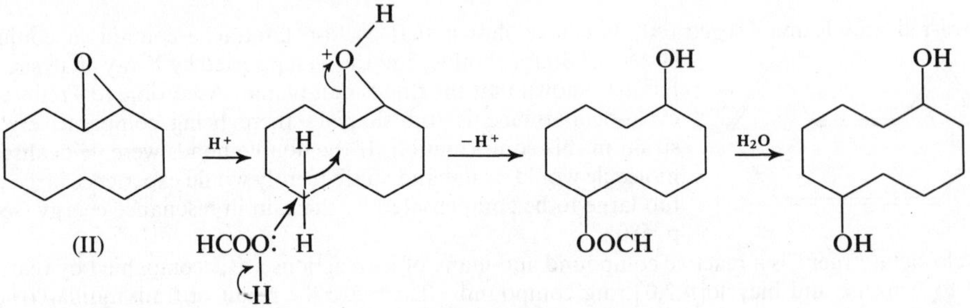

Another example is the catalytic reduction of cyclodecane-1,6-dione to give a mixture of 1,6-dihydroxycyclodecane and 9,10-dihydroxydecalin.

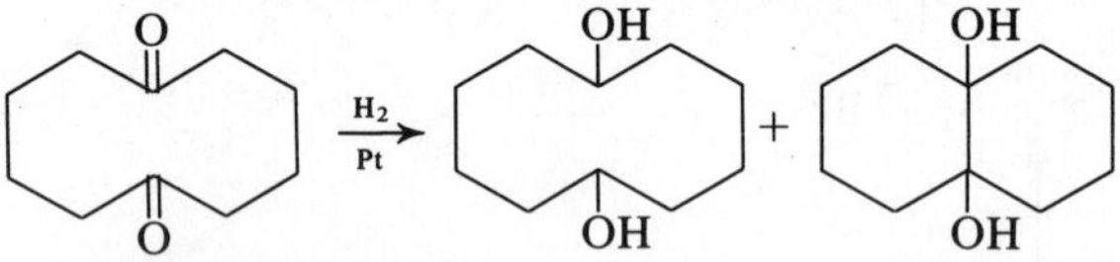

Large-ring compounds

Preparations of large-ring compounds. Up to 1926, the largest ring compound known contained eight carbon atoms. Ruzicka and his co-workers (1926, onwards) prepared large rings containing up to thirty-four carbon atoms. Their first method was to distil the calcium salts of dibasic acids, this method being limited to the preparation of cyclopentanone, cyclohexanone and cyclo-heptanone; the yields were poor. Ruzicka and his co-workers increased the yields, and also were able to obtain larger rings, by distilling *in vacuo* at about 300°C the thorium, cerium or yttrium salt mixed with copper powder (which aids heat conduction). Usually a number of products was obtained by the distillation of a particular acid, *viz.*, a cyclic hydrocarbon, a cyclic monoketone and a cyclic diketone; *e.g.*, the yttrium salt of the dibasic acid $(CH_2)_{10}(CO_2H)_2$ gave a cyclic hydrocarbon, the cyclic monoketone (I), and the cyclic diketone (II).

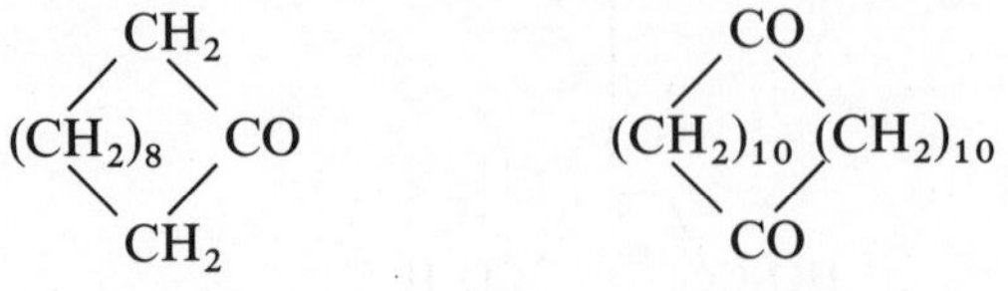

These ketones were then converted into the corresponding cycloalkanes (see Method 2, p. 533). A point of interest about this method is that 9- and 10-rings cannot be prepared this way. The structure of the ring compound (the cyclic ketone) was established by oxidation to the α,ω-dicarboxylic acid.

Ziegler *et al.* (1933) made use of the *high-dilution principle* (p. 471), obtaining large rings by the intramolecular condensation of α,ω-normal aliphatic dicyanides in the presence of alkali derivatives of secondary amines. The mechanism of the reaction is not certain; it may be as follows:

$$
\begin{array}{c}
\text{CH}_2\text{CN} \\
(\text{CH}_2)_n \\
\text{CH}_2\text{CN}
\end{array}
\xrightarrow{\text{LiN(C}_6\text{H}_5)(\text{C}_2\text{H}_5)}
\begin{array}{c}
\overset{\text{Li}}{\underset{}{\text{CHCN}}} \\
(\text{CH}_2)_n \\
\text{CH}_2\text{CN}
\end{array}
+\text{C}_6\text{H}_5\text{NHC}_2\text{H}_5 \longrightarrow
\begin{array}{c}
\text{CHCN} \\
(\text{CH}_2)_n \quad \text{C}=\text{NLi} \\
\text{CH}_2
\end{array}
\xrightarrow{\text{acid}}
$$

$$
\begin{array}{c}
\text{CHCN} \\
(\text{CH}_2)_n \quad \text{C}=\text{NH} \\
\text{CH}_2
\end{array}
\longrightarrow
\begin{array}{c}
\text{CHCO}_2\text{H} \\
(\text{CH}_2)_n \quad \text{CO} \\
\text{CH}_2
\end{array}
\xrightarrow{\text{heat}}
\begin{array}{c}
\text{CH}_2 \\
(\text{CH}_2)_n \quad \text{CO} \\
\text{CH}_2
\end{array}
$$

This method may be regarded as an extension of the Thorpe nitrile condensation (p. 357).

Hunsdiecker (1942) prepared large rings starting from ω-bromoacyl chlorides and E.A.A. (*cf.* p. 293). This method also requires the high-dilution technique, but the yields of medium rings are poor.

$$\text{Br(CH}_2)_n\text{COCl}+[\text{CH}_3\text{COCHCO}_2\text{C}_2\text{H}_5]^-\text{Na}^+ \longrightarrow$$

$$
\text{Br(CH}_2)_n\overset{\overset{\text{COCH}_3}{|}}{\text{COCHCO}_2\text{C}_2\text{H}_5}+\text{NaCl}
\xrightarrow[\text{CH}_3\text{OH}]{\text{CH}_3\text{ONa}}
\text{Br(CH}_2)_n\text{COCH}_2\text{CO}_2\text{CH}_3
\xrightarrow[\text{acetone}]{\text{NaI}}
$$

$$
\text{I(CH}_2)_n\text{COCH}_2\text{CO}_2\text{CH}_3
\xrightarrow[\text{butanone}]{\text{K}_2\text{CO}_3}
\begin{array}{c}
\text{CO} \\
(\text{CH}_2)_n \quad | \\
\text{CHCO}_2\text{CH}_3
\end{array}
\xrightarrow{\text{H}_2\text{SO}_4}
\begin{array}{c}
\text{CO} \\
(\text{CH}_2)_n \quad | \\
\text{CH}_2
\end{array}
$$

Leonard *et al.* (1958), using the high-dilution technique, extended the Dieckmann method (p. 534) to the preparation of large rings.

$$
\begin{array}{c}
\text{CO}_2\text{Et} \\
(\text{CH}_2)_n \\
\text{CO}_2\text{Et}
\end{array}
\xrightarrow{t-\text{BuOK}}
\begin{array}{c}
\text{CO} \\
(\text{CH}_2)_n \quad | \\
\text{CHCO}_2\text{Et}
\end{array}
\xrightarrow{\text{H}_2\text{SO}_4}
\begin{array}{c}
\text{CO} \\
(\text{CH}_2)_n \quad | \\
\text{CH}_2
\end{array}
$$

The most useful method of preparing large rings is the intramolecular acyloin condensation of α,ω-dicarboxylic esters (Prelog, 1947; *cf.* p. 253). This method does not require the high-dilution technique:

$$
\begin{array}{c}
\text{CO}_2\text{R} \\
(\text{CH}_2)_n \\
\text{CO}_2\text{R}
\end{array}
\xrightarrow[\text{Et}_2\text{O; N}_2]{\text{Na}}
\begin{array}{c}
\text{CO} \\
(\text{CH}_2)_n \quad | \\
\text{CHOH}
\end{array}
$$

The corresponding cycloalkane is obtained by reduction of the acyloin by the Clemmensen method.

Large rings have also been prepared by the reduction of cyclopoly-ynes (see below).

Large-ring compounds are very stable and those containing more than 20 carbon atoms consist of two approximately parallel chains joined at both ends by methylene bridges.

Cyclophanes. These are ring systems which contain benzene rings, *e.g.*,

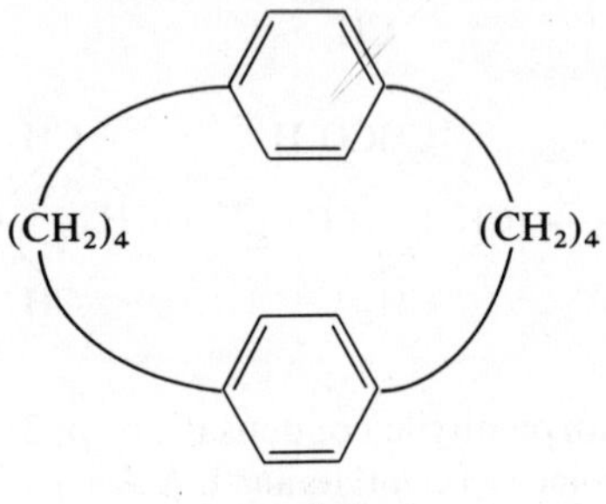

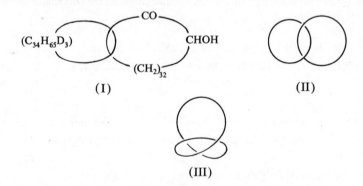

$(CH_2)_4$ $(CH_2)_4$

paracyclophane

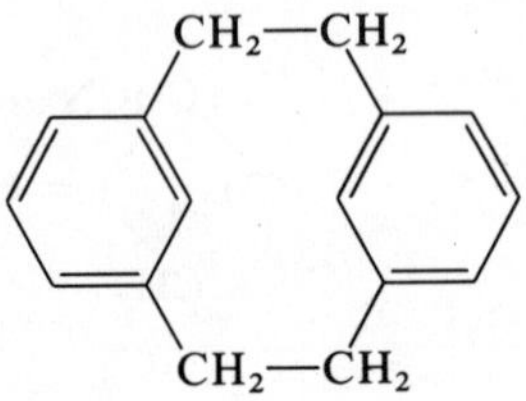

CH_2—CH_2

CH_2—CH_2

metacyclophane

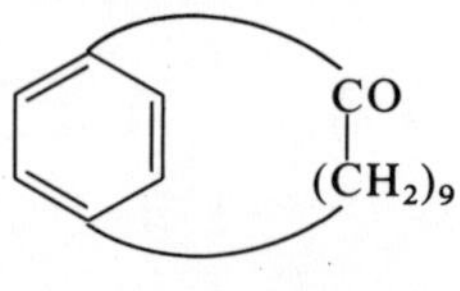

CO
$(CH_2)_9$

benzocyclenone

$(CH_2)_{10}$

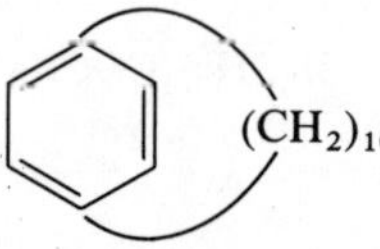

cyclophane

Benzocyclenones, which contain one benzene ring, have also been prepared and have been converted into the corresponding cyclophanes.

Catenanes. These are large-ring compounds consisting of interlocked rings. Wasserman (1960) prepared the ketone (I) which, in the form of its hydrocarbon, may be represented as (II). (III), which is a *trefoil*, is an example of a single chain that is knotted. No trefoil has yet been synthesised.

$(C_{34}H_{65}D_3)$ CO   CHOH $(CH_2)_{32}$

(I)   (II)

(III)

Cyclopoly-ynes and cyclopolyenes. Although large *saturated* rings are relatively strainless, rings containing multiple bonds present a different picture. So far, the smallest ring prepared that contains a *cis*-double bond is (I), a *trans*-double bond, (II), and a triple bond, (III).

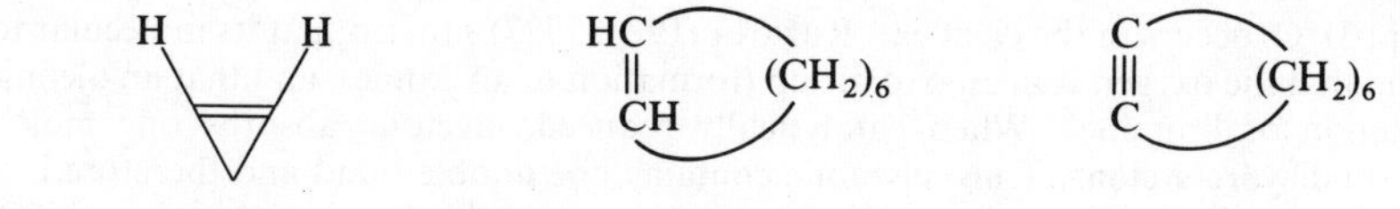

(I)   (II)   (III)

All are highly strained and, in consequences, show abnormal reactivity. Strain in (II) is due to the fact that the molecule is twisted about the double bond.

Large-ring cyclopoly-ynes and cyclopolyenes. Many of these have been prepared and they are relatively stable. The following example illustrates the method used. Tetradeca-1,13-di-yne, on treatment with excess of cupric acetate in methanolic pyridine, in high dilution (see p. 471), forms cyclotetradeca-1,3-di-yne and also some of the cyclic dimer cyclo-octacosa-1,3,15,17-tetra-yne.

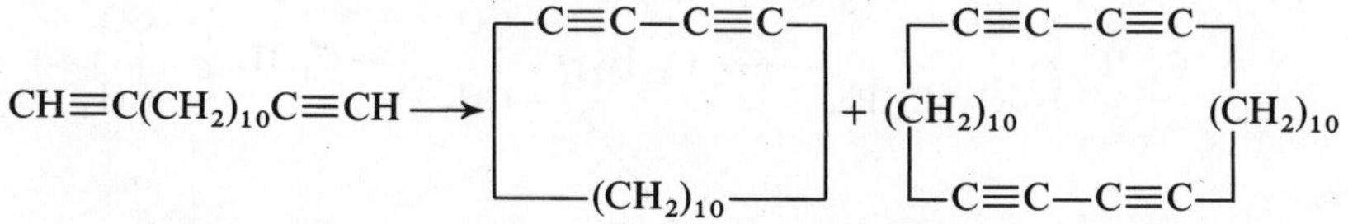

$$CH{\equiv}C(CH_2)_{10}C{\equiv}CH \longrightarrow$$

These, on catalytic hydrogenation (platinum), gave respectively cyclotetradecane $(C_{14}H_{28})$ and cyclo-octacosane $(C_{28}H_{56})$. Sondheimer *et al* (1959) have carried out oxidative coupling with octa-1,7-di-yne and obtained 16-, 24-, 32- and 40-membered polyalkynes which, on hydrogenation, give cycloalkanes. These authors have now prepared up to a 54-membered ring system; the largest ring hitherto contained 34 carbon atoms.

The above method was used by Sondheimer to prepare **annulenes** (cyclopolyenes; see also p. 580), *e.g.*, [18]annulene was prepared by the oxidative coupling of hexa-1,5-di-yne to give the cyclic trimer (plus the tetramer and pentamer), and this, on heating with potassium t-butoxide in t-butanol, isomerises (by prototropic rearrangement) to a conjugated enyne which, on hydrogenation in the presence of Lindlar's catalyst (p. 134), gave [18]annulene.

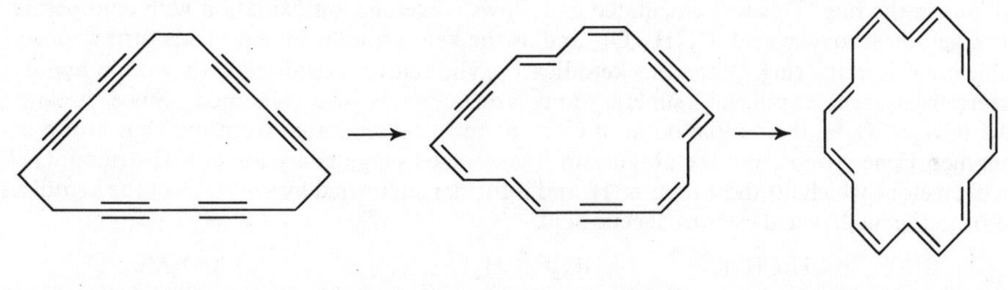

Factors influencing the ease of ring formation. Inspection of Table 19.1 shows that if strain were the only factor, ease of ring-closure would be greatest for the six-membered ring and large rings. In practice, the ease of formation is greatest for common rings (5–7), and it therefore follows that factors other than strain must be operating. In view of the fact that rings are formed by reaction between the functional groups *at each end of the chain*, it is reasonable to suppose that the distance between these two groups will play a part in deciding how readily ring-closure will occur. A detailed examination of cyclic compounds prepared by different methods, *e.g.*, pyrolysis of thorium salts, cyclisation of α,ω-dicyanides, acyloin condensation, etc., has shown that the yields are a function of the length of the chain. The curves obtained (*yield against chain-length*) are similar but *not* identical. Thus, there is a second factor, the *distance* or *probability factor*, which operates in addition to the strain factor.

Some large-ring compounds occur in nature, two of which are *civetone* and *muscone*. The following account of the elucidation of their structures is typical of the 'classical' methods used in the examination of natural products.

Civetone, m.p. 31°C, occurs in the civet cat. Ruzicka (1926, 1927) showed that its molecular formula is $C_{17}H_{30}O$, and that the oxygen is in an oxo group (formation of an oxime), and that an olefinic bond is present (addition of bromine). When catalytically reduced, civetone absorbs one molecule of hydrogen to form **dihydrocivetone**. Thus, civetone contains one double bond and therefore its parent (saturated) hydrocarbon is $C_{17}H_{34}$ (add two hydrogens for one double bond, and replace the O of the $C=O$ by two hydrogens). This formula corresponds to a *monocyclic* hydrocarbon.

Dihydrocivetone condenses with benzaldehyde; this shows the presence of a $-CH_2-CO-$ group (see p. 738). Also, since the oxidation of dihydrocivetone with chromic acid gives a dicarboxylic acid $C_{17}H_{32}O_4$, *without* loss of carbon, the oxo group is confirmed to be a keto group *which must be in the ring*. Furthermore, the same dicarboxylic acid is obtained by the Clemmensen reduction of civetone to **civetene** (in which the double bond remains intact), followed by ozonolysis. Thus, the double bond is also in the ring. The facts obtained so far may be formulated:

$$C_{15}H_{28}\left\{\begin{array}{l}-CO \\ | \\ -C{=}CHPh\end{array}\right. \xleftarrow{PhCHO} C_{15}H_{28}\left\{\begin{array}{l}-CO \\ | \\ -CH_2\end{array}\right. \xrightarrow[Ni]{H_2} C_{15}H_{30}\left\{\begin{array}{l}-CO \\ | \\ -CH_2\end{array}\right.$$

$$\downarrow CrO_3$$

$$C_{15}H_{30}\left\{\begin{array}{l}-CO_2H \\ -CO_2H\end{array}\right.$$

$$\Big/\!\!\Big/\!\!\Big/$$

$$\left.\begin{array}{l}CH- \\ \| \\ CH-\end{array}\right\}C_{13}H_{26}\left\{\begin{array}{l}-CH_2 \\ | \\ -CH_2\end{array}\right. \xrightarrow{O_3} \begin{array}{l}HO_2C- \\ \\ HO_2C-\end{array}\Big\}C_{13}H_{26}\left\{\begin{array}{l}-CH_2 \\ | \\ -CH_2\end{array}\right.$$

(via Zn/Hg, HCl)

Ruzicka now showed that the dicarboxylic acid, $C_{17}H_{32}O_4$, is identical with pentadecane-1,15-dicarboxylic acid (by comparison with a *synthetic* specimen). Hence, civetone is a seventeen-membered ring containing a keto group and one double bond, but there is no information which gives their relative positions in the ring. This was elucidated as follows. Civetone, on oxidation with cold permanganate, gave a ketodicarboxylic acid, $C_{17}H_{30}O_5$, and as the keto group is intact, this is further proof that the double bond is in the ring. When this ketodicarboxylic acid was oxidised with sodium hypobromite, a mixture of succinic, pimelic, suberic, and azelaic acids was obtained. Since azelaic acid is $HO_2C(CH_2)_7CO_2H$, the double bond in civetone must be separated from the keto group by at *least* seven methylene groups, and the absence of higher acids suggests a symmetrical structure. The structure of civetone which fits these facts is (I), and is further supported by synthesis of the ketodicarboxylic acid from the methyl acid ester of azelaic acid:

$$\begin{array}{l}CH(CH_2)_7 \\ \| \\ CH(CH_2)_7\end{array}\!\!\Big\rangle CO \xrightarrow{KMnO_4} \begin{array}{l}HO_2C(CH_2)_7 \\ \\ HO_2C(CH_2)_7\end{array}\!\!\Big\rangle CO \xleftarrow[\text{(ii) hydrolysis}]{\text{(i) Fe; heat}} 2\begin{array}{l}CO_2Me \\ (CH_2)_7 \\ CO_2H\end{array}$$

(I)

Final confirmation of the structure of civetone comes from its synthesis. Several syntheses have been carried out; that given here is due to Stoll (1948), and points to note are the use of the acyloin synthesis, protection of the keto group, and the formation of *cis*- and *trans*-isomers.

$$OC\!\!\left\langle\begin{array}{l}(CH_2)_7CO_2Me \\ \\ (CH_2)_7CO_2Me\end{array}\right. \xrightarrow[H_3O^+]{HOCH_2CH_2OH} \left[\begin{array}{l}O \\ \\ O\end{array}\right]\!\!C\!\!\left\langle\begin{array}{l}(CH_2)_7CO_2Me \\ \\ (CH_2)_7CO_2Me\end{array}\right. \xrightarrow[xylene]{Na}$$

$$\left[\begin{array}{l}O \\ \\ O\end{array}\right]\!\!C\!\!\left\langle\begin{array}{l}(CH_2)_7-CO \\ | \\ (CH_2)_7-CHOH\end{array}\right. \xrightarrow{H_2/Ni} \left[\begin{array}{l}O \\ \\ O\end{array}\right]\!\!C\!\!\left\langle\begin{array}{l}(CH_2)_7-CHOH \\ | \\ (CH_2)_7-CHOH\end{array}\right. \xrightarrow[AcOH]{HBr}$$

cis and *trans*

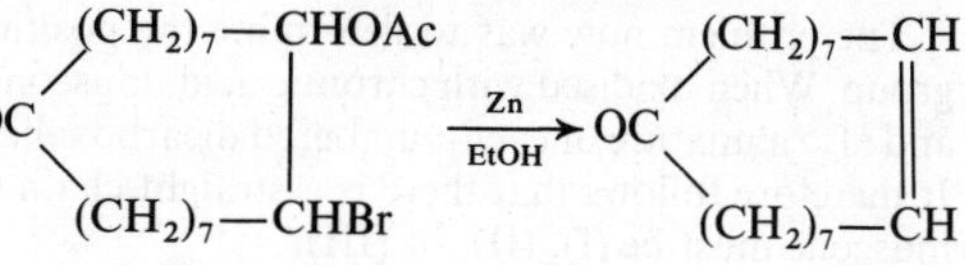

$$\begin{array}{c}
\text{(CH}_2)_7\text{---CHOAc} \\
\text{OC} \\
\text{(CH}_2)_7\text{---CHBr}
\end{array}
\xrightarrow[\text{EtOH}]{\text{Zn}}
\begin{array}{c}
\text{(CH}_2)_7\text{---CH} \\
\text{OC} \quad\quad \| \\
\text{(CH}_2)_7\text{---CH}
\end{array}$$

cis and *trans*   *cis* and *trans*

Because of the large ring, *cis*- and *trans*-isomers are possible, and both were present in the product.

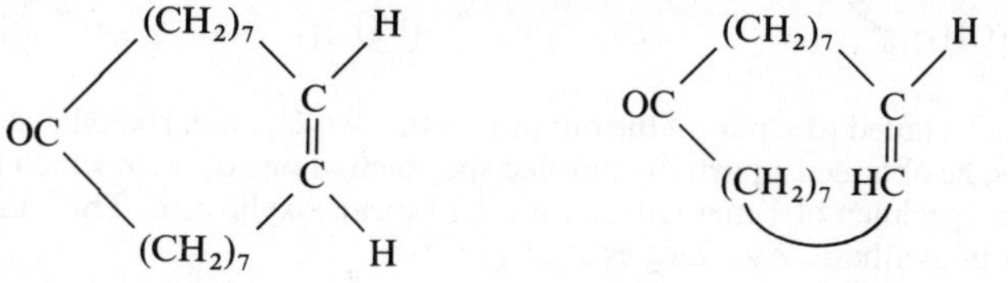

cis   *trans*

The two forms, referred to as α- and β-forms, were separated by fractional crystallisation of their dioxalans, and the β-form was then treated as shown:

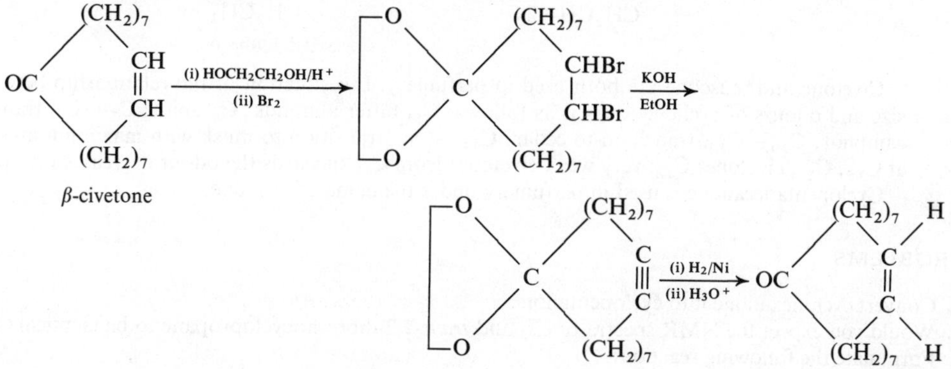

β-civetone

cis-civetone

Since catalytic hydrogenation of a triple bond normally gives *cis*-addition, the configuration of the product is *cis* (*i.e.*, α-civetone). This was shown to be identical with natural civetone.

Muscone (*muskone*) occurs in natural musk (from the musk deer). It is a thick colourless oil and is optically active. Its molecular formula is $C_{16}H_{30}O$, and it was shown to be a ketone. Since it was also shown to be saturated, it must therefore be cyclic, since if it were an open-chain compound, its formula would be $C_{16}H_{32}O$; if it had contained more than one ring, the number of hydrogen atoms would have been less than thirty.

The investigation of the odours of cyclic compounds was then used to elucidate the structure of muscone. Ruzicka was led to adopt this procedure because he had found that civetone and dihydrocivetone had about the same odour, *i.e.*, he assumed that similar structures gave rise to similar odours (this might be termed a *physiologico-chemical* method). Ruzicka found that the odour of muscone was identical with that of synthetic cyclopentadecanone and its methyl derivatives (see also below). He believed that muscone was therefore a methyl derivative of the fifteen-membered cyclic ketone. This was proved by preparing cyclopentadecanone (by distillation of the thorium salt of tetradecane-1,14-dicarboxylic acid), treating it with methylmagnesium iodide, and dehydrating the tertiary alcohol so produced to the alkene which was then catalytically reduced to methylcyclopentadecane:

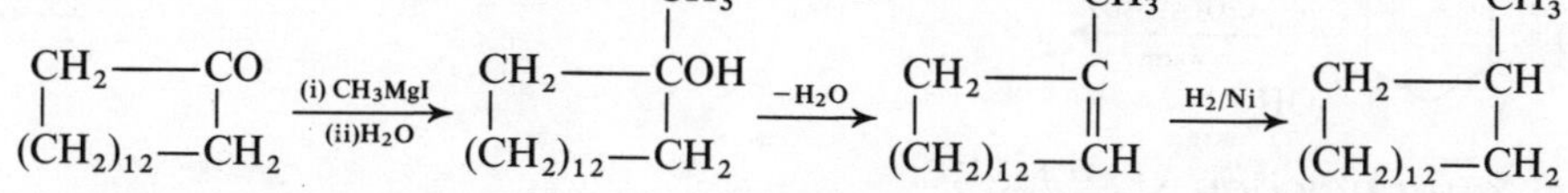

Methylcyclopentadecane is also obtained when muscone is reduced by the Clemmensen method.

The problem now was to determine the position of the methyl group with respect to the carbonyl group. When oxidised with chromic acid, muscone yields two acids with the formula $C_{14}H_{28}(CO_2H)_2$, and also a mixture of even-numbered dicarboxylic acids from succinic to decane-1,10-dicarboxylic acid. It therefore follows that there is a straight chain of at least ten methylene groups in muscone. Hence, muscone must be (I), (II), or (III).

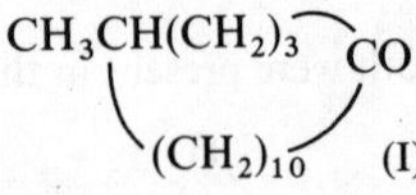

$$CH_3CH(CH_2)_3 \overgroup{CO}$$
$$(CH_2)_{10} \quad (I)$$

$$CH_3CH(CH_2)_2 \overgroup{CO}$$
$$(CH_2)_{11} \quad (II)$$

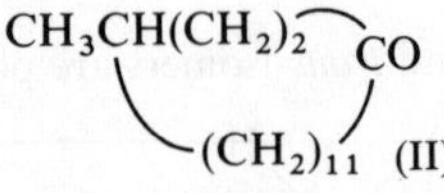

$$CH_3CHCH_2 \overgroup{CO}$$
$$(CH_2)_{12} \quad (III)$$

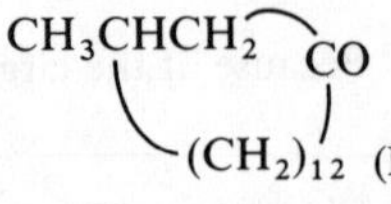

Although Ruzicka failed to separate the mixture of the two C_{14} dicarboxylic acids (see above) into two pure specimens, he obtained a partially purified specimen of one of them which he showed was identical with a *synthetic* specimen of 1-methyltridecane-1,13-dicarboxylic acid. Thus, muscone is (III), and this was confirmed by synthesis, *e.g.*, Ziegler *et al.* (1934):

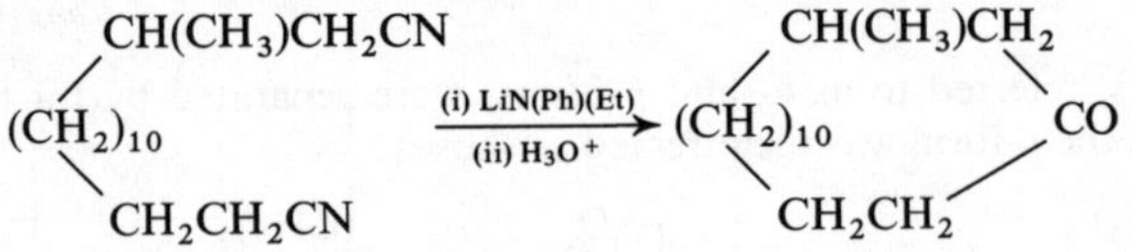

$$CH(CH_3)CH_2CN$$
$$(CH_2)_{10}$$
$$CH_2CH_2CN$$
$$\xrightarrow[\text{(ii) } H_3O^+]{\text{(i) LiN(Ph)(Et)}}$$
$$CH(CH_3)CH_2$$
$$(CH_2)_{10} \qquad CO$$
$$CH_2CH_2$$
$$(\pm)\text{-muscone}$$

Civetone and muscone are both used in perfumery. Investigation of the relationship between ring size and odours of cyclic ketones is as follows: C_5, bitter almonds; C_6, mint; C_7—C_9, transition to camphor; C_{10}—C_{13}, transition to cedar: C_{14}—C_{16}, transition to musk with maximum musk odour at C_{15}; C_{17}, civetone; C_{18}, very weak civetone; from C_{19} onwards the odour decreases very rapidly. Cyclopentadecanone is used in perfumery under the name of *exaltone*.

PROBLEMS

1. Convert cyclohexanone into cyclopentanone.
2. Would you expect the NMR spectra of *cis*- and *trans*-1,2-dibromocyclopropane to be identical? Explain.
3. Formulate the following reaction:

$$CH_3CO_2Et + (CH_2)_5(MgBr)_2 \xrightarrow[\text{(ii) } H^+]{\text{(i) } Et_2O} (A) \ C_7H_{14}O \xrightarrow{H^+} (B) \ C_7H_{12} \xrightarrow[\text{reagent}]{\text{Lemieux}} (C) \ C_7H_{12}O_3$$

4. Complete the following equations:

(i) 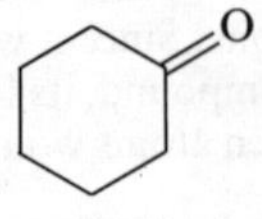 $\xrightarrow{SeO_2}$?

(ii) 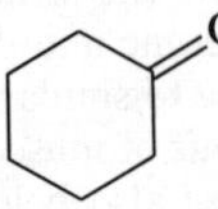 $+ HCO_2Et \xrightarrow{EtONa}$?

(iii) $ClCH_2CH_2CH_2CN \xrightarrow{NaOH}$ (cyclic compound)?

(iv) $PhCHBrCH_2CH_2Br \xrightarrow{Zn-Cu}$

(v) 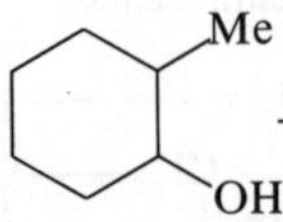 $\xrightarrow[\text{warm}]{CrO_3-H_2SO_4}$?

(vi) $C_3H_7CHClCOCH_2Me \xrightarrow{EtONa}$?

(vii) $MeCH(CH_2CO_2Et)CH_2CH_2CO_2Et \xrightarrow{EtONa}$?

(viii) $MeCO(CH_2)_3CO_2Et \xrightarrow{EtONa}$?

(ix) 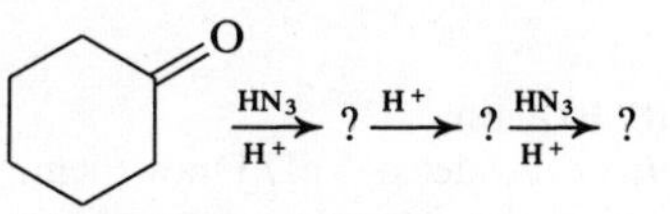 $\xrightarrow[H^+]{HN_3}$? $\xrightarrow{H^+}$? $\xrightarrow[H^+]{HN_3}$?

(x) 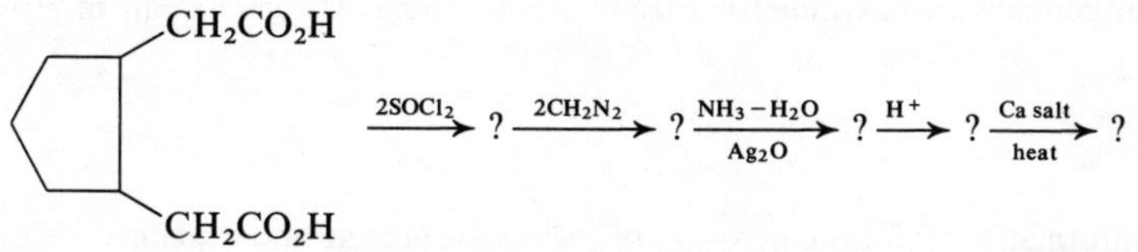 $\xrightarrow{2SOCl_2}$? $\xrightarrow[Ag_2O]{2CH_2N_2}$? $\xrightarrow{NH_3-H_2O}$? $\xrightarrow{H^+}$? $\xrightarrow[heat]{Ca\ salt}$?

(xi) 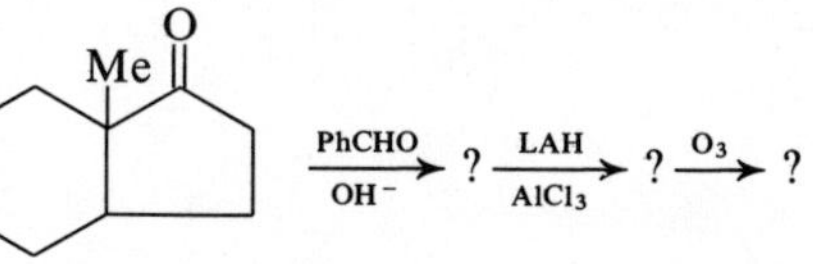 $\xrightarrow[OH^-]{PhCHO}$? $\xrightarrow[AlCl_3]{LAH}$? $\xrightarrow{O_3}$?

5. Synthesise

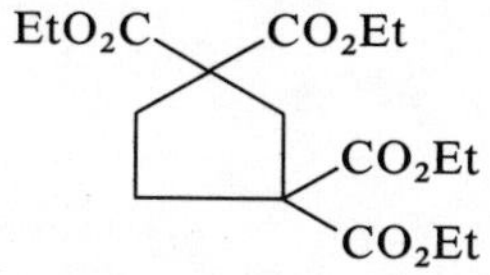

6. Treatment of $Me_2C{=}CHCH_2CH_2Cl$ with $CaCO_3{-}H_2O$ gives cyclopropyldimethylmethanol. Suggest a mechanism.

6. Discuss:

(i) 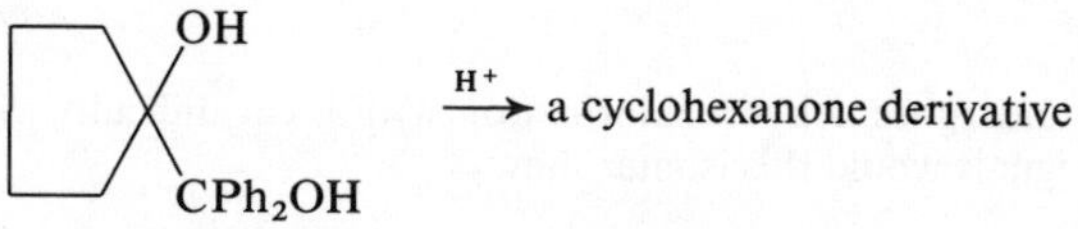 $\xrightarrow{H^+}$ a cyclohexanone derivative

(ii) 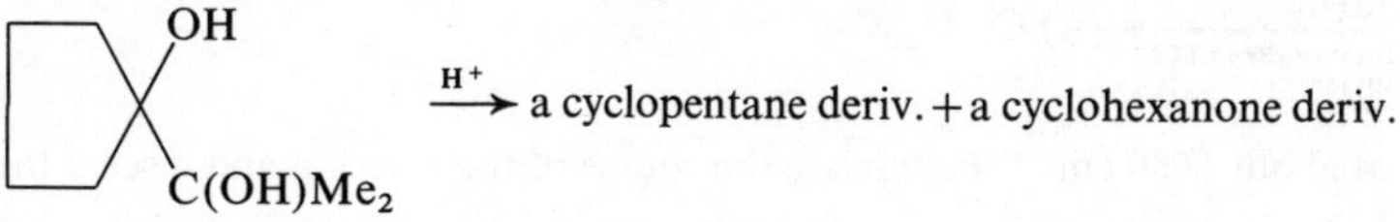 $\xrightarrow{H^+}$ a cyclopentane deriv. + a cyclohexanone deriv.

(iii) 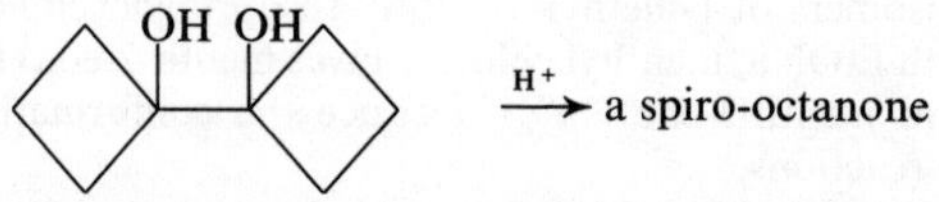 $\xrightarrow{H^+}$ a spiro-octanone

8. Using i.r. spectroscopy, differentiate between 2-methylcyclopentanone, cyclohexanone, cyclohex-2-enone, and cyclohex-3-enone.

9. Given the conformation shown, state: (*a*) the nature (*e* or *a*) of each group, and (*b*) the geometrical relationship (*cis* or *trans*) of: (i) Cl and Br; (ii) I and Et; (iii) Cl and Et; (iv) Et and Br; (v) Me and F; (vi) Cl and Me; (vii) F and Br; (viii) Et and Me; (ix) Cl and I; (x) I and Me.

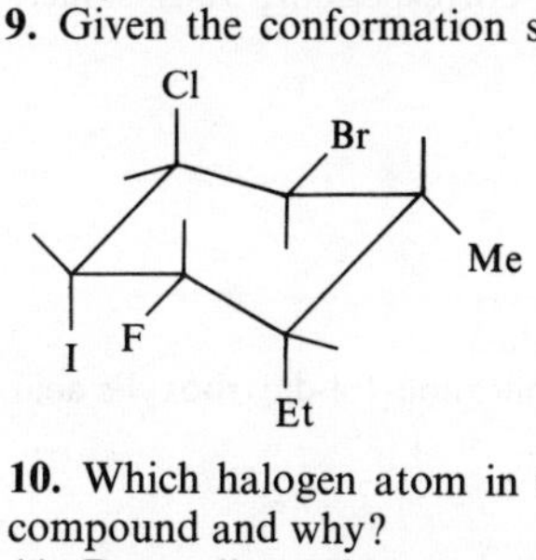

10. Which halogen atom in Q.9 would be most readily eliminated by the E2 mechanism to form the ene compound and why?

11. Draw all possible conformations for the disubstituted cyclohexanes:

(i) 1-Br-2-Cl; (ii) 1,3-Br$_2$; (iii) 1-Cl-4-I; (iv) 1,3-(OH)$_2$.

Which conformation is the most stable one in each set? Which are optically active compounds?

12. *cis-* and *trans*-Cyclohexane-1,2-diol are each treated with:

(i) 1 molecule of Ac_2O; (ii) a second molecule of Ac_2O.

Are the rates of the first and second reactions the same or different? Explain.

13. How many conformations are possible for: (i) *cis*-decal-2-ol; (ii) *trans*-decal-2-ol? Draw them.

14. 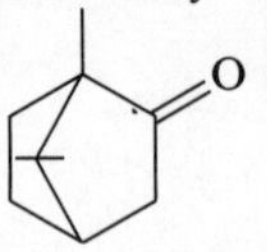 Camphor contains *two* asymmetric carbon atoms, and exists as *one* pair of enantiomers. Explain.

15. Draw the equation (conformational) for the pyrolysis of cyclohexyl acetate and explain.

16. Complete the equations:

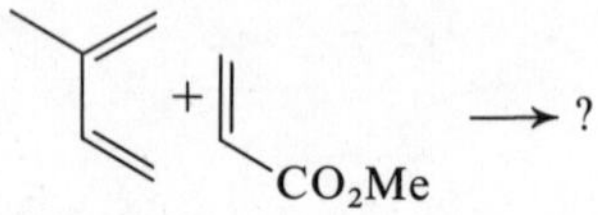

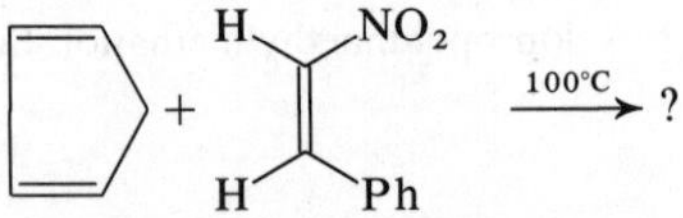

17. *One* isomer of hexachlorocyclohexane undergoes dehydrochlorination with great difficulty. Suggest its conformation. Explain. How many NMR signals would this isomer show?

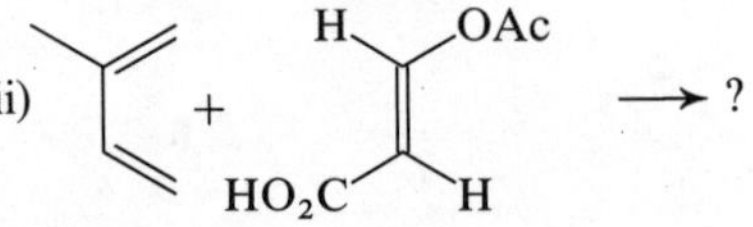

18.

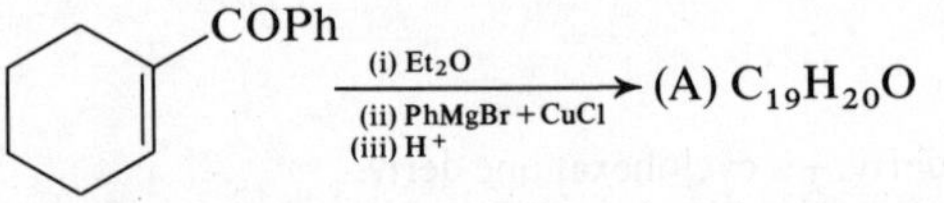

(**A**) showed an i.r. band at about 1680 cm^{-1}. Formulate the course of the reaction and discuss the geometry of the product.

19. Menthyl and neomenthyl chlorides are stereoisomers of 1-methyl-3-chloro-4-isopropylcyclohexane and in both the alkyl groups are *trans*. On treatment with EtONa, menthyl chloride gives menth-2-ene (100%) and neomenthyl chloride gives menth-2-ene (25%) and menth-3-ene (75%). Deduce the conformation of the chlorine atom in each chloride, and formulate the reactions.

20. Discuss the data of the heats of formation ($kJ \text{ mol}^{-1}$) of the following methylcyclohexanols: *cis*-2-Me-, $-396 \cdot 6$; *trans*-2-Me-, $-422 \cdot 2$; *cis*-3-Me-, $-422 \cdot 6$; *trans*-3-Me-, $-400 \cdot 8$.

21. *cis*-Cyclohexane-1,2-diol is oxidised by periodic acid more rapidly than the corresponding *trans*-isomer. Discuss.

22. (**A**)

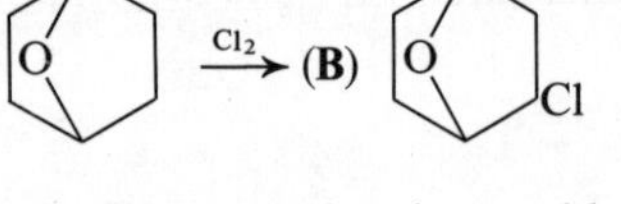

Two forms of (**B**) are produced, one with $\mu = 1 \cdot 07D$ and the other $3 \cdot 06D$. Discuss.

23. Starting from D.E.M. and any other compounds you like, synthesise cyclohexane-1,4-dicarboxylic acid.

24. Convert 1,2,4-trihydroxycyclohexane into 3,4-dihydroxycyclohexanone.

25. Suggest a method for the conversion:

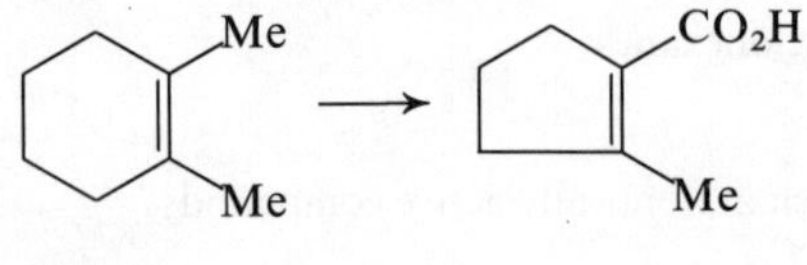

26. Expoxidation of 4-methylcyclopentene gives two isomeric epoxides, with one predominating. Discuss.

27. Convert cyclohexanol into *cis*- and *trans*-cyclohexane-1,2-diol.

28. Discuss: *cis*-2-chlorocyclohexanol with aqueous alkali gives cyclohexanone, whereas the *trans* isomer gives cyclohexene oxide.

29. Complete the equations:

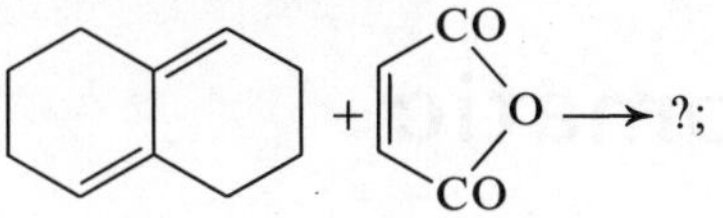

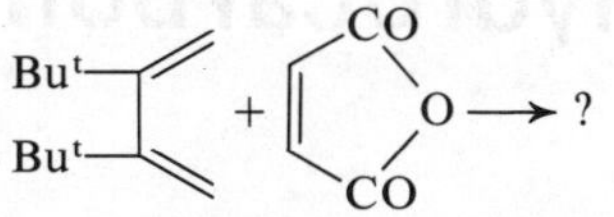

30. The deamination (with nitrous acid) of *trans*-2-aminocyclohexanol gives a cyclopentane derivative, $C_6H_{10}O$, as the only product and that of the *cis*-isomer gives two products, one the same as that from the *trans*-isomer and the other a cyclohexane derivative, $C_6H_{10}O$. Formulate the reactions and explain them.

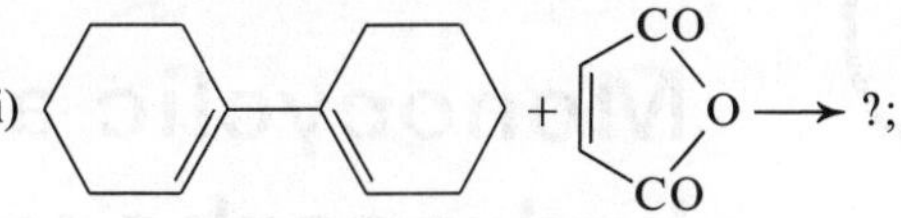

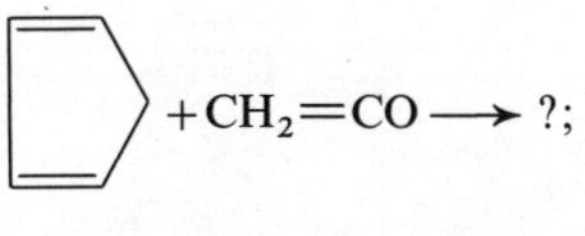

REFERENCES

Organic Reactions, Wiley, Vol. IV (1948), Chs. 1 and 2: Vol. V (1949), Ch. 3, 'The Diels–Alder Reaction': Vol. XI (1960), Ch. 2, 'The Demjanov and Tiffeneau–Demjanov Ring Expansions'; Ch. 4, 'The Favorsky Rearrangement of Haloketones': Vol. XII (1962), Ch. 1, 'Cyclobutane Derivatives from Thermal Cyclo-addition Reactions': Vol. XV (1967), Ch. 1, 'The Dieckmann Condensation': Vol. XVII (1969), Ch. 1, 'The Synthesis of Substituted Ferrocenes and Other π-Cyclopentadienyl-Transition Metal Compounds'.

SLOCUM *et al.*, 'Metalation of Metallocenes', *J. Chem. Educ.*, 1969, **46**, 144.

WASSERMANN, *Diels–Alder Reactions*, Elsevier (1965).

PATAI (ed.), *The Chemistry of Alkenes*, Interscience (1964), Ch. 11, 'Cycloaddition Reactions of Alkenes'.

NEWMAN (ed.), *Steric Effects in Organic Chemistry*, Wiley (1956), Ch. 1. 'Conformational Analysis'.

BARTON and COOKSON, 'The Principles of Conformational Analysis', *Quart. Rev.*, 1956, **10**, 44.

ELIEL, *Stereochemistry of Carbon Compounds*, McGraw-Hill (1962).

LLOYD, *Alicyclic Compounds*, Arnold (1963).

WHITHAM, *Alicyclic Chemistry*, Oldbourne Press (1963).

DE LA MARE and KLYNE (ed.), *Progress in Stereochemistry*, Butterworths, Vol. 3 (1962), Ch. 6, 'Stereochemistry of Many-Membered Rings'.

WILSON and GOLDHAMER, 'Cyclobutane Chemistry', *J. Chem. Educ.*, 1963, **40**, 504, 599.

COPE *et al.*, 'Transannular Reactions in Medium-sized Rings', *Quart. Rev.*, 1966, **20**, 119.

GUTSCHE and REDMORE, *Carbocyclic Ring Expansion Reactions*, Academic Press (1968).

FERGUSON, 'Ring Strain and Reactivity of Alicycles', *J. Chem. Educ.*, 1970, **47**, 46.

TURRO *et al.*, 'Favorskii Rearrangement of Some α-Bromo-ketones', *Chem. Comm.*, 1969, 270

20 Monocyclic aromatic hydrocarbons

Early in the development of organic chemistry, organic compounds were arbitrarily classified as either aliphatic or aromatic. The aliphatic compounds were so named because the first members of this class to be studied were the fatty acids (see p. 68). The term *aliphatic* is now reserved for any compound that has an *open-chain* structure.

In addition to the aliphatic compounds, there was a large number of compounds which were obtained from natural sources, *e.g.*, resins, balsams, 'aromatic' oils, etc., which comprised a group of compounds whose structures were unknown but had one thing in common: a pleasant odour. Thus these compounds were arbitrarily classified as *aromatic* (Greek: *aroma*, fragrant smell). Careful examination of these compounds showed that they contained a higher percentage carbon content than the corresponding aliphatic hydrocarbons, and that most of the simple aromatic compounds contained at least *six* carbon atoms. Furthermore, it was shown that when aromatic compounds were subjected to various methods of treatment, they often produced benzene or a derivative of benzene. If attempts were made to convert aromatic compounds into compounds with fewer carbon atoms than six (as in benzene), the whole molecule generally disrupted. It became increasingly evident that aromatic compounds were related to benzene, and this led to reserving the term aromatic for benzene and its derivatives. Thus aromatic compounds are *benzenoid* compounds; these are cyclic, but their properties are totally different from those of the alicyclic compounds.

Benzene (*phene*), C_6H_6, was first isolated by Faraday (1825) from cylinders of compressed illuminating gas obtained from the pyrolysis of whale oil. In 1845, benzene was found in coal-tar by Hofmann, and this is still a source of benzene and its derivatives.

When coal is destructively distilled, four fractions are obtained: coal-gas, coal-tar, ammoniacal liquors and coke. The low temperature carbonisation process (550–600°C) is used for the production of chemicals and smokeless fuels, and the high temperature process (1 000–1 300°C) is used for the production of coal-gas and coke.

> **Coal-tar.** The number of fractions taken from coal-tar varies, *e.g.*, one sample is shown in Table 20.1; another sample is three fractions: middle oil, heavy oil, and pitch. Subsequent treatment depends on the number of fractions taken. The light-oil fraction is washed with sodium hydroxide (to remove acids), then with sulphuric acid (to remove bases), and finally with water. The washed light oil is now distilled.

Table 20.1

Number of fraction	Temperature range	Name of fraction	Density	Percentage by volume
1	Up to 170°C	Crude light oil	0·970	2·25
2	170–230°C	Middle oil or Carbolic oil	1·005	7·5
3	230–270°C	Heavy oil or Creosote oil	1·033	16·5
4	270–360°C	Green oil or Anthracene oil	1·088	12
5		Pitch (left in retort)		about 56

Various fractions may be taken, *e.g.*, the fraction collected up to 110°C is known as '90 per cent *benzol*' (70 per cent benzene, 24 per cent toluene and some xylene). Pure xylene is obtained from the fraction between 110–140°C. The distillate between 140–170°C is known as '*solvent naphtha*' or *benzine* (consists mainly of xylenes, cumenes, etc.); it is used as a solvent for resins, rubber, paints, etc. The fraction '90 per cent benzol' gives pure benzene, toluene and xylene on careful fractionation; about 9·5 per cent benzene and 8·7 per cent toluene are obtained from the crude light-oil fraction.

Benzene was first synthesised by Berthelot (1870) by passing acetylene through a red-hot tube:

$$3C_2H_2 \longrightarrow C_6H_6 + \text{other products}$$

Benzene may be prepared in the laboratory by many methods, most of which depend on the decarboxylation of aromatic acids, *e.g.*, by heating benzoic or phthalic acid with calcium oxide:

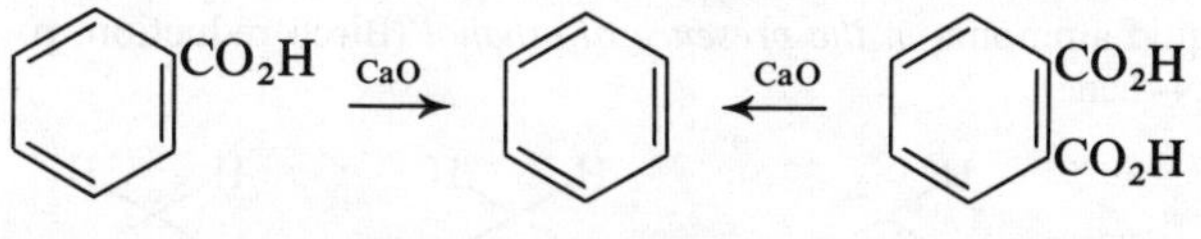

Preparation of benzene and its homologues from petroleum

Aromatic compounds can be extracted from petroleum in which they occur naturally. They are also prepared from the non-aromatic constituents of petroleum, which is now the main source.

(i) **Hydroforming or catalytic reforming.** This method is based on dehydrogenation, cyclisation and isomerisation reactions, and the aromatic compounds obtained contain the same number of carbon atoms as the aliphatic starting materials. Hydroforming is carried out under pressure at 480–550°C in the presence of, *e.g.*, chromium oxide on an alumina support. Many mechanisms have been proposed but none is certain. The following are the most important examples of hydroforming:

$$\underset{\text{n-hexane}}{CH_3(CH_2)_4CH_3} \longrightarrow \underset{\text{benzene}}{C_6H_6} + 4H_2$$

$$\underset{\text{n-heptane}}{CH_3(CH_2)_5CH_3} \longrightarrow \underset{\text{toluene}}{C_6H_5CH_3} + 4H_2$$

$$\underset{\text{n-octane}}{CH_3(CH_2)_6CH_3} \longrightarrow \underset{\substack{\text{xylene (three} \\ \text{isomers)}}}{C_6H_4(CH_3)_2} + \underset{\substack{\text{ethyl-} \\ \text{benzene}}}{C_6H_5C_2H_5}$$

The various hydrocarbons are separated by a selective solvent process, but since benzene is obtained in far less amount than toluene and the xylenes, these are converted into benzene by heating with hydrogen under pressure in the presence of a metal oxide catalyst. This process is called **hydro-dealkylation**; *e.g.*,

$$C_6H_5CH_3 + H_2 \longrightarrow C_6H_6 + CH_4$$

(ii) **High-temperature cracking in the presence of a catalyst.** The charging stock may be cracked at about 650–680°C in tubes packed with metallic dehydrogenation catalysts (which are the same as those used in hydroforming). By this means the following aromatic hydrocarbons have been isolated from the cracked alkanes: benzene, toluene, xylenes, naphthalene, anthracene and many other polynuclear hydrocarbons.

Properties of benzene

Benzene is a colourless liquid, m.p. 5·5°, b.p. 80°C, with a peculiar smell. It is inflammable, burning with a smoky flame, a property which is characteristic of most aromatic but not of most aliphatic compounds, and is due to the high carbon content of the former. It is a very good solvent for fats, resins, sulphur, iodine, etc., and is used in dry cleaning. It is also used as a motor fuel ('benzol') and for the manufacture of nitrobenzene, dyes, drugs, etc.

Benzene, obtained from coal-tar, always contains thiophen, which is removed (commercially) by heating benzene with hydrogen under pressure at about 400°C in the presence of a catalyst (the products from thiophen are n-butane and hydrogen sulphide; see also p. 834).

Benzene is a very stable compound; it is very slowly attacked by a solution of chromic acid or acid permanganate (both powerful oxidising agents) to form carbon dioxide and water. It can be reduced catalytically to cyclohexane, but the partially hydrogenated products, dihydro- and tetrahydrobenzene, have not been isolated in this reaction. Nickel and platinum are the best catalysts for hydrogenation of the benzene ring, whereas copper chromite or copper-barium chromite is useful when it is desired to reduce only side-chains.

Although benzene is not reduced by metals and acid, nor by sodium in ethanol, it is reduced by sodium in liquid ammonia *in the presence of ethanol* (Birch reduction, p. 101) to 1,4-dihydrobenzene (cyclohexa-1,4-diene).

Lithium in anhydrous ethylamine, however, reduces benzene to cyclohexene and cyclohexane (Benkeser *et al.*, 1955).

When heated with hydrogen iodide at 250°C, benzene is converted into a mixture of cyclohexane and methylcyclopentane. The action of chlorine and bromine on benzene depends on the conditions. In bright sunlight, halogen forms *addition* products with benzene, *e.g.*, chlorine adds on to form benzenehexachloride, $C_6H_6Cl_6$. In the absence of direct sunlight, benzene undergoes *substitution* with halogen; the reaction is slow, but in the presence of a halogen carrier, *e.g.*, iron or iodine, substitution is rapid. Thus, with chlorine, benzene forms chlorobenzene, C_6H_5Cl, dichlorobenzenes, $C_6H_4Cl_2$, etc. When benzene is heated with concentrated nitric acid or concentrated sulphuric acid, substitution products of benzene are obtained:

$$C_6H_5NO_2 \xleftarrow{\text{HNO}_3} C_6H_6 \xrightarrow{\text{H}_2\text{SO}_4} C_6H_5SO_3H$$

nitrobenzene   benzenesulphonic acid

Isomerism of benzene derivatives. Since all the six hydrogen atoms in benzene are equivalent (p. 575) only one monobromobenzene, mononitrobenzene, etc., is possible. In many compounds the univalent group C_6H_5— is known as **phenyl**, and is represented by the abbreviation Ph or

ϕ, *e.g.*, chlorobenzene may be written PhCl or ϕCl. When dealing with *any* univalent aromatic group, the symbol Ar is used.

When two hydrogen atoms in benzene are replaced by two univalent groups (which may be the same or different), three isomers are possible (position isomerism):

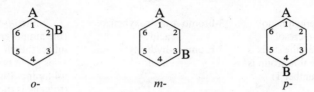

     o-   *m-*   *p-*

Since the six hydrogen atoms in benzene are equivalent, the positions 1,2- and 1,6- are equivalent. The 1,2-(1,6-)disubstituted benzene derivative is known as the *ortho-(o)*-compound. The 1,3- and 1,5-positions are equivalent and a 1,3-(1,5)-disubstituted derivative is known as the *meta-(m)*-compound. The 1,4-disubstituted derivative is known as the *para-(p)*-compound. The bivalent group $C_6H_4<$ is known as the **phenylene** group, *e.g.*, *m*-phenylenediamine.

In the case of trisubstitution derivatives of benzene, the number of isomers depends on the nature of the substituent groups.

  (i) If the three substituent groups are identical, then three isomers are possible:

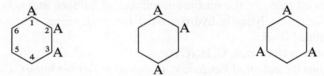

  (ii) If two substituent groups are identical and the third different, then six isomers are possible.

  (iii) If all three substituent groups are different, then ten isomers are possible.

Nomenclature

When two or more *functional groups* are present, number 1 is to be given to the *principal function*. The usual order for choosing the principal function is given on p. 297. For convenience of fixing the orientation (position of groups), the ring should be oriented with position 1 at the top and with the numbers proceeding in a *clockwise* direction. If a letter (*o*, *m*, *p*) is used, the principal function is still given position 1. The general rule is: **if the functional group is named as a suffix, this group is given number 1.** In many cases, however, it may be better to name aromatic compounds by using the *trivial* names of the *simple* (or *parent*) hydrocarbon. When this scheme is used, *the root name decides where numbering starts*. The following examples show the application of the above rules:

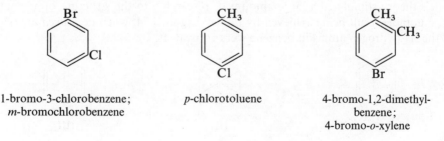

1-bromo-3-chlorobenzene;   *p*-chlorotoluene   4-bromo-1,2-dimethyl-
m-bromochlorobenzene   benzene;
                                              4-bromo-*o*-xylene

| 4-iodo-2-methylphenol; *p*-iodo-*o*-cresol (in both names, the suffix is -ol; therefore this group is given number 1) | 3-bromo-4-hydroxybenzoic acid; 3-bromo-*p*-hydroxybenzoic acid | 2-nitro-*p*-toluidine (the amino-group is in the suffix; therefore this group is given position 1); 4-amino-3-nitrotoluene (toluene is the parent, and so numbering starts at the *methyl* group |

Structure of benzene

Analysis and molecular-weight determinations show that the molecular formula of benzene is C_6H_6. The corresponding alkane is hexane, C_6H_{14}, and since the number of hydrogen atoms in benzene is eight less, it is to be expected that benzene would exhibit marked 'unsaturated reactions'. This is found to be so in practice, *e.g.*,

(i) Benzene adds on halogen, the maximum number of halogen atoms being six.

(ii) Benzene may be catalytically hydrogenated to cyclohexane, the maximum number of hydrogen atoms added is six.

(iii) Benzene forms a *triozonide*, $C_6H_6(O_3)_3$.

All these reactions indicate that benzene contains *three* double bonds. Further examination, however, shows that these double bonds behave in a most remarkable manner in comparison with double bonds in aliphatic compounds, *e.g.*,

(iv) Alkaline permanganate has no action on benzene in the cold, but on prolonged boiling, benzene is broken down into carbon dioxide and water.

(v) In the absence of sunlight (and preferably in the presence of a halogen carrier), benzene undergoes substitution when treated with halogen (*cf.* however, isobutene, p. 124).

(vi) Halogen acids do *not* add on to benzene.

Reactions (i)–(vi) lead to the conclusion that benzene contains three double bonds, but that these double bonds are different from aliphatic double bonds. This difference gives rise to 'aromatic properties' (unusual degree of saturation, stability, etc.; see also later).

Kekulé (1865) was the first to suggest a ring structure for benzene. He proposed formula (I), and believed (*but did not prove*) that this formula satisfied the following points:

(i) That benzene contains three double bonds.

(ii) That all the six hydrogen atoms in benzene are equivalent; consequently there is only one possible mono-substituted derivative, and there are three possible disubstitution products of benzene.

Kekulé's theory stimulated a large amount of research into the structure of aromatic compounds, a notable result being achieved by Ladenburg in 1874, when he *proved experimentally* that all the six hydrogen atoms in benzene were equivalent.

| (I) | (II) | (III) |

Kekulé's formula, however, did not explain the peculiar behaviour of the three double bonds in benzene. Claus (1867) therefore introduced his diagonal formula, (II), to overcome this difficulty. This formula also appeared to suggest a reason for the simultaneous formation of *o*- and *p*-compounds (see later). In 1882, Claus modified his formula, now postulating that the *para*-bonds were not like ordinary bonds, but could be easily ruptured.

Dewar (1867) suggested a number of formulæ for benzene, one being (III). This is not perfectly symmetrical and therefore was unacceptable (but see later). Ladenburg (1869) attacked Kekulé's formula on the grounds that it should give *four* disubstituted derivatives:

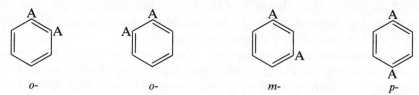

| *o*- | *o*- | *m*- | *p*- |

Thus there should be *two ortho*-derivatives, 1,2- and 1,6-; these have never been observed in practice. Ladenburg therefore proposed his prism formula, (IV). The six carbon atoms are at the corners of a regular prism, the edges of which denote linkages. This formula does not contain any double bonds, and therefore does not account for the addition products of benzene.

V. Meyer (1870) opposed Ladenburg; he pointed out that the 1,2- and 1,6-derivatives differed only in the position of double bonds and believed that this difference would be too slight to be noticeable. V. Meyer thus supported Kekulé's formula.

Kekulé (1872) felt that too much importance was being attached to the possible difference between the 1,2- and 1,6- positions, and pointed out that the difficulty arose from the inadequate representation of molecules by structural formulæ. According to Kekulé, the carbon atoms in

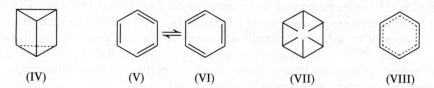

| (IV) | (V) | (VI) | (VII) | (VIII) |

benzene were continually in a state of vibration, and due to this vibration, each C—C pair had a single bond half of the time and a double bond the other half. This amounts to an oscillation between the two forms (V) and (VI), each molecule spending half its time in (V) and the other half in (VI). Thus neither (V) nor (VI) represents the benzene molecule satisfactorily; benzene is a 'combination' of the two, and so all bonds will be identical (neither single nor double), and hence there is no real difference between 1,2- and 1,6-disubstituted benzenes.

Baeyer (from about 1884–92) carried out a detailed investigation of benzene and some of its derivatives in order to decide between the formulæ of Kekulé and Ladenburg. One of the first things Baeyer proved was that cyclohexane and hexahydrobenzene were identical, thus establishing the ring structure of benzene (which Kekulé had assumed). Baeyer observed that as soon as one double bond was removed from benzene, the 'saturation' properties were lost, and that dihydrobenzene (cyclohexadiene) behaved as would be expected of an alkadiene. Baeyer showed that no *para* linkage was apparently present and hence rejected Claus' diagonal formula; he also rejected Dewar's formula for similar reasons. Baeyer also showed, by investigation of various benzene derivatives, that Ladenburg's prism formula was untenable and concluded that benzene contained six carbon atoms in a ring, but did not accept Kekulé's formula.

He adopted a suggestion of Armstrong (1887) and proposed what is known as the Armstrong–Baeyer centric formula (VII). According to this, the fourth valency of each carbon atom is represented as directed towards the centre of the ring but not actually linked to its opposite neighbour as in the Claus formula. This centric bond is not real but potential; by mutual action the power of each is rendered latent and there is a condition of equilibrium. Such a centric formula is unknown in aliphatic chemistry, and thus this formula could account for 'aromatic properties'.

The centric formula, however, is unsatisfactory for several reasons, *e.g.*, it did not explain the stability of the ring or the behaviour of the polynuclear hydrocarbons. The outcome of Baeyer's work was that 'aromatic properties' depend on the peculiar symmetrical arrangement of the fourth valency of each carbon atom in the ring.

Physico-chemical methods were also applied to the benzene problem, *e.g.*, Thomsen (1880) and Stohmann (1893) came to the conclusion that the heat of combustion of benzene was incompatible with the existence of three double bonds. On the other hand, Bruhl (1880) believed that the value of the refractive index of benzene proved their existence. These methods are particularly interesting in that they are some of the first examples of the application of physico-chemical methods to the elucidation of the structure of organic compounds.

In 1899, Thiele applied his theory of partial valency (p. 127) to the benzene problem, and suggested formula (VIII). This formula dispenses with Kekulé's oscillation hypothesis, and in this formula there is no real difference between 'single' and 'double' bonds, and so accounts for there being no difference between the 1,2- and 1,6- positions. At the same time it also accounts for the 'saturation' of benzene.

If Thiele's formula is correct, *i.e.*, 'aromatic character' is due to the symmetrical conjugation of the benzene ring, then cyclo-octatetraene should also exhibit 'aromatic character'. Willstätter prepared this compound with a view to testing Thiele's formula for benzene, and found that it had typical unsaturated properties (see p. 557), whereas, owing to its symmetrical conjugation (structure IX), it would have been expected to resemble benzene (see also p. 580). This led to a revival of Kekulé's oscillation formulæ and some of the modifications mentioned (other than Thiele's formula). The difficulty with the Kekulé formula is that it represents benzene with three double bonds, and the oscillation does not account for the difference in behaviour between these and olefinic bonds.

(IX)

Structural representation of aromatic compounds. According to the *Chemical Society*, the Kekulé type of structure should, in general, be used for benzenoid compounds. On the other hand, large circles representing *six delocalised π-electrons in cyclic systems* (with or without positive or negative signs as appropriate) may be used for certain types of compounds. Cyclic systems having more or fewer than *six* delocalised π-electrons may be represented by circles comprised of broken lines (see below).

Aromaticity (aromatic character)

Originally, the term *aromaticity* was used to describe all compounds that had the properties of benzene, and was confined, in consequence, only to compounds which contained benzene rings or a condensed system of benzene rings. The reason for this description was that compounds which had 'aromatic character' exhibited properties which were very much different from those of the analogous aliphatic and alicyclic compounds, *e.g.*, the ease of substitution (although the benzene ring was 'unsaturated'), the stability of the benzene ring, the weaker basic properties of aromatic amines, the acidic properties of phenols (as compared with alcohols), etc. These differences in chemical properties led chemists to seek an explanation, and in consequence, to define *aromaticity*.

Aromatic sextet theory. This was proposed by Robinson (1925), its essential feature being that there are six electrons more than necessary to link together the six carbon atoms (of the benzene ring). These six electrons, one being contributed by each carbon atom, form a 'closed group', and it is this closed group which gives rise to 'aromatic properties'. This closed group is not possible in aliphatic and alicyclic compounds, but is possible in heterocyclics, and so these also exhibit aromatic properties:

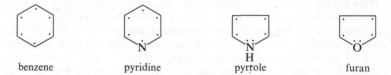

benzene	pyridine	pyrrole	furan

It should be noted that the hetero-atom in five-membered rings contributes *two* electrons to the aromatic sextet.

Valence bond theory of aromaticity. The V.B. method starts from the experimental results that benzene is a regular flat hexagon (angle 120°), with all six hydrogen atoms lying in the same plane (of the ring) and each C—C—H valency angle being 120°. This geometry of benzene has been established by X-ray analysis, which has shown that the benzene ring is planar and that all C—C bond lengths are the same, *viz.*, 1·397 Å. This value lies between that of a single bond (1·54 Å) and that of a double bond (1·33 Å). Thus, *all* the bonds in benzene have double-bond character. This regular hexagon structure accounts for all the bonding electrons in benzene except six. The procedure now is to correlate the spins of these six electrons in pairs, *i.e.*, to form bonding pairs. If this is done, then there are *five* possible canonical structures (X–XIV):

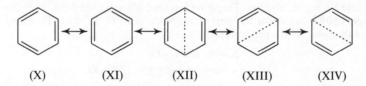

(X)	(XI)	(XII)	(XIII)	(XIV)

(X) and (XI) correspond to the Kekulé 'structures', and (XII–XIV) to the Dewar 'structures'. At first sight it might be thought that the Kekulé structure is cyclohexa-1,3,5-triene. However, consideration of this compound shows that it would not be a regular hexagon since it contains three ethylenic bonds (1·33 Å) and three single bonds (1·54 Å). As we have seen, one of the conditions for resonance is that the positions of the nuclei in each structure must be the same. Thus the Kekulé structures given above are *not* cyclohexatriene. Also, the Dewar structures do not contain ethylenic and single bonds since the presence of such bonds would not permit a structure corresponding to a regular hexagon. It therefore means that the 'double bonds' in the Kekulé and Dewar structures are not ethylenic double bonds. Also, because of the large distance between the *para* carbon atoms, these *p*-bonds are *formal* bonds, and their binding energy is very small in comparison with the other bonds present in the molecule. Calculations for weighting the contributions of the five hypothetical resonating structures show that the Kekulé structures contribute about 80 per cent and the Dewar structures 20 per cent to the benzene resonance hybrid.

It might be noted here that 'Dewar benzene' (bicyclo[2,2,0]hexadiene) has been prepared (Tamelen *et al.*, 1963); it is not planar.

For most purposes we use only the two equivalent Kekulé structures as the canonical forms of benzene, and on this basis it can be shown that the resonance (stabilisation) energy of benzene is about $150 \cdot 6 \, kJ \, mol^{-1}$ (see below). Resonance therefore makes benzene more stable than the corresponding conjugated acyclic triene (hexa-1,3,5-triene); hence its *aromatic character*.

So far, we have established, in a *qualitative* way, that benzene is very stable relative to acyclic and alicyclic compounds. Thus, its resonance energy is much greater than 'expected', and this raises the problem: If the Kekulé structures are fictitious, how are we going to calculate the resonance energy of benzene? One way is as follows. The heat of hydrogenation of cyclohexene is $-119 \cdot 7 \, kJ \, mol^{-1}$, and so the *calculated* value for *hypothetical* cyclohexatriene is $3 \times -119 \cdot 7 = -359 \cdot 1 \, kJ \, mol^{-1}$. Cyclohexene, and not ethylene, is chosen as the reference compound to minimise any steric effects. The *observed* heat of hydrogenation of benzene is $-208 \cdot 5 \, kJ \, mol^{-1}$. Thus, with cyclohexatriene as the reference standard, benzene is more stable than this 'compound' by $150 \cdot 6 \, kJ$. This value, then, is the resonance energy of the benzene molecule.

Alternatively, the resonance energy may be obtained from the heat of combustion of cyclohexatriene *calculated from average bond energies* (from Table 2.2). Since bond energies are estimated for the gaseous state, it is therefore necessary, when using bond-energy tables, to correct for latent heats. The stoichiometric equation for the combustion of cyclohexatriene (at 25°C) is:

$$C_6H_6(l) + 15/2O_2(g) \longrightarrow 6CO_2(g) + 3H_2O(l)$$

Let us assume that cyclohexatriene would be a liquid if it existed (*cf.* the physical properties of cyclohexene and cyclohexadiene; both are liquids), and that its heat of vaporisation would be about the same as that of benzene, *viz.*, $30 \cdot 5 \, kJ \, mol^{-1}$. Also, we shall take the heat of vaporisation of water as $43 \cdot 9 \, kJ \, mol^{-1}$. In order to evaluate ΔH from bond energies, reactants and products must be considered in the gaseous phase. Thus the energy required to break the bonds in the reactants is:

$$3C \!-\! C + 3C\!=\!C + 6C\!-\!II + 15/2 O\!=\!O$$
$$3 \times 347 \cdot 3 \quad 3 \times 606 \cdot 7 \quad 6 \times 414 \cdot 2 \quad 15/2 \times 497 \cdot 9 = 9 \, 081 \cdot 5 \, kJ$$

Since the triene is a liquid, the heat of vaporisation ($30 \cdot 5 \, kJ$) must be added. Therefore the total energy *absorbed* is $9 \, 112 \cdot 0 \, kJ$, *i.e.*, H for this part of the reaction is $+9 \, 112 \cdot 0 \, kJ$. The energy *liberated* in the formation of the bonds in the products is:

$$12C\!=\!0 + 6H\!-\!O$$
$$12 \times 803 \cdot 3 \quad 6 \times 464 \cdot 4 = 12 \, 426 \, kJ$$

Because the H_2O is finally produced as *liquid* water, to this sum must be added three times the heat of vaporisation mol^{-1}, since this amount of heat is *evolved* in the condensation. Therefore the total energy liberated is: $12 \, 426 + 3 \times 43 \cdot 9 = 12 \, 558 \, kJ$, *i.e.*, H for this part of the reaction is $-12 \, 558 \, kJ$. Thus, ΔH for the combustion of cyclohexatriene is $+9 \, 112 \cdot 0 - 12 \, 558 = -3 \, 446 \, kJ \, mol^{-1}$. The experimental value for the combustion of benzene is $-3 \, 301 \cdot 6 \, kJ \, mol^{-1}$. Therefore the resonance energy of benzene is $144 \cdot 4 \, kJ \, mol^{-1}$. This value compares favourably with that obtained from heats of hydrogenation.

In the above discussion, the reference model is the *hypothetical* cyclohexa-1,3,5-triene (XV). The geometry of this molecule, however, is entirely different from that of one Kekulé structure (X), in which

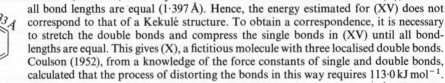 all bond lengths are equal ($1 \cdot 397$ Å). Hence, the energy estimated for (XV) does not correspond to that of a Kekulé structure. To obtain a correspondence, it is necessary to stretch the double bonds and compress the single bonds in (XV) until all bond-lengths are equal. This gives (X), a fictitious molecule with three localised double bonds. Coulson (1952), from a knowledge of the force constants of single and double bonds,

(XV)   calculated that the process of distorting the bonds in this way requires $113 \cdot 0 \, kJ \, mol^{-1}$. Therefore the energy content of (X) is greater than that of (XV) by $113 \cdot 0 \, kJ \, mol^{-1}$. Thus, the resonance (stabilisation) energy of the *actual* benzene molecule (*i.e.*, the resonance hybrid of X and XI) is $150 \cdot 6 + 113 \cdot 0 = 263 \cdot 6 \, kJ \, mol^{-1}$, with (X) as the reference molecule. The R.E. of benzene obtained in this way is called the **vertical resonance (stabilisation) energy** of benzene, and we now have a different definition of resonance (stabilisation) energy, *i.e.*, vertical resonance (stabilisation) energy is estimated by comparing a hypothetical molecule with the actual molecule, both molecules having the *same bond lengths* (*cf.* p. 34).

Molecular orbital theory of aromaticity. This also starts from the fact that benzene is a regular flat hexagon. *Thus each carbon atom is in a state of trigonal hybridisation.* Hence, in benzene, there are six σ C—H bonds, six σ C—C bonds and six $2p_z$ electrons (one on each carbon atom) which are all parallel and perpendicular to the plane of the ring (Fig. 20.1*a*). These electrons can be paired in two ways, both being equally good (*b* and *c*). Each $2p_z$ electron, however, overlaps its neighbours equally, and therefore all six can be treated as forming *an M.O.*

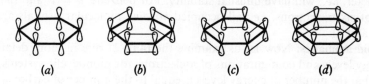

    (*a*)   (*b*)   (*c*)   (*d*)

Fig. 20.1

embracing all six carbon atoms, and so are completely delocalised (Fig. 20.1*d*; *cf.* p. 129). Since six $2p_z$ electrons are involved, six M.O.s are possible, three bonding and three antibonding (p. 36). These are shown in Fig. 20.2 ((I), (II), (III) are bonding, and (IV), (V), (VI) are antibonding); all have a node in the plane of the ring; but (II) and (III) have one node, (IV) and (V) two, and (VI) three nodes perpendicular to the plane of the ring. Now, as we have seen, no more

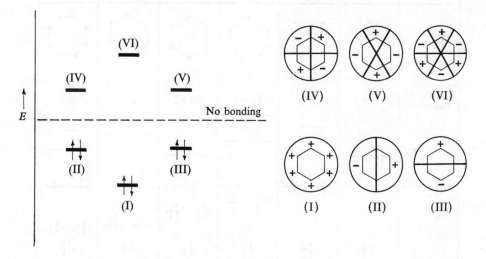

Fig. 20.2

than two electrons can occupy any particular M.O. Furthermore, in the ground state, the six $2p_z$ electrons of benzene will occupy, in pairs, the M.O.s of lowest energy. These three are (I), (II) and (III); these are the M.O.s in which the number of nodes are fewer than in (IV), (V) and (VI). When benzene is in an excited state, one or more of the π-electrons will occupy the higher energy level M.O.s.

    In the ground state, the total energy of the three pairs of *delocalised* π-electrons (Fig. 20.1*d*) is less than that of three pairs of *localised* π-electrons ((*b*) or (*c*)), and hence the benzene molecule is stabilised by delocalisation (resonance).

    Hückel (4*n* + 2) rule for aromaticity. Hückel (1937) carried out M.O. calculations on *monocyclic* systems $C_n H_n$ containing *n* π-electrons and each carbon atom providing one π-electron, and as a result connected *aromatic stability* (high delocalisation energy or high resonance (stabilisation) energy) with the presence of (4*n* + 2) π-electrons *in a closed shell,* where *n* is an integer. Thus, to be *aromatic*, a molecule must have 2 ($n = 0$), 6 ($n = 1$), 10 ($n = 2$), . . . π-electrons. In this description of aromaticity, no mention is made of the *number* of carbon atoms in the ring; the essential requirement is the presence of ($4n + 2$) π-electrons. Another requirement, however, for aromaticity is (reasonable) planarity of the ring. If the ring is not planar, overlap of the *p*-orbitals is diminished or absent. Thus, if a molecule is a monocyclic (reasonably) planar system and contains ($4n + 2$) π-electrons, that molecule will exhibit aromatic character, *i.e.*, will have unusual stability. For benzene, $n = 1$, and the molecule has a *closed* shell of six π-electrons, *i.e.*, all occupied *bonding* molecular orbitals are *doubly filled* (see Figs. 20.2 and 3).

Non-benzenoid aromatics. Now let us examine the ($4n + 2$) rule in more detail. Figure 20.3 shows the energy levels and configurations of π-electrons for monocyclic systems, C_nH_n (where n is 3–8; note that the number of energy levels is equal to the number of carbon atoms and that M.O. energy levels which are the *same* for a given molecule are said to be *degenerate*). Inspection of Fig. 20.3 (the molecules are in the ground state) shows that only benzene is a ($4n + 2$) π-electron molecule ($n = 1$) and consequently is the only molecule which has a closed shell electron configuration. Thus, for the 3–8 rings, only benzene is aromatic.

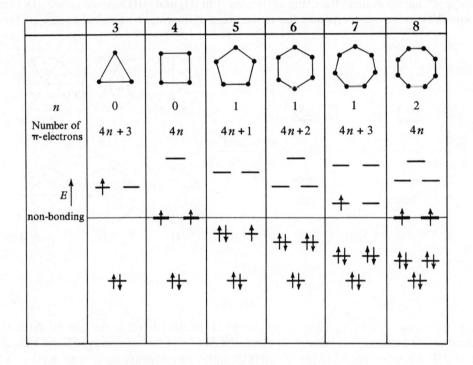

Fig. 20.3

Cyclopropenium salts. 3 (Fig. 20.3) has a doubly filled bonding and a singly occupied antibonding M.O., but if the latter electron is lost, the resulting cation becomes a closed-shell ($4n + 2$) π-electron molecule ($n = 0$). This is the *cyclopropenyl cation*, and may be represented as a resonance hybrid (V.B. method). In the M.O. method, since there is an empty $2p_z$ orbital, this can overlap with the two singly-occupied $2p_z$ orbitals to form a closed M.O. containing two π-electrons.

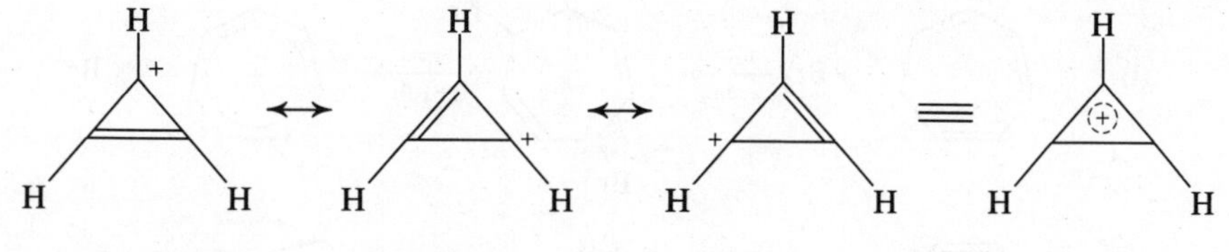

Hence the cyclopropenyl cation should be stable. Many cyclopropenium salts have actually been prepared, *e.g.*, (i) hydroxydiphenylcyclopropenyl bromide (in this case, the π-electrons of the C=O bond are 'lost' to the oxygen atom); (ii) cyclopropenyl hexachloroantimonate:

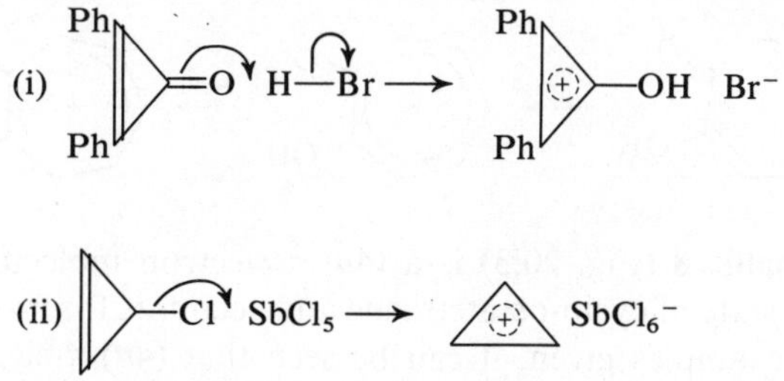

Cyclobutenium salts. 4 (Fig. 20.3) is a $(4n)$ π-electron molecule, and is in the triplet state, *i.e.*, each of a pair of degenerate orbitals contains one electron (see also p. 118). This is the ground state of cyclobutadiene and accounts for its high instability (see p. 543). On the other hand, loss of two π-electrons in the separate non-bonding M.O.s produces the closed-shell $(4n+2)$ π-electron molecule $(n = 0)$, *i.e.*, the *cyclobutenyl di-cation*, *e.g.*, tetraphenylcyclobutenyl fluoroborate:

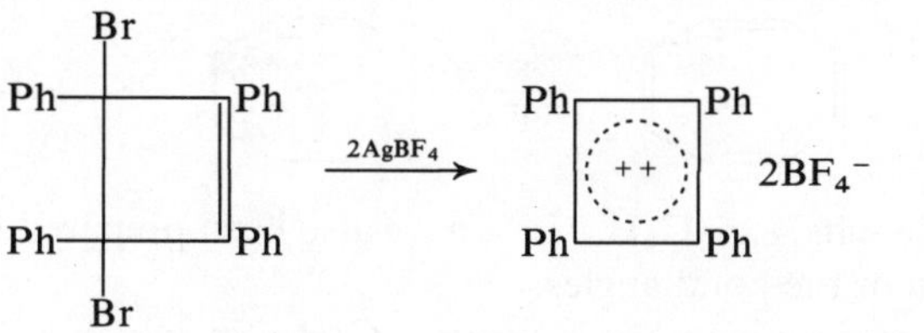

Cyclopentadienide salts. 5 (Fig. 20.3) is a $(4n+1)$ π-electron molecule, and the gain of one π-electron results in the formation of the *cyclopentadienyl anion*, which now has three doubly filled M.O.s. This anion is a closed-shell $(4n+2)$ π-electron molecule $(n = 1)$, *e.g.*, potassium cyclopentadienide:

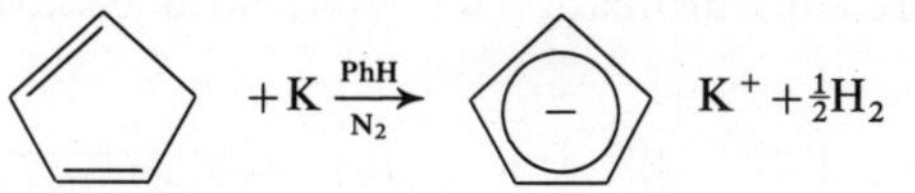

Ferrocene is dicyclopentadienyliron (p. 545); it has aromatic properties, but it is better described as an organometallic compound rather than a non-benzenoid aromatic (see later).

Tropylium (tropenium) salts. 7 (Fig. 20.3), by loss of the single π-electron in the anti-bonding M.O., gives the *tropylium cation*, a $(4n+2)$ π-electron molecule $(n = 1)$, *e.g.*, (i) tropylium bromide; (ii) hydroxytropylium chloride (from tropone); (iii) 1,2-dihydroxytropylium chloride (from tropolone):

(i)   cycloheptatriene $+ Br_2 \xrightarrow[\text{addn.}]{1,4^-}$ 3,7-dibromocycloheptadiene $\xrightarrow[-HBr]{\text{heat}}$ tropylium cation $+ \; Br^-$

(ii)   tropone $\longleftrightarrow$ tropylium $-O^- \xrightarrow{HCl}$ tropylium $-OH \quad Cl^-$

(iii)   hydroxytropone $\longleftrightarrow$ tropylium $O^-, OH \xrightarrow{HCl}$ tropylium $OH, OH \quad Cl^-$

Cyclo-octatetraenide salts. 8 (Fig. 20.3) is a $(4n)$ π-electron molecule, and is in the triplet state. This is the ground state of cyclo-octatetraene and accounts for its high (olefinic) reactivity (see p. 557). From the examples given, it can be seen that $(4n)$ π-electron molecules do not possess a closed-shell planar configuration in the ground state. This is the reason for the instability or reactivity of these compounds. In fact, monocyclic $(4n)$ π-electron molecules are less stable than their acyclic analogues. This decreased stability in $(4n)$ molecules has been called **anti-aromaticity** (Breslow, 1967).

Gain of two π-electrons (one to each non-bonding M.O.) by 8 gives the cyclo-octatetraenyl di-anion, a closed-shell $(4n+2)$ π-electron molecule $(n = 2)$, *e.g.*, dipotassium cyclo-octatetraenide:

$$\text{cyclo-octatetraene} \xrightarrow[\text{THF}]{2K} [\text{C}_8\text{H}_8]^{2-} \quad 2K^+ + H_2$$

Cyclononatetraenide salts, *e.g.*, $C_9H_9^- Li^+$, have also been prepared. The ion is planar in spite of the large distortion of the bond angles.

Annulenes. Conjugated monocyclic polyenes, C_nH_n, in which $n \geqslant 10$, are usually called *annulenes*, and were prepared by Sondheimer *et al.* (1962) to test the Hückel $(4n+2)$ rule, their aromatic character being investigated by NMR spectroscopy (see later). The annulenes prepared have $n = 12, 14, 16, 18, 20, 24,$ and 30. Of these, only [14]-, [18]-, and [30]annulenes are $(4n+2)$ π-electron molecules, whereas the rest are $(4n)$ molecules. The latter have olefinic properties (*cf.* above), but it was found that although [14]- and [18]annulene had the magnetic properties required for aromatic character, both behave *chemically* like an alkene (see later).

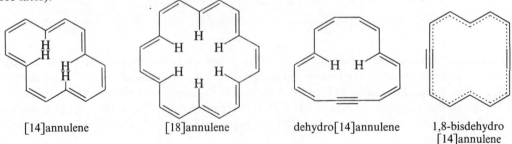

    [14]annulene   [18]annulene   dehydro[14]annulene   1,8-bisdehydro [14]annulene

The smallest annulene of the $(4n+2)$ type is [10]annulene. van Tamelen *et al.* (1971) have prepared it by the photolysis of *trans*-9,10-dihydronaphthalene.

Dehydro[14]annulene, $C_{14}H_{12}$, is a $(4n)$ molecule, but shows the magnetic properties required for aromaticity. In this molecule, the two inner hydrogen atoms are too far apart to experience trans-annular interaction, and consequently the molecule is *planar*. 1,8-Bisdehydro[14]annulene, $C_{14}H_{10}$, is a $(4n+2)$ molecule; it has aromatic magnetic property (note that it is not possible to represent this molecule with conjugated double bonds).

Polycyclic systems. As we have seen, Hückel's $(4n+2)$ rule applies to *monocyclic systems*. The rule, however, is not necessarily successful for *polycyclic* conjugated systems; it is in some cases, *e.g.*, naphthalene (10 π-electrons), but not in others, *e.g.*, pyrene (16 π-electrons). Both of these are aromatic *chemically* (both are *benzenoid* compounds). It has been suggested that

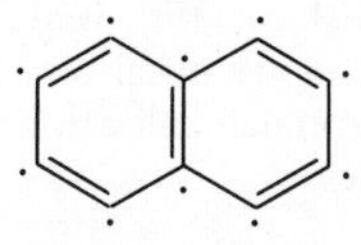

naphthalene

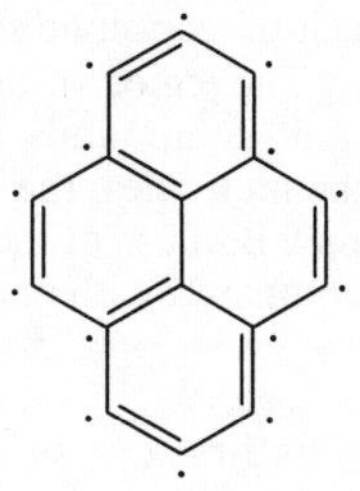

pyrene

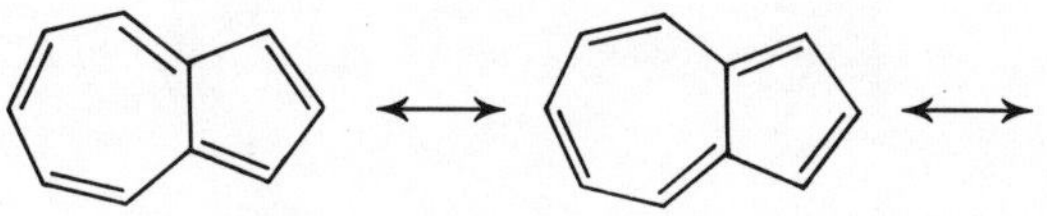

azulene

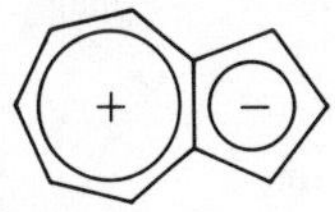

dipolar structure

the $(4n+2)$ rule should be applied to the *peripheral* (conjugated) π-electrons of polycyclic systems: naphthalene, 10 $(n = 2)$ peripheral π-electrons; pyrene, 14 $(n = 3)$ peripheral π-electrons.

Azulene (intensely blue solid, m.p. 99°C). There are two Kekulé resonating structures containing 10 π-electrons $(n = 2;$ and 10 peripheral π-electrons). The five-membered ring has five and the seven-membered ring has seven π-electrons (two π-electrons are common to both rings). If one π-electron is transferred from the seven- to the five-ring, each ring will now have a closed shell of six π-electrons (*cf.* naphthalene). In this condition, the molecule will have a *dipolar structure*, and this has been shown to be the case from dipole-moment measurements. Azulene also behaves chemically as an aromatic compound, *e.g.*, it undergoes bromination, nitration, and the Friedel–Crafts reaction. Furthermore, electrophilic substitution occurs preferentially in the five-membered ring, since this is more electron-rich than the seven-membered ring.

Conjugated cyclic systems (monocyclic and polycyclic) which are $(4n+2)$ π-electron molecules and not benzenoid compounds, are referred to as **non-benzenoid aromatic compounds.**

Definition of aromaticity. From the foregoing account, it can be seen that aromaticity (aromatic character) has been defined in different ways: (i) 'unexpected' chemical properties (particularly substitution rather than addition); (ii) unusual stability (large resonance or delocalisation energy); (iii) a $(4n+2)$ π-electron molecule with a (reasonably) planar cyclic structure. The tendency now, however, is to combine these criteria, and hence to consider a monocyclic compound to be aromatic if it has a reasonably planar cyclic structure, has $(4n+2)$ π-electrons, and has unusual stability due to π-electron delocalisation. It must be admitted that this definition of aromaticity includes many compounds which have very little resemblance *chemically* to benzene.

Experimental criteria for aromaticity. Current practice is to determine aromatic character by physical properties which depend on the extent of delocalisation of the π-electrons in the molecule.

> **Dipole-moment measurements.** In a number of cases, the experimental values can be readily explained on the basis of a dipolar structure, *e.g.*, tropone, azulene.
>
> **X-ray analysis.** If the compound is aromatic, the molecule will be planar and regular, *e.g.*, the X-ray analysis of the copper salt of tropolone showed it to be a planar, almost, regular, heptagon.
>
> **Infra-red spectroscopy.** The region of the C—H (stretch) in many non-benzenoid compounds is very similar to that in benzenoid compounds, *e.g.*, tropone exhibits this similarity. Furthermore, because of the high symmetry of the molecules, the infra-red spectra are often very simple.
>
> **Ultraviolet spectroscopy.** Because of the large π-electron delocalisation, aromatic compounds absorb at longer wavelengths compared with analogous alkenes.

NMR spectroscopy. Let us first consider the case of benzene placed in a magnetic field. The delocalised π-electrons in the ring can move in either direction, but under the influence of a

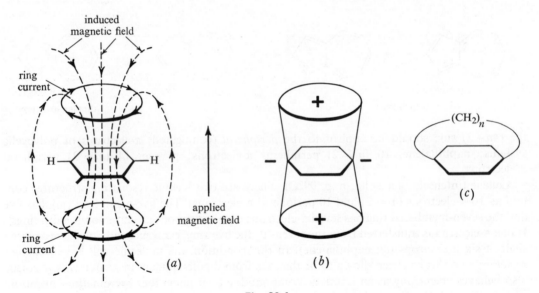

Fig. 20.4

magnetic field applied perpendicular to the molecular plane, circulation takes place in one direction, thereby producing a *ring current* which induces a magnetic field perpendicular to the molecular plane (see Fig. 20.4*a*). This induced magnetic field assists the applied field *outside* the ring and opposes the applied field *inside* the ring (and volume above and below bounded by the area of the ring; Fig. 20.4*b*). Thus, there are volumes in which deshielding (negative) and shielding (positive) occur (see also Fig. 4.4). Hence, for the hydrogen atoms of benzene, since they lie in the plane of the ring, their chemical shift occurs at lower field than had the deshielding effect been absent: τ-values of aromatic protons lie between 1·0 and 3·0; those of olefinic protons are 4·3 to 5·3. It is therefore usually assumed that a compound is aromatic if the NMR absorption peaks of the hydrogen atoms attached to the carbon atoms of the ring are at a lower field than that expected for olefinic hydrogen atoms.

For protons lying in the shielding cones (inside and above and below the ring), the chemical shift occurs at a higher field than had the shielding effect been absent, *i.e.*, the τ-value is greater than that of an olefinic proton.

In benzene, no protons lie inside the ring. In simple cyclophanes (Fig. 20.4c), some methylene groups lie directly above the plane of the benzene ring (*i.e.*, in a shielding cone); the τ-values of these methylene protons are higher than those of the others. In [18]annulene, which is planar, six protons lie inside the ring and the remaining twelve are peripheral (see p. 580). The NMR spectrum of this molecule shows two broad bands, one at τ 1·1 (due to the twelve peripheral protons) and the other at τ 11·8 (due to the six inner protons). Furthermore, integration of these peaks gives a ratio 2:1, thereby confirming the assignments.

The NMR spectrum of [14]annulene in a fresh solution shows a strong band at τ 4·42 and a very weak band at τ 3·93. The former is very close to the value for olefinic protons, and so the compound does not appear to be aromatic. However, at $-60°C$, the NMR spectrum shows two bands, one at τ 2·4 and the other at τ 10·0. The former value corresponds to the ten peripheral protons and the latter to the four inner protons. The explanation for this behaviour is that [14]annulene exists in two conformations (these have been separated), one being almost exclusive in the solid state, but in solution at room temperature, forms an equilibrium mixture of the two. In this case, the signals have coalesced to form a peak at some intermediate value. A point of particular interest is that [14]annulene is not completely planar owing to the transannular interactions of the four inner hydrogen atoms. Nevertheless, the molecule has a completely cyclic conjugated system.

NMR spectroscopy has been extremely useful for demonstrating the aromaticity of non-benzenoid aromatic compounds which are ions, *e.g.*, cyclopentadienyl cation, tropylium cation, cyclo-octatetraenyl di-anion. All show single sharp peaks which occur in the aromatic region. Thus, all the protons are equivalent and hence the rings are symmetrical.

It should also be noted that an aromatic compound has been defined as *a compound which has the ability to sustain an induced ring current* (Elvidge *et al.*, 1961; see also p. 829).

Orientation

The problem of assigning positions of substituents in disubstituted and higher substituted derivatives of benzene is known as *orientation*. When the substituent is a carbon group joined to the *benzene ring* or *nucleus* it is known as a *side-chain*.

(i) **Körner's absolute method** (1874). The term *absolute* is used because no assumption is made about the structures of the *initial* compounds (except that of benzene itself). The method is based on the principle that the introduction of a third substituent into the *p*-isomer gives *one* trisubstituted product, the *o*-isomer *two*, and the *m*-isomer *three* trisubstituted products. Körner applied this principle to establish the orientation of the isomeric dibromobenzenes. He nitrated each isomer and examined the number of nitrated products. One isomer gave *one* dibromonitrobenzene; this isomer is therefore the *p*-compound. Another gave *two* dibromonitrobenzenes; this is therefore the *o*-compound; and the third gave *three* and is therefore the *m*-compound:

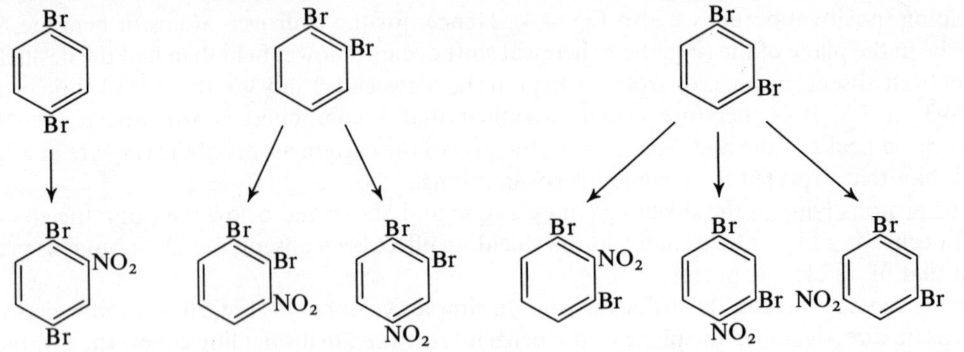

Körner also introduced a third bromine atom instead of a nitro-group (the number of isomers produced is independent of the nature of the third substituent).

The reverse procedure to Körner's method is often useful, *i.e.*, one group of the isomeric trisubstituted derivatives is removed, *e.g.*, Griess (1874) distilled the six diaminobenzoic acids (all known) with soda-lime. He obtained three phenylenediamines; three acids gave the same diamine, which is therefore the *m*-isomer; two acids gave the same diamine, which is therefore the *o*-isomer; and one acid gave one diamine, which is therefore the *p*-isomer.

(ii) **The relative method.** In this method the compound in question is converted into or synthesised from a substance of *previously determined orientation*; *e.g.*, reduction of one or both nitro-groups in *m*-dinitrobenzene gives rise to *m*-nitroaniline and *m*-phenylenediamine respectively:

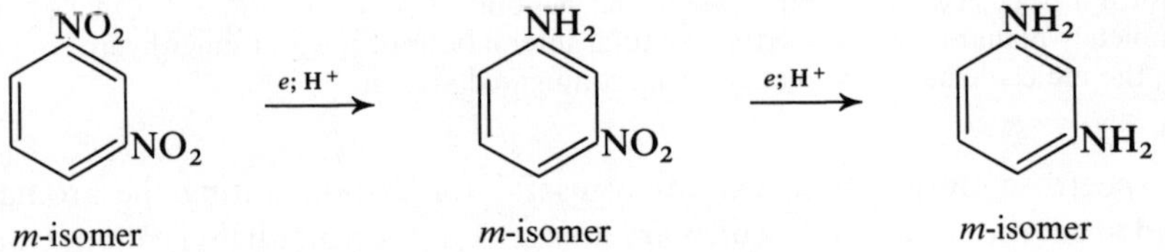

$$NO_2 \xrightarrow{e; \, H^+} NH_2 \xrightarrow{e; \, H^+} NH_2$$

m-isomer   *m*-isomer   *m*-isomer

Another example is the *replacement* of one group by another, *e.g.*, *m*-benzenedisulphonic acid on fusion with sodium hydroxide gives *m*-dihydroxybenzene:

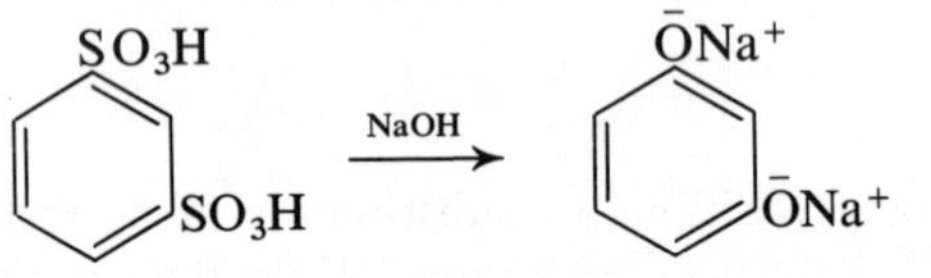

$$SO_3H \xrightarrow{NaOH} \bar{O}Na^+$$

A classical example of the relative method of orientation is the case of the three benzene-dicarboxylic acids. This starts with mesitylene. Mesitylene may be prepared by distilling acetone with sulphuric acid, and Baeyer *assumed* it to be 1,3,5-trimethylbenzene, arguing that the reaction

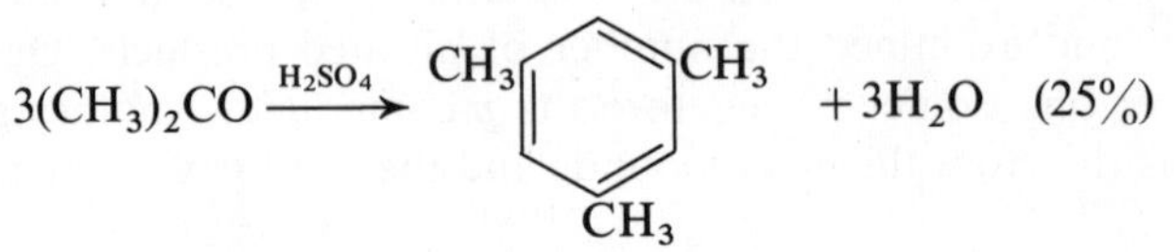

$$3(CH_3)_2CO \xrightarrow{H_2SO_4} CH_3\text{-}C_6H_3\text{-}(CH_3)_2 + 3H_2O \quad (25\%)$$

can be formulated only if the symmetrical structure of mesitylene is assumed. This assumption was later proved correct by Ladenburg (1874). Mesitylene may be converted into dimethylbenzene which, in turn, may be converted into a benzenedicarboxylic acid:

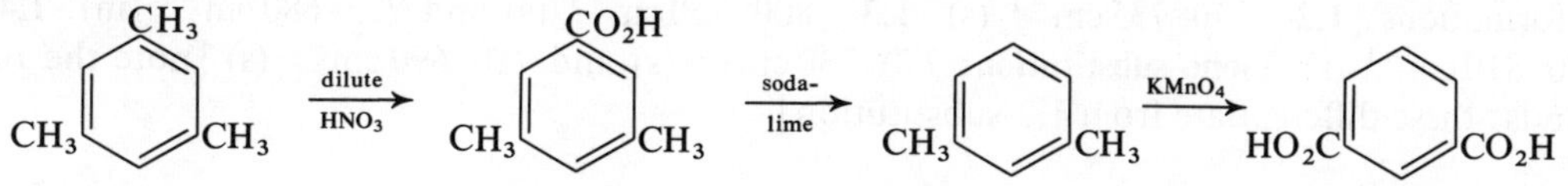

mesitylene

mesitylenic
acid

isophthalic
acid

The dimethylbenzene must be the *m*-isomer whichever methyl group in mesitylene is oxidised. Hence the benzenedicarboxylic acid must be the *m*-isomer, which is known as isophthalic acid. Of the three isomeric benzenedicarboxylic acids only one, phthalic acid, forms the anhydride; the other two, isophthalic acid and the one known as terephthalic acid, do not form anhydrides at all. Thus phthalic acid is the *o*-isomer (by analogy with succinic acid; *cf.* p. 450); and terephthalic acid is consequently the *p*-isomer (see also (iii) below):

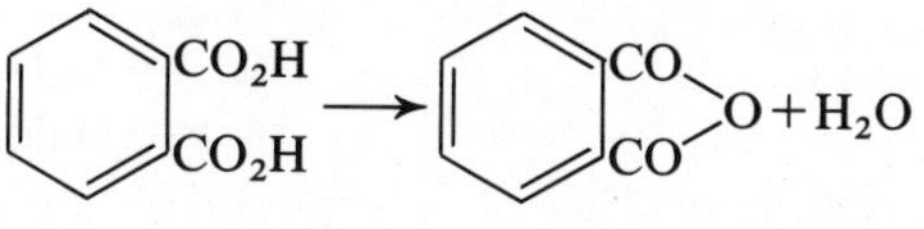

Another example of the relative method is the nitration of *o*- and *p*-nitrotoluenes; both give the *same* dinitrotoluene derivative which, therefore, must be the *m*-dinitro-compound:

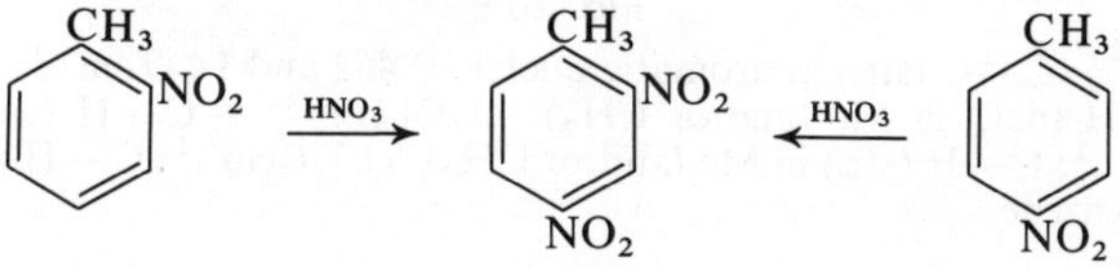

The relative method of orientation is based on the assumption that atoms or groups remain in the same positions or exchange positions with the incoming groups. It is also based on the assumption as to the structure of the *starting* material. Sometimes one or both of these assumptions are right, and sometimes they are wrong. Baeyer's assumption for the mesitylene formula was correct. On the other hand, *o*-, *m*- and *p*-bromobenzenesulphonic acids all give *m*-dihydroxybenzene (resorcinol) when fused with sodium hydroxide. These fusions are carried out at about 300°C, and since the conditions are vigorous, the interpretation of the results always contains an element of doubt. Thus Körner's absolute method is more satisfactory theoretically, but unfortunately it is often difficult to carry out in practice (see below).

(iii) In certain cases, the *o*-compound may be distinguished from the *m*- and *p*-isomers by reactions which produce ring compounds, *e.g.*, *o*-phthalic acid forms a cyclic anhydride (*cf.* (ii) above), *o*-diamines form quinoxalines with 1,2-diketones (p. 666).

(iv) **Infra-red absorption spectra.** Aromatic compounds produce a large number of characteristic bands in the infra-red region. The following regions are particularly useful for recognising the presence of benzene (and polynuclear aromatics) and benzene derivatives: the $=C-H$ (stretch) absorption region is $3\,080-3\,030\,cm^{-1}$ (w), and the bands for $C=C$ (in-plane vibration) are: $1\,625-1\,600\,cm^{-1}$ (v), $1\,590-1\,575\,cm^{-1}$ (v) and $1\,525-1\,475\,cm^{-1}$ (v).

Substituent groups have very little effect on the above bands, but new bands are introduced according to the orientation of the groups. However, because of a great deal of overlapping of the various bands in the region $1\,225$–$970\,cm^{-1}$, this region is not very useful. On the other hand, these isomers may be readily distinguished by the bands due to C—H (out-of-plane) deformations: **1,2-**: 770–$735\,cm^{-1}$ (s); **1,3-**: 800–$750\,cm^{-1}$ (s) and 725–$680\,cm^{-1}$ (m); **1,4-**: 840–$810\,cm^{-1}$ (s); **mono-substitution**: 770–$730\,cm^{-1}$ (s) and 710–$690\,cm^{-1}$ (s) [note the *two* bands; these differentiate from 1,2-substitution].

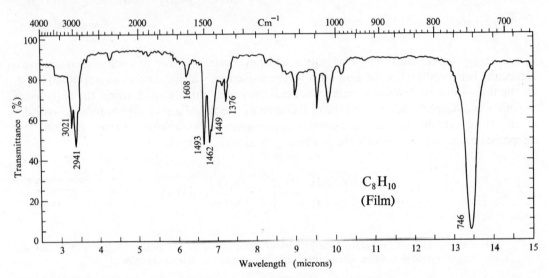

Fig. 20.5

1608 and $1493\,cm^{-1}$: C=C (str.) in aromatic nuclei. 1462 and $1449\,cm^{-1}$: C=C (str.) in aromatic nuclei and/or C—H (def.) in Me (and/or CH_2). $3021\,cm^{-1}$: =C—H (str.) in aromatics (and/or alkenes). $2941\,cm^{-1}$: C—H (str.) in Me (and/or CH_2). $1376\,cm^{-1}$: C—H (def.) in Me. $746\,cm^{-1}$: 1,2-disubstituted benzenes.

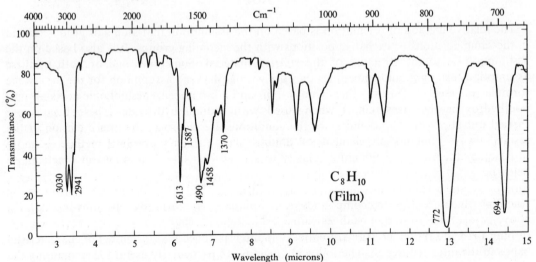

Fig. 20.6

1613, 1587 and $1490\,cm^{-1}$: C=C (str.) in aromatic nuclei. $1458\,cm^{-1}$: C=C (str.) in aromatic nuclei and/or C—H (def.) in Me (and/or CH_2). $3030\,cm^{-1}$: =C—H (str.) in aromatics (and/or alkenes). $2941\,cm^{-1}$: C—H (str.) in Me (and/or CH_2). $1370\,cm^{-1}$: C—H (def.) in Me. 772 and $694\,cm^{-1}$: 1,3-disubstituted benzenes.

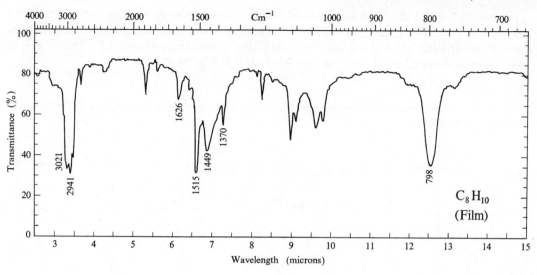

Fig. 20.7

1 626 and 1 515 cm^{-1}: C=C (str.) in aromatic nuclei. 1 449 cm^{-1}: C=C (str.) in aromatic nuclei and/or C—H (def.) in Me (and/or CH$_2$). 3 021 cm^{-1}: =C—H (str.) in aromatics (and/or alkenes). 2 941 cm^{-1}: C—H (str.) in Me (and/or CH$_2$). 1 370 cm^{-1}: C—H (def.) in Me. 798 cm^{-1}: 1,4-disubstituted benzenes.

Figures 20.5, 6 and 7 are the infra-red spectra of *o*-, *m*- and *p*-xylene, respectively (as films).

It can be seen from the legends that there could be different interpretations for some of the bands. However, when all the information is considered together, there is no doubt that the compound is aromatic, and the molecular formula indicates ethylbenzene or a xylene. The former is excluded because of the pattern present in the spectrum is that of a disubstituted benzene.

The NMR spectrum of benzene is a single line at τ 2·73 (the reason for this low field value has already been discussed; p. 582). In general, the τ-value of a nuclear hydrogen in substituted benzenes is 1·0 to 3·0, and depends on the position of the hydrogen atom and the nature of the substituent group. Let us consider a monosubstituted benzene, C$_6$H$_5$Z. If Z is an electron-withdrawing group (−I and/or −R), then because of the small positive charges on the *o*- and *p*-carbon atoms, *o*- and *p*-protons are deshielded, and consequently their resonance absorption occurs at a lower field than that of a proton in benzene. At the same time, the *m*-proton is also deshielded, but much less so than the *o*- and *p*-protons. In general, the effect of deshielding is *o* > *p* > *m*. On the other hand, if Z is an electron-donating group (+I and/or +R), shielding now occurs, and consequently *o*-, *p*-, and *m*-protons absorb upfield with respect to a proton in benzene, the effect of shielding being also *o* > *p* > *m*. If, in both cases, the chemical shifts of the *o*-, *p*-, and *m*-protons are sufficiently different, then coupling between a vicinal (*ortho*) pair of protons can occur, and also long range coupling between protons *meta* and *para* to each other. In these circumstances, the signal of the five protons in C$_6$H$_5$Z is a complicated multiplet. For benzene derivatives, coupling constants are: *ortho*, 7–10; *meta*, 2–3; *para*, 0–1 Hz. If, however, the spectrum is measured at low resolution, then, unless the differences in chemical shifts (*o*, *m* and *p*) are relatively large, the aromatic ring signal is essentially a singlet, *e.g.*, this is usually the case for Z = Cl, Br, R, CH$_2$Y (Y = R, Cl, OH, NH$_2$).

Figures 20.8–20.12 illustrate the types of NMR spectra given by various monosubstituted benzenes. It should be noted that protons attached to α- and β-carbon atoms (in side-chains) occur at lower fields than in the corresponding alkyl compounds (Table 3.2). This is due to their being partially deshielded by the benzene ring (see Fig. 20.4).

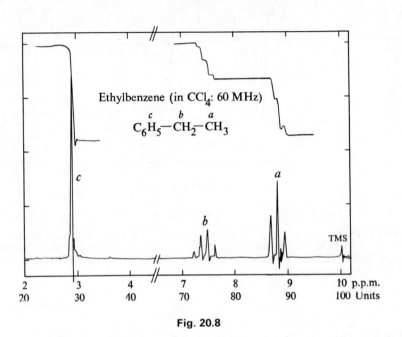

Fig. 20.8

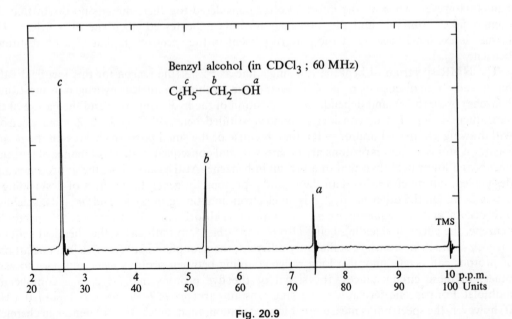

Fig. 20.9

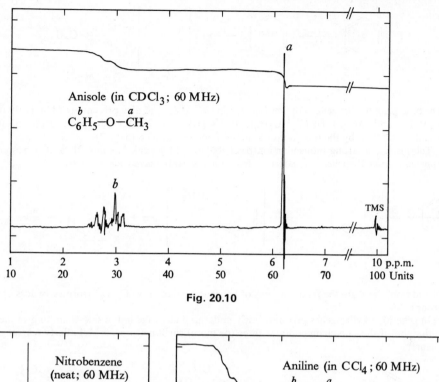

Fig. 20.10

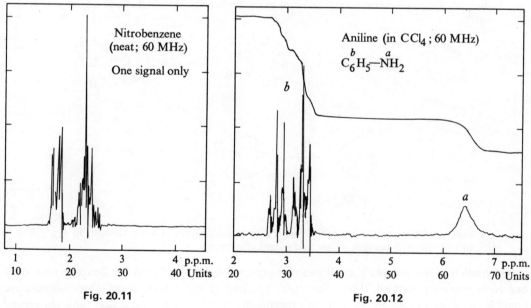

Fig. 20.11   **Fig. 20.12**

Mass spectra

Molecular ions of alkylbenzenes (and benzene) are strong, and are usually accompanied by $(M+1)$ and $(M+2)$ peaks (due to ^{13}C and/or D). **Benzene** shows a large number of peaks, *e.g.*, $C_6H_6^+$ (78), $C_6H_5^+$ (77), $C_4H_5^+$ (53), $C_4H_3^+$ (51), $C_4H_2^+$ (50), and $C_3H_3^+$ (39). All of these usually occur in the mass spectra of *all* benzene derivatives, but here we shall discuss the derivation of the most important peaks.

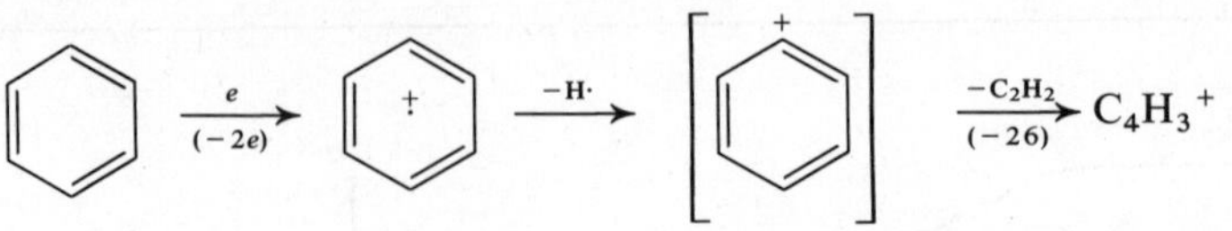

$$M^+ ; 78 \qquad m/e\ 77 \qquad m/e\ 51$$

The base peak for benzene is the molecular ion, M^+ (78), and also present are $(M+1)$, 79 $(C_5{}^{13}CH_6)$, (C_6H_5D), and $(M+2)$, 80 $(C_4{}^{13}C_2H_6)$, $(C_5{}^{13}CH_5D)$, $(C_6H_4D_2)$. The peak at m/e 51 $(C_4H_3{}^+)$ is usually confirmed by the presence of a metastable peak at 33·8 $(51^2/77)$.

Toluene has a strong molecular ion peak, but the base peak has m/e 91, and corresponds to the *tropylium cation* (believed to be derived from the less stable *benzylium cation*).

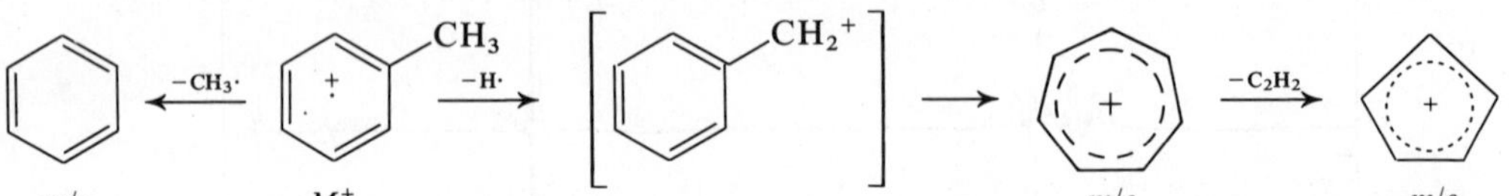

$$m/e \atop 77 \qquad M^+ \atop 92 \qquad\qquad m/e \atop 91 \qquad m/e \atop 65$$

In addition, there are the fragment ions of benzene derived from $C_6H_5{}^+$ (formed by loss of the methyl group).

In general, alkylbenzenes predominantly undergo β-cleavage in the side-chain to give the tropylium cation, but if the side-chain contains three (or more) carbon atoms, a McLafferty rearrangement also occurs:

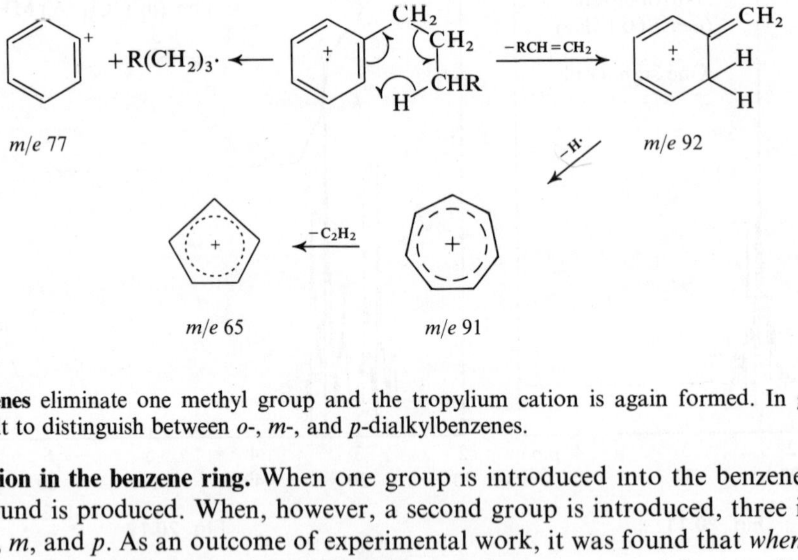

$$m/e\ 77 \qquad\qquad\qquad\qquad m/e\ 92$$

$$m/e\ 65 \qquad\qquad m/e\ 91$$

Xylenes eliminate one methyl group and the tropylium cation is again formed. In general, it is difficult to distinguish between *o*-, *m*-, and *p*-dialkylbenzenes.

Substitution in the benzene ring. When one group is introduced into the benzene ring, only one compound is produced. When, however, a second group is introduced, three isomers are possible: *o*, *m*, and *p*. As an outcome of experimental work, it was found that *when the second group enters the benzene nucleus, the main product is either a mixture of the* o- *and* p-*isomers or the* m-*isomer.* Pure *o*-, *p*- or *m*-substitution is rare, all three isomers being obtained simultaneously; but since the rates of their formation are very different, the slowest one results in the formation of very little of that derivative (hence the difficulty of Körner's absolute method of orientation). Experimental work also led to the conclusion that usually *o*-*p*-substitution is associated with *activation* of the benzene nucleus, *i.e.*, reaction is faster than in benzene itself, whereas *m*-substitution is associated with *deactivation* of the nucleus, *i.e.*, reaction is slower than in benzene.

Rules of orientation. Experience shows that the nature of group **A** already present in the nucleus determines the position taken by the incoming group (see also later).

Class I directs the incoming group mainly to the *o*- and *p*-positions. In this class, **A** may be any one of the following: R, OH, OR, NH_2, NHR, NR_2, $NHCOCH_3$, Cl, Br, I, F, CH_2Cl, SH, Ph, etc.

Class II directs the incoming group mainly to the *m*-position. In this class, **A** may be any one of the following: NO_2, CHO, CO_2H, CO_2R, SO_3H, SO_2Cl, $COCH_3$, CN, CCl_3, NH_3^+, NR_3^+, etc.

Introduction of a third group into the benzene ring. The position taken up by a third group entering the ring depends on the nature of the two groups already present. Experiments have shown that:

(i) When both groups belong to class I, the directive power of each group is generally in the following order:

$$O^- > NH_2 > NR_2 > OH > OMe, NHAc > Me > X$$

In the following examples, the number of arrow heads indicates (qualitatively) the amount of substitution, and the number below indicates the number of isomers:

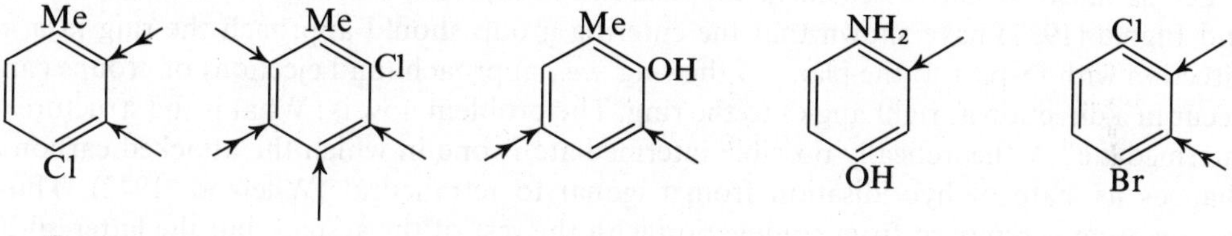

(ii) When both groups belong to class II, then it is difficult to introduce a third group, and the directive power of each group is generally:

$$Me_3N^+ > NO_2 > CN > SO_3H > CHO > COMe > CO_2H$$

(iii) When the two groups direct differently, *i.e.*, belong to classes I and II, then class I takes precedence. Furthermore, if the orientations reinforce each other, the third group enters almost entirely one position, *e.g.*,

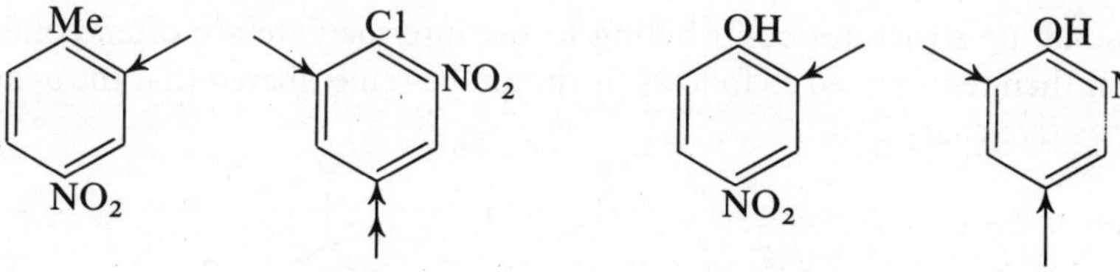

The above scheme is purely qualitative. A quantitative assessment of the isomers may be obtained from a knowledge of partial rate factors (see p. 601).

Separation of isomers. This usually means the separation of the *o*- and *p*-isomers.

(i) Steam distillation may be used, since the *o*-isomer is often steam-volatile whereas the *p*-is not. This method of separation is particularly applicable to *o*-hydroxy-compounds.

(ii) The boiling points of *o*- and *p*-isomers are often very close together and so it is difficult to separate them by fractional distillation. Their melting points, however, are usually very different, that of the *p*-isomer being much higher than that of the *o*-. Thus these isomers may be separated by filtration; it may be necessary to cool the mixture in order to get the *p*-isomer in the solid form.

(iii) Chemical methods of separation may be used in certain cases (see, *e.g.*, p. 691).

(iv) Chromatography. This is the most widely applicable method.

Mechanism of aromatic substitution

There are three possible mechanisms for aromatic substitution: electrophilic, nucleophilic and free-radical. Of these the common substitution reactions are those which involve electrophilic reagents.

Electrophilic substitution

The widely held theory of electrophilic aromatic substitution is that it proceeds by a bimolecular mechanism via the formation of an intermediate, and that the formation of this intermediate is the rate-determining step.

Let us illustrate this mechanism for benzene itself. Theoretical considerations of Hughes and Ingold (1937) have shown that the entering group should approach the ring in a *lateral* direction with respect to the plane of the ring, *i.e.*, approach (and ejection) of groups can only occur in a direction at right angles to the ring. The problem now is: What is the structure of the intermediate? A theoretically possible intermediate is one in which the attacked carbon atom changes its state of hybridisation from trigonal to tetrahedral (Wheland, 1942). Thus this carbon atom is removed from conjugation with the rest of the system, but the latter, although no longer benzenoid, still has a conjugated system which exhibits resonance. This resulting positive ion has been referred to by various names: the pentadienyl cation, the Wheland intermediate, the *σ*-complex and the benzenonium cation.

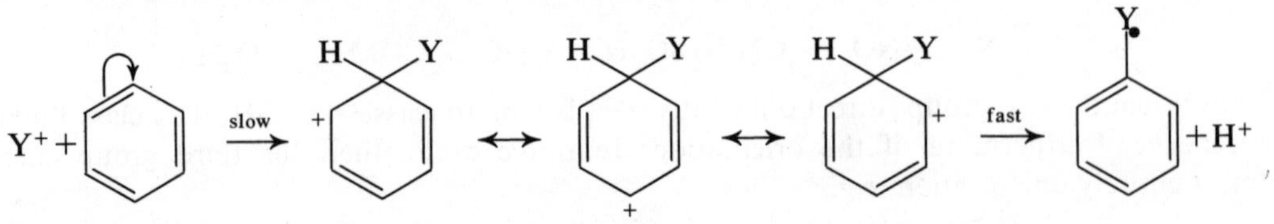

The three resonating structures contributing to the intermediate are often combined, and the intermediate is then represented as follows (it should be remembered that the *o*- and *p*-positions carry small positive charges):

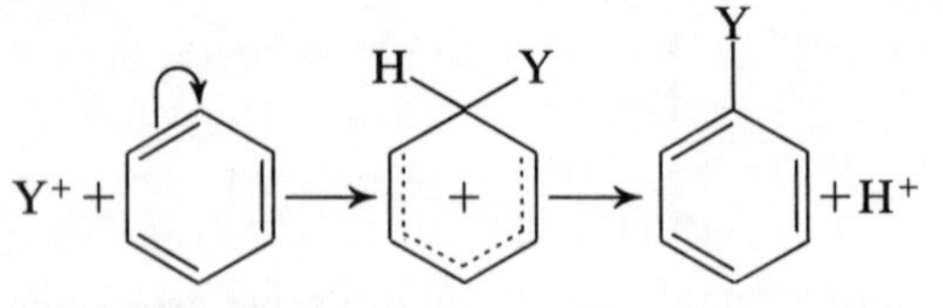

The resonance energy of this cation will be much smaller than that of benzene, but by the expulsion of a proton the molecule can revert to the benzenoid state. The proton is not set free as such, but is removed by some base present.

A more detailed picture of this mechanism is shown in Fig. 20.13, in which the electrophile is the nitronium ion, NO_2^+. The *σ*-complex, 2, is preceded by the formation of its transition

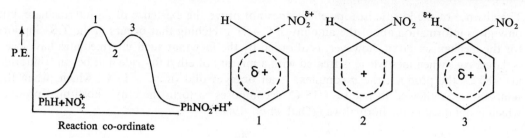

Fig. 20.13

state from the reactants. In this transition state, 1, the *new* covalent bond is *incompletely* formed (Fig. 20.13). The σ-complex, 2, now produces the products via the T.S. 3. The evidence that justifies this sequence is as follows. In nitration with mixed acid (nitric and sulphuric acids), the attacking species is the nitronium ion and this reaction can be a second-order reaction of the type (see also p. 640):

$$\text{rate} = k\,[\text{substrate}]\,[NO_2{}^+]$$

Two possible mechanisms would be in accord with second-order kinetics: (i) synchronous formation of the C—N bond and breaking of the C—H bond; (ii) addition of the nitronium ion to the nucleus, followed by the expulsion of a proton. In this two-step mechanism, either (*a*) the addition of the nitronium ion, or (*b*) the proton-expulsion may be the rate-determining step.

(i)

(ii)

It should be noted that the second step (*b*) will be second-order, since a proton acceptor is involved in the kinetics of this step.

To distinguish between these possibilities, Melander (1949, 1950) examined the nitration of, *e.g.*, toluene containing a tritium instead of a hydrogen atom in the 2-position, and showed that the product, 2,4-dinitrotoluene, contained 50 per cent of the original tritium content. Thus the rates of substitution in the 2-H and 2-T positions are equal. It therefore follows that, since there is no kinetic isotope effect, the expulsion of the proton does not occur in the rate-determining step. Hence mechanisms (i) and (ii*b*) are eliminated, *i.e.*, the nitration of benzene is:

The absence of the kinetic isotope effect does not prove the existence of an intermediate, but only gives information about the amount of C—H stretching that occurs in the T.S. Support for the mechanism given, however, is afforded by the fact that such intermediates have been isolated, *e.g.*, when mesitylene is treated with a mixture of ethyl fluoride and boron trifluoride at $-80°C$, the product, the σ-complex, is an orange solid (m.p. $-15°C$) which, when the temperature is allowed to rise to $-15°C$, decomposes to form the ethyl-substituted product which is obtained in the normal way (Olah *et al.*, 1958).

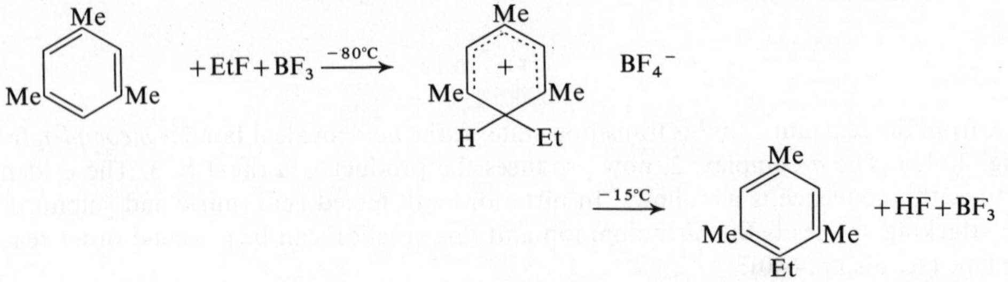

There is evidence to show that in some electrophilic aromatic substitutions, the first and rate-determining step involves the formation of a π-complex between the substrate and the electrophile, and this is then followed by fast steps leading to the formation of the various isomeric σ-complexes, *e.g.*,

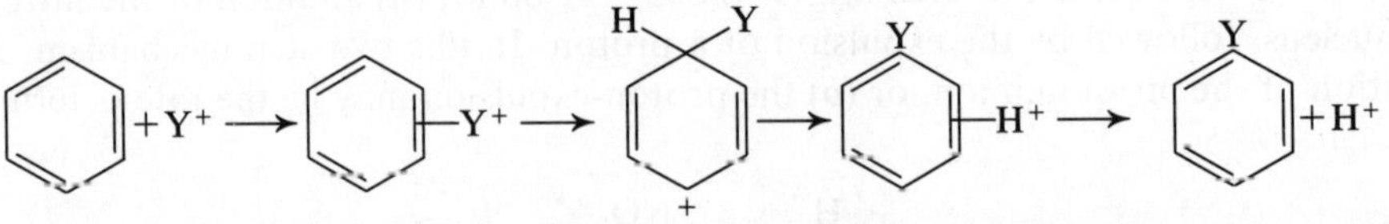

It appears that this mechanism tends to operate when the electrophile is highly reactive (and so shows little selectivity). On the other hand, the formation of the σ-complex in the r/d step is believed to operate for less reactive electrophiles (and so are more selective). This step may still be preceded (and followed) by a fast step in which a π-complex is formed.

The existence of π-complexes has been shown in various ways, *e.g.*, toluene dissolves in HCl at $-70°C$ to form a 1:1 complex which is electrically conducting. If DCl is used, similar results are obtained, but since, on decomposition, the starting materials are recovered, the complex cannot be a σ-complex. On the other hand, benzene dissolves in deuterosulphuric acid, D_2SO_4, to give hexadeuterobenzene. In this case, σ-complexes must have been formed as intermediates. These experiments also show that benzene, toluene, etc. have basic properties.

It is also of interest to consider the problem of substitution *v* addition in aromatic compounds. Let us examine the thermochemical changes for bromine substitution and addition with benzene as substrate. Using Table 2.2, we have (treating as a reasonable approximation the bonds in benzene as ethylenic bonds, etc.):

+414·2 +188·3 −284·5 −364·0 $\Delta H = -46{\cdot}0$ kJ

+606·7 +188·3 −347·3 −2(284·5) $\Delta H = 14{\cdot}9$ kJ
+150·6 (R.E.) −14·6 (R.E.)

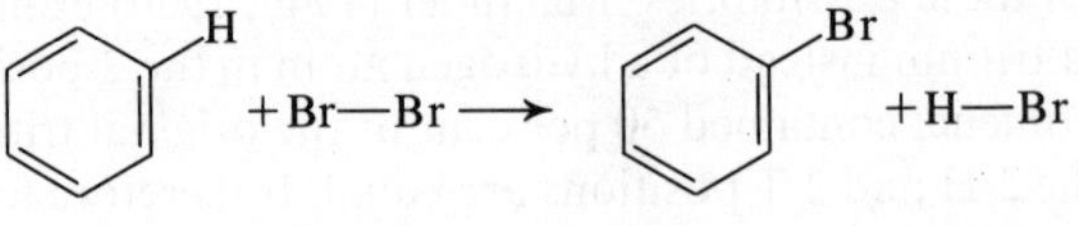

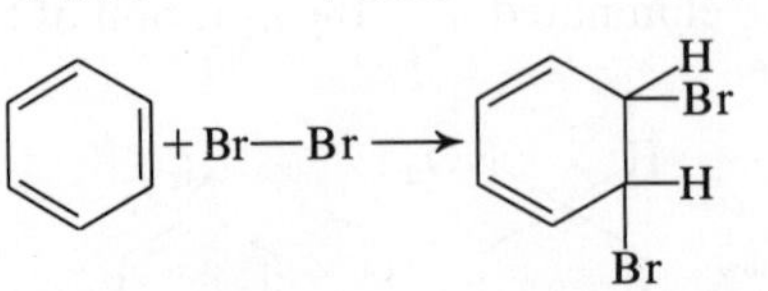

Substitution is therefore thermodynamically favourable, whereas addition is thermodynamically unfavourable. In substitution, there is no loss of R.E., but there is (136 kJ) in addition, and it is this loss that is the deciding factor.

Effect of substituents in electrophilic substitution. There are two ways of studying the relative reactivities of positions in substituted benzenes, one based on charge distribution and the other on the stabilities of the intermediate carbonium ions (σ-complexes; see also p. 105). In both approaches, the effects of substituents described below hold for kinetically controlled products; they do not always apply to the thermodynamically controlled products (see, *e.g.*, p. 690).

Orientation based on charge distribution. The charge distribution in a monosubstituted benzene depends on the polar effect (I and/or R) of the group. When the group present in the ring is OH, OR, NH_2, etc., the product of further substitution is mainly *o-p*. A property common to all these groups is that the atom adjacent to the nucleus—the 'key atom'— has at least one lone pair of electrons. The resonance effect gives rise to *increased* electron densities in the *o-p*-positions, *e.g.*,

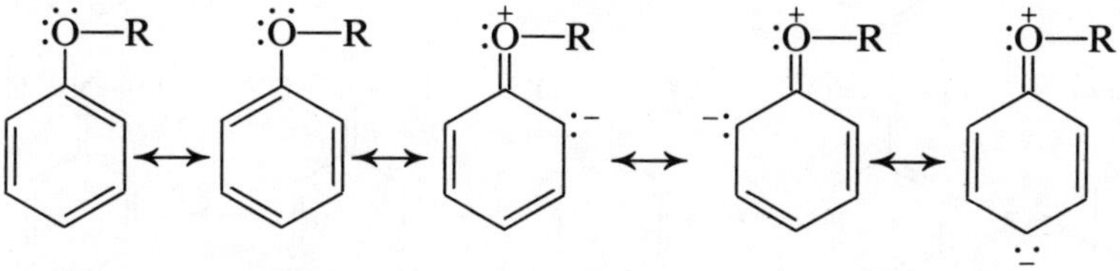

(I) and (II) are the normal structures; (III) and (IV) are *o*-quinonoid and (V) is *p*-quinonoid. (I–V) are the resonating structures, C_6H_5OR being a resonance hybrid of them. Hence the actual state of C_6H_5OR is a molecule having small negative charges at the two *o*-positions and at the *p*-position. Thus substitution with electrophilic reagents takes place in these positions, and owing to the *increased* electron-density in the *o-p*-positions, the substituent group present is associated with *activation* of the ring, *i.e.*, further substitution is *facilitated* by the presence of an *o-p*-orienting group. At the same time, it follows that the activation energy for *o-p*-substitution is lower than that for *m*-substitution (since we are dealing with kinetically controlled reactions).

According to Ingold's terminology of mesomerism, etc. (p. 39), these electronic displacements are +M. At the same time, since it is a +M effect that is present, then a strong +E effect will also operate because a +E effect is the requirement of the attacking (electrophilic) reagent.

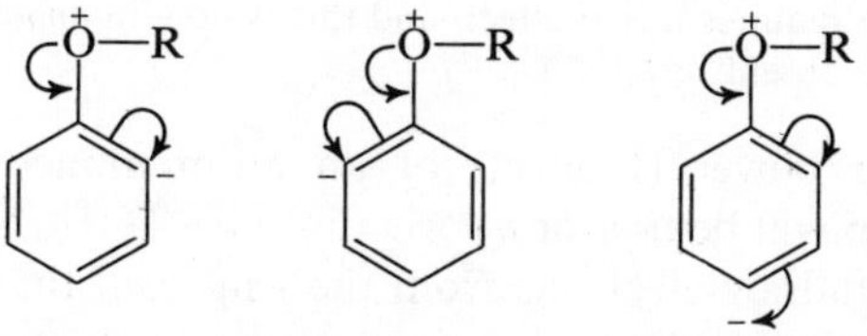

It therefore follows that *o-p*-substitution takes place by the +T effect (see p. 39).

When the group present in the ring is NO_2, CO_2H, COR, SO_3H, etc., the product of further substitution is mainly *m*-. All these groups, by virtue of having at least one strongly electron-attracting atom and a double or triple bond conjugated to the benzene ring, cause an electron displacement *away* from the nucleus and towards the group (*i.e.*, a $-$R effect), *e.g.*, C_6H_5COR, $C_6H_5NO_2$ and $C_6H_5SO_3H$. (VI–X) are resonating structures, and the actual states of C_6H_5COR, $C_6H_5NO_2$ and $C_6H_5SO_3H$ are molecules which have small positive charges in the *o*- and *p*-positions, *i.e.*, the *m*-positions have a *relatively* high electron density with respect to the *o-p*-positions. The above groups are therefore *m*-orienting to electrophilic reagents. It also follows that the activation energy for *m*-substitution is lower than that for *o-p*-substitution (*cf.* above).

The relative high electron density of the *m*-positions with respect to the *o-p*-positions is due to the *withdrawal* of electrons from the *o-p*-positions (*i.e.*, $-$R effect), and not due to a gain of electrons in the *m*-positions.

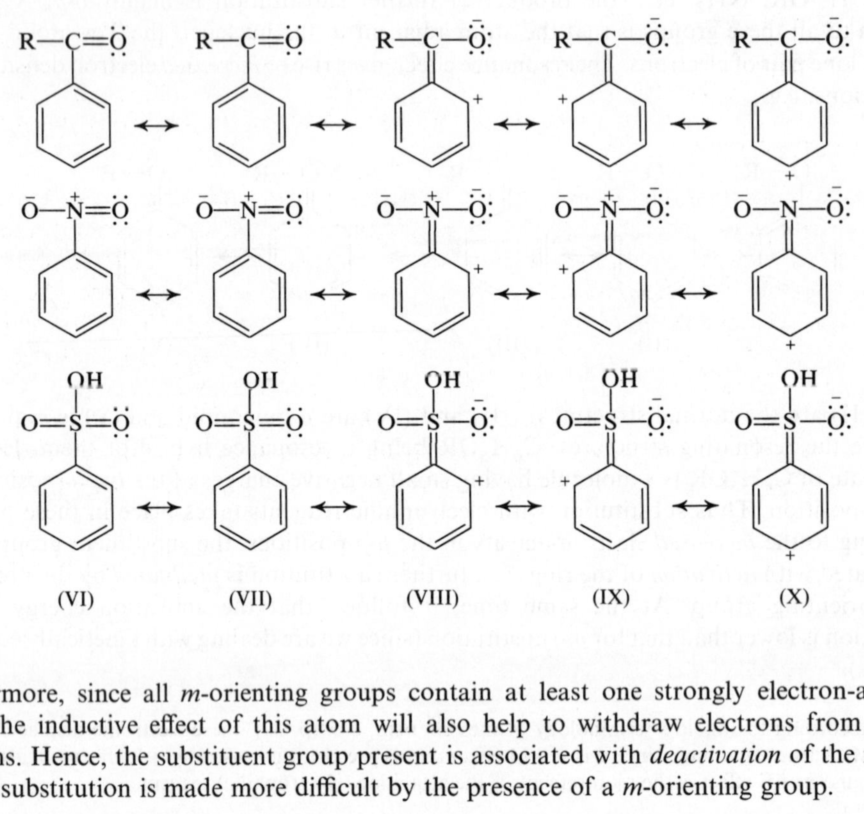

(VI)   (VII)   (VIII)   (IX)   (X)

Furthermore, since all *m*-orienting groups contain at least one strongly electron-attracting atom, the inductive effect of this atom will also help to withdraw electrons from the *o-p*-positions. Hence, the substituent group present is associated with *deactivation* of the ring, *i.e.*, further substitution is made more difficult by the presence of a *m*-orienting group.

Since an electrophile requires a $+$E effect, and this is now in *opposition* to the $-$M effect, the $+$E effect is very weak or is absent.

The electron-attracting power (I effect) of an atom alone cannot decide whether the substituent atom or group will be *o-p*- or *m*-orienting; *e.g.*, OH, NH_2, have a strong $-$I effect and therefore tend to withdraw electrons from the ring, *i.e.*, the $-$I effect tends to promote *m*-substitution, and not *o-p*- as is the case in practice. The 'key-atom' in these groups, however, has at least one lone pair of electrons, and consequently the $+$R effect is possible, and since this is stronger than the I effect, the above groups become *o-p*-orienting (with *activation* of the ring).

Using the Ingold terminology, we now have the $+$M effect *assisted* by the $+$E effect (since this is the requirement of the electrophile), and so *o-p*-substitution takes place by the $+$T effect (see above).

The orientation in chlorobenzene is a very interesting problem. Chlorine is $-I$ and $+R$, and since the direction of the C—Cl dipole is in the C⟶Cl direction, it follows that the $-I$ effect is greater than the $+R$ effect. That the $+R$ effect operates is indicated by the fact that the chlorine atom is less negative than expected (compared with RCl). On this basis, chlorine would be expected to be m-directing with deactivation of the ring. In practice, it has been found to be o-p-directing with *deactivation*, e.g., nitration of chlorobenzene gives the o- and p-disubstituted products, but the reaction is slower than that for benzene itself. This has been explained by Ingold as follows. The $-I$ effect deactivates the ring and the $+M$ effect is too small to be significant. When, however, the $+E$ effect is brought into play by the attacking reagent, the o- and p-positions are raised in electron density above the m-, but the increase is not so great as that in benzene itself. This o-p-orientation with deactivation is characteristic of aryl halides, styrene, cinnamic acid and benzyl chloride. In a number of reactions, however, *fluorine* behaves as an activating substituent towards p-substitution, e.g., the partial rate factor for the chlorination of fluorobenzene in 60 per cent aqueous acetic acid at 25°C is 3·93 (see also p. 601).

The above arguments can be extended to explain why the directive influence of an o-p-orienting group takes precedence over a m-orienting group when both are present in the ring. For the former, we have $+M$ and for the latter $-M$, but in the former there is also a strong $+E$, whereas in the latter the required $+E$ is very weak or absent. Hence, the electron distribution in the ring will be controlled by the o-p-orienting group.

As we have seen in the foregoing, the amino-group in aniline is o-p-orienting; e.g., bromination of aniline produces 2,4,6-tribromoaniline; nitration with dilute nitric acid produces a mixture of o- and p-nitroanilines. On the other hand, Ridd et al. (1964) have shown that nitration of aniline in 98 per cent sulphuric acid gives 38 per cent p-substitution, the rest being the m-isomer. These authors also showed that nitration of the tri-N-methylanilinium cation under the same conditions gives 11 per cent p-substitution, the rest being the m-isomer. Thus, it is the strong $-I$ effect of the NR_3^+ group which largely controls the orientation (m-).

The *carboxyl group* is m-orienting, but the *carboxylate ion* is o-p-orienting. This is attributed to the negative charge on each oxygen atom giving the carboxylate ion electron-repelling properties ($+I$) in contrast to the carboxyl group which is electron-attracting ($-I$).

Similarly, on the $-I$ effect of a carboxyl group, it would be expected that m-substitution would result in cinnamic acid. In practice, however, the main product is a mixture of the o- and p-compounds. This is difficult to explain on the charge-distribution method of orientation (see also p. 600).

The polar effect of an alkyl group in the alkylbenzenes is particularly interesting. Because alkyl groups are electron-repelling ($+I$), the o-p-positions become points of high electron

density, and consequently alkyl groups are o-p-orienting. Since the order of the inductive effect of alkyl groups is

methyl < ethyl < propyl < isopropyl < t-butyl

then the activating effect of an R group, if entirely due to the $+I$ effect, would be in the same order. Actually, in a number of cases the order is the reverse. One explanation offered for this reversal is hyperconjugation (p. 332), this being greatest in the methyl group and least in the t-butyl group; thus:

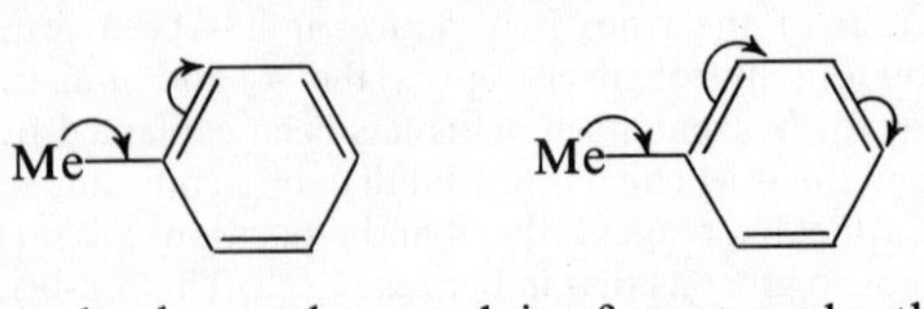

Hyperconjugation may also be used to explain, for example, the *m*-orienting power of the CCl_3 group (as well as by the $-I$ effect):

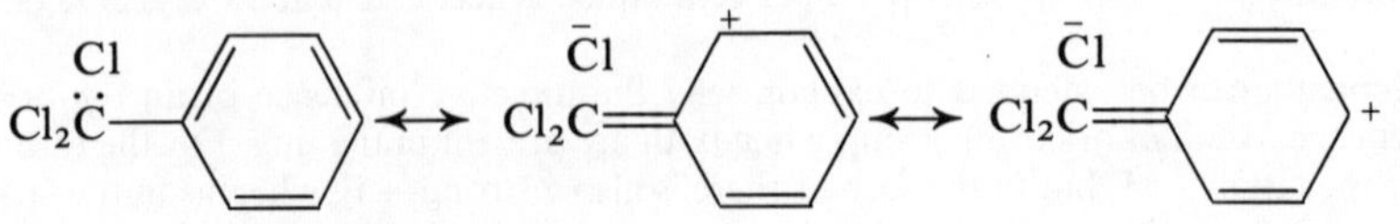

Summary of substituent polar effects on orientation
(i) $+I$: *o,p*-orienting; ring activated.
(ii) $+I$, $+T$: *o,p*-orienting; ring activated.
(iii) $-I$: *m*-orienting; ring deactivated.
(iv) $-I$, $-T$: *m*-orienting; ring deactivated.
(v) $-I > +T$: *o,p*-orienting; ring deactivated.
(vi) $-I < +T$: *o,p*-orienting; ring activated.

Orientation based on carbonium-ion (σ-complex) stability. If A is the substituent and Y the entering group, then the various carbonium ions for *o-*, *p-* and *m*-substitution are as follows (remember that the positions *o-* and *p-* to the attacked carbon atom carry small positive charges):

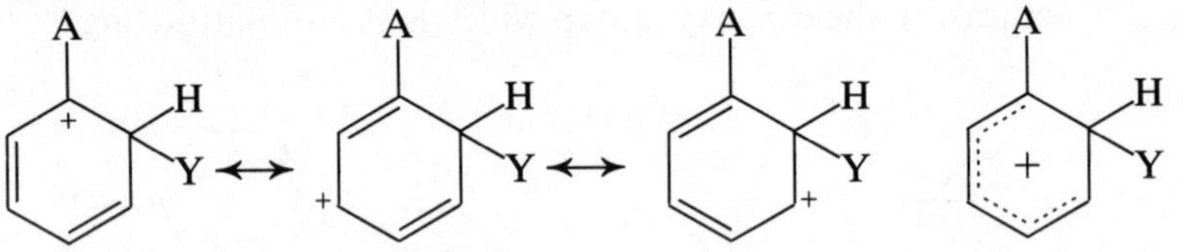

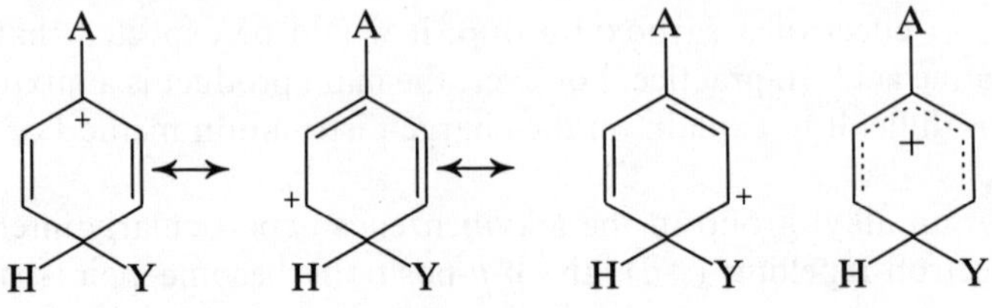

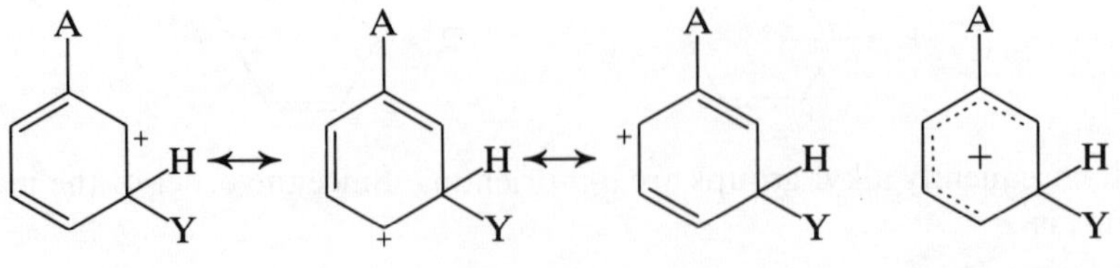

If A has a $+I$ effect (*e.g.*, Me) or is an electron-donating group (*e.g.*, OH), then the positive charge in 1. and 4. will tend to be neutralised, and consequently these ions are stabilised. Therefore the pentadienyl cations (I) and (II) each have a stabilised resonating structure contributing to the resonance hybrid. There is no such stabilised resonating structure contributing to the pentadienyl cation (III). Thus cations (I) and (II) will be more stable than (III); consequently, A will be *o,p*-orienting. If A has a $-I$ effect (*e.g.*, NR_3^+) or is an electron-withdrawing group (*e.g.*, NO_2) then these effects are opposed in 1. and 4., but not in 7., 8. or 9. (or to a far less extent). Thus, in this case cation (III) will be more stable than (I) or (II), and consequently A will be *m*-orienting.

In this method, because of the uncertainty of the structure of the T.S. in the r/d step, all transition states are taken to be σ-complexes.

Examples based on carbonium-ion stability.

(i) *Nitration of nitrobenzene*

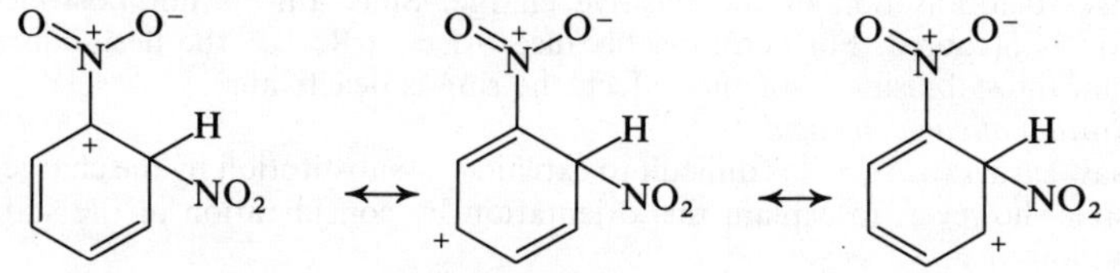

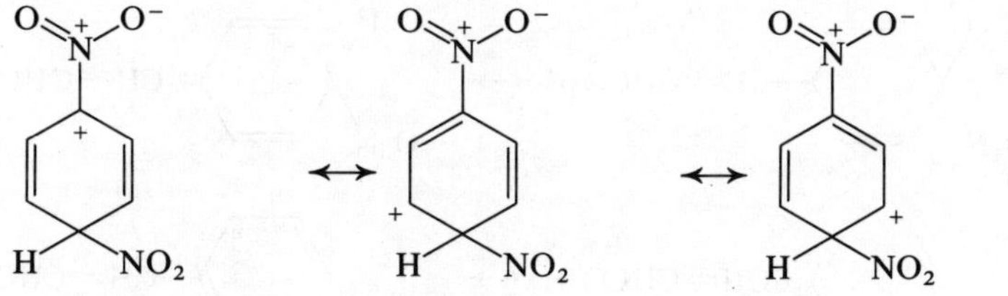

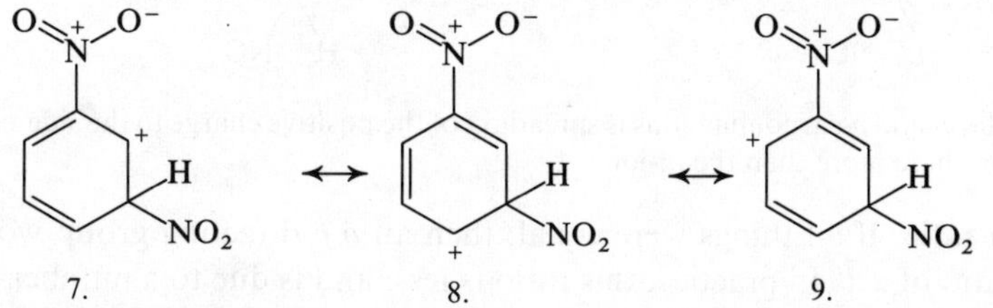

Both *o*- and *p*-carbonium ions each contribute one resonating structure carrying positive charges on *adjacent* atoms (1. and 4., respectively), and so both ions will be relatively unstable. On the other hand, there is no corresponding resonating structure for the *m*-carbonium ion and therefore this is more stable than the *o*- and *p*-ions, *i.e.*, *m*-substitution is favoured.

(ii) *Nitration of toluene*

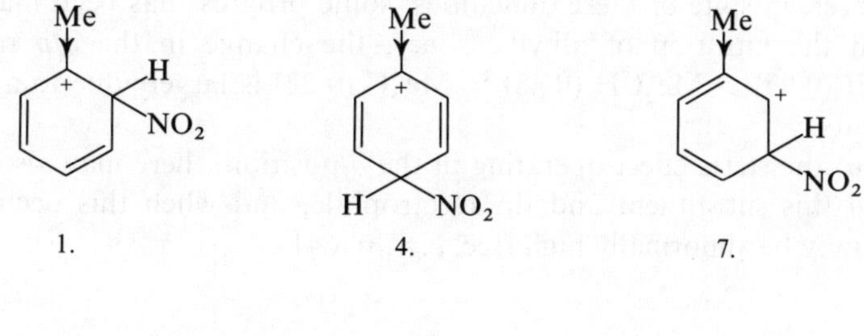

Ions 1. and 4. are stabilised by the $+I$ effect of the Me group due to partial neutralisation of the positive charge on C of C—Me and consequent spreading of the charge (the Me group now also acquires a small positive charge). Since there is no corresponding stabilisation in the m-ion, o,p-substitution is favoured.

(iii) *Nitration of chlorobenzene*

Let us consider the following structures:

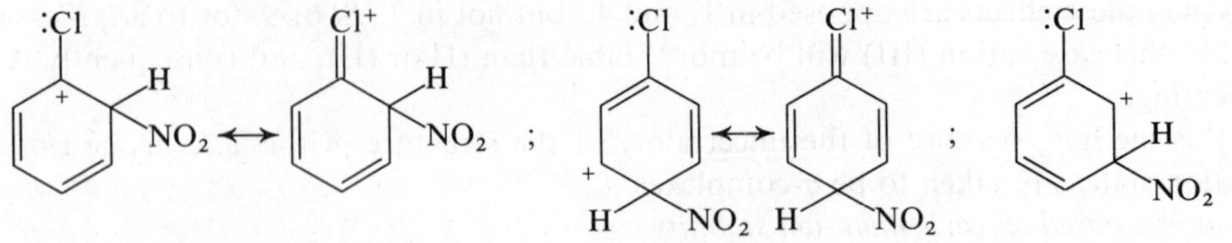

:Cl   Cl⁺   :Cl   Cl⁺   :Cl

H—NO₂ ↔ H—NO₂ ; H—NO₂ H—NO₂ ; H—NO₂

1.   5.   7.

For the o- and p-carbonium ions, resonance $(+R)$ involving a lone-pair of electrons on the chlorine atom causes delocalisation of the positive charge. Since this is not possible for the m-ion, o,p-substitution is favoured. Furthermore, because $-I > +R$, *i.e.*, the destabilising inductive effect is greater than the stabilising resonance effect, the ring is deactivated.

(iv) *Nitration of cinnamic acid*

As we saw earlier (p. 597), it is difficult to explain o,p-substitution by the charge distribution method. It is possible, however, to explain this orientation by consideration of the stabilities of the various carbonium ions, *e.g.*,

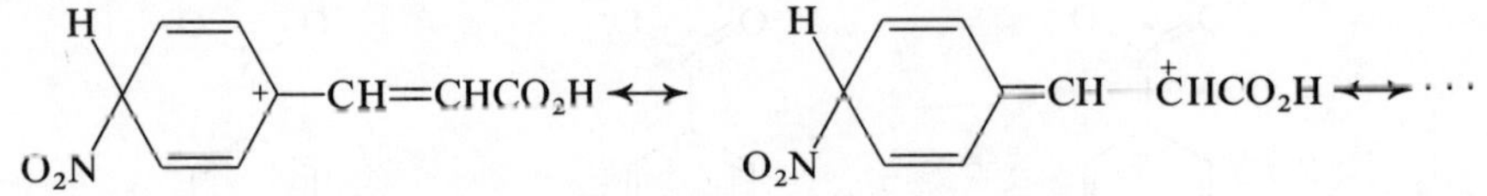

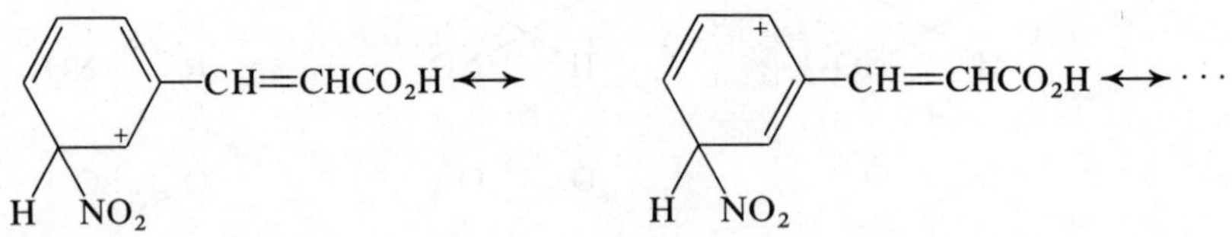

H—ring(O₂N)—CH=CHCO₂H ↔ H—ring(O₂N)=CH—ĊHCO₂H ↔ ···

H NO₂—ring—CH=CHCO₂H ↔ H NO₂—ring—CH=CHCO₂H ↔ ···

Only in the o- and p-carbonium ions is spreading of the positive charge to the side-chain possible, thereby stabilising these more than the m-ion.

Ortho-para ratio. If all things were equal, then an o,p-directing group would be expected to give an o/p ratio of 2/1. In practice, this ratio is less, and is due to a number of factors, *e.g.*, the polar effect of the substituent (I and R), its size (steric effect), the size of the entering group (steric effect), solvent effects, temperature effects, and electrostatic forces acting between substrate and electrophile as these approach each other. Much experimental work has been done to ascertain how these factors operate, but in many (apparently) analogous cases, the results may be different. Hence, predictions of orientation may not correspond with experimental results due to the fact that the relative importance of the various factors changes from one compound to another. However, in spite of these difficulties, some progress has been made, *e.g.*, it appears certain that in the nitration of alkylbenzenes, the change in the o/p ratio in the order: Me (1·57) > Et (0·93) > Me₂CH (0·48) > Me₃C (0·22) is largely due to a steric effect of the alkyl substituents.

In addition to the steric effect operating in the o-position, there may also be *chemical* interaction between this substituent and the electrophile, and when this occurs, the amount of o-substitution may be abnormally high (see, *e.g.*, p. 641).

Partial rate factors. The partial rate factor is defined as the rate of substitution at a given position in a substituted benzene relative to that at any one position in benzene itself (Ingold, 1953). Partial rate factors are calculated from experiments giving: (i) the proportions of the isomers produced from the substituted benzene under investigation; (ii) the relative rates of reaction of the substituted benzene and benzene. Formerly, (i) was carried out by actual isolation of the products. This often gave erroneous results, but now, with the use of methods such as chromatography and infra-red spectra, the results are extremely reliable. Various methods have been used to evaluate (ii); one of the best is by *competitive reactions*. An equimolecular mixture of the two substrates (substituted benzene and benzene) is treated with insufficient reagent to lead to complete reaction with both compounds. Thus the two substrates compete with each other, and analysis of the products will show which substrate has reacted with more of the reagent.

Let us consider the nitration of toluene to illustrate the calculation of partial rate factors. Ingold *et al.* (1931) showed that the nitration of toluene with nitric acid in acetic acid at 45°C is 24·5 times as fast as that of benzene under the same conditions, and the proportions of the isomers were *o*-, 57 per cent; *m*-, 3·2 per cent; and *p*-, 40 per cent. If the rate of nitration at any one of the six (equivalent) positions of benzene is taken as 1, then the total rate for benzene is $6 \times 1 = 6$, and the total relative rate for toluene is $6 \times 24·5 = 147$. Since there are two equivalent *o*-positions in toluene, the partial relative rate at each *o*-position $= 147 \times 0·57/2 = 42$. Similarly, at each *m*-position the partial relative rate is $147 \times 0·032/2 = 2·4$. Finally, for one *p*-position, the partial relative rate is $147 \times 0·4 = 59$.

If the partial rate factors are represented by the symbols o_f^R, m_f^R and p_f^R, where R is the substituent, then the foregoing account can be summarised as follows, where k is the rate constant:

$$o_f^R = \frac{k_{\mathrm{PhR}}}{k_{\mathrm{PhH}/6}} \times \frac{\%o\text{-}}{2 \times 100}$$

$$m_f^R = \frac{k_{\mathrm{PhR}}}{k_{\mathrm{PhH}/6}} \times \frac{\%m\text{-}}{2 \times 100}$$

$$p_f^R = \frac{k_{\mathrm{PhR}}}{k_{\mathrm{PhH}/6}} \times \frac{\%p\text{-}}{100}$$

Partial rate factors depend on the nature of the substituent group and on the nature of the entering group; they also depend on the conditions used: temperature and solvent. When the partial rate factor is greater than 1, then that position has been activated by the substituent; when less than 1, that position has been deactivated (see also below). Some partial rate factors for nitration are:

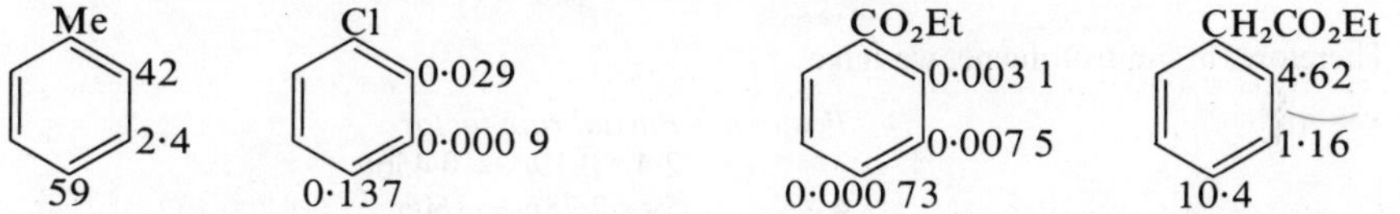

When dealing with several substituent groups in the ring then, as a first approximation, it is reasonably satisfactory to assume that the partial rate factor at a given position is the product of the partial rate factors corresponding to the substituents (Holleman's *product rule*, 1925), e.g.,

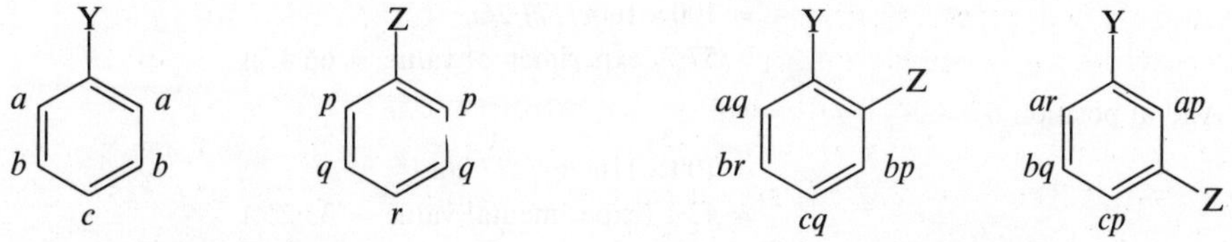

experimental values   predicted values

When dealing with three (or more) substituents, the same procedure is used. In this way, it is possible to assess (approximately) the amounts of isomeric substitution products in polysubstituted benzenes.

It is also possible to calculate the overall rate of substitution in a polysubstituted benzene relative to that of benzene. This rate is one-sixth of the sum of the partial rate factors for all the positions available for substitution (see equations given in the summary above). Thus:

$$k(\text{for } m^{YZ}) = 1/6(ar + bq + cp + ap)$$

If $Y = Z = Me$, then for bromination,

$$a = p = 600; b = q = 5.5; c = r = 2420$$

$$k = 1/6(600 \times 2420 + 5.5 \times 5.5 + 2420 \times 600 + 600 \times 600) = \frac{32.64 \times 10^5}{6} = 5.44 \times 10^5$$

The experimental result is 5.14×10^5 (Brown *et al.*, 1957).

It is also of interest to calculate the percentage yield of isomers from partial rate factors. The following example illustrates how this may be done even though the partial rate factors due to a particular substituent are unknown, but the proportions of the isomers produced are known. Let us consider the nitration of *o*-nitrotoluene. The values known for nitration are:

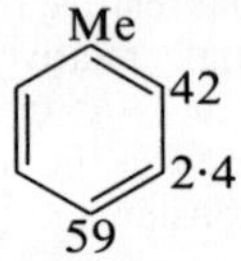

known partial rate factors	known isomer distribution	partial rate factors	

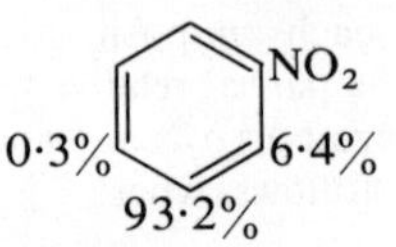

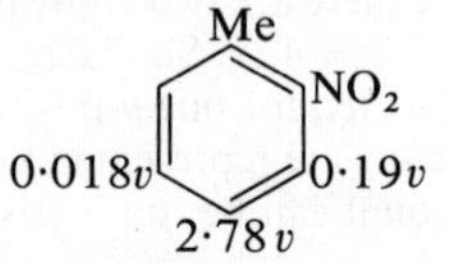

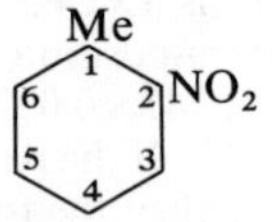

If $v = k_{PhNO_2}/k_{PhH}$, then the partial rate factors for nitrobenzene are:

$$o_f^{NO_2} = 6 \times v \times \frac{6.4}{2 \times 100} = 0.19v$$

$$m_f^{NO_2} = 6 \times v \times \frac{93.2}{2 \times 100} = 2.78v$$

$$p_f^{NO_2} = 6 \times v \times \frac{0.3}{100} = 0.018v$$

Therefore, in *o*-nitrotoluene, we have:

Position	Partial rate factor
3	$2.4 \times 0.19v = 0.456v$
4	$59 \times 2.78v = 160v$
5	$2.4 \times 0.018v = 0.043v$
6	$42 \times 2.78v = 116.76v$

$\therefore$ total rate $= 0.456v + 160v + 0.043v + 116.76v = 277.26v$

$\therefore$ per cent substitution at position 4

$$= 100 \times 160v/277.26v$$

$$= 57.7 \text{ (experimental value} = 66.4\%)$$

And at position 6

$$= 100 \times 116.76v/277.26v$$

$$= 42.1 \text{ (experimental value} = 33.2\%)$$

It might be noted that, provided the proportions of isomers produced by a substituent are known, then it is still possible to assess the orientation in derivatives containing two or more of the substituents; *e.g.*, in the above example, if the partial rate factors for toluene were unknown, then these could be evaluated in terms of *w*, the (unknown) rate of substitution. In the final step of the calculation, both *v* and *w* cancel out.

In a monosubstituted benzene, the total rate of substitution is one-sixth the sum of the partial rate factors of all positions (see above). However, since the o-position is subject to steric effects and the p-position is not, the partial rate factor for p-substitution may be taken as a measure of the ability of a reagent to choose between this substrate and benzene. Also, the ratio p_f/m_f is a measure of the ability of a reagent to choose between the p- and m-positions in the substrate. Brown *et al.* (1955–   ), from their study of the partial rate factors of these positions in toluene, found that for different types of reactions, *e.g.*, halogenation, nitration, etc.,

$$\log p_f{}^{Me} \propto \log (p_f{}^{Me}/m_f{}^{Me})$$
$$i.e.,\ \log p_f{}^{Me} = b \log (p_f{}^{Me}/m_f{}^{Me})$$

where b is a constant dependent on the nature of the substituent group. For toluene, a plot of $\log p_f{}^{Me}$ against $\log (p_f{}^{Me}/m_f{}^{Me})$ gave a very good straight-line graph, with $b = 1\cdot310$ (b is the slope of the line). Thus the selectivity of a reagent in choosing between the p- and m-positions in toluene is related to its selectivity in choosing between toluene and benzene.

Now, as we have already seen, the more reactive a reagent is, the less is its selectivity (p. 119). Hence, the more reactive the reagent is, the less selectivity it will show in competing reactions between the

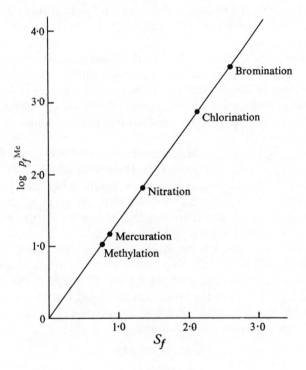

Fig. 20.14

two substrates toluene and benzene, and also less selectivity in choosing between the p- and m-positions in toluene. Thus, it can be anticipated that the selectivities $p_f{}^{Me}$ and $p_f{}^{Me}/m_f{}^{Me}$ both decrease with increasing activity of the reagent. In this way, the above equation becomes a Selectivity Relationship (Brown *et al.*, 1956, 1958). On this basis it follows that the order of selectivity is bromination > chlorination > nitration > mercuration > methylation (see Fig. 20.14). Hence, from the reactivity-selectivity principle, it follows that the order of reactivity of the reagents is the *reverse* of the above order (of selectivity).

A point of interest here is that a positively charged (electrophilic) reagent is less selective, and therefore more reactive, than the neutral reagent, *e.g.*, positive bromine (a brominium cation, Br^+; acid solution of hypobromous acid) is less selective than bromine (non-catalytic). Also, de la Mare *et al.*

(1956) have shown that m_f and p_f for the acid-catalysed bromination of toluene by hypobromous acid are about the same as those for nitration. Thus positive bromine and the positive nitronium ion have about the same selectivity (*cf.* the above order).

Hammett equation. In recent years, the data collected on rates and equilibria of reactions have led to the formulation of a number of empirical correlations. The introduction of a substituent may change the strength of an acid, *e.g.*, acetic and chloroacetic acids, *i.e.*, the position of ionisation *equilibrium* is changed. On the other hand, *rates* of reactions may be changed by the introduction of a substituent, *e.g.*, toluene is nitrated more rapidly than benzene and nitrobenzene less rapidly. These changes are due to a number of factors, one of which is the polar factor. If the other factors are considered to be essentially unchanged (by introduction of the substituent), it has then been found possible to derive an *empirical* linear relationship between the logarithms of rate or equilibrium constants for one reaction and those of another reaction which has been subjected to the same variations of reactant structure. For any *specified* reaction, there will be a set of rate or equilibrium constants; this set is termed a **reaction series**, *e.g.*, the hydrolysis of ethyl benzoate and its *m*- and *p*-substituted derivatives constitutes a specified reaction, and the various rate constants for hydrolysis under a given set of conditions form a reaction series.

The logarithm of a rate constant, according to the transition-state theory, is proportional to the standard free energy of activation, and the logarithm of the equilibrium constant is proportional to the change in the standard free energy of reaction (see p. 60). Hence, if there is a linear relationship between two sets of logarithms of rate or equilibrium constants, this is equivalent to a linear relationship between standard free energy changes, *i.e.*, the correlation is a **linear free-energy relationship**.

The linear free-energy relationship which describes the effects of polar factors (I and R) on reactivity in *aromatic* compounds is known as the **Hammett equation** (1937). It is important to note that this equation deals (as does the Selectivity Relationship, which is also a linear free-energy relationship) only with substituents *m*- and *p*- to the reaction site in *monosubstituted* benzenes. In these positions, steric effects are absent, and hence the correlation for polar factors is satisfactory, since other factors are essentially unchanged.

Let us now consider the reaction series of the ionisation constants (K) of *m*- and *p*-substituted benzoic acids and the reaction series of the ionisation constants (K') of the corresponding substituted phenylacetic acids. If the log K' values of the latter are plotted (as ordinates) against the log K values of the former (as abscissæ), the points will be found to lie on a good straight line. On the other hand, the corresponding values of *o*-substituents are very much scattered. This line is given by the equation

$$\log K' = \rho \log K + c$$

where ρ is the slope of the line and c is the intercept (on the vertical axis). For the special case where there is no substituent (*i.e.*, H is the 'substituent'), the equation may be written

$$\log K'_0 = \rho \log K_0 + c$$

Hence, by subtraction (which eliminates c), we have

$$\log (K'/K'_0) = \rho \log (K/K_0)$$

An equation of this type can be written for equilibrium or rate constants of almost any two types of side-chain reactions of benzene derivatives, *i.e.*, for almost any pair of reaction series. Suppose we now consider three reactions with equilibrium constants K, K', and K''. We can relate K and K', K and K'', or K' and K'' experimentally, or relate two pairs experimentally and *deduce* the third relation as follows:

$$\log(K/K_0) = \rho \log(K'/K_0')$$

and

$$\log(K''/K_0'') = \rho' \log(K'/K_0')$$

$$\therefore \ [\log(K/K_0)]/[\log(K''/K_0'')] = \rho/\rho'$$

or

$$\log(K/K_0) = \frac{\rho}{\rho'} \log(K''/K_0'')$$

From this it can be seen that it is possible to relate different reaction series to *one standard reaction*. Hammett chose the ionisation of substituted benzoic acids in water at 25°C as the standard reaction because many of these ionisation constants have been measured with great accuracy. Hammett then introduced a new constant, σ, defined by the equation

$$\sigma = \log[K(\text{substituted benzoic acid})/K(\text{benzoic acid})]$$

Thus, for *any* reaction involving an equilibrium constant (K) or a rate constant (k)

(i) $\log(K/K_0) = \sigma\rho$ *and* (ii) $\log(k/k_0) = \sigma\rho$

This is the Hammett equation in its two forms.

Since the set of ionisation constants of the substituted benzoic acids is *itself* a reaction series, then

$$\log(K/K_0) = \sigma\rho$$

However, *by definition*, for the substituted benzoic acids, $\log(K/K_0) = \sigma$. Hence, ρ for this reaction (*i.e.*, the ionisation of substituted benzoic acids) is *unity*.

As we have seen above, ρ is the slope of the line, and since its value is unity for the standard reaction, the line passes through the origin and is inclined at 45° to the axes. Hence, if we plot the *experimental* $\log(K/K_0)$ values (as ordinates) on this line, the corresponding abscissæ will be the σ values of the substituents. Alternatively, σ can be calculated from $\log(K/K_0)$.

The σ constant. σ is called the **substituent constant**. It is independent of the nature of the reaction (except in certain circumstances; see later), and is a quantitative measure of the polar effects in any reaction of a given *m*- or *p*-substituent relative to hydrogen, *i.e.*, it represents the electron-attracting (or withdrawing) or electron-repelling (or releasing) power of a substituent group. The more positive σ is, the more electron-withdrawing ($-$I and/or $-$R) is the substituent, and the more negative σ is, the more electron-releasing ($+$I and/or $+$R) is the substituent. It should be noted that, since electron-withdrawal or release is referred to hydrogen as standard, the σ value of hydrogen is zero.

Table 20.2

Group	σ_m	σ_p	Group	σ_m	σ_p
NH_2	-0.16	-0.66	F	$+0.34$	$+0.06$
Me	-0.07	-0.17	Cl	$+0.37$	$+0.23$
Et	-0.04	-0.15	Br	$+0.39$	$+0.23$
OH	$+0.12$	-0.37	I	$+0.35$	$+0.28$
H	0.000	0.000	COMe	$+0.38$	$+0.50$
OMe	$+0.12$	-0.27	NO_2	$+0.71$	$+0.78$

Some σ values are given in Table 20.2, and some points to note are:

(i) Polar effects are greater in the p-position than in the m-; the halogens are exceptions.

(ii) σ Values may be related to the orienting power of the group. Substituents whose σ values are more negative for the p- than for the m-position, e.g., NH_2, Me, increase the general reactivity of the ring for electrophilic reagents and hence are o- and p-directing. On the other hand, substituents whose σ values are more positive for the p- than for the m-position, e.g., COMe, NO_2, lower the general reactivity of the ring for electrophilic reagents and hence are m-orienting.

(iii) The positive σ values of the halogens indicate a lowering of the general reactivity of the ring, but in these cases the p-values are *more negative* than the m-values, and hence the halogens are o- and p-directing (see (ii) above).

(iv) When σ_m and σ_p have *different* signs, then I and R effects are acting in opposite directions (see also later), e.g., (a) OMe has $-$I from the m-position (no $+$R can operate from this position), and therefore σ_m is *positive*. On the other hand, OMe has $+$R from the p-position, and since this is much greater than the $-$I effect, σ_p is *negative*. (b) NO_2 has $-$I from the m-position and therefore σ_m is positive; but in the p-position, the effects are $-$I *and* $-$R, and consequently σ_p is more positive than σ_m.

The ρ constant. ρ is called the **reaction constant**, and is dependent on the nature of the reaction and on the conditions. It is a measure of the sensitivity of a given reaction series to the polar effects of the ring-substituents, *i.e.*, to changes in the σ value of the substituent.

The values of ρ are determined experimentally by choosing a number of benzene derivatives all containing the reaction centre under consideration, but each having a different ring substituent of *known* σ value. E.g., by plotting $\log(k/k_0)$ for the methanolysis of m- and p-substituted $(-)$-menthyl benzoates under the *same* conditions (MeONa—MeOH) as ordinates, against the σ values of the substituents as abscissæ, and drawing the best straight line through the points, the value of ρ is given by the slope of the line (Fig. 20.15). In the same way, one can plot $\log(K/K_0)$ against corresponding σ values for equilibria reactions.

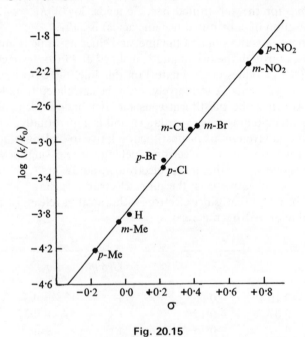

Fig. 20.15

It might be noted that if the value of ρ is known for a given reaction, it is possible to evaluate σ (from the Hammett equation) for various substituents if the rate (or equilibrium) constants are known for that reaction.

Table 20.3

Reaction	ρ
Ionisation of benzoic acids (water), 25°C	1·000
Ionisation of benzoic acids (40% aq. EtOH), 25°C	1·957
Ionisation of phenols (water), 25°C	2·113
Alkaline hydrolysis of methyl benzoates (60% aq. acetone), 25°C	2·460
Acid-catalysed esterification of benzoic acids (MeOH), 25°C	−0·521
Reaction between PhCOCl and substituted anilines (benzene), 25°C	−3·207

Some ρ values are given in Table 20.3, and points to note are:

(i) If σ is positive (electron-withdrawing), then $k > k_0$ (or $K > K_0$) if ρ is positive. Thus, for reactions in which rates or equilibria are facilitated by electron-withdrawing groups, ρ is positive; or alternatively, reactions for which ρ is positive are aided by electron-withdrawing groups.

(ii) If σ is positive, then $k < k_0$ (or $K < K_0$) if ρ is negative. Thus, for reactions in which rates or equilibria are hindered by electron-withdrawing groups, ρ is negative; or alternatively, reactions for which ρ is negative are hindered by electron-withdrawing groups.

(iii) By similar arguments, it can be shown that reactions with a positive ρ are hindered by electron-releasing groups (σ is negative), and those with a negative ρ are facilitated.

(iv) For a *given* positive σ value in different reactions, the more positive ρ is, the more is k (K) greater than k_0 (K_0), *i.e.*, the greater is the effect of the substituent on that reaction; or alternatively, the more sensitive is that reaction to polar effects. Similarly, the more negative ρ is, the more is k (K) smaller than k_0 (K_0).

Examples

(*a*) σ_m of NO_2 is $+0·71$; for the ionisation of the benzoic acids, $\rho = 1$, and for the alkaline hydrolysis of the methyl benzoates, $\rho = 2·460$. Thus, the latter reaction is much more sensitive to the effect of a *m*-nitro substituent than the former (see (iv)).

(*b*) Since the reaction of the alkaline hydrolysis of the methyl benzoates has a large positive ρ (2·460), this reaction is facilitated by electron-withdrawing groups (see (i)). This may be explained as follows. The reaction involves nucleophilic attack by the hydroxide ion on the

$$Z—Ar—\overset{\overset{\textstyle O}{\|}}{C}—OR \longleftrightarrow Z—\overset{+}{Ar}=\overset{\overset{\textstyle O^-}{|}}{C}—OR \qquad \qquad —\overset{\overset{\textstyle O}{\|}}{\underset{OH}{C}}—$$

$$(A) \qquad\qquad\qquad\qquad (B) \qquad\qquad\qquad\qquad (C)$$

carbon atom of the carbonyl group (see (C)). Thus, (A) is more readily attacked than (B). Also, (A) is favoured if Z is an electron-withdrawing group (σ is positive) and so reaction is facilitated. On the other hand, (B) is favoured if Z is an electron-releasing group (σ is negative) and therefore reaction is hindered.

(*c*) In the following bimolecular (S_N2) reaction, ρ is equal to $−0·99$.

$$ArO^- + EtI \xrightarrow[\text{EtOH}]{\text{OH}^-} [Ar\overset{\delta-}{O}\cdots Et \cdots \overset{\delta-}{I}] \longrightarrow ArOEt + I^-$$

Since ρ is negative, electron-releasing groups will increase the rate, whereas electron-withdrawing groups will decrease the rate (see (iii)). These results may be explained by the fact that ArO^- is a nucleophile, and it can be expected that the greater is the charge on the oxygen atom, the faster will be the reaction. Electron-releasing groups ($+I$ and/or $+R$) in Ar will increase the charge on the oxygen atom ((D) is more favoured than (E)) and so facilitate reaction, whereas

$$(D) \ Z-Ar-O^- \longleftrightarrow Z-Ar^-=O \ (E)$$

electron-withdrawing groups ($-I$ and/or $-R$) will decrease the charge on the oxygen atom ((E) is more favoured than (D)) and so the rate is retarded.

Deviations from the Hammett equation. (i) If in a *given* reaction series there is a deviation from the Hammett equation, this is a strong indication that this *particular* reaction is proceeding by a mechanism different from the expected one, *i.e.*, the mechanism of the rest of the series.

(ii) **Effect of resonance.** When the only polar connection between a substituent group and the reaction centre is an inductive effect, the Hammett equation holds good. When, however, resonance is possible between the substituent group and the reaction centre, this resonance affects the rate or equilibrium position in addition to the inductive effect. Resonance effects of this kind are one of the main causes of the failure of the Hammett equation. Thus, Hammett found that the σ value for the p-nitro group obtained from the ionisation of p-nitrobenzoic acid does not fit the equation when the ionisations of p-nitrophenol and p-nitroaniline are considered. This deviation has been explained by resonance involving the reaction centre and the substituent group, *e.g.*,

The increased positive charge on the oxygen atom (of the OH group) facilitates proton release, thereby increasing the acid strength of p-nitrophenol over that of phenol. Alternatively, the phenoxide ion produced from p-nitrophenol is more stabilised by resonance than is the phenoxide ion itself (*cf.* p. 701). From the Hammett equation, with the value of $\rho = 2\cdot1$ for this reaction, p-nitrophenol would be about forty-four times as strong an acid as phenol:

$$\log(K/K_0) = \sigma\rho = 0\cdot78 \times 2\cdot1 = 1\cdot64$$
$$K/K_0 = 44$$

The experimental value is about 680.

It is still possible to get good agreement with Hammett equation by giving the p-NO_2 group a *second substituent constant*. This second constant is designated as σ^- (sigma minus), the minus sign indicating that the p-substituent group is capable of a *resonance electron-withdrawal* effect. On the other hand, the second constant for a p-substituent group such as NH_2 is designated as σ^+ (sigma plus), the plus sign indicating a *resonance electron-donating* effect. Some σ^- and σ^+ values for p-substituents are given in Table 20.4 (*cf.* Table 20.2). These values have been obtained from a study of p-substituted phenols and anilines. The fact that the σ values of m-substituents may be used for these substrates indicates that the R effect of a given substituent is far stronger from the p-position than from the m-. Also, if substitution of a σ^+ (or σ^-) value gives a better fit in the Hammett equation, then this means that resonance is involved between the substituent group and the reaction centre.

It should be noted that since a second substituent constant has had to be introduced, σ values are *not* independent of the nature of the reaction.

Table 20.4

Group	σ^+	Group	σ^-
p-NH$_2$	$-1\cdot3$	p-COMe	$+0\cdot87$
p-Me	$-0\cdot31$	p-NO$_2$	$+1\cdot27$
p-OH	$-0\cdot92$	p-CO$_2$H	$+0\cdot73$
p-OMe	$-0\cdot78$	p-CO$_2$Me	$+0\cdot64$
p-F	$-0\cdot07$	p-CONH$_2$	$+0\cdot63$
p-Cl	$+0\cdot12$	p-CN	$+1\cdot00$
p-Br	$+0\cdot15$	p-CHO	$+1\cdot13$
p-I	$+0\cdot135$	p-CF$_3$	$+0\cdot74$

Hammett equation and electrophilic aromatic substitution. It appears that rates of m-substitution may be correlated satisfactorily using σ_m values (Table 20.2), but σ^+ values (Table 20.4) must be used for p-substitution. Since it is rates in these positions that are required, partial rate factors must be used. However, to make use of partial rate factors, m_f and p_f, instead of k/k_0 values in the Hammett equation, m_f is put equal to $3k/k_0$ and p_f is put equal to $6k/k_0$ (where $k \equiv k_{PhR}$ and $k_0 \equiv k_{PhH}$; see p. 601). Thus we have the equations:

$$\log m_f = \log(3k/k_0) = \sigma_m\rho$$
$$\log p_f = \log(6k/k_0) = \sigma^+\rho$$

Taft equation. Taft (1952, 1953) has examined *aliphatic* reactions, and has introduced the parameter σ^*, the *polar substituent constant*, which is believed to represent the I effect only. The Taft equation is:

$$\log(k/k_0) = \sigma^*\rho^*$$

A large number of aliphatic reaction rates can be correlated by means of this equation.
The Hammett equation fails with aliphatic compounds because of the different conformations the chains can assume and because substituents may interact sterically with the reaction centre.

General methods of preparation of the benzene homologues

Benzene homologues, *i.e.*, benzene derivatives in which the substituents are *alkyl* groups, are known collectively as **arenes**. This term also includes the alkyl derivatives of polynuclear aromatics such as naphthalene, anthracene, etc.

Friedel–Crafts reaction (1877). This reaction involves the introduction of an alkyl or acyl group into the benzene ring in the presence of a catalyst.

The aromatic compounds are usually hydrocarbons, aryl chlorides and bromides, mono- and polyhydric phenols or their ethers. In general, the Friedel–Crafts reaction fails with heterocyclic compounds containing a basic nitrogen atom, *e.g.*, pyridine, but is successful with pyrrole, furan and thiophen. The alkylating agents may be alkyl halides, aliphatic alcohols, alkenes, ethers and alkyl esters of organic and inorganic acids. From the point of view of convenience, the alkylating agent is usually confined to alkyl halides, alcohols and alkenes. The acylating agents may be acid chlorides or anhydrides, acids or esters. Aryl acid chlorides containing a ring amino-group may also be used provided it is protected; the tosyl (p-toluenesulphonyl) group is particularly good for this purpose. Since esters are not so reactive as acid chlorides, by using molecular proportions of EtO$_2$C(CH$_2$)$_n$COCl and substrate, then only the acid chloride end reacts. α,ω-Di-acid chlorides give different products according to the value of n. When $n > 3$, the product is a diketone; for $n = 2$ or 3, a cyclic product may also be obtained. Sulphonyl chlorides and cyclic anhydrides can also be used; the products are, respectively, sulphones and aryl keto-acids.

Two types of catalysts are used in the Friedel–Crafts reaction:

(i) Metal halides (Lewis acids); their general order of reactivity is as shown and usually, the more reactive the substrate, the weaker need be the Lewis acid used.

$$AlCl_3 > BF_3 > SbCl_5 > FeCl_3 > SnCl_4 > ZnCl_2$$

(ii) Proton acid catalysts: HF, H_2SO_4, H_3PO_4.

Aluminium chloride is the most widely used catalyst (this is the one originally used by Friedel and Crafts). It may be used with any alkylating or acylating agent. In the case of alkenes, it is necessary to have a small amount of halogen acid present. The amount of aluminium chloride required depends on the nature of the substrate and/or reagent. Thus, for *non-oxygenated substrates*, about $0·2$ molecule is required for alkyl halides and alkenes, $0·5$–$1·0$ molecule for alcohols ($ROAlCl_2$ is formed), 1 molecule for acid chlorides and 2 molecules for acid anhydrides ($RCOClAlCl_3$ and RCO_2AlCl_2 are formed). When the *substrate contains oxygen*, then large amounts of catalyst are required, *e.g.*, ArOH (forms $ArOAlCl_2$), $ArCO_2H$ (forms $ArCO_2AlCl_2$).

Solvents are generally used for Friedel–Crafts reactions, but if the substrate is a liquid, this may be used in excess (see p. 611). Two common solvents are nitrobenzene and carbon disulphide, and sometimes the orientation is affected by the nature of the solvent (see p. 800).

Although aluminium chloride can be used with any alkylating agent, the tendency is to use this catalyst with alkyl halides and alkenes; boron trifluoride or hydrogen fluoride with alkenes and alcohols; hydrogen fluoride with esters; and sulphuric acid with alkenes and alcohols.

The orienting influence of the catalyst in alkylation is interesting. $AlCl_3$ tends to give predominantly the *m*-product; BF_3 and H_2SO_4 at low temperatures give predominantly the *o*- and *p*-isomers.

The ease of alkylation with alkyl halides (with $AlCl_3$ as catalyst) depends on the nature of the alkyl group and on the halogen atom. For a given halogen atom, the order is t-R > s-R > p-R; this is also the order for alcohols, esters and ethers. For a given alkyl group, the order is RF > RCl > RBr > RI.

The Friedel–Crafts reaction has certain drawbacks:

(i) The structure of the alkyl group plays a part in the alkylation, *e.g.*, it is easy to introduce a methyl, ethyl or isopropyl group, but usually difficult to introduce n-propyl or n-butyl, since these tend to rearrange. Isobutyl halides very readily give a t-butyl substitution product. If the reaction is carried out in the cold with benzene and n-propyl chloride, the n-propyl group is introduced:

$$C_6H_6 + CH_3CH_2CH_2Cl \xrightarrow[\text{cold}]{AlCl_3} C_6H_5CH_2CH_2CH_3 + HCl$$

If, however, the reaction is carried out at higher temperatures, the isopropyl group is mainly introduced:

$$C_6H_6 + CH_3CH_2CH_2Cl \xrightarrow[\text{heat}]{AlCl_3} C_6H_5CH(CH_3)_2 + HCl$$

A very good means of introducing the n-propyl group is to use cyclopropane as the alkylating agent:

$$C_6H_6 + \overset{\displaystyle CH_2}{\underset{\displaystyle CH_2}{\overset{\diagup}{\underset{\diagdown}{CH_2}}}} \quad \xrightarrow{AlCl_3} C_6H_5CH_2CH_2CH_3 \quad (65\%)$$

Acid chlorides may be used to introduce long straight-chain groups (see below).

(ii) It is not always possible to stop the reaction at the required stage, *i.e.*, there is always a tendency to over-alkylate, particularly with Me and Et groups, *e.g.*,

$$\text{C}_6\text{H}_6 +\text{CH}_3\text{Cl} \xrightarrow{\text{AlCl}_3} \text{C}_6\text{H}_5\text{CH}_3 \xrightarrow{\text{CH}_3\text{Cl}} \text{C}_6\text{H}_4(\text{CH}_3)_2$$

It has now been shown that when an excess of the substrate is used as solvent, there is a separation of a dense catalyst layer in which the mono- and higher alkylbenzenes are soluble. The use of solvents, *e.g.*, nitrobenzene, or the addition of a small amount of, *e.g.*, ether, produces a *homogeneous* reaction mixture, and results in a very high yield of the monoalkylbenzene.

(iii) In the Friedel–Crafts reaction, an alkyl group may be removed from one molecule and transferred to another (disproportionation), or may migrate from one position to another in the ring. This renders the structure of the product uncertain in a number of cases (see also p. 615).

(iv) The presence of a *m*-orienting group in the ring hinders or inhibits the Friedel–Crafts reaction, *e.g.*, nitrobenzene and acetophenone do not undergo the Friedel–Crafts reaction. On the other hand, if a strongly activating group (*o-p*-orienting group) is present in either of the above two compounds, reaction can take place, *e.g.*, *o*-nitroanisole reacts with isopropanol in the presence of hydrogen fluoride to form 4-isopropyl-2-nitroanisole:

$$\text{(o-nitroanisole)} + \text{CH}_3\text{CHOHCH}_3 \xrightarrow{\text{HF}} \text{(4-isopropyl-2-nitroanisole)} + \text{H}_2\text{O} \quad (84\%)$$

This hindering effect can be used to advantage for preparing a monoalkylated benzene (free from dialkylated-product):

$$\text{C}_6\text{H}_6 + \text{CH}_3\text{COCl} \xrightarrow[\text{(1 mole)}]{\text{AlCl}_3} \text{C}_6\text{H}_5\text{COCH}_3 \xrightarrow{e;\, \text{H}^+} \text{C}_6\text{H}_5\text{CH}_2\text{CH}_3$$

The acetophenone (which is not further attacked) may be readily reduced to ethylbenzene by the Clemmensen method.

Mechanism of the Friedel–Crafts reaction

Alkyl halides. There appear to be at least two possible mechanisms that can operate with alkyl halides, one involving an alkyl halide–Lewis acid complex (bimolecular mechanism), and the other involving an alkyl carbonium ion (unimolecular mechanism). In either case, the Friedel–Crafts reaction can be described as a nucleophilic reaction in which the arene is the nucleophile.

It is well-known that alkyl halides form 1 : 1 adducts with Lewis acids, and there is evidence to show that this adduct reacts with the arene if the alkyl group is of the straight-chain type.

$$\text{RCl} + \text{AlCl}_3 \rightleftharpoons \text{R}-\overset{+}{\text{Cl}}-\overset{-}{\text{AlCl}}_3$$

$$\text{ArH} + \text{R}-\overset{+}{\text{Cl}}-\overset{-}{\text{AlCl}}_3 \longrightarrow \left[\text{Ar}\underset{\text{R}}{\overset{\text{H}}{<}}\right]^{+}\text{AlCl}_4^{-} \longrightarrow \text{ArR} + \text{HCl} + \text{AlCl}_3$$

On the other hand, because of the relative stability of s- and t-carbonium ions, the adducts with s- and t-alkyl halides ionise, and it is now the carbonium ion that is predominantly the active species:

$$\text{R}-\overset{+}{\text{Cl}}-\overset{-}{\text{AlCl}}_3 \rightleftharpoons \text{AlCl}_4^{-} + \text{R}^{+} \xrightarrow{\text{ArH}} \left[\text{Ar}\underset{\text{R}}{\overset{\text{H}}{<}}\right]^{+}\text{AlCl}_4^{-}$$

These mechanisms account for the fact that n-alkyl groups can be introduced (to a fair extent) without rearrangement at *low* temperatures (ionisation of the adduct is retarded), and that when carbonium ions are formed, rearrangement occurs (by hydride and/or methyl carbanion shift), *e.g.*, at higher temperatures (than those used above), n-propyl chloride gives isopropylbenzene:

$$MeCH_2CH_2Cl + AlCl_3 \rightleftharpoons AlCl_4^- + MeCH_2CH_2^+ \longrightarrow Me_2CH^+ \xrightarrow{PhH} PhCHMe_2$$

In the same way, isobutyl chloride gives t-butylbenzene.

These two mechanisms are *extremes*, MeX behaving as the adduct and t-RX as the carbonium ion. For all other cases, it would appear that the carbonium-ion route is increasingly favoured the greater is the branching in the alkyl halide, *i.e.*, many reactions proceed by a *dual* mechanism.

Alcohols. The mechanistic pattern is similar to that of the alkyl halides, but in this case it seems likely that the carbonium-ion path is followed even for a primary alkyl group (this mechanism also operates for ethers).

$$ROH + BF_3 \rightleftharpoons \begin{bmatrix} & H \\ & |_+ \\ R & -O-BF_3 \end{bmatrix} \rightleftharpoons R^+ + HOBF_3^-$$

Alkenes. With acids, the carbonium ion is formed first, and this then alkylates the substrate, *e.g.*, isobutene reacts with benzene in the presence of sulphuric acid to form t-butylbenzene:

$$Me_2C{=}CH_2 + H^+ \xrightarrow{H_2SO_4} Me_3C^+ \xrightarrow{PhH} PhCMe_3$$

Alkadienes may also be used, *e.g.* (note the 1,4-addition):

$$C_6H_6 + CH_2{=}CH{-}CH{=}CH_2 \xrightarrow[0°C]{HF} C_6H_5CH_2CH{=}CHCH_3$$

Many types of unsaturated compounds other than hydrocarbons may be used in the Friedel–Crafts reaction, *e.g.*, benzylideneacetophenone:

$$C_6H_6 + PhCH{=}CHCOPh \xrightarrow{AlCl_3} Ph_2CHCH_2COPh \quad (85\%)$$

Allyl halides react in the presence of sulphuric acid, *e.g.*,

$$C_6H_6 + CHMe{=}CHCH_2Cl \xrightarrow{H_2SO_4} C_6H_5CHMeCH_2CH_2Cl$$

Acyl chlorides. For many reactions involving acyl chlorides, there is evidence to show that the reaction proceeds through the intermediate formation of an acyl cation:

$$RCOCl + AlCl_3 \rightleftharpoons [RCO{-}\overset{+}{Cl}{-}\overset{-}{Al}Cl_3] \rightleftharpoons RCO^+ + AlCl_4^-$$

$$\xrightarrow{ArH} \begin{bmatrix} & H \\ & \diagup \\ Ar & \\ & \diagdown \\ & COR \end{bmatrix}^+ AlCl_4^- \longrightarrow ArCOR + HCl + AlCl_3$$

There is also evidence that the 1:1 adduct, $RCOClAlCl_3$, can be the active species, *i.e.*,

$$ArH + RCO{-}\overset{+}{Cl}{-}\overset{-}{Al}Cl_3 \longrightarrow ArCOR + HCl + AlCl_3$$

As with alkylation, it would appear that either mechanism could operate singly or both simultaneously according to circumstances.

Wurtz–Fittig reaction (1863). Homologues of benzene may be prepared by warming an ethereal solution of an alkyl and aryl halide with sodium (*cf.* Wurtz reaction):

$$C_6H_5Br + C_2H_5Br + 2Na \longrightarrow C_6H_5C_2H_5 + 2NaBr \quad (60\%)$$

Biphenyl and n-butane (and some other compounds) are obtained as by-products.

The advantages of the Wurtz–Fittig method over the Friedel–Crafts are that the structure of the product is known and that long n-side-chains can be easily introduced:

$$C_6H_5Br + CH_3(CH_2)_2CH_2Br \xrightarrow{Na} C_6H_5(CH_2)_3CH_3 \quad (62\text{--}72\%)$$

Two mechanisms have been suggested for the Wurtz–Fittig reaction, one via organometallics, and the other via the free radicals (*cf.* the Wurtz reaction). However, since there is more evidence to support the former mechanism, only this is given here.

(i) $C_6H_5Br + 2Na \longrightarrow C_6H_5{}^-Na^+ + NaBr$

(ii) $C_6H_5{}^-Na^+ + C_2H_5Br \longrightarrow C_6H_5{-}C_2H_5 + NaBr$

(iii) $C_2H_5Br + 2Na \longrightarrow C_2H_5{}^-Na^+ + NaBr$

(iv) $C_2H_5{}^-Na^+ + C_2H_5Br \longrightarrow C_2H_5{-}C_2H_5 + NaBr$

(v) $C_2H_5{}^-Na^+ + C_6H_6Br \longrightarrow C_6H_5{-}C_2H_5 + NaBr$

Since (i) proceeds very much faster than (iii), (ii) will be the main reaction ($C_6H_5{}^-$ is stabilised by resonance, whereas $C_2H_5{}^-$ is not).

Method using a Grignard reagent. Homologues of benzene may be prepared by the action of an alkyl halide on phenylmagnesium bromide:

$$C_6H_5Br \xrightarrow{Mg} C_6H_5MgBr \xrightarrow{CH_3I} C_6H_5CH_3 \quad (v.g.)$$

For the introduction of a methyl or ethyl group, the corresponding alkyl sulphate may be used instead of the alkyl halide (*cf.* p. 427).

Benzene homologues with branched side-chains may be prepared by the action of alkyl-magnesium halide on an aromatic ketone, *e.g.*, isopropylbenzene from acetophenone:

$$C_6H_5COCH_3 \xrightarrow[\text{(ii) } H_3O^+]{\text{(i) } CH_3MgI} C_6H_5C(CH_3)_2OH$$

$$\xrightarrow[-H_2O]{\text{heat}} C_6H_5C(CH_3){=}CH_2 \xrightarrow{H_2-Ni} C_6H_5CH(CH_3)_2$$

On the other hand, long side-chains may be produced by using allyl chloride followed by reduction (as above):

$$C_6H_5CH_2MgBr + CH_2{=}CHCH_2Br \longrightarrow C_6H_5CH_2CH_2CH{=}CH_2$$

Tosyl esters may also be used (Ts = p-MeC$_6$H$_4$SO$_2$—):

$$C_6H_5CH_2MgCl + 2TsOBu^n \longrightarrow C_6H_5CH_2Bu^n + Bu^nCl + Mg(OTs)_2 \quad (50\text{--}59\%)$$

Decarboxylation of aromatic acids. Many aromatic acids are readily decarboxylated when heated with soda-lime, *e.g.*, toluene from *p*-toluic acid (*cf.* p. 569).

Method of chloromethylation. This is the process whereby a hydrogen atom (generally of an aromatic compound) is replaced by a *chloromethyl group*, CH_2Cl. The reaction may be carried out by heating the aromatic hydrocarbon with formalin or paraformaldehyde and hydrochloric acid in the presence of zinc chloride as catalyst, *e.g.*, benzyl chloride from benzene:

$$C_6H_6 + CH_2O + HCl \xrightarrow{ZnCl_2} C_6H_5CH_2Cl + H_2O \quad (79\%)$$

Chloromethyl ether and dichloromethyl ether may be used instead of formalin or para-formaldehyde, but unlike the latter, do not require the presence of a catalyst. Many catalysts may be used with formalin (or paraformaldehyde), the most useful being zinc chloride, aluminium chloride, stannic chloride, acetic acid and sulphuric acid.

The introduction of a chloromethyl group is very useful, since this group is readily converted into other groups such as CH_3, CH_2OH, CHO, CH_2CN.

The presence of a halogen atom in the ring reduces the yield of the chloromethylated derivative. A nitro-group behaves in a similar manner, and the presence of two nitro-groups inhibits the reaction altogether, e.g., nitrobenzene gives only a small yield of the chloromethylated compound; m-dinitro-benzene gives none. Ketones of the type ArCOR give very poor yields, and those of the type ArCOAr do not react at all. Phenols, however, are very reactive, so much so that polymers are often obtained.

The mechanism of chloromethylation is uncertain; a possibility is:

$$2CH_2{=}O + 2HCl + ZnCl_2 \rightleftharpoons ZnCl_4^{2-} + 2\overset{+}{C}H_2OH \xrightarrow[-H^+]{ArH} ArCH_2OH \xrightarrow{HCl} ArCH_2Cl$$

General properties of the arenes

The benzene homologues are usually colourless liquids and their boiling points rise fairly regularly for each CH_2 added. In general, the boiling points of a group of isomers are also fairly close.

A characteristic chemical property of alkylbenzenes is their rearrangement and/or dis-proportionation in the presence of a Lewis acid and halogen acid. It appears that a methyl group is removed with great difficulty, if at all, but readily migrates. On the other hand, all other groups may be removed, the ease of disproportionation being in the order t-Bu > isoPr > Et, i.e., the more stable the alkyl carbonium ion derived from the alkyl substituent, the greater is the ease of disproportionation, e.g., ethylbenzene gives benzene and a mixture of diethyl-benzenes; t-butylbenzene gives benzene and 1,3,5-tri-t-butylbenzene.

The mechanism of these rearrangements and disproportionations is believed to involve the formation of an intermediate carbonium ion (cf. basicity of arenes, p. 594). Steinberg et al. (1962) showed that for toluene, no disproportionation occurs, but migration does. The latter was demonstrated by starting with the labelled compound, toluene-1-^{14}C. Thus:

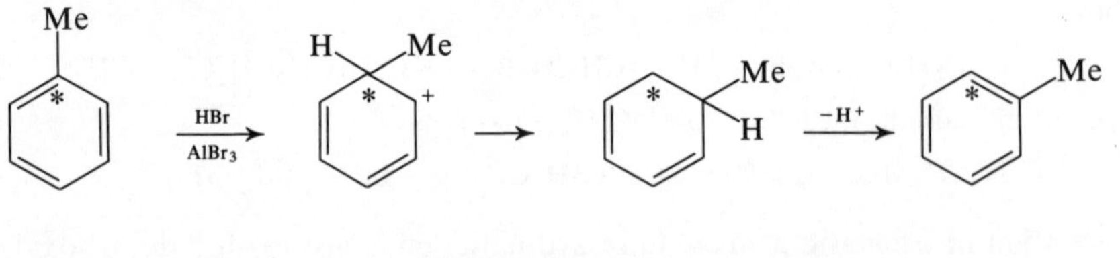

For the disproportionation of ethylbenzene, we may formulate the reaction as follows:

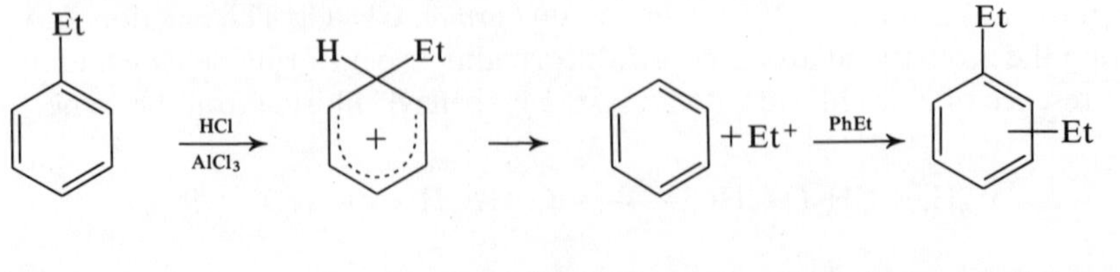

When each xylene is heated with $AlCl_3$—HCl, the product is a mixture of all three isomers, e.g. (Allen et al., 1959): o-, 17; m-, 62; p-, 21 per cent.

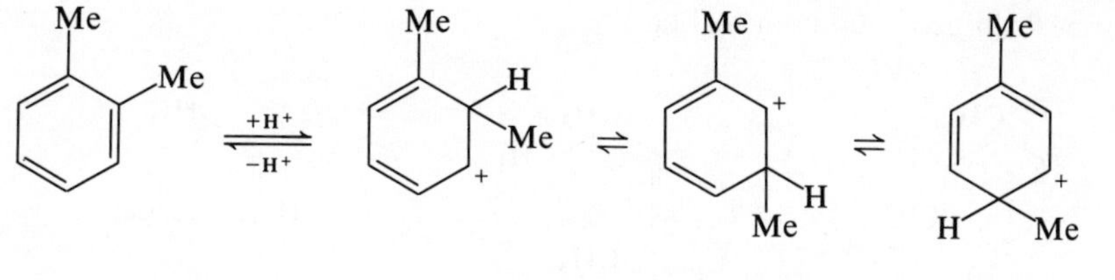

Everdell (1967) has estimated the thermodynamic data for all three xylenes and showed that the order of thermodynamic stability is $m > o > p$, and calculated that the composition of the equilibrium mixture at 50°C is o-, 20 ± 6; m-, 58 ± 10; p-, 22 ± 8 per cent. These values are in fair agreement with the experimental results, and can be explained on the basis that the m-isomer is the thermodynamically controlled product, whereas the other two isomers are kinetically controlled products (remember that these isomerisations are *reversible* under the conditions of the experiment).

These results have also been explained in terms of basicity. Rearrangement involving hydride shift (instead of Me) can also occur, *e.g.*,

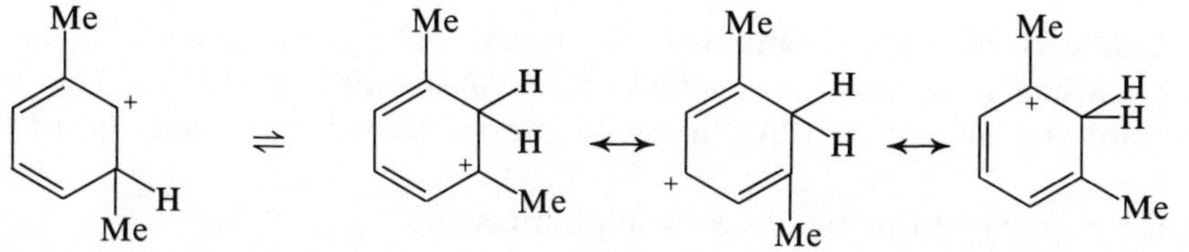

Other positions of protonation are also possible, but only for the m-carbonium ion are the *two* methyl groups ($+$I) in *one* canonical structure capable of partially neutralising the positive charges on ring carbon atoms. In the other two isomers, only one methyl group can act in this way. Thus the m-carbonium ion will be more stable than the other two, *i.e.*, is more basic, and so the equilibrium is shifted to the formation of the m-isomer.

Toluene (*methylbenzene*), b.p. 111°C, is obtained commercially from coal-tar, but mainly by catalytically dehydrogenating n-heptane or methylcyclohexane (both obtained from petroleum):

$$CH_3(CH_2)_5CH_3 \longrightarrow C_6H_5CH_3 + 4H_2$$

Xylenes, C_8H_{10}. Four isomers of the formula C_8H_{10} are known, and all are present in the

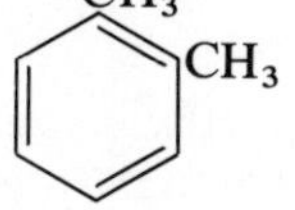

o-xylene
b.p. 144°C

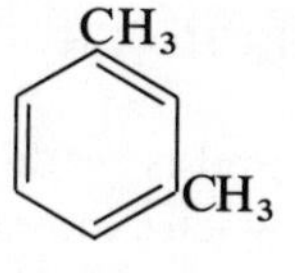

m-xylene
b.p. 139°C

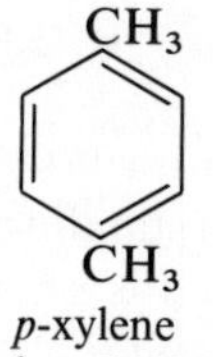

p-xylene
b.p. 138°C

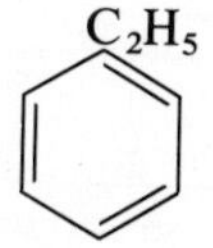

ethylbenzene
b.p. 136°C

light-oil fraction of coal-tar, but their main commercial source is the hydroforming process (p. 569).

Ethylbenzene is prepared industrially by the alkylation of benzene with ethylene using $AlCl_3$—HCl or H_3PO_4 as catalyst.

Hydrocarbons of formula C_9H_{12}. There are eight isomeric hydrocarbons of formula C_9H_{12}, *viz.*, three trimethylbenzenes, three ethylmethylbenzenes, one n-propylbenzene and one isopropylbenzene.

The three trimethylbenzenes occur in coal-tar:

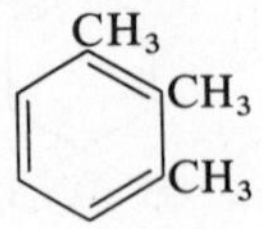

hemimellitene
b.p. 175°C

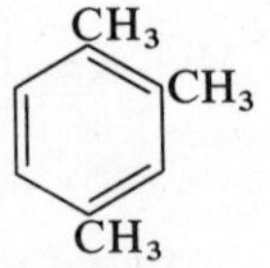

ψ-cumene
b.p. 169°C

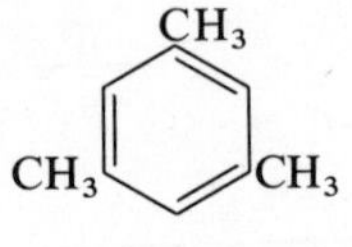

mesitylene
b.p. 164°C

Mesitylene is the most important of these, and is usually prepared by distilling acetone with sulphuric acid (p. 584).

Isopropylbenzene (*cumene*), b.p. 153°C, also occurs in coal-tar. It is manufactured from benzene and propene with sulphuric acid as catalyst (see also p. 699).

p-Cymene (*cymene*, *p*-isopropylmethylbenzene), b.p. 177°C, occurs in oil of thyme and eucalyptus oil, and is related to camphor and other terpenes.

Aromatic hydrocarbons with unsaturated side-chains. The only important unsaturated aromatic hydrocarbon is **styrene** (*phenylethylene*, *vinylbenzene*), b.p. 145°C. This occurs in storax (a balsam) and in coal-tar (distilling with the xylenes). Styrene may be prepared as follows:

(i) By heating phenylmethylmethanol with sulphuric acid:

$$C_6H_5MgBr + CH_3CHO \longrightarrow C_6H_5-\underset{\underset{CH_3}{|}}{C}HOMgBr \xrightarrow{H_2O}$$

$$C_6H_5CHOHCH_3 \xrightarrow[-H_2O]{H_2SO_4} C_6H_5CH=CH_2$$

(ii) The most convenient laboratory preparation is to heat cinnamic acid with a small amount of quinol:

$$C_6H_5CH=CHCO_2H \longrightarrow C_6H_5CH=CH_2 + CO_2 \quad (38-41\%)$$

A small amount of quinol is also placed in the receiving vessel to prevent the polymerisation of the styrene.

Styrene is manufactured by dehydrogenating ethylbenzene catalytically:

$$C_6H_5C_2H_5 \xrightarrow[600°C]{ZnO} C_6H_5CH=CH_2 + H_2$$

Styrene adds on bromine to form the dibromide, and is readily reduced to ethylbenzene. It polymerises slowly to a solid on standing, and rapidly when exposed to sunlight or in the presence of sodium. This polymer, which is known as *metastyrene*, $(C_8H_8)_n$, may be depolymerised by heating. Styrene is used for making plastics and synthetic rubbers.

Substitution in the side-chain of styrene gives rise to two possibilities:

$$C_6H_5CCl=CH_2 \qquad\qquad C_6H_5CH=CHCl$$

α-chlorostyrene β- or ω-chlorostyrene

Phenylacetylene, b.p. 142°C, may be prepared by decarboxylating phenylpropiolic acid, $C_6H_5C{\equiv}CCO_2H$, or by heating ω-bromostyrene with ethanolic potassium hydroxide (yield: 67%).

Phenylacetylene has acidic properties, *e.g.*, it forms metallic derivatives. It is reduced to styrene by zinc dust and acetic acid, and Budde *et al.* (1963) have shown that the hydration of phenylacetylene to acetophenone is effectively catalysed by mercuric perchlorate-perchloric acid in aqueous dioxan solution.

$$C_6H_5C\equiv CH + H_2O \longrightarrow C_6H_5COCH_3$$

Oxidation of arenes. The benzene ring is usually very resistant to oxidation, and so when benzene homologues are oxidised it is the side-chain which is attacked. Whatever the length of the side-chain, the ultimate oxidation product is benzoic acid; sometimes the intermediate products can be isolated, but it is usually difficult to control the oxidation. Initial attack is at the α-carbon, *e.g.* (with $KMnO_4$):

$$C_6H_5CH_2CH_3 \longrightarrow C_6H_5COCH_3 \longrightarrow C_6H_5CO_2H$$

If the side-chain is a t-R group, oxidation is difficult, and when it does occur, extensive ring-cleavage often results (Brandenberger *et al.*, 1961). When two side-chains are present, it is possible to oxidise them one at a time, *e.g.*, the xylenes may be oxidised first to a toluic acid and then to a phthalic acid.

The usual oxidising agents for side-chains are acid or alkaline permanganate, and less frequently, acid dichromate or dilute nitric acid. When an electron-withdrawing group ($-I$ and/or $-R$) is present, the ring is stable and the result of oxidation is a substituted benzoic acid. On the other hand, when the group present is **OH** or **NH$_2$** (strongly $+R$), the ring is very sensitive to oxidation and is largely broken down whatever the nature of the oxidising agent. Ring-rupture, however, can be prevented by protection of the group, *e.g.*, as tosyl derivative, followed by oxidation in acid media. Alternatively, when the ring contains a hydroxyl group, an alkaline solution of lead dioxide oxidises a methyl side-chain without ring-rupture.

Metalation of aromatic compounds. Metallic derivatives of aromatic compounds may be prepared by the general methods on p. 434, *e.g.*, phenylsodium, diphenylmercury, and phenyl-lithium:

(i) $C_6H_6 + R^-Na^+ \longrightarrow C_6H_5{}^-Na^+ + RH$

(ii) $2C_6H_5Br + Na_2Hg \longrightarrow (C_6H_5)_2Hg + 2NaBr$

(iii) $(C_6H_5)_2Hg + 2Na \longrightarrow 2C_6H_5{}^-Na^+ + Hg$

(iv) $LiBr + C_6H_5Li \xleftarrow{2Li} C_6H_5Br \xrightarrow{n\text{-}BuLi} C_6H_5Li + BuBr$

The lithium compounds, since they are readily prepared and undergo the Grignard-type of reactions, are useful for preparing various phenyl derivatives containing groups such as amino, hydroxyl, carboxyl, etc., *e.g.*,

Mercuration. Aromatic compounds containing almost all the common functional groups have been mercurated, and in general mercuration proceeds easily with the formation of mono-, di- and even polymercurated compounds.

Aromatic mercuration is carried out in two ways:

(i) **Direct mercuration.** This is the reaction in which a hydrogen atom of benzene, a polynuclear hydrocarbon, or a heterocyclic compound, is replaced by the *mercuri-acid group*, the most common one being the *acetoxymercuri-group*. This is usually introduced by heating the hydrocarbon with mercuric acetate, or with the equivalent of mercuric oxide in glacial acetic acid, at 90–160°C, for one or more hours, *e.g.*, *acetoxymercuribenzene* (*phenylmercury acetate*):

$$C_6H_6 + Hg(OCOCH_3)_2 \longrightarrow C_6H_5HgOCOCH_3 + CH_3CO_2H$$

Phenylmercury nitrate is formed by heating benzene with mercury nitrate in the presence of mercuric oxide. The latter compound is necessary to remove the nitric acid formed, since the reaction is reversible.

$$C_6H_6 + Hg(NO_3)_2 \xrightarrow{HgO} C_6H_5HgNO_3 + HNO_3$$

Mercuration of nitrobenzene with mercuric acetate in perchloric acid gives the expected orientation of the products, *i.e.*, predominantly *m*- (89%). In this case, the attacking species is an electrophile:

$$Hg(OOCCH_3)_2 + HClO_4 \longrightarrow CH_3CO_2H + CH_3CO_2Hg^+ClO_4^-$$

When the mercuration is carried out with mercuric acetate at 150°C, then the product is a mixture of *o*- and *p*- (57%) and *m*- (43%) acetoxymercuri-nitrobenzenes. In this case, the attacking species is the acetoxymercuri free radical.

$$(CH_3CO_2)_2Hg \rightleftharpoons CH_3CO_2Hg\cdot + CH_3COO\cdot$$

A characteristic of aromatic *homolytic* substitution is *o-p*-orientation, *irrespective* of the nature of the substituent present. This may be explained in terms of stabilisation of the pentadienyl free radicals formed $(Y\cdot = CH_3COOHg\cdot)$:

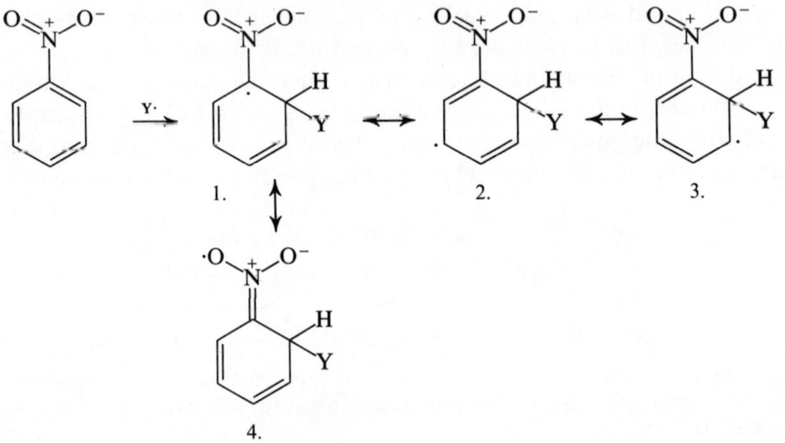

Similar structures may be drawn for the *p*-free radical, but not for the *m*-, for which the structure corresponding to 4. is *not* possible. Hence, since there are more resonating structures contributing to the *o*- and *p*-free radicals than to the *m*-, the former are more stable than the latter,

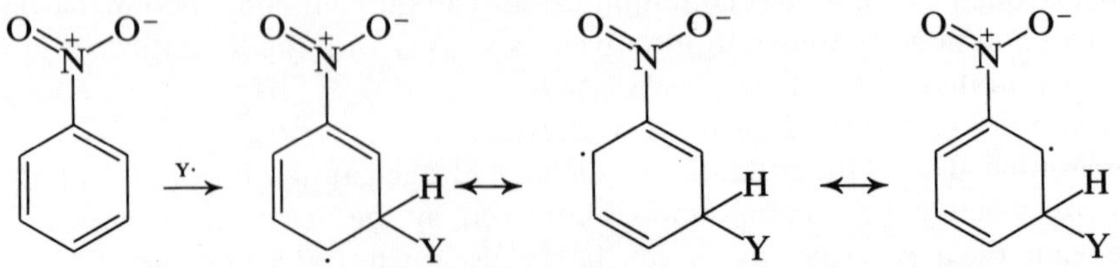

and consequently *o-p*-substitution is favoured. When the substituent group is *o-p*-orienting, *e.g.*, NH$_2$, OH, it can also be shown that the *o-p*-free radicals are more stable than the *m*-free radical.

The position of the acetoxymercuri-group can be found by treating the mercurated compound with halogen, whereupon the group is replaced by a halogen atom, probably by electrophilic attack on the nucleus:

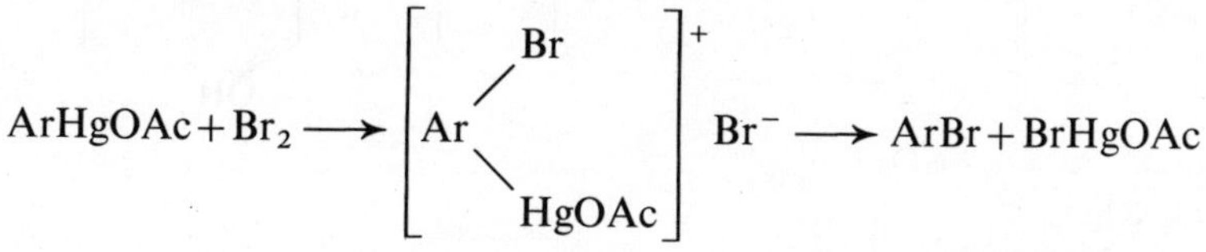

$$\text{ArHgOAc} + \text{Br}_2 \longrightarrow \left[\text{Ar} \underset{\text{HgOAc}}{\overset{\text{Br}}{\big<}} \right]^+ \text{Br}^- \longrightarrow \text{ArBr} + \text{BrHgOAc}$$

This reaction is the most characteristic reaction of mercurated compounds. On the other hand, if a mercurated compound is treated with sodium halide, the *acetyl group* is replaced by a halogen atom to form a *halogenomercuri*-compound (*phenylmercury halide*), probably by nucleophilic attack at the mercury atom:

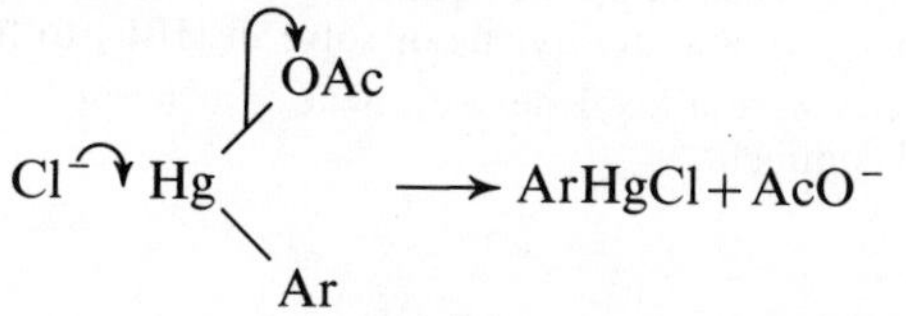

$$\text{Cl} \quad \text{Hg} \underset{\text{Ar}}{\overset{\text{OAc}}{\big<}} \quad \longrightarrow \quad \text{ArHgCl} + \text{AcO}^-$$

(ii) **Indirect mercuration.** This is the reaction in which a functional group is replaced by a mercuri-acid group, usually the *chloromercuri-group*. Mostly used for this purpose are the diazonium salts, sulphinic acids and aryl halides (see text).

The advantage of indirect over direct mercuration is that the former may be used to prepare a particular isomer, whereas the latter usually produces a mixture of isomers.

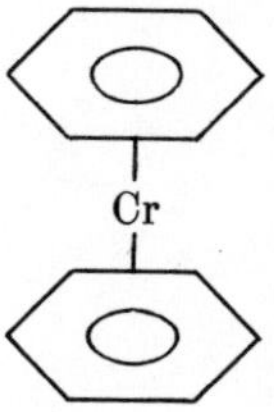

Benzene (and other aromatic compounds) form metal complexes which are similar to the ferrocenes (p. 545), *e.g.*, bis-benzenechromium (I) cation is readily prepared:

$$6\text{C}_6\text{H}_6 + 3\text{CrCl}_3 + \text{Al} + 2\text{AlCl}_3 \xrightarrow{\text{heat}} 3(\text{C}_6\text{H}_6)_2\text{Cr}^+\text{AlCl}_4^-$$

This cation, on reduction with sodium dithionite ($\text{Na}_2\text{S}_2\text{O}_4$), gives bis-benzenechromium (O). Many other transition metals behave in the same way, *e.g.*, Mo, V.

PROBLEMS

1. Write out the structures of the benzene isomers with the formula C_9H_{12}.
2. Assign orientations to:
(i) the three chlorotoluenes with dipole moments 1·3, 1·9, and 1·78D; (ii) the three chloronitrobenzenes with dipole moments 3·4, 2·5, and 4·3D.
3. Suggest oxidising agents for the conversions:

(i)

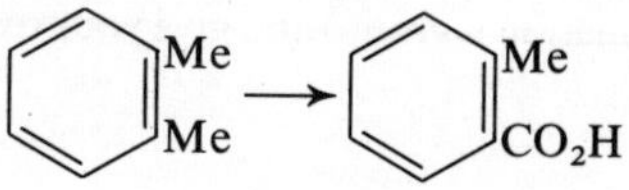

(ii)

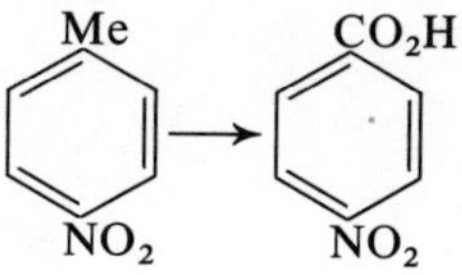

;   (iii)

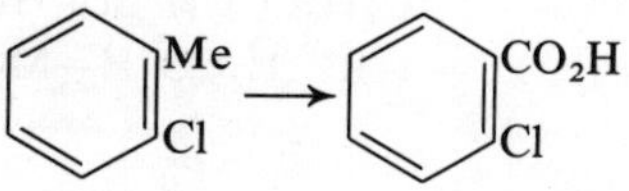

4. Indicate the positions of monosubstitution and, where possible, the predominant product:

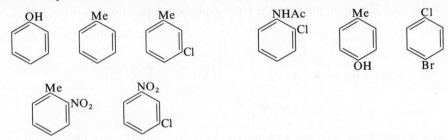

5. Suggest a reason for the catalytic hydrogenation of benzene giving cyclohexane and the inability to isolate intermediates.

6. What are the signs of the I and R effects for the following groups attached to the benzene ring: NO_2, NH_2, Br, COMe, NMe_3^+, SO_3H, OMe? When in opposition, which effect is greater?

7. Benzene, toluene, xylenes (o, m, p), and mesitylene dissolve in HBF_4 to form salts. Explain the order of basicity: mesitylene > m-xylene > o- and p-xylene > toluene > benzene.

8. Complete the equations and comment:

(i) $PhBr + Ac_2O \xrightarrow{AlCl_3} ?$

(ii) $PhCH_2CN \xrightarrow{HNO_3} ?$

(iii)

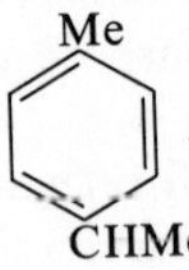

Me
⬡ CHMe₂ $\xrightarrow{M.A.} ?$

(iv)

Me
⬡ CHMe₂ $+ AcCl \xrightarrow{AlCl_3} ?$

(v) $Ph(CH_2)_3Me \xrightarrow{KMnO_4} ?$

(vi) $PhNO_2 + EtCl \xrightarrow{AlCl_3} ?$

(vii) $PhH + Me_3CCH_2Cl \xrightarrow{AlCl_3} ?$

9. Suggest a method for each of the following conversions:
(i) PhEt ⟶ PhH; (ii) PhH ⟶ p-Et$_2$C$_6$H$_4$; (iii) PhH ⟶ p-MeC$_6$H$_4$Pr; (iv) PhH ⟶ PhCH$_2$CO$_2$H;
(v) PhH ⟶ PhLi; (vi) 2-OH-3-Me-C$_6$H$_3$CO$_2$H ⟶ 2-OH-1,3-C$_6$H$_3$(CO$_2$H)$_2$.

10. Complete the equation and give your reasons:

$$PhCH{=}CHMe + HI \longrightarrow ?$$

11. Predict the sign of the Hammett σ constant for the following substituents and give your reasons: m-NO$_2$, p-MeO, p-Me$_2$S$^+$, m-Et, p-CF$_3$, m-OH, p-OH, m-NMe$_3^+$, p-CO$_2$H, m-CN, p-NMe$_2$, p-SO$_2$NH$_2$, m-IO$_2$.

12. Predict the sign of ρ for the following reactions (R is m or p to the functional group) and give your reasons:

(i) $RC_6H_4CO_2Me + OH^- \longrightarrow RC_6H_4CO_2^- + MeOH$

(ii) $ArH + NO_2^+ \longrightarrow ArNO_2 + H^+$

(iii) $R_2C_6H_3NH_2 + MeI \longrightarrow R_2C_6H_3NH_2Me^+I^-$

(iv) $RC_6H_4COMe + Br_2 \longrightarrow RC_6H_4COCH_2Br + HBr$

(v) $RC_6H_4CO_2Me + H_3O^+ \longrightarrow RC_6H_4CO_2H + MeOH$

(vi) $Ar_3CCl \rightleftharpoons Ar_3C^+ + Cl^-$

(vii) $ArH + Br_2 \longrightarrow ArBr + HBr$

(viii) $ArCOCl + EtOH \longrightarrow ArCO_2Et + HCl$

13. The acid-catalysed hydrolysis of ethyl benzoates has a ρ value of $+0.144$. What will be the effect of $-I$ groups on the rate? Explain.

14. Methyldiaryl chlorides (ArCHClAr) undergo solvolysis in EtOH by the S_N1 mechanism. This reaction has $\rho = -5.090$. What would be the effects of electron-withdrawing and releasing groups on the rate of reaction?

15. Given the following ionisation constants at 25°C (in H_2O): $PhCO_2H$, 6.3×10^{-5}; m-$CH_3COC_6H_4CO_2H$, 1.48×10^{-4}; p-$CH_3COC_6H_4CO_2H$, 2.0×10^{-4}, calculate σ_m and σ_p of the CH_3CO group.

16. The partial rate factor for the p-position for electrophilic substitution of fluorobenzene is greater than 1. Discuss.

17. Discuss the statement: In a monosubstituted benzene, the m/p ratio indicates the polar influences of the substituent at these two positions, whereas the o/m and o/p ratios indicate both polar and steric effects.

18. Discuss the reaction:

$$PhCH_2CH{=}CH_2 \xrightarrow[\text{heat}]{\text{KOH}} PhCH{=}CHMe$$

19. Which of the following ring compounds obey Hückel's rule: $C_{10}H_{10}$, $C_{12}H_{12}$, $C_{12}H_{12}{}^{2-}$, $C_{12}H_{12}{}^{2+}$, $C_{20}H_{20}$, $C_{20}H_{20}{}^{-}$, $C_{20}H_{20}{}^{2-}$, $C_{20}H_{20}{}^{+}$, $C_{20}H_{20}{}^{2+}$?

20. Arrange the following in order of increasing rate of nitration and give your reasons: PhH, PhMe, C_6D_6, $PhNO_2$, PhCl.

21. How many N.M.R. signals and their multiplicity (assume the benzene ring is a singlet) would be expected from:

(i) $PhCMe_3$; (ii) $PhCH_2CHMe_2$; (iii) $PhCH_2CClMe_2$; (iv) $PhCH_2OMe$; (v) $PhCHBrMe$; (vi) $PhCHBrCH_2Br$?

22. The following fragment ions are derived from n- and s-butylbenzene. Which is which? Give your reasons.

m/e	51	65	77	91	92	105	134	135
A (%)	5	10	6	100	60	9	30	3
B (%)	7	3	10	14	1	100	18	2

REFERENCES

Organic Reactions, Wiley, (i) Vol. I (1942), Ch. 3, 'Chloromethylation of Aromatic Compounds': (iii) Vol. III (1946), Ch. 1, 'The Alkylation of Aromatic Compounds by the Friedel–Crafts Method'.

NORMAN and TAYLOR, *Electrophilic Substitution in Benzenoid Compounds*, Elsevier (1965).

WILLIAMS, *Homolytic Aromatic Substitution*, Pergamon Press (1960).

NEWMAN (ed.), *Steric Effects in Organic Chemistry*, Wiley (1956), Ch. 3, 'Steric Effects in Aromatic Substitution'.

Advances in Physical Organic Chemistry, Academic Press, Vol. 1 (1963), p. 35. 'A Quantitative Treatment of Directive Effects in Aromatic Substitution'.

Progress in Physical Organic Chemistry, Interscience, Vol. 2 (1964), p. 253. 'Electrophilic Aromatic Substitution Reactions'.

FERGUSON, *The Modern Structural Theory of Organic Chemistry*, Prentice-Hall (1963).

OLAH (ed.), *Friedel–Crafts and Related Reactions*, Interscience, Vol. 1 (1963). Vols. 2 and 3 (1964). Vol. 4 (1965).

GOULD (ed.) *Kekulé Centennial, Amer. Chem. Soc.* Publications (1966).

LLOYD, *Carbocyclic Non-Benzenoid Aromatic Compounds*, Elsevier (1966).

WINSTEIN, 'Nonclassical Ions and Homoaromaticity', *Quart. Rev.*, 1969, **23**, 141.

WELLS, *Linear Free Energy Relationships*, Academic Press (1968).

SHORTER, 'The Separation of Polar, Steric, and Resonance Effects in Organic Reactions by the Use of Linear Free Energy Relationships', *Quart. Rev.*, 1970, **24**, 433.

EVERDELL, 'Some Remarks on the Thermodynamics of the Xylenes', *J. Chem. Educ.*, 1967, **44**, 538.

21

Aromatic halogen compounds

There are three types of halogen compounds, the *addition* compounds, the *nuclear substitution* products, and the *side-chain* substitution products.

Addition compounds

When treated with chlorine or bromine in the presence of sunlight, benzene forms the benzene hexa-halides, $C_6H_6Cl_6$ and $C_6H_6Br_6$, respectively.

$$Cl_2 \xrightarrow{h\nu} 2Cl\cdot$$

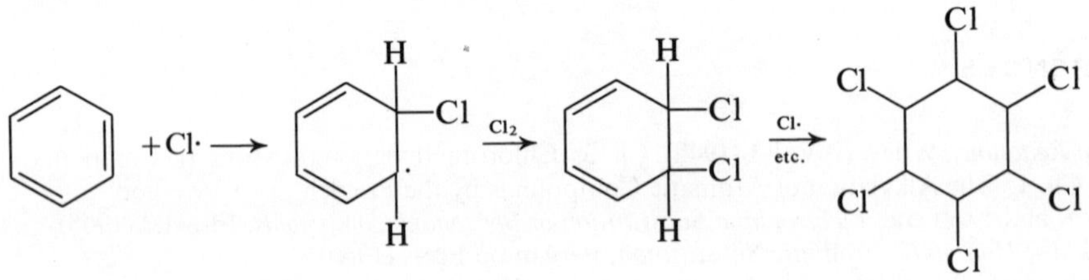

Benzene hexachloride, (1,2,3,4,5,6-hexachlorocyclohexane), theoretically can exist in eight stereo-isomeric forms (*cf.* inositol, p. 548), but only seven of these are known: α, β, γ, δ, ε, η and θ. The γ-isomer is a powerful insecticide; it is very stable and acts more quickly than D.D.T. All of the isomers have been shown to exist in the chair form, and the α-isomer has been identified as the ($\pm$)-form (Cristol, 1949). The following conformations have been assigned respectively to the α-, β- and γ-isomers: *aaeeee*, *eeeeee* and *aaaeee*.

Mullins (1955) has suggested that molecular size and shape are critical for the action of chlorinated hydrocarbon insecticides. According to Mullins, the molecule of γ-BHC is smaller than those of the other stereoisomers and so can penetrate more readily.

Nuclear substitution products (aryl halides)

These compounds may be prepared by the following methods:

1. Direct halogenation. *Low temperature and the presence of a halogen carrier favour nuclear substitution.* Chlorination and bromination may be very conveniently carried out at ordinary temperature in the presence of a Lewis acid, *e.g.*, the chlorides or bromides of Al, Fe, Sb. Iron is most commonly used, being converted to the Lewis acid ($FeCl_3$) by the halogen. The extent of the substitution depends on the amount of halogen used, *e.g.*, chlorobenzene is formed when benzene is treated with chlorine (1 molecule) in the presence of iron:

$$C_6H_6 + Cl_2 \xrightarrow{Fe} C_6H_5Cl + HCl \quad (90\%)$$

If 2 molecules of chlorine are used, then a mixture of *o*- and *p*-dichlorobenzenes is obtained, the latter predominating:

$$C_6H_5Cl + Cl_2 \xrightarrow{Fe} C_6H_4Cl_2 + HCl$$
$$o\text{-} + p\text{-}$$

When toluene is brominated in the presence of iron (using 1 molecule of bromine), a mixture of *o*- and *p*-bromotoluenes (*tolyl bromides*) is obtained:

$$C_6H_5CH_3 + Br_2 \xrightarrow{Fe} CH_3C_6H_4Br + HBr$$
$$o\text{-} + p\text{-}$$

To obtain the *o*-, *m*- and *p*-toluene derivatives pure, it is best to prepare them from the corresponding toluidines (*cf.* method 3).

The rate law for catalytic halogenation of benzene is often of the form (L.A. = Lewis acid):

$$\text{rate} = k[\text{PhH}][X_2][\text{L.A.}]$$

Hence, the mechanism of halogenation in the presence of a Lewis acid is probably similar to that of the Friedel–Crafts alkylation, *i.e.*, the active species is either the Lewis acid complex or positive halogen; *e.g.*,

$$Fe \xrightarrow{Cl_2} FeCl_3 \xrightarrow{Cl_2} Cl-\overset{+}{Cl}-\overset{-}{FeCl_3} \rightleftharpoons Cl^+ + FeCl_4^-$$

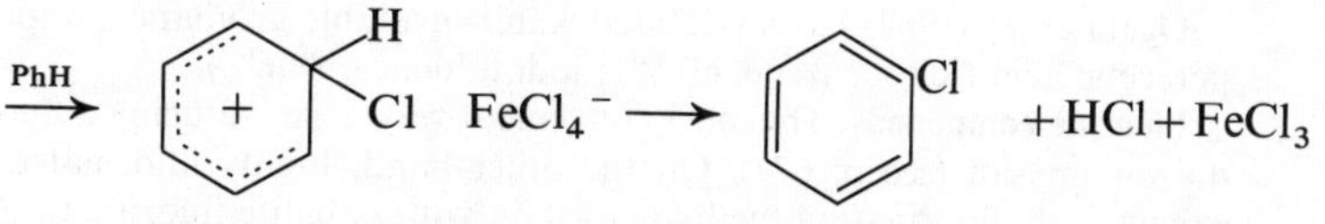

It also appears likely that both mechanisms can operate simultaneously, and that the actual path followed will depend on circumstances, *e.g.*, a strong Lewis acid will encourage the positive halogen mechanism. A similar dual mechanism can also operate when halogenation is carried out with hypochlorous or hypobromous acid. This reaction is acid-catalysed, and the active species could be H_2O^+—X or positive halogen (this is formed as follows: H_2O^+—X $\rightleftharpoons H_2O + X^+$).

2. Aromatic hydroxy-compounds cannot be converted into aryl halides by means of HX, PX_3, and $SOCl_2$ (which are commonly used with alcohols), but PCl_5 gives a very poor yield of chloro-compound, the main product being a triaryl phosphate.

Rydon *et al.* (1957), however, have shown that aryl chlorides may be prepared from phenols as follows:

$$3Ar^1OH + PCl_5 \xrightarrow{heat} 3HCl + (Ar^1O)_3PCl_2 \xrightarrow[heat]{Ar^2OH}$$

$$HCl + (Ar^1O)_3(Ar^2O)PCl \xrightarrow{heat} Ar^2Cl + (Ar^1O)_3PO$$

In these reactions Ar^2 contains a more powerful electron-attracting group than does Ar^1, *e.g.*, when $Ar^1 = C_6H_5$— and $Ar^2 = p\text{-}NO_2C_6H_4$—, the product is $p\text{-}NO_2C_6H_4Cl$ (91%).

Bromides and iodides may also be prepared as follows:

$$(Ar^1O)_3(Ar^2O)PCl + RX \xrightarrow{\text{heat}} RCl + (Ar^1O)_3(Ar^2O)PX \xrightarrow{\text{heat}} Ar^2X + (Ar^1O)_3PO$$

3. The decomposition of diazonium salts under suitable conditions is generally the most satisfactory method of preparing nuclear derivatives (see p. 673).

4. The silver salts of many aromatic acids react with bromine to form aryl bromides (*cf.* p. 151), *e.g.*,

$$C_6H_5CO_2Ag + Br_2 \longrightarrow C_6H_5Br + CO_2 + AgBr \quad (53\%)$$

$$p\text{-}NO_2C_6H_4CO_2Ag + Br_2 \longrightarrow p\text{-}NO_2C_6H_4Br + CO_2 + AgBr \quad (79\%)$$

Iodine compounds. Direct iodination presents some difficulties since iodine is the least active of the halogens but if, however, the iodination is carried out in the presence of an oxidising agent, *e.g.*, nitric acid, mercuric oxide, etc. the yield of iodo-compound is usually very good:

$$2C_6H_6 + I_2 \xrightarrow[\text{HNO}_3]{[\text{O}]} 2C_6H_5I + H_2O \quad (86\text{--}87\%)$$

The nature of the active species in these iodinations is still a matter of debate. According to Butler *et al.* (1971), in the presence of nitric acid the iodinating species appears to be HNO_2I^+.

Iodine compounds are generally most conveniently prepared via the diazonium salts (method 3), but if an activating group (hydroxyl or amino) is present in the ring, then iodine monochloride, or even iodine without the presence of an oxidising agent, may be used (see aniline, p. 658).

Ogata *et al.* (1964) have iodinated water-insoluble aromatic compounds with iodine in acetic acid–peracetic acid (iodobenzene, 60·5%; iodotoluenes, 84·6%).

Fluorine compounds. The most convenient way of introducing a fluorine atom into the ring is via the diazonium salt (see p. 674). On the other hand, highly fluorinated aromatics may be prepared by exhaustively fluorinating the hydrocarbon with cobaltic fluoride to give partially fluorinated alicyclic compounds. These are then converted into fluorinated aromatics by removal of hydrogen fluoride by means of, *e.g.*, potassium hydroxide, or by passing over heated metals, *e.g.*,

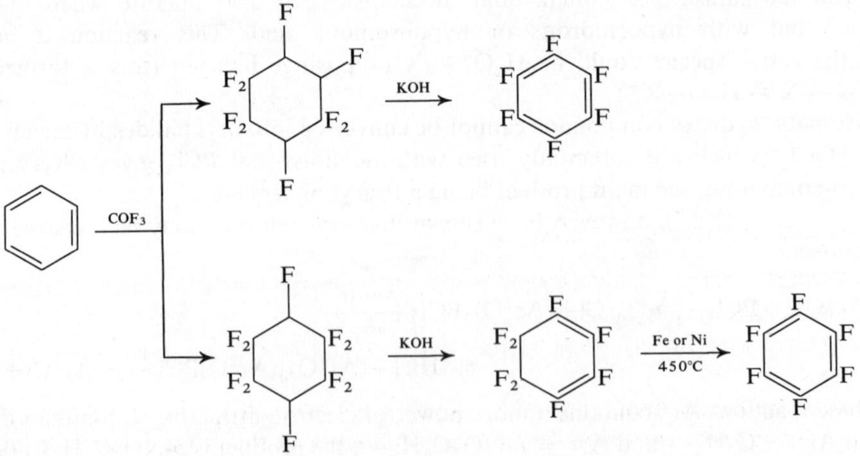

General properties of the nuclear halogen derivatives

Nuclear halogen derivatives are colourless oils or crystalline solids, and their boiling points and densities are greater than those of the parent arenes. Aryl halides form organo-aromatics (p. 617) and undergo hydrogenolysis with, *e.g.*, $Pd/C/H_2$ or Mg and isopropanol (Bryce-Smith *et al.*, 1963):

$$C_6H_5X + Mg + Me_2CHOH \longrightarrow C_6H_6 + Me_2COMgX \quad (89–96\%)$$

Aryl bromides and aryl iodides form Grignard reagents, and undergo the Ullmann reaction (p. 664). The aryl chlorides, however, form Grignard reagents in tetrahydrofuran or by the entrainment method (p. 424). The aryl halides do not react with sodioacetoacetic ester or with sodiomalonic ester, but they undergo the Wurtz–Fittig (p. 612) and Fittig reactions (p. 781). The aryl halides are difficult to mercurate, but they may be mercurated by refluxing them with mercuric acetate until a test portion gives no precipitate of mercuric oxide with aqueous sodium hydroxide. On the other hand, chloromercuri-derivatives may be prepared from aryl bromides or iodides as follows:

$$ArBr + Mg \longrightarrow ArMgBr \xrightarrow{HgCl_2} ArHgCl + MgClBr$$

Chlorobenzene (*phenyl chloride*), b.p. 132°C, is produced commercially by the Raschig process: a mixture of benzene vapour, air and hydrogen chloride is passed over a catalyst (copper chloride):

$$C_6H_6 + HCl + \tfrac{1}{2}O_2 \longrightarrow C_6H_5Cl + H_2O$$

It is used for the manufacture of aniline, phenol, and D.D.T. D.D.T. is a contraction for *p,p'*-dichlorodiphenyltrichloroethane[1,1,1-trichloro-2,2-bis(*p*-chlorophenyl)ethane]. It is a solid, m.p. 109–110°C, and is manufactured by heating chlorobenzene and chloral with concentrated sulphuric acid:

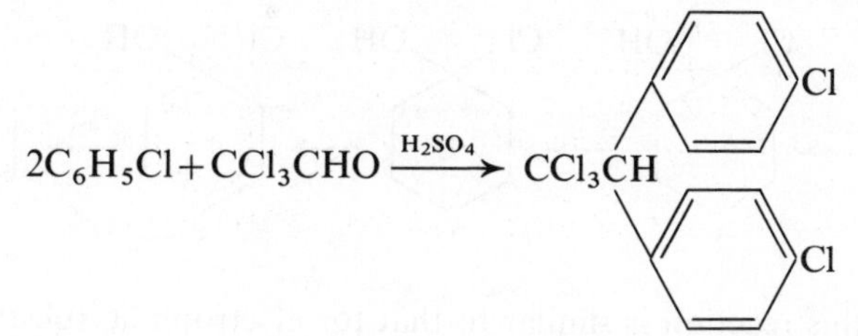

$$2C_6H_5Cl + CCl_3CHO \xrightarrow{H_2SO_4}$$

Good commercial D.D.T. contains about 75 per cent *p,p'*-D.D.T., 20 per cent *o,p'*-D.D.T. (m.p. 74–75°C), and 5 per cent of other substances as impurities; it is a powerful insecticide.

Nucleophilic aromatic substitution

This term covers the replacement of hydrogen or a substituent by a nucleophilic reagent, and also includes activated nucleophilic aromatic substitution (see below). In general, there are three types of mechanism: unimolecular (see p. 671), bimolecular, and the elimination-addition mechanism.

Bimolecular mechanism. Aryl halides usually undergo nucleophilic substitution with difficulty; vigorous conditions are generally required (thereby resembling vinyl halides), *e.g.*,

$$C_6H_5Cl \xrightarrow[\begin{array}{c}\text{aq. NaOH; 300°C}\end{array}]{} C_6H_5O^-Na^+$$

$$\xrightarrow[\begin{array}{c}\text{aq. NH}_3\text{; 200°C; Cu}_2\text{O}\end{array}]{} C_6H_5NH_2$$

$$\xrightarrow[\begin{array}{c}\text{CuCN in DMF; heat}\end{array}]{} C_6H_5CN$$

On the other hand, when a powerful electron-withdrawing group is present *o*- and/or *p*- to the halogen atom, replacement of the latter by nucleophilic reagents is now facilitated. This is *activated* nucleophilic substitution, and also includes the replacement of various types of groups such as NO_2, OMe, H (see also p. 644), *e.g.*,

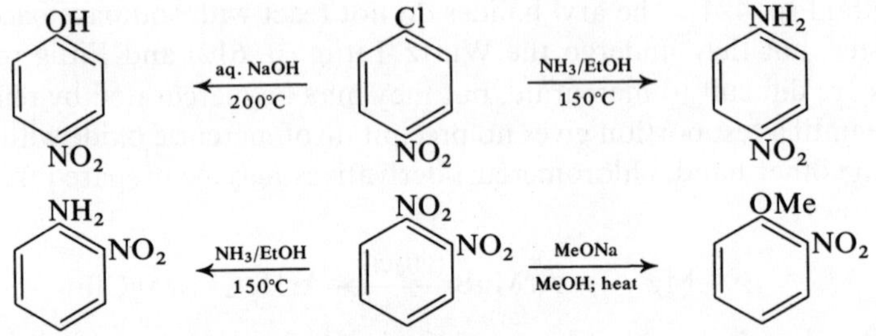

It is also interesting to note that in nucleophilic reactions involving activated halogen, the reactivity of the halogen atom is in the order F ≫ Cl > Br.

Evidence for the bimolecular mechanism comes from the fact that where kinetic studies have been made, the reaction is second-order (first order in both substrate and reagent). It is generally accepted that the reaction proceeds via an intermediate σ-complex, the *benzenonium carbanion* (or the *pentadienyl anion*), *e.g.*,

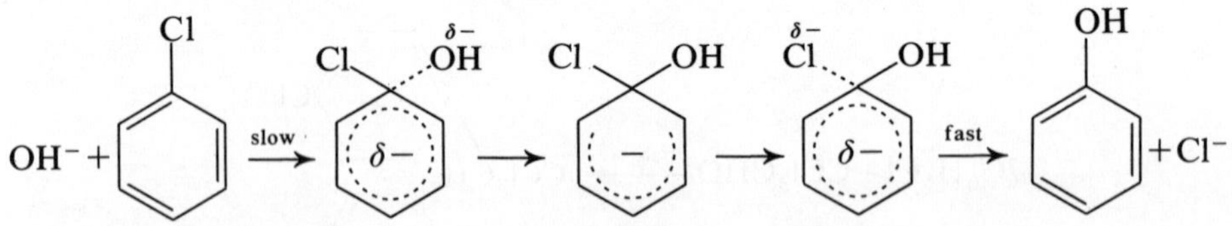

The energy profile of this reaction is similar to that for electrophilic substitution (Fig. 20.13). 2. is a resonance hybrid of three canonical forms, and it should be remembered that small negative charges are in the *o*- and *p*-positions (in 2.).

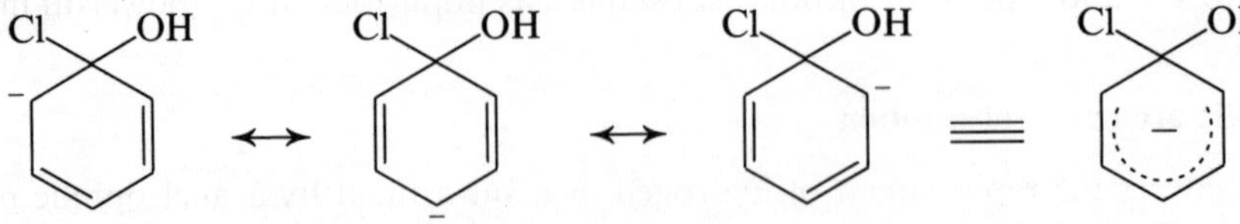

When we examine activated nucleophilic substitution then, as with electrophilic substitution, we can consider the problem from the point of view of charge distribution or the stabilities of the carbanions. In the former approach, we can draw canonical forms in which, for example, the C

of C—Cl in *o*-chloronitrobenzene carries a small positive charge:

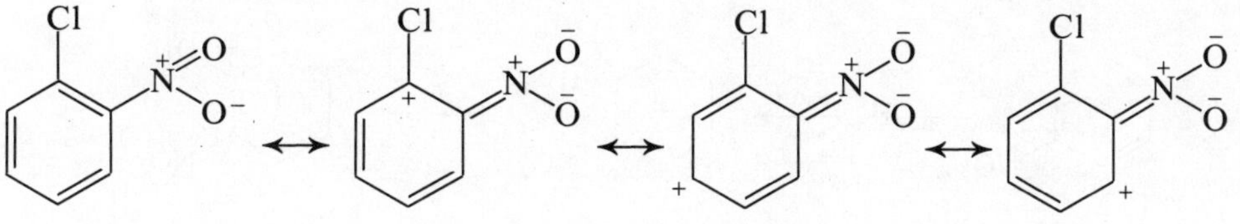

Because of the positive charge, attack at this carbon atom by the nucleophile is facilitated.

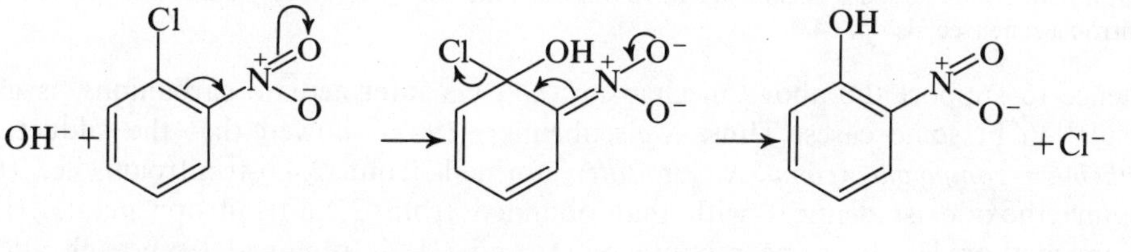

Now let us consider the stabilities of the various carbanions:

o

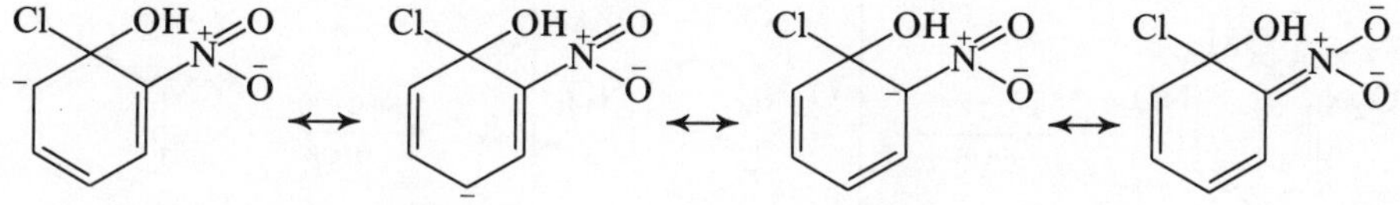

p

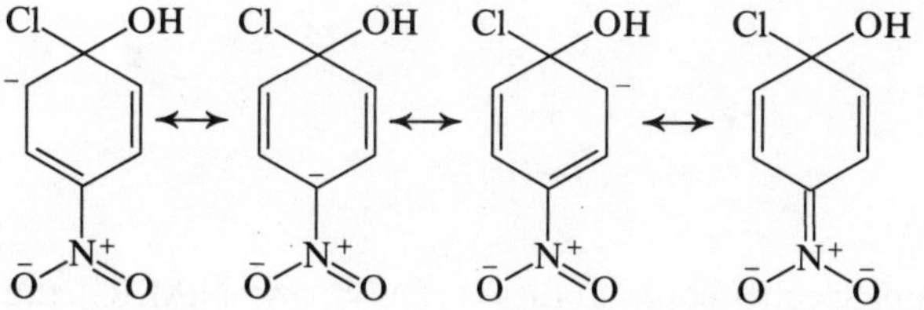

m

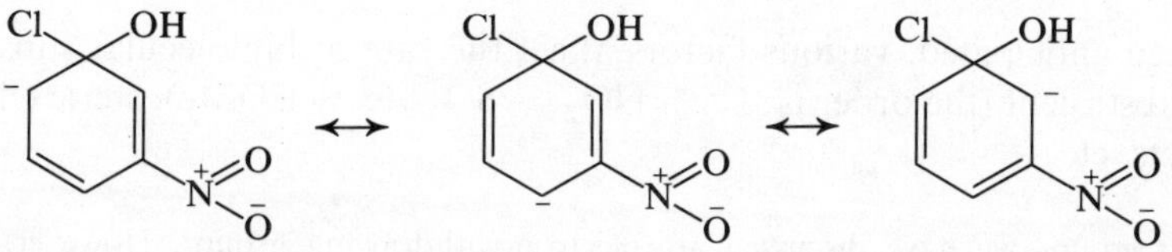

Inspection of these canonical structures shows that in both the *o*- and *p*-carbanions there are contributions from structures involving extended conjugation, resulting in spreading of the negative charge which resides on the most electronegative atom. These are the most stable contributing structures in *o*- and *p*-carbanions, and since no similar structure can be drawn for the *m*-carbanion, the latter is less stable than the former. Hence, an electron-withdrawing group activates *o*- and *p*-substitution, but not *m*-.

Substitution of *hydrogen* by a nucleophile does not occur in benzene itself, but the presence of one nitro group is sufficient to activate the *o*- and *p*-positions, *e.g.*,

$$\text{C}_6\text{H}_5\text{NO}_2 + \text{OH}^- \xrightarrow[\text{slow}]{\text{NaOH}} \quad \xrightarrow[\text{fast}]{-\text{H}^-} \quad o\text{-}O_2N\text{C}_6\text{H}_4\text{OH}$$

Removal of the hydride ion is more difficult than removal of X^-, OMe^-, etc. and hence, for completion of the reaction, it is necessary to have an oxidising agent present (this may then possibly be a *direct* oxidation of the carbanion). Nitrobenzene itself is a mild oxidising agent, and so reaction proceeds in the absence of an added oxidising agent. In this case, there is a decreased yield of product due to loss of nitrobenzene (see also p. 643).

Evidence to support the above mechanism, *i.e.*, via intermediate carbanions, is given by their isolation in some cases. Thus, Meisenheimer (1902) showed that the adduct (III), a *Meisenheimer compound* (*complex*, or *salt*), formed from 2,4,6-trinitroanisole, (I), and potassium ethoxide is identical with that obtained from 2,4,6-trinitrophenetole, (II), and potassium methoxide; the same mixture of (I) and (II) is obtained from each adduct on acidification:

$$\text{(I)} \xrightarrow[\text{EtOH}]{\text{EtO}^-\text{K}^+} \text{(III)} \xleftarrow[\text{MeOK}]{\text{MeO}^-\text{K}^+} \text{(II)}$$

(I)   (III)   (II)

$$\text{(III)} \xrightarrow{\text{HCl}} \text{(I)} + \text{(II)}$$

Further support comes from spectroscopic studies (i.r., u.v., and N.M.R.) and also from X-ray analysis.

Many groups are capable of activating the *o*- and *p*-positions, *e.g.*, CN, COMe, SO_2Me, etc., but the nitro group and the diazo group ($-N_2^+$) are by far the most powerful activators (see also p. 673).

As might have been anticipated, various factors affect the rate of bimolecular substitution: polar effect of the substituent (the order is: N_2^+, $NO_2 > SO_2Me > COMe$), steric effects (see p. 644), solvent effects, etc.

The activating groups we have discussed are electron-withdrawing groups. However, electron-releasing groups may also behave as activating groups, but in this case it is the *m*-position that is activated. This may be illustrated by the following example. When resorcinol is fused with sodium hydroxide in air (which is the oxidising agent), hydrogen is replaced by hydroxyl to give phloroglucinol (see also p. 717). The substrate is the resorcinol-oxide ion, and sites of high electron-density are the *o*- and *p*-positions. Hence, attack by the nucleophile, OH^-, is facilitated at the *m*-position. Alternatively, the *m*-carbanion is more stable than the *o*- and *p*-carbanions since, in the latter, contributing canonical structures carry negative charges on the C attached to O^-, thereby destabilising these ions (like charges are in opposition), *e.g.*,

Charge distribution

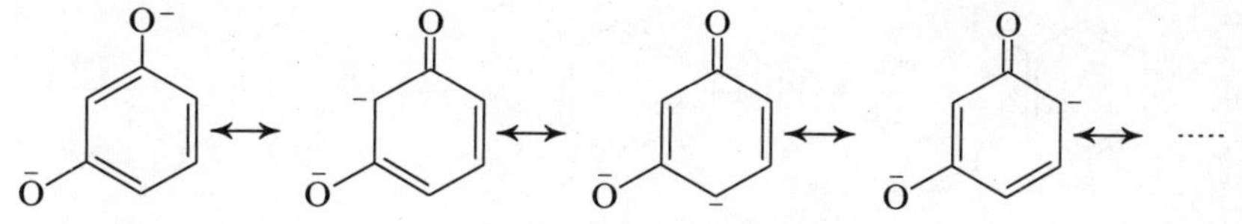

o-**carbanion**

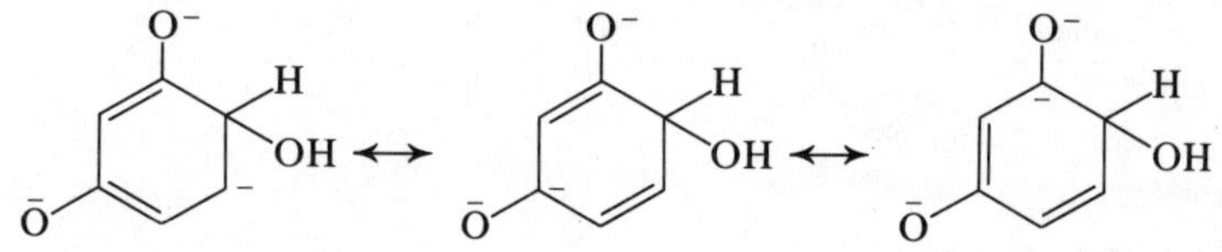

m-**carbanion**

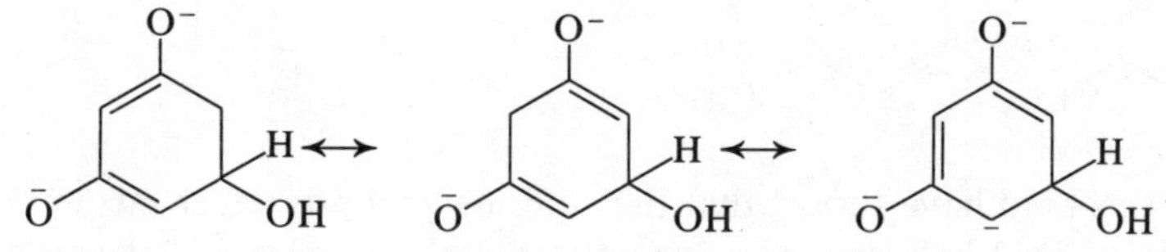

Elimination-addition mechanism. In a number of cases of nucleophilic aromatic substitution, the entering group does not occupy the position vacated by the expelled group. Reactions of this type have been called **cine-substitutions** (Bunnett *et al.*, 1951). Most of these rearrangement reactions involve the formation of a **benzyne** (1,2-dehydrobenzene) intermediate, and the mechanism is referred to as the **elimination-addition mechanism.** The majority of reactions of this type start with aryl halides as substrates and a strong base or an organolithium compound. Various mechanisms have been proposed for the formation of benzyne and for the subsequent steps leading to the products. It appears to be generally accepted that (i) benzyne is formed by a *stepwise* elimination, and (ii) benzyne undergoes *stepwise* addition to form the products. In the following account, only these mechanisms will be given.

When chlorobenzene labelled with ^{14}C at the carbon of the C—Cl group is treated with potassium in liquid ammonia, the amino-group enters partly at the labelled carbon atom and partly at the *ortho*-carbon atom.

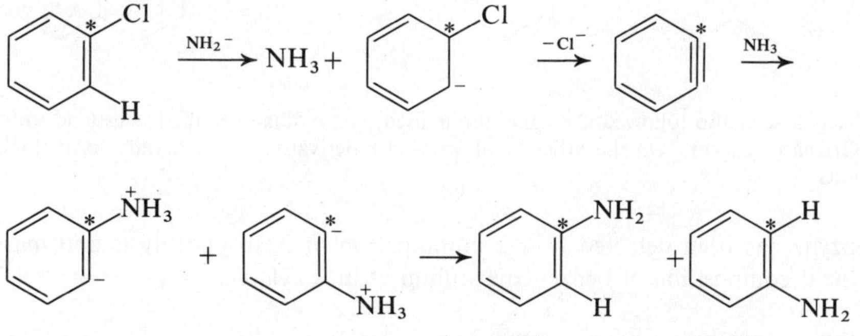

(47%)   (53%)

There is also evidence that the amide ion, NH_2^-, acts as an aminating agent as well as ammonia.

In the same way, when labelled fluorobenzene is treated with phenyl-lithium in ether solution, followed by acidification at the end of the reaction, the phenyl group enters partly at the labelled carbon atom and partly at the *ortho*-carbon atom.

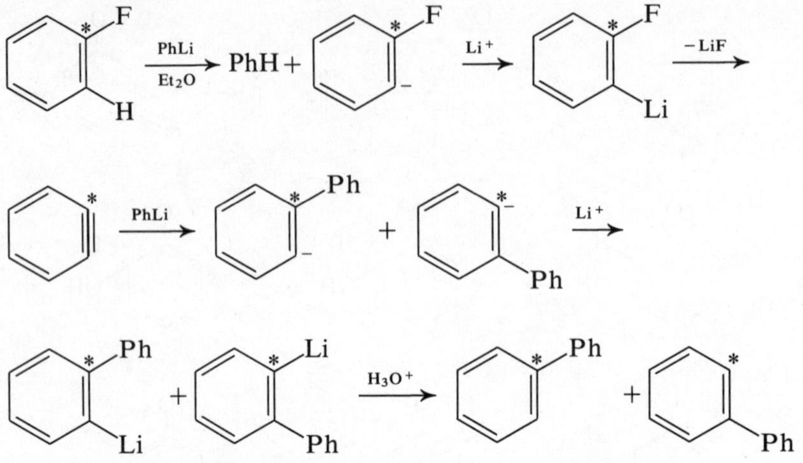

Experiments of this kind have shown that the entering group is never farther away than the *o*-position from the ejected halogen atom, and this led to the proposal that a benzyne intermediate is formed. This is amply supported by other experimental work, *e.g.*,

(i) Bromomesitylene does not react with potassamide in liquid ammonia. This is to be expected on the basis of the above mechanism, since there are no *o*-hydrogen atoms. In this connection, it might be noted that *o*-deuterochlorobenzene reacts more slowly than the protium analogue. Thus, there is a kinetic isotope effect, and therefore the formation of benzyne is the rate-determining step.

(ii) Benzyne intermediates have been detected by 'trapping' experiments, *e.g.*, in the presence of furan, *o*-bromofluorobenzene reacts with lithium amalgam to form 1,4-epoxy-1,4-dihydronaphthalene. This is the Diels–Alder reaction, and is one of the most important and useful reactions of benzyne.

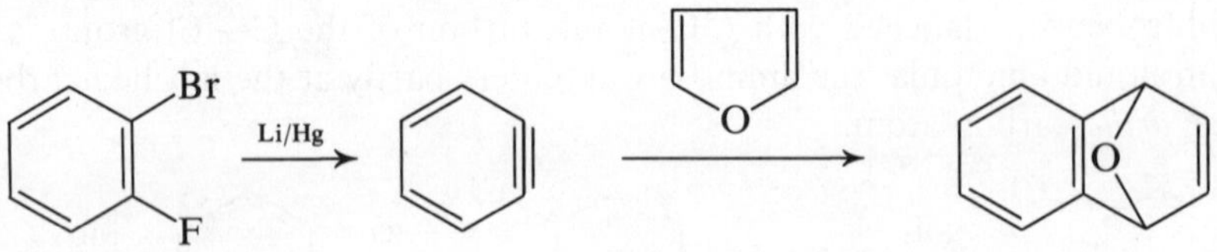

This course is also followed if magnesium is used, *i.e.*, *o*-dihalogenobenzenes tend to form arynes, *not* Grignard reagents. On the other hand, *m*- and *p*-derivatives do generally form stable Grignard reagents.

(iii) Benzyne has been detected by the combination of flash photolysis and mass spectrometry in the decomposition of benzenediazonium *o*-carboxylate:

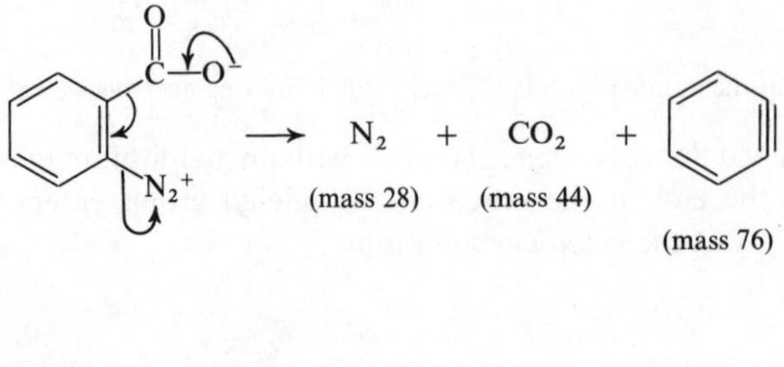

$$N_2 \quad + \quad CO_2 \quad +$$
(mass 28)   (mass 44)

(mass 76)

Further proof of the existence of benzyne has been given by Kettle *et al.* (1968), who have isolated it as a stable π-complex by heating *o*-di-iodobenzene in pentane with tetracarbonylnickel under pressure.

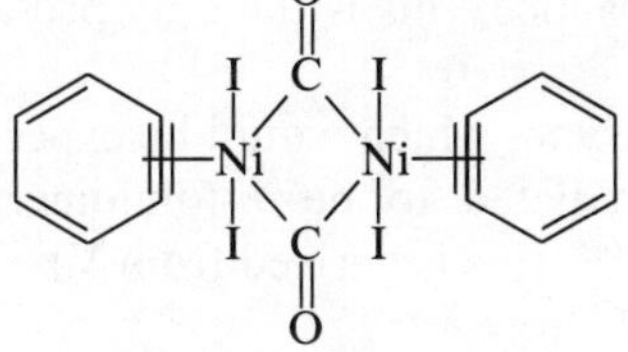

Ratio of isomeric products. If the benzyne intermediate is symmetrical, then attack by the nucleophile should occur equally well at either carbon atom of the 'triple' bond. This is reasonably so for labelled chlorobenzene (see above); the slightly less substitution at C* may possibly be due to a kinetic isotope effect.

When the aryne produced is not symmetrical, then two problems arise, the direction of elimination and the direction of addition.

Direction of elimination. Let us consider the reaction:

$$RC_6H_4X \xrightarrow[\text{liq. NH}_3]{\text{KNH}_2} RC_6H_4NH_2$$

The possible arynes are:

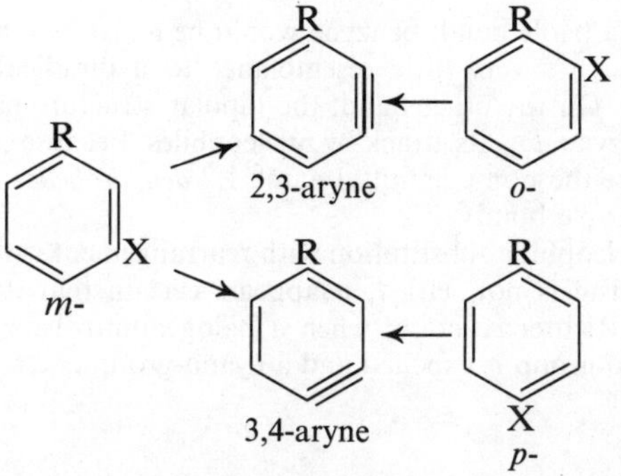

2,3-aryne · *o*-

3,4-aryne · *p*-

The *o*- and *p*-halides can each give only *one* aryne. The *m*-halide, however, can give *both*, and which predominates depends on the stability of the carbanion which precedes the aryne. This stability is dependent on the inductive effect of R. If R is electron-attracting ($-I$), then (I) is more stable than (II), since in the former the negative charge is closer to R. On the other hand, if R is $+I$, then (II) is more stable than (I), since in the former the negative charge is further from R.

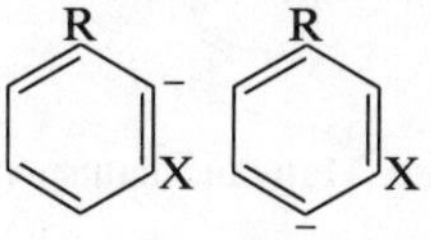

(I); $-I$ (II); $+I$

Direction of addition. Here also, the direction of addition depends largely on the I effect of R in the same way as before:

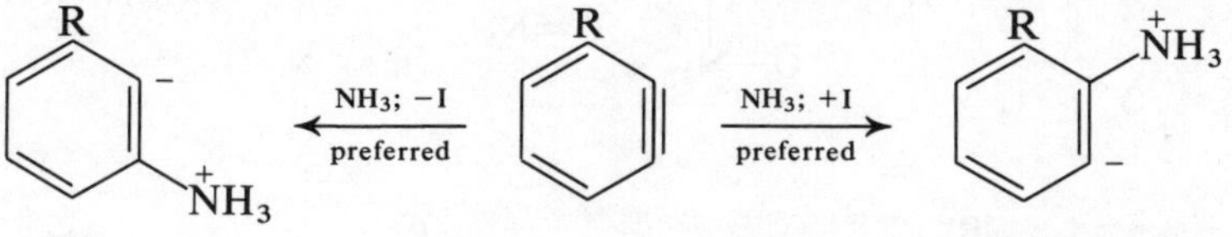

(III) · NH$_3$; $-I$ preferred ← → NH$_3$; $+I$ preferred · (IV)

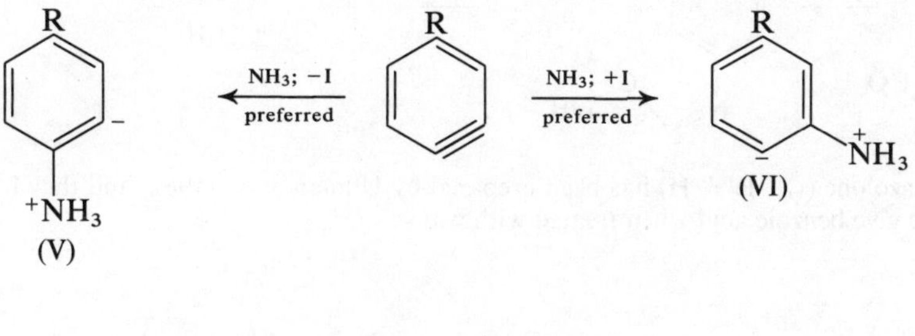

(V) · NH$_3$; $-I$ preferred ← → NH$_3$; $+I$ preferred · (VI)

Examples

(i) *m*-Bromoanisole (R = OMe; −I): this would be expected to give predominantly (III). Thus, the major product is predicted to be the *m*-amine. In practice, this is the only product (100%).

(ii) *p*-Chlorotoluene (R = Me; +I): this gives only the 3,4-aryne, which would be expected to give predominantly (VI). Thus, the major product is predicted to be *m*-toluidine. In practice, the products are *m*-toluidine (62%) and *p*-toluidine (38%; this is derived from V).

Structure of benzyne. This is still a matter of debate; three possibilities are:

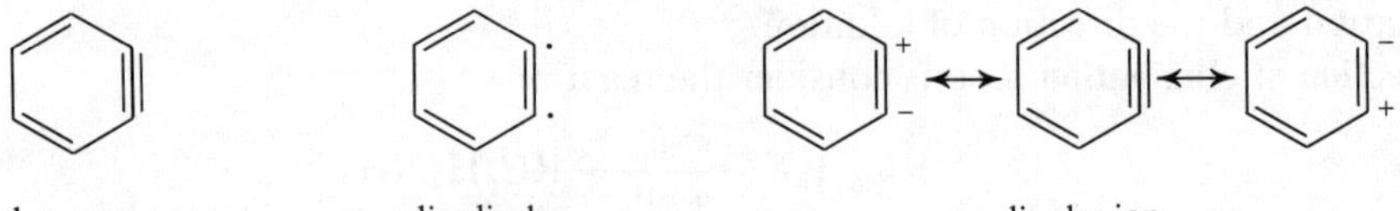

benzyne   diradical   dipolar ion

If the structure contained a triple bond, benzyne would be a *distorted* acetylene, and hence would be unstable. Since benzyne shows very little resemblance to a diradical in chemical behaviour, this structure is unsatisfactory. On the other hand, the dipolar structure has the merit of explaining the electrophilic nature of benzyne, *i.e.*, its attack by nucleophiles. Because of this uncertainty of structure, some workers prefer to use the non-committal name 1,2-*dehydrobenzene* rather than benzyne (which implies the presence of a triple bond).

Another example of nucleophilic substitution with rearrangement is the **von Richter reaction** (1871), but although the mechanism is not settled, it appears certain that it does *not* occur via an aryne intermediate. In the von Richter reaction, when a halogenonitrobenzene is heated with potassium cyanide at 150°C, the nitro-group is expelled and a cyano-group enters the ring *ortho* to the position occupied by the former, *e.g.*,

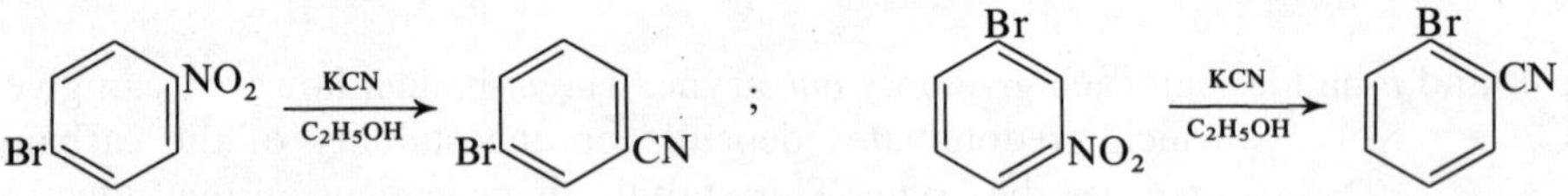

If the reaction is carried out in *aqueous* ethanol, the corresponding acid is obtained. Many mechanisms have been proposed; the following is that of Rosenblum (1960):

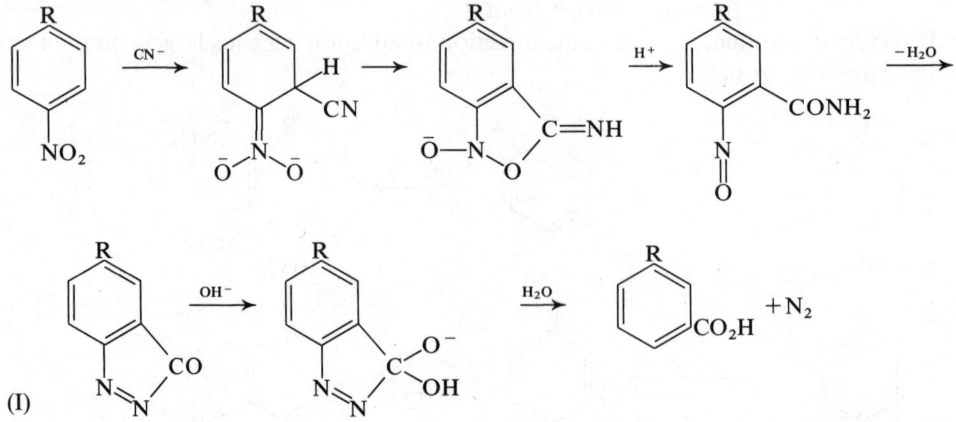

The unstable 3-indazolone ((I); R = H) has been prepared by Ullman *et al.* (1962) and they found it evolved nitrogen to give benzoic acid when treated with water.

An interesting application of the von Richter reaction is the preparation of 2-bromo-3-methyl-benzoic acid which, apparently, has not been prepared by any other means:

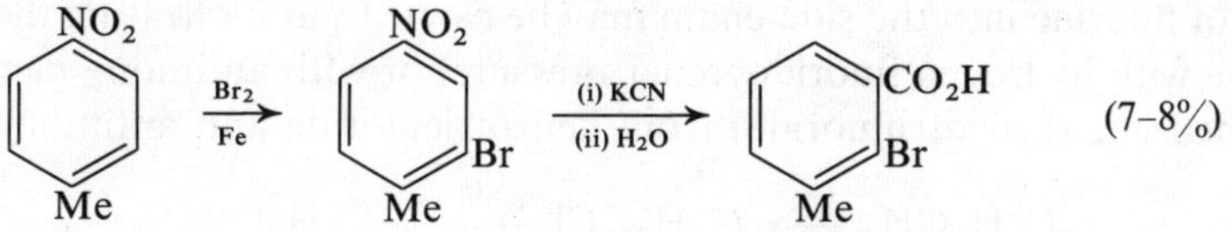

(7–8%)

Side-chain substituted compounds

Side-chain substitution is favoured by high temperature and light, and the *absence* of halogen carriers; *e.g.*, when chlorinated at its boiling point in the presence of light, toluene is converted into benzyl chloride.

$$Cl_2 \xrightarrow{h\nu} 2Cl\cdot$$
$$C_6H_5CH_3 + Cl\cdot \longrightarrow C_6H_5CH_2\cdot + HCl$$
$$C_6H_5CH_2\cdot + Cl_2 \longrightarrow C_6H_5CH_2Cl + Cl\cdot \; ; \text{ etc.}$$

When the side-chain is larger than a methyl group, its halogenation is more complicated, *e.g.*, chlorination of ethylbenzene at its boiling point in the presence of light first produces a mixture of α- and β-chloroethylbenzenes, and finally pentachloroethylbenzene:

$$C_6H_5CH_2CH_3 \xrightarrow{Cl_2} \underset{\text{α-isomer}}{C_6H_5CHClCH_3} + \underset{\text{β-isomer}}{C_6H_5CH_2CH_2Cl} \xrightarrow{Cl_2} C_6H_5CCl_2CCl_3$$

The monochloro-derivatives can be isolated (using 1 molecule of chlorine), but it is extremely difficult to isolate any of the other intermediates (when chlorine is used in excess). If, however, chlorination is carried out in the cold and in the presence of light, then the main product of substitution is the α-isomer (about 85%):

$$C_6H_5CH_2CH_3 \xrightarrow{Cl_2} C_6H_5CHClCH_3 \xrightarrow{Cl_2} C_6H_5CCl_2CH_3$$

The reason for this preferential substitution may be that the free radical produced at the α-carbon atom can enter into resonance with the benzene ring. This radical would be expected to be more stable than one at the β-carbon atom, since this atom is 'insulated' from the benzene ring and so resonance is not possible.

If two side-chains are present, both may be halogenated, *e.g.*,

$$CH_3C_6H_4CH_3 \xrightarrow{Cl_2} \underset{\text{methylbenzyl chloride}}{CH_3C_6H_4CH_2Cl} \xrightarrow{Cl_2} \underset{\text{xylylene chloride}}{ClCH_2C_6H_4CH_2Cl}$$

When a hydroxyl or an amino-group (activating groups) is present in the ring, side-chain halogenation is very difficult, if not impossible; *e.g.*, bromination of *p*-cresol gives 2,6-dibromo-*p*-cresol:

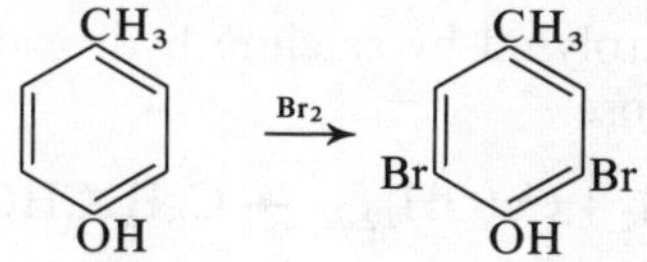

Side-chain chlorination may also be effected with sulphuryl chloride in the presence of a radical initiator, *e.g.*, peroxides or azobis-isobutyronitrile (see p. 151). t-Hypochlorites are also very useful reagents, *e.g.*, t-butyl hypochlorite readily generates free radicals on warming to 50°C.

$$t\text{-BuOCl} \longrightarrow t\text{-BuO}\cdot + Cl\cdot$$
$$C_6H_5CH_3 + t\text{-BuO}\cdot \longrightarrow C_6H_5CH_2\cdot + t\text{-BuOH}$$
$$C_6H_5CH_2\cdot + t\text{-BuOCl} \longrightarrow C_6H_5CH_2Cl + t\text{-BuO}\cdot \; ; \text{ etc.}$$

Side-chain bromination may be carried out with bromine in the presence of light, or with *N*-bromosuccinimide (p. 452).

Introduction of fluorine into the side-chain may be carried out by heating the corresponding chloro-derivative with hydrogen fluoride under pressure, or with antimony pentafluoride; *e.g.*, trifluoromethylbenzene (benzotrifluoride) from benzotrichloride and antimony pentafluoride:

$$C_6H_5CH_3 \xrightarrow{Cl_2} C_6H_5CCl_3 \xrightarrow{SbF_5} C_6H_5CF_3$$

The trifluoromethyl group is very resistant to attack, *e.g.*, oxidation of trifluoromethylbenzene results in the formation of trifluoroacetic acid:

$$C_6H_5CF_3 \xrightarrow{[O]} CF_3CO_2H$$

Benzyl chloride, b.p. 179°C, is prepared by passing chlorine into boiling toluene until the theoretical increase in weight for benzyl chloride is obtained:

$$C_6H_5CH_3 + Cl_2 \longrightarrow C_6H_5CH_2Cl + HCl \quad (60\%)$$

Benzyl chloride may also be prepared by the chloromethylation of benzene.

Benzyl chloride behaves like an alkyl halide, *e.g.*, when heated with aqueous alkali, aqueous potassium cyanide, or ethanolic ammonia, it is converted into benzyl alcohol, $C_6H_5CH_2OH$, benzyl cyanide, $C_6H_5CH_2CN$, and benzylamine, $C_6H_5CH_2NH_2$, respectively. Benzyl chloride is more reactive than alkyl chlorides in both S_N1 and S_N2 reactions. The increased S_N1 reactivity can be accounted for by the increased stability of the benzyl carbonium ion (over an R group) due to spreading of charge by resonance:

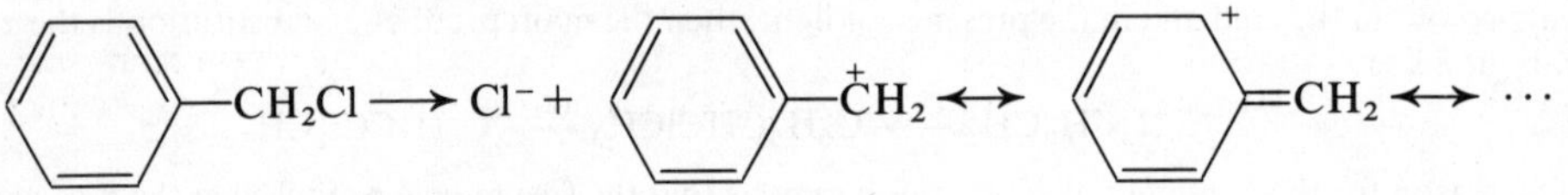

The reason for the increased S_N2 reactivity is uncertain.

Benzylidene chloride (*benzal chloride*), b.p. 207°C, may be prepared by passing chlorine into boiling toluene in the presence of light until the weight increase corresponds to $C_6H_5CHCl_2$:

$$C_6H_5CH_3 + 2Cl_2 \longrightarrow C_6H_5CHCl_2 + 2HCl \quad (g.)$$

It may also be prepared by the action of phosphorus pentachloride on benzaldehyde:

$$C_6H_5CHO + PCl_5 \longrightarrow C_6H_5CHCl_2 + POCl_3$$

Benzylidene chloride is hydrolysed by calcium hydroxide solution to benzaldehyde, and is used industrially for this purpose.

$$C_6H_5CHCl_2 + Ca(OH)_2 \longrightarrow C_6H_5CHO + CaCl_2 + H_2O$$

The conversion of toluene into benzaldehyde via benzylidene chloride is an example of *indirect* oxidation.

Benzotrichloride, b.p. 214°C, may be prepared by the continued action of chlorine on boiling toluene in the presence of light:

$$C_6H_5CH_3 + 3Cl_2 \longrightarrow C_6H_5CCl_3 + 3HCl$$

When heated with calcium hydroxide solution, it is converted into benzoic acid: it is used industrially for this purpose. This is another example of indirect oxidation:

$$C_6H_5CCl_3 + 2H_2O \xrightarrow{\text{Ca(OH)}_2} C_6H_5CO_2H + 3HCl$$

It can be seen from the foregoing discussion that the properties of the side-chain halogen derivatives are very much different from those of the nuclear halogen derivatives. The former closely resemble the alkyl halides but, in general, possess a pungent smell and are lachrymatory (provided the halogen atom is on the α-carbon); they have been used in warfare.

Mass spectra

For *nuclear* aromatic halides, the intensity of the molecular ion is usually strong. This is also the situation if methyl groups are present in the ring, but if an ethyl (or larger group) is present, then β-cleavage competes with the loss of the halogen atom (see also p. 589):

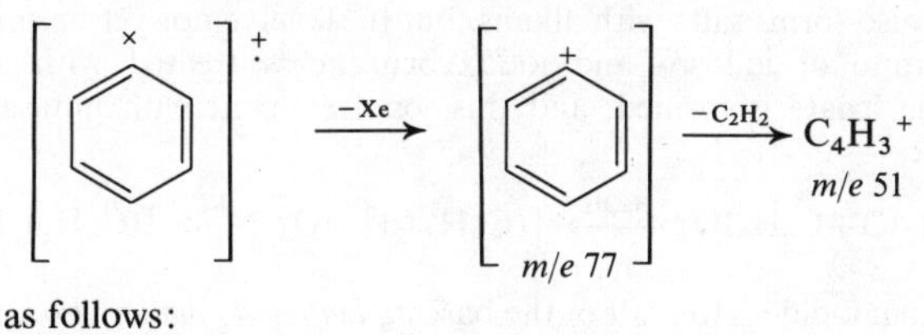

Benzyl halides behave as follows:

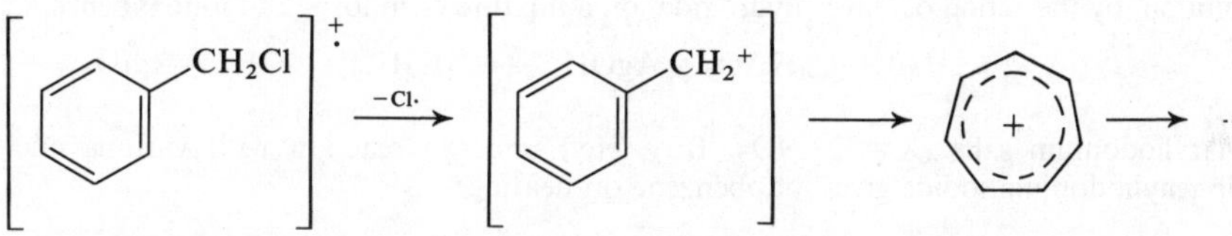

Chlorides and bromides give M (strong) and $M+2$ (fairly strong) molecular ions due to the two isotopes; fluorides and iodides are monoisotopic and only M is strong (other very weak molecular ions are due to ^{13}C and/or D).

Polyvalent iodine compounds

Iodine, in aryl iodides, differs from the other halogens in that it forms stable compounds in which it is polyvalent and has a valence shell of ten electrons. In tervalent compounds, there are two lone pairs ($5s^2$ and $5p^2$) on the iodine atom; in the oxygen compounds, iodosobenzene and iodoxybenzene, there are π_{d-p} bonds, and the structures are analogous to those of the sulphoxides and sulphones, respectively.

Iodobenzene dichloride is prepared as follows:

This is a yellow crystalline solid, and when exposed to light or heated to 110°C, it is converted into *p*-chloroiodobenzene:

$$C_6H_5ICl_2 \longrightarrow p\text{-}ClC_6H_4I + HCl$$

Iodobenzene dichloride is a useful reagent for the selective oxidation of sulphides (in aqueous pyridine solution) to sulphoxides (Montanari *et al.*, 1968). When treated with alkali, iodobenzene dichloride is converted into **iodosobenzene**.

$$C_6H_5ICl_2 + 2NaOH \longrightarrow C_6H_5\text{---}I{=}O + 2NaCl + H_2O \quad (60\text{--}62\%)$$

Iodosobenzene is a yellow solid which is basic, *e.g.*, with hydrochloric acid it forms iodobenzene dichloride, and with nitric acid, iodobenzene dinitrate:

$$C_6H_5IO + 2HNO_3 \longrightarrow C_6H_5I(NO_3)_2 + H_2O$$

When steam distilled, iodosobenzene is converted into **iodoxybenzene**:

$$2C_6H_5IO \longrightarrow C_6H_5-\overset{\overset{O}{\underset{}{\overset{+}{\parallel}}}}{\underset{O^-}{I}} \quad +C_6H_5I \quad (92\text{--}95\%)$$

This is also obtained when iodobenzene dichloride is treated with an alkaline solution of sodium hypochlorite:

$$C_6H_5ICl_2 + NaOCl + H_2O \longrightarrow C_6H_5IO_2 + NaCl + 2HCl \quad (87\text{--}92\%)$$

Iodoxybenzene is a solid which does not melt or vaporise when heated to 230°C; all iodoxy-compounds exhibit this refractory property. It forms salts with inorganic acids, *e.g.*, $C_6H_5IO_2 \cdot H_2SO_4$, but although these salts are fairly stable in the solid state, they are readily hydrolysed in solution. Iodoxybenzene also forms salts with alkalis, but these have not yet been isolated.

When a mixture of iodoso- and iodoxybenzene is treated with aqueous sodium hydroxide, **diphenyliodonium iodate** is formed, and this, on treatment with potassium iodide, forms **diphenyl-iodonium iodide**:

$$C_6H_5IO + C_6H_5IO_2 \xrightarrow{\text{NaOH}} [(C_6H_5)_2I]^+IO_3^- \xrightarrow{\text{KI}} [(C_6H_5)_2I]^+I^- \quad (70\text{--}72\%)$$

Diphenyliodonium iodide is the salt of the base *diphenyliodonium hydroxide* which may be prepared, in solution, by the action of 'silver hydroxide' on a mixture of iodoso- and iodoxybenzene:

$$C_6H_5IO + C_6H_5IO_2 + \text{`AgOH'} \longrightarrow [(C_6H_5)_2I]^+OH^- + AgIO_3$$

Diaryliodonium salts (X=X, NO_2, BF_4, etc.) undergo reaction with various nucleophiles, *e.g.*, diphenyliodonium iodide gives iodobenzene on heating.

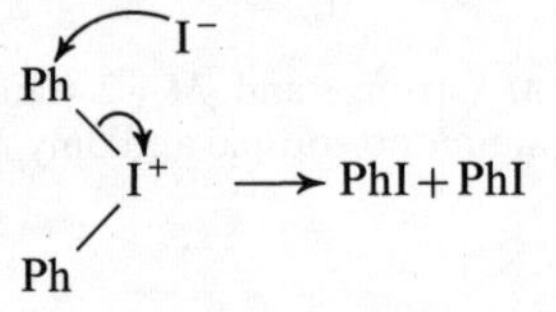

$$\begin{array}{c} \text{Ph} \\ \overset{\curvearrowright}{\underset{\text{Ph}}{I^+}} \end{array} \overset{I^-}{\curvearrowright} \longrightarrow \text{PhI} + \text{PhI}$$

This type of reaction has been used to prepare substituted diaryl ethers (Grundon *et al.*, 1963):

$$Ar_2^1I^+X^- + Ar^2O^-Na^+ \longrightarrow Ar^1OAr^2 + Ar^1I + NaX$$

Diaryliodonium salts have been found to arylate active methylene groups, *e.g.*, Beringer *et al.* (1963) have phenylated ethyl malonate (as potassium salt), and Kornblum *et al.* (1963) have phenylated nitroalkanes (as sodium salt) in dimethylformamide.

PROBLEMS

1. Write out the possible isomers of the aromatic compounds **A**, C_7H_7Cl. How could you distinguish between them? What would each give on oxidation?

2. Convert:

(i) 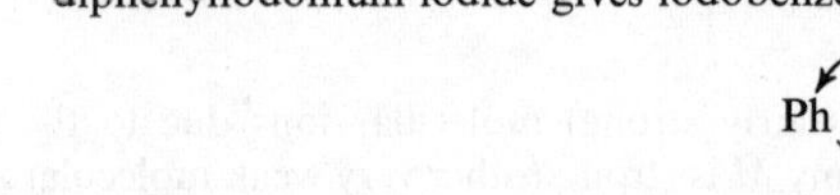 Cl—⟨benzene ring⟩—Cl $\longrightarrow$ Cl—⟨benzene ring⟩—CO_2H (ii) PhH $\longrightarrow$ Br—⟨benzene ring⟩—$COCH_2Br$

(iii) PhH $\longrightarrow$ *o*-bromotoluene.

3. Starting from benzene and any other compounds you like, prepare *p*-bromostyrene.

4. Complete the following equations:

(i)

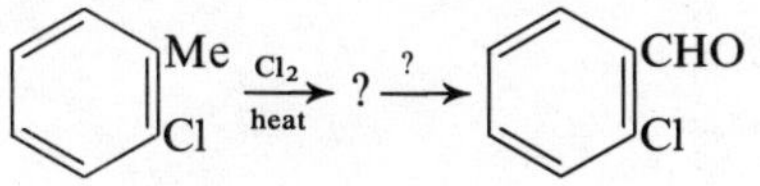

(ii) $2Ph_2CCl_2 \xrightarrow[PhH]{Cu}$?

(iii)

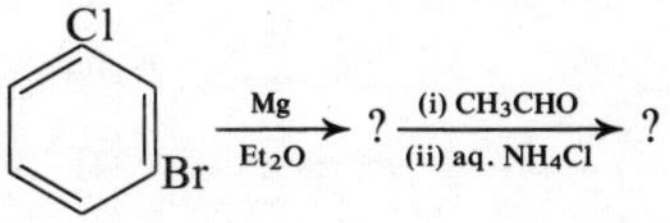

(iv)

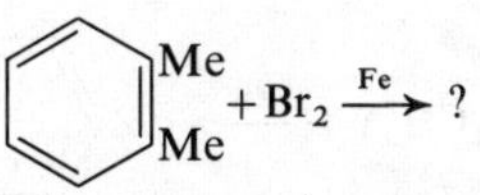

(v)

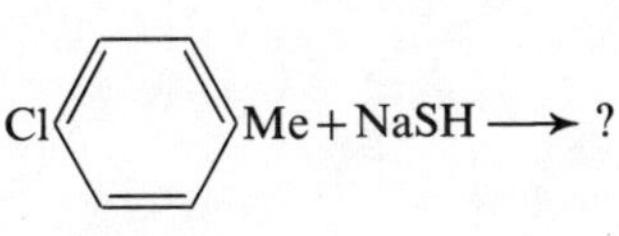

$+ NaSH \longrightarrow$?

(vi) $PhCH_3 + CBrCl_3 \xrightarrow{hv}$? (not ring substitution)

(vii)

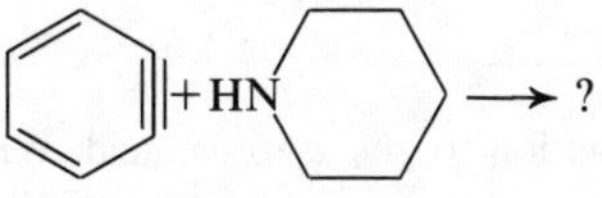

$+ HN$ $\longrightarrow$?

(viii) $PhBr \xrightarrow[liq.\ NH_3]{NaNH_2}$? $\xrightarrow{[EtCOCHMe]^- Na^+}$?

(ix)

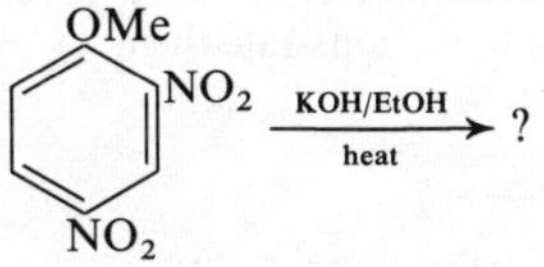

$\xrightarrow[heat]{KOH/EtOH}$?

(x)

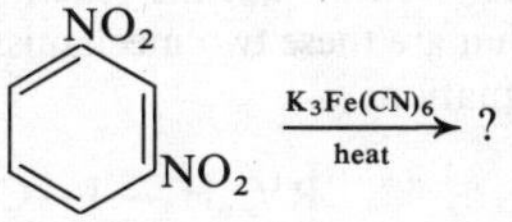

$\xrightarrow[heat]{K_3Fe(CN)_6}$?

(xi)

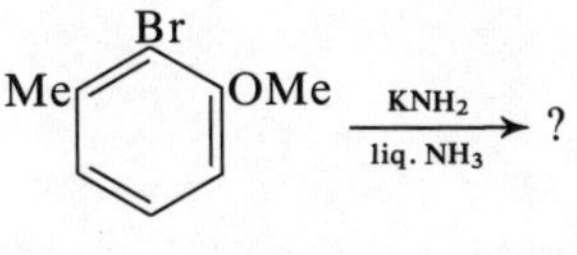

$\xrightarrow[liq.\ NH_3]{KNH_2}$?

(xii)

$\xrightarrow[Et_2O]{Mg}$?

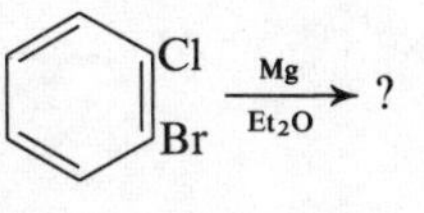

(xiii)

$+$ $\longrightarrow$?

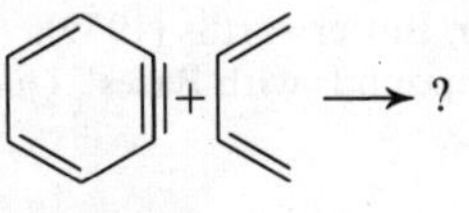

5. Bromine adds to styrene to give styrene dibromide only; no nuclear substitution occurs. Discuss.

6. Discuss the following reaction:

$$RC_6H_4X \xrightarrow[liq.\ NH_3]{NaNH_2} RC_6H_4NH_2$$

for: (i) R = o-OMe, X=Br; (ii) R = m-CF$_3$, X=Cl; (iii) R = m-Me, X=Br.

7. Complete and comment:

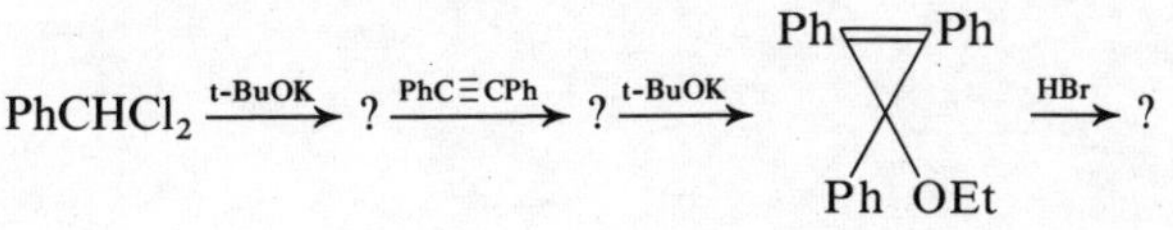

$$PhCHCl_2 \xrightarrow{t\text{-BuOK}} ? \xrightarrow{PhC \equiv CPh} ? \xrightarrow{t\text{-BuOK}}$$

$\xrightarrow{HBr}$?

8. Discuss the equation:

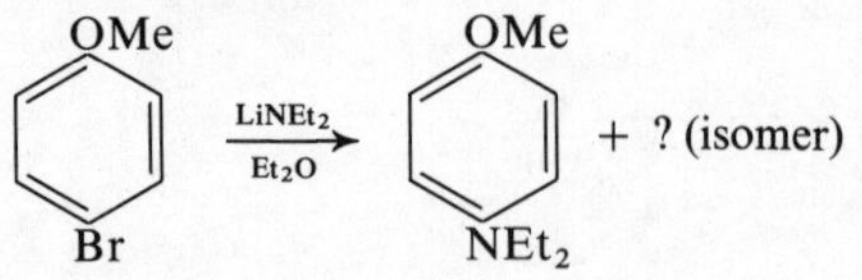

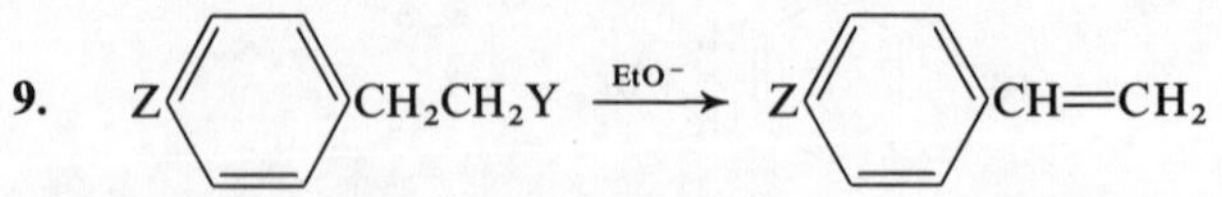

9.

This is an E2 reaction, and $\rho = 2\cdot1$ for $Y = Br$, and $\rho = 2\cdot7$ for $Y = Me_2S^+$. Discuss.

10. $PhMe + Br_2 \longrightarrow 0\cdot3\% m + 67\% p$

$PhMe + Cl_2 \longrightarrow 0\cdot5\% m + 40\% p$

No halogen carrier is used. Discuss.

11. Discuss the reaction:

$$Me_3CCHOHPh \xrightarrow[S_N1]{HCl} Me_3CHClPh$$

12. $PhCH_2MgCl + MeCHO \longrightarrow PhCH_2CHOHMe + A$

A is an isomer of the other product and undergoes the haloform reaction to give o-toluic acid. What is **A**? Suggest a mechanism for its formation.

13. The rates of bromination ($FeBr_3$ catalyst) of C_6H_6 and C_6D_6 are the same. Discuss.

14. There is evidence to show that the reaction between PhCl and aqueous NaOH at 300°C proceeds by a dual mechanism. What are these two mechanisms? Devise an experiment to demonstrate them.

15. Complete the equation:

$$R^1C_6H_4 - \overset{+}{I} - C_6H_4R^2\}X^- + ArO^-Na^+ \longrightarrow ?$$

where: (i) $R^1 = p\text{-}NO_2$, $R^2 = H$; (ii) $R^1 = p\text{-}NO_2$, $R^2 = m\text{-}NO_2$; (iii) $R^1 = m\text{-}MeO$, $R^2 = p\text{-}MeO$.

16. The mass spectrum of 1-chloro-4-ethylbenzene shows ions at 142, 140, 125, and 105. What are they? Suggest paths for their formation.

REFERENCES

STACEY *et al.* (eds.), *Advances in Fluorine Chemistry*, Butterworths (1960).

SHARTS, 'Organic Fluorine Chemistry', *J. Chem. Educ.*, 1968, **45**, 185.

DE LA MARE and RIDD, *Aromatic Substitution: Nitration and Halogenation*, Butterworths (1959).

BUNCEL, NORRIS, and RUSSELL, 'The interaction of Aromatic Nitro-compounds with Bases', *Quart. Rev.*, 1968, **22**, 123.

GILCHRIST and REES, *Carbenes, Nitrenes and Arynes*, Nelson (1969).

PIETRA, 'Mechanisms for Nucleophilic and Photonucleophilic Aromatic Substitution Reactions', *Quart. Rev.*, 1969, **23**, 504.

22 Aromatic nitro-compounds

Aromatic nitro-compounds are generally prepared by direct nitration, using one of the following reagents: concentrated or fuming nitric acid, mixed acid (mixtures of concentrated nitric and sulphuric acids), nitric acid in organic solvents (acetic acid, nitromethane, etc.), acyl nitrates (acetyl and benzoyl) in organic solvents, nitronium salts in organic solvents, and nitrosation followed by oxidation. The most common reagent is mixed acid, and the others are used according to circumstances, *e.g.*, acetyl nitrate is often useful for obtaining large amounts of the *o*-nitro-isomer (see p. 641).

When direct nitration is used, the orientation is determined by the nature of the substituent present. Hence, if the desired nitro-compound cannot be obtained directly, indirect methods are used (see, *e.g.*, pp. 646, 764).

As an outcome of a great deal of experimental work, it has been shown that for nitration with mixed acid, the active nitrating agent is the *nitronium cation*, NO_2^+, which is formed as follows (see also later):

$$HO-NO_2 + H_2SO_4 \underset{}{\overset{fast}{\rightleftharpoons}} HSO_4^- + H_2O^+ - NO_2 \overset{slow}{\rightleftharpoons}$$
$$NO_2^+ + H_2O \overset{H_2SO_4}{\rightleftharpoons} H_3O^+ + HSO_4^- \quad \dots (i)$$

The overall equation may therefore be written as:

$$HNO_3 + 2H_2SO_4 \rightleftharpoons NO_2^+ + H_3O^+ + 2HSO_4^-$$

Evidence for the existence of the nitronium cation has been obtained:

(i) By cryoscopic experiments in which it has been shown that i (the van't Hoff factor) is 4; this value satisfies the equation given above.

(ii) By the isolation of salts of the nitronium ion, *e.g.*, nitronium perchlorate, $(NO_2^+)(ClO_4^-)$, and nitronium tetrafluoroborate $(NO_2^+)(BF_4^-)$. These salts have been shown to be nitrating agents (see below).

(iii) By means of Raman spectroscopy (NO_2^+ has a band at $1\,400\,cm^{-1}$).

Nitration (with mixed acid) may therefore be formulated (see also p. 593):

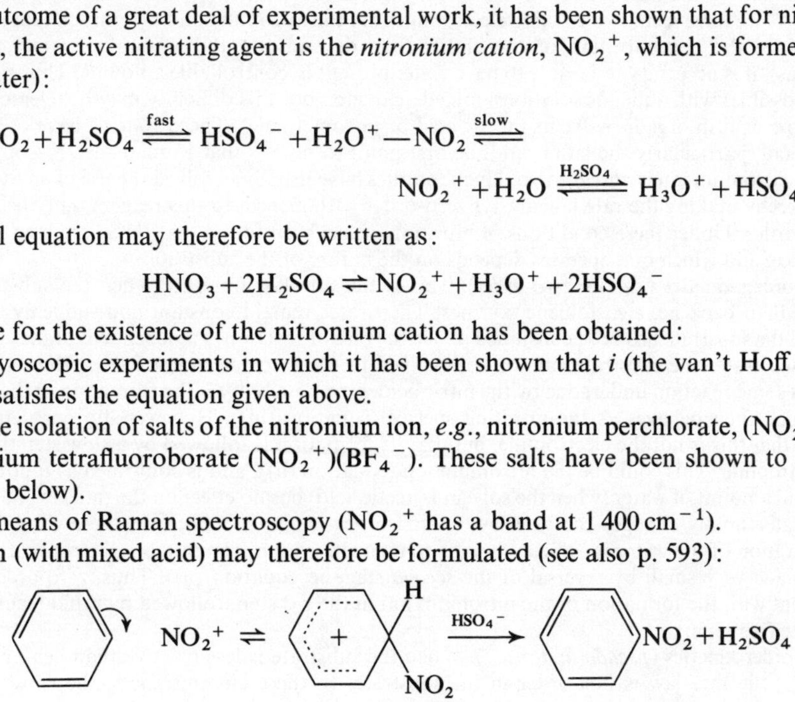

Nitration in concentrated aqueous nitric acid and in organic solvents also occurs via the nitronium ion:

$$2HNO_3 \overset{fast}{\rightleftharpoons} H_2O^+ \!-\! NO_2 + NO_3^- \overset{slow}{\rightleftharpoons} NO_2^+ + H_2O \quad \ldots \text{(ii)}$$

On the other hand, in more dilute nitric acid solutions, it is probably the *nitracidium cation* that is involved in nitration (this is less reactive than the nitronium ion):

$$2HNO_3 \rightleftharpoons H_2NO_3^+ + NO_3^- \quad \ldots \text{(iii)}$$

Evidence for the mechanism involving an intermediate σ-complex has been obtained by its isolation at $-80°C$ when trifluoromethylbenzene is treated with nitronium tetrafluoroborate (Olah *et al.*, 1958).

When the temperature was allowed to rise to $-15°C$, the σ-complex decomposed to form the nitro-compound which is obtained in the normal way (*cf.* p. 593).

Kinetics of nitration. In view of the fact that nitronium ions are present in mixed acid (equation (i); ionisation is virtually complete) and in concentrated aqueous nitric acid (equation (ii); ionisation is much less than for mixed acid), it is reasonable to suppose that the nitronium ion is the nitrating species. This is further supported by the fact that nitronium salts are effective nitrating agents. However, the existence of the nitronium ion in the nitrating solutions does not necessarily establish that this ion is, in fact, the nitrating species. Evidence that it is has been obtained mainly from kinetic studies, but the interpretation of the kinetics involves a knowledge of the nature and concentration of the nitrating species. As we have seen, two *possible* electrophiles are the nitronium and the nitracidium ion. Furthermore, since the concentration of the electrophile depends on the amount of water present (see equations), it is necessary to be able to have water present in controllable amounts. This is difficult (but not impossible) with aqueous solutions, mixed acid, etc., but this difficulty may be overcome by using pure nitric acid in organic solvents (acetic acid or nitromethane). These systems have been studied in great detail, particularly the latter, and the first point to note is that Raman spectroscopy has shown that the nitronium ion is *absent*. Also, kinetic studies have usually involved the use of an excess of nitric acid, thereby making the rate law always zero-order with respect to this reagent, *i.e.*, the rate law is a pseudo order. Under these conditions of nitration, it has been found that *three* rate laws are observed in practice, and which one operates depends on the nature of the substrate.

Zero-order kinetics (*pseudo zero-order*). This is the rate law obtained when the substrate is more reactive than benzene, *e.g.*, toluene, xylenes. Their rates remain constant and suddenly drop to zero when all the substrate has been consumed. Since the rate of reaction is independent of the concentration of the substrate, the substrate cannot be involved in the rate-determining step. Hence, the r/d step involves some reaction undergone by the nitric acid. One possibility is the formation of the nitracidium ion, which then behaves as the nitrating species (equation (iii)). However, further detailed studies showed that this is not the electrophile, but that its formation is followed by a slow step that generates the electrophile. This could be the nitronium ion (equation (ii)), and is confirmed as follows. Addition of a small amount of water (when the solvent is acetic acid) has no effect on the rate law, but addition of more water finally changes the rate law to first-order in the substrate. In these circumstances, the concentration of the water is sufficiently high to compete with nitronium ion (the concentration of this ion is made very small by reversal of the *second* stage in equation (ii)). Thus, zero-order kinetics is consistent with the formation of the nitronium ion in the r/d step, followed by rapid reaction with the substrate (see p. 56).

First-order kinetics (*pseudo first-order*). When the substrate is less reactive than benzene, *e.g.*, ethyl benzoate, the rate law is first-order in the substrate. In these circumstances, the substrate must be involved in the r/d step, and it follows that formation of the nitronium ion is faster than its rate of reaction with the substrate, *i.e.*, the concentration of the ion remains effectively constant. It should be noted that if the nitric acid was not present in large excess, then this reaction would be second-order (see p. 56).

Mixed-order kinetics. This is observed for substrates of 'intermediate' reactivity, *e.g.*, chlorobenzene. In these cases, the order is fractional, varying between zero- and first-order in substrate, the actual value depending on the concentrations of the substrate and nitric acid.

Nitrations with acetyl nitrate (HNO_3—Ac_2O). The mechanism of nitration involving acetyl nitrate appears to be an open question owing to the uncertainty of the nature of the nitrating species. The following have been proposed: CH_3COONO_2, $CH_3COOH^+NO_2$, N_2O_5, NO_2^+. An interesting feature of this reagent is the unusually high proportion of *o*-isomer that is obtained when the side-chain contains a hetero-atom, *e.g.*, nitroanisole with mixed acid gives *o*-, 31 per cent, *p*-, 67 per cent; with HNO_3—Ac_2O, *o*-, 71 per cent, *p*-, 28 per cent. One explanation offered is that the nitrating species is dinitrogen pentoxide, which reacts as follows (via a six-centre rearrangement):

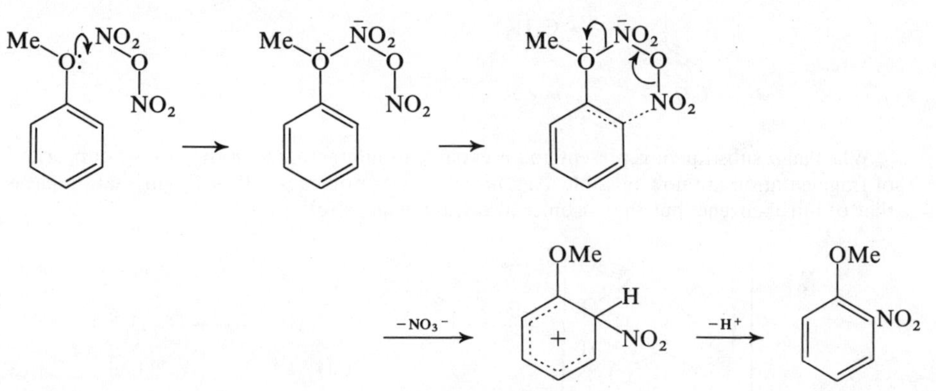

According to Schofield *et al.* (1969), however, the high yield of *p*-nitro-isomer with mixed acid is due to some of the nitration proceeding via *nitrosation* followed by oxidation (see also p. 710). In these circumstances, comparison of yields of *o*-isomer are unjustified, since the two reactions take place by different mechanisms.

General properties of the nitro-compounds

Most nitro-compounds are yellow crystalline solids; a few, including nitrobenzene, are yellow liquids. Many are steam-volatile, and except for a few mononitro-derivatives, cannot be distilled under atmospheric pressure because, when strongly heated, they decompose, often with explosive violence. All the nitro-compounds are denser than water, in which they are insoluble; they are, however, readily soluble in organic solvents. Their most important reaction is their reduction by various reducing agents. Since the nitro-group is attacked by Grignard reagents, nitro-substituted aryl halides cannot be used for the preparation of these reagents (*cf.* p. 434). 1,3-Dinitro- and 1,3,5-trinitro-compounds are different from the other polynitro-compounds in that they give colours with alkali due to the formation of Meisenheimer compounds (p. 628).

The infra-red absorption region of the nitro-group in aromatic compounds is 1 550–1 510 (s) and 1 365–1 335 cm^{-1} (s). Thus, aromatic nitro-compounds may be distinguished from aliphatic (p. 363). At the same time, the absorption regions of the aromatic nucleus will show its presence, with additional peaks according to the pattern of substitution.

Figure 22.1 is the infra-red spectrum of nitrobenzene (as a film), and Fig. 20.11 (p. 589) is the NMR spectrum of nitrobenzene.

Mass spectra

The molecular ion peak of aromatic mononitro-compounds is strong and has an odd mass number (see p. 373). For nitrobenzene, the fragmentation pattern is:

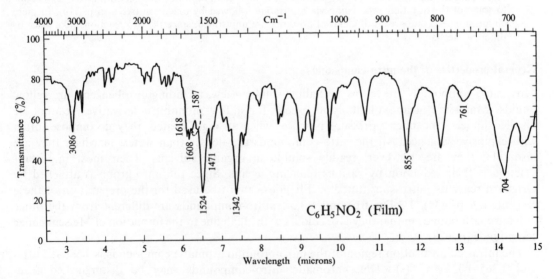

When an *o*-substituent is present and is capable of interacting with the nitro-group, additional paths of fragmentation are now possible, *e.g.*, *m*- and *p*-nitrotoluenes follow fragmentation paths similar to that of nitrobenzene, but the *o*-isomer also fragments as follows:

Fig. 22.1

1618, 1608, 1587 and 1471 cm⁻¹: C=C (str.) in aromatic nuclei.   3086 cm⁻¹: =C—H (str.) in aromatics. 1524 and 1342 cm⁻¹: NO₂ vibrations in aromatic nitro-compounds.   855 cm⁻¹: C—N (str.) or NO₂ (def.). 761 and 704 cm⁻¹: Monosubstituted benzenes.

Nitrobenzene (*oil of mirbane*), b.p. 211°C, may be prepared by the action of cold mixed acid on benzene, and then warming to complete the reaction:

$$C_6H_6 + HNO_3 \longrightarrow C_6H_5NO_2 + H_2O \quad (85\%)$$

When heated with solid potassium hydroxide, nitrobenzene produces a mixture of *o*- and *p*-nitrophenols and some azoxybenzene.

Dinitrobenzenes. When nitrobenzene is heated with mixed acid (fuming nitric and concentrated sulphuric acid), the main product is *m*-dinitrobenzene:

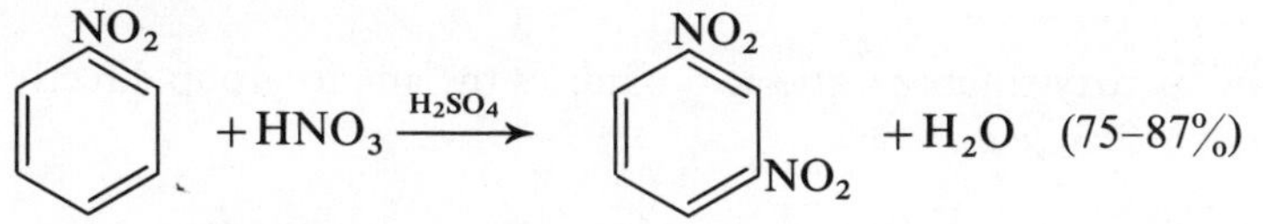

$$+ HNO_3 \xrightarrow{H_2SO_4} \cdots + H_2O \quad (75-87\%)$$

m-Dinitrobenzene, m.p. 90°C, may be reduced stepwise by, *e.g.*, sodium sulphide, to *m*-nitroaniline, and then to *m*-phenylenediamine:

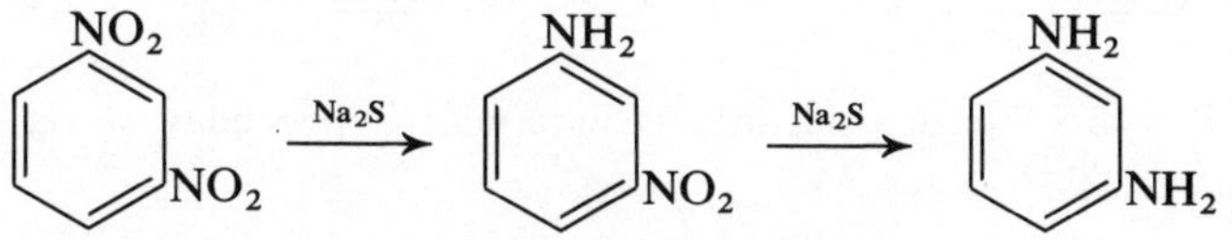

Hydrogen *ortho* to a nitro-group in *m*-dinitrobenzene undergoes ready activated nucleophilic substitution with sodium hydroxide in the presence of potassium ferricyanide to a mixture of 2,4-dinitrophenol (mainly) and 2,6-dinitrophenol (see p. 628).

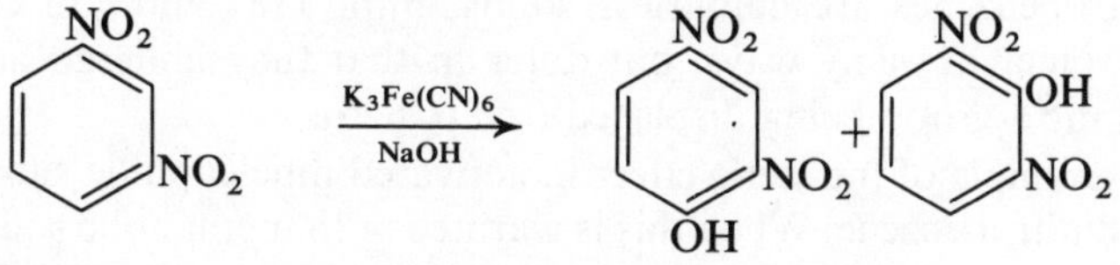

o- and *p*-Dinitrobenzene may be prepared from acetanilide as follows:

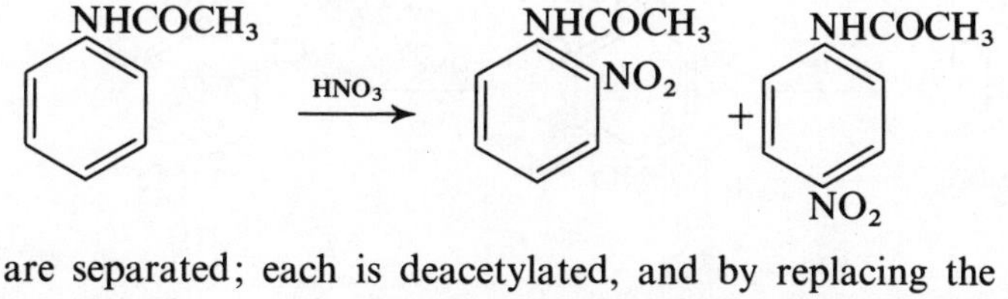

These isomers are separated; each is deacetylated, and by replacing the amino-group by a nitro-group (see p. 633), the *o*- and *p*-dinitrobenzenes may be obtained pure, *e.g.*,

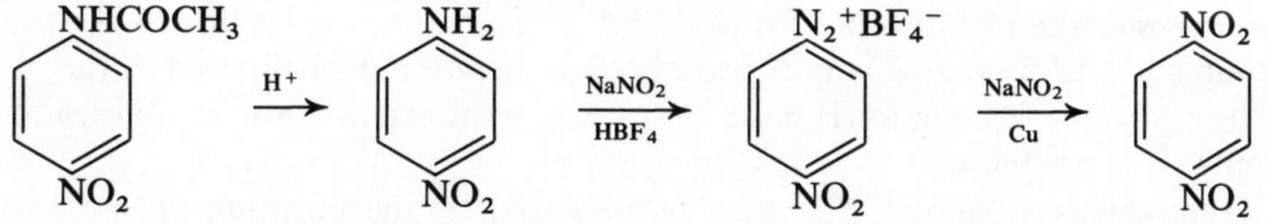

It is necessary to 'protect' compounds with a hydroxyl or amino-group in the ring, since these compounds are very easily oxidised instead of nitrated by nitric acid.

An alternative method for preparing *o*- and *p*-dinitrobenzenes is to oxidise the corresponding nitroaniline with Caro's acid, and then to oxidise the product, the nitronitroso-derivative, with a mixture of nitric acid and hydrogen peroxide:

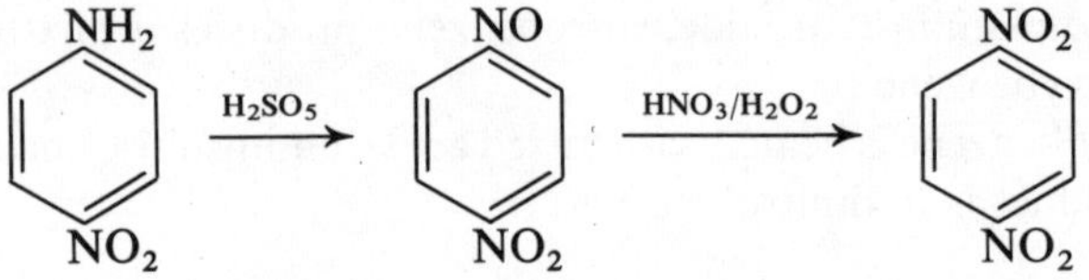

On the other hand, peroxytrifluoroacetic acid oxidises the amino-group directly to the nitro-group (Emmons *et al.*, 1953), *e.g.*,

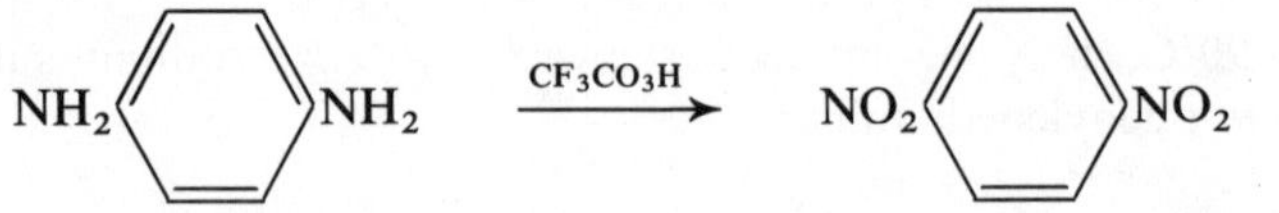

The mechanism of these oxidations is uncertain; a possibility is via the arylhydroxylamine (Gragerov *et al.*, 1961).

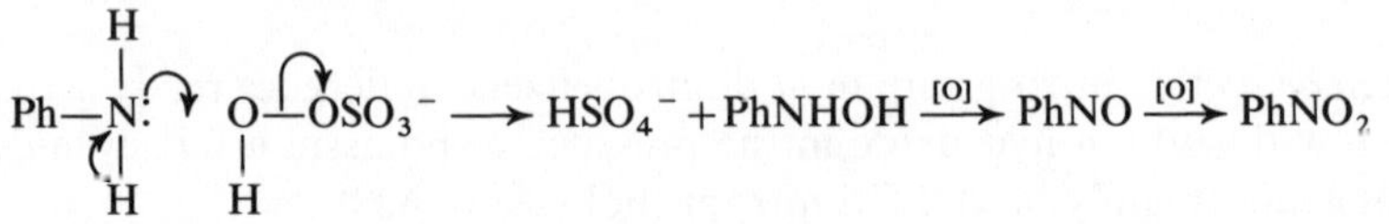

$$Ph-\overset{\overset{H}{|}}{\underset{\underset{H}{|}}{N}}: \quad O-\overset{|}{\underset{H}{O}}SO_3^- \longrightarrow HSO_4^- + PhNHOH \xrightarrow{[O]} PhNO \xrightarrow{[O]} PhNO_2$$

o- and *p*-Dinitrobenzenes are colourless solids, m.p. 118° and 173°C, respectively. They resemble the *m*-isomer in many ways, but differ in that they undergo activated nucleophilic substitution, one nitro-group being displaced (see p. 626).

An interesting example of the steric effect in activated mucleophilic substitution is shown by 1,3-dichloro-2,5-dinitrobenzene. When this is warmed with methanolic sodium methoxide, only *one* product is formed, 1,3-dichloro-5-methoxy-2-nitrobenzene (I). The 2-nitro-group cannot

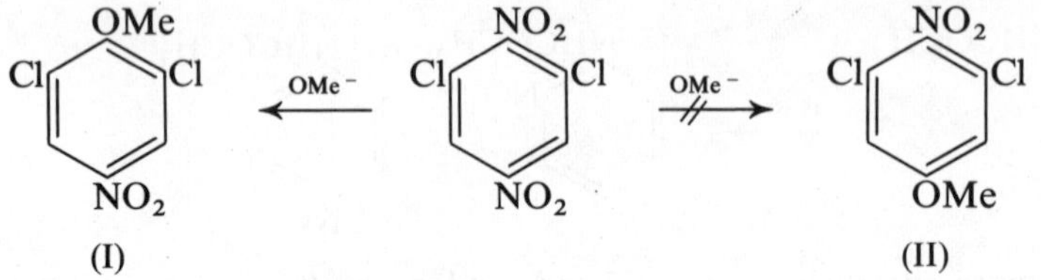

stabilise the carbanion leading to (II) since it is prevented from entering into resonance with the benzene ring because of the steric effects of the *o*-chlorine atoms. On the other hand, this **steric inhibition of resonance** (see also p. 656) is absent in the carbanion leading to (I) (the entering OMe group and the ejected NO$_2$ group are both *perpendicular* to the ring in the carbanion). Hence, the carbanion leading to (I) has a lower energy of activation than that leading to (II), and so only (I) is produced.

1,3,5-Trinitrobenzene, m.p. 122°C, may be prepared by the nitration of *m*-dinitrobenzene with mixed acid consisting of fuming nitric acid and fuming sulphuric acid. This reaction takes five days to complete; it is very difficult to introduce the third nitro-group, and it is impossible to introduce more than three nitro-groups by direct nitration. A better method of preparing

1,3,5-trinitrobenzene is to oxidise 2,4,6-trinitrotoluene and to decarboxylate the trinitro-benzoic acid so produced by heating it in acetic acid solution:

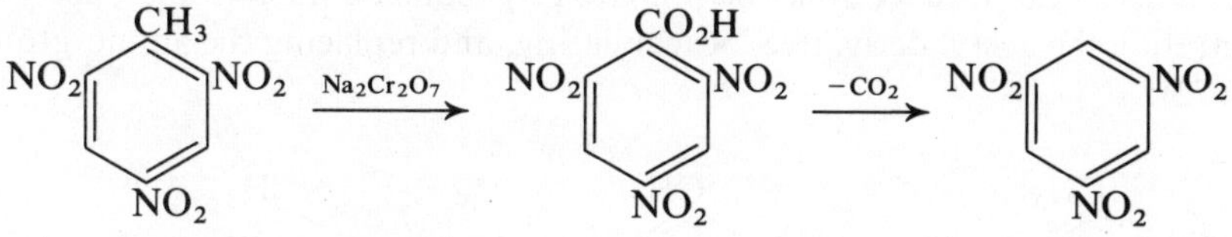

$$(43–46\%)$$

Charge-transfer complexes. Many di- and polynitro-compounds form complexes with various aromatic hydrocarbons, phenols, etc. These complexes usually contain the components in a 1:1 ratio, and are generally highly coloured. Their spectra are *not* the same as the *sum* of the spectra of the two individual components (see below). Complexes of this type are called **charge-transfer complexes**. Let us consider the charge-transfer complex formed from 1,3,5-trinitro-benzene and mesitylene. X-ray analysis has shown that the two rings lie in parallel planes and overlap (*complete* overlap does not always occur). One molecule behaves as an electron-donor

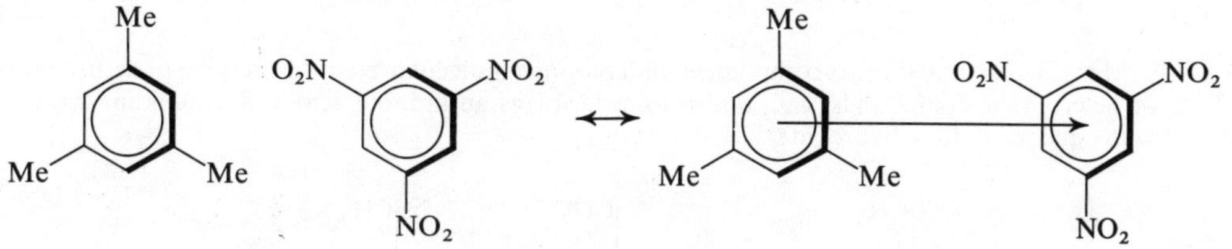

$$(D,A) \qquad\qquad (D^+A^-)$$

(*D*) and the other as an electron-acceptor (*A*). (*D*) has a low ionisation potential and (*A*) has a low-energy vacant orbital and a high electron affinity (see p. 32). Owing to their spatial arrangements, overlap occurs between the molecular orbitals, resulting in transfer of an electron from the highest occupied π-orbital of (*D*) to the lowest unoccupied π-orbital of (*A*). Alternatively, the complex may be regarded as a resonance hybrid of two contributing structures, non-bonded (*D*,*A*) held together by van der Waals forces, and (*D*$^+$*A*$^-$), the 'bonding' in which is very weak. It is this electron-transfer that produces the new band to give a **charge-transfer spectrum** (see also p. 873).

In the charge-transfer complex discussed, (*A*) is the polynitro-compound, since the ring becomes depleted of π-electron density due to the $-I$ and $-R$ effects of the nitro-groups. Thus, stability of the complex increases with the number of nitro-groups in (*A*) and with increasing basicity of (*D*), *e.g.*, 1,3,5-trinitrobenzene does not form a complex with benzene but does with mesitylene (see also p. 615). However, since the bonding in charge-transfer complexes is usually weak, they readily dissociate in solution.

Nitrotoluenes. Benzene homologues are more readily nitrated than benzene itself, due to the activating effect of the alkyl group. When nitrated with mixed acid, toluene forms a mixture of *o*- and *p*-nitrotoluenes:

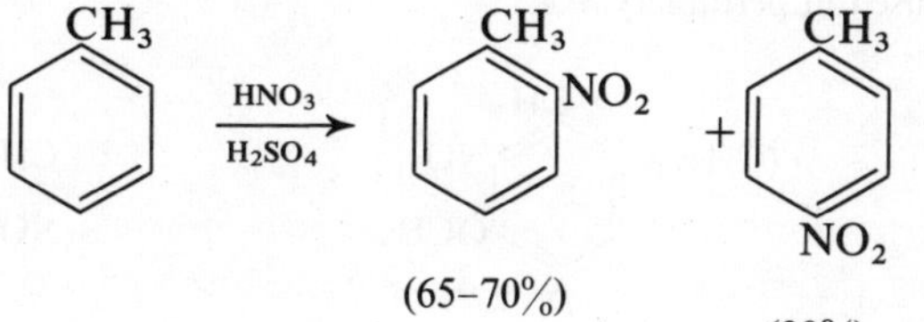

$$(65–70\%)$$

$$(30\%)$$

These isomers may readily be separated by fractional distillation under reduced pressure: *o*-isomer, m.p. −4°, b.p. 222°; *p*-isomer, m.p. 54°, b.p. 238°C.

m-Nitrotoluene, m.p. 16°, b.p. 227°C, may be prepared by reducing the *p*-isomer, acetylating the *p*-toluidine, nitrating the acetyl derivative, deacetylating, and replacing the amino-group by hydrogen (see p. 672):

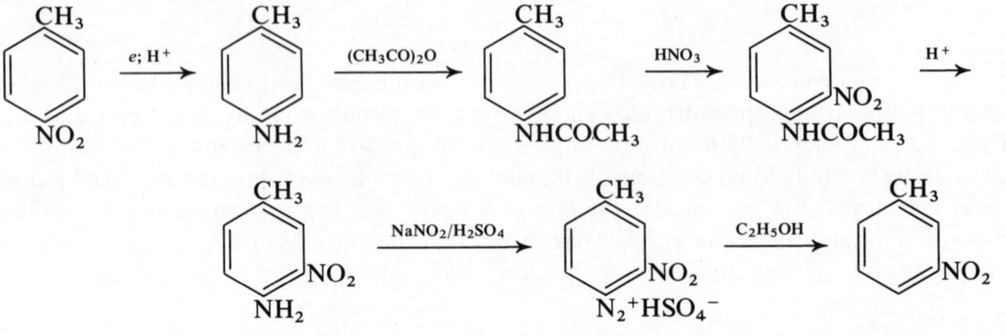

Many *o*-substituted nitro-compounds undergo intramolecular reactions to give uncyclised products or heterocyclic compounds, *e.g.*, *o*-nitrotoluene forms anthranilic acid and *o*-toluidine when heated with aqueous sodium hydroxide.

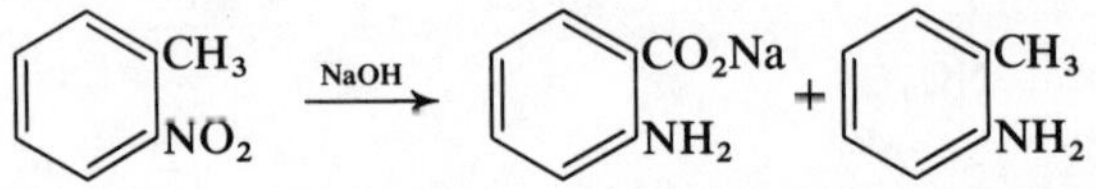

On the other hand, *o*-nitrobenzoylacetone gives isatin (p. 844) when heated with ethanolic potassium hydroxide:

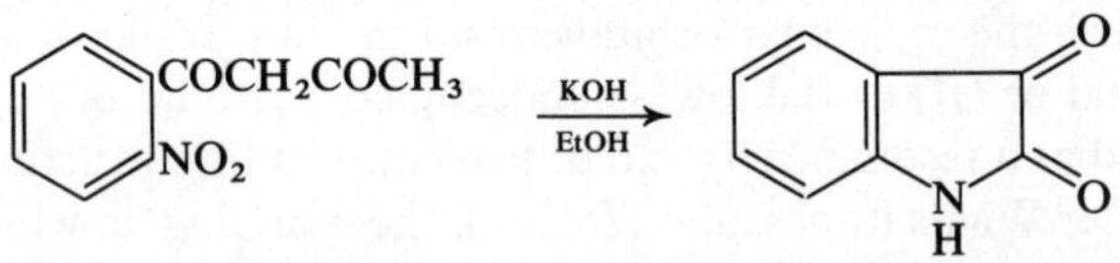

Dinitrotoluenes. Nitration of *p*-nitrotoluene gives almost exclusively the 2,4-dinitro-derivative; *o*-nitrotoluene gives a mixture of 2,4-dinitrotoluene (predominantly) and 2,6-dinitrotoluene; *m*-nitrotoluene gives a mixture of the 3,4-, 2,3-, and 2,5- dinitro-isomers, with the first predominating.

2,4,6-Trinitrotoluene (T.N.T.), m.p. 81°C, may be prepared by nitrating toluene with mixed acid consisting of fuming nitric and fuming sulphuric acids. This reaction takes place far more readily than with benzene because of the activation of the ring by the methyl group.

Trinitrotoluene is used as an explosive; mixed with ammonium nitrate, it forms the explosive *amatol*.

Artificial musks. Polynitro-derivatives of benzene containing a tertiary butyl group, possess odours resembling musk, and are used in perfumery, *e.g.*,

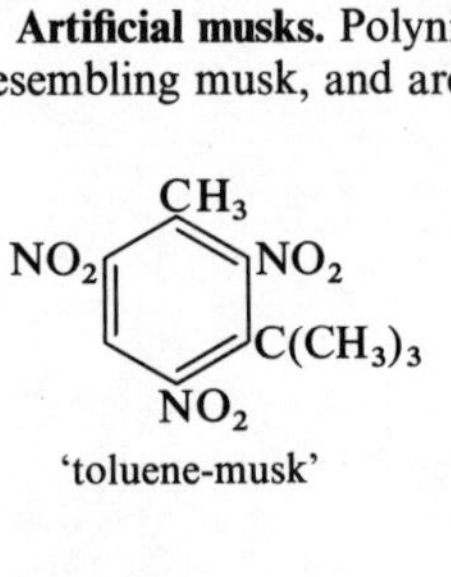

'toluene-musk' 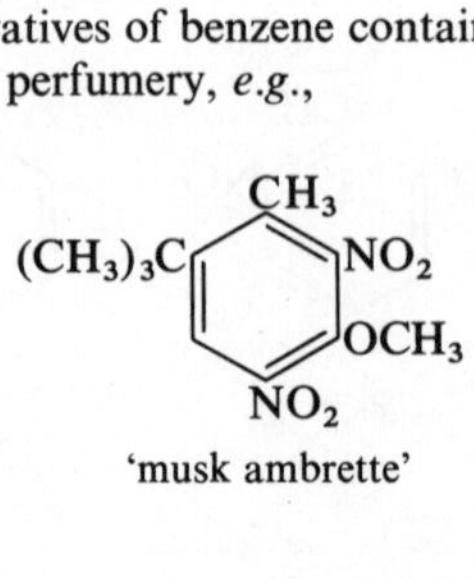   'musk ambrette' 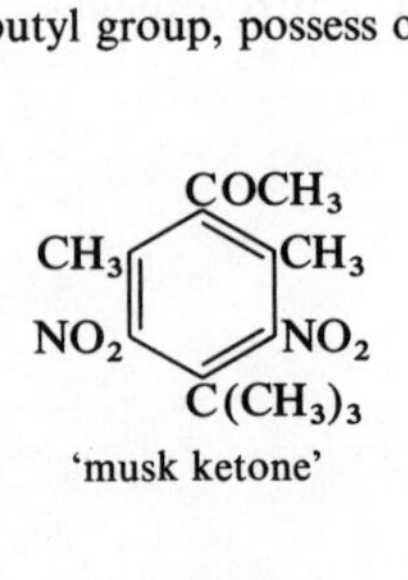   'musk ketone'

Halogenonitrobenzenes. Nitration of phenyl halides produces a mixture of the o- and p-halogenonitrobenzenes. Their characteristic property is that they undergo activated nucleophilic substitution (p. 626), reactivity increasing with the number of o- and/or p-nitro-groups, $e.g.$, picryl chloride forms picric acid merely by warming with water (which is a weak nucleophile).

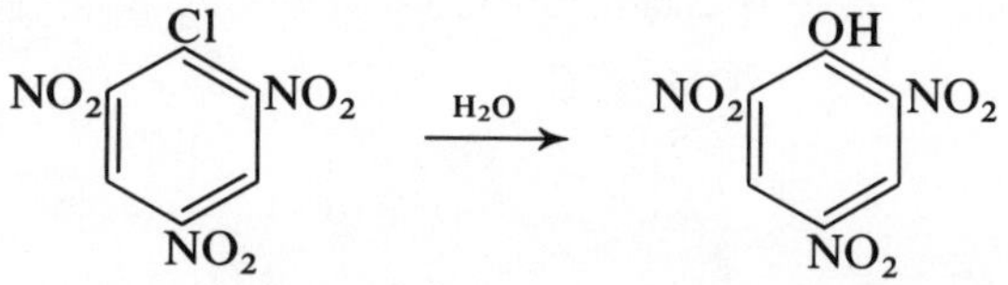

It is also interesting to note that the fluorine atom in 1-fluoro-2,4-dinitrobenzene is strongly activated. This compound condenses with amino-groups to form N-2,4-dinitrophenyl derivatives and this reaction has been used in protein and peptide studies (Sanger, 1945).

A methyl group is also made reactive by a nitro-group in the o- or p-position, $e.g.$, trinitrotoluene condenses with benzaldehyde in the presence of ethanolic potassium hydroxide to form *trinitrostilbene*:

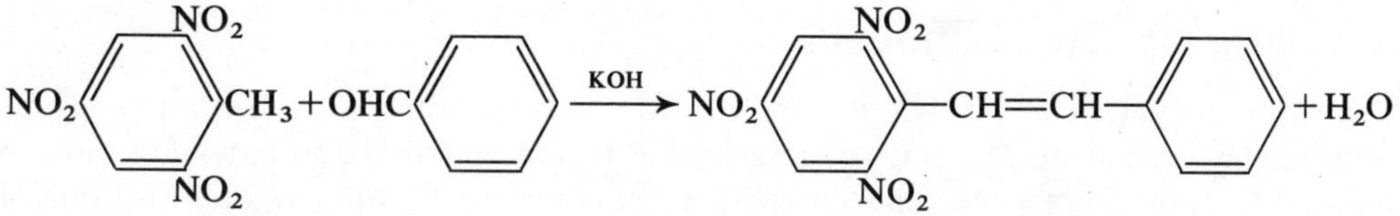

The activation of the methyl group may be explained in terms of the stability of the benzyl anion produced.

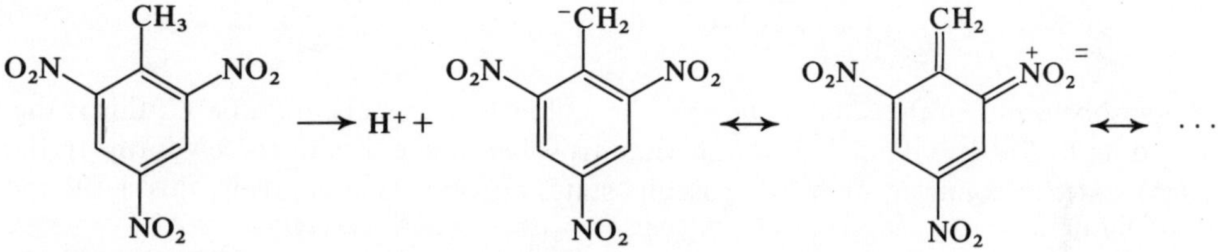

Phenylnitromethane may be prepared by heating toluene with dilute nitric acid in a sealed tube at 100°C, or by heating benzyl chloride with aqueous ethanolic silver nitrate:

$$C_6H_5CH_3 + HNO_3 \longrightarrow C_6H_5CH_2NO_2 + H_2O \quad (p.)$$
$$C_6H_5CH_2Cl + AgONO \longrightarrow C_6H_5CH_2NO_2 + AgCl \quad (f.g.)$$

Phenylnitromethane is best prepared as follows:

$$C_6H_5CH_2CN + CH_3ONO_2 + C_2H_5ONa \longrightarrow \underset{\underset{CN}{|}}{C_6H_5C}{=}NO_2Na + CH_3OH + C_2H_5OH \xrightarrow[\text{heat}]{\text{aq. NaOH}}$$

(I)

$$\underset{\underset{CO_2Na}{|}}{C_6H_5C}{=}NO_2Na \xrightarrow{HCl} \left[\underset{\underset{CO_2H}{|}}{C_6H_5C}{=}NO_2H\right] \xrightarrow{-CO_2}$$

(II)

$$C_6H_5CH{=}NO_2H \xrightarrow[\text{(3 days)}]{\text{isomerises}} C_6H_5CH_2NO_2 \quad (50\text{--}55\%)$$

(I) is sodio-phenyl*aci*nitroacetonitrile, and (II) is the sodium salt of phenylnitroacetic acid.

Phenylnitromethane behaves as a true primary aliphatic nitro-compound; both the nitro- and *aci*nitro-forms have been isolated (*cf.* p. 364):

yellow oil b.p. 226°C | solid m.p. 84°C

Abnormal nitration. In many cases of nitration, a substituent may be replaced by a nitro-group, *e.g.*,

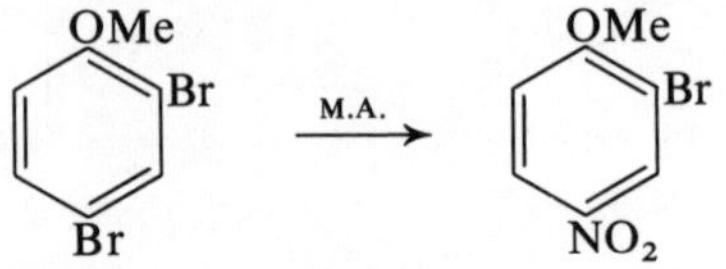

With polyalkylbenzenes, abnormal nitration usually occurs when s- and t-alkyl groups are present, *e.g.*,

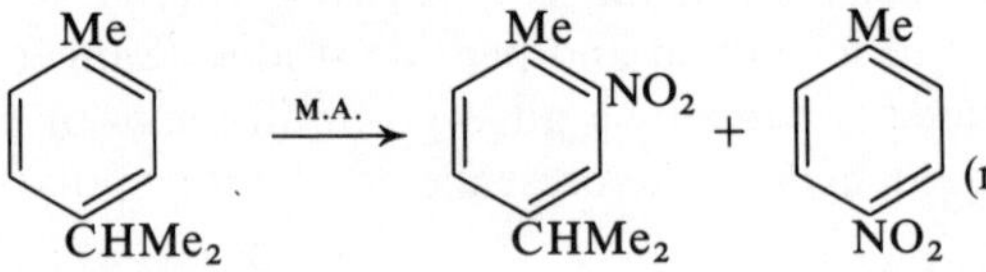

(minor product)

Aromatic nitroso-compounds

The direct reduction of nitro-compounds to the corresponding nitroso-compounds is not a satisfactory method of preparation, largely due to the fact that the latter are more readily reduced than the former. Nitrosobenzene is most conveniently prepared by oxidising phenylhydroxylamine with dichromate-sulphuric acid mixture, or by oxidising aniline with Caro's acid:

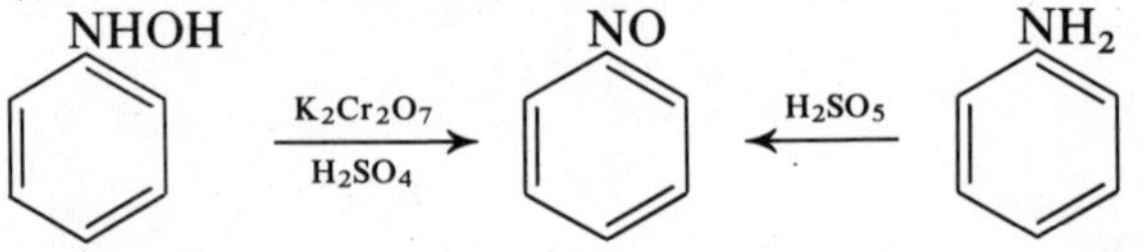

Nitrosobenzene, in the solid state, exists as colourless crystals, m.p. 68°C; this is the dimer. According to Lüttke (1956, 1957), dimeric nitrosobenzene exists in the *cis*-form. In the liquid (fused) state, in solution, or in the gaseous state, nitrobenzene is green; this is the monomer (*cf.* p. 366).

Nitrosobenzene is reduced by different reagents to a variety of products, *e.g.*, azobenzene (I), phenylhydroxylamine (II), aniline (III), or azoxybenzene (IV).

$$C_6H_5NO \xrightarrow{Pd/H_2} C_6H_5N{=}NC_6H_5 \quad (I)$$
$$C_6H_5NO \xrightarrow{HI/P} C_6H_5NHOH \quad (II)$$
$$C_6H_5NO \xrightarrow{Fe/HCl} C_6H_5NH_2 \quad (III)$$
$$C_6H_5NO \xrightarrow{Ph_3P} C_6H_5N{=}\overset{+}{N}(O^-)C_6H_5 \quad (IV)$$

Nitrosobenzene is readily oxidised to nitrobenzene by hydrogen peroxide-nitric acid mixture, and undergoes many condensation reactions in which the nitroso group behaves like a carbonyl group, *e.g.*,

$$C_6H_5N{=}NC_6H_5 \xleftarrow{C_6H_5NH_2} C_6H_5NO \xrightarrow{C_6H_5CH_2CN} C_6H_5C(CN){=}NC_6H_5$$
$$C_6H_5NO \xrightarrow{NH_2OH} C_6H_5N{=}NOH$$
$$C_6H_5NO \xrightarrow{(i)\ RMgX;\ (ii)\ H_2O} C_6H_5NR{-}OH$$

The reaction with an active methylene group affords a means of oxidising such a group to carbonyl (by hydrolysis of the imine; see p. 741).

Phenylhydroxylamine, m.p. 81°C, may be prepared by reducing nitrobenzene with zinc dust and aqueous ammonium chloride (yield: 62–68%), or by the electrolytic reduction of nitrobenzene in an aqueous solution of acetic acid containing sodium acetate.

Phenylhydroxylamine is a powerful reducing agent, *e.g.*, it reduces ammoniacal silver nitrate and Fehling's solution. It readily absorbs oxygen from the air to form nitrosobenzene; this can react with unchanged phenylhydroxylamine to give *azoxybenzene*:

$$C_6H_5NHOH + C_6H_5NO \longrightarrow C_6H_5\overset{\overset{\displaystyle O^-}{\displaystyle |}}{\underset{}{N}}{}^{+}=NC_6H_5 + H_2O$$

In dilute acid solution, phenylhydroxylamine rearranges to *p*-aminophenol. According to Hughes and Ingold (1951), the mechanism is possibly:

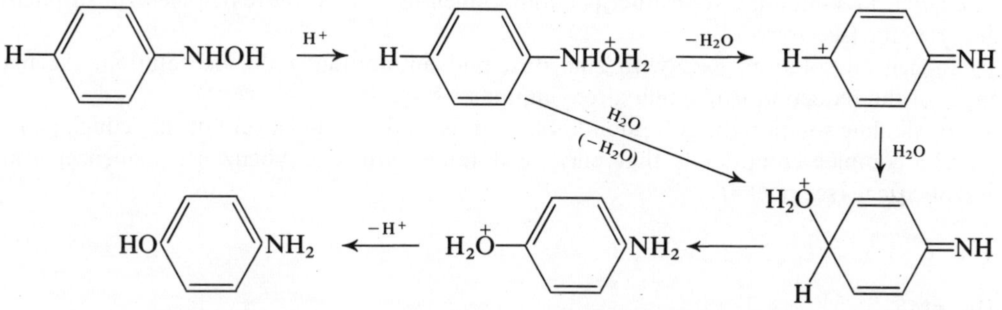

If the *p*-position is occupied by a methyl group, the rearrangement can still take place, but in this case ammonia is eliminated and a methylquinol is formed, which then readily rearranges to a quinol (see also p. 723):

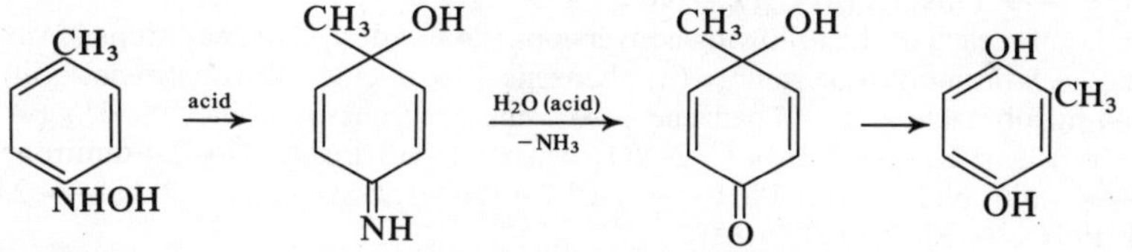

Phenylhydroxylamine is used to prepare **cupferron**; an ethereal solution of phenylhydroxylamine is treated with dry ammonia gas, and then n-butyl nitrite is added:

$$C_6H_5NHOH + C_4H_9ONO + NH_3 \longrightarrow C_6H_5N(NO)ONH_4 + C_4H_9OH \quad (85\text{--}90\%)$$

Cupferron is the ammonium salt of *N*-nitrosophenylhydroxylamine, and was originally used for the quantitative estimation of copper and iron (hence its name).

Reduction products of the nitro-compounds. The course of the reduction of nitro-compounds has been shown to take place through the following stages:

$$C_6H_5NO_2 \longrightarrow C_6H_5NO \longrightarrow C_6H_5NHOH \longrightarrow C_6H_5NH_2$$

The nature of the final product, however, depends mainly on the pH of the solution in which the reduction is carried out.

(i) In acid solution (metal and acid), aniline is obtained (via the nitrobenzene and nitroso-benzene free-radical anions):

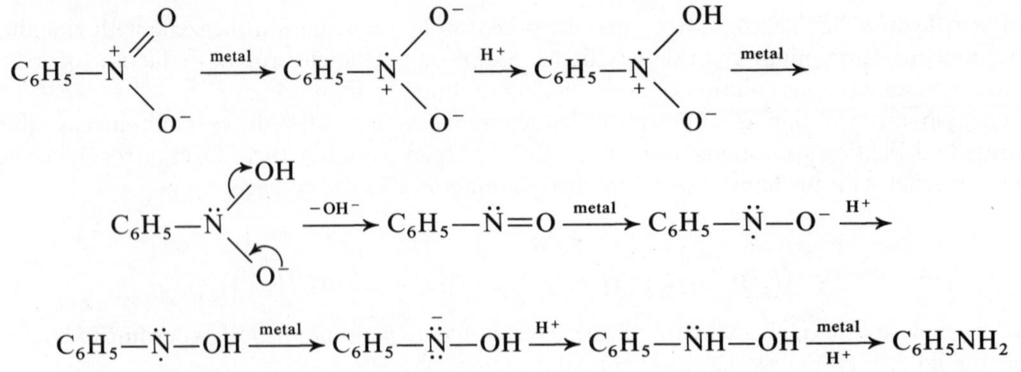

$$C_6H_5-\overset{+}{\underset{\overset{|}{O^-}}{N}}\overset{O}{\overset{\parallel}{\diagup}} \xrightarrow{\text{metal}} C_6H_5-\overset{+}{\underset{\overset{|}{O^-}}{\dot{N}}}\overset{O^-}{\diagup} \xrightarrow{H^+} C_6H_5-\overset{+}{\underset{\overset{|}{O^-}}{\dot{N}}}\overset{OH}{\diagup} \xrightarrow{\text{metal}}$$

$$C_6H_5-\ddot{N}\overset{\curvearrowright OH}{\underset{O^-}{\diagdown}} \xrightarrow{-OH^-} C_6H_5-\dot{N}=O \xrightarrow{\text{metal}} C_6H_5-\ddot{N}-O^- \xrightarrow{H^+}$$

$$C_6H_5-\ddot{N}-OH \xrightarrow{\text{metal}} C_6H_5-\overset{-}{\ddot{N}}-OH \xrightarrow{H^+} C_6H_5-\ddot{N}H-OH \xrightarrow[H^+]{\text{metal}} C_6H_5NH_2$$

In *strongly* acid solution, the product is *p*-aminophenol, formed by rearrangement of phenyl-hydroxylamine (see above).

(ii) In neutral solution, *e.g.*, with zinc dust and ammonium chloride solution, the main product of the reduction is phenylhydroxylamine.

(iii) In alkaline solution the compound obtained depends on the nature of the reducing agent used. The complex compounds that may be obtained are: azoxybenzene, azobenzene, and hydrazobenzene (see p. 684).

PROBLEMS

1. Arrange in order of relative rates the *o*-nitration of PhMe, 2-D-toluene, and 2,6-D$_2$-toluene. Give your reasons.

2. Draw the energy profiles for (and discuss):
(i) $PhH + NO_2^+ \longrightarrow PhNO_2$; (ii) $C_6D_6 + NO_2^+ \longrightarrow C_6D_5NO_2$.

3. Suggest a method for each of the following conversions (do not describe separations of isomers):
(i) benzene $\longrightarrow$ *p*-bromonitrobenzene; (ii) benzene $\longrightarrow$ *m*-chloronitrobenzene; (iii) benzene $\longrightarrow$ 1,2-dibromo-4-nitrobenzene; (iv) benzene $\longrightarrow$ *m*-nitrobenzenesulphonic acid; (v) benzene $\longrightarrow$ PhCH$_2$NO$_2$; (vi) benzene $\longrightarrow$ 1-Br-4-Cl-2-NO$_2$-benzene; (vii) PhMe $\longrightarrow$ 2,4-dinitrobenzoyl chloride; (viii) PhH $\longrightarrow$ *o*-nitrophenol; (ix) PhH $\longrightarrow$ 1-Cl-2,4-(NO$_2$)$_2$-benzene; (x) PhH $\longrightarrow$ 2,4,6-Br$_3$-1-NO$_2$-benzene; (xi) PhH $\longrightarrow$ PhN=C(CO$_2$Et)$_2$.

4. Discuss the following data:

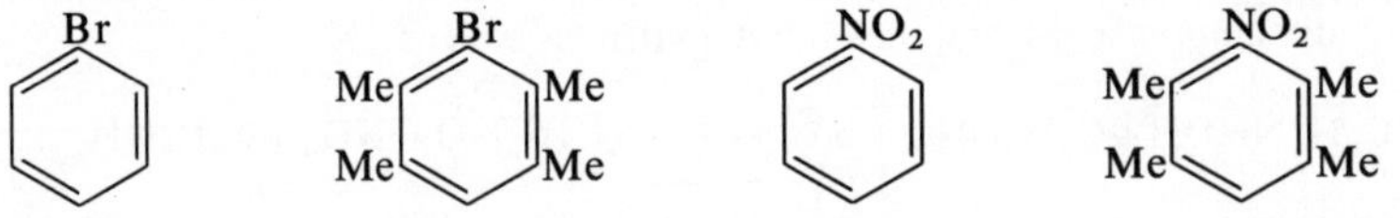

μ(D)1·53   1·55   3·9   3·39

5. The nitration of phenylboronic acid, PhB(OH)$_2$, with M.A. at $-20°C$ gives predominantly the *m*-derivative. With HNO$_3$—Ac$_2$O, the predominant product is the *o*-derivative. Suggest an explanation.

6. Complete the equations:

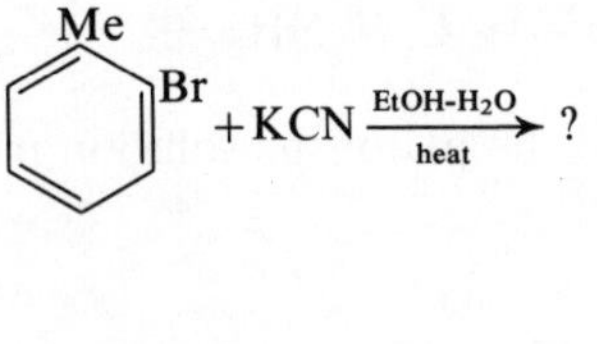

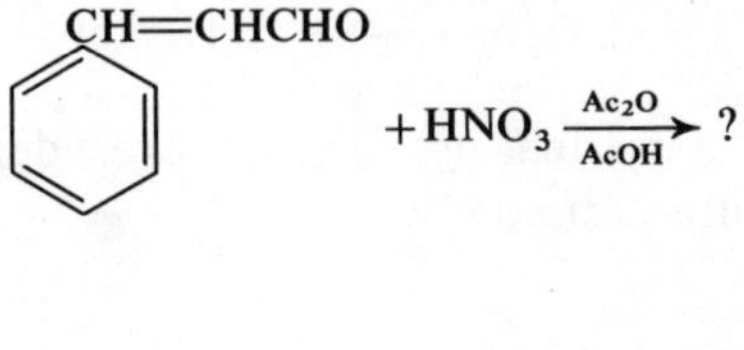

(iii) 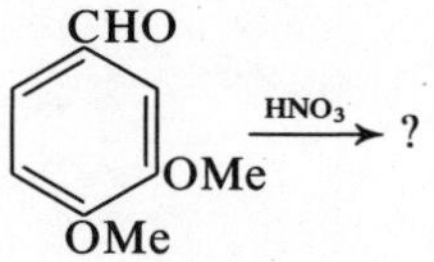 [benzene ring with CHO, OMe, OMe substituents] $\xrightarrow{HNO_3}$?

(iv) 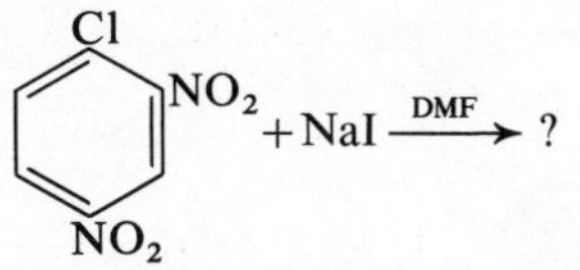 [benzene ring with Cl, NO_2, NO_2] $+ NaI \xrightarrow{DMF}$?

(v) 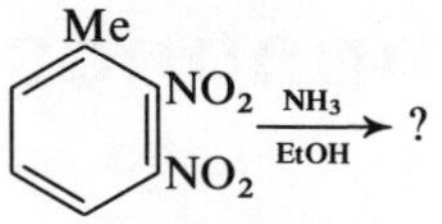 [benzene ring with Me, NO_2, NO_2] $\xrightarrow[EtOH]{NH_3}$?

(vi) 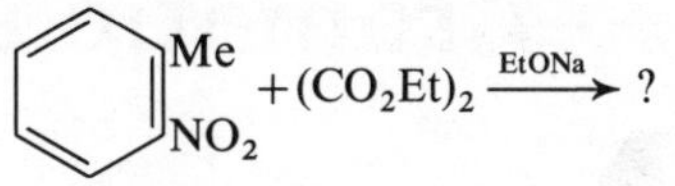 [benzene ring with Me, NO_2] $+ (CO_2Et)_2 \xrightarrow{EtONa}$?

(vii) 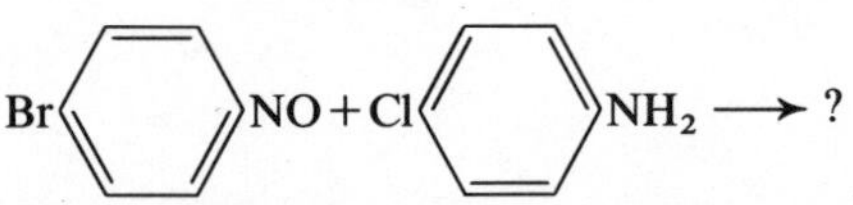 $Br\langle$ [ring] $\rangle NO + Cl\langle$ [ring] $\rangle NH_2 \longrightarrow$?

(viii) [benzene ring with NO_2, Me, O_2N, NO_2] $+ PhNO \longrightarrow$?

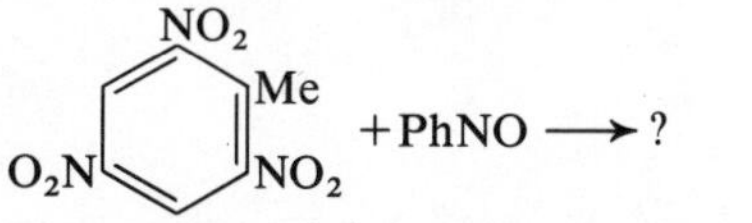

7. The aromatic compound **A**, C_8H_9Br, on treatment with DEM/EtONa, followed by reflux with H_2SO_4, gave **B**, $C_{10}H_{12}O_2$. **B**, on vigorous oxidation, gave **C**, $C_8H_6O_4$, which, on mononitration, gave only one mononitro derivative, **D**. What are **A–D**?

8. Offer an explanation for the following data of the mononitration of acetanilide (PhNHAc): M.A.: *o-*, 19%, *p-*, 79%; HNO_3—Ac_2O: *o-*, 68%; *p-*, 30%.

9. The rate of nitration of toluene in nitromethane with excess of HNO_3 is constant and the reaction is zero order with respect to toluene. If the rate is temperature-dependent, what conclusion can you draw?

10. Describe the fragmentation paths which could give the following ions (abundance ratio in brackets) from 1-chloro-3-nitrobenzene: 76(8), 87(2), 85(5·5), 113(32), 111(100), 129(2), 127(5), 157(73), 159(23). At least one ion that might have been expected is missing. Which is it?

REFERENCES

DE LA MARE and RIDD, *Aromatic Substitution: Nitration and Halogenation,* Butterworths (1959).

NORMAN and TAYLOR, *Electrophilic Substitution in Benzenoid Compounds,* Elsevier (1965).

SIDGWICK, *The Organic Chemistry of Nitrogen,* Oxford Press (1966, 3rd edn. by Millar and Springall).

MURRELL, 'The Theory of Charge-Transfer Spectra', *Quart. Rev.,* 1961, **15,** 191.

LOUDON and TENNANT, 'Substituent Interactions in *ortho*-Substituted Nitrobenzenes', *Quart. Rev.,* 1964, **18,** 389.

BUNCEL, NORRIS, and RUSSELL, 'The Interaction of Aromatic Nitro-compounds with Bases', *Quart Rev.,* 1968, **22,** 123.

23 Aromatic amino-compounds

Aromatic amines are of two types, one in which the amino-group is attached directly to the ring, and the other in which the amino-group is in the side-chain. The latter behave very much like the aliphatic amines, and so may be regarded as aryl-substituted aliphatic amines.

Nuclear amino-compounds fall into the same three classes as the aliphatic amines, *viz.*, *primary amines*, $ArNH_2$; *secondary amines*, Ar_2NH and $ArNHR$; and *tertiary amines*, Ar_3N, Ar_2NR and $ArNR_2$. Quaternary aromatic compounds containing 4, 3, or 2 aryl groups attached to nitrogen have not yet been prepared. This is possibly due partly to decreased basicity (see p. 664) and partly to steric effects.

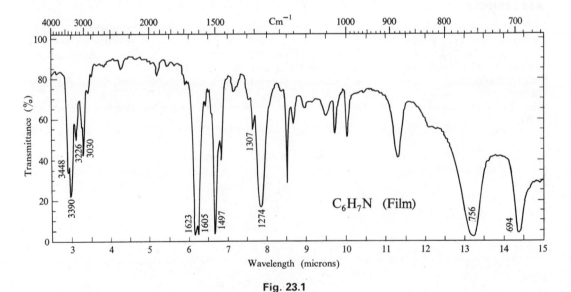

Fig. 23.1

(1 623), (1 605), and 1 497 cm^{-1}: C=C (str.) in aromatic nuclei.   3 030 cm^{-1}: =C—H (str.) in aromatics. 3 448 and 3 390 cm^{-1}: N—H (str.) in primary amines.   3 226 cm^{-1}: N—H (str.); association.   (1 623) and (1 605) cm^{-1}: N—H (def.) in primary amines.   1 307 and 1 274 cm^{-1}: C—N (str.) in primary aromatic amines. 756 and 694 cm^{-1}: Monosubstituted benzenes.

The N—H (stretch) infra-red absorption regions are the same as those for aliphatic amines (p. 372), but the C—N (stretch) region is different. Primary aromatic amines absorb in the region $1\,340$–$1\,250$ cm^{-1} (s); secondary, $1\,350$–$1\,280$ cm^{-1} (s); and tertiary, $1\,360$–$1\,310$ cm^{-1} (s). At the same time, the absorption regions of the aromatic nucleus will show its presence, with additional peaks according to the pattern of substitution.

Figure 23.1 is the infra-red spectrum of aniline (as a film), and Fig. 20.12 (p. 589) is the NMR spectrum of aniline.

Mass spectra

The peak for the molecular ion is strong and has an odd mass number. Primary amines lose a molecule of HCN, *e.g.*,

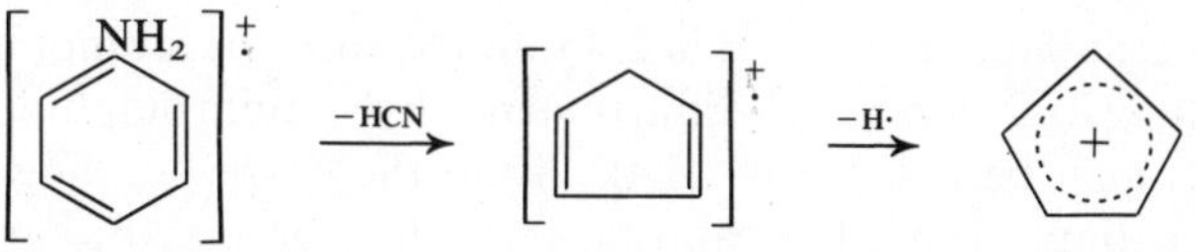

$$M^+ \; 93 \qquad m/e \; 66 \qquad m/e \; 65$$

Alkyl groups attached to the nucleus or to the nitrogen atom readily undergo β-cleavage, *e.g.*,

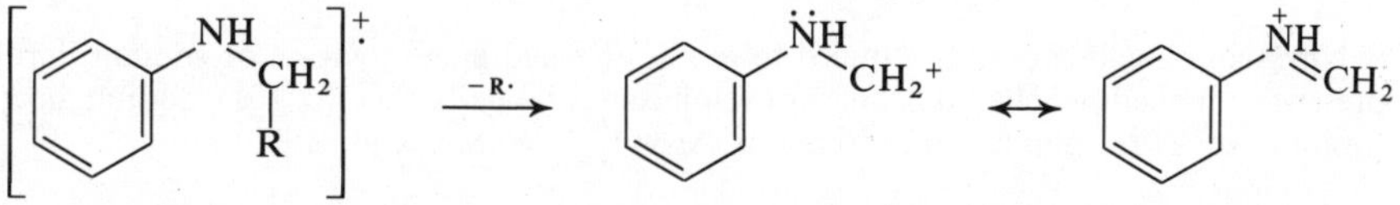

$$M^+ \qquad\qquad m/e \; 106$$

General methods of preparation of primary amines

1. Primary aromatic amines are usually prepared by the reduction of the corresponding nitro-compound. Various reducing reagents may be used, *e.g.*, tin, iron or zinc, with hydrochloric or acetic acid (see p. 649):

$$\text{ArNO}_2 \xrightarrow{\text{metal/acid}} \text{ArNH}_2 + 2\text{H}_2\text{O} \quad (v.g\text{–}ex.)$$

Zinc and hydrochloric acid often produce the chloroamino-compound as well as the desired amino-compound (see p. 664). Acetic acid is very useful for compounds which contain a group that is attacked by inorganic acid; *e.g.*, reduction of *p*-nitroacetanilide in hydrochloric acid solution leads to some hydrolysis of the acetyl group; use of acetic acid prevents this deacetylation.

Titanous chloride in hydrochloric acid reduces nitro-compounds to amines, and may be used for the quantitative estimation of nitro-groups in a compound:

$$\text{ArNO}_2 + 6\text{TiCl}_3 + 6\text{HCl} \longrightarrow \text{ArNH}_2 + 6\text{TiCl}_4 + 2\text{H}_2\text{O}$$

By means of aqueous ethanolic ammonium hydrogen sulphide, aqueous sodium sulphide, methanolic sodium hydrogen sulphide, or stannous chloride in hydrochloric acid, nitro-groups in a polynitro-compound can be reduced one at a time, *e.g.*, *m*-nitroaniline from *m*-dinitrobenzene:

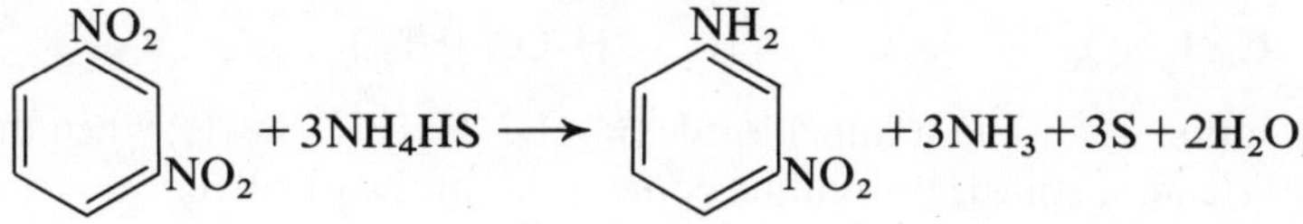

$$+\, 3\text{NH}_4\text{HS} \longrightarrow \qquad +\, 3\text{NH}_3 + 3\text{S} + 2\text{H}_2\text{O} \qquad (70\text{–}80\%)$$

It appears that ammonium hydrogen sulphide and sodium sulphide will not reduce a nitro-group in the *o*-position to an alkyl group, whereas stannous chloride in hydrochloric acid reduces this nitro-group preferentially:

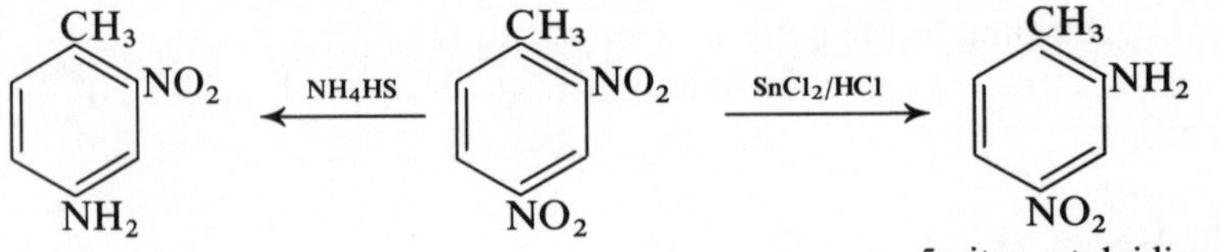

3-nitro-*p*-toluidine   5-nitro-*o*-toluidine

When the ring contains a group other than alkyl, then the course of reduction is less certain, *e.g.*, it is the *o*-nitro-group that is reduced in 2,4-dinitroaniline by ammonium hydrogen sulphide (52–58%), and the *o*-nitro-group in 2,4-dinitrophenol by sodium sulphide (58–61%).

Nitro-compounds may be reduced catalytically to the corresponding amine at 200–400°C; if Raney nickel or platinum is used as catalyst, the reduction may be carried out at room temperature. Catalytic reduction is particularly useful for compounds which contain a group that is hydrolysable in acid solution, *e.g.*, *p*-nitroacetanilide. Under milder conditions of catalytic reduction, nitro-compounds produce azo-compounds (see Table 6.1).

> Nitro-compounds are also reduced to amines by hydrazine hydrate or sodium borohydride in the presence of palladised charcoal, and by cyclohexene and palladium (p. 101). Sodium borohydride alone produces the corresponding radical anion, $ArNO_2^{-}\cdot$ (Waters *et al.*, 1970).

If a nitro-compound contains an unsaturated chain or an aldehyde group in the nucleus, alkaline ferrous sulphate may be used as the reducing agent to avoid reduction of these other groups. If only an aldehyde group is present, then stannous chloride may be used, *e.g.*, *m*-nitrobenzaldehyde may be reduced to *m*-aminobenzaldehyde:

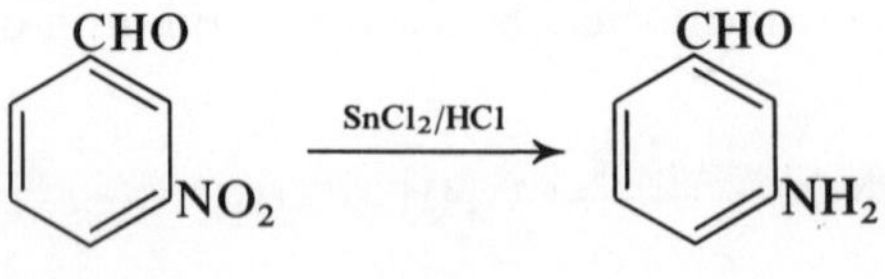

2. Amino-compounds may be prepared by reduction of aromatic nitroso-, azo-, azoxy-, and hydrazo-compounds, but since these are obtained from nitro-compounds, their use is limited to special cases (see, *e.g.*, p. 665).

3. Aromatic amines may be prepared by methods used for aliphatic amines, *e.g.*, amination of aryl halides, Hofmann amine preparation from amides (p. 264), Schmidt reaction with acids (p. 239), Curtius reaction (p. 268), etc. The Beckmann rearrangement of oximes (p. 755) is also useful in certain cases because of the ease of preparation of alkaryl ketones by the Friedel–Crafts reaction.

Aniline (*aminobenzene*). In the laboratory, aniline is prepared by the reduction of nitrobenzene with metal and acid, *e.g.*, tin and hydrochloric acid:

$$C_6H_5NO_2 \xrightarrow{e;\,H^+} C_6H_5NH_2 + 2H_2O \quad (80\%)$$

The aniline remains in solution as aniline stannichloride $(C_6H_5NH_2)_2\cdot H_2SnCl_6$; when the solution is made alkaline and steam distilled, the aniline comes over in the distillate.

Commercially, aniline is prepared by the reduction of nitrobenzene with iron, water and 30 per cent hydrochloric acid, or by catalytic reduction (copper catalyst).

Aniline (whose properties are characteristic of primary aromatic amines in general) is, when freshly prepared, a colourless liquid, b.p. 184°C; it has an unpleasant odour and is poisonous. When exposed to air, it rapidly darkens, since it is very sensitive to oxidation.

Aniline forms crystalline salts with strong inorganic acids, and these salts are considerably hydrolysed in solution. Aniline is a weaker base than the primary aliphatic amines, and this may be explained by resonance (which is not possible in aliphatic primary amines):

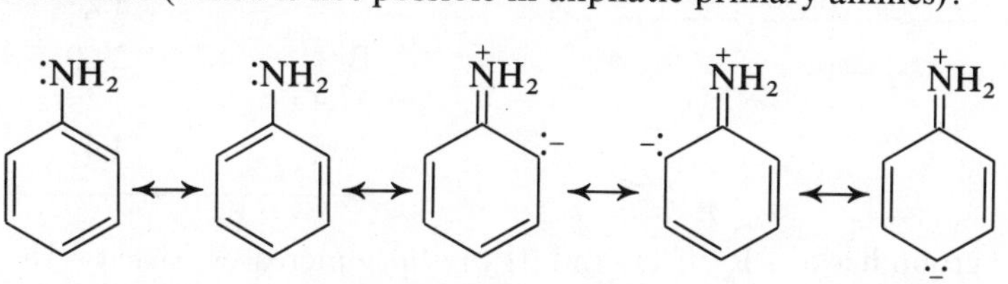

In aniline, owing to resonance, the lone pair of electrons on the nitrogen atom is less available for co-ordinating with a proton; at the same time, the small positive charge on the nitrogen atom would tend to repel a proton. Alternatively, since there are more resonating structures possible for aniline itself than for the cation $C_6H_5\overset{+}{N}H_3$, the former will be stabilised with respect to the latter.

The effect of a ring substituent on basicity depends on whether the substituent is electron-attracting or releasing, its ability to enter into resonance with the amino-group, and its position. All the nitroanilines are weaker bases than aniline (see Table 23.1). The nitro-group has a strong $-R$ effect, and o- and p-nitroaniline are therefore more resonance-stabilised than aniline itself.

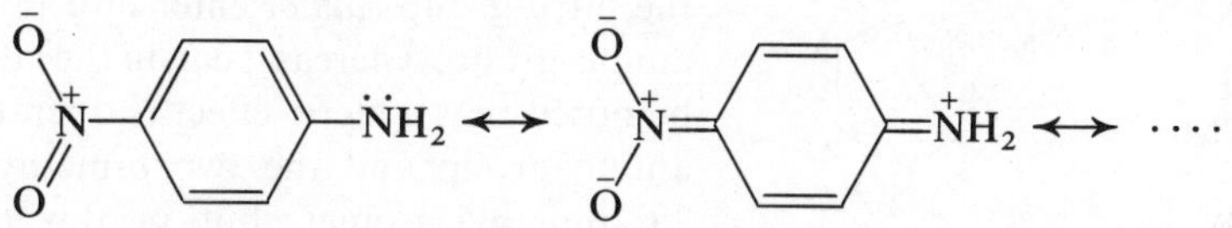

A m-nitro-group cannot enter into resonance with the amino-group, but nevertheless m-nitroaniline is a much weaker base than aniline. In this case, the explanation is that, in addition to exhibiting a $-R$ effect when possible, the nitro-group always has a strong $-I$ effect (due to the positive charge on the nitrogen atom). This $-I$ effect tends to draw into the ring the lone pair on the N atom of the NH_2 group, with consequent decrease in basicity. In the same way may be explained the decreased basicity due to halogen atoms in the ring (Table 23.1). The $+R$ effect of halogens is very small compared with their $-I$ effect (*cf.* p. 597). It can therefore be expected that increasing the number of nitro-groups or halogen atoms in the ring will decrease basicity.

When the ring contains an electron-releasing group, this reduces resonance of the NH_2 group with the ring, and consequently should increase basicity. The methyl group raises the electron density more at the o- and p-positions than at the m-. Thus, the carbon atom *para* to the methyl group has a high electron density, and the lone pair on the nitrogen atom is therefore prevented, to some extent, from entering into resonance with the ring. A methyl group raises the electron density at the m-position only to a very small extent. Hence resonance with the ring of the NH_2 group at this position is prevented less than for the p-position. Consequently, a methyl group in the ring increases the basicity of the aniline, more so from the p-position than from the m- (see Table 23.1).

Table 23.1

Base	pK_a	Base	pK_a
PhNH$_2$	4·58	PhNHEt	5·11
PhNHMe	4·85	PhNEt$_2$	6·56
PhNMe$_2$	5·06	Ph$_2$NH	0·9
		PhNHAc	0·4

Substituent in PhNH$_2$	pK_a		
	o-	m-	p-
Me	4·39	4·69	5·12
NO$_2$	−0·29	2·50	1·02
Cl	2·64	3·34	3·98
Br	2·60	3·51	3·91
MeO	4·49	4·20	5·29
NH$_2$	4·47	4·88	6·08

The p-OMe group has a +R effect, and its presence increases basicity. In the m-position, it can exert only a −I effect, and this decreases the basicity (see also Resonance and the Hammett equation, p. 608).

It can be seen from the foregoing account that when the substituent group can enter into resonance with the rest of the molecule, it has a marked effect on the basicity of the compound. It therefore follows that this effect will be diminished (or be completely absent) if this resonance can be diminished (or completely prevented). An essential requirement for resonance is that the resonating structures must be planar (or nearly planar). Thus, if planarity is partially reduced or completely prevented by the steric effect of a group, resonance is diminished or prevented. In these circumstances, the phenomenon is known as **steric inhibition of resonance**. This phenomenon may be used to explain the fact that 3,5-dimethyl-4-nitroaniline is a stronger base than the corresponding 2,6-dimethyl isomer. In the former, the nitro-group *cannot* enter into resonance with the amino-group, whereas it can in the latter. It might also be noted that a steric effect is operating between the amino group and the two o-methyl groups in the 2,6-dimethyl isomer, but in this case, since only hydrogen atoms (which are very small) are involved, the steric effect will be very much smaller than that due to oxygen atoms.

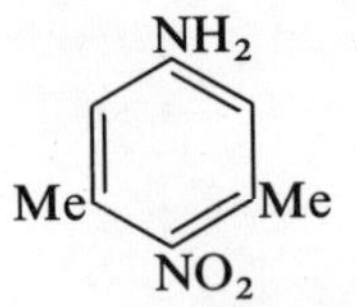

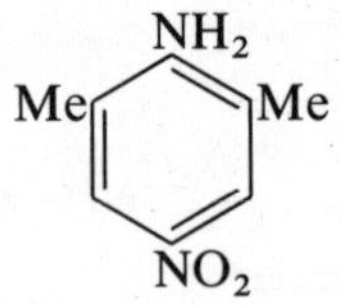

pK$_a$ 2·49   0·95

N-Alkylated anilines are stronger bases than aniline. At first sight one might have expected the opposite to be true, since alkyl groups have a +I effect and consequently increased resonance of the lone pair with the ring would be anticipated. Also, since an N-Et substituent increases the basic strength more than an N-Me, the observed results cannot be explained on the basis of polar effects. The explanation offered is that a steric effect operates (see also the *ortho* effect, p. 771). Since the ethyl group is larger than the methyl, the steric effect is greater for the former, and hence there is greater steric inhibition of resonance in the former. Thus, in the former, the lone pair on

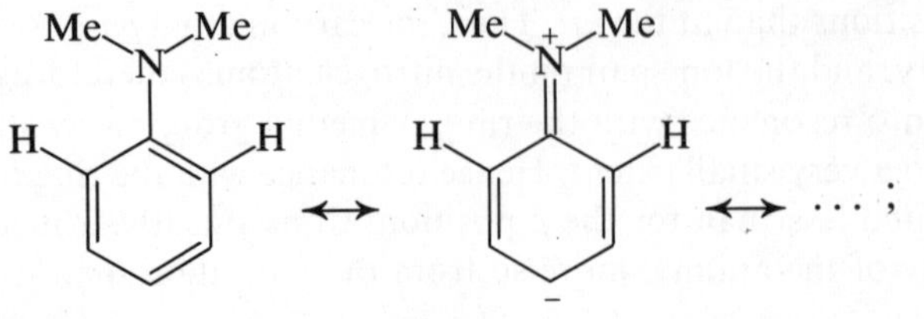

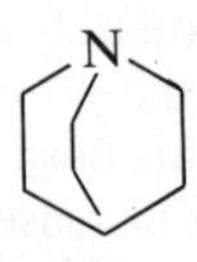

(I)

the nitrogen atom is more available for protonation, and consequently the basicity of N-ethyl-amine is greater than that of N-methylaniline. Steric inhibition of resonance will be further increased by the introduction of a second alkyl group on the nitrogen atom, and so basicity should·increase. This has been found to be the case (see Table 23.1). This argument for the steric effect is supported by other evidence, e.g., (i) N-t-butylaniline has a pK_a of 7·1 (the t-butyl group is very large). (ii) Quinuclidine ((I); pK_a of 10·58) is a very strong base; the bridge is perpendicular to the other ring, and a flat nitrogen atom is therefore not possible and so resonance is inhibited.

If we proceed with the above argument, then we can expect that o-methylaniline would be a stronger base than the m- and p-isomers, in which steric hindrance is absent. In practice, the reverse is true (see Table 23.1). Also, there are the basicities of the following compounds (in 50 per cent aqueous ethanol):

pK_a 5·07   4·69   4·09   (II)   (III)

The reason for these orders is not certain. A possible explanation is that angle α in the free base (II) is smaller than angle β in the protonated base (III). Thus protonation relieves crowding on the nitrogen atom but increases steric interaction with an o-methyl group. With one o-methyl group, nitrogen relief overcomes the steric effect of this methyl group, but with two o-methyl groups, their steric effect overcomes nitrogen relief; (but see also p. 374 and the ortho-effect, p. 771).

Hydrogen atoms of the amino-group are replaceable by halogen when aniline is treated with hypohalous acid and by sodium or potassium on heating:

$$C_6H_5NH_2 + K \longrightarrow [C_6H_5NH]^- K^+ + \tfrac{1}{2}H_2$$

Aniline combines with alkyl halides to give finally a quaternary ammonium compound, and gives the *isocyanide reaction* when heated with chloroform and ethanolic potassium hydroxide (see p. 367):

$$C_6H_5NH_2 + CHCl_3 + 3KOH \longrightarrow C_6H_5NC + 3KCl + 3H_2O$$

Aniline is readily acetylated, and condenses with aromatic aldehydes to form **anils** or **Schiff bases**; e.g., when warmed with benzaldehyde, it forms benzylideneaniline:

$$C_6H_5NH_2 + C_6H_5CHO \longrightarrow C_6H_5N{=}CHC_6H_5 + H_2O \quad (84\text{--}87\%)$$

These Schiff bases are easily hydrolysed to the free amine, and so their formation offers a means of 'protecting' an amino-group, e.g., during nitration. Schiff bases may also be used to prepare secondary amines:

$$C_6H_5N{=}CHAr \xrightarrow[\text{Ni}]{H_2} C_6H_5NHCH_2Ar$$

When refluxed with ethanolic carbon disulphide and solid potassium hydroxide, aniline forms **N,N'-diphenylthiourea** (*thiocarbanilide*), m.p. 154°C, which is used as a rubber accelerator:

$$CS_2 + 2C_6H_5NH_2 + 2KOH \longrightarrow S{=}C(NHC_6H_5)_2 + K_2S + 2H_2O \quad (70\%)$$

When treated with concentrated hydrochloric acid, diphenylthiourea is converted into phenyl isothiocyanate:

$$S{=}C(NHC_6H_5)_2 + HCl \longrightarrow C_6H_5NCS + C_6H_5NH_2{\cdot}HCl$$

Phenyl isothiocyanate may be conveniently prepared by running aniline into a cooled mixture of carbon disulphide and concentrated aqueous ammonium hydroxide, and decomposing the precipitated ammonium phenyldithiocarbamate with lead nitrate solution:

$$C_6H_5NH_2 + CS_2 + NH_4OH \longrightarrow C_6H_5NHCS_2NH_4 + H_2O \xrightarrow{Pb(NO_3)_2}$$
$$C_6H_5NCS + NH_4NO_3 + HNO_3 + PbS \quad (74\text{–}78\%)$$

The oxidation products of aniline are far more complex than those obtained from primary aliphatic amines (p. 377). Nitrosobenzene, nitrobenzene, phenylhydroxylamine, *p*-benzoquinone, azobenzene, azoxybenzene, aniline black (a dye), etc., may be isolated, the actual substance obtained depending on the nature of the oxidising agent used, *e.g.*, Caro's acid gives nitrosobenzene; peroxytrifluoroacetic acid, nitrobenzene (89%; Emmons, 1954); chromic acid, *p*-benzoquinone; sodium hypochlorite gives a purple coloration—this reaction is characteristic of aniline.

Aniline may be easily mercurated with an aqueous ethanolic solution of mercuric acetate in the cold or on warming.

The most important difference between aniline and primary aliphatic amines is their behaviour towards nitrous acid; aliphatic amines liberate nitrogen (see p. 376), whereas aniline (and all other primary aromatic monoamines) forms *diazonium salts*, which are stable in cold aqueous solution:

$$C_6H_5NH_2 + HNO_2 + HCl \longrightarrow C_6H_5N_2{}^+Cl^- + 2H_2O$$
$$\text{benzenediazonium chloride}$$

Acetanilide (N-*phenylacetamide*), m.p. 114°C, may be readily prepared, in very good yield, by heating aniline with acetyl chloride, acetic anhydride, or glacial acetic acid. It is hydrolysed by strong acids or alkalis to aniline. It is used in medicine, under the name of *antifebrin*, as a febrifuge. It is a useful intermediate in various reactions of aniline in which it is desirable to 'protect' the amino-group, *e.g.*, in nitration, halogenation, etc.; the acetamido-group is predominantly *p*-orienting. The *N*-phenyl derivative of the amide of any acid is known as an *anilide*.

The amino-group may also be protected by the formation of urethans with ethyl chloroformate (p. 459). Urethans are more readily hydrolysed than acetamido-compounds.

Halogenated anilines. The amino-group in aniline activates the *o*- and *p*-positions to a very large degree; *e.g.*, when aniline is treated with excess of chlorine- or bromine-water, an immediate precipitate of the 2,4,6-*trihalogeno-derivative* is obtained:

The yield is quantitative, and so this reaction is used to estimate aniline.

In order to introduce only one chlorine or bromine atom, the activating effect of the amino-group must be lowered; this may be done by acetylation (see also p. 664):

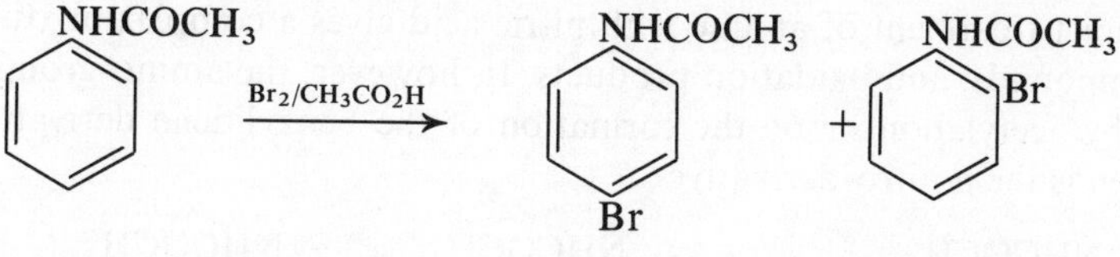

NHCOCH₃ → (Br₂/CH₃CO₂H) → NHCOCH₃ (with Br, 80%) + NHCOCH₃ (with Br, small amount)

These, on hydrolysis, yield the corresponding bromoanilines, which may be separated by steam distillation (the *o*-compound being steam volatile).

> The deactivating effect of the acetyl group may be explained as follows. The lone pair of the NH₂ group enters into resonance with the carbonyl group, and consequently resonance with the ring is

> strongly diminished. At the same time, this also explains the very weak basic character of acetanilide (see Table 23.1).

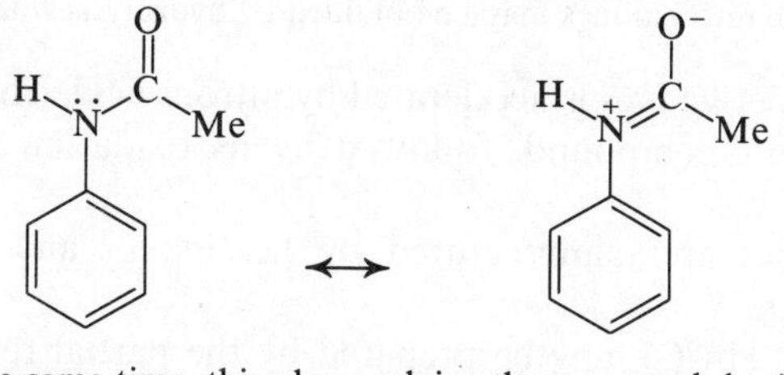

o-Bromoaniline may conveniently be prepared by first protecting the *p*-position by sulphonation (see also below);

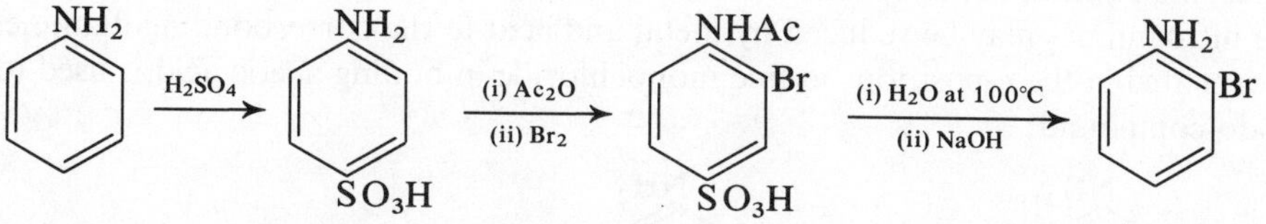

p-**Iodoaniline** may be prepared by the action of iodine on aniline in the presence of aqueous sodium hydrogen carbonate (*cf.* p. 624). The ring is highly activated by the amino-group and elemental iodine is now capable of direct iodination.

m-Chloro- and bromo-anilines may be prepared by reducing the corresponding halogeno-nitrobenzenes. They are also formed by the halogenation of aniline in concentrated sulphuric acid, the *m*-orienting effect being due to the presence of the $-\overset{+}{N}H_3$ group (see p. 597).

2,6-Dichloroaniline may be prepared by first blocking the *p*-position of aniline with a sulphonamide group, refluxing the sulphanilamide with hydrochloric acid and hydrogen peroxide, and subsequently refluxing the product, 3,5-dichlorosulphanilamide, with 70 per cent sulphuric acid, thereby removing the sulphonamide group:

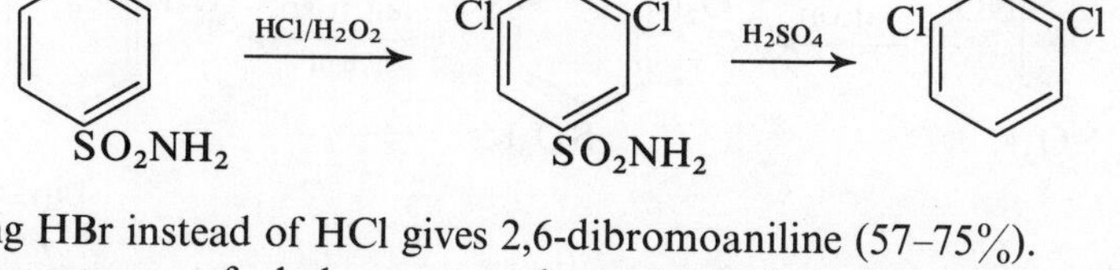

Using HBr instead of HCl gives 2,6-dibromoaniline (57–75%).

The presence of a halogen atom decreases the basic properties of aniline (see Table 23.1).

Nitroanilines. Direct treatment of aniline with nitric acid gives a complex mixture of mono-, di- and trinitro-compounds, and oxidation products. If, however, the amino-group is protected (and deactivated) by acetylation or by the formation of the benzylidene derivative, the main product on nitration is the *p*-nitro-derivative:

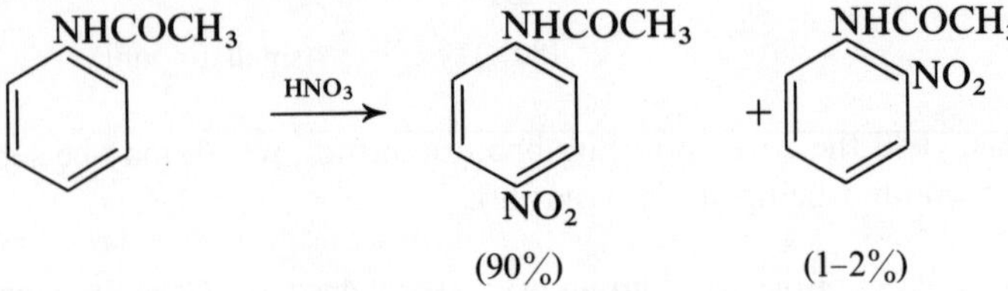

(90%)   (1–2%)

The nitroacetanilides may be separated by chromatographic adsorption, since the *o*-compound chelates. After separation, the nitroanilines may be obtained by hydrolysis with alkali.

Nitration of aniline derivatives is accelerated by nitrous acid. This is believed to be due to the formation of the nitroso-compound, followed by its oxidation to the nitro-compound (see phenol, p. 710).

o- and p-Nitroanilines are manufactured by heating *o*- and *p*-chloronitrobenzenes with ammonia under pressure.

m-Nitroaniline (m.p. 114°C) may be prepared by the partial reduction of *m*-dinitrobenzene with ammonium hydrogen sulphide (yield: 70–80%), sodium sulphide (yield: 70%), or stannous chloride in hydrochloric acid. *m*-Nitroaniline may also be prepared by the direct nitration of aniline *in the presence of concentrated sulphuric acid* (cf. *m*-chloroaniline, above). Industrially, it is prepared by reducing *m*-dinitrobenzene with the calculated quantity of iron (for one nitro-group) and a small amount of hydrochloric acid.

All three nitroanilines may be reduced by metal and acid to the corresponding diamines. All can be halogenated in the *o*-position; iodine monochloride in boiling acetic acid is used to prepare the iodo-compound:

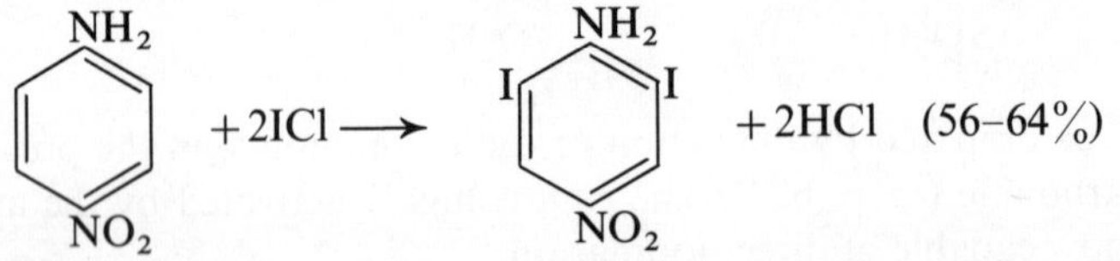

$+ 2ICl \longrightarrow$ $+ 2HCl$ (56–64%)

o- and *p*-Nitroanilines, but not *m*-, undergo activated nucleophilic substitution with boiling aqueous sodium hydroxide to give the corresponding nitrophenol. All the nitroanilines are weaker bases than aniline (see Table 23.1).

2,4-Dinitroaniline, m.p. 180°C, may be prepared by refluxing an ethanolic solution of 1-chloro-2,4-dinitrobenzene with concentrated ammonia solution. This is also an example of activated nucleophilic substitution.

2,6-Dinitroaniline may be prepared as follows (activated nucleophilic substitution):

(30–36%)

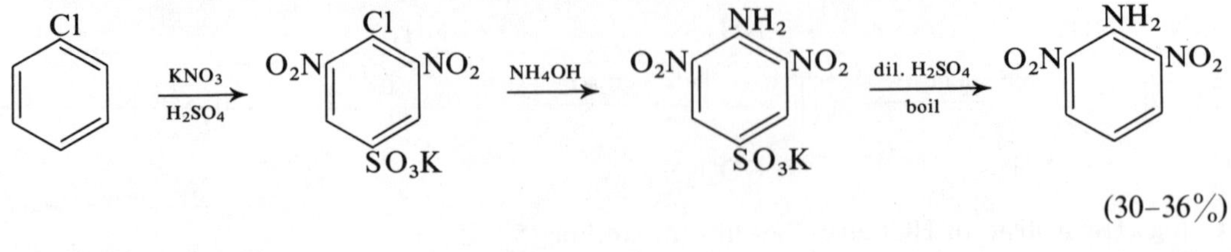

Aminobenzenesulphonic acids (see p. 696).

Toluidines (*tolylamines*). These may be obtained by reduction of the corresponding nitrotoluenes. *o*- and *m*-Toluidines are oils, b.p. 201° and 200°C, respectively; *p*-toluidine is a solid, m.p. 45°, b.p. 200°C.

N-Alkylanilines. The commonest *N*-alkylanilines are mono- and dimethylaniline. These may be prepared by heating aniline with methyl iodide:

$$C_6H_5NH_2 + CH_3I \longrightarrow C_6H_5NHCH_3 \cdot HI$$

$$C_6H_5NHCH_3 + CH_3I \longrightarrow C_6H_5N(CH_3)_2 \cdot HI$$

Commercially, these methylanilines are prepared by heating a mixture of aniline, methanol and sulphuric acid under pressure at 230°C. No quaternary compound is formed under these conditions (*cf.* p. 367). Monomethylaniline is the main product (50%) when about 1·2 molecules of methanol are used; dimethylaniline and unchanged aniline are also present. Dimethylaniline is obtained as the main product (95%) by using a large excess of methanol; monomethylaniline and unchanged aniline are also present, but these are removed by acetylating the mixture and distilling off the dimethylaniline.

To prepare higher *N*-alkyl derivatives of aniline, hydrochloric acid is used instead of sulphuric, since the latter tends to dehydrate the alcohol to alkene.

An interesting example of *reductive methylation* is the conversion of aniline into dimethylaniline; aniline and formaldehyde, in aqueous ethanolic sulphuric acid, is hydrogenated in the presence of platinum as catalyst (Pearson *et al.*, 1951).

$$C_6H_5NH_2 + 2CH_2O + 2H_2 \xrightarrow{\text{Pt}} C_6H_5N(CH_3)_2 + 2H_2O \quad (74\%)$$

Mono- and dimethylaniline are liquids, b.p. 196° and 193°C, respectively, and cannot be readily separated by fractional distillation. They can be separated, however, by acetylating the monomethyl derivative (as indicated above).

Pure monomethylaniline may be prepared by treating the sodium derivative of acetanilide with methyl iodide in toluene solution:

$$[C_6H_5NCOCH_3]^- Na^+ + CH_3I \longrightarrow C_6H_5N(CH_3)COCH_3 + NaI \xrightarrow{\text{KOH}} C_6H_5NHCH_3 \quad (90\%)$$

N-Alkylated-anilines are stronger bases than aniline (see p. 656).

Mono- and dimethylaniline resemble secondary and tertiary aliphatic amines in many ways, but differ in that they can also undergo substitution reactions (in the *o*- and *p*-positions) due to the presence of the benzene ring. Monoalkylanilines form pale yellow *N*-nitrosoamines with nitrous acid, and these give Liebermann's nitroso-reaction (p. 377):

$$C_6H_5NHCH_3 + HNO_2 \longrightarrow C_6H_5N(NO)CH_3 + H_2O \quad (87\text{--}93\%)$$

This may be converted into the original compound by heating with hydrochloric acid:

$$C_6H_5N(NO)CH_3 + H_2O \xrightarrow{\text{HCl}} C_6H_5NHCH_3 + HNO_2$$

The nitroso-compound may also be reduced to 1,1-methylphenylhydrazine:

$$C_6H_5N(NO)CH_3 \xrightarrow{\text{Zn/CH}_3\text{CO}_2\text{H}} C_6H_5N(CH_3)NH_2 + H_2O \quad (52\text{--}56\%)$$

N-Methylformanilide, b.p. 253°C, may be prepared by distilling a mixture of methylaniline, formic acid and toluene:

$$C_6H_5NHCH_3 + HCO_2H \longrightarrow C_6H_5N(CH_3)CHO + H_2O \quad (93\text{--}97\%)$$

It may also be prepared (in 80 per cent yield) by oxidising *N,N*-dimethylaniline with hydrated manganese dioxide (Henbest *et al.*, 1957).

It is used as a formylating agent in the preparation of aldehydes (see p. 744).

Dialkylanilines form *p*-nitroso-compounds with nitrous acid, *e.g.*, dimethylaniline forms *p*-nitrosodimethylaniline (green flakes):

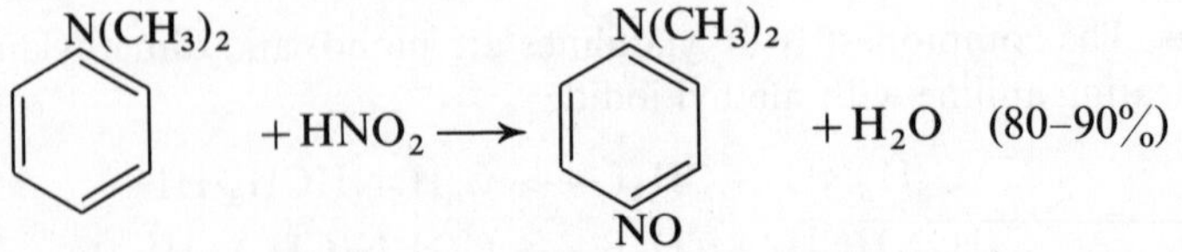

$$N(CH_3)_2 + HNO_2 \longrightarrow N(CH_3)_2\text{-}NO + H_2O \quad (80\text{--}90\%)$$

p-Nitrosodimethylaniline readily undergoes activated nucleophilic substitution, *e.g.*, it forms *p*-nitrosophenol with boiling aqueous sodium hydroxide. It is also readily oxidised by permanganate to *p*-nitrodimethylaniline, and may be reduced to *p*-aminodimethylaniline.

The dimethylamino-group activates the *p*-hydrogen atom so much that dimethylaniline will condense with formaldehyde and carbonyl chloride to form diphenylmethane derivatives, and with aromatic aldehydes to form triphenylmethane derivatives; *e.g.*, with carbonyl chloride, it forms *Michler's ketone*:

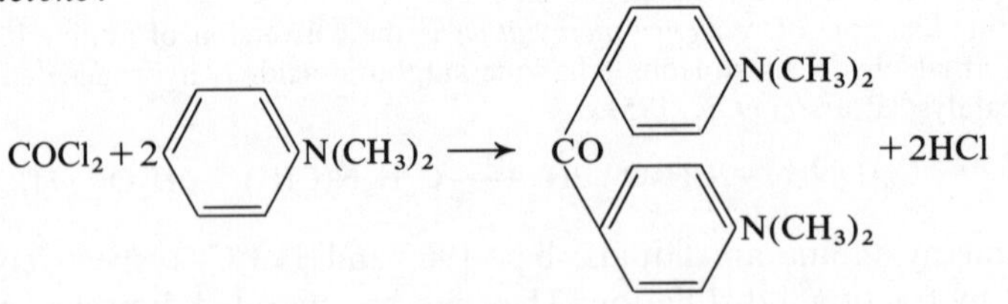

$$COCl_2 + 2\,C_6H_4N(CH_3)_2 \longrightarrow CO[C_6H_4N(CH_3)_2]_2 + 2HCl$$

Mono- and dialkylanilines (and the quaternary compounds) as their hydrochlorides (or hydrobromides) undergo rearrangement on strong heating, an alkyl group migrating from the nitrogen atom and entering preferentially the *p*-position or, if this is occupied, the *o*-; *e.g.*, when phenyltrimethylammonium chloride is heated under pressure, the following rearrangement takes place:

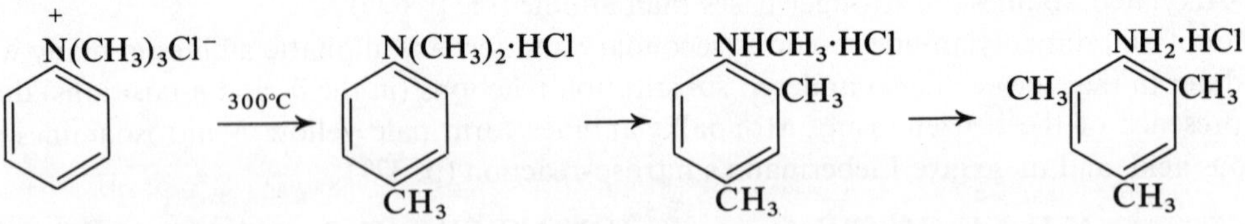

$$\overset{+}{N}(CH_3)_3Cl^- \xrightarrow{300°C} N(CH_3)_2\cdot HCl,\ CH_3 \longrightarrow NHCH_3\cdot HCl,\ CH_3 \longrightarrow NH_2\cdot HCl,\ CH_3,\ CH_3,\ CH_3$$

This reaction is known as the **Hofmann–Martius rearrangement** (1871); it may be used to prepare aniline homologues.

Many mechanisms have been proposed for the Hofmann–Martius rearrangement. Hughes and Ingold (1952) have proposed the following, based largely on the suggestion of Hickinbottom (1934) that the rearrangement occurs via the formation of an alkyl carbonium ion:

$$C_6H_5\overset{+}{N}H_2C_2H_5Cl^- \rightleftharpoons C_6H_5NH_2 + C_2H_5Cl$$
$$C_2H_5Cl \rightleftharpoons C_2H_5^+ + Cl^-$$
$$C_6H_5NH_2 + C_2H_5^+ \rightleftharpoons C_6H_5NH_3^+ + C_2H_4$$
$$2C_6H_5NH_2 + C_2H_5^+ \rightleftharpoons p\text{-}C_2H_5C_6H_4NH_2 + C_6H_5NH_3^+$$

It should be noted that the heterolytic fission of ethyl chloride is facilitated in the highly polar molten salt.

Rearrangements of this kind have been observed to take place with aniline derivatives of the type $C_6H_5N\text{—}Z$, where Z is R, X, NH_2, NO or NO_2, *e.g.*,

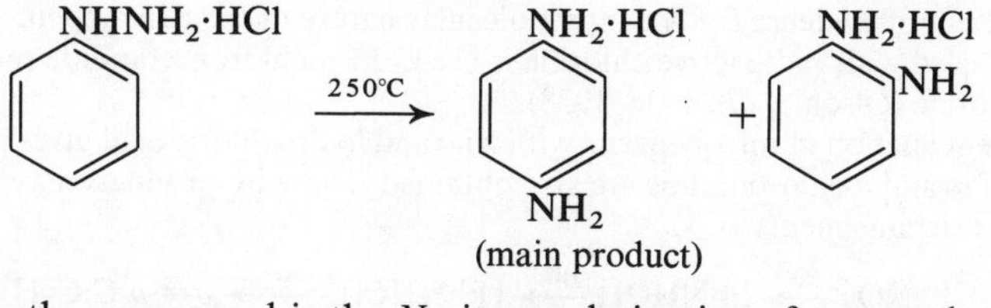

(main product)

When Z is NO, *i.e.*, the compound is the *N*-nitroso-derivative of a secondary aromatic amine, the rearrangement is known as the **Fischer–Hepp rearrangement** (1886).

The main product in the Fischer–Hepp rearrangement is the *p*-isomer. The mechanism of this reaction has long been *assumed* to be intermolecular (*cf.* the Orton rearrangement below), *i.e.*,

$$\text{PhNRNO} + \text{HCl} \rightleftharpoons \text{PhNHR} + \text{NOCl} \longrightarrow p\text{-RNHC}_6\text{H}_4\text{NO} + \text{HCl}$$

However, Williams *et al.* (1969), who carried out this rearrangement in the presence of ^{15}N-labelled sodium nitrite, showed that no ^{15}N was incorporated in the product. This suggests that the rearrangement is *intramolecular*, possibly by a mechanism similar to that for *N*-nitroanilines (see below).

Hughes *et al.* (1956, 1964) have studied the rearrangement of *N*-nitroaniline in sulphuric acid in the presence of enriched ^{15}N tracer. The products were *o*- and *p*-nitroaniline and they contained the normal amount of ^{15}N, thus showing that the rearrangement is *intramolecular*. This is consistent with the following six-centre mechanism for the nitramine rearrangement:

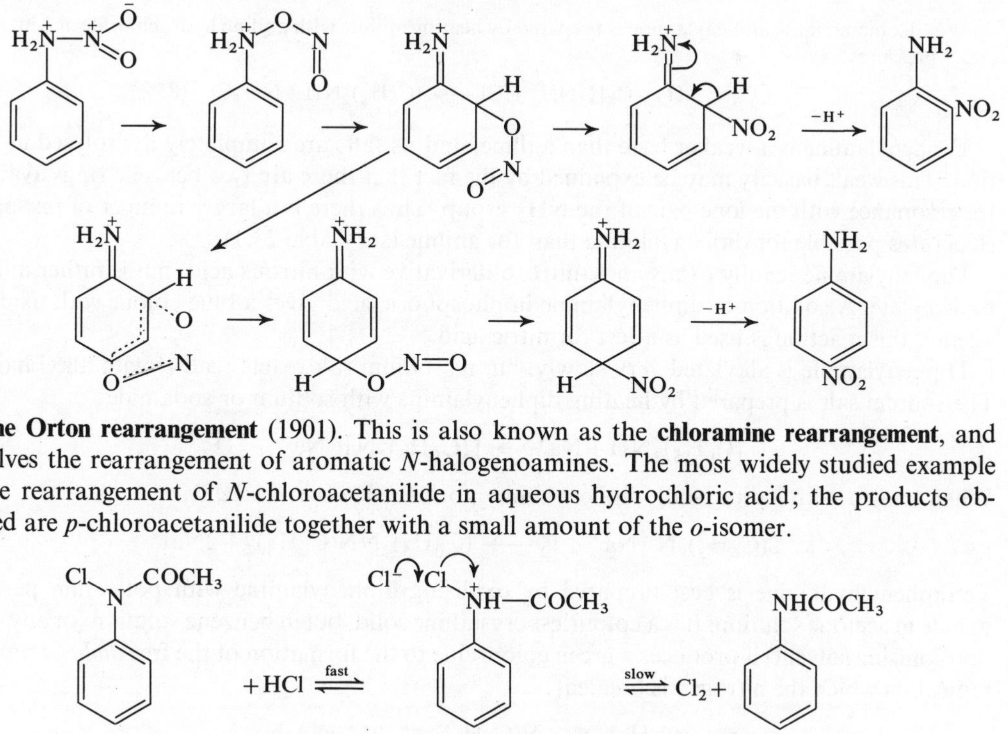

The Orton rearrangement (1901). This is also known as the **chloramine rearrangement**, and involves the rearrangement of aromatic *N*-halogenoamines. The most widely studied example is the rearrangement of *N*-chloroacetanilide in aqueous hydrochloric acid; the products obtained are *p*-chloroacetanilide together with a small amount of the *o*-isomer.

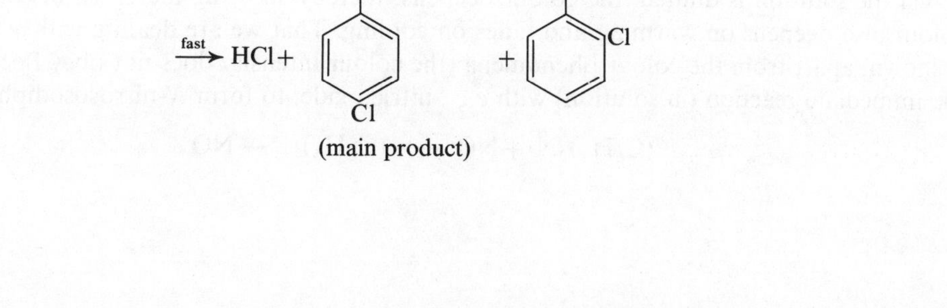

(main product)

One strong piece of evidence for the intermolecular nature of the rearrangement is that when hydrochloric acid labelled with radioactive chlorine is used, the *p*-chloroacetanilide produced contains some radioactive chlorine (Olson *et al.*, 1936–1938).

Although the reduction of nitrobenzene with zinc and hydrochloric acid gives predominantly aniline, fair amounts of *o*- and *p*-chloroaniline are also obtained. These by-products may possibly be formed via a chloroamine rearrangement:

$$PhNO_2 \xrightarrow[\text{HCl}]{\text{Zn}} PhNHOH \xrightarrow{\text{HCl}} [PhNHCl] \xrightarrow{\text{rearr.}} o\text{-}+p\text{-}ClC_6H_4NH_2$$

Diphenylamine, m.p. 54°C, may be prepared by heating phenol with aniline in the presence of zinc chloride at 260°C:

$$C_6H_5OH + C_6H_5NH_2 \xrightarrow{\text{ZnCl}_2} (C_6H_5)_2NH + H_2O$$

It may also be prepared by the **Ullmann reaction**; this is carried out by refluxing acetanilide, potassium carbonate, bromobenzene and a little copper powder in nitrobenzene solution (copper accelerates the reaction):

$$C_6H_5NHCOCH_3 + C_6H_5Br + K_2CO_3 \xrightarrow{\text{Cu}} (C_6H_5)_2NH + CO_2 + CH_3CO_2K + KBr \quad (60\%)$$

This is a useful general method for preparing substituted diphenylamines.

Commercially, diphenylamine is prepared by heating aniline with aniline hydrochloride at 140° under pressure:

$$C_6H_5NH_2 + C_6H_5NH_2 \cdot HCl \longrightarrow (C_6H_5)_2NH + NH_4Cl \quad (85\%)$$

Diphenylamine is a weaker base than aniline, and its salts are completely hydrolysed in solution. This weak basicity may be explained by the fact that there are two benzene rings available for resonance with the lone pair of the NH_2 group. Thus there is a larger number of resonating structures possible for diphenylamine than for aniline (see Table 23.1).

Diphenylamine readily forms an *N*-nitroso-derivative with nitrous acid, but is rather difficult to acetylate. A solution of diphenylamine in phosphoric acid gives a blue colour with oxidising agents; this reaction is used as a test for nitric acid.

Diphenylamine is alkylated very slowly, but its sodium salt reacts readily with alkyl halides. The sodium salt is prepared by heating diphenylamine with sodium or sodamide:

$$(C_6H_5)_2NH + Na \longrightarrow [(C_6H_5)_2N]^- Na^+ + \tfrac{1}{2}H_2$$

When treated with iodine, sodiodiphenylamine forms *tetraphenylhydrazine*:

$$2(C_6H_5)_2N^- Na^+ + I_2 \longrightarrow (C_6H_5)_2NN(C_6H_5)_2 + 2NaI$$

Tetraphenylhydrazine is best prepared by oxidising diphenylamine with potassium permanganate in acetone solution. It is a colourless crystalline solid, but in benzene solution (or any other non-ionising solvent) it produces a green colour due to the formation of the free *diphenylnitrogen radical*, in which the nitrogen is bivalent:

$$(C_6H_5)_2N\!-\!N(C_6H_5)_2 \rightleftharpoons 2(C_6H_5)_2N\cdot$$

When the solution is diluted, the colour deepens, thereby showing increased dissociation; the colour also deepens on warming and fades on cooling. That we are dealing with a free radical is shown, apart from the colour phenomena (the colour intensity does not obey Beer's law), by the immediate reaction (in solution) with *e.g.*, nitric oxide, to form *N*-nitrososodiphenylamine.

$$(C_6H_5)_2N\cdot + NO \longrightarrow (C_6H_5)_2N\!-\!NO$$

The stability of this free radical is attributed to its being a resonance hybrid (*cf.* triphenylmethyl, p. 788).

Triphenylamine, m.p. 127°C, may be prepared by the Ullmann reaction (in nitrobenzene solution):

$$2(C_6H_5)_2NH + 2C_6H_5I + K_2CO_3 \xrightarrow{\text{Cu}} 2(C_6H_5)_3N + CO_2 + 2KI + H_2O \quad (82\text{--}85\%)$$

Triphenylamine dissolves in concentrated sulphuric acid to give a blue solution. It is an extremely weak base, but its salts with HF, HBF_4, and $HClO_4$ have been prepared under special conditions, *e.g.*, triphenylammonium borofluorate, $Ph_3NH^+BF_4^-$, is precipitated when boron trifluoride is passed into a solution of triphenylamine in moist benzene (Sharp *et al.*, 1960).

Benzylamine (*α-aminotoluene*), b.p. 185°C, may be prepared by the reduction of phenyl cyanide or benzaldoxime:

$$C_6H_5CN \xrightarrow{\text{Na/C}_2\text{H}_5\text{OH}} C_6H_5CH_2NH_2 \xleftarrow{\text{Na/C}_2\text{H}_5\text{OH}} C_6H_5CH=NOH$$

It may also be prepared by heating benzyl chloride with ammonia under pressure, or by the Hofmann amine preparation on phenylacetamide:

$$C_6H_5CH_2CONH_2 \xrightarrow{\text{Br}_2/\text{KOH}} C_6H_5CH_2NH_2 + CO_2$$

Benzylamine resembles the aliphatic amines, and is best regarded (as far as the amino-group is concerned) as a phenyl-substituted methylamine; *e.g.*, it differs from the primary aromatic amines in that it is soluble in water, basic (more so than aniline), and with nitrous acid does not form a diazonium salt but gives benzyl alcohol:

$$C_6H_5CH_2NH_2 + HNO_2 \longrightarrow C_6H_5CH_2OH + N_2 + H_2O$$

Diamines

The aromatic diamines may be prepared by methods used for the monoamines, *e.g.*, by reduction of dinitro-compounds, nitroanilines, and in certain cases, aminoazo-compounds, *e.g.*, reduction of aminoazobenzene gives aniline and *p*-phenylenediamine:

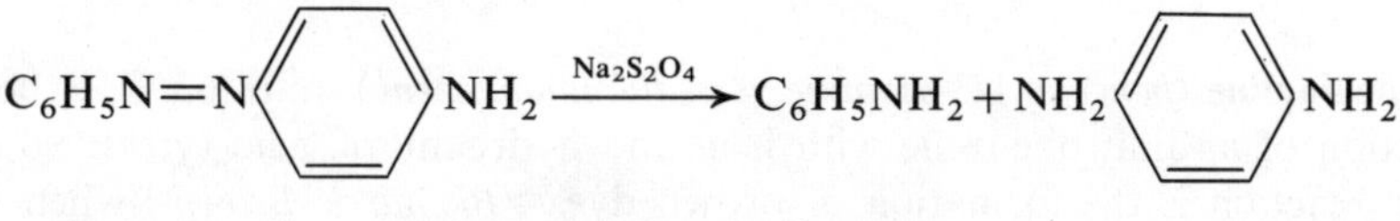

The diamines are colourless or white crystalline solids which turn brown when exposed to the air. They are 'di-acid' bases, and their reactions are characterised by those of the three phenylenediamines.

o-**Phenylenediamine** (*benzene-1,2-diamine*, *1,2-diaminobenzene*), m.p. 102°C, is best prepared by reducing *o*-nitroaniline with zinc dust and aqueous ethanolic sodium hydroxide (74–85%). The most characteristic property of *o*-diamines is the ease with which they form heterocyclic compounds.

(i) When *o*-phenylenediamine is treated with ferric chloride solution, a dark red colour is produced due to the formation of *2,3-diaminophenazine*:

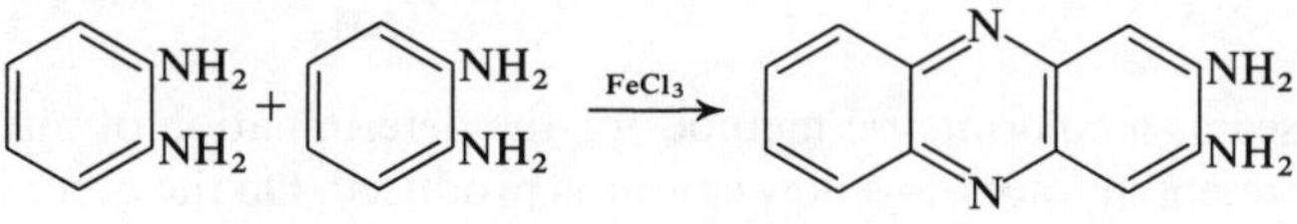

(ii) *Benzimidazoles* are formed when *o*-phenylenediamine is heated with organic acids, *e.g.*,

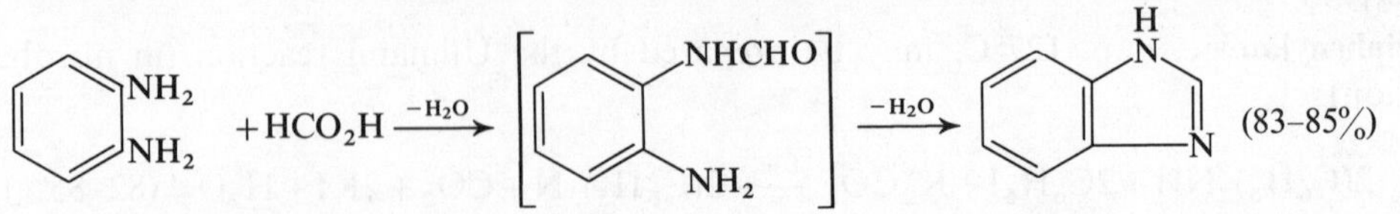

(iii) When *o*-phenylenediamine is treated with nitrous acid (a solution of the diamino-compound in acetic acid is treated with aqueous sodium nitrite), *benztriazole* is formed:

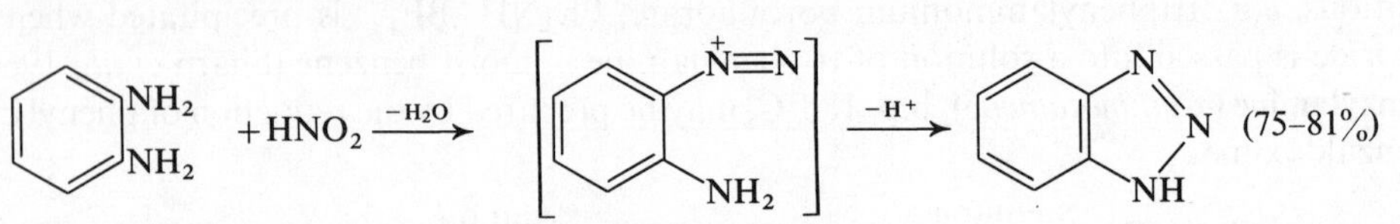

(iv) *o*-Phenylenediamine condenses with α-dicarbonyl compounds to form *quinoxalines*; *e.g.*, with glyoxal, quinoxaline is formed:

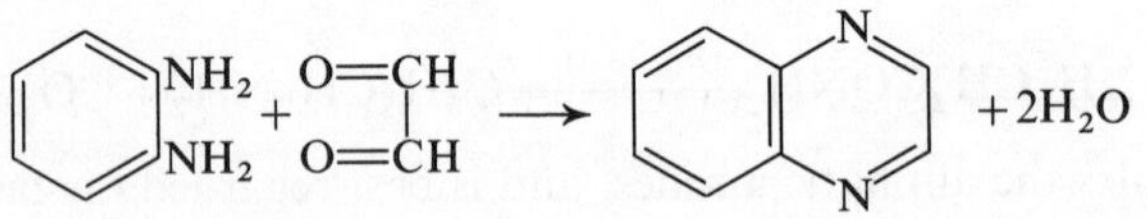

This reaction is used to identify *o*-diamines; the α-dicarbonyl compound employed for this purpose is phenanthraquinone, resulting in the formation of a sparingly soluble phenazine derivative:

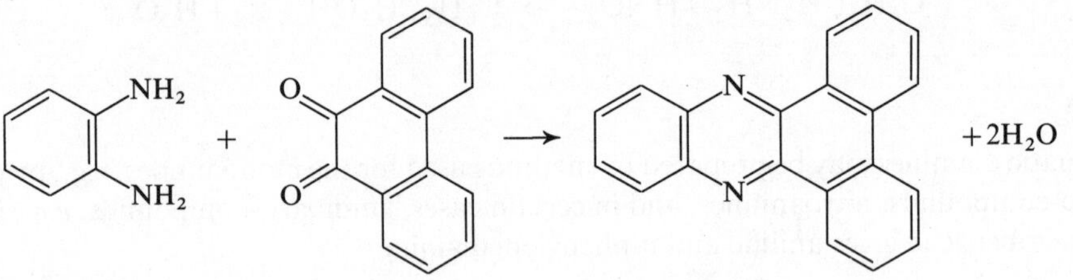

m-Phenylenediamine (*benzene-1,3-diamine*, *1,3-diaminobenzene*), m.p. 63°C, is best prepared by the reduction of *m*-dinitrobenzene with iron and hydrochloric acid (yield: 90%). Its most characteristic reaction is the formation of brown dyes—*Bismarck Brown*—when it is treated with nitrous acid; a monazo- and a bisazo-compound are formed:

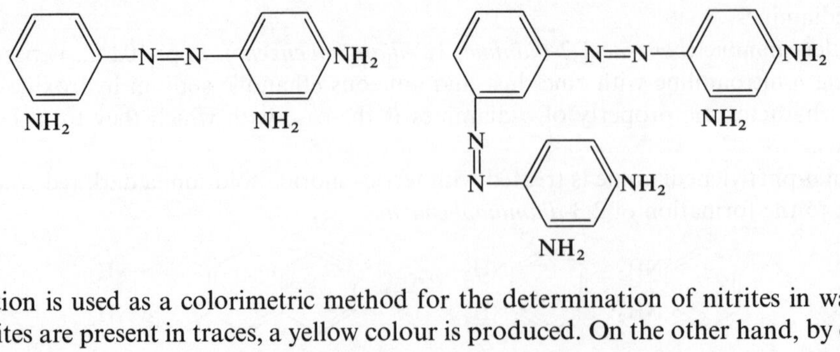

This reaction is used as a colorimetric method for the determination of nitrites in water; even when nitrites are present in traces, a yellow colour is produced. On the other hand, by dissolving *m*-phenylenediamine in concentrated hydrochloric acid, and by keeping the nitrous acid always in excess, both amino-groups may be diazotised to give the tetrazo-compound:

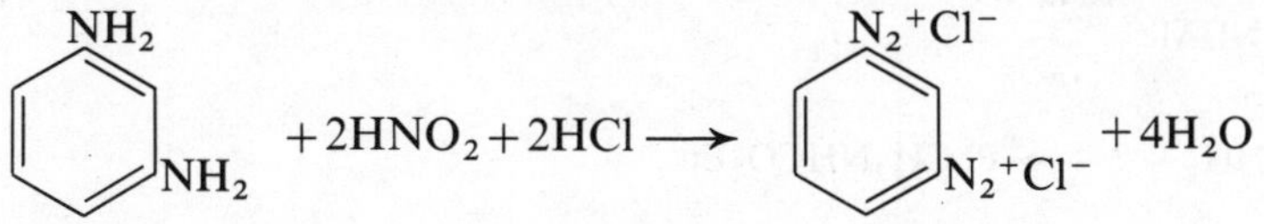
NH2
NH2
+2HNO2+2HCl
N2+Cl−
N2+Cl−
+4H2O

***p*-Phenylenediamine** (*benzene*-1,4-*diamine*, 1,4-*diaminobenzene*), m.p. 147°C, may be prepared by the reduction of *p*-nitroaniline or aminoazobenzene. On vigorous oxidation it forms *p*-benzoquinone:

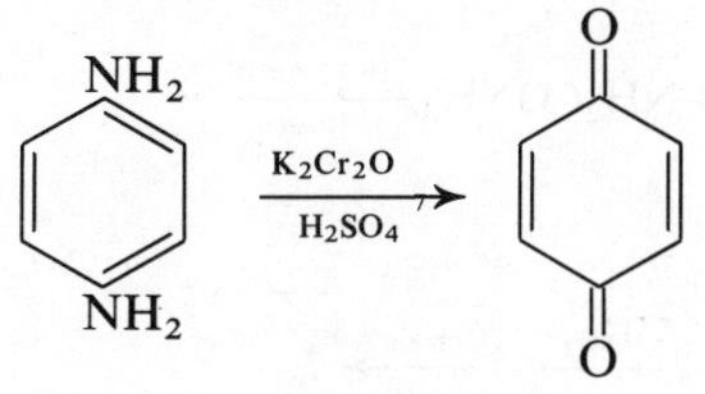
NH2
NH2
K2Cr2O
H2SO4
O
O

p-Phenylenediamine can be diazotised in the ordinary way, and is used in the preparation of dyes.

PROBLEMS

1. Draw the energy profiles for the reactions:

(i) $PhNH_2 + Z^+ \longrightarrow$ *o*-+*m*-+*p*-$ZC_6H_4NH_2$
(ii) $PhCl + Z^+ \longrightarrow$ *o*-+*m*-+*p*-ZC_6H_4Cl

2. Compare the polar effects in $Ph\overset{+}{N}Me_3$ and $Ph\overset{+}{P}Me_3$.

3. Discuss:

(i) $PhCH{=}CH_2 + Et_2NH \longrightarrow$ no reaction
(ii) *p*-$NO_2C_6H_4CH{=}CH_2 + Et_2NH \longrightarrow$ *p*-$NO_2C_6H_4CH_2CH_2NEt_2$

4. What would be the action of nitrous acid on each of the following: PhNHEt, $PhNMe_2$, *p*-MeC_6H_4NHMe, 2,5-$Me_2C_6H_3NH_2$?

5. What are the effects of the following groups on the basicity of aniline when in: (*a*) the *p*-position; (*b*) the *m*-position:

(i) NO_2; (ii) OR; (iii) CO_2R; (iv) NR_2?

6. Complete the equations:

(i) 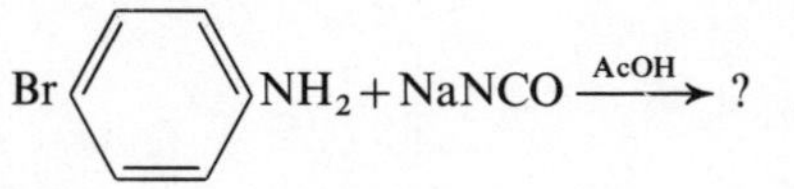
Br
NH2+NaNCO
AcOH
?

(ii) $PhNH_2 + \ ? \xrightarrow{Et_2O} PhNHCOCH{=}CHCO_2H$

(iii) 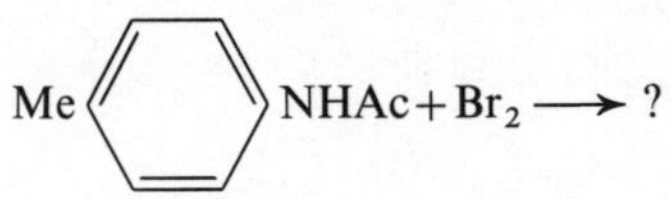
Me
NHAc+Br2
?

(iv) 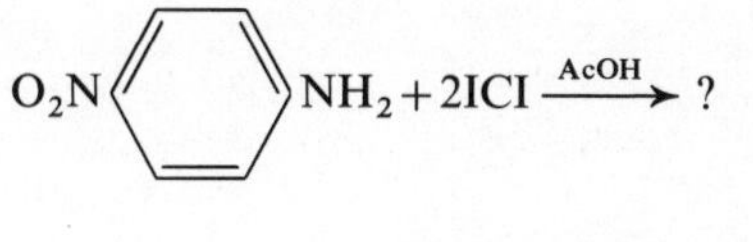
O2N
NH2+2ICl
AcOH
?

(v)
Me
NHMe
+NaNO2
HCl
?

(vi) $PhCH_2NH_2$ + ? ⟶ $PhCH_2NHCO_2Et$

(vii)
O2N
NH2+ ?
AcOEt
heat
O2N
NHCOCl

(viii)
EtO
NH2·HCl+NH2CONH2
HCl – AcOH
heat
?

(ix)
NH2
Ac2O
?
H2SO4
?
HNO3
?
steam
distil
?

(x) $Ph_2NH \xrightarrow{Li} ? \xrightarrow[CuI]{PhI} ?$

(xi)
Cl
NO2
Cl
+2NH3
heat
press.
?

(xii) $Ph_3N + MeI \longrightarrow ?$

(xiii)
COMe
M.A.
?
Zn – Hg
HCl
?

(xiv)
NMe3+Cl−
Me
Me
heat
?

(xv)
Br
CONH2
?
Br
NH2

(xvi)
CO2H
Cl
?
NH2
Cl
+N2+CO2

(xvii) $PhNH_2 + PhSO_2Cl \longrightarrow ? \xrightarrow[NaOH]{Me_2SO_4} ? \xrightarrow{H_3O^+} ?$

(xviii) $PhNH_2CH_2CHMe_2{}^+Br^- \xrightarrow{heat} ?$

(xix) $? \xleftarrow{LAH} PhNO_2 \xrightarrow[Pd-C]{NaBH_4} ?$

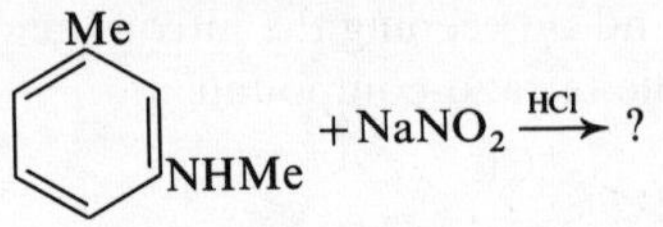

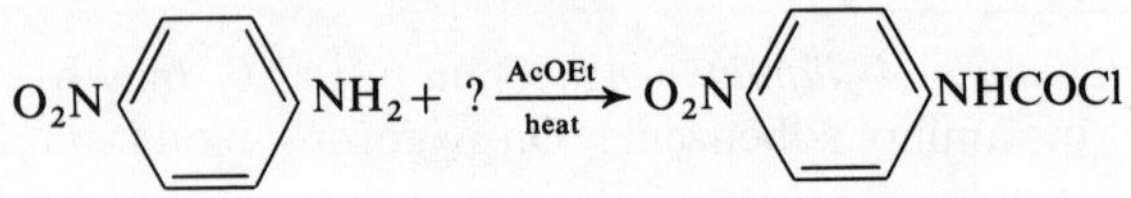

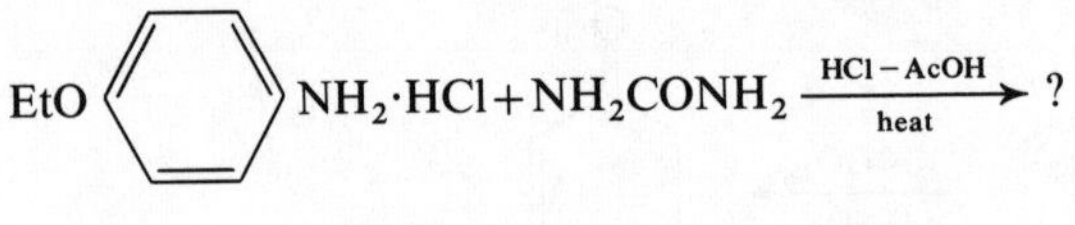

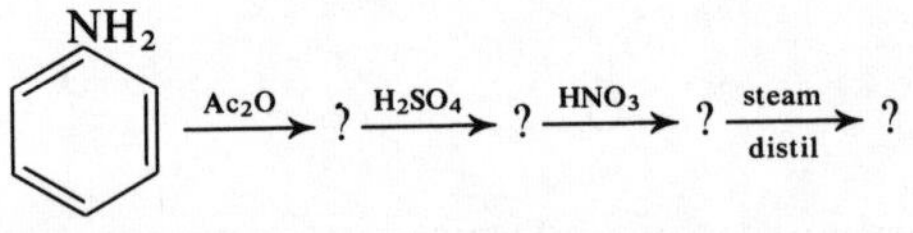

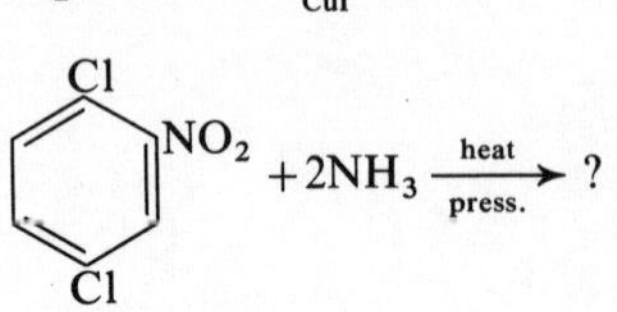

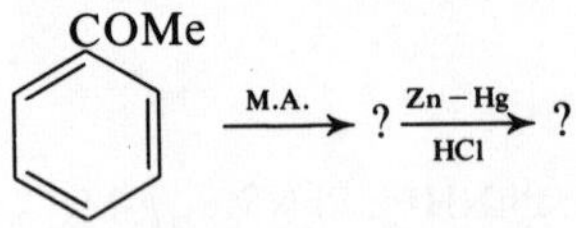

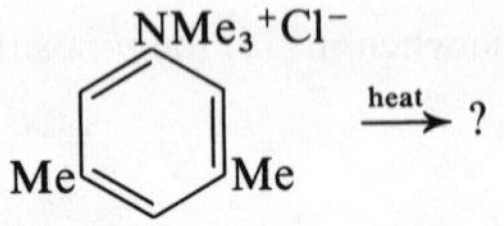

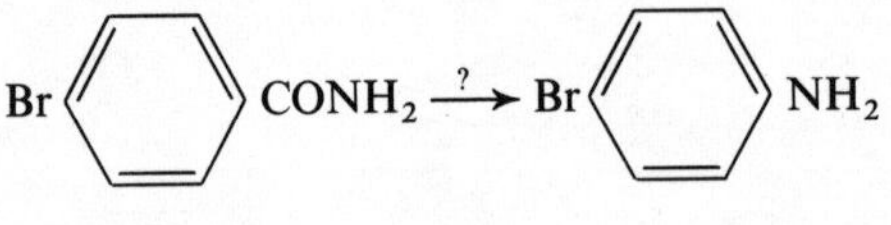

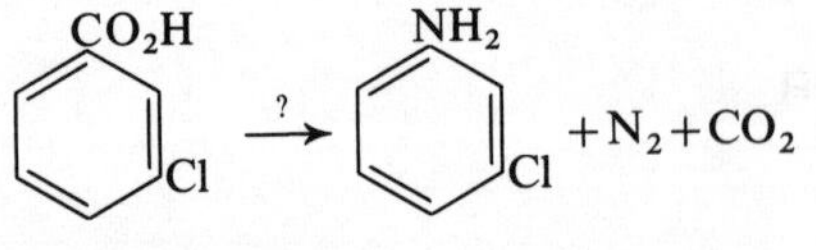

7. *N,N*-Dimethylaniline reacts with nitrous acid and so does the corresponding 2,6-dimethyl derivative, but the rate of reaction of the latter is much slower. Discuss.

8. Convert:

(i)   PhH $\longrightarrow$ *o*-ethylaniline;

(ii) PhMe $\longrightarrow$ 2-Br-4-Me-aniline;

(iii) PhH $\longrightarrow$ *m*-chloroaniline;

(iv) PhH $\longrightarrow$ *p*-iodoaniline;

(v)   PhH $\longrightarrow$ 2,4-dibromoaniline;

(vi) PhH $\longrightarrow$ 4-Br-3-Cl-aniline;

(vii) PhMe $\longrightarrow$ *p*-aminobenzoic acid;

(viii) PhH $\longrightarrow$ 2,5-dichloroaniline;

(ix) *p*-toluidine $\longrightarrow$ 4-NH_2-3-NO_2-toluene.

9. When a mixture of *N*-chloro-2,4-dichloroacetanilide and anisole (PhOMe) is dissolved in hydrochloric acid, among the products isolated are *o*- and *p*-chloroanisole. Discuss.

10. The N.M.R. spectrum of a compound, $C_8H_{11}N$, showed signals with τ-values at 8·6 (doublet), 8·4 (singlet), 5·9 (quartet), and 2·7 (singlet). Suggest a structure and give your reasons.

11. The mass spectrum of *N,N*-dimethylaniline contains the following ions: 121 (70%), 120 (100%), 105 (13%), 77 (25%), 51 (16%). Suggest paths by which they may be formed.

REFERENCES

SIDGWICK, *The Organic Chemistry of Nitrogen*, Oxford Press (1966, 3rd edn. by Millar and Springall), Ch. 5. 'Aromatic Amines'.

CLARK and PERRIN, 'Prediction of the Strengths of Organic Bases', *Quart. Rev.*, 1964, **18**, 295.

SHINE, *Aromatic Rearrangements*, Elsevier (1967).

24 Diazonium salts and their related compounds

When a primary aliphatic amine is treated with nitrous acid, the nature of the product depends on the amine used (see p. 376); nitrogen is always evolved:

$$RNH_2 + HNO_2 \longrightarrow ROH + N_2 + H_2O$$

When, however, a primary aromatic amine is treated with nitrous acid in a well-cooled solution, the product is an unstable compound known as a diazonium salt:

$$ArNH_2 + HNO_2 + HCl \longrightarrow ArN_2^+Cl^- + H_2O$$

Diazo-compounds may be obtained with primary aliphatic amines provided that the amino-group is attached to a carbon atom which is adjacent to a $-I$ group such as acyl, carbalkoxy, or cyano; *e.g.*, aminoacetic ester forms diazoacetic ester (p. 394).

The formation of a diazonium compound by the interaction of sodium nitrite, an inorganic acid (usually) and a primary aromatic amine, in ice-cold solution, is known as *diazotisation*. This reaction was discovered by Griess in 1858.

Diazonium salts have the structure $[Ar\overset{+}{N}\equiv N:]X^-$ (they are electrolytes in solution), and are salts of the strong base, *diazonium hydroxide*, $[Ar\overset{+}{N}\equiv N:]OH^-$, which has not yet been isolated, but is known in aqueous solution. Most diazonium salts of the inorganic acids are colourless solids, extremely soluble in water and many, particularly the nitrate, are explosive (in the solid state). They form complex salts with many metallic salts, of which one of the most important is zinc chloride, $(ArN_2)_2^{2+}ZnCl_4^{2-}$. These complex salts are stable in solution, and hence offer a means of stabilising a diazonium salt solution.

The stability of diazonium salts (as compared to RN_2^+) is attributed to resonance:

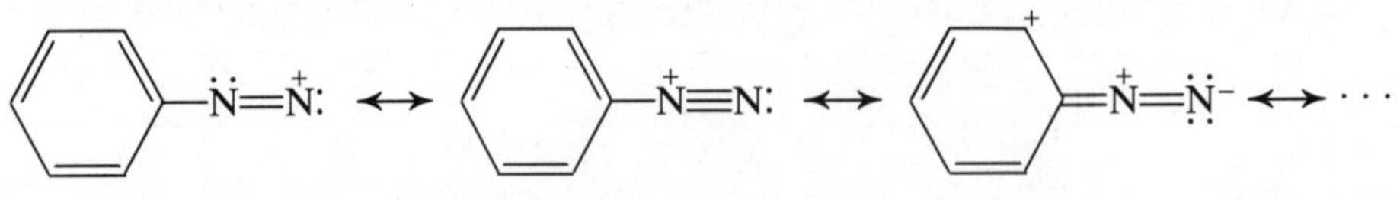

The diazonium salts are very important synthetic reagents, being the starting-point in the preparation of various aromatic compounds, dyes and drugs. As pointed out previously, the preparation of a diazonium salt in the solid state is usually a hazardous process. Fortunately, however, the aqueous solutions of the diazonium salts undergo practically all the necessary reactions (*cf.* the Grignard reagents).

There are various methods of preparing a solution of a diazonium salt, but the usual procedure is to dissolve (or suspend) the amine in excess of dilute inorganic acid (usually hydrochloric or sulphuric) cooled in ice and to add slowly, with stirring, a cooled aqueous solution of sodium nitrite, the addition of which is completed when the reaction mixture produces a blue colour with potassium iodide-starch paper, thereby showing the presence of free nitrous acid.

The mechanism of diazotisation is still not settled. Several factors affect the rate law of diazotisation, one being the acidity of the solution. In weakly aqueous acid solutions (and also in more strongly acid solutions), the rate-determining step is nitrosation of the free amine (to a primary nitrosamine), and a mechanism consistent with the kinetic studies is:

$$HO—NO + H^+ \xmroarrow{fast} H_2O^+—NO \xmroarrow{fast} H_2O + NO^+$$

$$ArNH_2 + NO^+ \xrightarrow{slow} ArNH_2{}^+NO \xrightarrow[fast]{-H^+} ArNHNO \xrightarrow[(rearr.)]{fast} Ar\ddot{N}{=}\ddot{N}—OH \xrightarrow[fast]{H^+}$$

$$Ar—\ddot{N}{=}\ddot{N}—OH_2{}^+ \xrightarrow{fast} Ar—\overset{+}{N}{\equiv}N{:} + H_2O$$

(I)

(II)

In the presence of certain substituents, diazotisation may lead to anomalous results. Thus, diazonium salts derived from aminocarboxylic and aminosulphonic acids are dipolar ions, *e.g.*, diazotised anthranilic (I) and sulphanilic (II) acids (see also p. 679).

Nomenclature

The name of the diazonium salt is obtained by adding diazonium chloride, diazonium sulphate, etc., to the name of the parent hydrocarbon, *e.g.*, $C_6H_5N_2{}^+Cl^-$, benzenediazonium chloride; $p\text{-}CH_3C_6H_4N_2{}^+HSO_4{}^-$, *p*-toluenediazonium sulphate.

Reactions of the diazonium salts

These may be divided into two groups; those which involve the liberation of nitrogen gas and the displacement of the diazo-group, $—N_2{}^+$, by another univalent group, and those in which the two nitrogen atoms are retained.

Replacement reactions

1. Replacement by hydroxyl. When a diazonium sulphate solution is boiled or steam distilled, the diazo-group is replaced by hydroxyl, *e.g.*,

$$C_6H_5N_2{}^+HSO_4{}^- + H_2O \longrightarrow C_6H_5OH + N_2 + H_2SO_4 \quad (g.)$$

This reaction is *nucleophilic* and occurs by the *unimolecular* mechanism involving an arene cation, and in the presence of halide ions (nucleophiles), aryl halides may also be formed.

$$Ar{-}N_2{}^+ \xrightarrow{slow} N_2 + Ar^+ \xrightarrow{fast} \begin{array}{l} \xrightarrow{H_2O} ArOH \\ \xrightarrow{X^-} ArX \end{array}$$

This mechanism has much to support it, *e.g.*, in water, the decomposition of benzenediazonium chloride is a first-order reaction. There are, however, some experimental results which are in apparent conflict with this mechanism. Electron-withdrawing groups, *e.g.*, CO_2H, in the *p*-position, retard the reaction. This is the anticipated result on the basis that heterolysis of the $Ar—N_2^+$ bond is the r/d step. By similar argument, the presence of a p-electron-releasing group, $e.g.$, OMe, would be expected to *accelerate* the decomposition. In practice, the opposite result is often observed. This has been explained as follows.

Resonance gives the $Ar—N_2^+$ bond some double bond character, with consequent increase in strength. Since the breaking of this bond is involved in the r/d step, the activation energy of this step is raised and so the reaction is retarded. This retardation is observed for groups which are electron-releasing by the +R effect (or by the hyperconjugative effect, $e.g.$, Me). On the other hand, electron-releasing groups in the m-position accelerate the reaction. In this case, the +R effect is absent; only the +I effect is operating.

It might be noted here that the arene cation cannot be stabilised by resonance, since a covalent pair of σ-electrons is lost, resulting in a vacant orbital *perpendicular* to the π-orbital ($i.e.$, p-orbitals) of the ring.

2. Replacement by hydrogen. Replacement of the diazo-group by hydrogen offers a means of preparing substituted aromatic compounds in which the substituents are in positions which they would not take up by direct substitution. According to Kornblum (1944, 1949), the most reliable method of replacing the diazo-group by hydrogen is by means of hypophosphorous acid. Also, this reaction is catalysed by cuprous salts.

A highly favoured mechanism for this reaction involves a free-radical chain:

Chain initiation

$$ArN_2^+Cl^- + H_2P(O)OH \longrightarrow HCl + Ar—N=N—PH(O)OH \longrightarrow Ar\cdot + N_2 + \cdot PH(O)OH$$

Chain propagation

$$Ar\cdot + H_2P(O)OH \longrightarrow ArH + \cdot PH(O)OH$$
$$ArN_2^+ + \cdot PH(O)OH \longrightarrow ArN_2\cdot + {}^+PH(O)OH$$

$$Ar\cdot + N_2 \qquad H_3PO_3 + H^+$$

Another common method of replacing the diazo-group by hydrogen is to dissolve the amine in a mixture of ethanol and concentrated sulphuric acid, add sodium nitrite and then warm (sometimes in the presence of copper as catalyst). Two products are usually obtained, one being the phenolic ether and the other the replacement of the diazo group by hydrogen.

$$AroEt \xleftarrow{\text{EtOH}} ArN_2^+HSO_4^- \xrightarrow{\text{EtOH}} ArH + CH_3CHO$$

The relative amounts of these two products depend predominantly on the nature of the diazo-compound and of the alcohol; also the pH of the solution and whether water is present affect the proportions of the products. If a −I group is present in the nucleus, and ethanol is used, the replacement of the diazo-group by hydrogen predominates. On the other hand, if methanol is used, the predominant product is an aryl methyl ether, so much so that this offers a means of preparing this type of ether.

The mechanism of this reaction is uncertain, but there is evidence to suggest that ethers are formed via phenyl cations which react with the alcohol ($cf.$ reaction 1; replace H_2O by ROH). On the other hand, the deamination reaction is believed to proceed via a free-radical mechanism involving aryl free radicals.

Some examples of replacement by hydrogen are:

m-*Bromotoluene*

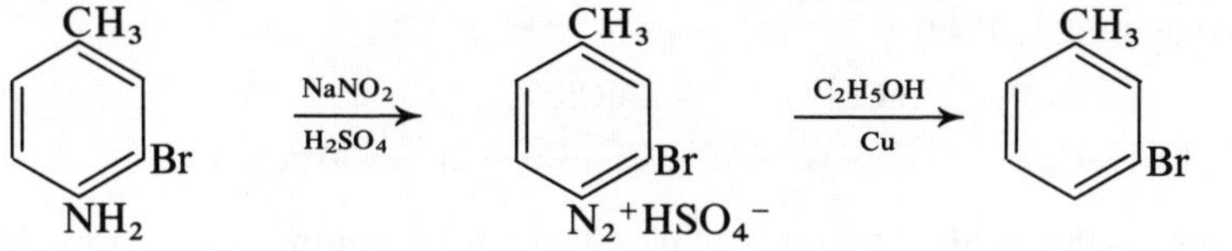

$$\text{CH}_3,\ \text{Br},\ \text{NH}_2 \xrightarrow[\text{H}_2\text{SO}_4]{\text{NaNO}_2} \text{CH}_3,\ \text{Br},\ \text{N}_2{}^+\text{HSO}_4{}^- \xrightarrow[\text{Cu}]{\text{C}_2\text{H}_5\text{OH}} \text{CH}_3,\ \text{Br} \qquad (54\text{–}59\%)$$

m-*Nitrotoluene*

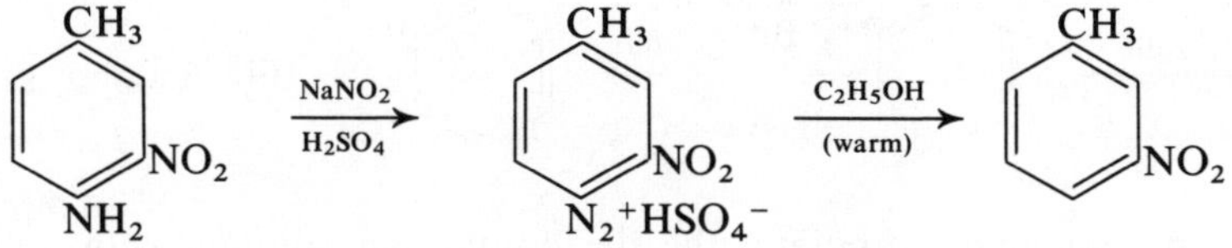

$$\text{CH}_3,\ \text{NO}_2,\ \text{NH}_2 \xrightarrow[\text{H}_2\text{SO}_4]{\text{NaNO}_2} \text{CH}_3,\ \text{NO}_2,\ \text{N}_2{}^+\text{HSO}_4{}^- \xrightarrow[\text{(warm)}]{\text{C}_2\text{H}_5\text{OH}} \text{CH}_3,\ \text{NO}_2 \qquad (62\text{–}72\%)$$

The diazo-group may be replaced by hydrogen in many other ways, *e.g.*, (i) decomposing the fluoroborate in ethanol in the presence of zinc dust (47–85%; Roe *et al.*, 1952); (ii) dissolving the hexafluorophosphate, $\text{ArN}_2{}^+\text{PF}_6{}^-$, in tetramethylurea (Rutherford *et al.*, 1963).

Apart from the factors described above that affect the nature of the deaminated product, there is also the fact that the diazo-group is a strong activator in nucleophilic substitution (p. 626). This is illustrated by the following example, in which replacement by hydrogen occurs together with activated nucleophilic substitution:

$$\text{Br},\ \text{N}_2{}^+\text{Cl}^-,\ \text{Br} \xrightarrow{\text{EtOH}} \text{Br},\ \text{Cl} \quad +\text{N}_2+\text{Br}^-$$

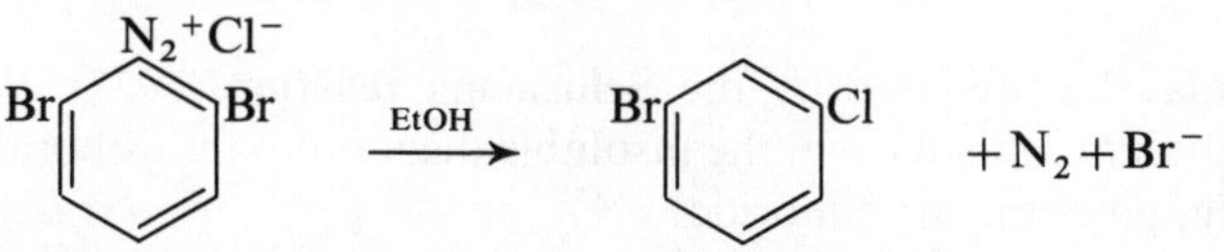

3. Replacement by halogen. (i) **Sandmeyer reaction** (1884). When a diazonium salt solution is run into a solution of cuprous halide dissolved in the corresponding halogen acid, the diazo-group is replaced by a halogen atom, *e.g.*,

m-*Chloronitrobenzene*

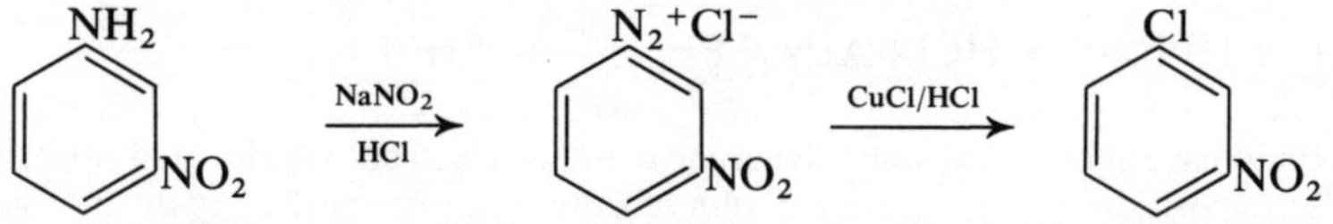

$$\text{NH}_2,\ \text{NO}_2 \xrightarrow[\text{HCl}]{\text{NaNO}_2} \text{N}_2{}^+\text{Cl}^-,\ \text{NO}_2 \xrightarrow{\text{CuCl/HCl}} \text{Cl},\ \text{NO}_2 \qquad (68\text{–}71\%)$$

p-*Bromotoluene*

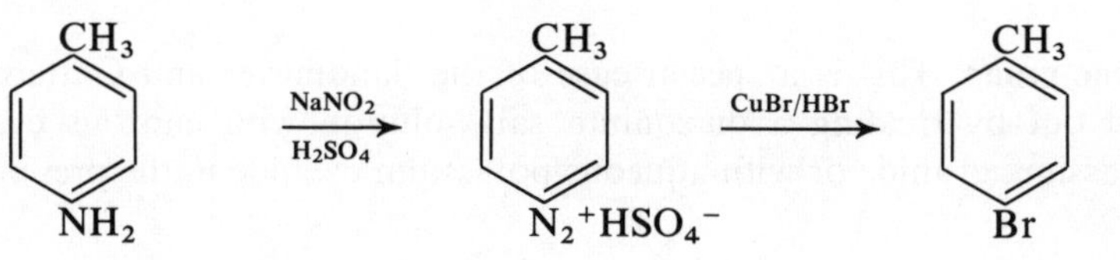

$$\text{CH}_3,\ \text{NH}_2 \xrightarrow[\text{H}_2\text{SO}_4]{\text{NaNO}_2} \text{CH}_3,\ \text{N}_2{}^+\text{HSO}_4{}^- \xrightarrow{\text{CuBr/HBr}} \text{CH}_3,\ \text{Br} \qquad (70\text{–}73\%)$$

The diazotisation may be carried out with hydrobromic acid, but in practice sulphuric acid is used since it is cheaper and only slightly affects the yield of aryl bromide.

The mechanism of the reaction is uncertain. A possibility is as follows. The 'cuprous chloride' is actually $CuCl_2^-$ (Cowdrey *et al.*, 1949).

$$CuCl + Cl^- \longrightarrow CuCl_2^-$$

$$ArN_2^+ + Cl—Cu—Cl^- \xrightarrow{slow} N_2 + Ar\cdot + Cl—Cu—Cl \xrightarrow{fast} ArCl + CuCl$$

(ii) **Gattermann reaction** (1890). This reaction is carried out by warming the diazonium salt solution in the presence of copper powder, *e.g.*,

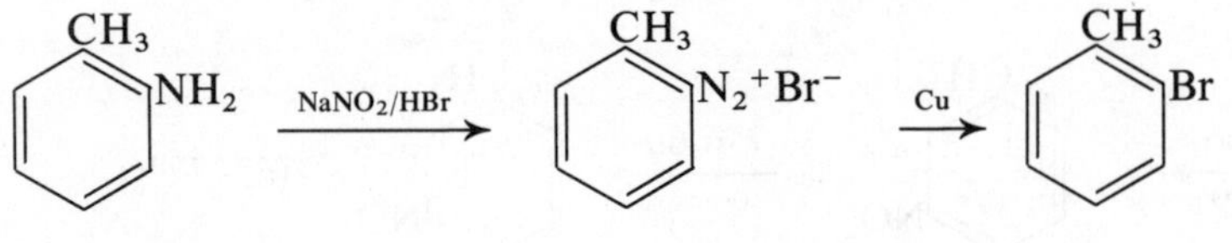

$$(43–47\%)$$

Iodo-compounds may be prepared by boiling the diazonium salt solution with aqueous potassium iodide:

$$C_6H_5N_2^+Cl^- + KI \longrightarrow C_6H_5I + KCl + N_2 \quad (74–76\%)$$

This is usually the best method for introducing iodine into the benzene ring.

The mechanism of this reaction is uncertain, but a possibility is that it occurs by a bimolecular mechanism, the displacement of the diazo-group being facilitated by the strong nucleophilic iodide ion (*cf.* p. 154). However, there is evidence that I_3^- ions are also the attacking species. These are produced by oxidation of I^- ions to I_2 by the diazonium ion or nitrous acid, followed by

$$I_2 + I^- \rightleftharpoons I_3^-$$

Fluoro-compounds may be prepared by the **Schiemann reaction** (1927). When fluoroboric acid is added to a diazonium salt solution, the insoluble diazonium fluoroborate is precipitated. This, on gentle heating, gives the aryl fluoride:

$$PhN_2^+Cl^- + HBF_4 \longrightarrow PhN_2^+BF_4^- \longrightarrow N_2 + Ph^+ + BF_4^- \longrightarrow PhF + BF_3 \quad (51–57\%)$$

Fluoroboric acid may be prepared by adding hydrochloric acid to sodium fluoroborate, or by dissolving boric acid in 50 per cent hydrofluoric acid (contained in a lead vessel); the latter usually gives better yields. A better method is to prepare the diazonium hexafluorophosphate; this is less water-soluble than the fluoroborate:

$$ArN_2^+Cl^- + HPF_6 \longrightarrow HCl + ArN_2^+PF_6^- \xrightarrow{heat} ArF + N_2 + PF_5$$

Suschitzky *et al.* (1969) have shown that when the arene contains electron-withdrawing substituents, *e.g.*, NO_2, CO_2H, decomposition of the arenediazonium hexafluoroantimonate gives the best yield of aryl fluoride.

$$ArN_2^+AsF_6^- \xrightarrow{heat} ArF + AsF_5 + N_2$$

4. Replacement by a cyano-group. This is a special case of the Sandmeyer and Gattermann reactions, and is carried out by treating a diazonium salt solution with cuprous cyanide dissolved in aqueous potassium cyanide or with aqueous potassium cyanide in the presence of copper powder:

$$C_6H_5N_2{}^+Cl^- \xrightarrow{K_3Cu(CN)_4} C_6H_5CN + N_2 \quad (65\%)$$

$$C_6H_5N_2{}^+HSO_4{}^- + KCN \xrightarrow{Cu} C_6H_5CN + KHSO_4 + N_2 \quad (60\%)$$

5. Replacement by a nitro-group. This may be carried out in several ways. A common method is to decompose the diazonium fluoroborate with aqueous sodium nitrite containing copper powder:

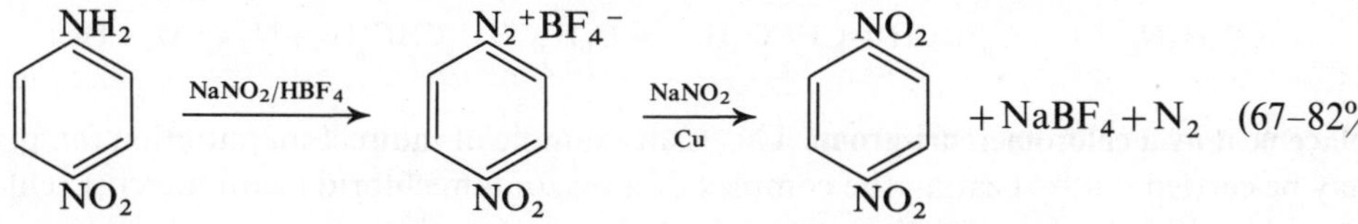

6. Replacement by an arsonic acid group, AsO_3H_2. One method is **Bart's reaction** (1910). This is carried out by decomposing a diazonium salt with sodium arsenite in the presence of a copper salt; *e.g.*, phenylarsonic acid:

$$C_6H_5N_2{}^+Cl^- + Na_3AsO_3 \xrightarrow{CuSO_4} C_6H_5AsO_3Na_2 + N_2 + NaCl \xrightarrow{HCl}$$

$$C_6H_5AsO_3H_2 \quad (39\text{--}45\%)$$

The yield may be increased by buffering the solution with sodium carbonate (Blas, 1940).

7. Replacement by an aryl group. This may be carried out by treating a diazonium sulphate with ethanol and copper powder, *e.g.*, biphenyl from benzenediazonium sulphate:

$$2C_6H_5N_2{}^+HSO_4{}^- \xrightarrow{C_2H_5OH} C_6H_5C_6H_5 + 2N_2 + 2H_2SO_4 \quad (25\%)$$

This is really a special case of the Gattermann reaction. Alternatively, a biaryl may be prepared by adding an aromatic compound to an *alkaline* solution of a diazonium salt, *e.g.*, *p*-bromobiphenyl from *p*-bromobenzenediazonium chloride and benzene (see also p. 781):

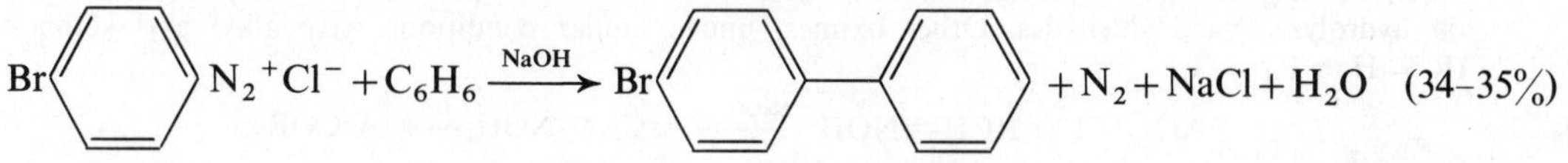

This method of preparation is known as the **Gomberg reaction** (1924), and experiment has shown that whatever is the nature of a substituent in the second component, *o*- or *p*-substitution always occurs predominantly, but some *m*-product is also formed; *e.g.*, benzenediazonium chloride forms *p*-nitrobiphenyl when treated with nitrobenzene:

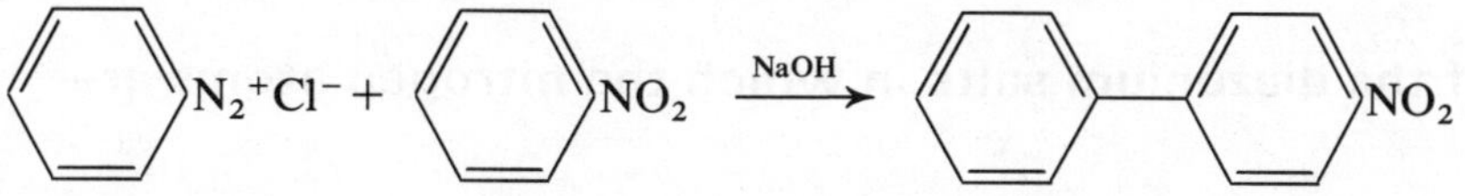

Thus, the reaction takes place by a free-radical mechanism (see also p. 818).

$$C_6H_5N_2{}^+Cl^- \xrightarrow{OH^-} C_6H_5-N=N-OH \longrightarrow C_6H_5\cdot + N_2 + \cdot OH$$

$$C_6H_5\cdot + C_6H_5NO_2 + \cdot OH \longrightarrow p\text{-}C_6H_5-C_6H_5NO_2 + H_2O$$

Meerwein reaction (1939). This is the reaction between diazonium salts and compounds containing an activated ethylenic bond to give arylated products. The reaction is carried out in acetone solution in the presence of a cupric salt, *e.g.*, benzenediazonium chloride adds to vinyl cyanide with elimination of nitrogen. The mechanism is uncertain, but it is generally accepted to be homolytic, addition occurring at the β-position (this free radical is more stable than the one formed from α-addition).

$$C_6H_5\cdot + CH_2{=}CHCN \longrightarrow C_6H_5{-}CH_2{-}\dot{C}HCN \xrightarrow{Cu^{2+}}$$

$$Cu^+ + C_6H_5{-}CH_2{-}\overset{+}{C}HCN \xrightarrow{Cl^-} C_6H_5CH_2CHClCN$$

When the alkene is an α,β-unsaturated acid, then arylation may occur with decarboxylation, e.g., benzenediazonium chloride and cinnamic acid give stilbene:

$$C_6H_5N_2{}^+Cl^- + C_6H_5CH{=}CHCO_2H \longrightarrow C_6H_5CH{=}CHC_6H_5 + N_2 + CO_2 + HCl$$

8. Replacement by a chloromercuri-group. This is an example of indirect mercuration (cf. p. 619), and may be carried out by heating the complex of a diazonium chloride with mercuric chloride in acetone or ethanol solution with copper powder:

$$ArN_2{\cdot}HgCl_3 + 2Cu \longrightarrow ArHgCl + N_2 + 2CuCl \quad (40\%)$$

The diazo-group may be replaced by many other groups, e.g.,

(i) $ArN_2{}^+Cl^- \xrightarrow{CuSCN} ArSCN$

(ii) $ArN_2{}^+Cl \xrightarrow{KNCO/Cu} ArNCO$

(iii) $2ArN_2{}^+Cl \xrightarrow{(NH_4)_2S} Ar_2S$

(iv) $ArN_2{}^+HSO_4{}^- + SO_2 + Cu \longrightarrow ArSO_2H + N_2 + CuSO_4$

(v) $ArN_2{}^+Cl^- + S{=}C\overset{\displaystyle SK}{\underset{\displaystyle OC_2H_5}{\big\langle}} \longrightarrow S{=}C\overset{\displaystyle SAr}{\underset{\displaystyle OC_2H_5}{\big\langle}} \xrightarrow{H^+} ArSH$

An alternative route to the thiol is:

$$ArN_2{}^+Cl^- \xrightarrow{NaSH} ArSH$$

(vi) Beech (1954) has shown that formaldoxime reacts with diazonium salts to give products which, on hydrolysis, yield aldehydes. Other oximes, under similar conditions, give alkyl aryl ketones (R = H or R):

$$ArN_2{}^+Cl^- + RCH{=}NOH \xrightarrow{NaOH} ArCR{=}NOH \xrightarrow{H^+} ArCOR$$

(vii) Cookson et al. (1962) have shown that the addition of acetic acid to diazonium fluoroborates in tetracarbonylnickel gives arenecarboxylic acids (25–75%).

In certain cases, diazotisation may lead to abnormal reactions; e.g.,

(i) Diazotisation of o-diamines produces triazoles (p. 666).

(ii) Treatment of m-phenylenediamine with nitrous acid gives Bismarck Brown (p. 666).

Reactions of the diazonium salts in which the nitrogen atoms are retained

1. When bromine and hydrobromic acid are added to a diazonium salt solution, a crystalline precipitate of *diazonium perbromide*, ArN_2Br_3, is obtained. This, on heating, forms the aryl bromide. On the other hand, if the perbromide is treated with aqueous ammonia, the azide, ArN_3, is produced.

Phenyl azide may also be prepared by the action of hydrazoic acid on nitrosobenzene (Maffei et al., 1954):

$$C_6H_5NO + 2HN_3 \longrightarrow C_6H_5N_3 + H_2O + 2N_2 \quad (92\%)$$

2. When reduced with stannous chloride and hydrochloric acid or with sodium sulphite, diazonium salts form phenylhydrazines (see p. 679). If vigorous reducing agents are used, *e.g.*, zinc and hydrochloric acid, the product is an aromatic amine:

$$ArN_2{}^+Cl^- \xrightarrow{Zn/HCl} [ArNHNH_2] \xrightarrow{Zn/HCl} ArNH_2 + NH_3$$

3. Diazonium salts readily undergo *coupling* reactions with phenols, naphthols, and aromatic amines, to form highly coloured azo-compounds, *e.g.*, benzenediazonium chloride couples with phenol in faintly alkaline solution to form *p*-hydroxyazobenzene, and with dimethylaniline in faintly acid solution to form *p*-dimethylaminoazobenzene (see also p. 682).

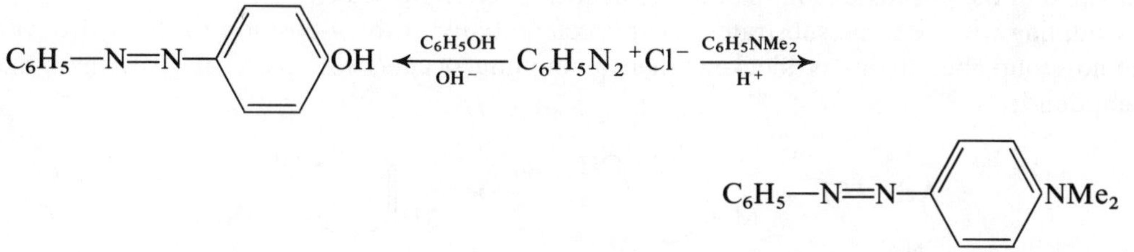

Coupling also occurs (in sodium acetate buffered solution) with compounds containing an active methylene group, *e.g.*, D.E.M., E.A.A.; the product is usually the phenylhydrazone, formed by the tautomerisation of the azo-compound to the more thermodynamically stable hydrazone.

$$C_6H_5N_2{}^+ + {}^-CH(CO_2Et)_2 \xrightarrow{AcONa} C_6H_5-N=N-CH(CO_2Et)_2 \rightleftharpoons C_6H_5NH-N=C(CO_2Et)_2$$

This coupling reaction is often used to detect the presence of enols, and is then referred to as the **Japp–Klingermann reaction** (1887).

A detailed study of the coupling reaction has shown that the mechanism is electrophilic substitution, the electrophile being the diazonium cation and the substrate being the free amine or the phenoxide ion:

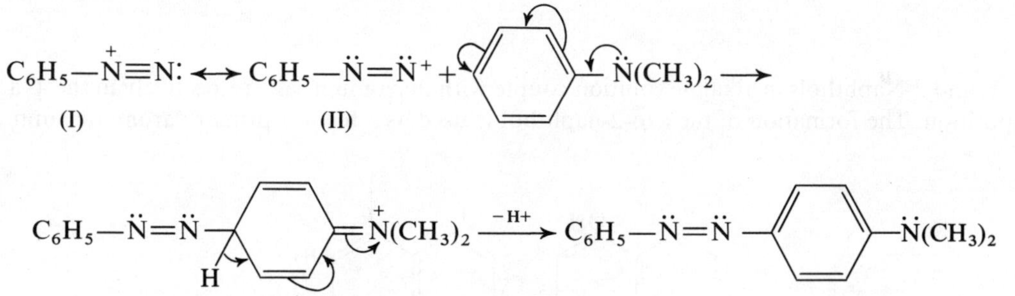

The phenoxide ion also undergoes the resonance effect in a similar manner.

Since (II) is the 'reactive' contributing resonating structure, any factor that increases its contribution will increase the reactivity of that diazonium cation. It can be anticipated that electron-withdrawing groups would favour (II) and electron-releasing groups would favour (I). This has been found to be so in practice, *e.g.*, the *p*-nitrobenzenediazonium cation is more than 10 000 times as reactive as the *p*-methoxy ion under the same conditions (Conant *et al.*, 1930). The 2,4,6-trinitrobenzenediazonium cation is so reactive that it couples with certain hydrocarbons, *e.g.*, mesitylene.

The mechanism described above is also in keeping with the following facts.

(i) Primary, secondary, and tertiary aromatic amines undergo coupling in weakly acid solutions but not in strongly acid solution; the rate of coupling decreases as the pH changes from 6 to 2. In the latter case, the substrate is the amine salt, and in this the ring is deactivated by the $-I$ effect of the ammonium cation group. Since m-substitution never occurs, this position is not sufficiently active to react even with highly reactive diazonium salts. It is because of this salt formation that diazotisation of primary amines is carried out in strongly acid solution in order to avoid coupling (see also below).

(ii) Phenols couple very readily in weakly alkaline solution (the rate of coupling increases as the pH changes from 5 to 8), and this is consistent with the fact that the ring is far more activated in the phenoxide ion than in the neutral phenol molecule.

Coupling with benzene substrates occurs preferentially in the p-position to the hydroxyl or amino-group, but if this is blocked, then o-coupling occurs, e.g., p-cresol gives the o-azo-compound:

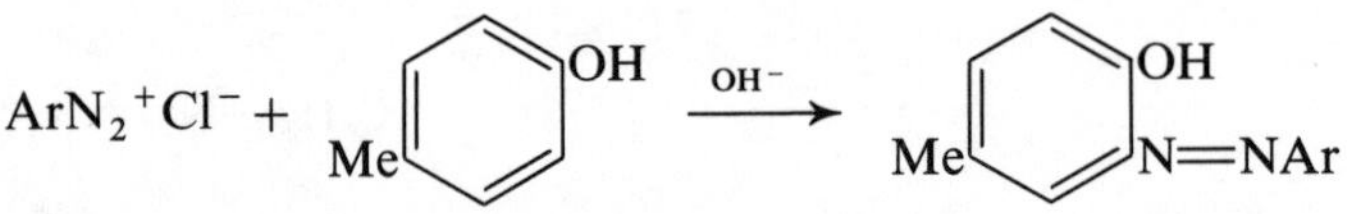

This is also exemplified by the fact that phenol forms bis-(2,4-) and tris-azo-compounds (2,4,6-) with an excess of diazonium salts (cf. below).

The presence of another activating group in the m-position of the phenol or amine accelerates coupling, e.g., resorcinol forms the mono-, bis-, and tris-azo-compounds more readily than does phenol:

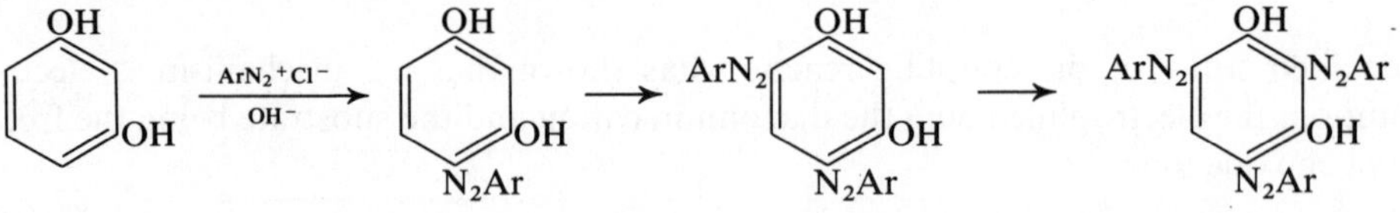

1- and 2-Naphthols in alkaline solution couple with diazonium salts respectively in the 4- and 1-position. The formation of the azo-2-naphthol is used as a test for primary aromatic amines.

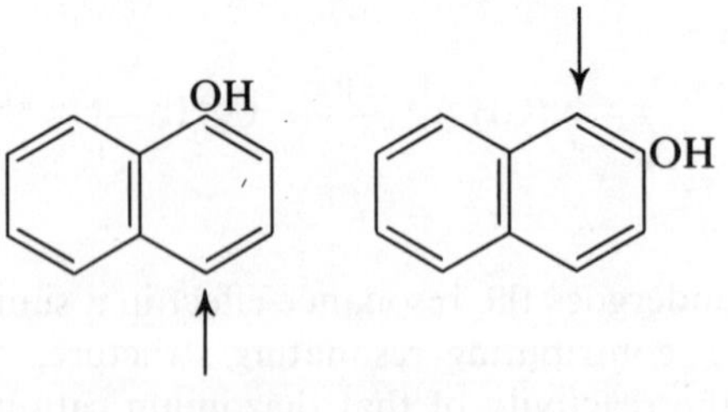

Diazo-compounds. Diazonium salts have the formula $[Ar-N\equiv N]^+X^-$, where X is Cl, Br, HSO_4, NO_3, etc. These are the salts of *diazonium hydroxide*, $[Ar-N\equiv N]^+OH^-$, which has never been isolated. When a diazonium salt solution is made alkaline, the diazonium hydroxide liberated rapidly rearranges to the *diazohydroxide*, the salts of which are called the *diazoates* (*diazotates*). The mechanism of these changes may be as follows:

$$[Ar\!-\!N\!\equiv\!N\!:]^+Cl^- \xrightarrow{\text{NaOH}} [Ar\!-\!N\!\equiv\!N\!:]^+OH^-$$

$$Ar\!-\!\overset{+}{N}\!\equiv\!N\!: \longleftrightarrow Ar\!-\!\overset{+}{\ddot{N}}\!=\!\ddot{N}\!: \underset{OH^-}{\overset{OH^-}{\rightleftharpoons}} Ar\!-\!\ddot{N}\!=\!\ddot{N}\!-\!OH \xrightarrow{\text{NaOH}} Ar\!-\!\ddot{N}\!=\!\ddot{N}\!-\!O^-Na^+$$

These diazoates exist in two geometrical isomeric forms, the *syn-* (*cis-*) and *anti-* (*trans-*):

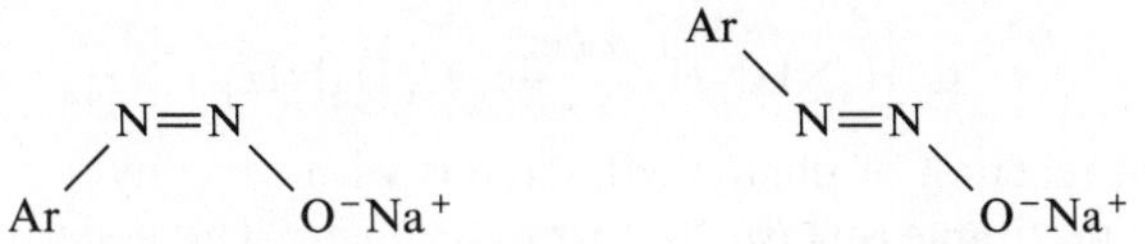

sodium n-diazoate	sodium isodiazoate
syn-form	*anti*-form

The *syn*-form, the n-diazoate, is produced in weakly alkaline solution and is unstable, slowly changing into the *anti*-form, the isodiazoate; this change is accelerated by making the solution strongly alkaline.

Other diazo-compounds that exhibit geometrical isomerism are the *diazosulphonates* ($X = SO_3H$) and the *diazocyanides* ($X = CN$):

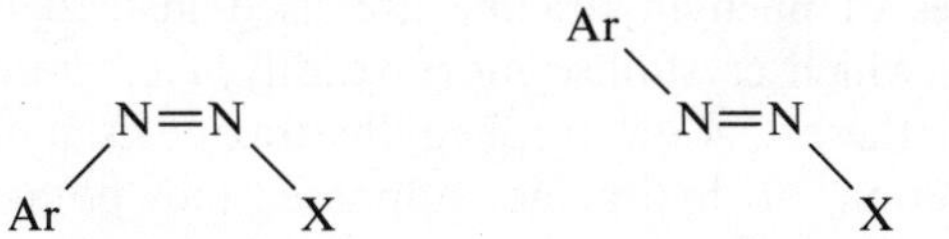

syn-form	*anti*-form

Diazo-oxides (diazo-phenols). When diazonium salts derived from *o-* and *p-*aminophenols are treated with alkali, the products are diazo-oxides, *e.g.*,

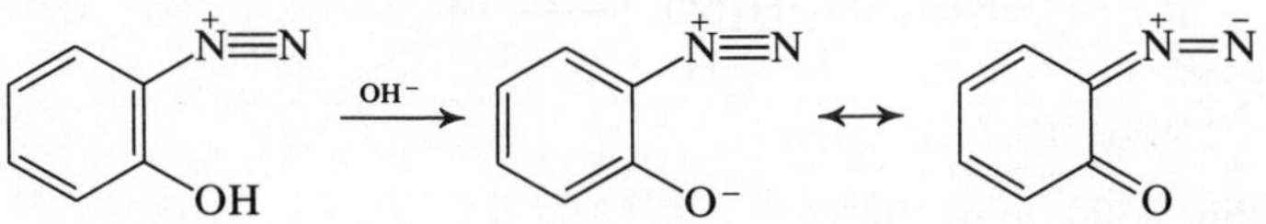

These compounds are usually more stable (through increased resonance stabilisation) than diazonium salts, and undergo the general reactions of diazonium salts.

Hydrazines. The most important substituted hydrazine is phenylhydrazine, b.p. 241°C. This may be prepared by the reduction of benzenediazonium chloride with stannous chloride and hydrochloric acid:

$$C_6H_5N_2{}^+Cl^- \xrightarrow{\text{SnCl}_2/\text{HCl}} C_6H_5NHNH_2\!\cdot\!HCl \quad (90\%)$$

The reduction is carried out commercially with sodium sulphite.

The mechanism of the reduction may be as follows, sodium benzene-*anti*-diazosulphonate (I), and the sodium salt of phenylhydrazine sulphonate (II), being intermediate products:

$$[C_6H_5N\!\equiv\!N]^+Cl^- \xrightarrow{\text{Na}_2\text{SO}_3} \underset{\substack{\| \\ N\!-\!SO_3Na \\ (I)}}{C_6H_5\!-\!N} \xrightarrow{\text{H}_2\text{SO}_4} \left[\underset{\substack{| \\ SO_3H}}{C_6H_5N\!-\!NHSO_3Na}\right] \xrightarrow{\text{H}_2\text{O}}$$

$$C_6H_5NHNHSO_3Na + H_2SO_4 \xrightarrow{\text{HCl}} C_6H_5NHNH_2\!\cdot\!HCl \xrightarrow{\text{NaOH}} C_6H_5NHNH_2 \quad (80\text{--}84\%)$$

$$(II)$$

Phenylhydrazine is readily oxidised when exposed to the air. It is strongly basic and forms well-defined salts, *e.g.*, phenylhydrazine hydrochloride, $C_6H_5NHNH_3{}^+Cl^-$; these salts are relatively stable to atmospheric oxidation. Phenylhydrazine is a powerful reducing agent, reducing

Fehling's solution in the cold:

$$C_6H_5NHNH_2 + 2CuO \longrightarrow Cu_2O + H_2O + C_6H_5N=NH \longrightarrow C_6H_6 + N_2$$

When reduced with vigorous reducing agents, *e.g.*, zinc and hydrochloric acid, phenyl-hydrazine gives aniline and ammonia:

$$C_6H_5NHNH_2 \xrightarrow{\text{Zn/HCl}} C_6H_5NH_2 + NH_3$$

The most important reaction of phenylhydrazine is with carbonyl compounds to form *phenyl-hydrazones*, and with simple sugars to form *osazones*. These derivatives are usually well-defined crystalline solids, and so are useful for the identification of carbonyl compounds. Many phenylhydrazones are decomposed by strong inorganic acids to reform the carbonyl compound. When reduced with zinc and acetic acid, phenylhydrazones form primary amines:

$$C_6H_5NHN=CHR \xrightarrow{e;\ H^+} C_6H_5NH_2 + RCH_2NH_2$$

Various derivatives of phenylhydrazine are used instead of phenylhydrazine, since these produce compounds which crystallise more readily, *e.g.*, *p*-nitro-, 2,4-dinitro- and *p*-bromo-phenylhydrazine. Of these, 2,4-dinitrophenylhydrazine is most commonly used, and this is prepared by the action of hydrazine sulphate and potassium acetate on 1-chloro-2,4-dinitrobenzene:

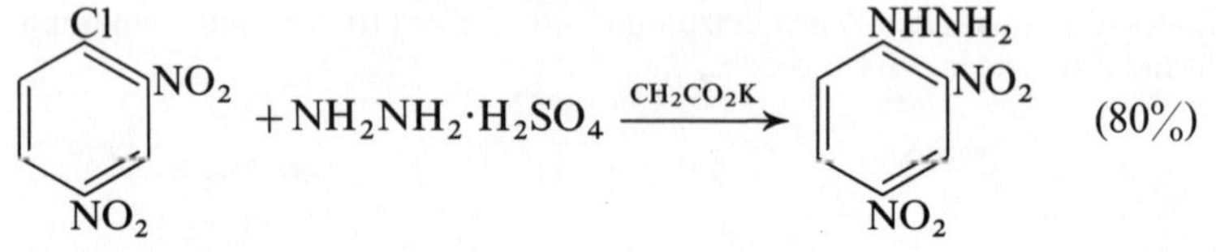

It is a red crystalline solid, m.p. 194°C.

Diazoamino- and aminoazo-compounds

Diazo-compounds have the structure Ar—N=N—Y, where Y is a group that is not attached to the nitrogen atom by carbon, **Azo-compounds**, on the other hand, have the structure Ar—N=N—Ar, each aryl group being attached to a nitrogen atom by carbon.

Coupling between diazonium salts and primary or secondary amines to form azo-compounds has already been discussed (p. 677). Under different conditions, however, particularly the pH of the solution, coupling occurs at the nitrogen atom to form diazoamino-compounds.

Diazoaminobenzene may be prepared by treating aniline hydrochloride solution with sodium nitrite just sufficient to diazotise *half* of the aniline, and then adding sodium acetate:

$$C_6H_5N_2{}^+Cl^- + C_6H_5NH_2 + CH_3CO_2Na \longrightarrow$$
$$C_6H_5N=NNHC_6H_5 + NaCl + CH_3CO_2H \quad (82\text{--}85\%)$$

Diazoaminobenzene is often formed during the diazotisation of aniline, especially when the solution is only weakly acid. It exists in two forms, one as golden-yellow prisms, m.p. 98°, and the other, yellow prisms, m.p. 80°C. It is feebly basic and does not form stable salts. When boiled with dilute sulphuric acid, diazoaminobenzene liberates nitrogen forming phenol and aniline:

$$C_6H_5NHN=NC_6H_5 + H_2SO_4 \longrightarrow C_6H_5NH_2 + C_6H_5N_2{}^+HSO_4{}^- \xrightarrow{H_2O}$$
$$C_6H_5OH + N_2 + H_2SO_4$$

When boiled with concentrated hydrobromic acid, diazoaminobenzene forms bromobenzene and aniline (the former being produced via benzenediazonium bromide), and when treated with sodium nitrite and hydrochloric acid, it forms benzenediazonium chloride.

A very important property of diazoaminobenzene is its tendency to rearrange to aminoazo-benzene (see below).

Tautomerism of the diazoamino compounds. Diazoamino-compounds are tautomeric, and this has been demonstrated in diazoaminobenzene itself by the use of this compound labelled with ^{15}N as shown in (I).

$$\overset{*}{PhN} = NNHPh \rightleftharpoons PhNHN = \overset{*}{N}Ph$$

(I)   (II)

When (I) was boiled with concentrated hydrochloric acid, aniline, chlorobenzene, and nitrogen were produced (Shemyakin *et al.*, 1957). If the structure were (I), then *all* the labelled nitrogen would be in the aniline (*cf.* the reaction above with HBr). In actual fact, the isotopic nitrogen was found to be distributed *between* the aniline and the nitrogen liberated. This is consistent with diazoaminobenzene being a tautomeric compound of forms (I) and (II).

This tautomeric system is known as the *diazoamino triad system*, the *three nitrogen system*, or the *triazene* system. Since such a system has never been separated into the two forms (as is the case when a mobile hydrogen atom is attached to a *carbon* atom), Hunter (1945) called this type of tautomerism *mesohydric tautomerism*.

p-**Aminoazobenzene** (orange-yellow solid, m.p. 126°C) is produced by the **diazoamino-aminoazo rearrangement** of diazoaminobenzene when this is warmed with hydrochloric acid.

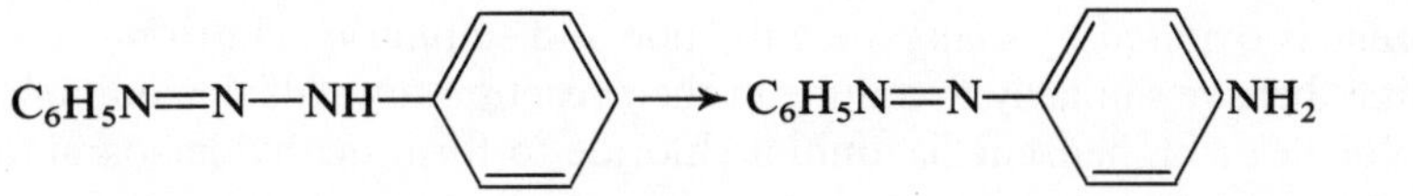

If the *p*-position in the anilino-group is occupied, then the isomeric change occurs more slowly, the azo-group taking up the *o*-position to the amino-group:

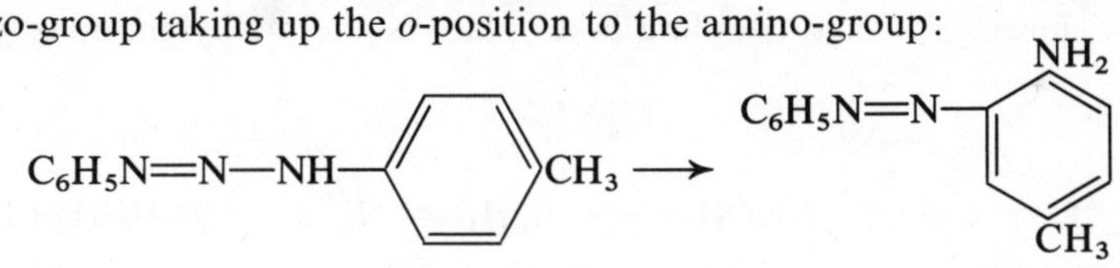

The rearrangement is better effected by aniline hydrochloride than by hydrochloric acid. This is in keeping with the following mechanism, the aniline hydrochloride serving both as a proton donor and as a source of aniline:

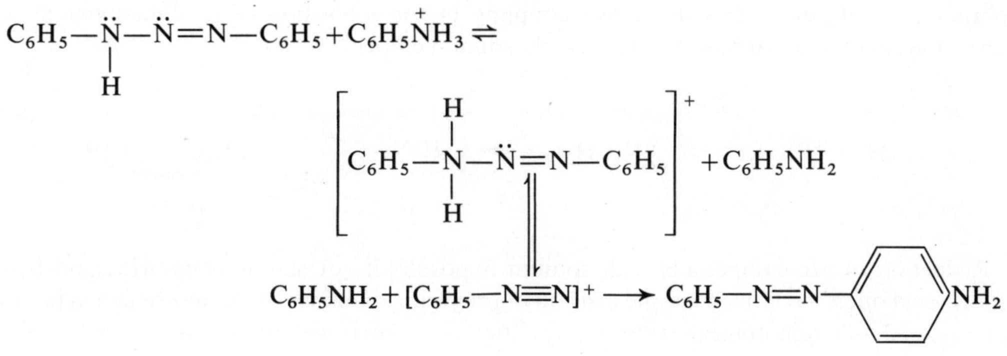

Aminoazobenzene forms two series of salts with acids. When one equivalent of hydrochloric acid is used, the salt formed is yellow and unstable, and probably has the structure (I) (since this has the same type of spectrum as azobenzene).

$$C_6H_5N=NC_6H_4\overset{+}{N}H_3Cl^-$$

(I)

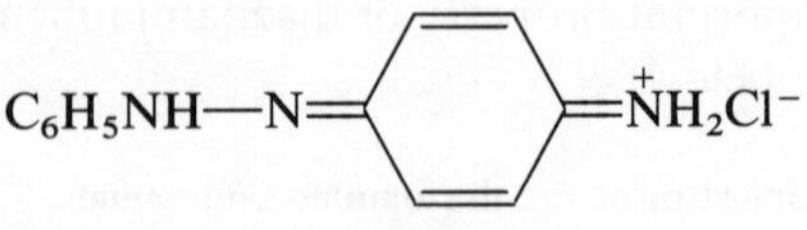

$$C_6H_5NH-N==\overset{+}{N}H_2Cl^-$$

(II)

When, however, a large excess of hydrochloric acid is used, the salt produced is dark violet and stable, and its structure is probably (II), *i.e.*, the salt of the quinone-iminohydrazone; the spectrum of this salt is different from that of azobenzene.

On vigorous reduction, aminoazobenzene is converted into aniline and *p*-phenylenediamine:

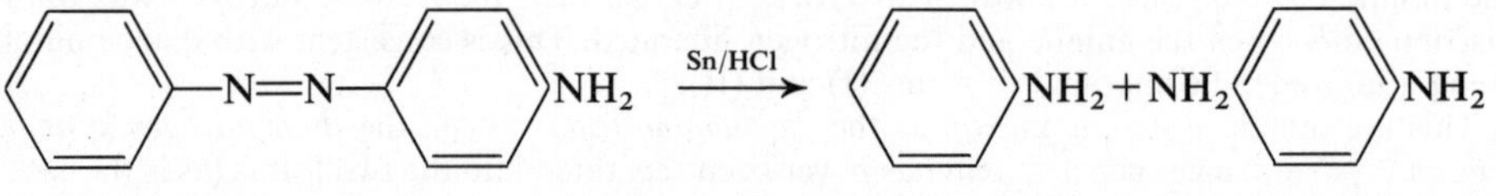

If titanous chloride is used as the reducing agent, the reaction is quantitative, and so may be used for the volumetric determination of aminoazo-dyes.

Aminoazobenzene is oxidised by manganese dioxide and sulphuric acid to *p*-benzoquinone.

Secondary amines behave similarly to aniline in their reaction towards diazonium salts; *e.g.*, *N*-methylaniline couples with benzenediazonium chloride to form methyldiazoaminobenzene:

$$C_6H_5N_2{}^+Cl^- + C_6H_5NHCH_3 \longrightarrow C_6H_5N=N-N(CH_3)C_6H_5 + HCl$$

At the same time, however, *C*-azo-coupling takes place, some methylaminoazobenzene being formed:

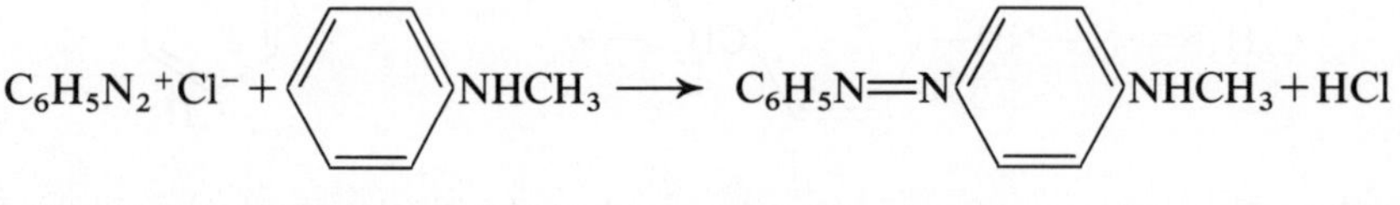

This is also formed by the rearrangement of methyldiazoaminobenzene.

With tertiary amines, the formation of a diazoamino-compound is impossible; the aminoazo-compound is always formed by direct coupling in the *p*-position; *e.g.*, dimethylaminoazobenzene from dimethylaniline and benzenediazonium chloride:

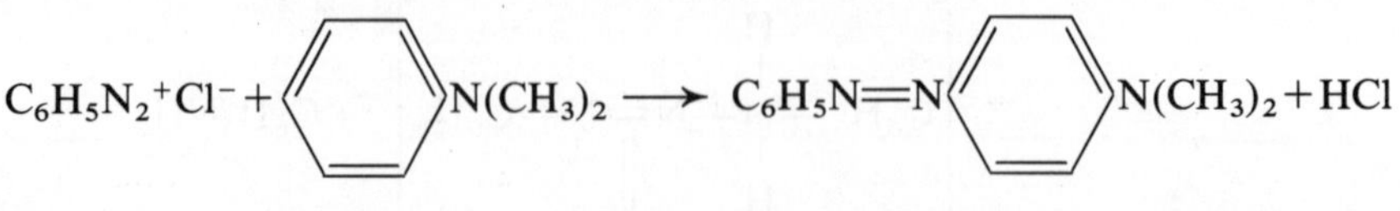

Reduction of azo-compounds with sodium hyposulphite or stannous chloride and hydrochloric acid offers a relatively simple method of preparing diamines (or amino-phenols) in a pure state; *e.g.*, 2,5-diaminotoluene from *m*-toluidine:

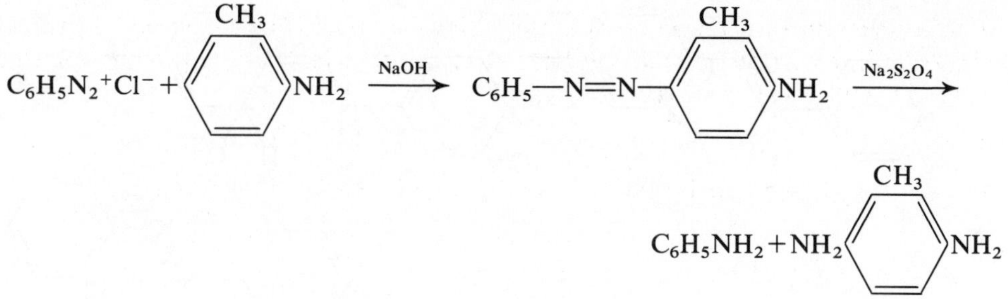

$$C_6H_5N_2{}^+Cl^- + \text{[ring: } CH_3, NH_2] \xrightarrow{\text{NaOH}} C_6H_5-N=N-\text{[ring: } CH_3, NH_2] \xrightarrow{\text{Na}_2\text{S}_2\text{O}_4}$$

$$C_6H_5NH_2 + NH_2\text{[ring: } CH_3, NH_2]$$

It should be noted that the azo-compound, 4-amino-2-methylazobenzene, is formed by direct coupling with the nucleus. Generally, the diazoamino-compound (*N*-coupling) is formed by reaction between diazonium salts and primary aromatic amines; *m*-toluidine and the naphthylamines always form the *C*-azo-compound.

Azoxybenzene may be prepared by reducing nitrobenzene with various reducing agents, *e.g.*, methanolic sodium hydroxide, alkaline sodium arsenite, or an alkaline solution of glucose. It is generally accepted that azoxybenzene is formed by interaction of the intermediates nitrosobenzene and phenylhydroxylamine (p. 649).

$$C_6H_5NO + C_6H_5NHOH \xrightarrow{-H_2O} C_6H_5N=N^+(O^-)C_6H_5$$

On the other hand, oxidation of suitable substrates with peroxyacetic acid gives azoxybenzene, *e.g.*, azobenzene. This method is useful for preparing unsymmetrical azoxybenzenes from unsymmetrical azobenzenes.

Azoxybenzene (yellow crystalline solid, m.p. 36°C), when warmed with iron filings, is reduced to azobenzene. When reduced with ammonium sulphide, azoxybenzene gives hydrazobenzene, and with metal and acid it gives aniline.

Azoxybenzene exists in two forms which are geometrical isomers; the *trans*-isomer is ordinary azoxybenzene.

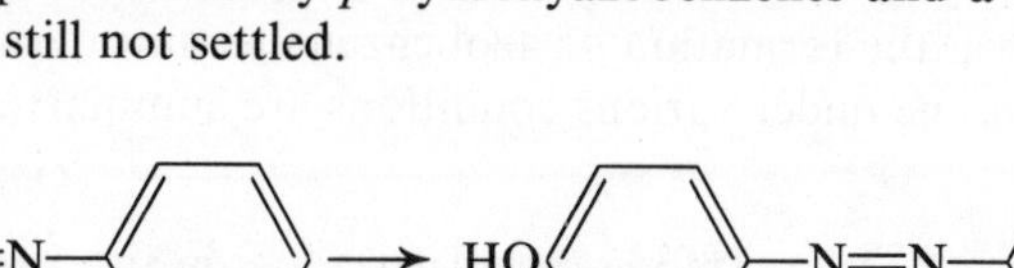

cis, m.p. 86°C   *trans*, m.p. 36°C

Wallach rearrangement (1880). When warmed with concentrated sulphuric acid, azoxybenzenes rearrange to give predominantly *p*-hydroxyazobenzenes and a small amount of the *o*-isomer. The mechanism is still not settled.

If, however, the rearrangement is effected by exposure to ultraviolet light, the product is exclusively the *o*-isomer (Badger *et al.*, 1954). These authors have proposed an intramolecular mechanism for this *o*-rearrangement:

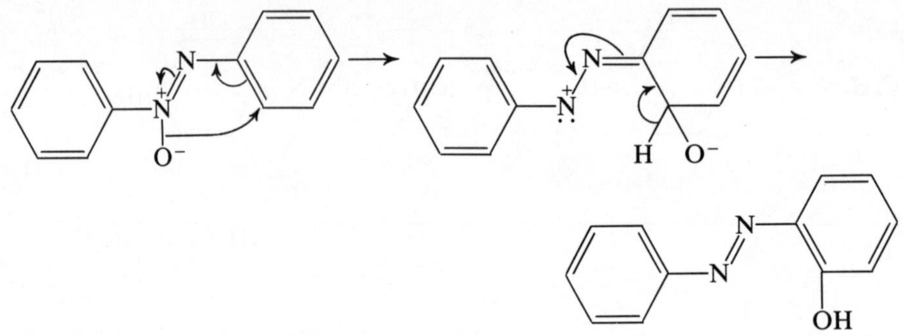

Azobenzene may be conveniently prepared by reducing nitrobenzene with iron filings, sodium amalgam, alkaline sodium stannite, or best with zinc dust and methanolic sodium hydroxide:

$$2C_6H_5NO_2 \xrightarrow{\text{Zn/NaOH}} C_6H_5N{=}NC_6H_5 + 4H_2O \quad (84\text{--}86\%)$$

Lithium aluminium hydride also reduces nitrobenzene to azobenzene. There is good evidence to suggest that these reductions occur via the intermediate formation of azoxybenzene.

General methods of preparing azo-compounds are the oxidation of hydrazo-compounds with, *e.g.*, mercuric oxide, and by the condensation between nitroso-compounds and anilines.

$$Ar^1NO + Ar^2NH_2 \longrightarrow Ar^1N{=}NAr^2 + H_2O$$

Azobenzene (orange-red plates) exists as geometrical isomers, the *trans*-form being 'ordinary' azobenzene:

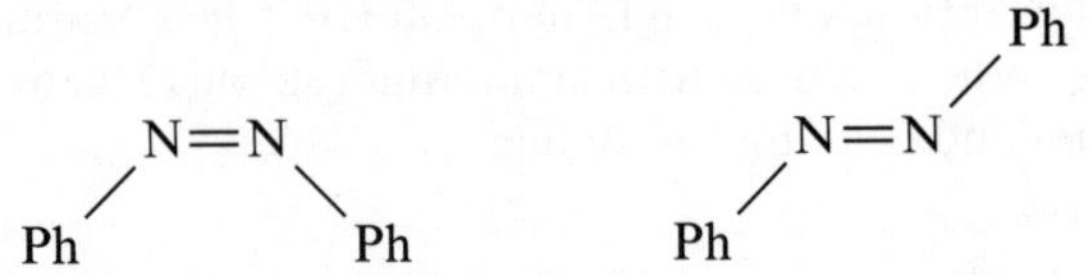

cis, m.p. 71·4°C   *trans*, m.p. 68°C

Azobenzene is reduced by zinc dust and aqueous sodium hydroxide, or by sodium borohydride in water or methanol in the presence of 10 per cent palladised charcoal (Wood *et al.*, 1962), to hydrazobenzene, and by stannous chloride or titanous chloride in acid solution, and by alkaline sodium hyposulphite, to aniline. The most outstanding property of the azo-compounds is their colour, and this is used to great advantage in the manufacture of azo-dyes.

Hydrazobenzene (s-*diphenylhydrazine*) may be prepared by reducing nitrobenzene or azobenzene with zinc dust and aqueous sodium hydroxide. It is also formed when azoxybenzene is reduced electrolytically. All of these reductions starting with oxygenated substrates are believed to proceed via the formation of azobenzene as an intermediate. The products by the reduction of nitrobenzene under various conditions are summarised in the chart.

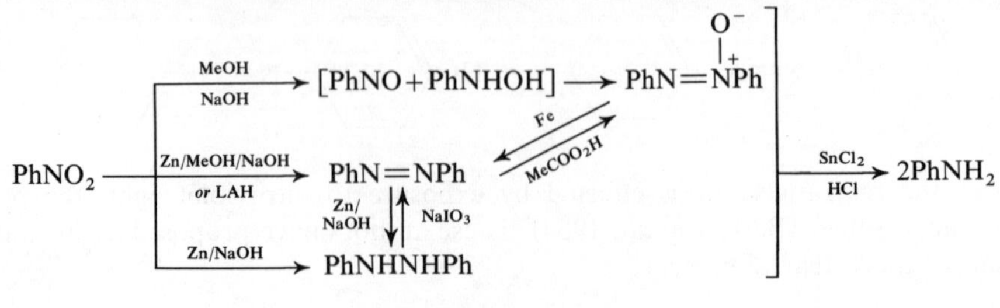

Hydrazobenzene, m.p. 126°C, is slowly oxidised by atmospheric oxygen to azobenzene; this oxidation is rapid with sodium hypobromite or periodate. Hydrazobenzene is reduced by stannous chloride and hydrochloric acid to aniline, and when heated, forms azobenzene and aniline.

Benzidine rearrangement. This term refers to the acid-catalysed conversions of hydrazoarenes into *diaminobiaryls* and *aminodiaryamines*. These rearrangements are usually carried out in aqueous or ethanolic solutions of hydrochloric or sulphuric acid. The types of products that may be produced are illustrated in the chart:

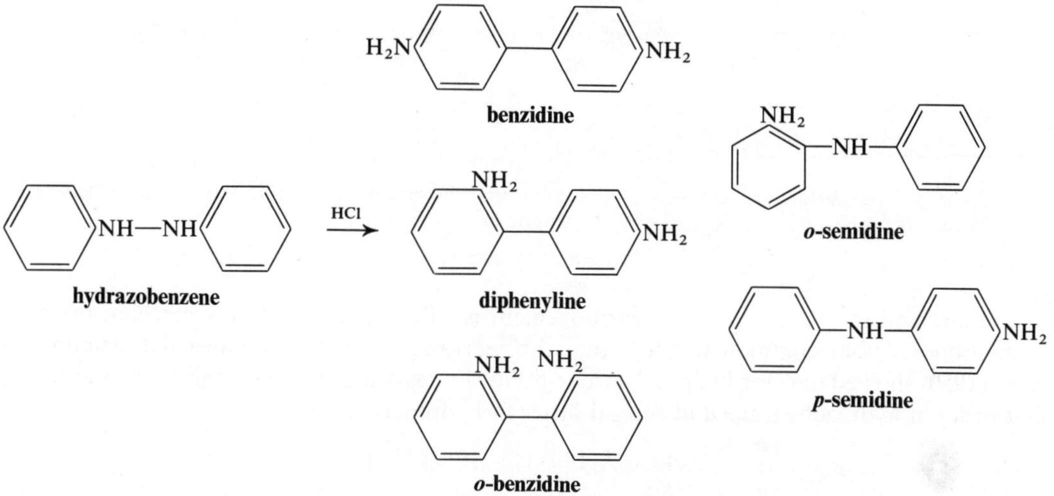

In addition to these rearrangements, products formed by *disproportionation* may also be obtained:

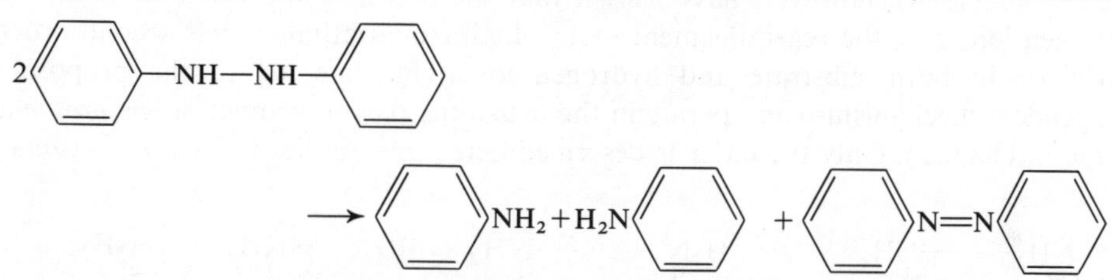

The actual amount of each product depends on the structure of the hydrazoarene under consideration, *e.g.*, hydrazobenzene gives benzidine ($\sim 70\%$), diphenyline ($\sim 30\%$), and other minor products; 2,2'-hydrazonaphthalene gives almost 100 per cent 2,2'-diamino-1,1'-binaphthyl (*o*-benzidine type).

When the *p*-position of one nucleus is occupied, the main product of the rearrangement depends on the nature of the substituent, *e.g.*, for OMe, OEt (strongly +R groups), an *o*-semidine is the main product together with a small amount of the *p*-semidine:

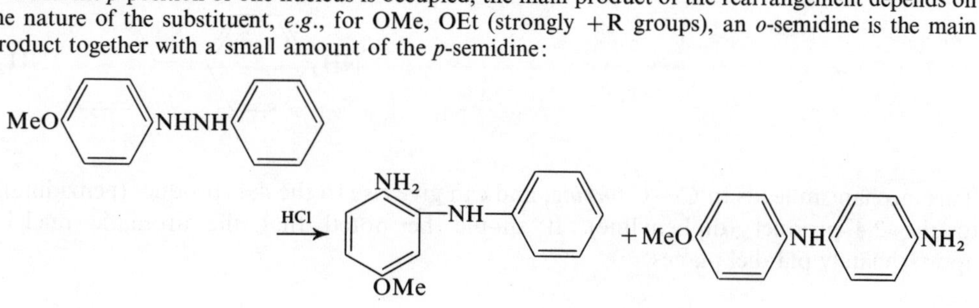

When the *p*-substituent is halogen (fairly strong −I groups), both a diphenyline and an *o*-semidine are formed:

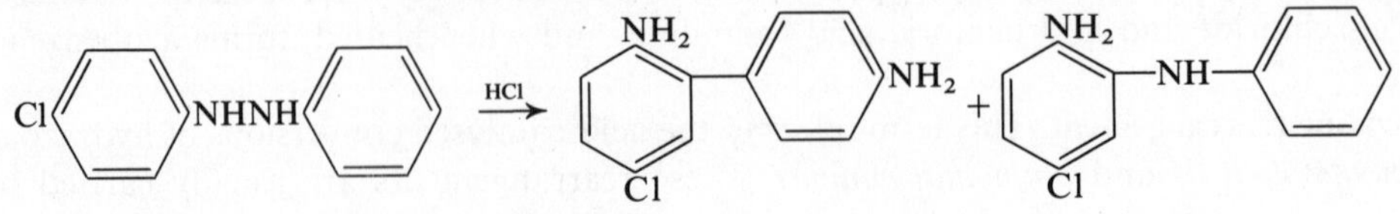

When both *p*-positions are occupied, the main product is an *o*-semidine:

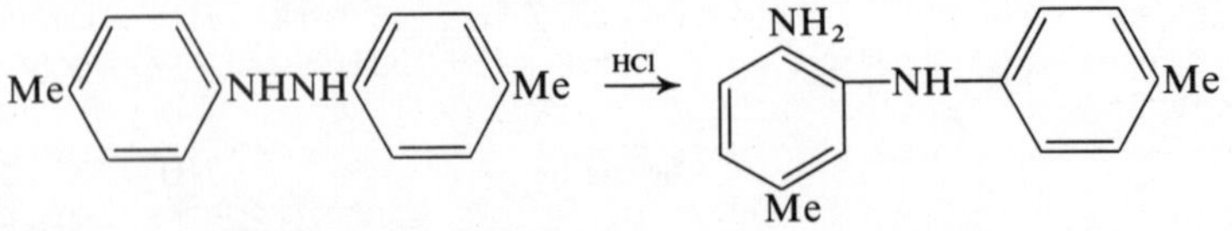

Some groups, particularly carboxyl and sulpho, may be eliminated during the rearrangement, *e.g.*, hydrazobenzene-4-carboxylic acid gives benzidine in very high yield.

The mechanism of the benzidine rearrangement is still not settled. Many mechanisms have been proposed, but common to all is that the rearrangement is intramolecular. Hammond *et al.* (1950) showed that for hydrazobenzene itself, the rearrangement has third-order kinetics; first-order in hydrazobenzene and second-order in hydrogen ion, *i.e.*,

$$\text{rate} = k[\text{PhNHNHPh}][\text{H}^+]^2$$

This led to the suggestion that the rearrangement proceeds via a diprotonated species. Ingold *et al.* (1962, 1964), however, have shown that there is a whole range of orders (1 to 2) in hydrogen ion, *e.g.*, the rearrangement of 1,1′-hydrazonaphthalene has second-order kinetics; first-order in both substrate and hydrogen ion. This has led to the proposal that two independent mechanisms can operate in the benzidine rearrangement, a *one-proton* and a *two-proton* mechanism. Only the latter is described here; this applies to hydrazobenzene.

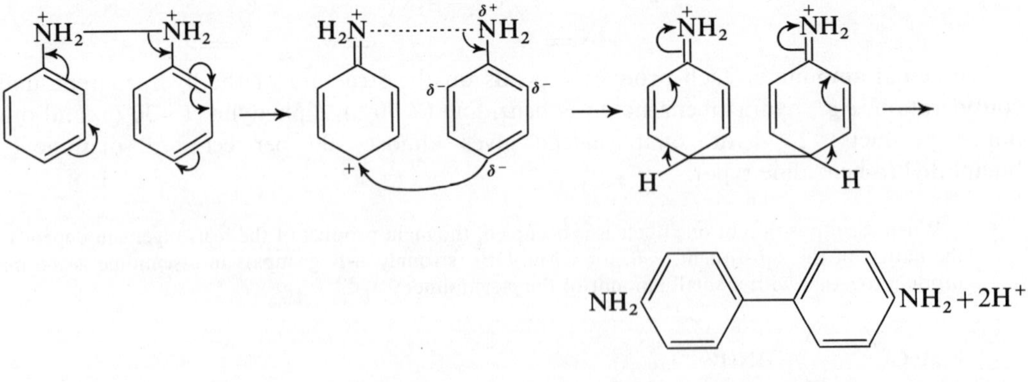

This mechanism leads to *C—C linking*, and can give rise to the 4,4′-product (benzidine) and/or to the 2,4′-product (diphenyline). It should be noted that the aromatic nuclei lie in approximately parallel planes.

PROBLEMS

1. $PhN_2^+Cl^-$ couples with $PhNMe_2$ but not with $2,6\text{-}Me_2C_6H_3NMe_2$. Explain.

2. Convert:

(i) 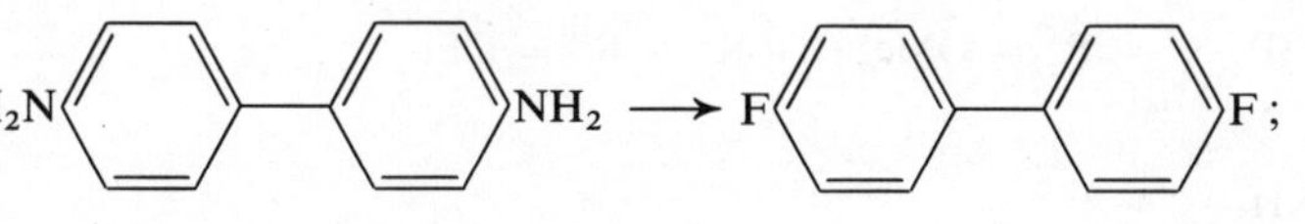 H_2N⟨benzene⟩-⟨benzene⟩NH_2 ⟶ F⟨benzene⟩-⟨benzene⟩F ;

(ii) *m*-dinitrobenzene ⟶ *m*-chloronitrobenzene;

(iii) benzaldehyde ⟶ *m*-chlorobenzaldehyde; (iv) *o*-chloronitrobenzene ⟶ *o*-bromochlorobenzene; (v) *p*-bromoaniline ⟶ *p*-bromoiodobenzene; (vi) 2-Br-4-Me-aniline ⟶ *m*-bromotoluene; (vii) *m*-aminobenzaldehyde ⟶ *m*-hydroxybenzaldehyde; (viii) *o*-toluidine ⟶ *o*-methylphenyl cyanide; (ix) *p*-nitroaniline ⟶ *p*-dinitrobenzene; (x) *o*-nitrotoluene ⟶ *o*-cresol; (xi) nitrobenzene ⟶ pentabromobenzene.

3. *p*-Toluidine reacts with benzenediazonium chloride to form a compound which, on boiling with H_2SO_4, gives *four* products (excluding nitrogen). Discuss.

4. Convert:
(i) PhH ⟶ *m*-dibromobenzene; (ii) PhMe ⟶ $3\text{-}Br\text{-}4\text{-}NO_2$-toluene; (iii) PhH ⟶ *m*-difluorobenzene; (iv) PhMe ⟶ 3,4-dinitrotoluene; (v) $PhNH_2$ ⟶ 1,3,5-tribromobenzene.

5. Discuss, including the use of thermochemical data, the reaction:

$$PhN_2^+Cl^- + CH_2(CN)CO_2Et \xrightarrow{\text{AcONa}} PhNHN{=}C(CN)CO_2Et$$

6. Complete the following equations:

(i)

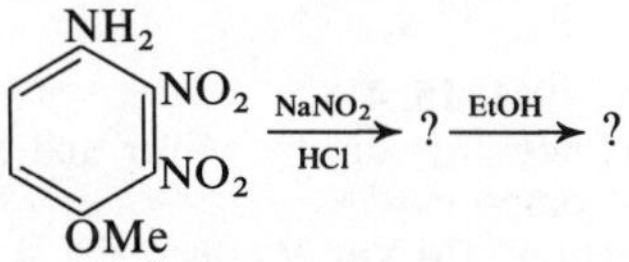

⟨benzene ring with NH_2, NO_2, NO_2, OMe substituents⟩ $\xrightarrow[\text{HCl}]{\text{NaNO}_2}$? $\xrightarrow{\text{EtOH}}$?

(ii) $PhN{=}NNHPh + PhOH \xrightarrow{\text{HCl}}$?

(iii)  EtO⟨benzene⟩$NHNH$⟨benzene⟩ $\xrightarrow{\text{HCl}}$?

(iv) 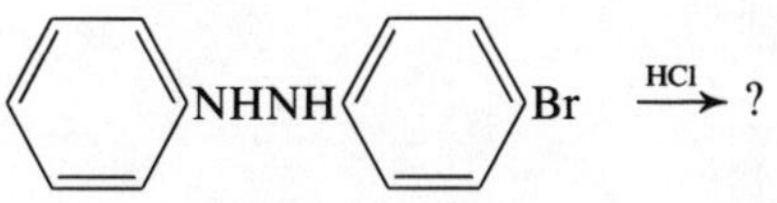   ⟨benzene⟩$NHNH$⟨benzene⟩Br $\xrightarrow{\text{HCl}}$?

(v) 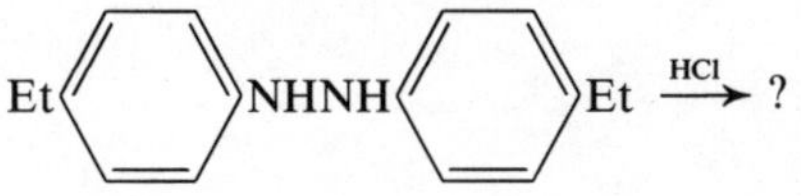 Et⟨benzene⟩$NHNH$⟨benzene⟩Et $\xrightarrow{\text{HCl}}$?

(vi) 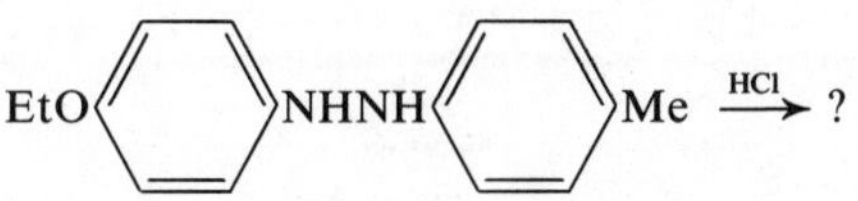 EtO⟨benzene⟩$NHNH$⟨benzene⟩Me $\xrightarrow{\text{HCl}}$?

(vii) 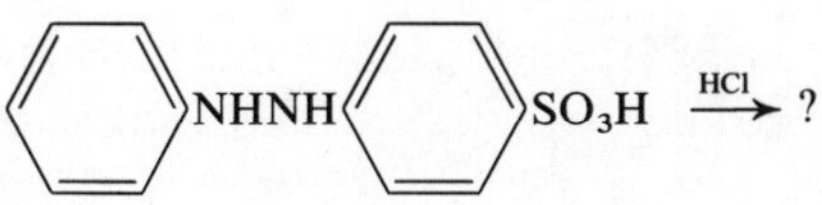   ⟨benzene⟩$NHNH$⟨benzene⟩SO_3H $\xrightarrow{\text{HCl}}$?

(viii) $\xrightarrow[\text{EtOH}]{\text{HCl}}$?

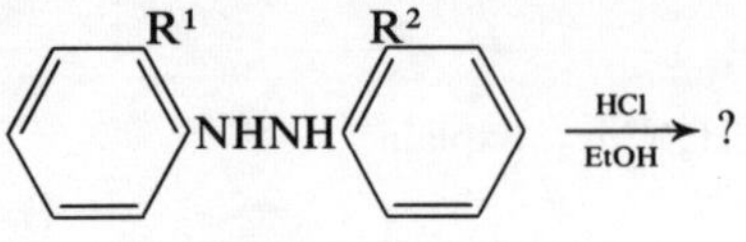

(I) + (II), where (I): $R^1 = R^2 = OMe$; (II): $R^1 = R^2 = OEt$

(ix) 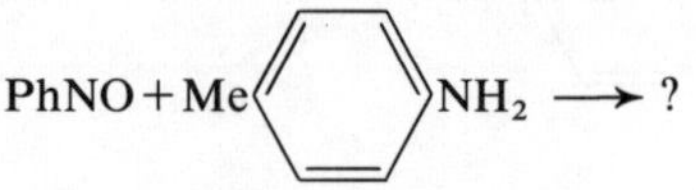 PhNO + Me NH$_2$ $\longrightarrow$?

(x) $PhN_2^+BF_4^- + PhNO_2 \xrightarrow{\text{heat}}$?

(xi) $p\text{-}R^1C_6H_4N{=}NC_6H_4R^2\text{-}p \xrightarrow{\text{MeCO}_3\text{H}}$?

(xii) 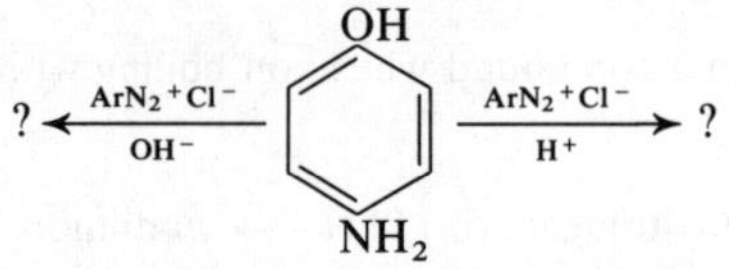 ? $\xleftarrow[\text{OH}^-]{\text{ArN}_2{}^+\text{Cl}^-}$ $\xrightarrow[\text{H}^+]{\text{ArN}_2{}^+\text{Cl}^-}$?

(xiii) $p\text{-}ClC_6H_4N_2^+Cl^- + CH_2{=}CHCHO \xrightarrow{\text{Cu}^{2+}}$?

REFERENCES

RIDD, 'Nitrosation, Diazotisation, and Deamination', *Quart. Rev.*, 1961, **15**, 418.

SIDGWICK, *The Organic Chemistry of Nitrogen*, Oxford Press (1966, 3rd edn. by Millar and Springall), Ch. 16, 'Aromatic Diazo-Compounds'; Ch. 17, 'Azoxy- and Azo-Compounds'.

Organic Reactions, Wiley, (i) Vol. II (1944), Ch. 10, 'The Bart Reaction': (ii) Vol. V (1949), Ch. 4, 'Preparation of Aromatic Fluorine Compounds from Diazonium Fluoroborates': (iii) Vol. X (1959), Ch. 1, 'The Coupling of Diazonium Salts with Aliphatic Carbon Atoms'; Ch. 2, 'The Japp–Klingermann Reaction'.

CADOGAN, 'Reduction of Nitro- and Nitroso-compounds by Tervalent Phosphorus Reagents', *Quart. Rev.*, 1968, **22**, 222.

SHINE, *Aromatic Rearrangements*, Elsevier (1967).

BANTHORPE, COOPER, and INGOLD, 'Acid Conversions of Hydrazoarenes', *Nature*, 1967, **216**, 232.

BANTHORPE *et al.*, 'Mechanism of Benzidine and Semidine Rearrangements', *J. Chem. Soc. (B)*, 1968, 627.

25 Aromatic sulphonic acids

One of the most characteristic properties of the aromatic hydrocarbons and their derivatives is the ease with which they can be sulphonated with concentrated or fuming sulphuric acid, or with chlorosulphonic acid. The alkanes are not so readily sulphonated, and so this reaction can be used to separate alkanes from aromatic hydrocarbons.

Aromatic sulphonic acids are usually prepared by direct sulphonation, since this is far more convenient than indirect methods. The usual sulphonating agents are: concentrated sulphuric acid, sulphur trioxide in sulphuric acid, *i.e.*, oleum (fuming sulphuric acid), or in organic solvents such as nitromethane, pyridine, etc.; chlorosulphonic acid in carbon tetrachloride to form the sulphonic acid (using one molecule of reagent) or the sulphonyl chloride (using excess of reagent).

Mechanism of sulphonation

This is still not settled, one reason being the difficulty in establishing the nature of the sulphonating species (*cf.* nitration). Thus, various possibilities exist for sulphonation with concentrated sulphuric acid, all arising from different equilibria, *e.g.*,

(i) $2H_2SO_4 \rightleftharpoons SO_3 + H_3O^+ + HSO_4^-$

(ii) $SO_3 + H_2SO_4 \rightleftharpoons H_2SO_4(SO_3)$ *or* $H_2S_2O_7$

(iii) $3H_2SO_4 \rightleftharpoons SO_3H^+ + H_3O^+ + 2HSO_4^-$

Experimental work based on kinetic studies in concentrated sulphuric acid and in oleum strongly favours the theory that the active species is sulphur trioxide. This is also believed to be the case with aqueous sulphuric acid. If we accept this, then it appears that the sequence of the steps and their relative rates depend on the conditions. Thus, in concentrated sulphuric acid, the mechanism proposed is:

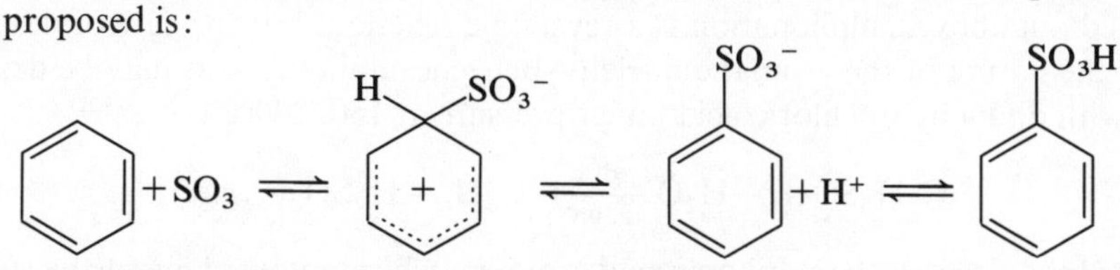

On the other hand, in oleum the mechanism proposed is:

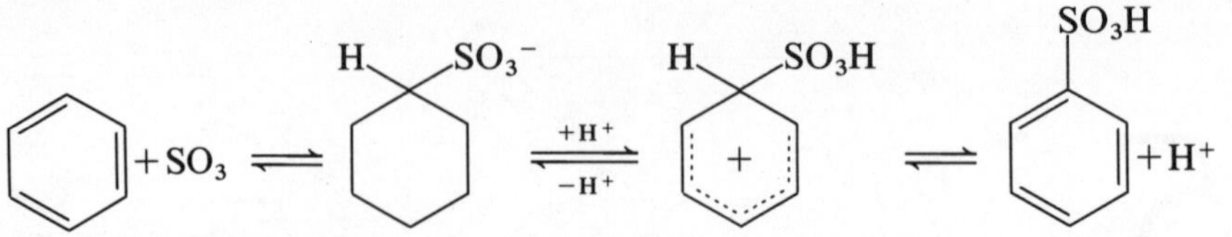

When sulphur trioxide is used in an organic solvent, the evidence is that the mechanism is different from those given above, e.g., according to Cerfontain et al. (1968), sulphonation of p-dichlorobenzene with sulphur trioxide in nitromethane proceeds as follows:

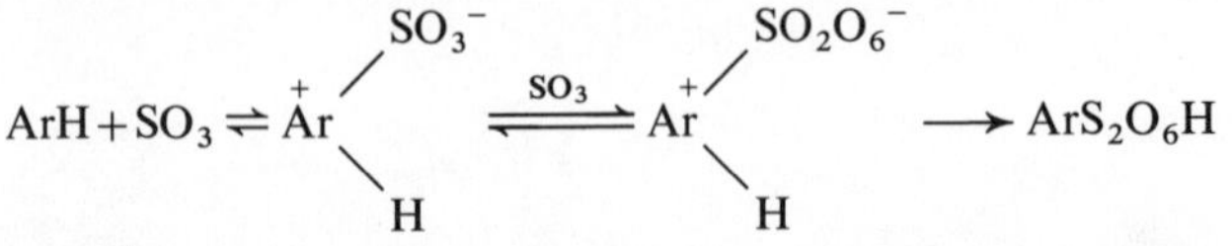

In addition to the difficulties mentioned above, there are also the complications that sulphonation is *reversible* and that isomerisation often occurs during the reaction. These factors affect the isomer distribution under different conditions. Thus, e.g., Wanders et al. (1963) have shown that toluene-*p*- and *m*-sulphonic both isomerise when heated in *aqueous* sulphuric acid to yield a mixture of mainly *m*- and *p*-sulphonic acid with only a small amount of the *o*-isomer. Also, the *o*-isomer rapidly isomerises to the *p*-isomer, but finally the same equilibrium mixture is obtained as before (i.e., all three isomers). From these results it appears that toluene-*m*-sulphonic acid is the most and the *o*-acid the least thermodynamically stable isomer.

Benzenesulphonic acid may be readily prepared by heating benzene with concentrated sulphuric acid at 80°C:

$$C_6H_6 + H_2SO_4 \rightleftharpoons C_6H_5SO_3H + H_2O \quad (75\text{-}80\%)$$

During sulphonation a small amount of sulphone is produced as a by-product, e.g., *diphenylsulphone*:

$$2C_6H_6 + H_2SO_4 \longrightarrow (C_6H_5)_2SO_2 + 2H_2O$$

Properties of benzenesulphonic acid

Benzenesulphonic acid is a colourless crystalline deliquescent solid, m.p. 44°C. It is very soluble in water and the solution is strongly acid (about as strong as sulphuric acid). Benzenesulphonic acid and other aromatic sulphonic acids are useful catalysts in esterification and dehydration, being better than sulphuric acid, since they attack the reaction constituents far less than does sulphuric acid. The sulphonic acids are valuable as synthetic reagents, and since the presence of a sulphonic acid group makes the compound soluble in water, sulphonation is an extremely important process in the preparation of dyes and drugs (converting them into soluble derivatives).

Reactions of benzenesulphonic acid

The following reactions are typical of all sulphonic acids.

1. As pointed out above, sulphonation is a reversible reaction, but the ease of desulphonation depends on the nature of the aromatic nucleus. Benzenesulphonic acid may be desulphonated by heating with dilute hydrochloric acid under pressure at 150–200°C:

$$C_6H_5SO_3H + H_2O \xrightarrow[150°C]{HCl} C_6H_6 + H_2SO_4 \quad (v.g.)$$

With some sulphonic acids the sulphonic acid group may be eliminated merely by steam distillation. This desulphonation is very useful for preparing certain isomers, e.g., *o*-chlorotoluene may be prepared as follows:

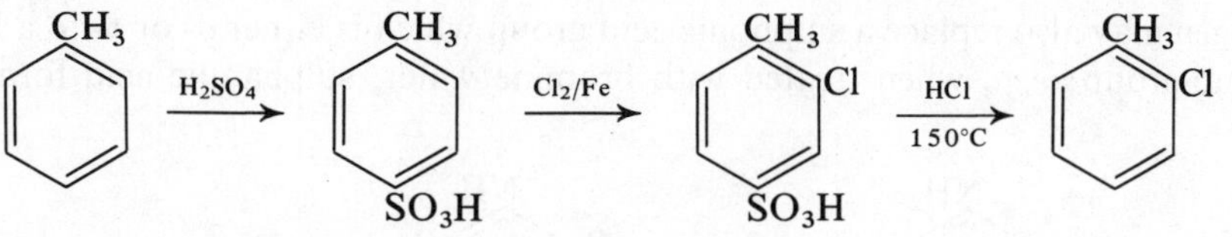

Desulphonation may also be used to separate certain isomers, *e.g.*, the three xylenes. The xylene fraction from coal tar may be treated with cold 80 per cent sulphuric acid. Under these conditions, the *m*-isomer is readily sulphonated to *m*-xylene-4-sulphonic acid, the *o*- and *p*-isomers remaining unaffected:

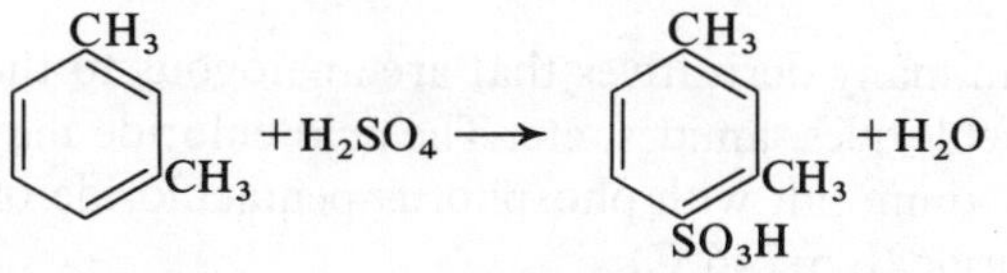

Thus only the *m*-isomer dissolves, and hence may be separated from the other two. The mixture of *o*- and *p*-xylenes is then sulphonated with concentrated sulphuric acid (98%), and the resulting *o*-xylene-4-sulphonic acid and *p*-xylene-2-sulphonic acid may be separated by fractional crystallisation from the diluted sulphonated mixture; the *p*-derivative is less soluble than the *o*-. The xylenes are then regenerated by heating their sulphonic acid derivatives with dilute hydrochloric acid under pressure.

2. When sodium benzenesulphonate is fused with sodamide, aniline is obtained:

$$C_6H_5SO_3^-Na^+ + NaNH_2 \longrightarrow C_6H_5NH_2 + Na_2SO_3 \quad (f.g.-g.)$$

3. Fusion with sodium hydroxide converts sodium benzenesulphonate into sodium phenoxide:

$$C_6H_5SO_3^-Na^+ + NaOH \longrightarrow Na_2SO_3 + C_6H_5OH \xrightarrow{\text{NaOH}} C_6H_5\bar{O}Na^+ \quad (g.-v.g.)$$

4. When sodium benzenesulphonate is fused with sodium cyanide, phenyl cyanide is formed:

$$C_6H_5SO_3^-Na^+ + NaCN \longrightarrow C_6H_5CN + Na_2SO_3 \quad (f.g.)$$

5. When the potassium salt of benzenesulphonic acid is fused with potassium hydrogen sulphide, thiophenol is formed:

$$C_6H_5SO_3^-K^+ + KSH \longrightarrow C_6H_5SH + K_2SO_3 \quad (f.g.)$$

6. The sulphonic acid group is often readily replaced by a nitro-group. This offers a means of preparing nitro-derivatives of compounds that are easily oxidised by nitric acid, since the sulphonic acid derivatives are not easily oxidised; *e.g.*, picric acid from phenol:

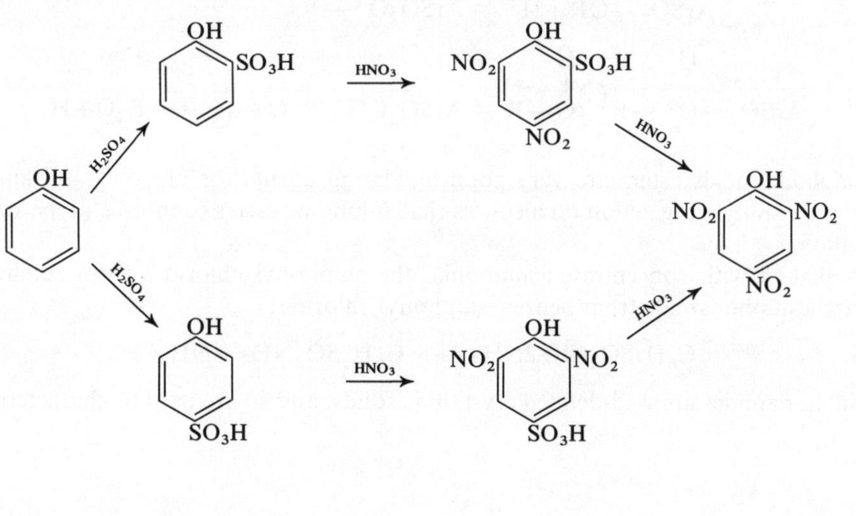

Halogen may also replace a sulphonic acid group which is either *o*- or *p*- to a hydroxyl or to an amino-group; *e.g.*, when treated with bromine water, sulphanilic acid forms *s*-tribromo-aniline:

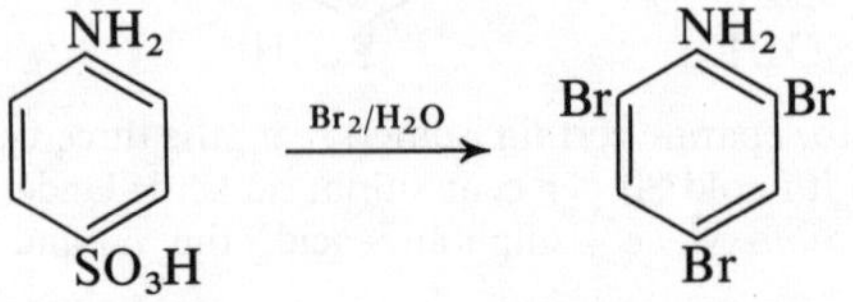

Sulphonic acids form many derivatives that are analogous to those of the carboxylic acids, *e.g.*, salts, esters, acid chlorides, amides, etc. The acid chloride may be prepared by treating a sulphonic acid or its sodium salt with phosphorus pentachloride or phorphoryl chloride, *e.g.*, benzenesulphonyl chloride (b.p. 246°C):

$$C_6H_5SO_3H + PCl_5 \longrightarrow C_6H_5SO_2Cl + HCl + POCl_3 \quad (g.-ex)$$

$$C_6H_5SO_3{}^-Na^+ + PCl_5 \longrightarrow C_6H_5SO_2Cl + NaCl + POCl_3 \quad (75-80\%)$$

$$2C_6H_5SO_3{}^-Na^+ + POCl_3 \longrightarrow 2C_6H_5SO_2Cl + NaCl + NaPO_3 \quad (74-87\%)$$

Thionyl chloride has no action on sulphonic acids (and some carboxylic acids), but in the presence of dimethylformamide, the sulphonyl chloride is obtained in excellent yield (Bosschard *et al.*, 1959). The reaction proceeds through the amide chloride Me_2NCHCl_2.

Sulphonyl chlorides may also be prepared by reaction between an arene and sulphuryl chloride in the presence of aluminium chloride (*cf.* the Friedel–Crafts reaction).

The aromatic sulphonyl chlorides, however, are usually best prepared by treating an aromatic compound with excess of chlorosulphonic acid:

$$ArH + ClSO_3H \longrightarrow HCl + ArSO_3H \overset{ClSO_3H}{\rightleftharpoons} ArSO_2Cl + H_2SO_4 \quad (75-77\%)$$

The sulphonyl chlorides are decomposed very slowly by water but rapidly by alkali; they react with alcohols in the presence of alkali to form esters:

$$ArSO_2Cl + ROH + NaOH \longrightarrow ArSO_3R + NaCl + H_2O$$

In the *absence* of alkali, the sulphonic ester produced alkylates the alcohol, possibly as follows:

$$\overset{\overset{\displaystyle H}{\displaystyle |}}{ArSO_2{-}OR + H^+ \rightleftharpoons ArSO_2O^+{-}R}$$

$$\overset{\overset{\displaystyle H}{\displaystyle |}}{ArSO_2{-}O^+}{-}R \quad \overset{\overset{\displaystyle H}{\displaystyle |}}{{:}O{-}R} \rightleftharpoons ArSO_2OH + R_2\overset{+}{O}{-}H \longrightarrow R_2O + H^+$$

Thus, sulphonic acids esters are very good alkylating agents for alcohols and amines. It is because of this alkylating action on alcohols that sulphonic esters cannot be prepared by direct esterification.

When shaken with concentrated ammonia, the sulphonyl chlorides form sulphonamides; *e.g.*, benzenesulphonamide from benzenesulphonyl chloride:

$$C_6H_5SO_2Cl + 2NH_3 \longrightarrow C_6H_5SO_2NH_2 + NH_4Cl$$

These sulphonamides are well-defined crystalline solids, and so are used to characterise the sulphonic acids. Better derivatives for characterising the sulphonic acids are their *S*-benzyliso-thiouronium salts:

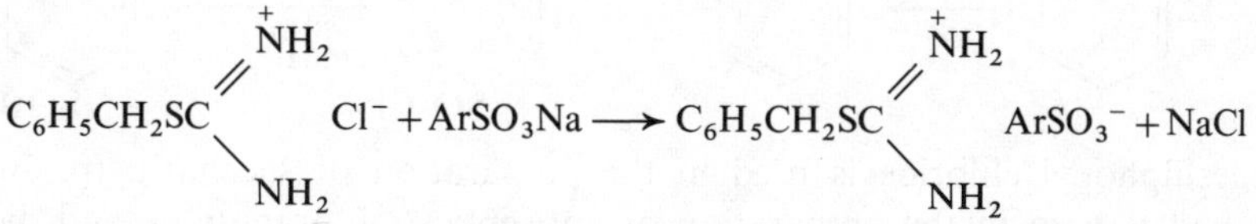

$$C_6H_5CH_2SC\overset{\overset{+}{N}H_2}{\underset{NH_2}{\diagdown}} Cl^- + ArSO_3Na \longrightarrow C_6H_5CH_2SC\overset{\overset{+}{N}H_2}{\underset{NH_2}{\diagdown}} ArSO_3^- + NaCl$$

Sulphonyl chlorides also react with primary and secondary amines to form *N*-substituted sulphonamides. Sulphonamides are more difficult to hydrolyse than carbonamides, and form salts with alkalis (see the Hinsberg separation, p. 368). Sulphonyl chlorides may be used in the Friedel–Crafts reaction to give sulphones.

Sulphonic acid anhydrides have been prepared by heating the acid with excess of phosphorus pentoxide (Field *et al.*, 1952), *e.g.*,

$$2C_6H_5SO_3H \xrightarrow[(-H_2O)]{P_2O_5} (C_6H_5SO_2)_2O \quad (70\%)$$

Benzenedisulphonic acids. When heated with excess of fuming sulphuric acid at 200°C, benzene forms benzene-*m*-disulphonic acid as the main product and the *p*-isomer in a small amount:

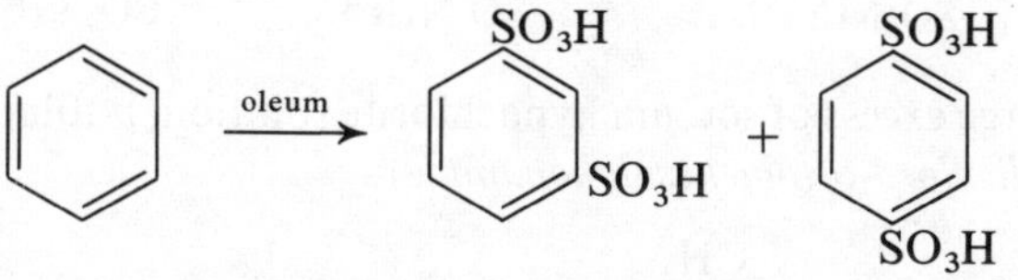

In the presence of mercuric sulphate as catalyst, the yield of *p*-disulphonic acid is increased (30–34%). Benzene-*o*-disulphonic acid may be prepared by sulphonating *m*-aminobenzene-sulphonic acid, and then replacing the amino-group by hydrogen (diazotising, etc.).

When fused with potassium hydroxide, *m*- and *p*-benzenedisulphonic acids both form resor-cinol (*m*-dihydroxybenzene); benzene-*o*-disulphonic acid forms catechol (*o*-dihydroxybenzene).

1,3,5-Benzenetrisulphonic acid may be prepared by heating benzene-*m*-disulphonic acid with fuming sulphuric acid.

Toluenesulphonic acids. When toluene is treated with concentrated sulphuric acid, the *o*- and *p*-toluenesulphonic acids are formed, low temperatures (below 100°C) favouring the formation of the *o*-isomer, and high temperatures (above 100°C) the *p*- (see also p. 690):

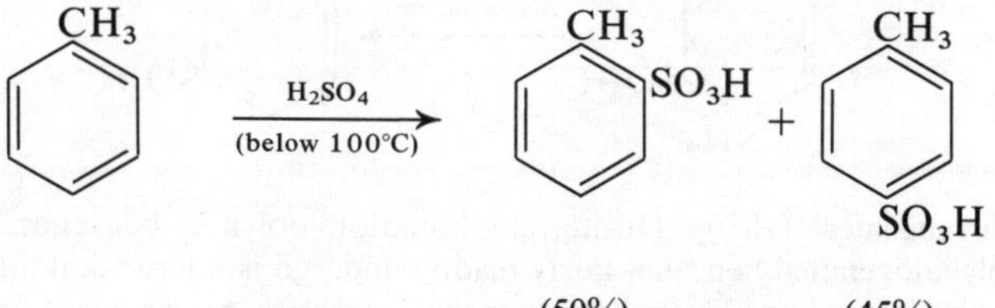

Both isomers are crystalline solids, the *o*- melting at 67·5°, and the *p*- at 106°C.

A much better yield of *o*-toluenesulphonic acid may be obtained by treating toluene with chlorosulphonic acid at low temperatures:

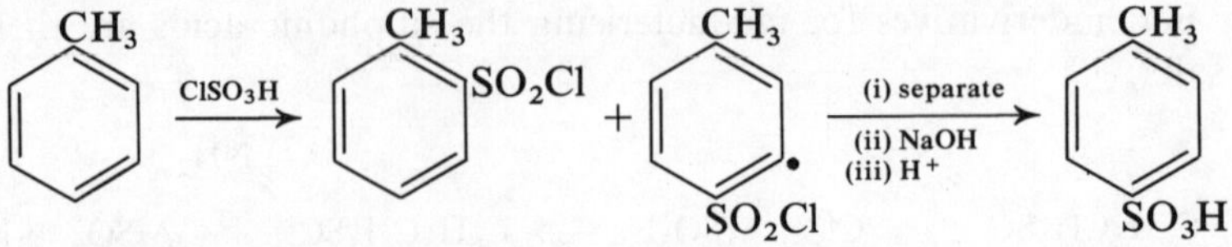

o-Toluenesulphonyl chloride is used in the preparation of saccharin (p. 765); *p*-toluenesulphonic acid is used in the preparation of antiseptics, chloramine T and dichloramine T. **Chloramine T** is the sodium salt of *N*-chloro-*p*-toluenesulphonamide and may be prepared as follows:

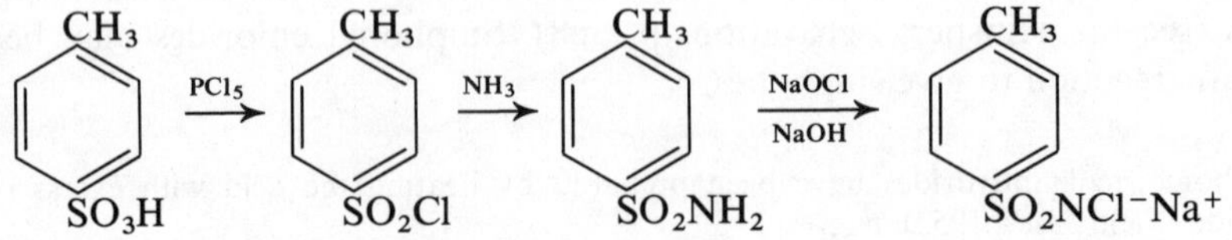

In addition to being used as an antiseptic, chloramine T is also used as a laboratory reagent instead of hypochlorite salts, since it is stable and liberates hypochlorous acid when acidified:

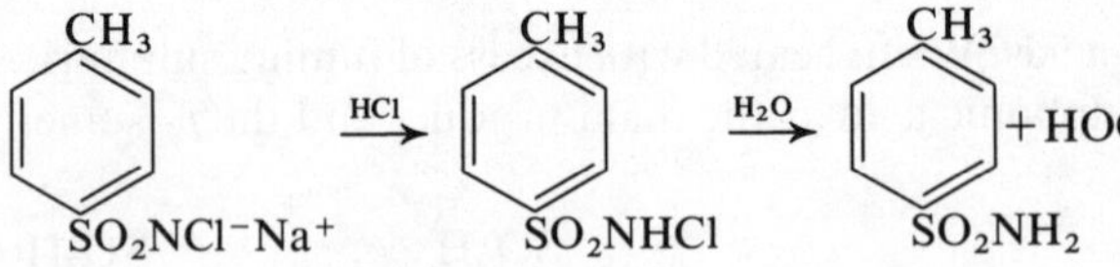

When treated with a large excess of sodium hypochlorite solution, *p*-toluenesulphonamide forms **dichloramine T** (N,N-*dichloro*-p-*toluenesulphonamide*):

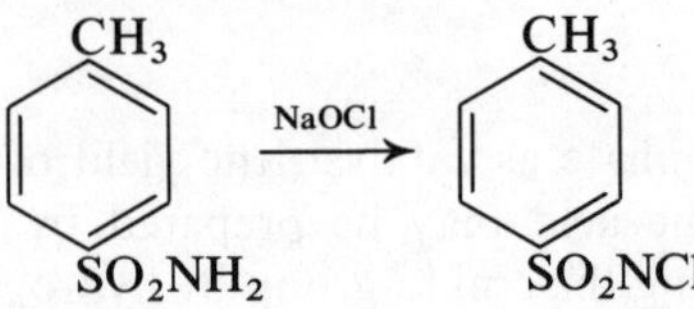

The *p*-toluenesulphonyl group is often referred to as the **tosyl** group, and is denoted by Ts. Tosylates are useful for preparing, *e.g.*, ethers (with ROH), cyanides (with KCN), thiols (with KSH) thioethers (with K_2S), thiocyanates (with KSCN), alkyl iodides (with NaI), alkanes (with Grignard reagents), etc. (see text).

m-*Toluenesulphonic acid* (an oil) may be prepared by replacing the amino-group in *p*-toluidine-*m*-sulphonic acid by hydrogen:

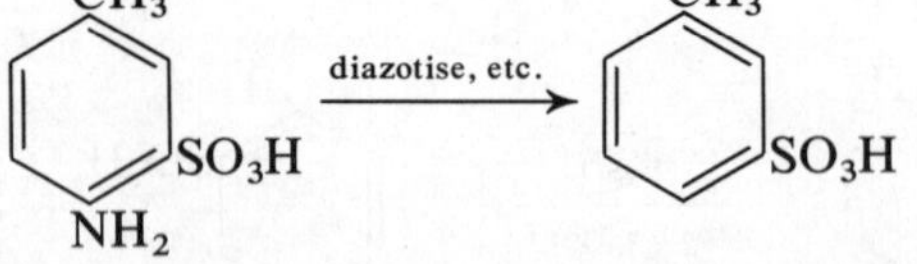

Jacobsen rearrangement (1886). During sulphonation polyalkylbenzenes, halogenated polyalkylbenzenes or polyhalogenated benzenes fairly readily undergo isomerisation due to the migration of an alkyl group or a halogen atom. This is known as the *Jacobsen rearrangement*, and in practice works best for tetra-alkylbenzenes. Furthermore, the rearrangement is mainly confined to the migration of methyl and ethyl groups; larger groups, particularly isopropyl and t-butyl, are eliminated as alkenes. In addition to rearrangement, disproportionation also occurs, *e.g.*, durene gives prehnitenesulphonic acid (2,3,4,5-tetramethylbenzenesulphonic acid) as the major product, together with small amounts of 2,4,5-trimethyl- and pentamethylbenzenesulphonic acid, and hexamethylbenzene.

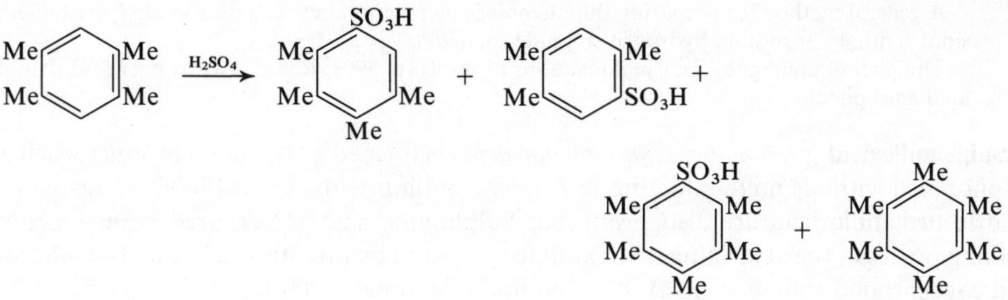

An example of halogen migration is:

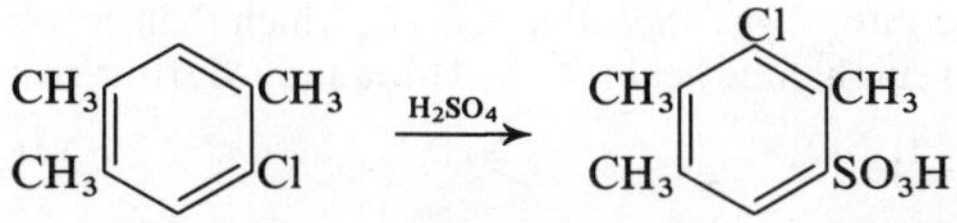

The mechanism of the Jacobsen rearrangement is uncertain. It is widely held that it is the sulphonic acid that rearranges, but a free-radical mechanism *not* involving the sulphonic acid has also been proposed.

Sulphinic acids. The general methods of preparing the aromatic sulphinic acids may be illustrated by the preparation benzenesulphinic acid:

(i) By the reduction of benzenesulphonyl chloride with zinc dust and water, aqueous sodium sulphide, or alkaline sodium sulphite, *e.g.*,

$$C_6H_5SO_2Cl \xrightarrow{Zn/H_2O} (C_6H_5SO_2)_2Zn \xrightarrow{HCl} 2C_6H_5SO_2H \quad (g.)$$

(ii) By the action of copper powder on a solution of a diazonium salt saturated with sulphur dioxide:

$$C_6H_5N_2{}^+HSO_4{}^- + SO_2 + Cu \longrightarrow C_6H_5SO_2H + N_2 + CuSO_4 \quad (g.)$$

(iii) By the action of sulphur dioxide on benzene in the presence of anhydrous aluminium chloride:

$$C_6H_6 + SO_2 \xrightarrow{AlCl_3} C_6H_5SO_2H \quad (g.)$$

(iv) By the action of sulphur dioxide on an arylmagnesium bromide.

$$C_6H_5MgBr + SO_2 \xrightarrow{-40°C} C_6H_5SO_2MgBr \xrightarrow{H_2O} C_6H_5SO_2H \quad (g.)$$

The usual by-product in this reaction is the sulphoxide.

$$C_6H_5SO_2MgBr + C_6H_5MgBr \longrightarrow (C_6H_5)_2SO + (MgBr)_2O$$

(v) Sulphinic acids may be prepared by fusion of sulphones with potassium hydroxide at 200°C (see p. 405).

The sulphinic acids are unstable solids which readily oxidise in the air to sulphonic acids. They are reduced by zinc and hydrochloric acid to thiols; *e.g.*, benzenesulphinic acid gives *thiophenol*:

$$C_6H_5SO_2H \xrightarrow{Zn/HCl} C_6H_5SH + 2H_2O$$

The most useful reaction of the sulphinic acids is the ease with which the sulphinic acid group is replaced by a chloromercuri-group (this is an example of indirect mercuration); *e.g.*, *p*-tolylmercuric chloride may be prepared by boiling an aqueous solution of sodium *p*-toluenesulphinate with mercuric chloride:

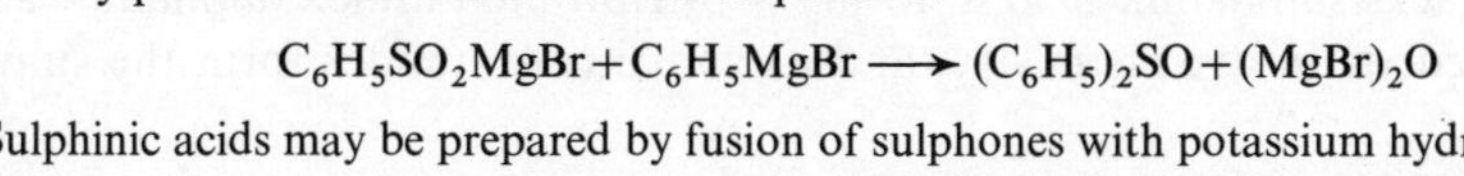

Thiophenol (b.p. 170°C, with a nauseating odour) may be prepared by reducing benzenesulphonyl chloride with zinc and sulphuric acid, or with lithium aluminium hydride, *e.g.*,

$$C_6H_5SO_2Cl \xrightarrow{e; H^+} C_6H_5SH + HCl + 2H_2O \quad (96\%)$$

A general method for preparing thiophenols is by reaction between diazonium salts and ethyl potassium xanthate or sodium hydrogen sulphide (p. 676).

Thiophenol undergoes the usual reactions of thiols (p. 399); it is also a stronger acid than its oxygen analogue, phenol.

Sulphanilic acid (p-*aminobenzenesulphonic acid*) is formed as the main product when aniline is sulphonated with oleum (containing 10 per cent sulphur trioxide) at 180°C; some metanilic and a little orthanilic acid are also produced. Sulphanilic acid is prepared commercially by the 'baking process'; the acid sulphate of aniline (prepared by mixing about equal weights of aniline and concentrated sulphuric acid) is heated for some time at 200°C.

The mechanism of the baking process is uncertain. According to Bamberger (1897), the acid sulphate of aniline is converted into *phenylsulphamic acid* (I), which then rearranges to *orthanilic acid* (II) which, in turn, rearranges to sulphanilic acid (*cf.* the Hofmann–Martius rearrangement, p. 662):

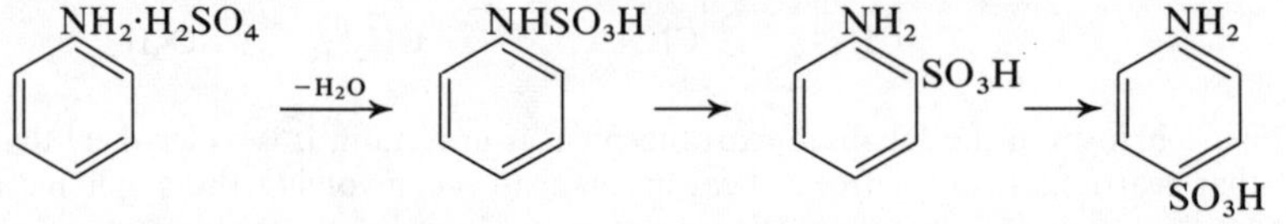

This rearrangement has been generally accepted as being intramolecular, but Scott *et al.* (1968) have shown that at least in dioxan-sulphuric acid solution, the rearrangement is intermolecular. These workers heated potassium phenylsulphamate in a mixture of dioxan and [^{35}S] sulphuric acid and obtained labelled sulphanilic acid. They also showed that orthanilic acid did not rearrange under these conditions. Thus, the rearrangement is intermolecular and does not proceed via orthanilic acid as an intermediate. The mechanism proposed (for the rearrangement under these conditions) is rapid hydrolysis to aniline and sulphuric acid followed by slow sulphonation. It might also be noted that phenylsulphamic acid has been postulated as, but never proved to be, an intermediate in the baking process.

Sulphanilic acid, m.p. 288°C (with decomp.), forms salts with bases but does not combine with acids. The latter is due to sulphanilic acid existing as a dipolar ion, $p\text{-}H_3\overset{+}{N}C_6H_4SO_3^-$ (sulphonic acids are as strong as the inorganic acids). When sulphanilic acid is treated with nitric acid, the sulphonic acid group is replaced by a nitro-group to form *p*-nitroaniline. Similarly, bromine water attacks sulphanilic acid to form 2,4,6-tribromoaniline. Sulphanilic acid is a very important intermediate in dye chemistry, and its substituted amides form the sulphanilamide drugs.

Orthanilic acid (o-*aminobenzenesulphonic acid*) may be prepared as follows:

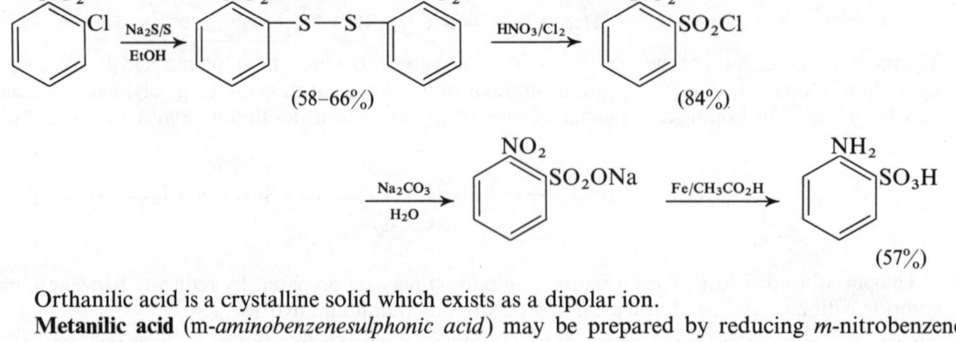

Orthanilic acid is a crystalline solid which exists as a dipolar ion.

Metanilic acid (m-*aminobenzenesulphonic acid*) may be prepared by reducing *m*-nitrobenzenesulphonic acid. It is a crystalline solid (dipolar ion structure) and is used in the manufacture of dyes.

PROBLEMS

1. Arenesulphonic esters are hydrolysed by aqueous alkali. There are two possible mechanisms. What are they? Devise an experiment to differentiate between them.

2. Complete the equations:

(i) $p\text{-MeC}_6H_4SO_2Cl \xrightarrow{?} p\text{-MeC}_6H_4SO_2H$ (ii) $PhSO_2Cl \xrightarrow{?} PhSH$

(iii)
S—S
NO2
O2N
Cl2
HCl/HNO3
?

(iv) $? + HgCl_2 \xrightarrow[\text{heat}]{H_2O} p\text{-MeC}_6H_4HgCl$

(v)
SO3H
SO3H
OH
+Br2
NaOH–H2O
?

(vi)
NH2
Me
+H2SO4
heat
?

(vii)
SO2NH2
Cl
Cl
NH2
aq. H2SO4
heat
?

(viii)
NH2
SO2NH2
+2HBr+2H2O2
H2O
heat
?

(ix)
Me
Br
Me
Br
(i) H2SO4
(ii) HCl; heat
?

(x) $PhSO_3^-Na^+ + PCl_5 \longrightarrow ?$

(xi) $p\text{-MeC}_6H_4SO_2Cl + PhH \xrightarrow{AlCl_3} ?$ (xii) $PhSO_3Et + 2MeNH_2 \longrightarrow ?$

(xiii)
NHAc+ ?
ClO2S
NHAc+ ?

(xiv)
CMe3
+H2SO4
?

(xv) $PhSO_3Et + EtMgI \xrightarrow{Et_2O} ?$

3. The sodium salt of sulphanilic acid can be readily acetylated with Ac_2O, but not the free acid. Explain.

4. Suggest a mechanism for the reaction

$$ArH + SO_2Cl_2 \xrightarrow{AlCl_3} ArSO_2Cl + HCl$$

5. Starting from *p*-chloronitrobenzene, prepare bis(4-aminophenyl)sulphone.

REFERENCES

SUTER, *Organic Compounds of Sulphur,* Wiley (1944).
GILBERT, *Sulphonation and Related Reactions,* Interscience (1965).
CERFONTAIN, *Mechanistic Aspects in Aromatic Sulphonation and Desulphonation,* Interscience (1968).

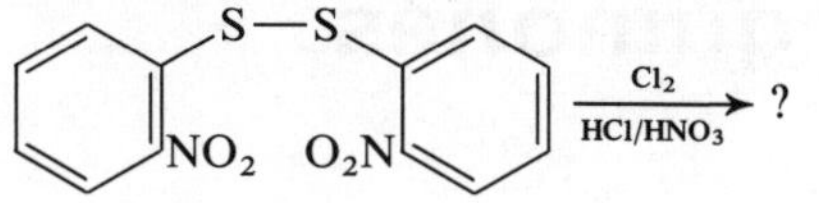

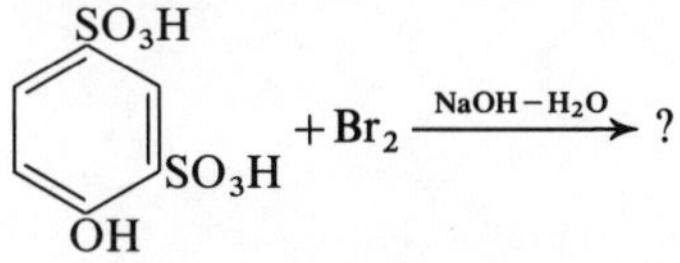

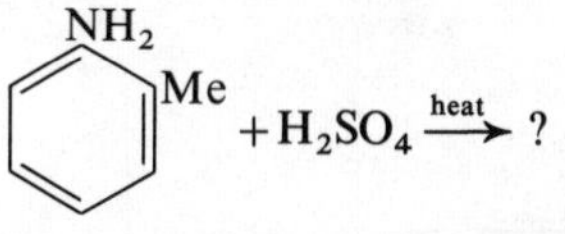

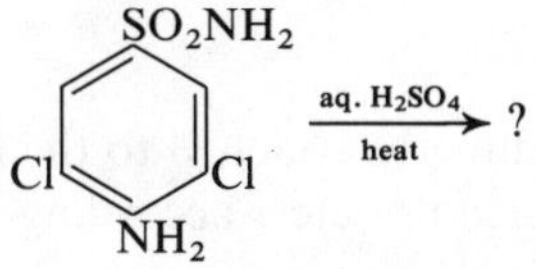

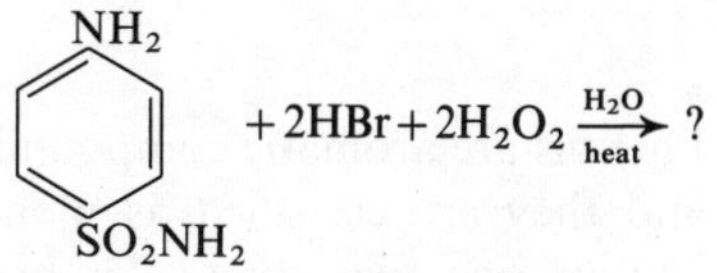

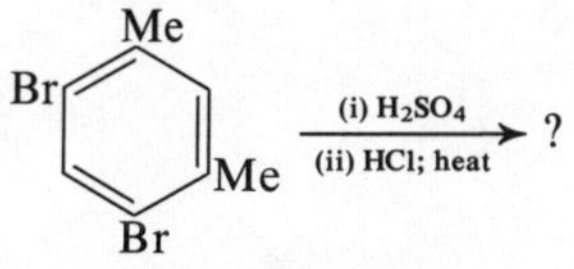

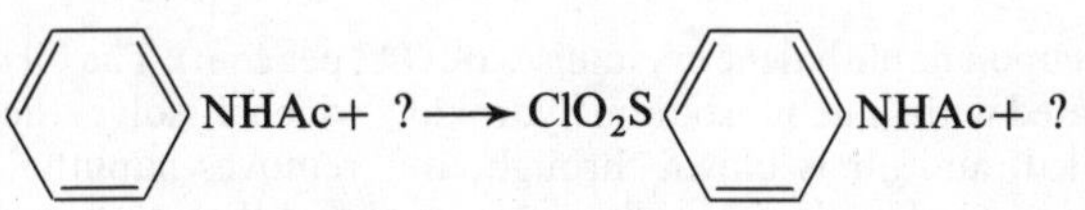

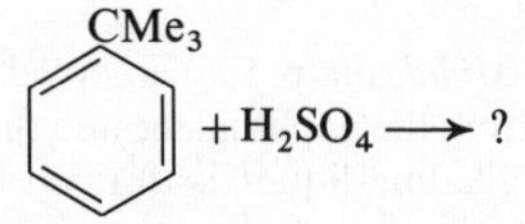

26 Phenols and quinones

Phenols are aromatic compounds containing hydroxyl groups directly attached to the nucleus, and they are classified as monohydric, dihydric, trihydric phenols, etc., according as they contain one, two, three, etc., hydroxyl groups.

Monohydric phenols

General methods of preparation

1. A number of monohydric phenols occur in coal-tar and their extraction from this source is very important commercially.

> (i) *Middle oil* (p. 568) is cooled, whereupon naphthalene crystallises out (43 per cent). The oil is pressed free from the naphthalene and then treated with aqueous sodium hydroxide, which dissolves the phenols. The alkaline liquor is drawn off, boiled, and air is blown through; this removes naphthalene (that remained after cooling the oil), pyridine, etc. The liquid is allowed to cool and then carbon dioxide is blown through, thereby decomposing the sodium phenoxides into the free phenols and sodium carbonate, the latter dissolving in the aqueous layer. This aqueous layer is drawn off, and the crude phenols (the yield of which is about 12 per cent of the middle oil) are fractionated. Three fractions are collected: *phenol*, b.p. 182°C (20%), *cresols*, b.p. 190–203°C (43%) and *xylenols*, b.p. 211–225°C (26%); the residue is *pitch*.
>
> (ii) *Heavy oil* is treated in the same manner as above. After being pressed to remove naphthalene, the residual oil contains cresols, higher phenols, naphthol, etc. (heavy oil contains about 7 per cent of the phenols).

2. Phenols may be prepared by fusion of sodium arylsulphonates with sodium hydroxide:

$$ArSO_3{}^-Na^+ + 2NaOH \longrightarrow ArO^-Na^+ + Na_2SO_3 + H_2O$$

3. When a diazonium sulphate solution is steam distilled, a phenol is produced:

$$ArN_2{}^+HSO_4{}^- + H_2O \longrightarrow ArOH + N_2 + H_2SO_4$$

4. Phenols are formed when compounds containing an activated halogen atom are heated with aqueous sodium hydroxide; *e.g.*, *p*-nitrophenol from *p*-chloronitrobenzene:

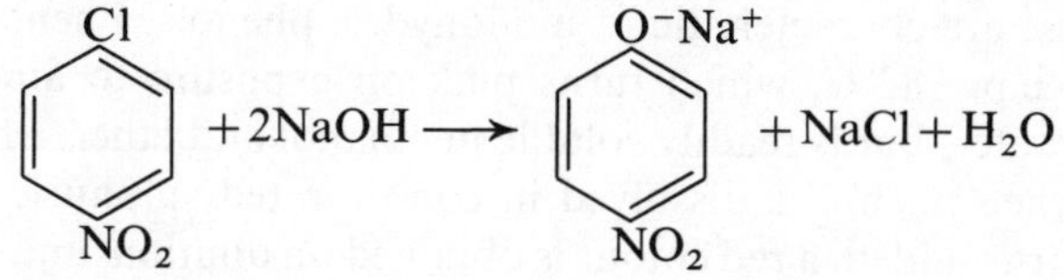

$$\text{(2,4-dinitrochlorobenzene with } Cl, NO_2) + 2NaOH \longrightarrow (\text{2,4-dinitrophenoxide with } O^-Na^+, NO_2) + NaCl + H_2O$$

5. Distillation of phenolic acids with soda-lime produces phenols; *e.g.*, sodium salicylate gives phenol:

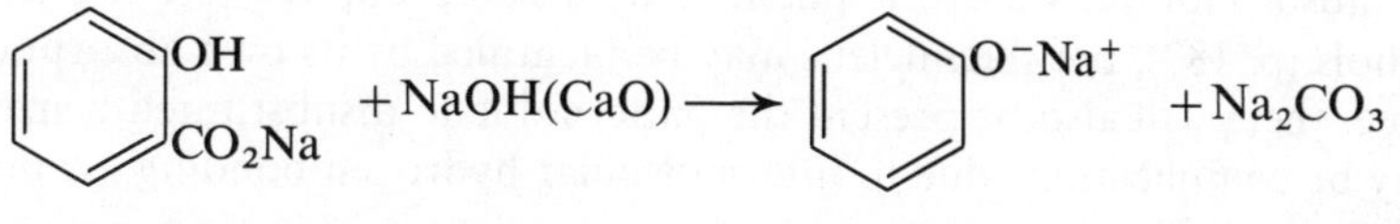

$$(\text{salicylate, } OH, CO_2Na) + NaOH(CaO) \longrightarrow (\text{phenoxide, } O^-Na^+) + Na_2CO_3$$

6. Phenols may be prepared by means of a Grignard reagent:

$$ArMgBr \xrightarrow{O_2} ArO_2MgBr \xrightarrow{ArMgBr} 2ArOMgBr \xrightarrow{H^+} 2ArOH$$

Alternatively, the Grignard reagent may be converted into a phenol via a boronic or borinic acid (these need not be isolated; Hawthorne, 1957):

$$ArMgX \xrightarrow[\text{(ii) } H_2O]{\text{(i) } B(OMe)_3} ArB(OH)_2 \text{ or } Ar_2BOH \xrightarrow{H_2O_2} ArOH \quad (60–70\%)$$

Phenol (*carbolic acid, hydroxybenzene*) may be prepared by any of the general methods; commercially, it is prepared from coal-tar (see above). The supply from this source is now insufficient to give the amount of phenol required for industry, and so it is also prepared synthetically. Various methods are used.

The oldest synthetic method is the alkaline fusion of sodium benzenesulphonate (see method **2** above). Two other methods start from chlorobenzene, one conversion being effected by heating with aqueous sodium carbonate or sodium hydroxide under pressure at about 300°C, and the other conversion being effected by heating with steam at 425°C in the presence of a catalyst (in the latter, the chlorobenzene is prepared by the Raschig process; see p. 625). Another method is the oxidation of toluene by air in the presence of manganous and cupric salts as catalyst:

$$C_6H_5CH_3 \xrightarrow[\text{cat.}]{O_2} C_6H_5OH + CO_2$$

The *cumene-phenol process* appears to be the most important commercial method and is carried out by the oxidation of cumene to its hydroperoxide, which is then decomposed by acid into phenol and acetone (which is sold as a by-product). The mechanism of the conversion of cumene hydroperoxide into phenol is not certain, but it most probably involves a 1,2-shift (*cf.* the Baeyer–Villiger oxidation, p. 223).

$$C_6H_6 + CH_3CH{=}CH_2 \xrightarrow[250°C;\text{ press.}]{H_3PO_4} C_6H_5CH(CH_3)_2 \xrightarrow[\text{OH}^-;\ 130°C]{O_2} (CH_3)_2\overset{\underset{\displaystyle |}{C_6H_5}}{C}{-}O{-}OH \xrightarrow[100°C]{H_2SO_4}$$

$$(CH_3)_2\overset{\underset{\displaystyle |}{C_6H_5}}{C}{-}O{-}\overset{+}{O}H_2 \xrightarrow{-H_2O} (CH_3)_2\overset{\underset{\displaystyle |}{C_6H_5}}{C}{-}\overset{+}{O} \longrightarrow (CH_3)_2\overset{+}{C}{-}O{-}C_6H_5 \xrightarrow[\text{(ii) }-H^+]{\text{(i) } H_2O}$$

$$(CH_3)_2\overset{\underset{\displaystyle |}{:OH}}{C}{-}OC_6H_5 \longrightarrow (CH_3)_2C{=}\overset{+}{O}H + C_6H_5O^- \longrightarrow (CH_3)_2C{=}O + C_6H_5OH$$

Properties of phenol—these are characteristic of monohydric phenols. Phenol is a colourless crystalline solid, m.p. 43°, b.p. 182°C, which turns pink on exposure to air and light. It is moderately soluble in cold water, but is readily soluble in ethanol and ether. Phenol undergoes the *Liebermann reaction*; when phenol is dissolved in concentrated sulphuric acid and a few drops of aqueous sodium nitrite added, a red colour is obtained on dilution, and turns blue when made alkaline with aqueous sodium hydroxide (see also p. 728).

Phenol is used as an antiseptic and disinfectant, and in the preparation of dyes, drugs, bakelite, etc.

The infra-red absorption region of the phenolic hydroxyl group (O—H stretch) is the same as that for alcohols (p. 187), but the nucleus may be identified by its own absorption spectrum. At the same time, there will also be present the pattern due to a substituted benzene. Furthermore, there may be complications due to intermolecular hydrogen bonding, or intramolecular hydrogen bonding in various *o*-derivatives.

Mass spectra

The molecular ion of phenols is usually very intense and fragments in various ways according to whether other substituent groups are present, *e.g.*, phenol itself:

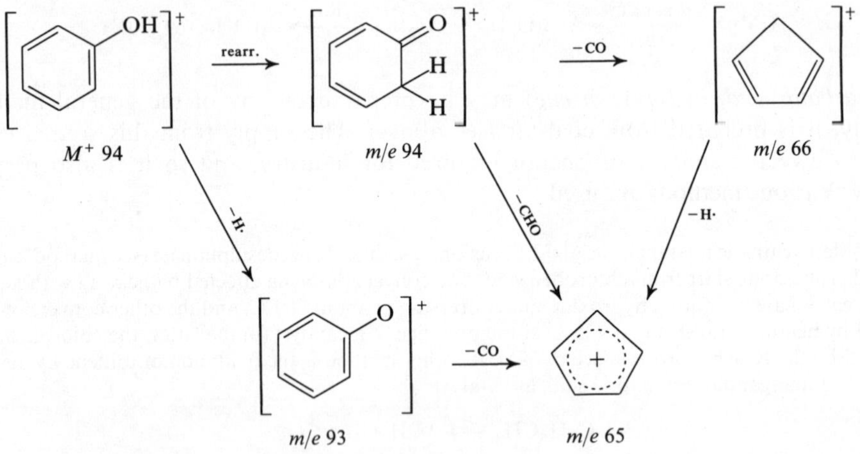

Cresols form the hydroxytropylium ion, *e.g.*,

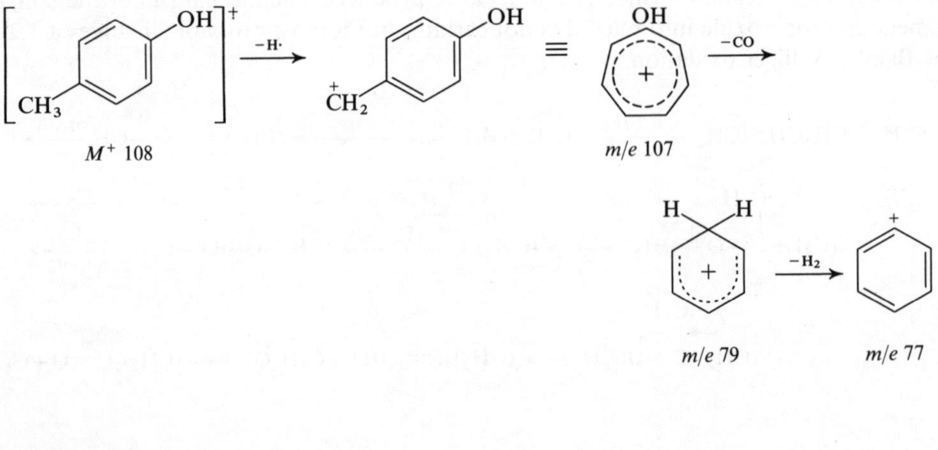

Phenolic ethers give a strong molecular ion peak and undergo several fragmentation patterns. One follows that of phenol, *e.g.*,

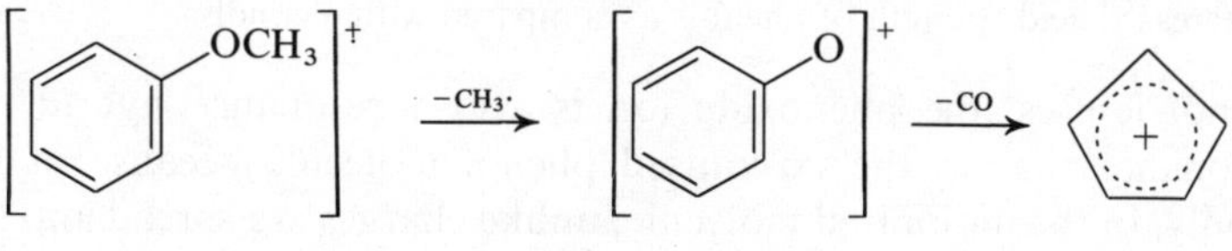

$$\left[\begin{array}{c}\text{OCH}_3\end{array}\right]^{+} \xrightarrow{-\text{CH}_3\cdot} \left[\begin{array}{c}\text{O}\end{array}\right]^{+} \xrightarrow{-\text{CO}} \left[\begin{array}{c}+\end{array}\right]$$

$$M^{+}\ 108 \qquad\qquad m/e\ 93 \qquad\qquad m/e\ 65$$

An alternative path involves a rearrangement:

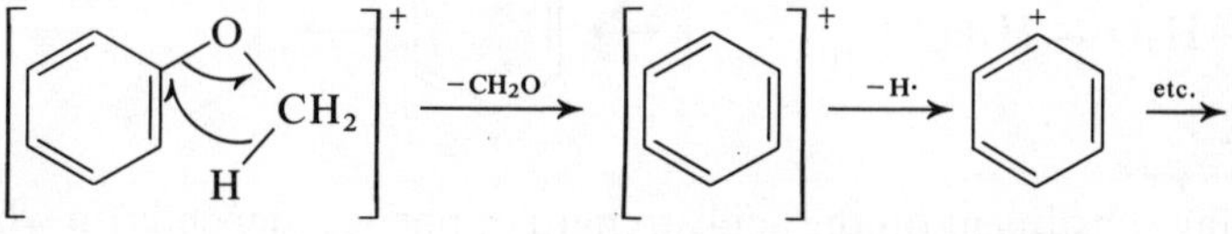

$$\left[\begin{array}{c}\text{O}\\\text{CH}_2\\\text{H}\end{array}\right]^{+} \xrightarrow{-\text{CH}_2\text{O}} \left[\begin{array}{c}\end{array}\right]^{+} \xrightarrow{-\text{H}\cdot} \left[\begin{array}{c}+\end{array}\right] \xrightarrow{\text{etc.}}$$

$$m/e\ 78 \qquad\qquad m/e\ 77$$

If the alkyl group contains two or more carbon atoms, then rearrangement can occur with elimination of an alkene (*cf.* alkyl ethers):

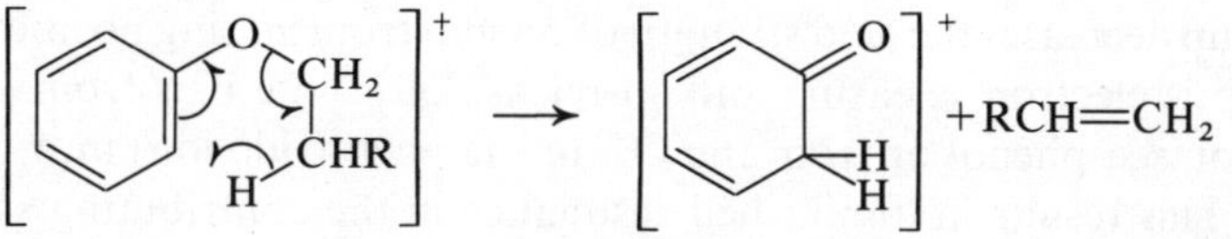

$$\left[\begin{array}{c}\text{O}\\\text{C}\quad\text{CH}_2\\\text{H}\quad\text{CHR}\end{array}\right]^{+} \longrightarrow \left[\begin{array}{c}\text{O}\\\text{H}\\\text{H}\end{array}\right]^{+} + \text{RCH}{=}\text{CH}_2$$

$$m/e\ 94$$

Reactions of phenols. The hydroxyl group in phenol behaves in some respects like that in alcohols, but the majority of the reactions of phenol are those connected with the benzene ring.

1. Phenol gives a violet colour with ferric chloride; this reaction is characteristic of all compounds containing the grouping —C(OH)=C (*cf.* enols, p. 280).

2. Phenol behaves as a weak acid, forming *phenoxides* with strong alkalis:

$$C_6H_5OH + NaOH \longrightarrow C_6H_5O^-Na^+ + H_2O$$

Since phenol is a weaker acid than carbonic acid, it may be separated from carboxylic acids by making the solution alkaline with sodium hydroxide, and then passing in carbon dioxide. Phenol is liberated from its sodium salt and so may be extracted with ether; the carboxylic acid salts are *not* decomposed by carbon dioxide.

Phenols are stronger acids than alcohols, and the explanations offered are similar to those for the carboxylic acids (p. 241).

(i) Because of resonance, the oxygen atom of the OH group acquires a positive charge and so proton release is facilitated:

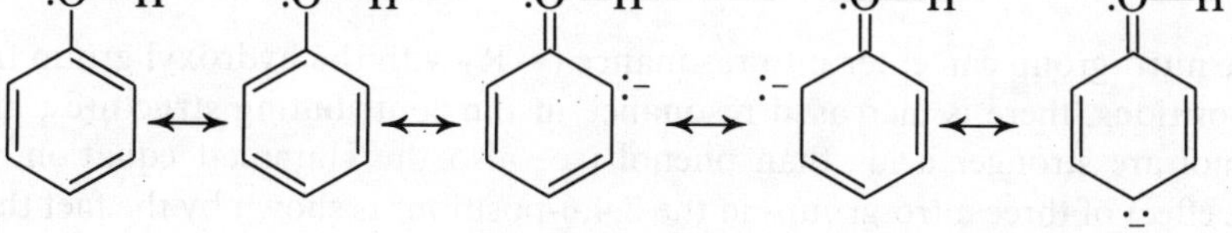

It might be noted that since the carbon atom of the C—OH group is sp^2 hybridised, it is more electron-attracting than an sp^3 carbon atom (as in alcohols), and consequently there is an increased $-I$ effect which facilitates proton release. This inductive effect *alone*, however, is not large enough to explain the increased acid strength of phenols as compared with alcohols.

(ii) When phenol ionises, the phenoxide ion is also a resonance hybrid, but it is more stabilised by resonance than is the un-ionised phenol molecule because of spreading of a negative charge only. In the un-ionised molecule, unlike charges are spread and so is less stable.

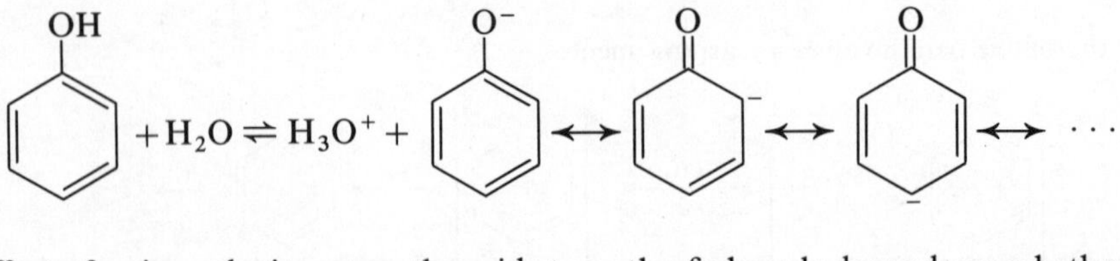

The effect of a ring substituent on the acid strength of phenols depends on whether the group is electron attracting or releasing, its ability to enter into resonance with the hydroxyl group, and its position (*cf.* the anilines). Effects in the *o*-position are similar to those in the *p*-position, but there are added complications due to the steric effect and to hydrogen bonding (when this is possible).

The methyl group decreases the acid strength of phenol from *all* ring positions (see Table 26.1). The methyl group is electron releasing and so release of a lone pair from oxygen (of the OH group in the un-ionised phenol or from the O^- in the phenoxide ion) into the ring is opposed (*cf.* the anilines). This results in diminished resonance in the contributing structures, and consequently the phenoxide ion is more resonance-stabilised with respect to phenol than is the methylphenoxide ion with respect to methylphenol. Hence phenol is a stronger acid than methylphenol.

Table 26.1

		pK_a	
Substituent	*o*-	*m*-	*p*-
None: 9·98			
Me	10·28	10·08	10·14
NO$_2$	7·23	8·40	7·15
F	8·81	9·28	9·95
Cl	8·48	9·02	9·38
MeO	9·98	9·65	10·21
OH	9·48	9·44	9·96
NH$_2$	9·71	9·87	10·30
2,4,6-tri-NO$_2$: 0·71			

Since the nitro-group can enter into resonance ($-R$) with the hydroxyl group from both the *o*- and *p*-positions, there is increased resonance in the contributing structures, and so *o*- and *p*-nitrophenol are stronger acids than phenol (see also the Hammett equation, p. 608). The cumulative effect of three nitro-groups in the 2,4,6-positions is shown by the fact that picric acid is a very strong acid (see Table 26.1). A *m*-nitro-group cannot enter into resonance with the hydroxyl group, but it can exert a −I effect from this position (*cf.* the anilines). Hence *m*-nitrophenol is a stronger acid than phenol, but is not so strong as the *o*- and *p*-isomers.

Groups with a +R effect, *e.g.*, NH$_2$, MeO, weaken the phenol when in the *p*-position, but 'strengthen it when in the *m*-position. In the latter case, only the −I effect can operate. All the halogenophenols are stronger acids than phenol, and so it follows that the −I effect of a halogen atom is greater than its +R effect (*cf.* p. 597).

A study of Table 26.1 shows that the effect of a given group in the *o*-position can be opposite to that when in the *p*-position. Intramolecular hydrogen bonding, when this is possible, may play a part; this would stabilise the un-ionised form. The steric effect, because of the small size of a hydrogen atom, would be expected to be very small or completely absent; if it is the phenoxide ion that is being considered, then the steric effect is absent. This line of argument is supported by the following observations:

pK_a 10·28	10·60	7·15	7·22	10·17	8·25
(I)	(II)	(III)	(IV)	(V)	(VI)

The decrease in acid strength from (I) to (II) can be explained by the additional +I effect in (II). (III) and (IV) have close pK_a values, and so the steric effect is absent, (IV) being slightly weaker due to the +I effect of the two *o*-methyl groups. In (V), the +I effect of the methyl group is operating, but in (VI) there is now steric inhibition of resonance. Thus (VI) is much weaker than (III); (*cf.* the anilines).

A point of interest in connection with the acid strength of phenols is that ring deuteration of phenol lowers the acidity by about 12 per cent (Kresge *et al.*, 1961). This is an example of the secondary isotope effect (p. 62), and it therefore appears that deuterium is more electron-releasing than protium.

The thermodynamic study of the ionisation of phenols may be carried out by considering the following equations (*cf.* carboxylic acids, p. 272):

$$ArOH \ (aq.) \rightleftharpoons ArO^- \ (aq.) + H^+ \ (aq.)$$

and

$$-RT \ln K_a = \Delta G^\ominus \ (\text{ionisation}) = \Delta H^\ominus \ (\text{ionisation}) - T\Delta S^\ominus \ (\text{ionisation})$$

These thermodynamic functions have been evaluated for a number of phenols (Biggs, 1956; Hepler *et al.*, 1959). Table 26.2 shows values for Me and NO$_2$ substituents.

Table 26.2

Substituent	$\Delta G^\ominus$ (kJ)	$\Delta H^\ominus$ (kJ)	$\Delta S^\ominus$ (J/K^{-1})
none	56·1	23·6	−111·7
o-Me	58·6	29·8	− 97·06
m-Me	57·7	20·5	−128·5
p-Me	58·6	17·7	−137·2
o-NO$_2$	41·0	19·5	− 72·8
m-NO$_2$	47·7	19·7	− 91·14
p-NO$_2$	40·6	19·7	− 70·7

3. Some other reactions involving mainly the hydroxyl group are:

(i) Alkali phenoxides react with alkyl halides to form phenolic ethers (see p. 718).

(ii) Phenol reacts with phosphorus pentachloride to give only a small amount of chloro-benzene; the main product is triphenyl phosphate. On the other hand, the presence of nitro-groups in the *o*- and/or *p*-positions leads to good yields of chloronitrobenzenes, *e.g.*, picric acid (2,4,6-trinitrophenol) gives picryl chloride. In this case, activated nucleophilic substitution occurs at one stage in the multistage reaction.

(iii) Phenol forms phenyl esters when warmed with acid chlorides or acid anhydrides, *e.g.*,

$$C_6H_5OH + (CH_3CO)_2O \xrightarrow{CH_3CO_2Na} CH_3CO_2C_6H_5 + CH_3CO_2H$$

These esters, however, because of the presence of the benzene ring, can undergo rearrangement under suitable conditions (*cf.* the allylic rearrangement, p. 334). This is known as the **Fries rearrangement** (1908), and is the conversion of a phenyl ester into an *o*- or *p*-hydroxyketone, or a mixture of both, by treatment with anhydrous aluminium chloride:

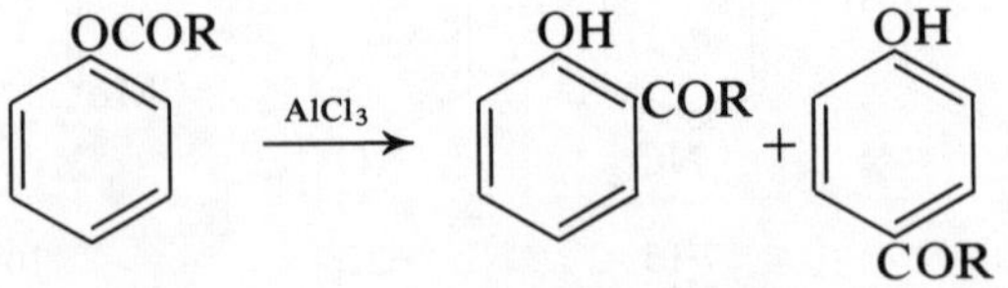

Generally low temperatures (60°C or less) favour the formation of the *p*-isomer, whereas high temperatures (above 160°C), favour the *o*-isomer. This rearrangement is a useful synthetic route to phenolic ketones (p. 753).

Many theories have been proposed for the Fries rearrangement, but none is certain. There is evidence for an intermolecular mechanism, (*e.g.*, Baltzly *et al.*, 1948), but more recently evidence has been obtained to show that the rearrangement is at least partly intramolecular (Crawford, 1959). It would appear that the rearrangement is best regarded as a combination of the two mechanisms occurring simultaneously:

(i) Intermolecular (*cf.* the Friedel–Crafts acylation):

$$PhOCOR + AlCl_3 \rightleftharpoons PhO^+\!-\!COR \rightleftharpoons PhO\bar{A}lCl_3 + RCO^+ \rightleftharpoons$$

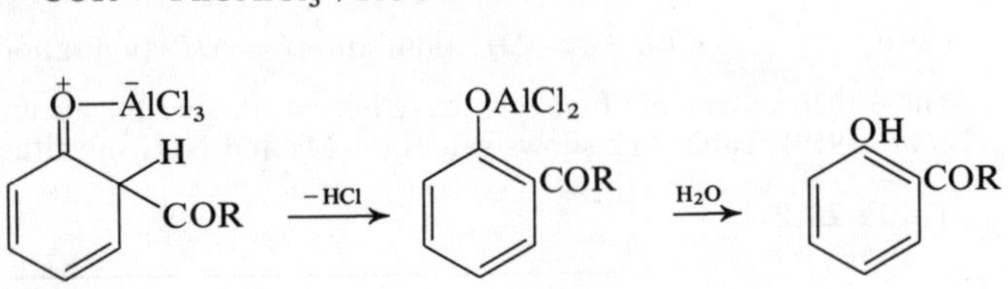

(ii) Intramolecular:

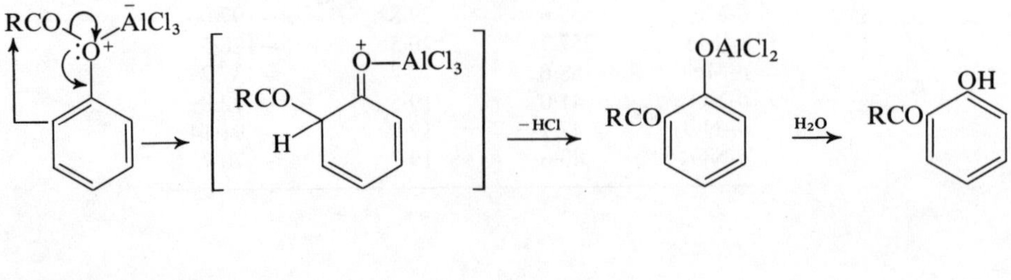

Just as alcohols undergo ammonolysis (p. 367), so can phenol be converted into aniline by ammonia either when heated under pressure or by heating in the presence of zinc chloride.

$$C_6H_5OH + NH_3 \xrightarrow{ZnCl_2} C_6H_5NH_2 + H_2O$$

4. Many of the reactions of phenol are those which are examples of electrophilic aromatic substitution, and since the hydroxyl group is *o-/p*-orienting, these are always the products formed (see below).

5. Phenol undergoes alkylation to form mainly the *p*-derivative and a small amount of the *o*- (see also p. 719):

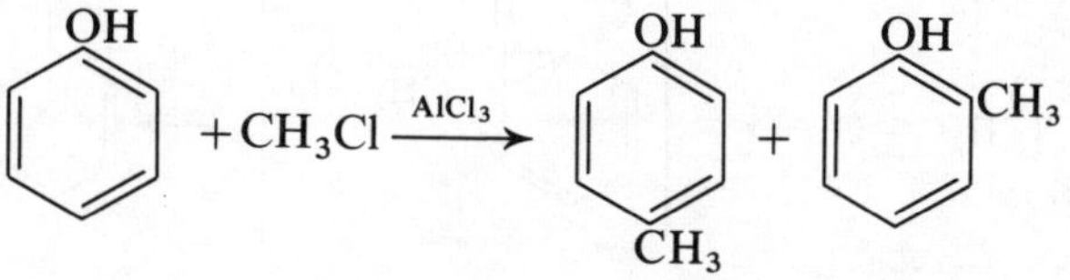

6. Phenol is chloromethylated (p. 613) so readily that usually polymers are obtained.

The presence of a $-I$ group, however, decreases the activating effect of the hydroxyl group; *e.g., p*-nitrophenol may be successfully chloromethylated to give 2-hydroxy-5-nitrobenzyl chloride.

On the other hand, phenol may be chloromethylated successfully by first converting it into an ester (usually the ethyl phenyl carbonate by means of chloroformic ester), and chloromethylating this.

7. Phenol can be readily mercurated, *e.g.*, when refluxed with aqueous mercuric acetate, *o*-acetoxymercuriphenol is formed (together with some dimercurated compound):

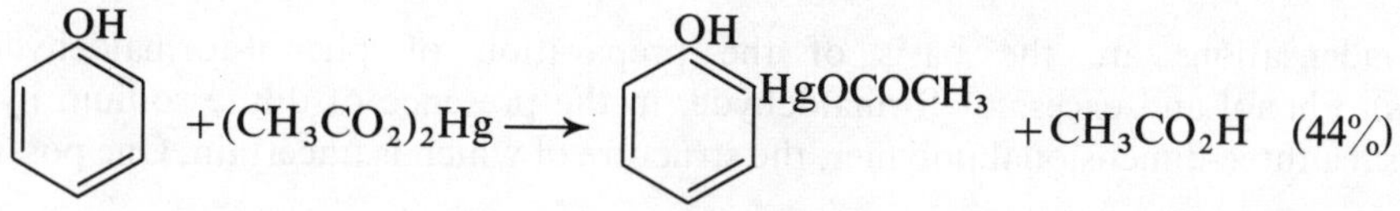

8. Phenol couples in the *p*-position with diazonium salts in alkaline solution to form hydroxyazo-compounds (p. 678).

9. Phenol condenses with aliphatic and aromatic aldehydes in the *o*- and *p*-positions, the most important example being the condensation with formaldehyde. At low temperature, in the presence of dilute acid or alkali, and using formalin (40 per cent aqueous formaldehyde), the main product is *p*-hydroxybenzyl alcohol, together with a small amount of the *o*-isomer.

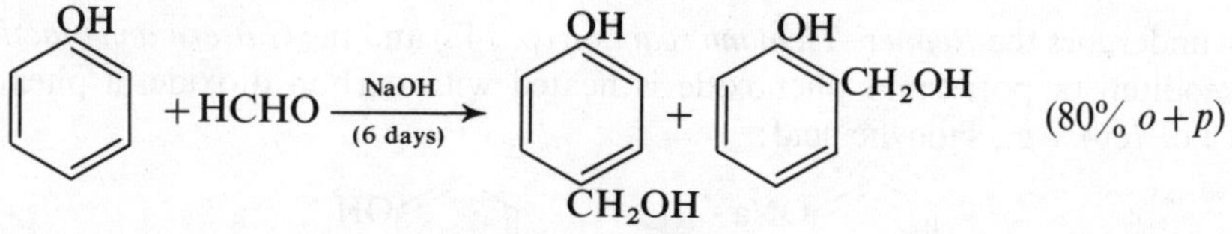

This is known as the **Lederer–Manasse reaction** (1894); its mechanism may be:

Acid catalysis

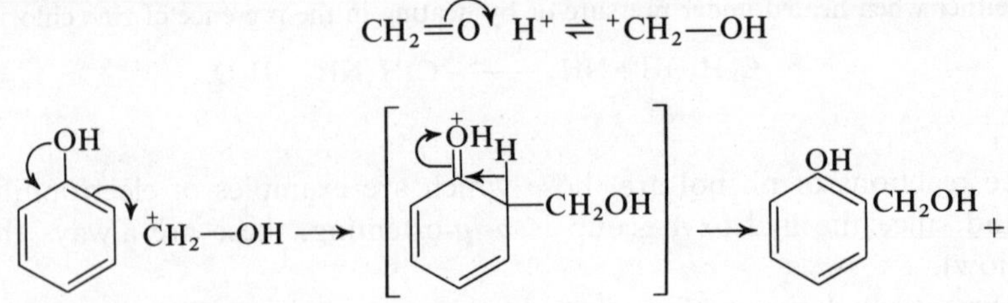

$$CH_2{=}O \quad H^+ \rightleftharpoons \ ^+CH_2{-}OH$$

Base catalysis

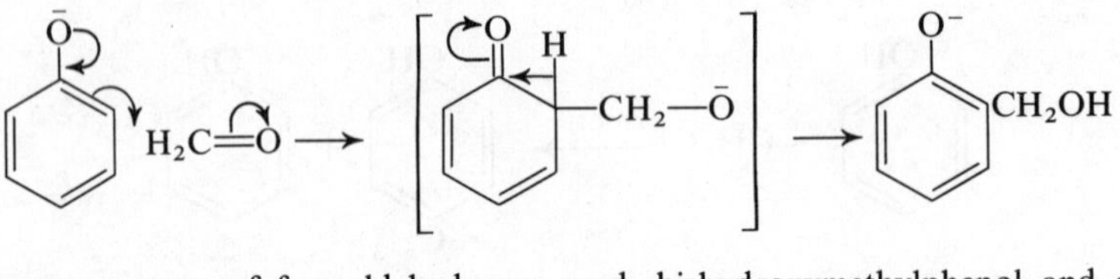

When larger amounts of formaldehyde are used, bishydroxymethylphenol and *p,p'*-dihydroxy-diphenylmethane are obtained:

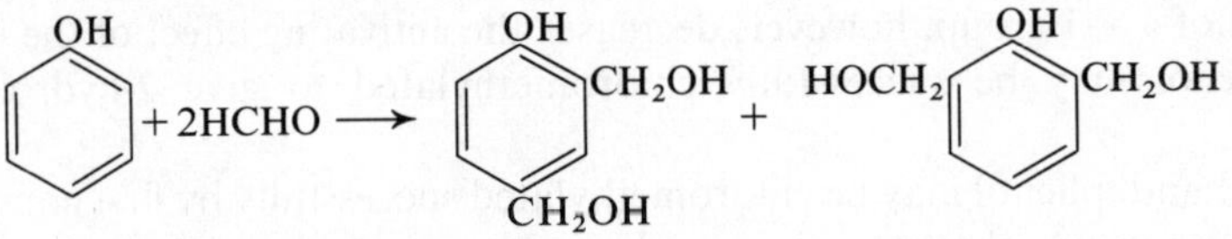

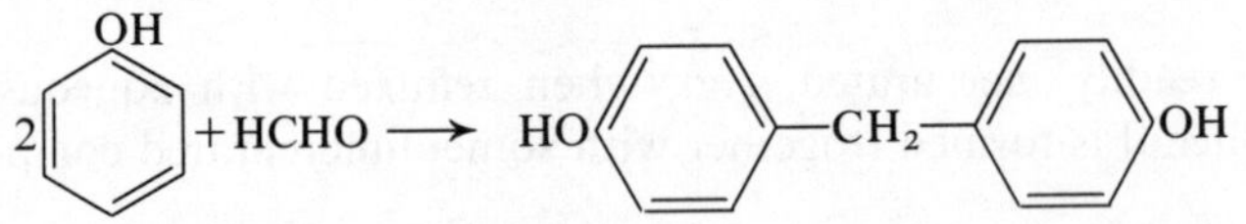

These condensations are the basis of the preparation of phenol-formaldehyde resins (**Bakelites**). Phenol and excess of formaldehyde, in the presence of dilute sodium hydroxide, slowly form a three-dimensional polymer, the structure of which is uncertain. One possibility is:

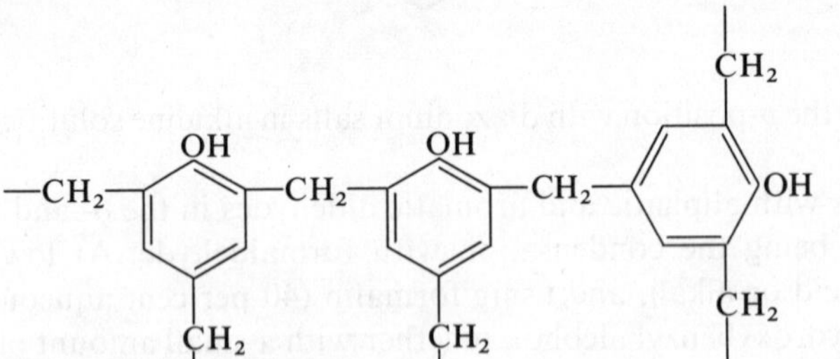

10. Phenol undergoes the *Reimer–Tiemann reaction* (p. 743) and the *Gattermann reaction* (p. 735).
11. When sodium or potassium phenoxide is heated with carbon dioxide, a phenolic acid is formed (see p. 766), *e.g.*, salicylic acid:

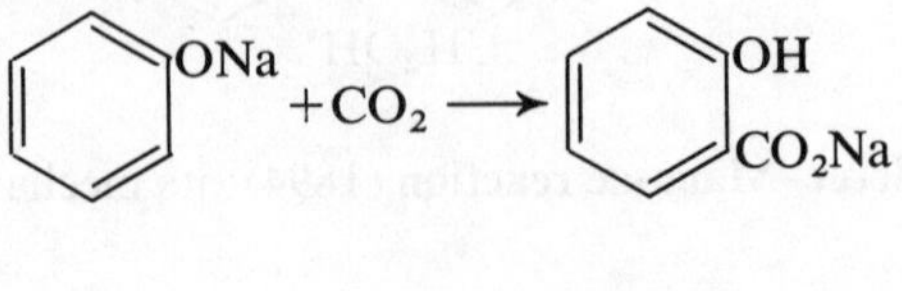

12. Phenol can be hydrogenated in the presence of a nickel catalyst at 160°C to cyclohexanol:

$$C_6H_5OH + 3H_2 \xrightarrow{Ni} C_6H_{11}OH \quad (ex.)$$

When distilled with zinc dust, phenol is converted into benzene. This is of very little preparative value, but is useful for the identification of various nuclei in natural products, e.g., alkaloids.

When phenol is oxidised with potassium permanganate, the ring is broken down (most phenols behave in a similar manner). Homologues of phenol can, however, be oxidised to the corresponding phenolic acid *provided the hydroxyl group is protected by alkylation or acylation*. The best means of protection is the formation of the benzenesulphonate, e.g. (see also p. 617):

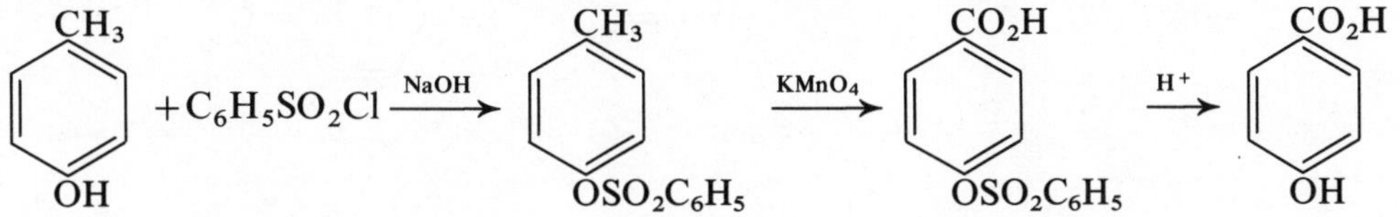

Hydrogen of the hydroxyl group of phenol is readily removed by 'one-electron transfer' oxidising agents, e.g., ferric ions (FeCl$_3$), alkaline potassium ferricyanide, etc.

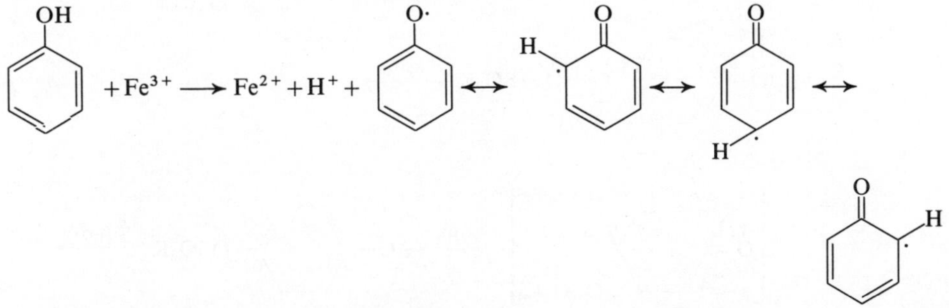

This phenoxyl radical, because of its high reactivity, quickly disappears, mainly by coupling (dimerisation), which is predominantly of the C—C type (o-o, p-p, o-p). The other possible modes of coupling, C—O and O—O, occur less frequently, particularly the latter. These combination reactions have useful synthetic applications and are also important in biosynthetic studies.

Coupling between aroxyl radicals is sterically hindered by the presence of large substituents, e.g., 2,4,6-tri-t-butylphenol forms a stable free radical when oxidised in benzene solution (the solution is dark blue). This radical has been isolated as a dark blue solid, m.p. 96°C. In the presence of oxygen, the colour of the solution becomes light yellow due to the formation of the peroxide.

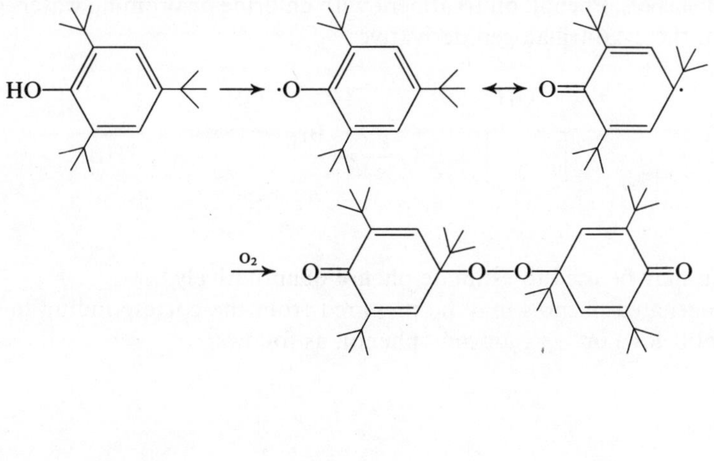

On the other hand, if the *p-t*-butyl group is absent, *p*-coupling occurs, and the dimer is then oxidised to a red compound.

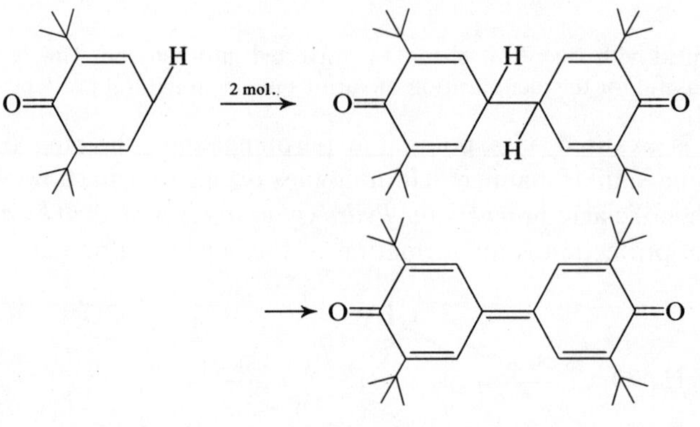

Phenols also undergo the **Elbs persulphate oxidation** (1890). In this reaction, monohydric phenols are oxidised to dihydric phenols by alkaline potassium persulphate. Hydroxylation normally occurs in the *p*-position but, if this is blocked, then in the *o*-position. According to Behrman *et al.* (1963), the mechanism is:

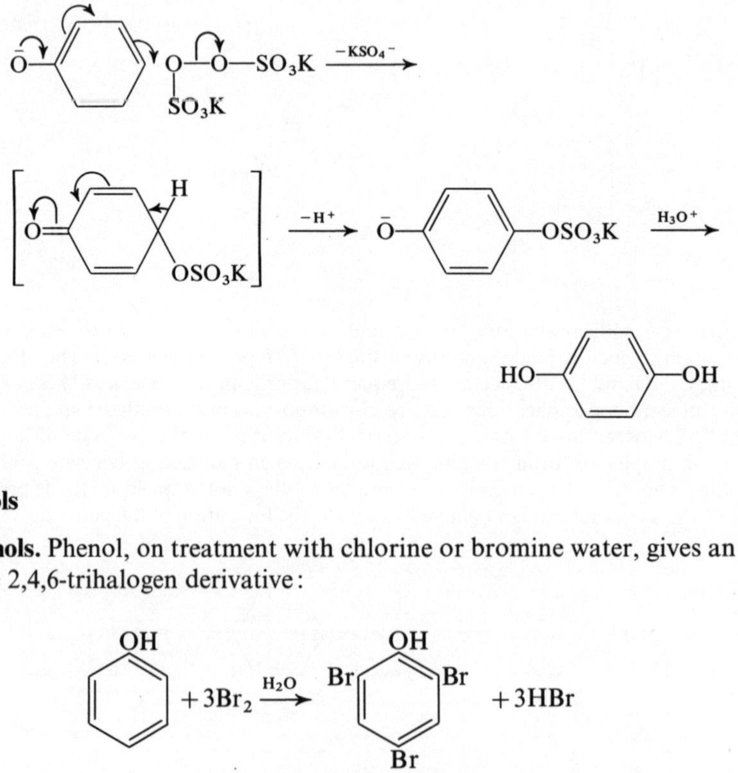

Substituted phenols

Halogenated phenols. Phenol, on treatment with chlorine or bromine water, gives an immediate precipitate of the 2,4,6-trihalogen derivative:

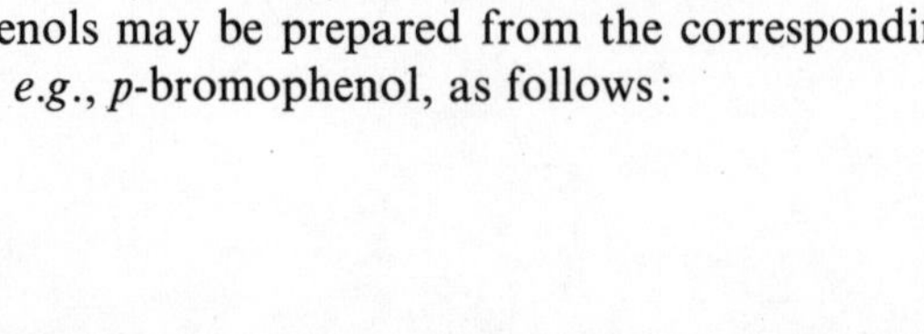

This reaction may be used to estimate phenol quantitatively.

Monohalogenated phenols may be prepared from the corresponding halogenoanilines (see *p*-iodophenol below) or, *e.g.*, *p*-bromophenol, as follows:

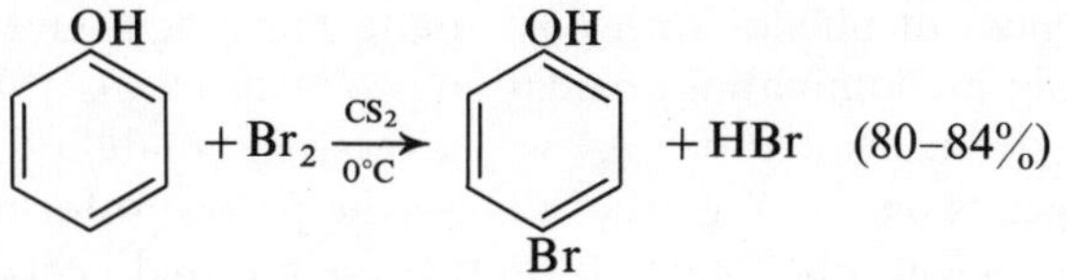

$$\text{OH-C}_6\text{H}_5 + Br_2 \xrightarrow[0°C]{CS_2} \text{o-Br-C}_6\text{H}_4\text{OH} + HBr \quad (80\text{--}84\%)$$

The reason for the formation of different products when water and carbon disulphide are used as solvents is believed to be due to the fact that in the former the substrate is the phenoxide ion. This is produced by ionisation in water, but is not formed in carbon disulphide (in the phenoxide ion, the benzene ring is more activated than in phenol itself).

o-Bromophenol may be prepared by protecting one *o*-position and the *p*-position by sulphonation:

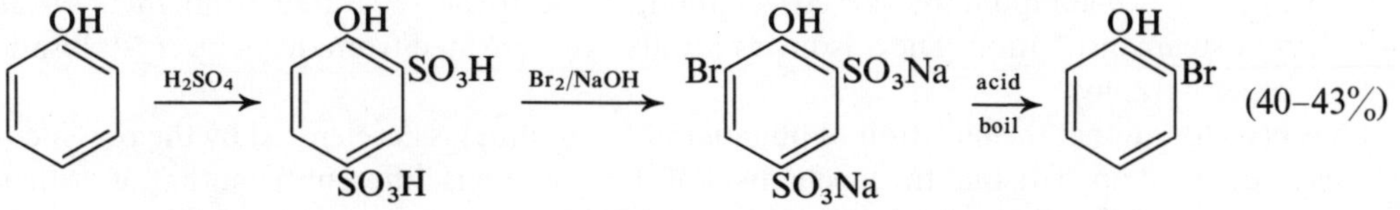

o-Iodophenol is best prepared by heating phenol with mercuric acetate, converting the *o*-acetoxymercuriphenol into the corresponding chloromercuri-derivative by heating with aqueous sodium chloride, and replacing the chloromercuri-group by treatment with iodine in chloroform solution:

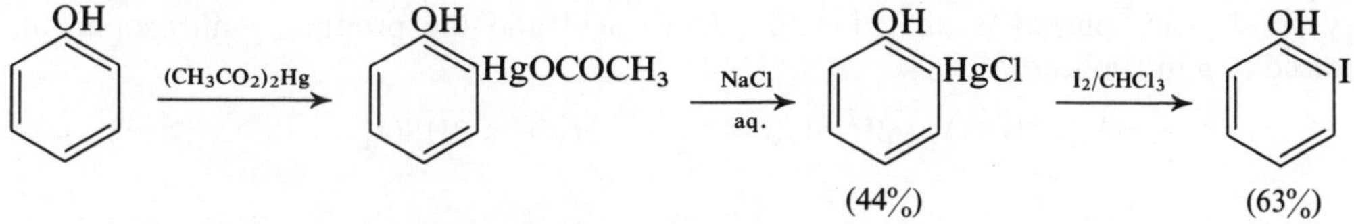

p-*Iodophenol* may be prepared as follows:

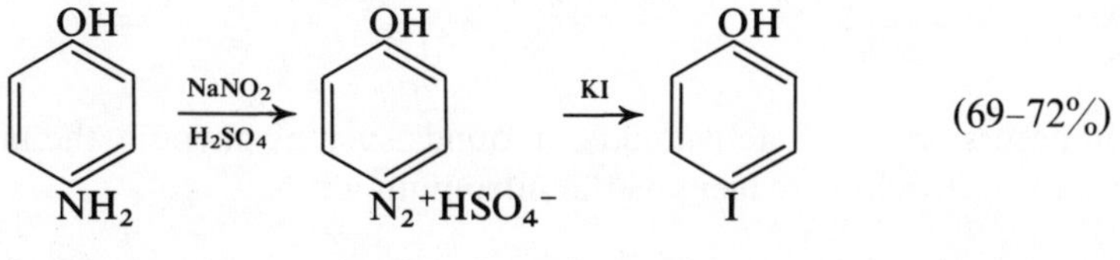

Dihalogenophenols are prepared by methods depending on the positions of the halogen atoms, *e.g.*, 2,6-dichlorophenol may be prepared from ethyl *p*-hydroxybenzoate:

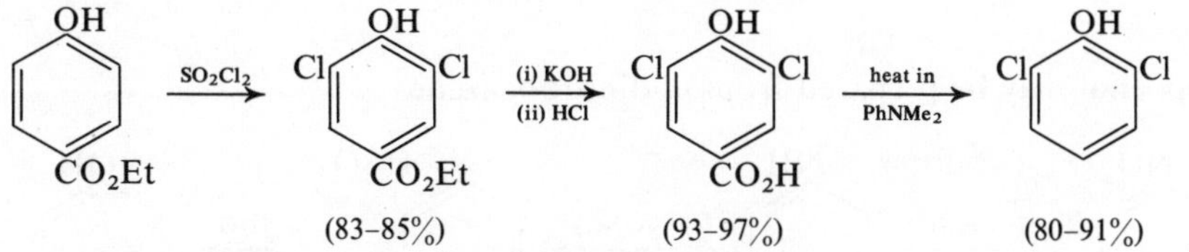

o-, *p-*, 2,4-Dichloro-, and 2,4,6-trichlorophenols are manufactured by the action of the requisite amount of chlorine on molten phenol. When monochlorination in the *p*-position is required, sulphuryl chloride is preferred as the chlorinating agent.

Nitrophenols. Treatment of phenol with cold dilute nitric acid gives a mixture of *o*- and *p*-nitrophenols, the latter predominating; oxidation products are also obtained. These isomers may be separated by steam distillation. As we have seen, solubility in hydroxylic solvents depends on, among other things, the power to form hydrogen bonds with the solvent. Phenol can form these bonds and hence a certain solubility in water can be expected. This argument also applies to substituted phenols since the hydroxyl group is still present, but in the *o*-compounds, however, because chelation is possible, hydrogen bonding with the solvent water molecules is hindered and hence the solubility is lowered. Furthermore, since chelation causes the *o*-compound to behave as a 'monomer', this isomer will be more volatile than the corresponding *m*- and *p*-isomers. Thus the effects of chelation are lower solubility and greater volatility in the *o*-compounds, thereby enabling these to be separated from their *m*- and *p*-isomers by steam distillation. The *o*-isomer may also be separated from the *p*- by crystallisation or by chromatography.

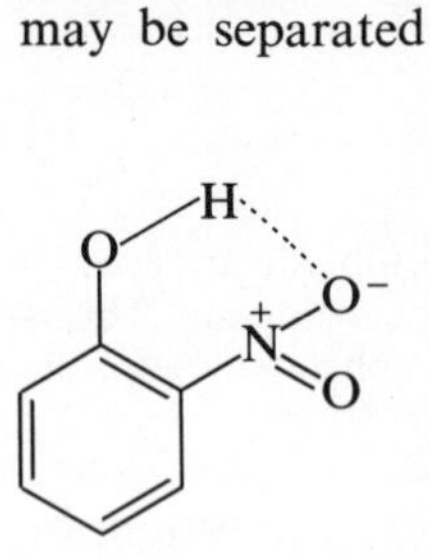

It has been found that the nitration of phenol (and of aniline) is accelerated by the presence of nitrous acid, and it appears that the mechanism of the reaction is different from that of ordinary nitration. Evidence for this is based on the observation that when phenol is nitrated in the presence of very little nitrous acid, *o*- and *p*-nitrophenols are formed in the ratio of 7:3. When nitrated in the presence of a large amount of nitrous acid, the ratio becomes 1:9, which is the ratio in which *o*- and *p*-nitrosophenols are formed if the nitric acid is omitted. It is therefore believed that the nitroso-compound is formed first, and this is then oxidised to the nitro-compound (Hughes *et al.*, 1946). This scheme may be used to prepare pure *p*-nitrophenol in very good yield; phenol is treated with nitrous acid and the product, *p*-nitrosophenol, is oxidised to *p*-nitrophenol.

$$HNO_2 + 2H_2SO_4 \rightleftharpoons NO^+ + H_3O^+ + 2HSO_4^-$$

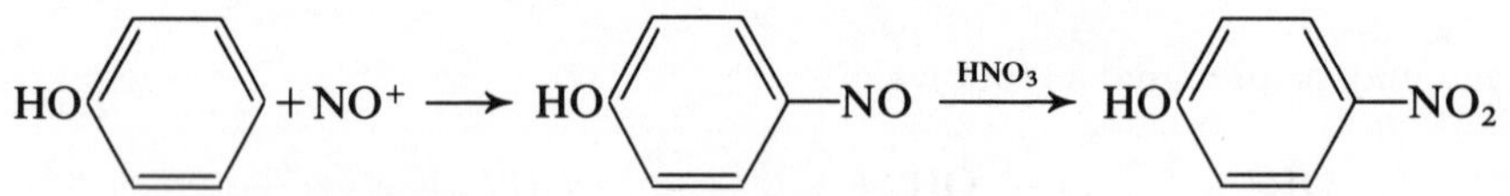

Since nitrosation occurs only with activated compounds such as phenols, the nitrosonium ion is not so reactive an electrophilic reagent as the nitronium ion.

o- and *p*-Nitrophenols are prepared commercially by heating *o*- and *p*-chloronitrobenzenes with aqueous sodium hydroxide under pressure.

o- and *p*-Nitrophenols are also formed when nitrobenzene is heated with solid potassium hydroxide (*cf.* p. 628).

m-Nitrophenol may be prepared from *m*-dinitrobenzene:

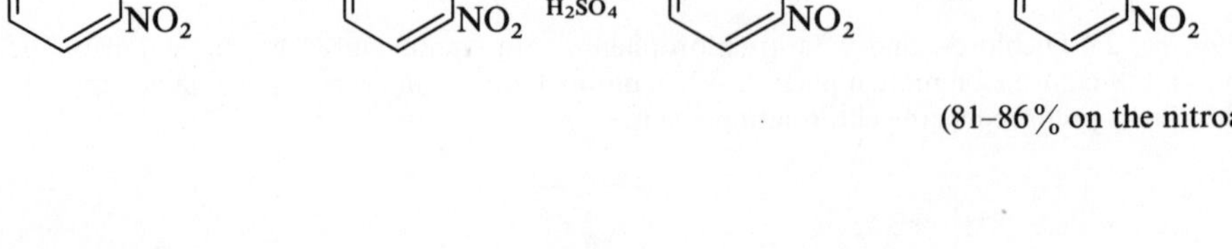

(81–86% on the nitroaniline)

o-Nitrophenol is a yellow solid, m.p. 45°C; the *m*- and *p*-isomers are colourless solids, m.p. 97° and 114°C, respectively. All three are readily reduced to the corresponding amino-phenols, and the nitro-group in the *o*- and *p*-compounds is displaced on treatment with bromine water, 2,4,6-tribromophenol being formed.

o- and *p*-Nitrophenols give rise to two series of ethers, one coloured and the other colourless. The colourless series contains the benzenoid structure, whereas the coloured series contains the chromo-phore quinone (see p. 887), *e.g.*,

colourless   coloured

The colourless ethers are stable but the coloured ethers are unstable and easily hydrolysed.

When *o*- and *p*-nitrophenols are nitrated, 2,4-dinitrophenol is formed and this, in turn, can be further nitrated to 2,4,6-trinitrophenol (*picric acid*). The yield of picric acid is poor due to large losses by oxidation. Picric acid is prepared commercially by first sulphonating phenol and then nitrating the product (see p. 691). Another commercial method is as follows:

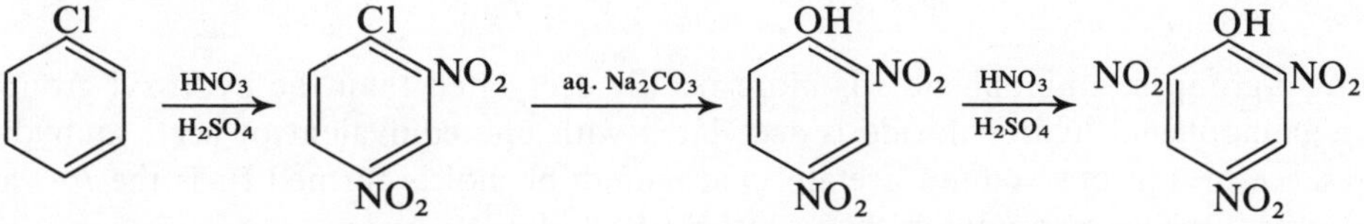

The interesting point to note about this method is that the presence of the nitro-group protects, to a large extent, the hydroxyl group from oxidation. It is in the nitration of phenol to the *o*- and *p*-nitrophenols that the loss by oxidation is greatest.

Picric acid may be obtained in the laboratory by oxidising 2,4,6-trinitrobenzene with potassium ferricyanide.

Picric acid is a yellow crystalline solid, m.p. 122°C, with a bitter taste (Greek: *pikros*, bitter), and is a strong acid, decomposing carbonates. It forms charge-transfer complexes known as *picrates*, with aromatic hydrocarbons, amines and phenols; their picrates are frequently used to identify these classes of compounds. When treated with bleaching powder, picric acid forms chloropicrin (p. 170) as one of the products and forms *picryl chloride* when treated with phosphorus pentachloride. The chlorine atom in picryl chloride is very reactive (owing to the presence of the three nitro-groups in the *p*- and two *o*-positions); *e.g.*, when boiled with water, picryl chloride forms picric acid, and when shaken with concentrated ammonia, *picramide*:

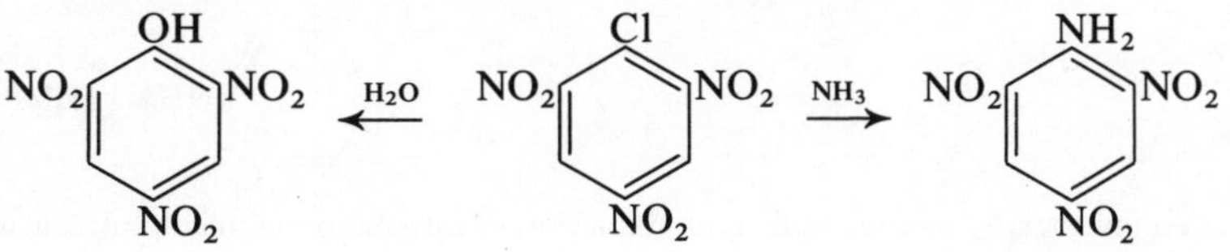

Picramic acid (2-*amino*-4,6-*dinitrophenol*) is formed when picric acid is reduced with sodium sulphide (*cf.* p. 653):

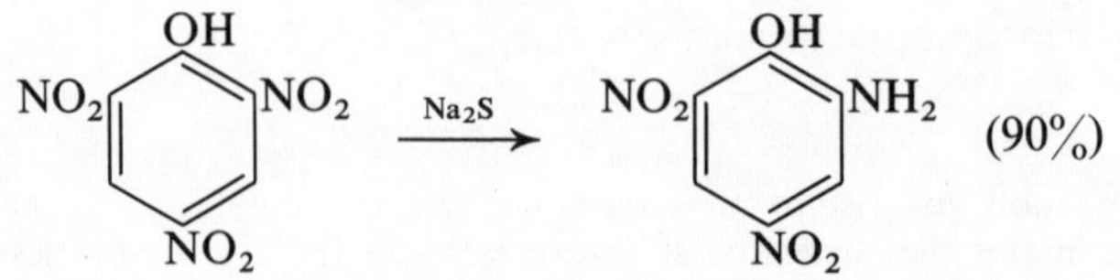

(90%)

Aminophenols. These may be prepared by reducing the corresponding nitrophenols with metal and acid, or catalytically. *m*-Aminophenol (used in the manufacture of dyes) is prepared commercially by fusion of metanilic acid (*m*-aminobenzenesulphonic acid) with sodium hydroxide or by the reduction of *m*-nitrophenol with Fe/HCl.

o- and *p*-Aminophenols are more weakly acidic than phenol; thus they do not form phenoxides with alkalis. On the other hand, they form salts with strong inorganic acids. The *o*- and *p*-derivatives are readily oxidised to the corresponding quinones; the *m*-compound is not easily oxidised (and does not give a quinone). *o*-Aminophenol has a marked tendency to form cyclic compounds (*cf.* *o*-phenylenediamines, p. 666); *e.g.*, with acetic acid it forms 2-methyl-benzoxazole:

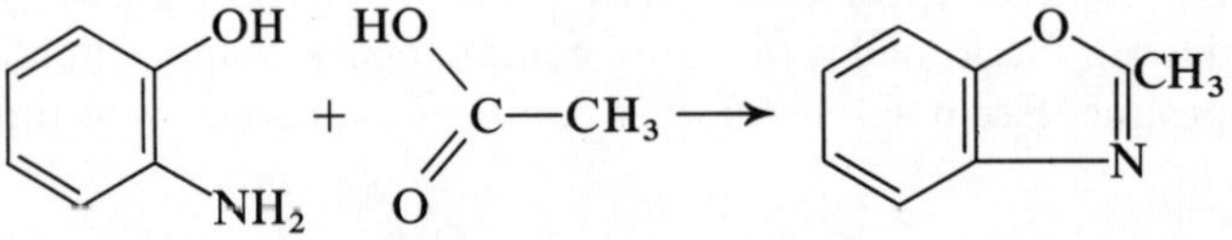

The amino-group in aminophenols is more readily acetylated than the hydroxyl group; *e.g.*, when *p*-aminophenol hydrochloride is acetylated with one equivalent of acetic anhydride in the presence of aqueous sodium acetate, *p*-acetamidophenol is formed (it is the *free* amino-group and not the ammonium ion that is attacked):

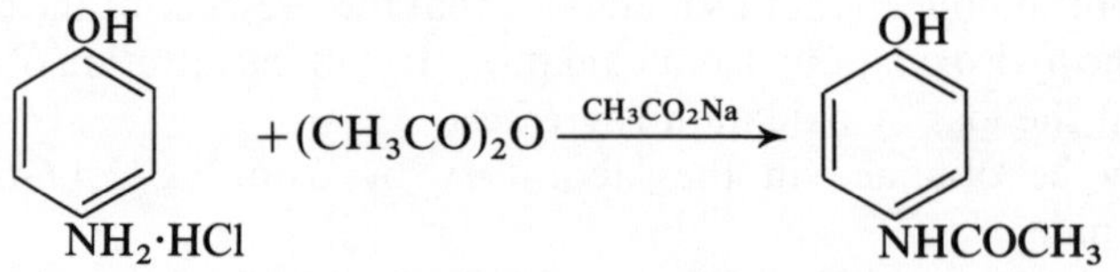

p-*Aminophenol*, m.p. 186°C, is very important as a photographic developer. It may be prepared by boiling phenylhydroxylamine with sulphuric acid (see p. 649), or by reducing hydroxyazobenzene with sodium dithionite:

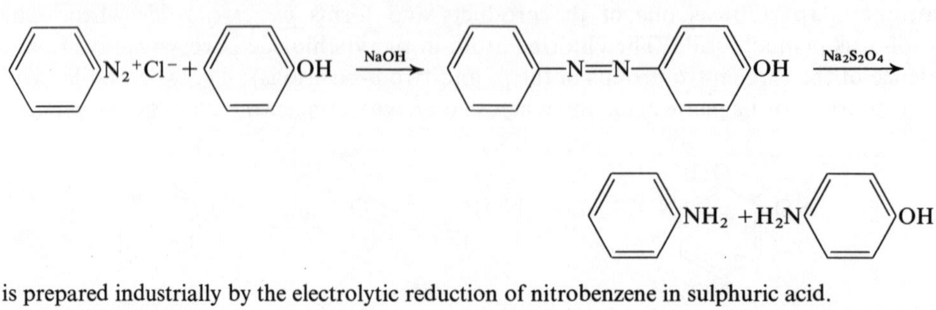

It is prepared industrially by the electrolytic reduction of nitrobenzene in sulphuric acid.

Two other important photographic developers are *amidol*, m.p. 78°C, and *metol*, m.p. 87°C:

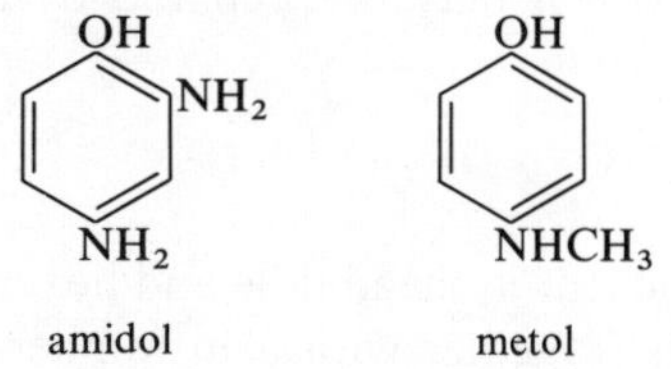

amidol   metol

Phenolsulphonic acids. When phenol is treated with concentrated sulphuric acid, *o*- and *p*-phenolsulphonic acids are formed, the former being the main product at ordinary temperatures, and the latter at higher temperatures (110°C); the *o*-compound rearranges to the *p*- on heating; thus, the *o*-isomer is the kinetically and the *p*-isomer is the thermodynamically controlled product. *m*-Phenolsulphonic may be obtained by the controlled potassium hydroxide fusion of benzene-*m*-disulphonic acid at about 180°C:

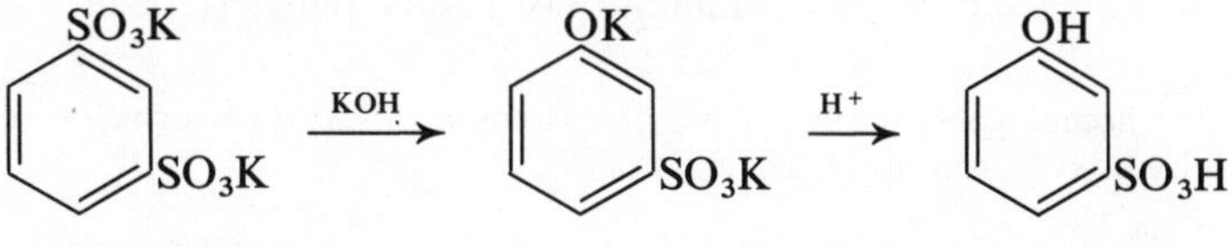

When *o*- or *p*- to a hydroxyl group (or an amino-group), a sulphonic acid group is often displaced by halogen when the sulphonic acid is halogenated in aqueous solution (*cf.* nitrophenols):

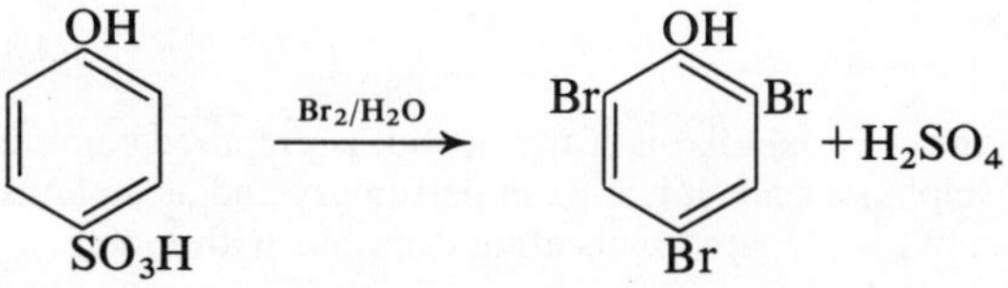

Nitrosophenols. When treated with nitrous acid, phenol forms mainly *p*-nitrosophenol, and a small amount of the *o*-compound:

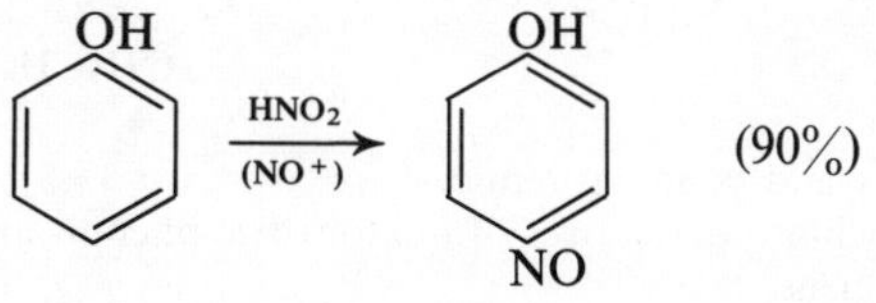

(90%)

p-Nitrosophenol is also formed when *p*-nitrosodimethylaniline is boiled with alkali (p. 662).

p-Nitrosophenol crystallises from hot water in pale yellow needles which readily turn brown. On the other hand, it crystallises from ether in brownish-green flakes. This colour suggests that *p*-nitrosophenol may have a quinonoid structure, *i.e.*, the following tautomeric system is present:

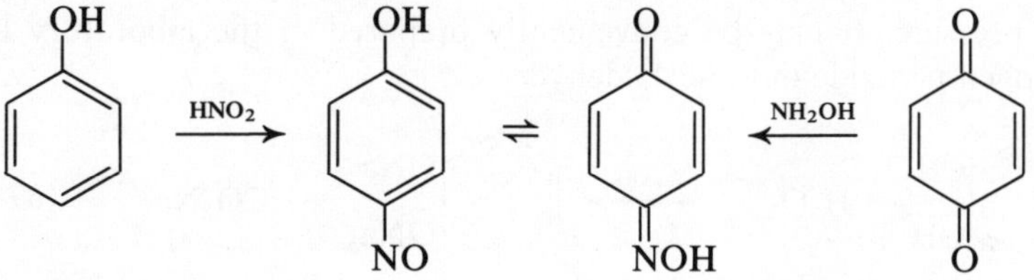

This is supported by the fact that *p*-nitrosophenol has been shown to be identical with the monoxime of *p*-benzoquinone, which may be prepared by the action of hydroxylamine on *p*-benzoquinone.

Homologues of phenol

Cresols (*hydroxytoluenes*). These occur in the middle and heavy oil fractions of coal-tar (p. 568). The mixture of the three cresols (together with a little phenol) is known as *cresylic acid* or *creosote*, and is used for preserving purposes, *e.g.*, timber, railway sleepers, etc. A solution of cresols in soapy water is known as *lysol*, which is used as a disinfectant.

The boiling points of the cresols are: *o*-, 191°; *m*-, 201°; *p*-, 202·5°C. By means of a very good fractionating column it is possible to separate the *o*-isomer from the other two. Each isomer can be obtained pure from the corresponding toluidine.

When either *o*- or *m*-chlorotoluene is heated with aqueous sodium hydroxide at about 300–320°C under pressure, a certain amount of *m*-cresol is obtained.

m- and *p*-Cresols are used in the manufacture of resins, plasticisers, etc.

Two higher homologues of phenol are the isomers *thymol* (3-hydroxy-4-isopropyltoluene), and *carvacrol* (2-hydroxy-4-isopropyltoluene):

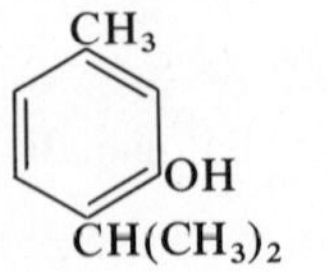

thymol, m.p. 51°C 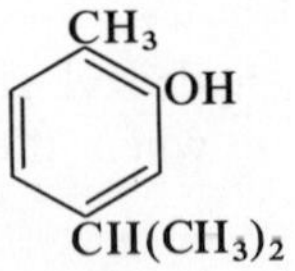   carvacrol, b.p. 237°C

Thymol occurs in the essential oil, oil of thyme, but is prepared commercially by heating *m*-cresol and isopropanol with sulphuric acid; it is used in perfumery and as an antiseptic. Carvacrol also occurs in some essential oils, but is prepared by heating camphor with iodine:

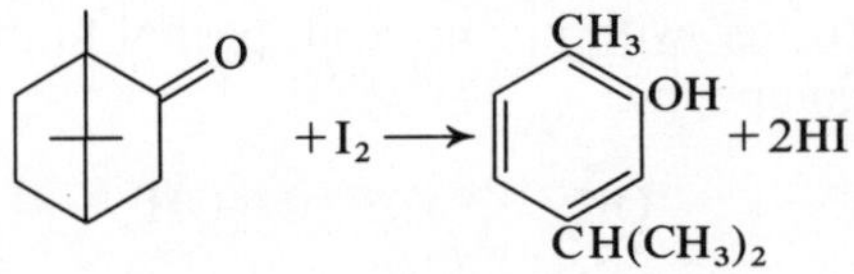

It is used in perfumery and as an antiseptic.

A number of monochloro-derivatives of monohydric phenols are also very good disinfectants, *e.g.*, 2-chloro-5-hydroxytoluene.

Dihydric phenols

Catechol (*o-dihydroxybenzene*) occurs in certain plants. It may be prepared by the alkaline fusion of *o*-phenolsulphonic acid; commercially, it is prepared by heating *o*-chlorophenol or *o*-dichlorobenzene with 20 per cent aqueous sodium hydroxide and a trace of copper sulphate at 190°C under pressure. It may be conveniently prepared in the laboratory by the action of alkaline hydrogen peroxide on salicylaldehyde:

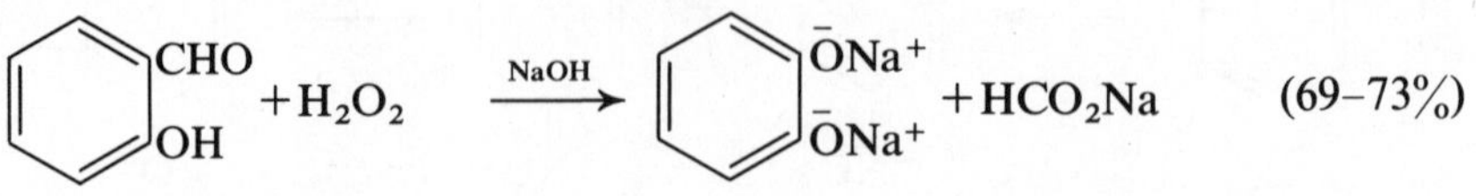

This reaction is characteristic of *o*- and *p*-hydroxyaldehydes; it is known as the **Dakin reaction** (1909).

The mechanism of the Dakin reaction is uncertain; a strong possibility is one that involves a 1,2-shift (*cf.* the cumene-phenol process, p. 699).

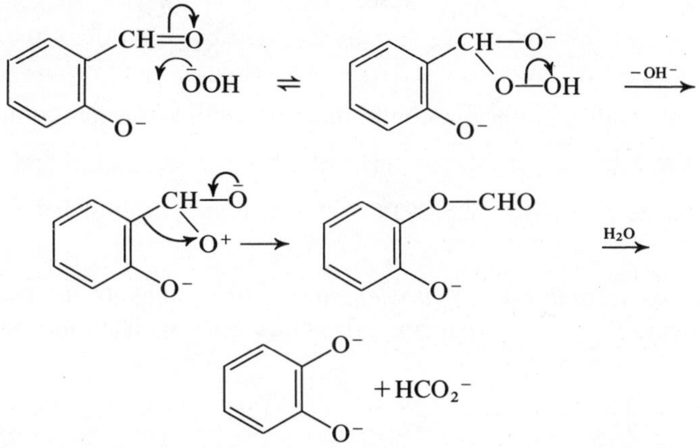

Catechol, m.p. 105°C, gives with ferric chloride a green colour which turns red on the addition of sodium carbonate. Catechol is a powerful reducing agent; its aqueous solution darkens on exposure to air due to oxidation; it reduces cold silver nitrate and warm Fehling's solution. It is used as a photographic developer. Catechol is oxidised by silver oxide in ether solution to *o*-benzoquinone, and it condenses with many compounds, *e.g.*, with phthalic anhydride in the presence of sulphonic acid to form alizarin (p. 898).

Important derivatives of catechol are *guaiacol* and *adrenaline* (the hormone secreted by the adrenal glands):

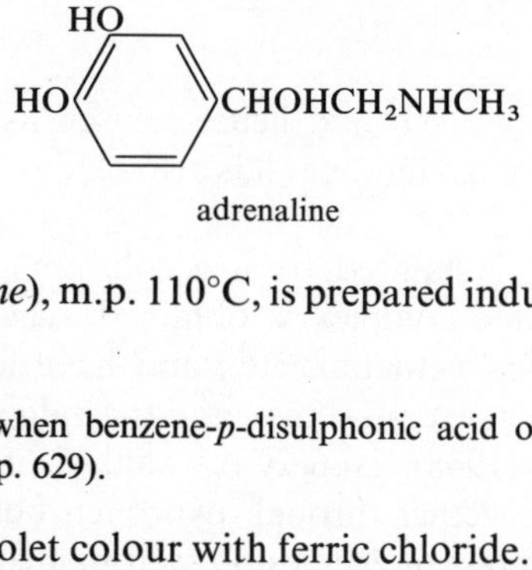

adrenaline

Resorcinol (m-*dihydroxybenzene*), m.p. 110°C, is prepared industrially by the alkaline fusion of benzene-*m*-disulphonic acid.

Resorcinol is also formed when benzene-*p*-disulphonic acid or all three bromobenzenesulphonic acids are fused with alkali (*cf.* p. 629).

Its aqueous solution gives a violet colour with ferric chloride. It is not so powerful a reducing agent as the *o*- and *p*-isomers, but it will reduce silver nitrate and Fehling's solution on warming. With nitrous acid it forms dinitrosoresorcinol:

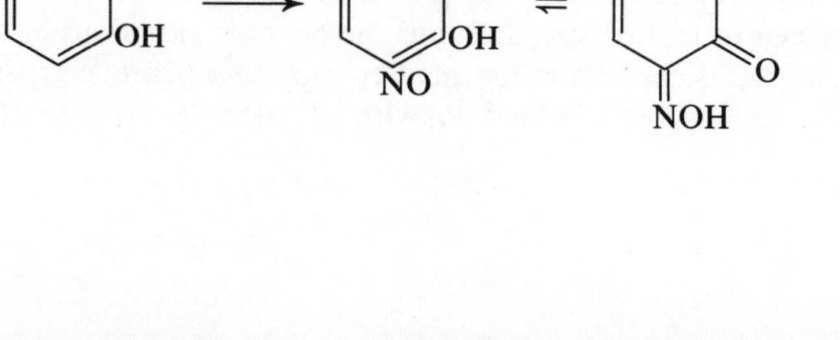

When nitrated, resorcinol forms **styphnic acid** (2,4,6-*trinitroresorcinol*), m.p. 180°C:

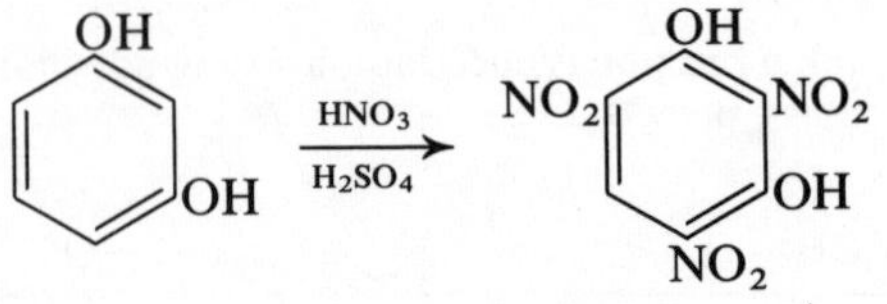

Styphnates are used to identify certain compounds by charge-transfer complex formation (*cf.* picrates; see also p. 645).

  Orcinol (3,5-*dihydroxytoluene*), m.p. 290°C, is found in many lichens, and is chemically related to litmus.

  Quinol (*hydroquinone*, p-*dihydroxybenzene*), m.p. 170°C, may be prepared by diazotising p-aminophenol (yield: 30%), or by reducing p-benzoquinone with sulphurous acid:

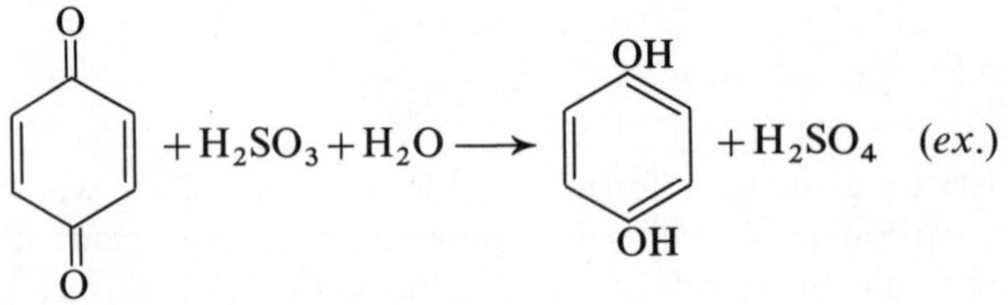

It is made commercially as follows:

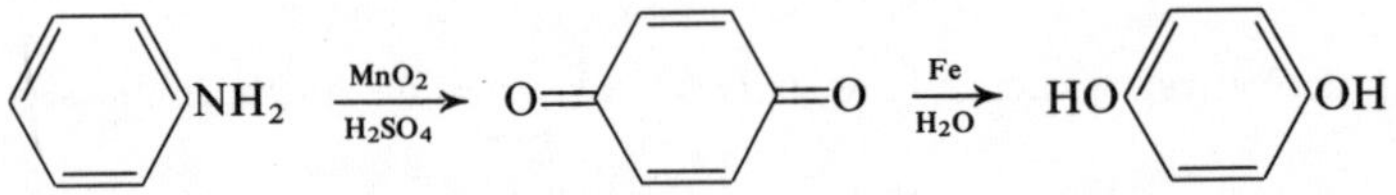

  Quinol is a powerful reducing agent and hence is used as a photographic developer. It is oxidised by ferric chloride to p-benzoquinone; it is also oxidised by diazonium salts, no coupling taking place at all (see p. 726).

  Clathrates. When quinol is crystallised from a solution in water saturated with sulphur dioxide, a quinol-sulphur dioxide complex is obtained. These inclusion complexes are also formed with hydrogen sulphide, methanol, etc., and have a molecular formula of the type $nC_6H_4(OH)_2 \cdot Z$ (where Z is one guest molecule), and the value of n (for the host molecules) can vary over a range. Powell *et al.* (1948), using X-ray analysis, found that in these complexes the quinol molecules were linked together through hydrogen bonds to form giant molecules the cavities in which enclosed the guest molecules. Powell named these complexes clathrates, and showed that the 'cages' must be large enough to contain the guest molecule and that they must be so arranged that the guest molecules do not escape. These clathrates are stable, but the imprisoned molecules may be released either by melting or by means of an organic solvent which dissolves quinol.

  Clathrates differ from channel complexes (p. 462) in that each guest molecule is enclosed in a separate molecular cage of limited size, and thus the possible variations of the guest molecules depend on *molecular size*. Hence clathrates may be used to separate certain homologues, *e.g.*, quinol forms a clathrate with methanol but not with any other homologue of this alcohol series.

Trihydric phenols

Pyrogallol (1,2,3-*trihydroxybenzene*) may be prepared by heating solid gallic acid in a stream of carbon dioxide, or by heating an aqueous solution of gallic acid at 210°C under pressure:

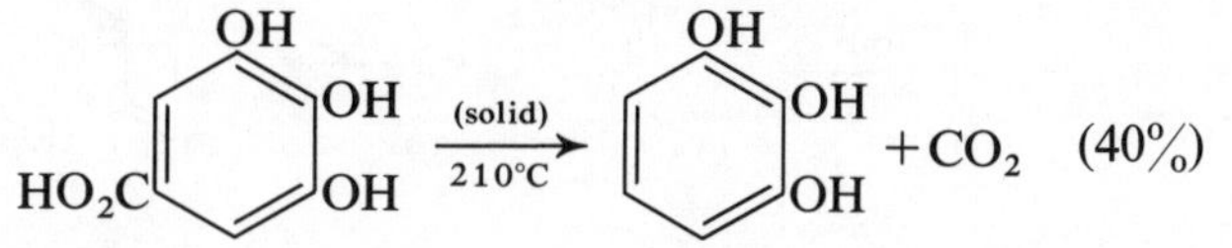

Pyrogallol, m.p. 133°C, in aqueous solution gives a red colour with ferric chloride. Alkaline solutions of pyrogallol oxidise very rapidly on exposure to air, and hence are used in gas analysis for the absorption of oxygen (and carbon dioxide); the pyrogallol is oxidised to a complex mixture containing, among other things, carbon monoxide, carbon dioxide, acetic and oxalic acids. Pyrogallol also reduces the salts of silver, gold, platinum and mercury to their metals. It is used as a photographic developer.

Hydroxyquinol (1,2,4-*trihydroxybenzene*), m.p. 140°C, may be prepared by the alkaline fusion of quinol in air (see p. 628). It is best prepared by hydrolysing its triacetate, which is obtained by heating *p*-benzoquinone with acetic anhydride and concentrated sulphuric acid (p. 726).

Phloroglucinol (1,3,5-*trihydroxybenzene*), m.p. 218°C, is obtained when many plant resins are fused with alkalis. It may be prepared by fusing resorcinol with sodium hydroxide in the air (p. 628), but a convenient laboratory method is to reduce 2,4,6-trinitrobenzoic acid, and heat the resulting amino-derivative with hydrochloric acid (yield: 46–53% as dihydrate):

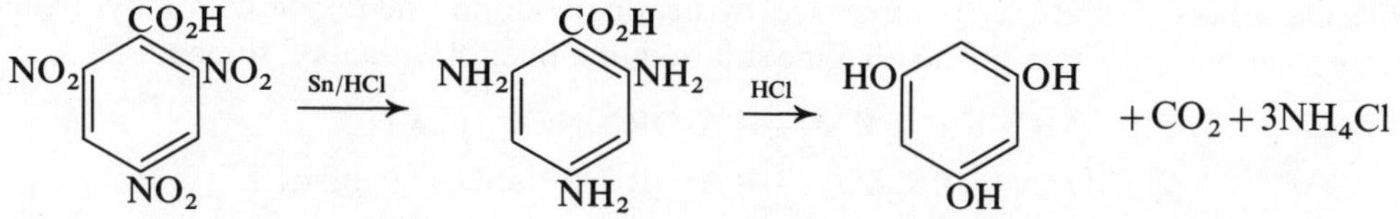

Alkaline solutions of phloroglucinol rapidly darken on exposure to air due to oxidation.

Since it reacts with certain reagents as if it were triketocyclohexane, it has been suggested that phloroglucinol is a tautomeric compound, *e.g.*,

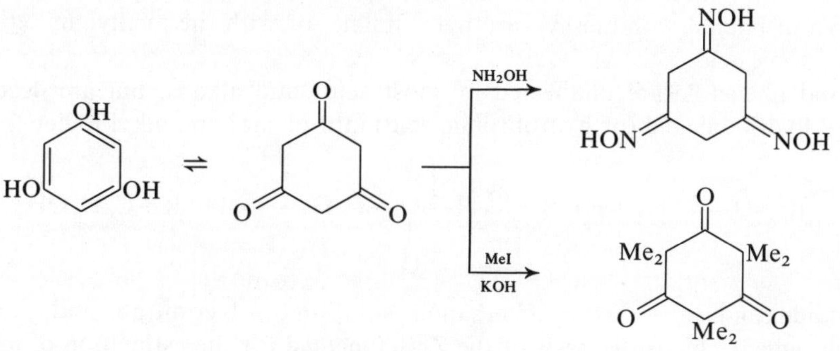

Infra-red studies, however, have shown that all phenols are entirely enolic in the solid state and also in solution (Thomson, 1956). This author has suggested that *C*-alkylation may be explained by attack of

alkyl carbonium ions at negative sites in the system. Thus, since alkylations are carried out in alkaline solution, the phenoxide ion may be regarded as a resonance hybrid:

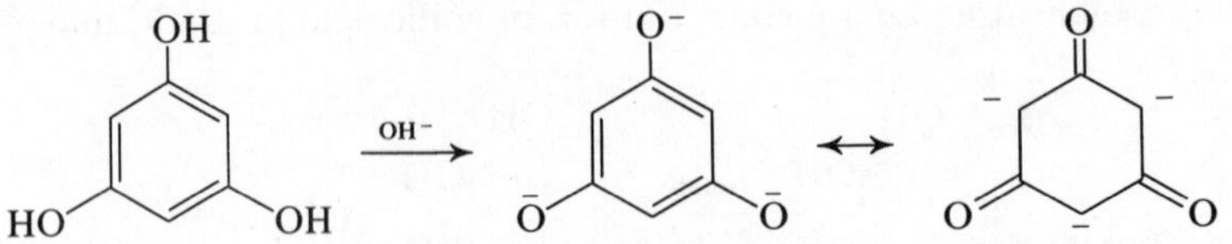

In protic solvents, *e.g.*, water, which are strongly hydrogen-bonding, the more electronegative oxygen atoms are solvated and so *C*-attack is favoured. For highly reactive halides in non-protic solvents (or weakly protic solvents, *e.g.*, ethanol), it appears that *C*-alkylation is favoured when the solvent is non-polar, *e.g.*,

$$\textit{o-}HOC_6H_4CH_2CH{=}CH_2 \xleftarrow[\text{C}_6\text{H}_6]{\text{CH}_2=\text{CHCH}_2\text{Br}} C_6H_5O^-Na^+ \xrightarrow[\text{Me}_2\text{CO}]{\text{CH}_2=\text{CHCH}_2\text{Br}} C_6H_5OCH_2CH{=}CH_2$$

Aromatic ethers

The aromatic ethers may be divided into two groups, the *phenolic ethers* (alkaryl ethers), which are of the type ArOR, and the ethers of the type Ar$_2$O (the diaryl ethers).

The infra-red absorption region of diaryl and alkaryl ethers ($=$C—O— stretch) is 1270–1230 cm^{-1} (s). Thus, these types of ethers may readily be distinguished from dialkyl ethers (p. 196), but not always from epoxides (p. 315; the molecular formula will help here; see p. 701 for mass spectra).

Phenolic ethers. These may be prepared by heating sodium phenoxide with alkyl halide in ethanol solution, or by treating an alkaline solution of a phenol with alkyl sulphate:

$$ArO^-Na^+ + RX \longrightarrow ArOR + NaX \quad (v.g.)$$
$$ArO^-Na^+ + R_2SO_4 \longrightarrow ArOR + RNaSO_4 \quad (v.g.-ex.)$$

Phenolic methyl ethers may be obtained in excellent yield by the action of diazomethane on a phenol:

$$ArOH + CH_2N_2 \longrightarrow ArOCH_3 + N_2 \quad (ex.)$$

Two important phenolic ethers are **anisole** (*methyl phenyl ether, methoxybenzene*), b.p. 155°C, and **phenetole** (*ethyl phenyl ether, ethoxybenzene*), b.p. 172°C. These are prepared industrially by the alkylation of phenol with methyl or ethyl sulphate, or with the methyl and ethyl esters of *p*-toluenesulphonic acid.

Anisole and phenetole are unaffected by most acids and alkalis, but are decomposed by concentrated hydriodic acid (or hydrobromic acid) into phenol and alkyl iodide (or bromide):

$$R{-}O{-}C_6H_5 + HI \rightleftharpoons I^- \quad R{-}\overset{+}{\underset{H}{O}}{-}C_6H_5 \longrightarrow I{-}R + C_6H_5OH$$

The alkyl iodide can be absorbed by an ethanolic solution of silver nitrate, and the silver iodide so formed, weighed. This is the basis of the **Zeisel method** for the estimation of methoxyl and ethoxyl groups.

Phenolic ethers may also be decomposed by heating with pyridine hydrochloride at 170–240°C. This method is particularly useful for ethers that are sensitive to other demethylating reagents.

$$PhOR + C_5H_5NH^+Cl^- \longrightarrow PhOH + C_5H_5\overset{+}{N}MeCl^- + H_2O$$

Formation of an ether provides a means of protecting the hydroxyl group in many reactions; e.g., in the Friedel–Crafts reaction, due to the acidic character of the hydroxyl group, aluminium chloride is attacked with the liberation of hydrogen chloride:

$$C_6H_5OH + AlCl_3 \longrightarrow C_6H_5OAlCl_2 + HCl$$

Thus, to carry out a successful Friedel–Crafts reaction with phenol, it is necessary to use a large amount of aluminium chloride. This, however, may be avoided by carrying out the reaction with the methyl ether, but the temperature must be kept as low as possible to prevent the catalyst decomposing the ether. Alternatively, alkylation of phenol may be carried out with an alkene or alcohol and hydrogen fluoride (p. 610).

In certain cases, phenolic ethers rearrange to the o- or p-alkyl compound:

$$C_6H_5OR \xrightarrow{\text{AlCl}_3} o\text{- or } p\text{-RC}_6H_4OH$$

Catechol ethers. Guaiacol (o-*hydroxyanisole*), m.p. 32°, b.p. 205°C, occurs in beech-wood tar from which it may be obtained by fractional distillation. Guaiacol is prepared synthetically from o-anisidine:

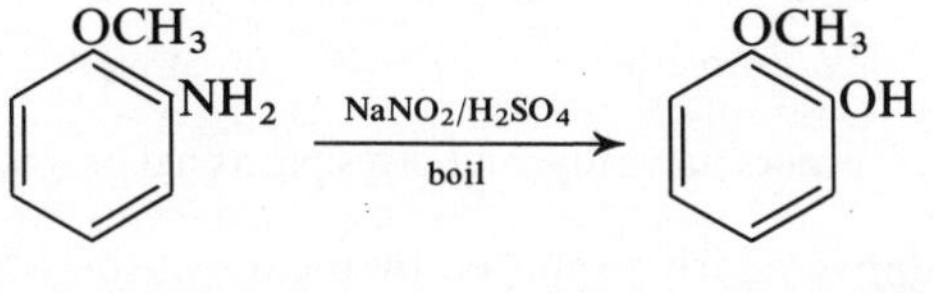

When refluxed with constant boiling hydrobromic acid, or heated with hydriodic acid at 130°C, guaiacol is converted into catechol.

Veratrole (1,2-*dimethoxybenzene*), b.p. 207°C, may be prepared by methylating catechol with methyl sulphate in alkaline solution:

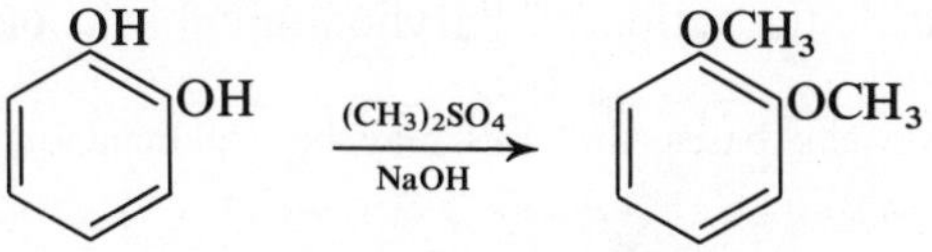

Eugenol (4-*allylguaiacol*), b.p. 254°C, occurs in oil of cloves and in many other essential oils. When heated with ethanolic potassium hydroxide, or better, with potassium hydroxide in diethylene glycol at 180°C, eugenol isomerises to **isoeugenol** (4-*propenylguaiacol*):

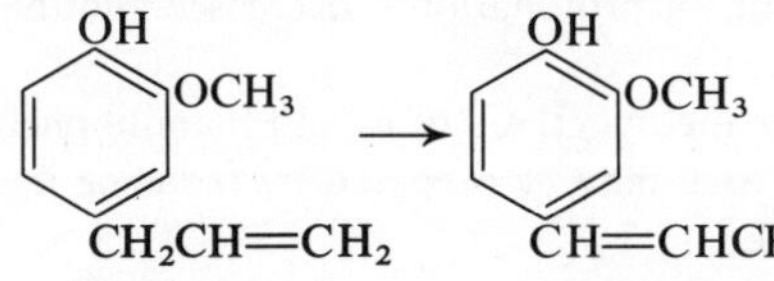

eugenol   isoeugenol

This migration of the double bond in the allyl side-chain, under the influence of alkali, is general. The rearranged product contains phenyl-vinyl conjugation, and this produces a more stable system (for mechanism, see p. 345).

Isoeugenol, b.p. 267·5°C, also occurs naturally, and gives vanillin on gentle oxidation (p. 747).

Anethole (p-*methoxypropenylbenzene*), m.p. 22–23°, b.p. 235°C, is one of the chief constituents of aniseed oil, from which it is obtained. It is also prepared synthetically by the interaction of *p*-methoxy-phenylmagnesium bromide and allyl bromide, and isomerising the product, **estragole** (*methylchavicol*), with alkali:

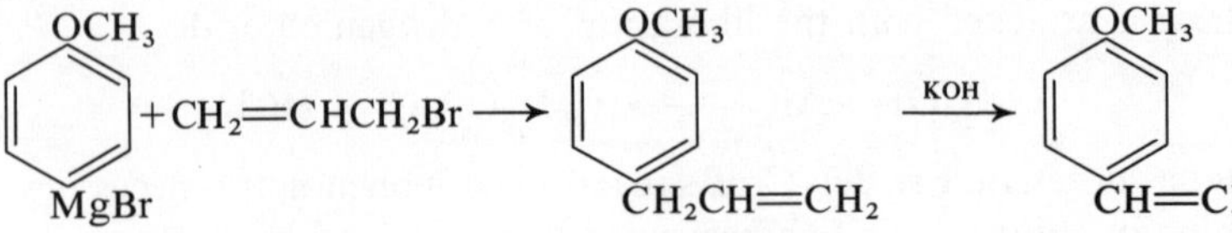

estragole, b.p. 215°C   anethole

Safrole (4-*allyl*-1,2-*methylenedioxybenzene*), b.p. 232°C, occurs in camphor oil and sassafras oil. When heated in ethanolic potassium hydroxide solution, or better, in a concentrated solution of potassium hydroxide in cellosolve, safrole isomerises to **isosafrole**:

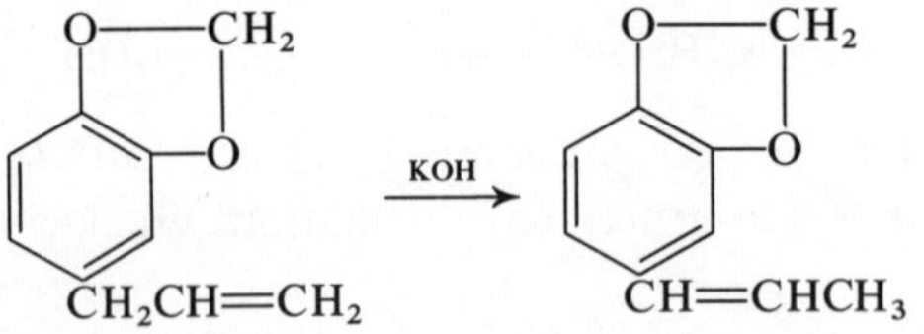

safrole   isosafrole

Isosafrole, b.p. 252°C, also occurs naturally, and gives piperonal on gentle oxidation (p. 747).

Diaryl ethers. These are conveniently prepared by means of the *Ullmann reaction* (*cf.* p. 664); *e.g.*, **diphenyl ether**:

$$C_6H_5Br + C_6H_5OH + KOH \xrightarrow{\text{Cu}} (C_6H_5)_2O + KBr + H_2O \quad (90\%)$$

Diphenyl ether, m.p. 28°C, is *not* decomposed by hydriodic acid, and is valuable as a high temperature heat transfer medium.

Diaryl ethers may be prepared by the action of diaryliodonium salts on phenoxides (p. 636).

Diaryl ethers are extremely weak bases, and this may be explained on the basis that they are resonance hybrids:

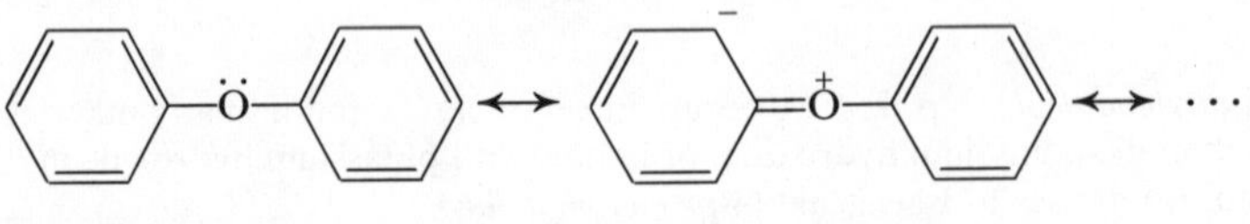

The availability of the lone pair for protonation is decreased and the extended resonance stabilisation is lost in the protonated molecule.

Ethers of aminophenols. The methyl ethers of *o*- and *p*-aminophenol are known as *o*-anisidine and *p*-anisidine, respectively, and each may be prepared by reducing its corresponding nitroanisole, *e.g.*, *o*-anisidine from *o*-nitroanisole:

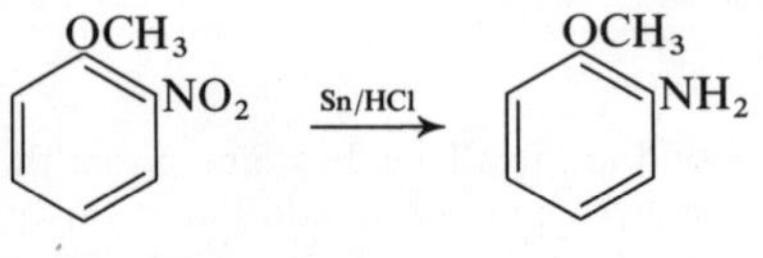

The anisidines are used in the preparation of azo-dyes.

The ethyl ethers of the aminophenols are known as **phenetidines**; these are also used in the preparation of azo-dyes. The acetyl derivative of *p*-phenetidine is known as **phenacetin**, and is prepared commercially as follows:

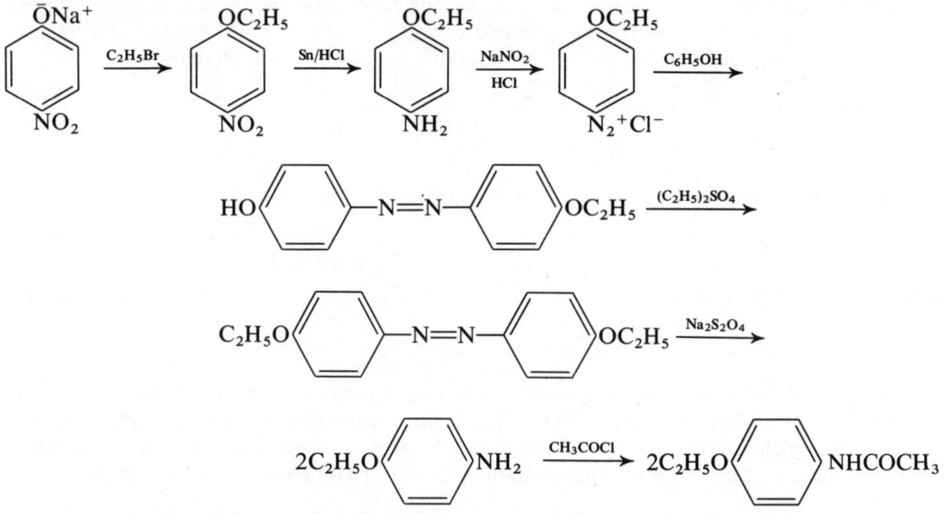

This preparation is a very good example of the *recycling procedure*; diazotisation of part of the *p*-phenetidine *produced* then permits *phenol* to be used as the starting material instead of *p*-nitrophenol (which is more expensive than phenol).

Phenacetin is replacing acetanilide in medicine; it is a very good analgesic (*i.e.*, promotes relief from pain) and antipyretic (*i.e.*, fever-reducing).

Dulcin is the carbamyl derivative of *p*-phenetidine, and is prepared by heating *p*-phenetidine with urea:

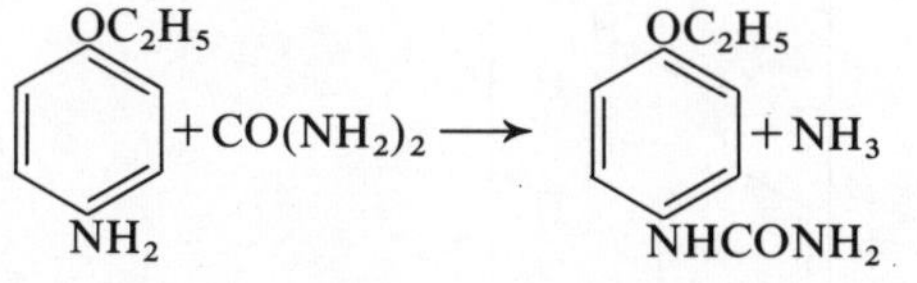

Dulcin, m.p. 171°C, is about 200 times as sweet as sugar; it is used commercially as a sweetening agent.

Claisen rearrangement (1912). This is the rearrangement of allyl aryl ethers to allylphenols. No catalyst is required and it occurs when the substrate is heated alone to about 200°C or in some inert solvent, *e.g.*, diphenyl ether. The allyl group migrates to the *o*-position if one is free, but if *both* *o*-positions are occupied, then migration occurs to the *p*-position, *e.g.*,

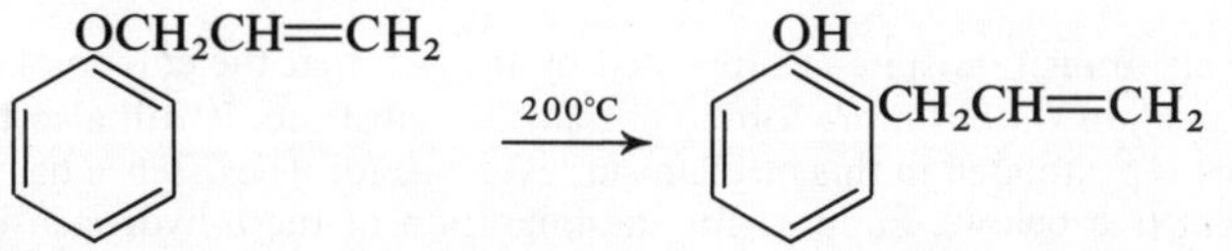

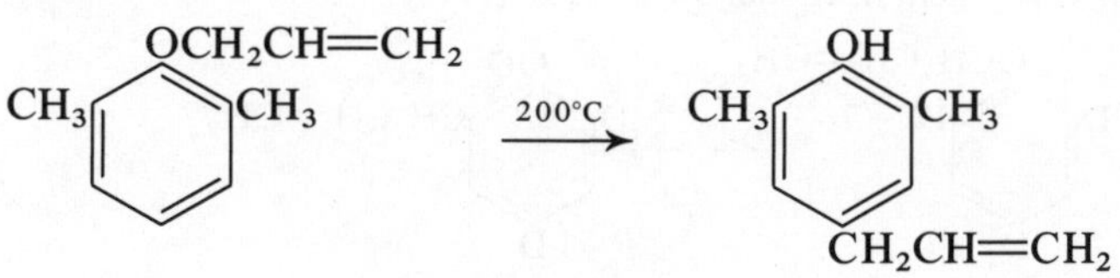

There are, however, some exceptions to these rules, *e.g.*, Schmid *et al.* (1963) found that crotyl 3,5-dimethylphenyl ether, $MeCH=CHCH_2OC_6H_3Me_2$, gave both the *o*- and *p*-rearrangement, with the former predominating.

The mechanism of the Claisen rearrangement has been the subject of a great deal of work. The intramolecular nature of the rearrangement has been demonstrated by the fact that if two different ethers are heated together, they rearrange independently, *e.g.*, Hurd *et al.* (1937) heated a mixture of allyl 2-naphthyl ether and cinnamyl phenyl ether and did *not* obtain cross products:

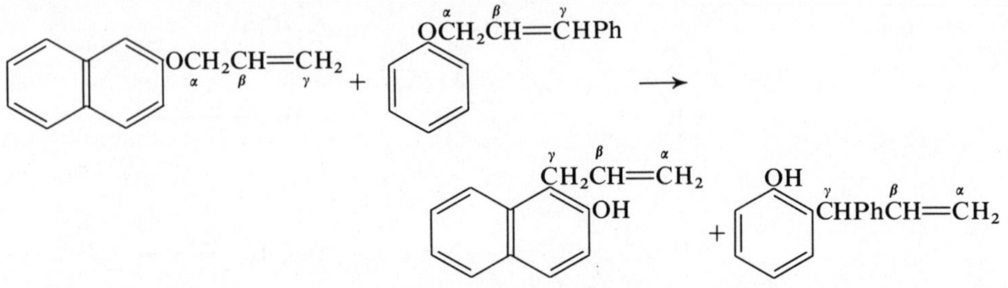

Kincaid *et al.* (1939) showed the rearrangement was of the first order, and other workers have shown that when the allyl group migrates to the ring in the *o*-rearrangement, the α,γ-bonding is reversed (see above example). Schmid *et al.* (1953), using the allyl group labelled with ^{14}C at the γ-carbon atom, also showed that the rearranged product contained the tracer atom attached to the nucleus:

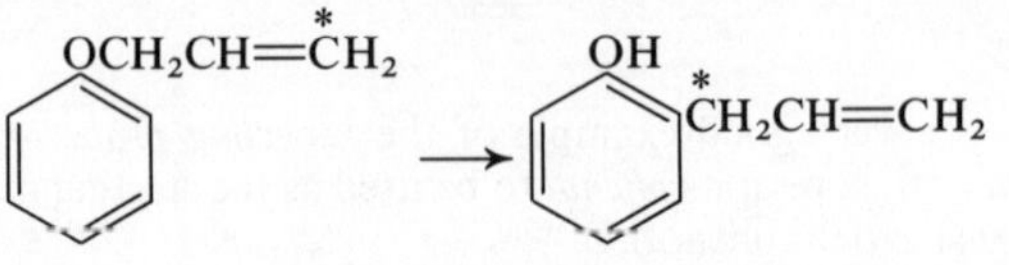

These examples suggest that the reversal of the bonding in the allyl group is a necessary requirement in the rearrangement. The following intramolecular *cyclic* mechanism is in agreement with the foregoing facts:

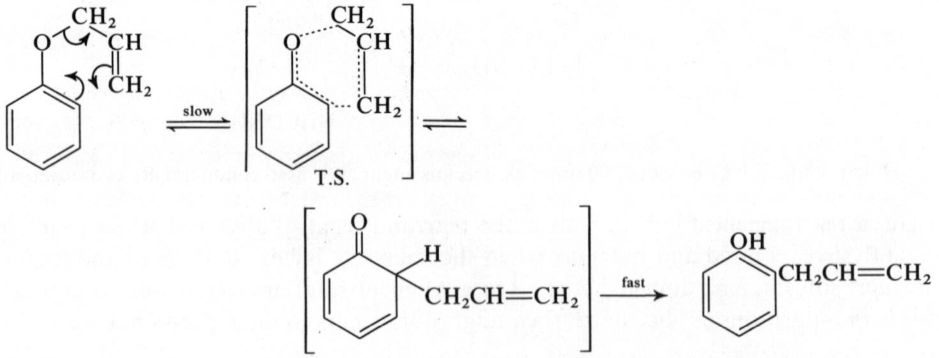

The formation of a cyclic transition state is supported by the fact that the entropies of activation are negative; thus, the transition state is more 'ordered' than the substrate. It will also be noted that the formation of a dienone is postulated in this mechanism; evidence for its existence has been obtained in the *para* rearrangement (see below). Support for the migration of the *o*-hydrogen in the dienone is given by the following experiment of Kistiakowsky *et al.* (1942):

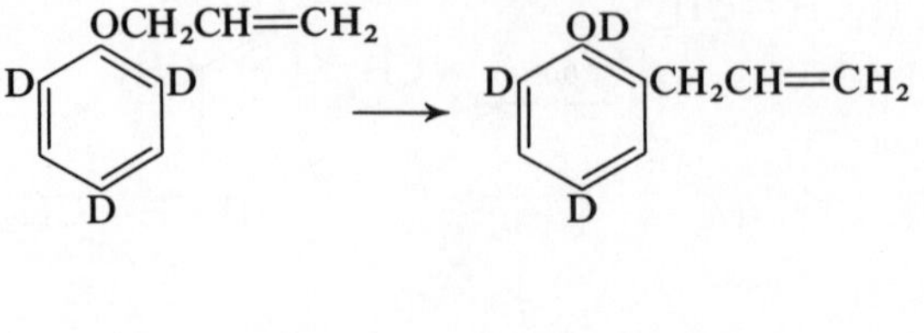

The mechanism of the *p*-rearrangement is believed to occur in two stages, each one with reversal, and thus the final product will *not* be reversed. Schmid *et al.* (1953) showed the following rearrangement occurred with labelled allyl 2,6-dimethylphenyl ether:

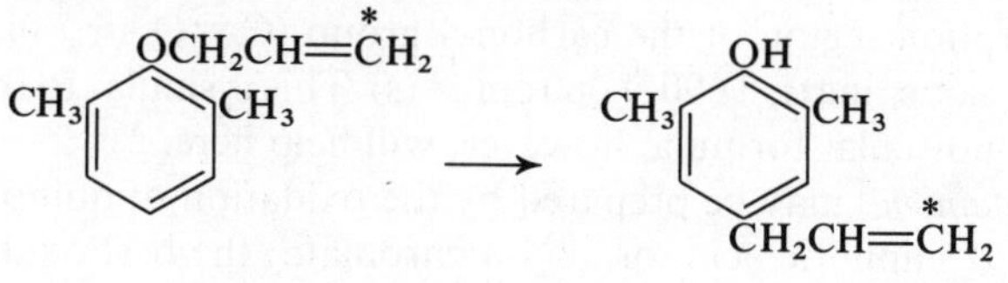

Conroy *et al.* (1954, 1956) have demonstrated the existence of the dienone by rearranging allyl 2,6-dimethylphenyl ether in the presence of maleic anhydride; a small amount of adduct was isolated. Thus a Diels–Alder reaction has occurred, and this is strong evidence for the formation of the dienone (the dienophile in this system). The reversibility of both stages in the *p*-arrangement has also been demonstrated by Schmid *et al.* (1956) using allyl groups labelled with ^{14}C:

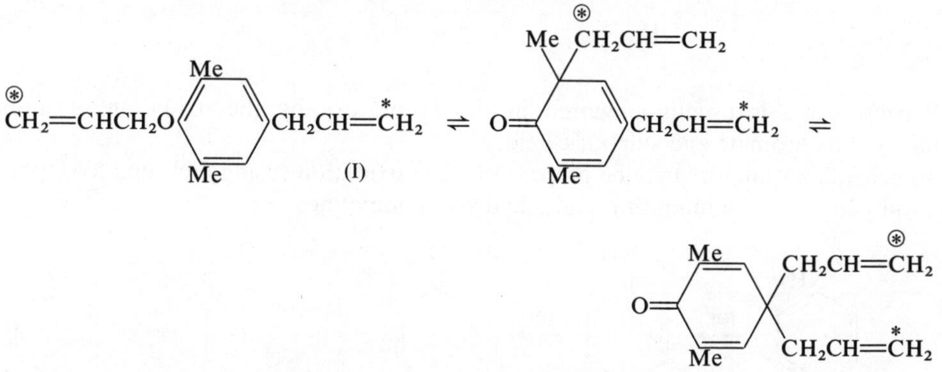

Compound (I), in which labelling was only in the *C*-allyl group $(\overset{*}{C})$, was recovered *chemically* unchanged after being heated to 160°C, but was now also labelled in the *O*-allyl group $(\overset{\circledast}{C})$, the isotope being equally distributed between the terminal methylene groups.

A two-stage mechanism for this *p*-rearrangement, each stage involving a cyclic transition state, accounts for all of the facts. In the second stage, *o*- to *p*-, the cyclic transition state will be highly deformed; this is probably the reason why the *o*-rearrangement is preferred when it is possible.

The Claisen rearrangement is the best known example of the class of reactions called **sigmatropic rearrangements**. These are uncatalysed intramolecular (unimolecular one-step) processes whereby a σ-bond, flanked by one or two π-electron systems, migrates to a new position. In the Claisen rearrangement, the migrating σ-bond is the O—C$_\alpha$ bond (in the ether) which forms the new σ-bond, *o*-C—C$_\gamma$, in the dienone (*o*-C is the *ortho*-carbon atom in the dienone).

Quinones

Quinones are cyclic unsaturated diketones in which the double bonds and the keto groups are conjugated. Thus they are not aromatic compounds, but they are readily prepared from, and readily converted into, aromatic compounds. Two quinones derived from benzene are possible, *o*- and *p*-benzoquinone:

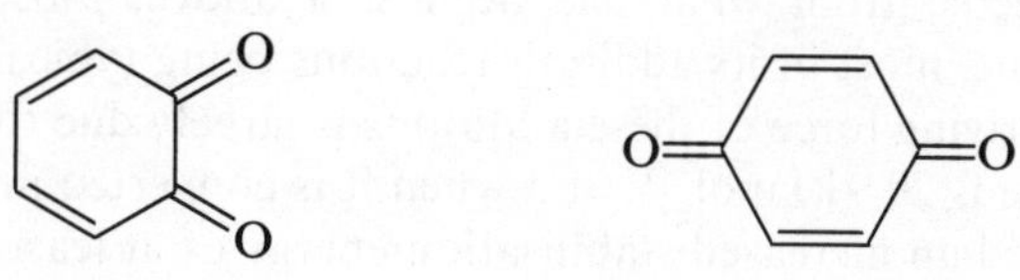

A characteristic property of quinones is their colour, the *p*-compounds being yellow and the *o*-compounds being (usually) red. These quinonoid structures are powerful chromophores (see p. 887).

The infra-red absorption region of the carbonyl group (C=O str.) in quinones (with both carbonyl groups in the same ring) is $1\,690\text{--}1\,660\,\text{cm}^{-1}$ (s). This is similar to that in α,β-unsaturated ketones (p. 339). The molecular formula, however, will help here.

p-**Benzoquinone** (p-*quinone*) may be prepared by the oxidation of quinol with ferric chloride, manganese dioxide and sulphuric acid, or acid dichromate; the best oxidising agent is sodium chlorate in dilute sulphuric acid in the presence of vanadium pentoxide as catalyst:

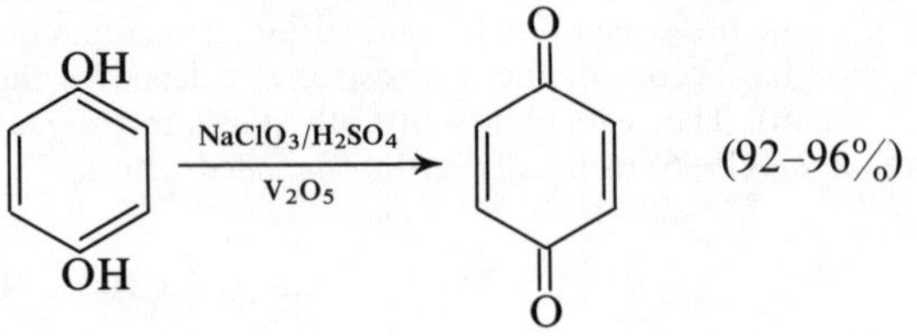

p-Benzoquinone is usually prepared in the laboratory by the oxidation of aniline with potassium dichromate and sulphuric acid.

In general, *p*-quinones may be prepared by the oxidation (using dichromate-sulphuric acid) of *p*-dihydroxy-, *p*-diamino- or *p*-aminohydroxy-compounds, *e.g.*,

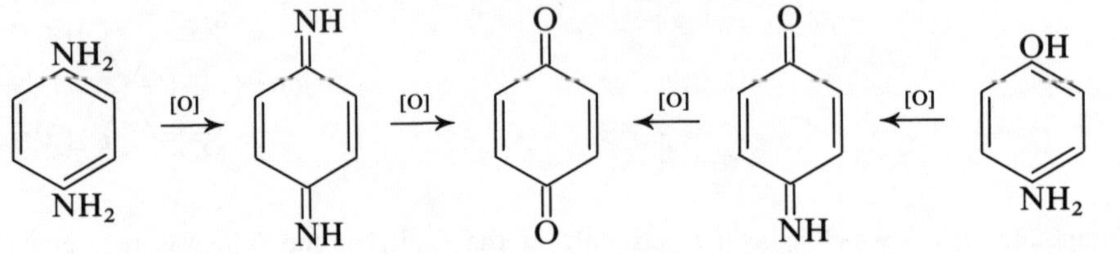

(I) (II)

The intermediate products (I) (p-*benzoquinonedi-imine*) and (II) (p-*benzoquinoneimine*) can be isolated under certain conditions (see p. 728).

p-Benzoquinone crystallises in yellow prisms, m.p. 116°C; it has a sharp smell, sublimes when heated, is slightly soluble in water, and is steam-volatile. On exposure to light, *p*-benzoquinone turns brown.

The structure of *p*-benzoquinone is known as a *crossed conjugated system*; it contains three (or more) conjugated double bonds which are not arranged in a continuous chain.

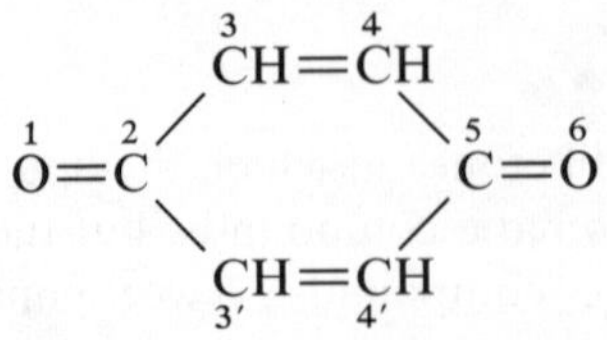

As might be expected from what has been said above, *p*-benzoquinone behaves as an α,β-unsaturated ketone, most of its addition reactions being typical 1,4-additions, followed by aromatisation. The driving force of these additions is largely due to the fact that the resonance energy of the quinone is $20\cdot9\,\text{kJ mol}^{-1}$, and when it is converted into the aromatic compound, the product has gained an increased stabilisation energy of at least $125\cdot5\,\text{kJ mol}^{-1}$.

p-Benzoquinone adds on hydrogen chloride to form mainly *chloroquinol* (*chlorohydroquinone*) m.p. 106°C:

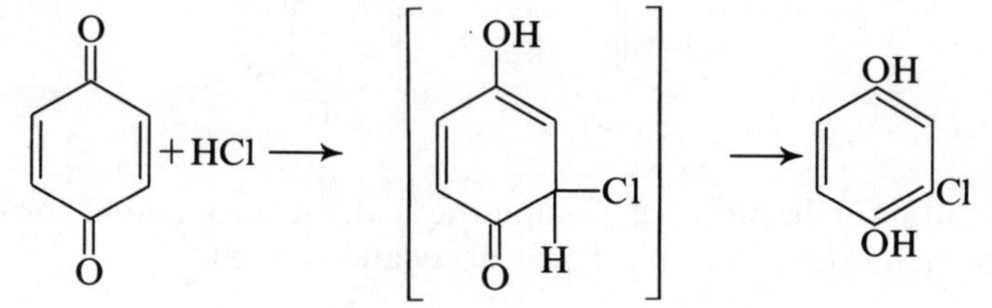

Chloroquinol, on oxidation, gives chloro-*p*-benzoquinone. By repeating the process of adding hydrogen chloride and then oxidising, the final product obtained is *tetrachloro*-p-*benzoquinone* (*chloranil*):

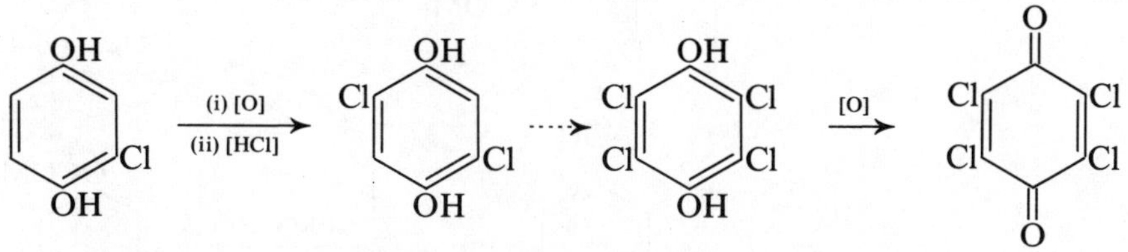

Chloranil is used as a fungicide and as an oxidising agent (see below).

In the same way, *p*-benzoquinone forms addition compounds with primary or secondary amines in ethanolic solution, but in this case the quinone and not the quinol derivative, is obtained, *e.g.*, with aniline, the final product is 2,5-dianilino-*p*-benzoquinone (see also below):

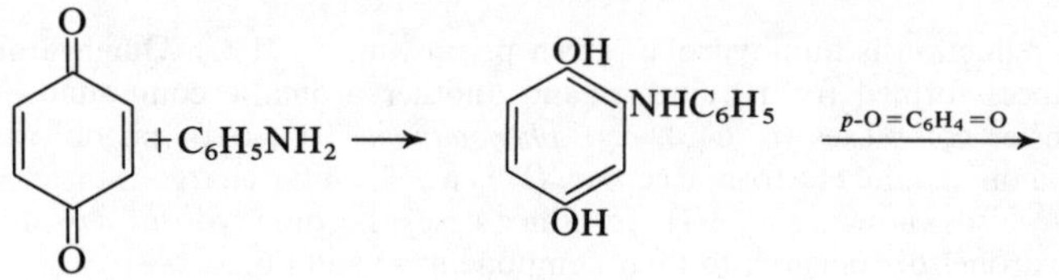

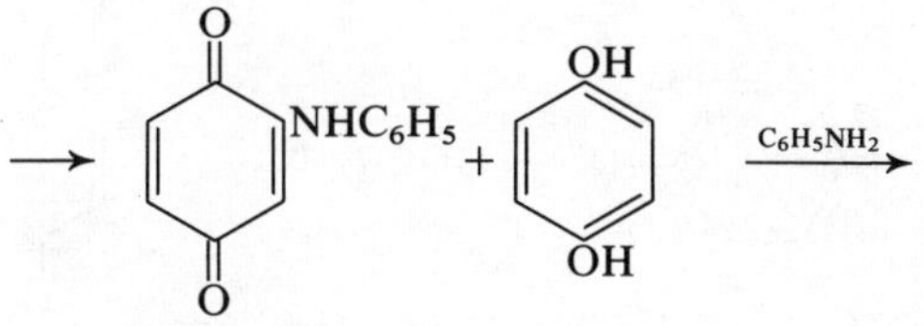

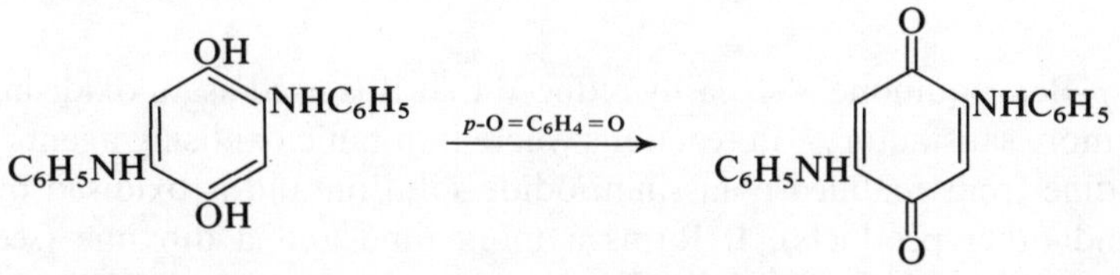

In a similar way, primary alcohols, in the presence of zinc chloride, form 2,5-dialkoxy-*p*-benzoquinones with *p*-benzoquinone.

When treated with aqueous potassium cyanide in the presence of sulphuric acid in ethanol, *p*-benzoquinone forms 2,3-dicyanoquinol (*cf.* the previous reaction):

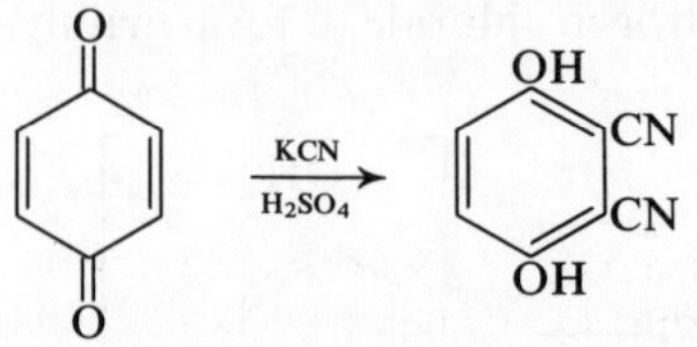

With acetic anhydride in the presence of sulphuric acid, *p*-benzoquinone forms hydroxyquinol triacetate; this reaction is known as the **Thiele acetylation** (1898).

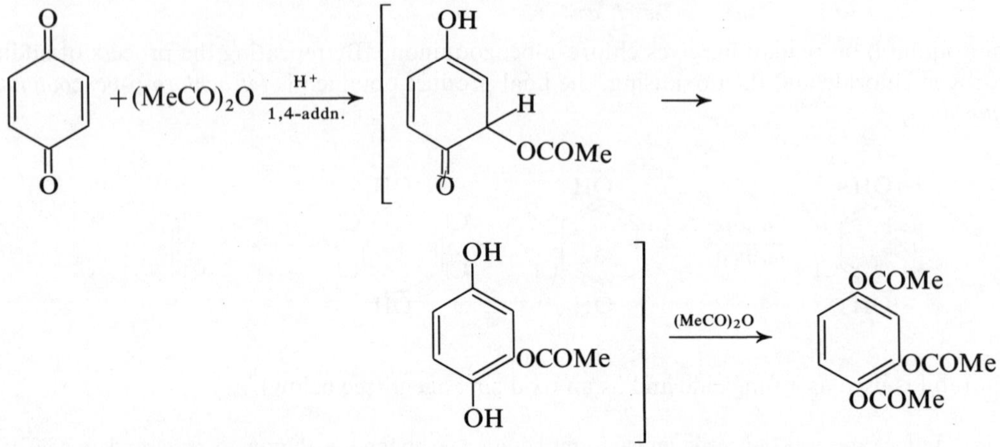

p-Benzoquinone is easily reduced to quinol (1,6-addition) by sulphurous acid, hydrogen sulphide, or sodium hyposulphite (yield: 80%).

> An intermediate in this reduction is quinhydrone (green prisms, m.p. 171°C). **Quinhydrones** are a group of coloured substances formed from quinones and another aromatic compound. They are believed to be charge-transfer complexes (p. 645), *e.g.*, *phenoquinone* ((I); red); phenol acts as the electron donor and the quinone as the electron acceptor. (I) is a 2:1 molar charge-transfer complex, but 1:1 molar adducts are also known, *e.g.*, (II). In either case, the quinhydrone is stabilised by hydrogen bonding. Quinhydrones dissociate into their components in solution.

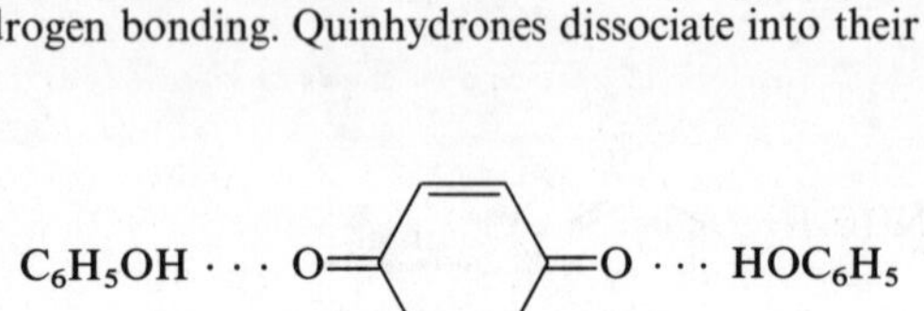

$C_6H_5OH \cdots O{=}{=}O \cdots HOC_6H_5$

(I) 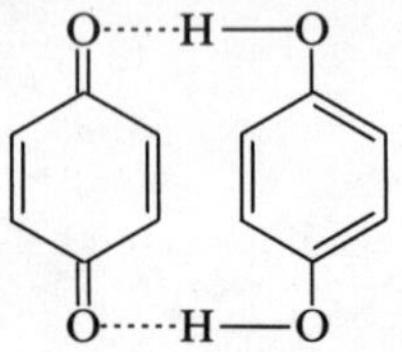   (II)

p-Benzoquinone is so easily reduced that it is used as an oxidising agent—generally chloranil is more satisfactory—in reactions where inorganic oxidising agents must be avoided. It liberates iodine from acidified potassium iodide solution and is oxidised by silver oxide to maleic acid (and other products). It forms a monoxime and a dioxime (see below), and can act as a dienophile in the Diels–Alder reaction, *e.g.*, with butadiene it forms a hydrogenated anthraquinone derivative (see p. 813).

> The oxidising property of benzoquinone is due to the fact that when the quinone adds on two electrons, the benzenoid structure (with a large resonance energy) is obtained:

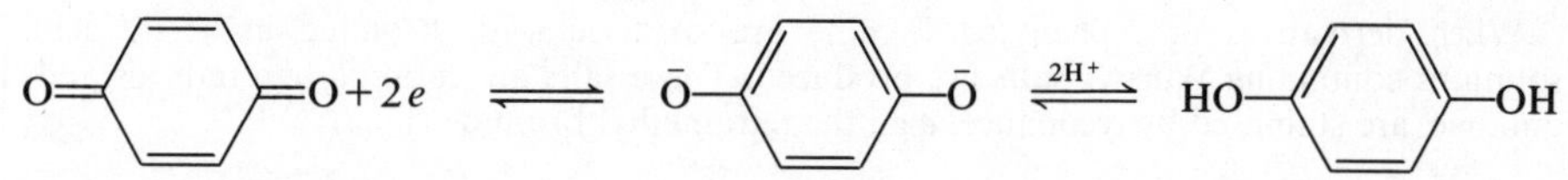

This *reversible* oxidation-reduction is characteristic of all quinones and a measure of the oxidising power is given by the *oxidation-reduction* or *redox potential* of that system. The system with the greater positive redox potential oxidises a system with a less positive redox potential, and the higher the redox potential the higher is the energy content of the quinone. Thus, the value of a redox potential depends on the number of nuclei in the quinone and on the relative positions of the two carbonyl groups. The potential of the *o*-quinone is higher than that of the corresponding *p*-isomer, and the larger the number of rings fused in a *linear* fashion, the lower is the redox potential, *e.g.*,

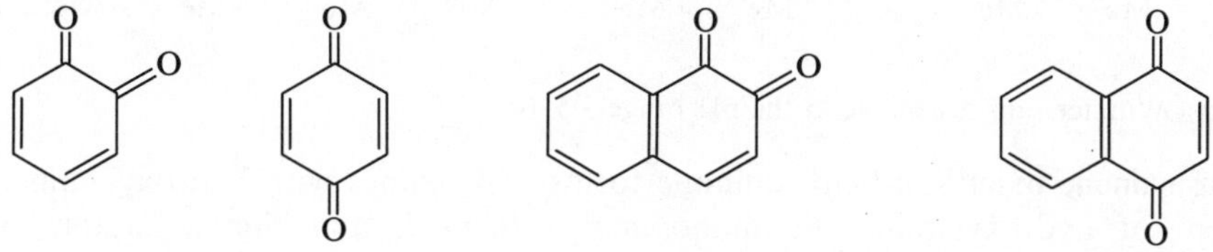

| volts | 0·79 | 0·70 | 0·56 | 0·47 |

| volts | 0·49 | 0·40 | 0·15 |

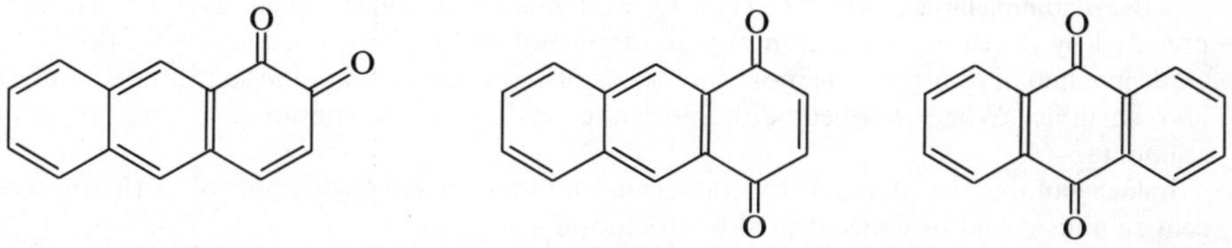

The property of being good electron-acceptors is generally due to the low energy level of the lowest empty M.O. in quinones. Apart from the factors mentioned above that control the oxidation-reduction potentials of quinones, it has also been found that substituent groups often have a marked effect. If the nucleus contains a strongly electron-attracting group, the quinone has a higher oxidation-reduction potential than that of the parent quinone. When the substituent is electron-releasing, the potential is lower than that of the parent quinone, *e.g.*, Cl (which has a −I effect) raises the potential of chloro-*p*-benzoquinone above that of *p*-benzoquinone. Thus the former quinone-quinol system will oxidise the latter system, *i.e.*, the latter will have its equilibrium position mainly on the quinone side. Hence in this case, when HCl adds to *p*-benzoquinone, the chloroquinol produced is *not* oxidised by unchanged quinone. On the other hand, the anilino and ethoxyl groups are electron-releasing (both have a +R effect). Therefore the oxidation-reduction potentials of these substituted *p*-benzoquinones are lower than that of *p*-benzoquinone. Hence, when aniline or ethanol adds to *p*-benzoquinone to form the corresponding quinol, this is then oxidised by unchanged quinone, and the reaction can proceed again (see above for the equations).

In the electrolytic reduction of benzoquinone, *one* electron is transferred to produce an anionic free radical, known as *semiquinone*; this is stabilised by resonance.

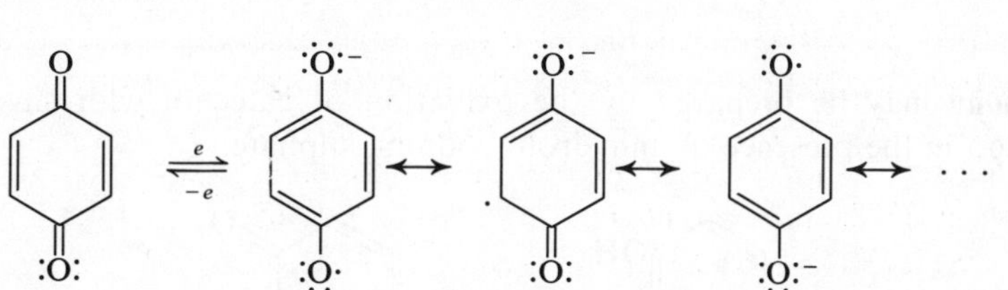

Corresponding anionic free radicals derived from *o*-benzoquinone are also known.

When derivatives of *p*-phenylenediamine are oxidised with bromine in acetic acid, coloured solutions containing **Wurster salts** are produced. These salts are cationic free radicals and, like semi-quinone, are stabilised by resonance, *e.g.*, the tetramethyl bromide (blue):

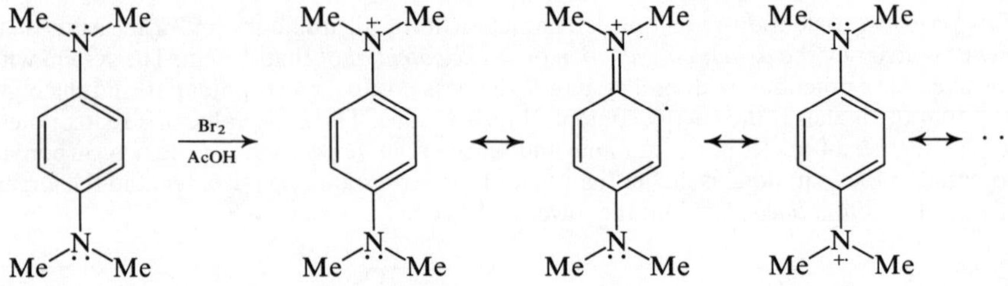

Most Wurster salts are stable in the pH range 3·5–6.

p-Benzoquinone oximes. *p*-Benzoquinone forms two oximes with hydroxylamine. With one equivalent of hydroxylamine the monoxime is formed, and this is tautomeric with *p*-nitrosophenol (p. 713). The dioxime is a yellow crystalline solid.

p-Benzoquinoneimine, $NH{=}C_6H_4{=}O$, and **p-benzoquinonedi-imine**, $NH{=}C_6H_4{=}NH$, may be prepared by the careful oxidation of *p*-aminophenol and *p*-phenylenediamine, respectively, with silver oxide in ethereal solution. The mono-imine is a bright yellow crystalline solid; the di-imine is colourless and unstable. When warmed with inorganic acids, both compounds form *p*-benzoquinone and ammonia.

Indophenol may be prepared by condensing *p*-nitrosophenol with phenol in the presence of 70 per cent sulphuric acid or concentrated hydrochloric acid:

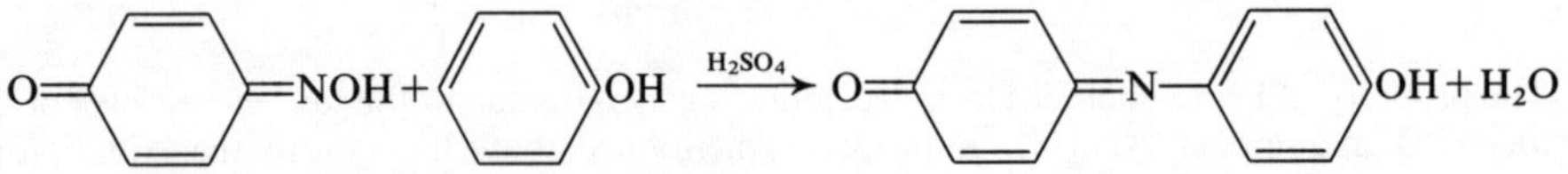

It is a brown solid, m.p. 160°C, giving a red solution in ethanol, and forms a blue solution in alkali; this is the basis of the Liebermann reaction (p. 377).

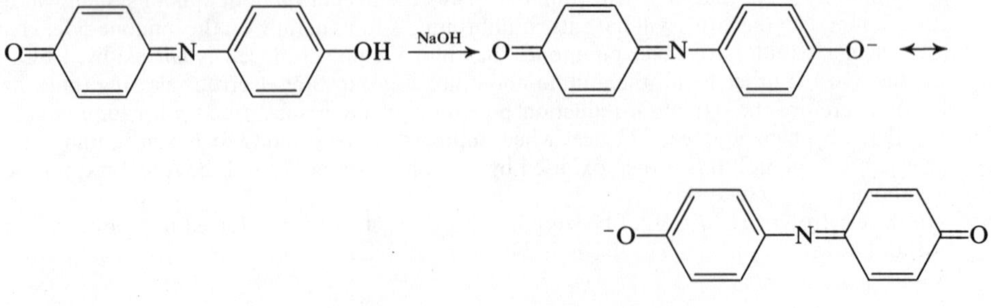

o-Benzoquinone may be prepared by the oxidation of catechol with silver oxide in dry ethereal solution in the presence of anhydrous sodium sulphate.

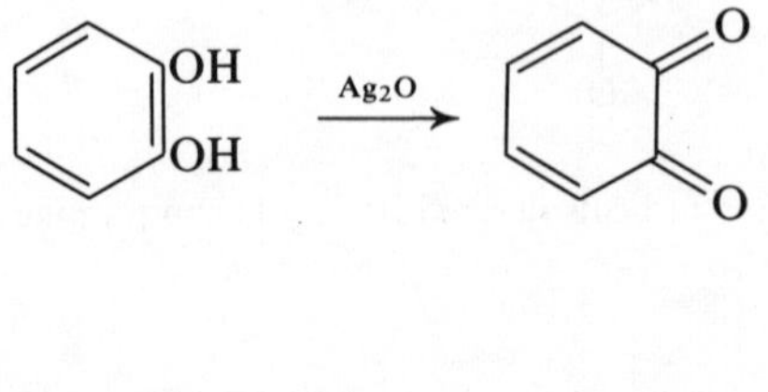

o-Benzoquinone exists in two forms, one as unstable green needles, and the other, stable light-red crystalline plates. It is odourless, not steam-volatile, and is reduced to catechol by sulphurous acid. It is a strong oxidising agent, *e.g.*, it liberates iodine from acidified potassium iodide.

Many benzoquinone compounds occur naturally, and are the cause of the colour in the pigments in which they are found.

Quinone methides. These occur naturally, and are formally derived from quinones by replacement of one of the carbonyl oxygen atoms by a methylene, or substituted methylene, group. Only *p*-quinone methides have been obtained in the pure state, *e.g.*, (I) is obtained by the oxidation of 2,6-di-t-butyl-4-isopropylphenol. The most characteristic property of quinone methides is their ready reaction with nucleophiles at the methylene group, *e.g.*,

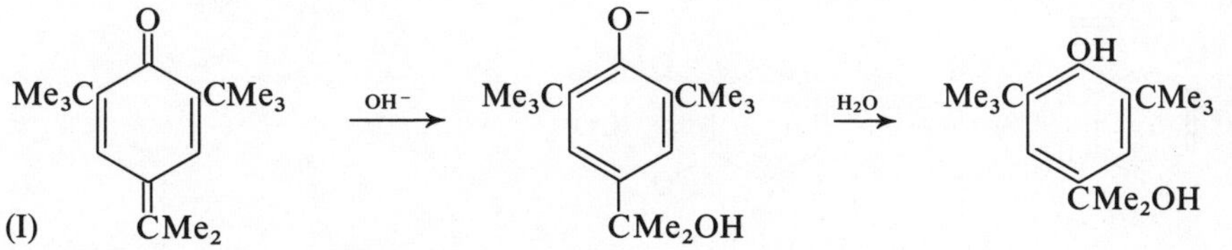

(I) CMe_2 CMe_2OH

Quinols (quinoles) may be regarded as derivatives of *p*-benzoquinone in which one carbonyl group has been converted into a tertiary alcohol. Quinols may be prepared by oxidising *p*-alkylphenols with Caro's acid; *e.g.*, *p*-cresol gives *p*-*toluquinol* (4-*methylquinol*):

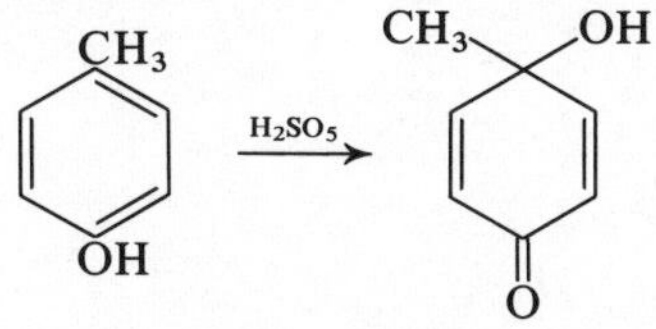

(*p*.)

The most characteristic reaction of quinols is their ready rearrangement to *p*-dihydric phenols (the **dienone-phenol rearrangement**), *e.g.*, *p*-toluquinol, when heated in aqueous methanol containing sulphuric acid, rearranges to 2-methylquinol (see also p. 649).

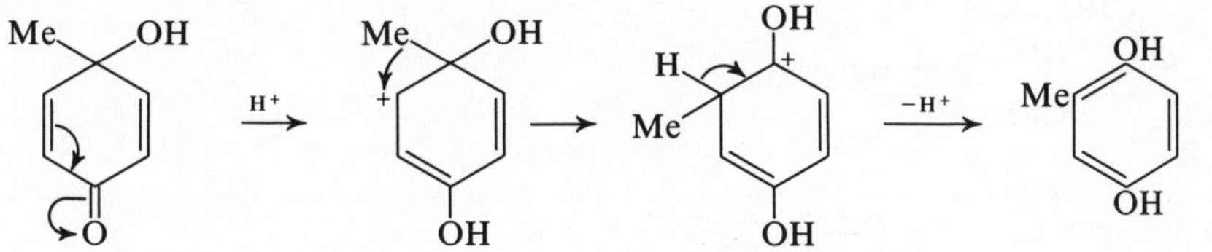

PROBLEMS

1. How would the i.r. spectra of *o*- and *p*-hydroxybenzaldehyde differ?
2. How could you separate chemically a mixture of benzene, aniline, phenol, and benzoic acid?
3. Alcohols are weaker acids than phenols but are stronger nucleophiles. Explain.
4. Which in each pair is the more stable and why?

(i) $PhCHMe$ and $PhCH_2CH_2{}^+$; (ii) PhO^- and $p\text{-}NO_2C_6H_4O^-$.
5. *o*-Nitrophenol has a lower b.p. and lower water-solubility than the *p*-isomer. Explain.
6. *m*-t-Butylphenol reacts with chlorine to give a trichloro-derivative, with bromine to give a dibromo-derivative, and with iodine to give a mono-iodo-derivative. What are these products? Explain.
7. 2,6-Di-t-butylphenol is a much weaker acid than phenol. Explain.

8. Discuss the following data:

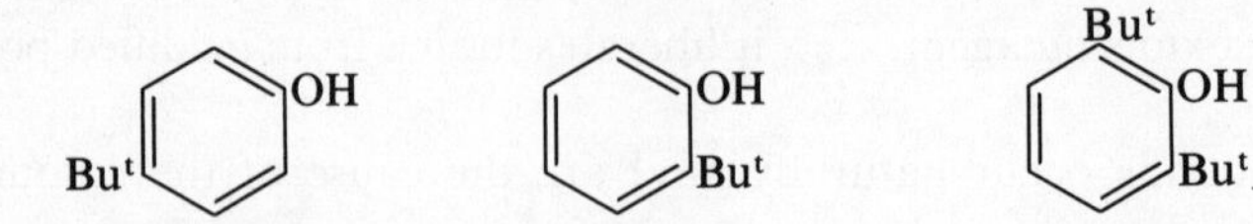

OH(cm^{-1})   3500   3500,   3640
                                   3640

9. Complete the equations:

(i) 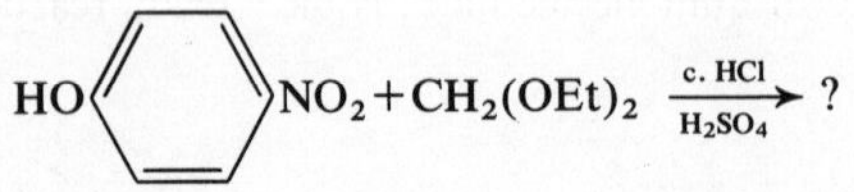   HO—⬡—NO$_2$ + CH$_2$(OEt)$_2$ $\xrightarrow[\text{H}_2\text{SO}_4]{\text{c. HCl}}$?

(ii) 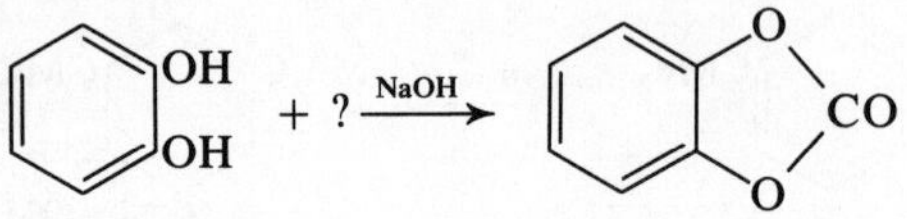   + ? $\xrightarrow{\text{NaOH}}$

(iii) PhOH + (AcO)$_2$Hg $\xrightarrow{\text{heat}}$? $\xrightarrow[\text{H}_2\text{O}]{\text{NaCl}}$?

(iv) 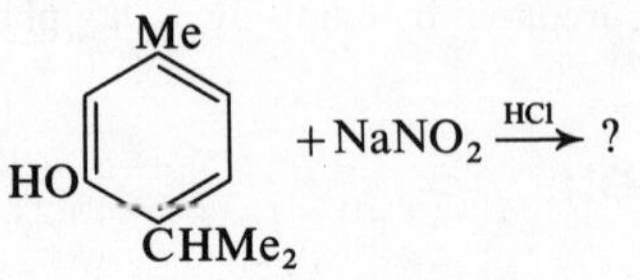   + NaNO$_2$ $\xrightarrow{\text{HCl}}$?

(v)   PhOH + ClCH$_2$CO$_2$H $\xrightarrow[\text{heat}]{\text{NaOH}}$?

(vi) 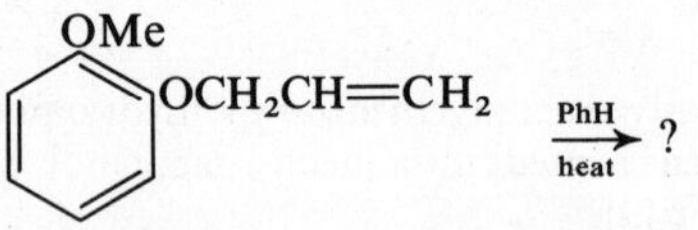 $\xrightarrow[\text{heat}]{\text{PhH}}$?

(vii) 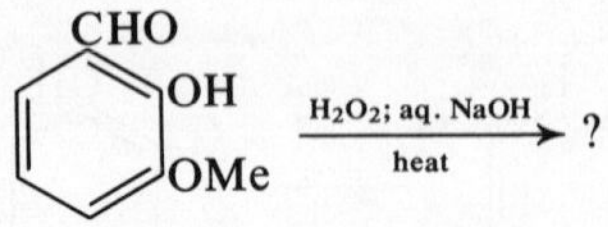 $\xrightarrow[\text{heat}]{\text{H}_2\text{O}_2;\ \text{aq. NaOH}}$?

(viii) 2,4,6-tri-t-butylphenol $\xrightarrow[\text{C}_6\text{H}_6]{\text{PbO}_2}$? $\xrightarrow{\text{Br}_2}$?

(ix)   PhOH + CH$_2$=CHCH$_2$Br $\xrightarrow{\text{K}_2\text{CO}_3}$?

(x) 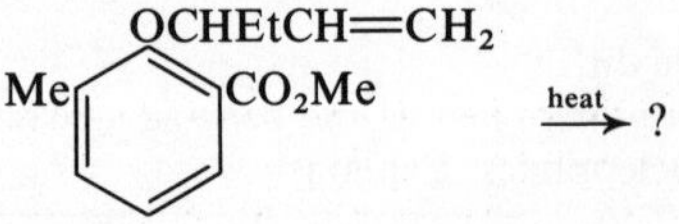 $\xrightarrow{\text{heat}}$?

(xi) 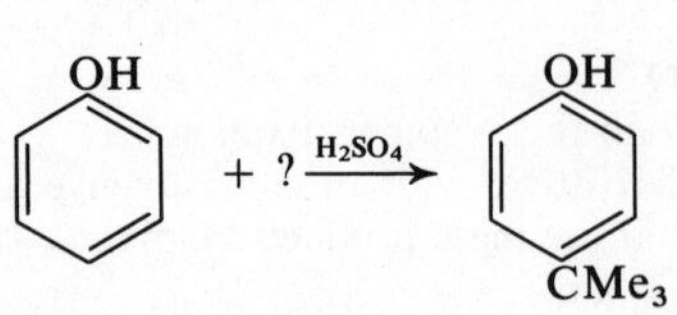   + ? $\xrightarrow{\text{H}_2\text{SO}_4}$

(xii)

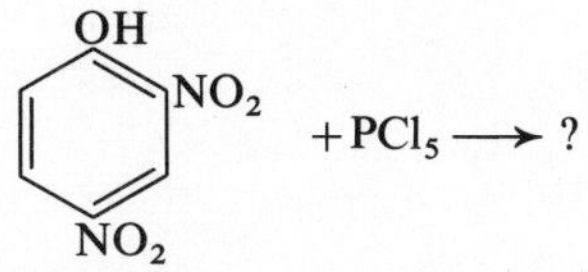

OH, NO$_2$, NO$_2$ + PCl$_5$ ⟶ ?

(xiii)

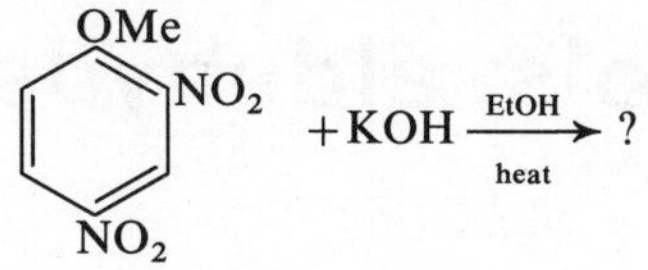

OMe, NO$_2$, NO$_2$ + KOH $\xrightarrow[\text{heat}]{\text{EtOH}}$?

(xiv)

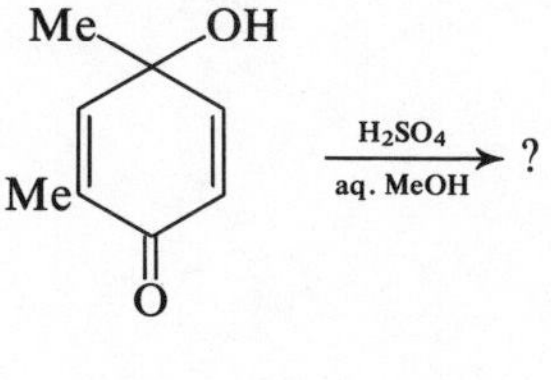

Me, OH, Me, O $\xrightarrow[\text{aq. MeOH}]{\text{H}_2\text{SO}_4}$?

(xv)

O, O + 2CH$_2$(CN)CO$_2$Et ⟶ ? $\xrightarrow[\text{boil}]{\text{aq. HCl}}$?

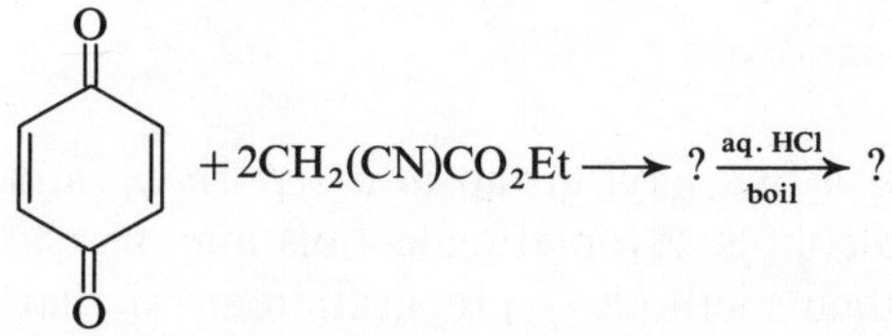

10. Phenyl triphenylmethyl ether, on treatment with acid, gives p-triphenylmethylphenol. Discuss.

11. How many N.M.R. signals would you expect from Ph—O—CH$_2$CH$_2$—O—Ph?

12. Convert:

(i) phenol ⟶ p-fluorophenol; (ii) PhH ⟶ m-HOC$_6$H$_4$NO$_2$; (iii) PhCl ⟶ 2,4-(NO$_2$)$_2$-phenol; (iv) PhOH ⟶ 2,6-Br$_2$-4-NO$_2$-phenol; (v) o-hydroxyacetophenone ⟶ 2,5-dihydroxyacetophenone; (vi) o-MeC$_6$H$_4$OH ⟶ o-HOC$_6$H$_4$CO$_2$H.

13. The Wurster salt derived from p-Me$_2$NC$_6$H$_4$NMe$_2$ is very stable, whereas that derived from the corresponding 2,3,5,6-tetramethyl compound (durene derivative) cannot be detected. Explain.

14. Account for the following ions in the mass spectra of:

(i) o-ethylphenol: 122 (40%), 107 (100%), 79 (9%), 77 (18%), 51 (5%); (ii) n-butyl phenyl ether: 150 (19%), 107 (2%), 94 (100%), 93 (1%), 66 (4%), 65 (4%).

REFERENCES

Organic Reactions, Wiley, Vol. I (1942), Ch. 2, 'The Fries Reaction': Vol. II (1944), Ch. 1, 'The Claisen Rearrangement'.

CHILD, 'Molecular Interactions in Clathrates: A Comparison with other Condensed Phases', *Quart. Rev.*, 1964, **18**, 321.

SCOTT, 'Oxidative Coupling of Phenolic Compounds', *Quart. Rev.*, 1965, **19**, 1.

TURNER, 'Quinone Methides', *Quart. Rev.*, 1964, **18**, 347.

SHINE, *Aromatic Rearrangements*, Elsevier (1967).

JEFFERSON and SCHEINMANN, 'Molecular Rearrangements Related to the Claisen Rearrangement', *Quart. Rev.*, 1968, **22**, 391.

FORRESTER, HAY, and THOMSON, *Organic Chemistry of Stable Free Radicals*, Academic Press (1968).

27

Aromatic alcohols, aldehydes and ketones

Aromatic alcohols

Aromatic alcohols are compounds containing a hydroxyl group in a *side-chain*, and may be regarded as aryl derivatives of the aliphatic alcohols. Aromatic alcohols may be classified as primary, secondary or tertiary alcohols, and their methods of preparation are similar to those used for aliphatic alcohols (p. 177). Only primary alcohols will be discussed in this section; secondary and tertiary aromatic alcohols are dealt with in the chapter on polynuclear hydrocarbons.

Benzyl alcohol (*phenylcarbinol*), b.p. 205°C, may be prepared:

(i) By hydrolysing benzyl chloride with aqueous sodium hydroxide:

$$C_6H_5CH_2Cl + NaOH \longrightarrow C_6H_5CH_2OH + NaCl \quad (70\%)$$

This method is used commercially.

(ii) By reducing benzaldehyde with zinc and hydrochloric acid:

$$C_6H_5CHO \xrightarrow{e;\, H^+} C_6H_5CH_2OH$$

(iii) By means of the Cannizzaro reaction:

$$2C_6H_5CHO + NaOH \longrightarrow C_6H_5CH_2OH + C_6H_5CO_2Na \quad (90\%)$$

It may also be prepared by means of a crossed Cannizzaro reaction (p. 226):

$$C_6H_5CHO + HCHO + NaOH \longrightarrow C_6H_5CH_2OH + HCO_2Na$$

The reactions of benzyl alcohol are similar to those of the primary aliphatic alcohols; *e.g.*, on oxidation, it forms benzaldehyde and finally benzoic acid:

$$C_6H_5CH_2OH \xrightarrow{[O]} C_6H_5CHO \xrightarrow{[O]} C_6H_5CO_2H$$

Benzyl alcohol forms esters, many of which are used in perfumery, *e.g.*, benzyl acetate (prepared from benzyl alcohol and acetic anhydride) occurs in oil of jasmine. Benzyl alcohol reacts with sodium to form sodium benzyl oxide.

In addition to the above reactions, benzyl alcohol exhibits aromatic properties due to the presence of the ring, *e.g.*, it can be nitrated, sulphonated, etc. Care, however, must be taken to avoid reaction with the hydroxyl group; it is therefore generally better, when preparing nuclear-substituted derivatives of benzyl alcohol, to use benzyl chloride (the CH_2Cl and CH_2OH groups are both *o-p*-orienting), and then hydrolyse to the alcohol.

Benzyl alcohol may be catalytically (palladium) reduced to toluene:

$$C_6H_5CH_2OH + H_2 \xrightarrow{Pd} C_6H_5CH_3 + H_2O$$

This reaction occurs when the benzyl group is attached to O, N or S, and may also be effected by sodium amalgam or lithium aluminium hydride. *Debenzylation* is a very useful reaction since a benzyl group may be introduced into a compound to protect a sensitive group and then removed at the end of the reaction.

Hydroxybenzyl alcohols may be prepared by the Lederer–Manasse reaction (p. 705). *o*-**Hydroxybenzyl alcohol, salicyl alcohol** (*saligenin*), $o\text{-}HOC_6H_4CH_2OH$, occurs in the glucoside *salicin*. It is a crystalline solid, m.p. 87°C, and is used in medicine as an antipyretic.

The NMR spectrum of benzyl alcohol is shown in Fig. 20.9 (p. 588).

Mass spectra. Benzyl alcohol shows an intense peak for the molecular ion, which undergoes the following fragmentations:

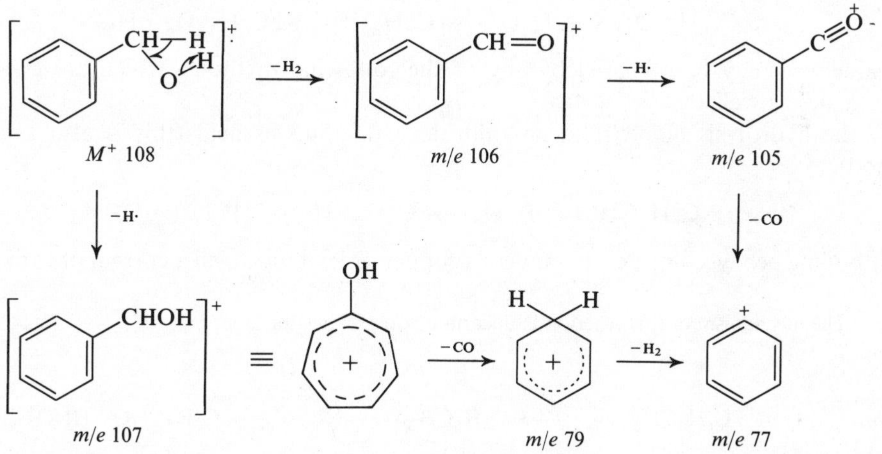

o-Substituted benzyl alcohols may behave differently from the *m*- and *p*-isomers in that they can also undergo elimination of a molecule of water by rearrangement (A = CH_2, O):

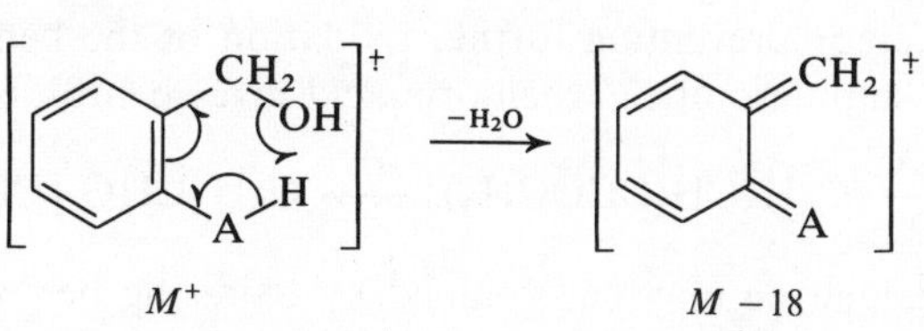

2-Phenylethanol (*β-phenylethyl alcohol*), b.p. 220°C, may be prepared by reducing phenylacetic ester (sodium and ethanol, LAH):

$$C_6H_5CH_2CO_2C_2H_5 \longrightarrow C_6H_5CH_2CH_2OH + C_2H_5OH$$

It is prepared industrially by reaction between ethylene oxide and benzene in the presence of aluminium chloride:

$$C_6H_6 + CH_2\!-\!\!-\!CH_2 \xrightarrow{AlCl_3} C_6H_5CH_2CH_2OH$$

When heated with alkali, it forms styrene.

Aromatic aldehydes

Aromatic aldehydes fall into two groups: those in which the aldehyde group is directly attached to the nucleus, and those in which it is attached to the side-chain. The former group comprises the aromatic aldehydes; the latter, which behave as aliphatic aldehydes, are best regarded as aryl-substituted aliphatic aldehydes.

The infra-red absorption region of the carbonyl group (C=O stretch) in aromatic aldehydes is $1\,715\text{--}1\,695\,\text{cm}^{-1}$ (s); this is lower than for aliphatic aldehydes. Also, the benzene nucleus and the pattern of orientation may be identified. There may be complications, however, due to intramolecular hydrogen bonding in certain *o*-derivatives.

Benzaldehyde (*benzenecarbonal*) is also known as *oil of bitter almonds*, since it is found in the glucoside *amygdalin* which occurs in bitter almonds. Amygdalin may be hydrolysed by dilute acids or the enzyme emulsin to benzaldehyde, glucose and hydrogen cyanide:

$$C_{20}H_{27}O_{11}N + 2H_2O \longrightarrow C_6H_5CHO + 2C_6H_{12}O_6 + HCN$$

Benzaldehyde may be prepared by any of the following methods, which are general for its homologues as well.

1. By the hydrolysis of benzylidene chloride with aqueous acid (this is also a commercial method).

$$C_6H_5CHCl_2 + H_2O \longrightarrow C_6H_5CHO + 2HCl \quad (76\%)$$

2. By boiling benzyl chloride with aqueous copper or lead nitrate in a current of carbon dioxide.

The mechanism of this reaction is uncertain; one suggestion is:

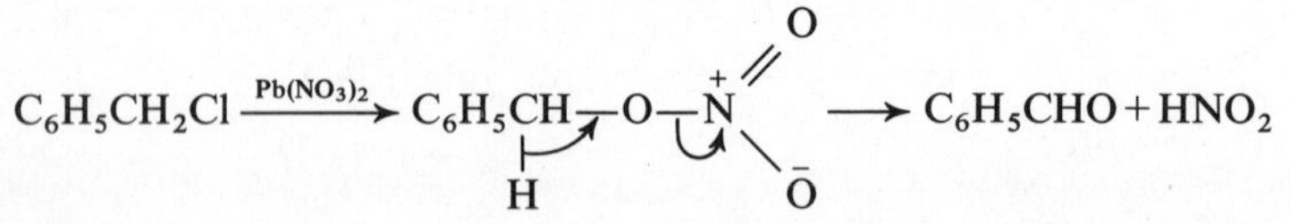

$$C_6H_5CH_2Cl \xrightarrow{Pb(NO_3)_2} C_6H_5CH\!-\!O\!-\!\overset{+}{N} \longrightarrow C_6H_5CHO + HNO_2$$

3. Benzaldehyde may be conveniently prepared in the laboratory by oxidising toluene with chromium trioxide in acetic anhydride. As the benzaldehyde is formed, it is converted into benzylidene acetate, thereby preventing further oxidation of the benzaldehyde. Hydrolysis of the acetate with dilute sulphuric or hydrochloric acid gives benzaldehyde:

$$C_6H_5CH_3 \xrightarrow[(CH_3CO)_2O]{CrO_3} C_6H_5CH(OCOCH_3)_2 \xrightarrow{H^+} C_6H_5CHO + 2CH_3CO_2H \quad (40\text{ }50\%)$$

A better yield of benzaldehyde may be obtained by oxidising benzyl alcohol with chromium trioxide in acetic anhydride (yield 90%) or with acid dichromate (yield 80–95%).

An interesting oxidising agent is chromyl chloride (**Étard's reaction**, 1877). In this method, toluene is treated with chromyl chloride in carbon tetrachloride solution and the complex, which is precipitated, is decomposed with water:

$$C_6H_5CH_3 + 2CrO_2Cl_2 \longrightarrow C_6H_5CH(OCrCl_2OH)_2 \xrightarrow{H_2O} C_6H_5CHO \quad (80\%)$$

Toluene is the main substrate that has been studied. The oxidation of n-propylbenzene gives propiophenone and benzyl methyl ketone (Miller *et al.*, 1890), and Wiberg *et al.* (1962) showed that the latter compound is produced by a rearrangement.

$$PhCH_2CH_2CH_3 \xrightarrow{CrO_2Cl_2} PhCOCH_2CH_3 + PhCH_2COCH_3$$

When more than one methyl group is present, only one is oxidised, *e.g.*, *m*-xylene gives *m*-tolualdehyde (80%).

Commercially, the oxidation of toluene to benzaldehyde is carried out in the vapour phase or in the liquid phase. In *vapour phase oxidation*, toluene is catalytically oxidised with air diluted with nitrogen to prevent complete oxidation of the hydrocarbon. The temperature may be as high as 500°C, and the catalyst is the oxide of metals such as manganese, molybdenum, zirconium, etc.:

$$C_6H_5CH_3 + O_2 \longrightarrow C_6H_5CHO + H_2O$$

In *liquid phase oxidation*, toluene is oxidised with manganese dioxide and 65 per cent sulphuric acid at 40°C.

4. Gattermann–Koch aldehyde synthesis (1897). Benzaldehyde may be synthesised by bubbling a mixture of carbon monoxide and hydrogen chloride through a solution of nitrobenzene or ether containing benzene and a catalyst consisting of aluminium chloride and a small amount of cuprous chloride:

$$C_6H_6 + CO + HCl \xrightarrow{AlCl_3} C_6H_5CHO + HCl$$

The mechanism of this reaction is uncertain, but it appears likely that the formyl cation is the active species.

$$CO + HCl + AlCl_3 \longrightarrow [H\overset{+}{C}{=}O \longleftrightarrow HC{\equiv}\overset{+}{O}] + AlCl_4^-$$

It also appears likely that the cuprous chloride forms a complex with the carbon monoxide, thereby increasing its local concentration.

When there are substituents in the ring, *e.g.*, a methyl group, the aldehyde group is introduced into the *p*-position only. The Gattermann–Koch aldehyde synthesis is not applicable to phenols and their ethers, or when the substituent is strongly deactivating, *e.g.*, nitrobenzene (*cf.* the Friedel–Crafts reaction).

5. Gattermann aldehyde synthesis (1906). When benzene is treated with a mixture of hydrogen cyanide and hydrogen chloride in the presence of aluminium chloride, and the complex so produced decomposed with water, benzaldehyde is produced (in low yield).

Several mechanisms have been proposed for this reaction. A widely accepted one is that imidoformyl chloride is formed as an intermediate:

$$HCl + HCN \xrightarrow{AlCl_3} HN{=}CHCl$$

$$C_6H_6 + HN{=}CHCl \xrightarrow{AlCl_3} HCl + C_6H_5CH{=}NH \xrightarrow{H_2O} C_6H_5CHO + NH_3$$
aryl imine

The Gattermann reaction is applicable to phenols and phenolic ethers, but not to nitrobenzene.

6. Sommelet's reaction (1913). Benzaldehyde is produced when benzyl chloride is refluxed with hexamethylenetetramine in aqueous ethanolic solution, followed by acidification and steam distillation:

$$C_6H_5CH_2Cl + (CH_2)_6N_4 \longrightarrow C_6H_5CHO \quad (60{-}70\%)$$

The mechanism of this reaction is uncertain. According to Angyal *et al.* (1953), the mechanism is:

(i) $C_6H_5CH_2Cl + C_6H_{12}N_4 \longrightarrow [C_6H_5CH_2C_6H_{12}N_4]^+Cl^- \longrightarrow$

$$C_6H_5CH_2N{=}CH_2 \underset{\rightleftharpoons}{\overset{H+}{}} C_6H_5CH_2\overset{+}{N}H{=}CH_2$$

(ii) $C_6H_5CH_2\overset{+}{N}H{=}CH_2 + C_6H_5\overset{..}{C}H{-}\overset{..}{N}H_2 \rightleftharpoons$

$$\underset{\overset{|}{H}}{}$$

$$C_6H_5CH_2NHCH_3 + C_6H_5CH{=}\overset{+}{N}H_2 \underset{\rightleftharpoons}{\overset{H_2O}{}} C_6H_5CHO + NH_3$$

Methylenebenzylamine, formed by hydrolysis of the quaternary compound, adds on a proton, and the ion thus formed reacts with benzylamine (formed as an intermediate) with transfer of a *hydride* ion from the latter to the former.

7. Rosenmund reduction (1918). Benzaldehyde is produced by the catalytic reduction of benzoyl chloride in the presence of a quinoline-sulphur poison (see p. 205).

$$C_6H_5COCl + H_2 \xrightarrow{Pd} C_6H_5CHO + HCl \quad (v.g.)$$

This method may be used to prepare hydroxybenzaldehydes provided the hydroxyl group is protected, *e.g.*, by acetylation.

8. Stephen's method (1925). When phenyl cyanide is reduced with stannous chloride and hydrochloric acid in ethereal solution, and then the product hydrolysed with water, benzaldehyde is formed (*cf.* p. 206):

$$C_6H_5CN \xrightarrow[\text{(ii) } H_2O]{\text{(i) } SnCl_2/HCl} C_6H_5CHO \quad (v.g.)$$

This method is a general one except for *o*-substituted cyanides, *e.g.*, *o*-tolyl cyanide, in which the yields are negligible due, probably, to steric hindrance.

Aryl cyanides may also be reduced to aldehydes by sodium triethoxyaluminium hydride (*cf.* p. 206), and sterically hindered cyanides by moist Raney nickel (Staskun *et al.*, 1964).

9. Benzaldehyde may be prepared by the reaction between phenylmagnesium bromide and ethyl formate, or better, ethyl orthoformate (*cf.* p. 430):

$$C_6H_5MgBr + HCO_2C_2H_5 \longrightarrow C_6H_5CHO + MgBr(OC_2H_5)$$

10. Benzaldehyde may be obtained from aniline via the diazonium salt and formaldoxime (Beech, 1954):

$$C_6H_5N_2{}^+Cl^- \xrightarrow{CH_2=NOH} C_6H_5CH{=}NOH \xrightarrow{H_2O} C_6H_5CHO \quad (40\%)$$

Benzaldehyde is a colourless liquid, b.p. 179°C, with a smell of almonds. It is used for flavouring purposes, in perfumery and in the dye industry (see also p. 748).

Benzaldehyde (and aromatic aldehydes in general) resembles aliphatic aldehydes in the following reactions:

(i) It gives the Schiff's reaction.

(ii) It is readily oxidised, *i.e.*, it is a strong reducing agent; *e.g.*, it reduces ammoniacal silver nitrate to silver, itself being oxidised to benzoic acid. Benzaldehyde oxidises to benzoic acid when exposed to air.

The mechanism of this autoxidation is not certain, but it is accepted that peroxy-acids are involved.

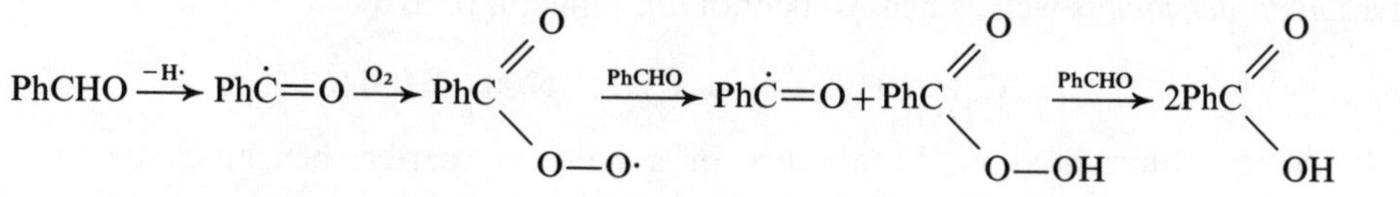

$$PhCHO \xrightarrow{-H\cdot} Ph\dot{C}=O \xrightarrow{O_2} PhC\overset{O}{\underset{O-O\cdot}{\diagup}} \xrightarrow{PhCHO} Ph\dot{C}=O + PhC\overset{O}{\underset{O-OH}{\diagup}} \xrightarrow{PhCHO} 2PhC\overset{O}{\underset{OH}{\diagup}}$$

(iii) Benzaldehyde forms a bisulphite compound, and may be prepared pure via this compound:

$$C_6H_5CHO + NaHSO_3 \longrightarrow C_6H_5CH(OH)SO_3Na$$

(iv) Benzaldehyde forms a cyanohydrin (*mandelonitrile*), $C_6H_5CH(OH)CN$, with hydrogen cyanide, and an oxime (and phenylhydrazone) with hydroxylamine (and phenylhydrazine). The latter compounds exist in two geometrical isomeric forms, *e.g.*, benzaldoxime:

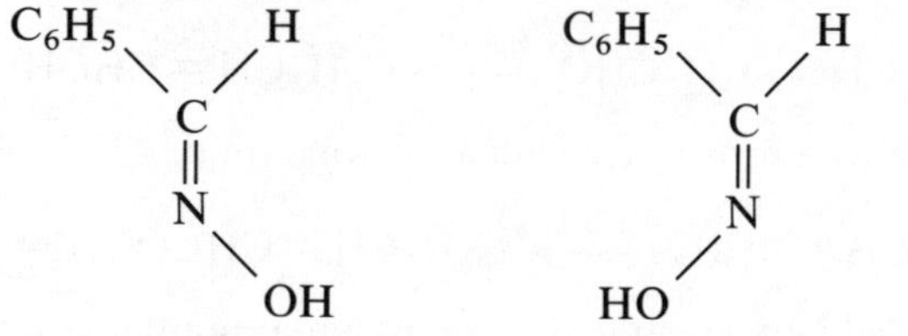

syn-form   *anti*-form

With hydrazine, benzaldehyde forms benzylideneazine, $C_6H_5CH=NN=CHC_6H_5$.

(v) Benzaldehyde reacts with phosphorus pentachloride to form benzylidene chloride:

$$C_6H_5CHO + PCl_5 \longrightarrow C_6H_5CHCl_2 + POCl_3$$

(vi) Benzaldehyde readily undergoes condensation with many aromatic and aliphatic compounds (see below).

Benzaldehyde (and other aromatic aldehydes) differs from aliphatic aldehydes in the following ways:

(i) It does not reduce Fehling's solution.

(ii) It does not readily polymerise; *e.g.*, it does not resinify with sodium hydroxide, but undergoes the Cannizzaro reaction due to its not having an α-hydrogen atom:

$$2C_6H_5CHO + NaOH \longrightarrow C_6H_5CH_2OH + C_6H_5CO_2Na$$

The mechanism involves hydride ion transfer, one possibility being as follows:

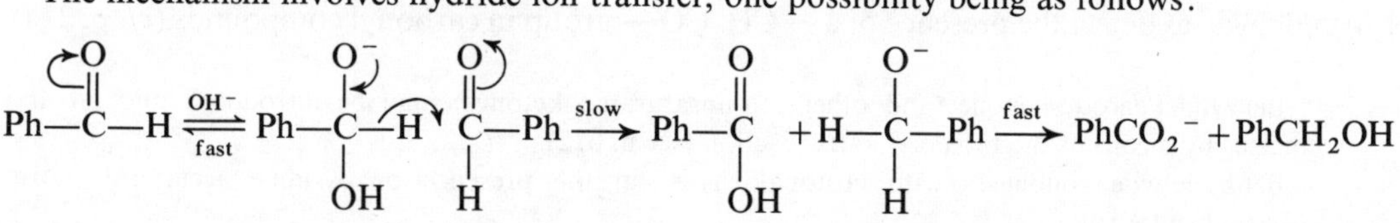

(iii) Benzaldehyde does not yield a simple addition product with ammonia, but forms a complex product, *hydrobenzamide* (90%). The mechanism is uncertain, but the reaction is believed to proceed via benzaldimine:

$$C_6H_5CHO \xrightarrow{NH_3} [C_6H_5CHOHNH_2] \xrightarrow{-H_2O} C_6H_5CH=NH \xrightarrow{3\ mol.}$$
$$C_6H_5CH=NCH(C_6H_5)N=CHC_6H_5 + NH_3$$

Benzaldehyde also reacts with primary aromatic amines to form *anils* or *Schiff bases* (p. 657).

(iv) Reduction of benzaldehyde with zinc and hydrochloric acid or with sodium amalgam gives *hydrobenzoin* as well as benzyl alcohol (*cf.* pinacol, p. 316):

$$2C_6H_5CHO \xrightarrow{e; H^+} C_6H_5CHOHCHOHC_6H_5$$

(v) When chlorinated in the absence of a halogen carrier, benzaldehyde forms benzoyl chloride (no α-hydrogen present):

$$C_6H_5CHO + Cl_2 \longrightarrow C_6H_5COCl + HCl$$

Condensation reactions of benzaldehyde

1. Claisen reaction (see also p. 219). Benzaldehyde, in the presence of dilute alkali, condenses with aliphatic aldehydes or ketones containing α-hydrogen; *e.g.*, with acetaldehyde it forms *cinnamaldehyde*:

$$C_6H_5CHO + CH_3CHO \xrightarrow{NaOH} C_6H_5CH=CHCHO + H_2O$$

With acetone, benzaldehyde forms benzylideneacetone (m.p. 42°C; used in perfumery):

$$C_6H_5CHO + CH_3COCH_3 \xrightarrow{NaOH} C_6H_5CH=CHCOCH_3 + H_2O \quad (65-78\%)$$

If the reaction is carried out in aqueous ethanolic sodium hydroxide, *dibenzylideneacetone* (m.p. 112°C) is produced by interaction of benzylideneacetone and another molecule of benzaldehyde:

$$C_6H_5CH=CHCOCH_3 + C_6H_5CHO \longrightarrow$$
$$C_6H_5CH=CHCOCH=CHC_6H_5 + H_2O \quad (90-94\%)$$

Benzaldehyde also condenses with acetophenone to form *phenyl styryl ketone* (*benzylidene-acetophenone, chalcone*):

$$C_6H_5CHO + CH_3COC_6H_5 \xrightarrow{NaOH} C_6H_5CH=CHCOC_6H_5 + H_2O \quad (85\%)$$

Phenyl styryl ketone may be reduced catalytically (platinum) in ethyl acetate solution or by zinc and acetic acid to *benzylacetophenone*:

$$C_6H_5CH=CHCOC_6H_5 + H_2 \xrightarrow{Pt} C_6H_5CH_2CH_2COC_6H_5 \quad (81-95\%)$$

The above condensation reactions, apart from their synthetic value, illustrate the use of benzaldehyde to detect the presence of a $-CH_2CO-$ group in carbonyl compounds (*cf.* p. 224).

> Benzylideneacetophenone (and other α,β-unsaturated ketones) may be introduced into an aryl nucleus by means of the Friedel–Crafts reaction (see p. 612).
> Benzaldehyde condenses with crotonaldehyde, in the presence of pyridine acetate, to form 5-phenylpentadienal:
>
> $$C_6H_5CHO + CH_3CH=CHCHO \longrightarrow C_6H_5CH=CHCH=CHCHO + H_2O$$
>
> This reaction is an example of vinylogy (p. 349), and was introduced by Kuhn (1929).

Benzaldehyde also condenses with nitromethane to form ω-nitrostyrene; the methylene group is made 'active' by the adjacent nitro-group:

$$C_6H_5CHO + CH_3NO_2 \xrightarrow{NaOH} C_6H_5CH=CHNO_2 + H_2O \quad (80-83\%)$$

2. Perkin reaction (1877). When benzaldehyde (or any other aromatic aldehyde) is heated with the anhydride of an aliphatic acid (containing two α-hydrogen atoms) in the presence of its sodium salt, condensation takes place to form a β-arylacrylic acid; *e.g.*, with acetic anhydride and sodium acetate, cinnamic acid is formed:

$$C_6H_5CHO + (CH_3CO)_2O \xrightarrow{CH_3CO_2Na} C_6H_5CH{=}CHCO_2H$$

With propionic anhydride and sodium propionate, α-methylcinnamic acid is formed:

$$C_6H_5CHO + (CH_3CH_2CO)_2O \xrightarrow{CH_3CH_2CO_2Na} C_6H_5CH{=}C(CH_3)CO_2H$$

The mechanism may be as follows (*cf.* the aldol condensation):

$$CH_3COOCOCH_3 + CH_3CO_2{}^- \rightleftharpoons \bar{C}H_2COOCOCH_3 + CH_3CO_2H$$

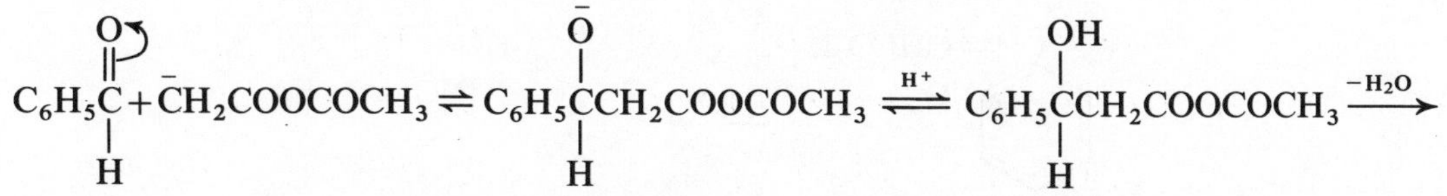

$$\underset{\underset{H}{|}}{\overset{\overset{O}{\|}}{C_6H_5C}} + \bar{C}H_2COOCOCH_3 \rightleftharpoons \underset{\underset{H}{|}}{\overset{\overset{O}{|}}{C_6H_5CCH_2COOCOCH_3}} \underset{\Longleftarrow}{\overset{H^+}{\Longrightarrow}} \underset{\underset{H}{|}}{\overset{\overset{OH}{|}}{C_6H_5CCH_2COOCOCH_3}} \xrightarrow{-H_2O}$$

$$C_6H_5CH{=}CHCOOCOCH_3 \xrightarrow{H_2O} C_6H_5CH{=}CHCO_2H + CH_3CO_2H$$

It should be noted that only the α-hydrogen atoms of the anhydride are involved in the condensation. The acyloxyl ion acts as the base, and in some cases triethylamine as base gives better yields.

The Perkin reaction does not usually take place with aliphatic aldehydes.

Benzaldehyde undergoes the Knoevenagel reaction with malonic acid in ethanolic ammonia to form cinnamic acid:

$$C_6H_5CHO + CH_2(CO_2H)_2 \longrightarrow C_6H_5CH{=}CHCO_2H + CO_2 + H_2O$$

A special case of the Perkin reaction is the condensation of benzaldehyde with cyclic anhydrides, *e.g.*, succinic anhydride (see p. 343).

3. Benzoin condensation. When refluxed with aqueous ethanolic potassium cyanide, benzaldehyde forms benzoin (see also p. 791):

$$2C_6H_5CHO \xrightarrow{KCN} C_6H_5CHOHCOC_6H_5 \quad (83\%)$$

4. Benzaldehyde condenses with phenols and tertiary aromatic amines in the presence of sulphuric acid or zinc chloride to form triphenylmethane derivatives; *e.g.*, with dimethylaniline it forms *malachite green* (see p. 886).

Derivatives of benzaldehyde

A large variety of substituted benzyl alcohols can be oxidised to the corresponding aldehydes by means of dinitrogen tetroxide solutions (yields: 91–98%). The benzyl alcohols (except the nitro-compounds) may be obtained by reduction of acid chloride or methyl esters with lithium aluminium hydride (Field *et al.*, 1955), *e.g.*,

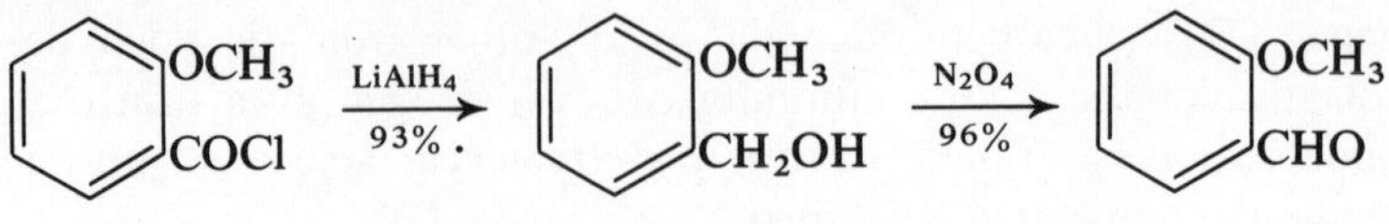

Various substituted benzaldehydes may also be obtained via diazonium salts (Beech, 1954), e.g.,

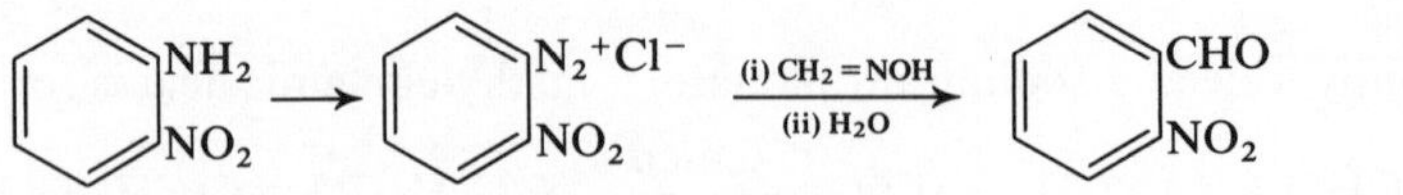

$$(33\%)$$

Nitrobenzaldehydes. Mixed acid nitrates benzaldehyde to give mainly the *m*-isomer, m.p. 58°C (about 50%), together with the *o*-isomer (about 20%).

o-*Nitrobenzaldehyde* may be prepared by oxidising *o*-nitrocinnamic acid with cold aqueous potassium permanganate:

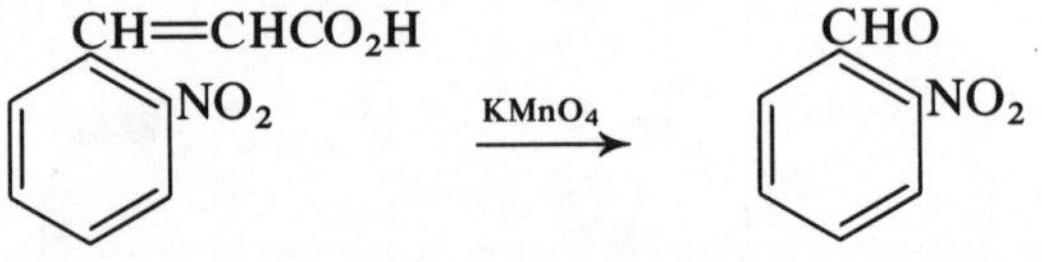

Alternatively, it may be prepared by dissolving *o*-nitrotoluene in glacial acetic acid containing acetic anhydride, and adding chromium trioxide and sulphuric acid. The product is the diacetate and this, on hydrolysis, gives the aldehyde:

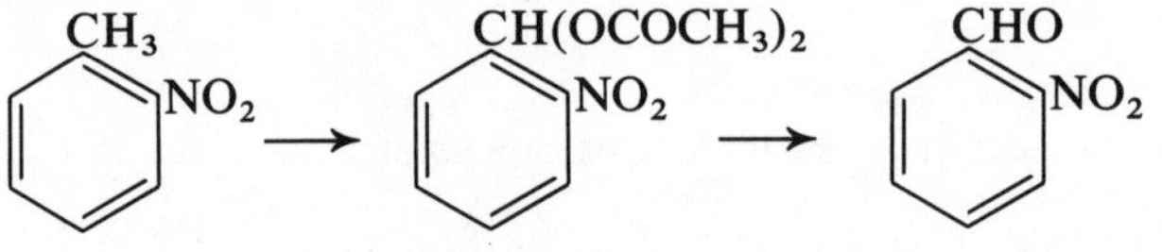

$$(23-24\%) \qquad (74\%)$$

A better yield may be obtained by oxidising the *o*-nitrotoluene with manganese dioxide and sulphuric acid.

o-Nitrobenzaldehyde is a yellow solid, m.p. 44°C. Its most important reaction is its conversion into indigotin when heated with acetone and sodium hydroxide (p. 895):

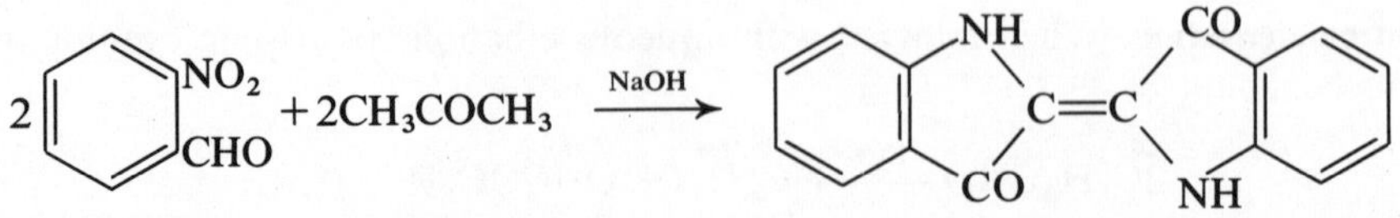

p-*Nitrobenzaldehyde* (m.p. 106°C) may be prepared by oxidising *p*-nitrocinnamic acid or *p*-nitrotoluene; it may also be prepared by oxidising *p*-nitrobenzyl chloride with aqueous lead nitrate.

All three nitrobenzaldehydes may be reduced to the corresponding aminobenzaldehydes with ferrous sulphate and ammonium hydroxide solution or by stannous chloride and hydrochloric acid. On the other hand, sodium borohydride, lithium aluminium hydride–aluminium chloride, and aluminium isopropoxide (MPV reduction) reduce nitrobenzaldehydes to the corresponding nitrobenzyl alcohols.

o-*Aminobenzaldehyde* readily condenses with compounds containing a —CH_2CO— group to form quinoline compounds; e.g., with acetaldehyde it forms *quinoline* (see p. 857). m- and p-*Amino-*

benzaldehydes are used in the preparation of dyes. *o*- and *p*-Aminobenzaldehydes are prepared commercially by the oxidation of the corresponding aminobenzyl alcohols (see method 3, p. 744).

2,4-*Dinitrobenzaldehyde* (m.p. 72°C) may be prepared as follows:

$$\text{N(CH}_3)_2\text{-C}_6\text{H}_5 \xrightarrow{\text{HNO}_2} \text{N(CH}_3)_2\text{-C}_6\text{H}_4\text{-NO}$$

$$(\text{CH}_3)_2\text{N-C}_6\text{H}_4\text{-NO} + \text{CH}_3\text{-C}_6\text{H}_3(\text{NO}_2)_2 \xrightarrow[\text{heat}]{\text{Na}_2\text{CO}_3}$$

$$(\text{CH}_3)_2\text{N-C}_6\text{H}_4\text{-N=CH-C}_6\text{H}_3(\text{NO}_2)_2 \xrightarrow[\text{heat}]{\text{HCl}} \text{N(CH}_3)_2\text{-C}_6\text{H}_4\text{-NH}_2 + \text{CHO-C}_6\text{H}_3(\text{NO}_2)_2 \quad (24\text{–}32\%)$$

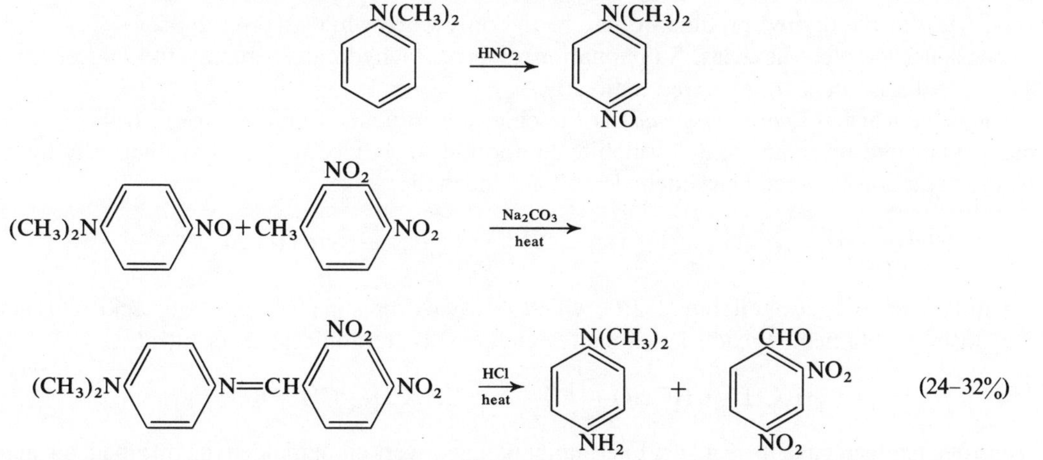

Halogen derivatives of benzaldehyde. *m*-Halogenobenzaldehydes may be prepared from *m*-nitrobenzaldehyde by reduction and subsequent replacement of the diazo-group by halogen (Sandmeyer reaction), *e.g.*, *m*-chlorobenzaldehyde:

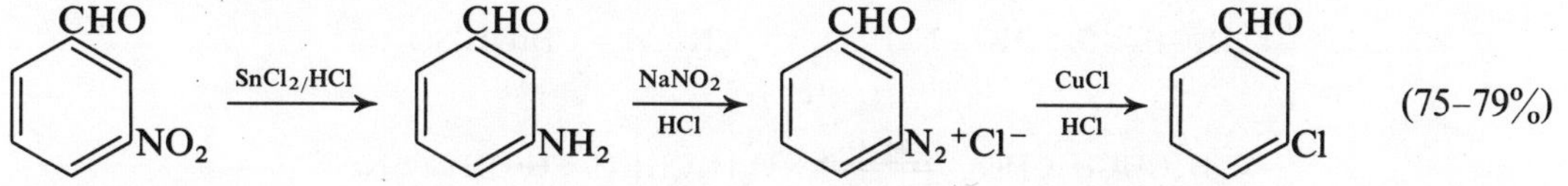

$$\text{CHO-C}_6\text{H}_4\text{-NO}_2 \xrightarrow{\text{SnCl}_2/\text{HCl}} \text{CHO-C}_6\text{H}_4\text{-NH}_2 \xrightarrow[\text{HCl}]{\text{NaNO}_2} \text{CHO-C}_6\text{H}_4\text{-N}_2{}^+\text{Cl}^- \xrightarrow[\text{HCl}]{\text{CuCl}} \text{CHO-C}_6\text{H}_4\text{-Cl} \quad (75\text{–}79\%)$$

o- and *p*-Halogenobenzaldehydes may be prepared by brominating the corresponding halogenotoluenes at their boiling points in the presence of light to give the benzylidene bromide derivative, and hydrolysing this with calcium hydroxide; *e.g.*, *p*-bromobenzaldehyde from *p*-bromotoluene:

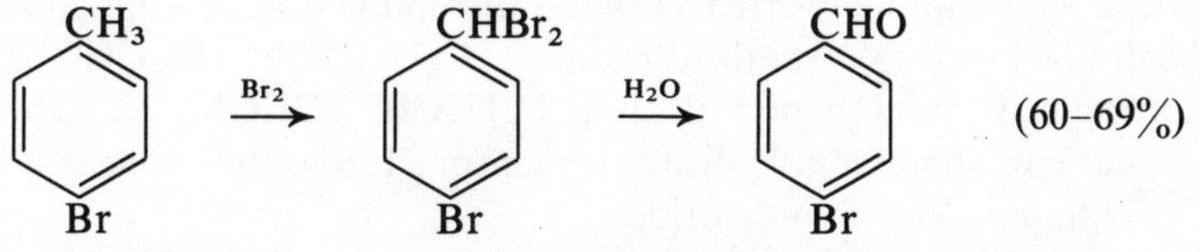

$$\text{CH}_3\text{-C}_6\text{H}_4\text{-Br} \xrightarrow{\text{Br}_2} \text{CHBr}_2\text{-C}_6\text{H}_4\text{-Br} \xrightarrow{\text{H}_2\text{O}} \text{CHO-C}_6\text{H}_4\text{-Br} \quad (60\text{–}69\%)$$

Alternatively, the halogenotoluene may be oxidised with manganese dioxide and sulphuric acid.

2,6-*Dichlorobenzaldehyde* (m.p. 72°C) is important in the preparation of triphenylmethane dyes; it may be prepared as follows:

$$\text{CH}_3\text{-C}_6\text{H}_4\text{-NO}_2 \xrightarrow[\text{Fe}]{\text{Cl}_2} \text{CH}_3\text{-C}_6\text{H}_3(\text{Cl})\text{NO}_2 \xrightarrow{\text{Fe/HCl}} \text{CH}_3\text{-C}_6\text{H}_3(\text{Cl})\text{NH}_2 \xrightarrow[\text{HCl}]{\text{NaNO}_2} \text{CH}_3\text{-C}_6\text{H}_3(\text{Cl})\text{N}_2{}^+\text{Cl}^- \xrightarrow[\text{HCl}]{\text{CuCl}}$$

$$\text{CH}_3\text{-C}_6\text{H}_3\text{Cl}_2 \xrightarrow[\text{(at b.p.)}]{\text{Cl}_2} \text{CHCl}_2\text{-C}_6\text{H}_3\text{Cl}_2 \xrightarrow[\text{(con. H}_2\text{SO}_4)]{\text{H}_2\text{O}} \text{CHO-C}_6\text{H}_3\text{Cl}_2$$

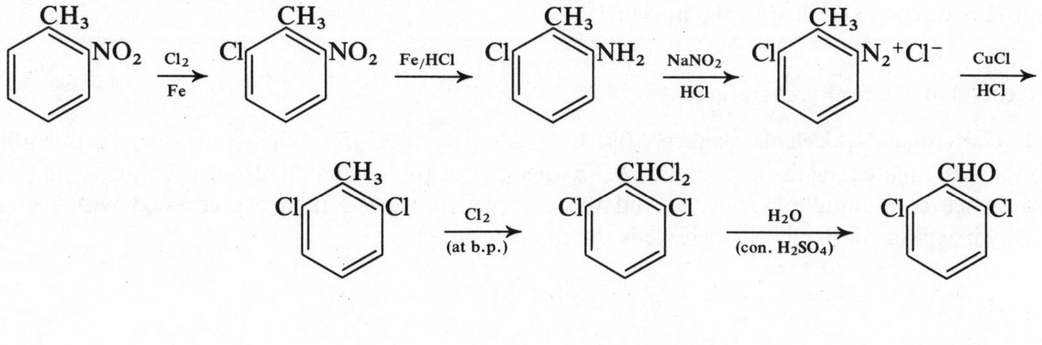

Concentrated sulphuric acid must be used to hydrolyse the *o,o'*-dichlorobenzylidene chloride; alkali has very little effect (this is an example of steric hindrance). The overall yield of 2,6-dichlorobenzaldehyde is very small; this is to be expected from the large number of steps involved. This method illustrates the important point that although a particular method gives a small yield of the desired product, it may be the only worthwhile method to use.

Benzaldehydesulphonic acids. Sulphonation of benzaldehyde gives mainly the *m*-derivative. The *o*- and *p*-isomers are prepared indirectly.

Cinnamaldehyde (3-*phenylpropenal*) is the chief constituent of cinnamon oil, from which it may be isolated by means of its bisulphite compound. It may be prepared synthetically by the Claisen reaction between benzaldehyde and acetaldehyde:

$$C_6H_5CHO + CH_3CHO \xrightarrow{\text{NaOH}} C_6H_5CH{=}CHCHO + H_2O$$

Cinnamaldehyde is an oil, b.p. 252°C, which slowly oxidises in air to cinnamic acid. This acid may also be obtained by oxidising cinnamaldehyde with ammoniacal silver nitrate:

$$C_6H_5CH{=}CHCHO + [O] \longrightarrow C_6H_5CH{=}CHCO_2H$$

Vigorous oxidising agents, *e.g.*, acid permanganate, convert cinnamaldehyde into benzoic acid.

Cinnamaldehyde forms the normal bisulphite compound with sodium hydrogen sulphite, but on prolonged treatment with this reagent, the sodium salt of a disulphonic acid is formed, possibly as follows (*cf.* p. 340):

$$C_6H_5CH{=}CHCHO + NaHSO_3 \rightleftharpoons C_6H_5CH{=}CH(OH)SO_3Na$$

$$\underset{\underset{\text{SO}_3\text{Na}}{|}}{C_6H_5CHCH_2CHO} \xrightarrow[\rightleftharpoons]{\text{NaHSO}_3} \underset{\underset{\text{SO}_3\text{Na}}{|}}{C_6H_5CHCH_2CH(OH)SO_3Na}$$

(1,4-addition)

The normal addition is reversible; the 1,4-addition is irreversible, and hence all the cinnamaldehyde is gradually converted into the disulphonic acid. Cinnamaldehyde adds on bromine to the double bond to form the dibromide. $C_6H_5CHBrCHBrCHO$, and is reduced by aluminium isopropoxide to cinnamyl alcohol, $C_6H_5CH{=}CHCH_2OH$. This reduction may also be carried out with aluminium hydride (Jorgenson, 1962); this reagent does *not* attack the double bond (*cf.* lithium aluminium hydride).

Phenolic aldehydes

Phenolic aldehydes (*hydroxyaldehydes*) contain an aldehyde group and one or more hydroxyl groups directly attached to the nucleus.

General methods of preparation

1. Gattermann's aldehyde synthesis (*cf.* benzaldehyde, p. 735). When phenol or a phenolic ether is treated with a mixture of hydrogen cyanide and hydrogen chloride in the presence of aluminium chloride, and the complex produced then decomposed with water, *p*-hydroxy-(or alkoxy)benzaldehyde is the main product:

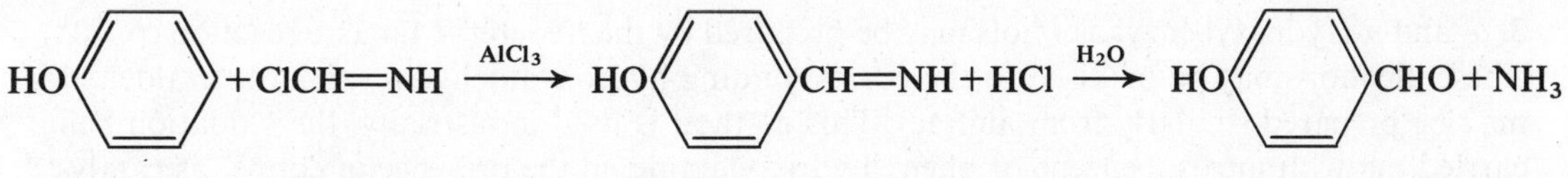

For more reactive compounds, *e.g.*, *m*-di- or trihydric phenols (*o*- and *p*-positions strongly activated), zinc chloride is a better catalyst. In either case, to avoid the use of hydrogen cyanide, zinc cyanide and hydrogen chloride are very convenient for generating the hydrogen cyanide *in situ*.

2. Reimer–Tiemann reaction (1876). This reaction is carried out by refluxing an alkaline solution of phenol with chloroform at 60°C, distilling off the excess of chloroform, acidifying the residual liquid with sulphuric acid, and then steam-distilling it. Unchanged phenol and *o*-hydroxy-benzaldehyde distil over, leaving behind *p*-hydroxybenzaldehyde.

The mechanism of the Reimer–Tiemann reaction is believed to involve the formation of dichloromethylene (produced from the chloroform; see p. 170):

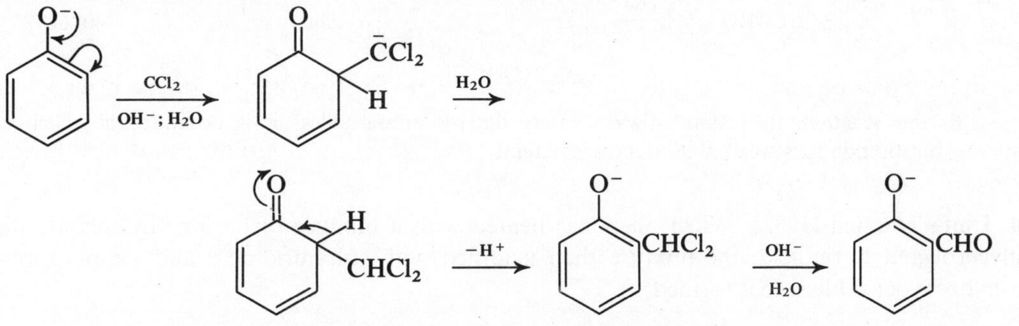

Phenols with blocked *p*-positions give cyclohexadienones containing the dichloromethyl group, *e.g.*,

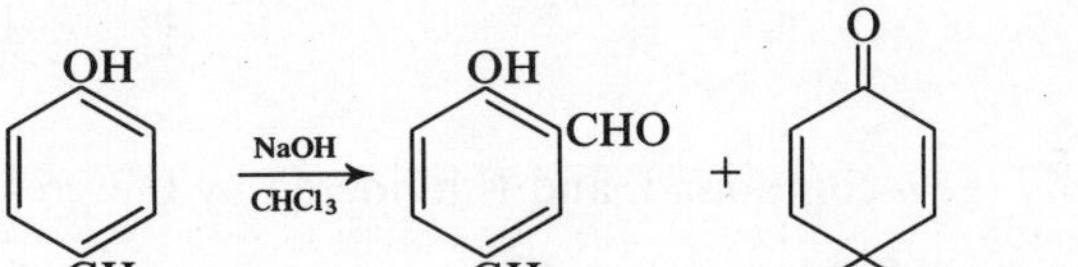

In the Reimer–Tiemann reaction, the *o*-isomer predominates, but if one of the *o*-positions is occupied, the aldehyde group tends to go to the *p*-position; *e.g.*, guaiacol forms vanillin:

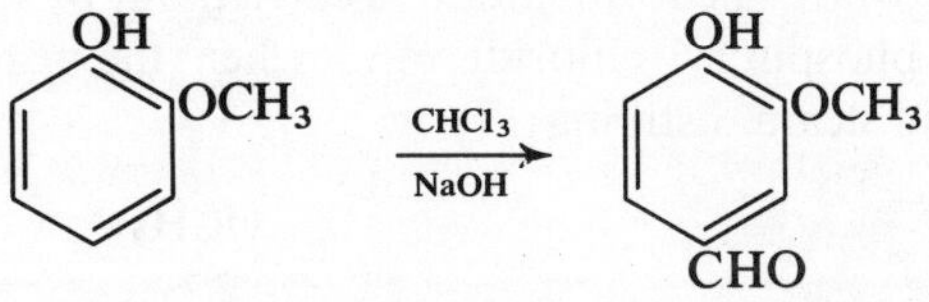

Phenols may be *selectively* formylated in the *o*-position by heating the phenoxymagnesium halide with ethyl orthoformate (Casnati *et al.*, 1965), *e.g.*,

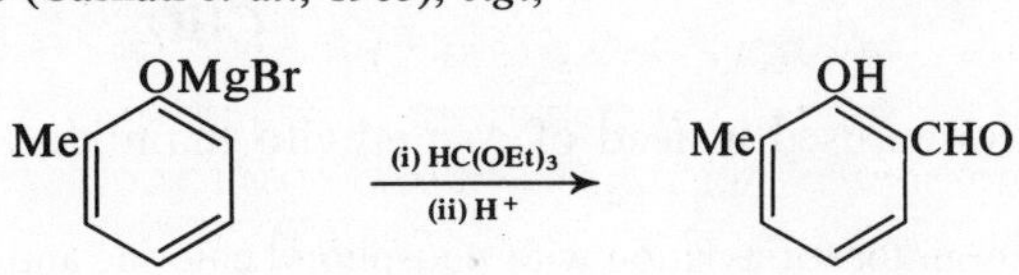

3. *o*- and *p*-Hydroxybenzyl alcohols may be prepared by the Lederer–Manasse reaction (p. 705). These alcohols may be oxidised to the corresponding hydroxyaldehydes (aminobenzaldehydes may be prepared similarly from aniline). This method is used industrially, the oxidation being carried out with nitrosobenzene or phenylhydroxylamine, in the presence of copper as catalyst:

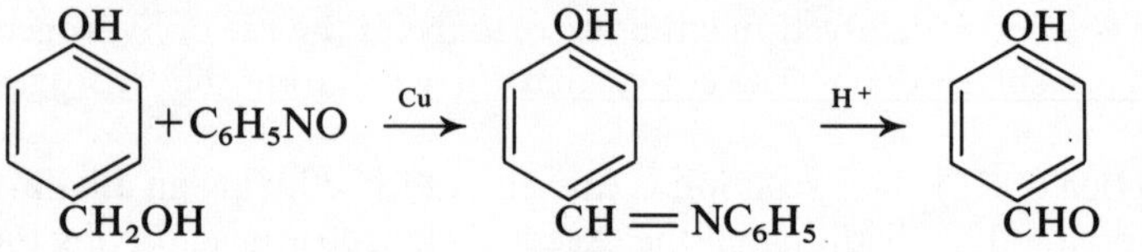

An extension of the above method is the *chloral condensation*. This is a modified Lederer–Manasse reaction, chloral being used instead of formaldehyde, *e.g.*,

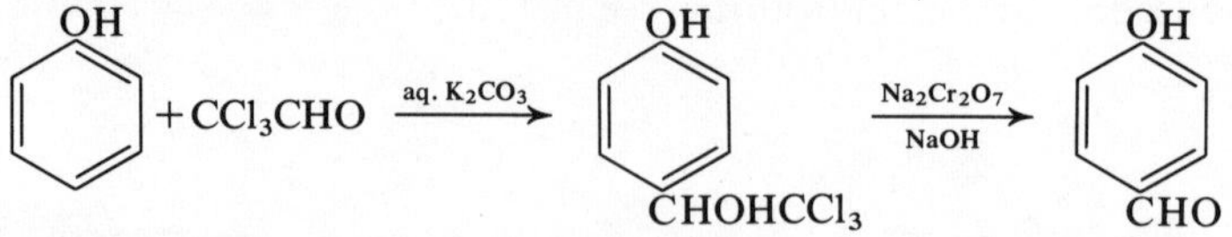

In this reaction, the chloral always enters the *p*-position unless it is occupied, in which case *o*-substitution takes place, but to a lesser extent.

4. Duff's reaction (1932). When phenol is heated with a mixture of hexamethylenetetramine, glycerol and boric acid, the mixture then acidified with sulphuric acid and steam distilled, *o*-hydroxybenzaldehyde is formed:

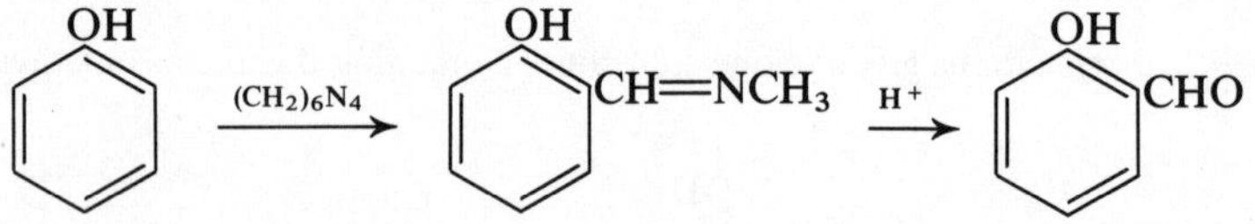

This method gives only the *o*-compound, and is hindered by the presence of a $-I$ group in the ring.

5. Formylation with *N*-methylformanilide (Vilsmeier and Haack, 1927). Provided the aromatic compound has an activated nucleus, it can be formylated by means of *N*-methylformanilide. The method works well for the *o*- and *p*-positions of phenolic ethers (and dialkyl anilines); it does not work with hydrocarbons, except with anthracene, in which case the aldehyde group enters the 9-position (see p. 814). The formylation is carried out by treating the compound with *N*-methylformanilide and phosphoryl chloride, and when the reaction is complete, adding aqueous sodium acetate and steam distilling:

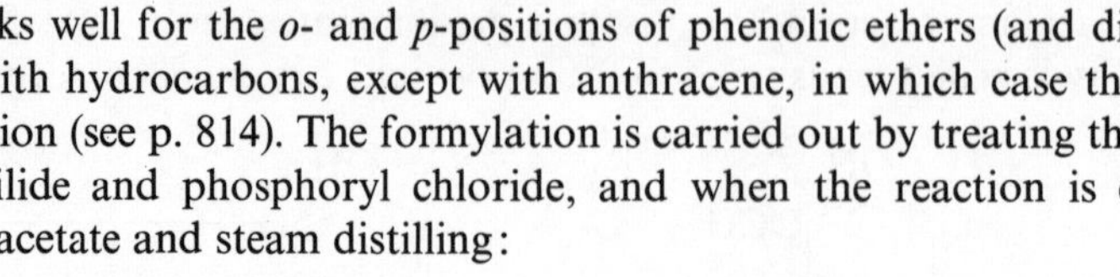

Dimethylformamide is often used instead of *N*-methylformanilide.

A possible mechanism for formylation with phosphoryl chloride and dimethylformamide is:

$$Me_2NCHO + POCl_3 \longrightarrow [Me_2N-\overset{+}{C}HCl \longleftrightarrow Me_2\overset{+}{N}=CHCl]OPOCl_2^-$$

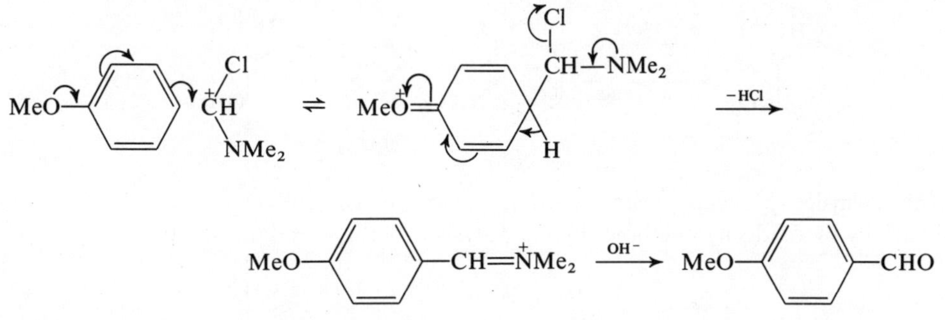

$$MeO-\!\!\!\!\bigcirc\!\!\!\!-CH=\overset{+}{N}Me_2 \xrightarrow{OH^-} MeO-\!\!\!\!\bigcirc\!\!\!\!-CHO$$

N-Methylformanilide (and, in general, disubstituted formamides) reacts with Grignard reagents to form aldehydes (Bouveault, 1904; Smith *et al.*, 1941), *e.g.*,

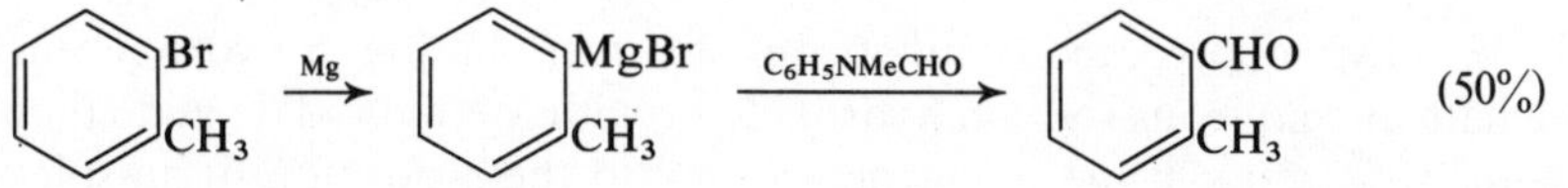

(50%)

6. By means of organolithium compounds. When a substituted phenyl-lithium compound is heated with ethyl orthoformate or *N*-methylformanilide, an intermediate compound is obtained which, on hydrolysis with acid, gives a substituted benzaldehyde in 70 per cent yield, *e.g.*,

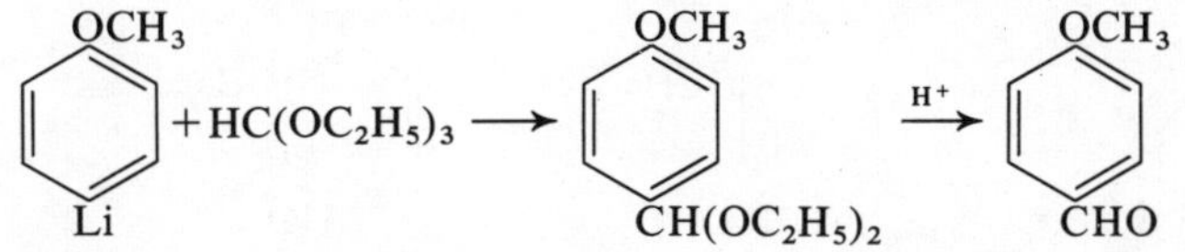

Salicylaldehyde (*o-hydroxybenzaldehyde*) occurs in certain essential oils. It may be prepared by any of the general methods applicable to *o*-hydroxyaldehydes. It is manufactured by the oxidation of *o*-hydroxybenzyl alcohol (obtained by method 3, above), and by the oxidation of *o*-methylphenyl benzenesulphonate with manganese dioxide and sulphuric acid:

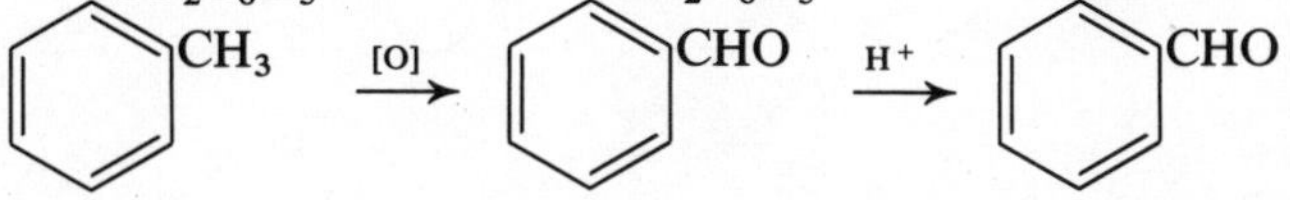

Salicylaldehyde is an oil, b.p. 197°C, and its aqueous solution gives a violet coloration with ferric chloride. Salicylaldehyde may be oxidised to salicylic acid and reduced to *o*-hydroxybenzyl alcohol. Oxidation with alkaline hydrogen peroxide converts it into catechol (see p. 714).

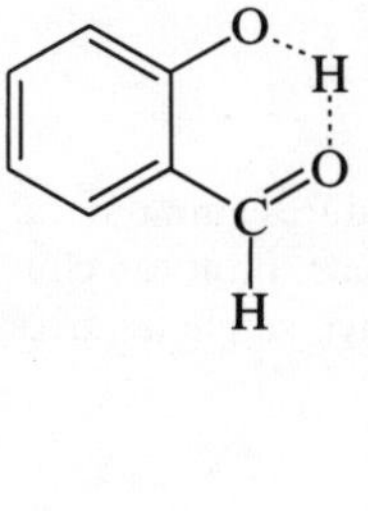

The hydroxyl group of salicylaldehyde is not so reactive as that in the *m*- and *p*-isomers. This is probably due to hydrogen bonding, which also accounts for the high volatility of salicylaldehyde (compared with the *m*- and *p*-compounds).

m-Hydroxybenzaldehyde may be prepared as follows:

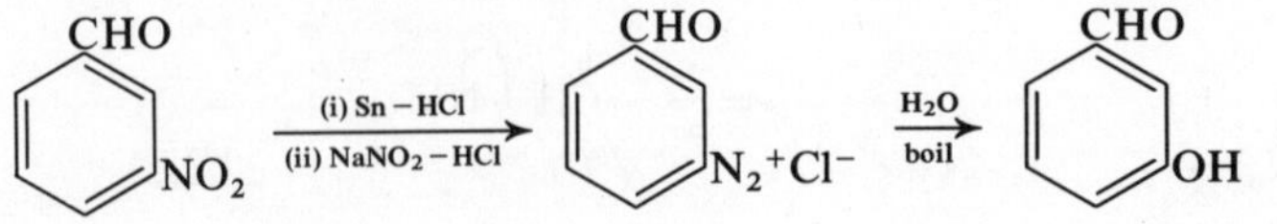

(51–56%)

Anisaldehyde (p-*methoxybenzaldehyde*) occurs in various essential oils. It is prepared industrially by oxidising anethole (by ozonolysis or with acid dichromate):

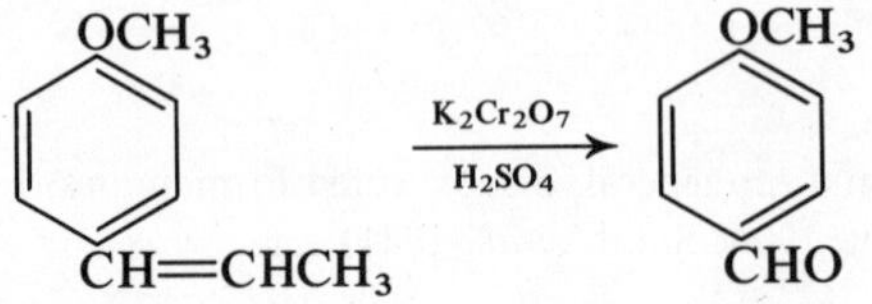

Anisaldehyde may be prepared synthetically by methylating *p*-hydroxybenzaldehyde with methyl sulphate and aqueous sodium hydroxide or with methyl iodide and ethanolic potassium hydroxide, or by introducing an aldehyde group into the *p*-position of anisole.

Anisaldehyde, b.p. 248°C, may be oxidised to anisic acid and reduced to anisyl alcohol.

Protocatechualdehyde (3,4-*dihydroxybenzaldehyde*) may be prepared by heating vanillin with hydrochloric acid:

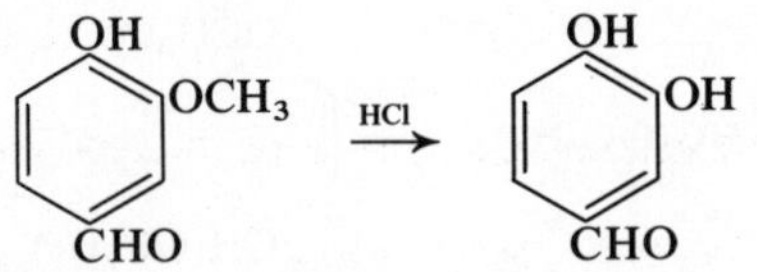

It may also be prepared by adding phosphorus pentachloride to piperonal, then treating the product with cold water, and finally boiling the solution:

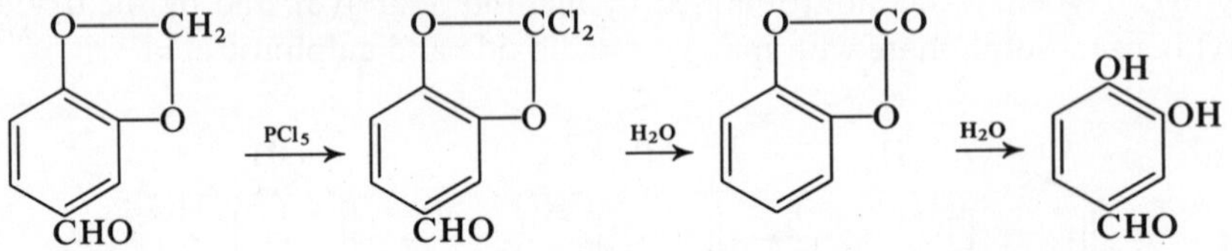

(61%)

Another preparation is by means of the Reimer–Tiemann reaction using catechol:

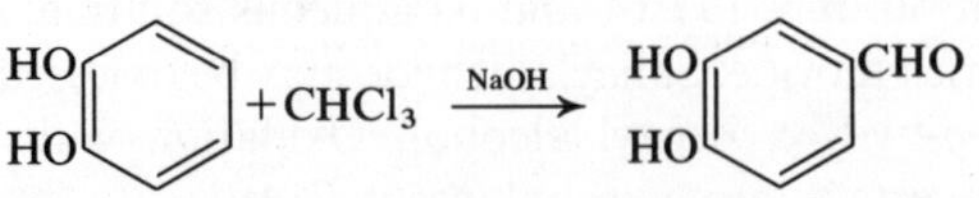

Protocatechualdehyde, m.p. 153°C, gives a green colour with ferric chloride, and reduces ammoniacal silver nitrate. When methylated with approximately one equivalent of methyl sulphate, protocatechu-aldehyde gives a mixture of vanillin (I) and isovanillin (II); excess of methyl sulphate gives veratraldehyde (III):

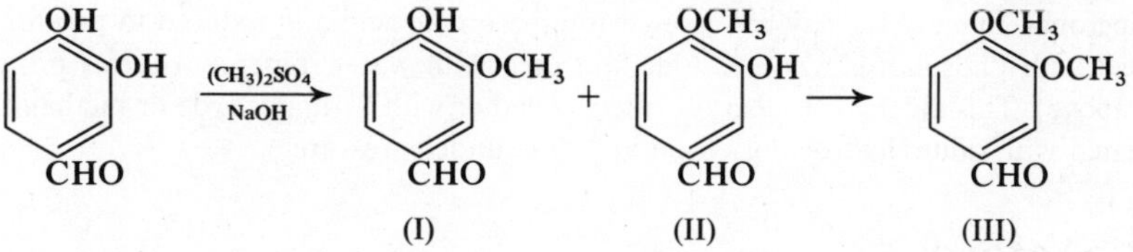

The reaction scheme shows phenolic aldehyde compounds (I), (II), and (III) with dimethyl sulphate and sodium hydroxide.

Vanillin (*p*-hydroxy-*m*-methoxybenzaldehyde) occurs in many substances of plant origin, *e.g.*, the vanilla bean. It may be prepared synthetically from guaiacol (p. 719) by means of the Reimer–Tiemann or the Gattermann reaction. Industrially, it is prepared:

(i) By oxidising isoeugenol (from eugenol) with nitrobenzene:

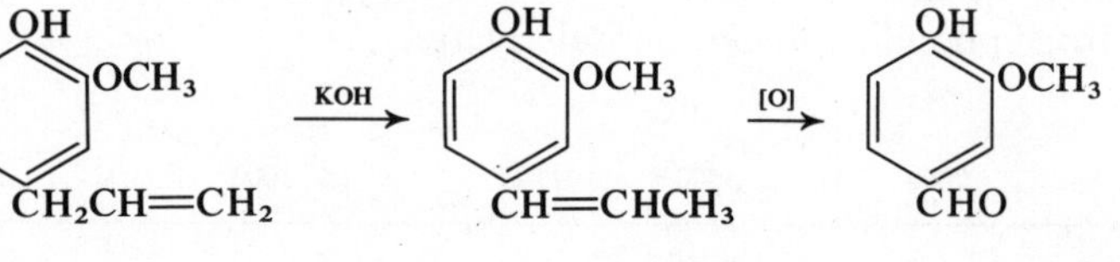

If the oxidation is carried out with acid dichromate, it is necessary to protect the hydroxyl group (by acetylation, etc.). On the other hand, ozonolysis may be used without protecting the hydroxyl group.

(ii) By the Lederer–Manasse reaction as follows:

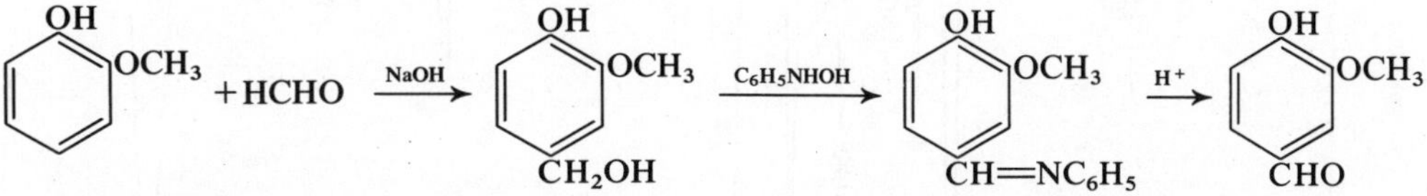

(iii) The liquors from the extraction of lignin from pulp-wood contain vanillin, and are now used as a source of vanillin.

Vanillin is a crystalline solid, m.p. 81°C. When heated with hydrochloric acid, it forms proto-catechualdehyde; with methyl sulphate it forms veratraldehyde. Oxidation by silver oxide in aqueous sodium hydroxide converts vanillin into vanillic acid (96–97%).

Piperonal (*heliotropin*, *3,4-methylenedioxybenzaldehyde*) is the methylene ether of protocatechu-aldehyde, and is obtained when piperic acid (from the alkaloid piperine) is oxidised. It may be prepared synthetically by treating protocatechualdehyde with methylene di-iodide and sodium hydroxide:

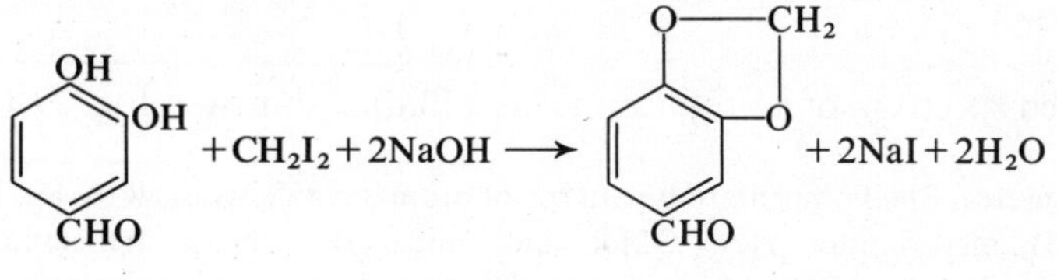

It is manufactured by the oxidation of isosafrole (from safrole) by ozonolysis or with acid dichromate:

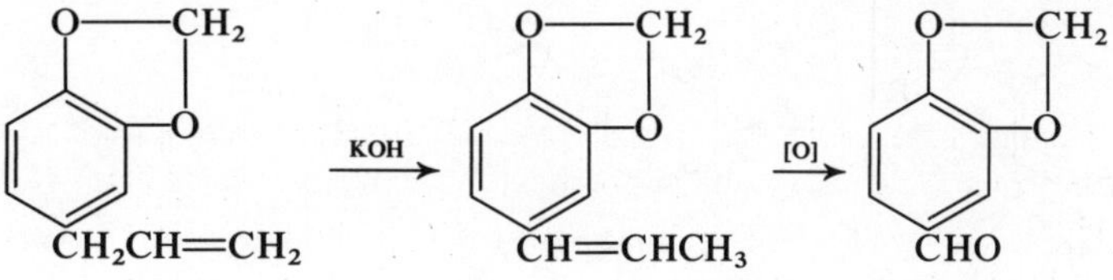

Piperonal, m.p. 37°C, may be oxidised to piperonylic acid and reduced to piperonyl alcohol. When treated with phosphorus pentachloride and then with water, piperonal forms protocatechualdehyde (see above). This aldehyde is also obtained, together with formaldehyde or methanol, when piperonal is heated with dilute hydrochloric acid at 200°C under pressure.

Aromatic ketones

Aromatic ketones may be either alkaryl ketones or diaryl ketones.

The infra-red absorption region of the carbonyl group (C=O stretch) in alkaryl ketones is $1\,700-1\,680\,\text{cm}^{-1}$ (s), and in diaryl ketones is $1\,670-1\,660\,\text{cm}^{-1}$ (s). In both cases, the frequencies are lower than for acyclic and cyclic ketones. At the same time, the benzene nucleus and the nature of the orientation may be identified. There may, however, be complications due to intra-molecular hydrogen bonding in certain o-derivatives.

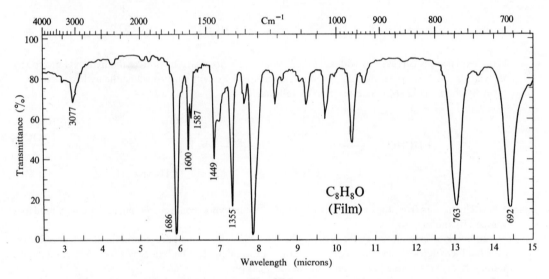

Fig. 27.1

$1\,686\,\text{cm}^{-1}$: C=O (str.) in alkaryl ketones. $1\,600$, $1\,587$, and $1\,449\,\text{cm}^{-1}$: C=C (str.) in aromatic nuclei. $3\,077\,\text{cm}^{-1}$: =C—H (str.) in aromatics. $1\,355\,\text{cm}^{-1}$: C—H (def.) in Me in MeCO. 763 and $692\,\text{cm}^{-1}$: Monosubstituted benzenes.

The infra-red spectrum of acetophenone (as a film) is shown in Fig. 27.1.

Mass spectra. The fragmentation patterns of aromatic aldehydes (R = H), ketones (R = R), acids (R = OH), methyl esters (R = OMe), and amides (R = NH_2) are similar in that all undergo β-cleavage with loss of R· and formation of the benzoyl cation. All give strong peaks for the molecular ion.

Esters in which the alkyl group is ethyl or higher also eliminate alkenes (*cf.* ethers, p. 701):

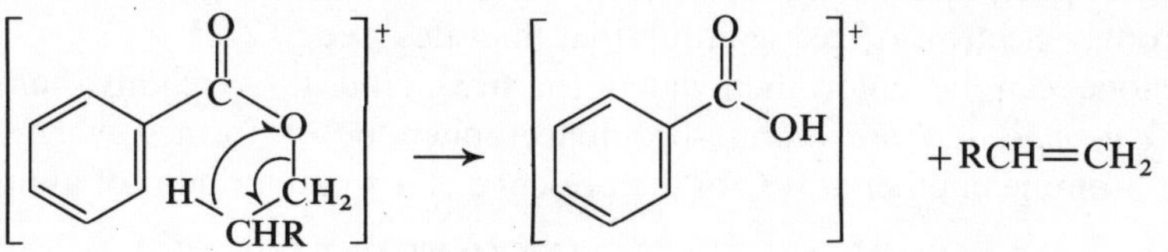

o-Substituted acids and esters (A = CH_2, O, NH), in addition to undergoing fragmentations described above, fragment by the McLafferty rearrangement:

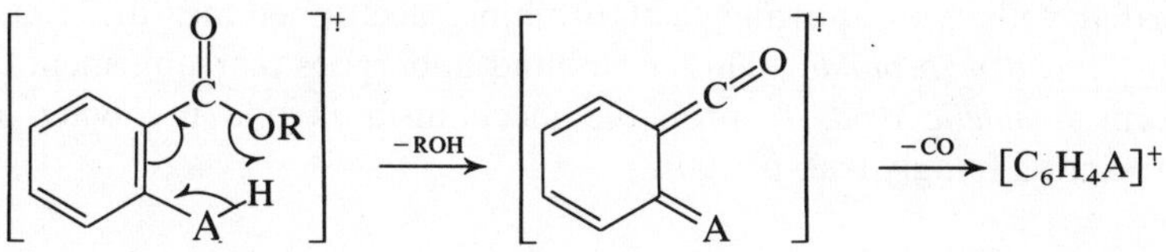

Acetophenone (*methyl phenyl ketone, acetylbenzene*) may be prepared by distilling a mixture of calcium benzoate and calcium acetate, or better, by heating a mixture of the corresponding iron salts (Granito *et al.*, 1963).

$$(C_6H_5CO_2)_2Ca + (CH_3CO_2)_2Ca \longrightarrow 2C_6H_5COCH_3 + 2CaCO_3$$

A convenient laboratory method is by means of the Friedel–Crafts reaction:

$$C_6H_6 + CH_3COCl \xrightarrow{\text{AlCl}_3} C_6H_5COCH_3 + HCl \quad (50\%)$$

Acetophenone is manufactured by the catalytic oxidation of ethylbenzene with air at 126°C under pressure in the presence of manganous acetate as catalyst (*cf.* p. 735).

Alkaryl ketones may be prepared from diazonium salts and aldoximes other than formaldoxime (see p. 676).

Acetophenone, m.p. 20°C, is used as an hypnotic (under the name of *hypnone*) and in perfumery. When reduced with sodium and ethanol, acetophenone gives *phenylmethylmethanol*, $C_6H_5CHOHCH_3$; reduction by Clemmensen's method gives ethylbenzene. Ethylbenzene is also produced when the reducing agent is lithium aluminium hydride containing aluminium chloride. Oxidation with cold potassium permanganate gives *phenylglyoxylic acid* (*benzoylformic acid*) which, on further oxidation, is converted into benzoic acid:

$$C_6H_5COCH_3 \xrightarrow{\text{[O]}} C_6H_5COCO_2H \xrightarrow{\text{[O]}} C_6H_5CO_2H$$

Phenylglyoxylic acid may be prepared as follows:

$$PhCOCl \xrightarrow{\text{CuCN}} PhCOCN \xrightarrow{\text{HCl}} PhCOCO_2H$$
$$(60–65\%) \qquad (73–77\%)$$

Acetophenone is oxidised by selenium dioxide to *phenylglyoxal*:

$$C_6H_5COCH_3 + SeO_2 \longrightarrow C_6H_5COCHO + Se + H_2O \quad (69–72\%)$$

Acetophenone undergoes the Baeyer–Villiger oxidation to form predominantly, with perbenzoic acid, phenyl acetate (63%). This is in keeping with the general observation that it is the more strongly electron-releasing group that migrates (see p. 223).

Acetophenone can be chloromethylated (p. 613), and it is readily halogenated in the ω-position; e.g., *phenacyl bromide* (ω-bromoacetophenone) is formed when acetophenone is treated with bromine in ether at 0°C in the presence of a small amount of aluminium chloride:

$$C_6H_5COCH_3 + Br_2 \longrightarrow C_6H_5COCH_2Br + HBr \quad (64–66\%)$$

Phenacyl bromide (m.p. 51°C) is used to identify acids, with which it forms well-defined crystalline esters. It is oxidised to phenylglyoxal by dissolving it in a large excess of dimethyl sulphoxide (Kornblum *et al.*, 1957).

When treated with two equivalents of bromine, acetophenone forms ω,ω-dibromoaceto-phenone (*phenacylidene bromide*). This compound undergoes rearrangement on treatment with alkali to form *mandelic acid*. A probable mechanism is one involving an *intramolecular* Cannizzaro reaction (*cf.* glyoxal, p. 300):

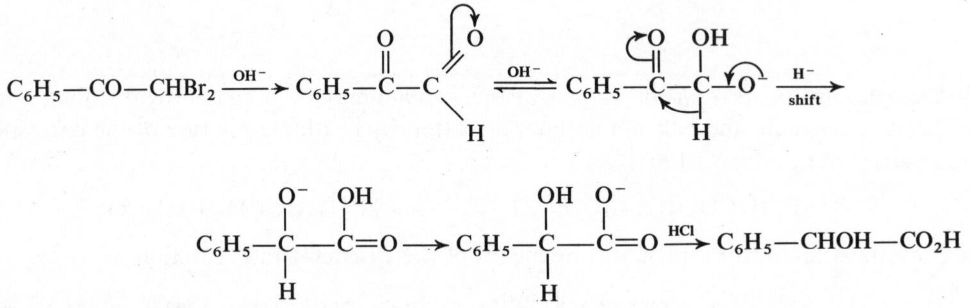

This rearrangement does not take place if both *o*-positions are occupied, *e.g.*, in the ω,ω-dibromo-derivative of acetomesitylene. This is probably due to a steric effect.

Phenacyl chloride, m.p. 59°C, may be conveniently prepared by the Friedel–Crafts condensation between chloroacetyl chloride and benzene:

$$C_6H_6 + ClCH_2COCl \xrightarrow{AlCl_3} C_6H_5COCH_2Cl + HCl$$

Acetophenone forms an oxime, phenylhydrazone and cyanohydrin, but does not form a bisulphite compound (*cf.* p. 211). It reacts with ammonia to form acetophenoneammonia, PhCMe=NCMePhN=CMePh (*cf.* benzaldehyde), and with Grignard reagents in the usual way. When, however, both *o*-positions in aceto-phenone are occupied, the compound exhibits steric hindrance; *e.g.*, acetomesitylene does not form an oxime, and does not react with Grignard reagents in the usual way.

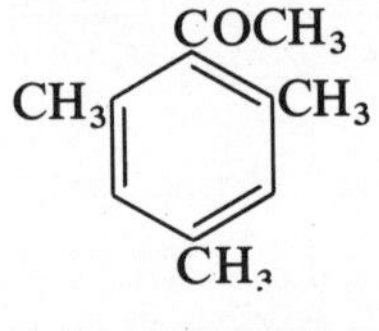

The acetyl group is *m*-orienting, and so when acetophenone undergoes nuclear substitution, the main product is the *m*-derivative.

Nuclear halogenation is difficult, since side-chain substitution occurs so readily. However, Pearson *et al.* (1958) have prepared *m*-bromoacetophenone (75%) by bromination in the presence of excess of aluminium chloride (which forms a complex with the carbonyl group; *cf.* Friedel–Crafts reaction). These conditions are referred to as 'swamping' catalytic conditions.

o- and *p*-Substituted acetophenones are prepared indirectly, *e.g.*,

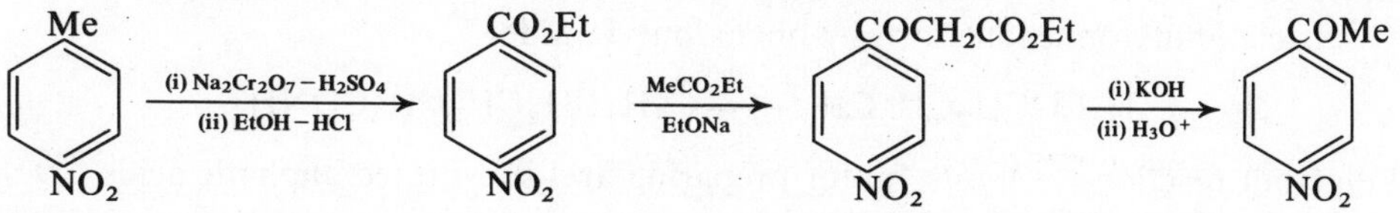

In the presence of aluminium t-butoxide, acetophenone undergoes condensation to form **dypnone** (b.p. 340–345°C):

$$\underset{\underset{CH_3}{|}}{C_6H_5C}{=}O + CH_3COC_6H_5 \longrightarrow \underset{\underset{CH_3}{|}}{C_6H_5C}{=}CHCOC_6H_5 + H_2O \quad (77\text{--}82\%)$$

In the presence of hydrochloric acid, 1,3,5-triphenylbenzene (m.p. 172°C) is formed:

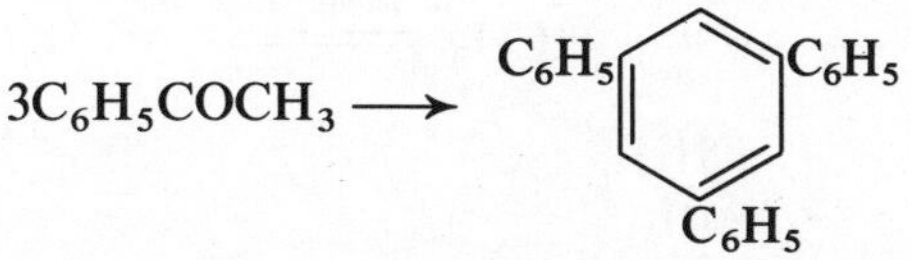

$$3C_6H_5COCH_3 \longrightarrow$$

Acetophenone condenses with acetic anhydride in the presence of boron trifluoride to form *benzoylacetone*, m.p. 61°C (Meerwein *et al.*, 1934):

$$C_6H_5COCH_3 + (CH_3CO)_2O \xrightarrow{BF_3} C_6H_5COCH_2COCH_2 + CH_3CO_2H \quad (83\%)$$

This is also formed by the condensation of acetophenone with ethyl acetate in the presence of sodium ethoxide:

$$C_6H_5COCH_3 + CH_3CO_2C_2H_5 \xrightarrow{C_2H_5ONa} C_6H_5COCH_2COCH_3 \quad (66\%)$$

Acetophenone condenses with ethyl benzoate in the presence of sodium ethoxide to form *dibenzoylmethane*, m.p. 77°C.

$$C_6H_5CO_2C_2H_5 + CH_3COC_6H_5 \xrightarrow{C_2H_5ONa} C_6H_5COCH_2COC_6H_5 \quad (62\text{--}71\%)$$

It also undergoes the Mannich reaction (see p. 375):

$$C_6H_5COCH_3 + HCHO + (CH_3)_2NH \cdot HCl \longrightarrow C_6H_5COCH_2CH_2N(CH_3)_2 \cdot HCl \quad (60\%)$$

Aromatic aldehydes and ketones generally behave as bases in concentrated sulphuric acid; cryoscopic experiments show that *i* (van't Hoff factor) is 2, and this value agrees with the following equation (R = H or R):

$$ArCOR + H_2SO_4 \rightleftharpoons [Ar\overset{+}{C}(=OH)R][HSO_4{}^-]$$

The carbonyl compound may be recovered by dilution with water.

Willgerodt reaction (1887). This is the name given to those reactions in which a carbonyl compound is converted into an amide with the same number of carbon atoms. The reaction was originally carried out by heating an aryl alkyl ketone with an aqueous solution of yellow ammonium polysulphide; *e.g.*, acetophenone forms the amide of phenylacetic acid, together with a small amount of the ammonium salt:

$$C_6H_5COCH_3 \xrightarrow{(NH_4)_2S_x} C_6H_5CH_2CONH_2 + C_6H_5CH_2CO_2NH_4$$

The amido-group is always formed at the end of the chain whatever the size of the R group in C_6H_5COR; *e.g.*, butyrophenone forms γ-phenylbutyramide:

$$C_6H_5COCH_2CH_2CH_3 \longrightarrow C_6H_5CH_2CH_2CH_2CONH_2$$

The Willgerodt reaction is very useful for preparing aryl-substituted aliphatic acids.

A modified technique is to use equimolecular amounts of sulphur and a secondary amine (Kindler, 1923, 1927), particularly morpholine, 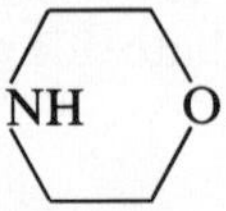. The final product is a thioamide and this, on acid or alkaline hydrolysis, gives an acid; or alternatively, on electrolytic reduction, the thioamide gives an amine.

According to Mayer *et al.* (1964), the reaction proceeds via enamine intermediates (I and II; see p. 383), *e.g.*,

$$ArCOCH_2CH_3 + R_2NH \longrightarrow \underset{(I)}{ArC(NR_2){=}CHCH_3} \xrightleftharpoons{\text{isomn. by S}} \underset{(II)}{ArCH_2C(NR_2){=}CH_2} \xrightleftharpoons{S}$$

$$ArCH_2C(NR_2){=}CHSH \xrightarrow{\text{isomn.}} ArCH_2CH_2C({=}S)NR_2$$

$$ArCH_2CH_2C({=}S)NR_2 \xrightarrow{H_3O^+} ArCH_2CH_2CO_2H \qquad ArCH_2CH_2C({=}S)NR_2 \xrightarrow{e;\ H^+} ArCH_2CH_2NR_2$$

It should be noted that isomerisation occurs along the side-chain until the amino-group reaches the *end* carbon atom.

Phenolic ketones may be prepared by the **Houben–Hoesch synthesis** (1927). This is the condensation of cyanides with polyhydric phenols, particularly *m*-compounds, in the presence of zinc chloride and hydrogen chloride; *e.g.*, phloroglucinol condenses with methyl cyanide to form *phloroacetophenone* (m.p. 219°C)·

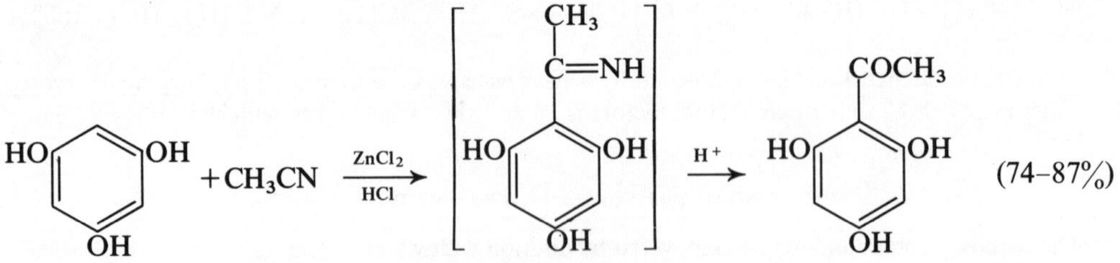

This method is really an extension of Gattermann's phenolic aldehyde synthesis. It is not applicable to phenol itself.

It is probable that the imido chloride is an intermediate:

$$MeC{\equiv}N + HCl \rightleftharpoons MeCCl{=}NH$$

Phenolic ketones may also be prepared by the Fries rearrangement (p. 704), and by heating polyhydric phenols with aliphatic acids in the presence of fused zinc chloride; *e.g.*, pyrogallol and acetic acid form *gallacetophenone* (m.p. 173°C):

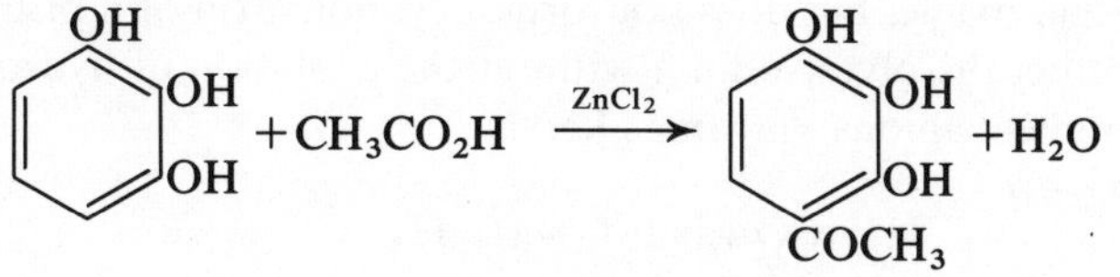

Benzophenone (*diphenyl ketone*) may be prepared in a similar manner to acetophenone, *e.g.*,

(i) By heating calcium benzoate:

$$(C_6H_5CO_2)_2Ca \longrightarrow C_6H_5COC_6H_5 + CaCO_3 \quad (30\%)$$

(ii) By the Friedel–Crafts condensation between benzoyl chloride and benzene:

$$C_6H_6 + C_6H_5COCl \xrightarrow{AlCl_3} C_6H_5COC_6H_5 + HCl \quad (80\%)$$

Carbonyl chloride may be used instead of benzoyl chloride:

$$2C_6H_6 + COCl_2 \xrightarrow{AlCl_3} C_6H_5COC_6H_5 + 2HCl$$

The Friedel–Crafts reaction may also be carried out with carbon tetrachloride, followed by steam distillation of the dichloro-compound produced (this method is used commercially):

$$2C_6H_6 + CCl_4 \xrightarrow{AlCl_3} (C_6H_5)_2CCl_2 \xrightarrow{H_2O} (C_6H_5)_2CO \quad (80–89\%)$$

Benzophenone exists in two solid forms, a stable, m.p. 49°C, and an unstable form, m.p. 26°C. It cannot be chloromethylated (*cf.* acetophenone). It is reduced by zinc and ethanolic potassium hydroxide to **benzhydrol** (*diphenylcarbinol*, m.p. 68°C) and by zinc and acetic acid to **benzopinacol** (m.p. 188°C):

$$(C_6H_5)_2C(OH)C(OH)(C_6H_5)_2 \xleftarrow[CH_3CO_2H]{Zn} (C_6H_5)_2CO \xrightarrow{Zn/KOH} (C_6H_5)_2CHOH$$
$$(90\%) \hspace{8cm} (90\%)$$

When benzophenone, dissolved in isopropanol to which a drop of acetic acid has been added, is exposed to bright sunlight, benzopinacol is formed (see also p. 904).

$$2(C_6H_5)_2CO + (CH_3)_2CHOH \xrightarrow{h\nu} (C_6H_5)_2C(OH)C(OH)(C_6H_5)_2 + (CH_3)_2CO \quad (93–94\%)$$

Benzopinacol is also obtained by the reduction of benzophenone with a mixture of magnesium and magnesium iodide.

$$2(C_6H_5)_2CO \xrightarrow{MgI_2} (C_6H_5)_2\overset{\displaystyle |}{\underset{\displaystyle {}^-O}{C}}\!-\!\overset{\displaystyle |}{\underset{\displaystyle O^-}{C}}(C_6H_5)_2 \xrightarrow{H^+} (C_6H_5)_2C(OH)C(OH)(C_6H_5)_2$$
$$Mg^{2+}$$

When heated in acetic acid solution in the presence of iodine as catalyst, benzopinacol undergoes the pinacol–pinacolone rearrangement (p. 229) to form *benzopinacolone* (m.p. 179°C):

$$(C_6H_5)_2C(OH)C(OH)(C_6H_5)_2 \longrightarrow (C_6H_5)_3CCOC_6H_5 + H_2O \quad (98–99\%)$$

When distilled with zinc dust, or reduced with a mixture of lithium aluminium hydride and aluminium chloride, benzophenone forms diphenylmethane, and when fused with potassium hydroxide, it forms benzene and potassium benzoate:

$$(C_6H_5)_2CO + KOH \longrightarrow C_6H_6 + C_6H_5CO_2K$$

Benzophenone forms an oxime, but does not form a cyanohydrin or a bisulphite compound.

Benzophenone in ether dissolves sodium without the evolution of hydrogen, forming a blue solution containing radical anions known as **ketyls**.

$$(C_6H_5)_2CO + Na \longrightarrow (C_6H_5)_2\dot{C}-\bar{O}Na^+$$

Aromatic ketyls are stabilised by resonance:

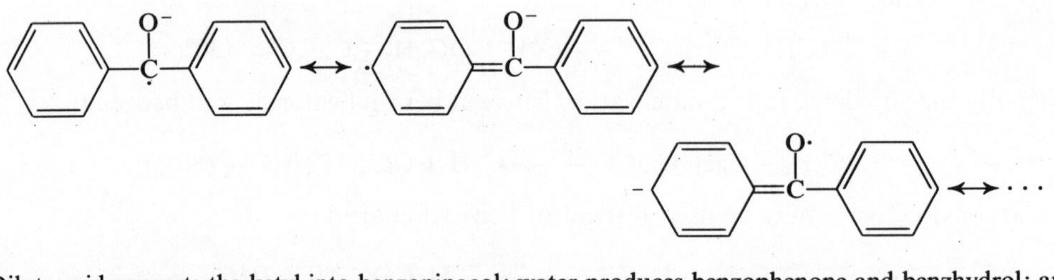

Dilute acid converts the ketyl into benzopinacol; water produces benzophenone and benzhydrol; and iodine regenerates benzophenone.

If an excess of sodium is added to benzophenone in ether, the violet dianion, $(C_6H_5)_2C^--O^-$, is produced by addition of a second electron.

A very important derivative of benzophenone is **Michler's ketone**, m.p. 172°C, which may be prepared by treating dimethylaniline with carbonyl chloride in the presence of zinc chloride.

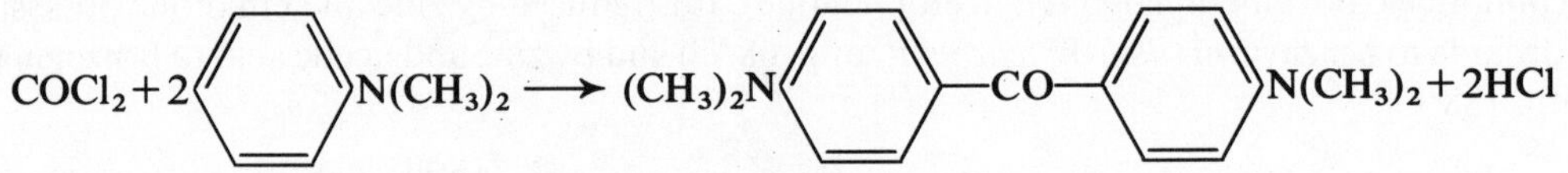

Michler's ketone is used in the preparation of certain triphenylmethane dyes (p. 887).

Stereochemistry of aldoximes and ketoximes

Many aromatic aldoximes and ketoximes exist in two isomeric forms, and this was explained by Hantzsch and Werner (1890) as being due to geometrical isomerism. According to these authors, nitrogen is tervalent (in oximes) and is situated at one corner of a tetrahedron with its three valencies directed towards the other three corners; consequently the three valencies are not coplanar. These authors also assumed that there is no free rotation about the C=N bond, and therefore proposed configurations (I) and (II) for the two isomers:

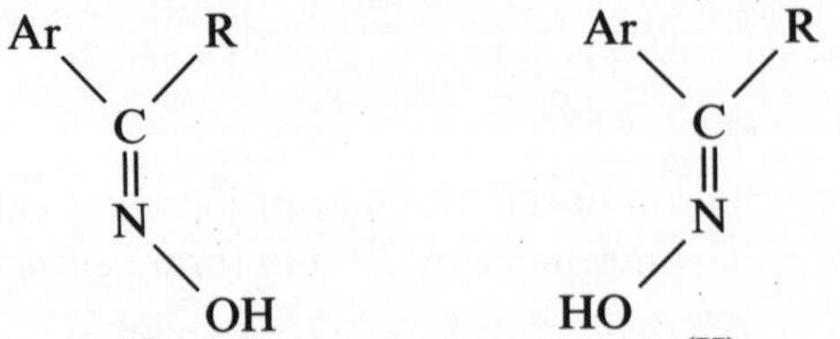

Many facts are in favour of geometrical isomerism, *e.g.*, (i) if Ar = R, then isomerism disappears; (ii) the absorption spectra show that both have identical structures.

Determination of configuration

Aldoximes. The two isomeric aldoximes may be distinguished by the behaviour of their acetyl derivatives towards aqueous carbonate; one gives the oxime back again, and the other forms the cyanide ArCN.

Brady *et al.* (1925) showed that the cyanide is formed by *anti*-elimination. These authors found that only one of the two isomers of 2-chloro-5-nitro-benzaldoxime readily gave ring closure on treatment with sodium hydroxide; this isomer is therefore the *anti-(trans-)*isomer. Furthermore, it was this isomer that gave the cyanide, thus showing that *anti*-elimination must have occurred.

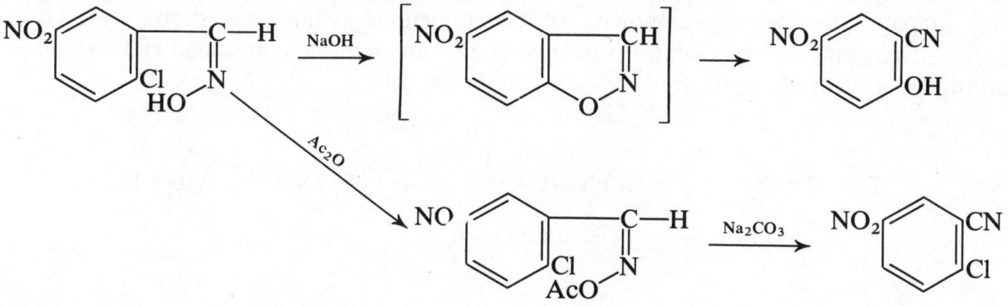

Actually, the ring compound produced, the 5-nitrobenzisoxazole, is unstable, and rearranges to nitrosalicylonitrile.

Thus the general reaction may be formulated as follows, the *anti*-oxime giving the cyanide and the *syn-(cis-)*oxime regenerating the oxime:

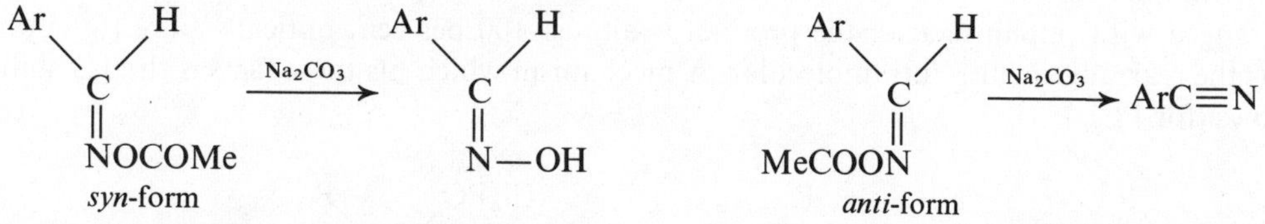

Ketoximes: Beckmann rearrangement (1886). When treated with reagents such as phosphorus pentachloride, sulphuric acid, polyphosphoric acid, etc., aromatic ketoximes undergo the Beckmann rearrangement to form an acid amide:

$$Ar_2C{=}NOH \longrightarrow ArCONHAr$$

Meisenheimer (1921, 1925) showed that the rearrangement occurs by *anti*-rearrangement, *e.g.*, he (1925) showed that the α-oxime of 2-bromo-5-nitroacetophenone is unaffected by sodium hydroxide, whereas the β-isomer undergoes ring closure to form 3-methyl-5-nitrobenzisoxazole; thus the α-oxime is the *syn*-methyl isomer (I), and the β-oxime the *anti*-methyl isomer (II). When treated with sulphuric acid or phosphorus pentachloride, the α-oxime underwent the Beckmann rearrangement to give the *N*-substituted acetamide; thus the exchange occurs in the *anti*-positions.

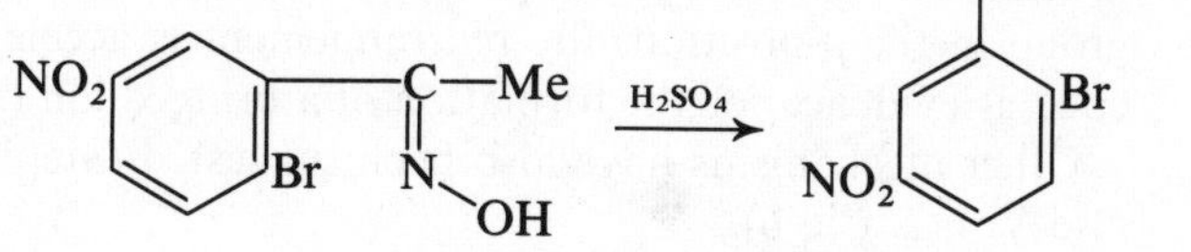

(II)

$$NO_2\text{-Ar}\cdots C(Me)=N\text{-OH (Br)} \xrightarrow{NaOH} NO_2\text{-Ar}\cdots C(Me)=N\text{-O}$$

Hence, by identifying the amide and using the fact that the rearrangement occurs in the *anti*-position, the configuration of the ketoxime can be determined.

The mechanism of the Beckmann rearrangement is still the subject of much discussion. Since the oxime itself does not rearrange, it is reasonable to suppose that some intermediate is formed between oxime and reagent, and it is this intermediate which rearranges. Kuhara *et al.* (1914, 1916) prepared the benzenesulphonate of benzophenone oxime and showed this readily underwent rearrangement in neutral solvents to give an isomeric compound (III) which gave benzanilide on hydrolysis; thus:

$$\underset{\underset{OSO_2Ph}{|}}{\overset{\overset{Ph}{|}}{Ph-C}}=N \longrightarrow \underset{\underset{OSO_2Ph}{|}}{\overset{\overset{Ph}{|}}{Ph-C}}=N \xrightarrow{H^+} PhCONHPh + PhSO_3H$$

(III)

The structure of (III) was assigned to it on the basis of the similarity of its absorption spectrum with that of synthetic compounds of similar structure. Chapman (1934–   ) showed the rate of rearrangement of benzophenone oxime picryl ester is more rapid in polar solvents than in non-polar. This is strong evidence that the rate-determining step is the ionisation of the intermediate. Furthermore, Kenyon *et al.* (1946) found that when (+)-α-phenylethyl methyl ketoxime is rearranged with sulphuric acid, the product is almost 100 per cent optically pure (retention). Thus the rearrangement is intramolecular. A mechanism which fits these facts is the 1,2-shift as follows (for PCl₅):

$$\underset{\underset{OPCl_4}{|}}{\overset{\overset{R^1\diagdown\diagup R^2}{C}}{N}} \longrightarrow \underset{\underset{R^1}{N}}{\overset{\overset{R^2}{^+C}}{||}} + OPCl_4^- \longrightarrow \underset{\underset{R^1}{N}}{\overset{\overset{Cl_4PO\diagdown\diagup R^2}{C}}{||}} \xrightarrow{H_2O}$$

$$\underset{\underset{R^1}{N}}{\overset{\overset{HO\diagdown\cdot R^2}{C}}{||}} \longrightarrow \underset{NHR^1}{\overset{\overset{O\diagdown\diagup R^2}{C}}{||}}$$

(When sulphuric acid is used, replace OPCl₄ in the above equation by OH_2^+.)

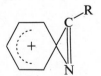

When the migrating group is aryl and contains an electron-releasing group in the *p*-position, the rearrangement is accelerated. This may be cited as evidence for the formation of a bridged ion (*cf.* p. 147).

Other mechanisms have also been suggested (Stephen *et al.*, 1956; see also Vol. 2, Ch. 6).

PROBLEMS

1. Distinguish between PhCOEt and p-MeC$_6$H$_4$COMe by:

(i) a chemical method; (ii) by a physical method.

2. Complete the following equations:

(i) 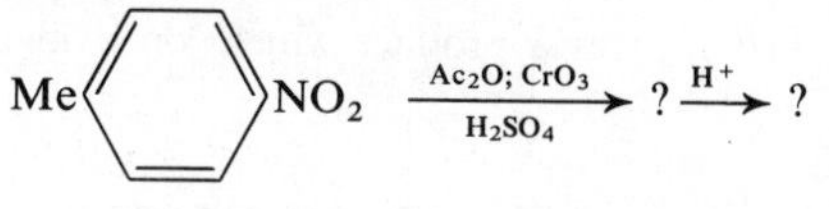 Me—⟨benzene ring⟩—NO$_2$ $\xrightarrow[\text{H}_2\text{SO}_4]{\text{Ac}_2\text{O; CrO}_3}$? $\xrightarrow{\text{H}^+}$?   (ii) Ph$_2$CO $\xrightarrow[\text{NaOH}]{\text{Zn}}$?

(iii) p-MeC$_6$H$_4$CHO $\xrightarrow[\text{KOH}]{\text{HCHO}}$?   (iv) 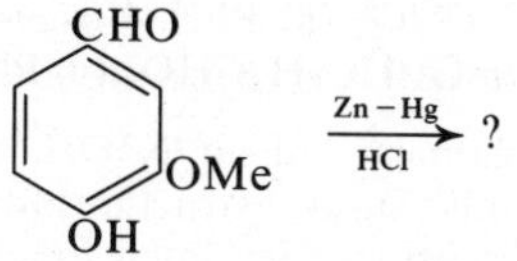 ⟨benzene ring with CHO, OMe, OH⟩ $\xrightarrow[\text{HCl}]{\text{Zn}-\text{Hg}}$?

(v) 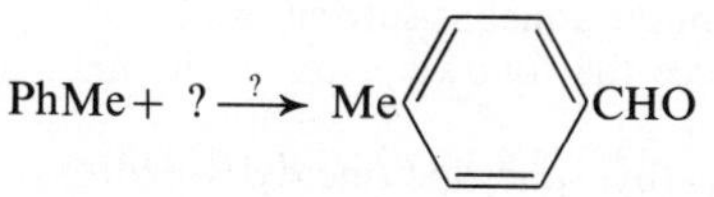 PhMe + ? $\xrightarrow{?}$ Me—⟨benzene ring⟩—CHO   (vi) PhCHO + NaCN + ? $\xrightarrow[\text{H}_2\text{O}]{\text{MeOH}}$ PhCH(CN)CH(CN)Ph

(vii) 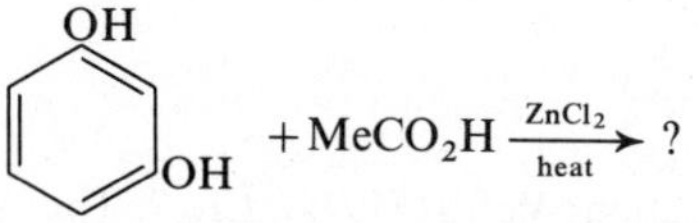 ⟨benzene ring with OH, OH⟩ + MeCO$_2$H $\xrightarrow[\text{heat}]{\text{ZnCl}_2}$?   (viii) ⟨benzene ring with OMe, OH, OMe⟩ + (CH$_2$)$_6$N$_4$ $\xrightarrow{\text{H}_3\text{BO}_3}$?

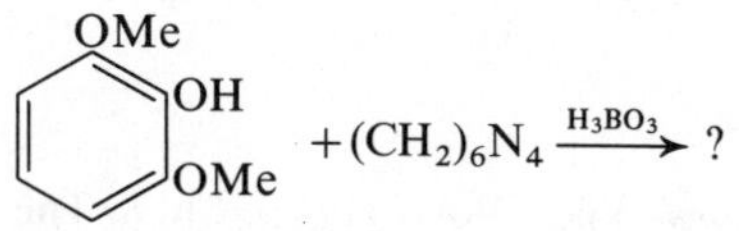

(ix) 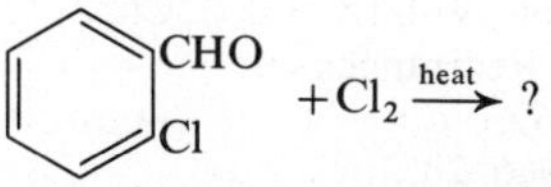 ⟨benzene ring with CHO, Cl⟩ + Cl$_2$ $\xrightarrow{\text{heat}}$?   (x) PhCH=CHCOPh + KCN $\xrightarrow{\text{AcOH}}$?

(xi) PhCH$_2$OH + ? $\longrightarrow$ PhCH$_2$OCOCl   (xii) PhCHBrCHBrCHO $\xrightarrow[\text{heat}]{\text{K}_2\text{CO}_3}$?

(xiii) PhCH(CN)COMe $\xrightarrow[\text{heat}]{\text{aq. H}_2\text{SO}_4}$?   (xiv) PhCOMe $\xrightarrow{?}$ PhCOCHO

(xv) 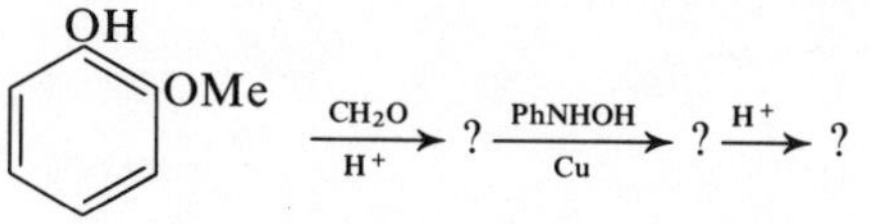 ⟨benzene ring with OH, OMe⟩ $\xrightarrow[\text{H}^+]{\text{CH}_2\text{O}}$? $\xrightarrow[\text{Cu}]{\text{PhNHOH}}$? $\xrightarrow{\text{H}^+}$?   (xvi) 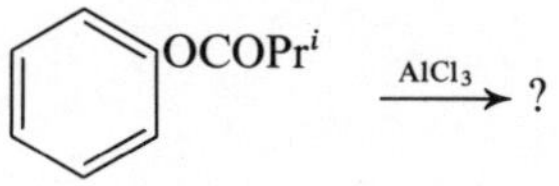 ⟨benzene ring with OCOPr^i⟩ $\xrightarrow{\text{AlCl}_3}$?

(xvii) PhCO$_2$H $\xrightarrow{\text{SOCl}_2}$? $\xrightarrow{\text{Me}_2\text{NH}}$? $\xrightarrow{?}$ PhCHO

3. Which of the carbonyl groups in p-MeOC$_6$H$_4$COMe and p-NO$_2$C$_6$H$_4$COMe protonates more readily in acid solution and why?

4. Starting with PhH, EtCO$_2$H and PhNH$_2$, synthesise PhCH$_2$CH$_2$CONHPh.

5. Convert PhCHO into PhCH=CHCOPh by *two* methods.

6. Convert Br(CH$_2$)$_3$CHO into PhCHOH(CH$_2$)$_3$CHO.

7. What are the products formed when the *syn*-phenyl and *anti*-phenyl monoximes of benzil undergo the acid-catalysed Beckmann rearrangement?

8. Suggest a mechanism for the rearrangement: Ph$_2$CHCHO $\xrightarrow{\text{H}^+}$ PhCH$_2$COPh.

9. Discuss the two reactions:

$$\text{PhCOMe} \xrightarrow[\text{NaOH}]{\text{NaOCl}} \text{PhCO}_2^- + \text{CHCl}_3$$

$$2,6\text{-Me}_2\text{C}_6\text{H}_3\text{COMe} \xrightarrow[\text{NaOH}]{\text{NaOCl}} 2,6\text{-Me}_2\text{C}_6\text{H}_3\text{COCCl}_3$$

10. Suggest what **A** and **B** could be in the reactions:

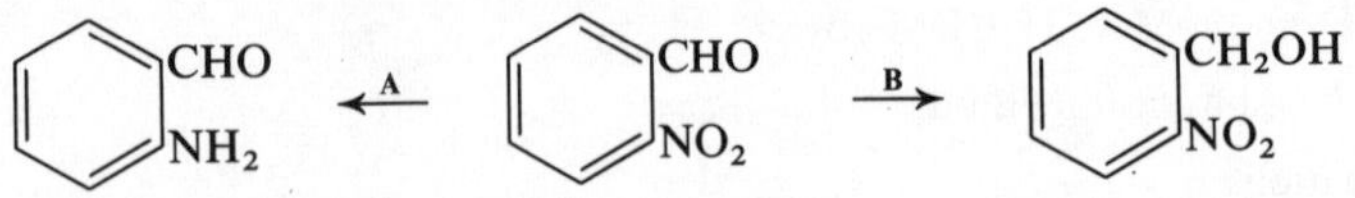

11. The deamination of $Ph_2C(OH)CH_2NH_2$ ($NaNO_2$—HCl) gives a product which, on oxidation, gives benzoic acid only. Formulate the steps involved.

12. Convert:

(i) *p*-xylene $\longrightarrow$ $2,5\text{-}Me_2C_6H_3CHO$; (ii) PhH $\longrightarrow$ *m*-nitroacetophenone; (iii) $PhNO_2 \longrightarrow$ *m*-bromo-benzaldehyde; (iv) PhH $\longrightarrow$ $p\text{-}OCHC_6H_4CHO$; (v) PhMe $\longrightarrow$ $p\text{-}MeC_6H_4CH{=}CHNO_2$.

13. Compound **A**, $C_8H_6O_2$, on treatment with aq. NaOH followed by acidification, gives **B**, $C_8H_8O_3$, which, on oxidation, gives benzoic acid only. Suggest structures for **A** and **B**, and formulate the reactions.

14. The mechanism of the Cannizzaro reaction involves addition of hydrogen to a benzaldehyde molecule. Devise an experiment to show whether the hydrogen comes from the aqueous solvent or not.

15. Suggest paths whereby *o*-methylbenzyl alcohol can produce the following ions in its mass spectrum: 122 (30%), 120 (1%), 119 (4%), 104 (100%), 91 (36%).

16. Account for the following ions in the mass spectrum of methyl salicylate (methyl *o*-hydroxybenzoate): 152 (41%), 121 (32%), 120 (100%), 93 (13%), 92 (58%), 65 (26%), 64 (17%).

REFERENCES

Organic Reactions, Wiley. Vol. I (1942), Ch. 8. The Perkin Reaction; Vol. II (1944), Ch. 2. The Willgerodt Reaction; Ch. 3. The Cannizzaro Reaction; Vol. V (1949), Ch. 6. The Gattermann–Koch Reaction; Ch. 9. The Hoesch Reaction; Vol. VIII (1954), Ch. 4. The Sommelet Reaction; Vol. IX (1957), Ch. 2. The Gattermann Synthesis of Aldehydes; Vol. XI (1960), Ch. 1. The Beckmann Rearrangement.

WYNBERG, 'The Reimer–Tiemann Reaction', *Chem. Rev.*, 1960, **60**, 169.

CARNDUFF, 'Recent Advances in Aldehyde Synthesis', *Quart. Rev.*, 1966, **20**, 169.

FINAR, *Organic Chemistry*, Vol. 2 (1968, 4th edn.), Ch. 6 (section 2d), 'Oximes'. Longmans.

28 Aromatic acids

Aromatic acids are compounds containing one or more carboxyl groups which are directly attached to the nucleus. Those acids in which the carboxyl group occurs in the side-chain may be regarded as aryl-substituted aliphatic acids, but they are also classified as aromatic acids, since they exhibit aromatic properties due to the presence of a benzene ring.

Monocarboxylic acids with the carboxyl group attached to the ring

General methods of preparation

1. By the oxidation of the corresponding alcohol or aldehyde; *e.g.*, benzaldehyde gives benzoic acid:

$$C_6H_5CHO \xrightarrow{\text{KMnO}_4} C_6H_5CO_2H \quad (v.g.)$$

2. By the hydrolysis of the corresponding cyanide, *e.g.*, *o*-tolyl cyanide gives *o*-toluic acid:

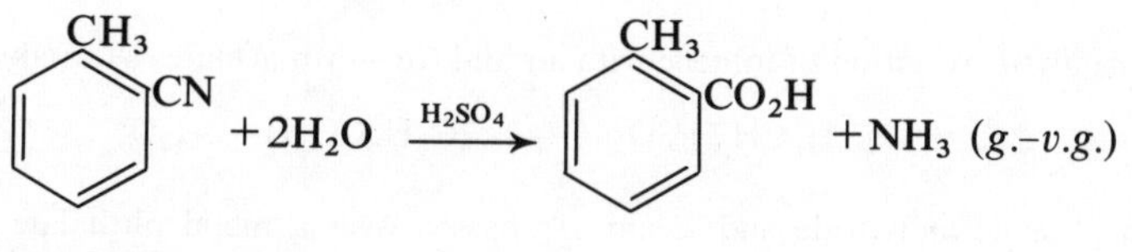

3. By means of a Grignard reagent, *e.g.*,

$$C_6H_5MgBr + CO_2 \longrightarrow C_6H_5CO_2MgBr \xrightarrow{\text{HCl}} C_6H_5CO_2H \quad (90\%)$$

4. By means of the Friedel–Crafts reaction; *e.g.*, treatment of benzene with carbonyl chloride in the presence of anhydrous aluminium chloride gives benzoyl chloride which, on hydrolysis, forms benzoic acid:

$$C_6H_6 + COCl_2 \xrightarrow{\text{AlCl}_3} C_6H_5COCl \longrightarrow C_6H_5CO_2H \quad (55\text{–}58\%)$$

Excess of carbonyl chloride must be used in this method; otherwise benzophenone will be the main product (p. 753).

5. By the oxidation of benzene homologues with dilute nitric acid, dichromate and sulphuric acid, alkaline permangante, or chromium trioxide in glacial acetic acid. In some cases it may be more convenient to chlorinate the hydrocarbon, and then oxidise the chloro-derivative, *e.g.*,

$$C_6H_5CH_3 \xrightarrow{Cl_2} C_6H_5CH_2Cl \xrightarrow[KMnO_4]{Na_2CO_3} [C_6H_5CH_2OH] \xrightarrow{[O]} C_6H_5CO_2H \quad (80\%)$$

In this way the oxidation of the side-chain is much easier, since the intermediate alcohol is much more readily oxidised than the hydrocarbon itself.

General properties

In general, the aromatic acids are slightly stronger acids than the aliphatic, less soluble in water, and less volatile. The reactions of the carboxyl group are very similar to those in the aliphatic acids, but apparently they differ in that they do not undergo coupling by the Kolbe electrolytic synthesis (p. 77). However, according to Fichter (1939), the electrolysis of a solution of benzoic acid in pyridine gave a number of products, among which was biphenyl.

The infra-red absorption region of the carbonyl group (C=O stretch) in aromatic acids is $1\,700–1\,680\ \text{cm}^{-1}$ (s), which is a lower frequency than for saturated aliphatic acids, and so can be distinguished from the latter. This region, however, lies within that of the α,β-unsaturated acids, and so the identification of a carboxyl group by infra-red studies will afford no certain distinction between aryl and the α,β-unsaturated acids (the molecular formula will help here). However, by identifying the presence of the benzene ring by its own characteristic frequencies, these two types of acids may then be distinguished. Also, the nature of the orientation may be identified.

Benzoic acid (*benzenecarboxylic acid*), m.p. 122°C, is present in certain resins, particularly gum-benzoin; it is also present in balsams. Benzoic acid is also found as *hippuric acid* (*benzoylglycine*) in the urine of horses.

Benzoic acid may be prepared in the laboratory by any of the general methods. Commercially, it is prepared as follows:

(i) By the hydrolysis of benzotrichloride with aqueous calcium hydroxide in the presence of iron powder as catalyst (this is an example of indirect oxidation):

$$C_6H_5CCl_3 + 2H_2O \xrightarrow[Fe]{Ca(OH)_2} C_6H_5CO_2H + 3HCl$$

(ii) By the catalytic oxidation of toluene with air and Co—Mn acetates as catalyst:

$$2C_6H_5CH_3 + 3O_2 \xrightarrow{170°C} 2C_6H_5CO_2H + 2H_2O$$

(iii) When phthalic anhydride and steam are passed over a metal phthalate catalyst, *e.g.*, zinc, chromium or nickel salt at 200–300°C, some of the resulting phthalic acid is decarboxylated to benzoic acid:

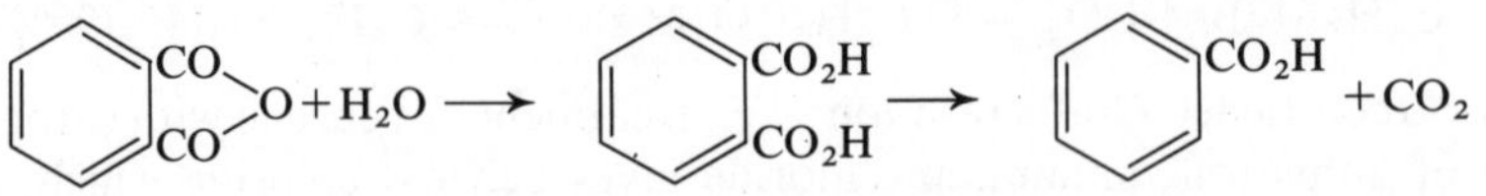

Benzoic acid readily forms esters when it is refluxed with an alcohol in the presence of a small amount of concentrated sulphuric acid or hydrogen chloride. Benzoic esters are pleasant-smelling liquids which are denser than water. Phenyl benzoate, which may be prepared by the action of benzoyl chloride on an alkaline solution of phenol, is a solid, m.p. 71°C. In general, it has been found that benzoic acid containing a substituent in the *o*-position does not esterify as easily as the *m*- or *p*-isomer. If both *o*-positions (with respect to the carboxyl group) are occupied, then esterification occurs with the greatest difficulty, if at all. Furthermore, once the ester is formed from these di-*ortho*-substituted benzoic acids, it is very difficult to hydrolyse them. The explanation for these abnormal reactions is not yet complete (see the *ortho*-effect, p. 771, for further details).

Benzoyl chloride (*benzenecarbonyl chloride*), b.p. 197°C, was the first acid chloride to be discovered. It may be readily prepared by distilling benzoic acid with phosphorus pentachloride or with thionyl chloride:

$$C_6H_5CO_2H + SOCl_2 \longrightarrow C_6H_5COCl + HCl + SO_2 \quad (75–80\%)$$

It is prepared commercially by chlorinating benzaldehyde in the cold (benzaldehyde contains no α-hydrogen atom):

$$C_6H_5CHO + Cl_2 \longrightarrow C_6H_5COCl + HCl \quad (ex.)$$

Another commercial method is to heat benzoic acid with benzotrichloride:

$$C_6H_5CO_2H + C_6H_5CCl_3 \longrightarrow 2C_6H_5COCl + HCl$$

Benzoyl chloride is only very slowly decomposed by water or by dilute sodium hydroxide (*cf.* acetyl chloride), and because of this, compounds containing an active hydrogen atom can be benzoylated in the presence of dilute aqueous sodium hydroxide. This method of benzoylation is known as the **Schotten–Baumann reaction**, *e.g.*,

$$C_6H_5COCl + C_6H_5NH_2 + NaOH \longrightarrow C_6H_5CONHC_6H_5 + NaCl + H_2O$$

p-Nitrobenzoyl and 3,5-dinitrobenzoyl derivatives of the alcohols are usually well-defined crystalline substances, and so are used to characterise alcohols.

Hippuric acid (*benzoylglycine*), m.p. 188°C, may be readily prepared by the action of benzoyl chloride on glycine:

$$C_6H_5COCl + NH_2CH_2CO_2H \longrightarrow C_6H_5CONHCH_2CO_2H + HCl$$

Its most important reaction is its condensation with aromatic aldehydes to form azlactones; *e.g.*, when heated with benzaldehyde, acetic anhydride and sodium acetate, hippuric acid forms benzoyl-α-aminocinnamic azlactone (3-*phenyl-4-benzylideneoxazol-5-one*):

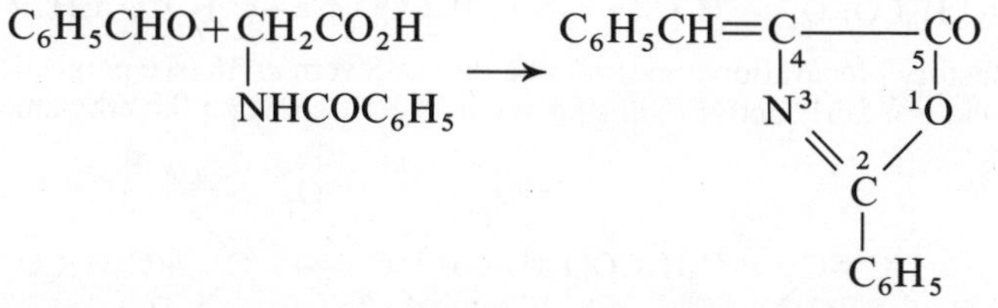

This reaction is usually referred to as the **Erlenmeyer azlactone synthesis** (1893). Azlactones are very important as intermediates in the preparation of amino- and keto-acids; *e.g.*, phenylalanine may be prepared from the above azlactone as follows:

$$C_6H_5CH{=}\underset{\underset{\underset{\displaystyle C_6H_5}{|}}{\underset{\displaystyle C}{\|}}{\overset{\displaystyle |}{\underset{\displaystyle N}{}}}C\underset{\underset{\displaystyle O}{\diagdown \diagup}}{\overset{\displaystyle |}{\underset{}{}}}CO \xrightarrow{\text{NaOH}} \underset{\underset{\displaystyle NHCOC_6H_5}{|}}{C_6H_5CH{=}CCO_2H} \xrightarrow{\text{Na/Hg}}$$

$$\underset{\underset{\displaystyle NHCOC_6H_5}{|}}{C_6H_5CH_2CHCO_2H} \xrightarrow{\text{HCl}} C_6H_5CH_2CH(NH_2)CO_2H + C_6H_5CO_2H$$

See also p. 769 for another synthetic use of azlactones.

Benzamide (*benzenecarbonamide*), m.p. 130°C, may be prepared by the action of concentrated aqueous ammonia on benzoyl chloride or by hydrolysing phenyl cyanide with warm alkaline 3 per cent hydrogen peroxide:

$$C_6H_5COCl + 2NH_3 \longrightarrow C_6H_5CONH_2 + NH_4Cl \quad (v.g.)$$

$$C_6H_5CN \xrightarrow{\text{H}_2\text{O}_2} C_6H_5CONH_2 \quad (ex.)$$

Benzamide undergoes most of the usual reactions of an aliphatic acid amide; *e.g.*, it is readily hydrolysed by dilute acids or alkalis to benzoic acid and ammonia; it forms mercury benzamide, $(C_6H_5CONH)_2Hg$, with mercuric oxide.

Benzoic anhydride, m.p. 42°C, may be prepared by heating a mixture of sodium benzoate and benzoyl chloride:

$$C_6H_5CO_2Na + C_6H_5COCl \longrightarrow (C_6H_5CO)_2O + NaCl \quad (g.)$$

Better yields are obtained by shaking an aqueous solution of sodium benzoate with benzoyl chloride at room temperature in the presence of pyridine (97.5%; see also p. 260).

An alternative method is to slowly distil a mixture of benzoic acid and acetic anhydride (*cf.* p. 260):

$$2C_6H_5CO_2H + (CH_3CO)_2O \longrightarrow (C_6H_5CO)_2O + 2CH_3CO_2H \quad (72\text{--}74\%)$$

McGookin *et al.* (1951) have prepared benzoic anhydride by the action of dry silver oxide or yellow mercuric oxide on benzoyl chloride in benzene:

$$2C_6H_5COCl + Ag_2O \longrightarrow (C_6H_5CO)_2O + 2AgCl$$

Benzoic anhydride is only very slowly decomposed by water (*cf.* acetic anhydride).

Benzoyl peroxide, m.p. 104°C, may be prepared by adding a mixture of benzoyl chloride and aqueous sodium hydroxide to cool hydrogen peroxide with vigorous shaking:

$$2C_6H_5COCl + 2NaOH + H_2O_2 \longrightarrow (C_6H_5CO)_2O_2 + 2NaCl + 2H_2O \quad (80\text{--}95\%)$$

It is widely used as a catalyst in polymerisation processes (it readily generates the phenyl free-radical).

Perbenzoic acid, m.p. 41°C, may be prepared as its sodium salt (from which the acid is obtained by acidification):

$$(C_6H_5CO)_2O_2 + CH_3ONa \longrightarrow C_6H_5COO_2Na + C_6H_5CO_2CH_3 \quad (82\%)$$

The most satisfactory preparation appears to be that of Swern *et al.* (see peracetic acid, p. 243).

Perbenzoic acid is a fairly active oxidising agent; *e.g.*, it converts the ethylenic bond quantitatively into the epoxide:

$$\overset{\diagdown}{\underset{\diagup}{}}C{=}C\overset{\diagup}{\underset{\diagdown}{}} + C_6H_5COO_2H \longrightarrow \overset{\diagdown}{\underset{\diagup}{}}C\overset{\overset{\displaystyle O}{\diagup \diagdown}}{\underset{}{}}C\overset{\diagup}{\underset{\diagdown}{}} + C_6H_5CO_2H$$

Perbenzoic acid is therefore used for the detection and estimation of ethylenic bonds (Prileschaiev reaction; *cf.* p. 111). It is far less troublesome to handle than peracetic acid, and so is used preferably (often as its sodium salt).

p-Nitroperbenzoic acid is better to use than perbenzoic acid, and has been prepared by treating *p*-nitrobenzoyl chloride with sodium peroxide in aqueous tetrahydrofuran (Vilkas, 1959):

$$2p\text{-}NO_2C_6H_4COCl + Na_2O_2 \longrightarrow (p\text{-}NO_2C_6H_4CO)_2O_2 + 2NaCl$$

Phenyl cyanide (*benzonitrile, benzenecarbonitrile*) may be prepared by heating benzamide with phosphorus pentoxide, by means of the Sandmeyer reaction with benzenediazonium chloride (p. 674), by fusing sodium benzenesulphonate with sodium cyanide (p. 691), or by refluxing benzaldehyde in formic acid containing hydroxylamine hydrochloride and sodium acetate (van Es, 1965); the yield with aromatic aldehydes is 81–97 per cent (*cf.* aliphatic aldehydes, p. 355). It is a colourless oil, b.p. 191°C, and behaves similarly to the aliphatic cyanides.

There are many homologues of benzoic acid, but only the **toluic acids** need be mentioned:

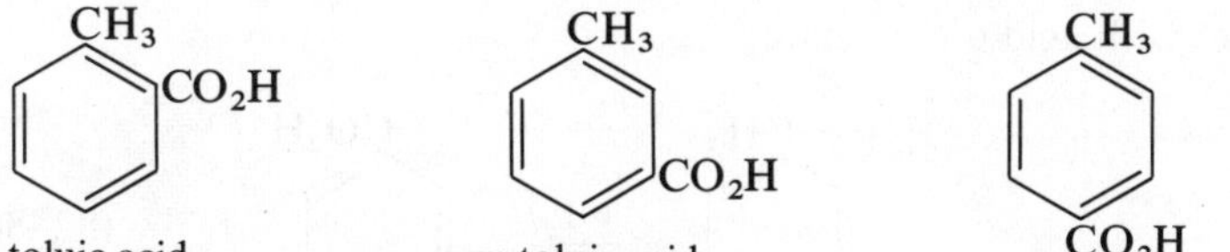

o-toluic acid,
m.p. 105°C

m-toluic acid,
m.p. 111°C

p-toluic acid,
m.p. 180°C

Each may be prepared by oxidising the corresponding xylene with dilute nitric acid, or from the corresponding toluidine (via the cyanide).

Substituted derivatives of benzoic acid. Benzoic acid is attacked by the usual electrophilic reagents chlorine, bromine, nitric and sulphuric acids, to give mainly the *m*-derivatives; *e.g.*, *m*-bromobenzoic acid may be obtained by heating benzoic acid, bromine and water under pressure:

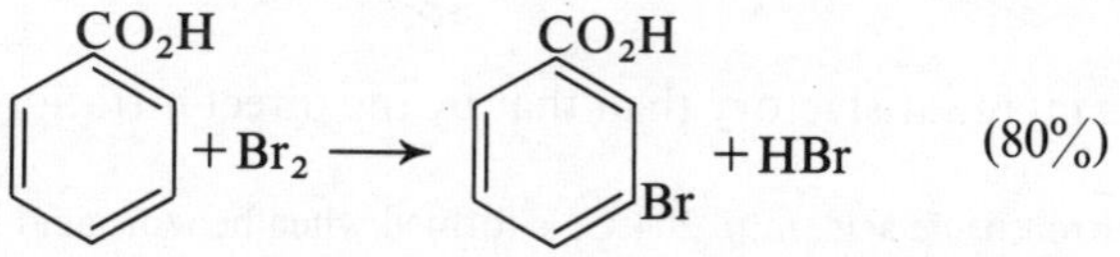

Waters *et al.* (1950) have prepared *m*-iodobenzoic acid (75% yield) by adding iodine to a solution of benzoic acid in concentrated sulphuric acid containing silver sulphate. The active species in this case is the iodinium ion (*cf.* p. 107):

$$I_2 + H_2O \rightleftharpoons HOI + H^+ + I^-$$
$$Ag_2SO_4 + 2I^- \longrightarrow 2AgI + SO_4^{2-}$$
$$HOI + H^+ \rightleftharpoons H_2OI^+ \rightleftharpoons H_2O + I^+$$

o- and *p*-Substituted benzoic acids may be obtained by oxidising the corresponding toluene derivatives, *e.g.*,

(i)

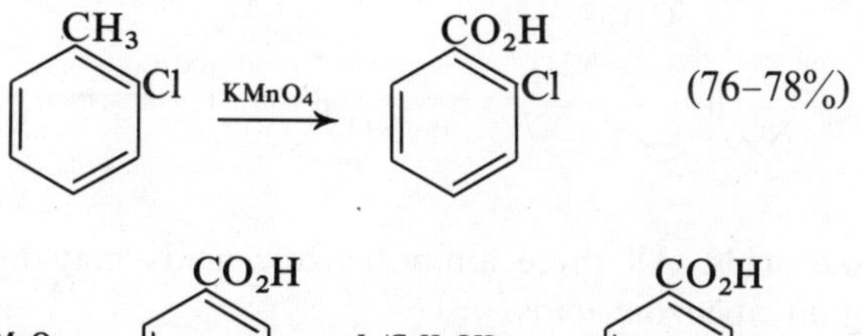

(76–78%)

(ii)

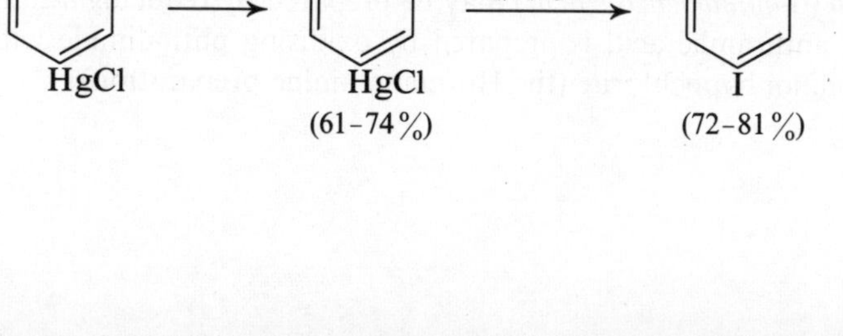

Many substituted benzoic acids may also be prepared from the corresponding aminobenzoic acids (via the diazonium salts). *o*- and *p*-Substituted benzoic acids may be obtained directly by using, *e.g.*, the sodium salt of benzoic acid (see p. 597).

2,4,6-Tribromobenzoic acid may be prepared as follows:

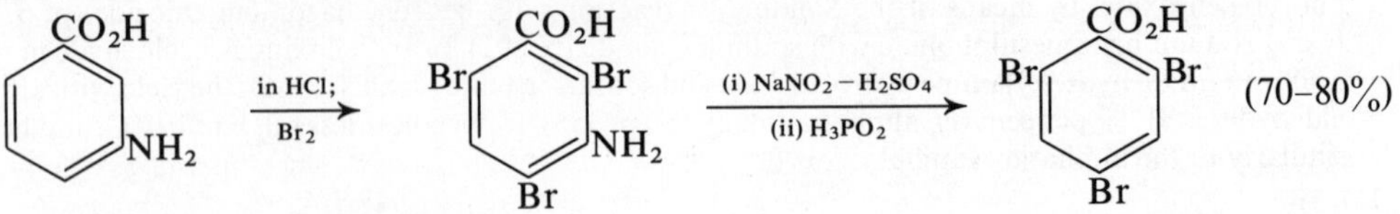

Nitrobenzoic acids. *o*- and *p*-Nitrobenzoic acids may be prepared by oxidising the corresponding nitrotoluenes with acid dichromate:

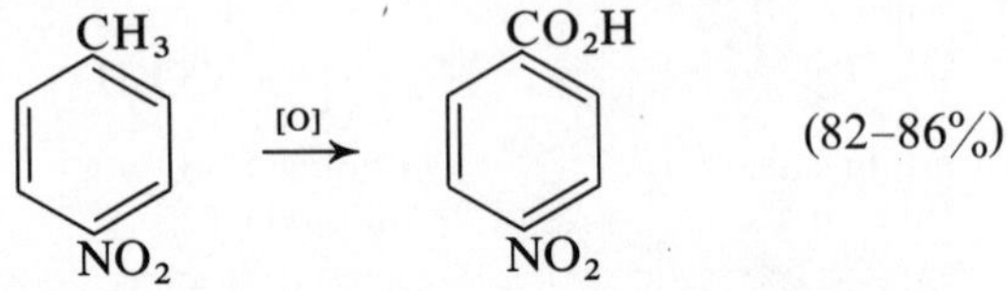

m-Nitrobenzoic acid may be prepared by direct nitration of methyl benzoate:

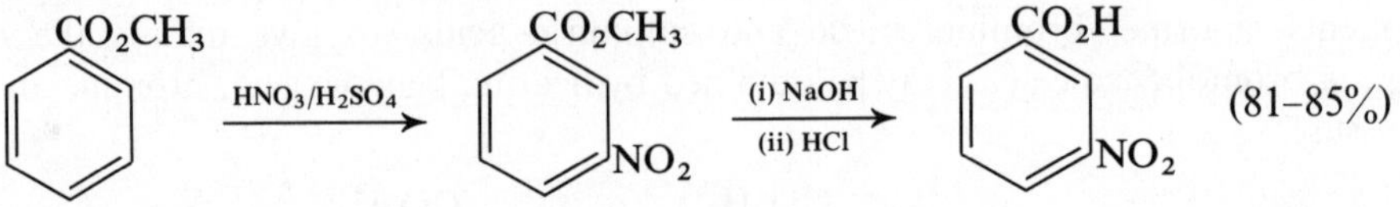

This method is more satisfactory than that by the direct nitration of benzoic acid.

3,5-Dinitrobenzoic acid, m.p. 204°C, is formed when benzoic acid is nitrated with a mixture of fuming nitric and sulphuric acids:

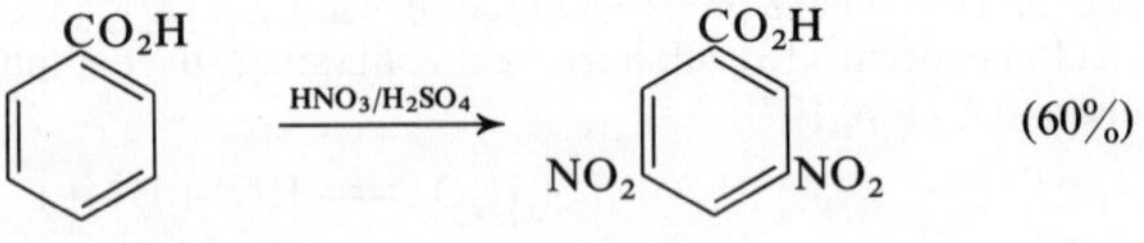

It is used to characterise alcohols.

2,5-Dinitrobenzoic acid may be prepared by treating 2-amino-5-nitrotoluene with potassium persulphate and sulphuric acid (*i.e.*, Caro's acid) and further oxidising the product, 5-nitro-2-nitrosotoluene, with potassium dichromate and sulphuric acid:

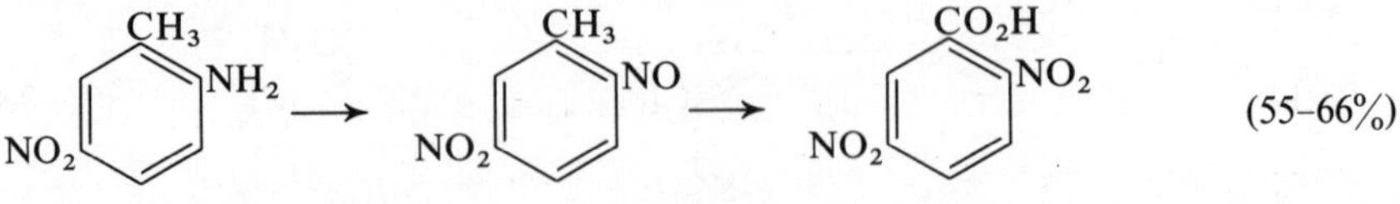

Aminobenzoic acids. All three aminobenzoic acids may be obtained by reduction of the corresponding nitrobenzoic acids.

Anthranilic acid (*o-aminobenzoic acid*) may be prepared by reducing *o*-nitrobenzoic acid.

Commercially, anthranilic acid is prepared by oxidising phthalimide with aqueous sodium hydroxide and sodium hypochlorite (the Hofmann amine preparation):

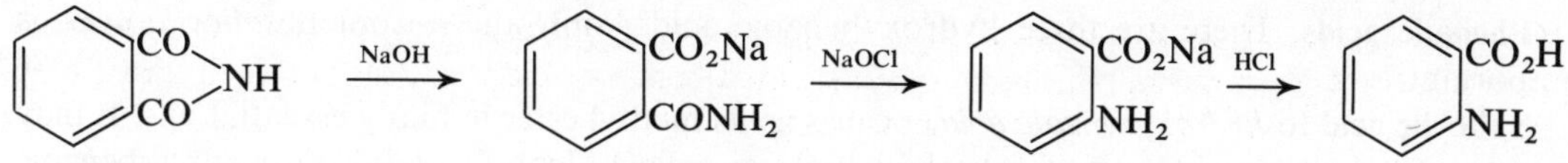

Anthranilic acid, m.p. 145°C, behaves as an acid and as an amine; it does not exist as a dipolar ion in the solid state (*cf.* sulphanilic acid). When distilled, anthranilic acid is decarboxylated to aniline.

Methyl anthranilate is used in perfumery, since it is the characteristic constituent of jasmine and orange blossoms. Anthranilic acid is used in the industrial preparation of indigotin (p. 895).

m-**Aminobenzoic acid**, m.p. 174°C, is used in the preparation of azo-dyes.

p-**Aminobenzoic acid**, m.p. 186°C, is one of the substances comprising the vitamin-B complex. Certain derivatives of *p*-aminobenzoic acid are used as local anaesthetics, *e.g.*, *novocaine*,

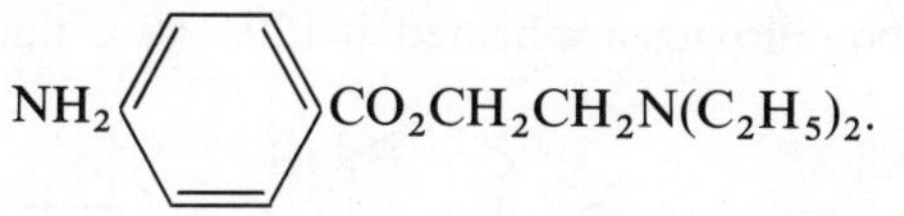

Sulphobenzoic acids. Sulphonation of benzoic acid gives *m*-sulphobenzoic acid. The *o*- and *p*-isomers may be prepared by oxidising the corresponding toluenesulphonic acids.

Saccharin (o-*sulphobenzoic imide*) may be prepared from *o*-toluenesulphonyl chloride (p. 694) as shown. If the solution is kept *neutral* during the permanganate oxidation, the product is saccharin. With *alkaline* permanganate, the salt of *o*-sulphamidobenzoic acid is obtained and this, on acidification, gives the free acid which, on heating, forms saccharin.

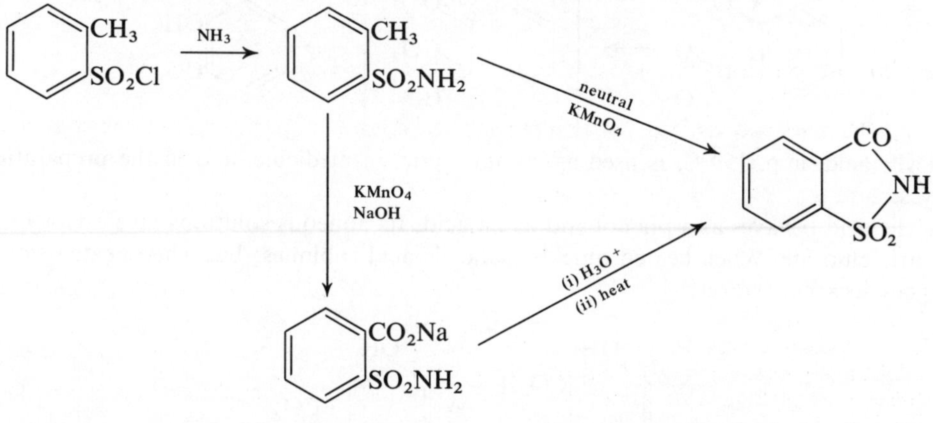

This was one of the first industrial methods, but others are also used, *e.g.*, starting with anthranilic acid:

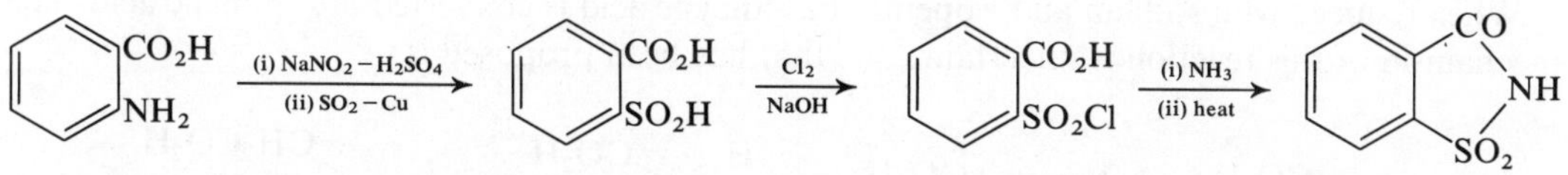

Saccharin, m.p. 224°C, is about 550 times sweeter than sugar. It is almost insoluble in water and hence is sold as its sodium salt, which is very soluble. Saccharin is very sweet in dilute but is bitter in concentrated solutions. It is used instead of sugar for many purposes, *e.g.*, sweetening preserves, drinks, etc. It is also used by diabetics and obese persons.

Phenolic acids. There are three hydroxybenzoic acids; only the *o*-isomer, *salicylic acid*, is important.

Salicylic acid (o-*hydroxybenzoic acid*) occurs as its methyl ester in many essential oils. It may be obtained by the oxidation of salicylaldehyde or salicyl alcohol; or fusing *o*-sulphobenzoic acid with sodium hydroxide. Salicylic acid may be prepared by replacing the amino-group in anthranilic acid by hydroxyl (via the diazonium salt), and also by the Reimer–Tiemann reaction using an alkaline solution of phenol and carbon tetrachloride (*cf.* p. 743):

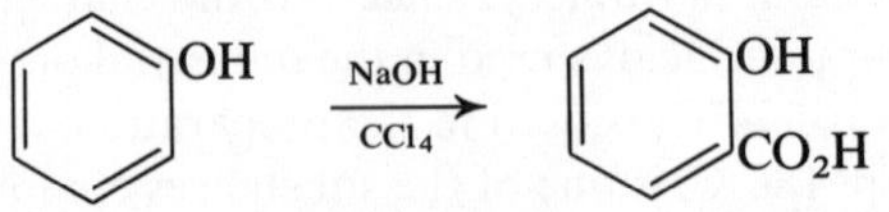

The industrial preparation of salicylic acid is carried out by the *Kolbe–Schmitt reaction* (1885); sodium phenoxide and carbon dioxide are heated at 120–140°C under pressure.

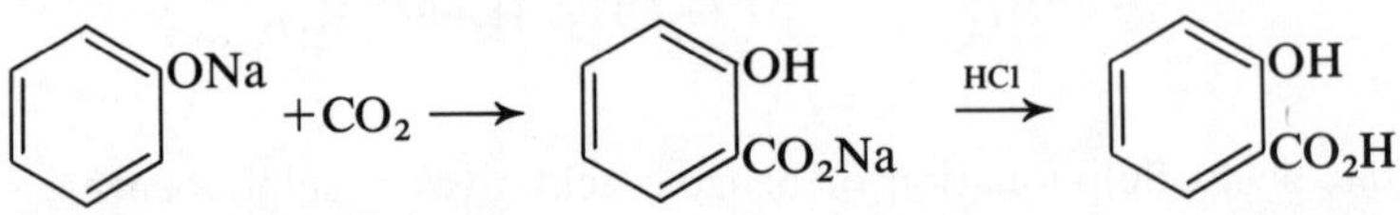

A small amount of the *p*-derivative is formed at the same time, and if the temperature rises above 140°C, the *p*-isomer is the main product.

The mechanism of the Kolbe–Schmitt reaction is uncertain; a possibility is:

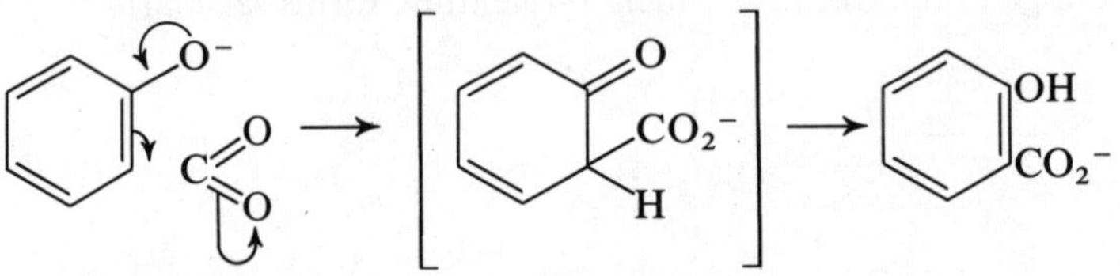

Salicylic acid, m.p. 159°C, is used as an antiseptic, in medicine, and in the preparation of azo-dyes.

Salicylic acid behaves as a phenol and as an acid. Its aqueous solutions give a violet colour with ferric chloride. When heated quickly, salicylic acid sublimes; but when heated slowly, it undergoes decarboxylation:

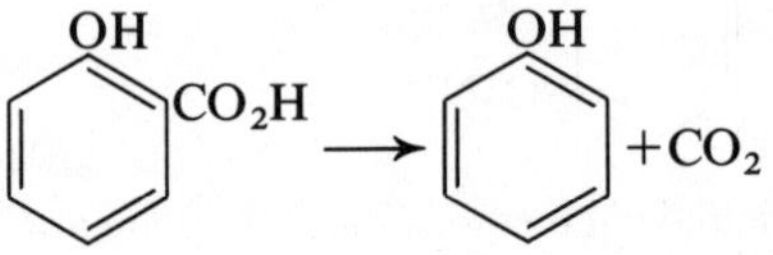

When reduced with sodium and isopentanol, salicylic acid is converted into pimelic acid. The mechanism of this reaction is uncertain; one that has been proposed is:

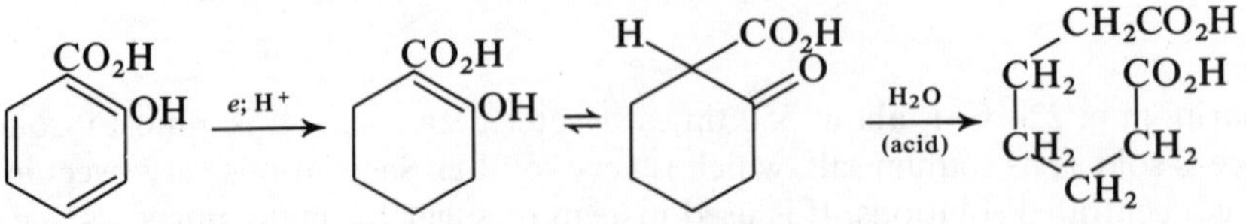

The final step is 'acid hydrolysis' of a β-keto-acid (*cf.* acetoacetic ester syntheses).

Bromination and nitration of salicylic acid cause displacement of the carboxyl group with formation of the corresponding 2,4,6-trisubstituted phenol (*cf.* p. 696).

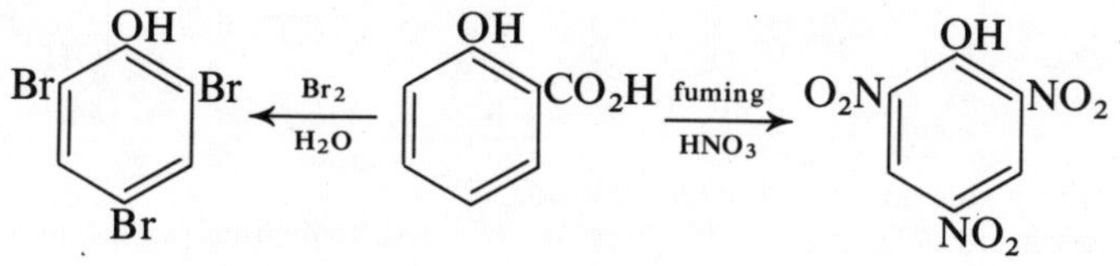

$$\text{2,4,6-tribromophenol} \xleftarrow[\text{H}_2\text{O}]{\text{Br}_2} \text{salicylic acid (OH, CO}_2\text{H)} \xrightarrow[\text{HNO}_3]{\text{fuming}} \text{2,4,6-trinitrophenol}$$

The reason for this ready displacement of the carboxyl group by bromine or by the nitro-group is not certain, but it may possibly occur by the following mechanism:

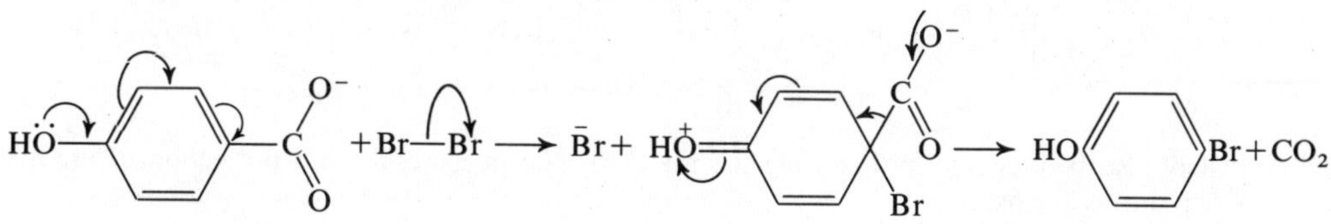

$$\text{HO–C}_6\text{H}_4\text{–CO}_2^- + \text{Br–Br} \longrightarrow \bar{\text{Br}} + \text{HO}=\!\!\text{(intermediate, Br)} \longrightarrow \text{HO–C}_6\text{H}_4\text{–Br} + \text{CO}_2$$

Decarboxylation in alkaline solution may be explained similarly (a proton from water is the attacking agent instead of Br^+).

These displacements of the carboxyl group occur when the reactions are carried out in aqueous solution; they do not occur in acetic acid solution, *e.g.*,

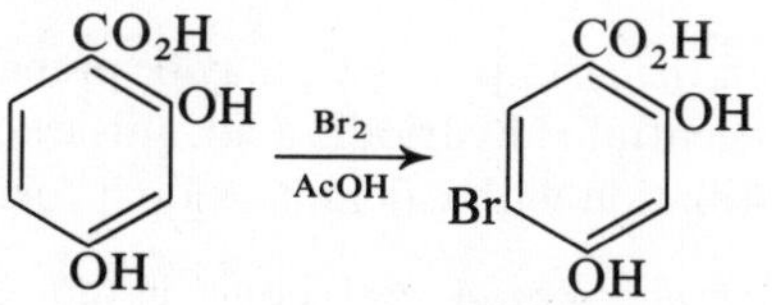

$$\text{2,3-dihydroxybenzoic acid (CO}_2\text{H, OH, OH)} \xrightarrow[\text{AcOH}]{\text{Br}_2} \text{bromo-2,3-dihydroxybenzoic acid (CO}_2\text{H, OH, OH, Br)}$$

When treated with carbonates, only the carboxyl group in salicylic acid forms salts; with alkali, both the carboxyl and hydroxyl groups form salts. Sodium salicylate is used in the treatment of rheumatism. Salicylic acid is a stronger acid than its *m*- and *p*-isomers (see the *ortho*-effect, below).

Methyl salicylate, o-$\text{HOC}_6\text{H}_4\text{CO}_2\text{CH}_3$, b.p. 224°C, is the principal constituent of oil of wintergreen. It may be prepared by direct esterification, and is used in perfumery and as a flavouring material. *Salol* (*phenyl salicylate*), o-$\text{HOC}_6\text{H}_4\text{CO}_2\text{C}_6\text{H}_5$, m.p. 43°C, may be prepared by heating salicylic acid with phenol in the presence of phosphoryl chloride. It is used as an internal antiseptic. *Aspirin* (*acetylsalicylic acid*), o-$\text{CH}_3\text{COOC}_6\text{H}_4\text{CO}_2\text{H}$, m.p. 135°C, may be prepared by acetylating salicylic acid with a mixture of acetic anhydride and glacial acetic acid.

An interesting point about the alkaline hydrolysis of aspirin is that it occurs by the following intramolecular mechanism (St. Pierre *et al.*, 1968; Kirby *et al.*, 1968):

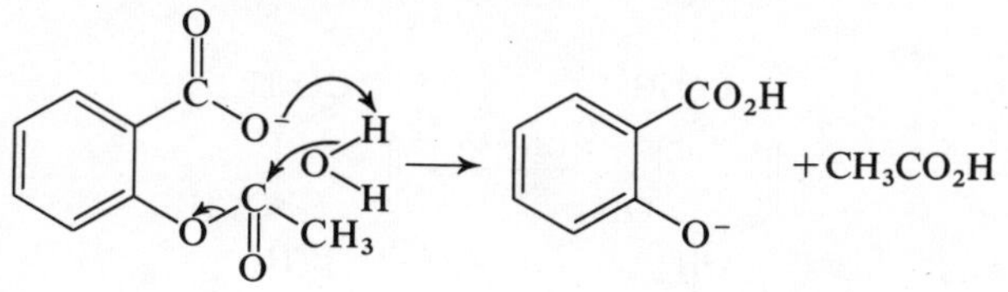

$$\text{aspirin anion intermediate} \longrightarrow \text{salicylate (CO}_2\text{H, O}^-) + \text{CH}_3\text{CO}_2\text{H}$$

m-Hydroxybenzoic acid, m.p. 201°C, may be prepared from *m*-aminobenzoic acid:

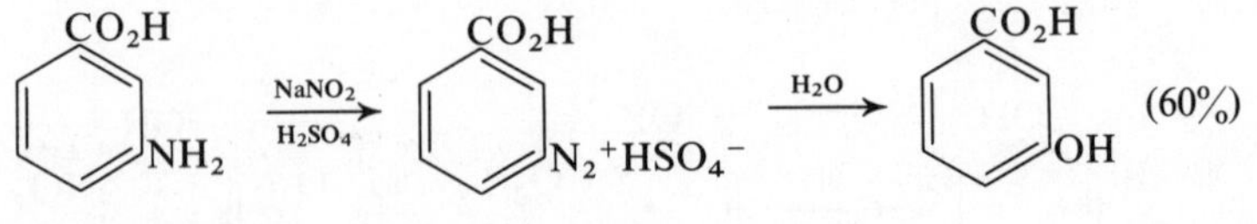

It does *not* give a coloration with ferric chloride.

p-Hydroxybenzoic acid, m.p. 214°C, may be prepared by heating potassium salicylate at 230°C:

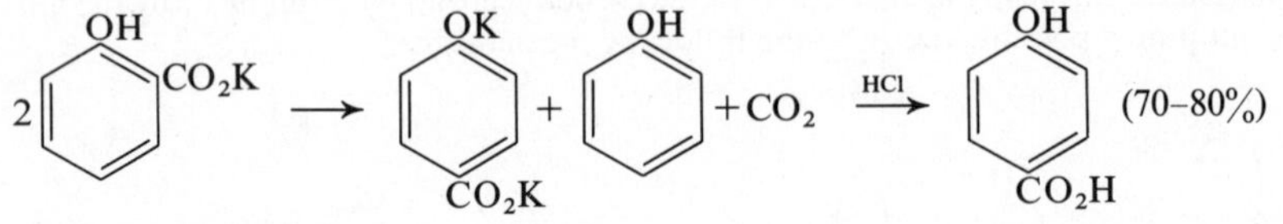

It may also be prepared by oxidising the methyl group in *p*-cresol (with protection of the hydroxyl group during the oxidation):

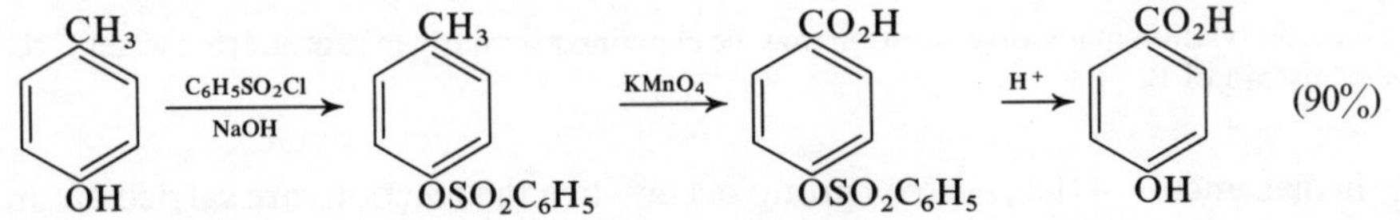

Many esters of *p*-hydroxybenzoic acid have antiseptic properties. The acid gives a red colour with ferric chloride.

Anisic acid (*p-methoxybenzoic acid*), m.p. 184°C, may be prepared by the oxidation of anethole (p. 720). When heated with concentrated hydriodic acid, anisic acid forms *p*-hydroxybenzoic acid and methyl iodide, and on distillation with calcium oxide it forms anisole.

Protocatechuic acid (3,4-*dihydroxybenzoic acid*) occurs in various plants, and is formed when certain resins (catechin, gum-benzoin, etc.) are fused with alkali. It may be prepared by heating catechol with aqueous ammonium carbonate at 140°C under pressure (p. 714), or from vanillin as follows (note the demethylation):

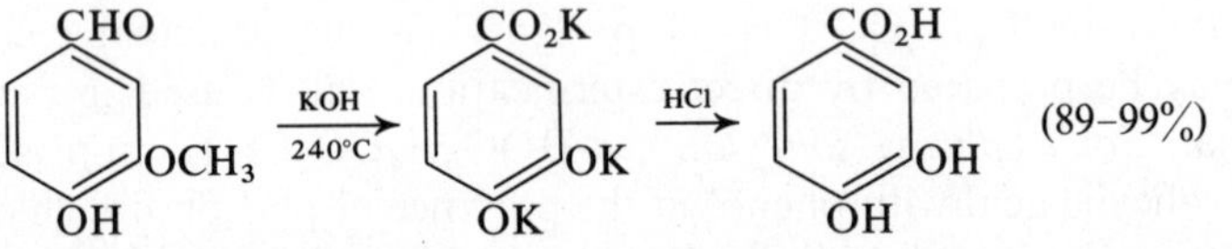

Protocatechuic acid crystallises with one molecule of water of crystallisation; the anhydrous acid melts at 199°C with decomposition into catechol and carbon dioxide.

Veratric acid (3,4-*dimethoxybenzoic acid*), m.p. 181°C, may be prepared by methylating proto-catechuic acid with methyl sulphate in the presence of aqueous sodium hydroxide:

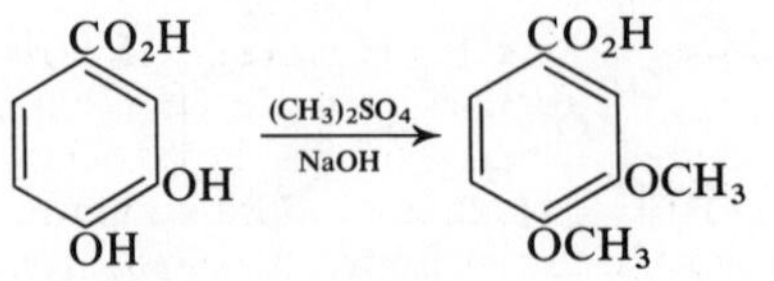

Homoprotocatechuic acid (3,4-*dihydroxyphenylacetic acid*), m.p. 127°C, may be prepared by treating veratraldehyde with hydrogen cyanide and boiling the product, 3,4-dimethoxymandelonitrile, with hydriodic acid (Pictet and Gams, 1909):

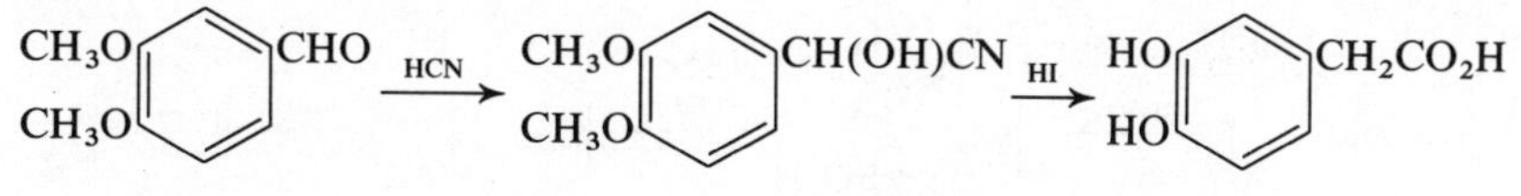

It should be noted that the prefix *homo* indicates that the compound contains one more carbon atom than the parent substance (*i.e.*, it is a *homo*logue).

Homoveratric acid (3,4-*dimethoxyphenylacetic acid*), m.p. 99°C, may be prepared by methylating homoprotocatechuic acid with methyl sulphate in the presence of aqueous sodium hydroxide. It may be prepared via the azlactone synthesis as follows:

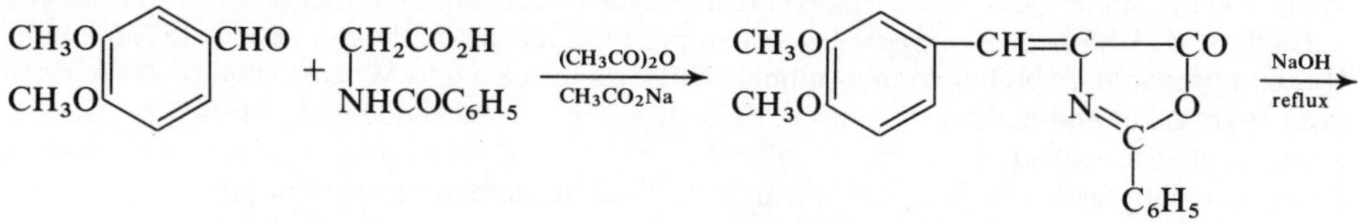

Homoveratric acid may also be prepared by chloromethylating veratrole, treating the product with potassium cyanide, and hydrolysing the homoveratronitrile so produced (Bide and Wilkinson, 1945);

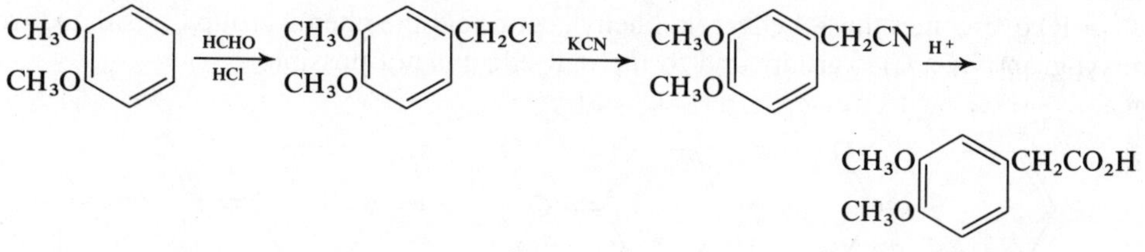

Piperonylic acid (3,4-*methylenedioxybenzoic acid*), m.p. 229°C, may be prepared by the oxidation of piperonal (p. 747), or by heating protocatechuic acid with methylene di-iodide and aqueous sodium hydroxide:

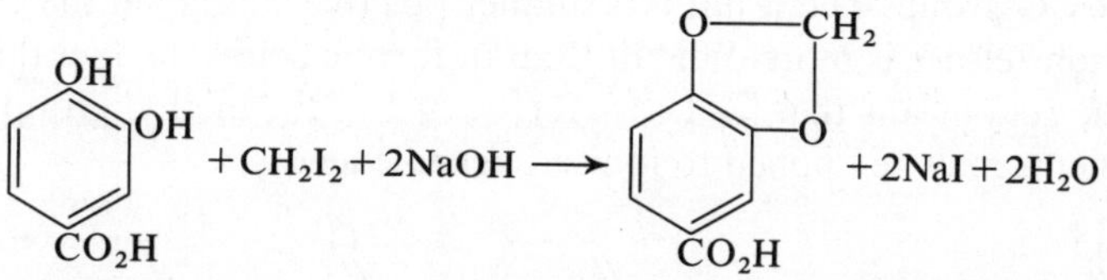

Piperonal and piperonylic acid are oxidation products of piperic acid, which occurs in the alkaloid piperine. Veratric acid is one of the products of oxidation of the alkaloid papaverine. The orientation of the groups in these acids (and in the corresponding homo-acids) may be shown as follows. Sulphonation of *p*-hydroxybenzoic acid (the orientation of which is known) produces a sulphonic acid derivative in which, according to the rules of directing power, the sulphonic acid group is *o*- to the hydroxyl group.

This derivative, on fusion with potassium hydroxide, produces the dihydroxybenzoic acid in which the two hydroxyl groups are *o*- to each other. This is confirmed by the fact that the acid, *viz.*, protocatechuic acid, gives catechol (orientation known) when heated to its melting point:

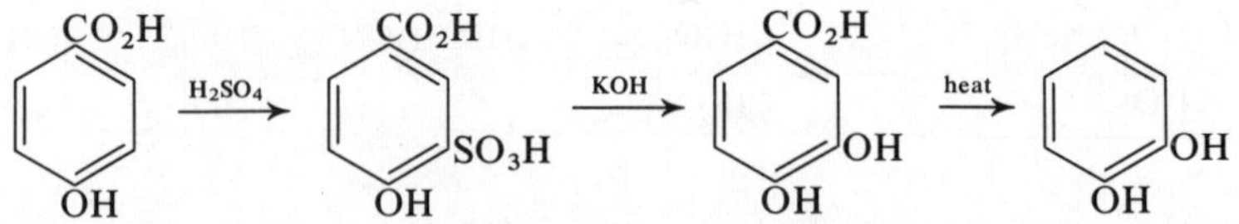

Thus protocatechuic acid must be 3,4-dihydroxybenzoic acid. Hence the dimethoxy-derivative, veratric acid, must be 3,4-dimethoxybenzoic acid. At the same time, the orientations of the corresponding aldehydes, protocatechualdehyde and veratraldehyde, are established. Similarly, since piperonylic acid is the methylene ether of protocatechuic acid (shown by analytical and synthetic evidence), its orientation is also established. Finally, since homoprotocatechuic acid (and its methylated derivative, homoveratric acid) can be synthesised from veratraldehyde (see above), its orientation is thus ascertained.

Gallic acid (3,4,5-*trihydroxybenzoic acid*), m.p. 253°C, occurs in the free state in tea and many plants. It is best prepared by boiling tannin (tannic acid) with dilute acids. When heated at its melting point, it forms pyrogallol and carbon dioxide (p. 717). It is a powerful reducing agent and hence is used as a photographic developer. It is readily soluble in water, and its aqueous solution gives a bluish-black precipitate with ferric chloride. This property is used in the manufacture of ink.

Strengths of aromatic acids

It has been pointed out (p. 241) that replacement of the hydrogen atom in formic acid by an alkyl group weakens the strength of the acid, and the greater the $+I$ effect of the R group, the weaker is the acid (see Table 9.1). Phenylacetic acid, $PhCH_2CO_2H$ (pK_a 4.31) is stronger than acetic acid (pK_a 4.76), and therefore the phenyl group has a $-I$ effect. On the other hand, benzoic acid (pK_a 4.17) is weaker than formic acid (pK_a 3.75). In this case, the phenyl group has a $+I$ effect (which is smaller than that of methyl group). These apparently contradictory results may be explained as follows. When the carboxyl group is *directly* attached to the nucleus, the resonance effect ($+R$) overcomes the $-I$ effect (in phenylacetic acid the phenyl group is insulated from the carboxyl group by a CH_2 group, and so the $+R$ effect is not possible):

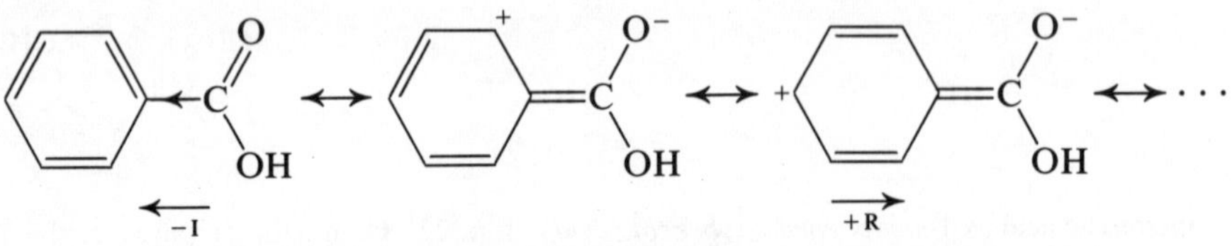

This prevents, to a large extent, the lone pair on the O atom of the OH group from entering into resonance with the CO group. The result is a smaller positive charge on the O atom of the OH group, and so proton release is more difficult than in formic acid. The fact that benzoic acid is stronger than acetic acid means that $[-I+(+R)] < +I$ of the methyl group.

The same arguments may be applied to ionised benzoic acid.

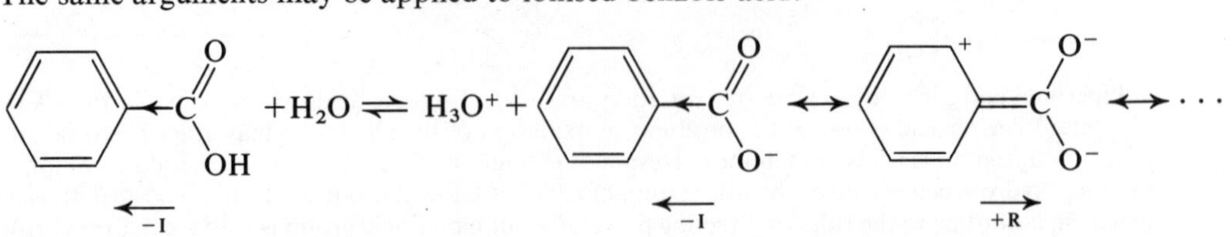

Now let us consider substituted benzoic acids (see Table 28.1). At this point we shall discuss effects from the *m*- and *p*-positions. The methyl group decreases the strength of benzoic acid from these positions. The methyl group has a $+I$ effect, and so the net result is to increase the $+R$ effect in the *m*- or *p*-tolyl group. Thus the two acids are weaker than benzoic acid. Resonance of the carboxyl group with the ring is from 'ring to the group' and not the other way round. Thus, any substituent group that reduces the electron density at the carbon atom attached to the carboxyl group will cause a withdrawal of electrons from the O atom of the OH group, and consequently facilitate proton release, thereby increasing the strength of the acid. Such a substituent is the nitro-group. This has a $-I$ effect, and this is greater from the *p*-position than from the *m*-. Therefore *p*-nitrobenzoic acid is stronger than the *m*-isomer.

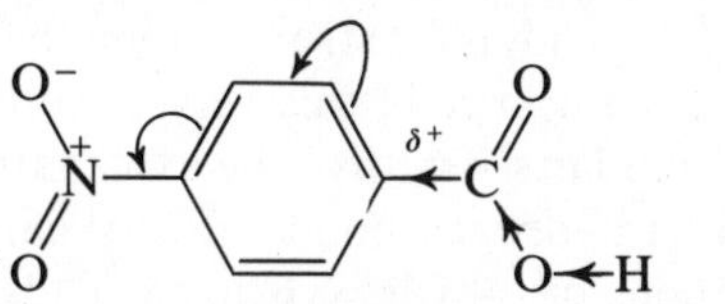

Groups such as MeO can enter into resonance with the carboxyl group in the *p*-position, but not from the *m*-, in which case it is the $-I$ effect that operates. Thus the *p*-substituted acid will be expected to be weaker and the *m*-substituted acid stronger than benzoic acid (see Table 28.1; see also the Hammett equation, p. 608). With chlorine as substituent, since the $+R$ effect is very small, the $-I$ effect controls acidity, and so *m*- and *p*-chlorobenzoic acid are stronger than benzoic acid.

The alternative way of explaining the strengths of these acids is to consider the stabilising effect of the substituent on the anion. The more stable the anion is with respect to the un-ionised acid, the stronger will be the acid. $+I$ and/or $+R$ groups will oppose, whereas $-I$ and/or $-R$ groups will assist (by dispersing the negative charge), the $+I$ effect of the *carboxylate ion group*. Hence, the former substituents weaken and the latter strengthen the acid (see also p. 597).

Table 28.1

Substituent	$o-$	pK_a $m-$	$p-$
None: 4.17			
Me	3.91	4.27	4.37
NO$_2$	2.17	3.49	3.43
F	3.27	3.87	4.14
Cl	2.94	3.83	3.98
OMe	4.09	4.09	4.47
OH	2.98	4.08	4.58
NH$_2$	4.98	4.79	4.92
2,4,6-tri-NO$_2$: 0.65			

The *ortho*-effect. This is a special effect that is shown by *o*-substituents, but is not necessarily just a steric effect, *e.g.*, the basicities of some *o*-substituted anilines were explained in terms of steric effects and differences in crowding round the nitrogen atom (see p. 657). This *ortho*-effect also operates with the benzoic acids. Examination of Table 28.1 shows that, irrespective of the polar type, nearly all *o*-substituted benzoic acids are stronger than benzoic acid. As we have seen, benzoic acid is a resonance hybrid, and so the carboxyl group is coplanar with the ring. An o-substituent tends to prevent this coplanarity. Thus resonance is diminished (or prevented), and so the O atom of the OH group has a greater positive charge, resulting in increased acid strength. It follows from this that the greater the steric inhibition of resonance, the stronger is the acid. Support for this is the following order of strengths of substituted benzoic acids:

$$2,6\text{-di-Me (p}K_a \text{ 3.21)} > 2\text{-t-Bu (p}K_a \text{ 3.46)} > 2\text{-Me (p}K_a \text{ 3.91).}$$

Here again, if we consider the stability of the anion, steric inhibition of resonance prevents the $+R$ effect of the *ring* coming into operation (see above), and since this weakens acid strength, its absence results in increased acid strength.

o-Hydroxybenzoic acid (salicylic acid) is far stronger than the corresponding *m*- and *p*-isomers. Steric inhibition of resonance cannot explain this very large increase, since the corresponding methoxybenzoic acids all have similar strengths. The explanation offered is hydrogen bonding; H of the o-OH group can form a hydrogen bond with the carboxyl group, whereas the Me of OMe cannot. The carboxylate ions of o-hydroxybenzoic acids are stabilised by intramolecular hydrogen bonding, and support for this is given by the following order of acid strength:

$$2,6\text{-di-OH (p}K_a \text{ 2.30)} > 2\text{-OH (p}K_a \text{ 2.98)} > \text{benzoic acid (p}K_a \text{ 4.17).}$$

It can be seen that two hydrogen bonds would be expected to bring about more stabilisation than one hydrogen bond:

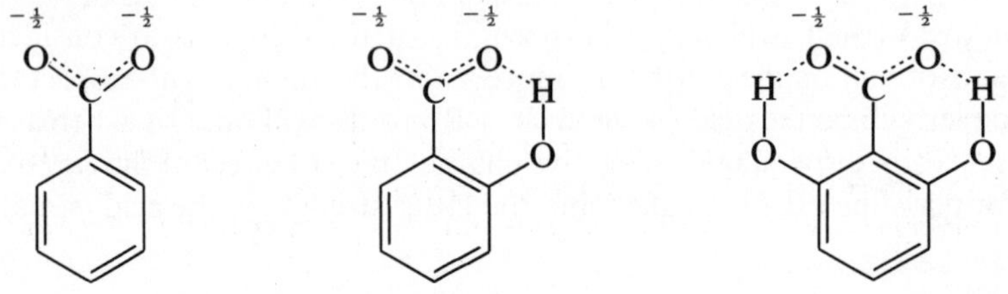

The much greater strength of o-nitrobenzoic acid as compared with that of its *m*- and *p*-isomers may also be explained in a similar manner:

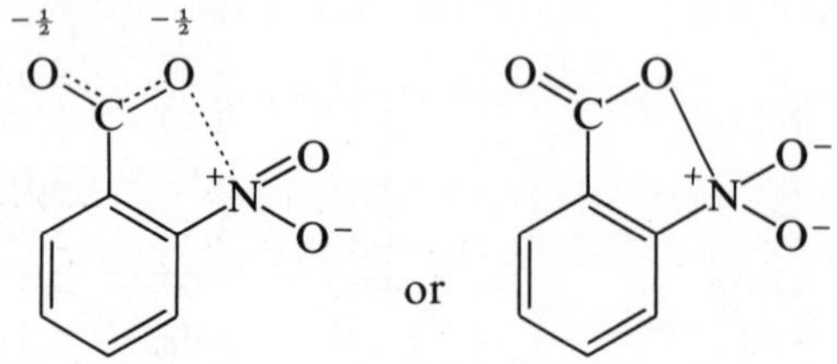

or

The actual geometry of these highly substituted benzenes is an interesting problem. In benzene the ring is flat and the hydrogen atoms are coplanar with this ring. Electron-diffraction investigations of polyhalogenobenzenes suggest that such molecules are non-planar (Hassel *et al.*, 1947), whereas X-ray studies indicate that in the solid state such molecules are very closely or even exactly planar (Tulinsky *et al.*, 1958; Gafner *et al.*, 1960). Ferguson *et al.* (1959, 1961) have examined, by X-ray analysis, various substituted benzoic acids, *e.g.*, o-chloro- and bromo-benzoic acid, and 2-chloro-5-nitrobenzoic acid. In all three molecules the steric strain is relieved by small out-of-plane displacements of the exocylic valency bonds in addition to the larger in-plane displacement of these bonds away from one another. This type of deformation is referred to as *molecular overcrowding* (see also p. 822).

Monocarboxylic acids with the carboxyl group in the side-chain

Phenylacetic acid (*α-toluic acid*), m.p. 77°C, may be prepared from benzyl chloride as follows:

$$C_6H_5CH_2Cl \xrightarrow{KCN} C_6H_5CH_2CN \xrightarrow[(H_2SO_4)]{H_2O} C_6H_5CH_2CO_2H \quad (60–70\%)$$

It may also be prepared by reducing mandelonitrile with hydriodic acid and red phosphorus:

$$C_6H_5CH(OH)CN \xrightarrow{HI/P} C_6H_5CH_2CO_2H \quad (90\%)$$

Phenylacetic acid contains two α-hydrogen atoms and so is reactive in the side-chain as well as in the ring, *e.g.*, chlorination in the cold in the presence of a halogen carrier, gives *o*- and *p*-nuclear substitution; chlorination at the boiling point of the acid, in the absence of a halogen carrier, gives replacement of the α-hydrogen (*cf.* toluene, p. 633). When oxidised with chromic acid, phenylacetic acid is converted into benzoic acid.

Mandelic acid (*phenylglycollic acid*) may be obtained from the glucoside amygdalin by regulated hydrolysis (*cf.* benzaldehyde, p. 734). It may be prepared as follows:

$$C_6H_5CHO \xrightarrow{NaHSO_3} C_6H_5CH(OH)SO_3Na \xrightarrow{NaCN}$$

$$C_6H_5CH(OH)CN \xrightarrow[(HCl)]{H_2O} C_6H_5CH(OH)CO_2H \quad (50–52\%)$$

Mandelic acid may also be prepared by the hydrolysis of phenacylidene chloride (*cf.* p. 750).

Mandelic acid behaves as a hydroxyacid, and is optically active. The acid obtained from amygdalin is lævorotatory, m.p. 133°C; the m.p. of the DL-acid is 118°C. On vigorous oxidation, mandelic acid gives benzoic acid.

Hydratropic acid (*α-phenylpropionic acid*), $C_6H_5CH(CH_3)CO_2H$, b.p. 265°C, may be prepared by the reduction of *atropic acid*, $C_6H_5C(=CH_2)CO_2H$, obtained by heating *tropic acid*, $C_6H_5CH(CH_2OH)CO_2H$, which occurs in the alkaloid atropine.

β-Phenylpropionic acid (*hydrocinnamic acid*), m.p. 47°C, may be prepared by reducing cinnamic acid with sodium amalgam:

$$C_6H_5CH=CHCO_2H \xrightarrow{e;\,H^+} C_6H_5CH_2CH_2CO_2H \quad (85\%)$$

It may also be synthesised from benzyl chloride and malonic ester:

$$C_6H_5CH_2Cl + [CH(CO_2C_2H_5)_2]^-Na^+ \longrightarrow C_6H_5CH_2CH(CO_2C_2H_5)_2 + NaCl \xrightarrow[(ii)\,boil]{\omega\,acid}$$

$$C_6H_5CH_2CH_2CO_2H + CO_2 + 2C_2H_5OH$$

β-Phenylpropionic acid is oxidised by chromic acid to benzoic acid.

Cinnamic acid is the *trans*-isomer, and hence is also known as ***trans-β*-phenylacrylic acid**. The *cis*-isomer (cis-*β-phenylacrylic acid*) has also been called allocinnamic acid:

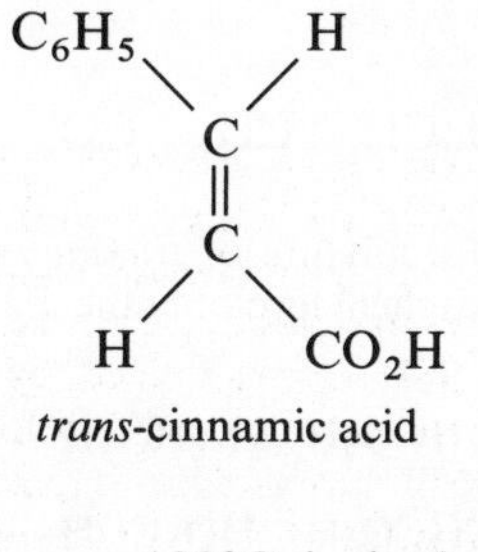

trans-cinnamic acid 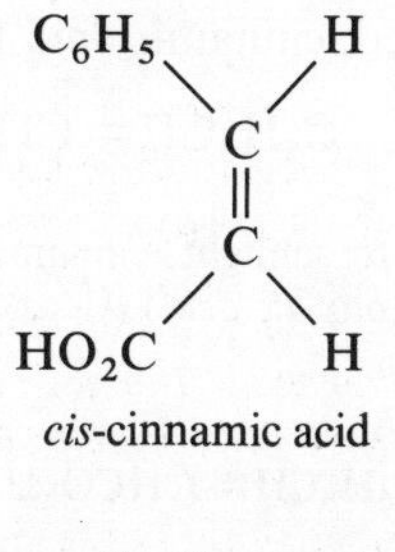   *cis*-cinnamic acid

trans-Cinnamic acid, m.p. 133°C, is the form that occurs naturally (free and as esters) in balsams and resins. It may be prepared:

(i) By *Perkin's reaction* (p. 739):

$$C_6H_5CHO + (CH_3CO)_2O \xrightarrow{CH_3CO_2Na} C_6H_5CH=CHCO_2H \quad (85\%)$$

(ii) By *Knoevenagel's reaction* (p. 342):

$$C_6H_5CHO + CH_2(CO_2H)_2 \xrightarrow[\text{heat}]{NH_3/C_2H_5OH} C_6H_5CH=CHCO_2H \quad (80\%)$$

(iii) By the *Claisen condensation* between benzaldehyde and ethyl acetate in the presence of sodium ethoxide (since benzaldehyde is one of the reactants, this is also referred to as a *Claisen reaction*; (see p. 219):

$$C_6H_5CHO + CH_3CO_2C_2H_5 \xrightarrow{C_2H_5ONa} C_6H_5CH=CHCO_2C_2H_5 \xrightarrow[\text{(acid)}]{H_2O}$$
$$C_6H_5CH=CHCO_2H + C_2H_5OH \quad (68-74\%)$$

(iv) By the oxidation of benzylideneacetone with sodium hypochlorite (*cf.* the haloform reaction):

$$C_6H_5CH=CHCOCH_3 \xrightarrow{NaOCl} C_6H_5CH=CHCO_2H$$

Methods (i) and (iv) are used industrially.

Cinnamic acid behaves as an α,β-unsaturated acid, and as a benzene derivative. It is reduced by sodium amalgam to γ-phenylpropionic acid; reduction of its ethyl ester, however, leads to the formation of the bimolecular product, ethyl β,β'-diphenyladipate (*cf.* p. 340):

$$2C_6H_5CH=CHCO_2C_2H_5 \xrightarrow{e;\,H^+} \begin{array}{l} C_6H_5CHCH_2CO_2C_2H_5 \\ | \\ C_6H_5CHCH_2CO_2C_2H_5 \end{array}$$

When a phenyl group is attached to the β-carbon atom of an α,β-unsaturated carbonyl compound, both the double bond and carbonyl group are reduced by lithium aluminium hydride (*cf.* pp. 337 and 742):

$$C_6H_5CH=CHCO_2H \xrightarrow{LiAlH_4} C_6H_5CH_2CH_2CH_2OH$$

By using the inverse addition at $-10°C$, the double bond is left intact, *e.g.*,

$$C_6H_5CH=CHCO_2C_2H_5 \xrightarrow[-10°C]{LiAlH_4} C_6H_5CH=CHCH_2OH$$

When oxidised with chromic acid, cinnamic acid forms a mixture of benzaldehyde and benzoic acid. Concentrated nitric acid nitrates it to a mixture of *o*- and *p*-nitrocinnamic acids, the latter predominating. *m*-Nitrocinnamic acid may be prepared by the Perkin reaction with *m*-nitrobenzaldehyde (74–77%).

When dry-distilled, cinnamic acid forms styrene:

$$C_6H_5CH=CHCO_2H \longrightarrow C_6H_5CH=CH_2 + CO_2$$

When exposed to sunlight, cinnamic acid dimerises to a mixture of *truxinic acid* (3,4-*diphenylcyclobutane*-1,2-*dicarboxylic acid*) (I), and *truxillic acid* (2,4-diphenylcyclobutane-1,3-dicarboxylic acid) (II) (see also p. 915).

$$2C_6H_5CH=CHCO_2H \longrightarrow \begin{array}{l} C_6H_5CH-CHCO_2H \\ | \quad\quad | \\ C_6H_5CH-CHCO_2H \end{array} + \begin{array}{l} C_6H_5CH-CHCO_2H \\ | \quad\quad | \\ HO_2CCH-CHC_6H_5 \end{array}$$
$$\quad\quad\quad\quad\quad\quad\quad\quad (I) \quad\quad\quad\quad\quad\quad\quad (II)$$

cis-**Cinnamic acid**, m.p. 68°C, may be prepared by exposing cinnamic acid in benzene to ultraviolet light, or by the partial reduction of phenylpropiolic acid with hydrogen and colloidal platinum as catalyst:

$$C_6H_5C{\equiv}CCO_2H + H_2 \xrightarrow{\text{Pt}} C_6H_5CH{=}CHCO_2H$$

cis-Cinnamic acid readily changes into *trans*-cinnamic acid, and under the influence of light dimerises to truxinic acid.

Phenylpropiolic acid (*phenylpropynoic acid*), m.p. 136°C, may be prepared by boiling the ethyl ester of cinnamic acid dibromide with ethanolic potassium hydroxide:

$$C_6H_5CH{=}CHCO_2C_2H_5 \xrightarrow[\text{(CCl}_4)]{\text{Br}_2} C_6H_5CHBrCHBrCO_2C_2H_5 \xrightarrow[\text{(ii) acid}]{\text{(i) KOH/C}_2\text{H}_5\text{OH}} C_6H_5C{\equiv}CCO_2H$$

Harley–Mason *et al.* (1961) have prepared phenylpropiolic acid by the action of *p*-bromobenzene-sulphonyl chloride (*brosyl chloride*), BsCl, on diethyl sodiobenzoylmalonate, and treating the brosylate with sodium hydroxide in aqueous dioxan at room temperature. This is one of the mildest methods for introducing a triple bond into a compound.

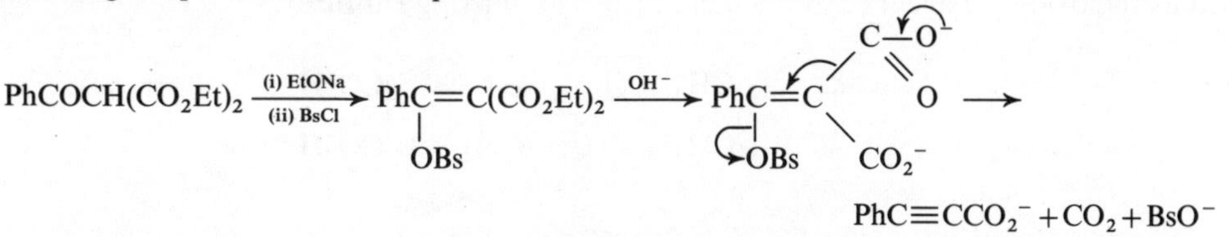

$$PhCOCH(CO_2Et)_2 \xrightarrow[\text{(ii) BsCl}]{\text{(i) EtONa}} \underset{\overset{|}{OBs}}{PhC{=}C(CO_2Et)_2} \xrightarrow{OH^-} \longrightarrow$$

$$PhC{\equiv}CCO_2^- + CO_2 + BsO^-$$

On catalytic reduction (using colloidal platinum), phenylpropiolic acid forms *cis*-cinnamic acid. Reduction with zinc and acetic acid gives *trans*-cinnamic acid, and with sodium amalgam, β-phenyl-propionic acid. This acid and *cis*-cinnamic acid are produced when di-imide is the reducing agent (p. 101). When refluxed with barium hydroxide solution, phenylpropiolic acid is converted into phenylacetylene.

$$C_6H_5C{\equiv}CCO_2H \longrightarrow C_6H_5C{\equiv}CH + CO_2$$

o-**Coumaric acid** (*trans-o-hydroxycinnamic acid*), m.p. 208°C, may be prepared by boiling coumarin (see below) with sodium ethoxide, or by diazotising *o*-aminocinnamic acid and then heating the diazonium sulphate solution. When exposed to ultraviolet light, *o*-coumaric acid (the stable *trans*-form) is converted into *coumarin* via *coumarinic acid* (the unstable *cis*-form of the acid).

Coumarinic acid (*cis-o-hydroxycinnamic acid*) is unstable, spontaneously forming its δ-lactone as soon as it is set free from its salts; the lactone is known as *coumarin* (1,2-benzopyrone), m.p. 69°C:

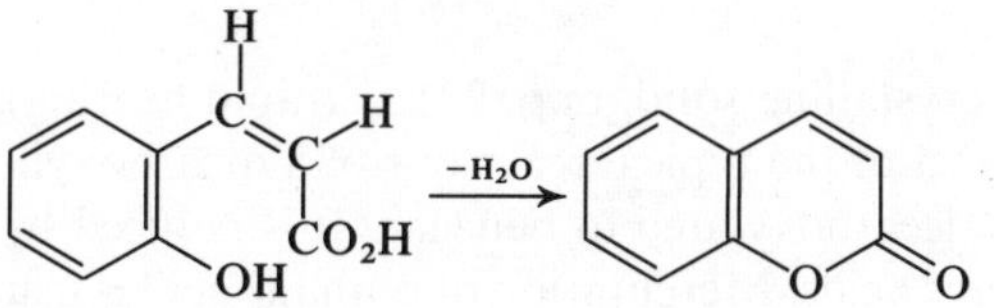

Coumarin may be prepared by the Perkin reaction.

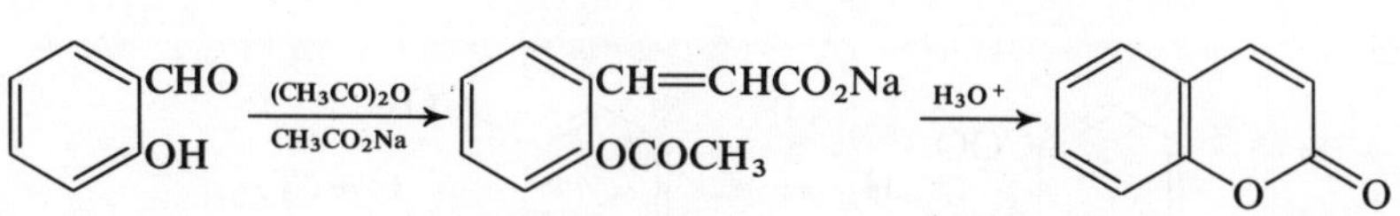

One of the most convenient methods for synthesising coumarins is the **Pechmann reaction** (1883) which is the condensation of a phenol with a β-keto-ester. The usual condensing reagents are sulphuric acid, phosphorus pentoxide, etc., but Koo (1955) has shown that polyphosphoric acid is very effective, *e.g.*, resorcinol and ethyl acetoacetate give 7-hydroxy-4-methylcoumarin:

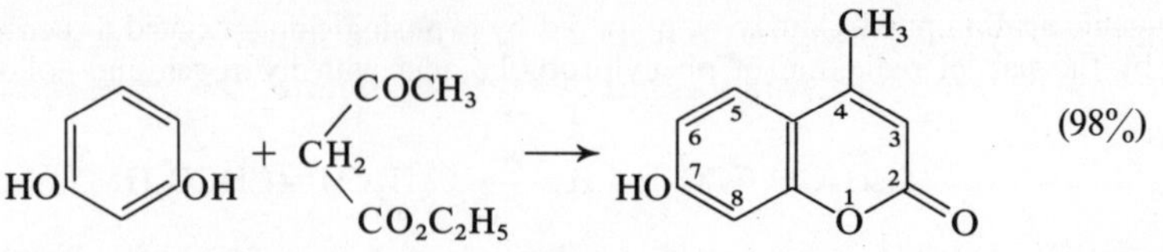

Coumarin is a natural perfume, and is also used as an artificial flavour. When heated with sodium ethoxide, it is converted into coumaric acid.

Benzenedicarboxylic acids

Phthalic acid (*benzene*-1,2-*dicarboxylic acid*) may be prepared by the oxidation of any benzene derivative having only two side-chains in the *o*-positions; permanganate or dilute nitric acid is used as the oxidising agent (chromic acid brings about ring rupture):

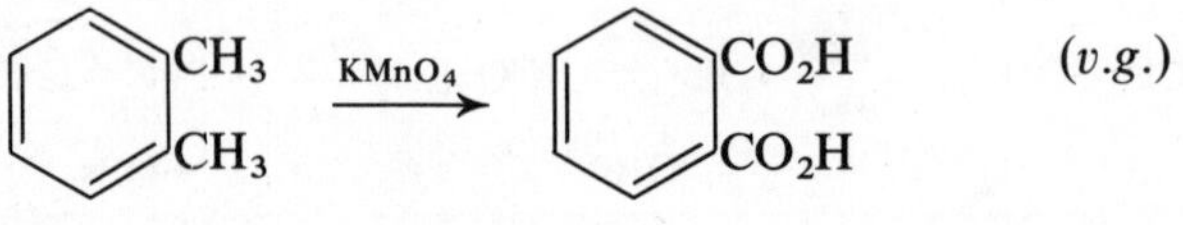

Industrially, phthalic acid is prepared by passing naphthalene vapour or *o*-xylene mixed with air over vanadium pentoxide as catalyst at 400–500°C:

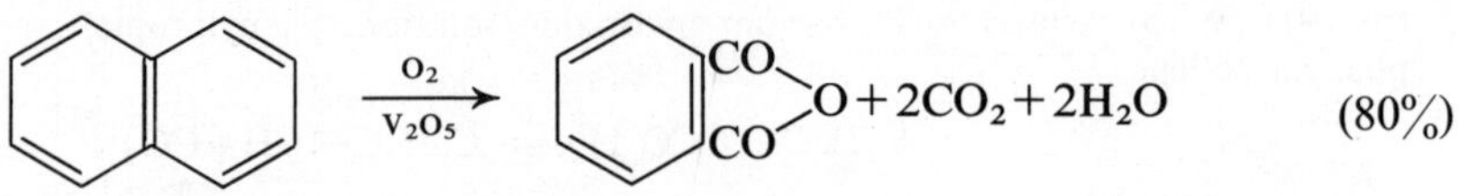

The anhydride is converted into the acid by heating it with alkali, and then acidifying.

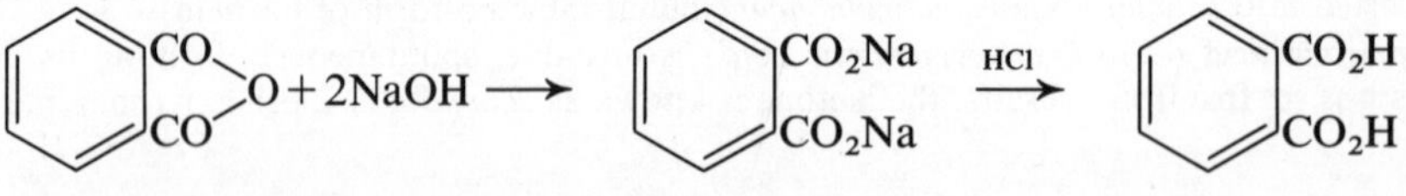

. Phthalic acid is a white crystalline solid, m.p. 231°C (rapid heating), with conversion into its anhydride. It undergoes most of the typical reactions of a dicarboxylic acid. When heated with potassium hydroxide, it is decarboxylated to benzene. It is reduced by sodium amalgam to di-, tetra- and hexahydrophthalic acids. Mercuration of phthalic acid is usually carried out by refluxing the sodium salt with aqueous mercuric acetate containing sodium acetate, until no ionic mercury remains in the solution. Phthalic acid may also be mercurated by fusing it with mercuric acetate, or by heating the mercury salt of phthalic acid until the mercury has been transferred to the nucleus:

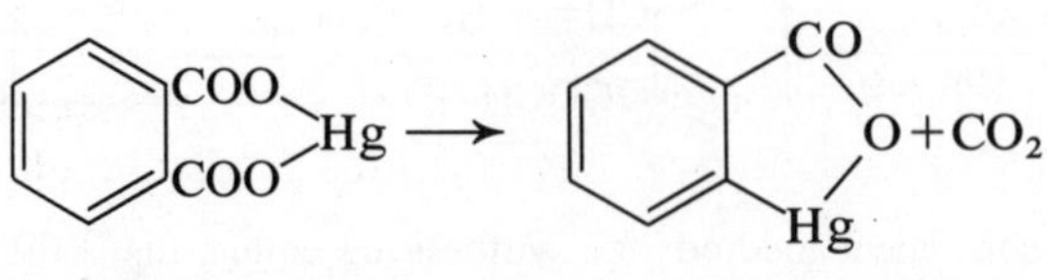

Phthalic anhydride, m.p. 128°C, is prepared industrially by the oxidation of naphthalene or *o*-xylene (see above). It is slowly hydrolysed by water but rapidly by alkalis or acids. When nitrated, phthalic anhydride forms a mixture of 3- and 4-nitrophthalic acids. When these are reduced, their corresponding aminophthalic acids spontaneously decarboxylate to form *m*-aminobenzoic acid:

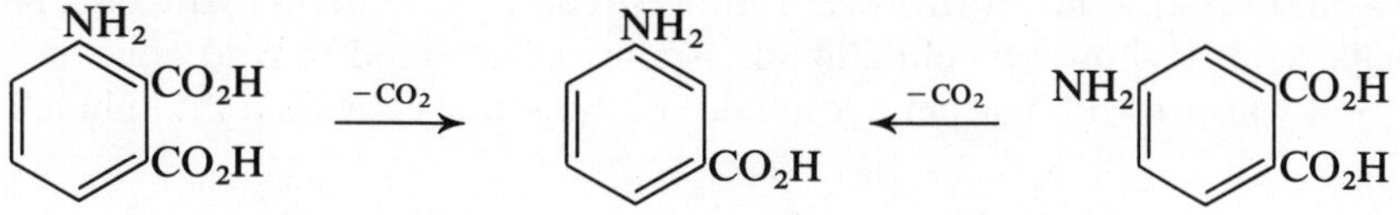

It is the carboxyl group *o*- or *p*- to the amino-group that is lost.

Phthalic anhydride undergoes a large number of condensation reactions (see text). Phthalic anhydride and phthalic acid are used industrially in the preparation of dyes, plastics (glyptals), plasticisers, benzoic acid, etc.

Phthalide may be prepared by reducing phthalic anhydride with zinc dust and aqueous sodium hydroxide:

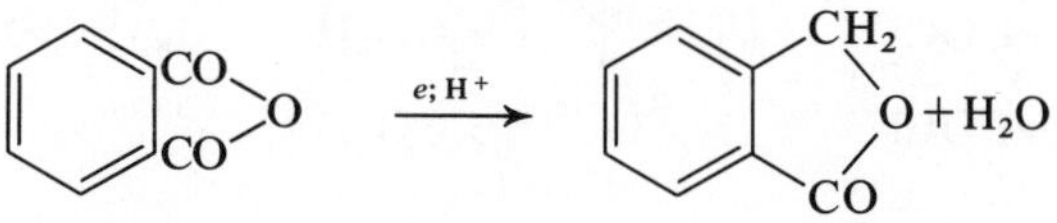

It may also be prepared by reducing phthalimide with a zinc-copper alloy and aqueous sodium hydroxide:

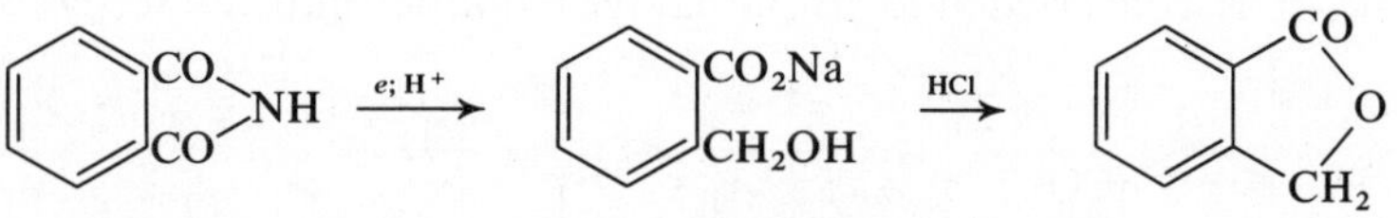

Phthalide, m.p. 75°C, may be converted into *homophthalic acid*, m.p. 175°C, by fusing it with potassium cyanide at 180°C and hydrolysing the product with boiling 50 per cent sulphuric acid:

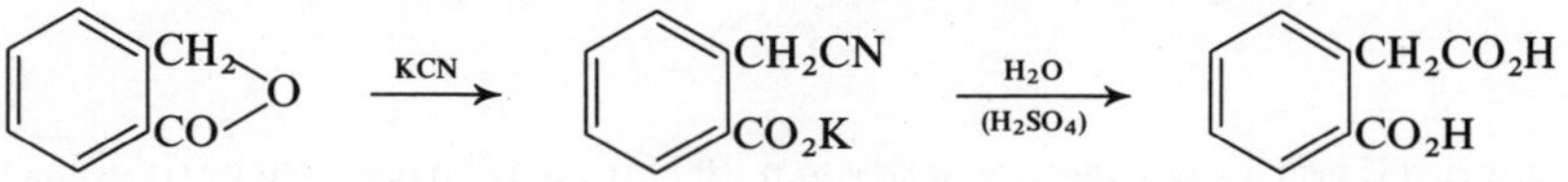

Phthalimide, m.p. 238°C, may be prepared by heating phthalic anhydride with aqueous ammonia:

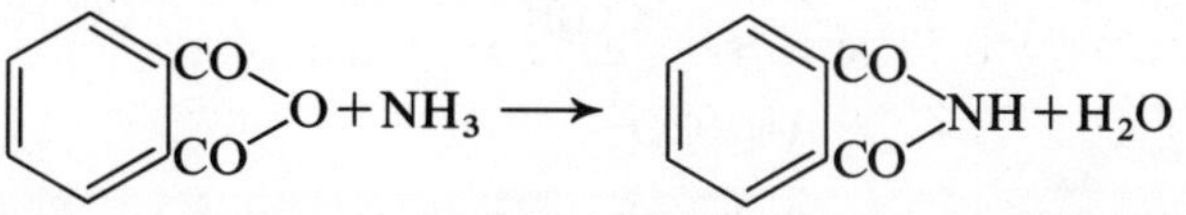

Alternatively, it may be prepared by fusing the anhydride with ammonium carbonate. It is weakly acidic; *e.g.*, with ethanolic potassium hydroxide it forms potassium phthalimide:

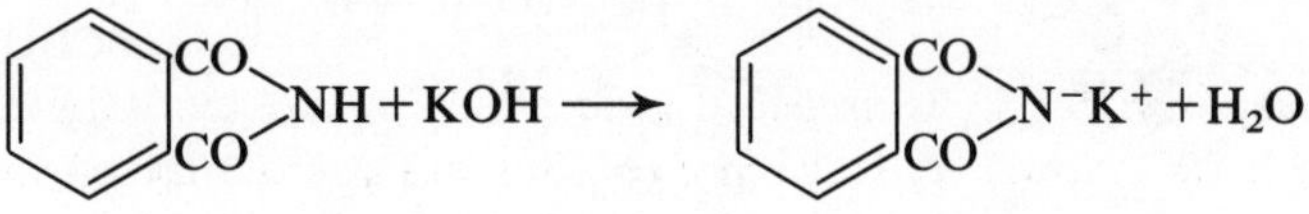

This salt is used in Gabriel's synthesis of primary amines (p. 370), and for preparing many α-aminoacids. The salt is decomposed into phthalimide by carbon dioxide.

Treatment of phthalimide with alkaline sodium hypochlorite results in the formation of anthranilic acid (p. 764). When hydrolysed with warm aqueous sodium hydroxide, phthalimide forms phthalic acid. If, however, phthalimide is allowed to stand in cold aqueous potassium hydroxide, or if warmed with barium hydroxide solution, it is converted into phthalamic acid:

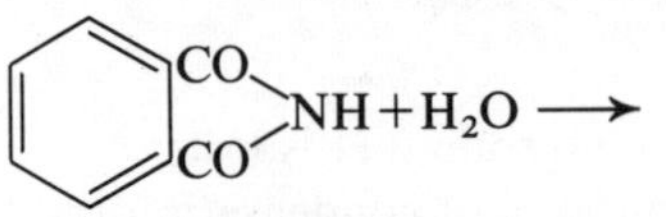

Phthaloyl chloride (phthalyl chloride) may be prepared by heating phthalic acid or phthalic anhydride with phosphorus pentachloride at 150°C. It may also be prepared by heating phthalic acid with thionyl chloride in the presence of zinc chloride at 220°C:

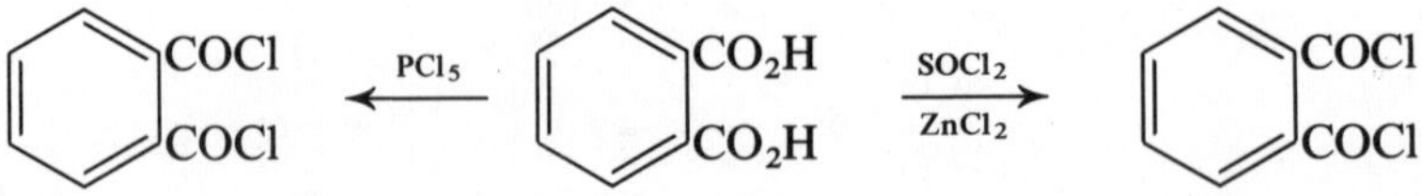

Phthaloyl chloride is a colourless oily liquid, m.p. 15–16°C. When heated with aluminium chloride for some time, it is converted into *as*-phthaloyl chloride, m.p. 89°C (*cf.* succinyl chloride):

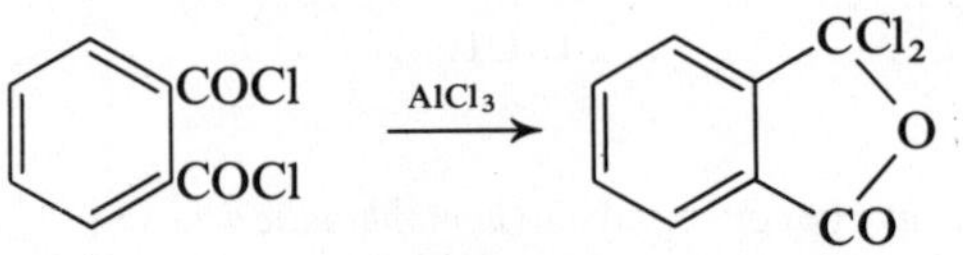

When reduced with zinc and hydrochloric acid, phthaloyl chloride is converted into phthalide.

Phthalaldehydic acid (o-*formylbenzoic acid*), m.p. 100·5°C, may be prepared by oxidising naphthalene with alkaline permanganate and decomposing the product, *phthalonic acid*, by boiling it in xylene solution:

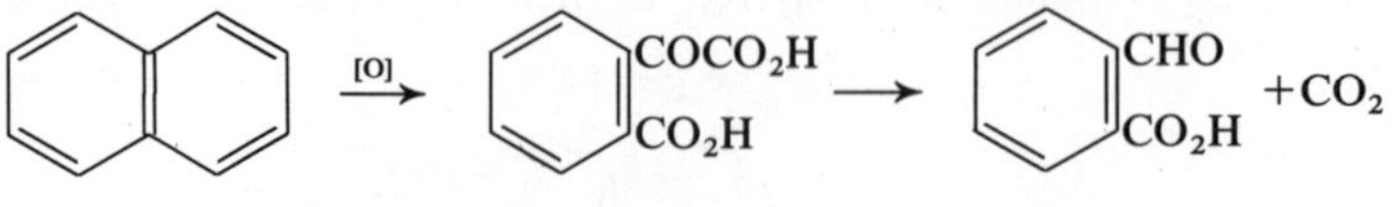

It may also be prepared by passing bromine vapour into phthalide and heating the product, 2-bromophthalide, with water:

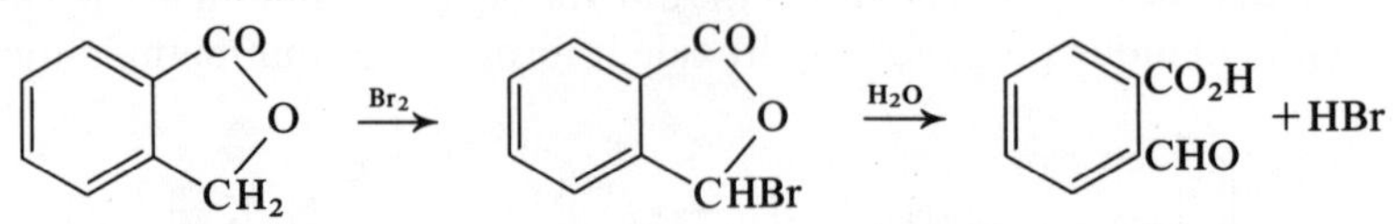

Phthalaldehyde, m.p. 56°C, may be prepared by ozonolysis of naphthalene (p. 758), or by the reduction of phthalobisdimethylamide with lithium aluminium hydride (Weygand *et al.*, 1951).

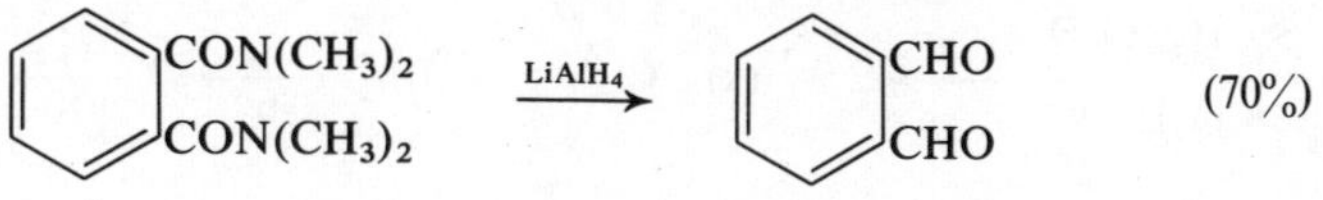

Monoperphthalic acid may be prepared as follows:

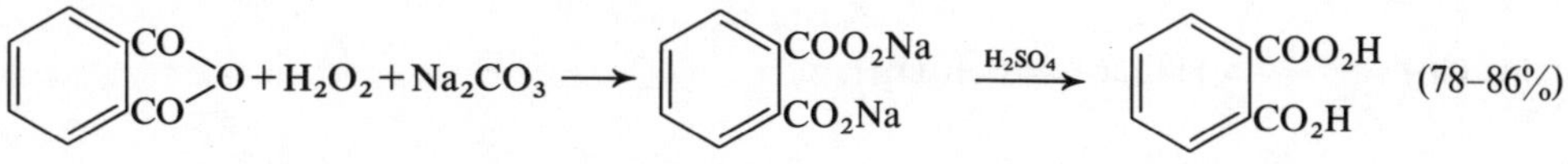

Monoperphthalic acid resembles perbenzoic acid in its properties, but is more stable.

Isophthalic acid (*benzene-1,3-dicarboxylic acid*), m.p. 346°C, may be prepared by oxidising *m*-xylene with permanganate. It does *not* form an anhydride.

Terephthalic acid (*benzene-1,4-dicarboxylic acid*) may be prepared by oxidising *p*-xylene with permanganate, or by oxidising *p*-methylacetophenone with concentrated nitric acid at about 300°C.

Commercially, it is prepared by the liquid phase oxidation of *p*-xylene at 140°C in the presence of a cobalt salt as catalyst.

Terephthalic acid sublimes without melting when heated above 300°C. It does *not* form an anhydride. It forms polyesters with glycol, and the plastic so obtained is known as *terylene*.

Polycarboxylic acids. Three isomeric tri- and three isomeric tetracarboxylic acids are known; the penta- and the hexacarboxylic acid are also known.

Mellitic acid (*benzenehexacarboxylic acid*) may be prepared by oxidising hexamethylbenzene with permanganate. It is also formed when graphite or wood-charcoal is oxidised with fuming nitric acid.

Mellitic acid is a stable solid, m.p. 288° (with decomposition). It occurs as its aluminium salt in peat and lignite.

PROBLEMS

1. Acetyl chloride is rapidly decomposed by water, but benzoyl chloride only very slowly. Suggest an explanation.

2. Complete the equations:

(i) $PhCO(CH_2)_2CO_2H \xrightarrow{\ ?\ } Ph(CH_2)_3CO_2H$ (ii) $Ph_2C(OH)CO_2H \xrightarrow[\text{heat}]{I_2/P}$?

(iii) 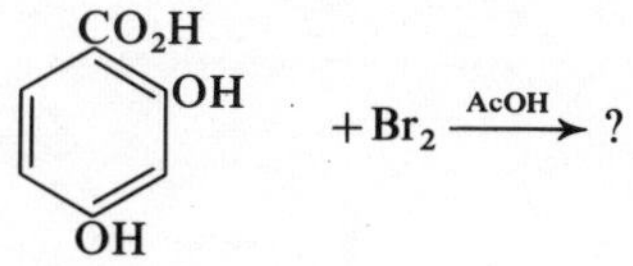 $+ Br_2 \xrightarrow{\text{AcOH}}$? (iv) 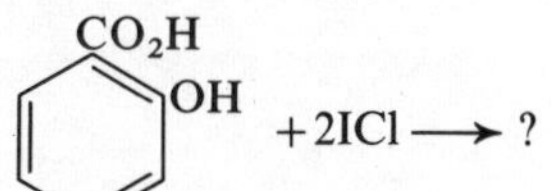 $+ 2ICl \longrightarrow$?

(v) 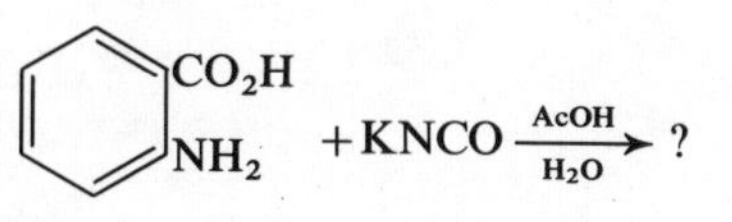 $+ KNCO \xrightarrow[\text{H}_2\text{O}]{\text{AcOH}}$? (vi) 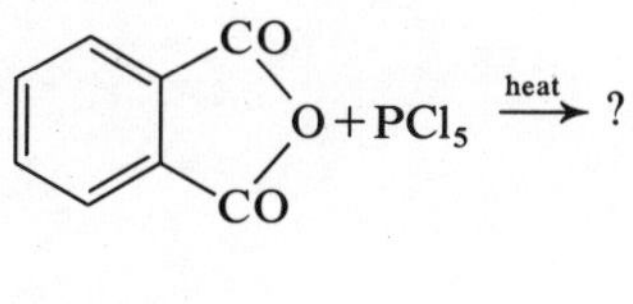 $+ PCl_5 \xrightarrow{\text{heat}}$?

(vii) $PhCOCl + HF$ (gas) $\longrightarrow$?   (viii) $PhCHO + ? \longrightarrow PhCHOHCH_2CO_2Et \xrightarrow{heat}$?

(ix) 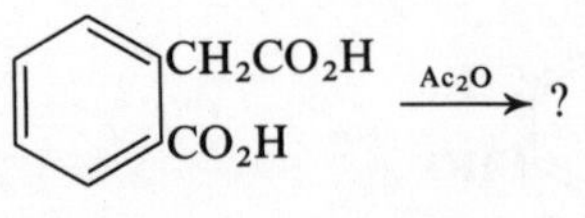 $\xrightarrow{Ac_2O}$?   (x) 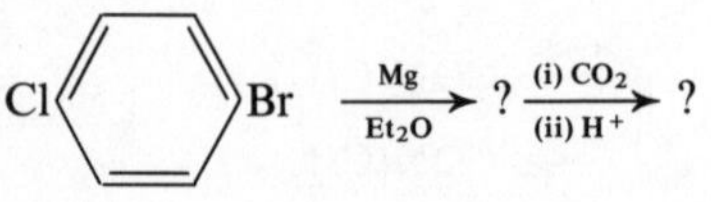 Cl——$Br \xrightarrow[Et_2O]{Mg}$? $\xrightarrow[\text{(ii) } H^+]{\text{(i) } CO_2}$?

(xi) $CH_2CH_2CH_2CO$—O (cyclic) $\xrightarrow[\text{(ii) } SOCl_2]{\text{(i) } HCl}$? $\xrightarrow[AlCl_3]{PhH}$? $\xrightarrow[MeOH]{KOH}$ cyclopropane derivative

(xii) $PhH + ? \xrightarrow{AlCl_3} PhCOCH{=}CHCO_2H$

(xiii) $\xrightarrow{?}$ $\xrightarrow{?}$ $\xrightarrow{?}$

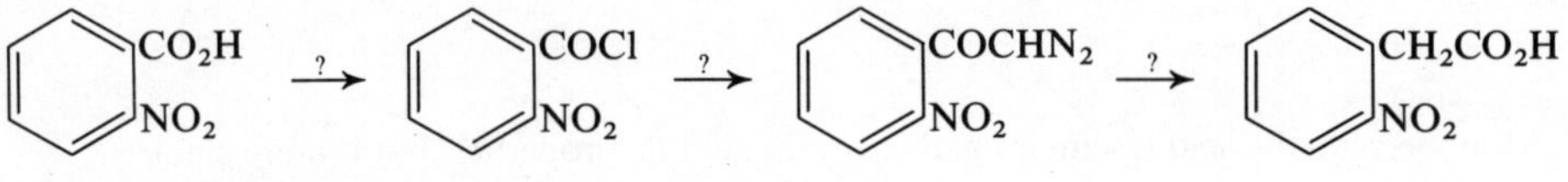

(xiv) $PhCOMe \xrightarrow{HNO_2}$? $\xrightarrow{Ac_2O}$? $\xrightarrow{H^+}$?

(xv) 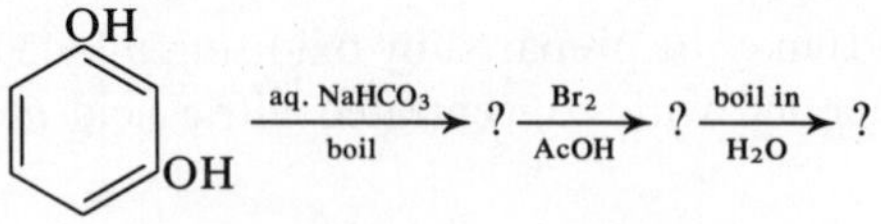 $\xrightarrow[\text{boil}]{\text{aq. } NaHCO_3}$? $\xrightarrow[AcOH]{Br_2}$? $\xrightarrow[H_2O]{\text{boil in}}$?

3. Deduce the structures of **A, B** and **C**:

A, $C_9H_{12} \xrightarrow{KMnO_4}$ **B**, $C_8H_6O_4 \xrightarrow[Fe]{Br_2}$ **C**, $C_8H_5BrO_4$ (1 product only)

4. Convert:

(i) $p\text{-}MeCOC_6H_4CO_2H \longrightarrow p\text{-}MeCHOHC_6H_4CO_2H$; (ii) $PhMe \longrightarrow o\text{-}BrC_6H_4COCl$;
(iii) $PhH \longrightarrow p\text{-}ClC_6H_4COC_6H_4Br\text{-}p$; (iv) $PhH \longrightarrow 4\text{-}NH_2\text{-}2\text{-}OH\text{-}$benzoic acid.

5. Convert toluene into *m*-toluic acid by *two* methods.
6. The sodium salt of 2-OH-*cis*-cinnamic acid (coumarinic acid), on acidification, gives coumarin, whereas the sodium salt of 2-OH-3-nitro-*cis*-cinnamic acid gives the free acid. Suggest an explanation.
7. Starting from benzene and any other compounds you like, synthesise:

(i) $Ph(CH_2)_3CO_2H$; (ii) $PhCH_2COPh$; (iii) 4-OH-3-MeO-cinnamic acid.

REFERENCES

Organic Reactions, Wiley. Vol. III (1946), Ch. 5. 'Azlactones'; Vol. VII (1953), Ch. 1. 'The Pechmann Reaction.'
INGOLD, 'Quantitative Study of Steric Hindrance', *Quart. Rev.*, 1957, **11**, 1.
NEWMAN (ed.), *Steric Effects in Organic Chemistry*, Wiley (1956).
LINDSEY ₋nd JESKY, 'The Kolbe–Schmitt Reaction', *Chem. Rev.*, 1957, **57**, 583.

29 Polynuclear hydrocarbons and their derivatives

Polynuclear hydrocarbons may be divided into two groups, those in which the rings are isolated, *e.g.*, biphenyl, diphenylmethane, etc.; and those in which two or more rings are fused together in the *o*-positions, *e.g.*, naphthalene, anthracene, etc.

Isolated systems

Biphenyl (diphenyl), m.p. 71°C, occurs in small quantities in coal-tar. It may be prepared:

(i) **By Fittig's reaction** (1863). This is carried out by treating bromobenzene with sodium in ethereal solution:

$$2C_6H_5Br + 2Na \longrightarrow C_6H_5C_6H_5 + 2NaBr \quad (20\text{--}30\%)$$

(ii) When bromobenzene in ethanolic solution is made alkaline and refluxed with hydrazine in the presence of a palladium catalyst on a calcium carbonate support, biphenyl is obtained (Busch *et al.*, 1936):

$$2C_6H_5Br \longrightarrow C_6H_5C_6H_5 + 2HBr \quad (80\%)$$

(iii) (*a*) By warming benzenediazonium sulphate in ethanol with copper powder (see also the Gomberg reaction, p. 675):

$$2C_6H_5N_2{}^+HSO_4{}^- \xrightarrow[\text{C}_2\text{H}_5\text{OH}]{\text{Cu}} C_6H_5C_6H_5 + 2N_2 + 2H_2SO_4 \quad (25\%)$$

The yield has been improved (to 50%) by warming a mixture of aniline, pentyl nitrite and benzene, and finally refluxing (Cadogan, 1962). This method may also be used for preparing biphenyl derivatives.

(b) By diazotising benzidine and allowing the diazonium salt solution to stand in contact with hypophosphorous acid:

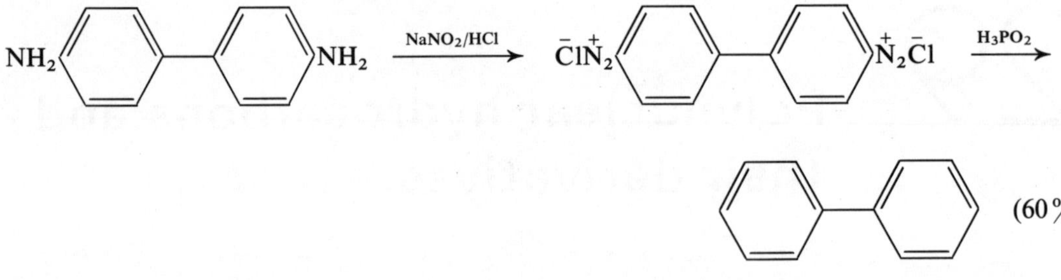

(iv) By the **Ullmann biaryl synthesis** (1903). Iodobenzene is heated with copper powder in a sealed tube:

$$2C_6H_5I + 2Cu \longrightarrow C_6H_5C_6H_5 + 2CuI \quad (80\%)$$

Aryl chlorides and bromides do not usually react unless there is a $-I$ group o- and/or p- to the halogen atom, e.g., o-chloronitrobenzene forms 2,2'-dinitrobiphenyl:

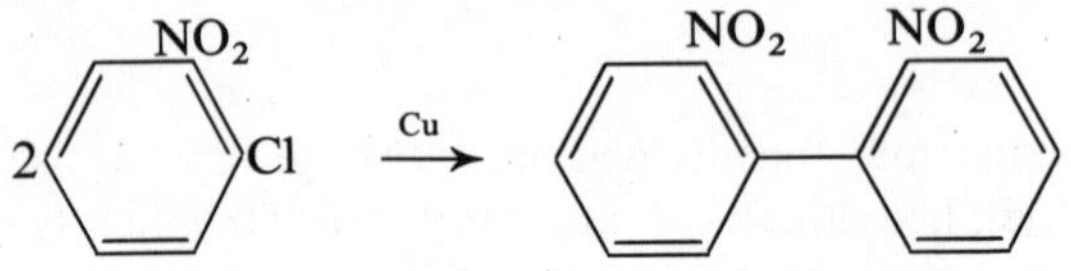

Kornblum et al. (1952) have shown that dimethylformamide is a good solvent for this synthesis, and the yields are higher.

There is still uncertainty about the mechanism of the Ullmann biaryl synthesis; both a free-radical and an ionic mechanism have been proposed.

(v) Biphenyl is formed by reaction between phenylmagnesium bromide and bromobenzene in the presence of a small amount of cobaltous chloride (see also p. 427).

$$C_6H_5MgBr + C_6H_5Br \xrightarrow{\text{CoCl}_2} C_6H_5C_6H_5 + MgBr_2 \quad (g.)$$

Industrially, biphenyl is prepared by passing benzene vapour through a red-hot tube, preferably packed with pumice (600–800°C):

$$2C_6H_6 \longrightarrow C_6H_5C_6H_5 + H_2$$

A more recent method is to mix benzene vapour (preheated to 650°C) with superheated steam (1 000–1 100°C), and passing the mixture into steel vessels coated internally with a film of Fe_3O_4.

Biphenyl is increasing in use as a heat transfer medium; chlorinated biphenyls are used as plasticisers.

Biphenyl undergoes the usual nuclear substitution reactions, the phenyl group being o- and p-orienting (one phenyl group behaving as an electron-releasing group and the other as an electron-acceptor). The first substituent enters mainly the 4-position, and to a lesser extent the 2-position. Whether this group is electron-releasing or withdrawing, introduction of a second substituent usually takes place in the *unsubstituted* ring in the p- and o-positions, e.g., on nitration, the main product is 4-nitrobiphenyl, together with a small amount of 2-nitrobiphenyl.

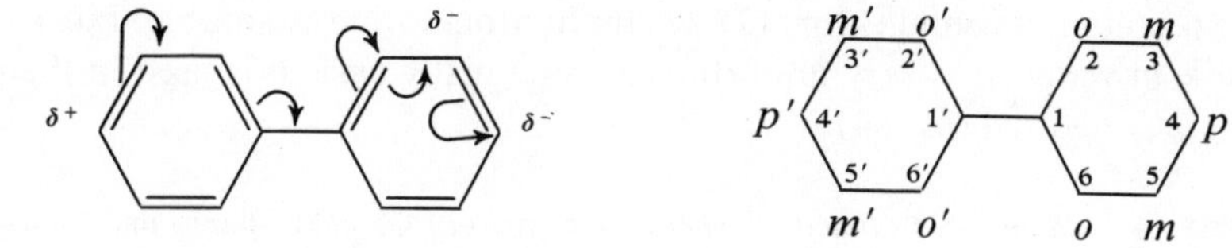

Further nitration gives 4,4'-dinitrobiphenyl (and some 2,4'- and 2,2'-dinitrobiphenyls). Biphenyl behaves similarly on halogenation and sulphonation.

According to the charge distribution approach, a *p*-nitro-group would have been expected to direct to the *m*-position in the other ring:

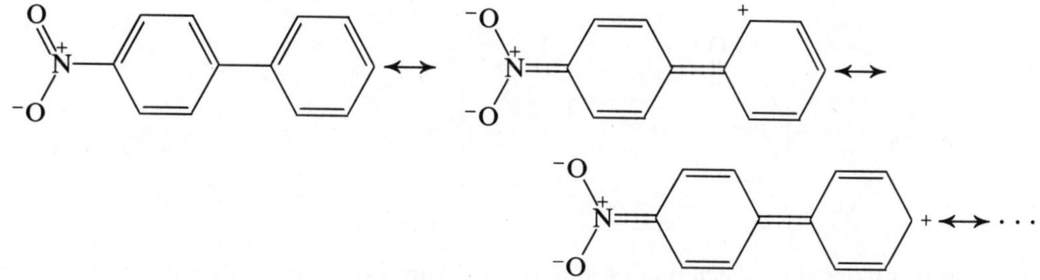

On the other hand, if the stabilities of the carbonium ion intermediates are examined, it will be seen that there is greater dispersal of charge in the *p*-(or *o*-)ion than in the *m*-. Consequently, the former is more stable and so this orientation is favoured.

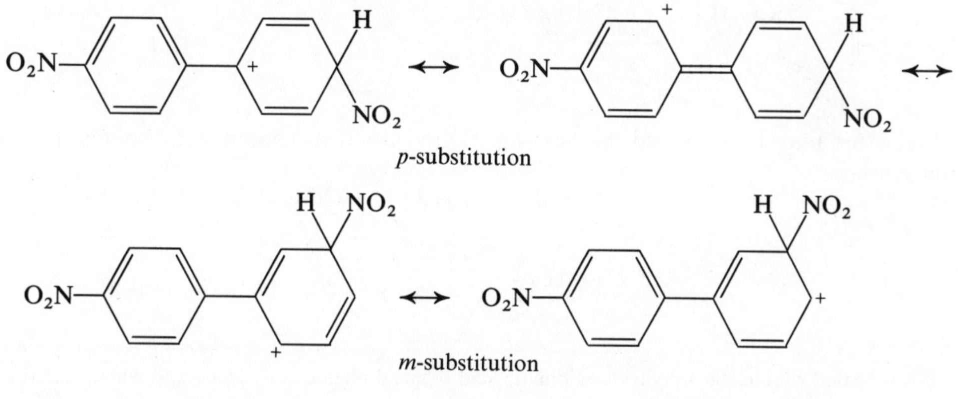

p-substitution

m-substitution

When oxidised with chromic acid, biphenyl forms a small amount of benzoic acid, most of the hydrocarbon being oxidised completely to carbon dioxide and water. On the other hand, ozonolysis of biphenyl at $-20°C$ gives an 86 per cent yield of benzoic acid (Copeland *et al.*, 1961).

Benzidine (4,4'-*diaminobiphenyl*) may be prepared by the *benzidine rearrangement* (p. 685); hydrazobenzene, on warming with hydrochloric acid, rearranges to benzidine:

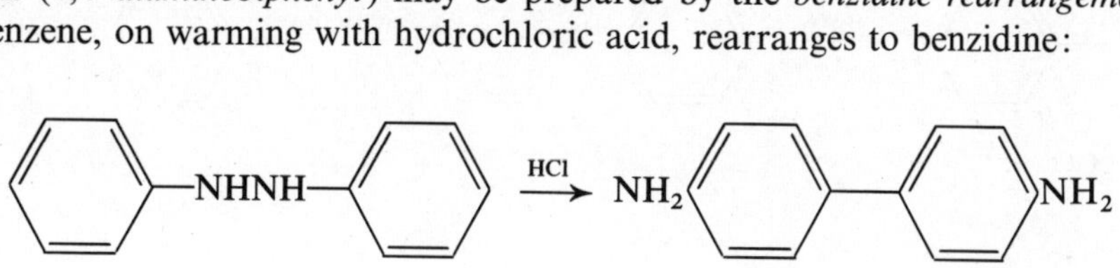

(90%)

Benzidine is a colourless solid, m.p. 127°C. Its hydrochloride is soluble, but its sulphate is sparingly soluble in water. It is very important commercially since it is used in the preparation of azo-dyes, *e.g.*, Congo red (p. 884).

o-**Tolidine** (4,4'-*diamino*-3,3'-*dimethylbiphenyl*), m.p. 129°C, and **dianisidine** (4,4'-*diamino*-3,3'-*dimethoxybiphenyl*), m.p. 138°C, are manufactured on a large scale as intermediates in the preparation of azo-dyes.

Diphenic acid (*biphenyl-2,2'-dicarboxylic acid*), m.p. 229°C, may be readily prepared by oxidising phenanthraquinone with potassium dichromate and sulphuric acid, or by the direct oxidation of phenanthrene with 50 per cent hydrogen peroxide in glacial acetic (68% yield).

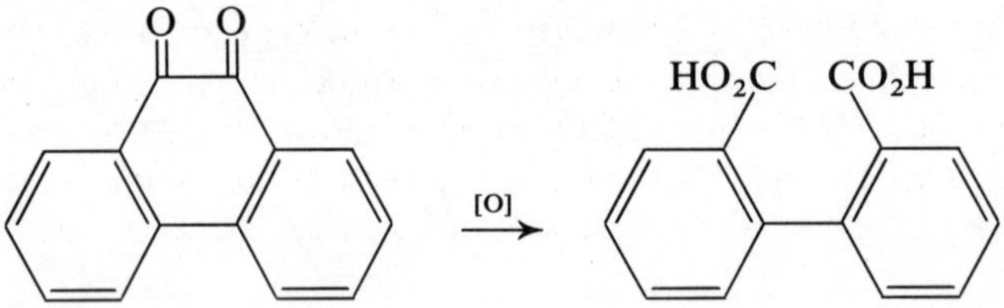

It may also be prepared by the action of ammoniacal cuprous oxide on diazotised anthranilic acid:

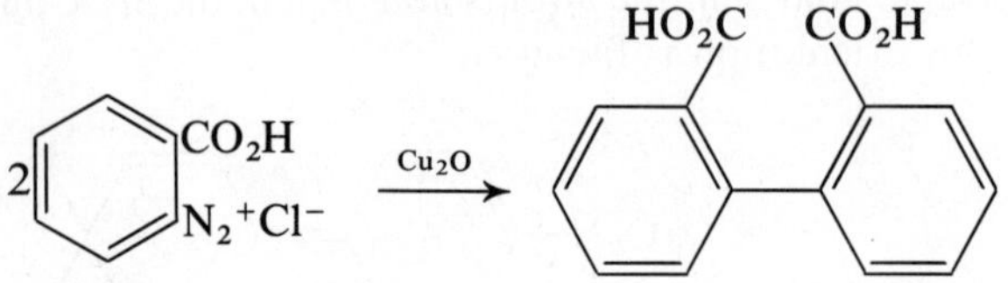

The methyl ester may be obtained by heating methyl *o*-iodobenzoate with copper powder (Ullmann synthesis):

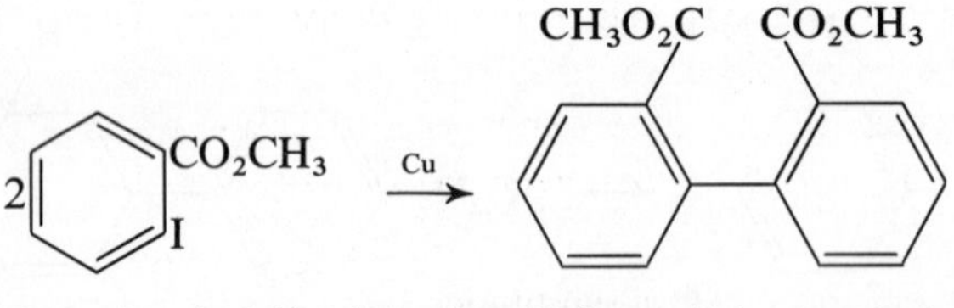

When heated with acetic anhydride, diphenic acid forms diphenic anhydride, and when heated with concentrated sulphuric acid, fluorenone-4-carboxylic acid:

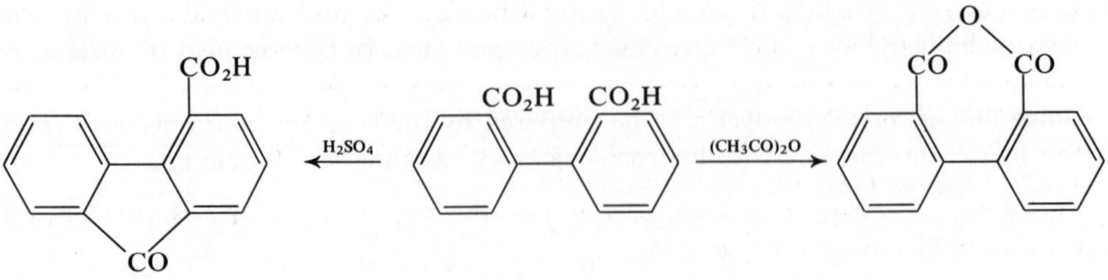

When the calcium salt of diphenic acid is distilled, fluorenone is obtained:

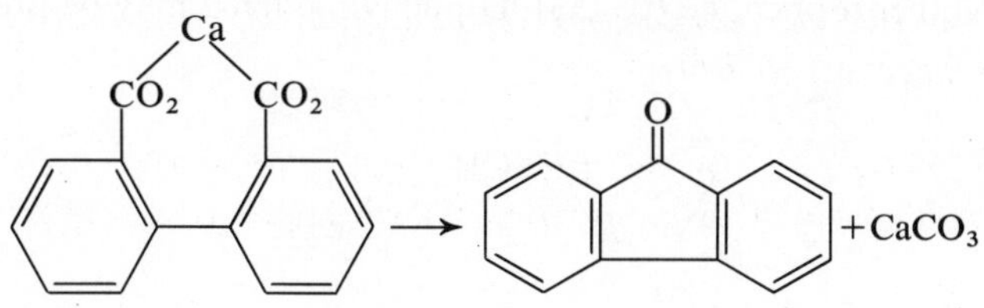

When oxidised with potassium permanganate, diphenic acid forms phthalic acid. Distillation with soda-lime gives biphenyl.

If at least three of the positions 2, 2′, 6 and 6′ are occupied by sufficiently large groups, free rotation about the single bond joining the two phenyl groups is no longer possible. Provided each ring has no vertical plane of symmetry, this restricted rotation gives rise to optical activity due to the molecule being *asymmetric as a whole* (p. 474); *e.g.*, *6-nitrobiphenyl-2,2′-dicarboxylic acid* (I), and *6,6′-diamino-2,2′-dimethylbiphenyl* (II), have been resolved. If the substituent group is large enough, then only two groups in the *o*- and *o′*-positions will cause restricted rotation, *e.g.*, *biphenyl-2,2′-disulphonic* acid (III), has been shown to be optically active.

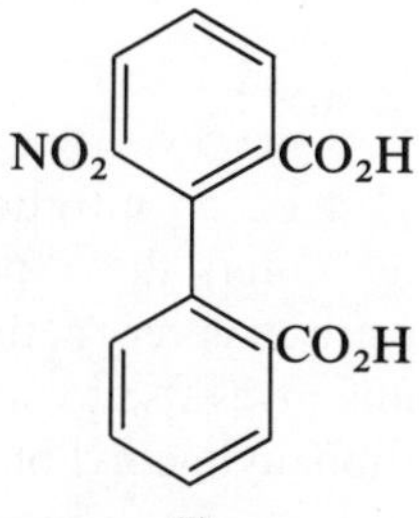

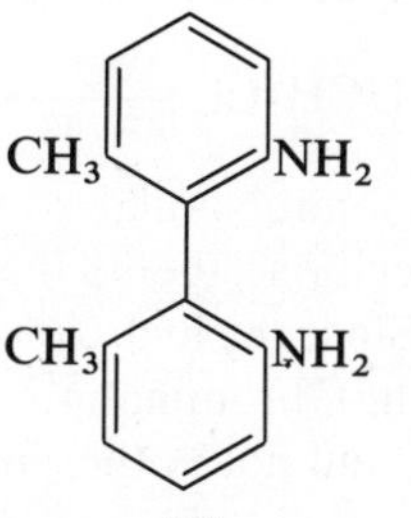

The cause of the restricted rotation is due mainly to the steric effects of the groups in the *o*- and *o′*-positions.

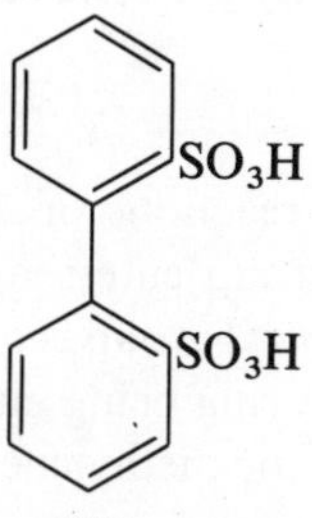

This type of stereoisomerism, arising from restricted rotation about a single bond and where the stereoisomers can be isolated, is called **atropisomerism**; the isomers are known as *atropisomers*.

Since the optical activity of biphenyl compounds arises from restricted rotation, it might be expected that racemisation would not be possible. In practice, however, many optically active biphenyls can be racemised, *e.g.*, by boiling in solution. The general theory of these racemisations is that heating increases the amplitude of vibrations of the substituents and also the amplitude of vibration of the interannular bond. In addition to these bond stretchings, the various valency angles between ring and substituents are also deformed. Thus the nuclei will pass through a common plane, and hence, when the molecule returns to its restricted state, it will do so equally in the (+)- and (−)-forms, *i.e.*, the molecules are racemised (see also Vol. 2, Ch. 5).

A number of polyphenyls and some of their derivatives have been prepared, *e.g.*, *terphenyl, quaterphenyl, quinquephenyl* and *sexiphenyl*.

Diphenylmethane, m.p. 26°C. Some related compounds of diphenylmethane have already been discussed, *e.g.*, benzophenone (p. 753). Diphenylmethane may be prepared:

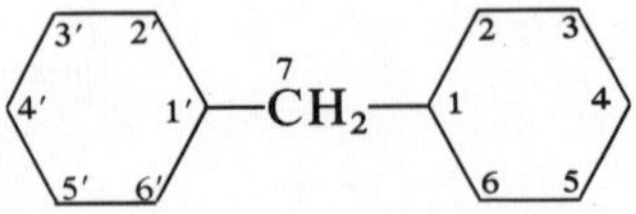

(i) By the Friedel–Crafts condensation between benzyl chloride and benzene:

$$C_6H_5CH_2Cl + C_6H_6 \xrightarrow{AlCl_3} C_6H_5CH_2C_6H_5 + HCl \quad (50\%)$$

(ii) By the condensation between one molecule of formaldehyde and two molecules of benzene in the presence of concentrated sulphuric acid:

$$2C_6H_6 + CH_2O \xrightarrow{H_2SO_4} C_6H_5CH_2C_6H_5 + H_2O$$

(iii) By heating benzophenone with hydriodic acid and red phosphorus at 160°C under pressure, by the Wolff–Kishner reduction, or by LAH—AlCl$_3$.

$$C_6H_5COC_6H_5 \xrightarrow{HI/P} C_6H_5CH_2C_6H_5 \quad (95\%)$$

(iv) By the following Grignard reaction:

$$C_6H_5MgBr + C_6H_5CH_2Cl \xrightarrow[60-70°C]{PhH} C_6H_5CH_2C_6H_5 \quad (ex.)$$

The reactions of diphenylmethane are similar to those of biphenyl, *e.g.*, on nitration, the nitro-group enters mainly the 4-position; the second nitro-group then enters the 4'-position (the benzyl group is *o-p*-orienting). Since the hydrogen of the methylene group is very active (due to each ring being electron-attracting), bromination of diphenylmethane gives rise to substitution at the methylene carbon atom, and not in the ring, to form, *e.g.*, diphenylmethyl bromide:

$$C_6H_5CH_2C_6H_5 + Br_2 \longrightarrow C_6H_5CHBrC_6H_5 + HBr$$

Diphenylmethyl bromide is readily hydrolysed by alkali to *benzhydrol* (*diphenylcarbinol*; p. 753). When oxidised with chromic acid, diphenylmethane forms benzophenone. When its vapour is passed through a red-hot tube, diphenylmethane forms fluorene:

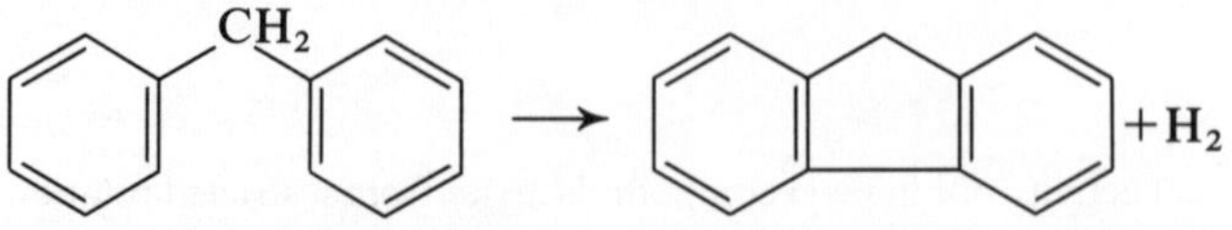

Triphenylmethane (tritane) may be prepared by the following methods:

$$2C_6H_5 + C_6H_5CHCl_2 \xrightarrow{AlCl_3} (C_6H_5)_3CH + 2HCl$$

$$3C_6H_6 + CHCl_3 \xrightarrow{AlCl_3} (C_6H_5)_3CH + 3HCl \quad (33\%)$$

$$C_6H_5CHO + 2C_6H_6 \xrightarrow{ZnCl_2} (C_6H_5)_3CH + H_2O$$

Triphenylmethane, m.p. 93°C, is the parent substance of the triphenylmethane dyes (p. 886). When brominated, it forms triphenylmethyl bromide, $(C_6H_5)_3CBr$ (*cf.* diphenylmethane), and may be oxidised directly to triphenylcarbinol (see below). Triphenylmethane is acidic (more so than diphenylmethane), the conjugate base being stabilised by resonance. Thus, when treated with sodium in ethereal solution (or in liquid ammonia), triphenylmethane forms triphenyl-methylsodium:

$$(C_6H_5)_3CH + Na \longrightarrow (C_6H_5)_3\bar{C}Na^+ + \tfrac{1}{2}H_2$$

Triphenylmethylsodium is also formed when triphenylmethyl chloride is treated with sodium amalgam.

Derivatives of triphenylmethane are best prepared synthetically (see, *e.g.*, p. 886).

Triphenylmethyl chloride (trityl chloride), m.p. 112°C, may be prepared from triphenyl-carbinol and acetyl chloride (see below), or by the Friedel–Crafts condensation between benzene and carbon tetrachloride; tetraphenylmethane is *not* formed:

$$3C_6H_6 + CCl_4 \xrightarrow{\text{AlCl}_3} (C_6H_5)_3CCl + 3HCl \quad (68\text{–}84\%)$$

The halogen atom is extremely reactive, *e.g.*, when boiled with water, triphenylmethyl chloride forms the corresponding alcohol, and with ethanol it forms the ethyl ether:

$$(C_6H_5)_3COC_2H_5 \xleftarrow{\text{C}_2\text{H}_5\text{OH}} (C_6H_5)_3CCl \xrightarrow{\text{H}_2\text{O}} (C_6H_5)_3COH$$

Triphenylmethyl chloride may be converted into triphenylmethane as follows:

$$(C_6H_5)_3CCl \xrightarrow{\text{Mg}} (C_6H_5)_3CMgCl \xrightarrow{\text{H}_2\text{O}} (C_6H_5)_3CH$$

This reaction is particularly interesting since the formation of the Grignard reagent from triphenylmethyl bromide was shown to occur via a free-radical mechanism (Gomberg and Bachmann, 1930). These authors showed that when half the total amount of magnesium had reacted, the solution contained *no Grignard reagent*:

$$2(C_6H_5)_3CBr + Mg \longrightarrow 2(C_6H_5)_3C\cdot + MgBr_2$$
$$Mg + MgBr_2 \rightleftharpoons 2\cdot MgBr$$
$$(C_6H_5)_3C\cdot + \cdot MgBr \longrightarrow (C_6H_5)_3CMgBr$$

When triphenylmethyl chloride is dissolved in liquid sulphur dioxide, it ionises:

$$Ph_3CCl \rightleftharpoons Ph_3C^+ + Cl^-$$

Triphenylcarbinol, m.p. 165°C, may be prepared by the oxidation of triphenylmethane with chromium trioxide:

$$(C_6H_5)_3CH \xrightarrow{\text{CrO}_3} (C_6H_5)_3COH \quad (85\%)$$

It may also be prepared from benzophenone or ethyl benzoate as follows:

$$(C_6H_5)_2CO \xrightarrow[\text{(ii) H}_3\text{O}^+]{\text{(i) C}_6\text{H}_5\text{MgBr}} (C_6H_5)_3COH \xleftarrow[\text{(ii) H}_3\text{O}^+]{\text{(i) 2C}_6\text{H}_5\text{MgBr}} C_6H_5CO_2C_2H_5$$

Triphenylcarbinol reacts extremely rapidly with hydrochloric acid or acetyl chloride to form triphenylmethyl chloride (*cf.* tertiary alcohols, p. 182):

$$(C_6H_5)_3COH + CH_3COCl \longrightarrow (C_6H_5)_3CCl + CH_3CO_2H \quad (93\text{–}95\%)$$

It also reacts rapidly with alcohols in the presence of a trace of acid or iodine to form ethers, *e.g.*,

$$(C_6H_5)_3COH \underset{}{\overset{\text{H}^+}{\rightleftharpoons}} (C_6H_5)_3C{-}OH_2{}^+ \rightleftharpoons H_2O + (C_6H_5)_3C^+$$
$$\xrightarrow{\text{C}_2\text{H}_5\text{OH}} (C_6H_5)_3COC_2H_5 + H^+$$

Triphenylcarbinol is reduced to triphenylmethane by formic acid or LAH—AlCl$_3$, and condenses with aniline hydrochloride to form *p*-aminotetraphenylmethane.

Tetraphenylmethane, m.p. 282°C, may be prepared as follows:

$$(C_6H_5)_3CCl + C_6H_5MgBr \longrightarrow (C_6H_5)_4C + MgClBr \quad (v.p.)$$

It may also be prepared by diazotising *p*-aminotetraphenylmethane in ethanolic solution and then boiling (see also triphenylcarbinol, above):

$$(C_6H_5)_3COH + C_6H_5NH_2 \cdot HCl \longrightarrow (C_6H_5)_3CC_6H_4NH_2 \cdot HCl \xrightarrow[C_2H_5OH]{NaNO_2/H_2SO_4} (C_6H_5)_4C$$

Tetraphenylmethane *cannot* be prepared by reaction between benzene and carbon tetrachloride in the presence of aluminium chloride (see triphenylmethyl chloride, above). The reason for this is not certain, but it is likely that the steric effect plays a large part in hindering the reaction.

When triphenylmethyl chloride, in benzene solution, is treated with silver *in the absence of air*, and the yellow benzene solution then evaporated, the residue is a white solid, m.p. 145–147°C. This was believed to be *hexaphenylethane*, $(C_6H_5)_6C_2$, the yellow colour being due to its dissociation into triphenylmethyl free radicals (Gomberg, 1900). Lankamp *et al.* (1968) and Guthrie *et al.* (1969), however, have now shown that the dimer of triphenylmethyl radical is not hexaphenylethane but is **1-diphenylmethylene-4-tritylcyclohexa-2,5-diene**:

$$2Ph_3CCl \xrightarrow{Ag} 2Ph_3C\cdot \rightleftharpoons Ph_2C=$$

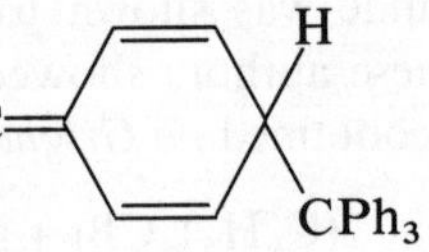

The stability of this free radical is attributed to resonance, a large number of resonating structures contributing to the resonance hybrid:

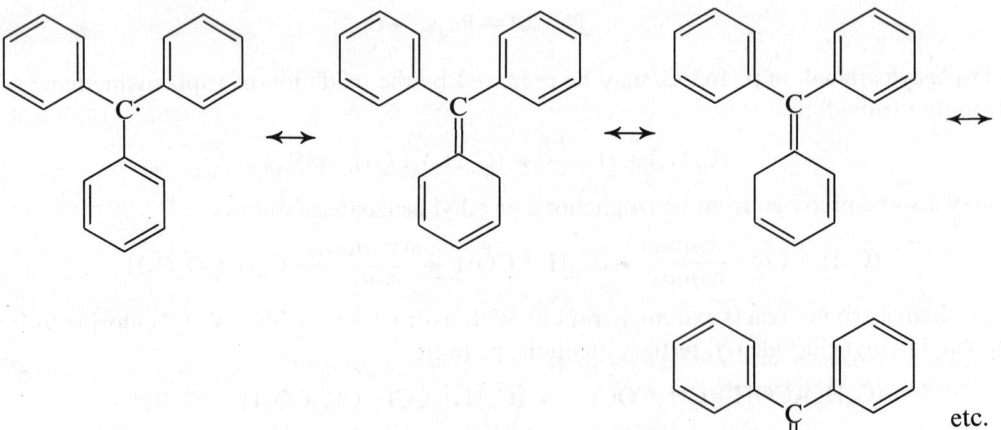

Triphenylmethyl free radical reacts immediately with a number of reagents to form triphenylmethyl derivatives, *e.g.*, with oxygen, it forms a colourless peroxide:

$$(C_6H_5)_3C\cdot \xrightarrow{O_2} (C_6H_5)_3C—O—O\cdot \xrightarrow{(C_6H_5)_6C_2} (C_6H_5)_3C—O—O—C(C_6H_5)_3 + (C_6H_5)_3C\cdot$$

With iodine it forms triphenylmethyl iodide, and with nitric oxide, nitrosotriphenylmethane (the formation of the latter compound appears to be uncertain):

$$2(C_6H_5)_3C\cdot + I_2 \rightleftharpoons 2(C_6H_5)_3CI$$
$$(C_6H_5)_3C\cdot + NO \longrightarrow (C_6H_5)_3CNO$$

It also combines with sodium to form triphenylmethylsodium, a brick-red solid and an electrical conductor:

$$(C_6H_5)_3C\cdot + Na \longrightarrow (C_6H_5)_3\overset{-}{C}Na^+$$

In addition to the above reactions, triphenylmethyl free radical can act as a powerful reducing agent, and will reduce the salts of silver, gold and mercury to the metals, *e.g.*,

$$(C_6H_5)_3C\cdot + AgCl \longrightarrow (C_6H_5)_3CCl + Ag$$

The triphenylmethyl free radical is an example of a long-life free radical (see also p. 440). The problem here is why does 'hexaphenylethane' dissociate so readily (as compared to ethane)? The answer is believed to be as follows. The free radical is stabilised by resonance, which therefore acts as a 'driving force' in the dissociation. At the same time, steric factors also operate. In 'hexaphenylethane' (sp^3 hybridisation) there is steric strain due to crowding, and this is relieved by dissociation to form a planar or almost planar free radical (sp^2 hybridisation; resonance requires complete or almost complete planarity). That the steric effect is operating is supported by the fact that methyl groups in the *o*-positions increase the dissociation. Further support for the steric effect is the case of Tschitschibabin's hydrocarbon:

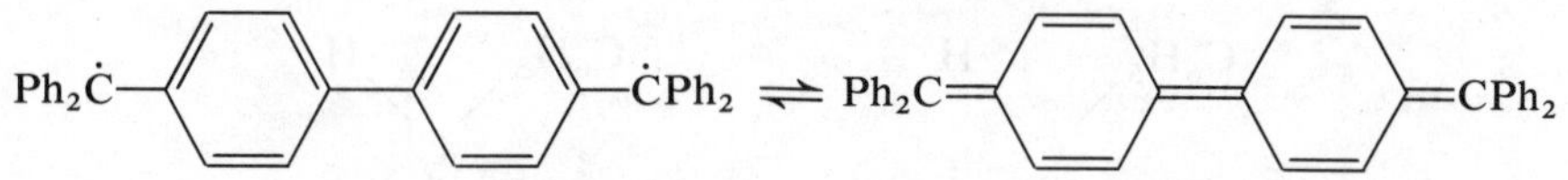

This compound contains about 4·5 per cent free radical, but when all four *o*-positions are occupied by chlorine the free-radical content is almost 100 per cent. In the chlorinated compound coplanarity of the two benzene rings is prevented, *i.e.*, this is an example of steric inhibition of resonance.

When dissolved in liquid sulphur dioxide, the solution shows a high conductivity (Walden, 1903), and this was explained by assuming that the 'hexaphenylethane' dissociates into positive and negative ions:

$$(C_6H_5)_6C_2 \rightleftharpoons (C_6H_5)_3C^+ + (C_6H_5)_3C^-$$

The existence of the triphenylmethyl carbonium ion has definitely been established by Dauben *et al.* (1960), who prepared the crystalline perchlorate, $Ph_3C^+ClO_4^-$, and the fluoroborate, $Ph_3C^+BF_4^-$. These carbonium ions have been shown to have the property of abstracting hydride ions from hydrocarbons of low nucleophilic power:

$$Ph_3C^+ClO_4^- + RH \rightleftharpoons Ph_3CH + R^+ClO_4^-$$

This property thus affords a means of preparing new carbonium salts.

Bibenzyl (1,2-*diphenylethane*), m.p. 52°C, may be prepared by the Friedel–Crafts condensation between benzene and ethylene dichloride:

$$2C_6H_6 + ClCH_2CH_2Cl \xrightarrow{AlCl_3} C_6H_5CH_2CH_2C_6H_5 + 2HCl$$

Another method of preparation is by the action of sodium or copper on benzyl bromide:

$$2C_6H_5CH_2Br \xrightarrow{Na} C_6H_5CH_2CH_2C_6H_5$$

Bibenzyl is oxidised by permanganate or chromic acid to benzoic acid.

***trans*-Stilbene** (*trans*-1,2-diphenylethylene) may be prepared:

(i) By treating benzylmagnesium bromide with benzaldehyde and dehydrating the product, benzylphenylmethanol:

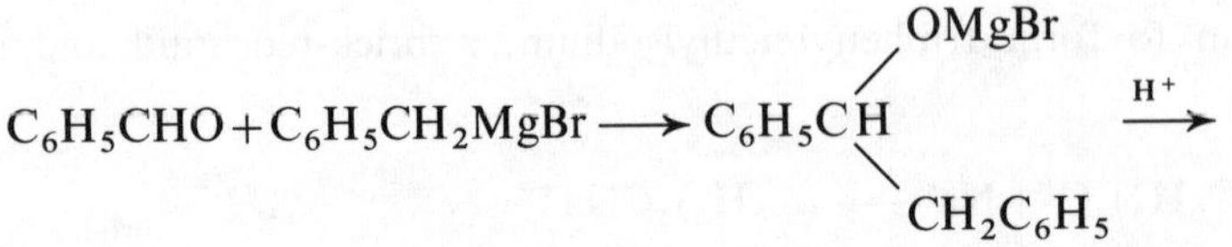

$$C_6H_5CHO + C_6H_5CH_2MgBr \longrightarrow C_6H_5\underset{\underset{CH_2C_6H_5}{|}}{\overset{\overset{OMgBr}{|}}{C}}H \xrightarrow{H^+}$$

$$C_6H_5CHOHCH_2C_6H_5 \xrightarrow[\text{(heat)}]{-H_2O} C_6H_5CH{=}CHC_6H_5$$

(ii) By reducing benzoin with amalgamated zinc and an ethanolic solution of hydrogen chloride:

$$C_6H_5CHOHCOC_6H_5 \longrightarrow C_6H_5CH{=}CHC_6H_5 \quad (53\text{--}57\%)$$

(iii) By heating α-phenylcinnamic acid in quinoline in the presence of copper chromite:

$$C_6H_5CHO + C_6H_5CH_2CO_2K \xrightarrow{(CH_3CO)_2O}$$

$$C_6H_5CH{=}C(C_6H_5)CO_2H \longrightarrow C_6H_5CH{=}CHC_6H_5$$

(iv) Stilbene may also be prepared by the Meerwein reaction (p. 675).

trans-Stilbene, m.p. 124°C, is the stable isomer; *cis*-stilbene (isostilbene), b.p. 145° (13 mm), is the unstable isomer:

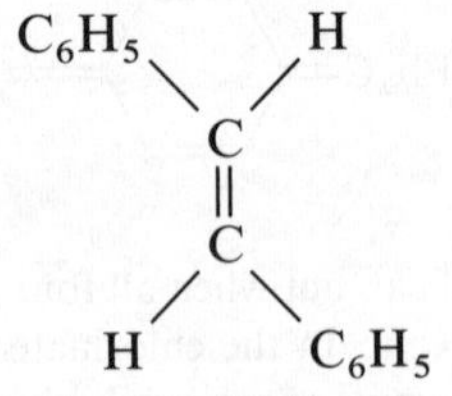

| *trans*-stilbene | *cis*-stilbene |

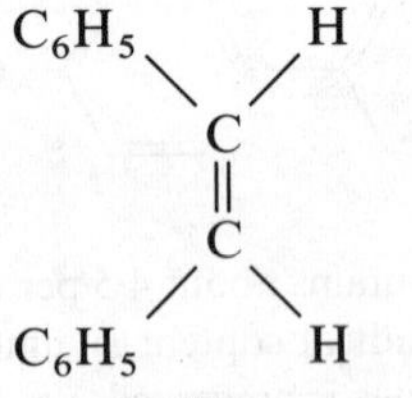

trans-Stilbene is reduced by sodium and ethanol to bibenzyl. It adds on bromine to form stilbene dibromide which, on heating with ethanolic potassium hydroxide, forms **tolan** (*diphenylacetylene*), m.p. 62°C:

$$C_6H_5CH{=}CHC_6H_5 \xrightarrow{Br_2} C_6H_5CHBrCHBrC_6H_5 \xrightarrow{KOH} C_6H_5C{\equiv}CC_6H_5$$

Stilbene dibromide reacts with silver acetate to form two isomeric diacetates. These, on hydrolysis, form hydrobenzoin and isohydrobenzoin (see later):

$$C_6H_5CH(OCOCH_3)CH(OCOCH_3)C_6H_5 \xrightarrow{H_2O} C_6H_5CHOHCHOHC_6H_5$$

cis-**Stilbene** may be prepared by reducing diphenylacetylene with zinc dust and ethanol, or by irradiating *trans*-stilbene with ultra-violet light.

$$C_6H_5C{\equiv}CC_6H_5 \xrightarrow{\;e;\,H^+\;} C_6H_5CH{=}CHC_6H_5$$

cis-Stilbene is readily converted into *trans*-stilbene under the catalytic influence of traces of hydrogen bromide and peroxides.

Benzoin, m.p. 137°C, is usually prepared by the *benzoin condensation*, this reaction being carried out by refluxing benzaldehyde with aqueous ethanolic potassium cyanide:

$$2C_6H_5CHO \xrightarrow{\;KCN\;} C_6H_5CHOHCOC_6H_5 \quad (83\%)$$

The generally accepted mechanism is

$$C_6H_5\overset{\displaystyle O}{\underset{\displaystyle H}{\overset{\|}{C}}}+:CN^- \rightleftharpoons \left[C_6H_5\overset{\displaystyle O^-}{\underset{\displaystyle H}{\overset{|}{C}}}{-}CN \right] \rightleftharpoons \left[C_6H_5{-}\overset{\displaystyle OH}{\underset{\displaystyle}{\overset{|}{\underset{..}{C}}}}{-}CN \right] \xrightarrow{\;C_6H_5CHO\;}$$

$$\left[C_6H_5\overset{\displaystyle OH}{\underset{\displaystyle CN}{\overset{|}{C}}}{-}\overset{\displaystyle O^-}{\underset{\displaystyle H}{\overset{|}{C}}}{-}C_6H_5 \right] \rightleftharpoons \left[C_6H_5\overset{\displaystyle O^-}{\underset{\displaystyle CN}{\overset{|}{C}}}{-}\overset{\displaystyle OH}{\underset{\displaystyle H}{\overset{|}{C}}}C_6H_5 \right] \rightleftharpoons C_6H_5\overset{\displaystyle O}{\overset{\|}{C}}{-}\overset{\displaystyle OH}{\underset{\displaystyle H}{\overset{|}{C}}}C_6H_5+:CN^-$$

When a mixture of aldehydes is treated with aqueous ethanolic potassium cyanide, 'mixed' benzoins (as well as the 'single' benzoins) are obtained:

$$Ar^1CHO+Ar^2CHO \xrightarrow{\;KCN\;} Ar^1CHOHCOAr^2+Ar^2CHOHCOAr^1$$

Furthermore, the reversibility of the benzoin condensation is indicated by the fact that when the benzoin $Ar^1CHOHCOAr^1$ is heated with the aldehyde Ar^2CHO in the presence of potassium cyanide, a mixed benzoin is obtained (Buck and Ide, 1933):

$$Ar^1CHOHCOAr^1+2Ar^2CHO \rightleftharpoons 2Ar^1CHOHCOAr^2$$

Aliphatic aldehydes do not undergo the benzoin condensation.

Benzoins (and acyloins) tautomerise to *enediols*:

$$-CHOHCO- \rightleftharpoons HO-\overset{|}{C}{=}\overset{|}{C}-OH$$

There is much evidence to support this, *e.g.*, acetylation produces the enediol diacetate. The enediol is usually less stable than its corresponding α-hydroxyketone; in certain cases the former has been isolated.

Benzoin is reduced to **deoxybenzoin** (*desoxybenzoin*), m.p. 60°C, which may also be prepared by the following Friedel–Crafts reaction:

$$C_6H_5CH_2COCl+C_6H_6 \xrightarrow{\;AlCl_3\;} C_6H_5CH_2COC_6H_5+HCl \quad (82\text{--}83\%)$$

The prefix *deoxy-* (formerly *desoxy-*) indicates the replacement of a hydroxyl group by hydrogen. Deoxybenzoin contains an active methylene group; *e.g.*, it forms a sodio-derivative, condenses with aldehydes in the presence of piperidine, and adds on to α,β-unsaturated carbonyl compounds (*cf.* ethyl malonate).

Benzoin is an α-hydroxyketone; hence it reduces Fehling's solution and forms an osazone.

When reduced by sodium amalgam, benzoin forms mainly **hydrobenzoin**, and a small amount of **isohydrobenzoin**:

$$C_6H_5CHOHCOC_6H_5 \xrightarrow{e; H^+} C_6H_5CHOHCHOHC_6H_5$$

This formula (tolylene glycol) contains two identical asymmetric carbon atoms, and hence the compound can exist in the *dextro*, *lævo* and *meso*-forms (*cf.* tartaric acid). The racemic modification is known as hydrobenzoin (m.p. 139°C), and the *meso*-form as isohydrobenzoin (m.p. 121°C). Both forms give benzoic acid on oxidation.

Benzoin is oxidised by chromic acid to a mixture of benzaldehyde and benzoic acid, and by nitric acid to **benzil**:

$$C_6H_5CHOHCOC_6H_5 \xrightarrow[\mathrm{HNO_3}]{[O]} C_6H_5COCOC_6H_5 \quad (96\%)$$

Purer benzil is obtained by oxidation with copper sulphate in aqueous pyridine (86% yield).

Benzil is a yellow crystalline solid, m.p. 95°C. It behaves as a typical α-diketone, *e.g.*, it is oxidised by hydrogen peroxide in acetic acid to benzoic acid, forms a monoxime and dioxime, etc.

> Two monoximes of benzil are known: α-(*cis*-), m.p. 134°C, and β-(*trans*-), m.p. 113°C. Three dioximes are possible and all are known: α-(*cis-cis*-), m.p. 237°C, β-(*trans-trans*-), m.p. 207°C, and γ- or *amphi*-(*cis-trans*-), m.p. 166°C (*cf.* p. 755).

Benzil condenses with *o*-phenylenediamine to form 2,3-*diphenylquinoxaline*:

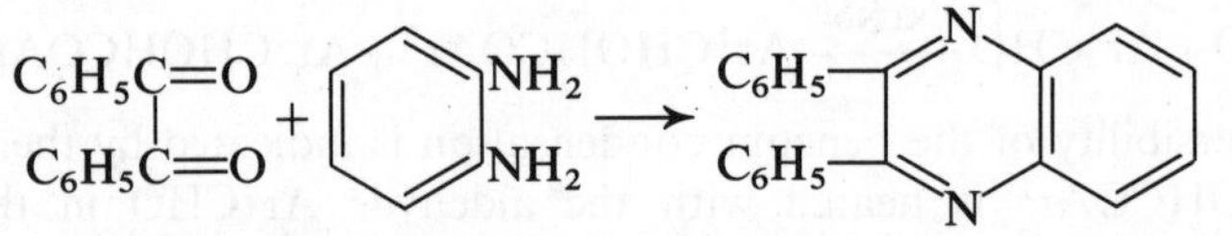

> The nature of the reduction products of benzil depends on the reducing agent used. Thus reduction with sodium amalgam gives hydrobenzoin (and some isohydrobenzoin); amalgamated zinc and hydrogen chloride in ethanol, stilbene; sodium hyposulphite in ethanol, benzoin; amalgamated tin and hydrogen chloride in ethanol, deoxybenzoin; and catalytic reduction using nickel at 230°C, bibenzyl.

When heated with ethanolic potassium hydroxide, benzil undergoes the **benzilic acid rearrangement**:

$$C_6H_5COCOC_6H_5 + KOH \longrightarrow (C_6H_5)_2C(OH)CO_2K \quad (77\text{--}79\%)$$

The rearrangement has been shown to be first order with respect to both benzil and hydroxide ion (Westheimer, 1936), and it has also been shown that when benzil is heated for a very short time with methanolic sodium hydroxide in water containing ^{18}O, the benzil recovered contained ^{18}O (Roberts and Urey, 1938). This is in keeping with the assumption that the first step is the rapid *reversible* addition of hydroxide ion to benzil.

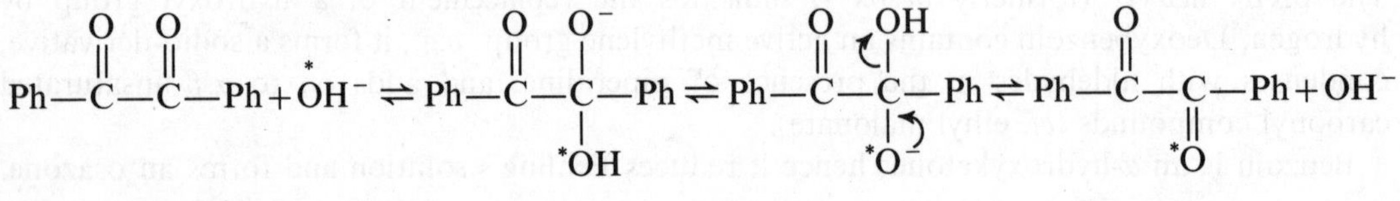

All of these results are in keeping with the mechanism proposed by Ingold in 1928:

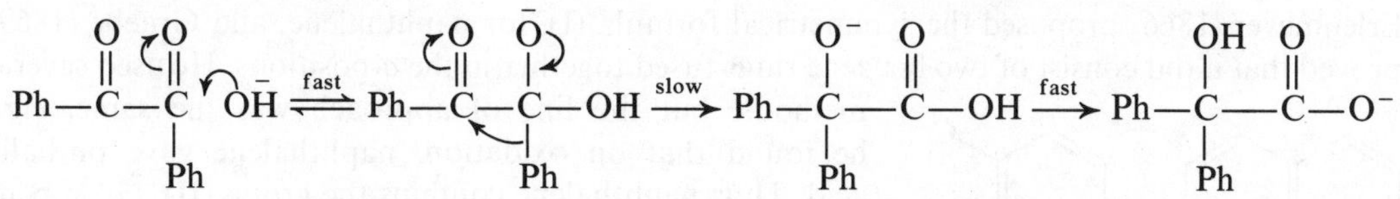

This is an example of the 1,2-shift, and since a phenyl group is involved, it is possible that a phenonium ion is formed as an intermediate. A prediction that can be made from the above mechanism is that when the two aryl groups are different, the one that is more electron-releasing (due to the presence of a suitable substituent, *e.g.*, *p*-MeO) will tend to neutralise the positive charge on the carbon atom to which it is attached when the C=O bond is polarised. Thus it will be the *other* CO group that will link with the hydroxide ion, and consequently it will be the aryl group attached to this 'other' CO group that migrates preferentially. This prediction is borne out in practice, *e.g.*, in *p*-methoxybenzil, the methoxyphenyl group migration was about 32 per cent. (Roberts *et al.*, 1951), and Eastham, *et al.* (1958), using labelled *p*-methylbenzil, showed that the phenyl group migrated almost exclusively.

Furthermore, it has also been shown that the hydroxide ion is *not* a specific catalyst; sodium methoxide and potassium t-butoxide produce the corresponding esters of benzilic acid (Doering *et al.*, 1956). Formation of these esters is strong evidence that the first step in the rearrangement is the addition of the nucleophilic reagent.

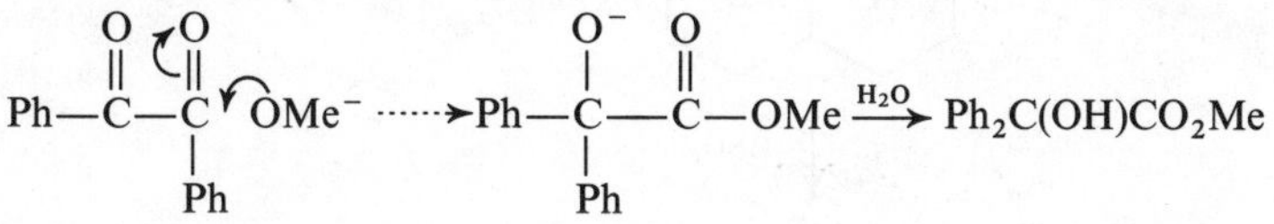

Benzilic acid (m.p. 150°C) forms diphenylacetic acid when refluxed in acetic acid solution with hydriodic acid and red phosphorus:

$$(C_6H_5)_2C(OH)CO_2H \xrightarrow{\text{HI/P}} (C_6H_5)_2CHCO_2H + H_2O \quad (94\text{--}97\%)$$

When oxidised with chromic acid, benzilic acid forms benzophenone.

Condensed systems†

Naphthalene

Naphthalene, $C_{10}H_8$, is the largest single constituent of coal-tar (6%). It is obtained by cooling the middle and heavy oiِs (p. 568), whereupon naphthalene crystallises out. The oil is pressed free from the naphthalene, the crude naphthalene cake melted, treated with concentrated sulphuric acid (to remove basic impurities), washed with water, and then treated with aqueous sodium hydroxide (to remove acidic impurities). Finally the naphthalene is distilled to give the pure product.

The modern tendency for isolating naphthalene is to replace the 'hot-pressing process' by the continuous washing or distillation processes. Naphthalene is also now being made synthetically from petroleum. Petroleum fractions are passed over a heated catalyst, *e.g.*, copper, at 680°C at atmospheric pressure; naphthalene and methylnaphthalenes are obtained, and the latter are hydrodealkylated to naphthalene (*cf.* benzene from toluene, p. 569).

† These systems are dealt with in more detail in Vol. 2, Ch. 10.

Structure of naphthalene

Erlenmeyer (1866) proposed the symmetrical formula (I), for naphthalene, and Graebe (1869) proved that it did consist of two benzene rings fused together in the *o*-positions. He used several methods, but the line of approach was the same, *e.g.*, he found that on oxidation, naphthalene gave phthalic acid. Thus naphthalene contains the group (II), *i.e.*, a benzene ring with two side-chains in the *o*-positions. When nitrated, naphthalene gave nitronaphthalene which, on oxidation, gave *o*-nitrophthalic acid. This indicates that the nitro-group is in the benzene ring, and that it is the side-chains which are oxidised. When nitronaphthalene was reduced and the corresponding aminonaphthalene oxidised, phthalic acid was obtained. As we have seen (p. 617), an amino-group attached to the nucleus renders the latter extremely sensitive to oxidation. Hence the inference is that the benzene ring in phthalic acid obtained by oxidation of aminonaphthalene is not the same ring as that originally containing the nitro-group in nitronaphthalene, *i.e.*, naphthalene contains *two* benzene rings. The above facts fit the following scheme:

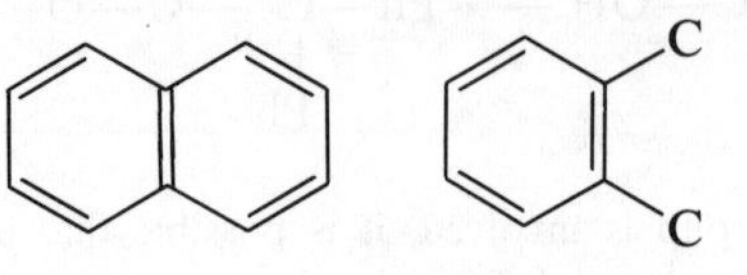

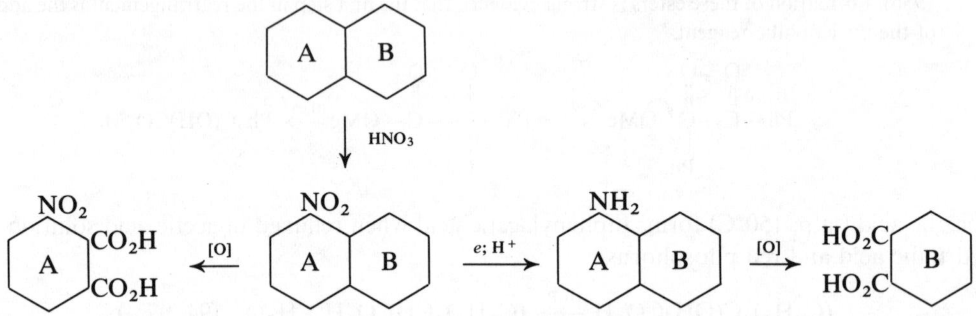

This structure for naphthalene has been confirmed by many syntheses, *e.g.*:

(i) When 4-phenylbut-1-ene is passed over red-hot calcium oxide, naphthalene is formed:

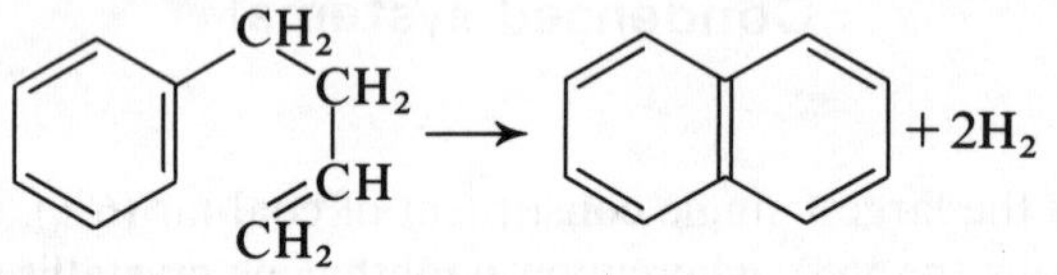

(ii) When 4-phenylbut-3-enoic acid is warmed with sulphuric acid, 1-naphthol is formed and this, on distillation with zinc dust, gives naphthalene (*cf.* p. 707):

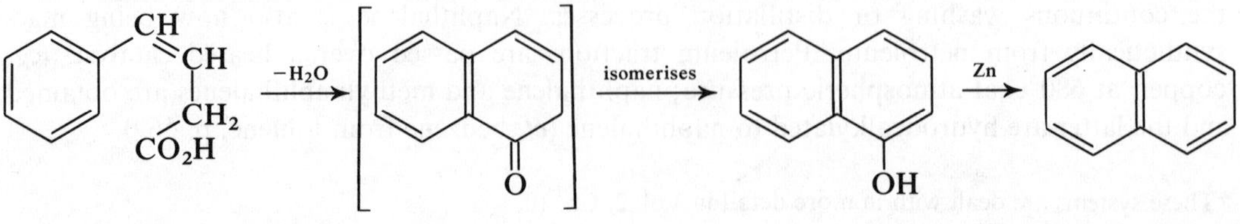

(iii) **Haworth synthesis** (1932). Benzene is treated with succinic anhydride in the presence of aluminium chloride, and the ketonic acid produced (Burcker, 1882) is reduced by the Clemmensen method. The ring is closed by heating with concentrated sulphuric acid and the product, α-tetralone, reduced to tetrahydronaphthalene by the Clemmensen method. This compound is then dehydrogenated to naphthalene by heating it with selenium or palladised charcoal:

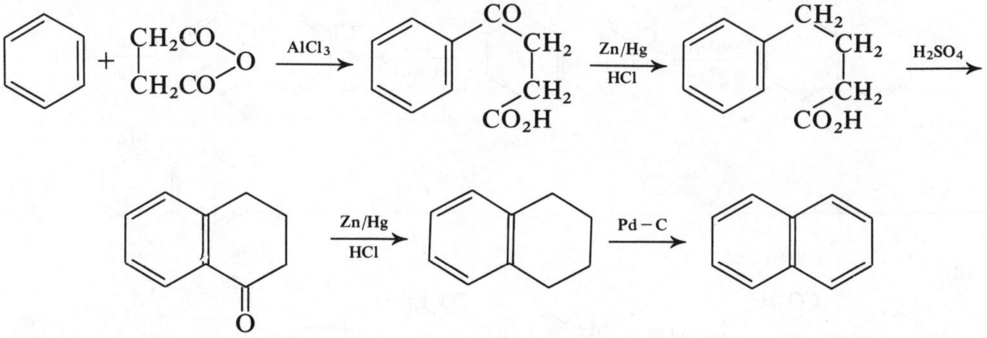

Ring closure may also be effected by means of a Friedel–Crafts reaction on the acid chloride as follows:

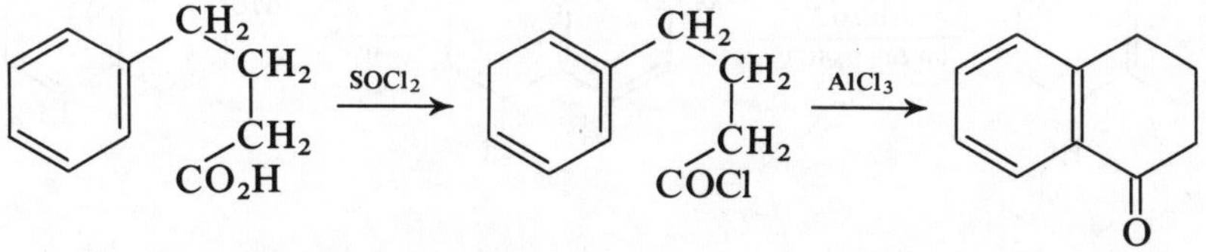

It should be noted here that the formation of cyclic ketones by *intramolecular* acylation is a very important synthetic process. Sulphuric, phosphoric, hydrofluoric acid, and especially polyphosphoric acid (PPA), are commonly used as catalysts for ring-closure of acids, and aluminium or stannic chlorides for acid chlorides.

An interesting point about the Haworth synthesis is that Birch *et al.* (1946) have shown that heating α-tetralone with a 1:1 mixture of NaOH—KOH at 220°C gives naphthalene (58%). The mechanism proposed is:

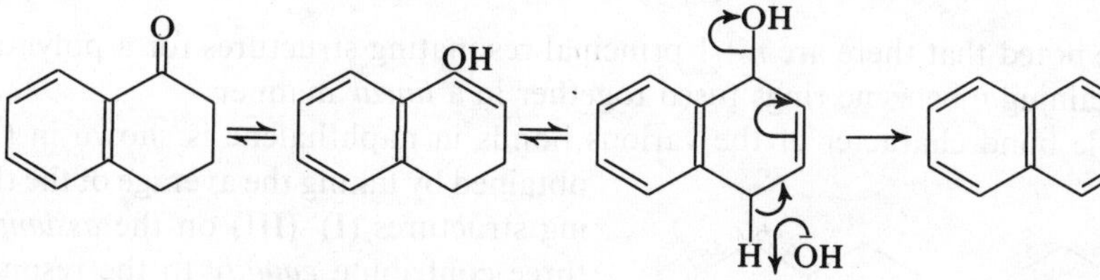

This may prove to be a better way of completing the Haworth synthesis.

Various substituted naphthalenes may also be prepared by the Haworth synthesis. With *o/p*-orienting groups present in the benzene ring, the main product from the Friedel–Crafts reaction is the *p*-compound, *e.g.*:

(i)

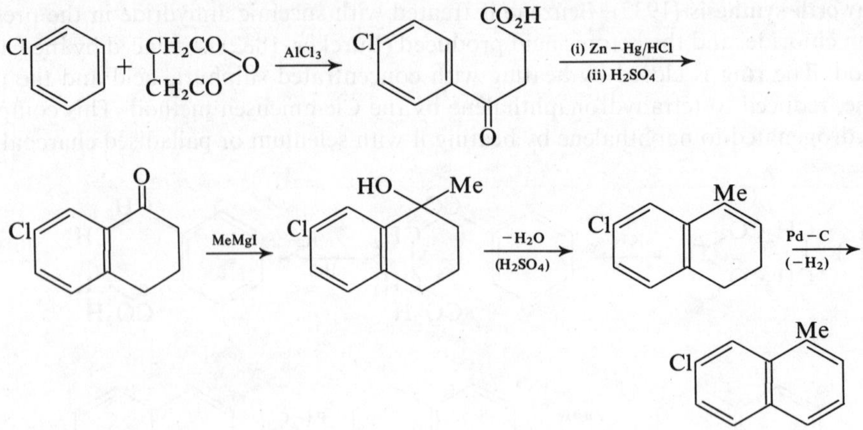

(ii)

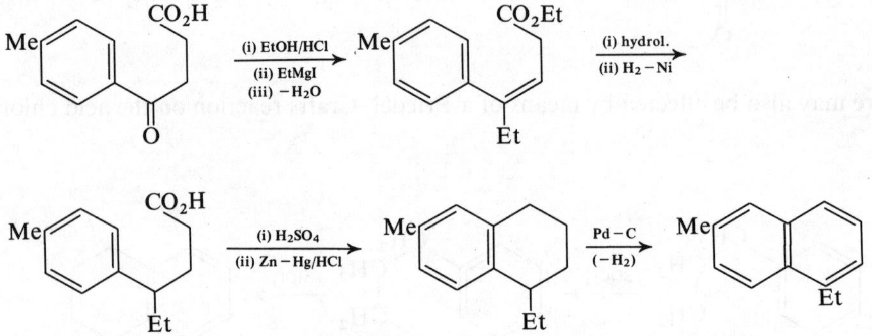

Positions of the double bonds in naphthalene

Physico-chemical evidence, *e.g.*, heat of combustion, etc., points towards naphthalene being a resonance hybrid of mainly three resonating structures, (I), (II) and (III):

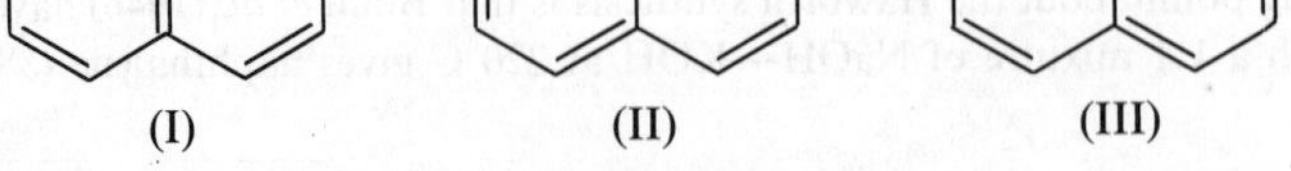

It should be noted that there are $n+1$ principal resonating structures for a polynuclear hydrocarbon containing n benzene rings fused together in a *linear* manner.

The double bond character of the various bonds in naphthalene is shown in (IV), and are

obtained by taking the average of the three resonating structures (I)–(III) on the *assumption* that all three contribute *equally* to the resonance hybrid. Thus, the 1,2-bond would be expected to be shorter than the 2,3-bond, and this has been shown to be the case by X-ray analysis of naphthalene (see V; in Å units).

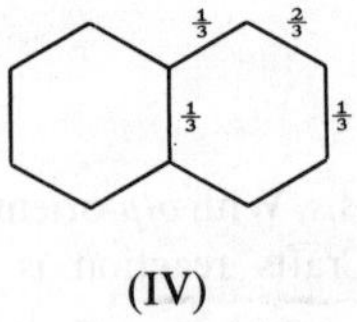

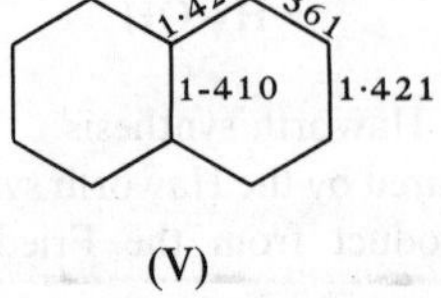

Fries rule (1935). Fries compared the possible arrangements of double bonds in polynuclear compounds with benzoquinones. Structures (II) and (III) have arrangements corresponding to

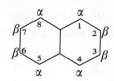

o-benzoquinone. Since quinones are far more reactive than an aromatic compound, Fries believed that the *stable* form of a polynuclear compound did not contain this quinonoid arrangement. He therefore formulated the following rule: *the most stable arrangement of a polynuclear compound is that form which has the maximum number of rings in the benzenoid condition*, i.e., *three double bonds in each individual ring*. Thus, according to the Fries rule, naphthalene tends to behave as structure (I) (with two benzenoid rings) rather than as (II) or (III) (with one benzenoid ring).

Isomerism and nomenclature of naphthalene derivatives

Positions 1, 4, 5 and 8 are identical (α-positions), as are positions 2, 3, 6 and 7 (β-positions). In the old literature, the positions 1,2- were known as *o*-; 1,3-, *m*-; 1,4-, *p*-; 1,5-, *ana*; 1,6-, *epi*; 1,7-, *kata*; 1,8-, *peri*; 2,6-, *amphi*; and 2,7-, *pros*. Some of these prefixes are still used, but it is best to use numbers; the Greek letters α- and β- are still frequently used to indicate the position of a single substituent.

Monosubstitution products	$C_{10}H_7X$	Two: 1- and 2-
Disubstitution products	$C_{10}H_6X_2$	10 isomers
	$C_{10}H_6XY$	14 isomers

There are 14 possible isomers for $C_{10}H_5X_3$, 22 for $C_{10}H_4X_4$, 14 for $C_{10}H_3X_5$, 10 for $C_{10}H_2X_6$, 2 for $C_{10}HX_7$, and 1 for $C_{10}X_8$.

Properties of naphthalene

Naphthalene, m.p. 80°C, has a characteristic odour and is very volatile. It is used as an insecticide and in the preparation of phthalic anhydride and dyes.

Since two benzene rings are present, it might have been expected that the resonance energy of naphthalene would be close to 301·2 (2×150·6) kJ mol^{-1}. Its RE is actually 255·2 kJ mol^{-1}, and hence naphthalene is less aromatic than benzene (each ring in naphthalene has an RE of 127·6 kJ). Consequently, naphthalene will have less aromatic character than benzene and so will be more reactive.

Addition compounds of naphthalene. A number of reduction products of naphthalene can be isolated, the nature of the product depending on the reducing agent used. When reduced with sodium and ethanol, naphthalene gives 1,4-*dihydronaphthalene* (1,4-*dialin*), m.p. 25°C (also see later):

$$\text{naphthalene} \xrightarrow{e; \text{H}^+} \text{1,4-dihydronaphthalene}$$

1,4-Dialin is unstable, readily isomerising to 1,2-dialin (m.p. −8°C) when heated with ethanolic sodium ethoxide. 1,2-Dialin is also unstable, fairly readily eliminating hydrogen to form naphthalene.

When reduced with sodium and isopentanol, naphthalene gives 1,2,3,4-*tetrahydronaphthalene* (*tetralin*), b.p. 206–208°C:

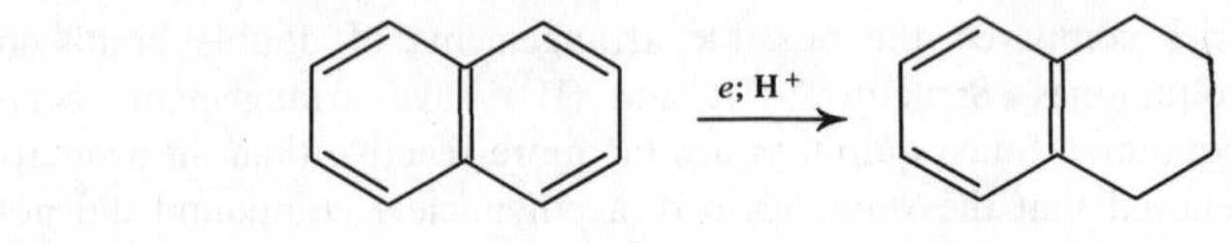

Tetralin is used as a solvent for varnishes, lacquers, etc. When treated with bromine in the presence of light, tetralin forms the mono- and dibromo-derivatives, substitution occurring in the *alicyclic* ring. When these bromo-derivatives are heated, hydrogen bromide is eliminated, the monobromo-compound forming dialin, and the dibromo-, naphthalene. In the absence of light and in the presence of iron as halogen carrier, tetralin undergoes substitution in the aromatic ring to form 5- and 6-bromo-1,2,3,4-tetrahydronaphthalenes. Catalytic dehydrogenation (Pd—C) converts tetralin into naphthalene.

α-Tetralone, b.p. 129·4°C/12 mm, may be prepared by synthesis (iii), p. 795, or by heating tetralin with air for 50 hours at 70°C and decomposing the peroxide with dilute sodium hydroxide:

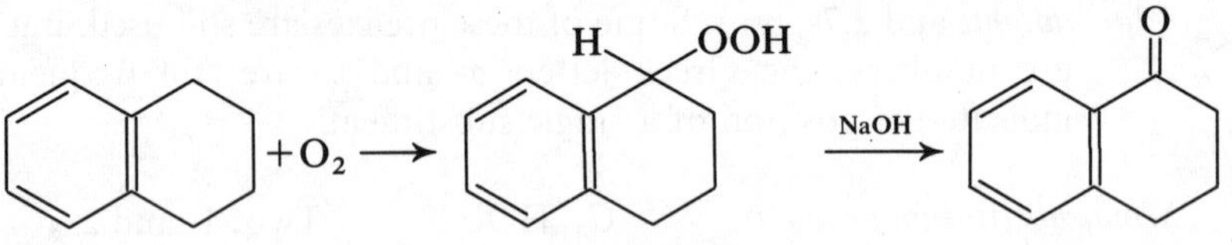

α-Tetralone may also be synthesised from γ-butyrolactone as follows:

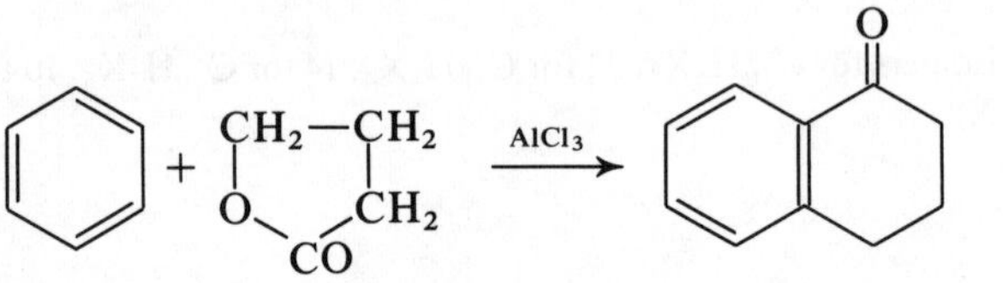

When naphthalene is catalytically reduced (under pressure) using nickel, tetralin and then *decahydronaphthalene* (*decalin*) are obtained:

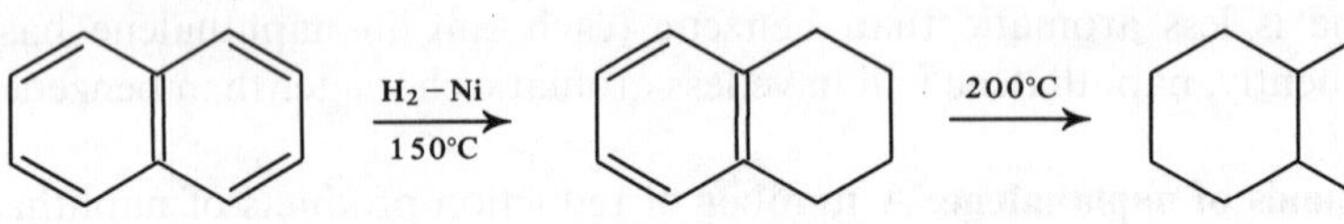

Decalin exists in two geometrical isomeric forms, *cis*- (b.p. 193°C) and *trans*- (b.p. 185°C). With nickel as catalyst, the main product is the *trans*-isomer; with platinum, *cis*- (see also p. 556). The commercial product is a mixture of the two forms, and is used as a solvent for varnishes, lacquers, etc.

Dry chlorine adds on to solid naphthalene to give naphthalene di- and tetrachlorides, $C_{10}H_8Cl_2$ and $C_{10}H_8Cl_4$. In both of these compounds the chlorine atoms are in the same ring (shown by the fact that on oxidation, both form phthalic acid). When naphthalene dichloride is heated at 40°C, hydrogen chloride is eliminated and 1-chloronaphthalene is produced. When naphthalene tetrachloride is treated with alkali, a mixture of dichloronaphthalenes is formed, the 1,3-isomer predominating.

Naphthalene, in dioxan solution, reacts with sodium to form a green-coloured solid. This is an anionic free radical, and is produced by the transfer of an electron from the sodium to the lowest vacant MO in naphthalene, and is stabilised by resonance:

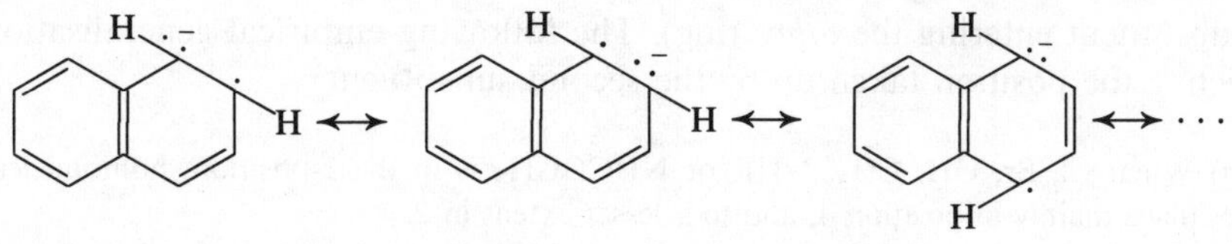

Sodium naphthalene reacts with water to give 1,4-dihydronaphthalene.

Oxidation of naphthalene

Naphthalene is oxidised by concentrated sulphuric acid and mercuric sulphate or by air in the presence of vanadium pentoxide, to phthalic anhydride (p. 776). It is oxidised by acid permanganate to phthalic acid, and by alkaline permanganate to phthalonic acid (p. 778). Chromic acid oxidises it to 1,4-naphthaquinone (p. 807). When treated with ozone, naphthalene forms the *diozonide* and this, on treatment with water, gives *phthalaldehyde*:

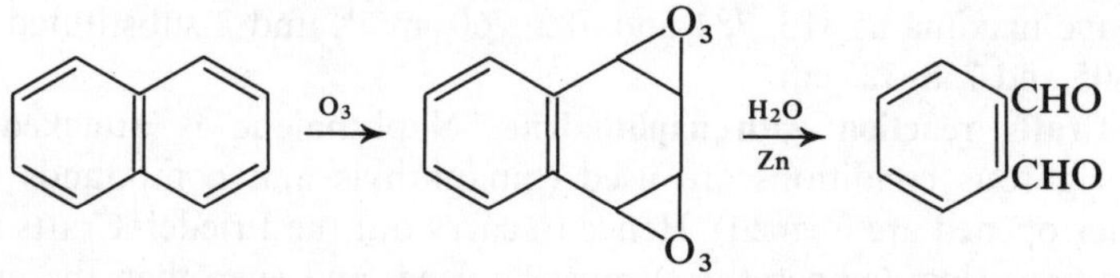

Substitution products of naphthalene

Orientation in naphthalene is more complicated than in benzene, due to the presence of two rings in the former. The first group always enters the 1-position in naphthalene except in two cases, when the 2-derivative is the main product; (i) sulphonation at high temperature, and (ii) in the Friedel–Crafts reaction.

The position of substitution may be obtained from a consideration of the nature of the intermediate carbonium ion, the one with the larger number of resonating structures being the more stable. The more important contributions will be by those structures in which one benzene ring is retained (benzene has a large resonance energy). It can be seen that 1-substitution gives rise to four such structures, whereas 2-substitution gives rise to only two. Thus, the former will be preferred (see also the naphthalenesulphonic acids, p. 801).

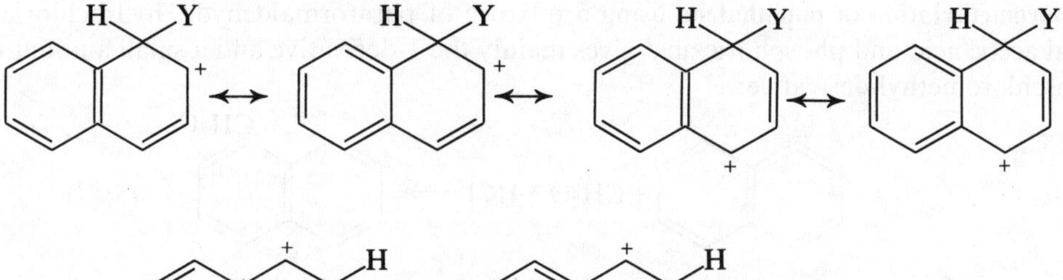

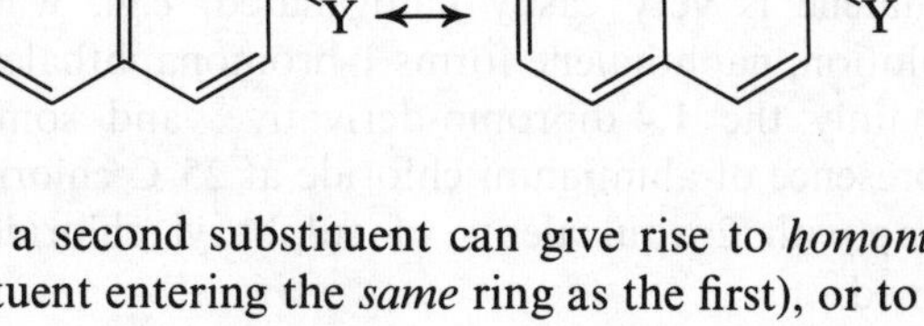

Introduction of a second substituent can give rise to *homonuclear* (*isonuclear*) substitution (the second substituent entering the *same* ring as the first), or to *heteronuclear* substitution (the second substituent entering the *other* ring). The following empirical generalisations are useful for predicting the position taken up by the second substituent:

>   (*a*) When Cl, Br, OH, CH_3, NHR or $NHCOCH_3$ is in the 1-position, homonuclear substitution takes place mainly in position 4, and to a lesser extent in 2.
>   (*b*) When OH, CH_3, NHR or $NHCOCH_3$ is in the 2-position, homonuclear substitution usually takes place in the 1-position.
>   It is worth noting that homonuclear substitution usually occurs when the group already present is *o-p*-orienting.
>   (*c*) When NO_2 or SO_3H is in the 1- or 2-position, heteronuclear substitution occurs in position 5 or 8; if halogen or NH_2 is in the 2-position, heteronuclear substitution also occurs in position 5 or 8.

The infra-red absorption regions of polynuclear hydrocarbons include many of those characteristic of the benzene compounds (see p. 585). Thus, naphthalenes have two bands near $1\,600\,cm^{-1}$ and bands in the regions $1\,520–1\,505$ and $1\,400–1\,390\,cm^{-1}$. The C—H deformation (out-of-plane) in the $800\,cm^{-1}$ region may often provide information, *e.g.*, 1-substituted naphthalenes have maxima at 813–784 and $782–760\,cm^{-1}$, and 2-substituted naphthalenes at 862–835, 830–805 and $758–737\,cm^{-1}$.

The Friedel–Crafts reaction with naphthalene. Naphthalene is attacked by aluminium chloride when vigorous conditions are used (binaphthyls and compounds with one of the naphthalene rings opened are formed). Hence to carry out the Friedel–Crafts reaction successfully, mild conditions (low temperatures) must be used, and even then the maximum yield is about 60 per cent. With methyl iodide, 1- and 2-methylnaphthalenes are formed; with ethyl bromide, only 2-ethylnaphthalene; and with n-propyl bromide, 2-isopropylnaphthalene. With alcohols and aluminium chloride, 2,6-dialkylnaphthalenes are obtained but with alcohols and boron trifluoride, 1,4-dialkylnaphthalenes. Introduction of an acyl group in the presence of aluminium chloride gives a mixture of 1- and 2-ketones, the nature of the solvent affecting the percentage of each; *e.g.*, with acetyl chloride in carbon disulphide as solvent, 1- and 2-naphthyl methyl ketone are formed in a ratio of 3:1, in nitrobenzene, 1:9:

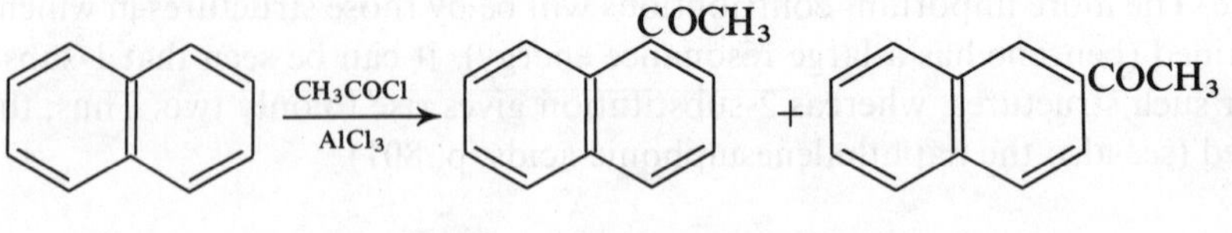

Chloromethylation of naphthalene using a mixture of paraformaldehyde, hydrochloric acid, glacial acetic acid and phosphoric acid gives mainly the 1-derivative and a small amount of the 1,5-bischloromethyl-derivative:

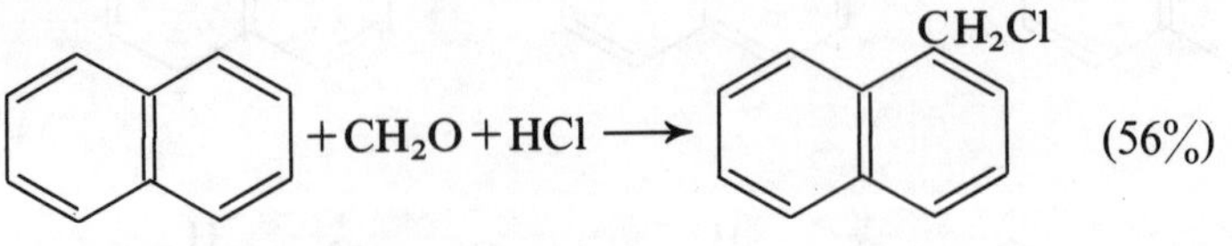

Halogen derivatives. Naphthalene is very easily halogenated, *e.g.*, when brominated in boiling carbon tetrachloride solution, naphthalene forms 1-bromonaphthalene (yield 72–75%). Further bromination gives mainly the 1,4-dibromo-derivative, and some 1,2-. Sulphuryl chloride (1 equivalent) in the presence of aluminium chloride at 25°C chlorinates naphthalene to give the 1-chloro-derivative; with 2 equivalents of sulphuryl chloride (at 100–140°C), 1,4-dichloronaphthalene is formed.

2-Halogenonaphthalenes are conveniently prepared from 2-naphthylamine by diazotisation, etc. *e.g.*,

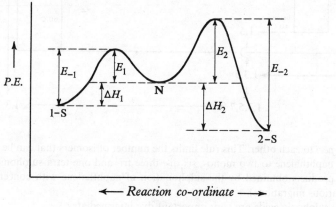

The halogen atom in naphthalene behaves similarly as in the benzene ring, but is more reactive.

Nitronaphthalenes. Nitric acid attacks naphthalene at room temperature to form 1-nitro-naphthalene. This is a yellow solid, m.p. 60°C, which behaves like nitrobenzene, but differs in that it forms 1-chloronaphthalene when treated with phosphorus pentachloride (nitrobenzene does not react).

Nitration of naphthalene at high temperature gives a mixture of 1,5- and 1,8-dinitro-naphthalenes. Other dinitro-derivatives are prepared by special means.

2-Nitronaphthalene, m.p. 79°C, may be prepared by heating 2-naphthalenediazonium fluoroborate with sodium nitrite and copper powder.

Naphthalenesulphonic acids. When naphthalene is treated with concentrated sulphuric acid at 40°C, the main product is the 1-derivative (m.p. 91°C); at 160°C, the main product is the 2-derivative (m.p. 102°C):

When heated with concentrated sulphuric acid, the 1-acid is converted into the 2-acid. Thus, the former is the kinetically controlled and the latter the thermodynamically controlled product. The reason for the greater stability of the 2-acid is uncertain; one suggestion is that there is steric repulsion in the 1-acid from the hydrogen atom in the 8-position.

Figure 29.1 is the energy profile of the sulphonation of naphthalene (**N**). Since 1-S is the kinetically controlled product, then $k_1 > k_2$, and therefore $E_2 > E_1$. Also, since 2-S is the

Fig. 29.1

thermodynamically controlled product, $\Delta H_2 > \Delta H_1$; It therefore follows that the activation energy for the reverse (desulphonation) reaction: 2-S $\rightarrow$ N, is much greater than that of the reverse reaction: 1-S $\rightarrow$ N, $i.e.$, $E_{-2}(= E_2 + H_2) \gg E_{-1}(= E_1 + \Delta H_1)$. Consequently, $k_{-1} \gg k_{-2}$.

At low temperatures, the equilibrium between the 1- and 2-acids is reached very slowly. Hence, if the reaction is stopped long before equilibrium is reached, 1-S > 2-S since $k_1 > k_2$. If one waited long enough for equilibrium to be reached, then 2-S $\gg$ 1-S. This is due to the fact that, since $k_{-1} \gg k_{-2}$, 1-S reforms N much faster than does 2-S, and hence 2-S accumulates at the expense of 1-S. Because raising the temperature speeds up a reaction, equilibrium is reached far more quickly at higher temperatures, and therefore to obtain 2-S (the thermodynamically controlled product), the reaction is carried out at some suitably elevated temperature.

1- and 2-Naphthalenesulphonic acids behave similarly to benzenesulphonic acid, but the sulphonic acid group in the former is more easily replaced; $e.g.$, when fused with phosphorus pentachloride, both acids give the corresponding chloronaphthalenes.

2-Naphthalenesulphonic acid is the starting-point of practically all 2-naphthalene derivatives.

Sulphonation of the 1-sulphonic acid with concentrated sulphuric acid below 40°C gives 1,5- (70%) and 1,6-disulphonic acids (25%); at 130°C, the main products are 1,6- and 1,7-disulphonic acids. Sulphonation of the 2-acid at 60°C gives 1,6- (80%) and 1,7-disulphonic acid (20%); above 140°C, the main product is 2,7-disulphonic acid, and a small amount of 2,6-. Armstrong and Wynne's rule (1890) for the orientation of naphthalenepolysulphonic acids is useful: two sulpho-groups never occupy

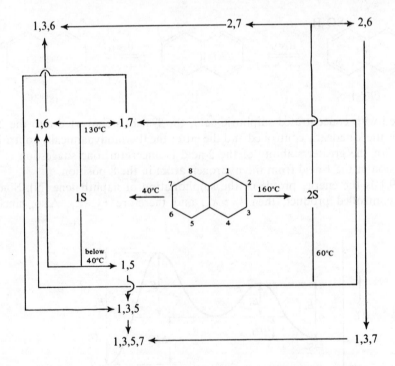

positions o, p or $peri$ to each other. This rule limits the number of isomers that can be formed by direct sulphonation of naphthalene to two mono-, six di-, three tri- and one tetra-sulphonic acid. The table summarises the products obtained by the sulphonation of naphthalene with concentrated sulphuric acid (note the various migrations).

Many of these sulphonic acids are very important dye-intermediates.

Naphthylamines. 1-Naphthylamine (α-naphthylamine) may be prepared by reducing 1-nitro-naphthalene with iron and hydrochloric acid (yield 80–85%). This method is used industrially. 1-Naphthylamine may also be prepared by the Bucherer reaction (see below) or by heating 1-naphthol with the double compound of zinc chloride and ammonia (this amination occurs more easily than with phenol):

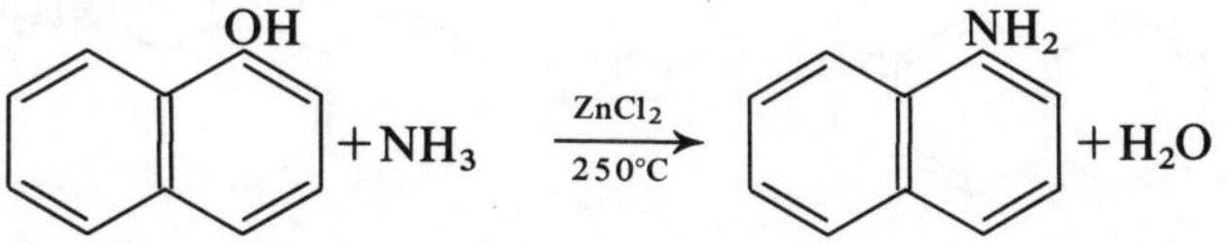

1-Naphthylamine, m.p. 50°C, has an unpleasant odour, and turns red on exposure to air. It reduces ammoniacal silver nitrate, and solutions of its salts give a blue precipitate with ferric chloride. Oxidation with boiling chromic acid gives α-naphthaquinone; with permanganate, phthalic acid is obtained. 1-Naphthylamine is reduced by sodium and isopentanol to *ar*-tetrahydro-1-naphthylamine; the prefix *ar*- is the abbreviation of *aromatic* and indicates that the four hydrogen atoms are *not* in the ring containing the amino-group:

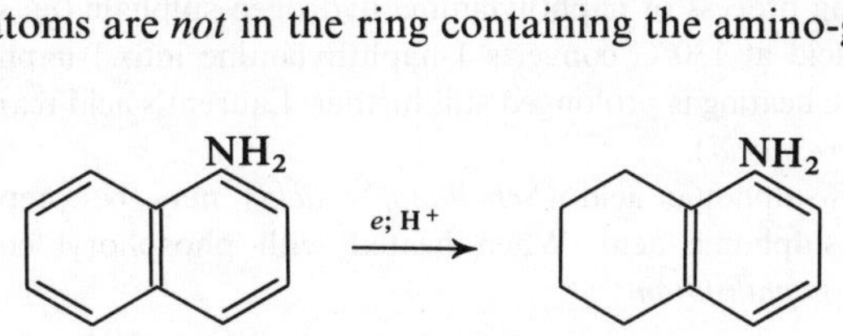

The systematic name is 5,6,7,8-tetrahydro-1-naphthylamine.

1-Naphthylamine couples with diazonium salts in the 4-position.

2-Naphthylamine (β-naphthylamine) is prepared industrially from 2-naphthol by **Bucherer reaction** (1904). This is the reversible conversion of a naphthol into a naphthylamine in the presence of an aqueous sulphite or hydrogen sulphite. According to Rieche *et al.* (1960), the mechanism of the formation of 2-naphthylamine from 2-naphthol, sodium hydrogen sulphite and ammonia is:

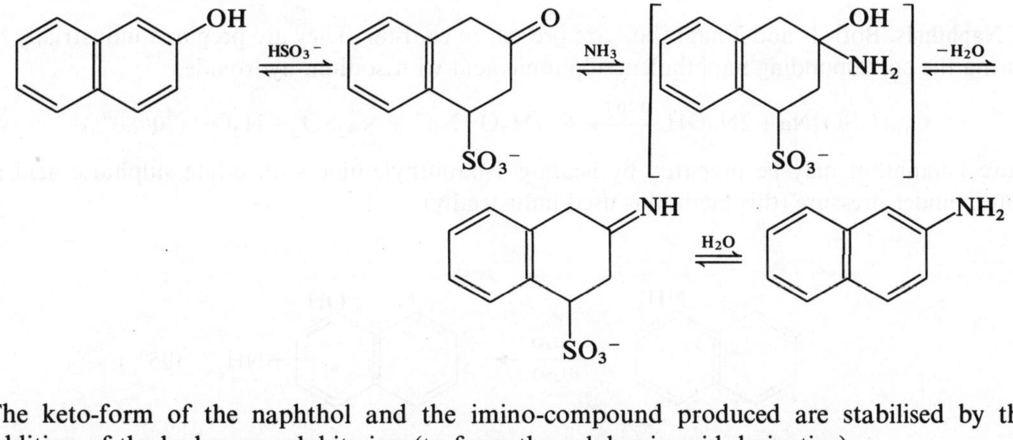

The keto-form of the naphthol and the imino-compound produced are stabilised by the addition of the hydrogen sulphite ion (to form the sulphonic acid derivative).

2-Naphthylamine is prepared commercially by heating 2-naphthol with aqueous ammonium hydrogen sulphite at 150°C under pressure (yield: 94–96%).

2-Naphthylamine, m.p. 112°C, is odourless, reduces ammoniacal silver nitrate, but gives *no* coloration with ferric chloride. It is oxidised by permanganate to phthalic acid and reduced by sodium and isopentanol to *ac*-tetrahydro-2-naphthylamine (1,2,3,4-tetrahydro-2-naphthylamine); the prefix *ac*- is the abbreviation of *alicyclic* and indicates that the four hydrogen atoms are in the ring containing the amino-group (*cf.* above):

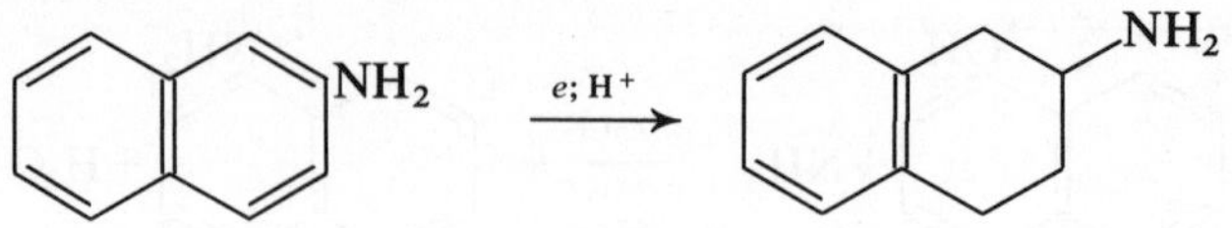

2-Naphthylamine couples with diazonium salts only in the 1-position; if this is occupied, no coupling occurs.

Naphthylaminesulphonic acids. These are very important industrially for making dyes. When heated with excess of concentrated sulphuric acid at 130°C, 1-naphthylamine forms 1-naphthylamine-4-sulphonic acid (*naphthionic acid*). This is used in the preparation of Congo red (p. 884), and is manufactured by the baking process of naphthylamine hydrogen sulphate (*cf.* sulphanilic acid). Prolonged action of sulphuric acid at 130°C converts 1-naphthylamine into 1-naphthylamine-5-sulphonic acid (*Laurent's acid*). If the heating is prolonged still further, Laurent's acid rearranges to 1-naphthylamine-6-sulphonic acid (*Cleve's acid*).

1-Naphthylamine-8-sulphonic acid (*Schollkopf's acid*) may be prepared by reduction of the corresponding nitro-sulphonic acid. When heated with phosphoryl chloride, 1-naphthylamine-8-sulphonic acid forms *naphthsultam*:

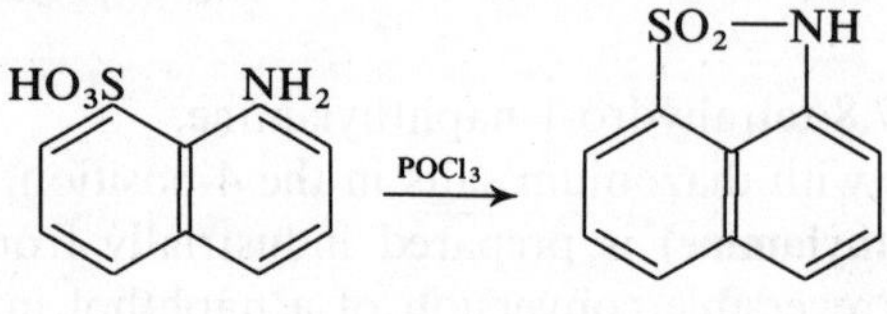

When heated with concentrated sulphuric acid, 2-naphthylamine gives four different sulphonic acids according to the temperature: 2-naphthylamine-5-sulphonic acid (*Dahl's acid*), 2-naphthylamine-6-sulphonic acid (*Bronner's acid*), 2-naphthylamine-7-sulphonic acid (*F-acid*) and 2-naphthylamine-8-sulphonic acid (*Badische's acid*).

Naphthols. Both 1- and 2-naphthols are present in coal-tar. They are prepared industrially by fusing the corresponding naphthalenesulphonic acid with sodium hydroxide:

$$C_{10}H_7SO_3Na + 2NaOH \xrightarrow{300°C} C_{10}H_7O^-Na^+ + Na_2SO_3 + H_2O \quad (70\text{-}80\%)$$

Pure 1-naphthol may be prepared by heating 1-naphthylamine with dilute sulphuric acid at 290°C under pressure (this method is used industrially).

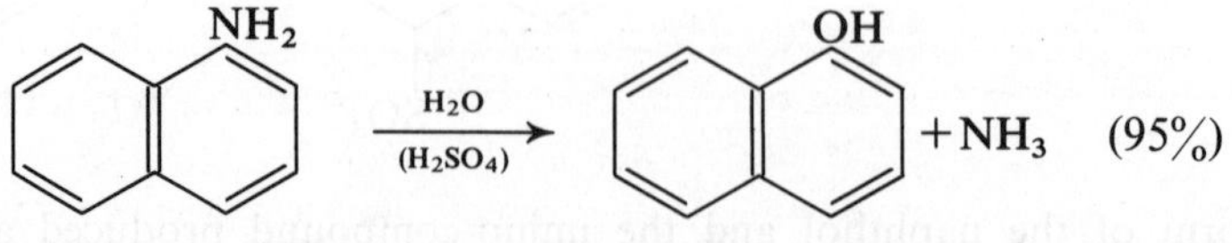

An interesting preparation of 1-naphthol is by the reaction between *o*-bromofluorobenzene and lithium amalgam in the presence of furan (Wittig *et al.*, 1955). The reaction proceeds via a benzyne intermediate (p. 629).

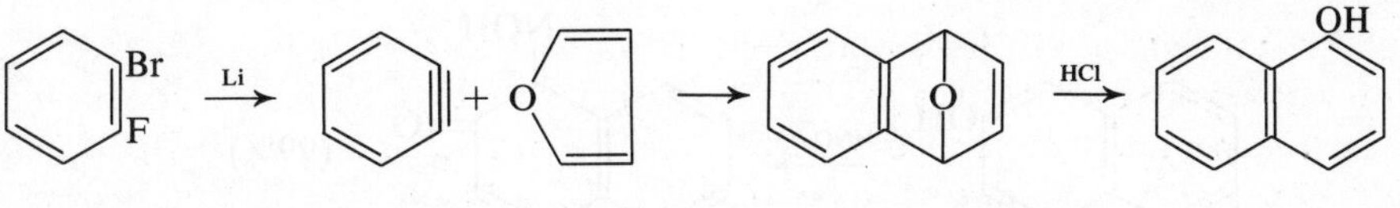

1-Naphthol (α-naphthol), m.p. 94°C, is readily soluble in alkalis to form naphthoxides, *e.g.*, $C_{10}H_7\bar{O}Na^+$. With ferric chloride it gives a violet-blue precipitate of α-binaphthol (4,4′-bis-1-naphthol):

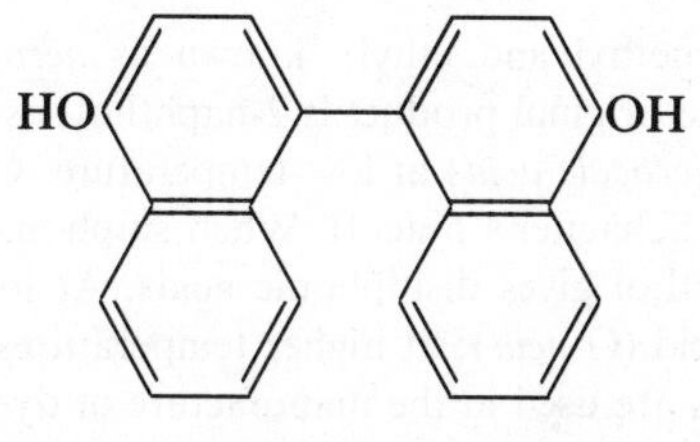

1-Naphthol reduces ammoniacal silver nitrate, is oxidised by alkaline permanganate to phthalonic acid and by chromic acid to α-naphthaquinone. It is reduced by sodium and isopentanol to *ar*-tetrahydro-1-naphthol (*ar-1-tetralol*).

Direct sulphonation of 1-naphthol under mild conditions gives a mixture of 1-naphthol-2-sulphonic acid (*Schaeffer's acid*) and 1-naphthol-4-sulphonic acid (*Nevile–Winther's acid*). More vigorous conditions result in the formation of 1-naphthol-2,4-disulphonic acid and finally 1-naphthol-2,4,7-trisulphonic acid.

1-Naphthol-8-sulphonic acid forms an *inner* ester, *naphthsultone*, when heated:

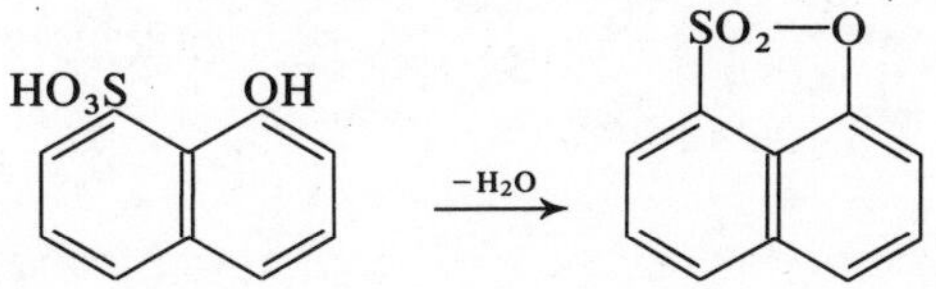

Treatment of 1-naphthol with nitrous acid gives mainly the 2-oxime of β-naphthaquinone (I), and a small amount of the 4-oxime of α-naphthaquinone (II) (*cf.* nitrosophenol):

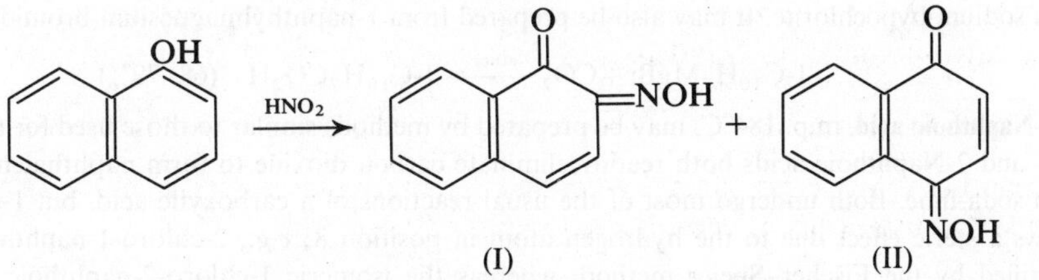

1-Naphthol couples with diazonium salts in the 4-position.

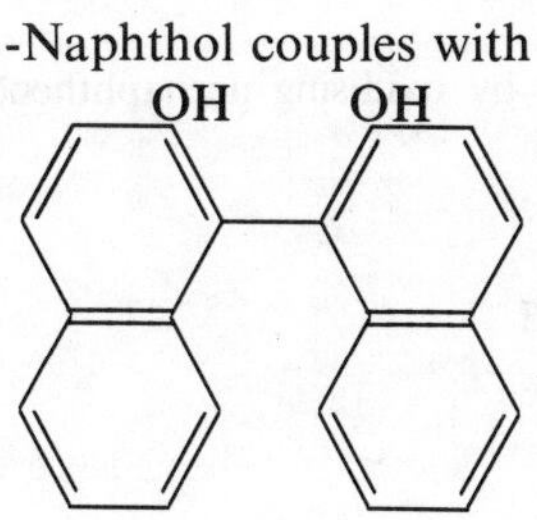

2-Naphthol (β-naphthol), m.p. 123°C, resembles 1-naphthol in most of its properties, but is more reactive. With ferric chloride it gives a green precipitate of *β-binaphthol* (1,1′-bis-2-naphthol). It reduces ammoniacal silver nitrate, and is oxidised by alkaline permanganate to phthalonic acid. It is reduced by sodium and isopentanol to mainly *ac*-tetrahydro-2-naphthol (*ac-2-tetralol*). With nitrous acid it forms the 1-oxime of β-naphthaquinone (nitroso-β-naphthol):

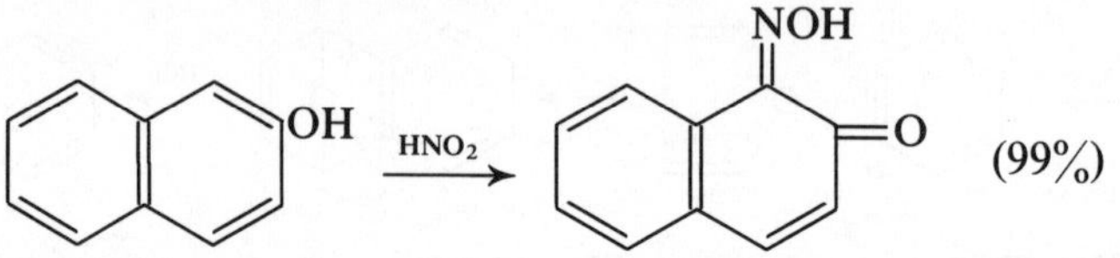

(99%)

2-Naphthol couples with diazonium salts only in the adjacent 1-position; if this is occupied, no coupling takes place.

Some 2-naphthol ethers (methyl and ethyl), known as *nerolins*, are used in perfumery. When 2-naphthol is sulphonated, the original product is 2-naphthol-1-sulphonic acid, but this rearranges to 2-naphthol-8-sulphonic acid (*croceic acid*) at low temperature. Croceic acid, at 100°C, rearranges to 2-naphthol-6-sulphonic acid (Schaeffer's β-acid). When sulphonated with larger amounts of concentrated sulphuric acid, 2-naphthol gives disulphonic acids. At low temperature the main product is 2-naphthol-6,8-disulphonic acid (*G-acid*); at higher temperatures, mainly 2-naphthol-3,6-disulphonic acid (R-*acid*). G- and R-acids are used in the manufacture of dyes.

Woodward (1940) prepared 10-methyldecal-2-one from *ar*-2-tetralol by means of the Reimer–Tiemann reaction as follows:

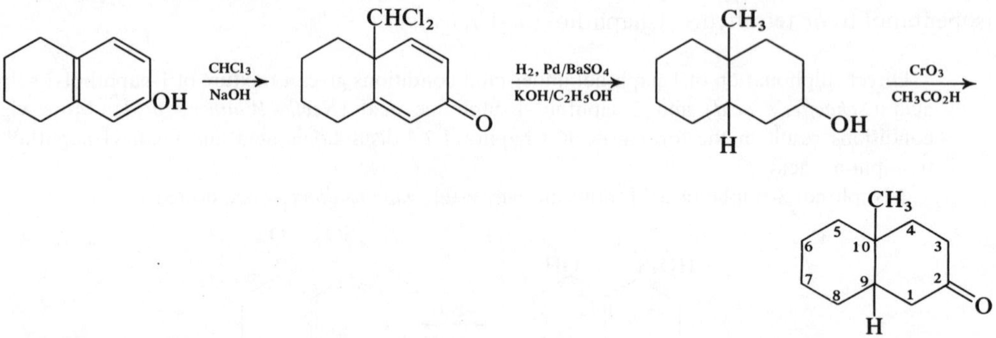

This synthesis is based on the fact that the intermediate product in the Reimer–Tiemann reaction can be isolated when the *p*-position to the hydroxyl group is occupied by, *e.g.*, a methyl group (p. 743).

Naphthalenecarboxylic acids. 1-Naphthoic acid (*naphthalene-1-carboxylic acid*), m.p. 161°C, may be prepared by the hydrolysis of the corresponding cyanide, or by the oxidation of 1-acetylnaphthalene with sodium hypochlorite. It may also be prepared from 1-naphthylmagnesium bromide as follows:

$$1\text{-}C_{10}H_7MgBr + CO_2 \xrightarrow{\text{ether}} 1\text{-}C_{10}H_7CO_2H \quad (68\text{-}70\%)$$

2-Naphthoic acid, m.p. 184°C, may be prepared by methods similar to those used for the 1-isomer.

1- and 2-Naphthoic acids both readily eliminate carbon dioxide to form naphthalene when heated with soda-lime. Both undergo most of the usual reactions of a carboxylic acid, but 1-naphthoic acid shows a steric effect due to the hydrogen atom in position 8; *e.g.*, 2-chloro-1-naphthoic acid is not esterified by the Fischer–Speier method, whereas the isomeric 1-chloro-2-naphthoic acid gives the ester under the same conditions.

Naphthalic acid (*naphthalene-1,8-dicarboxylic acid*) may be prepared by oxidising acenaphthene with acid dichromate:

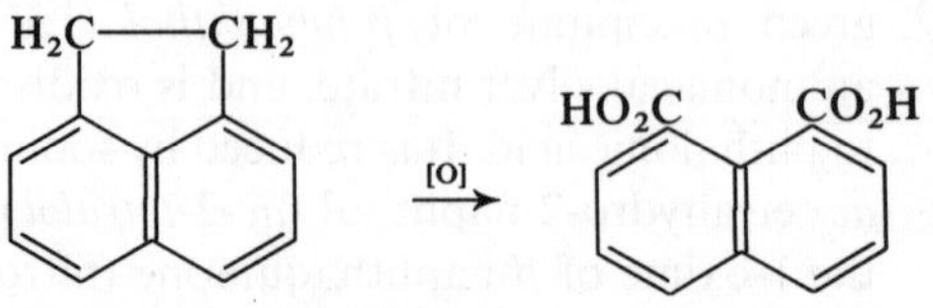

When heated at 180°C, it forms naphthalic anhydride:

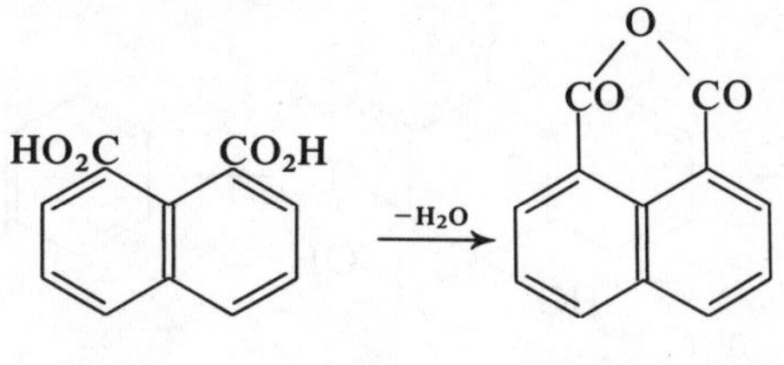

This is in keeping with the fact that all *peri*-(1,8)-substituents interact if possible (*cf.* sultams and sultones).

Naphthaquinones. Theoretically, six naphthaquinones are possible: 1,2-, 1,4-, 1,5-, 1,7-, 2,3- and 2,6-. Only three are known, the 1,2-, 1,4- and 2,6-, but it appears that derivatives of 2,3-naphthaquinone have been prepared (see also p. 727).

1,4-Naphthaquinone (α-naphthaquinone,, 1,4-*dihydronaphthalene*-1,4-*dione*) may be prepared by the oxidation of 1,4-diamino-, dihydroxy- or aminohydroxynaphthalene, *e.g.*,

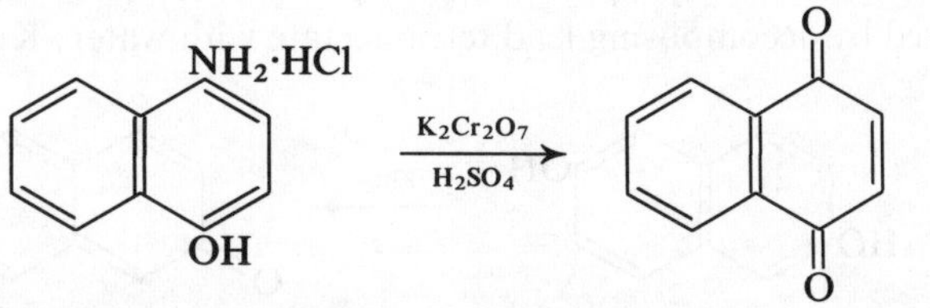

It may also be prepared by the direct oxidation of naphthalene with dichromate and sulphuric acid, or chromium trioxide in glacial acetic acid (yield 40%). Catalytic air-oxidation of naphthalene is the commercial method.

1,4-Naphthaquinone is a volatile yellow solid, m.p. 125°C, with a pronounced odour. It resembles *p*-benzoquinone in many ways chemically, but it is not reduced by sulphurous acid. It is reduced by metal and acid to 1,4-*dihydroxynaphthalene* (*naphthalene*-1,4-*diol*) and oxidised by nitric acid to phthalic acid. It forms a monoxime; this is tautomeric, and in the solid state it exists as the oxime, and in solution this form predominates in equilibrium with the nitroso-phenol form (Havinga *et al.*, 1955; Hadži, 1956). It is converted into *indane*-1,3-*dione* (1,3-*diketohydrindene*) on treatment with nitrous acid:

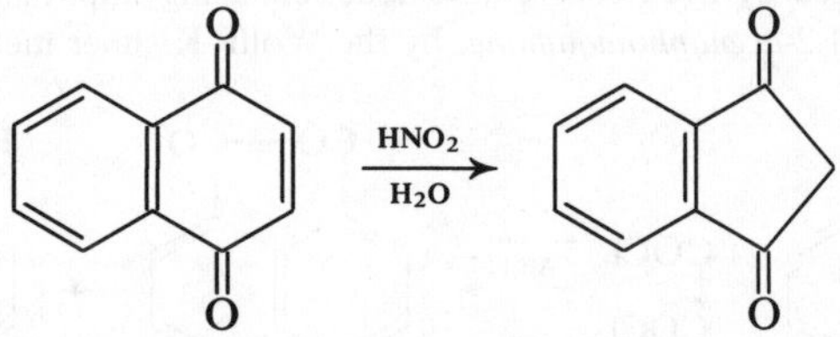

Vitamin K (the antihaemorrhagic factor) is a derivative of 1,4-naphthaquinone.

1,2-Naphthaquinone (β-naphthaquinone, 1,2-*dihydronaphthalene*-1,2-*dione*) may be prepared by oxidising 1-amino-2-naphthol hydrochloride with ferric chloride in hydrochloric acid (93–94%). It is a non-volatile, odourless, red solid which decomposes at 115–120°C. Fission of the quinone ring occurs with simultaneous hydroxylation when 1,2-naphthaquinone is treated with chlorine-water (or with hypochlorous acid); phenylglyceric-*o*-carboxylic lactone is the final product:

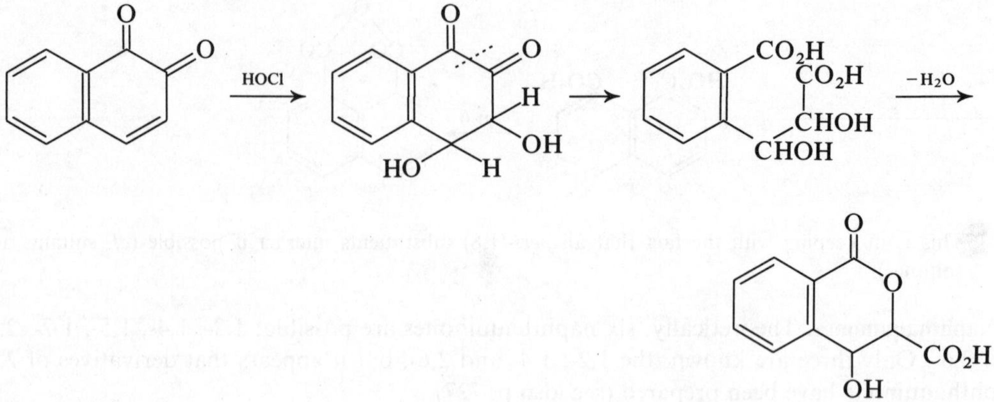

2,6-Naphthaquinone (*amphi*-naphthaquinone, 2,6-*dihydronaphthalene*-2,6-*dione*) may be prepared by oxidising 2,6-dihydroxynaphthalene in benzene solution with 'active' lead dioxide (this may be prepared by decomposing lead tetra-acetate with water; Kuhn *et al.*, 1950):

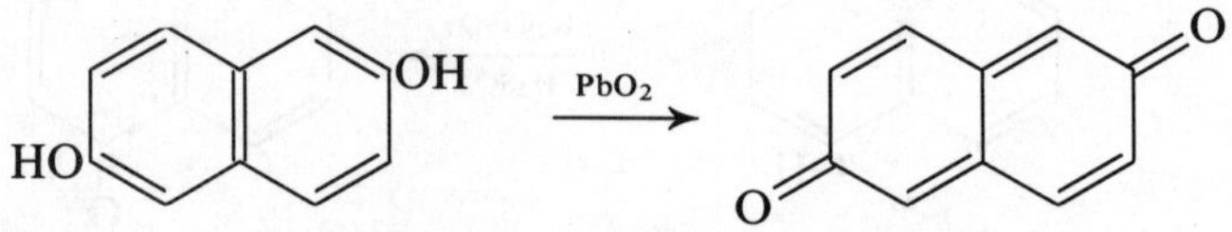

It is an orange, non-volatile, odourless solid, m.p. 135°C.

Acenaphthene, $C_{12}H_{10}$, occurs in coal-tar; it may be prepared by passing 1-ethylnaphthalene through a red-hot tube:

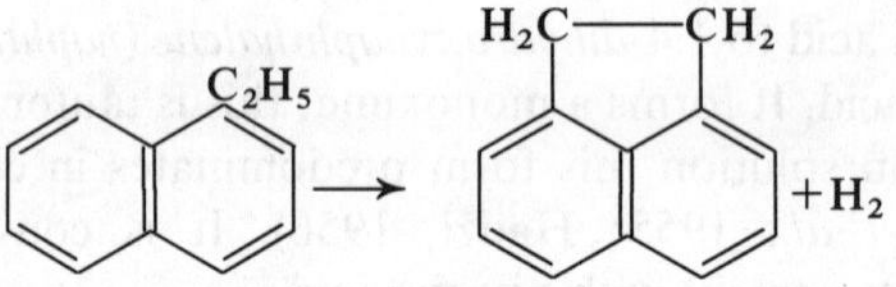

It may also be prepared by a Friedel–Crafts reaction using naphthalene and oxalyl chloride and reducing the product, 1,2-*acenaphthaquinone*, by the Wolff–Kishner method:

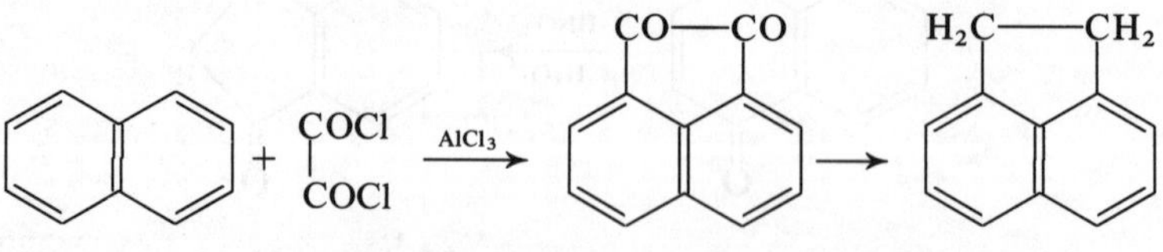

Acenaphthene, m.p. 96°C, on oxidation with dichromate and sulphuric acid, is converted into 1,2-acenaphthaquinone which, by further action of the oxidising agent, gives naphthalic acid. Substitution in acenaphthene occurs most readily in the 5-position, e.g., bromination, nitration and sulphonation give the corresponding 5-derivatives. When oxidised with lead dioxide or passed through a red-hot tube, acenaphthene forms *acenaphthylene*:

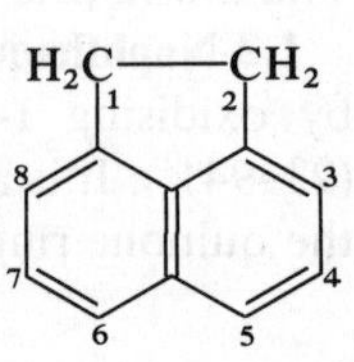

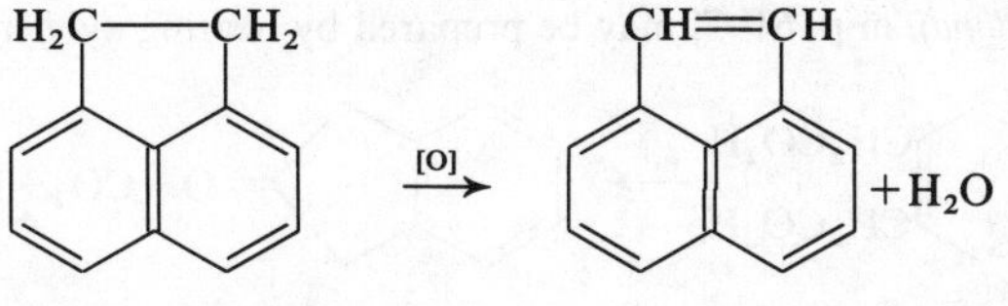

Indene (benzocyclopentadiene), C_9H_8, occurs in the coal-tar fraction boiling at 175–185°C. It may be obtained from this fraction by heating with sodium, separating the sodio-indene formed (solid) and steam-distilling it, whereupon indene distils over.

Indene, b.p. 182°C, readily polymerises (*cf.* cyclopentadiene). It is reduced by sodium and ethanol to 2,3-*dihydroindene* (*indane* or *hydrindene*), b.p. 177°C.

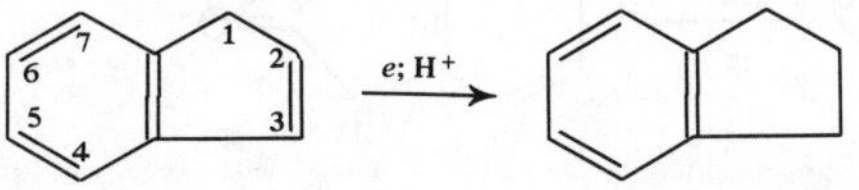

Oxidation of indene with acid dichromate gives homophthalic acid.

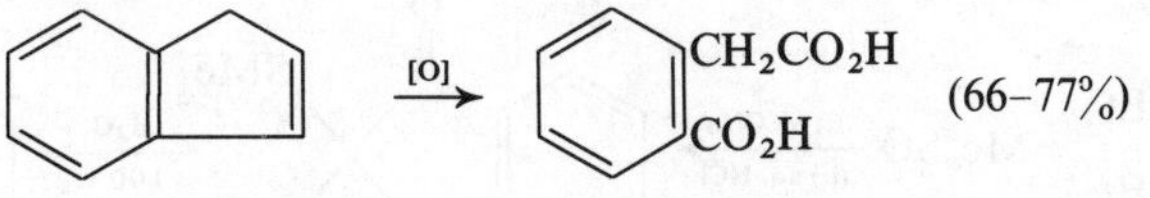

Indene combines with halogen to form 2,3-dihalogenoindane. It is acidic (*cf.* cyclopentadiene), *e.g.*, it forms a sodio-derivative:

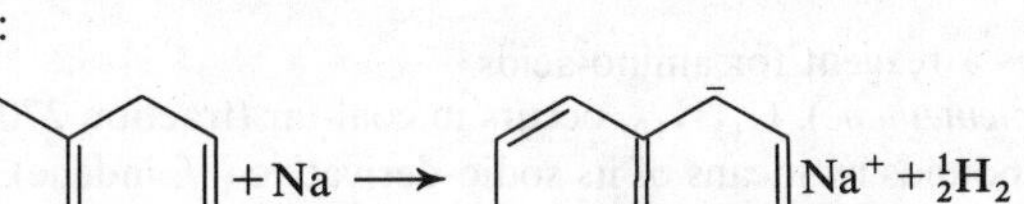

It condenses with ethyl oxalate in the presence of sodium ethoxide to form indene-oxalic ester, and with aldehydes or ketones in the presence of alkali to form benzofulvenes:

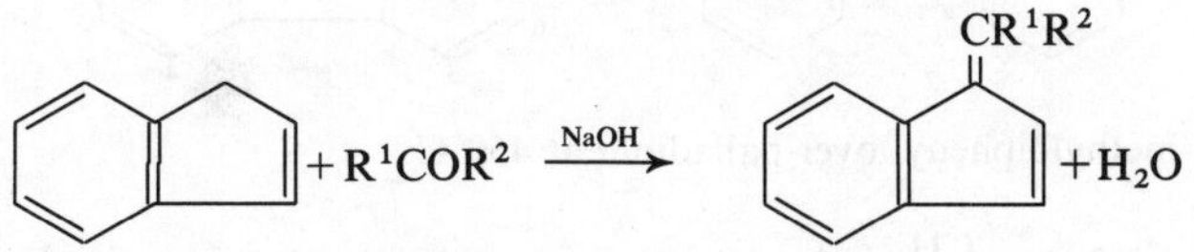

These benzofulvenes are highly coloured; *benzofulvene* has been prepared ($R^1 = R^2 = H$).

Indan-1-one (α-*hydrindone*), m.p. 42°C, may be synthesised by an internal Friedel–Crafts reaction on β-phenylpropionyl chloride:

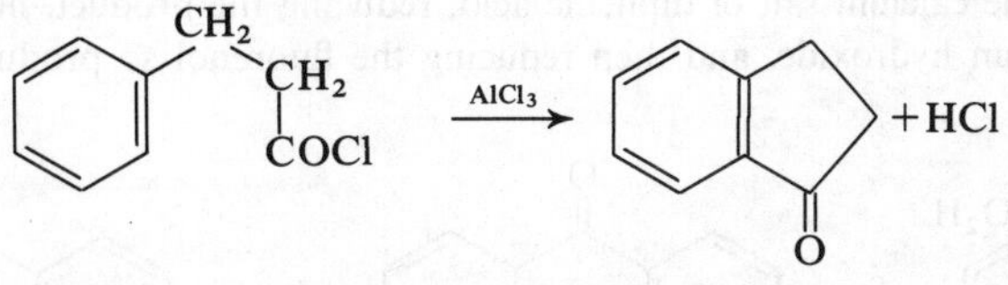

It may also be prepared by treating indene with hydrochloric acid and oxidising the product, 1-chloroindane, with chromium trioxide in acetic acid:

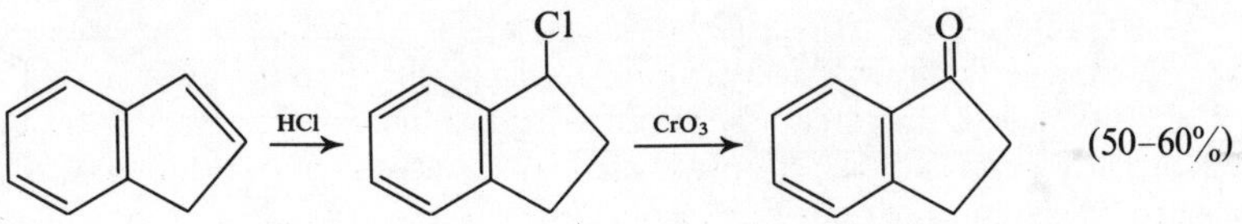

Indan-2-one (*β-hydrindone*), m.p. 61°C, may be prepared by heating xylylene-*o*-dicarboxylic acid:

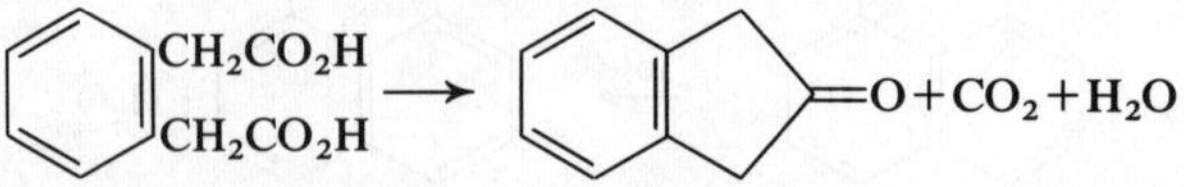

$$\text{(structure)} \quad \longrightarrow \quad \text{(structure)} =O + CO_2 + H_2O$$

Indane-1,3-dione may be prepared by the action of nitrous acid on 1,4-naphthaquinone (p. 807).
Indane-1,2,3-trione (ninhydrin) may be prepared as follows:

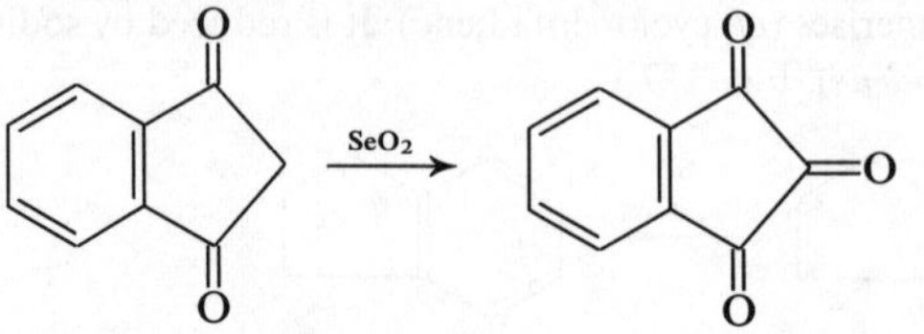

$$\text{(structure)} \xrightarrow{\text{SeO}_2} \text{(structure)} =O \qquad (35\%)$$

It has also been synthesised from ethyl phthalate and dimethyl sulphoxide (Becker *et al.*, 1963); note the water of constitution.

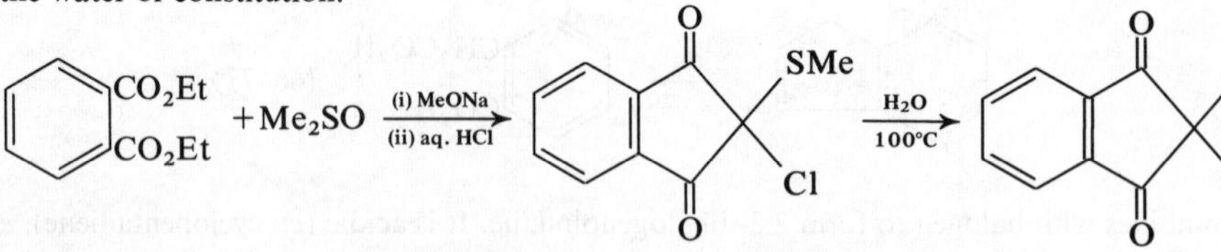

$$\text{(structure)} \begin{array}{c} CO_2Et \\ CO_2Et \end{array} + Me_2SO \xrightarrow[\text{(ii) aq. HCl}]{\text{(i) MeONa}} \text{(structure)} \begin{array}{c} SMe \\ Cl \end{array} \xrightarrow[\text{100°C}]{H_2O} \text{(structure)} \begin{array}{c} OH \\ OH \end{array} \quad (80\%)$$

Ninhydrin is used as a reagent for amino-acids.

Fluorene (*diphenylenemethane*), $C_{13}H_{10}$, occurs in coal-tar (fraction 270–300°C) and can be separated from the other compounds by means of its sodio-derivative (*cf.* indene). It may be prepared:

(i) By passing diphenylmethane through a red-hot tube:

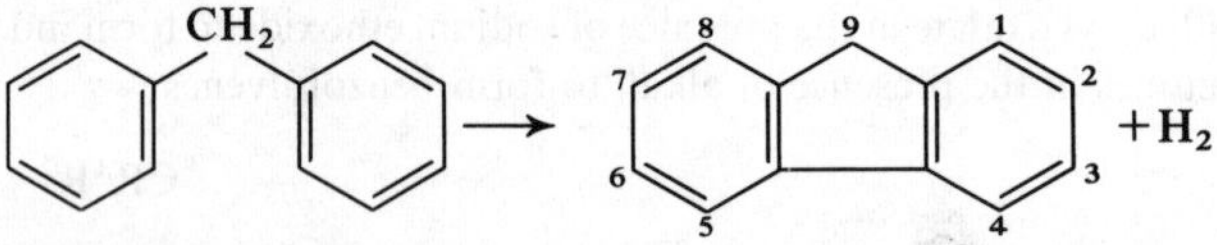

$$\text{(structure)} \longrightarrow \text{(structure)} + H_2$$

(ii) By passing *o*-methylbiphenyl over palladium at 450°C:

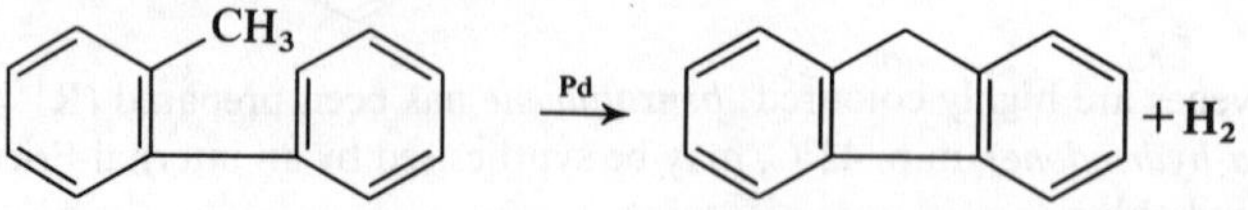

$$\text{(structure)} \xrightarrow{Pd} \text{(structure)} + H_2$$

(iii) By heating the calcium salt of diphenic acid, reducing the product, fluorenone, with zinc dust and ethanolic sodium hydroxide, and then reducing the fluorenol so produced by the Clemmensen method:

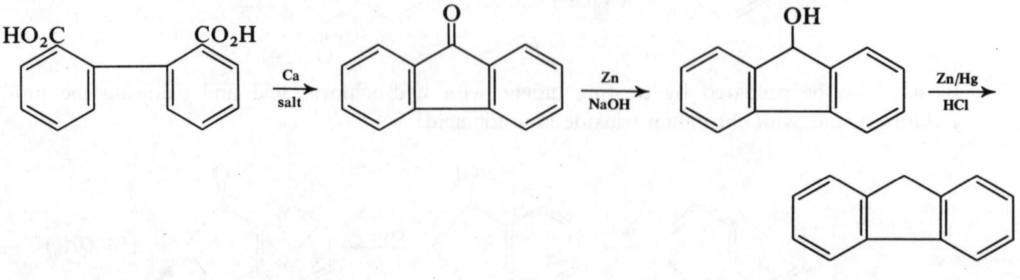

$$\text{(structure)} \xrightarrow[\text{salt}]{Ca} \text{(structure)} \xrightarrow[\text{NaOH}]{Zn} \text{(structure)} \xrightarrow[\text{HCl}]{Zn/Hg}$$

$$\text{(structure)}$$

(iv) Fluorene may be prepared by means of the Pschorr synthesis (p. 817), starting with *o*-aminodiphenylmethane.

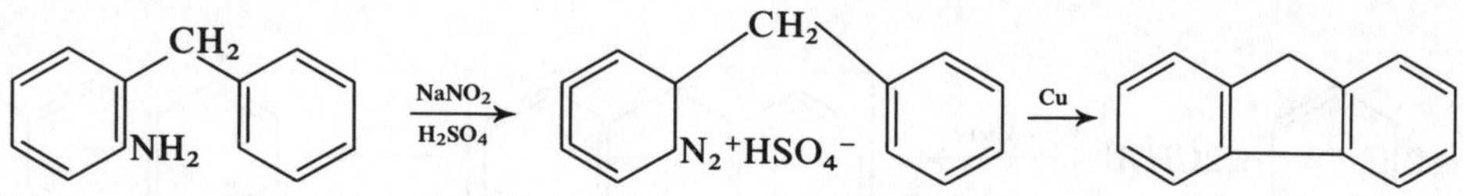

Alternatively, *o*-aminobenzophenone may be used as the starting material and the product, fluorenone, reduced to fluorene as in (iii).

Methods (iii) and (iv) are useful for preparing substituted fluorenes with the substituents in known positions.

Fluorene, m.p. 116°C, with a blue fluorescence, condenses with aldehydes and ketones in the presence of alkali. When fluorene is halogenated or nitrated, the first substituent enters the 2-position, and the second the 7-position. Fluorene gives 2-acetylfluorene (56–63%) when treated with acetic anhydride and aluminium chloride in carbon disulphide. Fluorene is oxidised by chromium trioxide in glacial acetic acid to *fluorenone* (m.p. 84°C). This is oxidised by permanganate to phthalic acid. When fused with potassium hydroxide, fluorenone is converted into biphenyl-2-carboxylic acid:

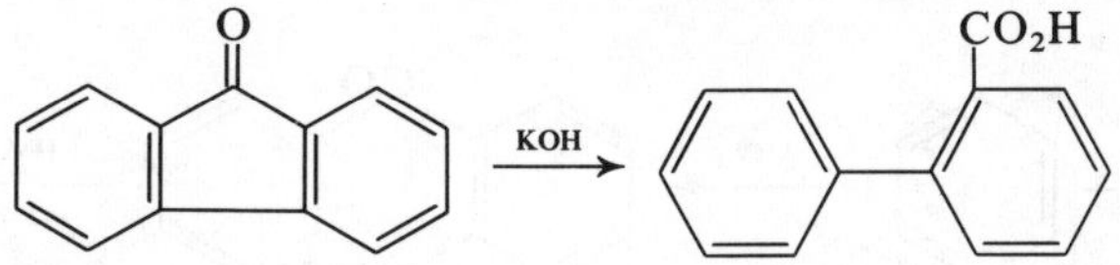

Anthracene

Anthracene, $C_{14}H_{10}$, is obtained from the anthracene oil fraction of coal-tar by cooling the latter and pressing the solid (which crystallises out) free from liquid. The crude anthracene contains phenanthrene and carbazole. The anthracene cake is powdered and washed with 'solvent naphtha' which dissolves the phenanthrene, and the remaining solid is then washed with pyridine which dissolves the carbazole. The anthracene is purified by sublimation. Alternatively, after removal of phenanthrene, the remaining solid is fused with potassium hydroxide, whereby potassio-carbazole is formed; unreacted anthracene is sublimed out of the melt and recovered.

Until recently, there has been very little use for carbazole, and thus the recovery of anthracene was expensive. Since anthracene is mainly used as the starting point of anthra-quinone, a cheaper method of isolating anthracene from coal-tar is to remove the phenanthrene first, and then catalytically oxidise the remaining mixture of anthracene and carbazole by air and vanadium pentoxide at 300–500°C. Under these conditions, anthracene is oxidised to anthraquinone and carbazole is completely oxidised (to carbon dioxide, etc.).

Synthesis of anthracene

(i) By a Friedel–Crafts reaction using benzyl chloride; 9,10-*dihydroanthracene*, which is first formed, readily eliminates two hydrogen atoms under the conditions of the experiment to form anthracene:

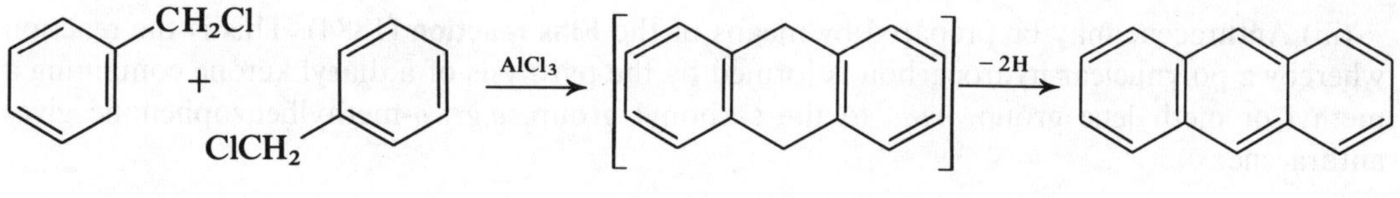

Anthracene is also formed by the Friedel–Crafts condensation between benzene and methylene dibromide, or between benzene and acetylene tetrabromide:

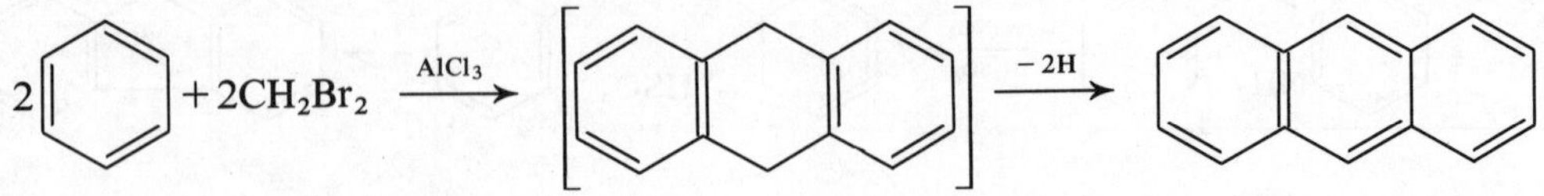

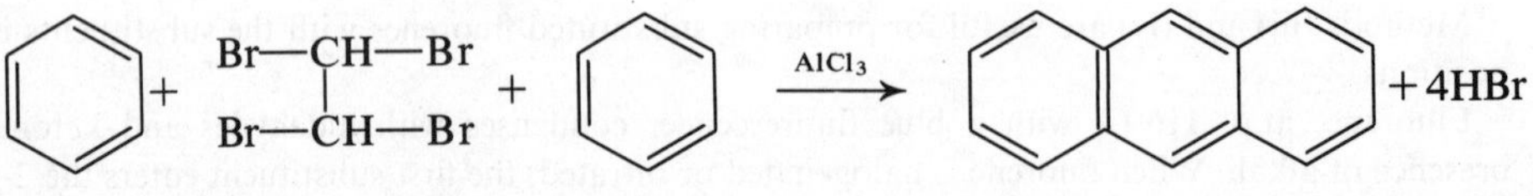

(ii) When phthalic anhydride in benzene solution is treated with aluminium chloride, o-benzoylbenzoic acid is formed. This, on heating with concentrated sulphuric acid or P.P.A. at 100°C, forms anthraquinone which, on distillation with zinc dust, gives anthracene:

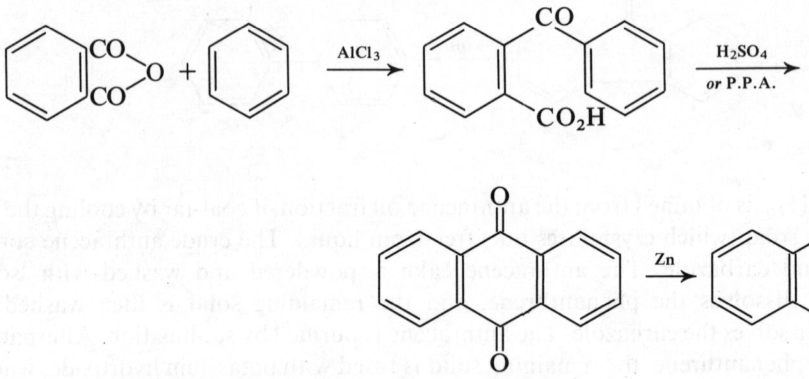

9-Alkylanthracenes may be prepared from o-benzoylbenzoic acid as follows (via anthrone):

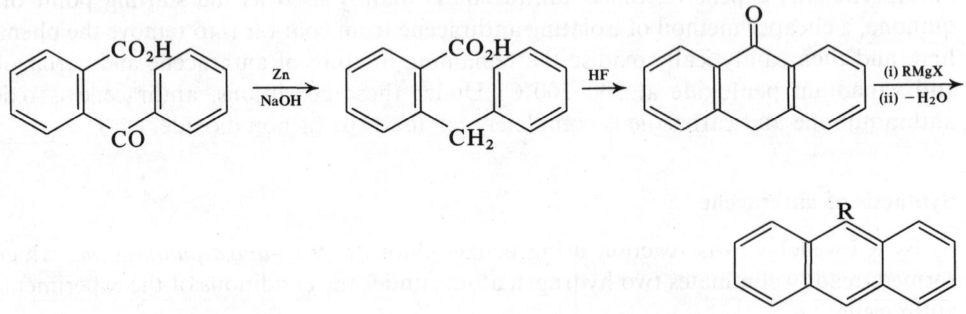

(iii) Anthracene may be prepared by means of the **Elbs reaction** (1884). This is the reaction whereby a polynuclear hydrocarbon is formed by the pyrolysis of a diaryl ketone containing a methyl or methylene group *ortho* to the carbonyl group, *e.g.*, o-methylbenzophenone gives anthracene:

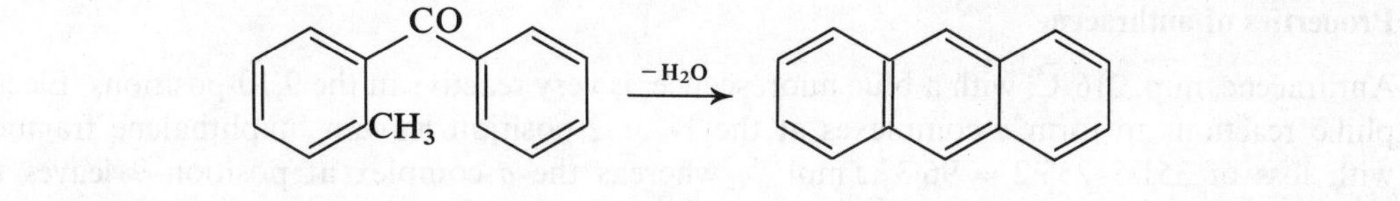

(iv) Anthracene may be synthesised by a Diels–Alder reaction involving 1,4-naphthaquinone and butadiene, followed by oxidation of the product with chromium trioxide in glacial acetic acid and then distillation with zinc dust.

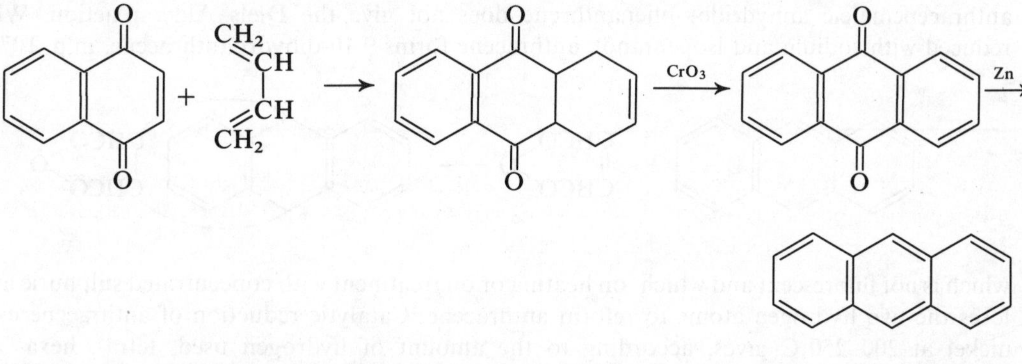

Positions of the double bonds in anthracene. Anthracene is best regarded as a resonance hybrid of four resonating structures:

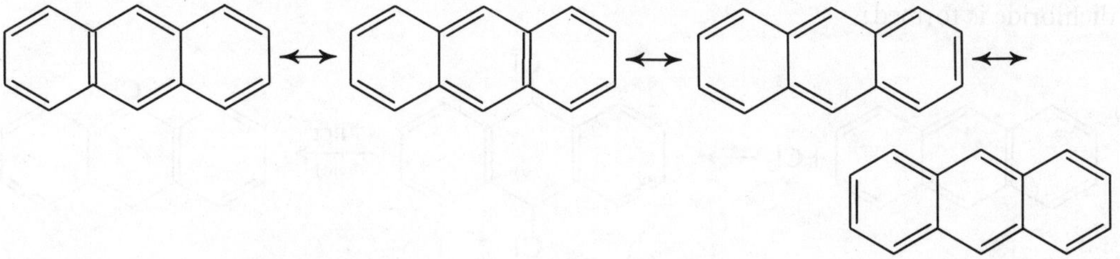

The resonance energy of anthracene is $351 \cdot 5 \, \text{kJ mol}^{-1}$

The double bond character of the various bonds is shown in (I) and bond lengths (in Å) are shown in (II).

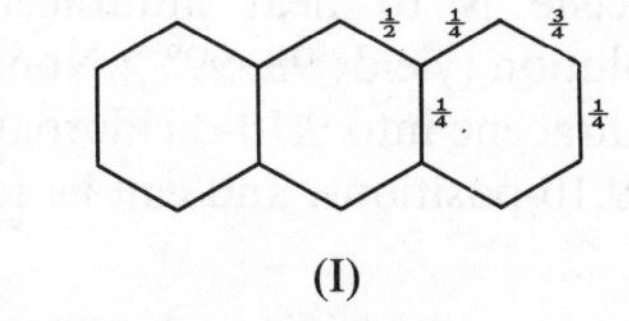

(I) 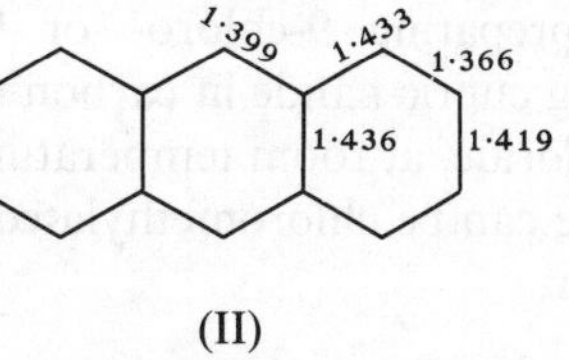   (II)

Isomerism of anthracene derivatives

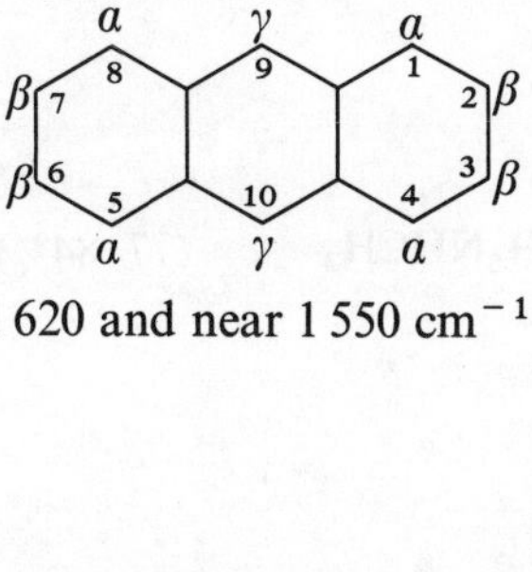

There are three possible mono-substitution products: 1- or α-, 2- or β- and 9- or γ- (or *meso-*). There are 15 possible disubstitution products if both substituents are identical; if the substituents are different, the number of isomers is larger.

The infra-red absorption regions of the anthracenes are 1 640–1 620 and near 1 550 cm^{-1} (see also phenanthrenes, p. 821).

Properties of anthracene

Anthracene, m.p. 216°C, with a blue fluorescence, is very reactive in the 9,10-positions. Electrophilic reactions to form σ-complexes at the 1- or 2-position leave a 'naphthalene fragment' with loss of $351\cdot5-255\cdot2 = 96\cdot3\,\text{kJ mol}^{-1}$, whereas the σ-complex at position 9 leaves two benzene rings intact, and so the loss is $351\cdot5-301\cdot2 = 50\cdot3\ \text{kJ mol}^{-1}$ (*cf.* p. 537). Hence, for polynuclear aromatic hydrocarbons (with three or more rings), the σ-complex that contains the largest number of *isolated* benzene rings will be the most stable.

Anthracene undergoes the Diels–Alder reaction in the 9,10-positions to form endo-anthracenemaleic anhydride; phenanthrene does not give the Diels–Alder reaction. When reduced with sodium and isopentanol, anthracene forms 9,10-dihydroanthracene, m.p. 107°C,

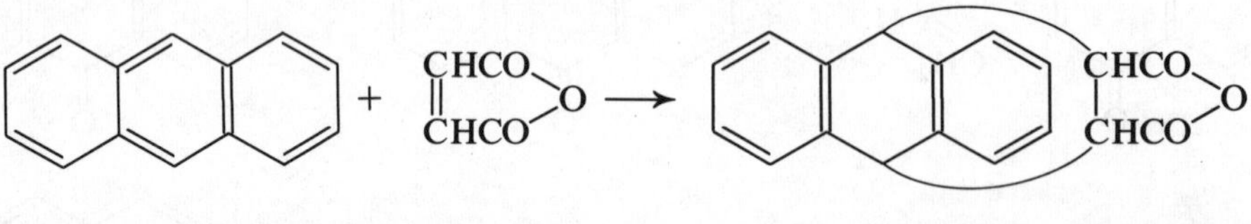

which is not fluorescent and which, on heating or on treatment with concentrated sulphuric acid, loses the two hydrogen atoms to reform anthracene. Catalytic reduction of anthracene using nickel at 200–250°C gives, according to the amount of hydrogen used, tetra-, hexa- and octahydroanthracene, and finally perhydroanthracene, $C_{14}H_{24}$ (the prefix *per* is often used to denote complete hydrogenation of a ring system).

When chlorine is passed into a cold solution of anthracene in carbon disulphide, anthracene dichloride is formed:

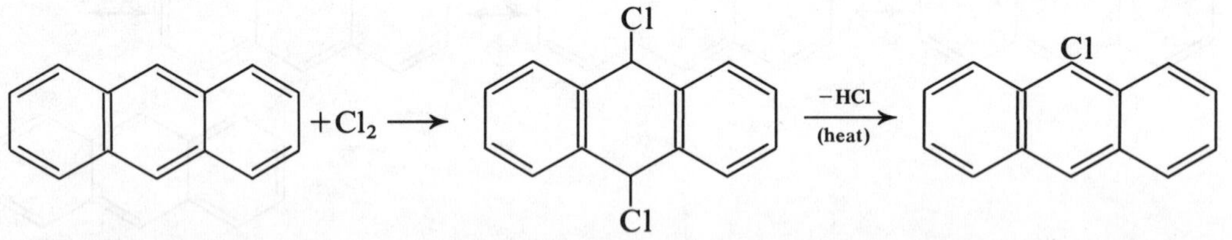

If this is heated or treated with alkali, hydrogen chloride is eliminated with the formation of 9-chloroanthracene. This is also obtained by chlorinating anthracene at 100°C, together with some 9,10-dichloroanthracene. Bromine reacts similarly, *e.g.*, bromination of anthracene in boiling carbon tetrachloride solution gives 9,10-dibromoanthracene (yield: 83–88%). The best method of preparing 9-chloro- or 9-bromoanthracene is to heat anthracene with the corresponding cupric halide in carbon tetrachloride solution (yield: 98–99%; Nonhebel, 1963). Sulphuryl chloride, at room temperature, converts anthracene into 9,10-dichloroanthracene.

Anthracene can be chloromethylated in the 9- and 9,10-positions, and can be formylated in the 9-position:

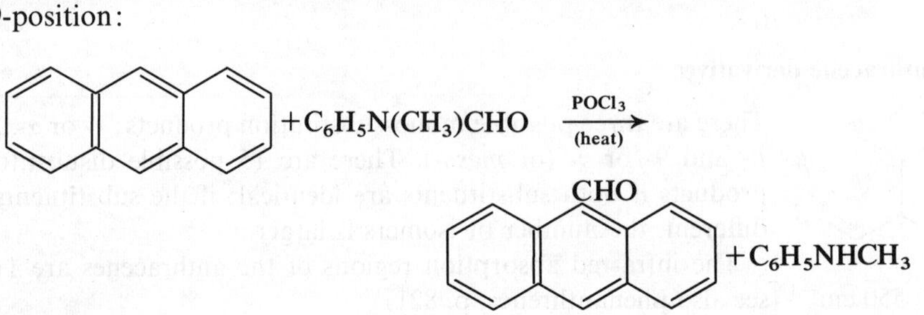

Attempts to nitrate anthracene with aqueous nitric acid lead to the formation of anthraquinone by oxidation. If, however, the nitration is carried out in acetic anhydride at 15–20°C, 9-nitroanthracene (m.p. 145°C) and 9,10-dinitroanthracene (m.p. 294°C) can be isolated. Anthracene is readily sulphonated to a mixture of the 1- and 2-sulphonic acids, some disulphonic acids also always being obtained; the 2-position is favoured at high temperature. If the sulphonation of anthracene is carried out in glacial acetic acid, a mixture of about equal amounts of 1- and 2-anthracenesulphonic acid is obtained. The 1-sulphonic acid shows no tendency to rearrange to the 2-acid (*cf.* naphthalenesulphonic acids, p. 801). With excess of concentrated sulphuric acid, anthracene gives disulphonic acids, the 1,8- at low temperatures, and the 2,7- at high temperatures.

According to Gore *et al.* (1966, 1967), the Friedel–Crafts acetylation (CH$_3$COCl and AlCl$_3$) of anthracene in benzene, nitrobenzene, etc. gives a complex mixture of mono- and di-acetyl derivatives. Furthermore, to obtain *pure* ketones, it was necessary to use elaborate chromatographic methods. In this way, it was shown that the main product in nitrobenzene as solvent is the 1-acetyl derivative, and in ethylene dichloride, the 9-acetyl derivative.

Hydroxyanthracenes. 1- and 2-*Hydroxyanthracenes* are known as **anthrols**. Each may be obtained from the corresponding anthracenesulphonic acid by alkaline fusion. 1-Anthrol is a yellow solid, m.p. 152°C; 2-anthrol is a brownish solid which decomposes at 200°C.

10-*Hydroxyanthracene*, also known as **anthranol**, is an unstable yellow solid, m.p. 120°C, and when quickly heated, forms **anthrone** (10-*keto*-9,10-*dihydroanthracene*—the keto-group is numbered last and consequently its enol isomer is known as 10- and *not* 9-hydroxyanthracene). Anthrone is the stable form and is a colourless solid, m.p. 154°C.

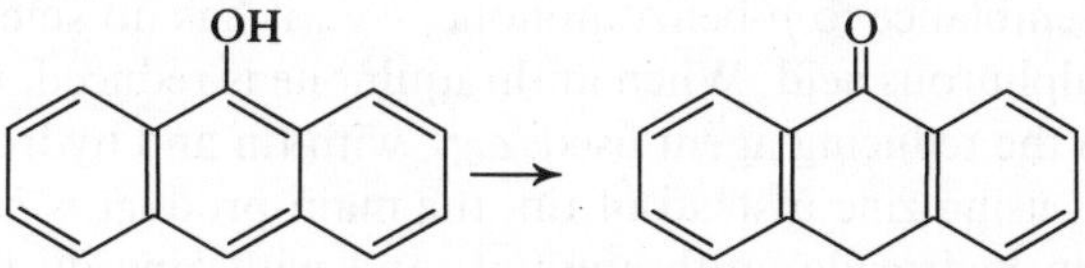

Since the hydrogen atom migrates across the ring, this type of tautomerism is called *trans-annular tautomerism*. Infra-red spectroscopy studies by Flett (1948) indicate that solid anthrone does not exist in equilibrium with the enol form (anthranol). Anthrone may be prepared by heating *o*-benzylbenzoic acid with hydrogen fluoride (see p. 812), or by reducing anthraquinone with tin and hydrochloric acid in glacial acetic acid. Anthrone dissolves in warm dilute alkalis, and these solutions, on acidification, precipitate the enol form.

Reduction of anthraquinone with zinc dust and aqueous sodium hydroxide gives *anthracene*-9,10-*diol* or *anthraquinol* (*anthrahydroquinone*). This is a brown solid, m.p. 180°C; its alkaline solutions (deep red) oxidise in air to give anthraquinone. When anthraquinol in alkaline solution is immediately acidified, it partly tautomerises to *oxanthranol*, m.p. 167°C (*trans*-annular tautomerism):

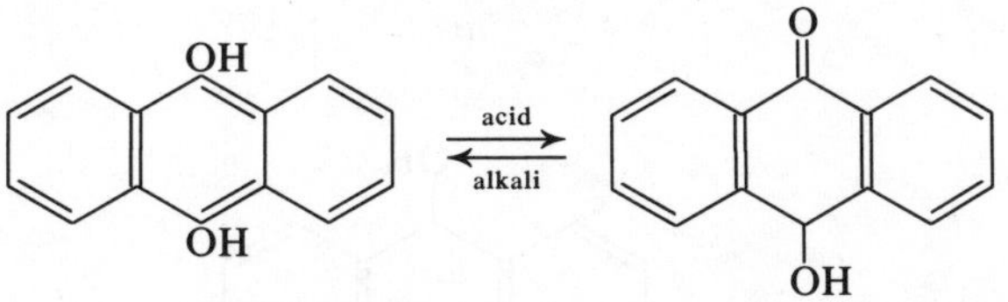

Anthracenecarboxylic acids. 1- and 2-**Anthroic acids** may be prepared by the hydrolysis of the corresponding cyanides (prepared by fusion of the sodium sulphonate with sodium cyanide). 9-Anthroic acid may be prepared by hydrolysing its acid chloride which may be prepared by heating anthracene with oxalyl chloride at 160°C. 9-Anthroic acid exhibits steric hindrance.

Anthraquinone (9,10-*dihydroanthracene*-9,10-*dione*). There are nine possible isomeric quinones of anthracene, but only three are known: 1,2-, 1,4- and 9,10-. The most important one is the 9,10-compound, and this is referred to simply as anthraquinone (see also p. 727).

Anthraquinone may be prepared by oxidising anthracene with sodium dichromate and sulphuric acid (yield: 90%). Commercially, it is prepared by the catalytic oxidation of anthracene (containing carbazole; see p. 811), but the cheapest method is its synthetic preparation (*cf.* p. 812):

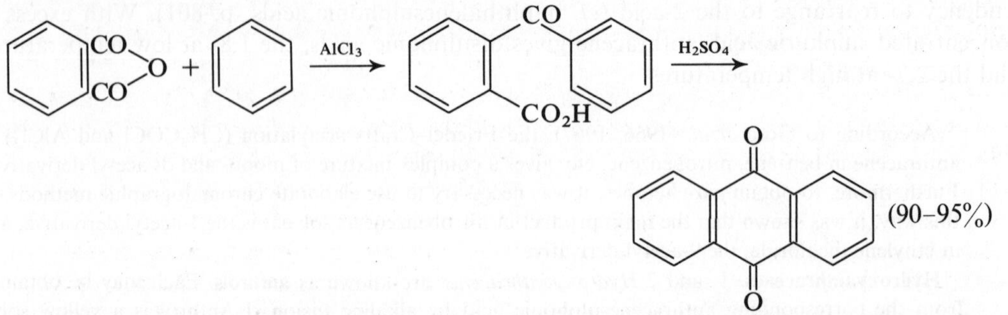

By using chlorobenzene or toluene instead of benzene, chloro- or methylanthraquinone is obtained; these are used in the manufacture of dyes.

Anthraquinone is a pale yellow solid which sublimes in needles that melt at 286°C. When distilled with zinc dust, anthraquinone forms anthracene. Anthraquinone is very stable and shows very little resemblance to *p*-benzoquinone, *e.g.*, it has no smell, is not very volatile, and is not reduced by sulphurous acid. When anthraquinone is reduced, the nature of the reduction product depends on the reducing agent used, *e.g.*, with tin and hydrochloric acid in acetic acid, anthrone is formed; using zinc instead of tin, the main product is *bianthryl* (I); with zinc dust and aqueous sodium hydroxide, anthraquinol; and with zinc dust and aqueous ammonium hydroxide, 9,10-dihydroanthrol (II):

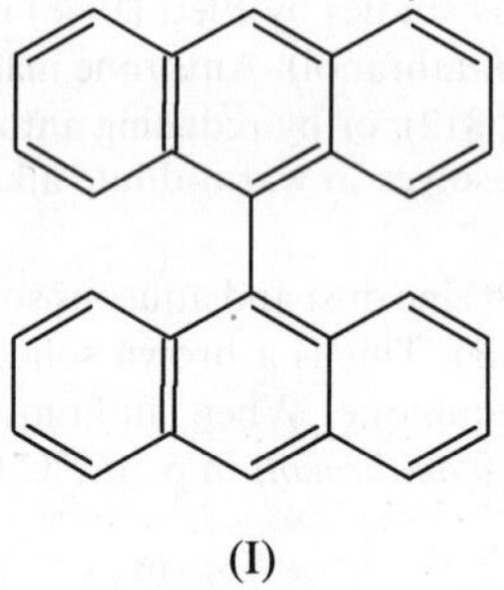

(I)

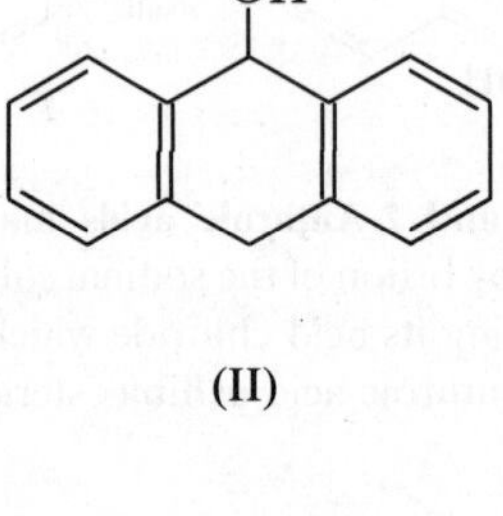

(II)

Nitration of anthraquinone with mixed acid gives 1-nitroanthraquinone; further nitration gives mainly 1,5- and 1,8-dinitroanthraquinones, and small amounts of the 1,6- and 1,7-dinitro-compounds. The nitro-group in the 1-position is very reactive, *e.g.*, it is replaced by an amino-group when 1-nitroanthraquinone is heated with ammonia.

Anthraquinone is very difficult to sulphonate with concentrated sulphuric acid, but it is readily sulphonated with oleum at 160°C. The first product is the 2-sulphonic acid and a small amount of the 1-isomer; prolonged heating gives a mixture of 2,6- and 2,7-anthraquinone-disulphonic acids in about equal amounts. If mercuric sulphate is used as a catalyst, the sulphonation takes an entirely different course. The first product now is the 1-sulphonic acid, and then a mixture of the 1,5- and 1,8-disulphonic acids. A possible explanation for this change in orientation is that the reaction now proceeds via a 1-mercurated derivative. The sulphonic acid group in the 1- or 2-position is easily displaced; *e.g.*, when 1- or 2-anthraquinonesulphonic acid is treated with chlorine, the corresponding chloroanthraquinone is obtained.

Anthraquinone does not undergo the Friedel–Crafts reaction, and is halogenated with very great difficulty; in fact, monohalogenoanthraquinones cannot be obtained directly.

2-Aminoanthraquinone, m.p. 304°C, is very important as an intermediate in the preparation of indanthrene dyes. It is prepared industrially by heating the sodium salt of anthraquinone-2-sulphonic acid with a solution of ammonia, ammonium chloride and sodium arsenate under pressure at 200°C. The sodium arsenate oxidises the liberated sulphite which otherwise would attack the amine produced (*cf.* p. 803):

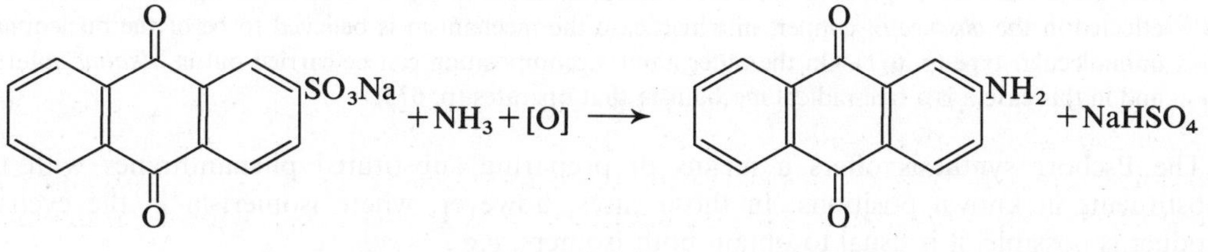

Alizarin (1,2-*dihydroxyanthraquinone*) is the most important dihydroxy-derivative of anthraquinone, and is used as a mordant dye (p. 898).

Phenanthrene

Phenanthrene, $C_{14}H_{10}$, is isomeric with anthracene; it is an example of an *angular* polynuclear hydrocarbon. It occurs in the anthracene oil fraction of coal-tar, and is separated from anthracene by means of solution in solvent naphtha (see p. 811). Phenanthrene is structurally related to certain alkaloids, *e.g.*, morphine, and to the steroids, *e.g.*, cholesterol.

Synthesis of phenanthrene

Many methods are available, but only two important ones will be described here.

(i) **Pschorr synthesis** (1896). This is carried out by heating *o*-nitrobenzaldehyde with sodium β-phenylacetate in the presence of acetic anhydride (Perkin's reaction), reducing and diazotising the product, α-phenyl-*o*-nitrocinnamic acid, and treating the diazonium salt with sulphuric acid and copper powder. Phenanthrene-9-carboxylic acid is produced and this, on strong heating, forms phenanthrene:

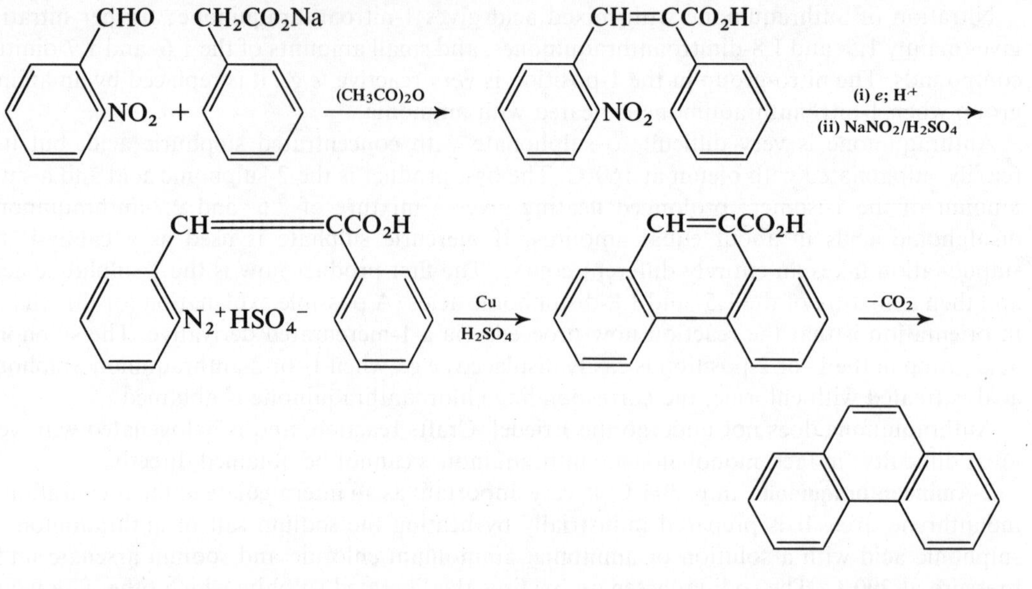

The Pschorr synthesis is really an example of intramolecular phenylation, and the decomposition can be effected in several ways. The method described above is the catalysed (copper) decomposition, and this is believed to proceed via a free-radical mechanism (*cf.* p. 781). Decomposition can also be effected in the *absence* of copper, in which case the mechanism is believed to be of the nucleophilic unimolecular type (p. 671). On the other hand, decomposition can be carried out in *alkaline* solution and in this case it is a free-radical mechanism that operates (p. 675).

The Pschorr synthesis offers a means of preparing substituted phenanthrenes with the substituents in known positions. In those cases, however, where isomerism in the cyclised product is possible, it is usual to obtain both isomers, *e.g.*,

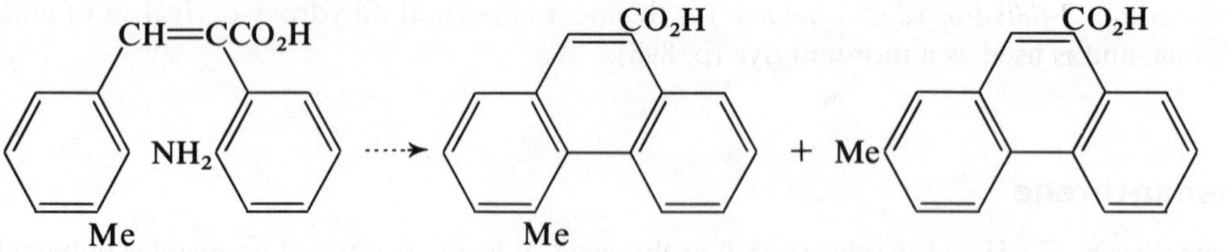

A further point to note is that since ring closure is effected between two rings, these rings must be in the *cis* position, *e.g.*, *cis-o*-aminostilbene gives phenanthrene, but the *trans* isomer does not.

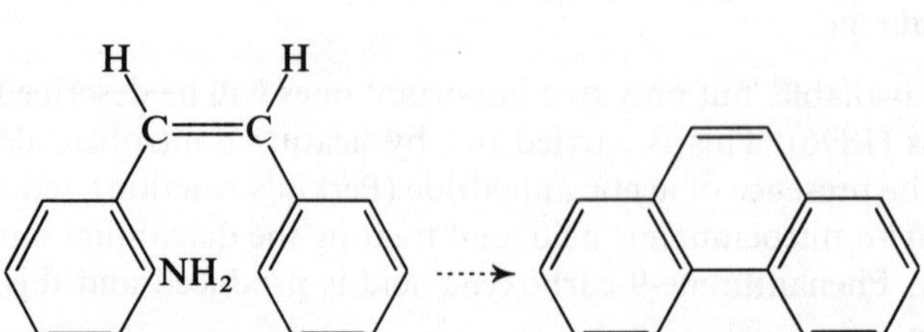

Kessar *et al.* (1969) have prepared phenanthrenes from cinnamic acids via what they believe to be benzyne intermediates, *e.g.*,

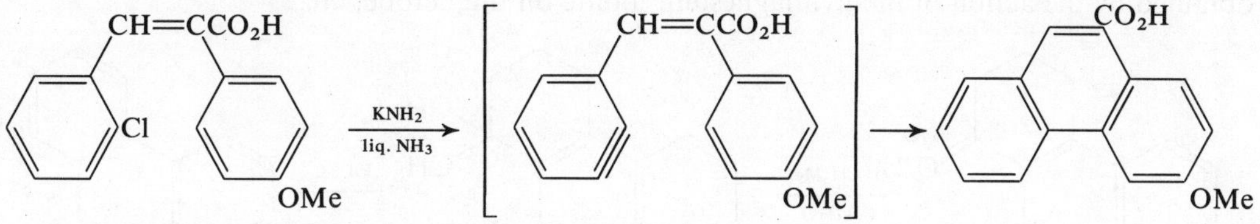

(ii) **Haworth synthesis** (*cf.* naphthalene):

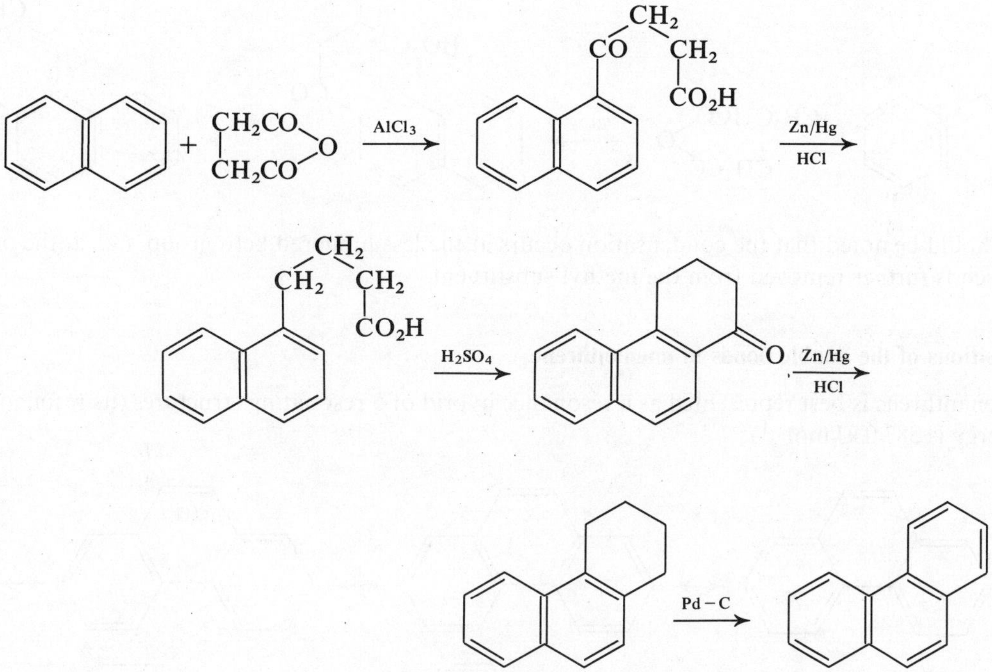

Succinoylation also occurs in the 2-position; the two keto-acids may be readily separated. The 2-compound also gives phenanthrene when treated as above; no anthracene is formed since ring closure occurs only in the 1-position of naphthalene, and not in 3.

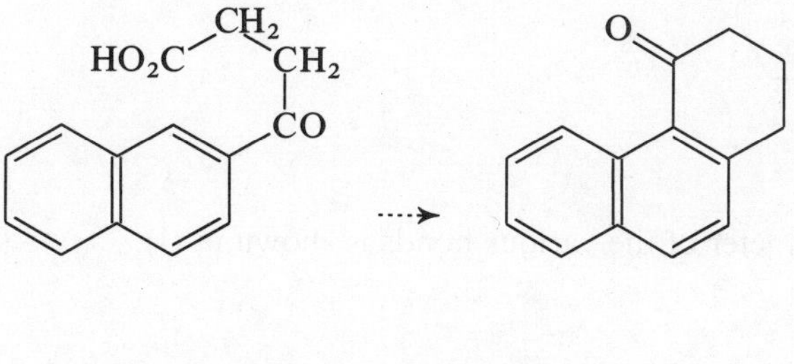

The Haworth synthesis is very useful for preparing alkylphenanthrenes with the alkyl groups in known positions; *e.g.*, after ring closure of the 1-derivative, 1-methylphenanthrene may be obtained by the action of methylmagnesium iodide on the ketone, etc.:

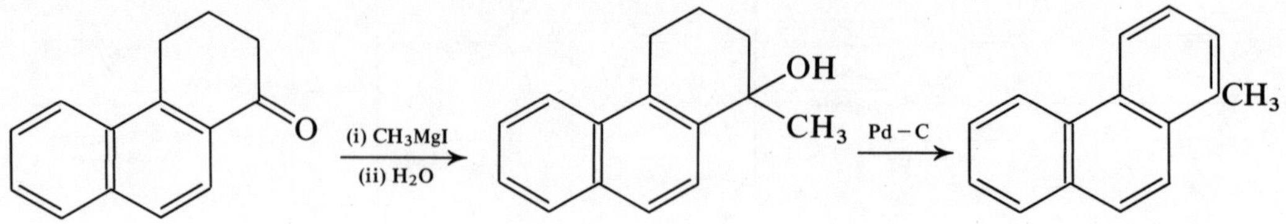

By using methylsuccinic anhydride instead of succinic anhydride, a methyl group can be introduced into the 2- or 3-position, *e.g.*, from the 2-derivative:

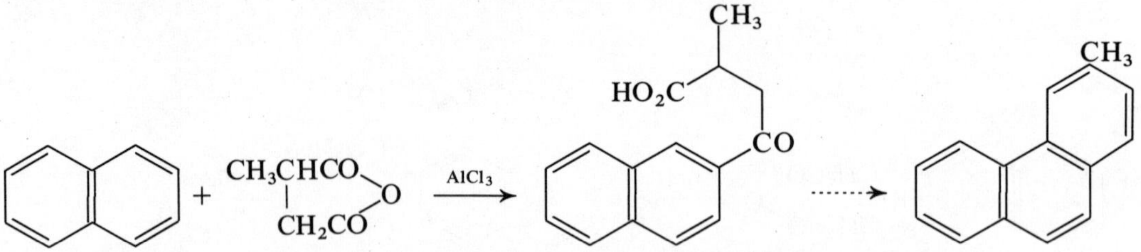

It should be noted that the condensation occurs at the less hindered keto group, *i.e.*, at the one which is further removed from the methyl substituent.

Positions of the double bonds in phenanthrene

Phenanthrene is best represented as a resonance hybrid of 5 resonating structures (its resonance energy is $387 \cdot 0\,\text{kJ mol}^{-1}$):

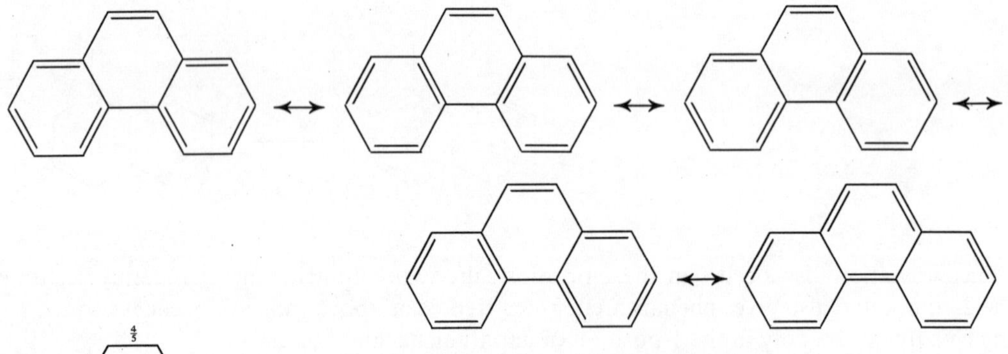

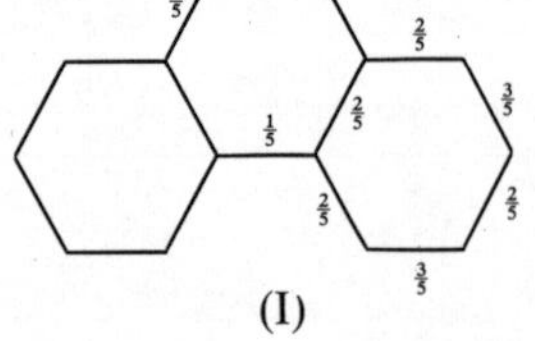

(I)

The double bond character of the various bonds is shown in (I).

Isomerism of phenanthrene derivatives

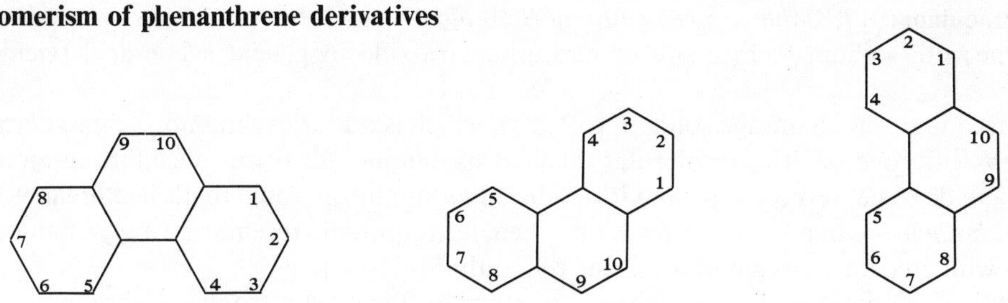

The formula of phenanthrene may be written in the three ways shown. There are 5 mono-substitution products possible: 1, 2, 3, 4 and 9. If the two substituents are identical, then 25 disubstitution products are possible. Owing to the great number of isomers, derivatives of phenanthrene are usually prepared synthetically and not by direct substitution in the phenanthrene nucleus.

Properties of phenanthrene

Phenanthrene, m.p. 99°C, in benzene shows a blue fluorescence. It is very reactive in the 9,10-positions (*cf.* anthracene, p. 814), *e.g.*, phenanthrene is readily catalytically reduced (using copper oxide-chromic oxide) to 9,10-dihydrophenanthrene, and it adds on bromine to form 9,10-phenanthrene dibromide. These addition reactions occur almost as easily as with a pure ethylenic bond. Ozonolysis (followed by oxidation) or oxidation with peracetic acid gives diphenic acid.

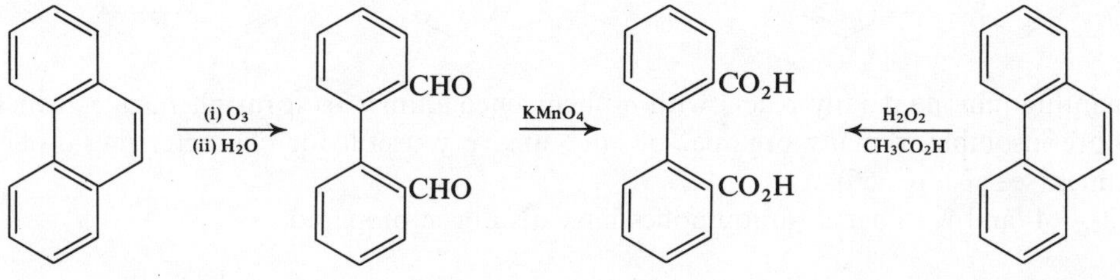

9-Bromophenanthrene (90–94%) is produced by adding bromine to a refluxing solution of phenanthrene in carbon tetrachloride. This is the starting-point of 9-substituted phenanthrenes, *e.g.*, when heated with cuprous cyanide at 260°C, 9-bromophenanthrene forms the corresponding cyano-compound; this may be hydrolysed to phenanthrene-9-carboxylic acid. Phenanthrene undergoes the Friedel–Crafts reaction mainly in the 3-, and to a small extent, in the 2-position. It is chloromethylated in the 9-position. When nitrated, phenanthrene gives a mixture of three mononitro-derivatives, the 3-isomer predominating. Sulphonation of phenanthrene gives a mixture of 1-, 2-, 3- and 9-phenanthrenesulphonic acids, and the ratio of these isomers depends on the temperature.

The infra-red absorption regions of phenanthrenes are $1\,600\,cm^{-1}$ (two bands) and near $1\,500\,cm^{-1}$. Hence, these compounds can be differentiated from anthracenes (p. 813).

Hydroxyphenanthrenes. Five *phenanthrols* are known: 1-, 2-, 3-, 4- and 9-.

Phenanthraquinone (9,10-*dihydrophenanthrene*-9,10-*dione*) may be prepared by oxidising phenanthrene with sodium dichromate or chromium trioxide in glacial acetic acid (yield is excellent).

Phenanthraquinone is an orange solid, m.p. 208°C, which is odourless and not steam-volatile. It combines with one or two molecules of hydroxylamine to form phenanthraquinone monoxime and dioxime, respectively, and it is reduced by sulphurous acid to **phenanthrene-9,10-diol** (*phenanthraquinol*). In all of these reactions phenanthraquinone resembles *o*-benzoquinone. Distillation with zinc dust gives phenanthrene (*cf.* anthraquinone).

When nitrated, phenanthraquinone gives a mixture of 2- and 4-nitrophenanthraquinones; more vigorous nitration produces a mixture of the 2,7- and 4,5-dinitro-compounds. Oxidation with sodium dichromate and sulphuric acid converts phenanthraquinone into diphenic acid. When warmed with alkali, phenanthraquinone undergoes the benzilic acid rearrangement (p. 792) to form 9-hydroxyfluorene-9-carboxylic acid:

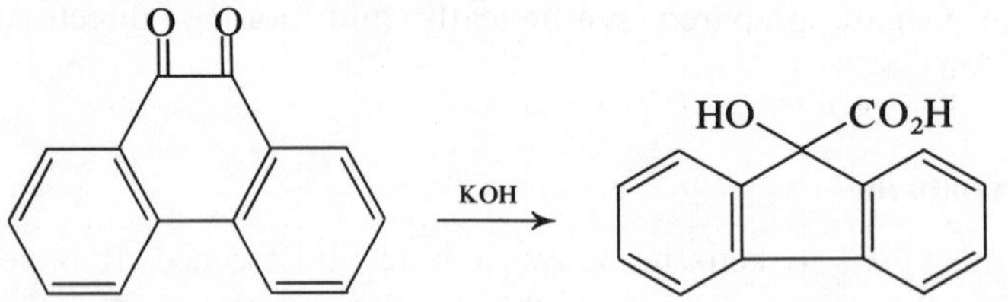

This, on heating in air, eliminates carbon dioxide and the product, fluorenol, is oxidised to fluorenone:

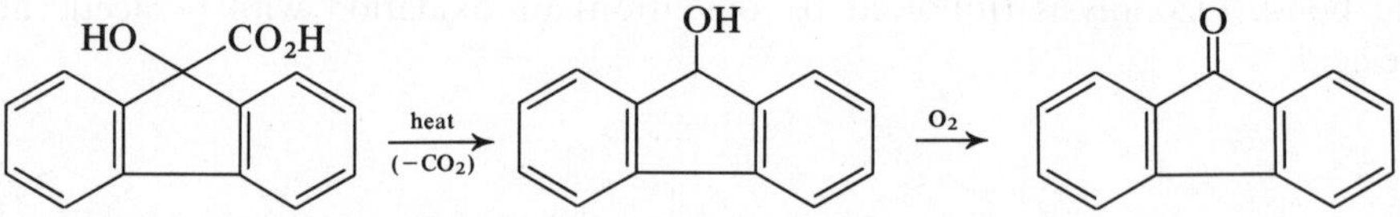

Phenanthraquinone readily reacts with *o*-phenylenediamines to form phenazines which, since they are insoluble in many organic solvents, are very useful for characterising *o*-phenylene-diamines (see also p. 666).

1,2-, 1,4- and 3,4-Phenanthraquinones have also been prepared.

Molecular overcrowding

A molecule becomes overcrowded when at least one pair of *non-bonded* atoms are closer to each other than the sum of their van der Waals radii. Two extremes are now possible. If the atoms were to remain very close to each other and the rest of the molecule remain unaltered, then there would be large steric repulsion. On the other hand, if the molecule were deformed so that the overcrowded atoms were separated to the sum of their van der Waals radii, there would be a large strain energy. In practice, a compromise is reached, the steric repulsion and the strain being relieved by bending of substituent bonds and buckling of the molecule as a whole. One example of this is the phenanthrene derivative (I). The phenanthrene nucleus is planar (but see below), and substituents lie in this plane. If, however, fairly large groups are in positions 4 and 5, there will not be enough room to accommodate both groups in the plane of the nucleus. Such a molecule will not be planar, and hence is asymmetric and consequently should be resolvable. Newman *et al.* (1940, 1947) have partially resolved (I), and also (in 1955) prepared the optically active forms of (II) (see also p. 772: and Vol. 2, Ch. 5). It is also interesting to note

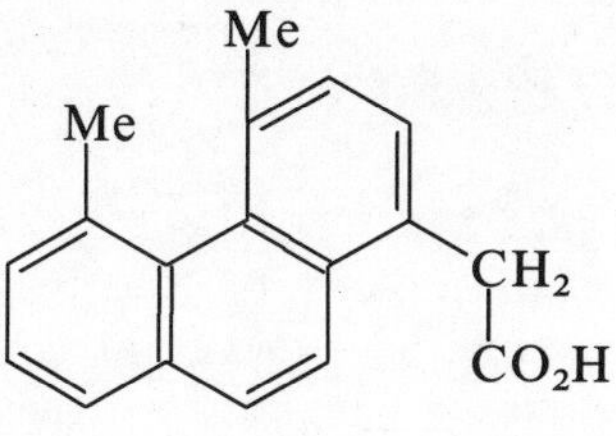

(I)

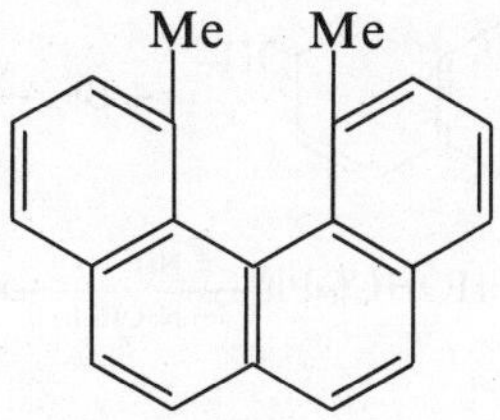

(II)

that, according to Trotter (1963), the structure of phenanthrene itself is not absolutely planar in the *crystalline* state, and this is probably due to molecular overcrowding.

PROBLEMS

1. Arrange the following in order of increasing stability and give your reasons: $PhCH_2^+$, Ph_3C^+, Me^+, Ph_2CH^+.

2. Using PhMgBr, what substrates would lead to Ph_3COH?

3. The rate of reaction of benhydryl chloride (Ph_2CHCl) in aqueous EtOH with KF to give benzhydryl fluoride is retarded by the addition of NaCl. Explain.

4. Devise an experiment to show whether the rearrangement of naphthalene-1-sulphonic acid to the 2-sulphonic acid is intramolecular or intermolecular.

5. Complete the equations:

(i) 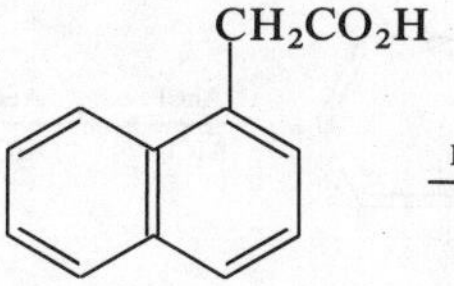 $\xrightarrow[H^+]{EtOH}$? $\xrightarrow[\text{(ii) MeI}]{\text{(i) NaH}}$? $\xrightarrow[\text{(ii) H}^+]{\text{(i) NaOH}}$? $\xrightarrow[\text{(ii) AlCl}_3]{\text{(i) SOCl}_2}$?

(ii) 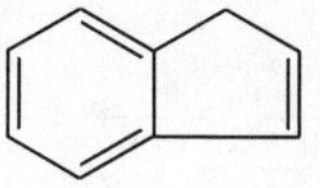 $\xrightarrow[\text{t-BuOK}]{CHBr_3}$?   (iii) $PhNH_2 \cdot HCl + Ph_3COH \xrightarrow{AcOH}$? (not an *N*-deriv.)

(iv) 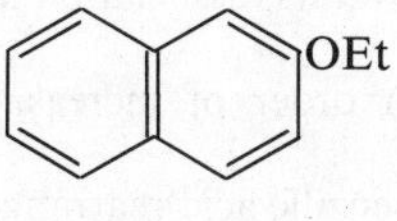 $+ PhNMeCHO \xrightarrow{POCl_3}$?

(v) 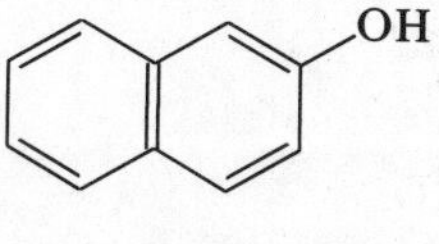 $+ CHCl_3 \xrightarrow[\text{NaOH}]{\text{aq. EtOH}}$?

(vi) 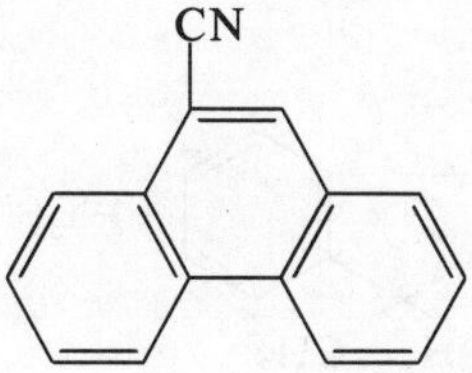 $+ MeMgI \xrightarrow[\text{(ii) H}^+]{\text{(i) Et}_2/PhH}$?

(vii) 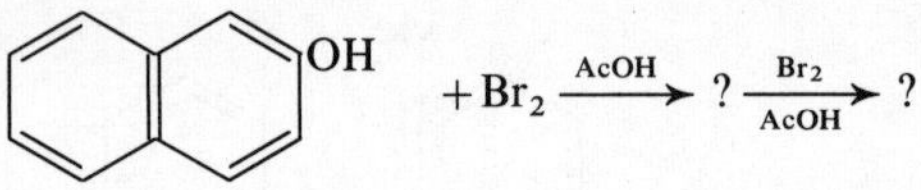 $+ Br_2 \xrightarrow{AcOH}$? $\xrightarrow[AcOH]{Br_2}$?

(viii) $PhCHOHCOPh \xrightarrow[\text{aq. NaOH; heat}]{NaBrO_3}$? (rearr.)

(ix) 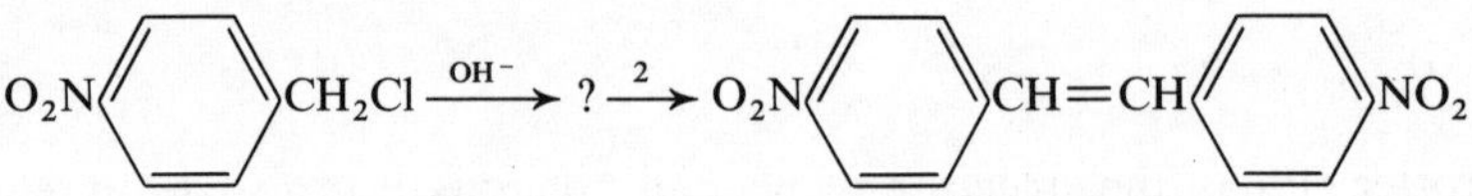 O_2N——$CH_2Cl \xrightarrow{OH^-}$? $\xrightarrow{2}$ O_2N——$CH=CH$——NO_2

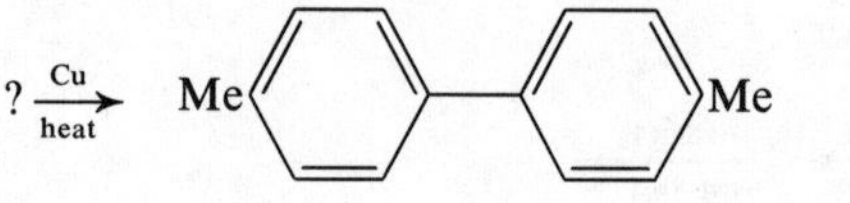

(x) 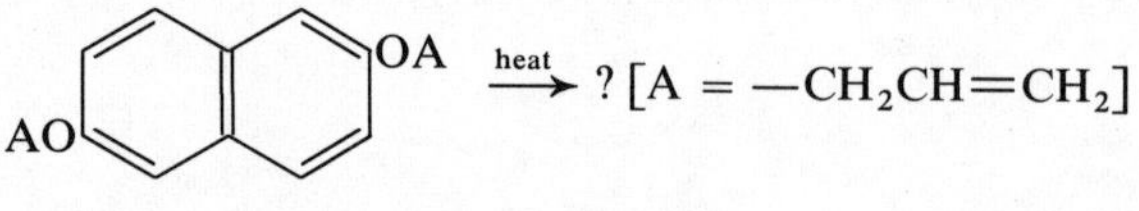 $\xrightarrow{heat}$? $[A = -CH_2CH=CH_2]$

(xi) 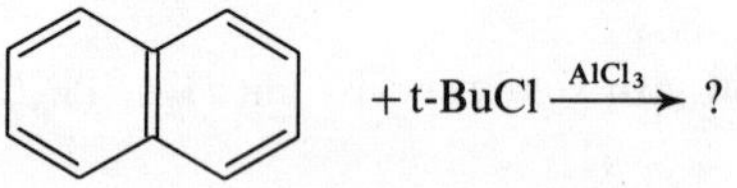 $+ t\text{-BuCl} \xrightarrow{AlCl_3}$?

(xii) 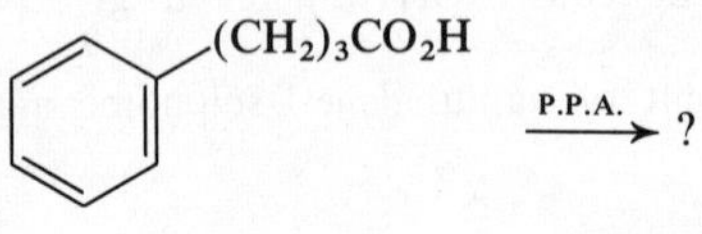 $\xrightarrow{P.P.A.}$?

(xiii) ? $\xrightarrow[\text{heat}]{Cu}$ Me——Me (xiv) 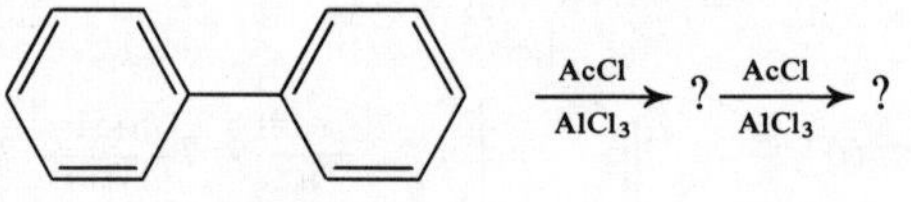 $\xrightarrow{AcCl}{AlCl_3}$? $\xrightarrow{AcCl}{AlCl_3}$?

(xv) 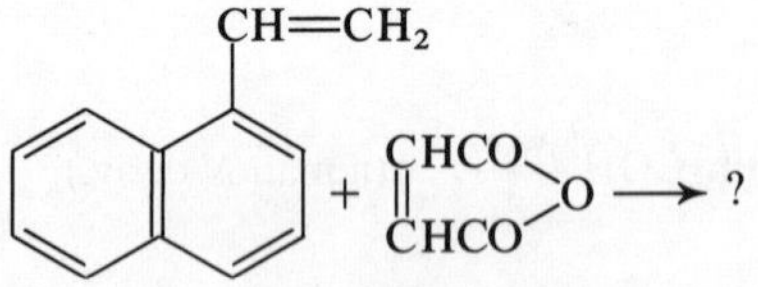 $+$ $\longrightarrow$?   (xvi) 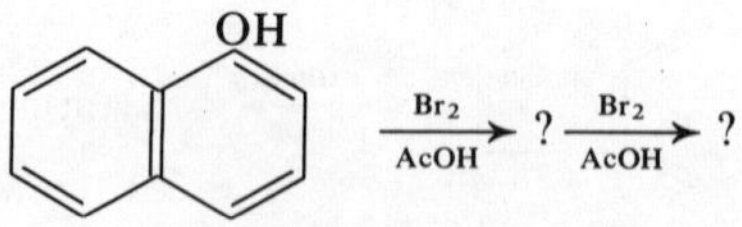 $\xrightarrow{Br_2}{AcOH}$? $\xrightarrow{Br_2}{AcOH}$?

6. Explain the effect of the introduction of *one p*-NO$_2$ or *p*-MeO group in Ph$_6$C$_2$ on its dissociation into free radicals.

7. Arrange Ph$_3$CH, (p-NO$_2$C$_6$H$_4$)$_3$CH, and (3,5-Me$_2$-4-NO$_2$—C$_6$H$_2$)$_3$CH in order of increasing acid strength and give your reasons.

8. What are the expected products when the following α-diketones undergo the benzilic acid rearrangement?

(i) $HO_2CCH_2COCOCH_2CO_2H$;   (ii) $MeCOCOCO_2Et$;   (iii)

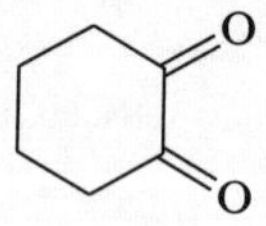

9. How could you convert naphthalene into 2-nitronaphthalene?
10. Complete the equation and suggest a mechanism for the last step.

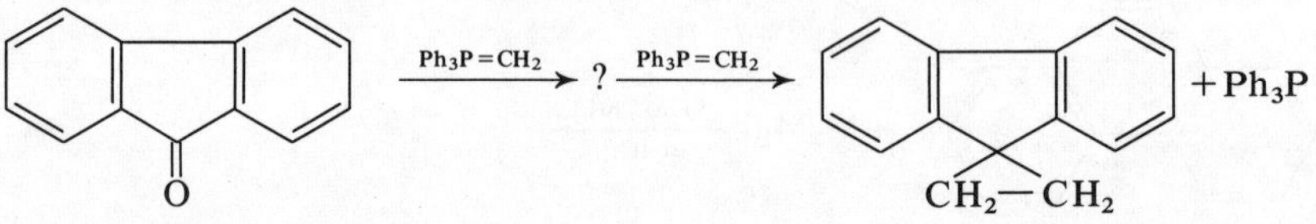

$\xrightarrow{Ph_3P=CH_2}$? $\xrightarrow{Ph_3P=CH_2}$ $+ Ph_3P$

11. Suggest a mechanism for the reaction:

$$2Ph_2CCl_2 + 2NaI \longrightarrow Ph_2C{=}CPh_2 + 2ICl + 2Cl^-$$

12. Synthesise phenanthrene, starting from naphthalene and butyric acid.

13. (i) The u.v. spectrum of a solution of anthracene in H_2SO_4 is similar to that of Ph_2CH^+; (ii) The NMR spectrum of anthracene in TFA indicates the presence of a CH_2 group. Offer an explanation.

14. The order of reactivity to electrophilic substitution is fluorene > biphenyl > benzene. Offer an explanation.

15. Convert: (*A*)

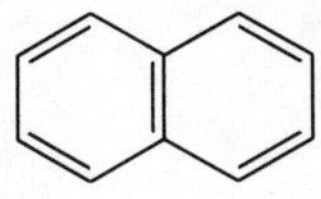

(i) 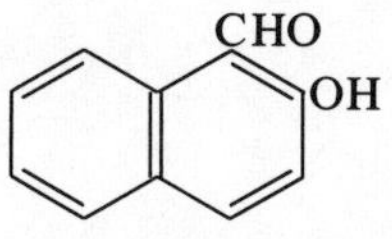 ; (ii) 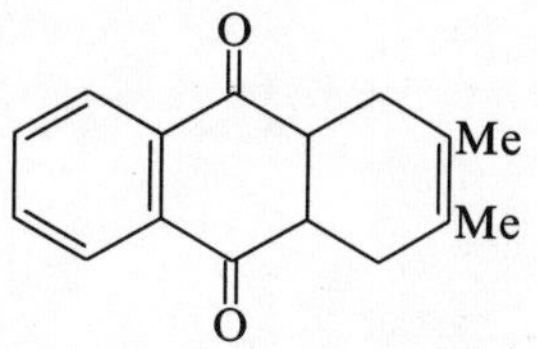 ;

(iii) 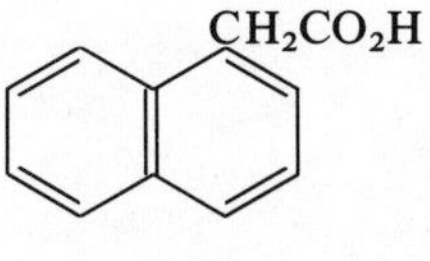 ; (iv)

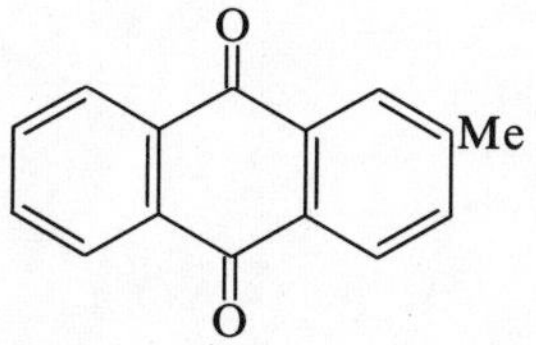

(*B*) (v) PhMe $\longrightarrow$ Ph$_2$CHCHPh$_2$; (vi) PhH $\longrightarrow$ Ph$_3$CCO$_2$H;

(vii) PhOMe $\longrightarrow$ (*a*) 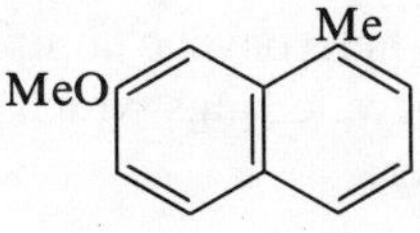 *and* (*b*) 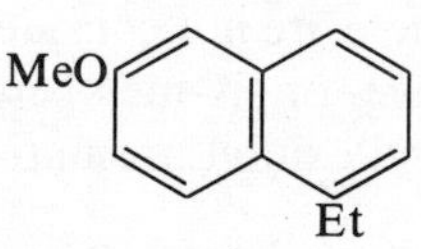 ;

(viii) phenanthrene $\longrightarrow$ phenanthrene-9-carboxylic acid.

16. Synthesise 2,9-dimethylanthracene starting with compounds containing one benzene ring.

REFERENCES

FANTA, 'The Ullmann Synthesis of Biaryls', *Chem. Rev.*, 1964, **64**, 613.

Organic Reactions, Wiley. Vol. IV (1948), Ch. 5. 'The Synthesis of Benzoins'; Vol. V (1949), Ch. 5. 'The Friedel–Crafts Reaction with Aliphatic Dibasic Anhydrides'; Vol. I (1942), (i) Ch. 5. 'The Bucherer Reaction.' (ii) Ch. 6. 'The Elbs Reaction'; Vol. VI (1951), Ch. 1. 'The Stobbe Condensation'; Vol. IX (1957), Ch. 7. 'The Pschorr Synthesis and Related Diazonium Ring Closure Reactions.'

STIRLING, *Radicals in Organic Chemistry*, Oldbourne Press (1965).

FORRESTER, HAY, and THOMSON, *Organic Chemistry of Stable Free Radicals*, Academic Press (1968).

FINAR, *Organic Chemistry*, Vol. 2, Longmans (1968, 4th edn.). Ch. 5. 'Stereochemistry of Diphenyl Compounds'; Ch. 10. 'Polycyclic Aromatic Hydrocarbons.'

DONALDSON, *The Chemistry and Technology of Naphthalene Compounds*, Arnold (1958).

BUEHLER, 'Hindered and Chelated 1,2-Enediols', *Chem. Rev.*, 1964, **64**, 7.

CLAR, *Polycyclic Hydrocarbons*, Academic Press, Vols. I and II (1964).

30 Heterocyclic compounds

Heterocyclic compounds are cyclic compounds with the ring containing carbon and other elements, the commonest being oxygen, nitrogen and sulphur. There are a number of heterocyclic rings which are easily opened and do not possess any aromatic properties, *e.g.*, ethylene oxide, γ- and δ-lactones, etc. These are not considered to be heterocyclic compounds. Heterocycles are those compounds with five- or six-membered heterocyclic rings which are stable, contain conjugated double bonds, and exhibit aromatic character.

Nomenclature

Many heterocyclic systems have trivial names (see text). The following is the systematic method of nomenclature.

(i) The names of monocyclic compounds are derived by a prefix (or prefixes) indicating the nature of the hetero-atoms present, and eliding the 'a' where necessary, *e.g.*, oxygen, **oxa**; sulphur, **thia**; nitrogen, **aza**; silicon, **sila**; phosphorus, **phospha**. When two or more of the same hetero-atoms are present, the prefixes di, tri, etc. are used, *e.g.*, dioxa, triaza. If the hetero-atoms are *different*, their order of citation starts with the hetero-atom of as high a group in the periodic table and as low an atomic number in that group. Thus, the order of naming will be O, S, N, P, Si, *e.g.*, thiaza (S then N).

(ii) The size of a monocyclic ring from 3 to 10 is indicated by a stem: 3, ir (*tri*); 4, et (*tetra*); 5, ol; 6, in; 7, ep (*hepta*); 8, oc (*octa*); 9, on (*nona*); 10, ec (*deca*) [see Table 30.1].

(iii) The state of hydrogenation is indicated in the suffix (see Table 30.1), or by the prefixes dihydro, tetrahydro, etc., or by prefixing the name of the parent unsaturated compound with the symbol *H* preceded by a number indicating the position of saturation.

(iv) (*a*) In a monocyclic compound containing only one hetero-atom, numbering starts at this atom.

(*b*) The ring is numbered to give substituents or other hetero-atoms the lowest numbers possible. If the hetero-atoms are different, then numbering starts at the atom cited first according to the rule in (i) and proceeds round the ring in order of precedence.

Table 30.1

No. of members in the ring	Rings containing nitrogen		Rings containing no nitrogen	
	Unsaturation	Saturation	Unsaturation	Saturation
	(a)		(a)	
3	-irine	-iridine	-iren	-iran
4	-ete	-etidine	-et	-etan
5	-ole	-olidine	-ole	-olan
6	-ine	(b)	-in	-ane
7	-epine	(b)	-epin	-epan
8	-ocine	(b)	-ocin	-ocan
9	-onine	(b)	-onin	-onan
10	-ecine	(b)	-ecin	-ecan

(a) Corresponding to the maximum number of non-cumulative double bonds.
(b) Expressed by prefixing 'perhydro' to the name of the corresponding unsaturated compound.

Examples (see text for the various trivial names).

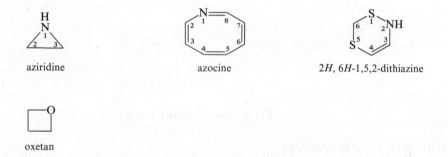

aziridine   azocine   2*H*, 6*H*-1,5,2-dithiazine

oxetan

Fused heterocyclic systems. Only a very elementary account is given here. When one heterocyclic ring is present, this is chosen as the parent compound. If more than one heterocyclic ring is present, the order of preference is given to the nitrogen-containing component (nitrogen rings are the most common). For a component containing a hetero-atom other than nitrogen, the order of preference is that in (i) above (O before S, etc.). When the parent compound has been chosen, its name is prefixed by the name of the fused ring attached, *e.g.*, benz(o), naphth(o). Also, the parent compound chosen is the component containing the largest number of rings and has a simple name. For the purpose of numbering, the structure is written with the greatest number of rings in a horizontal position and a maximum number of rings above and to the right of the horizontal row. Numbering is then carried out (usually) in a clockwise direction starting with the uppermost ring farthest to right and omitting atoms at ring junctions. To distinguish isomers, the *peripheral sides of the parent compound* are lettered *a*, *b*, *c*, etc., beginning with *a* for the side 1,2, *b* for 2,3, etc. To the letter as early in the alphabet as possible, denoting the side where fusion occurs, are prefixed, if necessary, the numbers indicating the positions of fusion of the other component; their order conforms to the *direction of lettering* of the base component. It should be noted that these numbers apply to the *prefixed component* (as a separate entity) and *not* to the combined system (which is numbered according to the usual rules). Two examples are:

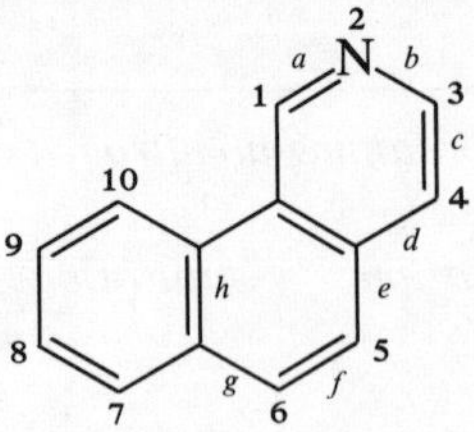

benzo[*h*]isoquinoline

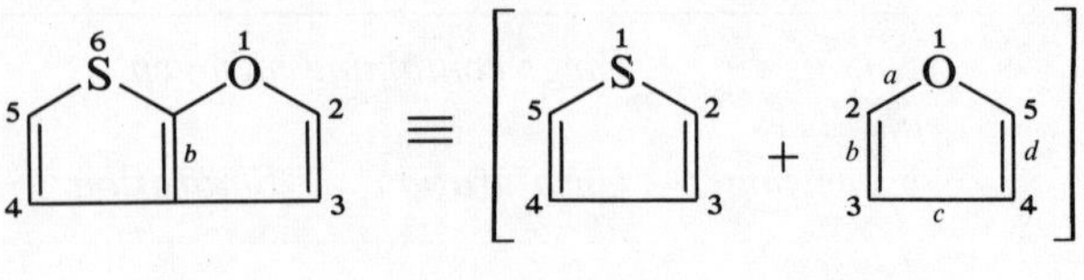

thieno[2,3-*b*]furan

In addition to the foregoing rules, there are the rules that the component chosen is the one containing the largest possible individual ring, or containing the greatest number or variety of hetero-atoms, etc. Some examples are:

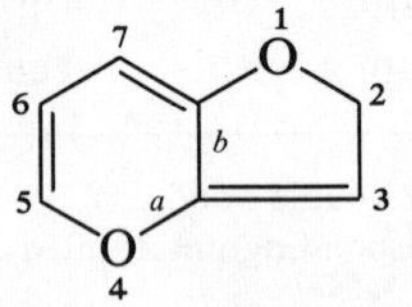

2*H*-furo[3,2-*b*]pyran

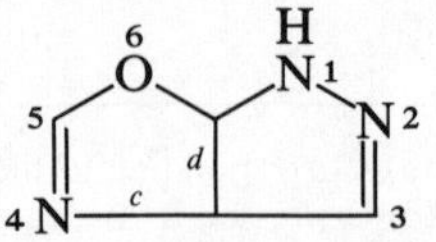

1*H*-pyrazolo[4,3-*d*]oxazole

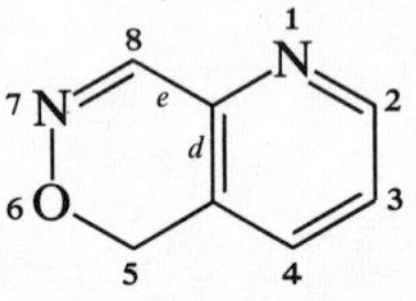

5*H*-pyrido[2,3-*d*]-*o*-oxazine

Five-membered rings

Furan and its derivatives

Furan (*furfuran*) contains one oxygen atom in its ring, and its structure is as shown.

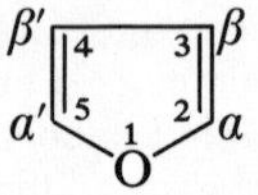

There are two monosubstituted derivatives of furan, 2 (or α) and 3 (or β); there are four disubstitution products: 2,3 (α,β), 2,4 (α,β′), 2,5 (α,α′) and 3,4 (β,β′).

Furan is obtained when wood, especially pine-wood, is distilled. It may be prepared by the dry distillation of mucic acid, and heating the product, *furoic acid*, at its b.p.:

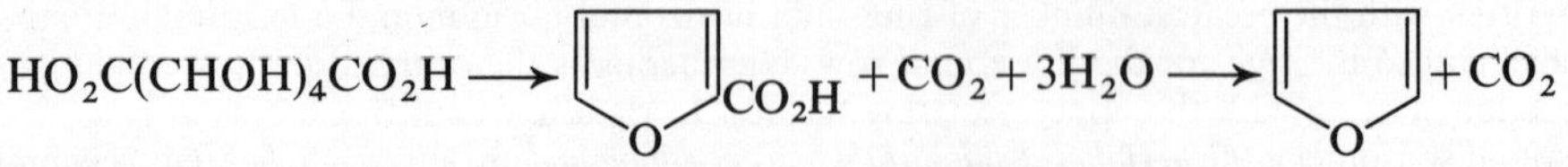

$$HO_2C(CHOH)_4CO_2H \longrightarrow \text{[furan-CO}_2\text{H]} + CO_2 + 3H_2O \longrightarrow \text{[furan]} + CO_2$$

Furan is most conveniently prepared by decarboxylating furoic acid in quinoline in the presence of copper powder.

It is manufactured by the catalytic decomposition of furfural in steam in the presence of an oxide catalyst:

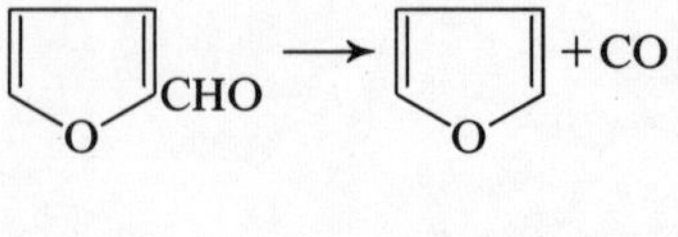

$$\text{[furan-CHO]} \longrightarrow \text{[furan]} + CO$$

A general method of preparing furan derivatives is to dehydrate 1,4-diketones or dialdehydes with, *e.g.*, phosphorus pentoxide, sulphuric acid, etc.:

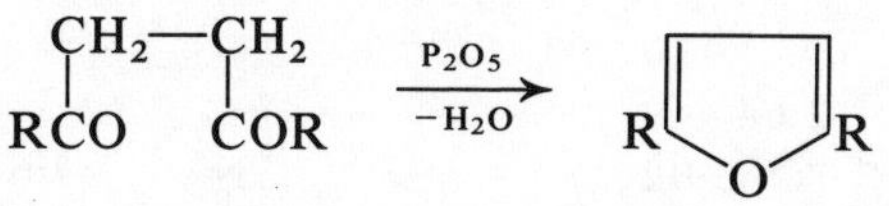

$$CH_2\text{---}CH_2 \quad \xrightarrow[-H_2O]{P_2O_5} \quad$$
$$RCO \quad COR$$

Alternatively, furan derivatives may be prepared from ethyl acetoacetate as follows:

$$2CH_3CO\overset{-}{C}HNaCO_2C_2H_5 + I_2 \longrightarrow 2NaI + \begin{array}{l} CH_3COCHCO_2C_2H_5 \\ | \\ CH_3COCHCO_2C_2H_5 \end{array}$$

When diacetosuccinic ester is heated with dilute sulphuric acid, 2,5-dimethylfuran-3,4-dicarboxylic acid is formed:

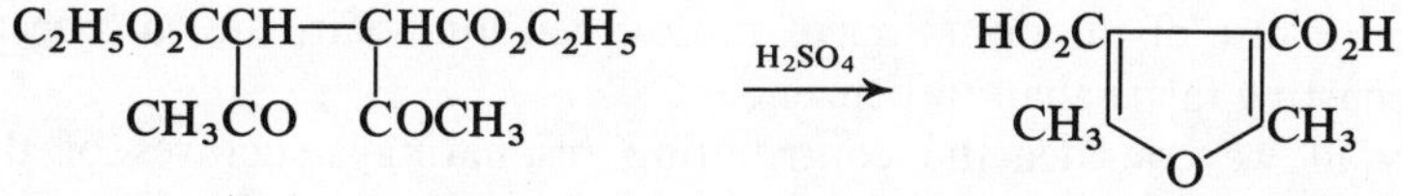

$$\begin{array}{l} C_2H_5O_2C\overset{}{C}H\text{---}\overset{}{C}HCO_2C_2H_5 \\ CH_3CO \quad COCH_3 \end{array} \quad \xrightarrow{H_2SO_4} \quad$$

If ammonia is used instead of sulphuric acid, 2,5-dimethylpyrrole-3,4-dicarboxylic ester is obtained.

Furan derivatives may also be prepared by the **Feist–Benary synthesis** (1902, 1911); an α-chloro-ketone is condensed with a β-keto-ester in the presence of pyridine (*cf.* the Hantzsch synthesis, p. 837, for a possible mechanism):

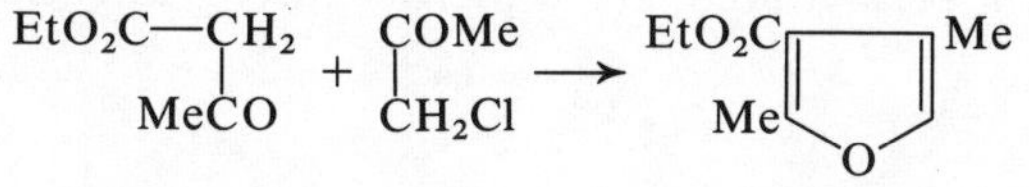

$$\begin{array}{l} EtO_2C\text{---}CH_2 \\ MeCO \end{array} + \begin{array}{l} COMe \\ CH_2Cl \end{array} \longrightarrow$$

Structure of furan

Furan has a resonance energy of 71·1–96·2 kJ mol^{-1}, and is best represented as a resonance hybrid of contributing structures (I)–(V). As we have seen (p. 577), to be aromatic, a monocyclic molecule must be a planar $(4n+2)$ π-electron molecule. Hence, to provide six π-electrons for the ring system, one of the lone pairs of electrons of the oxygen atom is involved.

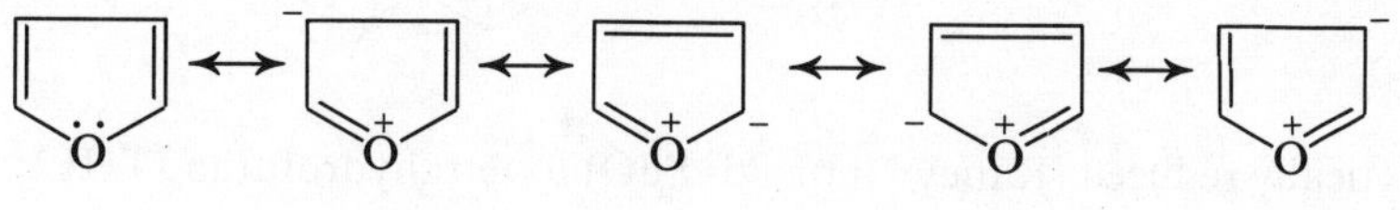

(I)   (II)   (III)   (IV)   (V)

The ring current in furan, evaluated from N.M.R. data (Elvidge, 1965; de Jongh *et al.*, 1965), is less than that of benzene and so furan is less aromatic than benzene (p. 583).

N.M.R. spectra of heterocyclic compounds. Because of the inductive effect of the hetero-atom (O, S, and N), ring protons have τ-values very much shifted downfield compared with the corresponding aromatic compound, and this downfield shift is more pronounced for hydrogen atoms in the α-positions, *e.g.*, (τ-values):

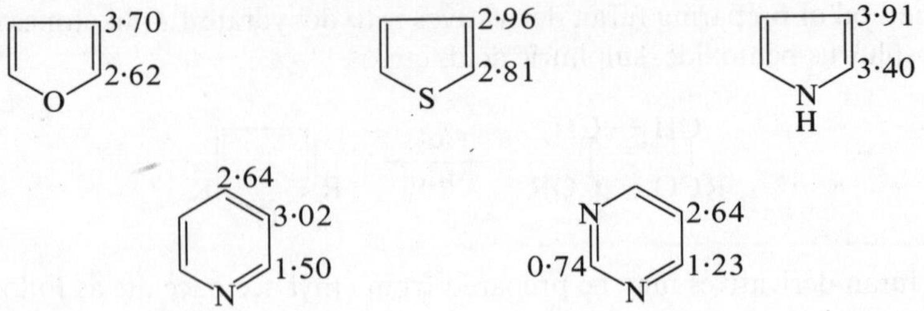

Substitution in furan

As we have seen (p. 38), canonical structures carrying unlike charges are not so stable as uncharged structures, and the larger the charge separation, the less stable is that structure. Hence, (II) and (V) contribute less than (III) and (IV), and consequently the resonance hybrid has a larger electron density at position 2 (or 5) than at 3 (or 4). Thus, on the basis of charge distribution, electrophilic substitution would be expected to occur at position 2 (or 5). This is the case in practice. Also, because of the donation of the oxygen lone-pair, the ring becomes activated, substitution occurring faster than in benzene.

Alternatively, if we consider the contributing resonating structures of the intermediate carbonium ion (σ-complex), then there are three for 2- and two for 3-substitution.

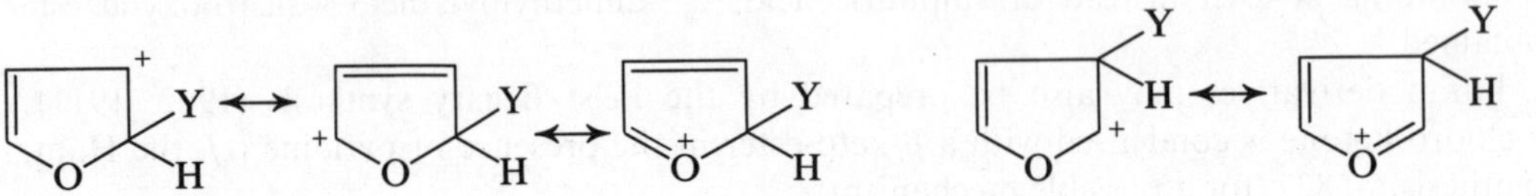

Hence 2-substitution will be favoured, since there is also greater spreading of charge in this resonance hybrid.

Reactions of furan

Furan, b.p. 32°C, turns green a pine splint moistened with hydrochloric acid. It behaves chemically as an aromatic compound, but in some ways it behaves as a 1,3-diene. Thus, it is less aromatic than thiophen and pyrrole, and is the only one of these to undergo the Diels–Alder reaction.

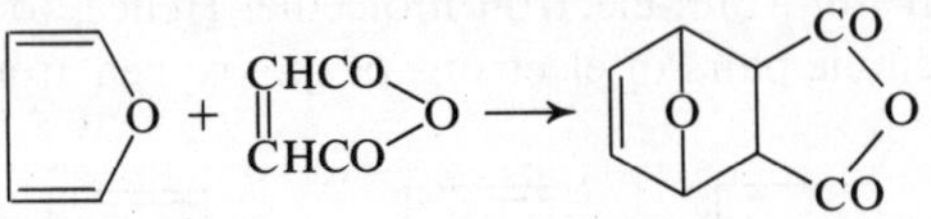

Furan is catalytically reduced (Raney Ni or Pd–PdO) to tetrahydrofuran (THF):

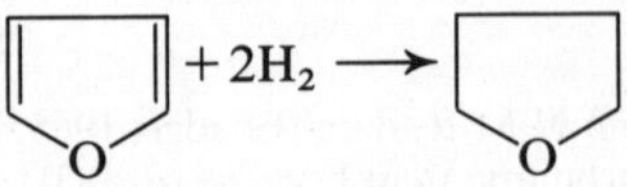

$$+ 2H_2 \longrightarrow \qquad (90\text{–}93\%)$$

Tetrahydrofuran is manufactured synthetically from butyne-1,4-diol:

$$HOCH_2C{\equiv}CCH_2OH \xrightarrow[\text{Ni}]{H_2} HOCH_2CH_2CH_2CH_2OH \xrightarrow[\text{(heat)}]{-H_2O}$$

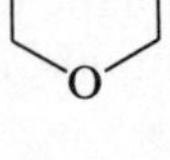

Tetrahydrofuran is a valuable solvent (*e.g.*, for the preparation of Grignard reagents). With ammonia it forms pyrrolidine, and with hydrogen chloride, tetramethylene chlorohydrin:

$$\text{(pyrrolidine)} \xleftarrow{\text{NH}_3} \text{(THF)} \xrightarrow{\text{HCl}} \text{ClCH}_2\text{CH}_2\text{CH}_2\text{OH} \quad (54\text{--}57\%)$$

Furan is readily polymerised by concentrated acids, and probably the oxonium salt is involved (furan is very weakly basic; *cf.* pyrrole, p. 838). Thus, direct nitration (mixed acid) and sulphonation result in resinified products. However, 2-nitrofuran may be prepared by nitrating furan with acetyl nitrate and the 2-sulphonic acid may be prepared by the action of pyridine-sulphur trioxide on furan. If a $-\text{I}$ group is present in the ring, then sulphonation can be carried out directly, *e.g.*, furoic acid gives furoic-5-sulphonic acid. Furan reacts readily with halogens, but the liberated halogen acid causes polymerisation. Because of this, halogenated furans are usually prepared by indirect methods, *e.g.*, by brominating furoic acid and decarboxylating the product, 5-bromofuroic acid, to 2-bromofuran by heating in quinoline in the presence of copper powder.

$$\text{(furoic acid)} \xrightarrow{\text{Br}_2} \text{(5-bromofuroic acid)} \xrightarrow{\text{heat}} \text{(2-bromofuran)} + \text{CO}_2$$

However, chlorination of furan at about $-40°\text{C}$ gives 2-chloro- and 2,5-dichlorofuran as the main products.

Furan undergoes the Gattermann reaction (p. 735) to form furfural (see also later). Since aluminium chloride attacks the ring, Friedel–Crafts reactions are best carried out with stannic chloride as catalyst. In any case, alkylation reactions with furan lead to polymers, but acylation with acid chlorides or anhydrides with stannic chloride as catalyst gives 2-acyl derivatives. A better catalyst for the reaction with anhydrides is boron trifluoride in ether (the reagent is the complex $\text{Et}_2\text{O}^+\text{—BF}_3{}^-$).

Furan is very readily mercurated in the 2-position, *e.g.*, when heated with mercuric chloride in aqueous sodium acetate, 2-chloromercurifuran is obtained. Since the mercuri-group is readily replaced by bromine or iodine, and by an acyl group, these mercuri-compounds are useful intermediates:

$$\text{(furan)} \xrightarrow[\text{AcONa}]{\text{HgCl}_2} \text{(furan-HgCl)} \quad \begin{array}{c} \xrightarrow{\text{X}_2 \ (\text{Br}_2 \ \text{or} \ \text{I}_2)} \text{(furan-X)} \\[2ex] \xrightarrow{\text{RCOCl}} \text{(furan-COR)} \end{array}$$

Furan reacts with n-butyl-lithium to form the 2-lithium derivative. This undergoes many of the usual reactions of organolithium compounds, *e.g.*, with carbon dioxide, furoic acid is obtained:

$$\text{(furan)} + \text{C}_4\text{H}_9\text{Li} \longrightarrow \text{C}_4\text{H}_{10} + \text{(furan-Li)} \xrightarrow[\text{(ii) H}^+]{\text{(i) CO}_2} \text{(furan-CO}_2\text{H)}$$

Furan undergoes the Gomberg reaction (p. 675) when treated in alkaline solution with diazonium salts:

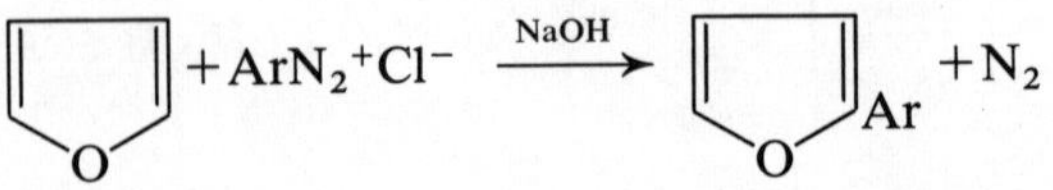

$$\text{(furan)} + ArN_2^+Cl^- \xrightarrow{\text{NaOH}} \text{(aryl furan)} + N_2$$

The furan ring is readily opened under suitable conditions to give 1,4-dioxo-compounds, *e.g.*,

(i) 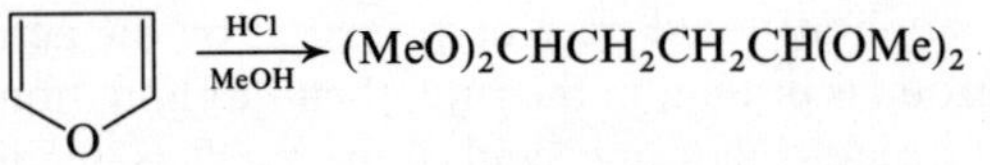 $\quad \text{(furan)} \xrightarrow[\text{MeOH}]{\text{HCl}} (MeO)_2CHCH_2CH_2CH(OMe)_2$

(ii) 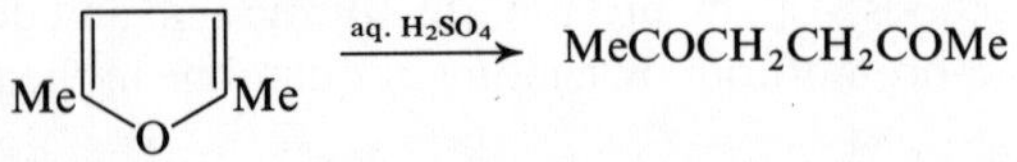 $\quad \text{Me(furan)Me} \xrightarrow{\text{aq. H}_2\text{SO}_4} MeCOCH_2CH_2COMe$

Furfural (*furfuraldehyde*) may be prepared by distilling pentoses with dilute sulphuric acid:

$$HOCH_2(CHOH)_3CHO \longrightarrow \text{(furan-CHO)} + 3H_2O \quad (100\%)$$

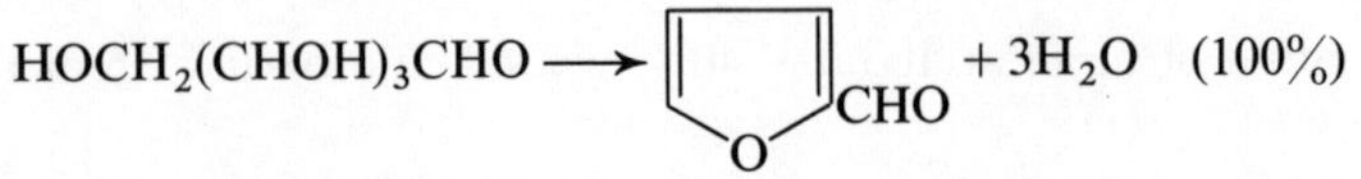

It is manufactured by treating oat husks, cotton-seed hulls or corn cobs with dilute sulphuric acid followed by steam distillation (the starting materials are rich in pentoses).

Furfural, b.p. 162°C, is chemically very similar to benzaldehyde, *e.g.*, with aqueous sodium hydroxide furfural forms *furfuryl alcohol* and *furoic acid* (Cannizzaro reaction):

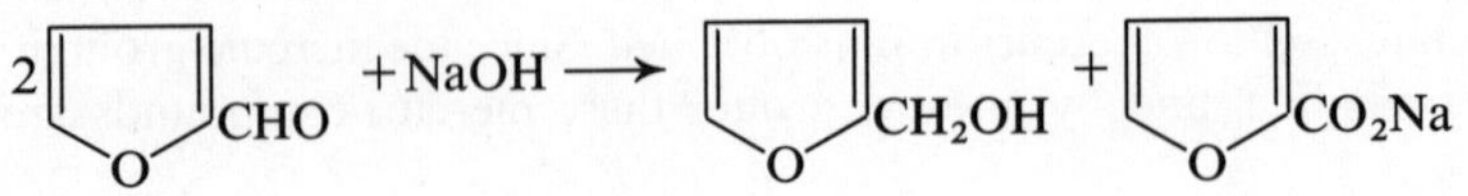

$$2\,\text{(furan-CHO)} + NaOH \longrightarrow \text{(furan-CH}_2\text{OH)} + \text{(furan-CO}_2\text{Na)} \quad (60\text{--}63\%)$$

With ethanolic potassium cyanide *furoin* is formed (benzoin condensation) and this, on oxidation, gives *furil*:

$$2\,\text{(furan-CHO)} \xrightarrow{\text{KCN}} \text{(furan-CH(OH)CO-furan)} \xrightarrow{\text{[O]}} \text{(furan-COCO-furan)}$$

When heated with aqueous potassium hydroxide, furil gives *furilic acid* (benzilic acid rearrangement):

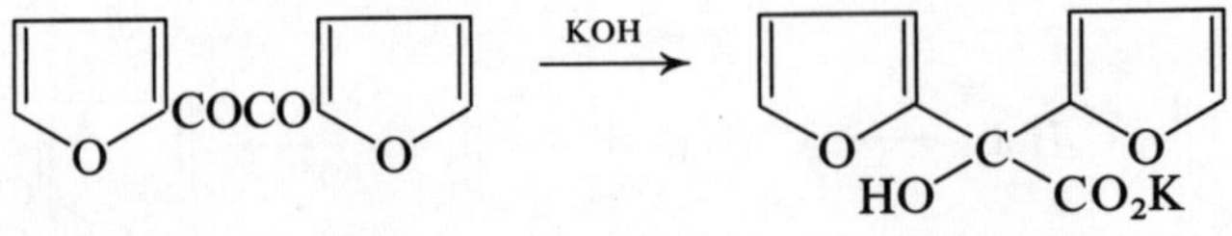

$$\text{(furan-COCO-furan)} \xrightarrow{\text{KOH}} \text{(furan-C(OH)(CO}_2\text{K)-furan)}$$

Furfural reacts with aniline to form an anil, and can undergo the Perkin reaction and the Claisen reaction. It is easily oxidised by silver oxide to the corresponding acid; it is oxidised by sodium chlorate to fumaric acid. Furfural condenses with dimethylaniline in the presence of zinc chloride to form *furfuraldehyde green* (analogous to malachite green, p. 886). A characteristic reaction of furfural is the red colour it gives with aniline and hydrochloric acid; it also turns green a pine splint moistened with hydrochloric acid (*cf.* furan, above).

Furfural is used for the preparation of dyes, plastics and fumaric acid. It is also used as a solvent in synthetic rubber manufacture, and as an extraction liquid in petroleum refining.

Furfuryl alcohol, b.p. 170°C, may be prepared by reducing furfural, or by means of the Cannizzaro reaction on furfural (see above). It is manufactured by the catalytic reduction (copper chromite) of furfural.

Furoic acid (furan-2-carboxylic acid, *pyromucic acid*), m.p. 133°C, may be prepared by the dry distillation of mucic acid, or by the oxidation of furfural with cuprous oxide-silver oxide in aqueous sodium hydroxide (yield: 86–90%). It behaves more like an unsaturated aliphatic acid rather than benzoic acid (*cf.* furfural); *e.g.*, furoic acid is readily oxidised by alkaline permanganate, brominated by bromine vapour (it adds on four bromine atoms), and oxidised by bromine-water to fumaric acid. The sodium salt of furoic acid is decarboxylated by mercuric chloride to give the mercuri-chloride (this does not occur with furan-3-carboxylic acid):

$$\text{(furan)}\text{CO}_2\text{Na} + \text{HgCl}_2 \longrightarrow \text{(furan)}\text{HgCl} + \text{CO}_2 + \text{NaCl}$$

On the other hand, when heated with mercuric acetate, furoic acid forms 3-acetoxymercurifuran.

$$\text{(furan)}\text{CO}_2\text{H} \xrightarrow{\text{Hg(OAc)}_2} \text{(furan)}\text{CO}_2\text{HgOAc} \xrightarrow{-\text{CO}_2} \text{(furan)}\text{HgOAc}$$

It has already been pointed out that the presence of a $-$I group stabilises the furan ring. Another example of this is that on nitration with mixed acids, furoic acid gives first the 2-nitro derivative, and then 2,5-dinitrofuran, the latter being produced by displacement of the carboxyl group.

$$\text{(furan)}\text{CO}_2\text{H} \xrightarrow[\text{H}_2\text{SO}_4]{\text{HNO}_3} \text{O}_2\text{N}\text{(furan)}\text{CO}_2\text{H} \longrightarrow \text{O}_2\text{N}\text{(furan)}\text{NO}_2$$

Furan-3-carboxylic acid occurs naturally.

Benzofuran (benzfuran, *coumarone*), b.p. 174°C, occurs in coal-tar. It is used in the manufacture of plastics.

Dibenzofuran (dibenzfuran, *diphenylene oxide*), m.p. 87°C, may be prepared by heating phenol with lead oxide at 150°C:

$$2\,\text{(phenol)OH} \xrightarrow{\text{PbO}} \text{(dibenzofuran)} \quad (28\%)$$

It may also be prepared by passing diphenyl ether through a red-hot tube.

Thiophen and its derivatives

Thiophen occurs in coal-tar and shale oils. Its b.p. (84°C) is close to that of benzene and hence it is difficult to separate from the benzene fraction obtained from coal-tar. Thiophen can be sulphonated more readily than benzene, and this property is used to separate the two compounds by repeatedly shaking benzene (from coal-tar) with cold concentrated sulphuric acid, whereby the water-soluble thiophensulphonic acid is formed. A better means of separation is to reflux the benzene with aqueous mercuric acetate whereupon thiophen is mercurated and benzene is not. Thiophen may be recovered from its mercurated derivative by distilling the latter with hydrochloric acid. The presence of thiophen in benzene may be detected by the *indophenin reaction*. This is the development of a blue colour when benzene is treated with isatin and sulphuric acid.

Thiophen may be prepared by passing a mixture of acetylene and hydrogen sulphide through a tube containing alumina at 400°C.

$$2C_2H_2 + H_2S \longrightarrow C_4H_4S + H_2$$

This method is used commercially. It is also manufactured by reaction between n-butane and sulphur in the vapour phase.

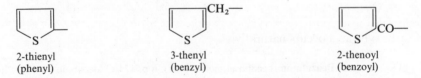

Thiophen may also be prepared by heating sodium succinate with phosphorus trisulphide:

$$\begin{array}{c}CH_2CO_2Na \\ | \\ CH_2CO_2Na\end{array} \xrightarrow{\text{P}_2\text{S}_3} \quad \text{[thiophene ring]} \quad (30\%)$$

Derivatives of thiophen may be prepared by heating 1,4-diketones with phosphorus trisulphide (*cf.* furan derivatives).

Thiophen is easily sulphonated, nitrated or chlorinated. It was this close chemical similarity to benzene that masked the presence of thiophen in benzene from coal-tar. V. Meyer (1882) found that a sample of benzene prepared by heating sodium benzoate with soda-lime did not give the indophenin test. Subsequently he showed that it was a sulphur-containing compound, which he called *thiophene*, that was responsible for the indophenin reaction.

This similarity to benzene has resulted in a similar nomenclature:

2-thienyl   3-thenyl   2-thenoyl
(phenyl)   (benzyl)   (benzoyl)

Structure of thiophen

Thiophen is a resonance hybrid (R.E. 117–130 kJ mol^{-1}), the sulphur atom contributing two electrons to form a $(4n+2)$ π-electron molecule (*cf.* furan). However, there are some complications in elucidating the electronic structure of thiophen. Sulphur is less electronegative than oxygen (or nitrogen) and can also use $3d$ orbitals (oxygen and nitrogen cannot). Hence, more canonical forms are possible for thiophen than for furan and pyrrole. In group (*a*), the sulphur atom uses *p*-orbitals; in group (*b*) it also uses *d*-orbitals.

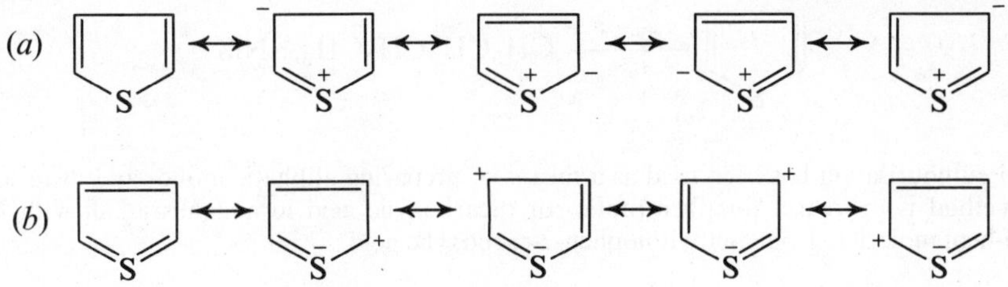

(a)

(b)

The estimation of the ring current in thiophen has shown it to be more aromatic than furan.

Substitution in thiophen

As with furan, electrophilic substitution would be expected, on the basis of charge distribution and stabilities of the carbonium ions, to take place at position 2 (or 5). This occurs in practice.

Reactions of thiophen

Thiophen shows no basic properties and is stable to aqueous acids. It can be nitrated by fuming nitric acid in acetic anhydride to give mainly 2-nitrothiophen, and is readily sulphonated in the 2-position with *cold* concentrated sulphuric acid. Chlorination of thiophen results in the formation of both substitution and addition products, but 2-chloro- and 2,5-dichlorothiophen are the main products when thiophen is chlorinated at $-30°C$. On the other hand, 2-bromothiophen is prepared by the action of *N*-bromosuccinimide on thiophen, and the 2-iodo-compound by the action of iodine in the presence of yellow mercuric oxide. Thiophen is readily acylated in the 2-position by acid chlorides in the presence of stannic chloride, or better, by acid anhydrides in the presence of phosphoric acid, *e.g.*, methyl 2-thienyl ketone:

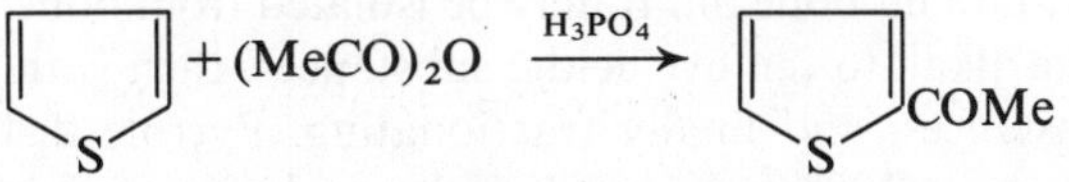

Thiophen may be chloromethylated (formaldehyde and HCl) and formylated (dimethylformamide and $POCl_3$) in the 2-position. Thiophen-2-aldehyde (b.p. 198°C) undergoes the Cannizzaro reaction and the benzoin condensation (*cf.* furfural). Mercuration with mercuric chloride in the presence of a small amount of sodium acetate produces the 2-mercuri-chloride as the main product, and when treated with n-butyl-lithium in ether, thiophen forms the 2-lithium derivative. These compounds may be used to prepare various 2-substituted thiophens.

Various derivatives of thiophen may also be prepared from the monobromo-derivative, *e.g.*,

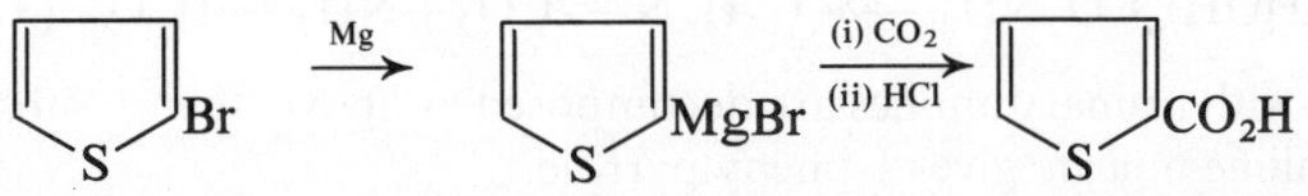

Thiophen does not undergo the Diels–Alder reaction. Reduction of thiophen with sodium in liquid ammonia produces 2,3-dihydro-(2-*thiolen*) and 2,5-dihydrothiophen (3-*thiolen*) (Birch *et al.*, 1951). Catalytic reduction to tetrahydrothiophen (*thiophan, thiolan*) may be carried out provided a large amount of the catalyst, palladium, is used to overcome the poisoning effect of the sulphur. On the other hand, catalytic reduction of thiophen with Raney nickel as catalyst results in the removal of sulphur, the main product being n-butane (*cf.* thiols, p. 400):

$$\text{Raney Ni} \atop H_2 \longrightarrow CH_3CH_2CH_2CH_3 + NiS$$

This desulphurisation has been used as a means of preparing aliphatic mono- and dicarboxylic acids. The method is to treat a thiophen mono- or dicarboxylic acid in aqueous alkali with Ni-Al alloy, *e.g.*, n-heptanoic acid from 5-ethylthiophen-2-carboxylic acid:

$$C_2H_5 \quad CO_2H \xrightarrow[\text{aq. NaOH}]{Ni-Al} CH_3(CH_2)_5CO_2H$$

Thiophen does not form sulphonium salts and cannot be oxidised to a sulphoxide or sulphone; hydrogen peroxide *opens* the thiophen ring, the sulphur being oxidised to sulphuric acid.

A number of condensed thiophen systems are known, *e.g.*, *benzothiophen* (I) (also known as *thionaphthen* because it closely resembles naphthalene), *dibenzothiophthen* (II), and *thiophthen* (III)

(I)   (II)   (III)

Pyrrole and its derivatives

Pyrrole is a very important five-membered heterocyclic ring because its nucleus occurs in many natural compounds, *e.g.*, alkaloids, chlorophyll, hæmatin, etc.

Pyrrole occurs in coal-tar and bone oil. It may be isolated from bone oil by washing the latter with dilute alkali to remove acidic substances, then with acid to remove strongly basic substances, and finally fractionating. Pyrrole distils over in the fraction boiling between 100° and 150°C, and may be purified by fusing with potassium hydroxide. Solid potassiopyrrole is formed and this, on steam distillation, gives pure pyrrole.

Pyrrole may be synthesised by passing a mixture of acetylene and ammonia through a red-hot tube:

$$2C_2H_2 + NH_3 \longrightarrow C_4H_5N + H_2$$

It is conveniently prepared by distilling a mixture of ammonium mucate and glycerol at 200°C:

$$H_4NO_2C(CHOH)_4CO_2NH_4 \longrightarrow C_4H_5N + 2CO_2 + NH_3 + 4H_2O \quad (37–40\%)$$

If salts of mucic acid with primary amines are decomposed as above, then 1-substituted pyrroles are obtained, *e.g.*, aniline mucate gives 1-phenylpyrrole.

Pyrrole is also formed when succinimide is distilled with zinc dust:

$$\begin{array}{c} CH_2-CH_2 \\ | \qquad | \\ CO \quad CO \\ \diagdown N \diagup \\ H \end{array} \xrightarrow{Zn} \begin{array}{c} \\ N \\ H \end{array}$$

Pyrrole is manufactured by passing a mixture of furan, ammonia and steam over heated alumina as catalyst:

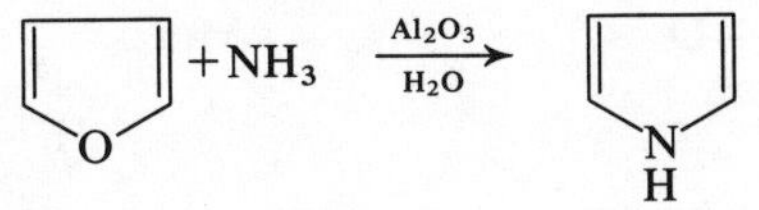

Many methods are available for synthesising pyrrole derivatives, *e.g.*,

(i) **Paal–Knorr synthesis** (1885). This is carried out by treating a 1,4-diketone with ammonia, primary amines, hydrazines, etc. The mechanism of this reaction (and others of a similar nature) is not fully understood, but there is evidence that it proceeds in a similar fashion to the reaction between oxo compounds and compounds of the type Z—NH_2 (see p. 212), *e.g.*, (proton transfers have been combined in one step).

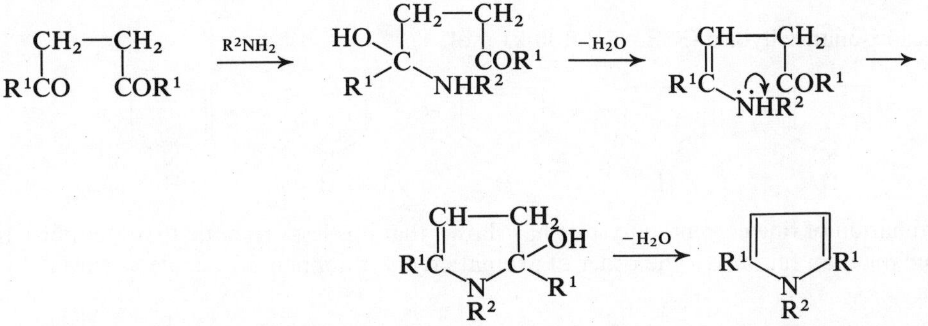

$$(g.)$$

Possibly the two molecules of water are eliminated from the ring-closed intermediate and not stepwise as shown.

If succinaldehyde is used as the 1,4-dicarbonyl compound, pyrrole itself is obtained (yield is poor).

(ii) **Knorr pyrrole synthesis** (1884; 1886). This is the most general method, and involves the condensation between an α-aminoketone and a β-diketone or β-keto-ester (*cf.* with (iii) below).

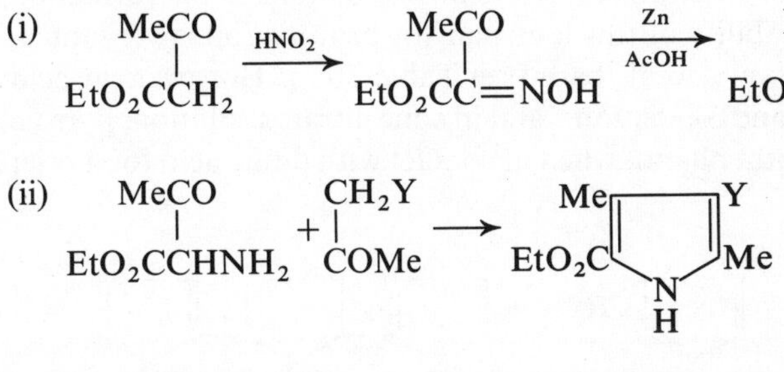

Thus, if Y is CO_2Et, the product is 3,5-dimethylpyrrole-2,4-dicarboxylic ester.

(iii) **Hantzsch synthesis** (1890). This is the condensation between chloroacetone, a β-ketoester and a primary amine, *e.g.*, (*cf.* p. 288):

(ii)

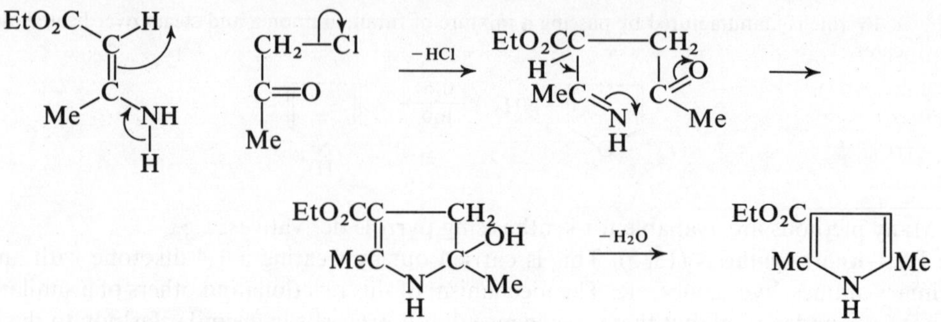

Some furan derivative is also formed.

Structure of pyrrole

Pyrrole is a resonance hybrid (R.E. 87·8–130 kJ mol^{-1}).

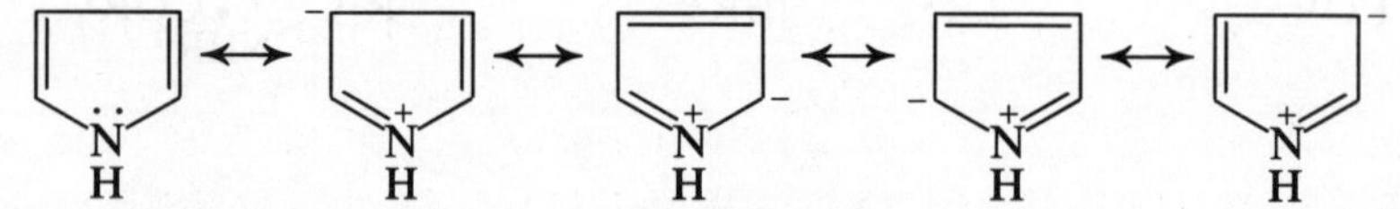

The estimation of ring current in pyrrole has shown that it is less aromatic than thiophen and more aromatic than furan, *i.e.*, the order of aromaticity is: thiophen > pyrrole > furan.

Substitution in pyrrole

Electrophilic substitution in pyrrole occurs predominantly in the 2- (or 5-) position, and arguments to explain this (charge distribution and carbonium ion stabilities) are similar to those used for furan (p. 830).

Basicity of pyrrole

Because of the contribution of the lone pair of electrons of the nitrogen atom to the formation of a $(4n+2)$ π-electron molecule, the availability of this lone pair for protonation is very much decreased, and consequently pyrrole is a very weak base (see Table 30.2). However, in acid solution, protonation also occurs at the 2- and 3-positions, and in concentrated solution, pyrrole polymerises (to form *pyrrole-red*). On the other hand, when in contact with dilute acid for a brief period, pyrrole forms the trimer (I).

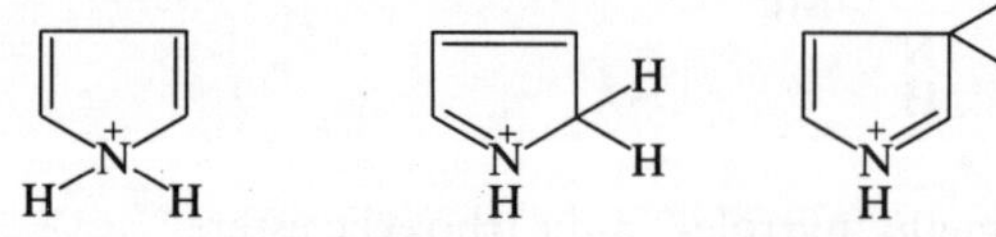

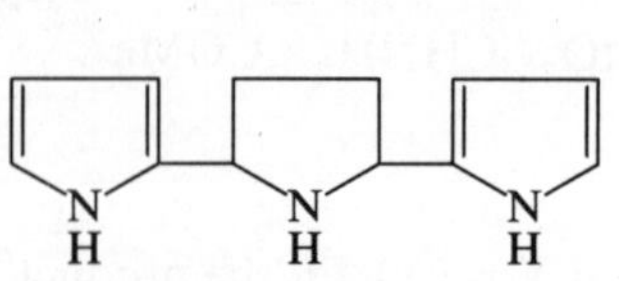

Properties of pyrrole

Pyrrole, b.p. 131°C, rapidly darkens on exposure to air. A characteristic reaction is the turning red of a pine splint moistened with hydrochloric acid when exposed to the vapour of pyrrole (and many of its derivatives). Pyrrole gives 2-nitropyrrole when nitrated with nitric acid in acetic anhydride at low temperatures, and sulphonation with pyridine-sulphur trioxide in ethylene dichloride gives the 2-sulphonic acid. Pyrrole reacts rapidly with halogens to form 2,3,4,5-tetrahalogenopyrroles, but with sulphuryl chloride, 2,5-dichloropyrrole is the main product.

Pyrrole shows a number of resemblances to phenols and aromatic amines. The imino-hydrogen of pyrrole is replaceable by potassium, alkyl or acyl groups. When pyrrole is heated with solid potassium hydroxide, potassiopyrrole is formed (thus pyrrole is acidic):

$$C_4H_4NH + KOH \longrightarrow C_4H_4\bar{N}K^+ + H_2O$$

Potassiopyrrole reacts with carbon dioxide and with chloroform as do phenols in the Kolbe–Schmitt and Reimer–Tiemann reactions, to form 2- and 3-pyrrolecarboxylic acids and pyrrole-2-aldehyde, respectively. In the latter case, however, some 3-chloropyridine is also formed (see also below). Potassiopyrrole also reacts with acyl chlorides, acid anhydrides, and alkyl halides to form the 1- (N-) derivatives. At higher temperature, the 2-derivative is the main product, presumably by rearrangement of the 1-derivative formed first (cf. Hofmann–Martius rearrangement, p. 662). 2-Ketones may also be prepared by the Houben–Hoesh synthesis (p. 752). Pyrrole undergoes the Gattermann reaction to form the 2-aldehyde, but this compound is most conveniently prepared by formylation with dimethylformamide and phosphoryl chloride.

Pyrrole is mercurated with great difficulty. Experience has shown that the ease of mercuration of heterocyclic compounds varies considerably with the nature of the hetero-atom. Generally, those containing oxygen or sulphur are easily mercurated (cf. thiophen), whereas those containing nitrogen are usually mercurated with greater difficulty. Pyrrole couples with diazonium salts in the 2-position in weakly acid solution, and in the 2- and 5-positions (to give the bisazo-compound) in alkaline solution. If the 2- and 5-positions are occupied by, e.g., methyl groups, coupling takes place in the 3-position.

When pyrrole is treated with methylmagnesium bromide, pyrrolylmagnesium bromide is formed, and this reacts with a series of alkyl halides to give both 2- and 3-alkylpyrroles in comparable amounts (Skell et al., 1962). Reinecke et al. (1964), from their NMR spectra studies of the indole Grignard reagent, have concluded that the structure of the indole and pyrrole reagents are resonance hybrids. This would explain attack at positions 2 and 3.

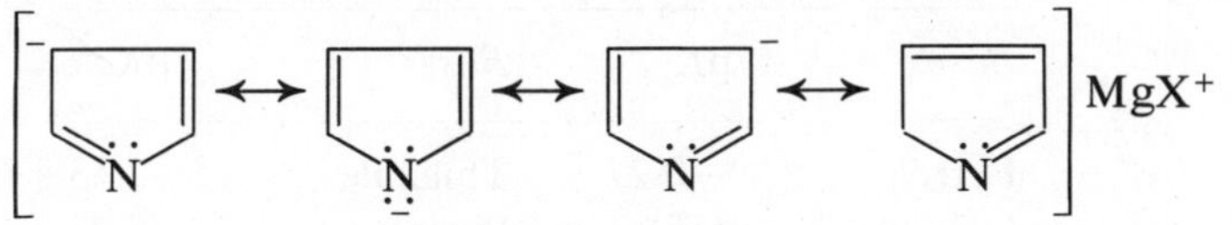

Pyrrole is oxidised by chromium trioxide in acetic acid to maleinimide:

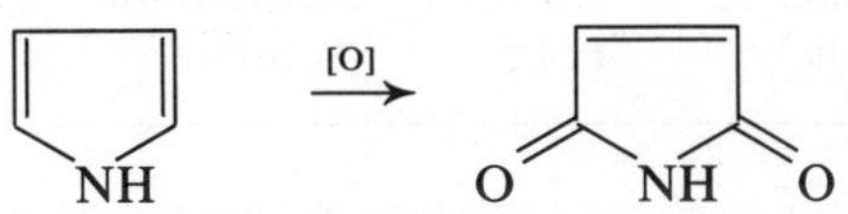

When pyrrole is refluxed with an ethanolic solution of hydroxylamine, *the ring is opened* and succinaldehyde dioxime is formed:

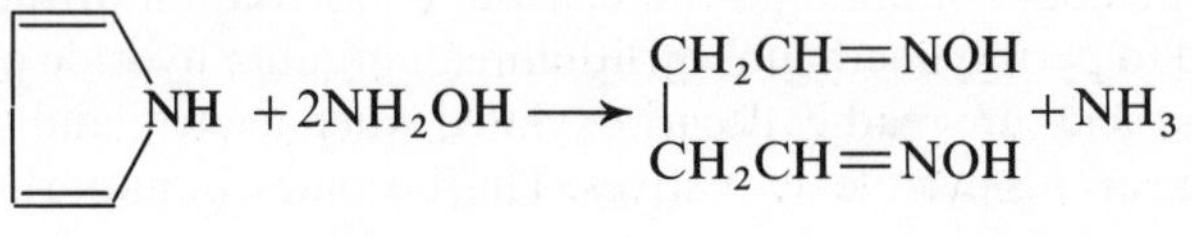

When potassiopyrrole is heated with chloroform and sodium ethoxide, *ring expansion* takes place, the product being 3-chloropyridine:

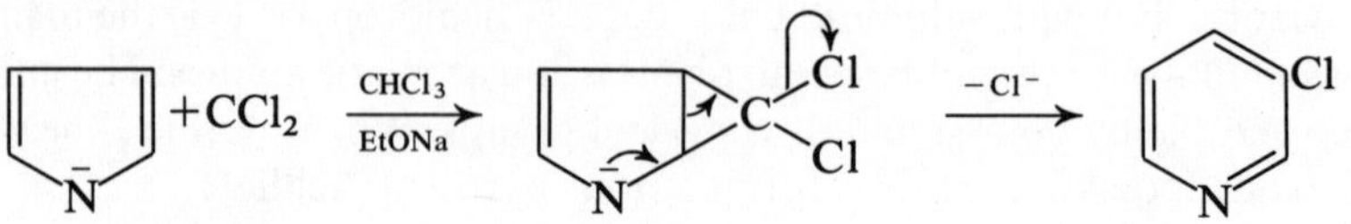

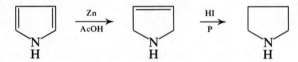

Reduction products of pyrrole

Pyrrole is reduced by zinc and acetic acid to pyrroline (2,5-*dihydropyrrole*), b.p. 91°C. This, on heating with hydriodic acid and red phosphorus, gives pyrrolidine (*tetrahydropyrrole*), b.p. 88°C:

Pyrrolidine may also be prepared by catalytically reducing pyrrole using nickel at 200°C, or by the electrolytic reduction of succinimide.

Pyrroline and pyrrolidine are both strong bases, and do not show any tendency to polymerise. Pyrrolidine is used in the preparation of enamines (p. 383).

2-Pyrrolidone, m.p. 25°C, may be prepared by the electrolytic reduction of succinimide:

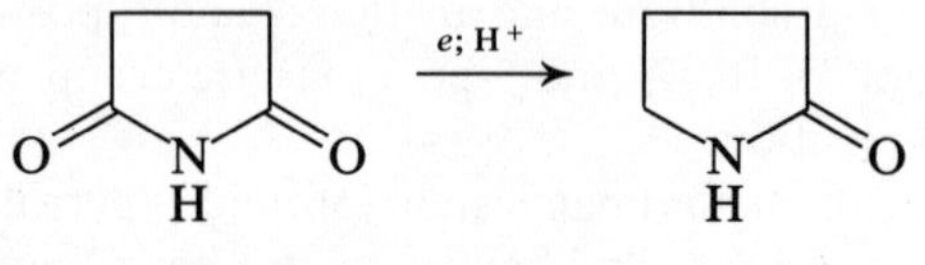

2-Pyrrolidone is the *lactam* (*cf.* lactones) of γ-aminobutyric acid.

3-*Pyrrolidone* is also known.

Table 30.2

Base	pK_a	Base	pK_a
Pyrrole	−0·27	Thiazole	2·53
(acidic)	(15)	Pyridine	5·23
Pyrazole	2·53	Piperidine	11·22
(acidic)	(14)	Quinoline	4·94
Imidazole	7·03	Isoquinoline	5·2
(acidic)	(14·5)	Acridine	5·60

Pyrrole derivatives

Methods of preparation have already been described; here will be mentioned some of their reactions.

Pyrrole-2-aldehyde does not undergo the benzoin condensation or the Cannizzaro reaction; it is readily reduced to pyrrolylmethanol by lithium aluminium hydride or sodium borohydride.

Pyrrolecarboxylic acids are readily decarboxylated when heated, and so this reaction offers a means of preparing various pyrrole derivatives. This becomes an important method when taken in conjunction with the fact that the 3-ester group is often hydrolysed by sulphuric acid at 40–60°C, whereas the 2-ester group is unaffected. Furthermore, the 2-ester group is hydrolysed by cold dilute sodium hydroxide solution, but not the 3-group.

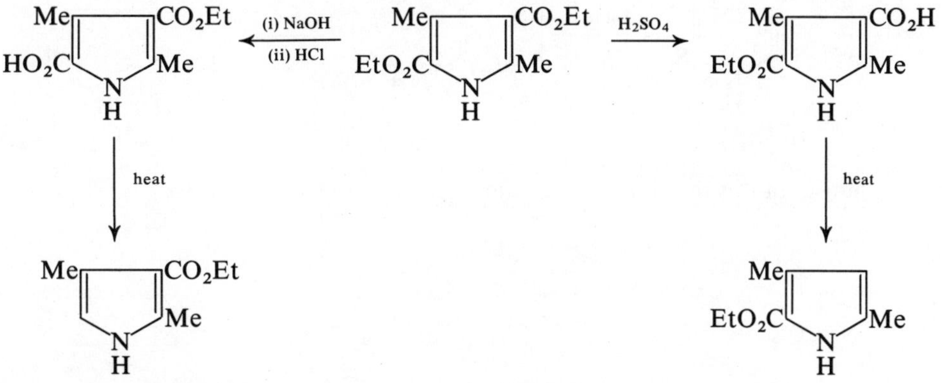

Complete hydrolysis with hot alkali, followed by decarboxylation of the free acid, leads to alkylpyrroles. Alkylated pyrroles may also be prepared from pyrrole–Grignard reagents (see above).

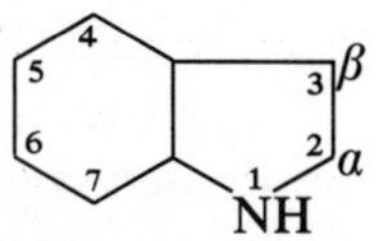

Indole (*benzopyrrole*), m.p. 52°C, occurs in coal-tar, jasmine flowers and orange blossoms. Indole is the parent substance of indigotin (p. 895).

Indoles may be synthesised in various ways:

(i) **Fischer's indole synthesis** (1886). This is the most important method of preparing indole derivatives, and is carried out by heating the phenylhydrazone or substituted phenylhydrazone of an appropriate aldehyde, ketone or ketonic acid with zinc chloride, polyphosphoric acid, sulphuric acid in ethanol, etc.

Various mechanisms have been proposed for this reaction; a widely accepted one is that suggested by Robinson (1918) and modified by later workers:

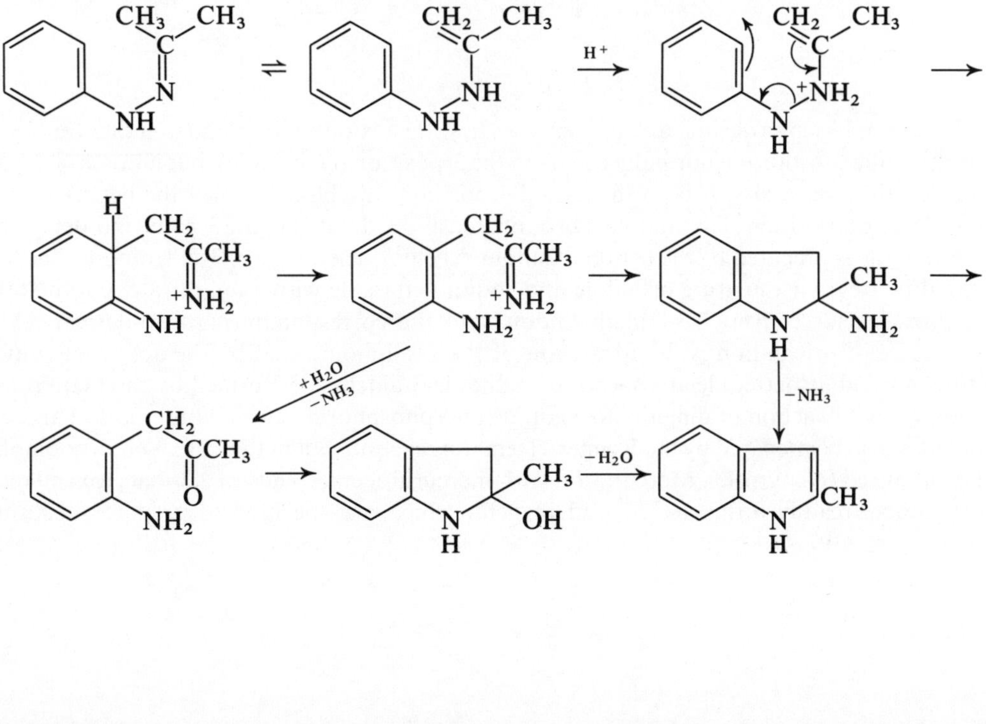

(ii) **Madelung synthesis** (1912). This is the cyclisation of an *o*-acylamidotoluene by means of a strong base, *e.g.*, indole (from *o*-formamidotoluene) and 2-methylindole (from *o*-acetamido-toluene):

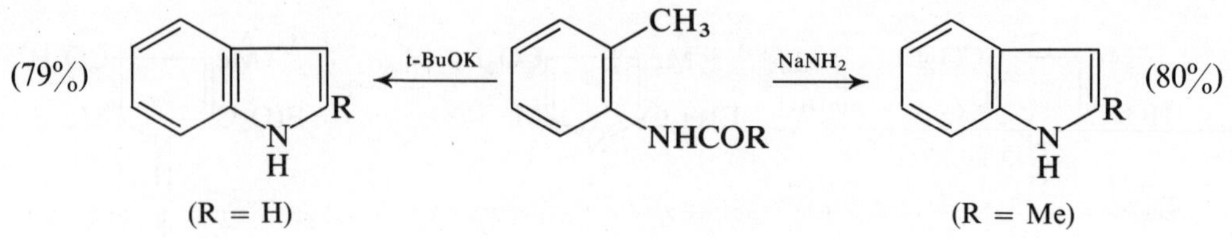

(79%)   t-BuOK ←   →   NaNH₂   (80%)

(R = H)   (R = Me)

The mechanism of this reaction is uncertain; a possibility (simplified) is:

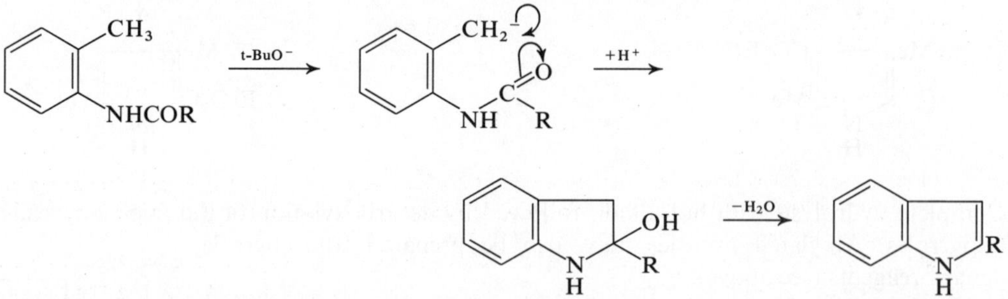

t-BuO⁻   +H⁺

−H₂O

(iii) **Reissert synthesis** (1897). This is a very good method, and is carried out with *o*-nitrotoluene (or its substituted derivatives) and ethyl oxalate as follows:

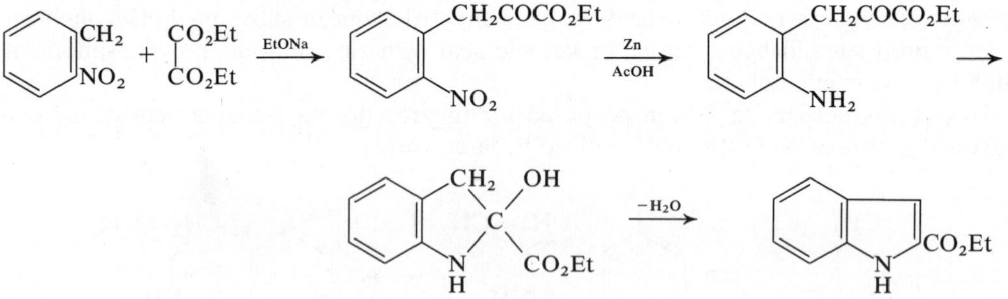

Indole resembles pyrrole in many of its properties. It is oxidised by ozone to indigotin.

Electrophilic substitution normally occurs in the 3-position (*cf.* pyrrole), but if this is occupied, then 2-substitution occurs. If both the 2- and 3-positions are blocked, then the benzene ring is attacked at the 6-position. Thus indole is brominated and iodinated to the 3-halogeno-derivative; 3-chloroindole is prepared by the action of sulphuryl chloride on indole. 3-Nitroindole has been prepared by treating a mixture of indole and sodium ethoxide with ethyl nitrate. On the other hand, mixed acids convert 2,3-dimethylindole into the corresponding 6-nitro-derivative. An unusual case of orientation is the formation of the 2-sulphonic acid by the action of sulphur trioxide in pyridine on indole at about 120°C. Indole-3-aldehyde is formed by the Gattermann reaction or by the action of dimethylformamide and phosphoryl chloride on indole. This compound may also be prepared by the Reimer–Tiemann reaction, but in this case 3-chloroquinoline is also produced (*cf.* pyrrole). Mercuration with mercuric acetate leads to 2,3-diacetoxymercuri-indole. Indole readily forms a Grignard reagent, which may be used to prepare substituted indoles (see p. 839).

Indole also undergoes the Mannich reaction with formaldehyde and dimethylamine to form *gramine* (3-dimethylaminomethylindole):

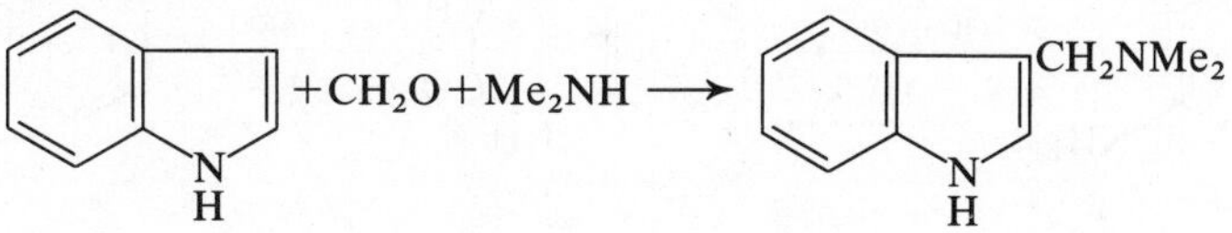

The dimethylamino-group is easily displaced by certain reagents, particularly when in the form of the trimethylammonium ion, *e.g.*,

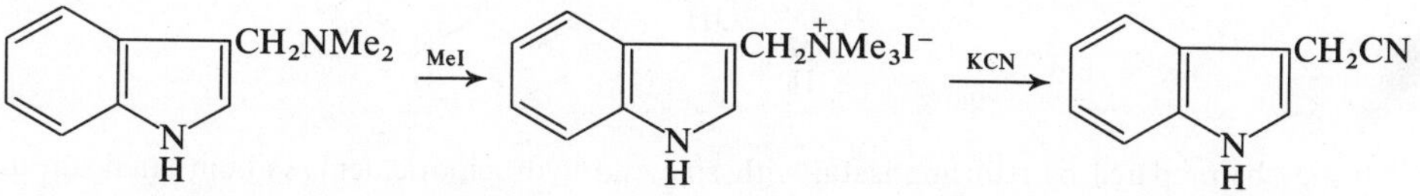

Indole is reduced electrolytically, by tin and hydrochloric acid, or by zinc dust and phosphoric acid to 2,3-dihydroindole (*indoline*). This is also produced by catalytic reduction using copper chromite, but if Raney nickel is used, the product is octahydroindole.

Indole behaves in some ways as a tautomeric substance, the tautomer being known as *indolenine*. Indolenine itself has never yet been isolated, but its derivatives are known, and these may be prepared, *e.g.*, by condensation between indoles and aldehydes.

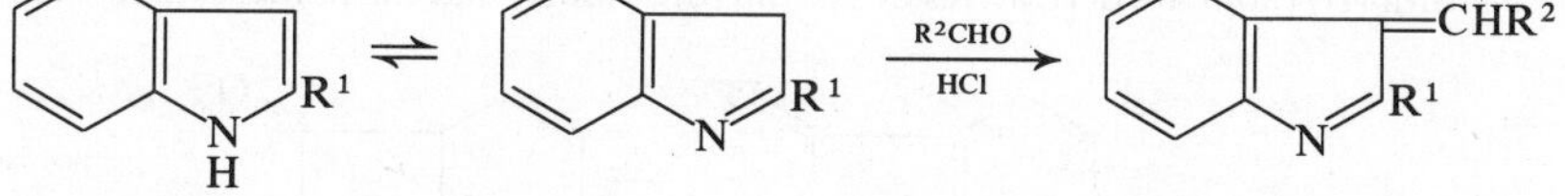

Indoxyl is the term usually applied to the keto-form of **3-hydroxyindole** (the enolic form):

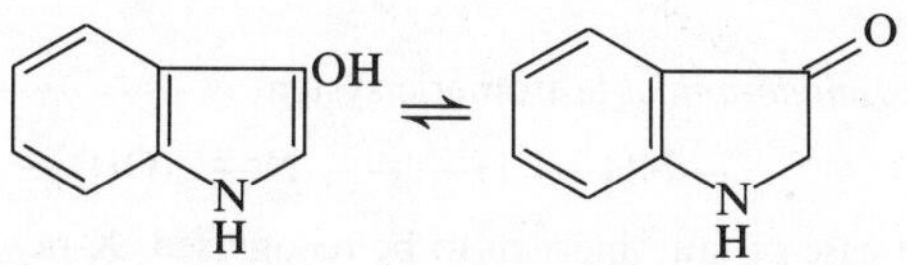

Derivatives of both are known (see p. 895 for its preparation). Indoxyl is a bright yellow solid, m.p. 85°C, and is readily oxidised in alkaline solution by air to indigotin. It gives a red colour with ferric chloride, and reacts with diazonium salts. These are typical reactions of an enol, but since it also reacts with aldehydes, this has been taken as evidence for the presence of the keto-form (aldehydes condense with an active methylene group).

Oxindole, m.p. 126°C. There are three possible formulae for oxindole:

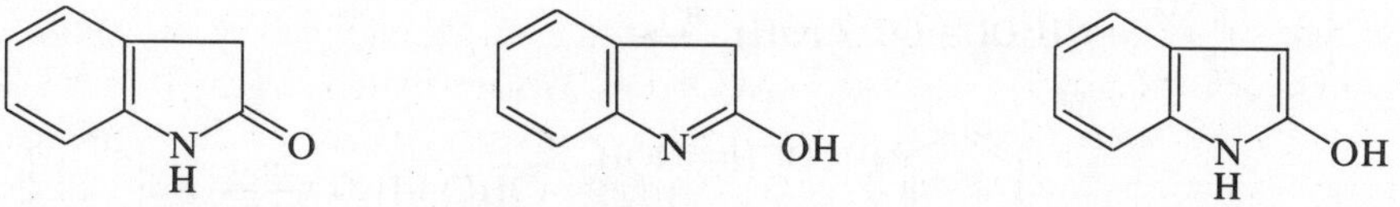

Evidence obtained from u.v. and i.r. spectra studies indicates that oxindole is almost completely the lactam (amide) form (I).

Oxindole is most conveniently prepared from aniline and chloroacetyl chloride as follows:

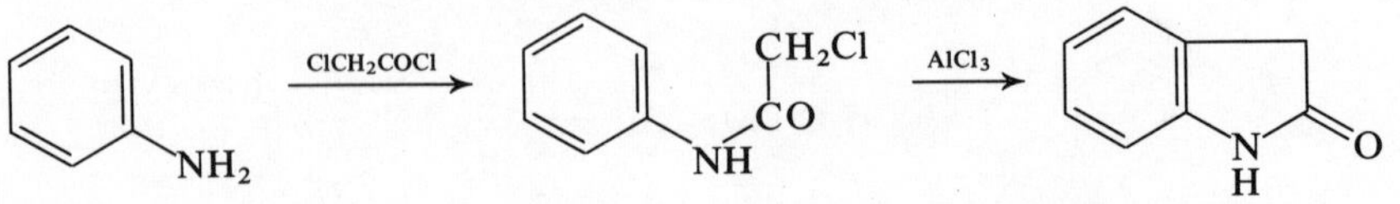

$$\text{(aniline, NH}_2\text{)} \xrightarrow{\text{ClCH}_2\text{COCl}} \text{(NHCOCH}_2\text{Cl)} \xrightarrow{\text{AlCl}_3} \text{(oxindole)}$$

Dioxindole (3-*hydroxyoxindole*), m.p. 180°C, is tautomeric with **2,3-dihydroxyindole**:

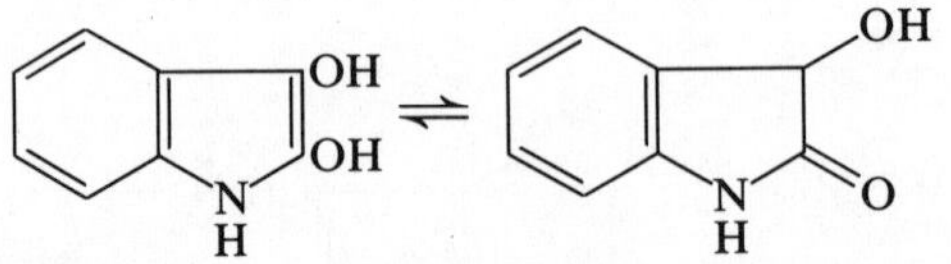

$$\rightleftharpoons$$

It may be prepared by reducing isatin with zinc and hydrochloric acid, sodium amalgam in alkaline solution, or with sodium dithionite.

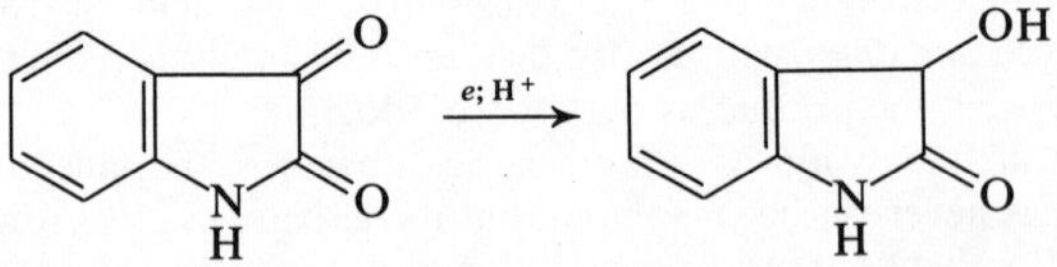

$$\xrightarrow{e;\,\text{H}^+}$$

Isatin exists in two forms, the term ψ-isatin being applied to the *lactam* form (I), and isatin to the *lactim* form (II) (both are derivatives of 2,3-dihydroindole; and the name isatin is often applied to I):

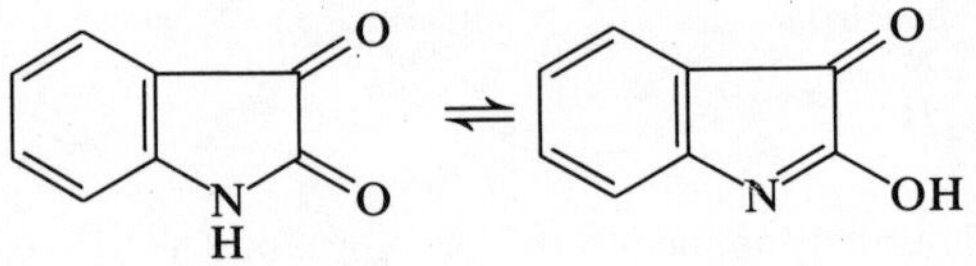

$$\rightleftharpoons$$

(I)   (II)

This is an example of the *amido-imidol* tautomeric system:

$$-\text{NH}-\text{CH}- \rightleftharpoons -\text{N}=\text{C(OH)}-$$

It appears to be the first case of tautomerism to be recognised. X-ray analysis of crystalline isatin has shown that it is predominantly in the amide (lactam) form. In aqueous solution, however, the species present depends on the pH, but in chloroform solution, i.r. spectra studies have shown that only the amide form is present.

Isatin was first obtained by the oxidation of indigotin with nitric acid. It is best prepared by heating a solution of concentrated hydrochloric acid containing aniline, chloral hydrate, hydroxylamine and sodium sulphate, 'Isonitrosoacetanilide' (oximinoacetanilide) crystallises out and this, on treatment with concentrated sulphuric acid, forms isatin by ring closure:

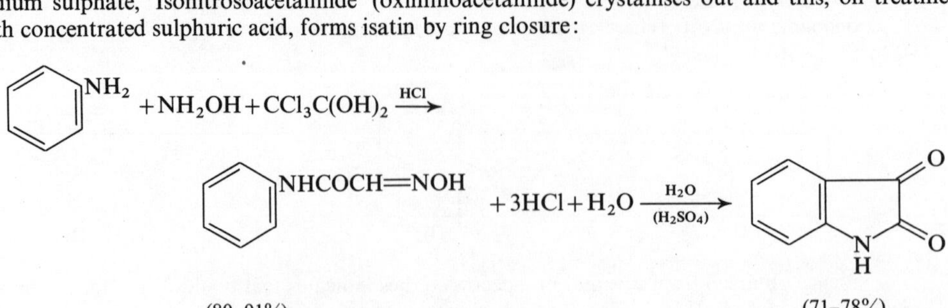

$$\text{(C}_6\text{H}_5\text{NH}_2) + \text{NH}_2\text{OH} + \text{CCl}_3\text{C(OH)}_2 \xrightarrow{\text{HCl}}$$

$$\text{(C}_6\text{H}_5\text{NHCOCH}=\text{NOH)} + 3\text{HCl} + \text{H}_2\text{O} \xrightarrow[\text{(H}_2\text{SO}_4)]{\text{H}_2\text{O}} \text{(isatin)}$$

(80–91%)   (71–78%)

The following synthesis from *o*-nitrobenzoyl chloride clearly shows the structure of isatin (Claisen *et al.*, 1879):

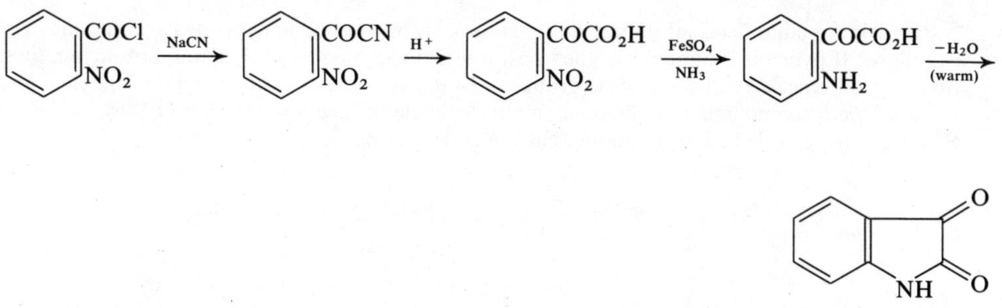

Isatin is a red solid, m.p. 200°C. With phosphorus pentachloride it forms *isatin chloride*, and with warm sodium hydroxide, *isatinic acid*.

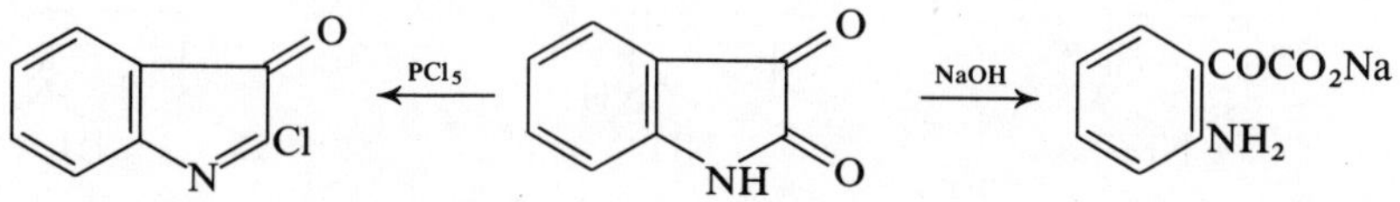

Carbazole (*dibenzopyrrole*), m.p. 245°C, may be isolated from the anthracene fraction of coal-tar (see p. 811). It may be synthesised by passing diphenylamine through a red-hot tube, or better, by heating 2,2'-diaminobiphenyl at 200°C with concentrated phosphoric acid (the yield is almost quantitative; Leditschke, 1953).

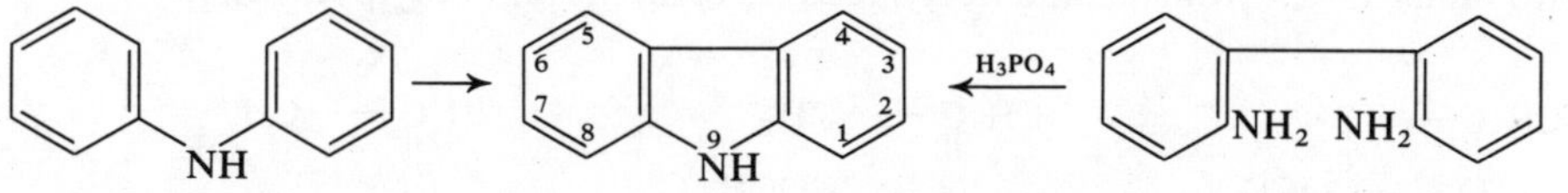

A convenient preparation is the reduction of 2-nitrobiphenyl with triethyl phosphite (82.5%; Cadogan *et al.*, 1962). Silver oxide converts carbazole into N,N'-*dicarbazyl*, which is a colourless compound but gives coloured solutions due to its dissociation into carbazyl free radical (*cf.* p. 664). Carbazole is used in the preparation of polyvinylcarbazole plastics.

Azoles†

Azole is the suffix used for five-membered rings containing two or more hetero-atoms, at least one of which is nitrogen.

Pyrazoles. Pyrazole may be prepared by passing acetylene into a cold ethereal solution of diazomethane; this is a 1,3-dipolar addition (p. 394):

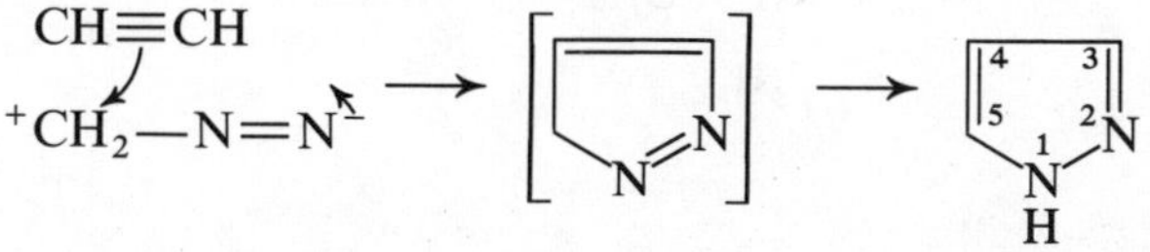

†These compounds are dealt with in more detail in Vol. 2, Ch. 12.

Pyrazole, m.p. 70°C, has aromatic properties, readily undergoing substitution (with the usual electrophilic reagents) in the 4-position.

Pyrazole is a tautomeric substance; this cannot be demonstrated in pyrazole itself, but may be shown as follows. If pyrazole is tautomeric, then positions 3 and 5 are identical; if not tautomeric, these positions are different. When the phenyl group in 3-methyl-1-phenyl- and 5-methyl-1-phenylpyrazole is removed, *both* compounds give the *same* methylpyrazole. Hence positions 3 and 5 are equivalent, and this can only be explained by assuming that pyrazole is tautomeric.

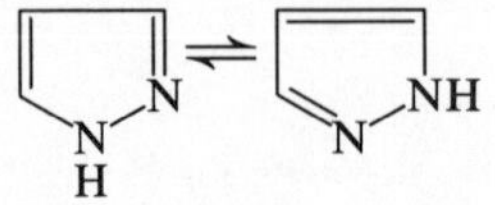

Pyrazole may be catalytically reduced to pyrazoline (I) and then to pyrazolidine (II). Both are stronger bases than pyrazole. 5-Ketopyrazoline or *pyrazol-5-one* is (III) (see below).

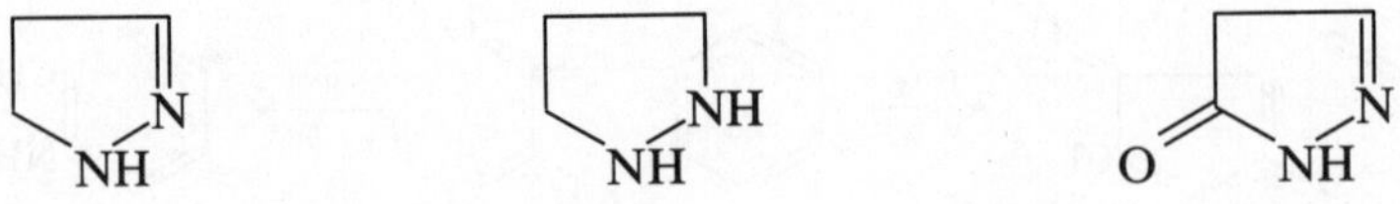

(I)   (II)   (III)

Pyrazole derivatives

One of the chief methods for preparing these is by reaction between hydrazines and 1,3-dicarbonyl compounds, *e.g.*, hydrazine and acetylacetone form 3,5-dimethylpyrazole.

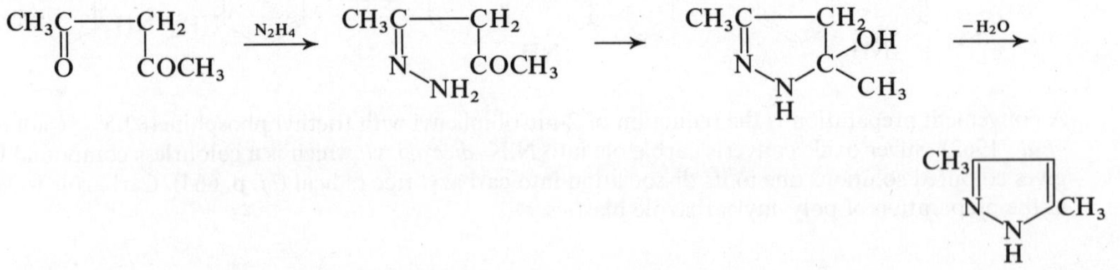

5-Pyrazolones are formed by reaction between hydrazines and β-ketonic esters, *e.g.*, 3-methyl-1-phenylpyrazolone from phenylhydrazine and ethyl acetoacetate. This, on methylation, gives *antipyrine* (*phenazone*, 2,3-dimethyl-1-phenylpyrazol-5-one), which is used in medicine as a febrifuge.

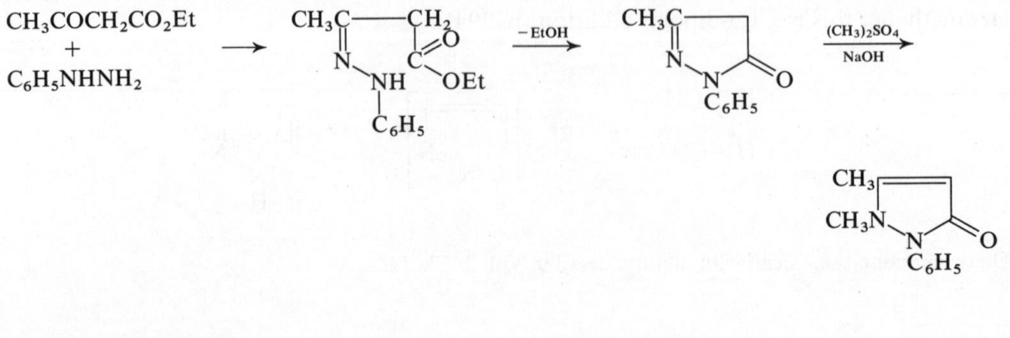

$$CH_3COCH_2CO_2Et$$
$$+$$
$$C_6H_5NHNH_2$$

Imidazoles (glyoxalines, iminazoles). Imidazole is isomeric with pyrazole, and occurs in the purine nucleus and in histidine.

A general method for preparing imidazoles is by reaction between α-dicarbonyl compounds, ammonia and an aldehyde.

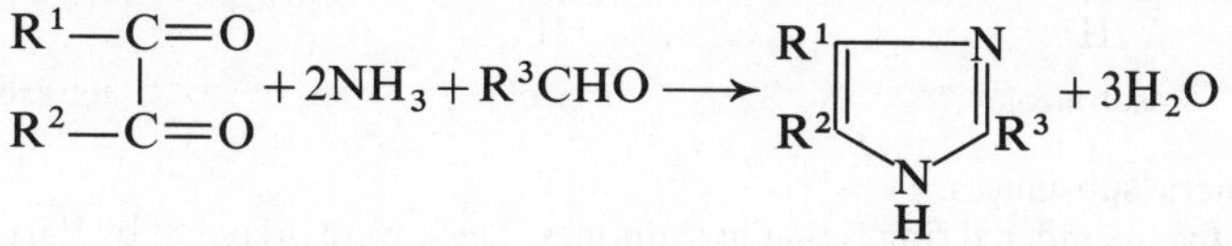

$$R^1-C=O \atop R^2-C=O \quad +2NH_3+R^3CHO \longrightarrow \quad {R^1 \atop R^2}\!\!\underset{N \atop H}{\overset{N}{\diagdown}}\!\!R^3 \quad +3H_2O$$

Glyoxal, ammonia and formaldehyde produce imidazole itself. This, however, is best prepared as follows: tartaric acid is treated with a mixture of concentrated nitric and sulphuric acids, the product, tartaric acid dinitrate ('dinitrotartaric acid'), then treated with formalin and ammonia solution, and the imidazole-4,5-dicarboxylic acid formed is then decarboxylated by heating with copper oxide.

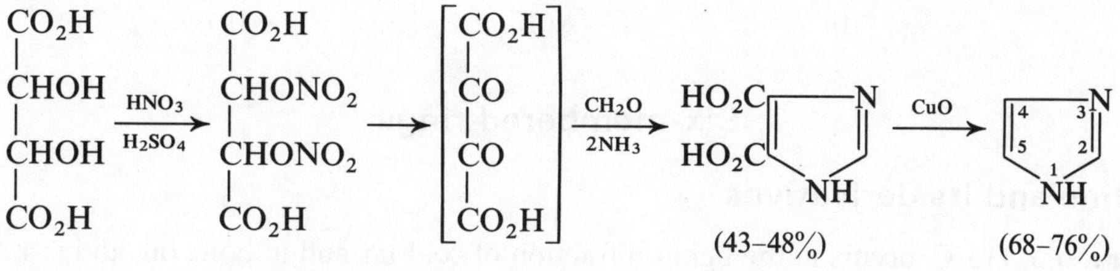

$$\begin{array}{c} CO_2H \\ | \\ CHOH \\ | \\ CHOH \\ | \\ CO_2H \end{array} \xrightarrow[H_2SO_4]{HNO_3} \begin{array}{c} CO_2H \\ | \\ CHONO_2 \\ | \\ CHONO_2 \\ | \\ CO_2H \end{array} \longrightarrow \left[\begin{array}{c} CO_2H \\ | \\ CO \\ | \\ CO \\ | \\ CO_2H \end{array}\right] \xrightarrow[2NH_3]{CH_2O} \begin{array}{c} HO_2C \\ HO_2C \end{array}\!\!\!\underset{NH}{\overset{N}{\diagdown}} \xrightarrow{CuO} \underset{NH}{\overset{N}{\diagdown}}$$

(43–48%)   (68–76%)

Imidazole, m.p. 90°C, is tautomeric (positions 4 and 5 are identical).

Oxazoles are formed by reaction between acid amides and α-halogenoketones, e.g., acetamide and bromoacetone form 2,4-dimethyloxazole.

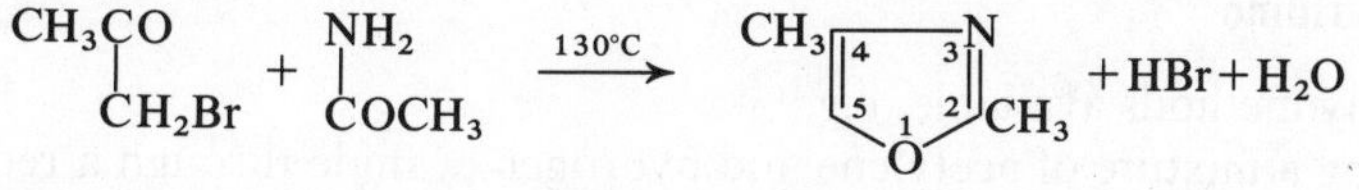

$$\begin{array}{c} CH_3CO \\ | \\ CH_2Br \end{array} + \begin{array}{c} NH_2 \\ | \\ COCH_3 \end{array} \xrightarrow{130°C} CH_3\!\!\underset{O}{\overset{N}{\diagdown}}\!\!CH_3 \quad +HBr+H_2O$$

Isoxazoles are formed by warming the monoximes of 1,3-diketones, e.g., 3,5-dimethylisoxazole from acetylacetonemonoxime.

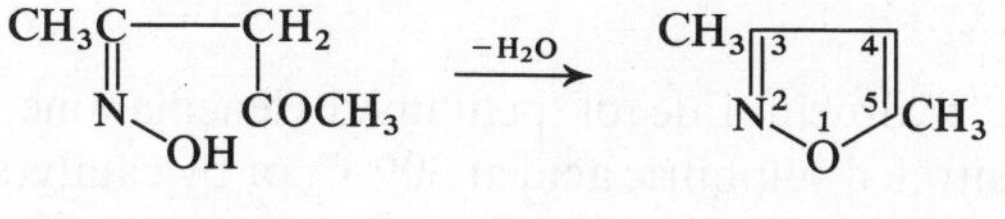

$$\begin{array}{c} CH_3C{\overset{}{-\!\!-}}CH_2 \\ \diagdown \quad | \\ N \quad COCH_3 \\ \diagdown \\ OH \end{array} \xrightarrow{-H_2O} CH_3\!\!\underset{O}{\overset{N}{\diagdown}}\!\!CH_3$$

Thiazoles may be prepared by reaction between an α-chlorocarbonyl compound and a thioacid amide.

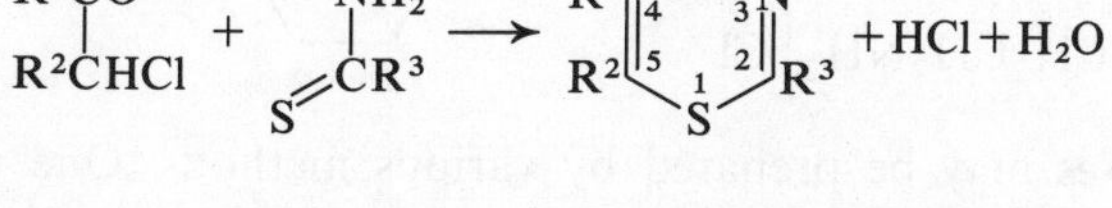

$$\begin{array}{c} R^1CO \\ | \\ R^2CHCl \end{array} + \begin{array}{c} NH_2 \\ | \\ S{\diagdown}CR^3 \end{array} \longrightarrow \begin{array}{c} R^1 \\ R^2 \end{array}\!\!\underset{S}{\overset{N}{\diagdown}}\!\!R^3 \quad +HCl+H_2O$$

Thiazole, b.p. 117°C, may be prepared from chloroacetaldehyde and thioformamide. Vitamin B_1 contains the thiazole nucleus.

Triazoles and **tetrazoles** are also known.

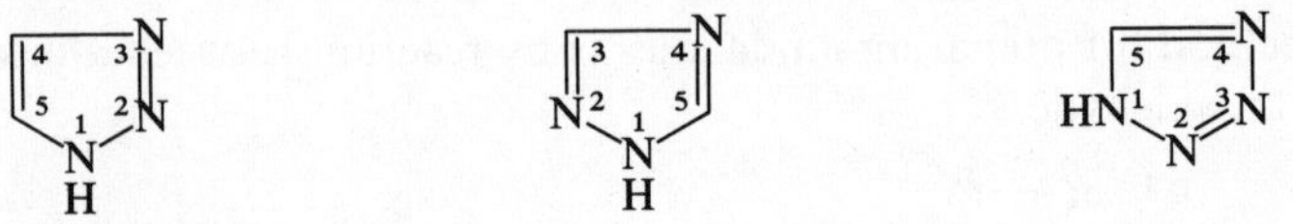

osotriazole   triazole   tetrazole

All are tautomeric substances.

An unusual five-membered ring is that in sydnones. These were prepared by Earl and Mackney (1935), and the peculiar feature of these compounds is that it is not possible to give them a structure which represents covalent bonds by the usual paired electrons, *e.g.*, sydnone ((I), (II), (III)). Dipole and infra-red studies indicate this compound is a resonance hybrid; it has been named a *meso-ionic structure*.

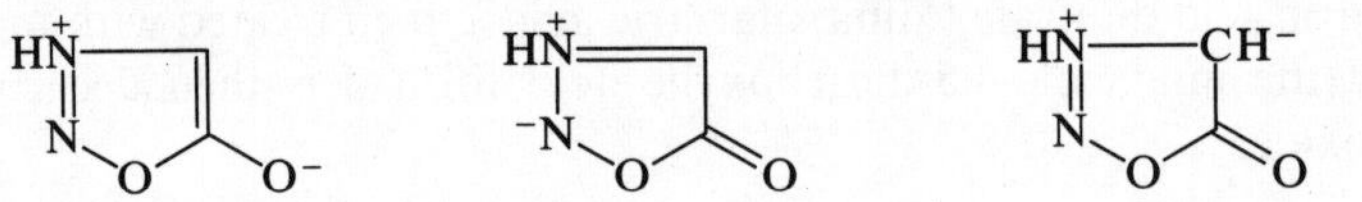

(I)   (II)   (III)

Six-membered rings

Pyridine and its derivatives

Pyridine, b.p. 115°C, occurs in the light oil fraction of coal-tar and in bone oil, and is a decomposition product of several alkaloids. It is obtained from light oil by treating the latter with dilute sulphuric acid. This dissolves pyridine and other basic substances. The acid layer is neutralised with sodium hydroxide and the liquid repeatedly fractionated.

Pyridine is basic, but it resembles benzene in many of its properties and this partly led Körner (1864) to adopt the ring structure for pyridine, a structure which is confirmed by synthesis.

Synthesis of pyridine

There are many methods available, *e.g.*,

(i) By passing a mixture of acetylene and hydrogen cyanide through a red-hot tube:

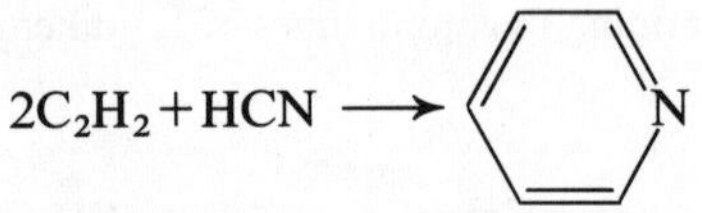

$$2C_2H_2 + HCN \longrightarrow$$

(ii) By heating the hydrochloride of pentamethylenediamine and oxidising the product, *piperidine*, with concentrated sulphuric acid at 300°C, or by catalytic dehydrogenation (Pd—C).

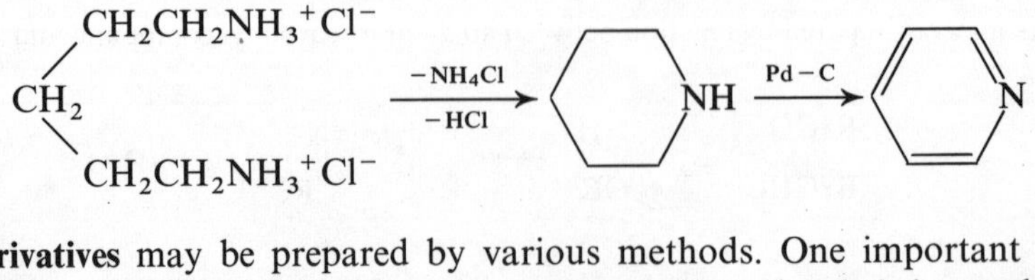

Pyridine derivatives may be prepared by various methods. One important method is the **Hantzsch synthesis** (1882). A β-dicarbonyl compound (2 molecules) is condensed with an aldehyde (1 molecule) and ammonia (1 molecule). The dihydropyridine derivative is obtained, and this gives the pyridine derivative on oxidation with nitric acid. The mechanism of the reaction is not fully understood, but intermediates proposed have been shown to give pyridines under the conditions of the experiment (proton transfers are given in one step):

(i) $\mathrm{MeCOCH_2CO_2Et} \xrightarrow{\mathrm{NH_3}} \mathrm{MeC(NH_2)}{=}\mathrm{CHCO_2Et}$

(ii) $\mathrm{MeCHO} + \mathrm{MeCOCH_2CO_2Et} \xrightarrow{\mathrm{NH_3}} \mathrm{MeCH}{=}\mathrm{C} \begin{smallmatrix} \mathrm{COMe} \\ \mathrm{CO_2Et} \end{smallmatrix}$

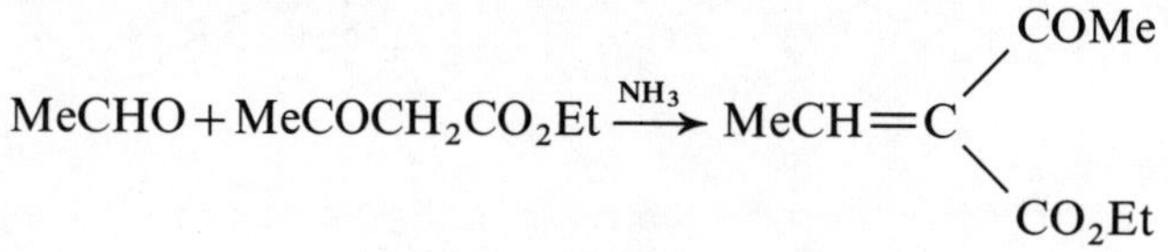

(iii)

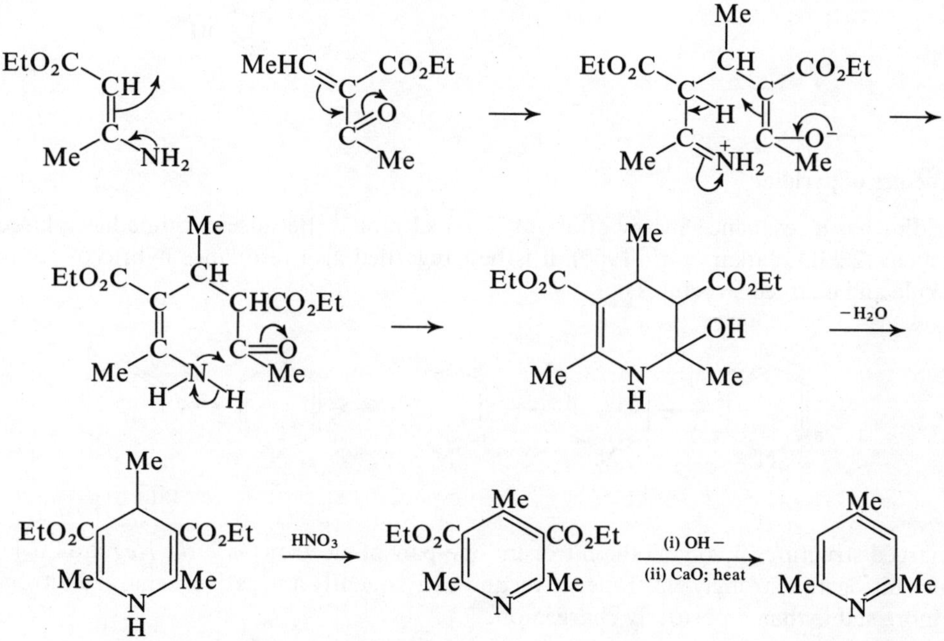

On the other hand, aldehydes (or ketones) alone (*i.e.*, no β-dicarbonyl compound need be present) form pyridines when heated with ammonia, *e.g.*,

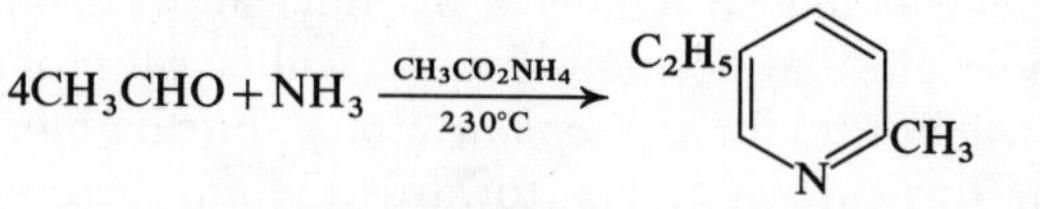

$$4\mathrm{CH_3CHO} + \mathrm{NH_3} \xrightarrow[230^{\circ}\mathrm{C}]{\mathrm{CH_3CO_2NH_4}} \text{(pyridine ring with } \mathrm{C_2H_5} \text{ and } \mathrm{CH_3}\text{)}$$

Pyridones may be prepared by the action of ammonia on 2- or 4-pyrones:

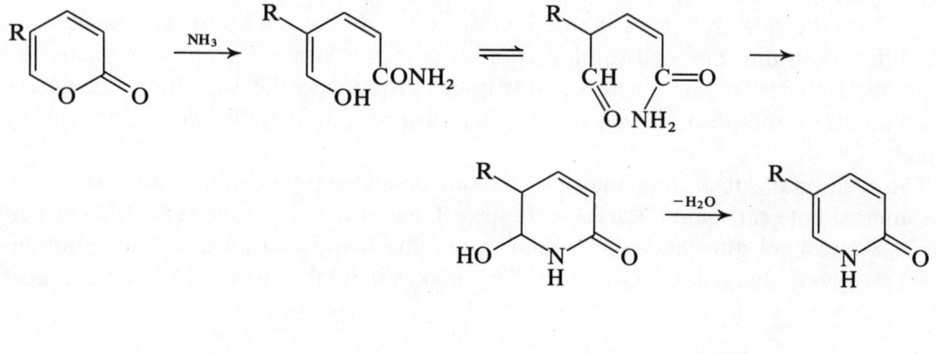

2-Furyl ketones may be used as starting material; when heated with ammonia and ammonium chloride, the product is a 3-hydroxypyridine (Leditschke, 1953):

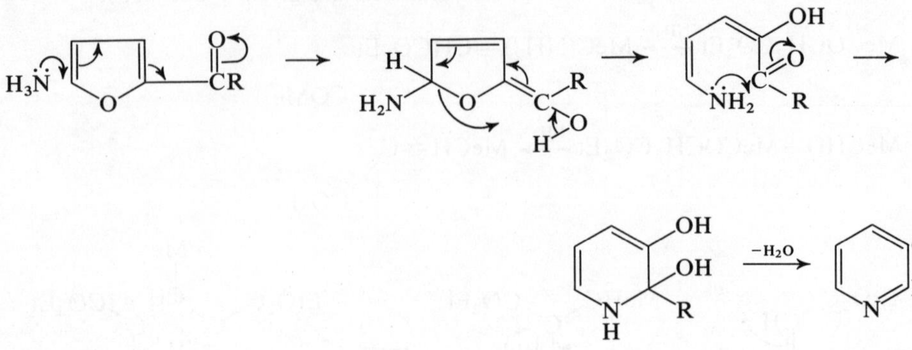

Structure of pyridine

Pyridine has a resonance energy of about $125 \cdot 5 \text{ kJ mol}^{-1}$. Because pyridine has a large dipole moment ($2 \cdot 23\text{D}$; Jatkar *et al.*, 1960), it is best regarded as a resonance hybrid of contributing Kekulé and charged structures:

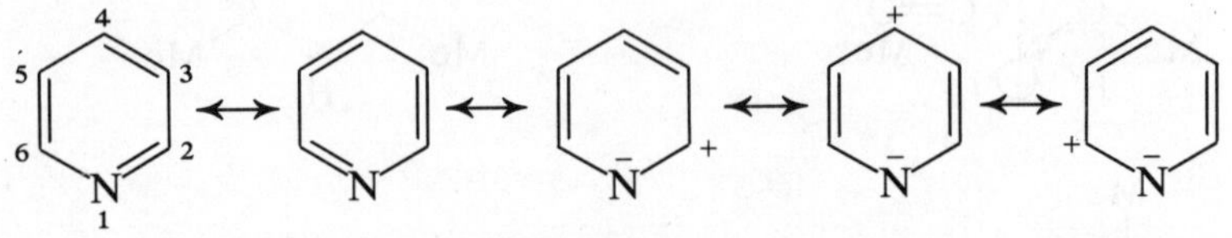

Charged structures involving the nitrogen lone-pair of electrons *into the ring* are far less likely since nitrogen is strongly electronegative, and consequently a negatively charged nitrogen atom is more stable than a positively charged one.

Substitution in pyridine

Examination of the contributing structures of pyridine shows, from the point of view of charge distribution in the molecule, that positions 3 and 5 will be sites for *electrophilic* attack, whereas positions 2, 4, and 6 will be sites for *nucleophilic* attack. Furthermore, because of the withdrawal of electrons from the ring-carbon atoms (towards the nitrogen atom), the ring is deactivated towards electrophilic reagents, thereby resembling the benzene ring in nitrobenzene. Also, in strongly acid solution, pyridine is protonated, and so the positively charged nitrogen atom deactivates the ring even more than the unprotonated nitrogen atom. This is clearly shown by the difficulty in nitration and sulphonation of pyridine. If a $+I$ group is present, electrophilic reactions are accelerated. When the substituent group is in the 3-position, the entering group takes up the 2- or 6-position; and if in the 2- or 4-position, then the 3- or 5-position is the site of attack.

The position of substitution may also be considered from the point of view of the stabilities of the intermediate carbonium ions (σ-complexes), *e.g.*, there are three resonating structures each for 2-, 3- and 4-substitution (electrophilic), but in the 2- and 4-cations, one contributing structure has a positively charged nitrogen atom. This makes these ions less stable than the 3-cation.

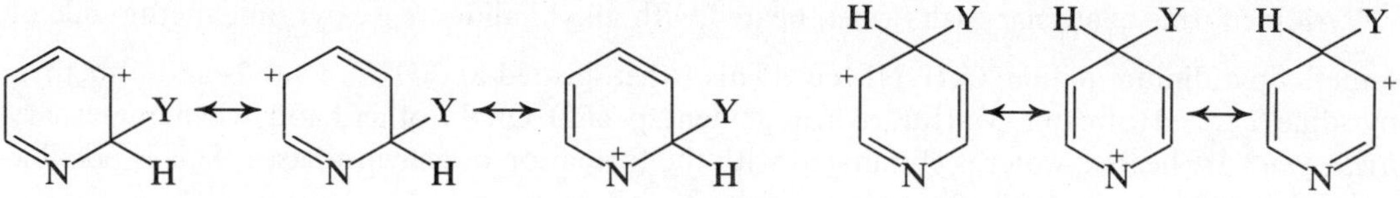

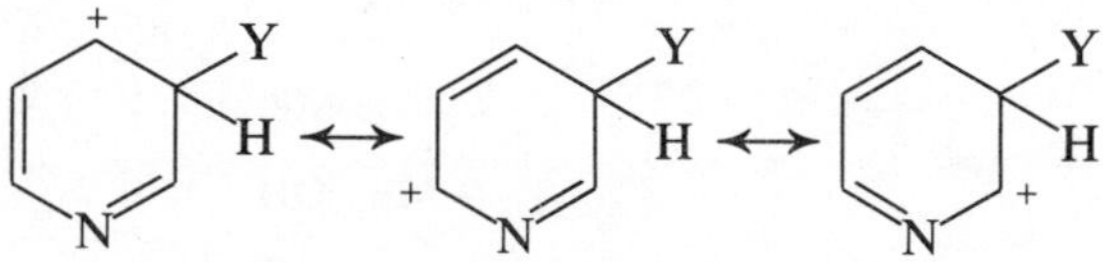

In a similar way, consideration of the anions in nucleophilic substitution shows that the 2-, and 4-anions have a nitrogen carrying a negative charge, and so these ions will be more stable than the 3-anion:

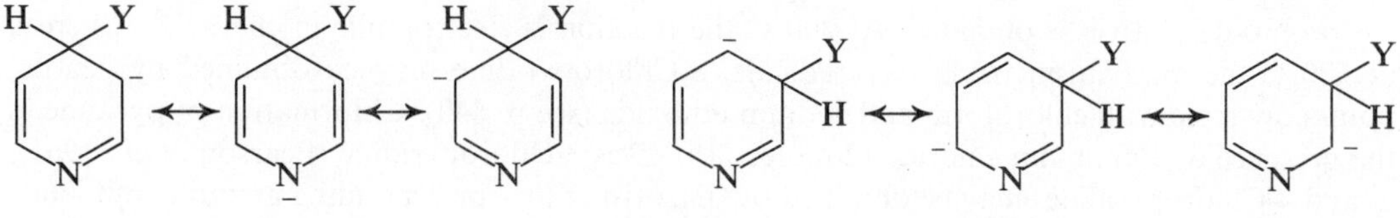

Basicity of pyridine

Since the nitrogen atom carries a negative charge (and the lone pair is 'completely' available for protonation; *cf.* pyrrole), pyridine would be expected to be a very strong base. In actual fact, pyridine is far less basic than expected (see Table 30.2), being less basic than the alkylamines (see Table 13.1). It is difficult to explain this, but a possibility is as follows. Since the nitrogen atom in pyridine is sp^2 hybridised, it is more electron-attracting than nitrogen that is sp^3 hybridised (as in RNH_2). Consequently, the lone pair is more tightly held and so is less available for protonation.

Reactions of pyridine

A number of reactions of pyridine occur at the nitrogen atom. The basic property of pyridine has already been discussed, and it is this property that is used in many reactions where it is desired to carry out dehydrohalogenation, *e.g.*, removal of HBr from bromosuccinic ester (KOH/EtOH would hydrolyse the ester at the same time).

Pyridine also reacts with halogen in the cold to form 1-halogenopyridinium halides, *e.g.*, $C_5H_5\overset{+}{N}Br\ Br^-$, and with acyl chlorides to form 1-acylpyridinium chlorides. This type of compound is believed to be the active species in the acylation of hydroxy- and amino-compounds with acyl halides in the presence of pyridine. This type of quaternary compound has also

$$C_5H_5N \xrightarrow{\ R^1COCl\ } C_5H_5\overset{+}{N}COR^1Cl^- \xrightarrow{\ R^2OH\ } R^1CO_2R^2 + C_5H_5NH^+Cl^-$$

been believed to be the active species when acetic anhydride is used as the acylating agent, but according to Bonner *et al.* (1968), it is hydrogen bonding between the substrate and pyridine which appears to be a prerequisite for reaction.

Pyridine forms quaternary salts when heated with alkyl halides, *e.g.*, pyridine methiodide or 1-methylpyridinium iodide, $C_5H_5\overset{+}{N}MeI^-$. This, when heated at 300°C, gives 2- and 4-methyl-pyridine (*cf.* the Hofmann–Martius rearrangement, p. 662). On the other hand, when quaternary *hydroxides* are heated, water is eliminated with the formation of **anhydro-bases**. This is possible only when there is a suitable alkyl group in the 2- or 4-position, *e.g.*,

Anhydro-bases, on treatment with aqueous inorganic acids, regenerate the quaternary hydroxide.

Pyridine is readily oxidised by per-acids (see later), and complexes with many metal ions and Lewis acids. It is this complex-formation that inhibits the Friedel–Crafts reaction with pyridine.

When pyridine and bromine are passed over a catalyst of charcoal at 300°C, a mixture of 3-bromopyridine and 3,5-dibromopyridine is obtained; at 500°C a mixture of the 2- and 2,6-bromo-derivatives is obtained. At 300°C the reaction is electrophilic in character, whereas at 500°C the mechanism is via free radicals. 3-Chloropyridine may be obtained by heating potassiopyrrole with chloroform and sodium ethoxide (see p. 840). Chlorination of pyridine in the presence of aluminium chloride (2 molecules) gives 3-chloropyridine (Pearson *et al.*, 1961). 2- and 4-Chloropyridines may be obtained by diazotising the corresponding amino-compounds in concentrated hydrochloric acid and treating with cuprous chloride.

Halogen in the 2- or 4-position is reactive, being fairly readily replaced by OH, CN, NH_2, etc. The bromopyridines form Grignard reagents in the presence of ethyl bromide (*cf.* p. 424).

Pyridine reacts with nitric acid fairly readily only if a hydroxyl or an amino-group is present in the ring. Pyridine itself is nitrated to 3-nitropyridine by heating with concentrated sulphuric acid and fuming nitric acid at 300°C. 2- and 4-Nitropyridines may be obtained by oxidising the corresponding amino-compounds with hydrogen peroxide in sulphuric acid.

Sulphonation of pyridine is difficult, but when heated with 20 per cent oleum (containing a little mercuric sulphate) at 220°C, pyridine gives pyridine-3-sulphonic acid (70%). 2- and 4-Pyridinesulphonic acids may be prepared by oxidation of the corresponding thiols (prepared by the action of potassium hydrogen sulphide on the chloropyridine).

The sulphonic acid group can be replaced by hydroxyl or by the cyano-group.

Pyridine may be fairly easily mercurated in the 3-position with aqueous mercuric acetate.

Sodium and ethanol, electrolytic reduction, or catalytic reduction using nickel convert pyridine into piperidine, but reduction with sodium in liquid ammonia in the presence of ethanol (Birch reduction) gives 1,4-dihydropyridine and with LAH, 1,2-dihydropyridine. On the other hand, when pyridine is heated with hydriodic acid at 300°C, the ring is opened with the formation of n-pentane and ammonia.

In addition to electrophilic reactions, pyridine undergoes nucleophilic reactions, but it does so only with strong nucleophiles. When heated with sodamide in toluene solution, pyridine forms 2-aminopyridine; excess of sodamide produces 2,6-diaminopyridine (**Tschitschibabin (Chichibabin) reaction**, 1914). Actually, the sodium salts are formed, but on treatment with water, are hydrolysed to the amine.

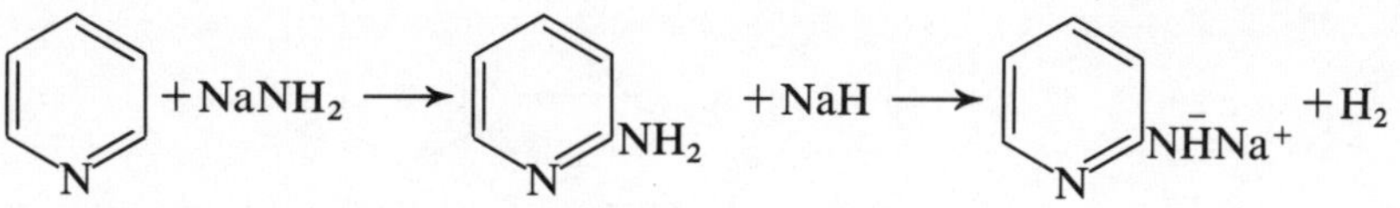

$$\text{(pyridine)} + NaNH_2 \longrightarrow \text{(2-aminopyridine)} + NaH \longrightarrow \text{(sodium salt)} + H_2$$

The three monoaminopyridines can be obtained by means of the Hofmann amine preparation on the amides of the pyridine monocarboxylic acids.

3-Aminopyridine can be diazotised easily; 2- and 4-aminopyridines are difficult to diazotise. Angyal *et al.* (1952), from an examination of the infra-red absorption spectra, have concluded that 2- and 4-aminopyridine (and aminoquinolines) are mainly in the amino form (*cf.* the corresponding hydroxy-compounds).

Pyridine forms 2-hydroxypyridine when heated with potassium hydroxide at 320°C.

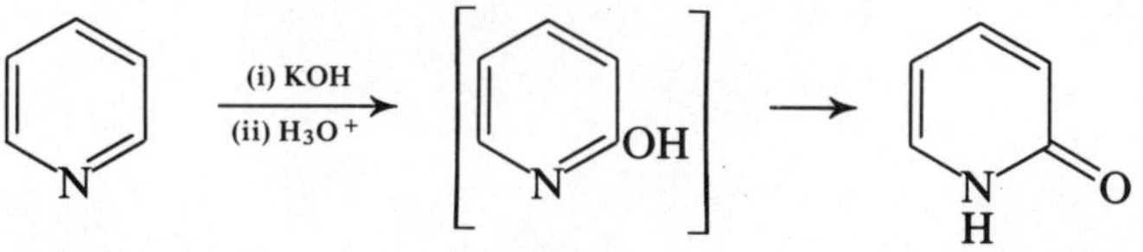

$$\text{(pyridine)} \xrightarrow[\text{(ii) } H_3O^+]{\text{(i) KOH}} \left[\text{(2-hydroxypyridine)} \right] \longrightarrow \text{(2-pyridone)}$$

However, the properties of 2- and 4-hydroxypyridine are not phenolic. X-ray and infra-red studies have shown the hydroxy-form (*pyridol*) is present only in small amount, the amide form (*pyridone*) being the predominant structure. On the other hand, 3-hydroxypyridine exhibits typical phenolic properties.

When heated with n-butyl-lithium, pyridine forms 2-n-butylpyridine (Ziegler *et al.*, 1930).

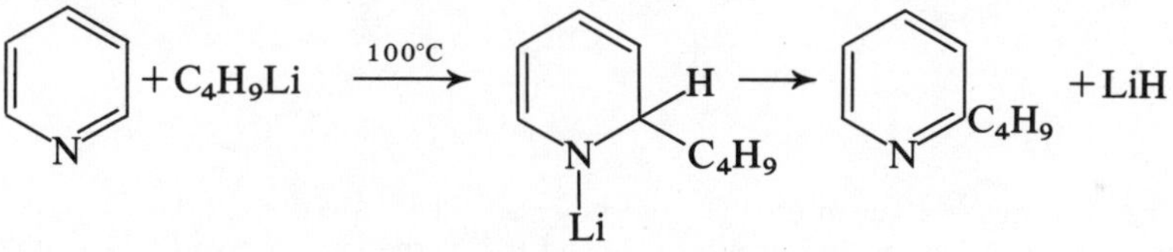

$$\text{(pyridine)} + C_4H_9Li \xrightarrow{100°C} \text{(intermediate)} \longrightarrow \text{(2-n-butylpyridine)} + LiH$$

Pyridine-1-oxide. Pyridine is oxidised by per-acids (peracetic, perbenzoic, etc.) to *pyridine oxide* which, on the basis of dipole moment studies, is formulated as a resonance hybrid.

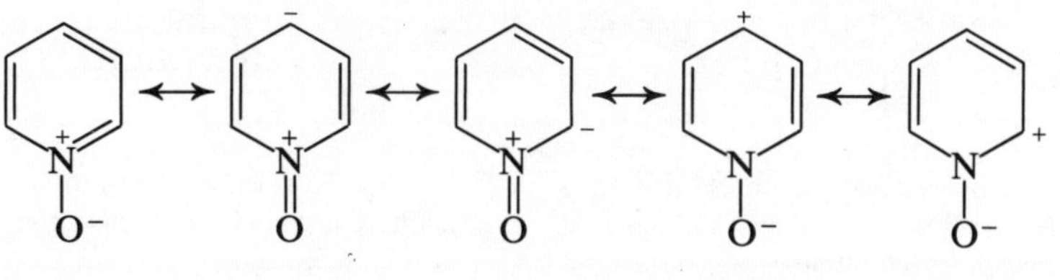

$$\text{(resonance structures of pyridine-1-oxide)}$$

Examination of these canonical forms shows that in pyridine oxide, high or low charge densities are produced at positions 2 and 4. Thus, pyridine oxide is more reactive towards both electrophilic and nucleophilic reagents than pyridine itself. In practice, substitution occurs predominantly at the 4-position, and since the oxygen atom may be readily removed, this offers a means of preparing pyridine derivatives that are difficult to prepare by other routes (see Chart).

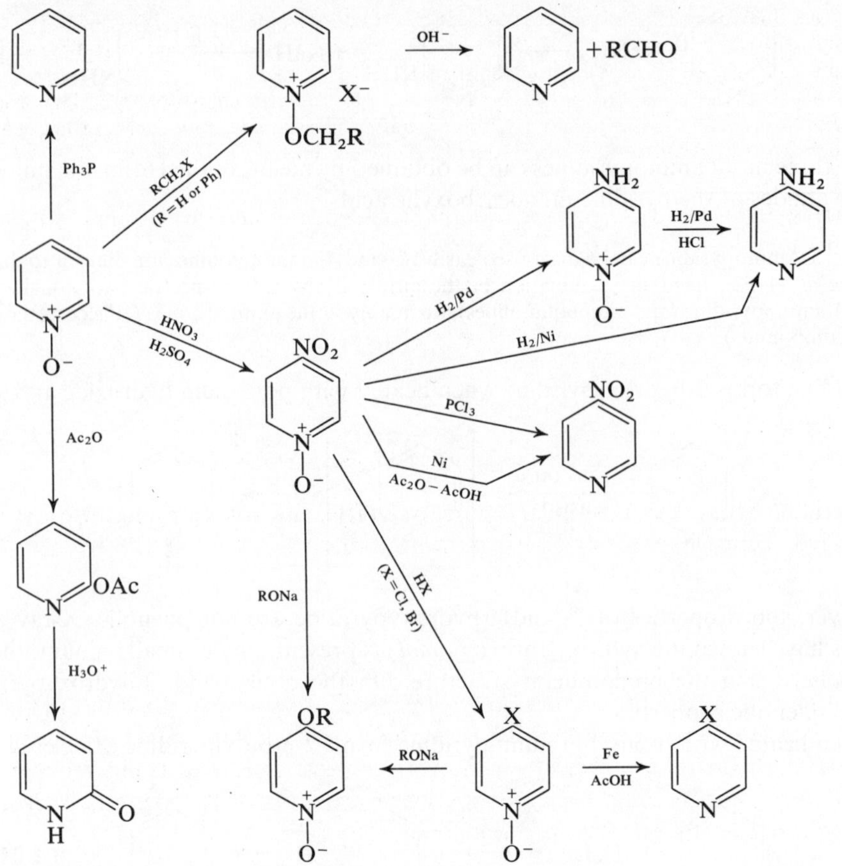

Pyridine homologues occur in coal-tar and bone oil. There are three methylpyridines and these are known as **picolines**. The methyl group in 2- and 4-picolines is reactive (*cf.* nitrotoluenes, p. 647); *e.g.*, 2-picoline condenses with acetaldehyde in the presence of warm aqueous sodium hydroxide to form 2-propenylpyridine:

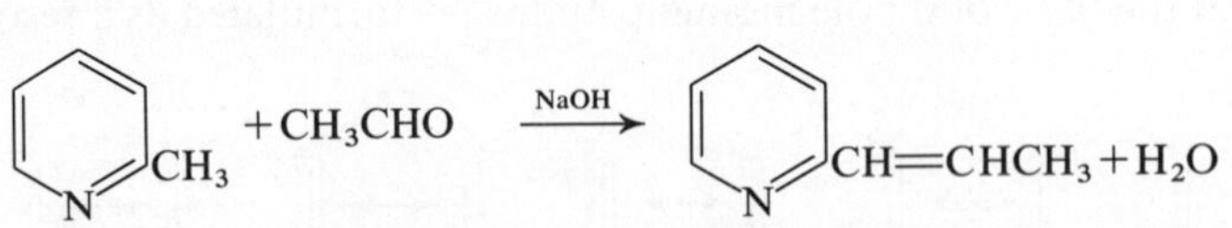

Oxidation of gaseous picolines with air at 380°C over mixed vanadium–molybdenum oxides gives mainly the pyridine–aldehydes (Mathes *et al.*, 1955). There are six *dimethylpyridines* (**lutidines**) and six *trimethylpyridines* (**collidines**).

Pyridinecarboxylic acids. There are three monocarboxylic acids, and each may be obtained by oxidising the corresponding picoline with alkaline permangante.

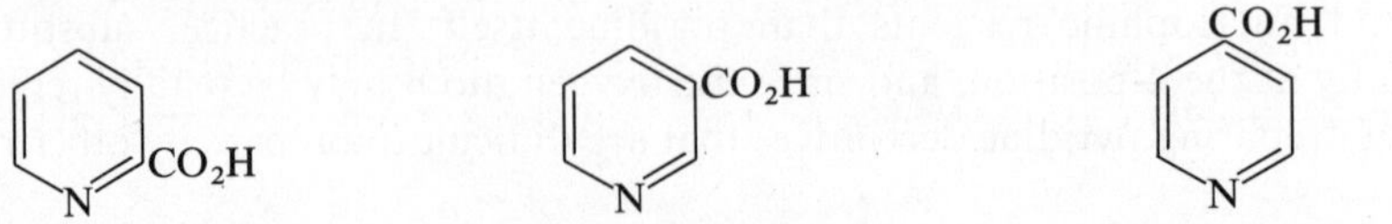

picolinic acid   nicotinic acid   isonicotinic acid

All three acids may be reduced to the corresponding piperidinecarboxylic acids by means of sodium and ethanol, or better, catalytically. Nicotinic acid is the most important one of the three isomers; this and its amide (nicotinamide) occur in the vitamin-B complex.

There are six pyridinedicarboxylic acids, but only two are important: **quinolinic acid** (*pyridine-2,3-dicarboxylic acid*), which is an oxidation product of quinoline, and **cinchomeronic acid** (*pyridine-3,4-dicarboxylic acid*), an oxidation product of isoquinoline.

Piperidine, b.p. 106°C, occurs in the alkaloid piperine. It may be prepared by reducing pyridine or by heating the hydrochloride of pentamethylenediamine:

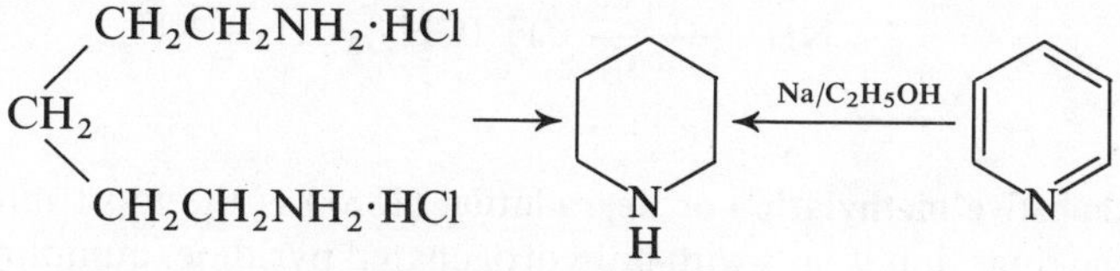

$$CH_2 \begin{cases} CH_2CH_2NH_2 \cdot HCl \\ \\ CH_2CH_2NH_2 \cdot HCl \end{cases} \longrightarrow \quad \underset{\overset{|}{H}}{N} \quad \xleftarrow{Na/C_2H_5OH} \quad N$$

Piperidine is manufactured by the catalytic hydrogenation of pyridine.

Piperidine behaves as a secondary aliphatic amine, and concentrated sulphuric acid at 300°C or catalytic dehydrogenation convert it into pyridine.

Methods of ring fission

Many heterocyclic compounds (containing nitrogen) occur naturally, and an extremely important step in the determination of their structure is to ascertain the disposition of the carbon atoms. A common procedure is to reduce the heterocyclic compounds and then open the ring of the product by the following methods (in which piperidine is used as the example):

(i) *Secondary* cyclic amines may be opened by treatment with 3 per cent hydrogen peroxide, *e.g.*, piperidine gives δ-aminovaleraldehyde:

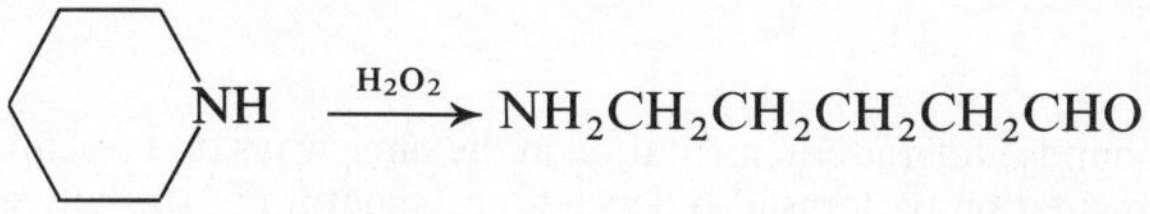

$$NH \xrightarrow{H_2O_2} NH_2CH_2CH_2CH_2CH_2CHO$$

(ii) *Von Braun's method* (1910) may be used to open *secondary* cyclic amines:

$$NH \xrightarrow[NaOH]{PhCOCl} NCOPh \xrightarrow[0°C]{PBr_3/Br_2} \left[NCBr_2Ph \right] \longrightarrow$$

$$Br(CH_2)_5Br + PhCN$$

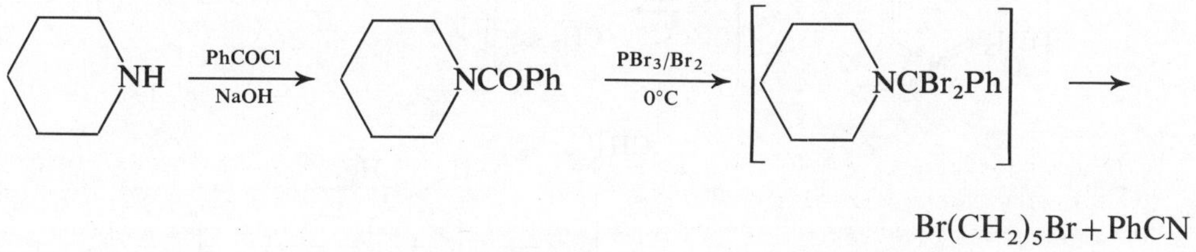

It might be noted that this provides a good method of preparing 1,5-dibromopentane.

(iii) *von Braun cyanogen bromide reaction* (1900). This is the reaction of a tertiary amine with cyanogen bromide to form an alkyl bromide and a disubstituted cyanamide (p. 378).

$$R_3N + BrCN \longrightarrow RBr + R_2NCN$$

This reaction has been used in alkaloid chemistry for opening tertiary *cyclic* amines.

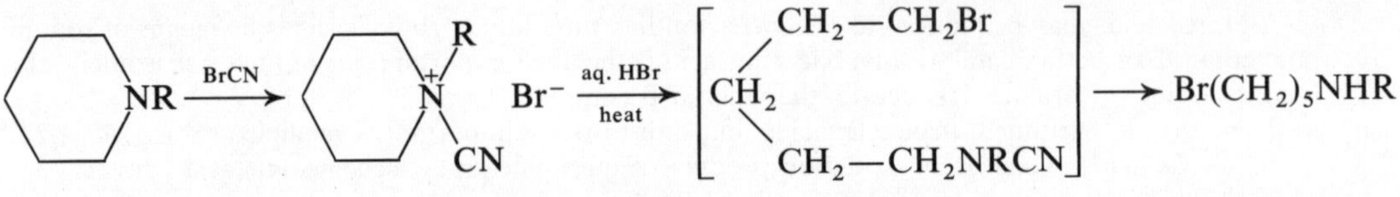

(iv) Piperidine may be converted into n-pentane by heating with hydriodic acid at 300°C:

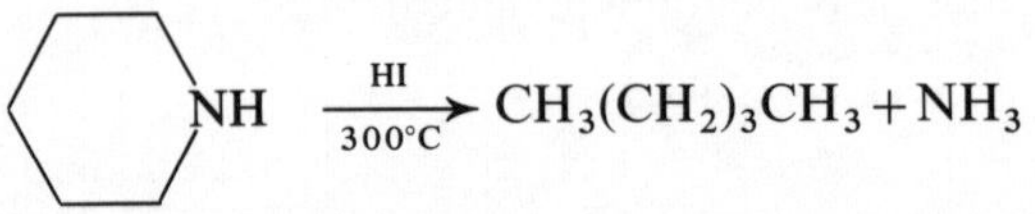

(v) **Hofmann exhaustive methylation or degradation** (1881) is the most important method of opening heterocyclic rings, but it fails with unhydrogenated pyridine, quinoline and isoquinoline derivatives, and with hydrogenated quinolines. Consider piperidine as our example. The process is repeated *twice*, the first step opening the ring, and the second step then eliminating trimethyl-amine to form penta-1,4-diene (see p. 379 for mechanism). This, however, isomerises to *piperylene* (penta-1,3-diene). This isomerisation is general, an isolated double bond system rearranging, if possible, to form a conjugated system.

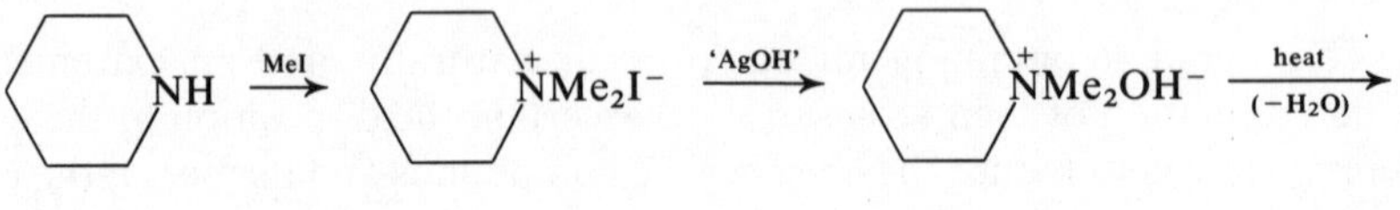

$$Me_2NCH_2CH_2CH_2CH=CH_2 \xrightarrow[\text{(ii) 'AgOH'}]{\text{(i) MeI}} HO^-Me_3\overset{+}{N}CH_2CH_2CH_2CH=CH_2 \xrightarrow[(-H_2O)]{\text{heat}}$$

$$Me_3N + [CH_2=CHCH_2CH=CH_2] \longrightarrow CH_3CH=CHCH=CH_2$$

Mass spectra

Heterocyclic compounds undergo fragmentation in the same ways that resemble the benzene analogues. However, the fragmentation patterns also depend on the nature of the hetero-atom, *e.g.*,

(i)

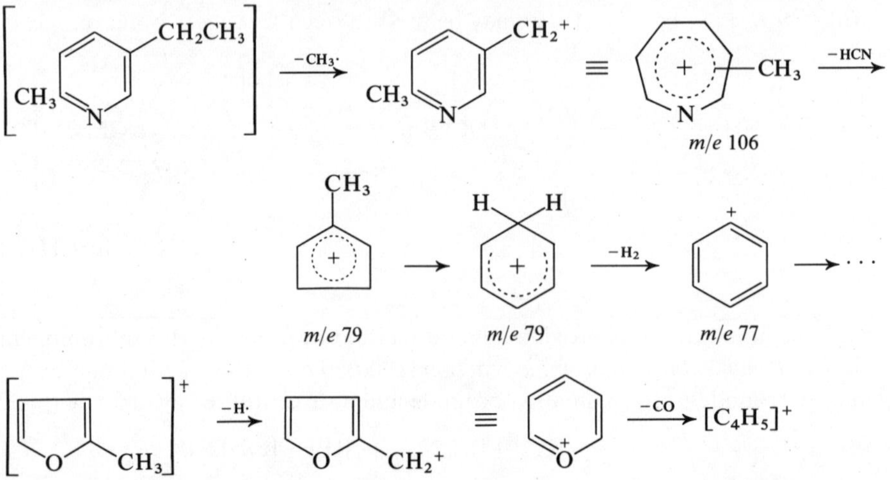

(ii)

Quinoline is present in coal-tar and bone oil, and was first obtained from the alkaloid *quinine* by alkaline decomposition. It is obtained commercially from coal-tar, or prepared synthetically.
(i) **Skraup synthesis** (1880) is a very important method, and may be carried out by heating a mixture of aniline, nitrobenzene, glycerol, concentrated sulphuric acid and ferrous sulphate. Nitrobenzene acts as an oxidising agent, and ferrous sulphate makes the reaction less violent. Arsenic acid may be used instead of nitrobenzene and the former is better since the reaction is less violent. The mechanism of the Skraup synthesis is not certain. It is generally believed that the first step is the conversion of glycerol into acraldehyde, which then undergoes 1,4-addition. Acraldehyde itself is not used since it polymerises under the conditions of the experiment.

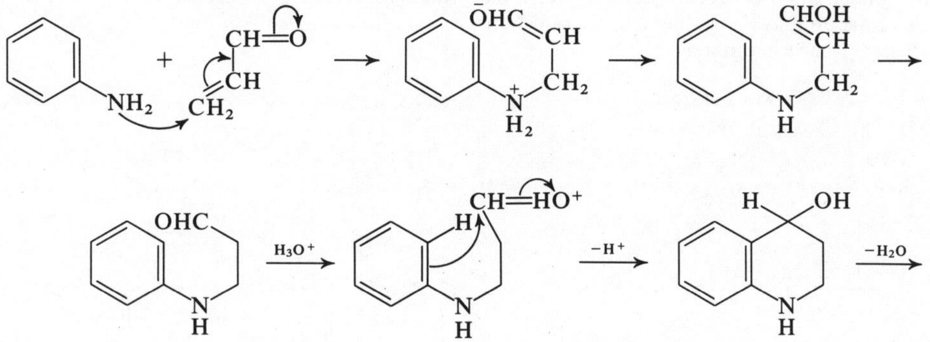

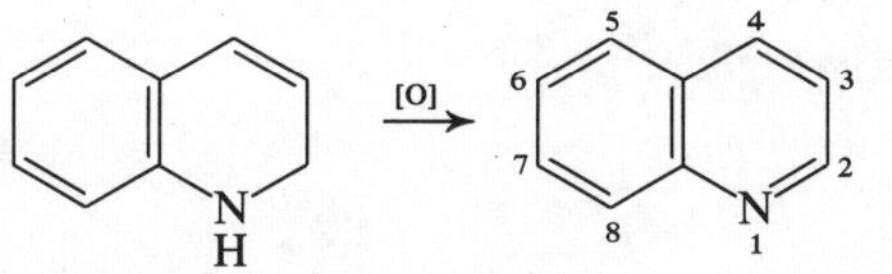

$(84-91\%)$

In general, the Skraup synthesis may be carried out with any primary aromatic amine in which at least one position *ortho* to the amino-group is vacant. If a strong *o/p*-directing group is present in the *m*-position, *e.g.*, OMe, then the 7-substituted quinoline is formed; if weakly *o/p*-directing, *e.g.*, Cl, then both 5- and 7-substituted quinolines are formed. If the *m*-substituent is *m*-directing, the predominant product is the 5-substituted quinoline.

(ii) **Friedländer's synthesis** (1882) is another important method for synthesising quinoline and many of its derivatives; *e.g.*, quinoline is formed when *o*-aminobenzaldehyde is condensed with acetaldehyde in aqueous sodium hydroxide:

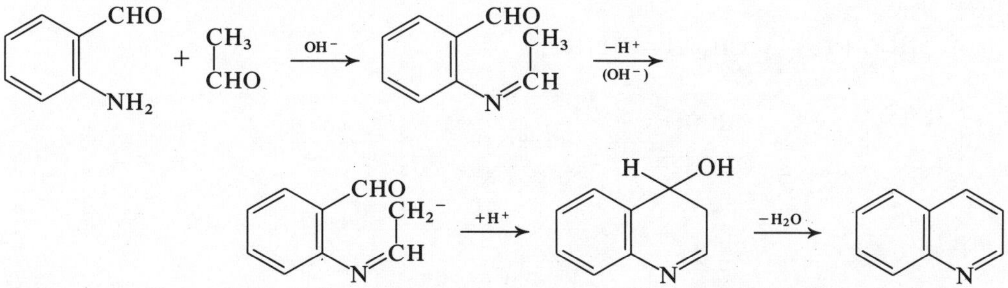

Acids are also effective catalysts for this reaction.

Derivatives of quinoline may be prepared by condensing *o*-aminoaldehydes or ketones with any aliphatic aldehyde or ketone containing the grouping —CH_2CO—:

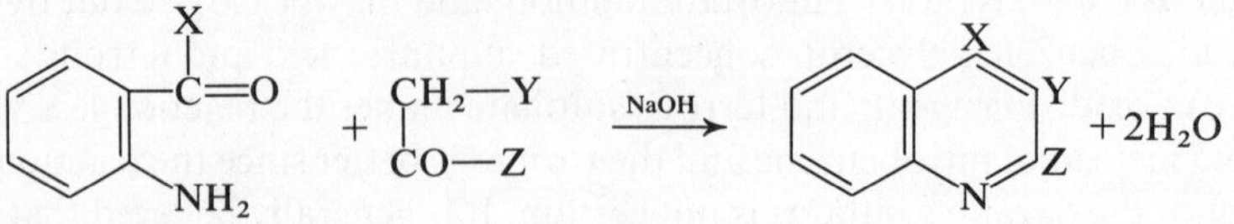

e.g., if X is CH_3, Y is $CO_2C_2H_5$ and Z is CH_3 (*i.e.*, the compound YCH_2COZ is ethyl acetoacetate), the product is 2,4-dimethylquinoline-3-carboxylic ester.

(iii) Condensation between *β*-ketonic esters and primary aromatic amines produces quinolines, the nature of the product depending on the conditions, *e.g.*, aniline and E.A.A.

(*a*) **Conrad–Limpach synthesis** (1887).

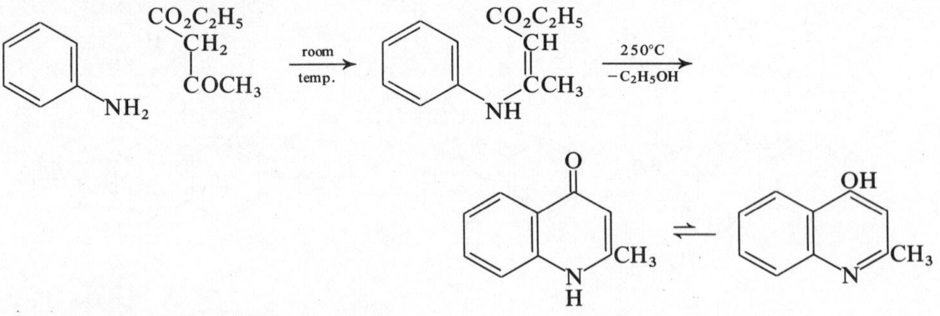

(*b*) **Knorr quinoline synthesis** (1886, 1888).

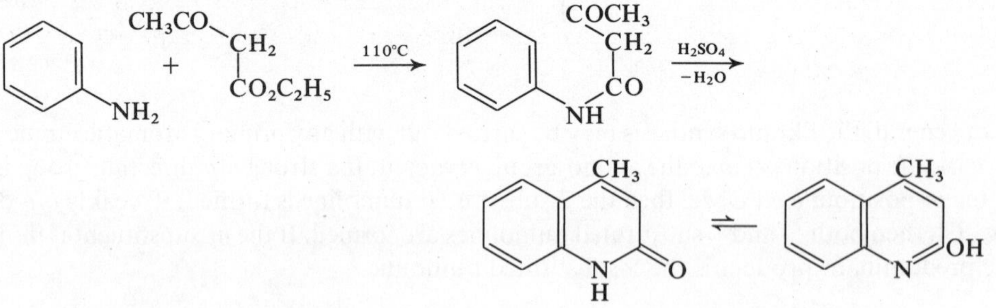

(iv) Homologues of quinoline may be prepared by the **Doebner–Miller synthesis** (1881); *e.g.*, aniline and paraldehyde heated with sulphuric acid form *quinaldine* (2-methylquinoline). Since, in general, the aldehyde may be any α,β-unsaturated aldehyde, the mechanism may be formulated (*cf.* the Skraup synthesis):

(i) $2CH_3CHO \xrightarrow{H_3O^+} CH_3CH=CHCHO$

(ii)

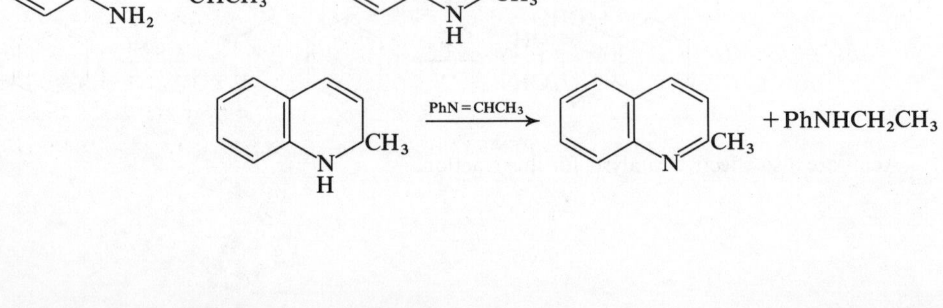

In this synthesis, no oxidising agent is added, and the final dehydrogenation is believed to occur by the action of the Schiff base produced from aniline and acetaldehyde.

(v) **Pfitzinger reaction** (1886). This is carried out by heating isatin with alkali in the presence of a ketone, e.g.,

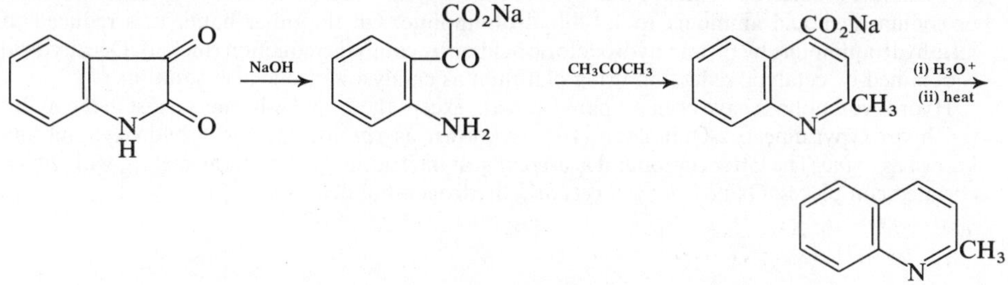

(vi) When methyl-lithium is added to indole in methylene dichloride solution, quinoline is produced by ring expansion via the addition of chloromethylene (Closs et al., 1961):

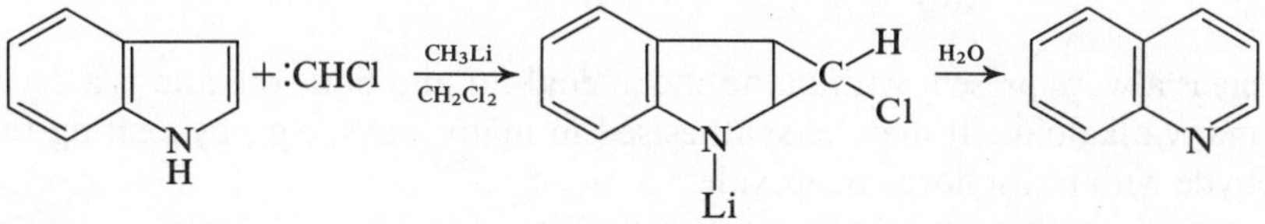

Quinoline, b.p. 238°C, is a tertiary base and forms salts with inorganic acids; with alkyl halides it forms *quinolinium salts* (quaternary salts), e.g., with methyl iodide it forms *quinoline methiodide* or 1-*methylquinolinium iodide* (I); with methyl sulphate it forms *quinoline methylmethosulphate* or 1-*methylquinolinium methyl sulphate* (II):

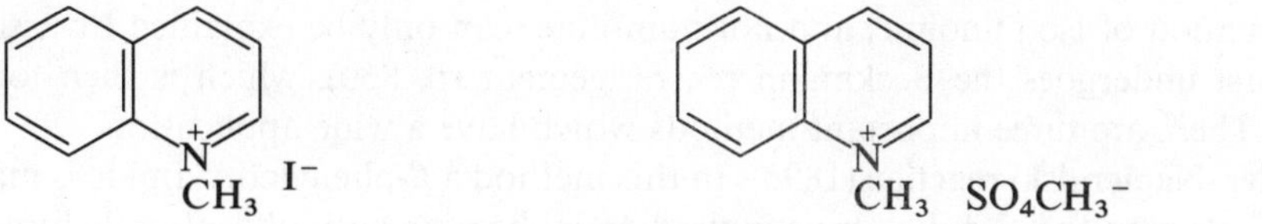

Quinoline is oxidised by permanganate to quinolinic acid (p. 855). The methyl group in the 2-position (quinaldine) and in the 4-position (lepidine) is very active and undergoes many condensation reactions (see cyanine dyes, p. 893).

Quinoline (R.E. 264·9 kJ mol^{-1}) resembles 1-nitronaphthalene in many ways. It is attacked preferentially at position 8 by electrophilic reagents, and positions 2 and 4 by nucleophilic reagents. It should also be remembered that in acid solution, it is the quinolinium cation that is present (cf. pyridine).

Vapour-phase bromination of quinoline produces 3-bromoquinoline, but at 500°C the product is 2-bromoquinoline (cf. pyridine). However, under conditions of high acidity, brominations occur in the 5- and 8-positions (de la Mare et al., 1958). Nitration with mixed acid gives a mixture of 5- and 8-nitroquinolines, but nitration with nitric acid-acetic anhydride forms the 3-nitro-compound. Sulphonation at 220°C gives predominantly the 8-sulphonic acid, together with a small amount of the 5-isomer. When the 8-sulphonic acid is heated to 300°C, it rearranges to the 6-acid.

Quinoline undergoes the Tschitschibabin reaction when heated with sodamide to form 2-aminoquinoline, and with n-butyl-lithium it forms the 2-butyl derivative.

Quinoline is oxidised to the 1-oxide by perbenzoic acid, and is reduced by lithium aluminium hydride or sodium in liquid ammonia to 1,2-dihydroquinoline. On the other hand, it is reduced to 1,2,3,4-tetrahydroquinoline by tin and hydrochloric acid or by catalytic reduction (nickel). Decahydroquinoline is obtained by catalytic reduction using platinum as catalyst in acetic acid solution.

Hydroxyquinolines have been prepared and all, except the 2- and 4-isomers, exist in the hydroxy-form (*cf.* hydroxypyridines). 2-Quinolone (III), is known as *carbostyril*, and 8-hydroxyquinoline (IV), is known as *oxine*. The latter compound is used in gravimetric analysis of many metals, with which it forms chelated complexes. Oxine is *intramolecularly* hydrogen-bonded.

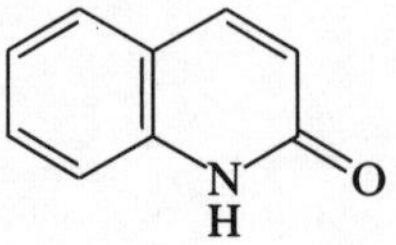

(III) 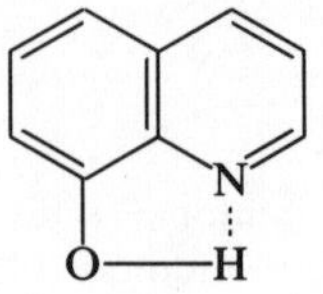   (IV)

Isoquinoline is always present with quinoline in coal-tar and bone oil, and is a decomposition product of many alkaloids. It may be synthesised in many ways, *e.g.*, by heating the oxime of cinnamaldehyde with phosphorus pentoxide:

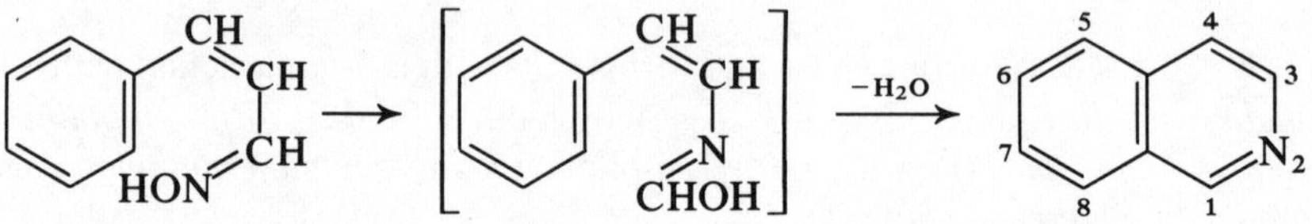

The formation of isoquinoline, and not quinoline, can only be explained by assuming that the oxime first undergoes the Beckmann rearrangement (p. 860), which is then followed by ring closure. There are three important methods which have a wide application.

1. Bischler–Napieralski reaction (1893). In this method a β-phenylethylamide is made to undergo cyclodehydration to a 3,4-dihydroisoquinoline by heating with phosphoryl chloride.

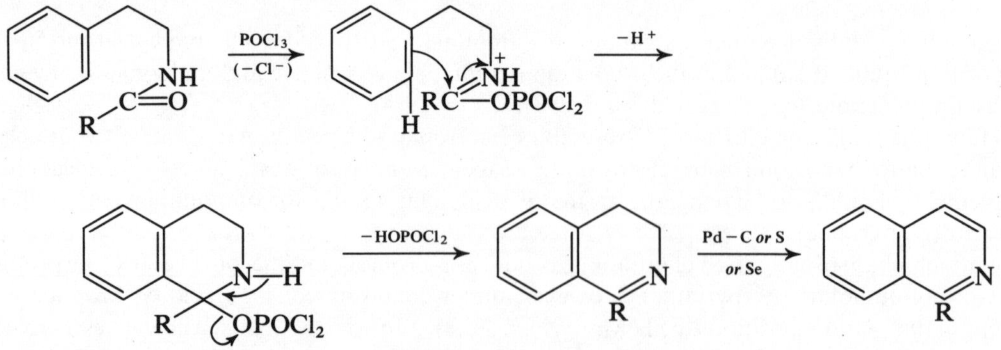

2. Pictet–Spengler reaction (1911). Condensation between a β-arylethylamine and an aldehyde in the presence of a large excess of hydrochloric acid at 100°C produces a 1,2,3,4-tetrahydroisoquinoline.

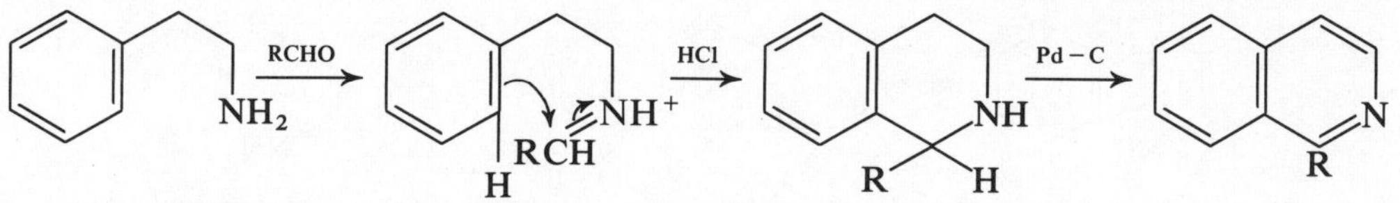

Pomeranz–Fritsch reaction (1893). This is carried out by condensing an aromatic aldehyde with an aminoacetal and then cyclising the product with sulphuric acid, *e.g.*,

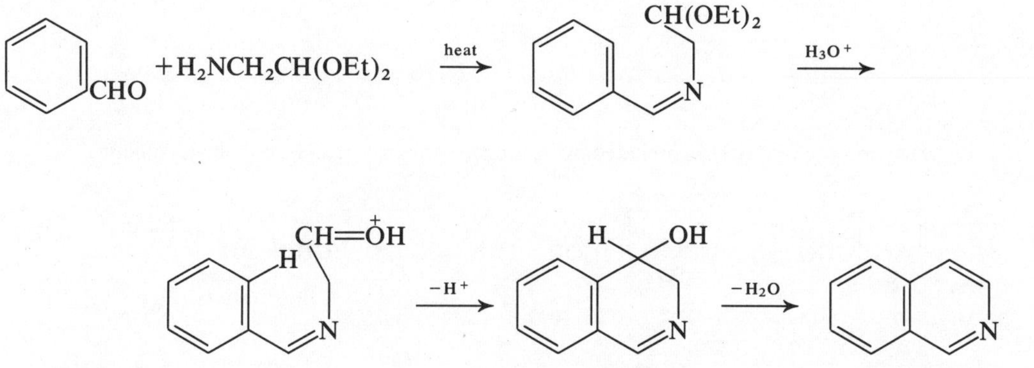

If an aromatic ketone is used instead of the aldehyde, then the product is a 1-substituted isoquinoline, *e.g.*,

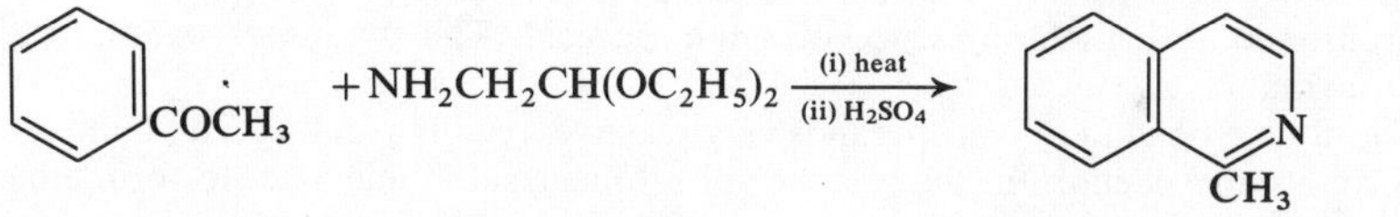

Isoquinoline, m.p. 26°C, is oxidised by permanganate to a mixture of phthalic and cinchomeronic acids (p. 855).

Isoquinoline resembles 2-nitronaphthalene in many of its properties. Electrophilic attack occurs at the 5- (predominantly) and 8-positions, *e.g.*, nitration and sulphonation (cation present) produce mainly the 5-derivative, together with a small amount of the 8-. On the other hand, bromination and mercuration (with mercuric acetate) give predominantly the 4-derivative. Nucleophilic substitution occurs at position 1, *e.g.*, 1-aminoisoquinoline is formed by the action of sodamide. Isoquinoline forms quaternary salts, and is oxidised by perbenzoic acid to the *N*-oxide. It is reduced by sodium in liquid ammonia to the 1,2-dihydro-compound, by tin and hydrochloric acid to the 1,2,3,4-tetrahydro-compound, and catalytic hydrogenation gives octahydroisoquinoline.

Acridine occurs in the anthracene fraction of coal-tar. It may be synthesised by passing the vapour of benzylaniline or *o*-aminodiphenylmethane through a red-hot tube (the outside numbering is to be used now):

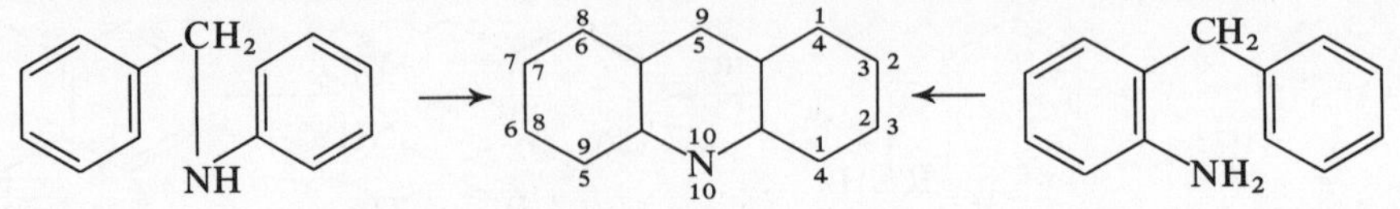

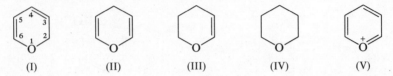

Acridine, m.p. 110°C, is a tertiary base, but weaker than quinoline. It is the parent substance of a number of dyes and antiseptics (p. 892).

Six-membered rings with one oxygen atom

The simplest six-membered rings with one oxygen are 2-*H*- (or α-)**pyran** (I), and 4-*H*- (or γ-)**pyran** (II).

(I)   (II)   (III)   (IV)   (V)

2,3-Dihydro-4-pyran, (III), is prepared by heating tetrahydrofurfuryl alcohol with alumina:

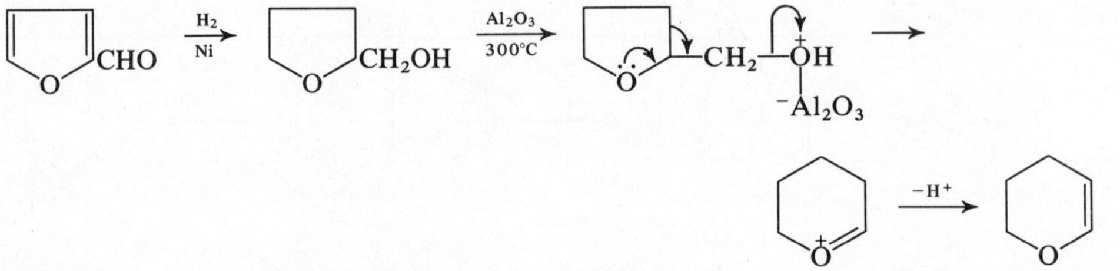

This, on catalytic reduction (Raney nickel), gives tetrahydropyran (IV). This compound is oxidised by nitric acid to glutaric acid, and is converted into pimelic acid by means of carbon monoxide and steam in the presence of a catalyst. Also, tetrahydropyran is readily converted into 1,5-dibromopentane by hydrobromic acid in the presence of sulphuric acid. The pyranose sugars are derivatives of tetrahydropyran.

The double bond in dihydropyran is very reactive, readily reacting with hydroxylic compounds such as alcohols and phenols, in the presence of *p*-toluenesulphonic acid, to form ethers. These ethers are stable to alkali, but are easily split by dilute acids. Thus, this reaction offers a very good means of protecting a hydroxyl group in reactions involving alkaline media:

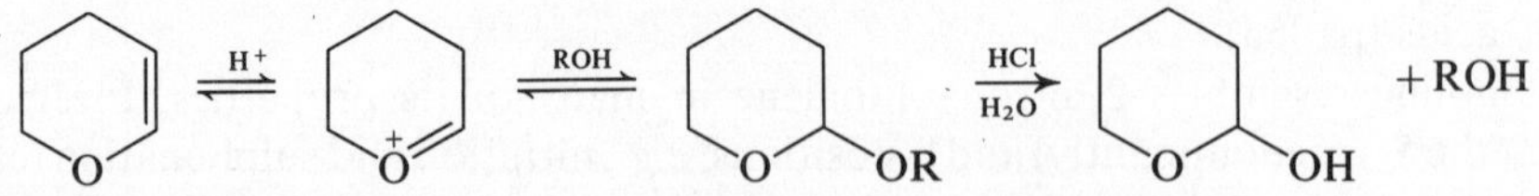

(V) is the aromatic **pyrylium cation**, and some salts have been isolated, *e.g.*, the perchlorate.

The corresponding keto-derivatives of the pyrans are known as **pyrones**:

2- or α-pyrone   4- or γ-pyrone

4-Pyrone may be prepared by heating *chelidonic acid* just above its m.p. (262°C). Chelidonic acid (a naturally occurring substance) may be prepared from acetone and ethyl oxalate.

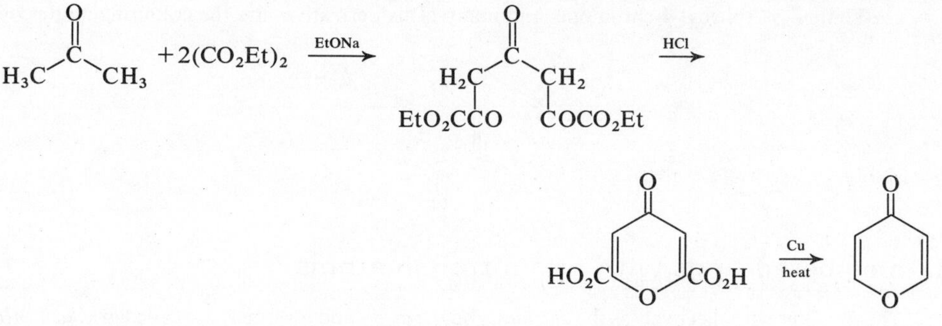

chelidonic acid

4-Pyrone, m.p. 32·5°C, is basic and shows some aromatic properties, and so is a resonance hybrid with contributing structures such as (III). 4-Pyrone does *not* form an oxime or phenylhydrazone. Similarly, the structure of 4-pyrone salts may be (VI) rather than (IV). According to Brown (1951), however, calculation of general charge distribution suggests (V) as the structure of the oxonium salts (this corresponds to (II) for 4-pyrone itself).

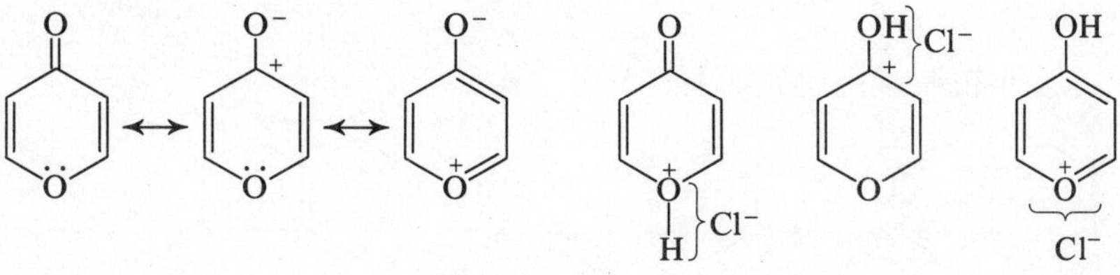

(I)	(II)	(III)	(IV)	(V)	(VI)

2,6-Dimethyl-4-pyrone is a very important derivative of 4-pyrone from the theoretical point of view, since its salt with hydrochloric acid was the first *oxonium* salt to be prepared (Collie and Tickle, 1890); the structure of this salt corresponds to that of (V) or (VI). The dimethyl-4-pyrone may be prepared from the copper salt of ethyl acetoacetate as follows:

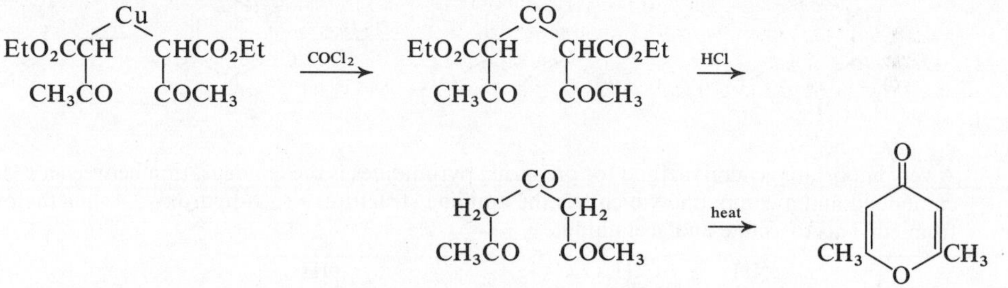

Condensed pyrone systems are important since many occur naturally:

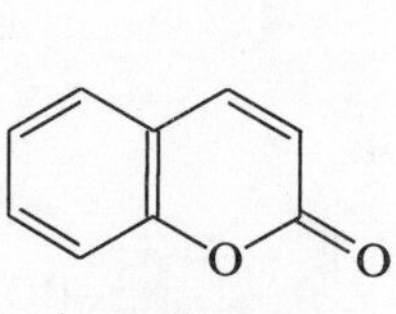

benzo-2-pyrone
or coumarin
(α-chromone)

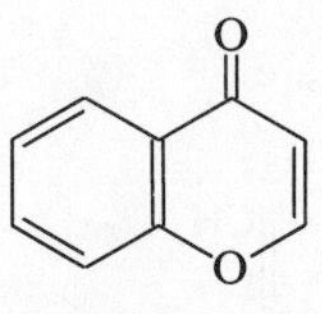

benzo-4-pyrone
(γ-chromone)

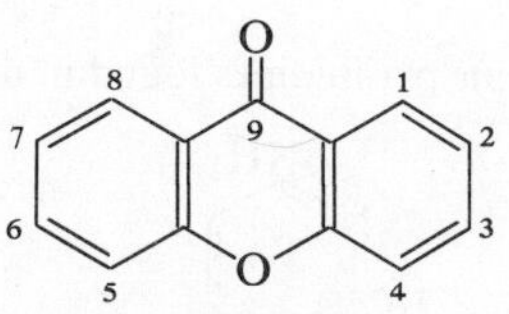

xanthone

α-Flavone is 2-phenyl-4-chromone, and many of its derivatives are the colouring matter of flowers:

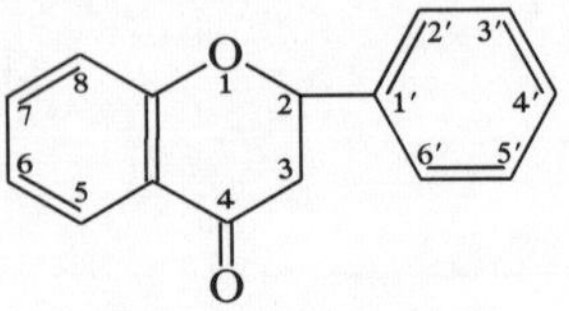

Six-membered rings with two nitrogen atoms

These are known collectively as the *diazines*, and the *o*-, *m*- and *p*-isomers are called *pyridazines* (*oiazines*), *pyrimidines* (*miazines*) and *pyrazines* (*piazines*), respectively (see also Vol. 2, Ch. 12).

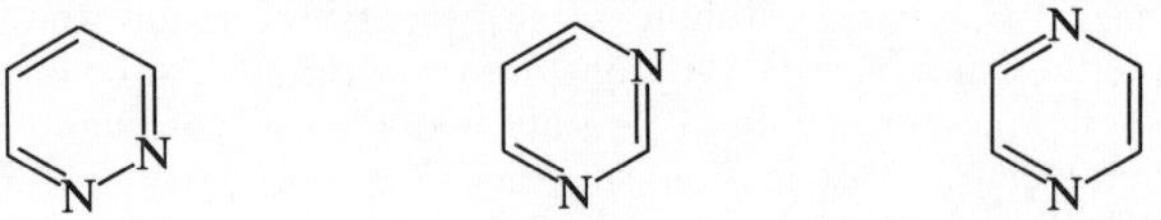

pyridazine
b.p. 208°C

pyrimidine
m.p. 22°C

pyrazine
m.p. 53°C

The corresponding *benzodiazines* are:

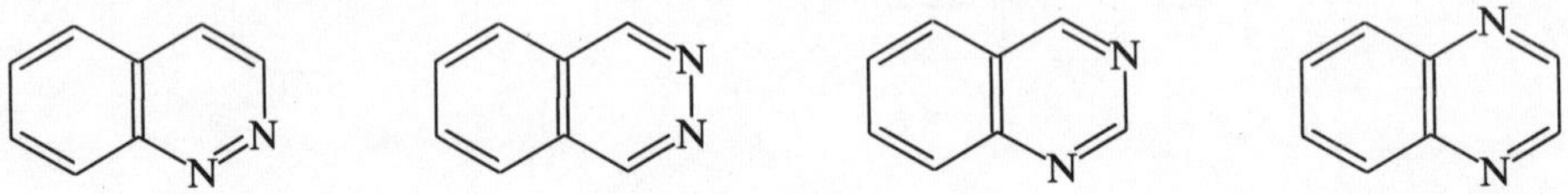

cinnoline   phthalazine   quinazoline   quinoxaline

The pyrimidines are a particularly important group of compounds, since the pyrimidine nucleus occurs in purines, nucleic acids and synthetic barbiturates. Pyrimidine may be prepared from barbituric acid as follows:

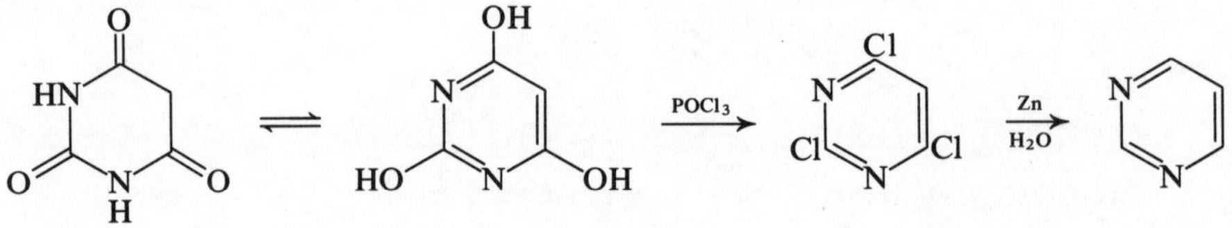

A very important general method for preparing pyrimidines is the condensation between a *β*-dicarbonyl compound and a compound which has the amidine structure, *e.g.*, 6-hydroxy-2,4-dimethylpyrimidine from ethyl acetoacetate and acetamidine.

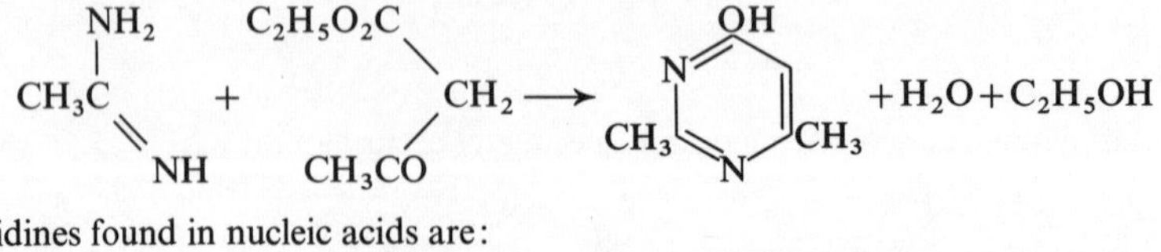

Some pyrimidines found in nucleic acids are:

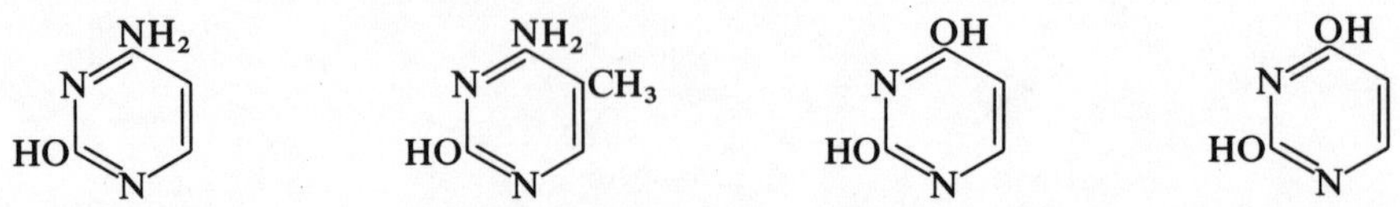

cytosine   5-methylcytosine   thymine   uracil

PROBLEMS

1. Pyrrole is more reactive than furan. Suggest a reason.

2. Complete the following equations:

(i) 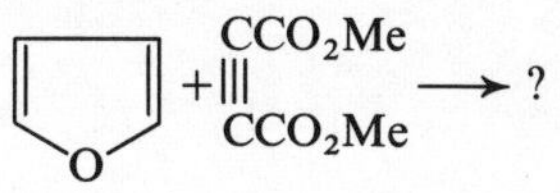 $+ \underset{CCO_2Me}{\overset{CCO_2Me}{\underset{\|}{|||}}} \longrightarrow$?

(ii) 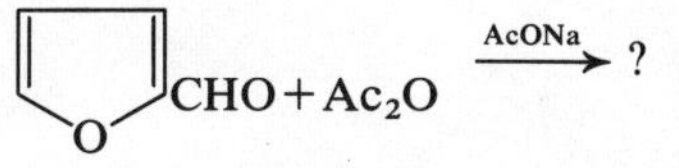 $CHO + Ac_2O \xrightarrow{AcONa}$?

(iii) 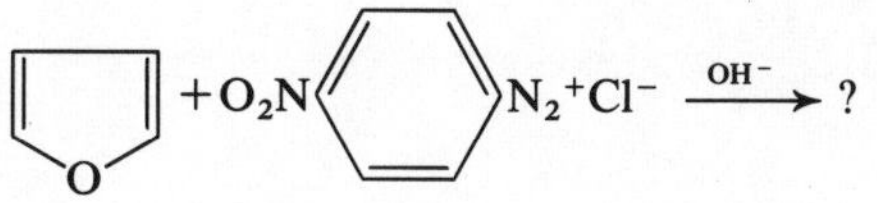 $+ O_2N \langle \rangle N_2^+Cl^- \xrightarrow{OH^-}$?

(iv) 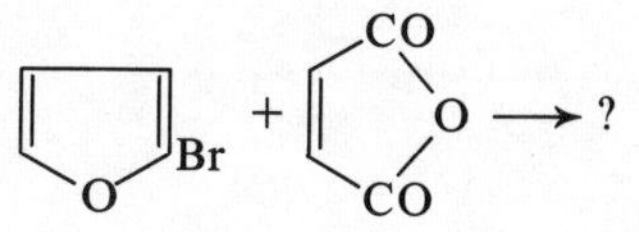 $\underset{Br}{}$ + $\longrightarrow$?

(v) 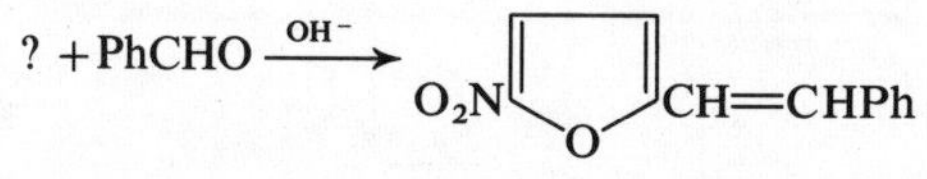 ? $+ PhCHO \xrightarrow{OH^-}$

(vi) 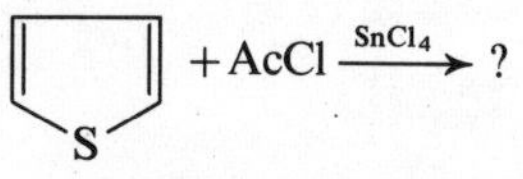 $+ AcCl \xrightarrow{SnCl_4}$?

(vii) 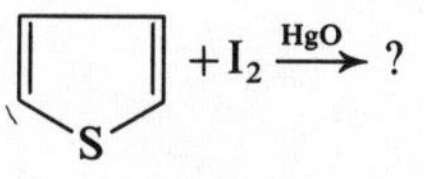 $+ I_2 \xrightarrow{HgO}$?

(viii) $+ Ph^-Na^+ \longrightarrow ? \xrightarrow[(ii) H^+]{(i) CO_2}$?

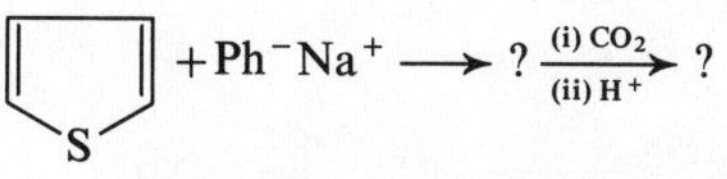

(ix) 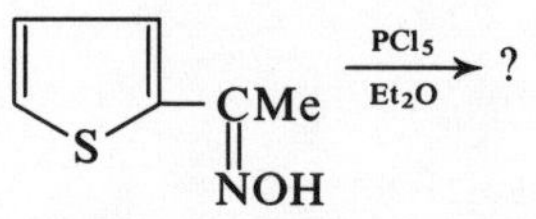 $\underset{\underset{NOH}{\|}}{CMe} \xrightarrow[Et_2O]{PCl_5}$?

(x) 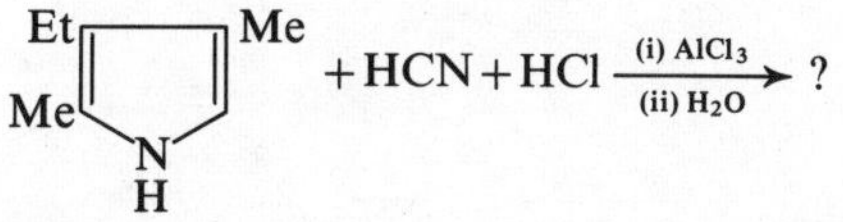 $+ HCN + HCl \xrightarrow[(ii) H_2O]{(i) AlCl_3}$?

(xi) 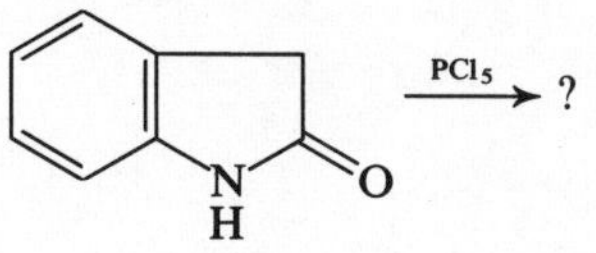 $\xrightarrow{PCl_5}$?

(xii) 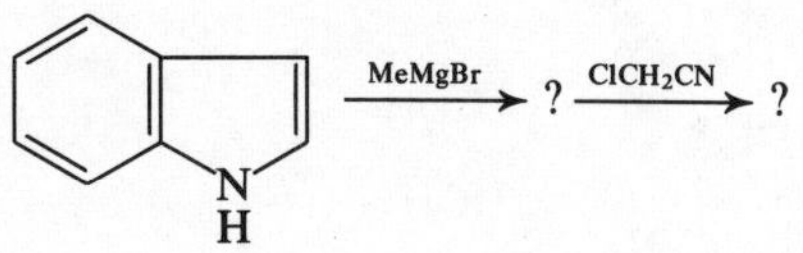 $\xrightarrow{MeMgBr} ? \xrightarrow{ClCH_2CN}$?

(xiii) $\xrightarrow[CH_2Cl_2]{MeLi}$?

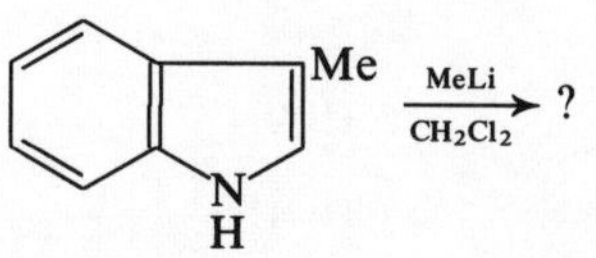

(xiv) 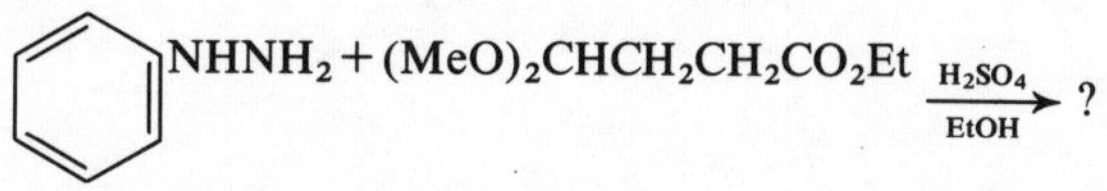 $NHNH_2 + (MeO)_2CHCH_2CH_2CO_2Et \xrightarrow[EtOH]{H_2SO_4}$?

3. Suggest a mechanism for the reaction

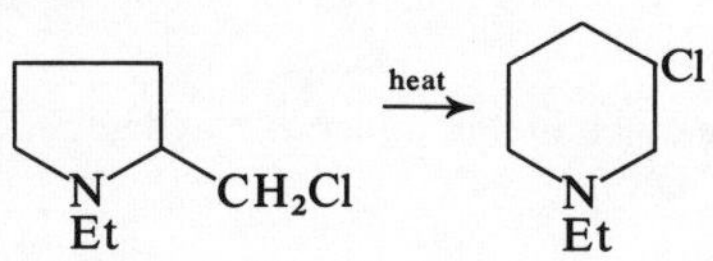

4. Suggest a mechanism for the reaction

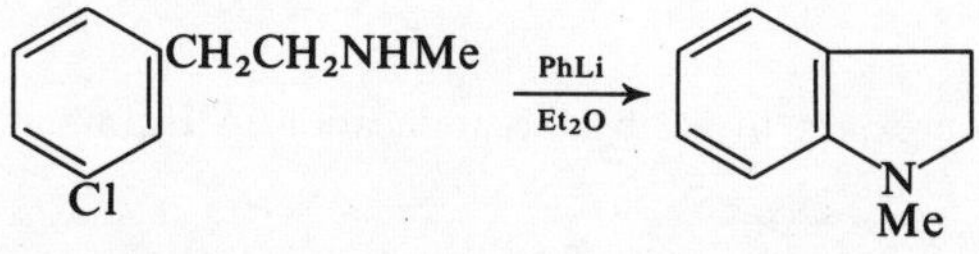

5. The nitration of furan with acetyl nitrate proceeds via a 2,5-addition intermediate with the reagent. Formulate the complete reaction.

6. Pyrrole is much more acidic than s-alkylamines. Suggest a reason.

7. no action $\xleftarrow{\text{furan}}$ CH$_2$O + Me$_2$NH $\xrightarrow{\text{2-Me-furan}}$ A

$\xrightarrow{\text{pyrrole}}$ B

What are **A** and **B**? Explain.

8. Complete the equations:

(i) 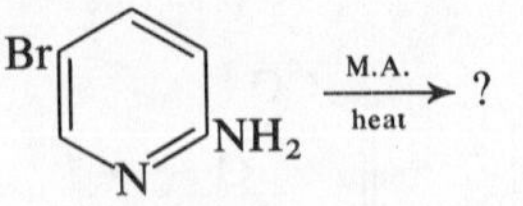 $\xrightarrow[\text{heat}]{\text{M.A.}}$?   (ii) 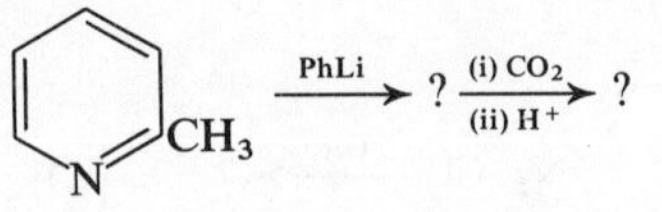 $\xrightarrow{\text{PhLi}}$? $\xrightarrow[\text{(ii) H}^+]{\text{(i) CO}_2}$?

(iii) 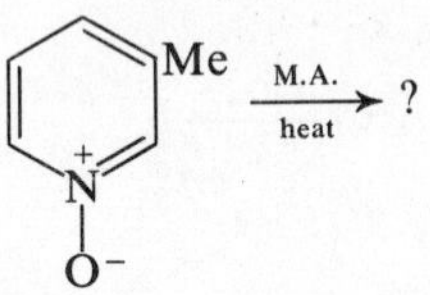 $\xrightarrow[\text{heat}]{\text{M.A.}}$?   (iv) 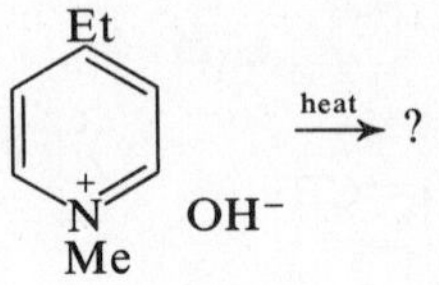 $\xrightarrow{\text{heat}}$?

(v) 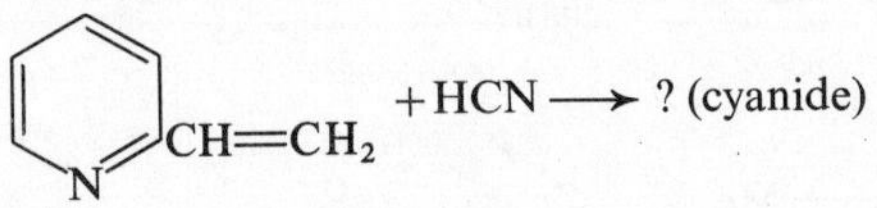 + HCN $\longrightarrow$? (cyanide)   (vi) 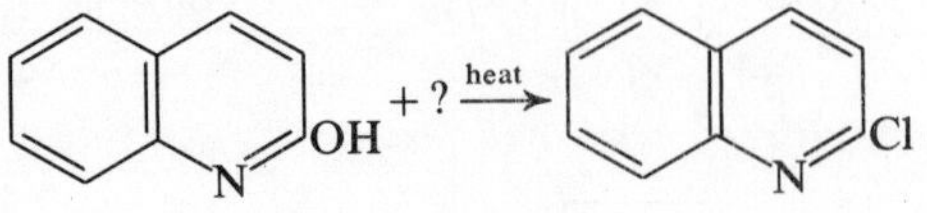 + ? $\xrightarrow{\text{heat}}$

(vii) 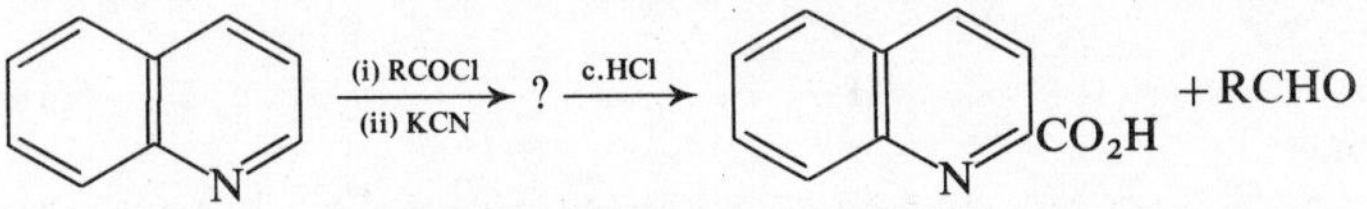 $\xrightarrow[\text{(ii) KCN}]{\text{(i) RCOCl}}$? $\xrightarrow{\text{c.HCl}}$ + RCHO

(viii) + CHOH(CH$_2$OH)$_2$ $\xrightarrow[\text{As}_2\text{O}_5]{\text{H}_2\text{SO}_4}$?

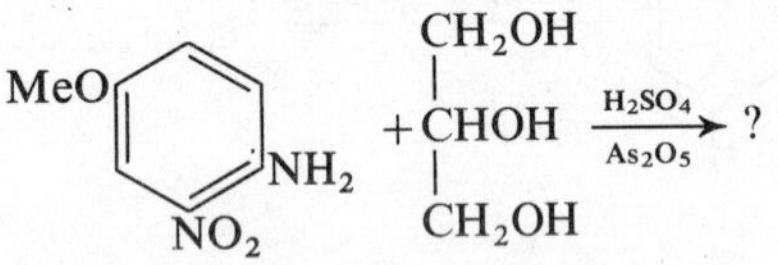

(ix) 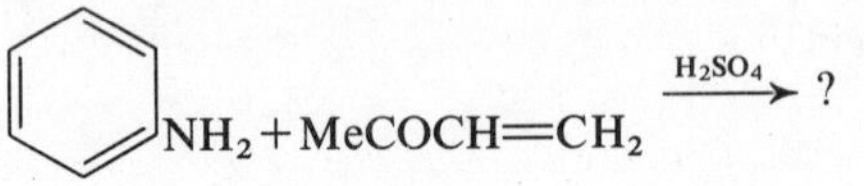 + MeCOCH=CH$_2$ $\xrightarrow{\text{H}_2\text{SO}_4}$?   (x) 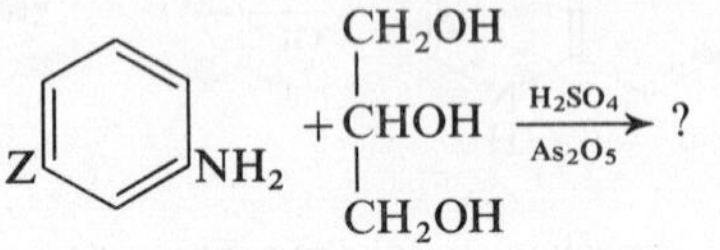 + CHOH(CH$_2$OH)$_2$ $\xrightarrow[\text{As}_2\text{O}_5]{\text{H}_2\text{SO}_4}$?

(Z = Cl, NO$_2$, OMe)

9. Discuss the behaviour of the three picolines in aqueous NaOH containing D$_2$O.
10. Suggest a mechanism for the reaction

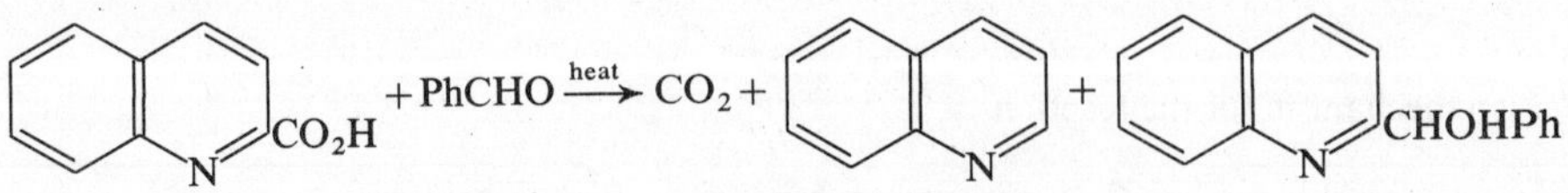

+ PhCHO $\xrightarrow{\text{heat}}$ CO$_2$ + +

11. What products would you expect to get by the application of Hofmann's exhaustive methylation to:

(i) 3-methylpyrrolidine; (ii) 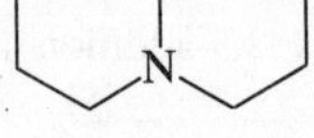 ; (iii) morpholine?

12. Complete:

(i) $2CH_2\!=\!CHCO_2Et + MeNH_2 \longrightarrow ? \xrightarrow{EtONa} ? \xrightarrow[heat]{H_3O^+} ?$

(ii) $\xrightarrow{EtONa} ? \xrightarrow[heat]{H_3O^+} ? \xrightarrow[redn.]{Wolff-Kishner} ?$

13. Suggest a synthesis for each of the following:
(i) 4H-imidazo[4,5-d]thiazole; (ii) 7H-pyrazino[2,3-c]carbazole; (iii) 4H-1,3-oxathiolo[5,4-b]pyrrole:
(iii) 3-aminopyridine from β-picoline; (iv) 2,6-dimethylpyridine from E.A.A.; (v) 4-nitroquinoline from
quinoline; (vi) 6-methoxyisoquinoline from p-methoxybenzaldehyde; (vii) 5-nitroquinoline from benzene;
(viii) 1-methylisoquinoline from benzene.
14. Arrange the following in order of increasing basicity and give your reasons: pyridine, 4-amino-, 4-methyl-,
and 4-cyano-pyridine.
15. Name the following:

16. Write out the structures (with numbering) of:
(i) 4H-imidazo[4,5-d]thiazole; (ii) 7H-pyrazino[2,3-c]carbazole; (iii) 4H-1,3-oxathiolo[5,4-b]pyrrole;
(iv) thiazolo[5,4-b]pyridine.
17. Pyridine methiodide readily reacts with aqueous KOH, and the product **A**, C_6H_9NO, on oxidation with
potassium ferricyanide, gives **B**, C_6H_7NO. What are **A** and **B**? Formulate the reaction.
18. The Skraup synthesis on 2-NH₂-1-Me-naphthalene gives the heterocycle **A**, $C_{14}H_{11}N$. The same synthesis
on 2-NH₂-1-NO₂-naphthalene gives the heterocycle **B**, $C_{13}H_9N$. What are **A** and **B**?

REFERENCES

PATTERSON et al., The Ring Index, Amer. Chem. Soc. (1960, 2nd edn.). Supplements 1 (1963) and 2 (1964).
Handbook for Chemical Society Authors, Special Publication No. 14 (1960).
ACHESON, An Introduction to the Chemistry of Heterocyclic Compounds, Interscience (1967, 2nd edn.).
BADGER, Chemistry of Heterocyclic Compounds, Academic Press (1961).
FINAR, Organic Chemistry, Vol. 2. Longmans (1968, 4th edn.). Ch. 12. 'Heterocyclic Compounds Containing
    Two or more Hetero-atoms.'
Organic Reactions, Wiley, Vol. VI (1951), Ch. 2, 3, 4. 'The Synthesis of Isoquinolines'; Vol. VII (1953). Ch. 2.
    'The Skraup Synthesis of Quinolines.' Ch. 4. 'The von Braun Cyanogen Bromide Reaction.'
SIDGWICK, The Organic Chemistry of Nitrogen, Oxford Press (1966, 3rd edn. by Millar and Springall).
KATRITZKY and LAGOWSKI, The Principles of Heterocyclic Chemistry, Methuen (1967).
PALMER, The Structure and Reactions of Heterocyclic Compounds, Arnold (1967).
ALBERT, Heterocyclic Chemistry, Athlone Press (1968, 2nd edn.).
PAQUETTE, Principles of Modern Heterocyclic Chemistry, Benjamin (1968).
ELDERFIELD (ed.), Heterocyclic Compounds, Wiley (1950–   ).
KATRITZKY and BOLTON (eds.), Advances in Heterocyclic Chemistry, Academic Press (1963–   ).
WEISBERGER (ed.), The Chemistry of Heterocyclic Compounds, Interscience (1950–   ).
OGATA et al., 'Mechanism of the Doebner–Miller Lepidine Synthesis', J. Chem. Soc. (B), 1969, 805.
JANSSEN (ed.), Organosulphur Chemistry, Interscience (1967).
SALMOND, 'Valence-shell Expansion in Sulphur Heterocycles', Quart. Rev., 1968, 22, 253.

31 Dyes and photochemistry

Colour. When white light (750–400 nm) falls on a substance, the light may be totally reflected or totally absorbed. In the former case, the substance appears white; in the latter, black. If a certain proportion of the light is absorbed and the rest reflected, the substance has the colour of the *reflected* light. If only a *single* band is absorbed, the substance has the *complementary* colour (of the absorbed band).

Table 31.1

nm (mμ)	Colour absorbed	Visible (complementary) colour
400–435	violet	yellow-green
435–480	blue	yellow
480–490	green-blue	orange
490–500	blue-green	red
500–560	green	purple
560–580	yellow-green	violet
580–595	yellow	blue
595–605	orange	green-blue
605–750	red	blue-green

If a substance absorbs all visible light except one band, which it reflects, the substance will have the colour of that reflected band. Thus, a substance can appear blue because it absorbs the yellow portion of the spectrum only; or because it absorbs *all the visible spectrum except blue*. The shades, however, will be different. Apparently no dye gives a pure shade, *i.e.*, does not reflect only one band of wave-lengths; *e.g.*, malachite green reflects green light, but also, to a small extent, red, blue and violet.

Many substances which appear to be colourless nevertheless have absorption spectra, but in these cases, absorption takes place in the infra-red or ultra-violet, and not in the region of the visible spectrum.

Relation between colour and constitution

The early theories of colour gave rise to a terminology which, with some modifications, is used in ultra-violet and visible spectroscopy (see p. 875). Witt (1876) was the first to show that colour usually appeared in a compound when it contained a group with multiple bonds. These groups were called **chromophores**, *e.g.*, NO_2, NO, N$=$N, quinonoid structure, etc. Witt also observed that certain groups which, although not chromophores, deepen colour when introduced into a coloured molecule. These groups were called **auxochromes**, *e.g.*, OH, NH_2, Cl, CO_2H. Other terms introduced later in connection with dyes are **bathochromic** and **hypsochromic groups**. These are, respectively, groups that bring about deepening and lightening of colour. Deepening of colour in *dye chemistry* is usually taken to mean the following changes: yellow → orange → red → purple → blue → green → black. This is effectively the order of the visible *complementary* colour of the colour *absorbed* (Table 31.1). Hence, we may say that the colour of one compound is deeper than that of another if the wavelength of maximum absorption of the former is longer than that of the latter. Because visible colour is the complementary colour of the absorbed band, bathochromic groups are said to have a red shift and hypsochromic groups a blue shift.

There are two approaches to the modern theories of colour, the *valence bond* and the *molecular orbital approach*. Both have some aspects in common, and these will be discussed first. Light consists of electromagnetic waves which are composed of a magnetic and an electric component, both oscillating in planes perpendicular to each other and perpendicular to the direction of propagation. It is only the electric component which is usually involved in interaction with matter, and this interaction is conveniently described in terms of light behaving as a train of discrete energy packets called *photons* or *quanta*. When light is absorbed by a molecule, the molecule undergoes transition from a state of lower energy (E_0) to a state of higher energy (E_1), and the *change* in energy, ΔE, is given by: $\Delta E = (E_1 - E_0)$. In a monatomic molecule, the energy absorbed can only be used to raise the *electronic* energy level from the lower (*ground state*) to a higher (an *excited state*) level. Since electrons occupy *definite* energy levels, changes from one level to another occur in discrete packets which are therefore not continuous, *i.e.*, the energy levels are *quantised*. Hence, in a monatomic molecule, absorption can occur only if the photon has an energy which is equal to the energy difference $E_1 - E_0$. Since the energy of a photon, E, is given by $E = h\nu$, it follows that

$$\Delta E = E_1 - E_0 = h\nu = hc/\lambda$$

where h is Planck's constant and c is the velocity of light. By insertion of the appropriate values, the value of ΔE is given by

$$\Delta E = 11 \cdot 97 \times (10^4/\lambda) \text{ kJ mol}^{-1} (2 \cdot 86 \times (10^4/\lambda) \text{ kcal mol}^{-1})$$

where λ is in nm (mμ). The value of ΔE for any particular wavelength is known as an *einstein* (for that given wavelength).

In polyatomic molecules, the total energy state is the sum of the electronic, vibrational, and rotational components. Since electronic transitions are associated with large amounts of energy relative to vibrational and rotational transitions (which are also quantised), ΔE is large for the former, *i.e.*, the frequency of the photon required to bring about electronic transitions is high (and consequently the wavelength is short). Thus, electronic transitions occur in the visible and ultra-violet parts of the spectrum, whereas vibrational and rotational transitions occur respectively in the near and far infra-red (see also p. 8).

If, after absorption of light, only electronic transitions occurred, each transition would be associated with some particular line. However, since an electronic level is associated with a large number of vibrational and rotational levels, a transition from one electronic level to another may be accompanied by vibrational and rotational transitions (see also below). This results in the absorption of a range of photons, and consequently the spectrum consists of a family of lines which are usually so closely spaced that they merge (for low resolution) into a broad band.

Let us now consider this problem in more detail. Figure 31.1 (b) represents (diagrammatically) two electronic energy states of a diatomic molecule, E_0 being the ground state and E_1 the *first* excited state. Excitation from the ground state (bonding M.O.) to the first excited state (anti-bonding M.O.) can occur in two possible ways: (i) with retention of electron spin, *i.e.*, the electron spins are still anti-parallel; this is represented as an $S_0 \rightarrow S_1$ transition. (ii) with spin inversion, *i.e.*, the electron spins are now parallel; this is represented as an $S_0 \rightarrow T_1$ transition.

In the ground state of a diatomic molecule, the M.O. is a σ-bonding type, and when excited, the M.O. is the σ^*-anti-bonding type (see Fig. 2.8). Their energy levels are represented diagrammatically in Fig. 31.1 (a). In the ground state, normal covalent compounds have all their electrons paired, *i.e.*, the molecule is in the *singlet state*, S_0. When the molecule is excited, the promoted electron can still be paired to give the first excited *singlet* state, S_1, or can be unpaired (diradical) to give the first excited *triplet* state, T_1 (see Fig. 31.1 (a)). These two states are said to have different *multiplicities* (see p. 118), and since the energy levels for S_1 and T_1 are both obtained from the relationship $\Delta E = E_1 - E_0$, they have the *same* levels independent of spin assignment. However, according to **Hund's rule of multiplicity**, *a system of highest multiplicity has the lowest energy*. Hence, the energy of a triplet state (multiplicity of 3) is lower than that of its *corresponding* singlet state (multiplicity of 1); *i.e.*, $T_1 < S_1$, $T_2 < S_2$, etc. By taking into account the different spin relationships, it is possible to develop the individual orbital states shown in Fig. 31.1 (b) but, in practice, the **Jablonski diagram**, Fig. 31.1 (c), which is a simplified version of Fig. 31.1 (b), is generally used.

For polyatomic molecules, the orbital level diagram (type Fig. 31.1 (a)) is basically the same, but naturally contains far more levels (see Fig. 31.3).

Another problem connected with electronic transitions is that theoretical considerations have shown that some of these transitions are *allowed*, *i.e.*, have high probability, or are *forbidden*, *i.e.*, have low probability. The essential condition for an electronic transition to take place is that the *transition dipole* (which is related to the actual dipole moment of the molecule) is not zero. Thus, there are *selection rules*, which are the conditions for which the transition dipole is not zero. One transition that would lead to a zero transition dipole is that in which there is electron spin inversion on excitation. This is the basis of the **selection rule of multiplicity**: *no change in spin multiplicity occurs during a transition*. Thus, singlet-singlet and triplet-triplet transitions are allowed, but singlet-triplet and triplet-singlet transitions are forbidden. This rule, however, is based on the assumption that a molecule has a fixed shape which remains unchanged on excitation. In practice, there are always some vibrations in a polyatomic molecule which periodically change the shape of that molecule, *e.g.*, bending vibrations. This enables a forbidden transition to become allowed (see also below).

It should also be noted that the molar absorptivity (ε_{max}) is a measure of the probability of a transition taking place, the higher the probability of the transition, the larger is the ε_{max}.

We now return to the problem of what happens to the molecule when it is in an excited state. In general, a molecule in the ground state will also be in the lowest vibrational state (rotational states, which lie between vibrational states, have been omitted in Figs. 31.1 (b) and (c)). Now suppose light is absorbed. If the energy of absorption is above that required for

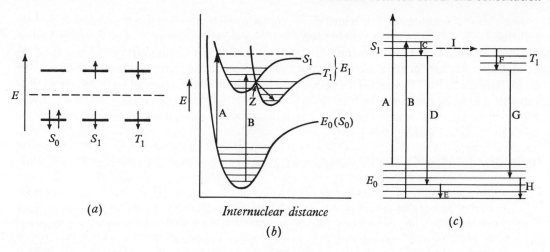

(a)

Internuclear distance

(b)

(c)

Fig. 31.1

dissociation ($>E_1$) for a diatomic molecule, the molecule splits into free atoms (path A). If the absorption of energy raises the molecule to the first excited state, the transition is from ground state to the first excited singlet state, S_1 (path B), since $S_0 \to T_1$ is forbidden. It should also be noted that when the electron is promoted (from the ground state molecule), it moves from a given vibrational level (usually the lowest one) to some (allowable) vibrational level in S_1. When excitation occurs, it does so extremely rapidly (10^{-13} s) and consequently the nuclei do not have time to change their positions **(Franck–Condon principle)**, *i.e.*, although the vibrational level has also been raised, the internuclear distance remains unchanged *immediately* after the excitation. Thus, the excited molecule is formed in a 'compressed' condition and *then* begins to vibrate normally at this higher vibrational level.

In the excited state, the molecule can lose its excess of energy (*i.e.*, $S_1 \to E_0$) in a number of ways. If the excitation raises the promoted electron to a vibrational level in S_1 which is *above* the lowest vibrational level in S_1, the electron rapidly drops to this lowest level, the excess of vibrational energy being lost by collisions with other molecules (path C). Loss of energy by this process of **vibrational relaxation** occurs so rapidly (10^{-12}–10^{-11} s) that the vibrationally relaxed excited state is reached before any other loss process can operate. This state can now lose energy by further collisions to reach the original ground state (S_0; paths D and E). When the molecule drops from S_1 to the ground state, the vibrational level in the latter is higher (in this level) than that in the former. In this series of processes, *all* the absorbed energy is lost through collisions, and the molecule is said to return from its excited state to the ground state by a **non-radiative process**, the 'lost' energy appearing as heat. This type of return is called **internal conversion**. On the other hand, the vibrationally relaxed excited state can lose energy by emission of radiation *before* collisions occur (path D). Loss of energy by this **radiative process**, however, results in the wavelength of the emitted light being longer than that of the absorbed light (ΔE for the former is smaller than that of the latter). This type of transition is known as **fluorescence** (emission between two states of the *same* multiplicity). When fluorescence occurs, it does so about 10^{-8} s after the light source is shut off (this is the life-time of a singlet state). Furthermore, had the original excitation been to the S_2 state, this decays extremely rapidly to the S_1 state, which then returns to S_0 by either of the two processes described.

There is still another way in which $S_1 \to S_0$. Although $S \to T$ and $T \to S$ are forbidden transitions, $S_1 \to T_1$ can occur. The reason for this is that usually the S_1 and T_1 energy curves intersect (point Z in Fig. 31.1 (*b*)), and at this point the nuclear configuration and energy of the S_1 and T_1 states are identical. At this instant the molecule may cross, *without radiation*, from S_1 to T_1 (path I; Fig. 31.1 (*c*)). This process is called **intersystem crossing** (*spin inversion*). The transition $T_1 \to S_0$ can now take place by a non-radiative (internal conversion; paths F, G, and H), or by a radiative process (path G). The latter process is called **phosphorescence** (emission between states of *different* multiplicity), and the wavelength of the emission is longer than that which occurs in fluorescence. Since $T_1 \to S_0$ is a transition of low probability, the T_1 state has a life-time of 10^{-5}–10^{-3} s, and so phosphorescence lasts longer than fluorescence.

> **Chemiluminescence** is the emission of light from chemical reactions at ordinary temperatures in the absence of flame. This light is emitted from an excited molecule which has been produced in the chemical reaction, and it is believed that most chemiluminescent reactions involve the singlet excited state, which decays to the ground state by fluorescing. The energy required for the excitation is obtained as a result of an *exothermic* reaction. A common type of reaction that exhibits chemiluminescence is the oxidation by oxygen of many organic compounds in solution.

Valence bond approach to colour

An extremely important difference between V.B. and M.O. theories is that in the former, electrons are dealt with in *pairs*, whereas in the latter they can be dealt with *singly*.

According to the V.B. theory, the electron pairs of a molecule in the ground state are in a state of oscillation, and when placed in the path of a beam of light, the amplitude of the oscillation is increased when a photon of the appropriate energy is absorbed. The molecule is now in an excited state, and the wavelength of the photon absorbed depends on the energy difference between the excited and ground states, the smaller the difference, the longer being the wavelength.

Let us first consider the case of ethylene. This, from the point of view of the V.B. method, may be regarded as a resonance hybrid of (I) and (II).

$$CH_2{=}CH_2 \longleftrightarrow \overset{+}{C}H_2{-}\overset{-}{C}H_2$$
$$\text{(I)} \qquad\qquad \text{(II)}$$

Also, to a first approximation, the ground state is represented predominantly by (I) and the excited state by (II). The energy difference between these two is very large and so the energy of the photon required to excite ethylene is very high, *i.e.*, its wavelength is very short. This is the general situation for *short* molecules, and it has been shown that: (i) Resonance among *charged* structures *lowers* the energies of both ground and excited states; (ii) Charged structures contribute more to the excited state than to the ground state; (iii) The larger the number of electrons involved in resonance, the smaller is the energy difference between the ground and excited states. Hence, the more extended the conjugation in a molecule and the greater the contribution of charged structures, the longer is the wavelength of the photon required to excite the molecule, *e.g.*,

$$CH_2{=}CH_2 \qquad\qquad CH_2{=}CH{-}CH{=}CH_2 \qquad\qquad CH_2{=}CH{-}CH{=}CH{-}CH{=}CH_2$$

| λ_{nm} | 175 | 217 | 258 |

From what has been said, it can be expected that the presence of $+I/+R$ and/or $-I/-R$ groups at the *ends* of a conjugated system will both *extend* the conjugation and *increase* the contributions of charged structures to the resonance hybrid. Therefore, in these circumstances, the wavelengths should become progressively longer, *e.g.*,

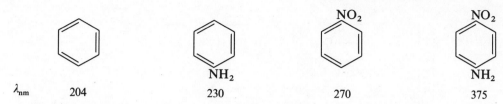

λ_{nm} 204   230   270   375

Benzene itself is a chromophore; it is considered to be a resonance hybrid of two Kekulé structures and a small amount of charged canonical structures (dipolar ions).

Steric effects and colour. As we have seen, resonance in a conjugated system is a maximum when the system is completely (or almost completely) planar. Hence, if resonance is restricted or inhibited, *i.e.*, steric inhibition of resonance is operating, depth of colour will diminish or the compound may even become colourless. It should be remembered that the colour of a compound is that of the complementary colour (this excludes reflection, etc.; see p. 868; also see later).

Molecular orbital approach to colour

In M.O. theory, an atom or molecule is excited when *one* electron is transferred from a *bonding* to an *anti-bonding* orbital. Electronic transitions, however, can occur in different ways. A transition in which a bonding σ-electron is excited to an anti-bonding σ-orbital is referred to as a $\sigma \longrightarrow \sigma^*$ transition. In the same way, $\pi \longrightarrow \pi^*$ represents the transition of a bonding π-electron to an anti-bonding π-orbital. An $n \longrightarrow \pi^*$ transition represents the transition of *one* electron of a *lone pair*, *i.e.*, a non-bonding pair

σ^*	Anti-Bonding
π^*	Anti-Bonding
n	{Lone Pair; Non-Bonding
π	Bonding
σ	Bonding

E

Fig. 31.2

of electrons, to an anti-bonding π-orbital. This type of transition occurs with compounds containing double bonds involving hetero-atoms, *e.g.*, $\diagdown C{=}O$, $\diagdown C{=}S$, $\diagdown C{=}N{-}$, etc., and may be represented as follows:

$$\diagdown\!\!C \overset{\cdot}{\underset{\cdot}{-}} \overset{\cdot\cdot}{O} \longleftarrow \diagdown\!\!C{=}\overset{\cdot\cdot}{O} \longrightarrow \diagdown\!\!C \overset{\cdot}{=} \overset{\cdot}{O}$$
$$(\pi \to \pi^*) \qquad\qquad\qquad (n \to \pi^*)$$

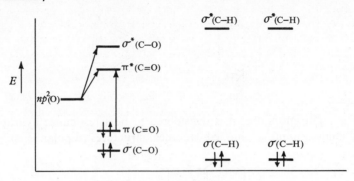

Fig. 31.3

Figure 31.2 shows diagrammatically the general pattern of the energy levels, and it can be seen that the transitions are brought about by the absorption of different amounts of energy. Of the large number of possibilities, only the following transitions are allowed, and their general order of energy difference is:

$$\sigma \longrightarrow \sigma^* > n \longrightarrow \sigma^* > \pi \longrightarrow \pi^* > n \longrightarrow \pi^*$$

The principles discussed so far are illustrated in Fig. 31.3, which represents the relative energies of the various types of orbitals for formaldehyde. This compound shows three bands: 285 ($n \longrightarrow \pi^*$), 190 ($\pi \longrightarrow \pi^*$), and 160 nm (this appears to be $n \longrightarrow \sigma^*$).

Now let us consider some examples. In alkanes, only $\sigma \longrightarrow \sigma^*$ transitions (C—H and C—C) are possible. The energy of a photon required to effect this transition is very high, e.g., ethane has a λ_{max} 135 nm (C—C), and since this is in the far ultra-violet region, ethane is colourless. In ethylene, two types of electronic transitions are now possible, $\sigma \longrightarrow \sigma^*$ and $\pi \longrightarrow \pi^*$, e.g., ethylene has λ_{max} 175 nm and ε (molar absorptivity), 5 000. This absorption band is due to a $\pi \longrightarrow \pi^*$ transition, and again, since it is in the far ultra-violet region, ethylene is colourless. Butadiene has λ_{max} 217 nm (ε, 21 000); this is also a $\pi \longrightarrow \pi^*$ transition. Thus, conjugation has shifted the absorption band to a longer wavelength (but still not into the visible region), and the intensity has greatly increased. These results can be explained as follows. If we refer to Fig. 4.3, we see that the two bonding orbitals (b) and (c) represent the molecule in the ground state; (d) and (e) represent excited states, that of (e) being the higher state. Excitation of butadiene can therefore cause transfer of one electron from (b) or (c) to (d) or (e). If all of these occurred, *four* absorption bands would be produced. However, since the energy difference between the highest occupied π-orbital level and lowest unoccupied level (i.e., $c \longrightarrow d$) is *smaller* than any other electronic transition, the *longest* wavelength in the absorption spectrum corresponds to this transition, and the shorter wavelengths correspond to the other transitions.

Calculation (and experimental work) has shown that as conjugation increases, the energy difference between the highest occupied and lowest unoccupied π-orbitals decreases (cf. V.B. theory). Thus, as conjugation increases, the wavelength of the absorption band increases, and when it reaches the visible region, colour (complementary) will appear in the compound, e.g., in the polyenes $CH_3(CH\!=\!CH)_nCH_3$, when $n = 6$ the absorption band occurs in the blue region and so the compound is yellow (complementary colour).

Now let us consider benzene. This is a symmetrical molecule and consequently has no transition dipole. However, since some bending vibrations distort its shape, benzene does show absorption. Introduction of substituents into the benzene ring destroys its symmetry and thereby gives rise to increased intensity of absorption, and by extending the conjugation, increases the wavelength of the absorption band. Thus, since nitrobenzene has a large transition dipole, its absorption is more intense than that of benzene, and because of extended conjugation the wavelength is shifted to the longer region.

So far, we have discussed $\sigma \longrightarrow \sigma^*$ and $\pi \longrightarrow \pi^*$ transitions. Now let us consider the case of methanol; this has λ_{max} 150 and 183 nm. As we have seen earlier, ethane has λ_{max} 135 nm. Hence, the introduction of the auxochrome, OH, has produced a bathochromic shift. The shorter wavelength is attributed to a $\sigma \longrightarrow \sigma^*$ transition (C—H) and the longer one to an $n \longrightarrow \sigma^*$ transition (this involves one electron of a lone pair on the oxygen atom). Crotonaldehyde has λ_{max} 220 (ε, 16 000) and 321 nm (ε, 20); the former is due to a $\pi \longrightarrow \pi^*$ transition, and the latter to an $n \longrightarrow \pi^*$ transition.

Steric effects on colour may be explained as in the V.B. approach, *i.e.*, decreased orbital overlap raises the energy difference between the highest occupied and lowest unoccupied M.O.s, thereby causing the absorption to shift to a shorter wavelength.

The u.v. (and i.r.) spectra of free radicals differ from those of the corresponding neutral molecules. This is due to the fact that the presence of the unpaired electron changes the electronic energy levels from those in the parent molecule. In general, the energy difference between the highest occupied and lowest unoccupied M.O.s is lowered, and consequently the absorption wavelength is lengthened and may enter the visible region. Hence, most free radicals are coloured.

Ultra-violet and visible spectroscopy

The principles of electronic absorption have been described. However, before discussing their applications, we shall restate some of the terms previously mentioned (p. 869), but with their modifications. A *chromophore* is any structural feature which produces light absorption in the ultra-violet region or colour in the visible region. An *auxochrome* is any group which, although not a chromophore, brings about a red shift when attached to a chromophore. Thus, the combination of chromophore and auxochrome behaves as a *new* chromophore. A *bathochromic effect* (*red shift*) and a *hypsochromic effect* (*blue shift*) are the shifting of the absorption band to the longer and shorter wavelengths, respectively. A *hyperchromic effect* and *hypochromic effect* are those which respectively increase and decrease the intensity of absorption.

As we have seen, isolated double bonds do not give strong bands, but when conjugated systems are present, the bands are usually strong and in the longer wavelength region. Thus, one particularly important application of ultra-violet (and visible) spectroscopy is the detection and elucidation of the nature of conjugated systems (including aromatics). This is often carried out with the use of 'model' molecules, *i.e.*, a 'simple' molecule that differs from the compound under investigation in a way that should have no effect on the chromophore.

Alkanes absorb in the region of 140–150 nm ($\sigma \longrightarrow \sigma^*$), and when combined with various auxochromes, new absorption bands ($n \longrightarrow \sigma^*$) are produced at longer wavelengths, *e.g.*, RCl, 170–175; RBr, 200–210; RI, 255–260; ROH and R_2O, 180–185; RNH_2, 190–200 nm. On the other hand, the carbonyl chromophore (in various functional groups) absorbs, in general, above 200 nm, *e.g.*, aldehydes, 180 nm ($\pi \longrightarrow \pi^*$) and 290–295 ($n \longrightarrow \pi^*$); ketones 190 ($\pi \longrightarrow \pi^*$) and 270–280 ($n \longrightarrow \pi^*$); saturated monocarboxylic acids, 200–210 ($n \longrightarrow \pi^*$); esters, 200–205 ($n \longrightarrow \pi^*$); amides, 205–220 ($n \longrightarrow \pi^*$). It is this absorption at wavelengths longer than 200 nm that permits the identification of many chromophores in compounds.

When ethylenic double bonds are in conjugation or conjugated with a carbonyl group, the absorption moves to longer wavelengths, *e.g.*, for crotonaldehyde, there is one band at 220 nm ($\pi \longrightarrow \pi^*$) and another at 321 nm ($n \longrightarrow \pi^*$).

An interesting point about conjugated systems is that the geometry of conjugated dienes and trienes affects both λ_{max} and ε. In general, the acyclic compound absorbs at a shorter wavelength and has a greater intensity than the corresponding cyclic compound, *e.g.*, butadiene, 217 (ε, 21 000) and cyclohexa-1,3-diene, 257 (ε, 8 000). λ_{max} and ε are also affected by strain in a molecule, the greater the strain, the shorter the wavelength, *e.g.*, cyclobutanone, 281 nm; cyclohexanone, 290 nm. Steric effects, when operating, decrease conjugation, and so the *trans*-isomer will absorb at a longer wavelength than the *cis*-.

Aromatic compounds show a number of bands, *e.g.*, benzene absorbs at 184 (ε, 60 000), 204 (ε, 7 400) and 254 (ε, 200) nm. All are $\pi \longrightarrow \pi^*$ transitions, and the 254 nm band is called

Table 31.2

Compound*	λ_{max} nm (mμ) (ε)	Compound	λ_{max} nm (mμ) (ε)
Ethylene	175 (5 000)	Resorcinol	277 (2 200)
Butadiene	217 (21 000)	Quinol	225 (5 000)
Hexatriene	258 (35 000)		293 (2 700)
Acetaldehyde	180 (10 000)	o-Nitroaniline	222 (16 000)
	290 (15)		275 (5 000)
Acetone	190 (900)	m- ,,	235 (16 000)
	280 (12)		373 (1 500)
Crotonaldehyde	220 (16 000)	p- ,,	229 (5 000)
	321 (20)		375 (16 000)
Benzene	204 (7 400)	Naphthalene	220 (100 000)
	254 (200)		275 (5 700)
Toluene	206·5 (7 000)	Anthracene	253 (200 000)
	261 (225)		375 (8 000)
Chlorobenzene	210 (7 400)	Phenanthrene	252 (50 000)
	264 (200)		293 (16 000)
Aniline	230 (8 600)	Furan	205 (6 000)
	280 (1 400)		250 (2)
Nitrobenzene	270 (7 800)	Thiophen	235 (4 500)
Phenol	210 (6 200)	Pyrrole	210 (10 000)
	271 (1 450)		240 (400)
Catechol	214 (6 000)	Pyridine	252 (2 000)
	278 (2 600)	Quinoline	313 (2 500)
		Stilbene (trans)	295 (27 000)
		Stilbene (cis)	280 (10 500)

*In most cases, ethanol is the solvent.

the *benzenoid band* and is characterised by a large degree of fine structure (Fig. 31.4). For benzene derivatives, this benzenoid band generally occurs between 250 and 280 nm, but for polynuclear aromatics it moves to the longer wavelengths as the number of rings increases (see Table 31.2).

All substituents in benzene have a bathochromic effect (see Table 31.2). For disubstituted benzenes, the positions of the absorption maxima depend on their orientation, and for *para* disubstitution, whether the substituents electronically assist, *e.g.*, NH$_2$ and NO$_2$, or whether they electronically oppose each other, *e.g.*, NH$_2$ and OMe. In the latter case, the absorption maximum is usually close to that of the 'stronger' chromophore.

The case of aniline is worth further consideration. In ethanol, λ_{max} is 230 nm (Fig. 31.5), but in dilute aqueous acid, λ_{max} is 203 nm. In the free base, the nitrogen lone pair of electrons can enter into conjugation with the benzene ring. Thus, there is increased delocalisation in aniline and consequently the absorption maximum is shifted to the lower wavelength. In the anilinium cation, the lone pair is no longer available for conjugation with the ring, and so the molecule now behaves like benzene itself.

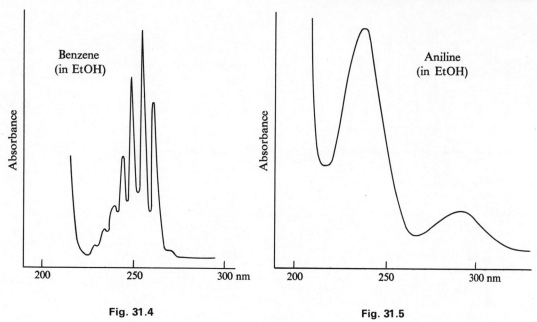

Fig. 31.4

Fig. 31.5

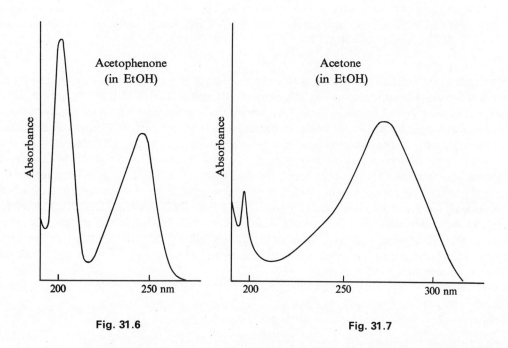

Fig. 31.6

Fig. 31.7

Figures 31.6–8 are the u.v. spectra of acetophenone, acetone, and *m*-nitrophenol, respectively. Heterocyclic compounds, to a large extent, have u.v. spectra similar to those of the analogous benzenoid compounds (Table 31.2).

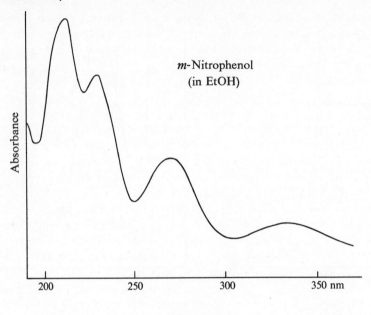

Fig. 31.8

Woodward and Fieser have developed empirical rules for calculating $\pi \longrightarrow \pi^*$ maxima of a given chromophore associated with unsaturation in conjugation and the type and position of substituents in the conjugated system (see Vol. 2, Ch. 8).

The final point we shall make here is the effect of solvents. Ethanol is most commonly used since it is a good solvent for many organic compounds and is transparent above 200 nm. However, polar solvents (and those which can form hydrogen bonds) tend to interact electrostatically (and form hydrogen bonds) with various chromophores, e.g., the carbonyl group. This changes the charge distribution in the molecule and results, in effect, in increased delocalisation. For $\pi \longrightarrow \pi^*$ transitions, both ground and excited states are stabilised, and the absorption moves to longer wavelengths. On the other hand, for $n \longrightarrow \pi^*$ transitions, the ground state is, e.g., hydrogen-bonded to a lone pair of electrons, whereas in the excited state, hydrogen bonding involves only *one* electron of the lone pair (the other having been promoted to an upper energy state). In these circumstances, the ground state is more stabilised than the excited state, and consequently absorption shifts to the *shorter* wavelengths. This blue shift with increasing polarity of solvent, e.g., cyclohexane $\longrightarrow$ EtOH $\longrightarrow$ H_2O, is a useful means of recognising $n \longrightarrow \pi^*$ transitions.

Dyes

For a substance to act as a dye, certain conditions must be fulfilled:
   (i) It must have a suitable colour.
   (ii) It must be able to 'fix' itself or be capable of being 'fixed' to the fabric.
   (iii) The fixed dye must have fastness properties: (*a*) fastness to light; (*b*) resistance to the action of water, dilute acids and alkalis, various organic solvents used in dry cleaning, etc.

Many natural dyes have been known for a long time. These were obtained from animal and vegetable sources. Today, however, practically all dyes are synthetic and are prepared from aromatic compounds, the only source of which was originally coal-tar; hence the name **coal-tar dyes**.

Nomenclature of dyes

There is no systematic nomenclature of dyes. Many have names that have been given to them by the manufacturers, and so it is not unusual to find a given dye having several names. Generally, each dye has a trade name (or names), and the shade is indicated by a letter, *e.g.*, Y or G = yellow (*gelb*); O = orange; R = red; B = blue. Sometimes the letter is repeated, the number of letters indicating roughly the intensity of the colour, *e.g.*, methyl violet 6B is a very deep purple (close to blue). Sometimes the letters have other meanings, *e.g.*, alizarin blue D; here the D means that this dye is a *direct* cotton colour; fuchsine S, the S indicating that the dye is an *acid* (*sauer*) colour. The letter F is often used to indicate that the dye is fast to light.

To avoid difficulties, the Society of Dyers and Colourists have compiled a Colour Index in which each dye is assigned its individual colour number (**C.I. no.**).

Classification of dyes. Dyes are classified according to their chemical constitution or by their application to the fibre. The former is of theoretical value to the chemist but of little importance to the dyer who is mainly concerned with the reaction of dyes towards the fibre being used.

Chemical classification. The chemical constitutions of dyes are so varied that it is difficult to classify them into distinct groups. The Colour Index classifies dyes as shown; in some cases a particular dye could be placed in one or other group.

Nitroso; nitro; azo (monazo, etc.); azoic; stilbene; diarylmethane; triarylmethane; xanthen; quinoline; methine; acridine; sulphur; thiazole; thiazine; indamine; azine; oxazine; lactone; anthraquinone; indigoid; phthalocyanine.

Classification according to application

The method of application of a dye depends on the nature of both the fibre and the dye. **Natural fibres** are of two types: *vegetable fibres* are cellulosic, *e.g.*, cotton, linen, flax, hemp, and jute; *animal fibres* are proteins, *e.g.*, wool, silk, leather, and fur. **Synthetic fibres** are high polymers built up of various types of monomers, *e.g.*, polyamides, polyesters, polyacrylonitrile, etc. Apart from the nature of the fibre, there is also the problem of the treatment of the fibre before dyeing, *e.g.*, natural and synthetic fibres are treated chemically to make them crease-resistant, shrink-resistant, water-repellent, etc. This chemical treatment alters the structure of the fibre, and these alterations can be detected by spectroscopy (u.v., visible, and particularly i.r.).

Dyeing has always been carried out in aqueous solutions, but recent work appears to show that various solvents, *e.g.*, 1,1,1-trichloroethane, may be used.

As pointed out above, dyes have diverse chemical structures, but from the point of view of their application, there are four ways in which the dye molecule may be bound to the fibre: (i) covalent bonds; (ii) hydrogen bonds; (iii) ionic bonds; (iv) van der Waals forces. The type of binding for a given dye will thus depend largely on the chemical nature of the fibre.

1. **Acid dyes** are the sodium salts of sulphonic and carboxylic acids, and are applied to wool, silk, polyamides and acrylic fibres. **Basic dyes** are salts of the colour bases (p. 886).

2. **Azoic dyes** are insoluble azo-dyes which are prepared *in situ*; they are mainly applied to cellulosic fibres. The term **ingrain dyes** has also been used with the same meaning, *i.e.*, produced on the fibre.

3. **Mordant dyes** do not dye a fibre directly; they require a mordant. This is, for acidic dyes, a metal hydroxide, and for basic dyes, tannin (tannic acid). The metal salts have been referred to as **lakes**.

4. **Metal complex dyes** differ from mordant dyes in that they are prepared as complexes which are *then* applied to the fibre.

5. **Vat dyes** are those which are used in their reduced state (*leuco-compounds*), and after application to the fibre, are oxidised to the dye. They are used mainly with cotton fibres.

6. **Sulphur dyes** are used mainly with cellulosic fibres.

7. **Disperse dyes** are water-insoluble dyes which are dispersed by suitable reagents before application to synthetic fibres.

8. **Reactive dyes** contain a reactive group which combines directly with cellulosic fibres.

9. **Organic pigments** are water-insoluble compounds which are used for colouring paints, etc., and when made water-soluble, are used as dyes.

Owing to the very large number of dyes in use, it is only possible to give a bare outline of dye chemistry. (For further information, see the appropriate references.)

NITRO-DYES   Two examples of this class are **Naphthol Yellow S**, (I), and **Amido Yellow E**, (II).

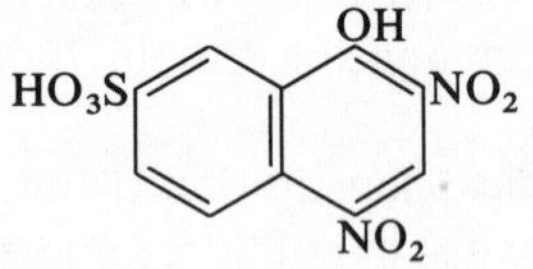

(I) 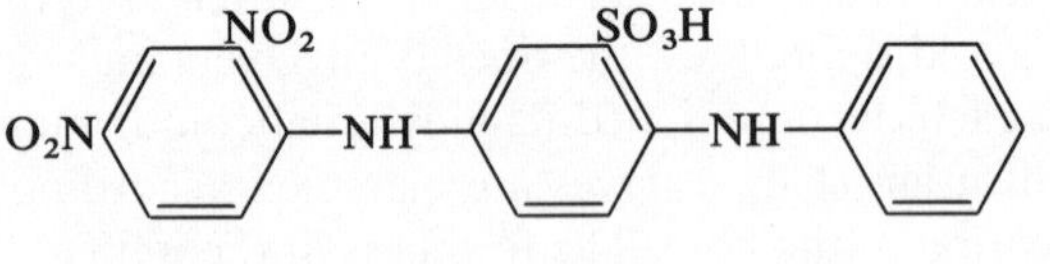   (II)

(III)

(I) is made by nitrating 1-naphthol-2,4,7-trisulphonic acid, and (II) by reaction between 1-chloro-2,4-dinitrobenzene and 4-aminodiphenylamine-2-sulphonic acid.

Lithol Yellow G, (III), is a pigment, and is used as a non-poisonous substitute for *Chrome Yellow* (lead chromate). It is prepared by condensing *p*-chloro-*o*-nitroaniline with formaldehyde.

NITROSO-DYES   Most nitroso-dyes have one hydroxyl group *ortho* to the nitroso-group, and their iron salts (lakes) are green, *e.g.*, the iron complex of **Gambine Y** (1-nitroso-2-naphthol).

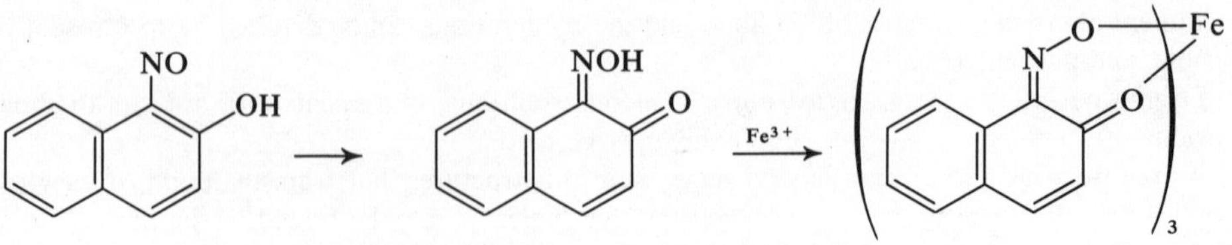

Azo-dyes

Azo-dyes (and pigments) are the largest class of synthetic dyes. Their chromophore is an aromatic system joined to the azo-group, and the common auxochromes are NH_2, NR_2, OH. The usual method of preparation of azo-dyes is direct coupling between a diazonium salt and a phenol or an amine.

The structure of an azo-dye is readily found by reduction with stannous chloride and hydro-chloric acid, or with sodium hyposulphite (dithionite), whereupon the azo-group is ruptured with the formation of primary amines which are then identified:

$$Ar^1N{=}NAr^2 \xrightarrow{e;\, H^+} Ar^1NH_2 + Ar^2NH_2$$

Azo-dyes are classified according to the number of azo-groups in the molecule: monazo, disazo, trisazo, etc., and are also sub-classified as basic (cationic) dyes, acid (anionic) dyes, etc.

Monazo dyes

Cationic dyes. Chrysoidine, (I), is a cationic dye, and is prepared by coupling benzenediazonium chloride with *m*-phenylenediamine. It is used for dyeing paper, leather, and jute. Another cationic dye is **Acid Red**, (II; from diazotised *p*-nitroaniline and naphthionic acid).

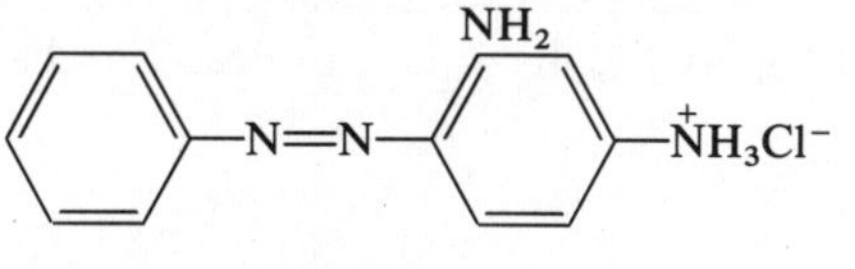

(I) 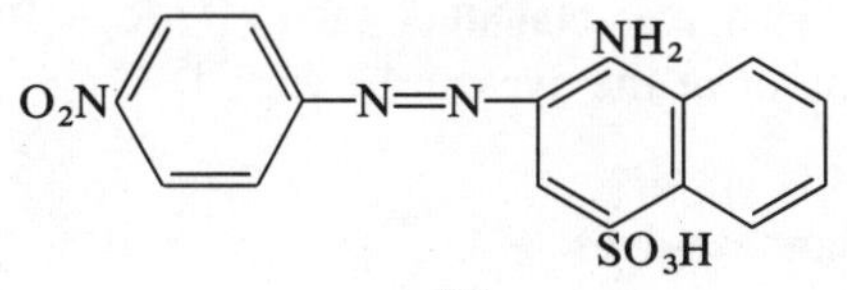   (II)

Anionic dyes

These contain a sulphonic or a carboxylic group, and also belong to the class of *soluble dyes*. **Methyl Orange** (*Helianthin*), (III), prepared by coupling diazotised sulphanilic acid with dimethylaniline, is not used as a dye. It is used as an indicator, being orange in alkaline solution and red in acid solution (*cf.* Congo Red; see later).

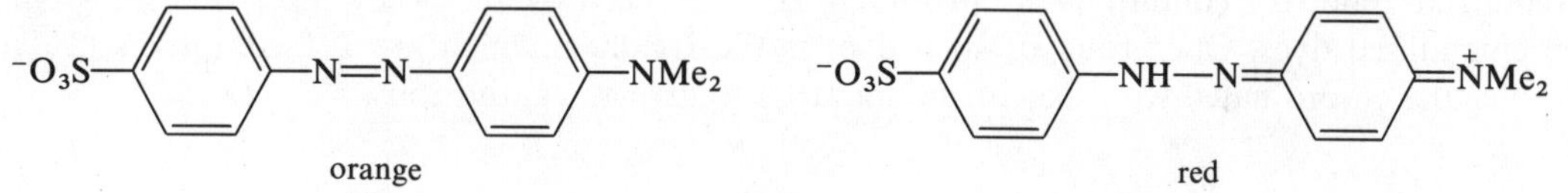

orange   red

(III)

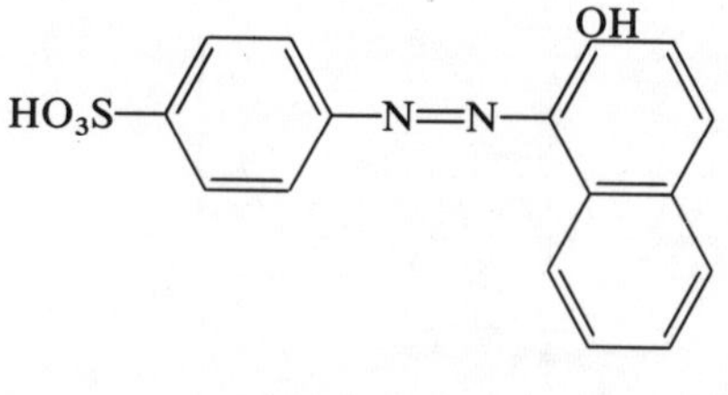

(IV) 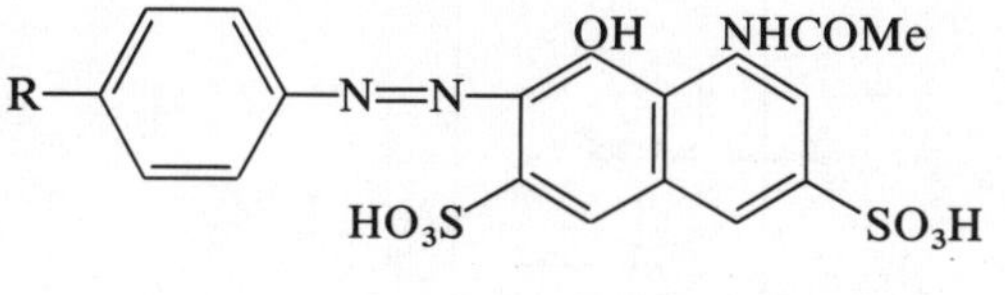   (V)

Orange II, (IV) (from diazotised sulphanilic acid and 2-naphthol) is used for dyeing paper.

A serious disadvantage with anionic azo-dyes is their relative easy removal from the fibre by water. This has been overcome by the introduction of the **Carbolan dyes**, (V). R is a long-chain alkyl group, *e.g.*, $C_{12}H_{25}$—. Because of their hydrophobic nature, these large groups prevent the removal of the dye by water.

Azoic dyes (*ingrain azo-dyes*)

These are water-insoluble azo-dyes which are formed on the fibre. They are prepared by 'padding' the fibre with an alkaline solution of 2-naphthol containing Turkey-red oil, drying, and then immersing in a solution of the diazotised amine, *e.g.*, **Para Red**, (VI) (diazotised *p*-nitroaniline).

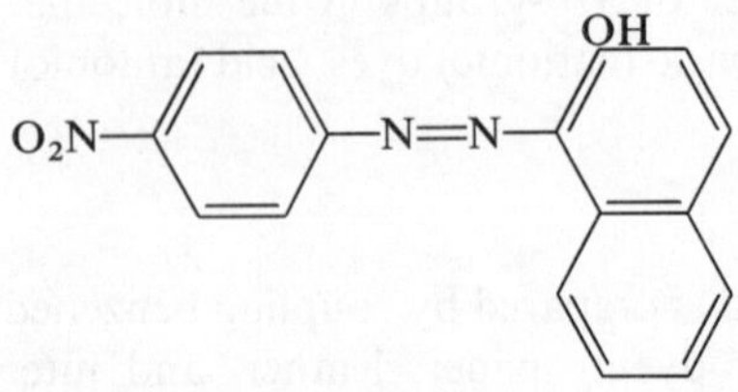

(VI) 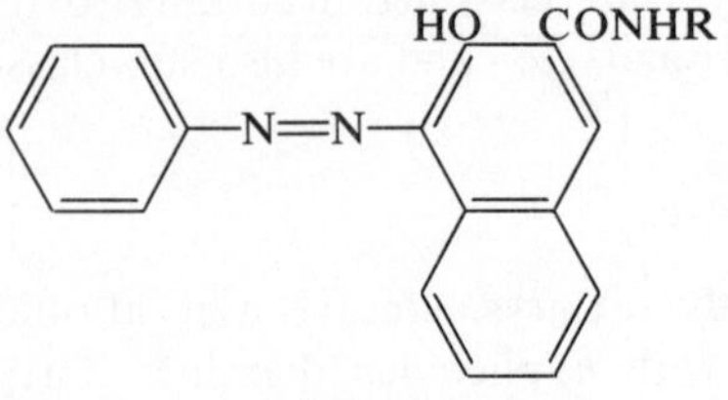 (VII)

Since this type of molecule has poor affinity for the fibre, 3-carbonamido-2-naphthols were introduced, *e.g.*, **Naphthol AS**, (VII; R = Ph). By changing the diazotised molecule and also R, the colour of the dye can be varied.

Mordant azo-dyes

Their characteristic structural feature is the presence of a hydroxyl group *ortho* to the azo-group. The most important mordanting metal is chromium. The fibre is mordanted by boiling with potassium dichromate solution, usually with a reducing agent such as formic, lactic, or oxalic acid (the dichromate is converted into chromium hydroxide). Metal salts (lakes) are then formed by reaction with various azo-dyes, *e.g.*, **Mordant Brown** (VIII; formed by coupling diazotised *p*-aminophenol with pyrogallol).

Metal azo-dye complexes. One of the difficulties with mordant dyes is that the fibre tends to be damaged under the conditions of chroming. This has been overcome by using neutral-dyeing premetallised dyes. One group of such dyes is the **Irgalans**, which are 2:1 complexes (2 dye molecules co-ordinated with the chromium atom to give a negative ion), *e.g.*, (IX):

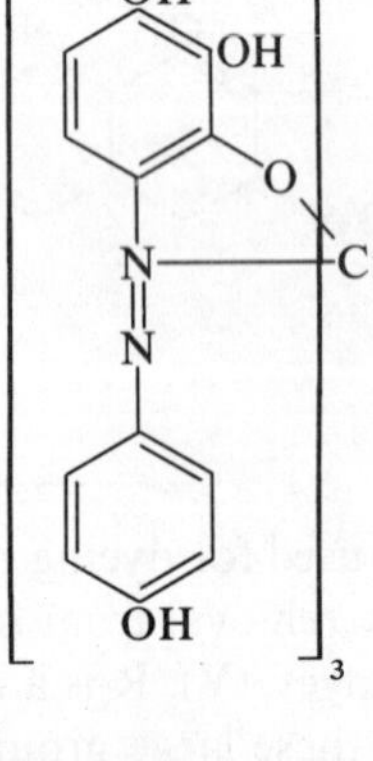

(VIII) 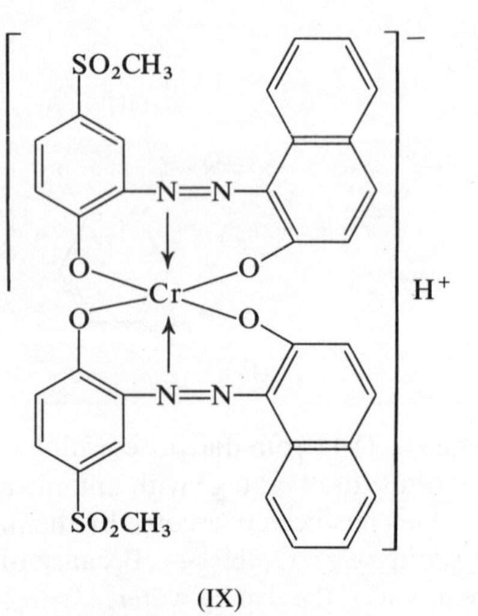 (IX)

These complexes are anionic, and dye wool in neutral solution. They have excellent fastness to light and to washing.

o-Amino- or carboxyl groups can be used instead of *o*-hydroxyl groups, and cobalt can replace chromium. Furthermore, a wider range of shades can be obtained by using two different azo-compounds to form the 2:1 complex.

Disperse azo-dyes

These are used for synthetic fibres, and the method of application is their preparation, by means of dispersing reagents, in a state of uniform fine dispersions in the dye bath. Terylene is dyed mainly with disperse dyes, *e.g.*, (X; a pyrazolone derivative; yellow), and (XI; red):

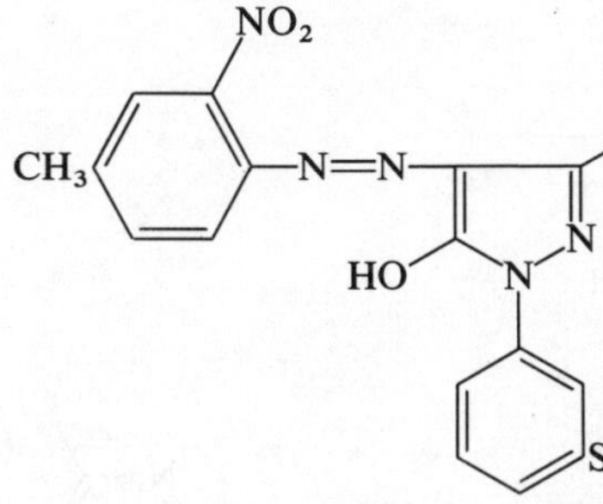

(X)

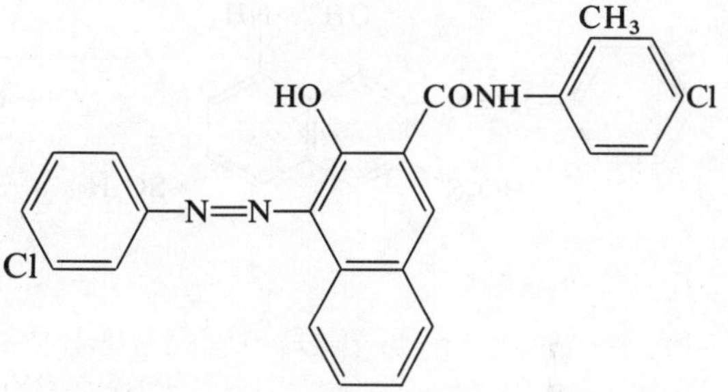

(XI)

Nylon may be dyed with acid dyes (Nylon is basic), with 2:1 metal complexes, and with disperse dyes. A typical acid dye is (XII), and a typical disperse dye is (XIII):

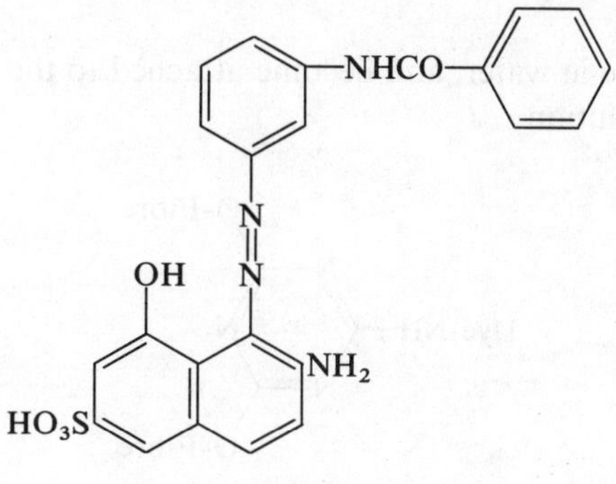

(XII)

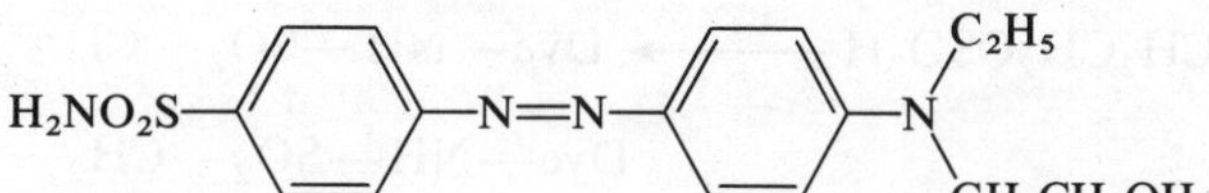

(XIII)

Polyacrylonitrile may be dyed with basic dyes (since it contains free carboxyl groups), with cationic dyes, *e.g.*, **Astrazone Red GTL**, (XIV), and with disperse dyes:

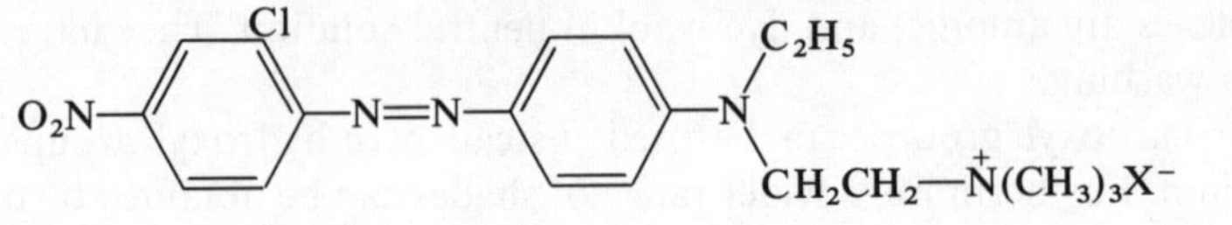

(XIV)

Reactive azo-dyes

These contain groups which react directly with the fibre. Thus, in **Procion dyes** (water-soluble dyes used for cellulosic fibres), the dye is attached to a chloro-1,3,5-triazine, *e.g.*,

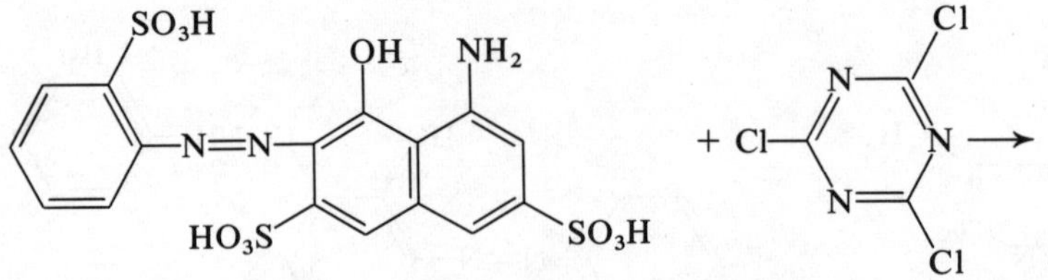

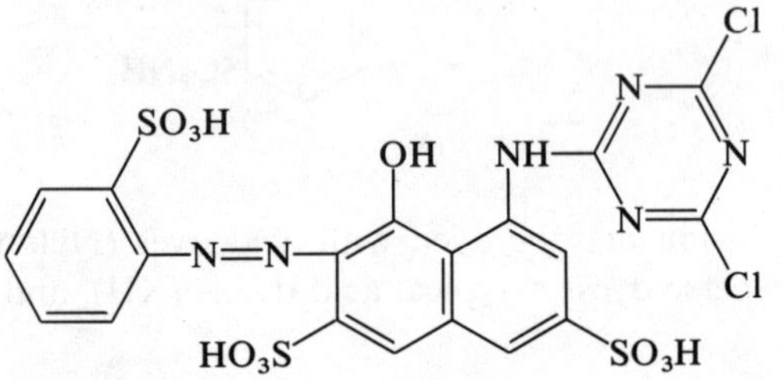

These reactive dyes are stable in water, and become attached to the cellulose fibre via hydroxyl groups in weakly alkaline solution.

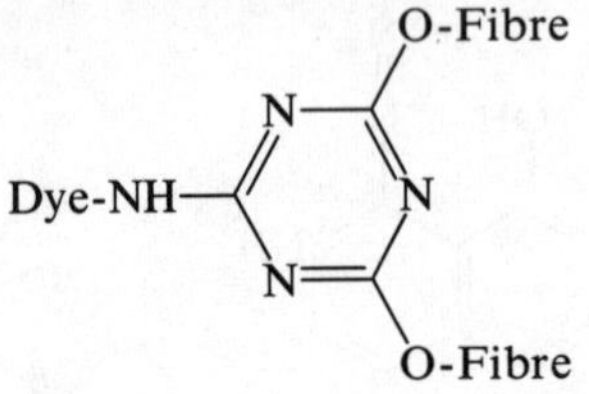

There are also other reactive dyes, *e.g.*, **Ramazol**:

$$\text{Dye—NH—SO}_2\text{CH}_2\text{CH}_2\text{OSO}_3\text{H} \xrightarrow{-\text{H}_2\text{SO}_4} \text{Dye—NH—SO}_2\text{—CH=CH}_2 \xrightarrow[\text{alkali}]{\text{HO-Fibre}}$$

$$\text{Dye—NH—SO}_2\text{—CH}_2\text{—CH}_2\text{—O—Fibre}$$

Disazo-dyes

Congo Red is an example of this group; it is prepared by coupling tetrazotised benzidine with two molecules of naphthionic acid:

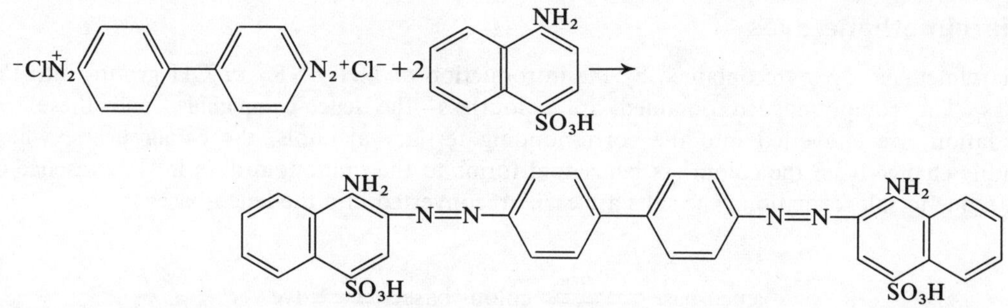

This is red in alkaline solution, and its sodium salt dyes cotton a full red. Congo Red was the first synthetic dye produced that was capable of dyeing cotton directly. It is very sensitive to acids, the colour changing from red to blue in the presence of inorganic acids. This blue colour, *i.e.*, the deepening of colour, may be attributed to resonance among charged canonical structures.

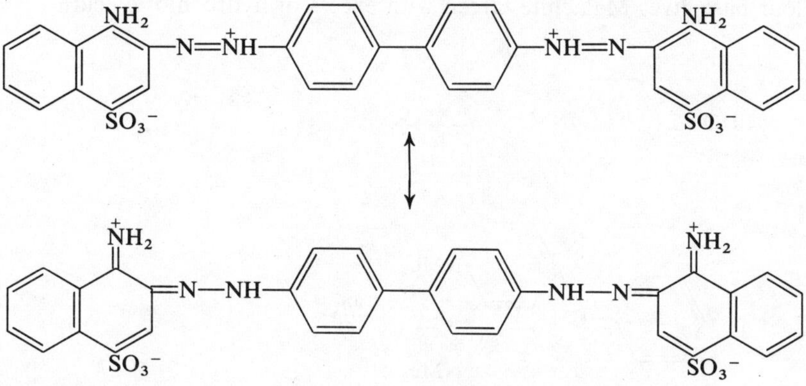

Diarylmethane dyes

Auramine O is prepared by fusing Michler's ketone (p. 754) with ammonium and zinc chlorides at 150–160°C, or by heating *p,p'*-tetramethyldiaminodiphenylmethane with sulphur, ammonium chloride and a large excess of sodium chloride in a current of ammonia at about 200°C. This produces *Auramine Base* which, with hydrochloric acid forms the hydrochloride, *Auramine O*, a yellow basic dye.

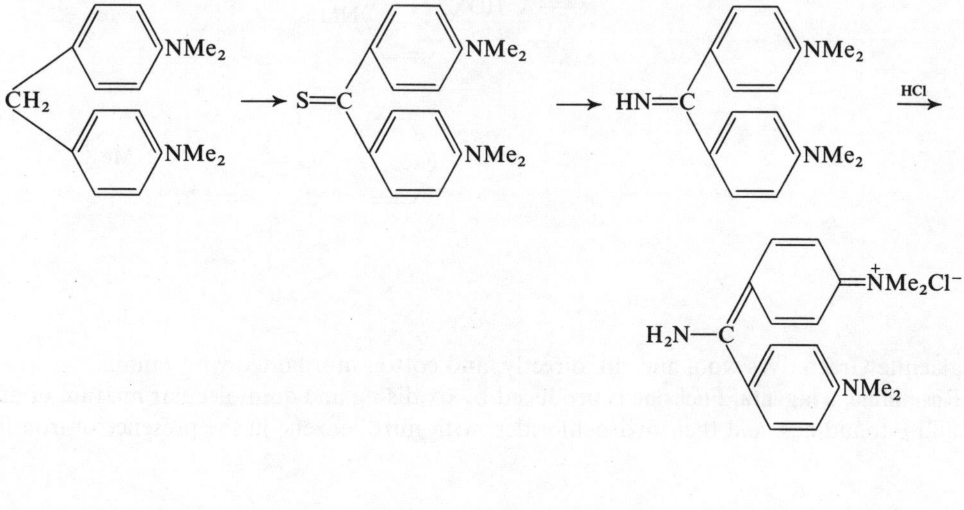

Triarylmethane dyes

Triarylmethane dyes are obtained by the introduction of NH_2, NR_2 or OH groups into the rings. The compounds so obtained are colourless—the **leuco-compounds**—and these, on oxidation, are converted into the corresponding tertiary alcohols, the **colour bases**, which readily change from the colourless benzenoid forms to the quinonoid dyes in the presence of acid, due to salt formation. The salts are easily reconverted into the leuco-base:

$$\text{leuco-base} \underset{\text{reduction}}{\overset{\text{oxidation}}{\rightleftarrows}} \text{colour base} \underset{\text{alkali}}{\overset{\text{acid}}{\rightleftarrows}} \text{dye}$$
$$\text{(colourless)} \qquad\qquad \text{(colourless)} \qquad\qquad \text{(coloured)}$$

Malachite Green is prepared by condensing dimethylaniline (2 molecules) with benzaldehyde (1 molecule) at 100°C in the presence of concentrated sulphuric acid. The leuco-base produced is oxidised with lead dioxide in a solution of acetic acid containing hydrochloric acid; the resulting colour base gives Malachite Green with excess of hydrochloric acid:

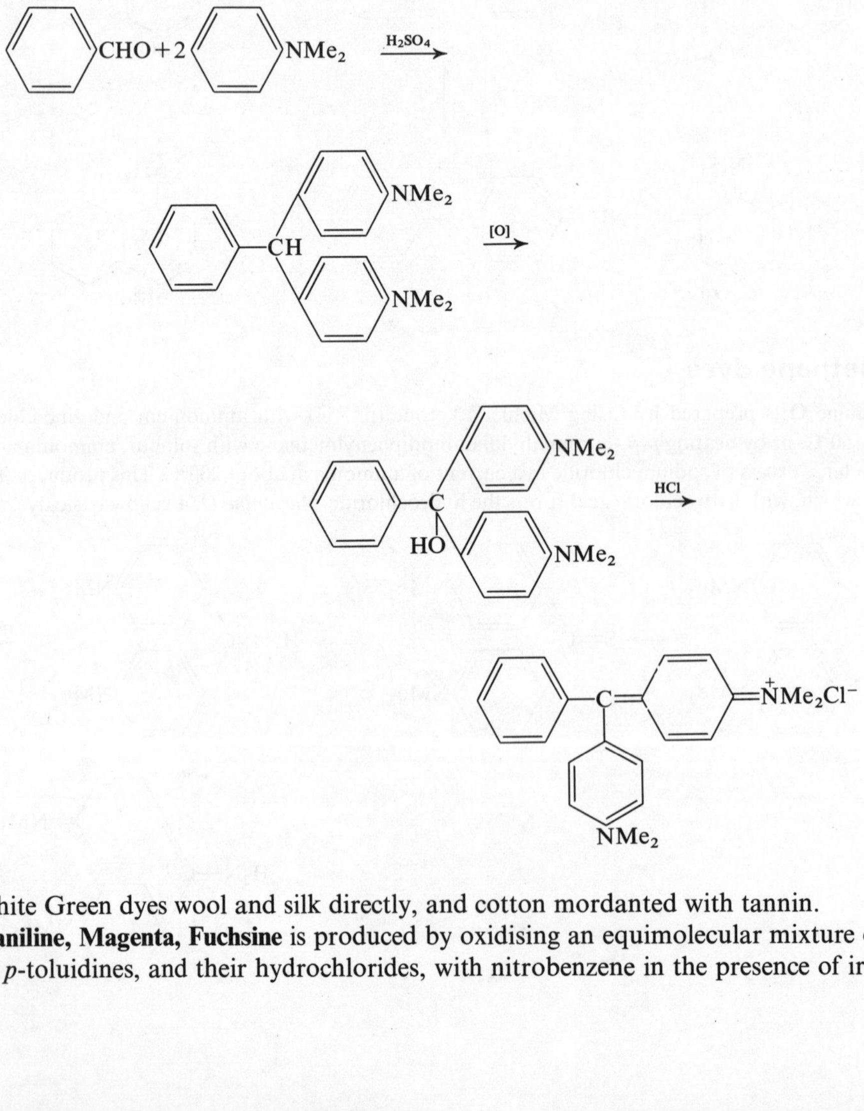

Malachite Green dyes wool and silk directly, and cotton mordanted with tannin.

Rosaniline, Magenta, Fuchsine is produced by oxidising an equimolecular mixture of aniline, o- and p-toluidines, and their hydrochlorides, with nitrobenzene in the presence of iron filings.

The product is a mixture of rosaniline and pararosaniline (no methyl group present), in which the former predominates:

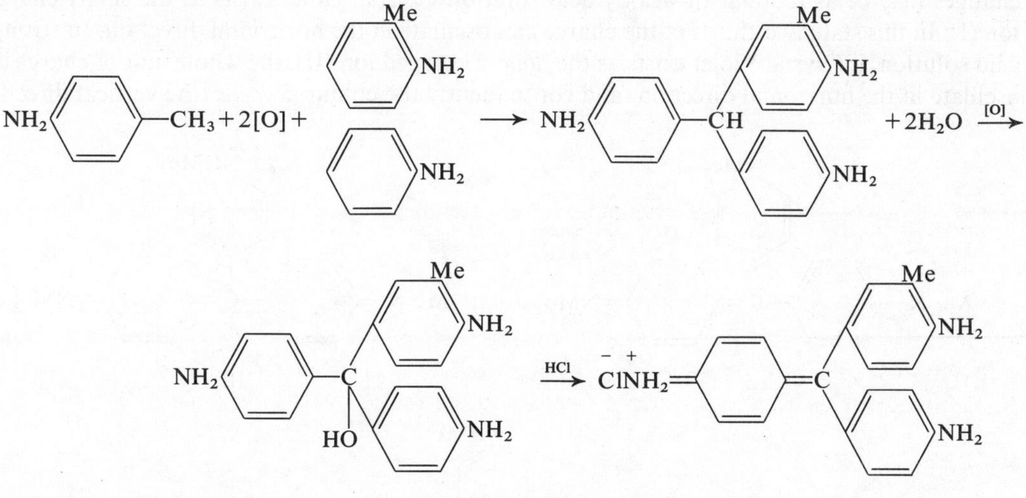

$$NH_2C_6H_4{-}CH_3 + 2[O] + (MeC_6H_3NH_2) + (C_6H_4NH_2) \longrightarrow NH_2C_6H_4{-}CH(C_6H_3(Me)NH_2)(C_6H_4NH_2) + 2H_2O \xrightarrow{[O]}$$

$$NH_2C_6H_4{-}\underset{HO}{C}(C_6H_3(Me)NH_2)(C_6H_4NH_2) \xrightarrow{HCl} Cl\overset{-}{N}H_2{=}C_6H_4{=}\overset{+}{C}(C_6H_3(Me)NH_2)(C_6H_4NH_2)$$

Crystals of rosaniline show a green metallic lustre, and dissolve in water to form a deep-red solution. This solution is decolorised by sulphur dioxide and is then known as Schiff's reagent, which is used as a test for aldehydes.

Rosaniline (and pararosaniline) dyes wool and silk directly, producing a violet-red colour; cotton must first be mordanted with tannin.

Crystal Violet may be prepared by heating *Michler's ketone* with dimethylaniline in the presence of phosphoryl chloride or carbonyl chloride. If the latter compound is used, then Crystal Violet may be prepared directly by heating carbonyl chloride and dimethylaniline:

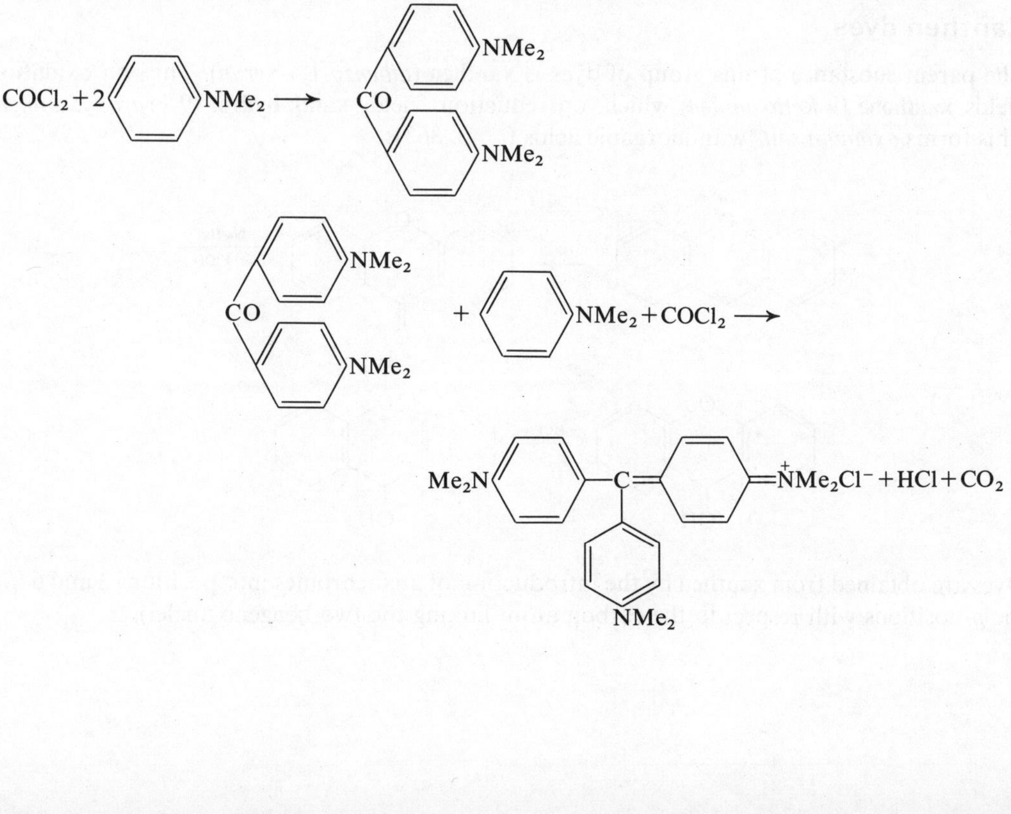

$$COCl_2 + 2(C_6H_4NMe_2) \longrightarrow CO(C_6H_4NMe_2)_2$$

$$CO(C_6H_4NMe_2)_2 + (C_6H_5NMe_2) + COCl_2 \longrightarrow$$

$$Me_2N{-}C_6H_4{-}\underset{\underset{C_6H_4NMe_2}{|}}{C}{=}C_6H_4{=}\overset{+}{N}Me_2Cl^- + HCl + CO_2$$

A weakly acid solution of Crystal Violet is purple. In *strongly* acid solution the colour is green, and in still more strongly acid solution, the colour is yellow. The explanation of these colour changes may be as follows. In weakly acid solution Crystal Violet exists as the singly charged ion (I). In this state two-thirds of the charge can oscillate in the horizontal direction. In strongly acid solution, if Crystal Violet exists as the *doubly* charged ion (II), the whole unit of charge can oscillate in the horizontal direction, and consequently the colour *deepens* (the vertical direction

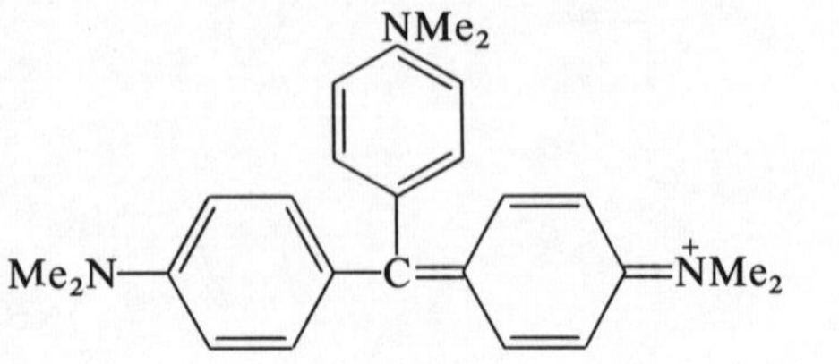

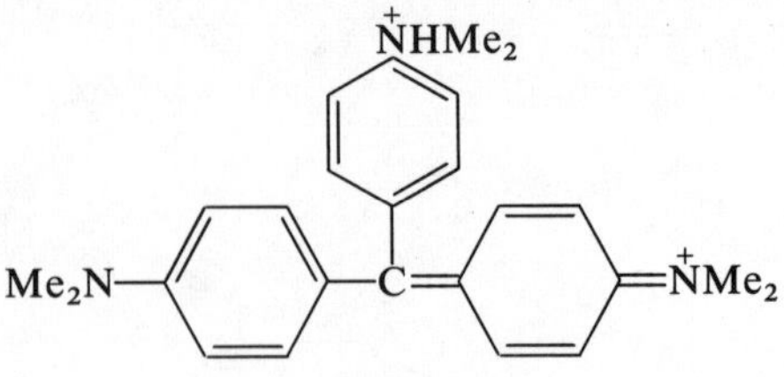

of oscillation is inhibited due to the fixation of the lone pair by proton addition). In very strongly acid solution another proton is added to form the ion (III) with three charges. In this ion relatively little resonance (with oscillation of charge) is possible, and consequently the colour lightens.

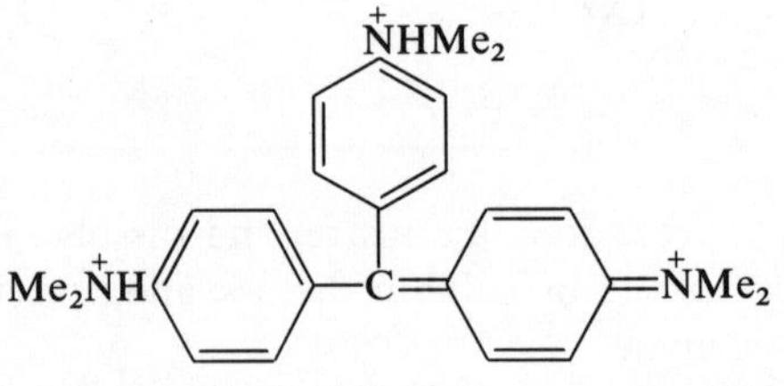

Xanthen dyes

The parent substance of this group of dyes is **xanthen** (*dibenzo*-1,4-*pyran*). This, on oxidation, yields **xanthone** (9-*ketoxanthen*) which, on reduction, yields **xanth-hydrol** (9-*hydroxyxanthen*). This forms *oxonium salts* with inorganic acids (*cf.* p. 863):

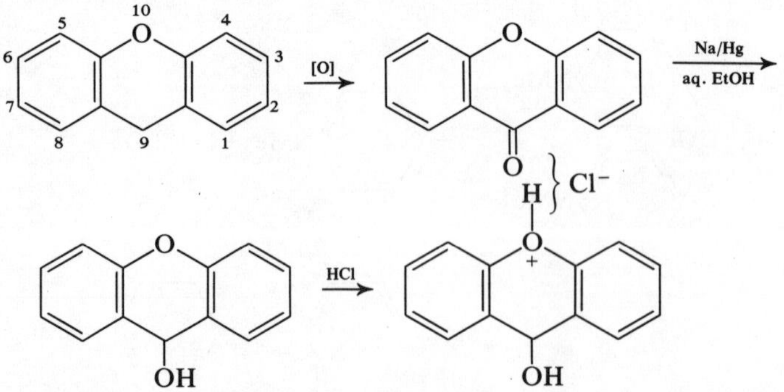

Dyes are obtained from xanthen by the introduction of auxochromes into positions 3 and 6 (*i.e.*, the *p*-positions with respect to the carbon atom linking the two benzene nuclei).

Pyronines. Pyronine G is prepared by condensing formaldehyde (1 molecule) with *m*-dimethyl-aminophenol (2 molecules) in the presence of concentrated sulphuric acid, and then oxidising the leuco-compound with ferric chloride. Pyronine G dyes silk and cotton mordanted with tannin a crimson red.

Since two electronic structures are possible in which the conjugation is totally different, more information is necessary to say which form predominates, or even whether both forms are present. However, because nitrogen is less electronegative than oxygen, the former can support a positive charge more easily than the latter, and therefore the structure with positive nitrogen is the more likely one.

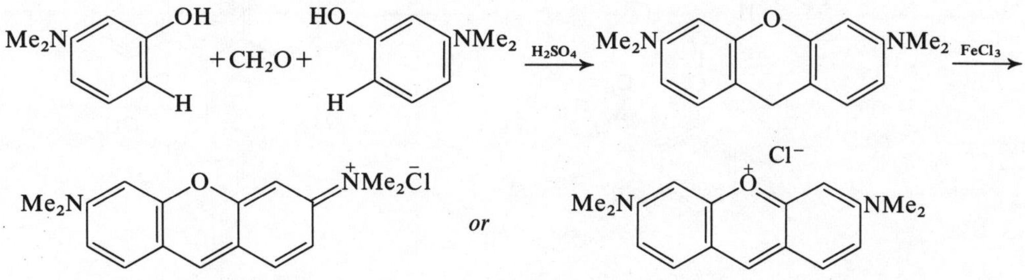

Phthaleins. These are obtained by condensing phenols with phthalic anhydride in the presence of certain dehydrating agents, *viz.*, concentrated sulphuric acid, zinc chloride or anhydrous oxalic acid.

Phenolphthalein is not a xanthen derivative but a triphenylmethane derivative. Since, however, it is prepared from phenol and phthalic anhydride, it is more closely allied to the phthaleins (in preparation and properties), and hence is here considered in the phthalein group. Phenolphthalein is prepared by heating phthalic anhydride (1 molecule) with phenol (2 molecules) in the presence of concentrated sulphuric acid:

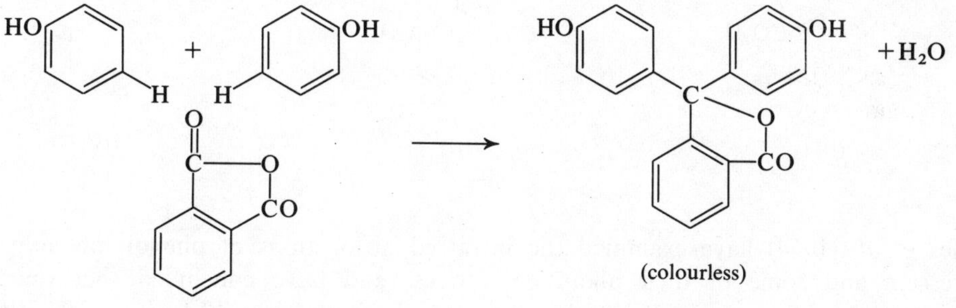

It is a white crystalline solid, insoluble in water, but soluble in alkalis to form deep red solutions:

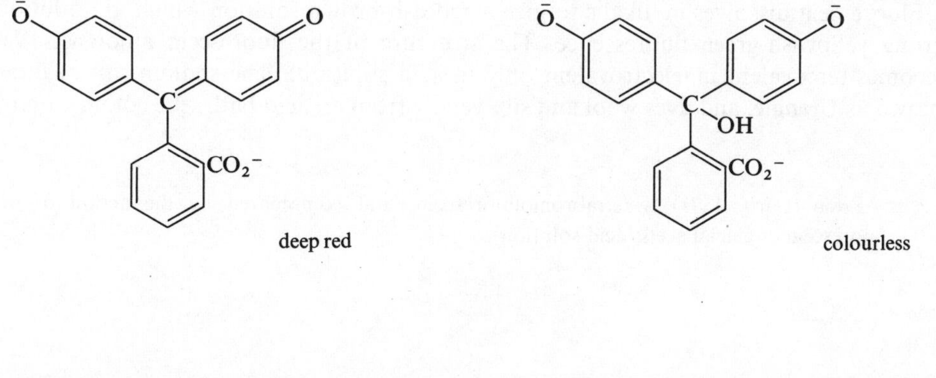

deep red   colourless

In the presence of *excess* of strong alkali, the solution of phenolphthalein becomes colourless
again due to the loss of the quinonoid structure and resonance (but see fluorescein, below).

Fluorescein is a xanthen derivative, and is prepared by heating phthalic anhydride (1 molecule)
with resorcinol (2 molecules) at 200°C, or at 110–120°C with anhydrous oxalic acid:

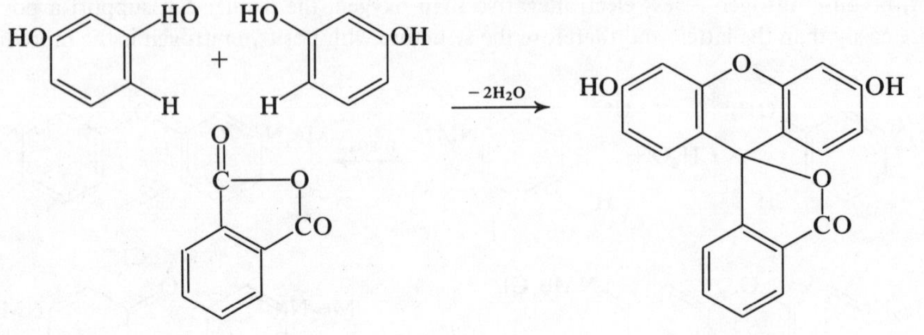

(I)

Fluorescein is a red powder insoluble in water. Since it is coloured, the non-quinonoid
uncharged structure (I) has been considered unsatisfactory. Two quinonoid structures are
possible in which the conjugation is totally different, one having the *p*-quinonoid structure (II),
and the other the *o*-quinonoid (III) (which contains tercovalent oxygen):

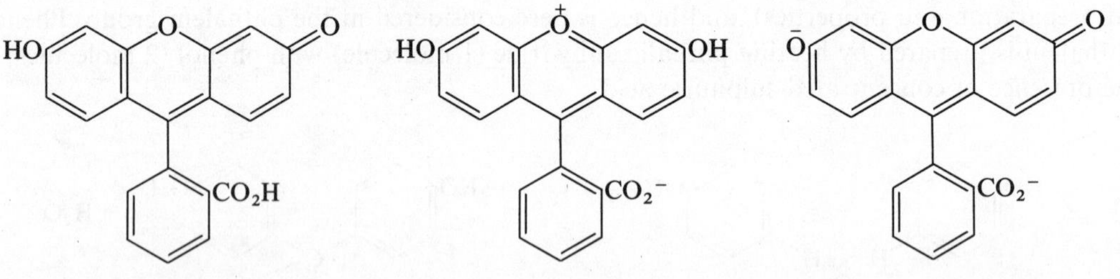

(II)   (III)   (IV)

Davies *et al.* (1954) have examined the infra-red absorptions of phenol, phenolphthalein,
fluorescein and some of their alkali derivatives, and have concluded that the classical
representation of fluorescein (I) is to be preferred. (II) is eliminated because of the absence of
the characteristic absorption of the carboxyl group, and similarly, the frequencies of the
carboxylate ion in (III) are also absent.

Fluorescein dissolves in alkalis to give a reddish-brown solution which, on dilution, gives a
strong yellowish-green fluorescence. The structure of the fluorescein anion is (IV) (oxygen
becomes tercovalent unielectrovalent only in acid solution). The sodium salt of fluorescein is
known as **Uranine**, and dyes wool and silk yellow from an acid bath; the colours are fugitive.

Eosin (Caro, 1871) is tetrabromofluorescein, and is prepared by the action of bromine on
fluorescein in glacial acetic acid solution:

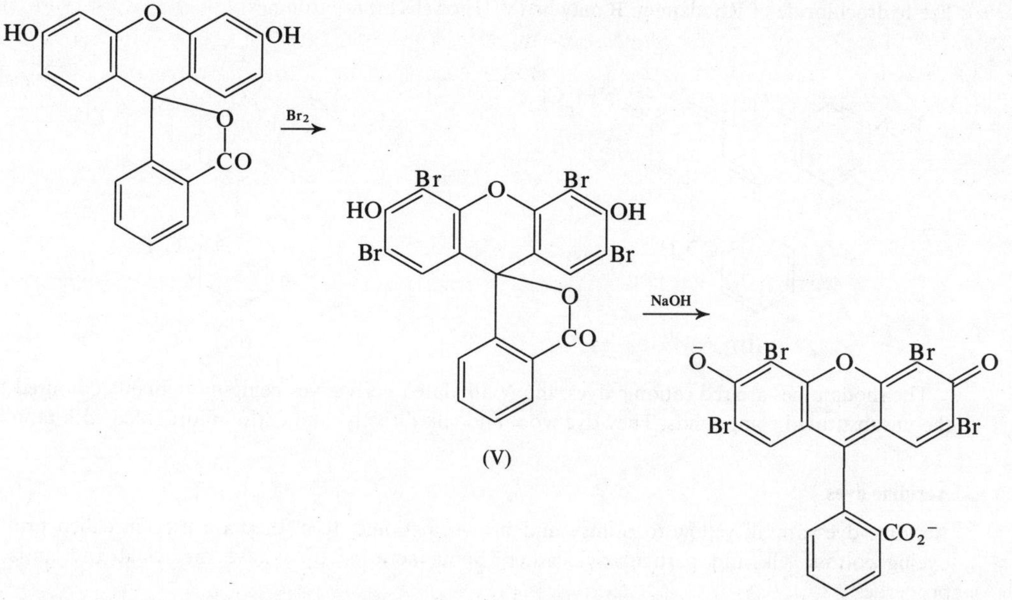

(V)

Eosin itself is a red powder and so its structure may not be (V), but may be (VI) or (VII) (*cf.* fluorescein):

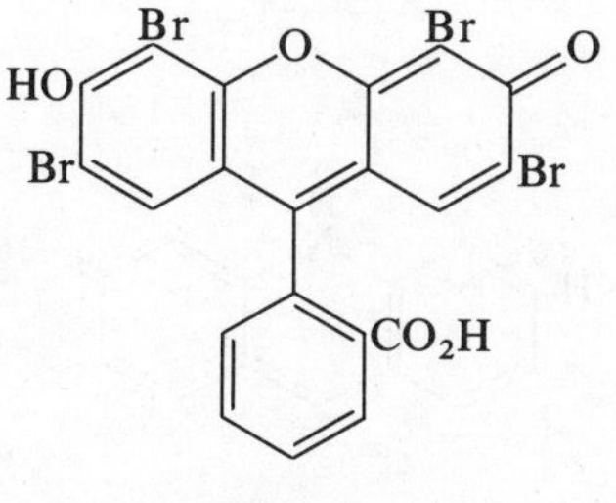

(VI) 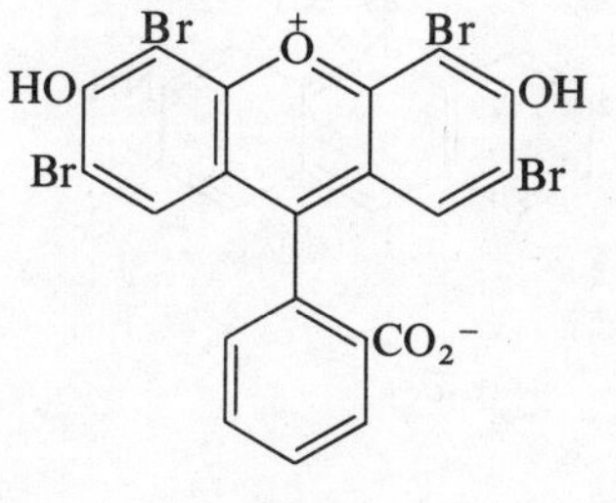   (VII)

Eosin dyes wool and silk a pure red, with a yellow fluorescence in the case of silk. Eosin is also used as the lead lake *Vermilionette* for poster printing. Most red inks are dilute solutions of eosin.

Rhodamines are prepared by condensing phthalic anhydride (1 molecule) with *m*-aminophenol (2 molecules) or its derivatives; *e.g.*, **Rhodamine B**:

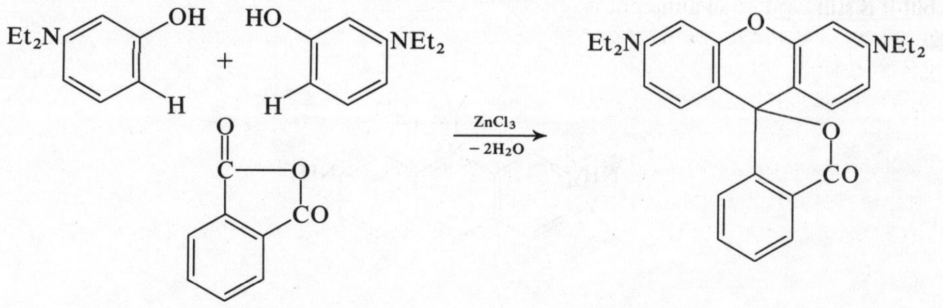

The hydrochloride of Rhodamine B may be (VIII) or (IX) (*cf.* pyronines):

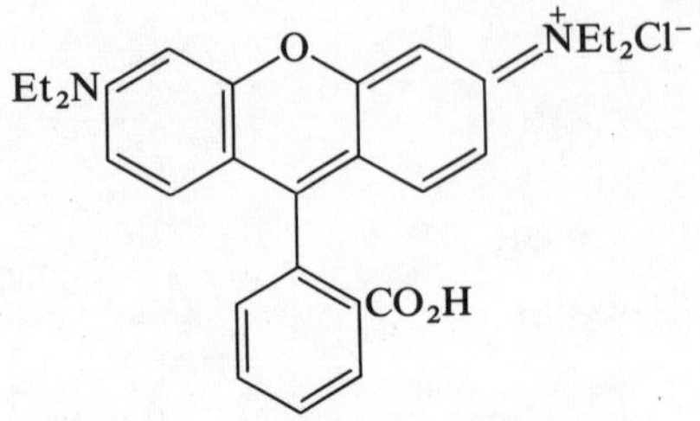

(VIII) 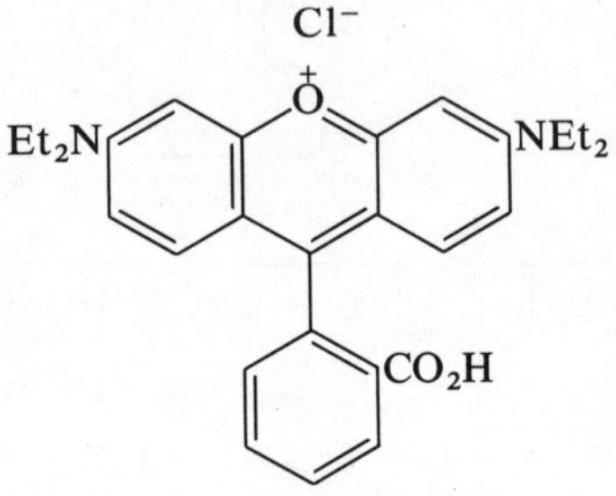   (IX)

The rhodamines are red cationic dyes, the *N*-alkylated derivatives being more highly coloured than the unsubstituted compounds. They dye wool and silk directly, and cotton mordanted with tannin.

Acridine dyes

Acridine dyes are all yellow to orange and brown, cationic dyes; they are used in calico printing, dyeing cotton, silk and particularly leather. Some acridine dyes have medicinal and antiseptic properties.

3,6-Diaminoacridine is made by heating a mixture of *m*-phenylenediamine, glycerol, oxalic acid and zinc chloride, and oxidising the leuco-compound (see p. 862 for numbering):

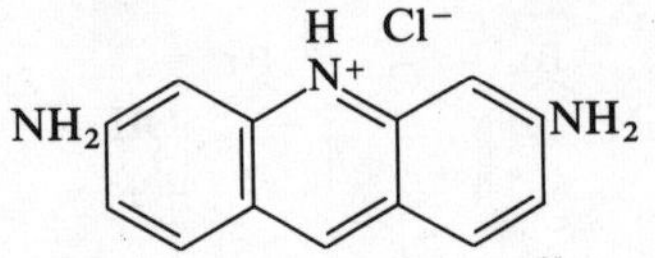

or

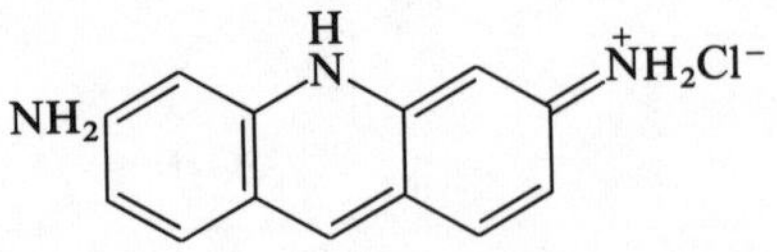

The corresponding sulphate is the antiseptic **proflavine**. On acetylating 3,6-diaminoacridine, methylating the 10-*N* atom by means of methyl *p*-toluenesulphonate, and then hydrolysing the product with dilute hydrochloric acid, 3,6-diamino-10-methylacridinium chloride is produced. This is known as **acriflavine**, and possesses trypanocidal action (*i.e.*, the power to kill trypanosomes which are micro-organisms causing sleeping sickness and other diseases). It has now been replaced by more potent trypanocides, but it is still used as an antiseptic:

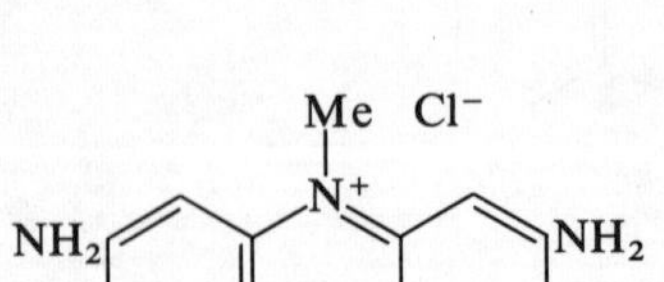

Acridine Yellow G is one of the more important acridine dyes, and is prepared from 2,4-diamino-toluene and formaldehyde:

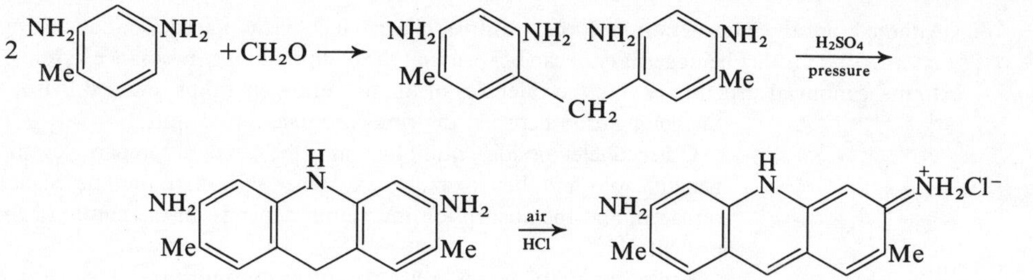

Quinoline dyes

Very few of these have any value as dyes because of their lack of fastness to light. However, most of them are very important photographic sensitisers (see also p. 901). *Cyanines* consist of two heterocyclic nuclei (containing nitrogen) linked by an *odd-numbered* conjugated carbon chain, and are in the form of a mono-acid salt. The term *cyanine*, when not preceded by a prefix, is used for the compounds in which *two quinoline nuclei* are linked through the 4,4′-positions. In other cases, numbers are used to designate other positions of attachment. Since the odd-conjugated chain can be of different lengths, a common practice is to indicate this by the prefix methin (=CH—), trimethin (=CH—CH=CH—), etc.

Cyanine (*methincyanine*; blue) may be prepared by heating a mixture of 1-ethylquinolinium iodide and 1-ethyl-4-methylquinolinium iodide in ethanolic sodium hydroxide solution.

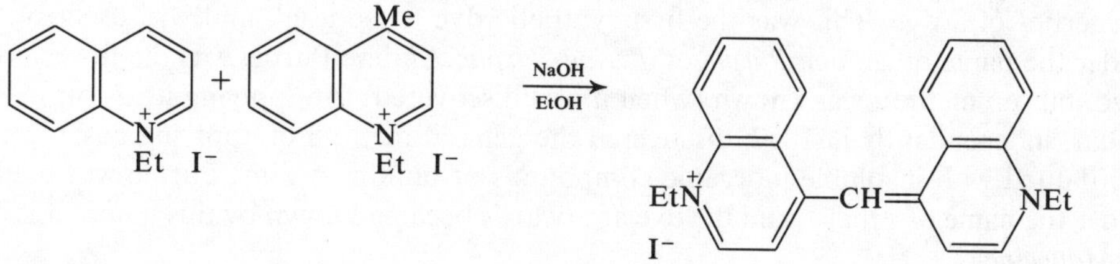

Ethyl Red (2,4′-*methincyanine*) is prepared in a similar manner from quinoline and quinaldine ethiodides.

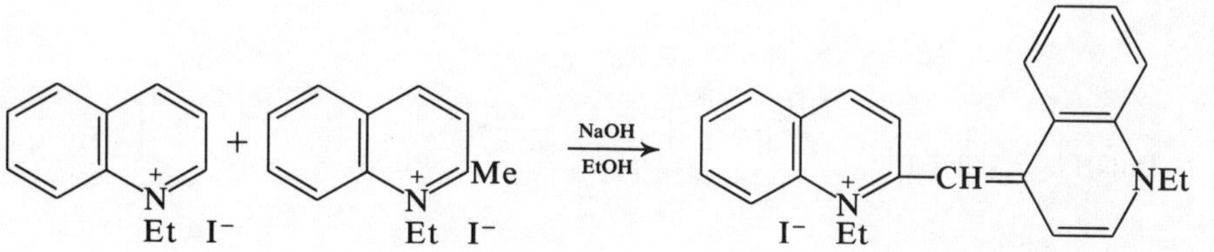

Pinacyanol (2,2′-*trimethincyanine*) may be prepared by heating a mixture of quinaldine ethiodide and ethyl orthoformate in pyridine.

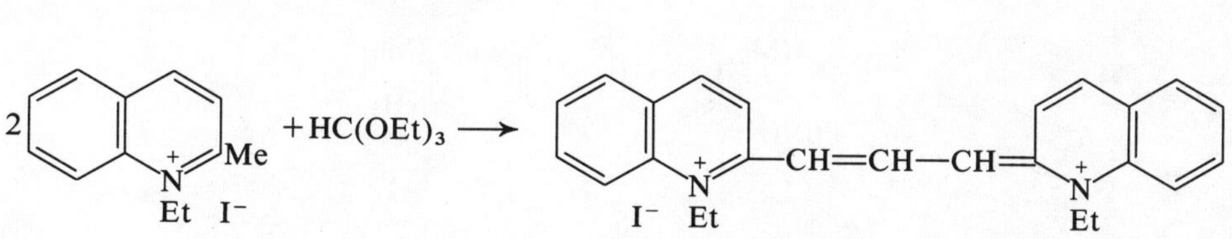

4,4′-Trimethincyanines are also known, and are often called the *kryptocyanines*.

Although in the above cyanines the positive charge has been placed on one nitrogen atom (quaternary), the other being tertiary, it can be seen that the compounds are resonance hybrids, the other 'extreme' canonical structure having the valencies of the two nitrogen atoms reversed. Also, extending the conjugation between the rings produces a red shift.

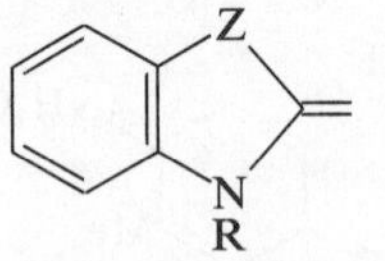

Other nuclei besides quinoline may be used to prepare 'cyanines', *e.g.*, benzthiazole and benzoxazole, etc. One or both quinoline nuclei may be replaced, and the absorption maximum depends on the nature of the nucleus present.

Z=S; benzthiazole
Z=O; benzoxazole

Photographic plates of silver chloride emulsions are sensitive from 350 to 450 nm; those of silver bromide, from 350 to 530 nm. On the other hand, silver bromide emulsions are sensitised to the green region (350 to 600 nm) by cyanine, but the emulsion is fogged. Pinacyanol sensitises the emulsion to the red region (350 to 710 nm), and there is no fogging. Hence, it is possible to make photographic emulsions sensitive to *all* parts of the visible spectrum; they are then said to be *panchromatic*.

Kryptocyanines sensitise emulsions to infra-red radiation.

Azine dyes

Perkin (1856), in search for the synthesis of the alkaloid quinine, oxidised crude aniline sulphate with potassium dichromate, and obtained a black precipitate from which he extracted a mauve colouring matter with methylated spirit. Perkin found that this material had the properties of a dye. This was the first synthetic dye to be used, and was used for dyeing silk under the name of *Aniline Purple* or *Tyrian Purple*. Aniline Purple was the first basic dye to be used, but no method was known (when it was discovered) for applying it to cotton. Perkin and Pullar, independently in 1857, discovered the general method of applying basic dyes to cotton mordanted with insoluble inorganic compounds of tannin. Aniline Purple was used in France under the name of *Mauve*, and the dye afterwards became known by this name; it is also known as *Mauveine*.

Perkin found that his material contained two products, one from pure aniline—*pseudo-Mauveine*, and the other from aniline and *o*-toluidine (the latter was an impurity in crude aniline)—*Mauveine*. The structure of pseudo-mauveine was established by Nietzki who prepared it by oxidising a mixture of *p*-phenylenediamine and diphenyl-*m*-phenylenediamine:

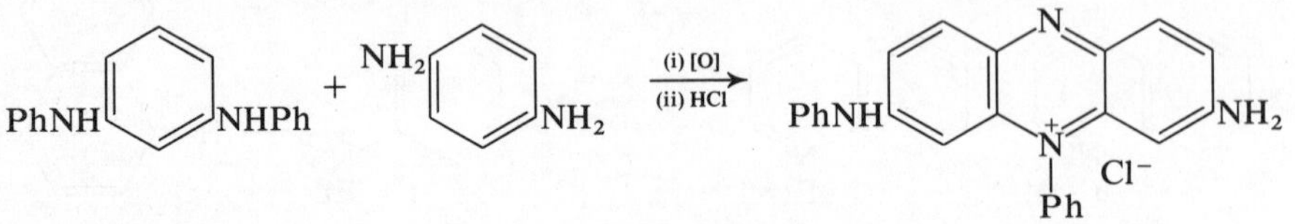

The structure may be the *p*-quinonoid instead of the *o*- shown above (*cf.* fluorescein, etc.).

Mauveine is a mixture of compounds of the type

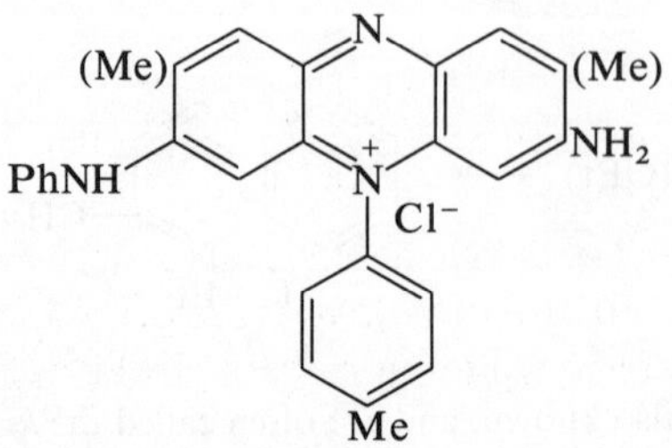

More useful dyes than Mauveine are the **Safranines**—these are more strongly basic than the mauveines. **Safranine T** is prepared by oxidising a mixture of 2,5-diaminotoluene, *o*-toluidine and aniline with sodium dichromate and hydrochloric acid (an indamine is formed as an intermediate):

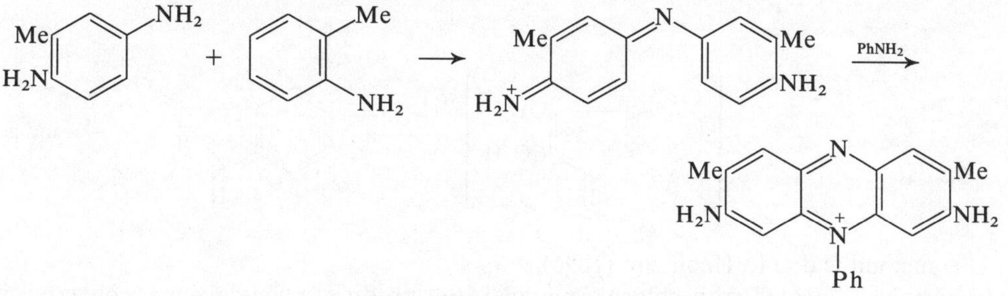

It is a red basic dye, dyeing wool, silk and 'tanned' cotton red shades.

Vat dyes

Indigoid Group. India appears to be the birth-place of indigo—indigotin is its official name. It is the oldest known dye—mummy cloths, 5 000 years old, have been found that were dyed with indigotin. *Woad*, which is an impure form of indigotin (and is obtained from the plant *Isatis tinctoria*), contains a small amount of *indican* which is the glucoside of indoxyl:

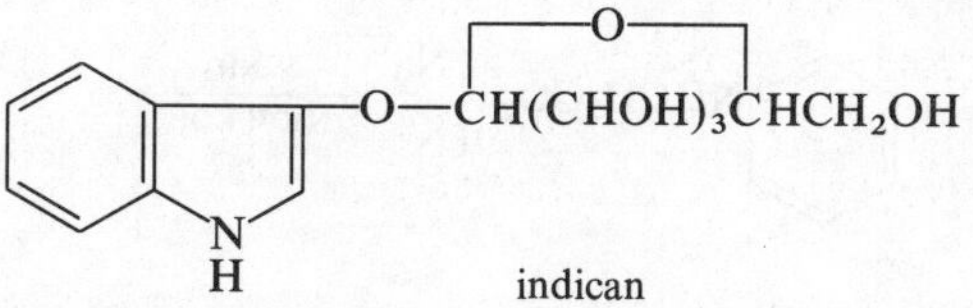

indican

Indigotin used to be obtained from plants of the *indigofera* group, but is now prepared synthetically. These plants do not contain indigotin but indican which, on hydrolysis with hydrochloric acid or by enzymes which occur in the crushed plant, is converted into indoxyl and this, on oxidation by atmospheric oxygen, gives indigotin:

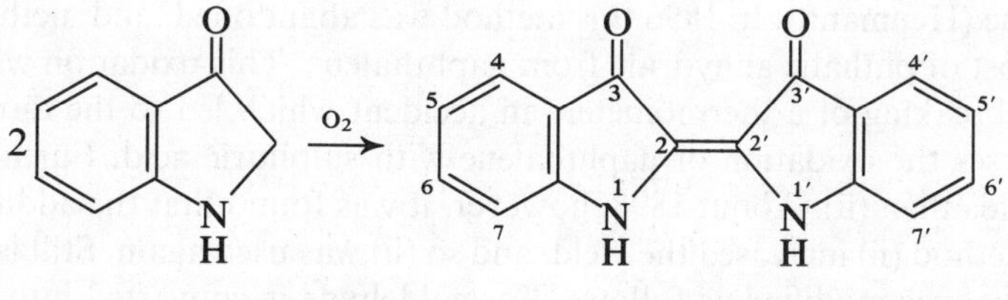

Synthesis of indigotin. Many syntheses of indigotin are known; only the commercial methods are given here.

(i) Anthranilic acid is heated with chloroacetic acid and the product, phenylglycine-*o*-carboxylic acid, is heated with a mixture of potassium hydroxide and sodamide. Indoxylic acid is thereby produced, and this decarboxylates into indoxyl which, on exposure to air, is oxidised to indigotin:

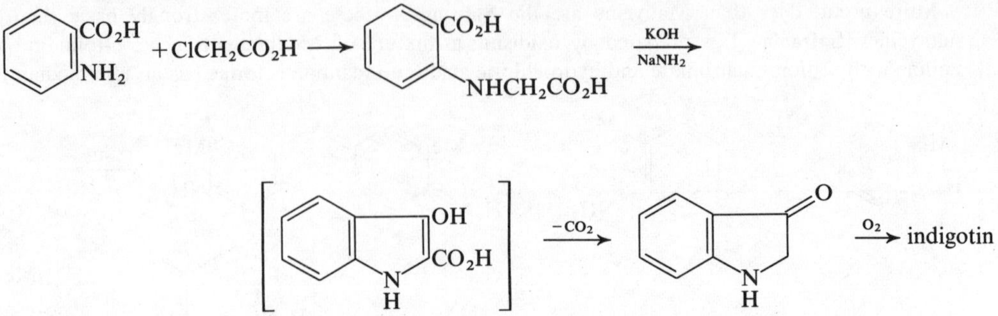

This method is due to Heumann (1896).

(ii) Aniline is heated with chloroacetic acid, the product, phenylglycine, converted into a mixture of its sodium and potassium salts, and these fused with sodamide and a mixture of sodium and potassium hydroxides at 220–240°C. Indoxyl which is thereby obtained is converted into indigotin by atmospheric oxygen (the 'mixed salt' requires a lower temperature of fusion, and consequently the yield of indoxyl is better):

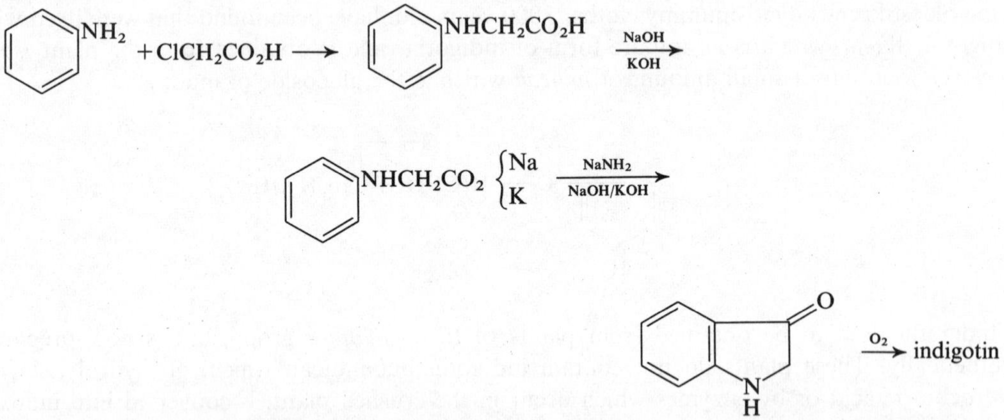

This method was first used in 1890, the fusion, however, being carried out with potassium hydroxide alone (Heumann). In 1896 this method was abandoned, and method (i) was used due to the cheap cost of phthalic anhydride from naphthalene. This oxidation was successful due to the accidental breaking of a thermometer, an accident which led to the discovery in 1893 that mercury catalyses the oxidation of naphthalene with sulphuric acid. Furthermore, method (i) gave better yields than (ii). About 1898, however, it was found that the addition of sodamide in the fusion in method (ii) increased the yield, and so (ii) was used again. Still later the preparation of phenylglycine was modified as follows. Formaldehyde is converted into its bisulphite compound on treatment with an aqueous saturated solution of sodium hydrogen sulphite at 50–80°C. This compound is warmed with aniline at 50–70°C, then treated with aqueous sodium cyanide, and the product hydrolysed with water at 70°C:

$$CH_2O + NaHSO_3 \longrightarrow CH_2(OH)SO_3Na \xrightarrow{C_6H_5NH_2} C_6H_5NHCH_2SO_3Na$$

$$\xrightarrow{NaCN} C_6H_5NHCH_2CN \xrightarrow{H_2O} C_6H_5NHCH_2CO_2H$$

(iii) Aniline, on treatment with ethylene oxide, forms *N*-phenyl-2-hydroxyethylamine and this, on fusion with a mixture of sodium and potassium hydroxides at 200°C, gives the sodium (and potassium) alkoxide. When this sodium salt is heated with sodamide and a mixture of sodium and potassium hydroxides at 200°C, the *N*-sodio-derivative is produced. This is dehydrogenated on heating rapidly to 300°C, and on rapid cooling to 235°C, ring-closure takes place with the formation of the sodium salt of indoxyl. This, when treated with water and allowed to stand exposed to the air, gives indigotin:

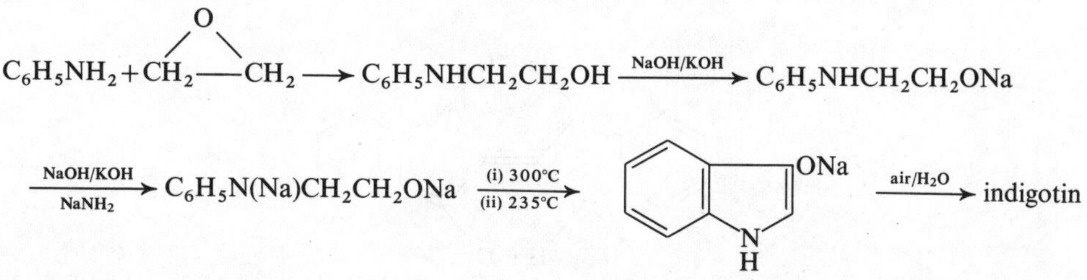

$$C_6H_5NH_2 + CH_2 \overset{O}{-\!\!\!-\!\!\!-} CH_2 \longrightarrow C_6H_5NHCH_2CH_2OH \xrightarrow{\text{NaOH/KOH}} C_6H_5NHCH_2CH_2ONa$$

$$\xrightarrow[\text{NaNH}_2]{\text{NaOH/KOH}} C_6H_5N(Na)CH_2CH_2ONa \xrightarrow[\text{(ii) 235°C}]{\text{(i) 300°C}}$$

$$\xrightarrow{\text{air/H}_2\text{O}} \text{indigotin}$$

Properties of indigotin. Indigotin is a dark blue powder with a coppered lustre. It is insoluble in water, but when its paste is agitated with alkaline sodium hyposulphite in large vats, the insoluble indigotin is reduced to the soluble leuco-compound, *indigotin-white*:

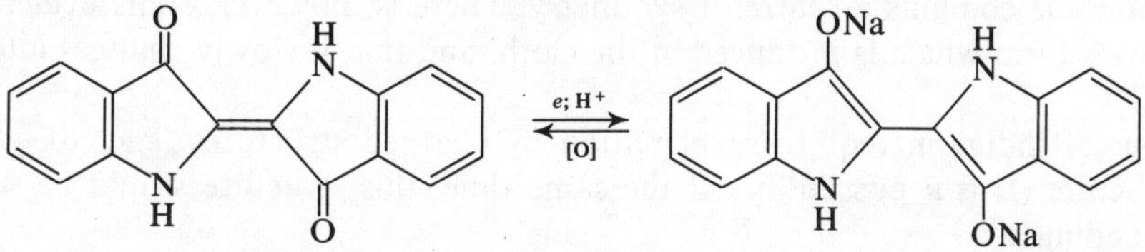

The material to be dyed is soaked in this alkaline solution and then exposed to the air, whereupon the original blue dye is regenerated in the cloth.

In vat-colour printing (of calico), 'local dyeing' is achieved by making the dye pigment into a thick paste (by means of a 'thickener', *e.g.*, starch), and applying this locally in the presence of alkali and a reducing agent, *e.g.*, sodium hydroxymethanesulphinate (*Rongalite, Formosul*), $CH_2(OH)SO_2Na$, which is prepared by reducing formaldehyde bisulphite with zinc and acetic acid. The colour is finally developed by means of an oxidising agent, *e.g.*, sodium nitrite.

In textile printing, *e.g.*, shirtings, dress materials, etc., indigotin-white is not sufficiently stable to be used. Hence textile printing (and ordinary dyeing) may be effected by first converting indigotin into **Indigosol O** by treating indigotin-white with chlorosulphonic acid in the presence of pyridine. This produces the disulphuric ester which, by means of sodium hydroxide, is converted into its sodium salt, Indigosol O. This is readily applied to animal or vegetable fibres by soaking the fabric in the solution, and then oxidising the Indigosol O in *acid* solution (with sodium nitrite) to the original insoluble vat dye. Here we have an example of a vat dye being applied without the vatting process. Indigosol O is stable in neutral or alkaline solution, but is hydrolysed to indigotin-white in acid solution.

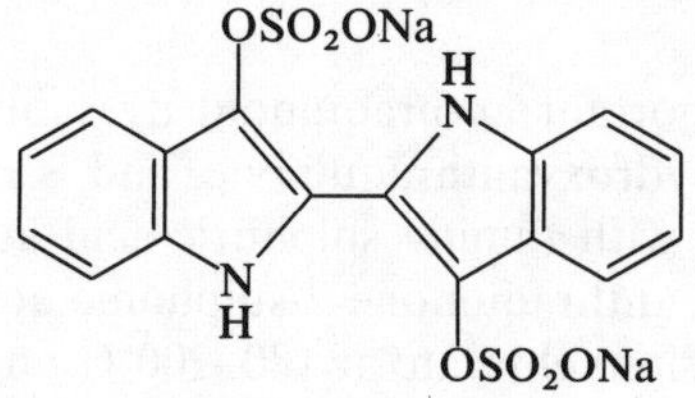

Two geometrical isomers are possible for indigotin:

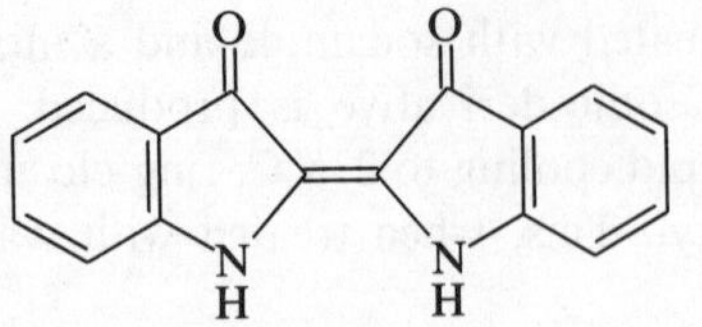

cis-form 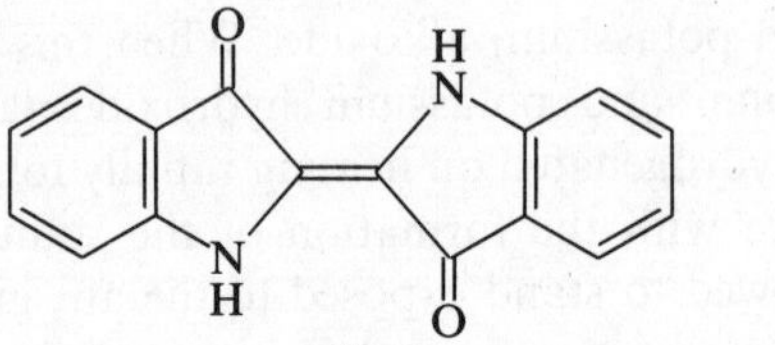   trans-form

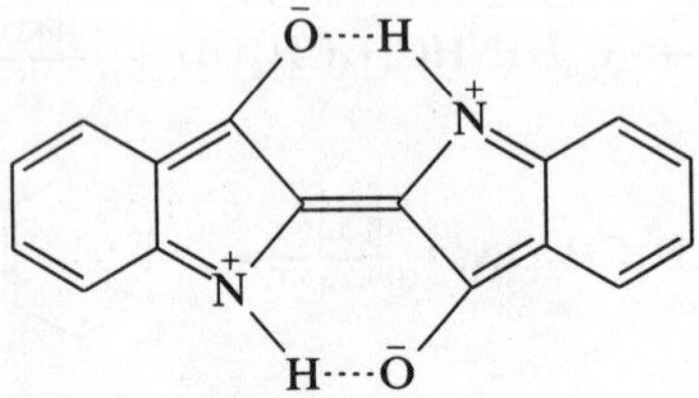

(I)

Derivatives of both are known, but the stable form is the *trans*; X-ray analysis shows that the indigotin molecule contains a centre of symmetry. There is, however, some evidence to show that it is the *cis*-form which is produced in the cloth, and that it slowly changes into the stable *trans*-form.

The colour of indigotin requires explanation by charged structures, *e.g.*, the *o*-quinonoid charged structure (I) is a possibility; at the same time, this structure would be stabilised by hydrogen bonding.

Derivatives of indigotin. Indigotin can be chlorinated or brominated directly in the 5-, 5′-, 7- and 7′-positions. The tetrachloro-derivative, **Brilliant Indigo B**, and the tetrabromo-derivative, **Brilliant Indigo 2B**, are both used commercially; both are brighter than indigotin. **Tyrian Purple** (‘*Purple of the Ancients*’) is 6,6′-dibromoindigotin, and was originally obtained from a species of molluscs in the Mediterranean. It is not commercially important since much cheaper dyes of similar colour are available. **Indigo Carmine** is the sodium salt of indigotin 5,5′-disulphonic acid. **Thioindigo** and many of its derivatives, *e.g.*, **Hydron Pink 2F** (6,6′-dichloro-4,4′-dimethylthioindigo), are important vat dyes. As in the case of indigotin, two geometrical isomers are possible, but since hydrogen bonding is very weak in the sulphur compounds, neither form is stabilised, and in solution both forms appear to exist in equilibrium (Wyman *et al.*, 1951).

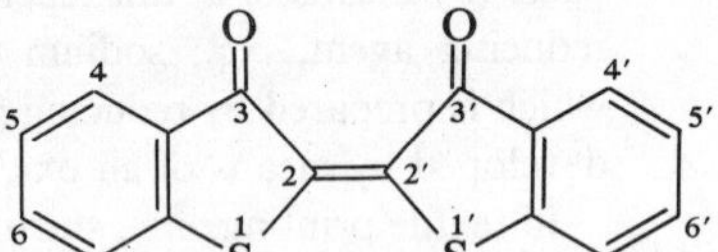

thioindigo

Anthraquinoid dyes

Alizarin is one of the most important anthraquinoid dyes; it is the chief constituent of the madder root. Alizarin is 1,2-dihydroxyanthraquinone, and is now manufactured synthetically by sulphonating anthraquinone with fuming sulphuric acid at high temperatures, and fusing the sodium salt of the product, anthraquinone-2-sulphonic acid, with sodium hydroxide and the calculated quantity of potassium chlorate at 180–200°C under pressure:

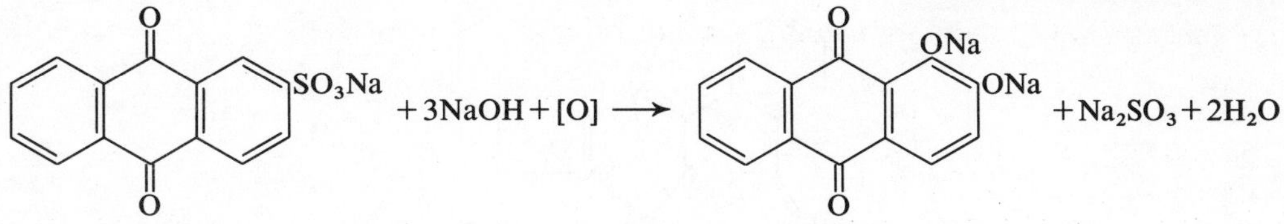

$$+ 3NaOH + [O] \longrightarrow$$

$$+ Na_2SO_3 + 2H_2O$$

The function of the potassium chlorate is to provide oxygen for the oxidation of the 1-carbon atom to COH (see also p. 628).

Properties of alizarin. Alizarin forms ruby red crystals, m.p. 290°C, and dissolves in alkalis to give purple solutions. It is a mordant dye, and the colour of the lake depends on the metal used. Aluminium gives a red lake (*Turkey red*), iron (ferric) a violet-black, and chromium a brown-violet. Aluminium and iron lakes are usually employed for cotton dyeing and for printing, and aluminium and chromium lakes for wood dyeing.

> In addition to the aluminium compound, the lake contains calcium salts, Turkey-red oil and sodium phosphate. These additional compounds affect the brightness of the lake and increase its stability towards acids and alkalis. It appears that aluminium hydroxide alone does not yield a lake with alizarin, but only after the addition of a calcium salt.
>
> Some anthraquinoid dyes of various types are:

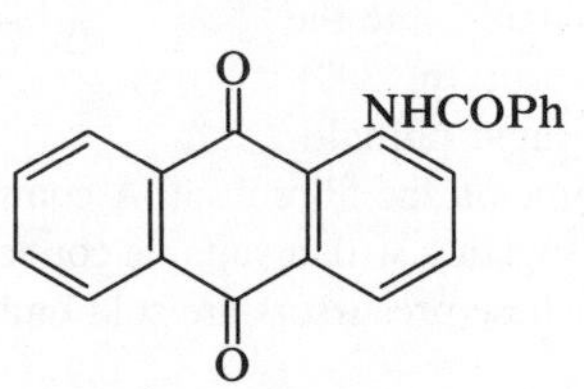

Indanthrene Yellow GK
(vat dye)

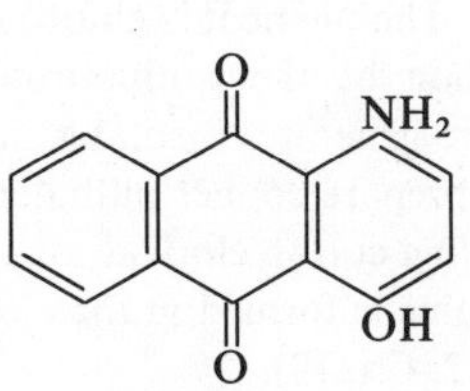

Duranol Red 2B
(disperse dye)

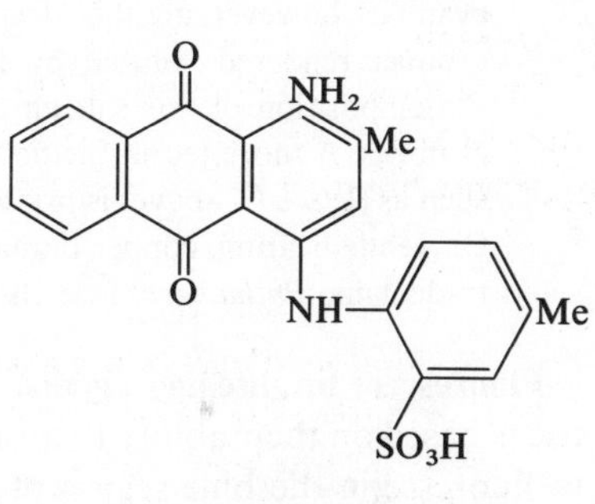

Kiton Fast Blue CR
(acid dye)

Phthalocyanine pigments and dyes

The **phthalocyanines** are a very important class of organic dyes and pigments; they are coloured blue to green. They are very fast to light, heat, acids and alkalis, and are very useful for paints, printing inks, synthetic plastics and fibres, rubber, etc.

The phthalocyanine dyes were discovered by accident at the works of Scottish Dyes Ltd. in 1928. It was there observed that some lots of phthalimide, manufactured by the action of ammonia on molten phthalic anhydride in an iron vessel, were contaminated with a blue pigment. The structure and method of formation of this iron compound were established by Linstead and his co-workers (1934).

The first commercial phthalocyanine dye was **Monastral Fast Blue BS**. This is copper phthalocyanine, the copper atom taking place of the hydrogen atoms in the two NH groups of phthalocyanine itself and being co-ordinated with the other two nitrogen atoms. The colour depends on the metal (copper, magnesium, lead, etc.), and the greener shades are obtained by direct chlorination or bromination.

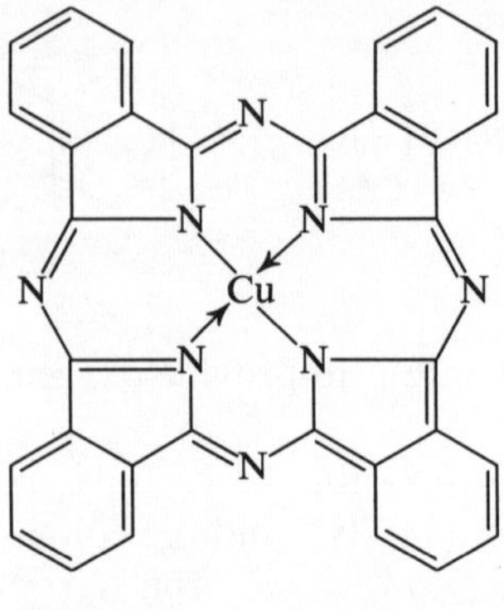

Metal phthalocyanines may be prepared:

(i) By passing ammonia into molten phthalic anhydride or phthalimide in the presence of a metal salt.

(ii) By heating *o*-cyanoarylamides or phthalonitriles with metals or metallic salts.

(iii) By heating phthalic anhydride or phthalimide with urea and a metallic salt, preferably in the presence of a catalyst such as boric acid.

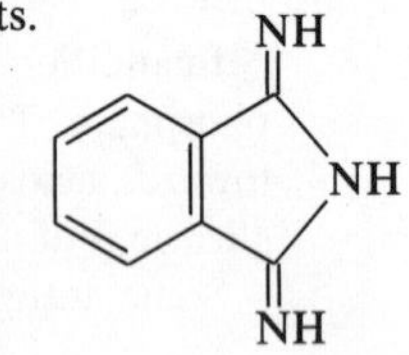

Phthalocyanine may be prepared by heating phthalonitrile with a little triethanolamine.

Phthalocyanines are made water soluble by sulphonation, and these soluble salts are used as direct dyes. The most successful water-soluble copper phthalocyanines, however, are the **Alcian dyes**. The pigment is chloromethylated, and the product rendered soluble by formation of the isothiouronium salt (p. 399). Decomposition of this salt on the fibre generates insoluble substituted phthalocyanines. A more recent method is to prepare copper phthalocyanine on the fibre itself. A compound such as (I; see (i) above) is printed on to a cotton cloth as a paste, together with a suitable copper salt. On gentle heating, copper phthalocyanine is formed in the cloth. These precursors are sold under the trade name *Phthalogens* (see also Vol. 2, Ch. 19).

(I)

Fluorescent brightening agents. These compounds are often called **optical bleaches**, and their use is based on their ability to absorb in the region 330–390 nm (hence they are 'colourless') and to fluoresce in the blue region of the spectrum (430–490 nm). Fabrics, after repeated washing, tend to develop a yellow colour. Since the complementary colour of yellow is blue (see Table 31.1), 'addition' of blue light will cause the fabric to appear white. Hence the use of detergents containing 'whiteners', many of which are stilbene derivatives (as sodium salts), *e.g.*,

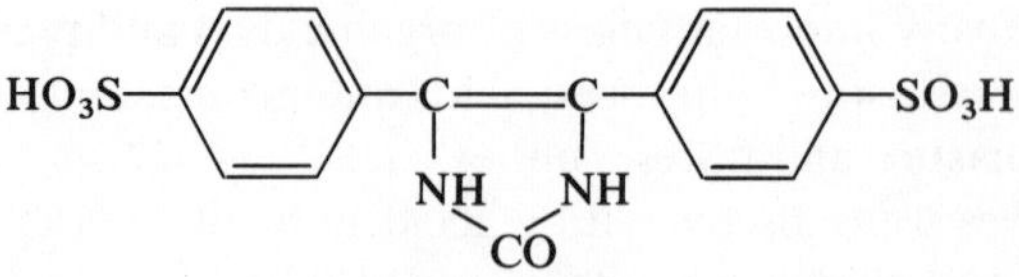

Blankophor WT

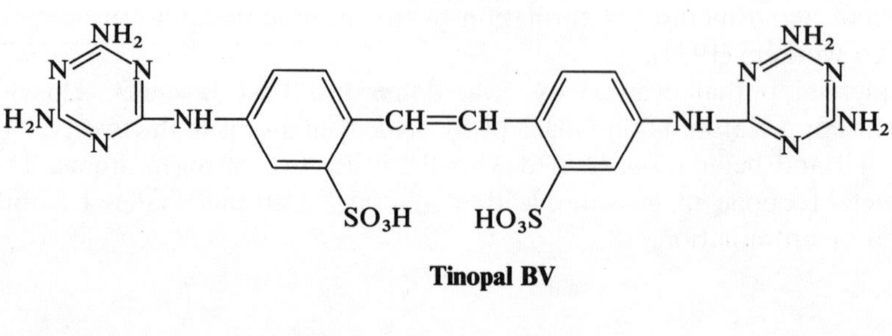

Tinopal BV

Organic photochemistry

Introduction

For chemical reaction to occur, the reacting molecules must be activated, and as we have seen, molecules usually acquire this activation energy by increasing their kinetic energy. In *photochemical reactions*, however, the activation energy is obtained by the absorption of radiation (light energy). In order to study the mechanisms of photochemical reactions, a necessary requirement is the nature of the electronic transition, *i.e.*, whether it is $\sigma \to \sigma^*$, $\pi \to \pi^*$, or $n \to \pi^*$, and also whether the state is singlet or triplet. As we have seen (p. 870), a molecule in the ground state is raised to an excited singlet state by absorption of a quantum of the appropriate energy. We have also seen how an excited molecule may lose this excess of energy to return to the ground state (loss by collision, fluorescence, and phosphorescence). However, because of their excitation, the molecules are more reactive (than ground-state molecules) and so can undergo chemical reaction.

In view of what has been said, it might be expected that *one* molecule would react for *each* quantum absorbed. This is the **law of photochemical equivalence** (Einstein, 1912), but is rarely observed in practice. The reason for this has been explained as follows. Photochemical reactions consist of *two* processes. The first is the **primary process**, in which the light is absorbed to produce the excited molecule. If the energy is high enough, dissociation occurs to give free radicals. Since this often involves fission of carbon-carbon single bonds (bond energy = $348\,\text{kJ mol}^{-1}$), a quantum of $\lambda = 300\,\text{nm}$ ($\Delta E = \sim 398\,\text{kJ mol}^{-1}$) will bring about homolytic fission. The primary process (which obeys the law of photochemical equivalence) is then followed by **secondary processes**, in which 'products' of the primary process undergo further reactions. The (total) **quantum yield**, Φ, *i.e.*, the *efficiency of the process*, is then defined as

$$\Phi = \frac{\text{Number of molecules reacting or produced}}{\text{Number of quanta absorbed}}$$

Φ is rarely unity in practice; its value varies from 10^{-2} to 10^4 (or more).

Photochemical secondary processes are of various types, *e.g.*, substitution, addition, dimerisation, elimination, rearrangement, etc.

Photochemical reactions are usually divided into two main types, **direct light-induced** or **non-sensitised reactions**, and **photosensitised reactions**. In non-sensitised reactions, the molecule that actually absorbs the light undergoes various secondary processes to give the final products of the reaction. On the other hand, in photosensitised reactions, one type of molecule, the *donor* (D) that is present absorbs the light (which is of the wavelength characteristic of the donor), becomes excited, and then transfers its excitation energy to another type of molecule, the *acceptor* (A), that is also present. In this way, the donor molecule returns to the ground state and the acceptor molecule is raised to an excited state, which then undergoes emission or photochemical reaction. This transfer of energy from D to A, followed by emission of light or reaction of A is called **sensitisation**.

For the donor-acceptor relationship to function, certain conditions must be fulfilled: (i) The donor-excited state must have a sufficiently long life-time to be able to transfer its energy to the acceptor molecule. Since the triplet state has a longer life-time than the singlet state, sensitisation involving the former is much more favourable than the latter. It therefore follows that sensitisers will be more effective the more efficient is their intersystem crossing. (ii) The transfer of energy from D to A will occur only if the excitation energy of the former is higher than that of the latter. A common sensitiser is benzophenone; the energies of its S_1 and T_1 states are, respectively, $346\,\text{kJ}$ and $291\,\text{kJ}$ above the S_0 state. (iii) In general, the multiplicities of the D and A excited states are the same.

Since molecules excited above the S_1 state rapidly decay to this state, it is the first excited triplet state that is usually involved in photochemical reactions. The actual transfer of energy from the donor to the acceptor molecule occurs through collisions, and occurs on *every* collision if the above conditions are fulfilled. The most usual types of energy transfer are (the asterisk represents the excited state):

$$D(S_0) \xrightarrow{h\nu} D^*(S_1) \longrightarrow D^*(T_1) \xrightarrow{A} D(S_0) + A^*(T_1)$$
$$\downarrow A$$
$$D(S_0) + A^*(S_1)$$

Photochemical reactions are being increasingly used for the synthesis of organic compounds, and in many cases, for the preparation of new compounds that cannot be obtained by the usual methods.

Many photochemical reactions have been mentioned in the text (see Index), but no details of mechanisms have been described. These will now be discussed, and since the most widely studied reactant molecules are the carbonyl compounds, they will be largely used to describe the various types of photochemical reactions. However, before proceeding further, let us first define two additional terms commonly used in photochemistry. **Photolysis** is the term used to describe the breaking of chemical bonds by light energy. The bonds are broken homolytically to form free radicals, and sometimes small molecules are eliminated (*cf.* mass spectrometry). In relatively few cases, photolysis produces ions. **Flash photolysis** is the process in which a flash of light of very high intensity, lasting a number of microseconds (10^{-6} s), irradiates the reaction mixture. In this way, reactive intermediates are formed in high concentration and their rate of decay may be measured by various means, particularly spectroscopically. Porter *et al.* (1968), by means of a laser pulse, have shortened the flash duration to 18 nanoseconds (10^{-9} s).

Photochemical elimination reactions. For ketones, these reactions usually involve the loss of carbon monoxide. An extensively studied reaction is the photolysis of acetone in the vapour phase (v.p.), and it has been found that the products depend on the temperature of the experiment:

(*a*) *Elevated temperature:* the products are carbon monoxide and ethane.

$$CH_3\overset{O}{\overset{\|}{C}}CH_3 \xrightarrow[(v.p.)]{h\nu} [CH_3\overset{O}{\overset{\|}{C}}CH_3]^* \longrightarrow CH_3\overset{O}{\overset{\|}{C}}\cdot + CH_3\cdot$$

$$CH_3\overset{O}{\overset{\|}{C}}\cdot \longrightarrow CH_3\cdot + CO$$
$$2CH_3\cdot \longrightarrow C_2H_6$$

The excited state of acetone could be S_1 or T_1, but it has been shown that the decomposition via the triplet state is the more efficient, the quantum yield (Φ) being almost unity. It should also be noted that of all the bonds present in acetone, the C—C bond is the weakest.

(*b*) *Room temperature:* the products are biacetyl, acetaldehyde, methane, and possibly hexane-2,5-dione.

$$CH_3\overset{O}{\overset{\|}{C}}CH_3 \xrightarrow{h\nu} [CH_3\overset{O}{\overset{\|}{C}}CH_3]^* \longrightarrow CH_3\overset{O}{\overset{\|}{C}}\cdot + CH_3\cdot$$

$$CH_3\overset{O}{\overset{\|}{C}}\cdot + CH_3\overset{O}{\overset{\|}{C}}CH_3 \longrightarrow CH_3\overset{O}{\overset{\|}{C}}H + \cdot CH_2\overset{O}{\overset{\|}{C}}CH_3$$

$$CH_3\cdot + CH_3\overset{\overset{\displaystyle O}{\|}}{C}CH_3 \longrightarrow CH_4 + CH_3\overset{\overset{\displaystyle O}{\|}}{C}CH_2\cdot$$

$$2CH_3\overset{\overset{\displaystyle O}{\|}}{C}\cdot \longrightarrow CH_3-\overset{\overset{\displaystyle O}{\|}}{C}-\overset{\overset{\displaystyle O}{\|}}{C}-CH_3$$

$$2CH_3\overset{\overset{\displaystyle O}{\|}}{C}CH_2\cdot \longrightarrow CH_3\overset{\overset{\displaystyle O}{\|}}{C}CH_2CH_2\overset{\overset{\displaystyle O}{\|}}{C}CH_3$$

Acetone is a symmetrical molecule, but when the ketone is unsymmetrical, preferential fission occurs to give the more stable alkyl radical, *e.g.*, photodissociation of ethyl methyl ketone gives ethyl and methyl radicals, with the former predominating:

$$CH_3\overset{\overset{\displaystyle O}{\|}}{C}\cdot + C_2H_5\cdot \overset{h\nu}{\longleftarrow} CH_3\overset{\overset{\displaystyle O}{\|}}{C}C_2H_5 \overset{h\nu}{\longrightarrow} CH_3\cdot + \cdot C C_2H_5$$

Photochemical reactions of ketones (and aldehydes) which result in fission of the C—CO bond followed by elimination of carbon monoxide, are classified as the **Norrish type I process**.

As the length and complexity of the alkyl groups increase, the products of photolysis become more complicated. Thus, when a hydrogen atom in the alkyl group can enter into the formation of a six-membered cyclic T.S. involving the oxygen atom as one member of the ring, *γ-hydrogen abstraction* and fission of the molecule occur to give an alkene and a ketone (via the enol form), *e.g.*, hexan-2-one gives mainly propene and acetone, the elimination of carbon monoxide occurring to a lesser extent (*cf.* mass spectra of ketones, p. 210).

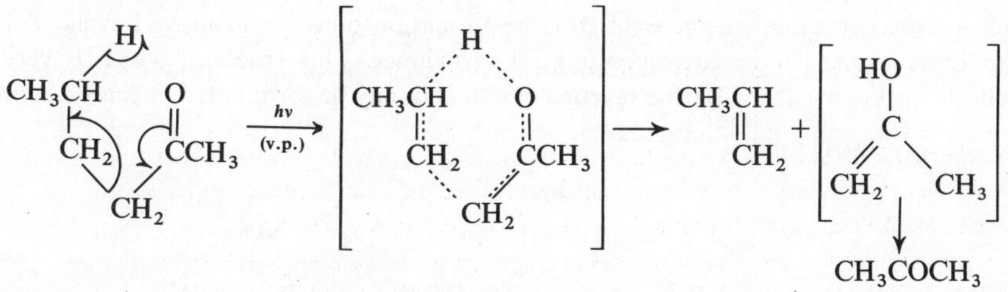

This type of photolysis undergone by ketones (and aldehydes) is called the **Norrish type II process**.

α,β-Unsaturated ketones, provided the light energy is high enough, undergo photolysis with elimination of carbon monoxide (see also later).

Cyclic ketones undergo elimination of carbon monoxide and also ring opening with the formation of an unsaturated aldehyde, etc., *e.g.*, cyclopentanone:

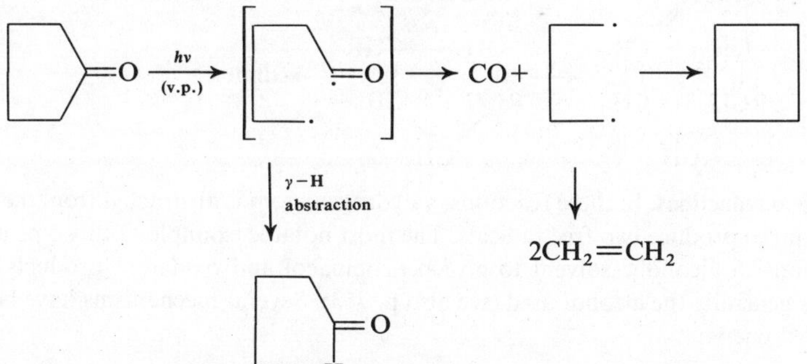

The photolysis of aldehydes is more complicated than that of ketones, the usual primary reaction resulting in the formation of the *formyl radical*, *i.e.*,

$$RCHO \xrightarrow[\text{(v.p.)}]{hv} R\cdot + \cdot CHO$$

This is then generally followed by the secondary process:

$$R\cdot + \cdot CHO \longrightarrow RH + CO$$

All of the foregoing decarbonylations are vapour phase reactions, and the products are the same if the photochemical reactions are carried out in a *chemically inert solvent*. In protic solvents, the products are usually different (see later).

Nitrogen is another type of molecule that is readily eliminated from compounds containing nitrogen-nitrogen double bonds, particularly azo-compounds, *e.g.*, azomethane in the gas phase (g.p.) ($\Phi = 1$).

$$CH_3-N{=}N-CH_3 \xrightarrow[\text{(g.p.)}]{hv} N_2 + 2CH_3\cdot \longrightarrow C_2H_6$$

In the same way, diazomethane gives nitrogen and methylene (*cf.* p. 119):

$$CH_2N_2 \xrightarrow[\text{(g.p.)}]{hv} :CH_2 + N_2$$

The photolysis of diazonium salts in alcohols gives aromatic hydrocarbons and aryl ethers, and the mechanism is believed to be via aryl free-radicals (see also p. 672).

The photolysis of nitrites which do not contain γ-hydrogen atoms usually results in the elimination of nitric oxide and the formation of hydroxy and carbonyl compounds, *e.g.*, [NO $\equiv$ NO$\cdot$]:

$$\overset{\beta}{C}H_3\overset{\alpha}{C}H_2CH_2ONO \xrightarrow{hv} NO + CH_3CH_2CH_2O\cdot$$

$$\begin{array}{l} CH_3CH_2CH-O\cdot \\ \quad \vert \\ \quad H \cdot OCH_2CH_2CH_3 \end{array} \longrightarrow CH_3CH_2CH{=}O + HOCH_2CH_2CH_3$$

When γ-hydrogen atoms are present (for the formation of a six-membered cyclic T.S.), then the product is an oxime or the corresponding nitroso dimer. This reaction, which is very important synthetically, is known as the **Barton reaction** (1960) and may be formulated in general terms as shown (note the nitroso-compound intermediate).

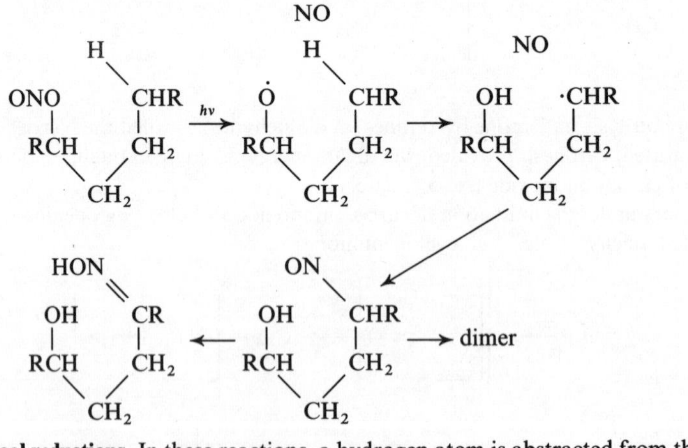

Photochemical reductions. In these reactions, a hydrogen atom is abstracted from the solvent or from another reactant to produce two free radicals. The most notable example of this type is the irradiation of benzophenone in alcoholic solvent to give benzopinacol and oxidation products of the alcohol; isopropanol is generally the alcohol used (see also p. 763). Several mechanisms have been proposed; a widely accepted one is:

(i) $Ph_2C{=}O \xrightarrow{hv} [Ph_2C{=}O]^*(S_1) \longrightarrow [Ph_2C{=}O]^*(T_1)$

(ii) $[Ph_2C{=}O]^*(T_1) + Me_2CHOH \longrightarrow Ph_2\dot{C}-OH + Me_2\dot{C}-OH$

(iii) $Me_2\dot{C}{-}OH + Ph_2C{=}O \longrightarrow Me_2C{=}O + Ph_2\dot{C}{-}OH$

(iv) $2Ph_2\dot{C}{-}OH \longrightarrow Ph_2C(OH)C(OH)Ph_2$

There is much evidence for this mechanism. *Two* molecules of benzophenone are consumed; one *excited* molecule (step i) is converted into the benzhydryl radical (step ii), and another *ground-state* benzophenone molecule reacts with the hydroxyisopropyl radical to give a benzhydryl radical (step iii). Thus, the quantum yield would be expected to be 2, and this is the value obtained in practice (under ideal experimental conditions). There is also much evidence to show that it is the triplet state of benzophenone that is involved. Benzophenone, in the presence of benzhydrol, also gives benzopinacol on irradiation:

(v) $[Ph_2C{=}O]^*(T_1) + Ph_2CHOH \longrightarrow 2Ph_2\dot{C}OH \longrightarrow Ph_2C(OH)C(OH)Ph_2$

When the photoreduction is carried out in the presence of a low concentration of benzhydrol, the quantum yield is still appreciable, thereby implying that the excited state has a relatively long life-time (hence triplet rather than singlet; p. 872). Also, fluorescence has never been observed for benzophenone solutions, and this shows that the singlet state decays rapidly, too rapidly to react with benzhydrol molecules in solution. Calculations have shown that benzophenone singlets undergo intersystem crossing to benzophenone triplets extremely rapidly; this explains the failure of benzophenone to fluoresce and supports the argument that it is benzophenone triplets which are the excited molecules involved in hydrogen abstraction.

At first sight it might have been expected from step (ii) that pinacol would be produced by dimerisation of the hydroxyisopropyl radicals. The absence of this product may be explained on the basis that the life-time of the hydroxyisopropyl radical is too short for it to combine with another radical to form the dimer; it transfers its hydroxylic proton too rapidly. On the other hand, the benzhydryl radical has a relative long life-time due to resonance stabilisation, and is sufficiently stable not to attack Me_2CHOH.

One other point about this reduction is the structure of the benzophenone triplet state. There are two possibilities, (n,π^*) and (π,π^*) triplets. The former, however, has more free-radical character than

(n,π^*) (π,π^*)

the latter, and can therefore abstract hydrogen atoms relatively easily. On the other hand, aromatic ketones which are not readily reduced are believed to be in the (π,π^*) triplet state.

Photochemical oxidations. Many photochemical reactions, when carried out in the presence of oxygen, result in the formation of peroxides, hydroperoxides, oxidation of a functional group, etc. Although oxygen is a triplet (diradical) in its ground state, it does not add to cyclic dienes. In the presence of a sensitiser, *e.g.*, fluorescein, eosin, the oxygen is raised to its S_1 state (all electrons paired), and it is the oxygen in this state that is the reactive species. Many linear polynuclear aromatic compounds, however, add oxygen in the absence of a sensitiser. In these cases, the substrate molecule is believed to behave as the sensitiser. Thus, the photo-oxidation of anthracene to form the transannular (9,10-) peroxide may possibly proceed as follows:

Alkenes which possess an allylic hydrogen atom tend to produce the 'allylic isomerised' hydroperoxide when they undergo photosensitised oxidation. This involves hydrogen abstraction, and it appears that a cyclic T.S. is involved, *e.g.*, 4-methylpent-2-ene:

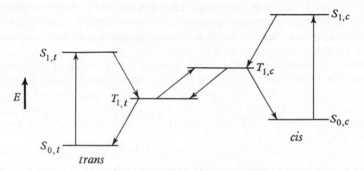

Photochemical *cis-trans* isomerisations. *cis-trans*-Isomerisation can be carried out (usually in solution) by irradiation alone or in the presence of a sensitiser or a catalyst (see also p. 498). In general, an equilibrium mixture is reached, the *cis-trans* ratio remaining constant no matter how much longer the irradiation is continued. This condition, called a **photostationary state**, is independent of which isomer is the starting material and always contains predominantly the *cis*-isomer. The actual ratio of the *cis-trans* forms, however, depends on a number of factors, *e.g.*, solvent, temperature, nature of the sensitiser, etc.

The *cis-trans* isomerisation of the stilbenes has been examined in great detail, and so we shall use this as our example. The evidence is strongly in favour that the reaction proceeds via a triplet state (Ph—ĊH—ĊH—Ph), but since the *cis-trans* ratio depends on the temperature, several theories have been proposed to explain the isomerisation. According to one theory, the *cis*- and *trans*-forms give

Fig. 31.9

different excited states, and the temperature effect is due to the existence of an energy barrier between these excited states (Fig. 31.9). Because of steric effects, the *cis*-isomer has a higher energy content than the *trans* in the ground state, and this is also (usually) the case for the corresponding excited states. Also, because of this larger steric effect, the molar absorptivity of the *cis*-isomer is (usually) lower than that of the *trans*-isomer. Hence, the population of the *trans* excited state is greater than that of the *cis*. These excited states can return to their respective ground states (Fig. 31.9) or, by rotation, the *cis** (T_1) and *trans** (T_1) become interconvertible. However, in this interconversion, only the *cis** → *trans** is energetically favourable (Fig. 31.9), but nevertheless, the overall process favours the *trans* → *cis* interconversion. The reason for this is that the *trans* population is greater than that of the *cis*, and the *cis* → *cis** is more difficult than the *trans* → *trans** (see above).

In *photosensitised cis-trans* isomerisation, the sensitiser excitation energy (of the T state) must be higher than that of the *cis*- and *trans*-isomers for energy transfer to occur. Hence, in practice, as long as the sensitiser has a triplet energy state above a certain minimum, the *cis-trans* ratio is very little affected by the nature of the sensitiser. Thus, if benzophenone is the sensitiser (donor) and stilbene (acceptor) may be written (*cf.* p. 904):

(i) $Ph_2CO\ (S_0) \xrightarrow{hv} [Ph_2CO]^*\ (S_1) \longrightarrow [Ph_2CO]^*\ (T_1)$

(ii) $[Ph_2CO]^* (T_1) + cis\text{-}Ph_2C_2H_2 \longrightarrow Ph_2CO (S_0) + \begin{bmatrix} H \quad\quad H \\ \diagdown\dot{C}\text{---}\dot{C}\diagup \\ \diagup \quad\quad \diagdown \\ Ph \quad\quad Ph \end{bmatrix}^* (T_1)$

(iii) $[Ph_2CO]^* (T_1) + trans\text{-}Ph_2C_2H_2 \longrightarrow Ph_2CO (S_0) + \begin{bmatrix} Ph \quad\quad H \\ \diagdown\dot{C}\text{---}\dot{C}\diagup \\ \diagup \quad\quad \diagdown \\ H \quad\quad Ph \end{bmatrix}^* (T_1)$

(iv) $\begin{bmatrix} H \quad\quad H \\ \diagdown\dot{C}\text{---}\dot{C}\diagup \\ \diagup \quad\quad \diagdown \\ Ph \quad\quad Ph \end{bmatrix} (T_1) \rightleftharpoons \begin{bmatrix} Ph \quad\quad H \\ \diagdown\dot{C}\text{---}\dot{C}\diagup \\ \diagup \quad\quad \diagdown \\ H \quad\quad Ph \end{bmatrix} (T_1)$

↓ spin inversion   ↓ spin inversion

$\begin{matrix} H \quad\quad H \\ \diagdown C{=}C \diagup \\ \diagup \quad\quad \diagdown \\ Ph \quad\quad Ph \end{matrix}$ (S_0) $\begin{matrix} Ph \quad\quad H \\ \diagdown C{=}C \diagup \\ \diagup \quad\quad \diagdown \\ H \quad\quad Ph \end{matrix}$ (S_0)

Actually, the mechanism given is an over-simplification; there is evidence to show that another triplet state of stilbene (intermediate in energy content between the *cis* and *trans* triplet states) may also be produced. This triplet state, because it can be reached *directly* from the ground state (normally a forbidden transition) has been called a **phantom triplet** (Hammond, 1960).

The structures of the *cis* and *trans* triplet states of stilbene are not clear. Both would be expected to be resonance hybrids (of the canonical forms given in equations ii and iii, above), but because rotation can occur, the CH—CH bond in both would be expected to be predominantly single in character. However, because of the different steric effects operating, the double bond character in the two forms will be different, presumably greater in the *trans* than in the *cis*, since the former is more stable than the latter (Fig. 31.9).

Photochemical *cis-trans* isomerisation can also be effected in the presence of, *e.g.*, bromine or iodine. A probable mechanism is:

$$Br_2 \xrightarrow{h\nu} 2Br\cdot$$

$\begin{matrix} Me \quad\quad Me \\ \diagdown C{=}C \diagup \\ \diagup \quad\quad \diagdown \\ H \quad\quad H \end{matrix} \xrightarrow{Br\cdot} \begin{matrix} Me \quad\quad Me \\ \diagdown\dot{C}\text{---}C\text{---}Br \\ \diagup \quad\quad \diagdown \\ H \quad\quad H \end{matrix} \rightleftharpoons \begin{matrix} H \quad\quad Me \\ \diagdown\dot{C}\text{---}C\text{---}Br \\ \diagup \quad\quad \diagdown \\ Me \quad\quad H \end{matrix}$

$\downarrow -Br\cdot$ $\downarrow -Br\cdot$

$\begin{matrix} Me \quad\quad Me \\ \diagdown C{=}C \diagup \\ \diagup \quad\quad \diagdown \\ H \quad\quad H \end{matrix}$ $\begin{matrix} H \quad\quad Me \\ \diagdown C{=}C \diagup \\ \diagup \quad\quad \diagdown \\ Me \quad\quad H \end{matrix}$

In addition to photochemical *cis-trans* isomerisations of alkenes, other systems, *e.g.*, oximes, azo-compounds, can be isomerised under similar conditions.

Photochemical rearrangements. α,β-Unsaturated ketones (acyclic and cyclic enones and dienones) undergo various types of rearrangements when irradiated with light. Acyclic α,β-unsaturated ketones containing a γ-hydrogen atom often undergo double-bond migration provided the light energy is not high enough to cause dissociation. The presence of the γ-hydrogen is necessary for the formation of a six-membered cyclic T.S., leading to hydrogen abstraction and transfer to the oxygen atom (*cf.* p. 906), *e.g.*, 5-methylhex-3-en-2-one.

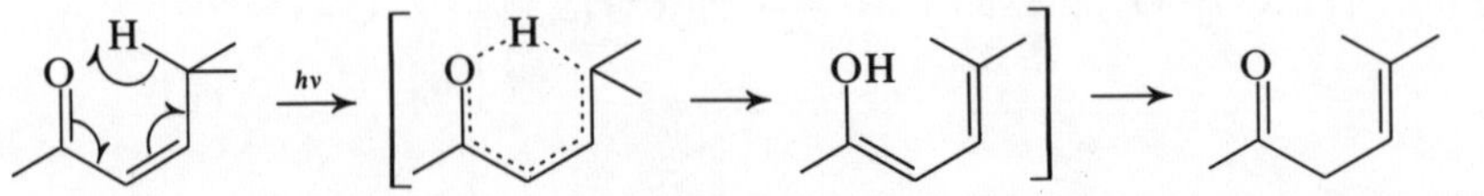

It should be noted that the system changes from conjugation to non-conjugation, and there is *no* equilibrium (as is the case with *cis-trans* isomerisation). The reason for this is that irradiation is carried out with a wavelength absorbed by the *conjugated* substrate, and since this is longer than that absorbed by the non-conjugated product, the latter remains unaffected once it is formed. Thus, isomerisation gives completely the less stable isomer. This type of isomerisation is an example of **optical pumping**.

The most widely studied compounds that undergo photochemical rearrangements are the dienones. These undergo the dienone-phenol rearrangement in acid solution (see p. 729), and also undergo this rearrangement under the influence of light, but in this case the reaction is more complex, *e.g.*, 4,4-diphenylcyclohexa-2,5-dienone (a cross-conjugated dienone), in aqueous dioxan solution, gives four products:

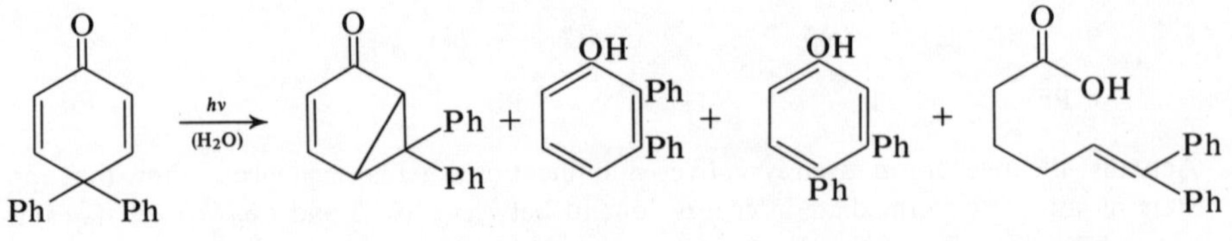

One mechanism that has been proposed (Zimmerman *et al.*, 1964) is given here to illustrate the variety of steps that can be involved in a photochemical reaction. (I) is the compound that leads to the formation of all other products, and it should be noted that the steps involve first a series of free radicals and then a series of carbonium ions, the latter undergoing rearrangements characteristic of such ions.

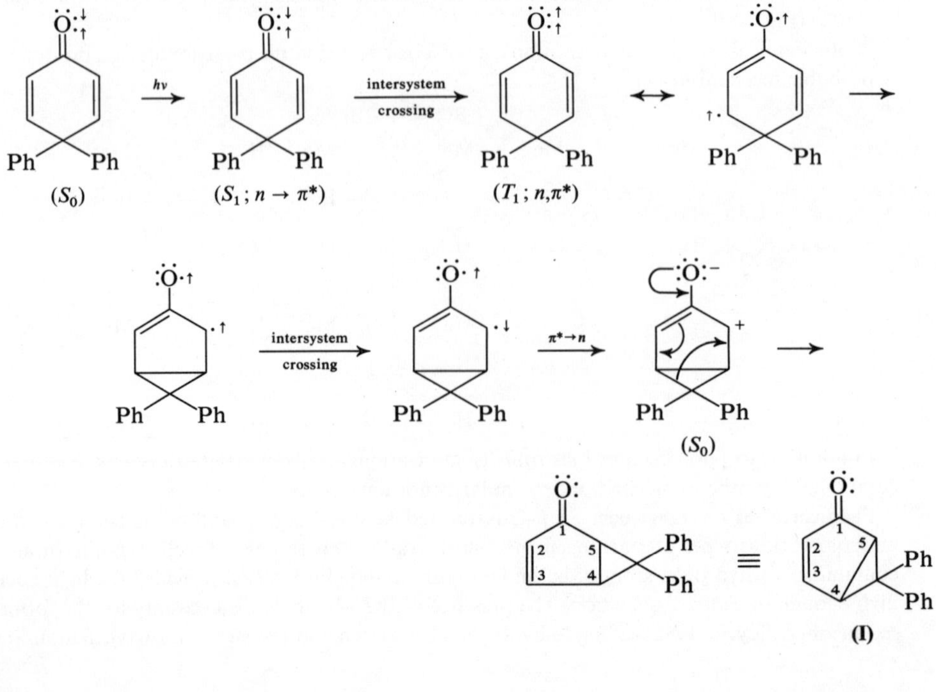

On the assumption of a similar mechanism, the formation of the other products can be explained as follows:

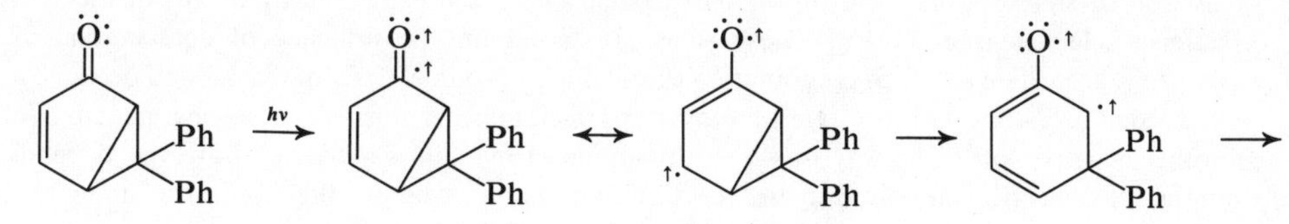

(I)

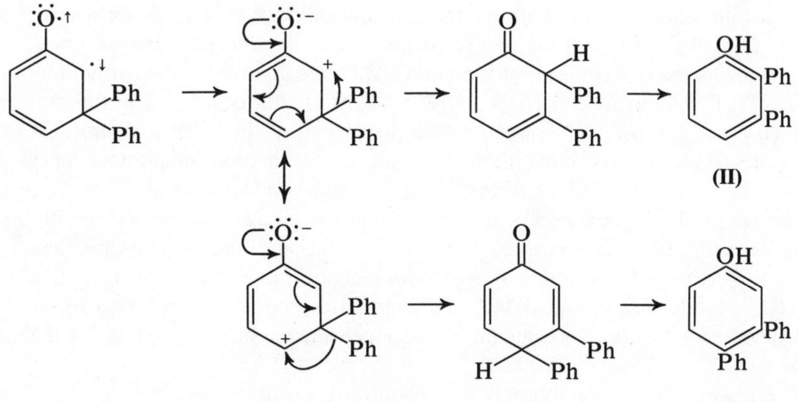

(II)

(III)

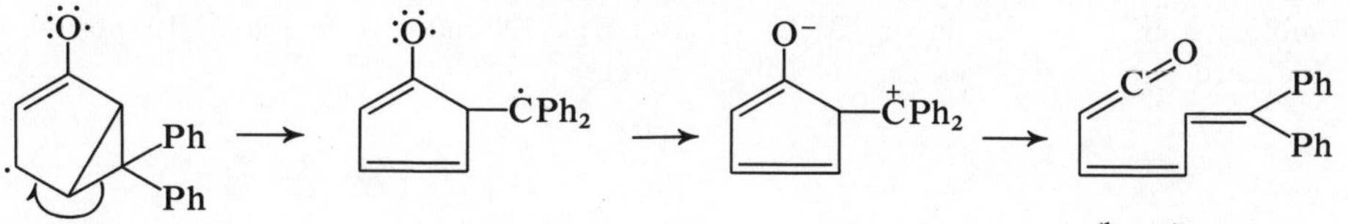

(keten)

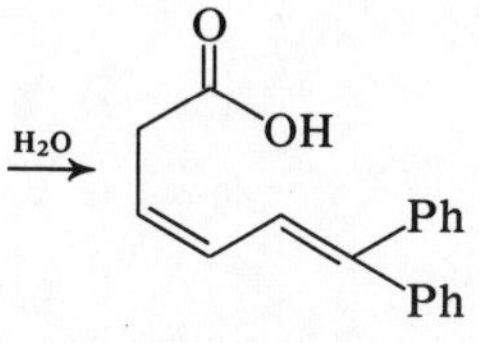

(IV)

Many rearrangements that occur thermally or in the presence of catalysts, can also be effected photo-chemically, *e.g.*, Orton, Fries, benzidine rearrangements.

Photochemical cyclisations and intermolecular cyclo-additions. Although we are here concerned with reactions induced by light, *i.e.*, *photochemical control*, nevertheless, since these reactions can also be induced by heat, *i.e.*, *thermal control*, both types will be discussed. Intramolecular cyclisations are **electrocyclic reactions**, which may be defined as the formation of a σ-bond between the terminal carbon atoms of a polyene to form a ring system, and the converse process.

Intermolecular cyclo-additions are of two types, *concerted* and *non-concerted*. Here we shall deal with only the concerted additions, which may be defined as a reaction in which *both new σ-bonds are formed simultaneously*. The Diels–Alder reaction is one example of this type (p. 536), and another is the dimerisation of alkenes (see below).

Both intramolecular cyclisation and concerted intermolecular addition are characterised by having a negative entropy of activation, which is in keeping with formation of a cyclic T.S. These reactions may also be stereospecific, and this stereospecificity has been explained by means of the **Woodward–Hoffman selection rules** (1965). These rules are based on the **principle of conservation of orbital symmetry**, according to which combination (involving p_z orbitals) to form the σ-bond occurs by overlap of *like-sign orbitals*, *i.e.*, a positive lobe overlaps with another positive lobe, or a negative lobe with another negative lobe. The way in which this mode of overlap is achieved, however, depends on the number of π-electrons involved in the reaction and also on whether the reaction is under thermal or photochemical control. For *thermally controlled* intramolecular cyclisations, the reactant molecule is assumed to be in a *vibrationally excited ground state*, and that the two π-electrons in the highest occupied M.O. are involved in the formation of the new σ-bond, and vice versa. For *photochemically controlled* intramolecular cyclisations, (i) the reactant molecule is in its *first electronic excited state, i.e.*, the (π,π^*) excited state, and the two electrons, one in the highest (occupied) bonding M.O. and the other in the lowest (unoccupied) anti-bonding M.O. are involved in the formation of the new σ-bond. In this case it is the symmetry of the *anti-bonding* M.O. that is involved in the conservation of orbital symmetry, and because this symmetry differs from that of the highest occupied M.O. (in the ground state), the stereochemistry of the products is different for thermal and photochemical control.

For thermally controlled concerted intermolecular cyclo-additions, it is assumed that the two π-electrons in the highest occupied M.O. overlap with the two π-electrons of the other reactant molecule in the lowest unoccupied M.O., *i.e.*, in the first electronic excited state. For the photochemically controlled reaction, however, *both* reactant molecules are in their first electronic excited states, *i.e.*, the highest occupied M.O. for thermal control is now replaced by the M.O. in the first electronic excited state. Once again, the stereochemistry of the products is different for the two processes.

We are now in a position to apply the Woodward–Hoffmann selection rules and, at the same time, introduce some other terms necessary for our discussion. We shall be concerned mainly with ethylene and butadiene as linear reactant molecules and so, for ease of reference, the M.O.s of these two molecules are given here in Fig. 31.10 (see also Figs. 2.9 and 4.3). In Fig. 31.10, positive lobes are indicated by blacked circles, negative lobes by circles.

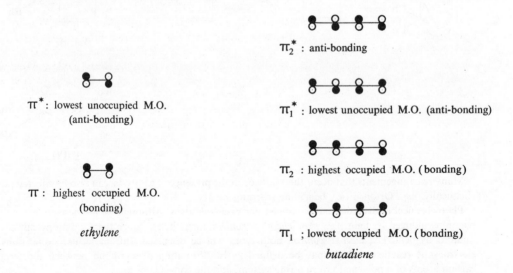

Fig. 31.10

Intramolecular electrocyclic reactions. (i) **Thermal control.** The most common examples are butadiene-cyclobutene interconversions, but ring fission is far more common than ring closure. We shall first consider the ring fission of 3,4-dimethylcyclobutene, and it should be noted that the terms *cis* or *trans* refer to the positions of the two hydrogen atoms with respect to a given double bond, *e.g.*, a diene described as *cis, trans* has the *cis* configuration for *one* double bond and the *trans* configuration for the *other* double bond. The product obtained from *cis*-3,4-dimethylcyclobutene, (I), on heating, is *cis,trans*-1,4-dimethylbutadiene, (II). The formation of this stereospecific product is a consequence of

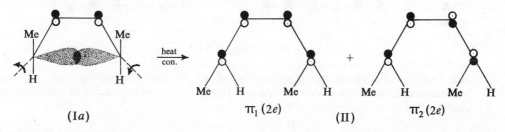

conservation of orbital symmetry, and is explained as follows. When the ring in (I) opens, (II) contains four π-electrons, two of which occupy the lowest occupied M.O. (orbital π_1; see Fig. 31.10) and the other two the highest occupied M.O. (orbital π_2). On the assumption that the two π-electrons derived from the σ-bond enter the highest occupied M.O., these enter orbital π_2, whereas the two π-electrons in the π-bond originally present in (I) enter orbital π_1. This sequence may be represented as $(Ia) \to \pi_1 + \pi_2$ (the combination of these being the ground state of II). However, when the σ-bond undergoes fission,

in order that the two individual π-electrons may enter orbital π_2 (in which the terminal lobes have opposite signs), it is necessary that in the 'de-overlapping' process, one lobe must rotate upwards and the other downwards. This is achieved by *both* rotating simultaneously in the same direction (anti-clockwise) as shown in (Ia). This mode of rotation, *i.e.*, both rotating in the *same* direction, is said to be **conrotatory**. It should be noted that if the mechanism was not concerted, *i.e.*, did not proceed via a cyclic T.S., *e.g.*, proceeded via a diradical, it is then very difficult to explain the stereospecificity of the reaction. It should also be noted that a *clockwise* conrotatory motion would still give the same product, *cis, trans* (Ib $\to \pi_2(a) \equiv \pi_2$; this is so, since + and − signs are purely relative).

If the conversion of *cis,trans*-1,4-dimethylbutadiene into *cis*-3,4-dimethylcyclobutene were thermally possible, then this reaction would also occur by a conrotatory motion, since both positive (or negative) lobes must be aligned to give maximum overlap for the formation of the σ-bond. This, of course, could also have been anticipated from the principle of microscopic reversibility (p. 49).

If we now consider the ring fission of *trans*-3,4-dimethylcyclobutene, (III), it then follows from the selection rules and on the basis that for every reaction there are two conrotatory motions, that the predicted products will be *cis, cis*- (IV*a*) and *trans,trans*-1,4-dimethylbutadiene (IV*b*).

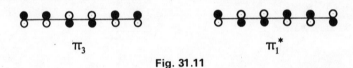

| (III) | (IV*a*) | (IV*b*) | III |

In practice, the *trans,trans* product (IV*b*) is obtained exclusively, since the strain in the T.S. leading to this configuration is far less than that leading to the *cis,cis* configuration (steric effects).

Now let us consider the thermal conversion of hexa-1,3,5-triene into cyclohexa-1,3-diene. The triene contains three bonding and three anti-bonding M.O.s, but since we shall be concerned with only the highest occupied M.O. (π_3; bonding) and the lowest unoccupied M.O. (π_1^*; anti-bonding), these two are given here (Fig. 31.11). Let us take as our example the thermal cyclisation of *trans,cis,trans*-

$$\pi_3 \qquad\qquad \pi_1^*$$

Fig. 31.11

octa-2,4,6-triene (V) to give the cyclohexadiene (VI). If we apply the selection rules, the predicted product is *cis*-5,6-dimethylcyclohexa-1,3-diene (VI); this is obtained in practice. In this case, however,

(V)   (VI)

to obtain maximum overlap in the formation of the σ-bond, the terminal groups must rotate in opposite directions (see V*a*), *i.e.*, they undergo a **disrotatory** motion. This is represented by (V*a*) → (VI*a*). In the same way, (VI*a*) is converted into (V*a*) by thermal control, also by a disrotatory motion.

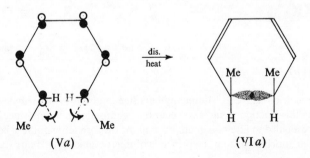

(V*a*)   (VI*a*)

(ii) **Photochemical control.** As we have seen, in thermally controlled cyclisations, the π-electrons forming the σ-bond are in the highest occupied M.O. For photochemically controlled reactions, the situation is different, the reactant molecule now being in its first excited electronic state. This also results in stereospecific reactions, but now the stereochemistry of the products is different, and the conrotatory and disrotatory motions are the reverse of those for thermal control. As our first example, let us consider the photochemical conversion of *trans*-3,4-dimethylcyclobutene (VII) into the butadiene derivative (VIII). (VII), by thermal control, undergoes a conrotatory motion (in the ground state) to give the *trans,trans* product (IV*b*). Under photochemical control, however, the predicted product is the *cis,trans* isomer (VIII). On irradiation, one π-electron in the reactant molecule is promoted to the π*-orbital (VII*a*), then the excited molecule undergoes fission to give the butadiene derivative in its first excited state (VIII), in which two π-electrons are in orbital π_1 (see Fig. 31.10), one in orbital π_2, and one in orbital π_1^*. Since it has been assumed that the two π-electrons derived from the σ-bond enter the higher energy M.O.s, to achieve this the σ-bond undergoes fission by a *disrotatory* motion, one π-electron entering the orbital π_2 and the other the orbital π_1^*. The excited molecule now loses energy, the π-electron in the orbital π_1^* returning to orbital π_2, thereby producing the butadiene derivative in its ground state $[\pi_1(2e) + \pi_2(2e)]$. This sequence is represented as shown in the diagram (VII) $\rightarrow$ (VII*a*) $\rightarrow$ (VIII).

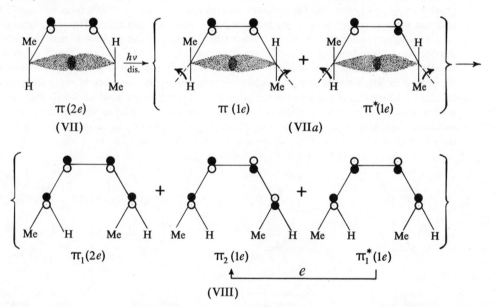

Now let us consider the photochemical cyclisation of *trans,cis,trans*-octa-2,4,6-triene (V). The predicted product is the *trans*-isomer, (IX) (*cis*, (VI), for the thermal cyclisation), and using similar arguments to those above, ring closure (and ring opening of the product) takes place by a conrotatory motion.

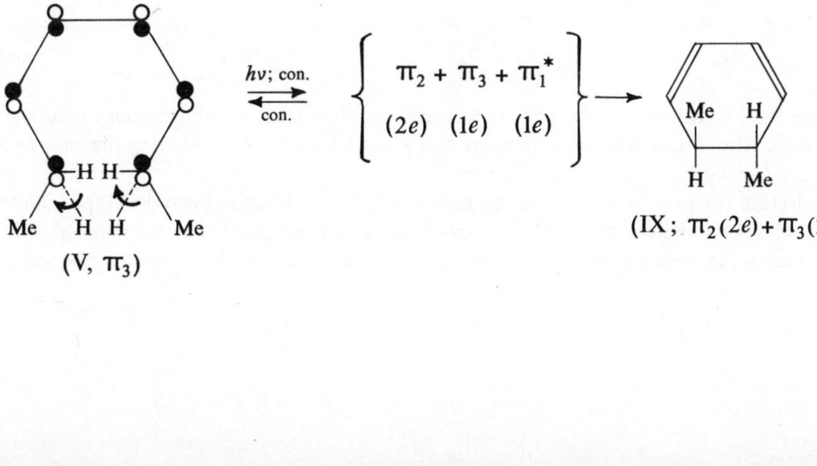

The foregoing discussion of the Woodward–Hoffmann rules for intramolecular electrocyclic reactions is summarised in Table 31.3, where n is the number of π-electrons in the *acyclic polyene* (and hence $n - 2$ π-electrons in the *cyclic isomer*). It should also be noted that if the polyene is *part* of a ring system (usually a medium ring), then ring closure may occur to give a bicyclic product, the stereochemistry of which will, unless isomerisations have occurred, be in accordance with the selection rules.

Table 31.3

Acyclic polyene ($n = 0, 1, 2, \ldots$)	Thermal	Photochemical
$4n$	Conrotatory	Disrotatory
$4n+2$	Disrotatory	Conrotatory

Concerted intermolecular cyclo-additions. The most common type of reaction is *dimerisation*, but another important type is the Diels–Alder reaction (p. 536). The selection rules also apply to these reactions which may, as before, be thermally or photochemically controlled. However, some complications arise in cyclo-additions owing to the fact that the reaction can proceed in four stereochemically distinct ways. Woodward and Hoffmann have introduced the following terms to indicate the geometrical relationship of the bonding (or fission) mode: (i) **suprafacial process** (*supra*; subscript s) is the process in which bonds that are made or broken lie on the *same* face of the system undergoing reaction; (ii) **antarafacial process** (*antara*; subscript a) is the process in which bonds that are made or broken lie on the *opposite* faces of the system. These terms are intended to replace the customary terms *cis* (*supra*) and *trans* (*antara*). The four combinations are illustrated by reactions leading to products (I–IV).

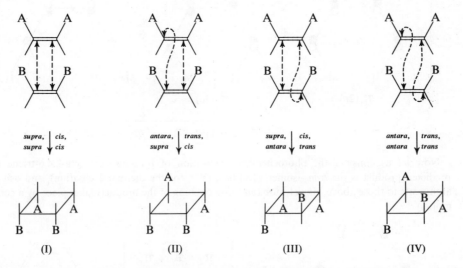

When the two reactant molecules are identical, *supra,antara* and *antara,supra* reactions give the same product. Also, for each process there are two possible cases, one producing the *endo* product, the other the *exo* (see later).

Selection rules for predicting the geometry of the product have been developed, and these are given in Table 31.4, where n is the total number of π-electrons participating in the reaction. In practice, *supra,antara* and *antara,antara* cyclo-additions are uncommon; they appear to be sterically difficult.

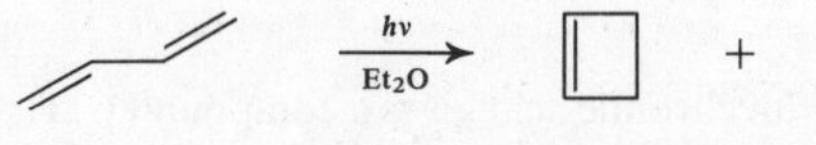

of butadiene is carried out in the *presence* of a sensitiser, three products are obtained, *trans*-(XIII) and *cis*-1,2-divinylcyclobutane (XIV), and 4-vinylcyclohexane (IX). (XIII) and (XIV) arise by 1,2-cyclo-addition (2+2) and (IX) by 1,4-cyclo-addition (4+2), and their relative amounts depend on the nature of the sensitiser, *e.g.*,

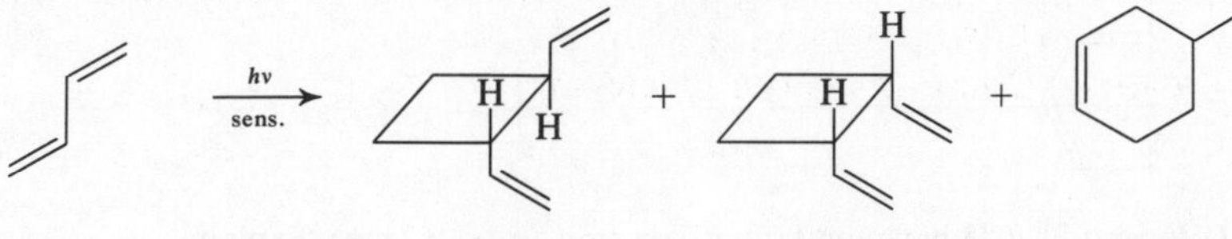

		(XIII)	(XIV)	(IX)
Sens.	acetophenone	82	14	4%
	benzil	50	8	42%

Since there is dependence on the sensitiser, it is reasonable to accept that the excited triplet state is involved; thus:

$$[\text{sens.}]\,(S_0) \xrightarrow{hv} [\text{sens.}]^*\,(S_1) \longrightarrow [\text{sens.}]^*(T_1)$$

$$\xrightarrow{CH_2=CH-CH=CH_2} [\text{sens.}]\,(S_0) + [CH_2=CH-CH=CH_2]^*\,(T_1)$$

If we also accept that butadiene in its triplet state can be represented as a resonance hybrid like butadiene in its ground state, then the central bond has double-bond character, with the *s-trans*-form being more stable than the *s-cis* (see p. 130).

$$\overset{\bullet\uparrow}{C}H_2-\overset{\bullet\uparrow}{C}H-CH=CH_2 \longleftrightarrow \overset{\bullet\uparrow}{C}H_2-CH=CH-\overset{\bullet\uparrow}{C}H_2 \longleftrightarrow CH_2=CH-\overset{\bullet\uparrow}{C}H-\overset{\bullet\uparrow}{C}H_2$$

and

Furthermore, since one reacting species is in the triplet state, the cyclo-addition is not a concerted reaction, and hence the selection rules do not apply (diradicals are in the triplet state). In the case under discussion, the geometry of the reacting species plays a part in the course of the reaction, the *s-trans**-form undergoing 1,2-addition and the *s-cis**-form undergoing both 1,2- and 1,4-additions with ground-state *s-trans*-butadiene. This is represented as shown:

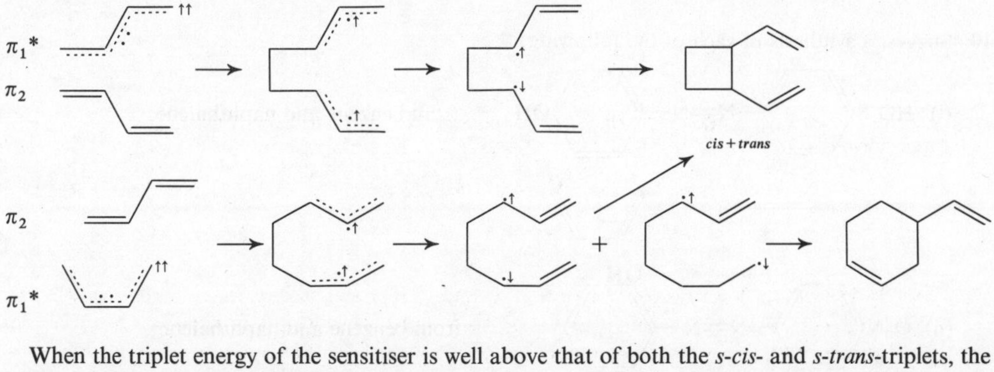

When the triplet energy of the sensitiser is well above that of both the *s-cis*- and *s-trans*-triplets, the same ratio of products is obtained. In these circumstances it is the *s-trans*-triplet that predominates (see *cis-trans* isomerisation, p. 906). As the triplet energy of the sensitiser decreases, the amount of the *s-cis*-triplet increases, and consequently the amount of cyclohexene derivative increases.

PROBLEMS

1. The oxidation of cyclohexene with chromic acid gives a compound C_6H_8O which shows a strong u.v. band at 225 nm. What is this compound and what is the 'expected' product that is also obtained?

2.

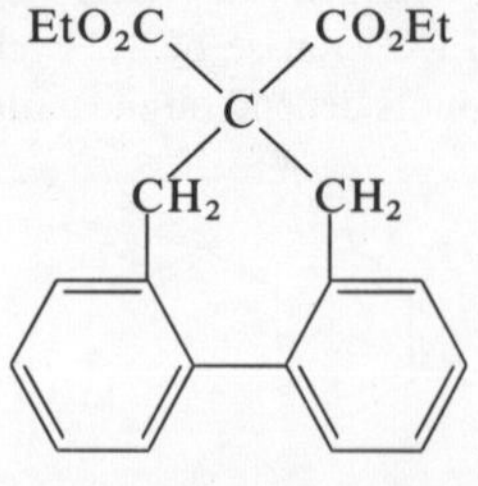

This biphenyl derivative has been resolved and has λ_{max} 249 nm (ε 17 000). Biphenyl has λ_{max} 251·5 nm (ε 18 300). Discuss these observations.

3. PhMe, Ph_2CH_2, and Ph_3CH have similar u.v. spectra ($\lambda_{max} \sim 262$ nm). Explain.

4. Discuss the following spectral data (nm):

(i) PhOH (210), PhOMe (217), PhO$^-$ (235).

(ii)

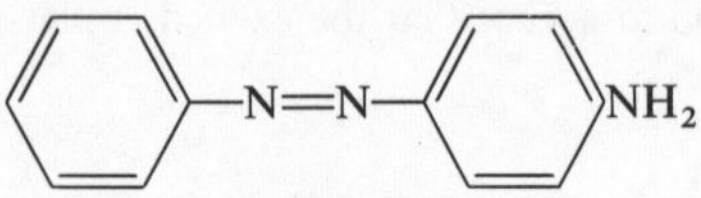

5. In 4-OH-2,6-Me$_2$—C$_6$H$_2$—N=N—C$_6$H$_2$-2,4,6-Me$_3$, the *cis* isomer is colourless and the *trans* isomer is coloured. Offer an explanation.

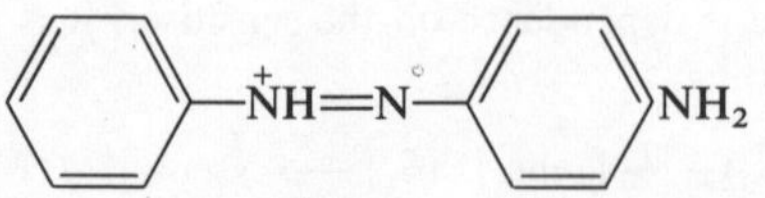

6. What effect would you expect the polarity of the solvent to have on the position of the absorption maximum of alkenes?

7. Offer an explanation for the following data: λ_{max} (nm): methane, 125; ethane, 135; cyclopropane, 190.

8. Draw diagrams for the relative energies of the various types of π-orbitals of butadiene (which has two bonding and two anti-bonding π-orbitals) to illustrate the ground state and the S_1, T_1, S_2, and T_2 excited states.

9. Complete the equations:

(i)   diazotised picramic acid + p-OH-acetanilide $\longrightarrow$?
(ii) diazotised 3-NH$_2$-4-OH-benzenesulphonic acid + m-phenylenediamine $\longrightarrow$?
(iii) benzenediazonium chloride + resorcinol $\longrightarrow$?

(iv)

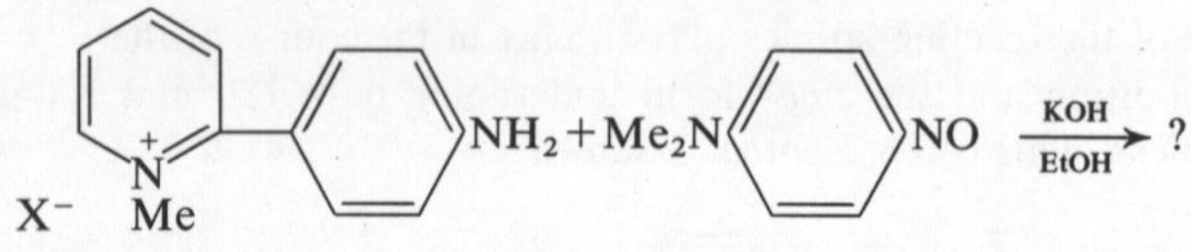

10. Suggest a synthesis of each of the following:

(i)

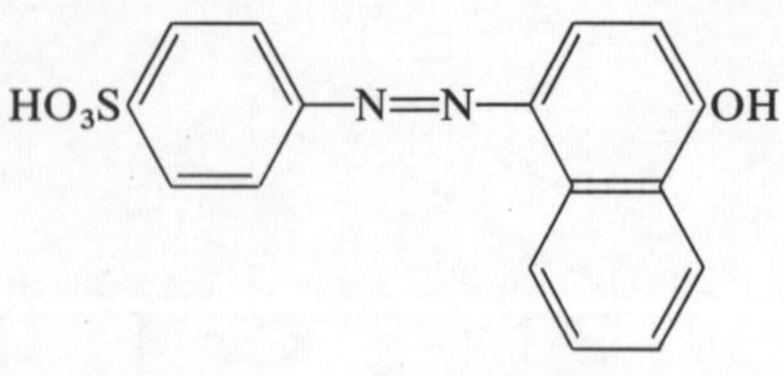

from benzene and naphthalene;

(ii)

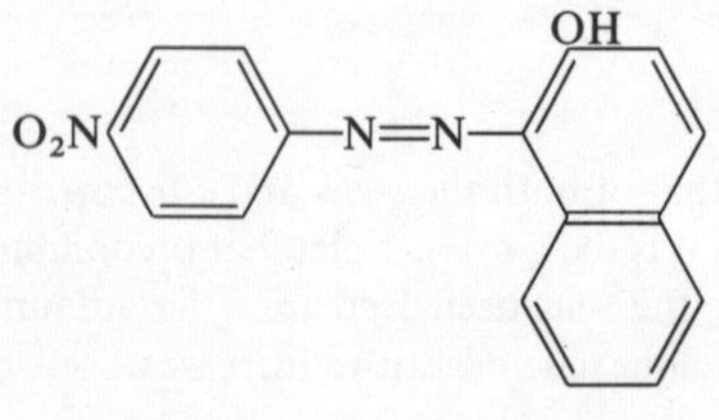

from benzene and naphthalene;

(iii) 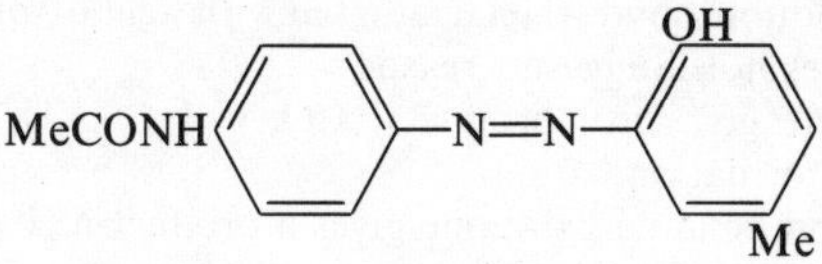 MeCONH⟨ ⟩—N=N—⟨ ⟩   from aniline and *p*-cresol;

(with OH at top and Me at bottom on the right ring)

(iv) 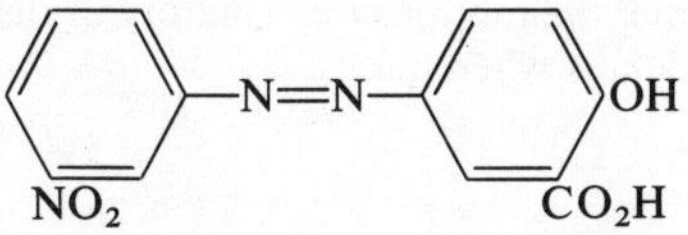 ⟨ ⟩—N=N—⟨ ⟩OH   from benzene;

(with NO₂ on left ring and CO₂H on right ring)

(v) 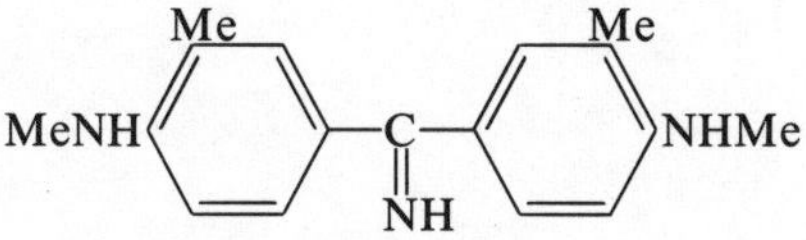 MeNH⟨ ⟩—C—⟨ ⟩NHMe   from *N*-Me-*o*-toluidine;

(with Me groups on both rings and NH below the central C)

(vi) 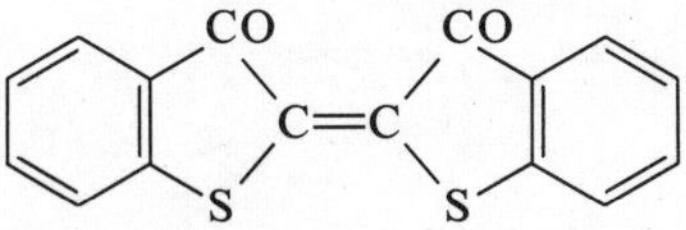 (structure with two benzene rings bearing CO and S groups linked by C=C)   from benzene.

11. What are the major products in the photochemical decomposition of methyl neopentyl ketone? Suggest a mechanism.

12. Complete the following equations:

(i) 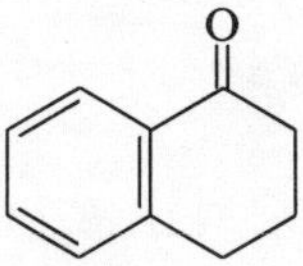 (α-tetralone structure) $\xrightarrow[h\nu]{Me_2CHOH}$?

(ii) $Me_2CHONO \xrightarrow{h\nu}$?

(iii) 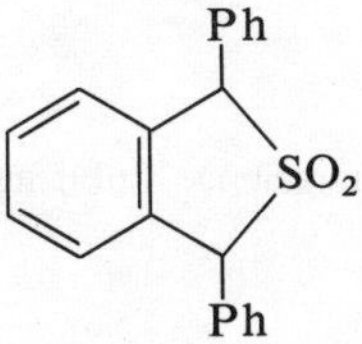 (2,3-diphenyl-2,3-dihydrobenzothiophene-1,1-dioxide with SO₂ and two Ph groups) $\xrightarrow{h\nu}$?

(iv) 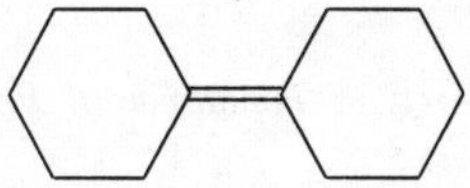 (bicyclohexylidene structure) $+ O_2 \xrightarrow{h\nu}$?

(v) $MeCH_2CH_2COMe \xrightarrow{h\nu}$? (a cyclobutane is major product)

(vi) 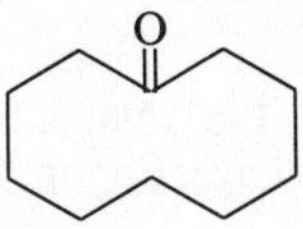 (cyclodecanone structure) $\xrightarrow{h\nu}$? (a decalin)

(vii) 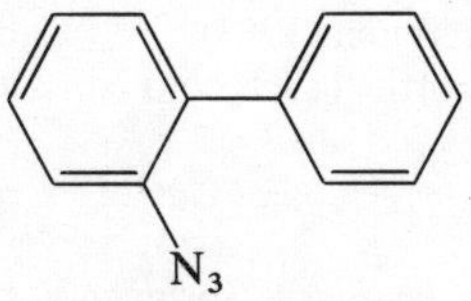 (2-azidobiphenyl structure with N₃) $\xrightarrow{h\nu}$? (a heterocycle)

(viii) $CH_3CH_2CH_2CH_2N_3 \xrightarrow{h\nu}$? (a cyclic compound)

13. 5-Methylhex-3-en-2-one undergoes migration of the double bond when irradiated with light of *low* energy, whereas pent-3-en-2-one, under the same conditions, does not. Suggest a reason.

14. The photochemical decomposition of cyclohexanone gives five products: CO, C_2H_4, C_3H_6, C_5H_{10}, $CH_2 = CH(CH_2)_3CHO$. Outline a mechanism for their formation.

15. The photochemical reaction between cyclohexane and pentane-2,4-dione gives a product **A**, $C_{11}H_{18}O_2$. **A**, on oxidation with alkaline hypobromite, gives **B**, $C_9H_{14}O_4$. **B**, on heating with soda-lime, gives **C**, C_7H_{14}, a saturated compound. What are **A**, **B**, and **C**? Formulate the reactions.

16. By means of the Woodward–Hoffmann selection rules, predict the motions (conrotatory or disrotatory) for the following reactions, and complete with stereochemical details where necessary.

(i) $\xrightarrow{\text{heat}}$? (ii) $\xrightarrow{\text{heat}}$?

(iii) $\xrightarrow{h\nu}$? (iv) $\xrightarrow{h\nu}$? $\xrightarrow{\text{heat}}$?

(v) $\xrightarrow{\text{heat}}$? (monocyclic) (vi) $\xrightarrow{h\nu}$? (bicyclic)

17. Suggest a mechanism for the reaction

$$2PhCH_2CH_2ONO \xrightarrow{h\nu} (PhCH_2NO)_2 + 2CH_2O$$

18. The Norrish type I process is not important for the photolysis of diaryl ketones. Suggest a reason.

REFERENCES

FIERZ-DAVID and BLANGEY, *Fundamental Processes of Dye Chemistry*, Interscience Publishers (1949; translated by Vittum).

LUBS (ed.), *The Chemistry of Synthetic Dyes and Pigments*, Reinhold (1955).

ABRAHART, *Dyes and Their Intermediates*, Pergamon (1968).

BRADLEY, 'Recent Progress in the Chemistry of Dyes and Pigments', *Roy. Inst. Chem., Lectures, Monographs and Reports*, 1958, No. 5.

STALLMANN, 'Use of Metal Complexes in Organic Dyes and Pigments', *J. Chem. Educ.*, 1960, **37**, 220.

STEAD, 'Recent Advances in Dyestuffs Chemistry', *Chem. in Britain*, 1965, **1**, 361.

INMAN, 'Organic Pigments', *Roy. Inst. Chem., Lecture Series*, 1967, No. 1.

GOODMAN, 'Synthetic Fibre-Forming Polymers', *Roy. Inst. Chem., Lecture Series*, 1967, No. 3.

FRODYMA and FREI, 'Reflectance Spectroscopic Analysis of Dyes Separated by Thin Layer Chromatography', *J. Chem. Educ.*, 1969, **46**, 522.

JAFFE and ORCHIN, *Theory and Applications of Ultraviolet Spectroscopy*, Wiley (1962).

MURRELL, *The Theory of the Electronic Spectra of Organic Molecules*, Methuen (1963).

RAO, *Ultraviolet and Visible Spectroscopy; Chemical Applications*, Butterworths (1968, 2nd edn).

TURRO, 'The Triplet State', *J. Chem. Educ.*, 1969, **46**, 2.

KAN, *Organic Photochemistry*, McGraw-Hill (1966).

NECKERS, *Mechanistic Organic Photochemistry*, Reinhold (1967).

SCHONBERG, SCHENCK, and NEUMULLER, *Preparative Organic Photochemistry*, Springer-Verlag (1968, 2nd edn).

WOODWARD and HOFFMANN, *The Conservation of Orbital Symmetry*, Verlag Chemie Academic Press (1970).

GILL, 'The Applications of the Woodward–Hoffmann Orbital Rules to Concerted Organic Reactions', *Quart. Rev.*, 1968, **22,** 338.

VOLLMER and SERVIS, 'Woodward–Hoffmann Rules: Cycloaddition Reactions', *J. Chem. Educ.,* 1970, **47,** 491.

Appendix

Nomenclature

In dealing with the systematic methods of nomenclature, no attempt has been made to indicate whether a particular system is described in the Geneva system, or whether it is described in the I.U.P.A.C. rules. Irrespective of its origin the method of nomenclature has been referred to in the text as the I.U.P.A.C. system. In any case, the reader should appreciate that organic nomenclature is in a state of flux. This is unfortunate, but inevitable. What is more unfortunate is that there is no complete agreement among the various publishing societies.

Many of the I.U.P.A.C. rules have been given in the text. The reader should find the following remarks of some assistance in nomenclature, and it should be noted that the I.U.P.A.C. name is *always* correct, even if it is rarely used.

(i) The names of substituents cited as prefixes are arranged alphabetically, regardless of the number of each, and commas are used to separate the numerals, *e.g.*,

3-**a**mino-4-**c**hloro-2-naphthol

ethyl **m**ethyl ketone

5-**n**itro-1,4-diphenylnaphthalene

1-**a**cetyl-2-**a**mino-4-**h**ydroxy-3,5,6-trimethylanthracene

methyl **p**ropyl succinate

N.B.   methyl **h**ydrogen phthalate

(ii) Compound group names are treated as *units*, and when several prefixes begin with the same letter, short names precede longer ones, *e.g.*,

1-(2,4-**d**initrophenyl)-3-**m**ethylnaphthalene

1-**methyl**-4-**methylamino**anthracene

1-**d**imethylamino-2-ethylnaphthalene

(iii) Isomeric substituents are arranged in alphabetical order of the prefixes, *e.g.*, butyl, ethyl, isobutyl, pentyl, s-butyl, t-butyl.

Prefixes which define the positions of named substituents or which are used to define stereoisomers, are italicised, *e.g.*, *anti, cis, endo, meso, peri, syn, trans*.

(iv) When a compound contains one principal function, then this is expressed in the ending of the name, *e.g.*,

3-ethyl-2-methylcyclohexane-1-carboxylic acid

This rule also applies to aromatic compounds.

(v) When a compound contains more than one functional group, the compound is named by using the suffix of the principal function and the prefixes of all the other functions (p. 297). This rule also applies to aromatic compounds.

(vi) In some cases the order may be changed to avoid ambiguity, *e.g.*, phenyldichloroarsine is used for $C_6H_5AsCl_2$, as dichlorophenylarsine might be taken as $Cl_2C_6H_3AsH_2$.

Use of Greek letters

The Greek alphabet is:

A	α	alpha	Ξ	ξ	xi
B	β	beta	O	o	omicron
Γ	γ	gamma	Π	π	pi
Δ	δ	delta	P	ρ	rho
E	ε	epsilon	Σ	σ	sigma
Z	ζ	zeta	T	τ	tau
H	η	eta	Y	υ	upsilon
Θ	θ	theta	Φ	ϕ	phi
I	ι	iota	X	χ	chi
K	κ	kappa	Ψ	ψ	psi
Λ	λ	lambda	Ω	ω	omega
M	μ	mu			
N	ν	nu			

The use of Greek letters to indicate the positions of substituents in a chain is to be avoided; numerals should be used. Greek letters, however, may be used (and are preferable) to indicate positions of substituents in compounds described by trivial names, *e.g.*, α-hydroxypropionic acid (lactic acid). Greek letters are also used to indicate the *class* of compound, *e.g.*, α, β, γ-, . . . diketones; γ- and δ-lactones, etc.

Writing of names

The general practice is to 'run on' the names of groups replacing hydrogen atoms in a compound, *e.g.*, chlorobenzene, 1-nitropropane, ethylmethylcarbinol, etc. If the compound is named by any *general term*, then each part of the name is written separately, *e.g.*, ethyl alcohol, ethyl methyl ether, acetic acid, methyl propyl ketone, butyraldehyde oxime (but butyraldoxime), acetaldehyde phenylhydrazone, etc.

Organic chemistry publications

Before using any book, consult the preface for information about the contents. The preface usually gives all the information necessary. Also note the date of the publication.

Chemical dictionaries

Thorpe's Dictionary of Applied Chemistry. This contains short articles on a very wide field of subjects, and also contains a large bibliography of original literature.

Encyclopædia of Chemical Technology. Similar in approach to Thorpe's Dictionary, but more up-to-date.

Heilbron's Dictionary of Organic Compounds. This gives a concise account of the physical and chemical properties of organic compounds, and references to good methods of preparation.

Physical constants

Dictionary of Organic Compounds.
International Critical Tables.
KAYE and LABY, *Tables of Physical and Chemical Constants.*
LANDOLT and BORNSTEIN, *Physikalisch-Chemische Tabellen.*
Chemist's Year Book (British).
Handbook of Chemistry and Physics (American).
LANGE'S *Handbook of Chemistry.*

Reference works

There are many excellent single volume texts available on Organic Chemistry. The following, however, largely consist of reviews of various topics, each review being published as a series.
ELDERFIELD (ed.), *Heterocyclic Compounds* (1950–).
Chemistry of the Carbon Compounds (Edited by Rodd), Elsevier.
RADT (ed.), *Elsevier's Encyclopædia of Organic Chemistry* (1940–).
GOWAN and WHEELER, *Name Index of Organic Reactions*, Longmans, Green (1960).

Accounts of Chemical Research.
Advances in: Carbohydrate Chemistry, Chemical Physics, Free Radical Chemistry, Heterocyclic Chemistry, Magnetic Resonance, Mass Spectrometry, Molecular Spectroscopy, Organic Chemistry (Methods and Results), Organometallic Chemistry, Photochemistry, Physical Organic Chemistry, Protein Chemistry.
Annual Reports of the Chemical Society.
Annual Review of: Biochemistry, NMR Spectroscopy.
Organic Reaction Mechanisms.
Progress in: Infrared Spectroscopy, Organic Chemistry, Physical Organic Chemistry, Reaction Kinetics, Stereochemistry.
Specialist Periodical Reports (The Chemical Society). Alkaloids, Terpenes and Steroids, etc. These are regular reviews.

There are also a number of reference books on practical organic chemistry, *e.g.*,
HOUBEN-WEYL, *Methoden der Organischen Chemie.*
    Organic Synthesis. These are published annually. There are now also four collected volumes.
    Organic Reactions. These contain detailed discussions of a large number of organic reactions, and also include an account of the practical methods.
WEISSBERGER, *Technique of Organic Chemistry.*
THEILHEIMER, *Synthetic Methods of Organic Chemistry.*
MIGRDICHIAN, *Organic Synthesis*, Reinhold (Vol. I and II).
HICKINBOTTOM, *Reactions of Organic Compounds.*
HOUSE, *Modern Synthetic Reactions.*
VOGEL, *Practical Organic Chemistry.*
FIESER and FIESER, *Reagents for Organic Synthesis* (Vols. 1 and 2).
SANDLER and KARO, *Organic Functional Group Preparations.*

Literature

Some important periodicals are:
American: **Journal of the American Chemical Society** (*J. Amer. Chem. Soc.*).
Belgian: **Bulletin de la Société chimique de Belgique** (*Bull. Soc. chim. belges*).
British: **Journal of the Chemical Society** (*J. Chem. Soc.*). Since 1966, the Journal has been divided into three sections: A, Inorganic, Physical and Theoretical, B, Physical Organic, C, Organic. This journal is undergoing reorganisation (1972–).
Dutch: **Recueil des Travaux chimiques des Pays-Bas** (*Rev. Trav. chim.*).
French: **Bulletin de la Société chimique de France** (*Bull. Soc. chim. France*).

German: **Berichte der deutschen chemischen Gesellschaft** (*Ber.*). This has been changed to *Chem. Ber.* since 1946.

Italian: **Gazzetta chimica italiana** (*Gazzetta*).

Russian: **Journal of General Chemistry** (U.S.S.R.) (*J. Gen. Chem.* (*U.S.S.R.*)). This is a U.S. translation of *Zhurnal obshchei Khimii.*

Swiss: **Helvetica Chimica Acta** (*Helv. Chim. Acta*).

The above journals are general in that they publish research on all branches of chemistry. Besides these, however, are a large number of specialised journals which deal with only Organic, Physical, Analytical, Biological, Industrial chemistry, etc.

Other useful sources of Organic Chemistry are:

Justus Leibig's Annalen der Chemie (*Annalen*).

Journal of Heterocyclic Chemistry (*J. Heterocyclic Chem.*)

Journal of Organic Chemistry (*J. Org. Chem.*).

Journal of Organometallic Chemistry (*J. Organometallic Chem.*).

Journal für praktische Chemie (*J. prakt. Chem.*).

Monatschefte für Chemie und verwandte Teile anderer Wissenschaften (*Monatsh.*).

Tetrahedron (*Tetrahedron*).

Transactions of the Faraday Society (*Trans. Faraday Soc.*).

In addition to journals containing original papers, there are also reviews published by various societies, *e.g.*, **Angewandte Chemie, International Edition in English, Chemical Reviews, Journal of Chemical Education, Education in Chemistry, Quarterly Reviews, R.I.C. Reviews.**

Many journals also give critical surveys, or publish lectures that have been given by specialists. The periodicals **Nature** and **Chemistry and Industry** are used as a means of correspondence as well as for publications. There are also the **Tetrahedron Letters** (begun 1959). **The Proceedings of the Chemical Society** (begun 1957) is now published separately as **Chemical Communications** (1965–), and other matter published in the former has been merged with the **Journal of the Royal Institute of Chemistry** to produce the new journal, **Chemistry in Britain** (1965–).

Owing to the large number of journals, it is impossible to read everything published. Hence abstracting journals have appeared for the benefit of the chemist. The most important abstracting journals are:

British Chemical Abstracts (ceased after 1953).

Chemical Abstracts. Five decennial indexes have appeared: 1917, 1927, 1937, 1947 and 1957, and two five-year indexes, 1962 and 1967.

Chemisches Zentrallblatt.

Chemical Titles. This is a publication of the *Amer. Chem. Soc.* Key words in the titles of chemical publications are listed alphabetically, and the names of authors are also listed.

Index Chemicus. This is a service of the *Institute for Scientific Information* (American). It is a twice-monthly register and index of *new* chemical compounds. It reports and indexes new chemical compounds within 30 to 60 days after their appearance in the primary journals.

The abstracts have the following three indexes: author, subject and formula.

Author index. This is arranged alphabetically. To search the author index it is necessary to know the name of the author and also how to spell it; care must be exercised with some surnames, *e.g.*, Tschitschibabin or Chichibabin.

Subject index. This is arranged alphabetically. When searching the literature about a particular subject, the word (name) and related words (names) should be looked up, *e.g.*, suppose we wish to ascertain the work that has been done on *fats*. The procedure would be to look up:

(i) Fats. (ii) Oils. (iii) Esters. (iv) Individuals, *e.g.*, olein, palmitin, etc. (v) Saponification. (vi) Soaps. (vii) Hydrolysis. (viii) Edible fats and oils. (ix) The author index of the various papers, since it is quite likely that the author has published more than one paper on the subject.

When using the subject index, it is advisable to work backwards chronologically, since recent papers will give references to earlier ones.

When searching for a particular compound, the reader should bear in mind alternative names for that particular compound, and also the possibility of alternative methods of numbering.

Formula index should offer no difficulties. It is probably best to search this.

Searching the literature

Some indication of searching the literature has already been given in the foregoing.

Finding a paper given the reference. Original papers are given as author, name of periodical, year of publication, volume number (in bold print, or underlined in written work), and page, *e.g.*,

E. F. Armstrong, *J. Chem. Soc.*, 1903, **83**, 1305. This is British practice. American practice is slightly different, *viz.*

E. F. Armstrong, *J. Chem. Soc.*, **83**, 1305 (1903).

Sometimes mistakes are made in giving a reference.

Wrong author. No difficulty will be encountered here, provided the nature of the article is known.

Wrong Journal. In this case it is best to consult the author or subject indexes of the abstracting journals.

Wrong year. If the volume number is correct, there should be no difficulty.

Wrong volume. If the year is correct, there should be no difficulty.

Wrong page. In this case one should look up the annual index of the authors or subjects, or both.

When consulting a reference in an abstracting journal, the reader should look up the same year and at least two years afterwards. The more efficient the abstracting journal, the sooner will an abstract of a publication appear.

Finding information about a compound of known formula (and structure). This is usually done by looking up Beilstein, and then the Chemical Abstracts.

Beilstein's Handbuch. There is now a fourth edition, and this is intended to give a complete survey of organic chemistry up to Jan. 1st, 1910. The plan of the arrangement is described in volume I, pp. 1–46 and pp. 939–944. The subject matter is divided into four main divisions:

I.	*Acyclic compounds*	Volumes I–IV
II.	*Isocyclic compounds*	Volumes V–XVI
III.	*Heterocyclic compounds*	Volumes XVII–XXVII
IV.	*Natural products*	Volumes XXX and XXXI

There is a first supplement to Beilstein's fourth edition, in most cases one to each volume; otherwise, the supplements may be combined, *e.g.*, the supplements for volumes III and IV are combined into one volume. The literature survey of the first supplement is from Jan. 1st, 1910 to Jan. 1st, 1920. There is also a second supplement giving the literature survey from Jan. 1st, 1920 to Jan. 1st, 1930. A third supplement has now been begun; this covers the period up to Jan. 1st, 1950.

Volume XXVIII is the general names index and Volume XXIX is the general formulæ index, and covers the main work and the first and second supplements (volumes I–XXVII). The names index is arranged alphabetically. In the formula index, the compounds are listed by the number of carbon atoms, the number of hydrogen atoms and according to the alphabetical order of the other elements, *e.g.*, $CBrCl_3O_2S$, $C_2H_5NO_2S$. Alongside the compound will be found various numbers, *e.g.*, **18**, 548 (this is the main work, Vol. 18, p. 548); **19**, 402, II 413 (this is the main work, Vol. 19, p. 402; and the second supplement, Vol. 19, which is not repeated, p. 413); **5**, I 357; **3**, 69, I 31, II 55.

How to use Beilstein. Each volume and supplement has a table of contents and an index consisting partly of names and partly of formulæ. The simplest way to find a compound is to look up the collected indexes of names and formulæ. Having found the compound in the main volumes, the searcher will then find later information (if any) in the corresponding supplements. System numbers are used for cross-reference purposes.

In Beilstein will be found an account of methods of preparation, properties, derivatives, etc., and references.

Each main division of Beilstein is subdivided into functional group compounds in the following order:

Acyclic and Isocyclic. Hydrocarbons, hydroxy-compounds, carbonyl compounds, hydroxy-carbonyl compounds, carboxylic acids, hydroxy-acids, carbonyl acids, hydroxy-carbonyl acids, sulphinic acids, sulphonic acids, hydroxy-sulphonic acids, carbonyl-sulphonic acids, carboxy-sulphonic acids, amines, hydroxyamines, carbonyl-amines, hydroxy-carbonyl-amines, aminoacids, hydroxylamines, hydrazines, azo-compounds, diazo-compounds, azoxy-compounds, metallic compounds.

Heterocyclic. 1 cyclic oxygen (S, Se, Te); 2 cyclic oxygen; 3 cyclic oxygen; 4 cyclic oxygen; 5 cyclic oxygen.

1 cyclic nitrogen; 2 cyclic nitrogen; . . . 8 cyclic nitrogen.

1 cyclic nitrogen and 1 cyclic oxygen; 1 cylic nitrogen and 2 cyclic oxygen; etc.

2 cyclic nitrogen and 1 cyclic oxygen; 2 cyclic nitrogen and 2 cyclic oxygen; etc.

Natural products. Hydrocarbons (petroleum), ethereal oils, sterols, fats, waxes, carbohydrates, alkaloids, proteins.

The division of a particular compound is determined by its *stem-nucleus*. This is obtained by replacing in the formula all the atoms or groups attached to carbon by the equivalent number of hydrogen atoms, except where replacement involves the breaking of a cyclic chain. When the stem-nucleus has been obtained, the following general rule is applied; when the compound is derived from two or more compounds, or contains two or more functional groups, that compound is discussed under the parent compound to be found *last* in the classification.

Example 1.

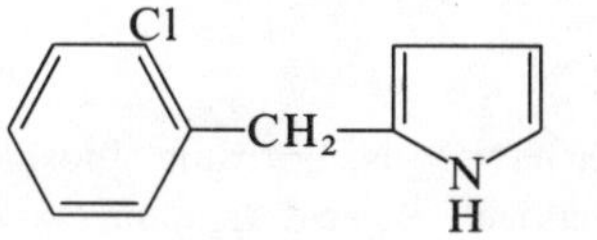

The stem nuclei are:

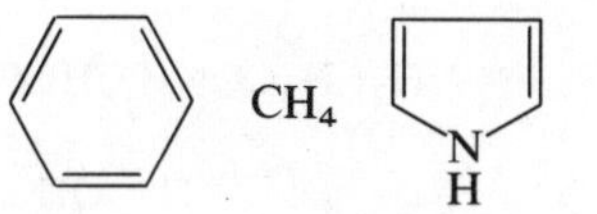

Since pyrrole is described last in the classification of these three compounds, the compound under consideration will be found under pyrrole.

Example 2.

$$ClCH_2CH_2OH$$

This is a derivative of ethanol, and will therefore be found under the substitution products of ethanol (described immediately after ethanol).

It is important to note that X, NO, NO_2 and N_3 are *not* functional groups, and that their order (of discussion) is F, Cl, Br, I, NO, NO_2, N_3 (immediately after the parent compound).

Difficulty may be encountered when more than one functional group, or one functional group and one or more *apparent* functional groups are attached to the *same* carbon atom. The following examples show how to find the compound:

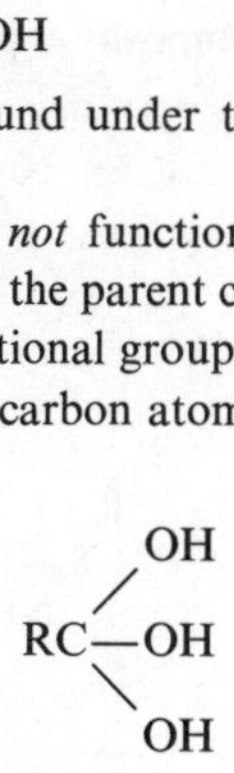

(i)

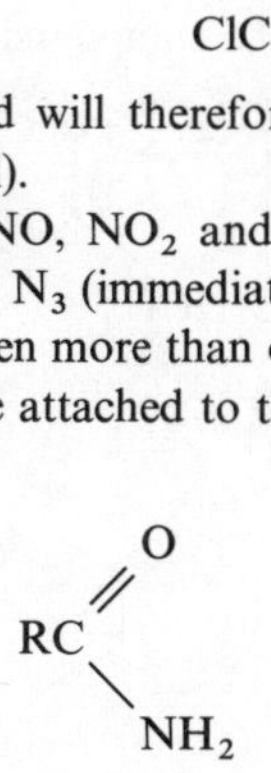

Replace the functional groups in (I) by hydroxyl groups. This gives (II), which is an acid (ortho-acid). (I) will therefore be found under the derivatives of the acid RCO_2H:

(ii)

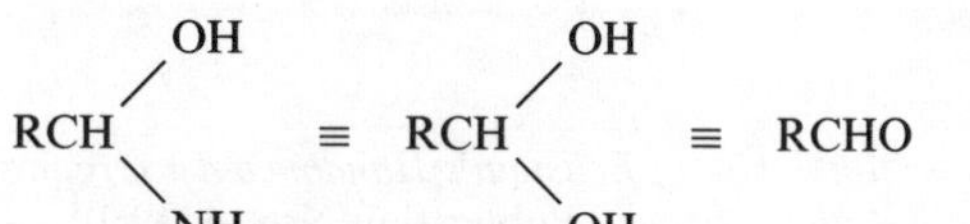

(iii)
$$NH_2C(=NH)NH_2 \equiv HOC(OH)_2OH \equiv H_2CO_3$$

(iv) $RCCl_3$. Since there is no functional group present, this compound will be found under RCH_3:

(v)
$$CH_3 + O-C_2H_5 \equiv CH_3OH + C_2H_5OH$$
$$HO \mid H$$

This compound will be found under the derivatives of ethanol (ethanol occurs later than methanol):

(vi)
$$R^1CO \mid OR^2 \equiv R^1CO_2H + R^2OH$$
$$HO \mid H$$

This will be found under R^1CO_2H, since this is discussed later than R^2OH.

Note however:

$$RO \mid NO_2 \equiv ROH + HNO_3$$
$$H \mid OH$$

This compound will be described under ROH, since HNO_3 is inorganic. Inorganic acid derivatives are discussed in the following order: hydrogen peroxide, halogen-oxyacids, sulphur-oxyacids, nitrogen-oxyacids, phosphorus oxyacids, arsenic-oxyacids, silicon-oxyacids, halogen acids, . . .

(vii)

$$\underset{HO \mid H}{NH_2CH_2CO \mid NHCHCO_2H} \overset{CH_3}{} \qquad \underset{H \mid OH}{NH_2CH_2CONH \mid CHCO_2H} \overset{CH_3}{}$$

(III)   (IV)

This is an example where a compound may be broken up in two (or more) ways. In such cases, the decomposition is carried out in the way which gives the stem-nucleus described later than any obtained in any other way. (III) gives glycine and alanine; (IV) gives glycine (amide derivative) and lactic acid. Since alanine is described later than any other decomposition product, the compound will be found under alanine.

(viii)

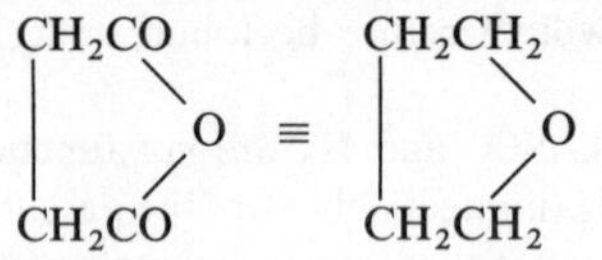

$$\underset{CH_2CO}{CH_2CO} \diagdown O \equiv \underset{CH_2CH_2}{CH_2CH_2} \diagdown O$$

(V)

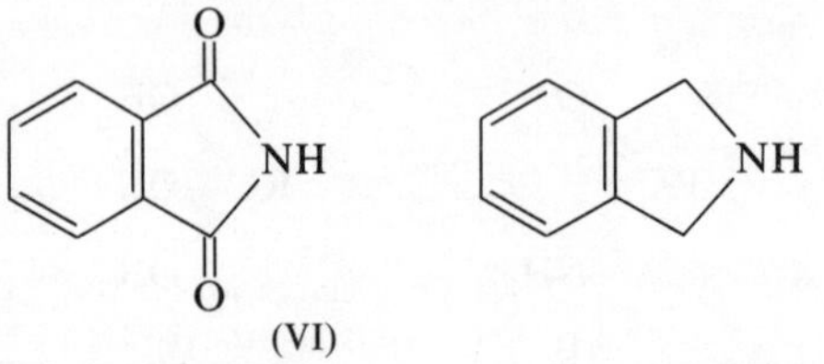

(VI)

(V) and (VI) are both to be found under *heterocyclic* compounds, the former under the division of 1 cyclic oxygen atom, and the latter, 1 cyclic nitrogen atom.

REFERENCES

HUNTRESS, *A Brief Introduction to the Use of Beilstein's Handbuch der Organischen Chemie*, Wiley (1930).
Handbook for Chemical Society Authors, Special Publication No. 14 (1960).
GOULD (ed.), *Searching the Chemical Literature*. Advances in Chemistry Series, No. 30, American Chemical Society (1961).
YESCOMBE, *Sources of Information on the Rubber, Plastics and Allied Industries*, Pergamon (1968).
BOTTLE (ed.), *The Use of Chemical Literature*, Butterworths (1969, 2nd edn).
CAHN, *An Introduction to Chemical Nomenclature*, Butterworths (1969, 3rd edn).

Index

Pages printed in bold type are the more important references. Letters marking the pages have the following meanings: **d** indicates definitions (of reactions, etc.); **n**, nomenclature; **p**, preparations; **r**, reactions; **s**, spectrum; **t**, table. These letters are generally not used for single page references. Derivatives of a parent substance will be found under the initial letter of the name, *e.g.*, nitrobenzene will be found under N. Names with prefixes will be found under the initial letter, *e.g.*, isobutanes will be found under I; cyclopropane under C. The alphabetical order has been used to name prefixes. Also the suffix *methanol* has been used as well as *carbinol* (p. 176). Salts are not listed separately, but are included with the references of the corresponding acids or bases. Names of authors are indexed only when they are associated with a particular reaction, condensation, etc. Dyestuffs are listed under Dyes; individual dyes are listed separately.

A

A_{AC} reactions, 249t, 250, 251
A_{AL} reactions, 249t, 252
Absolute alcohol, 190
Absolute asymmetric synthesis, 489
Absolute configuration, 474, 480–482
Absorbance, 6
Absorptivity (molar), 6
ac = alicyclic, 804
Accelerators, 407
Accessibility of compounds, 76
Acenaphthaquinone, 808
Acenaphthene, 808
Acenaphthylene, 808–809
Acetaldehyde, 91, 136, 190, 216, **217–218**, 219, 221, **227p–228r**, 240, 277–278, 298, 305, 312, 327, 336, 337, 342, 388, 471, 738, 849, 854, 857, 876t
Acetals, 185, **222pr**, 430
    cyclic, 311, 320
Acetamide, 351, 847
p-Acetamidophenol, 712
Acetanilide, **658**, 660, 661, 663
Acetic acid, 91, 136, 173, **240p–243r**, 259, 260, 305, 350, 712, 749, 753
Acetic anhydride, 199, 243, **259p–260r**, 261, 303, 339, 355, 504, 528, 739, 751, 835, 838, 839
Acetoacetaldehyde, 301
Acetoacetamides, 352
Acetoacetic acid, 282

Acetoacetic ester, **272–280**, 282, 284, 285s, **285p– 293r**, 341, 347, 468
Acetoacetic ester syntheses
    dicarboxylic acids, 293
    1,3-diketones, 292, 302
    1,4-diketones, 303–304
    heterocyclic compounds, 775, 829, 837, 846, 849, 858, 864
    homocyclic compounds, 534, 559
    keto-acids, 293, 306, 559
    ketones, 289–291
    monocarboxylic acids, 289, 290, 291, 292
Acetoin, 299, 302
Acetomesitylene, 750
Acetone, 91, 125, 136, 168, 169, 170, 210, 211, 215, 216, 218, 219, 222, 224, 225, **228p–229r**, 278, 279, 281, 287, 301, 302, 303, 316, **317p**, 339, 343, 346, 350, **506**, 584, 699, 738, 841, 863, 876t, 902–903
Acetonedicarboxylic acid, 492–493
Acetonitrile, *see* Methyl cyanide
Acetonylacetone, 303–304
Acetophenone, **292**, 375, 611, 617, 738, **748p–752r**, 917
Acetoxycrotonic ester, 292
Acetoxymercuribenzene, 617
o-Acetoxymercuriphenol, 705, 709
Acetylacetoacetic ester, 302
Acetylacetone, 213, 282, **302p–303r**
N-Acetylaminophenol, *see p*-Acetamidophenol